Y0-DXH-427

# HANDBOOK of MICROBIOLOGY

## Volume II
## Microbial Composition

EDITORS

**Allen I. Laskin, Ph.D.**
Esso Research and Engineering Company
Linden, New Jersey

**Hubert A. Lechevalier, Ph.D.**
Institute of Microbiology
Rutgers University
New Brunswick, New Jersey

Published by

A DIVISION OF
THE CHEMICAL RUBBER CO.
18901 Cranwood Parkway • Cleveland, Ohio 44128

HANDBOOK OF MICROBIOLOGY

Volume II: Microbial Composition





International Standard Book Number (ISBN)

Complete Set 0-87819-580-7
Volume II 0-87819-582-3

Library of Congress Catalog Card Number 72-88766
Second printing 1974

## PREFACE

In this second volume of the CRC Handbook of Microbiology, an effort has been made to put in the hands of microbiologists basic information about the constituents of microbial cells. The subject matter is covered under sections on amino acids and proteins, carbohydrates, lipids, nucleic acids and minerals. In addition, information is given about methods used to separate microbial components, as well as a wide variety of other pertinent information.

The Editors have been greatly helped in their task by the members of the Advisory Board and by the contributors. Especially helpful has been Dr. F. Persico, who kindly took charge of the section on nucleic acids.

It is the aim of the Publisher and of the Editors of the Handbook to periodically bring the material presented here up to date. In order to furnish the microbiologists with the information they need, the Editors will need all the feedback they can receive. It is thus the Editors' hope that the users of this Handbook will take the time to bring to our attention all the shortcomings of the first edition.

The Editors especially wish to thank Mrs. Lisbeth Hammer and Miss Beryl Tuffyas for their excellent editorial work and Mrs. Verna Lepping for her devoted assistance.

**A. I. Laskin**
**H. A. Lechevalier**
**New Jersey, 1973**

**Herman J. Phaff, Ph.D.**
Department of Food Technology
University of California
Davis, California

**Thomas B. Platt, Ph.D.**
Bioanalytical Section
The Squibb Institute of Medical Research
New Brunswick, New Jersey

**Otto J. Plescia, Ph.D.**
Institute of Microbiology
Rutgers University
New Brunswick, New Jersey

**G. Pontecorvo, Ph.D.**
Department of Cell Genetics
Imperial Cancer Research Fund
London, England

**Chase Van Baalen, Ph.D.**
Marine Science Institute
University of Texas
Port Aransas, Texas

**Claude Vezina, Ph.D.**
Microbiology Department
Ayerst Laboratories
St. Laurent, P. Q., Canada

**L. C. Vining, Ph.D.**
National Research Council
Atlantic Regional Laboratory
Halifax, N. S., Canada

**E. D. Weinberg, Ph.D.**
Department of Microbiology
Indiana University
Bloomington, Indiana

**Burton I. Wilner, Ph.D.**
Orinda, California

## CONTRIBUTORS

**Claude J. Abshire, Ph.D.**
Department of Chemistry
University of Quebec at Montreal
Montreal, P. Q., Canada

**Solomon Bartnicki-Garcia, Ph.D.**
Department of Plant Pathology
University of California
Riverside, California

**J. P. Bouley**
Universite de Strasbourg
Strasbourg, France

**Eugene H. Cota-Robles, Ph.D.**
Department of Microbiology
The Pennsylvania State University
University Park, Pennsylvania

**Cecil S. Cummins, Sc.D.**
Anerobe Laboratory
Virginia Polytechnic Institute and State University
Blacksburg, Virginia

**Ronald J. Elin, Ph.D., M.D.**
Institute of Allergy and Infectious Diseases
National Institute of Health
Bethesda, Maryland

**Douglas E. Eveleigh, Ph.D.**
Department of Biochemistry and Microbiology
Rutgers University
New Brunswick, New Jersey

**William Firshein, Ph.D.**
Department of Biology
Wesleyan University
Middletown, Connecticut

**R. Guay**
Laval University
Quebec, P.Q., Canada

**Hubert A. Lechevalier, Ph.D.**
Institute of Microbiology
Rutgers University
New Brunswick, New Jersey

**James C. MacDonald, Ph.D.**
Prairie Regional Laboratory
National Research Council of Canada
Saskatoon, Saskatchewan, Canada

**John B. Neilands, Ph.D.**
Department of Biochemistry
University of California
Berkeley, Calfornia

**Walter J. Nickerson, Ph.D.**
Institute of Microbiology
Rutgers University
New Brunswick, New Jersey

**William M. Normore, Ph.D.**
Biology Department
Staten Island Community College
Staten Island, New York

**William M. O'Leary, Ph.D.**
Department of Microbiology
Cornell University Medical College
New York, New York

**Elizabeth Percival, Ph.D., D.Sc.**
Chemistry Department
University of London
Royal Holloway College, Egham Hill
Egham, Surrey, England

**Carl A. Price, Ph.D.**
Department of Biochemistry and Microbiology
Rutgers University
New Brunswick, New Jersey

**Gail Carol Rodgers, A.B.**
University of California
San Francisco, California

**M. A. Q. Siddiqui, Ph.D.**
Roche Institute of Molecular Biology
Nutley, New Jersey

**Sylvia M. Stein, Ph.D.**
Department of Microbiology
The Pennsylvania State University
University Park, Pennsylvania

**David H. Strumeyer, Ph.D.**
Department of Biochemistry and Microbiology
Rutgers University
New Brunswick, New Jersey

**Clarence H. Suelter, Ph.D.**
Department of Biochemistry
Michigan State University
East Lansing, Michigan

**James R. Turvey, Ph.D., D.Sc.**
School of Physical and Molecular Sciences
University College of North Wales
Bangor, Caernarvonshire, England

**Eugene D. Weinberg, Ph.D.**
Department of Microbiology
Indiana University
Bloomington, Indiana

**Sheldon M. Wolff, M.D.**
National Institute of Allergy and Infectious Diseases
National Institute of Health
Bethesda, Maryland

## TABLE OF CONTENTS

**AMINO ACIDS AND PROTEINS** . . . . . **1**

AMINO ACIDS FOUND IN PROTEINS . . . . . 3
STRUCTURES OF AMINO ACIDS OCCURRING IN PROTEINS . . . . . 5
OPTICAL ISOMERS AND IONIZATION CONSTANTS OF AMINO ACIDS . . . . . 9
AMINO ACID COMPOSITION OF SELECTED PROTEINS . . . . . 15
AMINO ACID SEQUENCES OF MICROBIAL PROTEINS . . . . . 31
SOLUBILITIES OF THE AMINO ACIDS IN WATER AT VARIOUS TEMPERATURES . . . . . 53
AMINO ACID ANTAGONISTS . . . . . 57
PROPERTIES OF THE αKETO ACID ANALOGS OF AMINO ACIDS . . . . . 65
FAR-ULTRAVIOLET ABSORPTION SPECTRA OF AMINO ACIDS . . . . . 67
ULTRAVIOLET ABSORPTION CHARACTERISTICS OF N-ACETYL METHYL ESTERS OF THE AROMATIC AMINO ACIDS, CYSTINE AND N-ACETYLCYSTEINE . . . . . 69
ABSORBANCE VALUES OF THE AROMATIC AMINO ACIDS IN NEUTRAL, ALKALINE AND ACID SOLUTIONS . . . . . 71
SPECIFIC ROTATIONS OF AMINO ACIDS . . . . . 75
NUMBERING AND CLASSIFICATION OF ENZYMES . . . . . 83

**CARBOHYDRATES** . . . . . **87**

MICROBIAL MONOSACCHARIDES AND POLYSACCHARIDES . . . . . 89
POLYSACCHARIDES OF ALGAE . . . . . 147
BACTERIAL CELL WALL STRUCTURE . . . . . 167
FUNGAL CELL WALL COMPOSITIONS . . . . . 201
BACTERIAL ENDOTOXINS . . . . . 215

**LIPIDS** . . . . . **241**

LIPIDS . . . . . 243
THE NOMENCLATURE OF LIPIDS . . . . . 245
CHEMICAL AND PHYSICAL CHARACTERISTICS OF FATTY ACIDS . . . . . 248
STRUCTURES OF REPRESENTATIVE LIPID TYPES FOUND IN MICROORGANISMS . . . . . 261
STEROLS . . . . . 265
LIPOIDAL CONTENTS OF SPECIFIC MICROORGANISMS . . . . . 275
NMR AND MASS SPECTRA . . . . . 323

**NUCLEIC ACIDS** . . . . . **329**

PHYSICAL CONSTANTS AND SPECTRAL PROPERTIES OF PURINES, PYRIMIDINES, NUCLEOSIDES AND NUCLEOTIDES . . . . . 331
NATURAL OCCURRENCE OF THE MODIFIED NUCLEOSIDES . . . . . 427
NUCLEOSIDE ANTIBIOTICS . . . . . 437
SPECTROPHOTOMETRIC CONSTANTS OF RIBONUCLEOTIDES . . . . . 539
GENERAL FEATURES OF TRANSFER RIBONUCLEIC ACID STRUCTURE . . . . . 543
CONTROLLED PARTIAL HYDROLYSIS OF RNA . . . . . 553
NUCLEOTIDE SEQUENCES OF RIBONUCLEIC ACIDS . . . . . 557
DNA BASE COMPOSITIONS OF EUKARYOTIC PROTISTS . . . . . 561
DISTRIBUTION OF PURINES AND PYRIMIDINES IN DEOXYRIBONUCLEIC ACIDS . . . . . 567

GUANINE-PLUS-CYTOSINE (GC) COMPOSITION OF THE DNA OF BACTERIA, FUNGI, ALGAE AND PROTOZOA . . . 585
THE BASE COMPOSITION OF BACTERIOPHAGE NUCLEIC ACIDS . . . 741
VIRAL DNA MOLECULES . . . 751
DNA CONTENT PER CELL IN VIRUSES . . . 761
BUOYANT DENSITIES, MELTING TEMPERATURES AND GC CONTENT OF VIRAL DNA . . . 765
DEOXYRIBONUCLEIC ACID CONTENT PER CELL IN FUNGI, ALGAE AND PROTOZOA . . . 771
BUOYANT DENSITIES OF NUCLEIC ACIDS AND POLYNUCLEOTIDES . . . 775
MELTING TEMPERATURES ($T_m$) OF SYNTHETIC POLYNUCLEOTIDES . . . 781
CONTENT OF 6-METHYLAMINOPURINE AND 5-METHYLCYTOSINE IN DNA . . . 789
METABOLISMS OF NUCLEOSIDES . . . 791

**MINERALS . . . 803**

TRACE METALS . . . 805
INORGANIC ENZYME COFACTORS . . . 811
MICROBIAL IRON TRANSPORT COMPOUNDS . . . 823

**MISCELLANEOUS INFORMATION . . . 831**

CELL BREAKAGE . . . 833
Bacterial Cell Breakage of Lysis . . . 833
Isolation of Fungal Cell Walls . . . 844
CENTRIFUGATION . . . 845
TABLE OF ATOMIC WEIGHTS . . . 881
PERIODIC TABLE OF THE ELEMENTS . . . 883
DISSOCIATION AND IONIZATION CONSTANTS . . . 885
VISCOSITY AND DENSITY OF SUCROSE IN WATER . . . 899
DENSITY OF CESIUM CHLORIDE SOLUTIONS AS A FUNCTION OF REFRACTIVE INDEX AT 25°C . . . 903
BUFFER SOLUTIONS, pH INDICATORS, AND FLUORESCENT INDICATORS . . . 909
ANALYSES OF MEDIA CONSTITUENTS . . . 919
A NOMOGRAM FOR AMMONIUM SULFATE SOLUTIONS . . . 927
RECIPROCALS OF NUMBERS . . . 929
SQUARES, SQUARE ROOTS, CUBES, AND CUBE ROOTS . . . 953
DISTRIBUTION OF $t$ . . . 973
DISTRIBUTION OF $x^2$ . . . 975
RANDOM PERMUTATIONS OF TWENTY NUMBERS . . . 977
COMPONENTS OF ATMOSPHERIC AIR . . . 979
CONSTANT HUMIDITY . . . 981
CONVERSION OF TRANSPARENCY TO OPTICAL DENSITY . . . 983
PROPERTIES OF VARIOUS LABORATORY MATERIALS . . . 989
MISCIBILITY OF ORGANIC SOLVENT PAIRS . . . 1005
REFRACTIVE INDICES OF LIQUIDS . . . 1009
SPECIFIC GRAVITY OF LIQUIDS . . . 1013
INFRARED CORRELATION . . . 1017
FAR-INFRARED CORRELATION . . . 1025
CHARACTERISTIC NMR SPECTRAL POSITIONS FOR HYDROGEN IN ORGANIC STRUCTURES . . . 1037
**INDEX . . . 1039**

# AMINO ACIDS AND PROTEINS

# THE AMINO ACIDS FOUND IN PROTEINS

DR. D. STRUMEYER

I. Neutral Amino Acids

A. Aliphatic side chain

Glycine
Alanine
Valine
Isoleucine
Leucine

B. Polar side chain

1. Hydroxyl-containing

Threonine
Serine

2. Amide-containing

Asparagine
Glutamine

C. Sulfur-containing

Cysteine
Cystine
Methionine

D. Imino amino acids

Proline
Hydroxyproline

E. Aromatic

Phenylalanine
Tyrosine
Tryptophan

II. Acidic Amino Acids

Aspartic
Glutamic

III. Basic Amino Acids

Histidine
Lysine
Arginine

## ABBREVIATIONS AND MOLECULAR WEIGHTS OF AMINO ACIDS FOUND IN PROTEINS

| Amino Acid | 3-Letter Symbol | 1-Letter Symbol | M.W. | Amino Acid | 3-Letter Symbol | 1-Letter Symbol | M.W. |
|---|---|---|---|---|---|---|---|
| Alanine | Ala | A | 89.09 | Leucine | Leu | L | 131.17 |
| Arginine | Arg | R | 179.20 | Lysine | Lys | K | 146.19 |
| Asparagine | Asn | N | 132.12 | Methionine | Met | M | 141.21 |
| Aspartic acid | Asp | D | 133.10 | Phenylalanine | Phe | F | 165.19 |
| Cysteine | Cys | C | 121.15 | Proline | Pro | P | 115.13 |
| Cystine | $(Cys)_2$ | | 240.29 | Serine | Ser | S | 105.09 |
| Glutamic acid | Glu | E | 147.13 | Threonine | Thr | T | 119.12 |
| Glutamine | Gln | Q | 146.15 | Tryptophan | Trp | W | 204.24 |
| Glycine | Gly | G | 75.07 | Tyrosine | Tyr | Y | 181.19 |
| Histidine | His | H | 155.16 | Valine | Val | V | 117.15 |
| Isoleucine | Ile | I | 131.17 | "Other" | | X | |

Complete rules affecting amino acid and other biochemical nomenclature may be found in *Handbook of Biochemistry,* 2nd ed., Section A, H. A. Sober, Ed. The Chemical Rubber Co., Cleveland, Ohio (1970).

# STRUCTURES OF AMINO ACIDS OCCURRING IN PROTEINS

DR. D. STRUMEYER

## NEUTRAL AMINO ACIDS

### Aliphatic Side Chain

$$\boxed{H}-\overset{H}{\underset{NH_3^{\oplus}}{C}}-COO^{\ominus}$$

Glycine

$$\boxed{CH_3}-\overset{H}{\underset{NH_3^{\oplus}}{C}}-COO^{\ominus}$$

Alanine

$$\boxed{(CH_3)_2CH}-\overset{H}{\underset{NH_3^{\oplus}}{C}}-COO^{\ominus}$$

Valine

$$\boxed{(CH_3)_2CH-CH_2}-\overset{H}{\underset{NH_3^{\oplus}}{C}}-COO^{\ominus}$$

Leucine

$$\boxed{CH_3-CH_2-\overset{CH_3}{CH}}-\overset{H}{\underset{NH_3^{\oplus}}{C}}-COO^{\ominus}$$

Isoleucine

### Hydroxyl-Containing

$$\boxed{CH_3-\overset{OH}{CH}}-\overset{H}{\underset{NH_3^{\oplus}}{C}}-COO^{\ominus}$$

Threonine

$$\boxed{HO-CH_2}-\overset{H}{\underset{NH_3^{\oplus}}{C}}-COO^{\ominus}$$

Serine

### Amide-Containing

$$H_2N-\overset{O}{\overset{\|}{C}}-CH_2-\overset{H}{\underset{NH_3^{\oplus}}{C}}-COO^{\ominus}$$

Asparagine

$$H_2N-\overset{O}{\overset{\|}{C}}-CH_2-CH_2-\overset{H}{\underset{NH_3^{\oplus}}{C}}-COO^{\ominus}$$

Glutamine

**Sulfur-Containing**

Cysteine: $HS{-}CH_2{-}\overset{H}{\underset{NH_3^{\oplus}}{C}}{-}COO^{\ominus}$

Cystine:
$S{-}CH_2{-}\overset{H}{\underset{NH_3^{\oplus}}{C}}{-}COO^{\ominus}$
$|$
$S{-}CH_2{-}\overset{H}{\underset{NH_3^{\oplus}}{C}}{-}COO^{\ominus}$

Methionine: $CH_3{-}S{-}CH_2{-}CH_2{-}\overset{H}{\underset{NH_3^{\oplus}}{C}}{-}COO^{\ominus}$

Cysteine    Cystine    Methionine

**Imino Amino Acids**

H   H
C ——— $CH_2$
|   |   H
$CH_2$ — $N^{\oplus}$ — C — $COO^{\ominus}$
$H_2$

Proline

HO   H
C ——— $CH_2$
|   |   H
$CH_2$ — $N^{\oplus}$ — C — $COO^{\ominus}$
$H_2$

Hydroxyproline

**Aromatic**

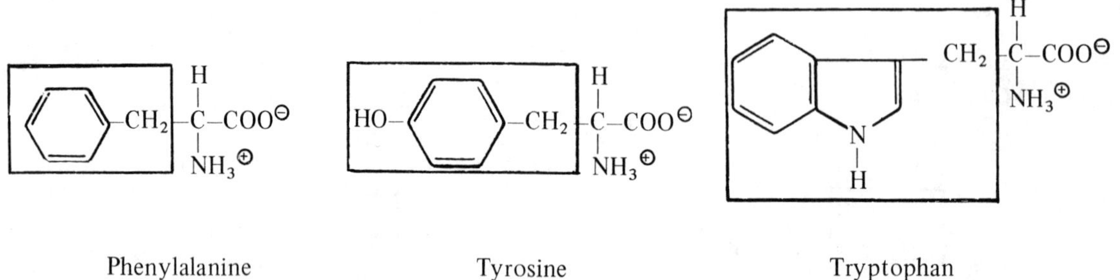


Phenylalanine    Tyrosine    Tryptophan

**ACIDIC AMINO ACIDS**

Aspartic acid: $\overset{O}{\underset{\ominus O}{\gtrless}}C{-}CH_2{-}\overset{H}{\underset{NH_3^{\oplus}}{C}}{-}COO^{\ominus}$

Glutamic acid: $\overset{O}{\underset{\ominus O}{\gtrless}}C{-}CH_2{-}CH_2{-}\overset{H}{\underset{NH_3^{\oplus}}{C}}{-}COO^{\ominus}$

Aspartic acid    Glutamic acid

**BASIC AMINO ACIDS**

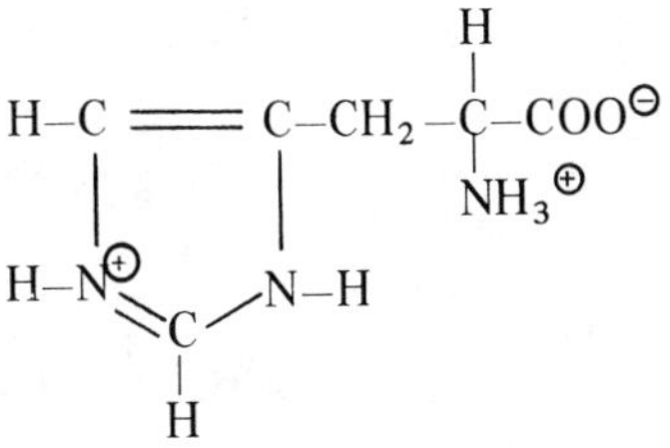

Histidine

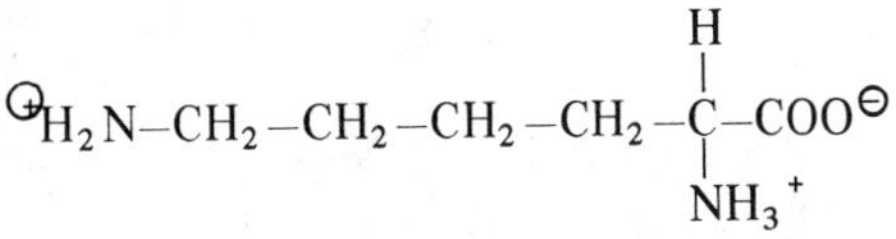

Lysine

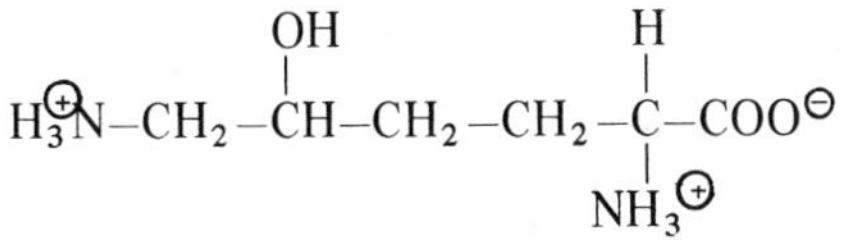

Hydroxylysine

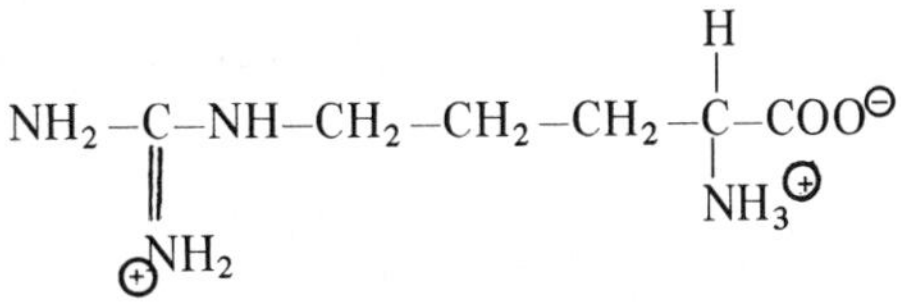

Arginine

**AMINO ACID DERIVATIVES**

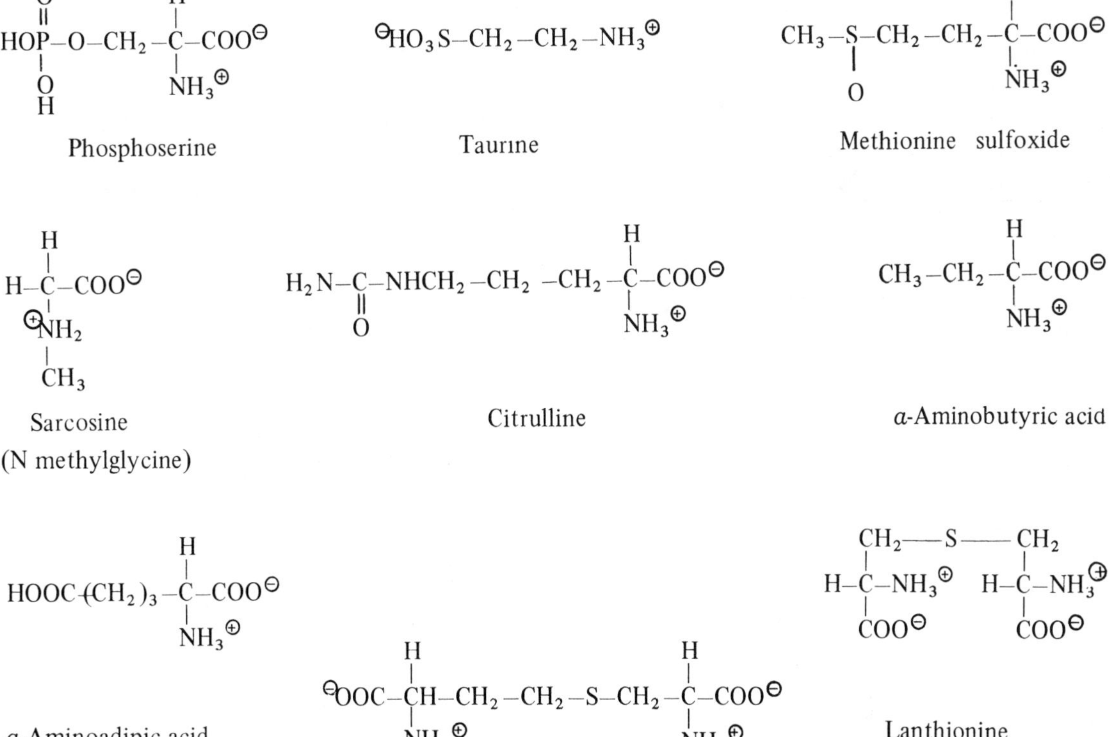

Phosphoserine

Taurıne

Methionine sulfoxide

Sarcosine
(N methylglycine)

Citrulline

$\alpha$-Aminobutyric acid

$\alpha$-Aminoadipic acid

Lanthionine

Cystathionine

$^{\oplus}H_3N{-}CH_2{-}CH_2{-}COO^{\ominus}$

β-Alanine

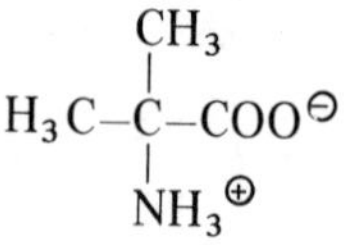

$H_3C{-}C(CH_3)(NH_3^{\oplus}){-}COO^{\ominus}$

α-Aminoisobutyric acid

$^{\oplus}NH_3{-}CH_2{-}CH_2{-}CH_2{-}COO^{\ominus}$

γ-Aminobutyric acid

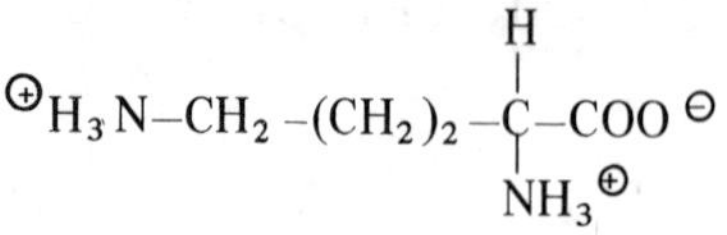

$^{\oplus}H_3N{-}CH_2{-}(CH_2)_2{-}CH(NH_3^{\oplus}){-}COO^{\ominus}$

Ornithine

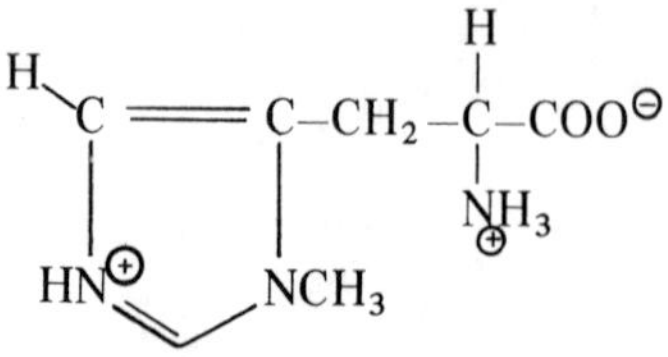

H–C = C–$CH_2$–CH($NH_3^{\oplus}$)–$COO^{\ominus}$ (ring: $HN^{\oplus}$, $NCH_3$)

1-Methylhistidine

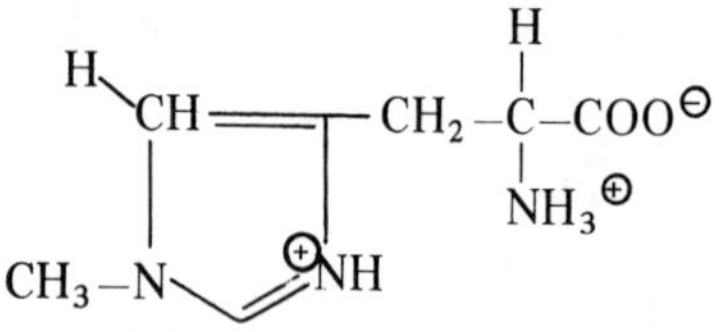

H–CH = C–$CH_2$–CH($NH_3^{\oplus}$)–$COO^{\ominus}$ (ring: $CH_3{-}N$, $^{\oplus}NH$)

3-Methylhistidine

$^{\ominus}OOC{-}CH(NH_3^{\oplus}){-}CH_2{-}CH_2{-}CH_2{-}CH(NH_3^{\oplus}){-}COO^{\ominus}$

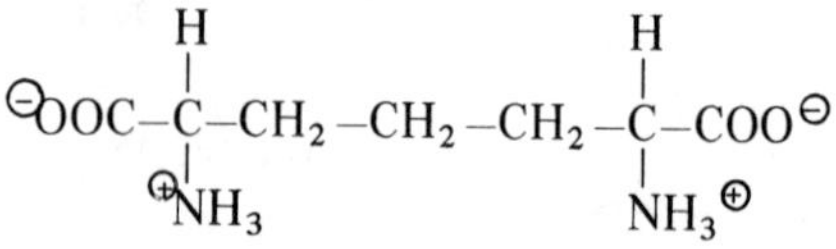

α,ε-Diaminopimelic acid

**Antibiotics**

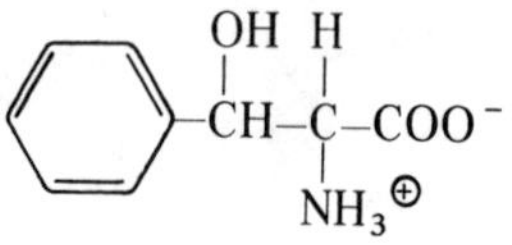

$C_6H_5{-}CH(OH){-}CH(NH_3^{\oplus}){-}COO^{-}$

L-Phenylserine

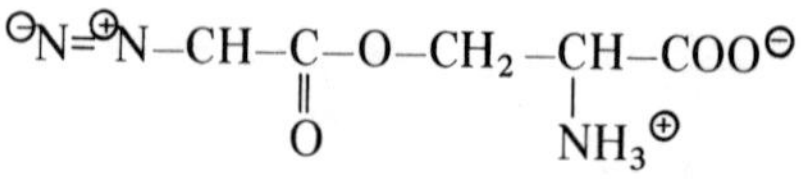

$^{\ominus}N{=}^{\oplus}N{-}CH{-}C({=}O){-}O{-}CH_2{-}CH(NH_3^{\oplus}){-}COO^{\ominus}$

Azaserine

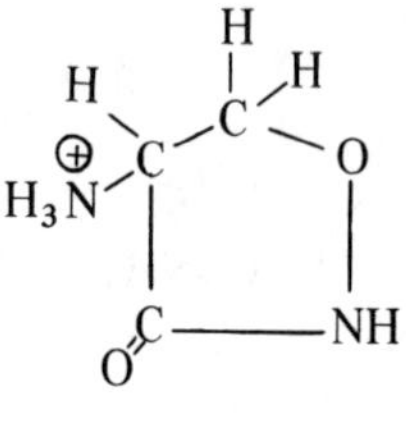

H, $^{\oplus}H_3N$–C–$CH_2$–O–NH–C(=O) (ring)

Cycloserine

$O_2N{-}C_6H_4{-}CH(OH){-}CH(CH_2OH){-}NH{-}C{=}O{-}CH{-}Cl_2$

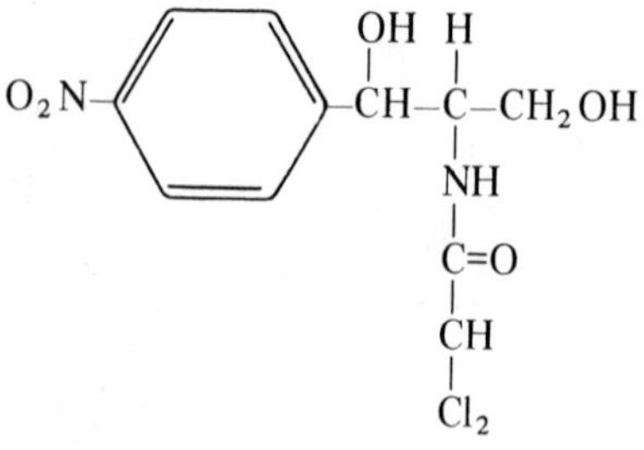

Chloramphenicol

# OPTICAL ISOMERS AND IONIZATION CONSTANTS OF AMINO ACIDS

DR. D. STRUMEYER

The proteins of all microorganisms, plants and animals contain the same twenty amino acids, except for a few additional amino acids that occur in some structural proteins, such as collagen. Upon acid hydrolysis of all proteins, free amino acids are released exclusively in the L-enantiomeric form relative to L-glyceraldehyde. Glycine is the only amino acid without an asymmetric α-carbon. D-amino acids are incorporated in cell wall structures and in several antibiotics. Except for proline, all protein amino acids are primary α-amino, α-carboxylic acid derivatives. The isomeric structures of amino acids are shown in Figure 1; alternate forms are shown in Figure 2. The structures are represented as dipolar ions or zwitter ions rather than as the uncharged species, since the amino acids exist in the fully ionized state in the crystal and in neutral solution.

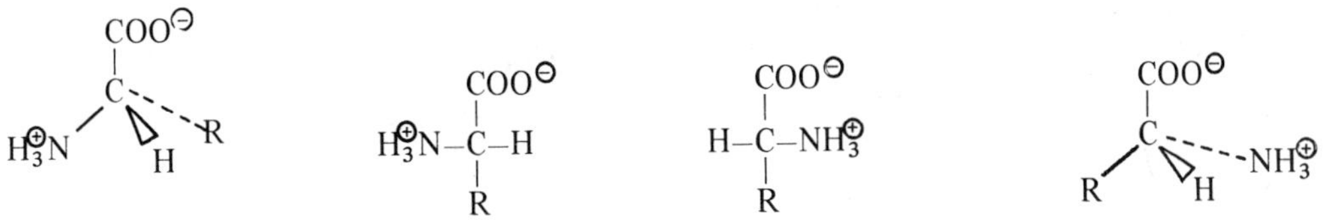

L-Amino acid D-Amino acid

FIGURE 1. Representations of optical isomers of an amino acid.

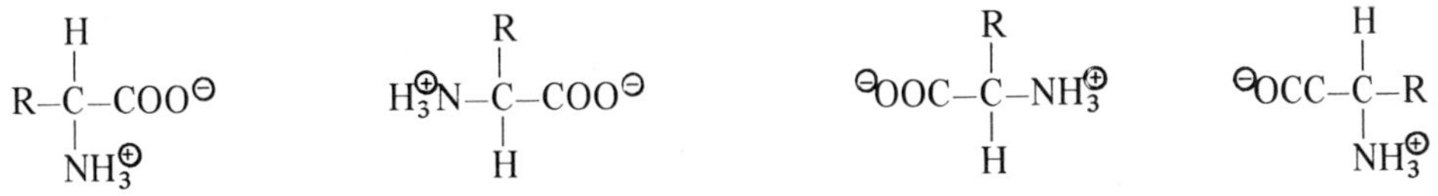

L-Amino acid D-Amino acid

FIGURE 2. Additional representations of amino acid isomers.

The titration by acid or base of a mono-amino, mono-carboxylic acid, such as alanine ($R = CH_3$), dissolved in water may be attributed to the reactions of the isolectric forms as shown in Figures 3 and 4.

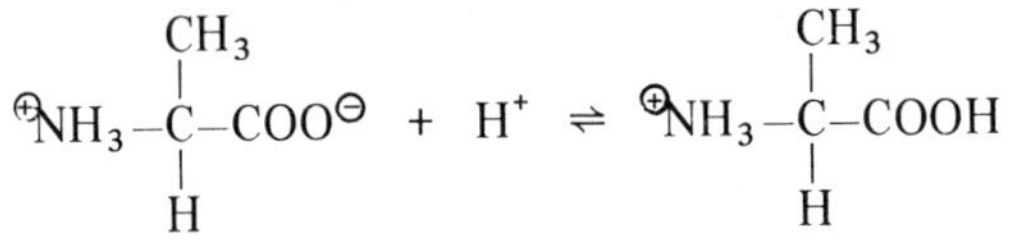

$$^{\oplus}NH_3-C(CH_3)(H)-COO^{\ominus} + H^+ \rightleftharpoons {}^{\oplus}NH_3-C(CH_3)(H)-COOH \qquad K_1 = 10^{-2.35}, pK_1 = 2.35$$

Isoelectric form (net charge = 0) Net charge = +1

FIGURE 3. Titration by acid of alanine dissolved in water.

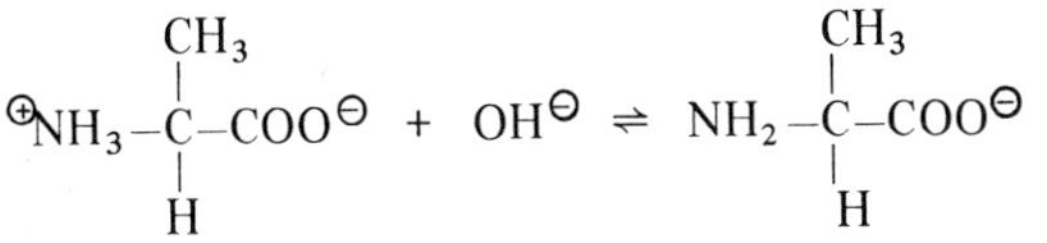

$$^{\oplus}NH_3-C(CH_3)(H)-COO^{\ominus} + OH^{\ominus} \rightleftharpoons NH_2-C(CH_3)(H)-COO^{\ominus} \qquad K_2 = 10^{-9.76}, \ pK_2 = 9.87$$

Isoelectric form (net charge = 0) Net charge = −1

FIGURE 4. Titration by base of alanine dissolved in water.

Amino acids carry a net charge, which is positive, zero or negative, depending on the pH of the solution. The terms "amphoteric" and "ampholytes" are used in referring to such substances, of which glycine is the simplest. Calculation of the pH at which amino acids carry a net charge of zero and will not migrate in an electric field, the *isoelectric point* ($pH_I$), can be made from the dissociation constants:

$$pH_I = \frac{pK_1 + pK_2}{2}$$

Applied to alanine, the calculation is made as follows:

$$pH_I = \frac{(2.35 + 9.87)}{2} = 6.11.$$

The same type of treatment may be extended to more complex ampholytes, which contain more than one carboxyl group or more than one amino group. For such substances (e.g., aspartic acid, glutamic acid, lysine, or arginine) the pK values are listed in the same sequence as the order of dissociation of the carboxyl and amino groups.

For a dicarboxylic amino acid, such as aspartic acid, the ionization sequence, beginning with the most acidic group, is shown in Figure 5.

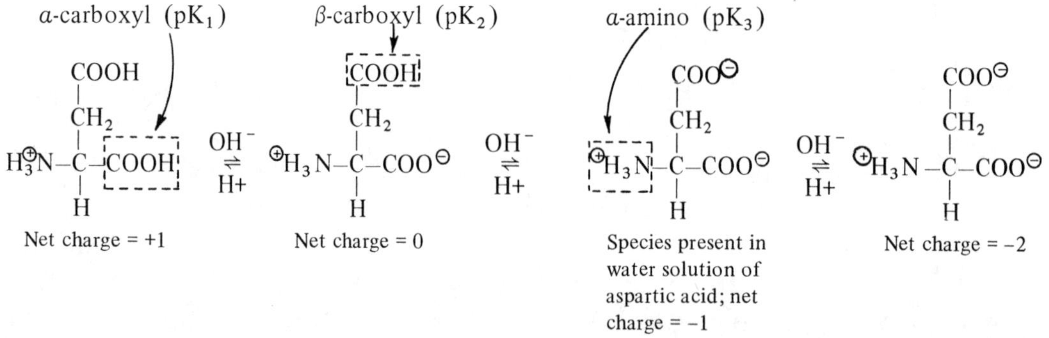

α-carboxyl ($pK_1$) β-carboxyl ($pK_2$) α-amino ($pK_3$)

$$H_3\overset{\oplus}{N}-\underset{H}{\overset{\overset{COOH}{|}\atop CH_2}{C}}-COOH \underset{H+}{\overset{OH^-}{\rightleftharpoons}} {}^{\oplus}H_3N-\underset{H}{\overset{\overset{COOH}{|}\atop CH_2}{C}}-COO^{\ominus} \underset{H+}{\overset{OH^-}{\rightleftharpoons}} {}^{\oplus}H_3N-\underset{H}{\overset{\overset{COO^{\ominus}}{|}\atop CH_2}{C}}-COO^{\ominus} \underset{H+}{\overset{OH^-}{\rightleftharpoons}} {}^{\ominus}H_3N-\underset{H}{\overset{\overset{COO^{\ominus}}{|}\atop CH_2}{C}}-COO^{\ominus}$$

Net charge = +1 | Net charge = 0 | Species present in water solution of aspartic acid; net charge = −1 | Net charge = −2

FIGURE 5. Ionization sequence for aspartic acid.

The pK values for the various groups are as follows:

$pK_1$ (α-carboxyl) = 2.10
$pK_2$ (β-carboxyl) = 3.86
$pK_3$ (α-amino) = 9.82

The isoelectric point, which must be at an acidic pH, is calculated from the dissociation constants of the two carboxyl groups:

$$pI = \frac{(2.10 + 3.86)}{2} = 2.98.$$

For a diamino (dibasic) amino acid, such as lysine, the ionization sequence also begins with the most acidic group, as shown in Figure 6.

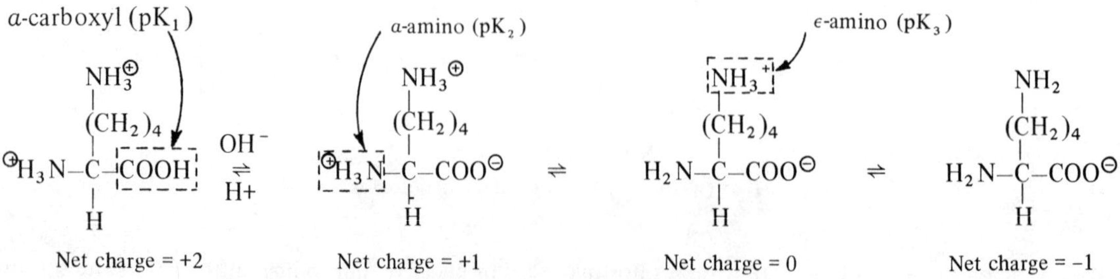

α-carboxyl ($pK_1$) α-amino ($pK_2$) ε-amino ($pK_3$)

$${}^{\oplus}H_3N-\underset{H}{\overset{\overset{NH_3^{\oplus}}{|}\atop (CH_2)_4}{C}}-COOH \underset{H+}{\overset{OH^-}{\rightleftharpoons}} {}^{\oplus}H_3N-\underset{H}{\overset{\overset{NH_3^{\oplus}}{|}\atop (CH_2)_4}{C}}-COO^{\ominus} \rightleftharpoons H_2N-\underset{H}{\overset{\overset{NH_3^{+}}{|}\atop (CH_2)_4}{C}}-COO^{\ominus} \rightleftharpoons H_2N-\underset{H}{\overset{\overset{NH_2}{|}\atop (CH_2)_4}{C}}-COO^{\ominus}$$

Net charge = +2 | Net charge = +1 | Net charge = 0 | Net charge = −1

FIGURE 6. Ionization sequence for lysine.

The pK values for the various groups are as follows:

$$pK_1\ (\alpha\text{-carboxyl}) = 2.18$$
$$pK_2\ (\alpha\text{-amino}) = 8.95$$
$$pK_3\ (\epsilon\text{-amino}) = 10.53$$

The isoelectric point is at a basic pH and is calculated from the dissociation constants of the amino groups:

$$pI = \frac{(8.95 + 10.53)}{2} = 9.74.$$

## IONIZATION CONSTANTS

### TABLE 1
### IONIZATION CONSTANTS AND ISOELECTRIC POINTS OF AMINO ACIDS AT 25°C

| Amino Acid | Acidic | | | | pI | References |
|---|---|---|---|---|---|---|
| | $pK_1$ | $pK_2$ | $pK_3$ | $pK_4$ | | |
| DL-Alanine | 2.348 | 9.866 | | | 6.107 | 1 |
| L-Arginine | 2.01 | 9.04 | 12.48 | | 10.76 | 2 |
| L-Aspartic acid | 2.10 | 3.86 | 9.82 | | 2.98 | 3 |
| L-Cystine | 1.04 | 2.05 | 8.00 | 10.25 | 5.02 | 4 |
| Diiodo-L-tyrosine | 2.12 | 6.48 | 7.82 | | 4.29 | 5, 6 |
| L-Glutamic acid | 2.10 | 4.07 | 9.47 | | 3.08 | 7 |
| Glycine | 2.350 | 9.778 | | | 6.064 | 8 |
| L-Histidine | 1.77 | 6.10 | 9.19 | | 7.64 | 3 |
| Hydroxy-L-proline | 1.92 | 9.73 | | | 5.82 | 9 |
| DL-Isoleucine | 2.318 | 9.758 | | | 6.038 | 1 |
| DL-Leucine | 2.328 | 9.744 | | | 6.036 | 1 |
| L-Lysine | 2.18 | 8.95 | 10.53 | | 9.47 | 2 |
| DL-Methionine | 2.28 | 9.21 | | | 5.74 | 10 |
| DL-Phenylalanine | 2.58 | 9.24 | | | 5.91 | 11 |
| L-Proline | 2.00 | 10.60 | | | 6.3 | 12 |
| DL-Serine | 2.21 | 9.15 | | | 5.68 | 9 |
| L-Tryptophan | 2.38 | 9.39 | | | 5.88 | 13 |
| L-Tyrosine | 2.20 | 9.11 | | | 5.63 | 6 |
| DL-Valine | 2.286 | 9.719 | | | 5.002 | 1 |

### TABLE 2
### IONIZATION CONSTANTS OF THE AMINO ACIDS IN AQUEOUS ETHANOL SOLUTIONS

| Amino Acid | Volume % Ethanol | Temperature, °C | $pK_1$ | $pK_2$ | $pK_3$ | Reference |
|---|---|---|---|---|---|---|
| Alanine | 72 | 25 | 3.55 | 10.02 | | 1 |
| Arginine | 72 | 25 | 3.34 | 9.40 | 14.1 | 1 |
| Aspartic acid | 72 | 25 | 2.85 | 5.20 | 10.51 | 1 |
| Glutamic acid | 72 | 25 | 3.16 | 5.63 | 10.75 | 2 |
| Glycine | 10 | 19.5 | 2.66 | 9.82 | | 2 |
| Glycine | 40 | 19.5 | 2.96 | 9.76 | | 2 |
| Glycine | 72 | 25 | 3.46 | 9.82 | | 1 |
| Glycine | 90 | 19.5 | 3.79 | 9.99 | | 2 |
| Histidine | 72 | 25 | 3.00 | 5.85 | 9.45 | 1 |
| Isoleucine | 72 | 25 | 3.69 | 9.81 | | 1 |
| Lysine | 48 | 25 | 2.75 | 8.95 | 10.53 | 1 |
| Lysine | 84 | 25 | 3.56 | 8.95 | 10.49 | 1 |
| Proline | 72 | 25 | 3.04 | 10.55 | | 1 |
| Valine | 72 | 25 | 3.60 | 9.73 | | 1 |

## TABLE 3
## IONIZATION CONSTANTS OF THE AMINO ACIDS IN AQUEOUS FORMALDEHYDE SOLUTIONS

| Amino Acid | Mole % Formaldehyde[a1] | | | | |
|---|---|---|---|---|---|
| | 0.99 | 3.95 | 5.60 | 10.0 | 17.9 |
| DL-Alanine | 8.36 | 7.42 | 6.96[a2] | 6.56 | 6.10 |
| L-Arginine | | 3.45[b3] | 3.40[b4] | | |
| L-Aspartic acid | | | 7.21[b4] | $\leqq$3.8[c5] | |
| L-Aspartic acid | | | | 6.85[d5] | |
| L-Glutamic acid | | | 6.91[b4] | $\leqq$4.2[c5] | |
| L-Glutamic acid | | | | 6.8[d5] | |
| Glycine | 7.16 | 6.08 | 5.92[a2] | 5.34 | 5.04 |
| L-Histidine | | 7.90[d3] | 7.90[d4] | | |
| Hydroxy-L-proline | | | 7.19[b4] | | |
| L-Leucine | 8.44 | 7.50 | 6.92[b4] | 6.62 | 6.20 |

| Amino Acid | Mole % Formaldehyde[a1] | | | | |
|---|---|---|---|---|---|
| | 0.99 | 3.95 | 5.60 | 10.0 | 17.9 |
| DL-Leucine | 8.44 | 7.48 | | 6.60 | 6.20 |
| L-Lysine | | 7.35[d3] | 7.15[d4] | | |
| L-Phenylalanine | | | 6.62[b4] | 5.9[c5] | |
| DL-Phenylalanine | 8.09 | 7.16 | 6.80[a2] | 6.35 | 6.13 |
| L-Proline | | | 7.78[b4] | | |
| DL-Serine | 6.66 | 5.74 | 5.64[a2] | | 4.94 |
| L-Tryptophan | | | 6.88[b4] | | |
| L-Tyrosine | | | 7.50[b4] | 6.2[c5] | |
| L-Tyrosine | | | | 9[d5] | |
| DL-Valine | 8.52 | 7.65 | 7.47[a2] | | 6.52 |

[a] $pK_2$ at 22°C.
[b] $pK_2$ at 30°C.
[c] $pK_2$ at 25°C.
[d] $pK_3$ at 30°C.

Data taken from: *Handbook of Chemistry and Physics,* 53rd ed., p. C-741, R. C. Weast, Ed. Copyright 1972, The Chemical Rubber Co., Cleveland, Ohio.

## REFERENCES


### Table 1

1. Smith, P. K., Taylor, A. C., and Smith, E. R. B., *J. Biol. Chem.*, *122*, 109 (1937–38).
2. Schmidt, C. L. A., Kirk, P. L., and Appleman, W. K., *J. Biol. Chem.*, *88*, 285 (1930).
3. Greenstein, J. P., *J. Biol. Chem.*, *93*, 479 (1931).
4. Borsook, H., Ellis, E. L., and Huffman, H. M., *J. Biol. Chem.*, *117*, 281 (1937).
5. Dalton, J. B., Kirk, P. L., and Schmidt, C. L. A., *J. Biol. Chem.*, *88*, 589 (1930).
6. Winnek, P. S., and Schmidt, C. L. A., *J. Gen. Physiol.*, *18*, 889 (1935).
7. Simms, H. S., *J. Gen. Physiol.*, *11*, 629 (1928); *12*, 231 (1929).
8. Owen, B. B., *J. Amer. Chem. Soc.*, *56*, 24 (1934).
9. Kirk, P. L., and Schmidt, C. L. A., *J. Biol. Chem.*, *81*, 237 (1929).
10. Emerson, O. H., Kirk, P. L., and Schmidt, C. L. A., *J. Biol. Chem.*, *92*, 449 (1931).
11. Miyamoto, S., and Schmidt, C. L. A., *J. Biol. Chem.*, *90*, 165 (1931).
12. McCay, C. M., and Schmidt, C. L. A., *J. Gen. Physiol.*, *9*, 333 (1926).
13. Schmidt, C. L. A., Appleman, W. K., and Kirk, P. L., *J. Biol. Chem.*, *85*, 137 (1929–30).

### Table 2

1. Jukes, T. H., and Schmidt, C. L. A., *J. Biol. Chem.*, *105*, 359 (1934).
2. Michaelis, L., and Mizutani, M., *Z. Phys. Chem.*, *116*, 135 (1925).

### Table 3

1. Dunn, M. S., and Weiner, J. G., *J. Biol. Chem.*, *117*, 381 (1937).
2. Dunn, M. S., and Loshakoff, A., *J. Biol. Chem.*, *113*, 691 (1936).
3. Levy, M., *J. Biol. Chem.*, *109*, 365 (1935).
4. Levy, M., and Silberman, D. E., *J. Biol. Chem.*, *118*, 723 (1937).
5. Harris, L. J., *Proc. Roy. Soc. London Ser. B*, *95*, 440 (1923–24).

# AMINO ACID COMPOSITIONS OF SELECTED PROTEINS

DR. GERALD REECK*

Table 1 (Complete Data) presents the amino acid compositions of polypeptides of at least fifty residues and proteins judged to be mutually non-homologous. Thus, within the limitations of our knowledge of the structure of these proteins, only one member of a set of homologous proteins is included in the table. Values for amino acid residues that result from the chemical modification of encoded residues are included in the values for the encoded residues from which they are derived; e.g., hydroxyproline is included in proline. In addition, the values for aspartic acid include asparagine and the values for glutamic acid include glutamine. Only those compositions in which half-cystine has been determined as cysteic acid following performic acid oxidation or as modified cysteine, e.g., carboxymethylcysteine, following reduction and chemical modification are included. Compositions in which half-cystine has been determined by extrapolation to time zero of acid hydrolysis or as cysteic acid following air oxidation will be found in Table 2 (Incomplete Data). Such half-cystine values are deemed unreliable. Unless the authors have done so, the reported values for threonine and serine are corrected for destruction during acid hydrolysis, and values for valine and isoleucine are corrected for incomplete liberation at times of hydrolysis less than 72 hours. The corrections factors used were determined by the compiler by visually averaging the time-course amino acid compositions of crotonase,[43] erythrocuprein,[59] lysine monooxygenase,[126] and nicotinamide dehydrogenase.[140] The correction factors by which a reported value for a given time of hydrolysis is multiplied are as follows:

| | Hours of Acid Hydrolysis | | | | | | |
|---|---|---|---|---|---|---|---|
| **Amino Acid** | **18** | **20** | **22** | **24** | **30** | **32** | **48** |
| Threonine | 1.011 | 1.014 | 1.016 | 1.018 | 1.027 | 1.031 | 1.060 |
| Serine | 1.045 | 1.053 | 1.060 | 1.068 | 1.093 | 1.101 | 1.175 |
| Valine | 1.182 | 1.156 | 1.136 | 1.121 | 1.086 | 1.074 | 1.026 |
| Isoleucine | 1.250 | 1.181 | 1.143 | 1.111 | 1.066 | 1.057 | 1.015 |

Correction factors for other times of hydrolysis were determined by linear interpolation, using the above factors. For the sake of uniformity, all values in both tables are given to 0.01 mole percent, regardless of the number of significant figures reported by the authors.

Table 2. (Incomplete Data) presents the amino acid compositions of polypeptides and proteins where a reliable value for half-cystine or tryptophan or both is lacking. A dash (–) is placed in every position for which a reliable value is unavailable; this is not to be confused with the value of 0.00 mole percent. The data in the table are calculated as mole percent of those amino acids listed for a given protein.

*Data previously published in *Handbook of Biochemistry,* 2nd ed., pp. C-281 – C-293, H. A. Sober, Ed. Copyright 1970, The Chemical Rubber Co., Cleveland, Ohio.

# TABLE 1
# COMPLETE DATA

| Protein | Amino Acid, mole % [a] | | | | | | | | | | | | | | | | | | References |
|---|---|---|---|---|---|---|---|---|---|---|---|---|---|---|---|---|---|---|---|
| | Asp[c] | Thr | Ser | Glu[d] | Pro[e] | Gly | Ala | Cys | Val | Met | Ile | Leu | Tyr | Phe | Trp | Lys | His | Arg | |
| **Acetoacetate decarboxylase,** *Clostridium acetobutylicum* | 10.47 | 5.56 | 5.16 | 7.15 | 7.11 | 5.45 | 7.51 | 1.08 | 6.64 | 3.00 | 7.11 | 9.60 | 5.70 | 3.00 | 0.36 | 7.29 | 3.29 | 4.51 | *1* |
| ***O*-Acetylserine sulfhydrylase A,** *Salmonella typhimurium* | 7.93 | 7.32 | 5.19 | 11.60 | 4.88 | 10.07 | 9.46 | 0.21 | 6.10 | 1.53 | 7.32 | 9.77 | 1.83 | 2.14 | 1.53 | 7.93 | 0.92 | 4.27 | *2* |
| **Actin,** rabbit skeletal muscle | 9.38 | 7.88 | 6.66 | 10.89 | 5.11 | 7.67 | 8.17 | 1.28 | 5.09 | 3.38 | 7.51 | 6.95 | 3.94 | 3.17 | 0.99 | 5.05 | 2.00 | 4.87 | *3* |
| **Acyl carrier protein,** *E. coli*[b] | 11.69 | 7.79 | 3.90 | 23.38 | 1.30 | 5.19 | 9.09 | 0.00 | 9.09 | 1.30 | 9.09 | 6.49 | 1.30 | 2.60 | 0.00 | 5.19 | 1.30 | 1.30 | *4* |
| **Adenosine deaminase,** *Aspergillus oryzae* | 13.50 | 4.92 | 6.63 | 12.78 | 4.18 | 6.93 | 9.56 | 0.72 | 6.19 | 1.79 | 3.88 | 9.24 | 1.74 | 4.57 | 2.31 | 3.09 | 2.97 | 5.00 | *5* |
| **Adenylate kinase,** bovine liver mitochondria | 8.77 | 4.59 | 5.44 | 9.78 | 6.19 | 7.55 | 8.43 | 1.70 | 6.52 | 2.48 | 4.35 | 9.64 | 2.32 | 3.33 | 4.50 | 7.57 | 2.01 | 4.84 | *6* |
| **Adenylosuccinase,** *Neurospora* | 9.15 | 5.87 | 5.60 | 10.56 | 4.28 | 10.06 | 11.97 | 2.00 | 6.96 | 1.96 | 5.01 | 7.97 | 2.64 | 3.64 | 1.32 | 6.19 | 1.32 | 3.50 | *7* |
| **Adrenal iron-sulfur protein,** bovine adrenal cortex | 15.02 | 7.26 | 6.26 | 11.02 | 1.08 | 7.18 | 6.42 | 4.39 | 5.08 | 4.42 | 7.23 | 9.95 | 0.81 | 2.97 | 0.00 | 5.01 | 2.82 | 3.09 | *8* |
| **Aequorin,** *Aequorea* (jellyfish) | 13.09 | 4.96 | 4.86 | 12.31 | 4.03 | 8.00 | 6.89 | 1.95 | 4.99 | 2.33 | 5.26 | 7.02 | 3.75 | 4.14 | 2.71 | 7.57 | 2.31 | 3.82 | *9* |
| **Alanine aminotransferase,** rat liver | 6.94 | 2.95 | 4.56 | 15.78 | 6.94 | 7.98 | 9.89 | 3.23 | 8.37 | 2.95 | 3.90 | 10.36 | 3.04 | 4.37 | 0.38 | 2.09 | 0.95 | 5.32 | *10* |
| **Aldehyde dehydrogenase,** bakers' yeast | 11.79 | 4.82 | 5.42 | 10.04 | 4.92 | 8.55 | 8.95 | 1.09 | 8.20 | 1.24 | 7.91 | 7.01 | 2.98 | 4.13 | 0.55 | 7.06 | 1.99 | 3.33 | *11* |
| **Alkaline phosphatase,** *E. coli* | 10.87 | 4.93 | 8.93 | 10.39 | 4.63 | 10.10 | 14.61 | 0.94 | 5.05 | 1.69 | 3.25 | 8.86 | 2.38 | 1.88 | 0.93 | 5.91 | 1.93 | 2.71 | *12* |
| **Allergen 1-B,** common rye grass pollen | 11.25 | 7.34 | 5.15 | 8.63 | 5.54 | 12.07 | 7.69 | 2.66 | 5.88 | 0.86 | 4.34 | 3.91 | 3.65 | 3.43 | 2.53 | 11.25 | 1.29 | 2.53 | *13* |
| **Allergen (antigen K),** ragweed pollen | 13.47 | 6.55 | 5.07 | 8.70 | 4.85 | 9.55 | 6.34 | 2.64 | 7.04 | 2.85 | 6.79 | 5.25 | 1.94 | 4.00 | 2.67 | 4.70 | 3.18 | 4.40 | *14* |
| **D-Amino acid oxidase,** hog kidney | 7.57 | 6.65 | 3.94 | 10.22 | 3.65 | 9.68 | 7.94 | 3.51 | 7.65 | 1.25 | 4.18 | 10.69 | 3.39 | 3.94 | 2.27 | 4.39 | 2.80 | 6.27 | *15* |
| **L-Amino acid oxidase,** *Crotalus adamanteus* venom | 9.64 | 6.48 | 5.72 | 9.69 | 4.27 | 6.96 | 7.29 | 1.59 | 5.18 | 1.53 | 6.25 | 6.73 | 5.34 | 5.76 | 1.37 | 7.30 | 2.61 | 6.30 | *16* |
| **α-Amylase,** rat pancreas | 13.16 | 5.91 | 7.46 | 7.73 | 4.19 | 9.27 | 5.45 | 2.62 | 7.88 | 1.66 | 5.37 | 6.49 | 3.29 | 5.32 | 1.80 | 5.83 | 2.48 | 4.09 | *17* |
| **Apoferritin,** horse spleen | 11.32 | 3.64 | 5.10 | 15.02 | 1.67 | 6.51 | 9.06 | 1.25 | 4.50 | 1.82 | 2.21 | 15.49 | 3.49 | 4.55 | 0.55 | 5.08 | 3.20 | 5.54 | *18* |
| **L-Arabinose isomerase,** *E. coli* | 11.37 | 6.45 | 3.38 | 10.74 | 4.07 | 7.92 | 8.42 | 1.22 | 6.83 | 2.99 | 4.73 | 9.71 | 2.25 | 4.45 | 1.92 | 4.67 | 4.13 | 4.76 | *19* |
| **Arginine decarboxylase,** *E. coli* | 11.81 | 5.91 | 5.63 | 10.55 | 4.92 | 7.17 | 8.30 | 0.98 | 7.17 | 3.52 | 4.50 | 7.88 | 3.94 | 4.22 | 1.83 | 3.94 | 3.23 | 4.50 | *20* |
| **Arginine kinase,** lobster tail muscle | 11.15 | 4.95 | 4.84 | 11.57 | 3.56 | 8.34 | 7.33 | 1.60 | 6.59 | 2.43 | 4.89 | 8.37 | 2.70 | 5.73 | 0.92 | 7.89 | 2.49 | 4.66 | *21* |
| **Asparaginase,** *E. coli* | 14.63 | 9.76 | 4.88 | 6.83 | 3.90 | 8.78 | 9.76 | 0.49 | 9.76 | 1.95 | 3.90 | 6.83 | 3.90 | 2.93 | 0.98 | 6.83 | 0.98 | 2.93 | *22* |
| **L-Aspartate β-decarboxylase,** *Alcaligenes faecalis* | 11.33 | 3.77 | 6.28 | 10.79 | 4.69 | 6.65 | 9.83 | 0.48 | 5.73 | 2.33 | 4.97 | 11.70 | 4.10 | 4.90 | 0.81 | 4.38 | 1.13 | 6.15 | *23* |
| **Aspartate transcarbamylase catalytic subunit,** *E. coli* | 11.46 | 5.67 | 6.05 | 8.98 | 4.17 | 4.94 | 10.45 | 0.44 | 7.21 | 2.56 | 4.81 | 11.87 | 2.84 | 3.84 | 0.69 | 5.28 | 3.72 | 5.01 | *24* |
| **Aspartate transcarbamylase regulatory subunit,** *E. coli*[b] | 13.16 | 4.61 | 7.24 | 10.53 | 4.61 | 3.95 | 6.58 | 2.63 | 7.89 | 1.32 | 7.89 | 9.87 | 1.97 | 3.29 | 0.00 | 6.58 | 2.63 | 5.26 | *25* |
| **Aspartokinase-homoserine dehydrogenase,** *E. coli* | 9.86 | 3.85 | 6.83 | 10.89 | 4.82 | 7.75 | 11.05 | 1.34 | 8.02 | 2.37 | 5.42 | 10.30 | 2.15 | 3.74 | 0.36 | 3.97 | 1.65 | 5.64 | *26* |
| **Azurin,** *Pseudomonas fluorescens*[b] | 14.06 | 7.81 | 7.81 | 7.81 | 3.13 | 8.59 | 5.47 | 2.34 | 7.81 | 4.69 | 3.13 | 7.81 | 1.56 | 4.69 | 0.78 | 8.59 | 3.13 | 0.78 | *27* |
| **Bacteriochlorophyll-protein complex,** *Chloropseudomonas ethylicum* | 10.93 | 4.09 | 7.53 | 9.23 | 4.91 | 11.26 | 6.02 | 0.62 | 9.55 | 0.98 | 6.58 | 5.73 | 2.65 | 4.88 | 1.73 | 5.20 | 2.26 | 5.86 | *28* |
| **Basic plasma protein,** human | 7.62 | 3.32 | 6.39 | 16.09 | 7.13 | 5.53 | 9.95 | 7.25 | 4.30 | 2.58 | 1.60 | 6.02 | 1.23 | 3.81 | 0.00 | 7.49 | 0.00 | 9.71 | *29* |
| **Carbamate kinase,** *Streptococcus faecalis* | 9.94 | 5.14 | 4.57 | 13.64 | 4.98 | 8.82 | 11.15 | 0.94 | 8.50 | 1.93 | 7.16 | 8.56 | 1.61 | 1.90 | 0.68 | 7.54 | 1.93 | 1.00 | *30* |
| **Carbonic anhydrase B,** horse erythrocytes | 12.16 | 4.55 | 10.50 | 9.56 | 6.59 | 8.69 | 5.61 | 0.94 | 7.45 | 0.75 | 3.46 | 8.05 | 3.01 | 4.14 | 1.96 | 6.96 | 3.65 | 1.96 | *31* |
| **Carboxypeptidase $A_{\alpha}^{Val}$,** bovine pancreas[b] | 9.51 | 8.52 | 10.49 | 8.20 | 3.28 | 7.54 | 6.56 | 0.66 | 5.25 | 0.98 | 6.56 | 7.54 | 6.23 | 5.25 | 2.30 | 4.92 | 2.62 | 3.61 | *32* |
| **κ-Casein component A-1,** bovine milk | 7.31 | 8.57 | 7.49 | 16.18 | 11.74 | 1.86 | 8.09 | 1.14 | 6.29 | 1.08 | 6.71 | 4.61 | 4.67 | 2.46 | 1.20 | 5.63 | 1.92 | 3.06 | *33* |
| **Catalase,** rat liver | 12.75 | 4.81 | 4.68 | 10.29 | 7.84 | 6.64 | 7.53 | 0.89 | 6.60 | 2.32 | 3.74 | 6.11 | 3.57 | 5.70 | 1.47 | 5.84 | 3.61 | 5.61 | *34* |
| **Cellulase,** *Myrothecium verrucaria* | 11.81 | 9.12 | 8.63 | 7.18 | 5.45 | 10.28 | 8.39 | 3.05 | 5.38 | 0.93 | 3.05 | 5.72 | 4.23 | 3.58 | 3.51 | 4.31 | 1.65 | 3.74 | *35* |
| **Ceruloplasmin,** human plasma | 11.34 | 7.23 | 5.80 | 10.92 | 4.79 | 7.14 | 4.71 | 1.26 | 6.47[h] | 2.18 | 5.55[h] | 6.64 | 5.97 | 4.54 | 2.27 | 5.71 | 3.61 | 3.87 | *36* |
| **Chorionic gonadotropin,** human pregnancy urine | 7.33 | 7.31 | 9.50 | 7.86 | 12.19 | 5.30 | 5.33 | 8.58 | 7.86 | 1.71 | 2.41 | 6.33 | 2.62 | 2.42 | 0.49 | 4.37 | 1.83 | 6.57 | *37* |
| **Cilia outer-fiber protein,** *Tetrahymena pyriformis* | 10.66 | 5.22 | 6.12 | 13.27 | 4.42 | 9.07 | 6.35 | 1.47 | 6.01 | 2.95 | 5.56 | 7.48 | 3.29 | 4.42 | 0.79 | 5.78 | 2.49 | 4.65 | *38* |
| **Cobramine B,** cobra venom | 9.62 | 5.77 | 3.85 | 0.00 | 7.69 | 3.85 | 3.85 | 11.54 | 11.54 | 3.85 | 1.92 | 9.62 | 5.77 | 1.92 | 0.00 | 15.38 | 0.00 | 3.85 | *39* |

# TABLE 1 (Continued)
## COMPLETE DATA

| Protein | Amino Acid, mole % [a] | | | | | | | | | | | | | | | | | References |
|---|---|---|---|---|---|---|---|---|---|---|---|---|---|---|---|---|---|---|
| | Asp[c] | Thr | Ser | Glu[d] | Pro[e] | Gly | Ala | Cys | Val | Met | Ile | Leu | Tyr | Phe | Trp | Lys | His | Arg | |
| **α1-Collagen,** rat skin | 4.48 | 1.96 | 3.81 | 7.25 | 22.11 | 33.95 | 11.25 | 0.00 | 2.04 | 0.63 | 0.76 | 1.96 | 0.18 | 1.16 | 0.00 | 3.27[f] | 0.21 | 4.96 | *40* |
| **Concanavalin A,** jack bean | 15.19 | 7.80 | 11.92 | 4.92 | 4.29 | 6.78 | 7.59 | 0.00 | 6.94 | 0.80 | 6.15 | 7.39 | 2.90 | 4.50 | 1.89 | 5.00 | 3.28 | 2.66 | *41* |
| **Corticosteroid-binding globulin,** rabbit serum | 9.43 | 7.30 | 7.86 | 11.37 | 6.02 | 7.08 | 7.92 | 2.27 | 6.78 | 1.28 | 3.64 | 9.79 | 2.97 | 4.34 | 2.01 | 3.95 | 2.32 | 3.69 | *42* |
| **Crotonase,** bovine liver | 7.30 | 4.93 | 4.61 | 12.88 | 3.52 | 9.35 | 13.39 | 2.05 | 6.53 | 2.88 | 5.96 | 7.94 | 1.92 | 4.48 | 0.58 | 8.14 | 0.38 | 3.14 | *43* |
| **β-Crustacyanin,** lobster shell | 12.98 | 6.51[i] | 6.80[i] | 8.80 | 4.74 | 5.05 | 8.64 | 3.35 | 7.17 | 0.44 | 4.27 | 4.28 | 7.96 | 7.63 | 0.99 | 5.93 | 1.27 | 3.17 | *44* |
| **$β_s$-Crystallin,** bovine lens | 8.97 | 2.89 | 6.23 | 12.46 | 3.84 | 8.04 | 5.63 | 3.65 | 4.76 | 3.21 | 4.04 | 5.65 | 6.58 | 5.67 | 1.99 | 4.94 | 4.05 | 7.41 | *45* |
| **Cystathionase,** rat liver | 7.61 | 5.87[i] | 7.09[i] | 9.93 | 4.77 | 6.60 | 9.14 | 5.16 | 5.85[i] | 1.97 | 4.25[i] | 10.88 | 1.80 | 4.42 | 0.86 | 6.21 | 3.38 | 4.21 | *46* |
| **Cystathionine γ-synthetase,** *Salmonella typhimurium* | 8.87 | 5.57[j] | 5.91[j] | 9.09 | 4.96 | 9.32 | 10.70 | 1.55 | 6.21[j] | 1.11 | 4.49[j] | 12.92 | 2.44 | 2.94 | 1.66 | 3.55 | 3.05 | 5.66 | *47* |
| **Cysteamine to hypotaurine enzyme,** horse kidney | 11.44 | 5.11 | 6.06 | 11.57 | 3.50 | 5.79 | 9.69 | 1.62 | 5.65 | 2.96 | 3.90 | 12.79 | 3.36 | 2.42 | 1.88 | 4.58 | 3.63 | 4.04 | *48* |
| **Cytochrome $b_{562}$,** *E. coli*[b] | 17.27 | 3.64 | 1.82 | 13.64 | 3.64 | 2.73 | 15.45 | 0.00 | 3.64 | 2.73 | 2.73 | 9.09 | 1.82 | 1.82 | 0.00 | 14.55 | 1.82 | 3.64 | *49* |
| **Cytochrome c,** human[b] | 7.69 | 6.73 | 1.92 | 9.62 | 3.85 | 12.50 | 5.77 | 1.92 | 2.88 | 2.88 | 7.69 | 5.77 | 4.81 | 2.88 | 0.96 | 17.31 | 2.88 | 1.92 | *50* |
| **Deoxyribonuclease,** bovine pancreas | 12.28 | 5.60 | 11.20 | 7.52 | 3.50 | 3.56 | 8.37 | 1.48 | 10.00 | 1.52 | 4.44 | 8.70 | 5.90 | 4.31 | 1.46 | 3.42 | 2.24 | 4.52 | *51* |
| **Deoxyribonuclease inhibitor II,** calf spleen | 9.09 | 6.91 | 6.41 | 9.95 | 4.89 | 8.16 | 7.74 | 1.60 | 6.74 | 3.66 | 6.33 | 7.79 | 3.44 | 3.51 | 1.25 | 5.65 | 2.28 | 4.58 | *52* |
| **Diphosphopyridine nucleosidase,** pig brain | 12.61 | 7.56 | 5.04 | 13.03 | 4.20 | 10.92 | 10.08 | 0.00 | 5.46 | 2.10 | 7.56 | 7.14 | 0.00 | 0.42 | 0.00 | 10.08 | 2.10 | 1.68 | *53* |
| **DNA polymerase,** T4-infected *E. coli* | 11.97 | 3.07 | 6.89 | 11.97 | 4.24 | 6.46 | 5.83 | 1.59 | 5.83 | 3.71 | 9.11 | 6.25 | 4.98 | 1.27 | 1.27 | 9.00 | 1.69 | 4.87 | *54* |
| **Egg white cross-reacting protein,** chicken eggs | 9.47 | 6.73[j] | 7.80[j] | 11.39 | 5.32 | 5.09 | 5.76 | 1.68 | 7.88 | 2.08 | 6.44 | 9.11 | 3.90 | 5.03 | 0.73 | 5.99 | 1.82 | 3.79 | *55* |
| **Encephalitogenic protein,** guinea pig brain | 6.77 | 4.15[j] | 9.84[j] | 6.90 | 7.52 | 14.81 | 8.19 | 0.00 | 1.55[j] | 0.99 | 1.68[j] | 5.88 | 2.28 | 4.80 | 0.98 | 7.55 | 5.92 | 10.19 | *56* |
| **Enolase,** white skeletal muscle of rainbow trout | 11.61 | 3.92 | 5.82 | 10.45 | 3.67 | 9.50 | 10.84 | 0.92 | 6.53 | 1.41 | 6.53 | 7.84 | 2.61 | 3.32 | 0.35 | 9.42 | 2.44 | 2.82 | *57* |
| **Enterotoxin C,** *Staphylococcus* | 17.87 | 6.05 | 6.08 | 8.01 | 2.57 | 6.05 | 3.01 | 0.81 | 7.57 | 2.80 | 4.16 | 6.66 | 6.93 | 4.19 | 0.61 | 12.97 | 2.43 | 1.25 | *58* |
| **Erythrocuprein,** human erythrocytes | 12.45 | 5.23 | 6.13 | 8.70 | 3.50 | 16.16 | 6.63 | 1.58 | 10.15[h] | 0.00 | 5.85[h] | 6.10 | 0.00 | 2.66 | 0.00 | 7.09 | 5.11 | 2.66 | *59* |
| **Ethanolamine deaminase,** choline-fermenting clostridium | 10.43 | 5.73 | 4.83 | 8.98 | 3.02 | 8.76 | 8.53 | 1.40 | 7.95 | 2.66 | 6.32 | 9.07 | 3.21 | 3.02 | 1.81 | 8.53 | 1.63 | 4.11 | *60* |
| **Exo-β-D-(1 → 3)-Glucanase,** *Basidiomycte* species QM 806 | 11.31 | 8.84 | 12.16 | 6.44 | 4.37 | 12.26 | 9.09 | 0.95 | 7.07 | 0.80 | 5.39 | 4.62 | 4.52 | 3.07 | 3.42 | 1.67 | 1.57 | 2.45 | *61* |
| **Factor X,** bovine plasma | 9.20 | 6.77 | 6.86 | 14.09 | 4.30 | 9.10 | 6.94 | 5.00 | 5.78 | 1.15 | 2.61 | 6.75 | 2.16 | 4.76 | 2.59 | 4.34 | 2.27 | 5.35 | *62* |
| **Ferredoxin,** alfalfa[b] | 9.28 | 6.19 | 8.25 | 16.49 | 3.09 | 7.22 | 9.28 | 5.15 | 9.28 | 0.00 | 4.12 | 6.19 | 4.12 | 2.06 | 1.03 | 5.15 | 2.06 | 1.03 | *63* |
| **Fetuin,** calf serum | 9.12 | 6.88 | 7.24 | 9.37 | 9.32 | 6.69 | 9.23 | 3.35 | 11.14 | 0.00 | 4.12 | 7.41 | 1.88 | 2.99 | 0.61 | 4.56 | 2.82 | 3.26 | *64* |
| **Flagellin,** *Proteus vulgaris* flagella | 17.31 | 8.24 | 8.52 | 11.26 | 0.27 | 8.24 | 10.44 | 0.00 | 7.97 | 0.55 | 6.04 | 8.79 | 0.82 | 2.20 | 0.00 | 6.04 | 0.00 | 3.30 | *65* |
| **Flavodoxin,** *Clostridium pasteurianum* | 13.66 | 2.44 | 9.37 | 12.04 | 2.54 | 11.40 | 9.36 | 0.96 | 10.25 | 2.57 | 3.23 | 8.03 | 1.58 | 2.29 | 2.46 | 6.57 | 0.00 | 1.27 | *66* |
| **Follicle-stimulating hormone,** human pituitary | 11.39 | 4.71 | 6.00 | 8.58 | 5.55 | 7.18 | 8.30 | 6.00 | 6.39 | 0.95 | 2.19 | 8.47 | 3.76 | 4.60 | 1.23 | 8.41 | 2.64 | 3.65 | *67* |
| **Fraction 1 protein,** spinach beet chloroplasts | 8.12 | 6.48 | 4.97 | 9.58 | 5.38 | 8.99 | 8.30 | 1.82 | 8.30 | 1.82 | 4.61 | 8.71 | 4.11 | 4.61 | 1.37 | 5.47 | 2.69 | 4.65 | *68* |
| **Fructose 1,6-diphosphatase,** rabbit liver | 10.71 | 5.86 | 6.61 | 6.78 | 4.35 | 8.37 | 8.87 | 1.67 | 8.70 | 2.85 | 5.86 | 7.87 | 4.35 | 3.35 | 0.00 | 9.79 | 1.17 | 2.85 | *69* |
| **Fructose 1,6-diphosphate aldolase,** baker's yeast | 11.15 | 5.24 | 5.47 | 10.98 | 4.62 | 8.46 | 9.49 | 1.30 | 6.32 | 1.95 | 6.86 | 6.40 | 3.43 | 4.12 | 1.29 | 7.17 | 3.07 | 2.69 | *70* |
| **Fructose 1,6-diphosphate aldolase A,** rabbit muscle | 7.88 | 5.63 | 5.27 | 11.25 | 5.49 | 7.88 | 10.83 | 0.56 | 6.33 | 0.84 | 5.84 | 9.85 | 3.09 | 2.04 | 0.84 | 8.51 | 3.16 | 4.71 | *70* |
| **Fumurase,** swine heart muscle | 10.04 | 5.67 | 5.22 | 9.87 | 4.54 | 8.11 | 11.34 | 0.68 | 7.43 | 3.52 | 5.50 | 8.34 | 2.27 | 3.57 | 0.45 | 7.32 | 3.06 | 3.06 | *71* |
| **Galactokinase,** *E. coli* | 9.99 | 3.30 | 3.71 | 13.87 | 3.90 | 8.94 | 10.67 | 2.79 | 9.78 | 2.71 | 5.96 | 8.23 | 2.55 | 3.47 | 0.51 | 4.28 | 1.79 | 3.55 | *72* |
| **β-Galactosidase,** *E. coli* | 10.50 | 5.52 | 5.69 | 12.11 | 5.70 | 7.22 | 7.96 | 1.64 | 6.35 | 2.08 | 4.07 | 9.40 | 3.09 | 3.80 | 3.00 | 2.48 | 3.11 | 6.30 | *73* |
| **2S $γ_2$-Globulin,** human plasma | 10.33 | 8.65 | 7.88 | 8.26 | 3.75 | 10.71 | 7.19 | 2.68 | 8.26 | 0.00 | 3.14 | 8.49 | 3.60 | 3.21 | 1.30 | 7.27 | 1.68 | 3.60 | *74* |
| **Glucose oxidase,** *Penicillium amagasakiense* | 12.65 | 6.12[k] | 5.56[k] | 9.67 | 3.55 | 8.94 | 9.99 | 0.56 | 7.90[k] | 2.42 | 4.67[k] | 9.35 | 3.14 | 4.35 | 2.01 | 4.83 | 1.45 | 2.82 | *75* |
| **Glucose 6-phosphate dehydrogenase,** human erythrocytes | 10.85 | 3.97 | 4.07 | 13.80 | 5.82 | 3.89 | 3.88 | 1.68 | 5.21 | 2.86 | 4.37 | 9.07 | 5.45 | 6.46 | 2.29 | 6.22 | 3.06 | 7.02 | *76* |
| **Glutamate-aspartate transaminase,** pig heart mitochondria | 9.39 | 3.81 | 5.27 | 10.10 | 4.45 | 8.51 | 8.84 | 1.18 | 6.67 | 2.93 | 6.55 | 7.19 | 2.86 | 4.63 | 2.36 | 7.35 | 2.86 | 5.04 | *77* |

## TABLE 1 (Continued)
## COMPLETE DATA

| Protein | Amino Acid, mole %[a] | | | | | | | | | | | | | | | | | | References |
|---|---|---|---|---|---|---|---|---|---|---|---|---|---|---|---|---|---|---|---|
| | Asp[c] | Thr | Ser | Glu[d] | Pro[e] | Gly | Ala | Cys | Val | Met | Ile | Leu | Tyr | Phe | Trp | Lys | His | Arg | |
| **Glutamate dehydrogenase,** bovine liver | 10.23 | 5.47 | 6.11 | 9.19 | 4.59 | 9.29 | 7.44 | 2.52 | 6.09 | 2.57 | 6.79 | 6.05 | 3.41 | 4.43 | 0.89 | 6.56 | 2.67 | 5.70 | *78* |
| **Glutamine phosphoribosyl pyrophosphate amidotransferase,** pigeon liver | 8.91 | 5.00 | 5.65 | 8.26 | 5.43 | 8.91 | 9.13 | 3.91 | 6.96 | 2.17 | 4.13 | 10.87 | 2.83 | 3.04 | 1.30 | 5.87 | 2.17 | 5.43 | *79* |
| **Glutamine synthetase,** *E. coli* | 9.56 | 8.10 | 5.71 | 8.31 | 5.19 | 7.65 | 9.18 | 0.87 | 6.19 | 3.37 | 5.36 | 6.90 | 3.28 | 4.74 | 0.83 | 5.38 | 5.40 | 3.97 | *80* |
| **Glyceraldehyde 3-phosphate dehydrogenase,** lobster tail muscle[b] | 9.61 | 6.01 | 7.51 | 7.21 | 3.60 | 9.01 | 9.61 | 1.50 | 11.41 | 3.00 | 5.41 | 5.41 | 2.70 | 4.50 | 0.90 | 8.41 | 1.50 | 2.70 | *81* |
| **L-Glycerol 3-phosphate dehydrogenase,** rat skeletal muscle | 8.53 | 5.09 | 5.09 | 9.50 | 4.47 | 9.78 | 8.26 | 3.21 | 9.16 | 0.73 | 7.69 | 9.24 | 1.14 | 4.44 | 0.74 | 7.39 | 2.53 | 3.01 | *82* |
| **Glycogen phosphorylase,** rabbit skeletal muscle | 11.26 | 3.47 | 2.22 | 12.48 | 3.51 | 2.76 | 4.53 | 0.96 | 6.19 | 2.87 | 5.67 | 9.24 | 5.98 | 5.78 | 2.34 | 6.76 | 3.32 | 10.65 | *83* |
| **$\beta_1$-Glycoprotein,** human plasma | 11.25 | 7.86 | 4.15 | 15.89 | 4.36 | 4.95 | 5.12 | 2.69 | 4.74 | 0.38 | 5.06 | 8.19 | 4.95 | 5.22 | 1.78 | 7.49 | 1.62 | 4.31 | *84* |
| **Glycoprotein 1,** kidney bean seeds | 15.40 | 8.56 | 14.43 | 7.52 | 3.82 | 5.34 | 5.95 | 0.06 | 8.68 | 0.98 | 4.72 | 5.28 | 3.14 | 5.40 | 1.96 | 3.78 | 1.39 | 3.58 | *85* |
| **Growth hormone,** human pituitary[b] | 10.64 | 5.32 | 9.57 | 13.83 | 4.26 | 4.26 | 3.72 | 2.13 | 3.72 | 1.60 | 4.26 | 13.30 | 4.26 | 6.91 | 0.53 | 4.79 | 1.60 | 5.32 | *86* |
| **Haptoglobin alpha 2,** human serum[b] | 16.20 | 3.52 | 2.11 | 11.27 | 7.75 | 8.45 | 5.63 | 4.23 | 8.45 | 0.00 | 3.52 | 4.23 | 7.04 | 0.00 | 1.41 | 10.56 | 2.82 | 2.82 | *87* |
| **Hemagglutinin,** *Helix pomatia* (snail) | 11.44 | 5.88 | 10.89 | 7.62 | 6.51 | 4.66 | 5.42 | 2.92 | 7.35 | 1.50 | 5.98 | 6.07 | 4.63 | 2.12 | 4.48 | 5.25 | 1.34 | 5.92 | *88* |
| **Hemerythrin,** *Golfingia gouldii* (Sipunculid worm)[b] | 15.04 | 3.54 | 3.54 | 8.85 | 3.54 | 6.19 | 4.42 | 0.88 | 3.54 | 0.88 | 7.96 | 7.08 | 4.42 | 7.96 | 3.54 | 9.73 | 6.19 | 2.65 | *89* |
| **Hemocyanin,** *Octopus vulgaris* | 11.96 | 5.27 | 4.93 | 9.54 | 5.64 | 5.16 | 6.64 | 2.34 | 6.16 | 2.60 | 5.15 | 9.03 | 4.33 | 5.75 | 1.60 | 4.71 | 5.32 | 3.86 | *90* |
| **Hemoglobin alpha chain,** human[b] | 8.51 | 6.38 | 7.80 | 3.55 | 4.96 | 4.96 | 14.89 | 0.71 | 9.22 | 1.42 | 0.00 | 12.77 | 2.13 | 4.96 | 0.71 | 7.80 | 7.09 | 2.13 | *91* |
| **Hirudin,** *Hirudo medicinalis* (leech) | 14.81 | 5.93 | 5.93 | 19.26 | 4.44 | 13.33 | 1.48 | 8.89 | 4.44 | 0.00 | 2.96 | 5.93 | 2.96 | 2.96 | 0.00 | 5.19 | 1.48 | 0.00 | *92* |
| **L-Histidine ammonia-lyase,** *Pseudomonas* | 8.04 | 3.54 | 6.00 | 10.03 | 4.58 | 9.12 | 16.27 | 1.06 | 8.89 | 2.26 | 4.88 | 11.12 | 1.26 | 2.14 | 0.37 | 2.17 | 2.91 | 5.37 | *93* |
| **Histidine decarboxylase,** *Lactobacillus* 30a | 12.69 | 5.08 | 6.67 | 10.27 | 6.31 | 9.21 | 8.97 | 0.60 | 5.33 | 3.22 | 6.02 | 5.96 | 4.63 | 2.45 | 1.77 | 6.55 | 0.61 | 3.66 | *94* |
| **Histidinol dehydrogenase,** *Salmonella typhimurium* | 8.45 | 7.18 | 8.14 | 10.63 | 6.11 | 5.80 | 14.38 | 1.54 | 7.81 | 0.96 | 5.19 | 9.29 | 1.61 | 2.74 | 0.50 | 3.55 | 1.04 | 5.09 | *95* |
| **Histone IV,** calf thymus[b] | 4.90 | 6.86 | 1.96 | 5.88 | 0.98 | 16.67 | 6.86 | 0.00 | 8.82 | 0.98 | 5.88 | 7.84 | 3.92 | 1.96 | 0.00 | 10.78 | 1.96 | 13.73 | *96* |
| **Hormone-binding polypeptide,** pig posterior pituitary | 5.72 | 2.27 | 8.20 | 14.79 | 7.05 | 15.59 | 7.81 | 12.96 | 2.26 | 1.09 | 2.24 | 8.08 | 1.10 | 3.21 | 0.00 | 2.20 | 0.00 | 5.43 | *97* |
| **Hyaluronidase,** bovine testes | 11.20 | 5.42 | 8.28 | 8.35 | 6.08 | 6.14 | 5.76 | 3.06 | 7.26 | 1.30 | 4.18 | 8.92 | 4.02 | 4.03 | 3.08 | 6.12 | 2.19 | 4.61 | *98* |
| **β-Hydroxydecanoyl thioester dehydrase,** *E. coli* | 10.42 | 4.47 | 2.42 | 8.73 | 4.32 | 12.90 | 7.17 | 1.07 | 8.21 | 3.25 | 3.86 | 10.79 | 2.30 | 5.09 | 1.53 | 7.20 | 1.35 | 4.90 | *99* |
| **Imidazoylacetolphosphate: L-glutamate aminotransferase,** *Salmonella typhimurium* | 10.03 | 6.29 | 4.58 | 13.79 | 5.69 | 5.84 | 9.71 | 2.43 | 8.23 | 0.69 | 4.56 | 11.17 | 3.14 | 2.88 | 1.16 | 2.86 | 0.77 | 6.19 | *100* |
| **Immobilizing antigen,** *Paramecium aurelia* | 11.76 | 15.00 | 8.44 | 6.15 | 2.37 | 7.35 | 12.31 | 9.34 | 4.99 | 0.41 | 2.82 | 4.01 | 2.81 | 1.65 | 3.50 | 5.09 | 0.70 | 1.31 | *101* |
| **γG Immunoglobulin kappa chain,** human myeloma plasma[b] | 7.94 | 8.41 | 14.95 | 11.68 | 4.67 | 6.07 | 6.07 | 2.34 | 7.01 | 1.40 | 2.80 | 7.01 | 4.21 | 3.74 | 1.40 | 7.01 | 0.93 | 2.34 | *102* |
| **Indole-3-glycerol phosphate synthetase,** *E. coli* | 10.36 | 3.10 | 5.36 | 12.32 | 3.60 | 6.89 | 13.15 | 1.48 | 8.12 | 1.07 | 5.19 | 10.79 | 3.41 | 3.36 | 0.50 | 3.88 | 2.38 | 5.03 | *103* |
| **Invertase (external),** *Saccharomyces* | 14.91 | 7.04 | 9.55 | 9.63 | 5.44 | 5.95 | 5.70 | 0.42 | 5.78 | 1.76 | 3.35 | 6.95 | 5.44 | 6.70 | 2.76 | 5.03 | 1.34 | 2.26 | *104* |
| **Isocitrate dehydrogenase,** *Azobacter vinelandii* | 10.77 | 6.00 | 6.46 | 9.38 | 5.08 | 6.92 | 11.69 | 0.46 | 6.46 | 1.23 | 6.00 | 9.23 | 2.92 | 2.62 | 1.08 | 8.00 | 2.00 | 3.69 | *105* |
| **Isocitrate lyase,** *Pseudomonas indigofera* | 9.65 | 5.87 | 4.39 | 11.77 | 3.94 | 8.62 | 13.03 | 1.03 | 7.64 | 2.10 | 4.06 | 6.86 | 4.10 | 3.46 | 1.05 | 5.58 | 2.28 | 4.61 | *106* |
| **Isomerase of histidine biosynthesis,** *Salmonella typhimurium* | 9.46 | 4.64 | 4.50 | 11.93 | 3.95 | 10.04 | 9.79 | 1.52 | 10.48 | 0.98 | 5.55 | 9.86 | 2.50 | 2.18 | 1.05 | 4.31 | 1.78 | 5.47 | *107* |
| **β-Isopropylmalate dehydrogenase,** *Salmonella typhimurium* | 10.55 | 3.46 | 5.75 | 10.46 | 5.13 | 9.12 | 11.51 | 1.33 | 5.13 | 2.35 | 6.61 | 9.12 | 2.69 | 3.46 | 1.08 | 4.41 | 2.02 | 5.85 | *108* |
| **α-Isopropylmalate synthetase,** *Neurospora* | 11.00 | 5.15[i] | 6.77 | 10.77[i] | 5.77 | 7.46 | 8.23 | 1.62 | 8.08[i] | 1.23 | 5.31[i] | 7.69 | 2.69 | 3.62 | 1.38 | 6.08 | 1.69 | 5.46 | *109* |
| **Keratin,** sheep wool[b] | 9.97 | 4.20 | 8.22 | 17.48 | 0.87 | 5.42 | 8.92 | 4.02 | 3.67 | 0.35 | 3.50 | 13.29 | 3.32 | 2.62 | 0.00 | 5.77 | 0.70 | 7.69 | *110* |
| **β-Ketoacyl acyl carrier protein synthetase,** *E. coli* | 7.31 | 5.22 | 6.89 | 10.02 | 3.97 | 10.86 | 10.65 | 1.67 | 7.52 | 4.38 | 5.85 | 6.68 | 3.34 | 3.13 | 0.42 | 5.43 | 2.51 | 4.18 | *111* |

## TABLE 1 (Continued)
## COMPLETE DATA

| Protein | Amino Acid, mole %[a] | | | | | | | | | | | | | | | | | | References |
|---|---|---|---|---|---|---|---|---|---|---|---|---|---|---|---|---|---|---|---|
| | Asp[c] | Thr | Ser | Glu[d] | Pro[e] | Gly | Ala | Cys | Val | Met | Ile | Leu | Tyr | Phe | Trp | Lys | His | Arg | |
| **$\Delta^5$-3-Ketosteroid isomerase,** *Pseudomonas testosteroni* | 9.73 | 5.43 | 4.05 | 9.61 | 3.93 | 7.40 | 17.19 | 0.00 | 11.42 | 2.03 | 3.01 | 6.57 | 2.25 | 6.33 | 0.00 | 3.22 | 2.32 | 5.50 | *112* |
| **D-Lactate dehydrogenase,** *E. coli* | 9.90 | 5.28 | 5.09 | 10.94 | 3.77 | 7.17 | 8.58 | 1.13 | 7.17 | 2.64 | 5.37 | 10.28 | 3.77 | 3.77 | 0.80 | 5.66 | 2.73 | 5.94 | *113* |
| **L-Lactate dehydrogenase $M_4$,** bovine muscle | 10.23 | 4.63 | 8.95 | 8.59 | 4.03 | 8.15 | 5.31 | 1.01 | 10.20 | 3.03 | 7.33 | 9.47 | 1.96 | 2.08 | 1.92 | 8.25 | 2.32 | 2.56 | *114* |
| **β-Lactoglobulin A,** bovine milk[b] | 9.88 | 4.94 | 4.32 | 15.43 | 4.94 | 1.85 | 8.64 | 3.09 | 6.17 | 2.47 | 6.17 | 13.58 | 2.47 | 2.47 | 1.23 | 9.26 | 1.23 | 1.85 | *115* |
| **Lactollin,** bovine milk | 11.15 | 2.03 | 8.28 | 12.10 | 9.43 | 3.09 | 1.10 | 2.23 | 5.13 | 0.00 | 6.13 | 8.30 | 6.10 | 4.16 | 2.26 | 9.30 | 4.05 | 5.17 | *116* |
| **Lactoperoxidase,** bovine milk | 11.58 | 4.57 | 4.89 | 9.79 | 6.85 | 6.69 | 6.53 | 2.61 | 4.73 | 1.96 | 4.57 | 11.09 | 2.45 | 5.06 | 2.61 | 5.38 | 2.28 | 6.36 | *117* |
| **Leucine aminopeptidase,** bovine lens | 9.50 | 4.95 | 5.81 | 10.70 | 5.55 | 8.51 | 10.65 | 2.62 | 6.55 | 2.08 | 5.45 | 7.77 | 1.82 | 3.93 | 1.33 | 6.90 | 1.62 | 4.25 | *118* |
| **Leucine binding protein,** *E. coli* | 11.58 | 4.85[i] | 3.72[i] | 10.90 | 4.22 | 10.01 | 12.86 | 0.44 | 7.91[i] | 1.19 | 6.11[i] | 6.65 | 3.73 | 2.93 | 0.94 | 8.55 | 1.30 | 2.12 | *119* |
| **Levansucrase,** *Bacillus subtilis* | 13.93 | 8.74[i] | 9.84[i] | 8.74 | 4.10 | 8.47 | 4.92 | 0.55 | 5.74[i] | 1.37 | 4.64[i] | 6.56 | 4.64 | 4.37 | 1.09 | 8.74 | 1.91 | 1.64 | *120* |
| **$\beta_1$-Lipoprotein (low density),** human serum | 10.64 | 6.36 | 8.15 | 12.40 | 3.76 | 4.82 | 6.12 | 0.66 | 6.13 | 1.81 | 6.12 | 11.81 | 2.97 | 5.27 | 0.56 | 6.73 | 2.51 | 3.19 | *121* |
| **Lipoprotein R-Gln (high density),** human serum | 4.06 | 8.19 | 8.23 | 19.76 | 5.23 | 4.25 | 6.81 | 1.50 | 7.68 | 1.48 | 1.50 | 10.32 | 4.52 | 4.99 | 0.00 | 11.50 | 0.00 | 0.00 | *122* |
| **Lipoprotein R-Thr (high density),** human serum | 9.80 | 3.80 | 6.25 | 18.10 | 3.69 | 4.43 | 7.58 | 0.00 | 5.13 | 1.50 | 0.00 | 14.81 | 2.84 | 2.29 | 3.00 | 7.77 | 2.26 | 6.76 | *122* |
| **β-Lipotropin,** sheep pituitary[b] | 4.44 | 4.44 | 5.56 | 17.78 | 5.56 | 8.89 | 14.44 | 0.00 | 2.22 | 2.22 | 1.11 | 6.67 | 3.33 | 3.33 | 1.11 | 11.11 | 2.22 | 5.56 | *123* |
| **Lipoyl dehydrogenase,** pig heart | 8.73 | 4.94[l] | 4.29[l] | 9.19 | 3.16 | 10.81 | 10.20 | 2.17 | 9.55[l] | 2.17 | 7.84[l] | 6.82 | 1.56 | 2.90 | 0.43 | 7.58 | 2.10 | 5.55 | *124* |
| **Luciferase chain α,** *Photobacterium fischeri* | 12.28 | 8.07 | 5.00 | 12.53 | 3.07 | 7.25 | 5.85 | 1.96 | 7.25 | 3.07 | 6.11 | 5.85 | 4.18 | 5.29 | 1.39 | 4.18 | 2.50 | 4.18 | *125* |
| **L-Lysine monooxygenase,** *Pseudomonas fluorescens* | 9.80 | 5.55 | 5.03 | 8.43 | 6.21 | 9.73 | 9.47 | 0.85 | 6.86 | 3.20 | 4.57 | 7.90 | 2.48 | 5.23 | 2.16 | 4.70 | 3.46 | 4.38 | *126* |
| **Lysozyme,** chicken egg white[b] | 16.28 | 5.43 | 7.75 | 3.88 | 1.55 | 9.30 | 9.30 | 6.20 | 4.65 | 1.55 | 4.65 | 6.20 | 2.33 | 2.33 | 4.65 | 4.65 | 0.78 | 8.53 | *127* |
| **Lysozyme,** T4 bacteriophage[b] | 13.41 | 6.71 | 3.66 | 7.93 | 1.83 | 6.71 | 9.15 | 1.22 | 5.49 | 3.05 | 6.10 | 9.76 | 3.66 | 3.05 | 1.83 | 7.93 | 0.61 | 7.93 | *128* |
| **Malic dehydrogenase (mitochondrial),** chicken heart | 7.91 | 6.70 | 6.39 | 9.13 | 6.24 | 9.28 | 10.05 | 2.13 | 7.31 | 1.98 | 6.09 | 8.68 | 1.37 | 4.57 | 0.00 | 7.46 | 1.67 | 3.04 | *129* |
| **Malic dehydrogenase (supernatant),** chicken heart | 10.29 | 4.54 | 5.90 | 9.08 | 4.39 | 8.93 | 9.08 | 1.21 | 8.02 | 2.12 | 6.35 | 8.93 | 2.42 | 3.63 | 1.82 | 8.62 | 1.66 | 3.03 | *129* |
| **Metapyrocatechase,** *Pseudomonas arvilla* | 11.62 | 5.40 | 3.19 | 9.08 | 4.01 | 8.10 | 7.94 | 0.98 | 7.53 | 3.27 | 3.27 | 11.46 | 2.13 | 4.91 | 1.23 | 4.91 | 4.83 | 6.14 | *130* |
| **β-Methlyaspartase,** *Clostridium tetanomorphum* | 12.08 | 5.03 | 2.13 | 10.07 | 2.68 | 9.40 | 9.62 | 2.24 | 8.28 | 3.58 | 6.71 | 6.82 | 3.02 | 3.58 | 0.45 | 7.38 | 1.79 | 5.15 | *131* |
| **Microtubule protein,** *Pseudocentrotus depressus* sperm flagella | 10.05 | 5.81[i] | 5.52[i] | 11.82 | 4.73 | 7.09 | 6.90 | 2.86 | 6.50[i] | 3.65 | 5.32[i] | 7.59 | 2.86 | 3.84 | 1.67 | 6.40 | 2.27 | 5.12 | *132* |
| **Mitochondrial structural protein,** beef heart | 7.15 | 4.69[j] | 4.81[j] | 2.47 | 6.78 | 8.51 | 9.12 | 8.38 | 6.78[j] | 2.22 | 6.17[j] | 8.38 | 3.45 | 4.81 | 3.58 | 6.91 | 1.36 | 4.44 | *133* |
| **Monoamine oxidase,** bovine plasma | 8.44 | 5.45 | 6.84 | 11.74 | 8.48 | 8.57 | 7.51 | 0.92 | 7.72 | 1.81 | 2.87 | 8.41 | 3.34 | 5.71 | 1.75 | 2.70 | 3.06 | 4.67 | *134* |
| **Multienzyme complex in tryptophan pathway,** *Neurospora crassa* | 9.49 | 5.33 | 8.52 | 11.47 | 5.68 | 8.02 | 9.84 | 1.01 | 5.83 | 1.37 | 4.01 | 9.79 | 1.98 | 3.55 | 1.01 | 5.38 | 2.23 | 5.48 | *135* |
| **Myeloperoxidase,** human leucocytes | 12.59 | 5.68 | 5.19 | 9.14 | 7.65 | 6.42 | 6.42 | 3.21 | 4.20 | 2.96 | 3.95 | 10.62 | 2.22 | 4.20 | 3.21 | 2.72 | 0.99 | 8.64 | *136* |
| **Myosin,** bovine heart | 10.40 | 4.97 | 4.52 | 19.10 | 2.60 | 4.29 | 8.70 | 0.84 | 4.29 | 2.60 | 4.18 | 10.51 | 1.81 | 3.50 | 0.40 | 10.28 | 1.70 | 5.31 | *137* |
| **Neurotoxin II,** *Androctonus australis* venom | 12.61 | 4.76 | 3.29 | 6.41 | 4.44 | 10.89 | 4.84 | 12.29 | 6.33 | 0.00 | 1.52 | 2.71 | 10.92 | 1.54 | 2.02 | 7.76 | 3.04 | 4.64 | *138* |
| **Neurotoxin α,** Egyptian cobra venom[b] | 11.48 | 11.48 | 6.56 | 11.48 | 6.56 | 8.20 | 0.00 | 13.11 | 1.64 | 0.00 | 4.92 | 1.64 | 1.64 | 0.00 | 1.64 | 9.84 | 3.28 | 6.56 | *139* |
| **Nicotinamide deamidase,** rabbit liver | 7.92 | 5.35 | 5.96 | 9.92 | 6.13 | 8.19 | 7.92 | 0.78 | 7.02 | 3.01 | 3.73 | 11.82 | 3.12 | 4.07 | 2.17 | 7.08 | 2.45 | 3.34 | *140* |
| **Nuclease,** *Staphylococcus aureaus* V8[b] | 9.40 | 6.71 | 3.36 | 12.08 | 4.03 | 6.71 | 9.40 | 0.00 | 6.04 | 2.68 | 3.36 | 8.05 | 4.70 | 2.01 | 0.67 | 15.44 | 2.01 | 3.36 | *141* |
| **Ornithine aminotransferase,** rat liver | 7.49 | 4.55 | 4.00 | 23.68 | 5.09 | 7.37 | 6.79 | 1.21 | 5.79 | 3.06 | 4.82 | 8.10 | 3.21 | 2.61 | 1.12 | 5.06 | 2.03 | 4.00 | *142* |
| **Parathyroid hormone,** bovine[b] | 12.05 | 0.00 | 8.43 | 14.46 | 3.61 | 4.82 | 7.23 | 0.00 | 9.64 | 2.41 | 3.61 | 8.43 | 1.20 | 2.41 | 1.20 | 10.84 | 3.61 | 6.02 | *143* |
| **Parathyroid polypeptide 2,** bovine | 5.84 | 3.87 | 4.40 | 9.63 | 4.17 | 10.10 | 12.92 | 0.00 | 6.38 | 0.00 | 4.38 | 11.69 | 1.85 | 1.05 | 0.00 | 11.31 | 2.96 | 9.46 | *144* |
| **Penicillinase,** *Staphylococcus aureus*[b] | 15.18 | 5.06 | 7.39 | 7.00 | 3.50 | 4.67 | 7.00 | 0.00 | 6.23 | 1.17 | 7.39 | 8.56 | 5.06 | 2.72 | 0.00 | 16.73 | 0.78 | 1.56 | *145* |
| **Pepsinogen,** bovine stomach mucosa | 11.14 | 7.39 | 13.83 | 8.91 | 4.27 | 9.68 | 4.42 | 1.63 | 6.92 | 1.05 | 8.76 | 6.95 | 4.89 | 4.17 | 1.60 | 2.23 | 0.55 | 1.62 | *146* |
| **Peroxidase B,** pineapple | 16.31 | 6.37 | 11.53 | 5.61 | 3.75 | 7.68 | 6.21 | 1.91 | 7.22 | 1.75 | 4.84 | 7.73 | 1.75 | 6.89 | 0.41 | 5.22 | 1.25 | 3.58 | *147* |

## TABLE 1 (Continued)
## COMPLETE DATA

| Protein | Amino Acid, mole %[a] | | | | | | | | | | | | | | | | | References |
|---|---|---|---|---|---|---|---|---|---|---|---|---|---|---|---|---|---|---|
| | Asp[c] | Thr | Ser | Glu[d] | Pro[e] | Gly | Ala | Cys | Val | Met | Ile | Leu | Tyr | Phe | Trp | Lys | His | Arg | |
| **Phosphoenolpyruvate carboxykinase,** pig liver mitochondria | 7.96 | 4.65 | 4.65 | 10.54 | 8.47 | 9.69 | 8.35 | 2.19 | 6.70 | 3.00 | 4.28 | 8.82 | 1.74 | 4.00 | 1.97 | 3.96 | 1.75 | 7.29 | *148* |
| **Phosphoenolpyruvate carboxylase,** *Salmonella typhimurium* | 10.88 | 4.76 | 2.49 | 14.17 | 5.78 | 5.89 | 8.05 | 0.23 | 10.20 | 0.14 | 5.21 | 14.17 | 0.23 | 2.15 | 0.12 | 6.57 | 1.93 | 7.03 | *149* |
| **Phosphofructokinase,** rabbit skeletal muscle | 8.92 | 7.02 | 4.61 | 9.22 | 3.15 | 10.41 | 8.55 | 2.00 | 7.53 | 2.83 | 5.78 | 8.18 | 2.04 | 3.74 | 1.59 | 5.20 | 2.45 | 6.78 | *150* |
| **Phosphoglucomutase,** rabbit skeletal muscle | 11.11 | 5.31 | 5.97 | 8.79 | 4.48 | 8.96 | 8.79 | 1.00 | 6.80 | 1.99 | 7.46 | 7.46 | 2.82 | 5.47 | 0.66 | 6.80 | 1.82 | 4.31 | *151* |
| **6-Phosphogluconate dehydrogenase,** *Candida utilis* | 11.19 | 3.64 | 4.43 | 11.99 | 4.37 | 12.38 | 8.21 | 0.83 | 5.55 | 1.40 | 6.12 | 8.01 | 3.66 | 3.49 | 1.07 | 8.62 | 1.36 | 3.69 | *152* |
| **Phosphoglycerate dehydrogenase,** *E. coli* | 9.72 | 4.38 | 5.30 | 9.83 | 4.32 | 8.38 | 10.95 | 0.87 | 7.07 | 1.68 | 6.84 | 12.03 | 2.16 | 3.31 | 0.32 | 5.82 | 3.25 | 3.79 | *153* |
| **Phosphoglyceric acid mutase,** bakers' yeast | 10.87 | 3.74 | 6.09 | 10.31 | 6.50 | 6.18 | 9.93 | 0.10 | 6.76 | 0.51 | 4.37 | 11.20 | 3.50 | 2.33 | 1.82 | 9.23 | 1.63 | 4.93 | *154* |
| **Phospholipase A,** *Crotalus adamanteus* venom | 11.28 | 4.89 | 4.89 | 9.02 | 6.02 | 9.02 | 5.64 | 11.28 | 4.14 | 0.75 | 4.14 | 4.14 | 6.02 | 3.76 | 2.63 | 6.02 | 1.88 | 4.51 | *155* |
| **Phosvitin fraction II,** ling roes | 2.66 | 1.39 | 65.51[g] | 3.76 | 1.23 | 0.74 | 3.90 | 0.00 | 2.26 | 0.00 | 1.28 | 4.29 | 0.00 | 0.38 | 0.00 | 5.50 | 0.13 | 6.97 | *156* |
| **Prealbumin,** human plasma | 7.05 | 9.06 | 8.59 | 9.92 | 6.36 | 7.80 | 9.57 | 0.86 | 8.72 | 0.54 | 3.90 | 5.40 | 3.79 | 4.13 | 1.61 | 6.45 | 3.28 | 3.00 | *157* |
| **Proinsulin,** porcine pancreas[b] | 4.76 | 2.38 | 3.57 | 16.67 | 4.76 | 13.10 | 8.33 | 7.14 | 5.95 | 0.00 | 2.38 | 13.10 | 4.76 | 3.57 | 0.00 | 2.38 | 2.38 | 4.76 | *158* |
| **Proteinase zymogen,** Group A streptococcus | 12.67 | 4.04 | 9.08 | 9.45 | 3.45 | 10.93 | 7.73 | 0.25 | 6.94[j] | 1.64 | 6.10[j] | 5.17 | 5.64 | 4.04 | 1.11 | 7.24 | 1.87 | 2.68 | *159* |
| **Protocatechuate 3,4-dioxygenase,** *Pseudomonas aeruginosa* | 12.75 | 5.08 | 3.64 | 9.11 | 7.54 | 8.17 | 8.47 | 1.69 | 4.70 | 0.82 | 6.46 | 8.15 | 3.34 | 4.38 | 2.80 | 3.23 | 3.44 | 6.23 | *160* |
| **Psoriatic scale protein,** human psoriasis patients | 15.14 | 5.22 | 8.65 | 8.82 | 3.04 | 6.15 | 6.17 | 1.57 | 1.62 | 3.85 | 5.71 | 7.46 | 3.23 | 6.03 | 0.72 | 10.52 | 3.10 | 3.00 | *161* |
| **Purine nucleoside phosphorylase,** *Bacillus cereus* vegatative cells | 9.18 | 4.59 | 4.59 | 12.62 | 2.29 | 12.62 | 8.03 | 1.36 | 8.03 | 2.29 | 4.59 | 6.88 | 3.44 | 4.59 | 1.15 | 5.74 | 4.59 | 3.44 | *162* |
| **Pyrocatechase,** *Pseudomonas fluorescens* | 11.85 | 5.59 | 4.31 | 13.50 | 4.11 | 9.45 | 10.27 | 0.36 | 6.73 | 0.72 | 4.11 | 9.31 | 2.96 | 4.47 | 0.60 | 1.65 | 3.25 | 6.74 | *163* |
| **Pyruvate kinase,** bakers' yeast | 10.76 | 12.21 | 7.81 | 7.48 | 4.71 | 6.48 | 8.40 | 0.99 | 7.78 | 1.51 | 5.67 | 7.27 | 2.61 | 2.96 | 0.64 | 6.82 | 1.45 | 4.45 | *164* |
| **R17 coat protein,** bacteriophage R17[b] | 10.85 | 6.98 | 10.08 | 8.53 | 4.65 | 6.98 | 10.85 | 1.55 | 10.85 | 1.55 | 6.20 | 5.43 | 3.10 | 3.10 | 1.55 | 4.65 | 0.00 | 3.10 | *165* |
| **Red protein,** bovine milk | 9.68 | 5.13 | 5.99 | 9.91 | 4.75 | 7.12 | 9.50 | 5.48 | 6.64 | 0.73 | 2.33 | 9.34 | 3.06 | 3.92 | 2.20 | 7.31 | 1.53 | 5.37 | *166* |
| **Ribonuclease,** bovine pancreas[b] | 12.10 | 8.06 | 12.10 | 9.68 | 3.23 | 2.42 | 9.68 | 6.45 | 7.26 | 3.23 | 2.42 | 1.61 | 4.84 | 2.42 | 0.00 | 8.06 | 3.23 | 3.23 | *167* |
| **Ribonuclease T1,** *Aspergillus oryzae*[b] | 14.42 | 5.77 | 14.42 | 8.65 | 3.85 | 11.54 | 6.73 | 3.85 | 7.69 | 0.00 | 1.92 | 2.88 | 8.65 | 3.85 | 0.96 | 0.96 | 2.88 | 0.96 | *168* |
| **Ribulose 1,5-diphosphate carboxylase,** *Hydrogenomonas facilis* | 11.44 | 5.54 | 4.43 | 9.22 | 4.79 | 8.87 | 9.22 | 2.57 | 7.01 | 3.02 | 4.79 | 7.76 | 3.06 | 4.43 | 1.51 | 3.55 | 2.48 | 6.30 | *169* |
| **L-Ribulose 5-phosphate 4-epimerase,** *E. coli* | 8.56 | 7.27 | 4.81 | 11.34 | 4.92 | 8.13 | 9.63 | 1.28 | 8.02 | 2.99 | 5.56 | 7.17 | 3.64 | 2.67 | 1.28 | 4.71 | 4.28 | 3.74 | *170* |
| **RNA polymerase α chain,** *E. coli* | 9.07 | 5.43 | 5.01 | 15.31 | 4.99 | 6.25 | 7.05 | 1.02 | 9.03 | 1.43 | 7.12 | 11.54 | 1.41 | 1.22 | 0.35 | 4.54 | 2.35 | 6.88 | *171* |
| **Rubredoxin,** *Peptostreptococcus elsdenii*[b] | 19.23 | 3.85 | 1.92 | 5.77 | 3.85 | 9.62 | 13.46 | 7.69 | 5.77 | 3.85 | 3.85 | 1.92 | 5.77 | 3.85 | 1.92 | 7.69 | 0.00 | 0.00 | *172* |
| **Serum albumin, bovine** | 9.36 | 5.65 | 4.59 | 13.25 | 4.95 | 2.83 | 8.13 | 6.18 | 6.18 | 0.71 | 2.30 | 10.78 | 3.36 | 4.59 | 0.35 | 9.89 | 3.00 | 3.89 | *173* |
| **Soybean trypsin inhibitor $F_3$** | 15.08 | 2.51 | 5.53 | 24.62 | 3.02 | 5.03 | 2.51 | 6.03 | 1.01 | 5.03 | 3.52 | 6.53 | 0.00 | 1.01 | 1.51 | 10.55 | 2.51 | 4.02 | *174* |
| **Streptokinase,** ***Streptococcus*** | 16.11 | 7.13 | 5.75 | 11.10 | 4.80 | 4.99 | 5.37 | 0.00 | 5.49 | 0.61 | 5.31 | 9.61 | 4.84 | 3.59 | 0.26 | 7.99 | 2.14 | 4.91 | *175* |
| **Subtilisin Carlsberg,** ***Bacillus subtilis***[b] | 10.22 | 6.93 | 11.68 | 4.38 | 3.38 | 12.77 | 14.96 | 0.00 | 11.31 | 1.82 | 3.65 | 5.84 | 4.74 | 1.46 | 0.36 | 3.28 | 1.82 | 1.46 | *176* |
| **Sulfate-binding protein,** ***Salmonella*** *typhimurium* | 13.39 | 4.91 | 5.80 | 9.37 | 4.02 | 7.59 | 10.27 | 0.00 | 7.59 | 0.45 | 5.36 | 7.14 | 4.02 | 4.02 | 2.23 | 8.48 | 1.79 | 3.57 | *177* |
| **Tamm-Horsfall mucoprotein,** human urine | 11.77 | 7.69 | 8.01 | 9.23 | 4.74 | 8.30 | 7.91 | 7.14 | 5.91 | 1.89 | 2.55 | 7.66 | 3.15 | 3.37 | 1.11 | 3.11 | 2.54 | 3.92 | *178* |
| **Tartrate dehydrogenase,** *Pseudomonas putida* | 10.64 | 3.81 | 5.81 | 9.82 | 5.23 | 8.64 | 10.56 | 0.97 | 6.69 | 2.44 | 6.77 | 8.40 | 2.39 | 4.76 | 1.35 | 3.47 | 2.46 | 5.78 | *179* |
| **Tartrate epoxidase,** *Pseudomonas putida* | 7.58 | 5.23 | 3.41 | 8.98 | 5.28 | 9.17 | 15.05 | 1.34 | 4.94 | 2.87 | 3.31 | 10.34 | 1.78 | 3.79 | 2.43 | 2.68 | 4.48 | 7.35 | *180* |
| **Tartronic semialdehyde reductase,** *Pseudomonas putida* | 9.01 | 5.75 | 8.34 | 10.85 | 2.56 | 11.90 | 11.52 | 1.42 | 8.06 | 1.50 | 4.94 | 6.16 | 1.74 | 3.48 | 0.80 | 6.76 | 1.76 | 3.45 | *181* |
| **Thiogalactoside transacetylase,** *E. coli* | 12.40 | 5.61[j] | 5.30[j] | 7.79 | 5.55 | 8.08 | 4.73 | 0.89 | 10.92[j] | 2.90 | 8.36[j] | 5.06 | 3.86 | 3.62 | 0.93 | 5.03 | 3.86 | 5.10 | *182* |
| **Thioredoxin,** *E. coli*[b] | 13.89 | 5.56 | 2.78 | 7.41 | 4.63 | 8.33 | 11.11 | 1.85 | 4.63 | 0.93 | 8.33 | 12.04 | 1.85 | 3.70 | 1.85 | 9.26 | 0.93 | 0.93 | *183* |

## TABLE 1 (Continued)
## COMPLETE DATA

| Protein | Amino Acid, mole %[a] | | | | | | | | | | | | | | | | | | References |
|---|---|---|---|---|---|---|---|---|---|---|---|---|---|---|---|---|---|---|---|
| | Asp[c] | Thr | Ser | Glu[d] | Pro[e] | Gly | Ala | Cys | Val | Met | Ile | Leu | Tyr | Phe | Trp | Lys | His | Arg | |
| **Thioredoxin reductase,** *E. coli* | 11.14 | 7.39 | 4.43 | 10.32 | 2.84 | 10.73 | 9.95 | 1.38 | 5.53 | 2.08 | 6.75 | 9.08 | 2.72 | 3.08 | 0.35 | 4.00 | 3.19 | 5.04 | *184* |
| **Threonine deaminase,** *E. coli* | 11.01 | 5.08 | 6.70 | 8.96 | 3.18 | 9.88 | 8.75 | 1.69 | 7.97 | 2.61 | 9.53 | 6.85 | 2.33 | 2.68 | 0.21 | 5.58 | 1.91 | 5.08 | *185* |
| **Tobacco mosaic virus (vulgare) coat protein**[b] | 11.39 | 10.13 | 10.13 | 10.13 | 5.06 | 3.80 | 8.86 | 0.63 | 8.86 | 0.00 | 5.70 | 7.59 | 2.53 | 5.06 | 1.90 | 1.27 | 0.00 | 6.96 | *186* |
| **Toxin,** diphtheria | 11.42 | 6.23 | 9.68 | 11.01 | 3.83 | 7.39 | 6.97 | 0.75 | 7.63 | 1.43 | 6.22 | 7.08 | 3.16 | 3.09 | 0.62 | 7.52 | 3.10 | 2.86 | *187* |
| **Toxin (type A),** *Clostridium botulinum* | 13.83 | 5.52 | 5.47 | 12.07 | 3.26 | 10.00 | 9.79 | 0.82 | 7.18 | 1.35 | 6.06 | 7.51 | 1.40 | 3.86 | 0.00 | 7.69 | 1.40 | 2.77 | *188* |
| **Transferrin,** human serum | 11.80 | 4.35 | 5.63 | 8.89 | 5.15 | 7.33 | 8.50 | 5.76 | 5.96 | 1.42 | 2.12 | 8.66 | 3.75 | 4.11 | 1.13 | 8.66 | 2.78 | 4.00 | *189* |
| **Transglutaminase,** guinea pig liver | 10.34 | 4.95 | 6.70 | 11.58 | 4.97 | 7.47 | 6.51 | 2.16 | 7.87 | 1.95 | 4.65 | 9.83 | 3.63 | 3.36 | 2.13 | 4.28 | 1.86 | 5.77 | *190* |
| **Triose phosphate isomerase,** rabbit muscle | 8.58 | 6.29[m] | 5.22[m] | 10.33 | 2.94 | 10.76 | 9.94 | 1.87 | 9.53[m] | 0.87 | 5.43[m] | 6.64 | 1.88 | 3.31 | 0.99 | 9.65 | 2.14 | 3.63 | *191* |
| **Tripeptide synthetase,** bakers' yeast | 11.06 | 4.46 | 6.79 | 11.74 | 5.04 | 7.47 | 8.54 | 0.78 | 6.60 | 1.16 | 5.33 | 9.89 | 2.91 | 3.49 | 0.78 | 7.76 | 2.23 | 3.98 | *192* |
| **Trypsin inhibitor,** *Ascaris*[b] | 7.58 | 6.06 | 1.52 | 16.67 | 9.09 | 9.09 | 7.58 | 15.15 | 3.03 | 0.00 | 4.55 | 0.00 | 0.00 | 3.03 | 1.52 | 10.61 | 0.00 | 4.55 | *193* |
| **Trypsin inhibitor (acidic),** porcine pancreatic juice[b] | 7.14 | 10.71 | 10.71 | 12.50 | 8.93 | 7.14 | 1.79 | 10.71 | 7.14 | 0.00 | 5.36 | 3.57 | 3.57 | 0.00 | 0.00 | 7.14 | 0.00 | 3.57 | *194* |
| **Trypsin inhibitor (basic),** bovine pancreas[b] | 8.62 | 5.17 | 1.72 | 5.17 | 6.90 | 10.34 | 10.34 | 10.34 | 1.72 | 1.72 | 3.45 | 3.45 | 6.90 | 6.90 | 0.00 | 6.90 | 0.00 | 10.34 | *195* |
| **Trypsinogen,** bovine pancreas[b] | 11.35 | 4.37 | 14.41 | 6.11 | 3.93 | 10.92 | 6.11 | 5.24 | 7.86 | 0.87 | 6.55 | 6.11 | 4.37 | 1.31 | 1.75 | 6.55 | 1.31 | 0.87 | *196* |
| **Tryptophanase,** *E. coli* | 8.18 | 5.52 | 4.33 | 11.66 | 4.16 | 7.66 | 9.27 | 1.24 | 6.25 | 3.21 | 5.62 | 7.71 | 4.81 | 4.91 | 0.43 | 7.13 | 2.14 | 5.78 | *197* |
| **Tryptophan oxygenase,** *Pseudomonas acidovorans* | 7.62 | 5.02 | 5.83 | 11.16 | 5.10 | 7.67 | 13.28 | 0.00 | 6.45 | 3.11 | 3.65 | 11.64 | 2.81 | 3.16 | 1.71 | 2.35 | 2.89 | 6.54 | *198* |
| **Tryptophan synthetase A protein,** *E. coli*[b] | 8.24 | 3.37 | 4.12 | 10.86 | 7.12 | 7.12 | 14.98 | 1.12 | 6.37 | 1.87 | 7.12 | 10.11 | 2.62 | 4.49 | 0.00 | 4.87 | 1.50 | 4.12 | *199* |
| **Tryptophanyl tRNA synthetase,** bovine pancreas | 10.26 | 4.65 | 6.28 | 10.72 | 6.13 | 9.02 | 8.27 | 1.49 | 5.08 | 1.95 | 5.30 | 7.42 | 3.50 | 5.23 | 1.27 | 6.56 | 2.17 | 4.71 | *200* |
| **Tyrosinase,** *Neurospora crassa* | 10.89 | 5.62 | 11.13 | 7.94 | 7.49 | 6.59 | 8.22 | 0.00 | 5.86 | 0.66 | 3.26 | 8.19 | 4.20 | 5.90 | 2.91 | 4.16 | 2.08 | 4.89 | *201* |
| **Tyrosine aminotransferase,** rat liver | 10.63 | 3.80 | 7.61 | 12.86 | 0.67 | 6.94 | 7.83 | 3.13 | 6.71 | 2.80 | 6.15 | 10.51 | 3.02 | 3.69 | 1.01 | 5.48 | 2.01 | 5.15 | *202* |
| **Uridine diphosphate galactose 4-epimerase,** *Candida pseudotropicalis* | 12.81 | 7.54 | 6.06 | 8.53 | 5.72 | 7.74 | 6.44 | 1.92 | 6.26 | 0.99 | 5.26 | 8.79 | 4.44 | 4.98 | 0.38 | 7.25 | 1.97 | 2.94 | *203* |
| **Visual pigment$_{500}$,** bovine retina | 6.38 | 7.23 | 5.11 | 8.94 | 5.53 | 6.81 | 8.51 | 2.13 | 8.51 | 3.40 | 5.53 | 8.51 | 4.68 | 8.09 | 2.13 | 4.26 | 1.70 | 2.55 | *204* |
| **Xanthine oxidase,** bovine milk | 8.37 | 7.11 | 6.48 | 10.22 | 5.47 | 8.24 | 7.53 | 2.61 | 6.85 | 1.98 | 5.05 | 8.66 | 2.57 | 4.96 | 0.38 | 6.81 | 2.31 | 4.42 | *205* |

# TABLE 2
# INCOMPLETE DATA

| Protein | Amino Acid, mole %[a] | | | | | | | | | | | | | | | | | | References |
|---|---|---|---|---|---|---|---|---|---|---|---|---|---|---|---|---|---|---|---|
| | Asp[c] | Thr | Ser | Glu[d] | Pro[e] | Gly | Ala | Cys | Val | Met | Ile | Leu | Tyr | Phe | Trp | Lys | His | Arg | |
| **Acetyl cholinesterase,** *Electrophorus electricus* | 9.53 | 4.50[i] | 7.47[i] | 9.44 | 8.27 | 7.94 | 5.57 | — | 7.94[i] | 2.95 | 4.15[i] | 9.05 | 3.87 | 5.23 | 2.09 | 4.40 | 2.26 | 5.34 | *206* |
| **α$_1$-Acid glycoprotein B,** human serum | 11.66 | 8.67 | 3.31 | 17.39 | 3.87 | 4.29 | 4.52 | — | 5.50[i] | 0.70 | 6.06[i] | 8.21 | 5.73 | 5.27 | — | 7.93 | 1.86 | 5.03 | *207* |
| **Acidic brain protein,** rat | 10.82 | 4.28 | 5.65 | 13.77 | 4.76 | 8.34 | 9.96 | 1.39 | 7.01 | 1.95 | 5.34 | 9.01 | 2.36 | 2.98 | — | 6.62 | 1.59 | 4.17 | *208* |
| **Acyl phosphatase,** bovine brain | 8.84 | 5.89[i] | 5.79[i] | 14.06 | 3.75 | 9.35 | 4.71 | — | 10.02[i] | 1.27 | 4.71[i] | 6.36 | 2.80 | 4.58 | 2.54 | 9.03 | 2.80 | 3.50 | *209* |
| **Antigen NN,** human erythrocytes | 6.17 | 10.17[j] | 10.13[j] | 8.01 | 7.11 | 5.46 | 6.97 | — | 8.34[j] | 4.62 | 6.12[j] | 8.71 | 2.73 | 3.16 | 0.00 | 4.43 | 3.77 | 4.10 | *210* |
| **Arginase,** rat liver | 9.08 | 7.11 | 6.26 | 9.08 | 7.42 | 9.39 | 6.70 | — | 8.95 | 1.34 | 5.32 | 7.42 | 2.50 | 3.24 | 1.27 | 9.11 | 2.41 | 3.39 | *211* |
| **Ascorbic acid oxidase,** crookneck squash | 11.21 | 5.93 | 6.21 | 8.49 | 7.77 | 7.95 | 6.48 | — | 6.71 | 1.43 | 6.11 | 7.68 | 4.25 | 4.46 | 4.25 | 4.64 | 3.28 | 3.18 | *212* |
| **Basic brain protein,** pig | 6.14 | 4.36[i] | 8.81[i] | 6.53 | 7.72 | 14.95 | 9.11 | 0.00 | 1.88[i] | 1.29 | 2.08[i] | 5.74 | 2.87 | 4.85 | — | 8.02 | 5.84 | 9.80 | *213* |
| **Bradykininogen,** bovine blood | 10.51 | 6.74 | 8.14 | 12.46 | 7.36 | 5.58 | 7.94 | — | 8.87 | 0.76 | 4.38 | 7.68 | 3.24 | 3.74 | 1.05 | 6.30 | 2.42 | 2.83 | *214* |
| **Calcium-binding protein,** chick intestinal mucosa | 14.03 | 4.14[j] | 4.47[j] | 15.71 | 1.45 | 7.05 | 7.38 | — | 3.10[j] | 1.92 | 5.29[j] | 11.97 | 4.18 | 5.51 | — | 9.76 | 1.57 | 2.47 | *215* |
| **Carcinogen-binding protein,** rat liver | 10.08 | 3.46[i] | 4.37[i] | 11.35 | 4.71 | 5.26 | 7.48 | — | 5.15[n] | 4.07 | 5.40[n] | 13.15 | 3.93 | 4.93 | — | 9.41 | 1.44 | 5.81 | *216* |
| **Chloroperoxidase,** *Caldariomyces fumago* | 14.94 | 6.51 | 12.64 | 9.96 | 8.81 | 5.36 | 9.20 | 0.77 | 4.60 | 0.77 | 3.45 | 7.66 | 3.83 | 4.98 | — | 1.53 | 2.68 | 2.30 | *217* |
| **Citrate oxaloacetate lyase,** *Aerobacter aerogenes* | 10.90 | 6.59 | 5.75 | 9.42 | 3.15 | 12.46 | 8.15 | — | 9.66 | 1.56 | 7.03 | 8.51 | 2.00 | 3.87 | 0.64 | 3.51 | 1.16 | 5.63 | *218* |
| **Cortisol metabolite binding large protein,** rat liver | 8.59 | 3.68 | 3.68 | 11.04 | 4.91 | 6.13 | 7.36 | — | 6.13 | 3.68 | 6.13 | 13.50 | 3.68 | 4.91 | 0.61 | 8.59 | 1.23 | 6.13 | *219* |
| **Deoxyribose 5-phosphate aldolase,** *Lactobacillus plantarum* | 9.98 | 5.50[o] | 5.50[o] | 9.57 | 1.83 | 8.15 | 13.65 | 0.61 | 10.59 | 2.44 | 6.31 | 6.72 | 1.83 | 2.24 | — | 5.91 | 4.07 | 5.09 | *220* |
| **Dipeptidase,** hog kidney cortex | 9.80 | 5.53 | 7.17 | 10.75 | 5.26 | 6.32 | 8.06 | — | 6.69 | 1.84 | 3.81 | 10.34 | 3.00 | 4.54 | 2.83 | 7.23 | 2.51 | 4.31 | *221* |
| **Disulfide-exchange enzyme,** beef liver | 11.47 | 4.23 | 4.47 | 15.85 | 3.85 | 6.47 | 9.41 | 0.93 | 5.16 | 0.83 | 4.47 | 9.06 | 2.26 | 7.01 | — | 10.38 | 2.16 | 1.98 | *222* |
| **Erabutoxin a,** sea snake venom | 9.53 | 9.36 | 13.24 | 15.32 | 7.82 | 9.55 | 0.00 | — | 3.75 | 0.00 | 6.85 | 1.99 | 1.58 | 3.92 | 2.46 | 7.35 | 1.71 | 5.59 | *223* |
| **Fatty acid synthetase,** pigeon liver | 8.57 | 4.69[j] | 6.91[j] | 11.60 | 4.61 | 8.09 | 8.43 | — | 9.19[j] | 1.74 | 6.41[j] | 11.46 | 2.56 | 3.34 | — | 5.09 | 2.70 | 4.61 | *224* |
| **Formyltetrahydrofolate synthetase,** *Clostridium* | 11.93 | 5.97[h] | 3.82[h] | 8.83 | 3.46 | 9.43 | 11.10 | — | 8.95[h] | 2.27 | 6.80[h] | 9.31 | 1.31 | 2.74 | 0.36 | 8.71 | 2.03 | 2.98 | *225* |
| **Galactose oxidase,** *Dactylium dendroides* | 10.73 | 7.29[j] | 8.96[j] | 7.42 | 5.86 | 14.18 | 8.63 | 1.54 | 5.21[j] | 1.98 | 3.88[j] | 4.78 | 1.28 | 3.67 | — | 6.88 | 2.05 | 5.65 | *226* |
| **Galactose 1-phosphate uridylyltransferase,** *E. coli* | 11.17 | 6.05[j] | 3.99[j] | 12.20 | 6.71 | 4.95 | 10.17 | — | 7.83[j] | 1.88 | 2.63[j] | 8.77 | 2.96 | 4.47 | 3.21 | 4.10 | 3.81 | 5.12 | *227* |
| **Gastrin,** porcine gastric secretion | 12.47 | 4.33[j] | 3.39[j] | 13.22 | 5.72 | 2.82 | 2.25 | — | 3.94[j] | 2.84 | 9.43[j] | 7.97 | 10.02 | 4.98 | — | 7.20 | 6.75 | 2.66 | *228* |
| **α-Gliadin,** wheat | 2.96 | 1.50 | 5.46 | 38.78 | 14.10 | 2.64 | 2.64 | — | 4.89 | 0.79 | 4.61 | 8.39 | 3.36 | 3.76 | 1.25 | 0.54 | 2.26 | 2.06 | *229* |
| **β-Glucuronidase,** bovine liver | 9.37 | 5.14 | 6.17 | 11.09 | 5.83 | 7.89 | 5.83 | — | 7.89 | 2.17 | 4.46 | 9.60 | 5.60 | 4.80 | 2.40 | 3.77 | 2.97 | 5.03 | *230* |
| **Glutamate mutase component S,** *Clostridium tetanomorphum* | 11.70 | 4.90 | 5.27 | 10.78 | 3.25 | 11.33 | 7.23 | — | 9.55 | 2.51 | 8.27 | 6.92 | 1.71 | 4.04 | 0.61 | 7.96 | 1.84 | 2.14 | *231* |
| **Glutathione reductase,** yeast | 13.10 | 1.78[j] | 0.44[j] | 12.66 | 3.46 | 10.79 | 8.34 | 1.07 | 13.25[j] | 2.60 | 8.21[j] | 8.96 | 2.06 | 4.31 | — | 5.20 | 1.89 | 1.88 | *232* |
| **Glycoprotein,** human aorta | 9.96 | 5.47[n] | 7.18[n] | 11.29 | 6.96 | 8.27 | 7.03 | — | 6.57[n] | 2.07 | 5.12[n] | 8.60 | 4.25 | 4.20 | 3.66 | 2.12 | 5.58 | 1.66 | *233* |
| **Glycoprotein M-1,** porcine plasma | 12.75 | 4.67[j] | 4.44[j] | 15.13 | 5.28 | 5.28 | 7.24 | — | 3.50 | 1.07 | 6.35 | 8.78 | 3.74 | 4.39 | — | 10.09 | 3.97 | 3.32 | *234* |
| **Group specific protein,** human serum | 10.90 | 5.96 | 7.93 | 13.17 | 6.30 | 4.49 | 7.12 | 4.22 | 5.29 | 1.32 | 2.05 | 9.98 | 3.95 | 4.11 | — | 8.22 | 1.52 | 3.48 | *235* |
| **Hageman factor,** human serum | 8.78 | 8.13[i] | 11.14[i] | 10.92 | 5.71 | 13.13 | 7.79 | — | 5.94[i] | 0.98 | 4.03[i] | 6.41 | 2.01 | 3.32 | — | 6.25 | 2.11 | 3.36 | *236* |
| **Hemagglutinin,** *Limulus polyphemus* (crab) | 11.46 | 6.13[i] | 6.93[i] | 13.67 | 4.42 | 4.72 | 3.22 | 2.81 | 5.03[i] | 1.41 | 5.03[i] | 10.25 | 5.13 | 5.03 | — | 6.13 | 5.13 | 3.52 | *237* |
| **Hemolysin,** *Staphylococcus aureus* | 12.28 | 6.84 | 5.02 | 7.83 | 0.44 | 3.66 | 3.98 | — | 5.07[j] | 4.78 | 10.88[j] | 7.26 | 1.65 | 10.20 | 2.56 | 16.22 | 0.40 | 0.92 | *238* |
| **Hexokinase,** bovine brain | 10.67 | 6.14[i] | 6.44[i] | 11.37 | 2.50 | 9.24 | 5.68 | — | 7.57[i] | 3.56 | 5.53[i] | 10.31 | 2.13 | 4.63 | — | 6.76 | 2.13 | 5.33 | *239* |
| **Hydroxypyruvate reductase,** *Pseudomonas acidovorans* | 10.19 | 4.72 | 5.44 | 10.22 | 4.02 | 10.58 | 13.44 | — | 7.59 | 2.00 | 4.41 | 8.39 | 1.83 | 2.62 | 0.61 | 6.75 | 3.37 | 3.83 | *240* |
| **Hypobranchial mucin,** *Buccinum undatum* | 11.33 | 6.53[i] | 6.92[i] | 11.70 | 5.44 | 8.53 | 8.06 | — | 7.06[i] | 1.38 | 4.29[i] | 6.95 | 2.28 | 3.72 | 3.31 | 5.66 | 2.27 | 4.58 | *241* |
| **Incisor eruption accelerating protein,** mouse submaxillary glands | 14.96 | 4.44[j] | 10.51[j] | 8.89 | 4.48 | 12.60 | 0.90 | — | 5.07[j] | 1.45 | 4.90[j] | 8.65 | 9.57 | 0.00 | 3.99 | 0.00 | 1.82 | 7.76 | *242* |

# TABLE 2 (Continued)
## INCOMPLETE DATA

| Protein | Asp[c] | Thr | Ser | Glu[d] | Pro[e] | Gly | Ala | Cys | Val | Met | Ile | Leu | Tyr | Phe | Trp | Lys | His | Arg | References |
|---|---|---|---|---|---|---|---|---|---|---|---|---|---|---|---|---|---|---|---|
| | Amino Acid, mole %[a] | | | | | | | | | | | | | | | | | | |
| **Interstitial cell-stimulating hormone,** sheep pituitary glands | 6.40 | 8.32 | 7.25 | 7.78 | 13.33 | 6.72 | 8.53 | — | 6.93 | 2.67 | 3.62 | 6.40 | 3.20 | 3.52 | **0.00** | **7.04** | 2.99 | 5.33 | *243* |
| **Intrinsic factor,** hog pyloric mucosa | 12.20 | 7.02 | 6.95 | 11.96 | 4.92 | 2.90 | 4.88 | — | 5.58 | 1.84 | 5.88 | 10.34 | 4.90 | **5.79** | **1.49** | **6.14** | 2.93 | 4.27 | *244* |
| **β-Lactamase,** *Enterobacter cloacae* | 8.33 | 6.17 | 5.83 | 9.33 | 7.33 | 7.83 | 12.00 | — | 8.67[i] | 3.83 | 4.83[i] | 8.00 | 4.00 | **2.50** | — | **6.50** | 1.50 | 3.33 | *245* |
| **Lipoprotein,** human erythrocyte stroma | 9.07 | 6.74 | 7.77 | 13.32 | 5.38 | 8.49 | 9.71 | 0.66 | 7.09 | 0.20 | 5.27 | 13.52 | 0.00 | **3.28** | — | **4.17** | 0.20 | 5.11 | *246* |
| **Lysostaphin,** *Staphylococcus aureus* | 12.46 | 6.66[j] | 6.08[j] | 9.53 | 5.07 | 11.50 | 8.18 | — | 7.18[j] | 3.22 | 5.19[j] | 4.89 | 6.03 | **3.44** | — | **4.63** | 3.64 | 2.29 | *247* |
| **Maleate isomerase,** *Pseudomonas fluorescens* | 9.28 | 4.24 | 7.87 | 9.68 | 4.79 | 4.93 | 8.95 | 1.94 | 10.30 | 5.15 | 6.38 | 10.75 | 2.65 | **1.89** | — | **4.33** | 2.50 | 4.39 | *248* |
| **β-Mannase,** *Aspergillus* | 12.40 | 9.61 | 10.85 | 8.21 | 2.47 | 9.69 | 7.93 | — | 5.81 | 1.20 | 4.75 | 5.85 | 5.51 | 3.25 | 6.89 | 2.50 | 1.51 | 1.58 | *249* |
| **$\beta_2$-Microglobulin,** urine of human tubular proteinurias patients | 12.39 | 5.02 | 10.24 | 11.44 | 4.91 | 3.20 | 2.12 | — | 7.20 | 0.83 | 5.16 | 7.24 | 5.95 | 5.02 | 2.12 | 8.00 | 4.07 | 5.09 | *250* |
| **Mitotic apparatus 22S protein,** *Strongylocentrotus purpuratus* | 14.36 | 4.08[j] | 1.71[j] | 12.25 | 5.80 | 5.67 | 5.01 | — | 10.14[j] | 2.90 | 7.11[j] | 8.43 | 3.43 | 5.01 | 0.40 | 7.64 | 1.58 | 4.48 | *251* |
| **Mucin (major),** porcine submaxillary gland | 1.91 | 13.68 | 23.44 | 5.73 | 5.43 | 19.82 | 14.59 | — | 6.64 | 0.00 | 2.92 | 1.11 | 0.60 | 0.00 | — | 1.01 | 0.40 | 2.72 | *252* |
| **Myxovirus hemagglutination inhibitor,** human erythrocytes | 5.78 | 13.59[h] | 13.82[h] | 9.74 | 6.46 | 6.57 | 6.57 | — | 8.61[h] | 1.36 | 5.10[h] | 4.42 | 3.51 | 3.40 | — | 3.40 | 3.74 | 3.96 | *253* |
| **Nerve growth factor fraction A,** mouse submaxillary gland | 12.63 | 6.33[p] | 8.25[p] | 9.56 | 7.20 | 9.18 | 5.69 | — | 6.40[p] | 1.60 | 4.09[p] | 9.72 | 3.36 | 2.72 | — | 6.78 | 3.29 | 3.20 | *254* |
| **5′-Nucleotidase,** *E. coli* | 8.77 | 4.22[j] | 3.46[j] | 8.33 | 3.40 | 6.24 | 7.54 | — | 6.32[j] | 12.20 | 3.51[j] | 5.96 | 1.84 | 2.68 | — | 5.75 | 1.05 | 18.71 | *255* |
| **Old yellow enzyme,** brewers' yeast | 11.68 | 4.66 | 6.16 | 11.68 | 5.95 | 8.98 | 9.30 | 0.16 | 5.15 | 1.11 | 6.00 | 7.75 | 4.85 | 4.57 | — | 5.82 | 1.80 | 4.38 | *256* |
| **Ovoinhibitor,** chicken egg white | 12.88 | 8.91 | 7.41 | 10.52 | 4.92 | 8.83 | 5.36 | — | 7.24 | 1.07 | 4.70 | 5.99 | 4.65 | 1.69 | 0.00 | 6.31 | 3.91 | 5.60 | *257* |
| **Oxaloglycolate reductive decarboxylase,** *Pseudomonas putida* | 7.79 | 6.37 | 7.87 | 10.53 | 4.13 | 11.04 | 10.24 | — | 7.73 | 1.99 | 4.38 | 8.61 | 2.20 | 3.66 | 0.64 | 7.01 | 1.67 | 4.13 | *258* |
| **Phenylpyruvate tautomerase,** hog thyroid | 10.11 | 2.55 | 7.06 | 9.06 | 6.44 | 9.29 | 9.70 | — | 8.30 | 3.30 | 5.25 | 10.23 | 4.09 | 3.84 | 0.54 | 3.20 | 2.01 | 5.04 | *259* |
| **N-1-(5′-Phosphoribosyl) adenosine triphosphate: pyrophosphate phosphoribosyl transferase,** *Salmonella typhimurium* | 9.42 | 4.63 | 5.26 | 12.56 | 4.58 | 7.77 | 9.49 | 1.23 | 6.77 | 2.96 | 6.46 | 11.12 | 1.98 | 2.31 | — | 5.24 | 1.74 | 6.49 | *260* |
| **Phycocyanin,** *Phormidium luridum* | 12.24 | 5.07 | 5.75 | 6.77 | 2.96 | 9.12 | 17.39 | — | 5.70 | 3.81 | 5.52 | 7.93 | 4.37 | 3.16 | 0.42 | 4.04 | 0.51 | 5.26 | *261* |
| **Phytochrome,** oat seedlings | 10.85 | 4.37 | 7.30 | 9.89 | 6.62 | 6.68 | 9.21 | — | 6.48 | 0.68 | 4.64 | 9.82 | 3.27 | 4.50 | 1.23 | 5.93 | 3.14 | 5.39 | *262* |
| **Polyol dehydrogenase,** *Candida utilis* | 13.23 | 6.22[j] | 6.14[j] | 11.06 | 5.89 | 7.55 | 7.58 | 0.58 | 4.37[j] | 0.45 | 9.17[j] | 10.08 | 2.68 | 3.67 | — | 5.62 | 2.26 | 3.43 | *263* |
| **Poricin,** *Poria corticola* | 17.04 | 7.72[j] | 7.54[j] | 5.44 | 4.15 | 4.05 | 10.32 | — | 5.34[j] | 0.67 | 3.81[j] | 10.65 | 5.40 | 6.24 | 2.66 | 4.72 | 0.72 | 3.52 | *264* |
| **Propionyl coenzyme A carboxylase,** pig heart | 11.27 | 5.14[j] | 7.18[j] | 3.87 | 5.14 | 8.87 | 9.86 | — | 9.58[j] | 2.75 | 7.68[j] | 7.46 | 2.82 | 4.51 | — | 5.99 | 2.25 | 5.63 | *265* |
| **Pyrophosphatase,** *Bacillus subtilis* | 12.30 | 4.84[j] | 4.55[j] | 14.16 | 2.56 | 5.08 | 11.23 | — | 9.31[j] | 1.32 | 7.39[j] | 10.35 | 2.28 | 2.69 | — | 8.18 | 1.30 | 2.46 | *266* |
| **Rhodopsin,** bovine retina | 6.91 | 7.51 | 6.31 | 7.81 | 7.21 | 7.21 | 8.11 | 2.10 | 7.51 | 3.00 | 5.11 | 8.41 | 4.20 | 7.81 | — | 4.80 | 2.40 | 3.60 | *267* |
| **Ribonucleotide reductase,** *Lactobacillus leichmannii* | 10.03 | 5.44[i] | 15.95[i] | 15.76 | 1.15 | 15.00 | 8.21 | — | 2.67[i] | 1.15 | 3.44[i] | 5.92 | 0.00 | 2.01 | — | 9.36 | 2.48 | 1.43 | *268* |
| **30S Ribosomal protein 1,** *E. coli* | 11.23 | 4.16[q] | 4.45[q] | 13.65 | 1.84 | 8.62 | 9.00 | — | 12.97[q] | 0.97 | 6.39[q] | 7.94 | 1.06 | 3.00 | — | 7.94 | 1.45 | 5.32 | *269* |
| **L-Ribulokinase,** *E. coli* | 7.68 | 3.74[j] | 5.66[j] | 20.07 | 4.67 | 16.54 | 12.98 | — | 0.00[j] | 0.00 | 4.94[j] | 7.36 | 1.29 | 2.97 | 3.18 | 2.88 | 1.96 | 4.07 | *270* |
| **S-100 brain protein,** bovine brain | 11.02 | 2.97[i] | 3.81[i] | 18.64 | 1.27 | 5.51 | 6.36 | — | 15.68[i] | 0.00 | 3.81[i] | 8.90 | 2.12 | 6.36 | 0.00 | 8.47 | 3.81 | 1.27 | *271* |
| **Serine transacetylase,** *Salmonella typhimurium* | 8.09 | 4.55[i] | 5.51[i] | 9.96 | 5.86 | 9.59 | 12.57 | 1.49 | 6.73[i] | 2.57 | 7.40[i] | 8.00 | 2.55 | 2.34 | — | 5.50 | 3.09 | 4.20 | *272* |
| **Serotypic antigen 51A,** *Paramecium aurelius* | 13.27 | 18.50 | 11.19 | 6.88 | 2.12 | 7.73 | 13.62 | — | 5.19 | 0.58 | 2.81 | 4.35 | 3.08 | 1.73 | 1.69 | 5.23 | 0.69 | 1.35 | *273* |
| **Stellacyanin,** *Rhus vernicifera* (Japanese lac tree) | 19.78 | 9.68[h] | 5.81[h] | 3.87 | 4.08 | 8.87 | 3.57 | — | 0.00 | 0.00 | 6.12[h] | 3.57 | 7.14 | 5.50 | 3.06 | 11.01 | 3.87 | 4.08 | *274* |

## TABLE 2 (Continued)
## INCOMPLETE DATA

| Protein | Asp[c] | Thr | Ser | Glu[d] | Pro[e] | Gly | Ala | Cys | Val | Met | Ile | Leu | Tyr | Phe | Trp | Lys | His | Arg | References |
|---|---|---|---|---|---|---|---|---|---|---|---|---|---|---|---|---|---|---|---|
| | Amino Acid, mole %[a] | | | | | | | | | | | | | | | | | | |
| **Sulfatase A,** ox liver | 7.41 | 5.86[j] | 6.48[j] | 9.05 | 9.26 | 11.21 | 9.67 | — | 5.76[j] | 1.65 | 1.85[j] | 13.68 | 2.88 | 5.14 | 1.13 | 1.44 | 4.01 | 3.50 | *275* |
| **Sweet-sensitive protein,** bovine taste buds | 9.50 | 5.68[h] | 6.70[h] | 14.02 | 7.57 | 7.54 | 6.55 | — | 8.52[h] | 1.46 | 4.95[h] | 8.05 | 2.58 | 3.28 | 0.00 | 7.72 | 2.07 | 3.82 | *276* |
| **Thyroglobulin,** rat thyroid gland | 8.00 | 5.24 | 10.79 | 13.93 | 7.19 | 8.51 | 9.13 | — | 6.23 | 0.96 | 2.66 | 9.47 | 2.50 | 5.12 | — | 2.77 | 1.21 | 6.28 | *277* |
| **Thyroxine-binding globulin,** human plasma | 9.34 | 6.14[h] | 8.85[h] | 12.78 | 7.37 | 6.14 | 7.62 | — | 6.63[h] | 1.23 | 2.70[h] | 10.81 | 2.46 | 4.42 | 0.98 | 6.14 | 2.46 | 3.93 | *278* |
| **Transcortin,** human plasma | 10.39 | 6.48[h] | 8.29[h] | 10.22 | 5.63 | 6.76 | 7.41 | — | 8.88[h] | 1.98 | 5.10[h] | 12.11 | 2.53 | 4.97 | — | 3.38 | 2.44 | 3.44 | *279* |
| **Urease,** jack bean | 11.16 | 7.03 | 5.49 | 9.43 | 4.65 | 9.18 | 9.13 | — | 6.63 | 2.82 | 8.36 | 7.79 | 2.46 | 2.60 | 1.14 | 5.39 | 2.65 | 4.11 | *280* |
| **Urokinase,** human urine | 7.42 | 7.38 | 8.47 | 10.99 | 6.37 | 8.70 | 3.92 | — | 4.23 | 1.75 | 6.06 | 8.00 | 4.97 | 3.61 | 1.90 | 6.72 | 3.96 | 5.55 | *281* |
| **Uridine diphosphate galactose: lipopolysaccharide α-3-galactosyl transferase,** *Salmonella typhimurium* | 9.03 | 5.22[i] | 4.48[i] | 12.71 | 3.61 | 7.97 | 9.58 | — | 6.96[i] | 1.68 | 6.82[i] | 8.73 | 2.37 | 3.06 | — | 6.82 | 2.30 | 8.66 | *282* |

[a] The actual number of amino acid residues in a given protein may be obtained by multiplying the listed value by Total Residues/100.
[b] Composition calculated from sequence data. See pp. C-226 to C-280.
[c] Includes asparagine.
[d] Includes glutamine.
[e] Includes hydroxyproline.
[f] Includes hydroxylysine.
[g] Includes phosphoserine.
[h] Corrected 20-hour hydrolysis value. See text.
[i] Corrected 24-hour hydrolysis value. See text.
[j] Corrected 22-hour hydrolysis value. See text.
[k] Corrected 32-hour hydrolysis value. See text.
[l] Corrected 30-hour hydrolysis value. See text.
[m] Corrected 23-hour hydrolysis value. See text.
[n] Corrected 48-hour hydrolysis value. See text.
[o] Corrected 19-hour hydrolysis value. See text.
[p] Corrected 21-hour hydrolysis value. See text.
[q] Corrected 18-hour hydrolysis value. See text.

Data taken from: Reeck, G., in *Handbook of Biochemistry,* 2nd ed., pp. C-282–C-290, H. A. Sober, Ed. Copyright 1970, The Chemical Rubber Co., Cleveland, Ohio.

## REFERENCES


1. Lederer, Coults, Laursen and Westheimer, *Biochemistry, 5,* 823 (1966).
2. Becker, Kredich and Tomkins, *J. Biol. Chem., 244,* 2418 (1969).
3. Carsten and Katz, *Biochim. Biophys. Acta, 90,* 534 (1964).
4. Vanaman, Wakil and Hill, *J. Biol. Chem., 243,* 6420 (1968).
5. Wolfenden, Tomozawa and Bamman, *Biochemistry, 7,* 3965 (1968).
6. Markland and Wadkins, *J. Biol. Chem., 241,* 4124 (1966).
7. Woodward and Braymer, *J. Biol. Chem., 241,* 580 (1966).
8. Kimura, Suzuki, Padmanabhan, Samejima, Tarutani and Ui, *Biochemistry, 8,* 4027 (1969).
9. Shimomura and Johnson, *Biochemistry, 8,* 3991 (1969).
10. Matsuzawa and Segal, *J. Biol. Chem., 243,* 5929 (1968).
11. Steinman and Jakoby, *J. Biol. Chem., 243,* 730 (1968).
12. Simpson, Vallee and Tait, *Biochemistry, 7,* 4336 (1968).
13. Johnson and Marsh, *Immunochemistry, 3,* 91 (1966).
14. King, Norman and Lichtenstein, *Biochemistry, 6,* 1992 (1967).
15. Kotaki, Harada and Yagi, *J. Biochem. (Tokyo), 61,* 598 (1967).
16. DeKok and Rawitch, *Biochemistry, 8,* 1405 (1969).
17. Vandermeers and Christophe, *Biochim. Biophys. Acta, 154,* 110 (1968).
18. Harrison, Hofmann and Mainwaring, *J. Mol. Biol., 4,* 251 (1962).
19. Patrick and Lee, *J. Biol. Chem., 244,* 4277 (1969).
20. Boeker, Fischer and Snell, *J. Biol. Chem., 244,* 5239 (1969).
21. Blethen and Kaplan, *Biochemistry, 6,* 1413 (1967).
22. Whelan and Wriston, *Biochemistry, 8,* 2386 (1969).
23. Tate and Meister, *Biochemistry, 7,* 3240 (1968).
24. Weber, *J. Biol. Chem., 243,* 543 (1968).
25. Weber, *Nature, 218,* 1116 (1968).
26. Truffa-Bachi, van Rapenbusch, Janin, Gros and Cohen, *Eur. J. Biochem., 5,* 73 (1968).
27. Ambler and Brown, *J. Mol. Biol., 9,* 825 (1964).
28. Thornber and Olson, *Biochemistry, 7,* 2242 (1968).
29. Iwasaki and Schmid, *J. Biol. Chem., 242,* 5247 (1967).
30. Marshall and Cohen, *J. Biol. Chem., 241,* 4197 (1966).
31. Furth, *J. Biol. Chem., 243,* 4832 (1968).
32. Bradshaw, Ericsson, Walsh and Neurath, *Proc. Nat. Acad. Sci. U.S.A., 63,* 1389 (1969).
33. Woychik, Kalan and Noelken, *Biochemistry, 5,* 2276 (1966).
34. Higashi and Shibata, *J. Biochem. (Tokyo), 56,* 361 (1964).
35. Datta, Hanson and Whitaker, *Can. J. Biochem. Physiol., 41,* 697 (1963).
36. Kasper and Deutsch, *J. Biol. Chem., 238,* 2325 (1963).
37. Bahl, *J. Biol. Chem., 244,* 567 (1969).
38. Renaud, Rowe and Gibbons, *J. Cell Biol., 36,* 79 (1968).
39. Larsen and Wolff, *J. Biol. Chem., 243,* 1283 (1968).
40. Butler, Piez and Bornstein, *Biochemistry, 6,* 3771 (1967).
41. Olson and Liener, *Biochemistry, 6,* 105 (1967).
42. Chader and Westphal, *J. Biol. Chem., 243,* 928 (1968).
43. Hass and Hill, *J. Biol. Chem., 244,* 6080 (1969).
44. Buchwald and Jencks, *Biochemistry, 7,* 844 (1968).
45. Van Dam, *Exp. Eye Res., 5,* 255 (1966).
46. Loiselet and Chatagner, *Biochim. Biophys. Acta, 130,* 180 (1966).
47. Kaplan and Flavin, *J. Biol. Chem., 241,* 5781 (1966).
48. Cavallini, De Marco, Scanduna, Dupré and Graziani, *J. Biol. Chem., 241,* 3189 (1966).
49. Itagaki and Hager, *Biochem. Biophys. Res. Commun., 32,* 1013 (1968).
50. Matsubara and Smith, *J. Biol. Chem., 237,* PC 3575 (1962).
51. Price, Liu, Stein and Moore, *J. Biol. Chem., 244,* 917 (1969).
52. Lindberg, *Biochemistry, 6,* 323 (1967).
53. Swislocki and Kaplan, *J. Biol. Chem., 242,* 1083 (1967).
54. Goulian, Lucas and Kornberg, *J. Biol. Chem., 243,* 627 (1968).
55. Miller and Feeney, *Biochemistry, 5,* 952 (1966).
56. Chao and Einstein, *J. Biol. Chem., 243,* 6050 (1968).
57. Cory and Wold, *Biochemistry, 5,* 3131 (1966).
58. Huang, Shih, Borja, Avena and Bergdoll, *Biochemistry, 6,* 1480 (1967).
59. Stansell and Deutsch, *J. Biol. Chem., 240,* 4306 (1965).
60. Kaplan and Stadtman, *J. Biol. Chem., 243,* 1794 (1968).

61. Huotari, Nelson, Smith and Kirkwood, *J. Biol. Chem., 243,* 952 (1968).
62. Jackson and Hanahan, *Biochemistry, 7,* 4506 (1968).
63. Keresztes-Nagy, Perini and Margoliash, *J. Biol. Chem., 244,* 981 (1969).
64. Spiro and Spiro, *J. Biol. Chem., 237,* 1507 (1962).
65. Chang, Brown and Glazer, *J. Biol. Chem., 244,* 5196 (1969).
66. Knight and Hardy, *J. Biol. Chem., 242,* 1370 (1967).
67. Papkoff, Mahlmann and Li, *Biochemistry, 6,* 3976 (1967).
68. Ridley, Thornber and Bailey, *Biochim. Biophys. Acta, 140,* 62 (1967).
69. Fernando, Pontremoli and Horecker, *Arch. Biochem. Biophys., 129,* 370 (1969).
70. Harris, Kobes, Teller and Rutter, *Biochemistry, 8,* 2442 (1969).
71. Kanarek and Hill, *J. Biol. Chem., 239,* 4202 (1964).
72. Wilson and Hogness, *J. Biol. Chem., 244,* 2137 (1969).
73. Craven, Steers and Anfinsen, *J. Biol. Chem., 240,* 2468 (1965).
74. Iwasaki and Schmid, *J. Biol. Chem., 242,* 2356 (1967).
75. Nakamura and Fujiki, *J. Biochem. (Tokyo), 63,* 51 (1968).
76. Yoshida, *J. Biol. Chem., 241,* 4966 (1966).
77. Martines-Carrion and Tiemeier, *Biochemistry, 6,* 1715 (1967).
78. Appella and Tomkins, *J. Mol. Biol., 18,* 77 (1966).
79. Rowe and Wyngaarden, *J. Biol. Chem., 243,* 6373 (1968).
80. Woolfolk, Shapiro and Stadtman, *Arch. Biochem. Biophys., 116,* 177 (1966).
81. Davidson, Sajgo, Noller and Harris, *Nature, 216,* 1181 (1967).
82. Fondy, Levin, Sollohub and Ross, *J. Biol. Chem., 243,* 3148 (1968).
83. Appleman, Yunis, Krebs and Fischer, *J. Biol. Chem., 238,* 1358 (1963).
84. Labat, Ishiguro, Fujisaki and Schmid, *J. Biol. Chem., 244,* 4975 (1969).
85. Pusztai, *Biochem. J., 101,* 379 (1966).
86. Li, Dixon and Chung, *Biochim. Biophys. Acta, 160,* 472 (1968).
87. Black and Dixon, *Nature, 218,* 736 (1968).
88. Hammarstrom and Kabat, *Biochemistry, 8,* 2696 (1969).
89. Klippenstein, Holleman and Klotz, *Biochemistry, 7,* 3868 (1968).
90. Ghiretti-Magaldi, Nuzzolo and Ghiretti, *Biochemistry, 5,* 1943 (1966).
91. Hill and Konigsberg, *J. Biol. Chem., 237,* 3151 (1962).
92. Markwardt and Walsmann, *Z. Physiol. Chem., 348,* 1381 (1967).
93. Rechler, *J. Biol. Chem., 244,* 551 (1969).
94. Chang and Snell, *Biochemistry, 7,* 2012 (1968).
95. Loper, *J. Biol. Chem., 243,* 3264 (1968).
96. DeLange, Fambrough, Smith and Bonner, *J. Biol. Chem., 244,* 319 (1969).
97. Wuu and Saffran, *J. Biol. Chem., 244,* 482 (1969).
98. Borders and Raftery, *J. Biol. Chem., 243,* 3756 (1968).
99. Helmkamp and Bloch, *J. Biol. Chem., 244,* 6014 (1969).
100. Martin, Voll and Appella, *J. Biol. Chem., 242,* 1175 (1967).
101. Steers, *Biochemistry, 4,* 1896 (1965).
102. Cunningham, Gottlieb, Konigsberg and Edelman, *Biochemistry, 7,* 1983 (1968).
103. Creighton and Yanofsky, *J. Biol. Chem., 241,* 4616 (1966).
104. Gascon and Lampen, *J. Biol. Chem., 243,* 1573 (1968).
105. Chung and Franzen, *Biochemistry, 8,* 3175 (1969).
106. Shiio, Shiio and McFadden, *Biochim. Biophys. Acta, 96,* 114 (1965).
107. Margolies and Goldberger, *J. Biol. Chem., 242,* 256 (1967).
108. Parsons and Burns, *J. Biol. Chem., 244,* 996 (1969).
109. Webster, Nelson and Gross, *Biochemistry, 4,* 2319 (1965).
110. Corfield and Fletcher, in *Atlas of Protein Sequence and Structure,* Vol. 4, p. D-183, Dayhoff, Ed. National Biomedical Research Foundation, Silver Spring, Maryland (1969).
111. Greenspan, Alberts and Vagelos, *J. Biol. Chem., 244,* 6477 (1969).
112. Boyer and Talalay, *J. Biol. Chem., 241,* 180 (1966).
113. Tarmy and Kaplan, *J. Biol. Chem., 243,* 2579 (1968).
114. Pesce, Fondy, Stolzenbach, Castillo and Kaplan, *J. Biol. Chem., 242,* 2151 (1967).
115. Frank and Braunitzer, *Z. Physiol. Chem., 348,* 1691 (1967).
116. Groves, Basch and Gordon, *Biochemistry, 2,* 814 (1963).
117. Rombauts, Schroeder and Morrison, *Biochemistry, 6,* 2965 (1967).
118. Kettmann, Kretschmer and Hanson, *Z. Physiol. Chem., 349,* 1537 (1968).
119. Anraku, *J. Biol. Chem., 243,* 3123 (1968).
120. Rapoport, *C. R. Acad. Sci. (Paris), 260,* 1016 (1965).
121. Margolis and Langdon, *J. Biol. Chem., 241,* 469 (1969).

122. Shore and Shore, *Biochemistry, 7,* 3396 (1968).
123. Li, Barnafi, Chretien and Chung, *Nature, 208,* 1093 (1965).
124. Massey, Hofmann and Palmer, *J. Biol. Chem., 237,* 3820 (1962).
125. Hastings, Weber, Friedland, Eberhard, Mitchell and Gunsalus, *Biochemistry, 8,* 4681 (1969).
126. Takeda, Yamamoto, Kojima and Hayaishi, *J. Biol. Chem., 244,* 2935 (1969).
127. Canfield, *J. Biol. Chem., 238,* 2698 (1963).
128. Tsugita and Inouye, *J. Mol. Biol., 37,* 201 (1968).
129. Kitto and Kaplan, *Biochemistry, 5,* 3966 (1966).
130. Nozaki, Ono, Nakazawa, Kotani and Hayaishi, *J. Biol. Chem., 243,* 2682 (1968).
131. Hsiang and Bright, *J. Biol. Chem., 242,* 3079 (1967).
132. Mohri, *Nature, 217,* 1053 (1968).
133. Criddle, Bock, Green and Tisdale, *Biochemistry, 1,* 827 (1962).
134. Yamada, Gee, Ebata and Yasunobu, *Biochim. Biophys. Acta, 81,* 165 (1964).
135. Gaertner and DeMoss, *J. Biol. Chem., 244,* 2716 (1969).
136. Schultz and Shmukler, *Biochemistry, 3,* 1234 (1964).
137. Tada, Bailin, Barany and Barany, *Biochemistry, 8,* 4842 (1969).
138. Rochat, Rochat, Miranda and Lissitzky, *Biochemistry, 6,* 578 (1967).
139. Botes and Strydom, *J. Biol. Chem., 244,* 4147 (1969).
140. Su, Albizati and Chaykin, *J. Biol. Chem., 244,* 2956 (1969).
141. Cusumano, Taniuchi and Anfinsen, *J. Biol. Chem., 243,* 4769 (1968).
142. Peraino, Bunville and Tahmisian, *J. Biol. Chem., 244,* 2241 (1969).
143. Potts, Keutmann, Niall, Deftos, Brewer and Aurbach, in *Parathyroid Hormone and Thyrocalcitonin,* p. 44, Talmage and Belanger, Eds. Excerpta Medica Foundation, Princeton, New Jersey (1968).
144. Hawker, Glass and Rasmussen, *Biochemistry, 5,* 344 (1966).
145. Ambler, in *Atlas of Protein Sequence and Structure,* Vol. 4, p. D-139, Dayhoff, Ed. National Biomedical Research Foundation, Silver Spring, Maryland (1969).
146. Chow and Kassell, *J. Biol. Chem., 243,* 1718 (1968).
147. Beaudreau and Yasunobu, *Biochemistry, 5,* 1405 (1966).
148. Chang and Lane, *J. Biol. Chem., 241,* 2413 (1966).
149. Maeba and Sanwal, *J. Biol. Chem., 244,* 2549 (1969).
150. Parmeggiani, Luft, Love and Krebs, *J. Biol. Chem., 241,* 4625 (1966).
151. Yankeelov, Horton and Koshland, *Biochemistry, 3,* 349 (1964).
152. Rippa, Signorini and Pontremoli, *Eur. J. Biochem., 1,* 170 (1967).
153. Rosenbloom, Sugimoto and Pizer, *J. Biol. Chem., 243,* 2099 (1968).
154. Sugimoto, Sasaki and Chiba, *Agr. Biol. Chem., 27,* 222 (1963).
155. Wells and Hanahan, *Biochemistry, 8,* 414 (1969).
156. Mano and Lipmann, *J. Biol. Chem., 241,* 3822 (1966).
157. Raz and Goodman, *J. Biol. Chem., 244,* 3230 (1969).
158. Chance, Ellis and Bromer, *Science, 161,* 165 (1968).
159. Liu, Neumann, Elliott, Moore and Stein, *J. Biol. Chem., 238,* 251 (1963).
160. Fujisawa and Hayashi, *J. Biol. Chem., 243,* 2673 (1968).
161. Liss, *Biochemistry, 4,* 2705 (1965).
162. Engelbrecht and Sadoff, *J. Biol. Chem., 244,* 6228 (1969).
163. Nakazawa, Kojima and Taniuchi, *Biochim. Biophys. Acta, 147,* 189 (1967).
164. Hunsley and Suelter, *J. Biol. Chem., 244,* 4815 (1969).
165. Weber, *Biochemistry, 6,* 3144 (1967).
166. Gordon, Groves and Basch, *Biochemistry, 2,* 817 (1963).
167. Smyth, Stein and Moore, *J. Biol. Chem., 238,* 227 (1963).
168. Takahashi, *J. Biol. Chem., 240,* PC 4117 (1965).
169. Kuehn and McFadden, *Biochemistry, 8,* 2403 (1969).
170. Lee, Patrick and Masson, *J. Biol. Chem., 243,* 4700 (1968).
171. Burgess, *J. Biol. Chem., 244,* 6168 (1969).
172. Bachmeyer, Yasunobu, Peel and Mayhew, *J. Biol. Chem., 243,* 1022 (1968).
173. Peters and Hawn, *J. Biol. Chem., 242,* 1566 (1967).
174. Frattali and Steiner, *Biochemistry, 7,* 521 (1968).
175. DeRenzo, Siiteri, Hutchings and Bell, *J. Biol. Chem., 242,* 533 (1967).
176. Smith, DeLange, Evans, Landon and Markland, *J. Biol. Chem., 243,* 2184 (1968).
177. Pardee, *Science, 162,* 632 (1968).
178. Friedmann and Johnson, *Biochim. Biophys. Acta, 130,* 355 (1966).
179. Kohn, Packman, Allen and Jakoby, *J. Biol. Chem., 243,* 2479 (1968).
180. Allen and Jakoby, *J. Biol. Chem., 244,* 2078 (1969).
181. Kohn, *J. Biol. Chem., 243,* 4426 (1968).

182. Zabin, *J. Biol. Chem., 238,* 3300 (1963).
183. Holmgren, *Eur. J. Biochem., 6,* 475 (1968).
184. Thelander and Baldesten, *Eur. J. Biochem., 4,* 420 (1968).
185. Shizuta, Nakazawa, Tokushige and Hayaishi, *J. Biol. Chem., 244,* 1883 (1969).
186. Funatsu, Tsugita and Fraenkel-Conrat, *Arch. Biochem. Biophys., 105,* 25 (1964).
187. Raynaud, Bizzini and Relyveld, *Bull. Soc. Chim. Biol., 47,* 261 (1965).
188. VanAlstyne, Gerwing and Tremain, *J. Bacteriol., 92,* 796 (1966).
189. Parker and Bearn, *J. Exp. Med., 115,* 83 (1962).
190. Folk and Cole, *J. Biol. Chem., 241,* 5518 (1966).
191. Burton and Waley, *Biochem. J., 100,* 702 (1966).
192. Mooz and Meister, *Biochemistry, 6,* 1722 (1967).
193. Fraefel and Acher, *Biochim. Biophys. Acta, 154,* 615 (1968).
194. Greene, Dicarlo, Sussman, Bartlett and Roark, *J. Biol. Chem., 243,* 1804 (1968).
195. Kassell and Laskowski, *Biochem. Biophys. Res. Commun., 20,* 463 (1965).
196. Walsh and Neurath, *Proc. Nat. Acad. Sci. U.S.A., 52,* 884 (1964).
197. Morino and Snell, *J. Biol. Chem., 242,* 5602 (1968).
198. Poillon, Maeno, Koike and Feigelson, *J. Biol. Chem., 244,* 3447 (1969).
199. Guest, Drapeau, Carlton and Yanofsky, *J. Biol. Chem., 242,* 5442 (1967).
200. Preddie, *J. Biol. Chem., 244,* 3958 (1969).
201. Fling, Horowitz and Heinemann, *J. Biol. Chem., 238,* 2045 (1963).
202. Valeriote, Auricchio, Tomkins and Riley, *J. Biol. Chem., 244,* 3618 (1969).
203. Darrow and Rodstrom, *Biochemistry, 7,* 1645 (1968).
204. Heller, *Biochemistry, 7,* 2906 (1968).
205. Bray and Malmstrom, *Biochem. J., 93,* 633 (1964).
206. Leuzinger and Baker, *Proc. Nat. Acad. Sci. U.S.A., 57,* 446 (1967).
207. Marshall, *J. Biol. Chem., 241,* 4731 (1966).
208. Bennett and Edelman, *J. Biol. Chem., 243,* 6234 (1968).
209. Diederich and Grisolia, *J. Biol. Chem., 244,* 2412 (1969).
210. Springer, Nagai and Tegtmeyer, *Biochemistry, 5,* 3254 (1966).
211. Hirsch-Kolb and Greenberg, *J. Biol. Chem., 243,* 6123 (1968).
212. Stark and Dawson, *J. Biol. Chem., 237,* 712 (1962).
213. Tomasi and Kornguth, *J. Biol. Chem., 242,* 4933 (1967).
214. Nagasawa, Mizushima, Sato, Iwanaga and Suzuki, *J. Biochem. (Tokyo), 60,* 643 (1966).
215. Wasserman, Corradino and Taylor, *J. Biol. Chem., 243,* 3978 (1968).
216. Ketterer, Ross-Mansell and Whitehead, *Biochem. J., 103,* 316 (1967).
217. Morris and Hager, *J. Biol. Chem., 241,* 1763 (1966).
218. Bowen and Rogers, *Nature, 205,* 1316 (1965).
219. Morey and Litwack, *Biochemistry, 8,* 4813 (1969).
220. Hoffee, Rosen and Horecker, *J. Biol. Chem., 240,* 1512 (1965).
221. Rene and Campbell, *J. Biol. Chem., 244,* 1445 (1969).
222. DeLorenzo, Goldberger, Steers, Givol and Anfinsen, *J. Biol. Chem., 241,* 1562 (1966).
223. Tamiya and Arai, *Biochem. J., 99,* 624 (1966).
224. Hsu, Wasson and Porter, *J. Biol. Chem., 240,* 3736 (1965).
225. Himes and Rabinowitz, *J. Biol. Chem., 237,* 2903 (1962).
226. Kelley-Falcoz, Greenberg and Horecker, *J. Biol. Chem., 240,* 2966 (1965).
227. Saito, Ozutsumi and Kurahashi, *J. Biol. Chem., 242,* 2362 (1967).
228. Tauber and Madison, *J. Biol. Chem., 240,* 645 (1965).
229. Bernardin, Kasarda and Mecham, *J. Biol. Chem., 242,* 445 (1967).
230. Plapp and Cole, *Arch. Biochem. Biophys., 116,* 193 (1966).
231. Switzer and Barker, *J. Biol. Chem., 242,* 2658 (1967).
232. Massey and Williams, *J. Biol. Chem., 240,* 4470 (1965).
233. Barnes and Partridge, *Biochem. J., 109,* 883 (1968).
234. Grant, Martin and Anastassiadis, *J. Biol. Chem., 242,* 3912 (1967).
235. Bowman, *Biochemistry, 8,* 4327 (1969).
236. Speer, Ridgway and Hill, *Thromb. Diath. Haemorrh., 14,* 1 (1965).
237. Marchalonis and Edelman, *J. Mol. Biol., 32,* 453 (1968).
238. Yoshida, *Biochim. Biophys. Acta, 71,* 544 (1963).
239. Schwartz and Basford, *Biochemistry, 6,* 1070 (1967).
240. Kohn and Jakoby, *J. Biol. Chem., 243,* 2494 (1968).
241. Hunt and Jevons, *Biochem. J., 97,* 701 (1965).
242. Cohen, *J. Biol. Chem., 237,* 1555 (1962).
243. Papkoff, Gospodarowicz, Candiotti and Li, *Arch. Biochem. Biophys., 111,* 431 (1965).

244. Highley, Davies and Ellenbogen, *J. Biol. Chem., 242,* 1010 (1967).
245. Hennessey and Richmond, *Biochem. J., 109,* 469 (1968).
246. Morgan and Hanahan, *Biochemistry, 5,* 1050 (1966).
247. Schindler and Schuhardt, *Biochim. Biophys. Acta, 97,* 242 (1965).
248. Scher and Jakoby, *J. Biol. Chem., 244,* 1878 (1969).
249. Eriksson and Winell, *Acta Chem. Scand., 22,* 1924 (1968).
250. Berggard and Bearn, *J. Biol. Chem., 243,* 4095 (1968).
251. Stephens, *J. Cell Biol., 32,* 255 (1967).
252. DeSalegui and Plonska, *Arch. Biochem. Biophys., 129,* 49 (1969).
253. Kathan and Winzler, *J. Biol. Chem., 238,* 21 (1963).
254. Schenkein and Bueker, *Ann. N.Y. Acad. Sci., 118,* 171 (1964).
255. Neu, *J. Biol. Chem., 242,* 3896 (1967).
256. Matthews and Massey, *J. Biol. Chem., 244,* 1779 (1969).
257. Davis, Zahnley and Donovan, *Biochemistry, 8,* 2044 (1969).
258. Kohn and Jakoby, *J. Biol. Chem., 243,* 2486 (1968).
259. Blasi, Fragomele and Covelli, *J. Biol. Chem., 244,* 4864 (1969).
260. Voll, Appella and Martin, *J. Biol. Chem., 242,* 1760 (1967).
261. Cope, Smith, Crespi and Katz, *Biochim. Biophys. Acta, 133,* 446 (1967).
262. Mumford and Jenner, *Biochemistry, 5,* 3657 (1966).
263. Scher and Horecker, *Arch. Biochem. Biophys., 116,* 117 (1966).
264. Schillings and Ruelius, *Arch. Biochem. Biophys., 127,* 672 (1968).
265. Kaziro, Grossman and Ochoa, *J. Biol. Chem., 240,* 64 (1965).
266. Tono and Kornberg, *J. Biol. Chem., 242,* 2375 (1967).
267. Shields, Dinovo, Henriksen, Kimbel and Millar, *Biochim. Biophys. Acta, 147,* 238 (1967).
268. Goulian and Beck, *J. Biol. Chem., 241,* 4233 (1966).
269. Craven, Voynow, Hardy and Kurland, *Biochemistry, 8,* 2906 (1969).
270. Lee and Bendet, *J. Biol. Chem., 242,* 2042 (1967).
271. Boore, *Biochem. Biophys. Res. Commun., 19,* 739 (1965).
272. Kredich, Becker and Tomkins, *J. Biol. Chem., 244,* 2428 (1969).
273. Reisner, Rowe and Sleigh, *Biochemistry, 8,* 4637 (1969).
274. Peisach, Levine and Blumberg, *J. Biol. Chem., 242,* 2847 (1967).
275. Nichol and Roy, *Biochemistry, 4,* 386 (1965).
276. Dastoli, Lopiekes and Price, *Biochemistry, 7,* 1160 (1968).
277. Salvatore, Vecchio, Salvatore, Cahmann and Robbins, *J. Biol. Chem., 240,* 2935 (1965).
278. Giorgio and Tabachnick, *J. Biol. Chem., 243,* 2247 (1968).
279. Slaunwhite, Schneider, Wissler and Sandberg, *Biochemistry, 5,* 3527 (1966).
280. Reithel and Robbins, *Arch. Biochem. Biophys., 120,* 158 (1967).
281. White, Barlow and Mozen, *Biochemistry, 5,* 2160 (1966).
282. Endo and Rothfield, *Biochemistry, 8,* 3500 (1969).

## TABLE 1
## COAT PROTEINS

| | |
|---|---|
| **Virus,** Tobacco Mosaic | 1 ... 10 ... 20 ... 30 |
| *Vulgare (1)* | Ac-Ser- Tyr-Ser- Ile- Thr-Thr-Pro-Ser- Gln-Phe-Val- Phe-Leu-Ser- Ser- Ala-Trp-Ala-Asp-Pro-Ile- Glu-Leu-Ile- Asn-Leu-Cys-Thr-Asn-Ala- |
| *Dahlemense (2)* | Ac — — — — — **Ser** — — — — — — — — **Val** — — — — — — — **Leu** — **Val** — — **Ser- Ser-** |
| *OM (8)* | Ac — — — — — — — — — — (Val, Phe, Leu, Ser, Ser, Ala, Trp, Ala, Asp, Pro, Ile, Glu, Leu, Ile, Asn, Leu)— — — — |
| **Bacteriophage** | 1 ... 10 ... 20 ... 30 |
| f2 (3) | H-Ala-Ser- Asn-Phe-Thr-Gln-Phe-Val- Leu-Val- Asn-Asp-Gly-Gly-Thr-Gly-Asn-Val-Thr-Val- Ala-Pro-Ser- Asn-Phe-Ala-Asn-Gly-Val- Ala- |
| R17 (4) | — — — — — — — — — — — — — — — — — — — — — — — — — — — — — — |
| MS2 (5) | — — — — — — — — — — — — — — — — — — — — — — — — — — — — — — |
| fr (6) | — — (Asn, Phe, **Glx**)**Glu** — — — — — (Asp, Gly, Gly, Thr, Gly)**Asp** — **Lys** — — — — — — — — — — — |
| **Bacteriophage** | 1 ... 10 ... 20 ... 30 |
| fd (7) | H-Ala-Glu-Gly-Asp-Asp-Pro-Ala-Lys-Ala-Ala-Phe-Asp-Ser- Leu-Glu-Ala-Ser- Ala-Thr-Glu-Tyr-Ile- Gly-Tyr (Ala, Trp, Gly, Val, Val, Val, |

| | |
|---|---|
| **Virus,** Tobacco Mosaic | 40 ... 50 ... 60 |
| *Vulgare (1)* | Leu-Gly-Asn-Gln-Phe-Gln-Thr-Gln-Gln-Ala- Arg-Thr-Val- Val- Gln-Arg-Gln-Phe-Ser- Gln-Val- Trp-Lys-Pro-Ser- Pro-Gln-Val- Thr-Val- |
| *Dahlemense (2)* | — — — — — — — — — — — — **Thr** — — **Gln** — — — **Glu** — — — — **Phe** — — **Ser** — — |
| *OM (8)* | — — — (Gln, Phe) — — — — — — — — — — — — — — — — — — — — — — — |
| **Bacteriophage** | 40 ... 50 ... 60 |
| f2 (3) | Glu-Trp-Ile- Ser- Ser- Asn-Ser- Arg-Ser- Gln-Ala-Tyr-Lys-Val- Thr-Cys-Ser- Val- Arg-Gln-Ser- Ser- Ala-Gln-Asn-Arg-Lys-Tyr-Thr-Ile- |
| R17 (4) | — — — — — — — — — — — — — — — — — — — — — — — — — — — — — — |
| MS2 (5) | — — — — — — — — — — — — — — — — — — — — — — — — — — — — — — |
| fr (6) | — — — — — — — — — — — — — — — — — — — — — — **Asn** — — — — — **Val** |
| **Bacteriophage** | 40 ... 49 ... 60 |
| fd (7) | Val, Met-Ile) Ala-Thr-Ile- Gly-Ile- Lys-Leu-Phe-Lys-Lys-Phe-Thr-Ser- Lys-Ala-Ser- OH |

| | |
|---|---|
| **Virus,** Tobacco Mosaic | 70 ... 80 ... 90 |
| *Vulgare (1)* | Arg-Phe-Pro-Asp-Ser- Asp-Phe-Lys- Val- Tyr- Arg-Tyr- Asn-Ala-Val- Leu-Asp-Pro-Leu-Val- Thr-Ala- Leu-Leu-Gly-Ala- Phe-Asp-Thr-Arg- |
| *Dahlemense (2)* | — — — **Gly-Asp-Val-Tyr** — — — — — — — — — — — — **Ile** — — — — — — **Thr** — — — — |
| *OM (8)* | — — — — — — — — — — — — — — — — — — — — — — — — — — — — — — |
| **Bacteriophage** | 70 ... 80 ... 90 |
| f2 (3) | Lys- Val- Glu-Val- Pro- Lys- Val- Ala- Thr-Gln-Thr-Val- Gly-Gly-Val- Glu-Leu-Pro- Val- Ala- Ala- Trp-Arg-Ser- Tyr-Leu-Asn-Leu Glu-Leu- |
| R17 (4) | — — — — — — — — — — — — — — — — — — — — — — — — — — — **Met** — — |
| MS2 (5) | — — — — — — — — — — — — — — — — — — — — — — — — — — — **Met** — — |
| fr (6) | — — — — — — — — — — **Gln** — — — — — — — — — — — — — — — **Met** — **Met** — — |

## TABLE 1 (Continued)
## COAT PROTEINS

| | |
|---|---|
| **Virus,** Tobacco Mosaic | 100 (Ala) · 110 (Ala) · 120 (Ala) |
| *Vulgare (1)* | Asn-Arg-Ile- Ile- Glu-Val- Glu-Asn-Gln-Ala-Asn-Pro- Thr-Thr-Ala- Glu-Thr-Leu-Asp-Ala- Thr-Arg-Arg-Val- Asp-Asp-Ala- Thr- Val- Ala- |
| *Dahlemense (2)* | — — — — — — — — — **Gln-Ser-** — — — — — — — — — — — — — — — — — — — |
| *OM (8)* | — — — — — — — — — — — — — — — — — — — — — — — — — — — — — — |
| **Bacteriophage** | 100 (Asp) · 110 (Gly) · 120 (Ser) |
| f2 *(3)* | Thr-Ile- Pro-Ile- Phe-Ala- Thr-Asn-Ser- Asp-Cys-Glu-Leu-Ile- Val- Lys- Ala- Met-Gln-Gly-Leu-Leu-Lys- Asp-Gly-Asn-Pro-Ile- Pro- Ser- |
| R17 *(4)* | — — — — — — — — — — — — — — — — — — — — — — — — — — — — — — |
| MS2 *(5)* | — — — — — — — — — — — — — — — — — — — — — — — — — — — — — — |
| fr *(6)* | — (**Val,** Pro, Ile) — — — — **Asx** — — **Ala** — — — — — **Leu** — — **Thr- Phe** — Thr (Gly, Asn, Pro, Ile, **Ala, Thr,** |

| | |
|---|---|
| **Virus,** Tobacco Mosaic | 130 (Val) · 140 (Asn) · 150 (Leu) |
| *Vulgare (1)* | Ile- Arg-Ser- Ala- Ile- Asn-Asn-Leu-Ile- Val- Glu-Leu-Ile- Arg-Gly-Thr- Gly-Ser- Tyr- Asn-Arg-Ser- Ser- Phe- Glu-Ser- Ser- Ser- Gly- Leu- |
| *Dahlemense (2)* | — — — — — — — — **Val- Asn** — — **Val** — — — — **Leu** — — **Gln- Asn- Thr** — — — **Met** — — — |
| *OM (8)* | — — — — — — — — **Val-** — — — — — — — — — — — — — — — — — — — — — |
| **Bacteriophage** | 129 (Tyr) |
| f2 *(3)* | Ala- Ile- Ala- Ala- Asn-Ser- Gly- Ile- Tyr- OH |
| R17 *(4)* | — — — — — — — — — |
| MS2 *(5)* | — — — — — — — — — |
| fr *(6)* | Ala) — — — — — — — — |

| | |
|---|---|
| **Virus,** Tobacco Mosaic | 158 (Thr) |
| *Vulgare (1)* | Val- Trp-Thr-Ser- Gly-Pro- Ala- Thr- OH |
| *Dahlemense (2)* | — — — — **Ala** — — **Ser-** OH |
| *OM (8)* | — — **Asn** — — — — — — |

Data taken from: Wojciech and Margoliash, in *Handbook of Biochemistry,* 2nd ed., pp. C-226–C-227, H. A. Sober, Ed. Copyright 1970, The Chemical Rubber Co., Cleveland, Ohio.

# TABLE 2
## EUKARYOTIC CYTOCHROMES *c*

| **Mammals** | | | | | | | | | | | 10 | | | | | | | | | | 20 | | | | | | | | | | 30 |
|---|---|---|---|---|---|---|---|---|---|---|---|---|---|---|---|---|---|---|---|---|---|---|---|---|---|---|---|---|---|---|---|
| Human, chimpanzee (*1, 2*) | Acetyl- | Gly- | Asp- | Val- | Glu- | Lys- | Gly- | Lys- | Lys- | Ile- | Phe- | Ile- | Met- | Lys- | Cys- | Ser- | Gln- | Cys- | His- | Thr- | Val- | Glu- | Lys- | Gly- | Gly- | Lys- | His- | Lys- | Thr- | Gly- | Pro- |
| | | | | | | | | | | | | | | | | HEME | | | | | | | | | | | | | | | |
| Rhesus Monkey (*Macaca mulatta*) (*3*) | — | — | — | — | — | — | — | — | — | — | — | — | — | — | — | — | — | — | — | — | — | — | — | — | — | — | — | — | — | — | — |
| Horse[b] (*4*) | — | — | — | — | — | — | — | — | — | — | — | **Val-** | **Gln** | — | — | **Ala** | — | — | — | — | — | — | — | — | — | — | — | — | — | — | — |
| Donkey[b] (*5*) | — | — | — | — | — | — | — | — | — | — | — | **Val-** | **Gln** | — | — | **Ala** | — | — | — | — | — | — | — | — | — | — | — | — | — | — | — |
| Cow, pig[c], sheep (*6, 7, 8*) | — | — | — | — | — | — | — | — | — | — | — | **Val-** | **Gln** | — | — | **Ala** | — | — | — | — | — | — | — | — | — | — | — | — | — | — | — |
| Dog (*9*) | — | — | — | — | — | — | — | — | — | — | — | **Val-** | **Gln** | — | — | **Ala** | — | — | — | — | — | — | — | — | — | — | — | — | — | — | — |
| Rabbit (*10*) | — | — | — | — | — | — | — | — | — | — | — | **Val-** | **Gln** | — | — | **Ala** | — | — | — | — | — | — | — | — | — | — | — | — | — | — | — |
| California grey whale (*Rhachianectes glaucus*) (*11*) | — | — | — | — | — | — | — | — | — | — | — | **Val-** | **Gln** | — | — | **Ala** | — | — | — | — | — | — | — | — | — | — | — | — | — | — | — |
| Great grey kangaroo (*Macropus canguru*) (*12*) | — | — | — | — | — | — | — | — | — | — | — | **Val-** | **Gln** | — | — | **Ala** | — | — | — | — | — | — | — | — | — | — | — | — | — | — | — |

| **Mammals** | | | | | | | | | | 40 | | | | | | | | | | 50 | | | | | | | | | | 60 |
|---|---|---|---|---|---|---|---|---|---|---|---|---|---|---|---|---|---|---|---|---|---|---|---|---|---|---|---|---|---|---|
| Human, chimpanzee (*1, 2*) | Asn- | Leu- | His- | Gly- | Leu- | Phe- | Gly- | Arg- | Lys- | Thr- | Gly- | Gln- | Ala- | Pro- | Gly- | Tyr- | Ser- | Tyr- | Thr- | Ala- | Ala- | Asn- | Lys- | Asn- | Lys- | Gly- | Ile- | Ile- | Trp- | Gly- |
| Rhesus Monkey (*Macaca mulatta*) (*3*) | — | — | — | — | — | — | — | — | — | — | — | — | — | — | — | — | — | — | — | — | — | — | — | — | — | — | — | — | — | — |
| Horse[b] (*4*) | — | — | — | — | — | — | — | — | — | — | — | — | — | — | — | **Phe-** | **Thr** | — | — | **Asp** | — | — | — | — | — | — | — | **Thr** | — | **Lys** |
| Donkey[b] (*5*) | — | — | — | — | — | — | — | — | — | — | — | — | — | — | — | **Phe** | — | — | — | **Asp** | — | — | — | — | — | — | — | **Thr** | — | **Lys** |
| Cow, pig[c], sheep (*6, 7, 8*) | — | — | — | — | — | — | — | — | — | — | — | — | — | — | — | **Phe** | — | — | — | **Asp** | — | — | — | — | — | — | — | **Thr** | — | — |
| Dog (*9*) | — | — | — | — | — | — | — | — | — | — | — | — | — | — | — | **Phe** | — | — | — | **Asp** | — | — | — | — | — | — | — | **Thr** | — | — |
| Rabbit (*10*) | — | — | — | — | — | — | — | — | — | — | — | — | — | **Val** | — | **Phe** | — | — | — | **Asp** | — | — | — | — | — | — | — | **Thr** | — | — |
| California grey whale (*Rhachianectes glaucus*) (*11*) | — | — | — | — | — | — | — | — | — | — | — | — | — | — | — | **Phe** | — | — | — | **Asp** | — | — | — | — | — | — | — | **Thr** | — | — |
| Grey grey kangaroo (*Macropus canguru*) (*12*) | — | — | **Asn** | — | **Ile** | — | — | — | — | — | — | — | — | — | — | **Phe-** | **Thr** | — | — | **Asp** | — | — | — | — | — | — | — | — | — | — |

| **Mammals** | | | | | | | | | | 70 | | | | | | | | | | 80 | | | | | | | | | | 90 |
|---|---|---|---|---|---|---|---|---|---|---|---|---|---|---|---|---|---|---|---|---|---|---|---|---|---|---|---|---|---|---|
| Human, chimpanzee (*1, 2*) | Glu- | Asp- | Thr- | Leu- | Met- | Glu- | Tyr- | Leu- | Glu- | Asn- | Pro- | Lys- | Lys- | Tyr- | Ile- | Pro- | Gly- | Thr- | Lys- | Met- | Ile- | Phe- | Val- | Gly- | Ile- | Lys- | Lys- | Lys- | Glu- | Glu- |
| Rhesus Monkey (*Macaca mulatta*) (*3*) | — | — | — | — | — | — | — | — | — | — | — | — | — | — | — | — | — | — | — | — | — | — | — | — | — | — | — | — | — | — |
| Horse[b] (*4*) | — | **Glu** | — | — | — | — | — | — | — | — | — | — | — | — | — | — | — | — | — | — | — | — | **Ala** | — | — | — | — | — | **Thr** | — |
| Donkey[b] (*5*) | — | **Glu** | — | — | — | — | — | — | — | — | — | — | — | — | — | — | — | — | — | — | — | — | **Ala** | — | — | — | — | — | **Thr** | — |
| Cow, pig,[c] sheep (*6, 7, 8*) | — | **Glu** | — | — | — | — | — | — | — | — | — | — | — | — | — | — | — | — | — | — | — | — | **Ala** | — | — | — | — | — | **Gly** | — |
| Dog (*9*) | — | **Glu** | — | — | — | — | — | — | — | — | — | — | — | — | — | — | — | — | — | — | — | — | **Ala** | — | — | — | — | **Thr-** | **Gly** | — |
| Rabbit (*10*) | — | — | — | — | — | — | — | — | — | — | — | — | — | — | — | — | — | — | — | — | — | — | **Ala** | — | — | — | — | — | **Asp** | — |
| California grey whale (*Rhachianectes glaucus*) (*11*) | — | **Glu** | — | — | — | — | — | — | — | — | — | — | — | — | — | — | — | — | — | — | — | — | **Ala** | — | — | — | — | — | **Gly** | — |
| Great grey kangaroo (*Macropus canguru*) (*12*) | — | — | — | — | — | — | — | — | — | — | — | — | — | — | — | — | — | — | — | — | — | — | **Ala** | — | — | — | — | — | **Gly** | — |

# TABLE 2 (Continued)
## EUKARYOTIC CYTOCHROMES *c*

| **Mammals** | | | | | | | | | | 100 | | | | 104 | |
|---|---|---|---|---|---|---|---|---|---|---|---|---|---|---|---|
| Human, chimpanzee (*1, 2*) | Arg- | Ala- | Asp- | Leu- | Ile- | Ala- | Tyr- | Leu- | Lys- | Lys- | Ala- | Thr- | Asn- | Glu- | OH |
| Rhesus Monkey (*Macaca mulatta*) (*3*) | — | — | — | — | — | — | — | — | — | — | — | — | — | — | — |
| Horse[b] (*4*) | — | **Glu** | — | — | — | — | — | — | — | — | — | — | — | — | — |
| Donkey[b] (*5*) | — | **Glu** | — | — | — | — | — | — | — | — | — | — | — | — | — |
| Cow, pig[c], sheep (*6, 7, 8*) | — | **Glu** | — | — | — | — | — | — | — | — | — | — | — | — | — |
| Dog (*9*) | — | — | — | — | — | — | — | — | — | — | — | — | — | **Lys** | — |
| Rabbit (*10*) | — | — | — | — | — | — | — | — | — | — | — | — | — | — | — |
| California grey whale (*Rhachianectes glaucus*) (*11*) | — | **Glu** | — | — | — | — | — | — | — | — | — | — | — | — | — |
| Great grey kangaroo (*Macropus canguru*) (*12*) | — | — | — | — | — | — | — | — | — | — | — | — | — | — | — |

| **Birds, reptiles and fish**[a] | | | | | | | | | | | 10 | | | | | | | | | | 20 | | | | | | | | | | 30 |
|---|---|---|---|---|---|---|---|---|---|---|---|---|---|---|---|---|---|---|---|---|---|---|---|---|---|---|---|---|---|---|---|
| Chicken, turkey (*13, 14*) | Acetyl- | Gly- | Asp- | Ile- | Glu- | Lys- | Gly- | Lys- | Lys- | Ile- | Phe- | Val- | Gln- | Lys- | Cys- | Ser- (HEME) | Gln- (HEME) | Cys- (HEME) | His- | Thr- | Val- | Glu- | Lys- | Gly- | Gly- | Lys- | His- | Lys- | Thr- | Gly- | Pro- |
| Pigeon (*14*) | ——— | — | — | — | — | — | — | — | — | — | — | — | — | — | — | — | — | — | — | — | — | — | — | — | — | — | — | — | — | — | — |
| Pekin duck (*14*) | ——— | — | — | **Val** | — | — | — | — | — | — | — | — | — | — | — | — | — | — | — | — | — | — | — | — | — | — | — | — | — | — | — |
| Snapping turtle (*Chelydra serpentina*) (*15*) | ——— | — | — | **Val** | — | — | — | — | — | — | — | — | — | — | — | **Ala** | — | — | — | — | — | — | — | — | — | — | — | — | — | — | — |
| Rattlesnake (*Crotalus adamanteus*) (*16*) | ——— | — | — | **Val** | — | — | — | — | — | — | — | **Thr** | **Met** | — | — | — | — | — | — | — | — | — | — | — | — | — | — | — | — | — | — |
| Bullfrog[d] (*Rana catesbiana*) (*17*) | ——— | — | — | **Val** | — | — | — | — | — | — | — | (—, | —, | —, | —, | **Ala,** | —, | —, | —, | —, | **Cys** | —, | —, | —, | —, | —, | —,) | — | **Val** | — | — |
| Tuna (*18, 19*) | ——— | — | — | **Val-** | **Ala** | — | — | — | — | **Thr** | — | — | — | — | — | **Ala** | — | — | — | — | — | — | **Asn** | — | — | — | — | — | **Val** | — | — |
| Dogfish (*Squalus sucklii*) (*20*) | ——— | — | — | **Val** | — | — | — | — | — | **Val** | — | — | — | — | — | **Ala** | — | — | — | — | — | — | **Asn** | — | — | — | — | — | — | — | — |
| Pacific lamprey (*Entosphenus tridentatus*) (*29*) | ——— | — | — | **Val** | — | — | — | — | — | **Val** | — | — | — | — | — | — | — | — | — | — | — | — | — | **Ala** | — | — | — | — | — | — | — |

| **Birds, reptiles and fish**[a] | | | | | | | | | | 40 | | | | | | | | | | 50 | | | | | | | | | | 60 |
|---|---|---|---|---|---|---|---|---|---|---|---|---|---|---|---|---|---|---|---|---|---|---|---|---|---|---|---|---|---|---|
| Chicken, turkey (*13, 14*) | Asn- | Leu- | His- | Gly- | Leu- | Phe- | Gly- | Arg- | Lys- | Thr- | Gly- | Gln- | Ala- | Glu- | Gly- | Phe- | Ser- | Tyr- | Thr- | Asp- | Ala- | Asn- | Lys- | Asn- | Lys- | Gly- | Ile- | Thr- | Trp- | Gly- |
| Pigeon (*14*) | — | — | — | — | — | — | — | — | — | — | — | — | — | — | — | — | — | — | — | — | — | — | — | — | — | — | — | — | — | — |
| Pekin duck (*14*) | — | — | — | — | — | — | — | — | — | — | — | — | — | — | — | — | — | — | — | — | — | — | — | — | — | — | — | — | — | — |
| Snapping turtle (*Chelydra serpentina*) (*15*) | — | — | **Asn** | — | — | **Ile** | — | — | — | — | — | — | — | — | — | — | — | — | — | **Glu** | — | — | — | — | — | — | — | — | — | — |
| Rattlesnake (*Crotalus adamanteus*) (*16*) | — | — | — | — | — | — | — | — | — | — | — | — | — | **Val** | — | **Tyr** | — | — | — | **Ala** | — | — | — | — | — | — | — | **Ile** | — | — |
| Bullfrog[d] (*Rana catesbiana*) (*17*) | — | — | **Tyr** | — | — | **Ile** | — | — | — | — | — | — | — | **Ala** | — | — | — | — | — | — | — | — | — | — | — | — | — | — | — | — |
| Tuna (*18, 19*) | — | — | **Trp** | — | — | — | — | — | — | — | — | — | — | — | — | **Tyr** | — | — | — | — | — | — | — | **Ser** | — | — | — | **Val** | — | **Asn-** |
| Dogfish (*Squalus sucklii*) (*20*) | — | — | **Ser** | — | — | — | — | — | — | — | — | — | — | — | — | — | — | — | — | — | — | — | — | **Ser** | — | — | — | — | — | **Gln-** |
| Pacific lamprey (*Entosphenus tridentatus*) (*29*) | — | — | **Gln** | — | — | — | — | — | — | — | — | — | — | **Pro** | — | — | — | — | — | — | — | — | — | **Ser** | — | — | — | **Val** | — | **Asn-** |

# TABLE 2 (Continued)
## EUKARYOTIC CYTOCHROMES *c*

| | | | | | | | | | | 70 | | | | | | | | | | 80 | | | | | | | | | | 90 |
|---|---|---|---|---|---|---|---|---|---|---|---|---|---|---|---|---|---|---|---|---|---|---|---|---|---|---|---|---|---|---|
| **Birds, reptiles and fish**[a] | | | | | | | | | | | | | | | | | | | | | | | | | | | | | | |
| Chicken, turkey (*13, 14*) | Glu- | Asp- | Thr- | Leu- | Met- | Glu- | Tyr- | Leu- | Glu- | Asn- | Pro- | Lys- | Lys- | Tyr- | Ile- | Pro- | Gly- | Thr- | Lys- | Met- | Ile- | Phe- | Ala- | Gly- | Ile- | Lys- | Lys- | Lys- | Ser- | Glu- |
| Pigeon (*14*) | — | — | — | — | — | — | — | — | — | — | — | — | — | — | — | — | — | — | — | — | — | — | — | — | — | — | — | — | **Ala** | — |
| Pekin duck (*14*) | — | — | — | — | — | — | — | — | — | — | — | — | — | — | — | — | — | — | — | — | — | — | — | — | — | — | — | — | — | — |
| Snapping turtle (*Chelydra serpentina*) (*15*) | — | **Glu** | — | — | — | — | — | — | — | — | — | — | — | — | — | — | — | — | — | — | — | — | — | — | — | — | — | — | **Ala** | — |
| Rattlesnake (*Crotalus adamanteus*) (*16*) | **Asp** | — | — | — | — | — | — | — | — | — | — | — | — | — | — | — | — | — | — | — | **Val** | — | **Thr** | — | **Leu-** | **Ser** | — | — | **Lys** | — |
| Bullfrog[d] (*Rana catesbiana*) (*17*) | (—, | —, | —, | —, | —, | —, | —) | — | — | — | — | — | — | — | — | — | — | — | — | — | — | — | — | — | — | — | — | — | **(Glu, Gly, Gln,** | |
| Tuna (*18, 19*) | **Asn** | — | — | — | — | — | — | — | — | — | — | — | — | — | — | — | — | — | — | — | — | — | — | — | — | — | — | — | **Gly** | — |
| Dogfish (*Squalus sucklii*) (*20*) | **Gln-** | **Glu** | — | — | **Arg-** | **Ile** | — | — | — | — | — | — | — | — | — | — | — | — | — | — | — | — | — | — | **Leu** | — | — | — | — | — |
| Pacific lamprey (*Entosphenus tridentatus*) (*29*) | **Gln-** | **Glu** | — | — | **Phe-** | **Val** | — | — | — | — | — | — | — | — | — | — | — | — | — | — | — | — | — | — | — | — | — | **Glu-** | **Gly** | — |

| | | | | | | | | | | 100 | | | | 104 | |
|---|---|---|---|---|---|---|---|---|---|---|---|---|---|---|---|
| **Birds, reptiles and fish**[a] | | | | | | | | | | | | | | | |
| Chicken, turkey (*13, 14*) | Arg- | Val- | Asp- | Leu- | Ile- | Ala- | Tyr- | Leu- | Lys- | Asp- | Ala- | Thr- | Ser- | Lys- | OH |
| Pigeon (*14*) | — | **Ala** | — | — | — | — | — | — | — | **Gln** | — | — | **Ala** | — | — |
| Pekin duck (*14*) | — | **Ala** | — | — | — | — | — | — | — | — | — | — | **Ala** | — | — |
| Snapping turtle (*Chelydra serpentina*) (*15*) | — | **Ala** | — | — | — | — | — | — | — | — | — | — | — | — | — |
| Rattlesnake (*Crotalus adamanteus*) (*16*) | — | **Thr-** | **Asn** | — | — | — | — | — | — | **Glu-** | **Lys** | — | **Ala-** | **Ala-** | OH |
| Bullfrog[d] (*Rana catesbiana*) (*17*) | —, | **Lys)** | — | — | — | — | — | (—, | —, | **Ser,** | —, | **Cys,** | —, | —) | OH |
| Tuna (*18, 19*) | — | **Gln** | — | — | **Val** | — | — | — | — | **Ser** | — | — | 103 | OH | |
| Dogfish (*Squalus sucklii*) (*20*) | — | **Gln** | — | — | — | — | — | — | — | **Lys-** | **Thr-** | **Ala-** | **Ala-** | **Ser-** | OH |
| Pacific lamprey (*Entosphenus tridentatus*) (*29*) | — | **Lys** | — | — | — | — | — | — | — | **Lys-** | **Ser** | — | — | **Glu-** | OH |

| | | −8 | | −6 | −5 | −4 | | | −1 | 1 | | | | | | | | | 10 | | | | | | | | | | 20 |
|---|---|---|---|---|---|---|---|---|---|---|---|---|---|---|---|---|---|---|---|---|---|---|---|---|---|---|---|---|---|
| **Invertebrates, fungi and plants**[a] | | | | | | | | | | | | | | | | | | | | | | | | | | | | | |
| Samia cynthia[e] (*21*) | | | | | H- | Gly- | Val- | Pro- | Ala- | Gly- | Asn- | Ala- | Glu- | Asn- | Gly- | Lys- | Lys- | Ile- | Phe- | Val- | Gln- | Arg- | Cys- | Ala- | Gln- | Cys- | His- | Thr- | Val- |
| | | | | | | | | | | | | | | | | | | | | | | | └ | HEME | | ┘ | | | |
| Tobacco horn worm moth[e] (*22*) | | | | | — | — | — | — | — | — | — | — | **Asp** | — | — | — | — | — | — | — | — | — | — | — | — | — | — | — | — |
| Screw worm fly[e] (*23*) (*Haematobia irritans*) | | | | | — | — | — | — | — | — | **Asp-** | **Val** | — | **Lys** | — | — | — | — | — | — | — | — | — | — | — | — | — | — | — |
| *Drosophila melanogaster*[e] (*24*) | | | | | — | — | — | — | — | — | **Asp-** | **Val** | — | **Lys** | — | — | — | **Leu** | — | — | — | — | — | — | — | — | — | — | — |
| Baker's yeast iso-1-cytochrome *c*[e] (*Saccharomyces cerevisiae*) (*25*) | | | | H- | **Thr-** | **Glu-** | **Phe-** | **Lys** | — | — | **Ser** | — | **Lys-** | **Lys** | — | **Ala-** | **Thr-** | **Leu** | — | **Lys-** | **Thr** | — | — | **Glu-** | **Leu** | — | — | — | — |
| *Candida krusei*[e] (*26*) | | | H- | **Pro-** | **Ala-** | **Pro-** | **Phe-** | **Glu-** | **Gln** | — | **Ser** | — | **Lys-** | **Lys** | — | **Ala-** | **Thr-** | **Leu** | — | **Lys-** | **Thr** | — | — | — | **Glu** | — | — | — | **Ile-** |
| *Neurospora crassa* (*27*) | | | | | H | — | **Phe-** | **Ser** | — | - | **Asp-** | **Ser-** | **Lys-** | **Lys** | — | **Ala-** | **Asn-** | **Leu** | — | **Lys-** | **Thr** | — | — | — | **Glu** | — | — | **Gly-** | **Glu-** |
| Wheat germ (*28*) | Acetyl- | **Ala-** | **Ser-** | **Phe-** | **Ser-** | **Glu-** | **Ala** | — | **Pro** | — | — | **Pro-** | **Asp-** | **Ala** | — | **Ala** | — | — | — | **Lys-** | **Thr-** | **Lys** | — | — | — | — | — | — | — |

# TABLE 2 (Continued)
## EUKARYOTIC CYTOCHROMES *c*

| **Invertebrates, fungi and plants**[a] | (30 at Pro; 40 at Thr; 50 at Asn) |
|---|---|
| Samia cynthia[e] (*21*) | Glu-Ala-Gly-Gly-Lys-His-Lys-Val-Gly-Pro-Asn-Leu-His-Gly-Phe-Tyr-Gly-Arg-Lys-Thr-Gly-Gln-Ala-Pro-Gly-Phe-Ser-Tyr-Ser-Asn- |
| Tobacco horn worm moth[e] (*22*) | — — — — — — — — — — — — — — — **Phe** — — — — — — — — — — — — — — |
| Screw worm fly[e] (*23*) (*Haematobia irritans*) | — — — — — — — — — — — — — — **Leu-Phe** — — — — — — **Ala** — — **Ala** — **Thr** — |
| *Drosophila melanogaster*[e] (*24*) | — — — — — — — — — — — — — — **Leu-Ile** — — — — — — **Ala** — — **Ala**— **Thr** — |
| Baker's yeast iso-1-cytochrome *c*[e] (*Saccharomyces cerevisiae*) (*25*) | — **Lys** — — **Pro** — — — — — — — — — **Ile- Phe** — — **His-Ser** — — — **Gln** — **Tyr** — — **Thr-Asp-** |
| *Candida krusei*[e] (*26*) | — — — — **Pro** — — — — — — — — — **Ile- Phe-Ser** — **His-Ser** — — — **Gln** — **Tyr** — — **Thr-Asp-** |
| *Neurospora crassa* (*27*) | **Gly-Gly-Asn-Leu-Thr-Gln** — **Ile** — — **Ala** — — — **Leu-Phe** — — — — — **Ser- Val- Asp** — **Tyr-Ala** — **Thr-Asp-** |
| Wheat germ (*28*) | **Asp** — — **Ala-Gly** — — **Gln** — — — — — — **Leu-Phe** — — **Gln-Ser** — **Thr-Thr-Ala** — **Tyr** — — — **Ala-** |

| **Invertebrates, fungi and plants**[a] | (60 at Gly; 70 at Asn; 80 at Met) |
|---|---|
| Samia cynthia[e] (*21*) | Ala-Asn-Lys-Ala-Lys-Gly-Ile- Thr-Trp-Gly-Asp-Asp-Thr-Leu-Phe-Glu-Tyr-Leu-Glu-Asn-Pro-Lys-Lys-Tyr-Ile- Pro-Gly-Thr-Lys-Met- |
| Tobacco horn worm moth[e] (*22*) | — — — — — — — — — **Gln** — — — — — — — — — — — — — — — — — — — — |
| Screw worm fly[e] (*23*) (*Haematobia irritans*) | — — — — — — — — — **Gln** — — — — — — — — — — — — — — — — — — — — |
| *Drosophila melanogaster*[e] (*24*) | — — — — — — — — — **Gln** — — — — — — — — — — — — — — — — — — — — |
| Baker's yeast iso-1-cytochrome *c*[e] (*Saccharomyces cerevisiae*) (*25*) | — — **Ile- Lys** — **Asn-Val-Leu** — **Asp-Glu-Asn-Asn-Met-Ser** — — — **Thr** — — — — — — — — — — — |
| *Candida krusei*[e] (*26*) | — — — **Arg-Ala-**— **Val-Glu** — **Ala-Glu-Pro** — **Met-Ser-Asp** — — — — — — — — — — — — — — |
| *Neurospora crassa* (*27*) | — — — **Gln** — — — — — **Asp-Glu-Asn** — — — — — — — — — — — — — — — — — — |
| Wheat germ (*28*) | — — — **Asn** — **Ala-Val-Glu** — **Glu-Glu-Asn** — — **Tyr-Asp** — — **Leu** — — — — — — — — — — — |

| **Invertebrates, fungi and plants**[a] | (90 at Glu; 100 at Glu; 104 at Lys) |
|---|---|
| Samia cynthia[e] (*21*) | Val-Phe-Ala-Gly-Leu-Lys-Lys-Ala-Asn-Glu-Arg-Ala-Asp-Leu-Ile- Ala-Tyr-Leu-Lys-Glu-Ser- Thr- D -Lys-OH |
| Tobacco horn worm moth[e] (*22*) | — — — — — — — — — — — — — — — — — — — **Gln-Ala** — D — |
| Screw worm fly[e] (*23*) (*Haematobia irritans*) | **Ile** — — — — — — **Pro** — — — **Gly** — — — — — — — **Ser-Ala** — D — |
| *Drosophila melanogaster*[e] (*24*) | **Ile** — — — — — — **Pro** — — — **Gly** — — — — — — — **Ser-Ala** — D — |
| Baker's yeast iso-1-cytochrome *c*[e] (*Saccharomyces cerevisiae*) (*25*) | **Ala** — **Gly** — — — — **Glu-Lys-Asp** — **Asn** — — — **Thr** — — — **Lys-Ala-** D- **Cys-Glu-**OH |
| *Candida krusei*[e] (*26*) | **Ala** — **Gly** — — — — — **Lys-Asp** — **Asp-Asn** — **Val-Thr** — **Met-Leu** — **Ala-** D- **Ser** — |
| *Neurospora crassa* (*27*) | **Ala** — **Gly** — — — — **Asp-Lys-Asp** — **Asn** — **Ile** — **Thr-Phe-Met** — — **Ala** — **Ala**(103)-OH |
| Wheat germ (*28*) | — — **Pro** — — — — **Pro-Gln-Asp** — — — — — — — — — **Lys-Ala** — **Ser-Ser** —OH |

## TABLE 2 (Continued)
## EUKARYOTIC CYTOCHROMES *c*

*a* The solid lines represent residues that are identical to those in the corresponding position in the amino acid sequence at the top of the table.
*b* Approximately 50% of the cytochrome *c* carried by mules and hinnies is identical to the horse protein, and 50% is identical to the donkey protein.[5]
*c* The following tissues of the pig have identical cytochromes *c*: kidney, liver, skeletal muscle, heart muscle, and brain.[7]
*d* The residues shown enclosed in parentheses have been aligned by analogy to chicken cytochrome *c* on the basis of the amino acid compositions of the corresponding peptides.
*e* The missing residues ("deletions") near the carboxyl-terminal ends of the amino acid sequences have been marked D when followed by one or more residues, and have been placed to provide maximal similarity between the different proteins.

Data taken from: Wojciech, Nolan and Margoliash, in *Handbook of Biochemistry,* 2nd ed., pp. C-228–C-233, H. A. Sober, Ed. Copyright 1970, The Chemical Rubber Co., Cleveland, Ohio.

# TABLE 3
# ENZYMES

**Aspartate Transcarbamylase, R chain**
*Escherichia coli* (*14*)

H-$^{1}$Met-Thr-His-Asn-Asp-Lys-Leu-Gln-Val-$^{10}$Ala-Glu-Ile-Lys-Arg-Gly-Thr-Val-Ile-Asn-$^{20}$His-Ile-Pro-Ala-Glu-Ile-Glu-Phe-Lys-Leu-$^{30}$Leu-
Ser-Leu-Phe-Lys-Leu-Thr-Glu-Thr-Gln-$^{40}$Asp-Arg-Ile-Thr-Ile-Gly-Leu-Asn-Leu-Pro-$^{50}$Ser-Gly-Glu-Met-Gly-Arg-Lys-Asp-Leu-Ile-$^{60}$Lys-
Ile-Glu-Asn-Thr-Phe-Leu-Ser-Glu-Asx-$^{70}$Glx-Val-Asx-Glx-Leu-Ala-Leu-Tyr-Ala-Pro-$^{80}$Gln-Ala-Thr-Val-Asn-Arg-Ile-Asn-Asp-Tyr-$^{90}$Glu-
Val-Val-Gly-Lys-Ser-Arg-Pro-Ser-Leu-$^{100}$Pro-Glu-Arg-Asn-Ile-Asp-Val-Leu-Val-Cys-$^{110}$Pro-Asp-Ser-Asn-Cys-Ile-Ser-His-Ala-Glu-$^{120}$Pro-
Val-Ser-Ser-Ser-Phe-Ala-Val-Arg-Arg-$^{130}$Ala-Asx-Asx-Ile-Ala-Leu-Lys-Cys-Lys-Tyr-$^{140}$Cys-Glu-Lys-Glu-Phe-Ser-His-Asn-Val-Val-$^{150}$Leu-
Ala-$^{152}$Asn-OH

**Catalase**, bovine (*15*)

H-($^{1}$Ala,Asx)-Asx-Arg-Asx-Pro-Ala-Ser-Asp-$^{10}$Gln-Met-Lys-His-Trp-Lys-Glu-Gln-Arg-Ala-$^{20}$Ala-Gln-Lys-Pro-Asp-Val-Leu-Thr-Thr-Gly-$^{30}$Gly-
Gly-Asn-Pro-Val-Gly-Asp-Lys-Leu-Asn-$^{40}$Ser-Leu-Thr-Val-Gly-Pro-Arg-Gly-Pro-Leu-$^{50}$Leu-Val-Gln-Asp-Val-Val-Phe-Thr-Asp-Glu-$^{60}$Met-
Ala-His-Phe-Asp-Arg-Glu-Arg-Ile-Pro-$^{70}$Glu-Arg-Val-Val-His-Ala-Lys-Gly-Ala-Gly-$^{80}$Ala-Phe-Gly-Tyr-Phe-Glu-Val-Thr-His-Asp-$^{90}$Ile-
Thr-Arg-Tyr-Ser-Lys-Ala-Lys-Val-Phe-$^{100}$Glu-His-Ile-Gly-Lys-Arg-Thr-Pro-Ile-Ala-$^{110}$Val-Arg-Phe-Ser-Thr-Val-Ala-Gly-Glu-Ser-$^{120}$Gly-
Ser-Ala-Asp-Thr-Val-Arg-Asp-Pro-Arg-$^{130}$Gly-Phe-Ala-Val-Lys-Phe-Tyr-(Asx,Asx,Asx,$^{140}$Asx,Thr, Thr, Glx, Pro, Gly, Gly, Val, Ile, Leu,$^{150}$Phe,
Phe,Trp)-Ile-Arg-Asp-Ala-Leu-Leu-Phe-$^{160}$Pro-Ser-Phe-Ile-His-Ser-Gln-Lys-Arg-Asn-$^{170}$Pro-Gln-Thr-His-Leu-Lys-(Asx,Asx,Pro)-Met-$^{180}$Val-
Trp-Asp-Phe-Trp-Ser-Leu-Arg-Pro-Glu-$^{190}$Ser-Leu-His-Gln-Val-Ser-Phe-Leu-Phe-Ser-$^{200}$Asp-Arg-Gly-Ile-Pro-Asp-Gly-His-Arg-His-$^{210}$Met-
(His,Asx,Thr,Ser,Gly,Gly,Tyr,Phe)-Lys-$^{220}$Leu-Val-Asn-Ala-Asp-Gly-Glu-Ala-Val-Tyr-$^{230}$Cys-Lys-Phe-His-Tyr-Lys-Thr-Asp-Gln-Gly-$^{240}$Ile-
Lys-Asn-Leu-Ser-Val-Glu-Asp-Ala-Ala-$^{250}$Arg-Leu-Ala-His-Glu-Asp-Pro-Asp-Tyr-Gly-$^{260}$Leu-Arg-Asp-Leu-Phe-(Asx,Thr,Ala,Ala,Ile)-$^{270}$Gly-
Asn-Tyr-Pro-Ser-Trp-Thr-Leu-Tyr-Ile-$^{280}$Gln-Val-Met-Thr-Phe-Ser-Glu-Ala-Glu-Ile-$^{290}$Phe-Pro-Phe-Asn-Pro-Phe-Asp-Leu-Thr-Lys-$^{300}$Val-
Trp-Pro-His-Gly-Asp-Tyr-Pro-Leu-Ile-$^{310}$Pro-Val-Gly-Lys-Leu-Val-Leu-Asn-Arg-Asn-$^{320}$Pro-Val-Asn-Tyr-Phe-Ala-Glu-Val-Glu-Gln-$^{330}$Leu-
Ala-Phe-Asp-Pro-Ser-Asn-Met-Pro-Pro-$^{340}$Gly-Ile-Glu-Pro-Ser-Pro-Asp-Lys-Met-Leu-$^{350}$Gln-Gly-Arg-Leu-Phe-Ala-Tyr-Pro-Asp-Thr-$^{360}$His-
Arg-His-Arg-Leu-Gly-Pro-Asn-Tyr-Leu-$^{370}$Gln-Ile-Pro-Val-Asn-Cys-Pro-Tyr-Arg-Ala-$^{380}$Arg-Val-Ala-Asn-Tyr-Gln-Arg-Asp-Gly-Pro-$^{390}$Met-
Cys-(Asx,Asx,Asx,Asx,Glx,Pro,Pro,Gly,$^{400}$Gly,Ala,Met,Met,Tyr,Tyr)-Ser-Phe-Ser-Ala-$^{410}$Pro-Glu-His-Gln-Pro-Ser-Ala-Leu-Glu-His-$^{420}$Arg-
Thr-His-Phe-Ser-Gly-Asp-Val-Gln-Arg-$^{430}$Phe-Asn-Ser-Ala-Asn-Asp-Asp-Asn-Val-Thr-$^{440}$Gln-Val-Arg-Thr-Phe-Tyr-Leu-Lys-Val-Leu-$^{450}$Asn-

# TABLE 3 (Continued)
## ENZYMES

**Catalase**, bovine—**(Continued)**

460 / 470 / 480
Glu-Glu-Gln-Arg-Lys-Arg-Leu-Cys-Glu-Asn-Ile- Ala-Gly-His-Leu-Lys-Asp-Ala-Gln-Leu-Phe-(Glx,Ile)-Lys-Lys-Ala-Val-Lys-Asn-Phe-

490 / 500 / 505
Ser-Asp-Val-His-Pro-Glu-Tyr-Gly-Ser-Arg-Ile- Gln-Ala-Leu-Leu-Asp-Lys-Tyr-Asn-Glu-Glu-Lys-Pro-Lys-Asn-OH

**Chymotrypsinogen** A, bovine (*1, 2, 3*)

1 / 10 / 20 / 30
H-Cys-Gly-Val-Pro-Ala-Ile- Gln-Pro-Val-Leu-Ser- Gly-Leu-Ser- Arg-Ile- Val-Asn-Gly-Glu-Glu-Ala-Val-Pro-Gly-Ser- Trp-Pro-Trp-Gln-

40 / 50 / 60
Val-Ser- Leu-Gln-Asp-Lys-Thr-Gly-Phe-His- Phe-Cys-Gly-Gly-Ser- Leu-Ile- Asn-Glu-Asn-Trp-Val- Val-Thr-Ala-Ala-His-Cys-Gly-Val-

70 / 80 / 90
Thr-Thr-Ser- Asp-Val-Val- Val-Ala-Gly-Glu-Phe-Asp-Gln-Gly-Ser- Ser- Ser- Glu-Lys-Ile- Gln-Lys-Leu-Lys-Ile- Ala-Lys-Val-Phe-Lys-

100 / 110 / 120
Asn-Ser- Lys-Tyr-Asn-Ser- Leu-Thr-Ile- Asn-Asn-Asn-Ile- Thr-Leu-Leu-Lys-Leu-Ser- Thr-Ala-Ala-Ser- Phe-Ser- Gln-Thr-Val-Ser- Ala-

130 / 140 / 150
Val-Cys-Leu-Pro-Ser- Ala-Ser- Asp-Asp-Phe-Ala-Ala-Gly-Thr-Thr-Cys-Val- Thr-Thr-Gly-Trp-Gly-Leu-Thr-Arg-Tyr-Thr-Asn-Ala-Asn-

160 / 170 / 180
Thr-Pro- Asp-Arg-Leu-Gln-Gln-Ala-Ser- Leu-Pro-Leu-Leu-Ser- Asn-Thr-Asn-Cys-Lys-Lys-Tyr-Trp-Gly-Thr-Lys-Ile- Lys-Asp-Ala-Met-

190 / 200 / 210
Ile- Cys-Ala-Gly-Ala-Ser- Gly-Val-Ser- Ser- Cys-Met-Gly-Asp-Ser- Gly-Gly-Pro-Leu-Val- Cys-Lys-Lys-Asn-Gly-Ala-Trp-Thr-Leu-Val-

220 / 230 / 240
Gly-Ile- Val-Ser- Trp-Gly-Ser- Ser- Thr-Cys-Ser- Thr-Ser- Thr-Pro-Gly-Val-Tyr-Ala-Arg-Val-Thr-Ala-Leu-Val-Asn-Trp-Val-Gln-Gln-

245
Thr-Leu-Ala-Ala-Asn-OH

**Glyceraldehyde 3-Phosphate Dehydrogenase**

1 / 10 / 20 / 30
Pig (*16*): Val-Lys-Val-Gly-Val-Asp-Gly-Phe-Gly-Arg-Ile- Gly-Arg-Leu-Val-Thr-Arg-Ala-Ala-Phe-Asn-Ser- Gly-Lys-Val-Asp-Ile- Val-Ala-Ile-
Lobster (*17*): **Ac.-Ser** — **Ile** — **Ile** — — — — — — — — — — **Leu** — — — **Leu-Ser-Cys** — — **Ala-Gln-Val** — — **Val-**

40 / 50 / 60
Pig: Asn-Asp-Pro-Phe-Ile- Asp-Leu-His-Tyr-Met-Val-Tyr-Met-Phe-Glu-Tyr-Asp-Ser- Thr-His- Gly-Lys-Phe-His- Gly-Thr-Val- Lys-Ala-Glu-
Lobster: — — — — — **Ala** — **Glu** — — — — — — **Lys** — — — — — — **Val** — **Lys** — **Glu** — — **Met** —

70 / 80 / 90
Pig: Asp-Gly-Lys-Leu-Val-Ile- Asp-Gly-Lys-Ala-Ile- Thr-Ile- Phe-Gln-Glu-Arg-Asp-Pro-Ala-Asn-Ile- Lys-Trp-Gly-Asp-Ala-Gly-Thr-Ala-
Lobster: — — **Ala** — — **Val** — — — **Lys** — — **Val** — **Asn** — **Met-Lys-** — **Glu** — — **Pro** — **Ser-Lys** — — **Ala-Glu-**

100 / 110 / 120
Pig: Tyr-Val-Val-Glu-Ser- Thr-Gly-Val-Phe-Thr-Thr-Met-Glu-Lys-Ala-Gly-Ala-His-Leu-Lys-Gly-Gly-Ala-Lys-Arg-Val-Ile- Ile- Ser- Ala-
Lobster: — **Ile** — — — — — — — — — **Ile** — — — **Ser** — — **Phe** — — — — — **Lys** — **Val** — — —

130 / 140 / 150
Pig: Pro-Ser- Ala-Asp-Ala-Pro-Met-Phe-Val-Met-Gly-Val-Asn-His-Glu-Lys-Tyr-Asp-Asn-Ser- Leu-Lys-Ile- Val-Ser- Asn-Ala-Ser- Cys-Thr-
Lobster: — — — — — — — — — **Cys** — — — **Leu** — — — **Ser- Lys-Asp-Met-Thr-Val** — — — — — — —

160 / 170 / 180
Pig: Thr-Asn-Cys-Leu-Ala-Pro-Leu-Ala-Lys-Val-Ile- His-Asp-His-Phe-Gly-Ile- Val-Glu-Gly-Leu-Met-Thr-Thr-Val-His-Ala-Ile- Thr-Ala-
Lobster: — — — — — — **Val** — — — **Leu** — **Glu-Asn** — **Glu** — — — — — — — — — — — **Val** — —

# TABLE 3 (Continued)
## ENZYMES

**Glyceraldehyde 3-Phosphate dehydrogenase—(Continued)**

| | |
|---|---|
| Pig | Thr-Gln-Lys-Thr-Val-Asp-Gly-Pro-Ser-$^{190}$Gly-Lys-Leu-Trp-Arg-Asp-Gly-Arg-Gly-Ala-$^{200}$Ala-Gln-Asn-Ile-Ile-Pro-Ala-Ser-Thr-Gly-$^{210}$Ala- |
| Lobster | — — — — — — — — — **Ala** — **Asp** — — **Gly** — — — — — — — — — — **Ser** — — — — |
| Pig | Ala-Lys-Ala-Val-Gly-Lys-Val-Ile-Pro-$^{220}$Glu-Leu-Asp-Gly-Lys-Leu-Thr-Gly-Met-Ala-$^{230}$Phe-Arg-Val-Pro-Thr-Pro-Asn-Val-Ser-Val-$^{240}$Val- |
| Lobster | — — — — — — — — — — — — — — — — — — — — — — — — — **Asp** — — — — |
| Pig | Asp-Leu-Thr-Cys-Arg-Leu-Glu-Lys-Pro-$^{250}$Ala-Lys-Tyr-Asp-Asp-Ile-Lys-Lys-Val-Val-$^{260}$Lys-Gln-Ala-Ser-Glu-Gly-Pro-Leu-Lys-Gly-$^{270}$Ile- |
| Lobster | — — — **Val** — — **Gly** — **Glu-Cys-Ser-** — — — — — **Ala- Ala- Met** — **Thr** — — — — — — **Gln** — **Phe** |
| Pig | Leu-Gly-Tyr-Thr-Glu-Asp-Gln-Val-Val-$^{280}$Ser-Cys-Asp-Phe-Asn-Asp-Ser-Thr-His-Ser-$^{290}$Ser-Thr-Phe-Asp-Ala-Gly-Ala-Gly-Ile-Ala-$^{300}$Leu- |
| Lobster | — — — — — — — **Asp** — — — **Ser** — — **Ile- Gly- Asp-Asn-Arg** — — **Ile** — — — **Lys** — — — **Gln** — |
| Pig | Asn-Asp-His-Phe-Val-Lys-Leu-Ile-Ser-$^{310}$Trp-Tyr-Asp-Asn-Glu-Phe-Gly-Tyr-Ser-Asn-$^{320}$Arg-Val-Val-Asp-Leu-Met-Val-His-Met-Ala-$^{330}$Ser- |
| Lobster | **Ser- Lys- Thr** — — — **Val- Val** — — — — — — — — — — **Gln** — — **Ile** — — **Leu-Lys** — — **Gln- Lys-** |
| Pig | Lys-$^{332}$Glu-OH |
| Lobster | **Val- Asp-Ser-** $^{334}$**Ala**-OH |

**Lysozyme**

| | |
|---|---|
| Bacteriophage T4 (*4, 18*) | H-$^{1}$Met-Asn-Ile-Phe-Glu-Met-Leu-Arg-Ile-$^{10}$Asp-Glu-Gly-Leu-Arg-Leu-Lys-Ile-Tyr-Lys-$^{20}$Asp-Thr-Glu-Gly-Tyr-Tyr-Thr-Ile-Gly-Ile-$^{30}$Gly- |
| Bacteriophage T2 (*19*) | — — — — — — — — — — — — — — — — — — — — — — — — — — — — — — |
| Bacteriophage T4 | His-Leu-Leu-Thr-Lys-Ser-Pro-Ser-Leu-$^{40}$Asn-Ala-Ala-Lys-Ser-Glu-Leu-Asp-Lys-Ala-$^{50}$Ile-Gly-Arg-Asn-Cys-Asn-Gly-Val-Ile-Thr-$^{60}$Lys- |
| Bacteriophage T2 | — — — — — — — — — **Ser- Val** — — — — — — — — — — — — — — — — — — — |
| Bacteriophage T4 | Asp-Glu-Ala-Glu-Lys-Leu-Phe-Asn-Gln-$^{70}$Asp-Val-Asp-Ala-Ala-Val-Arg-Gly-Ile-Leu-$^{80}$Arg-Asn-Ala-Lys-Leu-Lys-Pro-Val-Tyr-Asp-$^{90}$Ser- |
| Bacteriophage T2 | — — — — — — — — — — — — — — — — — — — — — — — — — — — — — — |
| Bacteriophage T4 | Leu-Asp-Ala-Val-Arg-Arg-Cys-Ala-Leu-$^{100}$Ile-Asn-Met-Val-Phe-Gln-Met-Gly-Glu-Thr-$^{110}$Gly-Val-Ala-Gly-Phe-Thr-Asn-Ser-Leu-Arg-$^{120}$Met- |
| Bacteriophage T2 | — — — — — — — — — — — — — — — — — — — — — — — — — — — — — — |
| Bacteriophage T4 | Leu-Gln-Gln-Lys-Arg-Trp-Asp-Glu-Ala-$^{130}$Ala-Val-Asn-Leu-Ala-Lys-Ser-Arg-Trp-Tyr-$^{140}$Asn-Gln-Thr-Pro-Asn-Arg-Ala-Lys-Arg-Val-$^{150}$Ile- |
| Bacteriophage T2 | — — — — — — — — — — — — — — — — — — — — — — — — — — — — — — |
| Bacteriophage T4 | Thr-Thr-Phe-Arg-Thr-Gly-Thr-Trp-Asp-$^{160}$Ala-Tyr-Lys-Asn-$^{164}$Leu-OH |
| Bacteriophage T2 | **Ala** — — — — — — — — — — — — — |

## TABLE 3 (Continued)
## ENZYMES

| | |
|---|---|
| **Lysozyme,** chicken egg white (*5, 6*) | H-$\overset{1}{\text{Lys}}$-Val-Phe-Gly-Arg-Cys-Glu-Leu-Ala-$\overset{10}{\text{Ala}}$-Ala-Met-Lys-Arg-His-Gly-Leu-Asp-Asn-$\overset{20}{\text{Tyr}}$-Arg-Gly-Tyr-Ser-Leu-Gly-Asn-Trp-Val-$\overset{30}{\text{Cys}}$-<br>Ala-Ala-Lys-Phe-Glu-Ser-Asn-Phe-Asn-$\overset{40}{\text{Thr}}$-Gln-Ala-Thr-Asn-Arg-Asn-Thr-Asp-Gly-$\overset{50}{\text{Ser}}$-Thr-Asp-Tyr-Gly-Ile-Leu-Gln-Ile-Asn-$\overset{60}{\text{Ser}}$-<br>Ar-Trp-Trp-Cys-Asn-Asp-Gly-Arg-Thr-$\overset{70}{\text{Pro}}$-Gly-Ser-Arg-Asn-Leu-Cys-Asn-Ile-Pro-$\overset{80}{\text{Cys}}$-Ser-Ala-Leu-Leu-Ser-Ser-Asp-Ile-Thr-$\overset{90}{\text{Ala}}$-<br>Ser-Val-Asn-Cys-Ala-Lys-Lys-Ile-Val-$\overset{100}{\text{Ser}}$-Asp-Gly-Asp-Gly-Met-Asn-Ala-Trp-Val-$\overset{110}{\text{Ala}}$-Trp-Arg-Asn-Arg-Cys-Lys-Gly-Thr-Asp-$\overset{120}{\text{Val}}$-<br>Gln-Ala-Trp-Ile-Arg-Gly-Cys-Arg-$\overset{129}{\text{Leu}}$-OH |
| Disulfide bridges: 6–127, 30–115, 64–80, 76–94. | |
| **Nuclease-T**<br>*Staphylococcus aureus V8* (*7*) | H-$\overset{1}{\text{Ala}}$-Thr-Ser-Thr-Lys-Lys-Leu-His-Lys-$\overset{10}{\text{Glu}}$-Pro-Ala-Thr-Leu-Ile-Lys-Ala-Ile-Asp-$\overset{20}{\text{Gly}}$-Asp-Thr-Val-Lys-Leu-Met-Tyr-Lys-Gly-$\overset{30}{\text{Gln}}$-<br>Pro-Met-Thr-Phe-Arg-Leu-Leu-Leu-Val-$\overset{40}{\text{Asp}}$-Thr-Pro-Gln-Thr-Lys-His-Pro-Lys-Lys-$\overset{50}{\text{Gly}}$-Val-Glu-Lys-Tyr-Gly-Pro-Glu-Ala-Ser-$\overset{60}{\text{Ala}}$-<br>Phe-Thr-Lys-Lys-Met-Val-Glu-Asn-Ala-$\overset{70}{\text{Lys}}$-Lys-Ile-Glu-Val-Glu-Phe-Asn-Lys-Gly-$\overset{80}{\text{Gln}}$-Arg-Thr-Asp-Lys-Tyr-Gly-Arg-Gly-Leu-$\overset{90}{\text{Ala}}$-<br>Tyr-Ile-Tyr-Ala-Asp-Gly-Lys-Met-Val-$\overset{100}{\text{Asn}}$-Glu-Ala-Leu-Val-Arg-Gln-Gly-Leu-Ala-$\overset{110}{\text{Lys}}$-Val-Ala-Tyr-Val-Tyr-Lys-Pro-Asn-Asn-$\overset{120}{\text{Thr}}$-<br>His-Glu-Gln-Leu-Leu-Arg-Lys-Ser-Glu-$\overset{130}{\text{Ala}}$-Gln-Ala-Lys-Lys-Glu-Lys-Leu-Asn-Ile-$\overset{140}{\text{Trp}}$-Ser-Glu-Asn-Asp-Ala-Asp-Ser-Gly-$\overset{149}{\text{Gln}}$-OH |

## TABLE 3 (Continued)
## ENZYMES

**Papain,** papaya (*8*)

H-[1]Ile- Pro-Glu-Tyr-Val- Asp-Trp-Arg-Gln-[10]Lys-Gly-Ala-Val-Thr-Pro-Val-Lys-Asn-Gln-[20]Gly-Ser- Cys-Gly-Ser- Cys-Trp(Ala, Phe)(Ile, [30]Ile)
Arg-Asn-Thr-Pro-Tyr-Tyr-Glu-Gly-Val-[40]Gln-Arg-Tyr-Cys-Arg-Ser- Arg-Glu-Lys-Gly-[50]Pro-Tyr-Ala-Ala-Lys-Thr-Asp-Gly-Val- Arg-[60]Gln-
Val-Gln-Pro-Tyr-Asn-Gln-Gly-Ala-Leu-[70]Leu-Tyr-Ser- Ile- Ala-Asn-Gln-Pro-Ser- Val-[80]Val-Leu-Gln-Ala-Ala-Gly-Lys-Asp-Phe-Gln-[90]Leu-
Tyr-Arg-Gly-Gly-Ile- Phe-Val-Gly-Pro-[100]Cys-Gly-Asn-Lys-Val-Asp-His-Ala-Val-Ala-[110]Ala-Val-Gly-Tyr-Asn-Pro-Gly-Tyr- Ile- Leu-[120]Ile-
Lys-Asn-Ser- Trp-Gly-Thr-Gly-Trp-Gly-[130]Glu-Asn-Gly-Tyr- Ile- Arg-Ile- Lys-Thr-Gly-[140]Asn-Leu-Asn-Gln-Tyr-Ser- Glu-Gln-Glu-Leu-[150]Leu-
Asp-Cys-Asp-Arg-Arg-Ser- Tyr-Gly-Cys-[160]Tyr-Pro-Gly-Asp-Gly-Trp(Ser, Ala, Leu)Val-[170]Ala-Gln-Tyr-Gly-Ile- His-Tyr-Arg-Gly-Thr-[180]Gly-
Asn-Ser- Tyr-Gly-Val- Cys-Gly-Leu-Tyr-[190]Thr-Ser- Ser- Phe-Tyr-Pro-Val-Lys-[198]Asn-OH

Disulfide bridges: 43–152, 100–186, 22–159, active sulfhydryl-group: 25.

**Penicillinase**
*Staphylococcus aureus PC1* (*20*)

H-[1]Lys-Glu-Leu-Asn-Asp-Leu-Glu-Lys-Lys-[10]Tyr-Asn-Ala-His-Ile- Gly-Val-Tyr-Ala-Leu-[20]Asp-Thr-Lys-Ser- Gly-Lys-Glu-Val- Lys-Phe-[30]Asn-
Ser- Asp-Lys-Arg-Phe-Ala-Tyr-Ala-Ser- [40]Thr-Ser- Lys-Ala-Ile- Asn-Ser- Ala-Ile- Leu-[50]Leu-Glu-Gln-Val- Pro-Tyr-Asn-Lys-Leu-Asn-[60]Lys-
Lys-Val-His-Ile- Asn-Lys-Asp-Asp-Ile- [70]Val-Ala-Tyr-Ser- Pro-Ile- Leu-Glu-Lys-Tyr-[80]Val-Gly-Lys-Asp-Ile- Thr-Leu-Lys-Ala-Leu-[90]Ile-
Glu-Ala-Ser- Met-Thr-Tyr-Ser- Asp-Asn-[100]Thr-Ala-Asn-Asn-Lys-Ile- Ile- Lys-Glu-Ile- [110]Gly-Gly-Ile- Lys-Val-Lys-Lys-Gln-Arg-Leu-[120]Lys-
Glu-Leu-Gly-Asp-Lys-Val-Thr-Asn-Pro-[130]Val-Arg-Tyr-Glu-Ile- Glu-Leu-Asn-Tyr-Tyr-[140]Ser- Pro-Lys-Ser- Lys-Lys-Asp-Thr-Ser- Thr-[150]Pro-
Ala-Ala-Phe-Gly-Lys-Thr-Leu-Asn-Lys-[160]Leu-Ile- Ala-Asn-Gly-Lys-Leu-Ser- Lys-Glu-[170]Asn-Lys-Lys-Phe-Leu-Leu-Asp-Leu-Met-Leu-[180]Asn-
Asn-Lys-Ser- Gly-Asp-Thr-Leu-Ile- Lys-[190]Asp-Gly-Val- Pro-Lys-Asp-Tyr-Lys-Val- Ala-[200]Asp-Lys-Ser- Gly-Gln-Ala-Ile- Thr-Tyr-Ala-[210]Ser-
Arg-Asn-Asp-Val-Ala-Phe-Val-Tyr-Pro-[220]Lys-Gly-Gln-Ser- Glu-Pro-Ile- Val-Leu-Val-[230]Ile- Phe-Thr-Asn-Lys-Asp-Asn-Lys-Ser- Asp-[240]Lys-
Pro-Asn-Asp-Lys-Leu-Ile- Ser- Glu-Thr-[250]Ala-Lys-Ser- Val-Met-Lys-Glu-[257]Phe-OH

# TABLE 3 (Continued)
## ENZYMES

**Ribonuclease,** pancreatic
Beef (*9*)
Rat (*21*)

−3 1 5 10 15 20 25 30
Beef: H-Lys-Glu-Thr-Ala-Ala-Ala-Lys-Phe-Glu-Arg-Gln-His-Met-Asp-Ser-Ser-Thr-Ser-Ala-Ala-Ser-Ser-Ser-Asn-Tyr-Cys-Asn-Gln-Met-Met-
Rat: **H-Gly-Glu-Ser-Arg** — **Ser-Ser** — **Asp** — — **Lys** — — — — — **Thr-Glu-Gly-Pro-Ser-Lys** — — **Pro-Thr** — — — — — — —

35 40 45 50 55 60
Beef: Lys-Ser-Arg-Asn-Leu-Thr-Lys-Asp-Arg-Cys-Lys-Pro-Val-Asn-Thr-Phe-Val-His-Glu-Ser-Leu-Ala-Asp-Val-Gln-Ala-Val-Cys-Ser-Gln-
Rat: — **Arg-Gln-Gly-Met** — — **Gly-Ser** — — — — — — — — — — **Pro** — **Glu** — — — — **Ile** — — —

65 70 75 80 85 90
Beef: Lys-Asn-Val-Ala-Cys-Lys-Asn-Gly-Gln-Thr-Asn-Cys-Tyr-Gln-Ser-Tyr-Ser-Thr-Met-Ser-Ile-Thr-Asp-Cys-Arg-Glu-Thr-Gly-Ser-Ser-
Rat: **Gly-Gln** — **Thr** — — — — **Arg-Asp** — — **His-Lys** — **Ser** — — **Leu-Arg** — — — — — **Leu-Lys** — — —

95 100 105 110 115 120
Beef: Lys-Tyr-Pro-Asn-Cys-Ala-Tyr-Lys-Thr-Thr-Gln-Ala-Asn-Lys-His-Ile-Ile-Val-Ala-Cys-Glu-Gly-Asn-Pro-Tyr-Val-Pro-Val-His-Phe-
Rat: — — — — — **Thr** — **Asn** — — **Asn-Ser-Glu** — — — — **Ile** — — **Asp** — — — — — — — — — —

124
Beef: Asp-Ala-Ser-Val-OH
Rat: — — — — —

Disulfide bridges (beef ribonuclease): 26–84, 40–95, 58–110 and 65–72.

**Ribonuclease T$_1$** (*10*)

1 10 20 30
H-Ala-Cys-Asp-Tyr-Thr-Cys-Gly-Ser-Asn-Cys-Tyr-Ser-Ser-Ser-Asp-Val-Ser-Thr-Ala-Gln-Ala-Ala-Gly-Tyr-Gln-Leu-His-Glu-Asp-Gly-

40 50 60
Glu-Thr-Val-Gly-Ser-Asn-Ser-Tyr-Pro-His-Lys-Tyr-Asn-Asn-Tyr-Glu-Gly-Phe-Asp-Phe-Ser-Val-Ser-Ser-Pro-Tyr-Tyr-Glu-Trp-Pro-

70 80 90
Ile-Leu-Ser-Ser-Gly-Asp-Val-Tyr-Ser-Gly-Pro-Gly-Ser-Gly-Ala-Asp-Arg-Val-Val-Phe-Asn-Glu-Asn-Asn-Gln-Leu-Ala-Gly-Val-Ile-

100 104
Thr-His-Thr-Gly-Ala-Ser-Gly-Asn-Asn-Phe-Val-Glu-Cys-Thr-OH

Disulfide bridges: 2–10 and 6–103.

## TABLE 3 (Continued)
## ENZYMES

| | |
|---|---|
| **Subtilisins** (*11*) | |
| BPN′ | H-[1]Ala-Gln-Ser-Val-Pro-Tyr-Gly-Val-Ser-[10]Gln-Ile-Lys-Ala-Pro-Ala-Leu-His-Ser-Gln-[20]Gly-Tyr-Thr-Gly-Ser-Asn-Val-Lys-Val-Ala-[30]Val- |
| Carlsberg | — — **Thr** — — — — **Ile- Pro-Leu** — — — **Asp-Lys-Val-Gln-Ala** — — **Phe-Lys** — **Ala** — — — — — — |
| BPN′ | Ile-Asp-Ser-Gly-Ile-Asp-Ser-Ser-His-[40]Pro-Asn-Leu-Lys-Val-Ala-Gly-Gly-Ala-Ser-[50]Met-Val-Pro-Ser-Glu-Thr-Pro-Asn-Phe-Gln-[60]Asp- |
| Carlsberg | **Leu** — **Thr** — — **Gln-Ala** — — — — — **Asn** — **Val** — — — — **Phe** — **Ala-Gly-Gln-Ala** — **Tyr-Asn-Thr** — |
| BPN′ | Asp-Asn-Ser-His-Gly-Thr-His-Val-Ala-[70]Gly-Thr-Val-Ala-Ala-Leu-Asn-Asn-Ser-Ile-[80]Gly-Val-Leu-Gly-Val-Ala-Pro-Ser-Ser-Ala-[90]Leu- |
| Carlsberg | **Gly** — **Gly** — — — — — — — — — — — — — — **Thr-Thr** — — — — — — — — **Val-Ser** — |
| BPN′ | Tyr-Ala-Val-Lys-Val-Leu-Gly-Asn-Ala-[100]Gly-Ser-Gly-Gln-Tyr-Ser-Trp-Ile-Ile-Asn-[110]Gly-Ile-Gln-Trp-Ala-Ile-Ala-Asn-Asn-Met-[120]Asp- |
| Carlsberg | — — — — — — **Asn-Ser-Ser** — — — **Ser** — — **Gly** — **Val-Ser** — — — — — — **Thr-Thr** — **Gly** — — |
| BPN′ | Val-Ile-Asn-Met-Ser-Leu-Gly-Gly-Pro-[130]Ser-Gly-Ser-Ala-Ala-Leu-Lys-Ala-Ala-Val-[140]Asp-Lys-Ala-Val-Ala-Ser-Gly-Val-Val-Val-[150]Val- |
| Carlsberg | — — — — — — — — — **Ala** — — — **Thr** — **Met** — **Gln** — — — **Asn** — **Tyr** — **Arg** — — — — — |
| BPN′ | Ala-Ala-Ala-Gly-Asn-Gln-Gly-Ser-Thr-[160]Gly-Ser-Ser-Ser-Thr-Val-Gly-Tyr-Pro-Gly-[170]Lys-Tyr-Pro-Ser-Val-Ile-Ala-Val-Gly-Ala-[180]Val- |
| Carlsberg | — — — — — **Ser** — **Asn-Ser** — — **Thr-Asn** — **Ile** — — — **Ala** — — **Asp** — — — — — — — — |
| BPN′ | Asp-Ser-Ser-Asn-Gln-Arg-Ala-Ser-Phe-[190]Ser-Ser-Val-Gly-Pro-Glu-Leu-Asp Val-Met-[200]Ala-Pro-Gly-Val-Ser-Ile-Gln-Ser-Thr-Leu-[210]Pro- |
| Carlsberg | — — **Asn-Ser-Asn** — — — — — — — — **Ala** — — **Glu** — — — — — **Ala-Gly-Val-Tyr** — — **Tyr** — |
| BPN′ | Gly-Asn-Lys-Tyr-Gly-Ala-Tyr-Asn-Gly-[220]Thr-Ser-Met-Ala-Ser-Pro-His-Val-Ala-Gly-[230]Ala-Ala-Ala-Leu-Ile-Leu-Ser-Lys-His-Pro-[240]Asn- |
| Carlsberg | **Thr** — **Thr** — **Ala-Thr-Leu** — — — — — — — — — — — — — — — — — — — — — — |
| BPN′ | Trp-Thr-Asn-Thr-Gln-Val-Arg-Ser-Ser-[250]Leu-Gln-Asn-Thr-Thr-Thr-Lys-Leu-Gly-Asp-[260]Ser-Phe-Tyr-Tyr-Gly-Lys-Gly-Leu-Ile-Asn-[270]Val- |
| Carlsberg | **Leu-Ser-Ala-Ser** — — — **Asn-Arg** — **Ser-Ser** — **Ala** — **Tyr** — — **Ser** — — — — — — — — — — — |
| BPN′ | Gln-Ala-Ala-Ala-[275]Gln-OH |
| Carlsberg | — — — — — |
| **Trypsinogen**, bovine (*12*) | H-[1]Val-Asp-Asp-Asp-Asp-Lys-Ile-Val-Gly-[10]Gly-Tyr-Thr-Cys-Gly-Ala-Asn-Thr-Val-Pro-[20]Tyr-Gln-Val-Ser-Leu-Asn-Ser-Gly-Tyr-His-[30]Phe- |
| | Cys-Gly-Gly-Ser-Leu-Ile-Asn-Ser-Gln-[40]Trp-Val-Val-Ser-Ala-Ala-His-Cys-Tyr-Lys-[50]Ser-Gly-Ile-Gln-Val-Arg-Leu-Gly-Gln-Asp-[60]Asn- |
| | Ile-Asn-Val-Val-Glu-Gly-Asn-Gln-Gln-[70]Phe-Ile-Ser-Ala-Ser-Lys-Ser-Ile-Val-His-[80]Pro-Ser-Tyr-Asn-Ser-Asn-Thr-Leu-Asn-Asn-[90]Asp- |
| | Ile-Met-Leu-Ile-Lys-Leu-Lys-Ser-Ala-[100]Ala-Ser-Leu-Asn-Ser-Arg-Val-Ala-Ser-Ile-[110]Ser-Leu-Pro-Thr-Ser-Cys-Ala-Ser-Ala-Gly-[120]Thr- |
| | Gln-Cys-Leu-Ile-Ser-Gly-Trp-Gly-Asn-[130]Thr-Lys-Ser-Ser-Gly-Thr-Ser-Tyr-Pro-Asp-[140]Val-Leu-Lys-Cys-Leu-Lys-Ala-Pro-Ile-Leu-[150]Ser- |
| | Asn-Ser-Ser-Cys-Lys-Ser-Ala-Tyr-Pro-[160]Gly-Gln-Ile-Thr-Ser-Asn-Met-Phe-Cys-Ala-[170]Gly-Tyr-Leu-Glu-Gly-Gly-Lys-Asn-Ser-Cys-[180]Gln- |
| | Gly-Asp-Ser-Gly-Gly-Pro-Val-Val-Cys-[190]Ser-Gly-Lys-Leu-Gln-Gly-Ile-Val-Ser-Trp-[200]Gly-Ser-Gly-Cys-Ala-Gln-Lys-Asn-Lys-Pro-[210]Gly- |
| | Val-Tyr-Thr-Lys-Val-Cys-Asn-Tyr-Val-[220]Ser-Trp-Ile-Lys-Gln-Thr-Ile-Ala-Ser-[229]Asn-OH |

## TABLE 3 (Continued)
## ENZYMES

**Tryptophan synthetase A protein,** *E. coli* (*13*)

H-$^{1}$Met-Gln-Arg-Tyr-Glu-Ser-Leu-Phe-Ala-$^{10}$Gln-Leu-Lys-Glu-Arg-Lys-Glu-Gly-Ala-Phe-$^{20}$Val-Pro-Phe-Val-Thr-Leu-Gly-Asp-Pro-Gly-$^{30}$Ile-
Glu-Gln-Ser-Leu-Lys-Ile-Asp-Thr-Leu-$^{40}$Ile-Glu-Ala-Gly-Ala-Asp-Ala-Leu-Glu-Leu-$^{50}$Gly-Ile-Pro-Phe-Ser-Asp-Pro-Leu-Ala-Asp-$^{60}$Gly-
Pro-Thr-Ile-Gln-Asn-Ala-Thr-Leu-Arg-$^{70}$Ala-Phe-Ala-Ala-Gly-Val-Thr-Pro-Ala-Gln-$^{80}$Cys-Phe-Glu-Met-Leu-Ala-Leu-Ile-Arg-Gln-$^{90}$Lys-
His-Pro-Thr-Ile-Pro-Ile-Gly-Leu-Leu-$^{100}$Met-Tyr-Ala-Asn-Leu-Val-Phe-Asn-Lys-Gly-$^{110}$Ile-Asp-Glu-Phe-Tyr-Ala-Gln-Cys-Glu-Lys-$^{120}$Val-
Gly-Val-Asp-Ser-Val-Leu-Val-Ala-Asp-$^{130}$Val-Pro-Val-Gln-Glu-Ser-Ala-Pro-Phe-Arg-$^{140}$Gln-Ala-Ala-Leu-Arg-His-Asn-Val-Ala-Pro-$^{150}$Ile-
Phe-Ile-Cys-Pro-Pro-Asn-Ala-Asp-Asp-$^{160}$Asp-Leu-Leu-Arg-Gln-Ile-Ala-Ser-Tyr-Gly-$^{170}$Arg-Gly-Tyr-Thr-Tyr-Leu-Leu-Ser-Arg-Ala-$^{180}$Gly-
Val-Thr-Gly-Ala-Glu-Asn-Arg-Ala-Ala-$^{190}$Leu-Pro-Leu-Asn-His-Leu-Val-Ala-Lys-Leu-$^{200}$Lys-Glu-Tyr-Asn-Ala-Ala-Pro-Pro-Leu-Gln-$^{210}$Gly-
Phe-Gly-Ile-Ser-Ala-Pro-Asp-Gln-Val-$^{220}$Lys-Ala-Ala-Ile-Asp-Ala-Gly-Ala-Ala-Gly-$^{230}$Ala-Ile-Ser-Gly-Ser-Ala-Ile-Val-Lys-Ile-$^{240}$Ile-
Glu-Gln-His-Asn-Ile-Glu-Pro-Glu-Lys-$^{250}$Met-Leu-Ala-Ala-Leu-Lys-Val-Phe-Val-Gln-$^{260}$Pro-Met-Lys-Ala-Ala-Thr-Arg-$^{267}$Ser-OH

Data taken from: Wojciech and Margoliash, in *Handbook of Biochemistry*, 2nd ed., pp. C-234–C-241, H. A. Sober, Ed. Copyright 1970, The Chemical Rubber Co., Cleveland, Ohio.

# TABLE 4
# FERREDOXINS

| Organism | Sequence |
|---|---|
| **Bacteria** | |
| *Clostridium pasteurianum* (*1*) | H-Ala[1]-Tyr-Lys-Ile- Ala-Asp-Ser- Cys-Val- Ser[10]- Cys-Gly-Ala-Cys-Ala-Ser- Glu-Cys-Pro- Val[20]- Asn-Ala-Ile- Ser- Gln-Gly-Asp-Ser- Ile- Phe[30]- |
| *Clostridium butyricum* (*2*) | — **Phe-Val** — **Asn** — — — — — — — — — — **Gly** — — — — **Ser** — — **Thr** — — — **Thr-Gln** — |
| *Clostridium acidi-urici* (*3*) | — — **Val** — **Asn-Glu-Ala** — **Ile** — — — — — **Asp-Pro** — — — — **Asp** — — — — — — — **Arg-Tyr-** |
| *Micrococcus aerogenes* (*4*) | — — **Val** — **Asn** — — — **Ile- Ala** — — — — **Lys-Pro** — — — — — □ — **Gln** — — □ — — **Tyr-** |
| **Bacteria** | |
| *Clostridium pasteurianum* | Val- Ile- Asp-Ala-Asp-Thr-Cys-Ile- Asp-Cys[40]-Gly-Asn-Cys-Ala- Asn-Val-Cys-Pro- Val- Gly[50]-Ala-Pro-Val- Gln-Glu[55]-OH |
| *Clostridium butyricum* | — — — — — — — — — — — — — — — — — — — — — — — **Asn** — — |
| *Clostridium acidi-urici* | — — — — — — — — — — — **Ala** — — **Gly** — — — — **Asp** — — — — **Ala-** |
| *Micrococcus aerogenes* | **Ala** — — — — **Ser** — — — — — **Ser** — — **Ser** — — — — — — — **Asn-Pro** — **Asp**-OH |
| **Plants** | |
| Spinach (*5*) | H-Ala[1]-Ala-Tyr-Lys-Val-Thr-Leu-Val-Thr-Pro[10]-Thr-Gly-Asn-Val-Glu-Phe-Gln-Cys-Pro-Asp[20]-Asp-Val-Tyr-Ile- Leu-Asp-Ala-Ala-Glu-Glu[30]- |
| Alfalfa (*6*) | — **Ser** — — — **Lys** — — — — **Glu** — **Thr-Gln** — — — — — — — — — — — — **His** — — — |
| *Scenedesmus* (*7*) | — **Thr** — — — — — **Lys** — — **Ser** — **Asp-Gln-Thr-Ile- Glu** — — — — **Thr** — — — — — — — — |
| *Leucaena glauca* (*8*) | □ — Phe — — Lys-**Leu**/**Val**-Leu — — **Asp** — **Pro-Lys** — — **Glu** — — — — — — — — — — **Gln** — — — |
| **Plants** | |
| Spinach | Glu-Gly-Ile- Asp-Leu-Pro-Tyr-Ser- Cys-Arg[40]-Ala-Gly-Ser- Cys-Ser- Ser- Cys-Ala-Gly-Lys[50]-Leu-Lys-Thr-Gly-Ser-Leu-Asn-Gln-Asp-Asp[60]- |
| Alfalfa | — — — **Val** — — — — — — — — — — — — — — — — — **Val·Ala-Ala** — **Glu-Val** — — **Ser** — |
| *Scenedesmus* | **Ala** — **Leu** — — — — — — — — — **Ala** — — — — — — — **Val-Glu-Ala** — **Thr-Val-Asp** — **Ser** — |
| *Leucaena glauca* | **Leu** — — **Asp**/**Glu** — — — — — — — — — — — — — — — — **Val-Glu** — **Asp** — **Asp** — **Ser** — |
| **Plants** | |
| Spinach | Gln-Ser-Phe-Leu-Asp-Asp-Asp-Gln-Ile- Asp[70]-Glu-Gly-Trp-Val- Leu-Thr-Cys-Ala-Ala-Tyr[80]-Pro-Val-Ser- Asp-Val-Thr-Ile- Glu-Thr-His[90]- |
| Alfalfa | **Gly** — — — — — — — — **Glu** — — — — — — — **Val** — — **Ala-Lys** — — — — — — — — |
| *Scenedesmus* | — — — — — — **Ser** — **Met** — **Gly** — **Phe** — — — — **Val** — — — **Thr** — — **Cys** — — **Ala** — — |
| *Leucaena glauca* | — — — — — — **Glu** — — **Glu** — — — — — — — — — — **Arg** — — — **Val** — — — — |

## TABLE 4 (Continued)
## FERREDOXINS

| | |
|---|---|
| **Plants—(Continued)** | |
| Spinach | Lys-Glu-Glu-Glu-Leu-Thr-Ala-OH (Ala = 97) |
| Alfalfa | — — — — — — — |
| *Scenedesmus* | — — — **Asp** — **Phe**-OH (Phe = 96) |
| *Leucaena glauca* | — — — — — — **Gly**/**Ala**-OH |

Data taken from: *Handbook of Biochemistry,* 2nd ed., pp. C-242–C-243, H. A. Sober, Ed. Copyright 1970, The Chemical Rubber Co., Cleveland, Ohio.

# TABLE 5
# MISCELLANEOUS

| | |
|---|---|
| **Acyl Carrier Protein** | |
| *Escherichia coli* (*7*) | H-Ser$^{1}$- Thr-Ile- Glu-Glu-Arg-Val- Lys- Lys-Ile$^{10}$- Ile- Gly-Glu-Gln-Leu-Gly-Val- Lys-Gln-Glu$^{20}$-Glu-Val- Thr-Asp-Asn-Ala-Ser- Phe-Val- Glu$^{30}$- |
| | Asp-Leu-Gly-Ala- Asp-Ser- Leu-Asp-Thr-Val$^{40}$- Glu-Leu-Val- Met-Ala-Leu-Glu-Glu-Glu-Phe$^{50}$-Asp-Thr-Glu-Ile- Pro- Asp-Glu-Glu-Ala-Glu$^{60}$- |
| | Lys- Ile- Thr-Thr-Val- Gln-Ala- Ala- Ile- Asp$^{70}$-Tyr-Ile- Asn-Gly-His- Gln-Ala$^{77}$- OH |
| **Azurin** | |
| *Pseudomonas fluorescens* (*1*) | H-Ala$^{1}$- Glu-Cys-Ser- Val- Asp-Ile- Gln-Gly-Asn$^{10}$-Asp-Gln-Met-Gln-Phe-Asn-Thr-Asn-Ala-Ile$^{20}$- Thr-Val- **Asp-Lys-Ser- Cys-Lys**- Gln-Phe-Thr$^{30}$- |
| | Val- Asn-Leu-Ser- His- Pro- Gly-Asn-Leu-Pro$^{40}$-Lys- Asn-Val-Met-Gly-His- Asn-Trp-Val- Leu$^{50}$-Ser- **Thr-Ala- Ala- Asp-Met**-Gln-Gly-Val- Val$^{60}$- |
| | Thr-Asp-Gly-Met-Ala-Ser- Gly-Leu-Asp-Lys$^{70}$-Asp-Tyr- Leu-Lys- Pro- Asp-Asp-Ser- Arg-Val$^{80}$- Ile- **Ala- His- Thr-Lys- Leu**-Ile- Gly-Ser- Gly$^{90}$- |
| | Glu-Lys- Asp-Ser- Val- Thr-Phe-Asp-Val-Ser$^{100}$- Lys-Leu-Lys- Glu-Gly-Glu-Gln-Tyr- Met- Phe$^{110}$-Phe-Cys-Thr-Phe-Pro- Gly-His-Ser- Ala- Leu$^{120}$- |
| | Met-Lys- Gly-Thr-Leu-Thr-Leu-Lys$^{128}$-OH |
| **Bee venom protein** | |
| Apamine (*2*) | H-Cys$^{1}$-Asn-Cys-Lys- Ala- Pro- Glu-Thr-Ala- Leu$^{10}$-Cys-Ala- Arg-Arg-Cys-Gln-Gln-His$^{18}$- OH |
| Melittin (*3*) | H-Gly$^{1}$-Ile- Gly-Ala- Val- Leu-Lys- Val- Leu-Thr$^{10}$-Thr-Gly-Leu-Pro- Ala- Leu-Ile- Ser- Trp-Ile$^{20}$- Lys- Arg-Lys- Arg-Gln-Gln$^{26}$-OH |
| **Clupeine,** Pacific herring ( ) | |
| Clupeine Z (*4*) | H-Ala$^{1}$- Arg-Arg-Arg-Arg-Ser- Arg-Arg-Ala- Ser$^{10}$- Arg-Pro- Val- Arg-Arg-Arg-Arg-Pro- Arg-Arg$^{20}$-Val- Ser- Arg-Arg-Arg-Arg-Ala- Arg-Arg-Arg$^{30}$-Arg-OH |
| Clupeine YII (*4*) | **H-Pro**$^{1}$ — — — **Thr** — — — — — —$^{10}$ — — — — — — — — — —$^{20}$ — — — — — — — — — —$^{30}$ |
| **Hemerythrin** | |
| *Golfingia gouldii* (*8*) | H-Gly$^{1}$-Phe-Pro- Ile- Pro- Asp-Pro- Tyr- Val- Asp$^{10}$-Trp-Pro- Ser- Phe- Arg-Thr-Phe- Tyr- Ser- Ile$^{20}$- Ile- Asp-Asp-Glu-His-Lys- Thr- Leu-Phe-Asn$^{30}$- |
| | Gly-Ile- Phe-His- Leu-Ala- Ile- Asp-Asp-Asn$^{40}$-Ala-Asp-Asn-Leu-Gly-Glu-Leu-Arg-Arg-Cys$^{50}$-Thr-Gly-Lys- His- Phe-Leu-Asn-Gln-Glu-Val$^{60}$- |
| | Leu-Met-Gln-Ala-Ser- Gln-Tyr- Gln-Phe-Tyr$^{70}$- Asp-Glu-His- Lys- Lys- Glu-His- Glu-Gly-Phe$^{80}$-Ile- His- Ala- Leu-Asp-Asn-Trp-Lys-Gly-Asp$^{90}$- |
| | Val- Lys- Trp-Ala- Lys-Ser- Trp-Leu-Val- Asn$^{100}$-His- Ile- Lys- Thr-Ile- Asp-Phe-Lys- Tyr- Lys$^{110}$- Gly-Lys- Ile- OH |

# TABLE 5 (Continued)
## MISCELLANEOUS

| | |
|---|---|
| **Histone IV**<br>Calf thymus (*9*) | Ac-[1]Ser- Gly-Arg-Gly-Lys-Gly-Gly-Lys-Gly-[10]Leu-Gly-Lys-Gly-Gly-Ala-Lys(Ac)-Arg-His-Arg-[20]Lys(Me)-Val-Leu-Arg-Asp-Asn-Ile- Gln-Gly-Ile- [30]Thr-<br>Lys-Pro-Ala-Ile- Arg-Arg-Leu-Ala-Arg-[40]Gly-Gly-Val-Lys-Arg-Arg-Ile- Ser- Gly-Leu-[50]Ile- Tyr-Glu-Glu-Thr-Arg-Gly-Val-Leu-Lys-[60]Val-<br>Phe-Leu-Glu-Asn-Val-Ile- Arg-Asp-Ala-[70]Val-Thr-Tyr-Thr-Glu-His- Ala-Lys-Arg-Lys-[80]Thr-Val-Thr-Ala-Met-Asp-Val-Val-Tyr-Ala-[90]Leu-<br>Lys-Arg-Gln-Gly-Arg-Thr-Leu-Tyr-Gly-[100]Phe-Gly-Gly-OH |
| **α-Lactalbumin,** bovine (*5*) | H-[1]Glu-Glu-Leu-Thr-Lys-Cys-Glu-Val- Phe-[10]Arg-Glu-Leu-Lys-Asp-Leu-Lys-Gly-Tyr-Gly-[20]Gly-Val-Ser- Leu-Pro-Glu-Trp-Val-Cys-Thr-[30]Thr-<br>Phe-His-Thr-Ser- Gly-Tyr-Asx-Thr-Glx([40]Ala, Ile, Val, Glx)Asx-Asx(Glx,Ser, Thr,Asx)[50]Tyr-Gly-Leu-Phe(Glx, Ile, Asx,Asx)Lys-Ile- [60]Trp-<br>Cys-Lys-Asx-Asx-Glx-Asx-Pro-His-Ser- [70]Ser- Asx-Ile- Cys-Asn-Ile- Ser- Cys-Asp-Lys-[80]Phe-Leu-Asx-Asx-Asx-Leu-Thr(Asx,Asx,Ile) [90]Met-<br>Cys-Val- Lys-Lys-Ile- Leu-Asp-Lys-Val-[100]Gly-Ile- Asn-Tyr-Trp-Leu-Ala-His- Lys-Ala-[110]Leu-Cys-Ser- Glu-Lys-Leu-Asp-Gln-Trp-Leu-[120]Cys-<br>Glu-Lys-[123]Leu-OH |
| **β-Lipotropin (β-LPH)**, sheep (*6*) | H-[1]Glu-Leu-Gly-Thr-Glu-Arg-Leu-Glu-Gln-[10]Ala-Arg-Gly-Pro-Glu-Ala-Ala-Glu-Glu-Ser- [20]Ala-Ala-Ala-Arg-Ala-Glu-Leu-Glu-Tyr-Gly-[30]Leu-<br>Val- Ala-Glu-Ala-Gln-Ala-Ala-Glu-Lys-[40]Lys-Asp-Ser- Gly-Pro-Tyr-Lys-Met-Glu-His-[50]Phe-Arg-Trp-Gly-Ser- Pro-Pro-Lys-Asp-Lys-[60]Arg-<br>Tyr-Gly-Gly-Phe-Met-Thr-Ser- Glu-Lys-[70]Ser- Gln-Thr-Pro-Leu-Val-Thr-Leu-Phe-Lys-[80]Asn-Ala-Ile- Lys-Lys-Asn-His-Ala-Lys-Gly-[90]Gln-OH |
| **Neurotoxin α, Cobra venom** (*10*)<br>*Naja haje haje*<br>*Naja nigricollis* | <br>H-[1]Leu-Gln-Cys-His- Asn-Gln-Gln-Ser- Ser- [10]Gln-Pro-Pro-Thr-Thr-Lys-Thr-Cys-Pro-Gly-[20]Glu-Thr-Asn-Cys-Tyr-Lys-Lys-Arg-Trp-Arg-[30]Asp-<br>— Glu — — — (Glx, Glx, Ser, Ser, Glx, Pro, Pro, Thr, Thr)— — — (Pro, Gly, Glx, Thr, Asx, Cys)— — — Val — — — |
| *Naja haje haje*<br>*Naja nigricollis* | His- Arg-Gly-Ser- Ile- Thr-Glu-Arg-Gly-[40]Cys-Gly-Cys-Pro-Ser- Val-Lys-Lys-Gly-Ile- [50]Glu-Ile- Asn-Cys-Cys-Thr-Thr-Asp-Lys-Cys-[60]Asn-<br>— — — Thr — Ile — — — — (Gly, Cys, Pro, Thr, Val, Lys, Pro, Gly, Ile) Lys-Leu — — (Cys, Thr, Thr, Asx)— — — |
| *Naja haje haje*<br>*Naja nigricollis* | [61]Asn-OH<br>— — |

## TABLE 5 (Continued)
## MISCELLANEOUS

| | |
|---|---|
| **Thioredoxin** | |
| *Escherichia coli* B (*11*) | H-$^{1}$Ser- Asp-Lys-Ile- Ile- His-Leu-Thr-Asp-$^{10}$Asp-Ser-Phe-Asp-Thr-Asp-Leu-Val-Lys-Ala-$^{20}$Asp-Gly-Ala-Ile- Leu-Val-Asp-Phe-Trp-Ala-$^{30}$Glu-<br>Trp-Cys-Gly-Pro-Cys-Lys-Met-Ile- Ala-$^{40}$Pro-Ile- Leu-Asp-Glu-Ile- Ala-Asp-Glu-Tyr-$^{50}$Gln-Gly-Lys-Leu-Thr-Val-Ala-Lys-Leu-Asn-$^{60}$Ile-<br>Asp-Gln-Asn-Pro-Gly-Thr-Ala-Pro-Lys-$^{70}$Tyr-Ile- Gly-Arg-Gly-Ile- Pro-Thr-Leu-Leu-$^{80}$Leu-Phe-Lys-Asn-Gly-Glu-Val-Ala-Ala-Thr-$^{90}$Lys-<br>Val-Gly-Ala-Leu-Ser- Lys-Gly-Gln-Leu-$^{100}$Lys-Glu-Phe-Leu-Asp-$^{105}$Ala-Asn-Leu-Ala-OH<br>(disulfide bridge: Cys–Cys, residues 32–35) |
| **Trypsin inhibitors** | |
| Ascaris (*Ascaris lumbricoides* var. suum)(*13*) | H-$^{1}$Glu-Ala-Glu-Lys-Cys(Asx,Glx,Glx,Pro,$^{10}$Gly,Trp)Thr-Lys-Gly-Gly-Cys-Glu-Thr-Cys-$^{20}$Gly-Cys-Ala-Gln-Lys-Ile- Val-Pro-Cys-Thr-$^{30}$Arg-<br>Glu-Thr-Lys-Pro-Asn-Pro-Gln-Cys-Pro-$^{40}$Arg-Lys-Gln-Cys-Cys-Ile- Ala-Ser- Ala-Gly-$^{50}$Phe-Val-Arg-Asp-Ala-Gln-Gly-Asn-Cys-Ile- $^{60}$Lys-<br>Phe-Glu-Asp-Cys-Pro-$^{66}$Lys-OH |
| Bovine pancreatic (Kazal's)(*12*) | H-$^{1}$Asn-Ile- Leu-Gly-Arg-Glu-Ala-Lys-Cys-$^{10}$Thr-Asn-Glu-Val-Asn-Gly-Cys-Pro-Arg-Ile- $^{20}$Tyr-Asn-Pro-Val-Cys-Gly-Thr-Asp-Gly-Val-$^{30}$Thr-<br>Tyr-Ser- Asn-Glu-Cys-Leu-Leu-Cys-Met-$^{40}$Glu-Asn-Lys-Glu-Arg-Gln-Thr-Pro-Val- Leu-$^{50}$Ile- Gln-Lys-Ser- Gly-Pro-$^{56}$Cys-OH |
| Bovine pancreatic (Kunitz') (*15, 16*) (Kallikrein inhibitor) | H-$^{1}$Arg-Pro-Asp-Phe-Cys-Leu-Glu-Pro-Pro-$^{10}$Tyr-Thr-Gly-Pro-Cys-Lys-Ala-Arg-Ile- Ile- $^{20}$Arg-Tyr-Phe-Tyr-Asn-Ala-Lys-Ala-Gly-Leu-$^{30}$Cys-<br>Gln-Thr-Phe-Val-Tyr-Gly-Gly-Cys-Arg-$^{40}$Ala-Lys-Arg-Asn-Asn-Phe-Lys-Ser- Ala-Glu-$^{50}$Asp-Cys-Met-Arg-Thr-Cys-Gly-Gly-$^{58}$Ala-OH |
| Disulfide bridges: 5–55, 14–38, 30–51. | |
| **Viscotoxin A3** | |
| *Viscum album L*, Loranthaceae (*14*) | H-$^{1}$Lys-Ser- Cys($HO_3S$)-Cys($SO_3H$)-Pro-Asn-Thr-Thr-Gly-$^{10}$Arg-Asn-Ile- Tyr-Asn-Ala-Cys($SO_3H$)-Arg-Leu-Thr-$^{20}$Gly-Ala-Pro-Arg-Pro-Thr-Cys($SO_3H$)-Ala-Lys-Leu-$^{30}$Ser-<br>Gly-Cys($SO_3H$)-Lys-Ile- Ile- Ser- Gly-Ser- Thr-$^{40}$Cys($SO_3H$)-Pro-Ser- Tyr-Pro-Asp-$^{46}$Lys-OH |

Data taken from: *Handbook of Biochemistry*, 2nd ed., pp. C-278–C-280, H. A. Sober, Ed. Copyright 1970, The Chemical Rubber Co., Cleveland, Ohio.

## REFERENCES


### Coat Proteins

1. Anderer, Wittmann-Liebold and Wittmann, *Z. Naturforsch. Teil B, 20,* 1203 (1965).
2. Wittmann-Liebold and Wittmann, *Z. Naturforsch. Teil B, 18,* 1032 (1963).
3. Weber and Konigsberg, *J. Biol. Chem., 242,* 3563 (1967).
4. Weber, *Biochemistry, 6,* 3144 (1967).
5. Wallis and Naughton, Unpublished Data, cited in Reference 3.
6. Wittmann-Liebold, *Z. Naturforsch. Teil B, 21,* 1249 (1966).
7. Braunitzer, Asbeck, Beyreuther, Kohler and von Wettstein, *Hoppe-Seyler's Z. Physiol. Chem., 348,* 725 (1967).
8. Nozu and Okada, *J. Mol. Biol., 35,* 643 (1968).

### Eukaryotic Cytochromes *c*

1. Matsubara and Smith, *J. Biol. Chem., 238,* 2732 (1963).
2. Needleman and Margoliash, Unpublished Data.
3. Rothfus and Smith, *J. Biol. Chem., 240,* 4277 (1965).
4. Margoliash, Smith, Kreil and Tuppy, *Nature, 192,* 1125 (1961).
5. Walasek and Margoliash, Unpublished Data.
6. Nakashima, Higa, Matsubara, Benson and Yasunobu, *J. Biol. Chem., 241,* 1166 (1966).
7. Stewart and Margoliash, *Can. J. Biochem., 43,* 1187 (1965).
8. Chan, Needleman, Stewart and Margoliash, Unpublished Data.
9. McDowall and Smith, *J. Biol. Chem., 240,* 4635 (1965).
10. Needleman and Margoliash, *J. Biol. Chem., 241,* 853 (1966).
11. Goldstone and Smith, *J. Biol. Chem., 241,* 4480 (1966).
12. Nolan and Margoliash, *J. Biol. Chem., 241,* 1049 (1966).
13. Chan and Margoliash, *J. Biol. Chem., 241,* 507 (1966).
14. Chan, Tulloss and Margoliash, Unpublished Data.
15. Chan, Tulloss and Margoliash, *Biochemistry, 5,* 2586 (1966).
16. Bahl and Smith, *J. Biol. Chem., 240,* 3585 (1965).
17. Chan, Walasek, Barlow and Margoliash, Unpublished Data.
18. Kreil, *Hoppe-Seyler's Z. Physiol. Chem., 334,* 154 (1963).
19. Kreil, *Hoppe-Seyler's Z. Physiol. Chem., 340,* 86 (1965).
20. Goldstone and Smith, *J. Biol. Chem., 242,* 4702 (1967).
21. Chan and Margoliash, *J. Biol. Chem., 241,* 335 (1966).
22. Chan, Unpublished Data.
23. Chan, Tulloss and Margoliash, Unpublished Data.
24. Nolan, Weiss, Adams and Margoliash, Unpublished Data.
25. Yaoi, Titani and Narita, *J. Biochem. (Tokyo), 59,* 247 (1966).
26. Narita and Titani, *Proc. Jap. Acad., 41,* 831 (1965).
27. Heller and Smith, *J. Biol. Chem., 241,* 3165 (1966).
28. Stevens, Glazer and Smith, *J. Biol. Chem., 242,* 2764 (1967).
29. Nolan, Uzzell, Fitch, Weiss and Margoliash, Unpublished Data.

### Enzymes

1. Hartley, *Nature, 201,* 1284 (1964).
2. Hartley and Kaufmann, *Biochem. J., 101,* 229 (1966).
3. Hartley, Brown, Kaufmann and Smillie, *Nature, 207,* 1157 (1965).
4. Inouye and Tsugita, *J. Mol. Biol., 22,* 193 (1966).
5. Canfield, *J. Biol. Chem., 238,* 2698 (1963).
6. Canfield and Liu, *J. Biol. Chem., 240,* 1997 (1965).
7. Taniuchi, Cusumano, Anfinsen and Cone, *J. Biol. Chem., 243,* 4775 (1968), revised.
8. Light, Frater, Kimmel and Smith, *Proc. Nat. Acad. Sci. U.S.A., 52,* 1276 (1964).
9. Smyth, Stein and Moore, *J. Biol. Chem., 238,* 227 (1963).
10. Takahashi, *J. Biol. Chem., 240,* 4117 (1965).
11. Smith, Markland, Kasper, DeLange, Landon and Evans, *J. Biol. Chem., 241,* 5974 (1966); Markland and Smith, *J. Biol. Chem., 242,* 5198 (1967).
12. Mikes, Holeysovsky, Tomasek and Sörm, *Biochem. Biophys. Res. Commun., 24,* 346 (1966).
13. Yanofsky, Drapeau, Gest and Carlton, *Proc. Nat. Acad. Sci. U.S.A., 57,* 296 (1967).
14. Weber, *Nature, 218,* 1116 (1968).

15. Schroeder, Shelton, Shelton, Robberson and Apell, *Arch. Biochem. Biophys., 131,* 653 (1969).
16. Harris and Perham, *Nature, 219,* 1025 (1968).
17. Davidson, Sajgo, Noller and Harris, *Nature, 216,* 1181 (1967).
18. Tsugita and Inouye, *J. Mol. Biol., 37,* 201 (1968).
19. Inouye and Tsugita, *J. Mol. Biol., 37,* 213 (1968).
20. Ambler and Meadway, *Nature, 222,* 24 (1969).
21. Beintema and Gruber, *Biochim. Biophys. Acta, 147,* 612 (1967).

**Ferredoxins**

1. Tanaka, Nakashima, Benson, Mower and Yasunobu, *Biochemistry, 5,* 1666 (1966).
2. Benson, Mower and Yasunobu, *Proc. Nat. Acad. Sci. U.S.A., 55,* 1532 (1966).
3. Rall, Bolinger and Cole, *Biochemistry, 8,* 2486 (1969).
4. Tsunoda, Yasunobu and Whiteley, *J. Biol. Chem., 243,* 6262 (1968).
5. Matsubara, Sasaki and Chain, *Proc. Nat. Acad. Sci. U.S.A., 57,* 439 (1967).
6. Keresztes-Nagy, Perini and Margoliash, *J. Biol. Chem., 244,* 981 (1969).
7. Sugena and Matsubara, *J. Biol. Chem., 244,* 2979 (1969).
8. Benson and Yasunobu, *J. Biol. Chem., 244,* 955 (1969).

**Miscellaneous**

1. Ambler and Brown, *J. Mol. Biol., 9,* 825 (1964).
2. van Haux, Sawerthal and Habermann, *Hoppe-Seyler's Z. Physiol. Chem., 348,* 737 (1967).
3. Habermann and Jentsch, *Hoppe-Seyler's Z. Physiol. Chem., 348,* 37 (1967).
4. Ando and Suzuki, *Biochim. Biophys. Acta, 121,* 427 (1966).
5. Brew, Vanaman and Hill, *J. Biol. Chem., 242,* 3747 (1967).
6. Li, Barnafi, Chretien and Chung, *Nature, 208,* 1093 (1965).
7. Vanaman, Wakil and Hill, *J. Biol. Chem., 243,* 6420 (1968).
8. Klippenstein, Holleman and Klotz, *Biochemistry, 7,* 3868 (1968).
9. DeLange, Fambrough, Smith and Bonner, *J. Biol. Chem., 244,* 319 (1969).
10. Botes and Strydom, *J. Biol. Chem., 244,* 4147 (1969).
11. Holmgren, *Eur. J. Biochem., 6,* 475 (1968).
12. Green and Bartelt, *J. Biol. Chem., 244,* 2646 (1969).
13. Fraefel and Acher, *Biochim. Biophys. Acta, 154,* 615 (1968).
14. Samuelson, Seger and Olson, *Acta Chem. Scand., 22,* 2624 (1968).
15. Kassell, Radicevic, Ansfield and Laskowski, Sr., *Biochem. Biophys. Res. Commun., 18,* 255 (1965).
16. Kassell and Laskowski, Sr., *Biochem. Biophys. Res. Commun., 20,* 463 (1966).

## SOLUBILITIES OF THE AMINO ACIDS IN WATER AT VARIOUS TEMPERATURES*

The values given below represent solubilities in grams of amino acid per kilogram of water. None of the experimental observations were made at temperatures above 70°C. In most cases the experimental values were obtained both from a colder, originally unsaturated solution and a warmer, originally saturated solution. As judged by agreement between investigators, many of the values are not reliable to better than 1 to 2%, even where four significant figures are given. Absence of a value at a given temperature indicates that, in the opinion of the compiler, extrapolation of the curve from lower temperatures could not be justified.

| 0°C | 10°C | 20°C | 30°C | 40°C | 50°C | 60°C | 70°C | 80°C | 90°C | 100°C | Reference |
|---|---|---|---|---|---|---|---|---|---|---|---|
| **L-Alanine** | | | | | | | | | | | |
| 127.3 | 141.7 | 157.8 | 175.7 | 195.7 | 217.9 | 242.6 | 270.2 | 300.8 | 335.0 | 373.0 | 1, 2 |
| **DL-Alanine** | | | | | | | | | | | |
| 121.1 | 137.8 | 156.7 | 178.3 | 202.9 | 230.9 | 262.7 | 299.0 | 340.1 | 387.0 | 440.4 | 1, 2, 3 |
| **L-Arginine·HCl** | | | | | | | | | | | |
| 400.0 | 553.0 | 718.0 | 931.0 | 1240.0 | – | – | – | – | – | – | 2 |
| **L-Asparagine·$H_2O$** | | | | | | | | | | | |
| 8.49 | 14.29 | 23.5 | 37.79 | 59.37 | 91.18 | 136.8 | 200.6 | 287.7 | 403.0 | 551.7 | 4 |
| **L-Aspartic Acid** | | | | | | | | | | | |
| 1.72 | 2.82 | 4.18 | 5.94 | 8.38 | 11.99 | 17.01 | 24.14 | 34.25 | 48.59 | 68.93 | 1, 3 |
| **DL-Aspartic Acid** | | | | | | | | | | | |
| 2.62 | 4.12 | 6.33 | 9.50 | 14.0 | 20.0 | 28.0 | 38.4 | 51.4 | 67.3 | 85.9 | 1, 3 |
| **L-Cystine** | | | | | | | | | | | |
| 0.0502 | 0.0686 | 0.0938 | 0.1281 | 0.1751 | 0.2394 | 0.3272 | 0.4472 | 0.612 | 0.836 | 1.142 | 4 |

* Data taken from: Hutchens, J. O., in *Handbook of Biochemistry,* 2nd ed., p. B-65, H. A. Sober, Ed. Copyright 1970, The Chemical Rubber Co., Cleveland, Ohio.

| 0°C | 10°C | 20°C | 30°C | 40°C | 50°C | 60°C | 70°C | 80°C | 90°C | 100°C | Reference |
|---|---|---|---|---|---|---|---|---|---|---|---|
| **L-Diiodotyrosine** | | | | | | | | | | | |
| 0.204 | 0.318 | 0.494 | 0.769 | 1.197 | 1.862 | 2.897 | 4.508 | 7.015 | 10.91 | 16.98 | 1 |
| **L-Glutamic Acid** | | | | | | | | | | | |
| 3.29 | 4.98 | 7.20 | 10.19 | 14.70 | – | – | – | – | – | – | 2 |
| **DL-Glutamic Acid** | | | | | | | | | | | |
| 8.55 | 12.13 | 17.22 | 24.47 | 34.75 | 49.34 | 70.06 | 99.50 | 141.3 | 200.8 | 284.9 | 1, 3 |
| **Glycine** | | | | | | | | | | | |
| 141.8 | 180.4 | 225.2 | 275.9 | 331.6 | 391.0 | 452.6 | 513.9 | 572.7 | 626.2 | 671.7 | 1, 3 |
| **L-Hydroxyproline** | | | | | | | | | | | |
| 288.6 | 315.6 | 345.2 | 377.6 | 413.0 | 451.7 | 494.1 | 540.4 | 591.0 | 646.5 | 706.9 | 5 |
| **L-Isoleucine** | | | | | | | | | | | |
| –[a] | 32.0 | 33.6 | 35.4 | 37.5 | – | – | – | – | – | – | 2 |
| **DL-Isoleucine** | | | | | | | | | | | |
| 18.26 | 19.52 | 21.23 | 23.50 | 26.47 | 303. | 35.39 | 42.01 | 50.75 | 62.37 | 78.02 | 1, 3 |
| **L-Leucine** | | | | | | | | | | | |
| 22.70 | 23.01 | 23.74 | 24.90 | 26.58 | 28.87 | 31.89 | 35.84 | 40.98 | 47.65 | 56.38 | 1 |
| **DL-Leucine** | | | | | | | | | | | |
| 7.97 | 8.56 | 9.39 | 10.51 | 12.03 | 14.06 | 16.78 | 20.46 | 25.46 | 32.38 | 42.06 | 1, 3 |
| **L-Lysine·HCl** | | | | | | | | | | | |
| 462.0 | 556.0 | 666.0 | 799.0 | 965.0 | – | – | – | – | – | – | 2 |

| 0°C | 10°C | 20°C | 30°C | 40°C | 50°C | 60°C | 70°C | 80°C | 90°C | 100°C | Reference |
|---|---|---|---|---|---|---|---|---|---|---|---|
| **L-Methionine** | | | | | | | | | | | |
| 36.8 | 43.7 | 51.4 | 60.4 | 70.9 | – | – | – | – | – | – | 2 |
| **DL-Methionine** | | | | | | | | | | | |
| 18.18 | 23.41 | 29.95 | 38.13 | 48.23 | 60.70 | 75.95 | 94.52 | 116.9 | 143.9 | 176.0 | 4 |
| **DL-Norleucine** | | | | | | | | | | | |
| 8.43 | 9.43 | 10.71 | 12.36 | 14.49 | 17.27 | 20.88 | 25.7 | 32.02 | 40.60 | 52.29 | 1, 3 |
| **L-Phenylalanine** | | | | | | | | | | | |
| 19.83 | 23.29 | 27.35 | 32.13 | 37.73 | 44.31 | 52.04 | 61.11 | 71.78 | 84.29 | 99.00 | 4 |
| **DL-Phenylalanine** | | | | | | | | | | | |
| 9.97 | 11.33 | 13.07 | 15.29 | 18.15 | 21.87 | 26.71 | 33.12 | 41.66 | 53.16 | 68.8 | 1, 3 |
| **L-Proline** | | | | | | | | | | | |
| 1274.0 | 1403.0 | 1545.0 | 1703.0 | 1876.0 | 2066.0 | 2277.0 | 2508.0 | 2764.0 | 3045.0 | 3355.0 | 5 |
| **L-Serine** | | | | | | | | | | | |
| 133.0 | 247.0 | 362.0 | 476.0 | 592.0 | – | – | – | – | – | – | 2 |
| **DL-Serine** | | | | | | | | | | | |
| 22.04 | 31.03 | 42.95 | 58.52 | 78.42 | 103.4 | 134.1 | 171.1 | 214.8 | 265.4 | 322.4 | 4 |
| **Taurine** | | | | | | | | | | | |
| 39.31 | 59.92 | 87.84 | 123.8 | 167.8 | 218.8 | 274.2 | 330.5 | 383.1 | 427.0 | 457.6 | 4 |
| **L-Tryptophan** | | | | | | | | | | | |
| 8.23 | 9.27 | 10.57 | 12.23 | 14.35 | 17.06 | 20.57 | 25.14 | 31.16 | 39.14 | 49.87 | 4 |

| 0°C | 10°C | 20°C | 30°C | 40°C | 50°C | 60°C | 70°C | 80°C | 90°C | 100°C | Reference |
|---|---|---|---|---|---|---|---|---|---|---|---|
| **L-Tyrosine** | | | | | | | | | | | |
| 0.196 | 0.274 | 0.384 | 0.537 | 0.752 | 1.052 | 1.473 | 2.1 | 2.884 | 4.036 | 5.7 | 1, 3 |
| **DL-Tyrosine** | | | | | | | | | | | |
| 0.147 | 0.208 | 0.294 | 0.417 | 0.743 | 0.836 | 1.2 | 1.7 | 2.4 | 3.4 | 4.8 | 6 |
| **L-Valine** | | | | | | | | | | | |
| –[b] | 53.7 | 56.5 | 59.7 | 63.6 | – | – | – | – | – | – | 2, 7 |
| **DL-Valine** | | | | | | | | | | | |
| 59.6 | 63.3 | 68.1 | 74.2 | 81.7 | 91.1 | 102.8 | 117.4 | 135.8 | 158.9 | 188.1 | 1, 3 |

[a] Experimental value of 33 g/kg water at 1°C.[2]

[b] Experimental value of 53.4 g/kg water at 1°C.[2]

## REFERENCES

1. Dalton and Schmidt, *J. Biol. Chem., 103,* 549 (1933).
2. Hade, Jr., Thesis. University of Chicago, Chicago, Illinois (1962).
3. Dunn, Ross and Read, *J. Biol. Chem., 103,* 579 (1933).
4. Dalton and Schmidt, *J. Biol. Chem., 109,* 241 (1935).
5. Tomiyama and Schmidt, *J. Gen. Physiol., 19,* 379 (1936).
6. Winnick and Schmidt, *J. Gen. Physiol., 18,* 889 (1935).
7. Dalton and Schmidt, *J. Gen. Physiol., 19,* 767 (1936).

# AMINO ACID ANTAGONISTS

| Amino acid | Antagonists | System | Reference |
|---|---|---|---|
| **α-Alanine** | α-Aminoethanesulfonic acid | Bacteria | *1* |
| | | Mouse tumor | *2* |
| | Glycine | Bacteria | *3* |
| | α-Aminoisobutyric acid | Bacteria | *4* |
| | Serine | Bacteria | *3* |
| **D-Alanine** | D-Cycloserine | Bacterial cell wall | *5–9, 10* |
| | *O*-Carbamyl-D-serine | Bacterial cell wall | *11* |
| | D-α-Aminobutyric acid | Bacterial cell wall | *12* |
| **β-Alanine** | β-Aminobutyric acid | Yeast | *13* |
| | Propionic acid | Bacteria | *14* |
| | Asparagine | Yeast | *15* |
| | D-Serine | Bacteria | *16* |
| **Arginine** | Canavanine | Yeast, *Neurospora*, Bacteria | *17, 18–22* |
| | | Carcinosarcoma | *23* |
| | | Animals | *24, 25* |
| | | Plants | *26, 27* |
| | | Tissue Culture | *28, 29* |
| | Lysine | Arginase | *30* |
| | Ornithine | Arginase | *31* |
| | Homoarginine | Bacteria | *22, 32* |
| **Aspartic acid** | Cysteic acid | Bacteria | *1, 33, 34* |
| | | Bacteria | *35* |
| | β-Hydroxyaspartic acid | Bacteria | *36, 34, 37* |
| | Diaminosuccinic acid | Bacteria | *36* |
| | Aspartophenone | Bacteria, yeast | *4* |
| | α-Aminolevulinic acid | Bacteria, yeast | *4* |
| | α-Methylaspartic acid | Bacteria | *38* |
| | β-Aspartic acid hydrazide | Bacteria | *38* |
| | *S*-Methylcysteine sulfoxide | Bacteria | *39* |
| | β-Methylaspartic acid | Bacteria | *40* |
| | Hadicidin | Purine biosynthesis | *41* |
| **Asparagine** | 2-Amino-2-carboxyethane-sulfonamide | *Neurospora* | *42* |
| **Cysteine** | Allylglycine | Bacteria, yeast | *43* |
| **α, ε-Diaminopimelic acid** | α,α-Diaminosuberic acid | Bacteria | *44* |
| | α,α-Diaminosebacic acid | Bacteria | *44* |
| | β-Hydroxy-α,ε-diaminopimelic acid | *Escherichia coli* | *45* |
| | γ-Methyl-α,ε-diaminopimelic acid | *E. coli* | *46* |
| | Cystine | *E. coli* | *47* |
| **Glutamic acid** | Methionine sulfoxide | Bacteria | *48, 49* |
| | | Glutamine synthesis | *50* |
| | γ-Glutamylethylamide | Bacteria | *51* |
| | β-Hydroxyglutamic acid | Bacteria | *52, 53* |
| | Methionine sulfoximine | Bacteria | *54, 55* |
| | α-Methylglutamic acid | Enzymes | *56, 57* |
| | γ-Phosphonoglutamic acid | Glutamine synthesis | *58* |
| | *P*-Ethyl-γ-phosphonoglutamic acid | Glutamine synthesis | *58* |
| | γ-Fluoroglutamic acid[a] | Glutamine synthesis | *59* |
| **Glutamine** | *S*-Carbamylcysteine | Bacteria | *60* |
| | | Ascites cells | *61* |
| | *O*-Carbamylserine | Bacteria | *62* |
| | *O*-Carbazylserine | Bacteria | *63* |
| | 3-Amino-3-carboxypropane-sulfonamide | *E. coli* | *64* |
| | *N*-Benzylglutamine | *Streptococcus lactis* | *65* |
| | Azaserine | Enzymes | *66* |
| | 6-Diazo-5-oxonorleucine | Enzymes | *66* |
| | γ-Glutamylhydrazide | *S. faecalis* | *67* |
| **Glycine** | α-Aminomethanesulfonic acid | Bacteriophage | *68* |
| | | Vaccinia virus | *69* |
| | | Bacteria | *1* |
| | | *E. coli* | *4* |
| **Histidine** | D-Histidine | Histidase | *70* |
| | Imidazole | | *70* |
| | 2-Thiazolealanine | *E. coli* | *71* |
| | 1,2,4-Triazolealanine | *E. coli* | *71, 72* |
| | | Salmonella | *73* |

| Amino acid | Antagonists | System | Reference |
|---|---|---|---|
| **Isoleucine** | Leucine | Bacteria | *74* |
| | | Rats | *75* |
| | Methallylglycine | Bacteria, yeast | *4, 43* |
| | ω-Dehydroisoleucine | Bacteria | *76* |
| | 3-Cyclopentene-1-glycine | Bacteria | *77* |
| | Cyclopentene glycine | Bacteria | *78* |
| | 2-Cyclopentene-1-glycine | Bacteria | *79* |
| | *O*-Methylthreonine | Tumor cells | *80* |
| | β-Hydroxyleucine | Bacteria | *37* |
| **Leucine** | D-Leucine | Bacteria | *81* |
| | α-Aminoisoamylsulfonic acid | Bacteria | *1, 33* |
| | | Mouse tumor | *2* |
| | Norvaline | Bacteria | *4* |
| | Norleucine | Bacteria | *1, 4, 82* |
| | Methallylglycine | Yeast, bacteria | *4* |
| | α-Amino-β-chlorobutyric acid | Yeast, bacteria | *4* |
| | Valine | Bacteria | *83* |
| | δ-Chloroleucine | *Neurospora* | *84* |
| | Isoleucine | Bacteria | *85* |
| | β-Hydroxynorleucine | Bacteria | *86* |
| | β-Hydroxyleucine | Bacteria | *86* |
| | Cyclopentene alanine | Bacteria | *87* |
| | 3-Cyclopentene-1-alanine | Bacteria | *88* |
| | 2-Amino-4-methylhexanoic acid | Bacteria | *89* |
| | 5′,5′,5′-Trifluoroleucine | *E. coli* | *90* |
| | 4-Azaleucine | *E. coli* | *91* |
| **Lysine** | α-Amino-ε-hydroxycaproic acid | Rat | *92* |
| | Arginine | *Neurospora* | *93* |
| | 2,6-Diaminoheptanoic acid | Bacteria | *94* |
| | Oxalysine | Bacteria | *95* |
| | 3-Aminomethylcyclohexane glycine | Bacteria | *96* |
| | 3-Aminocyclohexane alanine | Bacteria | *97* |
| | *trans*-4-Dehydrolysine | Bacteria | *98* |
| | *S*-(β-Aminoethyl)-cysteine | Bacteria | *99* |
| | 4-Azalysine | Bacteria | |
| **Methionine** | 2-Amino-5-heptenoic acid (crotylalanine) | *E. coli* | *100* |
| | 2-Amino-4-hexenoic acid (crotylglycine) | *E. coli* | *101* |
| | Methoxinine | Bacteria | *102* |
| | | Vaccinia virus | *69* |
| | | Rats | *103* |
| | Norleucine | Bacteria | *82, 104, 105* |
| | | Animal tissues | *106* |
| | | Casein | *107* |
| | Ethionine | Bacteria, animals | *28, 102, 105, 108–120* |
| | | Amylase | *121, 122* |
| | | Yeast | *123* |
| | | Tumors | *124* |
| | | Pancreatic proteins | *125* |
| | Methionine sulfoximine | Bacteria | *126* |
| | Threonine | *Neurospora* | *127* |
| | Selenomethionine | *Chlorella* | *128* |
| | | *E. coli*, yeast | *129–132* |
| **Ornithine** | α-Amino-δ-hydroxyvaleric acid | Bacteria | *21* |
| | Canaline | Bacteria | *21* |
| **Phenylalanine** | α-Amino-β-phenylethanesulfonic acid | Mouse tumor | *2* |
| | Tyrosine | Bacteria | *133* |
| | β-Phenylserine | Bacteria | *134, 135, 136* |
| | Cyclohexylalanine | Rats | *137* |
| | *o*-Aminophenylalanine | *E. coli* | *138* |
| | *p*-Aminophenylalanine | Bacteria | *139, 140* |
| | Fluorophenylalanines | Fungi, bacteria | *141–146, 140, 147, 148* |
| | | Lysozyme, albumin | *149* |
| | | Muscle enzymes | *150* |
| | | Amylase | *151* |
| | | Hemoglobin | *152* |
| | | Rats | *153* |

| Amino acid | Antagonists | System | Reference |
|---|---|---|---|
| **Phenylalanine—(Continued)** | Chlorophenylalanines | Fungi | *147* |
| | Bromophenylalanines | Fungi | *147* |
| | β-2-Thienylalanine | Rat, bacteria, yeast | *136, 154–164* |
| | | β-Galactosidase | *165* |
| | β-3-Thienylalanine | Bacteria, yeast | *166* |
| | β-2-Furylalanine | Bacteria, yeast | *4* |
| | β-3-Furylalanine | Bacteria, yeast | *4* |
| | β-2-Pyrrolealanine | Bacteria, yeast | *167* |
| | 1-Cyclopentene-1-alanine | Bacteria | *87* |
| | 1-Cyclohexene-1-alanine | Bacteria | *87* |
| | 2-Amino-4-methyl-4-hexenoic acid | Bacteria | *89* |
| | *S*-(1,2-Dichlorovinyl)-cysteine | *E. coli* | *168* |
| | β-4-Pyridylalanine | Bacteria | *169* |
| | Tryptophan | Bacteria | *170* |
| | β-2-Pyridylalanine | Bacteria | *171* |
| | β-4-Pyrazolealanine | Bacteria | *171* |
| | β-4-Thiazolealanine | Bacteria | *171* |
| | *p*-Nitrophenylalanine | Bacteria | *140* |
| **Proline** | Hydroxyproline | Fungi | *172* |
| | 3,4-Dehydroproline | Bacteria, beans | *173, 174* |
| | Azetidine-2-carboxylic acid | Bacteria, beans | *175, 176* |
| | | Actinomycin | *177, 178* |
| **Serine** | α-Methylserine | Bacteria | *4* |
| | Homoserine | Bacteria | *4* |
| | Threonine | Bacteria | *179, 180* |
| | Isoserine | Enzymes | *181* |
| **Threonine** | Serine | Bacteria | *179, 180, 182* |
| | β-Hydroxynorvaline | Bacteria | *86, 183* |
| | β-Hydroxynorleucine | Bacteria | *86, 183* |
| **Thyroxine** | Ethers of 3,5-diiodotyrosine | Tadpoles | *184* |
| **Tryptophan** | Methyltryptophans | Bacteria | *141, 185, 186–188* |
| | | Bacteriophage | *189* |
| | Naphthylalanines | Bacteria | *190, 191* |
| | | Rat | *4* |
| | Indoleacrylic acid | Bacteria | *192* |
| | Naphthylacrylic acid | Bacteria | *193* |
| | β-(2-Benzothienyl)alanine | Bacteria | *194* |
| | Styrylacetic acid | Bacteria | *193* |
| | Indole | Bacteriophage | *195* |
| | α-Amino-β-3(indazole)-propionic acid (Tryptazan) | Yeast | *196* |
| | | Enzyme | *197* |
| | | *E. coli* | *198* |
| | 5-Fluorotryptophan | Enzyme | *199, 200* |
| | 6-Fluorotryptophan | Enzyme | *201* |
| | 7-Azatryptophan | *E. coli* | *202–204, 205* |
| **Tyrosine** | Fluorotyrosines | Fungi | *147, 206* |
| | | Rat | *206* |
| | *p*-Aminophenylalanine | Fungi | *139* |
| | *m*-Nitrotyrosine | Bacteria | *140* |
| | β-(5-Hydroxy-2-pyridyl)-alanine | Bacteria | *207* |
| **Valine** | α-Aminoisobutanesulfonic acid | Bacteria | *1, 2, 33* |
| | | Vaccinia | *69* |
| | α-Aminobutyric acid | Bacteria | *182, 4* |
| | Norvaline | Bacteria | *4* |
| | Leucine, isoleucine | Bacteria | *182, 83* |
| | Methallylglycine | Bacteria, yeast | *4* |
| | β-Hydroxyvaline | Bacteria | *86, 183* |
| | ω-Dehydroalloisoleucine | Bacteria | *76* |

[a] Some of the observed inhibition may be due to fluoride ion present in the amino acid preparation or formed during incubation.[208]

Data taken from: Meister, in *Biochemistry of the Amino Acids,* 2nd ed., Vol. 1, p. 233 (1965). Reproduced by permission of Academic Press, New York.

## REFERENCES

1. McIlwain, *Brit. J. Exp. Pathol.*, *22*, 148 (1941).
2. Greenberg and Schulman, *Science*, *106*, 271 (1947).
3. Snell and Guirard, *Proc. Nat. Acad. Sci. U.S.A.*, *29*, 66 (1943).
4. Dittmer, *Ann. N.Y. Acad. Sci.*, *52*, 1274 (1950).
5. Bondi, Kornblum and Forte, *Proc. Soc. Exp. Biol. Med.*, *96*, 270 (1957).
6. Zygmunt, *J. Bacteriol.*, *84*, 154 (1962); *85*, 1217 (1963).
7. Strominger, Threnn and Scott, *J. Amer. Chem. Soc.*, *81*, 3803 (1959).
8. Strominger, Ito and Threnn, *J. Amer. Chem. Soc.*, *82*, 998 (1960).
9. Neuhaus and Lynch, *Biochem. Biophys. Res. Commun.*, *8*, 377 (1962).
10. Moulder, Novosel and Officer, *J. Bacteriol.*, *85*, 707 (1963).
11. Tanaka, *Biochem. Biophys. Res. Commun.*, *12*, 68 (1963).
12. Snell, Radin and Ikawa, *J. Biol. Chem.*, *217*, 803 (1955).
13. Nielsen, *Naturwissenschaften*, *31*, 146 (1943).
14. Wright and Skeggs, *Arch. Biochem.*, *10*, 383 (1946).
15. Sarett and Cheldelin, *J. Bacteriol.*, *49*, 31 (1945).
16. Durham and Milligan, *Biochem. Biophys. Res. Commun.*, *7*, 342 (1962).
17. Richmond, *Biochem. J.*, *73*, 261 (1959).
18. Horowitz and Srb, *J. Biol. Chem.*, *174*, 371 (1948).
19. Teas, *J. Biol. Chem.*, *190*, 369 (1951).
20. Miller and Harrison, *Nature*, *166*, 1035 (1950).
21. Volcani and Snell, *J. Biol. Chem.*, *174*, 893 (1948).
22. Walker, *J. Biol. Chem.*, *212*, 207, 617 (1955).
23. Kruse, White, Carter and McCoy, *Cancer Res.*, *19*, 122 (1959).
24. Owaga, *J. Agr. Chem. Soc. Jap.*, *10*, 225 (1934); *Chem. Abstr.*, *28*, 4458 (1934).
25. Owaga, *J. Agr. Chem. Soc. Jap.*, *11*, 11, 559 (1935); *Chem. Abstr.*, *29*, 740, 3379 (1935).
26. Steward, Pollard, Patchett and Witkop, *Biochim. Biophys. Acta*, *28*, 308 (1958).
27. Bonner, *Amer. J. Bot.*, *36*, 323, 429 (1949).
28. Gros and Gros-Doulcet, *Exp. Cell Res.*, *14*, 104 (1958).
29. Morgan, Morton and Pasieka, *J. Biol. Chem.*, *233*, 664 (1958).
30. Hunter and Downs, *J. Biol. Chem.*, *157*, 427 (1945).
31. Gross, *Z. Physiol. Chem.*, *112*, 236 (1921).
32. Walker, *J. Biol. Chem.*, *212*, 617 (1955).
33. McIlwain, *J. Chem. Soc.*, p. 75 (1941).
34. Ifland and Shive, *J. Biol. Chem.*, *223*, 949 (1956).
35. Shive, Ackerman and Ravel, *J. Amer. Chem. Soc.*, *69*, 2567 (1947).
36. Shive and Macow, *J. Biol. Chem.*, *162*, 452 (1946).
37. Otani, *Arch. Biochem. Biophys.*, *101*, 131 (1963).
38. Roberts and Hunter, *Proc. Soc. Exp. Biol. Med.*, *83*, 720 (1953).
39. Arnold, Morris and Thompson, *Nature*, *186*, 1051 (1960).
40. Woolley, *J. Biol. Chem.*, *235*, 3238 (1960).
41. Shigeura and Gordon, *J. Biol. Chem.*, *237*, 1932, 1937 (1962).
42. Heymann, Ginsberg, Gulick, Konopka and Mayer, *J. Amer. Chem. Soc.*, *81*, 5125 (1959).
43. Dittmer, Goering, Goodman and Cristol, *J. Amer. Chem. Soc.*, *70*, 2499 (1948).
44. Simmonds, *Biochem. J.*, *58*, 520 (1954).
45. Rhuland, *J. Bacteriol.*, *73*, 778 (1957).
46. Rhuland and Hamilton, *Biochim. Biophys. Acta*, *51*, 525 (1961).
47. Meadow, Hoare and Work, *Biochem. J.*, *66*, 270 (1957).
48. Borek, Miller, Sheiness and Waelsch, *J. Biol. Chem.*, *163*, 347 (1946).
49. Borek and Waelsch, *Arch. Biochem.*, *14*, 143 (1947).
50. Elliot and Gale, *Nature*, *161*, 129 (1948).
51. Lichtenstein and Grossowicz, *J. Biol. Chem.*, *171*, 387 (1947).
52. Borek and Waelsch, *J. Biol. Chem.*, *177*, 135 (1949).
53. Ayengar and Roberts, *Proc. Soc. Exp. Biol. Med.*, *79*, 476 (1952).
54. Pace and McDermott, *Nature*, *169*, 415 (1952).
55. Heathcote and Pace, *Nature*, *166*, 353 (1950).
56. Roberts, *J. Biol. Chem.*, *202*, 359 (1953).
57. Lichtenstein, Ross and Cohen, *Nature*, *171*, 45 (1953); *J. Biol. Chem.*, *201*, 117 (1953).
58. Mastalerz, *Arch. Immunol. Ter. Dos.*, *7*, 201 (1959).
59. Provided by Pattison [Buchanan, Dean and Pattison, *Can. J. Chem.*, *40*, 1571 (1962)] and studied in the author's laboratory.

60. Ravel, McCord, Skinner and Shive, *J. Biol. Chem.*, *232*, 159 (1958).
61. Rabinovitz and Fisher, *J. Nat. Cancer Inst.*, *28*, 1165 (1962).
62. Skinner, McCord, Ravel and Shive, *J. Amer. Chem. Soc.*, *78*, 2412 (1956).
63. McCord, Ravel, Skinner and Shive, *J. Amer. Chem. Soc.*, *80*, 3762 (1958).
64. Reisner, *J. Amer. Chem. Soc.*, *78*, 5102 (1956).
65. Edelson, Skinner and Shive, *J. Med. Pharm. Chem.*, *1*, 165 (1959).
66. Meister, in *The Enzymes*, Vol. 4, p. 247, Boyer, Lardy and Myrback, Eds. Academic Press, New York (1962).
67. McIlwain, Roper and Hughes, *Biochem. J.*, *42*, 492 (1948).
68. Spizizen, *J. Infect. Dis.*, *73*, 212 (1943).
69. Thompson, *J. Immunol.*, *55*, 345 (1947).
70. Edlbacher, Baur and Becker, *Z. Physiol. Chem.*, *265*, 61 (1940).
71. Moyed, *J. Biol. Chem.*, *236*, 2261 (1961).
72. Jones and Ainsworth, *J. Amer. Chem. Soc.*, *77*, 1538 (1955).
73. Levin and Hartman, *J. Bacteriol.*, *86*, 820 (1963).
74. Doudoroff, *Proc. Soc. Exp. Biol. Med.*, *53*, 73 (1943).
75. Harper, Benton, Winje and Elvehjem, *Arch. Biochem. Biophys.*, *51*, 523 (1954).
76. Parker, Skinner and Shive, *J. Biol. Chem.*, *236*, 3267 (1961).
77. Edelson, Fissekis, Skinner and Shive, *J. Amer. Chem. Soc.*, *80*, 2698 (1958).
78. Harding and Shive, *J. Biol. Chem.*, *206*, 401 (1954).
79. Dennis, Plant, Skinner, Sutherland and Shive, *J. Amer. Chem. Soc.*, *77*, 2362 (1955).
80. Rabinovitz, Olson and Greenberg, *J. Amer. Chem. Soc.*, *77*, 3109 (1955).
81. Fox, Fling and Bollenback, *J. Biol. Chem.*, *155*, 465 (1944).
82. Harding and Shive, *J. Biol. Chem.*, *174*, 743 (1948).
83. Brickson, Henderson, Solhjell and Elvehjem, *J. Biol. Chem.*, *176*, 517 (1948).
84. Ryan, *Arch. Biochem. Biophys.*, *36*, 487 (1952).
85. Hirsh and Cohen, *Biochem. J.*, *53*, 25 (1953).
86. Buston and Bishop, *J. Biol. Chem.*, *215*, 217 (1955).
87. Pal, Skinner, Dennis and Shive, *J. Amer. Chem. Soc.*, *78*, 5116 (1956).
88. Edelson, Skinner, Ravel and Shive, *Arch. Biochem. Biophys.*, *80*, 416 (1959).
89. Edelson, Skinner, Ravel and Shive, *J. Amer. Chem. Soc.*, *81*, 5150 (1959).
90. Rennert and Anker, *Biochemistry*, *2*, 471 (1963).
91. Smith, Bayliss and McCord, *Arch. Biochem. Biophys.*, *102*, 313 (1963).
92. Page, Gaudry and Gringras, *J. Biol. Chem.*, *171*, 831 (1948).
93. Daermann, *Arch. Biochem.*, *5*, 373 (1944).
94. McLaren and Knight, *J. Amer. Chem. Soc.*, *73*, 4478 (1951).
95. McCord, Ravel, Skinner and Shive, *J. Amer. Chem. Soc.*, *79*, 5693 (1957).
96. Davis, Skinner and Shive, *Arch. Biochem. Biophys.*, *87*, 88 (1960).
97. Davis, Ravel, Skinner and Shive, *Arch. Biochem. Biophys.*, *76*, 139 (1958).
98. Davis, Skinner and Shive, *J. Amer. Chem. Soc.*, *83*, 2279 (1961).
99. Shiota, Folk and Tietze, *Arch.Biochem. Biophys.*, *77*, 372 (1958).
100. Goering, Cristol and Dittmer, *J. Amer. Chem. Soc.*, *70*, 3314 (1948).
101. Skinner, Edelson and Shive, *J. Amer. Chem. Soc.*, *83*, 2281 (1961).
102. Robin, Lampen, English, Cole and Vaughan, *J. Amer. Chem. Soc.*, *67*, 290 (1945).
103. Shaffer and Critchfield, *J. Biol. Chem.*, *174*, 489 (1948).
104. Reisner, *J. Amer. Chem. Soc.*, *78*, 2132 (1956).
105. Porter and Meyers, *Arch. Biochem.*, *8*, 169 (1945).
106. Rabinovitz, Olson and Greenberg, *J. Biol. Chem.*, *210*, 837 (1954).
107. Black and Kleiber, *J. Amer. Chem. Soc.*, *77*, 6082 (1955).
108. Dyer, *J. Biol. Chem.*, *124*, 519 (1938).
109. Harris and Kohn, *J. Pharmacol. Exp. Ther.*, *73*, 383 (1941).
110. Halvorson and Spiegelman, *J. Bacteriol.*, *64*, 207 (1952).
111. Simpson, Farber and Tarver, *J. Biol. Chem.*, *182*, 81 (1950).
112. Simmonds, Keller, Chandler and du Vigneaud, *J. Biol. Chem.*, *183*, 191 (1950).
113. Stekol and Weiss, *J. Biol. Chem.*, *179*, 1049 (1949).
114. Stekol and Weiss, *J. Biol. Chem.*, *185*, 577, 585 (1950).
115. Farber, Simpson and Tarver, *J. Biol. Chem.*, *182*, 91 (1950).
116. Keston and Wortis, *Proc. Soc. Exp. Biol. Med.*, *61*, 439 (1946).
117. Levine and Tarver, *J. Biol. Chem.*, *192*, 835 (1951).
118. Tarver, in *The Proteins*, Vol. 2, p. 1199, Neurath and Bailey, Eds. Academic Press, New York (1954).
119. Swendseid, Swanson and Bethell, *J. Biol. Chem.*, *201*, 803 (1953).
120. Levy, Montanez, Murphy and Dunn, *Cancer Res.*, *13*, 507 (1953).
121. Yoshida, *Biochim. Biophys. Acta*, *29*, 213 (1958).

122. Yoshida and Yamasaki, *Biochim. Biophys. Acta, 34,* 158 (1959).
123. Parks, *J. Biol. Chem., 232,* 169 (1958).
124. Rabinovitz, Olson and Greenberg, *J. Biol. Chem., 227,* 217 (1957).
125. Hansson and Garzo, *Biochim. Biophys. Acta, 61,* 121 (1962).
126. Heathcote, *Lancet, 257,* 1130 (1949).
127. Doudney and Wagner, *Proc. Nat. Acad. Sci. U.S.A., 38,* 196 (1952); *39,* 1043 (1953).
128. Shrift, *Amer. J. Bot., 41,* 345 (1954).
129. Cowie and Cohen, *Biochim. Biophys. Acta, 26,* 252 (1957).
130. Cohen and Cowie, *C. R. Acad. Sci. (Paris), 244,* 680 (1957).
131. Blau, *Biochim. Biophys. Acta, 49,* 389 (1961).
132. Tuve and Williams, *J. Amer. Chem. Soc., 79,* 5830 (1957).
133. Beerstecher and Shive, *J. Biol. Chem., 167,* 527 (1947).
134. Beerstecher and Shive, *J. Biol. Chem., 164,* 53 (1946).
135. Fox and Warner, *J. Biol. Chem., 210,* 119 (1954).
136. Miller and Simmonds, *Science, 126,* 445 (1957).
137. Baltes, Elliott, Doisy and Doisy, *J. Biol. Chem., 194,* 627 (1952).
138. Davis, Lloyd, Fletcher, Bayliss and McCord, *Arch. Biochem. Biophys., 102,* 48 (1963).
139. Burckhalter and Stephens, *J. Amer. Chem. Soc., 73,* 56 (1951).
140. Bergmann, Sicher and Volcani, *Biochem. J., 54,* 1 (1953).
141. Munier and Cohen, *Biochim. Biophys. Acta, 21,* 592 (1956).
142. Munier and Cohen, *Biochim. Biophys. Acta, 31,* 378 (1959).
143. Cohen and Munier, *Biochim. Biophys. Acta, 31,* 347 (1959).
144. Cowie, Cohen, Bolton and De Robichon-Szulmajster, *Biochim. Biophys. Acta, 34,* 39 (1959).
145. Cohen, Halvorson and Spiegelman, Microsomal Particles in Protein Synthesis, in *Papers of the First Symposium of the Biophysical Society, Cambridge, Massachusetts,* p. 100. Pergamon Press, New York (1958).
146. Baker, Johnson and Fox, *Biochim. Biophys. Acta, 28,* 318 (1958).
147. Mitchell and Niemann, *J. Amer. Chem. Soc., 69,* 1232 (1947).
148. Atkinson, Melvin and Fox, *Arch. Biochem. Biophys., 31,* 205 (1951).
149. Vaughan and Steinberg, *Biochim. Biophys. Acta, 40,* 230 (1960).
150. Boyer and Westhead, *Abstr. Meet. Amer. Chem. Soc.,* p. 2C (1958).
151. Yoshida, *Biochim. Biophys. Acta, 41,* 98 (1960).
152. Kruh and Rose, *Biochim. Biophys. Acta, 34,* 561 (1959).
153. Armstrong and Lewis, *J. Biol. Chem., 188,* 91 (1951).
154. du Vigneaud, McKennis, Simmonds, Dittmer and Brown, *J. Biol. Chem., 159,* 385 (1945).
155. Dittmer, Ellis, McKennis and du Vigneaud, *J. Biol. Chem., 164,* 761 (1946).
156. Dittmer, Hertz and Chambers, *J. Biol. Chem., 166,* 541 (1946).
157. Garst, Campaigne and Day, *J. Biol. Chem., 180,* 1013 (1949).
158. Ferger and du Vigneaud, *J. Biol. Chem., 179,* 61 (1949).
159. Drea, *J. Bacteriol., 56,* 257 (1948).
160. Dunn and Dittmer, *J. Biol. Chem., 188,* 263 (1951).
161. Kihara and Snell, *J. Biol. Chem., 212,* 83 (1955).
162. Ferger and du Vigneaud, *J. Biol. Chem., 174,* 241 (1948).
163. Dunn, *J. Biol. Chem., 233,* 411 (1958).
164. Dunn, Ravel and Shive, *J. Biol. Chem., 219,* 809 (1956).
165. Janeček and Rickenberg, *Biochim. Biophys. Acta, 81,* 108 (1964).
166. Dittmer, *J. Amer. Chem. Soc., 71,* 1205 (1949).
167. Herz, Dittmer and Cristol, *J. Amer. Chem. Soc., 70,* 504 (1948).
168. Dickie and Schultz, *Arch. Biochem. Biophys., 100,* 279, 285 (1963).
169. Elliott, Fuller and Harington, *J. Chem. Soc.,* p. 85 (1948).
170. Beerstecher and Shive, *J. Amer. Chem. Soc., 69,* 461 (1947).
171. Lansford and Shive, *Arch. Biochem. Biophys., 38,* 347 (1952).
172. Robbins and McVeigh, *Amer. J. Bot., 33,* 638 (1946).
173. Fowden, Neale and Tristram, *Nature, 199,* 35 (1963).
174. Smith, Ravel, Skinner and Shive, *Arch. Biochem. Biophys., 99,* 60 (1962).
175. Fowden and Richmond, *Biochim. Biophys. Acta, 71,* 459 (1963).
176. Fowden, *J. Exp. Biol., 14,* 387 (1963).
177. Katz and Gross, *Biochem. J., 73,* 458 (1959).
178. Katz, *Ann. N.Y. Acad. Sci., 89,* 304 (1960).
179. Meinke and Holland, *J. Biol. Chem., 173,* 535 (1948).
180. Holland and Meinke, *J. Biol. Chem., 178,* 7 (1949).
181. Leibman and Fellner, *J. Biol. Chem., 237,* 2213 (1962).
182. Gladstone, *Brit. J. Exp. Pathol., 20,* 189 (1939).

183. Buston, Churchman and Bishop, *J. Biol. Chem., 204,* 665 (1953).
184. Woolley, *J. Biol. Chem., 164,* 11 (1946).
185. Trudinger and Cohen, *Biochem. J., 62,* 488 (1956).
186. Anderson, *Science, 101,* 565 (1945).
187. Fildes and Rydon, *Brit. J. Exp. Pathol., 28,* 211 (1947).
188. Marshall and Woods, *Biochem. J., 51,* ii (1952).
189. Cohen and Anderson, *J. Exp. Med., 84,* 525 (1946).
190. Erlenmeyer and Grubemann, *Helv. Chim. Acta, 30,* 297 (1947).
191. Dittmer, Herz and Cristol, *J. Biol. Chem., 173,* 323 (1948).
192. Fildes, *Brit. J. Exp. Pathol., 22,* 293 (1941).
193. Block and Erlenmeyer, *Helv. Chim. Acta, 25,* 694, 1063 (1942).
194. Avakian, Mars and Martin, *J. Amer. Chem. Soc., 70,* 3075 (1948).
195. Delbrück, *J. Bacteriol., 56,* 1 (1948).
196. Halvorson, Spiegelman and Hinman, *Arch. Biochem. Biophys., 55,* 512 (1956).
197. Durham and Martin, *Biochim. Biophys. Acta, 71,* 481 (1963).
198. Brawerman and Yŭcas, *Arch. Biochem. Biophys., 68,* 112 (1957).
199. Moyed, *J. Biol. Chem., 235,* 1098 (1960).
200. Bergmann, Eschinazi, Sicher and Volcani, *Bull. Res. Soc. Israel, 2,* 308 (1952).
201. Moyed and Friedmann, *Science, 129,* 968 (1959).
202. Robison and Robison, *J. Amer. Chem. Soc., 77,* 457 (1955).
203. Pardee, Shore and Prestidge, *Biochim. Biophys. Acta, 21,* 406 (1956).
204. Pardee and Prestidge, *Biochim. Biophys. Acta, 27,* 330 (1958).
205. Kidder and Dewey, *Biochim. Biophys. Acta, 17,* 288 (1955).
206. Niemann and Rapport, *J. Amer. Chem. Soc., 68,* 1671 (1946).
207. Norton, Skinner and Shive, *J. Org. Chem., 26,* 1495 (1961).
208. Kagan, Unpublished Data.

# PROPERTIES OF THE α-KETO ACID ANALOGS OF AMINO ACIDS

| α-Keto acid | α-Amino acid analog | 2,4-Dinitrophenylhydrazone | | | Reduction by lactic dehydrogenase[b,g] | Decarboxylation by yeast decarboxylase[c] |
|---|---|---|---|---|---|---|
| | | Crystallization | | Amino acids after hydrogenation (*14*) | | |
| | | M.p. (°C) | Solvent[a] | | | |
| **Pyruvic** | Alanine | 216 | h | Alanine | 26,800 | 1,200 |
| **α-Ketoadipamic** | α-Aminoadipamic acid (homoglutamine) | | | | | |
| **α-Ketoadipic** | α-Aminoadipic acid | 208 | h | α-Aminoadipic acid | | |
| **α-Ketobutyric** | α-Aminobutyric acid | 198 | h | α-Aminobutyric acid | 21,000 | |
| **α-Ketoheptylic** | α-Aminoheptylic acid | 130 | e, l | α-Aminoheptylic acid | 483 | |
| **α-Keto-ε-hydroxycaproic** | α-Amino-ε-hydroxy-caproic acid | 183 | h | α-Amino-ε-hydroxy-caproic acid | 181 | 25 |
| **Mesoxalic** | α-Aminomalonic acid | 205 | hc | α-Aminomalonic acid, glycine | | |
| **α-Ketophenylacetic** | α-Aminophenyl-acetic acid | 193 | h | α-Aminophenyl-acetic acid, cyclohexyl-glycine | 2.6 | 0 |
| **DL-Oxalosuccinic** | α-Aminotri-carballylic acid | | | | | |
| **α-Keto-δ-guanidinovaleric** | Arginine | 216, 267 (*1*) | | Arginine | 9.0 | 0 |
| **α-Ketosuccinamic** | Asparagine | 183 | | Asparagine, aspartic acid | 8,930 | |
| **Oxalacetic** | Aspartic acid | 218 | h | Aspartic acid, alanine, β-alanine | 12.8 | |
| **α-Keto-δ-carbamidovaleric** | Citrulline | 190 | h | Citrulline | 4.3 | 0 |
| **β-Cyclohexylpyruvic** | β-Cyclohexyl-alanine | 189 | h | β-Cyclohexyl-alanine | 14.6 | 0 |
| **β-Sulfopyruvic** | Cysteic acid | 210 | a | Cysteic acid, alanine | 89.6 | |
| **β-Mercaptopyruvic** | Cysteine | 195–200 (*2*) 161–162 (*3*) | | Alanine | 27,000 | 750 |
| **α-Keto-γ-ethiolbutyric** | Ethionine | 131 | h | Ethionine | 1,650 | 121 |
| **α-Ketoglutaric** | Glutamic acid | 220 | h | Glutamic acid | 9.2 | 0 |
| **α-Ketoglutaric γ-ethyl ester** | Glutamic acid γ-ethyl ester | | | | 49.0 | 721 |
| **α-Ketoglutaramic** | Glutamine | | | Glutamine, glutamic acid | | |
| **Glyoxylic** | Glycine | 203 | h | Glycine | 21,100 | 0 |
| **β-Imidazolylpyruvic** | Histidine | 190–192, 240 (*1*) | hc, e, l | Histidine | | |
| **α-Keto-γ-hydroxybutyric** | Homoserine | | | | | |
| **DL-α-Keto-β-methylvaleric** | DL-Isoleucine (or DL-alloisoleucine) | 169 | h | Isoleucine | | |
| **L-α-Keto-β-methylvaleric** | L-Isoleucine (or D-alloisoleucine) | 176 | h | Isoleucine | 5.0 | 1,000 |
| **D-α-Keto-β-methylvaleric** | D-Isoleucine (or L-alloisoleucine) | 176 | h | Isoleucine | 1.9 | 280 |
| **α-Ketoisocaproic** | Leucine | 162 | h | Leucine | 3.2 | 306 |
| **Trimethylpyruvic** | *tert*-Leucine | 180 | h | *tert*-Leucine | | |
| **α-Keto-ε-aminocaproic** | Lysine | 212 | h | Lysine, pipecolic acid | | |
| **α-Keto-γ-methiolbutyric** | Methionine | 150 | h | Methionine | 1,550 | 125 |
| **α-Keto-γ-methylsulfonylbutyric** | Methionine sulfone | 175 | h | Methione sulfone | | |
| **α-Keto-δ-nitroguanidinovaleric** | Nitroarginine | 225 | ac | Nitroarginine, arginine | 42.6 | 0 |
| **α-Ketocaproic** | Norleucine | 153 | h | Norleucine | 560 | |
| **α-Ketovaleric** | Norvaline | 167 | h | Norvaline | 1,470 | |

| α-Keto acid | α-Amino acid analog | 2,4-Dinitrophenylhydrazone: Crystallization M.p. (°C) | Solvent[a] | Amino acids after hydrogenation (*14*) | Reduction by lactic dehydro-genase[b,g] | Decar-boxylation by yeast decarboxy-lase[c] |
|---|---|---|---|---|---|---|
| **α-Keto-δ-aminovaleric** | Ornithine, proline | 232–242 (*4*)<br>219 (*5*)<br>211–212 (*6*) | | Ornithine, proline, pentahomoserine[d] | | |
| **Phenylpyruvic** | Phenylalanine | 162–164, 192–194 (*7*) | | Phenylalanine | 755 | 0 |
| ***S*-Benzyl-β-mercaptopyruvic** | *S*-Benzylcysteine | 150 | a | | | |
| **β-Hydroxypyruvic** | Serine | 162 | e | Serine, alanine | 26,000 | 0[h] |
| ***N*-Succinyl-α-amino-ε-ketopimelic** (*15*) | *N*-Succinyl-α,ε-diaminopimelic acid | 137–143 | h | *N*-Succinyl-α,ε-diaminopimelic acid | | |
| **DL-α-Keto-β-hydroxybutyric** | DL-Threonine (or DL-allothreonine) | 157–158 (*8*) | | Threonine, α-aminobutyric acid | 20,000 | |
| **β-[3,5-Diiodo-4-**(3′,5′-diiodo 4′-hydroxyphenoxy)phenyl] | Thyroxine | | | | | |
| **β-Indolylpyruvic** | Tryptophan | 169 (*1*) | | Tryptophan | 670 | 0 |
| ***p*-Hydroxyphenylpyruvic** | Tyrosine | 178 | h | Tyrosine | 345 | 0 |
| **α-Ketoisovaleric** | Valine | 196 | h | Valine | 103 | 922 |

[a] h = water; e = ethyl acetate; l = ligroin; hc = hydrochloric acid; ac = glacial acetic acid; a = ethanol.
[b] Mole x $10^{-8}$ of DPNH oxidized per mg of enzyme per minute at 26°C.[9,10]
[c] μl $CO_2$ per hour.[10]
[d] α-Amino-δ-hydroxy-*n*-valeric acid.
[e] Originally designated *d*.[11] Originally designated *l*.[11]
[g] Additional data have been published on the reduction of α-keto acids by lactic dehydrogenase.[12]
[h] This keto acid has been reported to be decarboxylated by yeast preparations; the reaction is much more rapid at pH 6.5 than at pH 5.[13]

Data taken from: Meister, in *Biochemistry of the Amino Acids,* 2nd ed., Vol. 1, pp. 162–164 (1965). Reproduced by permission of Academic Press, New York.

## REFERENCES


1. Stumpf and Green, *J. Biol. Chem., 153,* 387 (1944).
2. Schneider and Reinefeld, *Biochem. Z., 318,* 507 (1948).
3. Meister, Fraser and Tice, *J. Biol. Chem., 206,* 561 (1954).
4. Krebs, *Enzymologia, 7,* 53 (1939).
5. Blanchard, Green, Nocito and Ratner, *J. Biol. Chem., 155,* 421 (1944).
6. Meister, *J. Biol. Chem., 206,* 579 (1954).
7. Fones, *J. Org. Chem., 17,* 1534 (1952).
8. Sprinson and Chargaff, *J. Biol. Chem., 164,* 417 (1947).
9. Meister, *J. Biol. Chem., 197,* 309 (1953).
10. Meister, *J. Biol. Chem., 184,* 117 (1950).
11. Meister, *J. Biol. Chem., 190,* 269 (1951).
12. Czok and Büchler, *Advan. Protein Chem., 15,* 315 (1960).
13. Dickens and Williamson, *Nature, 178,* 1349 (1956).
14. Meister and Abendschein, *Anal. Chem., 28,* 171 (1956).
15. Gilvarg, *J. Biol. Chem., 236,* 1429 (1961).

## FAR-ULTRAVIOLET ABSORPTION SPECTRA OF AMINO ACIDS

The absorption of a chromophoric amino acid in this spectral region is due to the combined absorptions of the side chain chromophore and of the carboxylate group. Because the carboxylate group is consumed in polymerizing amino acids to polypeptides, an amino acid residue absorbs less intensely than a free amino acid. The magnitude of this difference can be estimated from the spectra of the non-chromophoric amino acids – leucine, proline, alanine, serine, and threonine – whose total absorption is due only to the carboxylate group. The variations between the absorptions of these amino acids reflect the variability of carboxylate absorption in slightly different environments.

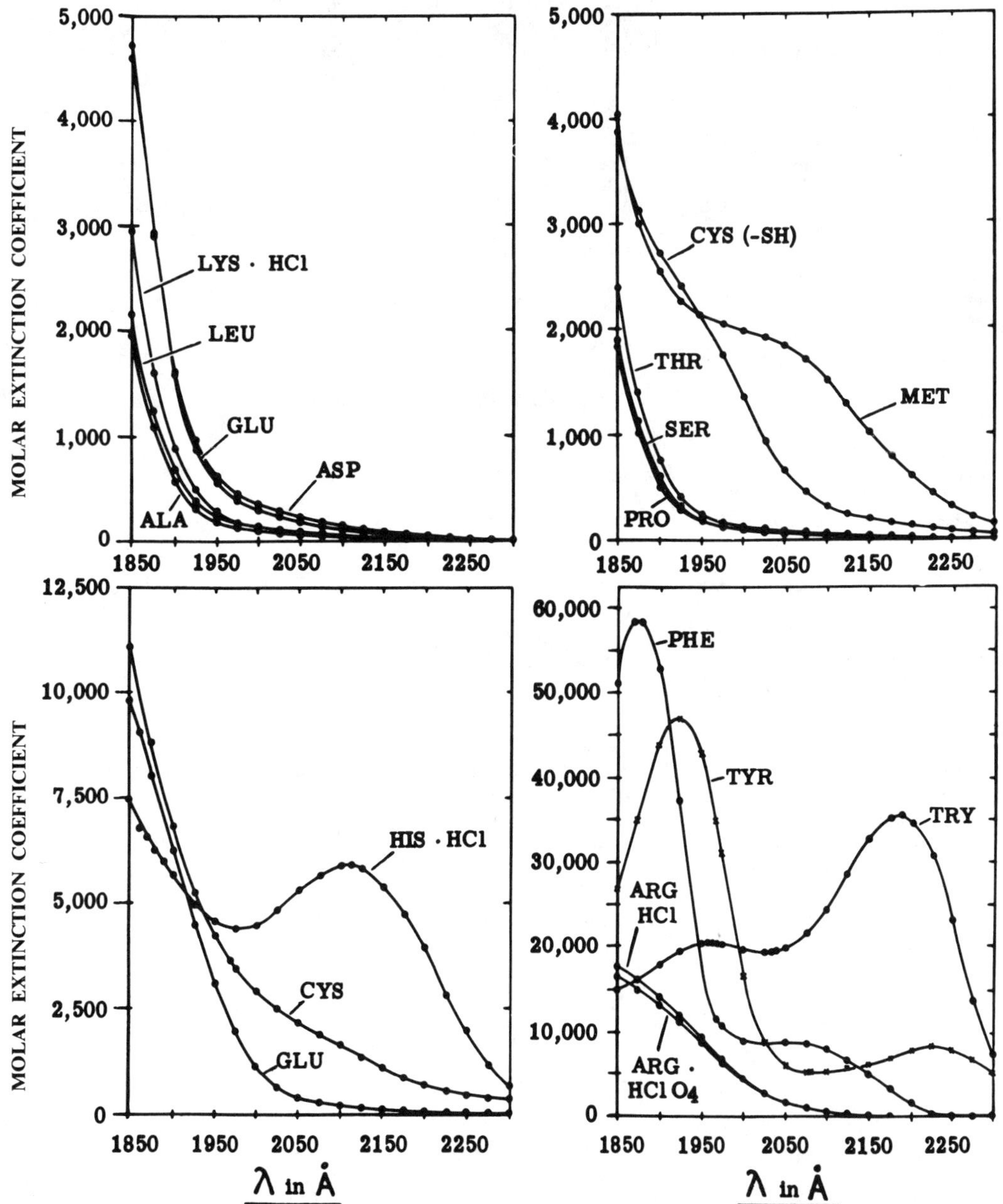


FIGURE 1. Far-ultraviolet spectra of amino acids. All amino acids were in aqueous solutions at pH 5, except cystine, which was at pH 3. The dibasic amino acids were measured as hydrochlorides and absorbances corrected by subtracting the absorbance contribution of chloride ion. Taken from: Wetlaufer, *Advan. Protein Chem.*, *17*, 320 (1962) and reproduced by permission of Academic Press, New York.

## TABLE 1
## FAR-ULTRAVIOLET ABSORPTION SPECTRA OF AMINO ACIDS

| Amino acid | $\lambda_{190.0}$ | $\lambda_{197.0}$ | $\lambda_{205.0}$ | Feature[a] | | | | | |
|---|---|---|---|---|---|---|---|---|---|
| | | | | Maxima | | Minima | | Shoulder | |
| | | | | λ | ε | λ | ε | λ | ε |
| **IN NEUTRAL WATER** | | | | | | | | | |
| **Tryptophan** | 17.60 | 20.50 | 19.60 | 196.7 | 20.60 | 203.3 | 19.40 | — | — |
| | — | — | — | 218.6 | 46.70 | — | — | — | — |
| **Tyrosine** | 42.80 | 35.50 | 5.60 | 192.5 | 47.50 | 208.0 | 4.88 | — | — |
| | — | — | — | 223.2 | 8.26 | — | — | — | — |
| **Phenylalanine** | 54.50 | 12.30 | 9.36 | 187.7 | 59.60 | 202.5 | 8.96 | — | — |
| | — | — | — | 206.0 | 9.34 | — | — | — | — |
| **Histidine**[b] | 5.57 | 4.35 | 5.17 | 211.3 | 5.86 | 198.4 | 4.22 | — | — |
| **Cysteine**[c] | 2.82 | 1.94 | 0.730 | — | — | — | — | 195.2 | 2.18 |
| **1/2 Cystine** | 3.25 | 1.76 | 1.05 | — | — | — | — | 207.0 | 0.96 |
| **Methionine** | 2.69 | 2.11 | 1.86 | — | — | — | — | 204.7 | 1.89 |
| **Arginine**[c] | 13.1 | 6.61 | 1.36 | — | — | — | — | — | — |
| **Acids**[d] | 1.61 | 0.460 | 0.230 | — | — | — | — | — | — |
| **Amides**[d] | 6.38 | 2.06 | 0.400 | — | — | — | — | — | — |
| **Lysine**[b] | 0.890 | 0.200 | 0.110 | — | — | — | — | — | — |
| **Leucine**[d] | 0.670 | 0.190 | 0.100 | — | — | — | — | — | — |
| **Alanine**[d] | 0.570 | 0.150 | 0.070 | — | — | — | — | — | — |
| **Proline**[d] | 0.540 | 0.150 | 0.070 | — | — | — | — | — | — |
| **Serine**[d] | 0.610 | 0.160 | 0.080 | — | — | — | — | — | — |
| **Threonine**[d] | 0.750 | 0.180 | 0.100 | — | — | — | — | — | — |
| **IN 0.1 *M* SODIUM DODECYL SULFATE**[d] | | | | | | | | | |
| **Tryptophan** | 16.70 | 19.70 | 19.00 | 197.3 | 19.80 | 204.1 | 18.90 | — | — |
| | — | — | — | 220.0 | 46.60 | — | — | — | — |
| **Tyrosine** | 39.10 | 36.60 | 5.87 | 193.2 | 45.10 | 208.5 | 4.88 | — | — |
| | — | — | — | 223.7 | 7.86 | — | — | — | — |
| **Phenylalanine** | 54.10 | 11.30 | 8.45 | 188.3 | 57.0 | 201.5 | 7.79 | — | — |
| | — | — | — | 207.2 | 8.66 | — | — | — | — |
| **Histidine**[b] | 5.88 | 4.48 | 5.03 | 212.0 | 5.88 | 199.0 | 4.25 | — | — |
| **Cysteine**[c] | 2.66 | 1.79 | 0.650 | — | — | — | — | 194.5 | 2.15 |
| **1/2 Cystine** | 2.62 | 1.61 | 0.92 | — | — | — | — | — | — |
| **Methionine** | 2.67 | 2.10 | 1.84 | — | — | — | — | 204.1 | 1.89 |
| **Arginine**[c] | 12.50 | 5.70 | 0.94 | — | — | — | — | — | — |

[a] Molar extinctions, $\epsilon \times 10^{-3}$; wavelength, λ, in nm.

[b] The absorptions of lysine and histidine were determined for the hydrochlorides and corrected for the absorption of the chloride ion: $\epsilon_{Cl} = 0.740$ ($\lambda = 190.0$); $\epsilon_{Cl} = 0.050$ ($\lambda = 197.0$); $\epsilon_{CL} = 0$ ($\lambda = 205.0$).

[c] The absorptions of cystein and arginine were determined for the $HClO_4$ salt.

[d] The spectra of the carboxylic acids, the amides, the aliphatic and hydroxy amino acids and lysine are unchanged from those obtained in neutral water.

Data taken from: McDiarmid, in *Handbook of Biochemistry,* 2nd ed., p. B-73, H. A. Sober, Ed. Copyright 1970, The Chemical Rubber Co., Cleveland, Ohio.

# ULTRAVIOLET ABSORPTION CHARACTERISTICS OF N-ACETYL METHYL ESTERS OF THE AROMATIC AMINO ACIDS, CYSTINE AND N-ACETYLCYSTEINE

| | Water[a] | | Ethanol[a] | |
|---|---|---|---|---|
| | λ | ε | λ | ε |
| **Phenylalanine** | | | | |
| Inflection | (208) | 10.20 | (208) | 10.40 |
| Inflection | (217) | 5.00 | (217) | 5.30 |
| Minimum | 240 | 0.080 | 244 | 0.088 |
| Maximum | 241.2 | 0.086 | 242.0 | 0.093 |
| Maximum | 246.5 | 0.115 | 247.3 | 0.114 |
| Maximum | 251.5 | 0.157 | 252.3 | 0.158 |
| Maximum | **257.4** | **0.197** | **258.3** | **0.195** |
| Inflection | (260.7) | — | (261.2) | — |
| Maximum | 263.4 | 0.151 | 264.2 | 0.155 |
| Maximum | 267.1 | 0.091 | 267.8 | 0.096 |
| **Tyrosine** | | | | |
| Maximum | 193 | 51.70 | — | — |
| Minimum | 212 | 7.00 | 212 | 6.20 |
| Maximum | 224 | 8.80 | 227 | 10.20 |
| Minimum | 247 | 0.176 | 246 | 0.174 |
| Maximum | **274.6** | **1.420** | **278.4** | **1.790** |
| Inflection | 281.9 | — | 285.7 | — |
| **Tryptophan** | | | | |
| Minimum | 205 | 21.40 | 206 | 21.30 |
| Maximum | 219 | 35.00 | 221 | 37.20 |
| Minimum | 245 | 1.900 | 245 | 1.560 |
| Maximum | **279.8** | **5.600** | **282.0** | **6.170** |
| Maximum | 288.5 | 4.750 | 290.6 | 5.330 |
| **Cystine** | | | | |
| Inflection | (250) | 0.360 | (253) | 0.372 |
| Inflection | 260 | 0.280 | 260 | 0.320 |
| Inflection | 280 | 0.110 | 280 | 0.135 |
| Inflection | 300 | 0.025 | 300 | 0.035 |
| Inflection | 320 | 0.006 | 320 | 0.007 |
| ***N*-Acetylcysteine** | | | | |
| | 250 | 0.015 | 250 | 0.020 |
| | 280 | 0.005 | 280 | 0.005 |
| | 320 | (nil) | 320 | (nil) |

[a] Wavelength, λ, in nm; molar extinctions, $\epsilon \times 10^{-3}$; inflection denotes unresolved inflection.

Data taken from: Gratzer, W. B., in *Handbook of Biochemistry,* 2nd ed., p. B-74, H. A. Sober, Ed. Copyright 1970, The Chemical Rubber Co., Cleveland, Ohio.

## ABSORPTION SPECTRA OF THE AROMATIC AMINO ACIDS AT pH 6

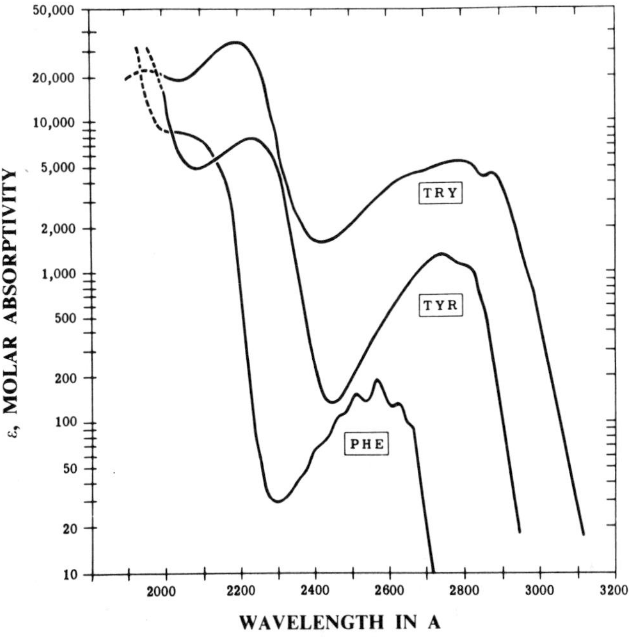


Taken from: Wetlaufer, D. B., *Advan. Protein Chem., 17,* 310 (1962). Reproduced by permission of Academic Press, New York.

# ABSORBANCE VALUES OF THE AROMATIC AMINO ACIDS IN NEUTRAL, ALKALINE AND ACID SOLUTIONS

## TABLE 1
## MOLECULAR ABSORBANCES OF TYROSINE

| mμ[d] | Neutral[a] | Alkaline[b] | mμ[d] | Neutral[a] | Alkaline[b] |
|---|---|---|---|---|---|
| 230 | 4980 | 7752 ± 108 | 276 | 1367 ± 0 | 1206 ± 4 |
| 232 | 3449 | 8667 ± 38 | 278 | 1260 ± 2 | 1344 ± 5 |
| 234 | 1833 ± 14 | 9634 ± 19 | 280 | 1197 ± 0 | 1507 ± 5 |
| 236 | 1014 ± 43 | 10440 ± 20 | 282 | 1112 ± 2 | 1675 ± 5 |
| 238 | 571 ± 36 | 11000 ± 10 | 284 | 845 ± 8 | 1850 ± 6 |
| 240 | 349 ± 34 | 11300 ± 20 | 286 | 506 ± 7 | 2024 ± 4 |
| 240.5 ↑ | — | 11340 ± 30 | 288 | 248 ± 8 | 2179 ± 5 |
| 242 | 252 ± 20 | 11230 ± 40 | 290 | 113 ± 0 | 2300 ± 7 |
| 244 | 209 ± 18 | 10760 ± 50 | 292 | 50 ± 1 | 2367 ± 5 |
| 245.3 ↓ | 202 ± 20 | — | 293.2 ↑ | — | 2381 ± 6 |
| 246 | 205 ± 17 | 9918 ± 78 | 294 | 23 ± 1 | 2377 ± 8 |
| 248 | 218 ± 15 | 8734 ± 72 | 296 | 13 ± 0 | 2317 ± 10 |
| 250 | 246 ± 14 | 7382 ± 56 | 298 | 8 ± 1 | 2195 ± 16 |
| 252 | 287 ± 13 | 5844 ± 77 | 300 | 6 ± 0 | 2006 ± 23 |
| 254 | 341 ± 14 | 4471 ± 55 | 302 | 5 ± 1 | 1747 ± 29 |
| 256 | 401 ± 12 | 3360 ± 46 | 304 | 3 ± 0 | 1445 ± 27 |
| 258 | 485 ± 10 | 2476 ± 20 | 306 | 2 ± 1 | 1107 ± 35 |
| 260 | 582 ± 9 | 1883 ± 17 | 308 | 1 ± 0 | 800 ± 27 |
| 262 | 693 ± 13 | 1467 ± 7 | 310 | 1 ± 0 | 547 ± 21 |
| 264 | 821 ± 13 | 1204 ± 14 | 312 | — | 346 ± 15 |
| 266 | 960 ± 14 | 1054 ± 16 | 314 | — | 206 ± 12 |
| 268 | 1083 ± 13 | 985 ± 13 | 316 | — | 118 ± 9 |
| 269.3 ↓ | — | 974 ± 8 | 318 | — | 67 ± 5 |
| 270 | 1197 ± 9 | 979 ± 9 | 320 | — | 32 ± 3 |
| 272 | 1310 ± 9 | 1019 ± 8 | 322 | — | 15 ± 3 |
| 274 | 1394 ± 6 | 1094 ± 8 | 324 | — | 6 ± 2 |
| 274.8 ↑ | 1405 ± 7 | — | 326 | — | 1 ± 1 |

## TABLE 2
## MOLECULAR ABSORBANCES OF TRYPTOPHAN

| mμ[d] | Neutral[a] | Alkaline[b] | mμ[d] | Neutral[a] | Alkaline[b] |
|---|---|---|---|---|---|
| 230 | 6818 | 13200 | 279.0 ↑ | 5579 ± 14 | — |
| 232 | 4037 ± 60 | 7470 | 280 | 5559 ± 12 | 5377 ± 43 |
| 234 | 2772 ± 71 | 4354 ± 81 | 280.4 ↑ | — | 5385 ± 34 |
| 236 | 2184 ± 64 | 2951 ± 50 | 282 | 5323 ± 10 | 5302 ± 34 |
| 238 | 1904 ± 55 | 2282 ± 29 | 284 | 4762 ± 11 | 4962 ± 42 |
| 240 | 1764 ± 52 | 1959 ± 30 | 285.8 ↓ | 4471 ± 6 | — |
| 242.0 ↓ | 1737 ± 49 | 1813 ± 25 | 286 | 4482 ± 11 | 4596 ± 22 |
| 244 | 1772 ± 48 | 1773 ± 29 | 286.8 ↓ | — | 4565 ± 27 |
| 244.4 ↓ | — | 1763 ± 29 | 287.8 ↑ | 4650 ± 12 | — |
| 246 | 1869 ± 40 | 1792 ± 27 | 288 | 4646 ± 16 | 4634 ± 19 |
| 248 | 2018 ± 35 | 1877 ± 23 | 288.3 ↑ | — | 4639 ± 28 |
| 250 | 2217 ± 32 | 2013 ± 25 | 290 | 3935 ± 5 | 4393 ± 32 |
| 252 | 2462 ± 19 | 2187 ± 37 | 292 | 2732 ± 5 | 3551 ± 46 |
| 254 | 2760 ± 27 | 2410 ± 38 | 294 | 1824 ± 5 | 2666 ± 27 |
| 256 | 3087 ± 20 | 2664 ± 25 | 296 | 1211 ± 10 | 1990 ± 24 |
| 258 | 3422 ± 18 | 2953 ± 39 | 298 | 797 ± 4 | 1472 ± 19 |
| 260 | 3787 ± 17 | 3261 ± 34 | 300 | 510 ± 1 | 1064 ± 19 |
| 262 | 4142 ± 14 | 3586 ± 46 | 302 | 314 ± 3 | 755 ± 16 |
| 264 | 4472 ± 10 | 3895 ± 32 | 304 | 184 ± 2 | 517 ± 10 |
| 266 | 4777 ± 14 | 4212 ± 48 | 306 | 112 ± 4 | 333 ± 6 |
| 268 | 5020 ± 15 | 4481 ± 46 | 308 | 55 ± 9 | 217 ± 4 |
| 270 | 5220 ± 8 | 4742 ± 37 | 310 | 27 ± 11 | 129 ± 5 |
| 272 | 5331 ± 5 | 4933 ± 45 | 312 | 11 ± 8 | 84 ± 8 |
| 272.1 ↑ | 5344 ± 5 | — | 314 | 3 ± 2 | 53 ± 7 |
| 273.6 ↓ | 5329 ± 10 | — | 316 | — | 31 ± 7 |
| 274 | 5341 ± 8 | 5025 ± 34 | 318 | — | 17 ± 4 |
| 274.5 ~ | — | 5062 ± 38 | 320 | — | 8 ± 2 |
| 276 | 5431 ± 8 | 5108 ± 39 | 322 | — | 3 ± 4 |
| 278 | 5554 ± 12 | 5275 ± 46 | | | |

## TABLE 3
## MOLECULAR ABSORBANCES OF PHENYLALANINE

| mμ[d] | Neutral[a] | mμ[d] | Alkaline[b] | mμ[d] | Neutral[a] | mμ[d] | Alkaline[b] |
|---|---|---|---|---|---|---|---|
| 230 | 32.8 ± 1.5 | 230 | 161.9 ± 1.9 | 257.6 ↑ | 195.1 ± 1.5 | 257 | 188.4 ± 2.8 |
| 232 | 32.1 ± 1.6 | 232 | 99.2 ± 1.9 | 258 | 193.4 ± 1.3 | 258 | 209.1 ± 0.3 |
| 234 | 35.6 ± 2.1 | 234 | 70.7 ± 2.4 | 259 | 171.9 ± 1.0 | 258.2 ↑ | 209.6 ± 0.2 |
| 236 | 42.8 ± 2.1 | 236 | 63.3 ± 2.7 | 260 | 147.0 ± 0.6 | 260 | 184.2 ± 1.0 |
| 238 | 48.5 ± 2.3 | 238 | 62.3 ± 2.6 | 261.9 ↓ | 127.7 ± 1.5 | 260.7 ~ | 178.6 ± 0.3 |
| 240 | 59.4 ± 2.0 | 240 | 68.9 ± 3.2 | 262 | 128.1 ± 1.4 | 262 | 157.8 ± 0.9 |
| 242 ~ | 72.2 ± 2.3 | 242 | 83.0 ± 2.8 | 263.7 ↑ | 151.5 ± 0.6 | 262.7 ↓ | 105.5 ± 1.3 |
| | | 243 ~ | 85.4 ± 2.9 | 264 | 148.7 ± 0.4 | 263.9 ↑ | 161.2 ± 1.0 |
| 244 | 80.1 ± 2.1 | 244 | 89.0 ± 3.0 | 265 | 119.8 ± 1.3 | 264 | 160.0 ± 2.1 |
| 246 | 102.0 ± 0.6 | 246 | 108.9 ± 2.8 | 266 | 91.8 ± 1.4 | 266 | 114.3 ± 1.6 |
| 247.4 ↑ | 110.7 ± 2.2 | 247 | 120.9 ± 1.5 | 266.8 ~ | 85.6 ± 1.5 | 266.5 ↓ | 109.7 ± 1.8 |
| 248 | 109.8 ± 1.9 | 248.0 ↑ | 126.1 ± 1.4 | | | 267.7 ↑ | 117.7 ± 1.8 |
| 248.3 ↓ | 109.5 ± 2.0 | 248.7 ↓ | 125.1 ± 1.7 | 268 | 74.7 ± 1.0 | 268 | 115.0 ± 1.0 |
| 250 | 123.5 ± 2.6 | 250 | 132.7 ± 1.8 | 270 | 30.0 ± 1.8 | 270 | 50.2 ± 2.0 |
| 251 | 143.0 ± 2.8 | 251 | 149.3 ± 1.9 | 272 | 14.3 ± 1.0 | 272 | 18.7 ± 1.1 |
| 252 | 153.9 ± 1.0 | 252 | 167.0 ± 1.1 | 274 | 5.4 ± 0.3 | 274 | 7.4 ± 0.3 |
| 252.2 ↑ | 154.1 ± 1.0 | 252.9 ↑ | 171.5 ± 1.3 | 276 | 2.2 ± 0.4 | 276 | 2.6 ± 0.4 |
| 254 | 139.6 ± 1.0 | 254 | 166.3 ± 0.8 | 278 | 1.1 ± 0.5 | 278 | 0.7 ± 0.3 |
| 254.5 ↓ | 138.5 ± 1.4 | 254.9 ↓ | 162.8 ± 1.7 | 280 | 0.7 ± 0.3 | 280 | 0.4 ± 0.2 |
| 256 | 156.5 ± 2.2 | 256 | 168.4 ± 1.9 | | | | |

## TABLE 4
## DIFFERENCE OF ALKALINE[b] VERSUS NEUTRAL[a] SPECTRA FOR TYROSINE, TRYPTOPHAN AND PHENYLALANINE

| mμ | Tyrosine | Tryptophan | Phenylalanine | mμ | Tyrosine | Tryptophan | Phenylalanine |
|---|---|---|---|---|---|---|---|
| 230 | 3041 | 4135 | 123.9 | 280 | 315 ± 1 | −191 ± 18 | — |
| 232 | 5440 | 3213 | 66.0 | 282 | 558 ± 1 | − 24 ± 4 | — |
| 234 | 7608 | 1621 | 35.4 | 284 | 994 ± 10 | 194 ± 3 | — |
| 236 | 9415 | 732 ± 35 | 20.7 | 286 | 1513 ± 11 | 110 ± 10 | — |
| 238 | 10490 | 345 ± 23 | 13.9 | 288 | 1936 ± 1 | 11 ± 3 | — |
| 240 | 11060 | 149 ± 21 | 9.9 | 290 | 2196 ± 14 | 467 ± 8 | — |
| 242 | 11090 | 45 ± 15 | 11.1 | 292 | 2331 ± 15 | 802 ± 3 | — |
| 244 | 10660 | − 40 ± 15 | 8.8 | 294 | 2357 ± 7 | 830 ± 8 | — |
| 246 | 9844 | −104 ± 11 | 9.0 | 296 | 2307 ± 6 | 755 ± 10 | — |
| 248 | 8567 | −172 ± 10 | 16.4 | 298 | 2194 ± 7 | 652 ± 13 | — |
| 250 | 7205 | −233 ± 9 | 8.3 | 300 | 2002 ± 3 | 527 ± 5 | — |
| 252 | 5671 | −298 ± 16 | 15.3 | 302 | 1754 ± 2 | 413 ± 11 | — |
| 254 | 4344 | −371 ± 10 | 25.8 | 304 | 1437 ± 2 | 300 ± 14 | — |
| 256 | 3127 | −435 ± 14 | 11.3 | 306 | 1097 ± 9 | 205 ± 8 | — |
| 258 | 2142 | −490 ± 19 | 19.1 | 308 | 792 ± 14 | 137 ± 5 | — |
| 260 | 1368 ± 37 | −535 ± 13 | 37.8 | 310 | 526 ± 13 | 88 ± 3 | — |
| 262 | 820 ± 36 | −564 ± 8 | 26.9 | 312 | 334 ± 9 | 55 ± 8 | — |
| 264 | 420 ± 25 | −580 ± 10 | 16.1 | 314 | 221 ± 19 | 22 ± 14 | — |
| 266 | 125 ± 20 | −573 ± 12 | 22.4 | 316 | 101 ± 7 | 16 ± 10 | — |
| 268 | − 78 ± 18 | −539 ± 14 | 42.8 | 318 | 62 ± 2 | 5 ± 2 | — |
| 270 | −225 | −486 ± 12 | 17.4 | 320 | 28 ± 4 | 0 | — |
| 272 | −296 | −394 ± 12 | 4.0 | 322 | 12 ± 5 | — | — |
| 274 | −299 | −308 ± 7 | 1.7 | 324 | 3 ± 4 | — | — |
| 276 | −158 | −312 ± 16 | 0.3 | 326 | 1 ± 1 | — | — |
| 278 | 89 ± 5 | −278 ± 13 | 0 | 328 | 0 | — | — |

# TABLE 5
# DIFFERENCE OF ACID[c] VERSUS NEUTRAL[a] SPECTRA FOR TYROSINE, TRYPTOPHAN AND PHENYLALANINE

| mμ | Tyrosine | Tryptophan | Phenylalanine | mμ | Tyrosine | Tryptophan | Phenylalanine |
|---|---|---|---|---|---|---|---|
| 230 | — | — | 46.7 | 276 | −40 | 128 ± 16 | 0 |
| 232 | 576 | 421 ± 49 | 34.1 | 278 | −45 | 110 ± 10 | — |
| 234 | 441 | 610 ± 37 | 23.3 | 280 | −34 | 71 ± 11 | — |
| 236 | 346 | 590 ± 31 | 16.0 | 282 | −46 | − 23 ± 7 | — |
| 238 | 218 | 512 ± 29 | 10.8 | 284 | −73 | − 92 ± 9 | — |
| 240 | 108 | 432 ± 23 | 6.9 | 286 | −71 | − 3 ± 9 | — |
| 242 | 40 | 358 ± 19 | 3.9 | 288 | −49 | − 26 ± 4 | — |
| 244 | 4 | 305 ± 20 | 3.3 | 290 | −31 | −250 ± 5 | — |
| 246 | −13 | 263 ± 21 | 1.2 | 292 | −20 | −317 ± 9 | — |
| 248 | −18 | 240 ± 16 | −0.3 | 294 | −16 | −276 ± 9 | — |
| 250 | −20 | 223 ± 14 | 2.5 | 296 | −14 | −227 ± 5 | — |
| 252 | −16 | 216 ± 17 | −1.8 | 298 | −12 | −177 ± 7 | — |
| 254 | −12 | 219 ± 15 | −1.9 | 300 | −10 | −131 ± 9 | — |
| 256 | − 7 | 222 ± 11 | 4.0 | 302 | − 9 | − 88 ± 6 | — |
| 258 | − 5 | 223 ± 14 | −3.3 | 304 | − 7 | − 59 ± 8 | — |
| 260 | − 3 | 225 ± 14 | −4.3 | 306 | − 6 | − 40 ± 10 | — |
| 262 | 0 | 232 ± 13 | 2.5 | 308 | − 4 | − 24 ± 10 | — |
| 264 | 0 | 232 ± 18 | −3.1 | 310 | − 3 | − 13 ± 10 | — |
| 266 | − 4 | 224 ± 15 | −3.4 | 312 | − 2 | − 7 ± 7 | — |
| 268 | − 8 | 214 ± 7 | −4.4 | 314 | − 1 | − 4 ± 4 | — |
| 270 | −11 | 190 ± 9 | −2.0 | 316 | − 1 | 0 | — |
| 272 | −13 | 159 ± 12 | −0.7 | 318 | 0 | — | — |
| 274 | −20 | 127 ± 15 | −0.3 | | | | |

[a] 0.1*M* phosphate buffer, pH 7.1.
[b] 0.1*N* potassium hydroxide.
[c] 0.1*N* hydrochloric acid.
[d] Maxima are indicated by ↑, minima by ↓, and inflection points by ~.

Data taken from Mihalyi, E., *J. Chem. Eng. Data, 13,* 179 (1969). Reproduced by permission of the American Chemical Society, Washington, D.C.

# SPECIFIC ROTATIONS OF AMINO ACIDS

## SPECIFIC OPTICAL ROTATIONS (25° ± 1) OF AMINO ACIDS*

| Amino Acids | $[\alpha]_D(H_2O)$ | $[\alpha]_D(5N\ HCl)$ |
|---|---|---|
| **Amino Acids Commonly Found in Protein Hydrolyzates** | | |
| L-Alanine[a] | + 1.8 | + 14.6 |
| L-Arginine[a] | + 12.5 | + 27.6 |
| L-Asparagine | – | + 33.2[b] |
| L-Aspartic acid[a] | + 5.0 | + 25.4 |
| L-Cysteine[a] | – 16.5 | + 6.5 |
| L-Cystine[a] | – | –232.0 |
| LGlutamic acid[a] | + 12.0 | + 31.8 |
| L-Glutamine | + 6.3 | + 31.8[c] |
| L-Histidine[a] | – 38.5 | + 11.8 |
| L-4-Hydroxyproline | – 76.0 | – 50.5 |
| L-Isoleucine[a] | + 12.4 | + 39.5 |
| L-Leucine[a] | – 11.0 | + 15.1[d] |
| L-Lysine[a] | + 13.5 | + 26.0 |
| L-Methionine[a] | – 10.0 | + 23.2 |
| L-Phenylalanine[a] | – 34.5 | – 4.5 |
| L-Proline[a] | – 86.2 | – 60.4 |
| L-Serine[a] | – 7.5 | + 15.1 |
| L-Threonine[a] | – 28.5 | – 15.0 |
| L-Tryptophan[a] | – 33.7 | + 2.8[c] |
| L-Tyrosine[a] | – | – 10.0 |
| L-Valine[a] | + 5.6 | + 26.6[d] |
| **Other Natural Amino Acids** | | |
| L-Albizziine | – 67 | – |
| (+)-*S*-Allyl-L-cysteine sulfoxide | + 62.8 | – |
| L-*α*-Aminoadipic acid[a] | + 3.2 | + 25.0 |
| L-*α*-Aminobutyric acid[a] | + 9.3 | + 20.6 |
| *β*-Aminoisobutyric acid | – 13 | – |
| L-Canavanine | + 7.9 | – |
| *O*-Carbamyl-D-serine | + 2.0 | – 19.6[c] |
| *m*-Carboxyphenyl-L-alanine | + 17.0 | – |
| L-Citrulline[a] | + 4.0 | + 24.2 |
| D-Cycloserine | +116 | – |
| L-Cystathionine | – | + 23.7[c] |
| L-*α*,*γ*-Diaminobutyric acid[a] | + 7.2 | + 31.7 |
| L-*α*,*ε*-Diaminopimelic acid[a] | + 8.1 | + 45.1 |
| L-*α*,*β*-Diaminopropionic acid[a] | 0 | – 34.0 |
| 3,5-Diiodo-L-tyrosine | – | + 2.9[e] |
| L-Djenkolic acid | – | – 44.5[f] |
| $N^4$-Ethyl-L-asparagine | – 5 | – |
| L-Homoserine[a] | – 8.8 | + 18.3 |
| *γ*-Hydroxy-L-arginine•HCl | – | + 5.4 |
| $N^4$-2-Hydroxyethyl-L-asparagine | – 2.5 | – |
| 5-Hydroxy-L-lysine[a] | + 9.2 | + 17.8 |
| 5-Hydroxy-L-pipecolic acid | –23.1 | – |
| allo-Hydroxy-L-proline | – 59.5 | – 18.8 |
| L-*α*-Kainic acid | – 15 | – |
| L-*α*-allo-Kainic acid | + 8 | – |
| L-Kynurenine | – 30.5 | – |
| threo-*β*-Methyl-L-aspartic acid | – 12.4 | + 13.4 |

## SPECIFIC OPTICAL ROTATIONS (25° ± 1) OF AMINO ACIDS* (Continued)

| Amino Acids | $[\alpha]_D(H_2O)$ | $[\alpha]_D(5N\ HCl)$ |
|---|---|---|
| **Other Natural Amino Acids (continued)** | | |
| (–) *S*-Methyl-L-cysteine | – 26 | – |
| (+) *S*-Methyl-L-cysteine sulfoxide | +125 | +168[c] |
| γ-Methylene-L-glutamic acid | – | + 14[b] |
| α-Methylserine[a] | + 4.5 | + 2.0 |
| L-Ornithine[a] | + 12.1 | + 28.4 |
| D-Penicillamine | – 56.0[g] | – |
| L-Pipecolic acid | – 24.6 | – |
| β-(N-Pyrazolyl)-L-alanine | – 73 | – |
| L-Theanine | + 6.3 | – |
| L-Thyroxine | – 4.4[h] | – |
| **Other Amino Acids** | | |
| L-Allylglycine[a] | – 37.1 | – 5.7 |
| L-α-Aminocaprylic acid[a] | + 9.1 | + 23.0 |
| L-α-Aminocyclohexylacetic acid[a] | + 6.7 | + 35.5 |
| L-α-Aminocyclohexylpropionic acid[a] | – 9.0 | + 15.0 |
| L-α-Aminoheptylic acid[a] | + 6.8 | + 23.3 |
| L-α-Amino-ε-hydroxy-*n*-caproic acid[a] | + 4.0 | + 23.7 |
| L-α-Amino-δ-hydroxy-*n*-valeric acid[a] | + 6.0 | + 28.8 |
| L-α-Aminophenylacetic acid[a] | +114 | +168 |
| L-α-Aminotricarballylic acid (A)[a] | + 7.5 | + 36.4 |
| L-α-Aminotricarballylic acid (B)[a] | – 32.8 | – 48.0 |
| L-allo-Cystathionine | – | – 25.0[c] |
| L-Ethionine[a] | – 9.2 | + 23.7 |
| L-Homocystine[a] | – | + 78.0 |
| L-Homoglutamine | + 2.6 | + 21.0[e] |
| L-Homolanthionine | – | + 37.3 |
| L-allo-5-Hydroxylysine[a] | + 10.9 | + 31.4 |
| L-allo-Isoleucine[a] | + 15.9 | + 39.6 |
| L-Isovaline[a] | + 11.2 | + 6.7 |
| L-Lanthionine | + 43.8[i] | – |
| L-*tert*-Leucine[a] | – 9.7 | + 7.4 |
| L-Norleucine[a] | + 4.7 | + 24.5 |
| L-Norvaline[a] | + 7.0 | + 24.1 |
| L-β-Phenylserine[a] | – 33.1 | – 50.3 |
| L-allo-β-Phenylserine[a] | + 8.2 | + 81.3 |
| L-β-2-Thienylalanine | – 31.7 | – |
| L-allo-Threonine[a] | + 10.0 | + 31.7 |

* Revised by Dr. J.C. MacDonald.

[a] Determinations made on isomers obtained by enzymatic resolution as per Greenstein (*Advan. Protein Chem.*, *9*, 121, 1954).
[b] 3*N* HCl.
[c] 1*N* HCl.
[d] Data taken from MacDonald, J.C., *Can. J. Chem.*, *47*, 2739 (1969). C = 2.
[e] 1.1*N* HCl.
[f] 1% HCl.
[g] 0.93*N* NaOH.
[h] 0.13*N* NaOH in 70% ethanol.
[i] 2% aqueous solution containing 1 equivalent of NaOH.

Data taken from: Meister, A., *Biochemistry of the Amino Acids*, 2nd ed., pp. 141, 142, and 144, Academic Press, New York (1965), and combined into one table. Reproduced by permission of the copyright owners.

# SPECIFIC ROTATIONS OF THE AMINO ACIDS, USING SODIUM LIGHT OF 5893 Å WAVELENGTH*

CODE

Sources:

A – prepared from a protein or other naturally occurring material
B – prepared by resolution of the inactive synthetic form
C – prepared by resolution of the inactive racemized form
D – prepared from the inactive synthetic form by a biological method
E – prepared from the inactive racemized form by a biological method
? – source not given
Column 1: grams of solute per 100 ml of solution, *c*
Column 2: solvent
Column 3: density of the solution, *d*
Column 4: grams of solute per 100 grams of solution, *p*
Column 5: moles acid or base per mole amino acid
Column 6: length of the tube, *l* (decimeters)
Column 7: temperature (°C)
Column 8: observed rotation, *a* (angular degrees)
Column 9: specific rotation, [*a*] (angular degrees), calculated from $[a]^t_\lambda = \frac{a \times 100}{c \times l} = \frac{a \times 100}{p \times d \times l}$,

where *t* is the temperature (°C) and λ the wavelength of the incident light (Ångstroms)

| Source | 1 | 2 | 3 | 4 | 5 | 6 | 7 | 8 | 9 | Reference |
|---|---|---|---|---|---|---|---|---|---|---|
| **L-Alanine** | | | | | | | | | | |
| A | 10 | HCl, 6*N* | – | – | 5 | 2 | 20 | – | + 14.7 | 1 |
| A | 5.790 | HCl, 0.97*N* | 1.033 | 5.605 | 1.5 | 2 | 15 | +1.70 | + 14.7 | 2 |
| A | 10.3 | $H_2O$ | 1.03 | 10.00 | 0 | 2 | 22 | +0.55 | + 2.7 | 3 |
| A | 1.781 | NaOH, 3*N* | – | – | 15 | 2 | 20 | – | + 3.0 | 4 |
| **D-Alanine** | | | | | | | | | | |
| B | 1.344 | HCl, 6*N* | – | – | 39.4 | 2 | 30.4 | –0.392 | – 14.6 | 5 |
| **L-Arginine** | | | | | | | | | | |
| A | 6.61 | HCl, 6*N* | – | – | 17 | 2 | 20 | – | + 27.2 | 1[a] |
| A | 1.653 | HCl, 6*N* | – | – | 63 | 4.001 | 23.4 | +1.777 | + 26.9 | 6 |
| A | 3.48 | $H_2O$ | – | – | 0 | 2 | 20 | – | + 12.5 | 7 |
| A | 0.87 | NaOH, 0.5*N* | – | – | 10 | 2 | 20 | – | + 11.8 | 7 |
| **L-Aspartic Acid** | | | | | | | | | | |
| A | 10 | HCl, 3*N* | – | – | 4 | 2 | 20 | – | + 26.2 | 1 |
| A | 1.3300 | NaOH, 3*N* | – | – | 30 | 3 | 18 | – | – 1.7 | 4 |
| A | 1.3300 | $H_2O$ | – | – | 0 | 3 | 18 | – | + 4.7 | 4 |
| A | 2.002 | HCl, 6*N* | – | – | 39 | 4.001 | 24.0 | +1.972 | + 24.6 | 8 |
| **D-Aspartic Acid** | | | | | | | | | | |
| C | 4.289 | HCl, 0.97*N* | 1.032 | 4.156 | 3 | 1 | 20 | –1.09 | – 25.5 | 9 |

Data taken from: *Handbook of Chemistry and Physics,* 53rd ed., p. C-742, R. C. Weast, Ed. Copyright 1972, The Chemical Rubber Co., Cleveland, Ohio. Revised by Dr. J. C. MacDonald.

## SPECIFIC ROTATIONS OF THE AMINO ACIDS, USING SODIUM LIGHT OF 5893 Å WAVELENGTH (Continued)

| Source | 1 | 2 | 3 | 4 | 5 | 6 | 7 | 8 | 9 | Reference |
|---|---|---|---|---|---|---|---|---|---|---|
| **L-Citrulline** | | | | | | | | | | |
| A | 8 | HCl, 6*N* | – | – | 13 | 2 | 20 | – | + 25.8 | 1 |
| **L-Cystine** | | | | | | | | | | |
| A | 0.400 | NaOH, 0.2*N* | – | – | 12 | 2 | 18.5 | – | – 70.0 | 4 |
| A | 0.9974 | HCl, 1.02*N* | 1.0181 | 0.9797 | 24.6 | 2 | 24.35 | –4.277 | –214.40 | 10 |
| **D-Cystine** | | | | | | | | | | |
| C | – | HCl, 1*N* | – | 1 | 24 | – | 20 | – | +223 | 11 |
| **3,5-Diiodo-L-tyrosine** | | | | | | | | | | |
| A | 5.08 | HCl, 1.1*N* | 1.05 | 4.84 | 9.4 | 1 | 20 | +0.15 | + 2.89 | 12 |
| A | 4.41 | $NH_4OH$, 13.4*N* | 0.9779 | 4.51 | 132 | 1 | 20 | +0.10 | + 2.27 | 12 |
| **L-Glutamic Acid** | | | | | | | | | | |
| A | 10 | HCl, 2*N* | – | – | 3 | 2 | 20 | – | + 32.0 | 1 |
| A | 1.471 | NaOH, 1*N* | – | – | 10 | 2 | 18 | – | + 10.96 | 4 |
| A | 1.471 | $H_2O$ | – | – | 0 | 2 | 18 | – | + 11.5 | 4 |
| A | 1.002 | HCl, 6*N* | – | – | 87 | 4.001 | 22.4 | +1.25 | + 31.2 | 13 |
| **D-Glutamic Acid** | | | | | | | | | | |
| C | 5.425 | HCl, 0.37*N* | 1.0233 | 5.3011 | 1 | 1 | 20 | –1.63 | – 30.05 | 9 |
| **L-Glutamine** | | | | | | | | | | |
| A | 4 | $H_2O$ | – | – | 0 | 2 | 20 | – | + 6.5 | 1 |
| **L-Histidine** | | | | | | | | | | |
| A | 7.40 | HCl, 6*N* | – | – | 13 | 2 | 20 | – | + 12.6 | 1[a] |
| A | 0.775 | NaOH, 0.5*N* | – | – | 10 | 2 | 20 | – | – 10.9 | 7 |
| A | 1.480 | HCl, 6*N* | – | – | 63 | 4.001 | 22.7 | +0.766 | + 13.0 | 8 |
| A | 1.128 | $H_2O$ | 1.0012 | 1.127 | 0 | 4 | 25.00 | –1.714 | – 39.01 | 14 |
| **D-Histidine** | | | | | | | | | | |
| ? | 4.000 | HCl, 1*N* | – | – | 4 | 1 | 20 | –0.407 | – 10.2 | 15 |
| B | 2.66 | $H_2O$ | – | – | 0 | 2 | 23 | +2.11 | + 39.8 | 15 |
| **Hydroxy-L-proline** | | | | | | | | | | |
| A | 4 | $H_2O$ | – | – | 0 | 2 | 20 | – | – 75.1 | 1 |
| A | 1.31 | HCl, 1*N* | – | – | 10 | 2 | 20 | – | – 47.3 | 7 |
| A | 0.655 | NaOH, 0.5*N* | – | – | 10 | 2 | 20 | – | – 70.6 | 7 |
| A | 1.001 | $H_2O$ | – | – | 0 | 4.001 | 22.5 | –3.009 | – 75.2 | 8 |
| **Hydroxy-D-proline** | | | | | | | | | | |
| B | 4.48 | $H_2O$ | 1.03 | 4.35 | 0 | 1 | 21 | +3.37 | + 75.2 | 16 |

## SPECIFIC ROTATIONS OF THE AMINO ACIDS, USING SODIUM LIGHT OF 5893 Å WAVELENGTH (Continued)

| Source | 1 | 2 | 3 | 4 | 5 | 6 | 7 | 8 | 9 | Reference |
|---|---|---|---|---|---|---|---|---|---|---|
| **allo-Hydroxy-L-proline** | | | | | | | | | | |
| B | 2.617 | $H_2O$ | 1.014 | 2.581 | 0 | 1 | 18 | −1.52 | − 58.1 | 16 |
| **allo-Hydroxy-D-proline** | | | | | | | | | | |
| B | 2.530 | $H_2O$ | 1.013 | 2.998 | 0 | 1 | 17 | +1.48 | + 58.5 | 16 |
| **L-Isoleucine** | | | | | | | | | | |
| A | 4 | HCl, 6*N* | – | – | 20 | 2 | 20 | – | + 40.7 | 1 |
| B | 5.09 | HCl, 6.1*N* | 1.098 | 4.64 | 15 | 1 | 20 | +2.07 | + 40.61 | 17 |
| B | 3.10 | $H_2O$ | 1.008 | 3.08 | 0 | 2 | 20 | +0.70 | + 11.29 | 17 |
| A | 3.34 | NaOH, 0.33*N* | 1.017 | 3.28 | 1.3 | 2 | 20 | +0.74 | + 11.09 | 18 |
| **D-Isoleucine** | | | | | | | | | | |
| B | 4.53 | HCl, 6.1*N* | 1.083 | 4.18 | 17 | 1 | 20 | −1.85 | − 40.86 | 17 |
| B | 3.12 | $H_2O$ | 1.006 | 3.10 | 0 | 2 | 20 | −0.66 | − 10.55 | 17 |
| **L-allo-Isoleucine** | | | | | | | | | | |
| B | 3.97 | HCl, 6*N* | – | – | 20 | 1 | 20 | +1.50 | + 38.1 | 19 |
| B | 2.00 | $H_2O$ | – | – | 0 | 1 | 20 | +0.28 | + 14.0 | 19 |
| **D-allo-Isoleucine** | | | | | | | | | | |
| D | 5.14 | HCl, 6*N* | 1.094 | 4.70 | 15.0 | 2 | 20 | −3.80 | − 36.95 | 20 |
| B | 2.00 | $H_2O$ | – | – | 0 | 1 | 20 | −0.285 | − 14.2 | 19 |
| **L-Leucine** | | | | | | | | | | |
| A | 4 | HCl, 6*N* | – | – | 20 | 2 | 20 | – | + 15.1 | 1 |
| A | 1.31 | NaOH, 3*N* | – | – | 30 | 2 | 20 | – | + 7.6 | 4 |
| A | 1.999 | HCl, 6*N* | – | – | 38 | 4.001 | 25.9 | +1.212 | + 15.1 | 6 |
| A | 2.001 | $H_2O$ | – | – | 0 | 4.001 | 24.7 | −0.863 | − 10.8 | 6 |
| **D-Leucine** | | | | | | | | | | |
| ? | 4.0 | HCl, 6*N* | 1.1 | 3.664 | 19 | 2 | 20 | +1.26 | − 15.6 | 21 |
| ? | – | $H_2O$ | – | 2.08 | 0 | 2 | 20 | +0.43 | + 10.34 | 22 |
| **L-Lysine** | | | | | | | | | | |
| A | 6.40 | HCl, 6*N* | – | – | 14 | 2 | 20 | – | + 26.5 | 1[a] |
| A | 2.00 | HCl, 6*N* | – | – | 43 | 4 | 22.9 | +1.652 | + 25.9 | 6 |
| A | 6.496 | $H_2O$ | – | – | 0 | 2 | 20 | +1.90 | + 14.6 | 23 |
| **D-Lysine** | | | | | | | | | | |
| B | 2.00 | HCl, 0.27*N* | – | – | 2 | 2 | 20 | −0.939 | − 23.48 | 24 |
| **L-Methionine** | | | | | | | | | | |
| A | 8 | HCl, 6*N* | – | – | 11 | 2 | 20 | – | + 24.2 | 1 |
| B | 0.80 | $H_2O$ | – | – | 0 | 2 | 25 | −0.13 | − 8.11 | 25 |

## SPECIFIC ROTATIONS OF THE AMINO ACIDS, USING SODIUM LIGHT OF 5893 Å WAVELENGTH (Continued)

| Source | 1 | 2 | 3 | 4 | 5 | 6 | 7 | 8 | 9 | Reference |
|---|---|---|---|---|---|---|---|---|---|---|
| **D-Methionine** | | | | | | | | | | |
| B | 0.80 | HCl, 0.2001*N* | – | – | 4 | 2 | 25 | −0.34 | – 21.18 | 25 |
| B | 0.80 | $H_2O$ | – | – | 0 | 2 | 25 | +0.13 | + 8.12 | 25 |
| B | 0.80 | $NaHCO_3$, 0.6*N* | – | – | 11 | 2 | 25 | −0.12 | – 7.47 | 25 |
| **L-Phenylalanine** | | | | | | | | | | |
| A | 2 | $H_2O$ | – | – | 0 | 2 | 20 | – | – 34.3 | 1 |
| B | 1.936 | $H_2O$ | 1.0040 | 1.928 | 0 | 2 | 20 | −1.36 | – 35.14 | 26 |
| **D-Phenylalanine** | | | | | | | | | | |
| B | 2.043 | $H_2O$ | 1.0045 | 2.034 | 0 | 2 | 20 | +1.43 | + 35.0 | 26 |
| B | 3.814 | HCl, 5.4*N* | 1.0895 | 3.501 | 23 | 2 | 20 | +0.54 | + 7.07 | 27 |
| **L-Proline** | | | | | | | | | | |
| A | 4 | $H_2O$ | – | – | 0 | 2 | 20 | – | – 84.8 | 1 |
| A | 0.575 | HCl, 0.50*N* | – | – | 10 | 2 | 20 | – | – 52.6 | 7 |
| A | 1.001 | $H_2O$ | – | – | 0 | 4.001 | 23.4 | −3.402 | – 85.0 | 8 |
| B | 2.42 | KOH, 0.6*N* | 1.031 | 2.35 | 3 | 1 | 20 | −2.25 | – 93.0 | 28 |
| **D-Proline** | | | | | | | | | | |
| B | 3.90 | $H_2O$ | 1.01 | 3.865 | 0 | 1 | 20 | +3.18 | + 81.5 | 28 |
| **L-Serine** | | | | | | | | | | |
| A | 10 | HCl, 2*N* | – | – | 2 | 2 | 20 | – | + 15.0 | 1 |
| B | 9.344 | HCl, 1*N* | 1.0465 | 8.929 | 1 | 1 | 25 | +1.35 | + 14.45 | 29 |
| B | 10.414 | $H_2O$ | 1.0414 | 9.997 | 0 | 2 | 20 | −1.42 | – 6.83 | 29 |
| **D-Serine** | | | | | | | | | | |
| B | 9.359 | HCl, 1*N* | 1.0465 | 8.943 | 1 | 1 | 25 | −1.34 | – 14.32 | 29 |
| B | 10.412 | $H_2O$ | 1.0414 | 9.998 | 0 | 2 | 20 | +1.43 | + 6.87 | 29 |
| **L-Threonine** | | | | | | | | | | |
| A | 6 | $H_2O$ | – | – | 0 | 2 | 20 | – | – 28.5 | 1 |
| B | – | $H_2O$ | – | 1.092 | 0 | 2 | 26 | −0.625 | – 28.3 | 30 |
| **D-Threonine** | | | | | | | | | | |
| B | – | $H_2O$ | – | 1.331 | 0 | 2 | 26 | +0.780 | + 28.4 | 30 |
| **L-allo-Threonine** | | | | | | | | | | |
| B | – | $H_2O$ | – | 1.643 | 0 | 2 | 26 | +0.320 | + 9.6 | 30 |
| **D-allo-Threonine** | | | | | | | | | | |
| B | – | $H_2O$ | – | 1.634 | 0 | 2 | 26 | −0.302 | – 9.1 | 30 |

## SPECIFIC ROTATIONS OF THE AMINO ACIDS, USING SODIUM LIGHT OF 5893 Å WAVELENGTH (Continued)

| Source | 1 | 2 | 3 | 4 | 5 | 6 | 7 | 8 | 9 | Reference |
|---|---|---|---|---|---|---|---|---|---|---|
| **L-Thyroxine** | | | | | | | | | | |
| A | – | NaOH, 0.13*N*, in 70% EtOH by weight | – | 3 | 3 | 1 | – | –0.147 | – 4.4 | 31 |
| **L-Tryptophan** | | | | | | | | | | |
| A | 1 | $H_2O$ | – | – | 0 | 2 | 20 | – | – 31.5 | 1 |
| A | 1.02 | HCl, 0.50*N* | – | – | 10 | 2 | 20 | – | + 2.4 | 7 |
| A | 1.004 | $H_2O$ | – | – | 0 | 4.001 | 22.7 | –1.266 | – 31.5 | 8 |
| A | 2.426 | NaOH, 0.5*N* | 1.0243 | 2.368 | 4.2 | 1 | 20 | +0.15 | + 6.17 | 32 |
| **D-Tryptophan** | | | | | | | | | | |
| C | 0.5024 | $H_2O$ | – | – | 0 | 2 | 25 | +0.326 | + 32.45 | 33 |
| **L-Tyrosine** | | | | | | | | | | |
| A | 5 | HCl, 1*N* | – | – | 3.6 | 2 | 20 | – | – 12.0 | 1 |
| A | 0.906 | NaOH, 3*N* | – | – | 60 | 3 | 18 | – | – 13.2 | 4 |
| B | 4.40 | HCl, 6.3*N* | 1.116 | 3.94 | 28 | 2 | 20 | –0.76 | – 8.64 | 34 |
| **D-Tyrosine** | | | | | | | | | | |
| B | 5.1484 | HCl, 6.3*N* | 1.1175 | 4.6071 | 24 | 2 | 20 | +0.89 | + 8.64 | 34 |
| **L-Valine** | | | | | | | | | | |
| A | 8 | HCl, 6*N* | – | – | 9 | 2 | 20 | – | + 28.0 | 1 |
| B | 3.4 | HCl, 6*N* | 1.1 | 3.05 | 20 | 2 | 20 | +1.93 | + 28.8 | 35 |
| B | 3.58 | $H_2O$ | 1.007 | 3.56 | 0 | 2 | 20 | +0.46 | + 6.42 | 35 |
| **D-Valine** | | | | | | | | | | |
| B | 3.2 | HCl, 6.0*N* | 1.1 | 2.91 | 21 | 2 | 20 | –1.86 | – 29.04 | 35 |
| E | 6.24 | $H_2O$ | 1.00 | 6.24 | 0 | 1 | 20 | –0.37 | – 6.06 | 36 |

[a] Data recalculated in terms of free base.

### REFERENCES

1. Hayashi, K., Fujii, Y., Saito, R., Kanao, H., and Hino, T., *Agr. Biol. Chem. (Japan), 30,* 1221 (1966).
2. Clough, G. W., *J. Chem. Soc., 113,* 526 (1918).
3. Fischer, E., and Raske, K., *Ber., 40,* 3717 (1907).
4. Lutz, O., and Jirgensons, B., *Ber., 63,* 448 (1930).
5. Dunn, M. S., Butler, A. W., and Naiditch, M. J., *Unpublished data.*
6. Dunn, M. S., and Courtney, G., *Unpublished data.*
7. Lutz, O., and Jirgensons, B., *Ber., 64,* 1221 (1931).
8. Dunn, M. S., and Stoddard, M. P., *Unpublished data.*
9. Fischer, E., *Ber., 32,* 2451 (1899).
10. Toennies, G., and Lavine, T. F., *J. Biol. Chem., 89,* 153 (1930).

11. Loring, H. S., and du Vigneaud, V., *J. Biol. Chem., 107,* 267 (1934).
12. Abderhalden, E., and Guggenheim, M., *Ber., 41,* 1237 (1908).
13. Dunn, M. S., and Sexton, E. L., *Unpublished data.*
14. Dunn, M. S., and Frieden, E. H., *Unpublished data.*
15. Cox, G. J., and Berg, C. P., *J. Biol. Chem., 107,* 497 (1934).
16. Leuchs, H., and Bormann, K., *Ber., 52,* 2086 (1919).
17. Locquin, R., *Bull. Soc. Chim. (4) 1,* 601 (1907).
18. Ehrlich, F., *Ber., 37,* 1809 (1904).
19. Abderhalden, E., and Zeisset, W., *Z. Physiol. Chem., 196,* 121 (1931).
20. Ehrlich, F., *Ber., 40,* 2538 (1907).
21. Fischer, E., and Warburg, O., *Ber., 38,* 3997 (1905).
22. Ehrlich, F., *Biochem. Z., 1,* 8 (1906).
23. Vickery, H.B., *Private communication* (1940).
24. Berg, C. P., *J. Biol. Chem., 115,* 9 (1936); *Private communication* (1940).
25. Windus, W., and Marvel, C. S., *J. Am. Chem. Soc., 53,* 3490 (1931).
26. Fischer, E., and Schoeller, W., *Ann., 357,* 1 (1907).
27. Fischer, E., and Mouneyrat, A., *Ber., 33,* 2383 (1900).
28. Fischer, E., and Zemplen, G., *Ber., 42,* 2989 (1909).
29. Fischer, E., and Jacobs, W. A., *Ber., 39,* 2942 (1906).
30. West, H. D., and Carter, H. E., *J. Biol. Chem., 119,* 109 (1937); Carter, H. E., *Private communication* (1940).
31. Foster, G. L., Palmer, W. W., and Leland, J. P., *J. Biol. Chem., 115,* 467 (1936).
32. Abderhalden, E., and Baumann, L., *Z. Physiol. Chem., 55,* 412 (1908).
33. Berg. C. P., *J. Biol. Chem., 100,* 79 (1933); *Private communication* (1940).
34. Fischer, E., *Ber., 32,* 3638 (1899).
35. Fischer, E., *Ber., 39,* 2320 (1906).
36. Ehrlich, F., and Wendel, A., *Biochem. Z., 8,* 399 (1908).

# NUMBERING AND CLASSIFICATION OF ENZYMES

DR. HUBERT A. LECHEVALIER

Six major classes of enzymes are recognized:

1. *Oxidoreductases* catalyzing the following reaction:

$$AH_2 + B \rightleftharpoons A + BH_2$$

Oxidases and dehydrogenases are enzymes of this type.

2. *Transferases* permitting the transfer of a functional group from one compound to another:

$$AR + B \rightleftharpoons BR + A$$

Transaminases and transacetylases are examples. Kinases that transfer phosphate from a nucleoside di- or triphosphate to an acceptor are also classified as transferases.

3. *Hydrolases* or hydrolyzing enzymes:

$$AB + H_2O \rightarrow AOH + BH$$

Peptidases, esterases, glycosidases, and phosphatases are examples of enzymes of this type.

4. *Lyases* are enzymes removing groups from substrates nonhydrolytically:

$$AB \rightarrow A + B$$

Decarboxylases, aldolases, and deaminases are examples.

5. *Isomerases:*

$$AB \rightleftharpoons BA$$

These enzymes are called *racemases* when they permit a change only in optical configuration. If there is more than one center of asymmetry in the substrate, the enzyme is called an *epimerase.*

6. *Ligases* or *synthetases* mediate multistep reactions in which two compounds are linked together. During the course of the reaction one of the two substrates is temporarily bound to AMP, ADP, or the active group of an enzyme. The overall reaction is of the general type:

$$A + B + ATP \rightarrow AB + ADP + \text{phosphate}$$

Electron carriers such as nicotinamide adenine dinucleotide (NAD) and flavine mononucleotide (FMN), phosphorylating agents such as ATP, and vitamins such as thiamine, nicotinamide, folic acid, biotin and pantothenic acid, all play a major role in enzymatic reactions.

# KEY TO NUMBERING AND CLASSIFICATION OF ENZYMES

## 1. OXIDOREDUCTASES

### 1.1. Acting on the CH–OH group of donors

1.1.1. With NAD or NADP as acceptor
1.1.2. With a cytochrome as acceptor
1.1.3. With $O_2$ as acceptor
1.1.99. With other acceptors

### 1.2. Acting on the aldehyde or keto group of donors

1.2.1. With NAD or NADP as acceptor
1.2.2. With a cytochrome as acceptor
1.2.3. With $O_2$ as acceptor
1.2.4. With lipoate as acceptor
1.2.99. With other acceptors

### 1.3. Acting on the CH–CH group of donors

1.3.1. With NAD or NADP as acceptor
1.3.2. With a cytochrome as acceptor
1.3.3. With $O_2$ as acceptor
1.3.99. With other acceptors

### 1.4. Acting on the CH–$NH_2$ group of donors

1.4.1. With NAD or NADP as acceptor
1.4.3. With $O_2$ as acceptor

### 1.5. Acting on the C–NH group of donors

1.5.1. With NAD or NADP as acceptor
1.5.3. With $O_2$ as acceptor

### 1.6. Acting on reduced NAD or NADP as donor

1.6.1. With NAD or NADP as acceptor
1.6.2. With a cytochrome as acceptor
1.6.4. With a disulfide compound as acceptor
1.6.5. With a quinone or related compound as acceptor
1.6.6. With a nitrogenous group as acceptor
1.6.99. With other acceptors

### 1.7. Acting on other nitrogenous compounds as donors

1.7.3. With $O_2$ as acceptor
1.7.99. With other acceptors

### 1.8 Acting on sulfur groups of donors

1.8.1. With NAD or NADP as acceptor
1.8.3. With $O_2$ as acceptor
1.8.4. With a disulfide compound as acceptor
1.8.5. With a quinone or related compound as acceptor
1.8.6. With a nitrogenous group as acceptor

### 1.9 Acting on heme groups of donors

1.9.3. With $O_2$ as acceptor
1.9.6. With a nitrogenous group as acceptor

### 1.10. Acting on diphenols and related substances as donors

1.10.3. With $O_2$ as acceptor

### 1.11. Acting on $H_2O_2$ as acceptor

### 1.12. Acting on hydrogen as donor

### 1.13. Acting on single donors with incorporation of oxygen (oxygenases)

### 1.14. Acting on paired donors with incorporation of oxygen into one donor (hydroxylases)

1.14.1. Using reduced NAD or NADP as one donor
1.14.2. Using ascorbate as one donor
1.14.3. Using reduced pteridine as one donor

## 2. TRANSFERASES

### 2.1. Transferring one-carbon groups

2.1.1. Methyltransferases
2.1.2. Hydroxymethyl-, formyl-, and related transferases
2.1.3. Carboxyl- and carbamoyltransferases
2.1.4. Amidinotransferases

### 2.2. Transferring aldehydic or ketonic residues

### 2.3. Acyltransferases

2.3.1. Acyltransferases
2.3.2. Aminoacyltransferases

### 2.4. Glycosyltransferases

2.4.1. Hexosyltransferases
2.4.2. Pentosyltransferases

### 2.5. Transferring alkyl or related groups

### 2.6. Transferring nitrogenous groups

2.6.1. Aminotransferases
2.6.3. Oximinotransferases

### 2.7. Transferring phosphorus-containing groups

2.7.1. Phosphotransferases with an alcohol group as acceptor

# KEY TO NUMBERING AND CLASSIFICATION OF ENZYMES (Continued)

**Transferring phosphorus-containing groups (continued)**

2.7.2. Phosphotransferases with a carboxyl group as acceptor
2.7.3. Phosphotransferases with a nitrogenous group as acceptor
2.7.4. Phosphotransferases with a phospho group as acceptor
2.7.5. Phosphotransferases, apparently intramolecular
2.7.6. Pyrophosphotransferases
2.7.7. Nucleotidyltransferases
2.7.8. Transferases for other substituted phospho groups

**2.8. Transferring sulfur-containing groups**

2.8.1. Sulfurtransferases
2.8.2. Sulfotransferases
2.8.3. CoA-transferases

3. HYDROLASES

**3.1. Acting on ester bonds**

3.1.1. Carboxylic ester hydrolases
3.1.2. Thiolester hydrolases
3.1.3. Phosphoric monoester hydrolases
3.1.4. Phosphoric diester hydrolases
3.1.5. Triphosphoric monoester hydrolases
3.1.6. Sulfuric ester hydrolases

**3.2. Acting on glycosyl compounds**

3.2.1. Glycoside hydrolases
3.2.2. Hydrolyzing N-glycosyl compounds
3.2.3. Hydrolyzing S-glycosyl compounds

**3.3. Acting on ether bonds**

3.3.1. Thioether hydrolases

**3.4. Acting on peptide bonds (peptide hydrolases)**

3.4.1. α-Aminoacyl–peptide hydrolases
3.4.2. Peptidyl–amino acid hydrolases
3.4.3. Dipeptide hydrolases
3.4.4. Peptidyl–peptide hydrolases

**3.5. Acting on C–N bonds other than peptide bonds**

3.5.1. In linear amides
3.5.2. In cyclic amides
3.5.3. In linear amidines
3.5.4. In cyclic amidines
3.5.5. In cyanides
3.5.99. In other compounds

**3.6. Acting on acid–anhydride bonds**

3.6.1. In phosphoryl-containing anhydrides

**3.7. Acting on C–C bonds**

3.7.1. In ketonic substances

**3.8. Acting on halide bonds**

3.8.1. In C-halide compounds
3.8.2. In P-halide compounds

**3.9. Acting on P–N bonds**

4. LYASES

**4.1. Carbon–carbon lyases**

4.1.1. Carboxyl lyases
4.1.2. Aldehyde lyases
4.1.3. Ketoacid lyases

**4.2. Carbon–oxygen lyases**

4.2.1. Hydro lyases
4.2.99. Other carbon–oxygen lyases

**4.3. Carbon–nitrogen lyases**

4.3.1. Ammonia lyases
4.3.2. Amidine lyases

**4.4. Carbon–sulfur lyases**

**4.5. Carbon–halide lyases**

**4.99. Other lyases**

5. ISOMERASES

**5.1. Racemases and epimerases**

5.1.1. Acting on amino acids and derivatives
5.1.2. Acting on hydroxyacids and derivatives
5.1.3. Acting on carbohydrates and derivatives
5.1.99. Acting on other compounds

**5.2. *Cis-trans* isomerases**

**5.3. Intramolecular oxidoreductases**

5.3.1. Interconverting aldoses and ketoses
5.3.2. Interconverting keto and enol groups
5.3.3. Transposing C=C bonds

## KEY TO NUMBERING AND CLASSIFICATION OF ENZYMES (Continued)

**5.4. Intramolecular transferases**

5.4.1. Transferring acyl groups
5.4.2. Transferring phosphoryl groups
5.4.99. Transferring other groups

**5.5. Intramolecular lyases**

**5.99. Other isomerases**

**6. LIGASES**

**6.1. Forming C–O bonds**

6.1.1. Amino acid–RNA ligases

**6.2. Forming C–S bonds**

6.2.1. Acid–thiol ligases

**6.3. Forming C–N bonds**

6.3.1. Acid–ammonia ligases (amide synthetases)
6.3.2. Acid–amino acid ligases (peptide synthetases)
6.3.3. Cyclo ligases
6.3.4. Other C–N ligases
6.3.5. C–N ligases with glutamine as N donor

**6.4. Forming C–C bonds**

Printed by permission of Harland G. Wood, Secretary General of the International Union of Biochemistry, Case Western Reserve University, Cleveland, Ohio.

# CARBOHYDRATES

# MICROBIAL MONOSACCHARIDES AND POLYSACCHARIDES

DR. DOUGLAS E. EVELEIGH

Carbohydrates are a major class of compounds composed of free sugars (monosaccharides) or of sugars in polymeric association (polysaccharides). Their essential roles in the metabolism of the protist cell include the following:

1. The backbone of DNA and RNA.
2. Intermediary compounds, especially sugar phosphates used in the energy transfer process; this process includes both the release of energy and the harnessing of energy by aerobic (algal and blue-green algal) and anaerobic (bacterial) photosynthesis.
3. Energy storage reserves: amylopectins, glycogen, laminaran.
4. Cell wall components: bacterial mureins, fungal and algal cellulose, and fungal chitins.
5. Extracellular polysaccharides: bacterial cellulose, yeast mannans, fungal glucans.

Carbohydrates often occur in association with other classes of compound, e.g., glycosides, glycolipids, and glycoproteins. The role of such heteropolymers has not been elucidated in detail. However, glycoproteins are recognized as integral components of microbial cell walls, and certain enzymes are glycoproteins (*Aspergillus niger* glucamylase and glucose oxidase; *Caldariomyces fumago* chloroperoxidase; *Saccharomyces cerevisiae* invertase). Lipopolysaccharides are important for their pyrogenic activity and as antigens.

Monosaccharides include polyhydroxy aldehydes, ketones, alcohols, and acids. Those monosaccharides containing an aldehyde group are termed "aldoses", and those with a keto group are termed "ketoses" (Figure 1). Suffixes denoting these forms are *-ose* and *-ulose,* respectively. The monosaccharides are further classified by the number of carbon atoms they contain, e.g., triose, tetrose, pentose, hexose, etc. (Figure 1). Monosaccharides containing up to nine carbon atoms have been recorded, e.g., N-acetyl neuraminic acid, found in *Escherichia coli* K 235 colominic acid.

**ALDOSES**

| biose | triose | tetrose | pentose | hexose |
|---|---|---|---|---|
| | | | | CHO |
| | | | CHO | *CHOH |
| | | CHO | *CH·OH | *CHOH |
| | CHO | *CH·OH | *CH·OH | *CHOH |
| CHO | *CHOH | *CH·OH | *CH·OH | *CHOH |
| $CH_2OH$ | $CH_2OH$ | $CH_2$·OH | $CH_2$·OH | $CH_2OH$ |

**KETOSES**

| triose | tetrose | pentose | hexose |
|---|---|---|---|
| $CH_2OH$ | $CH_2OH$ | $CH_2OH$ | $CH_2OH$ |
| C=O | C=O | C=O | C=O |
| $CH_2OH$ | CHOH | $(CHOH)_2$ | $(CHOH)_3$ |
| | $CH_2OH$ | $CH_2OH$ | $CH_2OH$ |

*Asymmetric carbon atom

FIGURE 1. Aldoses and ketoses.

Le Bel and van't Hoff (1874) independently proposed that asymmetric carbon atoms promoted optical activity. It thus follows that trioses with one asymmetric carbon atom have two possible isomers, tetroses with two asymmetric carbon atoms have four isomers, and hexoses with four asymmetric carbon atoms have sixteen isomers. It is thus remarkable that only four aldoses occur commonly in nature: D-glucose, D-mannose, and D- and L-galactose. The only abundant ketose is D-fructose. Microbial aldonic acids, uronic acids, glycitols, and amino sugars are well documented (Figure 2). However, the protista are exceptional in that they produce an extremely diverse range of monosaccharides that are not encountered in the plant and animal kingdoms; e.g., dideoxysugars, paratose, tyvelose, and abequose (Figure 2) occur in lipopolysaccharide complexes, and branched chain sugars, including cladinose, mycarose and streptose, occur in antibiotics (Figure 3). Several substituted microbial sugars have been recorded and are noteworthy. Thus, O-acetylation of the bacterial murein renders it intractable to lysozyme,[5] while 3-O-methyl-D-galactose found in the cell wall of actinomycetes has been used as a taxonomic criterion.[6]

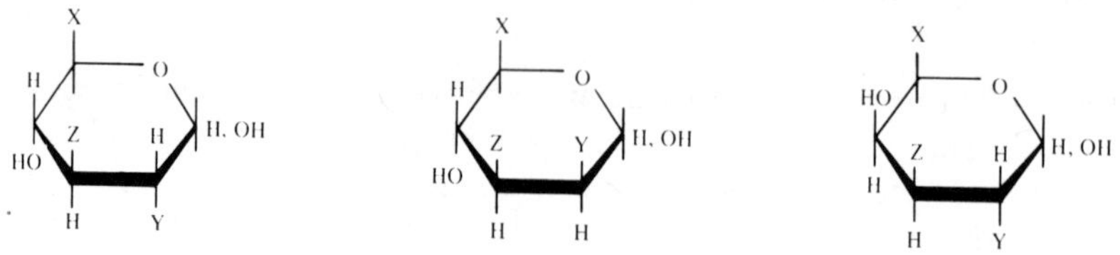

| When | | | |
|---|---|---|---|
| X = $-CH_2OH$<br>Y = $-OH$<br>Z = $-OH$ | D-Glucose | D-Mannose | D-Galactose |
| X = $-COOH$<br>Y = $-OH$<br>Z = $-OH$ | D-Glucuronic acid | D-Mannuronic acid | D-Galacturonic acid |
| X = $-H$<br>Y = $-OH$<br>Z = $-OH$ | D-Xylose | D-Lyxose[a] | L-Arabinose |
| X = $-CH_2OH$<br>Y = $-N(H)-C(=O)-CH_3$<br>Z = $-OH$ | N-Acetyl-D-glucosamine | N-Acetyl-D-mannosamine | N-Acetyl-D-galactosamine |
| X = $-CH(OH)-CH_2OH$<br>Y = $-OH$<br>Z = $-OH$ | D-Glycero- and L-glycero-D-glucoheptose | L-Glycero-D-mannoheptose, D-glycero-D-mannoheptose | D-Glycero-D-galactoheptose |
| X = $-CH_3$<br>Y = $-OH$<br>Z = $-OH$ | D-Quinovose[b] | D-Rhamnose | D-Fucose[c] |
| X = $-CH_3$<br>Y = $-OH$<br>Z = $-H$ | D-Paratose | D-Tyvelose | D-Abequose |

[a] Only L-lyxose has been found in protista, e.g., in the antibiotic curamycin.
[b] Not found in the protista to date.
[c] D-Fucose has not been reported of microbial origin; L-fucose occurs in lipopolysaccharides of Gram-negative bacteria; 3,6-anhydro-D-galactose occurs in algal carrageenans.

FIGURE 2. Microbial monosaccharides.

Rosanoff (1906) suggested that sugars should be related to glyceraldehyde. He designated the conformation (Figure 4A) as D-glyceraldehyde; this was confirmed in 1951 by Bijvoet, using X-ray crystallography. Sugars synthesized from it, retaining the asymmetric carbon atom as the lowest asymmetric carbon atom, form the D series. The L-sugar is analogously based on L-glyceraldehyde (Figure 4B). The prefixes D- and L- refer to configuration; the actual optical rotation is designated by (+) or (−).

Sugars generally exist in ring forms via a hemiacetal bridge between C1 and C4 (furanose) or C1 and C5 (pyranose). Ring closure creates a new asymmetric center at the C1 reducing carbon atom. These isomers are designated $\alpha$ and $\beta$ and termed anomers (Figure 5). Thus a D-sugar has two dextrorotary anomers. Hudson (1909) designated the more dextrorotary member as $\alpha$-D-, and the other $\beta$-D-.

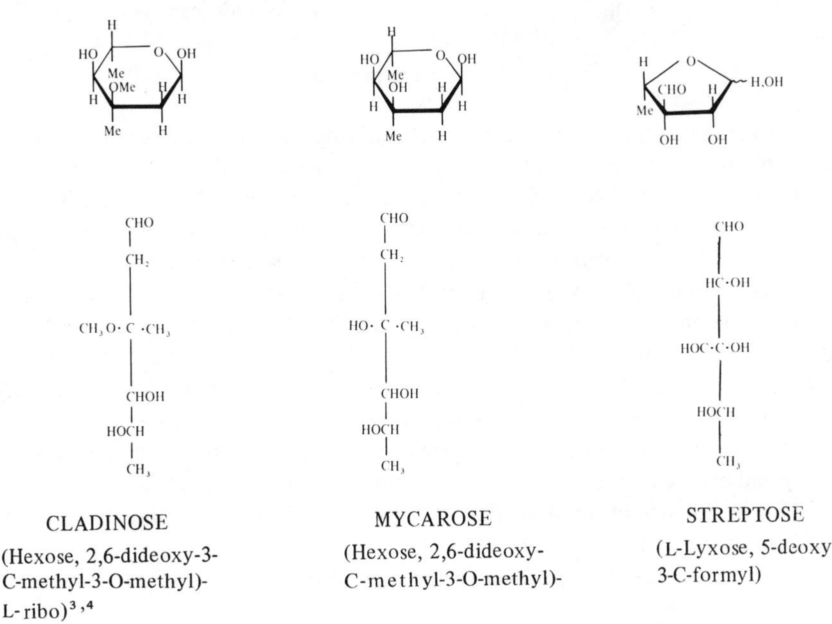


FIGURE 3. Branched-chain sugars.

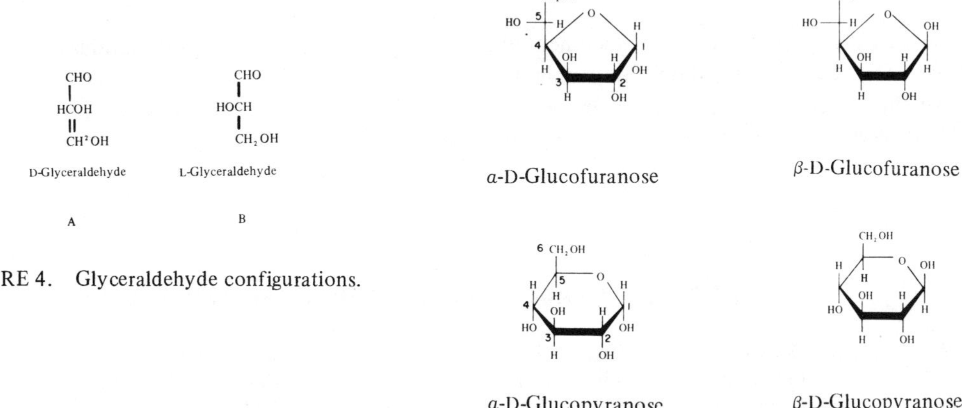


FIGURE 4. Glyceraldehyde configurations.

FIGURE 5. Anomers.

Ring structures may be depicted in three ways (Figure 6).

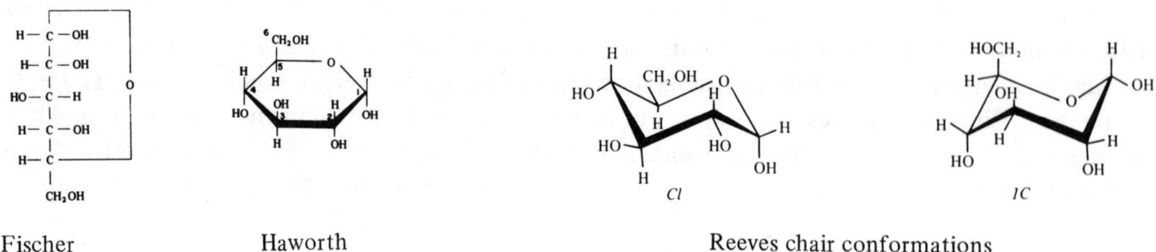


α-D-(+)-Glucopyranoside

FIGURE 6. Pyranose ring projections.

The Fischer and Haworth projections are planar. Furanoid rings can form almost planar strained rings, as depicted in the Haworth formulae (Figure 5). However, as the C-O-C bond is 111° (c.f. the C valence angle of 109.5°), the pyranose ring is under torsional strain, and to minimize this, the ring becomes puckered (e.g., Figure 6, C1 conformation). There are at least eight stereoisomeric structures possible; two are "chair forms" (Figure 6), and six are boat forms. Chair forms are preferred in that they have a minimal net energy e.g., any axial substituent other than hydrogen introduces some instability into the pyranose ring. Thus, pyranose sugars assume the chair form in which a larger number of the bulky substituent groups tend to occupy the equatorial positions and produce minimal interaction. The C1 α-D-glucopyranose conformation with equatorial substituents is more stable than the 1C conformation (Figure 6). Indeed, bulky equatorial C5 substituents stabilize the conformation, and hence, hexa- and heptapyranosides increase the stability of equivalent rings. The conformation will also be determined by the proximity of bulky substituents on the same side of the ring and by the interaction between the ring oxygen and hydroxyl groups, especially those of carbon 1 and 2. It is important to note that the properties of a sugar and its reactive groups are determined by the form of the ring.

Polysaccharides can be considered in relation to function and, on this basis, can be divided into three major groups:

1. Cell wall components.
2. Energy reserves.
3. Extracellular slimes and capsules of uncertain function.

Cell wall components are fully documented in subsequent chapters (Polysaccharides of Algae, Bacterial Cell Wall Structure, and Bacterial Endotoxins).

A listing of carbohydrate reserve materials is given in Table 1. Procaryotes are unique in also having polyhydroxybutyric acid as an energy reserve material.

## TABLE 1
## MICROBIAL CARBOHYDRATE STORAGE MATERIALS

| | |
|---|---|
| Starch | Occurs in a wide variety of algae. Various starch-like materials have been found in *Clostridium, Corynebacterium*, coliforms, yeasts (*Cryptococcus albidus*), fungi (*Aspergillus, Penicillium*) and protozoa (*Polytomella coeca*). Floridean starch occurs in the Rhodophyta. |
| | Amylopectins are found in blue-green algae (*Oscillatoria*) and in the Chlorophyta. |
| | α-Schardinger dextrans (6- to 8-D-glucose units linked α-(1→4): cycloamyloses) are formed by *Bacillus macerans*. |
| | *Pullularia pullulans* yields pullulan, and α-(1→4), α-(1→6) glucan (ratio 2:1). |

## TABLE 1 (Continued)
## MICROBIAL CARBOHYDRATE STORAGE MATERIALS

| | |
|---|---|
| Glycogen | Occurs widely in the protista (*Mycobacterium, Neisseria, Nostoc, Pneumococcus;* slime molds, yeasts, *Neurospora,* basidiomycetes; *Tetrahymena, Trichomonas;* Xanthophyta). |
| Sucrose | Found in Chlorophyta. |
| β-(1→3)-D-Glucans | May occur as paramylon (Euglenophyta, Xanthophyta), chrysolaminaran (Bacillariophyta, Chrysophyta), and laminaran (Phaeophyta). A questionable reserve, branched laminaran occurs in fungi and yeasts (e.g., pachyman in basidiomycetes). *Cetraria islandica, Perinidium westii,* and *Monodus subterraneus* yield a mixed-linkage β-(1→4), β-(1→3)-D-glucan. |

Extracellular polysaccharides are ill-defined on a conceptual basis, because they include extracellular slimes and capsules that may or may not be released from the major structural entity of the cell wall. Bacterial extracellular materials have proved most useful in serological analysis (*Pneumococcus, Salmonella,* and *Streptococcus*). The composition of extracellular polysaccharides has also been used in the classification of *Rhizobia.* NMR spectra of yeast mannans have proved to be a valuable taxonomic aid.

A list of extracellular microbial polysaccharides is given in Table 2. Several are considered of importance in soil fertility (*Arthrobacter, Azotobacter, Corynebacterium, Chlorobacterium, Myxobacterium,* and *Rhizobium*) or in phytopathology (*Agrobacterium* and *Xanthomonas*). Others have potential industrial importance for such uses as paper binders, food-packaging films and oil well drilling muds (*Sclerotium* glucan, *Hansenula holstii* Y-2448 phosphomannan, and *Xanthomonas campestris* B-1459 heteropolymer). Dextrans from *Leuconostoc mesenteroides* have been widely utilized as plasma expanders and also for forming gels for gel-filtration. The dextrans can vary from 50 to 90% α-(1→6) linkages with α-(1→3) and α-(1→4) cross links. Chitin has been isolated in the form of extracellular fibers from diatoms (*Thallasiosira*). This polysaccharide differs markedly in crystal structure from other crustacean or fungal chitins and has been termed chitan.

## TABLE 2
## EXTRACELLULAR MICROBIAL POLYSACCHARIDES

| Species | Basic Molecular Structure* | Trivial Name | Reference |
|---|---|---|---|
| *Acetobacter* | | | |
| *acetigenum* | β-(1 → 4)-glc | Cellulose | 1 |
| *capsulatum* | α-(1 → 6)-glc | Dextran | 2 |
| *viscosum* | α-(1 → 6)-glc | Dextran | 2,3 |
| *xylinum* | β-(1 → 4)-glc | | 4 |
| *Acinetobacter* | | | |
| *calco-aceticus* | rha, glc (4:1) in (1 → 3) linkages | | 5 |
| *Aerobacter* | | | |
| *levanicum* | 2 → 6′β-fru | Levan | 6 |
| *Agrobacterium* | | | |
| *tumefaciens* | mainly β-(1 → 2)-glc | | 7 |
| *Alcaligenes* | | | |
| *faecalis* var. *myxogenes* | glc, gal, succinic acid (7:1:1.5); glc has β-1 → 3, 1 → 4 and 1 → 6 linkages | Succinoglucan | 8 |
| *Alternaria* | | | |
| *solani* | 1. glc, gal, glcN (2.4:1:1)<br>2. glc, gal, man (6.4:1:1.2) | | 9 |
| *Anabaena* | | | |
| *flos-aquae* | glc, xyl, rha, glcUA | | 10, 11 |
| *Arthrobacter* | | | |
| *viscosus* B-1973 | glc, rha, manUA (1:1:1) | | 12 |

## TABLE 2 (Continued)
## EXTRACELLULAR MICROBIAL POLYSACCHARIDES

| Species | Basic Molecular Structure* | Trivial Name | Reference |
|---|---|---|---|
| *Aspergillus* | | | |
| *fumigatus* | gal, man (1:1); man core α-1 → 2, 1→ 3, 1 → 6 with gal furanose residues | | 13 |
| *nidulans* | α-(1 → 4)-gal, α-(1 → 4)-galNAc (1.8:1) | | 14 |
| *niger* (and other fungi) | α-(1 → 3), α-(1 → 4)-glc | Nigeran | 15, 16 |
| *Azotobacter* | | | |
| *chrococcum* | glc, gal, glcUA | | 17 |
| *vinelandii* | manUA, and relatively little gulUA | | 18 |
| *Bacillus* | | | |
| *megaterium* | (2 → 6) ketosidic fru | Levan | 19 |
| *mesentericus* | (2 → 6) ketosidic fru | Levan | 20 |
| *subtilis* | (2 → 6) ketosidic fru | Levan | 20 |
| species | glc, uronic acids | | 21 |
| | glc, man, uronic acids | | 21 |
| *Candida* | | | |
| species | β-(1 → 6)-glc (80% of linkage) | | 22 |
| *Cetraria* | | | |
| *islandicum* | β-(1 → 4), β-(1 → 3)-glucan (2:1) | | 23, 24 |
| *Chromobacterium* | | | |
| *violaceum* | glc, uronic acid, amino sugar, heptose | | 25 |
| *Claviceps* | | | |
| *fusiformis* | β-(1 → 3)-glucan with single β-(1 → 6)-glc substituents | | 26 |
| species | β-(1 → 3)-glucan with single β-(1 → 6)-glc substituents | | 27 |
| *Corynebacterium* | | | |
| *diphtheriae* | ara, gal, man (3:1:1) | | 28 |
| *insidiosum* | glc, gal, fuc (2:3:4), as glc-β-(1 → 4) fuc and glc, β-(1 → 4)-gal, and (1 → 3)-fuc, and with 4,6-O-(1′-carboxyethylide)-gal | | 29 |
| *Cryptococcus* | | | |
| *laurentii* | man, xyl, glcUA | | 30 |
| *Dictyostelium* | | | |
| *discoideum* | gal, glc | | 31 |
| | glc | Starch-like | 31 |
| | gal, galN, galUA | | 32 |
| *Escherichia* | | | |
| *coli* | N-acetylneuraminic acid, (2 → 8) ketosidic | Colominic acid | 33 |
| *coli* and Enterobacteriaceae | gal, glc, fuc, glcUA | | 34 |
| *Hansenula* | | | |
| *capsulata* NRRL Y-1842 | phosphomannans<br>–[– P → 6 β-D-Man*p* 1 → 2 α-D-Man*p* 1 –]$_4$ – P →<br>6 β-D-Man*p* 1 → 2 α-D-Man*p* 1 –<br>2<br>↑<br>1<br>α-D-Man*p*<br>– P – = orthophosphate in diester linkage | | 35 |
| *holstii* NRRL Y-2448 | phosphomannans | | 36 |
| *Helminthosporium* | | | |
| *sativum* | galN | | 37 |
| *Helotium* | | | |
| species NRRL 3129 | β-(1 → 3)-glucan with individual β-(1 → 6)-glc residues | | 38 |
| *Lactobacillus* | | | |
| *bifidus* | gal, β-(1 → 6)-glcNAc | | 39 |
| species | gal, glc, man | | 40 |
| *Lentinus* | | | |
| *edodes* | β-(1 → 3)-glucan, β-(1 → 4), β-(1 → 6)-glucan | | 41 |

## TABLE 2 (Continued)
## EXTRACELLULAR MICROBIAL POLYSACCHARIDES

| Species | Basic Molecular Structure* | Trivial Name | Reference |
|---|---|---|---|
| *Leuconostoc* | | | |
| *mesenteroides* B-512 | α-(1 → 6)-glucan, 95% [minor α-(1 → 3) and α-(1 → 6)] | Dextran | 42–44 |
| *mesenteroides* | (2 → 6)-β-fructan furanosyl | Fructan | 45 |
| *Microsporum* | | | |
| *quinckeanum* | (1 → 6)-, (1 → 3)-glucans (1.0:0.4) | | 46 |
| | galactomannans [α-(1 → 2), α-(1 → 6)] | | 47 |
| *Monilia* | | | |
| *fructicola* | β-(1 → 3)-, β-(1 → 4)-glucan | | 48 |
| *Myxobacterium* | | | |
| 402 | glc, man, rha, 3Merha, 2Merha, glcN | | 49 |
| *Neisseria* | | | |
| *sicca* | galN-acetyl-phosphorus (1.0:1.0:0.35) | | 50 |
| *Penicillium* | | | |
| *charlesii* | (1 → 5)-gal (furanose) | Galactocarolose | 51–53 |
| | α-man<br>α-D-Man*p* 1 → 2 α-D-Man*p* 1 → 2 α-D-Man*p* 1 → 2 α-D-Man*p*<br>1<br>↓<br>6<br>α-D-Man*p* 1 → 2 α-D-Man*p* 1 → 2 α-D-Man*p* 1 → 2 D-Man*p* | Mannan | 54 |
| | phosphogalactomannan | | 55 |
| *luteum* | β-(1 – 6)-glc + malonic acid | Lutean | 56 |
| *Physarum* | | | |
| *polycephalum* | gal with $SO_4$ (1:14) and $PO_4$ (1:21) | | 57 |
| *Plectania* | | | |
| *occidentalis* NRRL 3137 | β-(1 → 3)-glucan with single β-(1 → 6)-glc substituents | | 38 |
| *Polyporus* | | | |
| *tumulosus* | α-(1 → 3)-glc | | 58 |
| *Poria* | | | |
| *cocos* | β-(1 → 3)-glc | Pachyman | 59 |
| *Pullularia* | | | |
| *pullulans (Aureobasidium pullulans)* | α-(1 → 4), α-(1 → 6)-glucan (2:2:1) | Pullulan | 60, 61 |
| | β-(1 → 3)-glucan with β-(1 → 6)-glc substituents | | 60, 62 |
| | gal, glc, hexuronic acid | | 60 |
| *Rhizobium* | | | |
| *japonicum* | β-(1 → 2)-glc | | 63 |
| | glc, rha, 4MeglcUA | | 64 |
| *melioti* | glc, gal, glcUA (85:14:0.8) + 4MeglcUA | | 65 |
| *radicicola* | glc, glcUA (67:33) | | 66 |
| *Rhodotorula* | | | |
| *glutinis* | β-(1→3), (1→4)-man | | 67 |
| *Saccharomyces* | | | |
| *cerevisiae* | β-(1 → 6)-glucan | | 68 |
| species | α-(1 → 6)-man with α-(1 → 2) and α-(1 → 3) linkages | Mannan | 69 |
| *Schizophyllum* | | | |
| *commune* | β-(1 → 3)-glucan with single β-(1 → 6)-glc residues | Schizophyllan | 70, 71 |
| *Sclerotium* | | | |
| *glutanicum* | β-(1 → 3)-glucan with single β-(1 → 6)-glc substituents | | 72 |
| *rolfsii* | β-(1 → 3)-glucan with single β-(1 → 6)-glc residues | | 73 |
| *Serratia* | | | |
| *marcescens* | glc (major), man, heptose, fuc, rha | | 74 |
| | glucomannan | | 75 |
| | rhamnoglucan | | 75 |
| | glucoheptan | | 75 |
| | rhamnoheptoglucan | | 75 |

## TABLE 2 (Continued)
## EXTRACELLULAR MICROBIAL POLYSACCHARIDES

| Species | Basic Molecular Structure* | Trivial Name | Reference |
|---|---|---|---|
| *Sphaerotilus natans* | glcN, glc | | 76 |
| *Streptococcus* species | α-(1 → 6)-glc | Dextran | 42 |
| *Thallasiosira fluviatilis* | β-(1 → 4)-glcNAc | Chitan | 77 |
| *Tolypothrix tenuis* | fructofuranosyl-(2 → 4)-fructofuranosyl-(2 → 1)-α-glucopyranoside | | 78 |
| *Trichophyton granulosum* | man (84%) with α-(1 → 2) and α-(1 → 6) linkages, gal (16%) | | 47, 79 |
| *interdigitale* | 2 galactomannans, glucan | | 47, 80 |
| species | gal (furanose), man 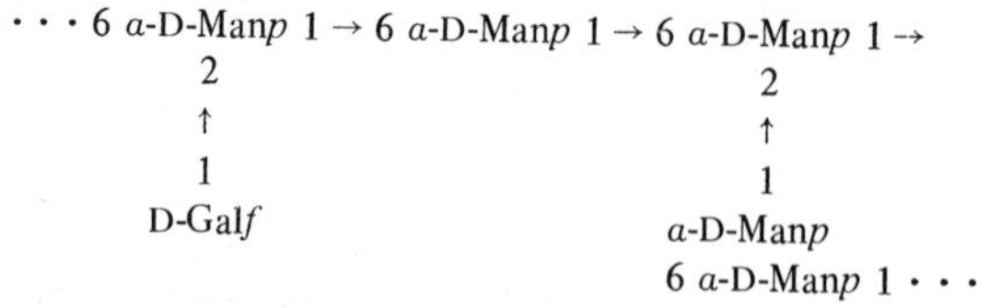<br>· · · 6 α-D-Man*p* 1 → 6 α-D-Man*p* 1 → 6 α-D-Man*p* 1 →<br>(first unit: 2 ↑ 1 D-Gal*f*; third unit: 2 ↑ 1 α-D-Man*p*)<br>6 α-D-Man*p* 1 · · · | | 47, 81, 82 |
| *Trichosporum cutaneum* | α-(1 → 3)-man with xyl, man and ara | | 83 |
| *Xanthomonas campestris* | glc, man, glcUA (branched), pyruvate | | 84 |
| *hyacinthi* | glc, man glcUA | | 85 |
| *maculofoliigardeniae* | glc, man, glcUA | | 85 |
| *stewartii* | glc, gal, glcUA | | 86 |
| *translucens* | glc, man, glcUA | | 85 |
| Yeasts | β-man, highly branched, with α-(1 → 6) and lesser amounts of α-(1 → 2) and α-(1 → 3) linkages 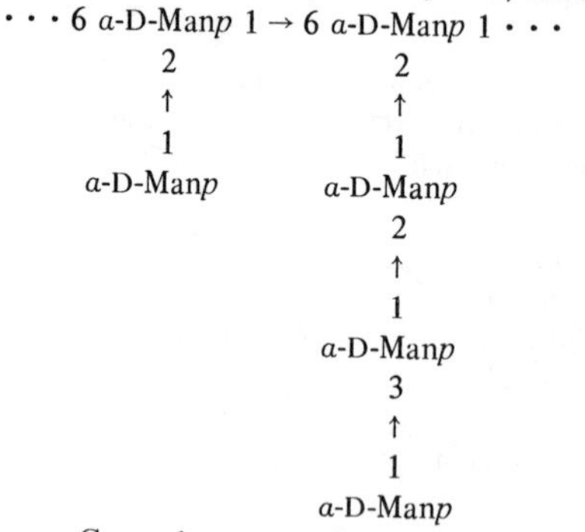<br>· · · 6 α-D-Man*p* 1 → 6 α-D-Man*p* 1 · · ·<br>(first unit: 2 ↑ 1 α-D-Man*p*; second unit: 2 ↑ 1 α-D-Man*p* 2 ↑ 1 α-D-Man*p* 3 ↑ 1 α-D-Man*p*)<br>General structure for yeast mannan. | | 69 |

*Abbreviations; glc = glucose; gal = galactose; man = mannose, fru = fructose; ara = arabinose; xyl = xylose, rib = ribose; fuc = fucose; rha = rhamnose; glcUA = glucuronic acid; 3 Merha = 3-O-methylrhamnose, etc.; galUA = galacturonic acid; glcN = 2-amino-2-deoxyglucose; glcNAc = 2-acetamino-2-deoxyglucose; gulUA = guluronic acid. Other sugars are given by their full names.

2-acetamido-2-deoxyglucose;

The chemistry and nomenclature of monosaccharides[7] have recently been reviewed.[8] Microbial polysaccharides[9-16] are discussed in subsequent chapters. Properties of natural sugars and their derivatives are given in Tables 3 to 7.

# TABLE 3
# NATURAL ALDITOLS AND INOSITOLS, INOSES, AMINO ALDITOLS AND INOSAMINES

| Substance[a] (synonym) | Derivative | Chemical formula | Melting point °C | Specific rotation[b] $[\alpha]_D$ | Reference |
|---|---|---|---|---|---|
| | | **ALDITOLS** | | | |
| **Glycerol** | | $C_3H_8O_3$ | 20 | None | *1* |
| | Tris-(*p*-nitrobenzoate) | $C_{24}H_{17}N_3O_{12}$ | 190–192 | — | *2* |
| **1-Deoxyglycerol**[c] (1,2-propanediol) | | $C_3H_8O_2$ | Oil, b.p. 188–189 | None (racemic) | *1,3* |
| | *Bis*(*p*-nitrobenzoate) | $C_{17}H_{14}N_2O_8$ | 125–126 | — | *2* |
| **Erythritol** | | $C_4H_{10}O_4$ | 118–120 | None (meso) | *4* |
| | Tetraacetate | $C_{12}H_{18}O_8$ | 85 | — | *4* |
| | Tetrakis-(*p*-nitrobenzoate) | $C_{32}H_{22}N_4O_{16}$ | 251–252 | — | *2* |
| **1,4-Dideoxy-erythritol** (2,3-butyleneglycol) | | $C_4H_{10}O_2$ | 25, 34 | None (meso) | *5* |
| | Dibenzoate | $C_{18}H_{18}O_4$ | 77 | — | *6* |
| **1,4-Dideoxy-D-threitol** | | $C_4H_{10}O_2$ | 19 | −13 | *5* |
| | Diacetate | $C_8H_{14}O_4$ | Oil, b.p. 192–194 (745 mm) | +1.4 | *7* |
| **L-Threitol** | | $C_4H_{10}O_4$ | 88–89 | −4.5 | *8* |
| | Di-*O*-benzylidene-L-threitol | $C_{18}H_{18}O_4$ | 218–220 | +87.2 (*c* 0.4, acetone) | *9* |
| **1,4-Dideoxy-L-threitol** | | $C_4H_{10}O_2$ | Oil, b.p. ca 170 | +12.4 | *10* |
| | *Bis*(*p*-nitrobenzoate) | $C_{18}H_{16}N_2O_8$ | 141–143 | +52 ($CHCl_3$) | *10* |
| **1,4-Dideoxy-D,L-threitol** | | $C_4H_{10}O_2$ | 7.6 | None (racemic) | *11* |
| | Diacetate | $C_8H_{14}O_4$ | 41–41.5 | — | *11* |
| **D-Arabinitol** (D-arabitol) | | $C_5H_{12}O_5$ | 103 | +7.8 (*c* 8, borax solution) | *12* |
| | Pentaacetate | $C_{15}H_{22}O_{10}$ | 76 | +37.2 ($CHCl_3$) | *12,13* |
| **L-Arabinitol** | | $C_5H_{12}O_5$ | 101–102 | −7.2 (*c* 9, borax solution) | *14* |
| | | — | — | −32 (*c* 0.4, 5% molybdate) | *15* |
| | Pentaacetate | $C_{15}H_{22}O_{10}$ | 72–73 | — | *16* |
| **Ribitol** (adonitol) | | $C_5H_{12}O_5$ | 102 | None (meso) | *17* |
| | Pentaacetate | $C_{15}H_{22}O_{10}$ | 51 | — | *18* |
| **Xylitol** | | $C_5H_{12}O_5$ | 61, 94 | None (meso) | *19, 20* |
| | Pentaacetate | $C_{15}H_{22}O_{10}$ | 62.5–63 | — | *19* |
| | Pentabenzoate | $C_{40}H_{35}O_{10}$ | 106 | — | *14* |
| **Galactitol** (dulcitol) | | $C_6H_{14}O_6$ | 186–188 | None (meso) | *21* |
| | Hexaacetate | $C_{18}H_{26}O_{12}$ | 168–169 | — | *21* |
| **D-Glucitol** (sorbitol) | | $C_6H_{14}O_6$ | 112 | −1.8 (15°) | *22* |
| | Hexaacetate | $C_{18}H_{26}O_{12}$ | 99 | +12.5 (*c* 0.8, $CHCl_3$) | *23, 24* |
| **1,5-Anhydro-D-glucitol** (polygalitol) | | $C_6H_{12}O_5$ | 140–141 | +42.4 | *25* |
| | Tetraacetate | $C_{14}H_{20}O_9$ | 73–74 | +38.9 ($CHCl_3$) | *25* |
| **L-Iditol** | | $C_6H_{14}O_6$ | 73.5 | −3.5 (*c* 10) | *26* |
| | Hexaacetate | $C_{18}H_{26}O_{12}$ | 121.5 | −25.7 ($CHCl_3$) | *26* |
| **D-Mannitol** | | $C_6H_{14}O_6$ | 166 | −0.21 | *27* |
| | | | | +16 (5% molybdate) | *15* |
| | Hexaacetate | $C_{18}H_{26}O_{12}$ | 126 | +18.8 (acetic acid) | *28, 29* |
| **1,5-Anhydro-D-mannitol** (Styrachitol) | | $C_6H_{12}O_5$ | 157 | −49.9 | *30* |
| | Tetraacetate | $C_{14}H_{20}O_9$ | 66–67 | −20.9 ($C_2H_5OH$) | *31* |
| **D-Mannitol 1-acetate** | | $C_8H_{16}O_7$ | 124–125 | +4 | *32* |
| **D-*glycero*-D-*galacto*-Heptitol** (L-*glycero*-D-*manno*-heptitol, perseitol) | | $C_7H_{16}O_7$ | 183–185, 188 | −1.1 | *33, 34* |
| | | | | +24 (5% molybdate) | *15* |
| | Heptaacetate | $C_{21}H_{30}O_{14}$ | 119–120.5 | −14 ($CHCl_3$) | *33* |
| **D-*glycero*-D-*gluco*-Heptitol** (L-*glycero*-D-*talo*-heptitol, β-sedoheptitol) | | $C_7H_{16}O_7$ | 131–132 | +46 (5% molybdate) | *35* |
| | Tri-*O*-methylene-β-sedoheptitol | $C_{10}H_{16}O_7$ | Sublimes 130, 276–278 d | −23.3 (*c* 0.4, $CHCl_3$) | *36* |

## TABLE 3 (Continued)
## NATURAL ALDITOLS AND INOSITOLS, INOSES, AMINO ALDITOLS AND INOSAMINES

| Substance[a] (synonym) | Derivative | Chemical formula | Melting point °C | Specific rotation[b] $[\alpha]_D$ | Reference |
|---|---|---|---|---|---|
| | | **Alditols—(Continued)** | | | |
| **D-*glycero*-D-*ido* Heptitol** | | $C_7H_{16}O_7$ | — | 0 | *37* |
| | Heptabenzoate | $C_{56}H_{44}O_{14}$ | 180–181 | +24 ($CHCl_3$) | *37* |
| **L-*glycero*-D-*ido*-Heptitol** | | $C_7H_{16}O_7$ | — | None (meso) | *37* |
| | Heptaacetate | $C_{21}H_{30}O_{14}$ | 175–176 | — | *37* |
| **D-*glycero*-D-*manno*-Heptitol** (D-*glycero*-D-*talo*-heptitol, volemitol) | | $C_7H_{16}O_7$ | 153 | +2.6 | *38* |
| | | — | — | +55 (5% molybdate) | *15* |
| | Heptaacetate | $C_{21}H_{30}O_{14}$ | 62 | +36.1 ($CHCl_3$) | *38, 39* |
| **D-*erythro*-D-*galacto*-octitol** | | $C_8H_{18}O_8 \cdot H_2O$ | 169–170 | −11 (5% molybdate) | *35* |
| | Octaacetate | $C_{24}H_{34}O_{16}$ | 99–100 | +2 ($CHCl_3$) | *35* |
| | | **INOSITOLS** | | | |
| **Asteritol** (an inositol monomethyl ether) | | $C_7H_{14}O_6$ | Sublimes, melts 164 | +157 (*c* 0.01) | *40* |
| **Betitol** (a dideoxyinositol) | | $C_6H_{12}O_4$ | 224 | — | *41* |
| **D-Bornesitol** (D-*myo*-inositol monomethyl ether) | | $C_7H_{14}O_6$ | 201–202 | +31.4 | *42* |
| | Pentaacetate | $C_{17}H_{24}O_{11}$ | 138–139 | +11.8 (*c* 0.8, acetone) | *42* |
| **L-Bornesitol** (1-*O*-methyl-L-*myo*-inositol) | | $C_7H_{14}O_6$ | 205–206 | −32.1 | *43, 63* |
| | Pentaacetate | $C_{17}H_{24}O_{11}$ | 142–143, 157 | −11.2 ($CHCl_3$) | *43, 44* |
| **Conduritol** (a 2,3-dehydro-2,3-dideoxy-inositol) | | $C_6H_{10}O_4$ | 142–143 | None (meso) | *45* |
| | Tetraacetate | $C_{14}H_{18}O_4$ | b.p. 165 (0.6 mm) | — | *46* |
| | Dihydroconduritol | $C_6H_{12}O_4$ | 204 | — | *46* |
| **Dambonitol** (1,3-di-*O*-methyl-*myo*-inositol) | | $C_8H_{16}O_6$ | 206, 210 | None (meso) | *47, 48* |
| | Tetraacetate | $C_{16}H_{24}O_{10}$ | 202 | — | *49* |
| **D-Inositol** (*d*-inositol) | | $C_6H_{12}O_6$ | 246–247 d. | +60, +65 | *50, 51* |
| | Hexabenzoate | $C_{48}H_{36}O_{12}$ | 252–253 | +64.5 | *52* |
| **L-Inositol** (*l*-inositol) | | $C_6H_{12}O_6$ | 247 | −64.1 | *1, 53* |
| | Hexaacetate | $C_{18}H_{24}O_{12}$ | 96 | — | *1* |
| **D,L-Inositol** | | $C_6H_{12}O_6$ | 253 | None (racemic) | *54* |
| | Hexaacetate | $C_{18}H_{24}O_{12}$ | 111 | — | *54* |
| ***muco*-Inositol methyl ether** | | $C_7H_{14}O_6$ | Gum | — | *55* |
| | Pentabenzoate | $C_{42}H_{34}O_{11}$ | Amorphous 95–100 | — | *55* |
| ***myo*-Inositol** (*meso*-inositol) | | $C_6H_{12}O_6$ | 225–227 | None (meso) | *50* |
| | Hexaacetate | $C_{18}H_{24}O_{12}$ | 206–208 | — | *56* |
| ***neo*-Inositol** | | $C_6H_{12}O_6$ | 314 | | *106, 107* |
| | Hexaacetate | $C_{18}H_{24}O_{12}$ | 252–253 | | *107* |
| **Laminitol** (a *C*-methyl *myo*-inositol) | | $C_7H_{14}O_6$ | 266–269 | −3 | *57* |
| | Hexaacetate | $C_{19}H_{26}O_{12}$ | 153 | −19.6 ± 1 ($CHCl_3$) | *58* |
| **Leucanthemitol** (a dehydro dideoxy inositol) | | $C_6H_{10}O_4$ | 131–132 | +101.5 | *59, 73* |
| | Dihydroleucan-themitol | $C_6H_{12}O_4$ | 161 | −40 | *73* |

# TABLE 3 (Continued)
# NATURAL ALDITOLS AND INOSITOLS, INOSES, AMINO ALDITOLS AND INOSAMINES

| Substance[a] (synonym) | Derivative | Chemical formula | Melting point °C | Specific rotation[b] $[\alpha]_D$ | Reference |
|---|---|---|---|---|---|
| | | **Inositols—(Continued)** | | | |
| **L-Liriodendritol** (1,4-di-*O*-methyl-*myo*-inositol) | | $C_8H_{16}O_6$ | 224 | −25 | *60, 62* |
| | Tetraacetate | $C_{16}H_{24}O_{10}$ | 139 | −24 | *62* |
| **Mytilitol** (a *C*-methyl *scyllo*-inositol) | | $C_7H_{14}O_6$ | 259 | None (meso) | *61* |
| | Hexaacetate | $C_{19}H_{26}O_{12}$ | 180–181 | — | *61* |
| ***d*-Ononitol** (4-*O*-methyl-L-*myo*-inositol) | | $C_7H_{14}O_6$ | 172 | +6.6 | *44* |
| | Pentaacetate | $C_{17}H_{24}O_{11}$ | 121 | −11.1 (*c* 0.8, $CHCl_3$) | *63* |
| ***d*-Pinitol** (a *dextro*-inositol monomethyl ether) | | $C_7H_{14}O_6$ | 186 | +65.5 | *64, 65* |
| | Pentaacetate | $C_{17}H_{24}O_{11}$ | 98 | +8.6 ($C_2H_5OH$) | *66* |
| ***l*-Pinitol** | | $C_7H_{14}O_6$ | 186 | −65 | *67, 68* |
| | Di-*O*-isopropylidene-pinitol | $C_{13}H_{22}O_6$ | 102–104 | +45 ($CHCl_3$) | *69* |
| ***l*-Quebrachitol** (a *levo*-inositol monomethyl ether) | | $C_7H_{14}O_6$ | 190–191 | −80.2 (28°) | *70, 71* |
| | Pentaacetate | $C_{17}H_{24}O_{11}$ | 96–98 | −25.1 (29°) ($CHCl_3$) | *70, 71* |
| ***d*-Quercitol** (a deoxy *dextro*-inositol) | | $C_6H_{12}O_5$ | 235 | +24.2 | *72* |
| ***d*-Quinic acid** (a dideoxy carboxy-*dextro*-inositol) | | $C_7H_{12}O_6$ | 164 | +44 (*c* 10) | *41* |
| ***l*-Quinic acid** | | $C_7H_{12}O_6$ | 162 | −42.1 | *74* |
| | Lactone | $C_7H_{10}O_5$ | 198 | None (racemate) | *80* |
| | Lactone triacetate | $C_{13}H_{16}O_8$ | 103 | −24 | *81* |
| | Quinic acid tetra-acetate | $C_{15}H_{20}O_{10}$ | 130–136 | −22.5 ($C_2H_5OH$) | *75* |
| **5-Dehydroquinic acid** | | $C_7H_{10}O_6$ | 138s, 140–142 | −82.4 (28°) ($C_2H_5OH$) | *76, 77* |
| | Protocatechuic acid | $C_7H_6O_2$ | 201–202 | — | *76* |
| **Scyllitol** (*scyllo*-inositol, cocositol) | | $C_6H_{12}O_6$ | 352–353 | None (meso) | *78, 79* |
| | Hexaacetate | $C_{18}H_{24}O_{12}$ | 299–300 | — | *78, 79* |
| **Sequoyitol** (5-*O*-methyl-*myo*-inositol) | | $C_7H_{14}O_6$ | 234–235 | None (meso) | *82* |
| | Pentaacetate | $C_{17}H_{24}O_{11}$ | 198 | — | *82* |
| **Shikimic acid** (a 3,4-anhydroquinic acid) | | $C_7H_{10}O_5$ | 183–184 | −200 (16°) | *80, 83* |
| | Methyl shikimate | $C_8H_{12}O_5$ | 115 | −136 ($CH_3OH$) | *84* |
| | Triacetate | $C_{13}H_{16}O_8$ | Syrup | −60 | *85* |
| **5-Dehydroshikimic acid** | | $C_7H_8O_5$ | 150–152 | −57.5 (28°) ($C_2H_5OH$) | *86* |
| | Methyl ester | $C_8H_{10}O_5$ | 124–126 | −47.1 ± 3 (*c* 0.2, $C_2H_5OH$) | *86* |
| **Viburnitol** (a deoxy *levo*-inositol, *l*-quercitol)[d] | | $C_6H_{12}O_5$ | 174 | −73.9 | *87* |
| | | — | 179–180 | −49.5 ± 1 | *88, 89* |
| | | — | 158–159 | — | *90* |
| | Pentaacetate | $C_{16}H_{22}O_{12}$ | 112–113 | — | *90* |
| | | — | 122, 124, 126 | — | *87–89* |
| | | **INOSOSES** | | | |
| **Bioinosose** (*myo*-inosose-2,*scyllo*-inosose, a deoxy keto or dehydro inositol) | | $C_6H_{10}O_6$ | 198–200 | None (meso) | *91, 92* |
| | Phenylhydrazone | $C_{12}H_{16}N_2O_5$ | 220–222 | — | *91* |
| | Pentaacetate | $C_{16}H_{20}O_{11}$ | 212–213 | — | *93* |
| ***myo*-Inosose-1** | | $C_6H_{10}O_6$ | 138–139 | +19.6 | *94* |
| | *Bis*(phenylhydrazone) | $C_{18}H_{20}N_4O_4$ | 196–197 | +62 | *94* |
| **2,3-Didehydro-4-deoxy-inositol** | | $C_6H_8O_5$ | No constants known | — | *95* |

# TABLE 3 (Continued)
# NATURAL ALDITOLS AND INOSITOLS, INOSES, AMINO ALDITOLS AND INOSAMINES

| Substance[a] (synonym) | Derivative | Chemical formula | Melting point °C | Specific rotation[b] $[\alpha]_D$ | Reference |
|---|---|---|---|---|---|
| | | **AMINO ALDITOLS AND INOSAMINES** | | | |
| **2-Aminoethanol** (ethanolamine) | | $C_2H_7NO$ | Oil, b.p. 171 | — | *1* |
| | Hydrochloride | $C_2H_7NOHCl$ | 100 | — | *1* |
| | Picrate | $C_8H_{10}N_4O_8$ | 158 | — | *1* |
| **Actinamine** (a 1,3-dideoxy-1,3-dimethyl-amino-*myo*-inositol) | | $C_8H_{18}N_2O_4$ | 135–136 | — | *96, 97* |
| | Dihydrochloride | $C_8H_{18}N_2O_4 \cdot 2HCl$ | >300 | — | *96* |
| | *N,N'*-Diacetyl tetraacetate | $C_{20}H_{30}N_2O_{10}$ | 205–206 | — | *96* |
| **Bluensidine** (3-*O*-carbamoyl-1-deoxy-1-guanidino-*scyllo*-inositol) | | $C_8H_{16}N_4O_6$ | No constants known | — | *98* |
| | Hydrochloride | $C_8H_{16}N_4O_6 \cdot HCl$ | 190–194 d. | +1 ± 0.5 | *98* |
| | *N,N'*-Diacetyl tetraacetate | $C_{20}H_{28}N_4O_{12}$ | 250–251 d. | +5 (*c* 0.9, $CHCl_3$) | *98* |
| ***neo*-Inosamine-2** (aminodeoxy-*neo*-inositol) | | $C_6H_{13}NO_5$ | 239–241 d. | None (meso) | *99* |
| | Hydrochloride | $C_6H_{13}NO_5 \cdot HCl$ | 217–221 d. | — | *100* |
| | *N*-acetyl-penta-acetate | $C_{18}H_{25}NO_{11}$ | 277–278 | — | *99* |
| ***scyllo*-Inosamine-2** | | $C_6H_{13}NO_5$ | No constants known | — | *108* |
| | *N*-acetyl pentaacetate | $C_{18}H_{25}NO_{11}$ | 299–301 | — | *109* |
| **Streptidine** (1,3-dideoxy-1,3-diguanidino-*scyllo*-inositol) | | $C_8H_{18}N_6O_4$ | — | None (meso) | *101* |
| | *N,N'*-Diacetyl tetraacetate | $C_{20}H_{30}N_6O_{10}$ | 342–345 | — | *101* |
| | Dipicrate | $C_{20}H_{24}N_{12}O_{18}$ | 284–285 d. | — | *102* |
| **Streptamine** (1,3-di-amino-1,3-dideoxy-*scyllo*-inositol) | | $C_6H_{14}N_2O_4$ | >290 (205 s.) | None (meso) | *101* |
| | Dihydroiodide | $C_6H_{14}N_2O_4 \cdot 2HI$ | >280 d. | — | *101* |
| | *N,N'*-Diacetyl tetraacetate | $C_{18}H_{26}N_2O_{10}$ | 342–345 d. | — | *101* |
| **2-Deoxystreptamine** | | $C_6H_{14}N_2O_3$ | 221–223 | None (meso) | *103* |
| | *N,N'*-Diacetyl triacetate | $C_{16}H_{24}N_2O_8$ | 340–350 | — | *103* |
| | Dihydrochloride | $C_6H_{14}N_2O_3 \cdot 2HCl$ | 283–286 d. | — | *104* |
| **Deoxy-*N*-methyl-streptamine** (hyosamine) | | $C_7H_{16}N_2O_3$ | 130–133 d. | −17.8 | *105* |
| | *N,N'*-Diacetyl triacetate | $C_{17}H_{26}N_2O_8$ | 204 | — | *103* |

[a] In order of increasing carbon chain length in the parent compounds, grouped according to class.
[b] Unless otherwise specified, $[\alpha]_D$ is given for 1 to 5 g solute, *c* representing the amount of solvent per 100 ml aqueous solution, at 20 to 25°C.
[c] Said to exist also as a phosphate ester.[3]
[d] Not the enanthiomorph of *d*-quercitol; other isomeric relations are involved.

Data taken from: Maher, G. G., and Wolfrom, M. L., in *Handbook of Biochemistry,* 2nd ed., pp. D-3–D-6, H. A. Sober, Ed. Copyright 1970, The Chemical Rubber Co., Cleveland, Ohio.

# TABLE 4
# NATURAL ACIDS OF CARBOHYDRATE DERIVATION

| Substance[a] (synonym) | Derivative | Chemical formula | Melting point °C | Specific rotation[b] $[\alpha]_D$ | Reference |
|---|---|---|---|---|---|
| | | **ALDONIC ACIDS** | | | |
| **D-Glyceric acid** | | $C_3H_6O_4$ | Gum | dextro | *1* |
| | Amide | $C_3H_7NO_3$ | 99.5–100 | −63.1 ($CH_3OH$) | *2* |
| | Calcium salt | $CaC_6H_{10}O_8$ | — | +10.9 | *3* |
| **L-Glyceric acid** | | $C_3H_6O_4$ | Gum | levo | *1* |
| | Calcium salt·$2H_2O$ | $CaC_6H_{10}O_8 \cdot 2H_2O$ | 134–135 | −12 (30°) | *4* |
| **D-Arabinonic acid** (arabonic acid) | | $C_5H_{10}O_6$ | 114–116 | +10.5 (*c* 6) | *5* |
| | Phenylhydrazide | $C_{11}H_{16}N_2O_5$ | 208–209 | −13 | *6* |
| **3-Deoxy-D-*glycero*-2-pentulosonic acid** (2-keto-3-deoxy-D-arabonic acid) | | $C_5H_8O_5$ | No constants known | — | *7* |
| | 2,4-Dinitrophenyl-hydrazone | $C_{11}H_{12}N_4O_8$ | 163 | — | *7* |
| **D-*threo*-4-Pentulosonic acid** (4-keto-D-arabonic acid) | | $C_5H_8O_6$ | — | −10.3 | *8* |
| | Brucine salt | $C_{28}H_{34}N_2O_{10}$ | 154–155 | −29.4 | *8* |
| **L-Arabinonic acid** | | $C_5H_{10}O_6$ | 118–119 | −9.6 →41.7[c] | *9* |
| | Phenylhydrazide | $C_{11}H_{16}N_2O_5$ | 215 | — | *10* |
| | 1,4-Lactone (γ-lactone) | $C_5H_8O_5$ | 95–98 | −71.6 | *11, 13* |
| | Amide | $C_5H_{11}NO_5$ | 135–136 | +37.2 | *12* |
| **3-Deoxy-L-*glycero*-2-pentulosonic acid** (2-keto-3-deoxy-L-arabonic acid) | | $C_5H_8O_5$ | No constants known | — | *14* |
| **L-Lyxonic acid**[e] | | $C_5H_{10}O_6$ | Syrup | — | *15* |
| | Phenylhydrazide | $C_{11}H_{16}N_2O_5$ | 163 | +13.7 | *16* |
| **D-Ribonic acid** | | $C_5H_{10}O_6$ | 112–113 | −17 | *17* |
| | Amide | $C_5H_{11}NO_5$ | 136–137 | +16.5 | *18, 18a* |
| | 1,4-Lactone | $C_5H_8O_5$ | 77 | +17 → +8 | *19, 20* |
| **D-Xylonic** | | $C_5H_{10}O_6$ | Syrup | −2.9 → +20.1[c] | *21* |
| | Amide | $C_5H_{11}NO_5$ | 81–82 | +44.5 → +23.8 | *22* |
| | 1,4-Lactone | $C_5H_8O_5$ | 98–101 | +91.8 → +86.7 | *11* |
| | Brucine salt | $C_{28}H_{36}N_2O_{10}$ | 170–172 | −37.4 | *23, 128* |
| **L-Xylonic acid** | | $C_5H_{10}O_6$ | No constants known | — | *16* |
| | Brucine salt | $C_{28}H_{36}N_2O_{10}$ | 177–178 | +24.3 | *16* |
| | 1,4-Lactone | $C_5H_8O_5$ | 97 | −82.2 | *24* |
| **D-Altronic acid**[f] | | $C_6H_{12}O_7$ | Syrup | +8 | *3* |
| | Phenylhydrazide | $C_{12}H_{18}N_2O_6$ | 150–152 | — | *25* |
| | Calcium salt·$3.5H_2O$ | $CaC_{12}H_{22}O_{14} \cdot 3.5H_2O$ | — | +11.5 →24.8[c] (*N* HCl) | *26* |
| | 1,4-Lactone | $C_6H_{10}O_6$ | — | +35 | *3* |
| **D-Galactonic acid** | | $C_6H_{12}O_7$ | 122 | −112. → −57.6[c] | *27, 28* |
| | | — | 148 | −13.6 → −17 | *11* |
| | Phenylhydrazide | $C_{12}H_{18}N_2O_6$ | 203 | +10.4 | *10* |
| | 1,4-Lactone | $C_6H_{10}O_6$ | 110–112 | −73 → −63.7 | *11, 29* |
| | | — | 132–133 | — | *30* |
| | Amide | $C_6H_{13}NO_6$ | 175 | +31.5 | *31* |
| **D-*lyxo*-2-Hexulosonic acid** (2-keto-D-galactonic acid) | | $C_6H_{10}O_7$ | 169 | −5 | *32* |
| | Brucine salt | $C_{29}H_{36}N_2O_{11}$ | 172 | −22.5 (50% EtOH) | *32* |

## TABLE 4 (Continued)
## NATURAL ACIDS OF CARBOHYDRATE DERIVATION

| Substance[a] (synonym) | Derivative | Chemical formula | Melting point °C | Specific rotation[b] $[\alpha]_D$ | Reference |
|---|---|---|---|---|---|
| | | **Aldonic Acids—(Continued)** | | | |
| **3-Deoxy-D-*threo*-2-hexulosonic acid** (2-keto-3-deoxy-D-galactonic acid) | | $C_6H_{10}O_6$ | — | +15 | *33, 109* |
| | Calcium salt | $CaC_{12}H_{18}O_{12}$ | 150–151 | +7.9 | *109* |
| | Phenylhydrazone phenylhydrazide | $C_{18}H_{22}N_4O_4$ | 204–205 | +13.9 ($C_5H_5N$) | *109* |
| | Potassium salt | $KC_6H_9O_6$ | 159–163d | — | *34* |
| | Lactone phenyl-hydrazone | $C_{12}H_{14}N_2O_4$ | 213–214 | −270 (*c* 0.5, $C_5H_5N$) | *109* |
| **L-Galactonic acid** | | $C_6H_{12}O_7$ | No constants known | — | *3* |
| | Amide | $C_6H_{13}NO_6$ | 175 | −30 | *35* |
| | 1,4-Lactone | $C_6H_{10}O_6$ | 110, 134–135 | +77 | *36, 37* |
| **D-*arabino*-5-Hexulosonic acid** (5-keto-L-galactonic acid or D-tagaturonic acid) | | $C_6H_{10}O_7$ | 108–109 | — | *38* |
| | Calcium salt·$5H_2O$ | $CaC_{12}H_{18}O_{14}$·$5H_2O$ | — | −14 | *39* |
| | Brucine salt | $C_{29}H_{36}N_2O_{11}$ | 148–149 | — | *38* |
| **D-Gluconic acid** | | $C_6H_{12}O_7$ | 120–131 | −6.9 → +7.3 | *11* |
| | Phenylhydrazide | $C_{12}H_{18}N_2O_6$ | 200 | +12 | *10* |
| | 1,4-Lactone | $C_6H_{10}O_6$ | 133–135 | +68 → +17.7 | *3, 11* |
| | Amide | $C_6H_{13}NO_6$ | 143–144 | +31.2 | *12, 18a* |
| | 1,5-Lactone | $C_6H_{10}O_6$ | 150–152 | +66 → +8.8 | *11* |
| **6-*O*-(*N,N*-dimethylglycyl)-D-gluconic acid** (pangamic acid, vitamine $B_{15}$) | | $C_{10}H_{19}NO_8$ | No constants known | — | *129* |
| | Lactone·HCl | $C_{10}H_{17}NO_7$·HCl | 69–73 | +32.3 ($CH_3OH$) | *129* |
| | Amide·HCl | $C_{10}H_{20}N_2O_7$·HCl | 92–95d | +20.9 ($CH_3OH$) | *129* |
| **2-Deoxy-D-*arabino*-hexonic acid** (2-deoxy-D-gluconic acid) | | $C_6H_{12}O_6$ | 142–144 | +2 | *40, 41* |
| | 1,4-Lactone | $C_6H_{10}O_5$ | 93–95 | +68 | *42* |
| | Phenylhydrazide | $C_{12}H_{18}N_2O_5$ | 156 | — | *42* |
| **D-*arabino*-2-Hexulosonic acid** (2-keto-D-gluconic acid) | | $C_6H_{10}O_7$ | — | −99.6 | *43* |
| | Methyl ester | $C_7H_{12}O_7$ | 175–176 | −76.8 | *39* |
| | Phenylhydrazone salt | $C_{18}H_{24}N_4O_6$ | 121 | −36.1 ($H_2O$, $C_5H_5N$) | *43* |
| **3-Deoxy-D-*erythro*-2-hexulosonic acid** (2-keto-3-deoxy-D-gluconic acid) | | $C_6H_{10}O_6$ | — | −34.5 | *109* |
| | Calcium salt·$1/2H_2O$ | $CaC_{12}H_{18}O_{12}$·$1/2H_2O$ | — | −29.2 (*c* 6) | *44* |
| | Phenylhydrazone | $C_{12}H_{16}N_2O_5$ | 229d | +168 ($C_5H_5N$) | *109* |
| **D-*xylo*-5-Hexulosonic acid** (5-keto-D-gluconic acid) | | $C_6H_{10}O_7$ | — | −14.5 | *45, 46* |
| | Calcium salt·$3H_2O$ | $CaC_{12}H_{18}O_{14}$·$3H_2O$ | — | −11.7 | *46* |
| | *p*-Nitrophenyl-hydrazone | $C_{12}H_{15}N_3O_8$ | 200–202 | — | *47* |
| **D-*threo*-Hex-2,5-diulosonic acid** (2,5-diketo-D-gluconic acid) | | $C_6H_8O_7$ | No constants known | — | *48* |
| | Calcium salt·$3H_2O$ | $CaC_{12}H_{14}O_{14}$ $3H_2O$ | — | −51 ± 5 | *48* |
| | *Bis*(2,4-dinitro-phenylhydrazone) | $C_{18}H_{16}N_4O_{13}$ | 156–157d | +57.2 ($C_5H_5N$) | *48* |

# TABLE 4 (Continued)
# NATURAL ACIDS OF CARBOHYDRATE DERIVATION

| Substance[a] (synonym) | Derivative | Chemical formula | Melting point °C | Specific rotation[b] $[\alpha]_D$ | Reference |
|---|---|---|---|---|---|
| | | **Aldonic Acids—(Continued)** | | | |
| **L-Gluconic acid** | | $C_6H_{12}O_7$ | No constants known | — | *3* |
| | Barium salt | $BaC_{12}H_{22}O_{14}$ | — | −6.4 (*c* 8) | *49* |
| | Phenylhydrazide | $C_{12}H_{18}N_2O_6$ | 203–204 | −11.7 | *50, 51* |
| | 1,4-Lactone | $C_6H_{10}O_6$ | 134–135 | −68.7 → −13.7 | *51* |
| | Brucine salt | $C_{29}H_{38}N_2O_{11}$ | 181–182 | −25.4 | *51* |
| **L-Gulonic acid** | | $C_6H_{12}O_7$ | Syrup | 0 | *52* |
| | 1,4-Lactone | $C_6H_{10}O_6$ | 183–185 | +55 | *53* |
| **L-*xylo*-2-Hexulosonic acid** (2-keto-L-gulonic acid or 2-keto-L-idonic acid) | | $C_6H_{10}O_7$ | 170–171 | −48.8 | *39* |
| | Sodium salt | $NaC_6H_9O_7$ | 145 | −24.4 | *54* |
| | Brucine salt·$H_2O$ | $C_{29}H_{36}N_2O_{11}$ $H_2O$ | 114 | — | *54* |
| **L-*xylo*-3-Hexulosonic acid** (3-keto-L-gulonic acid) | | $C_6H_{10}O_7$ | No constants known | — | *55* |
| **D-*lyxo*-5-Hexulosonic acid** (5-keto-L-gulonic acid or D-fructuronic acid) | | $C_6H_{10}O_7$ | No constants known | — | *56* |
| | Brucine salt | $C_{29}H_{36}N_2O_{11}$ | 195–197 | −15.5 | *38, 130* |
| | Potassium salt | $KC_6H_9O_7$ | 160–165 | +11 | *131* |
| **L-*threo*-Hex-2,3-diulosonic acid** (2,3-diketo-L-gulonic acid) | | $C_6H_8O_7$ | No constants known | — | *57* |
| | *Bis*(2,4-dinitrophenylhydrazone) | $C_{18}H_{16}N_4O_{13}$ | 281 | — | *57* |
| **4-Deoxy-D-*threo*-5-hexulosonic acid** (4-deoxy-5-keto-D-idonic acid) | | $C_6H_{10}O_6$ | No constants known | — | *58* |
| | Sodium salt | $NaC_6H_9O_6$ | 67–68 | +5.5 | *58* |
| | Phenylosazone | $C_{18}H_{22}N_4O_4$ | 113d | — | *58* |
| **L-Idonic acid** | | $C_6H_{12}O_7$ | No constants known | — | *59* |
| | Brucine salt | $C_{29}H_{38}N_2O_{11}$ | 190–192 | −17 | *3* |
| | 1,4-Lactone | $C_6H_{10}O_6$ | — | +50, +4.5 | *3, 60* |
| | Phenylhydrazide | $C_{12}H_{18}N_2O_6$ | 115–117 | +12.5 | *60* |
| **D-Mannonic acid** | | $C_6H_{12}O_7$ | — | −15.6 | *61* |
| | Phenylhydrazide | $C_{12}H_{18}N_2O_6$ | 212–214 | +15.8 | *25, 62* |
| | Brucine salt | $C_{29}H_{38}N_2O_{11}$ | 203 | −27.8 | *63* |
| | 1,4-Lactone | $C_6H_{10}O_6$ | 151–152 | +51.5 | *3, 11* |
| | 1,5-Lactone | $C_6H_{10}O_6$ | 158–160 | +114 → +30.3 | *11* |
| **3-Deoxy-D-*glycero*-hexo-2,5-diulosonic acid** (3-deoxy-2,5-diketo-D-mannonic acid) | | $C_6H_8O_6$ | No constants known | — | *64* |
| **6-Deoxy-L-mannonic acid** (L-rhamnonic acid) | | $C_6H_{12}O_6$ | — | +4 → −32.7 | *21, 50* |
| | Amide | $C_6H_{13}NO_5$ | 141–142 | +27.5 | *65* |
| | 1,4-Lactone | $C_6H_{10}O_5$ | 149–151 | −39.2 | *11* |
| | 1,5-Lactone | $C_6H_{10}O_5$ | 172–182 | −100 → −35.1 | *11* |
| **3-Deoxy-2-heptulosonic acid** | | $C_7H_{12}O_7$ | No constants known | — | *66, 67* |
| **3-Deoxy-D-*arabino*-2-heptulosonic acid** | | $C_7H_{12}O_7$ | No constants known | — | *68* |
| | 1,5 Lactone | $C_7H_{10}O_6$ | — | +33 | *68* |
| | Methyl glycoside methyl ester | $C_9H_{14}O_6$ | 148 | +78.2 ($CH_3OH$) | *68* |

## TABLE 4 (Continued)
## NATURAL ACIDS OF CARBOHYDRATE DERIVATION

| Substance[a] (synonym) | Derivative | Chemical formula | Melting point °C | Specific rotation[b] $[\alpha]_D$ | Reference |
|---|---|---|---|---|---|
| | | **Aldonic Acids—(Continued)** | | | |
| **3-Deoxy-2-octulosonic acid** | | $C_8H_{14}O_8$ | No constants known | — | *69, 70* |
| | $NH_4$ salt | $C_8H_{17}NO_8$ | 125–126 | +41.3 | *132* |
| | Lactone | $C_8H_{12}O_7$ | 192–194 | +31.8 | *132* |
| | Pentaacetate | $C_{18}H_{24}O_{13}$ | 98–103 | — | *133* |
| ***O*-β-D-Galactopyranosyl-(1 → 4)-D-gluconic acid** (lactobionic acid) | | | | | |
| | Calcium salt | $CaC_{24}H_{42}O_{24}$ | — | +25.1 | *71* |
| ***O*-α-D-Glucopyranosyl-(1 → 4)-D-gluconic acid** (maltobionic acid) | | | | | |
| | Calcium salt | $CaC_{24}H_{42}O_{24}$ | — | +105 | *71* |
| | Brucine salt | $C_{35}H_{48}N_2O_{16}$ | 155–157 | — | *71* |
| | | **URONIC ACIDS** | | | |
| **D-Lyxuronic acid** | | $C_5H_8O_6$ | No constants known | — | *72* |
| | Calcium salt·$2H_2O$ | $CaC_{10}H_{14}O_{12}\cdot 2H_2O$ | — | −23 → −53 | *72* |
| | Phenylosazone salt | $C_{23}H_{28}N_6O_4$ | 164d | — | *73* |
| **α-D-Galacturonic acid·$H_2O$** | | $C_6H_{10}O_7\cdot H_2O$ | 159–160 (110–115s) | +97.9 → +50.9 | *74* |
| | *p*-Bromophenylhydrazone salt | $Br_2C_{18}H_{24}N_4O_6$ | 145–146 | +9 ± 2 (*c* 0.7, $CH_3OH$) | *75* |
| **β-D-Galacturonic acid** | | $C_6H_{10}O_7$ | 160 | +27 → +55.6 | *74* |
| | *p*-Bromophenylhydrazone | $BrC_{12}H_{15}N_2O_6$ | 150–151 | +11.5 ± 2 ($CH_3OH$) | *75* |
| | Brucine salt | $C_{29}H_{36}N_2O_{11}$ | 180 | −7.7 | *74* |
| **β-D-Glucuronic acid** | | $C_6H_{10}O_7$ | 156 | +11.7 → +36.3 | *76, 77* |
| | 3,6-Lactone (γ-lactone) | $C_6H_8O_6$ | 163, 180 | +18.6 | *78, 91* |
| | Phenylhydrazone phenylhydrazide | $C_{18}H_{22}N_4O_5$ | 182 | — | *79* |
| | Brucine salt·$H_2O$ | $C_{29}H_{36}N_2O_{11}\cdot H_2O$ | 156–157 | −15.1 | *80* |
| **D-Glucuronic acid, 3-*O*-methyl-** | | $C_7H_{12}O_7$ | Syrup | +6 | *81, 82* |
| | *p*-Bromophenylosazone salt | $Br_3C_{25}H_{29}N_6O_5$ | 157 | −104 → −14 | *83* |
| **D-Glucuronic acid, 4-*O*-methyl-** | | $C_7H_{12}O_7$ | Syrup | +48, +82 | *84, 85* |
| | Methyl 4-*O*-methyl-α-D-glucopyranosiduronamide | $C_8H_{15}NO_6$ | 232 | +143 (*c* 0.7) | *86* |
| | Methyl 4-*O*-methyl-D-glucuronate | $C_8H_{14}O_7$ | 123–124 | +41 | *87* |
| **L-Guluronic acid** | | $C_6H_{10}O_7$ | No constants known | — | *47* |
| **L-Gulurono-3,6-lactone** | | $C_6H_8O_6$ | 141–142 | +81.7 | *88* |
| **L-Iduronic acid** | | $C_6H_{10}O_7$ | 131–133 | +37 → +33 | *89, 90* |
| | 3,6-Lactone | $C_6H_8O_6$ | Syrup | +30 | *91* |
| **4-Deoxy-L-*threo*-5-hexulosuronic acid** (4-deoxy-5-keto-L-iduronic acid) | | $C_6H_8O_6$ | No constants known | — | *64* |
| **α-D-Mannuronic acid·$H_2O$** | | $C_6H_{10}O_7\cdot H_2O$ | 120–130 (110s) | +16 → −6.1 (*c* 6.8) | *92* |
| | 3,6-Lactone | $C_6H_8O_6$ | 140–141 | +89.3 | *92* |
| **β-D-Mannuronic acid** | | $C_6H_{10}O_7$ | 165–167 | −47.9 → −23.9 | *92* |
| | *p*-Bromophenylhydrazone salt | $Br_2C_{18}H_{24}N_4O_6$ | 143–144d | +48.5 ± 1 ($CH_3OH$) | *75* |

## TABLE 4 (Continued)
## NATURAL ACIDS OF CARBOHYDRATE DERIVATION

| Substance[a] (synonym) | Derivative | Chemical formula | Melting point °C | Specific rotation[b] $[\alpha]_D$ | Reference |
|---|---|---|---|---|---|
| | | **ALDARIC ACIDS** | | | |
| **D-Threaric acid** (*l*-tartaric) | | $C_4H_6O_6$ | 170 | −15 | *93* |
| **L-Threaric acid** (*d*-tartaric) | | $C_4H_6O_6$ | 170 | +15 (15°) | *94, 95* |
| | Dimethyl ester | $C_6H_{10}O_6$ | 48, 61.5 | +2.7 | *96–98* |
| | Diamide | $C_4H_8N_2O_4$ | 195 | +106.5 | *99* |
| **L-Malic acid** (deoxy-L-tartaric) | | $C_4H_6O_5$ | 100 | −2.3 (*c* 8.4) | *100, 101* |
| | Malic acid acetate | $C_6H_8O_6$ | 132 | — | *102* |
| | Diamide | $C_4H_8N_2O_3$ | 156–157 | −37.9 | *103* |
| **Allaric acid** (allomucic) | | $C_6H_{10}O_8$ | 188–192d | — | *135* |
| **D-Glucaric acid** (saccharic) | | $C_6H_{10}O_8$ | 125–126 | +6.9 → +20.6 | *104* |
| | Potassium H salt | $KC_6H_9O_8$ | — | +10[d] | *104a* |
| | *Bis*(phenylhydrazide) | $C_{18}H_{22}N_4O_6$ | 209–210 | — | *105* |
| | Diamide | $C_6H_{12}N_2O_6$ | 172–173 | +13.3 | *12* |
| **D-Mannaric acid** (D-mannosaccharic) | | $C_6H_{10}O_8$ | 128.5 | +3.5 → +48.7 | *106* |
| | *Bis*(phenylhydrazide) | $C_{18}H_{22}N_4O_6$ | 214–216d | — | *107* |
| | Diamide | $C_6H_{12}N_2O_6$ | 188–189.5 | −24.4 | *12* |
| | | **AMINO SUGAR ACIDS** | | | |
| **3-Amino-3-deoxy-D-alluronic acid** | | $C_6H_{11}NO_6$ | No constants known | — | *108* |
| | 3-Acetamido-3-deoxy-D-allose | $C_8H_{13}NO_7$ | 182–183 | — | *108* |
| **2-Amino-2-deoxy-D-galactonic acid** (D-galactosaminic acid, chondrosaminic acid) | | $C_6H_{13}NO_6$ | 198–203d | −5 (*c* 0.6) | *109* |
| | | — | — | −11.3 → −31 (2*N* HCl) | *109* |
| | 2-Acetamido-2-deoxy-D-galactono-1,4-lactone | $C_8H_{13}NO_6$ | 165 | — | *110* |
| **2-Amino-2-deoxy-D-galacturonic acid** (D-galactosaminuronic acid) | | $C_6H_{11}NO_6$ | 160d | +84 (pH 2, HCl) | *111, 112* |
| | 2-Amino-2-deoxy-α-D-galactose·HCl | $C_6H_{13}NO_5{\cdot}HCl$ | 185 | +121 → +80 | *113* |
| | 2-Amino-2-deoxy-β-D-galactose·HCl | $C_6H_{13}NO_5{\cdot}HCl$ | 187 | +44 → +80 | *113* |
| **2-Amino-2-deoxy-D-glucuronic acid** (D-glucosaminuronic acid) | | $C_6H_{11}NO_6$ | 120–172d | +55 | *114, 115* |
| | Methyl glycopyranoside | $C_7H_{13}NO_6$ | 203–207 (196s) | +126.3 | *115* |
| | 2-Amino-2-deoxy-α-D-glucose | $C_6H_{13}NO_5$ | 88 | +100 → +47 | *116* |
| | 2-Amino-2-deoxy-β-D-glucose | $C_6H_{13}NO_5$ | 110–111 | +28 → +47 | *116* |
| **4-Amino-4-deoxy-D-*erythro*-hex-2-eneuronic acid** | | $C_6H_9NO_4$ | No constants known | — | *136* |
| **2-Amino-2-deoxy-D-mannuronic acid** (D-mannosaminuronic acid) | | $C_6H_{11}NO_6$ | No constants known | — | *117* |
| **3-*O*-(D-1-Carboxyethyl)-2-amino-2-deoxy-D-glucose** (muramic acid) | | $C_9H_{17}NO_7$ | 155 | +165 → +123 | *118–120* |
| | *N*-Acetyl deriv. | $C_{11}H_{19}NO_8$ | 122–124 | +59 → +39 | *121* |

## TABLE 4 (Continued)
## NATURAL ACIDS OF CARBOHYDRATE DERIVATION

| Substance[a] (synonym) | Derivative | Chemical formula | Melting point °C | Specific rotation[b] $[\alpha]_D$ | Reference |
|---|---|---|---|---|---|
| | | **Amino Sugar Acids—(Continued)** | | | |
| **5-Acetamido-3,5-dideoxy-D-*glycero*-D-galacto-nonulosonic acid** (*N*-acetylneuraminic acid, gynaminic acid, lactaminic acid, sialic acid) | | $C_{11}H_{19}NO_9$ | 183–186d | −31.7 | *122, 123* |
| | Quinoxaline deriv. | $C_{17}H_{23}N_3O_7$ | 204–205 | −100 (*c* 0.3, 1 : 1 $CH_3SOCH_3$, $H_2O$) | *124, 125* |
| ***N*-Glycolylneuraminic acid** | | $C_{11}H_{19}NO_{10}$ | — | −33.6 | *126* |
| **Destomic acid** | | $C_7H_{15}NO_7$ | 207–209d | +1.9 | *127* |
| | | — | — | −12.1 → −30.6 (2*N* HCl) | *127* |
| | Methyl destomate·HCl | $C_8H_{17}NO_7\cdot HCl$ | 150–151d | — | *127* |

[a] In order of increasing carbon chain length in the parent compounds, grouped according to class.
[b] Unless otherwise specified, $[\alpha]_D$ is given for 1 to 5 g solute, *c* representing the amount of solvent per 100 ml aqueous solution, at 20 to 25°C.
[c] Equilibrates with the lactone.
[d] As dipotassium salt.
[e] See References 21, 35 and 53 for data on the D-isomer.
[f] See Reference 134 for data on the L-isomer.

Data taken from: Maher, G. G., and Wolfrom, M. L., in *Handbook of Biochemistry,* 2nd ed., pp. D-8–D-13, H. A. Sober, Ed. Copyright 1970, The Chemical Rubber Co., Cleveland, Ohio.

## TABLE 5
## NATURAL ALDOSES

| Substance[a] (synonym) | Derivative | Chemical formula | Melting point °C | Specific rotation[b] $[\alpha]_D$ | Reference |
|---|---|---|---|---|---|
| **D-Glyceraldehyde** | | $C_3H_6O_3$ | Syrup | +13.5 ± 0.5 | *1* |
| | 2,4-Dinitrophenylhydrazone | $C_9H_{10}N_4O_6$ | 155–156 | — | *2* |
| | Dimethone | $C_{19}H_{28}O_6$ | 199–201 | +197.5 (*c* 0.7, $C_2H_5OH$) | *2* |
| **D-Glyceraldehyde, 3,3-*bis*(*C*-hydroxymethyl)-** (D-apiose) | | $C_5H_{10}O_5$ | 138–139 | −29 ($CH_3OH$)<br>+5.6 (*c* 10) (15°) | *3*<br>*4, 5* |
| | Benzylphenylhydrazone | $C_{18}H_{22}N_2O_4$ | 137–138 | −78.5 ($C_5H_5N$) $[\alpha]_{579}$ | *4, 5* |
| **D-Glyceraldehyde, 3,3-*bis*-(*C*-hydroxymethyl)-3-deoxy-** (Cordycepose) | | $C_5H_{10}O_4$ | Syrup | −26 (*c* 0.6, $C_2H_5OH$) | *6* |
| | Cordyceponic acid phenyl hydrazide | $C_{11}H_{16}N_2O_4$ | 151 | +26 ± 3 (*c* 0.3, $C_2H_5OH$) | *6* |
| **β-D-Arabinose** | | $C_5H_{10}O_5$ | 155 | −175 → −103 | *7* |
| | Benzylphenylhydrazone | $C_{18}H_{22}N_2O_4$ | 177–178 | +14.4 ($CH_3OH$) (16°) | *8* |
| **D-Arabinose, 2-*O*-methyl-** | | $C_6H_{12}O_5$ | Syrup | −102 ± 3 | *9, 10* |
| | Phenylhydrazone | $C_{12}H_{18}N_2O_4$ | 113 | — | *11* |
| | 2-*O*-Methyl-D-arabinitol | $C_6H_{14}O_5$ | 98–99 | −11 ($CH_3OH$) | *12* |

# TABLE 5 (Continued)
# NATURAL ALDOSES

| Substance[a] (synonym) | Derivative | Chemical formula | Melting point °C | Specific rotation[b] $[\alpha]_D$ | Reference |
|---|---|---|---|---|---|
| **α-L-Arabinose** | | $C_5H_{10}O_5$ | Amorphous 158 | +55.4 → +105 | *13* |
| **β-L-Arabinose** | | $C_5H_{10}O_5$ | 160 | +190.6 → +104.5 | *14* |
| | Tetra-*O*-acetyl-α-L-arabinopyranoside | $C_{13}H_{18}O_9$ | 97 | +42.5 ($CHCl_3$) | *15* |
| | Tetra-*O*-acetyl-β-L-arabinopyranoside | $C_{13}H_{18}O_9$ | 86 | +147.2 ($CHCl_3$) | *15* |
| | *p*-Nitrophenylhydrazone | $C_{11}H_{15}N_3O_6$ | 181 | +22.6 (1 : 1, $C_5H_5N$, $C_2H_5OH$) | *16* |
| **L-Arabinose 3-sulfate** | | $C_5H_{10}O_8S$ | — | +75 (*c* 0.6) | *17* |
| **α-L-Lyxose** | | $C_5H_{10}O_5$ | 105 | +5.8 → +13.5 | *18* |
| | Tetra-*O*-acetyl-α-L-lyxopyranoside | $C_{13}H_{18}O_9$ | 93–94 | +25 ($CHCl_3$)[d] | *19* |
| | *p*-Nitrophenylhydrazone | $C_{11}H_{15}N_3O_6$ | 172 | — | *18* |
| | *p*-Bromophenylhydrazone | $BrC_{11}H_{15}N_2O_4$ | 157–158 | −30.1 → −10 ($C_5H_5N$) | *20* |
| **L-Lyxose, 5-deoxy-3-*C*-formyl-** (streptose) | | $C_6H_{10}O_5$ | Syrup | −18 | *21* |
| | Streptosonolactone | $C_6H_8O_5$ | 146–148 | −37 (*c* 0.7) | *22* |
| **L-Lyxose, 5-deoxy-3-*C*-hydroxymethyl-** (Dihydrostreptose) | | $C_6H_{12}O_5$ | Syrup | −24 | *21, 23* |
| | Dihydrostreptosonolactone | $C_6H_{10}O_5$ | 140.5–142.5 | −32 | *21* |
| **L-Lyxose, 3-*C*-formyl-** (Hydroxystreptose) | | $C_6H_{10}O_6$ | No constants known | — | *24* |
| | 3-Hydroxy-2-hydroxymethyl-1,4-pyrone | $C_6H_6O_4$ | 152–153 | — | *24* |
| **Pentose, 4,5-anhydro-5-deoxy-D-*erythro*-** | | $C_5H_8O_3$ | No constants known | — | *25* |
| **Pentose, 2-deoxy-D-*erythro*-** (2-Deoxy-D-ribose) | | $C_5H_{10}O_4$ | 96–98 | −91 → −58 | *26* |
| | 1,3,4-Tri-*O*-acetyl-2-deoxy-D-*erythro*-pentopyranoside | $C_{11}H_{16}O_7$ | 98 | −171.8 (*c* 0.5, $CHCl_3$) | *27* |
| | *p*-Nitrophenylhydrazone | $C_{11}H_{15}N_3O_5$ | 160 | −11.1 (*c* 0.1, $C_2H_5OH$) (14°) | *27* |
| **D-Ribose** | | $C_5H_{10}O_5$ | 87 | −23.7 | *28* |
| | Tetra-*O*-acetyl-α-D-ribopyranoside | $C_{13}H_{18}O_9$ | Syrup | +46.1 ($CH_3OH$) | *29* |
| | Tetra-*O*-acetyl-β-D-ribopyranoside | $C_{13}H_{18}O_9$ | 110 | −54.5 ($CH_3OH$) | *29* |
| | *p*-Bromophenylhydrazone | $BrC_{11}H_{15}N_2O_4$ | 164–165 | +10.3 ($C_2H_5OH$) | *30* |
| **D-Ribose, 2-*C*-hydroxymethyl-** (D-Hamamelose) | | $C_6H_{12}O_6$ | 110–111 | +7.7 → −7.0 | *31* |
| | *p*-Nitrophenylhydrazone | $C_{12}H_{17}N_3O_7$ | 165–166 | +144 ($C_5H_5N$) $[\alpha]_{578}$ | *31, 32* |
| **α-D-Xylose** | | $C_5H_{10}O_5$ | 145 | +93.6 → +18.8 | *33, 34* |
| | Tetra-*O*-acetyl-α-D-xylopyranoside | $C_{13}H_{18}O_9$ | 59 | +89.3 ($CHCl_3$) | *35* |
| | Tetra-*O*-acetyl-β-D-xylopyranoside | $C_{13}H_{18}O_9$ | 128 | −24.7 ($CHCl_3$) | *35* |
| | Benzylphenylhydrazone | $C_{18}H_{22}N_2O_4$ | 99 | −20.3 ($CH_3OH$) | *36, 37* |
| | *p*-Bromophenylhydrazone | $BrC_{11}H_{15}N_2O_4$ | 128 | −20.7 | *38, 39* |
| **D-Xylose, 5-deoxy-** | | $C_5H_{10}O_4$ | — | +16 | *40* |
| | Triacetate | $C_{11}H_{16}O_7$ | — | +60.9 ($CHCl_3$) | *41* |
| | *p*-Bromophenylhydrazone | $BrC_{11}H_{15}N_2O_3$ | 69–70 d. | −26.1 ($C_5H_5N$) | *41* |
| **D-Xylose, 2-*O*-methyl-** | | $C_6H_{12}O_5$ | 132–133 | −21 → +34 (*c* 0.7) | *42, 45* |
| | Methyl 2-*O*-methyl-β-D-xylopyranoside | $C_7H_{14}O_5$ | 111–112 | −67.7 ($CHCl_3$) | *43* |

## TABLE 5 (Continued)
## NATURAL ALDOSES

| Substance[a] (synonym) | Derivative | Chemical formula | Melting point °C | Specific rotation[b] $[\alpha]_D$ | Reference |
|---|---|---|---|---|---|
| **D-Xylose, 3-*O*-methyl-** | | $C_6H_{12}O_5$ | 95 | +45 → +19 | *44, 45* |
| | 3-*O*-Methyl-*N*-phenyl-D-xylopyranosylamine | $C_{12}H_{17}NO_4$ | 136 | +120 (*c* 0.3, ethyl acetate) | *46* |
| | 3-*O*-Methyl-D-xylonolactone | $C_6H_{10}O_5$ | 90 | +72 → +40 (*c* 0.9) | *47* |
| **Aldgarose A** (a dideoxy-*C*-hydroxyethylhexose carbonate) | | $C_9H_{16}O_6$ | No constants known | — | *48* |
| | Methyl aldgaroside A | $C_{10}H_{18}O_6$ | 91–94 | — | *48* |
| **Aldgarose B** (a dideoxy-*C*-hydroxyethylhexose carbonate) | | $C_9H_{16}O_6$ | No constants known | — | *48* |
| | Methyl aldgaroside B[c] | $C_{10}H_{18}O_6$ | 175–177 | −41 ($CH_3OH$) | *48* |
| **D-Allose** | | $C_6H_{12}O_6$ | 128–128.5 | +0.6 → +14.4 | *49, 50* |
| | *p*-Bromophenylhydrazone | $BrC_{12}H_{17}N_2O_5$ | 145–147 | −6.7 ($C_2H_5OH$) | *51* |
| | β-Allopyranoside pentaacetate | $C_{16}H_{22}O_{11}$ | 93–93.5 | −13.7 ($CHCl_3$) | *52* |
| **D-Allose, 6-deoxy-** | | $C_6H_{12}O_5$ | 132–135, 146–148 | −4.7 → 0 | *53, 54* |
| | *p*-Bromophenylhydrazone | $BrC_{12}H_{17}N_2O_4$ | 138–140, 145–146 | −21.9 → −11.8 ($C_5H_5N$) | *55, 56* |
| | 6-Deoxy-D-allopyranoside tetraacetate | $C_{14}H_{20}O_9$ | 109–110 | +10.4 ($CHCl_3$) | *57, 58* |
| **D-Allose, 6-deoxy-2-*O*-methyl-** (javose) | | $C_7H_{14}O_5$ | 112–114 | −54 → −40, −8.2 | *54, 59* |
| | Methyl 6-deoxy-2-*O*-methyl-α-D-allopyranoside | $C_8H_{16}O_5$ | Syrup | +90 ± 3 ($CHCl_3$) | *59* |
| | Methyl-β-glycoside | | 97–98 | −82.8 ± 1 ($CH_3OH$) | *245* |
| **D-Allose, 6-deoxy-3-*O*-methyl-** | | $C_7H_{14}O_5$ | 122–123 | +9 (30°) | *60* |
| | Methyl 6-deoxy-3-*O*-methyl-α-D-allopyranoside | $C_8H_{16}O_5$ | 110–111 | +195 ± 5 (*c* 0.7, $CH_3OH$) | *60* |
| | Methyl 6-deoxy-3-*O*-methyl-β-D-allopyranoside | $C_{18}H_{16}O_5$ | 153–154 | −37.1 ± 2 ($CH_3OH$) | *246* |
| **D-Allose, 6-deoxy-2,3-di-*O*-methyl-** (mycinose) | | $C_8H_{16}O_5$ | 102–106 | −46 → −29 | *53* |
| | Methyl 6-deoxy-2,3-di-*O*-methyl-α-D-allopyranoside | $C_9H_{18}O_5$ | — | +140 (*c* 0.7, $CHCl_3$) | *247* |
| | Methyl 6-deoxy-2,3-di-*O*-methyl-β-D-allopyranoside | $C_9H_{18}O_5$ | 88–88.5 | −36 ($CHCl_3$) (27°) | *53, 61* |
| **D-Altrose, 6-deoxy-** | | $C_6H_{12}O_5$ | Syrup | +16.2 | *62* |
| | *p*-Bromophenylhydrazone | $BrC_{12}H_{17}N_2O_4$ | 155, 177 | — | *63, 64* |
| | Methyl 6-deoxy-α-D-altropyranoside | $C_7H_{14}O_5$ | Syrup | +118 ($CH_3OH$) (16°) | *63* |
| **D-Altrose, 6-deoxy-3-*O*-methyl-** (D-vallarose) | | $C_7H_{14}O_5$ | 111–113 | +8.6 → +22.3 (*c* 0.6) (18°) | *65, 66* |
| | Tri-*O*-acetyl-6-deoxy-3-*O*-methyl-α-D-altropyranoside | $C_{13}H_{20}O_8$ | 112–113 | +14.8 ± 2 (15°) | *66* |
| | Tri-*O*-acetyl-6-deoxy-3-*O*-methyl-β-D-altropyranoside | $C_{13}H_{20}O_8$ | 79 | −96.5 ± 2 (15°) | *66* |
| | 6-Deoxy-3-*O*-methyl-D-altronic acid phenylhydrazide | $C_{13}H_{20}N_2O_5$ | 144–145 | −7.1 ($CH_3OH$) (13°) | *66* |

## TABLE 5 (Continued)
## NATURAL ALDOSES

| Substance[a] (synonym) | Derivative | Chemical formula | Melting point °C | Specific rotation[b] $[\alpha]_D$ | Reference |
|---|---|---|---|---|---|
| **L-Altrose, 6-deoxy-3-*O*-methyl-** (L-vallarose) | | $C_7H_{14}O_5$ | 106–110 | −17.2 ± 2 (*c* 0.9) | *67* |
| **Antiarose** | | $C_6H_{12}O_5$ | Syrup | levo | *68, 69* |
| | Antiaronolactone | $C_6H_{12}O_6$ | Syrup | −30 | *68, 69* |
| | Antiaronic acid phenylhydrazide | $C_{12}H_{18}N_2O_5$ | 143–145 | — | *68, 69* |
| **α-D-Galactose** | | $C_6H_{12}O_6$ | 167 | +150.7 → +80.2 | *70* |
| **β-D-Galactose** | | $C_6H_{12}O_6$ | 143–145 | +52.8 → +80.2 | *70, 71* |
| | Penta-*O*-acetyl-α-D-galactopyranoside | $C_{16}H_{22}O_{11}$ | 96 | +106.7 ($CHCl_3$) | *72* |
| | Penta-*O*-acetyl-β-D-galactopyranoside | $C_{16}H_{22}O_{11}$ | 142 | +25 ($CHCl_3$) | *72* |
| | *p*-Nitrophenylhydrazone | $C_{12}H_{17}N_3O_7$ | 192 | +70 (*c* 0.3, $C_5H_5N : C_2H_5OH$) | *16* |
| **D-Galactose, 3,6-anhydro-** | | $C_6H_{10}O_5$ | — | +12, +21.3 (10°) | *73–75* |
| | Diphenylhydrazone | $C_{18}H_{20}N_2O_4$ | 153–155 | +34.5 → +23.6 ($CH_3OH$) (14°) | *74* |
| | Methyl 3,6-anhydro-α-D-galactopyranoside | $C_7H_{12}O_5$ | 109, 139–140 | +80, +175 (10°) | *73, 74* |
| **D-Galactose, 4,6-*O*-(1-carboxyethylidene)-** | | $C_9H_{14}O_8$ | No constants known | — | *76* |
| | Ethanolate | $C_9H_{14}O_8 \cdot C_2H_5OH$ | — | +49 | *76* |
| | 4,6-*O*-(1-carboxyethylidene methyl ester)-D-galactitol | $C_{10}H_{18}O_8$ | 104–105 | −18 (*c* 0.6, $CH_3OH$) | *76* |
| | Methyl 4,6-*O*-(1-carboxyethylidene methyl ester)-α-D-galacto-pyranoside | $C_{11}H_{18}O_8$ | Syrup | +133 ($CHCl_3$) | *77* |
| **α-D-Galactose, 6-deoxy-** (D-fucose, rhodeose) | | $C_6H_{12}O_5$ | 140–145 | +120 → +76.3 (*c* 10) | *78* |
| | Benzylphenylhydrazone | $C_{19}H_{24}N_2O_4$ | 178–179 | −14.9 (*c* 0.4, $CH_3OH$) | *78* |
| | Tetra-*O*-acetyl-6-deoxy-α-D-galactopyranoside | $C_{14}H_{20}O_9$ | 92–93 | +129 ($CHCl_3$) | *79* |
| **D-Galactose, 6-deoxy-2-*O*-methyl-** | | $C_7H_{14}O_5$ | 155–161 | +73 → +87 | *80* |
| | Methyl 6-deoxy-2-*O*-methyl-α-D-galactopyranoside | $C_8H_{16}O_5$ | Syrup | +173.6 | *81, 82* |
| | Methyl 6-deoxy-2-*O*-methyl-β-D-galactopyranoside | $C_8H_{16}O_5$ | 98.5–99.5 | +3.5 ($CH_3OH$) | *81* |
| **D-Galactose, 6-deoxy-3-*O*-methyl-** (digitalose) | | $C_7H_{14}O_5$ | 106, 119 | +106 | *83* |
| | Digitalonolactone | $C_7H_{14}O_6$ | 137–138 | −83 | *83* |
| | Methyl 6-deoxy-3-*O*-methyl-α-D-galactopyranoside | $C_8H_{16}O_5$ | 98.5–100 | +198 (*c* 0.7, $CH_3OH$) | *81* |
| | Methyl 6-deoxy-3-*O*-methyl-β-D-galactopyranoside | $C_8H_{16}O_5$ | 108–110 | +9.9 (*c* 0.3, $CH_3OH$) | *81* |
| **D-Galactose, 6-deoxy-4-*O*-methyl-** (curacose) | | $C_7H_{14}O_5$ | 131–132 | +82 | *20* |
| | *p*-Tolylsulfonylhydrazone | $C_{14}H_{22}N_2O_6S$ | 134 | −16 → −3 ($C_5H_5N$) | *20* |
| **D-Galactose, 6-deoxy-2,3-di-*O*-methyl-** | | $C_8H_{16}O_5$ | 75–76 | +73, +105 | *81, 84* |
| | Methyl 6-deoxy-2,3-di-*O*-methyl-α-D-galactopyranoside | $C_9H_{18}O_5$ | Syrup | +190 (acetone) | *81, 82* |
| | Methyl 6-deoxy-2,3-di-*O*-methyl-β-D-galactopyranoside | $C_9H_{18}O_5$ | Syrup | +0.7 (acetone) | *81, 82* |

## TABLE 5 (Continued)
## NATURAL ALDOSES

| Substance[a] (synonym) | Derivative | Chemical formula | Melting point °C | Specific rotation[b] $[\alpha]_D$ | Reference |
|---|---|---|---|---|---|
| **D-Galactose, 6-deoxy-2,4-di-*O*-methyl-** (labilose) | | $C_8H_{16}O_5$ | 129 | +82 (*c* 0.5) (27°) | *85* |
| | Methyl 6-deoxy-2,4-di-*O*-methyl-α-D-galactopyranoside | $C_9H_{18}O_5$ | 85 | +176 ($CHCl_3$) (30°) | *85* |
| | Methyl 6-deoxy-2,4-di-*O*-methyl-β-D-galactopyranoside | $C_9H_{18}O_5$ | 111 | −20.9 ($CHCl_3$) (30°) | *85* |
| **D-Galactose, 2-*O*-methyl-** | | $C_7H_{14}O_6$ | 148–149 | +84.9 (*c* 0.5)(16°) | *248* |
| | 2-*O*-Methyl-*N*-phenyl-galactosylamine | $C_{13}H_{19}NO_5$ | 164–165 | — | *248* |
| **D-Galactose, 3-*O*-methyl-** | | $C_7H_{14}O_6$ | 144–147 | +150 → +108 | *86* |
| | Methyl 3-*O*-methyl-β-D-galactopyranoside | $C_8H_{16}O_6$ | Syrup | +31.9 | *86* |
| **D-Galactose, 4-*O*-methyl-** | | $C_7H_{14}O_6$ | 207, 218 | +62 → +92 | *87–89* |
| | 4-*O*-methyl-*N*-phenyl-D-galactosylamine | $C_{13}H_{19}NO_5$ | 167–168 | −84 → −39 ($CH_3OH$) | *88* |
| **D-Galactose, 6-*O*-methyl-** | | $C_7H_{14}O_6$ | 122–123 | +117 → +77.3 | *90* |
| | Phenylhydrazone | $C_{13}H_{20}N_2O_5$ | 181.5–182.5 | — | *91* |
| | Methyl 6-*O*-methyl-α-D-galactopyranoside | $C_8H_{16}O_6$ | 137–138 | +165 | *92* |
| **D-Galactose 2-sulfate** | | $C_6H_{12}O_9S$ | — | +52 | *249* |
| **D-Galactose 4-sulfate** | | $C_6H_{12}O_9S$ | — | +64 (*c* 0.5) | *93* |
| | Sodium salt | $C_6H_{11}NaO_9S$ | — | +58.4 (16°) | *94* |
| | Methyl α-glycoside tribenzoate | $C_{28}H_{26}O_{15}S$ | — | +93 | *250* |
| **D-Galactose, 6-sulfate** | | $C_6H_{12}O_9S$ | — | +49 (*c* 0.3) | *93* |
| | Sodium salt | $C_6H_{11}NaO_9S$ | — | +47 (16°) | *94* |
| **L-Galactose** | | $C_6H_{12}O_6$ | 163–165 | −78 | *95* |
| | Phenylhydrazone | $C_{12}H_{18}N_2O_5$ | 158–160 | +21.6 | *96* |
| **L-Galactose, 3,6-anhydro-** | | $C_6H_{10}O_5$ | — | −39.4 → −25.2 | *97* |
| | Diphenylhydrazone | $C_{18}H_{20}N_2O_4$ | 154 | −34.3 → −24.1 ($CH_3OH$) (15°) | *97* |
| | Methyl 3,6-anhydro-α-L-galactopyranoside | $C_7H_{12}O_5$ | 138.5–140 | −77 (*c* 0.9) (19°) | *90* |
| | Methyl 3,6-anhydro-β-L-galactopyranoside | $C_7H_{12}O_5$ | 118 | −113.5 | *98* |
| **L-Galactose, 3,6-anhydro-2-*O*-methyl-** | | $C_7H_{12}O_5$ | — | −14.3 (*c* 0.6)(12°) | *249, 251* |
| | Onic acid | $C_7H_{12}O_6$ | 141–142 | −70.3 (*c* 0.8)(12°) | *251* |
| **α-L-Galactose, 6-deoxy-** (L-fucose) | | $C_6H_{12}O_5$ | 145 | −124.1 → −76.4 | *99* |
| | *p*-Tolylsulfonylhydrazone | $C_{13}H_{20}N_2O_6S$ | 167–170 | — | *100* |
| | Tetra-*O*-acetyl-6-deoxy-β-L-galactopyranoside | $C_{14}H_{20}O_9$ | 172 | −39 ($CHCl_3$) | *101* |
| | Methyl 6-deoxy-α-L-galactopyranoside | $C_7H_{14}O_5$ | 158–159 | −191 (18°) | *102* |
| | Methyl 6-deoxy-β-L-galactopyranoside | $C_7H_{14}O_5$ | 126–127 | +10.5 (18°) | *102* |

# TABLE 5 (Continued)
# NATURAL ALDOSES

| Substance[a] (synonym) | Derivative | Chemical formula | Melting point °C | Specific rotation[b] $[\alpha]_D$ | Reference |
|---|---|---|---|---|---|
| **L-Galactose, 6-deoxy-2-*O*-methyl-** | | $C_7H_{14}O_5$ | 149–150 | −85 (18°) | *102, 103* |
| | Methyl 6-deoxy-2-*O*-methyl-α-L-galactopyranoside | $C_8H_{16}O_5$ | Syrup | −179 (26°) | *81* |
| | Methyl 6-deoxy-2-*O*-methyl-*β*-L-galactopyranoside | $C_8H_{16}O_5$ | 98–99 | +17.2 ($CH_3OH$) | *81* |
| **L-Galactose, 2-*O*-methyl-** | | $C_7H_{14}O_6$ | — | −75 (*c* 0.5)(18°) | *249* |
| | Onic acid lactone | $C_7H_{12}O_6$ | — | +17 (18°) | *249* |
| **L-Galactose, 4-*O*-methyl-** | | $C_7H_{14}O_6$ | 203–206 | −84 (17°) | *252* |
| **L-Galactose 6-sulfate** | | $C_6H_{12}O_9S$ | — | −47 (*c* 0.2) | *104* |
| **β-D-Glucopyranoside, benzoyl-** (periplanetin) | | $C_{13}H_{16}O_7$ | 193 | −26.8 | *105, 106* |
| | Tetraacetate | $C_{21}H_{24}O_{11}$ | 140–141 | — | *105* |
| **β-D-Glucopyranoside, methyl-** | | $C_7H_{14}O_6$ | 102–104 | −32 | *107* |
| | Tetraacetate | $C_{15}H_{22}O_{10}$ | 104–105 | −18.7 ($CHCl_3$) | *108, 109* |
| | Methyl tetra-*O*-methyl-*β*-D-glucopyranoside | $C_{11}H_{22}O_6$ | 40–41 | −17.3 | *110* |
| **α-D-Glucose** | | $C_6H_{12}O_6$ | 146 | +112 → +52.7 | *111* |
| | Monohydrate | $C_6H_{12}O_6 \cdot H_2O$ | 83 | — | *111* |
| **β-D-Glucose** | | $C_6H_{12}O_6$ | 148–150 | +18.7 → +52.7 | *111* |
| | Penta-*O*-acetyl-*aldehydo*-D-glucose diethyl dithioacetal | — | 45–47 | +11 ($CHCl_3$) | *111a* |
| | Penta-*O*-acetyl-α-D-glucopyranoside | $C_{16}H_{22}O_{11}$ | 114 | +101.6 ($CHCl_3$) | *108, 112* |
| | Penta-*O*-acetyl-*β*-D-glucopyranoside | $C_{16}H_{22}O_{11}$ | 135 | +3.8 (*c* 7, $CHCl_3$) | *108, 113* |
| | *p*-Nitrophenylhydrazone | $C_{12}H_{17}N_3O_7$ | 189 | +21.5 ($C_5H_5N$, $C_2H_5OH$) | *39, 114* |
| | | — | 188 | −88 ($C_5H_5N$, $C_2H_5OH$) | *16* |
| **D-Glucose 6-acetate** | | $C_7H_{14}O_7$ | 135, 146 | +48, +53 | *115, 116* |
| | Tetrabenzoate | $C_{35}H_{30}O_{11}$ | 183–184 | — | *117* |
| **D-Glucose 6-benzoate** (vaccinin) | | $C_{13}H_{16}O_7$ | Amorphous | +48 ($C_2H_5OH$) | *118* |
| | Tetra-*O*-acetyl-6-*O*-benzoyl-*β*-D-glucopyranoside | $C_{21}H_{24}O_{11}$ | 132 | +32.9 ($CHCl_3$) | *119* |
| **α-D-Glucose, 6-deoxy-** (chinovose, quinovose, epirhamnose) | | $C_6H_{12}O_5$ | 139–140 | +73.3 → +29.7 (*c* 8) | *120* |
| | Tetra-*O*-acetyl-6-deoxy-D-glucopyranoside | $C_{14}H_{20}O_9$ | 145 | +23 ($CHCl_3$) | *121* |
| | D-Glucomethylonic acid lactone | $C_6H_{12}O_6$ | 151–152 | +66.9 → +5.4 | *122* |
| **D-Glucose, 6-deoxy-2,3-di-*O*-methyl-** | | $C_8H_{16}O_5$ | — | +40.4 ± 2 | *253* |
| | Methyl *β*-glycoside | $C_9H_{18}O_5$ | 76–78 | −49 ($CHCl_3$) | *253* |
| **α-D-Glucose, 6-deoxy-3-*O*-methyl** (D-thevetose) | | $C_7H_{14}O_5$ | 116, 126 | +84 → +33 | *123, 124* |
| | Tri-*O*-acetyl-6-deoxy-3-*O*-methyl-α-D-glucopyranoside | $C_{13}H_{20}O_8$ | 105 | +122 (acetone) | *123* |
| | Tri-*O*-acetyl-6-deoxy-3-*O*-methyl-*β*-D-glucopyranoside | $C_{13}H_{20}O_8$ | 121 | +6 (acetone) | *123* |

## TABLE 5 (Continued)
## NATURAL ALDOSES

| Substance[a] (synonym) | Derivative | Chemical formula | Melting point °C | Specific rotation[b] $[\alpha]_D$ | Reference |
|---|---|---|---|---|---|
| α-D-Glucose, 6 deoxy-3-*O*-methyl (*cont.*) | | | | | |
| | Methyl 6-deoxy-3-*O*-methyl-α-D-glucopyranoside | $C_8H_{16}O_5$ | 86–87 | +148 ± 2 | *125* |
| | Methyl 6-deoxy-3-*O*-methyl-β-D-glucopyranoside | $C_8H_{16}O_5$ | 116–117 | −44 ± 2 | *125* |
| **D-Glucose, 6-sulfonic acid, 6-deoxy-** (6-sulfoquinovose) | | $C_6H_{12}O_8S$ | No constants known | — | *126* |
| **D-Glucose, 6-sulfonic acid, 6-deoxy-** (6-sulfoquinovose) **(Continued)** | Methyl 6-deoxy-α-D-glucopyranoside 6-sulfonic acid cyclohexylamine salt | $C_{13}H_{27}NO_8S$ | 173–174 | +87 | *126* |
| | Glycerol 6-deoxy-α-D-glucopyranoside 6-sulfonic acid cyclohexylamine salt | $C_{15}H_{31}NO_{10}S$ | 191–193 | — | *126* |
| **D-Glucose, 3-malonate** | | $C_9H_{14}O_9$ | No constants known | — | *127* |
| **α-D-Glucose, 3-*O*-methyl-** | | $C_7H_{14}O_6$ | 162–167 | +98 → +59.5 | *128* |
| **β-D-Glucose, 3-*O*-methyl-** | | $C_7H_{14}O_6$ | 130–132 | +31.9 → +55.1 | *129* |
| | Methyl 3-*O*-methyl-α-D-glucopyranoside·1/2$H_2O$ | $C_8H_{16}O_6 \cdot 1/2H_2O$ | 80–81 | +164 ± 2 | *130* |
| | Methyl 3-*O*-methyl-β-D-glucopyranoside | $C_8H_{16}O_6$ | Syrup | −26 (*c* 5.5) | *131* |
| | 3-*O*-Methyl-*N*-phenyl-D-glucopyranosylamine | $C_{13}H_{19}NO_5$ | 152–153 | −108 → −46 ± 2 (*c* 0.5, $CH_3OH$) | *132* |
| | Tetra-*O*-acetyl-3-*O*-methyl-β-D-glucopyranoside | $C_{15}H_{22}O_{10}$ | 95–96 | −5.2 ($CHCl_3$) | *133* |
| **D-Glucose, 6-*O*-methyl-** | | $C_7H_{14}O_6$ | 139–141 | +57.5 | *134* |
| | Methyl-6-*O*-methyl-α-D-glucopyranoside | $C_8H_{16}O_6$ | Syrup | +127.9 | *135* |
| | Methyl 6-*O*-methyl-β-D-glucopyranoside | $C_8H_{16}O_6$ | 133–135 | −27 | *136* |
| | Tetra-*O*-acetyl-6-*O*-methyl-α-D-glucopyranoside | $C_{15}H_{22}O_{10}$ | 119–120 | +111.8 ($CHCl_3$) | *137* |
| | Tetra-*O*-acetyl-6-*O*-methyl-β-D-glucopyranoside | $C_{15}H_{22}O_{10}$ | 91–93 | +20.9 ($CHCl_3$) | *137* |
| **α-D-Glucose, 2,3-di-*O*-methyl-** | | $C_8H_{16}O_6$ | 85–87 | +81.9 → +48.3 (acetone) | *82, 129* |
| **β-D-Glucose, 2,3-di-*O*-methyl-** | | $C_8H_{16}O_6$ | 108–110, 121 | +5.9 → +50.9 (acetone) | *129, 138* |
| | 2,3-Di-*O*-methyl-*N*-phenyl-D-glucopyranosylamine | $C_{14}H_{21}NO_5$ | 134 | −83 ($CHCl_3$) | *139* |
| | Methyl 2,3-di-*O*-methyl-α-D-glucopyranoside | $C_7H_{18}O_6$ | 80–82 | +142.6 | *129* |
| | Methyl 2,3-di-*O*-methyl-β-D-glucopyranoside | $C_7H_{18}O_6$ | 62–64 | −36.6 | *131* |

# TABLE 5 (Continued)
## NATURAL ALDOSES

| Substance[a] (synonym) | Derivative | Chemical formula | Melting point °C | Specific rotation[b] $[\alpha]_D$ | Reference |
|---|---|---|---|---|---|
| **D-Glucose tri-β-nitropropionate** (karakin) | | $C_{15}H_{21}N_3O_{15}$ | 120–122 | +4.5 | *140, 141* |
| **α-L-Glucose** | | $C_6H_{12}O_6$ | 141–143 | −95.5 → −51.4 | *142* |
| **L-Glucose, 6-deoxy-3-*O*-methyl-** (L-thevetose) | | $C_7H_{14}O_5$ | 126–129 | −36.9 ± 2 | *143* |
| | Tri-*O*-acetyl-6-deoxy-3-*O*-methyl-α-L-glucopyranoside | $C_{13}H_{20}O_8$ | 103–104 | −113 ($CH_3OH$) | *144, 145* |
| **L-Glucose, 6-deoxy-3-*O*-methyl-** (L-thevetose) **(Continued)** | Tri-*O*-acetyl-6-deoxy-3-*O*-methyl-β-L-glucopyranoside | $C_{13}H_{20}O_8$ | 118–119 | −7.5 ± 2 (acetone) | *143* |
| **D-Gulose, 6-deoxy-** | | $C_6H_{12}O_5$ | 130–131 | −38 | *41, 146* |
| | *p*-Bromophenylhydrazone | $BrC_{12}H_{17}N_2O_4$ | 135–136 | −49 → −34.7 ($C_5H_5N$) | *41* |
| | 6-Deoxy-D-gulonolactone | $C_6H_{10}O_5$ | 180–181 | — | *41* |
| **Hexose, 1,5-anhydro-2,6-dideoxy-4-*C*-(1-hydroxyethyl)-** | | $C_8H_{14}O_4$ | 153–154 | −144 | *210* |
| | 3,5-Dinitrobenzoate | $C_{15}H_{18}O_5$ | 167–168 | — | *210* |
| **Hexose, 2-deoxy-D-*arabino*-** | | $C_6H_{12}O_5$ | 146 | +38.3 → +45.9 (*c* 0.5) (18°) | *147* |
| | 2-Deoxy-*N*-phenyl-D-*arabino*-hexopyranosylamine | $C_{12}H_{17}NO_4$ | 193–194 | −138 → −106 ($C_5H_5N$) | *147* |
| | Tetra-*O*-acetyl-2-deoxy-α-D-*arabino*-hexopyranoside | $C_{14}H_{20}O_9$ | 91 | +12.3 (*c* 0.3, $C_2H_5OH$) | *147* |
| | | | 109.7–110.7 | +107.7 ($CHCl_3$) | *254* |
| | Tetra-*O*-acetyl-2-deoxy-β-D-*arabino*-hexopyranoside | $C_{14}H_{20}O_9$ | 75–78 | +30 (*c* 0.2, $C_2H_5OH$) | *147* |
| | | | 92.2–93.2 | −2.8 ($CHCl_3$) | *254* |
| **Hexose, 2,6-dideoxy-D-*arabino*-** (chromose C, canarose, olivose) | | $C_6H_{12}O_4$ | 86–98, 100–103 | +19.6, +25 | *148, 149* |
| | | — | — | +95.9, +110 (acetone) | *148, 149* |
| | | — | — | +45 (*c* 0.5) | *150* |
| | 2,4-Dinitrophenylhydrazone | $C_{12}H_{16}N_4O_7$ | 132–132.5 | — | *148* |
| | Methyl 2,6-dideoxy-α-D-*arabino*-hexopyranoside | $C_7H_{14}O_4$ | Syrup | +87 | *151* |
| | | — | — | +131 (*c* 0.8, $C_2H_5OH$) | *150* |
| | Methyl 2,6-dideoxy-β-D-*arabino*-hexopyranoside | $C_7H_{14}O_4$ | 84 | −85 ($C_2H_5OH$) | *255* |
| **Hexose, 3,6-dideoxy-D-*arabino*-** (tyvelose) | | $C_6H_{12}O_4$ | 143–144 | +24 ± 2 | *152, 153* |
| | 3,6-Dideoxy-D-*arabino*-hexitol | $C_6H_{12}O_4$ | 113–115 | −35 ± 2 (*c* 0.7) | *153* |
| | Methyl 3,6-dideoxy-α-D-*arabino*-hexopyranoside | $C_7H_{14}O_4$ | 84.5–85.5 | +137 ± 2 ($CH_3OH$) $[\alpha]_{5461}$ | *154* |
| | Methyl 3,6-dideoxy-β-D-*arabino*-hexopyranoside | $C_7H_{14}O_4$ | Syrup | −72 ± 2 ($CH_3OH$) $[\alpha]_{5461}$ | *154* |

## TABLE 5 (Continued) NATURAL ALDOSES

| Substance[a] (synonym) | Derivative | Chemical formula | Melting point °C | Specific rotation[b] $[\alpha]_D$ | Reference |
|---|---|---|---|---|---|
| **Hexose, 2,6-dideoxy-3-*O*-methyl-D-*arabino*-** (D-oleandrose) | | $C_7H_{14}O_4$ | 62–63 | −12.5 | *155, 156* |
| | 2,6-Dideoxy-3-*O*-methyl-D-*arabino*-hexonolactone | $C_7H_{12}O_4$ | Syrup | +12.8 ± 2 (14°) (acetone) | *155* |
| | 2,6-Dideoxy-3-*O*-methyl-D-*arabino*-hexonic acid phenylhydrazide | $C_{13}H_{20}N_2O_4$ | 134–135 | −20.6 ± 2 (16°) (*c* 0.8, $CH_3OH$) | *155* |
| **Hexose, 4-*O*-acetyl-2,6-dideoxy-3-*C*-methyl-L-*arabino*-** (chromose B) | | $C_9H_{16}O_5$ | Syrup | −24 | *151, 157* |
| **Hexose, 3,6-dideoxy-L-*arabino*-** (ascarylose) | | $C_6H_{12}O_4$ | Syrup | −25 | *153, 158* |
| | 3,6-Dideoxy-L-*arabino*-hexitol | $C_6H_{14}O_4$ | 112–113 | +38 ± 3 ($CH_3OH$) | *158* |
| **Hexose, 2,6-dideoxy-3-*C*-methyl-L-*arabino*-** (olivomycose) | | $C_7H_{14}O_4$ | 103–106 | −13 → −22 (26°) | *150* |
| | Methyl 2,6-dideoxy-3-*C*-methyl-α-L-*arabino*-hexopyranoside | $C_8H_{16}O_4$ | Syrup | −147 ($C_2H_5OH$) | *150* |
| | Methyl 2,6-dideoxy-3-*C*-methyl-β-L-*arabino*-hexopyranoside | $C_8H_{16}O_4$ | 93–94 | +50 ($C_2H_5OH$) | *150* |
| **Hexose, 2,6-dideoxy-3-*O*-methyl-L-*arabino*-** (L-oleandrose) | | $C_7H_{14}O_4$ | 62–63 | +11.9 ± 2.5 | *159* |
| | 2,4-Dinitrophenylhydrazone | $C_{13}H_{18}N_4O_7$ | 155–160 | — | *160* |
| | Methyl 2,6-dideoxy-3-*O*-methyl-α-L-*arabino*-hexopyranoside | $C_8H_{16}O_4$ | Syrup | −125.6 ($C_2H_5OH$) | *161* |
| | Methyl 2,6-dideoxy-3-*O*-methyl-β-L-*arabino*-hexopyranoside | $C_8H_{16}O_4$ | 74–78 | +71.5 ($C_2H_5OH$) | *161* |
| **Hexose, 2,3,6-trideoxy-D-*erythro*-** (*erythro*-amicitose) | | $C_6H_{12}O_3$ | oil, b.p. 65–70 | +28.6 ($CHCl_3$) | *162* |
| | 2,4-Dinitrophenylhydrazone | $C_{12}H_{16}N_4O_6$ | 137.5–138, 152–153 | −10 (*c* 0.9, $C_5H_5N$) | *162, 163* |
| | Methyl 2,3,6-trideoxy-D-*erythro*-hexopyranoside | $C_7H_{14}O_3$ | Syrup | +75.1 (*c* 0.9) | *162* |
| **Everinose** (an *O*-methyl-deoxyhexose) | | $C_7H_{14}O_5$ | 186–188 | −69 | *164* |
| | *p*-Tolylsulfonylhydrazone | $C_{14}H_{22}N_2O_6S$ | 135–137 | +36.2 ($C_5H_5N$) | *164* |
| **Evernitrose** (a 3-*C*-methyl-4-*O*-methyl-3-nitro-2,3,6-trideoxy-L-hexose) | | $C_8H_{15}NO_5$ | 88–93 | −4.9 → −19.4 ($C_2H_5OH$) | *256* |
| | Monoacetate | $C_{10}H_{17}NO_6$ | 58–59 | −20.5 ($C_2H_5OH$) | *256* |
| | Evernitronolactone | $C_8H_{13}NO_5$ | 63–64 | −70 | *256* |
| **Hexo-, *galacto*-, dialdose** | | $C_6H_{10}O_6$ | No constants known | — | *165* |
| | Tetraacetate | $C_{14}H_{18}O_{10}$ | 184 d. | — | *166* |

# TABLE 5 (Continued)
# NATURAL ALDOSES

| Substance[a] (synonym) | Derivative | Chemical formula | Melting point °C | Specific rotation[b] $[\alpha]_D$ | Reference |
|---|---|---|---|---|---|
| **Hexose, 3-*O*-acetyl-2-6-dideoxy-D-*lyxo*-** (chromose D) | | $C_8H_{14}O_5$ | 115–116.5 | +100 → +78 (29°) | *167, 168* |
| | Methyl 3-*O*-acetyl-2,6-dideoxy-α-D-*lyxo*-hexopyranoside | $C_9H_{16}O_5$ | Syrup | +142 (16°) ($CHCl_3$) | *167* |
| **Hexose, 2,6-dideoxy-D-*lyxo*-** (oliose) | | $C_6H_{12}O_4$ | Syrup | +46, +53 | *151, 168* |
| **Hexose, 2,6-dideoxy-D-lyxo-** (oliose) **(Continued)** | Methyl α-D-olioside | $C_7H_{14}O_4$ | 70–72 | +122(16°)($CHCl_3$) | *257* |
| | *p*-Amino-phenyl-α-D-olioside | $C_{12}H_{17}NO_4$ | 129–130 | +201 ($CH_3OH$) | *258* |
| **Hexose, 2,6-dideoxy-4-*O*-methyl-D-*lyxo*-** (chromose A, olivomose) | | $C_7H_{14}O_4$ | 158–162 | +98.5 → +89 (*c* 0.5) | *150, 169* |
| | 2,4-Dinitrophenylhydrazone | $C_{13}H_{18}N_4O_7$ | 146–147 | — | *170* |
| | Methyl 2,6-dideoxy-4-*O*-methyl-α-D-*lyxo*-hexopyranoside | $C_8H_{16}O_4$ | 98 | +150 (*c* 0.4, $C_2H_5OH$) (26°) | *150, 170* |
| | Methyl 2,6-dideoxy-4-*O*-methyl-*β*-D-*lyxo*-hexopyranoside | $C_8H_{16}O_4$ | 152–153 | −37.5 (*c* 0.4, $C_2H_5OH$) (26°) | *150, 170* |
| **Hexose, 2,6-dideoxy-3-*O*-methyl-D-*lyxo*-** (diginose) | | $C_7H_{14}O_4$ | 90–92 | +56 ± 4 | *171, 172* |
| | Diginonolactone | $C_7H_{12}O_4$ | Syrup | −30 (14°) (acetone) | *171* |
| **Hexose, 2,6-dideoxy-L-*lyxo*-** | | $C_6H_{12}O_4$ | 103–106 | −90.4 → −61.6 | *173, 174* |
| | 2,6-Dideoxy-L-*lyxo*-hexonic acid phenylhydrazide | $C_{12}H_{18}N_2O_4$ | 167–169 | −8.5 ± 2 | *259* |
| **Hexose, 2,6-dideoxy-3-*O*-methyl-L-*lyxo*-** (L-diginose) | | $C_7H_{14}O_4$ | 78–85 | −65 | *175* |
| **Hexose, 2,6-dideoxy-D-*ribo*-** (digitoxose) | | $C_6H_{12}O_4$ | 110 | +46.4 | *176* |
| | Phenylhydrazone | $C_{12}H_{18}N_2O_3$ | 204–209 | +215 ($C_2H_5OH$, $C_5H_5N$) | *177* |
| | Oxime | $C_6H_{13}NO_4$ | 102 | — | *178* |
| | Digitoxonolactone | $C_6H_{10}O_4$ | — | −29.5 ± 2 (*c* 0.6, acetone) | *259* |
| **Hexose, 3,6-dideoxy-D-*ribo*-** (paratose) | | $C_6H_{12}O_4$ | Syrup | +10 ± 2 (*c* 0.9) | *158, 179* |
| | 3,6-Dideoxy-D-*ribo*-hexitol | $C_6H_{14}O_4$ | 67–68 | −18 ± 2 | *179* |
| **Hexose, 2,6-dideoxy-3-*O*-methyl-D-*ribo*-** (cymarose) | | $C_7H_{14}O_4$ | 83–90, 93 | +55 | *180, 181* |
| | Methyl 2,6-dideoxy-3-*O*-methyl-α-D-*ribo*-hexopyranoside | $C_8H_{16}O_4$ | 34–36 | +210 (14°) ($CH_3OH$) | *182* |
| | Cymaronic acid phenylhydrazide | $C_{13}H_{20}N_2O_4$ | 155–156 | +1.4 (16°) (*c* 0.7, $CH_3OH$) | *183, 259* |

## TABLE 5 (Continued)
## NATURAL ALDOSES

| Substance[a] (synonym) | Derivative | Chemical formula | Melting point °C | Specific rotation[b] $[\alpha]_D$ | Reference |
|---|---|---|---|---|---|
| **Hexose, 4,6-dideoxy-3-*O*-methyl-D-*ribo*-** (chalcose, lancavose) | | $C_7H_{14}O_4$ | 96–99 | +120 → +76 | *184, 185* |
| | Methyl 4,6-dideoxy-3-*O*-methyl-D-*ribo*-hexopyranoside | $C_8H_{16}O_4$ | 101–102 | −21 (27°) ($CHCl_3$) | *184* |
| **Hexose, 2,6-dideoxy-3-*O*-methyl-L-*ribo*-** (L-cymarose) | | $C_7H_{14}O_4$ | 87–91 | −53.6 ± 2 | *180* |
| | L-Cymaronic acid phenylhydrazide | $C_{13}H_{20}N_2O_4$ | 153–154 | 0.3 ± 3 (*c* 0.7, $CH_3OH$) | *180* |
| **Hexose, 2,3,6-trideoxy-D-*threo*-** (*threo*-amicitose) | | $C_6H_{12}O_3$ | Syrup | −0.2 (*c* 0.9) | *186* |
| | | — | — | +10.2 (acetone) | *186* |
| | 2,4-Dinitrophenyl-hydrazone | $C_{12}H_{16}N_4O_6$ | 105–106, 121–122 | +13.7 (*c* 0.9, $C_5H_5N$) | *163, 186* |
| **Hexose, 2,3,6-trideoxy-D-threo-** (L-*threo*-amicitose, rhodinose) | | $C_6H_{12}O_3$ | Syrup | −11 ± 1.6 | *187* |
| | 2,4-Dinitrophenyl-hydrazone | $C_{12}H_{16}N_4O_6$ | 121–122 | −14.9 (*c* 0.5, $C_5H_5N$) | *186* |
| **Hexose, 2-deoxy-D-*xylo*-** | | $C_6H_{12}O_5$ | Syrup | +12 ± 2 | *188, 189* |
| | 2-Deoxy-D-*xylo*-hexonic acid lactone | $C_6H_{10}O_5$ | Syrup | −56.6 ± 2 (acetone) | *189* |
| | 2-Deoxy-D-*xylo*-hexonic acid phenylhydrazide | $C_{12}H_{18}N_2O_5$ | 124–126 | −8.1 ± 2 ($CH_3OH$) | *189* |
| **Hexose, 2,6-dideoxy-D-*xylo*-** (boivinose) | | $C_6H_{12}O_4$ | 96–98 | −3.9 → +3.9 ± 2 (18°) | *190* |
| | Methyl 2,6-dideoxy-α-D-*xylo*-hexopyranoside | $C_7H_{14}O_4$ | Syrup | +108.7 ± 2 ($CH_3OH$) | *190* |
| **Hexose, 3,6-dideoxy-D-*xylo*-** (abequose) | | $C_6H_{12}O_4$ | 138–139 | −3.2 ± 0.6 | *152, 153* |
| | Methyl 3,6-dideoxy-α-D-*xylo*-hexopyranoside | $C_7H_{14}O_4$ | Syrup | +102 ± 5 $[\alpha]_{5461}$ ($CH_3OH$) | *154* |
| | Methyl 3,6-dideoxy-β-D-*xylo*-hexopyranoside | $C_7H_{14}O_4$ | Syrup | −90 ± 3 $[\alpha]_{5461}$ ($CH_3OH$) | *154* |
| | 3,6-Dideoxy-D-*xylo*-hexitol | $C_6H_{14}O_4$ | 92–93 | +51 ± 2 | *153* |
| **Hexose, 3,6-dideoxy-3-*O*-methyl-D-*xylo*-** (sarmentose) | | $C_7H_{14}O_4$ | 78–79 | +12 → +15.8 | *191* |
| | Methyl 3,6-dideoxy-3-*O*-methyl-α-D-*xylo*-hexopyranoside | $C_8H_{16}O_4$ | 33–36 | +156 ± 1 (acetone) | *192* |
| | Methyl 3,6-dideoxy-3-*O*-methyl-β-D-*xylo*-hexopyranoside | $C_8H_{16}O_4$ | 40–45 | −39.4 ± 1.5 (acetone) | *192* |
| | Sarmentonic acid *S*-benzylthiuronium salt | $C_{15}H_{24}N_2O_5S$ | 148–149 | +7.3 ± 1 ($CH_3OH$) | *192* |
| **Hexose, 3,6-dideoxy-L-*xylo*-** (colitose) | | $C_6H_{12}O_4$ | Syrup | +4 | *193* |
| | | — | — | −51 ± 2 ($CH_3OH$) | *193* |
| | 3,6-Dideoxy-L-*xylo*-hexitol | $C_6H_{14}O_4$ | 92–94 | — | *193* |
| | *p*-Nitrophenyl-sulfonylhydrazone | $C_{18}H_{21}N_3O_7S$ | 141 | — | *193* |

## TABLE 5 (Continued)
## NATURAL ALDOSES

| Substance[a] (synonym) | Derivative | Chemical formula | Melting point °C | Specific rotation[b] $[\alpha]_D$ | Reference |
|---|---|---|---|---|---|
| **Hexose, 2,6-dideoxy-3-*C*-methyl-L-*xylo*-** (mycarose)[e] | | $C_7H_{14}O_4$ | 128–129 | −31.1 | *194* |
| | | — | 132–134, 140–142 | — | *195* |
| | Methyl 2,6-dideoxy-3-*C*-methyl-α-L-*xylo*-hexopyranoside | $C_8H_{16}O_4$ | Syrup | +54, +22 ($CHCl_3$) | *194, 195* |
| **Hexose, 2,6-dideoxy-3-*C*-methyl-L-xylo-** (mycarose)[e] **(Continued)** | Methyl 2,6-dideoxy-3-*C*-methyl-β-L-*xylo*-hexopyranoside | $C_8H_{16}O_4$ | 62 | −155 ($CHCl_3$) | *195* |
| | Mycaronolactone | $C_7H_{12}O_4$ | 108–109 | −35 | *194* |
| **Hexose, 2,6-dideoxy-3-*C*-methyl-L-*xylo*-, 4-isovalerate** | | $C_{12}H_{22}O_5$ | No constants known | — | *197* |
| **Hexose, 2,6-dideoxy-3-*C*-methyl-3-*O*-methyl-L-*xylo*-** (cladinose)[e] | | $C_8H_{16}O_4$ | oil, b.p. 120–132 (0.25 mm) | −23.1 | *198* |
| | 2,6-Dideoxy-3-*C*-methyl-3-*O*-methyl-L-*xylo*-hexitol | $C_8H_{18}O_4$ | Syrup | −25 (27°) ($C_2H_5OH$) | *199* |
| | Cladinonolactone 3,5-dinitrobenzoate | $C_{14}H_{16}N_2O_9$ | 123–125 | — | *200* |
| **Hexose, 2,6-dideoxy-3-*C*-methyl-3-*O*-methyl-L-*xylo*-** (arcanose) | | $C_8H_{16}O_4$ | 96–98 | −20.9 ($C_2H_5OH$) | *185, 201* |
| | 2,6-Dideoxy-3-*C*-methyl-3-*O*-methyl-L-*xylo*-hexitol | $C_8H_{18}O_4$ | Syrup | −2.0 ($C_2H_5OH$) | *185* |
| | 2,6-Dideoxy-3-*C*-methyl-3-*O*-methyl-L-*xylo*-hexose 4-acetate | $C_{10}H_{18}O_5$ | Syrup | −52 ($C_2H_5OH$) | *185* |
| **Nogalose (a deoxy 3-*C*-methyl-tri-*O*-methyl-hexose)** | | $C_{10}H_{20}O_5$ | 115–121 | +15.5 | *260* |
| | Nogalonolactone | $C_{10}H_{18}O_5$ | — | +6.7 ($CH_3OH$) | *260* |
| **Vinelose (a deoxy 3-*C*-methyl-2-*O*-methyl-hexose)** | | $C_8H_{16}O_5$ | — | +12[ ]$_{546}$ | *261* |
| | Vinelitol | $C_8H_{18}O_5$ | — | −6.4[ ]$_{546}$ | *261* |
| **D-Idose**[f] | | $C_6H_{12}O_6$ | Syrup | +15.8 ± 1 | *202, 243* |
| | D-Idopyranose pentaacetate | $C_{16}H_{22}O_{11}$ | 91–92 | +54.3 ± 2 ($CHCl_3$) | *243* |
| | Methyl α-D-idopyranoside | $C_7H_{14}O_6$ | 67–68 | +99.8 ± 1 | *243* |
| **L-Idose, 1,6-anhydro-** | | $C_6H_{10}O_5$ | 128–129 | +113 (acetone) | *242* |
| | 1,6-anhydro-2,3,4-tri-*O*-methyl-β-L-idopyranose | $C_9H_{16}O_5$ | 39–40 | +88 ($CHCl_3$) | *203* |
| **α-D-Mannose** | | $C_6H_{12}O_6$ | 133 | +29.3 → +14.5 | *204* |
| **β-D-Mannose** | | $C_6H_{12}O_6$ | 132 | −16.3 → +14.5 | *205* |
| | *p*-Nitrophenylhydrazone | $C_{12}H_{17}N_3O_7$ | 202–203 | +56 ($C_2H_5OH$, $C_5H_5N$) | *38, 39, 114, 206* |
| | Penta-*O*-acetyl-α-D-mannopyranoside | $C_{16}H_{22}O_{11}$ | 64 | +55 ($CHCl_3$) | *108* |
| | Penta-*O*-acetyl-β-D-mannopyranoside | $C_{16}H_{22}O_{11}$ | 117–118 | −25.3 ($CHCl_3$) | *108, 207* |
| **D-Mannose, 3-*O*-methyl-** | | $C_7H_{14}O_6$ | 133–134 | +14 → +3 (*c* 0.6) | *208, 209* |
| **D-Mannose, 6-deoxy-** (D-rhamnose) | | $C_6H_{12}O_5$ | 86–90 | −7.0 | *211* |

## TABLE 5 (Continued)
## NATURAL ALDOSES

| Substance[a] (synonym) | Derivative | Chemical formula | Melting point °C | Specific rotation[b] $[\alpha]_D$ | Reference |
|---|---|---|---|---|---|
| **D-Mannose, 6-deoxy-2-*O*-methyl-** | | $C_7H_{14}O_5$ | See L-isomer | −22 | *262* |
| **D-Mannose, 6-deoxy-3-*O*-methyl-** | | $C_7H_{14}O_5$ | See L-isomer | −27 | *262* |
| **D-Mannose, 6-deoxy-2,3-di-*O*-methyl-** | | $C_8H_{16}O_5$ | — | [g] | *263* |
| **D-Mannose, 6-deoxy-3,4-di-*O*-methyl-** | | $C_8H_{16}O_5$ | 86–88 | [g] | *263* |
| **α-L-Mannose, 6-deoxy-** (L-rhamnose) | | $C_6H_{12}O_5{\cdot}H_2O$ | 93–94 | −8.6 → +8.2 | *34, 212* |
| **β-L-Mannose, 6-deoxy-** | | $C_6H_{12}O_5$ | 123–125 | +38.4 → +8.9 | *213* |
| | *p*-Nitrophenylhydrazone | $C_{12}H_{17}N_3O_6$ | 190–191 | −50 → −8.5 ($C_5H_5N$, $C_2H_5OH$) | *38* |
| | Tetra-*O*-acetyl-6-deoxy-*β*-L-rhamnopyranoside | $C_{14}H_{20}O_9$ | 98–99 | +13.9 (*c* 15) ($C_2H_2Cl_4$) | *214* |
| **L-Mannose, 6-deoxy-2-*O*-methyl-** | | $C_7H_{14}O_5$ | 113–114 | +31 (27°) | *80, 215, 217* |
| | 6-Deoxy-2-*O*-methyl-*N*-phenyl-L-mannopyranosylamine | $C_{13}H_{19}NO_4$ | 152 | +43 ($C_5H_5N$) | *216* |
| | 6-Deoxy-2-*O*-methyl-L-mannonolactone | $C_7H_{12}O_5$ | 116–117 | −62 | *216* |
| **L-Mannose, 6-deoxy-3-*O*-methyl-** (acofriose) | | $C_7H_{14}O_5$ | 112–116 | +37.3 ± 2 | *67* |
| | 6-Deoxy-3-*O*-methyl-L-mannonolactone | $C_7H_{12}O_5$ | Syrup | −20 (15°) | *218* |
| | Phenylosazone | $C_{19}H_{26}N_4O_3$ | 128–130 | +57 ($C_5H_5N$, $C_2H_5OH$) | *219* |
| **L-Mannose, 6-deoxy-2,3-di-*O*-methyl** | | $C_8H_{16}O_5$ | — | +47.6 | *219, 264* |
| | 6-Deoxy-2,3-di-*O*-methyl-*N*-phenyl-L-mannopyranosylamine | $C_{14}H_{21}NO_4$ | 136–137 | +147.8 → +42.8 (70 hr) (*c* 0.4, $C_2H_5OH$) | *265, 268* |
| **L-Mannose, 6-deoxy-2,4-di-*O*-methyl-** | | $C_8H_{16}O_5$ | 82 | −19 (16°) | *215, 220* |
| | 6-Deoxy-2,4-di-*O*-methyl-*N*-phenyl-L-mannopyranosylamine | $C_{14}H_{21}NO_4$ | 141–142 | +110 → +7 (*c* 0.4, $C_2H_5OH$) | *220* |
| | 6-Deoxy-2,4-di-*O*-methyl-L-mannonolactone | $C_8H_{14}O_5$ | Syrup | +47 (15°) (*c* 0.9) | *220* |
| **L-Mannose, 6-deoxy-3,4-di-*O*-methyl** | | $C_8H_{16}O_5$ | 98–99 | +18.5 (*c* 0.5) | *266, 267* |
| | 2,4-Dinitrophenylhydrazone | $C_{14}H_{20}N_4O_8$ | 170 | −75.6 (dioxane) | *268* |
| **L-Mannose, 6-deoxy-5-*C*-methyl-4-*O*-methyl-** (noviose) | | $C_8H_{16}O_5$ | 133–134 | +22.6 (50% $C_2H_5OH$) | *221* |
| | Novionolactone | $C_8H_{14}O_5$ | 111–113 | −35 (0.1 N HCl) | *222* |
| | Methyl 6-deoxy-5-*C*-methyl-4-*O*-methyl-α-L-mannopyranoside | $C_9H_{18}O_5$ | 68–70 | −62 ± 2 ($C_2H_5OH$) | *223* |
| | Methyl 6-deoxy-5-*C*-methyl-4-*O*-methyl-*β*-L-mannopyranoside | $C_9H_{18}O_5$ | 61–68 | +113.8 | *224* |

## TABLE 5 (Continued)
## NATURAL ALDOSES

| Substance[a] (synonym) | Derivative | Chemical formula | Melting point °C | Specific rotation[b] $[\alpha]_D$ | Reference |
|---|---|---|---|---|---|
| **L-Mannose, 6-deoxy-5-*C*-methyl-4-*O*-methyl-** (noviose)—**(Continued)** | | | | | |
| | 3-*O*-carbamoyl-6-deoxy-5-*C*-methyl-4-*O*-methyl-L-mannose | $C_9H_{17}NO_6$ | 124–126 | +45.3 ($C_2H_5OH$) | *224* |
| **D-Talose** | | $C_6H_{12}O_6$ | 128–132 | +16.9 | *225* |
| | Methylphenyl-hydrazone | $C_{13}H_{20}N_2O_5$ | 143–144 | — | *225* |
| | Penta-*O*-acetyl-α-D-talopyranoside | $C_{16}H_{22}O_{11}$ | 106–107 | +70.2 ($CHCl_3$) | *226* |
| **D-Talose, 6-deoxy-** (D-talomethylose) | | $C_6H_{12}O_5$ | 129–131 | +20.6 | *211* |
| | Methyl α-D-taloside triacetate | $C_{13}H_{20}O_8$ | 91–91.5 | +76 ($CH_3OH$) | *269* |
| **L-Talose, 6-deoxy-** (L-talomethylose) | | $C_6H_{12}O_5$ | 126–127 | −20.5 ± 1.4 | *227, 228* |
| | Methyl 6-deoxy-α-L-talopyranoside | $C_7H_{14}O_5$ | 63–64 | −104 | *228* |
| | *p*-Bromophenyl-hydrazone | $BrC_{12}H_{17}N_2O_4$ | 145–147 | −10 → +4 (16°) (*c* 0.8, $C_2H_5OH$) | *229* |
| **L-Talose, 6-deoxy-3-*O*-methyl-** (L-acovenose) | | $C_7H_{14}O_5$ | Syrup | −19.4 | *230, 263* |
| | 6-Deoxy-2-*O*-methyl-L-talonolactone | $C_7H_{12}O_5$ | 167–168 | +29.4 ± 2 (16°) ($CH_3OH$) | *230, 270* |
| **Heptose, D-*glycero*-D-*galacto*-** | | $C_7H_{14}O_7$ | 139–140 | +47 → +64 (*c* 0.5) | *231* |
| | 2,5-Dichlorophenyl-hydrazone | $C_{13}Cl_2H_{18}N_2O_6$ | 203–204 | — | *231* |
| | β-Hexaacetate | $C_{19}H_{26}O_{13}$ | 109–110 | +30.4 ($CHCl_3$) | *271* |
| **Heptose, D-*glycero*-D-*gluco*-** | | $C_7H_{14}O_7$ | 156–157 | +17 → +46.2 | *244* |
| | α-Hexaacetate | $C_{19}H_{26}O_{13}$ | 180–182 | +105 ($CCl_2H_2$) | *244* |
| | β-Hexaacetate | $C_{19}H_{26}O_{13}$ | 133–134 | +19.6 ($CHCl_3$) | *244* |
| **Heptose, D-*glycero*-D-*manno*-** | | $C_7H_{14}O_7$ | — | +21 ($CH_3OH$) | *232–234* |
| | *p*-Nitrophenyl-hydrazone | $C_{13}H_{19}N_3O_8$ | 176–177 | — | *233* |
| | Hexa-*O*-acetyl-D-*glycero*-D-*manno*-heptose | $C_{19}H_{26}O_{13}$ | 139–140 | +65 ($CHCl_3$) | *233* |
| | D-*glycero*-D-*manno*-Heptonolactone | $C_7H_{12}O_7$ | 164–165 | +48 (*c* 0.2) | *235* |
| **Heptose, L-*glycero*-D-*manno*-** | | $C_7H_{14}O_7$ | — | +14 (*c* 0.9) | *234, 236* |
| | Hexabenzoate | $C_{49}H_{38}O_{13}$ | 100 | −32 ($CHCl_3$) | *234* |
| | L-*glycero*-D-*manno*-Heptose diethyl dithioacetal | $C_{11}H_{24}O_6S_2$ | 201–202 | +9.9 ($C_5H_5N$) | *237* |
| **Heptose,** *unidentified*— | | — | No constants known | — | *238–240* |

[a] In alphabetical order of parent sugar names within groups of increasing carbon chain length in the parent compounds.
[b] Unless otherwise specified, $[\alpha]_D$ is given for 1 to 5 g solute, *c* representing the amount of solvent per 100 ml aqueous solution, at 20 to 25°C.
[c] Not anomeric with the aldgaroside A.
[d] By inference from the rotation of the D-isomer.
[e] Some authors claim that the configuration is L-*ribo*- (see Reference 196).
[f] Though never proved, some evidence exists for this sugar, or L-altrose, in the polysaccharide varianose.
[g] No rotations are given for these isolates by the authors; therefore, see the constants of the more commonly known L-isomer.

Data taken from: Maher, G. G., and Wolfrom, M. L., in *Handbook of Biochemistry,* 2nd ed., pp. D-16–D-29, H. A. Sober, Ed. 

## TABLE 6
## NATURAL KETOSES

| Substance[a] (synonym) | Derivative | Chemical formula | Melting point °C | Specific rotation[b] $[\alpha]_D$ | Reference |
|---|---|---|---|---|---|
| **Dihydroxyacetone** | | $C_3H_6O_3$ | 80 (dimer) | — | *1* |
| | *p*-Nitrophenylhydrazone | $C_9H_{11}N_3O_4$ | 160 | — | *1* |
| | Diacetate | $C_7H_{10}O_6$ | 46–47 | — | *1* |
| **Tetrulose, L-*glycero*-** (L-erythrulose, keto-erythritol, L-threulose) | | $C_4H_8O_4$ | Syrup | +12 | *2, 3* |
| | *o*-Nitrophenylhydrazone | $C_{10}H_{13}N_3O_5$ | 152–153 | +48 (18°) ($C_2H_5OH$) | *4* |
| **Pentulose, D-*erythro*-** (adonose, D-ribulose) | | $C_5H_{10}O_5$ | Syrup | −15 | *5, 6* |
| | *o*-Nitrophenylhydrazone | $C_{11}H_{15}N_3O_6$ | 165–166.5 | −52 ± 5 ($CH_3OH$) | *7* |
| **Pentulose, L-*erythro*-** (L-ribulose) | | $C_5H_{10}O_5$ | Syrup | +16.6 | *7, 8* |
| | *o*-Nitrophenylhydrazone | $C_{11}H_{15}N_3O_6$ | 162–163 | +47.4 (*c* 0.3, $CH_3OH$) | *9* |
| **Pentulose, *erythro*-3-** | | $C_5H_{10}O_5$ | No constants known | — | *10* |
| **Pentulose, D-*threo*-** (D-xylulose) | | $C_5H_{10}O_5$ | Syrup | −33 | *5* |
| | *p*-Bromophenylhydrazone | $BrC_{11}H_{15}N_2O_4$ | 126–128 | +24.1 → −31 ($C_5H_5N$, 7 days) | *5, 11* |
| **Pentulose, 5-deoxy-D-*threo*-** | | $C_5H_{10}O_4$ | Syrup | −5 ± 1 ($CH_3OH$) | *12* |
| **Pentulose, L-*threo*-** (L-xylulose, L-lyxulose, xyloketose) | | $C_5H_{10}O_5$ | Syrup | +33.1 | *6, 13, 14* |
| | *p*-Bromophenylhydrazone | $BrC_{11}H_{15}N_2O_4$ | 128–129 | −26 → +31.9 ($C_5H_5N$) | *13* |
| **Hexodiulose, 6-deoxy-D-*erythro*-2,5-** | | $C_6H_{10}O_5$ | No constants known | — | *15* |
| **Hexodiulose, D-*threo*-2,5-** | | $C_6H_{10}O_6$ | 157–159, 172–174 | −85 | *16* |
| | *Bis*(phenylhydrazone) | $C_{18}H_{22}N_4O_4$ | 133–135 | −164 ($C_5H_5N$) | *16* |
| **Hexos-, 4,6-dideoxy-, 2,3-diulose** (actinospectose) | | $C_6H_8O_4$ | No constants known | — | *17* |
| **Hexosulose, D-*arabino*-** (D-glucosone) | | $C_6H_{10}O_6$ | Syrup | −10.6 → +7.9 (15°) (*c* 8.5) | *18, 19* |
| | Tetraacetate·$H_2O$ | $C_{14}H_{18}O_{10}\cdot H_2O$ | 112 | +14.7 → +53.7 (20% $C_2H_5OH$) | *20* |
| **Hexosulose, 3-deoxy-D-*erythro*-** (3-deoxy-2-keto-D-glucose) | | $C_6H_{10}O_5$ | No constants known | — | *21* |
| | 2,4-Dinitrophenylosazone | $C_{18}H_{18}N_8O_{11}$ | 251d | +860 (*c* 0.09, DMSO) | *21, 64* |
| **Hexos-5-ulose, 6-deoxy-D-*arabino*-** | | $C_6H_{10}O_5$ | — | −4.3 (12°) $CH_3OH$ | *22, 61* |
| | *Bis*(*p*-nitrophenylhydrazone) | $C_{18}H_{20}N_6O_8$ | 211d | +1.1 (15°) (*c* 0.6, $C_5H_5N$) | *61* |
| **Hexos-5-ulose, D-*lyxo*-** | | $C_6H_{10}O_6$ | 157–158 | −86.6 | *23* |
| | *Bis*(*p*-nitrophenylhydrazone) | $C_{18}H_{20}N_6O_9$ | 173–174 | — | *23* |
| **Hexos-2-ulose, 6-deoxy-L-*lyxo*-** (angustose) | | $C_6H_{10}O_5$ | 115–116 | +18 ($C_2H_5OH$) | *37* |
| | Methyl angustoside dimethyl acetal | $C_9H_{18}O_6$ | — | −19.3 ($CH_3OH$) | *37* |
| **Hexos-3-ulose, D-*ribo*-** (3-keto-D-glucose) | | $C_6H_{10}O_6\cdot 5H_2O$ | 58–60 | +14.8 (26°) | *24* |
| **Hexulose, β-D-*arabino*-** (β-D-fructose, levulose) | | $C_6H_{12}O_6$[c] | 102–104 | −133.5 → −92 | *25, 26* |
| | Tetra-*O*-benzoyl-β-D-fructopyranose | $C_{34}H_{28}O_{10}$ | 174–175 | −165 ($CHCl_3$) | *27* |
| | Tetra-*O*-acetyl-β-D-fructopyranose | $C_{14}H_{20}O_{10}$ | 131–132 | −91.6 ($CHCl_3$) | *28, 28*a |

## TABLE 6 (Continued)
## NATURAL KETOSES

| Substance[a] (synonym) | Derivative | Chemical formula | Melting point °C | Specific rotation[b] $[\alpha]_D$ | Reference |
|---|---|---|---|---|---|
| | *p*-Nitrophenylhydrazone | $C_{12}H_{17}N_3O_7$ | 176 | — | *29* |
| **Hexulose, 6-deoxy-D-*arabino*-** (D-rhamnulose) | | $C_6H_{12}O_5$ | Syrup | $-6 \pm 1$, $-13 \pm 2$ | *30, 31* |
| | *o*-Nitrophenylhydrazone | $C_{12}H_{17}N_3O_6$ | 136–137 | $+40 \pm 3$ ($C_2H_5OH$) | *30* |
| **Hexulose, D-*lyxo*-** (D-tagatose) | | $C_6H_{12}O_6$ | 131–132 | $+2.7 \rightarrow -4$ | *32* |
| | Penta-*O*-acetyl-D-tagatopyranose | $C_{16}H_{22}O_{11}$ | 132 | +30.2 ($CHCl_3$) $[\alpha]_{578}$ | *33, 34* |
| | D-*arabino*-5-Hexulosonic acid (tagaturonic acid) | $C_6H_{10}O_7$ | 106–108 | −12.5 | *32* |
| **Hexulose, 6-deoxy-L-*lyxo*-** (L-fuculose) | | $C_6H_{12}O_5$ | 68–69 | $+3.4 \pm 1$ | *35* |
| | *o*-Nitrophenylhydrazone | $C_{12}H_{17}N_3O_6$ | 162–163 | — | *36* |
| **Hexulose, D-*ribo*-** (D-psicose, D-allulose) | | $C_6H_{12}O_6$ | Amorph. | +3.2, +4.7 | *38, 39* |
| | *keto*-Pentaacetate | $C_{16}H_{22}O_{11}$ | 63–65 | −21.5 (29°) ($CHCl_3$) | *40* |
| | Phenylosazone | $C_{18}H_{24}N_4O_4$ | 159–163 | $-74 \rightarrow -68$ ($C_5H_5N$) | *41* |
| **Hexulose, L-*xylo*-** (L-sorbose) | | $C_6H_{12}O_6$ | 159–161 | −43.1 | *42* |
| | Penta-*O*-acetyl-α-L-sorbopyranose | $C_{16}H_{22}O_{11}$ | 97 | −56.5 ($CHCl_3$) | *43* |
| | Penta-*O*-acetyl-β-L-sorbopyranose | $C_{16}H_{22}O_{11}$ | 113.8 | +74.4 ($CHCl_3$) | *43* |
| **Hexulose, 6-deoxy-L-*xylo*-** | | $C_6H_{12}O_5$ | 88 | $-25 \pm 2$ (*c* 0.7) | *31* |
| | Phenylosazone | $C_{18}H_{24}N_4O_3$ | 184–185 | — | *44* |
| **Maltol, 5-hydroxy-** | | $C_6H_6O_4$ | 184–184.5 | — | *45* |
| | Dimethyl ether | $C_8H_{10}O_4$ | 98 | — | *45* |
| **Heptulose, D-*allo*-** | | $C_7H_{14}O_7$ | 128–130 | — | *62* |
| | *Bis*(phenylhydrazone) | $C_{19}H_{26}N_4O_5$ | 164–167 | — | *62* |
| **Heptulose, D-*altro*-** (sedoheptulose, sedoheptose) | | $C_7H_{14}O_7$ | Amorph. | +2.5 (*c* 10) | *46, 62* |
| | 2,7-Anhydro-β-D-*altro*-heptulopyranose (sedoheptulosan) | $C_7H_{12}O_6$ | 155–156 | −145 | *62* |
| | monohydrate | $C_7H_{12}O_6 \cdot H_2O$ | 101–102 (91s) | −134 | *47* |
| **Heptulose, D-*altro*-3-** (coriose) | | $C_7H_{14}O_7$ | 169–171 | +20 | *62, 65* |
| | Pentabenzoate | $C_{42}H_{34}O_{12}$ | 100–102 | — | *65* |
| **Heptulose, L-*galacto*-** (perseulose) | Hemihydrate | $C_7H_{14}O_7 \cdot 1/2H_2O$ | 110–115 | −90, −80 | *48, 49* |
| | Hexaacetate | $C_{19}H_{26}O_{13}$ | 112 | −113 ($CHCl_3$) $[\alpha]_{578}$ | *50* |
| | Phenylosazone | $C_{19}H_{26}N_4O_5$ | 198–199 | −90, −45 | *51* |
| **Heptulose, L-*gulo*-** | | $C_7H_{14}O_7$ | — | −28 | *52* |
| **Heptulose, D-*ido*-** | | $C_7H_{14}O_7$ | — | −20 | *47* |
| | Idoheptulosan | $C_7H_{12}O_6$ | 172 | $-34 \pm 8$ (*c* 0.3) | *53* |
| **Heptulose, D-*manno*-** | | $C_7H_{14}O_7$ | 152 | +29.4 | *54, 55* |
| | *p*-Bromophenylhydrazone | $BrC_{13}H_{19}N_2O_6$ | 179 | — | *54* |
| | Hexaacetate | $C_{19}H_{26}O_{13}$ | 110 | +39 ($CHCl_3$) | *56* |
| **Heptulose, D-*talo*-** | | $C_7H_{14}O_7$ | No constants known | — | *57* |
| **Octulose, D-*glycero*-L-*galacto*-** | | $C_8H_{16}O_8$ | — | $-57$, $-43.4 \rightarrow -13.4$ | *58, 59* |
| | 2,5-Dichlorophenylhydrazone | $Cl_2C_{14}H_{20}N_2O_7$ | 178–180 | — | *58* |

## TABLE 6 (Continued)
## NATURAL KETOSES

| Substance[a] (synonym) | Derivative | Chemical formula | Melting point °C | Specific rotation[b] $[\alpha]_D$ | Reference |
|---|---|---|---|---|---|
| **Octulose, D-*glycero*-D-*manno*-** | | $C_8H_{16}O_8$ | — | +20 ($CH_3OH$) | *57* |
| | 2,5-Dichlorophenyl-hydrazone | $Cl_2C_{14}H_{20}N_2O_7$ | 169–170 | — | *57* |
| **Nonulose, D-*erythro*-L-*gluco*-** | | $C_9H_{18}O_9$ | — | −40 (*c* 0.6) | *60* |
| | 2,5-Dichlorophenyl-osazone | $Cl_4C_{21}H_{24}N_4O_7$ | 248–250d | — | *60* |
| **Nonulose, D-*erythro*-L-*galacto*-** | | $C_9H_{18}O_9$ | — | −9.7 ($H_2O$)<br>−36.2 ($CH_3OH$) | *63* |
| | 2,5-Dichlorophenyl-osazone | $Cl_4C_{21}H_{24}N_4O_7$ | 238–240 | — | *63* |

[a] In alphabetical order by parent sugar names within groups of increasing carbon chain length in the parent compounds.
[b] Unless otherwise specified, $[\alpha]_D$ is given for 1 to 5 g solute, *c* representing the amount of solvent per 100 ml aqueous solution, at 20 to 25°C.
[c] The ½$H_2O$ and $2H_2O$ forms also exist.

Data taken from: Maher, G. G., and Wolfrom, M. L., in *Handbook of Biochemistry,* 2nd ed., pp. D-33–D-35, H. A. Sober, Ed. Copyright 1970, The Chemical Rubber Co., Cleveland, Ohio.

## TABLE 7
## THE NATURALLY OCCURRING AMINO SUGARS

| Compound and formula | Source | Physical constants[a] | | $R_{GlcN}$ values on paper solvent systems[b] | | | | | $R_{GlcN}$ on ion exchange system[c] | | References |
|---|---|---|---|---|---|---|---|---|---|---|---|
| | | MP | $[\alpha]_D$ | A | B | C | D | E | I | II | |
| **I. 2-AMINO SUGARS** | | | | | | | | | | | |
| **Glucosamine (chitosamine): 2-amino-2-deoxy-D-glucose**<br>$CH_2OH$, O, H,OH, OH, HO, $NH_2$ | Polysaccharides of bacteria, fungi, invertebrates, chitin, antibiotics, higher plants, vertebrates, UDP-complexes | 88<br>190–210<br>110–111 | (α) FB<br>+100 → +47.5<br>(α) HCl<br>+100 → +72<br>(β) FB<br>+28 → +47.5<br>(β) HCl<br>+25 — +72.6 | 1.00 | 1.00 | 1.00 | 1.00 | 1.00 | 1.00 | 1.00 | *1–4* |
| **Galactosamine (chondrosamine): 2-amino-2-deoxy-D-galactose**<br>$CH_2OH$, HO, O, OH, H,OH, $NH_2$ | Polysaccharides of bacteria, fungi, invertebrates, antibiotics, vertebrates, UDP-complexes | 185 | (α) HCl : W–HCl<br>+121 → +80<br>(β) HCl : W–HCl<br>+44 → +80 | 0.90<br>$0.80^{\Delta}$ | 0.90 | 1.04 | 1.05 | 0.94 | 1.17<br>1.20 | 1.03 | *1–4* |
| **Mannosamine: 2-amino-2-deoxymannose**<br>$CH_2OH$, O, OH, $NH_2$, H,OH, HO | Pneumococcus type XIX polysaccharide *E. coli*, *Salmonella*, animal metabolite (*N*-acetyl-D-isomer) | 178–180 | HCl −3 | 1.05<br>$1.13^{\Delta}$<br>$1.17^{\Delta}$ | — | 1.19 | 1.05 | — | 1.06 | 1.12 | *1–4, 6[e], 7[e], 8* |
| **D-Gulosamine: 2-amino-2-deoxy-D-gulose**<br>$CH_2OH$, HO, O, H,OH, OH, $NH_2$ | Antibiotics: streptolin B, streptothricin | 150–170 | (α) HCl<br>+6.1 → −17.9 | 1.00<br>1.01 | — | — | 1.05 | — | 1.21 | 1.20 | *1–4* |
| **2-Amino heptose: D-glycero-2-desoxy-2-amino-gulo-(or ido-) heptose**[j] | *Anacystis nidulans* cell wall | — | — | — | — | — | — | — | 2.19 | — | *46* |

## TABLE 7 (Continued)
## THE NATURALLY OCCURRING AMINO SUGARS

| Compound and formula | Source | Physical constants[a] | | $R_{GlcN}$ values on paper solvent systems[b] | | | | | $R_{GlcN}$ on ion exchange system[c] | | References |
|---|---|---|---|---|---|---|---|---|---|---|---|
| | | MP | $[\alpha]_D$ | A | B | C | D | E | I | II | |
| **I. 2-Amino Sugars—(Continued)** | | | | | | | | | | | |
| **Quinovosamine: 2-amino-2,6-dideoxy-D-glucose** 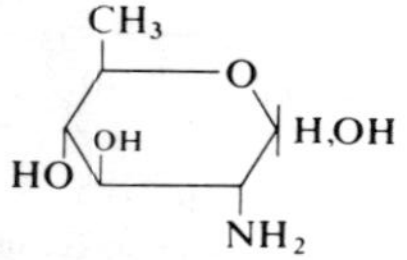 | *Achromobacter georgio politanum*<br>*Salmonella, Proteus vulgaris*<br>*Arizona, Neurospora crassa* | 165–170 | HCl +55.5 | 1.85 | 2.23 | 2.5 | — | 1.40 | 1.43 | 1.26 | *3, 4, 9[d], 10, 45, 47* |
| **D-Fucosamine: 2-amino-2,6-dideoxy-D-galactose** 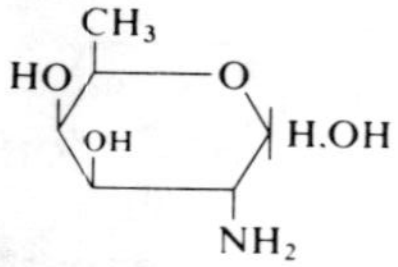 | *C. violaceum*<br>*B. licheniformis*<br>*B. cereus*<br>*Erysipelothrix insidiosa*<br>*Pseudomonas aeruginosa* | 155 chars<br>170–175<br>decomp. | HCl<br>+91 ± 2 | 1.32 | 1.94 | 2.4 | 1.31 | 1.20 | 1.73<br>1.75<br>1.95 | 1.24 | *1, 2, 4, 10, 11, 12 13[d], 14[d], 48[d], 49[d], 53[d]* |
| **L-Fucosamine: 2-amino-2,6-dideoxy-L-galactose** 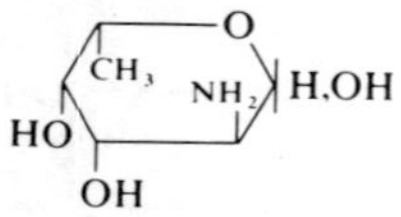 | Polysaccharide of pneumococcus type V<br>*Citrobacter freundii* 05: H30 mucopolysaccharide | 155 chars | HCl<br>−93.4 ± 2 | — | — | — | — | — | — | — | *2, 15[d], 16[d]* |
| **Pneumosamine: 2-amino-2,6-dideoxy-L-talose** 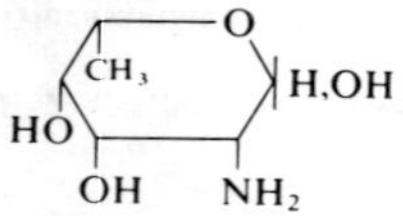 | Pneumococcus type V | 162–163 | HCl<br>+6.9 → +10.4 | — | — | — | — | — | — | 1.35 | *2, 4, 16[d]* |
| **L-Rhamnosamine: 2-amino-2,6-dideoxy-L-rhamnose** 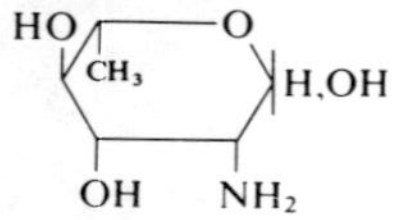 | *E. coli* 03:K2ab(L):H2 | — | HCl + 22.5 | — | — | — | — | — | — | 1.48 | *50[d]* |

# TABLE 7 (Continued)
## THE NATURALLY OCCURRING AMINO SUGARS

| Compound and formula | Source | Physical constants[a] | | $R_{GlcN}$ values on paper solvent systems[b] | | | | | $R_{GlcN}$ on ion exchange system[c] | | References |
|---|---|---|---|---|---|---|---|---|---|---|---|
| | | MP | $[\alpha]_D$ | A | B | C | D | E | I | II | |
| **II. 3-AMINO SUGARS** | | | | | | | | | | | |
| **Kanosamine: 3-amino-3-deoxy-D-glucose** 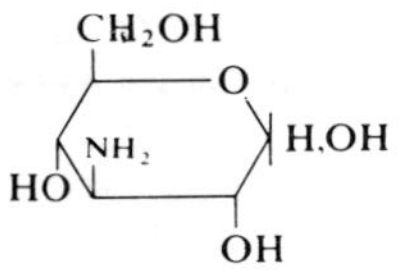 | Antibiotics: kanamycin group | — | — | — | — | — | — | — | — | — | *2*[e], *17*[d], *18*, *21* |
| **3-Amino-3,6-dideoxy-D-glucose** 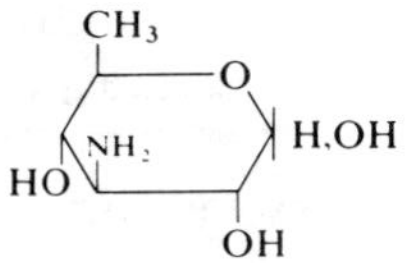 | Lipopolysaccharides, *Citrobacter freundii*, *Salmonella*, *E. coli* | Syrup | — | 1.6 | 2.0<br>2.2 | — | — | 1.36 | 1.34<br>1.37 | 129 | *10*[e], *19*[d], 20[e], 21[e] |
| **3-Amino-3,6-dideoxy-D-galactose** 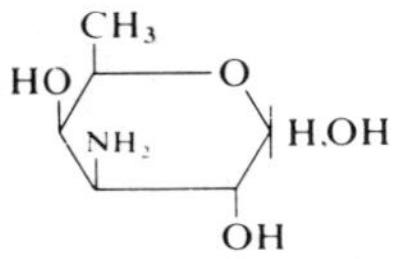 | Lipopolysaccharides, *Xanthomonas campestris*, *Salmonella*, *E. coli Arizona* | — | — | 1.2 | 1.6<br>1.8 | — | — | — | 1.54 | 120 | 20[e], *21*[e], *22*[d] |
| **Mycosamine: 3-amino-3,6-dideoxy-D-mannose** 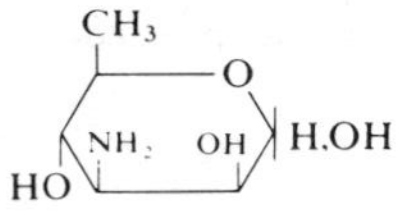 | Antibiotics: amphotericin B, nystatin pimaricin | — | — | 1.4<br>1.8 | 1.8<br>2.2 | — | — | — | 1.28 | 1.26 | *2*, *17*[d], *20*, *21* |

## TABLE 7 (Continued)
## THE NATURALLY OCCURRING AMINO SUGARS

| Compound and formula | Source | Physical constants[a] MP | Physical constants[a] $[\alpha]_D$ | $R_{GlcN}$ values on paper solvent systems[b] A | B | C | D | E | $R_{GlcN}$ on ion exchange system[c] I | II | References |
|---|---|---|---|---|---|---|---|---|---|---|---|
| **II. 3-Amino Sugars—(Continued)** | | | | | | | | | | | |
| **3-Aminoribose: 3-amino-3-deoxy-D-ribose** 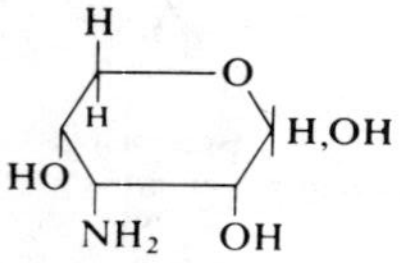 | 3′-amino-3′-deoxy-adenosine from *Helminthosporium* sp. and *Cordyceps militaris*, antibiotic: puromycin | 154–155<br>157–158<br>161 | HCl<br>$-37 \rightarrow -24.0$<br>Ac $-25$ | — | — | — | — | — | — | — | *2, 17, 23, 24* |
| **III. 4-AMINO SUGARS** | | | | | | | | | | | |
| **Viosamine: 4-amino-4,6-dideoxy-D-glucose** 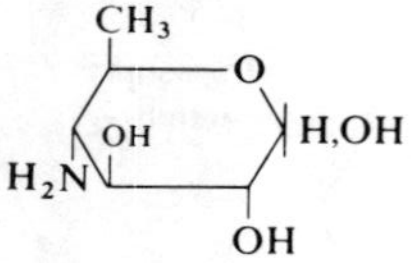 | Lipopolysaccharide *Chromobacterium violaceum*, *E. coli*, TDP-nucleotides in *Echerichia* and *Salmonella* | 132–138 | HCl<br>$-9 \rightarrow +21$ | 1.47 | 1.45 | 1.3 | — | — | 1.43 | — | *2, 3, 11*[d], *12, 25*[d], *51* |
| | | | | $R_{rhamnose}$ | | | | | | | |
| | | | | J | K | L | M | N | | | |
| | | | | 1.38 | 0.42 | 0.94 | 0.5<br>0.7 | 1.0<br>1.5 | | | *26*[d] |
| **4-Amino-4,6-dideoxy-D-galactose** 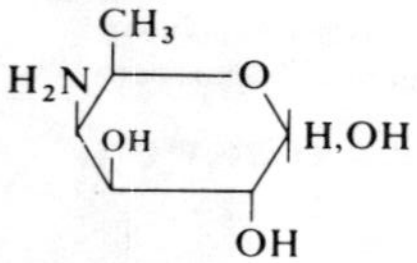 | TDP-nucleotide in *Escherichia*, *Salmonella* and *Pasteurella*, Lipopolysaccharide of *E. coli*. | — | — | 1.25 | 0.88 | 1.16 | 0.5<br>0.7 | 1.0<br>1.5 | — | — | *26*[d], *51* |
| **IV. 6-AMINO SUGARS** | | | | | | | | | | | |
| **6-Amino-6-deoxy-D-glucose** 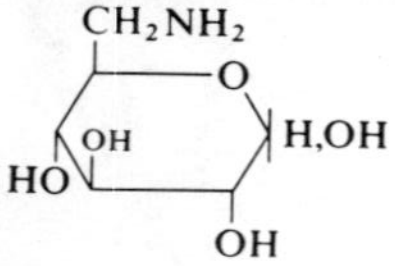 | Antibiotics: kanamycin A | 161–162 | HCl<br>$+23.0 \rightarrow +50.1$ | — | — | — | — | — | — | — | *2, 17*[d], *18*[d] |

## TABLE 7 (Continued)
## THE NATURALLY OCCURRING AMINO SUGARS

| Compound and formula | Source | Physical constants[a] | | $R_{GlcN}$ values on paper solvent systems[b] | | | | | $R_{GlcN}$ on ion exchange system[c] | | References |
|---|---|---|---|---|---|---|---|---|---|---|---|
| | | MP | $[\alpha]_D$ | A | B | C | D | E | I | II | |
| **V. DIAMINO-SUGARS** | | | | | | | | | | | |
| **Neosamine B: 2,6-diamino-2,6-dideoxy-L-idose** | Antibiotics: neomycin, paronomycin, zygomycin | — | HCl +17 | — | — | — | — | — | — | — | *2, 27, 28* |
| 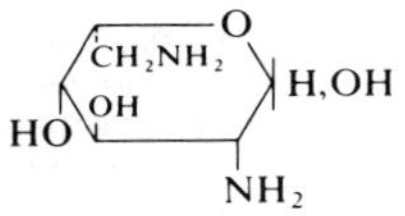 **Neosamine C: 2,6-diamino-2,6-dideoxy-D-glucose** | Antibiotics: neomycin B, neomycin A, zygomycin A | — | HCl +67 | — | — | — | — | — | — | 4.7[g] | *2, 27, 28* |
| | | | | $R_{GlcN}$ | | | | | | | |
| | | | | N | S | | | | | | |
| 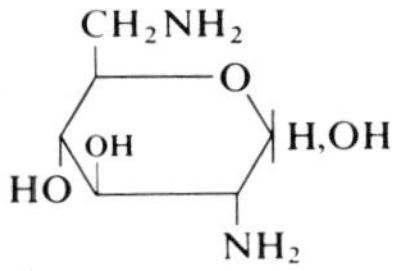 **2,4-Diamino-2,4,6-trideoxy-hexose** | Polysaccharide from *Bacillus lichenformis* | — | — | 0.92 | 1.09 | — | — | — | — | — | *2, 29[d], 30* |
| 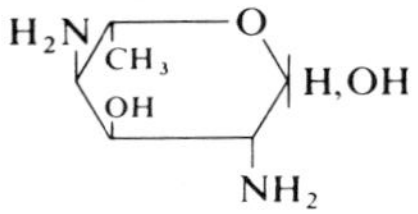 **2,4-Diamino-2,4,6-trideoxy hexose** | UDP-nucleotide synthesized by *D. pneumoniae* Type XIV | — | — | — | — | — | — | — | — | — | *31* |

## TABLE 7 (Continued)
## THE NATURALLY OCCURRING AMINO SUGARS

| Compound and formula | Source | Physical constants[a] | | $R_{GlcN}$ values on paper solvent systems[b] | | | | | $R_{GlcN}$ on ion exchange system[c] | | References |
|---|---|---|---|---|---|---|---|---|---|---|---|
| | | MP | $[\alpha]_D$ | A | B | C | D | E | I | II | |
| **VI. *N*-ACETYLATED[h] AND *N*-METHYLATED AMINO SUGARS** | | | | | | | | | | | |
| | | | | $R_{GlcN}$ | | | | | | | |
| | | | | A | B | C | D | E | | | |
| **2-Acetamido-D-glucose** | — | 205 | (α) +64 → +40.9 | 1.71<br>1.84[i] | — | — | 1.10 | — | — | — | *1, 3* |
| | | 182–184 | (β) −21.5 → +40.9 | 0.96[Δ] | — | — | — | — | — | — | — |
| **2-Acetamido-D-galactose** | — | 172–173 | (α) +115 → +86 | 1.59<br>1.70[i]<br>0.75[Δ] | — | — | 1.19 | — | — | — | *1, 3* |
| **2-Acetamido-D-mannose** | — | 105–108 | (β) −21 → +10 | 1.79<br>1.84[i]<br>1.05[Δ] | — | — | 1.12 | — | — | — | *1, 2, 3* |
| **2-Acetamido-D-gulose** | — | — | −55 → −59 | 1.87 | — | — | 1.18 | — | — | — | *1[e], 3* |
| **2-Acetamido-2,6-dideoxy-D-glucose** | — | — | — | — | — | — | — | — | — | — | — |
| **2-Acetamido-2,6-dideoxy-D-galactose** | — | — | — | 2.12 | — | — | 1.35 | — | — | — | *3* |
| **2-Acetamido-2,6-dideoxy-L-galactose** | — | 195–198 | −79 | — | — | — | — | — | — | — | *16[d]* |
| **2-Acetamido-2,6-dideoxy-L-talose** | — | — | — | | | | | | | | |
| | | | | $R_{fucose}$ | | | | | | | |
| | | | | A | B | F | G | H | I | | |
| **3-Acetamido-3-deoxy-D-glucose** | — | 199–202 | −43 | — | — | 0.71 | 1.13 | 0.90 | 0.92 | — | *18, 32* |
| **3-Acetamido-3,6-dideoxy-D-glucose** | — | — | — | 1.2<br>1.3 | 1.3<br>1.5 | — | — | — | — | — | *20, 21* |
| **3-Acetamido-3,6-dideoxy-D-galactose** | — | 174–176 | +114 | 1.1<br>1.2 | 1.4 | 1.20 | 1.31 | 1.24 | 1.18 | — | *20, 21, 22[d], 32* |
| **3-Acetamido-3,6-dideoxy-D-mannose** | — | 191–192 | −46E | 1.4 | 1.6 | 1.88 | 1.31 | 1.37 | 1.37 | — | *20, 21, 32, 33* |
| **3-Acetamido-3-deoxy-D-ribose** | — | — | — | — | — | 1.71 | 1.33 | 1.28 | 1.29 | — | *32* |

## TABLE 7 (Continued)
## THE NATURALLY OCCURRING AMINO SUGARS

| Compound and formula | Source | Physical constants[a] | | $R_{GlcN}$ values on paper solvent systems[b] | | | | | $R_{GlcN}$ on ion exchange system[c] | | References |
|---|---|---|---|---|---|---|---|---|---|---|---|
| | | MP | $[\alpha]_D$ | A | B | C | D | E | I | II | |
| **VI. *N*-Acetylated[h] and *N*-Methylated Amino Sugars—(Continued)** | | | | | | | | | | | |
| | | | | $R_{rhamnose}$ | | | | | | | |
| | | | | A | J | G | K | L | M | | |
| **4-Acetamido-4,6-dideoxy-D-glucose** | — | — | — | 0.97 | 1.20 | 1.27 | 0.61 | 1.02 | 1.09 | — | *26*[d] |
| **4-Acetamido-4,6-dideoxy-D-galactose** | — | — | — | 0.94 | 1.30 | 1.39 | 0.56 | 0.99 | 1.07 | — | *26*[d] |
| | | | | $R_{GlcN}$ | | | | | | | |
| | | | | A | N | Q | R | S | | | |
| **6-Acetamido-6-deoxy-D-glucose** | — | 196–198 | +44.0 → +34.9 | — | — | — | — | — | — | — | *18*[d] |
| **2-Amino-4-acetamido-2,4,6-trideoxy-L-altrose-HCl** | — | 216–219 | +115 → +94 | 1.70 | 2.33 | 1.38 | 1.8 | 1.38 | — | — | *23, 30* |
| **2,4-Diacetamido-2,4,6-trideoxy-hexose** | — | 262–264 | +67 W-E | — | 7.50 | — | — | — | — | — | *29, 30* |
| ***N*-methyl-L-glucosamine: 2-deoxy-2-methylamino-L-glucose**<br>HO, O, $CH_2OH$, $CH_3NH$, H,OH, OH | Antibiotics: Streptomycin group | 160–163<br>Gum<br>165–166 | HCl<br>−103 → −88<br>FB<br>−65 M<br>*N*-acetyl<br>−51 | — | — | — | — | — | — | — | *2, 17, 34*[d] |
| **Desosamine: 3-dimethylamino-3,4,6-trideoxy-D-xylohexose**<br>$CH_3$, H, O, $N(CH_3)_2$, H,OH, H, OH | Antibiotics: erythromycin, griseomycin, methymycin, narbomycin, neomethymycin, oleandomycin, picromycin, plicacetin | 189–191 | HCl<br>+49.5<br>+53.4 E | — | — | — | — | — | — | — | *2, 17*[d]<br>*35* |

## TABLE 7 (Continued)
## THE NATURALLY OCCURRING AMINO SUGARS

| Compound and formula | Source | Physical constants[a] | | $R_{GlcN}$ values on paper solvent systems[b] | | | | | $R_{GlcN}$ on ion exchange system[c] | | References |
|---|---|---|---|---|---|---|---|---|---|---|---|
| | | MP | $[\alpha]_D$ | A | B | C | D | E | I | II | |
| **VI. *N*-Acetylated[h] and *N*-Methylated Amino Sugars—(Continued)** | | | | | | | | | | | |
| **Mycaminose: 3-dimethyl-amino-3,6-dideoxy-D-glucose** | Antibiotics: carbomycin, spiromycin, leucomycin | 115–116 | HCl +31 | — | — | — | — | — | — | — | *2, 17[d], 36* |
| **Rhodosamine: 3-dimethyl-amino-2,3,6-trideoxy-L-lyxo-hexose** | Antibiotics: cinerubins, pyrromycin rhodomycins | 152–153 | HCl −65.2 | — | — | — | — | — | — | — | *2, 37* |
| **Amosamine: 4-dimethylamino-4,6-dideoxy-D-glucose** | Antibiotics: amicetin | 192–193 | HCl +45.5 | — | — | — | — | — | — | — | *2, 17[d], 38* |
| **4-Dimethylamino-2,3,4,6-tetradeoxy-hexose** | Antibiotics: spiramycins | 75 | +62.6, +83.9 M | — | — | — | — | — | — | — | *2, 39[d]* |

## TABLE 7 (Continued)
## THE NATURALLY OCCURRING AMINO SUGARS

| Compound and formula | Source | Physical constants[a] | | $R_{GlcN}$ values on paper solvent systems[b] | | | | | $R_{GlcN}$ on ion exchange system[c] | | References |
|---|---|---|---|---|---|---|---|---|---|---|---|
| | | MP | $[\alpha]_D$ | A | B | C | D | E | I | II | |
| **VII. ACIDIC AMINO SUGARS** | | | | | | | | | | | |
| **Glucosaminuronic acid: 2-amino-2-deoxy-D-glucuronic acid** (COOH; O; OH; H,OH; HO; $NH_2$) | Polysaccharide of *Haemophilus influenza* type d, *Staphylococcus* | 172 | +55 | 0.30<br>0.35 | — | — | 0.40 | 0.46 | 0.70 | — | *1*[f], *2*, *3*, *4*[e], *40*[d] |
| **D-Galactosaminuronic acid: 2-amino-2-deoxy-D-galacturonic acid** (COOH; HO; O; OH; H,OH; $NH_2$) | Vi antigens: *E. coli*, *Paracolobacterium ballerup* and *S. typhosa* | 160 | W-HCl, pH2<br>+84.5 | 0.10 | 0.83 | 0.58 | — | 0.54 | 1.00 | — | *2–4* |
| **Mannosaminuronic acid: 2-amino-2-deoxymannuronic acid** (COOH; O; OH; $H_2N$; H,OH; HO) | Polysaccharide of *Micrococcus lysodeikticus*, K7 antigen of *E. coli* (D-configuration) | —<br>— | —<br>— | —<br>0.42 | —<br>— | —<br>— | —<br>— | —<br>0.48 | —<br>— | —<br>— | *2*<br>*52* |
| **Muramic acid** ($CH_2OH$; O; H; $CH_3$—C—COOH; O; H,OH; OH; $NH_2$) | Bacterial cell walls, spores, nucleotide complexes, cell walls of blue-green algae | —<br>— | +109 | 1.00<br>1.02 | 1.86 | 2.10 | 0.83 | 1.11 | 1.1<br>1.2 | — | *2*, *3*, *41*[d] |

## TABLE 7 (Continued)
## THE NATURALLY OCCURRING AMINO SUGARS

| Compound and formula | Source | Physical constants[a] | | $R_{GlcN}$ values on paper solvent systems[b] | | | | | $R_{GlcN}$ on ion exchange system[c] | | References |
|---|---|---|---|---|---|---|---|---|---|---|---|
| | | MP | $[\alpha]_D$ | O | | | | | I | II | |
| **VII. Acidic Amino Sugars—(Continued)** | | | | | | | | | | | |
| **Neuraminic acid: 5-amino-3,5-dideoxy-D-glycero-D-galacto nonulosonic acid** 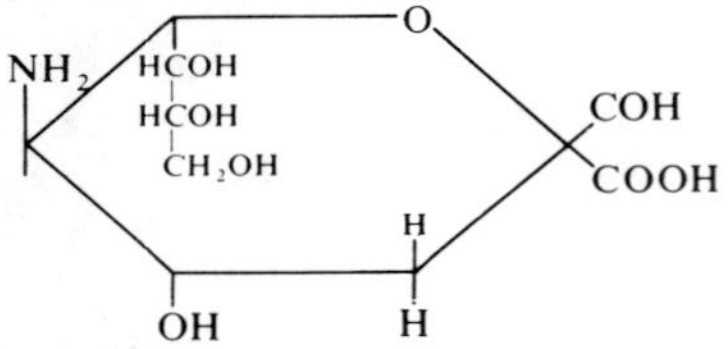<br>NH₂ HCOH HCOH CH₂OH O COH COOH H OH H | Polysaccharide of bacteria, invertebrates, vertebrates | — | — | — | — | — | — | — | — | — | 2 |
| **Occurs in Nature as:** | | | | | | | | | | | |
| *N*-acetyl-neuraminic acid | | 185–187 | −31 ± 2 | 0.48 | — | — | — | — | — | — | — |
| *N*-glycolyl-neuraminic acid | | 185–187 | −32 ± 2 | 0.33 | — | — | — | — | — | — | *8, 42* |
| *N*-4-*O*-diacetyl-neuraminic acid | | 200 | −61 ± 1 | — | — | — | — | — | — | — | — |
| *N*-7-*O*-diacetyl-neuraminic acid | | 138–140 | +6 ± 2 | — | — | — | — | — | — | — | — |
| Bovine *N*-acetyl-*O*-diacetylneuraminic acid | | 130–131 | +9 ± 2 | — | — | — | — | — | — | — | — |
| **VIII. KETO-AMINO SUGARS** | | | | | | | | | | | |
| **2-acetamido-4-keto 2,6-dideoxyhexose** 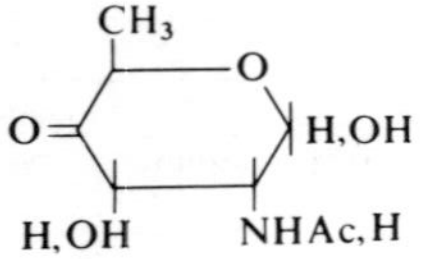<br>CH₃ O O H,OH H,OH NHAc,H | UDP-nucleotide made by enzyme of *D. pneumoniae* type XIV and *Citrobacter freundii* ATCC 10053 | — | — | — | — | — | — | — | — | — | *31, 43* |

## TABLE 7 (Continued)
## THE NATURALLY OCCURRING AMINO SUGARS

[a] MP = melting point. $[\alpha]_D$ = specific rotation; unless otherwise indicated, the rotation solvent is water; W–E = water:ethanol (1:1), W–HCl = water–hydrochloric acid, E = ethanol, M = methanol; FB = free base, HCl = hydrochloride, Ac = acetate; ($\alpha$) or ($\beta$) indicate the anomer.

[b] Solvent systems:

| | | | |
|---|---|---|---|
| A | *n*-Butanol-pyridine–water (6:4:3) | J | Isobutyric acid–N-ammonia (10:0.6 |
| A$^{\Delta}$ | Solvent system A on paper treated with 0.1*M* $BaCl_2$ or $BaAc_2$ | K | Pyridine–ethyl acetate–water (10:3.0:1.5 upper layer) |
| B | *n*-Butanol–glacial acetic acid–water (5:1:2) | L | *n*-Butanol–ethanol–water (13:8:4) |
| C | Phenol–water (70:30) | M | *n*-Butanol–acetic acid–water (3:1:1) |
| D | Phenol–water (80:20), ammonia atmosphere | N | *n*-Butanol–ethanol–water (4:1:1) |
| E | Ethylacetate–pyridine–water–acetic acid (5:5:3:1) | O | *n*-Butanol–pyridine–0.1*N* HCl (5:3:2) |
| F | Pyridine–ethyl acetate–water (10:36:11.5) | P | *n*-Butanol–acetic acid–water (25:6:25) |
| G | Phenol–water (80:20) | Q | Ethylacetate–pyridine–water (2:1:2) |
| H | *n*-Butanol–acetic acid–water (5:1.2:2.5) | R | *n*-Butanol–ethanol–water (5:1:4) |
| I | Ethylacetate–pyridine–*n*-butanol–butyric acid–water (10:10:5:1:5) | S | *n*-Propanol–1% ammonia (7:3) |

[c] System I: Dowex 50 $H^+$ column, 1 x 50 cm, packed according to Gardell and eluted with 0.33*N* HCl.[44]
System II: Technicon Amino Acid Analyzer, modified by Brendel *et al.*[4] and eluted with 0.133*M* pyridine–acetic acid (0.82*M*) buffer, pH 3.85.

[d] Reference describing the characterization of the compound in question.

[e] Chromatographic identification.

[f] The recorded physical properties are for the synthetic compound.

[g] Brendel *et al.* system, using 3.1*M* pyridine–acetic acid buffer at pH 4.5.[4]

[h] Not all of the N-acetamide compounds are necessarily found in nature, but they are listed here for convenience.

[i] N-Acetylmannosamine and N-acetylgalactosamine have an $R_N$-acetylglucosamine of 0.4 to 0.5 when run on borate-treated paper with solvent A.

[j] The C-2 carbon configuration is in doubt.

Data taken from: Raff, R. A., and Wheat, R. W., in *Handbook of Biochemistry,* 2nd ed., pp. D-71–D-80, H. A. Sober, Ed. Copyright 1970, The Chemical Rubber Co., Cleveland, Ohio.

## REFERENCES

### General References

1. Hofheinz, W., and Grisebach, H., *Berichte, 96,* 2867 (1963).
2. Celmer, W. D., *J. Amer. Chem. Soc., 87,* 1799 (1965).
3. Foster, A. B., *et al., Proc. Chem. Soc.,* p. 254 (1962).
4. Omura, S., *et al., J. Amer. Chem. Soc., 91,* 3401 (1969).
5. Brumfitt, W., *Brit, J. Exp. Pathol., 40,* 441 (1959).
6. Lechevalier, M. P., and Lechevalier, H. A., *Int. J. Syst. Bacteriol., 20,* 435 (1970).
7. Committee for Rules of Carbohydrate Nomenclature, *J. Org. Chem., 28,* 281 (1963).
8. Pigman, W., and Horton, D. (Eds.), *The Carbohydrates,* Vols. 1A, 1B, 2A, 2B. Academic Press, New York (1972).
9. Rogers, H. J., and Perkins, H. R., *Cell Walls and Membranes.* E. and F. N., Spon, London, England (1968).
10. Salton, M. J. R., *The Bacterial Cell Wall.* American Elsevier Publishing Co., New York (1964).
11. Osborn, M. J., *Ann. Rev. Biochem., 38,* 501 (1969).
12. Schleifer, K. H., and Kandler, O., *Bacteriol. Rev., 36,* 407 (1972).
13. Percival, E. E., and McDowell, R., *The Chemistry and Enzymology of Marine Algal Polysaccharides.* Academic Press, London, England (1967).
14. Bartnicki-Garcia, S., *Ann. Rev. Microbiol., 22,* 87 (1968).
15. Gorin, P. A. J., and Spencer, J. F. T., *Adv. Carbohydr. Chem., 23,* 367 (1968).
16. Stacey, M., and Barker, S. A., *Polysaccharides of Microorganisms.* Oxford University Press, London, England (1960).

### Table 2

1. Barclay, K. S., Bourne, E. J., Stacey, M., and Webb, M., *J. Chem. Soc.,* p. 1501 (1954).
2. Hehre, E. J., and Hamilton, P. M., *Proc. Soc. Exp. Biol. Med., 71,* 336 (1949).
3. Barker, S. A., Bourne, E. J., Bruce, G. T., and Stacey, M., *J. Chem. Soc.,* p. 4414 (1958).
4. Neufeld, E. F., and Hassid, W. Z., *Adv. Carbohydr. Chem., 18,* 309 (1963).
5. Heidelburger, M., Das, A., and Juni, E., *Proc. Nat. Acad. Sci. U.S.A., 63,* 47 (1969).
6. Feingold, D. S., and Gehatia, M., *J. Polymer Sci., 23,* 783 (1957).
7. Gorin, P. A. J., Spencer, J. F. T., and Westlake, D. W. S., *Can. J. Chem., 39,* 1067 (1961).
8. Misaki, A., Saito, H., Ito, T., and Harada, T., *Biochemistry, 8,* 4645 (1969).
9. Goatley, J. L., *Can. J. Microbiol., 14,* 1063 (1968).
10. Moore, B. G., and Tischer, R. G., *Science, 145,* 586 (1964).
11. Bishop, C. T., and Adams, G. A., *Can J. Chem., 32,* 999 (1954).
12. Jeanes, A., Knutson, C. A., Pittsley, J. E., and Watson, P. R., *J. Polymer Sci., 9,* 627 (1965).
13. Sakaguchi, O., Yokota, K., and Susuki, M., *Yakugaku Zasshi., 87,* 1268 (1967).
14. Gorin, P. A. J., and Eveleigh, D. E., *Biochemistry, 9,* 5023 (1970).
15. Barker, S. A., Bourne, E. J., O'Mant, D. M., and Stacey, M., *J. Chem. Soc.,* p. 2448 (1957).
16. Reese, E. T., and Mandels, M., *Can. J. Microbiol., 10,* 103 (1964).
17. Lawson, G. J., and Stacey, M., *J. Chem. Soc.,* p. 1925 (1954).
18. Gorin, P. A. J., and Spencer, J. F. T., *Can. J. Chem., 44,* 993 (1966).
19. Beijerinck, M. W., *Proc. K. Ned. Akad. Wet. Ser. C Biol. Med. Sci., 12,* 365 (1910).
20. Stacey, M., and Barker, S. A., in *Polysaccharides of Microorganisms,* p. 148. Oxford University Press, London, England (1960).
21. Stacey, M., and Barker, S. A., in *Polysaccharides of Microorganisms,* p. 150. Oxford University Press, London, England (1960).
22. Yu, R. J., *et al., Can. J. Chem., 45,* 2264 (1967).
23. Chanda, N. B., Hirst, E. L., and Manners, D. J., *J. Chem. Soc.,* p. 1951 (1957).
24. Perlin, A. S., and Suzuzi, S., *Can. J. Chem., 40,* 50 (1962).
25. Corpe, W. A., *J. Bacteriol., 88,* 1433 (1964).
26. Dickerson, A. G., Mantle, P. G., and Czczyrbak, C. A., *J. Gen. Microbiol., 60,* 403 (1970).
27. Perlin, A. S., and Taber, W. A., *Can. J. Chem., 41,* 2278 (1963).
28. Stacey, M., and Barker, S. A., in *Polysaccharides of Microorganisms,* p. 144. Oxford University Press, London, England (1960).
29. Gorin, P. A., and Spencer, J. F. T., *Can. J. Chem., 42,* 1230 (1964).
30. Jeanes, A., Pittsley, J. E., and Watson, P. R., *J. Appl. Polymer Sci., 8,* 2775 (1964).
31. White, G. J., and Sussman, M., *Biochim. Biophys. Acta, 74,* 173, (1963).
32. White, G. J., and Sussman, M., *Biochim. Biophys. Acta, 74,* 179, (1963).
33. McGuire, E. J., and Binkley, S. B., *Biochemistry, 3,* 247 (1964).
34. Grant, W. D., Sutherland, I. W., and Wilkinson, J. F., *J. Bacteriol., 100,* 1187 (1969).

35. Slodki, M. E., *Biochim. Biophys. Acta, 62,* 96 (1963).
36. Jeanes A., *et al., Can. J. Chem., 40,* 2256 (1962).
37. Applegarth, D. A., and Bozoian, G., *Arch. Biochem. Biophys., 134,* 285 (1969).
38. Wallen, L. L., Rhodes, R. A., and Shulke, H. R., *Appl. Microbiol., 13,* 272 (1965).
39. Zilliken, F., *et al., Arch. Biochem. Biophys., 54,* 398 (1955).
40. Stacey, M., and Barker, S. A., in *Polysaccharides of Microorganisms,* p. 143. Oxford University Press, London, England (1960).
41. Chihara, C., *et al., Nature, 222,* 687 (1969).
42. Jeanes, A., *et al., J. Amer. Chem. Soc., 76,* 5041 (1954).
43. Neely, W. B., *Adv. Carbohydr. Chem., 15,* 341 (1960).
44. Lindberg, B., and Svensson, S., *Acta Chem. Scand., 22,* 1907 (1968).
45. Lewis, B. A., St. Cyr, M. J., and Smith, F., *Carbohydr. Res., 5,* 194 (1967).
46. Grappel, S. F., Blank, F., and Bishop, C. T., *J. Bacteriol., 93,* 1001 (1967).
47. Grappel, S. F., Blank, F., and Bishop, C. T., *J. Bacteriol., 95,* 1238 (1967).
48. Feather, M. S., and Malek, A., *Biochim. Biophys. Acta, 264,* 103 (1972).
49. Morrison, I. M., Young, R., Perry, M. B., and Adams, G. A., *Can. J. Chem., 45,* 1987 (1967).
50. Adams, G. A., and Chaudhari, A. S., *Can. J. Biochem., 50,* 345 (1972).
51. Haworth, W. N., Raistrick, H., and Stacey, M., *Biochem. J., 31,* 640 (1937).
52. Gorin, P. A. J., and Spencer, J. F. T., *Can. J. Chem., 37,* 499 (1959).
53. Preston, J. F., and Gander, J. E., *Arch. Biochem. Biophys., 124,* 504 (1968).
54. Hough, L., and Perry, M. B., *J. Chem. Soc.,* p. 2801 (1962).
55. Preston, J. F., Lapis, E., and Gander, J. E., *Arch. Biochem. Biophys., 134,* 324 (1969).
56. Anderson, C. G., Haworth, W. N., Raistrick, H., and Stacey, M., *Biochem. J., 33,* 272 (1939).
57. Farr, D. R., Amister, H., and Horisberger, M., *Carbohydr. Res., 24,* 207 (1972).
58. Ralph, B. J., and Bender, V. J., *Chem. Ind. (London),* p. 1181 (1965).
59. Clarke, A. E., and Stone, B. A., *Pure Appl. Chem., 13,* 134 (1964).
60. Bouveng, H. O., Kiessling, H., Lindberg, B., and McKay, J., *Acta Chem. Scand., 17,* 1351 (1963).
61. Catley, B. J., *FEBS (Fed. Eur. Biochem. Soc.) Lett., 20,* 174 (1972).
62. Wallenfels, K., *et al., Angew, Chem. Int. Ed. Engl., 2,* 515 (1963).
63. Dedonder, R. A., and Hassid, W. Z., *Biochem. Biophys. Acta, 90,* 239 (1964).
64. Graham, P. H., *Antonie van Leeuwenhoek J. Microbiol. Serol., 31,* 349 (1965).
65. Amarger, N., Obaton, M., and Blachère, H., *Can. J. Microbiol., 13,* 99 (1967).
66. Cooper, E. A., Daker, W. D., and Stacey, M., *Biochem. J., 32,* 1732 (1938).
67. Gorin, P. A. J., Horitsu, K., and Spencer, J. F. T., *Can. J. Chem., 43,* 950 (1965).
68. Bacon, J. S. D., Farmer, V. C., Jones, D., and Taylor, I. F., *Biochem. J., 114,* 557 (1969).
69. Gorin, P. A. J., and Spencer, J. F. T., *Adv. Appl. Microbiol., 13,* 25 (1970).
70. Kikumoto, S., *et al., J. Agr. Chem., 45,* 162 (1971).
71. Wessels, J. G. H., *Biochim. Biophys. Acta, 178,* 191 (1969).
72. Johnson, J., *et al., Chem. Ind. (London),* p. 820 (1963).
73. Batra, K. K., Nordin, J. H., and Kirkwood, S., *Carbohydr. Res., 9,* 221 (1969).
74. Adams, G. A., and Martin, S. M., *Can. J. Biochem., 42,* 1403 (1964).
75. Adams, G. A., and Young, R., *Can. J. Biochem., 43,* 1499 (1965).
76. Romano, A. H., and Peloquin, J. P., *J. Bacteriol., 86,* 252 (1963).
77. Falk, M., *et al., Can. J. Chem., 44,* 2269 (1966).
78. Tsusue, Y., and Yamakawa, T. *J. Biochem. (Tokyo), 58,* 587 (1965).
79. Bishop, C. T., Blank, F., and Hranisavljevic-Jakovljevic, M., *Can. J. Chem., 40,* 1816 (1962).
80. Blank, F., and Perry, M. B., *Can. J. Chem., 42,* 2862 (1964).
81. Barker, S. A., Cruickshank, C. N. D., Morris. H. H., and Wood, S. R., *Immunology, 5,* 627 (1962).
82. Bishop, C. T., *et al., Can. J. Chem., 43,* 30 (1965).
83. Gorin, P. A. J., and Spencer, J. F. T., *Can. J. Chem., 45,* 1543 (1967).
84. Sloneker, J. H., and Jeanes, A., *Can. J. Chem., 40,* 2256 (1962).
85. Gorin, P. A. J., and Spencer, J. F. T., *Can. J. Chem., 41,* 2357 (1963).
86. Gorin, P. A. J., and Spencer, J. F. T., *Can. J. Chem., 34,* 2282 (1961).

**Table 3**

1. Heilbron and Bunbury, *Dictionary of Organic Compounds,* Vols. 2 and 3. Oxford University Press, New York (1943).
2. Dutton and Unrau, *Can. J. Chem., 43,* 924, 1738 (1965).
3. Lindberg, *Ark. Kemi Mineral. Geol., 23,* A2 (1946–1947).
4. Bamberger and Landsiedl, *Monatsh. Chem., 21,* 571 (1900).
5. Ward, Pettijohn, Lockwood and Coghill, *J. Amer. Chem. Soc., 66,* 541 (1944).
6. Ciamician and Silber, *Ber. Deut. Chem. Ges., 44,* 1280 (1911).

7. Morell and Auernheimer, *J. Amer. Chem. Soc., 66,* 792 (1944).
8. Bertrand, *Compt. Rend., 130,* 1472 (1900).
9. Hu, McComb and Rendig, *Arch. Biochem. Biophys., 110,* 350 (1965).
10. Rubin, Lardy and Fischer, *J. Amer. Chem. Soc., 74,* 425 (1952).
11. Wilson and Lucas, *J. Amer. Chem. Soc., 58,* 2396 (1936).
12. Asahina and Yanagita, *Ber. Deut. Chem. Ges., 67,* 799 (1934).
13. Frèrejacque, *Compt. Rend., 208,* 1123 (1939).
14. Onishi and Suzuki, *Agr. Biol. Chem. (Tokyo), 30,* 1139 (1966).
15. Richtmyer and Hudson, *J. Amer. Chem. Soc., 73,* 2249 (1951).
16. Touster and Harwell, *J. Biol. Chem., 230,* 1031 (1958).
17. Wessely and Wang, *Monatsh. Chem., 72,* 168 (1938).
18. Binkley and Wolfrom, *J. Amer. Chem. Soc., 70,* 2809 (1948).
19. Wolfrom and Kohn, *J. Amer. Chem. Soc., 64,* 1739 (1942).
20. Wolfrom, Hann and Hudson, *J. Amer. Chem. Soc., 74,* 1105 (1952).
21. Rogerson, *J. Chem. Soc. (London),* p. 1040 (1912).
22. Von Lippmann, *Ber. Deut. Chem. Ges., 60,* 161 (1927).
23. Haas and Hill, *Biochem. J., 26,* 987 (1932).
24. Jeger, Norymberski, Szpilfogel and Prelog, *Helv. Chim. Acta, 29,* 684 (1946).
25. Richtmyer, Carr and Hudson, *J. Amer. Chem. Soc., 65,* 1477 (1943).
26. Bertrand, *Bull. Soc. Chim. Fr. Ser. 3, 33,* 166 (1905).
27. Braham, *J. Amer. Chem. Soc., 41,* 1707 (1919).
28. Patterson and Todd, *J. Chem. Soc. (London),* p. 2876 (1929).
29. Iwate, *Chem. Zentralbl., 2,* 177 (1929).
30. Zervas, *Ber. Deut. Chem. Ges., 63,* 1689 (1930).
31. Asahina, *Ber. Deut. Chem. Ges., 45,* 2363 (1912).
32. Lindberg, *Acta Chem. Scand., 7,* 1119, 1123 (1953).
33. Jones and Wall, *Nature, 189,* 746 (1961).
34. Maquenne, *Ann. Chim. Phys. (Paris) Ser. 6, 19,* 5 (1890).
35. Charlson and Richtmyer, *J. Amer. Chem. Soc., 82,* 3428 (1960).
36. Buck, Foster, Richtmyer and Zissis, *J. Chem. Soc. (London),* p. 3633 (1961).
37. Onishi and Perry, *Can. J. Microbiol., 11* 929 (1965).
38. Bougault and Allard, *Compt. Rend., 135,* 796 (1902).
39. Maclay, Hann and Hudson, *J. Org. Chem., 9,* 293 (1944).
40. Ackerman, *Hoppe-Seyler's Z. Physiol. Chem., 336,* 1 (1964).
41. Von Lippmann, *Ber. Deut. Chem. Ges., 34,* 1159 (1901).
42. King and Jurd, *J. Chem. Soc. (London),* p. 1192 (1953).
43. Bien and Ginsburg. *J. Chem. Soc. (London),* p. 3189 (1958).
44. Plouvier, *Compt. Rend., 241,* 983 (1955).
45. Kübler, *Arch. Pharm., 246,* 620 (1908).
46. Dangschat and Fischer, *Naturwissenschaften, 27,* 756 (1939).
47. DeJong, *Rec. Trav. Chim. Pays-Bas, 27,* 257 (1908).
48. Kiang and Loke, *J. Chem. Soc. (London),* p. 480 (1956).
49. Angyal, Gilham and MacDonald, *J. Chem. Soc. (London),* p. 1417 (1957).
50. Ballou and Anderson, *J. Amer. Chem. Soc., 75,* 648 (1953).
51. Umezawa, Okami, Hashimoto, Suhara, Hamada and Takeuchi, *J. Antibiot. (Tokyo) Ser. A, 18,* 101 (1965).
52. Foxall and Morgan, *J. Chem. Soc. (London),* p. 5573 (1963).
53. Smith, *Biochem. J., 57,* 140 (1954).
54. Tanret, *Compt. Rend., 145,* 1196 (1907).
55. Adhikari, Bell and Harvey, *J. Chem. Soc. (London),* p. 2829 (1962).
56. Lindberg, *Acta Chem. Scand., 9,* 1093 (1955).
57. Lindberg and Wickberg, *Ark, Kemi, 13,* 447 (1959).
58. Posternak and Falbriard, *Helv. Chim. Acta, 44,* 2080 (1961).
59. Kindl, Kremlicka and Hoffman-Ostenhof, *Monatsh. Chem., 97,* 1783 (1966).
60. Angyal and Bender, *J. Chem. Soc. (London),* p. 4718 (1961).
61. Ackermann, *Ber. Deut. Chem. Ges., 54,* 1938 (1921).
62. Plouvier, *Compt. Rend., 241,* 765 (1955).
63. Post and Anderson, *J. Amer. Chem. Soc., 84,* 478 (1962).
64. Maquenne, *Ann. Chim. Phys. (Paris) Ser. 6, 22,* 264 (1891).
65. Anderson, Fischer and MacDonald, *J. Amer. Chem. Soc., 74,* 1479 (1952).
66. Pease, Reider and Elderfield, *J. Org. Chem., 5,* 198 (1940).
67. Plouvier, *Compt. Rend., 243,* 1913 (1956).
68. Anderson, Takeda, Angyal and McHugh, *Arch. Biochem. Biophys., 78,* 518 (1958).

69. Angyal, MacDonald and Matheson, *J. Chem. Soc. (London)*, p. 3321 (1953).
70. DeJong, *Rec. Trav. Chim. Pays-Bas, 25,* 48 (1906).
71. Adams, Pease and Clark, *J. Amer. Chem. Soc., 62,* 2194 (1940).
72. Prunier, *Ann. Chim. Phys. (Paris) Ser. 5, 15,* 5 (1878).
73. Plouvier, *Compt. Rend., 255,* 360 (1962).
74. Gorter, *Ann. Chem. (Justus Liebigs), 359,* 221 (1908).
75. Ervig and Koenigs, *Ber. Deut. Chem. Ges., 22,* 1457 (1889).
76. Weiss, Davis and Mingioli, *J. Amer. Chem. Soc., 75,* 5572 (1953).
77. Adlersberg and Sprinson, *Biochemistry, 3,* 1855 (1964).
78. Muller, *J. Chem. Soc. (London)*, p. 1767 (1907).
79. Posternak, *Helv. Chim. Acta, 25,* 746 (1942).
80. Eijkman, *Ber. Deut. Chem. Ges., 24,* 1278 (1891).
81. Grewe, Lorenzen and Vining, *Ber. Deut. Chem. Ges., 87,* 793 (1954).
82. Sherrard and Kurth, *J. Amer. Chem. Soc., 51,* 3139 (1929).
83. Eijkman, *Rec. Trav. Chim. Pays-Bas, 4,* 32 (1885).
84. Grewe, Buttner and Burmeister, *Angew. Chem., 69,* 61 (1957).
85. McCrindle, Overton and Raphael, *J. Chem. Soc. (London)*, p. 1560 (1960).
86. Salamon and Davis, *J. Amer. Chem. Soc., 75,* 5567 (1953).
87. Power and Tutin, *J. Chem. Soc. (London)* p. 624 (1904).
88. Angyal, Gorin and Pittman, *J. Chem. Soc. (London)*, p. 1807 (1965).
89. Posternak and Schopfer, *Helv. Chim. Acta, 33,* 343 (1950).
90. Nakajima and Kurihara, *Ber. Deut. Chem. Ges., 94,* 515 (1961).
91. Posternak, *Helv. Chim. Acta, 19,* 1333 (1936).
92. Kluyver and Boezaardt, *Rec. Trav. Chim. Pays-Bas, 58,* 958 (1939).
93. Stanacev and Kates, *J. Org. Chem., 26,* 912 (1961).
94. Magasanik and Chargaff, *J. Biol. Chem., 175,* 929 (1948).
95. Berman and Magasanik, *J. Biol. Chem., 241,* 800 (1966).
96. Johnson, Gourlay, Tarbell and Autrey, *J. Org. Chem., 28,* 300 (1963).
97. Nakajima, Kurihara, Hasegawa and Kurokawa, *Ann. Chem. (Justus Liebigs), 689,* 243 (1965).
98. Bannister and Argoudelis, *J. Amer. Chem. Soc., 85,* 119 (1963).
99. Allen, *J. Amer. Chem. Soc., 78,* 5691 (1956).
100. Patrick, Williams, Waller and Hutchings, *J. Amer. Chem. Soc., 78,* 2652 (1956).
101. Peck, Hoffhine, Peel, Graber, Holly, Mozingo and Folkers, *J. Amer. Chem. Soc., 68,* 776 (1946).
102. Peck, Graber, Walti, Peel, Hoffhine and Folkers, *J. Amer. Chem. Soc., 68,* 29 (1946).
103. Nakajima, Hasegawa and Kurihara, *Ann. Chem. (Justus Liebigs), 689,* 235 (1965).
104. Maeda, Murase, Mawatari and Umezawa, *J. Antibiot. (Tokyo) Ser. A, 11,* 73 (1958).
105. Kondo, Sezaki, Koika and Akita, *J. Antibiot. (Tokyo) Ser. A, 18,* 192 (1965).
106. Sherman, Goodwin and Stewart, *Abstracts of the 156th Meeting,* Paper 37. Biological Chemistry Division, American Chemical Society, Washington, D.C. (1968).
107. Allen, *J. Amer. Chem. Soc., 84,* 3128 (1962).
108. Walker and Walker, *Biochim. Biophys. Acta, 170,* 219 (1968).
109. Carter, Clark, Lytle and McCasland, *J. Biol. Chem., 175,* 683 (1948).

**Table 4**

1. Frankland and McGregor, *J. Chem. Soc. (London)*, p. 513 (1893).
2. Frankland, Wharton and Aston, *J. Chem. Soc. (London)*, p. 26 (1901).
3. Isherwood, Chen and Mapson, *Biochem. J., 56,* 1, 15 (1954).
4. Wolfrom and DeWalt, *J. Amer. Chem. Soc., 70,* 3148 (1948).
5. Robbins and Upson, *J. Amer. Chem. Soc., 62,* 1074 (1940).
6. Hardegger, Kreiss and El Khadem, *Helv. Chim. Acta, 35,* 618 (1952).
7. Palleroni and Doudoroff, *J. Biol. Chem., 223,* 499 (1956).
8. Liebster, Kulhanek and Tadra, *Chem. Listy, 47,* 1075 (1953).
9. Rehorst, *Ber. Deut. Chem. Ges., 63,* 2280 (1930).
10. Bates, in *Polarimetry, Saccharimetry and the Sugars (National Bureau of Standards Circular C440)*, pp. 709–760. United States Government Printing Office, Washington, D.C. (1942).
11. Isbell and Frush, *J. Res. Nat. Bur. Stand., 11,* 649 (1933).
12. Hudson and Komatsu, *J. Amer. Chem. Soc., 41,* 1141 (1919).
13. Assarson, Lindberg and Borbrueygen, *Acta Chem. Scand., 13,* 1395 (1959).
14. Weimberg, *J. Biol. Chem., 234,* 727 (1959).
15. Gardner and Wenis, *J. Amer. Chem. Soc., 73,* 1855 (1951).
16. Kanfer, Ashwell and Burns, *J. Biol. Chem., 235,* 2518 (1960).
17. Ladenberg, Tishler, Wellman and Babson, *J. Amer. Chem. Soc., 66,* 1217 (1944).

18. Wolfrom, Thompson and Evans, *J. Amer. Chem. Soc., 67,* 1793 (1945).
18a. Wolfrom, Bennett and Crum, *J. Amer. Chem. Soc., 80,* 944 (1958).
19. Steiger, *Helv. Chim. Acta, 19,* 189 (1936).
20. Hough, Jones and Mitchell, *Can. J. Chem., 36,* 1720 (1958).
21. Rehorst, *Ann. Chem. (Justus Liebigs), 503,* 143, 154 (1933).
22. Weerman, *Rec. Trav. Chim. Pays-Bas, 37,* 15, 40 (1917).
23. Lockwood and Nelson, *J. Bacteriol., 52,* 581 (1946).
24. Heyns and Stein, *Ann. Chem. (Justus Liebigs), 558,* 194 (1947).
25. Hickman and Ashwell, *J. Biol. Chem., 235,* 1566 (1960).
26. Richtmyer, Hann and Hudson, *J. Amer. Chem. Soc., 61,* 343 (1939).
27. Kiliani, *Ber. Deut. Chem. Ges., 55,* 75 (1922).
28. Pryde, *J. Chem. Soc. (London),* p. 1808 (1923).
29. Levene and Meyer, *J. Biol. Chem., 46,* 307 (1921).
30. Hudson and Isbell, *J. Amer. Chem. Soc., 51,* 2225 (1929).
31. Glattfield and MacMillan, *J. Amer. Chem. Soc., 56,* 2841 (1934).
32. Ettel, Liebster and Tadra, *Chem. Listy, 46,* 45 (1952).
33. Claus, *Biochem. Biophys. Res. Commun., 20,* 745 (1965).
34. Ley and Doudoroff, *J. Biol. Chem., 227,* 745 (1957).
35. Wolfrom, Berkebile and Thompson, *J. Amer. Chem. Soc., 71,* 2360 (1949).
36. Fukunaga and Kubata, *Bull. Chem. Soc. Jap., 13,* 272 (1938).
37. Wolfrom and Anno, *J. Amer. Chem. Soc., 74,* 5583 (1952).
38. Ashwell, Wahba and Hickman, *J. Biol. Chem., 235,* 1559 (1960).
39. Regna and Caldwell, *J. Amer. Chem. Soc., 66,* 243, 244, 246 (1944).
40. Hughes, Overend and Stacey, *J. Chem. Soc. (London),* p. 2846 (1949).
41. Bauer and Biely, *Collect. Czech. Chem. Commun., 33,* 1165 (1968).
42. Fischer and Dangschat, *Helv. Chim. Acta, 20,* 705 (1937).
43. Oble and Berend, *Ber. Deut. Chem. Ges., 60,* 1159 (1927).
44. Merrick and Roseman, *J. Biol. Chem., 235,* 1274 (1960).
45. Boutroux, *Ann. Chim. Phys. (Paris) Ser. 6, 21,* 565 (1890).
46. Barch, *J. Amer. Chem. Soc., 55,* 3656 (1933).
47. Bernhauer and Irrgang, *Biochem. Z., 280,* 360 (1935).
48. Wakisaka, *Agr. Biol. Chem., 28,* 819 (1964).
49. Hudson, *J. Amer. Chem. Soc., 73,* 4498 (1951).
50. Barber and Hassid, *Bull. Res. Counc. Isr. Sect. A, 11,* 249 (1963).
51. Upson, Sands and Whitnah, *J. Amer. Chem. Soc., 50,* 519 (1928).
52. Burns, *J. Amer. Chem. Soc., 79,* 1257 (1957).
53. Isbell, *J. Res. Nat. Bur. Stand., 29,* 227 (1942).
54. Heyns, *Ann. Chem. (Justus Liebigs), 558,* 177 (1947).
55. Smiley and Ashwell, *J. Biol. Chem., 236,* 357 (1961).
56. Kilgore and Starr, *Biochim. Biophys. Acta, 30,* 652 (1958).
57. Penney and Zilva, *Biochem. J., 37,* 403 (1943).
58. Berman and Magasanik, *J. Biol. Chem., 241,* 807 (1966).
59. Takagi, *Agr. Biol. Chem., 26,* 717 (1962).
60. Hamilton and Smith, *J. Amer. Chem. Soc., 76,* 3543 (1954).
61. Levene, *J. Biol. Chem., 59,* 123 (1924).
62. Gakhokidge and Gvelukashvili, *J. Gen. Chem. U.S.S.R., 22,* 143 (1952).
63. Pervozvanski, *Mikrobiologyia, 8,* 915 (1939).
64. Preiss and Ashwell, *J. Biol. Chem., 238,* 1571, 1577 (1963).
65. Kuhn, Bister and Dafeldecker, *Ann. Chem. (Justus Liebigs), 617,* 115 (1958).
66. Weissbach and Hurwitz, *J. Biol. Chem., 234,* 705, 710 (1959).
67. Srinivasan and Sprinson, *J. Biol. Chem., 234,* 716 (1959).
68. Adlersberg and Sprinson, *Biochemistry, 3,* 1855 (1964).
69. Heath and Ghalambor, *Biochem. Biophys. Res. Commun., 10,* 340 (1963).
70. Levin and Racker, *J. Biol. Chem., 234,* 2532 (1959).
71. Stodola and Lockwood, *J. Biol. Chem., 171,* 213 (1947).
72. Ameyama and Kondo, *Bull. Agr. Chem. Soc. Jap., 22,* 271, 380 (1958).
73. Bergmann, *Ber. Deut. Chem. Ges., 54,* 1362 (1921).
74. Ehrlich and Schubert, *Ber. Deut. Chem. Ges., 62,* 1987, 2022 (1929).
75. Nieman, Schoeffal and Link, *J. Biol. Chem., 101,* 337 (1933).
76. Winmann, *Ber. Deut. Chem. Ges., 62,* 1637 (1929).
77. Ehrlich and Rehorst, *Ber. Deut. Chem. Ges., 58,* 1989 (1925).
78. Goebel and Babers, *J. Biol. Chem., 100,* 573, 743 (1933).

79. Bergmann and Wolff, *Ber. Deut. Chem. Ges., 56,* 1060 (1923).
80. Ehrlich and Rehorst, *Ber. Deut. Chem. Ges., 62,* 628 (1929).
81. Das Gupta and Sarkar, *Text. Res. J., 24,* 705, 1071 (1954).
82. Marsh, *J. Chem. Soc. (London),* p. 1578 (1952).
83. Levene and Meyer, *J. Biol. Chem., 60,* 173 (1924).
84. Currie and Timell, *Can. J. Biochem., 37,* 922 (1959).
85. Jones and Painter, *J. Chem. Soc. (London),* p. 669 (1957).
86. Jones and Nunn, *J. Chem. Soc. (London),* p. 3001 (1955).
87. Wacek, Leitinger and Hochbahn, *Monatsh. Chem., 90,* 562 (1959).
88. Fischer and Dorfel, *Hoppe-Seyler's Z. Physiol. Chem., 302,* 186 (1955).
89. Cifonelli, Ludowieg and Dorfman, *J. Biol. Chem., 233,* 541 (1958).
90. Shafizadeh and Wolfrom, *J. Amer. Chem. Soc., 77,* 2568 (1955).
91. Fischer and Schmidt, *Ber. Deut. Chem. Ges., 92,* 2184 (1954).
92. Schoeffel and Link, *J. Biol. Chem., 100,* 397 (1933).
93. Pasteur, *Ann. Chim. Phys. (Paris) Ser. 3, 28,* 71 (1850).
94. Walden, *Ber. Deut. Chem. Ges., 29,* 1701 (1896).
95. Fandolt, *Ber. Deut. Chem. Ges., 6,* 1075 (1873).
96. Anschütz and Pictet, *Ber. Deut. Chem. Ges., 13,* 1176 (1880).
97. Patterson, *J. Chem. Soc. (London), 85,* 765 (1904).
98. Frankland and Wharton, *J. Chem. Chem. Soc. (London), 69,* 1310 (1896).
99. Frankland and Slater, *J. Chem. Soc. (London), 83,* 1354 (1903).
100. Pasteur, *Ann. Chem. (Justus Liebigs), 82,* 331 (1852).
101. Schneider, *Ber. Deut. Chem. Ges., 13,* 620 (1880).
102. Anschütz and Bennert, *Ann. Chem. (Justus Liebigs), 254,* 165 (1889).
103. Lutz, Dissertation, Rostock (1899); *Chem. Zentralbl., 2,* 1013 (1900).
104. Rehorst, *Ber. Deut. Chem. Ges., 61,* 163 (1928).
104a. Wolfrom and Rice, *J. Amer. Chem. Soc., 68,* 532 (1946).
105. Marsh, *Biochem. J., 86,* 77 (1963); *87* 82 (1963); *89,* 108 (1963).
106. Rehorst, *Ber. Deut. Chem. Ges., 65,* 1476 (1932).
107. Matsui, Okada and Ishidata, *J. Biochem. (Tokyo), 57,* 715 (1965).
108. Iwasaki, *Yakugaku Zasshi, 82,* 1380 (1962); *Chem. Abstr., 59,* 758 (1963).
109. Kuhn, Weiser and Fischer, *Ann. Chem. (Justus Liebigs), 628* 207 (1959).
110. Karrer and Mayer, *Helv. Chim. Acta, 20,* 407 (1937).
111. Heyns, Kiessling, Lindenberg and Paulsen, *Ber. Deut. Chem. Ges., 92,* 2435 (1959).
112. Heyns and Beck, *Ber. Deut. Chem. Ges., 90,* 2443 (1957).
113. Levene, *J. Biol. Chem., 57,* 337 (1923).
114. Williamson and Zamenhof, *J. Biol. Chem., 238,* 2255 (1963).
115. Heyns and Paulsen, *Ber. Deut. Chem. Ges., 88,* 188 (1955).
116. Westphal and Holzmann, *Ber. Deut. Chem. Ges., 75,* 1274 (1942).
117. Perkins, *Biochem. J., 86,* 475 (1963); *89,* 104P (1963).
118. Crumpton, *Biochem. J., 69,* 25P (1958).
119. Strange and Kent, *Biochem. J., 71,* 333 (1959).
120. Lambert and Zilliken, *Ber. Deut. Chem. Ges., 93,* 2915 (1960).
121. Flowers and Jeanloz, *J. Org. Chem., 28,* 1564, 2983 (1963).
122. Faillard, *Hoppe-Seyler's Z. Physiol. Chem., 307,* 62 (1957).
123. Zilliken and McGlick, *Naturwissenschaften, 43,* 536 (1956).
124. Kuhn and Baschang, *Ann. Chem. (Justus Liebigs), 659,* 156 (1962).
125. Cornforth, Daines and Gottschalk, *Proc. Chem. Soc. (London),* p. 25 (1957).
126. Faillard and Blohm, *Hoppe-Seyler's Z. Physiol. Chem., 341,* 167 (1965).
127. Kondo, Akita and Sezaki, *J. Antibiot. (Tokyo) Ser. A, 19,* 137 (1966).
128. Menzinsky, *Ber. Deut. Chem. Ges., 68,* 822 (1935).
129. Yurkevich, Verenikina, Dolgikh and Preobrazhenskii, *J. Gen. Chem., U.S.S.R., 37,* 1201 (1967).
130. Hart and Everett, *J. Amer. Chem. Soc., 61,* 1822 (1939).
131. Okazaki, Kanyaki, Doi, Nara and Motizuki, *Agr. Biol. Chem., 32,* 1250 (1968).
132. Hershberger and Binkley, *J. Biol. Chem., 243,* 1578 (1968).
133. Hershberger, Davis and Binkley, *J. Biol. Chem., 243,* 1585 (1968).
134. Humoller, McManus and Austin, *J. Amer. Chem. Soc., 58,* 2479 (1936).
135. Bond, *Chem. Commun.,* p. 338 (1969).
136. Yonehara and Otake, *Tetrahedron Lett.,* p. 3785 (1966).

Table 5


1. Wohl and Momber, *Ber. Deut. Chem. Ges., 50,* 456 (1917).
2. Fischer and Baer, *Helv. Chim. Acta, 17,* 622 (1934).
3. Williams and Jones, *Can. J. Chem., 42,* 69 (1964).
4. Schmidt, *Ann. Chem. (Justus Liebigs), 483,* 115 (1930).
5. Vongerichten, *Ann. Chem. (Justus Liebigs), 321,* 71 (1902).
6. Bentley, Cunningham and Spring, *J. Chem. Soc. (London),* p. 2301 (1951).
7. Hockett and Hudson, *J. Amer. Chem. Soc., 56,* 1632 (1934).
8. Fischer, Bergmann and Schotts, *Ber. Deut. Chem. Ges., 53,* 522 (1920).
9. Halliburton and McIlroy, *J. Chem. Soc. (London),* p. 299 (1949).
10. Lynch, Olney and Wright, *J. Sci. Food Agr., 9,* 56 (1958).
11. Jones, Kent and Stacey, *J. Chem. Soc. (London),* p. 1341 (1947).
12. Sowden, Oftedahl and Kirkland, *J. Org. Chem., 27,* 1791 (1962).
13. Vogel, *Helv. Chim. Acta, 11,* 1210 (1928).
14. Montgomery and Hudson, *J. Amer. Chem. Soc., 56,* 2074 (1934).
15. Hudson and Dale, *J. Amer. Chem. Soc., 40,* 995 (1918).
16. Whistler and Kirby, *J. Amer. Chem. Soc., 78,* 1755 (1956).
17. Mackie and Percival, *Biochem. J., 91,* 5P (1964).
18. Alberda van Ekenstein, *Chem. Weekbl., 11,* 189 (1914).
19. Levene and Wolfrom, *J. Biol. Chem., 78,* 525 (1928).
20. Galmarini and Deulofeu, *Tetrahedron, 15,* 76 (1961).
21. Dyer, McGonigal and Rice, *J. Amer. Chem. Soc., 87,* 654 (1965).
22. Kuehl, Jr., Flynn, Brink and Folkers, *J. Amer. Chem. Soc., 68,* 2679 (1946).
23. Tatsuoka, Kusaka, Miyake, Inone, Hitomi, Shiraishi, Iwasaki and Imanishi, *Pharm. Bull. (Tokyo), 5,* 343 (1957).
24. Stodola, Shotwell, Borud, Benedict and Riley, Jr., *J. Amer. Chem. Soc., 73,* 2290, 5912 (1951).
25. Hogenkamp and Barker, *J. Biol. Chem., 236,* 3097 (1961).
26. Deriaz, Overend, Stacey, Teece and Wiggins, *J. Chem. Soc. (London),* p. 1879 (1949).
27. Allerton and Overend, *J. Chem. Soc. (London),* p. 1480 (1951).
28. Phelps, Isbell and Pigman, *J. Amer. Chem. Soc., 56,* 747 (1934).
29. Zinner, *Ber. Deut. Chem. Ges., 86,* 817 (1953).
30. Levene and Tipson, *J. Biol. Chem., 115,* 731 (1936).
31. Overend and Williams, *J. Chem. Soc. (London),* p. 3446 (1964).
32. Freudenberg and Blummel, *Ann. Chem. (Justus Liebigs), 440,* 45 (1924).
33. Hudson and Yanovsky, *J. Amer. Chem. Soc., 39,* 1013 (1917).
34. Isbell and Pigman, *J. Res. Nat. Bur. Stand., 18,* 141 (1937).
35. Hudson and Johnson, *J. Amer. Chem. Soc., 37,* 2748 (1915).
36. Ruff and Ollendorf, *Ber. Deut. Chem. Ges., 32,* 3234 (1899).
37. Votoček, Valentin and Leminger, *Collect. Czech. Chem. Commun., 3,* 252 (1931).
38. Alberda van Ekenstein and Blanksma, *Rec. Trav. Chim. Pays-Bas, 22,* 434 (1903).
39. Reclaire, *Ber. Deut. Chem. Ges., 41,* 3665 (1908).
40. Gorin, Hough and Jones, *J. Chem. Soc. (London),* p. 2140 (1953).
41. Levene and Compton, *J. Biol. Chem., 111,* 325 (1935).
42. Andrews and Hough, *Chem. Ind. (London),* p. 1278 (1956).
43. Robertson and Speedie, *J. Chem. Soc. (London),* p. 824 (1934).
44. Aspinall and McKay, *J. Chem. Soc. (London),* p. 1059 (1958).
45. Laidlaw, *J. Chem. Soc. (London),* p. 752 (1954).
46. Glaudemans and Timell, *J. Amer. Chem. Soc., 80,* 1209 (1958).
47. Laidlaw and Percival, *J. Chem. Soc. (London),* p. 528 (1950).
48. Kunstmann, Mitscher and Bohonos, *Tetrahedron Lett.,* p. 839 (1966).
49. Kauss, *Z. Pflanzenphysiol., 53,* 58 (1965).
50. Phelps and Bates, *J. Amer. Chem. Soc., 56,* 1250 (1934).
51. Levene and Jacobs, *Ber. Deut. Chem. Ges., 43,* 3141 (1910).
52. Lerner and Kohn, *J. Med. Chem., 7,* 655 (1964).
53. Dion, Woo and Bartz, *J. Amer. Chem. Soc., 84,* 880 (1962).
54. Muhlradt, Weiss and Reichstein, *Ann. Chem. (Justus Liebigs), 685,* 253 (1965).
55. Levene and Compton, *J. Biol. Chem., 116,* 169 (1936).
56. Keller and Reichstein, *Helv. Chim. Acta, 32,* 1607 (1949).
57. Levene and Compton, *J. Biol. Chem., 117,* 37 (1937).
58. Iselin and Reichstein, *Helv. Chim. Acta, 27,* 1203 (1944).
59. Brimacombe and Husain, *Chem. Commun.,* p. 630 (1966).
60. Brimacombe and Portsmouth, *J. Chem. Soc. Sect. C,* p. 499 (1966).
61. Brimacombe, Stacey and Tucker, *J. Chem. Soc. (London),* p. 5391 (1964).
62. Jager, Dissertation. Basel, Switzerland (1959).

63. Gut and Prins, *Helv. Chim. Acta, 29,* 1555 (1946).
64. Iwadare, *Bull. Chem. Soc. Jap., 17,* 296 (1942).
65. Krauss, Dissertation. Basel, Switzerland (1959).
66. Grob and Prins, *Helv. Chim. Acta, 28,* 840 (1945).
67. Kaufmann, *Helv. Chim. Acta, 48,* 83 (1965).
68. Kiliani, *Ber. Deut. Chem. Ges., 46,* 667 (1913).
69. Kiliani, *Arch. Pharm., 234,* 449 (1896); *Chem. Zentralbl., 67,* 591 (1896).
70. Ruber, Minsaas and Lyche, *J. Chem. Soc. (London),* p. 2173 (1929).
71. Wolfrom, Schlamowitz and Thompson, *J. Amer. Chem. Soc., 76,* 1198 (1954).
72. Hudson and Parker, *J. Amer. Chem. Soc., 37,* 1589 (1915).
73. O'Neill, *J. Amer. Chem. Soc., 77,* 2837 (1955).
74. Araki and Hirase, *Bull. Chem. Soc. Jap., 29,* 770 (1956).
75. Clingman and Nunn, *J. Chem. Soc. (London),* p. 493 (1959).
76. Gorin and Spencer, *Can. J. Chem., 42,* 1230 (1964).
77. Gorin and Ishikawa, *Can. J. Chem., 45,* 521 (1967).
78. Votoček and Valentin, *Collect. Czech. Chem. Commun., 2,* 36 (1930).
79. Levy and McAllan, *Biochem. J., 80,* 433 (1961).
80. MacPhillamy and Elderfield, *J. Org. Chem., 4,* 150 (1939).
81. Springer, Desai and Kolechi, *Biochemistry, 3,* 1076 (1964).
82. Khare, Schindler and Reichstein, *Helv. Chim. Acta, 45,* 1534 (1962).
83. Lamb and Smith, *J. Chem. Soc. (London),* p. 442 (1936).
84. Schmidt and Wernicke, *Ann. Chem. (Justus Liebigs), 556,* 179 (1944). 85.
85. Akita, Maeda and Umezawa, *J. Antibiot. (Tokyo) Ser. A, 17,* 200 (1964).
86. Reber and Reichstein, *Helv. Chim. Acta, 28,* 1164 (1945).
87. Hirst and Jones, *J. Chem. Soc. (London),* p. 506 (1946).
88. Jeanloz, *J. Amer. Chem. Soc., 76,* 5684 (1954).
89. Itasaka, *J. Biochem. (Tokyo), 60,* 52 (1966).
90. Nunn and von Holdr, *J. Chem. Soc. (London),* p. 1094 (1947).
91. Hassid and Su, *Biochemistry, 1,* 468 (1962).
92. Goldstein, Hamilton and Smith, *J. Amer. Chem. Soc., 79,* 1190 (1957).
93. Love and Percival, *J. Chem. Soc. (London),* p. 3338 (1964).
94. Turvey and Williams, *J. Chem. Soc. (London),* p. 2119 (1962).
95. Anderson, *J. Biol. Chem., 100,* 249 (1933).
96. Fischer and Hertz, *Ber. Deut. Chem. Ges., 25,* 1247 (1892).
97. Araki and Hirase, *Bull Chem. Soc. Jap., 26,* 463 (1953).
98. Duff and Percival, *J. Chem. Soc. (London),* p. 830 (1941).
99. Minsaas, *Rec. Trav. Chim. Pays-Bas, 50,* 424 (1933).
100. Aspinall, Jamieson and Wilkinson, *J. Chem. Soc. (London),* p. 3483 (1956).
101. Westphal and Feier, *Ber. Deut. Chem. Ges., 89,* 582 (1956).
102. Gardiner and Percival, *J. Chem. Soc. (London),* p. 1414 (1958).
103. Anderson, Andrews and Hough, *Chem. Ind. (London),* p. 1453 (1957).
104. Turvey and Rees, *Nature, 189,* 831 (1961).
105. Quilico, Piozzi, Pavan and Mantia, *Tetrahedron, 5,* 10 (1959).
106. Zervas, *Ber. Deut. Chem. Ges., 64,* 2289 (1931).
107. Plouvier, *Compt. Rend., 256,* 1397 (1963).
108. Hudson and Dale, *J. Amer. Chem. Soc., 37,* 1264, 1280 (1915).
109. Harris, Hirst and Wood, *J. Chem. Soc. (London),* p. 2108 (1932).
110. Purdie and Irvine, *J. Chem. Soc. (London),* p. 1049 (1904).
111. Bates, in *Polarimetry, Saccharimetry and the Sugars (National Bureau of Standards Circular C440),* p. 728. U.S. Government Printing Office, Washington, D.C. (1942).
111a. Wolfrom and Karabinos, *J. Amer. Chem. Soc., 67,* 500 (1945).
112. Georg, *Helv. Chim. Acta, 12,* 261 (1929).
113. Brigl and Scheyer, *Hoppe Seyler's Z. Physiol. Chem., 160,* 214 (1926).
114. Alberda van Ekenstein and Blanksma, *Rec. Trav. Chim. Pays-Bas, 24,* 33 (1905).
115. Duff, *J. Chem. Soc. (London),* p. 4730 (1957).
116. Frohwein and Leibowitz, *Nature, 186,* 153 (1960).
117. Josephson, *Ber. Deut. Chem. Ges., 62,* 317 (1929).
118. Ohle, *Biochem. Z., 131,* 611 (1922).
119. Brigl and Grüner, *Ann. Chem. (Justus Liebigs), 495,* 60 (1932).
120. Fischer and Lieberman, *Ber. Deut. Chem. Ges., 26,* 2415 (1893).
121. Stanek and Tajmr, *Chem. Listy, 52,* 551 (1958).
122. Fischer and Zach, *Ber. Deut. Chem. Ges., 45,* 3761 (1902).

123. Frèrejacque, *Compt. Rend., 230,* 127 (1950).
124. Korte, *Ber. Deut. Chem. Ges., 88,* 1527 (1955).
125. Reyler and Reichstein, *Helv. Chim. Acta, 35,* 195 (1956).
126. Miyano and Benson, *J. Amer. Chem. Soc., 84,* 59 (1962).
127. Ebert and Zenk, *Arch. Mikrobiol., 54,* 276 (1966).
128. Chanley, Ledeen, Wax, Nigrelli and Sobotka, *J. Amer. Chem. Soc., 81,* 5180 (1959).
129. Irvine and Scott, *J. Chem. Soc. (London),* pp. 571, 575, 582 (1913).
130. Jeanloz and Gut, *J. Amer. Chem. Soc., 76,* 5793 (1954).
131. Oldham, *J. Amer. Chem. Soc., 56,* 1360 (1934).
132. Jeanloz, Rapin and Hakomori, *J. Org. Chem., 26,* 3939 (1961).
133. Levene and Raymond, *J. Biol. Chem., 88,* 513 (1930).
134. Lee and Ballou, *J. Biol. Chem., 239,* 3602 (1964).
135. Helferich, Klein and Schafer, *Ber. Deut. Chem. Ges., 59,* 79 (1926).
136. Helferich and Himmen, *Ber. Deut. Chem. Ges., 62,* 2136, 2141 (1929).
137. Helferich and Günther, *Ber. Deut. Chem. Ges., 64,* 1276 (1931).
138. White and Rao, *J. Amer. Chem. Soc., 75,* 2617 (1953).
139. Christensen and Smith, *J. Amer. Chem. Soc., 79,* 4492 (1957).
140. Finnegan, Mueller and Morris, *Proc. Chem. Soc. (London),* p. 182 (1963).
141. Carter, *J. Sci. Food Agr., 2,* 54 (1951).
142. Fischer, *Ber. Deut. Chem. Ges., 23,* 2618 (1890).
143. Blindenbacher and Reichstein, *Helv. Chim. Acta, 31,* 1669 (1948).
144. Frèrejacque and Hasenfratz, *Compt. Rend., 222,* 815 (1946).
145. Frèrejacque and Durgeat, *Compt. Rend., 228,* 1310 (1949).
146. Doebel, Schlittler and Reichstein, *Helv. Chim. Acta, 31,* 688 (1948).
147. Overend, Stacey and Stanek, *J. Chem. Soc. (London),* p. 2841 (1949).
148. Zorbach and Ciaudelli, *J. Org. Chem., 30,* 451 (1965).
149. Studer, Panavaram, Gavilanes, Linde and Meyer, *Helv. Chim. Acta, 46,* 23 (1963).
150. Berlin, Esipov, Kolosov, Shemyakin and Brazhnikova, *Tetrahedron Lett.,* p. 1323 (1964).
151. Miyamoto, Kawamatsu, Shinohara, Nakadaira and Nakanishi, *Tetrahedron, 22,* 2784 (1966).
152. Westphal, Lüderitz, Framme and Joseph, *Angew. Chem., 65,* 555 (1953).
153. Fouquey, Lederer, Lüderitz, Polonsky, Staub, Stirm, Tirelli and Westphal, *Compt. Rend., 246,* 2417 (1958).
154. Stirm, Lüderitz and Westphal, *Ann. Chem. (Justus Liebigs), 696,* 180 (1966).
155. Vischer and Reichstein, *Helv. Chim. Acta, 27,* 1332 (1944).
156. Tschesche and Buschauer, *Ann. Chem. (Justus Liebigs), 603,* 59 (1957).
157. Brimacombe and Portsmouth, *Carbohydr. Res., 1,* 128 (1965).
158. Davies, *Nature, 191,* 43 (1961).
159. Blindenbacher and Reichstein, *Helv. Chim. Acta, 31,* 2061 (1948).
160. Hesse, *Ber. Deut. Chem. Ges., 70,* 2264 (1937).
161. Celmer and Hobbs, *Carbohydr. Res., 1,* 137 (1965).
162. Stevens, Nagarajan and Haskell, *J. Org. Chem., 27,* 2991 (1962).
163. Stevens, Cross and Toda, *J. Org. Chem., 28,* 1283 (1963).
164. Herzog, Meseck, Delorenzo, Murawski, Charney and Rosselet, *Appl. Microbiol., 13,* 515 (1965).
165. Avigad, Amaral, Asensio and Horecker, *J. Biol. Chem., 237,* 2736 (1962).
166. Wolfrom and Usdin, *J. Amer. Chem. Soc., 75,* 4318 (1953).
167. Brimacombe and Portsmouth, *Chem. Ind. (London),* p. 468 (1965).
168. Berlin, Esipov, Kolosov and Shemyakin, *Tetrahedron Lett.,* p. 1431 (1966).
169. Brimacombe, Portsmouth and Stacey, *J. Chem. Soc. (London),* p. 5614 (1965).
170. Miyamoto, Kawamatsu, Shinohara, Asahi, Nakedaira, Kakisawa, Nakanishi and Bhacca, *Tetrahedron Lett.,* p. 693 (1963).
171. Shoppe and Reichstein, *Helv. Chim. Acta, 25,* 1611 (1942).
172. Tamm and Reichstein, *Helv. Chim. Acta, 31,* 1630 (1948).
173. Iselin and Reichstein, *Helv. Chim. Acta, 27,* 1200 (1944).
174. Brockmann and Waehneldt, *Naturwissenschaften, 48,* 717 (1961).
175. Renkonen, Schindler and Reichstein, *Helv. Chim. Acta, 42,* 182 (1959).
176. Kiliani, *Arch. Pharm., 234,* 486 (1896).
177. Micheel, *Ber. Deut. Chem. Ges., 63,* 347 (1930).

178. Kiliani, *Ber. Deut. Chem. Ges., 31,* 2454 (1898).
179. Fouquey, Polonsky, Lederer, Westphal and Lüderitz, *Nature, 182,* 944 (1958).
180. Krasso, Weiss and Reichstein, *Helv. Chim. Acta, 46,* 1691 (1963).
181. Jacobs, *J. Biol. Chem., 88,* 519 (1930).
182. Prins, *Helv. Chim. Acta, 29,* 378 (1946).
183. Bolliger and Ulrich, *Helv. Chim. Acta, 35,* 93 (1952).
184. Woo, Dion and Bartz, *J. Amer. Chem. Soc., 83,* 3352 (1961).
185. Keller-Schierlein and Roncari, *Helv. Chim. Acta, 45* 138 (1962).
186. Stevens, Blumbergs and Wood, *J. Amer. Chem. Soc., 86,* 3592 (1964).
187. Brockmann and Waehneldt, *Naturwissenschaften, 50,* 43 (1963).
188. Kowalewski, Schindler, Jager and Reichstein, *Helv. Chim. Acta, 43,* 1280 (1960).
189. Golab and Reichstein, *Helv. Chim. Acta, 44,* 616 (1961).
190. Bolliger and Reichstein, *Helv. Chim. Acta, 36,* 302 (1953).
191. Jacobs and Bigelow, *J. Biol. Chem., 96,* 355 (1932).
192. Hauenstein and Reichstein, *Helv. Chim. Acta, 33,* 446 (1950).
193. Lüderitz, Staub, Stirm and Westphal, *Biochem. Z., 330,* 193 (1958).
194. Regna, Hochstein, Wagner and Woodward, *J. Amer. Chem. Soc., 75,* 4625 (1953).
195. Paul and Tchelitcheff, *Bull. Soc. Chim. Fr.,* p. 443 (1957).
196. Flaherty, Overend and Williams, *J. Chem. Soc. (London),* p. 398 (1966).
197. Watanabe, Nishida and Satake, *Bull. Chem. Soc. Jap., 34,* 1285 (1961).
198. Flynn, Sigal, Wiley and Gerzon, *J. Amer. Chem. Soc., 76,* 3121 (1954).
199. Corcoran, *J. Biol. Chem., 236,* PC27 (1961).
200. Wiley and Weaver, *J. Amer. Chem. Soc., 77,* 3422 (1955); *78,* 808 (1956).
201. Roneari and Keller-Schierlein, *Helv. Chim. Acta, 49,* 705 (1966).
202. Haworth, Raistrick and Stacey, *Biochem. J., 29,* 2668 (1935).
203. Baggett, Stoffyn and Jeanloz, *J. Org. Chem., 28,* 1041 (1963).
204. Levene, *J. Biol. Chem., 57,* 329 (1923); *59,* 129 (1924).
205. Rüber and Minsaas, *Ber. Deut. Chem. Ges., 60,* 2402 (1927).
206. Butler and Cretcher, *J. Amer. Chem. Soc., 53,* 4358, 4363 (1931).
207. Fischer and Oetker, *Ber. Deut. Chem. Ges., 46,* 4039 (1913).
208. Caudy and Baddiley, *Biochem. J., 98,* 15 (1966).
209. Aspinall and Zweifel, *J. Chem. Soc. (London),* p. 2271 (1957).
210. Webb, Broschard, Cosulick, Mowat and Lancaster, *J. Amer. Chem. Soc., 84,* 3183 (1962).
211. Markovitz, *J. Biol. Chem., 237,* 1767 (1962).
212. Behrend, *Ber. Deut. Chem. Ges., 11,* 1353 (1878).
213. Fischer, *Ber. Deut. Chem. Ges., 29,* 324 (1896).
214. Fischer, Bergmann and Rabe, *Ber. Deut. Chem. Ges., 53,* 2362 (1920).
215. MacLennan, Smith and Randell, *Biochem. J., 74,* 3P (1960).
216. Andrews, Hough and Jones, *J. Amer. Chem. Soc., 77,* 125 (1955).
217. Young and Elderfield, *J. Org. Chem., 7,* 241 (1942).
218. Hirst, Percival and Williams, *J. Chem. Soc. (London),* p. 1942 (1958).
219. Schmidt, Plankenhorn and Kübler, *Ber. Deut. Chem. Ges., 75,* 579 (1942).
220. Charalambous and Percival, *J. Chem. Soc. (London),* p. 2443 (1954).
221. Vaterlaus, Kiss and Spieglberg, *Helv. Chim. Acta, 47,* 381 (1964).
222. Walton, Rodin, Stammer, Holly and Folkers, *J. Amer. Chem. Soc., 80,* 5168 (1958).
223. Barker, Homer, Keith and Thomas, *J. Chem. Soc. (London),* p. 1538 (1963).
224. Hinman, Caron and Hoeksema, *J. Amer. Chem. Soc., 79,* 3789 (1957).
225. Wiley and Sigal, *J. Amer. Chem. Soc., 80,* 1010 (1958).
226. Pigman and Isbell, *J. Res. Nat. Bur. Stand., 19,* 189 (1937).
227. MacLennan, *Biochim. Biophys. Acta, 48,* 600 (1961).
228. Collins and Overend, *J. Chem. Soc. (London),* p. 1912 (1965).
229. Schmitz, *Helv. Chim. Acta, 31,* 1719 (1948).
230. Von Euw and Reichstein, *Helv. Chim. Acta, 33,* 485 (1950).
231. Sephton and Richtmyer, *J. Org. Chem., 28,* 1691 (1963).
232. Palleroni and Doudoroff, *J. Biol. Chem., 218,* 535 (1956).
233. Hulyalkar, Jones and Perry, *Can. J. Chem., 41,* 1490 (1963).
234. Young and Adams, *Can. J. Chem., 43,* 2929 (1965).
235. Richtmyer and Charlson, *J. Amer. Chem. Soc., 82,* 3428 (1960).
236. Slein and Schnell, *Proc. Soc. Exp. Biol. Med., 82,* 734 (1953).
237. Weidell, *Hoppe-Seyler's Z. Physiol. Chem., 299,* 253 (1955).
238. Davies, *Nature, 180,* 1129 (1957).
239. Kuriki and Kurahashi, *J. Biochem. (Tokyo), 58,* 308 (1965).

240. Fraenkel, Osborn, Horecker and Smith, *Biochem. Biophys. Res. Commun., 11,* 423 (1963).
241. Missale, Colajacomo and Bologna, *Bull. Soc. Ital. Biol. Sper., 36,* 1885 (1960); *Chem. Abstr., 55,* 24869 (1961).
242. Stoffyn and Jeanloz, *J. Biol. Chem., 235,* 2507 (1960).
243. Sorkin and Reichstein, *Helv. Chim. Acta, 28,* 1, 662 (1945).
244. Begbie and Richtmyer, *Carbohydr. Res., 2,* 272 (1966).
245. Hoffman, Weiss and Reichstein, *Helv. Chim. Acta, 49,* 2209 (1966).
246. Krasso and Weiss, *Helv. Chim. Acta, 49,* 1113 (1966).
247. Brimacombe, Ching and Stacey, *J. Chem. Soc. Sect. C,* p. 197 (1969).
248. Nunn and Parolis, *Carbohydr. Res., 6,* 1 (1968).
249. Bowker and Turvey, *J. Chem. Soc. Sect. C,* p. 983 (1968).
250. Harris and Turvey, *Carbohydr. Res., 9,* 397 (1969).
251. Araki and Hirase, *Bull. Chem. Soc. Jap., 33,* 291 (1960).
252. Nunn and Parolis, *Carbohydr. Res., 8,* 361 (1968).
253. Allgeier, Weiss and Reichstein, *Helv. Chim. Acta, 50,* 456 (1967).
254. Bonner, *J. Org. Chem., 26,* 908 (1961).
255. Berlin, Esipov, Kiselava and Kolosov, *Chem. Natur. Compounds, 3,* 280 (1967).
256. Ganguly, Sarre and Reimann, *J. Amer. Chem. Soc., 90,* 7129 (1968).
257. Brimacombe and Portsmouth, *Carbohydr. Res., 1,* 128 (1965).
258. Siewert and Westphal, *Ann. Chem. (Justus Liebigs), 720,* 188 (1968).
259. Allgeier, *Helv. Chim. Acta, 51,* 668 (1968).
260. Wiley, Mackellar, Carron and Kelly, *Tetrahedron Lett.,* p. 663 (1968).
261. Okuda, Suzuki and Suzuki, *J. Biol. Chem., 242,* 958 (1967); *243,* 6353 (1968).
262. Morrison, Young, Barry and Adams, *Can. J. Chem., 45,* 1987 (1967).
263. MacLennan, *Biochem. J., 82,* 394 (1962).
264. Brown, Hough and Jones, *J. Chem. Soc. (London),* p. 1125 (1950).
265. Percival and Percival, *J. Chem. Soc. (London),* p. 690 (1950).
266. Chaput, Michel and Lederer, *Experientia, 17,* 107 (1961).
267. Hirst, Hough and Jones, *J. Chem. Soc. (London),* p. 3145 (1949).
268. Butler, Lloyd and Stacey, *J. Chem. Soc. (London),* p. 1531 (1955).
269. Stevens, Glinski and Taylor, *J. Org. Chem., 33,* 1586 (1968).
270. Kapur and Allgeier, *Helv. Chim. Acta, 51,* 89 (1968).
271. Strobach and Szabo, *J. Chem. Soc. (London),* p. 3970 (1963).

**Table 6**

1. Heilbron and Bunbury, *Dictionary of Organic Compounds,* Vol. 1. Oxford University Press, New York (1943).
2. Bertrand, *Bull. Soc. Chim. Fr. Ser. 3, 23,* 681 (1904).
3. Hu, McComb and Rendig, *Arch. Biochem. Biophys., 110,* 350 (1965).
4. Müller, Montigel and Reichstein, *Helv. Chim. Acta, 20,* 1468 (1937).
5. Hickman and Ashwell, *J. Amer. Chem. Soc., 78,* 6209 (1956).
6. Futterman and Roe, *J. Biol. Chem., 215,* 257 (1955).
7. Horecker, Smyrniotis and Seegmiller, *J. Biol. Chem., 193,* 383 (1951).
8. Reichstein, *Helv. Chim. Acta, 17,* 996 (1934).
9. Simpson, Wolin and Wood, *J. Biol. Chem., 230,* 457 (1958).
10. Ashwell and Hickman, *J. Amer. Chem. Soc., 77,* 1062 (1955).
11. Ashwell and Hickman, *J. Biol. Chem., 226,* 65 (1957).
12. Gorin, Hough and Jones, *J. Chem. Soc. (London),* p. 2140 (1953).
13. Levene and LaForge, *J. Biol. Chem., 18,* 319 (1914).
14. Wolfrom and Bennett, *J. Org. Chem., 30,* 458 (1965).
15. Chassy, Sugimori and Suhadolnik, *Biochim. Biophys. Acta, 130,* 12 (1966).
16. Avigad and Englard, *J. Biol. Chem., 240,* 2290 (1965).
17. Hoeksema, Argoudelis and Wiley, *J. Amer. Chem. Soc., 84,* 3212 (1962).
18. Bean and Hassid, *Science, 124,* 171 (1956).
19. Bayne, Collie and Fewster, *J. Chem. Soc. (London),* p. 2766 (1952).
20. Maurer, *Ber. Deut. Chem. Ges., 63,* 25 (1930).
21. Kato, *Agr. Biol. Chem., 27,* 461 (1963).
22. Mann and Woolf, *J. Amer. Chem. Soc., 79,* 120 (1957).
23. Weidenhagen and Bernsee, *Angew. Chem., 72,* 109 (1960).
24. Fukui and Hochster, *J. Amer. Chem. Soc., 85,* 1697 (1963).
25. Hudson and Brauns, *J. Amer. Chem. Soc., 38,* 1216 (1916).
26. Hudson and Yanovsky, *J. Amer. Chem. Soc., 39,* 1025 (1917).

27. Brigl and Schinle, *Ber. Deut. Chem. Ges., 66,* 325 (1933).
28. Pacsu and Rich, *J. Amer. Chem. Soc., 55,* 3018 (1933).
28a. Brauns, *Proc. Roy Acad. Amsterdam, 10,* 563 (1907–1908); Barry and Honeyman, *Adv. Carbohydr. Chem., 7,* 60, 85 (1952).
29. Reclaire, *Ber. Deut. Chem. Ges., 41,* 3665 (1908).
30. Morgan and Reichstein, *Helv. Chim. Acta, 21,* 1023 (1938).
31. Hough and Jones, *J. Chem. Soc. (London),* p. 4052 (1952).
32. Reichstein and Bosshard, *Helv. Chim. Acta, 17,* 753 (1934).
33. Khouvine and Tomoda, *Compt. Rend., 205,* 736 (1937).
34. Khouvine, Arragon and Tomoda, *Bull. Soc. Chim. Fr. Ser. 5, 6,* 354 (1939).
35. Barnett and Reichstein, *Helv. Chim. Acta, 21,* 913 (1938).
36. Green and Cohen, *J. Biol. Chem., 219,* 557 (1956).
37. Yüngsten, *J. Antibiot. (Tokyo) Ser. A, 11,* 77, 233 (1958).
38. Yüngsten, *J. Antibiot. (Tokyo) Ser. A, 11,* 244 (1958).
39. Wolfrom, Thompson and Evans, *J. Amer. Chem. Soc., 67,* 1793 (1945).
40. Binkley and Wolfrom, *J. Amer. Chem. Soc., 70,* 3940 (1948).
41. Strecker, Gouret and Montreuil, *Compt. Rend., 260,* 999 (1965).
42. Schlubach and Vorwerk, *Ber. Deut. Chem. Ges., 66,* 1251 (1933).
43. Schlubach and Graefe, *Ann. Chem. (Justus Liebigs), 532,* 211 (1937).
44. Müller and Reichstein, *Helv. Chim. Acta, 21,* 263 (1938).
45. Terada, Suzuki and Kinoshita, *Agr. Biol. Chem., 25,* 939 (1961).
46. LaForge and Hudson, *J. Biol. Chem., 30,* 61 (1917).
47. Pratt, Richtmyer and Hudson, *J. Amer. Chem. Soc., 74,* 2203, 2210 (1952).
48. Bertrand, *Bull. Soc. Chim. Fr. Ser. 4, 51,* 629 (1909).
49. Hann and Hudson, *J. Amer. Chem. Soc., 61,* 336 (1939).
50. Khouvine and Arragon, *Compt. Rend., 206,* 917 (1938).
51. McComb and Rendig, *Arch. Biochem. Biophys., 95,* 316 (1961).
52. Stewart, Richtmyer and Hudson, *J. Amer. Chem. Soc., 74,* 2206 (1952).
53. Gorin and Jones, *J. Chem. Soc. (London),* p. 1537 (1953).
54. LaForge, *J. Biol. Chem., 28,* 511 (1917).
55. Bevenne, White, Secor and Williams, *J. Assoc. Offic. Agr. Chem., 44,* 265 (1961).
56. Montgomery and Hudson, *J. Amer. Chem. Soc., 61,* 1654 (1939).
57. Charlson and Richtmyer, *J. Amer. Chem. Soc., 82,* 3428 (1960).
58. Sephton and Richtmyer, *J. Org. Chem., 28,* 1691 (1963).
59. Jones and Sephton, *Can. J. Chem., 38,* 753 (1960).
60. Sephton and Richtmyer, *J. Org. Chem., 28,* 2388 (1963).
61. Takahashi and Nakajima, *Tetrahedron Lett.,* p. 2285 (1967).
62. Begbie and Richtmyer, *Carbohydr. Res., 2,* 272 (1966).
63. Sephton and Richtmyer, *Carbohydr. Res., 2,* 289 (1966).
64. Fodor, Sachetto, Szent-Györgyi and Együd, *Proc. Nat. Acad. Sci. U.S.A., 57,* 1644 (1967).
65. Okuda and Konishi, *Tetrahedron, 24,* 6907 (1968).

**Table 7**

1. Horton, *Adv. Carbohydr. Chem., 15,* 159 (1960).
2. Sharon, in *The Amino Sugars,* Vol. 2A, p. 1, Balazo and Jeanloz, Eds. Academic Press, New York (1965).
3. Wheat, *Methods Enzymol., 8,* 60 (1966).
4. Brendel, Roszel, Wheat and Davidson, *Anal. Biochem., 18,* 147 (1967); Brendel, Steele, Wheat and Davidson, *Anal. Biochem., 18,* 161 (1967).
5. Sharbarova, Buchanan and Baddiley, *Biochim. Biophys. Acta, 57,* 146 (1962).
6. Lüderitz, Jann and Wheat, *Comp. Biochem., 26,* (in press, 1970).
7. Rüde and Goebel, *J. Exp. Med., 116,* 73 (1962).
8. Neuberger, Marshall and Gottschalk, in *Glycoproteins,* p. 158, Gottschalk, Ed. Elsevier, Amsterdam, The Netherlands (1966).
9. Smith, *Biochem. Biophys. Res. Commun., 15,* 593 (1964); Colwell, Smith and Chapman, *Can. J. Microbiol., 14,* 165 (1968).
10. Raff and Wheat, *J. Biol. Chem., 242,* 4610 (1967).
11. Wheat, Rollins and Leatherwood, *Biochem. Biophys. Res. Commun., 9,* 120 (1962).
12. Smith, Leatherwood and Wheat, *J. Bacteriol., 84,* 100 (1962).
13. Crumpton and Davies, *Biochem. J., 70,* 729 (1958).
14. Wheat, Rollins and Leatherwood, *Nature, 202,* 492 (1965).
15. Barry and Roark, *Nature, 202,* 493 (1965).

16. Barker, Brimacombe, Horn and Stacey, *Nature, 189,* 303 (1961).
17. Dutcher, *Adv. Carbohydr. Chem., 18,* 259 (1965).
18. Cron, Forbig, Johnson, Schmitz, Whitehead, Hooper and Lemieux, *J. Amer. Chem. Soc., 80,* 2342 (1958).
19. Raff and Wheat, *Fed. Proc., 26,* 281 (1967).
20. Lüderitz, Ruschmann, Westphal, Raff and Wheat, *J. Bacteriol., 94,* 5 (1967).
21. Jann, Jann and Müller-Seitz, *Nature, 215,* 170 (1967).
22. Ashwell and Volk, *J. Biol. Chem., 240,* 4549 (1965).
23. Baker, Schaub and Kissman, *J. Amer. Chem. Soc., 77,* 5911 (1955).
24. Baer and Fischer, *J. Amer. Chem. Soc., 81,* 5184 (1959).
25. Stevens, Blumberg, Daniker, Wheat, Kujomoto and Rollins, *J. Amer. Chem. Soc., 85,* 3061 (1963).
26. Matsuhashi and Strominger, *J. Biol. Chem., 239,* 2454 (1964).
27. Rinehart, Jr., Woo and Argoudelis, *J. Amer. Chem. Soc., 80,* 6461 (1958).
28. Hitchens and Rinehart, Jr., *J. Amer. Chem. Soc., 85,* 1547 (1963).
29. Zehavi and Sharon, *Isr. J. Chem., 2,* 322 (1964).
30. Sharon and Jeanloz, *J. Biol. Chem., 235,* 1 (1960).
31. Distler, Kauffman and Roseman, *Arch. Biochem. Biophys., 116,* 466 (1966).
32. Ashwell, Brown and Volk, *Arch. Biochem. Biophys., 112,* 648 (1965).
33. Walters, Dutcher and Wintersteiner, *J. Amer. Chem. Soc., 79,* 5076 (1957).
34. Kuehl, Jr., Flynn, Holly, Mozingo and Folkers, *J. Amer. Chem. Soc., 68,* 536 (1946); *69,* 3032 (1947).
35. Brockman, Konig and Oster, *Chem. Ber., 87,* 856 (1954).
36. Hochstein and Regna, *J. Amer. Chem. Soc., 77,* 3353 (1955).
37. Brockmann, Spohler and Waehneldt, *Chem. Ber., 96,* 2925 (1963).
38. Stevens, Gasser, Mukherjee and Haskell, *J. Amer. Chem. Soc., 78,* 6212 (1956).
39. Paul and Tchelitcheff, *Bull. Soc. Chim. Fr.,* p. 734 (1957).
40. Hanession and Haskell, *J. Biol. Chem., 239,* 2758 (1964).
41. Strange and Kent, *Biochem. J., 11,* 333 (1959).
42. Bourillon and Michon, *Bull. Soc. Chim. Biol., 41,* 267 (1959).
43. Raff and Wheat, *Fed. Proc., 24,* 478 (1965).
44. Gardell, *Acta Chem. Scand., 7,* 207 (1953).
45. Lüderitz, Gmeiner, Kickhofen, Mayer, Westphal and Wheat, *J. Bacteriol., 95,* 490 (1968).
46. Weise, Drews, Jann and Jann, Personal Communication from Dr. Drews (1969).
47. Livington, *J. Bacteriol., 99,* 85 (1969).
48. Erler, *Arch. Exp. Veterinaermed., 22,* 1155 (1968).
49. Suziki, *Biochim. Biophys. Acta, 177,* 371 (1969).
50. Jann and Jann, *Eur. J. Biochem., 5,* 173 (1967).
51. Jann and Jann, *Eur. J. Biochem., 2,* 26 (1967).
52. Mayer, *Eur. J. Biochem., 8,* 139 (1969).
53. Sharon, Shif and Zehavi, *Biochem. J., 93,* 210 (1964).

# POLYSACCHARIDES OF ALGAE

DR. E. E. PERCIVAL and DR. J. R. TURVEY

## INTRODUCTION

The algae cover a wide range of organisms with great diversity in size, morphology and metabolism. The majority are photosynthetic organisms, fairly low on the evolutionary scale, that is, lacking in the specialization encountered in bryophytes and vascular plants. In addition, certain colorless organisms are included on the basis of structural and morphological similarities to photosynthetic forms.

In size, algae vary from seaweeds several meters in length to unicellular microorganisms. In this survey macroalgae are described only in general terms, but microalgae, insofar as they have been examined, are covered in detail under individual species. Certain types that exist in filamentous or other forms but have a close relationship to unicellular species are also described in detail.

For the taxonomic classification, the authors have used essentially the system suggested by Silva.[1] Where a particular group of organisms is not covered in this system, other authorities, such as Fritsch,[2] have been used, and these are indicated, in parentheses, after the taxonomic group.

### Abbreviations and Symbols

In the following tables, monosaccharide constituents of a polysaccharide are indicated by the following abbreviations:

ara = arabinose　　fru = fructose　　fuc = fucose
gal = galactose　　glc = glucose　　man = mannose
rha = rhamnose　　rib = ribose　　xyl = xylose

galUA = galacturonic acid　　glcN = 2-amino-2-deoxyglucose
glcNAc = 2-acetamido-2-deoxyglucose　　glcUA = glucuronic acid
4MeglcUA = 4-O-methylglucuronic acid　　$SO_4$ = ester sulfate

Other sugars are given by their full names. The proportion of monosaccharides present is in most cases expressed as percentages; where no unit designation is given, the proportion is molecular.

In addition, the following abbreviations and symbols are used:

(t) = trace; (+) = present, but no quantitative data given

The symbols = and > have their usual significance in terms of quantities present.

## DIVISION CYANOPHYTA (BLUE-GREEN ALGAE)

| Taxonomic Group and Species | Polysaccharide | Monosaccharides | Reference |
|---|---|---|---|
| **CHROOCOCCALES** | | | |
| **Chroococcaceae** | | | |
| *Anacystis nidulans* | Cell-wall lipo-polysaccharide | man > gal, glc, fuc, rha, glcN | 3 |

| Taxonomic Group and Species | Polysaccharide | Monosaccharides | Reference |
|---|---|---|---|
| **NOSTOCALES** | | | |
| **Nostocaceae** | | | |
| *Amorphonostoc* | | | |
| *paludosum* | Unidentified | gal = man = xyl > glc, glcUA (t), rha (t) | 4 |
| *Anabaena* | | | |
| *cylindrica* | Unidentified | glc > xyl = glcUA > gal = rha = ara | 5 |
| *flos-aquae* | Extracellular and intracellular | glc, xyl, rha, glcUA | 6 |
| *oscillarioides* | Unidentified | gal = man = xyl > glc, ara (t) | 4 |
| *variabilis* | Unidentified | gal = glc = man = xyl, ara (t), rha (t), rib (t) | 4 |
| *Nostoc* sp. | Extracellular | ara, fuc, glc, glcUA | 6 |
| *commune* | Amylopectin | glc | 7 |
| | Unidentified | galUA + glcUA = 30%, gal + glc = 35%, rha = 10%, xyl = 25% | 7 |
| *muscorum* | Glycogen | glc | 8 |
| | Unidentified | ara, xyl, rha, glc, gal | 9 |
| *Pseudonostoc* sp. | Unidentified | gal = glc = man = xyl > glcUA | 4 |
| *Sphaeronostoc* | | | |
| *coeruleum* | Unidentified | gal = glc = xyl > man = rha, glcUA (t) | 4 |
| *Statonostoc* | | | |
| *linckia* | Unidentified | gal = glc = man, rha, ara (t), xyl (t), rib (t) | 4 |
| **Oscillatoriaceae** | | | |
| *Oscillatoria* sp. | Amylopectin | glc | 7 |
| | Unidentified | hexuronic acid = 30%, gal = 35%, xyl = 25%, rha = 10% | 7 |
| *princeps* | Glucan | glc | 9 |
| **Rivulariaceae** | | | |
| *Calothrix* | | | |
| *elenkinii* | Unidentified | gal = glc = xyl > man | 4 |
| *pulvinata* | Unidentified | gal, man | 10 |
| *scopulorum* | Water-soluble | gal, pentose, $SO_4$ | 11 |
| *Rivularia* | | | |
| *bullata* | Unidentified | ara, glc | 5 |
| **Uncertain** | | | |
| *Agmenellum* | | | |
| *quadruplicatum* | Whole-cell | glc = 17.4%, gal = 3.2%, fuc = 3.5%, hexuronic acid (+) | 12 |

# DIVISION RHODOPHYTA (RED ALGAE)

The marine Rhodophyta consist predominantly of macroscopic species. In general these contain water-soluble mucilages based on galactan sulfates,[13] but in certain species (e.g., *Rhodymenia palmata*) water-soluble xylans are the preponderant polysaccharides. In some species the cell wall consists of cellulose, but in others β-1,4-linked mannan, β-1,4-linked xylans, or β-1,3-linked xylans may form part or all of the skeletal polysaccharides.[13,14] Callose[15] and chitin[16] have also been reported in certain species. The principal reserve polysaccharide is floridean starch, which resembles glycogen and amylopectin in structure.[13]

The galactan sulfates belong to at least two distinct classes: agars and carrageenans. Agars have a structure based on 3-linked D-galactose units and 4-linked L-galactose (or 3,6-anhydro-L-galactose) units, whereas carrageenans have 3-linked D-galactose units and 4-linked D-galactose (or 3,6-anhydro-D-galactose) units. In addition, both polysaccharides may have additional ester sulfate and O-methyl ether groups on the sugar units.

Some other species of interest are given in the table below.

| Taxonomic Group and Species | Polysaccharide | Monosaccharides | Reference |
|---|---|---|---|
| **Class Bangiophyceae** | | | |
| **PORPHYRIDIALES** | | | |
| **Porphyridiaceae** | | | |
| *Porphyridium* | | | |
| *cruentum* | Cell-wall | gal, glc, xyl | 17,18 |
| | Extracellular | gal, glc, xyl, uronic acid, $SO_4$ = 10%, protein = 7% | 17,18 |
| **Class Florideophyceae** | | | |
| **NEMALIONALES** | | | |
| **Batrachospermaceae** | | | |
| *Batrachospermum* sp. | Extracellular | gal = 32.7%, man = 22%, xyl = 15.6%, glcUA = 15.3%, glc = 8.4%, ara = 4.5%, rha = 1.0%, 3-O-methyl rha = 0.55%, 3-O-methyl gal (t) | 19 |
| *moniliforme* | Mucilage | gal, xyl | 20 |
| **Lemaneaceae** | | | |
| *Lemanea* | | | |
| *fucina* | Hot-water extract | gal, xyl, man | 20 |
| *nodosa* | Hot-water extract | gal, xyl, man | 20 |
| *Tuomeya* | | | |
| *fluviatilis* | Cellulose | | 20 |
| | Glycogen (starch) | glc | 20 |
| | Mucilage | gal, xyl, ara, uronic acid | 20 |
| **Thoreaceae** | | | |
| *Thorea* | | | |
| *ramosissima* | Hot-water extract | man, gal (t), xyl (t) | 20 |

## DIVISION CRYPTOPHYTA

| Taxonomic Group and Species | Polysaccharide | Monosaccharides | Reference |
|---|---|---|---|
| **Class Cryptophyceae** | | | |
| **CRYPTOMONADALES** | | | |
| **Cryptomonadaceae** | | | |
| *Chilomonas paramecium* | Starch; 45% amylose, 55% amylopectin | glc | 21,22 |
| *Cryptomonas* sp. | Starch | glc | 23 |
| **Uncertain** | | | |
| *Cyanidium caldarum* | Highly branched glucan ($\alpha$-1,4 and $\alpha$-1,6) | glc | 24 |

## DIVISION PYRROPHYTA

The cell envelopes of Pyrrophyta are reputed to contain cellulose.[25] Food is stored as either fat or starch.[26]

| Taxonomic Group and Species | Polysaccharide | Monosaccharides | Reference |
|---|---|---|---|
| **Class Desmophyceae** | | | |
| **PROROCENTRALES** | | | |
| **Prorocentraceae** | | | |
| *Exuviella* sp. | Whole-plant | glc = 26.8%, gal = 8.3%, xyl (+), ara (+), rha (+) | 12 |
| **Class Dinophycae** | | | |
| **GYMNODINIALES** | | | |
| **Gymnodiniaceae** | | | |
| *Gymnodinium kovalevskii* | Acidic heteropolysaccharide | | 27 |
| *Amphidinium carteri* | Whole-plant | glc = 19%, gal = 8.4%, rha (+) | 12 |
| | Alcohol-insoluble fraction | glc = 8%, pentose = 2% | 28 |

| Taxonomic Group and Species | Polysaccharide | Monosaccharides | Reference |
|---|---|---|---|
| **PERIDINIALES** | | | |
| **Peridiniaceae** | | | |
| *Peridinium* | | | |
| *umbonatum* | Starch | glc | 29 |
| *westii* | Cell-wall glucan ($\beta$-1,3 and $\beta$-1,4) | glc, man (t), fuc (t), rha (t) | 30 |
| **Others** | | | |
| **DESMONADALES (Fritsch)** | | | |
| **Desmonadaceae** | | | |
| *Haplodinium antjoliense* | Starch | glc | 31 |
| **DINOCAPSALES (Fritsch)** | | | |
| **Dinocapsaceae** | | | |
| *Gloeodinium montanum* | Starch | glc | 32 |

## DIVISION BACILLARIOPHYTA (DIATOMS)

Diatoms usually have a siliceous cell wall. They do not store food as starch, but may do so as oil or as the $\beta$-1,3-linked glucan chrysolaminarin (leucosin), which occurs in the dissolved state in vacuoles of the living cell.[33] Of interest is the presence of crystalline chitan in extracellular fibers attached to the diatom. The diatom frustules consist mainly of silica, but there is some evidence of pectin in certain frustules.[34]

| Taxonomic Group and Species | Polysaccharide | Monosaccharides | Reference |
|---|---|---|---|
| **Class Centrobacillariophyceae** | | | |
| **EUPODISCALES** | | | |
| **Coscinodiscaceae** | | | |
| *Coscinodiscus* sp. | Whole-plant | glc = 2.1%, rha = 0.7%, fuc = 0.5%, man = 0.41%, gal = 0.4%, hexuronic acid (+) | 12 |
| *oculus-iridis* (and others) | Water-soluble fraction | glc + gal = 5.13%, xyl = 1.5%, ara = 1.2%, rha = 1.2% | 35 |
| *Cyclotella cryptica* | Chitan fibers ($\beta$-1,4) | glcNAc | 36,37 |

| Taxonomic Group and Species | Polysaccharide | Monosaccharides | Reference |
|---|---|---|---|
| *Melosira* | | | |
| *varians* | Chrysolaminarin | glc | 38,39 |
| | Water-soluble fraction | glc, xyl | 40 |
| *Schroederella* | | | |
| *schroederi* | Chrysolaminarin | glc | 38 |
| *Skeletonema* | | | |
| *costatum* | Whole-plant | glc = 16.4%, gal = 1.8%, rha = 1.0%, fuc = 0.9%, man = 0.87%, hexuronic acid (+) | 12 |
| *Thallassiosira* | | | |
| *gravida* and *nordenskjoldii* (mixed) | Water extract precipitated on concentration | glc + gal = 9, xyl = 2, rha = 1, ara (+) | 35,41 |
| | Water extract precipitated with alcohol | glc + gal = 24, xyl = 2, rha = 1, ara (+) | 35,41 |
| *fluviatilis* | Chitan fibers (β-1,4) | glcNAc | 36,37, 42,43 |
| **RHIZOSOLENIALES** | | | |
| **Rhizosoleniaceae** | | | |
| *Rhizosolenia* | | | |
| *stotherfoltii* | Chrysolaminarin | glc | 38 |
| **BIDDULPHIALES** | | | |
| **Biddulphiaceae** | | | |
| *Biddulphia* | | | |
| *sinensis* | Chrysolaminarin | glc | 38 |
| *Chaetoceras* sp. | Whole-plant | glc = 3.3%, rha = 2.8%, gal = 1.5%, man = 0.8%, xyl = 0.4%, fuc (+), hexuronic acid (+) | 12 |
| *decipiens* | Water-soluble fraction | gal | 44 |
| | Hemicellulose | glc, gal, man, rha, tyvelose | 44 |
| | Cell-wall material | glc, ara, xyl | 44 |
| *furcellatus* | See *Coscinodiscus oculus-iridicus* | See *Coscinodiscus oculus-iridicus* | |

## Class Pennatibacillariophyceae

| Taxonomic Group and Species | Polysaccharide | Monosaccharides | Reference |
|---|---|---|---|
| **FRAGILARIALES** | | | |
| **Fragilariaceae** | | | |
| *Rhabdonema* | | | |
| *adriaticum* | Water-soluble fraction | glc | 44 |
| | Hemicellulose | glc, gal, man, rha, tyvelose | 44 |
| | Cell-wall material | glc, ara, xyl | 44 |

| Taxonomic Group and Species | Polysaccharide | Monosaccharides | Reference |
|---|---|---|---|
| **ACHNANTHALES** | | | |
| **Achnanthaceae** | | | |
| *Achnanthes* sp. | Chrysolaminarin | glc | 38 |
| **NAVICULALES** | | | |
| **Naviculaceae** | | | |
| *Amphipleura rutilans* | Mucilage tube | man, xyl, rha (t), $SO_4$ = 16% | 45 |
| *Navicula* sp. | Chrysolaminarin | glc | 38 |
| *pelliculosa* | Capsular polysaccharide | glcUA | 46 |
| | Cell-wall | gal, glc, man, xyl, fuc, rha, glcN, glcUA | 47 |
| *Pinnularia* sp. | Chrysolaminarin | glc | 38,39 |
| **Cymbellaceae** | | | |
| *Gomphonema olivaceum* | Extracellular stalk substance, acidic fraction | gal = 1, xyl = 1, $SO_4$ = 1, uronic acid = 3.5% | 48 |
| | Extracellular stalk substance, neutral fraction | glc, rha | 48 |
| **PHAEODACTYLALES** | | | |
| **Phaeodactylaceae** | | | |
| *Phaeodactylum tricornutum* | Water-soluble mucilage | gal, man, xyl, fuc | 49 |
| | Whole-plant | glc = 10.7%, man = 3.7%, gal = 2.7%, rha = 1.5%, xyl = 0.7%, hexuronic acid (+) | 12 |
| | Glucan (β-1,3 and β-1,6) | glc | 50 |
| | Cell-wall glucuronomannan | glcUA = 27%, man, $SO_4$ = 7.5% | 50 |
| **BACILLARIALES** | | | |
| **Nitzschiaceae** | | | |
| *Nitzschia sigmoidea* | Chrysolaminarin | glc | 39 |
| *closterium* | Excreted mucilage | glc, xyl (t), rha (t) | 51 |
| | Intracellular | glc | 51 |
| **SURIRELLALES** | | | |
| **Surirellaceae** | | | |
| *Cymatopleura solea* | Chrysolaminarin | glc | 38,39 |

# DIVISION EUGLENOPHYTA

The majority of species in the Euglenophyta are characterized by the presence of characteristic granules containing the reserve polysaccharide paramylon (paramylum) in their cells. However, two species, *Eutreptiella marina* and *Sphenomonas laevi,* are reported not to contain paramylon.[52] Mucins formed by Euglenophytes have received little attention.[53]

Paramylon is an essentially linear β-1,3-linked glucan with perhaps a very small amount of branching.[54-56] This polysaccharide has been identified in the species listed in the table below.

| Taxonomic Group and Species | Reference |
|---|---|
| **Euglenaceae** | |
| *Euglena* | |
| *acus* | 57 |
| var. *hyalina* | 52 |
| var. *longissima* | 58 |
| *brevicaudata* | 59 |
| *deses* | 57 |
| *ehrenbergii* | 52 |
| *fusca* | 58 |
| *geniculata* | 60 |
| *gracilis* | 56,58 |
| var. *bacillaris*[a] | 57 |
| strain Z | 57 |
| apochlorotic form | 57 |
| *granulata* | 58,63 |
| *limnophila* | 57 |
| *magnifica* | 52 |
| *mutabilis* | 52 |
| *obtusa* | 52 |
| *oxyuris* | 57,63 |
| *pisciformis* | 52 |
| *polymorpha* | 63 |
| *proxima* | 52,58 |
| *sanguinea* | 60 |
| *spirogyra* | 57,63 |
| *tripteris* | 57,63 |
| *velata* | 63 |
| *virides* | 57,60 |
| *Colacium* | |
| *cyclopicolum* | 52 |
| *Cyclidiopsis* | |
| *acus* | 52 |
| *Distigma* | |
| *proteus* | 58 |
| *Eutreptia* | |
| *pertyi* | 52 |
| *Khawkinea* sp. | 64 |
| *ocellata (Astacia ocellata)* | 55,57,60 |
| *pertyi* | 57 |
| *quartana* | 52 |
| *Lepocinclis* | |
| *steinii* | 58 |
| *Phacus* | |
| *curvicauda* | 58 |
| *longicauda* | 58 |
| *pleuronectes* | 57,58 |
| *hispidula* | 58 |
| *Trachelomonas* spp. | 65 |
| *grandis* | 52 |
| **Astasiaceae** | |
| *Astasia* | |
| *dangeardii* | 58 |
| *longa* | 57 |
| *sagittifera* | 58 |
| *Menoidium* | |
| *bibacillatum* | 52 |
| *incurvum* | 58 |
| **Peranemataceae** | |
| *Anisonema* sp. | 58 |
| *Entosiphon* sp. | 58 |
| *Peranema* | |
| *trichophorum* | 54 |
| *Petalomonas* sp. | 58 |
| **Rhizaspidaceae (Skuja)** | |
| *Rhizaspis* | |
| *simplex* | 52 |
| **Rhynchopodacea (Skuja)** | |
| *Rhynchopus* | |
| *amitus* | 52 |
| **Uncertain** | |
| *Protaspis* | |
| *abovata* | 52 |

[a] Mucin of *Euglena gracilis* var. *bacillaris* contains glc, gal, man, xyl, fuc, and rha.[61,62]

# DIVISION CHRYSOPHYTA

| Taxonomic Group and Species | Polysaccharide | Monosaccharides | Reference |
|---|---|---|---|
| **PHAEOTHAMNIALES** | | | |
| **Phaeothamniaceae** | | | |
| *Pleurochrysis scherfellii* | Cellulose | glc | 66 |
| | Pectin | gal, rib | 66 |
| **OCHROMONADALES** | | | |
| **Ochromonadaceae** | | | |
| *Ochromonas malhamensis* | Chrysolaminarin | glc | 54 |
| | Water-soluble | glc, gal, man | 54 |
| | Extracellular | rha, gal = glc, man = xyl, glcUA, 4MeglcUA | 67 |
| *Pavlova mesolychnon* | Chrysolaminarin | glc | 68 |
| **Coccolithophoraceae** | | | |
| *Pontosphaeria roscoffensis* | Chrysolaminarin | glc | 69 |
| *Syracosphaera carterae* | Whole-cell | glc = 9.2%, gal = 7.1%, ara = 1.9%, xyl = 0.8%, hexuronic acid (+) | 12 |
| **ISOCHRYSIDALES** | | | |
| **Isochrysidaceae** | | | |
| *Isochrysis galbana* | Extracellular | gal = glc > ara, xyl (t), rha (t) | 70 |
| | Intracellular | gal > ara = xyl = rib, glc (t) | 70 |
| *Prymnesium parvum* | Extracellular | gal = glc = ara = xyl, rib (t), rha (t) | 70 |
| | Intracellular | gal = glc > ara = xyl | 70 |
| **CHROMULINALES** | | | |
| **Hydruraceae** | | | |
| *Hydrurus foetidus* | Chrysolaminarin | glc | 71 |
| **Chrysocapsaceae** | | | |
| *Monochrysis lutheri* | Whole-cell | glc = 22.1%, gal = 4.4%, xyl = 3.5%, hexuronic acid (+) | 12 |

## DIVISION XANTHOPHYTA

| Taxonomic Group and Species | Polysaccharide | Monosaccharides | Reference |
|---|---|---|---|
| **CHLORAMOEBALES** | | | |
| **Chloramoebaceae** | | | |
| *Chloramoeba* sp. | Glycogen | glc | 72 |
| *Botryococcus braunii* | Starch | glc | 73 |
| *Heterochloris* sp. | Chrysolaminarin | glc | 73 |
| **MISCHOCOCCALES** | | | |
| **Chlorobotrydaceae** | | | |
| *Monodus subterraneus* | Cell-wall β-glucan (85% β-1,4 and 15% β-1,3) | glc | 74,75 |
| | Unidentified | glc, gal, man, xyl | 74 |
| **TRIBONEMATALES** | | | |
| **Tribonemataceae** | | | |
| *Tribonema* sp. | Cellulose | glc | 76,77 |
| | Hemicellulose | glc > xyl, rha (t), fuc (t) | 76 |
| | Water-soluble | gal = glc > rha, fuc, ara, xyl, man | 76 |
| **VAUCHERIALES** | | | |
| **Botrydiaceae** | | | |
| *Botrydium* sp. | Unidentified | gal, man, ara, xyl, rha, fuc | 78 |
| *granulatum* | Cellulose | glc | 77 |
| **Vaucheriaceae** | | | |
| *Vaucheria sessilis* | Cellulose | glc | 79 |
| | Unidentified | glc, ara, xyl, rib | 79 |
| *hamata* | Cellulose | glc | |

## DIVISION PHAEOPHYTA

This division comprises almost entirely macroalgae, and only a general survey will, therefore, be given. Reference 13 gives additional details on the structures and distribution of the polysaccharides discussed below.

All the species examined metabolize alginic acid, an essentially linear polysaccharide consisting of variable amounts of 1,4-linked β-D-mannuronic acid and 1,4-linked (probably α-) L-guluronic acid. The proportion of this polysaccharide in the algae varies seasonally and from species to species.

This group also synthesizes a smaller proportion of laminarin, which is considered to be the food reserve polysaccharide. Laminarin is a glucan consisting mainly of 1,3-linked β-D-glucose units, with some 1,6-linked D-glucose units either in the linear chains or at branch points. The proportion of 1,6-linked units depends on the species, but it never exceeds 30% of the total linkages. In addition, the laminarin from some species of *Laminaria* has a proportion of non-reducing chains, which are terminated by the alcohol mannitol, linked either at C-1 or C-6.

A family of polysaccharides, consisting of L-fucose units carrying ester sulfate groups and different proportions of D-galactose, D-xylose and D-glucuronic acid, is also synthesized by algae in this division. In addition, the cell walls of most species contain a small proportion of cellulose.

## DIVISION CHLOROPHYTA (GREEN ALGAE)

The Chlorophyta, the majority of which grow in fresh water, include all green algae except the Stoneworts. They comprise a variety of structural forms, and members of the same genus may vary in size from macroscopic to microscopic. Some Chlorophyta are filamentous, and these may be branched or unbranched. Many species are unicellular. Selection by size or complexity of vegetative form proved to be very difficult. For this reason, all species whose polysaccharides have been chemically investigated are included here.

It is probable that all species contain starch similar in composition to that of land plants, comprising both amylose and amylopectin; however, the starch granules in the algae are not as well organized as those in the higher plants. Some species also store reserve foods as oil. Many of the species metabolize a high proportion of complex water-soluble polysaccharides in which some of the hydroxyl groups of the sugar units have been substituted by sulfate.

| Taxonomic Group and Species | Polysaccharide | Monosaccharides | Reference |
|---|---|---|---|
| **VOLVOCALES** | | | |
| **Polyblepharidaceae** | | | |
| *Dunaliella* | | | |
| *bioculata* | Starch, 13% amylose | glc | 80 |
| *Polytomella* | | | |
| *coeca* | Starch, 85% amylopectin | glc | 81 |
| *agilis* | Starch, 8% amylose (9°), 25% amylose (18°) | glc | 82 |
| **Chlamydomonadaceae** | | | |
| *Chlamydomonas* sp. | Extracellular | gal, ara, fuc (t), man (t), rha (t), uronic acid (t) | 83 |
| *globosa* | Cell-wall (70%) | glc > gal, ara, xyl, rha, three unidentified sugars | 84 |
| *ulvaënsis* | Extracellular | glc, xyl, others (t) | 85 |
| *Platymonas* | | | |
| *subcortiformis* | Cell-wall | gal, uronic acid, ara (t), *no* glc | 86 |

| Taxonomic Group and Species | Polysaccharide | Monosaccharides | Reference |
|---|---|---|---|
| **CHLOROCOCCALES** | | | |
| **Chlorellaceae** | | | |
| *Chlorella* | | | |
| *ellipsoides* | Starch granules | glc | 87 |
| | Extracellular | glc, ara, fuc, glcUA, rha | 6 |
| *pyrenoidosa* | α-Cellulose | glc | 88 |
| | Starch, 7% amylose | glc | 89 |
| | Hemicellulose A | gal = 7, rha = 5, ara = 3, xyl = 3, man = 2, glc = 1 | 88,89 |
| | Hemicellulose B | glc = 25, gal = 13, rha = 12, man = 2, ara = 1, xyl = 1, unknown = 20 | 88,89 |
| *vulgaris* | Extracellular | glc, xyl, rha, glcUA | 6 |
| *Prothoteca* | | | |
| *zopfii* | α-Glucan (glycogen) | glc | 90 |
| | Mannogalactan (1,4 man, 1,6 gal) | gal = 96%, man = 4% | 90 |
| **Hydrodictyaceae** | | | |
| *Hydrodictyon* | | | |
| *africanum* | α-Cellulose, 69% | glc, man | 91 |
| (Yaman) | Hemicellulose, 16% | glc, man, ara (t), xyl (t), glcUA | 91 |
| *patenaeforme* | Mannan | ·man | 92 |
| *reticulatum* | Starch | glc | 93 |
| | Cellulose II | glc | 92 |
| | Mannan | man | 92 |
| *Pediastrum* | | | |
| *tetras* | Hemicellulose | glc, man | 94 |
| **Oocystaceae** | | | |
| *Oocystis* sp. | Extracellular | gal, ara, fuc, glcUA | 6 |
| **SIPHONOCLADALES** | | | |
| **Siphonocladaceae** | | | |
| *Siphonocladus* sp. | Cellulose (X-ray) | | 77 |
| **Valoniaceae** | | | |
| *Valonia* | | | |
| *ventricosa* | Cellulose | glc | 95,96 |
| | Hemicellulose | glc, gal, ara, xyl, glcUA | 96,97 |
| *Apjohnia* | | | |
| *laetevirens* | Cellulose I | glc | 98 |
| | Pectin-like galacto-araban | glc, gal, ara, uronic acid = 1.8%, xyl (t), rha (t) | 98 |
| **CODIALES** | | | |
| **Bryopsidaceae** | | | |
| *Bryopsis* | | | |
| *corticulans* | Xylan (β-1,3) | xyl | 99 |
| *maxima* | Xylan (β-1,3) | xyl, glc (t) | 100 |
| | Glucan (β-1,3) | glc | 100 |
| *plumosa* | Xylan (β-1,3) | xyl | 99 |

| Taxonomic Group and Species | Polysaccharide | Monosaccharides | Reference |
|---|---|---|---|
| **Codiaceae** | | | |
| *Codium* | | | |
| *adherens* | Mannan | man | 101 |
| *intricata* | Mannan | man | 101 |
| *latum* | Mannan | man | 101 |
| *fragile* | Starch, 15% amylose | glc | 102 |
| | Water-soluble | gal, xyl, ara, man, glc, rha (t), $SO_4$ | 103 |
| | Mannan (β-1,4) | man | 104,105 |
| *tormentosum* | Mannan (β-1,4) | man | 99,101 |
| *dichotomum* | Mannogalacto-araban | man, gal, ara | 106 |
| | Starch | | 106 |
| **DERBESIALES** | | | |
| **Derbesiaceae** | | | |
| *Derbesia* | | | |
| *lamourouxii* | Mannan | man, glc (t) | 107 |
| **CAULERPALES** | | | |
| **Udoteaceae** | | | |
| *Chlorodesmis* sp. | Cell-wall xylan (β-1,3) + glucan (β-1,3) | xyl, glc | 100 |
| *formosana* | Cell-wall xylan (β-1,3) | xyl | 100,101 |
| *Halimeda* | | | |
| *cuneata* | Cell-wall xylan (β-1,3) | xyl | 100,107 |
| *incrassata* | Cell-wall xylan | xyl | 99 |
| *opunta* | Cell-wall xylan | xyl | 99 |
| *Penicillus* sp. | Cell-wall xylan | xyl | 108 |
| *capitatus* | Cell-wall xylan | xyl | 99 |
| *dumetosus* | Cell-wall xylan | xyl | 99,109 |
| *Udotea* | | | |
| *flabellum* | Cell-wall xylan | xyl | 99,101 |
| *orientalis* | | | 100 |
| **Caulerpaceae** | | | |
| *Caulerpa* | | | |
| *cupressoides* | Cell-wall xylan | xyl | 99 |
| *anceps* | Cell-wall xylan | xyl | 100 |
| *brachypus* | Cell-wall xylan | xyl | 107 |
| *filiformis* | Amylopectin | glc | 110 |
| | Cell-wall xylan (β-1,3) | xyl | 111 |
| | Water-soluble | gal, man, xyl, ara | 112 |
| *prolifera* | Cell-wall xylan (β-1,3) | xyl | 99 |
| *racemosa* var. *laetevirens* | Starch | glc | 112 |
| | Water-soluble | gal, man, xyl, glc (t), rha (t) | 112 |
| *sertularioides* | Starch | glc | 112 |
| | Water-soluble | gal, man = xyl, glc (t), rha (t) | 112 |

| Taxonomic Group and Species | Polysaccharide | Monosaccharides | Reference |
|---|---|---|---|
| **DASYCLADALES** | | | |
| **Dasycladaceae** | | | |
| *Acetabularia* | | | |
| *calyculus* | Mannan (β-1,4) | man | 101 |
| *crenulata* | Inulin | fru = 33, glc = 1 | 113 |
| | Mannan (β-1,4) | man, glc (t) | 113 |
| | Water-soluble | gal, xyl, rha, 4Megal, uronic acid = 7%, $SO_4$ | 113 |
| *mediterranea* | Inulin | fru | 114 |
| | Mannan | man | 99 |
| *Cymopolia* | | | |
| *barbata* | Fructan | fru | 115 |
| | Mannan | man | 99 |
| *Dasycladus* | | | |
| *clavaeformis* | Mannan | man | 99 |
| *vermicularis* | Mannan | man | 99 |
| | Fructan | fru | 116 |
| *Halicoryne* | | | |
| *wrightii* | Mannan | man | 101 |
| *Neomeris* | | | |
| *annulate* | Fructan | fru | 116 |
| **ULOTRICHALES** | | | |
| **Microsporaceae** | | | |
| *Microspora* sp. | Starch | glc | 117 |
| | Water-soluble | glc > rha > xyl > gal = man, ara (t) | 117 |
| **Ulotrichaceae** | | | |
| *Chlorhormidium* sp. | Cellulose (X-ray) | | 118 |
| *Ulothrix* sp. | Unidentified | xyl = rha, man | 119 |
| **ULVALES** | | | |
| **Ulvaceae** | | | |
| *Enteromorpha* | | | |
| *compressa* | Starch, 22% amylose | glc | 102 |
| | Glucuronoxylorhamnan | rha = 45%, glcUA = 18%, $SO_4$ = 16%, xyl = 15%, glc = 6% | 120 |
| *intestinales* | Starch | glc | 121 |
| *Ulva* | | | |
| *expansa* | Starch | glc | 93,122 |
| *lactuca* | Cellulose | glc | 123 |
| | Starch, 27% amylose | glc | 102 |
| | Glucuronoxylorhamnan | rha, xyl, glc, man (t), gal (t), glcUA = 14%, $SO_4$ = 20% | 124,125, 126 |
| *pertusa* | Uronoxylorhamnan | | 127 |
| *conglobota* | Water-soluble | rha, xyl, glc, glcUA | 128 |

| Taxonomic Group and Species | Polysaccharide | Monosaccharides | Reference |
|---|---|---|---|
| **SCHIZOGONIALES** | | | |
| **Schizogoniaceae** | | | |
| *Prasiola* | | | |
| *japonica* | Cellulose | glc | 129 |
| | Cell-wall xylomannan (β-1,4) | man = 93%, xyl = 7% | 129,130 |
| | Polyuronide | Uronic acid, rha | 129 |
| *Halicystis* | | | |
| *osterhoutii* | Cellulose 2 | glc 2 | 131 |
| | Xylan 1 | xyl 1 | 131 |
| **CLADOPHORALES** | | | |
| **Cladophoraceae** | | | |
| *Acrosiphonia* | | | |
| *centralis* | Starch, 9% | glc | 132 |
| | Water-soluble | xyl = rha, > glc, man (t), glcUA = 20%, $SO_4$ | 132 |
| *Chaetomorpha* | | | |
| *linum* | Starch | glc | 133 |
| | Xylogalactoaraban | ara > gal > xyl, $SO_4$ | 133,134 |
| *capillaris* | Xylogalactoaraban | ara > gal > xyl, $SO_4$ | 102,103 |
| *melagonium* | Cellulose | glc | 135 |
| | Xylogalactoaraban | ara > gal > xyl, $SO_4$ | 134 |
| *Cladophora* | | | |
| *rupestris* | Inulin-type | fru, glc (t) | 134 |
| | Starch, 20% amylose | glc | 102 |
| | Cellulose | glc | 77 |
| | Xyloarabogalactan | gal = ara > xyl, $SO_4$ | 136,137 |
| *sericea* | Xyloarabogalactan | ara = 1.3, gal = 1.0, xyl = 10.8, $SO_4$ | 134 |
| *Rhizoclonium* sp. | Cellulose (X-ray) | | 77 |
| | Starch | glc | 138 |
| *implexum* | Xyloarabogalactan | gal > ara = xyl | 134 |
| *riparium* | Inulin-type | fru, glc (t) | 134 |
| *turtuosum* | Water-soluble | gal, glc = xyl = ara, rha (t), $SO_4$ | 139 |
| *Urospora* | | | |
| *bangioides* | Xylorhamnan | rha = 7, xyl = 4, man (t), $SO_4$, uronic acid = 7% | 117,134 |
| **OEDOGONIALES** | | | |
| **Oedogoniaceae** | | | |
| *Oedogonium* | | | |
| *cardiaceum* | Cellulose | glc | 140 |
| **ZYGNEMATALES** | | | |
| **Zygnemataceae** | | | |
| *Mougeotia* sp. | Cellulose (X-ray) | | 141 |
| *Spirogyra* sp. | Cellulose (X-ray) | | 141,142 |
| | Starch | glc | 93 |

| Taxonomic Group and Species | Polysaccharide | Monosaccharides | Reference |
|---|---|---|---|
| DICHOTOMOSIPHONALES | | | |
| Dichotomosiphonaceae (G. M. Smith) | | | |
| *Dichotomosiphon* sp. | Xylan (β-1,3) | xyl | 83 |
| *tuberosus* | Xylan | xyl | 99 |

## DIVISION CHAROPHYTA (STONEWORTS)

| Taxonomic Group and Species | Polysaccharide | Monosaccharides | Reference |
|---|---|---|---|
| CHARALES | | | |
| Characeae | | | |
| *Chara* sp. | Cellulose | glc | 143 |
| | Starch | glc | 144 |
| *australis* | Starch | glc | 145 |
| | Cellulose, 26–29% | glc | 145 |
| | Pectic acid, 19% | galUA | 145 |
| | Hemicellulose, 8.8% | gal, glc, ara, xyl, man (t), uronic acid = 16% | 145 |
| *Nitella* sp. | Cellulose | glc | 146 |
| *translucens* | Cellulose, 23.4% | | 147 |
| | Starch, 4% (12% amylose) | glc | 147,148 |
| | Pectic acid, 14.9% | galUA = 74%, ara = 6, gal = 4, xyl = 3, rha = 1 | 147,149 |
| | Hemicellulose | gal, glc, ara, man, xyl | 147 |

## APOCHROMATIC GROUPS

| Taxonomic Group and Species | Polysaccharide | Monosaccharides | Reference |
|---|---|---|---|
| Protomastigineae | | | |
| Trypanosomaceae | | | |
| *Trypanosoma* | | | |
| *galba* | Glycogen | glc | 150 |
| *loricatum* | Glycogen | glc | 150 |
| *montezumae* | Glycogen | glc | 150 |
| *chattoni* | Glycogen | glc | 150 |

## REFERENCES


1. Silva, P. C., in *Physiology and Biochemistry of Algae,* p. 827, R. A. Lewin, Ed. Academic Press, New York (1962).
2. Fritsch, F. E., *Structure and Reproduction of the Algae,* Vols. 1 and 2. Cambridge University Press, New York (1956).
3. Weise, G., Drews, G., Jann, B., and Jann, K., *Arch. Mikrobiol., 71,* 89 (1970).
4. Kosenko, L. V., *Mikrobiol. Zh. (Kiev), 30,* 33 (1968); *Chem. Abstr., 68,* 102869 (1968).
5. Bishop, C. T., Adams, G. A., and Hughes, E. O., *Can. J. Chem., 32,* 999 (1954).
6. Moore, B. G., and Tischer, R. G., *Science, 145,* 586 (1964).
7. Hough, L., Jones, J. K., N., and Wadman, W. H., *J. Chem. Soc.,* p. 3393 (1952).
8. Fredrick, J. F., *Physiol. Plant. (Denmark), 5,* 37 (1952); *6,* 100 (1953); *8,* 288 (1955).
9. Biswas, B. B., *Sci. Cult. (India), 22,* 696 (1957).
10. Payen, J., *Rev. Algol. (France), 11,* 1 (1938).
11. Kylin, H., *Kgl. Fysiograf. Salisk. Lund. Forh. (Sweden), 13,* 1 (1943).
12. Parsons, T. R., Stephens, K., and Strickland, J. D. H., *J. Fish. Res. Board Can., 18,* 1001 (1961).
13. Percival, E. E., and McDowell, R., *The Chemistry and Enzymology of Marine Algal Polysaccharides.* Academic Press, London, England (1967).
14. Williams, E. L., and Turvey, J. R., *Phytochemistry, 9,* 2383 (1970).
15. Eschrich, W., *Protoplasma, 47,* 487 (1956).
16. Quillet, M., and Priou, M-L., *C. R. Hebd. Séances Acad. Sci. (Paris), 254,* 2210 (1962).
17. Haxo, F. T., and Oh Eocha, C., in *Proceedings of the Second International Seaweed Symposium, Trondheim,* p. 23, T. Baarud and N. A. Sorensen, Eds. Pergamon Press, London, England (1955).
18. Jones, R. F., *J. Cell. Comp. Physiol., 60,* 61 (1962).
19. Griffiths, L. M., and Turvey, J. R., Unpublished Data.
20. Augier, J., *C. R. Hebd. Seances Acad. Sci. (Paris), 242,* 1069 (1956); *239,* 87 (1954).
21. Archibald, A. R., Hirst, E. L., Manners, D. J., and Ryley, J. F., *J. Chem. Soc.,* p. 556 (1960).
22. Hutchens, J. O., Podolsky, B., and Morales, M. F., *J. Cell. Comp. Physiol., 32,* 117 (1948).
23. Wehrmeyer, W., *Arch. Mikrobiol., 71,* 367 (1970).
24. Fredrick, J. F., *Phytochemistry, 7,* 1573 (1968).
25. Fritsch, F. E., *Structure and Reproduction of the Algae,* Vol. 1, p. 667, 680. Cambridge University Press, New York (1956).
26. Fritsch, F. E., *Structure and Reproduction of the Algae,* Vol. 1, p. 668. Cambridge University Press, New York (1956).
27. Khailov, K. M., Burlakova, Z. P., and Lanskaya, L. A., *Mikrobiologiya, 36,* 359 (1967).
28. Bidwell, R. G. S., *Can. J. Bot., 35,* 945 (1957).
29. Geitler, L., *Arch. Protistenk., 53,* 343 (1926).
30. Nevo, Z., and Sharon, N., *Biochim. Biophys. Acta, 173,* 161 (1969).
31. Fritsch, F. E., *Structure and Reproduction of the Algae,* Vol. 1, p. 671. Cambridge University Press, New York (1956).
32. Killiani, C., *Arch. Protistenk., 50,* 50 (1924).
33. Lewin, J. C., and Guillard, R. R. L., *Annu. Rev. Microbiol., 17,* 373 (1963).
34. Desikachary, T. V., and Dweltz, N. E., *Proc. Indian Acad. Sci. Sect. B, 53,* 157 (1961).
35. Barashkov, G. K., *Dokl. Akad. Nauk S.S.S.R., 111,* 148 (1956).
36. Falk, M. G., Smith, D. G., McLachlan, J., and McInnes, A. G., *Can. J. Chem., 44,* 2269 (1966).
37. Blackwell, J., Parker, K. D., and Rudall, K. M., *J. Mol. Biol., 28,* 383 (1967).
38. von Stosch, H. A., *Naturwissenschaften, 38,* 192 (1951).
39. Beattie, A., Hirst, E. L., and Percival, E., *Biochem. J., 79,* 531 (1961).
40. Kleinkauf, H., *Helgolaender Wiss. Meeresunters., 11,* 39 (1964).
41. Barashkov, G. K., *Tr. Murmansk. Morsk. Biol. Inst.,* No. 4, 27 (1962).
42. McLachlan, J., McInnes, A. G., and Falk, M., *Can. J. Bot., 43,* 707 (1965).
43. Dweltz, N. E., Colvin, J. R., and McInnes, A. G., *Can. J. Chem., 46,* 1513 (1968).
44. Serenkov, G. P., and Pachomova, M. V., *Vestn. Mosk. Univ.,* No. 2, 39 (1959).
45. Lewin, R. A., *Limnol. Oceanogr., 3,* 111 (1958).
46. Lewin, J. C., *J. Gen. Microbiol., 13,* 162 (1955).
47. Coombs, J., and Volcani, B. E., *Planta, 80,* 264 (1968); *80,* 280 (1968).
48. Huntsman, S. de R., *Diss. Abstr., B 27,* 3012 (1967).
49. Lewin, J. C., Lewin, R. A., and Philpott, D. E., *J. Gen. Microbiol., 18,* 418 (1958).
50. Ford, C. W., and Percival, E., *J. Chem. Soc.,* p. 7035 (1965); p. 7042 (1965).
51. Tokuda, H., *Rec. Oceanogr. Works Jap., 10,* 109 (1969).
52. Leedale, G. F., *Euglenoid Flagellates.* Prentice-Hall, Englewood Cliffs, New Jersey (1967).
53. Leedale, G. F., Meeuse, B. J. D., and Pringsheim, E. G., *Arch. Mikrobiol., 50,* 68 (1965).
54. Archibald, A. R., Cunningham, W. L., Manners, D. J., Stark, J. R., and Ryley, J. F., *Biochem. J., 88,* 444 (1963).
55. Manners, D. J., Ryley, J. F., and Stark, J. R., *Biochem. J., 101,* 323 (1966).

56. Clarke, A. E., and Stone, B. A., *Biochim. Biophys. Acta, 44,* 161 (1960).
57. Leedale, G. F., Meeuse, B. J. D., and Pringsheim, E. G., *Arch. Mikrobiol., 50,* 133 (1965).
58. Deflandre, G., *Bull. Biol. Fr. Belg., 68,* 382 (1934).
59. Kamptner, E., *Oesterr. Bot. Z., 99,* 556 (1952).
60. Kreger, D. R., and Meeuse, B. J. D., *Biochim. Biophys. Acta, 9,* 699 (1952).
61. Barras, D. R., and Stone, B. A., *The Biology of Euglena,* Vol. 2, p. 149, D. E. Buetow, Ed. Academic Press, New York (1968).
62. Barras, D. R., and Stone, B. A., *Biochem. J., 97,* 14P (1965).
63. Pringsheim, E. G., *Nova Acta Leopoldina, 18,* 1 (1956).
64. Nath, V., Dutta, G. P., and Dhillon, B., *Res. Bull. Panjab Univ. Sci. (India), 11,* 159 (1960).
65. Singh, K. P., *Amer. J. Bot., 43,* 274 (1956).
66. Brown, R. M., Franke, W. W., Kleinig, H., Falk, H., and Sitte, P., *J. Cell. Biol., 45,* 246 (1970).
67. Kauss, K., *Z. Pflanzenphysiol., 58,* 281 (1968).
68. Kreger, D. R., and van der Veer, J., *Acta Bot. Neerl., 19,* 401 (1970).
69. Chadefaud, M., and Feldmann, J., *Bull. Mus. Nat. Hist. Natur. (Paris), 21,* 617 (1949).
70. Marker, A. F. H., *J. Mar. Biol. Assn. U.K., 45,* 755 (1965).
71. Quillet, M., *C. R. Hebd. Séances Acad. Sci. (Paris), 240,* 1001 (1955).
72. Printz, H., in *Die Natürlichen Pflanzenfamilien,* 2nd ed., Vol. 3, p. 380, A. Engler and K. Prantl, Eds. Duncker and Humblot, Berlin, Germany (1927).
73. Fritsch, F. E., *Structure and Reproduction of the Algae,* Vol. 1, pp. 473, 476. Cambridge University Press, New York (1956).
74. Beattie, A., and Percival, E., *Proc. Roy. Soc. Edinburgh, B 68,* 171 (1962).
75. Ford, C. W., and Percival, E., *J. Chem. Soc.,* p. 3014 (1965).
76. Cleare, M., and Percival, E., *Brit. Phycol. J., 7,* 185 (1972).
77. Nicolai, E., and Preston, R. D., *Proc. Roy. Soc. (England), B 140,* 244 (1952).
78. Percival, E., and Young, M., Unpublished Data.
79. Parker, B. C., Preston, R. D., and Fogg, G. E., *Proc. Roy. Soc. (England), B 158,* 435 (1963).
80. Eddy, B. O., Fleming, I. D., and Manners, D. J., *J. Chem. Soc.,* p. 2827 (1958).
81. Bourne, E. J., Stacey, M., and Wilkinson, I. A., *J. Chem. Soc.,* p. 2694 (1950).
82. Sheeler, P., Moore, J., Cantor, M., and Granik, R., *Life Sci., 7,* 1045 (1968).
83. Maeda, M., Kuroda, J., Iriki, Y., Chihara, M., Nisizawa, K., and Miwa, T., *Bot. Mag. (Tokyo), 79,* 634 (1966).
84. Pakhomova, M. V., and Zaitseva, G. N., *Mikrobiologiya, 38,* 689 (1969).
85. Lewin, R. A., *Can. J. Microbiol., 2,* 665 (1956).
86. Lewin, R. A., *J. Gen. Microbiol., 19,* 87 (1958).
87. Murakami, S., Norimura, Y., and Takamiya, A., *Studies on Microalgae and Photosynthetic Bacteria,* p. 65, Japanese Society of Plant Physiologists, Eds. University of Tokyo Press, Tokyo (1963).
88. Northcote, D. H., Goulding, K. J., and Horne, R. W., *Biochem. J., 70,* 391 (1958).
89. Northcote, D. H., and Olaitan, S. A., *Biochem. J., 81,* 7P (1961); *82,* 509 (1962).
90. Manners, D. J., Pennie, I. R., and Ryley, J. F., *Biochem. J., 104,* 32P (1961).
91. Northcote, D. H., Goulding, K. J., and Horne, R. W., *Biochem. J., 77,* 503 (1960).
92. Kreger, D. R., *Proc. Kon. Ned. Akad. Wetensch. Ser. C, 63,* 613 (1960).
93. Meeuse, B. J. D., and Smith, B. N., *Planta, 57,* 624 (1962).
94. Parker, B. C., *Phycologia, 4,* 63 (1964).
95. Marx-Figini, M., *Biochim. Biophys. Acta, 177,* 27 (1969).
96. Preston, R. D., and Cronshaw, J., *Nature (London), 181,* 248 (1958).
97. Blackwell, J., Vasko, P. D., and Koenig, J. L., *J. Appl. Physiol., 41,* 4375 (1970).
98. Stewart, C. M., *Aust. J. Mar. Freshwater Res., 20,* 143 (1969).
99. Mackie, W., and Sellen, D. B., *Proceedings of the 6th International Seaweed Symposium, Santiago,* p. 529, R. Margalef, Ed. Pergamon Press, London, England (1969).
100. Iriki, Y., Suzuki, T., Nisizawa, K., and Miwa, T., *Nature (London), 187,* 82 (1960).
101. Iriki, Y., and Miwa, T., *Nature (London), 185,* 178 (1960).
102. Love, J., Mackie, W., McKinell, J. P., and Percival, E., *J. Chem. Soc.,* p. 4177 (1963).
103. Love, J., and Percival, E., *J. Chem. Soc.,* p. 3338 (1964).
104. Mackie, W., and Sellen, D. B., *Polymer (London), 10,* 621 (1969).
105. Love, J., and Percival, E., *J. Chem. Soc.,* p. 3345 (1964).
106. Augier, J., and du Merac, M. L. R., *Bull. Lab. Mar. Dinard, 40,* 25 (1954).
107. Miwa, T., Iriki, Y., and Suzuki, T., *Colloq. Int. Centre Nat. Rech. Sci., 103,* 135 (1960).
108. Preston, R. D., *Endeavour, 23,* 153 (1964).
109. Preston, R. D., *Sci. Amer., 218,* 102 (1968).
110. Mackie, I. M., and Percival, E., *J. Chem. Soc.,* p. 2381 (1960).
111. Mackie, I. M., and Percival, E., *J. Chem. Soc.,* p. 1151 (1959).
112. Mackie, I. M., and Percival, E., *J. Chem. Soc.,* p. 3010 (1961).

113. Percival, E., and Smestad, B., *Carbohydr. Res.*, *22,* 75 (1972); *25,* 299 (1972).
114. du Merac, M. L. R., *Rev. Gen. Bot., 60,* 689 (1953).
115. du Merac, M. L. R., *C. R. Hebd. Séances Acad. Sci. (Paris), 243,* 714 (1956).
116. du Merac, M. L. R., *C. R. Hebd. Séances Acad. Sci. (Paris), 241,* 88 (1955).
117. Megarry, M., and Percival, E., Unpublished Data.
118. Brandenberger, E., and Frey-Wyssling, A., *Experientia, 3,* 492 (1947).
119. Percival, E., Megarry, M., and Young, M., Unpublished Data.
120. McKinnell, J. P., and Percival, E., *J. Chem. Soc.*, p. 3141 (1962).
121. Meeuse, B. J. D., in *Physiology and Biochemistry of Algae,* p. 289, R. A. Lewin, Ed. Academic Press, New York (1962).
122. Meeuse, B. J. D., and Kreger, D. R., *Biochim. Biophys. Acta, 13,* 593 (1954); *35,* 26 (1959).
123. Cronshaw, J., Myers, A., and Preston, R. D., *Biochim. Biophys. Acta, 27,* 89 (1958).
124. Percival, E., and Wold, J. K., *J. Chem. Soc.*, p. 5459 (1963).
125. Haq, Q. N., and Percival, E., *Contemporary Studies in Marine Science,* p. 355. George Allen and Unwin Ltd., London, England (1966).
126. Haq, Q. N., and Percival, E., *Proceedings of the 5th International Seaweed Symposium, Halifax,* p. 261, E. G. Young and J. McLachlan, Eds. Pergamon Press, London, England (1965).
127. Miyaki, S., Hayashi, K., and Takino, Y., *J. Soc. Trop. Agri. Taihoku Imp. Uno., 10,* 232 (1938).
128. Mita, K., *Bull. Chem. Soc. Jap., 42,* 3496 (1969).
129. Takeda, H., Nisizawa, K., and Miwa, T., *Bot. Mag. (Tokyo), 80,* 109 (1967).
130. Takeda, H., Nisizawa, K., and Miwa, T., *Sci. Rep. Tokyo Kyoiku Daigaku, Sect. B, 13,* 183 (1968).
131. Roelofsen, P. A., Dalitz, V. C., and Wijnman, C. F., *Biochim. Biophys. Acta, 11,* 344 (1953).
132. O'Donnell, J. J., and Percival, E., *J. Chem. Soc.*, p. 2168 (1959).
133. Hirst, Sir E., Mackie, W., and Percival, E., *J. Chem. Soc.*, p. 2958 (1965).
134. Percival, E., and Young, M., *Phytochemistry, 10,* 807 (1971).
135. Cronshaw, J., Myers, A., and Preston, R. D., *Biochim. Biophys. Acta,* 27, 89 (1958).
136. Fisher, I. S., and Percival, E., *J. Chem. Soc.*, p. 2666 (1957).
137. Johnson, P. G., and Percival, E., *J. Chem. Soc.*, p. 906 (1969); p. 1561 (1970).
138. Meeuse, B. J. D., and Kreger, D. R., *Biochim. Biophys. Acta, 13,* 593 (1954); *35,* 26 (1959).
139. Percival, E., and Beattie, M., Unpublished Data.
140. Parker, B. C., *Phycologia, 4,* 63 (1964).
141. Frey, R., *Ber. Schweiz. Bot. Ges., 60,* 199 (1950).
142. Kreger, D. R., *Nature (London), 180,* 914 (1957).
143. Amin, S. E., *J. Chem. Soc.*, p. 281 (1955).
144. Meeuse, B. J. D., in *Physiology and Biochemistry of Algae,* p. 289, R. A. Lewin, Ed. Academic Press, New York (1962).
145. Anderson, D. M. W., and King, N. J., *Biochim. Biophys. Acta, 52,* 449 (1961).
146. Probine, M. C., and Preston, R. D., *J. Exp. Bot., 12,* 261 (1961).
147. Anderson, D. M. W., and King, N. J., *Biochim. Biophys. Acta, 52,* 441 (1961).
148. Anderson, D. M. W., and King, N. J., *J. Chem. Soc.*, p. 2914 (1961).
149. Anderson, D. M. W., and King, N. J., *J. Chem. Soc.*, p. 5333 (1961).
150. Perez-Reyes, R., and Streber, F., *Rev. Latinoamer. Microbiol. Parasitol., 10,* 127 (1968).

# BACTERIAL CELL WALL STRUCTURE

C. S. CUMMINS

## INTRODUCTION

The bacterial cell wall is not composed of a single substance; it is a complex of polymers which can be separated in various ways in the laboratory, although it is likely that they are bound together covalently in the wall of the living cell. The study of bacterial wall structure essentially started in 1951 when Salton and Horne[1] reported chemical studies on cell wall fractions which could be seen to be pure by electron microscopy. Since that classical study, a great deal of work has been done, and the subject has been reviewed at fairly frequent intervals.[2–8] Reviews on specialized aspects of the subject (e.g., biosynthesis, structure of O-antigens, etc.) are referred to in the appropriate section below.

## PREPARATION OF CELL WALL FRACTIONS FOR ANALYSIS

Detailed discussions of methods for cell disruption can be found in References 1 and 9 to 11 and elsewhere in this volume. The matter is considered here briefly because of the considerable influence which methods of preparation have on the final product. An outline of the essential features of the three commonly used procedures is given below.

### Ultrasonic Disruption

Ultrasonic vibrations at 20 kHz or higher, produced in either a magnetostrictive or a piezoelectric oscillator, are applied to the bacterial suspension, usually by a probe connected to the oscillator. The bacteria are disrupted by the great pressure changes produced by the alternate development and collapse of gas cavities in the liquid. Considerable heating of the fluid takes place, and some form of cooling is necessary.

### Shaking with Small Glass Beads

This method was introduced by Dawson in 1949.[12] It was used by Salton from the start of his classic investigations into cell wall composition[1] and has remained a standard procedure. For bacteria, glass beads (Ballotini) 0.1–0.12 mm in diameter are most satisfactory, although beads up to 0.2 mm have been used. The original device was the Mickle tissue disintegrator[13] which shook a mixture of glass beads and bacteria at 50 Hz. The principal disadvantage of this apparatus is the rather small amount of material which can be treated at one time; and other machines such as the Braun shaker,[14] the Bühler "cell mill",[15] or the Nossal apparatus[16] will disintegrate 1–2 grams dry weight of bacteria in a total shaking time of about 5 minutes, although the whole process may take longer because of the intervals for cooling. After shaking, the glass beads are removed by filtration through a coarse sintered glass filter.

### Extrusion at High Pressure

If a bacterial suspension is placed under high pressure, at 20,000 psi or more, and then extruded through a narrow orifice by slightly opening a fine ball or needle valve, the combination of release of pressure and shear at the orifice will disrupt the organisms. In other methods a paste of bacteria frozen at −20°C or below is forced through a narrow orifice by the impact of a fly press on a closely fitting piston. In this case the ice crystals in the frozen paste probably have an abrasive effect. Examples of commercial machines are the Aminco® French pressure cell,[17] the Sorvall®- Ribi refrigerated cell fractionator,[18] and the Hughes press.[19]

Of these three methods, the best for the preparation of cell wall fractions is probably shaking with small glass beads. The major disadvantage of ultrasonic methods is that the cell wall fragments first produced are soon disintegrated still further to almost colloidal dimensions so that by the time the majority of the cells have been ruptured, a very wide range of sizes of wall fragments has been produced. This makes the clean recovery of cell walls in the centrifuge very difficult. Extrusion at high pressure works well with Gram-negative rods, but is much less efficient for Gram-positive organisms. Gram-positive cocci are particularly resistant to rupture by this method.

## Preparation of Cell Wall Fractions from the Crude Disintegrate

After disintegration, an essential preliminary step in purification is centrifugation to separate unbroken cells from the crude cell wall fractions. Alternate low speed (*ca.* 3,000 x G) and high speed (*ca.* 30,000 x G) centrifugation in distilled water or buffer is a standard procedure. However, it is often possible after a single high speed centrifugation to get a clear separation in the deposit into a lower, more opaque layer of unbroken cells and an upper, more translucent layer of cell walls. The upper cell wall layer can then usually be removed by gentle washing with a wash bottle or Pasteur pipette; and a relatively pure cell wall preparation is achieved in a single step.

Subsequent treatment of the crude cell wall fractions will depend on the purpose for which the walls are being prepared. If it is desired to recover cell wall fragments in as nearly native a state as possible, centrifugation in a sucrose or other density gradient is the method of choice.[20-22] If, on the other hand, it is desired to prepare peptidoglycan for analysis, then treatment with proteolytic enzymes such as trypsin or Pronase will be followed by extraction with formamide at 165°C, or with trichloroacetic acid or phenol, depending on the nature of the organism. Cell walls are very liable to autolysis because of the lytic enzymes liberated at the time of cell disruption, and it is usual to inactivate these by a short treatment at 100°C (10 to 15 minutes) or by a wash in 0.1% formalin. Alternatively, autolysis may be prevented if the cultures to be disintegrated are killed by 2% formalin before being washed and shaken with glass beads. However, this may make the organisms somewhat more difficult to disintegrate.

## Estimation of Purity of Cell Wall Fragments

Electron microscopy is the ideal method for determining whether or not cell wall fragments are free from contamination with other cell elements. Undisrupted cells, ribosomes, and other cytoplasmic particles are almost all more electron-dense than the wall; and examination of shadowed or negatively stained preparations will show contamination with membrane fragments. In the absence of facilities for electron microscopy, it is possible to assess the degree of contamination in other ways. For example, ribose or deoxyribose should be absent from cell wall hydrolysates, and there should be no UV absorption in the region 260–270 nm. In many Gram-positive organisms, treatment with proteolytic enzymes will remove all sources of amino acids except the peptide of peptidoglycan so that the finding of a very simple amino acid pattern in hydrolysates, e.g., alanine, glutamic acid, and *meso*-diaminopimelic acid, or alanine, glutamic acid, and lysine, is evidence of purity because almost all contaminating material will contain substantial amounts of nonpeptidoglycan amino acids such as leucine and isoleucine. This test cannot of course be applied to cell walls of Gram-negative organisms since the protein fraction of the O-antigen is unaffected by proteolytic enzymes.

## Preservation of Cell Wall Samples

Probably the most convenient method of preservation is freeze-drying. In most cases, especially if proteolytic enzymes have been used in their preparation, cell wall fractions after lyophilization form a very light, flaky, white material. It may need a brief sonication or shaking with beads to redisperse the dry cell wall fragments in saline or water. Suspensions of cell wall fragments in distilled water or dilute buffer appear to be stable for years in the refrigerator, provided that autolytic enzymes have been inactivated during their preparation. An inhibitor such as 0.1% sodium azide should be present to prevent attack by bacterial contaminants, many of which will actively digest cell wall fragments.

# DIFFERENCES BETWEEN CELL WALLS FROM GRAM-POSITIVE AND GRAM-NEGATIVE BACTERIA: NATURE OF POLYMERS PRESENT

There are certain broad chemical and structural differences between the cell walls of Gram-positive and Gram-negative bacteria. The major physical and chemical differences are set out in Table 1, and the structural differences, as shown in electron-microscope profiles of thin sections of walls are illustrated diagrammatically in Figure 1. These broad differences in structure are reflected in the different kinds of polymers found in cell walls of Gram-positive and Gram-negative organisms, and the polymers are detailed in Table 2. The most important of them is undoubtedly peptidoglycan (mucopeptide, murein), since it is responsible for the structural integrity of the cell in the face of its internal osmotic pressure. Peptidoglycan is present in bacteria of all kinds, except the extreme halophiles where the osmotic pressure of the environment presumably makes it unnecessary. It is also present in *Spirillum, Spirochaeta, Rickettsia*, and blue-green algae, but not in viruses, yeasts, fungi, higher plants, or animals. It is therefore confined to the cell wall of prokaryotic cells.

Peptidoglycan forms a large proportion of the wall in Gram-positive organisms, where it may make up 50% or more of the wall's dry weight. In Gram-negatives, on the other hand, it forms usually less than 10% of the wall and sometimes as little as 1%. These differences are largely responsible for the higher proportion of amino sugars in the walls of Gram-positives.

In the electron microscope the difference in appearance of cross sections of the wall in Gram-positives and Gram-negatives is generally rather obvious (Figure 1). The typical Gram-positive wall shows a thick (200–500 Å), featureless structure inside which is the double-track (dark-light-dark) appearance of the cytoplasmic membrane. The typical Gram-negative wall, however, is generally thinner (100–150 Å) and shows two double-track membranes separated by a space in which an intermediate layer may or may not be visible (Figure 1, IIb). The inner of the two double-track structures is the cytoplasmic membrane, and the intermediate layer is almost certainly the peptidoglycan. These "typical" appearances may vary considerably in different organisms and with different methods of fixation.[23-26]

## PEPTIDOGLYCAN

Peptidoglycan is a polymer with a backbone of amino sugar chains which are cross-linked through tetrapeptide side chains. An idealized diagram of the structure is given in Figure 2. The backbone of amino sugar is composed of alternate residues of *N*-acetylglucosamine and *N*-acetylmuramic acid linked by β-1 > 4 glycosidic bonds, and the tetrapeptide is attached to the –COOH group of the lactic acid side chain on C-3 of muramic acid. Since muramic acid is a substituted glucosamine, the carbohydrate skeleton of peptidoglycan resembles that of chitin.

The tetrapeptide is almost always composed of four amino acids in the following order, starting at muramic acid: L-alanine, D-glutamic acid, L-diamino acid, D-alanine. One of several diamino acids may be present, as described below. Since the lactic acid side chain of muramic acid is in the D-configuration, this gives the configurational sequence 'D-L-D-L-D' down the peptide chain. This sequence is a constant feature of peptidoglycan structure. The D-glutamic acid residue is linked in the chain by the γ-COOH, and the α-COOH is frequently amidated. For a review of peptidoglycan structure see Reference 27.

### Variations in Peptidoglycan Structure

Of the three major elements in peptidoglycan structure, i.e., the amino sugar backbone, the tetrapeptide, and the linking amino acids, no major variation has so far been found in the backbone structure, although some aerobic actinomycetes are reported to have *N*-glycolylmuramic acid instead of *N*-acetylmuramic acid.[28] However, there is considerable variation both in the tetrapeptide and in the linking amino acids. The principal variations are given here in two tables, Table 3, which lists diamino acids which have been found in different genera of bacteria and some other microorganisms, and Table 4, which gives the principal variations in linking amino acids which detailed structural studies have so far revealed.

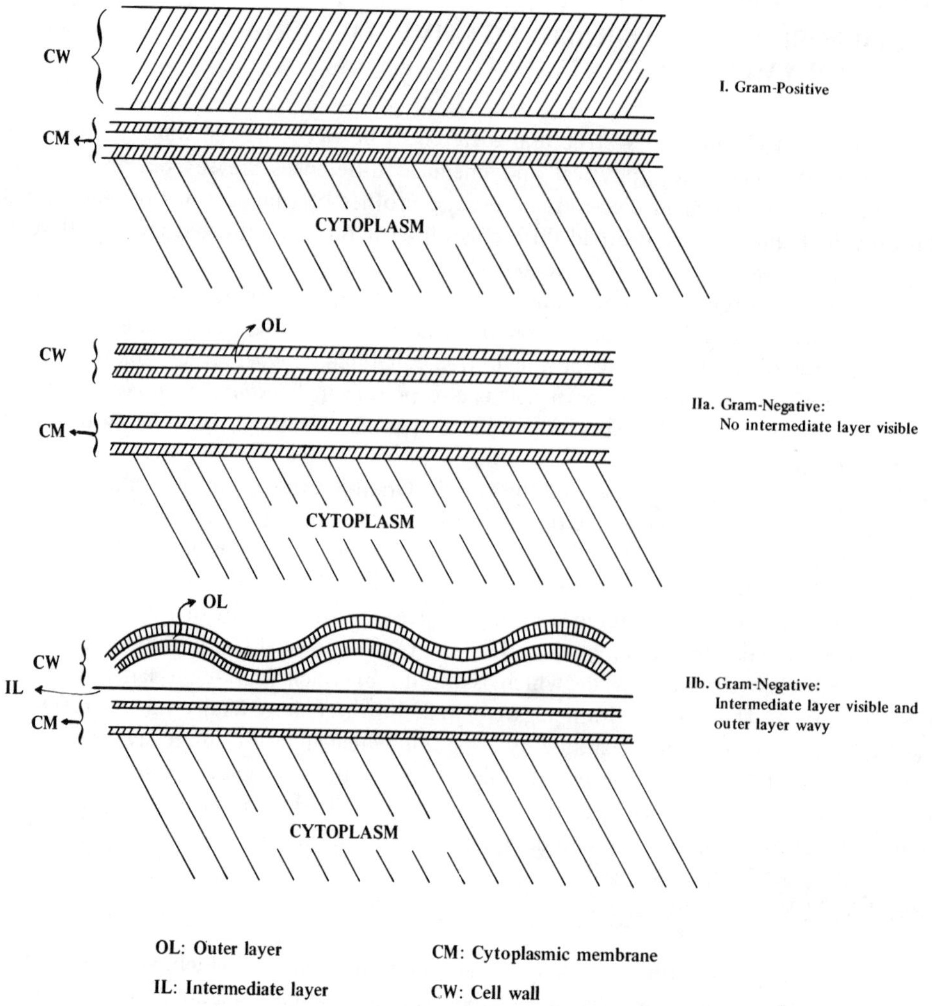


FIGURE 1. Electron-microscope profiles of bacterial cell walls.

## TABLE 1

## GENERAL DIFFERENCES IN CELL WALL COMPOSITIONS BETWEEN GRAM-POSITIVE AND GRAM-NEGATIVE ORGANISMS

| Gram-positives | Gram-negatives |
|---|---|
| Wall thick, about 200 Å; usually structureless. | Wall thinner, about 100–150 Å; double membrane structure usually obvious. |
| Amino sugar content high: 10–30%. | Amino sugar content low: 1–10%. |
| Lipid content generally low: 0–2% (except in mycobacteria, nocardias, and corynebacteria). | Lipid content high: 10–20%. |
| 3–4 major amino acids. | 14–18 amino acids (i.e., the full range found in protein, + diaminopimelic acid). |

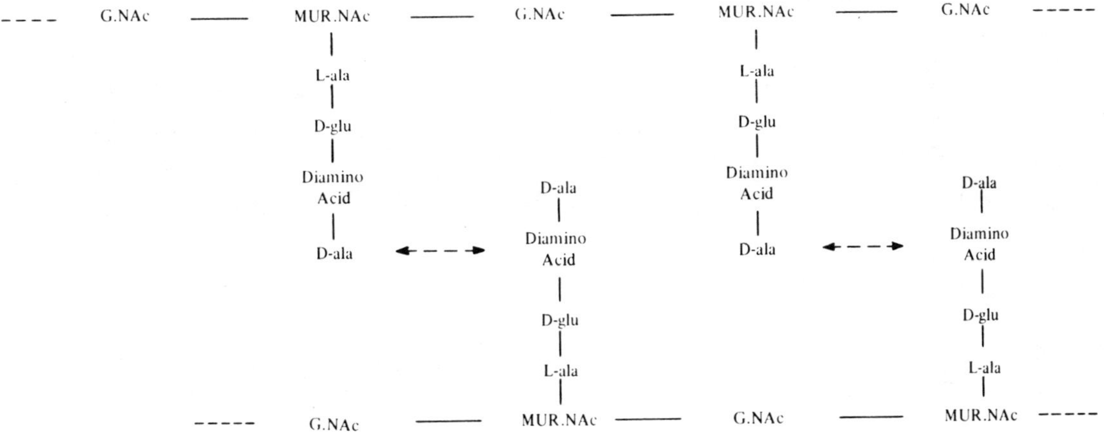


G.NAc = *N*-acetylglucosamine

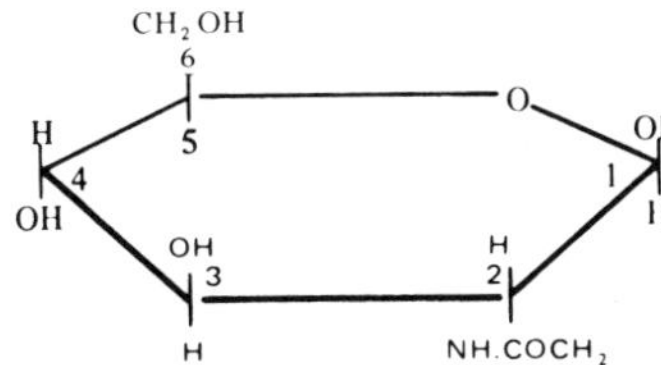


2-acetamido-2-deoxy-D-glucose

MUR.NAc = *N*-acetylmuramic acid

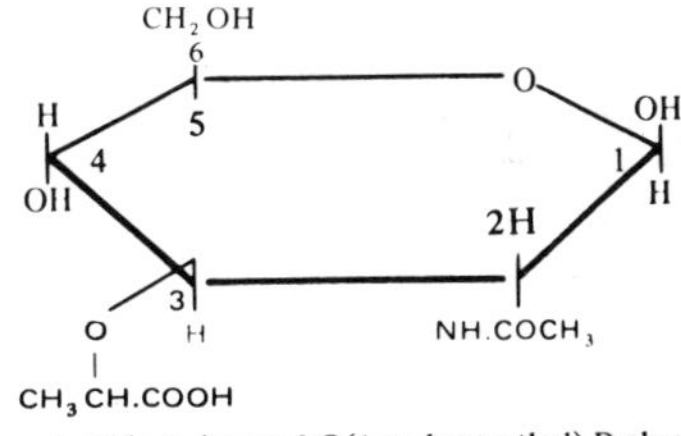


2-acetamido-2-deoxy-3-O(1-carboxyethyl)-D-glucose

FIGURE 2. A general structure for cell wall peptidoglycan.

## TABLE 2

## PRINCIPAL POLYMERS FOUND IN CELL WALLS OF GRAM-POSITIVE AND GRAM-NEGATIVE BACTERIA

### Gram-positives

Peptidoglycan
Polysaccharides
Teichoic acids
Teichuronic acids
Proteins

Glycolipids } mycobacteria, corynebacteria, nocardias
Mycosides }

### Gram-negatives

Peptidoglycan
O-antigen complex
Lipoprotein

## TABLE 3

### DISTRIBUTION OF DIAMINOPIMELIC ACID (DAP) AND OTHER DIAMINO ACIDS IN BACTERIA AND OTHER MICROORGANISMS

| Organism | DAP | | | Other Diamino Acids | | | | References |
|---|---|---|---|---|---|---|---|---|
| | *meso*/DD | LL | 3-Hydroxy | L-Lysine | L-Ornithine | L-DABA | L-Homoserine† | |
| **Gram-Positive Bacteria** | | | | | | | | |
| *Aerococcus* | | | | | | | | |
| *A. viridans* | | | | + | | | | 29* |
| *Aerobic Actinomycetes* | | | | | | | | |
| *Streptomyces* spp. | | + | | | | | | |
| *Micromonospora* spp., *Actinoplanes* spp., *Streptosporangium* spp. | + | Variable, generally small amount | + to – | | | | | 30,* 31* 32,* 33* 34* |
| *Nocardia* | + | | | | | | | |
| *Actinomyces, microaerophilic* | | | | | | | | |
| *A. israelii, A. viscosus, A. naeslundii* | | | | Lysine or ornithine (structure not known in detail) | | | | 35, 36 |
| *A. bovis* | | | | + | | | | |
| *Arachnia* | | | | | | | | |
| *A. propionica* | | + | | | | | | 36,* 37* |
| *Arthrobacter* | | | | | | | | |
| *A. albidus* and other strains | + | | | | | | | 38,* 39* |
| *A. simplex* and other strains | | + | | | | | | 40, 41* |
| *A. globiformis* and other strains | | | | + | | | | 42 |

TABLE 3 (Continued)

DISTRIBUTION OF DIAMINOPIMELIC ACID (DAP) AND OTHER DIAMINO ACIDS IN BACTERIA AND OTHER MICROORGANISMS

| | DAP | | | Other Diamino Acids | | | | |
|---|---|---|---|---|---|---|---|---|
| Organism | *meso*/DD | LL | 3-Hydroxy | L-Lysine | L-Ornithine | L-DABA | L-Homoserine† | References |
| **Gram-positive Bacteria (continued)** | | | | | | | | |
| *Bacillus* | | | | | | | | |
| Most spp. | + | | | | | | | 43,* 44, 45, 46,* 47,* 48, 49, 50, 51, 52, 53* |
| *B. sphaericus, B. pasteurii* | | | | + | | | | 47,* 53* |
| *Bacterionema* | | | | | | | | |
| *B. matruchottii* | + | | | | | | | 54 |
| *Bifidobacterium* | | | | | | | | |
| *B. infantis* and other spp. | | | | + | | | | 53 |
| *B. bifidum* and other spp. | | | | | + | | | |
| *Brevibacterium* | | | | | | | | |
| c. 50% of strains (15 named spp.) | + | | | | | | | |
| *B. acetylicum* | | | | + | | | | 41* |
| *B. helvolum* (1 strain) | | | | | | + | | |
| *B. albidum* (and some other strains) | | | | | | | + | |
| *Butyribacterium* | | | | | | | | |
| *B. rettgeri* | | | | + | | | | 55 |

## TABLE 3 (Continued)

## DISTRIBUTION OF DIAMINOPIMELIC ACID (DAP) AND OTHER DIAMINO ACIDS IN BACTERIA AND OTHER MICROORGANISMS

| Organism | DAP | | | Other Diamino Acids | | | | References |
|---|---|---|---|---|---|---|---|---|
| | *meso*/DD | LL | 3-Hydroxy | L-Lysine | L-Ornithine | L-DABA | L-Homoserine† | |
| **Gram-positive bacteria (continued)** | | | | | | | | |
| *Cellulomonas* | | | | | | | | |
| 6 spp. (1 strain each) | | | | + | | | | 38 |
| *Clostridium* | | | | | | | | |
| Most spp. | + | | | | | | | |
| *C. perfringens, C. pectinovorum* | | + | | | | | | 56 |
| *C. paraputrificum, C. tertium, C. innocuum* | | | | + | | | | |
| *Corynebacterium* | | | | | | | | |
| *C. diphtheriae* (and most human and animal pathogenic coryneforms) | + | | | | | | | 57 |
| *C. pyogenes, C. haemolyticum* | | | | + | | | | 58 |
| *C. acnes (Propionibacterium acnes)* and related organisms | | + | | | | | | 37* |
| *C. poinsettiae* and some other plant pathogens | | | | | | | + | 59 |
| *C. insidiosum, C. tritici, C. sepedonicum* | | | | | | + | | 59 |
| *Diplococcus* | | | | | | | | |
| *D. pneumoniae* | | | | + | | | | 60 |

TABLE 3 (Continued)

DISTRIBUTION OF DIAMINOPIMELIC ACID (DAP) AND OTHER DIAMINO ACIDS IN BACTERIA AND OTHER MICROORGANISMS

| Organism | DAP | | | Other Diamino Acids | | | | References |
|---|---|---|---|---|---|---|---|---|
| | *meso*/DD | LL | 3-Hydroxy | L-Lysine | L-Ornithine | L-DABA | L-Homoserine[†] | |
| **Gram-positive Bacteria (continued)** | | | | | | | | |
| *Erysipelothrix* | | | | | | | | |
| *E. rhusiopathiae* | | | | + | | | | 61 |
| *Gaffkya* | | | | | | | | |
| *G. homari* | | | | + | | | | 29* |
| *Lactobacillus* | | | | | | | | |
| Most spp. | | | | + | | | | |
| *L. plantarum, L. inulinus* | + | | | | | | | 53* |
| *L. cellobiosus* | | | | | + | | | |
| *Leuconostoc* | | | | | | | | |
| All spp. examined | | | | + | | | | 53, 62 |
| *Listeria* | | | | | | | | |
| *L. monocytogenes* | + | | | | | | | 61,* 63, 64 |
| *Microbacterium* | | | | | | | | |
| *M. flavum, M. thermosphactum* | + | | | | | | | |
| *M. lacticum* | | | | + | | | | 41,* 65* |
| *M. liquefaciens* | | | | | | + | | |

TABLE 3 (Continued)

DISTRIBUTION OF DIAMINOPIMELIC ACID (DAP) AND OTHER DIAMINO ACIDS IN BACTERIA AND OTHER MICROORGANISMS

| | DAP | | | Other Diamino Acids | | | | |
|---|---|---|---|---|---|---|---|---|
| Organism | *meso*/DD | LL | 3-Hydroxy | L-Lysine | L-Ornithine | L-DABA | L-Homoserine[†] | References |
| **Gram-positive Bacteria (continued)** | | | | | | | | |
| *Micrococcus* | | | | | | | | |
| Most spp. | | | | + | | | | 29 |
| *M. varians* | + | | | | | | | |
| *M. radiodurans* | | | | | + | | | 66, 67 |
| *Mycobacterium* | | | | | | | | |
| All spp. examined | + | | | | | | | 68,* 69* |
| *Planococcus* | | | | + | | | | 29 |
| *Pediococcus* | | | | | | | | |
| *P. cerevisiae* | | | | + | | | | 53,* 70 |
| *Peptostreptococcus* | | | | | | | | 71, 72* |
| *Propionibacterium** | | | | | | | | |
| Most spp. | | + | | | | | | 37* |
| *P. shermanii, P. freudenreichii* | + | | | | | | | |
| *Rothia* | | | | | | | | |
| *R. dentocariosa* | | | | + | | | | 73 |
| *Sarcina* | | | | | | | | |
| *S. ventricule, S. maxima* | | + | | | | | | 29* |

TABLE 3 (Continued)

DISTRIBUTION OF DIAMINOPIMELIC ACID (DAP) AND OTHER DIAMINO ACIDS IN BACTERIA AND OTHER MICROORGANISMS

| | DAP | | | Other Diamino Acids | | | | |
|---|---|---|---|---|---|---|---|---|
| Organism | *meso*/DD | LL | 3-Hydroxy | L-Lysine | L-Ornithine | L-DABA | L-Homoserine† | References |
| **Gram-positive Bacteria (continued)** | | | | | | | | |
| *Spororsarcina* | | | | | | | | |
| *S. ureae* | | | | + | | | | 29* |
| *Staphylococcus* | | | | | | | | |
| All spp. examined | | | | + | | | | 74, 75, 76* |
| *Streptococcus* | | | | | | | | |
| All spp. examined | | | | + | | | | 57, 77,* 78,* 79, 80 |
| **Gram-negative Bacteria** | | | | | | | | |
| **Gram-negative Bacilli** | | | | | | | | |
| *Aerobacter* | + | | | | | | | 81,*‡ 82 |
| *Agrobacterium* | + | | | | | | | 83 |
| *Brucella* | + | | | | | | | 81,*‡ 84,* 85 |
| *Citrobacter* | + | | | | | | | 86 |
| *Erwinia* | + | | | | | | | 87 |
| *Escherichia* | + | | | | | | | 48,* 86 |
| *Ferrobacillus* | + | | | | | | | 88 |
| *Haemophilus* | + | | | | | | | 81*‡ |

TABLE 3 (Continued)

**DISTRIBUTION OF DIAMINOPIMELIC ACID (DAP) AND OTHER DIAMINO ACIDS IN BACTERIA AND OTHER MICROORGANISMS**

| Organism | DAP | | | Other Diamino Acids | | | | References |
|---|---|---|---|---|---|---|---|---|
| | *meso*/DD | LL | 3-Hydroxy | L-Lysine | L-Ornithine | L-DABA | L-Homoserine† | |
| **Gram-negative Bacteria (continued)** | | | | | | | | |
| *Klebsiella* | + | | | | | | | 86 |
| *Proteus* | + | | | | | | | 86, 89 |
| *Pseudomonas* | + | | | | | | | 86, 89, 90,* 91* |
| *Salmonella* | + | | | | | | | 89, 92, 93 |
| *Serratia* | + | | | | | | | 86, 89 |
| *Thiobacillus* | + | | | | | | | 94 |
| **Gram-negative cocci** | | | | | | | | |
| *Neisseria, Veillonella* | + | | | | | | | 95 |
| *Vibrio* | | | | | | | | |
| *V. metschnikovi* | + | | | | | | | 92 |
| *V. fetus* | + | | | | | | | 86 |
| *V. comma* | + | | | | | | | 81*‡ |
| **Other Organisms** | | | | | | | | |
| Blue-green algae | + | | | | | | | 81,*‡ 97, 98 99 |
| Myxobacteria | + | | | | | | | 81,*‡ 100 |

## TABLE 3 (Continued)

## DISTRIBUTION OF DIAMINOPIMELIC ACID (DAP) AND OTHER DIAMINO ACIDS IN BACTERIA AND OTHER MICROORGANISMS

| | DAP | | | Other Diamino Acids | | | | |
|---|---|---|---|---|---|---|---|---|
| Organism | *meso*/DD | LL | 3-Hydroxy | L-Lysine | L-Ornithine | L-DABA | L-Homoserine† | References |
| **Other Organisms (continued)** | | | | | | | | |
| *Caryophanon* | | | | + | | | | 101 |
| *Spirillum* | + | | | | | | | 102,* 103, 104, 105 |
| *Spirochaeta* | Traces? | | | Probably lysine or ornithine | | | | 106, 107 |

† Homoserine is not a diaminoacid, but where it occurs it occupies the position in the tetrapeptide structure normally occupied by a diaminoacid.
* Isomer of DAP definitely identified, or Lysine, Ornithine, etc. definitely identified as L-isomer.
‡ Isomer of DAP identified in whole cell hydrolysate only.

## TABLE 4

## PEPTIDOGLYCAN LINKAGES ASSOCIATED WITH DIFFERENT DIAMINO ACIDS*

General Peptidoglycan Structure (from Figure 2)

```
– – – – M – G – M – G – M – – – –
              |
            L–ala
              |
            D–glu
              |            (linkage)
         Diamino Acid . . . . . . . . D-ala
              |                          |
            D–ala                  Diamino Acid
                                         |
                                       D-glu
                                         |
                                       L-ala
                                         |
                       – – – – M – G – M – G – M – – – –
```

| Diamino Acid | Organism | Linkage (Linking Amino Acids or Peptides) | References |
|---|---|---|---|
| L-Lysine | *Staphylococcus* | $-(Gly)_5-$<br>$-(Gly)_2-Ser-(Gly)_2-$<br>$-L-ala-(Gly)_4-$ | 29, 76 |
| | *Micrococcus*<br>*Sarcina* | $-L-ala-$<br>$-D-ala-L-lys-D-glu-L\text{-}ala-$<br>Gly<br>$-Gly-L-glu-$<br>$-(L-ala)_3-$ | 29, 108, 109, 110, 111 |
| | *Aerococcus*<br>*Gaffkya* | Direct peptide bond between L–lys and D–ala | 29 |
| | *Planococcus* | –D–glu– | 111 |
| | *Pediococcus* | –D–asp– | 29, 53 70 |
| | *Leuconostoc* | $-L-Ser-(L-ala)_2-$<br>$-L-Ser-L-ala-$<br>$-L-ala-L-ser-$ | 53, 112 |
| | *Lactobacillus* | $-D-asp-$<br>$-L-ser-(L-ala)_2-$<br>$-(L-ala)_2-$ | 53, 112, 113, 114, 115 |
| | *Streptococcus* | $-L-asp-$<br>$-Thr-L-ala-$<br>$-(L-ala)_3-$ | 77, 78, 114, 116 |
| | *Bifidobacterium* | –Gly– | 53, 117, 118 |

TABLE 4 (Continued)

PEPTIDOGLYCAN LINKAGES ASSOCIATED WITH DIFFERRENT DIAMINO ACIDS*

| Diamino Acid | Organism | Linkage | References |
|---|---|---|---|
| | | Linking Amino Acids or Peptides | |
| *meso*–DAP | *Bacillus, Arthrobacter, Corynebacterium, Clostridium,* etc.<br>Gram-negative bacteria | Direct peptide bond between *meso*–DAP and D–ala | See Table 3 for genera *meso*–DAP |
| L–DAP | *Propionibacterium, Streptomyces, Clostridium* | $-(Gly)_n-$ (n usually 1–4) | 119 |
| L–Ornithine | *Bifidobacterium* | –L–ser–L–ala–L–Thr–L–ala–<br>–L–lys–D–asp–<br>–L–lys–L–ser–$(L–ala)_2$–<br>–L–lys–$(L–ala)_2$– | 53 |
| L–DABA | *Corynebacterium tritici, C. insidiosum, C. michiganense*† | –D–DABA– | 41, 59 |
| L–Homoserine | *Corynebacterium betae, C flaccumfaciens, C. poinsettiae* | –D–orn– | 41, 115 |
| | Some strains in *Arthrobacter, Microbacterium, and Brevibacterium*‡ | –Gly–D–orn–<br>–$(Gly)_2$–L–lys– | 41 |

* The material in the table has been arranged broadly in terms of the linkages occurring in different genera, but it must be noted that (1) the same linkage may occur in several different genera and (2) organisms classified in a single genus may have different diamino acids and types of linkage.

† The peptidoglycan in these organisms is unusual because L-ala is replaced by Gly as the amino acid attached to muramic acid, and the cross-linkage is through the α-COOH of glutamic acid instead of through the diamino acid (L-DABA).

‡ Peptidoglycans with homoserine are unusual because Gly is the amino acid joined to muramic acid, because the cross-link is through the α-COOH of glutamic acid, and because homoserine is not a diamino acid: D-glu in these strains may be replaced by 3-threo-hydroxy-glutamic acid.

## VARIATIONS IN TETRAPEPTIDE

Any one of four diamino acids may be present: L-lysine, diaminopimelic acid (DAP), L-ornithine, or L-diaminobutyric acid (see Table 3). Rarely, no diamino acid is present, as in the homoserine-containing tetrapeptide of some plant pathogenic corynebacteria. In the case of DAP, the *meso-* or L-isomers may be present, but the end involved in the tetrapeptide has the L-configuration.[27] The D-isomer of DAP has also been found in the cell walls of some organisms (e.g., *Bacillus megaterium* and some aerobic actinomycetes[34,45]), but always in association with other DAP isomers, and its position in the structure is not certain. It is thought not to occur in the tetrapeptide, since this would violate the 'D-L-D-L-D' rule. In some actinomycetes, the 3-hydroxy derivative of DAP is found.[120]

Most variations in tetrapeptide structure concern the diamino acid, but rarely the L-alanine is replaced by glycine or L-serine (see Table 4). For a review of the distribution of diamino acids in cell walls, see Reference 121.

### VARIATION IN LINKING AMINO ACIDS

In some cases, e.g., with *meso*-DAP, the link between tetrapeptide chains is a direct peptide bond from diamino acid to D-alanine on another tetrapeptide. In most other cases, however, the link is composed of a single amino acid or a short peptide. In unusual types of peptidoglycan structure, the link may involve the $\alpha$-COOH of glutamic acid.

### RECOGNITION OF PEPTIDOGLYCAN COMPONENTS IN ACID HYDROLYSATES OF WHOLE CELLS

Two components of peptidoglycan, diaminopimelic acid and muramic acid, are so distinctive that their presence in hydrolysates of whole cells may be taken as evidence of their occurrence in the peptidoglycan of that particular strain, or more generally, as evidence that peptidoglycan is present in a particular group of organisms.

#### Diaminopimelic Acid (DAP)

DAP occurs as an intermediate in the more common of the two pathways of lysine biosynthesis and as such is found in small quantities in the cytoplasm of bacteria and other prokaryotic organisms, in green algae, and in plants. However, as a structural amino acid it is found only in peptidoglycan. In Gram-positive bacteria the cell wall may constitute 15–20% or more of the dry weight of the cell, and 50% or more of the wall may be peptidoglycan. Under these circumstances DAP, if present, is likely to be a major constituent of acid hydrolysates of whole cells and can readily be detected by paper chromatography. Occasional strains of lactobacilli with defective lysine metabolism may produce large amounts of free DAP, although the diamino acid of their peptidoglycan is lysine.[122]

In Gram-negative bacteria, however, the wall forms a smaller proportion of the cell (5–10%), and peptidoglycan may form less than 1% of the wall, so that it may be difficult to detect DAP in crude whole cell hydrolysates. (For recognition of DAP and its isomers and derivatives by paper chromatography, see References 45, 81, 120, and 123.)

#### Muramic Acid

This is a more reliable indicator substance for peptidoglycan than DAP, as it appears to be an invariable component of the backbone glycan chain and is not known to occur elsewhere. The presence of muramic acid in whole cell hydrolysates has been used to demonstrate the occurrence of peptidoglycan in *Leptospira* and *Borrelia*[124] and in *Rickettsia.*[125]

## OTHER CELL WALL POLYMERS

In most cases detailed knowledge about other cell wall polymers is confined to a few selected groups of organisms since they have not been investigated as thoroughly or as widely as has peptidoglycan. In the following sections, polymers occurring in Gram-positive and Gram-negative organisms are discussed separately.

### Gram-positive Organisms

#### TEICHOIC ACIDS AND OTHER ACIDIC POLYSACCHARIDES

The teichoic acids are complex polymers of polyols and phosphate, originally described by Baddiley and his co-workers.[126,127] Most commonly they have the general structure illustrated in Figure 3, which shows a backbone of repeating units of glycerol or ribitol phosphate with associated residues of sugars or amino sugars and D-alanine. However, teichoic acid from *Staphylococcus lactis* has alternate residues of *N*-acetylglucosamine and glycerol in the backbone structure,[128] and the C-substance from pneumococcal cell walls is a teichoic acid containing choline phosphate.[129]

```
OH              Sugar          OH          Sugar            OH   Sugar
|                 |            |             |              |      |
OH – P – O – ALDITOL – O  [ – P – O – ALDITOL – O ]  – P – ALDITOL
||                |            ||            |          n   ||     |
O             D-alanine        O         D-alanine          O   D-alanine
```

The ALDITOL may be either:

1. GLYCEROL or 2. RIBITOL

```
H2C.OH          H2C.OH
  |               |
H.C.OH          H.C.OH
  |               |
H2C.OH          H.C.OH
                  |
                H.C.OH
                  |
                H2C.OH
```

FIGURE 3. General structure of teichoic acids. (Taken from Archibald, A. R., and Baddiley, J., *Adv. Carbohydr. Chem.*, *21*, 323 (1966). Reproduced by permission of the Academic Press.)

Although originally named because of their association with cell walls (*teichos* is the Greek for wall), teichoic acids are also widely distributed in Gram-positive organisms in association with the cytoplasmic membrane, in which case they are often referred to as intracellular or cytoplasmic teichoic acids, although they probably accumulate between the wall and the membrane.

Cell wall teichoic acids appear to be especially prominent in certain genera such as *Staphylococcus, Micrococcus, Bacillus,* and *Lactobacillus,* where they may form up to 50% of the wall. They may also be important cell wall antigens (see below). The presence of appreciable amounts of phosphorus (i.e., 2–5%) in cell walls may be taken as presumptive evidence of the presence of teichoic acids, since in their absence the phosphorus content of bacterial walls is low (i.e., 0.2–0.5%).

Acidic polysaccharides containing uronic acids but no phosphate have been isolated from some organisms. For example, Janczura et al.[130] isolated from *Bacillus licheniformis* a polysaccharide consisting of equal parts of glucuronic acid and *N*-acetyglucosamine which thev called teichuronic acid: a similar substance composed of glucose and amino-mannuronic acid was found in the walls of *Micrococcus lysodeikticus* by Perkins.[131] Since uronic and (especially) amino-uronic acids are acid-labile, it is possible that they occur rather widely in bacterial walls but have remained unrecognized because they are destroyed by the usual methods of hydrolysis.

There appears to be a close relationship between synthesis of teichoic and teichuronic acids because in *Bacillus subtilis* teichuronic acid replaces teichoic acid if the organisms are deprived of phosphate.[132] However, NaCl concentration is also important.[133]

## Extraction of Teichoic Acid from Cell Walls

Teichoic acids are generally prepared from cell walls by extraction with dilute trichloroacetic acid (TCA: 5–10% aqueous) at 0–5°C. The teichoic acid is precipitated from the extract with ethanol (2 to 5 volumes). The extraction may need to be long-continued (2 to 3 days) and repeated several times to obtain the bulk of the material. The process can be followed by estimating the phosphorus content of the residual cell walls. Extraction is much faster at higher temperatures, i.e., 37°C or even 80–90°C, but this is usually undesirable because other polysaccharides, if present, will also be extracted readily and because degradation of teichoic acid may occur. Cell wall teichoic acids may also be obtained by extraction of intact bacterial cells but will then be contaminated by nucleic acids and intracellular teichoic acids from which it may be difficult to purify them.

Details of TCA extraction procedures as applied to walls of *Lactobacillus buchneri, Lactobacillus arabinosus, Staphylococcus albus,* and *Bacillus subtilis* may be found in References 134, 135, 136, and 137. Other extraction procedures using dilute alkali or dilute aqueous *N,N*-dimethylhydrazine will be found in References 138 and 139.

The detailed structures of a glycerol teichoic acid from *Staphylococcus albus* and a ribitol teichoic acid from *Bacillus subtilis* are shown in Figures 4 and 5, respectively.[136,140]

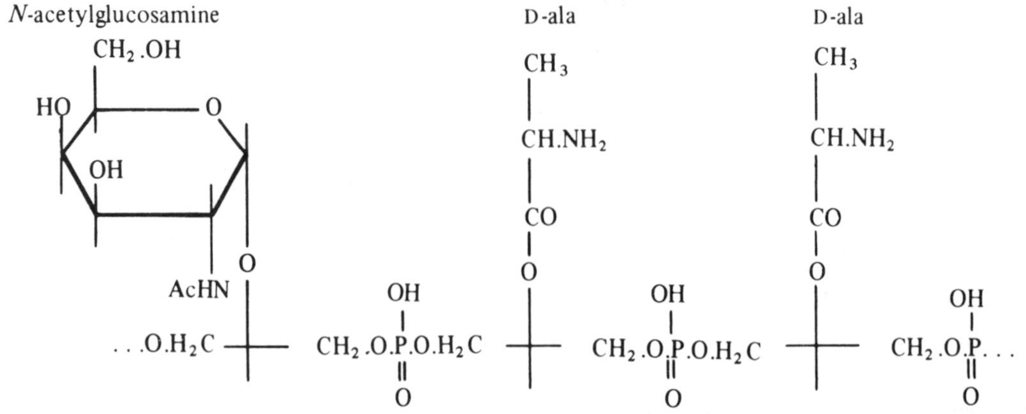

FIGURE 4. Repeating unit of the glycerol teichoic acid in cells walls of *Staphylococcus albus.* (Taken from Ellwood, D. C., Kelleman, M. V., and Baddiley, J., *Biochem. J., 86,* 213 (1963). Reproduced by permission of *The Biochemical Journal.*)

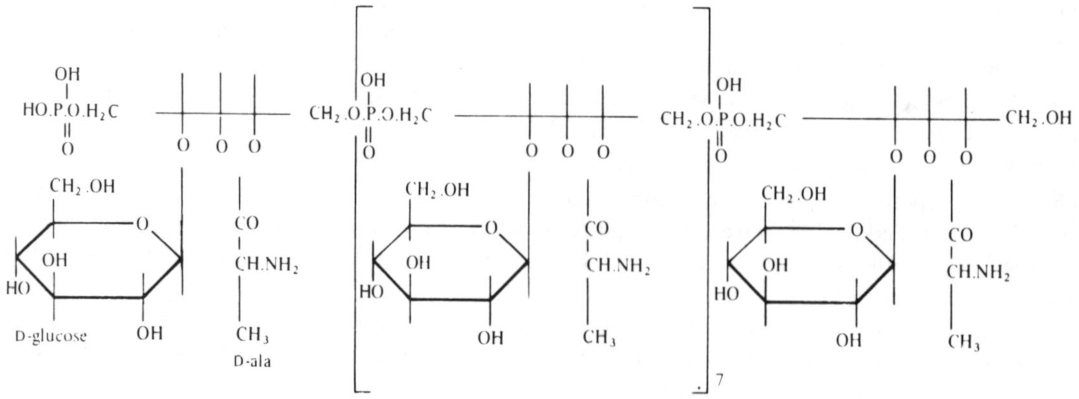

FIGURE 5. Ribitol teichoic acid from cell walls of *Bacillus subtilis.* (Taken from Armstrong, J. J., Baddiley, J., and Buchanan, J. G., *Biochem. J., 80,* 254 (1961). Reproduced by permission of *The Biochemical Journal.*)

TABLE 5

LOCATION AND CHEMICAL NATURE OF GROUP ANTIGENS IN DIFFERENT LACTOBACILLI

| Serological Group | Species Included in Group | Location of Antigen in Cell | Chemical Nature of Antigen | Sugar | Alkali Hydrolysis Product |
|---|---|---|---|---|---|
| A | *L. helveticus* | Membrane | GTA[a] | Glucose (trace) | |
| | *L. jugurti* | | | Ribose (trace) | Glucosyl glycerol |
| B | *L. casei* | Wall | Polysaccharide | Rhamnose | |
| C | *L. casei* | Wall | Polysaccharide | Glucose | |
| D | *L. plantarum* | Wall | RTA[b] | Glucose | Glucosyl ribitol |
| E | *L. lactis*<br>*L. bulgaricus*<br>*L. brevis*<br>*L. buchneri* | Wall | GTA[a] | Glucose | Glucosyl glycerol |
| F | *L. fermenti* | Membrane | GTA[b] | | |
| G | *L. salivarius* | ? | ? | | |

[a] Glycerol teichoic acid
[b] Ribitol teichoic acid

Taken from Sharpe, E., *Int. J. Syst. Bacteriol., 24,* 509 (1970). Reproduced by permission of the International Association of Microbiological Societies.

### Teichoic Acids as Antigens

Teichoic acids are good antigens and have been used as the basis for defining antigenic groups in several genera. The determinants involved are usually the sugars or amino sugars attached to the alditols. (See Figure 3.) For example, in *Staphylococcus aureus*[141] the walls contain ribitol teichoic acid which has *N*-acetylglucosamine attached to the ribitol by either an $\alpha$- or a $\beta$-linkage so that two serologically distinct teichoic acids may occur in the same organism. On the other hand, *Staphylococcus epidermidis* strains have glycerol teichoic acids which may have $\alpha$- or $\beta$-linked glucose, or $\alpha$-linked glucosamine.[142,143] As a further example of the use of teichoic acid antigens in classification, Table 5 shows the chemical nature and anatomical location of the group antigen in various serological groups of lactobacilli.[144] It may be noted that not all lactobacilli contain teichoic acids in the cell wall.

## NEUTRAL POLYSACCHARIDES

Apart from acidic polysaccharides such as teichoic and teichuronic acids, many Gram-positive organisms have cell wall polysaccharides composed of neutral sugars, usually hexoses and pentoses, and less commonly amino sugars. The polysaccharides of Gram-negative organisms will not be dealt with here since they are an essential part of the O-antigen complex which is considered in this article under the heading, Lipopolysaccharide–Protein Complexes: O-antigens, Endotoxins.

The cell wall polysaccharides of streptococci have been most thoroughly investigated, probably because of their economic importance and because of the pioneer work of Lancefield in developing an antigenic scheme for these organisms.[145,146] Qualitatively, most streptococcal polysaccharides consist of rhamnose and some combination of glucose, galactose, glucosamine, and galactosamine; mannose is found in some groups. However, a number of strains in Lancefield groups K, M, and O have polysaccharides without rhamnose.

The results of qualitative analysis of the sugars in cell wall polysaccharides of more than two hundred strains of streptococci can be found in papers by Slade and Slamp[147] and Colman and Williams.[79] The quantitative composition of polysaccharides from strains in several Lancefield groups is given in Table 6.

In other groups of Gram-positive organisms, knowledge of cell wall polysaccharide composition is fragmentary and is largely confined to reports of the sugar components released on hydrolysis. A list of the sugars found in some genera of Gram-positive organisms other than streptococci is given in Table 7. More than one polysaccharide may be identified in cell wall extracts. For example, Michel and Krause[179] found two polysaccharides in formamide extracts of Group F streptococci, and two antigenically distinct carbohydrates were found in strains of *Lactobacillus casei* by Knox.[180] Since in some cases teichoic acids and neutral polysaccharides may be found together in walls of the same organism, some of the sugars reported may come from teichoic acid.

Most methods of extraction of polysaccharides from cell walls were originally devised to get material from whole cells into solution for immunological testing. The classic method is that used by Lancefield for streptococci, which involved heating with *N*/20 HCl (pH 2.0) at 100°C for 15 minutes and then cooling and neutralizing the extract. Fuller in 1938[181] introduced extraction with neutral formamide at 160 – 165°C, and Maxted,[182] the use of muralytic enzymes from *Streptomyces.* References to the use of these and other methods are given in Table 8; for a general discussion of the action of hot formamide on cell walls, see Reference 183.

## TABLE 6

## PERCENT COMPOSITION OF CELL WALL POLYSACCHARIDES IN STREPTOCOCCI[a]

| Lancefield Serological Group | Rhamnose[b] | Glucose[b] | Galactose[b] | Mannose | N-Acetyl-Glucosamine | N-Acetyl-Galactosamine | Method of Preparation | Reference |
|---|---|---|---|---|---|---|---|---|
| A | 39 | | | | 24.3 | | TCA[c] | 148 |
| | 42–49 | | | | 23–28 | | Enzyme[d] | 149 |
| B | 50.2 | | 8.9 | | 12.3 | | | |
| | 50.5 | | 11.0 | | 11.4 | | Formamide | 151 |
| C | 43 | | | | 3.9 | 35.1 | Formamide | 152 |
| E | 44.2 | 22.0 | | | 2.2 | | | |
| | 36.4 | 19.4 | | | 2.3 | | TCA[c] | 148 |
| F | 48.2 | 20.6 | 3 | 0 | 1 | 27 | | |
| | 17.3 | 13 | 13 | 20.5 | 18.9 | 17.3 | | |
| | 55.3 | 4.5 | 14.6 | 5.9 | 4.5 | 11.9 | | |
| | 36.4 | 11.7 | 24.3 | 12.1 | 0.8 | 14.6 | Formamide | 152 |
| G | 40.7 | | 23.7 | | | 20.6 | | |
| | 38.7 | | 20.1 | | | 17.8 | | |
| | 36.8 | | 21.2 | | | 16.2 | Formamide | 153 |
| | 34.8 | | 16.0 | | | 18.3 | TCA[c] | 148 |
| T | 22.6 | 7.0 | 26.3 | | 15.9 | 2.8 | TCA[c] | 148 |

[a] During hydrolysis of the carbohydrates to liberate individual components, some sugar is destroyed; hence the figures for percentage composition, based on weight before hydrolysis, normally total 75–80% rather than 100%. The figures for Group F *Streptococci,* however, were calculated on the basis of 100% recovery. (See Reference 152.)
[b] Wherever the isomers have been determined, the sugars have been found to be L-rhamnose, D-glucose, and D-galactose.
[c] Trichloroacetic acid
[d] Muralytic enzyme

## TABLE 7

## SUGARS FOUND IN CELL WALL POLYSACCHARIDES FROM GRAM-POSITIVE BACTERIA

| Genus or Group | Sugars Present[a] | References |
|---|---|---|
| Aerobic Actinomycetes | Very variable; at least 4 patterns recognized: (1) Arabinose, galactose, (2) Madurose (3-O-methyl-D-galactose),[b] (3) No sugars, (4) Arabinose, xylose[b] | 154, 155 |
| | Teichoic acids may also occur in these strains. | 127 |
| *Actinomyces* | *A. israelii:* (1) Galactose, (2) Galactose and rhamnose | 156 |
| | *A. bovis, A. naeslundii:* rhamnose, glucose, fucose, 2-deoxytalose | 157, 158 |
| *Arachnia* | *A. propionica:* (1) galactose, (2) galactose and glucose | 37, 159 |
| *Arthrobacter* | Some combination of rhamnose, glucose, galactose, and mannose (also, rarely: fucose, unknown sugars). | 38, 160 |
| *Bacillus* | Glucose and/or galactose appear to be present in the walls of most strains, but are probably from teichoic acids, as in: | |
| | *B. subtilis* | 137 |
| | *B. licheniformis* | 130 |
| | *B. coagulans* | 43 |
| | *B. stearothermophilus* | 43 |
| | However, a polysaccharide composed of glucose, glucosamine, and galactosamine was obtained from *B. thuringiensis* var. *thuringiensis,* and one | 49,161 |
| | composed of galactose and glucosamine was obtained from *B. anthracis.* | 161 |
| *Bacterionema* | *Bacterionema matruchottii:* arabinose, galactose, glucose | 54, 162 |
| *Bifidobacterium* | Some combination of rhamnose, galactose, and glucose: considerable variation. (Some strains may have teichoic acids also.) | 163, 164, 165 |
| *Clostridium* | Usually either (1) glucose, (2) glucose, galactose, or (3) rhamnose, glucose, galactose, mannose | 56, 166, 167 |
| *Corynebacterium* | Human and animal pathogens | |
| | *C. diphtheriae* and most others: arabinose, galactose (may also be mannose and glucose) | 168, 169 |
| | *C. pyogenes:* rhamnose, glucose | 58 |
| | Plant pathogens | |
| | *C. poinsettiae, C. betae:* rhamnose, mannose, galactose, fucose | |
| | *C. tritici, C. insidosum, C. sepedonicum:* xylose, glucose, mannose | 170, 171, 172 |
| *Erysipelothrix* | Galactose, glucose, mannose | 61 |
| *Leuconostoc* | Rhamnose, glucose (+a glycerol teichoic acid) | 62 |
| *Lactobacillus* | *L. casei:* rhamnose, galactose, glucose | |
| | Other lactobacilli: some combination of glucose, galactose, and mannose probably in association with teichoic acids | 144, 169, 173 |
| *Listeria* | Rhamnose, galactose, glucose, mannose. (Some strains also reported to have fucose and xylose.) | 25, 41 |
| *Micrococcus* | Cell walls of most strains probably have teichoic or teichuronic acids, but some have polysaccharides, e.g., *M. roseus,* which has a polysaccharide containing glucose, galactose, and mannose. | 75, 174 |

## TABLE 7 (Continued)

## SUGARS FOUND IN CELL WALL POLYSACCHARIDES FROM GRAM-POSITIVE BACTERIA

| Genus or Group | Sugars Present[a] | References |
|---|---|---|
| *Mycobacterium* | Arabinose, galactose, glucose. (The polysaccharides of mycobacteria occur in combination with lipid.) | 175 |
| *Peptostreptococcus* | Some combination of rhamnose, glucose, mannose | 71 |
| *Propionibacterium* | *P. freudenreichii, P. shermanii:* rhamnose, galactose, mannose | |
| | Other propionibacteria, including *P. acnes* and related organisms: Some combination of glucose, galactose, and mannose | 37, 157, 176 |
| *Rothia* | *Rothia dentocariosa* (*Nocardia salivae*): (1) galactose, (2) galactose, glucose, ribose, fructose | 73, 177<br>178 |
| *Staphylococcus* | Teichoic acids | 75, 127, 174 |
| *Streptococcus* | Some combination of rhamnose, galactose, glucose, mannose, glucosamine, and galactosamine | 79, 147<br>(See also Table 6) |

[a] Amino sugars are not generally included in this survey: by analogy with the results for streptococci, it is likely that they are present in a considerable number of cell wall polysaccharides.

[b] These sugars have been found in whole cell hydrolysates, but it is not yet certain that they are from cell wall polysaccharides.

## TABLE 8

## EXTRACTION OF POLYSACCHARIDES FROM GRAM-POSITIVE BACTERIA

| Organism | Method | Reference |
|---|---|---|
| *Streptococcus* | *N*/20 HCl at 100°C, 15 min | 184 |
| *Streptococcus* | Formamide at 160–165°C | 181 |
| *Streptococcus* | Cell wall lytic enzymes from *Streptomyces* spp. | 182 |
| *Streptococcus* | Cell wall lytic enzymes from bacteriophage | 185 |
| *Streptococcus* | Activation of autolytic enzyme systems | 186 |
| *Streptococcus* | 5% Trichloracetic acid at 90°C | 148 |
| *Actinomyces* | (1) Formamide at 160–165°C<br>(2) 5% Trichloracetic acid at 55°C, 15 min | 156 |
| *Listeria monocytogenes* | Formamide at 145–150°C, 20 min | 64 |
| *Rothia dentocariosa* | 10% Perchloric acid in the cold, followed by column chromatography. | 178 |

The material extracted by these methods usually behaves antigenically as a haptene in that it reacts strongly in precipitin tests against suitable antisera but is not itself immunogenic. To prepare antisera against cell wall polysaccharides, animals are usually immunized with suspension of whole cells; but suspensions of crude or purified cell wall fragments can also be used.

In many organisms the wall polysaccharides are the main polysaccharide elements in the cell, and it may reasonably be assumed that the sugars found in whole cell hydrolysates are from the cell wall. Obviously the presence of capsular material or of intracellular polysaccharides is a source of error to be guarded against. Patterns of sugar components in whole cell hydrolysates may be of considerable value in identification, especially if unusual sugars such as arabinose are present.[187]

## CELL WALL LIPIDS

In most cases the walls of Gram-positive bacteria are characterized by a low lipid content (<5%). However, mycobacteria, nocardias, and most human and animal pathogenic corynebacteria are exceptional in having a high content of cell wall lipids, which may be over 50% of the dry weight of the wall in some mycobacteria. Many of these lipids are characterized by the presence of mycolic acids, which are $\alpha$-substituted, $\beta$-hydroxy long chain acids with the general formula:

$$\begin{array}{l} \text{R–CH–CH–COOH} \\ \quad\;\;| \qquad \backslash \\ \quad\text{OH} \qquad \text{R}_2 \end{array}$$

In mycobacteria, mycolic acids with high molecular weights occur with numbers of carbon atoms usually from $C_{79}$–$C_{85}$, while acids with lower molecular weights are found in nocardias ($C_{48}$–$C_{58}$) and corynebacteria ($C_{32}$–$C_{36}$).

In the cell wall, mycolic acids are found in combination in the form of glycolipids. Thus they may be esterified with the arabino-galactan polysaccharide to form *WaxD*, or with trehalose to form Cord factor (trehalose-6′-6-dimycolate). The cell walls of mycobacteria and related organisms may also contain other lipids in the form of mycosides or lipoproteins, but these have been little studied so far.

The lipids of the mycobacterial wall, especially the mycolic acids, are associated with acid-fastness, and this property is lost if cells are thoroughly delipidated. However, acid-fastness is also dependent on the integrity of the wall and cell membrane since only intact cells are acid-fast. For detailed information on mycobacterial cell wall lipids, see References 175 and 188.

The cell wall lipids of Gram-negative bacteria are associated with the endotoxin complex and are considered in the section on Gram-negative organisms.

For general information on bacterial lipids, see References 189, 190, and 191.

## PROTEIN COMPONENTS

Except in one or two genera, the protein components of the cell walls of Gram-positive organisms have been little investigated. As in the case of wall polysaccharides, the most detailed information on cell wall proteins is available for streptococci, in which numerous acid-soluble protein antigens have been recognized (M–, T–, and R–antigens). Immunological and electron-microscope studies agree in locating these antigens in the superficial layers of the cell wall:[192,193] they are not necessary for the integrity of the wall, since strains which lack them appear to grow as well as those in which they are present, and they can be removed by proteolytic enzymes without affecting the other wall components. In the case of a Group A, Type 12 streptococcus, acid-extractable protein forms about 12% of the wall.[194]

In *Staphylococcus aureus*, the ability of suspension of the organism to absorb agglutinins from antisera is completely abolished by treatment with Pronase,[195] suggesting that the antigens concerned are protein. A protein (Antigen A) which was isolated and purified from the walls of some strains was shown to be a surface component responsible for agglutination.[196]

There is immunological evidence for the presence of a protein in the cell walls of *Corynebacterium diphtheriae*,[197] and a trypsin-insoluble protein was found in the wall of *Bacillus licheniformis*.[198] In fact, it seems likely that cell wall proteins are present quite widely in Gram-positive bacteria, but it must be admitted that as yet little concrete evidence supports this statement.

## Gram-negative Organisms

### Lipopolysaccharide–Protein Complexes: O-antigens, Endotoxins

These substances form a major part of the cell wall of Gram-negative bacteria. In their most complete form, they are macromolecular complexes of lipid, polysaccharide, and protein, with a molecular weight of at least several million. The name endotoxin was given because the toxic principle was regarded as being firmly bound to the cell substance, although in fact variable amounts occur free in young cultures, especially in some strains. (See Reference 199.) The polysaccharide part of the complex carries antigenic groupings which determine the serological reactions of agglutinating suspensions of the organisms (O-antigens) unless the cell surface is covered by some other material such as a capsule, or unless flagellar antigens interfere.

One major characteristic which is shared by all high-molecular-weight O-antigen complexes is that they are toxic and pyrogenic in man and animals. An active preparation from, for example, *Escherichia coli* will cause a febrile reaction in a rabbit at a dosage level of 0.1 μg/kg or less. Other effects of endotoxin injection are the Schwartzmann phenomenon (hemorrhage at the site of intradermal injection in a sensitized animal), hemorrhagic necrosis of tumors, enhanced antigenicity of proteins, and increased nonspecific resistance to infection. (For a general review of the properties of endotoxin, see Reference 200.)

### Variation In Lipopolysaccharide Structure

It is not proposed to do more than summarize briefly the results of the extensive work on the polysaccharides of Gram-negative O-antigens which has been done over the past twenty years. Several detailed reviews are available by Westphal, Lüderitz, Simmonds, Nikaido, Kauffmann, and others which cover the subject both adequately and authoritatively, since the authors of the reviews have themselves been the principal investigators in the field. (See References 201–206.)

The general structure of the lipopolysaccharide portion of a typical O-antigen (from *Salmonella typhimurium*) is shown in Figure 6.[207] Apart from laboratory-induced mutants, the structure of the core polysaccharide appears to be uniform over a rather wide range of Gram-negative organisms. However, heptose may be absent in some groups, e.g., *Xanthomonas* and *Bacteroides*. (See References 208–211.) In contrast to the relative stability of the core, the polysaccharide side chains are very variable in composition, and these variations are responsible for the large number of serologically distinct O-antigens which exist. More than thirty different sugar components have been found in different combinations in the lipopolysaccharides from Gram-negative organisms: a list of the components so far identified is given in Table 9, which is taken from Reference 206. Details of the chemical properties of some of the more unusual of these constituents can be found in Reference 235.

**TABLE 9**

**SUGAR CONSTITUENTS OF LIPOPOLYSACCHARIDES**

| Structure | Configuration | | Trivial Name | Occurrence |
|---|---|---|---|---|
| Hexoses | Glucose | D | – | Frequent |
| | Galactose | D | – | Frequent |
| | Mannose | D | – | Frequent |
| 6-Deoxyhexoses | Galacto- | L | Fucose | Frequent |
| | Manno- | L | Rhamnose | Frequent |

## TABLE 9 (Continued)

## SUGAR CONSTITUENTS OF LIPOPOLYSACCHARIDES

| Structure | Configuration | | Trivial Name | Occurrence |
|---|---|---|---|---|
| 6-Deoxyhexoses (cont.) | Manno- | D | Rhamnose | *Xanthomonas campestris*[212] |
| | Talo- | L | – | *Escherichia coli* 045, 066, 084, 088[213] |
| 3,6-Dideoxyhexoses | Gluco- | D | Paratose | *Salmonella* group A[214]<br>*Pasteurella pseudotuberculosis* I, III[215] |
| | Galacto- | D | Abequose | *Salmonella* groups B, C2, C3[214]<br>*Citrobacter* 4.5[216]<br>*Pasteurella pseudotuberculosis* II[215] |
| | Galacto- | L | Colitose | *Salmonella* groups 35, 50[214]<br>*Escherichia coli* 055, 0111[216]<br>*Arizona* 9, 20[216] |
| | Manno- | D | Tyvelose | *Salmonella* group D[214]<br>*Pasteurella pseudotuberculosis* IV[215] |
| | Manno- | L | Ascarylose | *Pasteurella pseudotuberculosis* V[215] |
| Pentoses | Ribose | D | – | *Salmonella* groups 28, 52, 56[214,217] |
| | Ribose | – | – | *Escherichia coli* 0114[213] |
| | Xylose | – | – | *Citrobacter freundii* 08[218]<br><br>*Chromobacterium violaceum* Lewitus MWB[215] |
| Heptoses | | | | |
| L-Glycero- | Manno- | D | – | Frequent |
| D-Glycero- | Manno- | D | – | *Chromobacterium violaceum* NCTC 7917[215]<br>*Serratia marcescens*[219]<br>*Proteus marabilis*[220]<br>*Salmonella, Escherichia coli,* and others[208]<br>*Brucella melitensis*[221] |
| D-Glycero- | Galacto- | D | – | *Chromobacterium violaceum* Brown, Birch, BN[215] |
| 2-Amino-2-deoxyhexoses | Glu- | D | Glucosamine | Frequent |
| | Galacto- | D | Galactosamine | Frequent |
| | Manno- | D | Mannosamine | *Salmonella* groups 17, 42[222]<br>*Escherichia coli* K235,[223] 031[225]<br>*Arizona* 15 |
| 2-Amino-2-deoxy-D-glyceroheptose | Ido- or Gulo- | D | – | Blue-green alga *Anacystis nidulans*[224] |
| 2-Amino-2,6-dideoxy-hexoses | Gluco- | D | Quinovosamine | Unidentified strain no. COC21[225] |
| | Gluco- | – | Quinovosamine | *Salmonella groups* 41, 58[222]<br>*Arizona* 1, 33[222]<br>*Proteus vulgaris*[222] |

TABLE 9 (Continued)

SUGAR CONSTITUENTS OF LIPOPOLYSACCHARIDES

| Structure | Configuration | | Trivial Name | Occurrence |
|---|---|---|---|---|
| 2-Amino-2,6-dideoxyhexoses (cont.) | Galacto- | D | Fucosamine | *Chromobacterium violaceum* NCTC 7917, Frazier's Hill[226] |
| | Galacto- | L | Fucosamine | *Salmonella* group 48<br>*Citrobacter freundii* 5<br>*Arizona* 5, 29[227] |
| | Galacto- | D,L | Fucosamine | *Pseudomonas aeruginosa*[227] |
| | Galacto- | – | Fucosamine | *Escherichia coli* 04, 012, 015, 016, 025, 026, 029, 045, 057[213] |
| | Manno- | L | Rhamnosamine | *Escherichia coli* 03[229] |
| 3-Amino-3,6-dideoxy-hexoses | Gluco- | D | | *Citrobacter freundii* 8090[230] |
| | Gluco- | – | | *Salmonella* groups 28, 39[231]<br>*Citrobacter freundii* 896[231]<br>*Escherichia coli* 05, 065, 070, 071, 0114[213] |
| | Galacto- | D | | *Xanthomonas campestris*[212]<br>*Escherichia coli* 02, 074[213] |
| | Galacto- | – | | *Salmonella* groups 55[231]<br>*Arizona* 24[231]<br>*Brucella melitensis*[232] |
| 4-Amino-4,6-dideoxy-hexoses | Gluco- | D | Viosamine | *Chromobacterium violaceum* NCTC 7917[233]<br>*Escherichia coli* 07[234] |
| | Galacto- | D | Thomosamine | *Escherichia coli* 010[234] |

This table is taken from Lüderitz, O., Westphal, O., Staub, A. M., and Nikaido, H., *Microbial Toxins,* Vol. 4, p. 145, G. Weinbaum, S. Kadis and S. Ajl, Eds. Academic Press, New York (1971). Reproduced by permission of the Academic Press.

### Extraction of the O-Antigen Complex

Several different extraction methods can be used to obtain O-antigen complexes from Gram-negative bacteria. For convenience, intact cells are frequently extracted, but contamination with nucleic acid and other cell components is avoided if cell walls are used. The original classic method is that of Boivin and Mesrobeanu[236] in which acetone-dried cells are extracted with dilute trichloracetic acid (0.25$M$–0.5$M$) at 4°C. This gives an antigenic, endotoxic, lipo-protein-polysaccharide complex. Another classic method, described by Westphal et al.,[237] uses 45% aqueous phenol at 68°C, and gives a lipopolysaccharide fraction devoid of protein. The state of aggregation and other properties of the product obtained thus vary considerably according to the method of extraction used; the type of complex extracted by different methods is given in Table 10.

The lipid, protein, and polysaccharide elements of the O-antigen complex may be separated by various mild chemical treatments. For example, hydrolysis in 0.2$N$ acetic acid at 100°C will liberate lipid, polysaccharide, and protein components. For a thorough discussion of the composition and properties of O-antigens, see Reference 203 and various articles on bacterial endotoxins in Volume 4 of *Microbial Toxins.*[242]

## TABLE 10

## TYPE OF O-ANTIGEN COMPLEX EXTRACTED FROM GRAM-NEGATIVE BACTERIA BY DIFFERENT METHODS

| Method | Type of Complex Extracted | Antigenicity (Rabbits) | Endotoxic Activity | Molecular Weight | Ability to Sensitize RBC |
|---|---|---|---|---|---|
| TCA[242]<br>Diethylene glycol[244]<br>50% aqueous pyridine[245]<br>50% aqueous glycol[245]<br>EDTA at alkaline pH[246] | Lipid-polysaccharide protein | +++ | +++ | One to several millions | + |
| Phenol/water[247]<br>Aqueous ether[248] | Lipopolysaccharide (Lipid) | ± | +++ | One to several millions | ± |
| Dilute alkali (e.g., 0.25*N* NaoH at 56°)[205] | Lipopolysaccharide (Deacylated lipid A) | Only if fixed to RBC | ± | 200,000 | +++ |
| Dilute acid (e.g., 0.1*N* acetic acid at 90°)[205] | Polysaccharide | Only if coupled to protein | – | 20,000–30,000 | – |

This table is modified from Table 2 of Lüderitz, O., Staub, A. M., and Westphal, O., *Bacteriol. Rev., 30,* 192 (1966). Reproduced by permission of the American Society for Microbiology.

### Lipid Components

These are divided into lipid A (firmly bound) and lipid B (loosely bound). Lipid A is obtained by mild acid hydrolysis of lipopolysaccharide; polysaccharide is liberated at the same time. Lipid B is obtained from the complete antigen complex by extraction with neutral fat solvents.

It seems probable that lipid A has a backbone structure composed of *N*-β-hydroxymyristoyl-glucosamine-phosphate, in which the available –OH groups are esterified with fatty acids.[243,244] However, other hydroxylated fatty acids may occur, e.g., in *Pseudomonas aeruginosa.*[245] Lipid B, on the other hand, is a phospholipid of cephalin type with mainly palmitic and oleic acids.

### Protein Components

Compared with the lipid and polysaccharide components of endotoxin, little is known about the protein, which has been obtained in two different forms: conjugated protein, and simple amphoteric protein. The conjugated protein is released, together with lipid B and polysaccharide, when complete endotoxin is heated with 1% acetic acid at 100°C. Since lipid A is not liberated separately, it is assumed that conjugated protein is Protein + Lipid A. Simple protein is released when either conjugated protein or complete endotoxin is dissociated with phenol. Since lipid A is found associated with polysaccharide when the complete endotoxin is dissociated in this way, it is assumed that the simple protein does not contain lipid A.

There is some evidence that the protein part of the O-antigen or endotoxic complex may be associated with bacteroicine activity.[246–248]

### Smooth and Rough Forms: S → R Variation

Originally, S → R variation referred to the change in colony type and stability in saline which corresponded to the loss of antigens that characterized the smooth type. In terms of the structure shown in Figure 6,

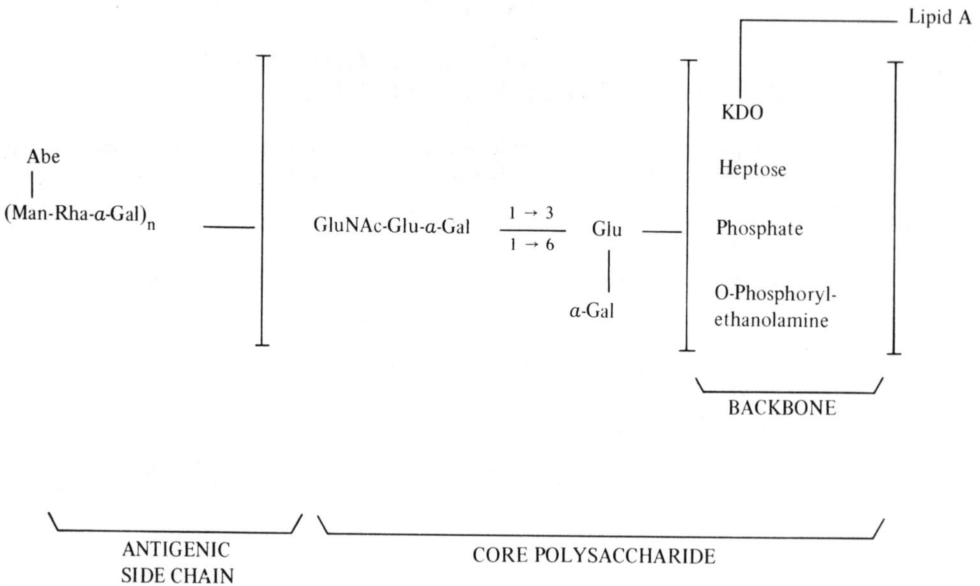


Abe = abequose = 3:6-dideoxy D-galactose.

KDO = 2-keto-3-deoxyoctanic acid.

FIGURE 6. Schematic structure of O-antigen in *Salmonella typhimurium*. (Taken from Weiner, I. M., Higuchi, T., Osborn, M. J., and Horecker, B. L., *Ann. N.Y. Acad. Sci., 133*, 391 (1966). Reproduced by permission of the New York Academy of Science.)

S-organisms have both the side chain and the core, while R-organisms have the core-polysaccharide only; and the S → R variation consists of loss in ability to synthesize the side chain. However, numerous intermediate mutants are known in which some or all of the sugars of both the side chain and the core are missing. (Since "S" and "R" have now come to have rather specialized meanings in immunochemical terms, it is probably better to describe the morphological appearance or stability of cultures by using the words "smooth" and "rough" written in full.)

## BACTERIAL CELL WALL SYNTHESIS

The basic synthetic and energy-yielding mechanisms of the cell are inside the cytoplasmic membrane, whereas the cell wall is outside it. The process of cell wall synthesis therefore essentially requires that the building-blocks for the wall be passed through the membrane at some stage. It has now become evident that this transfer is performed by carrier lipids and that some of the intermediate stages of the synthesis of several cell wall polymers are carried out while the growing polymers are attached to the cytoplasmic membrane.

The carrier lipid in peptidoglycan synthesis has been identified as a phosphorylated $C_{55}$ polyisoprenoid alcohol: similar or identical substances act as carriers in the synthesis of O-antigens and teichoic acids.

An outline of the probable course of peptidoglycan synthesis is given in Figure 7. The uridine-diphosphomuramyl-pentapeptide units are built up by the stepwise addition of amino acids, each step being catalyzed by a specific soluble enzyme. The dipeptide D-alanyl-D-alanine is added last as a single unit. This part of the process therefore differs from protein synthesis in that it is extra-ribosomal and t-RNA is not involved. In the formation of the linking peptides, however, which occurs after the UDP-muramyl-pentapeptide units have become linked to the carrier lipid, some at least of the amino acids (e.g., glycine) are incorporated via t-RNA, although others are not (e.g., D-aspartic acid). For detailed discussions of peptidoglycan synthesis, see Reference 8.

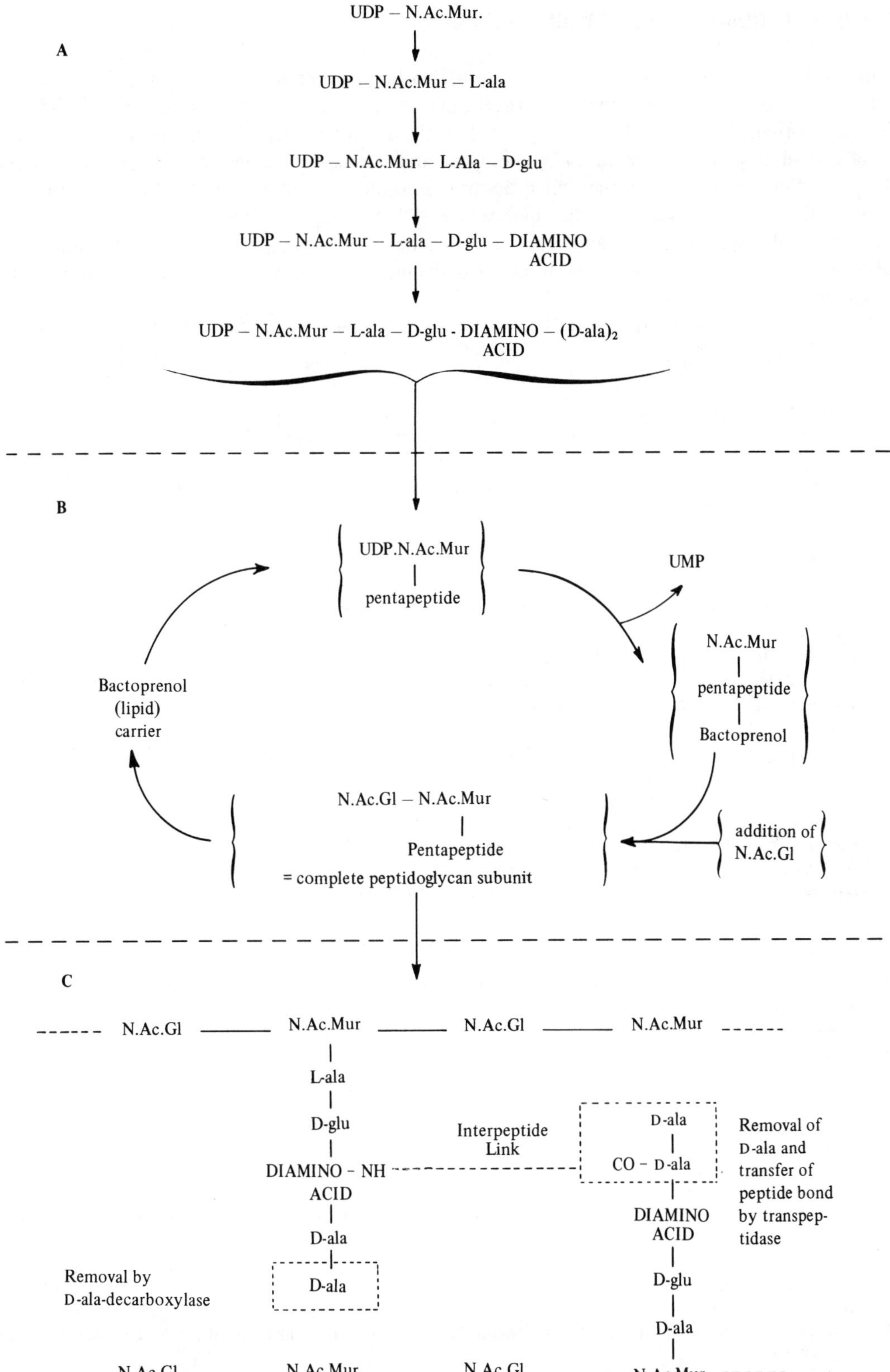


FIGURE 7. Outline of peptidoglycan synthesis. (A)Cytoplasm: Soluble enzymes and building blocks. (B)Cytoplasmic Membrane: Attachment to lipid carrier and transport through membrane. (C)Cell Wall: Complete unit incorporated and cross-linked in wall.

## Effects of Antibiotics on Cell Wall Synthesis

A number of antibiotics are inhibitory or lethal to bacterial cells because they affect some aspect of peptidoglycan synthesis. In the first stage (in the cytoplasm) D-cycloserine interferes with the formation of the D-alanyl-D-alanine dipeptide which is added to the developing peptide chain to form the pentapeptide. In the second stage (cell membrane), vancomycin and ristocetin become attached to the peptidoglycan subunits and prevent their subsequent incorporation into the cell wall. Bacitracin also acts at this stage since it inhibits the dephosphorylation of the lipid carrier so that the lipid cannot re-enter the cycle.

In the third stage, penicillins and cephalosporins are specific inhibitors of the cross-linking reaction in which the transpeptidase removes the terminal D-alanine residue and forms the peptide bond which links two chains.

(For the effects of antibiotics on wall synthesis, see References 8 and 249–252.)

**Note added in proof:** Two recent reviews, on wall and membrane composition in bacteria, and on the effect of cultural conditions on the composition of the cell wall, will be found in References 253, 254, and 255.

### REFERENCES


1. Salton, M. R. J., and Horne, R. W., *Biochim. Biophys. Acta, 7,* 177 (1951).
2. Cummins, C. S., *Int. Rev. Cytol., 5,* 25 (1957).
3. Perkins, H. R., *Bacteriol. Rev., 27,* 18 (1963).
4. Salton, M. R. J., *The Bacterial Cell Wall.* American Elsevier Co., New York (1964).
5. Martin, H. H., *Annu. Rev. Biochem., 35,* 457 (1966).
6. Ghuysen, J. M., *Bacteriol. Rev., 32,* 425 (1968).
7. Rogers, H. J., and Perkins, H. R., *Cell Walls and Membranes.* E. and F. N. Spon, London, England (1968).
8. Osborn, M. J., *Annu. Rev. Biochem., 38,* 501 (1969).
9. Hugo, W. B., *Bacteriol. Rev., 18,* 87 (1954).
10. Edebo, L., *Disintegration of Microorganisms.* Almqvist and Wiksell, Uppsala, Sweden (1961).
11. Ribi, E., and Milner, K. C., in *Methods in Immunology and Immunochemistry,* Vol. 1, p. 13, C. A. Williams and M. W. Chase, Eds. Academic Press, New York (1967).
12. Dawson, I. M., *Symp. Soc. Gen. Microbiol.,* No. 1, p. 119 (1949).
13. Mickle, H., *J. R. Microsc. Soc., 68,* 10 (1948).
14. B. Braun, Melsungen, Germany; Bronwill Scientific, Rochester, N.Y.
15. E. Bühler, Tübingen, Germany; RHO Scientific, Commack, N.Y.
16. Nossal, P. M., *Aust. J. Exp. Biol. Med. Sci., 31,* 583 (1953).
17. American Instrument Co., Silver Spring, Maryland.
18. Ivan Sorvall, Inc., Norwalk, Connecticut.
19. Colab Laboratories, Inc., Chicago Heights, Illinois.
20. Roberson, B. S., and Schwab, J. H., *Biochim. Biophys. Acta, 44,* 436 (1960).
21. Yoshida, A., Hedén, G. C., Cedergren, B., and Edebo, L., *J. Biochem. Microbiol. Technol. Eng., 3,* 151 (1961).
22. Hofsten, B. von, and Baird, G. D., *Biotechnol. Bioeng., 4,* 403 (1962).
23. Nermuth, M. V., and Murray, R. G. E., *J. Bacteriol., 93,* 1949 (1967).
24. Glauert, A. M., and Thornley, M. J., *Annu. Rev. Microbiol., 23,* 159 (1969).
25. Buckmire, F. L. A., *Int. J. Syst. Bacteriol., 20,* 345 (1970).
26. Freer, J. H., and Salton, M. R. J., in *Microbial Toxins,* Vol. 4, p. 67, G. Weinbaum, S. Kadis and S. Ajl, Eds. Academic Press, New York (1971).
27. Tipper, D. J., *Int. J. Syst. Bacteriol., 20,* 361 (1970).
28. Azuma, I., Thomas, D. W., Adam, A., Ghuysen, J. M., Bonaly, R., Petit, J. F., and Lederer, E., *Biochim. Biophys. Acta, 208,* 444 (1970).
29. Kandler, O., Schleifer, K. H., Niebler, E., Nakel, M., Zahradnik, H., and Reid, M., *Publ. Fac. Sci. Univ. Brno., 47,* 143 (1970).
30. Becker, B., Lechevalier, M. P., and Lechevalier, H. A., *Appl. Microbiol., 13,* 236 (1965).
31. Lechevalier, H. A., Lechevalier, M. P., and Becker, B., *Int. J. Syst. Bacteriol., 16,* 151 (1966).
32. Szaniszlo, P. J., and Gooder, H., *J. Bacteriol., 94,* 2037 (1967).
33. Yamaguchi, T., *J. Bacteriol., 89,* 444 (1965).

34. Hoare, D. S., and Work, E., *Biochem. J., 65,* 441 (1957).
35. Cummins, C. S., *Nature, 206,* 1272 (1965).
36. DeWeese, M. S., Gerencser, M. A., and Slack, J. M., *Appl. Microbiol., 16,* 1713 (1968).
37. Johnson, J. L., and Cummins, C. S., *J. Bacteriol., 109,* 1047 (1972).
38. Keddie, R. M., Leask, B. G. S., and Grainger, J. M., *J. Appl. Bacteriol., 29,* 17 (1966).
39. Krulwich, T. A., Ensign, J. C., Tipper, D. J., and Strominger, J. L., *J. Bacteriol., 94,* 741 (1967).
40. Gillespie, D. C., *Can. J. Microbiol., 9,* 515 (1963).
41. Fiedler, F., Schleifer, K. H., Cziharz, B., Interschick, E., and Kandler, O., *Publ. Fac. Sci. Univ. Brno., 47,* 111 (1970).
42. Cummins, C. S., and Harris, H., *Nature, 184,* 831 (1959).
43. Forrester, I. T., and Wicken, A. J., *J. Gen. Microbiol., 42,* 147 (1966).
44. Boylen, C. W., and Ensign, J. C., *J. Bacteriol., 96,* 421 (1968).
45. Bricas, E., Ghuysen, J. M., and Dezelée, P., *Biochemistry, 6,* 2598 (1967).
46. Hughes, R. C., *Biochem. J., 106,* 41 (1968).
47. Hungerer, K. D., and Tipper, D. J., *Biochemistry, 8,* 3577 (1969).
48. Van Heijenoort, J., Elbaz, L., Dezelée, P., Petit, J. F., Bricas, E., and Ghuysen, J. M., *Biochemistry, 8,* 207 (1969).
49. Kingan, S. L., and Ensign, J. C., *J. Bacteriol., 95,* 724 (1968).
50. Ratney, R. S., *Biochim. Biophys. Acta, 101,* 1 (1965).
51. Reynolds, P. E., *Biochim. Biophys. Acta, 237,* 239 (1971).
52. Sutow, A. B., and Welker, N. E., *J. Bacteriol., 93,* 1452 (1967).
53. Kandler, O., *Int. J. Syst. Bacteriol., 20,* 491 (1970).
54. Baboolal, R., *J. Gen. Microbiol., 58,* 217 (1969).
55. Guinand, M., Ghuysen, J. M., Schleifer, K. H., and Kandler, O., *Biochemistry, 8,* 200 (1969).
56. Cummins, C. S., and Johnson, J. L., *J. Gen. Microbiol., 67,* 33 (1971).
57. Cummins, C. S., and Harris, H., *J. Gen. Microbiol., 14,* 583 (1956).
58. Barksdale, W. L., Li, K., Cummins, C. S., and Harris, H., *J. Gen. Microbiol., 16,* 749 (1957).
59. Perkins, H. R., *Biochem. J., 121,* 417 (1971).
60. Mosser, J. C., and Tomasz, A., *J. Biol. Chem., 245,* 287 (1970).
61. Mann, S., *Zentralbl. Bakteriol. (Orig.), 209,* 510 (1969).
62. Harney, S. J., Simopoulos, N. D., and Ikawa, M., *J. Bacteriol., 93,* 273 (1967).
63. Keeler, R. F., and Gray, M. L., *J. Bacteriol., 80,* 683 (1960).
64. Ullman, W. W., and Cameron, J. A., *J. Bacteriol., 98,* 486 (1969).
65. Robinson, K., *J. Appl. Bacteriol., 29,* 616 (1966).
66. Work, E., and Griffiths, H., *J. Bacteriol., 95,* 641 (1968).
67. Work, E., *Nature, 201,* 1107 (1964).
68. Cummins, C. S., and Harris, H., *J. Gen. Microbiol., 18,* 173 (1958).
69. Acharya, P. V. N., and Goldman, D. W., *J. Bacteriol., 102,* 733 (1970).
70. White, P. J., *J. Gen. Microbiol., 50,* 107 (1968).
71. Bahn, A. N., Kung, P. C. Y., and Hayashi, J. A., *J. Bacteriol., 91,* 1672 (1966).
72. Schleifer, K. H., and Kandler, O., *Arch. Mikrobiol., 61,* 292 (1968).
73. Davis, G. H. G., and Freer, J. H., *J. Gen. Microbiol., 23,* 163 (1960).
74. Cummins, C. S., and Harris, H., *Int. Bull. Bacteriol. Nomencl. Taxon., 6,* 111 (1956).
75. Baird-Parker, A. C., *J. Gen. Microbiol., 38,* 363 (1965).
76. Tipper, D. J., and Berman, M. F., *Biochemistry, 8,* 2183 (1969).
77. Schleifer, K. H., and Kandler, O., *Arch. Mikrobiol., 57,* 335 (1967).
78. Schleifer, K. H., and Kandler, O., *Arch. Mikrobiol., 57,* 365 (1967).
79. Colman, G., and Williams, R. E. O., *J. Gen. Microbiol., 41,* 375 (1965).
80. Slade, H. D., and Slamp, W. C., *J. Bacteriol., 109,* 691 (1972).
81. Hoare, D. S., and Work, E., *Biochem. J., 61,* 562 (1955).
82. Jusic, D., Roy, C., and Watson, R. W., *Can. J. Biochem., 42,* 1553 (1964).
83. Manasse, R. J., and Corpe, W. A., *Can. J. Microbiol., 13,* 1591 (1967).
84. LaCave, C., and Roux, J., *C. R. Acad. Sci., 260,* 1514 (1965).
85. Mardarowitz, C., *Z. Naturforsch. B, 21,* 1006 (1966).
86. Mandelstam, J., *Biochem. J., 84,* 294 (1962).
87. Grula, E. A., Smith, G. L., and Grula, M. M., *Can. J. Microbiol., 11,* 605 (1965).
88. Wang, W. S., and Lundgren, D. G., *J. Bacteriol., 95,* 1851 (1968).
89. Braun, V., Rehn, K., and Wolff, H., *Biochemistry, 9,* 5041 (1970).
90. Clarkson, C. E., and Meadow, P. M., *J. Gen. Microbiol., 66,* 161 (1971).
91. Forsberg, C. W., Rayman, M. K., Costerton, J. W., and MacLeod, R. A., *J. Bacteriol., 109,* 895 (1972).
92. Salton, M. R. J., and Shafa, F., *Nature, 181,* 1321 (1958).
93. Colobert, L., and Creach, O., *Ann. Inst. Pasteur, 99,* 672 (1960).
94. Crum, E. H., and Siehr, D. J., *J. Bacteriol., 94,* 2069 (1967).

95. Graham, R. K., and May, J. W., *J. Gen. Microbiol., 41,* 243 (1965).
96. Winter, A. J., Katz, W., and Martin, H. H., *Biochim. Biophys. Acta, 244,* 58 (1971).
97. Drews, G., and Meyer, H., *Arch. Mikrobiol., 48,* 259 (1964).
98. Frank, H., Lefort, M., and Martin, H. H., *Z. Naturforsch. B, 17,* 262 (1962).
99. Höcht, H., Martin, H. H., and Kandler, O., *Z. Pflanzenphysiol., 53,* 39 (1965).
100. Verma, J. P., and Martin, H. H., *Arch. Mikrobiol., 59,* 355 (1967).
101. Becker, B., Worzel, E. M., and Nelson, J. H., *Nature, 213,* 300 (1967).
102. Kolenbrander, P. E., and Ensign, J. C., *J. Bacteriol., 95,* 201 (1968).
103. Martin, H. H., and Frank, H., *Zentralbl. Bakteriol. (Orig.), 184,* 306 (1962).
104. Newton, J. W., *Biochim. Biophys. Acta, 165,* 534 (1968).
105. Preusser, H. J., *Arch. Mikrobiol., 68,* 150 (1969).
106. Tinelli, R., and Pillot, J., *C.R. Acad. Sci., 263,* 739 (1966).
107. Cummins, C. S., and Smibert, R. M., *Unpublished Data.*
108. Campbell, J. N., Leyh-Bouille, M., and Ghuysen, J. M., *Biochemistry, 8,* 193 (1969).
109. Ghuysen, J. M., Bricas, E., Lache, M., and Leyh-Bouille, M., *Biochemistry, 7,* 1450 (1968).
110. Niebler, E., Schleifer, K. H., and Kandler, O., *Biochem. Biophys. Res. Commun., 34,* 560 (1969).
111. Schleifer, K. H., and Kandler, O., *Biochem. Biophys. Res. Commun., 28,* 965 (1967).
112. Kandler, O., Plapp, R., and Holzapfel, W., *Biochim. Biophys. Acta, 147,* 252 (1967).
113. Hungerer, K. D., Fleck, J., and Tipper, D. J., *Biochemistry, 8,* 3567 (1969).
114. Plapp, R., Schleifer, K. H., and Kandler, O., *Folia Microbiol., 12,* 205 (1967).
115. Plapp, R., and Kandler, O., *Arch. Mikrobiol., 58,* 305 (1967).
116. Kandler, O., Schleifer, K. H., and Dandl, R., *J. Bacteriol., 96,* 1935 (1968).
117. Koch, D., Schleifer, K. H., and Kandler, O., *Z. Naturforsch. B, 25,* 1294 (1970).
118. Kandler, O., Koch, D., and Schleifer, K. H., *Arch. Mikrobiol., 61,* 181 (1968).
119. Schleifer, K. H., Plapp, R., and Kandler, O., *FEBS Lett., 1,* 287 (1968).
120. Perkins, H. R., *Nature, 208,* 872 (1965).
121. Work, E., *Int. J. Syst. Bacteriol., 20,* 425 (1970).
122. Bottazzi, V., Weiss, N., and Kandler, O., *Arch. Mikrobiol., 58,* 35 (1967).
123. Rhuland, L. E., Work, E., Denman, R. F., and Hoare, D. S., *J. Am. Chem. Soc., 77,* 4844 (1955).
124. Ginger, C. D., *Nature, 199,* 159 (1963).
125. Perkins, H. R., and Allison, A. C., *J. Gen. Microbiol., 30,* 496 (1963).
126. Baddiley, J., *J. R. Inst. Chem., 86,* 366 (1962).
127. Archibald, A. R., and Baddiley, J., *Adv. Carbohydr. Chem., 21,* 323 (1966).
128. Archibald, A. R., Baddiley, J., and Button, D., *Biochem. J., 110,* 543 (1968).
129. Brundish, D. E., and Baddiley, J., *Biochem. J., 110,* 573 (1968).
130. Janczura, E., Perkins, H. R., and Rogers, H. J., *Biochem. J., 80,* 82 (1961).
131. Perkins, H. R., *Biochem. J., 86,* 475 (1963).
132. Ellwood, D. C., and Tempest, D. W., *Biochem. J., 111,* 1 (1969).
133. Ellwood, D. C., *Biochem. J., 121,* 349 (1971).
134. Shaw, N., and Baddiley, J., *Biochem. J., 93,* 317 (1964).
135. Archibald, A. R., Baddiley, J., and Buchanan, J. G., *Biochem. J., 81,* 124 (1961).
136. Ellwood, D. C., Kelleman, M. V., and Baddiley, J., *Biochem. J., 86,* 213 (1963).
137. Armstrong, J. J., Baddiley, J., and Buchanan, J. G., *Biochem. J., 76,* 610 (1960).
138. Archibald, A. R., Coapes, H. E., and Stafford, G. H., *Biochem. J., 113,* 899 (1969).
139. Hughes, R. C., *Biochem. J., 117,* 431 (1970).
140. Armstrong, J. J., Baddiley, J., and Buchanan, J. G., *Biochem. J., 80,* 254 (1961).
141. Davison, A. L., Baddiley, J., Hofstad, T., Losengard, N., and Oeding, P., *Nature, 202,* 872 (1964).
142. Davison, A. L., and Baddiley, J., *Nature, 202,* 874 (1964).
143. Oeding, P., Mykelstad, B., and Davison, A. L., *Acta Pathol Microbiol. Scand., 69,* 458 (1967).
144. Sharpe, E., *Int. J. Syst. Bacteriol., 24,* 509 (1970).
145. Lancefield, R. C., *Harvey Lect., 36,* 251 (1941).
146. McCarty, M., *Harvey Lect., 65,* 73 (1970).
147. Slade, H. D., and Slamp, W. C., *J. Bacteriol., 84,* 345 (1962).
148. Slade, H. D., *J. Bacteriol., 90,* 667 (1965).
149. McCarty, M., and Lancefield, R. C., *J. Exp. Med., 102,* 11 (1955).
150. Curtis, S. N., and Krause, R. M., *J. Exp. Med., 120,* 629 (1964).
151. Krause, R. M., and McCarty, M., *J. Exp. Med., 115,* 49 (1962).
152. Willers, J. M. N., Michel, M. F., Sysma, M. J., and Winkler, K. C., *J. Gen. Microbiol., 36,* 95 (1964).
153. Curtis, S. N., and Krause, R. M., *J. Exp. Med., 119,* 997 (1964).
154. Lechevalier, M. P., and Lechevalier, H., *Int. J. Syst. Bacteriol., 20,* 435 (1970).
155. Lechevalier, M. P., and Gerber, N. N., *Carbohydr. Res., 13,* 451 (1970).
156. Cummins, C. S., in *The Actinomycetales,* p. 29, H. Prauser, Ed. Gustav Fisher, Jena, Germany (1970).

157. Cummins, C. S., and Harris, H., *J. Gen. Microbiol.*, *18,* 173 (1958).
158. Pine, L., and Boone, C. J., *J. Bacteriol.*, *94,* 875 (1967).
159. Pine, L., and Georg, L. K., *Int. J. Syst. Bacteriol.*, *19,* 267 (1969).
160. Cummins, C. S., and Harris, H., *Nature, 184,* 831 (1959).
161. Smith, H., Strange, R. E., and Zwartouw, H. T., *Nature, 178,* 865 (1965).
162. Cummins, C. S., and Harris, H., *Unpublished Data.*
163. Cummins, C. S., Glendenning, O. M., and Harris, H., *Nature, 180,* 337 (1957).
164. Cummins, C. S., *Unpublished Data.*
165. Veerkamp, J. H., Lambert, R., and Saito, Y., *Arch. Biochem. Biophys., 112,* 120 (1965).
166. Cato, E. P., Cummins, C. S., and Smith, L. DS., *Int. J. Syst. Bacteriol., 20,* 305 (1970).
167. Haythornthwaite, S. U., *Ph.D. Thesis.* University of London, London, England (1968).
168. Holdsworth, E. S., *Biochim. Biophys. Acta, 8,* 110 (1952).
169. Cummins, C. S., and Harris, H., *J. Gen. Microbiol., 14,* 583 (1956).
170. Perkins, H. R., *Biochem. J., 97,* 3C (1965).
171. Perkins, H. R., *Biochem. J., 102,* 29C (1967).
172. Perkins, H. R., *Int. J. Syst. Bacteriol., 20,* 379 (1970).
173. Glastonbury, J., and Knox, K. W., *J. Gen. Microbiol., 31,* 73 (1963).
174. Baird-Parker, A. C., *Int. J. Syst. Bacteriol., 20,* 483 (1970).
175. Lederer, E., *Pure Appl. Chem., 25,* 135 (1971).
176. Allsop, J., and Work, E., *Biochem. J., 87,* 512 (1963).
177. Georg, L. K., and Brown, J. M., *Int. J. Syst. Bacteriol., 17,* 79 (1967).
178. Hammond, B. F., *J. Bacteriol., 103,* 634 (1970).
179. Michel, M. F., and Krause, R. M., *J. Exp. Med., 125,* 1075 (1967).
180. Knox, K. W., *J. Gen. Microbiol., 31,* 59 (1963).
181. Fuller, A. T., *Br. J. Exp. Pathol., 19,* 130 (1938).
182. Maxted, W. R., *Lancet, 2,* 255 (1948).
183. Perkins, H. R., *Biochem. J., 95,* 876 (1965).
184. Lancefield, R. C., *J. Exp. Med., 57,* 571 (1933).
185. Krause, R. M., *J. Exp. Med., 108,* 803 (1958).
186. Bleiweiss, A. S., Young, F. E., and Krause, R. M., *J. Bacteriol., 94,* 1381 (1967).
187. Becker, B., Lechevalier, M. P., Gordon, R. E., and Lechevalier, H. A., *Appl. Microbiol., 12,* 421 (1964).
188. Goren, M. B., *Bacteriol. Rev., 36,* 33 (1972).
189. Carter, H. E., Johnson, P., and Weber, E. J., *Annu. Rev. Biochem., 34,* 109 (1965).
190. Asselineau, J., *The Bacterial Lipids.* Holden-Day, San Francisco, California (1967).
191. O'Leary, W. M., *The Chemistry and Metabolism of Microbial Lipids.* World Publishing Co., Cleveland, Ohio (1967).
192. Hahn, J. J., and Cole, R. M., *J. Exp. Med., 118,* 659 (1963).
193. Swanson, J., Hsu, K. C., and Gotschlich, E. C., *J. Exp. Med., 130,* 1063 (1969).
194. Lange, C. F., Lee, R., and Merdinger, E., *J. Bacteriol., 100,* 1277 (1969).
195. Pillot, J., Rouyer, M., and Orta, B., *Ann. Inst. Pasteur, 88,* 662 (1955).
196. Yoshida, A., Mudd, S., and Lenhart, N. A., *J. Immunol., 91,* 777 (1963).
197. Cummins, C. S., *Br. J. Exp. Pathol., 35,* 166 (1956).
198. Hughes, R. C., *Biochem. J., 96,* 100 (1965).
199. Work, E., Knox, K. W., and Vesk, M., *Ann. N.Y. Acad. Sci., 133,* 438 (1966).
200. Milner, K. C., Rudbach, J. A., and Ribi, E., in *Microbial Toxins,* Vol. 4, p. 1, G. Weinbaum, S. Kadis, and S. Ajl, Eds. Academic Press, New York (1971).
201. Kauffmann, F., *Die Bakteriologie der Salmonella Species.* Munksgaard, Copenhagen, Denmark (1961).
202. Lüderitz, O., Staub, A. M., and Westphal, O., *Bacteriol. Rev., 30,* 192 (1966).
203. Lüderitz, O., Jann, K., and Wheat, R., in *Comprehensive Biochemistry,* Vol. 26A, p. 105, M. Florkin and E. H. Stotz, Eds. American Elsevier Co., New York (1968).
204. Nikaido, H., *Int. J. Syst. Bacteriol., 20,* 383 (1970).
205. Simmons, D. A. R., *Bacteriol. Rev., 35,* 117 (1971).
206. Lüderitz, O., Westphal, O., Staub, A. M., and Nikaido, H., in *Microbial Toxins,* Vol. 4, p. 145, G. Weinbaum, S. Kadis, and S. Ajl, Eds. Academic Press, New York (1971).
207. Weiner, I. M., Higuchi, T., Osborn, M. J., and Horecker, B. L., *Ann. N.Y. Acad. Sci., 133,* 391 (1966).
208. Adams, G. A., Quadling, C., and Perry, M. B., *Can. J. Microbiol., 13,* 1605 (1967).
209. Volk, W. A., *J. Bacteriol., 91,* 39 (1966).
210. Volk, W. A., *J. Bacteriol., 95,* 980 (1968).
211. Hofstad, T., and Kristoffersen, T., *Acta Pathol. Microbiol. Scand. B., 79,* 12 (1971).
212. Hickman, G., and Ashwell, J., *J. Biol. Chem., 241,* 1424 (1966).
213. Orskov, F., Orskov, I., Jann, B., Jann, K., Müller-Seitz, E., and Westphal, O., *Acta Pathol. Microbiol. Scand., 71,* 339 (1967).
214. Kauffmann, F., Lüderitz, O., Stierlin, H., and Westphal, O., *Zentralbl. Bakteriol. (Orig.), 178,* 442 (1960).

215. Davies, D. A. L., *Adv. Carbohydr. Chem., 15,* 271 (1960).
216. Westphal, O., Kauffmann, F., Lüderitz, O., and Stierlin, H., *Zentralbl. Bakteriol. (Orig.), 179,* 336 (1960).
217. Kauffmann, F., Jann, B., Krüger, L., Lüderitz, O., and Westphal, O., *Zentralbl. Bakteriol. (Orig.), 186,* 509 (1962).
218. Fromme, I., Lüderitz, O., and Westphal, O., *Z. Naturforsch. B, 9,* 303 (1954).
219. Adams, G. A., and Young, R., *Can. J. Biochem., 43,* 1499 (1965).
220. Bagdian, G., Dröge, W., Kotelko, K., Lüderitz, O., Westphal, O., Yamakawa, T., and Ueta, N., *Biochem. Z., 344,* 197 (1966).
221. Lacave, C., Asselineau, J., Serre, A., and Roux, J., *Eur. J. Biochem., 9,* 189 (1969).
222. Lüderitz, O., Gmeiner, J., Kickhöfen, B., Mayer, H., Westphal, O., and Wheat, R. W., *J. Bacteriol., 95,* 490 (1968).
223. Rüde, E., and Goebel, W. F., *J. Exp. Med., 116,* 73 (1962).
224. Weise, G., Drews, G., Jann, B., and Jann, K., *Arch. Mikrobiol., 71,* 89 (1970).
225. Smith, E. J., *Biochem. Biophys. Res. Commun., 15,* 593 (1964).
226. Crumpton, M. J., and Davies, D. A. L., *Biochem. J., 70,* 729 (1958).
227. Barry, G. T., *Bull. Soc. Chim. Biol., 47,* 529 (1964).
228. Suzuki, N., *Biochim. Biophys. Acta, 177,* 371 (1969).
229. Jann, B., and Jann, K., *Eur. J. Biochem., 5,* 173 (1968).
230. Raff, R. A., and Wheat, R. W., *J. Biol. Chem., 242,* 4610 (1967).
231. Lüderitz, O., Rusehmann, E., Westphal, O., Raff, R., and Wheat, R., *J. Bacteriol., 93,* 1681 (1967).
232. Lacave, C., *Thesis.* University of Toulouse, France (1969).
233. Stevens, C. L., Blumbergs, P., Daniher, F. A., Wheat, R. W., Kujomoto, A., and Rollins, E., *J. Am. Chem. Soc., 85,* 3061 (1963).
234. Jann, B., and Jann, K., *Eur. J. Biochem., 2,* 26 (1967).
235. Ashwell, G., and Hickman, J., in *Microbial Toxins,* Vol. 4, p. 235, G. Weinbaum, S. Kadis, and S. Ajl, Eds. Academic Press, New York (1971).
236. Boivin, A., and Mesrobeanu, L., *Rev. Immunol., 1,* 553 (1935).
237. Westphal, O., Lüderitz, O., and Bister, F., *Z. Naturforsch. B, 7,* 148 (1952).
238. Morgan, W. T. J., *Biochem. J., 31,* 2003 (1937).
239. Goebel, W. F., Binkley, F., and Perlman, E., *J. Exp. Med., 81,* 315 (1945).
240. Westphal, O., and Jann, K., in *Methods in Carbohydrate Chemistry,* Vol. 5, p. 83, R. L. Whistler, Ed. Academic Press, New York (1965).
241. Ribi, E., Anacker, R. L., Fukushi, K., Haskins, W. T., Landy, M., and Milner, K. C., in *Bacterial Endotoxins,* p. 16, M. Landy and W. Braun, Eds. Rutgers University Press, New Brunswick, New Jersey (1964).
242. G. Weinbaum, S. Kadis, and S. Ajl, Eds., *Microbial Toxins,* Vol. 4. Academic Press, New York (1971).
243. Nowotny, A., *J. Am. Chem. Soc., 83,* 501 (1961).
244. Burton, A. J., and Carter, H. E., *Biochemistry, 3,* 411 (1964).
245. Roberts, N. A., Gray, G. W., and Wilkinson, S. G., *Biochim. Biophys. Acta, 135,* 1068 (1967).
246. Goebel, W. F., and Barry, G. T., *J. Exp. Med., 107,* 185 (1958).
247. Homma, J. Y., and Suzuki, N., *Ann. N.Y. Acad. Sci., 133,* 508 (1966).
248. Mesrobeanu, L., *Ann. N.Y. Acad. Sci., 133,* 685 (1966).
249. Ghuysen, J. M., Strominger, J. L., and Tipper, D. J., in *Comprehensive Biochemistry,* Vol. 26A, p. 53, M. Florkin and E. H. Stoltz, Eds. American Elsevier Co., New York (1968).
250. Perkins, H. R., *Biochem. J., 111,* 195 (1969).
251. Siewert, G., and Strominger, J. L., *Proc. Natl. Acad. Sci. U.S.A., 57,* 767 (1967).
252. Rothfield, L., and Romeo, D., *Bacteriol. Rev., 35,* 14 (1971).
253. Reaveley, D. A., and Burge, R. E., in *Advances in Microbial Physiology,* Vol. 7, p. 1, A. H. Rose and D. W. Tempest, Eds. Academic Press, New York (1972).
254. Ellwood, D. C., and Tempest, D. W., in *Advances in Microbial Physiology,* Vol. 7, p. 83, A. H. Rose and D. W. Tempest, Eds. Academic Press, New York (1972).
255. Schliefer, K. H., and Kandler, O., *Bacteriol. Rev., 36,* 407 (1972).

# FUNGAL CELL WALL COMPOSITION

DR. SALOMON BARTNICKI-GARCIA

## MONOSACCHARIDES

Values are in % wall dry weight. Whenever feasible, values were recalculated from available data and expressed as anhydrosugar. Multiple and range values were averaged.

### Abbreviations

| | | | | | | | | |
|---|---|---|---|---|---|---|---|---|
| GlcNAc | = | N-acetylglucosamine | Man | = | mannose | Fuc | = | fucose |
| GlcN | = | glucosamine | Gal | = | galactose | Xyl | = | xylose |
| GalN | = | galactosamine | GluA | = | glucuronic acid | Ara | = | arabinose |
| Glu | = | glucose | Rha | = | rhamnose | Rib | = | ribose |

## AMINO ACIDS AND PROTEIN

Amino acid composition is expressed in mole % of the sum total of amino acids. Values in parentheses pertain to purified glycoprotein fractions isolated from cell walls. All other values correspond to unfractionated cell walls. Whenever necessary, values were recalculated from published data.

Protein content of isolated cell walls is expressed in % wall dry weight.

## TABLE 1
## MONOSACCHARIDE COMPOSITION OF FUNGAL CELL WALLS

| Fungus | Form | Hexos-amine | GlcNAc | GlcN | GalN | Neutral Sugar | Glu | Man | Gal | GluA | Rha | Fuc | Xyl | Ara | Rib | Ref. |
|---|---|---|---|---|---|---|---|---|---|---|---|---|---|---|---|---|
| **Acrasiales** | | | | | | | | | | | | | | | | |
| *Polysphondylium pallidum* | Microcyst | | | 0 | 0 | | 55 | | | | | | | | | 72 |
| **Myxomycetes** | | | | | | | | | | | | | | | | |
| *Physarum polycephalum* | Spherules | | | | 88.4 | | | | | | | | | | | 47 |
| | Spores | | | | 81.0 | | | | | | | | | | | 47 |
| **Trichomycetes** | | | | | | | | | | | | | | | | |
| *Amoebidium parasiticum* | Thallus | | | | 30 | | | | 10 | | | | 3 | | | 74 |
| **Oomycetes** | | | | | | | | | | | | | | | | |
| *Apodachlya brachynema* | Mycelium | | | 3.2 | | | 87 | tr | | | | | | | 0 | 67 |
| *Atkinsiella dubia* | Mycelium | | | 1.8 | 0.03 | | 78.4 | | | | | | | | | 5 |
| *Dictyuchus sterile* | Mycelium | | | 2.5 | | | 82 | tr | | | | | | | 0 | 67 |
| *Phytophthora cinnamomi* | Mycelium | | | 0.3 | tr | | 88 | 0.6 | | 0 | | | | | tr | 8 |
| *heveae* | Mycelium | | | 2.3 | | | 90 | + | + | | + | | | | + | 55 |
| *megasperma* | Oospores | | | 0.4 | | | 77.7 | 1.9 | 0.08 | | | | | | | 41 |
| *palmivora* | Mycelium | | | 0.6 | | | 90.1 | | | | | | | | | 73 |
| | Cysts | | | 0.1 | | | 92.9 | | | | | | | | | 73 |
| | Sporangia | | | 0.2 | | | 93.5 | | | | | | | | | 73 |
| *parasitica* | Mycelium | | | 0.3 | tr | | 86 | 0.7 | | 0 | | | | | tr | 8 |
| *Pythium* sp. | Mycelium | | | 1.2 | | | 79 | tr | | | | | | | 0 | 67 |
| *butleri* | Mycelium | | | 1.3 | | | 81 | + | + | | + | | | | 0 | 55 |
| *debaryanum* | Mycelium | | | 0.5 | | | 82.4 | + | | | | | | | 0 | 20 |

## TABLE 1 (Continued)
## MONOSACCHARIDE COMPOSITION OF FUNGAL CELL WALLS

| Fungus | Form | Hexos-amine | GlcNAc | GlcN | GalN | Neutral Sugar | Glu | Man | Gal | GluA | Rha | Fuc | Xyl | Ara | Rib | Ref. |
|---|---|---|---|---|---|---|---|---|---|---|---|---|---|---|---|---|
| **Oomycetes (continued)** | | | | | | | | | | | | | | | | |
| *Saprolegnia* | | | | | | | | | | | | | | | | |
| *ferax* | Mycelium | | | 2.7 | | | 84 | tr | | | | | | | 0 | 67 |
| | Mycelium | | | 1.7 | | | 93 | tr | tr | | + | | | | + | 55 |
| *Sapromyces* | | | | | | | | | | | | | | | | |
| *elongatus* | Mycelium | | | tr | | | 89.1 | | | | | | | | | 56 |
| **Chytridiomycetes** | | | | | | | | | | | | | | | | |
| *Allomyces* | | | | | | | | | | | | | | | | |
| *arbuscula* | | 45 | + | | | 12.25 | + | + | + | | | | | | | 28 |
| *macrogynus* | Mycelium | | 58 | | | | 16 | | | | | | | | | 6 |
| *macrogynus* | | | | | | | | | | | | | | | | |
| (gametophyte) | Mycelium | 38 | + | | | 12 | + | + | + | | | | | | | 28 |
| (sporophyte) | Mycelium | 33 | + | | | 11.6 | + | + | + | | | | | | | 28 |
| *neo-moniliformis* | Mycelium | 39 | + | | | 12.4 | + | + | + | | | | | | | 28 |
| *Blastocladiella* | | | | | | | | | | | | | | | | |
| *britannica* | | | + | | | 10.0 | + | + | + | | | | | | | 28 |
| *emersonii* | | 39 | + | | | | + | ? | + | | | | | | | 28 |
| **Zygomycetes** | | | | | | | | | | | | | | | | |
| *Mucor* | | | | | | | | | | | | | | | | |
| *erectus* | Mycelium | 19.2 | | | | | | | | | | | | | | 77 |
| *javanicus* | Mycelium | 24.3 | | | | | | | | | | | | | | 77 |
| *pusillus* | Mycelium | 21.2 | | | | | | | | | | | | | | 77 |
| *rouxii* | Mycelium | | 9.4 | 32.7 | | | 0 | 1.6 | 1.6 | 11.8 | | 3.8 | | | | 9, 12 |
| | Yeast | | 8.4 | 27.9 | | | 0 | 8.9 | 1.1 | 12.2 | | 3.2 | | | + | 9, 12 |
| | Sporangio-phore | | 18.0 | 20.6 | | | tr | 0.9 | 0.8 | 25.0 | | 2.1 | | | | 11 |
| | Sporangio-spore | | 2.1 | 9.5 | | | 42.6 | 4.8 | 0 | 1.9? | | 0 | | | | 10, 12 |
| *Zygorhynchus* | | | | | | | | | | | | | | | | |
| *vuilleminii* | Mycelium | | + | 31.5 | | | 0 | tr | 5.1 | 16.0 | | 6.8 | | | | 7 |

## TABLE 1 (Continued)
## MONOSACCHARIDE COMPOSITION OF FUNGAL CELL WALLS

| Fungus | Form | Hexos-amine | GlcNAc | GlcN | GalN | Neutral Sugar | Glu | Man | Gal | GluA | Rha | Fuc | Xyl | Ara | Rib | Ref. |
|---|---|---|---|---|---|---|---|---|---|---|---|---|---|---|---|---|
| **Hemiascomycetes** | | | | | | | | | | | | | | | | |
| *Hansenula saturnus* | Yeast | | | | | | 45 | 19.3 | | | | | | | | 76 |
| *Hanseniaspora uvarum* | Yeast | | 0.05 | | | | 30.4 | 34.9 | | | | | | | | 49 |
| *Nadsonia elongata* | Yeast | | | 0.39 | | | 45 | 31 | | | | | | | | 24 |
| *Pichia farinosa* | Yeast | | | | | | 54 | 42.4 | | | | | | | | 76 |
| *Saccharomyces cerevisiae* | | | | | | | | | | | | | | | | |
| baker's | Yeast | | | 1.4 | | | 29 | 31 | | | | | | | | 39, 53 |
| | Yeast | | | <1.0 | | | ~34 | ~34 | | | | | | | | 61 |
| | Yeast | | | 2.9 | | | ~42 | ~42 | | | | | | | | 16 |
| | Yeast | | | 2.7 | | 84.4 | + | + | | | | | | | | 26 |
| brewer's | Yeast | | 0.85 | | | 80 | ~58 | ~22 | | | | | | | | 25 |
| | Yeast[a] | | | 1.2 | | | 47 | 44 | | | | | | | | 48 |
| | Yeast[b] | | | 1.1 | | | 46 | 43 | | | | | | | | 48 |
| *carlsbergensis* | Yeast | | | | | | 43 | 38.1 | | | | | | | | 76 |
| *Saccharomycopsis guttulata* | Yeast | | | 1.7 | | 72 | ~48 | ~24 | | | | | | | | 16 |
| | Filaments | | | 2.3 | | 74 | ~50 | ~25 | | | | | | | | 16 |
| *Schizosaccharomyces pombe* | Yeast | | | | | | 64[x] | 10.6[x] | 0.7[x] | | | | | | | 21 |
| | Abnormal[c] | | | | | | 64.6[x] | 4.5[x] | 3.6[x] | | | | | | | 21 |
| **Euascomycetes** | | | | | | | | | | | | | | | | |
| *Aspergillus* sp. | Mycelium | | 16.5 | | | 48 | 36.7 | 4.3 | 4.8 | | | | | | | 62 |
| *carbonarius* | Mycelium | | | 6.38 | | | | | | | | | | | | 77 |
| *fumigatus* | Mycelium | | | 7.23 | | | | | | | | | | | | 77 |
| *nidulans* | | | | | | | | | | | | | | | | |
| wild type | Mycelium | | + | | + | 60.5 | + | + | + | 1.9 | | | | | | 17 |
| albino mutant | Mycelium | 25.1 | 2.3 | 12.0 | 10.8 | 82.8 | 28.9 | 2.8 | 3.8 | 3.5 | | | | | | 17 |
| wild type at 30°C | Mycelium | | | 16.9 | 0.28 | 38 | 32 | + | + | | | | | tr | | 37 |

## TABLE 1 (Continued)
## MONOSACCHARIDE COMPOSITION OF FUNGAL CELL WALLS

| Fungus | Form | Hexos-amine | GlcNAc | GlcN | GalN | Neutral Sugar | Glu | Man | Gal | GluA | Rha | Fuc | Xyl | Ara | Rib | Ref. |
|---|---|---|---|---|---|---|---|---|---|---|---|---|---|---|---|---|
| **Euascomycetes (continued)** | | | | | | | | | | | | | | | | |
| *Aspergillus* | | | | | | | | | | | | | | | | |
| *nidulans (cont.)* wild type at 41°C | Mycelium | | | 16.7 | 0.28 | 27 | 18.8 | + | + | | | | | tr | | 37 |
| Ts6 mutant at 30°C | Mycelium | | | 21.9 | 0.8 | 42.5 | 29.0 | + | + | | | | | tr | | 37 |
| Ts6 mutant at 41°C | Mycelium | | | 1.9 | 0.05 | 41.5 | 20.7 | + | + | | | | | tr | | 37 |
| *niger* | Mycelium | | | 12 | | | 52 | 1.0 | 4 | | | | | 1.0 | | 40 |
| | Mycelium | 11 | + | | + | 70 | 57[x] | 2.2[x] | 10.2[x] | | | | | tr | | 34 |
| *oryzae* | Mycelium | | | 48 | | 54 | + | + | + | | | | | | | 32 |
| | Conidia | | | 20 | | 27 | + | + | + | | | | | | | 32 |
| *phoenicis* | Mycelium | | 23.7 | | + | 61 | 58.5 | + | + | | | | | | | 13 |
| | Conidia | | 36.2 | | + | | 46 | + | + | | | | | | | 13 |
| *wentii* | Mycelium | | | 8.88 | | | | | | | | | | | | 77 |
| *Blastomyces dermatitidis* | Yeast | | 44 | | | | 42.5 | 0 | 0 | | | | | | | 36 |
| | Yeast | 34 | | | | | 36 | tr | tr | | | | | | | 35 |
| | Mycelium | | 13 | | | | 39 | + | + | | | | | | | 36 |
| | Mycelium | 22.8 | | | | | 44 | 8.8 | 4.4 | | | | | | | 35 |
| *Cordyceps militaris* | Mycelium | 11 | | 6.55 | | | 46 | 9 | 7 | | | | | | | 46 |
| *Fusarium* sp. | Mycelium | | | | | | | | | 0.5 | | | | | | 29 |
| *solani* f. *phaseoli* | Mycelium | | 47 | | | | 14 | + | + | + | | | | | | 68 |
| *Microsporum* | | | | | | | | | | | | | | | | |
| *canis* | Mycelium | | 26.6 | | 0.3[x] | 48.3 | 37.5 | 11.4 | tr | | | | | | | 66 |
| *gypseum* | Mycelium | | 31.2 | | 0.25[x] | 45.9 | 36.6 | 10.3 | tr | | | | | | | 66 |
| *Neurospora* | | | | | | | | | | | | | | | | |
| *crassa* | Mycelium | | 11.9 | | | | 48.6 | | | | | | | | | 57 |
| SYR 9-7a | Mycelium[d] | | | 7.1 | | | 56.4 | | | | | | | | | 22 |
| Perkins A | Mycelium[d] | | | 6.4 | | | 56.8 | | | | | | | | | 22 |
| | Mycelium[d] | | 8.0 | | | 14[e] | 48[f] | | | | | | | | | 45 |
| | Conidia[d] | | 7.4 | | | 30[e] | 40[f] | | | | | | | | | 45 |
| St. Lawrence | Mycelium[d] | | | 6.9 | | | 58.2 | | | | | | | | | 22 |
| B6 | Mycelium[g] | | | 11.3 | | | 50.1 | | | | | | | | | 22 |
| B28 | Mycelium[g] | | | 12.4 | | | 51.9 | | | | | | | | | 22 |
| colonial-1 | Mycelium[g] | | | 17.7 | | | 49.7 | | | | | | | | | 22 |
| melon-1 | Mycelium[g] | | | 12.3 | | | 51.2 | | | | | | | | | 22 |
| | Mycelium[g,h] | | | 9.4 | | | 43.4 | | | | | | | | | 22 |

## TABLE 1 (Continued)
## MONOSACCHARIDE COMPOSITION OF FUNGAL CELL WALLS

| Fungus | Form | Hexos-amine | GlcNAc | GlcN | GalN | Neutral Sugar | Glu | Man | Gal | GluA | Rha | Fuc | Xyl | Ara | Rib | Ref. |
|---|---|---|---|---|---|---|---|---|---|---|---|---|---|---|---|---|
| **Euascomycetes (continued)** | | | | | | | | | | | | | | | | |
| *Neurospora crassa (cont.)* | | | | | | | | | | | | | | | | |
| B132 | Mycelium[i] | | | 6.7 | | | 56.6 | | | | | | | | | 22 |
| B110 | Mycelium[i] | | | 6.7 | | | 54.0 | | | | | | | | | 22 |
| B4 | Mycelium[i] | | | 8.6 | | | 58.3 | | | | | | | | | 22 |
| B54 | Mycelium[j] | | | 6.8 | | | 62.4 | | | | | | | | | 22 |
| RL-3-8-A | Mycelium[d] | | 10.2–14.4 | | 1.5 | | 47.5 | + | | + | | | | | | 44 |
| RL | Mycelium[d] | | 10.0 | | | 16[e] | 50[f] | | | | | | | | | 45 |
| | Conidia[d] | | 9.0 | | | 28[e] | 35[f] | | | | | | | | | 45 |
| STL 74A | Mycelium[d] | | 5.1 | | 0.27 | 75.4 | 71.4 | | | | | | | | | 42 |
| os-4-NM201o | Mycelium[l] | | 2.6 | | 0.06 | 81.3 | 78.9 | | | | | | | | | 42 |
| os-3-S2 | Mycelium[l] | | 4.9 | | 0.15 | 79.6 | 58.2 | | | | | | | | | 42 |
| os-5-C24 | Mycelium[l] | | 4.7 | | 0.27 | 72.8 | 59.5 | | | | | | | | | 42 |
| cut-A49 | Mycelium[l] | | 10.6 | | 0.25 | 70.4 | 60.2 | | | | | | | | | 42 |
| os-1-B135 | Mycelium[l] | | 7.9 | | 0.37 | 82.4 | 58.9 | | | | | | | | | 42 |
| *Penicillium* | | | | | | | | | | | | | | | | |
| *album* | Mycelium | | | | | | | | | 0.2 | | | | | | 29 |
| *chrysogenum* | Mycelium | | | 19.5 | | | 38.9 | 4.4 | 12.0 | | 1.9 | | 2.1 | | | 31 |
| | Mycelium | | >42 | | | 52 | 40 | 8 | 4 | | | | | | | 75 |
| | Mycelium | | | 18.0 | | | 33.9 | 2.65 | 7.1 | | 0.9 | | 0 | | | 60 |
| | Conidia | | | 11.4 | | | 26.3 | 2.6 | 19.4 | | 0.3 | | 0 | | | 60 |
| *digitatum* | Mycelium | | | 5.7 | | | 45.4 | + | 3.8 | | tr | | tr | | | 30 |
| *italicum* | Mycelium | | | 9.0 | | | 51.6 | + | 3.8 | | tr | | tr | | | 30 |
| *notatum* | Mycelium | | | 18.5 | tr | | 43 | 1 | 7 | | 0 | | 0 | | | 1 |
| *patulum* | Mycelium | | | 12.3 | 0.5 | | | | | | | | | | | 2 |
| *roqueforti* | Mycelium | | | 13 | <0.6 | | 41 | 1 | 14 | | | | | | | 3 |
| *stoloniferum* | | | | | | | | | | | | | | | | |
| ATCC 14586 | Mycelium[m] | | | 18.5 | 20.1 | | | | | | | | | | | 15 |
| | Mycelium[n] | | | 15.2 | 0.8 | | | | | | | | | | | 15 |
| ATCC 10111 | Mycelium | | | 25.6 | 0.7 | | | | | | | | | | | 15 |
| CMI 31200 | Mycelium | | | 25.5 | 0.8 | | | | | | | | | | | 15 |
| CMI 960 | Mycelium | | | 25.1 | 1.1 | | | | | | | | | | | 15 |
| CMI 92219 | Mycelium | | | 18.2 | 0.45 | | | | | | | | | | | 15 |
| *Trychophyton mentagrophytes* | Mycelium | | 28.1 | | 0.4[x] | 55.5 | 45.9 | 7.8 | | | | | | | | 66 |

## TABLE 1 (Continued)
## MONOSACCHARIDE COMPOSITION OF FUNGAL CELL WALLS

| Fungus | Form | Hexos-amine | GlcNAc | GlcN | GalN | Neutral Sugar | Glu | Man | Gal | GluA | Rha | Fuc | Xyl | Ara | Rib | Ref. |
|---|---|---|---|---|---|---|---|---|---|---|---|---|---|---|---|---|
| **Loculoascomycetes** | | | | | | | | | | | | | | | | |
| *Cochliobolus miyabeanus* | Mycelium | | | 38 | + | 51.7 | + | + | + | | + | + | + | + | | 52 |
| *Helminthosporium sativum* | Mycelium | | | 8.6 | 8.3 | | 37.0 | 2.0 | 2.0 | | | | | | | 4 |
| *Leptosphaeria albopunctata* | Mycelium[o] | | | 4.8 | <0.1 | | 51.6 | 20.4 | 4.4 | | | | | | | 70 |
| *allorgei* | Mycelium[k] | | | 11.35 | <0.1 | | 45.9 | 18.0 | 8.0 | | | | | | | 70 |
| *discors* | Mycelium[o] | | | 20.6 | <0.1 | | 39.8 | 18.9 | 4.1 | | | | | | | 70 |
| *nitschkei* | Mycelium[k] | | | 14.5 | <0.1 | | 32.8 | 22.2 | 6.8 | | | | | | | 70 |
| *orae-maris* | Mycelium[o] | | | 10.75 | <0.1 | | 40.3 | 12.2 | 10.4 | | | | | | | 70 |
| *robusta* | Mycelium[k] | | | 17.4 | <0.1 | | 32.6 | 16.3 | 9.6 | | | | | | | 70 |
| *Venturia inaequalis* | Mycelium | | 7.3 | | | | + | + | | | | | | | | 33 |
| **Homobasidiomycetes** | | | | | | | | | | | | | | | | |
| *Rhizoctonia solani* | Mycelium | | + | + | | 31.6 | 22.4 | + | + | + | | | | | | 58 |
| *Schizophyllum commune* | Mycelium | | 5.0 | | | | 81.4 | | | | | | + | | | 80 |
| | Primordium | | 3.1 | | | | 86.8 | | | | | | + | | | 80 |
| 699 | Mycelium[d] | 8.7 | | | | | 54.1 | | | | | | | | | 79 |
| 1019 | Mycelium[d] | 8.2 | | | | | 67.8 | | | | | | | | | 79 |
| 699D | Mycelium[p] | 8.8 | | | | | 56.1 | | | | | | | | | 79 |
| 1737T | Mycelium[p] | 7.6 | | | | | 63.2 | | | | | | | | | 79 |
| 3532 | Mycelium[q] | 12.6 | | | | | 45.2 | | | | | | | | | 79 |
| 3535 | Mycelium[q] | 16.1 | | | | | 41.2 | | | | | | | | | 79 |
| *Sclerotium rolfsii* | Mycelium | | | 3.5 | | 67.5 | + | + | | | | | | | | 19 |
| | Sclerotia | | | 1.6 | | 39.5 | + | + | | | | | | | | 19 |
| | Mycelium | | | 61.0 | + | | 16.5 | + | + | | | | | | | 13 |
| **Deuteromycetes** | | | | | | | | | | | | | | | | |
| *Alternaria* sp. | Mycelium | | | | | | | | | 0.7 | | | | | | 29 |

**TABLE 1 (Continued)**
**MONOSACCHARIDE COMPOSITION OF FUNGAL CELL WALLS**

| Fungus | Form | Hexos-amine | GlcNAc | GlcN | GalN | Neutral Sugar | Glu | Man | Gal | GluA | Rha | Fuc | Xyl | Ara | Rib | Ref. |
|---|---|---|---|---|---|---|---|---|---|---|---|---|---|---|---|---|
| **Deuteromycetes (continued)** | | | | | | | | | | | | | | | | |
| *Candida* | | | | | | | | | | | | | | | | |
| *albicans* | Yeast[r] | | | 1.5[x] | | 42.8[x] | 25.4[x] | 17.5[x] | | | | | | | | 18 |
| | Yeast[s] | | | 1.7[x] | | 44.7[x] | 29.5[x] | 15.2[x] | | | | | | | | 18 |
| | Mycelium[t] | | | 6.5[x] | | 47.4[x] | 30.4[x] | 17.0[x] | | | | | | | | 18 |
| | Mycelium[u] | | | 6.5[x] | | 41.15[x] | 23.2[x] | 18.0[x] | | | | | | | | 18 |
| *krusei* | Yeast | 1.3 | | | | 41.2 | + | tr | 0 | | | | | | | 64 |
| *utilis* | Yeast | | | 0.35 | | 78 | 48 | 30 | | | | | | | | 54 |
| | Tubular[v] | | | 16 | | 50 | 48 | 1.5 | | | | | | | | 54 |
| *Dactylium* | | | | | | | | | | | | | | | | |
| *dendroides* | Mycelium | | | | | | | | | 2 | | | | | | 29 |
| *Epidermophyton* | | | | | | | | | | | | | | | | |
| *floccosum* | Mycelium | | 29.7 | | 0.43[x] | 56.6 | 45.8 | 6.7 | | | | | | | | 66 |
| *Histoplasma* | | | | | | | | | | | | | | | | |
| *capsulatum* | Yeast | | 25 | | | | 21.1 | 1.2 | | | | | | | | 23 |
| | Mycelium | | 4 | | | | 5.5 | 5.1 | | | | | | | | 23 |
| *Paracoccidioides* | | | | | | | | | | | | | | | | |
| *brasiliensis* | Yeast | | 37 | | | 33.5 | 33.5 | 0 | 0 | | | | | | | 36 |
| | Mycelium | | 11 | | | 34.2 | ~34 | + | + | | | | | | | 36 |
| *Piricularia* | | | | | | | | | | | | | | | | |
| *oryzae* | Mycelium | | + | 12 | | 71 | 62 | 4 | 0.5 | | | | | | | 51 |
| *Pithomyces* | | | | | | | | | | | | | | | | |
| *chartarum* | Mycelium | | | 9.5 | | 39 | + | | | | | | | | | 63 |
| | Spores | 3.6 | | + | | 45 | + | + | + | | | | | | | 69 |
| *Pityrosporum* | | | | | | | | | | | | | | | | |
| *ovale* | | | | | | | | | | | | | | | | |
| MRL-3074 | Yeast | | | | | | | | | | | | | | | |
| | 2 days old | | | 18 | | | 48 | 0 | 0 | | | | | | | 71 |
| | 9 days old | | | 15 | | | 57 | 0 | 0 | | | | | | | 71 |
| | 14 days old | | | 12.5 | | | 52 | 0 | 0 | | | | | | | 71 |
| ATCC 14521 | Yeast | | | | | | | | | | | | | | | |
| | 2 days old | | | 8 | | | 40 | 0 | 0 | | | | | | | 71 |
| | 9 days old | | | 10 | | | 47 | 0 | 0 | | | | | | | 71 |
| *Pullularia* | | | | | | | | | | | | | | | | |
| *(Aureobasidium)* | | | | | | | | | | | | | | | | |
| *pullulans* | Yeast | | | 1.9 | | | 54 | 9.3 | 4.2 | 1.8 | 0.9 | | | | | 14 |
| | Filaments | | | 2.7 | | | 70 | 7.1 | 3.8 | tr | 1.9 | | | | | 14 |

## TABLE 1 (Continued)
## MONOSACCHARIDE COMPOSITION OF FUNGAL CELL WALLS

| Fungus | Form | Hexos-amine | GlcNAc | GlcN | GalN | Neutral Sugar | Glu | Man | Gal | GluA | Rha | Fuc | Xyl | Ara | Rib | Ref. |
|---|---|---|---|---|---|---|---|---|---|---|---|---|---|---|---|---|
| **Deuteromycetes (continued)** | | | | | | | | | | | | | | | | |
| *Verticillium albo-atrum* | Yeast | | 8.8 | 2.6 | | | 49 | 6.2 | 8.2 | 1.3 | | | | | | 78 |
| *Trigonopsis variabilis* | Triangular | 1.9 | | | | 80.6 | + | + | | | | | | | | 65 |
| | Ellipsoidal | 1.9 | | | | 91.0 | + | | | | | | | | | 65 |

[a] Non-flocculent.
[b] Flocculent.
[c] Caused by isomytilitol.
[d] Wild type.
[e] Alkali- soluble fraction only.
[f] Alkali-insoluble $\beta$-1,3-glucan fraction only.
[g] Sorbose-type semicolonial
[h] Grown on sorbose.
[i] Semicolonial.
[j] Colonial.
[k] Terrestrial fungus.
[l] Osmotic mutant.
[m] Virus-infected.
[n] Virus-free.
[o] Marine fungus.
[p] Thin mutants.
[q] Puff mutants.
[r] In starch medium at 30°C.
[s] In glucose medium at 37°C.
[t] In starch medium at 40°C.
[u] In ox serum at 37°C.
[v] Regenerating protoplast.
[x] Calculated from published data.

## TABLE 2

## AMINO ACID COMPOSITION AND PROTEIN CONTENT OF FUNGAL CELL WALLS

| Fungus | Form | Protein | Ala | Arg | Asp | Cys | Glu | Gly | His | Ile | Leu |
|---|---|---|---|---|---|---|---|---|---|---|---|
| **Myxomycetes** | | | | | | | | | | | |
| *Physarum polycephalum* | Spherules | 2 | 12.2 | 2.2 | 7.8 | 0 | 9 | 11.3 | 1.3 | 6.3 | 8.9 |
| | Spores | 2 | 9.0 | 4.3 | 11.6 | 0 | 13.4 | 12.5 | 2.1 | 4.8 | 6.9 |
| **Oomycetes** | | | | | | | | | | | |
| *Atkinsiella dubia* | Mycelium | 13.7 | 6.1 | 1.3 | 8.7 | 2.5 | 7.3 | 5.1 | 2.9 | 1.5 | 1.9 |
| *Phytophthora cinnamomi* | Mycelium | 3.5 | 9.0 | 4.2 | 9.0 | 1.1 | 11.1 | 4.5 | 1.1 | 1.8 | 4.5 |
| *Sapromyces elongatus* | Mycelium | 3.7 | 4.3 | 3.5 | 23.6 | | 8.8 | 5.0 | 3.0 | 2.6 | 5.3 |
| **Hemiascomycetes** | | | | | | | | | | | |
| *Nadsonia elongata* | Yeast[a] | 8.9 | 10.3 | 1.0 | 9.7 | 3.0 | 11.9 | 2.9 | 1.45 | 4.35 | 4.9 |
| *Saccharomyces cerevisiae* | Yeast[c] | | (9.0) | (2.6) | (28.0) | (1.5) | (9.3) | (5.5) | | (4.3) | (4.4) |
| | Yeast[a] | 15.9 | 11.2 | 2.1 | 7.0 | 0.3 | 7.5 | 5.9 | 2.4 | 5.3 | 6.4 |
| | Abnormal[a,q] | 11.1 | 12.0 | 1.0 | 8.4 | 0.2 | 9.8 | 6.5 | 0.9 | 4.2 | 3.9 |
| *Schizosaccharomyces pombe* | Yeast | | 8.2 | 4.9 | 9.9 | 0.3 | 8.9 | 9.5 | 2.1 | 6.2 | 8.5 |
| | Abnormal[d] | | 8.1 | 5.2 | 9.8 | 0.45 | 8.75 | 9.5 | 2.4 | 6.6 | 8.8 |
| **Euascomycetes** | | | | | | | | | | | |
| *Aspergillus niger* | Mycelium | 1.15 | 9.3 | 2.25 | 8.95 | 3.0 | 5.95 | 11.0 | tr | 2.0 | 2.6 |
| *Blastomyces dermatitidis* | Mycelium | 26.8 | 7.3 | 2.1 | 6.1 | | 9.6 | 22.3 | 1.4 | 5.1 | 5.75 |
| | Yeast | 7.1 | 6.8 | 2.9 | 15.7 | | 10.3 | 9.9 | 11.2 | 2.7 | 2.3 |
| *Chaetomium globosum* | Mycelium | 5.0 | 8.5 | 3.8 | 10.4 | 1.7 | 12.2 | 9.8 | tr | 2.6 | 6.0 |
| *Cordyceps militaris* | Mycelium | 6.3 | 9.9 | 4.3 | 9.7 | | 8.9 | 8.1 | 1.8 | 3.7 | 6.4 |
| *Microsporum canis* | Mycelium | 6.8 | 4.4 | tr | 5.8 | tr | 7.5 | 8.5 | 1.4 | 2.0 | 19.4 |
| *gypseum* | Mycelium | 8.0 | 6.9 | tr | 6.3 | | 7.2 | 10.8 | 3.0 | 2.1 | 4.5 |
| *Neurospora crassa* | Mycelium | 6.0 | 13.9 | 3.5 | 8.5 | | 6.2 | 13.7 | 1.2 | 3.9 | 6.2 |
| *Penicillium expansum* | Conidia | 7.7 | 7.9 | 3.9 | 9.9 | 10.2 | 8.9 | 8.0 | 4.8 | 3.1 | 6.0 |
| *notatum* | Mycelium | 8.9 | 10.7 | 2.6 | 9.9 | | 9.1 | 11.4 | 1.7 | 3.1 | 7.5 |
| *roqueforti* | Mycelium | 9.7 | 11.4 | 2.4 | 8.6 | | 8.6 | 10.1 | 0.4 | 2.4 | 6.2 |
| *Trichophyton mentagrophytes* | Mycelium | 7.1 | 6.3 | tr | 4.8 | 3.3 | 6.3 | 9.2 | 1.7 | 2.1 | 14.6 |
| **Loculoascomycetes** | | | | | | | | | | | |
| *Helminthosporium sativum* | Mycelium | 18 | 11.2 | 2.0 | 9.9 | 1.4 | 10.3 | 8.9 | 0.7 | 4.1 | 7.5 |
| *Leptosphaeria albopunctata* | Mycelium | 4.8 | 12.6 | 0.7 | 5.3 | | 9.45 | 10.1 | 2.7 | 1.5 | 2.4 |

TABLE 2 (Continued)

AMINO ACID COMPOSITION AND PROTEIN CONTENT OF FUNGAL CELL WALLS

| Lys | Met | Phe | Pro | Hyp | Ser | Thr | Tyr | Val | Others | Ref. |
|---|---|---|---|---|---|---|---|---|---|---|
| 3.5 | 0 | 4.7 | 6.8 | | 6.9 | 7.4 | 2.2 | 9.4 | | 47 |
| 6.0 | 0 | 3.3 | 6.4 | | 5.7 | 5.9 | 1.3 | 6.9 | | 47 |
| 5.5 | 0.2 | 1.2 | 4.1 | 20.4 | 3.6 | 17.9 | 3.0 | 3.8 | | 5 |
| 4.7 | 1.1 | 2.1 | 5.5 | 5.0 | 9.8 | 14.8 | 2.6 | 5.0 | | 8 |
| 9.1 | 2.8 | 3.6 | 3.3 | 2.5 | 7.4 | 6.0 | 3.1 | 4.6 | | 56 |
| 4.2 | 0.4 | 3.0 | 4.0 | | 10.1 | 17.2 | 2.75 | 7.7 | *b* | 24 |
| (4.3) | | (2.7) | (8.5) | | (4.8) | (6.25) | (1.5) | (7.0) | | 38 |
| 5.9 | 0.3 | 2.8 | 4.9 | | 14.6 | 14.0 | 1.9 | 7.4 | | 59 |
| 4.2 | | 2.3 | 5.8 | | 15.8 | 17.3 | 1.5 | 6.2 | | 59 |
| 6.8 | 1.55 | 4.3 | 4.2 | | 7.8 | 6.2 | 2.9 | 7.6 | | 21 |
| 7.0 | 1.7 | 4.75 | 4.15 | | 6.5 | 5.2 | 3.0 | 7.9 | | 21 |
| 4.5 | 2.3 | 3.1 | 6.5 | | 19.3 | 11.0 | 2.9 | 4.8 | | 50 |
| 4.2 | 0.6 | 2.8 | 12.6 | | 6.3 | 8.5 | | 5.1 | | 36 |
| 13.2 | 1.2 | 2.5 | 5.6 | | 6.6 | 5.0 | | 3.9 | | 36 |
| 1.0 | 0.7 | 2.65 | 9.9 | | 9.2 | 14.0 | 2.2 | 5.4 | | 50 |
| 5.4 | 1.4 | 3.1 | 8.2 | | 9.9 | 11.0 | 2.3 | 5.7 | | 46 |
| 3.7 | 2.4 | 7.8 | 4.1 | | 7.5 | 8.8 | 4.8 | 5.8 | *e* | 66 |
| 2.4 | 2.4 | 8.1 | 9.6 | | 10.8 | 12.3 | 6.0 | 3.6 | *f* | 66 |
| 5.6 | | 2.5 | 8.1 | | 9.75 | 8.7 | 1.0 | 7.0 | | 42 |
| 2.4 | 1.0 | 2.4 | 6.0 | | 8.7 | 8.4 | 2.4 | 5.7 | | 27 |
| 3.4 | | 5.7 | 9.4 | | 12.8 | 9.7 | 2.9 | | | 1 |
| 3.4 | | 5.8 | 9.1 | | 14.9 | 11.4 | 4.3 | | | 3 |
| 4.0 | tr | 7.1 | 5.4 | | 7.9 | 8.8 | 6.4 | 6.9 | *g* | 66 |
| 4.6 | 0.7 | 2.0 | 9.0 | | 9.0 | 7.1 | 3.0 | 7.6 | | 4 |
| 5.7 | 1.3 | 1.3 | 17.8 | | 12.3 | 10.0 | 2.8 | 3.9 | | 70 |

## TABLE 2 (Continued)

## AMINO ACID COMPOSITION AND PROTEIN CONTENT OF FUNGAL CELL WALLS

| Fungus | Form | Protein | Ala | Arg | Asp | Cys | Glu | Gly | His | Ile | Leu |
|---|---|---|---|---|---|---|---|---|---|---|---|
| **Loculoascomycetes (Continued)** | | | | | | | | | | | |
| *allorgei* | Mycelium | 9.5 | 11.0 | 3.1 | 10.4 | | 10.6 | 10.0 | 1.6 | 2.7 | 5.1 |
| *discors* | Mycelium | 8.65 | 11.2 | 2.0 | 8.0 | | 8.85 | 10.9 | 1.4 | 2.4 | 3.8 |
| *nitschkei* | Mycelium | 12.8 | 10.6 | 2.2 | 8.0 | | 9.0 | 10.6 | 1.9 | 2.2 | 3.6 |
| *orae-maris* | Mycelium | 10.9 | 10.0 | 3.0 | 7.0 | | 10.6 | 10.5 | 1.75 | 2.2 | 5.3 |
| *robusta* | Mycelium | 11.2 | 10.2 | 2.55 | 8.6 | | 10.1 | 9.55 | 2.2 | 2.6 | 4.7 |
| *Venturia* | | | | | | | | | | | |
| *inaequalis* | Mycelium | 2.7 | 9.9 | 3.5 | 7.3 | 0.8 | 9.3 | 10.4 | 2.7 | 5.2 | 8.3 |
| **Deuteromycetes** | | | | | | | | | | | |
| *Candida* | | | | | | | | | | | |
| *albicans* | Yeast | 25.6 | 8.9 | 0 | 6.5 | | 6.55 | 6.9 | 11.2 | 5.3 | 6.3 |
| | Mycelium | 15.7 | 10.7 | 5.3 | 6.5 | | 9.1 | 7.4 | 0 | 5.9 | 7.5 |
| *Cladosporium* | | | | | | | | | | | |
| *werneckii* | Yeast[l] | | (12.8) | (0.7) | (6.8) | | (8.6) | (6.8) | (0.5) | (3.2) | (3.9) |
| *Epidermophyton* | | | | | | | | | | | |
| *floccosum* | Mycelium | 7.4 | 6.8 | tr | 5.4 | 2.2 | 6.8 | 10.2 | 2.2 | 2.2 | 5.4 |
| *Histoplasma* | | | | | | | | | | | |
| *capsulatum* | Mycelium | 10 | 7.9 | 2.2 | 7.9 | | 13.5 | 19.1 | 2.2 | 3.3 | 4.5 |
| | Yeast | 5 | 7.9 | 7.9 | 2.6 | | 13.15 | 15.8 | 2.6 | 7.9 | 10.5 |
| *Paracoccidioides* | | | | | | | | | | | |
| *brasiliensis* | Mycelium | 37 | 8.9 | 3.6 | 9.3 | | 11.8 | 14.8 | 1.0 | 3.9 | 5.3 |
| | Yeast | 11.2 | 8.35 | 2.3 | 8.9 | | 9.9 | 11.8 | 10.8 | 2.7 | 4.7 |
| *Piricularia* | | | | | | | | | | | |
| *oryzae* | Mycelium | 4.6 | 9.5 | 2.1 | 11.4 | 0.8 | 7.0 | 7.6 | 0.8 | 2.7 | 4.6 |
| *Verticillium* | | | | | | | | | | | |
| *albo-atrum* | Yeast | 12.5 | 7.9 | 0.9 | 8.2 | 1.9 | 8.5 | 10.0 | 2.2 | 2.7 | 2.9 |

[a] Excluding $NH_3$.
[b] Tryptophan, 0.9%.
[c] Composition of a glucan-protein fraction from baker's yeast.
[d] Caused by isomytilitol.
[e] Ornithine, 6.1%.
[f] Ornithine, 3.6%.
[g] Ornithine, 5.2%.
[h] Ornithine, 0.6%.
[i] Three unidentified components, amounting to 2.8%.
[j] Mixture of 3- and 4-hydroxyproline.
[k] Unidentified amino acid, amounting to 4.3%.
[l] Composition of a galactomannan-protein fraction.
[m] Ornithine, 8.5%.
[n] Ornithine, traces.
[p] Unresolved.
[q] Caused by inositol deficiency.

## REFERENCES


1. Applegarth, D. A., *Arch. Biochem. Biophys., 120,* 471 (1967).
2. Applegarth, D. A., and Bozoian, G., *J. Bacteriol., 94,* 1787 (1967).
3. Applegarth, D. A., and Bozoian, G., *Can. J. Microbiol., 14,* 489 (1968).
4. Applegarth, D. A., and Bozoian, G., *Arch. Biochem. Biophys., 134,* 285 (1969).
5. Aronson, J. M., and Fuller, M. S., *Arch. Mikrobiol., 68,* 295 (1969).
6. Aronson, J. M., and Machlis, L., *Amer. J. Bot., 46,* 292 (1959).
7. Ballesta, J.-P. G., and Alexander, M., *J. Bacteriol., 106,* 938 (1971).
8. Bartnicki-Garcia, S., *J. Gen. Microbiol., 42,* 57 (1966).
9. Bartnicki-Garcia, S., and Nickerson, W. J., *Biochim. Biophys. Acta, 58,* 102 (1962).
10. Bartnicki-Garcia, S., and Reyes, E., *Arch. Biochem. Biophys., 108,* 125 (1964).
11. Bartnicki-Garcia, S., and Reyes, E., *Biochim. Biophys. Acta, 165,* 32 (1968).

## TABLE 2 (Continued)

## AMINO ACID COMPOSITION AND PROTEIN CONTENT OF FUNGAL CELL WALLS

| Lys | Met | Phe | Pro | Hyp | Ser | Thr | Tyr | Val | Others | Ref. |
|---|---|---|---|---|---|---|---|---|---|---|
| 4.8 | 0 | 2.1 | 11.8 | | 10.7 | 8.9 | 2.2 | 4.9 | | 70 |
| 4.7 | 0.6 | 2.7 | 12.9 | | 10.8 | 10.7 | 2.9 | 6.0 | | 70 |
| 5.8 | 0.4 | 1.8 | 12.9 | | 11.1 | 10.7 | 3.5 | 5.6 | | 70 |
| 4.5 | 0 | 2.9 | 12.4 | | 9.4 | 12.3 | 3.3 | 4.8 | | 70 |
| 5.0 | 0.65 | 2.3 | 13.1 | | 8.3 | 11.7 | 3.2 | 5.2 | | 70 |
| 7.0 | 2.3 | 4.2 | 5.4 | | 6.9 | 5.0 | 3.0 | 6.0 | *h* | 33 |
| 2.6 | 7.9 | *p* | 6.3 | | 5.6 | 6.1 | 5.1 | 5.1 | *i* | 18 |
| 0.8 | 2.2 | *p* | 3.7 | 3.1[j] | 5.6 | 7.4 | 3.6 | 9.4 | *k* | 18 |
| (1.7) | tr | (2.1) | (5.4) | | (16.3) | (26.4) | tr | (4.71) | | 43 |
| 4.1 | tr | 5.85 | 10.0 | | 9.5 | 10.2 | 5.6 | 4.9 | *m* | 65 |
| 6.7 | tr | 1.1 | tr | | 11.2 | 12.3 | tr | 7.9 | *n* | 23 |
| 10.5 | tr | 7.9 | tr | | 5.3 | 5.3 | 2.6 | | | 23 |
| 2.8 | 2.0 | 3.1 | 11.7 | | 8.1 | 7.6 | | 6.0 | | 36 |
| 9.4 | 2.1 | 3.5 | 6.4 | | 7.6 | 8.0 | | 3.5 | | 36 |
| 5.3 | 0.8 | 3.6 | 9.5 | | 5.3 | 8.6 | 1.9 | 5.8 | | 51 |
| 4.8 | tr | 1.9 | 13.2 | | 8.6 | 14.0 | 3.8 | 8.4 | | 78 |


12. Bartnicki-Garcia, S., and Reyes, E., *Biochim. Biophys. Acta, 170,* 54 (1968).
13. Bloomfield, B. J., and Alexander, M., *J. Bacteriol., 93,* 1276 (1967).
14. Brown, R. G., and Nickerson, W. J., Unpublished Results, *Bacteriol. Proc.,* p. 26 (1965).
15. Buck, K. W., Chain, E. B., and Darbyshire, J. E., *Nature (London), 223,* 1273 (1969).
16. Buecher, E. J., and Phaff, H. J., in *Proceedings of the Second International Symposium on Yeast Protoplasts,* p. 165 (1968).
17. Bull, A. T., *J. Gen. Microbiol., 63,* 75 (1970).
18. Chattaway, F. W., Holmes, M. R., and Barlow, A. J. E., *J. Gen. Microbiol., 51,* 367 (1968).
19. Chet, I., Henis, Y., and Mitchell, R., *Can. J. Microbiol., 13,* 137 (1967).
20. Cooper, B. A., and Aronson, J. M., *Mycologia, 59,* 658 (1967).
21. Deshusses, J., Berthoud, S., and Posternak, T., *Biochim. Biophys. Acta, 176,* 803 (1969).
22. De Terra, N., and Tatum, E. L., *Amer. J. Bot., 50,* 669 (1963).
23. Domer, J. E., Hamilton, J. G., and Harkin, J. C., *J. Bacteriol., 94,* 466 (1967).
24. Dyke, K. G. H., *Biochim. Biophys. Acta, 82,* 374 (1964).
25. Eddy, A. A., *Proc. Roy. Soc. Ser. B, 149,* 425 (1958).
26. Falcone, G., and Nickerson, W. J., *Science, 124,* 272 (1956).
27. Fisher, D. J., and Richmond, D. V., *J. Gen. Microbiol., 64,* 205 (1970).
28. Fultz, S. A., Ph. D. Thesis, *Diss. Abstr.,* p. 2433 (1965).

29. Gancedo, J. M., Gancedo, C., and Asensio, C., *Biochem. Z., 346,* 328 (1966).
30. Grisaro, V., Sharon, N., and Barkai-Golan, R., *J. Gen. Microbiol., 51,* 145 (1968).
31. Hamilton, P. B., and Knight, S. G., *Arch. Biochem. Biophys., 99,* 282 (1962).
32. Horikoshi, K., and Iida, S., *Biochim. Biophys. Acta, 83,* 197 (1964).
33. Jaworski, E. G., and Wang, L. C., *Phytopathology, 55,* 401 (1965).
34. Johnston, I. R., *Biochem. J., 96,* 651 (1965).
35. Kanetsuna, F., and Carbonell, L. M., *J. Bacteriol., 106,* 946 (1971).
36. Kanetsuna, F., Carbonell, L. M., Moreno, R. E., and Rodriguez, J., *J. Bacteriol., 97,* 1036 (1969).
37. Katz, D., and Rosenberger, R. F., *Biochim. Biophys. Acta, 208,* 452 (1970).
38. Kessler, G., and Nickerson, W. J., *J. Biol. Chem., 234,* 2281 (1959).
39. Korn, E. D., and Northcote, D. H., *Biochem. J., 75,* 12 (1960).
40. Leopold, J., and Seichertova, O., *Folia Microbiol., 12,* 345 (1967).
41. Lippman, E., Erwin, D. C., and Bartnicki-Garcia, S., *Phytopathology, 61,* 901 (1971); Unpublished Data.
42. Livingston, L. R., *J. Bacteriol., 99,* 85 (1969).
43. Lloyd, K. O., *Biochemistry, 9,* 3446 (1970).
44. Mahadevan, P. R., and Tatum, E. L., *J. Bacteriol., 90,* 1073 (1965).
45. Mahadevan, P. R., and Mahadkar, U. R., *Indian J. Exp. Biol., 8,* 207 (1970).
46. Marks, D. B., Keller, B. J., and Guarino, A. J., *Biochim. Biophys. Acta, 183,* 58 (1969).
47. McCormick, J. J., Blomquist, J. C., and Rusch, H. P., *J. Bacteriol., 104,* 1119 (1970).
48. Mill, P. J., *J. Gen. Microbiol., 44,* 329 (1966).
49. Miller, M. W., and Phaff, H. J., *Antonie van Leeuwenhoek J. Microbiol. Serol., 24,* 225 (1958).
50. Mitchell, A. D., and Taylor, I. E. P., *J. Gen. Microbiol., 59,* 103 (1969).
51. Nakajima, T., Tamari, K., Matsuda, K., Tanaka, H., and Ogasawara, N., *Agr. Biol. Chem., 34,* 553 (1970).
52. Nanba, H., and Kuroda, H., *Chem. Pharm. Bull. (Tokyo), 19,* 252 (1971).
53. Northcote, D. H., and Horne, R. W., *Biochem. J., 51,* 232 (1952).
54. Novaes-Ledieu, M., and Garcia-Mendoza, C., *J. Gen. Microbiol., 61,* 335 (1970).
55. Novaes-Ledieu, M., Jimenez-Martinez, A., and Villanueva, J. R., *J. Gen. Microbiol., 47,* 237 (1967).
56. Pao, V. M., and Aronson, J. M., *Mycologia, 62,* 531 (1970).
57. Potgieter, H. J., and Alexander, M., *Can. J. Microbiol., 11,* 122 (1965).
58. Potgieter, H. J., and Alexander, M., *J. Bacteriol., 91,* 1526 (1966).
59. Power, D. M., and Challinor, S. W., *J. Gen. Microbiol., 55,* 169 (1969).
60. Rizza, V., and Kornfeld, J. M., *J. Gen. Microbiol., 58,* 307 (1969).
61. Roelofsen, P. A., *Biochim. Biophys. Acta, 10,* 477 (1953).
62. Ruiz-Herrera, J., *Arch. Biochem. Biophys., 122,* 118 (1967).
63. Russell, D. W., Sturgeon, R. J., and Ward, V., *J. Gen. Microbiol., 36,* 289 (1964).
64. Rzucidlo, L., Stachow, A., Nowakowska, A., and Kubica, J., *Bull. Acad. Pol. Sci. Ser. Sci. Biol., 6,* 15 (1958).
65. Sentheshanmuganathan, S., and Nickerson, W. J., *J. Gen. Microbiol., 27,* 451 (1962).
66. Shah, V. K., and Knight, S. G., *Arch. Biochem. Biophys., 127,* 229 (1968).
67. Sietsma, J. H., Eveleigh, D. E., and Haskins, R. H., *Biochim. Biophys. Acta, 184,* 306 (1969).
68. Skujins, J. J., Potgieter, H. J., and Alexander, M., *Arch. Biochem. Biophys., 11,* 358 (1965).
69. Sturgeon, R. J., *Nature (London), 209,* 204 (1966).
70. Szaniszlo, P. J., and Mitchell, R., *J. Bacteriol., 106,* 640 (1971).
71. Thompson, E., and Colvin, J. R., *Can. J. Microbiol., 16,* 263 (1970).
72. Toama, M. A., and Raper, K. B., *J. Bacteriol., 94,* 1150 (1967).
73. Tokunaga, J., and Bartnicki-Garcia, S., *Arch. Mikrobiol., 79,* 293 (1971).
74. Trotter, M. J., and Whisler, H. C., *Can. J. Bot., 43,* 869 (1965).
75. Troy, F. A., and Koffler, H., *J. Biol. Chem., 244,* 5563 (1969).
76. Wai, N., *Bull. Inst. Chem. Acad. Sinica, 17,* 41 (1970).
77. Wai, N., *Bull. Inst. Chem. Acad. Sinica, 17,* 1 (1970).
78. Wang, M. C., and Bartnicki-Garcia, S., *J. Gen. Microbiol., 64,* 41 (1970).
79. Wang, C.-S., Schwalb, M. N., and Miles, P. G., *Can. J. Microbiol., 14,* 809 (1968).
80. Wessels, J. G. H., *Wentia, 13,* 1 (1965).

# BACTERIAL ENDOTOXINS

DR. RONALD J. ELIN and DR. SHELDON M. WOLFF

Bacterial endotoxins are biologically active compounds present in the cell wall of Gram-negative bacteria. Endotoxin-like materials have been found in other organisms; however, this review will be concerned only with the endotoxins of Gram-negative bacteria. Using chemical methods of extraction and purification, bacterial endotoxins can be obtained as relatively pure chemical complexes that consist mainly of lipids and polysaccharides and contain a small amount of amino acids. The chemical name, lipopolysaccharides (LPS), is used synonymously with bacterial endotoxins.

The chemical and biological properties of lipopolysaccharides have been under investigation for more than a century. These investigations have been hampered by two intrinsic problems: (1) different bacterial strains produce chemically dissimilar LPS, which may modify their biological properties, and (2) different extraction and purification methods produce a heterogeneity of LPS preparations. Although endotoxins have a myriad of biological activities that affect all systems of the body, no clear-cut molecular-biological basis for these activities is known.

## CELLULAR LOCALIZATION

The term "endotoxin" (LPS) was initially used because it was thought that the toxic substances were present within the bacterial cell and could only be liberated by disruption or digestion of the cell. This is in contrast to classical toxins (exotoxins), which are usually proteins and are excreted by the bacterial cells. It is now known that lipopolysaccharides are located in the outer layer of the cell wall.[263] Electron microscopy of Gram-negative bacteria shows three distinct layers to the cell wall (Figure 1): (1) an inner layer, which is the cytoplasmic membrane; (2) a middle layer, which is composed of a rigid peptidoglycan; and (3) an outer layer containing LPS, phospholipid, and protein.[11,309,310] It is postulated that a large portion of the LPS molecule is exposed on the cell surface, since a carbohydrate moiety of the LPS molecule is responsible for the somatic O-antigenic determinants of the bacteria.[309]

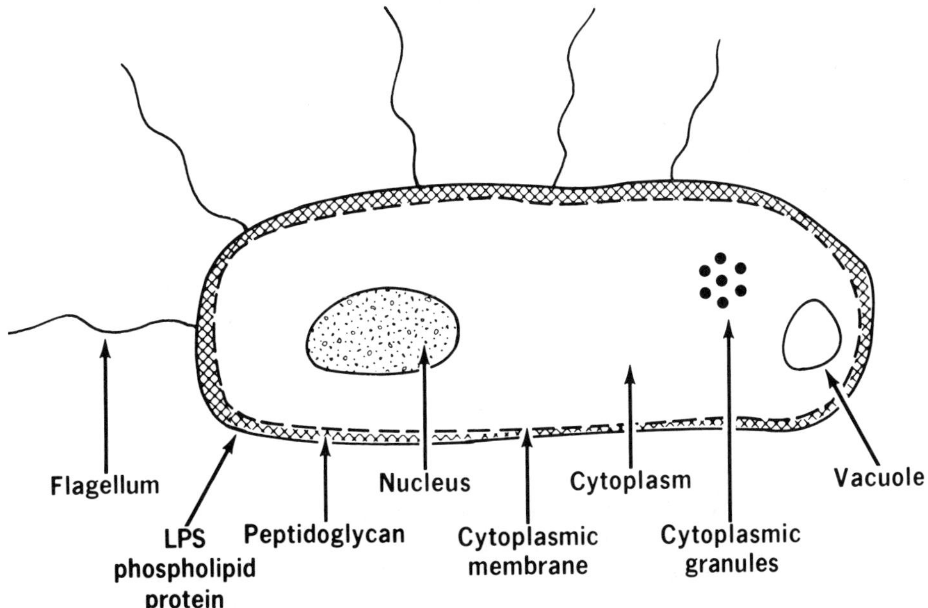


FIGURE 1. Diagrammatic structure of a Gram-negative bacterium, showing the location of LPS in the outer layer of the cell wall.

The attachment of the LPS to the cell wall is uncertain. It has been shown that LPS are synthesized by the cytoplasmic membrane. The molecules move to the surface of the bacterial cell wall, possibly because of their hydrophilic properties. At the cell wall surface, LPS are probably linked to protein granules that are covalently bound to the peptidoglycan layer. The treatment of Gram-negative bacterial cells with EDTA[196] or with inhibitors of protein synthesis[184,294] results in the liberation of chemical complexes composed of LPS, phospholipids and proteins. It is, therefore, postulated that LPS are found in the outer layer of the cell wall of Gram-negative bacteria, complexed to phospholipids and proteins.

## CHEMICAL PROPERTIES

### Methods of Extraction and Purification

As indicated previously, LPS preparations vary with the methods of extraction and purification. This has led to major difficulties in comparing the work of different investigators who used different methods to obtain LPS. Several methods of extraction and purification of LPS are listed in Table 1.

### Structure of Endotoxins (LPS)

Considerable work has been done on the structure of LPS, primarily using *Salmonella* mutants. On the basis of these studies, it is possible to divide the LPS molecule into three parts: (1) the lipid moiety, (2) the R-core, and (3) the O-polysaccharide (Figure 2 and Table 2). Chemical analyses of LPS isolated from several different strains of Gram-negative bacteria have shown a consistency of certain organic molecules for the three parts of the LPS molecule (Figures 2 and 3).

#### LIPID MOIETY

Initial studies of the structure of the lipid moiety of the LPS molecule were conducted by Boivin and co-workers.[33] They found that acid hydrolysis of LPS resulted in a phosphorus-containing lipid precipitate, designated as "Fraction A" (later called "Lipid A"), and a soluble "Fraction B", which contained polysaccharides. The term "Lipid A" has persisted through the years and refers to all chemical compounds that become insoluble after acid hydrolysis of LPS. Lipid A, therefore, is a mixture of several chemical compounds rather than one chemically defined molecule.

The chemical structure of the lipid moiety of LPS has been evaluated by careful dissection and analysis of the several components of the lipid A precipitate. These studies show that the lipid component of LPS is composed entirely of carboxylic acids (fatty acids), which have a range of chain length from at least $C_{10}$ to $C_{22}$.[40] Ten different carboxylic acids were found in a mutant of *Salmonella typhimurium;* the major component was β-hydroxmyristic acid, which appears to be unique to the LPS molecule.[203]

One of the basic structural components of the lipid A precipitate is a diglucosamine unit that contains fatty acids attached to amino and hydroxyl groups (Figure 4). Alhough there is no unanimity of opinion, it appears that the glucosamine units are bound together by 1,6-glycosidic linkages and 1,4-phosphodiester bridges.[2,3,40,107]

#### R-CORE

The R-core is the middle portion of the LPS molecule, linking the lipid moiety to the O-polysaccharide moiety. Investigation of the structure of the R-core has been greatly facilitated by the group of mutants termed R or rough forms, which are devoid of O-polysaccharide. These mutants are easily identified by their rough colony morphology, in contrast to the S or smooth forms, which contain the O-polysaccharide moiety.

Chemical analyses of the R-core have shown five basal sugars, phosphate, and O-phosphorylethanolamine.[143,204,249] For *Salmonella* the five sugars are the following: the amino sugar, N-acetylglucosamine;

the hexoses, glucose and galactose; the heptose, L-glycero-O-manno heptose; and the octose, 2-keto-3-deoxyoctonate (KDO). These basal polysaccharides seem to be common constituents in all *Salmonellae.*[143,266] Thus far, KDO has been found only with LPS (Figure 5).

The lipid moiety of the LPS molecule is covalently bound to the R-core by attachment of KDO (Figures 5 and 6). The O-phosphorylethanolamine is linked to a different KDO unit by its phosphate group. The outermost KDO unit serves as a link to the outer polysaccharide portion of the R-core (Figure 6). Molecular-weight determinations suggest that there is an average of three R-core units, which are probably linked by phosphodiester bridges, per LPS molecule.[369]

**TABLE 1**
**EXTRACTION AND PURIFICATION METHODS FOR LPS**

| Extraction | Purification | Reference |
|---|---|---|
| 0.25*M* Trichloroacetic acid at 4°C | Dialysis | 33 |
| Diethylene glycol | Alcohol or acetone precipitation | 238 |
| 2% Phenol and autolysis | Filtration + $(NH_4)_2SO_4$ precipitation | 232 |
| 95% Phenol | Precipitation with alcohol | 270 |
| 2.5*M* Urea | Acetone and alcohol precipitation | 352 |
| 45% Phenol | Precipitation with alcohol | 204 |
| Water and ethyl ether | Precipitation with alcohol | 285 |
| Detergent (Cetavlon) | Precipitation with alcohol | 262 |
| Dimethyl sulfoxide | Precipitation with acidic acetone | 1 |
| Phenol, chloroform, and petroleum ether | Precipitation with water | 98 |
| Ethylenediaminetetraacetic acid and tris(hydroxymethyl)aminomethane | Membrane partition chromatography | 300 |

**TABLE 2**
**CHEMICAL STRUCTURE OF LPS**

| Chemical Unit | Investigations | References |
|---|---|---|
| Lipid moiety | Structure | 2, 3, 184 |
| | Biosynthesis | 40, 162, 249, 341 |
| | Binding to cell wall | 107 |
| R-core | Structure | 72–74, 107, 108, 123, 127, 143, 257 |
| | Biosynthesis | 137, 249, 266 |
| O-Polysaccharide | Structure | 11, 76, 108, 127, 144, 145, 203, 205, 262 |
| | Biosynthesis | 63, 137, 208, 249, 252, 253, 267–269, 288 |
| | Binding to R-core | 250, 251 |
| | Review | 203, 204, 208, 288 |
| Protein | | 363 |
| | General reviews | 170, 266, 294, 338, 363, 369 |

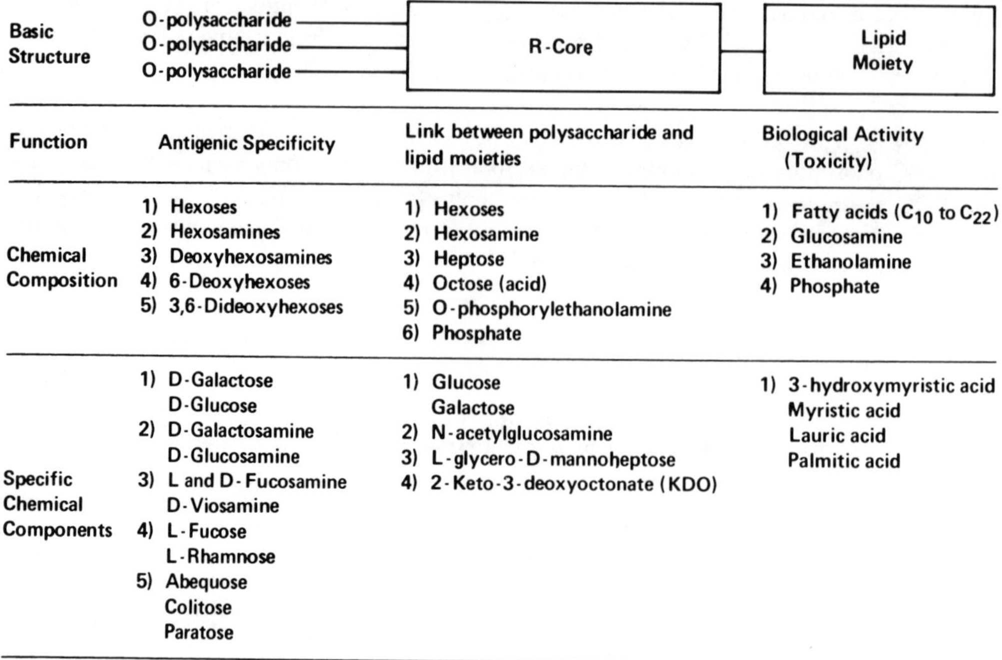

| Basic Structure | O-polysaccharide<br>O-polysaccharide<br>O-polysaccharide | R-Core | Lipid Moiety |
|---|---|---|---|
| Function | Antigenic Specificity | Link between polysaccharide and lipid moieties | Biological Activity (Toxicity) |
| Chemical Composition | 1) Hexoses<br>2) Hexosamines<br>3) Deoxyhexosamines<br>4) 6-Deoxyhexoses<br>5) 3,6-Dideoxyhexoses | 1) Hexoses<br>2) Hexosamine<br>3) Heptose<br>4) Octose (acid)<br>5) O-phosphorylethanolamine<br>6) Phosphate | 1) Fatty acids ($C_{10}$ to $C_{22}$)<br>2) Glucosamine<br>3) Ethanolamine<br>4) Phosphate |
| Specific Chemical Components | 1) D-Galactose<br>D-Glucose<br>2) D-Galactosamine<br>D-Glucosamine<br>3) L and D-Fucosamine<br>D-Viosamine<br>4) L-Fucose<br>L-Rhamnose<br>5) Abequose<br>Colitose<br>Paratose | 1) Glucose<br>Galactose<br>2) N-acetylglucosamine<br>3) L-glycero-D-mannoheptose<br>4) 2-Keto-3-deoxyoctonate (KDO) | 1) 3-hydroxymyristic acid<br>Myristic acid<br>Lauric acid<br>Palmitic acid |

FIGURE 2. Function and chemical composition of the three parts of the LPS molecule. The numbers of the specific chemical components are correlated with the numbers in the chemical composition above.

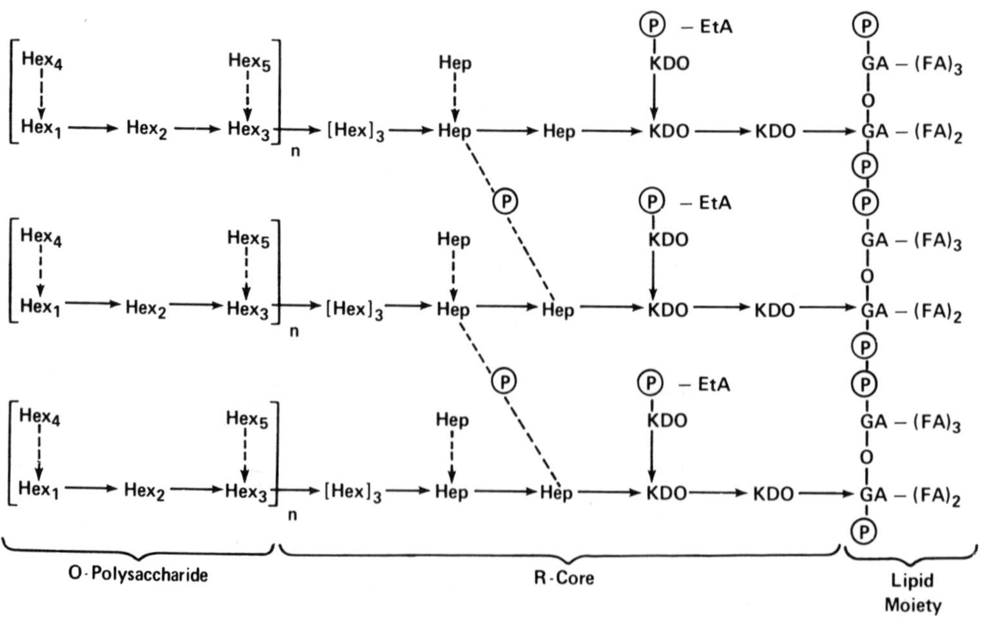



FIGURE 3. A general structure of a LPS molecule based upon the work of many different investigators.[72,74,143,203,204,249,266] Intermittent occurrence of a structure is indicated by the broken lines. Hex = hexose; Hep = heptose; P = phosphate; KDO = 2-keto-3-deoxyoctonate; EtA = ethanolamine; GA = glucosamine; FA = fatty acid.

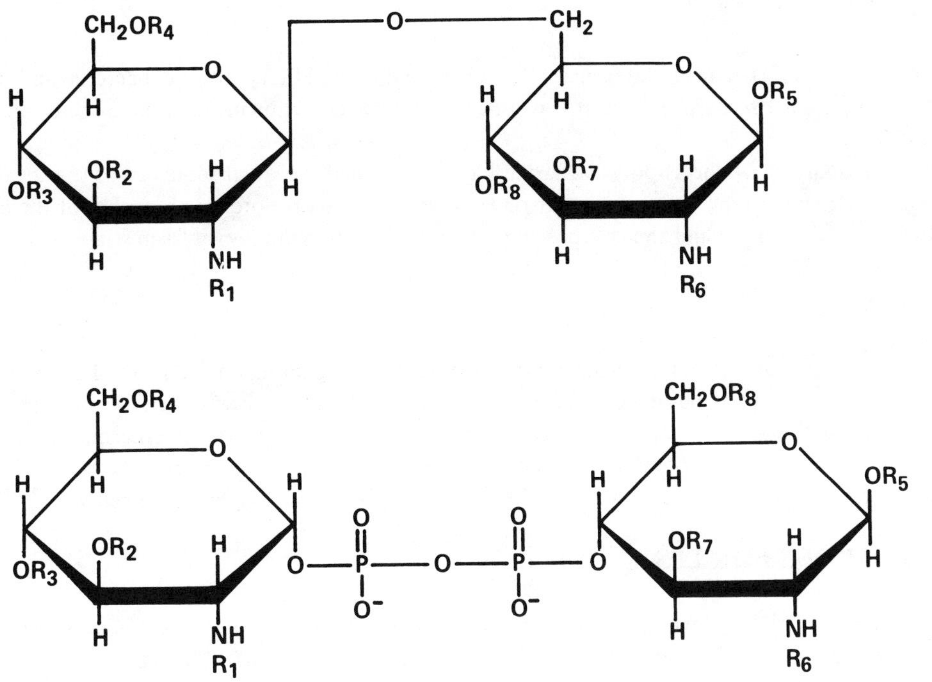

FIGURE 4. Structure of the diglucosamine unit of the lipid moiety of LPS.[2,3,40,107] The β-1,6-glycosidic linkage is shown above, and the 1,4-phosphodiester bridge is shown below. The R positions are the sites of attachment for carboxylic acid, KDO, a glycosidic linkage or a phosphodiester bridge.

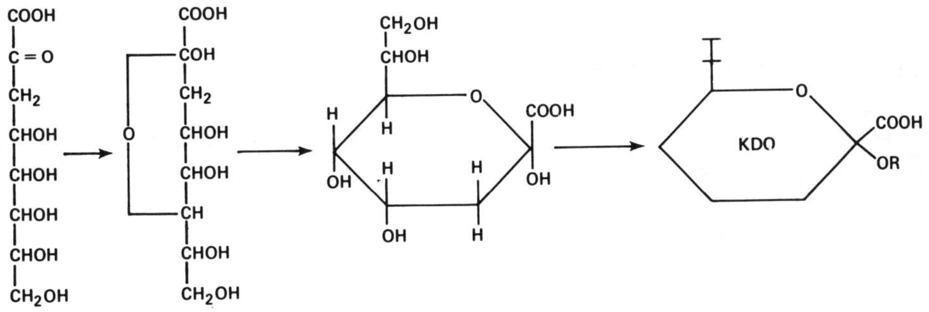

FIGURE 5. The structure of 2-keto-3-deoxyoctonate (KDO). This eight-carbon α-ketonic acid is present in the R-core of the LPS molecule as a 6-membered ring.

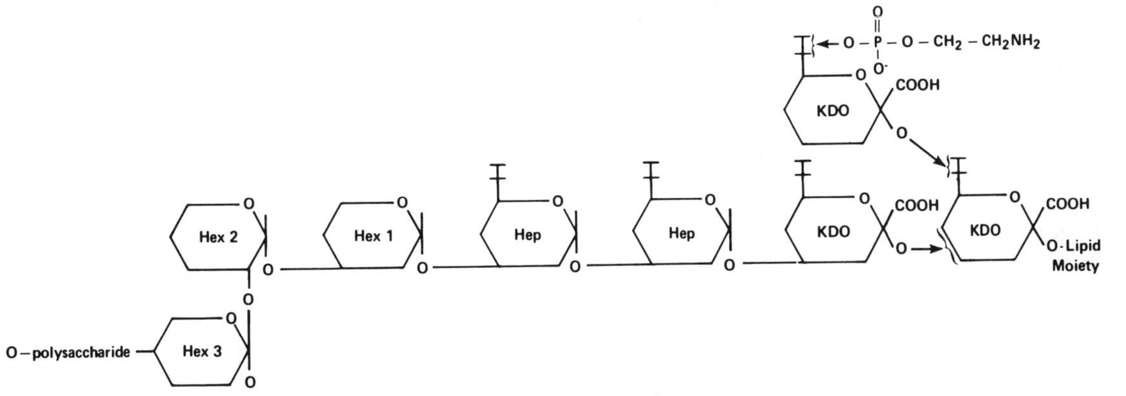

FIGURE 6. General structure of the R-core of the LPS molecules.[143,204,249,266] KDO = 2-keto-3-deoxyoctonate; Hep = heptose; Hex = hexose.

### O-POLYSACCHARIDE

The O-polysaccharide structure determines the O-antigenic specificity of the bacteria and their LPS. Members of a particular serogroup have at least one O-antigenic determinant in common with members of a different serogroup.

Biochemical analyses of the O-polysaccharide have established a regular sequence of the sugars.[144,145] The repeating sequences may either be linear or may contain one or more side branches of the component sugars (Figure 3). The length and the composition of these repeating units vary from strain to strain.[288]

### AMINO ACIDS

All LPS preparations contain amino acids in varying amounts, depending on the method of extraction. The lowest amino acid content is obtained by the phenol–water extraction or by proteolysis. The importance of amino acids to the structure and biological activity of LPS is unknown. Furthermore, it is uncertain whether the amino acids are covalently bound to the LPS molecule or are merely contaminants. It is possible that the amino acids serve to bind the LPS to the peptidoglycan layer of the cell wall.[363]

## ASSAY METHODS FOR LPS

Several methods have been devised to assay for LPS in tissues and body fluids (Table 3). Most of these procedures are based upon the toxicity of LPS *in vivo,* i.e., the ability of LPS to produce fever, abortion, dermal necrosis, or death. Recently, *in-vitro* methods of assay for LPS have been introduced, which appear to offer greater precision and sensitivity. The *Limulus* method for the detection of LPS, based upon the *in-vitro* reaction between LPS and a lysate prepared from amebocytes, the only circulating blood cell of the *Limulus* crab, seems to have great potential for the future because of its simplicity and sensitivity.[199,200] However, since the biological activities of each LPS are not a uniform gravimetric property but are unique, it may be virtually impossible for any method to measure LPS quantitatively with any meaningful accuracy.

## BIOLOGICAL PROPERTIES OF LPS

The biological and pathological effects of LPS in humans and in experimental animals are multiple and diverse. These biological changes are briefly summarized in the following paragraphs and in Table 4.

**TABLE 3**
**ASSAY METHODS FOR ENDOTOXIN**

| Method | End Point of Assay | Sensitivity, μg/ml | References |
|---|---|---|---|
| Chick embryo lethality | Lethality | 0.01 | 235, 326 |
| Intradermal injection in rabbits | Dermal inflammation | 0.15 | 194 |
| Fever in rabbits | 0.5°C rise in temperature | $10^{-3}$ | 176, 235, 365 |
| Actinomycin-treated mice | Lethality | 1 | 71, 274 |
| Adrenalectomized mice | Lethality | 0.5 | 48 |
| Hepatic enzyme induction | Inhibition of enzyme induction | 1 | 27 |
| Macrophage migration | Inhibition of migration | 0.1 | 142 |
| Rabbit epinephrine skin test | Dermal inflammation | 1 | 216, 217 |
| Hypoferremia | Decrease in serum iron concentration | 0.1 | 10 |
| *Limulus* lysate | Gelation | $10^{-6}$ | 59, 198–200, 281, 289, 373 |

## TABLE 4
## BIOLOGICAL PROPERTIES OF LPS

| LPS Reaction | Type of Study | References |
|---|---|---|
| Fever | Mechanism | 19, 41, 69, 82, 313, 349, 371 |
| | Relationship to endogenous pyrogen | 8, 9, 69, 126, 290 |
| | Tolerance | 12, 69, 240, 272, 353 |
| | Measurement | 62, 176, 365, 368 |
| | Review | 6, 7, 17, 18, 330 |
| Effect on blood | Leukocytes | 57, 105, 210, 239, 313, 367 |
| | Leukocytosis | 32, 130, 210 |
| | Leukopenia | 70, 239 |
| | Leukopoiesis | 130, 279, 329 |
| | Leukotaxis | 38, 94, 306 |
| | Leukotoxic effects | 105, 264, 306, 362 |
| | Other leukocyte changes | 94, 109, 264, 306, 362 |
| | Erythrocytes | 56, 95, 96 |
| | Platelets | 67, 68, 161, 219, 333 |
| | Coagulation | 61, 132, 199 |
| | Mechanism | 81, 374 |
| | Disseminated intravascular coagulation (DIC) | 81, 132, 374 |
| | Complement | 29, 99, 103, 188, 228, 229, 242 |
| Effect on the endocrine system | Pituitary | 93, 180, 187, 207, 236, 339, 357, 358 |
| | Adrenal | 180, 227, 236, 339, 358, 371 |
| | Reproduction-abortion | 47, 222, 286, 320 |
| Effect on metabolism | Carbohydrates | 28, 128, 167, 311, 312 |
| | Lipids | 100, 101, 157, 197, 199 |
| | Proteins and enzymes | 26, 237, 315, 323, 331 |
| | Minerals | |
| | Iron | 10, 171–174 |
| | Zinc | 271 |
| Effect on the reticuloendothelial system | Description | 5, 14, 16, 110, 117, 139, 347 |
| | Mechanism | 111, 112, 140, 141 |
| | Morphologic changes | 138, 141 |
| | Distribution of injected LPS | 34, 50, 125, 146, 298 |
| Effect on the vascular system | Shock | |
| | Mechanism | 15, 147, 188, 291, 346, 351 |
| | Therapy | 25, 85, 233, 278, 325, 342, 354 |
| | Experimental models | 20, 189 |
| | Primate | 45, 46, 150, 152, 220, 247 |
| | Dog | 39, 334, 348 |
| | Rat | 97 |
| | Biochemical mediators | 45, 149, 165, 186, 226, 291 |
| | Catecholamines | 115, 334, 372 |
| | Histamine | 64, 116, 152, 159, 244, 275 |
| | Kinins | 44, 79, 181, 247, 323 |
| | Serotonin | 64, 65 |
| | Steroids | 97, 155, 233, 342 |
| | Hemodynamic changes | 31, 37, 45, 52, 53, 104, 156, 189, 202 |
| | Heart | 30, 39, 148, 150, 151, 332, 343 |
| | Capillaries | 54, 133, 134, 218, 219 |
| | Lung | 39, 133, 134 |
| | Liver | 31, 84, 154, 226, 255 |
| | Other organs | 134, 153, 164, 318 |
| | Pathological changes | 149, 219, 220 |
| | Reviews | 55, 104, 149, 155, 202, 377 |

## TABLE 4 (Continued) BIOLOGICAL PROPERTIES OF LPS

| LPS Reaction | Type of Study | References |
|---|---|---|
| Immunological phenomena | Immunogenicity of LPS | 51, 124, 193, 204, 232, 234, 238, 259 |
| | Description | 292, 293, 301, 361, 364 |
| | Effect on endotoxicity | 178, 179, 224, 280, 287, 340 |
| | Immune response | 89, 223, 246, 273 |
| | Humoral | 86, 87, 124, 129, 191 |
| | Cellular | 77, 86, 88, 106, 121, 191 |
| | Adjuvant effect | 42, 89, 90, 113, 223, 359 |
| | Immunosuppression | 88–90, 113, 223, 359 |
| | Thymus | 42, 102, 295 |
| | Lymphocytes | 102, 129, 223 |
| | Reviews | 168, 243 |
| Detoxification | Biological | 62, 175, 211, 321, 324, 347 |
| | Plasma | 49, 183, 192, 302, 303, 307, 308, 375, 376 |
| | Leukocytes | 49, 84 |
| | Reticuloendothelial system | 84, 321, 324, 347 |
| | Immunochemical | 178, 179, 261, 280 |
| | Chemical | 212, 213, 245, 256, 258, 259, 282–284 |
| | Alkali | 56, 166, 211, 245, 254 |
| | Acids | 135, 284 |
| | Enzymes | 177, 304 |
| | Reduction of particle size | 135, 214, 265, 283, 284 |
| Endotoxin tolerance | Description | 12–14, 272, 353 |
| | Mechanism | 14, 43, 48, 113, 117–120, 122, 240, 272, 299, 366 |
| Non-specific resistance to infection | Description | 21, 35, 58 |
| | Bacteria | 23, 75, 215, 231, 296 |
| | Fungi | 78, 182, 185, 370 |
| | Parasites | 206, 319 |
| | Viruses | 277, 305, 350 |
| | Mechanism | 58, 78, 239, 261, 297, 360 |
| | Interferon | 158, 277, 305, 337 |
| | Comparison with other compounds | 22, 80, 356 |
| | Reviews | 35, 58, 297, 314 |
| Shwartzman phenomenon | Description | |
| | Localized | 316, 317 |
| | Generalized | 160, 195, 322, 345 |
| | Mechanism | 60, 61, 114, 160, 195, 209, 221, 241, 335, 344 |
| | Clinical correlation | 61, 241, 322 |
| Pathology | Cytotoxicity | 24, 36, 66, 91, 138, 141 |
| | Lymphocytes | 4, 223 |
| | Liver | 4, 66, 83, 119, 154, 157, 201, 336 |
| | Other organs | 92, 219, 223 |
| | Effect on tumor growth | 136, 163 |
| Protection against irradiation | Description | 131, 225, 328 |
| | Mechanism | 96, 327, 329 |
| General reviews | Books | 170, 190, 355 |
| | Journals | 260 |

## Fever

The fever produced by LPS may occur by direct LPS action on the thermoregulatory center in the brain or, more likely, by direct LPS action on blood leukocytes, on reticuloendothelial cells, or perhaps other tissues, to liberate an endogenous pyrogen.[69,290] The evidence for the direct action of LPS on the brain is based upon studies in which the injection of LPS into the central nervous system resulted in the abrupt onset of fever after three minutes.[19,349] After an intravenous injection of LPS, there is a lag period of ten to twenty minutes in the rabbit and of one to two hours in man before the temperature elevation begins. The length of the lag period is dose-dependent. During this period, LPS are rapidly removed from the blood by phagocytic cells. These cells are activated by LPS to release a fever-producing substance, which has been named "endogenous pyrogen."[368] This low-molecular-weight protein, which is heat-labile, then acts on the anterior hypothalamus to produce fever.[7] The fever is produced by alteration of either heat production or vasomotor control of heat redistribution and by heat loss at rates that are independent.[41]

## Effect on Blood

### LEUKOCYTES

The intravenous administration of LPS to humans and experimental animals produces transient leukopenia followed by marked granulocytosis. Depth and length of the leukopenia are dose-dependent.[313] The leukopenia has been shown to result predominantly from a decrease in granulocytes, due to an increased margination of circulating granulocytes to the endothelium of capillaries.[239,367] The granulocytosis seen after LPS stimulation is dependent on an adequate bone marrow reserve of granulocytes and on a normal granulocyte-releasing mechanism.[210] The release of granulocytes may be controlled by a humoral factor, which is increased by LPS injection.[32] In addition, LPS injection increases the production of a colony-stimulating factor, which may regulate granulocytopoiesis.[279] *In-vitro* studies have shown that LPS in the presence of complement are ingested by, or adhere to, the human granulocyte and induce a series of metabolic changes resembling those associated with phagocytosis.[57]

### ERYTHROCYTES

Studies in mice have shown that a single injection of LPS results in a significant shift in erythropoiesis from the bone marrow to the spleen.[95] The leukocytic hyperplasia of the bone marrow following LPS administration may cause this shift in erythropoiesis.[96]

### PLATELETS

Platelet aggregation *in vivo* following LPS injection has been demonstrated in rats and rabbits.[219,335] After aggregation, degranulation of platelets occurs, which results in an increased activity of platelet factor 3.[161] Electron microscopy investigations have shown that LPS adheres to platelets, causing them to aggregate, fuse and fragment.[333]

### DISSEMINATED INTRAVASCULAR COAGULATION (DIC)

Numerous cases of septicemia in man due to Gram-negative bacteria have been reported, which resulted in DIC.[374] The DIC is considered to be secondary to the LPS of the bacteria. There is experimental evidence that LPS may initiate DIC by two different pathways: (1) interaction with platelets to liberate platelet factor 3, and (2) activation of coagulation factor XII. LPS injection causes platelets to aggregate and fragment, with a resultant increase in the activity of platelet factor 3.[161] Platelet factor 3 in the presence of calcium and factor V acts on activated factor X to form blood thromboplastin, which then transforms prothrombin to thrombin. Also, *in-vitro* studies indicate that LPS can initiate coagulation by activating factor XII.[221]

### COMPLEMENT

LPS are efficient activators of the complement system. They do not activate the complement system via the classical pathway. Instead, an "alternate pathway" of complement activation has been demonstrated, involving consumption of the terminal components (C′3–C′9) and sparing the early components (C′1, C′4, and C′2).[229] The biologically active products generated due to complement-system activation by LPS effect changes in the clotting system, in vascular permeability, in smooth-muscle reactivity, and in neutrophil chemotaxis.[229] Indeed, several of the biological changes induced by endotoxin may be mediated by activated components of the complement system.

## Effects on the Endocrine System

LPS have a selective effect on the hypothalamic-pituitary system, causing the release of adrenocorticotropic hormone (ACTH) and growth hormone (GH) but having no influence on thyrotropin (TSH) and luteinizing hormone (LH) release.[93,187] The site of LPS action on the hypothalamic-pituitary axis has not been settled. There is evidence that LPS act upon or through the neuronal elements of the hypothalamus to cause the release of ACTH by the pituitary gland.[236] On the other hand, studies show LPS can bypass the medial hypothalamus and act directly upon the pituitary gland to release ACTH.[207,357] Also, there is a circadian rhythm in the pituitary response after LPS injection.[339]

A significant elevation of plasma cortisol can be demonstrated two hours after LPS injection.[180,227] Fever and the associated non-specific stress do not appear to be the stimuli for the release of ACTH by LPS.[180,371]

## Effect on Metabolism

### CARBOHYDRATES

The injection of LPS into experimental animals produces initial hyperglycemia, which is followed by a prompt decrease in blood sugar to hypoglycemic levels and a concomitant decrease in liver glycogen.[28,167] In the mouse, the hypoglycemia has been shown to be of sufficient magnitude to cause death.[311] The mechanism of the hypoglycemia appears to be inhibition of the synthesis of glucose from non-carbohydrate sources.[312]

### LIPIDS

The injection of LPS into rabbits produces hyperlipidemia, which is characterized by an early transitory rise in plasma free fatty acids and, at 24 hours, an elevation of serum cholesterol, serum phospholipids and plasma triglycerides.[157] Patients with Gram-negative infections have been shown to have a marked elevation of their plasma triglycerides or free fatty acids.[100]

### PROTEINS AND ENZYMES

Protein synthesis by the liver is stimulated by LPS injection.[315] However, the influence of LPS on specific enzyme activities in the liver indicates a selective effect, as increases and decreases have been demonstrated.[237,331] LPS injection causes an increase in several serum enzymes (lactic dehydrogenase, isocitric dehydrogenase, transaminases, and creatine phosphokinase) secondary to tissue damage.[323]

### MINERALS

Hypoferremia proportional to the dose of LPS has been demonstrated in rats, with the maximum decrease occurring eight to ten hours after injection.[171,174] The total iron-binding capacity of the serum also decreases following LPS administration.[172] Endogenous pyrogen liberated by LPS may in part mediate the hypoferremia, since endogenous pyrogen from rabbit polymorphonuclear leukocytes lowers the serum iron concentration of endotoxin-tolerant rats.[173] In addition, serum zinc concentrations decrease significantly in a dose-dependent response after LPS injection in the rat.[271]

## Effect on the Reticuloendothelial System (RES)

The RES was first implicated in the host response to LPS by Beeson in 1947.[14] While a variety of studies have supported this basic hypothesis, they also have indicated that the relationship between host RES activity and LPS toxicity is complex.[16,347] In general, LPS are potent stimulators of RES activity, as evidenced by enhanced blood clearance of colloidal carbon.[111,140] Such alterations in the activity of the RES have, in turn, been related to many of the biological responses of LPS, such as the enhancement of nonspecific resistance,[111] the adjuvant effects of LPS,[111] the development of endotoxin tolerance,[14] and biological detoxification of these substances. The importance of the RES in LPS metabolism is made evident by the fact that the major portion of an injected dose of LPS can be traced to the Kupffer cells of the liver and to the spleen.[34]

## Effect on the Vascular System

The complex interaction of LPS with the cardiovascular system and the production of shock have been intensively investigated since the syndrome of endotoxin shock in man was first described in 1951.[351] The progress in the understanding of the pathogenesis and of the hemodynamic and biochemical events of endotoxin shock is indicated by the still appallingly high mortality rate of this condition in man.

The syndrome of endotoxin shock begins with Gram-negative septicemia and resultant endotoxemia. The LPS reacts with leukocytes, platelets, complement system, and other serum proteins to increase blood levels of proteolytic enzymes and of certain vasoactive substances, such as histamine, kinins, and serotonin.[149] This results in pooling of the blood, primarily in the pulmonary and splanchnic vasculature.[202] Cardiac output is diminished by the direct myocardial depressant action of LPS and by a decrease in venous return secondary to pooling of blood. The decrease in cardiac output initiates a circular series of events, beginning with a drop in blood pressure and peripheral perfusion. Endogenous catecholamines are liberated to compensate for the falling blood pressure; however, the arteriolar-venular constriction produced by the catecholamines leads to tissue acidosis and anoxia.[291] This initiates capillary congestion and dilatation, which completes the circle by accentuating the decreased venous return and cardiac output. In time, the cellular hypoxia results in lysosomal disruption, leading to cell lysis, propagation of tissue injury, and death.[55]

A concomitant feature of this syndrome is disseminated intravascular coagulation (DIC). Whether DIC is a cause or an effect of endotoxin shock is unknown, but the localized or generalized hemorrhagic diathesis due to the consumption of clotting factors certainly intensifies the shock condition.

Current therapy for endotoxin shock attempts to disrupt the circle of events mentioned above by restoring and maintaining the integrity of the microcirculation. This treatment regimen begins with massive fluid administration and plasma expanders. Although of unproven benefit, *beta*-adrenergic stimulators, *alpha*-adrenergic blockade, and large doses of corticosteroids are frequently administered.[55,325] The DIC can be reversed in most cases with small doses of heparin.[85,278] In man, the most effective therapy is the use of specific antibiotic treatment of the underlying infection.

## Immunological Phenomena

In 1927, White established that the somatic polysaccharides of LPS on the surface of Gram-negative bacteria confer on these organisms their characteristic serologic specificities.[361] This was followed in the early 1930's by the pioneer work of Boivin, who extracted and isolated these LPS and showed that the same complex carried both antigenic and toxic attributes.[33] It has now been clearly established that the repeating oligosaccharide units of the O-polysaccharide portion of the LPS molecule represent, in part or completely, the antigenic determinant group.[204] The grouping of *Salmonella* by the Kauffmann-White

Schema depends on the O-polysaccharide antigen. In general, the O-polysaccharide of LPS is a potent antigen that can elicit antibody formation in animals in submicrogram quantities.[301]

The immune response to LPS in all species studied is quite characteristic.[293] At three to four days after intravenous injection of LPS, circulating antibodies to that LPS are measurable, reaching their maximum concentration at six to eight days.[292] The antibody response occurs after administration of nanogram quantities of LPS in man.[292] Although the initial antibody response is predominantly IgM, IgG and IgA antibodies are present within a few days.[292]

Extensive studies have unequivocally documented the adjuvant effect of LPS. Adjuvancy has been shown by increase of titers of circulating antibodies, increase in the number of antibodies, increase in the number of antibody-producing cells, accelerated antigen eliminations, shortening of the lag period of the immune response, and/or longer persistence of circulating antibodies.[243] This adjuvant effect has been demonstrated in several animal species, including mouse,[89] rat,[273] rabbit,[169] guinea pig,[243] and chicken.[243] On the other hand, LPS can act as an immunosuppressant by manipulation of the dose and time of injection relative to giving the antigen.[223]

## Detoxification

Several methods are known by which the toxic effects of LPS can be altered. These range from biological observations through chemical and physicochemical changes in the LPS molecule. The detoxifying properties of human serum for LPS have been studied in the greatest detail.[192] However, there is experimental evidence for the following two mechanisms of detoxification: (1) enzymatic degradation of the LPS molecule into non-toxic fragments,[192,308,375] and (2) reversible complex formation between LPS and serum proteins, which effects loss of toxicity.[302,303] In addition, acid and alkaline hydrolysis and reduction of LPS particle size result in detoxification.[260]

## Endotoxin Tolerance

Since the classical experiments of Beeson in the rabbit,[12] it has been known that the administration of LPS results in a progressive increase in the resistance to some of its biological effects. This phenomenon, termed endotoxin tolerance, can be readily induced by single or repetitive doses. Tolerance to the pyrogenic and lethal properties of LPS has been most widely studied.

The precise mechanism underlying the development of endotoxin tolerance is unknown. Early experiments suggested that enhanced RES activity effected a more rapid clearance of subsequent LPS injections, thereby producing tolerance.[14] However, other studies have shown that endotoxin tolerance persists following RES blockade,[118,365] that tolerance develops in man without a change in RES activity,[117,299] that passive transfer of tolerance readily occurs,[120,121] and that stimulation of the RES with zymosan or BCG (bacille Calmette Guérin) fails to induce a state of tolerance.[272] Another study has suggested that a 19S *gamma* globulin fraction of serum is responsible for tolerance.[353] Still other investigations have failed to demonstrate a specific antibody as the cause of endotoxin tolerance.[240,366] Recent evidence indicates that hepatic mechanisms dominate the second phase of the biphasic febrile response to LPS and that the tolerant state may be an inability of the Kupffer cells of the liver to produce or release endogenous pyrogen.[119]

## Non-Specific Resistance to Infection

LPS administration can alter resistance to infection by bacteria,[23,75,215,231,296] fungi,[78,182,185,370] viruses,[277,305,350] and parasites[206,319] in several animal species. This change in susceptibility to infection is characterized by a transient increased susceptibility (so-called "negative phase") followed by a prolonged increase in non-specific resistance to infection. The mechanism of the increased resistance to infection secondary to LPS injection is unclear; however, enhanced macrophage and reticulo-endothelial-system activity,[5] granulocytosis,[58] and changes in iron metabolism[78] may be factors.

## Shwartzman Phenomenon

A local tissue reactivity was described in 1928 by Shwartzman, who observed that, if a single injection of a culture filtrate of *Salmonella typhosa* was made into the skin of a rabbit and the same material was injected intravenously twenty-four hours later, a severe hemorrhagic, necrotic lesion developed at the site of the original skin injection.[316] The reaction was non-specific, since the filtrate used for the preparatory skin injection showed the same reaction after the intravenous injection of the filtrate from another organism. The component of the culture filtrate that produced the localized Shwartzman reaction was later shown to be LPS.

The classical generalized Shwartzman reaction (GSR) requires two properly spaced (usually twenty-four hours apart) intravenous injections of LPS into a rabbit. This results in bilateral renal cortical necrosis, due to fibrin deposition secondary to diffuse intravascular clotting. A precipitous drop in the leukocytes and platelets is essential for the reaction.[160] An attempt to define the mechanism of this phenomenon has implicated granulocyte products as the clot-promoting factor;[160] other studies implicate stimulation of $\alpha$-adrenergic receptor sites in the microcirculation as the cause of the formation of fibrin thrombi.[195,209] Heparin effectively blocks the GSR,[160] but the mechanism is unclear.[241]

## Pathology

The injection of a single dose of LPS causes a profound depletion of the lymphoid elements of the spleen and lymph nodes, which persists for several days.[4,223] If an antigen is given at the time of LPS injection, regeneration of the lymphoid tissue is evident after twenty-four hours.[223]

In the liver, LPS causes vesiculation of hepatocytes with progressive necrosis in the centrilobular zone and congestion of the central veins and sinusoids.[201] In addition, there is progressive increase in the number, size and variety of lysosomes, depletion of glycogen, swelling of the Kupffer cells, and fibrin deposition in the sinusoids.[66] In the lung there is hemorrhage into the alveolar wall with clumping of polymorphonuclear leukocytes in the pulmonary capillaries.[66] Examination of the myocardium by electron microscopy shows marked distortion of the intercalated discs.[66]

### REFERENCES


1. Adam, G. A., Extraction of Lipopolysaccharides from Gram-Negative Bacteria with Dimethyl Sulfoxide, *Can. J. Biochem., 45,* 422 (1967).
2. Adams, G. A., and Singh, P. P., The Chemical Constitution of Lipid A from *Serratia marcescens, Can. J. Biochem., 48,* 55 (1970).
3. Adams, G. A., and Singh, P. P., Structural Features of Lipid A Preparations Isolated from *Escherichia coli* and *Shigella flexneri, Biochim. Biophys. Acta, 202,* 553 (1970).
4. André-Schwartz, J., Rubenstein, H. S., and Coons, A. H., Electron Microscopy of Cellular Responses Following Immunization with Endotoxin, *Amer. J. Pathol., 53,* 331 (1968).
5. Arredondo, M. I., and Kampschmidt, R. F., Effect of Endotoxin on Phagocytic Activity of the Reticuloendothelial System of the Rat, *Proc. Soc. Exp. Biol. Med., 112,* 78 (1963).
6. Atkins, E. A., Pathogenesis of Fever, *Physiol. Rev., 40,* 580 (1960).
7. Atkins, E. A., and Bodel, P., Fever, *N. Engl. J. Med., 286,* 27 (1972).
8. Atkins, E. A., and Wood, W. B., Jr., Studies on the Pathogenesis of Fever, I, The Presence of Transferable Pyrogen in the Blood Stream Following Injection of Typhoid Vaccine, *J. Exp. Med., 101,* 519 (1955).
9. Atkins, E. A., and Wood, W. B., Jr., Studies on the Pathogenesis of Fever, II, Identification of an Endogenous Pyrogen in the Blood Stream Following the Injection of Typhoid Vaccine, *J. Exp. Med., 102,* 499 (1955).
10. Baker, P. J., and Wilson, J. B., Hypoferremia in Mice and Its Application to the Bioassy of Endotoxin, *J. Bacteriol., 90,* 903 (1965).

11. Bayer, M., and Anderson, T. F., The Surface Structure of *Escherichia coli, Proc. Nat. Acad. Sci. U.S.A., 54,* 1592 (1965).
12. Beeson, P. B., Development of Tolerance to Typhoid Bacterial Pyrogen and Its Abolition by Reticuloendothelial Blockade, *Proc. Soc. Exp. Biol. Med., 61,* 248 (1946).
13. Beeson, P. B., Tolerance to Bacterial Pyrogens, I, Factors Influencing Its Development, *J. Exp. Med., 86,* 29 (1947).
14. Beeson, P. B., Tolerance to Bacterial Pyrogens, II, Role of the Reticuloendothelial System, *J. Exp. Med., 86,* 39 (1947).
15. Bell, M. L., Herman, A. H., Smith, E. E., Egdahl, R. H., and Rutenburg, A. M., Role of Lysosomal Instability in the Development of Refractory Shock, *Surgery, 70,* 341 (1971).
16. Benacerraf, B., and Sebestyen, M. M., Effect of Bacterial Endotoxins on the Reticuloendothelial System, *Fed. Proc., 16,* 860 (1957).
17. Bennett, I. L., Jr., Pathogenesis of Fever, *Bull. N.Y. Acad. Med., 37,* 440 (1961).
18. Bennett, I. L., Jr., and Beeson, P. B., Pathogenesis of Fever, *Med. Sci. Law, 29,* 365 (1950).
19. Bennett, I. L., Jr., Petersdorf, R. G., and Keene, W. R., Pathogenesis of Fever: Evidence for Direct Cerebral Action of Bacterial Endotoxins, *Trans. Ass. Amer. Physicians Philadelphia, 70,* 64 (1957).
20. Berczi, I., Bertok, L., and Bereznay, T., Comparative Studies on the Toxicity of *Escherichia coli* Lipopolysaccharide Endotoxin in Various Animal Species, *Can. J. Microbiol., 12,* 1070 (1966).
21. Berger, F. M., The Effect of Endotoxin on Resistance to Infection and Disease, *Advan. Pharmacol., 5,* 19 (1967).
22. Berger, F. M., Fukui, G. M., Gustafson, R. H., and Rosselet, J. P., Studies on the Mechanism of Protodyne-Induced Protection against Microbial Infections, *Proc. Soc. Exp. Biol. Med., 138,* 391 (1971).
23. Berger, F. M., Fukui, G. M., Ludwig, B. J., and Rosselet, J. P., Increased Host Resistance to Infection Elicited by Lipopolysaccharides from *Brucella abortus, Proc. Soc. Exp. Biol. Med., 131,* 1376 (1969).
24. Bergman, S., and Nilsson, S. B., Effect of Endotoxin on Embryonal Chick Fibroblasts Cultured in Monolayer, *Acta Pathol. Microbiol. Scand., 59,* 161 (1963).
25. Berk, J. L., Hagen, J. F., Beyer, W. H., Gerber, M. J., and Dochat, G. R., The Treatment of Endotoxin Shock by Beta Adrenergic Blockade, *Ann. Surg., 169,* 74 (1969).
26. Berry, L. J., and Smythe, D. S., Effects of Bacterial Endotoxins on Metabolism, VII, Enzyme Induction and Cortisone Protection, *J. Exp. Med., 120,* 721 (1964).
27. Berry, L. J., Smythe, D. S., and Colwell, L. S., Inhibition of Hepatic Enzyme Induction as a Sensitive Assay for Endotoxin, *J. Bacteriol., 96,* 1191 (1968).
28. Berry, L. J., Smythe, D. S., and Young, L. G., Effects of Bacterial Endotoxin on Metabolism, I, Carbohydrate Depletion and the Protective Role of Cortisone, *J. Exp. Med., 110,* 389 (1959).
29. Bladen, H. A., Gewurz, H., and Mergenhagen, S. E., Interactions of the Complement System with the Surface and Endotoxin Lipopolysaccharide of *Veillonella alcalescens, J. Exp. Med., 125,* 767 (1967).
30. Blattberg, B., and Levy, M. N., Nature of Bradycardia Evoked by Bacterial Endotoxin, *Amer. J. Physiol., 216,* 249 (1969).
31. Blattberg, B., and Levy, M. N., Early Hepatic and Extraheptic Pooling in Response to Endotoxin, *Amer. J. Physiol., 219,* 460 (1970).
32. Boggs, D. R., Chervenick, P. A., Marsh, J. C., Cartwright, G. E., and Wintrobe, M. M., Neutrophil-Releasing Activity in Plasma of Dogs Injected with Endotoxin, *J. Lab. Clin. Med., 72,* 177 (1968).
33. Boivin, A., Mesrobeanu, I., and Mesrobeanu, L., Technique pour la Préparation des Polysaccharides Microbiens Spécifiques, *C. R. Séances Soc. Biol. Filiales, 113,* 490 (1933).
34. Braude, A. I., Absorption, Distribution and Elimination of Endotoxins and Their Derivatives, in *Bacterial Endotoxins,* p. 98, M. Landy and W. Braun, Eds. Rutgers University Press, New Brunswick, New Jersey (1964).
35. Braude, A. I., and Siemienski, J., The Influence of Endotoxin on Resistance to Infection, *Bull. N.Y. Acad. Med., 37,* 448 (1961).
36. Braun, W., and Kessel, R. W. I., Cytotoxicity of Endotoxins in Relation to Their Effects on Host Resistance, in *Bacterial Endotoxins,* p. 397, M. Landy and W. Braun, Eds. Rutgers University Press, New Brunswick, New Jersey (1964).
37. Brockman, S. K., Thomas, C. S., and Vasko, J. S., The Effect of *Escherichia coli* Endotoxin on the Circulation, *Surg. Gynecol. Obstet. Int. Abstr. Surg., 125,* 763 (1967).
38. Bryant, R. E., Des Prez, R. M., and Rogers, D. E., Studies on Human Leukocyte Mobility, II, Effects of Bacterial Endotoxin on Leukocyte Migration, Adhesiveness, and Aggregation, *Yale J. Biol. Med., 40,* 192 (1968).
39. Burch, G. E., Giles, T. D., Quiroz, A. C., and Shen, Y., Influence of *E. coli* Endotoxin on the Pulmonary Veins, Circulatory System and Work and Tone of the Heart of Intact Dog, *Cardiologia, 53,* 77 (1968).
40. Burton, A. J., and Carter, H. E., Purification and Characterization of the Lipid A Component of the Lipopolysaccharides from *Escherichia coli, Biochemistry, 3,* 411 (1964).
41. Buskirk, E. R., Thompson, R. H., Rubenstein, M., and Wolff, S. M., Heat Exchange in Men and Women Following Intravenous Injection of Endotoxin, *J. Appl. Physiol., 19,* 907 (1964).
42. Campbell, P. A., Rowlands, D. T., Jr., Harrington, M. J., and Kind, P. D., The Adjuvant Action of Endotoxin in Thymectomized Mice, *J. Immunol., 96,* 849 (1966).

43. Carey, F. J., Braude, A. I., and Zalesky, M., Studies with Radioactive Endotoxin, III, The Effect of Tolerance on the Distribution of Radioactivity after Intravenous Injection of *Escherichia coli* Endotoxin Labeled with $Cr^{51}$, *J. Clin. Invest., 37,* 441 (1958).
44. Carretero, O. A., Nasjletti, A., and Fasciolo, J. C., Kinins, and Kininogen in Endotoxin Shock, *Experientia, 26,* 63 (1970).
45. Cavanagh, D., and Rao, P. S., Endotoxin Shock in the Subhuman Primate, I, Hemodynamic and Biochemical Changes, *Arch. Surg., 99,* 107 (1969).
46. Cavanagh, D., Rao, S. P., Sutt, D. M., Bhagat, B. D., and Bachmann, F., Pathophysiology of Endotoxin Shock in the Primate, *Amer. J. Obstet. Gynecol., 108,* 705 (1970).
47. Chedid, L., Boyer, F., and Parant, M., Etude de l'Action Abortive des Endotoxines Injectées a la Souris Gravide Normale, Castrée ou Hypophysectomisée, *Ann. Inst. Pasteur (Paris), 102,* 77 (1962).
48. Chedid, L., Parant, M., Boyer, F., and Skarnes, R. C., Nonspecific Host Responses in Tolerance to the Lethal Effect of Endotoxins, in *Bacterial Endotoxins,* p. 500, M. Landy and W. Braun, Eds. Rutgers University Press, New Brunswick, New Jersey (1964).
49. Chedid, L., Lamensans, A., and Prixova, J., Comparison of the Effects of an Antibacterial Leukocytic Extract and the Serum Endotoxin-Detoxifying Component of Lipopolysaccharides Extracted from Rough and Smooth *Salmonella, J. Infect. Dis., 121,* 634 (1970).
50. Chedid, L., Skarnes, R. C., and Parant, M., Characterization of a $^{51}$Cr-Labelled Endotoxin and Its Identification in Plasma and Urine after Parenteral Administration, *J. Exp. Med., 117,* 561 (1963).
51. Chernokhvostova, E., Luxemburg, K. I., Starshinova, V., Andreeva, N., and German, G., Study on the Production of IgG, IgA, and IgM Antibodies to Somatic Antigens of *Salmonella typhi* in Humans, *Clin. Exp. Immunol., 4,* 407 (1969).
52. Chien, S., Chang, C., Dellenback, J., Usami, S., and Gregsen, M. I., Hemodynamic Changes in Endotoxin Shock, *Amer. J. Physiol., 210,* 1401 (1966).
53. Chien, S., Dellenback, J., Usami, S., Treitel, K., Chang, C., and Gregsen, M. I., Blood Volume and Its Distribution in Endotoxin Shock, *Amer. J. Physiol., 210,* 1411 (1966).
54. Chien, S., Sinclair, D. G., Dellenback, R. J., Chang, C., Peric, B., Usami, S., and Gregsen, M. I., Effect of Endotoxin on Capillary Permeability to Macromolecules, *Amer. J. Physiol., 207,* 518 (1964).
55. Christy, J. H., Pathophysiology of Gram-Negative Shock, *Amer. Heart J., 81,* 694 (1971).
56. Ciznar, I., and Shands, J. W., Jr., Effect of Alkali-Treated Lipopolysaccharide on Erythrocyte Membrane Stability, *Infect. Immun., 4,* 362 (1971).
57. Cline, M. J., Melmon, K. L., David, W. C., and Williams, H. E., Mechanism of Endotoxin Interaction with Human Leukocytes, *Brit. J. Haematol., 15,* 539 (1968).
58. Cluff, L. E., Effects of Endotoxins on Susceptibility to Infections, *J. Infect. Dis., 122,* 205 (1970).
59. Cooper, J. F., Levin, J., and Wagner, H. N., Jr., Quantitative Comparison of *in vitro* and *in vivo* Methods for the Detection of Endotoxin, *J. Lab. Clin. Med., 78,* 138 (1971).
60. Corrigan, J. J., Jr., Effect of Anticoagulating and Non-Anticoagulating Concentrations of Heparin on the Generalized Shwartzman Reaction, *Thromb. Diath. Haemorrh., 24,* 136 (1970).
61. Corrigan, J. J., Jr., Abildgaard, C. F., Vanderheiden, J. F., and Schulman, I., Quantitative Aspects of Blood Coagulation in the Generalized Shwartzman Reaction, I, Effects of Variation of Preparation and Provocative Dose of *E. coli* Endotoxin, *Pediat. Res., 1,* 39 (1967).
62. Cundy, K. R., and Nowotny, A., Quantitative Comparison of Toxicity Parameters of Bacterial Endotoxins, *Proc. Soc. Exp. Biol. Med., 127,* 999 (1968).
63. Dankert, M. A., Wright, A., Kelley, W. S., and Robbins, P. W., Isolation, Purification and Properties of the Lipid-Linked Intermediates of O-Antigen Biosynthesis, *Arch. Biochem. Biophys., 116,* 425 (1966).
64. Davis, R. B., Bailey, W. L., and Hanson, N. P., Modification of Serotonin and Histamine Release after *E. coli* Endotoxin Administration, *Amer. J. Physiol., 205,* 560 (1963).
65. Davis, R. B., Meeker, W. R., Jr., and Bailey, W. L., Serotonin Release by Bacterial Endotoxin, *Proc. Soc. Exp. Biol. Med., 108,* 774 (1961).
66. Depalma, R. G., Coil, J., David, J. H., and Holden, W. D., Cellular and Ultrastructural Changes in Endotoxemia: A Light and Electron Microscopic Study, *Surgery, 62,* 505 (1967).
67. Des Prez, R. M., Effects of Bacterial Endotoxin on Rabbit Platelets, III, Comparison of Platelet Injury Induced by Thrombin and by Endotoxin, *J. Exp. Med., 120,* 305 (1964).
68. Des Prez, R. M., and Bryant, R. E., Effects of Bacterial Endotoxin on Rabbit Platelets, IV, The Divalent Ion Requirements of Endotoxin Induced and Immunologically Induced Platelet Injury, *J. Exp. Med., 124,* 971 (1966).
69. Dinarello, C. A., Bodel, P. T., and Atkins, E., The Role of the Liver in the Production of Fever and in Pyrogenic Tolerance, *Trans. Ass. Amer. Physicians Philadelphia, 81,* 334 (1968).
70. Donald, W. D., Winkler, C. H., and Hare, K., Investigations on the Mechanisms of the Leukopenic Response to *Shigella* Endotoxin, *J. Clin. Invest., 37,* 1100 (1958).
71. Dowling, J. N., and Feldman, H. A., Quantitative Biological Assay of Bacterial Endotoxins, *Proc. Soc. Exp. Biol. Med., 134,* 861 (1970).

72. Dröge, W., Lehmann, V., Lüderitz, O., and Westphal, O., Structural Investigations on KDO Region of Lipopolysaccharides, *Eur. J. Biochem., 14,* 175 (1970).
73. Dröge, W., Lüderitz, O., and Westphal, O., Biochemical Studies on Lipopolysaccharides of *Samonella* R Mutants, III, The Linkage of the Heptose Units, *Eur. J. Biochem., 4,* 126 (1968).
74. Dröge, W., Ruschmann, E., Lüderitz, O., and Westphal, O., Biochemical Studies on Lipopolysaccharides of *Salmonella* R Mutants, IV, Phosphate Groups Linked to Heptose Units and Their Absence in Some R Lipopolysaccharides, *Eur. J. Biochem., 4,* 134 (1968).
75. Dubos, R. J., Schaedler, R. W., and Bohome, D., Effects of Bacterial Endotoxins on Susceptibility to Infection with Gram-Positive and Acid-Fast Bacteria, *Fed. Proc., 16,* 856 (1957).
76. Edstrom, R. D., and Heath, E. C., The Biosynthesis of Cell Wall Lipopolysaccharide in *Escherichia coli,* VI, Enzymatic Transfer of Galactose, Glucose, N-Acetyl-glucosamine and Colitose into the Polymer, *J. Biol. Chem., 242,* 3581 (1967).
77. Elekes, E., Meretey, K., and Kocsar, L., The Action of Aluminium Hydroxide or Endotoxin on Natural Sheep Haemolysin Producing Cells in Rats, *Pathol. Microbiol., 32,* 345 (1968).
78. Elin, R. J., and Wolff, S. M., Iron and Endotoxin Induced Nonspecific Resistance to Infection, *Fed. Proc., 31,* 802 (1972).
79. Erdos, E. G., and Miwa, I., Effect of Endotoxin Shock on the Plasma Kallikrein-Kinin System of the Rabbit, *Fed. Proc., 27,* 92 (1968).
80. Erlondson, A. L., Jr., The Induction of Nonspecific Resistance in Mice by Papain, *J. Infect. Dis., 116,* 297 (1966).
81. Evensen, S. A., Jeremic, M., and Hjort, P. F., The Effect of Endotoxin on Factor VII in Rats: *in vivo* and *in vitro* Observations, *Scand. J. Clin. Lab. Invest., 18,* 509 (1966).
82. Fekety, R. F., Heat Balance and Reactivity to Endotoxin, *Amer. J. Physiol., 204,* 719 (1963).
83. Filkins, J. P., Hepatic Vascular Response to Endotoxin, *Proc. Soc. Exp. Biol. Med., 131,* 1235 (1969).
84. Filkins, J. P., Comparison of Endotoxin Detoxification by Leukocytes and Macrophages, *Proc. Soc. Exp. Biol. Med., 137,* 1396 (1971).
85. Filkins, J. P., and DiLuzio, N. R., Heparin Protection in Endotoxin Shock, *Amer. J. Physiol., 214,* 1074 (1968).
86. Finger, H., Beneke, G., and Fresenius, H., Cellular Kinetics of 19S and 7S Hemolysin Production in Mice under the Influence of Bacterial Endotoxins, *Pathol. Microbiol., 35,* 324 (1970).
87. Finger, H., Emmerling, P., and Busse, M., Increased Priming for the Secondary Response in Mice to Sheep Erythrocytes by Bacterial Endotoxins, *Int. Arch. Allergy Appl. Immunol., 38,* 598 (1970).
88. Floersheim, G. L., Suppression of Cellular Immunity by Gram-Negative Bacteria, *Antibiot. Chemother., 15,* 407 (1969).
89. Franzl, R. E., and McMaster, P. D., The Primary Immune Response in Mice, I, The Enhancement and Suppression of Hemolysin Production by a Bacterial Endotoxin, *J. Exp. Med., 127,* 1087 (1968).
90. Freedman, H. H., Fox, A. E., and Schwartz, B. S., Antibody Formation at Various Times after Previous Treatment of Mice with Endotoxin, *Proc. Soc. Exp. Biol. Med., 125,* 583 (1967).
91. Fritz, H., and Nordenfelt, E., Endotoxin-Induced Cytotoxicity of Rabbit Serum, *Acta Pathol. Microbiol. Scand., 75,* 631 (1969).
92. Frohlich, E. D., Effect of *Salmonella typhosa* Endotoxin on Perfused Dog Spleen, *Proc. Soc. Exp. Biol. Med., 113,* 559 (1963).
93. Frohman, L. A., Horton, E. S., and Lebovitz, H. E., Growth Hormone Releasing Action of a *Pseudomonas* Endotoxin (Piromen), *Metabolism, 16,* 57 (1967).
94. Fruhman, G. J., Mobilization of Neutrophils into Peritoneal Fluid Following Intraperitoneal Injection of Bacterial Endotoxins, *Proc. Soc. Exp. Biol. Med., 102,* 423 (1959).
95. Fruhman, G. J., Bacterial Endotoxin: Effects on Erythropoiesis, *Blood, 27,* 363 (1966).
96. Fruhman, G. J., Endotoxin-Induced Shunting of Erythropoiesis in Mice, *Amer. J. Physiol., 212,* 1095 (1967).
97. Fukuda, T., and Hata, N., Mechanisms of Endotoxin Shock in Rats and the Anti-Endotoxic Effect of Glucocorticoids and Endotoxin-Conditioning, *Jap. J. Physiol., 19,* 509 (1969).
98. Galanos, C., Lüderitz, O., and Westphal, O., A New Method for the Extraction of R Lipopolysaccharides, *Eur. J. Biochem., 9,* 245 (1969).
99. Galanos, C., Rietschel, E. T., Lüderitz, O., and Westphal, O., Interaction of Lipopolysaccharides and Lipid A with Complement, *Eur. J. Biochem., 19,* 143 (1971).
100. Gallin, J. I., Kaye, D., and O'Leary, W. M., Serum Lipids in Infection, *N. Engl. J. Med., 281,* 1081 (1969).
101. Gallin, J. I., O'Leary, W. M., and Kaye, D., Serum Concentrations of Lipids in Rabbits Infected with *Escherichia coli* and *Staphylococcus aureus, Proc. Soc. Exp. Biol. Med., 133,* 309 (1970).
102. Gery, I., Krüger, J., and Spiesel, S. Z., Stimulation of B-Lymphocytes by Endotoxin. Reactions of Thymus-Deprived Mice and Karyotypic Analysis of Dividing Cells in Mice Bearing $T_6$ $T_6$ Thymus Grafts, *J. Immunol., 108,* 1088 (1972).
103. Gewurz, H., Mergenhagen, S. E., Nowotny, A., and Philips, J. K., Interactions of the Complement System with Native and Chemically Modified Endotoxins, *J. Bacteriol., 95,* 397 (1968).
104. Gilbert, R. P., Mechanisms of the Hemodynamic Effects of Endotoxin, *Physiol. Rev., 40,* 245 (1960).

105. Gimber, P. E., and Rafter, G. W., The Interaction of *Escherichia coli* Endotoxin with Leukocytes, *Arch. Biochem. Biophys., 135,* 14 (1969).
106. Gingold, J. L., and Freedman, H. H., Inhibition by Endotoxin of the Migration of Peritoneal Exudate Cells from Endotoxin Sensitive Mice, *Proc. Soc. Exp. Biol. Med., 128,* 599 (1968).
107. Gmeiner, J., Lüderitz, O., and Westphal, O., Biochemical Studies on Lipopolysaccharides of *Salmonella* R Mutants, VI, Investigations on the Structure of the Lipid A Component, *Eur. J. Biochem., 7,* 370 (1969).
108. Gmeiner, J., Simon, M., and Lüderitz, O., The Linkage of Phosphate Groups and of 2-Keto-3-deoxyoctonate to the Lipid A Component in a *Salmonella minnesota* Lipopolysaccharide, *Eur. J. Biochem., 21,* 355 (1971).
109. Goldstein, I. M., Wünschmann, B., Astrup, T., and Henderson, E. S., Effects of Bacterial Endotoxin on the Fibrinolytic Activity of Normal Human Leukocytes, *Blood, 37,* 447 (1971).
110. Golub, S., Groschel, D. H. M., and Nowotny, A., RES Uptake of Endotoxin in Mice, *Bacteriol. Proc.,* p. 79 (1967).
111. Golub, S., Groschel, D. H. M., and Nowotny, A., Factors Which Affect the Reticuloendothelial System Uptake of Bacterial Endotoxins, *J. Reticuloendothelial Soc., 5,* 324 (1968).
112. Golub, S., Groschel, D. H. M., and Nowotny, A., Studies on the Opsonization of a Bacterial Endotoxoid, *J. Reticuloendothel. Soc., 7,* 518 (1970).
113. Golub, E. S., and Weigle, W. O., Studies on the Induction of Immunologic Unresponsiveness, I, Effects of Endotoxin and Phytohemagglutinin, *J. Immunol., 98,* 1241 (1967).
114. Good, R. A., and Thomas, L., Studies on the Generalized Shwartzman Reaction, IV, Prevention of the Local and Generalized Shwartzman Reactions with Heparin, *J. Exp. Med., 97,* 871 (1953).
115. Gourzis, J. T., Hollenberg, M. W., and Nickerson, M., Involvement of Adrenergic Factors in the Effects of Bacterial Endotoxin, *J. Exp. Med., 114,* 593 (1961).
116. Greisman, S. E., Activation of Histamine-Releasing Factor in Normal Rat Plasma by *E. coli* Endotoxin, *Proc. Soc. Exp. Biol. Med., 103,* 628 (1960).
117. Greisman, S. E., Wagner, H. N., Iio, M., and Hornick, R. B., Mechanisms of Endotoxin Tolerance, II, Relationship Between Endotoxin Tolerance and Reticuloendothelial System Phagocytic Activity in Man, *J. Exp. Med., 119,* 241 (1964).
118. Greisman, S. E., Carozza, F. A., Jr., and Hill, J. D., Mechanisms of Endotoxin Tolerance, I, Relationship Between Tolerance and Reticuloendothelial System Phagocytic Activity in the Rabbit, *J. Exp. Med., 117,* 663 (1963).
119. Greisman, S. E., and Woodward, C. L., Mechanisms of Endotoxin Tolerance, VII, The Role of the Liver, *J. Immunol., 105,* 1468 (1970).
120. Greisman, S. E., and Woodward, W. E., Mechanisms of Endotoxin Tolerance, III, The Refractory State During Continuous Intravenous Infusions of Endotoxin, *J. Exp. Med., 121,* 911 (1965).
121. Greisman, S. E., and Young, E. J., Mechanisms of Endotoxin Tolerance, VI, Transfer of the "Anamnestic" Tolerance Response with Primed Spleen Cells, *J. Immunol., 103,* 1237 (1969).
122. Greisman, S. E., Young, E. J., and Carozza, F. A., Jr., Mechanisms of Endotoxin Tolerance, V, Specificity of the Early and Late Phases of Pyrogenic Tolerance, *J. Immunol., 103,* 1223 (1969).
123. Grollman, A. P., and Osborn, M. J., O-Phosphorylethanolamine: A Component of Lipopolysaccharide in Certain Gram-Negative Bacteria, *Biochemistry, 3,* 1571 (1964).
124. Gupta, J. D., and Reed, C. E., The Direct Reaction Between *Salmonella enteritidis* Endotoxin and Antibody – Measurement of 7S and 19S Antibody in Normal, Tolerant, and Immune Sera, *J. Immunol., 98,* 1093 (1967).
125. Gupta, J. D., and Reed, C. E., Distribution and Degradation of Sublethal Doses of $I^{125}$ Labeled Endotoxin from *Salmonella enteritidis* in Mice, *Proc. Soc. Exp. Biol. Med., 131,* 481 (1969).
126. Hahn, H., Char, D. C., Postel, W. B., and Wood, W. B., Studies on the Pathogenesis of Fever, XV, The Production of Endogenous Pyrogen by Peritoneal Macrophages, *J. Exp. Med., 126,* 385 (1967).
127. Hammerling, G., Lüderitz, O., and Westphal, O., Structural Investigations of the Core Polysaccharide of *Salmonella typhimurium* Mode of Attachment to the O-Specific Chains, *Eur. J. Biochem., 15,* 48 (1970).
128. Hamosh, M., and Shapiro, B., The Mechanism of Glycogenolytic Action of Endotoxin, *Brit. J. Exp. Pathol., 41,* 372 (1960).
129. Han, S. S., Johnson, A. G., and Han, I. H., The Antibody Response in the Rat, I, A Histometric Study of the Spleen following a Single Injection of Bovine Gamma Globulin with and without Endotoxin, *J. Infect. Dis., 115,* 149 (1965).
130. Handler, E. S., Varsa, E. E., and Gordon, A. S., Mechanisms of Leukocyte Production and Release, V, Studies on the Leukotosis-Inducing Factor in the Plasma of Rats Treated with Typhoid-Paratyphoid Vaccine, *J. Lab. Clin. Med., 67,* 398 (1966).
131. Harcourt, K. F., Robertson, R. D., and Fletcher, W. S., Effect of Pretreatment with Endotoxin on the Response of Skin to Irradiation in the Rabbit, *Cancer, 21,* 812 (1968).
132. Hardaway, R. M., and Johnson, D., Clotting Mechanism in Endotoxin Shock, *Arch. Intern. Med., 112,* 775 (1963).
133. Harrison, L. H., Beller, J. J., Hinshaw, L. B., Coalson, J. J., and Greenfield, L. J., Effects of Endotoxin on Pulmonary Capillary Permeability, Ultra-Structure and Surfactant, *Surg. Gynecol. Obstet., 129,* 723 (1969).
134. Harrison, L. H., Jr., Hinshaw, L. B., Coalson, J. J., and Greenfield, L. J., Effects of *E. coli* Septic Shock on Pulmonary Hemodynamics and Capillary Permeability, *J. Thorac. Cardiovasc. Surg., 61,* 795 (1971).
135. Haskins, W. T., Landy, M., Milner, K. C., and Ribi, E., Biological Properties of Parent Endotoxins and Lipoid Fractions with a Kinetic Study of Acid-Hydrolyzed Endotoxin, *J. Exp. Med., 114,* 665 (1961).

136. Havas, H. F., Groesbeck, M. E., and Donnelly, A. J., Mixed Bacterial Toxins in the Treatment of Tumors, I, Methods of Preparation and Effects on Normal and Sarcoma 37 Bearing Mice, *Cancer Res., 18,* 141 (1958).
137. Heath, E. C., Mayer, R. M., Edstrom, R. D., and Beaudreau, C. A., Structure and Biosynthesis of the Cell Wall Lipopolysaccharide of *Escherichia coli, Ann. N.Y. Acad. Sci., 133,* 315 (1971).
138. Heilman, D. H., Cellular Aspects of the Action of Endotoxin: The Role of the Macrophage, in *Bacterial Endotoxins,* p. 610, M. Landy and W. Braun, Eds. Rutgers University Press, New Brunswick, New Jersey (1964).
139. Heilman, D. H., *In vitro* Studies on Changes in the Reticuloendothelial System of Rabbits after an Injection of Endotoxin, *J. Reticuloendothelial Soc., 2,* 89 (1965).
140. Heilman, D. H., Effect of Dosage on Endotoxin-Induced Changes in the Reticuloendothelial System of Rabbits, *J. Reticuloendothelial Soc., 2,* 273 (1965).
141. Heilman, D. H., Selective Toxicity of Endotoxin for Phagocytic Cells of the Reticuloendothelial System, *Int. Arch. Allergy Appl. Immunol., 26,* 63 (1965).
142. Heilman, D. H., and Bast, R. C., Jr., *In vitro* Assay of Endotoxin by the Inhibition of Macrophage Migration, *J. Bacteriol., 93,* 15 (1967).
143. Hellerqvist, C. G., and Lindberg, A. A., Structural Studies on the Common Core Polysaccharide of the Cell Wall Lipopolysaccharide from *Salmonella typhimurium, Carbohyd. Res., 16,* 39 (1971).
144. Hellerqvist, C. G., Lindberg, B., Svensson, S., Holme, T., and Lindberg, A. A., Structural Studies on the O-Specific Side-Chains of the Cell Wall Lipopolysaccharide from *Salmonella typhimurium* 395 MS, *Carbohyd. Res., 8,* 43 (1968).
145. Hellerqvist, C. G., Lindberg, B., Svensson, S., Holme, T., and Lindberg, A. A., Structural Studies on the O-Specific Side-Chains of the Cell Wall Lipopolysaccharide from *Salmonella typhimurium* LT2, *Carbohyd. Res., 9,* 237 (1969).
146. Herring, W. B., Herion, J. C., Walker, R. I., and Palmer, J. G., Distribution and Clearance of Circulating Endotoxin, *J. Clin. Invest., 42,* 79 (1963).
147. Hildebrand, G. J., Ng, J., Seys, Y., and Madin, S. H., Differential between Pathogenic Mechanisms of Early and Late Phase of Endotoxin Shock, *Amer. J. Physiol., 210,* 1451 (1966).
148. Hinshaw, L. B., Archer, L. T., Greenfield, L. J., and Guenter, C. A., Effects of Endotoxin on Myocardial Hemodynamics, Performance, and Metabolism, *Amer. J. Physiol., 221,* 504 (1971).
149. Hinshaw, L. B., Brake, C. M., and Emerson, T. E., Jr., Biochemical and Pathologic Alterations in Endotoxin Shock, in *Shock and Hypotension: Pathogenesis and Treatment,* p. 431, L. J. Mills and J. H. Moyer, Eds. Grune and Stratton, New York (1965).
150. Hinshaw, L. B., Emerson, T. E., Jr., and Reins, D. A., Cardiovascular Responses of the Primate in Endotoxin Shock, *Amer. J. Physiol., 210,* 335 (1966).
151. Hinshaw, L. B., Greenfield, L. J., Archer, L. T., and Guenter, C. A., Effects of Endotoxin on Myocardial Hemodynamics, Performance, and Metabolism During Beta Adrenergic Blockade, *Proc. Soc. Exp. Biol. Med., 137,* 1217 (1971).
152. Hinshaw, L. B., Jordan, M. M., and Vick, J. A., Histamine Release and Endotoxin Shock in the Primate, *J. Clin. Invest., 40,* 1631 (1961).
153. Hinshaw, L. B., and Nelson, D. L., Venous Response of Intestine to Endotoxin, *Amer. J. Physiol., 203,* 870 (1962).
154. Hinshaw, L. B., Reins, D. A., and Hill, R. J., Response of Isolated Liver to Endotoxin, *Can. J. Physiol. Pharmacol., 44,* 529 (1966).
155. Hinshaw, L. B., Solomon, L. A., Freeny, P. C., and Reins, D. A., Endotoxin Shock, *Arch. Surgery, 94,* 61 (1967).
156. Hinshaw, L. B., Vick, J. A., Jordan, M. M., and Wittmers, L. E., Vascular Changes Associated with Development of Irreversible Endotoxin Shock, *Amer. J. Pathol., 202,* 103 (1962).
157. Hirsch, R. L., MacKay, D. G., and Travers, A. I., Hyperlipidemia, Fatty Liver, and Bromsulfophthalein Retention in Rabbits Injected Intravenously with Bacterial Endotoxins, *J. Lipid Res., 5,* 563 (1964).
158. Ho, M., Interferon-Like Viral Inhibitor in Rabbits after Intravenous Administration of Endotoxin, *Science, 146,* 1472 (1964).
159. Hook, W. A., Snyderman, R., and Mergenhagen, S. E., Histamine-Releasing Factor Generated by the Interaction of Endotoxin with Hamster Serum, *Infect. Immun., 2,* 462 (1970).
160. Horn, R. G., and Collins, R. D., Studies on the Pathogenesis of the Generalized Shwartzman Reaction, The Role of Granulocytes, *Lab. Invest., 18,* 101 (1968).
161. Horowitz, H. I., Des Prez, R. M., and Hook, E. W., Effect of Bacterial Endotoxin on Rabbit Platelets Factor 3 Activity *in vitro* and *in vivo, J. Exp. Med., 116,* 619 (1962).
162. Humphreys, G. O., and Meadow, P. M., The Biosynthesis of Lipid A in *Pseudomonas, J. Gen. Microbiol., 68,* 5 (1971).
163. Ikawa, M., Koepfli, J. B., Mudd, S. G., and Niemann, C., An Agent from *E. coli* Causing Hemorrhage and Regression of an Experimental Mouse Tumor, I, Isolation and Properties, *J. Nat. Cancer Inst., 13,* 157 (1952).
164. Jacobson, E. D., Dooley, E. S., Scott, J. B., and Frohlich, E. D., Effects of Endotoxin on the Hemodynamics of the Stomach, *J. Clin. Invest., 42,* 391 (1963).
165. Jacobson, E. D., Mehlman, B., and Kalas, J. P., Vasoactive Mediators as the "Trigger-Mechanism" of Endotoxin Shock, *J. Clin. Invest., 43,* 1000 (1964).
166. Jarvis, F. G., Mesenko, M. T., Martin, D. G., and Perrine, T. D., Physicochemical Properties of the $V_i$ Antigen Before and After Mild Alkaline Hydrolysis, *J. Bacteriol., 94,* 1406 (1967).

167. Jeffries, C. D., Liver Carbohydrate Levels in Mice Treated with Endotoxin, Cortisone, and Elipten, *Proc. Soc. Exp. Biol. Med., 132,* 540 (1969).
168. Johnson, A. G., The Adjuvant Action of Bacterial Endotoxins on the Primary Antibody Response, in *Bacterial Endotoxins,* p. 252, M. Landy and W. Braun, Eds. Rutgers University Press, New Brunswick, New Jersey (1964).
169. Johnson, A. G., Gaines, S., and Landy, M., Studies on the O-Antigen of *Salmonella typhosa,* V, Enhancement of Antibody Response to Protein Antigens by the Purified Lipopolysaccharide, *J. Exp. Med., 103,* 225 (1956).
170. Kadis, S., Weinbaum, G., and Ajl, S. J., Bacterial Endotoxins, in *Microbial Toxins,* Vol. 5, S. Kadis, G. Weinbaum and S. J. Ajl, Eds. Academic Press, New York (1971).
171. Kampschmidt, R. F., and Schultz, G. A., Hypoferremia in Rats Following Injection of Bacterial Endotoxin, *Proc. Soc. Exp. Biol. Med., 106,* 870 (1961).
172. Kampschmidt, R. F., and Upchurch, H. F., Effect of Endotoxin upon Total Iron-Binding Capacity of the Serum, *Proc. Soc. Exp. Biol. Med., 116,* 420 (1964).
173. Kampschmidt, R. F., and Upchurch, H. F., A Comparison of the Effects of Rabbit Endogenous Pyrogen on the Body Temperature of the Rabbit and Lowering of Plasma Iron in the Rat, *Proc. Soc. Exp. Biol. Med., 133,* 128 (1970).
174. Kampschmidt, R. F., Upchurch, H. F., and Eddington, C. L., Hypoferremia Produced by Plasma from Endotoxin-Treated Rats, *Proc. Soc. Exp. Biol. Med., 132,* 817 (1969).
175. Keene, W. R., Detoxification of Bacterial Endotoxin by Soluble Tissue Extracts, *J. Lab. Clin. Med., 60,* 433 (1962).
176. Keene, W. R., Silberman, H. S., and Landy, M., Observations on the Pyrogenic Response and Its Application to the Bioassay of Endotoxin, *J. Clin. Invest., 40,* 295 (1961).
177. Kim, Y. B., and Watson, D. W., Inactivation of Gram-Negative Bacterial Endotoxins by Papain, *Proc. Soc. Exp. Biol. Med., 115,* 140 (1964).
178. Kim, Y. B., and Watson, D. W., Modification of Host Responses to Bacterial Endotoxins, II, Passive Transfer of Immunity to Bacterial Endotoxin with Fractions Containing 19S Antibodies, *J. Exp. Med., 121,* 751 (1965).
179. Kim, Y. B., and Watson, D. W., Role of Antibodies in Reaction to Gram-Negative Bacterial Endotoxins, *Ann. N.Y. Acad. Sci., 133,* 727 (1966).
180. Kimball, H. R., Lipsett, M. B., Odell, W. D., and Wolff, S. M., Comparison of the Effect of the Pyrogens, Etiocholanolone and Bacterial Endotoxin on Plasma Cortisol and Growth Hormone in Man, *J. Clin. Endocrinol. Metab., 28,* 337 (1968).
181. Kimball, H. R., Melmon, K. L., and Wolff, S. M., Endotoxin-Induced Kinin Production in Man, *Proc. Soc. Exp. Biol. Med., 139,* 1078 (1972).
182. Kimball, H. R., Williams, T. W., and Wolff, S. M., Effect of Bacterial Endotoxin on Experimental Fungal Infections, *J. Immunol., 100,* 24 (1968).
183. Kimball, H. R., and Wolff, S. M., Febrile Responses of Rabbits to Bacterial Endotoxin Following Incubation in Homologous Serum and Plasma, *Proc. Soc. Exp. Biol. Med., 124,* 269 (1967).
184. Knox, K. W., Cullen, J., and Work, E., An Extracellular Lipopolysaccharide-Phospholipid-Protein Complex Produced by *Escherichia coli* Grown under Lysine Limiting Conditions, *Biochem. J., 103,* 192 (1967).
185. Kobayashi, H., Yasuhira, K., and Uesaka, I., Effect of *Escherichia coli* and Its Endotoxin on the Resistance of Mice to Experimental Cryptococcal Infection, *Jap. J. Microbiol., 13,* 223 (1969).
186. Kobold, E. E., Lovell, R., Katz, W., and Thal, A. P., Chemical Mediators Released by Endotoxin, *Surg. Gynecol. Obstet., 118,* 807 (1964).
187. Kohler, P. O., O'Malley, B. W., Rayford, P. L., Lipsett, M. B., and Odell, W. D., Effect of Pyrogen on Blood Levels of Pituitary Tropic Hormones: Observations of the Usefulness of the Growth Hormone Response in the Detection of Pituitary Disease, *J. Clin. Endocrinol. Metab., 27,* 219 (1967).
188. Kohler, P. F., and Spink, W. W., Complement in Endotoxin Shock: Effect of Decomplementation by Aggregated Gamma Globulin, *Proc. Soc. Exp. Biol. Med., 117,* 207 (1964).
189. Kuida, H., Gilbert, R. P., Hinshaw, L. B., Brunson, J. G., and Visscher, M. B., Species Differences in Effect of Gram-Negative Endotoxin on Circulation, *Amer. J. Physiol., 200,* 1197 (1961).
190. Landy, M., and Braun, W. (Eds.), *Bacterial Endotoxins.* Rutgers University Press, New Brunswick, New Jersey (1964).
191. Landy, M., Sanderson, R. P., and Jackson, A. L., Humoral and Cellular Aspects of the Immune Response to the Somatic Antigen of *Salmonella enteritidis, J. Exp. Med., 122,* 483 (1965).
192. Landy, M., Trapani, R. J., and Rosen, F. S., Inactivation of Endotoxin by a Humoral Component, VI, Two Separate Systems Required for Viable and Killed *Salmonella typhosa, J. Clin. Invest., 39,* 352 (1960).
193. Landy, M., and Weidanz, W. P., Natural Antibodies Against Gram-Negative Bacteria, in *Bacterial Endotoxins,* p. 275, M. Landy and W. Braun, Eds. Rutgers University Press, New Brunswick, New Jersey (1964).
194. Larson, C., Ribi, E., Milner, K., and Lieberman, J., A Method for Titrating Endotoxic Activity in the Skin of Rabbits, *J. Exp. Med., 111,* 1 (1960).
195. Latour, J. G., Prejean, J. B., and Margaretten, W., Corticosteroids and the Generalized Shwartzman Reaction, Mechanisms of Sensitization in the Rabbit, *Amer. J. Pathol., 65,* 189 (1971).
196. Leive, L., Shovlin, V. K., and Mergenhagen, S. E., Physical, Chemical, and Immunological Properties of Lipopolysaccharide Released from *Escherichia coli* by Ethylenediaminetetraacetate, *J. Biol. Chem., 243,* 6384 (1968).

197. Lequire, V. S., Hutcherson, J. D., Hamilton, R. L., and Gray, M. E., Effects of Bacterial Endotoxin on Lipid Metabolism, *J. Exp. Med., 110,* 293 (1959).
198. Levin, J., and Bang, F. B., Clottable Protein in *Limulus:* Its Localization and Kinetics of Its Coagulation by Endotoxin, *Thromb. Diath. Haemorrh., 19,* 186 (1968).
199. Levin, J., Poore, E., Young, N. S., Margolis, S., Zauber, N. P., Townes, A. S., and Bell, W. R., Gram-Negative Sepsis: Detection of Endotoxemia with the Limulus Test with Studies of Associated Changes in Blood Coagulation, Serum Lipids, and Complement, *Ann. Intern. Med., 76,* 1 (1972).
200. Levin, J., Poore, T. E., Zauber, N. P., and Oser, R. S., Detection of Endotoxin in the Blood of Patients with Sepsis Due to Gram-Negative Bacteria, *N. Engl. J. Med., 283,* 1313 (1970).
201. Levy, E., Path, F. C., and Ruebner, B. H., Hepatic Changes Produced by a Single Dose of Endotoxin in the Mouse, *Amer. J. Pathol., 51,* 269 (1967).
202. Lillehei, R. C., Longerbeam, J. K., Bloch, J. H., and Manax, W. G., Hemodynamic Changes in Endotoxin Shock, in *Shock and Hypotension: Pathogenesis and Treatment,* p. 442, L. J. Mills and J. H. Moyer, Eds. Grune and Stratton, Inc., New York (1965).
203. Lüderitz, O., Jann, K., and Wheat, R., Somatic and Capsular Antigens of Gram-Negative Bacteria, *Compr. Biochem., 26A,* 105 (1968).
204. Lüderitz, O., Staub, A. M., and Westphal, O., Immunochemistry of O and R Antigens of *Salmonella* and Related Enterobacteriaceae, *Bacteriol. Rev., 30,* 192 (1966).
205. Lüderitz, O., Galanos, C., Risse, H. J., Ruschmann, E., Schlecht, S., Schmidt, G., Schulte-Holthausen, H., Wheat, R., and Westphal, O., Structural Relationships of *Salmonella* O and R Antigens, *Ann. N.Y. Acad. Sci., 133,* 349 (1966).
206. MacGregor, R. R., Sheagren, J. N., and Wolff, S. M., Endotoxin Induced Modification of *Plasmodium berghei* Infection in Mice, *J. Immunol., 102,* 131 (1969).
207. MaKara, G. B., Stark, E., and Meszaros, T., Corticotrophin Release Induced by *E. coli* Endotoxin after Removal of the Medial Hypothalamus, *Endocrinology, 88,* 412 (1971).
208. Mäkelä, P. H., and Stocker, B. A. D., Genetics of Polysaccharide Synthesis, *Annu. Rev. Genet., 3,* 291 (1969).
209. Margaretten, W., and McKay, D. G., The Effect of Leukocyte Antiserum on the Generalized Shwartzman Reaction, *Amer. J. Pathol., 57,* 299 (1969).
210. Marsh, J. C., and Perry, S., The Granulocyte Response to Endotoxin in Patients with Hematologic Disorders, *Blood, 23,* 581 (1964).
211. Martin, A. R., The Toxicity for Mice of Certain Fractions Isolated from *Bact. aertrycke., Brit. J. Exp. Pathol., 15,* 137 (1934).
212. Martin, W. J., and Marcus, S., Detoxified Bacterial Endotoxins, I, Preparation and Biological Properties of an Acetylated Crude Endotoxin from *Salmonella typhimurium, J. Bacteriol., 91,* 1453 (1966).
213. Martin, W. J., and Marcus, S., Detoxified Bacterial Endotoxins, II, Preparation and Biological Properties of Chemically Modified Crude Endotoxins from *Salmonella typhimurium, J. Bacteriol., 91,* 1750 (1966).
214. Marx, A., Musetescu, M., Sendrea, M., and Mihalca, M., Relationship Between Particle Size and Biological Activity of *Salmonella typhimurium* Endotoxin, *Zentr. Bakteriol. Parasitenk. Infektionskr. Hyg. Abt. Orig., 207,* 313 (1968).
215. McCabe, W. R., Immunization with R Mutants of *S. minnesota,* I, Protection against Challenge with Heterologous Gram-Negative Bacilli, *J. Immunol., 108,* 601 (1972).
216. McGill, M. W., Porter, P. J., and Kass, E. H., The Use of a Bioassay for Endotoxin in Clinical Infections, *J. Infect. Dis., 121,* 103 (1970).
217. McGill, M. W., Porter, P. J., Vivaldi, E., and Kass, E. H., Use of a Bioassay for Endotoxin in Clinical Infections and in Experimental Vasomotor Collapse, in *Antimicrobial Agents and Chemotherapy – 1967,* p. 132, G. L. Hobby, Ed. American Society for Microbiology, Ann Arbor, Michigan (1968).
218. McGrath, B. J., and Stewart, G. J., The Effects of Endotoxin on Vascular Endothelium, *J. Exp. Med., 129,* 833 (1969).
219. McKay, D. G., Margaretten, W., and Csavossy, I., An Electron Microscope Study of the Effects of Bacterial Endotoxin on the Blood-Vascular System, *Lab. Invest., 15,* 815 (1966).
220. McKay, D. G., Margaretten, W., and Csavossy, I., An Electron Microscope Study of Endotoxin Shock in Rhesus Monkeys, *Surg. Gynecol. Obstet., 125,* 825 (1967).
221. McKay, D. G., and Müller-Berghaus, G., Hageman Factor (HF) and the Generalized Shwartzman Reaction (GSR), *Fed. Proc., 27,* 436 (1968).
222. McKay, D. G., and Wong, T. C., The Effect of Bacterial Endotoxin on the Placenta of the Rat, *Amer. J. Pathol., 42,* 357 (1963).
223. McMaster, P. D., and **Franzl,** R. E., The Primary Immune Response in Mice, II, Cellular Responses of Lymphoid Tissue Accompanying the Enhancement or Complete Suppression of Antibody Formation by a Bacterial Endotoxin, *J. Exp. Med., 121,* 1109 (1968).
224. Medearis, D. N., Camitta, B. M., and Heath, E. C., Cell Wall Composition and Virulence in *Escherichia coli, J. Exp. Med., 128,* 399 (1968).
225. Mefferd, R. B., Henkel, D. T., and Loefer, J. B., Effect of Piromen on Survival of Irradiated Mice, *Proc. Soc. Exp. Biol. Med., 83,* 54 (1953).

226. Mela, L., Bacalzo, L. V., Jr., and Miller, L. D., Defective Oxidative Metabolism of Rat Liver Mitochondria in Hemorrhagic and Endotoxin Shock, *Amer. J. Physiol., 220,* 571 (1971).

227. Melby, J. C., Egdahl, R. H., and Spink, W. W., Secretion and Metabolism of Cortisol after Injection of Endotoxin, *J. Lab. Clin. Med., 56,* 50 (1960).

228. Mergenhagen, S. E., Gewurz, H., Bladen, H. A., Nowotny, A., Kasai, N., and Lüderitz, O., Interactions of the Complement System with Endotoxins from *Salmonella minnesota* Mutant Deficient in O-Polysaccharide and Heptose, *J. Immunol., 100,* 227 (1968).

229. Mergenhagen, S. E., Snyderman, R., Gewurz, H., and Shin, H. S., Significance of Complement to the Mechanism of Action of Endotoxin, *Curr. Top. Microbiol. Immunol., 50,* 37 (1969).

230. Meritt, K., and Johnson, A. G., Studies on the Adjuvant Action of Bacterial Endotoxins on Antibody Formation, V, The Influence of Endotoxin and 5-Fluoro-2-deoxyuridine on the Primary Antibody Response of the Balb Mouse to a Purified Protein Antigen, *J. Immunol., 91,* 266 (1963).

231. Michael, J. G., and Massell, B. F., Factors Involved in the Induction of Non-Specific Resistance to Streptococcal Infection in Mice by Endotoxin, *J. Exp. Med., 116,* 101 (1962).

232. Miles, A. A., and Pirie, N. W., The Properties of Antigenic Preparations from *Brucella melitensis,* I, Chemical and Physical Properties of Bacterial Fractions, *Brit. J. Exp. Pathol., 20,* 83 (1939).

233. Mills, L. C., Corticosteroids in Endotoxin Shock, *Proc. Soc. Exp. Biol. Med., 138,* 507 (1971).

234. Milner, K. C., Anacker, R. L., Fukushi, K., Haskins, W. T., Landy, M., Malmgren, B., and Ribi, E., Symposium on Relationship of Structure of Microorganisms to their Immunological Properties, III, Structure and Biological Properties of Surface Antigens from Gram-Negative Bacteria, *Bacteriol. Rev., 27,* 352 (1963).

235. Milner, K. C., and Finkelstein, R. A., Bioassay of Endotoxin: Correlation Between Pyrogenicity for Rabbits and Lethality for Chick Embryos, *J. Infect. Dis., 116,* 529 (1966).

236. Moberg, G. P., Site of Action of Endotoxins on Hypothalamic-Pituitary-Adrenal Axis, *Amer. J. Physiol., 220,* 397 (1971).

237. Moon, R. J., and Berry, L. J., Role of Tryptophan Pyrrolase in Endotoxin Poisoning, *J. Bacteriol., 95,* 1247 (1968).

238. Morgan, W. T. J., Studies in Immunochemistry, II, The Isolation and Properties of a Specific Antigenic Substance from *B. Dysenteriae* (Shiga), *Biochem. J., 31,* 2003 (1937).

239. Mulholland, J. H., and Cluff, L. E., The Effect of Endotoxin upon Susceptibility to Infection: The Role of the Granulocyte, in *Bacterial Endotoxins,* p. 211, M. Landy and W. Braun, Eds. Rutgers University Press, New Brunswick, New Jersey (1964).

240. Mulholland, J. H., Wolff, S. M., Jackson, A. L., and Landy, M., Quantitative Studies of Febrile Tolerance and Levels of Specific Antibody Evoked by Bacterial Endotoxin, *J. Clin. Invest., 44,* 920 (1965).

241. Müller-Berghaus, G., and Schneberger, R., Hageman Factor Activation in the Generalized Shwartzman Reaction Induced by Endotoxin, *Brit. J. Haematol., 21,* 513 (1971).

242. Muschel, L. H., Schmoker, K., and Webb, P. M., Anticomplementary Action of Endotoxin, *Proc. Soc. Exp. Biol. Med., 117,* 639 (1964).

243. Neter, E., Endotoxins and the Immune Response, *Curr. Top. Microbiol. Immunol., 47,* 82 (1969).

244. Neter, E., and Vogt, W., Anaphylatoxin Formation by a Pyrogenic Lipopolysaccharide, *Arch. Pathol., 248,* 261 (1964).

245. Neter, E., Westphal, O., Lüderitz, O., Gorzynski, E. A., and Eichenberger, E., Studies on Enterobacterial Lipopolysaccharides; Effect of Heat and Chemicals on Erythrocyte Modifying Antigenic, Toxic and Pyrogenic Properties, *J. Immunol., 76,* 377 (1956).

246. Neter, E., Whang, H. Y., Lüderitz, O., Gorzynski, E. A., and Westphal, O., Immunological Priming without Production of Circulating Antibodies Conditioned by Endotoxin and Its Lipid A Component, *Nature (London), 212,* 420 (1966).

247. Nies, A. S., Forsyth, R. P., Williams, H. E., and Melmon, K. L., Contribution of Kinins to Endotoxin Shock in Unanesthetized Rhesus Monkeys, *Circ. Res., 22,* 155 (1968).

248. Nikaido, H., Studies on the Biosynthesis of Cell Wall Polysaccharide in Mutant Strains of *Salmonella,* I, *Proc. Natl. Acad. Sci. U.S.A., 48,* 1337 (1962).

249. Nikaido, H., Biosynthesis of Cell Wall Lipopolysaccharide in Gram-Negative Enteric Bacteria, *Advan. Enzymol., 31,* 77 (1969).

250. Nikaido, H., Structure of Cell Wall Lipopolysaccharide from *Salmonella typhimurium,* I, Linkage between O Side Chains and R-Core, *J. Biol. Chem., 244,* 2835 (1969).

251. Nikaido, H., Structure of Cell Wall Lipopolysaccharide from *Salmonella typhimurium.* Further Studies on the Linkage between O Side Chains and R-Core, *Eur. J. Biochem., 15,* 57 (1970).

252. Nikaido, H., and Nikaido, K., Biosynthesis of Cell Wall Polysaccharide in Mutant Strains of *Salmonella,* IV, Synthesis of S-Specific Side Chains, *Biochem. Biophys. Res. Commun., 19,* 322 (1965).

253. Nikaido, H., Nikaido, K., and Mäkelä, P. H., Genetic Determination of Enzymes Synthesizing O-Specific Sugars of *Salmonella* Lipopolysaccharides, *J. Bacteriol., 91,* 1126 (1966).

254. Niwa, M., Milner, K. C., Ribi, E., and Rudbach, J. A., Alteration of Physical, Chemical, and Biological Properties of Endotoxin by Treatment with Mild Alkali, *J. Bacteriol., 97,* 1069 (1969).

255. Nolan, J. P., and O'Connell, C. J., Vascular Response in the Isolated Rat Liver, I, Endotoxin, Direct Effects, *J. Exp. Med., 122,* 1063 (1965).

256. Noll, H., and Braude, A. I., Preparation and Biological Properties of a Chemically Modified *Escherichia coli* Endotoxin of High Immunogenic Potency and Low Toxicity, *J. Clin. Invest., 40,* 1935 (1961).
257. Nowotny, A., Chemical Structure of a Phosphomucopeptide and Its Occurrence in Some Strains of *Salmonella, J. Amer. Chem. Soc., 83,* 501 (1961).
258. Nowotny, A., Chemical Detoxification of Bacterial Endotoxins, in *Bacterial Endotoxins,* p. 29, M. Landy and W. Braun, Eds. Rutgers University Press, New Brunswick, New Jersey (1964).
259. Nowotny, A., Immunogenicity of Toxic Detoxified Endotoxin Preparations, *Proc. Soc. Exp. Biol. Med., 127,* 745 (1968).
260. Nowotny, A., Molecular Aspects of Endotoxin Reactions, *Bacteriol. Rev., 33,* 72 (1969).
261. Nowotny, A., Radvany, R., and Neale, N. L., Neutralization of Toxic Bacterial O-Antigens with O-Antibodies while Maintaining Their Stimulus on Non-Specific Resistance, *Life Sci., 4,* 1107 (1965).
262. Nowotny, A. M., Thomas, S., Duron, S., and Nowotny, A., Relation of Structure to Function in Bacterial O-Antigens, I, Isolation Methods, *J. Bacteriol., 85,* 418 (1963).
263. Ogura, M., High Resolution Electron Microscopy on the Surface Structure of *Escherichia coli, J. Ultrastruct. Res., 8,* 251 (1963).
264. Oppenheim, J. J., and Perry, S., Effects of Endotoxins on Cultured Leukocytes, *Proc. Soc. Exp. Biol. Med., 118,* 1014 (1965).
265. Oroszlan, S. I., and Mora, P. T., Dissociation and Reconstitution of an Endotoxin, *Biochem. Biophys. Res. Commun., 12,* 345 (1963).
266. Osborn, M. J., Structure and Biosynthesis of the Bacterial Wall, *Annu. Rev. Biochem., 38,* 501 (1969).
267. Osborn, M. J., Rosen, S. M., Rothfield, L., Zeleznick, L. D., and Horecker, B. L., Lipopolysaccharide of the Gram-Negative Cell Wall: Biosynthesis of a Complex Heteropolysaccharide Occurs by Successive Addition of Specific Sugar Residues, *Science, 145,* 783 (1964).
268. Osborn, M. J., and Weiner, I. M., Biosynthesis of a Bacterial Lipopolysaccharide, VI, Mechanism of Incorporation of Abequose into the O-Antigen of *Salmonella typhimurium, J. Biol. Chem., 243,* 2631 (1968).
269. Osborn, M. J., and Yuan Tze-Yeuen, R., Biosynthesis of Bacterial Lipopolysaccharide, VII, Enzymatic Formation of the First Intermediate in Biosynthesis of the O-Antigen of *Salmonella typhimurium, J. Biol. Chem., 243,* 5145 (1968).
270. Palmer, J. W., and Gerlough, T. D., Scientific Apparatus and Laboratory Methods; A Simple Method for Preparing Antigenic Substances from the Typhoid Bacillus, *Science, 92,* 155 (1940).
271. Pekarek, R. S., and Beisel, W. R., Effect of Endotoxin on Serum Zinc Concentrations in the Rat, *Appl. Microbiol., 18,* 482 (1969).
272. Petersdorf, R. G., and Shulman, J. A., The Role of Tolerance in the Action of Bacterial Endotoxins, in *Bacterial Endotoxins,* p. 487, M. Landy and W. Braun, Eds. Rutgers University Press, New Brunswick, New Jersey (1964).
273. Pierce, W. C., The Effects of Endotoxin on the Immune Response in the Rat, III, Elimination of $I^{125}$-Labeled Bovine γ-Globulin from the Circulation of Rats, *Lab. Invest., 17,* 380 (1967).
274. Pieroni, R. E., Broderick, E. J., Bundeally, A., and Levine L., A Simple Method for the Quantitation of Submicrogram Amounts of Bacterial Endotoxin, *Proc. Soc. Exp. Biol. Med., 133,* 790 (1969).
275. Pieroni, R. E., Broderick, E. J., and Levine, L., Endotoxin-Induced Hypersensitivity to Histamine in Mice, I, Contrasting Effects of Bacterial Lipopolysaccharides and the Classical Histamine-Sensitizing Factor of *Bordetella pertussis, J. Bacteriol., 91,* 2169 (1966).
276. Porter, P. J., Spievack, A. R., and Kass, E. H., Endotoxin-Like Activity of Serum from Patients with Severe Localized Infections, *N. Engl. J. Med., 271,* 445 (1964).
277. Postic, B., DeAngelis, C., Breinig, M. K., and Ho, M., Effect of Temperature on the Induction of Interferons by Endotoxin and Virus, *J. Bacteriol., 91,* 1277 (1966).
278. Priano, L. L., Wilson, R. D., and Traber, D. L., Lack of Significant Protection Afforded by Heparin during Endotoxin Shock, *Amer. J. Physiol., 200,* 901 (1971).
279. Quesenberry, P., Morley, A., Stohlman, F., Richard, K., Howard, D., and Smith, M., Effect of Endotoxin on Granulopoiesis and Colony-Stimulating Factor, *N. Engl. J. Med., 286,* 227 (1972).
280. Radvany, R., Neale, N. L., and Nowotny, A., Relation of Structure to Function in Bacterial O-Antigens, VI, Neutralization of Endotoxic O-Antigens by Homologous O-Antibody, *Ann. N.Y. Acad. Sci., 133,* 763 (1966).
281. Reinhold, R. B., and Fine, J., A Technique for Quantitative Measurement of Endotoxin in Human Plasma, *Proc. Soc. Exp. Biol. Med., 137,* 334 (1971).
282. Reitschel, E. T., Galanos, C., Tanaka, A., Ruschmann, E., Lüderitz, O., and Westphal, O., Biological Activities of Chemically Modified Endotoxins, *Eur. J. Biochem., 22,* 218 (1971).
283. Ribi, E., Anacker, L., Brown, R., Haskins, W. T., Malmgren, B., Milner, K. C., and Rudbach, J. A., Reaction of Endotoxin and Surfactants, I, Physical and Biological Properties of Endotoxins Treated with Sodium Desoxycholate, *J. Bacteriol., 92,* 1493 (1966).
284. Ribi, E., Haskins, W. T., Milner, K. C., Anacker, R. L., Ritter, D. B., Goode, G., Trapani, R. J., and Landy, M., Physicochemical Changes in Endotoxin Associated with Loss of Biological Potency, *J. Bacteriol., 84,* 803 (1962).
285. Ribi, E., Milner, K. C., and Perrine, T. D., Endotoxic and Antigenic Fractions from the Cell Wall of *Salmonella enteritidis.* Methods for Separation and Some Biologic Activities, *J. Immunol., 82,* 75 (1959).

286. Rieder, R. F., and Thomas, L., Studies of the Mechanisms Involved in the Production of Abortion by Endotoxin, *J. Immunol., 84,* 189 (1960).
287. Roantree, R. J., *Salmonella* O-Antigens and Virulence, *Annu. Rev. Microbiol., 21,* 443 (1967).
288. Robbins, P. W., and Wright, A., Biosynthesis of O-Antigens, in *Microbial Toxins,* Vol. 4, p. 351, G. Weinbaum, S. Kadis and S. J. Ajl, Eds. Academic Press, New York (1971).
289. Rojas-Corona, R. R., Skarnes, R., Tamakuma, S., and Fine, J., The *Limulus* Coagulation Test for Endotoxin. A Comparison with Other Assay Methods, *Proc. Soc. Exp. Biol. Med., 132,* 599 (1969).
290. Root, R. K., Nordlund, J. J., and Wolff, S. M., Factors Affecting the Quantitative Production and Assay of Human Leukocytic Pyrogen, *J. Lab. Clin. Med., 75,* 679 (1970).
291. Rosenberg, J. C., and Rush, B. F., Lethal Endotoxin Shock: Oxygen Deficit, Lactic Acid Levels, and Other Metabolic Changes, *J. Amer. Med. Ass., 196,* 767 (1966).
292. Rossen, R. D., Wolff, S. M., and Butler, W. T., The Antibody Response in Nasal Washings and Serum to *S. typhosa* Endotoxin Administered Intravenously, *J. Immunol., 99,* 246 (1967).
293. Rossen, R. D., Wolff, S. M., Butler, W. T., and Vannier, W. E., The Identification of Low Molecular Weight Bentonite Flocculating Antibodies in the Serum of the Rabbit, Monkey and Man, *J. Immunol., 98,* 764 (1967).
294. Rothfield, L., and Pearlman-Kothencz, M., Synthesis and Assembly of Bacterial Membrane Components; A Lipopolysaccharide-Phospholipid-Protein Complex Excreted by Living Bacteria, *J. Mol. Biol., 44,* 477 (1969).
295. Rowlands, D. T., Jr., Claman, H. N., and Kind, P. D., The Effect of Endotoxin on the Thymus of Young Mice, *Amer. J. Pathol., 46,* 165 (1965).
296. Rowley, D., Stimulation of Natural Immunity to *Escherichia coli* Infections, *Lancet, 1,* 232 (1955).
297. Rowley, D., Endotoxin-Induced Changes in Susceptibility to Infection, in *Bacterial Endotoxins,* p. 359, M. Landy and W. Braun, Eds. Rutgers University Press, New Brunswick, New Jersey (1964).
298. Rowley, D., Howard, J. G., and Jenkin, C. R., The Fate of $P^{32}$-Labeled Bacterial Lipopolysaccharide in Laboratory Animals, *Lancet, 1,* 366 (1956).
299. Rubenstein, M., Mulholland, J. H., Jeffery, G. M., and Wolff, S. M., Malaria Induced Endotoxin Tolerance, *Proc. Soc. Exp. Biol. Med., 118,* 283 (1965).
300. Rubio, N., and Lopez, R., Purification of *Pseudomonas aeruginosa* Endotoxin by Membrane Partition Chromatography, *Appl. Microbiol., 23,* 211 (1972).
301. Rudbach, J. A., Molecular Immunogenicity of Bacterial Lipopolysaccharide Antigens: Establishing a Quantitative System, *J. Immunol., 106,* 993 (1971).
302. Rudbach, J. A., and Johnson, A. G., Restoration of Endotoxin Activity Following Alteration by Plasma, *Nature (London), 202,* 811 (1964).
303. Rudbach, J. A., and Johnson, A. G., Alteration and Restoration of Endotoxin Activity after Complexing with Plasma Proteins, *J. Bacteriol., 92,* 892 (1966).
304. Rudbach, J. A., Ribi, E., and Milner, K. C., Reaction of Papain-Treated Endotoxin, *Proc. Soc. Exp. Biol. Med., 119,* 115 (1965).
305. Sauter, C., and Gifford, G. E., Interferon-Like Inhibitor and Lysosomal Enzyme Induced in Mice Injected with Endotoxin, *Nature (London), 212,* 626 (1966).
306. Schmidt, G., Eichenberger, E., and Westphal, O., Die Wirkung der Lipoid- und Polysaccharide-Komponente Endotoxischer Lipopolysaccharide Gram-Negativer Bakterien auf die Leukozyten-Kultur, *Experientia, 14,* 289 (1958).
307. Schultz, D. R., and Becker, E. L., The Alteration of Endotoxin by Postheparin Plasma and Its Purified Fractions, I, Comparison of the Ability of Guinea Pig Postheparin and Normal Plasma to Detoxify Endotoxin, *J. Immunol., 98,* 473 (1967).
308. Schultz, D. R., and Becker, E. L., The Alteration of Endotoxin by Postheparin Plasma and Its Purified Fractions, II, Relationship of the Endotoxin Detoxifying Activity of Euglobulin from Postheparin Plasma to Lipoprotein Lipase, *J. Immunol., 98,* 482 (1967).
309. Shands, J. W., Localization of Somatic Antigen on Gram-Negative Bacteria by Electron Microscopy, *J. Bacteriol., 90,* 266 (1965).
310. Shands, J. W., Localization of Somatic Antigen on Gram-Negative Bacteria using Ferritin Antibody Conjugates, *Ann. N.Y. Acad. Sci., 133,* 292 (1966).
311. Shands, J. W., Jr., Miller, V., and Martin, H., The Hypoglycemic Activity of Endotoxin, I, Occurrence in Animals Hyperactive to Endotoxin, *Proc. Soc. Exp. Biol. Med., 130,* 413 (1969).
312. Shands, J. W., Jr., Miller, V., Martin, H., and Senterfitt, V., Hypoglycemic Activity of Endotoxin, II, Mechanism of the Phenomenon in BCG-Infected Mice, *J. Bacteriol., 98,* 494 (1969).
313. Sheagren, J. N., Wolff, S. M., and Shulman, N. R., Febrile and Hematologic Responses of Rhesus Monkeys to Bacterial Endotoxin, *Amer. J. Physiol., 212,* 884 (1967).
314. Shilo, M., Non-Specific Resistance to Infections, *Annu. Rev. Microbiol., 13,* 255 (1959).
315. Shtasel, T. F., and Berry, L. J., Effect of Endotoxin and Cortisone on Synthesis of Ribonucleic Acid and Protein in Livers of Mice, *J. Bacteriol., 97,* 1018 (1969).
316. Shwartzman, G., Studies of *Bacillus typhosus* Toxic Substances, I, The Phenomenon of Local Skin Reactivity to *B. typhosus* Culture Filtrate, *J. Exp. Med., 48,* 247 (1928).

317. Shwartzman, G., and Michailovsky, N., Phenomenon of Local Skin Reactivity to Bacterial Filtrates in the Treatment of Mouse Sarcoma 180, *Proc. Soc. Exp. Biol. Med., 29,* 737 (1931).
318. Silk, M. R., The Effect of Endotoxin on Renal Hemodynamics, *Arch. Surg., 93,* 531 (1966).
319. Singer, I., Kimble, E. T., III, and Ritts, R. E., Jr., Alterations of the Host–Parasite Relationship by Administration of Endotoxin to Mice with Infections of Trypanosomes, *J. Infect. Dis., 114,* 243 (1964).
320. Skarnes, R. C., and Harper, M. J. K., Relationship between Endotoxin-Induced Abortion and the Synthesis of Prostaglandin F, *Prostaglandins, 1,* 191 (1972).
321. Skarnes, R., Rutenburg, S., and Fine, J., Fractionation of an Esterase from Calf Spleen Implicated in the Detoxification of Bacterial Endotoxin, *Proc. Soc. Exp. Biol. Med., 128,* 75 (1968).
322. Skjorten, F., Bilateral Renal Cortical Necrosis and the Generalized Shwartzman Reaction, I, Review of Literature and Report of Seven Cases, *Acta Pathol. Microbiol. Scand., 61,* 394 (1964).
323. Sleeman, H. K., Lamborn, P. B., Diggs, J. W., and Emery, C. E., Effects of Endotoxin and Histamine on Serum Enzyme Activity, *Proc. Soc. Exp. Biol. Med., 138,* 536 (1971).
324. Smith, E. E., Rutenburg, S. H., Rutenburg, A. M., and Fine, J., Detoxification of Endotoxin by Splenic Extracts, *Proc. Soc. Exp. Biol. Med., 113,* 781 (1963).
325. Smith, L. L., Muller, W., and Hinshaw, L. B., The Management of Experimental Endotoxin Shock, *Arch. Surg., 89,* 630 (1964).
326. Smith, R. T., and Thomas, L., The Lethal Effect of Endotoxins in the Chick Embryo, *J. Exp. Med., 104,* 217 (1956).
327. Smith, W. W., Alderman, I. M., and Cornfield, J., Granulocyte Release by Endotoxin in Normal and Irradiated Mice, *Amer. J. Physiol., 201,* 396 (1961).
328. Smith, W. W., Alderman, I. M., and Gillespie, R. E., Increased Survival in Irradiated Animals Treated with Bacterial Endotoxins, *Amer. J. Physiol., 191,* 124 (1957).
329. Smith, W. W., Marston, R. A., and Cornfield, J., Patterns of Hemopoietic Recovery in Irradiated Mice, *Blood, 14,* 737 (1959).
330. Snell, E. S., and Atkins, E. A., The Mechanisms of Fever, in *The Biological Basis of Medicine,* Vol. 2, p. 397, E. E. Bittar, Ed. Academic Press, London, England (1968).
331. Snyder, I. S., Deters, M., and Ingle, J., Effect of Endotoxin on Pyruvate Kinase Activity in Mouse Liver, *Infect. Immun., 4,* 138 (1971).
332. Solie, R. T., and Downing, S. E., Effects of *E. coli* Endotoxemia on Ventricular Performance, *Amer. J. Physiol., 211,* 307 (1966).
333. Spielvogel, A. R., An Ultrastructural Study of the Mechanisms of Platelet–Endotoxin Interaction, *J. Exp. Med., 126,* 235 (1967).
334. Spink, W. W., Reddin, J., Zak, S. J., Peterson, M., Starzecki, B., and Seljeskog, E., Correlation of Plasma Catecholamine Levels with Hemodynamic Changes in Canine Endotoxin Shock, *J. Clin. Invest., 45,* 78 (1966).
335. Stetson, C. A., Jr., Studies on the Mechanisms of the Shwartzman Phenomenon, *J. Exp. Med., 93,* 489 (1951).
336. Stewart, G. J., Effect of Endotoxin on the Ultrastructure of Liver and Blood Cells of Hamsters, *Brit. J. Exp. Pathol., 51,* 114 (1970).
337. Stinebring, W. R., and Youngner, J. S., Patterns of Interferon Appearance in Mice Injected with Bacteria or Bacterial Endotoxin, *Nature (London), 204,* 712 (1964).
338. Stocker, B. A. D., and Mäkelä, P. H., Genetic Aspects of Biosynthesis and Structure of *Salmonella* Lipopolysaccharide, in *Microbial Toxins,* Vol. 4, p. 369, G. Weinbaum, S. Kadis and S. J. Ajl, Eds. Academic Press, New York (1971).
339. Takebe, K., Setaishi, C., Hirama, M., Yamamoto, M., and Horiuchi, Y., Effects of a Bacterial Pyrogen on the Pituitary-Adrenal Axis at Various Times in 24 Hours, *J. Clin. Endocrinol. Metab., 26,* 437 (1966).
340. Tate, W. J., III, Douglas, H., Braude, A. I., and Wells, W. W., Protection against Lethality of *E. coli* Endotoxin with "O" Antiserum, *Ann. N.Y. Acad. Sci., 133.* 746 (1966).
341. Taylor, S. S., and Heath, E. C., The Incorporation of β-Hydroxy Fatty Acids into Phospholipid of *Escherichia coli* B., *J. Biol. Chem., 244,* 6605 (1969).
342. Thomas, C. S., Jr., and Brockman, S. K., The Role of Adrenal Corticosteroid Therapy in *Escherichia coli* Endotoxin Shock, *Surg. Gynecol. Obstet., 126,* 61 (1968).
343. Thomas, L., Possible New Mechanism of Tissue Damage in the Experimental Cardiovascular Effects of Endotoxin, *Amer. Heart J., 52,* 507 (1956).
344. Thomas, L., and Good, R. A., Bilateral Cortical Necrosis of Kidneys in Cortisone-Treated Rabbits Following Injection of Bacterial Toxins, *Proc. Soc. Exp. Biol. Med., 76,* 604 (1951).
345. Thomas L., and Good, R. A., Studies on the Generalized Shwartzman Reaction, I, General Observations Concerning the Phenomenon, *J. Exp. Med., 96,* 605 (1952).
346. Thomas, L., and Zweifach, B. W., Mechanisms in the Production of Tissue Damage and Shock by Endotoxin, *Trans. Ass. Amer. Physicians Philadelphia, 70,* 54 (1957).
347. Trejo, R. A., and DiLuzio, N. D., Influence of Reticuloendothelial System (RES) Functional Modification on Endotoxin Detoxification by Liver and Spleen, *J. Reticuloendothelial Soc., 10,* 515 (1971).
348. Tsagaris, T. J., Gani, M., and Lange, R. L., Central Blood Volume during Endotoxin Shock in Dogs, *Amer. J. Physiol., 212,* 498 (1967).

349. Villablanca, J., and Myers, R. D., Fever Produced by Microinjections of Typhoid Vaccine into Hypothalmus of Cats, *Amer. J. Physiol., 208,* 703 (1965).

350. Wagner, R. R., Snyder, R. M., Hook, E. W., and Luttrell, C. N., Effect of Bacterial Endotoxin on Resistance of Mice to Viral Encephalitides Including Comparative Studies of the Interference Phenomenon, *J. Immunol., 83,* 87 (1959).

351. Waisbren, B. A., Bacteremia due to Gram-Negative Bacilli Other than *Salmonella, Arch. Intern. Med., 88,* 467 (1951).

352. Walker, J., A Method for the Isolation of Toxic and Immunizing Fractions from Bacteria of the *Salmonella* Group, *Biochem. J., 34,* 325 (1940).

353. Watson, D. W., and Kim, Y. B., Modification of Host Responses to Bacterial Endotoxins, I, Specificity of Pyrogenic Tolerance and the Role of Hypersensitivity in Pyrogenicity, Lethality, and Skin Reactivity, *J. Exp. Med., 118,* 425 (1963).

354. Weil, M. H., Shubin, H., Udhoji, V. N., and Rossoff, L., Effects of Vasopressor Agents and Corticosteroid Hormones in Endotoxin Shock, in *Shock and Hypotension: Pathogenesis and Treatment,* p. 470, L. J. Mills and J. H. Moyer, Eds. Grune and Stratton, New York (1965).

355. Weinbaum, G., Kadis, S., and Ajl, S. J., Bacterial Endotoxins, in *Microbial Toxins,* Vol. 4, G. Weinbaum, S. Kadis and S. J. Ajl, Eds. Academic Press, New York (1971).

356. Weinstein, M. J., Waitz, J. A., and Came, P. E., Induction of Resistance to Bacterial Infections of Mice with Poly I Poly C, *Nature (London), 226,* 170 (1970).

357. Wexler, B., Effects of a Bacterial Polysaccharide (Piromen) on the Pituitary-Adrenal Axis: Modification of ACTH Release, *Metabolism, 12,* 49 (1963).

358. Wexler, B. C., Dolgin, A. E., and Tryczynski, E. W., Effects of a Bacterial Polysaccharide (Piromen) on the Pituitary-Adrenal Axis; Adrenal Ascorbic Acid, Cholesterol, and Histologic Alterations, *Endocrinology, 61,* 300 (1965).

359. Whang, H. Y., and Neter, E., Immunosuppression by Endotoxin and Its Lipoid A Component, *Proc. Soc. Exp. Biol. Med., 124,* 919 (1967).

360. Whitby, J. L., Michael, J. G., Woods, M. W., and Landy, M., Symposium on Bacterial Endotoxins, II, Possible Mechanism Whereby Endotoxins Evoke Increased Nonspecific Resistance to Infection, *Bacteriol. Rev., 25,* 437 (1961).

361. White, P. B., On the Relation of the Alcohol-Soluble Constituents of Bacteria to Their Spontaneous Agglutination, *J. Pathol. Bacteriol., 30,* 113 (1927).

362. Wiener, E., Beck, A., and Shilo, M., Effect of Bacterial Lipopolysaccharides on Mouse Peritoneal Leukocytes, *Lab. Invest., 14,* 475 (1965).

363. Wober, W., and Alaupovic, P., Studies on the Protein Moiety of Endotoxin from Gram-Negative Bacteria; Characterization of the Moiety Isolated by Phenol Treatment of Endotoxin from *Serratia marcescens* 08 and *Escherichia coli* 0141: K85 (B), *Eur. J. Biochem., 19,* 340 (1971).

364. Wolff, S. M., Mulholland, J. H., and Rubenstein, M., Suppression of the Immune Response to Bacterial Endotoxins, in *Bacterial Endotoxins,* p. 319, M. Landy and W. Braun, Eds. Rutgers University Press, New Brunswick, New Jersey (1964).

365. Wolff, S. M., Mulholland, J. H., and Ward, S. B., Quantitative Aspects of the Pyrogenic Response of Rabbits to Endotoxin, *J. Lab. Clin. Med., 65,* 268 (1965).

366. Wolff, S. M., Mulholland, J. H., Ward, S. B., Rubenstein, M., and Mott, P. D., Effect of 6-Mercaptopurine on Endotoxin Tolerance, *J. Clin. Invest., 44,* 1402 (1965).

367. Wolff, S. M., Rubenstein, M., Mulholland, J. H., and Alling, D. W., Comparison of Hematologic and Febrile Response to Endotoxin, *Blood, 26,* 190 (1965).

368. Wood, W. B., Jr., Studies on the Cause of Fever, *N. Engl. J. Med., 258,* 1023 (1958).

369. Wright, A., and Kanegasaki, S., Molecular Aspects of Lipopolysaccharides, *Physiol. Rev., 51,* 748 (1971).

370. Wright, L. J., Kimball, H. R., and Wolff, S. M., Alterations in Host Responses to Experimental *Candida albicans* Infections by Bacterial Endotoxin, *J. Immunol., 103,* 1276 (1969).

371. Wright, L. J., Lipsett, M. B., Ross, G. T., and Wolff, S. M., Effects of Dexamethasone and Aspirin on the Responses to Endotoxin in Man, *J. Clin. Endocrinol. Metab., 34,* 13 (1972).

372. Wright, R. C., and Winkelmann, R. K., The Epinephrine Response of Isolated Rabbit Vascular Strips after *in vivo* and *in vitro* Endotoxin Exposure, *Angiology, 22,* 495 (1971).

373. Yin, E. T., Galanos, C., Kinsky, S., Bradshaw, R. A., Wessler, S., Lüderitz, O., and Sariento, M. E., Picogram-Sensitive Assay for Endotoxin: Gelation of *Limulus polyphemus* Blood Cell Lysate Induced by Purified Lipopolysaccharides and Lipid A from Gram-Negative Bacteria, *Biochim. Biophys. Acta, 261,* 284 (1972).

374. Yoshikawa, T., Tanaka, K. R., and Guze, L. B., Infection and Disseminated Intravascular Coagulation, *Medicine (Baltimore), 50,* 237 (1971).

375. Yoshioka, M., and Konno, S., Characteristics of Endotoxin-Altering Fractions Derived from Normal Serum, III, Isolation and Properties of Horse Serum $a_2$-Macroglobulin, *Infect. Immun., 1,* 431 (1970).

376. Yoshioka, M., and Johnson, A. B., Characteristics of Endotoxin-Altering Fractions Derived from Normal Human Serum, *J. Immunol., 89,* 326 (1962).

377. Zweifach, B. W., Vascular Effects of Bacterial Endotoxin, in *Bacterial Endotoxins,* p. 110, M. Landy and W. Braun, Eds. Rutgers University Press, New Brunswick, New Jersey (1964).

# LIPIDS

# LIPIDS

DR. WILLIAM M. O'LEARY

In a remarkably few years our knowledge of microbial lipids has progressed from nil to sketchy to massive and complex (though still incomplete). The powerful aid of the latest methods of chromatography and instrumental analysis now makes possible an increasing avalanche of reports on the lipid composition of specific microorganisms and of the chemical nature of individual lipids. So profuse has this information become that, as all reviewers now lament, no format is sufficiently expansive to accommodate all of it. The information presented here is an admittedly selected but representative collection of data available at the time of publication.

For the reader who requires more information, discussion, or an idea of historical development in this field, the following reviews and monographs are recommended.

1. Asselineau, J., *The Bacterial Lipids.* Holden-Day, San Francisco, California (1966).
2. Kates, M., *Advan. Lipid Res., 2,* 17 (1964).
3. Kates, M., and Wassef, M. K., *Annu. Rev. Biochem., 39,* 323 (1970).
4. Law, J. H., in *The Specificity of Cell Surfaces,* B. D. Davis and L. Warren, Eds. Prentice-Hall, Englewood Cliffs, New Jersey (1967).
5. Lennarz, W. J., *Annu. Rev. Biochem., 39,* 359 (1970).
6. Lennarz, W. J., in *Lipid Metabolism,* S. J. Wakil, Ed. Academic Press, New York (1970).
7. O'Leary, W. M., *The Chemistry and Metabolism of Microbial Lipids.* World Publishing Co., Cleveland, Ohio (1967).
8. O'Leary, W. M., in *Comprehensive Biochemistry,* Vol. 18, M. Florkin and E. Stotz, Eds. Elsevier, Amsterdam, The Netherlands (1969).

# THE NOMENCLATURE OF LIPIDS

DR. W. M. O'LEARY

The following is an excerpt of a synopsis of the recommendations of the IUPAC-IUB Commission on Biochemical Nomenclature (CBN) prepared by Dr. Waldo E. Cohn, Director, NAS-NRC Office of Biochemical Nomenclature (CBN). Requests for reprints of CBN publications may be directed to Dr. Cohn at Oak Ridge National Laboratory, Box Y, Oak Ridge, TN 37830.

## GENERIC TERMS

| | |
|---|---|
| 1. Phosphoglyceride | Any derivative of glycerophosphoric acid (glycerol-*P*) that contains at least one *O*-acyl, *O*-alkyl, or *O*-alk-1′-en-1′-yl group attached to glycerol. glycerol. |
| 2. X phosphoglyceride | If X is the other ester component of a phosphoglyceride (e.g., X = choline). |
| 3. Phosphatidic acid | Both alcoholic OH groups of glycerophosphoric acid are esterified by fatty acids. |
| 4. Lecithin | Permitted, but not recommended, for 1,2-diacyl-*sn*-glycero-3-phosphoryl-cholines (see 5). |
| 5. 3-*sn*-phosphatidylcholine | Recommended for lecithin (see 4). |
| 6. Tri(di)acylglycerol | Replaces tri(di)glyceride (Figure 2). |
| 7. Phospholipid | Any lipid containing a radical of $H_3PO_4$ (Figures 1, 3, 4, 6, and 7). |
| 8. Phosphoinositide | Any lipid containing inositol and $H_3PO_4$ radicals. |
| 9. "Long chain base" | Used in this table to refer to sphinganine (see 20, below) and its homologues, stereoisomers, and hydroxy or unsaturated derivatives of these. |
| 10. Sphingolipid | Any lipid containing a "long chain base" (see 9). |
| 11. Glycosphingolipid | Any lipid containing a "9" and one or more sugars. |
| 12. Ceramide | Any *N*-acyl "9". |
| 13. Cerebroside | Any monoglycosyl ceramide (see 12). |
| 14. Ganglioside | Any cerebroside (13) containing neuraminic acid (see 16). |
| 15. Sphingomyelin | A ceramide 1-phosphorylcholine (see 12). |
| 16. Neuraminic acid | 5-Amino-3,5-dideoxy-D-*glycero*-D-*galacto*-nonulosonic acid; radicals = neuraminoyl,[a] neuraminosyl[b] (Figure 8). |
| 17. Sialic acid | *N*-Acylneuraminic acids, their esters, and other derivatives of the alcoholic hydroxyl groups (radicals = sialoyl,[a] sialosyl.[b] |

## SPECIFIC NOMENCLATURE OF LIPIDS

18. a. Substitution for H from alcoholic-OH by "group": name is *group X*.
    b. Esterification of alcoholic-OH of X by "acidate", the anion of "acidic acid": name is *X acidate.*

19. Glycerol derivatives may be numbered "stereospecifically", the C atom at top in that Fischer projection having the secondary hydroxyl to the left being C-1 (Figure 5). Such numbering takes the prefix *sn* (for stereospecifically numbered), which immediately precedes the "glycero" term. It is replaced by *rac* (which precedes the *full* name) if the product is an equal mixture of both antipodes, or by *X* (same position) if the configuration is unknown or unspecified.

    **Examples**

    The older terms

    | | |
    |---|---|
    | L-*α*-Glycerophosphoric acid | *sn*-Glycerol 3-(dihydrogen phosphate) |
    | D-Glycerol 1-phosphate | *sn*-Glycero-3-phosphoric acid (Figure 1) |

    1-Alk-1′-enyl-2-acyl-*sn*-glycerophosphoric ester
    *O*-(1-Acyl-*sn*-glycero-3-phosphoryl)ethanolamine
    Diacyl-*sn*-glycero-3-phosphoryl-1′-*sn*-glycerol = 3-*sn*-phosphatidyl-1′-*sn*-glycerol (Figure 7)

20. Sphingolipids may be named as sphinganine [2*D*-aminoöctadecane-1,3*D*-diol, or D-erythro-2-aminoöctadecane-1,3-diol, or (2*S*, 3*R*)-2-aminoöctadecane-1,3-diol] derivatives. The name implies identical configurations in the *D*/*L* system, not the *R*,*S* system, unless the contrary is specified in *D*/*L* terms. *D* and *L* refer to right or left position of functional groups written in the Fischer projection, vertically, with C-1 at the top; they follow the number of the substituted C atom, and are replaced by *X* when the configuration is unknown or by *rac* (before the name) for racemic mixtures.

    **Examples**

    | Old, non-recommended Terms: | Recommended equivalents: |
    |---|---|
    | "phytosphingosine" | 4*D*-hydroxysphinganine |
    | sphingosine | 4-sphingenine $^{18}CH_3(CH_2)_{12}CH = {}^{4}CH\text{-}CHOH\text{-}CHNH_2\text{-}{}^{1}CH_2OH$ |
    | "cis"-sphingosine | *cis*-4-sphingenine |
    | C-2 epimer of sphinganine | 2*L*-sphinganine 2*L*-$CH_3(CH_2)_{12}CH_2\text{-}CH_2\text{-}CHOH\text{-}CHNH_2\text{-}CH_2OH$ |
    | | 4*X*-hydroxy-2*X*,3*X*-eicosasphinganine |
    | | 4*X*-hydroxy-19-methyl-2*X*,3*X*-eicosasphinganine |

[a] If OH is deleted from carboxyl group.
[b] If OH is deleted from anomeric carbon of cyclic structure.

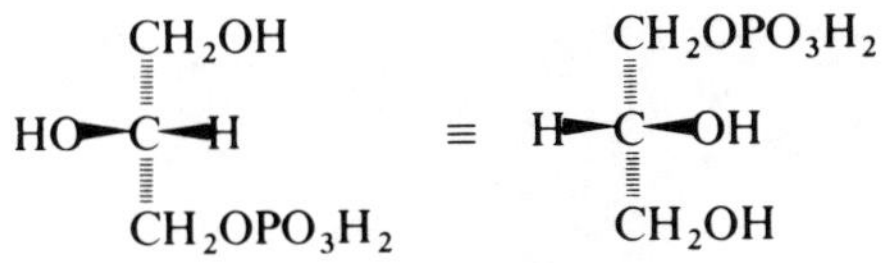

$CH_2OH$
HO—C—H ≡ H—C—OH
$CH_2OPO_3H_2$ $CH_2OH$

$CH_2OPO_3H_2$ (right structure, top)

FIGURE 1.
*sn*-Glycerol 3-phosphate
(cf. Figure 6).

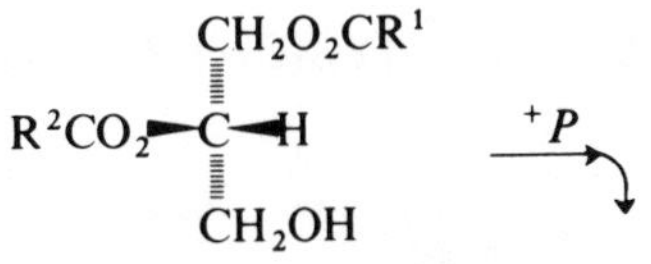

$CH_2O_2CR^1$
$R^2CO_2$—C—H $\xrightarrow{+P}$
$CH_2OH$

FIGURE 2.
(*S*)-1,2-Diacylglycerol or
1,2-diacyl-*sn*-glycerol
(see 6).

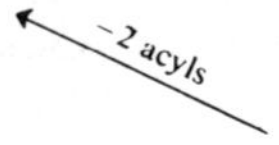

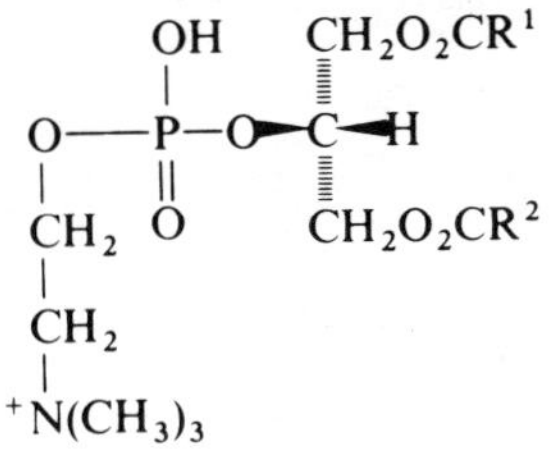

−2 acyls

$CH_2O_2CR^1$ $CH_2OPO_3H_2$
$R^2CO_2$—C—H ≡ H—C—$O_2CR^2$
$CH_2OPO_3H_2$ $CH_2O_2CR^1$

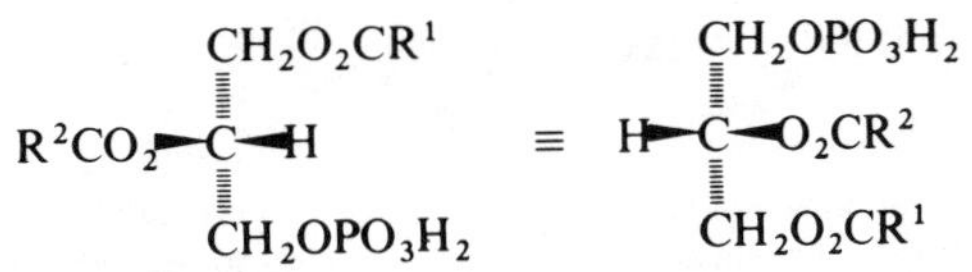

FIGURE 3.
(*R*)-Phosphatidic acid or 3-*sn*-phosphatidic acid
or 1,2-diacyl-*sn*-glycerol 3-phosphate:

OH $CH_2O_2CR^1$
O—P—O—C—H
$CH_2$ O $CH_2O_2CR^2$
$CH_2$
$^+N(CH_3)_3$

FIGURE 4.
1,3-Diacyl-2-*sn*-phosphatidylcholine.

$CH_2OH$ (1)
HO—C—H (3)
$CH_2OH$ (3)

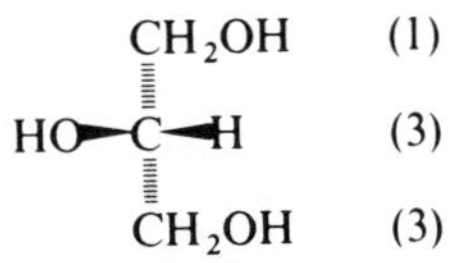

FIGURE 5.
Glycerol (*sn* numbering to right)

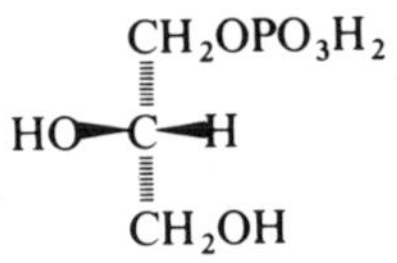

$CH_2OPO_3H_2$
HO—C—H
$CH_2OH$

FIGURE 6.
*sn*-Glycerol 1-phosphate
(cf. Figure 1).

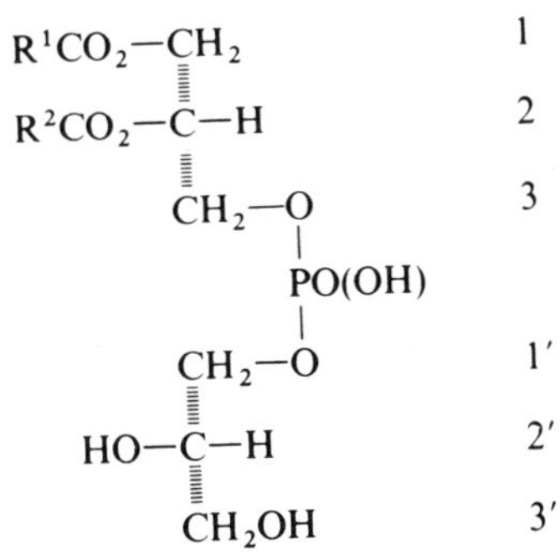

$R^1CO_2$—$CH_2$ 1
$R^2CO_2$—C—H 2
$CH_2$—O 3
PO(OH)
$CH_2$—O 1′
HO—C—H 2′
$CH_2OH$ 3′

FIGURE 7.
3-*sn*-Phosphatidyl-1′-*sn*-glycerol
(see 7).

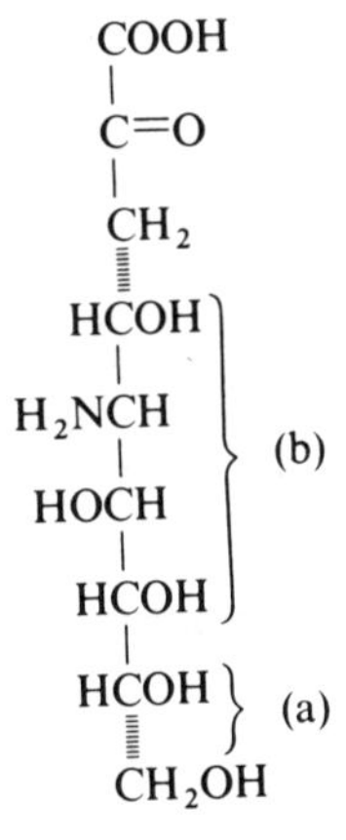

COOH
C=O
$CH_2$
HCOH
$H_2NCH$
HOCH (b)
HCOH
HCOH (a)
$CH_2OH$

FIGURE 8.
Neuraminic acid.
5-Amino-3,5-dideoxy-D-*glycero*-D-*galacto*-nonulosonic acid (see 16).

# CHEMICAL AND PHYSICAL CHARACTERISTICS OF FATTY ACIDS

| No. | Systematic Name | Common Name | Chemical Formula | Molecular Weight | Melting Point °C | Boiling point °C/mm[a] |
|---|---|---|---|---|---|---|
| **Saturated Fatty Acids** | | | | | | |
| 1. | Methanoic acid | Formic acid | HCOOH | 46.0 | 8.4 | 100.5 |
| 2. | Ethanoic acid | Acetic acid | $CH_3COOH$ | 60.1 | 16.7 | 118.2 |
| 3. | Propanoic acid | Propionic acid | $C_2H_5COOH$ | 74.1 | −22.0 | 141.1 |
| 4. | Butanoic acid | Butyric acid | $C_3H_7COOH$ | 88.1 | −7.9 | 163.5 |
| 5. | Pentanoic acid | Valeric acid | $C_4H_9COOH$ | 102.1 | −34.5 | 187 |
| 6. | Hexanoic acid | Caproic acid | $C_5H_{11}COOH$ | 116.2 | −3.4 | 205.8 |
| 7. | Heptanoic acid | Heptylic[h] acid | $C_6H_{13}COOH$ | 130.2 | −10.5 | 223.0 |
| 8. | Octanoic acid | Caprylic acid | $C_7H_{15}COOH$ | 144.2 | 16.7 | 239.7 |
| 9. | Nonanoic acid | Pelargonic acid | $C_8H_{17}COOH$ | 158.2 | 12.5 | 255.6 |
| 10. | Decanoic acid | Capric acid | $C_9H_{19}COOH$ | 172.3 | 31.6 | 270 |
| 11. | Undecanoic[i] acid | Undecylic acid | $C_{10}H_{21}COOH$ | 186.3 | 29.3 | 284 |
| 12. | Dodecanoic acid | Lauric acid | $C_{11}H_{23}COOH$ | 200.3 | 44.2 | 225/100 |
| 13. | Tridecanoic acid | Tridecylic acid | $C_{12}H_{25}COOH$ | 214.3 | 41.5 | 236/100 |
| 14. | Tetradecanoic acid | Myristic acid | $C_{13}H_{27}COOH$ | 228.4 | 53.9 | 250/100 |
| 15. | Pentadecanoic acid | Pentadecylic acid | $C_{14}H_{29}COOH$ | 242.2 | 52.3 | 202.5/10 |
| 16. | Hexadecanoic acid | Palmitic acid | $C_{15}H_{31}COOH$ | 256.4 | 63.1 | 268/100 |
| 17. | Heptadecanoic acid | Margaric acid | $C_{16}H_{33}COOH$ | 270.4 | 61.3 | 220/10 |
| 18. | Octadecanoic acid | Stearic acid | $C_{17}H_{35}COOH$ | 284.5 | 69.6 | 213/5 |
| 19. | Nonadecanoic acid | Nonadecyclic acid | $C_{18}H_{37}COOH$ | 298.5 | 68.6 | 299/10 |
| 20. | Eicosanoic acid | Arachidic acid | $C_{19}H_{39}COOH$ | 312.5 | 76.5 | 204/1 |
| 21. | Docosanoic acid | Behenic acid | $C_{21}H_{43}COOH$ | 340.6 | 81.5 | 306/60 |
| 22. | Tetracosanoic acid | Lignoceric acid | $C_{23}H_{47}COOH$ | 368.6 | 86.0 | 272/10 |
| 23. | Hexacosanoic acid | Cerotic acid | $C_{25}H_{51}COOH$ | 396.7 | 88.5 | |
| 24. | Octacosanoic acid | Montanic acid | $C_{27}H_{55}COOH$ | 424.7 | 90.9 | |
| 25. | Triacontanoic acid | Melissic acid | $C_{29}H_{59}COOH$ | 452.8 | 93.6 | |
| 26. | Dotriacontanoic acid | Lacceroic acid | $C_{31}H_{63}COOH$ | 480.0 | 96.2 | |
| 27. | Tetratriacontanoic acid | Gheddic acid | $C_{33}H_{67}COOH$ | 508.9 | 98.4 | |
| 28. | Pentatriacontanoic acid | Ceroplastic acid | $C_{34}H_{69}COOH$ | 522.9 | 98.4 | |
| **Unsaturated Fatty Acids, Monoethenoic** | | | | | | |
| 29. | *trans*-2-Butenoic acid | Crotonic acid | $C_4H_6O_2$ | 86.1 | 72 | 189.0 |
| 30. | *cis*-2-Butenoic acid | Isocrotonic acid | $C_4H_6O_2$ | 86.1 | 15.5 | 169.3 |
| 31. | 2-Hexenoic acid | Isohydrosorbic acid | $C_6H_{10}O_2$ | 114.1 | 32 | 217 |
| 31a. | *cis*-3-Decenoic acid | | $C_{10}H_{18}O_2$ | 170.2 | 10 | |
| 32. | 4-Decenoic acid | Obtusilic acid | $C_{10}H_{18}O_2$ | 170.2 | | 149/13 |
| 33. | 9-Decenoic acid | Caproleic acid | $C_{10}H_{18}O_2$ | 170.2 | | 142/4 |
| 33a. | *cis*-3-Dodecenoic acid | | $C_{12}H_{22}O_2$ | 198.3 | | |
| 33b. | *cis*-5-Dodecenoic acid | | $C_{12}H_{22}O_2$ | 198.3 | | |
| 34. | 4-Dodecenoic acid | Linderic acid | $C_{12}H_{22}O_2$ | 198.3 | 1.0–1.3 | 171/13 |
| 35. | 5-Dodecenoic acid | Denticetic acid | $C_{12}H_{22}O_2$ | 198.3 | | |
| 36. | 9-Dodecenoic acid | Lauroleic acid | $C_{12}H_{22}O_2$ | 198.3 | | 142/4 |
| 37. | 4-Tetradecenoic acid | Tsuzuic acid | $C_{14}H_{26}O_2$ | 226.4 | 18.0–18.5 | 185–188/13 |
| 38. | 5-Tetradecenoic acid | Physeteric acid | $C_{14}H_{26}O_2$ | 226.4 | | 190–195/15 |
| 38a. | *cis*-7-Tetradecenoic acid | | $C_{14}H_{26}O_2$ | 226.4 | 4.0–5.0 | |
| 39. | 9-Tetradecenoic acid | Myristoleic acid | $C_{14}H_{26}O_2$ | 226.4 | −4 | |
| 39a. | *cis*-7-Hexadecenoic acid | | $C_{16}H_{30}O_2$ | 254.4 | | |
| 40. | 9-Hexadecenoic acid | Palmitoleic acid | $C_{16}H_{30}O_2$ | 254.4 | −0.5 to +0.5 | 131/0.06 |
| 40a. | *cis*-10-Hexadecenoic acid | | $C_{16}H_{30}O_2$ | 254.4 | | |
| 40b. | *cis*-11-Hexadecenoic acid | | $C_{16}H_{30}O_2$ | 254.4 | | |

| No. | Specific Gravity[b] | Refractive Index[c] $n_D$ °C | Neutralization Value[d] | Iodine Value (Calculated)[e] | Solubility[f] | References |
|---|---|---|---|---|---|---|
| **Saturated Fatty Acids (continued)** | | | | | | |
| 1. | $1.220^{20}$ | $1.3714^{20}$ | 1,219 | | s w | 13, 28, 29 |
| 2. | $1.049^{20}$ | $1.3718^{20}$ | 934.2 | | s w | 15, 28, 29 |
| 3. | $0.992^{20}$ | $1.3874^{20}$ | 757.3 | | s al, chl, eth, w | 28, 29 |
| 4. | $0.9587^{20}$ | $1.33906^{20}$ | 636.8 | | s al, eth, w | 13, 29, 36 |
| 5. | $0.942^{20}$ | $1.4086^{20}$ | 549.3 | | s al, eth; sl s w | 29, 36 |
| 6. | $0.929^{20}$ | $1.41635^{20}$ | 483.0 | | s al, eth; sl s w | 29, 36 |
| 7. | $0.92215^{20}$ | $1.4230^{20}$ | 431.0 | | s al, eth; v sl s w | 29, 36 |
| 8. | $0.910^{20}$ | $1.4285^{20}$ | 389.1 | | s al, bz, eth; v sl s w | 13, 29, 36 |
| 9. | $0.907^{20}$ | $1.4322^{20}$ | 354.6 | | s al, chl, eth; v sl s w | 29, 36 |
| 10. | $0.8858^{40}$ | $1.42855^{40}$ | 325.7 | | s al, eth, pet eth; v sl s w | 13, 29, 36 |
| 11. | $0.9905^{25}$ | $1.4202^{70}$ | 301.2 | | s al, chl, eth, pet eth | 29, 36 |
| 12. | $0.8690^{50}$ | $1.4261^{60}$ | 280.1 | | s acet, al, eth, pet eth | 13, 29, 36 |
| 13. | $0.8458^{80}$ | $1.4286^{60}$ | 261.8 | | s acet, al, eth, pet eth | 14, 29, 31, 36 |
| 14. | $0.8622^{54}$ | $1.4273^{70}$ | 245.7 | | s acet, al, eth, pet eth | 13, 29, 36 |
| 15. | $0.8423^{80}$ | $1.4292^{70}$ | 231.5 | | s acet, al, eth, pet eth | 29, 36 |
| 16. | $0.8487^{70}$ | $1.4309^{70}$ | 218.8 | | s acet, h al, eth, pet eth | 29, 36 |
| 17. | $0.853^{60}$ | $1.4324^{70}$ | 207.5 | | s acet, h al, eth, pet eth | 29, 36 |
| 18. | $0.8390^{80}$ | $1.4337^{70}$ | 197.2 | | s acet, h al, eth, pet eth | 13, 29, 36 |
| 19. | $0.8771^{24}$ | $1.4512^{25}$ | 188.0 | | s acet, h al, eth, pet eth | 29 |
| 20. | $0.8240^{100}$ | $1.4250^{100}$ | 179.5 | | s bz, chl, eth, pet eth | 13, 29, 36 |
| 21. | $0.8221^{100}$ | $1.4270^{100}$ | 164.7 | | sl s al, eth | 13, 29, 36 |
| 22. | $0.8207^{100}$ | $1.4287^{100}$ | 152.2 | | s ac a, bz, $CS_2$, eth | 28, 29, 36, 39 |
| 23. | $0.8198^{100}$ | $1.4301^{100}$ | 141.4 | | s h acet, h chl, h me al | 13, 29, 36 |
| 24. | $0.8191^{100}$ | $1.4313^{100}$ | 132.1 | | s h ac a, h bz, h me al | 13, 29, 36 |
| 25. | | $1.4323^{100}$ | 123.9 | | s chl, $CS_2$, h me al | 13, 29, 36 |
| 26. | | | 116.7 | | s h acet, h bz, chl | 13, 17, 29, 36 |
| 27. | | | 110.2 | | s h acet, h bz, chl | 13, 29, 32, 36 |
| 28. | | | 107.3 | | s h acet, h bz, chl | 13, 29, 32, 36 |
| **Unsaturated Fatty Acids, Monoethenoic (continued)** | | | | | | |
| 29. | $0.964^{80}$ | $1.4228^{80}$ | 651.7 | 294.9 | s acet, al, tol, w | 29 |
| 30. | $1.0312^{15}$ | $1.4457^{20}$ | 651.7 | 294.9 | s al, pet eth, w | 29 |
| 31. | $0.965^{20}$ | $1.4460^{40}$ | 491.5 | 222.5 | s $CS_2$, eth | 29, 36 |
| 31a. | | | | | | 41 |
| 32. | $0.9197^{20}$ | $1.4497^{20}$ | 329.6 | 149.1 | s bz, eth | 2, 13, 29, 36 |
| 33. | $0.9238^{15}$ | $1.4507^{15}$ | 329.6 | 149.1 | s al, eth | 13, 29, 36, 37 |
| 33a. | | | | | | |
| 33b. | | | | | | |
| 34. | $0.9081^{20}$ | $1.4529^{20}$ | 282.9 | 128.0 | s bz, chl, eth | 13, 29, 36 |
| 35. | $0.9130^{15}$ | $1.4535^{15}$ | 282.9 | 128.0 | s bz, chl, eth | 13, 29, 36, 39 |
| 36. | | | 282.9 | 128.0 | s bz, chl, eth | 13, 29, 36 |
| 37. | $0.9024^{20}$ | $1.4557^{20}$ | 247.9 | 112.2 | s bz, pet eth | 2, 13, 29, 36 |
| 38. | $0.9046^{20}$ | $1.4552^{20}$ | 247.9 | 112.2 | s bz, eth, pet eth | 13, 29, 36, 39 |
| 38a. | | | | | | 43 |
| 39. | $0.9018^{20}$ | $1.4519^{20}$ | 247.9 | 112.2 | s bz, eth, pet eth | 2, 13, 29, 36 |
| 39a. | | | | | | |
| 40. | | | 220.5 | 99.8 | s bz, eth, pet eth | 2, 13, 18, 29 |
| 40a. | | | | | | |
| 40b. | | | | | | |

| No. | Systematic Name | Common Name | Chemical Formula | Molecular Weight | Melting Point °C | Boiling point °C/mm[a] |
|---|---|---|---|---|---|---|
| **Unsaturated Fatty Acids, Monoethenoic (continued)** | | | | | | |
| 40c. | 9-Heptadecenoic acid | | $C_{17}H_{32}O_2$ | 268.5 | 13.0–13.5 | |
| 41. | 6-Octadecenoic acid | Petroselinic acid | $C_{18}H_{34}O_2$ | 282.5 | 32–33 | 237.5/18 |
| 42. | *cis*-9-Octadecenoic acid | Oleic acid | $C_{18}H_{34}O_2$ | 282.5 | 13.4(α), 16.3(β) | 234/15 |
| 43. | *trans*-9-Octadecenoic acid | Elaidic acid | $C_{18}H_{34}O_2$ | 282.5 | 44.5 | 288/100 |
| 44. | *trans*-11-Octadecenoic acid | Vaccenic acid | $C_{18}H_{34}O_2$ | 282.5 | 44 | |
| 44a. | *cis*-11-Octadecenoic acid | *cis*-Vaccenic acid | $C_{18}H_{34}O_2$ | 282.5 | 13.0–14.0 | 44 |
| 45. | 9-Eicosenoic acid | Gadoleic acid | $C_{20}H_{38}O_2$ | 310.5 | 24–24.5 | 220/6 |
| 46. | 11-Eicosenoic acid | Gondoic acid | $C_{20}H_{38}O_2$ | 310.5 | 23.5–24 | 267/15 |
| 47. | 11-Docosenoic acid | Cetoleic acid | $C_{22}H_{42}O_2$ | 338.6 | 32.5–33 | |
| 48. | 13-Docosenoic acid | Erucic acid | $C_{22}H_{42}O_2$ | 338.6 | 34.7 | 242/5 |
| 49. | 15-Tetracosenoic acid | Nervonic[j] acid | $C_{24}H_{46}O_2$ | 366.6 | 42.5–43.0 | |
| 50. | 17-Hexacosenoic acid | Ximenic acid | $C_{26}H_{50}O_2$ | 394.7 | 45–45.5 | |
| 51. | 21-Triacontenoic acid | Lumequeic acid | $C_{30}H_{58}O_2$ | 450.8 | | |
| **Unsaturated Fatty Acids, Dienoic** | | | | | | |
| 52. | 2,4-Pentadienoic acid | β-Vinylacrylic acid | $C_5H_6O_2$ | 98.1 | 80 | 110 d. |
| 53. | 2,4-Hexadienoic acid | Sorbic acid | $C_6H_8O_2$ | 112.1 | 134.5 | 228 d. |
| 54. | 2,4-Decadienoic acid | Stillingic acid | $C_{10}H_{16}O_2$ | 168.2 | | |
| 55. | 2,4-Dodecadienoic acid | | $C_{12}H_{20}O_2$ | 196.3 | | |
| 56. | 9,12-Hexadecadienoic acid | | $C_{16}H_{28}O_2$ | 252.4 | | |
| 56a. | *cis*-5,9-Octadecadienoic acid | | $C_{18}H_{34}O_2$ | 280.5 | | |
| 56b. | *cis*-5,11-Octadecadienoic acid | | $C_{18}H_{34}O_2$ | 280.5 | | |
| 57. | *cis*-9,*cis*-12-Octadecadienoic acid | α-Linoleic acid | $C_{18}H_{32}O_2$ | 280.5 | -5.2 to -5.0 | 202/1.4 |
| 58. | *trans*-9,*trans*-12-Octadecadienoic acid | Linolelaidic acid | $C_{18}H_{32}O_2$ | 280.5 | 28–29 | |
| 59. | *trans*-10,*trans*-12-Octadecadienoic acid | | $C_{18}H_{32}O_2$ | 280.5 | 55.5–56 | |
| 60. | 11,14-Eicosadienoic acid | | $C_{20}H_{36}O_2$ | 308.4 | | |
| 61. | 13,16-Docosadienoic acid | | $C_{22}H_{40}O_2$ | 336.6 | | |
| 62. | 17,20-Hexacosadienoic acid | | $C_{26}H_{48}O_2$ | 392.7 | 61 | |
| **Unsaturated Fatty Acids, Trienoic** | | | | | | |
| 63. | 6,10,14-Hexadecatrienoic acid | Hiragonic acid | $C_{16}H_{26}O_2$ | 250.4 | | 180–190/15 |
| 64. | 7,10-13-Hexadecatrienoic acid | | $C_{16}H_{26}O_2$ | 250.4 | | |
| 65. | *cis*-6,*cis*-9,*cis*-12-Octadecatrienoic acid | γ-Linolenic acid | $C_{18}H_{30}O_2$ | 278.4 | | |
| 66. | *trans*-8,*trans*-10,*cis*-12-Octadecatrienoic acid | α-Calendic acid | $C_{18}H_{30}O_2$ | 278.4 | 40–40.5 | |
| 67. | *trans*-8,*trans*-10,*trans*-12-Octadecatrienoic acid | β-Calendic acid | $C_{18}H_{30}O_2$ | 278.4 | 77–78 | |
| 68. | *cis*-8,*trans*-10,*cis*-12-Octadecatrienoic acid | | $C_{18}H_{30}O_2$ | 278.4 | | |
| 69. | *cis*-9,*cis*-12,*cis*-15-Octadecatrienoic acid | α-Linolenic acid | $C_{18}H_{30}O_2$ | 278.4 | -10 to -11.3 | 157/0.001 |
| 70. | *trans*-9,*trans*-12,*trans*-15-Octadecatrienoic acid | Linolenelaidic acid | $C_{18}H_{30}O_2$ | 278.4 | 29–30 | |
| 71. | *cis*-9,*trans*-11,*trans*-13-Octadecatrienoic acid | α-Eleostearic acid | $C_{18}H_{30}O_2$ | 278.4 | 48–49 | 235/15 |
| 72. | *trans*-9,*trans*-11,*trans*-13-Octadecatrienoic acid | β-Eleostearic acid | $C_{18}H_{30}O_2$ | 278.4 | 71.5 | |

| No. | Specific Gravity$^{b}$ | Refractive Index$^{c}$ $n_D$ °C | Neutralization Value$^{d}$ | Iodine Value (Calculated)$^{e}$ | Solubility$^{f}$ | References |
|---|---|---|---|---|---|---|
| **Unsaturated Fatty Acids, Monoethenoic (continued)** | | | | | | |
| 40c. | | | | | | 41 |
| 41. | $0.8824^{35}$ | $1.4533^{40}$ | 198.6 | 89.9 | s al, eth, pet eth | 13, 21, 22, 29, 36 |
| 42. | $0.8905^{20}$ | 1.45823 20 | 198.6 | 89.9 | s acet, eth, me al | 13, 29, 36 |
| 43. | $0.851^{79}$ | $1.4468^{50}$ | 198.6 | 89.9 | s al, chl, eth, pet eth | 29 |
| 44. | $0.8563^{70}$ | $1.4406^{70}$ | 198.6 | 89.9 | s acet, me al | 13, 29, 36 |
| 44a. | | | | | | 44 |
| 45. | $0.8882^{25}$ | $1.4597^{25}$ | 180.7 | 81.8 | s acet, me al, pet eth | 13, 16, 21, 29 |
| 46. | | | 180.7 | 81.8 | s al, me al | 13, 16, 29 |
| 47. | | | 165.7 | 75.0 | s al | 13, 21, 29, 39 |
| 48. | $0.85321^{70}$ | $1.4444^{70}$ | 165.7 | 75.0 | v s eth, me al | 13, 29, 30, 36 |
| 49. | | | 153.0 | 69.2 | s acet, al, eth | 13, 21, 29, 36 |
| 50. | | | 142.2 | 64.3 | s bz, chl, eth, pet eth | 13, 29 |
| 51. | | | 124.5 | 56.3 | s bz, chl, eth, pet eth | 13, 29 |
| **Unsaturated Fatty Acids, Dienoic (continued)** | | | | | | |
| 52. | | | 572.0 | 517.5 | v s al, eth; s h w | 29, 36 |
| 53. | | | 500.4 | 452.7 | s al, eth; sl s w | 29, 36 |
| 54. | | | 333.5 | 301.7 | s acet, eth, hex | 29 |
| 55. | | | 285.8 | 258.6 | s acet, eth, pet eth | 29 |
| 56. | | | 222.3 | 201.1 | s acet, eth, pet eth | 7 |
| 56a. | | | | | | |
| 56b. | | | | | | |
| 57. | $0.9038^{18}$ | $1.4699^{20}$ | 200.1 | 181.0 | s acet, al, eth, pet eth | 13, 29, 36 |
| 58. | | | 200.1 | 181.0 | s al, eth, me al, pet eth | 29, 36 |
| 59. | | | 200.1 | 181.0 | s acet, cyc, eth | 23 |
| 60. | | | 181.9 | 164.5 | s acet, eth, pet eth | 29 |
| 61. | | | 166.7 | 150.8 | s acet, eth | 29 |
| 62. | | | 142.9 | 129.3 | s eth, pet eth | 29 |
| **Unsaturated Fatty Acids, Trienoic (continued)** | | | | | | |
| 63. | $0.9296^{20}$ | $1.4850^{50}$ | 224.1 | 304.1 | s al, eth | 13, 29, 36, 38 |
| 64. | | | 224.1 | 304.1 | s al, eth | 29 |
| 65. | | | 201.5 | 273.5 | s acet, eth, me al | 29, 36 |
| 66. | | | 201.5 | 273.5 | s acet, pent | 8, 29, 36 |
| 67. | | | 201.5 | 273.5 | s me al, pet eth | 8 |
| 68. | | | 201.5 | 273.5 | v s acet, al, pent, pet eth | 10, 29 |
| 69. | $0.914^{20}$ | $1.4678^{50}$ | 201.5 | 273.5 | s acet, al, eth, pet eth | 13, 29, 36 |
| 70. | | | 201.5 | 273.5 | s me al, pet eth | 29, 36 |
| 71. | | $1.5112^{50}$ | 201.5 | 273.5 | s al, cyc, eth, pet eth | 13, 28, 29, 36 |
| 72. | | $1.5002^{75}$ | 201.5 | 273.5 | s al, eth, me al, pet eth | 13, 29, 36 |

| No. | Systematic Name | Common Name | Chemical Formula | Molecular Weight | Melting Point °C | Boiling point °C/mm[a] |
|---|---|---|---|---|---|---|
| **Unsaturated Fatty Acids, Trienoic (continued)** | | | | | | |
| 73. | *cis*-9,*trans*-11,*cis*-13-Octadecatrienoic acid | Punicic acid | $C_{18}H_{30}O_2$ | 278.4 | 43.5–44 | |
| 74. | *trans*-9,*trans*-11,*trans*-13-Octadecatrienoic acid | | $C_{18}H_{30}O_2$ | 278.4 | | |
| 75. | 5,8,11-Eicosatrienoic acid | | $C_{20}H_{34}O_2$ | 306.5 | | |
| 76. | 8,11,14-Eicosatrienoic acid | | $C_{20}H_{34}O_2$ | 306.5 | | |
| **Unsaturated Fatty Acids, Tetranoic** | | | | | | |
| 77. | 4,8,11,14-Hexadecatetraenoic acid | | $C_{16}H_{24}O_2$ | 248.4 | | |
| 78. | 6,9,12,15-Hexadecatetraenoic acid | | $C_{16}H_{24}O_2$ | 248.4 | | |
| 79. | 4,8,12,15-Octadecatetraenoic acid | Moroctic acid | $C_{18}H_{28}O_2$ | 276.4 | | 208–213/15 |
| 80. | 6,9,12,15-Octadecatetraenoic acid | | $C_{18}H_{28}O_2$ | 276.4 | -57.4 to -56.6 | |
| 81. | 9,11,13,15-Octadecatetraenoic acid | α-Parinaric acid | $C_{18}H_{28}O_2$ | 276.4 | 85–86 | |
| 82. | 9,11,13,15-Octadecatetraenoic acid | β-Parinaric acid | $C_{18}H_{28}O_2$ | 276.4 | 95–96 | |
| 83. | 9,12,15,18-Octadecatetraenoic acid | | $C_{18}H_{28}O_2$ | 276.4 | | |
| 84. | 4,8,12,16-Eicosatetraenoic acid | | $C_{20}H_{32}O_2$ | 304.5 | | 217–220/10 |
| 85. | 5,8,11,14-Eicosatetraenoic acid | Arachidonic acid | $C_{20}H_{32}O_2$ | 304.5 | -49.5 | 163/1 |
| 86. | 6,10,14,18-Eicosatetraenoic acid? | | $C_{20}H_{32}O_2$ | 304.5 | | |
| 87. | 4,7,10,13-Docosatetraenoic acid | | $C_{22}H_{36}O_2$ | 332.5 | | |
| 88. | 7,10,13,16-Docosatetraenoic acid | | $C_{22}H_{36}O_2$ | 332.5 | | |
| 89. | 8,12,16,19-Docosatetraenoic acid | | $C_{22}H_{36}O_2$ | 332.5 | | |
| **Unsaturated Fatty Acids, Penta- and Hexaenoic** | | | | | | |
| 90. | 4,8,12,15,18-Eicosapentaenoic acid | Timnodonic acid? | $C_{20}H_{30}O_2$ | 302.5 | | |
| 91. | 5,8,11,14,17-Eicosapentaenoic acid | | $C_{20}H_{30}O_2$ | 302.5 | -54.4 to -53.8 | |
| 92. | 4,7,10,13,16-Docosapentaenoic acid | | $C_{22}H_{34}O_2$ | 330.5 | | |
| 93. | 4,8,12,15,19-Docosapentaenoic acid | Clupanodonic acid | $C_{22}H_{34}O_2$ | 330.5 | 207-212/2 | |
| 94. | 7,10,13,16,19-Docosapentaenoic acid | | $C_{22}H_{34}O_2$ | 330.5 | | |
| 95. | 4,7,10,13,16,19-Docosahexaenoic acid | | $C_{22}H_{32}O_2$ | 328.5 | -44.5 to -44.1 | |
| 96. | 4,8,12,15,18,21-Tetracosahexaenoic acid | Nisinic acid | $C_{24}H_{36}O_2$ | 356.6 | | |
| 96a. | 3-Hydroxybutanoic acid | | $C_4H_9O_3$ | 104.1 | 46.0–48.0 | |
| 96b. | 2-Methyl-3-hydroxypentanoic acid | | $C_6H_{12}O_3$ | 132.2 | | |
| 96c. | 3,5-Dihydroxy-3-methylpentanoic acid | Mevalonic acid | $C_6H_{12}O_4$ | 148.2 | 96.0–97.0 | |
| 96d. | 3-Hydroxyoctanoic acid | | $C_8H_{16}O_3$ | 160.2 | | |
| 96e. | 3-Hydroxydecanoic acid | | $C_{10}H_{20}O_3$ | 188.3 | | |
| **Hydroxyalkanoic Acids** | | | | | | |
| 97. | 2-Hydroxydodecanoic acid | 2-Hydroxylauric acid | $C_{12}H_{24}O_3$ | 216.3 | 73–74 | |
| 97a. | 3-Hydroxydodecanoic acid | β-Hydroxylauric acid | $C_{12}H_{24}O_3$ | 216.3 | 70.0–70.5 | |
| 98. | 12-Hydroxydodecanoic acid | Sabinic acid | $C_{12}H_{24}O_3$ | 216.3 | 84 | |
| 99. | 2-Hydroxytetradecanoic acid | 2-Hydroxymyristic acid | $C_{14}H_{28}O_3$ | 244.4 | 81.5-82 | |
| 99a. | 3-Hydroxytetradecanoic acid | β-Hydroxymyristic acid | $C_{14}H_{28}O_3$ | 244.4 | 72.0-73.0 | |
| 100. | 11-Hydroxypentadecanoic acid | Convolvulinolic acid | $C_{15}H_{30}O_3$ | 258.4 | 63.5–64 | |
| 101. | 2-Hydroxyhexadecanoic acid | 2-Hydroxypalmitic acid | $C_{16}H_{32}O_3$ | 272.4 | 86-87 | |

| No. | Specific Gravity[b] | Refractive Index[c] $n_D$ °C | Neutralization Value[d] | Iodine Value (Calculated)[e] | Solubility[f] | References |
|---|---|---|---|---|---|---|
| **Unsaturated Fatty Acids, Trienoic (continued)** | | | | | | |
| 73. | $0.9027^{50}$ | $1.5114^{50}$ | 201.5 | 273.5 | s al, pent, pet eth | 13, 29, 36 |
| 74. | | | 201.5 | 273.5 | s acet, al, $CS_2$, pent | 9, 29 |
| 75. | | | 183.1 | 248.3 | s $CS_2$, hept, me al | 29 |
| 76. | | | 183.1 | 248.3 | s $CS_2$, hept, me al | 29 |
| **Unsaturated Fatty Acids, Tetranoic (continued)** | | | | | | |
| 77. | | | 225.9 | 408.8 | s acet, al, eth, pet eth | 29 |
| 78. | | $1.4870^{29}$ | 225.9 | 408.8 | s acet, al, $CS_2$, eth, pent | 29 |
| 79. | $0.9297^{20}$ | $1.4911^{20}$ | 203.0 | 367.3 | s acet, al, eth, pet eth | 13, 29, 36, 38 |
| 80. | | $1.4888^{16}$ | 203.0 | 367.3 | s $CS_2$, me al | 29 |
| 81. | | | 203.0 | 367.3 | s acet, al, eth, pet eth | 13, 29, 36 |
| 82. | | | 203.0 | 367.3 | s eth, pet eth | 13, 29, 36 |
| 83. | | | 203.0 | 367.3 | s $CS_2$, me al | 29 |
| 84. | $0.9263^{20}$ | $1.4915^{20}$ | 184.3 | 333.4 | s acet, eth | 13, 29, 36 |
| 85. | $0.9082^{20}$ | $1.4824^{20}$ | 184.3 | 333.4 | s acet, eth, me al, pet eth | 13, 29, 36 |
| 86. | $0.9263^{20}$ | $1.4935^{20}$ | 184.3 | 333.4 | s acet, me al, pet eth | 29 |
| 87. | | | 168.7 | 305.4 | s acet, me al, pet eth | 24, 29 |
| 88. | | | 168.7 | 305.4 | s $CS_2$, hept, me al | 24, 29 |
| 89. | | | 168.7 | 305.4 | s acet, me al, pet eth | 1, 29 |
| **Unsaturated Fatty Acids, Penta- and Hexaenoic (continued)** | | | | | | |
| 90. | $0.9399^{15}$ | $1.5109^{15}$ | 185.5 | 419.6 | s bz, chl, eth, pet eth | 13, 29, 36 |
| 91. | | $1.4977^{23}$ | 185.5 | 419.6 | s hept, me al | 25 |
| 92. | | | 169.8 | 384.0 | s chl, hept, me al | 24, 29 |
| 93. | $0.9356^{20}$ | $1.5014^{20}$ | 169.8 | 384.0 | s acet, eth, pet eth | 13, 29, 36 |
| 94. | | | 169.8 | 384.0 | s bz, chl, me al, pet eth | 26, 29 |
| 95. | | $1.5017^{26}$ | 170.8 | 463.6 | s bz, chl, me al, pet eth | 19, 24, 29 |
| 96. | $0.9452^{20}$ | $1.5122^{20}$ | 157.4 | 427.1 | s bz, chl, eth, pet eth | 13, 29, 36 |
| 96a. | | | | | | 42 |
| 96b. | | | | | | |
| 96c. | | | | | | 42 |
| 96d. | | | | | | |
| 96e. | | | | | | |
| **Hydroxyalkanoic Acids (continued)** | | | | | | |
| 97. | | | 259.4 | | s al, me al | 29 |
| 97a. | | | | | | 41 |
| 98. | | | 259.4 | | s al, h bz | 3, 13, 29 |
| 99. | | | 229.1 | | s al, chl, eth | 29 |
| 99a. | | | | | | 41 |
| 100. | | | 217.1 | | s al, chl, eth | 13, 29, 34 |
| 101. | | | 206.0 | | s al, me al | 29, 40 |

| No. | Systematic Name | Common Name | Chemical Formula | Molecular Weight | Melting Point °C | Boiling point °C/mm[a] |
|---|---|---|---|---|---|---|
| **Hydroxyalkanoic Acids (continued)** | | | | | | |
| 101a. | 3-Hydroxyhexadecanoic acid | β-Hydroxypalmitic acid | $C_{16}H_{32}O_3$ | 272.4 | | |
| 102. | 11-Hydroxyhexadecanoic acid | Jalapinolic acid | $C_{16}H_{32}O_3$ | 272.4 | 68–69 | |
| 103. | 16-Hydroxyhexadecanoic acid | Juniperic acid | $C_{16}H_{32}O_3$ | 272.4 | 95 | |
| 104. | 2-Hydroxyoctadecanoic acid | 2-Hydroxystearic acid | $C_{18}H_{36}O_3$ | 300.5 | 91 | |
| 104a. | 3-Hydroxyoctadecanoic acid | β-Hydroxystearic acid | $C_{18}H_{36}O_3$ | 300.5 | | |
| 105. | 23-Hydroxydocosanoic acid | Phellonic acid | $C_{22}H_{44}O_3$ | 356.6 | 95–96 | |
| 106. | 2-Hydroxytetracosanoic acid | Cerebronic acid | $C_{24}H_{48}O_3$ | 384.6 | 99.5–100.5 | |
| 107. | 3,11-Dihydroxytetradecanoic acid | Ipurolic acid | $C_{14}H_{28}O_4$ | 260.4 | 100-101 | |
| 108. | 2,15-Dihydroxypentadecanoic acid | Dihydroxypentadecyclic acid | $C_{15}H_{30}O_4$ | 274.4 | 102-103 | |
| 109. | 15,16-Dihydroxyhexadecanoic acid | Ustilic acid A | $C_{16}H_{32}O_4$ | 288.4 | 112-113 | |
| 110. | 9,10-Dihydroxyoctadecanoic acid | 9,10-Dihydroxystearic acid | $C_{18}H_{36}O_4$ | 316.5 | 141 | |
| 111. | 9,10-Dihydroxyoctadecanoic acid | 9,10-Dihydroxystearic acid | $C_{18}H_{36}O_4$ | 316.5 | $90^{10}$ | |
| 112. | 11,12-Dihydroxyeicosanoic acid | 11,12-Dihydroxyarachidic acid | $C_{20}H_{40}O_4$ | 344.5 | $130^{9}$ | |
| 113. | 2,15,16-Trihydroxyhexadecanoic acid | Ustilic acid | $C_{16}H_{32}O_5$ | 304.4 | 140 | |
| 114. | 9,10,16-Trihydroxyhexadecanoic acid | Aleuritic acid | $C_{16}H_{32}O_5$ | 304.4 | 100 | |
| **Keto, Epoxy, and Cyclo Fatty Acids** | | | | | | |
| 115. | 4-Ketopentanoic acid | Levulinic acid | $C_5H_8O_3$ | 116.1 | 37.2 | 154/15 |
| 115a. | *cis*-9,10-Methylenehexadecanoic acid | | $C_{17}H_{32}O_2$ | 268.4 | | |
| 116. | 6-Ketooctadecanoic acid | Lactarinic acid | $C_{18}H_{34}O_3$ | 298.5 | 87 | |
| 117. | 4-Keto-9,11,13,octadecatrienoic acid | α-Licanic acid | $C_{18}H_{28}O_3$ | 292.4 | 74-75 | |
| 118. | 4-Keto-*trans*-9,-*trans*-11,-*trans*-13-octadecatrienoic acid | β-Licanic acid | $C_{18}H_{28}O_3$ | 292.4 | 99.5 | |
| 119. | *cis*-12,13-Epoxy-*cis*-9-octadecenoic acid | Vernolic acid | $C_{18}H_{32}O_3$ | 296.5 | 31–32 | |
| 120. | *cis*-9,10-Epoxyoctadecanoic acid | Epoxystearic acid | $C_{18}H_{34}O_3$ | 298.5 | 57.5-58 | |
| 120a. | *cis*-9,10-Methyleneoctadecanoic acid | Dihydrosterculic acid | $C_{19}H_{36}O_2$ | 296.5 | 39.7–40.5 | |
| 121. | ω-(2-*n*-Octylcycloprop-1-enyl)-octanoic acid | Sterculic acid | $C_{19}H_{34}O_2$ | 294.5 | 18 | |
| 122. | ω-(2-*n*-Octylcyclopropyl)-octanoic acid | Lactobacillic acid | $C_{19}H_{36}O_2$ | 296.5 | 28–29 | |
| 123. | 13-(2-Cyclopentenyl)-tridecanoic acid | Chaulmoogric acid | $C_{18}H_{32}O_2$ | 280.2 | 68.5 | 247.5/20 |
| 124. | 11-(2-Cyclopentenyl)-hendecanoic acid | Hydnocarpic acid | $C_{16}H_{28}O_2$ | 252.2 | 60.5 | |
| 125. | 9-(2-Cyclopentenyl)-nonanoic acid | Alepric acid | $C_{14}H_{24}O_2$ | 224.2 | 48.0 | |
| 126. | 7-(2-Cyclopentenyl)-heptanoic acid | Aleprylic acid | $C_{12}H_{20}O_2$ | 196.2 | 32.0 | |

| No. | Specific Gravity[b] | Refractive Index[c] $n_D$ °C | Neutralization Value[d] | Iodine Value (Calculated)[e] | Solubility[f] | References |
|---|---|---|---|---|---|---|
| **Hydroxyalkanoic Acids (continued)** | | | | | | |
| [illegible]1a. | | | | | | |
| 102. | | | 206.0 | | s al, eth | 13, 29, 33 |
| 103. | | | 206.0 | | s al, bz, eth | 13, 21, 29, 36 |
| 104. | | | 186.7 | | s al, me al | 29, 40 |
| 104a. | | | | | | |
| 105. | | | 157.3 | | s acet, chl, eth, glac ac a pvr | 5, 13, 29, 36 |
| 106. | | | 145.9 | | s acet, h al, eth, pyr | 13, 29, 36 |
| 107. | | | 215.5 | | s chl, eth | 13, 29, 33 |
| 108. | | | 204.5 | | s me al | 29 |
| 109. | | | 194.5 | | s me al | 29 |
| 110. | | | 177.3 | | s h al; sl s eth | 13, 29 |
| 111. | | | 177.3 | | s al, eth, h w | 29 |
| 112. | | | 162.9 | | s acet, eth | 13, 29 |
| 113. | | | 184.3 | | s me al | 29 |
| 114. | | | 184.3 | | s me al | 29 |
| **Keto, Epoxy, And Cyclo Fatty Acids (continued)** | | | | | | |
| 115. | $1.1395^{20}$ | $1.442^{15.8}$ | 483.2 | | v s al, eth, w | 29 |
| 115a. | | | | | | |
| 116. | | | 188.0 | | s h al, chl, eth | 13, 21, 29, 36 |
| 117. | | | 191.9 | 260.4 | s h pet eth | 4, 15, 29, 36 |
| 118. | | | 191.9 | 260.4 | s h pet eth | 4, 15, 29, 36 |
| 119. | | | 189.3 | 85.6 | s acet, al, hex | 27, 29 |
| 120. | | | 188.0 | | s acet, al, hex | 6 |
| 120a. | | | | | | 44 |
| 121. | | | 190.5 | 86.2 | s eth | 29 |
| 122. | | | 189.2 | | s acet, eth, pet eth | 29 |
| 123. | | | 200.1 | 90.5 | s acet, chl, eth | 12, 13, 21, 29, 36 |
| 124. | | | 222.3 | 100.6 | s al, chl, pet eth | 21, 29, 36 |
| 125. | | | 250.1 | 113.1 | s al, eth pet eth | 29 |
| 126. | | | 285.8 | 129.3 | s acet, eth, pet eth | 29 |

| No. | Systematic Name | Common Name | Chemical Formula | Molecular Weight | Melting Point °C | Boiling point °C/mm[a] |
|---|---|---|---|---|---|---|
| **Keto, Epoxy, and Cyclo Fatty Acids (continued)** | | | | | | |
| 127. | 5-(2-Cyclopentenyl)-pentanoic acid | Aleprestic acid | $C_{10}H_{16}O_2$ | 168.1 | Liquid | |
| 128. | 2-Cyclopentenyl-1-oic acid | Aleprolic acid | $C_6H_8O_2$ | 112.1 | Liquid | |
| 129. | 13-(2-Cyclopentenyl)-6-tridecenoic acid | Gorlic acid | $C_{18}H_{30}O_2$ | 278.2 | 6.0 | 232.5 |
| 129a. | 3-Hydroxy-5-dodecenoic acid | | $C_{12}H_{22}O_3$ | 214.3 | | |
| **Hydroxy Unsaturated Acids** | | | | | | |
| 130. | 16-Hydroxy-7-hexadecenoic acid | Ambrettolic acid | $C_{16}H_{30}O_3$ | 270.5 | 25 | |
| 131. | 9-Hydroxy-12-octadecenoic acid | | $C_{18}H_{34}O_3$ | 298.5 | | |
| 132. | *d*-12-Hydroxy-*cis*-9-octadecenoic acid | Ricinoleic acid | $C_{18}H_{34}O_3$ | 298.5 | 5, 7.7, & 16 | 225/10 |
| 133. | *d*-12-Hydroxy-*trans*-9-octadecenoic acid | Ricinelaidic acid | $C_{18}H_{34}O_3$ | 298.5 | 52–53 | |
| 134. | 2-Hydroxy-15-tetracosenoic acid | Hydroxynervonic acid | $C_{24}H_{46}O_3$ | 382.6 | 65 | |
| 135. | 9-Hydroxy-10,12-octadecadienoic acid | | $C_{18}H_{32}O_3$ | 296.5 | | |
| 136. | 13-Hydroxy-9,11-octadecadienoic acid | | $C_{18}H_{32}O_3$ | 296.5 | | |
| 137. | 18-Hydroxy-*cis*-9,*trans*-11,-*trans*-13-octadecatrienoic acid | α-Kamlolenic acid | $C_{18}H_{30}O_3$ | 294.4 | 77–78 | |
| 138. | 18-Hydroxy-*trans*-9,*trans*-11,-*trans*-13-octadecatrienoic acid | β-Kamlolenic acid | $C_{18}H_{30}O_3$ | 294.4 | 88–89 | |
| **Branched-Chain Fatty Acids** | | | | | | |
| 139. | 3-Methylbutanoic acid | Isovaleric acid | $C_5H_{10}O_2$ | 102.1 | –37.6 | 176.7 |
| 139a. | 2-Methyl-3-hydroxy-pentanoic acid | | $C_6H_{12}O_3$ | 132.2 | | |
| 139b. | 6-Methylheptanoic acid | | $C_8H_{16}O_3$ | 144.2 | 0 | |
| 140. | *d*-6-Methyloctanoic acid | | $C_9H_{18}O_2$ | 158.2 | | |
| 141. | 8-Methyldecanoic acid | | $C_{11}H_{22}O_2$ | 186.3 | –18.5 | |
| 142. | 10-Methylhendecanoic acid | Isolauric acid | $C_{12}H_{24}O_2$ | 200.3 | 41.2 | |
| 143. | *d*-10-Methyldodecanoic acid | | $C_{13}H_{26}O_2$ | 214.3 | 6.2-6.5 | |
| 144. | 11-Methyldodecanoic acid | Isoundecylic acid | $C_{13}H_{26}O_2$ | 214.3 | 39.4–40 | |
| 145. | 12-Methyltridecanoic acid | Isomyristic acid | $C_{14}H_{28}O_2$ | 228.4 | 53.6 | |
| 146. | *d*-12-Methyltetradecanoic | | $C_{15}H_{30}O_2$ | 242.4 | 25.8 | |
| 147. | 13-Methyltetradecanoic acid | Isopentadecylic acid | $C_{15}H_{30}O_2$ | 242.4 | 52.2 | |
| 148. | 14-Methylpentadecanoic acid | Isopalmitic acid | $C_{16}H_{32}O_2$ | 256.4 | 62.4 | |
| 149. | *d*-14-Methylhexadecanoic acid | | $C_{17}H_{34}O_2$ | 270.4 | 38.0 | |
| 150. | 15-Methylhexadecanoic acid | | $C_{17}H_{34}O_2$ | 270.4 | 60.5 | |
| 151. | 10-Methylheptadecanoic acid | | $C_{18}H_{36}O_2$ | 284.5 | 33.5 | |
| 152. | 16-Methylheptadecanoic acid | Isostearic acid | $C_{18}H_{36}O_2$ | 284.5 | 69.5 | |
| 152a. | 9-Methyloctadecanoic acid | | $C_{19}H_{38}O_2$ | 298.5 | 38.5–39.1 | |
| 153. | *l*-D-10-Methyloctadecanoic acid | Tuberculostearic acid | $C_{19}H_{38}O_2$ | 298.5 | 13.2 | 175–178/0.7 |
| 154. | *d*-16-Methyloctadecanoic acid | | $C_{19}H_{38}O_2$ | 298.5 | 49.9-50.7 | |
| 155. | 18-Methylnonadecanoic acid | Isoarachidic acid | $C_{20}H_{40}O_2$ | 312.5 | 75.3 | |
| 156. | *d*-18-Methyleicosanoic acid | | $C_{21}H_{42}O_2$ | 326.6 | 55.6 | |

| No. | Specific Gravity[b] | Refractive Index[c] $n_D$ °C | Neutralization Value[d] | Iodine Value (Calculated)[e] | Solubility[f] | References |
|---|---|---|---|---|---|---|
| **Keto, Epoxy, and Cyclo Fatty Acids (continued)** | | | | | | |
| 127. | | | 333.5 | 150.8 | s acet, eth, pet eth | 29 |
| 128. | | | 500.4 | 226.4 | s acet, eth, pet eth | 29 |
| 129. | $0.9436^{25}$ | $1.4782^{25}$ | 201.5 | 182.5 | s h al | 11, 13, 29, 36 |
| 129a. | | | | | | |
| **Hydroxy Unsaturated Acids (continued)** | | | | | | |
| 130. | | | 207.5 | 93.9 | s al, eth | 29, 36 |
| 131. | | | 188.0 | 85.0 | s acet, al, eth | 29 |
| 132. | $0.940^{27.4}$ | $1.4716^{20}$ | 188.0 | 85.0 | s acet, al, eth | 13, 29, 36 |
| 133. | | | 188.0 | 85.0 | s acet, al, eth | 13, 29, 36 |
| 134. | | | 146.6 | 66.3 | s acet, al, chl, eth, pyr; sl s pet eth | 13, 29, 36 |
| 135. | | | 189.2 | 171.2 | s acet, al, pent | 8 |
| 136. | | | 189.2 | 171.2 | s acet, al, pent | 8 |
| 137. | | | 190.5 | 258.6 | | 29 |
| 138. | | | 190.5 | 258.6 | | 29 |
| **Branched-Chain Fatty Acids (continued)** | | | | | | |
| 139. | $0.937^{15}$ | $1.40178^{22.4}$ | 549.3 | | s al, chl, eth; sl s w | 13, 28, 29, 36 |
| 139a. | | | | | | |
| 139b. | | | | | | 41 |
| 140. | | | 354.6 | | s acet, eth, me al, pet eth | 13, 29, 36, 40 |
| 141. | | | 301.2 | | s acet, eth, me al, pet eth | 13, 29, 36, 40 |
| 142. | | | 280.1 | | s acet, eth, me al, pet eth | 13, 29, 36, 40 |
| 143. | | $1.4424^{25}$ | 261.8 | | s bz, chl, me al, pet eth | 13, 29, 36, 40 |
| 144. | | $1.4293^{60}$ | 261.8 | | s acet, al, me al, pet eth | 29 |
| 145. | | | 245.7 | | s acet, me al, pet eth | 13, 29, 36, 40 |
| 146. | | $1.4327^{59}$ | 231.5 | | s chl, eth, me al, pet eth | 13, 29, 36, 40 |
| 147. | | $1.4312^{59}$ | 231.5 | | s me al, pet eth | 29 |
| 148. | | $1.4293^{70}$ | 218.8 | | s acet, eth, me al, pet eth | 13, 29, 36, 40 |
| 149. | | | 207.5 | | s acet, eth, me al, pet eth | 13, 29, 36, 40 |
| 150. | | $1.4315^{70}$ | 207.5 | | s acet, eth, pet eth | 13, 29, 36, 40 |
| 151. | | | 197.2 | | s acet, glac ac a | 20 |
| 152. | | | 197.2 | | s acet, eth, pet eth | 13, 29, 36, 40 |
| 152a. | | | | | | 41 |
| 153. | $0.887^{25}$ | $1.4512^{25}$ | 188.0 | | s acet, al, me al, pent | 13, 29, 36 |
| 154. | | | 188.0 | | s acet, me al, pet eth | 13, 29, 35, 36 40 |
| 155. | | | 179.5 | | s al, eth, pet eth | 13, 29, 36, 40 |
| 156. | | | 171.8 | | s acet, chl, pet eth | 13, 29, 36, 40 |

| No. | Systematic Name | Common Name | Chemical Formula | Molecular Weight | Melting Point °C | Boiling point °C/mm[a] |
|---|---|---|---|---|---|---|
| **Branched-Chain Fatty Acids (continued)** | | | | | | |
| 157. | 20-Methylheneicosanoic acid | Isobehenic acid | $C_{22}H_{44}O_2$ | 340.6 | 79.5 | |
| 158. | *d*-20-Methyldocosanoic acid | | $C_{23}H_{46}O_2$ | 354.6 | 62.1 | |
| 159. | 22-Methyltricosanoic acid | Isolignoceric acid | $C_{24}H_{48}O_2$ | 368.6 | 83.1 | |
| 160. | *d*-22-Methyltetracosanoic acid | | $C_{25}H_{50}O_2$ | 382.7 | 67.8 | |
| 161. | 24-Methylpentacosanoic acid | Isocerotic acid | $C_{26}H_{52}O_2$ | 396.7 | 86.9 | |
| 162. | *d*-24-Methylhexacosanoic acid | | $C_{27}H_{54}O_2$ | 410.7 | 72.9 | |
| 163. | 26-Methylheptacosanoic acid | Isomontanic acid | $C_{28}H_{56}O_2$ | 424.7 | 89.3 | |
| 164. | *d*-28-Methyltriacontanoic acid | | $C_{31}H_{62}O_2$ | 466.8 | 80.7 | |
| 165. | 2,4,6-(D)-Trimethyloctacosanoic acid | Mycoceranic acid[k] | $C_{31}H_{62}O_2$ | 466.8 | 27-28 | |
| 166. | 2-Methyl-*cis*-2-butenoic acid | Angelic acid | $C_5H_8O_2$ | 100.1 | 45 | 185 |
| 167. | 2-Methyl-*trans*-2-butenoic acid | Tiglic acid | $C_5H_8O_2$ | 100.1 | 65.5 | 198.5 |
| 168. | 4-Methyl-3-pentenoic acid | Pyroterebic acid | $C_6H_{10}O_2$ | 114.1 | | 207 |
| 169. | *d*-2,4(L),6(L)-Trimethyl-*trans*-2-tetracosenoic acid | $C_{27}$-Phthienoic acid[l] | $C_{27}H_{52}O_2$ | 408.7 | 39.5–41 | |

[a] At 760 mm of mercury (atmospheric pressure), unless otherwise specified; d = decomposition.
[b] At the temperature indicated in the superscript, referred to water at 4°C.
[c] Refractive index (n) given is for the sodium D-line at the temperature indicated in the superscript.
[d] Milligrams of KOH required to neutralize one gram of acid.
[e] Grams of iodine absorbed by 100 grams of acid.
[f] s = soluble; sl = slightly; v = very.
[g] a = acid; ac = acetic; acet = acetone; al = alcohol; bz = benzene; chl = chloroform; cyc = cyclohexane; eth = ether; glac = glacial; h = hot; hept = heptane; hex = hexane; me = methyl; pent = pentane; pet = petroleum; pyr = pyridine; tol = toluene; w = water

| No. | Specific Gravity[b] | Refractive Index[c] $n_D$ °C | Neutralization Value[d] | Iodine Value (Calculated)[e] | Solubility[f] | References |
|---|---|---|---|---|---|---|

**Branched-Chain Fatty Acids (continued)**

| No. | Specific Gravity[b] | Refractive Index[c] $n_D$ °C | Neutralization Value[d] | Iodine Value (Calculated)[e] | Solubility[f] | References |
|---|---|---|---|---|---|---|
| 157. | | | 164.7 | | s chl, eth, me al, pet eth | 13, 29, 36, 40 |
| 158. | | | 158.2 | | s acet, chl, eth, pet eth | 13, 29, 36, 40 |
| 159. | | | 152.2 | | s acet, chl, pet eth | 13, 29, 36, 40 |
| 160. | | | 146.6 | | s al, bz, chl, pet eth | 13, 29, 36, 40 |
| 161. | | | 141.4 | | s acet, chl, glac ac a | 13, 29, 36, 40 |
| 162. | | | 136.6 | | s bz, chl, glac ac a, pet eth | 13, 29, 36, 40 |
| 163. | | | 132.1 | | s bz, chl, glac ac a, pet eth | 13, 29, 36, 40 |
| 164. | | | 120.2 | | s bz, chl, glac ac a, pet eth | 13, 29, 36, 40 |
| 165. | | | 120.2 | | s ch, pet eth | 13, 29 |
| 166. | $0.983^{47}$ | $1.4434^{47}$ | 560.4 | 253.6 | v s eth; s al; sl s w | 29 |
| 167. | | $1.4342^{81}$ | 560.4 | 253.6 | v s h w; s al, eth | 29 |
| 168. | | | 491.6 | 222.4 | s al, chl, eth | 29 |
| 169. | | $1.4598^{25}$ | 137.3 | 62.1 | s acet, me al, pet eth | 29 |

*h* Also called enanthic acid.
*i* Also called hendecanoic acid.
*j* Also called selacholeic acid.
*k* Also called mycoserosic acid.
*l* Also called mycolipenic acid.

The majority of the data were originally compiled by Klare S. Markley for the *Biology Data Book,* pp. 370–381, and reproduced in modified form by permission of the copyright owners (© 1964), Federation of American Societies for Experimental Biology, Washington, D.C. Additional data were supplied by William M. O'Leary.

## REFERENCES

1. Baudert, *Bull. Soc. Chim. Fr. Ser. 5, 9,* 922 (1942).
2. Bosworth and Brown, *J. Biol. Chem., 103,* 115 (1933).
3. Bougault and Bourdier, *J. Pharm. Chim. Ser. 6, 30,* 10 (1909).
4. Brown and Farmer, *Biochem. J., 29,* 631 (1935).
5. Chibnall, Piper and Williams, *Biochem. J., 30,* 100 (1936).
6. Chisholm and Hopkins, *Chem. Ind. (London),* p. 1154 (1959).
7. Chisholm and Hopkins, *Can. J. Chem., 38,* 805 (1960).
8. Chisholm and Hopkins, *Can. J. Chem., 38,* 2500 (1960).
9. Chisholm and Hopkins, *J. Chem. Soc.,* p. 573 (1962).
10. Chisholm and Hopkins, *J. Org. Chem., 27,* 3137 (1962).
11. Cole and Cardoss, *J. Amer. Chem. Soc., 60,* 612 (1938).
12. Cole and Cardoss, *J. Amer. Chem. Soc., 61,* 2349 (1939).
13. Deuel, H. J., Jr., *The Lipids: Their Chemistry and Biochemistry.* Interscience Publications, John Wiley and Sons, New York (1951–1957).
14. Dorinson, McCorkle and Ralston, *J. Amer. Chem. Soc., 64,* 2739 (1942).
15. Dyson, *A Manual of Organic Chemistry,* Vol. 1. Longmans and Green, London, England (1950).
16. Foreman and Brown, *Oil Soap (Chicago), 21,* 183 (1944).
17. Francis and Piper, *J. Amer. Chem. Soc., 61,* 577 (1939).
18. Gupta, Grollman and Niyogy, *Proc. Nat. Inst. Sci. India, 19,* 519 (1953).
19. Hammond and Lundberg, *J. Amer. Oil Chem. Soc., 30,* 438 (1953).
20. Hansen, Shorland and Cooke, *Chem. Ind. (London),* p. 839 (1951).
21. Heilbron, *Dictionary of Organic Compounds.* Eyre and Spottiswood, London, England (1934).
22. Hilditch and Jones, *Biochem. J., 22,* 326 (1928).
23. Hopkins and Chisholm, *Chem. Ind. (London),* p. 2064 (1962).
24. Klenk and Bongard, *Z. Physiol. Chem., 291,* 104 (1952).
25. Klenk and Montag, *Ann. Chem., 604,* 4 (1957).
26. Klenk and Tomuschat, *Z. Physiol. Chem., 308,* 165 (1957).
27. Krewson, Ard and Riemenschneider, *J. Amer. Oil Chem. Soc., 39,* 334 (1962).
28. Lange, N. A., *Handbook of Chemistry,* 6th ed. Handbook Publications, Sandusky, Ohio (1946).
29. Markley, K. S., *Fatty Acids,* 2nd ed., Parts 1 and 2. Interscience Publications, John Wiley and Sons, New York (1960–1961).
30. Noller and Talbot, *Organic Synthesis Collection,* Vol. 12. John Wiley and Sons, New York (1943).
31. Nunn, *J. Chem. Soc.,* p. 313 (1952).
32. Piper *et al., Biochem. J., 28,* 2185 (1934).
33. Power and Rogerson, *J. Amer. Chem. Soc., 32,* 106 (1910).
34. Power and Rogerson, *J. Chem. Soc., 101(T),* 1 (1912).
35. Prout, Cason and Ingersoll, *J. Amer. Chem. Soc., 69,* 1233 (1947).
36. Ralston, *Fatty Acids and Their Derivations.* John Wiley and Sons, New York (1948).
37. Smedley, *Biochem. J., 6,* 451 (1912).
38. Teresi, J. D., Unpublished Data. U.S. Naval Radiological Defense Laboratory, San Francisco, California.
39. Warth, *The Chemistry and Technology of Waxes,* 2nd ed. Reinhold Books, New York (1956).
40. Weitkamp, *J. Amer. Chem. Soc., 67,* 447 (1945).
41. Doss, *Properties of the Principal Fats, Fatty Oils, Waxes and Fatty Acids.* Texas Co., New York (1952).
42. Stacher (Ed.), *The Merck Index,* 8th ed. Merck and Co., Rahway, New Jersey (1968).
43. Hofmann, O'Leary, Yoho and Liu, *J. Biol. Chem., 234,* 1672 (1959).
44. Hofmann, *Fatty Acid Metabolism in Microorganisms,* John Wiley and Sons, New York (1963).

# STRUCTURES OF REPRESENTATIVE LIPID TYPES FOUND IN MICROORGANISMS

DR. WILLIAM M. O'LEARY

## TYPES OF PHOSPHOLIPIDS

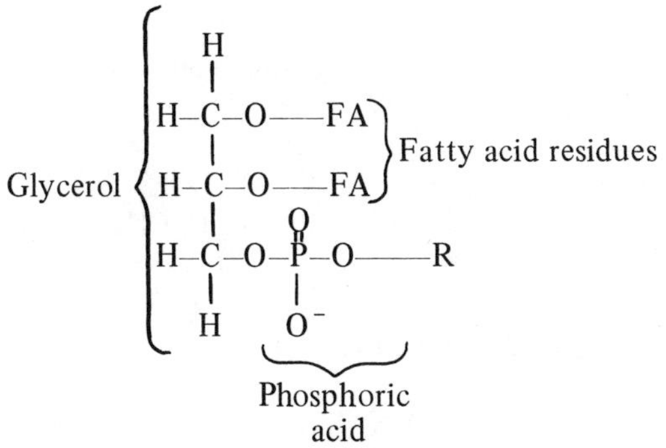

```
           ┌  H
           │  |
           │H–C–O——FA ┐
           │  |        } Fatty acid residues
  Glycerol { H–C–O——FA ┘
           │  |   O
           │  |   ‖
           │H–C–O–P–O——R
           │  |   |
           └  H   O⁻
                └──┬──┘
               Phosphoric
                 acid
```

| R | Compound |
|---|---|
| $-H$ | Phosphatidic acid |
| $-CH_2-CH_2-\overset{+}{N}-(CH_3)_3$ | Phosphatidyl choline |
| $-CH_2-CH_2-NH_3^+$ | Phosphatidyl ethanolamine |
| $-CH_2-CH(COOH)-NH_3^+$ | Phosphatidyl serine |
| 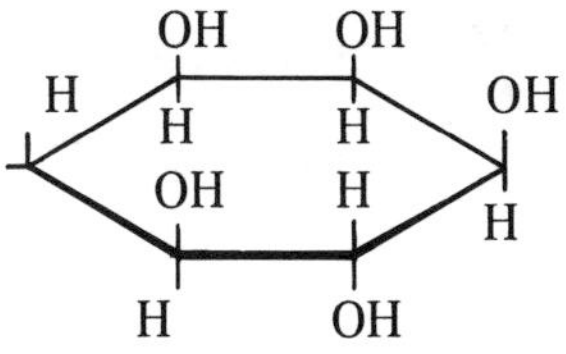 Inositol ring (substituents OH, OH, H, OH, H, H, OH, H, H, OH, H, OH) | Phosphatidyl inositol |
| $-CH_2-CHOH-CH_2OH$ | Phosphatidyl glycerol |
| 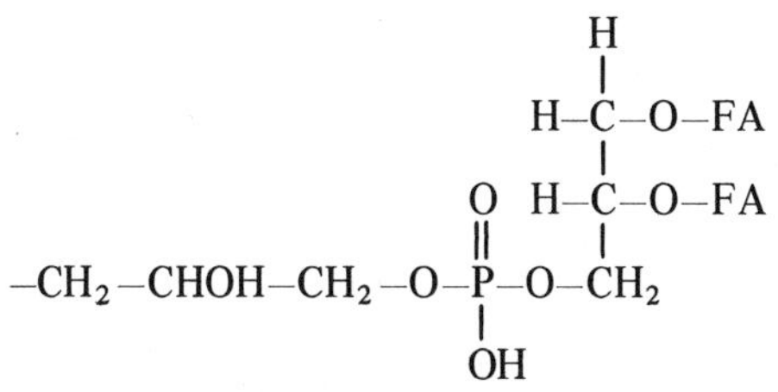 $-CH_2-CHOH-CH_2-O-P(=O)(OH)-O-CH_2-CH(O-FA)-CH_2(O-FA)$ (H–C–O–FA / H–C–O–FA / CH$_2$) | Diphosphatidyl glycerol (cardiolipin) |
| 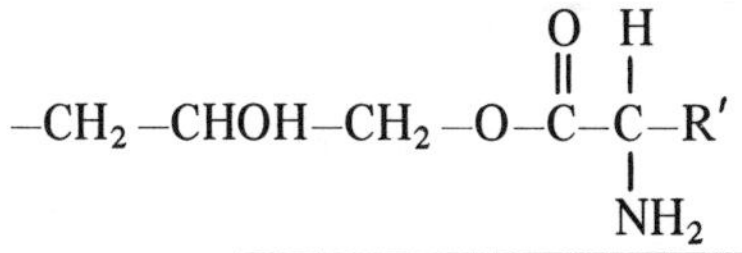 $-CH_2-CHOH-CH_2-O-C(=O)-CH(NH_2)-R'$ | 3′-O-Aminoacyl phosphatidyl glycerol |

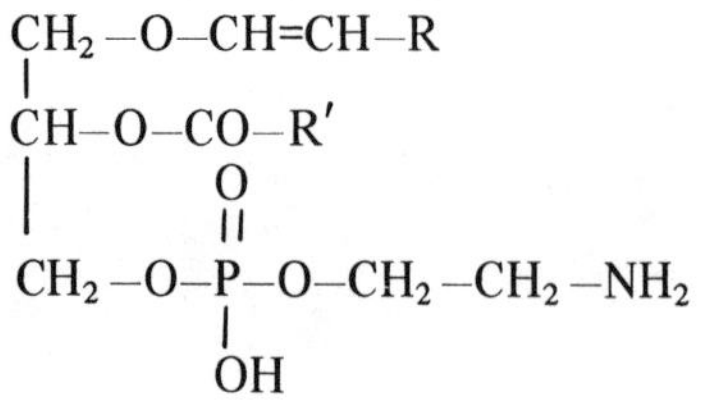

$$\begin{array}{l} CH_2-O-CH=CH-R \\ | \\ CH-O-CO-R' \\ | \qquad\quad O \\ | \qquad\quad \| \\ CH_2-O-P-O-CH_2-CH_2-NH_2 \\ \qquad\quad | \\ \qquad\quad OH \end{array}$$

Plasmalogen

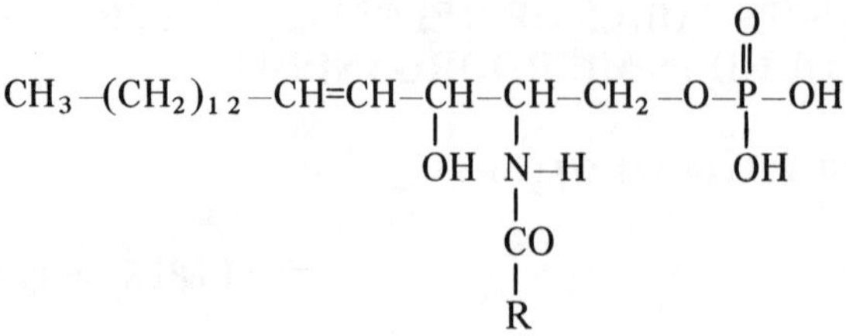

$$\begin{array}{l} \qquad\qquad\qquad\qquad\qquad\qquad\qquad\qquad\qquad O \\ \qquad\qquad\qquad\qquad\qquad\qquad\qquad\qquad\qquad \| \\ CH_3-(CH_2)_{12}-CH=CH-CH-CH-CH_2-O-P-OH \\ \qquad\qquad\qquad\qquad\quad | \quad\;\; | \qquad\qquad\quad | \\ \qquad\qquad\qquad\qquad\; OH \;\; N-H \qquad\quad OH \\ \qquad\qquad\qquad\qquad\qquad\;\; | \\ \qquad\qquad\qquad\qquad\qquad\; CO \\ \qquad\qquad\qquad\qquad\qquad\;\; | \\ \qquad\qquad\qquad\qquad\qquad\;\; R \end{array}$$

Ceramide phosphate

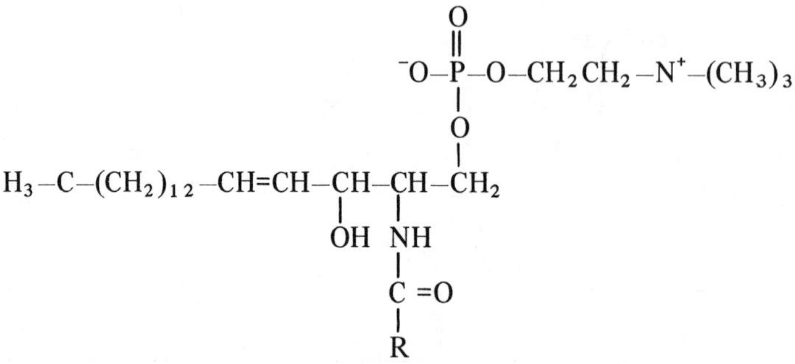

$$\begin{array}{l} \qquad\qquad\qquad\qquad\qquad\qquad\quad O \\ \qquad\qquad\qquad\qquad\qquad\qquad\quad \| \\ \qquad\qquad\qquad\qquad\qquad\; ^{-}O-P-O-CH_2CH_2-N^{+}-(CH_3)_3 \\ \qquad\qquad\qquad\qquad\qquad\qquad\quad | \\ \qquad\qquad\qquad\qquad\qquad\qquad\quad O \\ \qquad\qquad\qquad\qquad\qquad\qquad\quad | \\ H_3-C-(CH_2)_{12}-CH=CH-CH-CH-CH_2 \\ \qquad\qquad\qquad\qquad\quad | \quad\;\; | \\ \qquad\qquad\qquad\qquad\; OH \;\; NH \\ \qquad\qquad\qquad\qquad\qquad\;\; | \\ \qquad\qquad\qquad\qquad\qquad\; C=O \\ \qquad\qquad\qquad\qquad\qquad\;\; | \\ \qquad\qquad\qquad\qquad\qquad\;\; R \end{array}$$

Sphingomyelin

$$\begin{array}{l} CH_2OOCR \\ | \\ RCOOCH \qquad O \\ | \qquad\qquad\quad \| \\ CH_2-O-P-O-CH_2-CH-CH_2OH \\ \qquad\quad | \qquad\qquad\quad | \\ \qquad\quad O^{-} \qquad\qquad | \\ \qquad\qquad\qquad\qquad\;\; O \end{array}$$

CH₂OH, O, OH, HO, NH₂ (glucosamine ring)

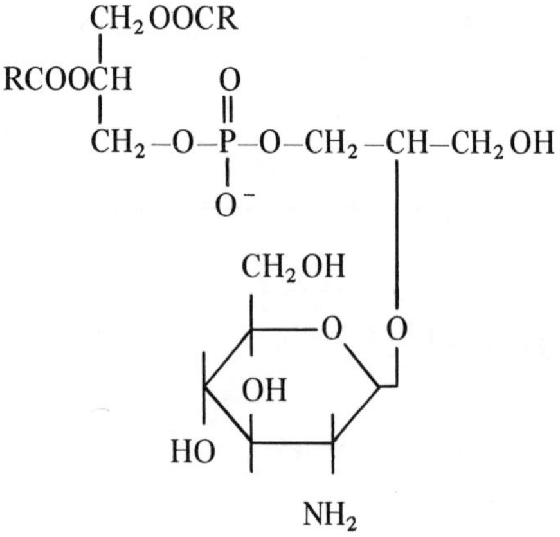

Glucosaminylphosphatidylglycerol

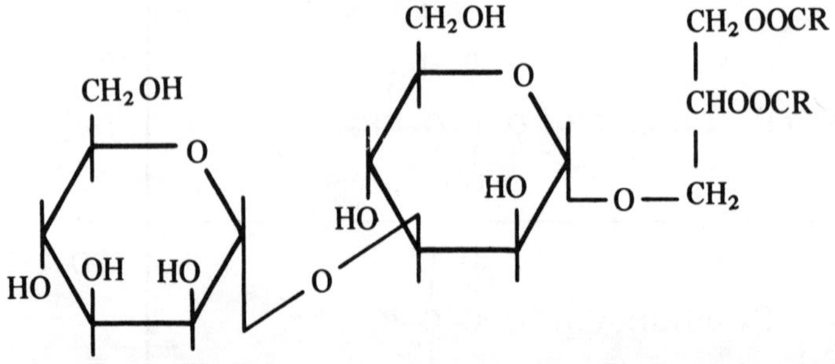

CH₂OH, O, O—CH₂, O, O—CH₂, HO, OH, OH, HO, OH, OH

$$\begin{array}{l} CH_2 \\ | \\ CHOOCR \\ | \\ CH_2OOCR' \end{array}$$

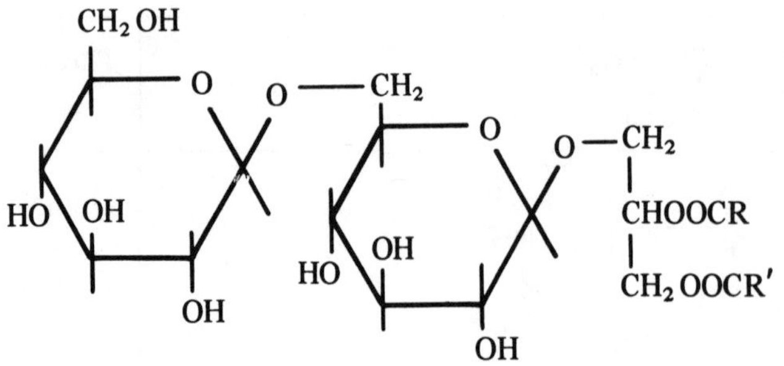

β-D-Glucosyl-(1→6)-β-D-glucosyl-(1→3)-diglyceride

CH₂OH, O, HO, OH, HO, O, CH₂OH, O, HO, HO

$$\begin{array}{l} CH_2OOCR \\ | \\ CHOOCR \\ | \\ O-CH_2 \end{array}$$

α-D-Mannosyl-(1→3)-α-D-mannosyl-(1→3)-diglyceride

HO, $CH_3$, O, O, HO, $CH_3$, O, OH, OH, OH

$OCHCH_2COOCHCH_2COOH$ with $(CH_2)_6CH_3$ on each CH

Rhamnolipid from *Pseudomonas aeruginosa*

$$HO-\underset{CH_3}{CH}-CH_2-\underset{O}{\overset{\|}{C}}\left[O-\underset{CH_3}{CH}-CH_2-\underset{O}{\overset{\|}{C}}\right]_n O-\underset{CH_3}{CH}-CH_2-\underset{O}{\overset{\|}{C}}-OH$$

Poly-$\beta$-hydroxybutyrate

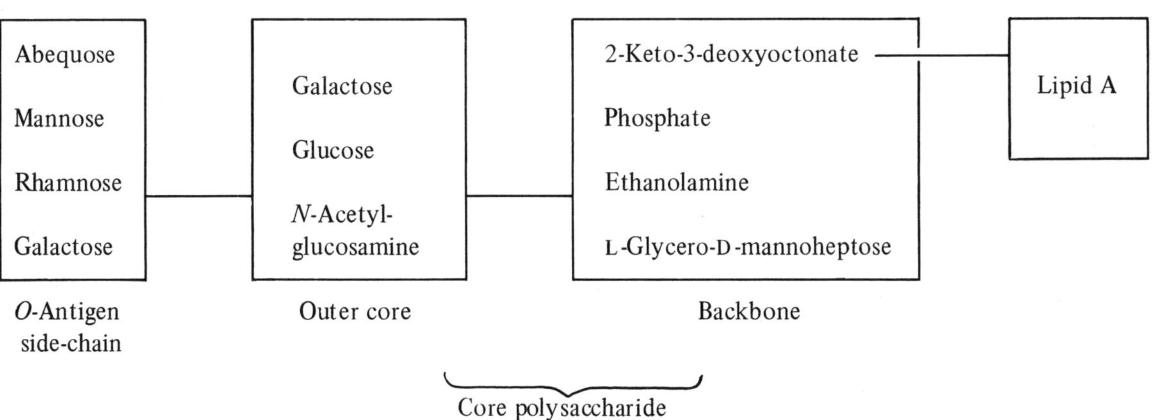


Schematic structure of a lipopolysaccharide (from *Salmonella typhimurium*)

Taken from: O'Leary, W. M., in *Comprehensive Biochemistry,* Vol. 18, p. 253, M. Florkin and E. Stotz, Eds. (1970). Reproduced by permission of the Federation of American Societies for Experimental Biology, Bethesda, Maryland.

# STEROLS

| Systematic Name[a] (Trivial Name) | Structure | Formula (Mol. Wt.) | Melting Point, °C | $[\alpha]^b$ |
|---|---|---|---|---|
| Cholest-5,7-diene-3β-ol (7-Dehydrocholesterol) | 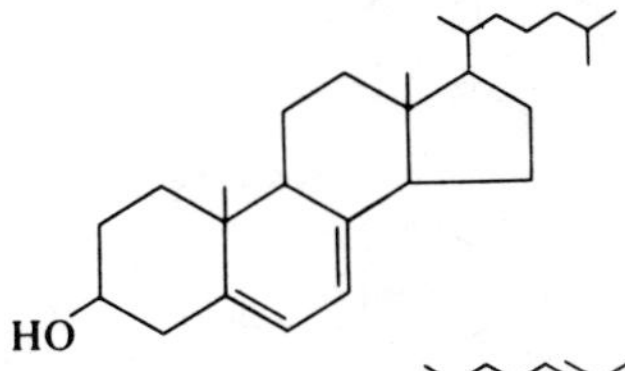 HO | $C_{27}H_{44}O$ (384.6) | 150 | –114° |
| Cholest-5,24(25)-diene-3β-ol (24-Dehydrocholestadione-3β-ol) | 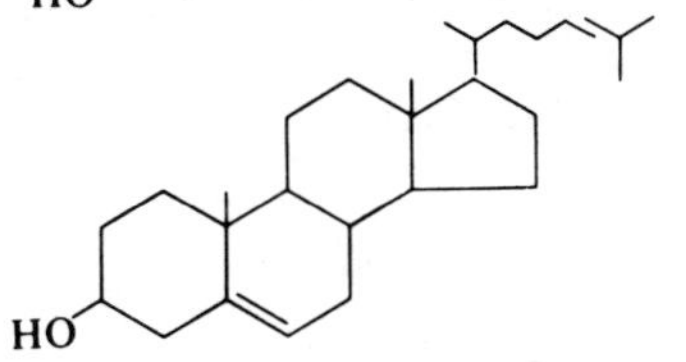 HO | $C_{27}H_{44}O$ (384.6) | 117 | –38° |
| Cholest-5-ene-3β-ol (Cholesterol) | 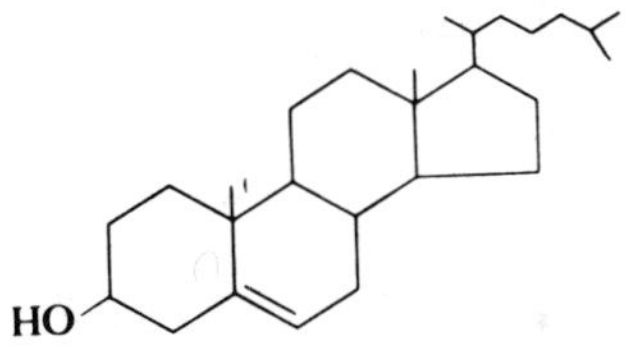 HO | $C_{27}H_{46}O$ (386.6) | 149 | –39° |
| Coprostan-3β-ol (Coprostanol) | 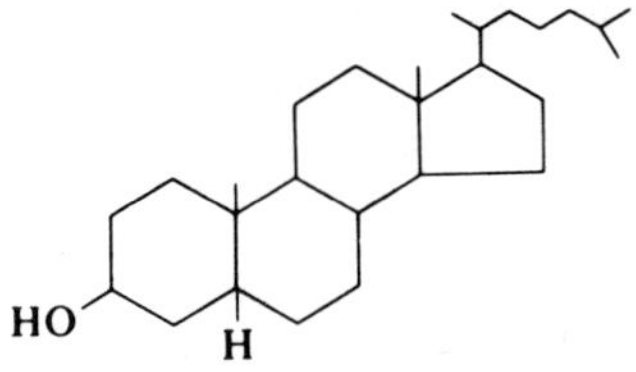 HO H | $C_{27}H_{48}O$ (388.4) | 101 | +28° |
| Cholestan-3β-ol (Cholestanol) | 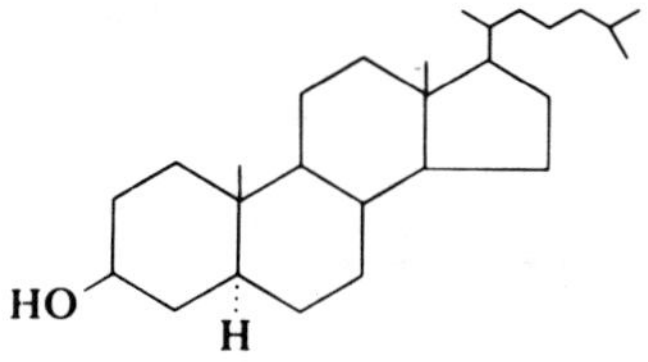 HO H | $C_{27}H_{48}O$ (388.4) | 140–142 | +23° |
| Ergosta-5,7,22-triene-3β-ol (Ergosterol) | 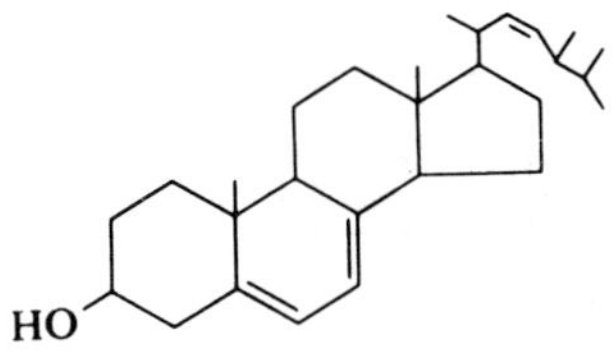 HO | $C_{28}H_{44}O$ (396) | 165 | –130° |
| Stigmasta-5,22-diene-3β-ol (Stigmasterol) | 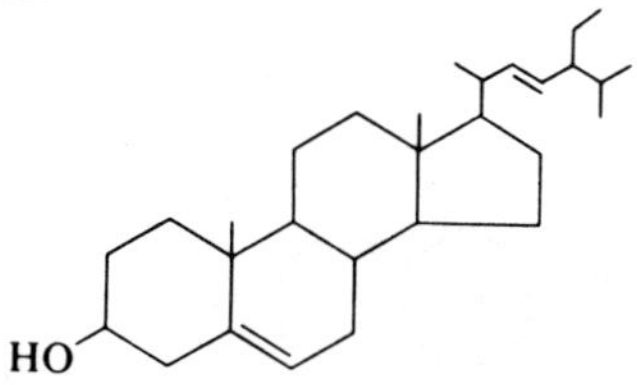 HO | $C_{29}H_{48}O$ (412) | 170 | –49° |

[a] Numbers after symbol Δ indicate the position of double bonds in the basic cyclopentano perhydrophenanthrene ring.
[b] Specific rotation at the sodium D line, with chloroform as the solvent.

Compiled by F. Edward Roberts and T. Windholz. Data were selected from *Biology Data Book*, pp. 385-386, P. L. Altman and D. S. Dittmer, Eds. (1964). Reproduced by permission of the Federation of American Societies for Experimental Biology, Washington, D.C.

| Systematic Name[a] (Trivial Name) | Structure | Formula (Mol. Wt.) | Melting Point, °C | [a] [b] |
|---|---|---|---|---|
| Stigmast-5-ene-3β-ol (β-Sitosterol) | 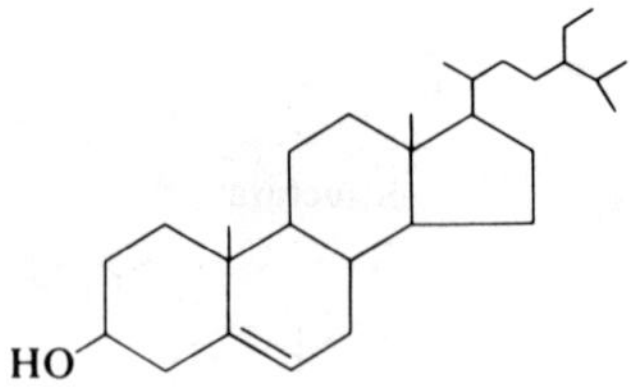 HO | $C_{29}H_{50}O$ (414) | 140 | −36° |
| $\Delta^{3,5}$-Cholestadien-7-one | 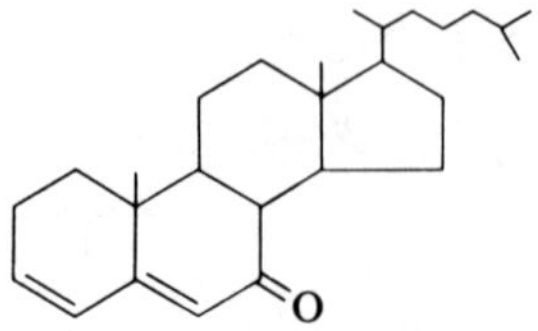 O | $C_{27}H_{42}O$ (382.6) | 112 | −305° |
| $\Delta^{4,6}$-Cholestadien-3-one | 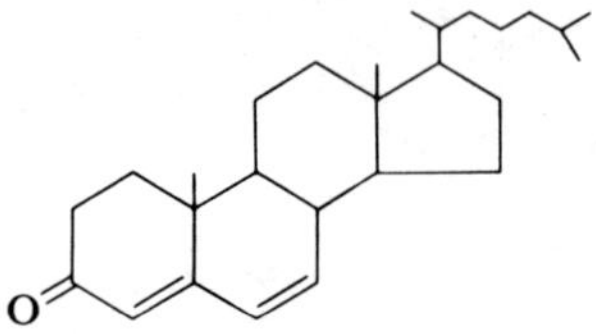 O | $C_{27}H_{242}O$ (382.6) | 80 | +35° |
| $\Delta^{5,7,22}$-Cholestatrien-3β-ol | 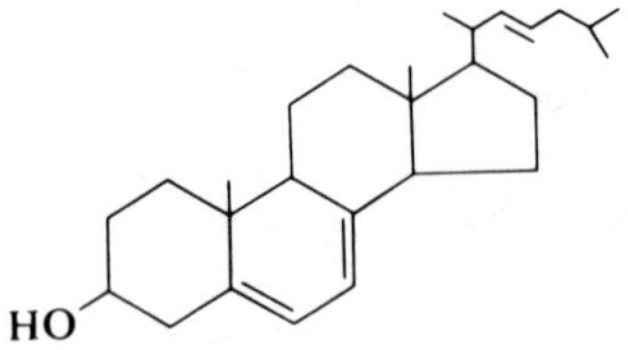 HO | $C_{27}H_{42}O$ (382.6) | | |
| $\Delta^{4}$-Cholesten-3-one | 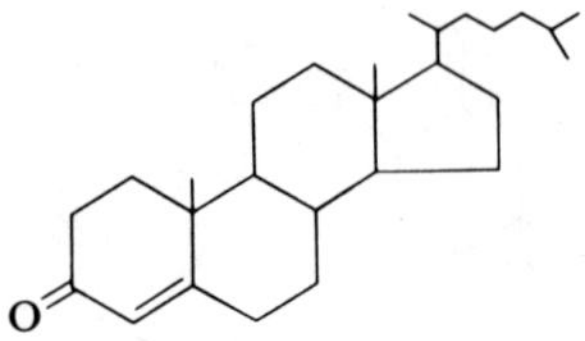 O | $C_{27}H_{44}O$ (384.6) | 81 | +89° |
| $\Delta^{5,22}$-Cholestadien-3β-ol (22-Dehydrocholesterol) | 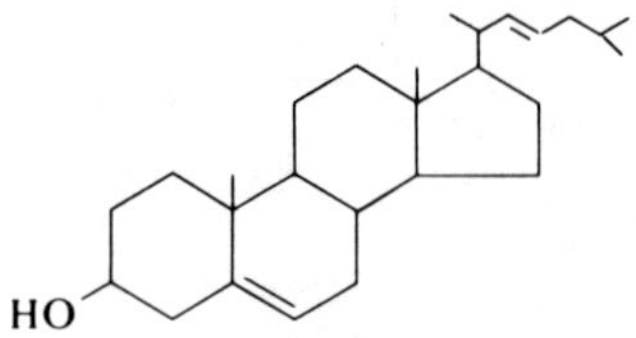 HO | $C_{27}H_{44}O$ (384.6) | 135 | −57° |
| $\Delta^{8,24}$-Cholestadien-3β-ol (Zymosterol) | 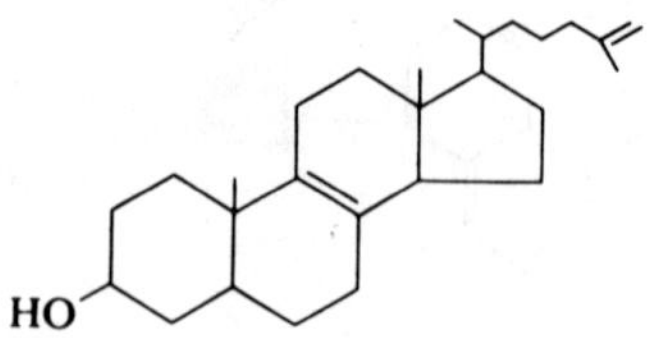 HO | $C_{27}H_{44}O$ (384.6) | 108 | +47° |
| Cholestane-3,6-dione | 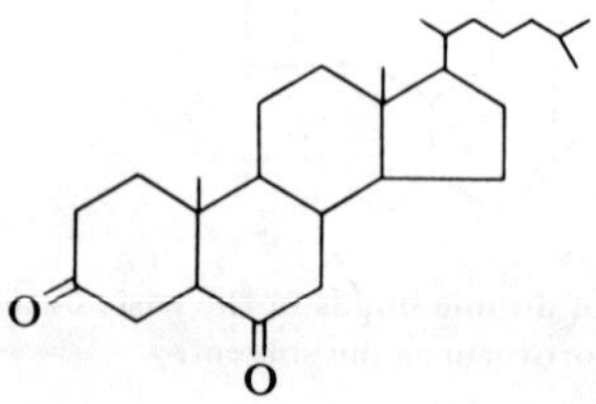 O O | $C_{27}H_{44}O_2$ (400.6) | 175 | |

| Systematic Name[a] (Trivial Name) | Structure | Formula (Mol. Wt.) | Melting Point, °C | $[\alpha]^b$ |
|---|---|---|---|---|
| $\Delta^5$-Cholesten-3β-ol-7-one (7-Ketocholesterol) | 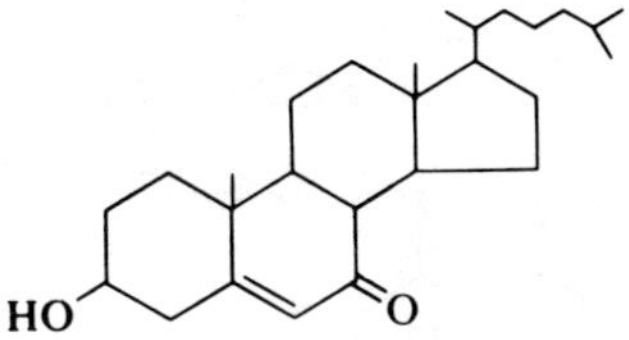 | $C_{27}H_{44}O_2$ (400.6) | 170 | −104° |
| $\Delta^7$-Cholesten-3β-ol (Lathosterol) | 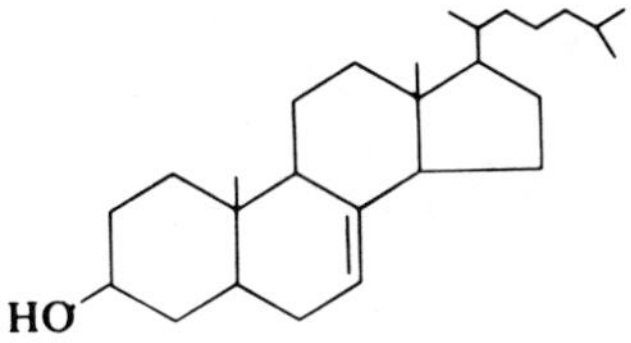 | $C_{27}H_{46}O$ (386.6) | 122 | +5.7° |
| Coprostan-3-one | 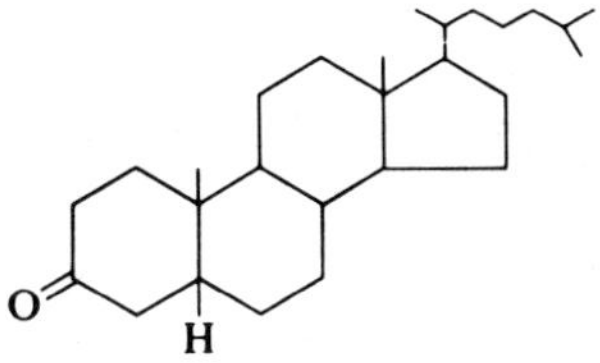 | $C_{27}H_{46}O$ (386.6) | 63 | +36° |
| $\Delta^4$-Cholestene-3β,6β-diol | 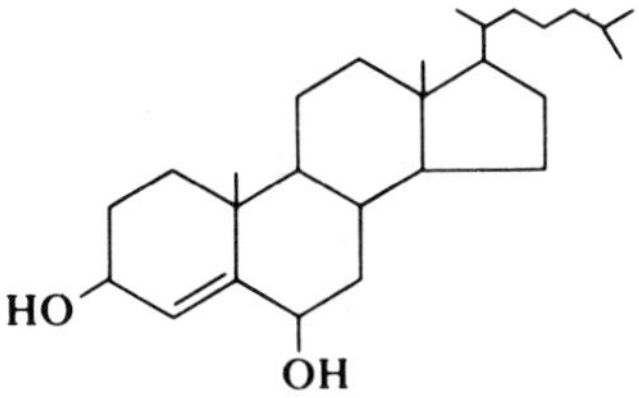 | $C_{27}H_{46}O_2$ (402.6) | 258 | +9° |
| $\Delta^5$-Cholestene-3β,7α-diol (7α-Hydroxycholesterol) | 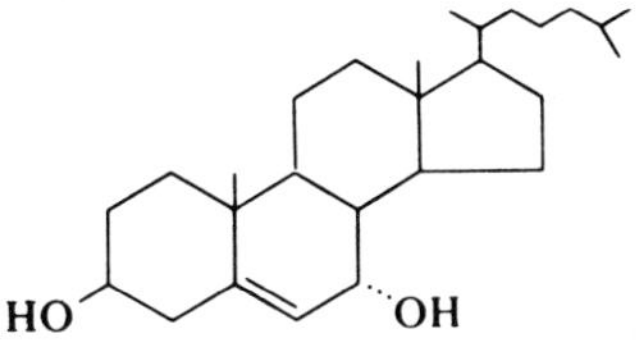 | $C_{27}H_{46}O_2$ (402.6) | 184 | −93° |
| $\Delta^5$-Cholestene-3β,7β-diol (7β-Hydroxycholesterol) | 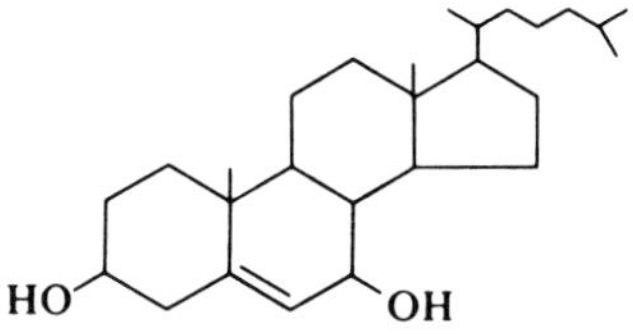 | $C_{27}H_{46}O_2$ (402.6) | 178 | +7° |
| $\Delta^5$-Cholestan-3β,20α-diol (22-Hydroxycholesterol) | 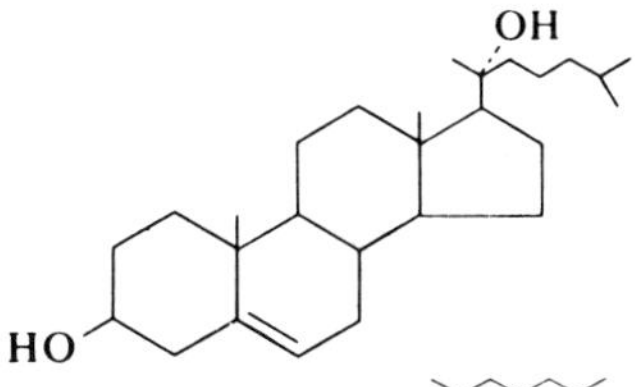 | $C_{27}H_{46}O_2$ (402.6) | 186 | −39° |
| Cholestan-3β-ol-6-one | 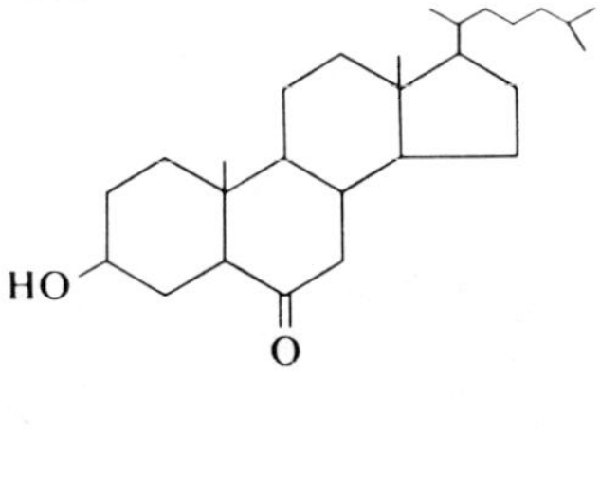 | $C_{27}H_{46}O_2$ (402.6) | 143 | |

| Systematic Name[a] (Trivial Name) | Structure | Formula (Mol. Wt.) | Melting Point, °C | $[\alpha]^b$ |
|---|---|---|---|---|
| Cholestane-3β,5α-diol-6-one | 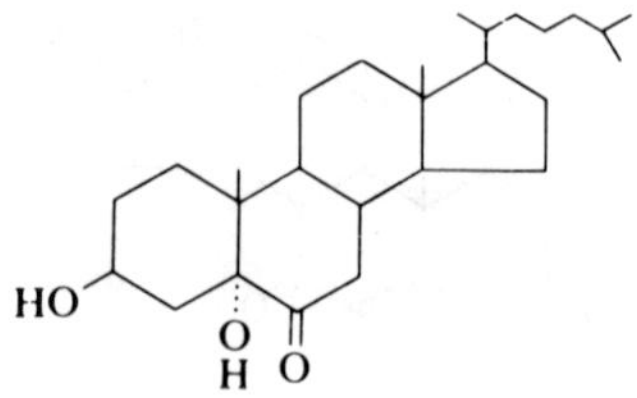 | $C_{27}H_{46}O_3$ (418.6) | 236 | |
| Cholestan-3β-ol (Cholestanol) | 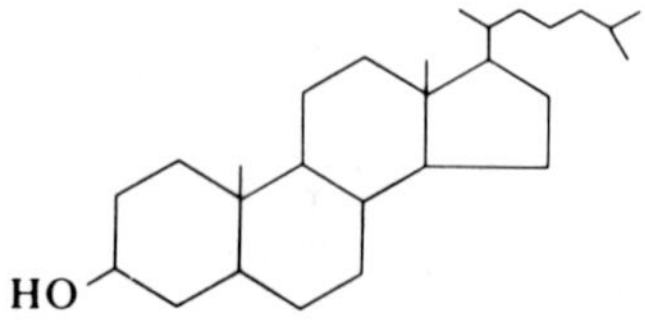 | $C_{27}H_{48}O$ (388.6) | 142 | +24° |
| Coprostan-3β-ol (Coprostanol) | 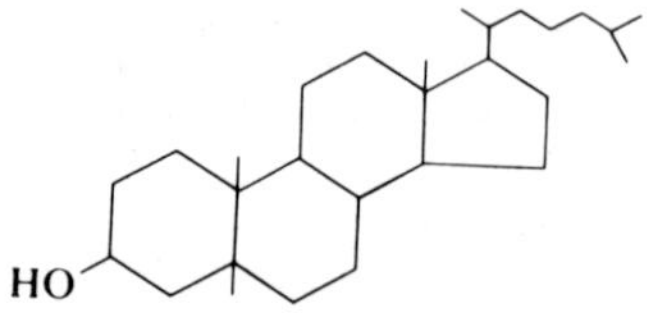 | $C_{27}H_{48}O$ (388.6) | 101 | +28° |
| Coprostan-3α-ol (Epicoprostanol) | 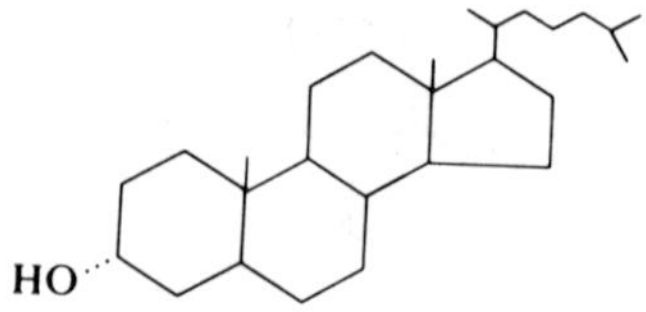 | $C_{27}H_{48}O$ (388.6) | 117 | +32° |
| Cholestane-3β,5,6β-triol | 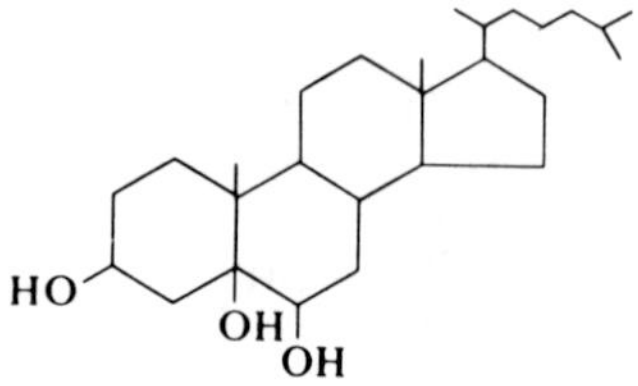 | $C_{27}H_{48}O_3$ (420.6) | 239 | +3° |
| $\Delta^{5,7,9(11),22}$-Ergostatetraen-3β-ol (Dehydroergosterol) | 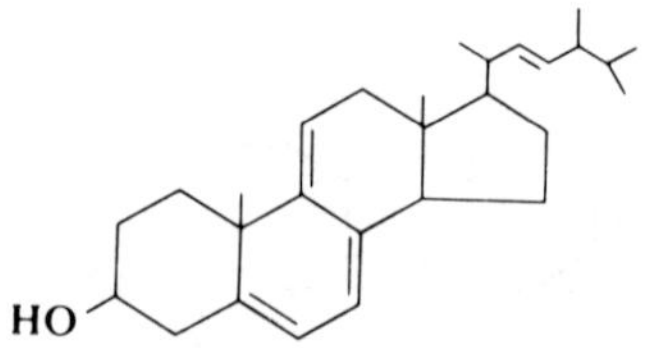 | $C_{28}H_{42}O$ (394.6) | 146 | +149° |
| $\Delta^{5,7,14,22}$-Ergostatetraen-3β-ol (14-dehydroergosterol) | 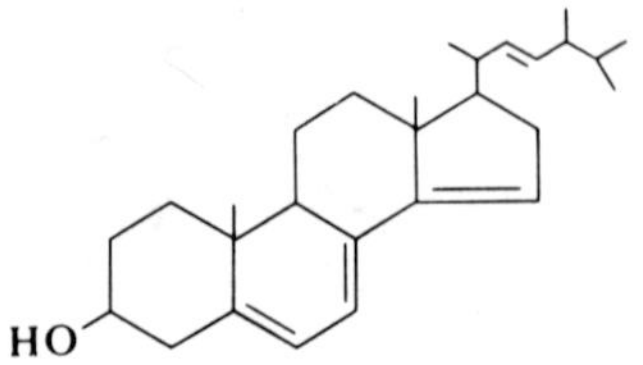 | $C_{28}H_{42}O$ (394.6) | 198 | –396° |
| $\Delta^{5,7,22,24(28)}$-Ergostatetraen-3β-ol (24-Dehydroergosterol) | 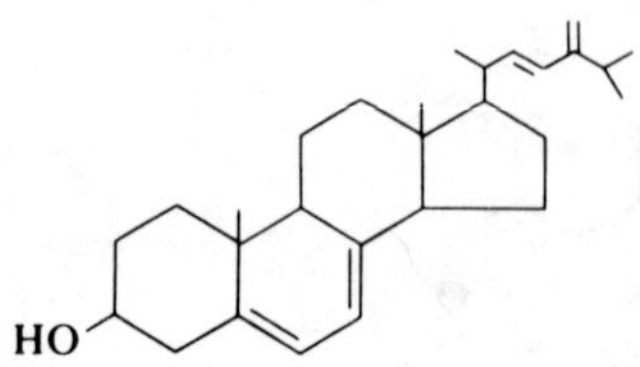 | $C_{28}H_{42}O$ (394.6) | 118 | –78° |

| Systematic Name[a] (Trivial Name) | Structure | Formula (Mol. Wt.) | Melting Point, °C | [α][b] |
|---|---|---|---|---|
| $\Delta^{7}$-Ergosten-3β-ol (Fungisterol) | 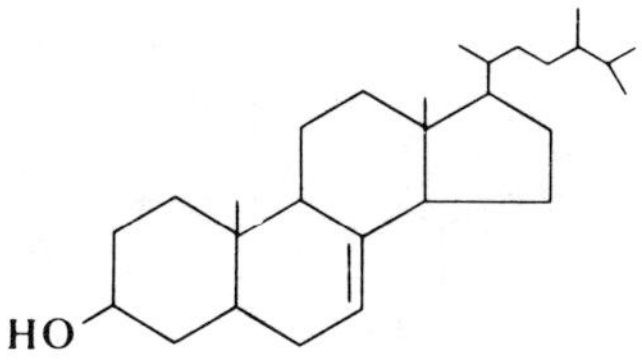 | $C_{28}H_{44}OO$ (396.6) | 148 | −0.2° |
| $\Delta^{5,7,22}$-Ergostatrie-3β-ol (Ergosterol) | 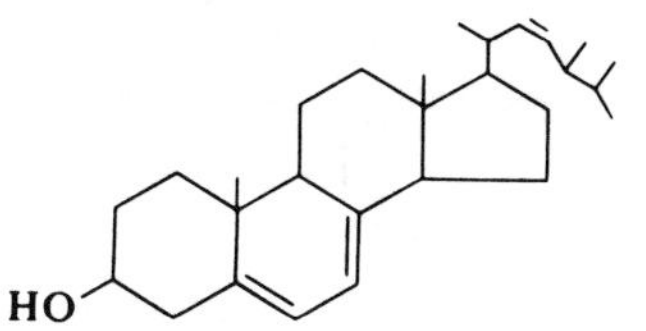 | $C_{28}H_{44}O$ (396.6) | 165 | −130° |
| $\Delta^{5,7}$-Ergosta-dien-3β-ol (22-Dihydroergosterol) | 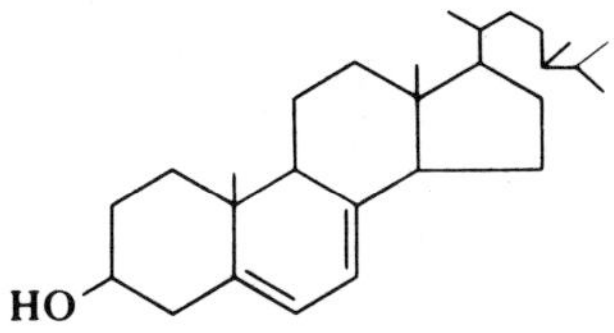 | $C_{28}H_{46}O$ (398.6) | 153 | −109° |
| $\Delta^{5,22}$-Ergostadien-3β-ol (Brassicasterol) | 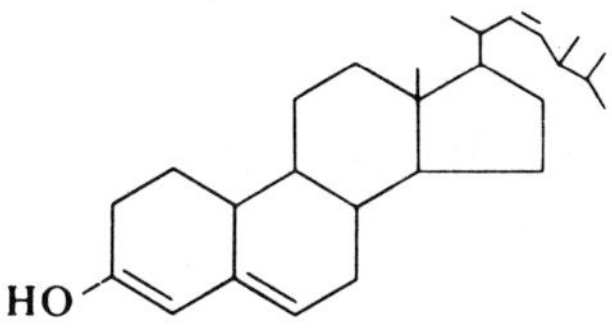 | $C_{28}H_{46}O$ (398.6) | 148 | −64° |
| $\Delta^{5,24(28)}$-Ergostadien-3β-ol (24-Methylenecholesterol) | 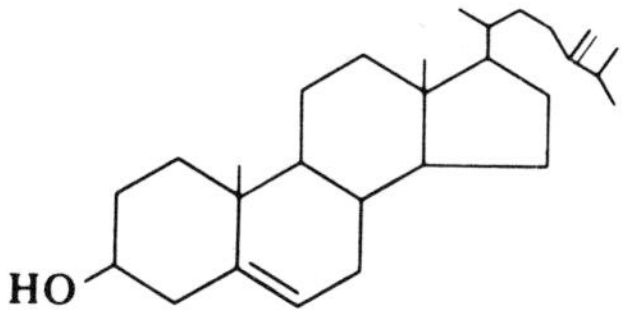 | $C_{28}H_{46}O$ (398.6) | 144 | −42° |
| $\Delta^{7,22}$-Ergosta-dien-3β-ol (5-Dihydroergosterol) | 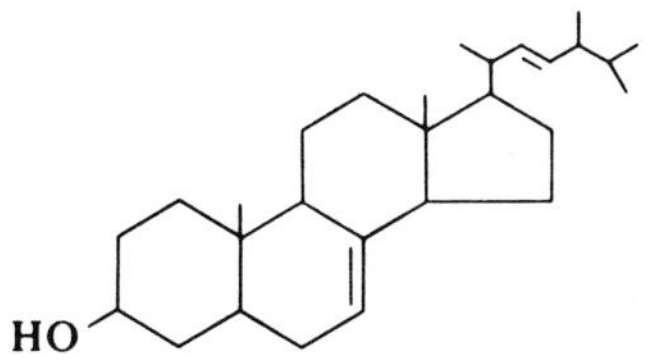 | $C_{28}H_{46}O$ (398.6) | 174 | −20° |
| $\Delta^{7,24(28)}$-Ergostadien-3β-ol | 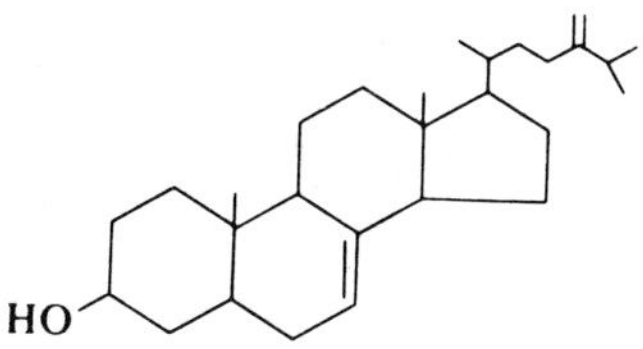 | $C_{28}H_{46}O$ (398.6) | 130 | +6.0° |
| $\Delta^{7,24(28?)}$-Ergostadien-3β-ol (Episterol) | 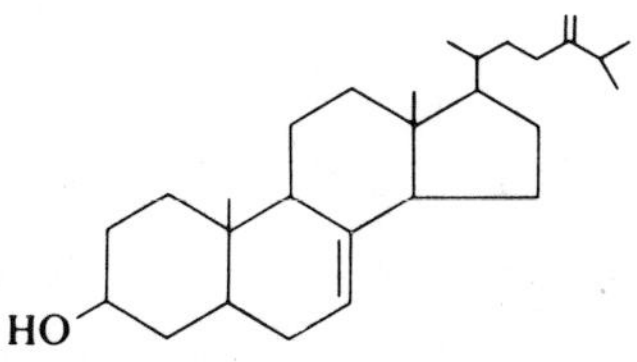 | $C_{28}H_{46}O$ (398.6) | 151 | −5° |

| Systematic Name[a] (Trivial Name) | Structure | Formula (Mol. Wt.) | Melting Point, °C | $[\alpha]$[b] |
|---|---|---|---|---|
| $\Delta^{8,23(?)}$-Ergostadien-3β-ol (Ascosterol) | 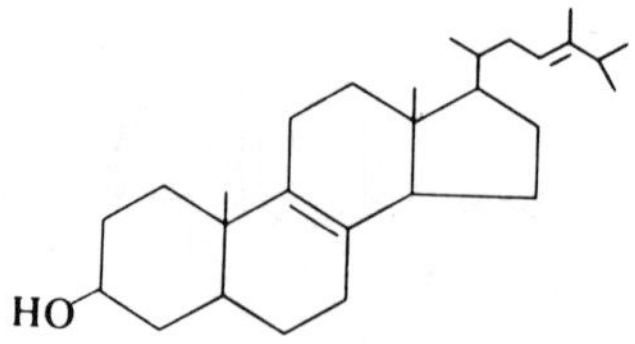 | $C_{28}H_{46}O$ (398.6) | 147 | +45° |
| $\Delta^{8,24(28)}$-Ergostadien-3β-ol (Fecosterol) | 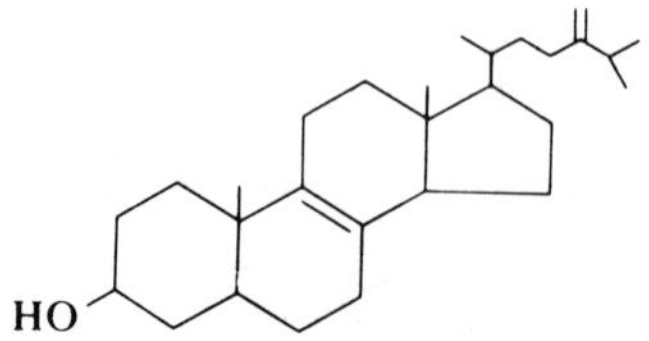 | $C_{28}H_{46}O$ (398.6) | 162 | +42° |
| $\Delta^{7,22}$-Ergostadiene-3β,5α,6β-triol (Cerevisterol) | 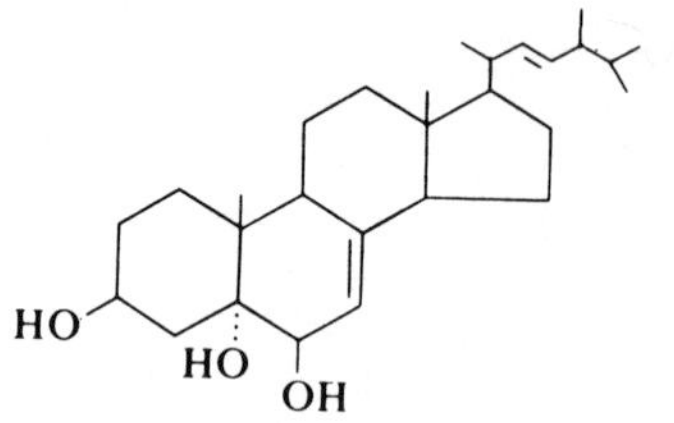 | $C_{28}H_{46}O_3$ (430.6) | 265 | –79° |
| (Haliclonasterol) | 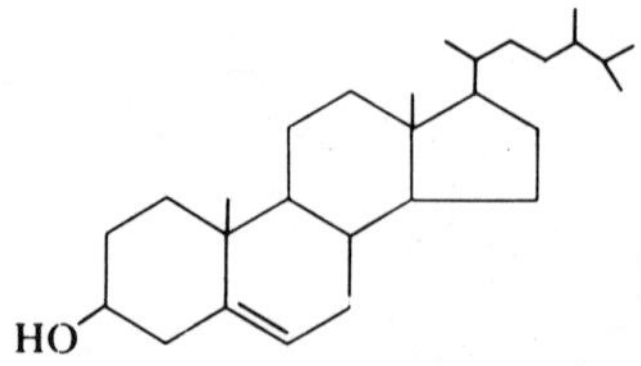 | $C_{28}H_{48}O$ (398.6) | 141 | –41.5° |
| $\Delta^{5}$-24-Isoergosten-3β-ol (Campesterol) | 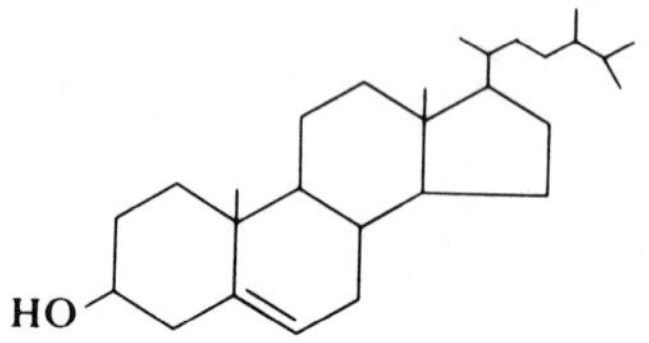 | $C_{28}H_{48}O$ (398.6) | 158 | –33° |
| $\Delta^{22}$-24-Isoergosten-3β-ol (Neospongosterol) | 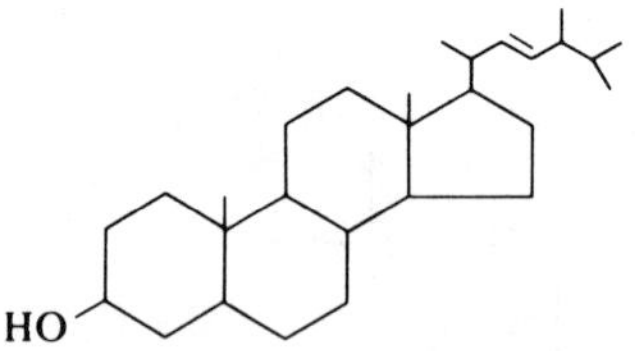 | $C_{28}H_{48}O$ (398.6) | 153 | +10° |
| $\Delta^{5,24(28)}$-Stigmastadien-3β-ol (Fucosterol) | 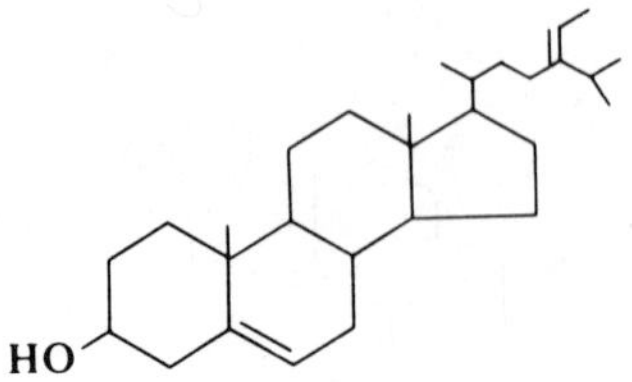 | $C_{29}H_{48}O$ (410.6) | 124 | –38° |
| (Aptostanol) | 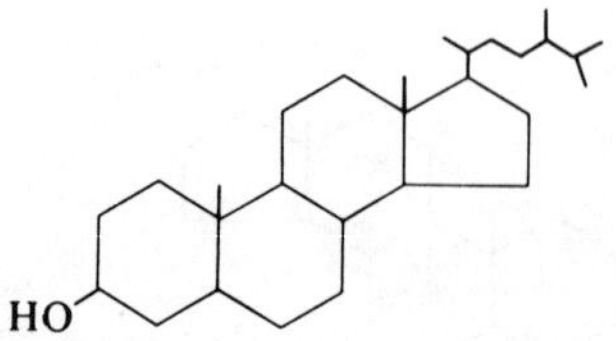 | $C_{28}H_{50}O$ (400.6) | 135 | +22° |

| Systematic Name[a] (Trivial Name) | Structure | Formula (Mol. Wt.) | Melting Point, °C | $[\alpha]^b$ |
|---|---|---|---|---|
| Ergostan-3β-ol (Ergostanol) | 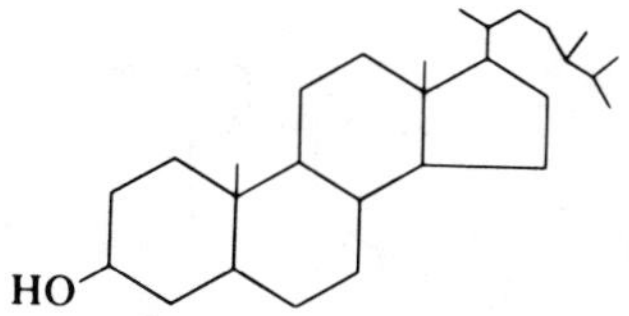 | $C_{28}H_{50}O$ (402.4) | 143 | +16° |
| $\Delta^5$-Stigmasten-3β-ol (β-Sitosterol) | 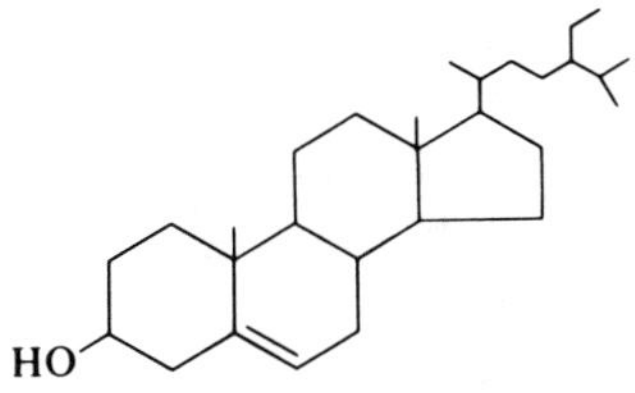 | $C_{29}H_{50}O$ (414.4) | 140 | −36° |
| $\Delta^{5,7,22}$-Stigmastadien-3β-ol (Corbisterol) | 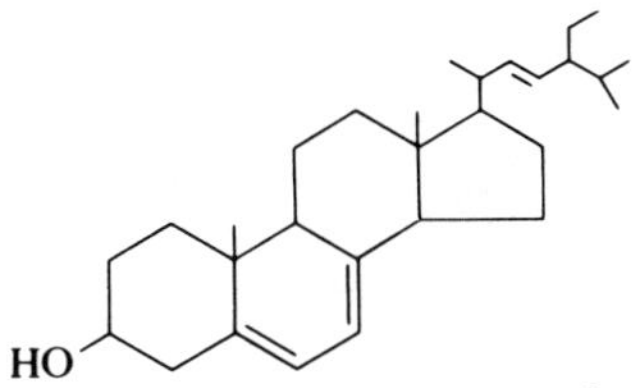 | $C_{29}H_{46}O$ (410.4) | 154 | −114° |
| $\Delta^{7,22}$-24-Isoergostadien-3β-ol (Chondrillasterol) | 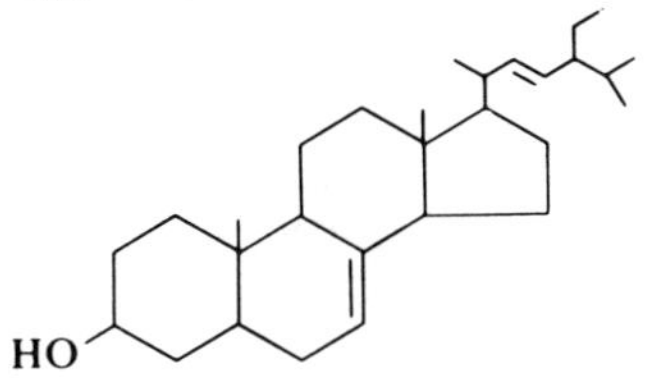 | $C_{29}H_{48}O$ (410.6) | 164 | −2° |
| $\Delta^{5,22}$-24-Isostigmastadien-3β-ol (Poriferasterol) | 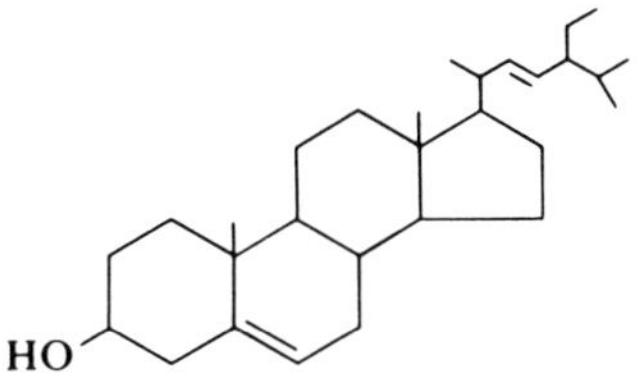 | $C_{29}H_{48}O$ (410.6) | 156 | −49° |
| $\Delta^{5,24(28)}$-20-Isostigmastadien-3β-ol (Sargasterol) | 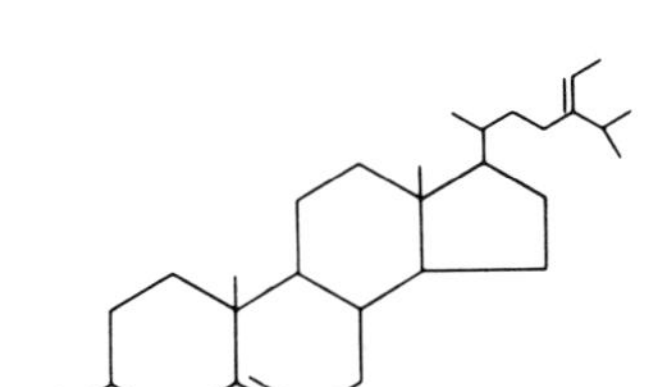 | $C_{29}H_{48}O$ (410.6) | 133.5 | −47.5° |
| $\Delta^{5,11}$(?)-Stigmastadien-3β-ol ($\Delta^5$-Avenasterol) | 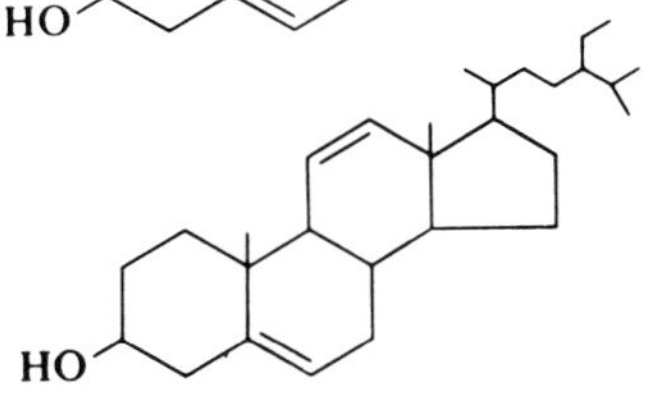 | $C_{29}H_{48}O$ (410.6) | 137 | −37° |

| Systematic Name[a] (Trivial Name) | Structure | Formula (Mol. Wt.) | Melting Point, °C | $[\alpha]^b$ |
|---|---|---|---|---|
| $\Delta^{7,11\,(?)}$-Stigmastadien-3β-ol ($\Delta^7$-Avenasterol) | 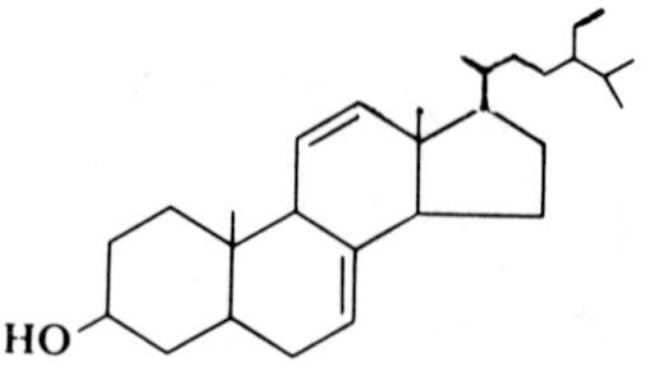 | $C_{29}H_{48}O$ (410.6) | 145 | +8.8° |
| $\Delta^{5,22}$-Stigmastadien-3β-ol (Stigmasterol) | 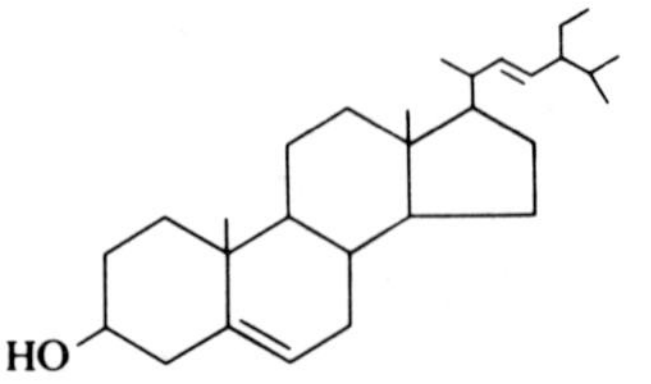 | $C_{29}H_{48}O$ (410.6) | 170 | –49° |
| $\Delta^{7,22}$-Stigmastadien-3β-ol (α-Spinasterol) | 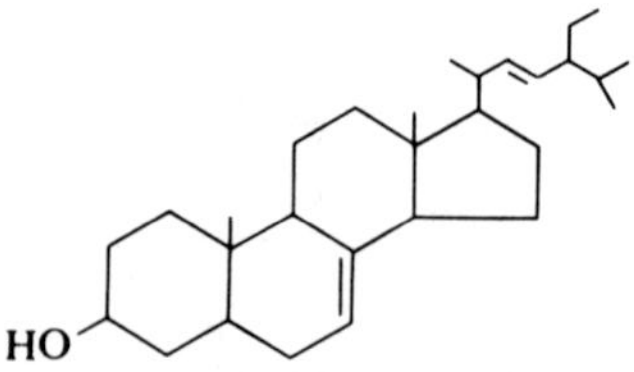 | $C_{29}H_{48}O$ (410.6) | 175 | –2.7° |
| (Palysterol) | 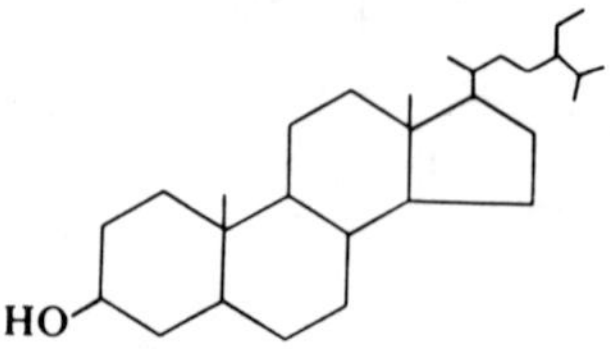 | $C_{29}H_{50}O$ (412.6) | 140 | –47° |
| $\Delta^5$-24-Isostigmasten-3β-ol? (Clionasterol) | 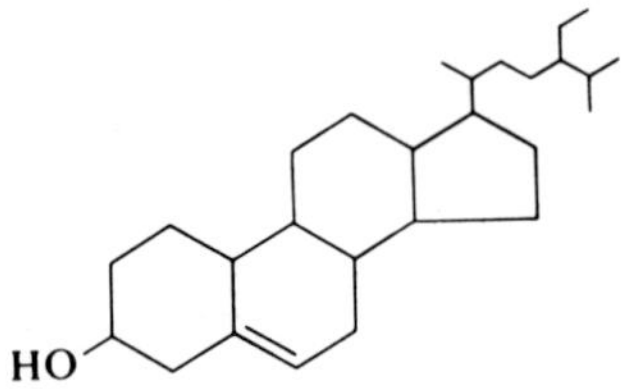 | $C_{29}H_{50}O$ (412.6) | 138 | –37° |
| $\Delta^5$-24-Isostigmasten-3β-ol (γ-Sitosterol) | 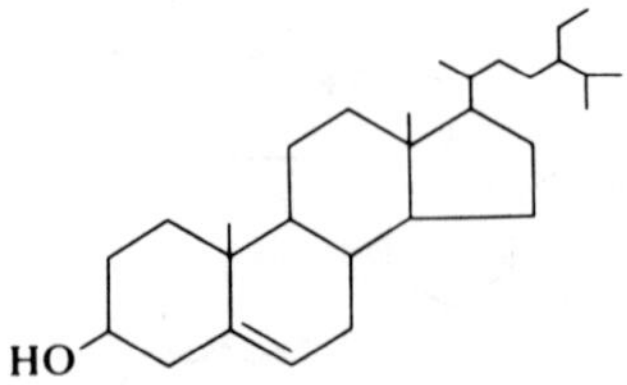 | $C_{29}H_{50}O$ (412.6) | 148 | –43° |
| $\Delta^7$-Stigmasten-3β-ol | 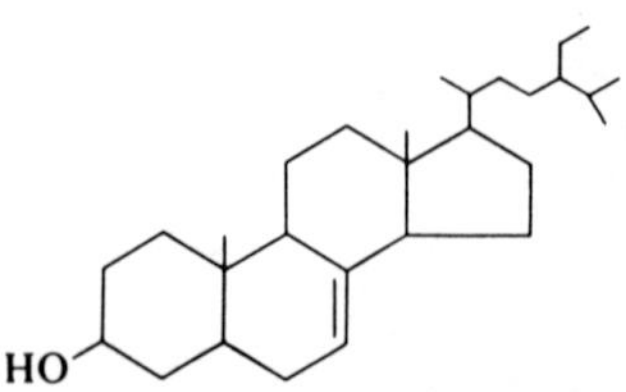 | $C_{29}H_{50}O$ (412.6) | 145 | +9° |

| Systematic Name[a] (Trivial Name) | Structure | Formula (Mol. Wt.) | Melting Point, °C | [α][b] |
|---|---|---|---|---|
| Stigmastan-3β-ol (Dihydrositosterol) | 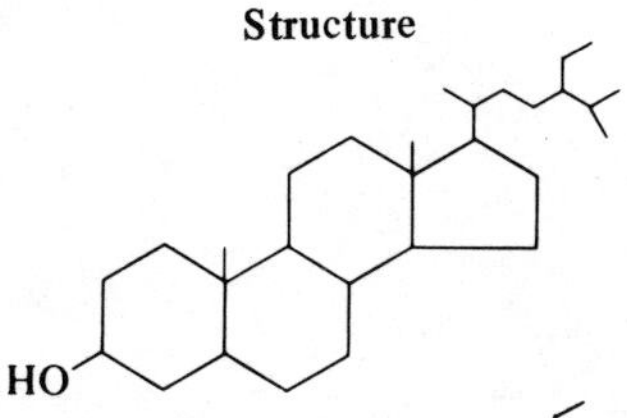 | $C_{29}H_{52}O$ (414.6) | 140 | +25° |
| 22-Stigmastan-3β-ol | 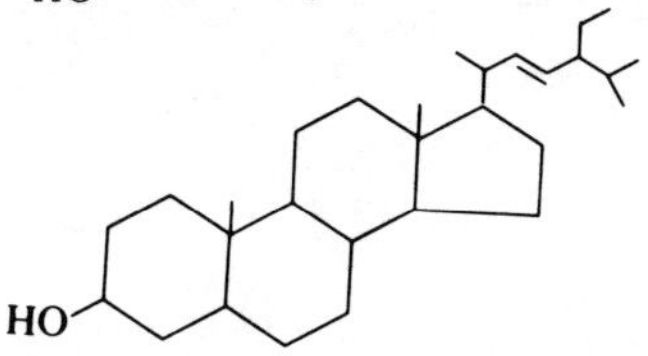 | $C_{29}H_{52}O$ (414.6) | 159 | +3.3° |
| (Dicholesterylether) | 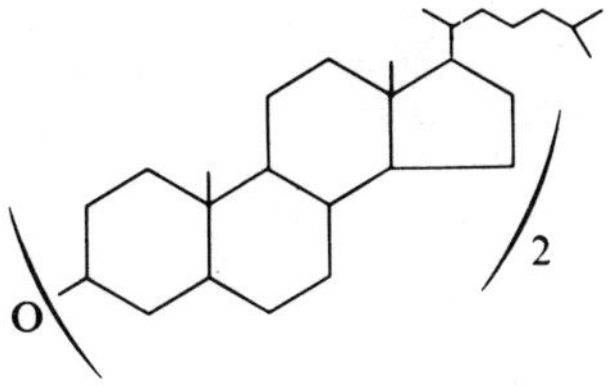 | $C_{54}H_{90}O$ (754) | 196 | −38° |

# LIPOIDAL CONTENTS OF SPECIFIC MICROORGANISMS

DR. WILLIAM M. O'LEARY

## MICROBIAL FATTY ACIDS

Those consulting this or any other tabulation of the fatty acid contents of microorganisms should be aware that there are a number of problems and limitations involved. Prominent among these are the following:

1. The fatty acid composition of a single species can vary considerably, depending on such factors as strain, culture medium, the conditions of incubation, age at harvest, and the techniques of extraction and analysis. For details on this important aspect, see O'Leary, W. M., *The Chemistry and Metabolism of Microbial Lipids,* World Publishing Co., Cleveland, Ohio (1967).

2. In some investigations total fatty acids are studied, and in others only those in a certain type of extract or a certain category of lipids, such as phospholipids. Often it is not clear just what fraction was studied.

3. The precision of identification varies greatly from one report to another. Earlier papers suffer unavoidably from limitations in the then available methods. Some papers report unequivocal identification by varied and rigorous procedures. Some papers report characterization based only on chromatographic behavior. Some papers report only the chain length of fatty acids encountered.

Consequently, for the especially interested reader, it is important to consult the original references for explicit data.

### Symbols and Abbreviations

Fatty acids are listed primarily according to chain length; e.g., $C_{10}$ = a straight 10-carbon chain. Multiple unsaturations are indicated by the number of double bonds following the chain length; e.g., $C_{18}$:2. Positions of double bonds are indicated by the conventional *delta* designation; e.g., $\Delta^9$. In branched acids, iso = $CH_3\underset{|}{\overset{CH_3}{C}}H-$ and ant (anteiso) = $CH_3CH_2\overset{CH_3}{\overset{|}{C}}H-$ . Positions of subgroupings, such as hydroxyls, are indicated by the number of the carbon atoms to which they are attached; e.g., 3 – OH– $C_{12}$. The figures given in the body of Table 1 are the percentage figures reported in each reference cited for the material analyzed; + = reported present but not quantitated, tr = trace, and unk = unknown.

## MICROBIAL LIPIDS

As is the case with fatty acids, the lipids of microorganisms are variously reported. Some papers give rigorously supported characterizations; others give only tentative or incomplete identifications. Some papers only intend to report on certain types of components without commenting on whether any others are present. Some papers report on total cellular lipids, and some deal only with certain fractions or types of extracts. Some papers give precise procedural details, others are more general. Earlier papers are necessarily less conclusive than more recent ones because of the limitations of past techniques. Finally, as with fatty acid composition, lipid composition even within a species is affected by strain differences, culture procedures and media compositions, and recovery methods. The reader with a special interest should, therefore, consult the original references for details that are too numerous to include in any manageable table.

**Symbols and Abbreviations**

MG = monoglyceride; DG = diglyceride; TG = triglyceride; PA = phosphatidic acid; PS = phosphatidyl serine; PE = phosphatidyl ethanolamine; MPE and DPE = mono- and dimethyl phosphatidyl ethanolamine respectively; PC = phosphatidyl choline (lecithin); PI = phosphatidyl inositol; PG = phosphatidyl glycerol; PGP (CL) = diphosphatidyl glycerol (cardiolipin); lyso = one rather than two fatty acids esterified to glycerol moiety. Figures given in the body of Table 2 are the percentage figures reported in each reference for the material analyzed. Where quantitation was not given, tr = trace and + = present in any amount.

## LIPOIDAL CONSTITUENTS OF CORYNEBACTERIA, MYCOBACTERIA AND NOCARDIA

The lipoidal constituents of these three groups of organisms are exceedingly unusual, perhaps unique, and exceedingly numerous. Here are found complex large-molecule fatty acids, such as diphtheric, corynolic, mycolic, phthioic, and nocardic acids, to name only a few, and equally distinctive lipids, including higher waxes, glycolipids, peptidolipids, sulfolipids, and so on, in astonishing variety. These unusual constituents show the interrelationship of these three genera and at the same time provide a means for taxonomic differentiation of the species. However, these lipids are so numerous, so complicated, and so limited in distribution that their compendious description would consume far more space than could be justified in a handbook such as this. Table 3 describes some of the more prominent categories of lipoidal substances found in these genera. Those who require more detail should consult the following reviews and monographs:

1. Asselineau, J., *The Bacterial Lipids.* Holden-Day, San Francisco, California (1966).
2. Kates, M. and Wassef, M. K., *Annu. Rev. Biochem., 39,* 323 (1970).
3. Lechevalier, M. P., Horan, A. C., and Lechevalier, H., *J. Bacteriol., 105,* 313 (1971).

# TABLE 1
# MICROBIAL FATTY ACIDS

| Organism | Saturated | | | | | | Monounsaturated | | | | | Cyclopropane | | | Branched | Hydroxy | Other Compounds | Ref. |
|---|---|---|---|---|---|---|---|---|---|---|---|---|---|---|---|---|---|---|
| | $C_{10}$ | $C_{12}$ | $C_{14}$ | $C_{16}$ | $C_{18}$ | $C_{20}$ | $C_{12}$ | $C_{14}$ | $C_{16}$ | $C_{18}$ | $C_{20}$ | $C_{15}$ | $C_{17}$ | $C_{19}$ | | | | |
| **Order Pseudomonadales** | | | | | | | | | | | | | | | | | | |
| **Suborder Rhodobacteriineae** | | | | | | | | | | | | | | | | | | |
| ATHIORHODACEAE | | | | | | | | | | | | | | | | | | |
| *Rhodopseudomonas* | | | | | | | | | | | | | | | | | | |
| *palustris* | | | | | | | | | | | | | | | | | | |
| light, anaerobic | | | | 12.7 | 6.2 | | | | 3.1 | 72.3 | | | | | | | | 32 |
| dark, aerobic | | | | 3.5 | tr | | | | 4.8 | 91.5 | | | | | | | | 32 |
| *palustris* 2 | | | | | | | | | | | | | | | | | | |
| 1-7 | | | 0.17 | 5.17 | 7.10 | | | | 1.57 | 78.0 | | | | | | | | 7 |
| 1-23 | | | 0.11 | 6.6 | 7.8 | | | | 1.97 | 75.0 | | | | | | | | 7 |
| *spheroides* | | | | | | | | | | | | | | | | | | |
| light, anaerobic | | | | 1.7 | 4.8 | | | | 1.0 | 90.8 | | | | | | | | 32 |
| dark, aerobic | | | | tr | tr | | | | tr | 99.0 | | | | | | | | 32 |
| *capsulata* | | | | | | | | | | | | | | | | | | |
| light, anaerobic | | | | 2.3 | 4.2 | | | | 2.4 | 90.3 | | | | | | | | 32 |
| dark, aerobic | | | | 0.9 | 0.2 | | | 2.4 | 11.8 | 84.0 | | | | | | | | 32 |
| *gelatinosa* | | | | | | | | | | | | | | | | | | |
| light, anaerobic | | | 2.9 | 33.4 | tr | | | | 51.0 | 6.2 | | | | | | | | 32 |
| dark, aerobic | | | 2.21 | 13.8 | tr | | | | 58.0 | 24.4 | | | | | | | | 32 |
| *Rhodospirillum* | | | | | | | | | | | | | | | | | | |
| *molischianum* | | | 0.28 | 21.9 | 0.62 | | | | 29.8 | 47.1 | | | | | | | | 7 |
| *rubrum* | | | 10.9 | 48.9 | 0.6 | | | | 9.2 | 14.1 | | | | | $C_{17}$ = 4.3 | | | 2 |
| light, anaerobic | | | 1.6 | 16.3 | 0.9 | | | | 37.6 | 37.3 | | | | | | | | 32 |
| dark, aerobic | | | 1.6 | 8.3 | tr | | | 1.4 | 51.0 | 35.3 | | | | | | | | 32 |

## TABLE 1 (Continued)
## MICROBIAL FATTY ACIDS

| Organism | Saturated | | | | | | Monounsaturated | | | | | Cyclopropane | | | Branched | Hydroxy | Other Compounds | Ref. |
|---|---|---|---|---|---|---|---|---|---|---|---|---|---|---|---|---|---|---|
| | $C_{10}$ | $C_{12}$ | $C_{14}$ | $C_{16}$ | $C_{18}$ | $C_{20}$ | $C_{12}$ | $C_{14}$ | $C_{16}$ | $C_{18}$ | $C_{20}$ | $C_{15}$ | $C_{17}$ | $C_{19}$ | | | | |
| **Order Pseudomonadales** | | | | | | | | | | | | | | | | | | |
| **Suborder Pseudomonadineae** | | | | | | | | | | | | | | | | | | |
| **PSEUDOMONADACEAE** | | | | | | | | | | | | | | | | | | |
| *Pseudomonas* sp. | | 2.6 | 3.5 | 49.5 | 8.6 | | | 2.9 | 18.3 | 2.3 | | | | | | | $C_{15}$ = 4.2<br>$C_{17}$ = 8.1 | 2 |
| | | | 3.0 | 42.6 | 1.0 | | | | 40.7 | 9.0 | | | | | | | $C_{17}$ = 1.3<br>$C_{18}$ = 2.4 | 2 |
| *aeruginosa* | | | | | | | | | | | | | | | | | | |
| 95% alcohol extract | | | 1 | 37 | 2 | | | | 7 | 33 | | | 5 | | | | | 3 |
| C/M 2:1, grown at 25°C | | | | 23.2 | | | | | 15.8 | 49.5 | | | | | | | $C_{18}$ or $C_{21}$ = 11.6 | 4 |
| in lipid A | | 11.8 | | 7.8 | | | | | | | | | | | | 3-OH-$C_{10}$ = 14.6<br>2-OH-$C_{12}$ = 18.5<br>3-OH-$C_{12}$ = 35.2 | | 134, 136 |
| *alcaligenes* | | 0.1 | 1.0 | 38.4 | tr | | | | 14.4 | 42.8 | | | | | | | | 1 |
| in lipopolysaccharide | | 35.0 | | | | | 10.5 | | | | | | | | | 3-OH-$C_{10}$ = 20.3<br>3-OH-$C_{12}$ = 34.2 | | 88 |
| in lipid A | | 39.6 | | | | | 8.9 | | | | | | | | | 3-OH-$C_{10}$ = 18.9<br>3-OH-$C_{12}$ = 32.6 | | 88 |
| *diminuta* | | | | | | | | | | | | | | | | | | |
| glycolipid | | | + | · 18.1 | + | | | + | 2.7 | 77.3 | | | | | | | $C_{15}$ | 135 |
| phospholipid | | | + | 11.6 | + | | | + | 2.7 | 82.8 | | | | | | | $C_{17}$ | 135 |

## TABLE 1 (Continued)
## MICROBIAL FATTY ACIDS

| Organism | Saturated | | | | | | Monounsaturated | | | | | Cyclopropane | | | Branched | Hydroxy | Other Compounds | Ref. |
|---|---|---|---|---|---|---|---|---|---|---|---|---|---|---|---|---|---|---|
| | $C_{10}$ | $C_{12}$ | $C_{14}$ | $C_{16}$ | $C_{18}$ | $C_{20}$ | $C_{12}$ | $C_{14}$ | $C_{16}$ | $C_{18}$ | $C_{20}$ | $C_{15}$ | $C_{17}$ | $C_{19}$ | | | | |
| **Order Pseudomonadales** | | | | | | | | | | | | | | | | | | |
| **Suborder Pseudomonadineae (continued)** | | | | | | | | | | | | | | | | | | |
| *Pseudomonas* | | | | | | | | | | | | | | | | | | |
| *fluorescens* | | | | 36.9 | 18.7 | | | | | 9.1 | | | | | | | $C_{18}$:2 = 22.6 | 4 |
| | | | 1.3 | 29.3 | 1.5 | | | | 12.7 | 42.3 | | | 3.1 | | | | $C_{15}$ = 1.0<br>$C_{17}$ = 1.5 | 5 |
| *Photobacterium* | | | | | | | | | | | | | | | | | | |
| *albensis* | | | 3.7 | 25.9 | 4.7 | | | | 41.10 | 22.2 | | | | | | | $C_{17}$ = 2.5 | 2 |
| **SPIRILLACEAE** | | | | | | | | | | | | | | | | | | |
| *Vibrio* | | | | | | | | | | | | | | | | | | |
| *cholerae* | | | 3.9 | 29.5 | 3.1 | | | tr | 34.2 | 22.9 | | | | | iso-$C_{16}$ = 2.0 | | $C_{15}$ = tr<br>$C_{17}$ = 1.2<br>$C_{17}$:1 or iso-$C_{18}$ = 1.8 | 5 |
| *cholerae* CA 72 | | | 6.1 | 36.8 | 2.7 | | | | 38.2 | 12.7 | | | | | | | $C_{17}$:1 = 1.0 | 78 |
| rugose strain | | | 1.8 | 29.3 | 2.8 | | | | 14.2 | 24.8 | | | 10.1 | 15.7 | | | $C_{17}$:1 = 1.5 | 78 |
| *cholerae* CA 113 | | | 4.7 | 37.2 | 2.5 | | | | 36.2 | 17.5 | | | | | iso-$C_{16}$ = tr<br>iso-$C_{18}$ = tr | | $C_{17}$:1 = tr | 78 |
| rugose strain | | | 1.5 | 26.5 | 2.4 | | | | 17.2 | 40.0 | | | 4.5 | 8.0 | | | $C_{17}$:1 = tr | 78 |
| *cholerae* NIH 35A3 | | | 4.4 | 37.7 | 3.1 | | | | 36.9 | 14.7 | | | | | | | $C_{17}$:1 = tr | 78 |
| rugose strain | | | 1.1 | 26.5 | 2.0 | | | | 17.2 | 38.4 | | | 5.3 | 9.3 | | | $C_{17}$:1 = tr | 78 |
| *marinus* | | | 18.1 | 41.7 | 22.8 | | | | | | | | | | $C_{12}$ = 1.1<br>$C_{13}$ = 0.7<br>$C_{17}$ = 1.5 | $C_{15}$ = 0.7 | $C_{14}$<br>$C_{16}$<br>$C_{19}$ | 71 |
| *metchnikovii* | | | 3.3 | 54.4 | 3.7 | | | | | | | | | | $C_{17}$ = 3.7 | | $C_{15}$ = 2.5 | 2 |

# TABLE 1 (Continued)
# MICROBIAL FATTY ACIDS

| Organism | Saturated | | | | | | Monounsaturated | | | | | Cyclopropane | | | Branched | Hydroxy | Other Compounds | Ref. |
|---|---|---|---|---|---|---|---|---|---|---|---|---|---|---|---|---|---|---|
| | $C_{10}$ | $C_{12}$ | $C_{14}$ | $C_{16}$ | $C_{18}$ | $C_{20}$ | $C_{12}$ | $C_{14}$ | $C_{16}$ | $C_{18}$ | $C_{20}$ | $C_{15}$ | $C_{17}$ | $C_{19}$ | | | | |
| **Order Hyphomicrobiales** | | | | | | | | | | | | | | | | | | |
| **HYPHOMICROBIACEAE** | | | | | | | | | | | | | | | | | | |
| *Rhodomicrobium* | | | | | | | | | | | | | | | | | | |
| *vanielli* | | | | | | | | | | | | | | | | | | |
| extractable lipids | 0.10 | 0.30 | 1.75 | 5.40 | 0.11 | | | 0.50 | 1.08 | 88.2 | | | | | | | $C_{18}$:2 = 2.70 | 85 |
| *Rhodomicrobium* | | | | | | | | | | | | | | | | | | |
| *vanielli* | | | | | | | | | | | | | | | | | | |
| bound lipids | | | | | | | | | | | | | | | | | | |
| fraction I | | | 10.3 | 3.7 | 2.5 | 0.9 | | | 1.1 | 35.1 | 1.6 | | | 4.4 | $C_{15}$ = 6.9<br>$C_{17}$ = 9.8 | | $C_{15}$:1 = 6.9<br>$C_{17}$ = 1.0<br>$C_{21}$ = 8.0<br>$C_{22}$ = 8.0<br>$C_{22}$:1 = 4.4 | 85 |
| fraction II | | | | | | | | | | | | | | | 2-OH-$C_{13}$ br. = 17.2 | 2-OH-$C_{12}$ = 24.0<br>3-OH-$C_{12}$ = 34.5<br>2-OH-$C_{13}$ = 10.3<br>2-OH-$C_{14}$ = 6.9<br>3-OH-$C_{14}$ = 6.9 | | 85 |
| **Order Eubacteriales** | | | | | | | | | | | | | | | | | | |
| **AZOTOBACTERIACEAE** | | | | | | | | | | | | | | | | | | |
| *Azotobacter* | | | | | | | | | | | | | | | | | | |
| *agilis* | | | | | | | | | | | | | | | | 3-OH-$C_{10}$<br>3-OH-$C_{12}$<br>2-OH-$C_{13}$ | | 81 |
| **RHIZOBIACEAE** | | | | | | | | | | | | | | | | | | |
| *Agrobacterium* | | | | | | | | | | | | | | | | | | |
| *tumefaciens* | | | 1.1 | 15.0 | | | | | 0.6 | 29.7 | | | [illegible] | [illegible] | | | | |

## TABLE 1 (Continued)
## MICROBIAL FATTY ACIDS

| Organism | Saturated | | | | | | Monounsaturated | | | | | Cyclopropane | | | Branched | Hydroxy | Other Compounds | Ref. |
|---|---|---|---|---|---|---|---|---|---|---|---|---|---|---|---|---|---|---|
| | $C_{10}$ | $C_{12}$ | $C_{14}$ | $C_{16}$ | $C_{18}$ | $C_{20}$ | $C_{12}$ | $C_{14}$ | $C_{16}$ | $C_{18}$ | $C_{20}$ | $C_{15}$ | $C_{17}$ | $C_{19}$ | | | | |
| **Order Eubacteriales (continued)** | | | | | | | | | | | | | | | | | | |
| *Chromobacterium* | | | | | | | | | | | | | | | | | | |
| *violaceum* ND | | 1.7 | 35.8 | 0.4 | | | | 52.7 | 9.4 | | | | | | | | | 2 |
| *violaceum* 9 | | 4.1 | 44.2 | 1.4 | | | | 44.2 | 6.1 | | | | | | | | | 2 |
| *Rhizobium* | | | | | | | | | | | | | | | | | | |
| *japonicum* | | | | | | | | | | | | | | | | | | |
| strain 83 (indifferent to biotin) | | 2.6 | 17.3 | 1.9 | | | | 0.3 | 52.2 | | | 1.0 | 24.1 | | | | | 119 |
| strain 508 (inhibited by biotin) | | 12.3 | 22.6 | 6.6 | | | | 0.9 | 43.3 | | | | 0.7 | | | | | 119 |
| strain 5633 (requires biotin) | | 9.6 | 19.3 | 4.8 | | | | 3.6 | 49.8 | | | | 1.8 | | | | | 119 |
| **ACHROMOBACTERACEAE** | | | | | | | | | | | | | | | | | | |
| *Achromobacter* sp. | 7.9 | 9.0 | 21.7 | | | | 7.8 | 9.3 | 23.6 | | | | | | | | $C_{18}$:3 or $C_{21}$ = 20.7 | 4 |
| *Flavobacterium* | | | | | | | | | | | | | | | | | | |
| *aquatile* | 17.6 | 15.0 | 29.2 | | | | | 11.0 | | | | | | | | | $C_{18}$:3 or $C_{21}$ = 9.0 | 4 |
| **ENTEROBACTERIACEAE** | | | | | | | | | | | | | | | | | | |
| **Escherichieae** | | | | | | | | | | | | | | | | | | |
| *Escherichia* | | | | | | | | | | | | | | | | | | |
| *coli* | | 3.8 | 43.8 | tr | | | | 2.1 | 7.2 | | | 20.9 | 18.8 | | | | $C_{15}$ = 1.2<br>$C_{17}$ = 1.4 | 5 |
| | <1 | 8 | 47 | | | | <1 | <1 | | | | 29 | 13 | | | | | 3 |
| *coli* E-26 | | | | | | | | | | | | | | | | | | |
| in synthetic medium | 3.72 | 6.22 | 45.70 | 1.23 | | | | 2.32 | 1.38 | | 2.70 | 16.50 | 9.78 | | | | $C_{16}$:7,8 = 4.10<br>*cis*-vaccenic = 3.92<br>unk = 2.46 | 8 |

## TABLE 1 (Continued)
## MICROBIAL FATTY ACIDS

| Organism | | Saturated | | | | | Monounsaturated | | | | | Cyclopropane | | | | | | |
|---|---|---|---|---|---|---|---|---|---|---|---|---|---|---|---|---|---|---|
| | $C_{10}$ | $C_{12}$ | $C_{14}$ | $C_{16}$ | $C_{18}$ | $C_{20}$ | $C_{12}$ | $C_{14}$ | $C_{16}$ | $C_{18}$ | $C_{20}$ | $C_{15}$ | $C_{17}$ | $C_{19}$ | Branched | Hydroxy | Other Compounds | Ref. |
| **Order Eubacteriales (continued)** | | | | | | | | | | | | | | | | | | |
| *Escherichia* | | | | | | | | | | | | | | | | | | |
| *coli* E-26 | | | | | | | | | | | | | | | | | | |
| in inducing medium | | 2.84 | 7.78 | 55.30 | 1.99 | | | | 1.25 | 1.46 | | 2.34 | 14.20 | 7.65 | | | $C_{16}$:7,8 = 3.32<br>*cis*-vaccenic = 1.48<br>unk = 1.33 | 8 |
| *coli* B | | | 6.8 | 42.0 | 1.7 | | | 2.1 | 2.3 | 25.7 | | | | | | | $C_{15}$ = 0.8<br>unk = 12.1, 2.3, 4.2 | 2 |
| in synthetic medium | | 3.27 | 10.00 | 44.40 | 0.40 | | | | 1.05 | 0.80 | | 3.84 | 18.10 | 8.95 | | | $C_{16}$:7,8 = 2.27<br>*cis*-vaccenic = 2.68<br>unk = 2.13 | 8 |
| in inducing medium | | 6.48 | 9.28 | 54.70 | 1.51 | | | | 1.25 | 2.20 | | 1.33 | 5.86 | 4.81 | | | $C_{16}$:7,8 = 2.90<br>*cis*-vaccenic = 3.55<br>unk = 3.18 | 8 |
| in phosphatydyl ethanolamine | 0.3 | 0.5 | 2.7 | 48.2 | 2.1 | | | | 10.8 | 26.3 | | | 5.7 | 0.5 | | | $C_{15}$ = 0.6<br>$C_{19}$ = 12.2 | 9 |
| FA in PG | tr | 2.4 | 2.1 | 18.0 | tr | | | tr | 2.7 | 70.6 | | | tr | | | | $C_{15}$ = 0.9<br>$C_{17}$ = 2.5<br>$C_{19}$ = 10.8 | 9 |
| FA in cardiolipin | tr | 0.7 | 3.2 | 32.4 | | | | | 10.4 | 26.2 | | 19.4 | 3.1 | 3.5 | | | $C_{19}$ = 10.6 | 9 |
| *alkalescens* | | | 5.3 | 46.0 | | | | | 5.3 | 1.5 | | | | | $C_{17}$ = 26.4<br>$C_{19}$ = 12.0 | | $C_{15}$ = 3.5 | 2 |
| *dispar* 4169 | | 16.4 | 10.0 | 36.6 | 2.3 | | | | 1.0 | 10.3 | | | | | $C_{18}$ = 1.5 | | $C_{13}$ = 10.4<br>$C_{15}$ = 10.2<br>$C_{17}$ = 1.3 | 2 |
| *Aerobacter* | | | | | | | | | | | | | | | | | | |
| *aerogenes* | | | 5.7 | 56.4 | | | | | 5.7 | 9.6<br>18.8 | | | | | | | $C_{17}$ = 3.8 | 2 |
| total cell | 0.1 | 5.9 | 1.5 | 15.6 | 5.8 | 0.7 | | 8.1 | 30.0 | 0.4 | | 4.2 | 10.1 | 1.5 | | 3-OH-$C_{14}$ = 12.9 | $C_{13}$:1 = 0.2<br>$C_{15}$ = 1.0<br>$C_{17}$ = 0.9 | 10 |

# TABLE 1 (Continued)
# MICROBIAL FATTY ACIDS

| Organism | Saturated | | | | | | Monounsaturated | | | | | Cyclopropane | | | Branched | Hydroxy | Other Compounds | Ref. |
|---|---|---|---|---|---|---|---|---|---|---|---|---|---|---|---|---|---|---|
| | $C_{10}$ | $C_{12}$ | $C_{14}$ | $C_{16}$ | $C_{18}$ | $C_{20}$ | $C_{12}$ | $C_{14}$ | $C_{16}$ | $C_{18}$ | $C_{20}$ | $C_{15}$ | $C_{17}$ | $C_{19}$ | | | | |
| **Order Eubacteriales (continued)** | | | | | | | | | | | | | | | | | | |
| *Aerobacter* | | | | | | | | | | | | | | | | | | |
| *aerogenes* | | | | | | | | | | | | | | | | | | |
| C/M extract lipid | tr | 0.05 | | 25.5 | 8.0 | | | 3.9 | 43.1 | 0.7 | | | 13.7 | 1.2 | | 3-OH-$C_{14}$ = 1.0 | $C_{13}$:1 = tr<br>$C_{15}$ = 1.3<br>$C_{17}$ = 1.2 | 10 |
| in lipid A | 4.4 | 20.2 | | 5.5 | 29.8 | | | 11.5 | 6.8 | | | | 5.9 | | | 3-OH-$C_{14}$ = 6.7 | $C_{15}$ = 9.8 | 10 |
| **Serratieae** | | | | | | | | | | | | | | | | | | |
| *Serratia* | | | | | | | | | | | | | | | | | | |
| *marcescens* | | | 10.5 | 55.5 | 6.1 | | | | 0.9 | 3.1<br>2.6 | | | 19.7 | 0.3 | | | $C_{15}$ = 1.3 | 2 |
| | | | 1.2 | 43.2 | 1.2 | | | | 5.2 | 11.0 | | | 31.4 | 5.4 | | | $C_{15}$ = tr<br>$C_{17}$ = 1.0 | 5 |
| | | | 7.9 | 33.9 | 27.1 | | | | | 12.1<br>10.8 | | | | | | | $C_{18}$:3 or $C_{21}$ = 8.4 | 4 |
| | | | 9 | 51 | <1 | | | | | <1 | | | 36 | 3 | | | | 3 |
| | | | | | | | | | | | | | | | | 3-OH-$C_{10}$ = 5.2<br>3-OH-dodec-anoic = 2.5<br>3-OH-5-dodec-anoic = 1.5 | | 10 |
| | | tr | 4 | 51 | 1 | | | | 1 | 2 | | | 28 | 12 | | | $C_{10}$:0:OH = tr<br>unk = 2 | 12 |
| at 10°C | | tr | 1 | 24 | 3 | | | 1 | 25 | 33 | | | 7 | 2 | $C_{14}$ = 0.4<br>$C_{15}$ = 1<br>$C_{16}$ = 0.5<br>$C_{17}$ = 2 | | $C_{15}$ = tr<br>$C_{17}$ = 2 | 86 |

# TABLE 1 (Continued)
## MICROBIAL FATTY ACIDS

| Organism | Saturated | | | | | | Monounsaturated | | | | | Cyclopropane | | | Branched | Hydroxy | Other Compounds | Ref. |
|---|---|---|---|---|---|---|---|---|---|---|---|---|---|---|---|---|---|---|
| | $C_{10}$ | $C_{12}$ | $C_{14}$ | $C_{16}$ | $C_{18}$ | $C_{20}$ | $C_{12}$ | $C_{14}$ | $C_{16}$ | $C_{18}$ | $C_{20}$ | $C_{15}$ | $C_{17}$ | $C_{19}$ | | | | |
| **Order Eubacteriales (continued)** | | | | | | | | | | | | | | | | | | |
| *Serratia* | | | | | | | | | | | | | | | | | | |
| *marcescens* | | | | | | | | | | | | | | | | | | |
| at 30°C | | 0.4 | 2 | 36 | tr | | | | 3 | 9 | | | 25 | 6 | $C_{14}$ = 2<br>$C_{15}$ = 6<br>$C_{16}$ = 2<br>$C_{17}$ = 4 | | $C_{15}$ = 1<br>$C_{17}$ = 4 | 86 |
| ethanol-soluble | | | 4.0 | 15.6 | | | | | | | | | 11.1 | tr | | 3-OH-$C_{10}$ = 44.7<br>3-OH-$C_{12}$ = 14.5<br>3-OH-5-$C_{12}$ = 9.6 | | 13 |
| light-petroleum-soluble | tr | tr | 6.0 | 43.8 | 0.5 | | | | 2.3 | 10.5 | | | 32.4 | 3.1 | | 3-OH-$C_{10}$ = 0.7<br>3-OH-$C_{12}$ = tr<br>3-OH-5-$C_{12}$ = tr | $C_{11}$ = tr<br>$C_{13}$ = tr<br>$C_{15}$ = tr | 13 |
| bound lipids | tr | tr | 6.0 | 46.0 | 0.8 | | | | 1.8 | 7.5 | | | 29.8 | 4.3 | | 3-OH-$C_{10}$ = 1.8<br>3-OH-$C_{12}$ = 0.4<br>3-OH-5-$C_{12}$ = 0.2 | $C_{11}$ = tr<br>$C_{13}$ = tr<br>$C_{15}$ = 0.6 | 13 |
| *marcescens* 2446 | | | 4.9 | 49.7 | | | | | 3.7 | 12.2<br>5.8 | | | | | ant-$C_{15}$ = 0.6 | | $C_{15}$ = 1.5<br>unk = 21.6 | 2 |
| *psychrophilis* | | | | | | | | | | | | | | | | | | |
| at 5°C | | 0.8 | 3.5 | 30.6 | | | | 6.6 | 52.5 | tr | | | | | $C_{14}$ = tr<br>$C_{15}$ = tr<br>$C_{16}$ = tr<br>$C_{17}$ = tr | | $C_{15}$ = 1.2<br>$C_{15}$:1 = 3.8<br>$C_{17}$:1 = 1.0<br>$C_{19}$ = tr | 86 |
| **Proteae** | | | | | | | | | | | | | | | | | | |
| *Proteus* | | | | | | | | | | | | | | | | | | |
| *vulgaris* | | | 3.6 | 35.6 | 19.8 | | | | 3.6 | 11.2 | | | | | | | $C_{18}$:2 = 22.8<br>$C_{18}$ = 3, or<br>$C_{21}$ = 3.4 | 4 |

## TABLE 1 (Continued)
## MICROBIAL FATTY ACIDS

| Organism | Saturated | | | | | | Monounsaturated | | | | | Cyclopropane | | | Branched | Hydroxy | Other Compounds | Ref. |
|---|---|---|---|---|---|---|---|---|---|---|---|---|---|---|---|---|---|---|
| | $C_{10}$ | $C_{12}$ | $C_{14}$ | $C_{16}$ | $C_{18}$ | $C_{20}$ | $C_{12}$ | $C_{14}$ | $C_{16}$ | $C_{18}$ | $C_{20}$ | $C_{15}$ | $C_{17}$ | $C_{19}$ | | | | |
| **Order Eubacteriales (continued)** | | | | | | | | | | | | | | | | | | |
| *Proteus* | | | | | | | | | | | | | | | | | | |
| *mirabilis* | | | | | | | | | | | | | | | | | | |
| wild type | | | | | | | | | | | | | | | | | | |
| log phase | | $C_{12} + C_{14} = 3$ | | 47 | | | | | 35 | 15 | | | | | | | | 72 |
| stationary phase | | $C_{12} + C_{14} = 2$ | | 7 | | | | | 3 | 12 | | | 7 | 23 | | | | 72 |
| mutant, polymyxin B-sensitive | | | | | | | | | | | | | | | | | | |
| log phase | | $C_{12} + C_{14} = 9$ | | 42 | | | | | 27 | 21 | | | | | | | | 72 |
| stationary phase | | $C_{12} + C_{14} = 2$ | | 20 | | | | | 3 | 18 | | | 6 | 16 | | | | 72 |
| wild SD, grown in sulfadiazine | | | | | | | | | | | | | | | | | | |
| log phase | | $C_{12} + C_{14} = 2$ | | 50 | | | | | 26 | 19 | | | | | | | | 72 |
| stationary phase | | $C_{12} + C_{14} = 11$ | | 71 | | | | | | 17 | | | | tr | | | | 72 |
| **Salmonelleae** | | | | | | | | | | | | | | | | | | |
| *Salmonella* | | | | | | | | | | | | | | | | | | |
| *typhimurium* | | | | | | | | | | | | | | | | | | |
| in phospholipid | | tr | 3.2 | 35.0 | 0.7 | | | 0.2 | 18.3 | 21.0 | | | 16.0 | 3.7 | | | $C_{15} = 0.4$<br>$C_{17} =$ tr | 14 |
| *gallinarum* | | 2.9 | 9.1 | 49.0 | 2.0 | | | 3.8 | 4.7 | 2.0 | | | | | | | $C_{13} = 5.9$<br>$C_{15} = 3.8$<br>unk = 4.9, 11.9 | 2 |
| **BRUCELLACEAE** | | | | | | | | | | | | | | | | | | |
| *Pasteurella* | | | | | | | | | | | | | | | | | | |
| *pestis* | | | + | + | | | | | | + | | | + | | $C_{19}$ | | | 138 |
| *tularensis* | + | | + | + | + | + | | | + | + | + | | | | | | $C_{15}$<br>$C_{22}$<br>$C_{24}$<br>$C_{26}$ | 139 |

## TABLE 1 (Continued)
## MICROBIAL FATTY ACIDS

| Organism | Saturated | | | | | | Monounsaturated | | | | | Cyclopropane | | | Branched | Hydroxy | Other Compounds | Ref. |
|---|---|---|---|---|---|---|---|---|---|---|---|---|---|---|---|---|---|---|
| | $C_{10}$ | $C_{12}$ | $C_{14}$ | $C_{16}$ | $C_{18}$ | $C_{20}$ | $C_{12}$ | $C_{14}$ | $C_{16}$ | $C_{18}$ | $C_{20}$ | $C_{15}$ | $C_{17}$ | $C_{19}$ | | | | |
| **Order Eubacteriales (continued)** | | | | | | | | | | | | | | | | | | |
| **BRUCELLACEAE** | | | | | | | | | | | | | | | | | | |
| *Brucella* | | | | | | | | | | | | | | | | | | |
| *abortus* | | | | | | | | | | | | | | | | | | |
| in agar I | | 0.3 | 0.5 | 9.5 | 14.3 | | | | 0.4 | 11.9 | | | | 58.0 | | | $C_{13}$ = 2.5<br>$C_{17}$ = 0.4<br>others = 2.2 | 15 |
| in agar II | | 1.7 | 3.5 | 8.7 | 14.2 | | | | 2.6 | 15.9 | | | | 30.7 | | unk OH's = 22.1 | $C_{13}$ = tr<br>$C_{17}$ = 0.6 | 15 |
| *melitensis* | | | 0.5 | 9.7 | 8.5 | | | | 1.2 | 42.5 | | | | 35.2 | | | $C_{17}$ = 0.3<br>others = 2.1 | 15 |
| *Haemophilus* | | | | | | | | | | | | | | | | | | |
| *parainfluenzae* | | 0.8 | 27.9 | 24.8 | 2.3 | | | 0.8 | 17.9 | 2.5 | 0.3 | | 0.4 | 0.1 | | OH-$C_{14}$ = 11.8 | $C_{13}$ = <0.1<br>$C_{13}$:1 = 0.1<br>$C_{15}$ = 1.0<br>$C_{15}$:1 = 7.4<br>$C_{17}$ = 0.1<br>$C_{19}$ = 0.2 | 16 |
| **BACTEROIDACEAE** | | | | | | | | | | | | | | | | | | |
| *Bacteroides* | | | | | | | | | | | | | | | | | | |
| *melaninogenicus* | | | 1.4 | 2.5 | 1.4 | | | | | | | | | | $C_{13}$ = 2.4<br>$C_{15}$ = 84.4<br>$C_{17}$ = 1.3<br>$C_{18}$ br. = 6.5 | | | 82 |
| *succinogenes* | | 3.2 | 5.4 | 5.6 | 3.4 | | | | | 3.6<br>1.4 | | | | | iso-$C_{14}$ = 7.3<br>$C_{15}$ = 1.2<br>iso-$C_{16}$ = 4.8 | | $C_{13}$ = 11.4<br>$C_{15}$ = 45.5<br>$C_{17}$ = 1.6<br>unk = 2.1, 2.4, 1.1 | 17 |

# TABLE 1 (Continued)
# MICROBIAL FATTY ACIDS

| Organism | Saturated | | | | | | Monounsaturated | | | | | Cyclopropane | | | Branched | Hydroxy | Other Compounds | Ref. |
|---|---|---|---|---|---|---|---|---|---|---|---|---|---|---|---|---|---|---|
| | $C_{10}$ | $C_{12}$ | $C_{14}$ | $C_{16}$ | $C_{18}$ | $C_{20}$ | $C_{12}$ | $C_{14}$ | $C_{16}$ | $C_{18}$ | $C_{20}$ | $C_{15}$ | $C_{17}$ | $C_{19}$ | | | | |
| **Order Eubacteriales (continued)** | | | | | | | | | | | | | | | | | | |
| MICROCOCCACEAE | | | | | | | | | | | | | | | | | | |
| *Micrococcus* | | | | | | | | | | | | | | | | | | |
| *lysodeikticus* | | 0.4 | 4.4 | 0.2 | tr | | | | | | | | | | $C_{13}$ = 1.2<br>$C_{15}$ = 85.4<br>$C_{16}$ = 5.0 | | $C_{17}$ = 2.6 | 2 |
| *halodenitrificans* | | + | | 21 | | | | | 24 | 52 | | | | | $C_{15}$<br>$C_{19}$ | | $C_{15}$<br>$C_{17}$ | 140 |
| *Staphylococcus* | | | | | | | | | | | | | | | | | | |
| *aureus* | | | | + | + | + | | | | | | | | | iso-$C_{15}$ = 13<br>ant-$C_{15}$ = 45 | | $C_{12}$ br.<br>$C_{17}$ | 141, 142 |
| | | | 0.71 | | 2.7 | | | 0.41 | 0.38 | 0.49 | | | | | iso-$C_{15}$ = 3.5<br>ant-$C_{15}$ = 36.2 | | iso-$C_{14}$ = 0.5<br>ant-$C_{14}$ = 0.1<br>iso-$C_{17}$ = 3.5<br>ant-$C_{17}$ = 8.5 | 18 |
| *epidermidis* | | tr | tr | 2 | 6 | 7 | | | | | | | | | $C_{13}$ = tr<br>$C_{14}$ = 1<br>$C_{15}$ = 37<br>$C_{16}$ = tr<br>$C_{17}$ = 12<br>$C_{18}$ = tr<br>$C_{19}$ = 20<br>$C_{20}$ = tr<br>$C_{21}$ = 11 | | | 3 |
| *Sarcina* | | | | | | | | | | | | | | | | | | |
| *lutea* | | | 1.8 | 1.5 | 2.5 | | | | 1.3 | 3.6 | | | | | $C_{14}$ = 6.0<br>ant-$C_{15}$ = 80.2<br>$C_{16}$ = 1.8 | | $C_{17}$ = 1.3 | 22 |
| in B-F, various concentrations | 8.10<br>5.2 | 8.1 | 5.2 | 10.1 | 5.7 | | | | 2.7 | | 1.9 | | | | $C_{14}$ = 0.9<br>$C_{16}$ = 3.2 | | $C_{13}$ = 1.8<br>$C_{15}$ = 1.5<br>$C_{19}$ = 2.5<br>$C_{20}$:4 = 3.4<br>$C_{21}$:1 = 2.6<br>$C_{22}$ = 0.9 | 19 |

## TABLE 1 (Continued)
## MICROBIAL FATTY ACIDS

| Organism | Saturated | | | | | | Monounsaturated | | | | | Cyclopropane | | | Branched | Hydroxy | Other Compounds | Ref. |
|---|---|---|---|---|---|---|---|---|---|---|---|---|---|---|---|---|---|---|
| | $C_{10}$ | $C_{12}$ | $C_{14}$ | $C_{16}$ | $C_{18}$ | $C_{20}$ | $C_{12}$ | $C_{14}$ | $C_{16}$ | $C_{18}$ | $C_{20}$ | $C_{15}$ | $C_{17}$ | $C_{19}$ | | | | |
| **Order Eubacteriales (continued)** | | | | | | | | | | | | | | | | | | |
| *Sarcina* | | | | | | | | | | | | | | | | | | |
| *lutea* | | | | | | | | | | | | | | | | | | |
| in trypticase | | 0.14 | 6.45 | 5.4 | 0.14 | | | | | 0.22 | | | | | iso-$C_{12}$ = 0.09<br>$C_{13}$ = 0.69<br>iso-$C_{13}$ = 1.13<br>iso-$C_{14}$ = 3.45 | | iso-$C_{15}$ = 25.74<br>ant-$C_{15}$ = 44.35<br>iso-$C_{16}$ = 5.97<br>iso-$C_{17}$ = 1.3<br>ant-$C_{17}$ = 2.84 | 20 |
| in nutrient broth | | | 1.45 | 0.63 | | | | | | | | | | | iso-$C_{12}$ = 0.41<br>iso-$C_{13}$ = 1.06<br>ant-$C_{13}$ = 4.63<br>iso-$C_{14}$ = 6.49 | | iso-$C_{15}$ = 10.88<br>ant-$C_{15}$ = 63.86<br>iso-$C_{16}$ = 4.00<br>iso-$C_{17}$ = 0.57<br>ant-$C_{17}$ = 1.28 | 20 |
| in defined medium | | 0.55 | 1.35 | 0.63 | | | | | | | | | | | iso-$C_{12}$ = 0.59<br>iso-$C_{13}$ = 0.75<br>ant-$C_{13}$ = 2.27<br>iso-$C_{14}$ = 4.6 | | iso-$C_{15}$ = 5.03<br>ant-$C_{15}$ = 69.58<br>iso-$C_{16}$ = 4.26<br>iso-$C_{17}$ = 0.64<br>ant-$C_{17}$ = 3.27 | 20 |
| *lutea* ATCC 533 | | $C_{11}$ + $C_{12}$ tr | 0.9 | 2.0 | 0.7 | | | | | | | | | | $C_{13}$ = 0.3<br>$C_{14}$ = 0.4<br>$C_{15}$ = 89.5 | | $C_{13}$ = tr<br>$C_{13}$ = tr<br>$C_{15}$ = tr<br>$C_{16}$ br. = 1.9<br>$C_{17}$ = 0.1<br>$C_{17}$ br. = 2.8 | 21 |
| | | | 6.4 | 5.4 | 0.1 | | | | | | | | | | $C_{12}$ = 0.1<br>$C_{13}$ = 1.8<br>$C_{14}$ = 4.1<br>$C_{15}$ = 70.1<br>$C_{16}$ = 6.0<br>$C_{17}$ = 4.1 | | | 71 |
| *lutea* FD 533 | | $C_{11}$ + $C_{12}$ tr | 0.2 | 0.5 | | | | | | | | | | | $C_{13}$ br. = 0.3<br>$C_{14}$ = 1.6<br>$C_{15}$ = 94.1 | | $C_{13}$ = tr<br>$C_{15}$ = 0.7<br>$C_{16}$ br. = 1.9<br>$C_{17}$ = tr<br>$C_{17}$ br. = 0.8 | 21 |
| | | 0.2 | 0.5 | | | | | | | | | | | | $C_{13}$ = 0.3<br>$C_{14}$ = 1.6<br>$C_{15}$ = 94.1<br>$C_{16}$ = 1.9<br>$C_{17}$ = 0.8 | | n-$C_{15}$ = 0.7 | 71 |

## TABLE 1 (Continued)
## MICROBIAL FATTY ACIDS

| Organism | $C_{10}$ | Saturated | | | | | Monounsaturated | | | | | Cyclopropane | | | Branched | Hydroxy | Other Compounds | Ref. |
|---|---|---|---|---|---|---|---|---|---|---|---|---|---|---|---|---|---|---|
| | | $C_{12}$ | $C_{14}$ | $C_{16}$ | $C_{18}$ | $C_{20}$ | $C_{12}$ | $C_{14}$ | $C_{16}$ | $C_{18}$ | $C_{20}$ | $C_{15}$ | $C_{17}$ | $C_{19}$ | | | | |
| **Order Eubacteriales (continued)** | | | | | | | | | | | | | | | | | | |
| **NEISSERIACEAE** | | | | | | | | | | | | | | | | | | |
| *Neisseria* | | | | | | | | | | | | | | | | | | |
| *catarrhalis* | | <1 | 1 | 23 | 1 | | <1 | | 17 | 55 | | | | | | | $C_{15}$ = 1.0<br>$C_{17}$ = 1.0 | 3 |
| | 16 | 11 | 6 | 12 | 10 | | | | 8 | 26 | | | | | | | | 79 |
| *flavia* | | 20 | 13 | 23 | tr | | | | 27 | 6 | | | | | | | | 79 |
| *flavescens* | | 16 | 10 | 18 | tr | | | | 19 | 20 | | | | | | | | 79 |
| *meningitidis* | | | | | | | | | | | | | | | | | | |
| Type A | | 15 | 11 | 24 | <5 | | | | 17 | 7 | | | | | | | | 79 |
| Type B | | 13 | 15 | 29 | tr | | | | 24 | 9 | | | | | | | | 79 |
| Type C | | 9 | 11 | 33 | tr | | | | 26 | 8 | | | | | | | | 79 |
| Type D | | 18 | 13 | 33 | tr | | | | 24 | 4 | | | | | | | | 79 |
| Type Slaterus X | | 17 | 14 | 24 | tr | | | | 20 | 9 | | | | | | | | 79 |
| Type Slaterus Y | | 20 | 18 | 22 | tr | | | | 12 | 7 | | | | | | | | 79 |
| Type Slaterus Z | | 14 | 17 | 38 | tr | | | | 23 | 8 | | | | | | | | 79 |
| Type Slaterus Z[1] | | 11 | 12 | 36 | tr | | | | 26 | 9 | | | | | | | | 79 |
| untypable | | 13 | 16 | 25 | tr | | | | 21 | 11 | | | | | | | | 79 |
| *mucosa* | | 20 | 12 | 26 | tr | | | | 22 | 8 | | | | | | | | 79 |
| *perflava* | | 26 | 16 | 23 | tr | | | | 21 | tr | | | | | | | | 79 |
| *sicca* | | 19 | 10 | 25 | tr | | | | 24 | 8 | | | | | | | | 79 |
| *subflava* | | 24 | 9 | 22 | tr | | | | 26 | 7 | | | | | | | | 79 |
| *Veillonella* | | | | | | | | | | | | | | | | | | |
| in lipopolysaccharides | | | | | | | | | | | | | | | | 3-OH-$C_{13}$<br>3-OH-$C_{15}$ | | 87 |

## TABLE 1 (Continued)
## MICROBIAL FATTY ACIDS

| Organism | Saturated | | | | | Monounsaturated | | | | | Cyclopropane | | | Branched | Hydroxy | Other Compounds | Ref. |
|---|---|---|---|---|---|---|---|---|---|---|---|---|---|---|---|---|---|
| | $C_{10}$ | $C_{12}$ | $C_{14}$ | $C_{16}$ | $C_{18} \cdot C_{20}$ | $C_{12}$ | $C_{14}$ | $C_{16}$ | $C_{18}$ | $C_{20}$ | $C_{15}$ | $C_{17}$ | $C_{19}$ | | | | |
| **Order Eubacteriales (continued)** | | | | | | | | | | | | | | | | | |
| **LACTOBACILLACEAE** | | | | | | | | | | | | | | | | | |
| **Streptococceae** | | | | | | | | | | | | | | | | | |
| *Diplococcus* | | | | | | | | | | | | | | | | | |
| *pneumoniae* | | | | | | | | | | | | | | | | | |
| in glycolipid | | 6.9 | 13.6 | 33.8 | 2.9 | | 6.5 | 25.9 | 10.4 | | | | | | | | 153 |
| **Streptococceae** | | | | | | | | | | | | | | | | | |
| *Streptococcus* | | | | | | | | | | | | | | | | | |
| *cremoris* | | 0.4 | 15.0 | 27.5 | 0.4 | | | 2.3 | 4.8 | | | | 44.7 | | | | 24 |
| *lactis* | | 7.7 | 17.1 | | | | | | 7.2 | | | | | | | $C_{18}$:2 = 18.5<br>$C_{18}$:3 or $C_{21}$ = 49.6 | 4 |
| var. *maltigenes* | | 0.5 | 16.6 | 34.5 | 0.7 | | | 2.7 | 16.8 | | | | 19.3 | $C_{13}$:1 or $C_{14}$ br. = 2.1<br>$C_{15}$:1 or $C_{16}$ br. = 3.5 | | unk = 3.3 | 24 |
| *pyogenes* | | | | | | | | | | | | | | | | | |
| L-forms | | 1.06 | 3.58 | 26.39 | 8.34 | | 0.48 | 6.92 | 34.87 | | | | | $C_{15}$ = tr<br>$C_{16}$ = tr<br>$C_{17}$ = 0.41 | | $C_{15}$ = 0.37<br>$C_{17}$ = 0.58<br>$C_{17}$ unsat. = 0.82<br>$C_{18}$:2 = 13.12 | 23 |
| whole cells | | 0.92 | 2.37 | 34.84 | 6.74 | | 1.15 | 18.05 | 25.10 | | | | | $C_{15}$ = 0.25<br>$C_{16}$ = 0.17<br>$C_{17}$ = 0.45 | | $C_{15}$ = 0.17<br>$C_{17}$ = 0.67<br>$C_{17}$ unsat. = 0.55<br>$C_{18}$:2 = 5.42 | 23 |
| **Lactobacilleae** | | | | | | | | | | | | | | | | | |
| *Lactobacillus* | | | | | | | | | | | | | | | | | |
| *arabinosus* | | | | 37 | 2 | | | | 20 | | | | 31 | | | | 34 |

## TABLE 1 (Continued)
## MICROBIAL FATTY ACIDS

| Organism | Saturated | | | | | | Monounsaturated | | | | | Cyclopropane | | | Branched | Hydroxy | Other Compounds | Ref. |
|---|---|---|---|---|---|---|---|---|---|---|---|---|---|---|---|---|---|---|
| | $C_{10}$ | $C_{12}$ | $C_{14}$ | $C_{16}$ | $C_{18}$ | $C_{20}$ | $C_{12}$ | $C_{14}$ | $C_{16}$ | $C_{18}$ | $C_{20}$ | $C_{15}$ | $C_{17}$ | $C_{19}$ | | | | |
| **Order Eubacteriales (continued)** | | | | | | | | | | | | | | | | | | |
| *Lactobacillus* | | | | | | | | | | | | | | | | | | |
| *arabinosus (plantarum)* | | | 0.7 | 7.0 | 1.0 | | | | 14.0 | 26.0 | | | | 48.0 | | | | 144 |
| *casei* | | | | 28 | 5 | | | | | 45 | | | | 19 | | | | 34 |
| | | | 0.7 | 9.0 | 1.0 | | | | 10.0 | 26.0 | | | | 49.0 | | | | 144 |
| *acidophilus* | | | 4.0 | 10.0 | 1.0 | | | | 11.0 | 40.0 | | | | 28.0 | | | | 144 |
| **PROPIONIBACTERIACEAE** | | | | | | | | | | | | | | | | | | |
| *Propionibacterium* | | | | | | | | | | | | | | | | | | |
| *freudenreichii* | | tr | 1 | 4 | 1 | 1 | | | 1 | 1 | | | | | $C_{17}$ = 20 | | $C_{15}$ = 2<br>iso-$C_{15}$ = 3<br>ant-$C_{15}$ = 47<br>$C_{17}$ = 2<br>$C_{19}$ = 2<br>$C_{21}$ = 1<br>$C_{22}$ = 10<br>$C_{22}$ = 11 | 96 |
| *shermanii* | | tr | 2 | 4 | tr | 14 | | | 1 | tr | | | | | $C_{17}$ = 2 | | $C_{15}$ = 1<br>iso-$C_{15}$ = 9<br>ant-$C_{15}$ = 35<br>$C_{17}$ = 1<br>$C_{19}$ = 1<br>$C_{21}$ = 8 | 96 |
| *arabinosum* | | tr | tr | 4 | 1 | 5 | | | 2 | tr | | | | | $C_{17}$ = 5 | | $C_{15}$ = 5<br>iso-$C_{15}$ = 25<br>ant-$C_{15}$ = 11<br>$C_{17}$ = 4<br>$C_{19}$ = 1<br>$C_{21}$ = 11<br>$C_{22}$ = 13 | 96 |

## TABLE 1 (Continued)
## MICROBIAL FATTY ACIDS

| Organism | Saturated | | | | | | Monounsaturated | | | | | Cyclopropane | | | Branched | Hydroxy | Other Compound | Ref. |
|---|---|---|---|---|---|---|---|---|---|---|---|---|---|---|---|---|---|---|
| | $C_{10}$ | $C_{12}$ | $C_{14}$ | $C_{16}$ | $C_{18}$ | $C_{20}$ | $C_{12}$ | $C_{14}$ | $C_{16}$ | $C_{18}$ | $C_{20}$ | $C_{15}$ | $C_{17}$ | $C_{19}$ | | | | |
| **Order Eubacteriales (continued)** | | | | | | | | | | | | | | | | | | |
| *Propionibacterium* | | | | | | | | | | | | | | | | | | |
| *jensenii* | | tr | 1 | 4 | 1 | 11 | | | tr | | | | | | $C_{17}$ = 1 | | $C_{15}$ = 1<br>iso-$C_{15}$ = 23<br>ant-$C_{15}$ = 15<br>$C_{17}$ = 1<br>$C_{19}$ = 9<br>$C_{21}$ = 13<br>$C_{22}$ = 10 | 96 |
| *pentosaceum* | | tr | 3 | 6 | 2 | 6 | | | 1 | 1 | | | | | $C_{17}$ = 1 | | $C_{15}$ = 11<br>iso-$C_{15}$ = 24<br>ant-$C_{15}$ = 15<br>$C_{17}$ = 1<br>$C_{19}$ = 4<br>$C_{21}$ = 7<br>$C_{22}$ = 18 | 96 |
| *thoenii* | | 1 | 3 | 7 | 3 | 2 | | | 1 | 3 | | | | | $C_{17}$ = 3 | | $C_{15}$ = 7<br>iso-$C_{15}$ = 34<br>ant-$C_{15}$ = 9<br>$C_{17}$ = 20<br>$C_{19}$ = 1<br>$C_{21}$ = 3<br>$C_{22}$ = 3 | 96 |
| *zeae* | | tr | 2 | 3 | tr | 10 | | | 1 | tr | | | | | $C_{17}$ = 4 | | $C_{15}$ = 6<br>iso-$C_{15}$ = 24<br>ant-$C_{15}$ = 15<br>$C_{17}$ = 1<br>$C_{19}$ = 1<br>$C_{21}$ = 13<br>$C_{22}$ = 12 | 96 |
| *Bifidobacterium* | | | | | | | | | | | | | | | | | | |
| *bifidum* | | | | | | | | | | | | | | | | | | |
| grown without human milk | | 1.5 | 10.1 | 23.9 | 7.9 | | | | 9.2 | 42.9 | | | + | 1.2 | iso-$C_{14}$ = 1.8<br>iso-$C_{16}$ = +<br>iso-$C_{18}$ = + | | $C_{15}$ = +<br>ant-$C_{15}$ = 1.0<br>$C_{17}$ = +<br>ant-$C_{17}$ = +<br>$C_{18}$:2 = 1.5<br>$C_{19}$ = + | 91 |

# TABLE 1 (Continued)
# MICROBIAL FATTY ACIDS

| Organism | Saturated | | | | | | Monounsaturated | | | | | Cyclopropane | | | Branched | Hydroxy | Other Compounds | Ref. |
|---|---|---|---|---|---|---|---|---|---|---|---|---|---|---|---|---|---|---|
| | $C_{10}$ | $C_{12}$ | $C_{14}$ | $C_{16}$ | $C_{18}$ | $C_{20}$ | $C_{12}$ | $C_{14}$ | $C_{16}$ | $C_{18}$ | $C_{20}$ | $C_{15}$ | $C_{17}$ | $C_{19}$ | | | | |
| **Order Eubacteriales (continued)** | | | | | | | | | | | | | | | | | | |
| *Bifidobacterium* | | | | | | | | | | | | | | | | | | |
| *bifidum* | | | | | | | | | | | | | | | | | | |
| grown with human milk | | + | 3.5 | 22.9 | 22.6 | | | | 4.9 | 33.4 | | | + | 2.8 | iso-$C_{14}$ = 1.1<br>iso-$C_{16}$ = +<br>iso-$C_{18}$ = + | | $C_{15}$ = +<br>ant-$C_{15}$ = +<br>$C_{17}$ = +<br>ant-$C_{17}$ = +<br>$C_{18}$:2 = 2.2<br>$C_{19}$ = +<br>ant-$C_{19}$ = + | 91 |
| **CORYNEBACTERIACEAE** | | | | | | | | | | | | | | | | | | |
| *Corynebacterium* | | | | | | | | | | | | | | | | | | |
| *liquefaciens* | | 1 | 2 | 10 | 2 | 2 | | | 2 | 2 | | | | | $C_{17}$ = 8 | | $C_{15}$ = 4<br>iso-$C_{15}$ = 55<br>ant-$C_{15}$ = 3<br>$C_{17}$ = 2<br>$C_{19}$ = tr<br>$C_{21}$ = 6<br>$C_{22}$ = tr | 96 |
| *granulosum* | | 2 | 1 | 11 | 2 | 7 | | | 2 | 2 | | | | | $C_{17}$ = 8 | | $C_{15}$ = 4<br>iso-$C_{15}$ = 55<br>ant-$C_{15}$ = 3<br>$C_{17}$ = 2<br>$C_{19}$ = tr<br>$C_{21}$ = tr<br>$C_{22}$ = tr | 96 |
| *anaerobium* | | 2 | 1 | 10 | 1 | 1 | | | 1 | 1 | | | | | $C_{17}$ = 14 | | $C_{15}$ = 4<br>iso-$C_{15}$ = 62<br>ant-$C_{15}$ = 2<br>$C_{17}$ = 1<br>$C_{19}$ = tr<br>$C_{21}$ = tr<br>$C_{22}$ = 1 | 96 |
| *diphtheroides* | | tr | 2 | 7 | 2 | 1 | | | 1 | 1 | | | | | $C_{17}$ = 17 | | $C_{15}$ = 13<br>iso-$C_{15}$ = 41<br>ant-$C_{15}$ = 2<br>$C_{17}$ = 2<br>$C_{19}$ = 2<br>$C_{21}$ = 1<br>$C_{22}$ = 7 | 96 |

## TABLE 1 (Continued)
## MICROBIAL FATTY ACIDS

| Organism | $C_{10}$ | Saturated | | | | | Monounsaturated | | | | | Cyclopropane | | | Branched | Hydroxy | Other Compounds | Ref. |
|---|---|---|---|---|---|---|---|---|---|---|---|---|---|---|---|---|---|---|
| | | $C_{12}$ | $C_{14}$ | $C_{16}$ | $C_{18}$ | $C_{20}$ | $C_{12}$ | $C_{14}$ | $C_{16}$ | $C_{18}$ | $C_{20}$ | $C_{15}$ | $C_{17}$ | $C_{19}$ | | | | |
| **Order Eubacteriales (continued)** | | | | | | | | | | | | | | | | | | |
| *Corynebacterium* | | | | | | | | | | | | | | | | | | |
| *acnes* | | tr | 3 | 15 | 1 | 1 | | | 1 | 1 | | | | | $C_{17}$ = 24 | | $C_{15}$ = 16<br>iso-$C_{15}$ = 32<br>ant-$C_{15}$ = 2<br>$C_{17}$ = 3<br>$C_{19}$ = 1<br>$C_{21}$ = tr<br>$C_{22}$ = tr | 96 |
| *pyogenes* | | 1 | 1 | 3 | 3 | 16 | | | 1 | 2 | | | | | $C_{17}$ = 2 | | $C_{15}$ = 2<br>iso-$C_{15}$ = 33<br>ant-$C_{15}$ = 5<br>$C_{17}$ = 1<br>$C_{19}$ = 7<br>$C_{21}$ = 17<br>$C_{22}$ = 4 | 96 |
| *Listeria* | | | | | | | | | | | | | | | | | | |
| *monocytogenes* | | 4 | 8 | 15 | tr | 2 | | | 3 | tr | | | | | $C_{15}$ = 40<br>$C_{17}$ = 16 | | $C_{15}$ = tr<br>$C_{17}$ = 1<br>$C_{19}$ = 1<br>$C_{21}$ = 1<br>$C_{22}$ = 6<br>$C_{23}$ = tr | 97 |
| **BACILLACEAE** | | | | | | | | | | | | | | | | | | |
| *subtilis* | | | 1 | 10 | | | | | | | | | | | iso-$C_{14}$ = 3<br>iso-$C_{15}$ + ant-<br>$C_{15}$ = 51<br>iso-$C_{16}$ = 10 | | iso-$C_{17}$ +<br>ant-$C_{17}$ = 25 | 25 |
| *anthracis* I | | 0.5 | 2.7 | 6.2 | | | | | 5.9 | | | | | | iso-$C_{12}$ = tr<br>iso-$C_{13}$ = 1.0<br>ant-$C_{13}$ = 0.7<br>iso-$C_{14}$ = 5.9 | | $C_{15}$ = 0.9<br>iso-$C_{15}$ = 19.4<br>ant-$C_{15}$ = 15.6<br>iso-$C_{16}$ = 13.5<br>iso-$C_{16}$:1 = 6.2<br>iso-$C_{17}$ = 3.5<br>ant-$C_{17}$ = 5.7<br>iso-$C_{17}$:1 = 2.3<br>ant-C[illegible]:1 = 4.6 | 26 |

# TABLE 1 (Continued)
# MICROBIAL FATTY ACIDS

| Organism | Saturated $C_{10}$ | Saturated $C_{12}$ | Saturated $C_{14}$ | Saturated $C_{16}$ | Saturated $C_{18}$ | Saturated $C_{20}$ | Monounsaturated $C_{12}$ | Monounsaturated $C_{14}$ | Monounsaturated $C_{16}$ | Monounsaturated $C_{18}$ | $C_{20}$ | Cyclopropane $C_{15}$ | Cyclopropane $C_{17}$ | Cyclopropane $C_{19}$ | Branched | Hydroxy | Other Compounds | Ref. |
|---|---|---|---|---|---|---|---|---|---|---|---|---|---|---|---|---|---|---|
| **Order Eubacteriales (continued)** | | | | | | | | | | | | | | | | | | |
| *Bacillus* | | | | | | | | | | | | | | | | | | |
| *anthracis* II | | 0.5 | 1.8 | 9.1 | | | | | 6.8 | | | | | | iso-$C_{13}$ = 0.9<br>iso-$C_{14}$ = 2.5 | | $C_{15}$ = 0.9<br>iso-$C_{15}$ = 31.4<br>ant-$C_{15}$ = 9.8<br>iso-$C_{16}$ = 11.6<br>iso-$C_{16}$:1 = 2.4<br>iso-$C_{17}$ = 10.0<br>ant-$C_{17}$ = 7.6<br>iso-$C_{17}$:1 = 2.4<br>ant-$C_{17}$:1 = 2.2 | 26 |
| *thuringiensis* | | 0.2 | 2.2 | 7.0 | | | | | 8.2 | | | | | | iso-$C_{12}$ = 1.1<br>iso-$C_{13}$ = 4.5<br>iso-$C_{14}$ = 1.7<br>iso-$C_{15}$ = 11.7 | | $C_{15}$ = 0.8<br>iso-$C_{15}$ = 15.2<br>ant-$C_{15}$ = 7.1<br>iso-$C_{16}$ = 15.3<br>iso-$C_{16}$:1 = 7.0<br>iso-$C_{17}$ = 6.4<br>ant-$C_{17}$ = 2.8<br>iso-$C_{17}$:1 = 1.9<br>ant-$C_{17}$:1 = 0.9 | 26 |
| *cereus* | | 1 | 7 | 19 | | | | | | | | | | | $C_{12}$ = <1<br>$C_{13}$ = 11<br>$C_{14}$ = 6<br>$C_{15}$ = 29<br>$C_{16}$ = 11<br>$C_{17}$ = 15 | | | 3 |
| | | 0.1 | 2.4 | 8.6 | 0.7 | | | | 4.7 | | | | | | $C_{12}$ = 0.2<br>$C_{13}$ = 8.7<br>$C_{14}$ = 2.0<br>$C_{18}$ = 0.9 | | $C_{13}$ = 0.1<br>$C_{15}$ = 0.3<br>$C_{15}$ br. = 44.0<br>$C_{16}$ br. = 4.4<br>$C_{17}$ = 0.2<br>$C_{17}$ br. = 16.9<br>$C_{17}$:1 = 4.8 | 27 |
| *megaterium* | | <1 | 4 | 18 | 2 | | | | | | | | | | $C_{12}$ = <1<br>$C_{13}$ = 7<br>$C_{14}$ = 3<br>$C_{15}$ = 37<br>$C_{16}$ = 6<br>$C_{17}$ = 20 | | | 3 |

# TABLE 1 (Continued)
# MICROBIAL FATTY ACIDS

| Organism | Saturated | | | | | | Monounsaturated | | | | | Cyclopropane | | | Branched | Hydroxy | Other Compounds | Ref. |
|---|---|---|---|---|---|---|---|---|---|---|---|---|---|---|---|---|---|---|
| | $C_{10}$ | $C_{12}$ | $C_{14}$ | $C_{16}$ | $C_{18}$ | $C_{20}$ | $C_{12}$ | $C_{14}$ | $C_{16}$ | $C_{18}$ | $C_{20}$ | $C_{15}$ | $C_{17}$ | $C_{19}$ | | | | |
| **Order Eubacteriales (continued)** | | | | | | | | | | | | | | | | | | |
| *Bacillus* | | | | | | | | | | | | | | | | | | |
| *licheniformis* | | | 4.2 | 3.9 | 0.2 | | | | 12.0 | 1.1 | | | | | ant-$C_{15}$ = 50.4<br>$C_{17}$ = 28.2 | | | 2 |
| *licheniformis* 9259 | | | | 25.0 | | | | | 8.5 | | | | | | $C_{13}$ = 9.1<br>$C_{14}$ = 1.4<br>$C_{15}$ = 17.1<br>$C_{16}$ = 5.4<br>$C_{17}$ = 22.4<br>others = 1.0 | | $C_{13}$:1 br. = 4.5<br>$C_{16}$:1 br. = 3.5 | 28 |
| *stearothermophilus* | | | 2.9 | 3.1 | 1.2 | | | 4.2 | 23.5 | 1.2 | | | | | ant-$C_{15}$ = 30.7<br>$C_{17}$ = 33.2 | | | 2 |
| | | | 2.3 | 42.1 | | | | | 10.9 | | | | | | iso-$C_8$ = 0.3<br>iso-$C_9$ = 0.4<br>iso-$C_{10}$ = 0.3 | | iso-$C_{11}$ = 0.2<br>iso-$C_{14}$ = 0.3<br>iso-$C_{15}$ = 19.8<br>ant-$C_{15}$ = 13.0<br>iso-$C_{16}$ = 3.1<br>$C_{17}$ = 0.9<br>ant-$C_{17}$ = 7.6 | 29 |
| *larvae* B2605 | | | 1.1 | 11.5 | | | | | | | | | | | iso-$C_{14}$ = 0.8<br>iso-$C_{15}$ = 6.6<br>iso-$C_{16}$ = 7.6 | | ant-$C_{15}$ = 38.7<br>iso-$C_{17}$ = 6.6<br>ant-$C_{17}$ = 25.0 | 30 |
| *larvae* B2610 | | | 1.0 | 12.2 | | | | | | | | | | | iso-$C_{14}$<br>iso-$C_{15}$ = 8.3<br>iso-$C_{16}$ = 4.1 | | ant-$C_{15}$ = 50.2<br>iso-$C_{17}$ = 4.2<br>ant-$C_{17}$ = 18.1 | 30 |
| *popilliae* B2309 | | | 3.9 | 18.8 | | | | | | | | | | | iso-$C_{14}$ = 1.8<br>iso-$C_{15}$ = 5.2<br>iso-$C_{16}$ = 3.0 | | ant-$C_{15}$ = 56.5<br>iso-$C_{17}$ = 1.6<br>ant-$C_{17}$ = 2.2 | 30 |
| *popilliae* B2519 | | | 4.8 | 9.8 | | | | | | | | | | | iso-$C_{14}$ = 3.5<br>iso-$C_{15}$ = 5.1<br>iso-$C_{16}$ = 3.8 | | ant-$C_{15}$ = 62.0<br>iso-$C_{17}$ = 1.0<br>ant-$C_{17}$ = 1.8 | 30 |
| *lentimorbus* B2522 | | | 5.3 | 26.6 | | | | | | | | | | | iso-$C_{15}$ = 5.9<br>iso-$C_{16}$ = 0.6<br>iso-$C_{17}$ = 1.9 | | ant-$C_{15}$ = 44.6<br>ant-$C_{17}$ = 1.5 | 30 |

# TABLE 1 (Continued)
## MICROBIAL FATTY ACIDS

| Organism | Saturated $C_{10}$ | Saturated $C_{12}$ | Saturated $C_{14}$ | Saturated $C_{16}$ | Saturated $C_{18}$ | Saturated $C_{20}$ | Monounsaturated $C_{12}$ | Monounsaturated $C_{14}$ | Monounsaturated $C_{16}$ | Monounsaturated $C_{18}$ | Monounsaturated $C_{20}$ | Cyclopropane $C_{15}$ | Cyclopropane $C_{17}$ | Cyclopropane $C_{19}$ | Branched | Hydroxy | Other Compounds | Ref. |
|---|---|---|---|---|---|---|---|---|---|---|---|---|---|---|---|---|---|---|
| **Order Eubacteriales (continued)** | | | | | | | | | | | | | | | | | | |
| *Bacillus* | | | | | | | | | | | | | | | | | | |
| *lentimorbus* B2530 | | | 1.3 | 32.5 | | | | | | | | | | | iso-$C_{15}$ = 2.4<br>iso-$C_{16}$ = 1.4<br>iso-$C_{17}$ = 4.8 | | ant-$C_{15}$ = 40.7<br>ant-$C_{17}$ = 7.3 | 30 |
| *Clostridium* | | | | | | | | | | | | | | | | | | |
| *perfringens* (Hobbs) | | 24 | 24 | 8 | 8 | 5 | | 3 | 5 | 4 | | | | | $C_{15}$ = 3<br>$C_{16}$ = 2 | | $C_{13}$ = tr<br>$C_{15}$ = 4<br>$C_{17}$ = 4<br>unk = 4, 2, tr | 31 |
| *perfringens* E98 | 3 | 33 | 20 | 8 | 9 | 3 | | 4 | 4 | 2 | | | | | $C_{15}$ = 3<br>$C_{16}$ = 4 | | $C_{13}$ = tr<br>$C_{15}$ = 2<br>$C_{17}$ = 5<br>unk = tr, 3, tr | 31 |
| *perfringens* KA137 | tr | 31 | 19 | 8 | 7 | 8 | | 1 | 1 | 2 | | | | | $C_{15}$ = 2<br>$C_{16}$ = tr | | $C_{13}$ = tr<br>$C_{15}$ = 2<br>$C_{17}$ = 5<br>unk = 3, 5, 6 | 31 |
| *sporogenes* | tr | tr | 17 | 24 | 2 | | | 3 | 5 | 7 | | | | | $C_{15}$ = 5<br>$C_{16}$ = 5 | | $C_{13}$ = tr<br>$C_{15}$ = 7<br>$C_{17}$ = 1<br>unk = 22, 1, 1 | 31 |
| *bifermentans* | | 2 | 8 | 13 | 2 | tr | | 6 | 11 | 5 | | | | | $C_{15}$ = 10<br>$C_{16}$ = tr | | $C_{13}$ = 7<br>$C_{17}$ = 8<br>unk = tr, 2, 14, tr | 31 |
| *histolyticum* | tr | 6 | 33 | 20 | 2 | tr | | 8 | 10 | 4 | | | | | $C_{15}$ = 6 | | $C_{15}$ = 8<br>$C_{17}$ = 1<br>unk = 1, 1 | 31 |
| *thermosaccharolyticum* TA-37 | | | | | | | | | | | | | | | | | | |
| spores | | 2.5 | 9.5 | 19.3 | 7.4 | | | 10.9 | 10.1 | 11.7 | | | | | | OH-$C_{18}$ = 19.6 | $C_{18}$:2 = tr<br>$C_{18}$:3 = tr | 33 |
| *thermosaccharolyticum* 3814 | | 1.8 | 19.6 | 15.7 | 6.7 | | | 16.7 | 16.6 | 6.2 | | | | | | OH-$C_{18}$ = 10.9 | $C_{18}$:2 = tr<br>$C_{18}$:3 = tr | 33 |

## TABLE 1 (Continued)
## MICROBIAL FATTY ACIDS

| Organism | Saturated | | | | | | Monounsaturated | | | | | Cyclopropane | | | Branched | Hydroxy | Other Compounds | Ref. |
|---|---|---|---|---|---|---|---|---|---|---|---|---|---|---|---|---|---|---|
| | $C_{10}$ | $C_{12}$ | $C_{14}$ | $C_{16}$ | $C_{18}$ | $C_{20}$ | $C_{12}$ | $C_{14}$ | $C_{16}$ | $C_{18}$ | $C_{20}$ | $C_{15}$ | $C_{17}$ | $C_{19}$ | | | | |
| **Order Eubacteriales (continued)** | | | | | | | | | | | | | | | | | | |
| *Clostridium* | | | | | | | | | | | | | | | | | | |
| *butyricum* | | 1.24 | 6.47 | 52.1 | 1.95 | | | 0.89 | $\Delta^{7}=$ 12.9<br>$\Delta^{9}=$ 6.03 | $\Delta^{4}=$ 1.29<br>$\Delta^{11}=$ 0.92 | | | $\Delta^{7,8}=$ 0.30<br>$\Delta^{9,10}=$ 9.15 | $\Delta^{9,10}=$ 1.07<br>$\Delta^{11,12}=$ 3.47 | | | | 92 |
| **Order Spirochaetales** | | | | | | | | | | | | | | | | | | |
| **TREPONEMATACEAE** | | | | | | | | | | | | | | | | | | |
| *Treponema* | | | | | | | | | | | | | | | | | | |
| *pallidum* | | 0.4 | 0.8 | 28.5 | 35.0 | | | | $\Delta^{9}=$ 3.6 | $\Delta^{9}=$ 1.80 | | | | | | | $C_{15}$ = 0.9<br>$C_{18}$:2-$\Delta^{9,12}$ = 13 | 93 |
| *zuelzerae* | | 1.0 | 10.9 | 5.0 | | | $\Delta^{5}=$ 2.5<br>$\Delta^{7}=$ 1.5 | $\Delta^{7}=$ 3.5<br>$\Delta^{9}=$ 6.5 | $\Delta^{9}=$ 6.0<br>$\Delta^{11}=$ 2.5 | | | | | | iso-$C_{14}$ = 17.6<br>iso-$C_{16}$ = 4.9 | | iso-$C_{13}$ = 8.1<br>$C_{15}$ = 13.4<br>iso-$C_{15}$ = 16.4 | 93 |
| *Leptospira* | | | | | | | | | | | | | | | | | | |
| *interrogans* | | | | 26.3 | 1.0 | | | | $\Delta^{9}=$ 3.0<br>$\Delta^{11}=$ 22.3 | $\Delta^{9}=$ 37.6 | | | | | | | | 98 |
| **Order Mycoplasmatales** | | | | | | | | | | | | | | | | | | |
| **MYCOPLASMATACEAE** | | | | | | | | | | | | | | | | | | |
| "PPLO" 07 strain | 1.1 | 1.3 | 3.0 | 56.5 | 15.5 | | | | 6.0 | 7.1 | | | | | | | $C_{15}$ = 2.4<br>$C_{17}$ = 5.3 | 110 |
| *Mycoplasma* sp. KHS | 0.21 | 19.56 | 40.44 | 29.91 | 0.97 | | | tr | $\Delta^{7,8}=$ tr<br>$\Delta^{9,10}=$ 0.32 | $\Delta^{9,10}=$ 3.31<br>$\Delta^{11,12}=$ 0.65 | | | | | $C_{13}$ = 0.21<br>$C_{14}$ = tr | | $C_{13}$ = 0.46<br>$C_{14}$ = 0.45<br>$C_{18}$:3 = 1.57<br>margaric = tr | 114 |
| | 0.42 | 18.78 | | 27.11 | 1.27 | | | $\Delta^{9,10}=$ tr | $\Delta^{7,8}=$ tr<br>$\Delta^{9,10}=$ 0.45 | $\Delta^{9,10}=$ 3.75<br>$\Delta^{11,12}=$ 1.38 | | | | | | | $C_{13}$ = 0.73<br>$C_{15}$ = 0.76<br>$C_{18}$:3 = 1.57<br>margaric = tr | 115 |

## TABLE 1 (Continued)
## MICROBIAL FATTY ACIDS

| Organism | Saturated | | | | | | Monounsaturated | | | | | Cyclopropane | | | Branched | Hydroxy | Other Compounds | Ref. |
|---|---|---|---|---|---|---|---|---|---|---|---|---|---|---|---|---|---|---|
| | $C_{10}$ | $C_{12}$ | $C_{14}$ | $C_{16}$ | $C_{18}$ | $C_{20}$ | $C_{12}$ | $C_{14}$ | $C_{16}$ | $C_{18}$ | $C_{20}$ | $C_{15}$ | $C_{17}$ | $C_{19}$ | | | | |
| **Order Mycoplasmatales (continued)** | | | | | | | | | | | | | | | | | | |
| *Mycoplasma* sp. KHS | | | | | | | | | | | | | | | | | | |
| *gallisepticum* | 3.8 | 3.0 | 1.4 | 36.2 | 21.2 | | | 4.6 | 12.6 | 12.0 | | | | | | | $C_{15}$ = 1.0<br>$C_{17}$ = 1.0<br>$C_{18}$:2 = 1.4 | 111 |
| *laidlawii* | | | | | | | | | | | | | | | | | | |
| in phospholipid in<br>glycolipid | | 3.3<br>0.9 | 20.3<br>2.4 | 55.3<br>55.7 | 4.1<br>2.1 | | | | | 1.9<br>39.0 | | | | | | | | 112<br>117 |
| *mycoides* | | | | | | | | | | | | | | | | | | |
| in glycolipid | | | 1 | 18 | 49 | | | | | 24 | | | | | | | $C_{18}$:2 = 3 | 111 |
| **Order Rickettsiales** | | | | | | | | | | | | | | | | | | |
| **CHLAMYDIADEAE** | | | | | | | | | | | | | | | | | | |
| *Chlamydia* | | | | | | | | | | | | | | | | | | |
| *psittaci* 6BC | | | 0.6 | 13.7 | 14.8 | 1.5 | | | | 19.8 | 6.9 | | | | $C_{15}$ = 9.9<br>$C_{16}$ = 2.8<br>$C_{17}$ = 2.9<br>$C_{18}$ = 0.5<br>$C_{19}$ = 1.3<br>$C_{20}$ = 1.1 | | $C_{18}$:2 = 5.6<br>$C_{20}$:4 = 2.2 | 70 |
| **Viruses** | | | | | | | | | | | | | | | | | | |
| **POX VIRUSES** | | | | | | | | | | | | | | | | | | |
| Fowlpox | | 0.3 | 1.4 | 35.3 | 12.9 | 3.1 | | | 6.2 | 24.0 | 4.1 | | | | | | $C_{15}$:2 = 4.6<br>Numerous other acids, in small amounts, from $C_{14}$ | 66 |

## TABLE 1 (Continued)
## MICROBIAL FATTY ACIDS

| Organism | Saturated | | | | | | Monounsaturated | | | | | Cyclopropane | | | Branched | Hydroxy | Other Compounds | Ref. |
|---|---|---|---|---|---|---|---|---|---|---|---|---|---|---|---|---|---|---|
| | $C_{10}$ | $C_{12}$ | $C_{14}$ | $C_{16}$ | $C_{18}$ | $C_{20}$ | $C_{12}$ | $C_{14}$ | $C_{16}$ | $C_{18}$ | $C_{20}$ | $C_{15}$ | $C_{17}$ | $C_{19}$ | | | | |
| **Viruses (continued)** | | | | | | | | | | | | | | | | | | |
| **MYXOVIRUSES** | | | | | | | | | | | | | | | | | | |
| Influenza Mel-1935 | | | | | | | | | | | | | | | | | | |
| calf kidney host | | | 0.7 | 23 | 21 | | | | 1.3 | 21 | | | | | $C_{15}$ = 0.6<br>$C_{17}$ = 0.9 | | $C_{15}$ = 0.3<br>$C_{17}$ = 0.9 | 58 |
| chick embryo host | | | 0.3 | 32 | 13 | | | | 1.2 | 29 | | | | | $C_{17}$ = tr | | $C_{15}$ = tr<br>$C_{17}$ = tr<br>$C_{22}$-polyenes = 7.5<br>$C_{28}$-triene = 0.9<br>arachidic = 1.6<br>arachidonic = 5.8<br>lindeic = 9.4 | 58 |
| Influenza PRB | | | 0.7 | 16.2 | 14.3 | | | | 6.7 | 13.2 | 10.0 | | | | | | $C_{18}$:2 = 2.8<br>$C_{18}$:3 = 1.1<br>$C_{20}$ = 15.5<br>$C_{22}$-polyene = 8.9<br>$C_{24}$ = 7.8 | 146 |
| **PARAMYXOVIRUSES** | | | | | | | | | | | | | | | | | | |
| Parainfluenza SV5/W3 | | | | | | | | | | | | | | | | | | |
| HaK host | | | 3.5 | 28.5 | 16.3 | | | | 4.1 | 33.9 | | | | | | | $C_{18}$:2 = 6.2<br>$C_{20}$:4 = 11.4 | 67 |
| Sendai | | tr | tr | 24.2 | 12.7 | | | | 1.0 | 22.3 | 3.9 | | | | | | $C_{18}$:2 = 5.1<br>$C_{20}$:4 = 12.2<br>$C_{22}$:6 = 8.0<br>$C_{24}$ = 5.1 | 145 |
| Newcastle | | tr | tr | 11.9 | 14.6 | | | | tr | 20.0 | 7.0 | | | | | | $C_{18}$:2 = 6.9<br>$C_{22}$ = 23.8<br>$C_{22}$:6 = 6.5<br>$C_{24}$ = 9.3 | 145 |

# TABLE 1 (Continued)
# MICROBIAL FATTY ACIDS

| Organism | Saturated | | | | | | Monounsaturated | | | | | Cyclopropane | | | Branched | Hydroxy | Other Compounds | Ref. |
|---|---|---|---|---|---|---|---|---|---|---|---|---|---|---|---|---|---|---|
| | $C_{10}$ | $C_{12}$ | $C_{14}$ | $C_{16}$ | $C_{18}$ | $C_{20}$ | $C_{12}$ | $C_{14}$ | $C_{16}$ | $C_{18}$ | $C_{20}$ | $C_{15}$ | $C_{17}$ | $C_{19}$ | | | | |
| **Yeasts** | | | | | | | | | | | | | | | | | | |
| *Candida* | | | | | | | | | | | | | | | | | | |
| *lipolytica* | | tr | tr | 32.60 | 3.25 | | | | 17.80 | 40.30 | | | | | | | $C_{11}$ = tr<br>$C_{13}$ = tr<br>$C_{15}$ = tr<br>$C_{17}$ = tr<br>$C_{17}$:1 =<br>$C_{18}$:2 = 4.03 | 106 |
| *utilis* | | | | 26.0 | 4.0 | | | | 14.1 | 39.0 | 5.5 | | | | | | $C_{18}$:2 = 5.0 | 107 |
| at 20°C | | | | 11.1 | + | | | | 4.5 | 27.0 | | | | | | | $C_{16}$:2 = +<br>$C_{17}$:1 = 2.6<br>$C_{18}$:2 = 42.9<br>$C_{18}$:3 = 11.9 | 108 |
| at 30°C | | | | 12.3 | 1.0 | | | | 2.3 | 33.4 | 33.4 | | | | | | $C_{16}$:2 = +<br>$C_{17}$:1 = 1.2<br>$C_{18}$:2 = 47.6<br>$C_{18}$:3 = 2.7 | 108 |
| *Debaryomyces* | | | | | | | | | | | | | | | | | | |
| *hansenii* | | | 0.4 | 23.7 | 8.2 | | | 2.5 | | | | | | | | | $C_{15}$ = 0.6<br>$C_{17}$ = 4.2<br>$C_{18}$:2 = 2.5<br>$C_{18}$:3 = 1.5<br>$C_{18}$:4 or $C_{20}$:1 = 31 | 122 |
| *Pullularia* | | | | | | | | | | | | | | | | | | |
| *pullulans* | | | | | | | | | | | | | | | | | | |
| DEGS | 0.12 | 0.12 | 0.12 | 30.80 | 8.7 | | | | 2.60 | 41.93 | | | | | | | $C_8$ or $C_8$:1 = 0.12<br>$C_{13}$ = 0.09<br>$C_{15}$ = tr<br>$C_{18}$:2 = 13.10<br>$C_{18}$:3 = 0.60<br>$C_{20}$:2 = 0.70 | 102 |
| SE-30 | 0.08 | 0.12 | 0.10 | 31.1 | 7.65 | | | | 2.9 | 45.48 | | | | | | | $C_8$ or $C_8$:1 = 0.08<br>$C_{12}$:1 = 0.10<br>$C_{18}$:2 = 11.20<br>$C_{18}$:3 = 0.29<br>$C_{20}$:2 = 0.25 | 102 |

## TABLE 1 (Continued)
## MICROBIAL FATTY ACIDS

| Organism | Saturated | | | | | | Monounsaturated | | | | | Cyclopropane | | | Branched | Hydroxy | Other Compounds | Ref. |
|---|---|---|---|---|---|---|---|---|---|---|---|---|---|---|---|---|---|---|
| | $C_{10}$ | $C_{12}$ | $C_{14}$ | $C_{16}$ | $C_{18}$ | $C_{20}$ | $C_{12}$ | $C_{14}$ | $C_{16}$ | $C_{18}$ | $C_{20}$ | $C_{15}$ | $C_{17}$ | $C_{19}$ | | | | |
| **Yeasts (continued)** | | | | | | | | | | | | | | | | | | |
| *Rhodotorula* | | | | | | | | | | | | | | | | | | |
| *graminis* | | | | | | | | | | | | | | | | | | |
| in fat | 0.3 | 0.4 | 3.9 | 31.9 | 3.2 | | | 1.1 | 0.3 | 52.0 | 2.5 | | | | | | $C_{10}$ = 0.1 wt % | 101 |
| *Saccharomyces* | | | | | | | | | | | | | | | | | | |
| *cerevisiae* | | + $C_{13}$ = 3 | 4 | 7 | $<1$ | | | 5 | 59 | 21 | | | | | | | $C_{15}$ = $<1$<br>$C_{17}$ = $<1$<br>$C_{18}$:2 + $C_{18}$:3 = $<1$ | 100 |
| at first stage of industrial production | | 0.9 | 0.4 | 23.7 | 5.1 | | | | 39.0 | 28.5 | | | | | | | $C_{13}$ = 0.2<br>$C_{15}$ = 0.4<br>$C_{17}$ = 1.1 | 105 |
| at last stage of industrial production | 0.4 | 0.8 | 4.0 | 11.8 | 3.3 | | | | 52.8 | 27.2 | | | | | | | $C_{13}$ = 0.1<br>$C_{15}$ = 1.2<br>$C_{17}$ = 0.6 | 105 |
| *Torula* | | | | | | | | | | | | | | | | | | |
| *utilis* | | | 0.3 | 7.9 | 3.8 | | | | 7.6 | 21.5 | | | | | | | $C_{18}$:2 = 49.7<br>$C_{18}$:3 = 0.4 | 167 |
| **Fungi** | | | | | | | | | | | | | | | | | | |
| *Agaricus* | | | | | | | | | | | | | | | | | | |
| *bisporus* (Lange) Sing., strain 310 | | | | | | | | | | | | | | | | | | |
| sporophore | | | | | | | | | | | | | | | | | | |
| in neutral lipid | tr | tr | tr | 13.2 | 4.1 | | | | tr | 5.4 | | | | | | | $C_{18}$:2 = 77.1 | 128 |
| in polar lipid | tr | tr | tr | 6.1 | 2.0 | | | | tr | tr | | | | | | | $C_{18}$:2 = 91.3 | 128 |
| mycelium | | | | | | | | | | | | | | | | | | |
| in neutral lipid | 4.1 | 5.2 | 8.3 | 22.1 | 10.0 | | | | 6.6 | 16.9 | | | | | | | $C_{17}$ = tr<br>$C_{18}$:2 = 26.3 | 128 |
| in polar lipid | tr | tr | tr | 18.8 | 3.7 | | | | 8.5 | 14.1 | | | | | | | $C_{17}$ = 5.1<br>$C_{18}$:2 = 49.8 | 128 |

# TABLE 1 (Continued)
## MICROBIAL FATTY ACIDS

| Organism | Saturated | | | | | | Monounsaturated | | | | | Cyclopropane | | | Branched | Hydroxy | Other Compounds | Ref. |
|---|---|---|---|---|---|---|---|---|---|---|---|---|---|---|---|---|---|---|
| | $C_{10}$ | $C_{12}$ | $C_{14}$ | $C_{16}$ | $C_{18}$ | $C_{20}$ | $C_{12}$ | $C_{14}$ | $C_{16}$ | $C_{18}$ | $C_{20}$ | $C_{15}$ | $C_{17}$ | $C_{19}$ | | | | |
| **Fungi (continued)** | | | | | | | | | | | | | | | | | | |
| *Blastomyces* | | | | | | | | | | | | | | | | | | |
| *dermatitidis* | | | | | | | | | | | | | | | | | | |
| yeast phase | | | | | | | | | | | | | | | | | | |
| cell sap | | | | 13.5 | 7.3 | | | | 1.8 | 59.1 | | | | | | | $C_{18}$:2 = 17.8 | 120 |
| cell wall | | | | 14.8 | 7.4 | | | | 1.3 | 61.8 | | | | | | | $C_{18}$:2 = 14.5 | 120 |
| mycelial phase | | | | | | | | | | | | | | | | | | |
| cell sap | | | | 13.8 | 3.8 | | | | 1.0 | 38.3 | | | | | | | $C_{18}$:2 = 13.2 | 120 |
| cell wall | | | | 16.2 | 7.9 | | | | 3.1 | 41.2 | | | | | | | $C_{18}$:2 = 31.7 | 120 |
| *Chaetomium* | | | | | | | | | | | | | | | | | | |
| *thermophile* | | | | 57.8 | 4.4 | | | | 3.1 | 8.0 | | | | | | | $C_{18}$:2 = 26.8 | 124 |
| *globosum* | | | 1.4 | 30.6 | 9.6 | | | | 10.8 | 9.7 | | | | | | | $C_{15}$ = 1.0<br>$C_{18}$:2 = 35.6<br>$C_{18}$:3 = 1.3 | 124 |
| *Claviceps* | | | | | | | | | | | | | | | | | | |
| *purpurea* | | | 0.1 | 19.9 | 4.3 | tr | | | 6.5 | 22.5 | | | | | | OH-$C_{18}$ = 32.3<br>2-OH-$C_{18}$ = tr | $C_{18}$:2 = 14.3<br>$C_{18}$:3 = tr | 132 |
| *paspali* | | | 0.2 | 19.0 | 2.1 | tr | | | 7.4 | 56.7 | | | | | | OH-$C_{18}$ = tr<br>2-OH-$C_{18}$ = tr | $C_{18}$:2 | 132 |
| *gigantea* | | | 0.2 | 17.2 | 2.7 | tr | | | 2.6 | 55.6 | | | | | | OH-$C_{18}$ = ~1<br>2-OH-$C_{18}$ = ~1 | $C_{18}$:2 = 19.6<br>$C_{18}$:3 = tr | 132 |
| *ex* Pennisetum | | | 0.3 | 34.1 | 7.3 | tr | | | 2.4 | 38.6 | | | | | | $C_{18}$ = ~1<br>2-OH-$C_{18}$ = ~2 | $C_{18}$:2 = 14.4<br>epoxy-$C_{18}$ = tr | 132 |
| *sulcata* | | | 0.1 | 11.1 | 4.9 | 1.4 | | | 0.8 | 12.8 | | | | | | $C_{18}$ = tr<br>2-OH-$C_{18}$ = 63.6 | $C_{18}$:2 = 5.1<br>$C_{18}$:3 = tr<br>epoxy-$C_{18}$ = tr | 132 |

# TABLE 1 (Continued)
# MICROBIAL FATTY ACIDS

| Organism | Saturated | | | | | | Monounsaturated | | | | | Cyclopropane | | | Branched | Hydroxy | Other Compounds | Ref. |
|---|---|---|---|---|---|---|---|---|---|---|---|---|---|---|---|---|---|---|
| | $C_{10}$ | $C_{12}$ | $C_{14}$ | $C_{16}$ | $C_{18}$ | $C_{20}$ | $C_{12}$ | $C_{14}$ | $C_{16}$ | $C_{18}$ | $C_{20}$ | $C_{15}$ | $C_{17}$ | $C_{19}$ | | | | |
| **Fungi (continued)** | | | | | | | | | | | | | | | | | | |
| *Clitocybe* | | | | | | | | | | | | | | | | | | |
| *illudens* | | | tr | 18.7 | 1.8 | | | | 1.5 | 42.8 | | | | | | | $C_{18}$:2 = 34.8 | 123 |
| *Dictyostelium* | | | | | | | | | | | | | | | | | | |
| *discoideum* | | | | 14 | 2 | | | | 7 | 23 | | | | | | | $C_{16}$:3 = 10<br>$C_{18}$:2 = 41 | 125 |
| *Histoplasma* | | | | | | | | | | | | | | | | | | |
| *capsulatum* | | | | | | | | | | | | | | | | | | |
| yeast phase | | | | | | | | | | | | | | | | | | |
| cell sap | | | | 15.5 | 0.9 | | | | 3.0 | 50.9 | | | | | | | $C_{18}$:2 = 26.8 | 120 |
| cell wall | | | | 20.8 | 2.1 | | | | 3.3 | 66.1 | | | | | | | $C_{18}$:2 = 7.6 | 120 |
| mycelial phase | | | | | | | | | | | | | | | | | | |
| cell sap | | | | 17.1 | 6.4 | | | | 1.4 | 33.7 | | | | | | | $C_{18}$:2 = 41.1 | 120 |
| cell wall | | | | 17.3 | 7.4 | | | | 1.3 | 32.8 | | | | | | | $C_{18}$:2 = 41.3 | 120 |
| *Humicola* | | | | | | | | | | | | | | | | | | |
| *grisea* | | | | 15.3 | 1.5 | | | | | 30.9 | | | | | | | $C_{18}$:2 = 33.9<br>$C_{18}$:3 = 18.5 | 124 |
| var. *termoidea* | | | | 28.8 | 2.2 | | | | | 40.4 | | | | | | | $C_{18}$:2 = 28.5 | 124 |
| *brevis* | | | | 28.8 | 3.7 | | | | 1.9 | 20.4 | | | | | | | $C_{18}$:2 = 41.3<br>$C_{18}$:3 = 4.0 | 124 |
| *nigrescens* | | | | 20.5 | 3.6 | | | | | 29.2 | | | | | | | $C_{18}$:2 = 34.3<br>$C_{18}$:3 = 12.2 | 124 |
| *insolens* | | | | 29.9 | 1.1 | | | | | 57.3 | | | | | | | $C_{18}$:2 = 31.8 | 124 |
| *lanuginosa* | | | | 21.4 | 4.5 | | | | | 65.2 | | | | | | | $C_{18}$:2 = 8.6 | 124 |
| *Malbranchea* | | | | | | | | | | | | | | | | | | |
| *pulchella* | | | | 11.3 | 11.4 | | | | | 26.6 | | | | | | | $C_{18}$:2 = 50.7 | 124 |
| var. *sulfurea* | | | | 26.2 | 7.5 | | | | | 35.0 | | | | | | | $C_{18}$:2 = 31.3 | 124 |

# TABLE 1 (Continued)
## MICROBIAL FATTY ACIDS

| Organism | Saturated $C_{10}$ | Saturated $C_{12}$ | Saturated $C_{14}$ | Saturated $C_{16}$ | Saturated $C_{18}$ | Saturated $C_{20}$ | Monounsaturated $C_{12}$ | Monounsaturated $C_{14}$ | Monounsaturated $C_{16}$ | Monounsaturated $C_{18}$ | Monounsaturated $C_{20}$ | Cyclopropane $C_{15}$ | Cyclopropane $C_{17}$ | Cyclopropane $C_{19}$ | Branched | Hydroxy | Other Compounds | Ref. |
|---|---|---|---|---|---|---|---|---|---|---|---|---|---|---|---|---|---|---|
| **Fungi (continued)** | | | | | | | | | | | | | | | | | | |
| *Mucor* | | | | | | | | | | | | | | | | | | |
| *globosus* | | 2.1 | 7.6 | 26.1 | 6.9 | | | | 7.7 | 25.8 | | | | | | | $C_{18}$:2 = 8.3<br>$C_{18}$:3 = 15.6 | 124 |
| *pusillus* | | | | 23.5 | 2.9 | 0.8 | | | 1.2 | 59.4 | | | | | | | $C_{18}$:2 = 11.2<br>$C_{18}$:3 = 1.1 | 124 |
| *Penicillium* | | | | | | | | | | | | | | | | | | |
| *atrovenetum* | | | 0.4 | 13.8 | 3.6 | 1.9 | | | 2.7 | 30.9 | | | | | | | $C_{15}$ = 0.6<br>$C_{17}$ = 0.8 | 126 |
| *chrysogenum* | | | | 12.2 | 5.5 | | | | | 10.9 | | | | | | | $C_{18}$:2 = 65.4<br>$C_{18}$:3 = 6.0 | 124 |
| on D-glucose | | tr | tr | 23.8 | 9.0 | 4.5 | | | 3.0 | 4.7 | | | | | | | $C_{15}$ = 1.8<br>$C_{17}$ = tr<br>$C_{18}$ = 48.0<br>$C_{20}$:4 = 4.2 | 129 |
| on sucrose | | tr | tr | 17.3 | 8.85 | 6.3 | | | 1.15 | 11.25 | | | | | | | $C_{15}$ = 1.2<br>$C_{17}$ = tr<br>$C_{18}$:2 = 53.0<br>$C_{20}$:4 = 0.65 | 129 |
| *duponti* | | | | 25.2 | 10.8 | | | | | 42.2 | | | | | | | $C_{18}$:2 = 21.8 | 124 |
| *pulvillorum* | | | | | | | | | | | | | | | | | | |
| mycelium, per unit dry weight | | | 0.14 | 1.34 | 0.69 | | | | 0.12 | 1.87 | | | | | | | $C_{15}$ = 0.19<br>iso-$C_{16}$ = 0.04<br>$C_{17}$ = 0.16<br>iso-$C_{18}$ = 0.10<br>$C_{18}$:2 = 4.91<br>$C_{18}$:3 = 0.10 | 130 |
| *Rhizopus* | | | | | | | | | | | | | | | | | | |
| *arrhizus* | | | | | | | | | | | | | | | | | | |
| mycelia | tr | tr | 1.2 | 18.4 | 11.0 | 16.2 | | | 3.7 | 29.4 | | | | | | | $C_{15}$ = 0.2<br>$C_{18}$:2 = 16.3<br>$C_{18}$:3 = 0.2<br>$C_{22}$ = 0.9<br>$C_{24}$ = 1.7 | 127 |

## TABLE 1 (Continued)
## MICROBIAL FATTY ACIDS

| Organism | Saturated | | | | | | Monounsaturated | | | | | Cyclopropane | | | Branched | Hydroxy | Other Compounds | Ref. |
|---|---|---|---|---|---|---|---|---|---|---|---|---|---|---|---|---|---|---|
| | $C_{10}$ | $C_{12}$ | $C_{14}$ | $C_{16}$ | $C_{18}$ | $C_{20}$ | $C_{12}$ | $C_{14}$ | $C_{16}$ | $C_{18}$ | $C_{20}$ | $C_{15}$ | $C_{17}$ | $C_{19}$ | | | | |
| **Fungi (continued)** | | | | | | | | | | | | | | | | | | |
| *Rhizopus* | | | | | | | | | | | | | | | | | | |
| *arrhizus* | | | | | | | | | | | | | | | | | | |
| spores | | | 1.1 | 16.8 | 19.9 | 6.9 | | | 1.7 | 42.4 | | | | | | | $C_{15}$ = 0.8<br>$C_{18}$:2 = 7.7<br>$C_{18}$:3 = 0.1 | 127 |
| *Scenedesmus* | | | | | | | | | | | | | | | | | | |
| *obliquus* | | | 1.8 | 22.3 | 1.1 | | | | 4.5 | | | | | | $C_{16}$:3 = 10.4 | | $C_{13}$ = 3.5<br>$C_{15}$ = 3.2<br>$C_{16}$:2 = 12.2<br>$C_{18}$:2 = 30.7<br>$C_{18}$:3 = 1.8 | 131 |
| *Sporotrichum* | | | | | | | | | | | | | | | | | | |
| *exile* | | 1.4 | 1.9 | 17.0 | 8.8 | 2.1 | | | 2.1 | 8.3 | | | | | | | $C_{18}$:2 = 58.4 | 124 |
| *thermophile* | | | | 28.4 | 6.8 | | | | | 2.7 | | | | | | | $C_{18}$:2 = 35.1 | 124 |
| *Stibella* sp. | | | | 19.5 | 2.3 | | | | 1.4 | 13.4 | | | | | | | $C_{18}$:2 = 58.3<br>$C_{18}$:3 = 5.0 | 124 |
| *thermophila* | | | 2.1 | 42.5 | 13.7 | | | | 1.9 | 25.4 | | | | | | | $C_{18}$:2 = 14.3 | 124 |
| *Trychophyton* | | | | | | | | | | | | | | | | | | |
| *rubrum* | | tr | 0.8 | 23.8 | 7.4 | | | | | 13.1 | | | | | | | $C_{15}$ = 2.1<br>$C_{22}$ = tr | 121 |

# TABLE 2
# MICROBIAL LIPIDS

| Organism | Glycerides | | | Phospholipids | | | | | | | | | | | | |
|---|---|---|---|---|---|---|---|---|---|---|---|---|---|---|---|---|
| | MG | DG | TG | PA | PS | PE | MPE | DPE | PC | PI | PG | PGP (CL) | Glycolipids | Amino Acid Lipids | Other Compounds | Ref. |
| **Order Pseudomonadales Suborder Rhodobacteriineae** | | | | | | | | | | | | | | | | |
| **THIORHODACEAE** | | | | | | | | | | | | | | | | |
| *Chromatium* | | | | | | | | | | | | | | | | |
| strain D | | | | | | 1.32 | | | | | 17.8 | 3.75 | 0.73<br>0.23<br>3.26 | | lyso-PE = 55.76 | 74 |
| **ATHIORHODACEAE** | | | | | | | | | | | | | | | | |
| *Rhodopseudomonas* | | | | | | | | | | | | | | | | |
| *gelatinosa* | | | | | | + | | | | | + | | | O-ornithyl-PG = + | | 32 |
| *spheroides* | | | | | | + | | | + | | + | + | | O-ornithyl-PG = + | sulfoquinovasyl diglyceride | 32 |
| | | | | 24 | | 40 | | | 22 | | 14 | | | | | 35 |
| *capsulata* | | | | | | + | | | + | | + | | | O-ornithyl-PG = + | | 32 |
| *palustris* | | | | | | + | | | + | | + | | | O-ornithyl-PG = tr | | 32 |
| *Rhodospirillum* | | | | | | | | | | | | | | | | |
| *rubrum* | | | | | | 12 | | | 30 | 15 | 42 | | | | | 35 |
| | | | | | | + | | | | | + | + | | O-ornithyl-PG = + | | 32 |
| *Chloropseudomonas* | | | | | | | | | | | | | | | | |
| *ethylicum* | | | | | | | | | | | | + | + | | (monogalactosyl diglyceride) | 63 |
| **THIOBACTERIACEAE** | | | | | | | | | | | | | | | | |
| *Thiobacillus* | | | | | | | | | | | | | | | | |
| *thioxidans* | | | | | | + | | | + | | + | | | | | 37 |
| **PSEUDOMONADACEAE** | | | | | | | | | | | | | | | | |
| *Pseudomonas* | | | | | | | | | | | | | | | | |
| *aeruginosa* | | | | | | + | | | + | | | + | | | | 38 |
| | | | | | | | | | | | | | | | rhamnolipid | 147 |

## TABLE 2 (Continued)
## MICROBIAL LIPIDS

| Organism | Glycerides | | | Phospholipids | | | | | | | | | | | | |
|---|---|---|---|---|---|---|---|---|---|---|---|---|---|---|---|---|
| | MG | DG | TG | PA | PS | PE | MPE | DPE | PC | PI | PG | PGP (CL) | Glycolipids | Amino Acid Lipids | Other Compounds | Ref. |
| **Order Pseudomonadales** | | | | | | | | | | | | | | | | |
| **Suborder Rhodobacteriineae (continued)** | | | | | | | | | | | | | | | | |
| *Pseudomonas* | | | | | | | | | | | | | | | | |
| *aeruginosa* | | | | | | | | | | | | | | | | |
| | | | | | | 48.7 | | | 4.7 | | | 45.4 | | | lyso-PE = 1.2 | 148 |
| | | | | 1.7 | | 69.4 | | | 1.4 | | | 9.4 | | | | 149 |
| *Halobacterium* | | | | | | | | | | | | | | | | |
| *cutirubrum* | | | | | | | | | 2.3 | 1.3 | 6.1 | | | | lyso-lecithin = 0.8<br>polyglycerol phosphatide = 73.0 | 41 |
| | | | | | | | | | | | | | | | *dl*-O-phytanil analogues of phosphatidyl glycerophosphate + PG and a glycolipid sulfate | 75 |
| | | | | | | | | | | | | | | | squalene and hydrosqualenes | 80 |
| **SIDEROCAPSACEAE** | | | | | | | | | | | | | | | | |
| *Ferrobacillus* | | | | | | | | | | | | | | | | |
| *ferrooxidans* | | | | | | 20 | | | 1.5 | | 23 | 13 | | | phosphatidyl monomethyl ethanolamine = 42<br>phosphatidyl dimethyl ethanolamine = 1 | 84 |
| **Order Hyphomicrobiales** | | | | | | | | | | | | | | | | |
| **HYPHOMICROBIACEAE** | | | | | | | | | | | | | | | | |
| *Rhodomicrobium* | | | | | | | | | | | | | | | | |
| *vanielli* | | | | 1.8 | | 4.5 | | | 26.5 | | 9.7 | | | 46.5 | biphosphatidic acid = 6.7<br>lyso-PG-*o*-ornithine ester = 3.2 | 36 |

## TABLE 2 (Continued)
## MICROBIAL LIPIDS

| Organism | Glycerides | | | Phospholipids | | | | | | | | | | | | |
|---|---|---|---|---|---|---|---|---|---|---|---|---|---|---|---|---|
| | MG | DG | TG | PA | PS | PE | MPE | DPE | PC | PI | PG | PGP (CL) | Glycolipids | Amino Acid Lipids | Other Compounds | Ref. |
| **Order Eubacteriales** | | | | | | | | | | | | | | | | |
| **AZOTOBACTERIACEAE** | | | | | | | | | | | | | | | | |
| *Azotobacter* | | | | | | | | | | | | | | | | |
| *vinelandii* | | | | | | + | | | | | | + | | | | 39 |
| *agilis* | | | | tr | | 5.68 | 4.9 | | 2.4 | | 13.3 | 22.6 | | | | 43, 149 |
| **RHIZOBIACEAE** | | | | | | | | | | | | | | | | |
| *Rhizobium* | | | | | | | | | | | | | | | | |
| *japonicum* (two of five strains listed) | | | | | | | | | | | | | | | | |
| strain 83 | | | | | 11.0 | 36.0 | | | 38.0 | | | 15.0 | | | | 89 |
| strain 508 | | | | | 13.4 | 35.7 | | | 37.5 | | | 13.4 | | | | 89 |
| *Rhizobium* | | | | | | | | | | | | | | | | |
| *japonicum* | | | | | + | + | | | + | | | + | | | | 115 |
| *Agrobacterium* | | | | | | | | | | | | | | | | |
| *tumefaciens* | | | | | | + | | | + | | | | | | | 43 |
| | | | | 4.1 | | 17.5 | 16.4 | | 27.7 | | 13.0 | 18.5 | | | | 149 |
| *Chromobacterium* | | | | | | | | | | | | | | | | |
| *violaceum* | | | | 0.9 | | 76.6 | | | | | 17.9 | 4.6 | | | | 149 |
| **ENTEROBACTERIACEAE** | | | | | | | | | | | | | | | | |
| *Escherichia* | | | | | | | | | | | | | | | | |
| *coli* | | | | | | | + | | | | | | | | | 43 |
| | | | | 0.9 | | | 52.1 | | | | 18.3 | 28.1 | | | | 149 |

## TABLE 2 (Continued)
## MICROBIAL LIPIDS

| Organism | Glycerides | | | Phospholipids | | | | | | | | | Glycolipids | Amino Acid Lipids | Other Compounds | Ref. |
|---|---|---|---|---|---|---|---|---|---|---|---|---|---|---|---|---|
| | MG | DG | TG | PA | PS | PE | MPE | DPE | PC | PI | PG | PGP (CL) | | | | |
| **Order Eubacteriales (continued)** | | | | | | | | | | | | | | | | |
| *Escherichia* | | | | | | | | | | | | | | | | |
| *coli* B | | | | | | + | + | | | | + | 5-12 | | | | 40 |
| | | | | | | | 18.6 | | | | 3.7 | 1.6 | | | | 44 |
| | | | | minimal | + | | + | | | | + | | | | | 9 |
| *coli* K-12 | | | | + | + | | 69 | | | | 19 | 6.5 | | | | 45 |
| *Aerobacter* | | | | | | | | | | | | | | | | |
| *aerogenes* | | | | 1.0 | | | 74.3 | | | | 20.8 | 3.2 | | | | 149 |
| *Serratia* | | | | | | | | | | | | | | | | |
| *marcescens* | | | | | | | 25 | | | | 4 | | | | lyso-PE = 0.7 | 95 |
| | | | | | 9 | | 42 | | | | 7 | | | | polyglycerol phosphatide = tr<br>3 unidentified spots = 42 | 12 |
| | | | | 2.3 | | 66.1 | | | | | 14.4 | 17.0 | | | | 149 |
| with streptomycin | | | | | | | 52 | | | | 1 | | | | lyso-PE = 3 | 95 |
| *Proteus* | | | | | | | | | | | | | | | | |
| *mirabilis* | | | | | | | | | | | | | | | | |
| wild type | | | | | | + | + | | | | + | + | | | | 72 |
| mutant | | | | | | + | + | | | | + | + | | | | 72 |
| wild SD | | | | | | + | + | | | | + | + | | | | 72 |
| *vulgaris* | | | | 0.9 | | 63.0 | 12.1 | | | | 5.8 | 14.8 | | | | 149 |
| *Salmonella* | | | | | | | | | | | | | | | | |
| *typhimurium* | | | | 0.2 | 0.2 | 75 | | | | | 18 | 4.5 | | | | 45 |
| | | | | | | 66.4 | | | | | 15.0 | 5.2 | | | | 94 |
| infected with bacteriophage | | | | | | 57.8 | | | | | 10.8 | 3.7 | | | | 94 |

# TABLE 2 (Continued)
# MICROBIAL LIPIDS

| Organism | Glycerides | | | Phospholipids | | | | | | | | | Glycolipids | Amino Acid Lipids | Other Compounds | Ref. |
|---|---|---|---|---|---|---|---|---|---|---|---|---|---|---|---|---|
| | MG | DG | TG | PA | PS | PE | MPE | DPE | PC | PI | PG | PGP (CL) | | | | |
| **Order Eubacteriales (continued)** | | | | | | | | | | | | | | | | |
| **BRUCELLACEAE** | | | | | | | | | | | | | | | | |
| *Brucella* | | | | | | + | | | + | | + | + | | | | |
| *abortus* | | | | | | | | | | | | | | | phosphatidyl-N-dimethylethanolamine<br>phosphatidyl-N-methylethanolamine | 46 |
| | | | | | + | + | | | + | | | | | | lyso-lecithin<br>3 unidentified | 47 |
| *abortus* Bang | | | | | | | | | | | | | | | | |
| Scherle II | | | | | + | + | | | 34.5 | | + | + | | | phosphatidyl-N-dimethylethanolamine<br>phosphatidyl-N-methylethanolamine | 76 |
| strain 1119 | | | | | | + | | | 37.5 | | + | + | | | phosphatidyl-N-dimethylethanolamine<br>phosphatidyl-N-methylethanolamine | 76 |
| *Haemophilus* | | | | | | | | | | | | | | | | |
| *parainfluenzae* | | | | + | + | + | | | | | + | + | | | | 48 |
| **BACTEROIDACEAE** | | | | | | | | | | | | | | | | |
| *Bacteroides* | | | | | | | | | | | | | | | | |
| *melaninogenicus* | | | | + | + | + | | | | | + | + | | | ceramide phosphorylglycerol<br>ceramide phosphorylethanolamine<br>ceramide phosphorylglycerol phosphate | 49 |
| | | | | 0.2 | 1.5 | 15.6 | | | | | +CPG 50.8 | +CPGP 4.7 | | | ceramide phosphorylethanolamine = 27.1 | 83 |
| *succinogenes* | | | | | | + | | | | | | | | | lyso-PE<br>plasmalogen | 17 |

## TABLE 2 (Continued)
## MICROBIAL LIPIDS

| Organism | Glycerides | | | Phospholipids | | | | | | | | | Glycolipids | Amino Acid Lipids | Other Compounds | Ref. |
|---|---|---|---|---|---|---|---|---|---|---|---|---|---|---|---|---|
| | MG | DG | TG | PA | PS | PE | MPE | DPE | PC | PI | PG | PGP (CL) | | | | |
| **Order Eubacteriales (continued)** | | | | | | | | | | | | | | | | |
| **MICROCOCCACEAE** | | | | | | | | | | | | | | | | |
| *Micrococcus* | | | | | | | | | | | | | | | | |
| *halodenitrificans* | | | | | | + | | | | | + | | | + | lyso-PE<br>4 unidentified | 50 |
| *lysodeikticus* | | | | | | | | | | | | | + | | | 51 |
| | | | | | | | | | | | | | | | polyphosphatidic acid | 151 |
| | | + | | | | | | | | + | + | | + | | | 152 |
| *Staphylococcus* | | | | | | | | | | | | | | | | |
| *aureus* | | | | + | | | | | | | | + | + | | lysyl-PG | 52 |
| | 2.31 | 9.60 | | | | | | | | | 31.9 | | 15.99 | 11.9 | lysyl-PG = 7.95 | 54 |
| | + | + | | + | | + | | | | | + | | + | | 3-sn-phosphatidyl-1′-glucose | 73 |
| at pH 7 | | | | | | | | | | | 94 | 4 | | + | | 53 |
| *Sarcina* | | | | | | | | | | | | | | | | |
| *lutea* | | | | 4.27 | 2.0 | | | | 1.0 | 5.5 | 13.8 | | | 8.4 + 6.7 = 15.1 | polyglycerol phosphatide = 17.0 | 55 |
| **LACTOBACILLACEAE** | | | | | | | | | | | | | | | | |
| *Diplococcus* | | | | | | | | | | | | | | | | |
| *pneumoniae* | | | | | | | | | | | | | + | | | 153, 154 |
| *Streptococcus* | | | | | | | | | | | | | | | | |
| *faecalis* | | | | | | | | | | | + | + | | + | | 57 |
| | | | | + | | | | | | | + | + | | + | | 159 |
| *Lactobacillus* | | | | | | | | | | | | | | | | |
| *casei* | | | | | | 1.2 | | | 45.4 | | | 8.0 | | | | 56 |

# TABLE 2 (Continued)
# MICROBIAL LIPIDS

| Organism | Glycerides | | | Phospholipids | | | | | | | | | | | | |
|---|---|---|---|---|---|---|---|---|---|---|---|---|---|---|---|---|
| | MG | DG | TG | PA | PS | PE | MPE | DPE | PC | PI | PG | PGP (CL) | Glycolipids | Amino Acid Lipids | Other Compounds | Ref. |
| **Order Eubacteriales (continued)** | | | | | | | | | | | | | | | | |
| *Bifidobacterium* | | | | | | | | | | | | | | | | |
| *bifidum* | | | | | | | | | | | | | | | | |
| var. *pennsylvanicus* | | | | | | | | | | | +[a] | | +[b] | | | 63 |
| **BACILLACEAE** | | | | | | | | | | | | | | | | |
| *Bacillus* | | | | | | | | | | | | | | | | |
| *subtilis* | | | | | | 30 | | | | | 36 | 12 | | | α-glucolipid lysyl-PG = 22 | 58 |
| *megaterium* | | | | | | + | | | | | + | + | | | glucosaminyl-PG | 59 |
| *stearothermophilus* | | | | | | 21-32 | | | | | 22-39 | 23.42 | | | | 60 |
| *polymyxa* | | | | + | | + | | | | | + | | | | lyso-lecithin<br>lyso-PE<br>lyso-PS | 61 |
| *cereus* | | + | | | | 40.8 | | | 4.78 | | 26.3 | | | | lyso-lecithin = 2.7<br>lyso-PE = 9.4 | 27 |
| | | + | | | | + | | | | | + | | | | alanine ester of PG<br>diphosphatidyl glycerol | 82 |
| *Clostridium* | | | | | | | | | | | | | | | | |
| *butyricum* | | | | | + | + | + | | | | | | | | lyso-phosphatide<br>plasmalogen | 155, 156 |
| | | | | | | | | | | | | | | | lyso-phosphatidic acid | 157 |
| *welchii* | | | | | | | | | | | | + | | + | | 158 |
| **Order Spirochaetales** | | | | | | | | | | | | | | | | |
| **TREPONEMATACEAE** | | | | | | | | | | | | | | | | |
| *Treponema* | | | | | | | | | | | | | | | | |
| *pallidum* | | | | | | | | | | | | | + | | | 90 |

[a] Major fraction: diphosphatidyl-PG.

[b] Minor fraction: alanyl-PG.

## TABLE 2 (Continued)
## MICROBIAL LIPIDS

| Organism | Glycerides | | | Phospholipids | | | | | | | | | Glycolipids | Amino Acid Lipids | Other Compounds | Ref. |
|---|---|---|---|---|---|---|---|---|---|---|---|---|---|---|---|---|
| | MG | DG | TG | PA | PS | PE | MPE | DPE | PC | PI | PG | PGP (CL) | | | | |
| **Order Spirochaetales (continued)** | | | | | | | | | | | | | | | | |
| *Treponema* | | | | | | | | | | | | | | | | |
| *pallidum* | | 18 + 5 = 23 | | | | | | | 41 | | 7 | 4 | | | lyso-PC = 1<br>lyso-CL = 2<br>cholesterol = 11<br>cholesterol ester = 2 | 93 |
| *zuelzerae* | | 27 + 7 = 34 | | | | | | | | | 21 | 16 | | | lyso-CL = 5 | 93 |
| **Order Mycoplasmatales** | | | | | | | | | | | | | | | | |
| *Mycoplasma* | | | | | | | | | | | | | | | | |
| *gallisepticum* | | 55 | 3.9 | 10.2 | | | | | + inosit-ides = 18.1 | | | | | | cholesterol = 13.3<br>cholesterol ester = 5.5<br>cephaline + inositides = 31.5<br>sphingomyelin = 0.8 | 111 |
| *mycoides* | | | | | | | | | | | | | + | | | 111 |
| *laidlawii* | | + | | | | | | | | | | | | | | 116 |
| strain Y | | + | | | | | | | | | + | | + | | | 118 |
| **Order Rickettsiales** | | | | | | | | | | | | | | | | |
| *Wolbachia* | | | | | | | | | | | | | | | | |
| *persica* | | | | + | | + | | | + | | | + | | | | 109 |
| *Chlamydia* | | | | | | | | | | | | | | | | |
| *psittaci* 6BC | | + | + | | 6.9 | 31.1 | | | 23.8 | 6.7 | 15.0 | 6.9 | | | sterol<br>sterol ester<br>sphingomyelin | 70 |
| **Viruses** | | | | | | | | | | | | | | | | |
| **POX VIRUSES** | | | | | | | | | | | | | | | | |
| Fowlpox, Doll | 1.9 | 2.4 | 23.2 | | | | | | | | | | | | phospholipid<br>cholesterol<br>cholesterol ester<br>squalene | 66 |

# TABLE 2 (Continued)
# MICROBIAL LIPIDS

| Organism | Glycolipids | | | Phospholipids | | | | | | | | | | | | |
|---|---|---|---|---|---|---|---|---|---|---|---|---|---|---|---|---|
| | MG | DG | TG | PA | PS | PE | MPE | DPE | PC | PI | PG | PGP (CL) | Glycolipids | Amino Acid Lipids | Other Compounds | Ref. |
| **Viruses (continued)** | | | | | | | | | | | | | | | | |
| **MYXOVIRUSES** | | | | | | | | | | | | | | | | |
| Influenza Mel-1935 | | | | | | | | | | | | | | | | |
| calf kidney host | | | | 64.2 | 3.4 | 7.2 | | | 11.2 | 1.7 | | | | | sphingomyelin<br>lyso-PE<br>polyglycerol phosphatide | 64, 65, 68 |
| chick embryo host | | | | | + | + | | | + | + | | | | | sphingomyelin<br>lyso-PE<br>lyso-PC<br>polyglycerol phosphatide | 64, 65, 68 |
| Influenza PR-8 | | | | 7.9 | 8.8 | 11.7 | | | 32.8 | 3.9 | | | | | cholesterol<br>cholesterol esters<br>glycerides<br>lyso-phosphatides<br>sphingomyelin | 146 |
| **PARAMYXOVIRUSES** | | | | | | | | | | | | | | | | |
| Parainfluenza SV5/W3 | | | | | | | | | | | | | | | | |
| HaK host | | | + | | + | + | | | + | + | | | | | sphingomyelin<br>cholesterol<br>cholesterol ester | 67 |
| Sendai | | | | 10.1 | 15.0 | 37.0 | | | 8.0 | 6.4 | | | | | phospholipids<br>sphingomyelin | 145 |
| Newcastle B1 | | | | 7.0 | 12.0 | 34.6 | | | 11.6 | 6.5 | | | | | cholesterol<br>cholesterol esters<br>glycerides<br>lyso-phosphatides<br>sphingomyelin | 145 |
| Mumps | | | | 10 | | 19 | | | 11 | | | | | | sphingomyelin | 160 |
| **ARBOVIRUSES** | | | | | | | | | | | | | | | | |
| Eastern equine encephalitis | | | | | | | | | | | | | | | phospholipid<br>cholesterol<br>neutral lipids | 161 |
| Venezuelan equine encephalitis | | | | | | + | | | + | | | | | | sphingomyelin<br>cholesterol | 162 |

# TABLE 2 (Continued)
## MICROBIAL LIPIDS

| Organism | Glycerides | | | Phospholipids | | | | | | | | | | | | |
|---|---|---|---|---|---|---|---|---|---|---|---|---|---|---|---|---|
| | MG | DG | TG | PA | PS | PE | MPE | DPE | PC | PI | PG | PGP (CL) | Glycolipids | Amino Acid Lipids | Other Compounds | Ref. |
| **Viruses (continued)** | | | | | | | | | | | | | | | | |
| **ARBOVIRUSES** | | | | | | | | | | | | | | | | |
| Sindbis | | | | | | + | | | + | | | | | | sphingomyelin<br>cholesterol | 163 |
| Murray Valley encephalitis | | | + | | | | | | | | | | | | phospholipid<br>cholesterol | 164 |
| Japanese B encephalitis | | | | | | | | | | | | | | | phospholipid<br>cholesterol | 165 |
| **BACTERIOPHAGES** | | | | | | | | | | | | | | | | |
| PM 2 | | | | | | + | | | | | | | | | | 69 |
| **Yeasts** | | | | | | | | | | | | | | | | |
| *Saccharomyces* | | | | | | | | | | | | | | | | |
| *carlsbergensis*[c] | | 3.8 | 36.0 | 2.0 | 2.2 | 9.0 | | | 13.1 | 7.9 | | 2.9 | | | | 99 |
| *cerevisiae* | | | | | + | + | | | + | + | + | + | | | lyso-PE<br>lyso-PC<br>lyso-PI<br>N-monomethyl-PE<br>N,N-dimethyl-PE | 103 |
| | + | + | + | | + | + | | | + | + | | | | | free sterols | 105 |
| % phosphorus | | | | | 7.2 | 22.6 | | | 42.5 | 20.9 | 1.9 | 2.5 | | | N,N-dimethyl-PE = 2.4 | 104 |
| % of total phosphorus in phospholipids | | | | <5 | + PI = 15 | 20 | | | 55 | | | <5 | | | | 100 |
| **Fungi** | | | | | | | | | | | | | | | | |
| *Agaricus* | | | | | | | | | | | | | | | | |
| *bisporus* (Lange) | | | | | | | | | | | | | | | | |
| sporophore | | | + | | + | + | | | + | + | | | | | sterol | 128 |
| mycelium | | | + | | | + | | | + | | | | | | | 128 |

## TABLE 2 (Continued) MICROBIAL LIPIDS

| Organism | Glycerides | | | Phospholipids | | | | | | | | | | Glycolipids | Amino Acid Lipids | Other Compounds | Ref. |
|---|---|---|---|---|---|---|---|---|---|---|---|---|---|---|---|---|---|
| | MG | DG | TG | PA | PS | PE | MPE | DPE | PC | PI | PG | PGP (CL) | | | | | |
| **Fungi (continued)** | | | | | | | | | | | | | | | | | |
| *Blastomyces* | | | | | | | | | | | | | | | | | |
| *dermatitidis* | | | | | | | | | | | | | | | | | |
| 2 days | | | 54 | | | 23 | | | 23 | | | | | | | | 120 |
| *Clitocybe* | | | | | | | | | | | | | | | | | |
| *illudens* | | | | | + | + | | | + | | | | | | | ergosterol | 123 |
| *Dictyostelium* | | | | | | | | | | | | | | | | | |
| *discoideum* | | | | | 4.5 | 39 | | | 37 | | | 6 | | | | monophosphoinositide | 125 |
| *Histoplasma* | | | | | | | | | | | | | | | | | |
| *capsulatum* | | | 56 | | | 16 | | | 28 | | | | | | | | 120 |
| *Neurospora* | | | | | | | | | | | | | | | | | |
| *crassa* | | | | | + | + | + | + | | + | | | | | | | 168, 169, 170 |

## TABLE 3

## LIPOIDAL CONSTITUENTS OF CORYNEBACTERIA, MYCOBACTERIA AND NOCARDIA

| Common Name | Carbon Atoms | Structural Formula |
|---|---|---|
| Tuberculostearic acid | 19 | $CH_3-(CH_2)_7-\underset{CH_3}{CH}-(CH_2)_8-COOH$ |
| Diphtheric acid | 35 | $(C_{25}H_{51})-CH{=}\underset{CH_3}{C}-(CH_2)_6-COOH$ |
| Corrinic acid | 35 | ? (empirical formula: $C_{35}H_{65}O_2$) |
| Corynomycolic acid | 32 | $CH_3-(CH_2)_{14}-\underset{OH}{CH}-\underset{C_{14}H_{29}}{CH}-COOH$ |
| Corynomycolenic acid | 32 | $CH_3-(CH_2)_5-CH{=}CH-(CH_2)_7-\underset{OH}{CH}-\underset{C_{14}H_{29}}{CH}-COOH$ |
| Corynolic acid | 52 | $CH_3-(CH_2)_{14}-\underset{CH_3}{CH}-(CH_2)_{17}CH-COOH$<br>$CH_3-\underset{OH}{CH}-(CH_2)_7-\underset{CH_3}{CH}-\underset{OH}{CH}-\underset{CH_3}{CH}-\underset{CH_3}{CH}$ |
| Phthioic acid | 26 | $CH_3-(CH_2)_3-\underset{CH_3}{CH}-(CH_2)_5-\underset{CH_3}{CH}-(CH_2)_9-\underset{CH_3}{CH}-CH_2-COOH$ |
| Phthienoic acids (several) | 27,29 | $CH_3-(CH_2)_6-\underset{C_5H_{11}}{CH}-(CH_2)_8-\underset{CH_3}{CH}-CH{=}\underset{COOH}{C}-CH_3$<br>($C_{27}$ acid) |
| Mycolipenic acids (several) | 25,27 | $CH_3-(CH_2)_{17}-\underset{CH_3}{CH}-CH_2-\underset{CH_3}{CH}-CH{=}\underset{CH_3}{C}-COOH$<br>($C_{27}$ acid) |
| Mycocerosic acid | 32 | $CH_3-(CH_2)_{19}-\underset{CH_3}{CH}-CH_2-\underset{CH_3}{CH}-CH_2-\underset{CH_3}{CH}-CH_2-\underset{CH_3}{CH}-COOH$ |
| Mycolic acids (many) | 87,88, etc. | $R-\underset{R'}{CH}-\underset{OH}{CH}-\underset{R''}{CH}-COOH$<br>where R, R′, and R″ may be of varying but appreciable length and may or may not contain groups, double bonds, etc. |

## TABLE 3 (Continued)

## LIPOIDAL CONSTITUENTS OF CORYNEBACTERIA, MYCOBACTERIA AND NOCARDIA

| Common Name | Carbon Atoms | Structural Formula |
|---|---|---|
| Cord factor | | OH<br>$CHC_{60}H_{120}(OH)$<br>$CH_2OOCCH$<br>$C_{24}H_{49}$<br>O<br>HO OH<br>OH<br>O<br>OH<br>OH<br>$H_2C$ OH<br>O<br>OH<br>$OOCCHCHC_{60}H_{120}(OH)$<br>$C_{24}H_{49}$ |

Taken from: O'Leary, W. M. (Ed.), in *The Chemistry and Metabolism of Microbial Lipids,* pp, 47, 48, and 80 (1967). Reproduced by permission of the publishers, World Publishing Company, Cleveland, Ohio.

## REFERENCES


1. Key, Gray, and Wilkinson, *Biochem. J., 117,* 721 (1970).
2. Cho and Salton, *Biochim. Biophys. Acta, 116,* 73 (1966).
3. Bergh, Webb, and McArthur, *Can. J. Biochem., 42,* 1141 (1964).
4. Steinhauer, Flentge, and Lechowich, *Appl. Microbiol., 15,* 826 (1967).
5. Brian and Gardner, *Appl. Microbiol., 16,* 549 (1968).
6. Kaneshiro and Marr, *J. Biol. Chem., 236,* 2615 (1961).
7. Constantopoulos and Bloch, *J. Bacteriol., 93,* 1788 (1967).
8. Weinbaum and Panos, *J. Bacteriol., 92,* 1576 (1966).
9. De Siervo, *J. Bacteriol., 100,* 1342 (1969).
10. Bishop and Still, *Biochem. Biophys. Res. Commun., 7,* 337, (1962).
11. Gallin and O'Leary, *J. Bacteriol., 96,* 660 (1968).
12. Kates, Adams, and Martin, *Can. J. Biochem., 42,* 461 (1964).
13. Bishop and Still, *J. Lipid Res., 4,* 81 (1963).
14. Gray, *Biochim. Biophys. Acta, 65,* 135 (1962).
15. Thiele, Lacave, and Asselineau, *Eur. J. Biochem., 7,* 393 (1969).
16. White and Cox, *J. Bacteriol., 93,* 1079 (1967).
17. Wegner and Foster, *J. Bacteriol., 85,*53 (1963).
18. White and Frerman, *J. Bacteriol., 95,* 2198 (1968).
19. Huston and Albro, *J. Bacteriol., 88,* 425 (1964).
20. Tornabene, Bennett, and Oro, *J. Bacteriol., 94,* 344 (1967).
21. Albro and Dittmer, *Biochemistry, 8,* 394 (1969).
22. Albro and Dittmer, *Biochemistry, 8,* 953 (1969).
23. Panos, Cohen, and Fagan, *Biochemistry, 5,* 1461 (1966).
24. MacLeod and Brown, *J. Bacteriol., 85,* 1056 (1963).
25. Kaneda, *J. Biol. Chem., 238,* 1222 (1963).
26. Kaneda, *J. Bacteriol., 95,* 2210 (1968).
27. Kates, Kushner, and James, *Can. J. Biochem., Physiol., 40,* 83 (1962).
28. Fulco, *J. Biol. Chem., 245,* 2985 (1970).
29. Yao, Walker, and Lillard, *J. Bacteriol., 102,* 877 (1970).
30. Kaneda, *J. Bacteriol., 98,* 143 (1969).
31. Moss and Lewis, *Appl. Microbiol., 15,* 390 (1967).
32. Wood, Nichols, and James, *Biochim. Biophys. Acta, 106,* 261 (1965).
33. Phiel, and Ordal, *J. Bacteriol., 93,* 1727 (1967).
34. Hofmann, *Rec. Chem. Progr., 14,* 7 (1953).
35. Ikawa, *Bacteriol., Rev., 31,* 54 (1967).
36. Park and Berger, *J. Bacteriol., 93,* 221 (1967).
37. Jones and Benson, *J. Bacteriol., 89,* 260 (1965).
38. Gordon and MacLeod, *Biochem. Biophys. Res. Commun., 24,* 684 (1966).
39. Jurtshuk and Schlech, *J. Bacteriol., 97,* 1507 (1969).
40. Kanemasa, Akamatsu, and Nojima, *Biochim. Biophys. Acta, 144,* 382 (1967).
41. Sehgal, Kates, and Gibbons, *Can. J. Biochem. Physiol., 40,* 69 (1962).
42. Kunsman, *J. Bacteriol., 103,* 104 (1970).
43. Kaneshiro and Marr, *J. Lipid Res., 3,* 184 (1962).
44. Okuyama, *Biochim. Biophys. Acta, 176,* 125 (1969).
45. Ames, *J. Bacteriol., 95,* 833 (1968).
46. Thiele and Busse, *Experientia, 24,* 112 (1968).
47. Wober, Thiele, and Urbaschek, *Biochim. Biophys. Acta, 84,* 376 (1964).
48. White and Tucker, *J. Bacteriol., 97,* 199 (1969).
49. White and Tucker, *Lipids, 5,* 56 (1969).
50. Kates, Sehgal, and Gibbons, *Can. J. Microbiol., 7,* 427 (1961).
51. Lennarz and Talamo, *J. Biol. Chem., 241,* 2707 (1966).
52. Gale and Folkes, *Biochim. Biophys. Acta, 144,* 452 (1967).
53. Houtsmuller and Van Deenen, *Biochim. Biophys. Acta, 106,* 564 (1965).
54. White and Frerman, *J. Bacteriol., 94,* 1854 (1967).
55. Huston, Albro, and Grindey, *J. Bacteriol, 89,* 768 (1965).
56. Thorne, *Biochim. Biophys. Acta, 84,* 350 (1964).
57. Kocun, *Biochim. Biophys. Acta, 202,* 277 (1970).
58. Op den Kamp, Redai, and Van Deenen, *J. Bacteriol., 99,*298 (1969).
59. Bertsch, Bonsen, and Kornberg, *J. Bacteriol., 98,* 75 (1969).
60. Card, Georgi, and Militzer, *J. Bacteriol., 97,* 186 (1969).

61. Matches, Walker, and Ayres, *J. Bacteriol., 87,* 16 (1964).
62. Yano, Furukawa, and Kusunose, *Biochim. Biophys. Acta, 210,* 105 (1970).
63. Exterkate and Veerkamp, *Biochim. Biophys. Acta, 176,* 65 (1969).
64. Kates, Allison, Tyrrell, and James, *Biochim. Biophys. Acta, 52,* 455 (1961).
65. Kates, Allison, Tyrell, and James, *Cold Spring Harbor Symp. Quant. Biol., 27,* 293 (1962).
66. White, Powell, Gafford, and Randall, *J. Biol. Chem., 243,* 4517 (1968).
67. Klenk and Choppin, *Virology, 40,* 939 (1970).
68. Fraenkel-Conrat, *The Chemistry and Biology of Viruses,* Academic Press, New York (1969).
69. Espejo and Canelo, *J. Virol., 2,* 1235 (1968).
70. Makino, Jenkin, Yu, and Townsend, *J. Bacteriol., 103,* 62 (1970).
71. Albro and Dittmer, *Lipids, 5,* 320 (1969).
72. Sud and Feingold, *J. Bacteriol., 104,* 289 (1970).
73. Short and White, *J. Bacteriol., 104,* 126 (1970).
74. Steiner, Conti, and Lester, *J. Bacteriol., 98,* 10 (1969).
75. Kates, Wassef, and Pugh, *Biochim. Biophys. Acta, 202,* 206 (1970).
76. Thiele, Busse, and Hoffmann, *Eur. J. Biochem., 5,* 513 (1968).
77. Bauman and Simmons, *J. Bacteriol., 98,* 528 (1969).
78. Brian and Gardner, *J. Bacteriol., 96,* 2181 (1968).
79. Lewis, Weaver, and Hollis, *J. Bacteriol., 95,* 1 (1968).
80. Tornabene, Kates, Gelpi, and Oro, *J. Lipid Res., 10,* 294 (1969).
81. Kaneshiro and Marr, *Biochim. Biophys. Acta, 70,* 271 (1963).
82. Lang and Lundgren, *J. Bacteriol., 101,* 483 (1970).
83. Rizza, Tucker, and White, *J. Bacteriol., 101,* 84 (1970).
84. Short, White, and Aleem, *J. Bacteriol., 99,* 142 (1969).
85. Park and Berger, *J. Bacteriol., 93,* 230 (1967).
86. Kates and Hagen, *Can. J. Biochem., 42,* 481 (1964).
87. Bishop, Hewett, and Knox, *Biochim. Biophys. Acta, 231,* 274 (1971).
88. Key, Gray, and Wilkinson, *Biochem. J., 120,* 559 (1970).
89. Bunn and Elkan, *Can. J. Microbiol., 17,* 291 (1971).
90. Livermore and Johnson, *Biochim. Biophys. Acta, 210,* 315 (1970).
91. Veerkamp, *Biochim. Biophys. Acta, 210,* 267 (1970).
92. Goldfine and Panos, *J. Lipid Res., 12,* 214 (1971).
93. Meyer and Meyer, *Biochim. Biophys. Acta, 231,* 93 (1971).
94. Knipprath, Cohen, and Allen, *Biochim. Biophys. Acta, 231,* 107 (1971).
95. Bermingham, Deol, and Still, *Biochem. J., 119,* 861 (1970).
96. Moss, Dowell, Farshtchi, Raines, and Cherry, *J. Bacteriol., 97,* 561 (1969).
97. Raines, Moss, Farshtchi, and Pittman, *J. Bacteriol., 96,* 2175 (1968).
98. Livermore, Johnson, and Jenkin, *Lipids, 4,* 166 (1968).
99. Paltauf and Johnston, *Biochim. Biophys. Acta, 218,* 424 (1970).
100. Suomalainen and Nurminen, *Chem. Phys. Lipids, 4,* 247 (1970).
101. Hartman, Hawke, Shorland, and di Menna, *Arch. Biochem. Biophys., 81,* 346 (1959).
102. Merdinger, Kohn, and McClain, *Can. J. Microbiol., 14,* 1021 (1968).
103. Letters and Brown, *Biochim. Biophys. Acta, 116,* 482 (1966).
104. Letters, *Biochim. Biophys. Acta, 116,* 489 (1966).
105. Suomalainen and Keranen, *Biochim. Biophys. Acta, 70,* 493 (1963).
106. Klug and Markovetz, *J. Bacteriol., 93,* 1847 (1967).
107. Dawson and Craig, *Can. J. Microbiol., 12,* 775 (1966).
108. Brown and Rose, *J. Bacteriol., 99,* 371 (1969).
109. Neptune, Weiss, Davies, and Suitor, *J. Infect. Dis., 114,* 39 (1964).
110. O'Leary, *Biochem. Biophys. Res. Commun., 8,* 87 (1962).
111. Tourtellotte, Jensen, Gander, and Morowitz, *J. Bacteriol., 86,* 370 (1963).
112. Razin, Tourtellotte, McElhaney, and Pollack, *J. Bacteriol., 91,* 609 (1966).
113. Plackett, *Biochemistry., 6,* 2746 (1967).
114. Panos and Henrikson, *Biochemistry, 8,* 652 (1969).
115. Henrikson and Panos, *Biochemistry, 8,* 646 (1969).
116. Smith, *J. Bacteriol., 99,* 480 (1969).
117. McElhaney and Tourtellotte, *Biochim. Biophys. Acta, 202,* 120 (1970).
118. Plackett and Rodwell, *Biochim., Biophys. Acta, 210,* 230 (1970).
119. Bunn, McNeill, and Elkan, *J. Bacteriol., 102,* 24 (1970).
120. Domer and Hamilton, *Biochim. Biophys. Acta, 231,* 465 (1971).
121. Wirth and Anand, *Can. J. Microbiol., 10,* 23 (1964).
122. Merdinger and Devine, *J. Bacteriol., 89,* 1488 (1965).

123. Bentley, Lavate, and Sweeley, *Comp. Biochem. Physiol., 11,* 263 (1964).
124. Mumma, Fergus, and Sekura, *Lipids, 5,* 100 (1969).
125. Davidoff and Korn, *Biochem. Biophys. Res. Commun., 9,* 54 (1962).
126. Van Etten and Gottlieb, *J. Bacteriol., 89,* 409 (1965).
127. Weete, Weber, and Laseter, *J. Bacteriol., 103,* 536 (1970).
128. Holtz and Schisler, *Lipids, 6,* 176 (1970).
129. Divakaran and Modak, *Experientia, 24,* 1102 (1968).
130. Nakajima and Tanenbaum, *Arch. Biochem. Biophys., 127,* 150 (1968).
131. Graff, Szczepanik, Klein, Chipault, and Holman, *Lipids, 5,* 786 (1969).
132. Morris, *Lipids, 3,* 260 (1967).
133. Yano, Furukawa, and Kusunose, *J. Bacteriol., 98,* 124 (1969).
134. Fensom and Gray, *Biochem. J., 114,* 185 (1969).
135. Wilkinson, *Biochim. Bíophys. Acta, 187,* 492 (1969).
136. Hancock, Humphreys, and Meadow, *Biochim. Biophys. Acta, 202,* 389 (1970)
137. Kaneshiro and Marr, *J. Lipid Res., 3,* 184 (1962).
138. Asselineau, *Ann. Inst. Pasteur (Paris), 100,* 109 (1961).
139. Abel, de Schmertzing, and Peterson, *J. Bacteriol., 85,* 1039 (1963).
140. Kates, Sehgal, and Gibbons, *Can. J. Microbiol., 7,* 427 (1961).
141. Macfarlane, *Biochem. J., 82,* 40P (1962).
142. Macfarlane, *Nature, 196,* 136 (1962).
143. Brundish, Shaw, and Baddiley, *Biochem. Biophys. Res. Commun., 18,* 308 (1965).
144. Thorne and Kodicek, *Biochim. Biophys. Acta, 59,* 306 (1962).
145. Blough and Lawson, *Virology, 36,* 286 (1968).
146. Blough, Weinstein, Lawson, and Kodicek, *Virology, 33,* 459 (1967).
147. Edwards and Hayashi, *Arch. Biochem. Biophys., 111,* 415 (1965).
148. Bobo and Eagon, *Can. J. Microbiol., 14,* 503 (1968).
149. Randle, Albro, and Dittmer, *Biochim. Biophys. Acta, 187,* 214 (1969).
150. Wegner and Foster, *J. Bacteriol., 85,* 53 (1963).
151. Gilby, Few, and McQuillen, *Biochim. Biophys. Acta, 29,* 21 (1958).
152. Macfarlane, *Biochem. J., 80,* 45P (1961).
153. Kaufman, Kundig, Distler, and Roseman, *Biochem. Biophys. Res. Commun., 18,* 312 (1965).
154. Panos, *Ann. N.Y. Acad. Sci., 26,* 954 (1964).
155. Goldfine, *Biochim. Biophys. Acta, 59,* 504 (1962).
156. Goldfine, *Fed. Proc., 22,* 415 (1963).
157. Goldfine, Ailhaud, and Vagelos, *J. Biol. Chem., 242,* 4466 (1967).
158. Macfarlane, *Advan. Lipid Res., 2,* 91 (1964).
159. Kocun, *Biochim. Biophys. Acta, 202,* 277 (1970).
160. Soule, Marinetti, and Morgan, *J. Exp. Med., 110,* 93 (1959).
161. Taylor, Sharp, Beard, and Beard, *J. Infect. Dis., 72,* 31 (1943).
162. Wachter and Johnson, *Fed. Proc., 21,* 461 (1962).
163. Pfefferkorn and Hunter, *Virology, 20,* 446 (1963).
164. Hardy and Arbiter, *J. Bacteriol., 89,* 1101 (1965).
165. Ada, Abbot, Anderson, and Collins, *J. Gen. Microbiol., 29,* 165 (1962).
166. Nozima, Mori, Minobe, and Yamamoto, *Acta Virol., 8,* 97 (1964).
167. Hilditch and Williams, *Chemical Constitution of Natural Fats,* 4th ed. John Wiley and Sons, New York (1964).
168. Ellman and Mitchell, *J. Amer. Chem. Soc., 76,* 4028 (1954).
169. Fuller and Tatum, *Amer. J. Botany, 43,* 361 (1956).
170. Crocken and Nyc, *J. Biol. Chem., 239,* 1727 (1964).

## NMR SPECTRA OF SOME UNSATURATED METHYL ESTERS[a]

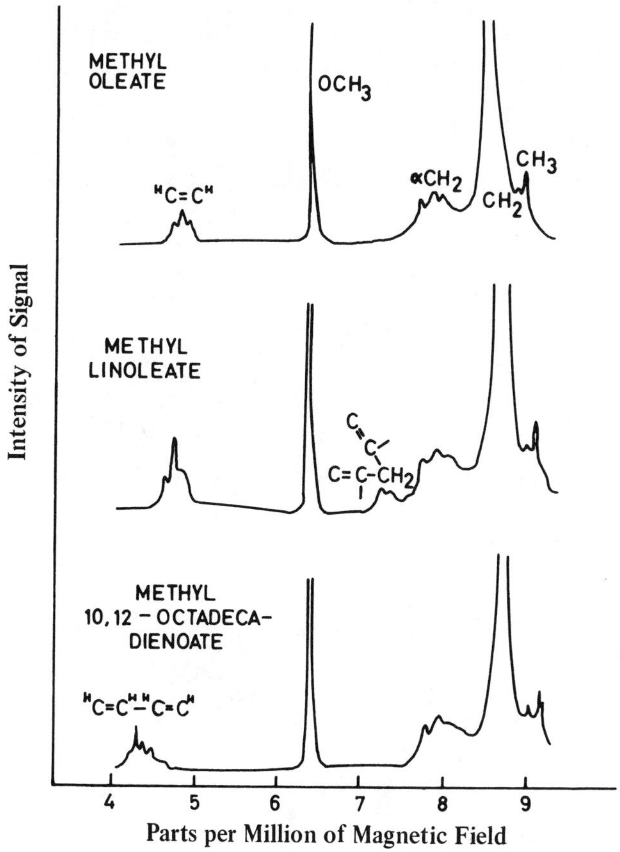


From: Chapman, D., *The Structure of Lipids,* p. 193 (1965). Reproduced by permission of the copyright owners, John Wiley and Sons, New York.

[a] Hopkins, *J. Amer. Oil Chem. Soc., 38,* 664, 1961.

## PROTON CHEMICAL SHIFTS, τ

| Groups | Resonance Lines of Protons[a] (τ) | Groups | Resonance Lines of Protons[a] (τ) |
|---|---|---|---|
| **Alkane Groups** | | | |
| $^{\Delta}CH_2$ | 9.78 | $-CH_2-$ | 8.52 to 8.75 |
| **β-Functional Groups** | | | |
| $-CH_2-C-N$ | 8.38 to 8.80 | $-CH_2-C-CO-R$ | 8.10 to 8.37 |
| $CH_3-C-N$ | 8.92 to 9.12 | $CH_3-C-N-CO-R$ | 8.80 |
| $CH_3-C-CO-R$ | 8.88 to 9.07 | $-CH_2-C-C=C-$ | 8.40 to 8.82 |
| $-CH_2-C-Ar$ | 8.22 to 8.40 | $-CH_2-C-O-R$ | 8.19 to 8.79 |
| $-CH_2-C-O-COR$ | 8.50 | $-CH_2-C-O-Ar$ | 8.50 |
| $-CH_2-C-Cl$ | 8.04 to 8.40 | $-CH_2-C-Br$ | 7.97 to 8.32 |
| $-CH_2-C-I$ | 8.14 to 8.35 | $-CH_2-C-SO_2-R$ | 7.84 |
| $-CH_2-C-NO_2$ | 7.93 | | |
| **α-Functional Groups** | | | |
| $-CH_2-C=C-$ | 7.69 to 8.17 | $CH_3-C=C-CO-R$ | 7.94 to 8.07 |
| $-CH_2-C=C-O-R$ | 8.07 | $CH_3-C=C-$ | 7.97 to 8.06 |
| | | COOR | |
| $CH_3-C=C-$ | 7.97 to 8.06 | $CH_3-C=C-$ | 8.09 to 8.13 |
| CN | | $O-CO-R$ | |
| $CH_3-C=C-$ | 8.17 | $-CH_2-Ar$ | 6.94 to 7.47 |
| $-C=C-$ | | | |
| $-CH_2-C=N$ | 7.42 | $CH_3-C=NOH$ | 8.19 |
| $-CH_2-CO-R$[b] | 7.61 to 7.98 | $CH_3-CO-Cl$ | 7.19 to 7.34 |
| $CH_3-CO-Br$ | 7.19 to 7.34 | $CH_3-CO-C=C-$ | 7.32 to 8.17 |
| $CH_3-CO-Ar$ | 7.32 to 8.17 | $CH_3-CO-Sr$ | 7.46 to 7.67 |
| $-CH_3-S-R$ | 7.47 to 7.61 | $-N-C-$ | 8.52 |
| | | $CH_2$ | |
| $-CH_2-N$ | 6.88 to 7.72 | $-CH_2-N-CO-R$ | 6.63 to 6.72 |
| $-CH_2-N-SO_2R$ | 6.63 to 6.72 | $-CH_2-N-Ar$ | 6.63 to 6.72 |
| $-CH_2-N^+$ | 6.60 | $-CH_3-N-N-$ | 7.67 |
| $-C-O$ | 7.71 | $-CH_2-O-R$ | 6.42 to 7.69 |
| $CH_2$ | | | |
| $-CH_3-O-CO-R$ | 5.71 to 6.08 | $-CH_2-O-Ar$ | 5.71 to 6.08 |
| $-CH_2-Cl$ | 6.43 to 6.65 | $-CH_2-Br$ | 6.42 to 6.75 |
| $-CH_2-I$ | 6.80 to 6.97 | $CH_3-SO-R$ | 7.50 |
| $-CH_2-SO_2-R$ | 7.08 | $CH_3-SO_2-Cl$ | 6.36 |
| $-CH_2-SO_2F$ | 6.72 | $CH_3-O-SO-OR$ | 6.42 |
| $CH_3-O-SO_2-OR$ | 6.06 | $CH_3-S-C=N$ | 7.37 |
| $-CH_2-N=C=S$ | 6.39 | $-CH_2-NO_2$ | 5.62 |
| $-CH_2(C=C-)_2$ | 6.95 to 7.10 | $Ar-CH_2-C=C-$ | 6.62 |

## PROTON CHEMICAL SHIFTS, $\tau$ (Continued)

| Groups | Resonance Lines of Protons[a] ($\tau$) | Groups | Resonance Lines of Protons[a] ($\tau$) |
|---|---|---|---|
| **Alkyl Groups $\alpha$ to Two or More Functional Groups** | | | |
| Ar–$CH_2$–Ar | 6.08 to 6.19 | Ar–$CH_2$–N | 6.68 |
| Ar–$CH_3$–OR | 5.51 to 5.64 | Ar–$CH_2$–Cl | 5.50 |
| Ar–$CH_2$–Br | 5.57 to 5.59 | Ar–$CH_2$–O–CO–R | 4.74 |
| –C=C–$CH_2$–O–R | 6.03 to 6.10 | –C=C–$CH_2$–Cl | 5.96 to 6.04 |
| –C=C–$CH_2$–OR | 5.82 | –C≡C–$CH_2$–Cl | 5.84 to 5.91 |
| –C≡C–$CH_2$–Br | 6.18 | Cl–$CH_2$–C≡N | 5.93 |
| Br–$CH_2$–C≡N | 6.30 | –CH$(OR)_2$ | 4.80 to 5.20 |
| **Acetylenic and Olefinic Protons** | | | |
| –C≡C–H | 7.07 to 7.67 | –C=C–C≡C–H | 7.13 |
| Ar–C≡C–H | 6.95 | –C=CH– | 4.87 |
| –C=$CH_2$ | 5.37 | –C=CH–CO–R | 3.95 to 4.32 |
| –CH=C–CO–R | 2.96 to 4.53 | –C=CH–O–R | 3.55 to 3.78 |
| –CH=C–O–R | 4.45 to 5.46 | C=CH–O–CO–$CH_3$ | 2.75 |
| $H_2C$=C–O–$CH_2$–C=C | 5.87 to 6.17 | R–CO–CH=C–CO–R | 3.87 to 3.97 |
| Br–CH=C– | 3.00 to 3.38 | –CH=C–C≡N | 4.25 |
| Ar–C=CH– | 4.60 to 4.72 | Ar–CH=C– | 3.72 to 3.77 |
| H–C=C–<br>H CO–R | 3.60 to 3.70 | Ar–CH–C–CO–R | 2.28 to 2.62 |
| **Aldehydes** | | | |
| R–CHO | 0.20 to 0.43 | >C=C–CHO | 0.32 to 0.57 |
| Ar–CHO | −0.08 to +0.35 | | |
| **Carboxylic Acids** | | | |
| R–COOH | −0.97 to −1.52 | >C=C–COOH | −1.57 to −2.18 |

### Hydroxyl Groups

The position of the hydroxyl proton is very dependent on concentration, temperature, and the presence of other easily exchanged protons ($H_2O$!). It is displaced strongly to lower $\tau$ values by intramolecular bonding (in some stable planar systems to $\tau = -5.0$), and to higher values by increased shielding.

### Amine Protons

The position of $-NH_2$ and –NH protons is dependent above all on the basicity of the nitrogen atom. Stronly basic amines show NH– absorption in the C-methyl region.

| | | | |
|---|---|---|---|
| R–$NH_2$ | 7.88 to 8.90 | R–NH– | 7.88 to 8.90 |
| R–CO–NH– | 2.3 to 3.9 | | 2.6 to 2.7[c] |

[a] Protons designated by bold-face type.
[b] R = H, alkyl, aryl, OH, OR, or $HN_2$.
[c] 2.68 $\tau$ is due to the symmetrically placed aromatic protons; this assignment is justified on the basis of the chemical shift as well as the known pattern for *p*- disubstituted benzene derivatives. Area measurements indicate that the quartet around 5.58 $\tau$ is one-proton signal. The multiplet centered at 6.75 $\tau$ must account for the remaining two protons. The first-degree approximation of the $n + 1$ rule is not applicable here. The three protons on the heterocyclic ring constitute a special grouping (ABX) and lead to a characteristic pattern.

# MASS SPECTRA OF METHYL OLEATE, METHYL LINOLEATE AND METHYL LINOLENATE

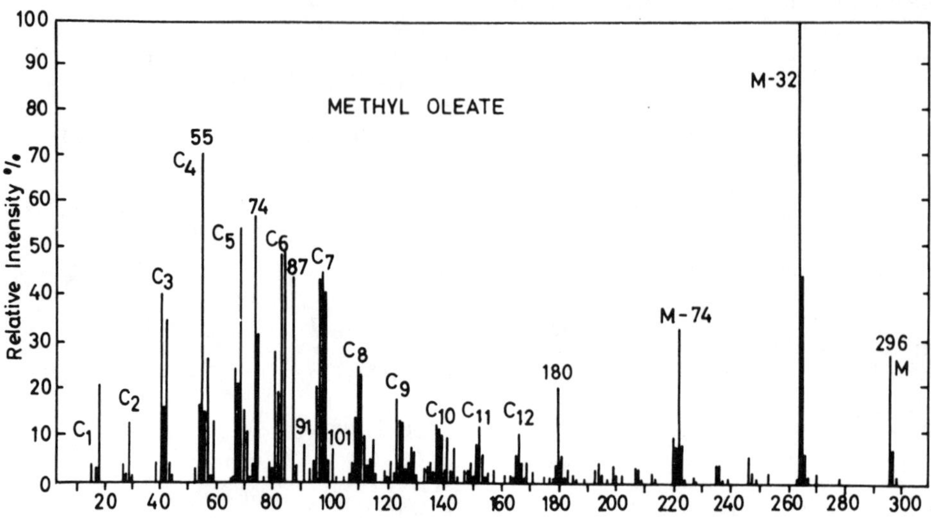


Methyl oleate (methyl $\Delta^{9:10}$-octadecenoate), M = 296.

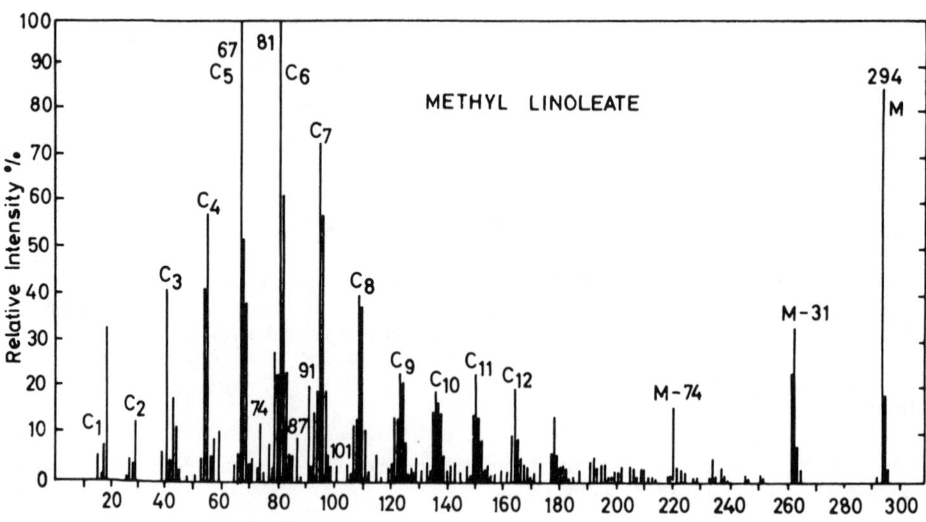


Methyl linoleate (methyl $\Delta^{9:10,12:13}$-octadecdienoate), M = 294.

# MASS SPECTRA OF METHYL OLEATE, METHYL LINOLEATE AND METHYL LINOLENATE (Continued)

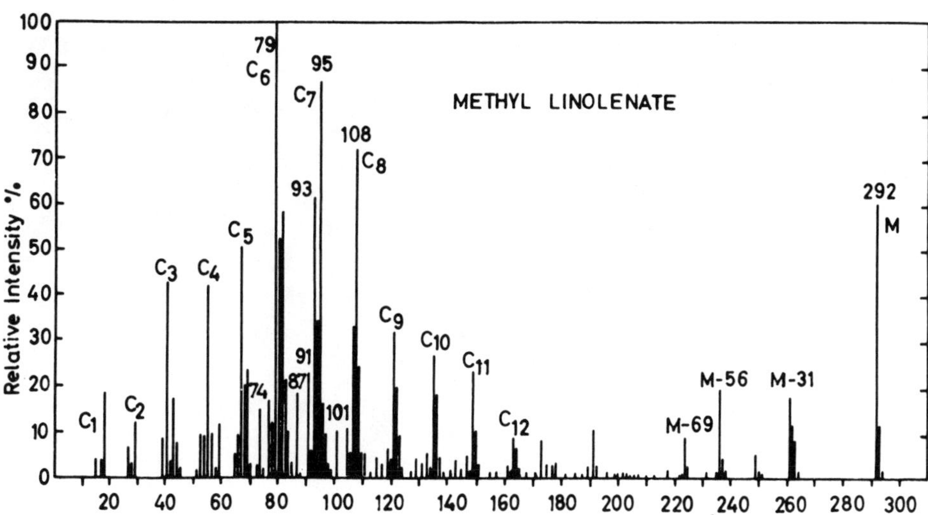


Methyl linolenate (methyl $\Delta^{9:10,12:13,15:16}$-octadectrienoate), M = 292. (Hallgren, Rvhage, and Stenhagen, *Acta Chem. Scand., 13,* 845, 1959.)

From: Chapman, D., *The Structure of Lipids,* p. 151 (1965). Reproduced by permission of the copyright owners, John Wiley and Sons, New York.

# NUCLEIC ACIDS

# PHYSICAL CONSTANTS AND SPECTRAL PROPERTIES OF PURINES, PYRIMIDINES, NUCLEOSIDES AND NUCLEOTIDES

DR. DAVID B. DUNN and DR. ROSS H. HALL

The data in the following table cover a total of 190 compounds, including most modified components of nucleic acids, some naturally occurring purine and pyrimidine compounds, and related synthetic compounds. Where possible, data were taken on chemically synthesized material.

Abbreviations were suggested by Dr. Waldo E. Cohn, Director, NAS-NRC Office of Biochemical Nomenclature (OBN), in accordance with the principles set out by the IUPAC-IUB Commission on Biochemical Nomenclature in Section 5 of "Abbreviations and Symbols for Chemical Names of Special Interest in Biological Chemistry" (1965 Revision), published in *J. Biol. Chem., 241,* 527 (1966), *Biochemistry, 5,* 1445 (1966), and elsewhere. The three-letter abbreviations are proposed for use in tables, illustrations and equations involving the monomeric units themselves; the one-letter symbols are proposed for polymer sequences. For deoxyribonucleosides in sequences, the letter d may precede the sequence and thus be eliminated from each residue.

The first reference for origin and synthesis gives the origin of the compound used to obtain the principal spectral data: C = chemical synthesis; E = prepared enzymically; R = isolated from RNA; D = isolated from DNA; N = isolated from natural products other than nucleic acids.

Melting point data were taken from the first reference to chemical synthesis, except where otherwise indicated (footnote[a]); dec signifies decomposition.

For $[a]_D^t$ the temperature is given as a superscript. Concentration and solvent used in obtaining the value are given in parentheses: W = water; E = ethanol.

pK values were taken from the first reference quoted, except where values differed by only 0.2 pH units when a mean value was used. References that give values deviating from those quoted by more than ±0.1 pH unit are marked with an asterisk (*); the pK values involved are similarly marked. The pK values for nucleotides are only those for the nucleoside moiety, not phosphate ionizations. Where pK values were determined from electrophoretic mobilities (footnote[p]), values were obtained from mobilities relative to the parent compound and pK values of these given by Jordan in *The Nucleic Acids,* Vol. 1, p. 447, Chargaff and Davidson, Eds. (Academic Press, New York, 1955).

Where possible, all spectral values are given at pH values away from pK values; exceptions to this are indicated by a footnote. Data obtained at a pH value where the compound was unstable were obtained soon after subjecting the material to this pH and differ from those of the decomposition products. The reference giving the largest amount of spectral data is cited first; references giving additional data are marked by footnotes, indicating that some – but not necessarily all – of the data mentioned come from the cited reference (footnotes [c,d,e]). The spectral ratios represent absorption at the wavelength given to absorption at 260 mμ. In the pH column of spectral data, W = water, E = ethanol, and M = methanol. References that give spectral data differing from the values quoted are marked with an asterisk (*), as are the values involved; such deviations are marked only where they exceed ±1 mμ for $\lambda_{max}$ or $\lambda_{min}$, ±5% for $\epsilon_{max}$, and ±10% for spectral ratios. The latter applies to all instances where spectral ratios are quoted in the additional reference, but not necessarily to all references giving spectra.

The authors are indebted to a number of collaborators, who supplied unpublished data, provided original spectra for calculation of the values or gave advice on the selection of the most reliable data. They also wish to thank particularly Mr. I. H. Flack, Mr. R. Thedford and Miss L. Csonka for their assistance in the preparation of the table.

| No. | Compound | Abbreviation | Structure | Formula (mol wt) | Melting point °C | $[\alpha]_D^t$ | pK Basic | pK Acidic |
|---|---|---|---|---|---|---|---|---|
| **PURINES AND PYRIMIDINES** | | | | | | | | |
| 1 | **Adenine** | Ade | 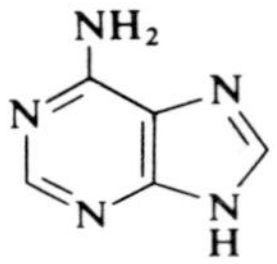 | $C_5H_5N_5$ (135.13) | 360° (dec) (sublimes 220°) | — | < 1,4.15 | 9.8 |
| 2 | **1-Methyladenine** | 1MeAde | 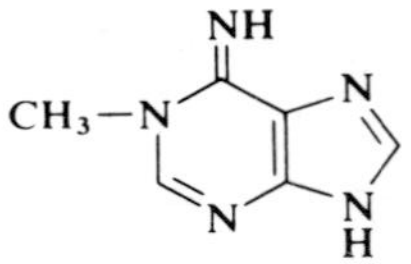 | $C_6H_7N_5$ (149.16) | 296–299° (dec) | — | 7.2 | 11.0 |
| 3 | **1-($\Delta^2$-Isopentenyl)adenine** (1-($\gamma\gamma$-Dimethylallyl)adenine) | 1iPeAde | 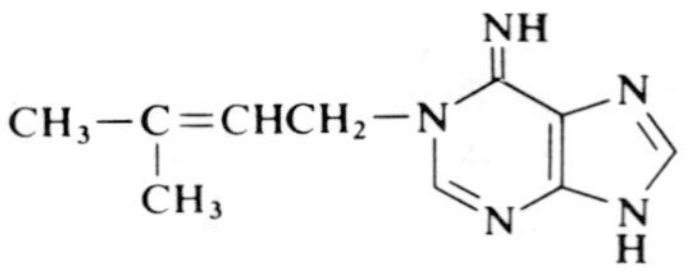 | $C_{10}H_{13}N_5$ (203.24) | 237–238° | — | 7.1* | 11.6[h] |

**Acidic spectral data**

| No. | pH | $\lambda_{max}$ | $\varepsilon_{max}$ ($\times 10^{-3}$) | $\lambda_{min}$ | 230 | 240 | 250 | 270 | 280 | 290 |
|---|---|---|---|---|---|---|---|---|---|---|
| 1 | 1 | 262.5 | 13.2 | 229 | — | — | 0.76 | — | 0.38 | 0.04 |
| 2 | 4 | 259 | 11.7 | 228* | 0.20 | 0.41 | 0.80 | 0.73 | 0.23* | 0.02* |
| 3 | 1 | 260 | 13.4 | 233 | — | — | — | — | — | — |

**Neutral spectral data**

| No. | pH | $\lambda_{max}$ | $\varepsilon_{max}$ ($\times 10^{-3}$) | $\lambda_{min}$ | 230 | 240 | 250 | 270 | 280 | 290 |
|---|---|---|---|---|---|---|---|---|---|---|
| 1 | 7 | 260.5 | 13.4 | 226 | — | — | 0.76 | — | 0.13 | 0.01 |
| 2 | 8.8 | 270 | 11.9 | 242 | 1.43 | 0.39 | 0.52 | 1.34 | 0.87 | 0.14 |
| 3 | E | 273 | 12.3 | 246 | — | — | — | — | — | — |

**Alkaline spectral data**

| No. | pH | $\lambda_{max}$ | $\varepsilon_{max}$ ($\times 10^{-3}$) | $\lambda_{min}$ | 230 | 240 | 250 | 270 | 280 | 290 |
|---|---|---|---|---|---|---|---|---|---|---|
| 1 | 12 | 269 | 12.3 | 237 | — | — | 0.57 | — | 0.60 | 0.03 |
| 2 | 13 | 270* | 14.4 | 239* | 0.64 | 0.29 | 0.55* | 1.27 | 0.85 | 0.35* |
| 3 | 13 | 274 | 15.2 | 242 | — | — | — | — | — | — |

(Columns 230–290 are spectral ratios.)

REFERENCES

| No. | Origin and synthesis | $[\alpha]_D^t$ | pK | Spectral data | $R_f$ |
|---|---|---|---|---|---|
| 1 | C: *231, 256*[a]*, 257* | — | *56, 22, 220, 51* | *232*[h]*. 205* | *258, 21, 18* |
| 2 | C: *8, 10*[a] R: *9* | — | *8* | *8*[b]*, 9*[b]*, 10, 11*, 12*, 265**[b] | *8–13* |
| 3 | C: *14* | — | *15, 16** | *14* | *14, 32* |

| No. | Compound | Abbreviation | Structure | Formula (mol wt) | Melting point °C | $[\alpha]_D^t$ | pH Basic | pH Acidic |
|---|---|---|---|---|---|---|---|---|
| 4 | **1-Methyl-$N^6$-methyladenine** (1-Methyl-6-methylaminopurine) | 1,6Me$_2$Ade | | $C_7H_9N_5$ (163.18) | 236° (picrate) | — | — | — |
| 5 | **2-Methyladenine** | 2MeAde | | $C_6H_7N_5$ (149.16) | > 300° | — | ~5.1[f] | — |
| 6 | **2-Hydroxyadenine** (Isoguanine) | 2(OH)Ade *iso*Gua | | $C_5H_5N_5O$ (151.13) | — | — | 4.5 | 9.0 |

Acidic spectral data

| No. | pH | $\lambda_{max}$ | $\varepsilon_{max}$ ($\times 10^{-3}$) | $\lambda_{min}$ | Spectral ratios 230 | 240 | 250 | 270 | 280 | 290 |
|---|---|---|---|---|---|---|---|---|---|---|
| 4 | 1 | 261 | 12.9 | 230* | — | — | — | — | — | — |
| 5 | 1 | 266 | 12.9 | 229 | 0.26 | 0.48 | 0.79 | 1.03 | 0.56 | 0.04 |
| 6 | 2 | 284* | 11.7* | 248 | — | — | 0.45 | — | 3.16 | — |

Neutral spectral data

| No. | pH | $\lambda_{max}$ | $\varepsilon_{max}$ ($\times 10^{-3}$) | $\lambda_{min}$ | Spectral ratios 230 | 240 | 250 | 270 | 280 | 290 |
|---|---|---|---|---|---|---|---|---|---|---|
| 4 | — | — | — | — | — | — | — | — | — | — |
| 5 | — | — | — | — | — | — | — | — | — | — |
| 6 | 7 | 240<br>286 | 7.8<br>8.0* | 210<br>255 | — | — | — | — | — | — |

Alkaline spectral data

| No. | pH | $\lambda_{max}$ | $\varepsilon_{max}$ ($\times 10^{-3}$) | $\lambda_{min}$ | Spectral ratios 230 | 240 | 250 | 270 | 280 | 290 |
|---|---|---|---|---|---|---|---|---|---|---|
| 4 | 11 | 274 | 12.7 | 245 | — | — | — | — | — | — |
| 5 | 13 | 271 | 10.7 | 238 | 0.77 | 0.40 | 0.61 | 1.28 | 0.84 | 0.04 |
| 6 | 12 | 284 | 12.3 | 253 | — | — | 0.80 | — | 3.47 | — |

REFERENCES

| No. | Origin and synthesis | $[\alpha]_D^t$ | pK | Spectral data | $R_f$ |
|---|---|---|---|---|---|
| 4 | C: *24, 12, 30* | — | — | *24, 12** | *12, 24* |
| 5 | C: *17*[g] R: *18* | — | *19* | *18*[b], *20*[d], *21* | *18, 21* |
| 6 | C: *282* | — | *22* | *28, 283**[b,c], *223**[e], *282*[b], *284*[b] | — |

| No. | Compound | Abbreviation | Structure | Formula (mol wt) | Melting point °C | $[\alpha]^t_D$ | pK Basic | pK Acidic |
|---|---|---|---|---|---|---|---|---|
| 7 | **3-($\Delta^2$-Isopentenyl)adenine** (Triacanthine) | 3iPeAde | 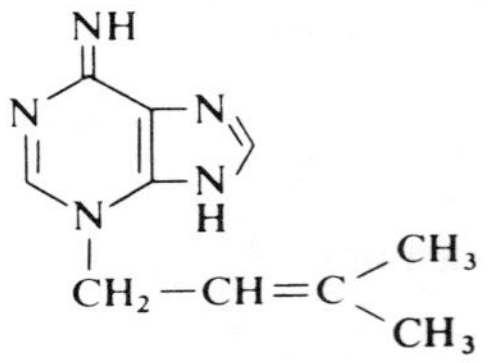 | $C_{10}H_{13}N_5$ (203.24) | 231–232° | — | — | 5.4 |
| 8 | **$N^6$-Methyladenine** (6-Methylaminopurine) | 6MeAde | 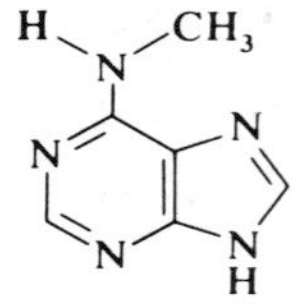 | $C_6H_7N_5$ (149.16) | 319–320° | — | <1,4.2 | 10.0 |
| 9 | **$N^6$-($\Delta^2$-Isopentenyl)adenine** ($N^6$-(γ,γ, Dimethylallyl) adenine, 6-(3-Methyl-2-butenylamino) purine) | 6iPeAde | 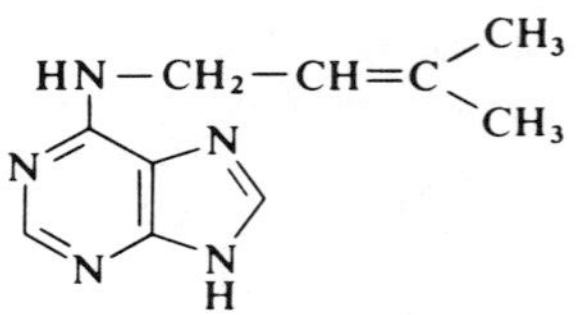 | $C_{10}H_{13}N_5$ (203.24) | 212–214° | — | 3.4[h] | 10.4[h] |

| No. | Acidic spectral data pH | $\lambda_{max}$ | $\varepsilon_{max}$ ($\times 10^{-3}$) | $\lambda_{min}$ | Spectral ratios 230 | 240 | 250 | 270 | 280 | 290 |
|---|---|---|---|---|---|---|---|---|---|---|
| 7 | 1 | 277 | 18.3 | 239 | — | — | — | — | — | — |
| 8 | 1 | 267 | 15.3* | 232* | 0.22 | 0.31 | 0.64 | 1.06 | 0.70 | 0.32* |
| 9 | 1 | 273* | 18.6 | 235 | 0.14 | 0.22 | 0.57 | 1.27 | 1.11 | 0.64 |

| No. | Neutral spectral data pH | $\lambda_{max}$ | $\varepsilon_{max}$ ($\times 10^{-3}$) | $\lambda_{min}$ | Spectral ratios 230 | 240 | 250 | 270 | 280 | 290 |
|---|---|---|---|---|---|---|---|---|---|---|
| 7 | 7 | 273 | 12.5 | 247 | — | — | — | — | — | — |
| 8 | 7 | 266 | 16.2 | 231 | — | — | — | — | — | — |
| 9 | 7 | 269 | 19.4 | 225 | 0.06 | 0.17 | 0.38 | 1.31 | 0.95 | 0.26 |

| No. | Alkaline spectral data pH | $\lambda_{max}$ | $\varepsilon_{max}$ ($\times 10^{-3}$) | $\lambda_{min}$ | Spectral ratios 230 | 240 | 250 | 270 | 280 | 290 |
|---|---|---|---|---|---|---|---|---|---|---|
| 7 | — | — | — | — | — | — | — | — | — | — |
| 8 | 13 | 273 | 15.9* | 239* | 0.77 | 0.39 | 0.55 | 1.48 | 1.19 | 0.25 |
| 9 | 13 | 275 | 18.1 | 240 | 0.86 | 0.28 | 0.49 | 1.70 | 1.60 | 0.54 |

## REFERENCES

| No. | Origin and synthesis | $[\alpha]^t_D$ | pK | Spectral data | $R_f$ |
|---|---|---|---|---|---|
| 7 | C: *287, 1* | — | *1* | *1* | — |
| 8 | C: *22, 23, 24*[a], *12*, D: *21* R: *18* | — | *22* | *21*[b], *18*[b], *25*[b,c], *28*[d], *8, 23, 12**, *24**, *265**[b] | *9, 10, 12, 13, 18, 21, 22, 24–27* |
| 9 | C: *31, 16* | — | *16* | *32, 14** | *32* |

| No. | Compound | Abbreviation | Structure | Formula (mol wt) | Melting point °C | $[\alpha]^t_D$ | pK Basic | pK Acidic |
|---|---|---|---|---|---|---|---|---|
| 10 | **$N^6$-($\Delta^2$-Isopentenyl)-2-methylthioadenine** | 6iPe2MeSAde | 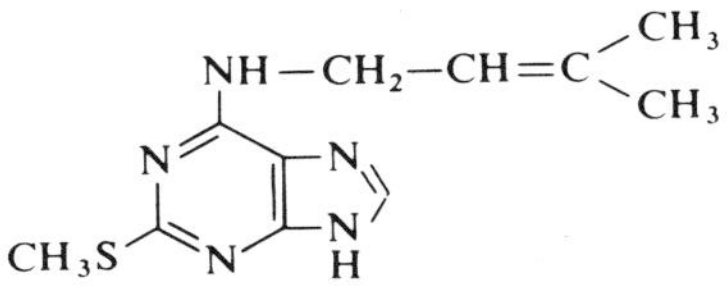 | $C_{11}H_{15}N_5S$ (249.32) | 259–260° | — | — | — |
| 11 | **$N^6$,$N^6$-Dimethyladenine** (6-Dimethylaminopurine) | $6Me_2Ade$ | 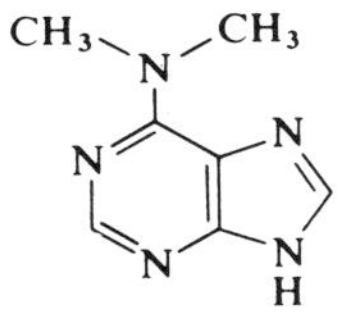 | $C_7H_9N_5$ (163.18) | 257° | — | <1,3.9 | 10.5 |
| 12 | **7-Methyladenine** | 7MeAde | 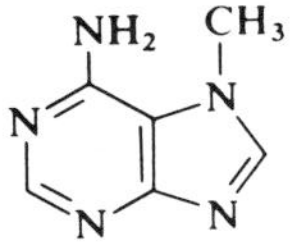 | $C_6H_7N_5$ (149.16) | 336° (dec) | — | 4.2 | — |

| No. | Acidic spectral data | | | | | | | | | | Neutral spectral data | | | | | | | | | | Alkaline spectral data | | | | | | | | | |
|---|---|---|---|---|---|---|---|---|---|---|---|---|---|---|---|---|---|---|---|---|---|---|---|---|---|---|---|---|---|---|
| | pH | $\lambda_{max}$ | $\varepsilon_{max}$ ($\times 10^{-3}$) | $\lambda_{min}$ | Spectral ratios 230 | 240 | 250 | 270 | 280 | 290 | pH | $\lambda_{max}$ | $\varepsilon_{max}$ ($\times 10^{-3}$) | $\lambda_{min}$ | Spectral ratios 230 | 240 | 250 | 270 | 280 | 290 | pH | $\lambda_{max}$ | $\varepsilon_{max}$ ($\times 10^{-3}$) | $\lambda_{min}$ | Spectral ratios 230 | 240 | 250 | 270 | 280 | 290 |
| 10 | — | — | — | — | — | — | — | — | — | — | — | — | — | — | — | — | — | — | — | — | — | — | — | — | — | — | — | — | — | — |
| 11 | 1 | 277 | 15.6 | 236 | 0.29* | 0.27 | 0.57 | 1.33 | 1.36* | 0.94* | 7 | 275 | 17.8 | — | — | — | — | — | — | — | 13 | 281 | 17.8 | 245 | 1.62 | 0.54 | 0.53 | 1.86 | 2.61 | 2.09* |
| 12 | 1 | 273* | 14.0* | 237 | 0.58 | 0.35 | 0.60 | 1.35 | 1.06 | 0.21 | — | — | — | — | — | — | — | — | — | — | 12[f] | 270* | 10.6* | 231 | 0.40 | 0.48 | 0.70 | 1.21 | 0.80 | 0.06 |

## REFERENCES

| No. | Origin and Synthesis | $[\alpha]^t_D$ | pK | Spectral data | $R_f$ |
|---|---|---|---|---|---|
| 10 | C: *274* | — | — | — | — |
| 11 | C: *29, 22*[a], *23* R: *18* | — | *22* | *18*[b], *23*[d], *28*[c,d], *21, 265**[b] | *18, 22, 32* |
| 12 | C: *33, 34* | — | *35* | *36*[b], *33**[d], *35*[b], *37**[b] | *26, 33, 36* |

| No. | Compound | Abbreviation | Structure | Formula (mol wt) | Melting point °C | $[\alpha]^t_D$ | pK Basic | pK Acidic |
|---|---|---|---|---|---|---|---|---|
| 13 | ***N*-(Purin-6-ylcarbamoyl)-threonine** | 6Thr(CO)Ade | 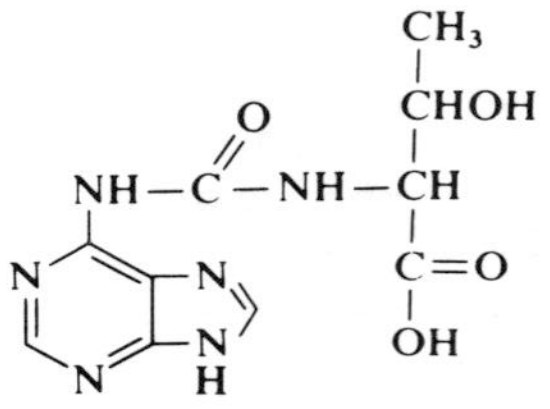 | $C_{10}H_{12}N_6O_4$ (280.24) | 219–221° | $+42^{25}$ (0.12,W) | — | ~3 |
| 14 | **Cytosine** | Cyt | 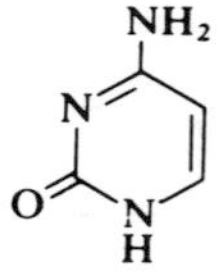 | $C_4H_5N_3O$ (111.10) | 312° (dec) | — | 4.45* | 12.2 |
| 15 | **3-Methylcytosine** | 3MeCyt | 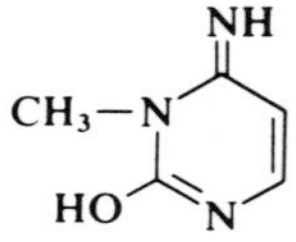 | $C_5H_7N_3O$ (125.13) | 242–245° (HCl salt) | — | 7.4 | >13 |

| No. | Acidic spectral data | | | | | | | | | |
|---|---|---|---|---|---|---|---|---|---|---|
| | pH | $\lambda_{max}$ | $\varepsilon_{max}$ ($\times 10^{-3}$) | $\lambda_{min}$ | Spectral ratios 230 | 240 | 250 | 270 | 280 | 290 |
| 13 | 1.5 | 279 | 20.0 | 239 | 0.60 | 0.43 | 0.62 | 1.76 | 2.17 | 1.14 |
| 14 | 1 | 276 | 10.0 | 239 | — | — | 0.48 | — | 1.53 | 0.78 |
| 15 | 4 | 274 | 9.4 | 240 | 0.52 | 0.27 | 0.50 | 1.47 | 1.33 | 0.56 |

| No. | Neutral spectral data | | | | | | | | | |
|---|---|---|---|---|---|---|---|---|---|---|
| | pH | $\lambda_{max}$ | $\varepsilon_{max}$ ($\times 10^{-3}$) | $\lambda_{min}$ | Spectral ratios 230 | 240 | 250 | 270 | 280 | 290 |
| 13 | 7.0 | 270<br>278 | 19.2<br>18.8 | 236 | 0.39 | 0.45 | 0.67 | 1.67 | 1.33 | 0.23 |
| 14 | 7 | 267 | 6.1 | 247 | — | — | 0.78 | — | 0.58 | 0.08 |
| 15 | — | — | — | — | — | — | — | — | — | — |

| No. | Alkaline spectral data | | | | | | | | | |
|---|---|---|---|---|---|---|---|---|---|---|
| | pH | $\lambda_{max}$ | $\varepsilon_{max}$ ($\times 10^{-3}$) | $\lambda_{min}$ | Spectral ratios 230 | 240 | 250 | 270 | 280 | 290 |
| 13 | 11.5 | 279 | 17.2 | 241 | 1.56 | 0.38 | 0.53 | 1:91 | 2.44 | 1.56 |
| 14 | 13[a]<br>14 | 281.5<br>282 | 7.1<br>7.9 | 251<br>251 | — | — | 0.60 | — | 3.28 | 2.6 |
| 15 | 12 | 294 | 11.9 | 250 | 5.10 | 1.60 | 0.53 | 2.60 | 5.90 | 9.30 |

REFERENCES

| No. | Origin and synthesis | $[\alpha]^t_D$ | pK | Spectral data | $R_f$ |
|---|---|---|---|---|---|
| 13 | C: *4*, R: *3* | *4* | *4* | *3* | — |
| 14 | C: *233*, *64*[a] | — | *66*, *80** | *66*[b], *232*[b], *205* | *258*, *21* |
| 15 | C: *57*, *30*[a] | — | *58*, *57* | *57*, *58*[b] | *13*, *57* |

| No. | Compound | Abbreviation | Structure | Formula (mol wt) | Melting point °C | $[\alpha]^t_D$ | pK Basic | pK Acidic |
|---|---|---|---|---|---|---|---|---|
| 16 | **$N^4$-Acetylcytosine** | 4AcCyt | 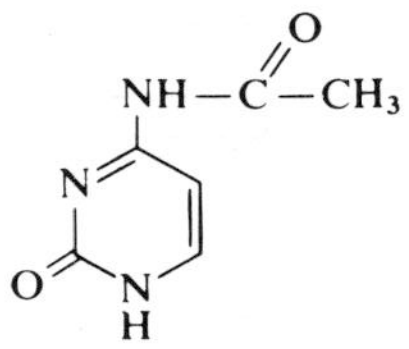 | $C_6H_7N_3O_2$ (153.14) | 326–328° | — | — | — |
| 17 | **$N^4$-Methylcytosine** (4-Methylaminopyrimidin-2-one) | 4MeCyt | 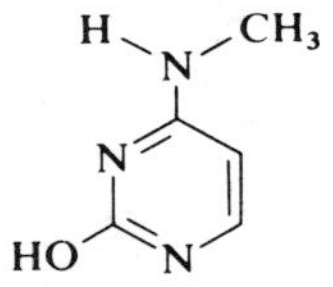 | $C_5H_7N_3O$ (125.13) | 275–278° (dec) | — | 4.5 | 12.7 |
| 18 | **5-Methylcytosine** | 5MeCyt | 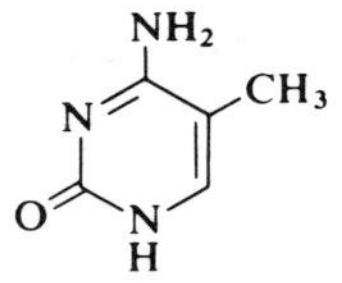 | $C_5H_7N_3O$ (125.13) | 299-301° (dec) (HCl salt) | — | 4.6 | 12.4 |

Acidic spectral data

| No. | pH | $\lambda_{max}$ | $\varepsilon_{max}$ ($\times 10^{-3}$) | $\lambda_{min}$ | Spectral ratios 230 | 240 | 250 | 270 | 280 | 290 |
|---|---|---|---|---|---|---|---|---|---|---|
| 16 | | | | — | — | — | | | | — |
| 17 | 1 | 277 | 10.5* | 240 | 0.60 | 0.29 | 0.50 | 1.56 | 1.66 | 1.03 |
| 18 | 1 | 283 | 9.8 | 242 | 0.97 | 0.26 | 0.40 | 1.90 | 2.62 | 2.43 |

Neutral spectral data

| No. | pH | $\lambda_{max}$ | $\varepsilon_{max}$ ($\times 10^{-3}$) | $\lambda_{min}$ | Spectral ratios 230 | 240 | 250 | 270 | 280 | 290 |
|---|---|---|---|---|---|---|---|---|---|---|
| 16 | 7 | 244.5<br>293 | 14.2<br>4.9 | 226<br>270 | — | — | — | — | — | — |
| 17 | 7 | 267 | 7.2* | 248 | 1.07 | 0.89 | 0.79 | 1.07 | 0.66 | 0.14 |
| 18 | 7 | 273 | 6.2 | 252 | 1.60 | 1.10 | 0.80 | 1.35 | 1.21 | 0.54 |

Alkaline spectral data

| No. | pH | $\lambda_{max}$ | $\varepsilon_{max}$ ($\times 10^{-3}$) | $\lambda_{min}$ | Spectral ratios 230 | 240 | 250 | 270 | 280 | 290 |
|---|---|---|---|---|---|---|---|---|---|---|
| 16 | — | — | — | — | — | — | — | — | — | — |
| 17 | 14 | 286 | 8.0* | 256 | 3.18 | 2.02 | 0.96 | 2.02 | 3.27 | 3.23 |
| 18 | 14 | 289 | 8.1 | 254 | 5.05 | 1.97 | 0.84 | 2.02 | 3.64 | 4.71 |

REFERENCES

| No. | Origin and synthesis | $[\alpha]^t_D$ | pK | Spectral data | $R_f$ |
|---|---|---|---|---|---|
| 16 | C: *280, 281* | — | — | *281* | — |
| 17 | C: *59, 60, 61*[a]*, 62* | — | *59, 63* | *59*[b]*, 60** | *59* |
| 18 | C: *64, 65* | — | *66* | *66*[b]*, 67*[b]*, 64*[b]*, 68*[b]*, 265*[b] | *67, 69, 70* |

| No. | Compound / Abbreviation | Structure | Formula (mol wt) | Melting point °C | $[\alpha]^t_D$ | pK Basic | pK Acidic |
|---|---|---|---|---|---|---|---|
| 19 | **5-Hydroxymethylcytosine**<br>5HmCyt | $NH_2$, $CH_2OH$, N, HO, N | $C_5H_7N_3O_2$ (141.13) | > 200° (dec) | — | 4.3 | ~13 |
| 20 | **2,6-Diamino-4-hydroxy-5-*N*-methyl-formamidopyrimidine**<br>(6-Amino-5-*N*-methylformamidoisocytosine) | OH, $CH_3$, N—CHO, N, $H_2N$, N, $NH_2$ | $C_6H_9N_5O_2$ (183.17) | — | — | 3.8 | 9.9[k] |
| 21 | **Guanine**<br>Gua | O, HN, N, $H_2N$, N, N, H | $C_5H_5N_5O$ (151.13) | > 350° | — | <0.3.2 | 9.6,*12.4 |

| No. | Acidic spectral data: pH | $\lambda_{max}$ | $\varepsilon_{max}$ ($\times 10^{-3}$) | $\lambda_{min}$ | Spectral ratios 230 | 240 | 250 | 270 | 280 | 290 |
|---|---|---|---|---|---|---|---|---|---|---|
| 19 | 1 | 279 | 9.7 | 241 | 0.63 | 0.25 | 0.45 | 1.68 | 1.97 | 1.37 |
| 20 | 1 | 263 | 17.8 | 232* | 0.37 | 0.39 | 0.59 | 0.92 | 0.34 | 0.04 |
| 21 | 1 | 248<br>276 | 11.4<br>7.35 | 224<br>267 | — | — | 1.37 | — | 0.84 | 0.50 |

| No. | Neutral spectral data: pH | $\lambda_{max}$ | $\varepsilon_{max}$ ($\times 10^{-3}$) | $\lambda_{min}$ | Spectral ratios 230 | 240 | 250 | 270 | 280 | 290 |
|---|---|---|---|---|---|---|---|---|---|---|
| 19 | 7 | 269 | 5.7 | 251 | 0.59 | 0.93 | 0.80 | 1.14 | 0.80 | 0.15 |
| 20 | 7 | 264 | 13.8 | 242 | — | — | — | — | — | — |
| 21 | 7 | 246<br>276 | 10.7<br>8.15 | 225<br>262 | — | — | 1.42 | — | 1.04 | 0.54 |

| No. | Alkaline spectral data: pH | $\lambda_{max}$ | $\varepsilon_{max}$ ($\times 10^{-3}$) | $\lambda_{min}$ | Spectral ratios 230 | 240 | 250 | 270 | 280 | 290 |
|---|---|---|---|---|---|---|---|---|---|---|
| 19 | 13 | 283 | 7.6 | 254 | 3.91 | 2.59 | 0.83* | 1.97 | 2.98 | 2.50 |
| 20 | 13 | 262 | 9.8 | 242 | 0.72 | 0.67 | 0.71 | 0.78 | 0.31 | 0.09 |
| 21 | 11 | 274 | 8.0 | 255 | — | — | 0.99 | — | 1.14 | 0.59 |
| | 14 | 274 | 9.9 | 238 | — | — | 0.81 | — | 1.24 | 0.61 |

## REFERENCES

| No. | Origin and Synthesis | $[\alpha]^t_D$ | pK | Spectral data | $R_f$ |
|---|---|---|---|---|---|
| 19 | C: *71* | — | *72* | *73*[b], *72*[b], *71*, *74*, *265**[b] | *69*, *70*, *73*, *75* |
| 20 | C: *41* | — | *41*, *76* | *90*, *41**[c,d] | *41* |
| 21 | C: *234*, *257*[a] | — | *170*, *56**, *232* | *232*[b], *205* | *258*, *21*, *40* |

| No. | Compound | Abbreviation | Structure | Formula (mol wt) | Melting point °C | $[\alpha]^t_D$ | pK Basic | pK Acidic |
|---|---|---|---|---|---|---|---|---|
| 22 | **1-Methylguanine** | 1MeGua | | $C_6H_7N_5O$ (165.16) | None (dec) | — | ~0,3.1 | 10.5 |
| 23 | ***N*²-Methylguanine** (6-Hydroxy-2-methylaminopurine) | 2MeGua | | $C_6H_7N_5O$ (165.16) | — | — | 3.3 | 8.9, 12.8 |
| 24 | ***N*²,*N*²-Dimethylguanine** (2-Dimethylamino-6-hydroxypurine) | $2Me_2Gua$ | | $C_7H_9N_5O$ (179.18) | — | — | — | — |

Acidic spectral data

| No. | pH | $\lambda_{max}$ | $\varepsilon_{max}$ ($\times 10^{-3}$) | $\lambda_{min}$ | 230 | 240 | 250 | 270 | 280 | 290 |
|---|---|---|---|---|---|---|---|---|---|---|
| 22 | 1 | 250<br>272 | 10.2<br>7.1 | 227<br>— | 0.47 | 0.93 | 1.28 | 0.87 | 0.81 | 0.50 |
| 23 | 1 | 250*<br>279 | 13.9*<br>6.2* | 228* | 0.51 | 0.95 | 1.34* | 0.62* | 0.64 | 0.54* |
| 24 | 1 | 256* | 19.0* | 233* | 0.36 | 0.48 | 0.92 | 0.52 | 0.37* | 0.39 |

Neutral spectral data

| No. | pH | $\lambda_{max}$ | $\varepsilon_{max}$ ($\times 10^{-3}$) | $\lambda_{min}$ | 230 | 240 | 250 | 270 | 280 | 290 |
|---|---|---|---|---|---|---|---|---|---|---|
| 22 | 7 | 248<br>272 | 10.0<br>7.9 | 227<br>264 | 0.54 | 1.01 | 1.24 | 0.96 | 0.93 | 0.46 |
| 23 | 7 | 249<br>277 | — | 227<br>266 | — | — | — | — | — | — |
| 24 | — | — | — | — | — | — | — | — | — | — |

Alkaline spectral data

| No. | pH | $\lambda_{max}$ | $\varepsilon_{max}$ ($\times 10^{-3}$) | $\lambda_{min}$ | 230 | 240 | 250 | 270 | 280 | 290 |
|---|---|---|---|---|---|---|---|---|---|---|
| 22 | 13 | 277 | 8.7 | 241* | 1.49 | 0.64 | 0.80 | 1.12 | 1.19 | 0.81 |
| 23 | 11 | 244<br>278* | 9.5*<br>7.2 | 263* | — | — | — | — | — | — |
| 24 | 11 | 282* | 7.5* | 265 | — | — | — | — | — | — |

(Spectral ratios are given under the column headings 230–290.)

REFERENCES

| No. | Origin and synthesis | $[\alpha]^t_D$ | pK | Spectral data | $R_f$ |
|---|---|---|---|---|---|
| 22 | C: *33*[g], *38*[a] | — | *39, 27* | *40*[b], *41*[c,d], *27*[b,e], *39*[b,c,e] *25, 33, 42, 265**[b] | *24, 25, 27, 33, 40, 42* |
| 23 | C: *43*[g], *44* | — | *2* | *40*[b], *44*[c,d], *43**, *2**, *27**, *25, 265**[b] | *25, 27, 33, 40* |
| 24 | C: *43*[g], *44* | — | — | *40*[b], *44, 43, 265**[b] | *40* |

| No. | Compound | Abbreviation | Structure | Formula (mol wt) | Melting point °C | $[\alpha]_D^t$ | pK Basic | pK Acidic |
|---|---|---|---|---|---|---|---|---|
| 25 | **7-Methylguanine** | 7MeGua | 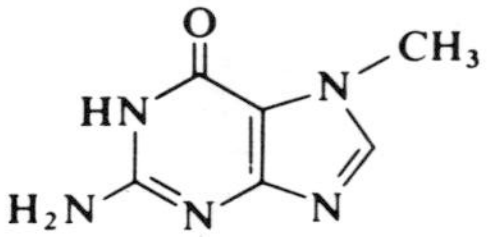 | $C_6H_7N_5O$ (165.16) | None (dec > 390°) | — | ~0,3.5 | 9.9* |
| 26 | **Hypoxanthine** | Hyp | 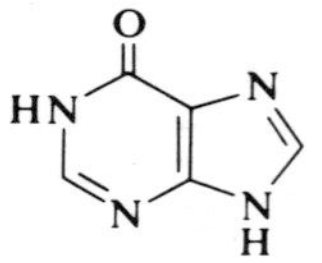 | $C_5H_4N_4O$ (136.11) | None (dec) | — | 2.0 | 8.9*, 12.1 |
| 27 | **1-Methylhypoxanthine** | 1MeHyp | 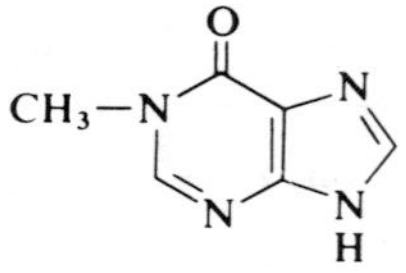 | $C_6H_6N_4O$ (150.14) | 311–312° | — | ~2 | 8.9*, ~13 |

| No. | Acidic spectral data | | | | | | | | | |
|---|---|---|---|---|---|---|---|---|---|---|
| | pH | $\lambda_{max}$ | $\varepsilon_{max}$ ($\times 10^{-3}$) | $\lambda_{min}$ | Spectral ratios 230 | 240 | 250 | 270 | 280 | 290 |
| 25 | 1 | 250<br>272 | 10.6<br>6.9 | 228 | 0.55 | 0.99 | 1.30 | 0.84 | 0.79 | 0.52 |
| 26 | 0 | 248 | 10.8 | 215 | — | — | 1.45 | — | 0.04 | 0.00 |
| 27 | 1 | 249 | 9.4 | — | 0.59 | 1.11 | 1.37 | 0.43 | 0.10 | 0.01 |

| No. | Neutral spectral data | | | | | | | | | |
|---|---|---|---|---|---|---|---|---|---|---|
| | pH | $\lambda_{max}$ | $\varepsilon_{max}$ ($\times 10^{-3}$) | $\lambda_{min}$ | Spectral ratios 230 | 240 | 250 | 270 | 280 | 290 |
| 25 | 7 | 248<br>283 | 5.7*<br>7.4* | 235<br>261 | 1.54 | 1.42 | 1.46 | 1.35 | 1.87 | 1.73 |
| 26 | 6 | 249.5 | 10.7 | 222 | — | — | 1.32 | — | 0.09 | 0.01 |
| 27 | 5 | 251 | 9.4 | — | 0.51 | 0.95 | 1.31 | 0.53 | 0.16 | 0.02 |

| No. | Alkaline spectral data | | | | | | | | | |
|---|---|---|---|---|---|---|---|---|---|---|
| | pH | $\lambda_{max}$ | $\varepsilon_{max}$ ($\times 10^{-3}$) | $\lambda_{min}$ | Spectral ratios 230 | 240 | 250 | 270 | 280 | 290 |
| 25 | 12 | 280* | 7.4 | 257 | 1.92 | 1.50 | 1.13 | 1.50 | 1.89 | 1.47 |
| 26 | 11<br>14 | 259<br>263 | 11.1<br>11.5 | 232<br>233 | — | — | 0.84<br>0.71 | — | 0.12<br>0.19 | 0.01<br>0.01 |
| 27 | 11 | 260 | 9.7 | — | — | — | — | — | — | — |

REFERENCES

| No. | Origin and synthesis | $[\alpha]_D^t$ | pK | Spectral data | $R_f$ |
|---|---|---|---|---|---|
| 25 | C: *45* | — | *39, 27** | *46*[b], *47*[b], *42**[b], *35*[b], *39, 41, 48* | *10, 13, 24, 27, 41, 42, 46* |
| 26 | C: *231* | — | *22, 170, 253, 51** | *232*[b], *205* | *258* |
| 27 | C: *33, 26* | — | *33, 27** | *49, 33*[c,d], *27, 121* | *24, 26, 27, 49, 33, 121* |

| No. | Compound | Abbreviation | Structure | Formula (mol wt) | Melting point °C | $[\alpha]^{t}_{D}$ | pK Basic | pK Acidic |
|---|---|---|---|---|---|---|---|---|
| 28 | **3-Methylhypoxanthine** | 3MeHyp | 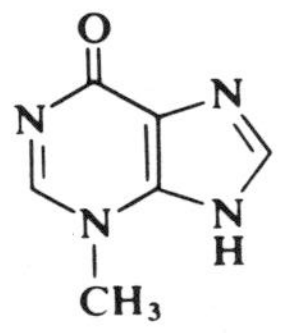 | $C_6H_6N_4O$ (150.14) | None (dec >280°) | — | 2.6 | 8.3 |
| 29 | **7-Methylhypoxanthine** | 7MeHyp | 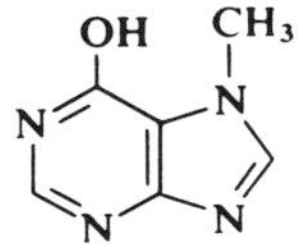 | $C_6H_6N_4O$ (150.14) | 355° (dec) | — | 2.1 | 8.9 |
| 30 | **Uracil** | Ura | 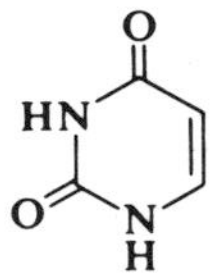 | $C_4H_4N_2O_2$ (112.09) | 335° (dec) | — | — | 9.5, >13 |

Acidic spectral data

| No. | pH | $\lambda_{max}$ | $\varepsilon_{max}$ ($\times 10^{-3}$) | $\lambda_{min}$ | 230 | 240 | 250 | 270 | 280 | 290 |
|---|---|---|---|---|---|---|---|---|---|---|
| 28 | 0 | 253 | 11.0 | — | — | — | — | — | — | — |
| 29 | 0 | 250 | 10.2 | 224 | — | — | — | — | — | — |
| 30 | 0[a] | 260 | 7.8 | 229 | — | — | 0.80 | — | 0.30 | 0.05 |
| | 4 | 259.5 | 8.2 | 227 | — | — | 0.84 | — | 0.17 | 0.01 |

Neutral spectral data

| No. | pH | $\lambda_{max}$ | $\varepsilon_{max}$ ($\times 10^{-3}$) | $\lambda_{min}$ | 230 | 240 | 250 | 270 | 280 | 290 |
|---|---|---|---|---|---|---|---|---|---|---|
| 28 | 5 | 264 | 14.0 | — | — | — | — | — | — | — |
| 29 | 5 | 256 | 9.5 | 229 | 0.42 | 0.81 | 0.97 | 0.55 | 0.07 | 0.00 |
| 30 | 7 | 259.5 | 8.2 | 227 | — | — | 0.84 | — | 0.17 | 0.01 |

Alkaline spectral data

| No. | pH | $\lambda_{max}$ | $\varepsilon_{max}$ ($\times 10^{-3}$) | $\lambda_{min}$ | 230 | 240 | 250 | 270 | 280 | 290 |
|---|---|---|---|---|---|---|---|---|---|---|
| 28 | 11 | 265 | 10.9 | — | — | — | — | — | — | — |
| 29 | 11 | 262 | 10.6 | 230 | 0.36 | 0.49 | 0.76 | 0.83 | 0.21 | 0.00 |
| 30 | 12 | 284 | 6.2 | 241 | — | — | 0.71 | — | 1.40 | 1.27 |

(Columns 230–290 are spectral ratios.)

REFERENCES

| No. | Origin and synthesis | $[\alpha]^{t}_{D}$ | pK | Spectral data | $R_f$ |
|---|---|---|---|---|---|
| 28 | C: *26, 50*[a] | — | *33* | *33, 121* | *26, 33, 121* |
| 29 | C: *45* | — | *33, 51* | *52, 33*[c,d] | *10, 26, 27, 33* |
| 30 | C: *231* | — | *66, 80* | *66*[b], *232*[b,c,d,e] | *258, 40* |

| No. | Compound | Abbreviation | Structure | Formula (mol wt) | Melting point °C | $[\alpha]^t_D$ | pK Basic | pK Acidic |
|---|---|---|---|---|---|---|---|---|
| 31 | **1-Methyluracil** | 1MeUra | | $C_5H_6N_2O_2$ (126.11) | 232–233° | — | — | 9.7 |
| 32 | **3-Methyluracil** | 3MeUra | | $C_5H_6N_2O_2$ (126.12) | 179° | — | — | 10.0 |
| 33 | **4-Thiouracil** | 4SUra | | $C_4H_4N_2OS$ (128.15) | 289–290° (dec) | — | — | — |

| No. | Acidic spectral data | | | | | | | | | |
|---|---|---|---|---|---|---|---|---|---|---|
| | pH | $\lambda_{max}$ | $\varepsilon_{max}$ ($\times 10^{-3}$) | $\lambda_{min}$ | Spectral ratios 230 | 240 | 250 | 270 | 280 | 290 |
| 31 | 2 | 272 | — | — | — | — | 0.60 | — | 0.64 | 0.08 |
| 32 | 3 | 259 | 7.3 | — | — | — | 0.86 | — | 0.14 | 0.02 |
| 33 | 1 | 327 | 17.5 | 277 | — | — | — | — | — | — |

| No. | Neutral spectral data | | | | | | | | | |
|---|---|---|---|---|---|---|---|---|---|---|
| | pH | $\lambda_{max}$ | $\varepsilon_{max}$ ($\times 10^{-3}$) | $\lambda_{min}$ | Spectral ratios 230 | 240 | 250 | 270 | 280 | 290 |
| 31 | 7 | 267 | 9.8 | 232 | 0.17 | 0.25 | 0.59 | 1.12 | 0.70 | 0.12 |
| 32 | 7 | 259 | 7.3 | 230 | 0.29 | 0.46 | 0.84 | 0.66 | 0.14 | 0.00 |
| 33 | 7 | 328 | 16.6 | 275 | — | — | — | — | — | — |

| No. | Alkaline spectral data | | | | | | | | | |
|---|---|---|---|---|---|---|---|---|---|---|
| | pH | $\lambda_{max}$ | $\varepsilon_{max}$ ($\times 10^{-3}$) | $\lambda_{min}$ | Spectral ratios 230 | 240 | 250 | 270 | 280 | 290 |
| 31 | 12 | 265* | 7.0 | 241 | 0.92 | 0.53 | 0.68 | 1.00 | 0.44 | 0.03 |
| 32 | 12 | 283 | 10.7 | 243 | 1.43 | 0.34 | 0.37 | 2.23 | 3.47 | 3.01 |
| 33 | 11 | 335 | 17.6 | 278 | — | — | — | — | — | — |

## REFERENCES

| No. | Origin and synthesis | $[\alpha]^t_D$ | pK | Spectral data | $R_f$ |
|---|---|---|---|---|---|
| 31 | C: *78, 79*[a] | — | *66, 80* | *66*[b], *81**[b,c,e] | *69, 81* |
| 32 | C: *82, 79*[a], *83* | — | *66, 80* | *66*[b], *81*[b,e] | *13, 69, 81* |
| 33 | C: *84, 85*[a] | — | — | *86*[b] | — |

| No. | Compound | Abbreviation | Structure | Formula (mol wt) | Melting point °C | $[\alpha]_D^t$ | pK Basic | pK Acidic |
|---|---|---|---|---|---|---|---|---|
| 34 | **Thymine** (5-Methyluracil) | Thy or 5MeUra | 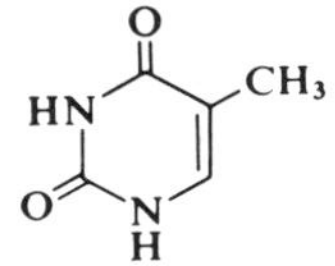 | $C_5H_6N_2O_2$ (126.11) | 325–335° (dec) | — | — | 9.9, >13 |
| 35 | **5-Methylaminomethyluracil** | 5MeNHMeUra | 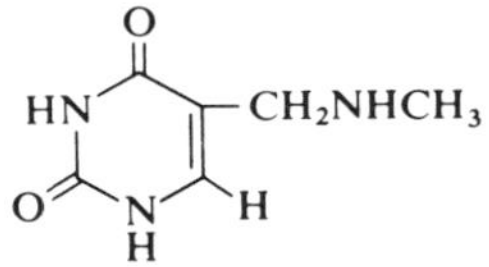 | $C_6H_9N_3O_2$ (155.16) | 230–232° | — | — | — |
| 36 | **5-Carboxymethyluracil** | 5CmUra | 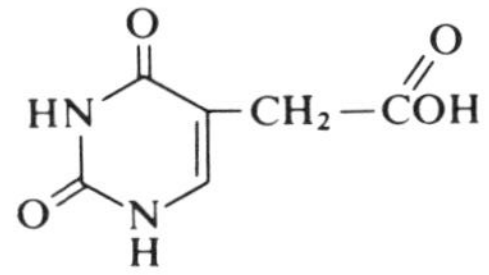 | $C_6H_6N_2O_4$ (170.12) | 315–320° (dec) | — | — | — |

Acidic spectral data

| No. | pH | $\lambda_{max}$ | $\varepsilon_{max}$ ($\times 10^{-3}$) | $\lambda_{min}$ | 230 | 240 | 250 | 270 | 280 | 290 |
|---|---|---|---|---|---|---|---|---|---|---|
| 34 | 4 | 264.5 | 7.9 | 233 | — | — | 0.67 | — | 0.53 | 0.09 |
| 35 | 1 | 262 | — | 230 | 0.01 | 0.07 | 0.70 | 0.85 | 0.39 | 0.13 |
| 36 | 1 | 262 | | 231 | 0.23 | 0.37 | 0.71 | 0.89 | 0.42 | 0.05 |

Neutral spectral data

| No. | pH | $\lambda_{max}$ | $\varepsilon_{max}$ ($\times 10^{-3}$) | $\lambda_{min}$ | 230 | 240 | 250 | 270 | 280 | 290 |
|---|---|---|---|---|---|---|---|---|---|---|
| 34 | 7 | 264.5 | 7.9 | 233 | — | — | 0.67 | — | 0.53 | 0.09 |
| 35 | 7.5 | 262 | — | 230 | 0.01 | 0.07 | 0.70 | 0.85 | 0.39 | 0.13 |
| 36 | 7 | 264 | 7.2 | 234 | 0.33 | 0.33 | 0.66 | 0.98 | 0.56 | 0.13 |

Alkaline spectral data

| No. | pH | $\lambda_{max}$ | $\varepsilon_{max}$ ($\times 10^{-3}$) | $\lambda_{min}$ | 230 | 240 | 250 | 270 | 280 | 290 |
|---|---|---|---|---|---|---|---|---|---|---|
| 34 | 12 | 291 | 5.4 | 244 | — | — | 0.65 | — | 1.31 | 1.41 |
| 35 | 10.1 | 287 | — | 244 | 3.88 | 0.42 | 0.39 | 1.69 | 2.18 | 2.32 |
| 36 | 13 | 290 | — | 246 | 1.95 | 0.78 | 0.63 | 1.40 | 1.77 | 2.00 |

REFERENCES

| No. | Origin and synthesis | $[\alpha]_D^t$ | pK | Spectral data | $R_f$ |
|---|---|---|---|---|---|
| 34 | C: *254, 231, 101* | — | *66, 80* | *66*[b], *232*[b] | *258, 21, 18* |
| 35 | C: *270* | — | — | *270* | *270* |
| 36 | C: *279, 278* | — | — | *278*[b] | *278* |

| No. | Compound | Abbreviation | Structure | Formula (mol wt) | Melting point °C | $[\alpha]^t_D$ | pK Basic | pK Acidic |
|---|---|---|---|---|---|---|---|---|
| 37 | **2-Thio-5-carboxymethyl-uracil, methyl ester** | 2S5(CmMe)Ura | 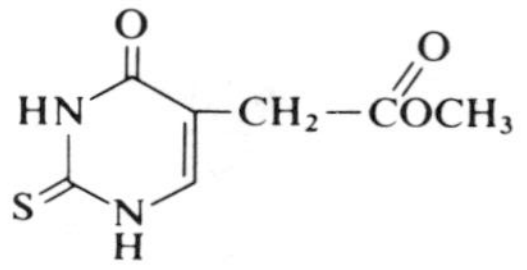 | $C_7H_8N_2O_3S$ (200.15) | 218–220° | — | — | — |
| 38 | **5-Hydroxyuracil** | 5(OH)Ura | 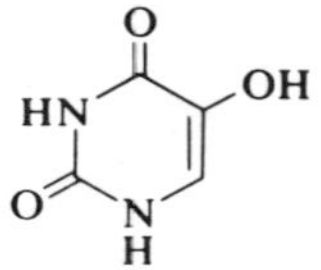 | $C_4H_4N_2O_3$ (128.09) | None (dec > 300°) | — | — | 8.0 |
| 39 | **5-Hydroxymethyluracil** | 5HmUra | 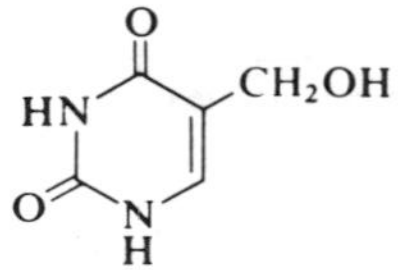 | $C_5H_6N_2O_3$ (142.11) | 260–300° (dec) | — | — | 9.4*, ~14 |

| No. | Acidic spectral data | | | | | | | | | | Neutral spectral data | | | | | | | | | | Alkaline spectral data | | | | | | | | | |
|---|---|---|---|---|---|---|---|---|---|---|---|---|---|---|---|---|---|---|---|---|---|---|---|---|---|---|---|---|---|---|
| | pH | $\lambda_{max}$ | $\varepsilon_{max}$ ($\times 10^{-3}$) | $\lambda_{min}$ | Spectral ratios 230 | 240 | 250 | 270 | 280 | 290 | pH | $\lambda_{max}$ | $\varepsilon_{max}$ ($\times 10^{-3}$) | $\lambda_{min}$ | Spectral ratios 230 | 240 | 250 | 270 | 280 | 290 | pH | $\lambda_{max}$ | $\varepsilon_{max}$ ($\times 10^{-3}$) | $\lambda_{min}$ | Spectral ratios 230 | 240 | 250 | 270 | 280 | 290 |
| 37 | 1 | 275<br>214 | — | — | 0.49 | 0.31 | 0.47 | 1.48 | 1.52 | 1.44 | 7 | 275<br>215 | — | — | 0.47 | 0.32 | 0.47 | 1.45 | 1.49 | 1.41 | 12 | 313<br>260<br>234 | — | — | 0.91 | 0.83 | 0.85 | 0.80 | 0.59 | 0.51 |
| 38 | 2 | 278* | 6.4 | 244 | 1.16 | 0.58 | 0.56 | 1.52 | 1.63 | 1.26 | 6 | 278 | 6.4 | 244 | 1.08 | 0.53 | 0.54 | 1.45 | 1.55 | 1.18 | 9[q] | 240<br>304* | 6.5*<br>5.1 | 272 | 1.49 | 1.65 | 1.45 | 0.75 | 0.87 | 1.09 |
| 39 | 2 | 261 | 8.0 | 231 | — | — | 0.77 | — | 0.32 | — | 7 | 261 | — | 231 | 0.27 | 0.43 | 0.77 | 0.80 | 0.33 | 0.05 | 12 | 286 | 7.4 | 245 | 1.77 | 0.75 | 0.67 | 1.39 | 1.80 | 1.75 |

REFERENCES

| No. | Origin and synthesis | $[\alpha]^t_D$ | pK | Spectral data | $R_f$ |
|---|---|---|---|---|---|
| 37 | R: *5* | — | — | *5* | — |
| 38 | C: *87, 88*[a]*, 89* | — | *90* | *87*[b]*, 91**[b] | — |
| 39 | C: *69, 92, 93* | — | *72, 94** | *94*[b]*, 69*[c,d,e]*, 72*[b]*, 95*[b] | *69, 96, 97* |

| No. | Compound | Abbreviation | Structure | Formula (mol wt) | Melting point °C | $[\alpha]_D^t$ | pK Basic | pK Acidic |
|---|---|---|---|---|---|---|---|---|
| 40 | **Dihydrouracil** | $H_2Ura$ | 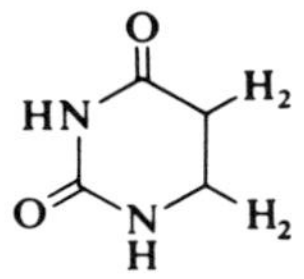 | $C_4H_6N_2O_2$ (114.10) | 275–276° | — | — | — |
| 41 | **Orotic Acid** (Uracil-6-carboxylic acid) | Oro | 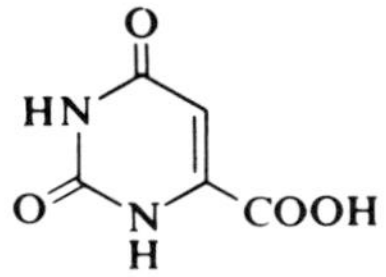 | $C_5H_4N_2O_4$ (156.10) | 345° (dec) | — | — | 2.4, 9.5, >13 |
| 42 | **Xanthine** | Xan | 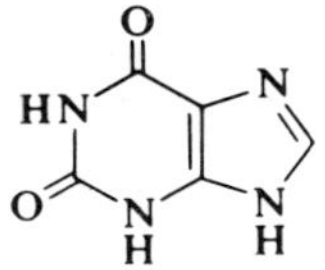 | $C_5H_4N_4O_2$ (152.11) | None (dec) | — | ~0.8 | 7.5*, 11.1* |

Acidic spectral data

| No. | pH | $\lambda_{max}$ | $\varepsilon_{max}$ ($\times 10^{-3}$) | $\lambda_{min}$ | 230 | 240 | 250 | 270 | 280 | 290 |
|---|---|---|---|---|---|---|---|---|---|---|
| 40 | — | — | — | — | — | — | — | — | — | — |
| 41 | 1 | 280 | 7.5 | 241 | 0.61 | 0.41 | 0.54 | 1.54 | 1.82 | 1.56 |
| 42 | 0[a] | 231<br>260 | 6.35<br>9.15 | 220<br>242 | — | — | 0.77 | — | 0.15 | 0.01 |

Neutral spectral data

| No. | pH | $\lambda_{max}$ | $\varepsilon_{max}$ ($\times 10^{-3}$) | $\lambda_{min}$ | 230 | 240 | 250 | 270 | 280 | 290 |
|---|---|---|---|---|---|---|---|---|---|---|
| 40 | — | — | — | — | — | — | — | — | — | — |
| 41 | 7 | 279 | 7.7 | 241 | 0.68 | 0.43 | 0.57 | 1.49 | 1.71 | 1.36 |
| 42 | 6 | 267 | 10.25 | 239 | — | — | 0.57 | — | 0.61 | 0.07 |

Alkaline spectral data

| No. | pH | $\lambda_{max}$ | $\varepsilon_{max}$ ($\times 10^{-3}$) | $\lambda_{min}$ | 230 | 240 | 250 | 270 | 280 | 290 |
|---|---|---|---|---|---|---|---|---|---|---|
| 40 | 13[e] | 230 | 8.2 | — | — | — | — | — | — | — |
| 41 | 12 | 286 | 6.0 | 244 | 1.36 | 0.80 | 0.80 | 1.38 | 1.71 | 1.72 |
| 42 | 10 | 240<br>277 | 8.9<br>9.3 | 222<br>257 | — | — | 1.29 | — | 1.71 | 0.92 |

(Spectral ratio columns 230–290 in each group are headed "Spectral ratios".)

REFERENCES

| No. | Origin and synthesis | $[\alpha]_D^t$ | pK | Spectral data | $R_f$ |
|---|---|---|---|---|---|
| 40 | C: *98, 85, 99–102* | — | — | *103*[b] | — |
| 41 | C: *104–106* | — | *66, 105* | *66*[b], *104*[b] | *69* |
| 42 | C: *231* | — | *22, 248, 170*, 53*, 55*, 253** | *232*[b], *205* | *258* |

| No. | Compound | Abbreviation | Structure | Formula (mol wt) | Melting point °C | $[\alpha]^t_D$ | pK Basic | pK Acidic |
|---|---|---|---|---|---|---|---|---|
| 43 | **7-Methylxanthine** | 7MeXan | | $C_6H_6N_4O_2$ (166.14) | ~380° (dec) | — | — | 8.4, ~13 |
| 44 | **Zeatin** (6-(*trans*-4-Hydroxy-3-methylbut-2-enylamino)purine, $N^6$-(*trans*-4-Hydroxy-3-methylbut-2-enyl)adenine) | Zea | | $C_{10}H_{13}N_5O$ (219.24) | 207–208° | — | — | — |

| No. | Acidic spectral data | | | | | | | | | |
|---|---|---|---|---|---|---|---|---|---|---|
| | pH | $\lambda_{max}$ | $\varepsilon_{max}$ ($\times 10^{-3}$) | $\lambda_{min}$ | Spectral ratios 230 | 240 | 250 | 270 | 280 | 290 |
| 43 | 2 | 268 | 9.3 | 241 | — | 0.38 | — | — | 0.75 | — |
| 44 | 1 | 207<br>275 | 14.5<br>14.65 | 235 | — | — | — | — | — | — |

| No. | Neutral spectral data | | | | | | | | | |
|---|---|---|---|---|---|---|---|---|---|---|
| | pH | $\lambda_{max}$ | $\varepsilon_{max}$ ($\times 10^{-3}$) | $\lambda_{min}$ | Spectral ratios 230 | 240 | 250 | 270 | 280 | 290 |
| 43 | 6 | 269 | 10.0 | 240 | — | — | — | — | — | — |
| 44 | 7 | 212<br>270 | 17.1<br>16.2 | 233 | — | — | — | — | — | — |

| No. | Alkaline spectral data | | | | | | | | | |
|---|---|---|---|---|---|---|---|---|---|---|
| | pH | $\lambda_{max}$ | $\varepsilon_{max}$ ($\times 10^{-3}$) | $\lambda_{min}$ | Spectral ratios 230 | 240 | 250 | 270 | 280 | 290 |
| 43 | 14 | 284 | 9.4 | 257 | — | — | 1.11 | — | 2.39 | 2.27 |
| | 12 | 290* | 9.1* | 255 | — | — | — | — | — | — |
| 44 | 13 | 220<br>276 | 15.9<br>14.65 | 242 | — | — | — | — | — | — |

REFERENCES

| No. | Origin and synthesis | $[\alpha]^t_D$ | pK | Spectral data | $R_f$ |
|---|---|---|---|---|---|
| 43 | C: *45* | — | *53–56* | *53*[b], *41*[c,d], *27*[e], *54** | *10, 27* |
| 44 | C: *227* N: *228* | — | — | *227* | *32, 227* |

| No. | Compound | Abbreviation | Symbol | Structure | Formula (mol wt) | Melting point °C | $[\alpha]_D^t$ | pK Basic | pK Acidic |
|---|---|---|---|---|---|---|---|---|---|
| **RIBONUCLEOSIDES** | | | | | | | | | |
| 45 | **Adenosine** | Ado | A | 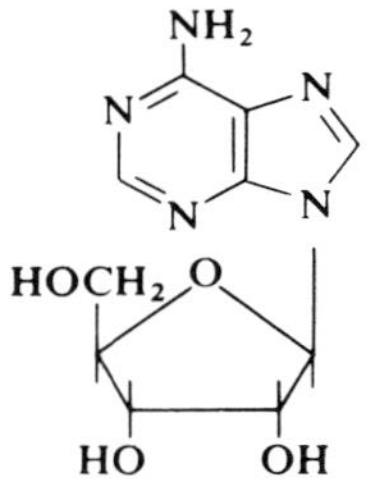 | $C_{10}H_{13}N_5O_4$ (267.24) | 234–235° | $-61.7^{11}$ (0.706, W) | 3.5* | 12.5 |
| 46 | **1-Methyladenosine** | 1MeAdo | $m^1A$ | 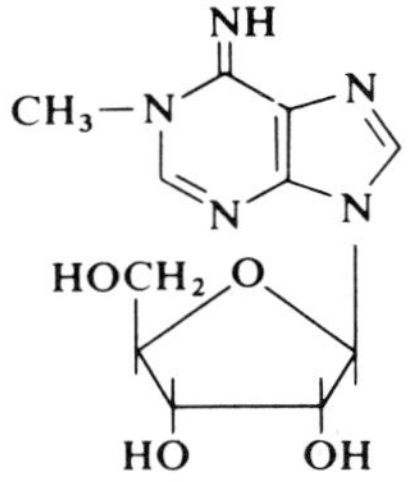 | $C_{11}H_{15}N_5O_4$ (281.27) | 214–217° (dec) | $-59^{26}$ (2.0, W) | ~7.6[l] | — |

Acidic spectral data

| No. | pH | $\lambda_{max}$ | $\varepsilon_{max}$ ($\times 10^{-3}$) | $\lambda_{min}$ | Spectral ratios 230 | 240 | 250 | 270 | 280 | 290 |
|---|---|---|---|---|---|---|---|---|---|---|
| 45 | 1 | 257 | 14.6 | 230 | — | — | 0.84 | — | 0.22 | 0.03 |
| 46 | 1 | 257 | 13.7 | 231 | 0.87 | 0.62 | 0.83 | 0.69 | 0.26 | 0.07 |

Neutral spectral data

| No. | pH | $\lambda_{max}$ | $\varepsilon_{max}$ ($\times 10^{-3}$) | $\lambda_{min}$ | Spectral ratios 230 | 240 | 250 | 270 | 280 | 290 |
|---|---|---|---|---|---|---|---|---|---|---|
| 45 | 6 | 260 | 14.9 | 227 | — | — | 0.78 | — | 0.14 | 0.00 |
| 46 | W | 257* | 14.6 | — | — | — | — | — | — | — |

Alkaline spectral data

| No. | pH | $\lambda_{max}$ | $\varepsilon_{max}$ ($\times 10^{-3}$) | $\lambda_{min}$ | Spectral ratios 230 | 240 | 250 | 270 | 280 | 290 |
|---|---|---|---|---|---|---|---|---|---|---|
| 45 | 11 | 259 | 15.4 | 227 | — | — | 0.79 | — | 0.15 | — |
| 46 | 11[o] | 257 | 14.6 | 223 | — | — | — | — | — | — |

REFERENCES

| No. | Origin and synthesis | $[\alpha]_D^t$ | pK | Spectral data | $R_f$ |
|---|---|---|---|---|---|
| 45 | C: *235, 237* | *235* | *250, 231*, 51, 220, 170* | *232*[b], *212*[c,d,e], *183*[b], *205* | *258, 110, 18* |
| 46 | C: *10, 30* | *10* | — | *10, 12*[b], *109*[e] | *10, 12* |

| No. | Compound | Abbreviation | Symbol | Structure | Formula (mol wt) | Melting point °C | $[\alpha]^t_D$ | pK Basic | pK Acidic |
|---|---|---|---|---|---|---|---|---|---|
| 47 | **1-($\Delta^2$-Isopentenyl)adenosine** (1-($\gamma\gamma$-dimethylallyl)adenosine) | 1iPeAdo | $i^1A$ | 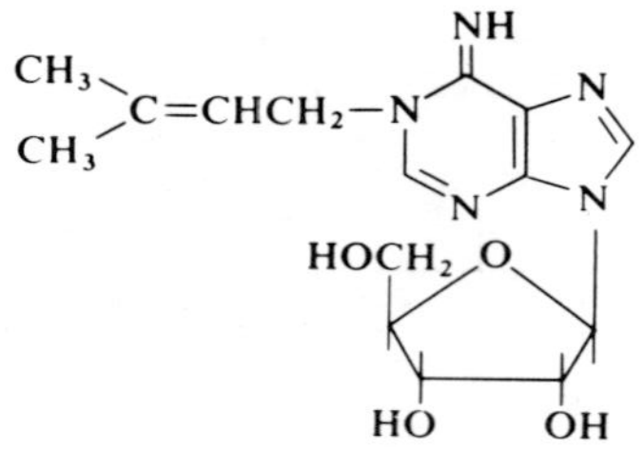  | $C_{15}H_{21}N_5O_4$ (335.36) | — | — | 8.5 | — |
| 48 | **1-Methyl-$N^6$-methyladenosine** (1-Methyl-6-methylamino-9-β-D-ribofuranosyl purine) | 1,6-$Me_2$Ado | $m_2^{1,6}A$<br>$m^1m^6A$ | 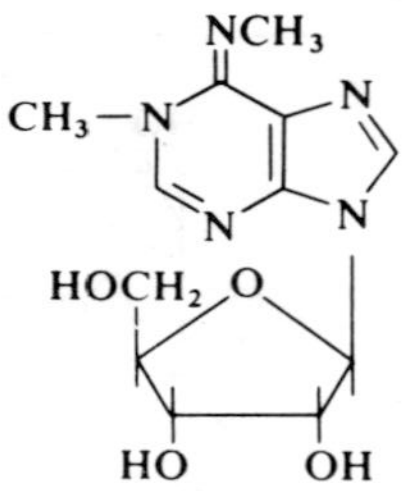  | $C_{12}H_{17}N_5O_4$ (295.30) | 206° | — | — | — |

| No. | Acidic spectral data | | | | | | | | | | Neutral spectral data | | | | | | | | | | Alkaline spectral data | | | | | | | | | |
|---|---|---|---|---|---|---|---|---|---|---|---|---|---|---|---|---|---|---|---|---|---|---|---|---|---|---|---|---|---|---|
| | pH | $\lambda_{max}$ | $\varepsilon_{max}$ ($\times 10^{-3}$) | $\lambda_{min}$ | Spectral ratios 230 | 240 | 250 | 270 | 280 | 290 | pH | $\lambda_{max}$ | $\varepsilon_{max}$ ($\times 10^{-3}$) | $\lambda_{min}$ | Spectral ratios 230 | 240 | 250 | 270 | 280 | 290 | pH | $\lambda_{max}$ | $\varepsilon_{max}$ ($\times 10^{-3}$) | $\lambda_{min}$ | Spectral ratios 230 | 240 | 250 | 270 | 280 | 290 |
| 47 | — | — | — | — | — | — | — | — | — | — | — | — | — | — | — | — | — | — | — | — | — | — | — | — | — | — | — | — | — | — |
| 48 | 1 | 261 | 14.2 | 234 | — | — | — | — | — | — | — | — | — | — | — | — | — | — | — | — | 14 | 262 | 14.9 | 234 | — | — | — | — | — | — |

REFERENCES

| No. | Origin and Synthesis | $[\alpha]^t_D$ | pK | Spectral data | $R_f$ |
|---|---|---|---|---|---|
| 47 | C: *14* | — | *15* | — | *32* |
| 48 | C: *24, 12* | — | — | *24, 12* | *24, 12* |

| No. | Compound | Abbreviation | Symbol | Structure | Formula (mol wt) | Melting point °C | $[\alpha]^{t}_{D}$ | pK Basic | pK Acidic |
|---|---|---|---|---|---|---|---|---|---|
| 49 | **2-Methyladenosine** | 2MeAdo | $m^2A$ | 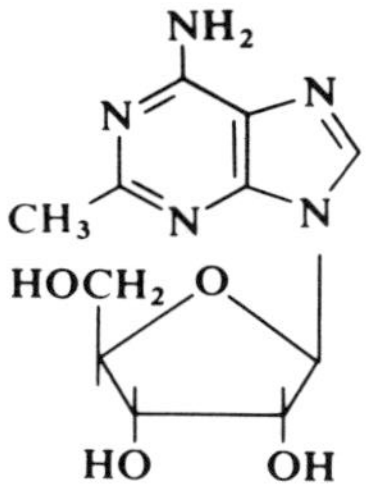 | $C_{11}H_{15}N_5O_4$ (281.27) | > 200° (dec) (picrate) | — | — | — |
| 50 | **2-Hydroxyadenosine** (Crotonoside, isoguanosine) | 2(OH)Ado *iso*Guo | *iso*G $o^2A$ | 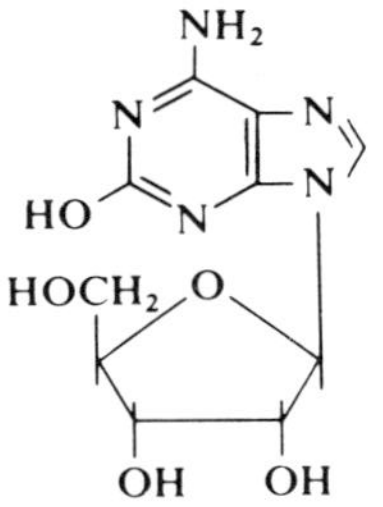 | $C_{10}H_{13}N_5O_5$ (283.25) | 237–252° (dec) | $-71^{26}$ (1.06,0.1 *N* NaOH) | — | — |

**Acidic spectral data**

| No. | pH | $\lambda_{max}$ | $\varepsilon_{max}$ ($\times 10^{-3}$) | $\lambda_{min}$ | 230 | 240 | 250 | 270 | 280 | 290 |
|---|---|---|---|---|---|---|---|---|---|---|
| 49 | 1 | 258 | 12.5[m] | 230 | 0.22 | 0.44 | 0.84 | 0.86 | 0.40 | 0.05 |
| 50 | 1.2 | 235<br>283* | 6.14<br>12.7* | — | — | — | — | — | — | — |

**Neutral spectral data**

| No. | pH | $\lambda_{max}$ | $\varepsilon_{max}$ ($\times 10^{-3}$) | $\lambda_{min}$ | 230 | 240 | 250 | 270 | 280 | 290 |
|---|---|---|---|---|---|---|---|---|---|---|
| 49 | — | — | — | — | — | — | — | — | — | — |
| 50 | W | 247*<br>293* | 8.9<br>11.1* | — | — | — | — | — | — | — |

**Alkaline spectral data**

| No. | pH | $\lambda_{max}$ | $\varepsilon_{max}$ ($\times 10^{-3}$) | $\lambda_{min}$ | 230 | 240 | 250 | 270 | 280 | 290 |
|---|---|---|---|---|---|---|---|---|---|---|
| 49 | 13 | 263 | 14.7[m]<br>13.5[n] | 230 | 0.24 | 0.41 | 0.75 | 0.84 | 0.17 | 0.01 |
| 50 | 12.8 | 251*<br>285* | 6.9*<br>10.55* | — | — | — | — | — | — | — |

REFERENCES

| No. | Origin and synthesis | $[\alpha]^{t}_{D}$ | pK | Spectral data | $R_f$ |
|---|---|---|---|---|---|
| 49 | E: *18* C: *111* | — | — | *18*[b], *20*[d] | *18* |
| 50 | C: *269*, N: *268* | *269* | — | *269*[b], *268**[b] | *269* |

| No. | Compound | Abbreviation | Symbol | Structure | Formula (mol wt) | Melting point °C | $[\alpha]_D^t$ | pK Basic | pK Acidic |
|---|---|---|---|---|---|---|---|---|---|
| 51 | **$N^6$-Methyladenosine** (6-Methylaminopurine ribonucleoside) | 6MeAdo | $m^6A$ | 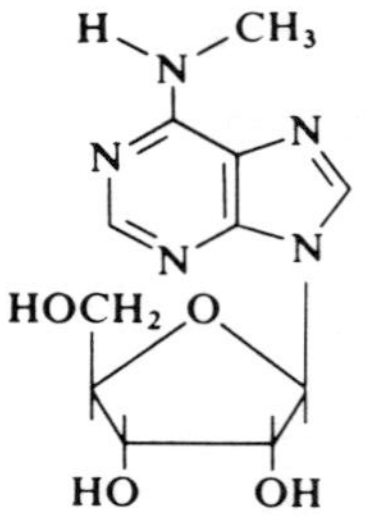 | $C_{11}H_{15}N_5O_4$ (281.27) | 219–221° | $-54^{26}$ (0.6, W) | 4.0 | — |
| 52 | **$N^6$-($\Delta^2$-Isopentenyl)adenosine** ($N^6$-($\gamma\gamma$-Dimethylallyl)adenosine, 6-($\gamma\gamma$-Dimethylallylamino)-9-$\beta$-D-ribofuranosyl purine) | 6iPeAdo | $i^6A$ | 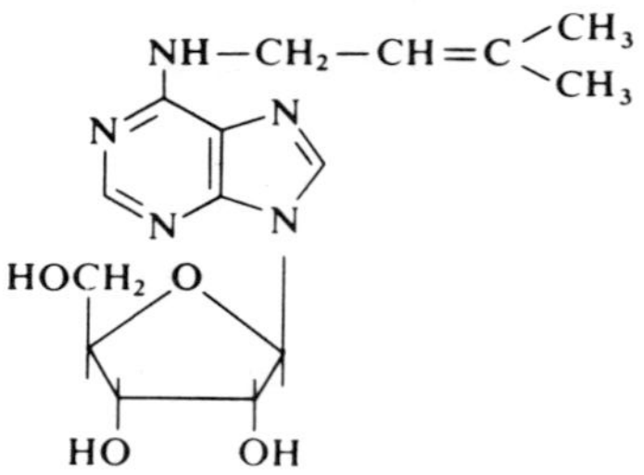 | $C_{15}H_{21}N_5O_4$ (335.36) | 145–147° | $-97^{25}_{546}$ (0.07, E) | 3.8 | — |

Acidic spectral data

| No. | pH | $\lambda_{max}$ | $\varepsilon_{max}$ ($\times 10^{-3}$) | $\lambda_{min}$ | Spectral ratios 230 | 240 | 250 | 270 | 280 | 290 |
|---|---|---|---|---|---|---|---|---|---|---|
| 51 | 1 | 262 | 16.6 | 231 | 0.20 | 0.31 | 0.66 | 0.88 | 0.41 | 0.13 |
| 52 | 1 | 265 | 20.4 | 232 | — | — | 0.57 | 1.05 | 0.69* | 0.18* |

Neutral spectral data

| No. | pH | $\lambda_{max}$ | $\varepsilon_{max}$ ($\times 10^{-3}$) | $\lambda_{min}$ | Spectral ratios 230 | 240 | 250 | 270 | 280 | 290 |
|---|---|---|---|---|---|---|---|---|---|---|
| 51 | 7 | 266 | 15.9 | 229 | — | — | — | — | — | — |
| 52 | 7 | 269 | 20.0 | 234* | — | — | 0.52 | 1.17 | 0.94* | 0.36* |

Alkaline spectral data

| No. | pH | $\lambda_{max}$ | $\varepsilon_{max}$ ($\times 10^{-3}$) | $\lambda_{min}$ | Spectral ratios 230 | 240 | 250 | 270 | 280 | 290 |
|---|---|---|---|---|---|---|---|---|---|---|
| 51 | 13 | 266 | 15.9* | 232* | 0.17 | 0.22 | 0.57 | 1.09 | 0.68 | 0.24 |
| 52 | 12 | 269 | 20.0 | 234 | — | — | 0.52* | 1.17 | 0.94 | 0.36 |

REFERENCES

| No. | Origin and synthesis | $[\alpha]_D^t$ | pK | Spectral data | $R_f$ |
|---|---|---|---|---|---|
| 51 | E: *18* C: *10, 112, 166, 30* | *112* | *15* | *18*[b], *112*[c,d], *10**[c], *8, 12* | *10, 12, 18, 110* |
| 52 | C: *31, 32, 14* R: *31, 116* | *31* | *15* | *32*[b], *116**[b], *14, 31, 265**[b] | *32, 116, 117* |

| No. | Compound | Abbreviation | Symbol | Structure | Formula (mol wt) | Melting point °C | $[\alpha]_D^t$ | pK Basic | pK Acidic |
|---|---|---|---|---|---|---|---|---|---|
| 53 | **$N^6$-(*cis*-4-Hydroxy-3-methylbut-2-enyl) adenosine** | | | 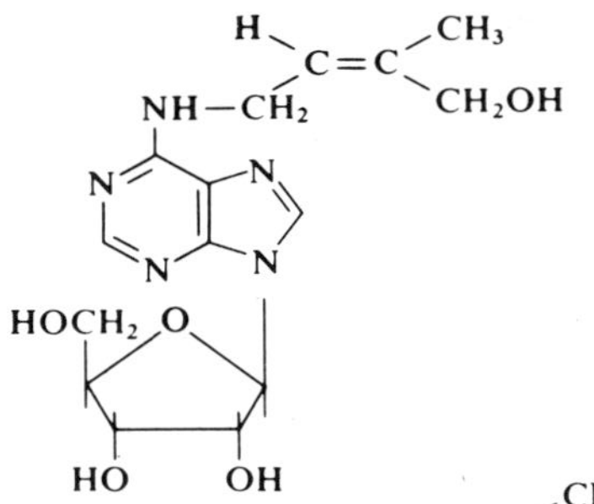 | $C_{15}H_{21}N_5O_5$ (351.36) | 206° | $-98_{546}^{27}$ (0.02, W) | — | — |
| 54 | **$N^6$-($\Delta^2$-Isopentenyl)-2-methylthioadenosine** [6(3-Methyl-2-butenyl-amino)-2-methylthio-9β-D-ribofuranosyl purine] | 2MeS6iPeAdo | $ms^{2}i^{6}A$ | 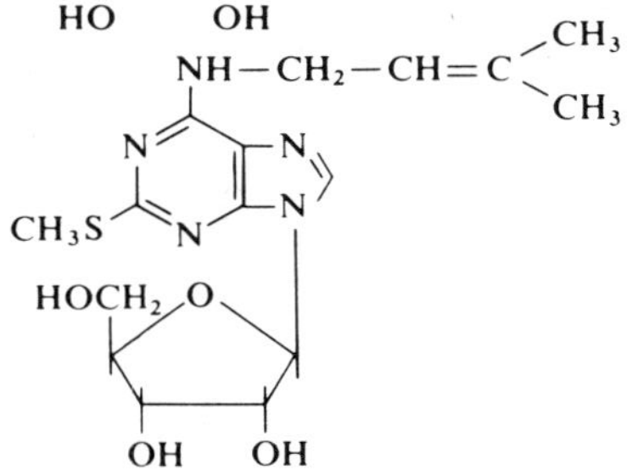 | $C_{16}H_{23}N_5O_4S$ (381.147) | 194–195° | — | — | — |

| No. | Acidic spectral data pH | $\lambda_{max}$ | $\varepsilon_{max}$ ($\times 10^{-3}$) | $\lambda_{min}$ | Spectral ratios 230 | 240 | 250 | 270 | 280 | 290 |
|---|---|---|---|---|---|---|---|---|---|---|
| 53 | 1 | 265 | 20.4 | — | 0.53 | 0.42 | 0.68 | 1.01 | 0.65 | 0.21 |
| 54 | 1<br>(E) | 246<br>286 | 18.6<br>16.1 | 265 | 0.83 | 1.12 | 1.27 | 1.00 | 1.12 | 1.13 |

| No. | Neutral spectral data pH | $\lambda_{max}$ | $\varepsilon_{max}$ ($\times 10^{-3}$) | $\lambda_{min}$ | Spectral ratios 230 | 240 | 250 | 270 | 280 | 290 |
|---|---|---|---|---|---|---|---|---|---|---|
| 53 | 7 | 268 | 20.0 | — | 0.36 | 0.27 | 0.53 | 1.16 | 0.79 | 0.27 |
| 54 | E | 244<br>283 | 25.3<br>18.0 | 259 | 1.39 | 2.39 | 1.78 | 1.39 | 1.85 | 1.67 |

| No. | Alkaline spectral data pH | $\lambda_{max}$ | $\varepsilon_{max}$ ($\times 10^{-3}$) | $\lambda_{min}$ | Spectral ratios 230 | 240 | 250 | 270 | 280 | 290 |
|---|---|---|---|---|---|---|---|---|---|---|
| 53 | 12 | 268 | 20.0 | — | 0.43 | 0.29 | 0.50 | 1.23 | 0.85 | 0.29 |
| 54 | 10<br>(E) | 243<br>283 | 24.9<br>18.0 | 258 | 1.39 | 2.40 | 1.79 | 1.40 | 1.85 | 1.68 |

REFERENCES

| No. | Origin and synthesis | $[\alpha]_D^t$ | pK | Spectral data | $R_f$ |
|---|---|---|---|---|---|
| 53 | R: *117*[a] | *117* | — | *117* | *117* |
| 54 | C: *274*, R: *274, 288* | — | — | *274, 288*[b] | *274, 288* |

| No. | Compound | Abbreviation | Symbol | Structure | Formula (mol wt) | Melting point °C | $[\alpha]^t_D$ | pK Basic | pK Acidic |
|---|---|---|---|---|---|---|---|---|---|
| 55 | **N-(Nebularin-6-ylcarbamoyl)-threonine** (N-(9-β-D-Ribofuranosylpurin-6-ylcarbamoyl)threonine) | 6Thr(CO)Ado | | 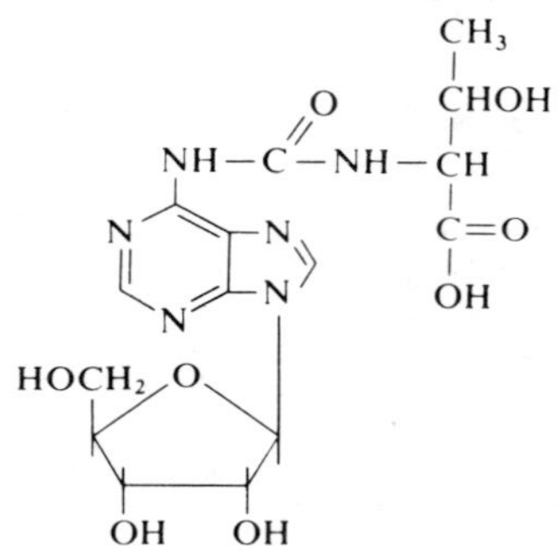 | $C_{15}H_{20}N_6O_8$ (412.36) | — | — | — | — |
| 56 | **$N^6$,$N^6$-Dimethyladenosine** (6-Dimethylaminopurine ribonucleoside) | 6Me$_2$Ado | $m^6_2A$ | 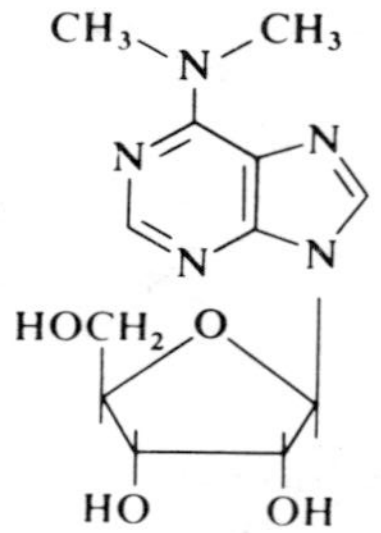 | $C_{12}H_{17}N_5O_4$ (295.30) | 183–184° | $-62.6^{25}$ (2.6, W) | 4.5 | — |

**Acidic spectral data**

| No. | pH | $\lambda_{max}$ | $\varepsilon_{max}$ ($\times 10^{-3}$) | $\lambda_{min}$ | Spectral ratios 230 | 240 | 250 | 270 | 280 | 290 |
|---|---|---|---|---|---|---|---|---|---|---|
| 55 | 1 | 276 | 23.4 | — | — | — | — | — | — | — |
| 56 | 1 | 268 | 18.4 | 234 | 0.24 | 0.27 | 0.60 | 1.20 | 0.94 | 0.42 |

**Neutral spectral data**

| No. | pH | $\lambda_{max}$ | $\varepsilon_{max}$ ($\times 10^{-3}$) | $\lambda_{min}$ | Spectral ratios 230 | 240 | 250 | 270 | 280 | 290 |
|---|---|---|---|---|---|---|---|---|---|---|
| 55 | 5 | 268 | 24.9 | — | — | — | — | — | — | — |
| 56 | 7 | 275 | 18.8 | 236 | — | — | — | — | — | — |

**Alkaline spectral data**

| No. | pH | $\lambda_{max}$ | $\varepsilon_{max}$ ($\times 10^{-3}$) | $\lambda_{min}$ | Spectral ratios 230 | 240 | 250 | 270 | 280 | 290 |
|---|---|---|---|---|---|---|---|---|---|---|
| 55 | 12 | 297 | 13.8 | — | — | — | — | — | — | — |
| 56 | 13 | 276 | 19.2 | 237 | 0.17 | 0.09 | 0.37 | 1.57 | 1.68 | 1.01 |

REFERENCES

| No. | Origin and synthesis | $[\alpha]^t_D$ | pK | Spectral data | $R_f$ |
|---|---|---|---|---|---|
| 55 | R: *4* | — | — | *6, 290* | — |
| 56 | E: *18* C: *113–115* | *113* | *15* | *18*[b], *114*[c,d], *113* | *18, 32, 117* |

| No. | Compound | Abbreviation | Symbol | Structure | Formula (mol wt) | Melting point °C | $[\alpha]_D^t$ | pK Basic | pK Acidic |
|---|---|---|---|---|---|---|---|---|---|
| 57 | **2′-*O*-Methyladenosine** | 2′MeAdo | Am | 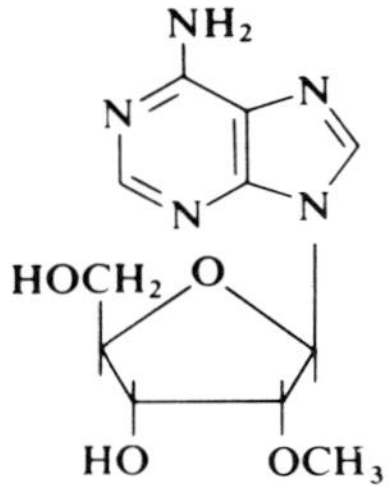 | $C_{11}H_{15}N_5O_4$ (281.27) | 201–202° | $-57.9^{23}$ (1.0, W) | — | — |
| 58 | **2′(3′)-*O*-Ribosyladenosine** | *O*RibAdo | | 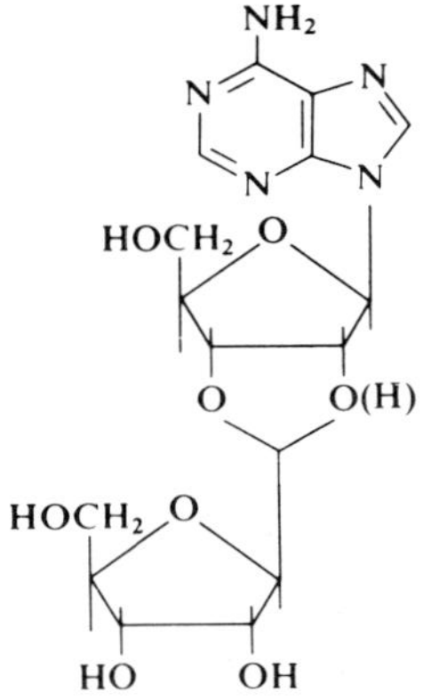 | $C_{15}H_{21}N_5O_8$ (399.36) | — | — | — | — |

| No. | Acidic spectral data | | | | | | | | | | Neutral spectral data | | | | | | | | | | Alkaline spectral data | | | | | | | | | |
|---|---|---|---|---|---|---|---|---|---|---|---|---|---|---|---|---|---|---|---|---|---|---|---|---|---|---|---|---|---|---|
| | pH | $\lambda_{max}$ | $\varepsilon_{max}$ ($\times 10^{-3}$) | $\lambda_{min}$ | Spectral ratios 230 | 240 | 250 | 270 | 280 | 290 | pH | $\lambda_{max}$ | $\varepsilon_{max}$ ($\times 10^{-3}$) | $\lambda_{min}$ | Spectral ratios 230 | 240 | 250 | 270 | 280 | 290 | pH | $\lambda_{max}$ | $\varepsilon_{max}$ ($\times 10^{-3}$) | $\lambda_{min}$ | Spectral ratios 230 | 240 | 250 | 270 | 280 | 290 |
| 57 | 1 | 257 | 13.8 | — | — | — | — | — | — | — | M | 259 | 13.7 | — | — | — | — | — | — | — | 11 | 259 | 13.9 | — | — | — | — | — | — | — |
| 58 | 1.5 | 257 | — | 230 | 0.36 | 0.47 | 0.84 | 0.72 | 0.32 | 0.12 | 7 | 259 | — | 227 | 0.35 | 0.47 | 0.82 | 0.67 | 0.21 | 0.04 | 11 | 259 | — | 227 | 0.34 | 0.47 | 0.79 | 0.66 | 0.20 | 0.05 |

## REFERENCES

| No. | Origin and synthesis | $[\alpha]_D^t$ | pK | Spectral data | $R_f$ |
|---|---|---|---|---|---|
| 57 | C: *107*, *108*[a] | *108* | — | *107* | *107*, *109*, *110* |
| 58 | C: *286* R: *110* | — | — | *110* | *110* |

| No. | Compound / Abbreviation | Symbol | Structure | Formula (mol wt) | Melting point °C | $[\alpha]^t_D$ | pK Basic | pK Acidic |
|---|---|---|---|---|---|---|---|---|
| 59 | **7-α-D-Ribofuranosyladenine** (Pseudovitamin $B_{12}$ nucleoside) 7αDAdo | | 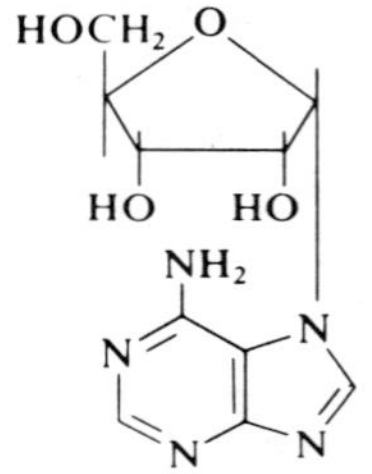 | $C_{10}H_{13}N_5O_4$ (267.24) | 220–222° | $0^{25}$ (0.4, W) | 3.9 | — |
| 60 | **2-Methyl-7-α-ribofuranosyladenine** (Factor A nucleoside) 2Me7αDAdo | | 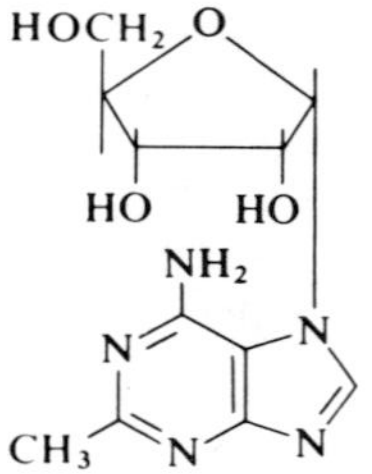 | $C_{11}H_{15}N_5O_4$ (281.27) | 219–220° | — | 4.8 | — |

| No. | Acidic spectral data | | | | | | | | | |
|---|---|---|---|---|---|---|---|---|---|---|
| | pH | $\lambda_{max}$ | $\varepsilon_{max}$ (× $10^{-3}$) | $\lambda_{min}$ | Spectral ratios 230 | 240 | 250 | 270 | 280 | 290 |
| 59 | 1 | 273 | 13.6 | 239 | 0.86 | 0.50 | 0.66 | 1.39 | 1.19 | 0.33 |
| 60 | 1 | 273 | 13.2 | 238 | 0.95 | 0.55 | 0.70 | 1.42 | 1.24 | 0.35 |

| No. | Neutral spectral data | | | | | | | | | |
|---|---|---|---|---|---|---|---|---|---|---|
| | pH | $\lambda_{max}$ | $\varepsilon_{max}$ (× $10^{-3}$) | $\lambda_{min}$ | Spectral ratios 230 | 240 | 250 | 270 | 280 | 290 |
| 59 | W | 271 | 9.8 | 233 | 0.59 | 0.65 | 0.78 | 1.26 | 0.97 | 0.25 |
| 60 | 7 | 276 | 9.1 | 233 | 0.85 | 0.90 | 0.89 | 1.40 | 1.35 | 0.51 |

| No. | Alkaline spectral data | | | | | | | | | |
|---|---|---|---|---|---|---|---|---|---|---|
| | pH | $\lambda_{max}$ | $\varepsilon_{max}$ (× $10^{-3}$) | $\lambda_{min}$ | Spectral ratios 230 | 240 | 250 | 270 | 280 | 290 |
| 59 | 13 | 271 | 9.8 | — | — | — | — | — | — | — |
| 60 | 12 | 244<br>276 | 5.7<br>9.1 | — | — | — | — | — | — | — |

## REFERENCES

| No. | Origin and synthesis | $[\alpha]^t_D$ | pK | Spectral data | $R_f$ |
|---|---|---|---|---|---|

| No. | Compound | Abbreviation | Symbol | Structure | Formula (mol wt) | Melting point °C | $[\alpha]_D^t$ | pK Basic | pK Acidic |
|---|---|---|---|---|---|---|---|---|---|
| 61 | **Cytidine** | Cyd | C | 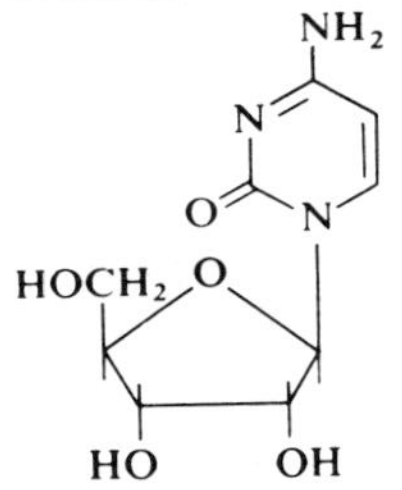 | $C_9H_{13}N_3O_5$ (243.22) | 224–225° (dec) (sulphate) | $+34.2^{16}$ (2.0, W) | 4.15 | 12.5 |
| 62 | **2-Thiocytidine** | 2Syd | $s^2C$ | 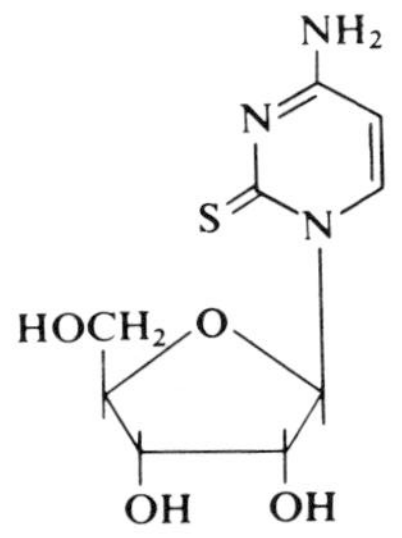 | $C_9H_{13}N_3O_4S$ (259.28) | 208–209° | $+64.2^{25}$ (1.8, W) | — | — |

| No. | Acidic spectral data | | | | | | | | | |
|---|---|---|---|---|---|---|---|---|---|---|
| | pH | $\lambda_{max}$ | $\varepsilon_{max}$ ($\times 10^{-3}$) | $\lambda_{min}$ | Spectral ratios 230 | 240 | 250 | 270 | 280 | 290 |
| 61 | 1 | 280 | 13.4 | 242 | — | — | 0.45 | — | 2.10 | 1.55 |
| 62 | 0 | 229<br>276 | 17.0<br>17.4 | 213<br>250 | 1.44* | 1.14 | 0.86 | 1.33* | 1.30* | 1.11 |

| No. | Neutral spectral data | | | | | | | | | |
|---|---|---|---|---|---|---|---|---|---|---|
| | pH | $\lambda_{max}$ | $\varepsilon_{max}$ ($\times 10^{-3}$) | $\lambda_{min}$ | Spectral ratios 230 | 240 | 250 | 270 | 280 | 290 |
| 61 | 7 | 229.5<br>271 | 8.3<br>9.1 | 226<br>250 | — | — | 0.86 | — | 0.93 | 0.28 |
| 62 | 7 | 249 | 22.3 | — | 0.59 | 0.95 | 1.08 | 0.78 | 0.50* | 0.42 |

| No. | Alkaline spectral data | | | | | | | | | |
|---|---|---|---|---|---|---|---|---|---|---|
| | pH | $\lambda_{max}$ | $\varepsilon_{max}$ ($\times 10^{-3}$) | $\lambda_{min}$ | Spectral ratios 230 | 240 | 250 | 270 | 280 | 290 |
| 61 | 13[q]<br>14 | 272.5<br>273 | 9.15<br>9.2 | 251<br>252 | — | — | — | — | — | — |
| 62 | 13 | 252 | — | 229 | 0.68 | 0.70 | 1.04 | 0.91 | 0.72 | 0.39 |

REFERENCES

| No. | Origin and synthesis | $[\alpha]_D^t$ | pK | Spectral data | $R_f$ |
|---|---|---|---|---|---|
| 61 | R: *236, 181* C: *251, 128, 252* | *236* | *163, 231* | *163*[b], *232*[b,e], *181*[b], *212*, *183*[b], *205* | *258, 110* |
| 62 | C: *272, 271*[a], *285* R: *270* | *285* | — | *271**, *270*[e] | *273* |

| No. | Compound | Abbreviation | Symbol | Structure | Formula (mol wt) | Melting point °C | $[\alpha]^t_D$ | pK Basic | pK Acidic |
|---|---|---|---|---|---|---|---|---|---|
| 63 | **3-Methylcytidine** | 3MeCyd | $m^3C$ | 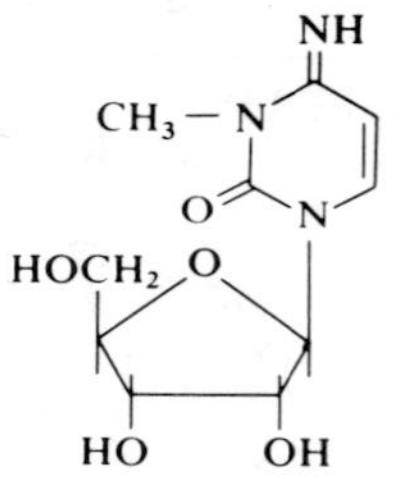 | $C_{10}H_{15}N_3O_5$ (257.24) | 193–194° (methosulphate) | — | 8.7 | >12 |
| 64 | **$N^4$-Acetylcytidine** (1-β-D-Ribofuranosyl-4-acetylamino-2-pyrimidinone) | 4AcCyd | $ac^4C$ | 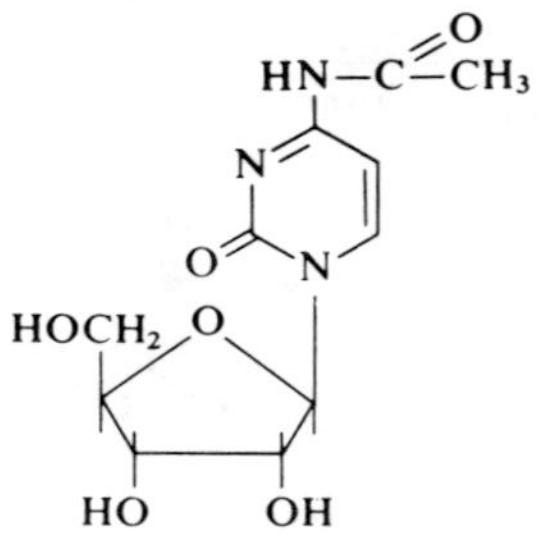 | $C_{11}H_{15}N_3O_6$ (285.25) | 208–209° | $+60.1^{23}$ (1.0,W) | <1.5 | — |

Acidic spectral data

| No. | pH | $\lambda_{max}$ | $\varepsilon_{max}$ ($\times 10^{-3}$) | $\lambda_{min}$ | 230 | 240 | 250 | 270 | 280 | 290 |
|---|---|---|---|---|---|---|---|---|---|---|
| 63 | 4 | 278* | 11.8 | 243* | 0.75 | 0.33 | 0.47 | 1.67 | 1.89* | 1.21 |
| 64 | $1^{o,q}$ | 241<br>308 | 12.4<br>13.8 | 267 | 2.4 | 2.2* | 1.7* | 0.78 | 1.4 | 2.4* |

Neutral spectral data

| No. | pH | $\lambda_{max}$ | $\varepsilon_{max}$ ($\times 10^{-3}$) | $\lambda_{min}$ | 230 | 240 | 250 | 270 | 280 | 290 |
|---|---|---|---|---|---|---|---|---|---|---|
| 63 | — | — | — | — | — | — | — | — | — | — |
| 64 | 7 | 247*<br>297* | 15.2<br>8.7 | 226<br>271* | 0.77 | 1.65 | 1.80 | 0.46 | 0.61* | 0.95 |

Alkaline spectral data

| No. | pH | $\lambda_{max}$ | $\varepsilon_{max}$ ($\times 10^{-3}$) | $\lambda_{min}$ | 230 | 240 | 250 | 270 | 280 | 290 |
|---|---|---|---|---|---|---|---|---|---|---|
| 63 | $12^o$ | 266 | 9.0 | 243* | 1.12 | 0.68 | 0.73 | 1.01 | 0.69* | 0.26 |
| 64 | $13^o$ | 302 | — | 241 | 0.88 | 0.97 | 0.72 | 1.34 | 1.57 | 1.84 |

(Columns 230–290 are spectral ratios.)

REFERENCES

| No. | Origin and synthesis | $[\alpha]^t_D$ | pK | Spectral data | $R_f$ |
|---|---|---|---|---|---|
| 63 | C: *57, 30, 125* | — | *58, 57, 59** | *57, 58*[b]*, 59*, 125*, 126* | *57, 59, 110, 125* |
| 64 | C: *127, 244, 245* R: *116* | *127* | *127* | *127, 116**[b,c,e]*, 244*, 245, 265**[b] | *116* |

| No. | Compound | Abbreviation | Symbol | Structure | Formula (mol wt) | Melting point °C | $[\alpha]^t_D$ | pK Basic | pK Acidic |
|---|---|---|---|---|---|---|---|---|---|
| 65 | **$N^4$-Methylcytidine** (1-β-D-Ribofuranosyl-4-methylamino-2-pyrimidinone) | 4MeCyd | $m^4C$ | 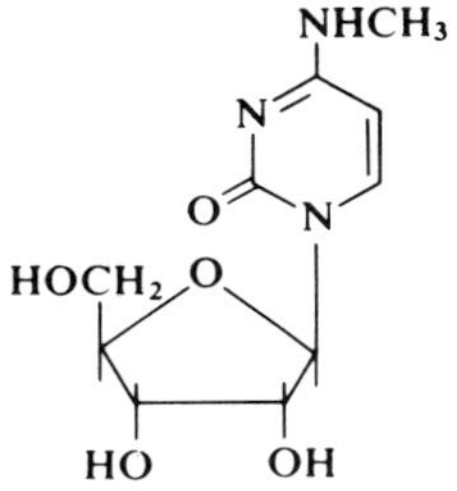 | $C_{10}H_{15}N_3O_5$ (257.24) | 237° (dec) | — | 3.9 | — |
| 66 | **$N^4$-Methyl-2-thiocytidine** | 4Me2Syd or 4Me2SCyd | $m^4s^2C$ | 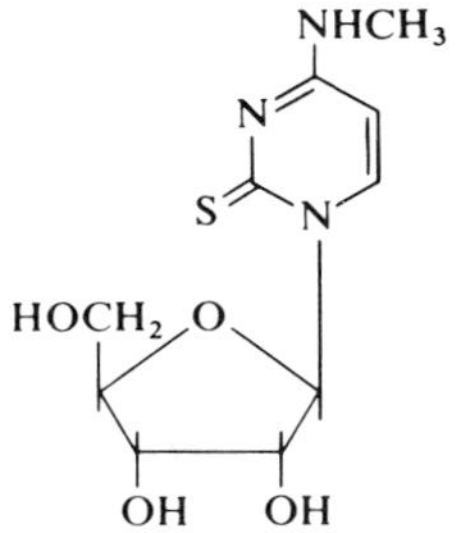 | $C_{10}H_{15}N_3O_4S$ (273.32) | 189–190° | — | — | — |

| No. | Acidic pH | Acidic $\lambda_{max}$ | Acidic $\varepsilon_{max}$ $(\times 10^{-3})$ | Acidic $\lambda_{min}$ | 230 | 240 | 250 | 270 | 280 | 290 | Neutral pH | Neutral $\lambda_{max}$ | Neutral $\varepsilon_{max}$ $(\times 10^{-3})$ | Neutral $\lambda_{min}$ | 230 | 240 | 250 | 270 | 280 | 290 | Alkaline pH | Alkaline $\lambda_{max}$ | Alkaline $\varepsilon_{max}$ $(\times 10^{-3})$ | Alkaline $\lambda_{min}$ | 230 | 240 | 250 | 270 | 280 | 290 |
|---|---|---|---|---|---|---|---|---|---|---|---|---|---|---|---|---|---|---|---|---|---|---|---|---|---|---|---|---|---|---|
| 65 | 1 | 281 | 14.3 | 243 | 0.85 | 0.37 | 0.49 | 1.72 | 2.13 | 1.73 | 7 | 237<br>271 | 9.2<br>11.6 | 227<br>250 | 0.88 | 0.91 | 0.84 | 1.15 | 0.92 | 0.38 | 14 | 236<br>273 | 8.9<br>11.6 | 251 | 0.93 | 0.93 | 0.84 | 1.22 | 1.06 | 0.51 |
| 66 | 0 | 238<br>276 | 18.5<br>18.1 | 216<br>257 | 1.31 | 1.50 | 1.05 | 1.39 | 1.40 | 0.86 | W | 224<br>262 | 14.9<br>27.4 | 236 | 0.49 | 0.51 | 0.64 | 0.92 | 0.65 | 0.37 | — | — | — | — | — | — | — | — | — | — |

(Columns 230–290 under each of Acidic, Neutral and Alkaline spectral data are Spectral ratios.)

REFERENCES

| No. | Origin and synthesis | $[\alpha]^t_D$ | pK | Spectral data | $R_f$ |
|---|---|---|---|---|---|
| 65 | C: *59, 128* | — | *128, 59* | *59*[b], *128* | *59, 129, 130* |
| 66 | C: *271* | — | — | *271* | — |

| No. | Compound | Abbreviation | Symbol | Structure | Formula (mol wt) | Melting point °C | $[\alpha]^t_D$ | pK Basic | pK Acidic |
|---|---|---|---|---|---|---|---|---|---|
| 67 | **5-Methylcytidine** | 5MeCyd | $m^5C$ | 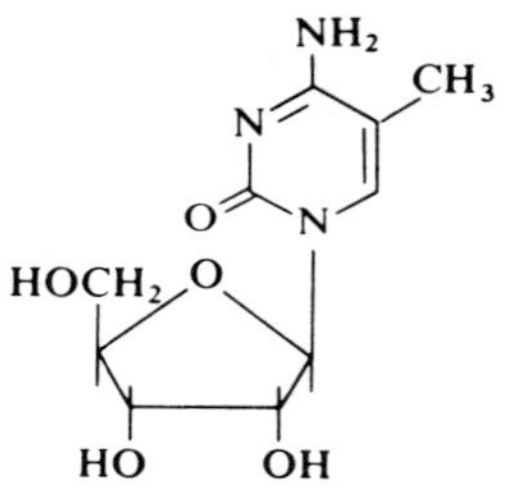 | $C_{10}H_{15}N_3O_5$ (257.24) | 210–211° (dec) | $-3^{23}$ (2.5, 1*N* NaOH) | 4.3 | >13 |
| 68 | **2′-*O*-Methylcytidine** | 2′MeCyd | Cm | 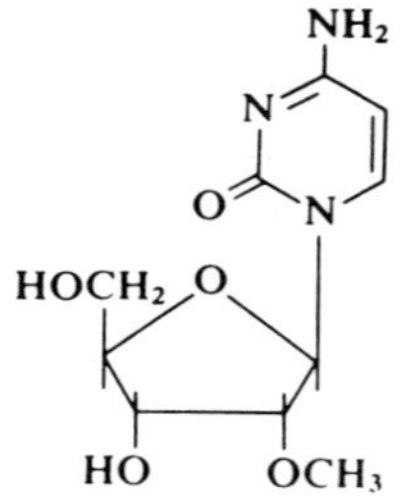 | $C_{10}H_{15}N_3O_5$ (257.24) | 220–222° (dec) (HCl salt) | $+54^{21}$ (1.1, W) | 4.2[p] | — |

Acidic spectral data

| No. | pH | $\lambda_{max}$ | $\varepsilon_{max}$ ($\times 10^{-3}$) | $\lambda_{min}$ | 230 | 240 | 250 | 270 | 280 | 290 |
|---|---|---|---|---|---|---|---|---|---|---|
| 67 | 0 | 287 | 12.6 | 245 | 1.72 | 0.42 | 0.33 | 2.24 | 3.59 | 3.96 |
| 68 | 1 | 281 | 12.9 | 241 | 0.62 | 0.03 | 0.43 | 1.75 | 2.13 | 1.61 |

Neutral spectral data

| No. | pH | $\lambda_{max}$ | $\varepsilon_{max}$ ($\times 10^{-3}$) | $\lambda_{min}$ | 230 | 240 | 250 | 270 | 280 | 290 |
|---|---|---|---|---|---|---|---|---|---|---|
| 67 | 7 | 277 | 8.9 | 255 | 1.53 | 1.34 | 1.03 | 1.33 | 1.45 | 0.96 |
| 68 | — | — | — | — | — | — | — | — | — | — |

Alkaline spectral data

| No. | pH | $\lambda_{max}$ | $\varepsilon_{max}$ ($\times 10^{-3}$) | $\lambda_{min}$ | 230 | 240 | 250 | 270 | 280 | 290 |
|---|---|---|---|---|---|---|---|---|---|---|
| 67 | 14[q] | 279 | 9.0 | 256 | 1.58 | 1.35 | 1.05 | 1.37 | 1.58 | 1.19 |
| 68 | 11 | 272 | 8.9 | 251 | 1.46 | 0.96 | 0.85 | 1.18 | 0.94 | 0.31 |

(Spectral ratios columns: 230–290.)

REFERENCES

| No. | Origin and synthesis | $[\alpha]^t_D$ | pK | Spectral data | $R_f$ |
|---|---|---|---|---|---|
| 67 | C: *128* | *128* | *128* | *128*[b], *132*[b] | *110, 132* |
| 68 | C: *124* | *124* | *124* | *124* | *109, 110, 124* |

| No. | Compound | Abbreviation | Symbol | Structure | Formula (mol wt) | Melting point °C | $[\alpha]_D^t$ | pK Basic | pK Acidic |
|---|---|---|---|---|---|---|---|---|---|
| 69 | **$N^4$,$O^{2'}$-Dimethylcytidine** | 2',4Me$_2$Cyd | m$^4$Cm | | $C_{11}H_{17}N_3O_5$ (271.27) | — | — | 3.9 | — |
| 70 | **2-Amino-4-hydroxy-5-*N*-methylformamido-6-ribosylaminopyrimidine**[r] (5-*N*-Methylformamido-6-ribosyl-amino-isocytosine) | | | | $C_{11}H_{17}N_5O_6$ (315.29) | None (dec > 180°) | $+32.5^{25}$ (1, W) | ~0.6 | 9.6 |

Acidic spectral data

| No. | pH | $\lambda_{max}$ | $\varepsilon_{max}$ ($\times 10^{-3}$) | $\lambda_{min}$ | 230 | 240 | 250 | 270 | 280 | 290 |
|---|---|---|---|---|---|---|---|---|---|---|
| 69 | 1 | 281 | — | 243 | 0.79 | 0.37 | 0.49 | 1.65 | 1.99 | 1.66 |
| 70 | 0.1[o]<br>1[oq] | 270<br>271 | 25.1[n]<br>22.3 | 239<br>246 | 0.49 | 0.44 | 0.54 | 1.41 | 0.85 | 0.14 |

Neutral spectral data

| No. | pH | $\lambda_{max}$ | $\varepsilon_{max}$ ($\times 10^{-3}$) | $\lambda_{min}$ | 230 | 240 | 250 | 270 | 280 | 290 |
|---|---|---|---|---|---|---|---|---|---|---|
| 69 | 7 | 238<br>270 | — | 227<br>250 | 0.87 | 0.88 | 0.83 | 1.15 | 0.92 | 0.34 |
| 70 | 7 | 273 | — | 247 | 0.96 | 0.51 | 0.46 | 1.76 | 1.27 | 0.24 |

Alkaline spectral data

| No. | pH | $\lambda_{max}$ | $\varepsilon_{max}$ ($\times 10^{-3}$) | $\lambda_{min}$ | 230 | 240 | 250 | 270 | 280 | 290 |
|---|---|---|---|---|---|---|---|---|---|---|
| 69 | — | — | — | — | — | — | — | — | — | — |
| 70 | 12 | 265 | 16.3 | 244 | 0.74 | 0.57 | 0.59[s] | 0.99 | 0.24 | 0.02 |

(Spectral ratios columns: 230, 240, 250, 270, 280, 290.)

REFERENCES

| No. | Origin and synthesis | $[\alpha]_D^t$ | pK | Spectral data | $R_f$ |
|---|---|---|---|---|---|
| 69 | R: *131* | — | *131* | *131*[b] | *131* |
| 70 | C: *137*, *41* | *137* | *52* | *52*, *137*[d], *13*[b], *41* | *41* |

| No. | Compound | Abbreviation | Symbol | Structure | Formula (mol wt) | Melting point °C | $[\alpha]^{t}_{D}$ | pK Basic | pK Acidic |
|---|---|---|---|---|---|---|---|---|---|
| 71 | **Guanosine** | Guo | G | 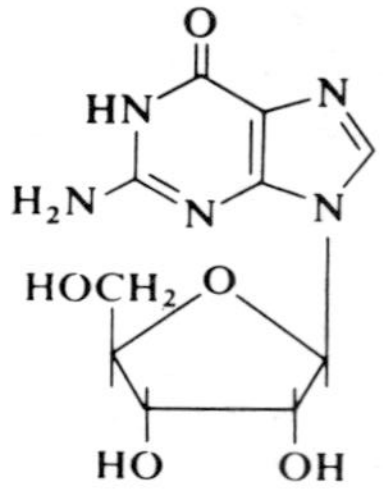 | $C_{10}H_{13}N_5O_5$ (283.24) | >235° (dec) | −72[26] (1.4, 0.1*N* NaOH) | 1.6* | 9.2*, 12.4 |
| 72 | **1-Methylguanosine** | 1MeGuo | $m^1G$ | 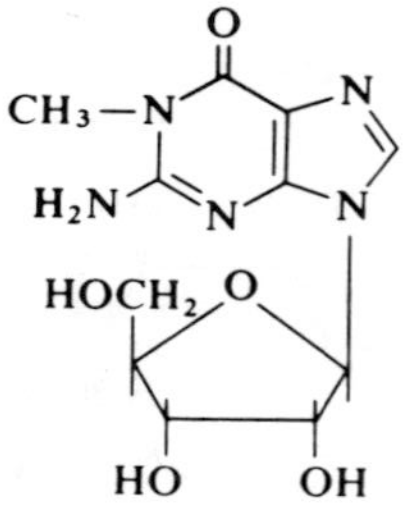 | $C_{11}H_{15}N_5O_5$ (297.27) | 225–227° (dec) | — | ~2.4[l] | — |

| No. | Acidic spectral data | | | | | | | | | | Neutral spectral data | | | | | | | | | | Alkaline spectral data | | | | | | | | | |
|---|---|---|---|---|---|---|---|---|---|---|---|---|---|---|---|---|---|---|---|---|---|---|---|---|---|---|---|---|---|---|
| | pH | $\lambda_{max}$ | $\varepsilon_{max}$ ($\times 10^{-3}$) | $\lambda_{min}$ | Spectral ratios 230 | 240 | 250 | 270 | 280 | 290 | pH | $\lambda_{max}$ | $\varepsilon_{max}$ ($\times 10^{-3}$) | $\lambda_{min}$ | Spectral ratios 230 | 240 | 250 | 270 | 280 | 290 | pH | $\lambda_{max}$ | $\varepsilon_{max}$ ($\times 10^{-3}$) | $\lambda_{min}$ | Spectral ratios 230 | 240 | 250 | 270 | 280 | 290 |
| 71 | 0.7[q] | 256 | 12.3 | 228 | — | — | 0.94 | — | 0.70 | 0.50 | 6 | 253 | 13.6 | 223 | — | — | 1.15 | — | 0.67 | 0.28 | 11.3 | 256–266 | 11.3 | 230 | — | — | 0.89 | — | 0.61 | 0.13 |
| 72 | 1 | 258 | 9.4* | 232* | 0.28 | 0.45 | 0.85 | 0.77 | 0.71 | 0.53 | M | 256 | 10.8 | — | — | — | — | — | — | — | 13 | 256* | 10.4 | 231 | 0.39 | 0.61 | 0.99 | 0.83 | 0.63 | 0.22* |

## REFERENCES

| No. | Origin and synthesis | $[\alpha]^{t}_{D}$ | pK | Spectral data | $R_f$ |
|---|---|---|---|---|---|
| 71 | C: *237, 231* | *237* | *231, 170** | *232*[b], *212, 183*[b], *205* | *258, 110, 40, 189* |
| 72 | E: *40* C: *24* | — | — | *40*[b], *24**[d,c], *265**[b] | *24, 40, 110* |

| No. | Compound / Abbreviation / Symbol | Structure | Formula (mol wt) | Melting point °C | $[\alpha]^{t}_{D}$ | pK Basic | pK Acidic |
|---|---|---|---|---|---|---|---|
| 73 | ***N*²-Methylguanosine** (2-Methylamino-9-$\beta$-ribofuranosyl-purin-6-one)<br>2MeGuo<br>$m^2G$ | 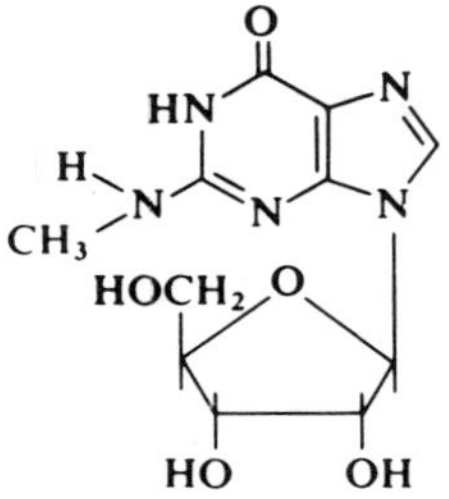 | $C_{11}H_{15}N_5O_5$ (297.27) | None (dec > 200°) | $-34.6^{26}$ (1.0 DMS/E[i]) | 2.3[p] | 9.7[p] |
| 74 | ***N*²,*N*²-Dimethylguanosine** (2-Dimethylamino-9-$\beta$-ribofuranosyl-purin-6-one)<br>2Me₂Guo<br>$m^2_2G$ | 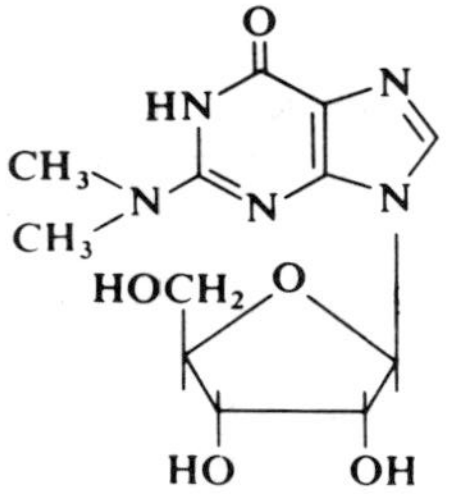 | $C_{12}H_{17}N_5O_5$ (311.30) | 242° (dec) | $-35.6^{26}$ (1.1 DMS/E[i]) | 2.5[p] | 9.7[p] |

| No. | Acidic pH | $\lambda_{max}$ | $\varepsilon_{max}$ ($\times 10^{-3}$) | $\lambda_{min}$ | 230 | 240 | 250 | 270 | 280 | 290 | Neutral pH | $\lambda_{max}$ | $\varepsilon_{max}$ ($\times 10^{-3}$) | $\lambda_{min}$ | 230 | 240 | 250 | 270 | 280 | 290 | Alkaline pH | $\lambda_{max}$ | $\varepsilon_{max}$ ($\times 10^{-3}$) | $\lambda_{min}$ | 230 | 240 | 250 | 270 | 280 | 290 |
|---|---|---|---|---|---|---|---|---|---|---|---|---|---|---|---|---|---|---|---|---|---|---|---|---|---|---|---|---|---|---|
| 73 | 1 | 258 | 14.3 | 231 | 0.52 | 0.49 | 0.85 | 0.73 | 0.56 | 0.52 | — | — | — | — | — | — | — | — | — | — | 11<br>13 | 254<br>258 | 14.8<br>12.3[n] | 228<br>237 | 0.98 | 0.70 | 0.93 | 0.91 | 0.82 | 0.48 |
| 74 | 1 | 264 | 12.8* | 236* | 0.45 | 0.32 | 0.63 | 0.92 | 0.56 | 0.48 | — | — | — | — | — | — | — | — | — | — | 11<br>13 | 262<br>262 | 12.2<br>10.6[n] | 240<br>242* | 1.37 | 0.75 | 0.85 | 0.88 | 0.78 | 0.65 |

(Columns 230–290 under each condition are Spectral ratios.)

REFERENCES

| No. | Origin and Synthesis | $[\alpha]^{t}_{D}$ | pK | Spectral data | $R_f$ |
|---|---|---|---|---|---|
| 73 | E: *40* C: *44* | *44* | *40* | *40*[b], *44*[c,d] | *40, 44, 110* |
| 74 | E: *40* C: *44* | *44* | *40* | *40*[b], *44*[c,d], *265**[b] | *40, 44, 110* |

| No. | Compound | Abbreviation | Symbol | Structure | Formula (mol wt) | Melting point °C | $[\alpha]^t_D$ | pK Basic | pK Acidic |
|---|---|---|---|---|---|---|---|---|---|
| 75 | **7-Methylguanosine** | 7MeGuo | $m^7G$ | 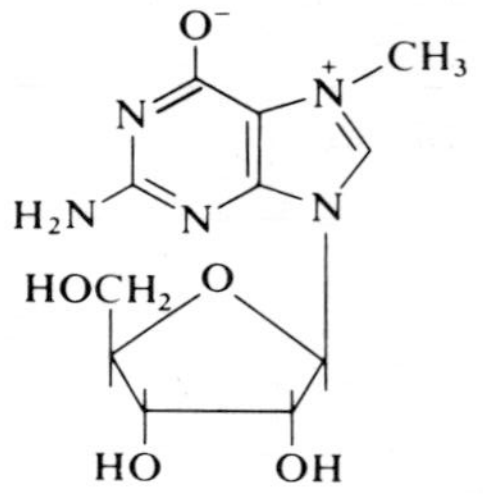 | $C_{11}H_{15}N_5O_5$ (297.27) | 165° (hemihydrate) | $-33.5^{27}$ (0.4, W) | [r] | 7.1 |
| 76 | ***N*²,*N*²,7-Trimethylguanosine** (2-Dimethylamino-7-methylguanosine) | 2,2,7Me$_3$Guo | $m_2^{2,7}G$ or $m^2m^7G$ | 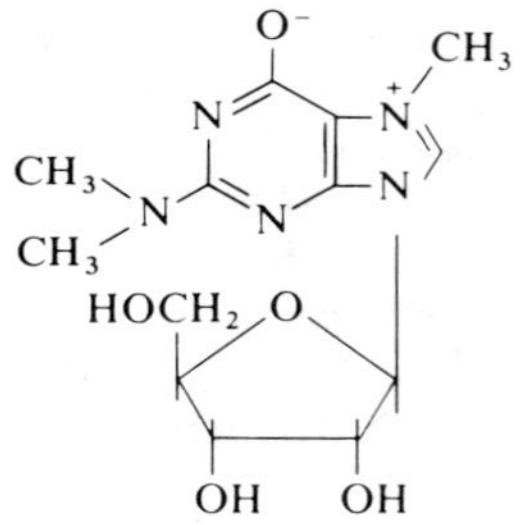 | $C_{13}H_{19}N_5O_5$ (325.33) | — | — | — | — |

Acidic spectral data

| No. | pH | $\lambda_{max}$ | $\varepsilon_{max}$ ($\times 10^{-3}$) | $\lambda_{min}$ | Spectral ratios 230 | 240 | 250 | 270 | 280 | 290 |
|---|---|---|---|---|---|---|---|---|---|---|
| 75 | 3 | 257 | 10.7* | 230 | 0.27 | 0.47 | 0.88 | 0.73 | 0.68 | 0.53 |
| 76 | 1 | 266<br>295 | — | 240 | 0.67 | 0.32 | 0.57 | 1.06 | 0.59 | 0.49 |

Neutral spectral data

| No. | pH | $\lambda_{max}$ | $\varepsilon_{max}$ ($\times 10^{-3}$) | $\lambda_{min}$ | Spectral ratios 230 | 240 | 250 | 270 | 280 | 290 |
|---|---|---|---|---|---|---|---|---|---|---|
| 75 | 7[q] | 258<br>281 | 8.5<br>7.4 | 238 | — | — | 0.89 | — | 1.04 | 0.90 |
| 76 | 5 | 266 | 10.3 | 239 | 0.67 | 0.32 | 0.57 | 1.08 | 0.63 | 0.51 |

Alkaline spectral data

| No. | pH | $\lambda_{max}$ | $\varepsilon_{max}$ ($\times 10^{-3}$) | $\lambda_{min}$ | Spectral ratios 230 | 240 | 250 | 270 | 280 | 290 |
|---|---|---|---|---|---|---|---|---|---|---|
| 75 | 9[o] | 282 | 8.0 | 242 | 2.10 | 0.84 | 0.90 | 1.16 | 1.46 | 1.30 |
| 76 | 10[o] | 234<br>302 | — | 280 | 2.04 | 1.93 | 1.31 | 0.84 | 0.62 | 0.73 |

REFERENCES

| No. | Origin and synthesis | $[\alpha]^t_D$ | pK | Spectral data | $R_f$ |
|---|---|---|---|---|---|
| 75 | C: *13, 41, 10* | *10* | *13, 41* | *52, 13*[b,d], *41**[c,d], *10**[s] | *10, 41* |
| 76 | C: *7* | — | — | *7*[b] | — |

| No. | Compound | Abbreviation | Symbol | Structure | Formula (mol wt) | Melting point °C | $[\alpha]_D^t$ | pK Basic | pK Acidic |
|---|---|---|---|---|---|---|---|---|---|
| 77 | **2′-*O*-Methylguanosine** | 2′MeGuo | Gm | 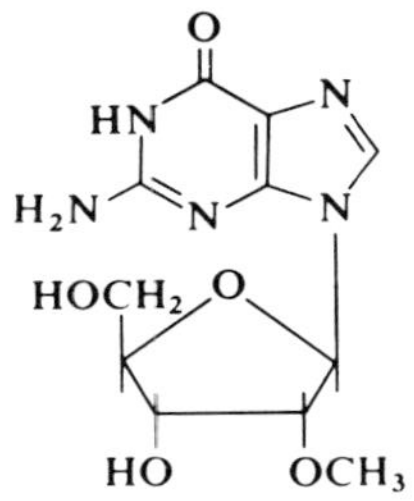 | $C_{11}H_{15}N_5O_5$ (297.27) | 218–220° | $-38.4^{22}$ (0.6, W) | — | — |
| 78 | **Inosine** | Ino | I | 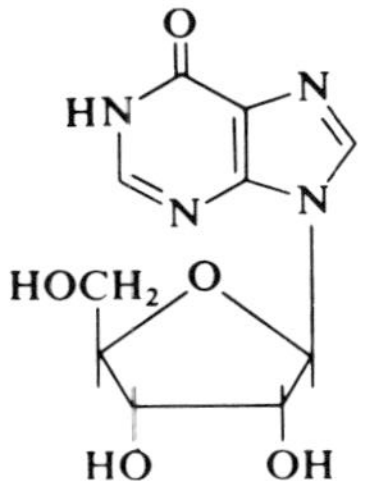 | $C_{10}H_{12}N_4O_5$ (268.23) | 218° | −58.8 (2.5, W) | 1.2 | 8.8, 12.3 |

| No. | Acidic spectral data | | | | | | | | | | Neutral spectral data | | | | | | | | | | Alkaline spectral data | | | | | | | | | |
|---|---|---|---|---|---|---|---|---|---|---|---|---|---|---|---|---|---|---|---|---|---|---|---|---|---|---|---|---|---|---|
| | pH | $\lambda_{max}$ | $\varepsilon_{max}$ ($\times 10^{-3}$) | $\lambda_{min}$ | Spectral ratios 230 | 240 | 250 | 270 | 280 | 290 | pH | $\lambda_{max}$ | $\varepsilon_{max}$ ($\times 10^{-3}$) | $\lambda_{min}$ | Spectral ratios 230 | 240 | 250 | 270 | 280 | 290 | pH | $\lambda_{max}$ | $\varepsilon_{max}$ ($\times 10^{-3}$) | $\lambda_{min}$ | Spectral ratios 230 | 240 | 250 | 270 | 280 | 290 |
| 77 | 1 | 256 | 10.7 | — | — | — | — | — | — | — | — | — | — | — | — | — | — | — | — | — | 11 | 258 | 9.8 | — | — | — | — | — | — | — |
| 78 | 0 | 251 | 10.9 | 221 | — | — | 1.21 | — | 0.11 | 0.00 | 6 | 248.5 | 12.3 | 223 | — | — | 1.68 | — | 0.25 | 0.03 | 11 | 253 | 13.1 | 224 | — | — | 1.05 | — | 0.18 | 0.01 |
| | 3 | 248 | 12.2 | 223 | — | — | 1.68 | — | 0.25 | 0.03 | | | | | | | | | | | | | | | | | | | | |

## REFERENCES

| No. | Origin and synthesis | $[\alpha]_D^t$ | pK | Spectral data | $R_f$ |
|---|---|---|---|---|---|
| 77 | C: *108* | *108* | — | *108* | *109*, *110* |
| 78 | R: *231*[a] | *255* | *170*, *231*, *51*, *250* | *232*[b], *205* | *110* |

| No. | Compound | Abbreviation | Symbol | Structure | Formula (mol wt) | Melting point °C | $[\alpha]^t_D$ | pK Basic | pK Acidic |
|---|---|---|---|---|---|---|---|---|---|
| 79 | **1-Methylinosine** | 1MeIno | m¹I | 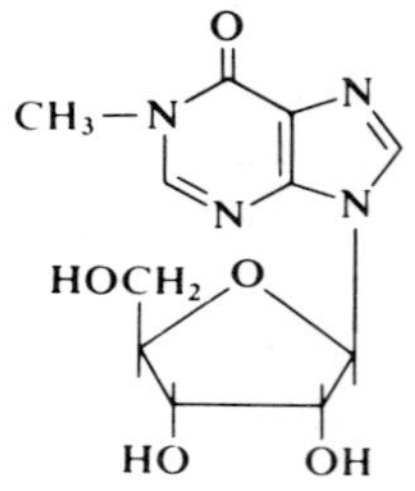 | $C_{11}H_{14}N_4O_5$ (282.25) | 210–212° | $-49.2^{28}$ (0.5, W) | — | — |
| 80 | **Nebularine** (9-β-Ribofuranosylpurine) | Neb | | 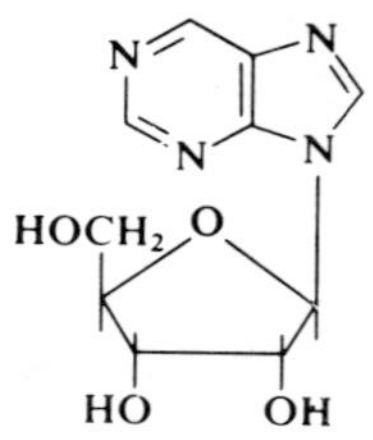 | $C_{10}H_{12}N_4O_4$ (252.23) | 181–182° | $-48.6^{25}$ (1.0, W) | 2.1 | — |

| No. | Acidic spectral data | | | | | | | | | |
|---|---|---|---|---|---|---|---|---|---|---|
| | pH | $\lambda_{max}$ | $\varepsilon_{max}$ ($\times 10^{-3}$) | $\lambda_{min}$ | Spectral ratios 230 | 240 | 250 | 270 | 280 | 290 |
| 79 | 2 | 250 | 10.4 | 223 | 0.59 | 1.14 | 1.42 | 0.57 | 0.23 | 0.03 |
| 80 | 1 | 262 | 5.9 | 235 | 0.51 | 0.49 | 0.74 | 0.85 | 0.29 | 0.07 |

| No. | Neutral spectral data | | | | | | | | | |
|---|---|---|---|---|---|---|---|---|---|---|
| | pH | $\lambda_{max}$ | $\varepsilon_{max}$ ($\times 10^{-3}$) | $\lambda_{min}$ | Spectral ratios 230 | 240 | 250 | 270 | 280 | 290 |
| 79 | 6 | 251* | 10.4 | 226 | 0.70 | 1.38 | 1.51 | 0.68 | 0.37 | 0.08 |
| 80 | W | 262 | 7.1 | 222 | 0.35 | 0.54 | 0.71 | 0.68 | 0.11 | 0.04 |

| No. | Alkaline spectral data | | | | | | | | | |
|---|---|---|---|---|---|---|---|---|---|---|
| | pH | $\lambda_{max}$ | $\varepsilon_{max}$ ($\times 10^{-3}$) | $\lambda_{min}$ | Spectral ratios 230 | 240 | 250 | 270 | 280 | 290 |
| 79 | 12 | 249 | 10.7 | — | 0.86 | 1.28 | 1.60 | 0.67 | 0.35 | 0.07 |
| 80 | 13 | 262 | 7.1 | 234 | 0.45 | 0.48 | 0.71 | 0.75 | 0.22 | 0.11 |

REFERENCES

| No. | Origin and synthesis | $[\alpha]^t_D$ | pK | Spectral data | $R_f$ |
|---|---|---|---|---|---|
| 79 | C: *10, 122* | *10* | — | ***49****, *10*[c,d], *122* | *10, 49, 110, 122* |
| 80 | C: *123* | *123* | *123* | *123*[b] | — |

| No. | Compound | Abbreviation | Symbol | Structure | Formula (mol wt) | Melting point °C | $[\alpha]^t_D$ | pK Basic | pK Acidic |
|---|---|---|---|---|---|---|---|---|---|
| 81 | **Uridine** | Urd | U | 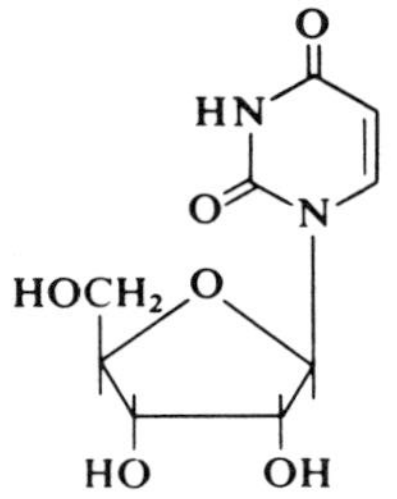 | $C_9H_{12}N_2O_6$ (244.20) | 163–165° | $+9.6^{16}$ (2.0, W) | — | 9.2, 12.5 |
| 82 | **2-Thiouridine** | 2Srd<br>2SUrd | $s^2U$ | 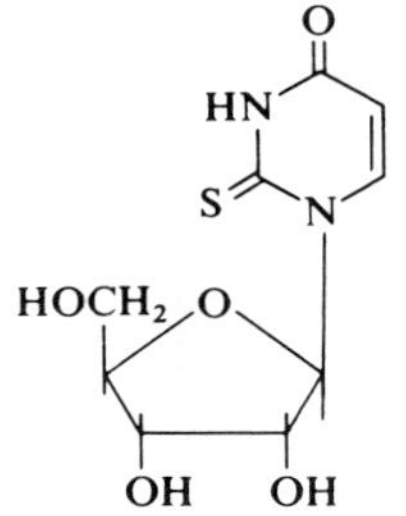 | $C_9H_{12}N_2O_5S$ (260.26) | 214° | $+39^{20}$ (1.2, W) | — | 8.8 |

| No. | Acidic spectral data | | | | | | | | | |
|---|---|---|---|---|---|---|---|---|---|---|
| | pH | $\lambda_{max}$ | $\varepsilon_{max}$ $(\times 10^{-3})$ | $\lambda_{min}$ | Spectral ratios 230 | 240 | 250 | 270 | 280 | 290 |
| 81 | 1 | 262 | 10.1 | 230 | — | — | 0.74 | — | 0.35 | 0.03 |
| 82 | 2 | 279 | 16.4 | 247 | 1.32 | 0.63 | 0.53 | 1.66 | 1.86 | 1.66 |

| No. | Neutral spectral data | | | | | | | | | |
|---|---|---|---|---|---|---|---|---|---|---|
| | pH | $\lambda_{max}$ | $\varepsilon_{max}$ $(\times 10^{-3})$ | $\lambda_{min}$ | Spectral ratios 230 | 240 | 250 | 270 | 280 | 290 |
| 81 | 7 | 262 | 10.1 | 230 | — | — | 0.74 | — | 0.35 | 0.03 |
| 82 | 7 | 222<br>278 | 15.5<br>16.1 | 247 | 1.16 | 0.70 | 0.62 | 1.47 | 1.57 | 1.40 |

| No. | Alkaline spectral data | | | | | | | | | |
|---|---|---|---|---|---|---|---|---|---|---|
| | pH | $\lambda_{max}$ | $\varepsilon_{max}$ $(\times 10^{-3})$ | $\lambda_{min}$ | Spectral ratios 230 | 240 | 250 | 270 | 280 | 290 |
| 81 | 12<br>14 | 262<br>264.5 | 7.45<br>7.5 | 243<br>243 | — | — | 0.83 | — | 0.29 | 0.02 |
| 82 | 9 | 241 | 21.8 | 261 | 1.07 | 1.43 | 1.22 | 1.02 | 0.88 | 0.59 |

## REFERENCES

| No. | Origin and synthesis | $[\alpha]^t_D$ | pK | Spectral data | $R_f$ |
|---|---|---|---|---|---|
| 81 | R: *236*[a], *181* | *236* | *163*, *231* | *163*[b], *232*[b,e], *181*[b],*212*, *183*[b] | *258*, *110*, *40* |
| 82 | C: *276*, *285*, E: *277* | *276* | *285* | *276*[b], *277*[b] | *276* |

| No. | Compound | Abbreviation | Symbol | Structure | Formula (mol wt) | Melting point °C | $[\alpha]_D^t$ | pK Basic | pK Acidic |
|---|---|---|---|---|---|---|---|---|---|
| 83 | **2,4-Dithiouridine** | 2,4Srd or 2,4$S_2$Urd | $s^2, {}^4U$ or $s^2s^4U$ | 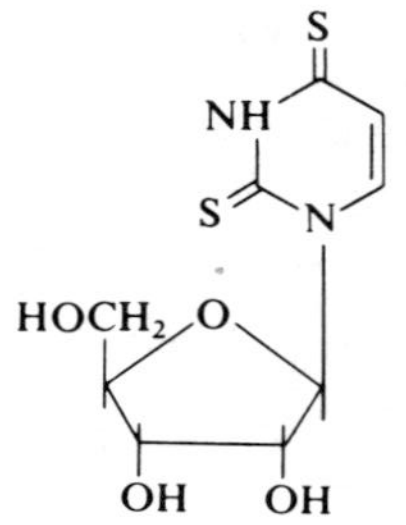 | $C_9H_{12}N_2O_4S_2$ (276.33) | 166–167° | — | — | 7.4 |
| 84 | **2-Thio-5-carboxymethyl-uridine, methyl ester** | 2S5(CmMe)Urd | $mcm^5s^2U$ | 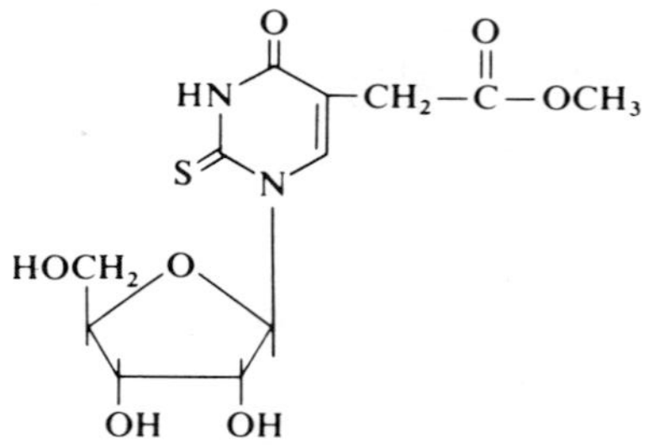 | $C_{12}H_{16}N_2O_7S$ (332.32) | 199° | $+9^{25}$ (0.20, E) | — | — |

Acidic spectral data

| No. | pH | $\lambda_{max}$ | $\varepsilon_{max}$ ($\times 10^{-3}$) | $\lambda_{min}$ | Spectral ratios 230 | 240 | 250 | 270 | 280 | 290 |
|---|---|---|---|---|---|---|---|---|---|---|
| 83 | — | — | — | — | — | — | — | — | — | — |
| 84 | 1 | 277* | 15.6 | 244 | 1.22 | 0.73 | 0.68 | 1.49 | 1.54 | 1.41 |

Neutral spectral data

| No. | pH | $\lambda_{max}$ | $\varepsilon_{max}$ ($\times 10^{-3}$) | $\lambda_{min}$ | Spectral ratios 230 | 240 | 250 | 270 | 280 | 290 |
|---|---|---|---|---|---|---|---|---|---|---|
| 83 | 5.8 | 283 | 22.5 | — | — | — | — | — | — | — |
| 84 | 7 | 277* | 15.8 | 244 | 1.18 | 0.84 | 0.76 | 1.42 | 1.45 | 1.24 |

Alkaline spectral data

| No. | pH | $\lambda_{max}$ | $\varepsilon_{max}$ ($\times 10^{-3}$) | $\lambda_{min}$ | Spectral ratios 230 | 240 | 250 | 270 | 280 | 290 |
|---|---|---|---|---|---|---|---|---|---|---|
| 83 | 9 | 280<br>320 | 16.9<br>24.8 | — | — | — | — | — | — | — |
| 84 | 12 | 242 | 22.4 | — | 1.06 | 1.37 | 1.16 | 1.02 | 0.86 | 0.46 |

REFERENCES

| No. | Origin and synthesis | $[\alpha]_D^t$ | pK | Spectral data | $R_f$ |
|---|---|---|---|---|---|
| 83 | C: *271* | — | *271* | *271* | *271* |
| 84 | C: *5* R: *275* | *5* | — | *5, 275**,b | *275* |

| No. | Compound | Abbreviation | Symbol | Structure | Formula (mol wt) | Melting point °C | $[\alpha]_D^t$ | pK Basic | pK Acidic |
|---|---|---|---|---|---|---|---|---|---|
| 85 | **3-Methyluridine** | 3MeUrd | $m^3U$ | 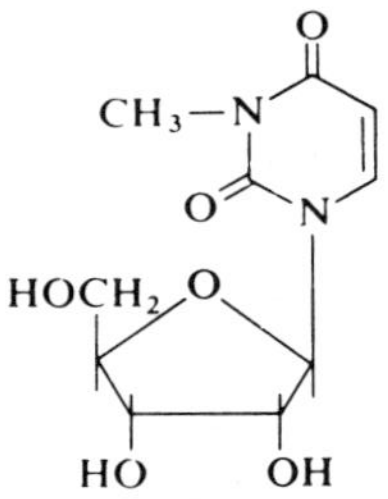 | $C_{10}H_{14}N_2O_6$ (258.23) | 119–120° | $+20.1^{26}$ (W) | — | — |
| 86 | **4-Thiouridine** | 4Srd or 4SUrd | $s^4U$ | 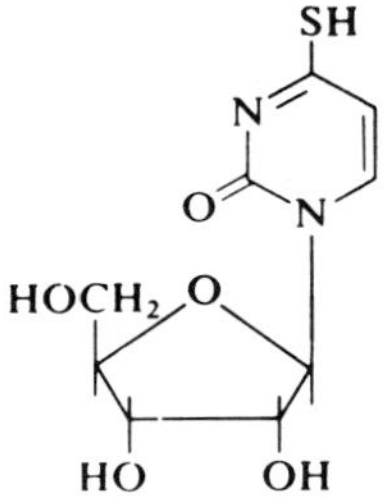 | $C_9H_{12}N_2O_5S$ (260.26) | — | — | — | 8.2 |

| No. | Acidic spectral data | | | | | | | | | | Neutral spectral data | | | | | | | | | | Alkaline spectral data | | | | | | | | | |
|---|---|---|---|---|---|---|---|---|---|---|---|---|---|---|---|---|---|---|---|---|---|---|---|---|---|---|---|---|---|---|
| | pH | $\lambda_{max}$ | $\varepsilon_{max}$ $(\times 10^{-3})$ | $\lambda_{min}$ | Spectral ratios 230 | 240 | 250 | 270 | 280 | 290 | pH | $\lambda_{max}$ | $\varepsilon_{max}$ $(\times 10^{-3})$ | $\lambda_{min}$ | Spectral ratios 230 | 240 | 250 | 270 | 280 | 290 | pH | $\lambda_{max}$ | $\varepsilon_{max}$ $(\times 10^{-3})$ | $\lambda_{min}$ | Spectral ratios 230 | 240 | 250 | 270 | 280 | 290 |
| 85 | $2^v$ | 262* | — | — | — | — | 0.76 | — | 0.30* | 0.02 | 5 | 262 | 9.0 | — | — | — | — | — | — | — | 11.6 | 263* | 9.0 | — | — | — | — | — | 0.34 | — |
| 86 | 2 | 245<br>331 | 5.2<br>17.0 | 285 | — | — | — | — | — | — | $7^q$ | 244<br>328 | — | 225<br>272 | — | — | — | — | — | — | 12 | 317 | 14.9 | 260 | — | — | — | — | — | — |

REFERENCES

| No. | Origin and synthesis | $[\alpha]_D^t$ | pK | Spectral data | $R_f$ |
|---|---|---|---|---|---|
| 85 | C: *139, 125, 140–142, 81* | *139* | — | *142, 13*[c,e], 81[e], 139, 126** | *13, 57, 81, 110, 125, 126, 142, 143, 173* |
| 86 | C: *128* R: *144* | — | *144* | *144[b], 128[c]* | *144* |

| No. | Compound | Abbreviation | Symbol | Structure | Formula (mol wt) | Melting point °C | $[\alpha]_D^t$ | pK Basic | pK Acidic |
|---|---|---|---|---|---|---|---|---|---|
| 87 | **4-Thiouridine disulphide** (bis(4-4′-Dithiouridine)) | $(S^4Urd)_2$ | | 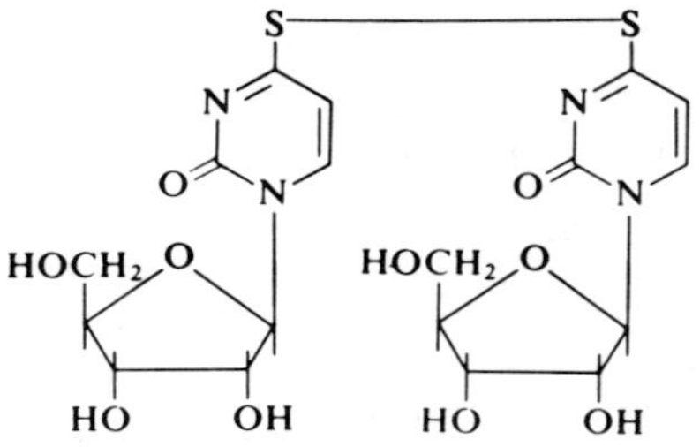 | $C_{18}H_{22}N_4O_{10}S_2$ (518.51) | 188–190° | — | — | — |
| 88 | **5-Methyluridine** (1-β-Ribofuranosylthymine. Ribosylthymine) | 5MeUrd<br>rThd, Thd | $m^5U$<br>T | 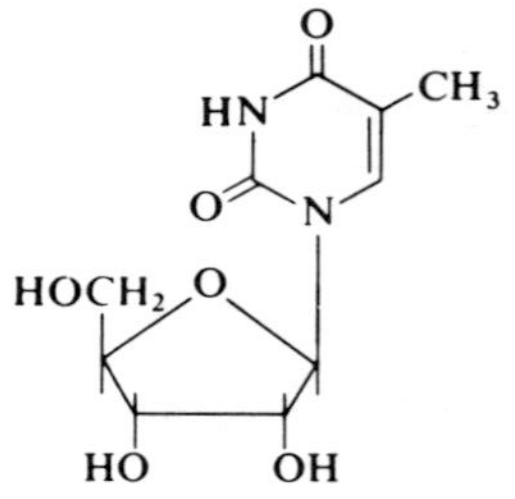 | $C_{10}H_{14}N_2O_6$ (258.23) | 183–185° | $-10^{31}$ (2.0, W) | — | 9.7 |

| No. | Acidic spectral data | | | | Spectral ratios | | | | | |
|---|---|---|---|---|---|---|---|---|---|---|
| | pH | $\lambda_{max}$ | $\varepsilon_{max}$ $(\times 10^{-3})$ | $\lambda_{min}$ | 230 | 240 | 250 | 270 | 280 | 290 |
| 87 | — | — | — | — | — | — | — | — | — | — |
| 88 | 1 | 267 | 9.9* | 235* | 0.44 | 0.39 | 0.67 | 1.07 | 0.74 | 0.27 |

| No. | Neutral spectral data | | | | Spectral ratios | | | | | |
|---|---|---|---|---|---|---|---|---|---|---|
| | pH | $\lambda_{max}$ | $\varepsilon_{max}$ $(\times 10^{-3})$ | $\lambda_{min}$ | 230 | 240 | 250 | 270 | 280 | 290 |
| 87 | 7 | 261<br>309 | — | 236<br>278 | — | — | — | — | — | — |
| 88 | 7 | 267 | 9.8 | 236 | — | — | — | — | — | — |

| No. | Alkaline spectral data | | | | Spectral ratios | | | | | |
|---|---|---|---|---|---|---|---|---|---|---|
| | pH | $\lambda_{max}$ | $\varepsilon_{max}$ $(\times 10^{-3})$ | $\lambda_{min}$ | 230 | 240 | 250 | 270 | 280 | 290 |
| 87 | — | — | — | — | — | — | — | — | — | — |
| 88 | 13 | 268 | 7.5 | 246 | 1.31 | 0.91 | 0.83 | 1.08 | 0.75 | 0.31* |

## REFERENCES

| No. | Origin and synthesis | $[\alpha]_D^t$ | pK | Spectral data | $R_f$ |
|---|---|---|---|---|---|
| 87 | C: *128* | — | — | *128, 208*[b] | ***144*** |
| 88 | E: *18, 149* C: *150, 69* | *150* | *150* | *18*[b], *150*[b,d], *149*[b,c,d], *69**,*265**[b] | *18, 69, 110, 142, 149* |

| No. | Compound | Abbreviation | Symbol | Structure | Formula (mol wt) | Melting point °C | $[\alpha]^t_D$ | pK Basic | pK Acidic |
|---|---|---|---|---|---|---|---|---|---|
| 89 | **5-Methyl-2-thiouridine** (2-Thioribosyl-thymine) | 2S5MeUrd | $m^5s^2U$ | 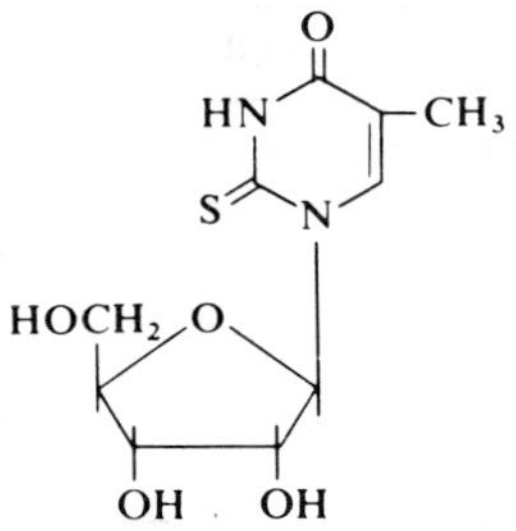 | $C_{10}H_{14}N_2O_5S$ (274.30) | 217° | $+31^{28}$ (1.23, W) | — | — |
| 90 | **2-Thio-5(*N*-methylamino-methyl)uridine** | 5MeNHMeSrd | | 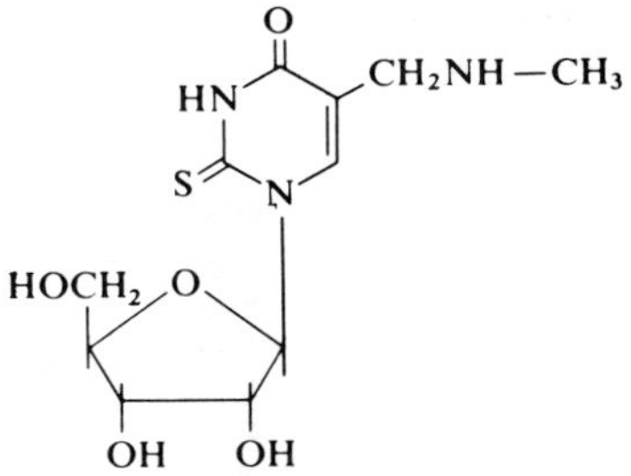 | $C_{11}H_{17}N_3O_5S$ (303.33) | — | — | — | — |

Acidic spectral data

| No. | pH | $\lambda_{max}$ | $\varepsilon_{max}$ ($\times 10^{-3}$) | $\lambda_{min}$ | 230 | 240 | 250 | 270 | 280 | 290 |
|---|---|---|---|---|---|---|---|---|---|---|
| 89 | 2 | 218<br>273 | 17.4<br>14.8 | 243 | 1.05 | 0.53 | 0.58 | 1.41 | 1.43 | 1.22 |
| 90 | 1 | 220<br>273 | — | 242 | 0.84 | 0.56 | 0.66 | 1.31 | 1.29 | 1.20 |

Neutral spectral data

| No. | pH | $\lambda_{max}$ | $\varepsilon_{max}$ ($\times 10^{-3}$) | $\lambda_{min}$ | 230 | 240 | 250 | 270 | 280 | 290 |
|---|---|---|---|---|---|---|---|---|---|---|
| 89 | 7 | 219<br>272 | 16.2<br>14.1 | 247 | 1.06 | 0.77 | 0.69 | 1.40 | 1.39 | 1.15 |
| 90 | 7 | 220<br>273 | — | 242 | 0.84 | 0.56 | 0.66 | 1.33 | 1.31 | 1.22 |

Alkaline spectral data

| No. | pH | $\lambda_{max}$ | $\varepsilon_{max}$ ($\times 10^{-3}$) | $\lambda_{min}$ | 230 | 240 | 250 | 270 | 280 | 290 |
|---|---|---|---|---|---|---|---|---|---|---|
| 89 | 9 | 239 | 21.0 | 259 | 1.35 | 1.64 | 1.20 | 1.10 | 1.04 | 0.60 |
| 90 | 13 | 243 | — | 227 | 1.00 | 1.33 | 1.21 | 1.01 | 0.87 | 0.49 |

(Spectral ratios columns: 230–290.)

REFERENCES

| No. | Origin and synthesis | $[\alpha]^t_D$ | pK | Spectral data | $R_f$ |
|---|---|---|---|---|---|
| 89 | C: *276* | *276* | — | *276*[b] | *276* |
| 90 | R: *270* | — | — | *270* | *270* |

| No. | Compound | Abbreviation | Symbol | Structure | Formula (mol wt) | Melting point °C | $[\alpha]_D^t$ | pK Basic | pK Acidic |
|---|---|---|---|---|---|---|---|---|---|
| 91 | **5-Hydroxymethyluridine** | 5HmUrd | $hm^5U$ | 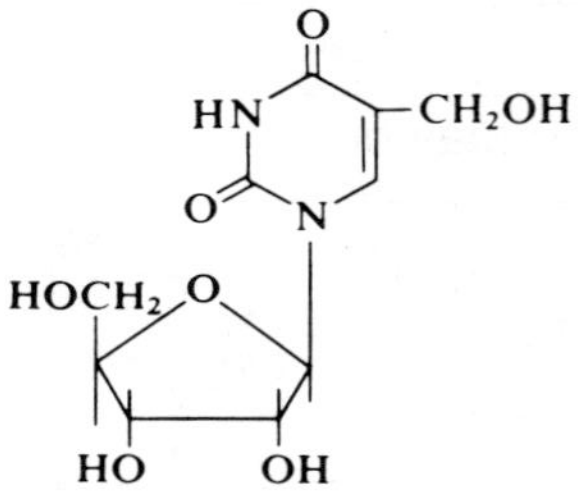 | $C_{10}H_{14}N_2O_7$ (274.23) | 167–168° | — | — | — |
| 92 | **5-Carboxymethyluridine** | 5CmUrd | $cm^5U$ | 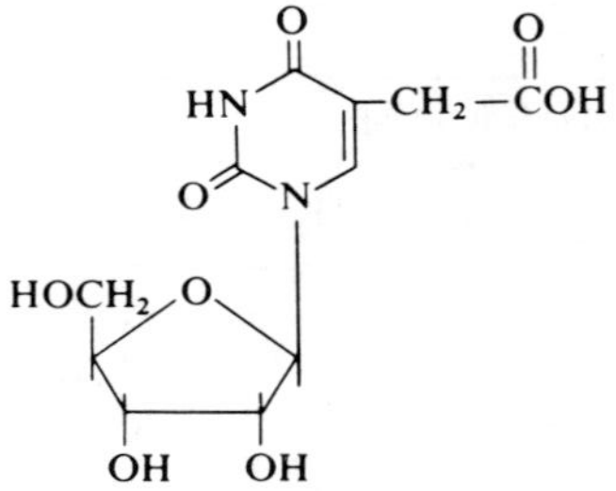 | $C_{11}H_{14}N_2O_8$ (302.24) | — | — | — | — |

| No. | Acidic spectral data | | | | | | | | | | Neutral spectral data | | | | | | | | | | Alkaline spectral data | | | | | | | | | |
|---|---|---|---|---|---|---|---|---|---|---|---|---|---|---|---|---|---|---|---|---|---|---|---|---|---|---|---|---|---|---|
| | pH | $\lambda_{max}$ | $\varepsilon_{max}$ ($\times 10^{-3}$) | $\lambda_{min}$ | Spectral ratios 230 | 240 | 250 | 270 | 280 | 290 | pH | $\lambda_{max}$ | $\varepsilon_{max}$ ($\times 10^{-3}$) | $\lambda_{min}$ | Spectral ratios 230 | 240 | 250 | 270 | 280 | 290 | pH | $\lambda_{max}$ | $\varepsilon_{max}$ ($\times 10^{-3}$) | $\lambda_{min}$ | Spectral ratios 230 | 240 | 250 | 270 | 280 | 290 |
| 91 | 2 | 264 | 9.5 | 233 | — | — | 0.70 | — | 0.52 | — | 7 | 263 | — | — | — | — | — | — | 0.53 | — | 12 | 263 | 7.0 | 243 | — | — | 0.79 | — | 0.45 | — |
| 92 | 1 | 265 | — | 234 | 0.39 | 0.40 | 0.69 | 1.02 | 0.64 | 0.19 | 7 | 266.5 | — | 236 | 0.50 | 0.41 | 0.68 | 1.07 | 0.75 | 0.27 | 13 | 266.5 | — | 245.5 | 1.29 | 0.84 | 0.80 | 1.05 | 0.71 | 0.25 |

## REFERENCES

| No. | Origin and synthesis | $[\alpha]_D^t$ | pK | Spectral data | $R_f$ |
|---|---|---|---|---|---|
| 91 | C: *69, 151* | — | — | *69, 94*[c,e] | *69* |
| 92 | R: *278* | — | — | *278*[b] | *278* |

| No. | Compound | Abbreviation | Symbol | Structure | Formula (mol wt) | Melting point °C | $[\alpha]^t_D$ | pK Basic | pK Acidic |
|---|---|---|---|---|---|---|---|---|---|
| 93 | **5-Hydroxyuridine** | 5(OH)Urd | ho⁵U | 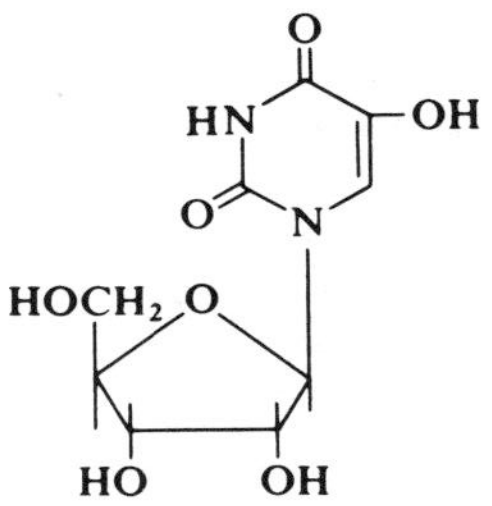 | $C_9H_{12}N_2O_7$ (260.20) | 242–245° | — | — | — |
| 94 | **2′-*O*-Methyluridine** | 2′MeUrd | Um | 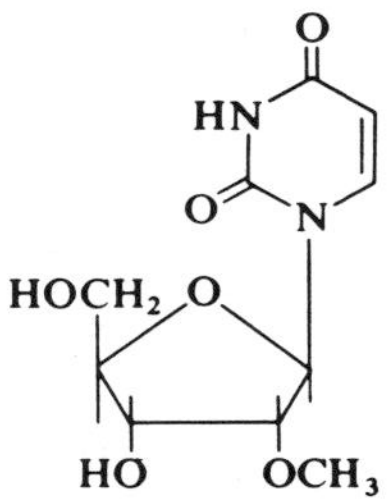 | $C_{10}H_{14}N_2O_6$ (258.23) | 159° | $+41^{20}$ (1.6, W) | — | ~9.3[p] |

Acidic spectral data

| No. | pH | $\lambda_{max}$ | $\varepsilon_{max}$ ($\times 10^{-3}$) | $\lambda_{min}$ | Spectral ratios 230 | 240 | 250 | 270 | 280 | 290 |
|---|---|---|---|---|---|---|---|---|---|---|
| 93 | 2 | 280 | — | 245 | 0.92 | 0.50 | 0.56 | 1.52 | 1.76 | 1.46 |
| 94 | 2 | 263* | 10.0 | 231 | 0.23 | 0.39 | 0.75 | 0.86 | 0.38 | 0.04 |

Neutral spectral data

| No. | pH | $\lambda_{max}$ | $\varepsilon_{max}$ ($\times 10^{-3}$) | $\lambda_{min}$ | Spectral ratios 230 | 240 | 250 | 270 | 280 | 290 |
|---|---|---|---|---|---|---|---|---|---|---|
| 93 | 7 | 280* | 8.2 | — | — | — | — | — | — | — |
| 94 | — | — | — | — | — | — | — | — | — | — |

Alkaline spectral data

| No. | pH | $\lambda_{max}$ | $\varepsilon_{max}$ ($\times 10^{-3}$) | $\lambda_{min}$ | Spectral ratios 230 | 240 | 250 | 270 | 280 | 290 |
|---|---|---|---|---|---|---|---|---|---|---|
| 93 | 12 | 306 | — | 267 | — | 1.60 | 1.29 | 0.70 | 0.89 | 1.16 |
| 94 | 12 | 262 | 7.4 | 243 | 0.98 | 0.76 | 0.83 | 0.82 | 0.30 | 0.03 |

REFERENCES

| No. | Origin and synthesis | $[\alpha]^t_D$ | pK | Spectral data | $R_f$ |
|---|---|---|---|---|---|
| 93 | C: *145–147* | — | — | *148*[b], *145*[c,d], *146** | *148* |
| 94 | C: *124* | *124* | *124, 138* | *124** | *109, 110, 124* |

| No. | Compound | Abbreviation | Symbol | Structure | Formula (mol wt) | Melting point °C | $[\alpha]_D^t$ | pK Basic | pK Acidic |
|---|---|---|---|---|---|---|---|---|---|
| 95 | **Dihydrouridine** | $H_2Urd$ | hU or D | 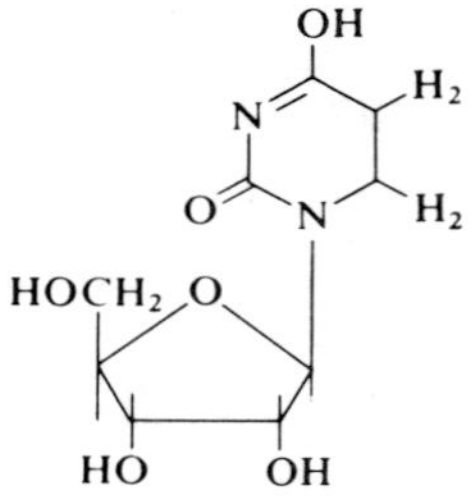 | $C_9H_{14}N_2O_6$ (246.22) | 275–276° | +39.1 | — | — |
| 96 | **Spongothymidine** (1-β-D-Arabinofuranosylthymine) | araThd aThd | aT | 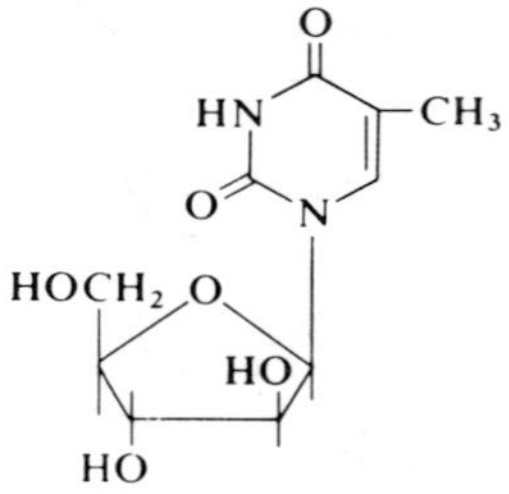 | $C_{10}H_{14}N_2O_6$ (258.23) | 238–242° | $+93^{24}_{589}$ (0.5, W) | — | 9.8 |

| No. | Acidic pH | Acidic $\lambda_{max}$ | Acidic $\varepsilon_{max}$ ($\times 10^{-3}$) | Acidic $\lambda_{min}$ | Acidic 230 | Acidic 240 | Acidic 250 | Acidic 270 | Acidic 280 | Acidic 290 | Neutral pH | Neutral $\lambda_{max}$ | Neutral $\varepsilon_{max}$ ($\times 10^{-3}$) | Neutral $\lambda_{min}$ | Neutral 230 | Neutral 240 | Neutral 250 | Neutral 270 | Neutral 280 | Neutral 290 | Alkaline pH | Alkaline $\lambda_{max}$ | Alkaline $\varepsilon_{max}$ ($\times 10^{-3}$) | Alkaline $\lambda_{min}$ | Alkaline 230 | Alkaline 240 | Alkaline 250 | Alkaline 270 | Alkaline 280 | Alkaline 290 |
|---|---|---|---|---|---|---|---|---|---|---|---|---|---|---|---|---|---|---|---|---|---|---|---|---|---|---|---|---|---|---|
| 95 | — | — | — | — | — | — | — | — | — | — | W | 208 | 6.6 | — | — | — | — | — | — | — | 13[a] | 235 | — | — | — | — | — | — | — | — |
| 96 | 1 | 268 | 10.0 | 236 | 0.36 | 0.31 | 0.61 | 1.15 | 0.85 | 0.36 | 7 | 268 | 10.0 | 236 | 0.36 | 0.31 | 0.61 | 1.15 | 0.85 | 0.36 | 12 | 269 | 7.9 | 245 | 1.17 | 0.70 | 0.69 | 1.15 | 0.77 | 0.22 |

REFERENCES

| No. | Origin and synthesis | $[\alpha]_D^t$ | pK | Spectral data | $R_f$ |
|---|---|---|---|---|---|
| 95 | C: *266, 99, 147* | *147* | — | *265*[b], *266*[c,d] | — |
| 96 | N: *161* C: *162* | *162* | *163* | *163*[b], *161*[b] | — |

| No. | Compound | Abbreviation | Symbol | Structure | Formula (mol wt) | Melting point °C | $[\alpha]_D^t$ | pK Basic | pK Acidic |
|---|---|---|---|---|---|---|---|---|---|
| 97 | **Orotidine** (Uridine 6-carboxylic acid) | Ord | O | | $C_{10}H_{12}N_2O_8$ (288.21) | 183–184° (CHA salt[y]) | — | — | — |
| 98 | **Pseudouridine C** (5-β-D-Ribofuranosyluracil) | Ψrd<br>Ψrd C | Ψ<br>$Ψ_C$ | | $C_9H_{12}N_2O_6$ (244.20) | 223–224° | −3 (1.0, W) | — | 8.9*, >13 |

| No. | Acidic spectral data | | | | | | | | | | Neutral spectral data | | | | | | | | | | Alkaline spectral data | | | | | | | | | |
|---|---|---|---|---|---|---|---|---|---|---|---|---|---|---|---|---|---|---|---|---|---|---|---|---|---|---|---|---|---|---|
| | pH | $\lambda_{max}$ | $\varepsilon_{max}$ ($\times 10^{-3}$) | $\lambda_{min}$ | Spectral ratios 230 | 240 | 250 | 270 | 280 | 290 | pH | $\lambda_{max}$ | $\varepsilon_{max}$ ($\times 10^{-3}$) | $\lambda_{min}$ | Spectral ratios 230 | 240 | 250 | 270 | 280 | 290 | pH | $\lambda_{max}$ | $\varepsilon_{max}$ ($\times 10^{-3}$) | $\lambda_{min}$ | Spectral ratios 230 | 240 | 250 | 270 | 280 | 290 |
| 97 | 1 | 267 | 9.8 | 234 | 0.39 | 0.41 | 0.66 | 1.12 | 0.81 | 0.37 | — | — | — | — | — | — | — | — | — | — | 13 | 266 | 7.8 | 245 | 1.07 | 0.83 | 0.83 | 1.04 | 0.71 | 0.29 |
| 98 | 2[r] | 262 | 7.9* | 233 | — | — | 0.74 | — | 0.42* | 0.06* | 7 | 263 | 8.1* | 233 | 0.33 | 0.42 | 0.74 | 0.90 | 0.44 | 0.08 | 12 | 286 | 7.7* | 245 | 2.06 | 0.73 | 0.62 | 1.51 | 2.06 | 2.16 |
| | | | | | | | | | | | | | | | | | | | | | 14[q] | 279 | 5.7[n] | 248 | 2.31 | 1.11 | 0.61 | 1.67 | 2.09 | 1.51 |

## REFERENCES

| No. | Origin and synthesis | $[\alpha]_D^t$ | pK | Spectral data | $R_f$ |
|---|---|---|---|---|---|
| 97 | N: *160* | — | — | *160*[b] | — |
| 98 | N: *153*, C: *154* | *95* | *155, 156, 94**, *95** | *155, 94**[b,c,d,e], *154*[c,d], *95**[b,e], *156*[b], *157, 265**[b] | *110, 157–159* |

| No. | Compound | Abbreviation | Symbol | Structure | Formula (mol wt) | Melting point °C | $[\alpha]^t_D$ | pK Basic | pK Acidic |
|---|---|---|---|---|---|---|---|---|---|
| 99 | **Pseudouridine B** (5-α-D-Ribofuranosyluracil) | Ψrd B | $\Psi_B$ | | $C_9H_{12}N_2O_6$ (244.20) | — | — | — | 9.2,* >13 |
| 100 | **Pseudouridine $A_S$** (5-β-D-Ribopyranosyluracil) | Ψrd $A_S$ | $\Psi_{AS}$ | | $C_9H_{12}N_2O_6$ (244.20) | — | — | — | 9.6, >13 |

| No. | Acidic pH | Acidic $\lambda_{max}$ | Acidic $\varepsilon_{max}$ ($\times 10^{-3}$) | Acidic $\lambda_{min}$ | Acidic ratio 230 | 240 | 250 | 270 | 280 | 290 | Neutral pH | Neutral $\lambda_{max}$ | Neutral $\varepsilon_{max}$ ($\times 10^{-3}$) | Neutral $\lambda_{min}$ | Neutral ratio 230 | 240 | 250 | 270 | 280 | 290 | Alkaline pH | Alkaline $\lambda_{max}$ | Alkaline $\varepsilon_{max}$ ($\times 10^{-3}$) | Alkaline $\lambda_{min}$ | Alkaline ratio 230 | 240 | 250 | 270 | 280 | 290 |
|---|---|---|---|---|---|---|---|---|---|---|---|---|---|---|---|---|---|---|---|---|---|---|---|---|---|---|---|---|---|---|
| 99 | [r] | — | — | — | — | — | — | — | — | — | 7 | 264 | — | 234 | 0.33 | 0.38 | 0.70 | 0.95 | 0.51 | 0.09 | 12 | 288 | — | 245 | 1.88 | 0.76 | 0.70 | 1.26 | 1.46 | 1.56 |
| | | | | | | | | | | | | | | | | | | | | | 14[q] | 279 | — | 248 | 2.20 | 1.02 | 0.61 | 1.56 | 1.81 | 1.37 |
| 100 | [r] | — | — | — | — | — | — | — | — | — | 7 | 262 | 8.3 | 231 | 0.25 | 0.41 | 0.75 | 0.83 | 0.34 | 0.04 | 12 | 286 | 9.2 | 244 | 2.10 | 0.60 | 0.56 | 1.69 | 2.50 | 2.66 |
| | | | | | | | | | | | | | | | | | | | | | 14[q] | 281 | 7.5[n] | 247 | 2.24 | 0.81 | 0.54 | 1.73 | 2.28 | 1.93 |

REFERENCES

| No. | Origin and synthesis | $[\alpha]^t_D$ | pK | Spectral data | $R_f$ |
|---|---|---|---|---|---|
| 99 | C: *154* R: *94* | — | *156, 94** | *155, 94*[b,c,e], *156*[b] | *158* |
| 100 | R: *94* C: *154* | — | *94* | *155, 94*[b,c,d,e] | *97* |

| No. | Compound | Abbreviation | Symbol | Structure | Formula (mol wt) | Melting point °C | $[\alpha]_D^t$ | pK Basic | pK Acidic |
|---|---|---|---|---|---|---|---|---|---|
| 101 | **Pseudouridine $A_F$** (5-α-D-Ribopyranosyluracil) | Ψrd $A_F$ | $\Psi_{AF}$ | 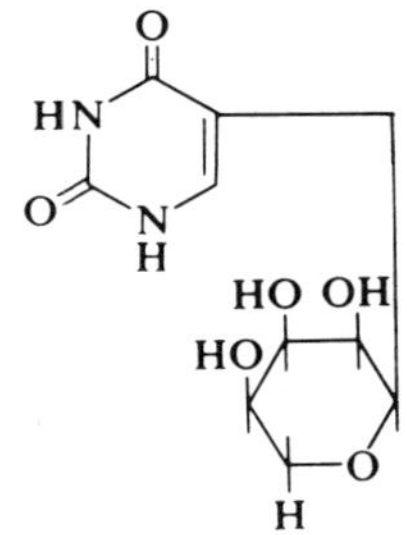 | $C_9H_{12}N_2O_6$ (244.20) | — | — | — | 9.6, >13 |
| 102 | **2′-*O*-Methylpseudouridine** | 2′MeΨrd | Ψm | 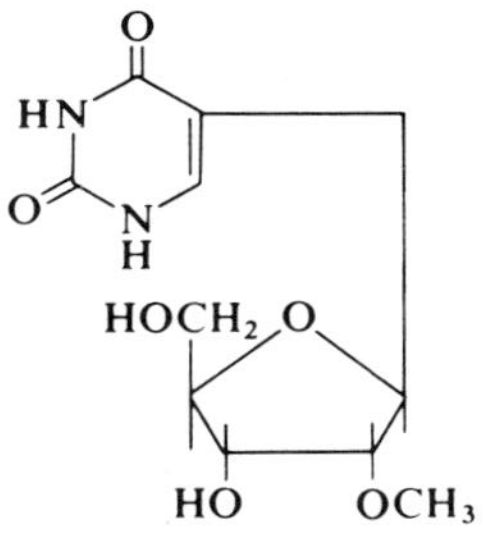 | $C_{10}H_{14}N_2O_6$ (258.23) | — | — | — | — |

| No. | Acidic pH | Acidic $\lambda_{max}$ | Acidic $\varepsilon_{max}$ ($\times 10^{-3}$) | Acidic $\lambda_{min}$ | Acidic 230 | Acidic 240 | Acidic 250 | Acidic 270 | Acidic 280 | Acidic 290 | Neutral pH | Neutral $\lambda_{max}$ | Neutral $\varepsilon_{max}$ ($\times 10^{-3}$) | Neutral $\lambda_{min}$ | Neutral 230 | Neutral 240 | Neutral 250 | Neutral 270 | Neutral 280 | Neutral 290 | Alkaline pH | Alkaline $\lambda_{max}$ | Alkaline $\varepsilon_{max}$ ($\times 10^{-3}$) | Alkaline $\lambda_{min}$ | Alkaline 230 | Alkaline 240 | Alkaline 250 | Alkaline 270 | Alkaline 280 | Alkaline 290 |
|---|---|---|---|---|---|---|---|---|---|---|---|---|---|---|---|---|---|---|---|---|---|---|---|---|---|---|---|---|---|---|
| 101 | [r] | — | — | — | — | — | — | — | — | — | 7 | 263 | — | 233 | 0.27 | 0.37 | 0.71 | 0.90 | 0.42* | 0.05 | 12 | 287 | — | 244* | 1.59 | 0.62 | 0.67 | 1.26 | 1.49 | 1.58 |
| | | | | | | | | | | | | | | | | | | | | | 14[q] | 278 | — | 248 | 2.09 | 1.04 | 0.65 | 1.52 | 1.75 | 1.32 |
| 102 | 1 | 261 | — | — | — | — | — | — | — | — | 7 | 261 | — | — | — | — | — | — | — | — | 13 | 281 | — | — | — | — | — | — | — | — |

## REFERENCES

| No. | Origin and synthesis | $[\alpha]_D^t$ | pK | Spectral data | $R_f$ |
|---|---|---|---|---|---|
| 101 | R: *94* C: *154* | — | *94* | *155*, *94*[b,c,e] | *97* |
| 102 | R: *118* | — | — | *118*[b] | *118* |

| No. | Compound | Abbreviation | Symbol | Structure | Formula (mol wt) | Melting point °C | $[\alpha]_D^t$ | pK Basic | pK Acidic |
|---|---|---|---|---|---|---|---|---|---|
| 103 | **Xanthosine** | Xao | X | 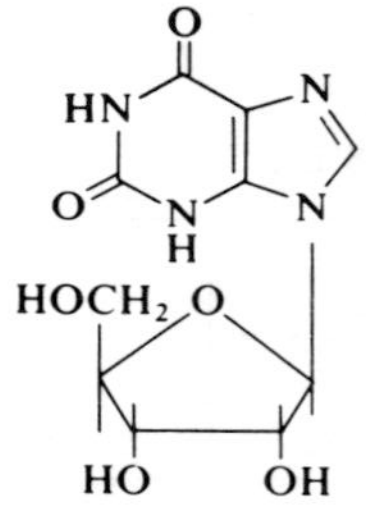 | $C_{10}H_{12}N_4O_6$ (284.23) | — | $-51.2^{30}$ (8, 0.3*N* NaOH) | <2.5 | 5.7* ~13.0 |
| 104 | **Ribosylzeatin** ($N^6$-(*trans*-4-Hydroxy-3-methyl-but-2-enyl)adenosine) | Zeo | Z | 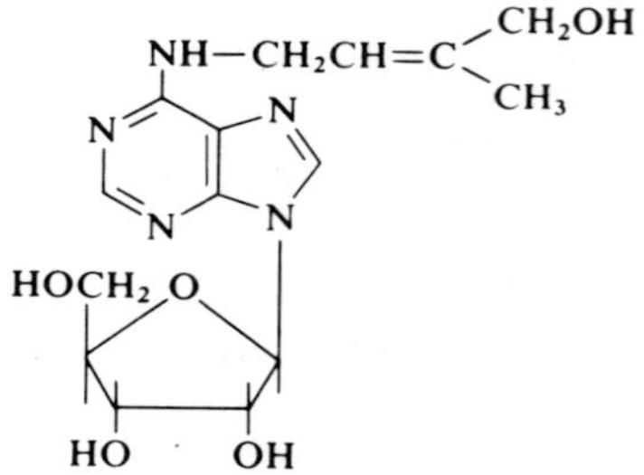 | $C_{15}H_{21}N_5O_5$ (351.36) | 180–182° | — | — | — |

Acidic spectral data

| No. | pH | $\lambda_{max}$ | $\varepsilon_{max}$ ($\times 10^{-3}$) | $\lambda_{min}$ | Spectral ratios 230 | 240 | 250 | 270 | 280 | 290 |
|---|---|---|---|---|---|---|---|---|---|---|
| 103 | 3 | 235<br>263 | 8.4<br>8.95 | 248 | — | — | 0.75 | — | 0.28 | 0.03 |
| 104 | 1 | 208<br>266 | 19.8<br>18.5 | 235 | — | — | — | — | — | — |

Neutral spectral data

| No. | pH | $\lambda_{max}$ | $\varepsilon_{max}$ ($\times 10^{-3}$) | $\lambda_{min}$ | Spectral ratios 230 | 240 | 250 | 270 | 280 | 290 |
|---|---|---|---|---|---|---|---|---|---|---|
| 103 | 8 | 248<br>278 | 10.2<br>8.9 | 223<br>264 | — | — | 1.30 | — | 1.13 | 0.61 |
| 104 | 7 | 211<br>270 | 19.3<br>17.8 | 233 | — | — | — | — | — | — |

Alkaline spectral data

| No. | pH | $\lambda_{max}$ | $\varepsilon_{max}$ ($\times 10^{-3}$) | $\lambda_{min}$ | Spectral ratios 230 | 240 | 250 | 270 | 280 | 290 |
|---|---|---|---|---|---|---|---|---|---|---|
| 103 | 14 | 252<br>276 | 8.6<br>9.3 | 230<br>262 | — | — | 1.12 | — | 1.16 | 0.59 |
| 104 | 11 | 215<br>270 | 18.1<br>18.3 | 235 | — | — | — | — | — | — |

REFERENCES

| No. | Origin and synthesis | $[\alpha]_D^t$ | pK | Spectral data | $R_f$ |
|---|---|---|---|---|---|
| 103 | C: *224* | *224* | *250*, *53**, *55**, *170* | *232*[b], *205* | — |
| 104 | C: *227* | — | — | *227* | *32*, *227* |

| No. | Compound | Abbreviation | Symbol | Structure | Formula (mol wt) | Melting point °C | $[\alpha]^t_D$ | pK Basic | pK Acidic |
|---|---|---|---|---|---|---|---|---|---|
| **DEOXYRIBONUCLEOSIDES** | | | | | | | | | |
| 105 | **Deoxyadenosine** | dAdo | dA | 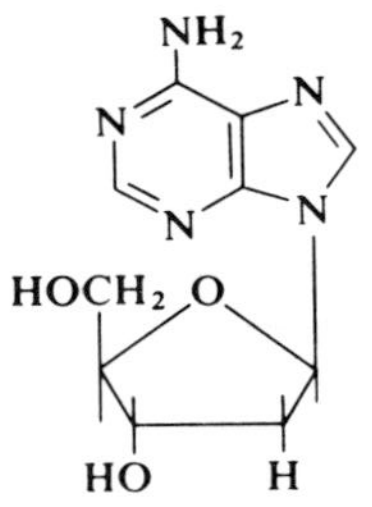 | $C_{10}H_{13}N_5O_3$ (251.24) | 191–192° | $-26.0^{21}$ (1.0, W) | 3.8 | — |
| 106 | **$N^6$-Methyldeoxyadenosine** (6-Methylaminopurinedeoxyribonucleoside) | 6MedAdo<br>6dMeAdo | m$^6$dA | 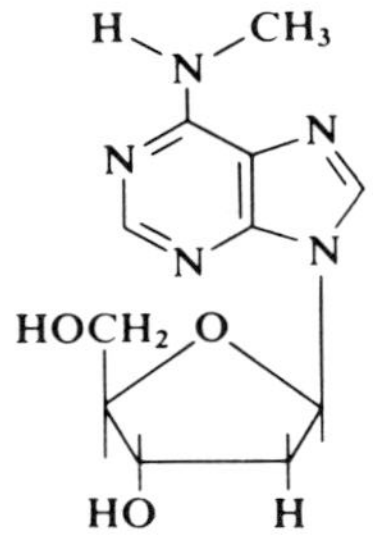 | $C_{11}H_{15}N_5O_3$ (265.27) | 206–208° | $-23.5^{26}$ (1.0, W) | — | — |

| No. | Acidic spectral data | | | | | | | | | | Neutral spectral data | | | | | | | | | | Alkaline spectral data | | | | | | | | | |
|---|---|---|---|---|---|---|---|---|---|---|---|---|---|---|---|---|---|---|---|---|---|---|---|---|---|---|---|---|---|---|
| | pH | $\lambda_{max}$ | $\varepsilon_{max}$ ($\times 10^{-3}$) | $\lambda_{min}$ | Spectral ratios 230 | 240 | 250 | 270 | 280 | 290 | pH | $\lambda_{max}$ | $\varepsilon_{max}$ ($\times 10^{-3}$) | $\lambda_{min}$ | Spectral ratios 230 | 240 | 250 | 270 | 280 | 290 | pH | $\lambda_{max}$ | $\varepsilon_{max}$ ($\times 10^{-3}$) | $\lambda_{min}$ | Spectral ratios 230 | 240 | 250 | 270 | 280 | 290 |
| 105 | 2 | 258 | 14.1 | 228 | — | — | 0.83 | — | 0.24 | — | 7 | 259 | 15.0 | 225 | — | — | 0.79 | — | 0.15 | <0.01 | 13[q] | 260 | 14.9 | — | — | — | — | — | — | — |
| 106 | 1[o] | 261 | 15.1 | 230 | — | — | — | — | — | — | 7 | 265 | 15.4 | 229* | 0.29 | 0.35 | 0.61 | 1.07 | 0.65 | 0.22 | 11 | 265 | 15.4 | 226* | 0.17 | 0.30 | 0.58 | 1.08 | 0.63 | 0.11 |

REFERENCES

| No. | Origin and synthesis | $[\alpha]^t_D$ | pK | Spectral data | $R_f$ |
|---|---|---|---|---|---|
| 105 | D: *238* C: *249, 267* | *239* | *246, 223* | *246, 223, 240, 267*[b] | *21* |
| 106 | C: *10* D: *21* | *10* | — | *10, 21**[b,e] | *10, 18, 21* |

| No. | Compound | Abbreviation | Symbol | Structure | Formula (mol wt) | Melting point °C | $[\alpha]_D^t$ | pK Basic | pK Acidic |
|---|---|---|---|---|---|---|---|---|---|
| 107 | **Deoxycytidine** | dCyd | dC | 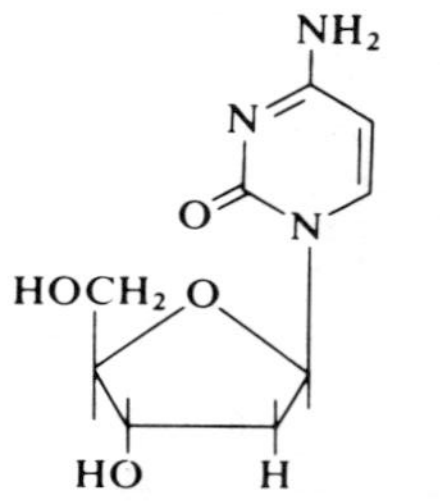 | $C_9H_{13}N_3O_4$ (227.22) | 200–201° | $+82.4^{19}$ (1.31, 1*N* NaOH) | 4.3 | >13 |
| 108 | **$N^4$-Methyldeoxycytidine** (1-β-2′-Deoxyribofuranosyl-4-methylamino-2-pyrimidinone) | 4MedCyd or 4dMeCyd | $m^4dC$ | 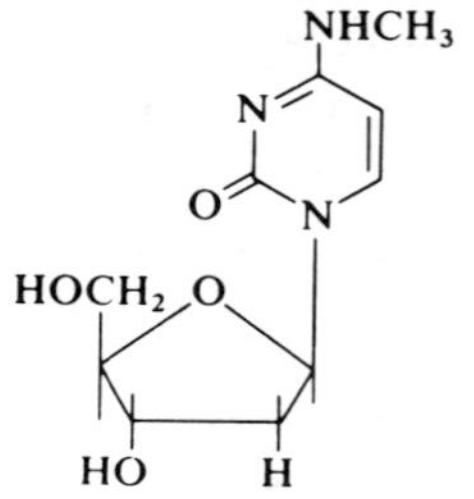 | $C_{10}H_{15}N_3O_4$ (241.24) | 191–193° | $+48^{28}$ (1.2, W) | 4.0 | — |

| No. | Acidic spectral data | | | | | | | | | |
|---|---|---|---|---|---|---|---|---|---|---|
| | pH | $\lambda_{max}$ | $\varepsilon_{max}$ ($\times 10^{-3}$) | $\lambda_{min}$ | Spectral ratios 230 | 240 | 250 | 270 | 280 | 290 |
| 107 | 1 | 280 | 13.2 | 241 | — | — | 0.42 | — | 2.15 | 1.61 |
| 108 | 1 | 282 | 14.6 | 242 | — | — | — | — | 1.98 | — |

| No. | Neutral spectral data | | | | | | | | | |
|---|---|---|---|---|---|---|---|---|---|---|
| | pH | $\lambda_{max}$ | $\varepsilon_{max}$ ($\times 10^{-3}$) | $\lambda_{min}$ | Spectral ratios 230 | 240 | 250 | 270 | 280 | 290 |
| 107 | 7 | 271 | 9.0 | 250 | — | — | 0.83 | — | 0.97 | 0.31 |
| 108 | 7 | 236<br>270 | 9.1<br>11.7 | 229 | — | — | — | — | — | — |

| No. | Alkaline spectral data | | | | | | | | | |
|---|---|---|---|---|---|---|---|---|---|---|
| | pH | $\lambda_{max}$ | $\varepsilon_{max}$ ($\times 10^{-3}$) | $\lambda_{min}$ | Spectral ratios 230 | 240 | 250 | 270 | 280 | 290 |
| 107 | 11<br>13[q] | 271<br>271.5 | 9.0<br>9.1 | 250<br>250 | — | — | 0.83 | — | 0.97 | 0.31 |
| 108 | 12 | 236<br>270 | 9.1<br>11.7 | 229 | — | — | — | — | — | — |

## REFERENCES

| No. | Origin and synthesis | $[\alpha]_D^t$ | pK | Spectral data | $R_f$ |
|---|---|---|---|---|---|
| 107 | D: *238, 241* C: *247, 129* | *241* | *163* | *163*[b], *232*[b,e] | *201* |
| 108 | C: *129* | *129* | *129* | *129* | *129* |

| No. | Compound | Abbreviation | Symbol | Structure | Formula (mol wt) | Melting point °C | $[\alpha]^t_D$ | pK Basic | pK Acidic |
|---|---|---|---|---|---|---|---|---|---|
| 109 | **5-Methyldeoxycytidine** | 5MedCyd or 5dMeCyd | $m^5dC$ | 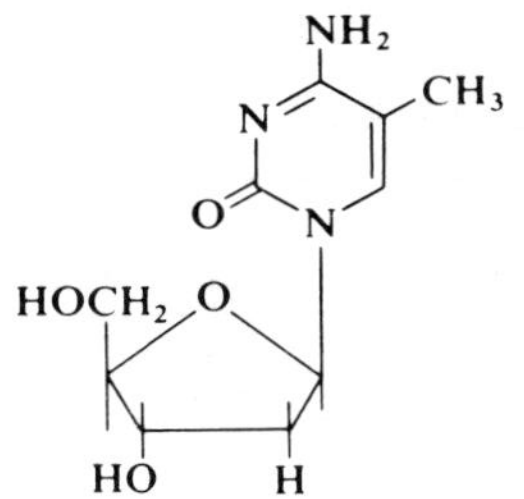 | $C_{10}H_{15}N_3O_4$ (241.24) | 154–155° (dec) (HCl salt) | $+62^{23}$ (1.0, 1*N* NaOH) | 4.4 | > 13 |
| 110 | **5-Hydroxymethyldeoxycytidine** | 5HmdCyd 5dHmCyd | $hm^5dC$ | 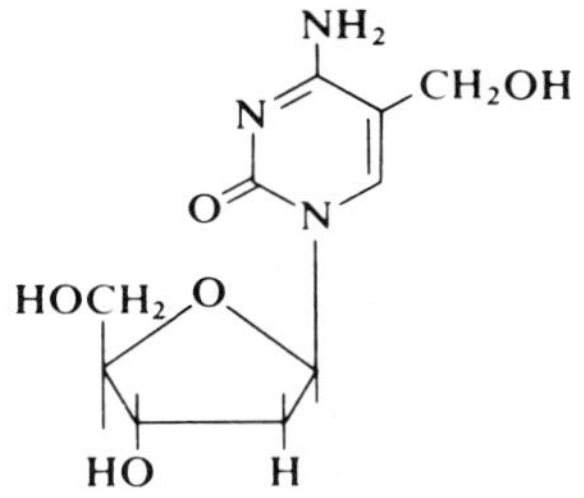 | $C_{10}H_{15}N_3O_5$ (257.24) | 203° (dec) | $+51^{20}$ (W) | 3.5 | — |

| No. | Acidic spectral data | | | | | | | | | | Neutral spectral data | | | | | | | | | | Alkaline spectral data | | | | | | | | | |
|---|---|---|---|---|---|---|---|---|---|---|---|---|---|---|---|---|---|---|---|---|---|---|---|---|---|---|---|---|---|---|
| | pH | $\lambda_{max}$ | $\varepsilon_{max}$ ($\times 10^{-3}$) | $\lambda_{min}$ | Spectral ratios 230 | 240 | 250 | 270 | 280 | 290 | pH | $\lambda_{max}$ | $\varepsilon_{max}$ ($\times 10^{-3}$) | $\lambda_{min}$ | Spectral ratios 230 | 240 | 250 | 270 | 280 | 290 | pH | $\lambda_{max}$ | $\varepsilon_{max}$ ($\times 10^{-3}$) | $\lambda_{min}$ | Spectral ratios 230 | 240 | 250 | 270 | 280 | 290 |
| 109 | 1 | 287 | 12.4* | 245 | 1.34 | 0.43 | 0.42 | 1.93 | 2.93 | 3.12 | 7 | 277 | 8.5 | 255 | 1.47 | 1.29 | 1.00 | 1.37 | 1.54 | 1.01 | 14 | 279 | 8.8 | 255 | 1.57 | 1.27 | 0.98 | 1.43 | 1.67 | 1.13 |
| 110 | 1 | 283 | 12.6 | 243 | 0.16 | 0.21* | 0.64* | 1.88* | 2.47* | 2.27* | 7 | 272 | | 247 | | 0.88 | 0.97 | 1.19 | 1.17 | 0.58* | 13 | 274 | — | 252 | — | 1.21 | 0.97 | 1.26 | 1.17* | 0.68* |

REFERENCES

| No. | Origin and Synthesis | $[\alpha]^t_D$ | pK | Spectral data | $R_f$ |
|---|---|---|---|---|---|
| 109 | C: *128* D: *133*, *134* | *128* | *128* | *128*[b], *133**[b], *134*[b] | *70* |
| 110 | D: *75*, *135* C: *136* | *136* | *75* | *75*[b], *135**[b,c,e], *136**[d] | *70*, *75* |

| No. | Compound | Abbreviation | Symbol | Structure | Formula (mol wt) | Melting point °C | $[\alpha]^t_D$ | pK Basic | pK Acidic |
|---|---|---|---|---|---|---|---|---|---|
| 111 | **Deoxyguanosine** | dGuo | dG | 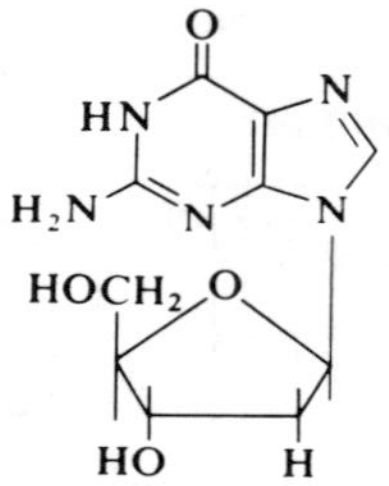 | $C_{10}H_{13}N_5O_4$ (267.24) | 250° | $-30.2^{23.5}$ (0.2, W) | 2.5 | — |
| 112 | **1-Methyldeoxyguanosine** | 1MedGuo or 1dMeGuo | $m^1dG$ | 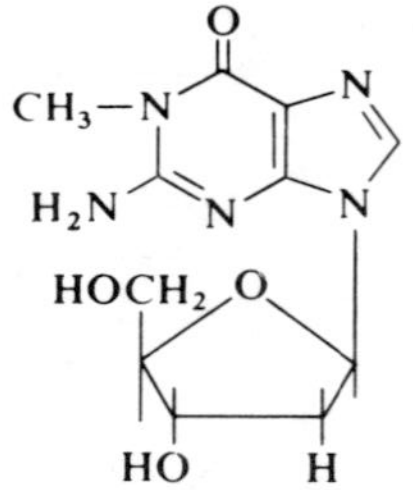 | $C_{11}H_{15}N_5O_4$ (281.27) | 249–250° (dec) | — | — | — |

| No. | Acidic spectral data | | | | | | | | | |
|---|---|---|---|---|---|---|---|---|---|---|
| | pH | $\lambda_{max}$ | $\varepsilon_{max}$ ($\times 10^{-3}$) | $\lambda_{min}$ | Spectral ratios 230 | 240 | 250 | 270 | 280 | 290 |
| 111 | 1° | 255 | 12.1 | — | — | — | — | — | — | — |
| 112 | 1° | 257 | 12.1 | — | — | — | — | — | — | — |

| No. | Neutral spectral data | | | | | | | | | |
|---|---|---|---|---|---|---|---|---|---|---|
| | pH | $\lambda_{max}$ | $\varepsilon_{max}$ ($\times 10^{-3}$) | $\lambda_{min}$ | Spectral ratios 230 | 240 | 250 | 270 | 280 | 290 |
| 111 | W | 254 | 13.0 | 223 | — | — | 1.16 | — | 0.65 | — |
| 112 | — | — | — | — | — | — | — | — | — | — |

| No. | Alkaline spectral data | | | | | | | | | |
|---|---|---|---|---|---|---|---|---|---|---|
| | pH | $\lambda_{max}$ | $\varepsilon_{max}$ ($\times 10^{-3}$) | $\lambda_{min}$ | Spectral ratios 230 | 240 | 250 | 270 | 280 | 290 |
| 111 | 12 | 260 | 9.2 | 230 | — | — | 0.99 | — | 0.61 | — |
| 112 | 11 | 254 | 13.6 | — | — | — | — | — | — | — |

REFERENCES

| No. | Origin and synthesis | $[\alpha]^t_D$ | pK | Spectral data | $R_f$ |
|---|---|---|---|---|---|
| 111 | D: *238* C: *267* | *267* | *246* | *242, 267*[b,c,d,e] | — |
| 112 | C: *24* | — | — | *24* | *24* |

| No. | Compound | Abbreviation | Symbol | Structure | Formula (mol wt) | Melting point °C | $[\alpha]^t_D$ | pK Basic | pK Acidic |
|---|---|---|---|---|---|---|---|---|---|
| 113 | **7-Methyldeoxyguanosine** | 7MedGuo or 7dMeGuo | $m^7dG$ | 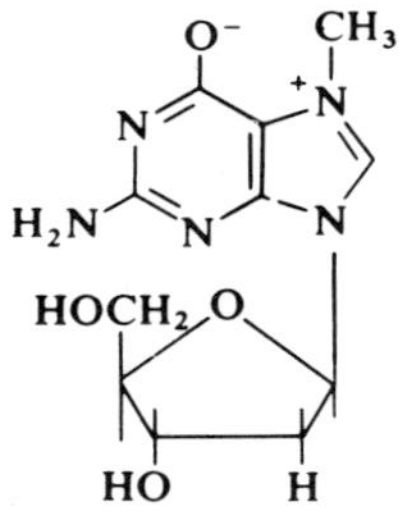 | $C_{11}H_{15}N_5O_4$ (281.27) | None (dec) | — | — | — |
| 114 | **Deoxyuridine** | dUrd | dU | 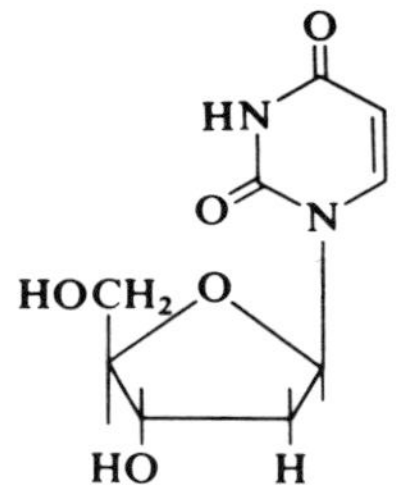 | $C_9H_{12}N_2O_5$ (228.20) | 163° | $+50.0^{22}$ (1.1, 1*N* NaOH) | — | 9.3, >13 |

| No. | Acidic spectral data | | | | | | | | | |
|---|---|---|---|---|---|---|---|---|---|---|
| | pH | $\lambda_{max}$ | $\varepsilon_{max}$ ($\times 10^{-3}$) | $\lambda_{min}$ | Spectral ratios 230 | 240 | 250 | 270 | 280 | 290 |
| 113 | 1[a] | 256 | 10.8 | 229 | — | — | — | — | — | — |
| 114 | 1 | 262 | 10.2 | 231 | — | — | 0.72 | — | 0.37 | — |

| No. | Neutral spectral data | | | | | | | | | |
|---|---|---|---|---|---|---|---|---|---|---|
| | pH | $\lambda_{max}$ | $\varepsilon_{max}$ ($\times 10^{-3}$) | $\lambda_{min}$ | Spectral ratios 230 | 240 | 250 | 270 | 280 | 290 |
| 113 | 6 | 257 | — | 235 | — | — | — | — | — | — |
| 114 | 7 | 262 | 10.2 | 231 | — | — | 0.72 | — | 0.37 | — |

| No. | Alkaline spectral data | | | | | | | | | |
|---|---|---|---|---|---|---|---|---|---|---|
| | pH | $\lambda_{max}$ | $\varepsilon_{max}$ ($\times 10^{-3}$) | $\lambda_{min}$ | Spectral ratios 230 | 240 | 250 | 270 | 280 | 290 |
| 113 | 9[a] | — | — | — | — | — | — | — | — | — |
| 114 | 12 | 262 | 7.6 | 242 | — | — | 0.81 | — | 0.31 | — |

## REFERENCES

| No. | Origin and synthesis | $[\alpha]^t_D$ | pK | Spectral data | $R_f$ |
|---|---|---|---|---|---|
| 113 | C: *10, 41* | — | — | *10*[s], *41*[c] | *10* |
| 114 | D: *238, 243*[a] | *243* | *163* | *163*[b], *232*[e] | *243* |

| No. | Compound | Abbreviation | Symbol | Structure | Formula (mol wt) | Melting point °C | $[\alpha]^t_D$ | pK Basic | pK Acidic |
|---|---|---|---|---|---|---|---|---|---|
| 115 | **Thymidine** (5-Methyldeoxyuridine) | dThd | dT | 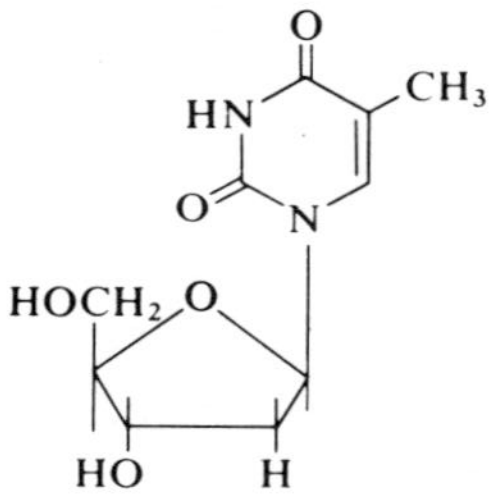 | $C_{10}H_{14}N_2O_5$ (242.23) | 183–184° | $+32.8^{16}$ (1.04, 1*N* NaOH) | — | 9.8, >13 |
| 116 | **5-Hydroxymethyldeoxyuridine** | 5HmdUrd | $hm^5dU$ | 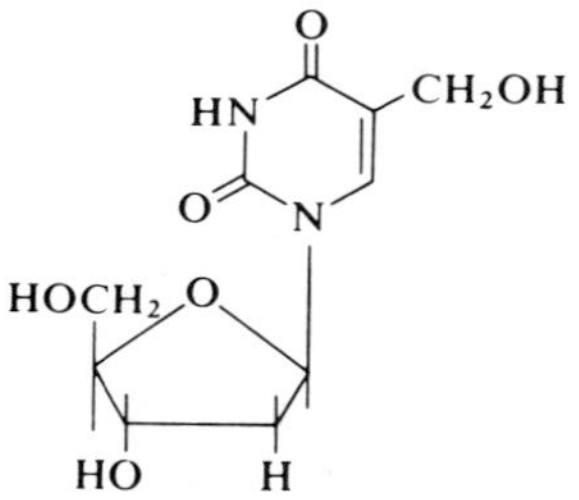 | $C_{10}H_{14}N_2O_6$ (258.23) | 180–182° | $+19^{20}$ (W) | — | — |

Acidic spectral data

| No. | pH | $\lambda_{max}$ | $\varepsilon_{max}$ ($\times 10^{-3}$) | $\lambda_{min}$ | Spectral ratios 230 | 240 | 250 | 270 | 280 | 290 |
|---|---|---|---|---|---|---|---|---|---|---|
| 115 | 1 | 267 | 9.65 | 235 | — | — | 0.65 | — | 0.72 | 0.24 |
| 116 | 2 | 264 | 9.6 | 233 | — | — | 0.68 | — | 0.52 | — |

Neutral spectral data

| No. | pH | $\lambda_{max}$ | $\varepsilon_{max}$ ($\times 10^{-3}$) | $\lambda_{min}$ | Spectral ratios 230 | 240 | 250 | 270 | 280 | 290 |
|---|---|---|---|---|---|---|---|---|---|---|
| 115 | 7 | 267 | 9.65 | 235 | — | — | 0.65 | — | 0.72 | 0.24 |
| 116 | — | — | — | — | — | — | — | — | — | — |

Alkaline spectral data

| No. | pH | $\lambda_{max}$ | $\varepsilon_{max}$ ($\times 10^{-3}$) | $\lambda_{min}$ | Spectral ratios 230 | 240 | 250 | 270 | 280 | 290 |
|---|---|---|---|---|---|---|---|---|---|---|
| 115 | 13 | 267 | 7.4 | 246 | — | — | 0.75 | — | 0.67 | 0.16 |
| 116 | 12 | 264 | 7.0 | 243 | — | — | 0.78 | — | 0.44 | — |

REFERENCES

| No. | Origin and Synthesis | $[\alpha]^t_D$ | pK | Spectral data | $R_f$ |
|---|---|---|---|---|---|
| 115 | D: *238* C: *128* | *241* | *163* | *163*[b], *232*[b,e] | *21, 18* |
| 116 | C: *69, 152*[a] | *152* | — | *69* | *69* |

| No. | Compound | Abbreviation | Symbol | Structure | Formula (mol wt) | Melting point °C | $[\alpha]_D^t$ | pK Basic | pK Acidic |
|---|---|---|---|---|---|---|---|---|---|
| **RIBONUCLEOTIDES** | | | | | | | | | |
| 117 | **Adenosine 2′-phosphate** | Ado-2′-P<br>2′-AMP | | | $C_{10}H_{14}N_5O_7P$ (347.22) | 183° (dec) | $-65.4^{22}$ (0.5, 0.5*M*, $Na_2HPO_4$) | 3.8 | — |
| 118 | **Adenosine 3′-phosphate** | Ado-3′-P<br>3′-AMP | Ap<br>A- | | $C_{10}H_{14}N_5O_7P$ (347.22) | 195° (dec) | $-45.4^{22}$ (0.5, 0.5*M* $Na_2HPO_4$) | 3.65 | — |

| No. | Acidic spectral data | | | | | | | | | | Neutral spectral data | | | | | | | | | | Alkaline spectral data | | | | | | | | | |
|---|---|---|---|---|---|---|---|---|---|---|---|---|---|---|---|---|---|---|---|---|---|---|---|---|---|---|---|---|---|---|
| | pH | $\lambda_{max}$ | $\varepsilon_{max}$ ($\times 10^{-3}$) | $\lambda_{min}$ | Spectral ratios 230 | 240 | 250 | 270 | 280 | 290 | pH | $\lambda_{max}$ | $\varepsilon_{max}$ ($\times 10^{-3}$) | $\lambda_{min}$ | Spectral ratios 230 | 240 | 250 | 270 | 280 | 290 | pH | $\lambda_{max}$ | $\varepsilon_{max}$ ($\times 10^{-3}$) | $\lambda_{min}$ | Spectral ratios 230 | 240 | 250 | 270 | 280 | 290 |
| 117 | 2 | 257[z] | 14.4[z] | 229[z] | — | — | 0.85 | — | 0.23 | 0.04 | 7 | 259[z] | 15.4[z] | — | — | — | 0.80 | — | 0.15 | 0.01 | 12 | 259[z] | 15.4[z] | — | — | — | 0.80 | — | 0.15 | — |
| 118 | 1 | 257 | 15.1 | 230 | — | — | 0.85 | 0.71 | 0.22* | 0.04* | 7 | — | — | — | — | — | 0.80 | — | 0.15 | 0.01 | 13 | 259 | 15.4 | 227 | — | — | 0.78 | 0.73 | 0.22 | 0.05 |

REFERENCES

| No. | Origin and synthesis | $[\alpha]_D^t$ | pK | Spectral data | $R_f$ |
|---|---|---|---|---|---|
| 117 | C: *191, 193, 219* R: *182, 221* | *221* | *220, 170, 218* | *179*[e], *223*[z] | *258, 219, 263* |
| 118 | C: *193, 190, 219* R: *182, 221, 194*[z] | *221* | *220, 170, 218* | *265**[b], *179*[e] | *258, 219, 263* |

| No. | Compound | Abbreviation | Symbol | Structure | Formula (mol wt) | Melting point °C | $[\alpha]_D^t$ | pK Basic | pK Acidic |
|---|---|---|---|---|---|---|---|---|---|
| 119 | **Adenosine 5′-phosphate** | Ado-5′-P<br>AMP | pA<br>-A | 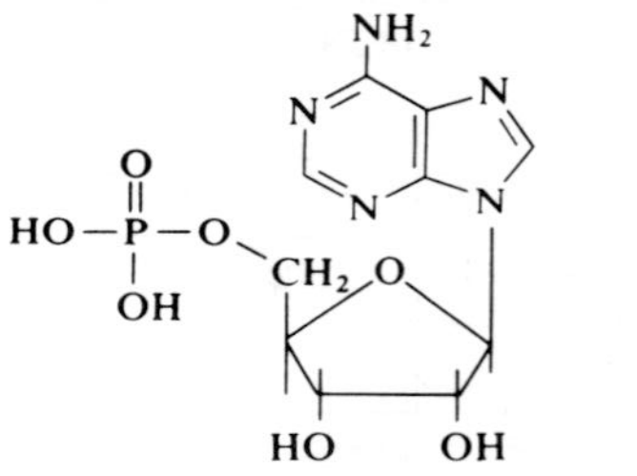 | $C_{10}H_{14}N_5O_7P$ (347.22) | 192° (dec) | $-26.0^{20}$ (1.0, 10% HCl) | 3.8 | — |
| 120 | **Adenosine 5′-diphosphate** | Ado-5′-$P_2$<br>ADP | ppA | 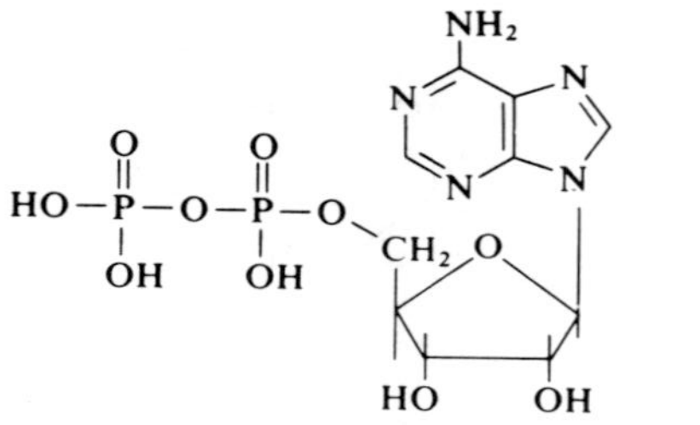 | $C_{10}H_{15}N_5O_{10}P_2$ (427.21) | — | — | 3.9 | — |

| No. | Acidic spectral data | | | | | | | | | |
|---|---|---|---|---|---|---|---|---|---|---|
| | pH | $\lambda_{max}$ | $\varepsilon_{max}$ ($\times 10^{-3}$) | $\lambda_{min}$ | Spectral ratios 230 | 240 | 250 | 270 | 280 | 290 |
| 119 | 2 | 257 | 15.0 | 230 | 0.23 | 0.43 | 0.84 | 0.68 | 0.22 | 0.04 |
| 120 | 2 | 257 | 15.0 | 230 | — | — | 0.85 | — | 0.21 | — |

| No. | Neutral spectral data | | | | | | | | | |
|---|---|---|---|---|---|---|---|---|---|---|
| | pH | $\lambda_{max}$ | $\varepsilon_{max}$ ($\times 10^{-3}$) | $\lambda_{min}$ | Spectral ratios 230 | 240 | 250 | 270 | 280 | 290 |
| 119 | 7 | 259 | 15.4 | 227 | 0.18 | 0.39 | 0.79 | 0.66 | 0.16 | 0.01 |
| 120 | 7 | 259 | 15.4 | 227 | — | — | 0.78 | — | 0.16 | — |

| No. | Alkaline spectral data | | | | | | | | | |
|---|---|---|---|---|---|---|---|---|---|---|
| | pH | $\lambda_{max}$ | $\varepsilon_{max}$ ($\times 10^{-3}$) | $\lambda_{min}$ | Spectral ratios 230 | 240 | 250 | 270 | 280 | 290 |
| 119 | 11 | 259 | 15.4 | 227 | — | — | 0.79 | — | 0.16 | — |
| 120 | 11 | 259 | 15.4 | 227 | — | — | 0.78 | — | 0.15 | — |

REFERENCES

| No. | Origin and synthesis | $[\alpha]_D^t$ | pK | Spectral data | $R_f$ |
|---|---|---|---|---|---|
| 119 | C: *191, 197* R: *195* N: *198* | *261* | *220, 170, 212* | *212, 179*[e]*, 184*[b]*, 183*[b] | *188, 219, 263* |
| 120 | C: *225, 211, 188, 196* E: *226* N: *198* | — | *212* | *212, 183*[b]*, 206* | *188* |

| No. | Compound | Abbreviation | Symbol | Structure | Formula (mol wt) | Melting point °C | $[\alpha]^{t}_{D}$ | pK Basic | pK Acidic |
|---|---|---|---|---|---|---|---|---|---|
| 121 | **Adenosine-5′-triphosphate** | Ado-5′-$P_3$ ATP | pppA | | $C_{10}H_{16}N_5O_{13}P_3$ (507.19) | — | — | 4.1 | — |
| 122 | **1-Methyladenosine 3′(2′)-phosphate** | 1-MeAdo-3′(2′)-P | $m^1Ap$ $m^1A$- for 3′ | | $C_{11}H_{16}N_5O_7P$ (361.25) | — | — | 7.6[p] | — |

**Acidic spectral data**

| No. | pH | $\lambda_{max}$ | $\varepsilon_{max}$ ($\times 10^{-3}$) | $\lambda_{min}$ | 230 | 240 | 250 | 270 | 280 | 290 |
|---|---|---|---|---|---|---|---|---|---|---|
| 121 | 2 | 257 | 14.7 | 230 | — | — | 0.85 | — | 0.22 | — |
| 122 | 2 | 258 | 13.3[w] | 230 | 0.24 | 0.44 | 0.83 | 0.67 | 0.26 | 0.07 |

**Neutral spectral data**

| No. | pH | $\lambda_{max}$ | $\varepsilon_{max}$ ($\times 10^{-3}$) | $\lambda_{min}$ | 230 | 240 | 250 | 270 | 280 | 290 |
|---|---|---|---|---|---|---|---|---|---|---|
| 121 | 7 | 259 | 15.4 | 227 | — | — | 0.80 | — | 0.15 | — |
| 122 | — | — | — | — | — | — | — | — | — | — |

**Alkaline spectral data**

| No. | pH | $\lambda_{max}$ | $\varepsilon_{max}$ ($\times 10^{-3}$) | $\lambda_{min}$ | 230 | 240 | 250 | 270 | 280 | 290 |
|---|---|---|---|---|---|---|---|---|---|---|
| 121 | 11 | 259 | 15.4 | 227 | — | — | 0.80 | — | 0.15 | — |
| 122 | — | — | — | — | — | — | — | — | — | — |

(Columns 230–290 are Spectral ratios.)

REFERENCES

| No. | Origin and synthesis | $[\alpha]^{t}_{D}$ | pK | Spectral data | $R_f$ |
|---|---|---|---|---|---|
| 121 | C: *211*, *214*, *188* N: *198* | — | *212* | *212*[p], *206*, *183*[b] | *188* |
| 122 | C: *90* R: 77 | — | *90* | *90* | *143* |

| No. | Compound | Abbreviation | Symbol | Structure | Formula (mol wt) | Melting point °C | $[\alpha]^{t}_{D}$ | pK Basic | pK Acidic |
|---|---|---|---|---|---|---|---|---|---|
| 123 | **1-Methyladenosine 5′-phosphate** | 1MeAdo-5′-P<br>1MeAMP | pm¹A<br>-m¹A | | $C_{11}H_{16}N_5O_7P$ (361.25) | — | — | — | — |
| 124 | **1-Methyladenosine 5′-diphosphate** | 1MeAdo-5′-$P_2$<br>1MeADP | ppm¹A | | $C_{11}H_{17}N_5O_{10}P_2$ (441.23) | — | — | — | — |

| No. | Acidic spectral data | | | | | | | | | |
|---|---|---|---|---|---|---|---|---|---|---|
| | pH | $\lambda_{max}$ | $\varepsilon_{max}$ ($\times 10^{-3}$) | $\lambda_{min}$ | Spectral ratios 230 | 240 | 250 | 270 | 280 | 290 |
| 123 | 2 | 258 | — | 232 | 0.34 | 0.46 | 0.81 | 0.74 | 0.32 | 0.10 |
| 124 | 2 | 257 | 11.9 | 234 | 0.37 | 0.44 | 0.84 | 0.66 | 0.23 | 0.04 |

| No. | Neutral spectral data | | | | | | | | | |
|---|---|---|---|---|---|---|---|---|---|---|
| | pH | $\lambda_{max}$ | $\varepsilon_{max}$ ($\times 10^{-3}$) | $\lambda_{min}$ | Spectral ratios 230 | 240 | 250 | 270 | 280 | 290 |
| 123 | W[q] | 259 | — | 233 | 0.22 | 0.34 | 0.77 | 0.74 | 0.27 | — |
| 124 | — | — | — | — | — | — | — | — | — | — |

| No. | Alkaline spectral data | | | | | | | | | |
|---|---|---|---|---|---|---|---|---|---|---|
| | pH | $\lambda_{max}$ | $\varepsilon_{max}$ ($\times 10^{-3}$) | $\lambda_{min}$ | Spectral ratios 230 | 240 | 250 | 270 | 280 | 290 |
| 123 | 12[o] | 259 | — | 230 | 0.22 | 0.36 | 0.75 | 0.71 | 0.36 | 0.30 |
| 124 | 12[o] | 259 | 12.5 | 232 | 0.32 | 0.40 | 0.75 | 0.74 | 0.36 | 0.31 |

## REFERENCES

| No. | Origin and synthesis | $[\alpha]^{t}_{D}$ | pK | Spectral data | $R_f$ |
|---|---|---|---|---|---|
| 123 | C: *164*, *8* | — | — | *164*, *8*[b] | *8*, *164* |
| 124 | C: *164* | — | — | *164*[b] | *164* |

| No. | Compound | Abbreviation | Symbol | Structure | Formula (mol wt) | Melting point °C | $[\alpha]^t_D$ | pK Basic | pK Acidic |
|---|---|---|---|---|---|---|---|---|---|
| 125 | **$N^6$-Methyladenosine 3′(2′)-phosphate** | 6MeAdo-3′(2′)-P | $m^6Ap$ for 3′, or $m^6A$- |  | $C_{11}H_{16}N_5O_7P$ (361.25) | — | — | — | — |
| 126 | **$N^6$-Methyladenosine 5′-phosphate** | 6MeAdo-5′-P<br>6MeAMP | $pm^6A$<br>$-m^6A$ |  | $C_{11}H_{16}N_5O_7P$ (361.25) | — | — | ~3.7[p] | — |

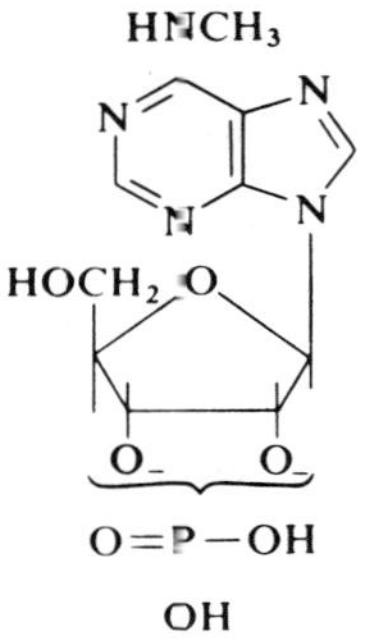

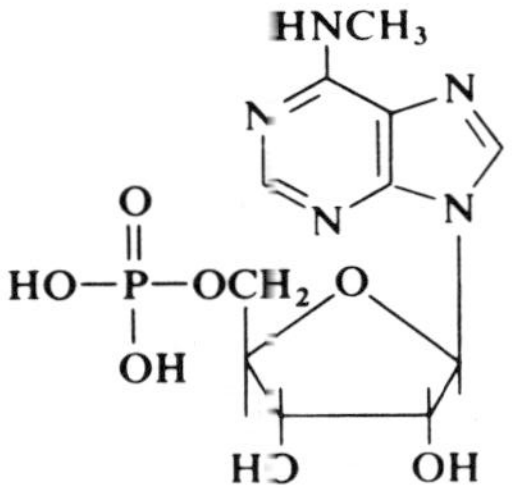

| No. | Acidic pH | $\lambda_{max}$ | $\varepsilon_{max}$ ($\times 10^{-3}$) | $\lambda_{min}$ | 230 | 240 | 250 | 270 | 280 | 290 | Neutral pH | $\lambda_{max}$ | $\varepsilon_{max}$ ($\times 10^{-3}$) | $\lambda_{min}$ | 230 | 240 | 250 | 270 | 280 | 290 | Alkaline pH | $\lambda_{max}$ | $\varepsilon_{max}$ ($\times 10^{-3}$) | $\lambda_{min}$ | 230 | 240 | 250 | 270 | 280 | 290 |
|---|---|---|---|---|---|---|---|---|---|---|---|---|---|---|---|---|---|---|---|---|---|---|---|---|---|---|---|---|---|---|
| 125 | 1 | 262 | 18.3 | 236 | — | — | 0.64 | 0.91 | 0.45 | 0.14 | — | — | — | — | — | — | — | — | — | — | 12 | 266 | — | — | — | — | — | — | — | — |
| 126 | 2 | 261 | 16.3 | 231 | 0.28 | 0.39 | 0.73 | 0.85 | 0.36* | 0.13 | W | 264 | 13.4 | 229 | 0.17 | 0.29 | 0.63 | 0.97 | 0.56 | 0.17 | 12 | 264* | 13.0 | 229 | 0.05 | 0.20 | 0.58 | 0.97 | 0.56* | 0.18 |

(Columns 230–290 under each of Acidic, Neutral and Alkaline spectral data are Spectral ratios.)

REFERENCES

| No. | Origin and synthesis | $[\alpha]^t_D$ | pK | Spectral data | $R_f$ |
|---|---|---|---|---|---|
| 125 | R : *165* | — | — | *265*[b], *165*[c] | *143*, *165* |
| 126 | C : *164*, *166* | — | *164* | *164*, *8**[b] | *164*, *166* |

| No. | Compound | Abbreviation | Symbol | Structure | Formula (mol wt) | Melting point °C | $[\alpha]^t_D$ | pK Basic | pK Acidic |
|---|---|---|---|---|---|---|---|---|---|
| 127 | **$N^6$-Methyladenosine 5′-diphosphate** | 6MeAdo-5′-$P_2$<br>6MeADP | | 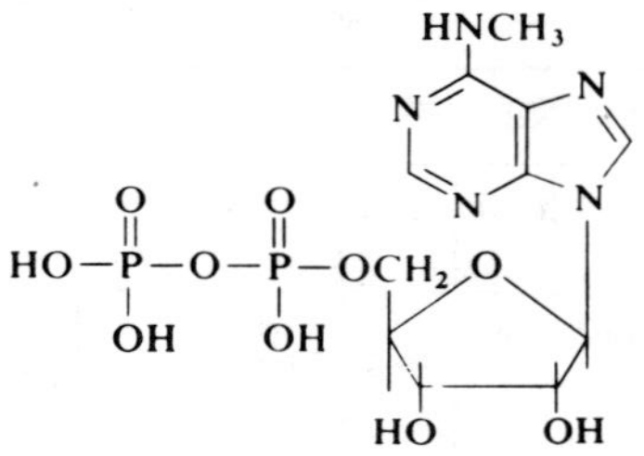 | $C_{11}H_{17}N_5O_{10}P_2$ (441.23) | — | — | ~3.7[p] | — |
| 128 | **$N^6$,$N^6$-Dimethyladenosine 5′-phosphate** | 6$Me_2$Ado-5′-P<br>6$Me_2$AMP | $pm^6_2A$ or $-m^6_2A$ | 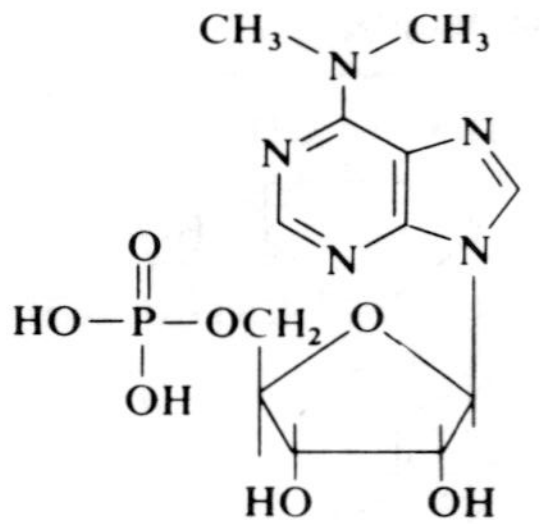 | $C_{12}H_{18}N_5O_7P$ (375.28) | 225° (dec) | $-51^{20}$ (2.0, W) | — | — |

Acidic spectral data

| No. | pH | $\lambda_{max}$ | $\varepsilon_{max}$ ($\times 10^{-3}$) | $\lambda_{min}$ | Spectral ratios 230 | 240 | 250 | 270 | 280 | 290 |
|---|---|---|---|---|---|---|---|---|---|---|
| 127 | 2 | 262 | 15.7 | 231 | 0.18 | 0.32 | 0.69 | 0.84 | 0.29 | 0.08 |
| 128 | — | 268 | 18.3 | — | — | — | — | — | — | — |

Neutral spectral data

| No. | pH | $\lambda_{max}$ | $\varepsilon_{max}$ ($\times 10^{-3}$) | $\lambda_{min}$ | Spectral ratios 230 | 240 | 250 | 270 | 280 | 290 |
|---|---|---|---|---|---|---|---|---|---|---|
| 127 | — | — | — | — | — | — | — | — | — | — |
| 128 | 7 | 274 | — | — | — | — | — | — | — | — |

Alkaline spectral data

| No. | pH | $\lambda_{max}$ | $\varepsilon_{max}$ ($\times 10^{-3}$) | $\lambda_{min}$ | Spectral ratios 230 | 240 | 250 | 270 | 280 | 290 |
|---|---|---|---|---|---|---|---|---|---|---|
| 127 | 12 | 265 | 15.4 | 229 | 0.14 | 0.26 | 0.60 | 0.99 | 0.57 | 0.18 |
| 128 | — | — | — | — | — | — | — | — | — | — |

REFERENCES

| No. | Origin and synthesis | $[\alpha]^t_D$ | pK | Spectral data | $R_f$ |
|---|---|---|---|---|---|
| 127 | C: *164* | — | *164* | *164*[b] | *164* |
| 128 | C: *115, 166* | *115* | — | *115, 166*[c] | *115, 166, 167* |

| No. | Compound | Abbreviation | Symbol | Structure | Formula (mol wt) | Melting point °C | $[\alpha]^{t}_{D}$ | pK Basic | pK Acidic |
|---|---|---|---|---|---|---|---|---|---|
| 129 | **$N^6$,$N^6$-Dimethyladenosine 5′-diphosphate** | 6Me$_2$Ado-5′-P$_2$<br>6Me$_2$ADP | | 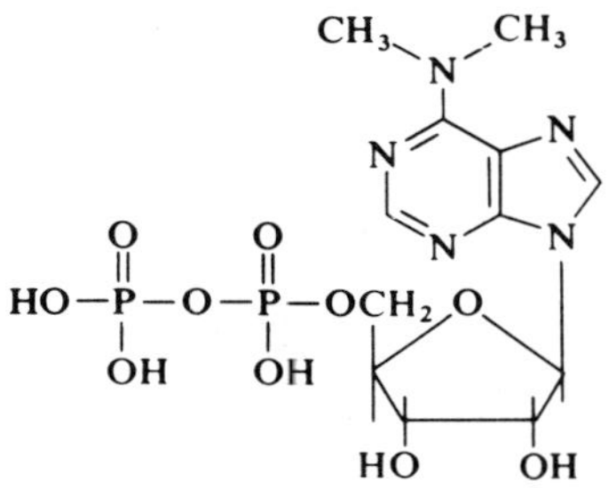 | $C_{12}H_{19}N_5O_{10}P_2$ (455.26) | — | — | — | — |
| 130 | **Cytidine 2′-phosphate** | Cyd-2′-P<br>2′-CMP | | 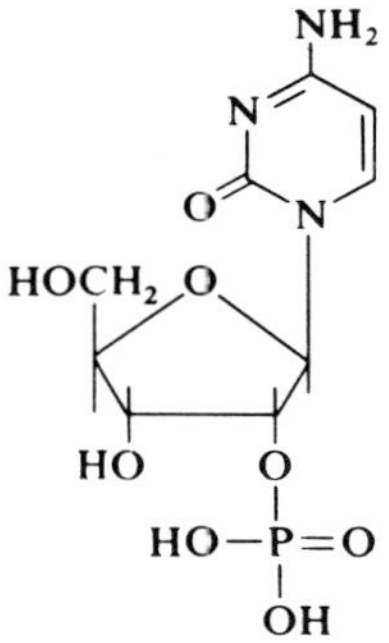 | $C_9H_{14}N_3O_8P$ (323.21) | 238–240° (dec) | +20.7[20] (1.0, W) | 4.4 | — |

| No. | Acidic pH | Acidic $\lambda_{max}$ | Acidic $\varepsilon_{max}$ ($\times 10^{-3}$) | Acidic $\lambda_{min}$ | Acidic 230 | Acidic 240 | Acidic 250 | Acidic 270 | Acidic 280 | Acidic 290 | Neutral pH | Neutral $\lambda_{max}$ | Neutral $\varepsilon_{max}$ ($\times 10^{-3}$) | Neutral $\lambda_{min}$ | Neutral 230 | Neutral 240 | Neutral 250 | Neutral 270 | Neutral 280 | Neutral 290 | Alkaline pH | Alkaline $\lambda_{max}$ | Alkaline $\varepsilon_{max}$ ($\times 10^{-3}$) | Alkaline $\lambda_{min}$ | Alkaline 230 | Alkaline 240 | Alkaline 250 | Alkaline 270 | Alkaline 280 | Alkaline 290 |
|---|---|---|---|---|---|---|---|---|---|---|---|---|---|---|---|---|---|---|---|---|---|---|---|---|---|---|---|---|---|---|
| 129 | — | — | — | — | — | — | — | — | — | — | — | — | — | — | — | — | — | — | — | — | — | — | — | — | — | — | — | — | — | — |
| 130 | 2 | 278 | 12.7 | 240 | — | — | 0.48 | — | 1.80 | 1.22 | 7 | — | — | — | — | — | 0.90 | — | 0.85 | 0.26 | 12 | 272 | 8.6 | 250 | — | — | 0.90 | — | 0.85 | 0.26 |

(Spectral ratios columns: 230, 240, 250, 270, 280, 290.)

REFERENCES

| No. | Origin and synthesis | $[\alpha]^{t}_{D}$ | pK | Spectral data | $R_f$ |
|---|---|---|---|---|---|
| 129 | C : *167* | — | — | — | *167* |
| 130 | R : *215*[a], *192*, *178* | *215* | *218*, *170*, *192* | *223*, *179*[b,e] | *258* |

| No. | Compound | Abbreviation | Symbol | Structure | Formula (mol wt) | Melting point °C | $[\alpha]^t_D$ | pK Basic | pK Acidic |
|---|---|---|---|---|---|---|---|---|---|
| 131 | **Cytidine 3′-phosphate** | Cyd-3′-P<br>3′-CMP | Cp<br>C- | | $C_9H_{14}N_3O_8P$ (323.21) | 232–234° (dec) | $+49.4^{20}$ (1.0, W) | 4.3 | — |
| 132 | **Cytidine 5′-phosphate** | Cyd-5′-P<br>CMP | pC<br>-C | | $C_9H_{14}N_3O_8P$ (323.21) | 233° (dec) | $+27.1^{14}$ (0.54, W) | 4.5 | — |

Acidic spectral data

| No. | pH | $\lambda_{max}$ | $\varepsilon_{max}$ $(\times 10^{-3})$ | $\lambda_{min}$ | Spectral ratios 230 | 240 | 250 | 270 | 280 | 290 |
|---|---|---|---|---|---|---|---|---|---|---|
| 131 | 2 | 279 | 13.0 | 240 | — | — | 0.45* | 1.51 | 2.00* | 1.43* |
| 132 | 2 | 280 | 13.2 | 241 | 0.56 | 0.25 | 0.44 | 1.73 | 2.09 | 1.55 |

Neutral spectral data

| No. | pH | $\lambda_{max}$ | $\varepsilon_{max}$ $(\times 10^{-3})$ | $\lambda_{min}$ | Spectral ratios 230 | 240 | 250 | 270 | 280 | 290 |
|---|---|---|---|---|---|---|---|---|---|---|
| 131 | 7 | 270[z] | 9.0[z] | 250[z] | — | — | 0.86 | — | 0.93 | 0.30 |
| 132 | 7 | 271 | 9.1 | 249 | 1.07 | 0.92 | 0.84 | 1.21 | 0.98 | 0.33 |

Alkaline spectral data

| No. | pH | $\lambda_{max}$ | $\varepsilon_{max}$ $(\times 10^{-3})$ | $\lambda_{min}$ | Spectral ratios 230 | 240 | 250 | 270 | 280 | 290 |
|---|---|---|---|---|---|---|---|---|---|---|
| 131 | 12 | 272 | 8.9 | 250 | — | — | 0.86 | 1.16 | 0.93 | 0.30* |
| 132 | 11 | 271 | 9.1 | 249 | — | — | 0.84 | — | 0.98 | 0.33 |

REFERENCES

| No. | Origin and synthesis | $[\alpha]^t_D$ | pK | Spectral data | $R_f$ |
|---|---|---|---|---|---|
| 131 | R : *215*[a], *192*, *178* | *215* | *218*, *170*, *192* | *223*, *179*,[b,e] *181*[z,b,c,d], *215*, *265**[b,e] | *258* |
| 132 | C : *190*, *196* R : *195* N : *198* | *190* | *212* | *212*, *183*[b,c], *184*[b], *179*, *205* | — |

| No. | Compound | Abbreviation | Symbol | Structure | Formula (mol wt) | Melting point °C | $[\alpha]_D^t$ | pK Basic | pK Acidic |
|---|---|---|---|---|---|---|---|---|---|
| 133 | **Cytidine 5′-diphosphate** | Cyd-5′-$P_2$ CDP | | 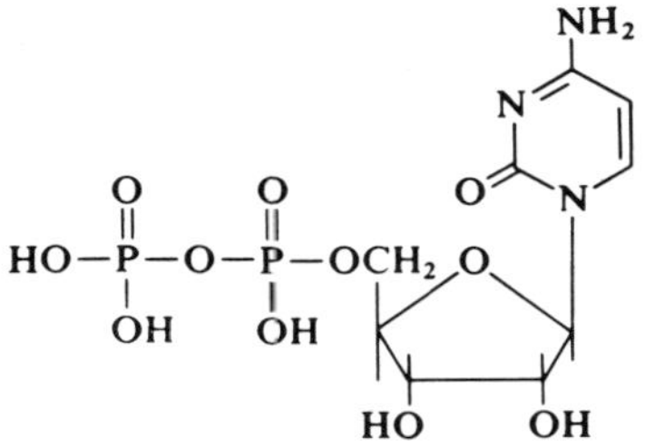 | $C_9H_{15}N_3O_{11}P_2$ (403.18) | — | — | 4.6 | — |
| 134 | **Cytidine 5′-triphosphate** | Cyd-5′-$P_3$ CTP | | 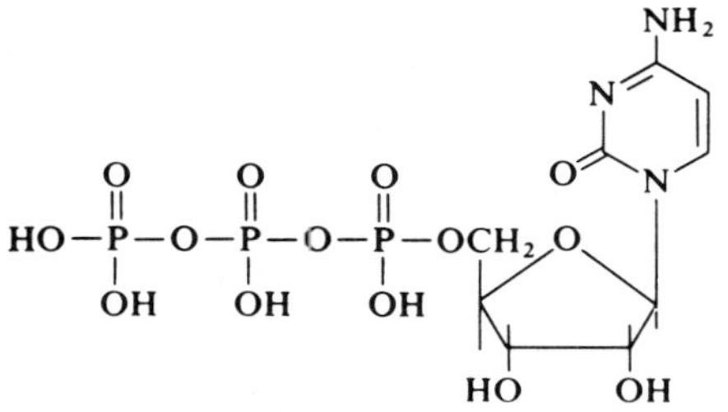 | $C_9H_{16}N_3O_{14}P_3$ (483.16) | — | — | 4.8 | — |

| No. | Acidic spectral data | | | | | | | | | | Neutral spectral data | | | | | | | | | | Alkaline spectral data | | | | | | | | | |
|---|---|---|---|---|---|---|---|---|---|---|---|---|---|---|---|---|---|---|---|---|---|---|---|---|---|---|---|---|---|---|
| | pH | $\lambda_{max}$ | $\varepsilon_{max}$ ($\times 10^{-3}$) | $\lambda_{min}$ | Spectral ratios 230 | 240 | 250 | 270 | 280 | 290 | pH | $\lambda_{max}$ | $\varepsilon_{max}$ ($\times 10^{-3}$) | $\lambda_{min}$ | Spectral ratios 230 | 240 | 250 | 270 | 280 | 290 | pH | $\lambda_{max}$ | $\varepsilon_{max}$ ($\times 10^{-3}$) | $\lambda_{min}$ | Spectral ratios 230 | 240 | 250 | 270 | 280 | 290 |
| 133 | 2 | 280 | 12.8 | 241 | — | — | 0.46 | — | 2.07 | 1.48 | 7 | 271 | 9.1 | 249 | — | — | 0.83 | — | 0.98 | 0.32 | 11 | 271 | 9.1 | 249 | — | — | 0.83 | — | 0.98 | — |
| 134 | 2 | 280 | 12.8 | 241 | — | — | 0.45 | — | 2.12 | — | 7 | 271 | 9.0 | 249 | — | — | 0.84 | — | 0.97 | — | 11 | 271 | 9.0 | 249 | — | — | 0.84 | — | 0.97 | — |

REFERENCES

| No. | Origin and synthesis | $[\alpha]_D^t$ | pK | Spectral data | $R_f$ |
|---|---|---|---|---|---|
| 133 | C : *196*, N : *198* | — | *212* | *212*, *179*[e], *183*[b] | — |
| 134 | N : *198* | — | *212* | *212*[b], *183*[b] | — |

| No. | Compound | Abbreviation | Symbol | Structure | Formula (mol wt) | Melting point °C | $[\alpha]^t_D$ | pK Basic | pK Acidic |
|---|---|---|---|---|---|---|---|---|---|
| 135 | **2-Thiocytidine 3′(2′)-phosphate** | 2 Syd-3′(2′)-P or 2SCyd-3′(2′)-P | $s^2Cp$ for 3′ | | $C_9H_{14}N_3O_7SP$ (339.27) | — | — | — | — |
| 136 | **3-Methylcytidine 3′(2′)-phosphate** | 3MeCyd-3′(2′)-P | $m^3Cp$ $m^3C$- for 3′ | | $C_{10}H_{16}N_3O_8P$ (337.22) | — | — | ~9.0[p] | — |

| No. | Acidic spectral data | | | | Spectral ratios | | | | | | Neutral spectral data | | | | Spectral ratios | | | | | | Alkaline spectral data | | | | Spectral ratios | | | | | |
|---|---|---|---|---|---|---|---|---|---|---|---|---|---|---|---|---|---|---|---|---|---|---|---|---|---|---|---|---|---|---|
| | pH | $\lambda_{max}$ | $\varepsilon_{max}$ ($\times 10^{-3}$) | $\lambda_{min}$ | 230 | 240 | 250 | 270 | 280 | 290 | pH | $\lambda_{max}$ | $\varepsilon_{max}$ ($\times 10^{-3}$) | $\lambda_{min}$ | 230 | 240 | 250 | 270 | 280 | 290 | pH | $\lambda_{max}$ | $\varepsilon_{max}$ ($\times 10^{-3}$) | $\lambda_{min}$ | 230 | 240 | 250 | 270 | 280 | 290 |
| 135 | 1 | 227<br>276 | — | 247 | 1.24 | 0.91 | 0.75 | 1.24 | 1.28 | 0.83 | W | 248 | — | 220 | 0.64 | 0.94 | 1.08 | 0.88 | 0.69 | 0.39 | 13 | 249 | — | 228 | 0.71 | 0.93 | 1.08 | 0.89 | 0.71 | 0.40 |
| 136 | 1 | 276 | 11.5 | 242 | — | — | — | — | — | — | 7 | 276 | 11.2 | 242 | — | — | — | — | — | — | — | — | — | — | — | — | — | — | — | — |

REFERENCES

| No. | Origin and synthesis | $[\alpha]^t_D$ | pK | Spectral data | $R_f$ |
|---|---|---|---|---|---|
| 135 | R: *270* | — | — | *270*[b] | *270* |
| 136 | C: *130* | — | *130* | *130* | *13, 130, 143* |

| No. | Compound | Abbreviation | Symbol | Structure | Formula (mol wt) | Melting point °C | $[\alpha]^t_D$ | pK Basic | pK Acidic |
|---|---|---|---|---|---|---|---|---|---|
| 137 | **3-Methylcytidine 5′-phosphate** | 3MeCyd-5′-P<br>3MeCMP | $pm^3C$<br>$-m^3C$ | 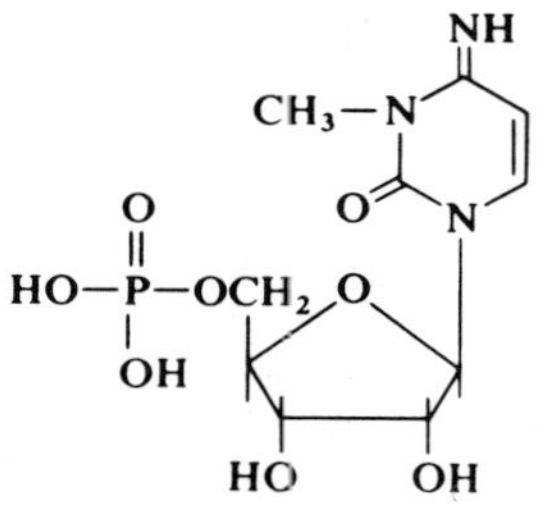 | $C_{10}H_{16}N_3O_8P$ (337.22) | — | — | — | — |
| 138 | **3-Methylcytidine 5′-diphosphate** | 3MeCyd-5′-$P_2$<br>3MeCDP | | 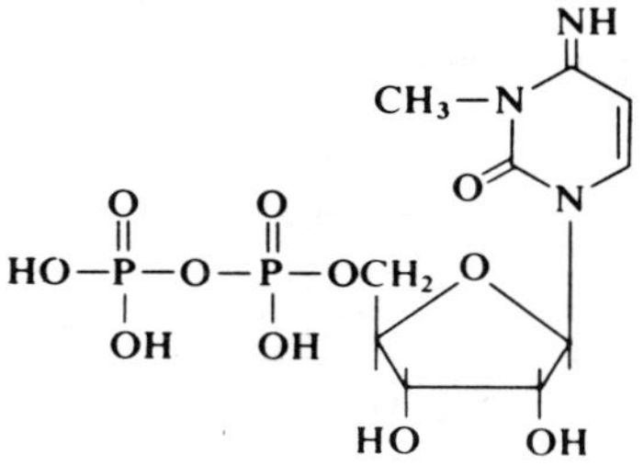 | $C_{10}H_{17}N_3O_{11}P_2$ (417.21) | — | — | ~9.0[p] | — |

| No. | Acidic spectral data | | | | | | | | | | Neutral spectral data | | | | | | | | | | Alkaline spectral data | | | | | | | | | |
|---|---|---|---|---|---|---|---|---|---|---|---|---|---|---|---|---|---|---|---|---|---|---|---|---|---|---|---|---|---|---|
| | pH | $\lambda_{max}$ | $\varepsilon_{max}$ $(\times 10^{-3})$ | $\lambda_{min}$ | Spectral ratios 230 | 240 | 250 | 270 | 280 | 290 | pH | $\lambda_{max}$ | $\varepsilon_{max}$ $(\times 10^{-3})$ | $\lambda_{min}$ | Spectral ratios 230 | 240 | 250 | 270 | 280 | 290 | pH | $\lambda_{max}$ | $\varepsilon_{max}$ $(\times 10^{-3})$ | $\lambda_{min}$ | Spectral ratios 230 | 240 | 250 | 270 | 280 | 290 |
| 137 | — | — | — | — | — | — | — | — | — | — | — | — | — | — | — | — | — | — | — | — | — | — | — | — | — | — | — | — | — | — |
| 138 | 1 | 278 | 11.0 | 241 | — | — | — | — | — | — | 7 | 277 | 11.0 | 241 | — | — | — | — | — | — | — | — | — | — | — | — | — | — | — | — |

REFERENCES

| No. | Origin and synthesis | $[\alpha]^t_D$ | pK | Spectral data | $R_f$ |
|---|---|---|---|---|---|
| 137 | C : *125* | — | — | — | *125* |
| 138 | C : *130* | — | *130* | *130* | *130* |

| No. | Compound / Abbreviation | Symbol | Structure | Formula (mol wt) | Melting point °C | $[\alpha]^t_D$ | pK Basic | pK Acidic |
|---|---|---|---|---|---|---|---|---|
| 139 | **$N^4$-Methylcytidine 3′(2′)-phosphate** 4MeCyd-3′(2′)-P | $m^4Cp$ $m^4C$- for 3′ | 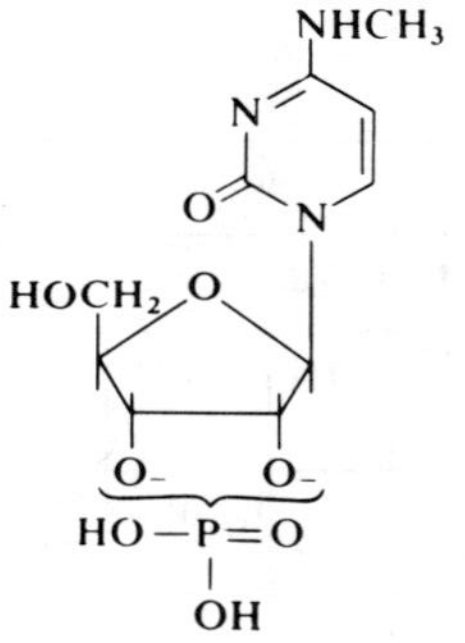 | $C_{10}H_{16}N_3O_8P$ (337.22) | — | — | — | — |
| 140 | **$N^4$-Methylcytidine 5′-phosphate** 4MeCyd-5′-P 4MeCMP | $pm^4C$ -$m^4C$ | 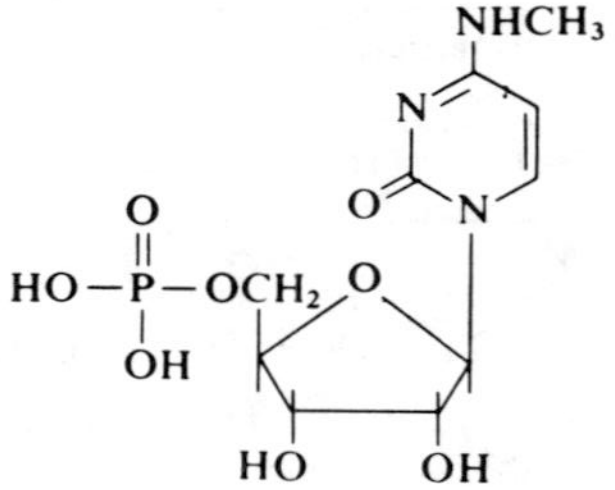 | $C_{10}H_{16}N_3O_8P$ (337.22) | — | — | — | — |

| No. | Acidic pH | Acidic $\lambda_{max}$ | Acidic $\varepsilon_{max}$ ($\times 10^{-3}$) | Acidic $\lambda_{min}$ | Acidic ratio 230 | 240 | 250 | 270 | 280 | 290 | Neutral pH | Neutral $\lambda_{max}$ | Neutral $\varepsilon_{max}$ ($\times 10^{-3}$) | Neutral $\lambda_{min}$ | Neutral ratio 230 | 240 | 250 | 270 | 280 | 290 | Alkaline pH | Alkaline $\lambda_{max}$ | Alkaline $\varepsilon_{max}$ ($\times 10^{-3}$) | Alkaline $\lambda_{min}$ | Alkaline ratio 230 | 240 | 250 | 270 | 280 | 290 |
|---|---|---|---|---|---|---|---|---|---|---|---|---|---|---|---|---|---|---|---|---|---|---|---|---|---|---|---|---|---|---|
| 139 | 1 | 281 | 12.9 | 242 | — | — | — | — | — | — | — | — | — | — | — | — | — | — | — | — | — | — | — | — | — | — | — | — | — | — |
| 140 | 1 | 280 | 14.8 | 242 | — | — | — | — | — | — | W | 237<br>272 | — | 227<br>248 | — | — | — | — | — | — | — | — | — | — | — | — | — | — | — | — |

## REFERENCES

| No. | Origin and synthesis | $[\alpha]^t_D$ | pK | Spectral data | $R_f$ |
|---|---|---|---|---|---|
| 139 | C: *130* | — | — | *130* | *130* |
| 140 | C: *130, 59, 168* | — | — | *130, 168*[c] | *130, 59* |

| No. | Compound | Abbreviation | Symbol | Structure | Formula (mol wt) | Melting point °C | $[\alpha]_D^t$ | pK Basic | pK Acidic |
|---|---|---|---|---|---|---|---|---|---|
| 141 | **$N^4$-Methylcytidine 5′-diphosphate** | 4MeCyd-5′-$P_2$<br>4MeCDP | | 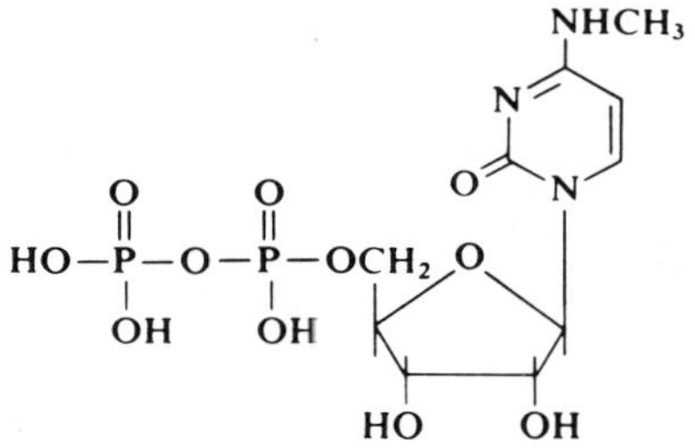 | $C_{10}H_{17}N_3O_{11}P_2$ (417.21) | — | — | — | — |
| 142 | **5-Methylcytidine 5′-phosphate** | 5MeCyd-5′-P<br>5MeCMP | $pm^5C$<br>$-m^5C$ | 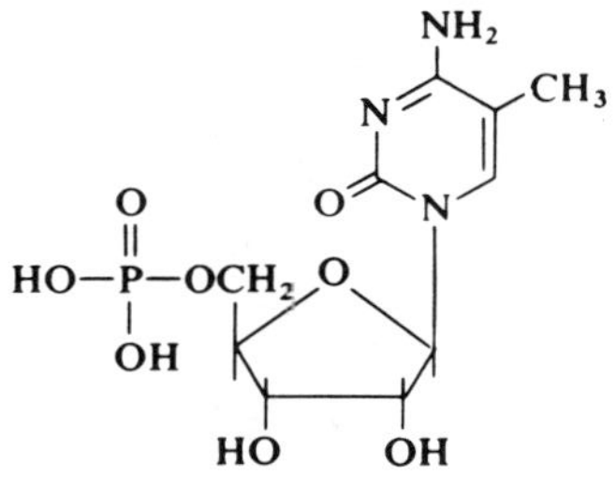 | $C_{10}H_{16}N_3O_8P$ (337.22) | — | — | — | — |

| No. | Acidic spectral data | | | | | | | | | | Neutral spectral data | | | | | | | | | | Alkaline spectral data | | | | | | | | | |
|---|---|---|---|---|---|---|---|---|---|---|---|---|---|---|---|---|---|---|---|---|---|---|---|---|---|---|---|---|---|---|
| | pH | $\lambda_{max}$ | $\varepsilon_{max}$ ($\times 10^{-3}$) | $\lambda_{min}$ | Spectral ratios 230 | 240 | 250 | 270 | 280 | 290 | pH | $\lambda_{max}$ | $\varepsilon_{max}$ ($\times 10^{-3}$) | $\lambda_{min}$ | Spectral ratios 230 | 240 | 250 | 270 | 280 | 290 | pH | $\lambda_{max}$ | $\varepsilon_{max}$ ($\times 10^{-3}$) | $\lambda_{min}$ | Spectral ratios 230 | 240 | 250 | 270 | 280 | 290 |
| 141 | 1 | 280 | 12.9 | 241 | — | — | — | — | — | — | — | — | — | — | — | — | — | — | — | — | — | — | — | — | — | — | — | — | — | — |
| 142 | 4[q] | 284 | 10.7 | — | — | — | — | — | — | — | 8 | 278 | 8.8 | — | — | — | — | — | — | — | — | — | — | — | — | — | — | — | — | — |

REFERENCES

| No. | Origin and synthesis | $[\alpha]_D^t$ | pK | Spectral data | $R_f$ |
|---|---|---|---|---|---|
| 141 | C: *130, 59* | — | — | *130* | *130, 59* |
| 142 | C: *59* | — | — | *169* | *59* |

| No. | Compound | Abbreviation | Symbol | Structure | Formula (mol wt) | Melting point °C | $[\alpha]^t_D$ | pK Basic | pK Acidic |
|---|---|---|---|---|---|---|---|---|---|
| 143 | **5-Methylcytidine 5′-diphosphate** | 5MeCyd-5′-$P_2$<br>5MeCDP | | 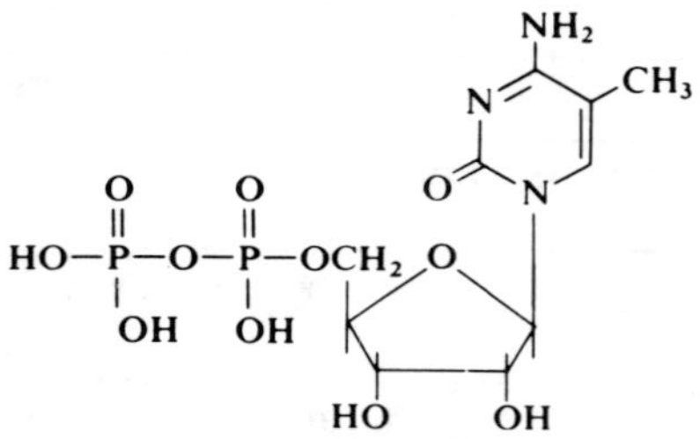 | $C_{10}H_{17}N_3O_{11}P_2$ (417.21) | — | — | — | — |
| 144 | **Guanosine 2′-phosphate** | Guo-2′-P<br>2′-GMP | | 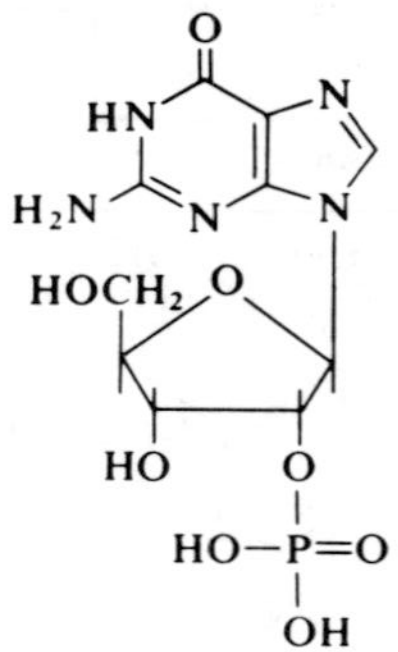 | $C_{10}H_{14}N_5O_8P$ (363.22) | 175–180° (dec)[z] (dihydrate) | $-57.0^{25}$[z] (1.0, 2% NaOH) | — | — |

| No. | Acidic pH | Acidic $\lambda_{max}$ | Acidic $\varepsilon_{max}$ ($\times 10^{-3}$) | Acidic $\lambda_{min}$ | Acidic ratio 230 | 240 | 250 | 270 | 280 | 290 | Neutral pH | Neutral $\lambda_{max}$ | Neutral $\varepsilon_{max}$ ($\times 10^{-3}$) | Neutral $\lambda_{min}$ | Neutral ratio 230 | 240 | 250 | 270 | 280 | 290 | Alkaline pH | Alkaline $\lambda_{max}$ | Alkaline $\varepsilon_{max}$ ($\times 10^{-3}$) | Alkaline $\lambda_{min}$ | Alkaline ratio 230 | 240 | 250 | 270 | 280 | 290 |
|---|---|---|---|---|---|---|---|---|---|---|---|---|---|---|---|---|---|---|---|---|---|---|---|---|---|---|---|---|---|---|
| 143 | — | — | — | — | — | — | — | — | — | — | — | — | — | — | — | — | — | — | — | — | — | — | — | — | — | — | — | — | — | — |
| 144 | 1 | — | — | — | — | — | 0.90 | — | 0.68 | 0.48 | 7 | — | — | — | — | — | 1.15 | — | 0.68 | 0.29 | 12 | — | — | — | — | — | 0.89 | — | 0.60 | 0.11 |

REFERENCES

| No. | Origin and synthesis | $[\alpha]^t_D$ | pK | Spectral data | $R_f$ |
|---|---|---|---|---|---|
| 143 | C : *59* | — | — | — | *59* |
| 144 | R : *222*[a,z], *170* | *222*[z] | — | *179* | — |

| No. | Compound | Abbreviation | Symbol | Structure | Formula (mol wt) | Melting point °C | $[\alpha]^t_D$ | pK Basic | pK Acidic |
|---|---|---|---|---|---|---|---|---|---|
| 145 | **Guanosine 3′-phosphate** | Guo-3′-P<br>3′-GMP | Gp<br>G- | 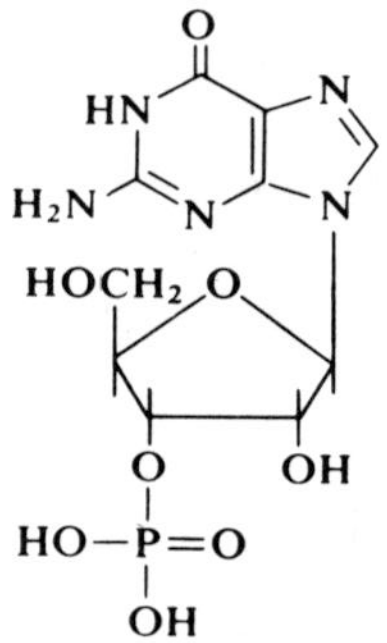 | $C_{10}H_{14}N_5O_8P$ (363.22) | 175–180° (dec)[z] (dihydrate) | $-57.0^{25}$[z] (1.0, 2% NaOH) | 2.3 | 9.7 |
| 146 | **Guanosine 5′-phosphate** | Guo-5′-P<br>GMP | pG<br>-G | 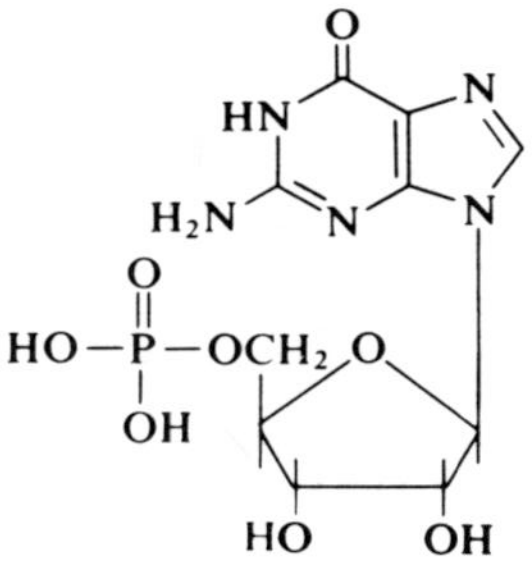 | $C_{10}H_{14}N_5O_8P$ (363.22) | 190–200° (dec) | — | 2.4 | 9.4 |

| No. | Acidic spectral data | | | | | | | | | | Neutral spectral data | | | | | | | | | | Alkaline spectral data | | | | | | | | | |
|---|---|---|---|---|---|---|---|---|---|---|---|---|---|---|---|---|---|---|---|---|---|---|---|---|---|---|---|---|---|---|
| | pH | $\lambda_{max}$ | $\varepsilon_{max}$ ($\times 10^{-3}$) | $\lambda_{min}$ | Spectral ratios 230 | 240 | 250 | 270 | 280 | 290 | pH | $\lambda_{max}$ | $\varepsilon_{max}$ ($\times 10^{-3}$) | $\lambda_{min}$ | Spectral ratios 230 | 240 | 250 | 270 | 280 | 290 | pH | $\lambda_{max}$ | $\varepsilon_{max}$ ($\times 10^{-3}$) | $\lambda_{min}$ | Spectral ratios 230 | 240 | 250 | 270 | 280 | 290 |
| 145 | 1 | 257 | 12.2 | 228 | — | — | 0.93 | 0.77 | 0.69 | 0.49 | 7 | 252 | 13.4 | 227 | — | — | 1.15 | — | 0.68 | 0.29 | 10.8[q] | 257 | 11.25 | 230 | — | — | 0.92 | — | 0.64 | 0.15 |
| 146 | 1 | 256 | 12.2 | 228 | 0.22 | 0.55 | 0.96 | 0.74 | 0.67 | 0.29 | 7 | 252 | 13.7 | 224 | 0.36 | 0.81 | 1.16 | 0.81 | 0.66 | 0.29 | 11 | 258 | 11.6 | 230 | 0.38 | 0.82 | 0.90 | 0.97 | 0.61 | 0.29 |

REFERENCES

| No. | Origin and synthesis | $[\alpha]^t_D$ | pK | Spectral data | $R_f$ |
|---|---|---|---|---|---|
| 145 | R : *222*[a,z], *170* | *222*[z] | *231* | *232*[b], *179*[e], *265*[b,e] | *258* |
| 146 | C : *190*, *199*, *189* N : *198* R : *195* | — | *212* | *212*, *183*[b], *184*[b], *198* | *189*, *213* |

| No. | Compound | Abbreviation | Symbol | Structure | Formula (mol wt) | Melting point °C | $[\alpha]^t_D$ | pK Basic | pK Acidic |
|---|---|---|---|---|---|---|---|---|---|
| 147 | **Guanosine 5′-diphosphate** | Guo-5′-$P_2$ GDP | ppG | 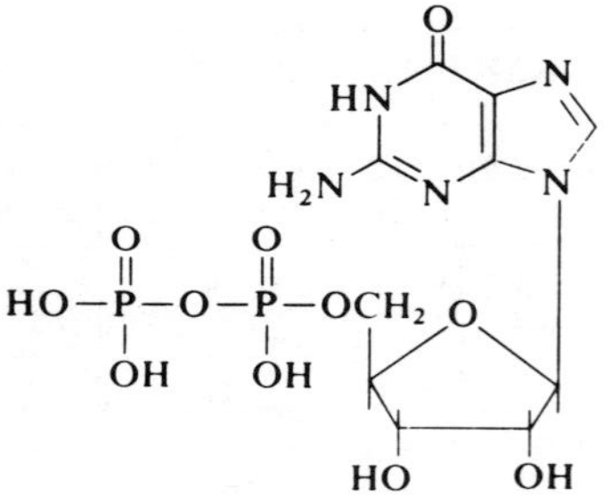 | $C_{10}H_{15}N_5O_{11}P_2$ (443.21) | — | — | 2.9 | 9.6 |
| 148 | **Guanosine 5′-triphosphate** | Guo-5′-$P_3$ GTP | pppG | 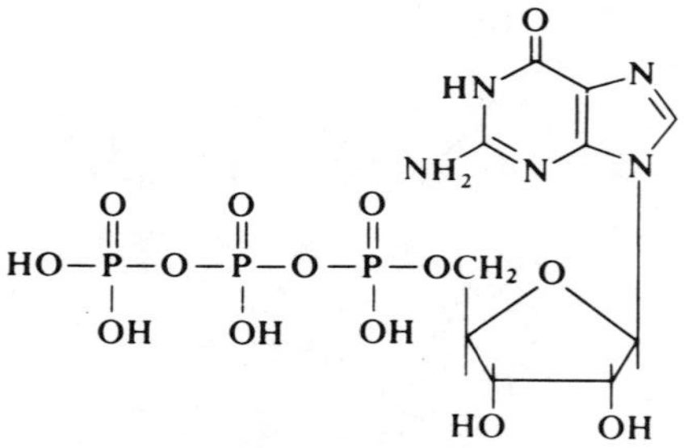 | $C_{10}H_{16}N_5O_{14}P_3$ (523.19) | — | — | 3.3 | 9.3 |

| No. | Acidic spectral data | | | | | | | | | | Neutral spectral data | | | | | | | | | | Alkaline spectral data | | | | | | | | | |
|---|---|---|---|---|---|---|---|---|---|---|---|---|---|---|---|---|---|---|---|---|---|---|---|---|---|---|---|---|---|---|
| | pH | $\lambda_{max}$ | $\varepsilon_{max}$ ($\times 10^{-3}$) | $\lambda_{min}$ | Spectral ratios | | | | | | pH | $\lambda_{max}$ | $\varepsilon_{max}$ ($\times 10^{-3}$) | $\lambda_{min}$ | Spectral ratios | | | | | | pH | $\lambda_{max}$ | $\varepsilon_{max}$ ($\times 10^{-3}$) | $\lambda_{min}$ | Spectral ratios | | | | | |
| | | | | | 230 | 240 | 250 | 270 | 280 | 290 | | | | | 230 | 240 | 250 | 270 | 280 | 290 | | | | | 230 | 240 | 250 | 270 | 280 | 290 |
| 147 | 1 | 256 | 12.3 | 228 | — | — | 0.95 | — | 0.67 | — | 7 | 253 | 13.7 | 224 | — | — | 1.15 | — | 0.66 | — | 11 | 258 | 11.7 | 230 | — | — | 0.91 | — | 0.61 | — |
| 148 | 1 | 256 | 12.4 | 228 | — | — | 0.96 | — | 0.67 | — | 7 | 253 | 13.7 | 223 | — | — | 1.17 | — | 0.66 | — | 11 | 257 | 11.9 | 230 | — | — | 0.92 | — | 0.59 | — |

## REFERENCES

| No. | Origin and synthesis | $[\alpha]^t_D$ | pK | Spectral data | $R_f$ |
|---|---|---|---|---|---|
| 147 | C: *213*, *196* N: *198* | — | *212* | *212*, *183*[b] | *213* |
| 148 | C: *213* N: *198* | — | *212* | *212*[b], *183*[b] | *213* |

| No. | Compound | Abbreviation | Symbol | Structure | Formula (mol wt) | Melting point °C | $[\alpha]^{t}_{D}$ | pK Basic | pK Acidic |
|---|---|---|---|---|---|---|---|---|---|
| 149 | **1-Methylguanosine 3′(2′)-phosphate** | 1MeGuo-3′(2′)-P | m[1]Gp for 3′ | 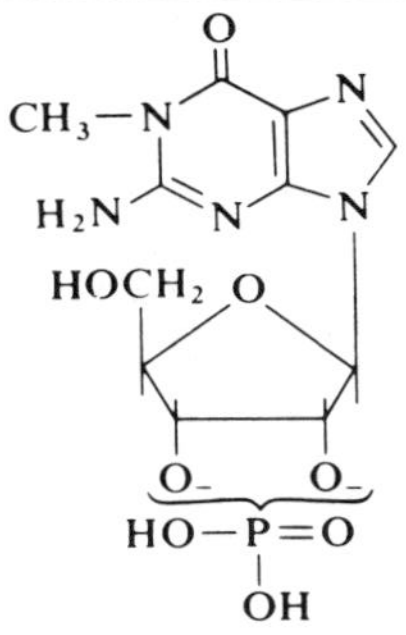 | $C_{11}H_{16}N_5O_8P$ (377.25) | — | — | 2.4[p] | — |
| 150 | ***N*[2]-Methylguanosine 3′(2′)-phosphate** | 2MeGuo-3′(2′)-P | m[2]Gp or m[2]G- for 3′ | 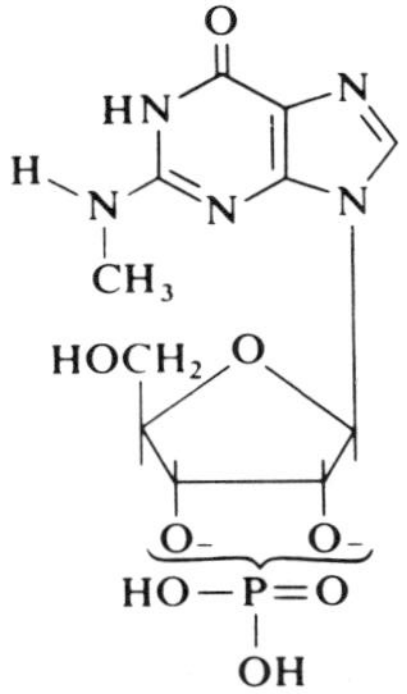 | $C_{11}H_{16}N_5O_8P$ (377.25) | — | — | 2.4[p] | — |

| No. | Acidic spectral data | | | | | | | | | | Neutral spectral data | | | | | | | | | | Alkaline spectral data | | | | | | | | | |
|---|---|---|---|---|---|---|---|---|---|---|---|---|---|---|---|---|---|---|---|---|---|---|---|---|---|---|---|---|---|---|
| | pH | $\lambda_{max}$ | $\varepsilon_{max}$ ($\times 10^{-3}$) | $\lambda_{min}$ | Spectral ratios 230 | 240 | 250 | 270 | 280 | 290 | pH | $\lambda_{max}$ | $\varepsilon_{max}$ ($\times 10^{-3}$) | $\lambda_{min}$ | Spectral ratios 230 | 240 | 250 | 270 | 280 | 290 | pH | $\lambda_{max}$ | $\varepsilon_{max}$ ($\times 10^{-3}$) | $\lambda_{min}$ | Spectral ratios 230 | 240 | 250 | 270 | 280 | 290 |
| 149 | 1 | 258 | 9.4[n] | 230* | 0.21 | 0.45 | 0.86 | 0.80 | 0.72 | 0.51* | W | 255 | 10.2[n] | 222 | 0.23 | 0.67 | 1.04 | 0.86 | 0.63 | 0.20 | 13 | 256 | 10.7[n] | 227 | 0.30 | 0.66 | 1.02 | 0.84 | 0.63 | 0.20* |
| 150 | 1 | 259 | 14.3[n] | 232 | 0.29 | 0.44 | 0.81 | 0.77 | 0.60 | 0.53 | W | 253 | 15.8[n] | 224 | 0.40 | 0.78 | 1.12 | 0.71 | 0.69 | 0.45 | 13 | 258 | 13.4[n] | 236 | 0.72 | 0.59 | 0.89 | 0.91 | 0.78 | 0.41 |

REFERENCES

| No. | Origin and synthesis | $[\alpha]^{t}_{D}$ | pK | Spectral data | $R_f$ |
|---|---|---|---|---|---|
| 149 | R : *165*, *40* | — | *40* | *90*, *265*[b], *165* | *165* |
| 150 | R : *40* | — | *40* | *90* | — |

| No. | Compound / Abbreviation | Symbol | Structure | Formula (mol wt) | Melting point °C | $[\alpha]_D^t$ | pK Basic | pK Acidic |
|---|---|---|---|---|---|---|---|---|
| 151 | **$N^2$,$N^2$-Dimethylguanosine 3′(2′)-phosphate**<br>2Me$_2$Guo-3′(2′)-P | $m_2^2$Gp or $m_2^2$G- for 3′ | 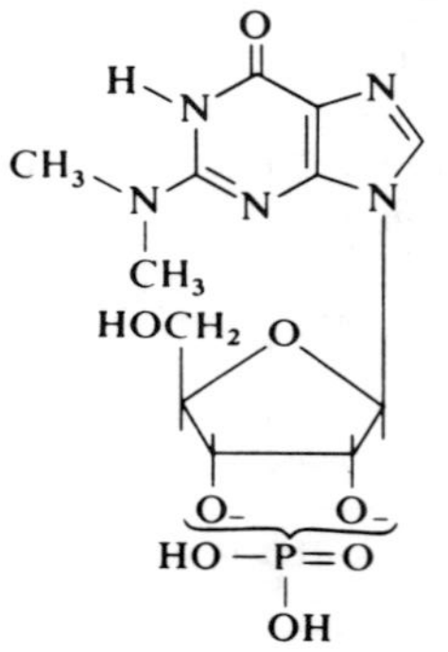 | $C_{12}H_{18}N_5O_8P$ (391.28) | — | — | 2.6[p] | — |
| 152 | **7-Methylguanosine 3′(2′)-phosphate**<br>7MeGuo-3′(2′)-P | $m^7$Gp or $m^7$G- for 3′ | 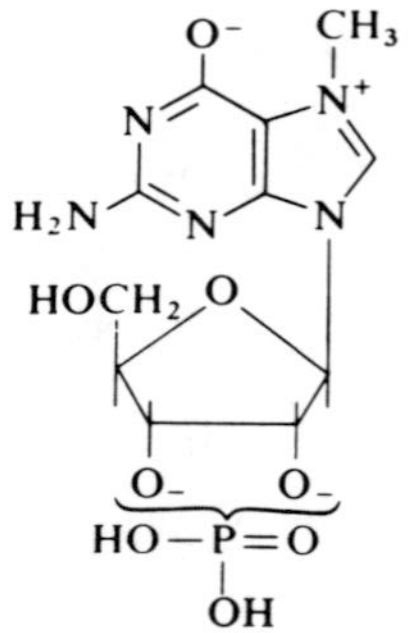 | $C_{11}H_{16}N_5O_8P$ (377.25) | — | — | | 7.1[p] |

| No. | Acidic pH | Acidic $\lambda_{max}$ | Acidic $\varepsilon_{max}$ ($\times 10^{-3}$) | Acidic $\lambda_{min}$ | Acidic 230 | Acidic 240 | Acidic 250 | Acidic 270 | Acidic 280 | Acidic 290 | Neutral pH | Neutral $\lambda_{max}$ | Neutral $\varepsilon_{max}$ ($\times 10^{-3}$) | Neutral $\lambda_{min}$ | Neutral 230 | Neutral 240 | Neutral 250 | Neutral 270 | Neutral 280 | Neutral 290 | Alkaline pH | Alkaline $\lambda_{max}$ | Alkaline $\varepsilon_{max}$ ($\times 10^{-3}$) | Alkaline $\lambda_{min}$ | Alkaline 230 | Alkaline 240 | Alkaline 250 | Alkaline 270 | Alkaline 280 | Alkaline 290 |
|---|---|---|---|---|---|---|---|---|---|---|---|---|---|---|---|---|---|---|---|---|---|---|---|---|---|---|---|---|---|---|
| 151 | 1 | 265 | 12.8[n] | 237* | 0.42 | 0.29 | 0.62 | 0.97 | 0.58* | 0.57 | W | 259 | 13.9[n] | 228 | 0.25 | 0.47 | 0.84 | 0.72 | 0.58 | 0.50 | 13 | 263 | 10.8[n] | 241* | 1.18 | 0.54 | 0.77 | 0.93 | 0 83 | 0.60* |
| 152 | 2 | 257 | 10.9 | 229 | 0.26 | 0.51 | 0.89 | 0.74 | 0.68 | 0.52 | — | — | — | — | — | — | — | — | — | — | — | — | — | — | — | — | — | — | — | — |

(Spectral ratios columns: 230, 240, 250, 270, 280, 290 under each of Acidic, Neutral and Alkaline spectral data.)

## REFERENCES

| No. | Origin and synthesis | $[\alpha]_D^t$ | pK | Spectral data | $R_f$ |
|---|---|---|---|---|---|
| 151 | R : *165*, *40* | — | *40* | *90*, *265*[b], *165* | *165* |
| 152 | C : *90* R : *77* | — | *90* | *90* | — |

| No. | Compound | Abbreviation | Symbol | Structure | Formula (mol wt) | Melting point °C | $[\alpha]^{t}_{D}$ | pK Basic | pK Acidic |
|---|---|---|---|---|---|---|---|---|---|
| 153 | **7-Methylguanosine 5′-phosphate** | 7MeGuo-5′-P<br>7MeGMP | $pm^7G$<br>$-m^7G$ | 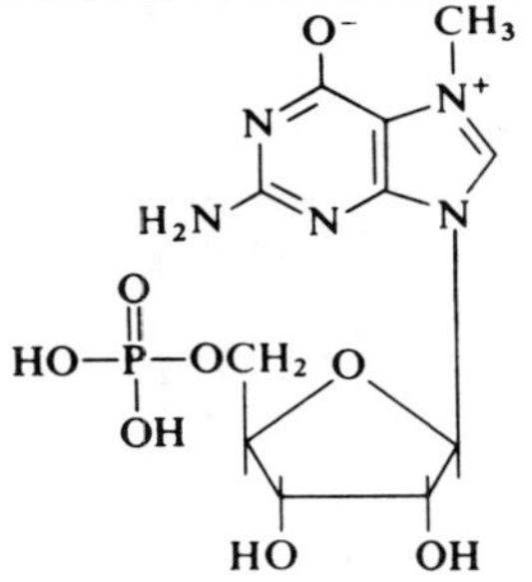 | $C_{11}H_{16}N_5O_8P$ (377.25) | — | — | — | — |
| 154 | **Uridine 2′-phosphate** | Urd-2′-P<br>2′-UMP | | 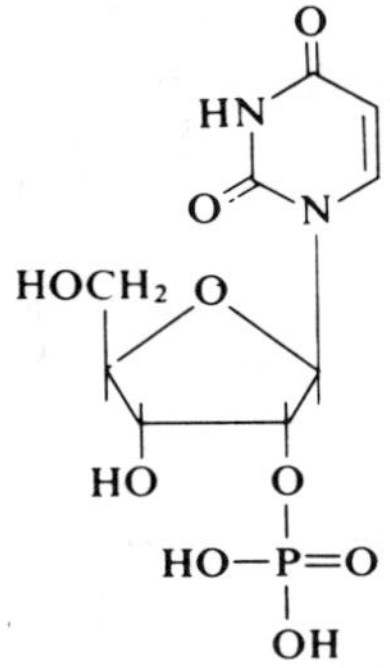 | $C_9H_{13}N_2O_9P$ (324.18) | 190–191°(dec)[z] (Diammonium salt) | $+22.3^{22}$ [z] (2.0, W) | — | — |

| No. | Acidic spectral data | | | | | | | | | | Neutral spectral data | | | | | | | | | | Alkaline spectral data | | | | | | | | | |
|---|---|---|---|---|---|---|---|---|---|---|---|---|---|---|---|---|---|---|---|---|---|---|---|---|---|---|---|---|---|---|
| | pH | $\lambda_{max}$ | $\varepsilon_{max}$ ($\times 10^{-3}$) | $\lambda_{min}$ | Spectral ratios 230 | 240 | 250 | 270 | 280 | 290 | pH | $\lambda_{max}$ | $\varepsilon_{max}$ ($\times 10^{-3}$) | $\lambda_{min}$ | Spectral ratios 230 | 240 | 250 | 270 | 280 | 290 | pH | $\lambda_{max}$ | $\varepsilon_{max}$ ($\times 10^{-3}$) | $\lambda_{min}$ | Spectral ratios 230 | 240 | 250 | 270 | 280 | 290 |
| 153 | — | — | — | — | — | — | — | — | — | — | — | — | — | — | — | — | — | — | — | — | — | — | — | — | — | — | — | — | — | — |
| 154 | 2 | 260[*,z] | 9.9[z] | 230[z] | — | — | 0.80 | — | 0.28 | 0.03 | 7 | 260[*,z] | 10.0[z] | 230[z] | — | — | 0.78 | — | 0.30 | 0.03 | 12 | 261[z] | 7.3[z] | 242[z] | — | — | 0.85 | — | 0.25 | 0.02 |

REFERENCES

| No. | Origin and synthesis | $[\alpha]^{t}_{D}$ | pK | Spectral data | $R_f$ |
|---|---|---|---|---|---|
| 153 | C : *125* | — | — | — | *125* |
| 154 | R : *178*, *170*, *216*[az] | *216*[z] | — | *181*[*,z,b], *179*[e], *90*[z,c] | *258* |

| No. | Compound | Abbreviation | Symbol | Structure | Formula (mol wt) | Melting point °C | $[\alpha]_D^t$ | pK Basic | pK Acidic |
|---|---|---|---|---|---|---|---|---|---|
| 155 | **Uridine 3′-phosphate** | Urd-3′-P<br>3′-UMP | Up<br>U- | 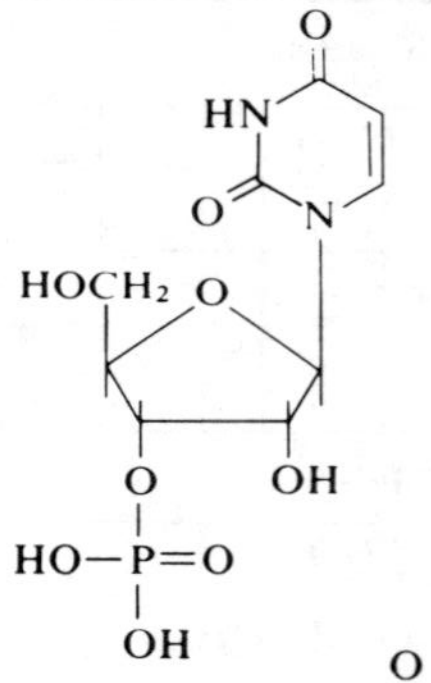 | $C_9H_{13}N_2O_9P$ (324.18) | 192° | $+22.3^{22z}$ (2.0, W) | — | 9.4 |
| 156 | **Uridine 5′-phosphate** | Urd-5′-P<br>UMP | pU<br>-U | 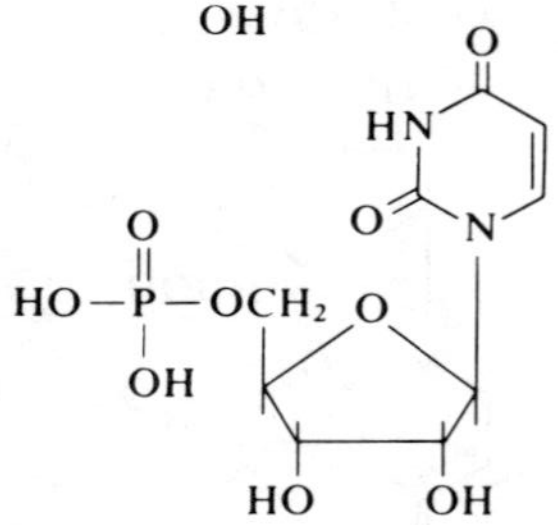 | $C_9H_{13}N_2O_9P$ (324.18) | 190–202° (dibrucine salt) | $+3.44^{28}$ (1.02, 10% HCl) | — | 9.5 |

| No. | Acidic spectral data | | | | | | | | | | Neutral spectral data | | | | | | | | | | Alkaline spectral data | | | | | | | | | |
|---|---|---|---|---|---|---|---|---|---|---|---|---|---|---|---|---|---|---|---|---|---|---|---|---|---|---|---|---|---|---|
| | pH | $\lambda_{max}$ | $\varepsilon_{max}$ ($\times 10^{-3}$) | $\lambda_{min}$ | Spectral ratios 230 | 240 | 250 | 270 | 280 | 290 | pH | $\lambda_{max}$ | $\varepsilon_{max}$ ($\times 10^{-3}$) | $\lambda_{min}$ | Spectral ratios 230 | 240 | 250 | 270 | 280 | 290 | pH | $\lambda_{max}$ | $\varepsilon_{max}$ ($\times 10^{-3}$) | $\lambda_{min}$ | Spectral ratios 230 | 240 | 250 | 270 | 280 | 290 |
| 155 | 1 | 262 | 10.0 | 230 | — | — | 0.76 | 0.82 | 0.32* | 0.03* | 7 | $262^z$ | $10.0^z$ | $230^z$ | — | — | 0.73 | — | 0.35 | 0.03 | 13 | 261 | 7.8 | 241 | — | — | 0.83 | 0.85 | 0.28* | 0.02* |
| 156 | 2 | 262 | 10.0 | 230 | — | — | 0.73 | — | 0.39 | 0.03 | 7 | 262 | 10.0 | 230 | 0.21 | 0.38 | 0.73 | 0.87 | 0.39 | 0.03 | 11 | 261 | 7.8 | 241 | 0.79 | 0.50 | 0.80 | — | 0.31 | 0.02 |

## REFERENCES

| No. | Origin and synthesis | $[\alpha]_D^t$ | pK | Spectral data | $R_f$ |
|---|---|---|---|---|---|
| 155 | C: *190*, R: *178*, *170*, *216*[z] | *216*[z] | *223* | *265*[*,b] *181*[z,b,c,d], *179*[e] | *258* |
| 156 | C: *264*, *190*[a]R: *195* N: *198* | *217* | *212* | *212*, *183*[b,e] *184*[b], *179*[e] | *210* |

| No. | Compound | Abbreviation | Symbol | Structure | Formula (mol wt) | Melting point °C | $[\alpha]_D^t$ | pK Basic | pK Acidic |
|---|---|---|---|---|---|---|---|---|---|
| 157 | **Uridine 5′-diphosphate** | Urd-5′-$P_2$ UDP | | 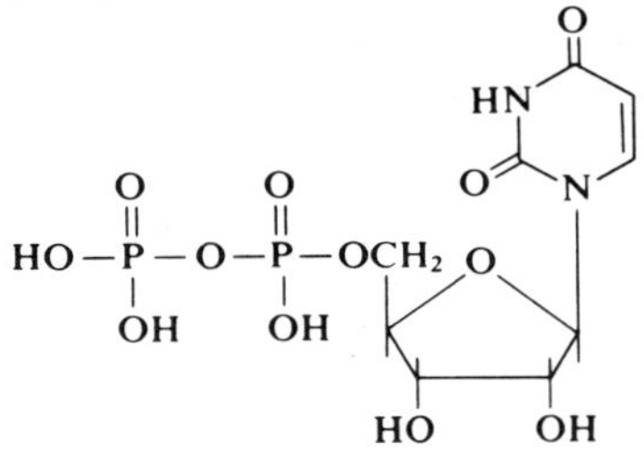 | $C_9H_{14}N_2O_{12}P_2$ (404.16) | — | — | — | 9.4 |
| 158 | **Uridine 5′-triphosphate** | Urd-5′-$P_3$ UTP | | 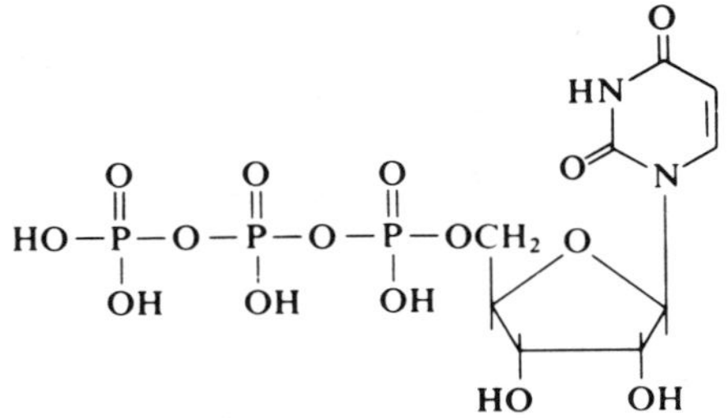 | $C_9H_{15}N_2O_{15}P_3$ (484.15) | — | — | — | 9.6 |

| No. | Acidic spectral data pH | $\lambda_{max}$ | $\varepsilon_{max}$ $(\times 10^{-3})$ | $\lambda_{min}$ | Spectral ratios 230 | 240 | 250 | 270 | 280 | 290 |
|---|---|---|---|---|---|---|---|---|---|---|
| 157 | 2 | 262 | 10.0 | 230 | — | — | 0.73 | — | 0.39* | 0.04 |
| 158 | 2 | 262 | 10.0 | 230 | 0.21 | 0.37 | 0.75 | 0.88 | 0.38 | — |

| No. | Neutral spectral data pH | $\lambda_{max}$ | $\varepsilon_{max}$ $(\times 10^{-3})$ | $\lambda_{min}$ | Spectral ratios 230 | 240 | 250 | 270 | 280 | 290 |
|---|---|---|---|---|---|---|---|---|---|---|
| 157 | 7 | 262 | 10.0 | 230 | — | — | 0.73 | — | 0.39 | — |
| 158 | 7 | 262 | 10.0 | 230 | 0.21 | 0.38 | 0.75 | 0.86 | 0.38 | — |

| No. | Alkaline spectral data pH | $\lambda_{max}$ | $\varepsilon_{max}$ $(\times 10^{-3})$ | $\lambda_{min}$ | Spectral ratios 230 | 240 | 250 | 270 | 280 | 290 |
|---|---|---|---|---|---|---|---|---|---|---|
| 157 | 11 | 261 | 7.9 | 241 | — | — | 0.80 | — | 0.32 | — |
| 158 | 11 | 261 | 8.1 | 239 | 0.79 | 0.65 | 0.81 | 0.78 | 0.31* | — |

REFERENCES

| No. | Origin and synthesis | $[\alpha]_D^t$ | pK | Spectral data | $R_f$ |
|---|---|---|---|---|---|
| 157 | C : *210* N : *198* | — | *212* | *212*, *183*[b], *179*[*e] | *210* |
| 158 | C : *210* N : *198*, *204* | — | *204*, *212* | *212*[b], *204*[*e], *183*[b] | *210* |

| No. | Compound | Abbreviation | Symbol | Structure | Formula (mol wt) | Melting point °C | $[\alpha]^t_D$ | pK Basic | pK Acidic |
|---|---|---|---|---|---|---|---|---|---|
| 159 | **3-Methyluridine 3′(2′)-phosphate** | 3MeUrd-3′(2′)-P | $m^3Up$ or $m^3U$- for 3′ | 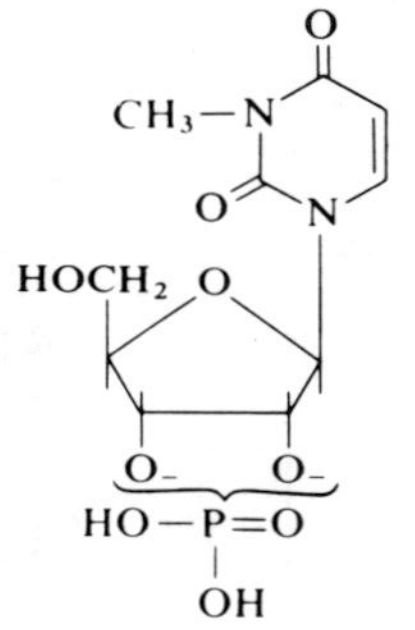 | $C_{10}H_{15}N_2O_9P$ (338.21) | — | — | — | — |
| 160 | **3-Methyluridine 5′-phosphate** | 3MeUrd-5′-P<br>3MeUMP | $pm^3U$<br>$-m^3U$ | 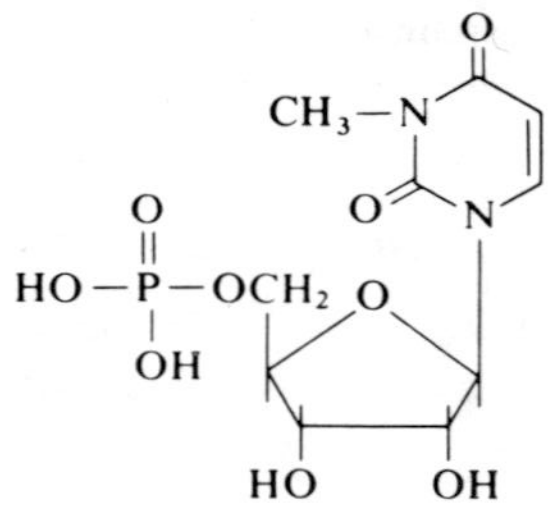 | $C_{10}H_{15}N_2O_9P$ (338.21) | — | — | — | — |

| No. | Acidic spectral data | | | | | | | | | | Neutral spectral data | | | | | | | | | | Alkaline spectral data | | | | | | | | | |
|---|---|---|---|---|---|---|---|---|---|---|---|---|---|---|---|---|---|---|---|---|---|---|---|---|---|---|---|---|---|---|
| | pH | $\lambda_{max}$ | $\varepsilon_{max}$ $(\times 10^{-3})$ | $\lambda_{min}$ | Spectral ratios 230 | 240 | 250 | 270 | 280 | 290 | pH | $\lambda_{max}$ | $\varepsilon_{max}$ $(\times 10^{-3})$ | $\lambda_{min}$ | Spectral ratios 230 | 240 | 250 | 270 | 280 | 290 | pH | $\lambda_{max}$ | $\varepsilon_{max}$ $(\times 10^{-3})$ | $\lambda_{min}$ | Spectral ratios 230 | 240 | 250 | 270 | 280 | 290 |
| 159 | 2r | 258 | — | 233 | — | — | — | — | — | — | W | 262 | 8.8* | — | — | — | 0.77 | — | 0.45 | — | 11.6 | 260 | 9.3 | 233 | — | — | — | — | — | — |
| 160 | — | — | — | — | — | — | — | — | — | — | — | — | — | — | — | — | — | — | — | — | — | — | — | — | — | — | — | — | — | — |

REFERENCES

| No. | Origin and synthesis | $[\alpha]^t_D$ | pK | Spectral data | $R_f$ |
|---|---|---|---|---|---|
| 159 | C: *173, 142, 174* | — | — | *173, 174*[c], *142**[c,d] | *13, 142, 143, 173* |
| 160 | — | — | — | — | *125* |

| No. | Compound | Abbreviation | Symbol | Structure | Formula (mol wt) | Melting point °C | $[\alpha]_D^t$ | pK Basic | pK Acidic |
|---|---|---|---|---|---|---|---|---|---|
| 161 | **4-Thiouridine 3′(2′)-phosphate** | Srd-3′(2′)-P or SUrd-3′(2′)-P 3′(2′)-SMP | $^4$Sp $^4$S- s$^4$Up s$^4$U- for 3′ | 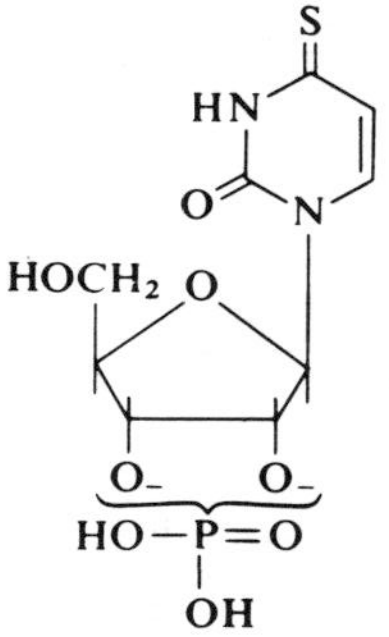 | $C_9H_{13}N_2O_8PS$ (340.25) | — | — | — | — |
| 162 | **Ribosylthymine 3′(2′)-phosphate** | Thd-3′(2′)-P or 5MeUrd-3′(2′)-P | Tp T- for 3′ | 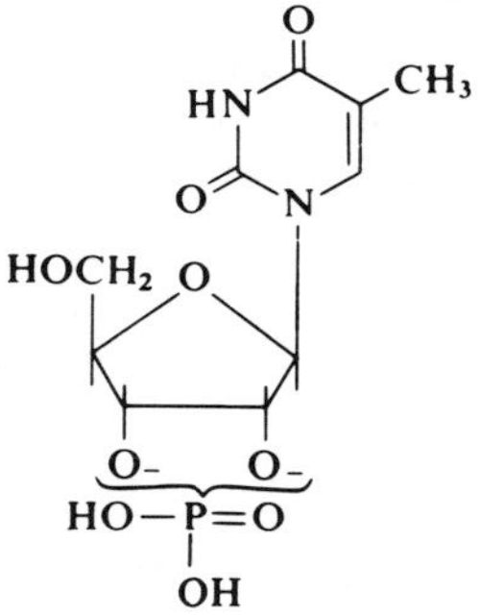 | $C_{10}H_{15}N_2O_9P$ (338.21) | — | — | — | — |

Acidic spectral data

| No. | pH | $\lambda_{max}$ | $\varepsilon_{max}$ $(\times 10^{-3})$ | $\lambda_{min}$ | Spectral ratios 230 | 240 | 250 | 270 | 280 | 290 |
|---|---|---|---|---|---|---|---|---|---|---|
| 161 | 2 | 331 | 17.0 | 279 | — | — | — | — | — | — |
| 162 | 1 | 267* | 9.8 | 235 | — | — | 0.68 | 1.05 | 0.66 | 0.23 |

Neutral spectral data

| No. | pH | $\lambda_{max}$ | $\varepsilon_{max}$ $(\times 10^{-3})$ | $\lambda_{min}$ | Spectral ratios 230 | 240 | 250 | 270 | 280 | 290 |
|---|---|---|---|---|---|---|---|---|---|---|
| 161 | 8.6[q] | 320 | — | — | — | — | — | — | — | — |
| 162 | — | — | — | — | — | — | — | — | — | — |

Alkaline spectral data

| No. | pH | $\lambda_{max}$ | $\varepsilon_{max}$ $(\times 10^{-3})$ | $\lambda_{min}$ | Spectral ratios 230 | 240 | 250 | 270 | 280 | 290 |
|---|---|---|---|---|---|---|---|---|---|---|
| 161 | 12 | 317 | 15.2 | 256 | — | — | — | — | — | — |
| 162 | 13 | 268 | — | 247 | — | — | 0.79 | 1.04 | 0.69 | 0.23 |

REFERENCES

| No. | Origin and synthesis | $[\alpha]_D^t$ | pK | Spectral data | $R_f$ |
|---|---|---|---|---|---|
| 161 | R : *144* | — | — | *144*[b] | — |
| 162 | R : *165* C : *142* | — | — | *265*[b] *165** | *165*, *142* |

| No. | Compound | Abbreviation | Symbol | Structure | Formula (mol wt) | Melting point °C | $[\alpha]_D^t$ | pK Basic | pK Acidic |
|---|---|---|---|---|---|---|---|---|---|
| 163 | **Ribosylthymine 5′-phosphate** | Thd-5′-P<br>TMP | pT<br>-T | | $C_{10}H_{15}N_2O_9P$ (338.21) | — | — | — | — |
| 164 | **Ribosylthymine 5′-diphosphate** | Thd-5′-$P_2$<br>TDP | | | $C_{10}H_{16}N_2O_{12}P_2$ (418.18) | — | — | — | — |

| No. | Acidic pH | Acidic $\lambda_{max}$ | Acidic $\varepsilon_{max}$ ($\times 10^{-3}$) | Acidic $\lambda_{min}$ | Acidic 230 | Acidic 240 | Acidic 250 | Acidic 270 | Acidic 280 | Acidic 290 |
|---|---|---|---|---|---|---|---|---|---|---|
| 163 | — | — | — | — | — | — | — | — | — | — |
| 164 | 2 | 268 | 10.0 | 234 | 0.37 | 0.34 | 0.64 | 1.10 | 0.77 | 0.27 |

| No. | Neutral pH | Neutral $\lambda_{max}$ | Neutral $\varepsilon_{max}$ ($\times 10^{-3}$) | Neutral $\lambda_{min}$ | Neutral 230 | Neutral 240 | Neutral 250 | Neutral 270 | Neutral 280 | Neutral 290 |
|---|---|---|---|---|---|---|---|---|---|---|
| 163 | — | — | — | — | — | — | — | — | — | — |
| 164 | — | — | — | — | — | — | — | — | — | — |

| No. | Alkaline pH | Alkaline $\lambda_{max}$ | Alkaline $\varepsilon_{max}$ ($\times 10^{-3}$) | Alkaline $\lambda_{min}$ | Alkaline 230 | Alkaline 240 | Alkaline 250 | Alkaline 270 | Alkaline 280 | Alkaline 290 |
|---|---|---|---|---|---|---|---|---|---|---|
| 163 | — | — | — | — | — | — | — | — | — | — |
| 164 | — | — | — | — | — | — | — | — | — | — |

REFERENCES

| No. | Origin and synthesis | $[\alpha]_D^t$ | pK | Spectral data | $R_f$ |
|---|---|---|---|---|---|
| 163 | — | — | — | — | *142* |
| 164 | C : *175* | — | — | *175* | *175* |

| No. | Compound | Abbreviation | Symbol | Structure | Formula (mol wt) | Melting point °C | $[\alpha]^t_D$ | pK Basic | pK Acidic |
|---|---|---|---|---|---|---|---|---|---|
| 165 | **5-Hydroxyuridine 5′-phosphate** | 5HOUrd-5′-P<br>5(OH)UMP | pho$^5$U<br>-ho$^5$U | 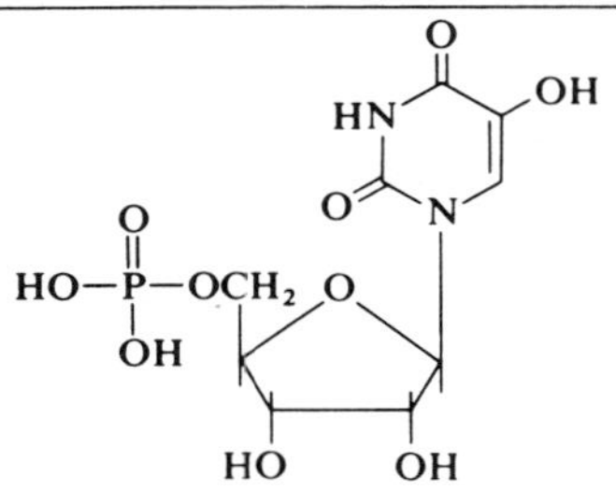 | $C_9H_{13}N_2O_{10}P$<br>(340.18) | — | — | — | — |
| 166 | **Pseudouridine C 3′(2′)-phosphate** | Ψrd-3′(2′)-P | Ψp<br>Ψ-<br>for 3′ | 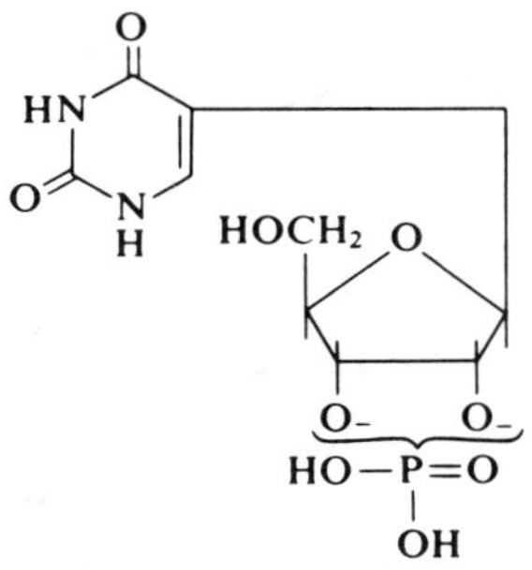 | $C_9H_{13}N_2O_9P$<br>(324.18) | — | — | — | 9.6 |

| No. | Acidic spectral data | | | | | | | | | | Neutral spectral data | | | | | | | | | | Alkaline spectral data | | | | | | | | | |
|---|---|---|---|---|---|---|---|---|---|---|---|---|---|---|---|---|---|---|---|---|---|---|---|---|---|---|---|---|---|---|
| | pH | $\lambda_{max}$ | $\varepsilon_{max}$ ($\times 10^{-3}$) | $\lambda_{min}$ | Spectral ratios 230 | 240 | 250 | 270 | 280 | 290 | pH | $\lambda_{max}$ | $\varepsilon_{max}$ ($\times 10^{-3}$) | $\lambda_{min}$ | Spectral ratios 230 | 240 | 250 | 270 | 280 | 290 | pH | $\lambda_{max}$ | $\varepsilon_{max}$ ($\times 10^{-3}$) | $\lambda_{min}$ | Spectral ratios 230 | 240 | 250 | 270 | 280 | 290 |
| 165 | — | — | — | — | — | — | — | — | — | — | 6 | 278 | — | 245 | — | — | — | — | — | — | 9 | 236<br>300 | — | 268 | — | — | — | — | — | — |
| 166 | 2[r] | 263 | 8.4 | 233 | 0.30 | 0.41 | 0.75 | 0.86 | 0.40* | 0.07* | 7 | 263 | — | 233 | 0.28 | 0.39 | 0.74 | 0.85 | 0.40 | 0.07 | 12 | 286 | 8.4 | 246 | 2.13 | 0.75 | 0.64* | 1.54* | 2.06* | 2.14* |

## REFERENCES

| No. | Origin and synthesis | $[\alpha]^t_D$ | pK | Spectral data | $R_f$ |
|---|---|---|---|---|---|
| 165 | C : *146* | — | — | *146* | — |
| 166 | R : *157*, *94* | — | *157* | *157*[b], *94*, *158*, *265*[*,b] | *165* |

| No. | Compound | Abbreviation | Symbol | Structure | Formula (mol wt) | Melting point °C | $[\alpha]^t_D$ | pK Basic | pK Acidic |
|---|---|---|---|---|---|---|---|---|---|
| 167 | **Pseudouridine C 5′-phosphate** | Ψrd-5′-P<br>ΨMP | pΨ<br>-Ψ | 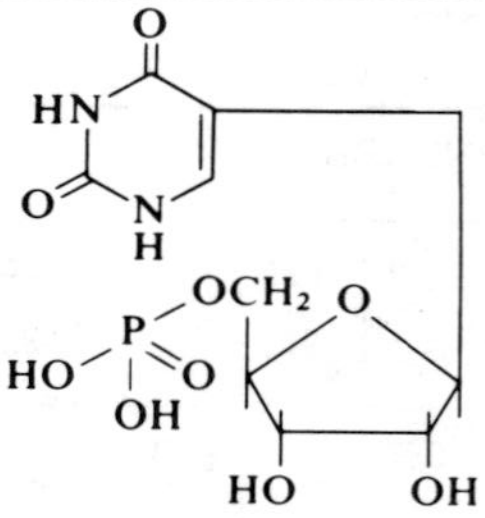 | $C_9H_{13}N_2O_9P$ (324.18) | — | — | — | — |
| 168 | **Pseudouridine C 5′-diphosphate** | Ψrd-5′-$P_2$<br>ΨDP | | 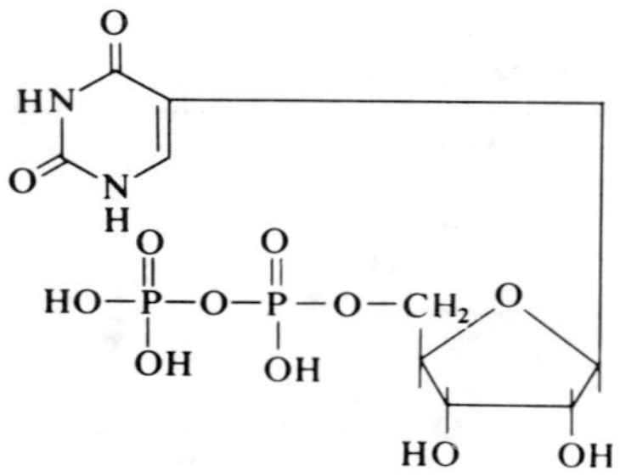 | $C_9H_{14}N_2O_{12}P_2$ (404.16) | — | — | — | — |

| No. | Acidic spectral data | | | | | | | | | | Neutral spectral data | | | | | | | | | | Alkaline spectral data | | | | | | | | | |
|---|---|---|---|---|---|---|---|---|---|---|---|---|---|---|---|---|---|---|---|---|---|---|---|---|---|---|---|---|---|---|
| | pH | $\lambda_{max}$ | $\varepsilon_{max}$ $(\times 10^{-3})$ | $\lambda_{min}$ | Spectral ratios 230 | 240 | 250 | 270 | 280 | 290 | pH | $\lambda_{max}$ | $\varepsilon_{max}$ $(\times 10^{-3})$ | $\lambda_{min}$ | Spectral ratios 230 | 240 | 250 | 270 | 280 | 290 | pH | $\lambda_{max}$ | $\varepsilon_{max}$ $(\times 10^{-3})$ | $\lambda_{min}$ | Spectral ratios 230 | 240 | 250 | 270 | 280 | 290 |
| 167 | — | — | — | — | — | — | — | — | — | — | — | — | — | — | — | — | — | — | — | — | 12 | — | — | — | — | — | — | — | 1.40 | — |
| 168 | — | — | — | — | — | — | — | — | — | — | — | — | — | — | — | — | — | — | — | — | 12 | — | — | — | — | — | — | — | 1.30 | — |

## REFERENCES

| No. | Origin and synthesis | $[\alpha]^t_D$ | pK | Spectral data | $R_f$ |
|---|---|---|---|---|---|
| 167 | C : *158, 176* | — | — | *158* | *158* |
| 168 | C : *158* | — | — | *158* | *158* |

| No. | Compound | Abbreviation | Symbol | Structure | Formula (mol wt) | Melting point °C | $[\alpha]^t_D$ | pK Basic | pK Acidic |
|---|---|---|---|---|---|---|---|---|---|
| 169 | **Orotidine 5′-phosphate** | Ord-5′-P<br>OMP | pO<br>-O | 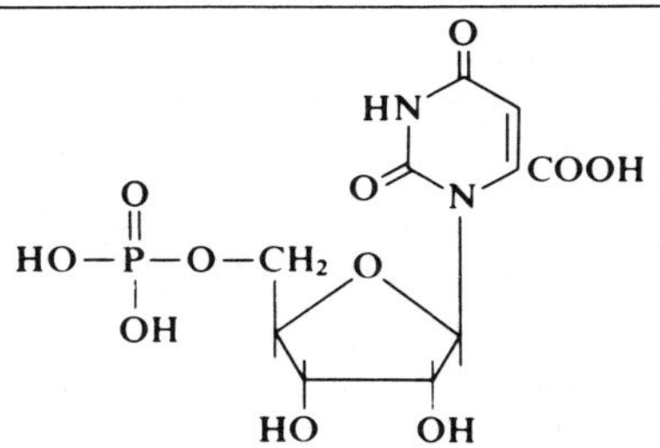 | $C_{10}H_{13}N_2O_{11}P$ (368.19) | — | — | — | — |
| 170 | **2-Amino-4-hydroxy-5-*N*-methylformamido-6-ribosylaminopyrimidine 3′(2′)-phosphate**[f] | | | 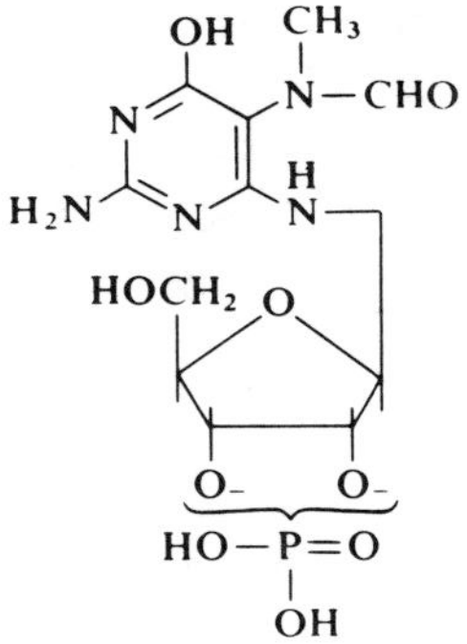 | $C_{11}H_{18}N_5O_9P$ (395.27) | — | — | — | — |

| No. | Acidic spectral data | | | | | | | | | | Neutral spectral data | | | | | | | | | | Alkaline spectral data | | | | | | | | | |
|---|---|---|---|---|---|---|---|---|---|---|---|---|---|---|---|---|---|---|---|---|---|---|---|---|---|---|---|---|---|---|
| | pH | $\lambda_{max}$ | $\varepsilon_{max}$ ($\times 10^{-3}$) | $\lambda_{min}$ | Spectral ratios 230 | 240 | 250 | 270 | 280 | 290 | pH | $\lambda_{max}$ | $\varepsilon_{max}$ ($\times 10^{-3}$) | $\lambda_{min}$ | Spectral ratios 230 | 240 | 250 | 270 | 280 | 290 | pH | $\lambda_{max}$ | $\varepsilon_{max}$ ($\times 10^{-3}$) | $\lambda_{min}$ | Spectral ratios 230 | 240 | 250 | 270 | 280 | 290 |
| 169 | — | — | — | — | — | — | — | — | — | — | 7 | 266 | — | — | — | — | — | — | 0.66 | — | — | — | — | — | — | — | — | — | — | — |
| 170 | 2 | 273 | 14.0[x] | 247 | 1.00 | 0.50 | 0.47 | 1.81 | 1.41 | 0.31 | — | — | — | — | — | — | — | — | — | — | 13 | 265 | 10.5[x] | 244 | 0.77 | 0.51 | 0.56 | 1.03 | 0.26 | 0.05 |

REFERENCES

| No. | Origin and synthesis | $[\alpha]^t_D$ | pK | Spectral data | $R_f$ |
|---|---|---|---|---|---|
| 169 | E : *177* | — | — | *177* | — |
| 170 | C : *90* | — | — | *90* | — |

| No. | Compound / Abbreviation | Symbol | Structure | Formula (mol wt) | Melting point °C | $[\alpha]^t_D$ | pK Basic | pK Acidic |
|---|---|---|---|---|---|---|---|---|
| **DEOXYRIBONUCLEOTIDES** | | | | | | | | |
| 171 | **Deoxyadenosine 3′-phosphate**<br>dAdo-3′-P<br>3′-dAMP | dAp<br>dA- | 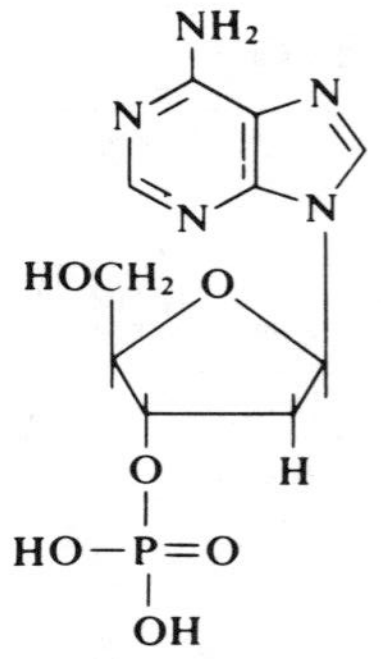 | $C_{10}H_{14}N_5O_6P$ (331.22) | — | — | — | — |
| 172 | **Deoxyadenosine 5′-phosphate**<br>dAdo-5′-P<br>dAMP | pdA<br>-dA | 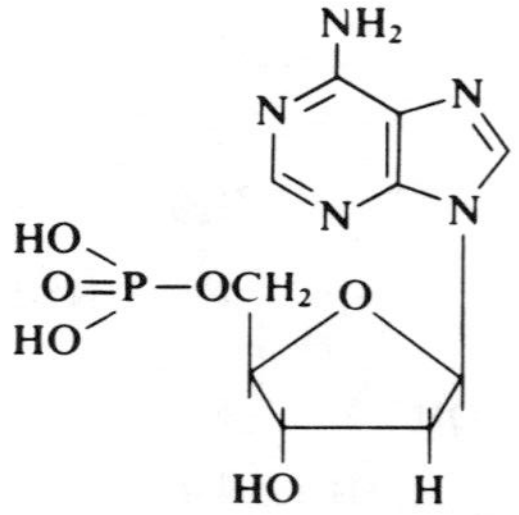 | $C_{10}H_{14}N_5O_6P$ (331.22) | 142° | $-38.0^{19}$ (0.23, W) | ~4.4 | — |

| No. | Acidic pH | Acidic $\lambda_{max}$ | Acidic $\varepsilon_{max}$ ($\times 10^{-3}$) | Acidic $\lambda_{min}$ | Acidic ratio 230 | 240 | 250 | 270 | 280 | 290 | Neutral pH | Neutral $\lambda_{max}$ | Neutral $\varepsilon_{max}$ ($\times 10^{-3}$) | Neutral $\lambda_{min}$ | Neutral ratio 230 | 240 | 250 | 270 | 280 | 290 | Alkaline pH | Alkaline $\lambda_{max}$ | Alkaline $\varepsilon_{max}$ ($\times 10^{-3}$) | Alkaline $\lambda_{min}$ | Alkaline ratio 230 | 240 | 250 | 270 | 280 | 290 |
|---|---|---|---|---|---|---|---|---|---|---|---|---|---|---|---|---|---|---|---|---|---|---|---|---|---|---|---|---|---|---|
| 171 | — | — | — | — | — | — | — | — | — | — | 7 | — | — | — | — | — | 0.79 | 0.68 | 0.14 | — | — | — | — | — | — | — | — | — | — | — |
| 172 | 2 | 258 | 14.3* | 230 | — | — | 0.82 | — | 0.23 | 0.04 | 7 | — | 15.3 | — | — | 0.42 | 0.80 | 0.66 | 0.14 | 0.01 | — | — | — | — | — | — | — | — | — | — |

## REFERENCES

| No. | Origin and synthesis | $[\alpha]^t_D$ | pK | Spectral data | $R_f$ |
|---|---|---|---|---|---|
| 171 | D: *263* C: *203* | — | — | *263* | *263* |
| 172 | D: *200*[a], *260* C: *202* | *260* | *180* | *186*, *185*[*e] *200*, *223*[c,d] | *263* |

| No. | Compound | Abbreviation | Symbol | Structure | Formula (mol wt) | Melting point °C | $[\alpha]_D^t$ | pK Basic | pK Acidic |
|---|---|---|---|---|---|---|---|---|---|
| 173 | **Deoxyadenosine 5′-triphosphate** | dAdo-5′-$P_3$ dATP | pppA | 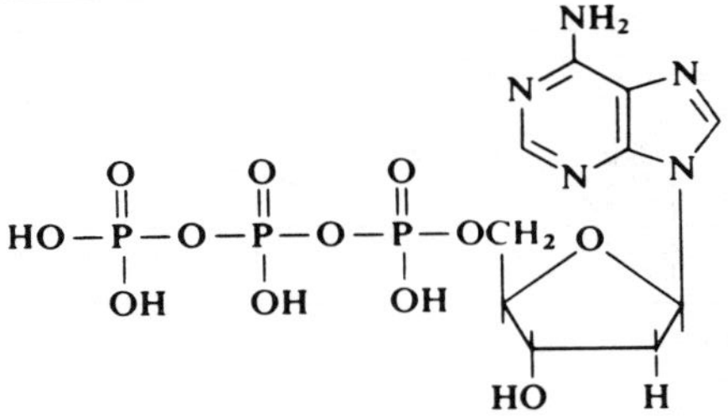 | $C_{10}H_{16}N_5O_{12}P_3$ (491.19) | — | — | — | — |
| 174 | **$N^6$-Methyldeoxyadenosine 5′-phosphate** | 6dMeAdo-5′-P 6dMeAMP | $pm^6dA$ $-m^6dA$ | 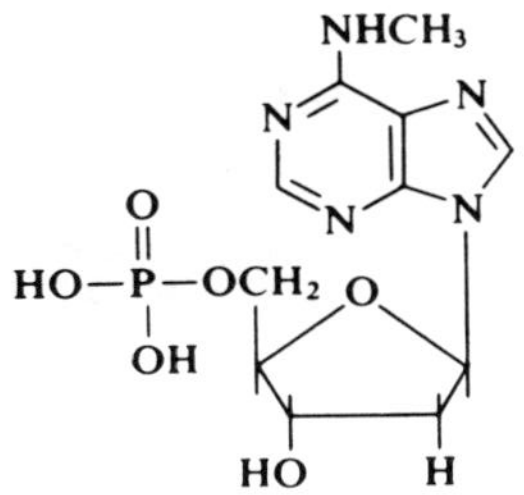 | $C_{11}H_{16}N_5O_6P$ (345.25) | — | — | 3.6[p] | — |

| No. | Acidic spectral data | | | | | | | | | |
|---|---|---|---|---|---|---|---|---|---|---|
| | pH | $\lambda_{max}$ | $\varepsilon_{max}$ ($\times 10^{-3}$) | $\lambda_{min}$ | Spectral ratios 230 | 240 | 250 | 270 | 280 | 290 |
| 173 | — | — | — | — | — | — | — | — | — | — |
| 174 | 4[q] | 266 | — | — | 0.18 | 0.28 | 0.63 | 1.09 | 0.66 | 0.25 |

| No. | Neutral spectral data | | | | | | | | | |
|---|---|---|---|---|---|---|---|---|---|---|
| | pH | $\lambda_{max}$ | $\varepsilon_{max}$ ($\times 10^{-3}$) | $\lambda_{min}$ | Spectral ratios 230 | 240 | 250 | 270 | 280 | 290 |
| 173 | 7 | — | — | — | — | — | 0.77 | — | 0.14 | — |
| 174 | — | — | — | — | — | — | — | — | — | — |

| No. | Alkaline spectral data | | | | | | | | | |
|---|---|---|---|---|---|---|---|---|---|---|
| | pH | $\lambda_{max}$ | $\varepsilon_{max}$ ($\times 10^{-3}$) | $\lambda_{min}$ | Spectral ratios 230 | 240 | 250 | 270 | 280 | 290 |
| 173 | — | — | — | — | — | — | — | — | — | — |
| 174 | 13 | 266 | — | 234 | 0.18 | 0.24 | 0.57 | 1.08 | 0.63 | 0.21 |

REFERENCES

| No. | Origin and synthesis | $[\alpha]_D^t$ | pK | Spectral data | $R_f$ |
|---|---|---|---|---|---|
| 173 | E : *209* | — | — | *209* | — |
| 174 | D : *21* | — | *21* | *21*[b] | — |

| No. | Compound | Abbreviation | Symbol | Structure | Formula (mol wt) | Melting point °C | $[\alpha]_D^t$ | pK Basic | pK Acidic |
|---|---|---|---|---|---|---|---|---|---|
| 175 | **Deoxycytidine 3′-phosphate** | dCyd-3′-P<br>3′-dCMP | dCp<br>dC- | 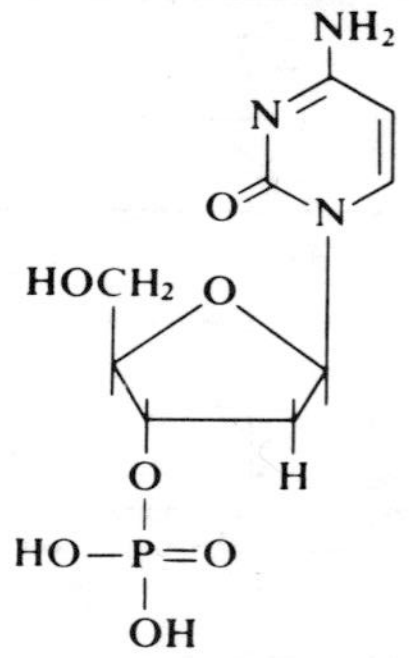 | $C_9H_{14}N_3O_7P$ (307.20) | 196–197° (dec) | $+57.0^{17}$ (1.35, W) | — | — |
| 176 | **Deoxycytidine 5′-phosphate** | dCyd-5′-P<br>dCMP | pdC<br>-dC | 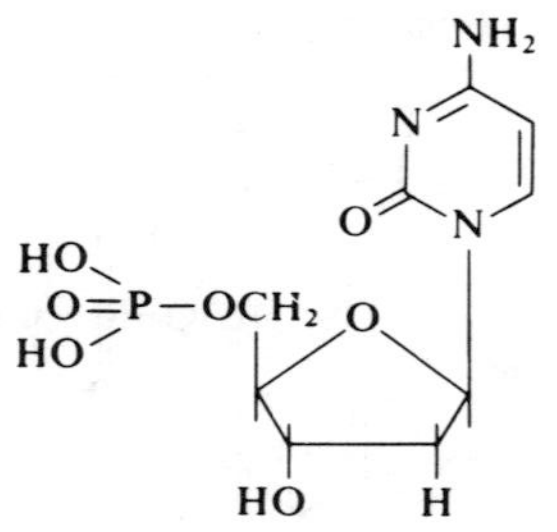 | $C_9H_{14}N_3O_7P$ (307.20) | 183–184° (dec) | $+35.0^{21}$ (0.2, W) | 4.6 | — |

| No. | Acidic spectral data | | | | | | | | | | Neutral spectral data | | | | | | | | | | Alkaline spectral data | | | | | | | | | |
|---|---|---|---|---|---|---|---|---|---|---|---|---|---|---|---|---|---|---|---|---|---|---|---|---|---|---|---|---|---|---|
| | pH | $\lambda_{max}$ | $\varepsilon_{max}$ ($\times 10^{-3}$) | $\lambda_{min}$ | Spectral ratios 230 | 240 | 250 | 270 | 280 | 290 | pH | $\lambda_{max}$ | $\varepsilon_{max}$ ($\times 10^{-3}$) | $\lambda_{min}$ | Spectral ratios 230 | 240 | 250 | 270 | 280 | 290 | pH | $\lambda_{max}$ | $\varepsilon_{max}$ ($\times 10^{-3}$) | $\lambda_{min}$ | Spectral ratios 230 | 240 | 250 | 270 | 280 | 290 |
| 175 | 3 | — | — | — | — | — | — | — | 2.0 | — | 7 | — | — | — | — | — | 0.84 | 1.19 | 0.93 | — | — | — | — | — | — | — | — | — | — | — |
| 176 | 2 | 280 | 13.5 | 239 | — | — | 0.43 | — | 2.12 | 1.55 | 7 | 271 | 9.3 | 249 | — | 0.91 | 0.82 | 1.25 | 0.99 | 0.30 | 12 | — | — | — | — | — | 0.82 | — | 0.99 | 0.30 |

REFERENCES

| No. | Origin and synthesis | $[\alpha]_D^t$ | pK | Spectral data | $R_f$ |
|---|---|---|---|---|---|
| 175 | D : *263* C : *201, 203, 202* | *201* | — | *263, 201*[e] | *203, 201, 263* |
| 176 | C : *201* D : *200* | *207* | *180* | *185*[b], *179*[e], *186*[e], *201* | *185, 201, 263* |

| No. | Compound | Abbreviation | Symbol | Structure | Formula (mol wt) | Melting point °C | $[\alpha]_D^t$ | pK Basic | pK Acidic |
|---|---|---|---|---|---|---|---|---|---|
| 177 | **Deoxycytidine 5′-triphosphate** | dCyd-5′-$P_3$<br>dCTP | | | $C_9H_{16}N_3O_{13}P_3$ (467.17) | — | — | — | — |
| 178 | **5-Methyldeoxycytidine 5′-phosphate** | 5dMeCyd-5′-P<br>5dMeCMP | $pm^5dC$<br>$-m^5dC$ | | $C_{10}H_{16}N_3O_7P$ (321.22) | — | — | 4.4 | — |

| No. | Acidic spectral data | | | | | | | | | | Neutral spectral data | | | | | | | | | | Alkaline spectral data | | | | | | | | | |
|---|---|---|---|---|---|---|---|---|---|---|---|---|---|---|---|---|---|---|---|---|---|---|---|---|---|---|---|---|---|---|
| | pH | $\lambda_{max}$ | $\varepsilon_{max}$ ($\times 10^{-3}$) | $\lambda_{min}$ | Spectral ratios 230 | 240 | 250 | 270 | 280 | 290 | pH | $\lambda_{max}$ | $\varepsilon_{max}$ ($\times 10^{-3}$) | $\lambda_{min}$ | Spectral ratios 230 | 240 | 250 | 270 | 280 | 290 | pH | $\lambda_{max}$ | $\varepsilon_{max}$ ($\times 10^{-3}$) | $\lambda_{min}$ | Spectral ratios 230 | 240 | 250 | 270 | 280 | 290 |
| 177 | 2 | — | — | — | — | — | 0.44 | — | 2.14 | — | — | — | — | — | — | — | — | — | — | — | — | — | — | — | — | — | — | — | — | — |
| 178 | 2 | 287 | — | 244 | 1.51 | 0.43 | 0.36 | 2.10 | 3.14 | 3.44 | 7 | 278 | — | 254 | 1.52 | 1.29 | 0.95 | 1.40 | 1.52 | 1.01 | 12 | 278 | — | — | — | — | — | — | — | — |

## REFERENCES

| No. | Origin and synthesis | $[\alpha]_D^t$ | pK | Spectral data | $R_f$ |
|---|---|---|---|---|---|
| 177 | E: *209* C: *214* | — | — | *209* | — |
| 178 | D: *68* | — | *170* | *68*[b], *186* | — |

| No. | Compound | Abbreviation | Symbol | Structure | Formula (mol wt) | Melting point °C | $[\alpha]^t_D$ | pK Basic | pK Acidic |
|---|---|---|---|---|---|---|---|---|---|
| 179 | **5-Hydroxymethyldeoxycytidine 5′-phosphate** | 5HmdCyd-5′-P<br>5HmdCMP | phm$^5$dC<br>-hm$^5$dC | 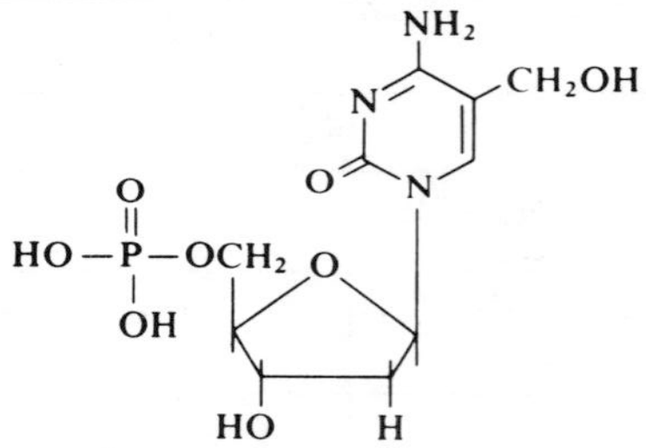 | $C_{10}H_{16}N_3O_8P$ (337.22) | — | — | — | — |
| 180 | **Deoxyguanosine 3′-phosphate** | dGuo-3′-P<br>3′-dGMP | dGp<br>dG- | 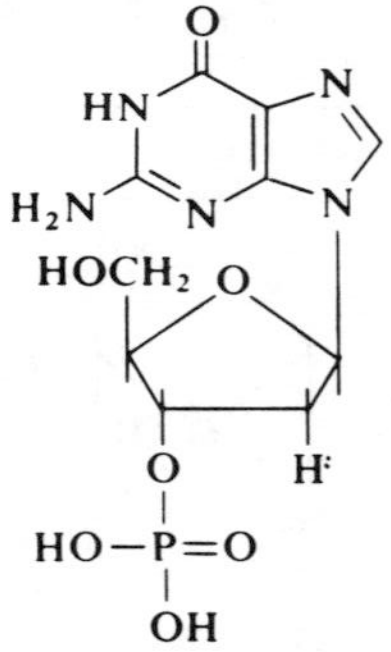 | $C_{10}H_{14}N_5O_7P$ (347.23) | — | — | — | — |

| No. | Acidic spectral data | | | | | | | | | | Neutral spectral data | | | | | | | | | | Alkaline spectral data | | | | | | | | | |
|---|---|---|---|---|---|---|---|---|---|---|---|---|---|---|---|---|---|---|---|---|---|---|---|---|---|---|---|---|---|---|
| | pH | $\lambda_{max}$ | $\varepsilon_{max}$ ($\times 10^{-3}$) | $\lambda_{min}$ | Spectral ratios 230 | 240 | 250 | 270 | 280 | 290 | pH | $\lambda_{max}$ | $\varepsilon_{max}$ ($\times 10^{-3}$) | $\lambda_{min}$ | Spectral ratios 230 | 240 | 250 | 270 | 280 | 290 | pH | $\lambda_{max}$ | $\varepsilon_{max}$ ($\times 10^{-3}$) | $\lambda_{min}$ | Spectral ratios 230 | 240 | 250 | 270 | 280 | 290 |
| 179 | 1 | 284 | 12.5 | 245 | 1.12* | 0.39 | 0.44 | 1.89 | 2.68 | 2.53 | 7 | 275 | 7.7 | 254 | 1.50 | 1.10 | 0.90 | 1.35 | 1.33 | 0.71 | 12 | 275 | 7.7 | 254 | 1.40* | 1.08 | 0.93 | 1.33 | 1.31 | 0.65 |
| 180 | — | — | — | — | — | — | — | — | — | — | 7 | — | — | — | — | — | 1.20 | 0.82 | 0.67 | — | — | — | — | — | — | — | — | — | — | — |

## REFERENCES

| No. | Origin and synthesis | $[\alpha]^t_D$ | pK | Spectral data | $R_f$ |
|---|---|---|---|---|---|
| 179 | C : *74* E : *70* D : *171*, *172* | — | — | *74*, *70**[b,c,d,e], *171*[d], *172** | — |
| 180 | D : *263* C : *203* | — | — | *263* | *263* |

| No. | Compound | Abbreviation | Symbol | Structure | Formula (mol wt) | Melting point °C | $[\alpha]^t_D$ | pK Basic | pK Acidic |
|---|---|---|---|---|---|---|---|---|---|
| 181 | **Deoxyguanosine 5′-phosphate** | dGuo-5′-P<br>dGMP | pdG<br>-dG | 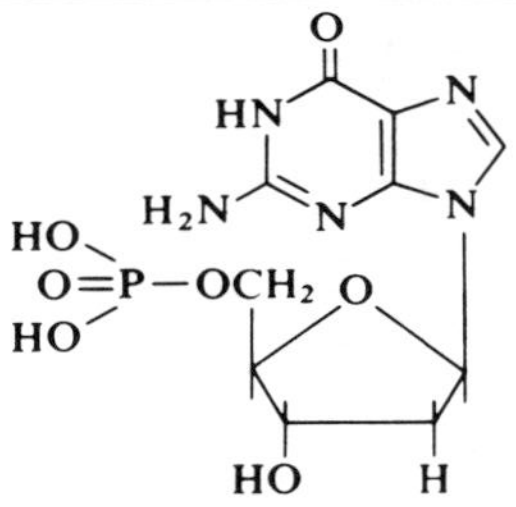 | $C_{10}H_{14}N_5O_7P$<br>(347.23) | 180–182° | $-31^{19}$<br>(0.43, W) | 2.9 | 9.7 |
| 182 | **Deoxyguanosine 5′-triphosphate** | dGuo-5′-$P_3$<br>dGTP | pppdG | 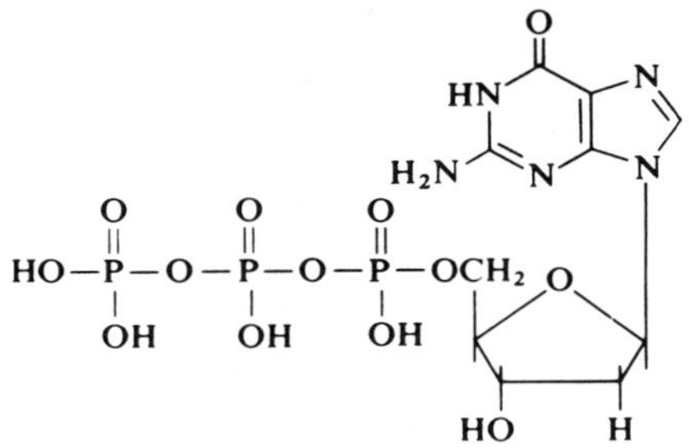 | $C_{10}H_{16}N_5O_{13}P_3$<br>(507.20) | — | — | — | — |

| No. | Acidic spectral data | | | | | | | | | |
|---|---|---|---|---|---|---|---|---|---|---|
| | pH | $\lambda_{max}$ | $\varepsilon_{max}$ ($\times 10^{-3}$) | $\lambda_{min}$ | Spectral ratios 230 | 240 | 250 | 270 | 280 | 290 |
| 181 | 1[e] | 255 | 11.8 | 228 | — | — | 1.02 | — | 0.70 | — |
| | 2[q] | — | — | — | — | — | 1.03 | — | 0.70 | 0.46 |
| 182 | — | — | — | — | — | — | — | — | — | — |

| No. | Neutral spectral data | | | | | | | | | |
|---|---|---|---|---|---|---|---|---|---|---|
| | pH | $\lambda_{max}$ | $\varepsilon_{max}$ ($\times 10^{-3}$) | $\lambda_{min}$ | Spectral ratios 230 | 240 | 250 | 270 | 280 | 290 |
| 181 | 7 | — | — | — | — | 0.79 | 1.13 | 0.81 | 0.67 | 0.27 |
| 182 | 7 | — | — | — | — | — | 1.14 | — | 0.66 | — |

| No. | Alkaline spectral data | | | | | | | | | |
|---|---|---|---|---|---|---|---|---|---|---|
| | pH | $\lambda_{max}$ | $\varepsilon_{max}$ ($\times 10^{-3}$) | $\lambda_{min}$ | Spectral ratios 230 | 240 | 250 | 270 | 280 | 290 |
| 181 | — | — | — | — | — | — | — | — | — | — |
| 182 | — | — | — | — | — | — | — | — | — | — |

## REFERENCES

| No. | Origin and synthesis | $[\alpha]^t_D$ | pK | Spectral data | $R_f$ |
|---|---|---|---|---|---|
| 181 | D : *200, 180[a], 259* | *259* | *180* | *186, 185[e], 200, 223[c,d,e]* | *185, 263* |
| 182 | E : *209* C : *214* | — | — | *209* | — |

| No. | Compound | Abbreviation | Symbol | Structure | Formula (mol wt) | Melting point °C | $[\alpha]^t_D$ | pK Basic | pK Acidic |
|---|---|---|---|---|---|---|---|---|---|
| 183 | **7-Methyldeoxyguanosine 5′-phosphate** | 7dMeGuo-5′-P<br>7dMeGMP | $pm^7dG$<br>$-m^7dG$ | | $C_{11}H_{16}N_5O_7P$ (361.24) | — | — | — | — |
| 184 | **Deoxyuridine 3′-phosphate** | dUrd-3′-P<br>3′-dUMP | dUp<br>dU- | | $C_9H_{13}N_2O_8P$ (308.18) | — | — | — | — |

| No. | Acidic spectral data | | | | | | | | | | Neutral spectral data | | | | | | | | | | Alkaline spectral data | | | | | | | | | |
|---|---|---|---|---|---|---|---|---|---|---|---|---|---|---|---|---|---|---|---|---|---|---|---|---|---|---|---|---|---|---|
| | pH | $\lambda_{max}$ | $\varepsilon_{max}$ ($\times 10^{-3}$) | $\lambda_{min}$ | Spectral ratios 230 | 240 | 250 | 270 | 280 | 290 | pH | $\lambda_{max}$ | $\varepsilon_{max}$ ($\times 10^{-3}$) | $\lambda_{min}$ | Spectral ratios 230 | 240 | 250 | 270 | 280 | 290 | pH | $\lambda_{max}$ | $\varepsilon_{max}$ ($\times 10^{-3}$) | $\lambda_{min}$ | Spectral ratios 230 | 240 | 250 | 270 | 280 | 290 |
| 183 | — | — | — | — | — | — | — | — | — | — | $7^{aq}$ | 256<br>283 | 9.8<br>7.8 | — | — | — | — | — | — | — | — | — | — | — | — | — | — | — | — | — |
| 184 | — | — | — | — | — | — | — | — | — | — | — | — | — | — | — | — | — | — | — | — | — | — | — | — | — | — | — | — | — | — |

REFERENCES

| No. | Origin and synthesis | $[\alpha]^t_D$ | pK | Spectral data | $R_f$ |
|---|---|---|---|---|---|
| 183 | C: *46* | — | — | *46* | *46* |
| 184 | — | — | — | — | — |

| No. | Compound | Abbreviation | Symbol | Structure | Formula (mol wt) | Melting point °C | $[\alpha]_D^t$ | pK Basic | pK Acidic |
|---|---|---|---|---|---|---|---|---|---|
| 185 | **Deoxyuridine 5′-phosphate** | dUrd-5′-P<br>dUMP | pdU<br>-dU | | $C_9H_{13}N_2O_8P$ (308.18) | — | — | — | — |
| 186 | **Deoxyuridine 5′-triphosphate** | dUrd-5′-$P_3$<br>dUTP | pppdU | | $C_9H_{15}N_2O_{14}P_3$ (468.15) | — | — | — | — |

| No. | Acidic spectral data | | | | | | | | | |
|---|---|---|---|---|---|---|---|---|---|---|
| | pH | $\lambda_{max}$ | $\varepsilon_{max}$ ($\times 10^{-3}$) | $\lambda_{min}$ | Spectral ratios 230 | 240 | 250 | 270 | 280 | 290 |
| 185 | 2 | 260 | 9.8 | 231 | — | — | — | — | — | — |
| 186 | 1 | 262 | — | — | — | — | 0.72 | — | 0.45 | — |

| No. | Neutral spectral data | | | | | | | | | |
|---|---|---|---|---|---|---|---|---|---|---|
| | pH | $\lambda_{max}$ | $\varepsilon_{max}$ ($\times 10^{-3}$) | $\lambda_{min}$ | Spectral ratios 230 | 240 | 250 | 270 | 280 | 290 |
| 185 | 7 | 260 | — | 230 | — | — | — | — | — | — |
| 186 | — | — | — | — | — | — | — | — | — | — |

| No. | Alkaline spectral data | | | | | | | | | |
|---|---|---|---|---|---|---|---|---|---|---|
| | pH | $\lambda_{max}$ | $\varepsilon_{max}$ ($\times 10^{-3}$) | $\lambda_{min}$ | Spectral ratios 230 | 240 | 250 | 270 | 280 | 290 |
| 185 | 12 | 261 | 7.6[a] | 241 | — | — | — | — | — | — |
| 186 | — | — | — | — | — | — | — | — | — | — |

## REFERENCES

| No. | Origin and synthesis | $[\alpha]_D^t$ | pK | Spectral data | $R_f$ |
|---|---|---|---|---|---|
| 185 | E : *229* | — | — | *230*[b] | — |
| 186 | C : *289* | — | — | *289* | — |

| No. | Compound | Abbreviation | Symbol | Structure | Formula (mol wt) | Melting point °C | $[\alpha]^t_D$ | pK Basic | pK Acidic |
|---|---|---|---|---|---|---|---|---|---|
| 187 | **Thymidine 3′-phosphate** | dThd-3′-P<br>3′-dTMP | dTp<br>dT- | 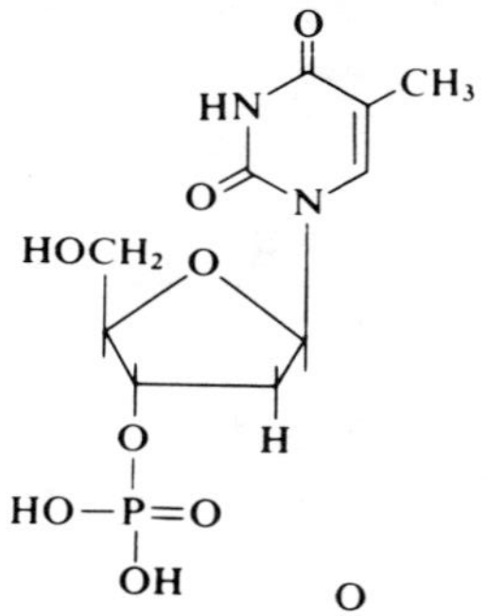 | $C_{10}H_{15}N_2O_8P$<br>(322.21) | 178°<br>(dibrucine salt) | $+7.3^{20}$<br>(1.5, W) | — | — |
| 188 | **Thymidine 5′-phosphate** | dThd-5′-P<br>dTMP | pdT<br>-dT | 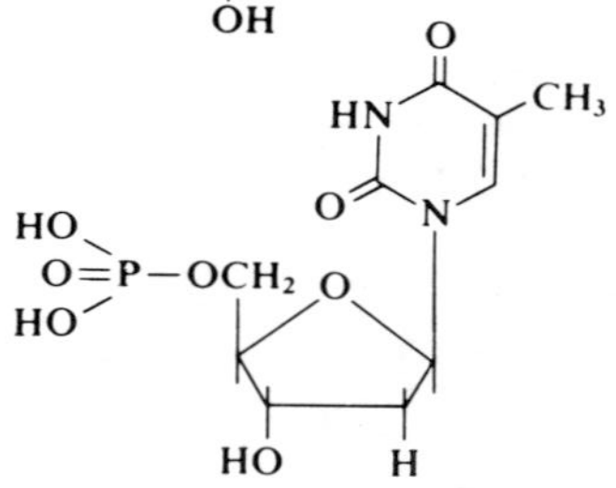 | $C_{10}H_{15}N_2O_8P$<br>(322.21) | 175°<br>(dibrucine salt) | $-4.4^{21}$<br>(0.4, W) | — | 10.0 |

| No. | Acidic spectral data | | | | | | | | | |
|---|---|---|---|---|---|---|---|---|---|---|
| | pH | $\lambda_{max}$ | $\varepsilon_{max}$ ($\times 10^{-3}$) | $\lambda_{min}$ | Spectral ratios 230 | 240 | 250 | 270 | 280 | 290 |
| 187 | 2 | — | — | — | — | — | — | — | 0.69 | — |
| 188 | 2 | 267 | 10.2 | — | — | — | 0.64 | — | 0.72 | 0.23 |

| No. | Neutral spectral data | | | | | | | | | |
|---|---|---|---|---|---|---|---|---|---|---|
| | pH | $\lambda_{max}$ | $\varepsilon_{max}$ ($\times 10^{-3}$) | $\lambda_{min}$ | Spectral ratios 230 | 240 | 250 | 270 | 280 | 290 |
| 187 | 7 | 267 | 9.5 | — | — | — | 0.65 | 1.08 | 0.71 | — |
| 188 | 7 | 267 | 10.2 | — | — | 0.34 | 0.65 | 1.10 | 0.73 | 0.24 |

| No. | Alkaline spectral data | | | | | | | | | |
|---|---|---|---|---|---|---|---|---|---|---|
| | pH | $\lambda_{max}$ | $\varepsilon_{max}$ ($\times 10^{-3}$) | $\lambda_{min}$ | Spectral ratios 230 | 240 | 250 | 270 | 280 | 290 |
| 187 | — | — | — | — | — | — | — | — | — | — |
| 188 | 12 | — | — | — | — | — | 0.74 | — | 0.67 | 0.17 |

REFERENCES

| No. | Origin and synthesis | $[\alpha]^t_D$ | pK | Spectral data | $R_f$ |
|---|---|---|---|---|---|
| 187 | C : *262, 202, 187*[a]*, 203* D : *263* | *187* | — | *262, 187*[e]*, 263*[e] | *187, 262, 263* |
| 188 | C : *187, 202* D : *200* | *207* | *180* | *185*[b]*, 186*[e]*, 179*[e]*, 223* | *187, 185, 263* |

| No. | Compound | Abbreviation | Symbol | Structure | Formula (mol wt) | Melting point °C | $[\alpha]^{t}_{D}$ | pK Basic | pK Acidic |
|---|---|---|---|---|---|---|---|---|---|
| 189 | **Thymidine 5′-triphosphate** | dThd-5′-$P_3$<br>dTTP | | | $C_{10}H_{17}N_2O_{14}P_3$ (482.18) | — | — | — | — |
| 190 | **5-Hydroxymethyldeoxyuridine 5′-phosphate** | 5HmdUrd-5′-P<br>5HmdUMP | phm$^5$dU<br>-hm$^5$dU | | $C_{10}H_{15}N_2O_9P$ (338.21) | — | — | — | — |

| No. | Acidic spectral data | | | | | | | | | | Neutral spectral data | | | | | | | | | | Alkaline spectral data | | | | | | | | | |
|---|---|---|---|---|---|---|---|---|---|---|---|---|---|---|---|---|---|---|---|---|---|---|---|---|---|---|---|---|---|---|
| | pH | $\lambda_{max}$ | $\varepsilon_{max}$ ($\times 10^{-3}$) | $\lambda_{min}$ | Spectral ratios 230 | 240 | 250 | 270 | 280 | 290 | pH | $\lambda_{max}$ | $\varepsilon_{max}$ ($\times 10^{-3}$) | $\lambda_{min}$ | Spectral ratios 230 | 240 | 250 | 270 | 280 | 290 | pH | $\lambda_{max}$ | $\varepsilon_{max}$ ($\times 10^{-3}$) | $\lambda_{min}$ | Spectral ratios 230 | 240 | 250 | 270 | 280 | 290 |
| 189 | 2 | — | — | — | — | — | 0.64 | — | 0.72 | — | — | — | — | — | — | — | — | — | — | — | — | — | — | — | — | — | — | — | — | — |
| 190 | 2 | 264 | 10.2 | 234 | 0.32 | 0.37 | 0.69 | 0.97 | 0.56 | 0.11 | — | — | — | — | — | — | — | — | — | — | 12 | 264 | — | 244* | 1.15 | 0.75 | 0.80 | 0.95 | 0.48 | 0.09 |

REFERENCES

| No. | Origin and synthesis | $[\alpha]^{t}_{D}$ | pK | Spectral data | $R_f$ |
|---|---|---|---|---|---|
| 189 | E : *209* C : *209* | — | — | *209* | — |
| 190 | C : *74* D : *96*, *171* | — | — | *74*, *171**[d], *96* | — |

## FOOTNOTES

[a] Reference giving melting point.
[b] Full spectrum given.
[c] Reference giving $\lambda_{max}$ and/or $\lambda_{min}$.
[d] Reference giving $\epsilon_{max}$.
[e] Reference giving spectral ratios.
[f] pK of 6-methylamino-2-methyladenine; compare the similar pK of adenine and 6-methylaminopurine.[15]
[g] Spectral data taken on material obtained by chemical synthesis and further purified by paper chromatography.
[h] In 50% dimethyl formamide.
[i] In 50% dimethyl sulfoxide/ethanol.
[k] pK of 6-amino-5-formamidoisocytosine.
[l] pK of nucleotide.
[m] $\epsilon$ of 2-methyl-9β-xylanosyladenine.
[n] Calculated from spectral data, using $\epsilon_{max}$ acid of nucleoside.
[o] Decomposes at this pH.
[p] Determined from electrophoretic mobility.
[q] Values very dependent on pH (near pK).
[r] Basic ionization at all pH values.
[s] Spectral data in water and pH 11 indicate decomposition.
[t] Alkaline degradation product of 7-methylguanosine or nucleotide.
[v] Acidic and neutral spectra are similar.
[w] $\epsilon$ estimated from conversion to $N^6$-methyladenosine 3′(2′)-phosphate, using $\epsilon$ of $15.3 \times 10^3$ for the latter.
[x] Based on $\epsilon$ of 7-methylguanosine 3′(2′)-phosphate, assuming quantitative conversion in alkali.
[y] Cyclohexamine salt.
[z] Data on mixed 2′ and 3′ phosphates.

Data taken from: Dunn, D. B., and Hall, R. H., in *Handbook of Biochemistry*, 2nd ed., pp. G-8–G-95, H. A. Sober, Ed. 

## REFERENCES

1. Leonard and Deyrup, *J. Amer. Chem. Soc., 84,* 2148 (1962).
2. Shapiro and Gordon, *Biochem. Biophys. Res. Commun., 17,* 160 (1964).
3. Chheda, Hall, Magrath, Mozejko, Schweizer, Stasiuk and Taylor, *Biochemistry, 8,* 3278 (1969).
4. Schweizer, Chheda, Baczynskyj and Hall, *Biochemistry, 8,* 3283 (1969).
5. Baczynskyj, Biemann, Fleysher and Hall, *Can. J. Biochem., 47,* 1202 (1969).
6. Hall, *Biochemistry, 3,* 769 (1964).
7. Saponara and Enger, *Nature, 223,* 1365 (1969).
8. Brookes and Lawley, *J. Chem. Soc.,* p. 539 (1960).
9. Dunn, *Biochim. Biophys. Acta, 46,* 198 (1961).
10. Jones and Robins, *J. Amer. Chem. Soc., 85,* 193 (1963); Unpublished Data.
11. Mandel, Srinivasan and Borek, *Nature, 209,* 586 (1966).
12. Wacker and Ebert, *Z. Naturforsch. Teil B, 14,* 709 (1959).
13. Lawley and Brookes, *Biochem. J., 89,* 127 (1963).
14. Leonard, Achmatowicz, Loeppky, Carraway, Grimm, Szweykowska, Hamzi and Skoog, *Proc. Nat. Acad. Sci. U.S.A., 56,* 709 (1966).
15. Martin and Reese, Unpublished Data.
16. Leonard and Fujii, *Proc. Nat. Acad. Sci. U.S.A., 51,* 73 (1964).
17. Baddiley, Lythgoe, McNeil and Todd, *J. Chem. Soc.,* p. 383 (1943).
18. Littlefield and Dunn, *Biochem. J., 70,* 642 (1958).
19. Lynch, Robins and Cheng, *J. Chem. Soc.,* p. 2973 (1958).
20. Baddiley, Lythgoe and Todd, *J. Chem. Soc.,* p. 318 (1944).
21. Dunn and Smith, *Biochem. J., 68,* 627 (1958).
22. Albert and Brown, *J. Chem. Soc.,* p. 2060 (1954).
23. Elion, Burgi and Hitchings, *J. Amer. Chem. Soc., 74,* 411 (1952).
24. Broom, Townsend, Jones and Robins, *Biochemistry, 3,* 494 (1964).
25. Adler, Weissman and Gutman, *J. Biol. Chem., 230,* 717 (1958).
26. Elion, in *The Chemistry and Biology of Purines,* p. 39, CIBA Symposium. Churchill, London, England (1957).
27. Weissman, Bromberg and Gutman, *J. Biol. Chem., 224,* 407 (1957).
28. Mason, *J. Chem. Soc.,* p. 2071 (1954).
29. Baker, Joseph and Schaub, *J. Org. Chem., 19,* 631 (1954).

30. Brederick, Haas and Martini, *Chem. Ber., 81,* 307 (1948).
31. Hall, Robins, Stasiuk and Thedford, *J. Amer. Chem. Soc., 88,* 2614 (1966).
32. Robins, Hall and Thedford, *Biochemistry, 6,* 1837 (1967).
33. Elion, *J. Org. Chem., 27,* 2478 (1962).
34. Fischer, *Chem. Ber., 31,* 104 (1898).
35. Pal, *Biochemistry, 1,* 558 (1962).
36. Lawley and Brookes, *Biochem. J., 92,* 19c (1964).
37. Gulland and Holiday, *J. Chem. Soc.,* p. 765 (1936).
38. Traube and Dudley, *Chem. Ber., 46,* 3839 (1913).
39. Pfleiderer, *Ann. Chem. (Justus Liebigs), 647,* 167 (1961).
40. Smith and Dunn, *Biochem. J., 72,* 294 (1959).
41. Haines, Reese and Todd, *J. Chem. Soc.,* p. 5281 (1962).
42. Reiner and Zamenhof, *J. Biol. Chem., 228,* 475 (1957).
43. Elion, Lange and Hitchings, *J. Amer. Chem. Soc., 78,* 217 (1956).
44. Gerster and Robins, *J. Amer. Chem. Soc., 87,* 3752 (1965).
45. Fischer, *Chem. Ber., 30,* 2400 (1897).
46. Lawley, *Proc. Chem. Soc.,* p. 290 (1957).
47. Gulland and Story, *J. Chem. Soc.,* p. 692 (1938).
48. Brookes and Lawley, *J. Chem. Soc.,* p. 3923 (1961).
49. Hall, *Biochem. Biophys. Res. Commun., 13,* 394 (1963); Unpublished Data.
50. Traube and Winter, *Arch. Pharm., 244,* 11 (1906).
51. Ogston, *J. Chem. Soc.,* p. 1713 (1936).
52. Cohn, Unpublished Data.
53. Cavalieri, Fox, Stone and Chang, *J. Amer. Chem. Soc., 76,* 1119 (1954).
54. Pfeiderer and Nübel, *Ann. Chem. (Justus Liebigs), 647,* 155 (1961).
55. Ogston, *J. Chem. Soc.,* p. 1376 (1935).
56. Taylor, *J. Chem. Soc.,* p. 765 (1948).
57. Brookes and Lawley, *J. Chem. Soc.,* p. 1348 (1962); Unpublished Data.
58. Ueda and Fox, *J. Amer. Chem. Soc., 85,* 4024 (1963).
59. Szer and Shugar, *Acta Biochim. Pol., 13,* 177 (1966).
60. Ueda and Fox, *J. Org. Chem., 29,* 1770 (1964).
61. Brown, *J. Appl. Chem., 5,* 358 (1955).
62. Johns, *J. Biol. Chem., 9,* 161 (1911).
63. Brown, *J. Appl. Chem., 9,* 203 (1959).
64. Hitchings, Elion, Falco and Russell, *J. Biol. Chem., 177,* 357 (1949).
65. Wheeler and Johnson, *Amer. Chem. J., 31,* 591 (1904).
66. Shugar and Fox, *Biochim. Biophys. Acta, 9,* 199 (1952).
67. Wyatt, *Biochem. J., 48,* 581 (1951).
68. Cohn, *J. Amer. Chem. Soc., 73,* 1539 (1951); Beaven, Holiday and Johnson, in *The Nucleic Acids,* Vol. 1, p. 520, Chargaff and Davidson, Eds., Academic Press, New York (1955).
69. Cline, Fink and Fink, *J. Amer. Chem. Soc., 81,* 2521 (1959).
70. Flaks and Cohen, *J. Biol. Chem., 234,* 1501 (1959).
71. Miller, *J. Amer. Chem. Soc., 77,* 752 (1955).
72. Fissekis, Myles and Brown, *J. Org. Chem., 29,* 2670 (1964).
73. Wyatt and Cohen, *Biochem. J., 55,* 774 (1953).
74. Alegria, *Biochim. Biophys. Acta, 149,* 317 (1967); Unpublished Data.
75. Loeb and Cohen, *J. Biol. Chem., 234,* 364 (1959).
76. Pohland, Flynn, Jones and Shive, *J. Amer. Chem. Soc., 73,* 3247 (1951).
77. Dunn, *Biochem. J., 86,* 14p (1963).
78. Hilbert and Johnson, *J. Amer. Chem. Soc., 52,* 2001 (1930).
79. Brown, Hoerger and Mason, *J. Chem. Soc.,* p. 211 (1955).
80. Levene, Bass and Simms, *J. Biol. Chem., 70,* 229 (1926).
81. Scannell, Crestfield and Allen, *Biochim. Biophys. Acta, 32,* 406 (1959).
82. Whitehead, *J. Amer. Chem. Soc., 74,* 4267 (1952).
83. Johnson and Heyl, *Amer. Chem. J., 37,* 628 (1907).
84. Wheeler and Liddle, *Amer. Chem. J., 40,* 547 (1908).
85. Fox and Van Praag, *J. Amer. Chem. Soc., 82,* 486 (1960).
86. Elion, Ide and Hitchings, *J. Amer. Chem. Soc., 68,* 2137 (1946).
87. Wang, *J. Amer. Chem. Soc., 81,* 3786 (1959).
88. Johnson and McCollum, *J. Biol. Chem., 1,* 437 (1906).
89. Behrend and Roosen, *Ann. Chem. (Justus Liebigs), 251,* 235 (1889).
90. Dunn and Flack, Unpublished Data.

91. Stimson, *J. Amer. Chem. Soc., 71,* 1470 (1949).
92. Johnson and Litzinger, *J. Amer. Chem. Soc., 58,* 1940 (1936).
93. Dornow and Petsch, *Ann. Chem. (Justus Liebigs), 588,* 45 (1954).
94. Cohn, *J. Biol. Chem., 235,* 1488 (1960).
95. Yu and Allen, *Biochem. Biophys. Acta, 32,* 393 (1959).
96. Kallen, Simon and Marmur, *J. Mol. Biol., 5,* 248 (1962).
97. Chambers and Kurkov, *Biochemistry, 3,* 326 (1964).
98. di Carlo, Schultz and Kent, *J. Biol. Chem., 199,* 333 (1952).
99. Green and Cohen, *J. Biol. Chem., 225,* 397 (1957).
100. Brown and Johnson, *J. Amer. Chem. Soc., 45,* 2702 (1923).
101. Fischer and Roeder, *Chem. Ber., 34,* 3751 (1901).
102. Lengfeld and Stieglitz, *Amer. Chem. J., 15,* 504 (1893).
103. Batt, Martin, Ploeser and Murray, *J. Amer. Chem. Soc., 76,* 3663 (1954).
104. Mitchell and Nyc, *J. Amer. Chem. Soc., 69,* 674 and 1382 (1947).
105. Bachstetz, *Chem. Ber., 63b,* 1000 (1930).
106. Johnson and Schroeder, *J. Amer. Chem. Soc., 54,* 2941 (1932).
107. Broom and Robins, *J. Amer. Chem. Soc., 87,* 1145 (1965).
108. Khwaja and Robins, *J. Amer. Chem. Soc., 88,* 3640 (1966).
109. Hall, *Biochim. Biophys. Acta, 68,* 278 (1963); Unpublished Data.
110. Hall, *Biochemistry, 4,* 661 (1965); Unpublished Data.
111. Davoll and Lowy, *J. Amer. Chem. Soc., 74,* 1563 (1952).
112. Johnson, Thomas and Schaeffer, *J. Amer. Chem. Soc., 80,* 699 (1958).
113. Kissman, Pidacks and Baker, *J. Amer. Chem. Soc., 77,* 18 (1955).
114. Townsend, Robins, Loeppky and Leonard, *J. Amer. Chem. Soc., 86,* 5320 (1964).
115. Andrews and Barber, *J. Chem. Soc.,* p. 2768 (1958).
116. Feldmann, Dütting and Zachau, *Hoppe-Seyler's Z. Physiol. Chem., 347,* 236 (1966).
117. Hall, Csonka, David and McLennan, *Science, 156,* 69 (1967); Unpublished Data.
118. Hall, *Biochemistry, 3,* 876 (1964).
119. Friedrich and Bernhauer, *Chem. Ber., 89,* 2507 (1956).
120. Montgomery and Thomas, *J. Amer. Chem. Soc., 85,* 2672 (1963).
121. Friedrich and Bernhauer, *Chem. Ber., 90,* 465 (1957).
122. Miles, *J. Org. Chem., 26,* 4761 (1961).
123. Brown and Weliky, *J. Biol. Chem., 204,* 1019 (1953).
124. Furukawa, Kobayashi, Kanai and Honjo, *Chem. Pharm. Bull. (Tokyo), 13,* 1273 (1965); Unpublished Data.
125. Haines, Reese and Todd, *J. Chem. Soc.,* p. 1406 (1964).
126. Hall, *Biochem. Biophys. Res. Commun., 12,* 361 (1963).
127. Van Montagu and Stockx, *Arch. Int. Physiol. Biochem., 73,* 158 (1965); Unpublished Data.
128. Fox, Van Praag, Wempen, Doerr, Cheong, Knoll, Eidinoff, Bendich and Brown, *J. Amer. Chem. Soc., 81,* 178 (1959).
129. Wempen, Duschinsky, Kaplan and Fox, *J. Amer. Chem. Soc., 83,* 4755 (1961).
130. Brimacombe and Reese, *J. Chem. Soc.,* p. 588c (1966).
131. Nichols and Land, *Can. J. Biochem., 44,* 1633 (1966).
132. Dunn, *Biochim. Biophys. Acta, 38,* 176 (1960).
133. Dekker and Elmore, *J. Chem. Soc.,* p. 2864 (1951).
134. Cohen and Barner, *J. Biol. Chem., 226,* 631 (1957).
135. Cohen, *Cold Spring Harbor Symp. Quant. Biol., 18,* 221 (1953).
136. Brossmer and Röhm, *Angew. Chem. Int. Ed. Engl., 2,* 742 (1963).
137. Townsend and Robins, *J. Amer. Chem. Soc., 85,* 242 (1963).
138. Smith and Dunn, *Biochim. Biophys. Acta, 31,* 573 (1959).
139. Miles, *Biochim. Biophys. Acta, 22,* 247 (1956).
140. Visser, Barron and Beltz, *J. Amer. Chem. Soc., 75,* 2017 (1953).
141. Levene and Tipson, *J. Biol. Chem., 104,* 385 (1934).
142. Thedford, Fleysher and Hall, *J. Med. Chem., 8,* 486 (1965).
143. Brimacombe, Griffin, Haines, Haslam and Reese, *Biochemistry, 4,* 2452 (1965).
144. Lipsett, *J. Biol. Chem., 240,* 3975 (1965).
145. Roberts and Visser, *J. Amer. Chem. Soc., 74,* 668 (1952).
146. Ueda, *Chem. Pharm. Bull. (Tokyo), 8,* 455 (1960).
147. Levene and La Forge, *Chem. Ber., 45,* 608 (1912).
148. Lis and Passarge, *Arch. Biochem. Biophys., 114,* 593 (1966).
149. Reichard, *Acta Chem. Scand., 9,* 1275 (1955).
150. Fox, Yung, Davoll and Brown, *J. Amer. Chem. Soc., 78,* 2117 (1956).
151. Farkas and Sorm, *Collect. Czech. Chem. Commun., 28,* 1620 (1963).

152. Brossmer and Röhm, *Angew. Chem. Int. Ed. Engl., 3,* 66 (1964).
153. Cohn, Kurkov and Chambers, *Biochem. Prep., 10,* 135 (1963).
154. Shapiro and Chambers, *J. Amer. Chem. Soc., 83,* 3920 (1961).
155. Chambers, *Progr. Nucleic Acid Res. Mol. Biol., 5,* 349 (1966); Shapiro, Reeves and Chambers, Unpublished Data.
156. Ofengand and Schaeffer, *Biochemistry, 4,* 2832 (1965).
157. Davis and Allen, *J. Biol. Chem., 227,* 907 (1957).
158. Chambers, Kurkov and Shapiro, *Biochemistry, 2,* 1192 (1963).
159. Michelson and Cohn, *Biochemistry, 1,* 490 (1962).
160. Michelson, Drell and Mitchell, *Proc. Nat. Acad. Sci. U.S.A., 37,* 396 (1951).
161. Bergmann and Feeney, *J. Org. Chem., 16,* 981 (1951).
162. Fox, Yung and Bendich, *J. Amer. Chem. Soc., 79,* 2775 (1957).
163. Fox and Shugar, *Biochim. Biophys. Acta, 9,* 369 (1952).
164. Griffin and Reese, *Biochim. Biophys. Acta, 68,* 185 (1963); Unpublished Data.
165. Davis, Carlucci and Roubein, *J. Biol. Chem., 234,* 1525 (1959).
166. Ikehara, Ohtsuka and Ishikawa, *Chem. Pharm. Bull. (Tokyo), 9,* 173 (1961).
167. Griffin, Haslam and Reese, *J. Mol. Biol., 10,* 353 (1964).
168. Ikehara, Ueda and Ikeda, *Chem. Pharm. Bull. (Tokyo), 10,* 767 (1962).
169. Szer, *Biochem. Biophys. Res. Commun., 20,* 182 (1965).
170. Cohn, in *The Nucleic Acids,* Vol. 1, p. 211, Chargaff and Davidson, Eds. Academic Press, New York (1955).
171. Kuno and Lehman, *J. Biol. Chem., 237,* 1266 (1962).
172. Lehman and Pratt, *J. Biol. Chem., 235,* 3254 (1960).
173. Szer and Shugar, *Acta Biochim. Pol., 7,* 491 (1960).
174. Letters and Michelson, *J. Chem. Soc.,* p. 71 (1962).
175. Griffin, Todd and Rich, *Proc. Nat. Acad. Sci. U.S.A., 44,* 1123 (1958); Unpublished Data.
176. Goldberg and Rabinowitz, *Biochim. Biophys. Acta, 54,* 202 (1961).
177. Lieberman, Kornberg and Simms, *J. Amer. Chem. Soc., 76,* 2844 (1954).
178. Cohn, *J. Amer. Chem. Soc., 72,* 2811 (1950).
179. Cohn, in *The Nucleic Acids,* Vol. 1, p. 513, Chargaff and Davidson, Eds. Academic Press, New York (1955).
180. Hurst, Marko and Butler, *J. Biol. Chem., 204,* 847 (1953).
181. Ploeser and Loring, *J. Biol. Chem., 178,* 431 (1949).
182. Cohn, *J. Cell. Comp. Physiol., 38 (Suppl. 1),* 21 (1951).
183. Anonymous, *Circular OR-10.* Pabst Laboratories, Milwaukee, Wisconsin (1956).
184. Steiner and Beers, in *Polynucleotides,* p. 155. Elsevier, New York (1961).
185. Shapiro and Chargaff, *Biochim. Biophys. Acta, 26,* 596 (1957).
186. Sinsheimer, *J. Biol. Chem., 208,* 445 (1954).
187. Michelson and Todd, *J. Chem. Soc.,* p. 951 (1953).
188. Clark, Kirby and Todd, *J. Chem. Soc.,* p. 1497 (1957).
189. Chambers, Moffatt and Khorana, *J. Amer. Chem. Soc., 79,* 3747 (1957).
190. Michelson and Todd, *J. Chem. Soc.,* p. 2476 (1949).
191. Brown, Fasman, Magrath and Todd, *J. Chem. Soc.,* p. 1448 (1954).
192. Loring, Bortner, Levy and Hammell, *J. Biol. Chem., 196,* 807 (1952).
193. Barker, *J. Chem. Soc.,* p. 3396 (1954).
194. Jones and Perkins, *J. Biol. Chem., 62,* 557 (1925).
195. Cohn and Volkin, *Arch. Biochem. Biophys., 35,* 465 (1952).
196. Chambers, Shapiro and Kurkov, *J. Amer. Chem. Soc., 82,* 970 (1960).
197. Brown, Haynes and Todd, *J. Chem. Soc.,* p. 3299 (1950).
198. Schmitz, Hurlbert and Potter, *J. Biol. Chem., 209,* 41 (1954).
199. Chambers, Moffatt and Khorana, *J. Amer. Chem. Soc., 77,* 3416 (1955).
200. Volkin, Khym and Cohn, *J. Amer. Chem. Soc., 73,* 1533 (1951).
201. Michelson and Todd, *J. Chem. Soc.,* p. 34 (1954).
202. Tener, *J. Amer. Chem. Soc., 83,* 159 (1961).
203. Schaller, Weimann, Lerch and Khorana, *J. Amer. Chem. Soc., 85,* 3821 (1963).
204. Lipton, Morell, Frieden and Bock, *J. Amer. Chem. Soc., 75,* 5449 (1953).
205. Volkin and Cohn, in *Methods in Biochemical Analysis,* Vol. 1, p. 287, Glick, Ed. Interscience Publications, John Wiley and Sons, New York (1954).
206. Morell and Bock, 126th Meeting of the American Chemical Society, New York, *Biol. Abstr.,* p. 44c (1954).
207. Klein and Thannhauser, *Z. Physiol. Chem., 231,* 96 (1935).
208. Lipsett, *Cold Spring Harbor Symp. Quant. Biol., 31,* 449 (1966).
209. Lehman, Bessman, Simms and Kornberg, *J. Biol. Chem., 233,* 163 (1958).
210. Hall and Khorana, *J. Amer. Chem. Soc., 76,* 5056 (1954).
211. Khorana, *J. Amer. Chem. Soc., 76,* 3517 (1954).
212. Bock, Nan-Sing Ling, Morell and Lipton, *Arch. Biochem. Biophys., 62,* 253 (1956).

213. Chambers and Khorana, *J. Amer. Chem. Soc.*, *79*, 3752 (1957).
214. Smith and Khorana, *J. Amer. Chem. Soc.*, *80*, 1141 (1958).
215. Loring and Luthy, *J. Amer. Chem. Soc.*, *73*, 4215 (1951).
216. Loring, Roll and Pierce, *J. Biol. Chem.*, *174*, /29 (1948).
217. Levene and Tipson, *J. Biol. Chem.*, *106*, 113 (1934).
218. Cavalieri, *J. Amer. Chem. Soc.*, *75*, 5268 (1953).
219. Brown and Todd, *J. Chem. Soc.*, p. 44 (1952).
220. Alberty, Smith and Bock, *J. Biol. Chem.*, *193*, 425 (1951).
221. Reichard, Takenaka and Loring, *J. Biol. Chem.*, *198*, 599 (1952).
222. Levene, *J. Biol. Chem.*, *41*, 483 (1920).
223. California Corporation for Biochemical Research, *Properties of Nucleic Acid Derivatives*, 4th revision (1961).
224. Levene and Jacobs, *Chem. Ber.*, *43*, 3150 (1911).
225. Chambers, Moffatt and Khorana, *J. Amer. Chem. Soc.*, *79*, 4240 (1957).
226. Le Page, *Biochem. Prep.*, *1*, 1 (1949).
227. Shaw, Smallwood and Wilson, *J. Chem. Soc.*, p. 921c (1966).
228. Carrington, Shaw and Wilson, *J. Chem. Soc.*, p. 6864 (1965).
229. Scarano, *Boll. Soc. Ital. Biol. Sper.*, *34*, 722 (1958).
230. Scarano, *Boll. Soc. Ital. Biol. Sper.*, *34*, 727 (1958).
231. Levene and Bass, in *The Nucleic Acids.* Chemical Catalog Co., New York (1931).
232. Beaven, Holiday and Johnson, in *The Nucleic Acids*, Vol. 1, p. 493, Chargaff and Davidson, Eds. Academic Press, New York (1955).
233. Hilbert and Johnson, *J. Amer. Chem. Soc.*, *52*, 1152 (1930).
234. Traube, *Chem. Ber.*, *33*, 1371 (1900).
235. Davoll, Lythgoe and Todd, *J. Chem. Soc.*, p. 967 (1948).
236. Elmore, *J. Chem. Soc.*, p. 2084 (1950).
237. Davoll and Lowy, *J. Amer. Chem. Soc.*, *73*, 1650 (1951).
238. Andersen, Dekker and Todd, *J. Chem. Soc.*, p. 2721 (1952).
239. Klein, *Z. Physiol. Chem.*, *224*, 244 (1934).
240. Deutsch, Unpublished Data.
241. Schindler, *Helv. Chim. Acta*, *32*, 979 (1949).
242. Hotchkiss, *J. Biol. Chem.*, *175*, 315 (1948).
243. Dekker and Todd, *Nature*, *166*, 557 (1950).
244. Watanabe and Fox, *Angew. Chem. Int. Ed. Engl.*, *78*, 589 (1966).
245. Mizuno, Itoh and Tagawa, *Chem. Ind. (London)*, p. 1498 (1965).
246. Anonymous, Schwartz Bioresearch Data (1966).
247. Hoffer, Duschinsky, Fox and Yung, *J. Amer. Chem. Soc.*, *81*, 4112 (1959).
248. Wood, *J. Chem. Soc.*, *89*, 1839 (1906).
249. Ness and Fletcher, *J. Amer. Chem. Soc.*, *81*, 4752 (1959).
250. Albert, *Biochem. J.*, *54*, 646 (1953).
251. Fox, Yung, Wempen and Doerr, *J. Amer. Chem. Soc.*, *79*, 5060 (1957).
252. Howard, Lythgoe and Todd, *J. Chem. Soc.*, p. 1052 (1947).
253. Bergmann and Dikstein, *J. Amer. Chem. Soc.*, *77*, 691 (1955).
254. Johnson and Mackenzie, *Amer. Chem. J.*, *42*, 353 (1909).
255. Levene, Simms and Bass, *J. Biol. Chem.*, *70*, 243 (1926).
256. Baddiley, Lythgoe and Todd, *J. Chem. Soc.*, p. 386 (1943).
257. Robins, Dille, Willits and Christensen, *J. Amer. Chem. Soc.*, *75*, 263 (1953).
258. Carter, *J. Amer. Chem. Soc.*, *72*, 1466 (1950).
259. Klein and Thannhauser, *Z. Physiol. Chem.*, *218*, 173 (1933).
260. Klein and Thannhauser, *Z. Physiol. Chem.*, *224*, 252 (1934).
261. Embden and Schmidt, *Z. Physiol. Chem.*, *181*, 130 (1929).
262. Turner and Khorana, *J. Amer. Chem. Soc.*, *81*, 4651 (1959).
263. Cunningham, *J. Amer. Chem. Soc.*, *80*, 2546 (1958).
264. Hall and Khorana, *J. Amer. Chem. Soc.*, *77*, 1871 (1955).
265. Venkstern and Baev, in *Absorption Spectra of Minor Bases, Their Nucleosides, Nucleotides and Selected Oligoribonucleotides.* Plenum Press Data Division, New York (1965).
266. Hanze, *J. Amer. Chem. Soc.*, *89*, 6720 (1967).
267. Venner, *Chem. Ber.*, *93*, 140 (1960).
268. Falconer, Gulland and Story, *J. Chem. Soc.*, p. 1784 (1939).
269. Davoll, *J. Amer. Chem. Soc.*, *73*, 3174 (1951).
270. Carbon, David and Studier, *Science*, *161*, 1146 (1968); Unpublished Data.
271. Ueda, Iida, Ikeda and Mizuno, *Chem. Pharm. Bull. (Tokyo)*, *16*, 1788 (1968); Unpublished Data.
272. Ueda and Nishino, *J. Amer. Chem. Soc.*, *90*, 1678 (1968).

273. Ueda, Iida and Mizuno, *Chem. Pharm. Bull. (Tokyo), 14,* 666 (1966).
274. Burrows, Armstrong, Skoog, Hecht, Boyle, Leonard and Occolowitz, *Science, 161,* 691 (1968); Unpublished Data.
275. Baczynskyj, Biemann and Hall, *Science, 159,* 1481 (1968).
276. Shaw, Warrener, Maguire and Ralph, *J. Chem. Soc.*, p. 2294 (1958).
277. Strominger and Friedkin, *J. Biol. Chem., 208,* 663 (1954).
278. Gray and Lane, *Biochemistry, 7,* 3441 (1968).
279. Johnson and Speh, *Amer. Chem. J., 38,* 602 (1907).
280. Codington, Fecher, Maguire, Thomson and Brown, *J. Amer. Chem. Soc., 80,* 5164 (1958).
281. Brown, Todd and Varadarajan, *J. Chem. Soc.*, p. 2384 (1956).
282. Bendich, Tinker and Brown, *J. Amer. Chem. Soc., 70,* 3109 (1948).
283. Cavalieri, Bendich, Tinker and Brown, *J. Amer. Chem. Soc., 70,* 3875 (1948).
284. Wyngaarden and Dunn, *Arch. Biochem. Biophys., 70,* 150 (1957).
285. Lee and Wigler, *Biochemistry, 7,* 1427 (1968).
286. Lis and Passarge, *Physiol. Chem. Phys., 1,* 68 (1969).
287. Fuji and Leonard, in *Synthetic Procedures in Nucleic Acid Chemistry,* Vol. 1, p. 13, Zorbach and Tipson, Eds. Interscience Publications, John Wiley and Sons, New York (1968).
288. Nishimura, Yamada and Ishikura, *Biochim. Biophys. Acta, 179,* 517 (1969).
289. Bessman, Lehman, Adler, Zimmerman, Simms and Kornberg, *Proc. Nat. Acad. Sci. U.S.A., 44,* 633 (1958).
290. Chheda, *Life Sci., 8,* 979 (1969).

# NATURAL OCCURRENCE OF THE MODIFIED NUCLEOSIDES

DR. ROSS H. HALL AND DR. DAVID B. DUNN

The following table lists the natural distribution of the modified nucleosides known to occur in RNA. Three compounds originally reported as components of RNA have not been included, because recent evidence suggests that they may be artifacts of the isolation procedure or that their structural assignment may be incorrect. Neoguanosine, isolated by Hemmens[1] and identified as $N^2$-ribosylguanine by Shapiro and Gordon,[2] appears to result from an acid-catalyzed interaction of ribose and guanine.[3] α-Cytidylic acid[4] may result from alkali-catalyzed anomerization during the hydrolysis procedure.[5] The isolated "1,5-diribosyluracil"[6] does not correspond to a chemically synthesized sample.[7]

The values reported for many of the analyses represent the actual amounts isolated and should be considered only as minimal amounts.

The RNA samples are listed as transfer (t) or ribosomal (r), but – particularly in earlier studies – these designations represent more accurately the soluble and high-molecular-weight RNA. If no designation is given, the sample consisted of a mixture of all classes of RNA isolated from the organism.

| RNA Source | Amount, moles/100 moles | Ref. |
|---|---|---|
| **1-Methyladenosine[a]** | | |
| Carp liver (t) | 1.0 | 17 |
| Mouse adenosarcoma (t) | 0.84[c] | 18 |
| Mouse liver (t) | 0.15[c] | 18 |
| Pig liver | 0.95[b,c] | 9 |
| Pigeon (t) | 0.6 | 17 |
| Rabbit liver (whole RNA) | 0.06[c] | 16 |
| Rat liver (r) | 0.10[b,c] | 16, 19 |
| Rat liver (t) | 1.6[c] | 19 |
| S-180 ascites tumor (t) | 0.70[c] | 18 |
| Chinese cabbage leaves (t) | 0.65[b,c] | 44 |
| Tobacco leaves (t) | 0.1[c], 0.8[b,c] | 18, 44 |
| Wheat germ | 0.7[b,c] | 16 |
| Wheat germ (t) | 0.91 | 21 |
| Yeast, commercial | 0.05[c] | 15, 16 |
| Yeast (r) | 0.03[c] | 23, 45 |
| Yeast (t) | 0.9[b,c] | 22 |
| *Euglena gracilis* | 1.2 | 20 |
| *Neurospora* (r) | 0.04[c] | 45 |
| *Neurospora* (t) | 0.8[c] | 45 |
| *Aerobacter aerogenes* | 0.06[c] | 16 |
| *Azotobacter vinelandii* (t) | 0.3 | 17 |
| *Escherichia coli* (t) | 0.1[d] | 24 |
| *Escherichia coli* (23S) | 0.13[e] | 27 |
| *Proteus vulgaris* (t) | 0.02 | 17 |
| *Sarcina flava* (t) | 0.4 | 17 |
| *Staphylococcus aureus* | 0.04[c] | 16 |
| **2-Methyladeonsine[f]** | | |
| Carp liver (t) | 0.02 | 17 |
| S-180 ascites | 0.48[c] | 18 |
| Tobacco leaves (t) | 0.07[c] | 18 |
| Wheat germ | 0.02[c] | 16 |
| Yeast, commercial | 0.02[c] | 16 |
| **2-Methyladeonsine (continued)** | | |
| *Aerobacter aerogenes* | 0.06[c] | 16 |
| *Azotobacter vinelandii* (t) | 0.14 | 17 |
| *Escherichia coli* (t) | 0.25 | 24 |
| *Escherichia coli* (23S) | 0.07[e] | 27 |
| *Proteus vulgaris* (t) | 0.06 | 17 |
| *Sarcina flava* (t) | 0.11 | 17 |
| *Staphylococcus aureus* | 0.01[c] | 16 |
| **$N^6$,$N^6$-Dimethyladenosine[g]** | | |
| Mouse adenosarcoma (t) | 0.23[c] | 18 |
| Mouse liver (t) | 0.11[c] | 18 |
| Pigeon liver(t) | 0.11 | 17 |
| Rat liver (r) | 0.02[c] | 16, 19 |
| Rat liver (t) | 0.02[c] | 19 |
| S-180 ascites (t) | 0.43[c] | 18 |
| Tobacco leaves (t) | 0.07[c] | 18 |
| Wheat germ | 0.01[c] | 16 |
| Yeast, commercial | 0.01[c] | 16 |
| *Aerobacter aerogenes* | 0.02[c] | 16 |
| *Azotobacter vinelandii* (t) | 0.24 | 17 |
| *Escherichia coli* (r) | 0.08 | 29 |
| *Escherichia coli* B | 0.15[c] | 16 |
| *Escherichia coli* (16S) | 0.27[e] | 27 |
| *Escherichia coli* (16S) | 0.14[h] | 28 |
| *Proteus vulgaris* (t) | 0.2 | 17 |
| **2′-O-Methyladenosine[i]** | | |
| Calf liver (t) | 0.19 | 41, 42 |
| Human liver (r) | 0.08 | 42 |
| Human liver (t) | 0.19 | 42 |
| Human tumors | 0.05 | 42 |
| Mouse Ehrlich ascites | 0.17 | 42 |

| RNA Source | Amount, moles/100 moles | Ref. |
|---|---|---|

**2′-O-Methyladeonsine (continued)**

| RNA Source | Amount, moles/100 moles | Ref. |
|---|---|---|
| Rat liver (r) | Present | 31 |
| Rat liver (t) | Present | 31 |
| Rat Murphy-Sturm lympho-sarcoma | 0.30 | 42 |
| Sheep heart | 0.26 | 42 |
| Sheep liver | 0.21 | 41, 42 |
| Chinese cabbage leaves (r) | ~0.4 | 43 |
| Tobacco leaves | Present | 31 |
| Wheat germ | Present | 31 |
| Wheat germ (r) | 0.54 | 35 |
| Wheat germ (t) | 0.005 | 21 |
| Yeast (r) | Present | 45 |
| Yeast (t) | 0.03 | 32, 42 |
| *Neuraspora* (r) | Present | 45 |
| *Neuraspora* (t) | Present | 45 |
| *Escherichia coli* (t) | 0.01 | 41 |

**$N^6$-($\Delta^2$-Isopentenyl)adenosine[j]**

| RNA Source | Amount, moles/100 moles | Ref. |
|---|---|---|
| Calf liver (t) | 0.05 | 52 |
| Chick embryo (r) | Not detected | 52 |
| Chick embryo (t) | 0.03 | 52 |
| Human liver (r) | Not detected | 52 |
| Human liver (t) | 0.05 | 52 |
| Rat liver (t) | Present | 49 |
| Immature corn kernels (t) | Not detected | 353 |
| Immature peas (t) | 0.003 | 53 |
| Spinach leaves (t) | 0.02 | 53 |
| Yeast (t) | 0.065 | 52 |
| *Escherichia coli* B (t) | Not detected | 51 |
| *Lactobacillus acidophilus* (t) | Present | 51 |
| *Lactobacillus plantarum* (t) | Present | 51 |

**$N^6$-(*cis*-4-Hydroxy-3-methylbut-2-enyl)adenosine[k]**

| RNA Source | Amount, moles/100 moles | Ref. |
|---|---|---|
| Calf liver (t) | Not detected | 32 |
| Chick embryo (t) | Not detected | 32 |
| Immature corn kernels | 0.01 | 53 |
| Immature peas (t) | 0.05 | 53 |
| Mature corn kernels (seed corn) | Present | 32 |
| Spinach leaves | 0.01 | 53 |
| Yeast (t) | Not detected | 32 |
| *Escherichia coli* B (t) | Not detected | 32 |
| *Lactobacillus acidophilus* (t) | Not detected | 32 |

**$N^6$-($\Delta^2$-Isopentenyl)-2-methylthioadenosine[l]**

| RNA Source | Amount, moles/100 moles | Ref. |
|---|---|---|
| *Escherichia coli* (t) | Present | 58–60, 108 |

**N-[9-($\beta$-D-Ribofuranosyl)pruin-6-ylcarbamoyl]-L-threonine[m]**

| RNA Source | Amount, moles/100 moles | Ref. |
|---|---|---|
| Calf liver (t) | 0.19 | 62 |
| Rat liver (t) | Present | 62 |

| RNA Source | Amount, moles/100 moles | Ref. |
|---|---|---|

**N-[9-($\beta$-D-Ribofuranosyl)pruin-6-ylcarbamoyl]-L-threonine[m] (continued)**

| RNA Source | Amount, moles/100 moles | Ref. |
|---|---|---|
| Yeast (r) | Not detected | 62 |
| Yeast (t) | 0.28 | 62 |
| *Escherichia coli* (r) | Not detected | 62 |
| *Escherichia coli* (t) | 0.07 | 62 |

**2′(3′)-O-Ribosyladenosine[n]**

| RNA Source | Amount, moles/100 moles | Ref. |
|---|---|---|
| Animal cells | Present | 65–67 |
| Yeast (t) | Present | 23 |

**3-Methylcytidine[o]**

| RNA Source | Amount, moles/100 moles | Ref. |
|---|---|---|
| Rat liver (t) | Present | 49 |
| Yeast (t) | 0.1 | 68 |

**$N^4$,$O^2$-Dimethylcytidine[p]**

| RNA Source | Amount, moles/100 moles | Ref. |
|---|---|---|
| *Escherichia coli* 16S | 0.05[r] | 70 |

**5-Methylcytidine[q]**

| RNA Source | Amount, moles/100 moles | Ref. |
|---|---|---|
| Pig liver (t) | 1.2[c] | 75 |
| Rat liver (r) | 0.02[c] | 10 |
| Rat liver (t) | 2.0[c] | 75 |
| Chinese cabbage leaf (r) | 0.03[c] | 43 |
| Chinese cabbage leaf (t) | 0.8[c] | 43, 44 |
| Tobacco leaf (t) | 0.75[c] | 44 |
| Wheat germ (t) | 1.37[c] | 21 |
| Yeast (r) | 0.02[c] | 45 |
| Yeast (t) | 1.2[c] | 22 |
| *Neurospora* (r) | 0.02[c] | 45 |
| *Neurospora* (t) | 0.8[c] | 45 |
| *Aerobacter aerogenes* | 0.03[c] | 75 |
| *Escherichia coli* $K_{12}$[q] | 1.5 | 74 |
| *Escherichia coli* (16S) | 0.13[e] | 27 |
| *Escherichia coli* (23S) | 0.07[e] | 27 |

2

**2′-O-Methylcytidine[r]**

| RNA Source | Amount, moles/100 moles | Ref. |
|---|---|---|
| Calf liver (t) | 0.11 | 41, 42 |
| Human liver (r) | 0.09 | 42 |
| Human liver (t) | 0.34 | 42 |
| Human tumor | 0.06 | 42 |
| Mouse Ehrlich ascites | 0.14 | 42 |
| Rat liver | Present | 77 |
| Rat liver (r) | Present | 19 |
| Rat liver (t) | Present | 19 |
| Rat Murphy-Sturm lympho-sarcoma | 0.20 | 42 |
| Sheep liver | 0.26 | 41 |
| Chinese cabbage leaf (r) | Present | 43 |
| Wheat germ (r) | 0.35 | 35 |
| Wheat germ (t) | 0.40 | 21 |
| Yeast (r) | Present | 45 |
| Yeast (t) | 0.28 | 32, 42, 77 |

| RNA Source | Amount, moles/100 moles | Ref. |
|---|---|---|
| **2′-O-Methylcytidine (continued)** | | |
| *Anacystis nidulans* | 0.05 | 34 |
| *Neurospora* (r) | Present | 45 |
| *Escherichia coli* (r) | 0.03 | 70 |
| *Escherichia coli* (t) | 0.06 | 41, 42 |
| **N⁴-Acetylcytidine**[s] | | |
| Rat liver (t) | Present | 49 |
| Yeast (t) | Present | 48 |
| **2-Thiocytidine**[t] | | |
| *Escherichia coli* (t) | Present | 78 |
| **1-Methylgauanosine**[u] | | |
| Carp liver (t) | 0.6 | 17 |
| Frog liver (t) | 0.6 | 17 |
| Mouse adenosarcoma (t) | 0.21[c] | 18 |
| Mouse liver (t) | 0.07–0.45[c] | 18 |
| Mouse spleen (t) | 0.13[c] | 18 |
| Pigeon liver (t) | 0.65 | 17 |
| Rabbit liver (r) | 0.03[c] | 18 |
| Rabbit liver (t) | 0.94[c] | 18 |
| Rat liver (r) | 0.02[c] | 19, 79 |
| Rat liver (t) | 0.8[c] | 19, 79 82 |
| S-180 ascites (t) | 0.6[c] | 18 |
| Chinese cabbage leaves (t) | 0.50[c] | 43, 44 |
| Sugar beet leaves | 0.05[c] | 79 |
| Tobacco leaves (t) | 0.50[c] | 44 |
| Wheat germ (t) | 0.73 | 21 |
| Yeast (t) | 1.00[c] | 22 |
| *Neurospora* (t) | 0.5[c] | 45 |
| *Aerobacter aerogenes* | 0.02[c] | 79 |
| *Azotobacter vinelandii* (t) | 0.18 | 17 |
| *Escherichia coli* (t) | 0.14 | 24 |
| *Escherichia coli* (23S) | 0.07[e] | 27 |
| *Proteus vulgaris* (t) | 0.32 | 17 |
| *Sarcina flava* (t) | 0.14 | 17 |
| **N²-Methylguanosine**[v] | | |
| Mouse adenosarcoma (t) | 2.5[c] | 18 |
| Mouse liver (t) | 0.7–1.6[c] | 18 |
| Mouse spleen (t) | 0.64[c] | 18 |
| Rabbit liver (r) | 0.005[c] | 18 |
| Rabbit liver (t) | 0.5[c] | 18 |
| Rat liver (r) | 0.02[c] | 19, 79 |
| Rat liver (t) | 0.4[c] | 19, 31, 82 |
| S-180 ascites (t) | 1.0[c] | 18 |
| Chinese cabbage leaves (t) | 0.2[c] | 43 |
| Tobacco leaves (t) | 0.38[c] | 18 |
| Wheat germ (t) | 0.29 | 21 |
| Yeast (t) | 0.35[c] | 22, 45 |
| **N²-Methylguanosine (continued)** | | |
| *Euglena gracilis* | 1.0[c] | 20 |
| *Neurospora* (t) | 0.10[c] | 45 |
| *Aerobacter aerogenes* | 0.01[c] | 79 |
| *Escherichia coli* (16S) | 0.27[e] | 27 |
| *Escherichia coli* (23S) | 0.13[e] | 27 |
| **N²,N²-Dimethylguanosine**[w] | | |
| Mouse adenosarcoma (t) | 0.64[c] | 18 |
| Mouse liver (t) | 0.14–0.28[c] | 18 |
| Mouse spleen (t) | 0.17[c] | 18 |
| Rabbit liver (r) | 0.01[c] | 18 |
| Rabbit liver (t) | 0.6[c] | 18 |
| Rat liver (r) | 0.02[c] | 19, 79 |
| Rat liver (t) | 0.65[c] | 19, 79, 82 |
| S-180 ascites (t) | 0.40[c] | 18 |
| Chinese cabbage leaves (t) | 0.55[c] | 44 |
| Sugar beet leaves | 0.025[c] | 79 |
| Tobacco leaves (t) | 0.55[c] | 44 |
| Wheat germ (t) | 0.55 | 21 |
| Yeast (t) | 0.70[c] | 22, 45 |
| *Neurospora* (t) | 0.65[c] | |
| **7-Methylguanosine**[x] | | |
| Pig liver (t) | 0.20[c] | 9 |
| Rat liver (r) | Present | 84 |
| Rat liver (t) | Present | 84 |
| Chinese cabbage leaves (t) | 0.30[c] | 43, 44 |
| Tobacco leaves (t) | 0.20[c] | 44 |
| Wheat germ (t) | 0.18[h] | 21 |
| Yeast (t) | 0.35[c] | 44, 45 |
| *Escherichia coli* (t) | 0.50[c] | 85 |
| *Escherichia coli* (16S) | 0.13[e] | 27 |
| *Escherichia coli* (23S) | 0.07[e] | 27 |
| **N²,N²,7-Trimethylguanosine**[y] | | |
| Hamster overy HeLa Cells | Present | 109 |
| **2′-O-Methylguanosine**[z] | | |
| Calf liver (t) | 0.16 | 41, 42 |
| Human liver (r) | 0.15 | 42 |
| Human liver (t) | 0.30 | 42 |
| Human tumor | 0.12 | 42 |
| Mouse Ehrlich ascites | 0.23 | 42 |
| Rat liver | Present | 77 |
| Rat liver (r) | Present | 19 |
| Rat liver (t) | Present | 19 |
| Rat Murphy-Sturm lympho-sarcoma | 0.30 | 42 |
| Sheep heart | 0.38 | 42 |
| Sheep liver | 0.27 | 41, 42 |

| RNA Source | Amount, moles/100 moles | Ref. |
|---|---|---|
| **2′-O-Methylguanosine (continued)** | | |
| Chinese cabbage leaves (r) | Present | 43 |
| Wheat germ (r) | 0.38 | 35 |
| Wheat germ (t) | 0.47 | 21 |
| Yeast (r) | Present | 45 |
| Yeast (t) | 0.25 | 32, 42, 77 |
| *Neurospora* (r) | Present | 45 |
| *Escherichia coli*[aa] | 0.10 | 41, 42 |
| *Escherichia coli* (r) | 0.02 | 70 |
| **Inosine**[bb] | | |
| Pig liver (t) | 0.15[c] | 9 |
| Tobacco leaves (t) | 0.1[c] | 44 |
| Wheat germ (t) | 0.09 | 21 |
| Yeast (t) | 0.33 | 87 |
| **1-Methylinosine**[cc] | | |
| Yeast (t) | 0.045 | 87 |
| **3-Methyluridine**[dd] | | |
| Human liver (t) | 0.03 | 68 |
| Yeast (t) | 0.01 | 68 |
| **5-Methyluridine**[ee] | | |
| Human kidney (t) | 0.09[ff] | 92 |
| Rat liver (t) | 0.04 | 92 |
| Chinese cabbage leaves (t) | 0.70[c] | 44 |
| Tobacco leaves (t) | 0.75[c] | 44 |
| Wheat germ | 0.75[c] | 16 |
| Wheat germ (t) | 0.64 | 21 |
| Yeast (t) | 1.2[c] | 22, 45 |
| *Neurospora* (t) | 1.1[c] | 45 |
| *Aerobacter aerogenes* | 0.24[c] | 16 |
| *Azotobacter vinelandii* (t) | 1.2 | 17 |
| *Escherichia coli* (t) | 1.1[c] | 24 |
| *Escherichia coli* (23S) | 0.13[e] | 27 |
| *Proteus vulgaris* (t) | 1.4 | 17 |
| *Sarcina flava* (t) | 1.4 | 17 |
| *Staphylococcus aureus* | 0.2[c] | 16 |
| **2′-O-Methyluridine**[gg] | | |
| Calf liver (t) | 0.21 | 41, 42 |
| Human liver (r) | 0.04 | 42 |
| Human liver (t) | 0.22 | 42 |
| Human tumor | 0.11 | 42 |
| Mouse Ehrlich ascites | 0.20 | 42 |
| Rat liver (r) | 0.18[c] | 19 |
| Rat liver (t) | 0.30[c] | 19 |
| Rat Murphy-Sturm lympho-sarcoma | 0.21 | 42 |
| Sheep liver | 0.31 | 41, 42 |

| RNA Source | Amount, moles/100 moles | Ref. |
|---|---|---|
| **2′-O-Methyluridine (continued)** | | |
| Chinese cabbage leaves (r) | 0.35[c] | 43 |
| Wheat germ (r) | 0.47 | 35 |
| Wheat germ (t) | 0.30 | 21 |
| Yeast (r) | 0.07 | 45 |
| *Neurospora* (r) | 0.16[c] | 45 |
| *Neurospora* (t) | 0.07[c] | 45 |
| *Escherichia coli* | 0.06 | 32, 41, 42 |
| *Escherichia coli* (23S) | 0.07 | 27, 70 |
| **5-β-D-Ribofuranosyluracil**[hh] | | |
| Carp liver (t) | 4.2 | 17 |
| Frog liver (t) | 4.3 | 17 |
| Pig liver (t) | 3.2[c] | 9 |
| Pigeon liver (t) | 3.5 | 17 |
| Rat liver (r) | 1.6[c] | 10 |
| Rat liver (t) | 4.3[c] | 19 |
| Chinese cabbage leaves (r) | 1.4[c] | 43 |
| Chinese cabbage leaves (t) | 2.9[c] | 44 |
| Tobacco leaves (t) | 2.6[c] | 44 |
| Wheat germ (r) | 1.8 | 35 |
| Wheat germ (t) | 2.7 | 21 |
| Yeast (r) | 0.8–0.5[c] | 11, 45 |
| Yeast (t) | 4.0–4.4[c] | 11, 45 |
| *Neurospora* (r) | 1.2[c] | 45 |
| *Neurospora* (t) | 4.3[c] | 45 |
| *Azotobacter vinelandii* (t) | 3.2–4.1 | 17 |
| *Escherichia coli* B (r) | 0.15[c] | 24, 70 |
| *Escherichia coli* B (t) | 2.0 | 24, 28 |
| *Escherichia coli* B (t) | 1.3 | 70 |
| *Escherichia coli* 16S | 0.12 | 28 |
| *Escherichia coli* (16S) | 0.6 | 70 |
| *Escherichia coli* (23S) | 0.30 | 28 |
| *Escherichia coli* (23S) | 0.15 | 70 |
| *Proteus vulgaris* (t) | 3.5–4.5 | 17 |
| *Sarcina flava* (t) | 3.1–3.8 | 17 |
| **5-(2′-O-Methylribosyl)uracil**[ii] | | |
| Rat liver (t) | Present | 49 |
| Wheat germ (t) | 0.12 | 21 |
| Yeast (t) | Present | 42 |
| **2-Thio-5-carboxymethyluridine Methyl Ester**[jj] | | |
| Yeast (2) | Present | 97 |
| **4-Thiouridine**[kk] | | |
| *Escherichia coli* B (t) | 0.9 | 100 |
| *Escherichia coli* B (t) | 0.7 | 98 |
| **5-Hydroxyuridine (Isobarbituridine)**[ll] | | |
| Yeast | Present | 101 |

| RNA Source | Amount, moles/100 moles | Ref. |
|---|---|---|
| **2-Thio-5-(N-Methylaminomethyl)uridine$^{mm}$** | | |
| *Escherichia coli* (t) | Present | 78 |
| **5-Carboxymethyluridine$^{nn}$** | | |
| Wheat embryo (r) | Not detected | 102 |
| Wheat embryo (t) | 0.15 | 102 |
| Yeast (r) | Not detected | 102 |
| Yeast (t) | 0.34 | 102 |
| *Escherichia coli* (r) | Not detected | 102 |
| *Escherichia coli* (t) | Not detected | 102 |

| RNA Source | Amount, moles/100 moles | Ref. |
|---|---|---|
| **5,6-Dihydrouridine$^{oo}$** | | |
| Rat liver (t) | 1.1 | 100 |
| Wheat embryo (t) | 1.9 | 102 |
| Yeast (t) | 3.0 | 102 |
| Yeast (t) | 3.6 | 104 |
| Yeast (t) | 3.9 | 100 |
| *Bacillus subtilis* (t) | 1.7 | 100 |
| *Escherichia coli* B (t) | 2.2 | 102, 104 |
| *Escherichia coli* B (t) | 3.3 | 100 |

*a* 1-Methyladenosine rearranges readily in neutral and alkaline solution to N$^6$-methyladenosine.[13] The kinetics of the rate of the rearrangement at different pH values has been studied by Macon and Wolfenden.[14] N$^6$-Methyladenosine or its base has often been isolated from nucleic acid samples under conditions that do not preclude the occurrence of the rearrangement during isolation. Therefore, analytical values for N$^6$-methyl- and N$^1$-methyladenosine are included under this heading. The compound was first isolated as N$^6$-methyladenine from commercial yeast RNA by Adler *et al.*[15] and from the RNA of several organisms by Littlefield and Dunn.[16] N$^6$-Methyladenine (but not the N$^1$-methyl derivative) occurs in the DNA of some organisms. This nucleoside appears in the primary sequence of several molecular species of yeast tRNA.

*b* The presence of 1-methyladenosine was confirmed by isolation of either 1-methyladenine[8] or 1-methyladenylic acid.[9]

*c* The value was originally reported relative to uridine. Since the uridylic acid content of most RNA samples, including tRNA of yeast and t- and rRNA of mammalian liver, is between 18 and 21 mole %,[11,12,105] the reported values are divided by 5. This does not apply, however, to results on tRNA from *Escherichia coli* and rRNA from yeast and *Neurospora,* which are corrected for uridylic acid contents of 15, 26, and 25% respectively.[11,24,106]

*d* The presence of N$^6$-methyladenine was established as a component by acid hydrolysis.[24]

*e* Values have been calculated assuming chain lengths for 16S and 23S RNA of 1500 and 3000 nucleotides respectively. These values were determined using data of Stanley and Bock.[71]

*f* 2-Methyladenosine was first isolated by Littlefield and Dunn[16] as the nucleotide and nucleoside from the RNA of several organisms. It also exists as a component of vitamin $B_{12}$ (the 7$a$ isomer) that occurs in *Escherichia coli.*[25,26]

*g* N$^6$,N$^6$-Dimethyladenosine was first isolated by Littlefield and Dunn[16] as the nucleotide and nucleoside from the RNA of several sources. It occurs exclusively in the 16S ribosomal RNA of *Escherichia coli* as the dinucleotide $m_2^6$Apm$_2^6$Ap.[27,28,29] The base occurs as a component of the antibiotic, puromycin.[30]

*h* Value probably low, due to alkaline decomposition.

*i* 2′-O-Methyladenosine and the corresponding methylated derivatives of the three other major ribonucleosides were first isolated by Smith and Dunn[31] and characterized as 2′(3′)-O-derivatives. Hall[32] identified this group of nucleosides as the 2′-isomer; this identification was confirmed by Honjo *et al.*[33] Biswas and Myers[34] identified the 2′-O-methylcytidine in the nucleic acids of the blue-green alga, *Anacystis nidulans.* Alkali hydrolysates of wheat germ rRNA and tRNA have yielded alkali-stable dinucleotides composed of all combinations of the four 2′-O-methylribonucleosides and the four major ribonucleosides.[21,35,36] The 2′-O-methylribose content of the nucleic acids of plant and mammalian tissues appears to be much higher than that of *Escherichia coli.* This difference is particularly striking in rRNA. The 16S + 28S RNA of L cells and HeLa cells contains about one mole % of 2′-O-methylribonucleosides, and these methyl groups account for 80 to 90% of the total methyl groups in the rRNA.[37,38,39] *Escherichia coli* rRNA, on the other hand, contains about the same proportion of methyl groups, but only about 10% of the methyl groups are attached to the O$^2$-position of the nucleoside residues. The presence of a methyl group at the 2′-position of the pyrimidine nucleosides blocks the action of pancreatic ribonucleases. Honjo *et al.*[33] found that the 5′-phosphate derivatives of these four 2′-O-methylribonucleosides are not dephosphorylated by the action of bull semen or snake venom 5′-nucleotidases. Gray and Lane[40] found that the 2′-O-methyl substituent adjacent to the 3′-phosphoester internucleotide linkage inhibits the action of snake venom diesterase.

*j* N$^6$-($\Delta^2$-Isopentenyl)adenosine was isolated from yeast tRNA and identified by Biemann *et al.*[46] and Hall *et al.*[47] It occurs in the primary sequence adjacent to the anticodon of yeast and rat liver tRNA$^{Ser}$[48,49] and yeast tRNA$^{Tyr}$.[50] The following data were obtained by means of hydrolysis of the tRNA by whole snake venom and bacterial alkaline phosphatase. More recently it has been found that this procedure does not give complete release of the N$^6$-($\Delta^2$-isopentenyl)adenosine.[51] Consequently, the values reported may be low.

*k* N$^6$-(*cis*-4-Hydroxy-3-methylbut-2-enyl)adenosine was isolated and identified as the nucleoside by Hall *et al.,*[54] starting from plant tissue tRNA. This compound has not been detected in the tRNA of yeast or mammalian tissue. The *trans* isomer occurs in plant tissue in a relatively unbound form (perhaps as the nucleotide or nucleoside).[55,56,57] The values reported here may be low for the same reason that the values for $N^6$-($\Delta^2$-isopentenyl)adenosine are low, although the authors have no specific data on the rate of release of this compound.

*l* $N^6$-($\Delta^2$-Isopentenyl)-2-methylthioadenosine has been isolated from *Escherichia coli* tRNA[58] and has been identified in a specific sequence of *E. coli* $tRNA^{Tyr}$ adjacent to the presumed anticodon.[59,60] It has also been detected in *E. coli* $tRNA^{Phe}$, $tRNA^{Ser}_{Ia}$, and $tRNA^{Ser}_{II}$.[108]

*m* N-[9-($\beta$-D-Ribofuranosyl)purin-6-ylcarbamoyl]-L-threonine was first isolated from yeast tRNA and characterized as an $N^6$-(aminoacyl)adenosine by Hall.[61] Subsequently, Chheda *et al.*[62] and Schweizer *et al.*[63] described the isolation and characterization of this nucleoside. This compound has been identified in the primary sequence of yeast $tRNA^{Ile}$.[64]

*n* 2′(3′)-O-Ribosyladenosine was isolated from samples of yeast tRNA and characterized by Hall.[23] More recently, Chambon *et al.*,[65] Sugimura *et al.*,[66] and Hasegawa *et al.*[67] described the presence of a polymer in animal cells that appears to consist of repeating units of the phosphate of this nucleoside. The polymer is insoluble in acid; it is reasonable to assume that if it were present in yeast, it would be co-isolated together with the tRNA. Therefore, the source of 2′(3′)-O-ribosyladenosine may not have been the tRNA, but rather this polymer.

*o* 3-Methylcytidine was first detected as the nucleoside in yeast tRNA.[68] The nucleotide is unstable to alkali[107] and is converted entirely to non-ultraviolet-absorbing compounds by the conditions normally used for alkaline hydrolysis. Under mild alkaline conditions, such as those used for hydrolysis with snake venom, partial conversion to 3-methyluridine can occur.[107] The nucleotide can be isolated from RNA hydrolyzed with acid at 28°C.[22] It has been identified in a specific sequence of rat liver $tRNA^{Ser}$.[49]

*p* $N^4$,$O^2$-Dimethylcytidine was isolated from an alkaline hydrolysate of *Escherichia coli* ribosomal RNA by Nichols and Lane[29,69] and represents the only identified methylated nucleoside containing a methyl group attached to both the sugar and base position. It occurs in the 16S RNA of *E. coli* at a level of about 0.05 mole %, which corresponds to about one per 2,000 nucleotides.[70] The chain length of the 16S component is about 1500 nucleotides.[71] In all probability, there is one such residue per 16S molecule.[27,72,73]

*q* 5-Methylcytidine was originally isolated from acid-insoluble polyribonucleotides of a strain of *Escherichia coli* by Amos and Korn[74] and was positively identified by Dunn[75] in the RNA of several species. It has been identified in the primary sequences of several molecular species of yeast tRNA.

*r* 2′-O-Methylcytidine. See 2′-O-Methyladenosine for general remarks. Methylcytidine has been identified in the primary sequence of yeast $tRNA^{Phe}$.[76]

*s* $N^4$-Acetylcytidine was identified in a specific sequence of yeast and rat liver $tRNA^{Ser}$.[48,49]

*t* 2-Thiocytidine has been detected in an alkaline hydrolysate of *Escherichia coli* tRNA.[78]

*u* 1-Methylguanosine was first isolated as the free base from commercial yeast RNA by Adler *et al.*,[15] and as the nucleoside from the RNA of several organisms by Smith and Dunn.[79] It has been identified in the primary sequence of yeast $tRNA^{Ala}$[80] and $tRNA^{Val}$.[81]

*v* $N^2$-Methylguanosine was first isolated as the free base from commercial yeast by Adler *et al.*,[15] and as the nucleoside from the RNA of several species by Smith and Dunn.[79] It has been identified in the primary sequence of yeast $tRNA^{Phe}$[76] and $tRNA^{Tyr}$.[50]

*w* $N^2$,$N^2$-Dimethylguanosine was first isolated from the RNA of several organisms by Smith and Dunn.[79]

*x* 7-Methylguanosine. The free base of this nucleoside was isolated from hot acid hydrolysates of the tRNA from *Escherichia coli*, and the nucleotide was isolated from cold acid hydrolysates of tRNA from pig liver, yeast and Chinese cabbage leaves by Dunn.[9] At neutral and alkaline pH values, this nucleoside undergoes ring scission to form 2-amino-4-hydroxy-5-(N-methyl) formamido-6-ribosylaminopyrimidine. Because of the ease of this reaction isolation, procedures involving enzymatic processes at slightly alkaline pH values result in conversion to the ring-opened form. Hall[83] first reported the isolation of the ring-opened form from an enzymatic hydrolysate of yeast tRNA. The conversion of 7-methylguanosine to the ring-opened form by treatment with alkali is not quantitative and, therefore, cannot be used as a reliable assay for 7-methylguanosine. Other products appear to be formed; analysis for 7-methylguanosine based on the ring-opened base gives results 25% lower than that based on isolation of 7-methylguanylic acid.[9] 7-Methylguanosine occurs in yeast $tRNA^{Phe}$.[76]

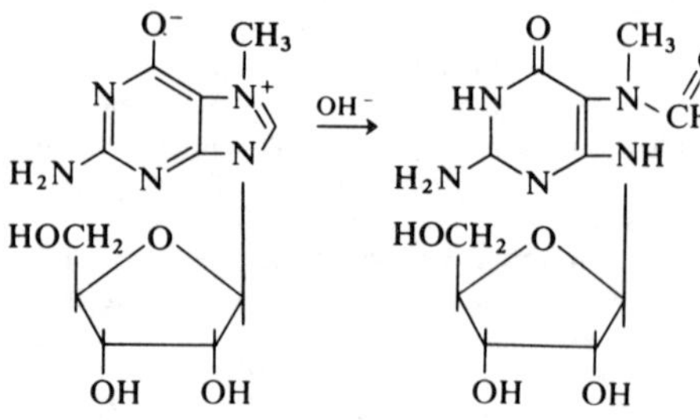

*y* $N^2$,$N^2$,7-Trimethylguanosine was isolated by Saponara and Enger[109] from a nuclear RNA fraction of hamster ovary cells and HeLa cells. They isolated it as the nucleoside from an enzymic hydrolysate. In alkali, the imidazole ring is readily cleaved, analogous to the alkali-catalyzed degradation of 7-methylguanosine. The nucleoside has been detected only by radioactive labeling, but the properties of the radioacive sample correspond to those of a synthetic sample.

*z* 2′-O-Methylguanosine has been identified in the primary sequence of several molecular species of yeast tRNA. See 2′-O-Methyladenosine for general remarks.

*aa* 2′-O-Methylguanosine is the most common $O^2$-methyl nucleoside in *Escherichia coli* RNA (see also 86).

*bb* Inosine was identified by yeast tRNA[45,87] and occurs as the third letter of the anticodon of yeast $tRNA^{Ala}$,[80] $tRNA^{Ser}$[48] and $tRNA^{Val}$.[81] It has not been found in any other position of the tRNA molecule, and this fact may be related to its ability to pair with more than one base residue (wobbling).

*cc* 1-Methylinosine occurs in yeast tRNA[87] and has been identified in the primary sequence of yeast $tRNA^{Ala}$.[80]

*dd* 3-Methyluridine was detected as the nucleoside in yeast tRNA and in human liver tRNA.[68] Since 3-methylcytidine readily undergoes deamination at slightly alkaline pH values, the possibility exists that this nucleoside could have been formed from 3-methylcytidine during the isolation procedure. Dunn and Flack[22] report the failure to detect any 3-methyluridylic acid in mild acid hydrolysates of yeast tRNA, although >0.003% should have been detectable. While more stable to alkali than 3-methylcytidine, 3-methyluridine and its nucleotide are converted to non-ultraviolet-absorbing compounds in alkali.[108] The conditions necessary for alkaline hydrolysis of RNA result in complete degradation of the nucleotide. The attachment of a methyl at the 3 position of uridylic acid derivatives markedly changes the properties of the compound. For example, 3-methyluridylyl-3′ derivatives are resistant to the action of pancreatic ribonuclease.[88,89] Methyl groups at position 3 of uridine residues of polyuridylic acid eliminate all vestiges of a secondary structure;[88] this fact is in distinct contrast to methyl groups at C-5.

*ee* 5-Methyluridine was originally identified in the RNA of several sources by Littlefield and Dunn.[16] A 5-methyl group attached to uridylic acid residues in polynucleotides can considerably influence the stacking pattern of adjacent nucleotides. This influence is indicated by the substantial increase in the hyperchromicity of poly 5-methyluridylic acid over that of polyuridylic acid,[90] and by a relatively high melting point, Tm = 36°.[91] The presence of the 5-methyl group in uridylyl-3′ oligoribonucleotide derivatives has no effect on the normal action of pancreatic ribonuclease.[89] 5-Methyluridine has been found in every molecular species of tRNA for which the sequence has been determined.

*ff* Value probably low, due to contamination of tRNA with degraded rRNA.[92]

*gg* 2′-O-Methyluridine occurs in yeast $tRNA^{Ser}_{III}$.[48] See 2′-O-Methyladenosine for general remarks.

*hh* 5-β-D-Ribofuranosyluracil appeared originally in the elution profile of an RNA hydrolysate[93] and later became known as the "fifth" ribonucleotide[94] or "pseudouridine".[95] It was identified by Cohn[95] as a 5-ribosyluracil; the β-D-ribofuranosyl configuration was established by Michelson and Cohn.[96] This nucleoside is the most predominant of the modified nucleosides. In yeast and rat liver tRNA, it accounts for about 25% of the uridine residues. 5-β-D-ribofuranosyluracil has been identified in every known sequence of tRNA.

*ii* 5-(2′-O-Methylribosyl)uracil was first isolated as the nucleoside from an enzymic digest of yeast tRNA;[42] however, the amount obtained did not permit a rigorous identification. Evidence for its existence rests solely on spectral and chromatographic behavior of the isolated sample. A similar compound was isolated as the nucleotide by Hudson *et al.*[21] from an alkaline digest of tRNA of wheat germ. Its chromatographic and spectral properties conform to those expected for 5′(2′-O-methylribosyl)uracil. It has been identified in a specific sequence of rat liver $tRNA^{Ser}$.[49]

*jj* 2-Thio-5-carboxymethyluridine methyl ester was detected in a snake venom hydrolysate of yeast tRNA.[97]

*kk* 4-Thiouridine was isolated from the tRNA of *Escherichia coli* and identified by Lipsett.[98] It has not been detected in the tRNA of yeast or mammalian tissue,[32] although the sulfur-containing nucleoside 2-thio-5-carboxymethyluridine methyl ester occurs in the tRNA of yeast.[97] Lipsett and Doctor[99] purified a species of *E. coli* $tRNA^{Tyr}$ and found two 4-thiouridylate residues per molecule.

*ll* 5-Hydroxyuridine was identified as the nucleoside in yeast total RNA by Lis and Passarge.[101]

*mm* 2-Thio-5-(N-methylaminomethyl)uridine has been detected in an alkaline hydrolysate of *Escherichia coli* tRNA.[78]

*nn* 5-Carboxymethyluridine was first detected in wheat embryo tRNA by Gray and Lane.[40,102] Gray and Lane[102] report that very little of this compound can be isolated from a snake venom diesterase digest of the tRNA; on the other hand, the compound can be readily obtained from an alkaline digest. This fact led the authors to suggest that in native tRNA the carboxymethyl group is esterified. There is precedence for the existence of an esterified carboxyl group in the tRNA; Baczynskyj *et al.*[97] identified 2-thio-5-carboxymethyluridine methyl ester in yeast tRNA.

*oo* 5,6-Dihydrouridine was first detected in pancreatic digests of yeast $tRNA^{Ala}$ by Madison and Holley.[103] From two to six 5,6-dihydrouridylic acid units occur in each of the yeast tRNA molecules of known sequences.

Data taken from: Hall, R. H., *The Modified Nucleosides in Nucleic Acids* (1971). Reproduced by permission of Columbia University Press, New York.

## REFERENCES


1. Hemmens, *Biochim. Biophys. Acta, 68,* 284 (1963).
2. Shapiro and Gordon, *Biochem. Biophys. Res. Commun., 17,* 160 (1964).
3. Hemmens, *Biochim. Biophys. Acta, 91,* 332 (1964).
4. Gassen and Witzel, *Biochim. Biophys. Acta, 95,* 244 (1965).
5. Dekker, *Symposium on Chemistry of Nucleosides and Nucleotides.* 150th Meeting of American Chemical Society at Atlantic City, New Jersey (1965).
6. Lis and Lis, *Biochim. Biophys. Acta, 61,* 799 (1962).
7. Dlugajczyk and Eiler, *Biochim. Biophys. Acta, 119,* 11 (1966).
8. Dunn, *Biochim. Biophys. Acta, 46,* 198 (1961).
9. Dunn, *Biochem. J., 86,* 14P (1963).
10. Dunn, Smith and Simpson, *Biochem. J., 76,* 24P (1960).

11. Monier, Stephenson and Zamecnik, *Biochim. Biophys. Acta, 43,* 1 (1960).
12. Osawa, *Biochim. Biophys. Acta, 43,* 110 (1960).
13. Brookes and Lawley, *J. Chem. Soc.*, p. 539 (1960).
14. Macon and Wolfenden, *Biochemistry, 7,* 3453 (1968).
15. Adler, Weissmann and Gutman, *J. Biol. Chem., 230,* 717 (1958).
16. Littlefield and Dunn, *Biochem. J., 70,* 642 (1958).
17. Zaitseva, Dmitriva, Chanfa and Belozerskii, *Dokl. Akad. Nauk. (SSSR), 147,* 1211 (1962).
18. Bergquist and Matthews, *Biochem. J., 85,* 305 (1962).
19. Dunn, *Biochim. Biophys. Acta, 34,* 286 (1959).
20. Brawerman, Hufnagel and Chargaff, *Biochim. Biophys. Acta, 61,* 340 (1962).
21. Hudson, Gray and Lane, *Biochemistry, 4,* 2009 (1965).
22. Dunn and Flack, *Abstr. 4th Fed. Eur. Biochem. Soc.*, p. 82 (1967).
23. Hall, *Biochemistry, 4,* 661 (1965).
24. Dunn, Smith and Spahr, *J. Mol. Biol., 2,* 113, (1960).
25. Brown, Cain, Gant, Parker and Smith, *Biochem. J., 59,* 82 (1955).
26. Dion, Calkins and Pfiffner, *J. Amer. Chem. Soc., 76,* 948 (1954).
27. Fellner and Sanger, *Nature, 219,* 236 (1968).
28. Dubin and Gunalp, *Biochim. Biophys. Acta, 134,* 106 (1967).
29. Nichols and Lane, *Biochim. Biophys. Acta, 119,* 649 (1966).
30. Waller, Fryth, Hutchings and Williams, *J. Amer. Chem. Soc., 75,* 2025 (1953).
31. Smith and Dunn, *Biochim. Biophys. Acta, 31,* 573 (1959).
32. Hall, *Biochim. Biophys. Acta, 68,* 278 (1963); Unpublished Data.
33. Honjo, Kanai, Furukawa, Mizuno and Sanno, *Biochim. Biophys. Acta, 87,* 696 (1964).
34. Biswas and Myers, *Nature, 186,* 238 (1960).
35. Lane, *Biochemistry, 4,* 212 (1965).
36. Singh and Lane, *Can. J. Biochem., 42,* 1011 (1964).
37. Tamaoki and Lane, *Biochemistry, 7,* 3431 (1968).
38. Wagner, Penman and Ingram, *J. Mol. Biol., 29,* 371 (1967).
39. Vaughan, Soeiro, Warner and Darnell, *Proc. Nat. Acad. Sci. U.S.A., 58,* 1527 (1967).
40. Gray and Lane, *Biochim. Biophys. Acta, 134,* 243 (1967).
41. Hall, *Biochem. Biophys. Res. Commun., 12,* 429 (1963).
42. Hall, *Biochemistry, 3,* 876 (1964).
43. Dunn, Hitchborn and Trim, *Biochem. J., 88,* 34P (1963).
44. Dunn and Flack, *Phytochemistry, 6,* 459 (1967).
45. Dunn, *Fifth Int. Congr. Biochem. Moscow, 10,* 68 (1961).
46. Biemann, Tsunakawa, Sönnenbichler, Feldmann, Dütting and Zachau, *Angew. Chem., 78,* 600 (1966).
47. Hall, Robins, Stasiuk and Thedford, *J. Amer. Chem. Soc., 88,* 2614 (1966).
48. Zachau, Dütting and Feldmann, *Z. Physiol. Chem., 347,* 212 (1966).
49. Staehelin, Rogg, Baguley, Ginsberg and Wehrli, *Nature, 219,* 1363 (1968).
50. Madison and Kung, *J. Biol. Chem., 242,* 1324 (1967).
51. Fittler, Kline and Hall, *Biochemistry, 7,* 940 (1968).
52. Robins, Hall and Thedford, *Biochemistry, 6,* 1837 (1967).
53. Hall, Csonka, David and McLennan, *Science, 156,* 69 (1967).
54. Hall, Mittelman, Horoszewicz and Grace, *Ann. N. Y. Acad. Sci., 143,* 799 (1967).
55. Letham, *Life Sci. 5,* 551 (1966).
56. Letham, *Life Sci., 5,* 1999 (1966).
57. Miller, *Proc. Nat. Acad. Sci. U.S.A., 54,* 1052 (1965).
58. Burrows, Armstrong, Skoog, Hecht, Boyle, Leonard and Occolowitz, *Science, 161,* 691 (1968).
59. Goodman, Abelson, Landy, Brenner and Smith, *Nature, 217,* 1019 (1968).
60. Harada, Gross, Kimura, Chang, Nishimura and RajBhandary, *Biochem. Biophys. Res. Commun., 33,* 299 (1968).
61. Hall, *Biochemistry, 3,* 769 (1964).
62. Chheda, Hall, Magrath, Mozejko, Schweizer, Stasiuk and Taylor, *Biochemistry, 8,* 3278 (1969).
63. Schweizer, Chheda, Baczynskyj and Hall, *Biochemistry, 8,* 3283 (1969).
64. Takemura, Murakami and Miyazaki, *J. Biochem. (Tokyo), 65,* 553 (1969).
65. Chambon, Weill, Doly, Strosser and Mandel, *Biochem. Biophys. Res. Commun., 25,* 638 (1966).
66. Sugimura, Fujimura, Hasegawa and Kawamura, *Biochim. Biophys. Acta, 138,* 438 (1967).
67. Hasegawa, Fujimura, Shimizu and Sugimura, *Biochim. Biophys. Acta, 149,* 369 (1967).
68. Hall, *Biochem. Biophys. Res. Commun., 12,* 361 (1963).
69. Nichols and Lane, *Can. J. Biochem., 44,* 1633 (1966).
70. Nichols and Lane, *J. Mol. Biol., 30,* 477 (1967).
71. Stanley and Bock, *Biochemistry, 4,* 1302 (1965).
72. Nichols and Lane, *Can. J. Biochem., 46,* 109 (1968).

73. Nichols and Lane, *Biochim. Biophys. Acta, 166,* 605 (1968).
74. Amos and Korn, *Biochim. Biophys. Acta, 29,* 444 (1958).
75. Dunn, *Biochim. Biophys. Acta, 38,* 176 (1960).
76. RajBhandary, Chang, Stuart Faulkner, Hoskinson and Khorana, *Proc. Nat. Acad. Sci. U.S.A., 57,* 751 (1967).
77. Morisawa and Chargaff, *Biochim. Biophys. Acta, 68,* 147 (1963).
78. Carbon, David and Studier, *Science, 161,* 1146 (1968).
79. Smith and Dunn, *Biochem. J., 72,* 294 (1959).
80. Holley, Apgar, Everett, Madison, Marquisee, Merrill, Penswick and Zamir, *Science, 147,* 1462 (1965).
81. Bayev, Venkstern, Mirzabekov, Krutilina, Li and Axelrod, *Mol. Biol. (U.S.S.R.), 1,* 754 (1967).
82. Sluyser and Bosch, *Biochim. Biophys. Acta, 55,* 479 (1962).
83. Hall, in *Methods in Enzymology,* Vol. 12, Nucleic Acids, Part A, p. 305, Grossman and Moldave, Eds. Academic Press, New York (1967).
84. Craddock, Villa-Trevino and Magee, *Biochem. J., 107,* 179 (1968).
85. Dunn and Flack, Unpublished Data.
86. Sanger, Brownlee and Barrell, *J. Mol. Biol., 13,* 373 (1965).
87. Hall, *Biochem. Biophys. Res. Commun., 13,* 394 (1963).
88. Szer and Shugar, *Acta Biochim. Pol., 8,* 235 (1961).
89. Thedford, Fleysher and Hall, *J. Med. Chem., 8,* 486 (1965).
90. Griffin, Todd and Rich, *Proc. Nat. Acad. Sci. U.S.A., 44,* 1123 (1958).
91. Shugar and Szer, *J. Mol. Biol., 5,* 580 (1962).
92. Price, Hinds and Brown, *J. Biol. Chem., 238,* 311 (1963).
93. Cohn and Volkin, *Nature, 167,* 483 (1951).
94. Davis and Allen, *J. Biol. Chem., 227,* 907 (1957).
95. Cohn, *J. Biol. Chem., 235,* 1488 (1960).
96. Michelson and Cohn, *Biochemistry, 1,* 490 (1962).
97. Baczynskyj, Biemann and Hall, *Science, 159,* 1481 (1968).
98. Lipsett, *J. Biol. Chem., 240,* 3975 (1965).
99. Lipsett and Doctor, *J. Biol. Chem., 242,* 4072 (1967).
100. Cerutti, Holt and Miller, *J. Mol. Biol., 34,* 505 (1968).
101. Lis and Passarge, *Arch. Biochem. Biophys., 114,* 593 (1966).
102. Gray and Lane, *Biochemistry, 7,* 3441 (1968).
103. Madison and Holley, *Biochem. Biophys. Res. Commun., 18,* 153 (1965).
104. Magrath and Shaw, *Biochem. Biophys. Res. Commun., 26,* 32 (1967).
105. Trim, Baker and Leah, *Biochem. J., 93,* 14 (1964).
106. Henney and Storck, *J. Bacteriol., 85,* 822 (1963).
107. Brookes and Lawley, *J. Chem. Soc.,* p. 1348 (1962).
108. Nishimura, Yamada and Ishikura, *Biochim. Biophys. Acta, 179,* 517 (1969).
109. Saponara and Enger, *Nature, 223,* 1365 (1969).

# NUCLEOSIDE ANTIBIOTICS

DR. LEROY TOWNSEND and DR. ROLAND K. ROBINS

The data on nucleoside antibiotics on the following pages were selected from the published literature to which reference is made. In the case of the more important nucleoside antibiotics, actual spectra as determined in the author's own laboratory have been reproduced for reference. The IR spectra were obtained in KBr, the UV spectra in methanol, and the PMR spectra with TMS as internal standard. The exceptions are the polyoxins, where some of the PMR spectra have dioxane as an internal standard and whose UV spectra were obtained to 0.05*N* HCl (solid line) and in 0.05*N* NaOH (dotted line). Additional physical data, such as melting point, optical rotation, chromatography, pK, method for purification, and solubility, have been included.

Biological and chemotherapeutic information includes the source or microorganism that gives rise to the antibiotic, a brief summary of the more important biological and chemotherapeutic activity, and reference to the mechanism of action if known. Other information, such as toxicity and any unusual activity, has also been included.

It should be noted that often a single nucleoside antibiotic has been isolated from several sources and has been given different names. In those instances where this identity has been verified, it has been noted. In certain cases some of the antibiotics are related or presumed to be identical to another, better-known antibiotic. When it is not known whether the products are identical, due to lack of data or sample, this has been noted. In some instances compounds have been included as nucleoside antibiotics even though insufficient data are available at this time to verify the type of structure; these compounds have been included for completeness and in the hope that these entries will stimulate further work in a class of compounds of considerable interest to biochemistry and molecular biology.

## INDEX


| Antibiotic Name | No. |
|---|---|
| **3′-Acetamido-3′-deoxyadenosine** | 1 |
| **Allomycin** (Amicetin) | 2 |
| **Amicetin** (Allomycin, Antibiotic D-13, Sacromycin) | 3 |
| **Amicetin B** (Plicacetin) | 4 |
| **3′-Amino-3′-deoxyadenosine** | 5 |
| **Angustmycin A** (Antibiotic A-14, Decoyinine) | 6 |
| **Angustmycin C** (Antibiotic U-9586, Psicofuranine) | 7 |
| **Antibiotic 1037** (Toyocamycin) | 8 |
| **Antibiotic C** (Plicacetin) | 9 |
| **Antibiotic D** (Bamicetin) | 10 |
| **Antibiotic D-13** (Amicetin) | 11 |
| **Antibiotic E-212** first substance (Toyocamycin) | 12 |
| **Antibiotic R-285** (Amicetin B or Plicacetin) | 13 |
| **Antibiotic T-3018** (Nucleocidin) | 14 |
| **9-β-D-Arabinofuranosyladenine** (Ara-A) | 15 |
| **Aristeromycin** | 16 |
| **BA-90912** (Sangivamycin) | 17 |
| **5-Azacytidine** (Antibiotic U, 18496) | 18 |
| **Bamicetin** (Antibiotic D, N-Nor-amicetin) | 19 |
| **Blasticidin S** | 20 |
| **Cordycepin** (3′-Deoxyadenosine) | 21 |
| **Cytomycin** | 22 |
| **Cytovirin** | 23 |
| **Decoyinine** (Angustmycin A) | 24 |
| **Formycin** | 25 |
| **Formycin B** (Laurusin, Oyamycin) | 26 |
| **Gougerotin** (No. 21544 Substance) | 27 |
| **Homocitrullyaminoadenosine** | 28 |
| **Laurusin** (Formycin B) | 29 |
| **Lysylaminoadenosine** | 30 |
| **Nariterashin** (Toyocamycin) | 31 |
| **Naritheracin** (Toyocamycin) | 31 |
| **Nebularine** (9-β-D-Ribofuranosylpurine) | 32 |
| **Nucleocidin** (Antibiotic T-3018) | 33 |
| **Oyamycin** (Formycin B) | 34 |
| **Plicacetin** (Amicetin B, Antibiotic C, Antibiotic R-285) | 35 |
| **Polyoxin A** | 36 |
| **Polyoxin B** | 37 |
| **Polyoxin C** | 38 |
| **Polyoxin D** | 39 |
| **Polyoxin E** | 40 |
| **Polyoxin F** | 41 |
| **Polyoxin G** | 42 |
| **Polyoxin H** | 43 |
| **Polyoxin I** | 44 |
| **Polyoxin J** | 45 |
| **Polyoxin K** | 46 |
| **Polyoxin L** | 47 |
| **Psicofuranine** (Angustmycin C) | 48 |
| **Puromycin** (Achromycin, *(260)*) | 49 |
| **Pyrazomycin** | 50 |
| **Sacromycin** (Amicetin) | 51 |
| **Sangivamycin** (BA-90912) | 52 |
| **Showdomycin** (MSD-125 A) | 53 |
| **Sparsomycin A** (Tubercidin) | 54 |
| **Tubercidin** (Sparsomycin A) | 55 |
| **Toyocamycin** (Antibiotic 9-27, Antibiotic 9-48, Antibiotic 1037, Antibiotic E-212 first substance, Naritheracin, Siromycin, Unamycin B, Vengicide) | 56 |
| **Unamycin B** (Toyocamycin) | 57 |
| **Vengicide** (Toyocamycin) | 58 |

| No. | Name | Structure | Formula (mol wt) | Melting point °C | $[\alpha_D]$ | Chromatography |
|---|---|---|---|---|---|---|
| 1 | **3′-Acetamido-3′-deoxyadenosine** (*1*) | 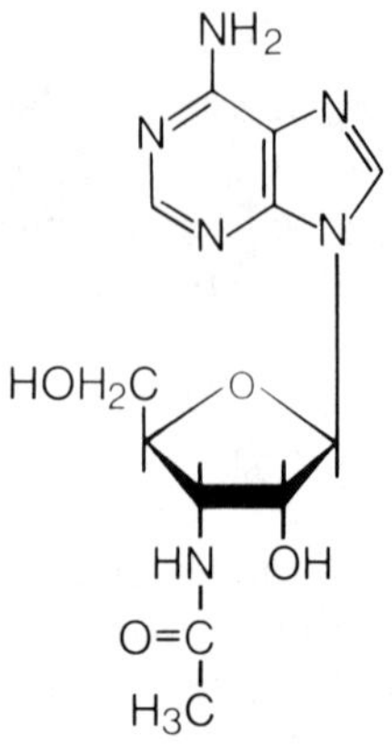 NH2, N, N, N, N, HOH2C, O, HN, OH, O=C, H3C | $C_{12}H_{16}N_6O_4$ (308.3) | 263–265 (*1*) | 8 ± 2° (0.1 *M* HCl) (*1*) | Column (*1*), Paper (*1*) |

| pK | Purification | Source | Solubility | Toxicity | Miscellaneous |
|---|---|---|---|---|---|
| — | Isolated by Column Chromatography and recrystallized from EtOH (*1*). | *Helminthosporium sp.* 215 (*1, 2*) | Soluble in $H_2O$ | — | uv (*1*), ir (*1*), pmr (*1*), Mass spectroscopy (*1*), uv($A_{280nm}$/$A_{260nm}$), Chemical synthesis (*1*) |

| Biological activity |
|---|
| Inactive against Ehrlich–Lettre ascitic tumor cells, *B. cereus*, *E. coli B*, *B. megaterium*, *S. lindegren*, and *S. showdoensis* (*1*). |

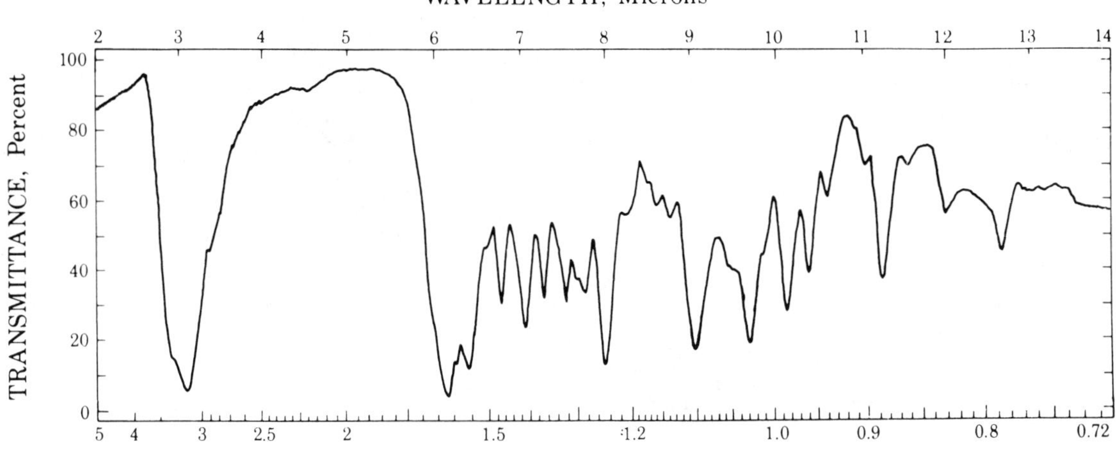
WAVELENGTH, Microns
TRANSMITTANCE, Percent
FREQUENCY (cm⁻¹) × 10⁻³

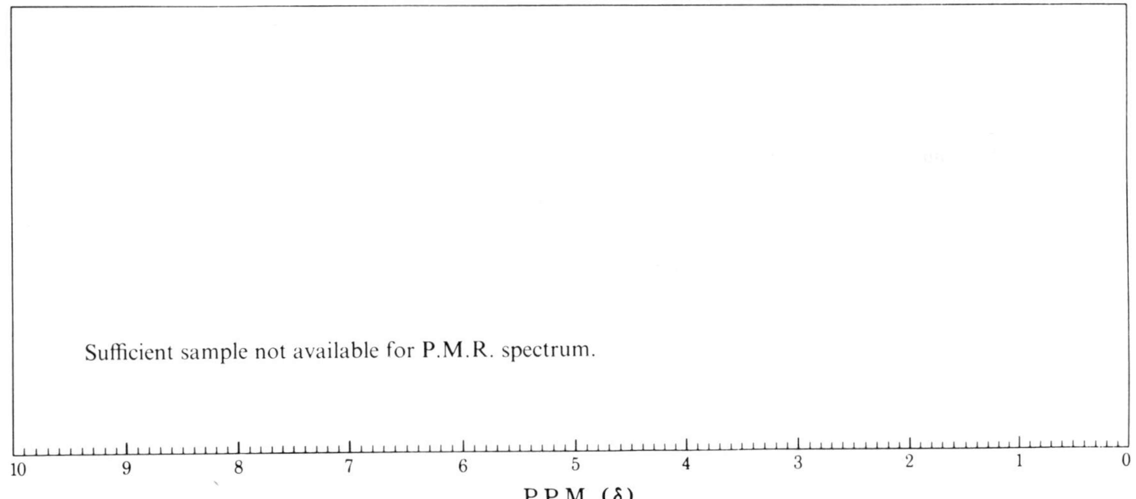
Sufficient sample not available for P.M.R. spectrum.
P.P.M. (δ)

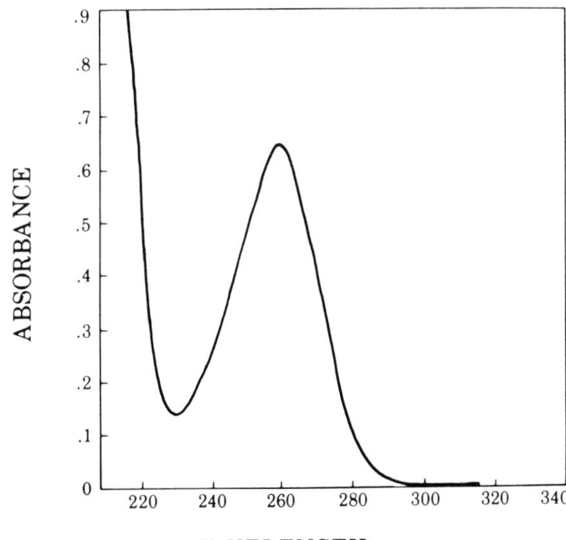
ABSORBANCE
WAVELENGTH, nm.

| No. | Name | Structure | Formula (mol wt) | Melting point °C | $[\alpha]_D$ | Chromatography |
|---|---|---|---|---|---|---|
| 2 | **Allomycin** | 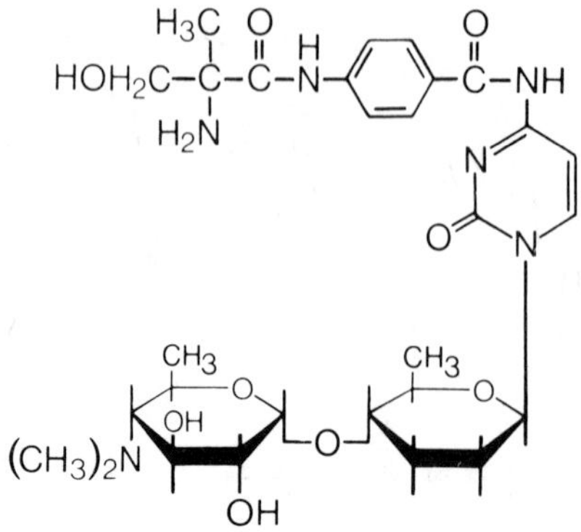 | $C_{29}H_{42}N_6O_9$ (636.7) | 220–225 dec. (3–5), 195 dec. (2 HCl) (*3*), 175 dec. (citrate) (*3*), 175 (benzoate) (3) | — | — |

| pK | Purification | Source | Solubility | Toxicity | Miscellaneous |
|---|---|---|---|---|---|
| — | Purified by ion-exchange and isolated as the HCl salt (*3*). | *Streptomyces sindenensis* (*3, 6, 7*) | Soluble in $H_2O$ | $LD_{50}$ 6.8 mg/kg (i.v.), 57 mg/kg (s.c.) (mice) | uv |

| Biological activity |
|---|
| Active against gram-positive bacteria including mycobacteria. |

| No. | Name | Structure | Formula (mol wt) | Melting point °C | $[\alpha]_D^{26}$ | Chromatography |
|---|---|---|---|---|---|---|
| 3 | **Amicetin** (*8*), (*9*) | 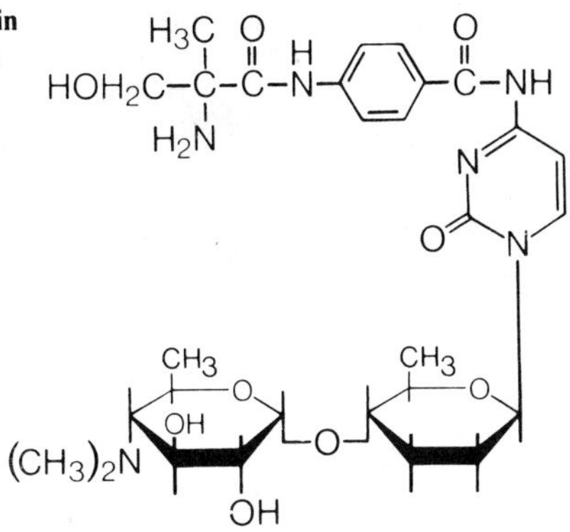 | $C_{29}H_{42}N_6O_9$ (636.7) | 252–253 dec. (granular), 166–169 dec. (needle), 190–192 (hydrochloride) | +116° (c = 0.5, 0, 0.1 *N* HCl) | Column, Paper (*11*) |

| pK | Purification | Source | Solubility | Toxicity | Miscellaneous |
|---|---|---|---|---|---|
| 1.1, 7.0, 10.4 | Recrystallized from $H_2O$ or Methanol. The hydrochloride from aqueous acetone (*17*). | *Streptomyces vinaceus-drappus*, *Streptomyces fasciculatis*, *Streptomyces plicatus* and *Streptomyces sacromyceticus* (*10–14*) | Cold $H_2O$ (slightly soluble), dilute acids, dilute alkali, organic solvents (very slightly), water saturated with butanol (soluble). | $LD_{50}$ 90 mg/kg (i.v.) (mice), 600 mg/kg (s.c.) (mice), 200 mg/kg (i.v.) (rat), 600 mg/kg (s.c.) (rat). | ir (*11*, *14*) |

Biological activity

Active against *Mycobacterium tuberculosis* (*14*), *E. coli* P-D 04863 (*waksman* 52) (*14*), *Mycobacterium phlei* (*15*), *Mycobacterium avium* (*11*) and *Mycobacterium tuberculosis* H37Rv (*12*). Antileukemic (*16*) activity and some activity against the KB strain of human epidermoid carcinoma cells (*19*) was demonstrated. Inhibits protein synthesis in *E. coli* cells and a possible mode of action (*18*). There was reported a complete inhibition of *Mycobacterium phlei* and the bacteriostatic effect consists of the inhibition of protein synthesis while RNA and DNA syntheses remain unaffected (*20*).

| No. | Name | Structure | Formula (mol wt) | Melting point °C | $[\alpha]_D$ | Chromatography |
|---|---|---|---|---|---|---|
| 4 | **Amicetin B** | 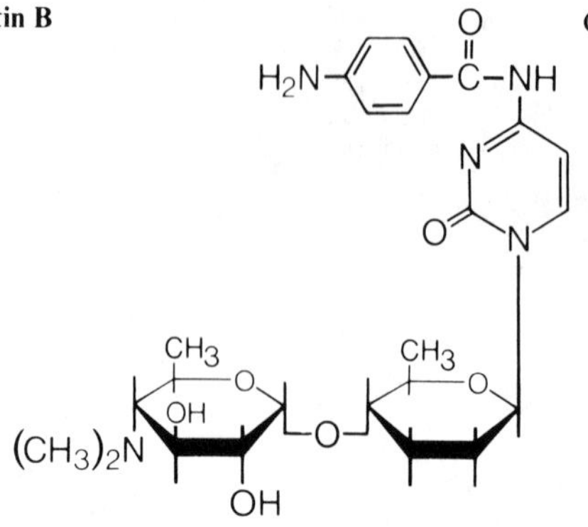 | $C_{25}H_{35}N_5O_7$ (517.59) | 160–161 (*21, 22*) | +122° (c = 1%, 0.1 *N* HCl) (*21, 22*) | — |

| pK | Purification | Source | Solubility | Toxicity | Miscellaneous |
|---|---|---|---|---|---|
| — | | *Streptomyces sp.* 285 (*21, 22*) | — | — | — |

Biological activity

Active against gram-positive microorganisms.

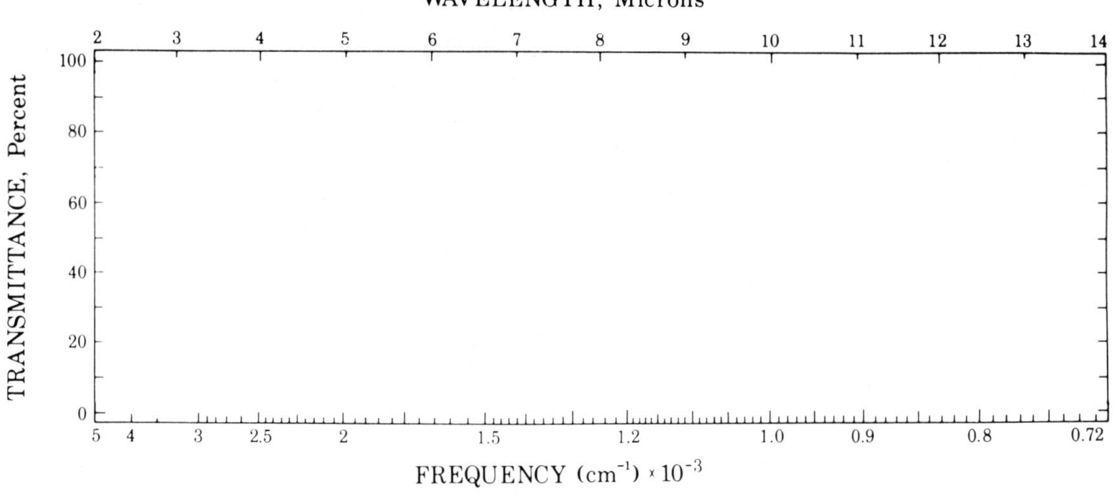
WAVELENGTH, Microns
2
3
4
5
6
7
8
9
10
11
12
13
14
TRANSMITTANCE, Percent
100
80
60
40
20
0
5
4
3
2.5
2
1.5
1.2
1.0
0.9
0.8
0.72
FREQUENCY (cm⁻¹) × 10⁻³

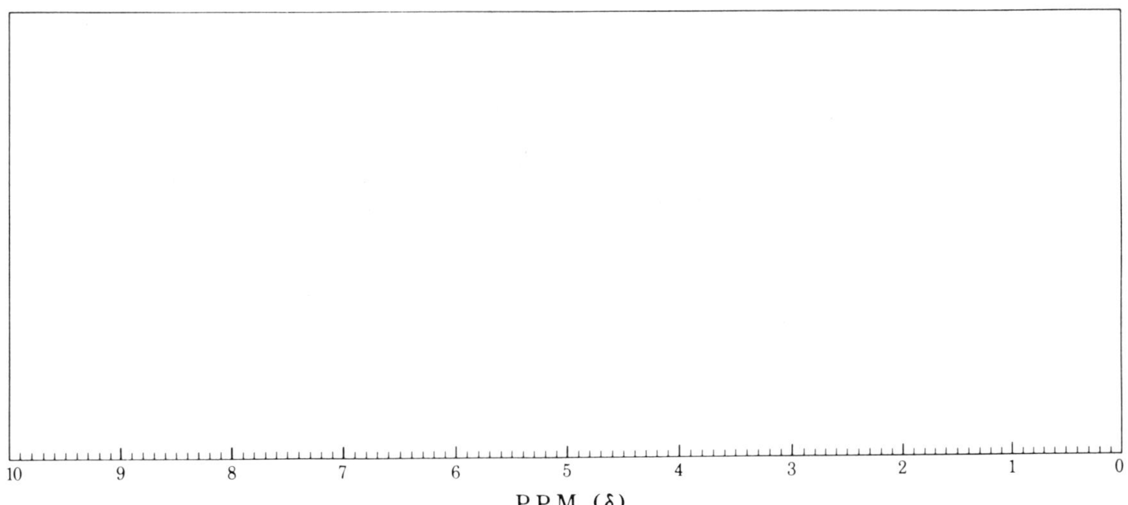
10
9
8
7
6
5
4
3
2
1
0
P.P.M. (δ)

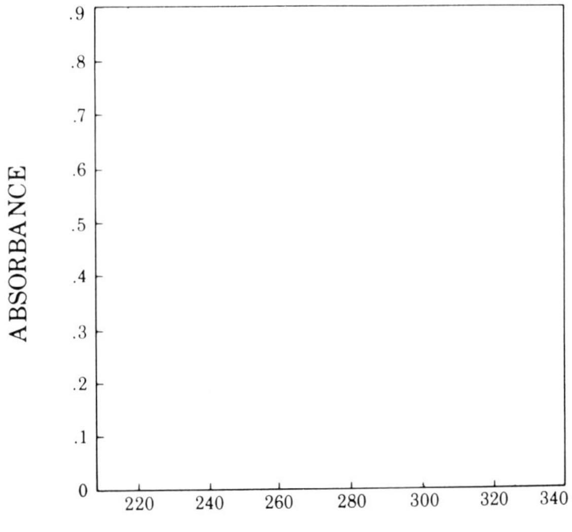
ABSORBANCE
.9
.8
.7
.6
.5
.4
.3
.2
.1
0
220
240
260
280
300
320
340
WAVELENGTH, nm.

| No. | Name | Structure | Formula (mol wt) | Melting point °C | $[\alpha]_D$ | Chromatography |
|---|---|---|---|---|---|---|
| 5 | **3′-Amino-3′-deoxyadenosine** | 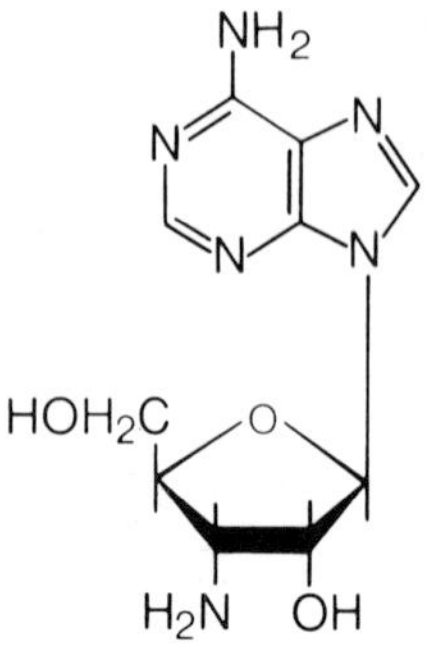 NH2, N, N, N, N, HOH2C, O, H2N, OH | $C_{10}H_{14}N_6O_3$ (266.26) | 271–273 dec. | −37° (0.1 *N* HCl) (*1*) | Paper (*1*) |

| pK | Purification | Source | Solubility | Toxicity | Miscellaneous |
|---|---|---|---|---|---|
| — | Recrystallized from $H_2O$ | *Helminthosporium sp.* (*23, 31*) *Cordyceps militaris* (*24, 1, 33*) | Soluble in $H_2O$ and MeOH | — | Chemical synthesis (*23, 39*) |

Biological activity

Active against *Candida albicans* and *Streptococcus faecalis* (*37*). A substrate for adenosine kinase (*25*), adenosine deaminase (*37*), for adenosine aminohydrolase (*26, 27*) but not for adenosine phosphorylase (*37*). Inhibits RNA and DNA synthesis in Ehrlich ascites tumor cells (*28–30*) (*32*). Antitumor activity against adenocarcinoma S3A (ascitic), sarcoma 180 (ascitic), Gardner lymphosarcoma (ascitic) and Ehrlich carcinoma (ascitic) (*34–37*).

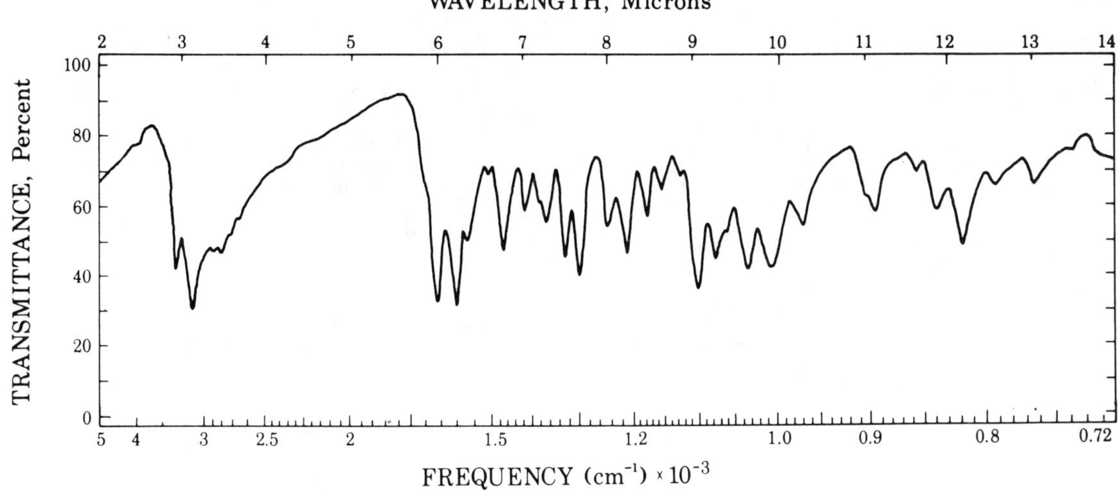
WAVELENGTH, Microns
2 3 4 5 6 7 8 9 10 11 12 13 14
TRANSMITTANCE, Percent
100 80 60 40 20 0
5 4 3 2.5 2 1.5 1.2 1.0 0.9 0.8 0.72
FREQUENCY (cm$^{-1}$) × $10^{-3}$

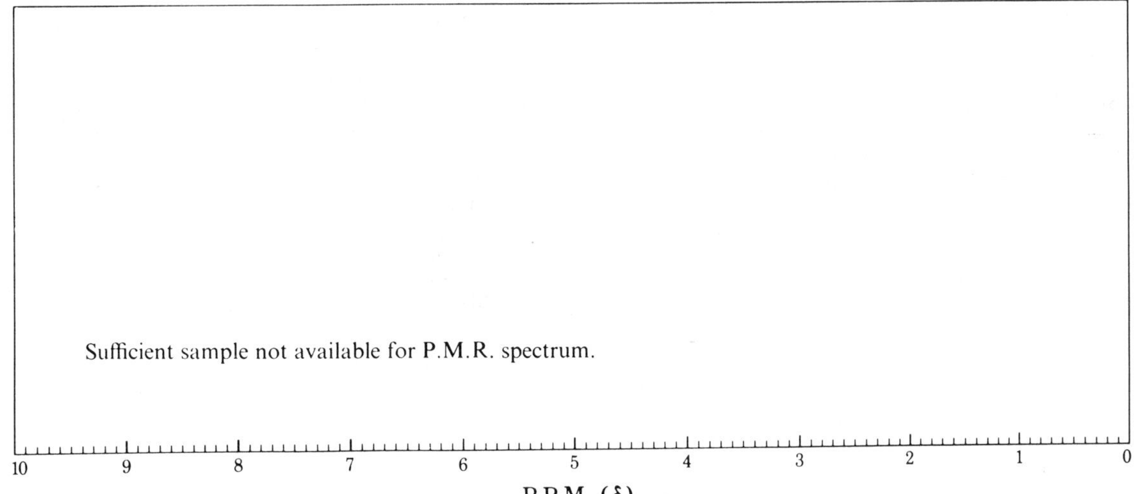
Sufficient sample not available for P.M.R. spectrum.
10 9 8 7 6 5 4 3 2 1 0
P.P.M. (δ)

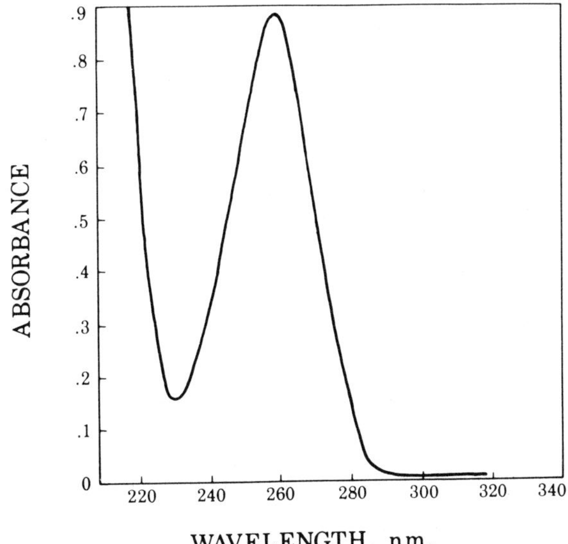
ABSORBANCE
.9 .8 .7 .6 .5 .4 .3 .2 .1 0
220 240 260 280 300 320 340
WAVELENGTH, nm.

| No. | Name | Structure | Formula (mol wt) | Melting point °C | $[\alpha]^{25}$ | Chromatography |
|---|---|---|---|---|---|---|
| 6 | **Angustmycin A** (*41*) | 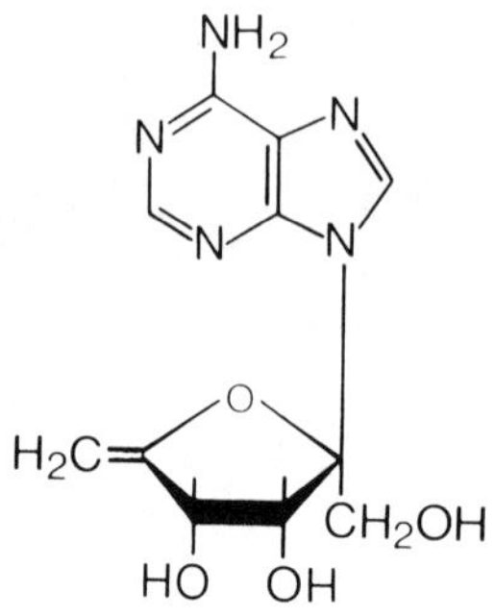 | $C_{11}H_{13}N_5O_4$ (279.3) | 164.5–165.5, 128–130 remelting at 164.5 with dec. (hydrochloride), 187–188 (tetraacetate) (*42*) | +17.02° (c = 1.4, DMF) (*40*) | Column, Paper (*42, 43, 44*) |

| pK | Purification | Source | Solubility | Toxicity | Miscellaneous |
|---|---|---|---|---|---|
| 9.8 | Column chromatography freeze-dried and then counter-current distribution. Recrystallized from $H_2O$. | *S. hygroscopicus* var. *angustmyceticus* (*40*), *S. hygroscopicus* var. *decoyicus* (*41*) | Soluble in $H_2O$, MeOH, EtOH, HOAc, DMF, Pyridine, and Phenol. Sparingly soluble in Dioxane and BuOH. | 2.5 g/kg (i.p.) in mice not toxic | ir (*42*), Chemical synthesis (*49*) |

| Biological activity |
|---|
| Inhibits incorporation of $^{32}P$ into nucleic acid of *B. subtilis* (*47*). Active against *Mycobacterium 607*, *Mycobacterium phlei*, *Staphylococcus aureus*, *Micrococcus varians*, *Micrococcus ureae*, *Bacillus subtilis*, *E. coli*, *Salmonella typhi* and *Salmonella enteritidis* (*46, 47*). *Mycobacterium tuberculosis* $H_{37}R_v$ (inactive at 100 mg/ml) (*47*) and does not function as a substrate for adenosine deaminase from *Aspergillis oryzae* (*48*). |

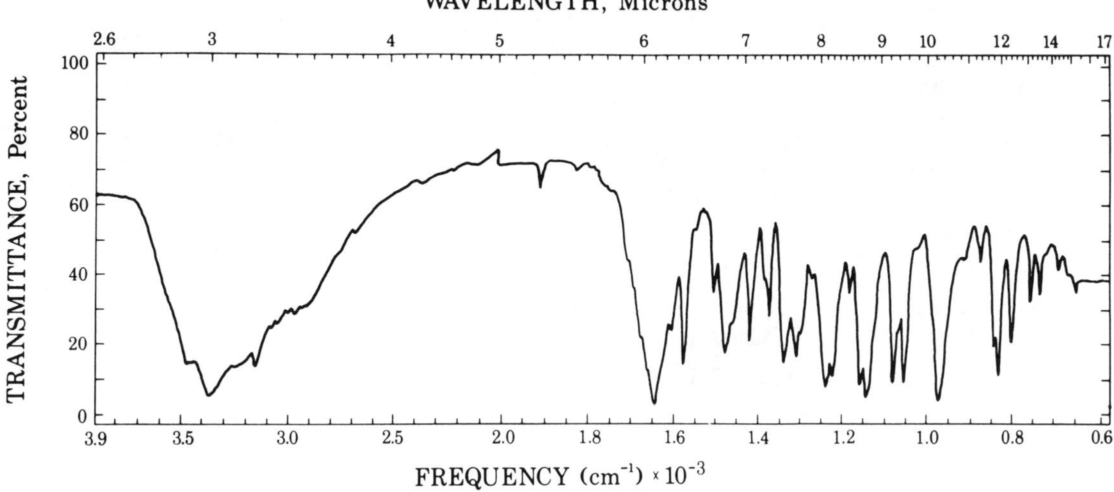


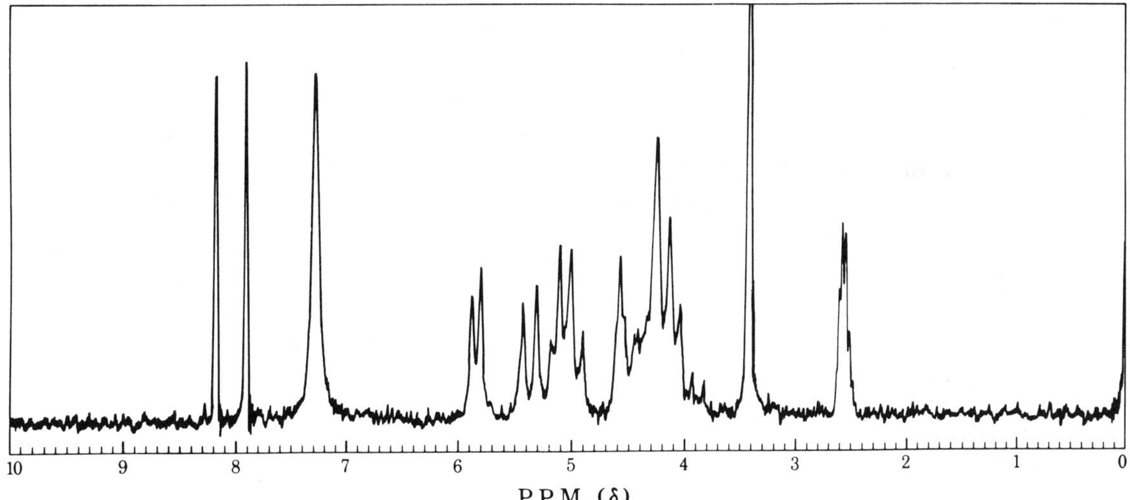


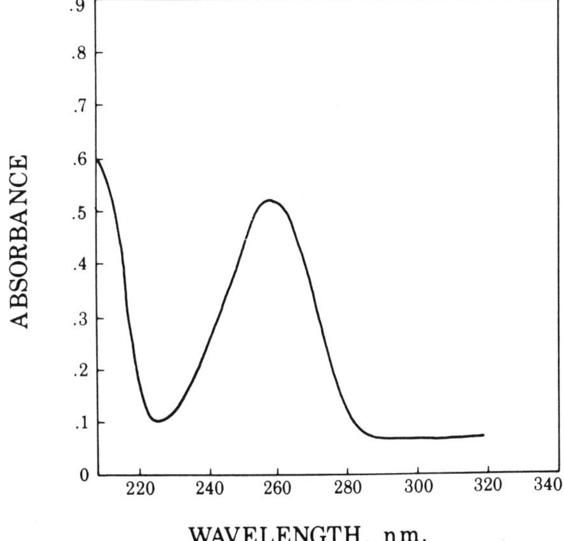


$NH_2$
N
N
N
N
O
$H_2C$
$CH_2OH$
HO
OH

| No. | Name | Structure | Formula (mol wt) | Melting point °C | $[\alpha]_D^{19}$ | Chromatography |
|---|---|---|---|---|---|---|
| 7 | **Angustmycin C** (*50*) | 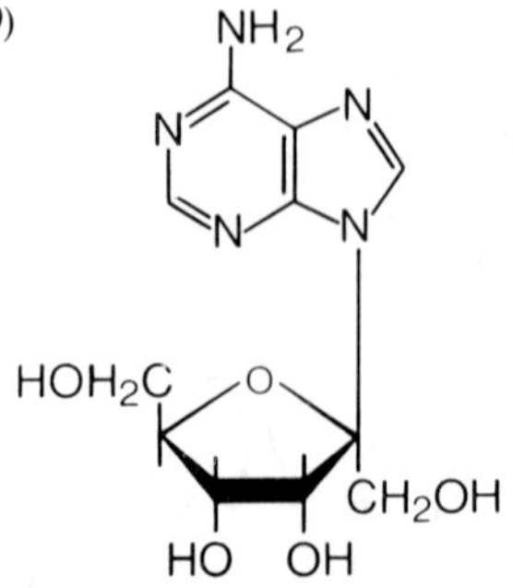 | $C_{11}H_{15}N_5O_5$ (297.3) | 202–204 dec. | −71.1° (c = 1.8, pyridine) | Column |

| pK | Purification | Source | Solubility | Toxicity | Miscellaneous |
|---|---|---|---|---|---|
| — | Column chromatography and recrystallized from $H_2O$. | *Streptomyces hygroscopicus* var. *angustmyceticus* (*40, 50*) | Soluble in $H_2O$ | — | ir (*56*), Chemical synthesis (*51, 57, 58*) |

Biological activity

Inhibition of DNA and RNA biosynthesis in *E. coli*, inhibition of xanthosine-5′-phosphate aminase (*52, 53*) and incorporated into RNA of *E. coli*. (*53*). Inhibits growth of *S. aureus* (*54*), inhibits 5′-XMP aminase in *Br. ammoniagenes* (*55*) and is not a substrate for adenosine deaminase from *Aspergillus oryzae* (*48*). Active against *Mycobacterium 607*, *Mycobacterium phlei*, *Staphylococcus aureus*, *Micrococcus varians*, *Micrococcus ureae*, *Bacillus subtilis*, *E. coli* and *Salmonella typhi* (*45, 46*).

| No. | Name | Structure | Formula (mol wt) | Melting point °C | $[\alpha]_D^{35}$ | Chromatography |
|---|---|---|---|---|---|---|
| 8 | **Antibiotic 1037** *(60)* | 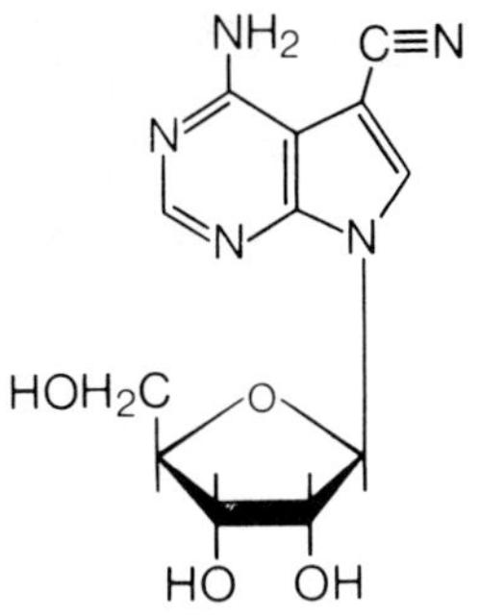 NH2 C≡N N N N HOH2C O HO OH | $C_{12}H_{13}N_5O_4$ (291.3) | 238–239 *(59)*, 244 *(60)*, 227–230 (picrate) *(60)*, 101 (acetate) *(60)* | −51 (c = 0.13, $H_2O$) *(59)*, −51.8 (0.1 *N* HCl) *(60)* | Column, Paper *(60)* |

| pK | Purification | Source | Solubility | Toxicity | Miscellaneous |
|---|---|---|---|---|---|
| — | Absorption on carbon, elution and then extracted *(59)*. Recrystallized from MeOH *(59)*. | *Streptomyces* sp. No. 1037 *(59, 60)* | Very soluble in dilute HCl and $H_2O$. | $LD_{50}$ 12 mg/kg (i.v.) (mice) *(59)* | uv *(59, 60)*, ir *(59, 60)* |

Biological activity

Strong activity against *Candida albicans*, *C. tropicalis* and *C. parakrusei* *(59)*. Moderate activity against phytopathogenic bacteria *(59)*, completely inhibits *Trichomonas vaginalis* No. 1099 *(59)*.

| No. | Name | Structure | Formula (mol wt) | Melting point °C | $[\alpha]_D^{26}$ | Chromatography |
|---|---|---|---|---|---|---|
| 9 | **Antibiotic C** (*21*, *22*), | 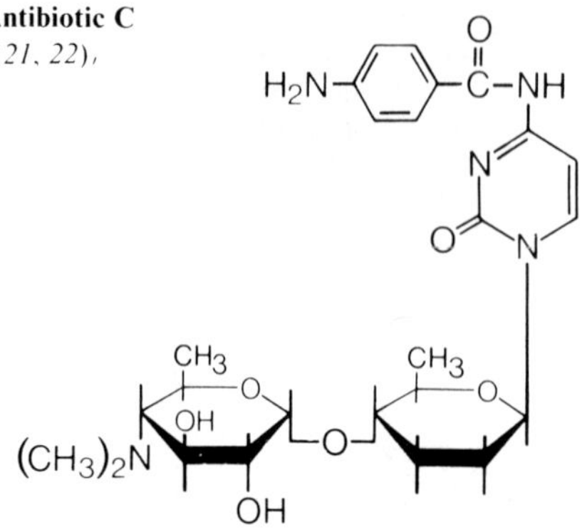 H2N–C6H4–C(=O)–NH; N; O; N; CH3; O; OH; (CH3)2N; O; OH; CH3; O | $C_{26}H_{37}N_5O_8$ (*21*) (547.6). If same as plicacetin it should be $C_{25}H_{35}N_5O_7$. | 180–182 (*21*) | +180.9 (MeOH) (*21*) | Paper (*21*) |

| pK | Purification | Source | Solubility | Toxicity | Miscellaneous |
|---|---|---|---|---|---|
| | Countercurrent extraction (*21*, *22*). Recrystallized from absolute EtOH (*21*, *22*) and from $H_2O$ (*21*). | *Streptomyces plicatus* n. sp. (*21*) | Soluble in alcohols, $CHCl_3$ and $CH_2Cl_2$. | — | — |

| Biological activity |
|---|
| Active against *E. coli*, various gram-negative and gram-positive bacteria (*21*). |

| No. | Name | Structure | Formula (mol wt) | Melting point °C | $[\alpha]_D$ | Chromatography |
|---|---|---|---|---|---|---|
| 10 | **Antibiotic D** | 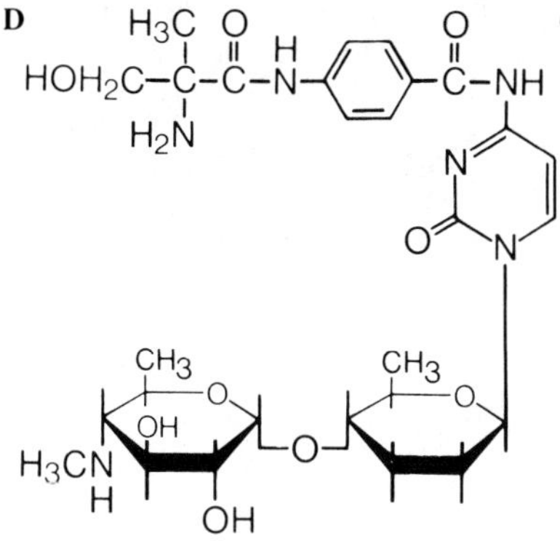 | $C_{28}H_{40}N_6O_9$ (504.63) | — | — | Paper (*22*) |

| pK | Purification | Source | Solubility | Toxicity | Miscellaneous |
|---|---|---|---|---|---|
| — | Purified by ion exchange chromatography and then freeze-dried. Recrystallized from a MeOH/AcOEt mixture (*22*). | *Streptomyces plicatus* n. sp. (*22*) | Soluble in EtOH, MeOH and $H_2O$. | — | uv (*22*), ir (*22*) |

| Biological activity |
|---|
| Active against various gram-negative and gram-positive bacteria including *Mycobacterium tuberculosis* (*22*). |

| No. | Name | Structure | Formula (mol wt) | Melting point °C | $[\alpha]_D$ | Chromatography |
|---|---|---|---|---|---|---|
| 11 | **Antibiotic D-13** | 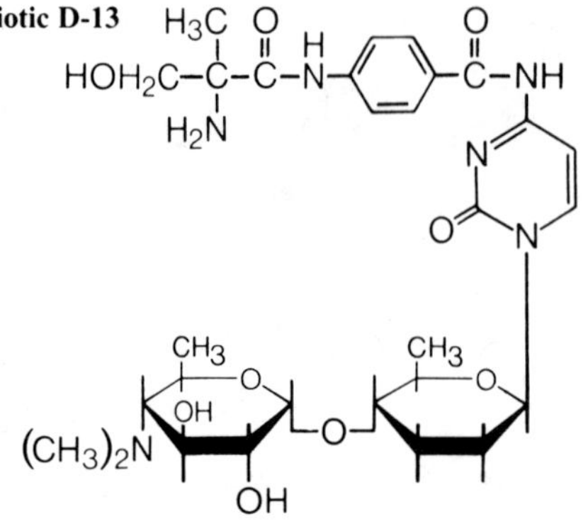 | $C_{29}H_{42}N_6O_9$ (636.69) | 243–244, 190–192 (HCl) (*61*) | +143° [c = 1.01, EtOH/MeOH (95:5)] (*61*) | Column |

| pK | Purification | Source | Solubility | Toxicity | Miscellaneous |
|---|---|---|---|---|---|
| 7.0 | Purified by extracting the fermentation broth with organic solvents (*62*). | *Streptomyces vinaceus-drappus* (*61, 62*) | Insoluble in $H_2O$. | — | ir |

Biological activity

Active against *Mycobacteria* and other gram-positive bacteria (*62*).

| No. | Name | Structure | Formula (mol wt) | Melting point °C | $[\alpha]_D$ | Chromatography |
|---|---|---|---|---|---|---|
| 12 | **Antibiotic E-212 First substance** (*65*) | 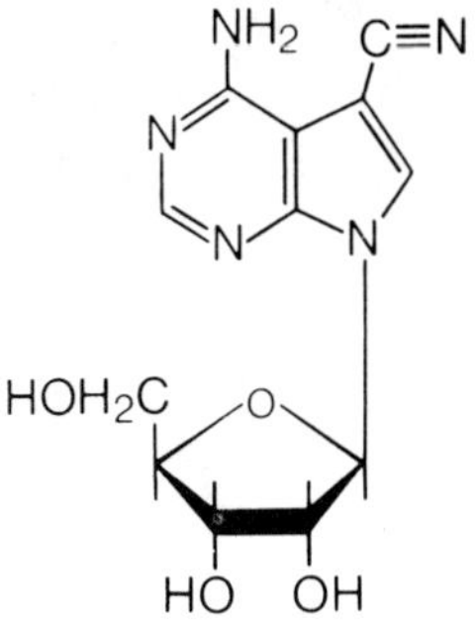 | $C_{12}H_{13}N_5O_4$ (291.3) | 233–234 (*63*) | — | — |



| pK | Purification | Source | Solubility | Toxicity | Miscellaneous |
|---|---|---|---|---|---|
| — | Purified by column chromatography (*63*), extracted into $H_2O$ and evaporated to dryness. Recrystallized from absolute MeOH (*63*). | *Streptomyces sp.* No. E-212 (*63*) | Slightly soluble in $H_2O$. Insoluble in benzene, $CHCl_3$, ether and pet. ether. Soluble in MeOH and EtOH (*63*). | No toxicity at 5 mg/kg (i.v.) (mice). A delayed death, accompanied by yellowish turbidites of the livers at 10 mg/kg (i.v.) (mice). Toxicity reduced by simultaneous addition of *dl*-methionine or glucuronic acid. Human red cells did not exhibit hemolysis at 100 mcg/ml (*63*). | uv (*63*), ir (*63*), Chemical synthesis (*65*) |

Biological activity

Active against Ehrlich ascites carcinoma, *C. albicans* and *C. parakrusei* (*63*).

| No. | Name | Structure | Formula (mol wt) | Melting point °C | $[\alpha]_D$ | Chromatography |
|---|---|---|---|---|---|---|
| 13 | **Antibiotic R-285** | 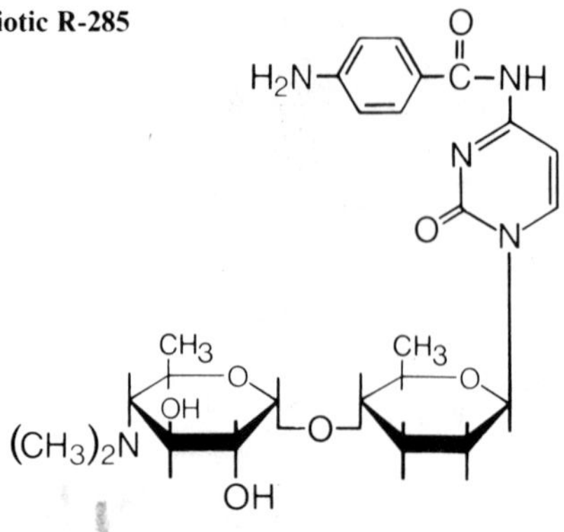 | $C_{25}H_{35}N_5O_7$ (*66*) (517.59) | 165–167 (*66, 67*) | +122 (c = 0.5, 0.1 *N* HCl) (*66, 67*) | — |

| pK | Purification | Source | Solubility | Toxicity | Miscellaneous |
|---|---|---|---|---|---|
| — | Extracted from the filtrate at pH 7.0–8.5 (*66*). | *Streptomyces sp.* R-285 (*66, 67*) | — | $LD_{50}$ 200 mg/kg (i.v.) (mice. (*66, 67*) | uv (*66, 67*). ir (*66, 67*) |

| Biological activity |
|---|
| Active against gram-positive bacteria including mycobacteria (*66, 67*). |

| No. | Name | Structure | Formula (mol wt) | Melting point °C | $[\alpha]_D$ | Chromatography |
|---|---|---|---|---|---|---|
| 14 | **Antibiotic T-3018** (*69*) | 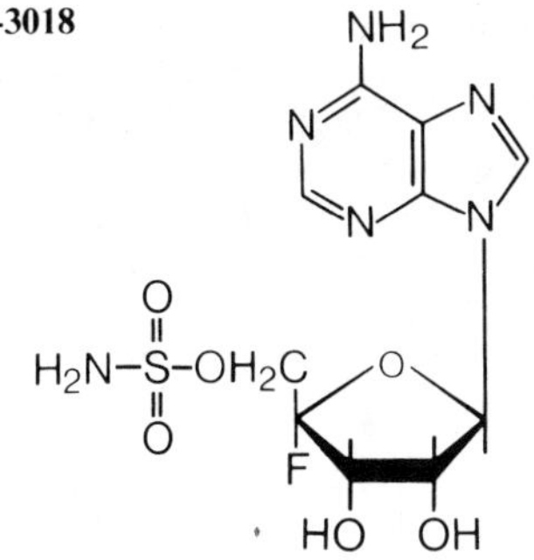 | $C_{11}H_{15-16}N_6O_8S$ (*69*) (391.34 *or* 392.34). However, if the same as nucleocidin then $C_{10}H_{13}FN_6O_6S$ (364.31). | 143–144 (picrate) (*69*) | −33.3 (c = 1.052, EtOH/0.1 *N* HCl) (*69*) | — |

| pK | Purification | Source | Solubility | Toxicity | Miscellaneous |
|---|---|---|---|---|---|
| — | Adsorbed on active carbon eluted and crystallized from $H_2O$ (*69*). | *Streptomyces sp.* T-3018 (*68, 69*) | Soluble in $H_2O$ (*69*) | — | uv (*69*) |

| Biological activity |
|---|
| Active against gram-positive, gram-negative bacteria and certain protozoa (*69*). |

| No. | Name | Structure | Formula (mol wt) | Melting point °C | $[\alpha]_D^{27}$ | Chromatography |
|---|---|---|---|---|---|---|
| 15 | **9-β-D-Arabinofuranosyladenine (Ara-A)** | 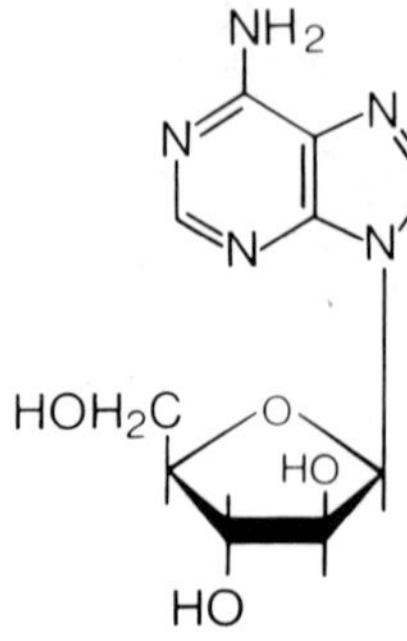 | $C_{10}N_{13}N_5O_4$ (267.2) | 257 (*90*), 258–260 (*92*), 257–257.5 (*91*) | −5° (c = 0.25%, $H_2O$) (*90, 91*), −1.7 (c = 0.54, pyridine) (*92*) | Paper (*90, 91*), Column (*92*), TLC (*93*) |

| pK | Purification | Source | Solubility | Toxicity | Miscellaneous |
|---|---|---|---|---|---|
| — | Recrystallization from $H_2O$ (*91, 92*) *or* MeOH/$H_2O$ (*93*). | An antibiotic concentrate derived from a microbial fermentation (*76*). | Soluble in $H_2O$, solubility is essentially the same as for adenosine (*89*). | Not lethal for any normal toxicity at 2500 mg/kg day preorally (*71*). Acute oral and i.p. to tolerance toxicity studies (*70, 71*). Acute $LD_{50}$ 4700 mg/kg (i.p.) (mice) (*71*). | uv (*91, 93*), Chemical synthesis (*90–93*) |

Biological activity

Active against DNA viruses with limited activity against RNA viruses (*70, 76, 71*). Active against herpes simplex virus in hamsters (*70, 72, 77, 71*) vaccinia (pox) virus in mice (*71*), central nervous system viral infections in mice, vaccinia virus *in vitro* and cell cultures (*72, 73*). Active against cytomegalovirus *in vitro* (*74*) vaccinia virus *in vivo* (*75*), encephalitis, intracebral herpes simplex virus (*78, 79, 88*), experimental vaccinial encephalitis in mice (*80*) with a therapeutic index (TI) = 8 against vaccinia virus (*80*). Reduces rate of deamination in comparison with adenosine in mouse tissue (*82*) and inhibits the rate of nucleoside cleavage for thioguanosine by a synergistic effect (*83*). A kinetic study on the rate of enzymatic deamination (*84*). Ara-A has been found to be a noncompetitive inhibitor of DNA polymerase at the triphosphate level (*85, 86, 87*).

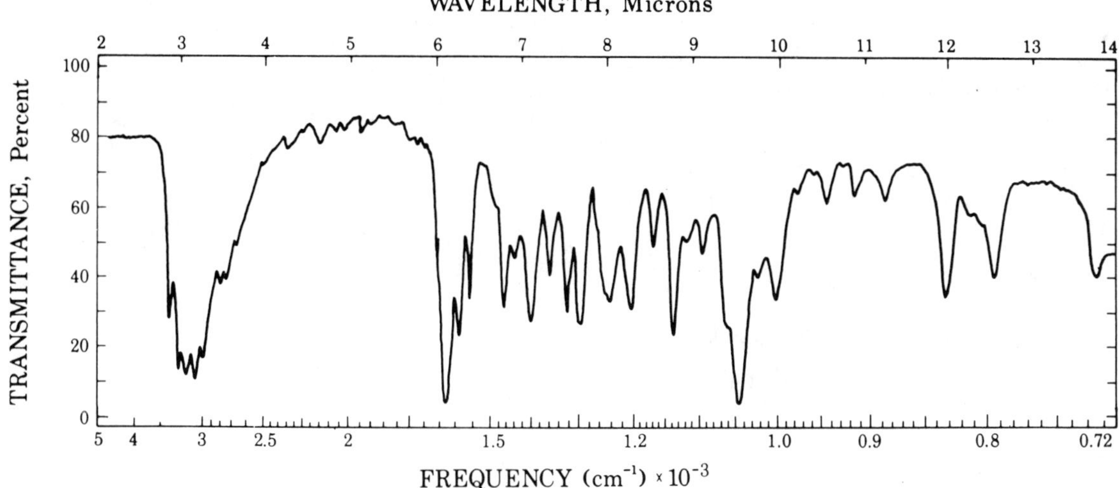


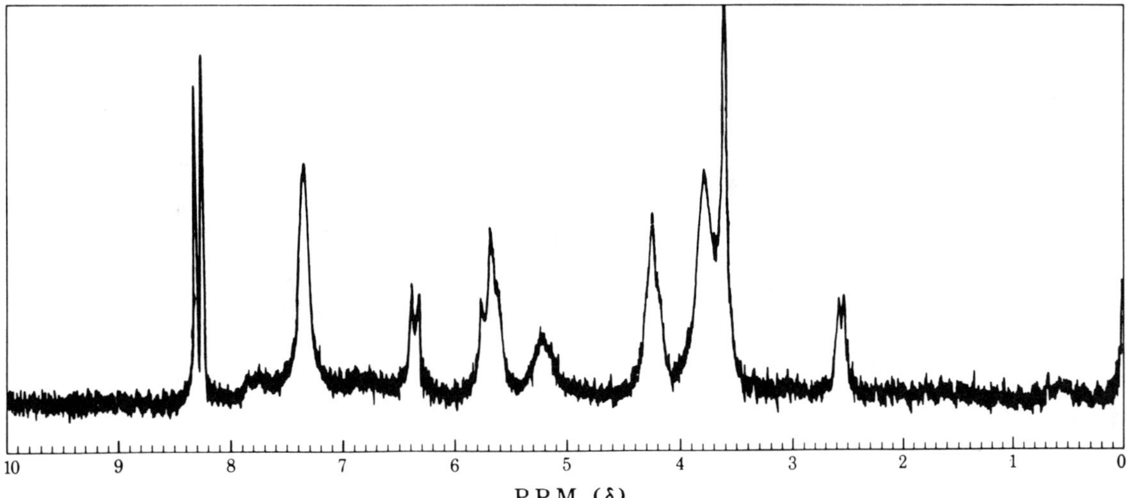


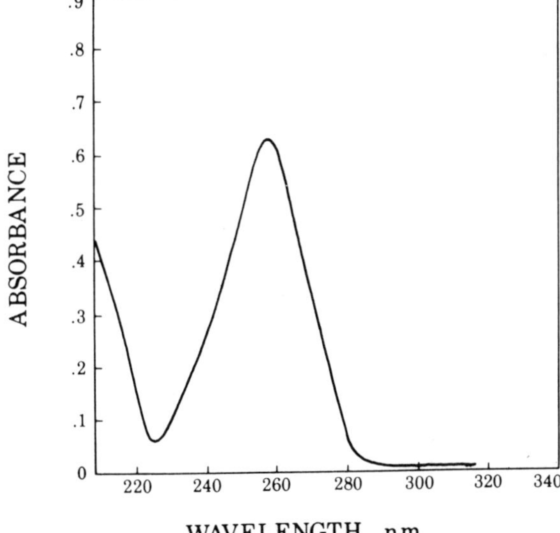


$NH_2$
N N
N N
$HOH_2C$ O
HO
HO

| No. | Name | Structure | Formula (mol wt) | Melting point °C | $[\alpha]_D^{25}$ | Chromatography |
|---|---|---|---|---|---|---|
| 16 | **Aristeromycin** | 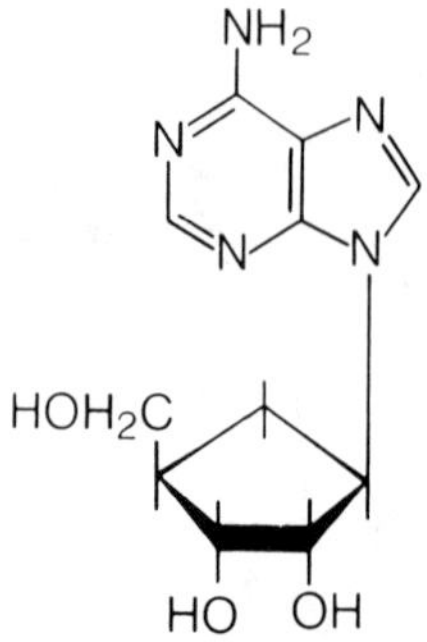 NH2 / HOH2C / HO OH | $C_{11}H_{15}N_5O_3$ (265.3) | 213–215 dec. | −52.5 (c = 1.0, DMF) | Column (*94*), TLC (*94*), Paper (*94*) |

| pK | Purification | Source | Solubility | Toxicity | Miscellaneous |
|---|---|---|---|---|---|
| — | Recrystallized from hot $H_2O$. | *Streptomyces citricolor Nov. sp.* (*94*) | Soluble in HoAc, DMF, DMSO, $H_2O$, ethylene-glycol, aqueous MeOH and aqueous acetone (*94*). Slightly soluble or in-soluble in EtOH, MeOH, EtOAc, $CHCl_3$, ether and $C_6H_6$ (*94*). | The synthetic mixture (*97*) of nucleosides was highly cytotoxic, toxic at 50 mg/kg/day and non-toxic but inactive at 25 mg/kg/day (*97*). | Mass spectroscopy (*94*), X-ray (*94*, *96*), uv and ir (*94*) |

Biological activity

Active against *Xanthomonas oryzae* and *Piricularia oryzae* (*in vitro*) and bacterial leaf bright and blast disease of rice plant (*in vivo*) (*94*). Showed activity against *B. subtilis*, *B. brevis*, *B. cereus*, *E. coli.*, *P. vulgaris*, *Aerobacter aerogenes*, *Micrococcus flavus*, *M. avium*, *M. phlei*, *Micobacterium ATCC607*, *M. smegmatis*, and numerous others (*94*). The following data were obtained on a mixture (*97*) of the *cis* and the naturally occurring *trans*-nucleoside. Inhibits cell cultures [H. Ep. No. 2/MemPR], *de novo* purine synthesis (*95*) and is a substrate for partially purified adenosine kinase, calf intestinal deaminase and a nucleotide kinase (*95*).

| No. | Name | Structure | Formula (mol wt) | Melting point °C | $[\alpha]_D^{20}$ | Chromatography |
|---|---|---|---|---|---|---|
| 17 | **BA-90912** | 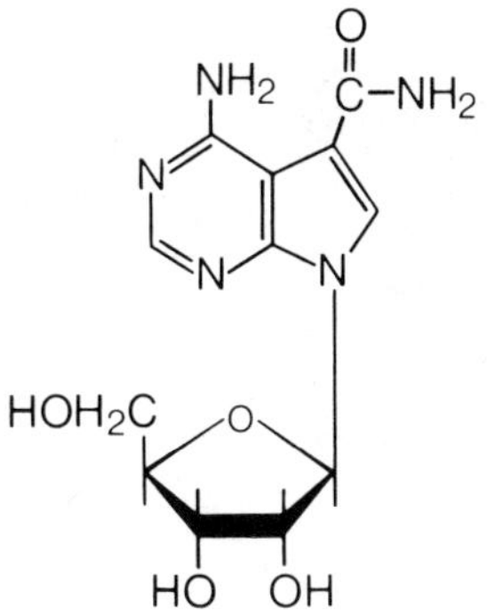 | $C_{12}H_{15}N_5O_5 \cdot H_2O$ (327.3) | 258–260 (*117*), 254–255 (HCl) (*117*), 271–272 (picrate) (*117*) | 96.7 (c = 1%, DMF) (*117*) | — |

| pK | Purification | Source | Solubility | Toxicity | Miscellaneous |
|---|---|---|---|---|---|
| — | Adsorption on carbon and then elution and counter-current distribution. Crystallizes from aqueous pyridine as the free base. (*117*). | Unidentified species of *Streptomyces* (*117*). | Slightly soluble in $H_2O$ and alcohols. Almost insoluble in acetone, EtOAc and $CHCl_3$. Very soluble in pyridine or DMF (*117*). | — | uv (*117*), ir (*117*), Chemical synthesis (*64, 65*) |

| Biological activity |
|---|
| Cytotoxic to HeLa cells with weak activity against sarcoma 180 and carcinoma 755 tumors in mice (*117*). Significant activity against leukemia 1210 (*117*). |

| No. | Name | Structure | Formula (mol wt) | Melting point °C | $[\alpha]_D$ | Chromatography |
|---|---|---|---|---|---|---|
| 18 | **5-Azacytidine** | 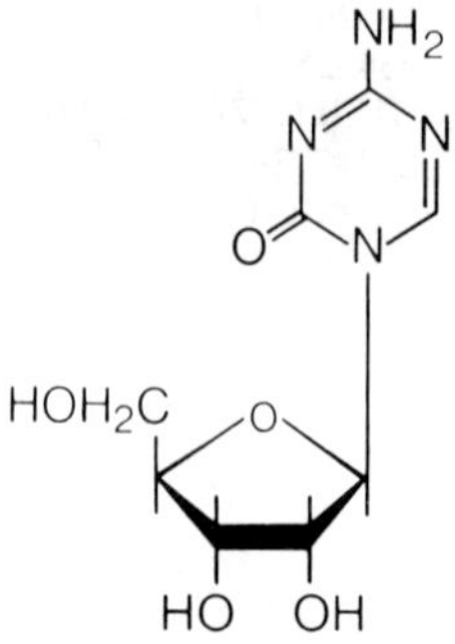 $NH_2$; N, N, N; O; $HOH_2C$; O; HO OH | $C_8H_{12}N_4O_5$ (244.2) | 230–231, dec. (*116*) | — | Paper (*116*) |

| pK | Purification | Source | Solubility | Toxicity | Miscellaneous |
|---|---|---|---|---|---|
| — | Recrystallized from absolute MeOH and aqueous MeOH (1:6) (*116*). | *Streptoverticillium ladakanus* var. *ladakanus* sp. n. (*98*) | Soluble in $H_2O$ and MeOH. | Optional radio-protective dose (1.5 mg/kg) (*114*). | ir (*116*), uv (*116*), Chemical synthesis (*116*) |

Biological activity

Active against *E. coli.* ATCC 26, *Salmonella gallinarium* USDA 8410, *S. schottmuelleri* ATCC 9149, *Pseudomonas mildenbergii* ATCC 795, *proteus vulgaris* ATCC 8427 and *Staphylococcus aureus* FDA 20clp (*98*). Active as an anticancer agent against leukemia L-1210, leukemia L-1210/c95 (triple resistant), T-4 lymphoma in mice (*98*) and also inhibits growth of KB cells (*98*). Antileukemic activity (*100, 101*), e.g. leukemia L-1210 (*102*) and has also been evaluated in the clinic (*102*). 5-Azacytidine has been incorporated *in vivo* into the RNA of Ehrlich ascitic, tumor bearing mice and the liver RNA of leukemic mice (*103, 104*). Inhibition of orotidylic decarboxylase by the 5′-phosphate derivative (formed enzymatically) (*105, 111*), inhibition of a kinase, (*106*) incorporation into RNA (*107, 108*) and DNA (*109*) and has demonstrated the ability to interrupt pregnancy in mice (*110*). 5-Azacytidine has also been found to inhibit the induction of tryptophan pyrrolase (*115*), liver regeneration in partially hepatectomized rats, (*113*) the phosphorolysis of thymidine by normal and T-4 phage-infected cells of *E. coli* (*115*). It has also demonstrated significant radioprotective activity for the stem-cells (*114*) and replication of phage T4 (*353*).

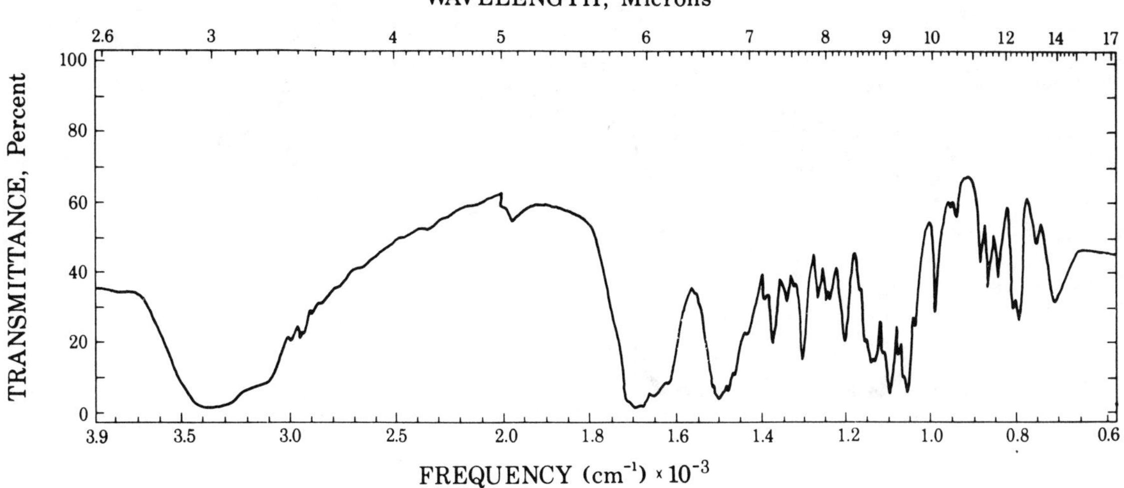
WAVELENGTH, Microns
2.6 3 4 5 6 7 8 9 10 12 14 17
TRANSMITTANCE, Percent
100 80 60 40 20 0
3.9 3.5 3.0 2.5 2.0 1.8 1.6 1.4 1.2 1.0 0.8 0.6
FREQUENCY (cm$^{-1}$) × 10$^{-3}$

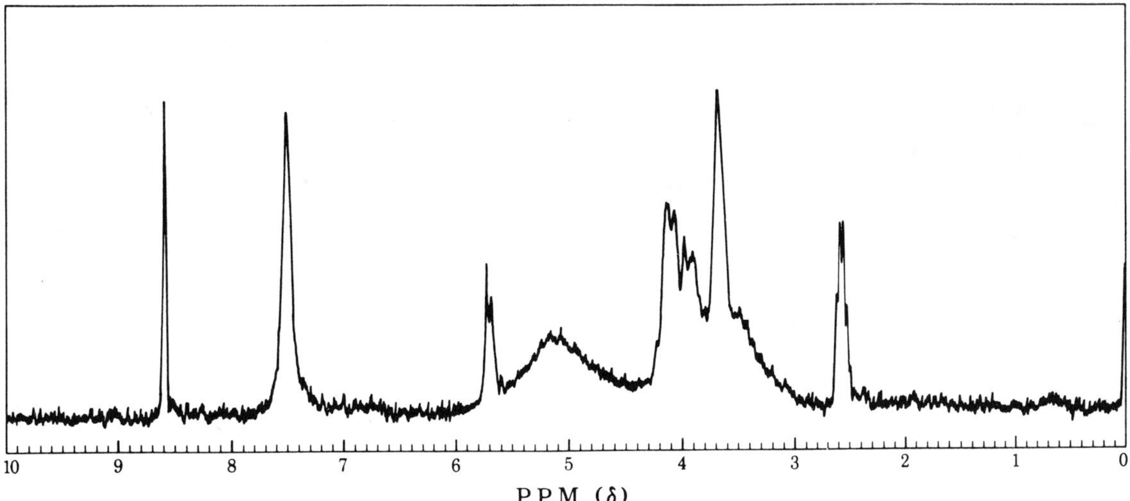
10 9 8 7 6 5 4 3 2 1 0
P.P.M. (δ)

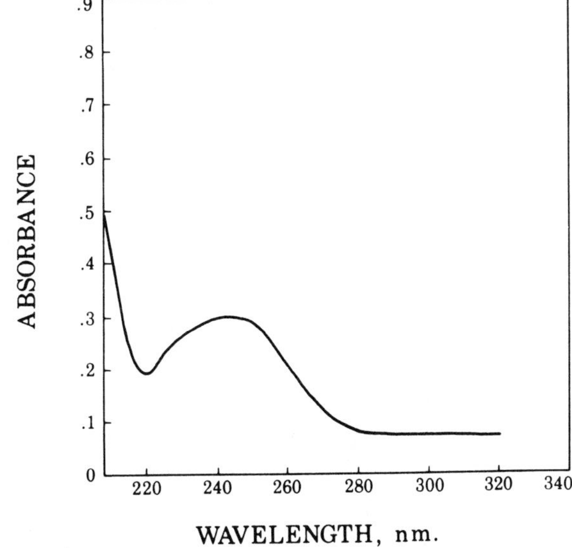
ABSORBANCE
.9 .8 .7 .6 .5 .4 .3 .2 .1 0
220 240 260 280 300 320 340
WAVELENGTH, nm.

$NH_2$
N
N
O
N
$HOH_2C$
O
HO
OH

| No. | Name | Structure | Formula (mol wt) | Melting point °C | $[\alpha]_D$ | Chromatography |
|---|---|---|---|---|---|---|
| 19 | **Bamicetin** Tentative (*120*) | 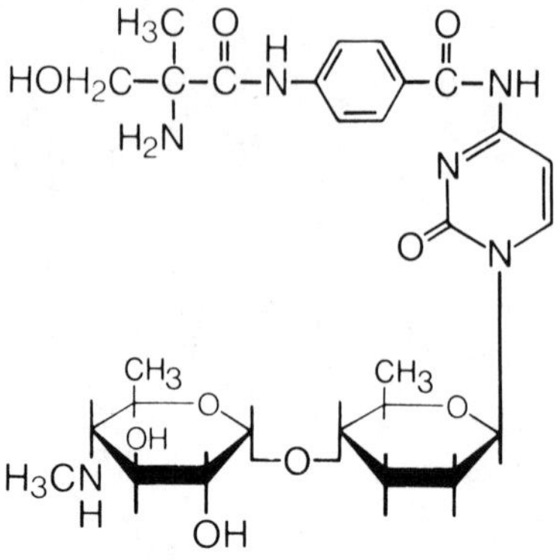 | $C_{28}H_{40}N_6O_9$ (504.63) | 240–241 dec. | +123 (c = 0.5, 0.1 *N* HCl) | Paper |

| pK | Purification | Source | Solubility | Toxicity | Miscellaneous |
|---|---|---|---|---|---|
| 6.6, 7.8, 10.9 | Counter-current distribution extraction. Recrystallized from absolute EtOH (*14*, *118*). | *Streptomyces plicatus* (*118*) | Soluble in $H_2O$, EtOH and MeOH. | Less toxic than Amicetin to dogs (s.c.) (*118*). | uv (*14*, *119*), ir (*14*, *119*) |

| Biological activity |
|---|
| Active against *E. Coli* (*118*). |

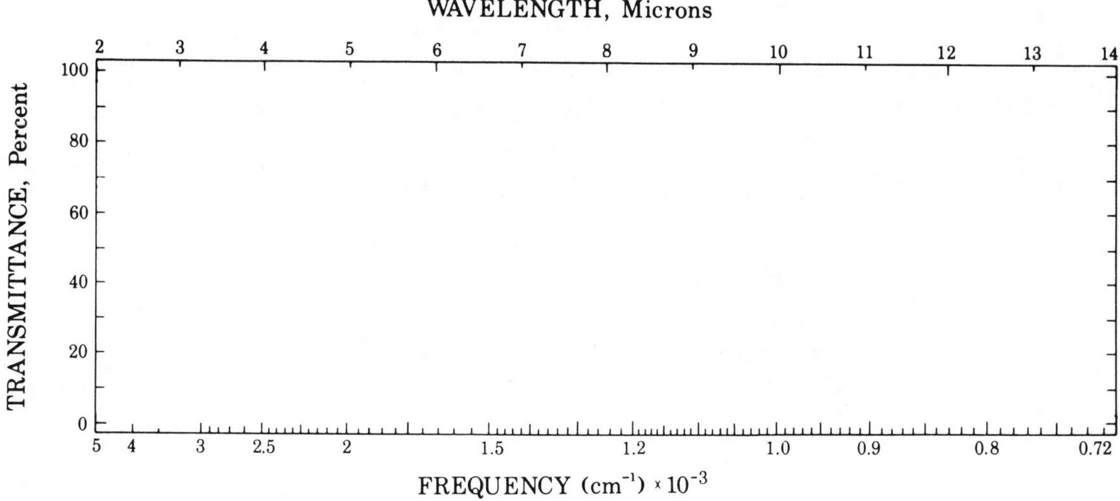
WAVELENGTH, Microns
2
3
4
5
6
7
8
9
10
11
12
13
14
TRANSMITTANCE, Percent
100
80
60
40
20
0
5
4
3
2.5
2
1.5
1.2
1.0
0.9
0.8
0.72
FREQUENCY (cm⁻¹) × 10⁻³

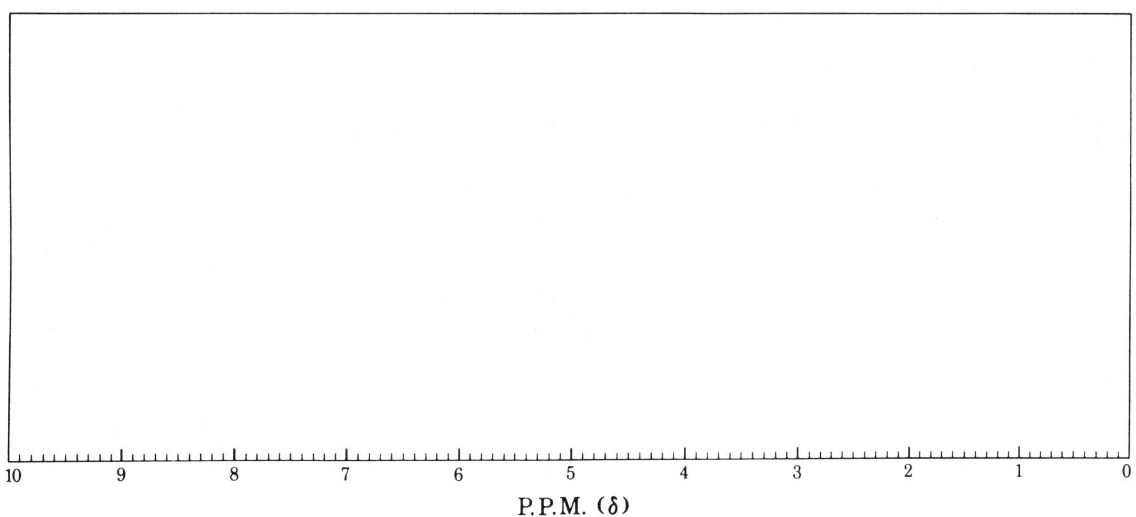
10
9
8
7
6
5
4
3
2
1
0
P.P.M. (δ)

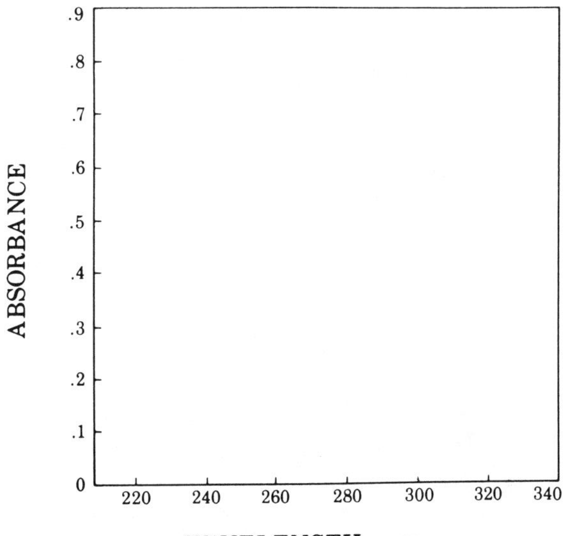
ABSORBANCE
.9
.8
.7
.6
.5
.4
.3
.2
.1
0
220
240
260
280
300
320
340
WAVELENGTH, nm.

| No. | Name | Structure | Formula (mol wt) | Melting point °C | $[\alpha]_D^{11}$ | Chromatography |
|---|---|---|---|---|---|---|
| 20 | **Blasticidin S** | 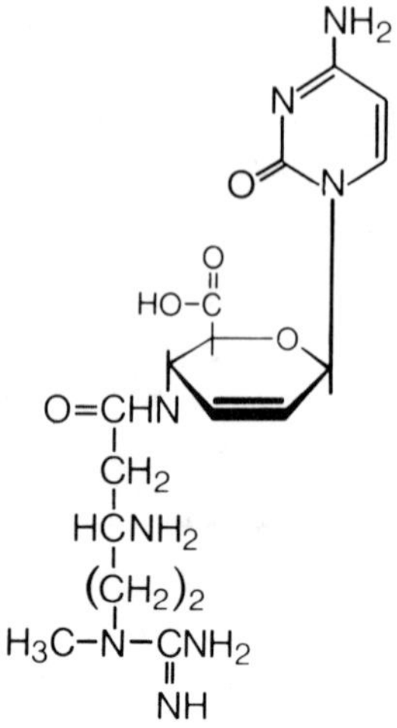 $NH_2$<br>N<br>O N<br>O<br>HO–C<br>O<br>O=CHN<br>$CH_2$<br>$HCNH_2$<br>$(CH_2)_2$<br>$H_3C–N–CNH_2$<br>NH | $C_{17}H_{26}N_8O_5$ (422.45) | 235.236 dec., 224–225 (HCl), 200–202 (picrate) | +108.4° (c = 1, $H_2O$) | Column (*121*) |

| pK | Purification | Source | Solubility | Toxicity | Miscellaneous |
|---|---|---|---|---|---|
| 2.4, 4.6, 8.0, 12.5 | Purified by column chromatography and recrystallized from aqueous acetone. | *Streptomyces griseochromogenus* (*121, 122*) | Soluble in $H_2O$, HOAc. Insoluble in the common organic solvents, MeOH, EtOH, acetone, benzene, ether, EtOAc, $CHCl_3$, $CCl_4$ and Methylethylketone. | 2.82 mg/kg (i.v.) (mice) | ir (*121*), uv (*121*), ORD (*126*), X-ray (*129*), Chemical synthesis (*132*) |

Biological activity

Active against *Piricularia oryzae* (*123*), antitumor (*124*), various antifungal and antibacterial microorganisms (*121*). Inhibition of protein synthesis in *E. coli* (*127, 128*), inhibition of rice stripe virus disease and TMV multiplication in the tissues of tobacco leaves (*130*).

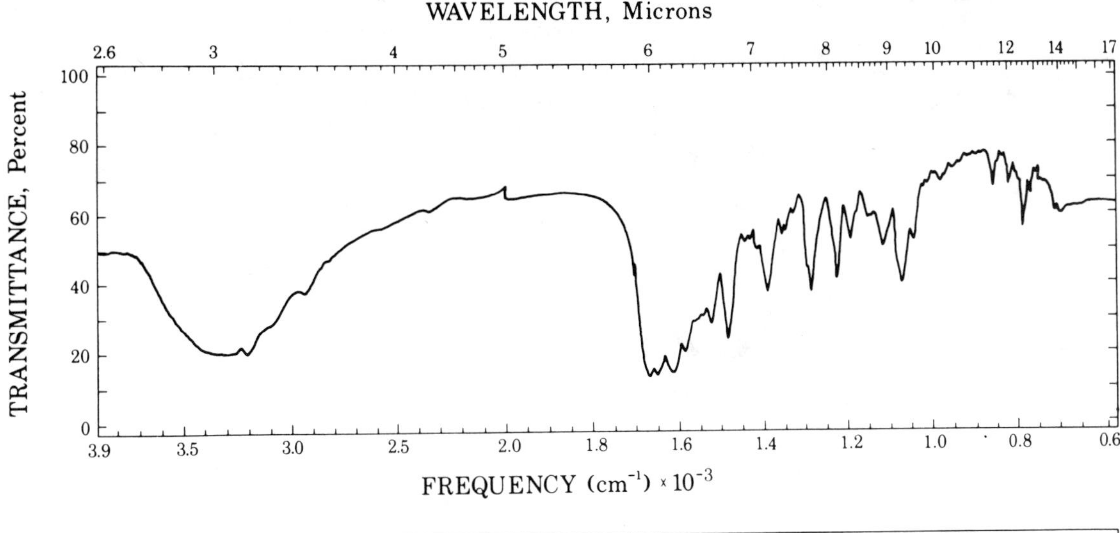


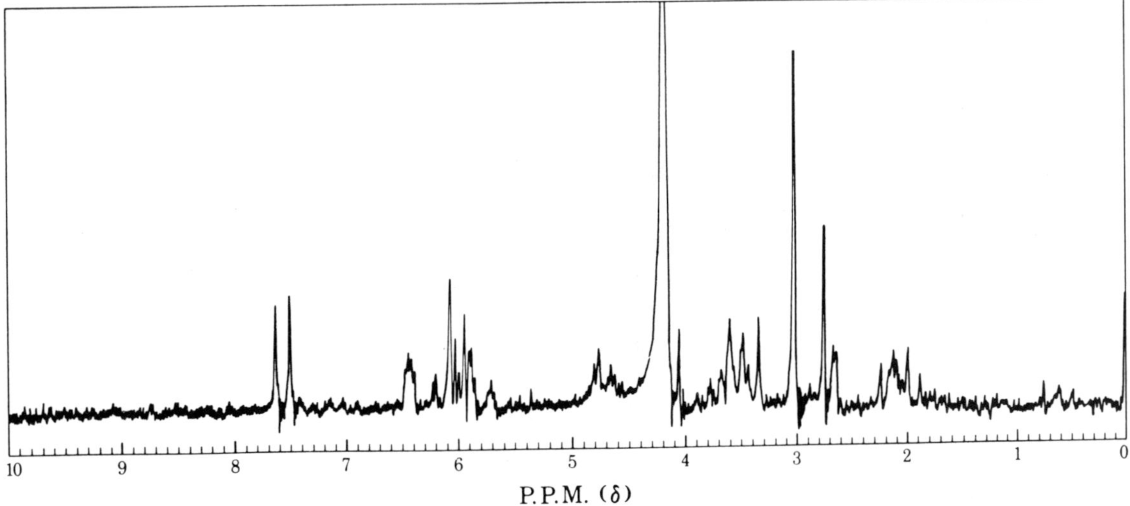


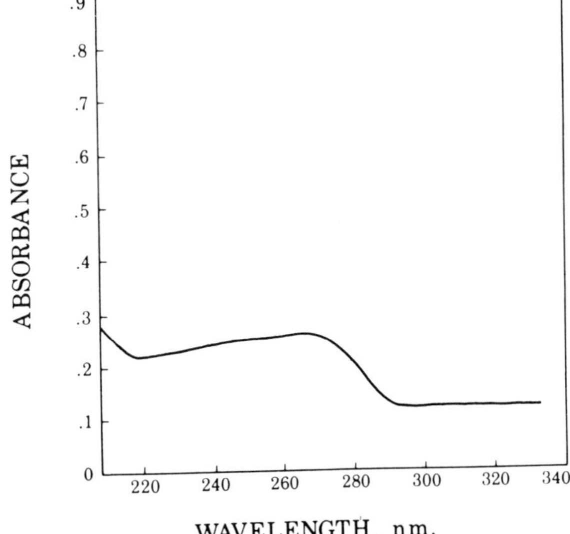


NH2
N
O N
O
HO–C
O
O=CHN
CH2
HCNH2
(CH2)2
H3C–N–CNH2
NH

| No. | Name | Structure | Formula (mol wt) | Melting point °C | $[\alpha]_D^{20}$ | Chromatography |
|---|---|---|---|---|---|---|
| 21 | **Cordycepin** | 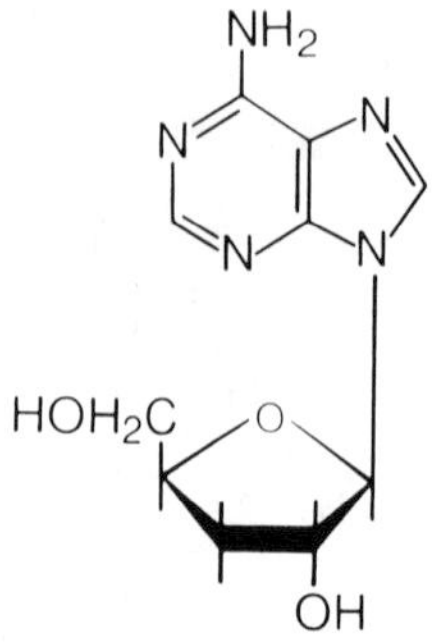 NH2 … HOH2C … OH | $C_{10}H_{13}N_5O_3$ (251.2) | 225–226, 195 (picrate) | −47.0° | — |

| pK | Purification | Source | Solubility | Toxicity | Miscellaneous |
|---|---|---|---|---|---|
| — | Purified by Soxhlet extraction and then recrystallized from EtOH. | *Cordyceps militaris* (*Linn.*), Link, *Aspergillus nidulans* (EIDAM) (*135*), *Cordeceps militaris* (L. Ex. Fr.) Link (*137*). | Soluble in $H_2O$ and possesses essentially the same solubility as 2′-deoxyadenosine. | — | ir (*136*), pmr (*136*) and mass spectrometry (*144*), Chemical synthesis (*147–149*) |

Biological activity

Active against *Bacillus subtilis* 6752 NCTC, *Avain-tubercle bacillus* and *Bovine-type tubercle bacillus* (133, *138*). Active against Ehrlich mouse ascites tumor (*in vivo* and *in vitro*) *139*, *37*), inhibits RNA synthesis in intact Ehrlich ascites cells (*140*) and the growth of human tumor cells (H. Ep. No. 1) in culture (*141*). Incorporation into RNA and DNA of growing H. Ep. No. 1 cells (*142*), the inhibition of KB cells (*37*), chick embryo fibroblast cells and inhibits *Trypanosoma rhodesiense* (*in vitro*), *T. congolense* (*in vivo*) (*145*) and *Streptococcus faecalis* (*37*). Does not inhibit cell growth of actively growing cultures of *C. militaris* (*146*). It was found to be a substrate for adenosine deaminase from *Aspergillis oryzae* (*48*) but not a substrate for adenosine phosphorylase (*37*). A kinetic study on the rate of enzymatic deamination (*84*) and the inhibition of ribosephosphate pyrophosphokinase in intact neoplastic cells has been reported (*150*).

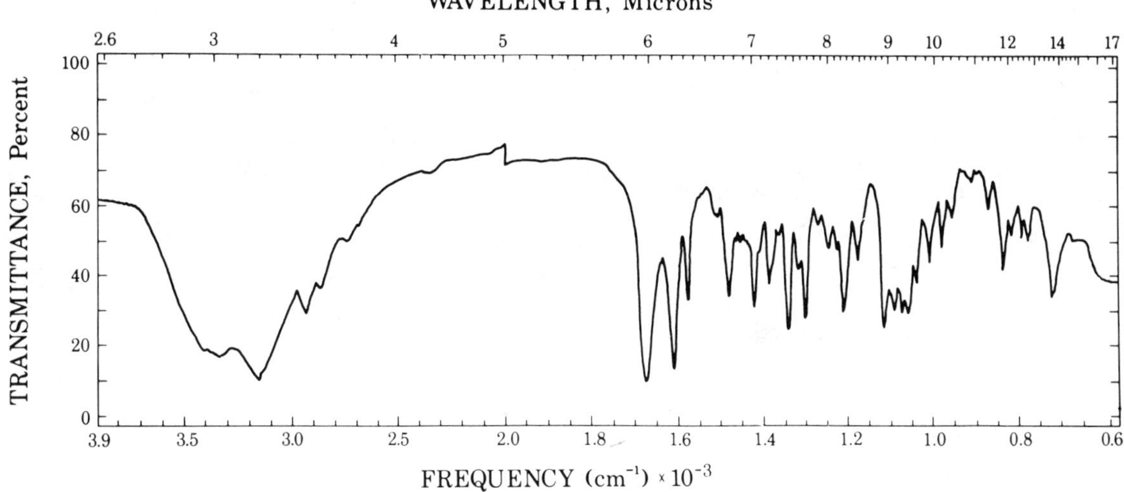


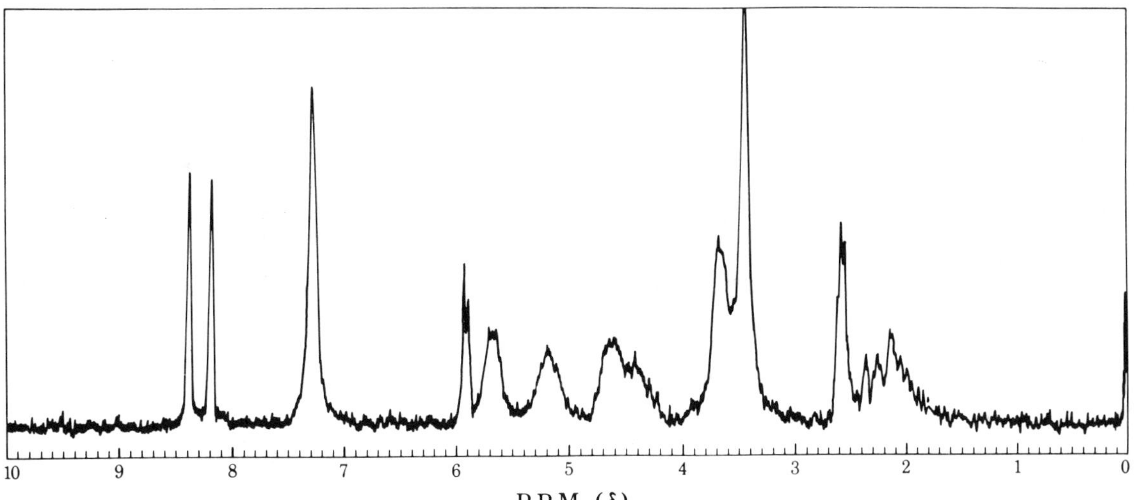


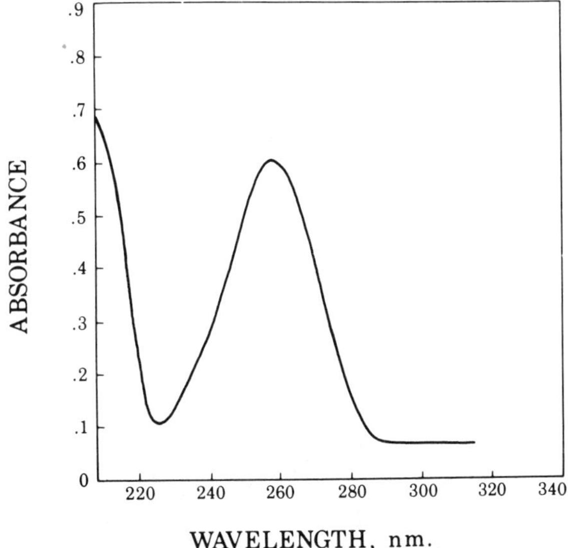


$NH_2$

N N N N

$HOH_2C$ O

OH

| No. | Name | Structure | Formula (mol wt) | Melting point °C | $[\alpha]_D$ | Chromatography |
|---|---|---|---|---|---|---|
| 22 | **Cytomycin** | 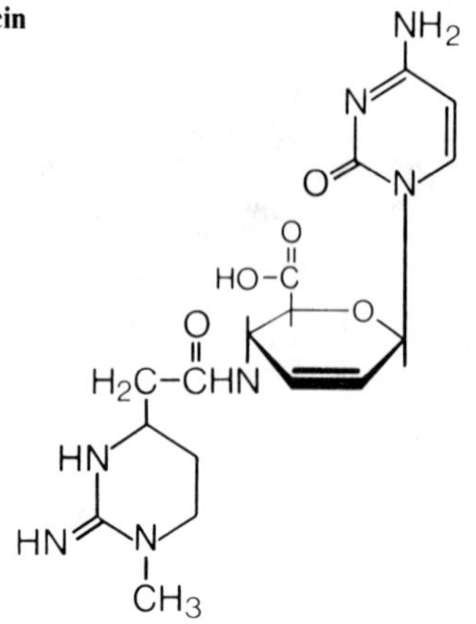 | $C_{17}H_{23}N_7O_5$ (225.42) | 236–237, 237–239 dec. (*156*), 268 dec. (monoacetate) (*156*) | +137, +84.36 (c = 2, 0.1 *N* HCl) (*156*) | Column (*156*), Paper (*156*) |

| pK | Purification | Source | Solubility | Toxicity | Miscellaneous |
|---|---|---|---|---|---|
| — | Essentially the same purification and isolation as Blasticidin S. | *Streptomyces griseochromogenes*(*154*), or by mild aklaline hydrolysis of Blasticidin S (*153, 156*). | Soluble in $H_2O$. | $LD_{50}$ 2 g/kg (i.p.) (mice) (*152*). Less toxic than Blasticidin S but less effective (*152*). | uv, ($\lambda$, 274, 0.1 *N* HCl and $\lambda$ 266, nm, 0.1 *N* NaOH) (*156*) |

| Biological activity |
|---|
| Active against Walker's adenocarcinoma 256 in rats, Ehrlich's ascites carcinoma and sarcoma 180 in mice (*152*). Activity also against solid forms of leukemia SN28, Ehrlich carcinoma and sarcoma 180 (*155*). |

| No. | Name | Structure | Formula (mol wt) | Melting point °C | $[\alpha]_D^{25}$ | Chromatography |
|---|---|---|---|---|---|---|
| 23 | **Cytovirin** ↓ $HO^-$ then HCl **Cytoviridine · HCl** $C_{10}H_{12}N_4O_4 \cdot HCl$ | Unknown. Presumably belongs to the same class as Cytomycin and Blasticidin S. | $C_{17}H_{24-28}N_8O_5$ (*158*) (420.43–424.43) | 245–250 dec. (*158*), 185–195 dec. (picrate) (*158*) | 85° (c = 0.077, $H_2O$) (*158*), 40 (c = 0.07, dil. HCl) (*158*), 82 (c = 0.058, pH 10–11) (*158*) | Column for removal of salts (*161*). Gas chromatography (*168*). |

| pK | Purification | Source | Solubility | Toxicity | Miscellaneous |
|---|---|---|---|---|---|
| — | Purified by column chromatography and precipitation (*158*). | *S. olivochromogenes* var. *cytovirinus* ATTC 12791 (*158*). Unidentified *Streptomyces* sp (*157*). | — | — | ir HCl (*158*), uv (275–276) (*158*) |

Biological activity

Active against tomato spotted with virus and tobacco mosaic virus (*157*), with antiviral activity for plants (*158*). Active against tobacco mosaic virus, stone fruit ring virus (*159*) and inhibits infection by potato virus Y(160). Inhibits egg-laying of *Tetranychus telarius* (*162*, *165*), *Panonychus ulmi* (*163*) and reproduction by *Myzus persicae* (*167*). Activity against vaccinia virus (*75*) and a more complete antiviral study (*164*) was reported.

| No. | Name | Structure | Formula (mol wt) | Melting point °C | $[\alpha]_D^{25}$ | Chromatography |
|---|---|---|---|---|---|---|
| 24 | **Decoyinine** (Antibiotic A-14) (*169, 173*) | 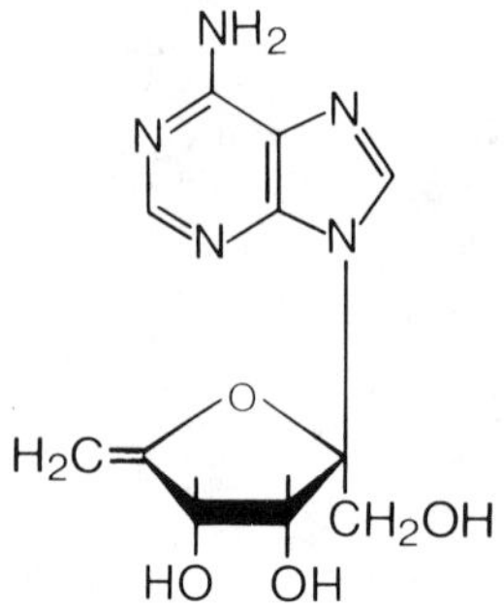 | $C_{11}H_{13}N_5O_4$ (279.3) | 183–186, 124–128 ($H_2O$), 188–190 (triacetate) (*169, 172, 174*), ca. 65 (tetraacetate) (*169, 172*), 124–126 ($H_2O$) (*172*), 124–125 ($H_2O$) (*174*) | −18.6° (DMF), −35.6 (c = 1, MeOH) (*172*) | Paper (*173*) |

| pK | Purification | Source | Solubility | Toxicity | Miscellaneous |
|---|---|---|---|---|---|
| — | Purified by column chromatography counter-current distribution and recrystallized from $H_2O$. | *S. hydroscopicus var. decoyicus* NRRL 2666 res. *S. endos* (*173, 174*), *S. hygroscopicus* CBS | Soluble in $H_2O$ and MeOH. | $LD_{50}$ 2,500 mg/kg (s.c.) (mice). Oral and parental $LD_{50}$ 2.5 g/kg and tolerated oral dose of 800 mg/kg daily for 4 days (mice). | pmr (*169*). Hydrogenated to the dihydroderivative (*169, 172*). Ozonolysis (*169*), uv (*173, 174*), ir (*173, 174*) |

Biological activity

Active against gram-positive and gram-negative bacteria, certain fungi and inhibits *Streptococcus faecalis* (*170, 37*), inhibits conversion of XMP to GMP (*171*). Active against *Streptomyces hemolyticus*, *Staphylococcus aureus*, *Diplococius pneumoniae*, *Proteus vulgaris*, *Pseudomonas aeruginosa*, *Salmonella typhiosa* and *E. coli* (*172*). Inhibits PRPP amido transferase activity in intact tumor cells (*175*). Does not function as a substrate for adenosine deaminase or adenosine phosphorylase (*37*). Antitumor activity against Walker adenocarcinoma 256, Adenocarcinoma 755, Ehrlich carcinoma (solid) and sarcoma 180 (solid).

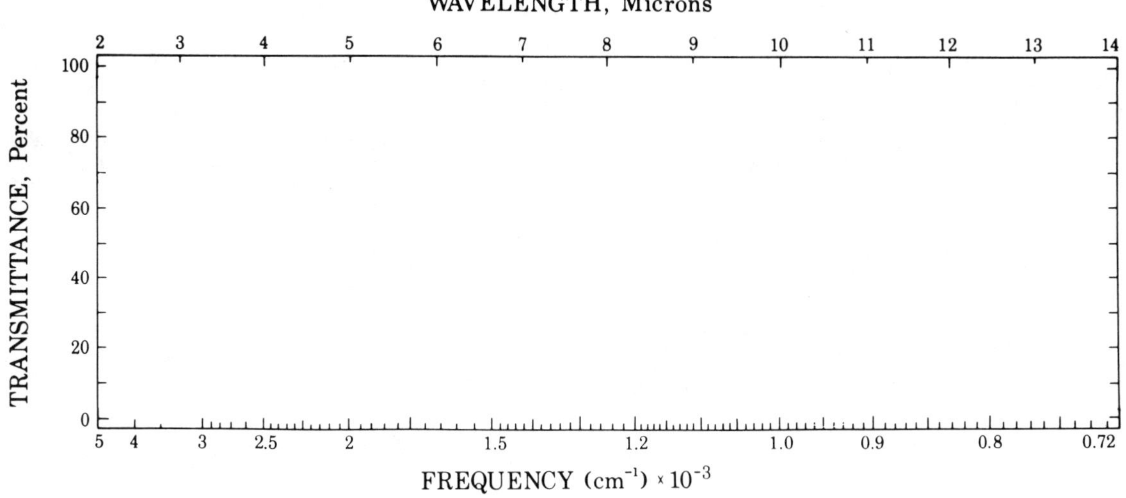
WAVELENGTH, Microns
2
3
4
5
6
7
8
9
10
11
12
13
14
TRANSMITTANCE, Percent
100
80
60
40
20
0
5
4
3
2.5
2
1.5
1.2
1.0
0.9
0.8
0.72
FREQUENCY (cm⁻¹) × 10⁻³

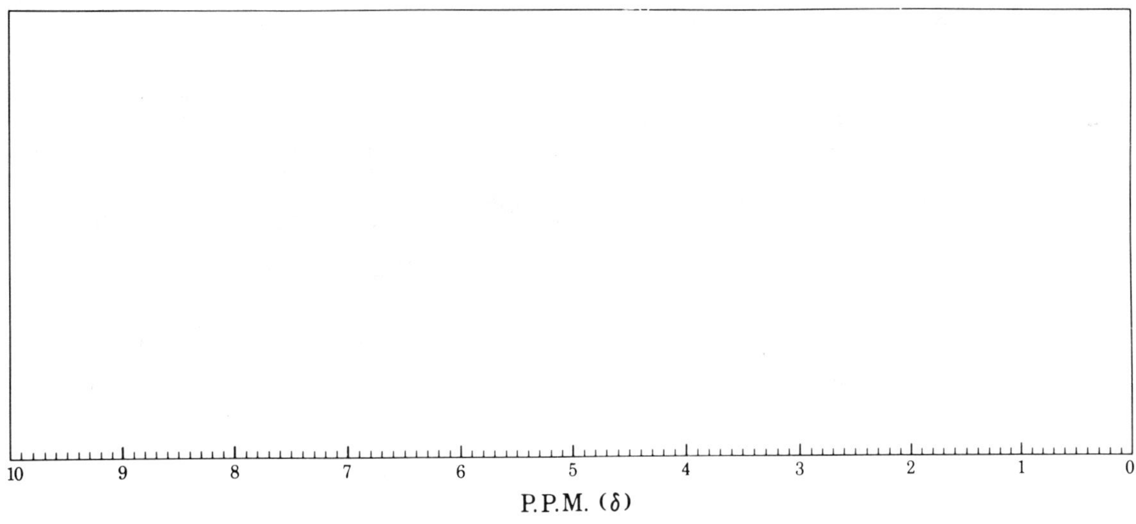
10
9
8
7
6
5
4
3
2
1
0
P.P.M. (δ)

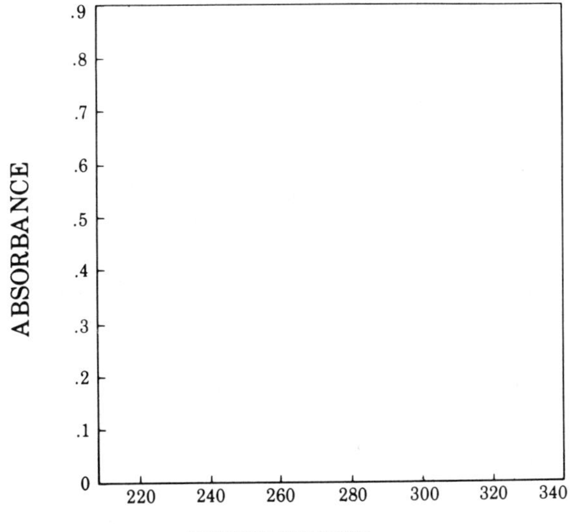
ABSORBANCE
.9
.8
.7
.6
.5
.4
.3
.2
.1
0
220
240
260
280
300
320
340
WAVELENGTH, nm.

| No. | Name | Structure | Formula (mol wt) | Melting point °C | $[\alpha]_D$ | Chromatography |
|---|---|---|---|---|---|---|
| 25 | **Formycin** (*177*) (*178*) | 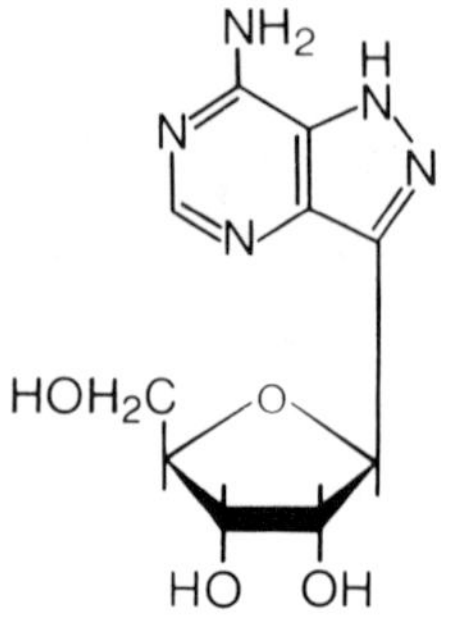 | $C_{10}H_{13}N_5O_4$ (267.2) | 141–144 dec. (*176*), 141–142 dec. and dec. above 250 (*182*), 172–182 isopropylidene) (*184*) | −35.5 (c = 1, 0.1 *N* HCl) (*182*) | tlc (*176*), Paper (*176*, *182*), Column (*176*) |

| pK | Purification | Source | Solubility | Toxicity | Miscellaneous |
|---|---|---|---|---|---|
| 4.4 9.7 (*182*) | Purified by column chromatography and recrystallized from $H_2O$ (*176*) | *Streptomyces lavendulae* (*181*), *Norcardia interforma* (*176*) | Soluble in $H_2O$ and MeOH. Slightly soluble in EtOH. Insoluble in acetone and ether. | No acute or delayed toxicity at 250 mg/kg, (i.v.) (mice) (*176*). Lethal at 50 mg/kg. Viral: cytotoxic effect at 2.5 mcg/ml in chick embryo cells (*179*). | Paper electrophoresis (*176*), Hydrobromide (*177*), X-ray (*177*), pmr (*177*, *178*), uv (*176*, *178*, *182*), ir (*182*), isopropylidene (*184*), mass spectrometry (*194*) |

Biological activity

Shows inhibition against Yoshida Rat Sarcoma cells (*176*, *180*), Ehrlich ascites carcinoma (*176*, *180*), *Xanthomonas oryzae* (*176*), *Mycobacterium 607*, HeLa cells (*176*, *180*), Vaccinia virus (*179*), Influenza virus (*179*). Deamination by an adenosine deaminase from takadiastase (*183*). *S. aureus*, *M. flavus*, *B. subtilis*, *S. latea*, *E. coli*, *P. aeruginosa*, *P. vulgaris OX19*, *Klebsiella pneumonia* and *S. flexneri* were not inhibited at 100 mcg/ml. Inhibition of *Pseudomonas dacunhae*, and *Pellicubria fiamentosa* (*183*). Substrate for adenosine deaminase from *Aspergillis oryzae* (*48*) and for a kinase from *Serratia marcescens* (*184*). Inhibits *de novo* purine biosynthesis by inhibition of PRPP synthesis and also inhibits incorporation of lysine and methionine into protein (*175*). Has shown some inhibition of purine-$\beta$-ribosidases (*185*). Formycin has demonstrated the ability to substitute for adenosine as a substrate for the enzymes, RNA polymerase, aminoacyl-transfer RNA synthetases, polynucleotide phosphorylase of tRNA (*186*). A study has been made on the mode of action against HeLa cells in synchronized culture (*187*) and on the deamination of formycin by several actinomycetes and bacteria (*188*). A report on the behavior of formycin and formycin B in cells Ehrlich carcinoma and E. coli was reported (*189*) as well as a distribution of formycin in various organs of mice (*190*). A study on the protein synthesis directed by ribopolynucleotide containing formycin (*191*), the fluorescent properties of phosphorylated formycins (*192*) and the hydrolysis of polyformycin by pancreatic ribonuclease (*193*) have been present.

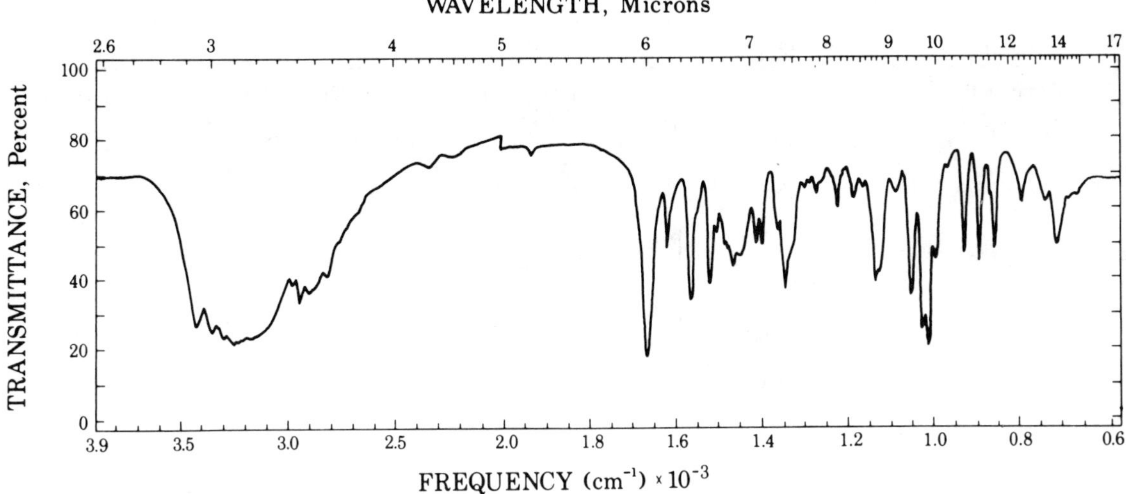


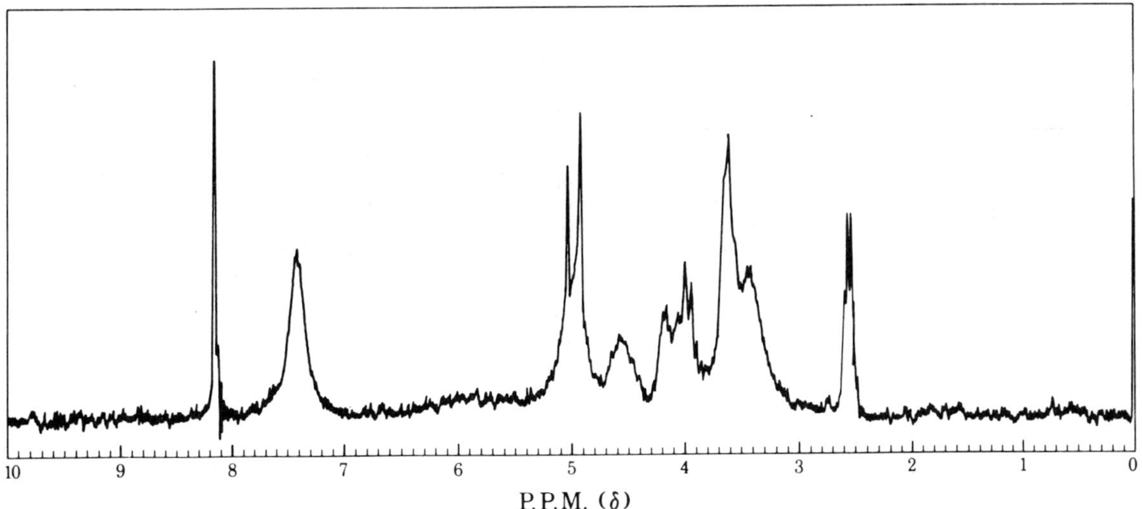


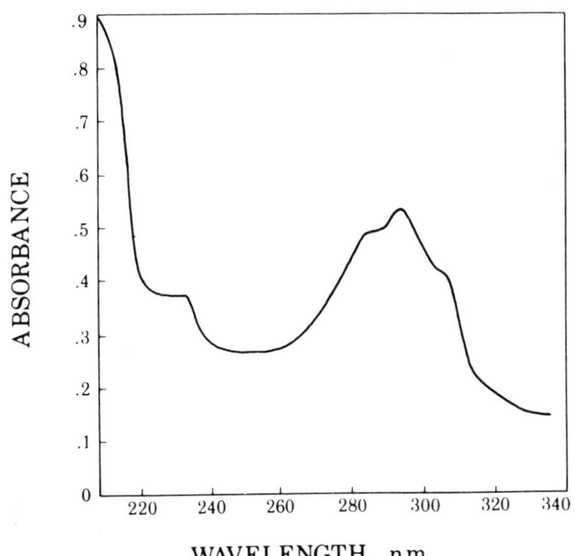


$NH_2$ H N N N N $HOH_2C$ O HO OH

| No. | Name | Structure | Formula (mol wt) | Melting point °C | $[\alpha]_D^{20}$ | Chromatography |
|---|---|---|---|---|---|---|
| 26 | **Formycin B.** (*177, 178*) | 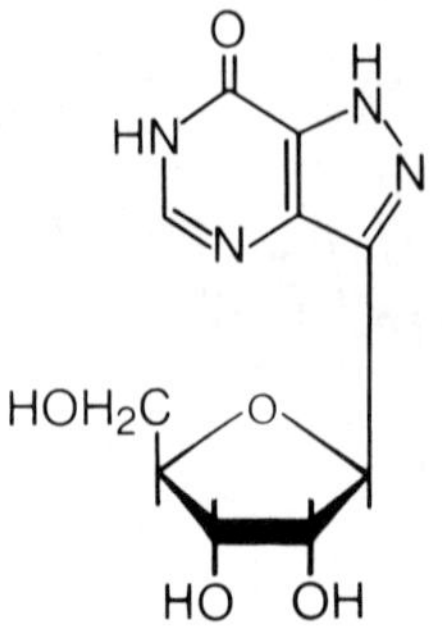 | $C_{10}H_{12}N_4O_5$ (268.2) | 247 dec. (*182*), 259 (isopropylidene) (*184*) | −51.5° (c = 1, $H_2O$) (*182*) | Column (*182*) |

| pK | Purification | Source | Solubility | Toxicity | Miscellaneous |
|---|---|---|---|---|---|
| 8.8 | Purified by column chromatography (*182*). Recrystallized from $H_2O$ (*182*). | *Nocardia interforma* (*182*), by chemical deamination of formycin (*183*) or enzymatic deamination of formycin (*183*). | Soluble in $H_2O$. Sparingly soluble in organic solvents. | $LD_{50}$ estimated 1000 mg/kg. Cytotoxicity against CEC and monkey kidney cells, 100 mcg/ml. | uv (*182*), ir (*182*), isopropylidene (*184*). Formation of oxoformycin (*190, 198, 199*), (196). Formation of oxoformycin and formycin from formycin B (*201*). Mass spectroscopy (*194*). |

Biological activity

Active against *Xanthomonas oryzae* (*182, 183*) and is a substrate for aldehyde oxidase in rabbit liver and is converted to oxoformycin (*196*). No inhibition of Yoshida rat sarcoma (*183*) but has shown some slight activity as an inhibitor of a purine $\beta$-ribosidase (*185*). Some inhibition of myxoviruses (*197*) and has been reported to be a potent inhibitor of purified erythrocytic purine nucleoside phosphorylase (*198*) as well as to inhibit degradation of nucleosides by hemolyzed and intact human erythrocytes or Sarcoma 180 ascites cells (*198*). A study on the distribution of formycin B in various organs of mice (*190*) and the biochemical effects of formycin B on *Xanthomonas oryzae* (*200*) were reported.

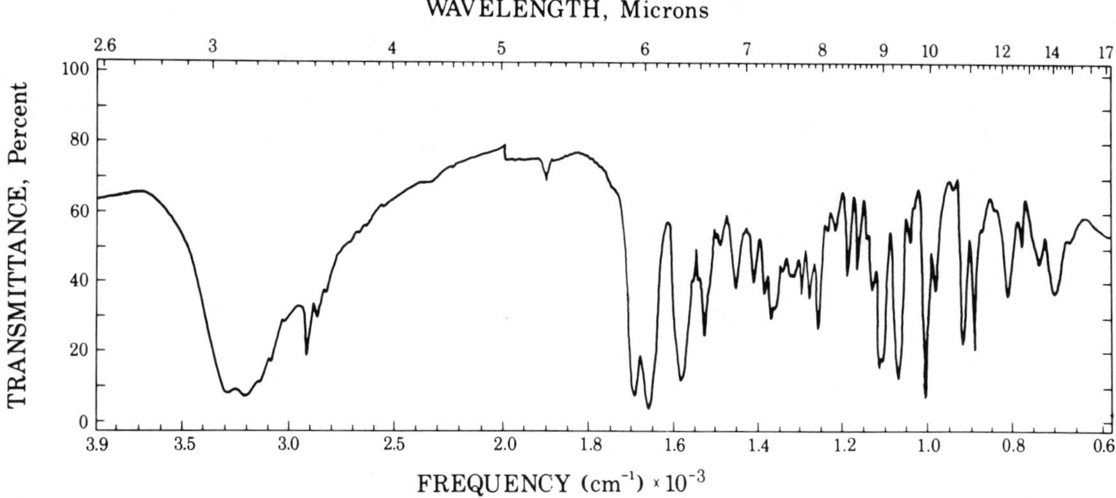
WAVELENGTH, Microns
2.6
3
4
5
6
7
8
9
10
12
14
17
TRANSMITTANCE, Percent
100
80
60
40
20
0
3.9
3.5
3.0
2.5
2.0
1.8
1.6
1.4
1.2
1.0
0.8
0.6
FREQUENCY (cm⁻¹) × 10⁻³

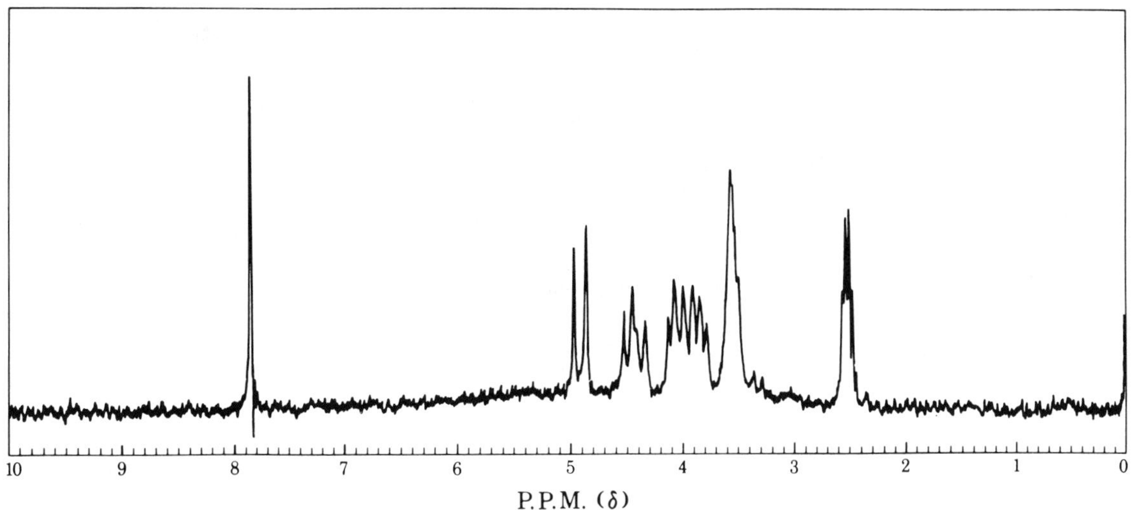
10
9
8
7
6
5
4
3
2
1
0
P.P.M. (δ)

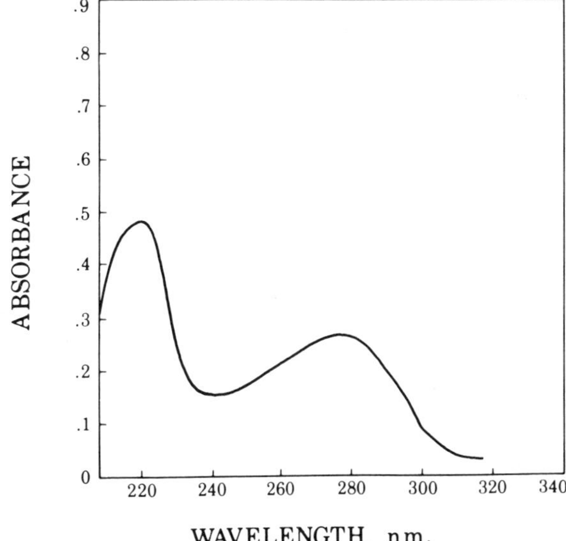
ABSORBANCE
.9
.8
.7
.6
.5
.4
.3
.2
.1
0
220
240
260
280
300
320
340
WAVELENGTH, nm.

| No. | Name | Structure | Formula (mol wt) | Melting point °C | $[\alpha]_D^{21}$ | Chromatography |
|---|---|---|---|---|---|---|
| 27 | **Gougerotin** (*203, 354*) | 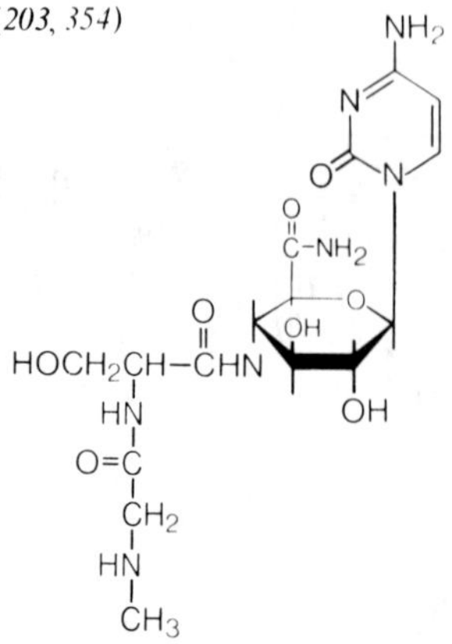 [structure: NH2; N; O; N; O; C-NH2; O; OH; HOCH2CH-CHN; HN; O=C; CH2; HN; CH3; OH] | $C_{16}H_{25}N_7O_8$ (442.41) | 200–215 dec., 200–210 dec. (picrate) | −45 (c = 1, $H_2O$) | Column |

| pK | Purification | Source | Solubility | Toxicity | Miscellaneous |
|---|---|---|---|---|---|
| 3.7, 8.05 | Purified by column chromatography. Recrystallized from aqueous methanol. | *S. gougerotti* No. 21544 (*202*) | Soluble in $H_2O$. Slightly soluble in MeOH and EtOH. Insoluble in acetone and other organic solvents. | $LD_{50}$ 57 mg/kg (i.v.) (mice) | uv, pmr (*203*) |

Biological activity

Active against *Micrococcus flavus* with a weak activity against gram-positive and gram-negative bacteria. Inhibits protein synthesis in *E. coli* (*204*) and inhibits the transfer of amino acids from amino acyl-S-RNA to protein (*205–207*). At tolerated doses there was observed essentially no antitumor activity (*203*).

| No. | Name | Structure | Formula (mol wt) | Melting point °C | $[\alpha]_D$ | Chromatography |
|---|---|---|---|---|---|---|
| 28 | **Homocitrullyl-aminoadenosine** (*208*) | 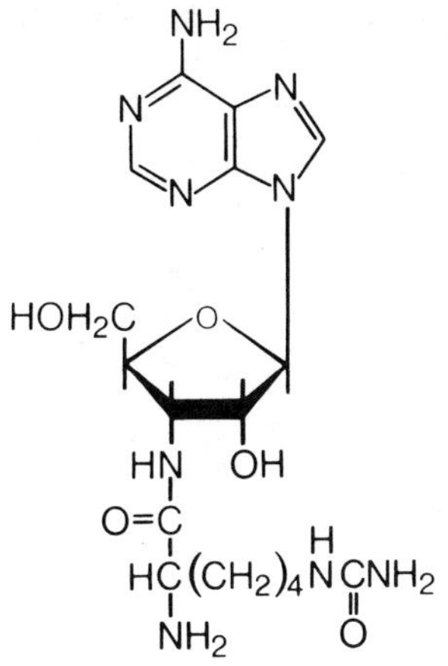 | $C_{17}H_{27}N_9O_5$ (421.46) | — | — | Paper (*208*) |

| pK | Purification | Source | Solubility | Toxicity | Miscellaneous |
|---|---|---|---|---|---|
| — | Purified by column chromatography (Dowex 50) and recrystallized from $H_2O$ (*208*). | *Cordeceps militaris* (Linn.) Link (*208*) | Soluble in $H_2O$ | $1.1 \times 10^{-4}$ *M* conc. required for inhibition of incorporation of amino acids into *E. coli* (*209*). | uv, hydrolyzed by acid (*208*). |

Biological activity

Inhibits the incorporation of amino acids into cell free systems derived from *E. coli* and rat livers and appears to be acting analogously to puromycin (*209*).

| No. | Name | Structure | Formula (mol wt) | Melting point °C | $[\alpha]_D^{20}$ | Chromatography |
|---|---|---|---|---|---|---|
| 29 | **Laurusin** (*210*) | 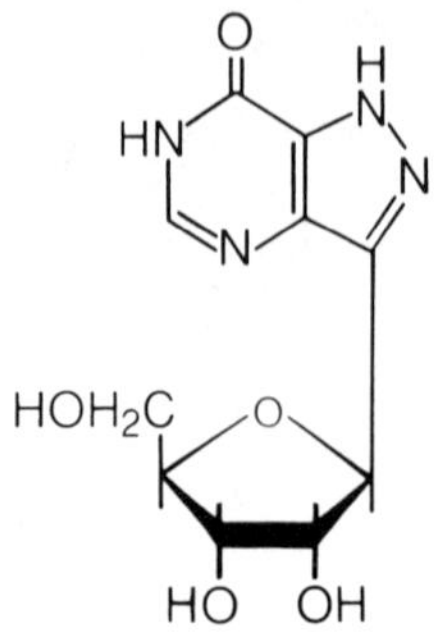 | $C_{10}H_{12}N_4O_5$ (268.2) | 254–255 dec. (*181, 195*) | 44.6 (c = 1, 0.1, *N* HCl) (*181*) | Paper, Column, Partition (*181, 195*) |

| pK | Purification | Source | Solubility | Toxicity | Miscellaneous |
|---|---|---|---|---|---|
| 9.2 (*181*) | Purified by column chromatography (*181*). Recrystallized from $H_2O$ (*181*). | *Streptomyces lavendulae* (*181*). Enzymatic deamination of formycin (*195*). Chemical deamination of formycin (*195*). | Moderately soluble in $H_2O$, MeOH, HOAc and pyridine. Sparingly soluble in EtOH and *n*-BuOH. Insoluble in ether, EtOAc, $CHCl_3$, acetone, *n*- hexane, benzene and tetrahydrofuran (*181*). | M.I.C. 0.75 mg/ml. No toxic symptoms observed at 500 mg/kg (i.v.) (mice) (*181*). | uv, ir (*181*) |

| Biological activity |
|---|
| Active against *Xanthomonas oryzae* but less active against other bacteria and inactive against fungi and yeasts (*181*). Showed inhibition of TMV-multiplication without phytotoxicity on the host plant and also inhibited local lesion formation on bean leaf (*210*). |

| No. | Name | Structure | Formula (mol wt) | Melting point °C | $[\alpha]_D$ | Chromatography |
|---|---|---|---|---|---|---|
| 30 | **Lysylaminoadenosine** | 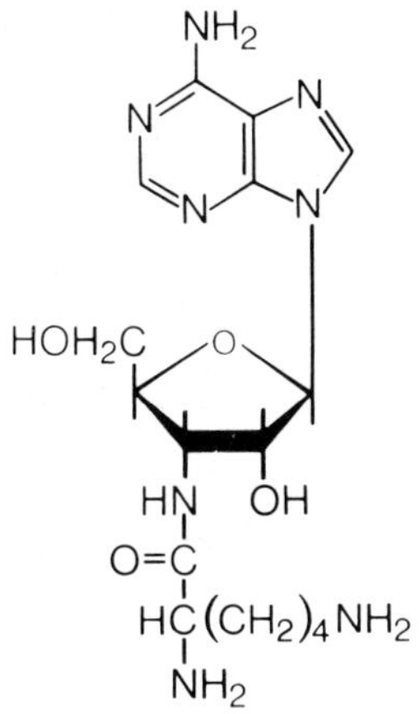 NH<sub></sub> | $C_{16}H_{26}N_8O_4$ (394.44) | — | — | Paper (*211*) |

| pK | Purification | Source | Solubility | Toxicity | Miscellaneous |
|---|---|---|---|---|---|
| — | Purified by column chromatography (Dowex-50) and paper chromatography (*211*). | *Cordeceps militaris* (Linn.) Link (*211*) | — | — | uv, alkaline hydrolysis furnished 3′-amino-3′-deoxyadenosine (*211*). |

| Biological activity |
|---|
| |

| No. | Name | Structure | Formula (mol wt) | Melting point °C | $[\alpha]_D^{20}$ | Chromatography |
|---|---|---|---|---|---|---|
| 31 | **Naritheracin** (*213*), **Nariterashin** (*212*) | 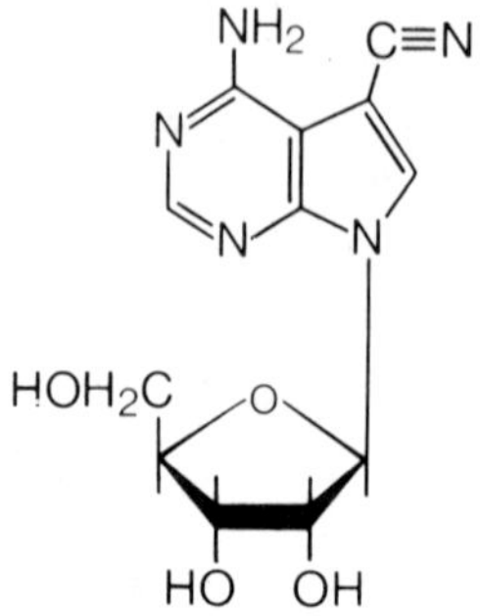 $NH_2$ C≡N N N N $HOH_2C$ O HO OH | $C_{12}H_{13}N_5O_4$ (suggested) (*212, 213*) (291.3) | 239–240 dec. (*212, 213*) | −151.7 (c = 1, DMF) (*212, 213*) | — |

| pK | Purification | Source | Solubility | Toxicity | Miscellaneous |
|---|---|---|---|---|---|
| — | Adsorption on activated charcoal, eluted, column chromatography. Recrystallized from EtOH (*212, 213*). | *Actinomycetales* A-399-Y4 (*212*), *S. sp.* A-392-Y4 res. *S. antimycoticus* (*213*) | — | — | uv (*212*), ir (*213*) |

| Biological activity |
|---|
| Inhibits the growth of Ehrlich sarcoma with no antibacterial activity (*212*). Active against *Candida albicans* and HeLa cells in tissue culture (*213*). |

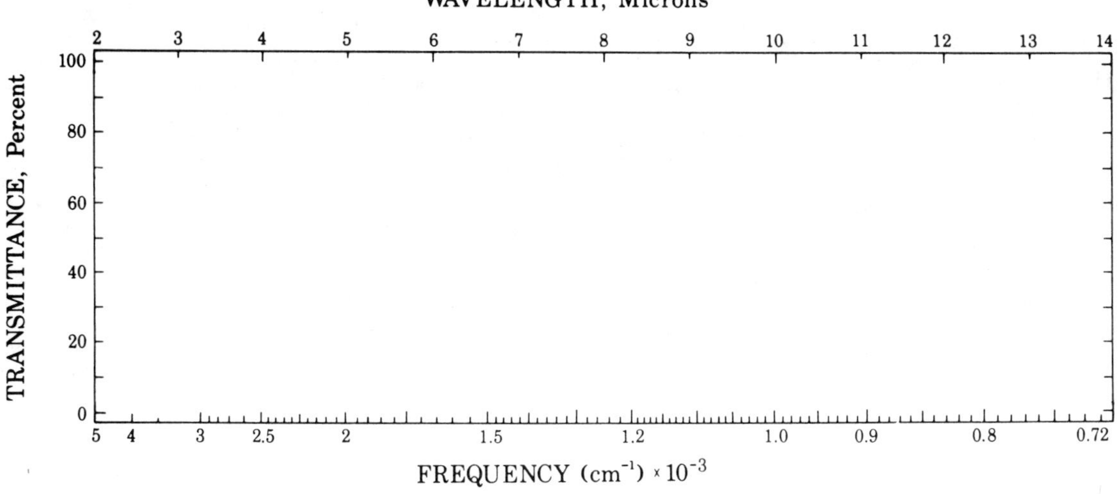
WAVELENGTH, Microns
2 3 4 5 6 7 8 9 10 11 12 13 14
TRANSMITTANCE, Percent
100 80 60 40 20 0
5 4 3 2.5 2 1.5 1.2 1.0 0.9 0.8 0.72
FREQUENCY (cm$^{-1}$) × $10^{-3}$

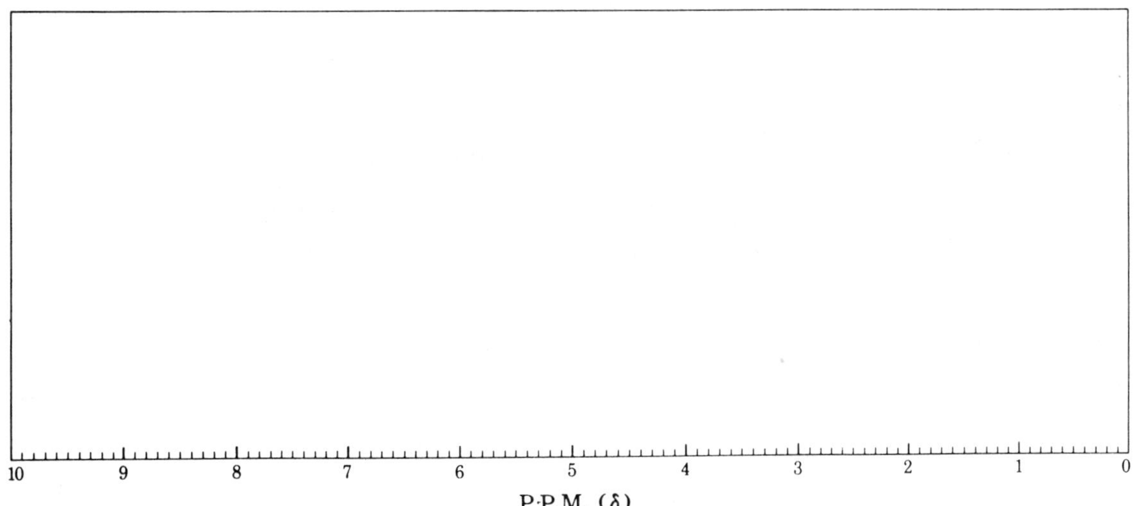
10 9 8 7 6 5 4 3 2 1 0
P.P.M. (δ)

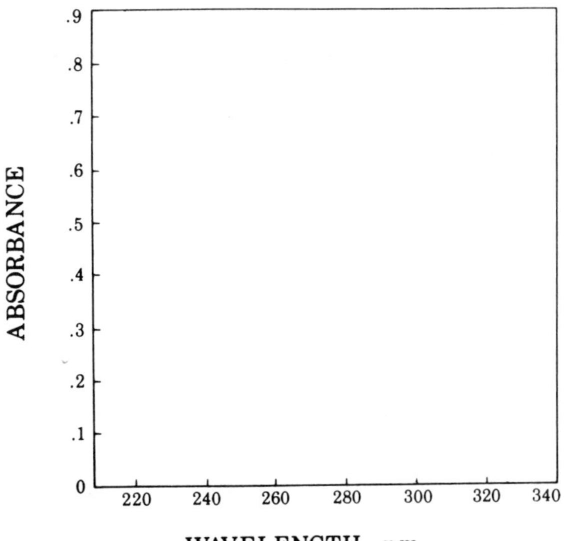
ABSORBANCE
.9 .8 .7 .6 .5 .4 .3 .2 .1 0
220 240 260 280 300 320 340
WAVELENGTH, nm.

| No. | Name | Structure | Formula (mol wt) | Melting point °C | $[\alpha]_D^{25}$ | Chromatography |
|---|---|---|---|---|---|---|
| 32 | **Nebularine** (*216, 217*) | 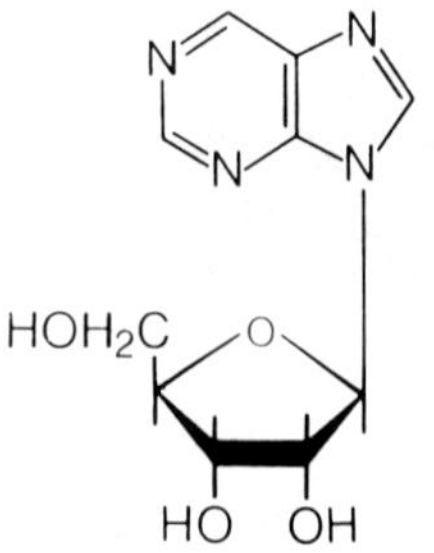 | $C_{10}N_{12}N_4O_4$ (252.23) | 180–181 | −48.6 (c = 1.0, $H_2O$) (*224*) | — |

| pK | Purification | Source | Solubility | Toxicity | Miscellaneous |
|---|---|---|---|---|---|
| 2.1 | Purified by column chromatography. Recrystallized from absolute EtOH (*225*). | *S. yokosukaensis* (*214, 215*), *Agaricus* (*Clitocybe*) *nebularis batsch* (*216, 217, 219*). | Soluble in $H_2O$. Insoluble in ether and $CHCl_3$. | $LD_{50}$ 200 mg/kg (s.c.) (mice). Very toxic in tissue cultures (*222, 223*). | uv (*224*), Chemical synthesis (*224–228*) |

Biological activity

Active against *Mycobacterium avium*, *Mycobacterium phlei*, *Mycobacterium tuberculosis BCG* and *Candida albicans*. (*214–217, 221*). There was also some antitumor activity reported (*218*).

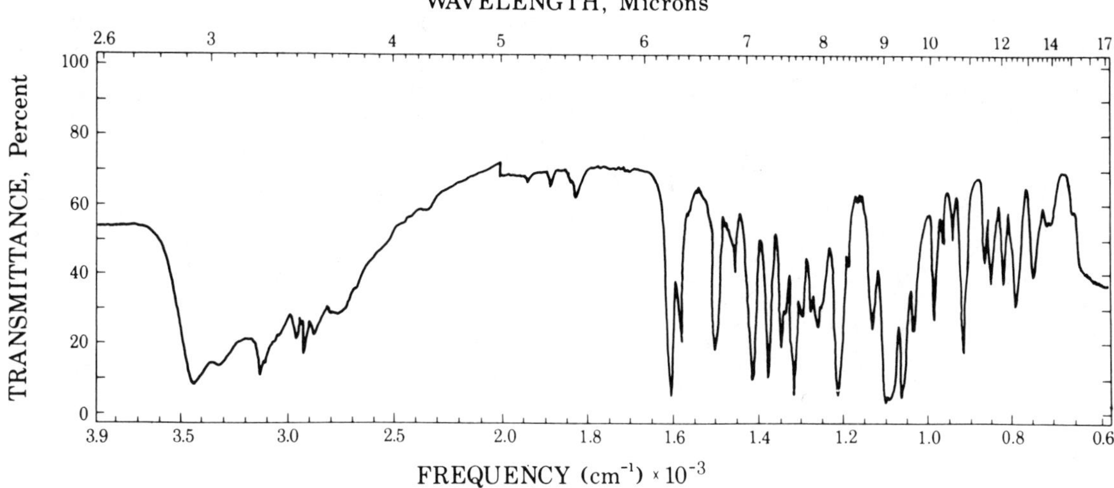
WAVELENGTH, Microns
2.6
3
4
5
6
7
8
9
10
12
14
17
TRANSMITTANCE, Percent
100
80
60
40
20
0
3.9
3.5
3.0
2.5
2.0
1.8
1.6
1.4
1.2
1.0
0.8
0.6
FREQUENCY (cm$^{-1}$) × 10$^{-3}$

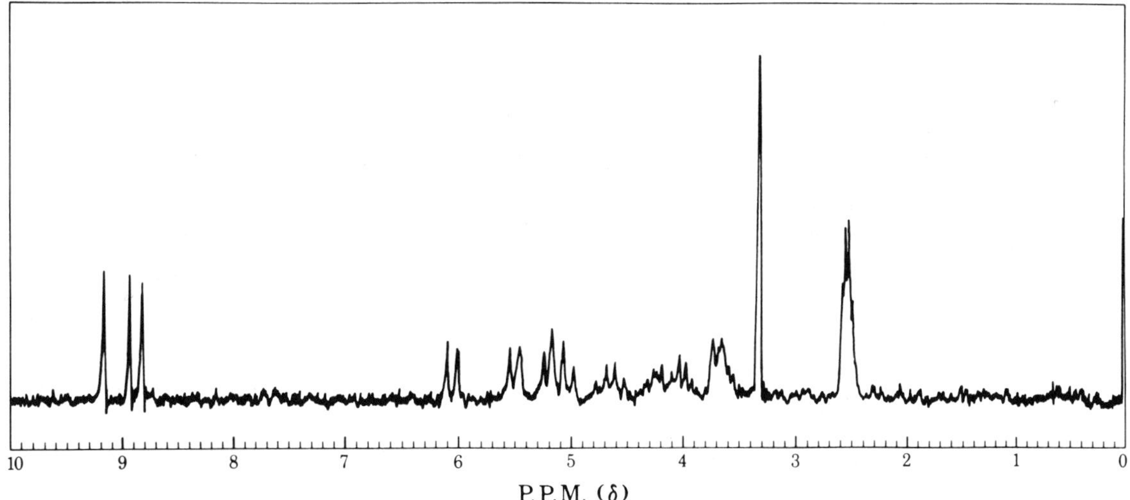
10
9
8
7
6
5
4
3
2
1
0
P.P.M. (δ)

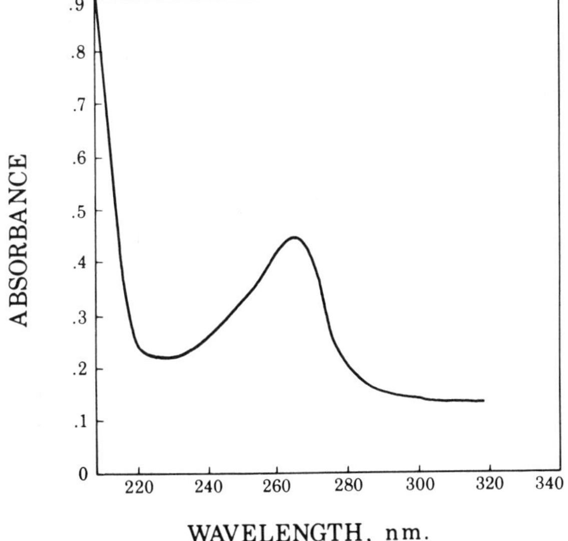
ABSORBANCE
.9
.8
.7
.6
.5
.4
.3
.2
.1
0
220
240
260
280
300
320
340
WAVELENGTH, nm.

| No. | Name | Structure | Formula (mol wt) | Melting point °C | $[\alpha]_D^{24.5}$ | Chromatography |
|---|---|---|---|---|---|---|
| 33 | **Nucleocidin** (*233*) | 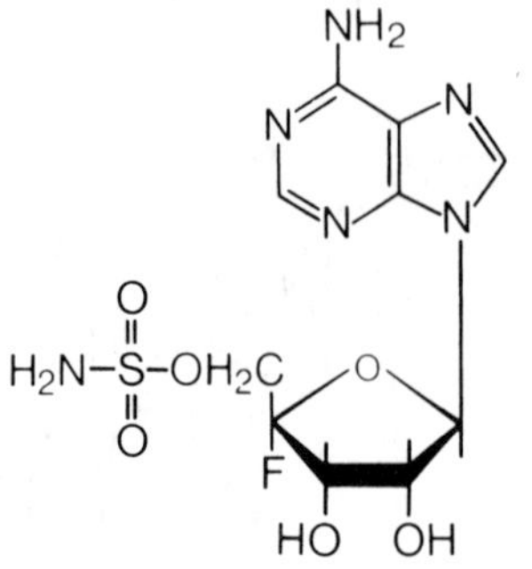 | $C_{11}H_{15-16}N_6O_8S$ (391.34–392.34) (*299*, *239*), $C_{10}H_{13}FN_6O_6S$ (364.31) (*233*) | 143–144° (picrate) (*238*, *239*), | −33.3 [c = 1.05, EtOH/0.1 *N* HCl (1:1)] | Column |

| pK | Purification | Source | Solubility | Toxicity | Miscellaneous |
|---|---|---|---|---|---|
| 9.3 | Purified by adsorption on charcoal and then column chromatography. Crystallized from dilute aqueous solution at pH 4 and at 5°. | *Streptomyces calvus T3018* (*229*, *239*), *Streptomyces calvus* (ATCC No. 13,382) (*238*) | Soluble at pH 9.2 and pH 3.2, only slightly soluble at pH 6.5. | $LD_{50}$ 0.2 mg/kg (i.p.) (mice) (*230*). Lethal at 0.8 mg/kg (i.p.) (rats) (*234*). Lethal at 0.05 mg/kg for young bovines (*230*). | pmr (*233*), $^{19}F$ nmr (*233*), mass spectroscopy (*233*), uv (*238*), HCl, sulfate (*238*), ir (*239*) |

| Biological activity |
|---|
| Has shown antitrypanosomal (*230*) activity and inhibition of protein biosynthesis (*234*, *237*). Active against gram-positive and gram-negative bacteria, *Endoamoeba*, *Trypanosoma equiperdum* (*238*), *T. congolense*, *T. equinum* and *T. gambiense* in rats and mice (*235*). Active against *T. vivax* in West African zebu cattle (*236*). |

| No. | Name | Structure | Formula (mol wt) | Melting point °C | $[\alpha]_D^{21}$ | Chromatography |
|---|---|---|---|---|---|---|
| 34 | **Oyamycin** (*197*) | 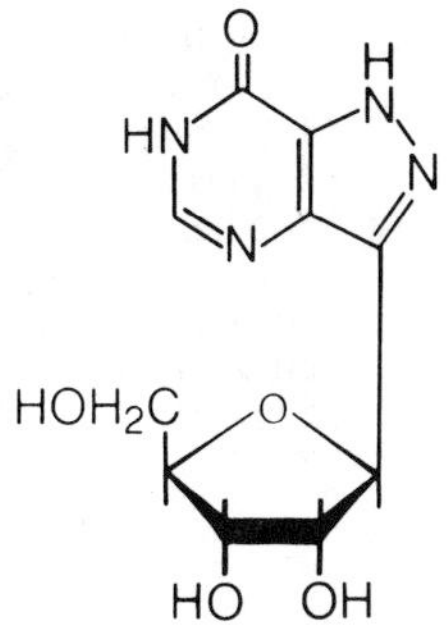 | $C_{10}H_{12}N_6O_{10}$ (*240*) (formula in patent and was revised at a later date to $C_{10}H_{12}N_4O_5$). | 244–246 dec. (*240*) | −44 (c = 1%, 0.1 *N* HOAc) | Column (*240*) |

| pK | Purification | Source | Solubility | Toxicity | Miscellaneous |
|---|---|---|---|---|---|
| — | Adsorbed on carbon, eluted, Sephadex G-25 column and freeze-dried. Recrystallized from an EtOH/hexane mixture (*240*). | *Streptomyces roseochromogenes* var. *oyaensis* (*240*) | Soluble in $H_2O$, AcOH, pyridine, MeOH, EtOH, methyl cellosolve, ethylene glycol and formamide. Slightly soluble in $Me_2CO$ and AcOBu, $CCl_4$, $C_6H_6$, hexane, $Et_2O$ and petroleum ether (*240*). | No toxicity at 300 mg/kg (i.v.) (mice) (*240*). | uv (*240*), ir (*240*) |

| Biological activity |
|---|
| Very weak antibacterial activity but strong inhibition of *Xanthomonas oryzae* (*240*). |

| No. | Name | Structure | Formula (mol wt) | Melting point °C | $[\alpha]_D$ | Chromatography |
|---|---|---|---|---|---|---|
| 35 | **Plicacetin** (Amicetin B) (Antibiotic C) (*21*, *22*) | 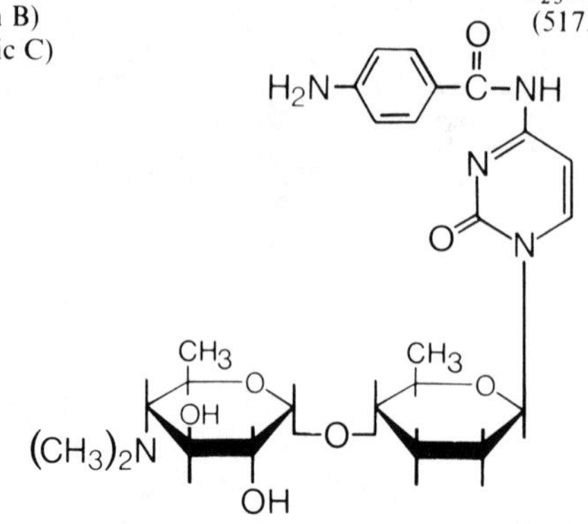 | $C_{25}H_{35}N_5O_7$ (517.59) | Recry. from $H_2O$ (*14*), 182–184; from ethyl acetate, 160–163; from ethanol (absolute), 222–225 | +181° (c = 2.7% in methanol) (*14*) | Column (*14*) |

| pK | Purification | Source | Solubility | Toxicity | Miscellaneous |
|---|---|---|---|---|---|
| 2.2, 7.0, 10.9 | Recrystallized from $H_2O$, dil. aqueous methanol, ethyl acetate, ethanol (absolute). | *Streptomyces plicatus* (*21*, *22*, *14*). | Soluble in lower alcohols, chloroform and methylene chloride. Sparingly soluble in ethyl acetate, ether and cold water. Insoluble in benzene and petroleum ether (*14*). | $LD_{50}$ 210 mg/kg (i.v.) (mice) (*14*) | M.W. by Rast procedure and titration. uv, ir (*14*). |

| Biological activity |
|---|

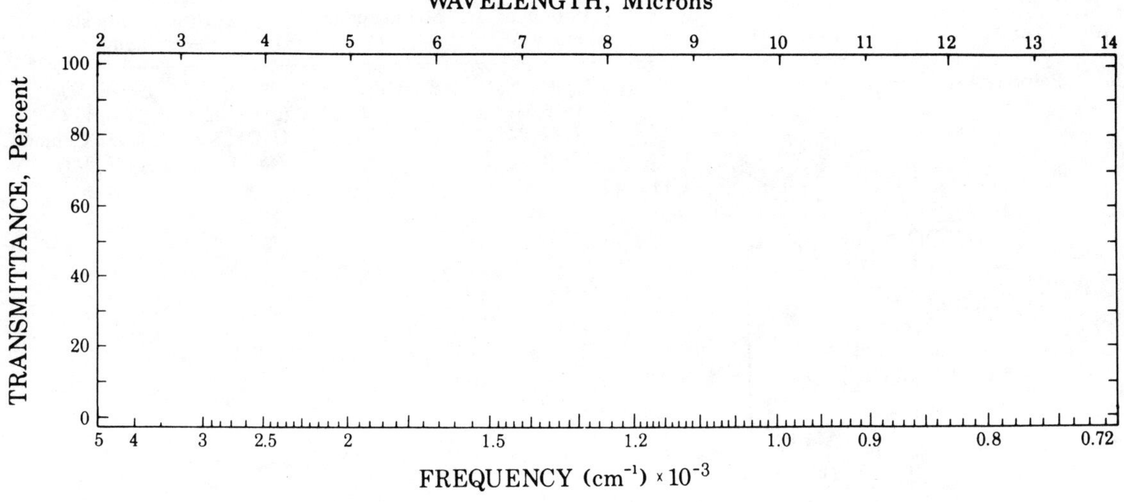
WAVELENGTH, Microns
2
3
4
5
6
7
8
9
10
11
12
13
14
TRANSMITTANCE, Percent
100
80
60
40
20
0
5
4
3
2.5
2
1.5
1.2
1.0
0.9
0.8
0.72
FREQUENCY (cm⁻¹) × 10⁻³

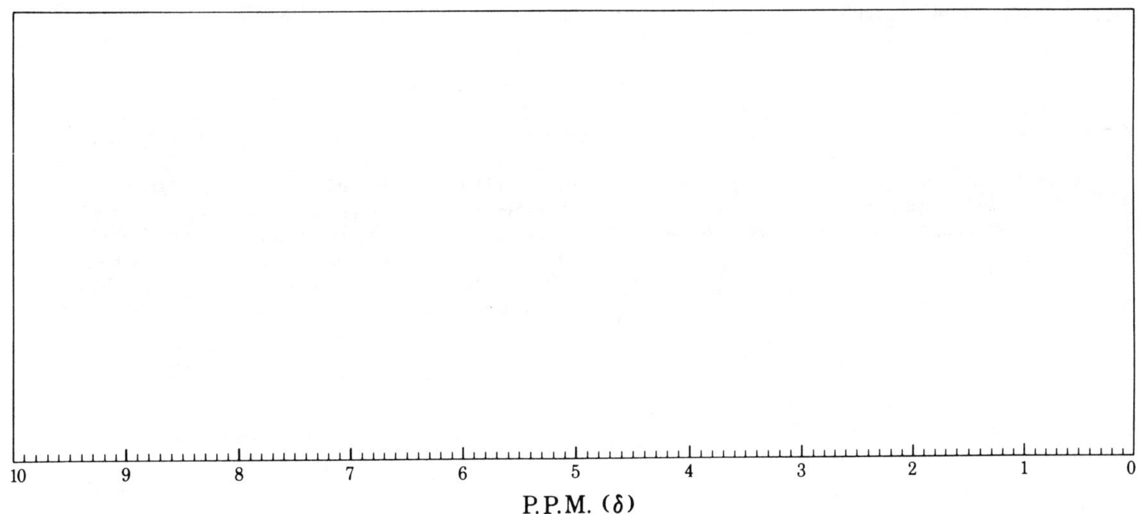
10
9
8
7
6
5
4
3
2
1
0
P.P.M. (δ)

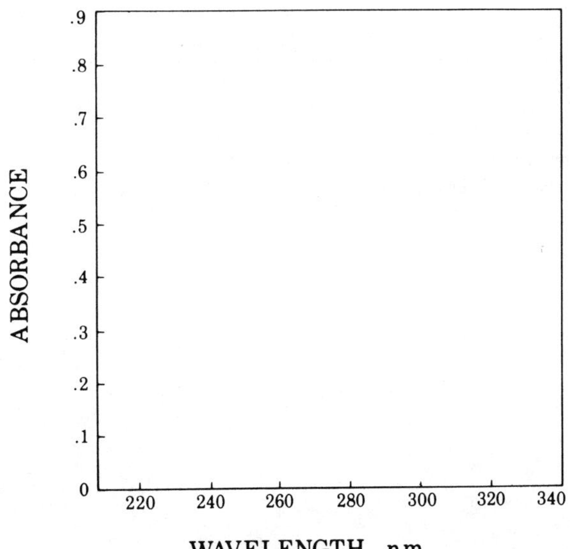
ABSORBANCE
.9
.8
.7
.6
.5
.4
.3
.2
.1
0
220
240
260
280
300
320
340
WAVELENGTH, nm.

| No. | Name | Structure | Formula (mol wt) | Melting point °C | $[\alpha]_D^{20}$ | Chromatography |
|---|---|---|---|---|---|---|
| 36 | **Polyoxin A** (*246, 248, 355*) | 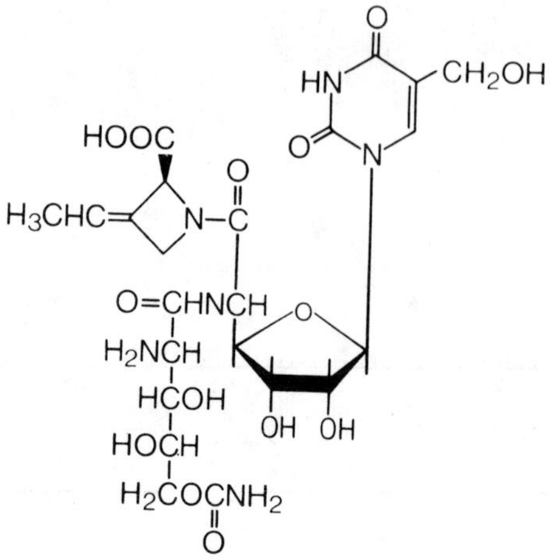 HOOC, $H_3CHC$=, N–C(=O), HN, $CH_2OH$, O=CHNCH, $H_2NCH$, HCOH, HOCH, $H_2COCNH_2$, OH OH | $C_{23}H_{32}N_6O_{14}$ (616.53), (*242, 246, 248*) | > 160 dec. | −30 (c = 1.02, $H_2O$) (*242, 248*) | Paper (*241, 243*), TLC (*243*), Column (*241, 243*) |

| pK | Purification | Source | Solubility | Toxicity | Miscellaneous |
|---|---|---|---|---|---|
| 3.0, 7.3, 9.6 (*241, 246, 248, 242*) | Recrystallized from aqueous ethanol (*241*). Purified by column chromatography (*241*). | *Streptomyces cacaoi* var. *asoensis* (*241*) | Soluble in $H_2O$. Insoluble in MeOH, EtOH, BuOH, $Me_2CO$, $CHCl_3$, $C_6H_6$ and $Et_2O$ (*241*). | Mice tolerated 500 mg/kg (i.v.). No toxicity against mice or killifish (up to 4 p.p.m.) (10 mg/20 g, i.v.). | uv (*248, 241, 242*), Paper electrophoresis (*241*), pmr data (*248*), ir (*241, 242*), MW detr. (*241*) |

Biological activity

Tested as antimicrobial and antifungal agents (*241, 244*) against *Piricularia oryzae*, *Cochliobolus miyabianus*, *Pellicularia sasakii*, *Alternia kikuchiana*, *Sclerotinia cinerera*, *Cladospotium fulvum*, *Corticium rolfsii* and *Guignardia laricina* (244).

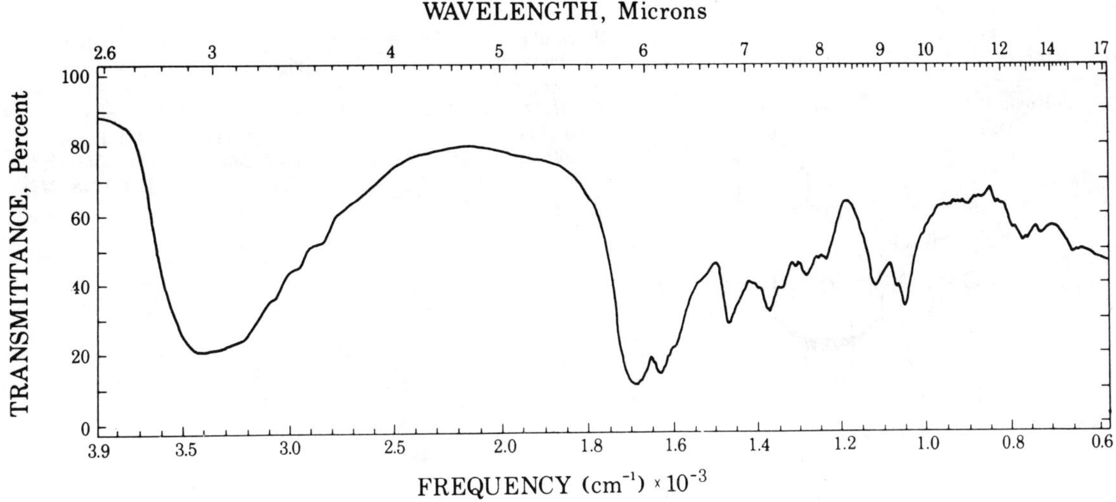


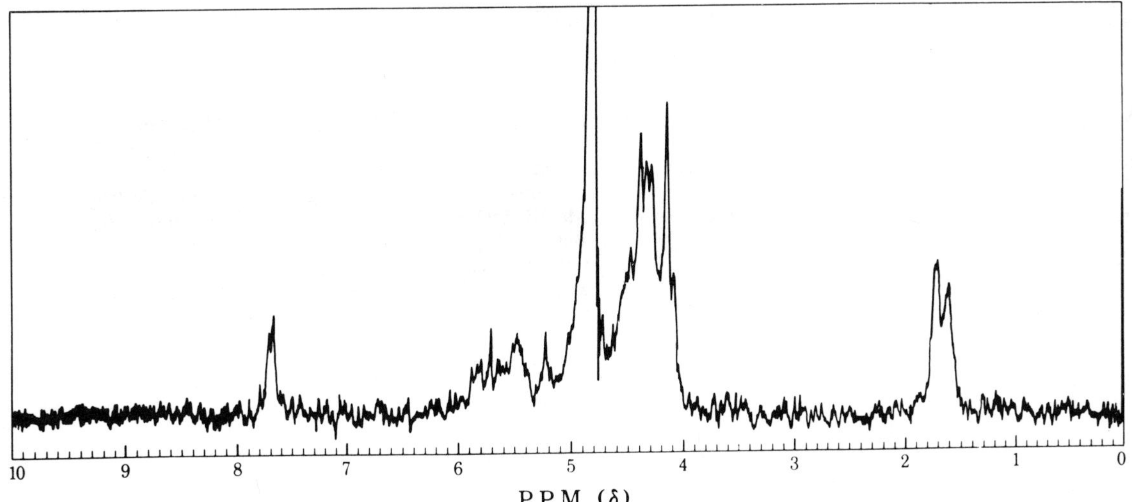


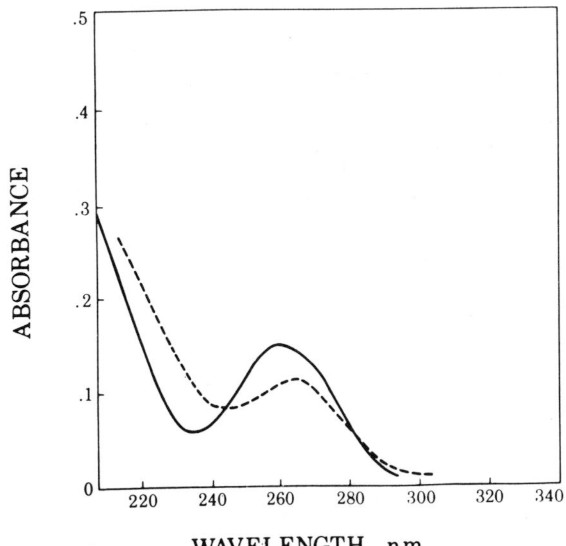


HOOC
$H_3CHC=$
N—C
O=CHNCH
$H_2NCH$
HCOH
HOCH
$H_2COCNH_2$
O
HN
$CH_2OH$
N
OH OH

| No. | Name | Structure | Formula (mol wt) | Melting point °C | $[\alpha]_D^{20}$ | Chromatography |
|---|---|---|---|---|---|---|
| 37 | **Polyoxin B** (*246, 355*) | 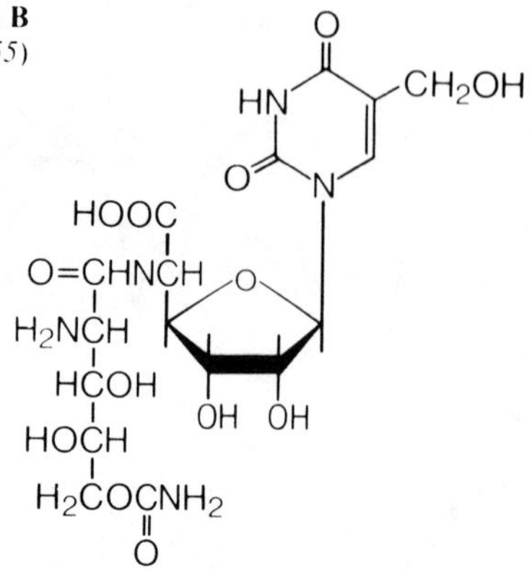 [structure: uracil-5-CH₂OH nucleoside with HOOC–CH(–NH–CO–CH(NH₂)–CH(OH)–CH(OH)–CH₂OCONH₂) side chain, ring OH OH] | $C_{17}H_{25}N_5O_{13}$ (507.41) (*246*) | 160 dec. (*241*) | +34° (c = 1, $H_2O$) (*241, 246, 248*) | Column (*241, 243*), Paper (*241, 243*), TLC (*248, 243*) |

| pK | Purification | Source | Solubility | Toxicity | Miscellaneous |
|---|---|---|---|---|---|
| 3.0, 6.9, 9.4, (*241, 242, 248*) | Recrystallized from aqueous ethanol (*241*). | *Streptomyces cacaoi* var. *asoensis* | Soluble in $H_2O$. Insoluble in MeOH, EtOH, BuOH, $Me_2CO$, $CHCl_3$, $C_6H_6$ and $Et_2O$ (*241*). | Mice tolerated 500 mg/kg (i.v.) No toxicity against mice or killifish (up to 4 p.p.m.) (10 mg/20 g, i.v.). | uv (*248, 242*), Paper electrophoresis (*241*), ir (*242*) |

| Biological activity |
|---|
| Active antimicrobial (*241*) and tested as antifungal and antimicrobial agents against *Piricularia oryzae*, *Cochliobolus miyabianus*, *Pellicularia sasakii*, *Alternia kikuchiana*, *Sclerotinia cinerera*, *Cladospotium fulvum*, *Corticium rolfsii* and *Guignardia laricina* (*244*). |

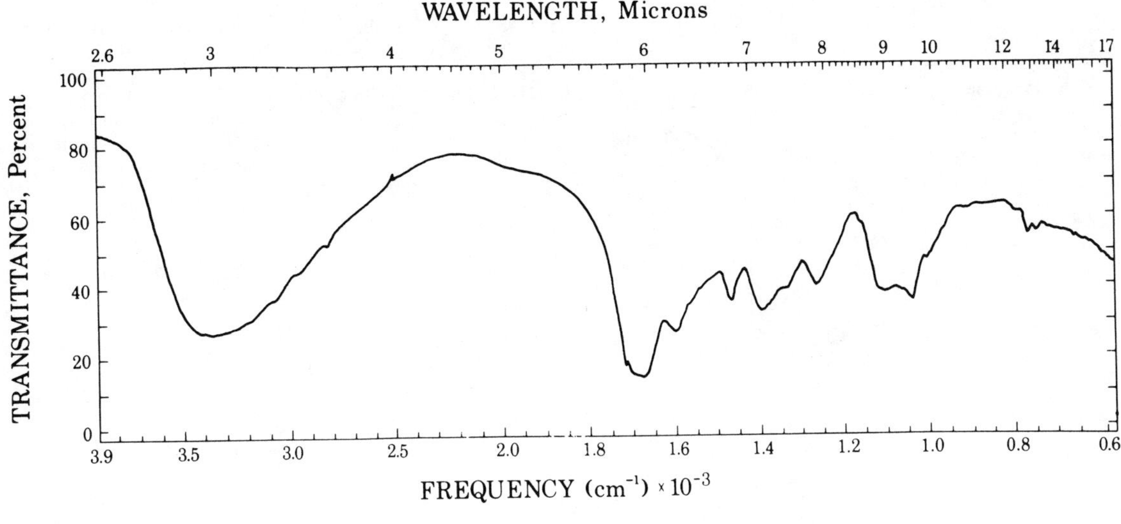
WAVELENGTH, Microns
2.6
3
4
5
6
7
8
9
10
12
14
17
TRANSMITTANCE, Percent
100
80
60
40
20
0
3.9
3.5
3.0
2.5
2.0
1.8
1.6
1.4
1.2
1.0
0.8
0.6
FREQUENCY (cm⁻¹) × 10⁻³

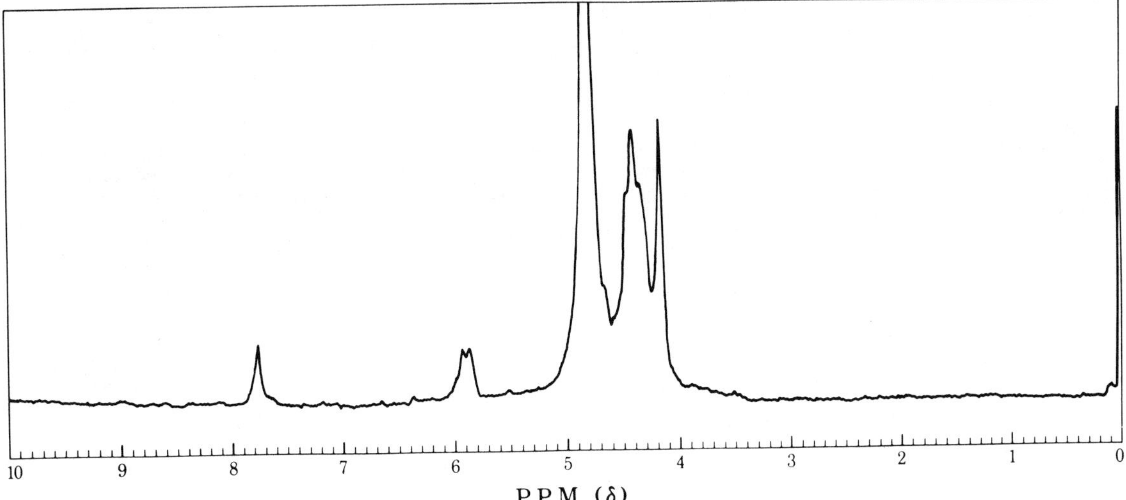
10
9
8
7
6
5
4
3
2
1
0
P.P.M. (δ)

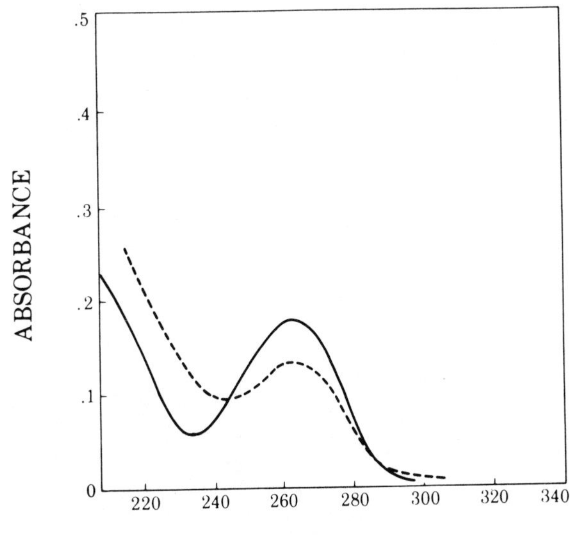
ABSORBANCE
.5
.4
.3
.2
.1
0
220
240
260
280
300
320
340
WAVELENGTH, nm.

| No. | Name | Structure | Formula (mol wt) | Melting point °C | $[\alpha]_D$ | Chromatography |
|---|---|---|---|---|---|---|
| 38 | **Polyoxin C** (*246, 247, 355*) | 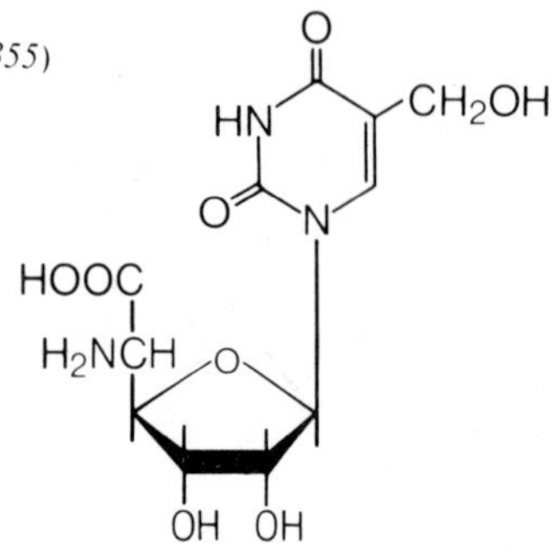 | $C_{11}H_{15}N_3O_8$ (317.25) (*246*) | 260–267 (*243*) | +11.2° (c = 0.5, $H_2O$) (*246, 247*) | Paper (*243*), TLC (*243*), Column (*242*) |

| pK | Purification | Source | Solubility | Toxicity | Miscellaneous |
|---|---|---|---|---|---|
| 2.4, 8.1, 9.4 (*246*) | Recrystallized from aqueous ethanol. | *Streptomyces cacaoi* var. *asoensis* (*247, 243*) | Soluble in $H_2O$, MeOH and EtOH. | Mice tolerated 500 mg/kg (i.v.). No toxicity against mice or killifish (up to 4 p.p.m.) (10 mg/20 g, i.v.). | CD, (*247*), Paper electrophoresis (*243, 242*), pmr of *N*-Acetyl deriv. (*247*), uv (*242*), ir (*242*), X-ray of *N*-Brosyl derivatives (*245*). Hydrogenation to afford deoxypolyoxin C (*246, 245*). |

| Biological activity |
|---|
| Phytopathogenic fungi (*242*), antifungal (*242*) and antimicrobial (*242*). |

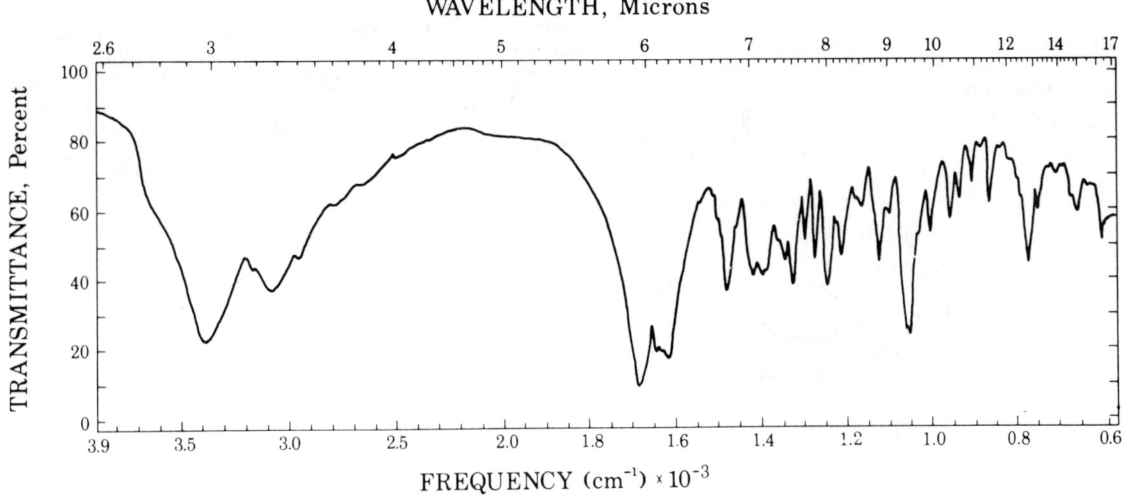


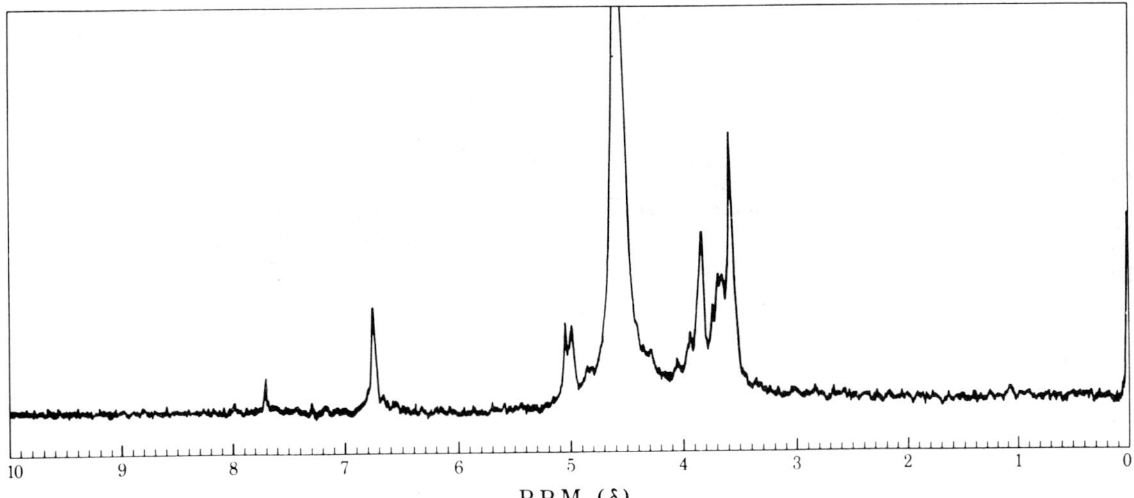


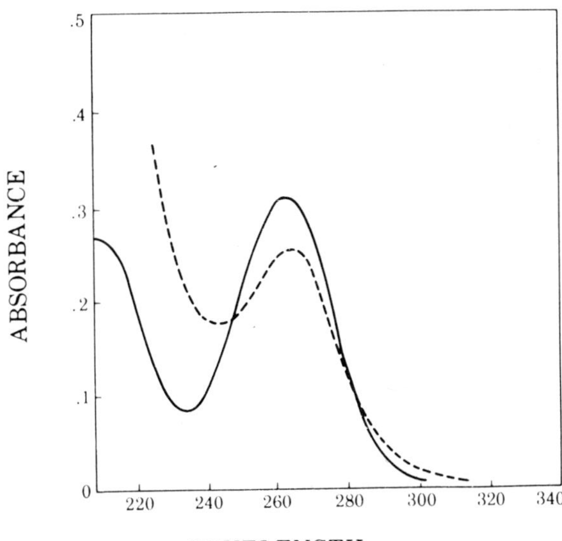


O
HN
$CH_2OH$
O
N
HOOC
$H_2NCH$
O
OH OH

| No. | Name | Structure | Formula (mol wt) | Melting point °C | $[\alpha]_D^{20}$ | Chromatography |
|---|---|---|---|---|---|---|
| 39 | **Polyoxin D** (*246, 355*) | 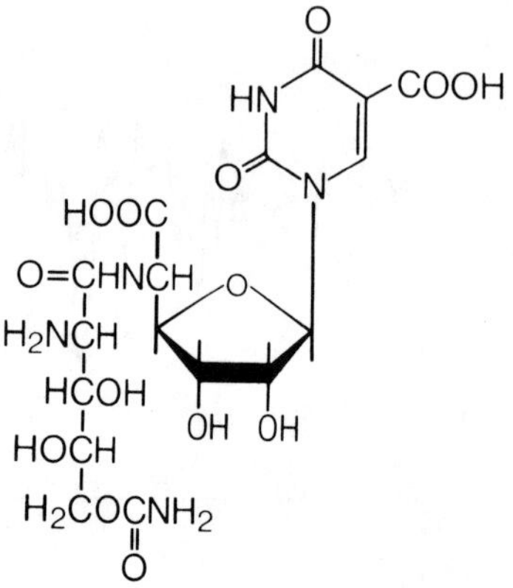 | $C_{17}H_{23}N_5O_{14}$ (521.39) (*246*) | 190 | 30 (c = 1, $H_2O$) (*242, 246*) | Column (*242, 243*) Paper (*242, 243*), TLC (*243*) |

| pK | Purification | Source | Solubility | Toxicity | Miscellaneous |
|---|---|---|---|---|---|
| 2.6, 3.7, 7.3, 9.4, (*242, 246*) | Recrystallized aqueous EtOH. | *Streptomyces cacaoi* var. *asoensis* (*242*) | Soluble in $H_2O$, MeOH and EtOH. | Mice tolerated 500 mg/kg (i.v.). No toxicity against mice or killifish (up to 4 p.p.m.) (10 mg/20 g, i.v.). | uv (*242*), ir (*242*), Paper electrophoresis (*243*) |

| Biological activity |
|---|
| Phytopathogenic fungi (*242*), antifungal (*243*) and antimicrobial (*243*). |

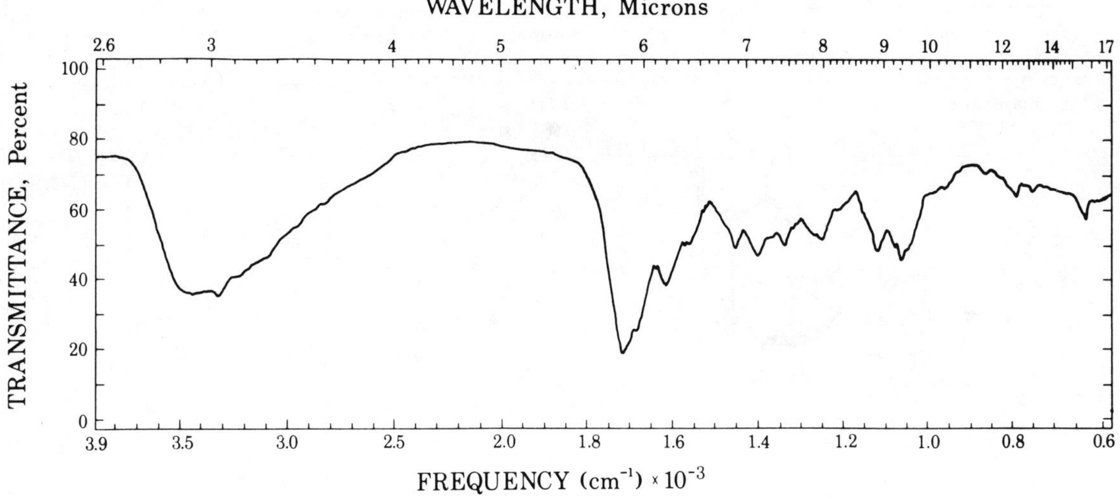
WAVELENGTH, Microns
2.6
3
4
5
6
7
8
9
10
12
14
17
TRANSMITTANCE, Percent
100
80
60
40
20
0
3.9
3.5
3.0
2.5
2.0
1.8
1.6
1.4
1.2
1.0
0.8
0.6
FREQUENCY (cm$^{-1}$) × 10$^{-3}$

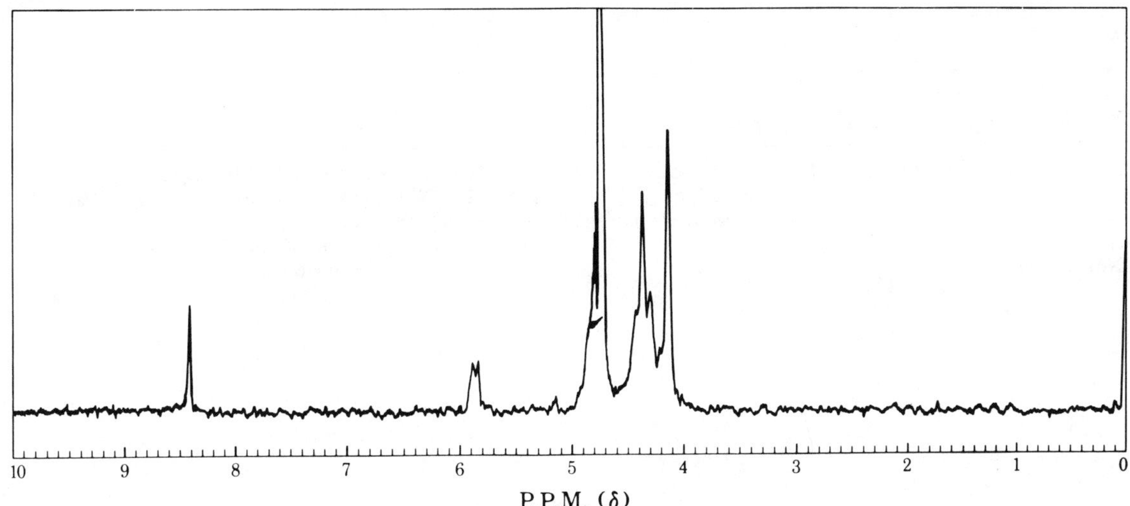
10
9
8
7
6
5
4
3
2
1
0
P.P.M. (δ)

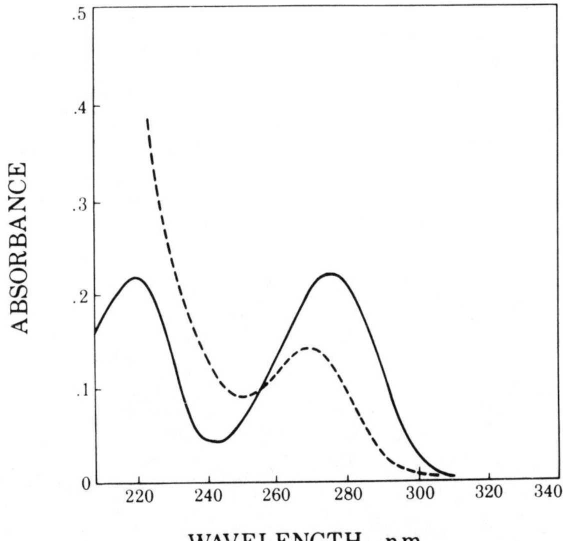
ABSORBANCE
.5
.4
.3
.2
.1
0
220
240
260
280
300
320
340
WAVELENGTH, nm.

O
HN
COOH
O
N
HOOC
O=CHNCH
O
H$_2$NCH
HCOH
OH
OH
HOCH
H$_2$COCNH$_2$
O

| No. | Name | Structure | Formula (mol wt) | Melting point °C | $[\alpha]_D^{20}$ | Chromatography |
|---|---|---|---|---|---|---|
| 40 | **Polyoxin E** (*246, 355*) | 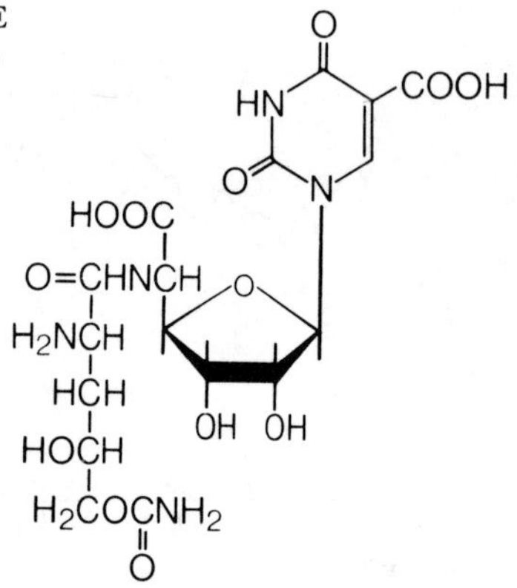 HOOC–CH(NHCH=O)–CH(NH₂)–CH₂–CH(OH)–CH₂OC(=O)NH₂ nucleoside with 5-carboxyuracil (HN, COOH, O, N), furanose ring with OH OH | $C_{17}H_{23}N_5O_{13}$ (505.39) (*246*) | > 180 (*242*) | + 19° (c = 1, $H_2O$) (*246*) | Column (*242, 243*), Paper (*242, 243*), TLC (*242, 243*) |

| pK | Purification | Source | Solubility | Toxicity | Miscellaneous |
|---|---|---|---|---|---|
| 2.8, 3.9, 7.4, 9.3 (*246*) | Aqueous alcohol (*242*) | *Streptomyces cacaoi* var. *asoensis* (*242*) | Soluble in $H_2O$, MeOH and EtOH. | Mice tolerated 500 mg/kg (i.v.). No toxicity against mice or killifish (up to 4 p.p.m.) (10 mg/20 g, i.v.). | uv (*242, 243*), ir (*242, 243*), Paper electrophoresis (*243*) |

| Biological activity |
|---|
| Phytopathogenic fungi (*242*), antifungal (*243*) and antimicrobial (*243*). |

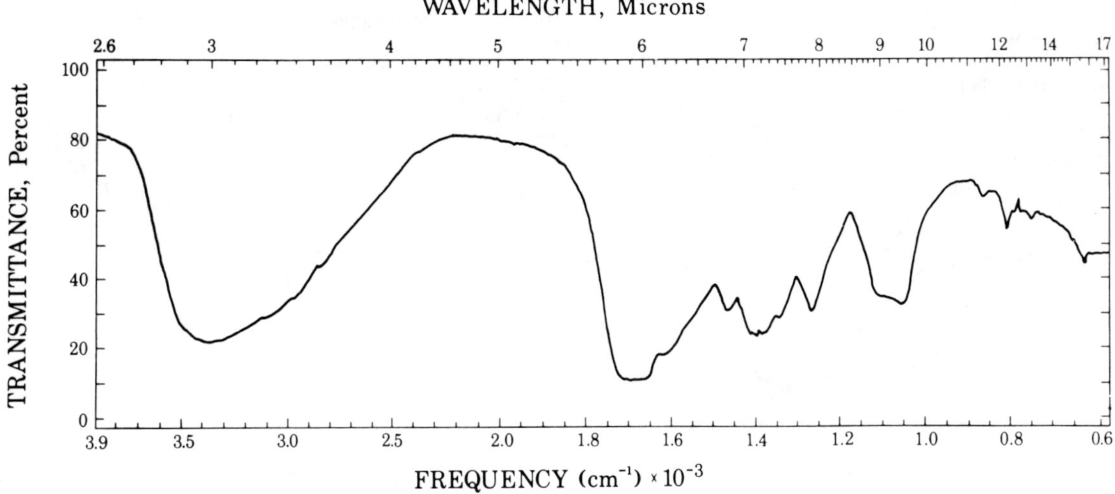
WAVELENGTH, Microns
TRANSMITTANCE, Percent
FREQUENCY (cm$^{-1}$) × 10$^{-3}$

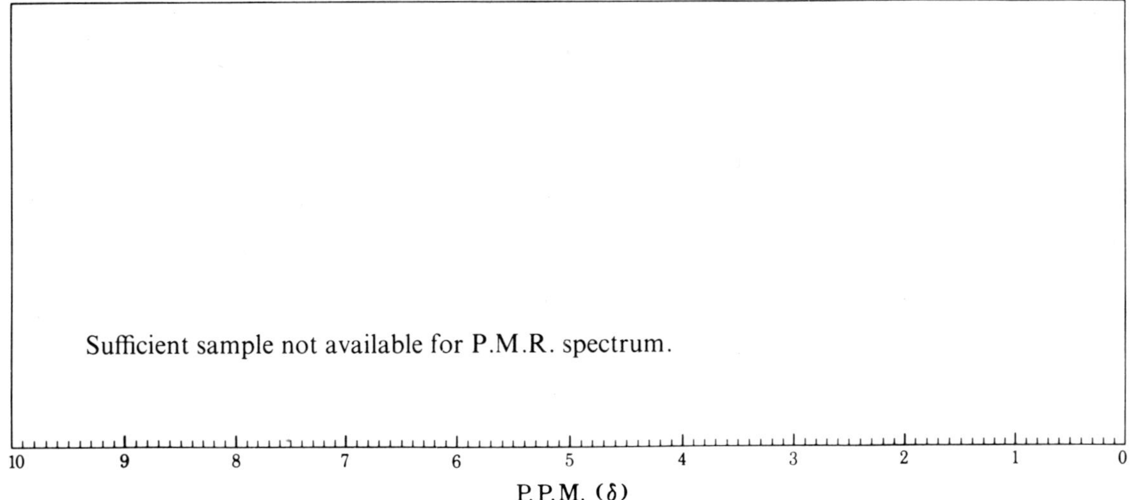
Sufficient sample not available for P.M.R. spectrum.
P.P.M. (δ)

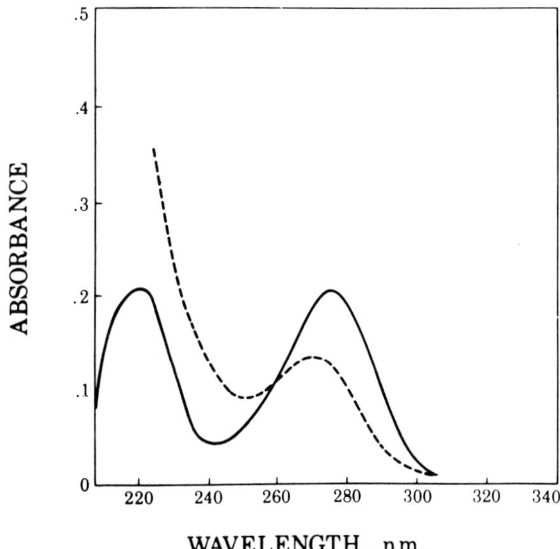
ABSORBANCE
WAVELENGTH, nm.

| No. | Name | Structure | Formula (mol wt) | Melting point °C | $[\alpha]_D^{20}$ | Chromatography |
|---|---|---|---|---|---|---|
| 41 | **Polyoxin F** (*246, 355*) | 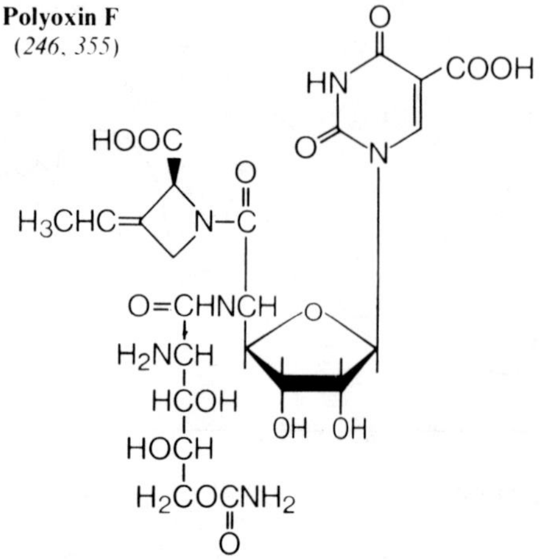 O, HN, COOH, O, N, HOOC, O, $H_3CHC$, N–C, O=CHNCH, O, $H_2NCH$, HCOH, OH OH, HOCH, $H_2COCNH_2$, O | $C_{23}H_{30}N_6O_{15}$ (630.52) (*246*) | > 190 (*241*) | − 18° (c = 1, $H_2O$) (*246*) | Paper (*242, 243*), Column (*242, 243*), TLC (*242, 243*) |

| pK | Purification | Source | Solubility | Toxicity | Miscellaneous |
|---|---|---|---|---|---|
| 2.7, 3.9, 7.2, 9.3 (*246*) | Aqueous EtOH (*242*) | *Streptomyces cacaoi* var. *asoensis* (*242*) | Soluble in $H_2O$, MeOH and EtOH. | Mice tolerated 500 mg/kg (i.v.). No toxicity against mice or killifish (up to 4 p.p.m.) (10 mg/20 g, i.v.). | uv (*242, 243*), ir (*242*), Paper electrophoresis (*243*) |

| Biological activity |
|---|
| Phytopathogenic fungi (*242*), antifungal (*243*) and antimicrobial (*243*). |

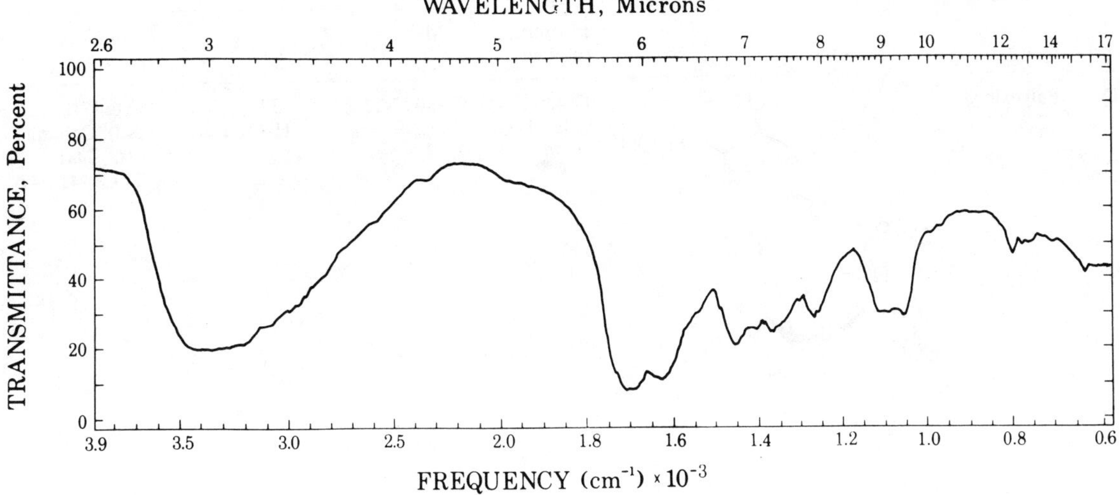
WAVELENGTH, Microns
2.6
3
4
5
6
7
8
9
10
12
14
17
TRANSMITTANCE, Percent
100
80
60
40
20
0
3.9
3.5
3.0
2.5
2.0
1.8
1.6
1.4
1.2
1.0
0.8
0.6
FREQUENCY (cm⁻¹) × 10⁻³

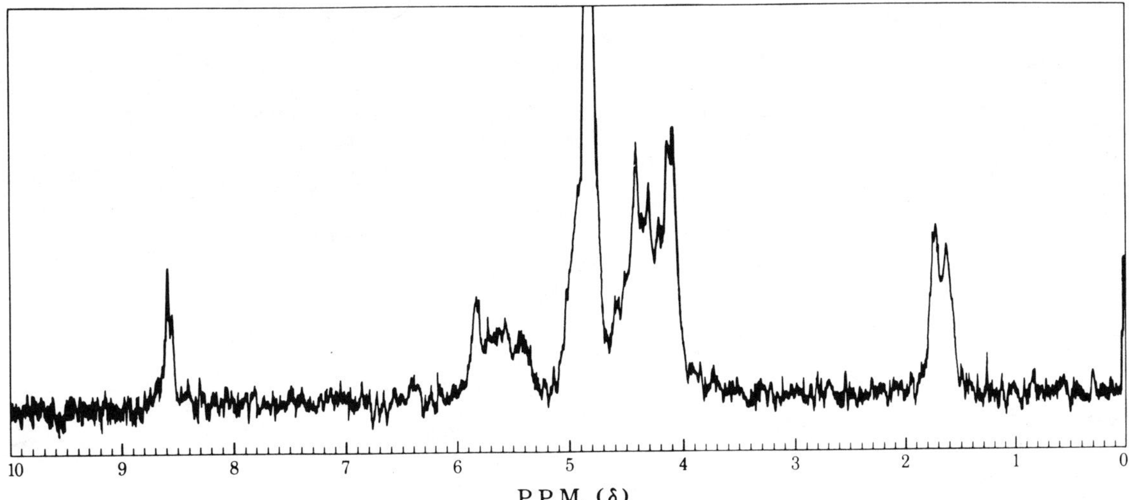
10
9
8
7
6
5
4
3
2
1
0
P.P.M. (δ)

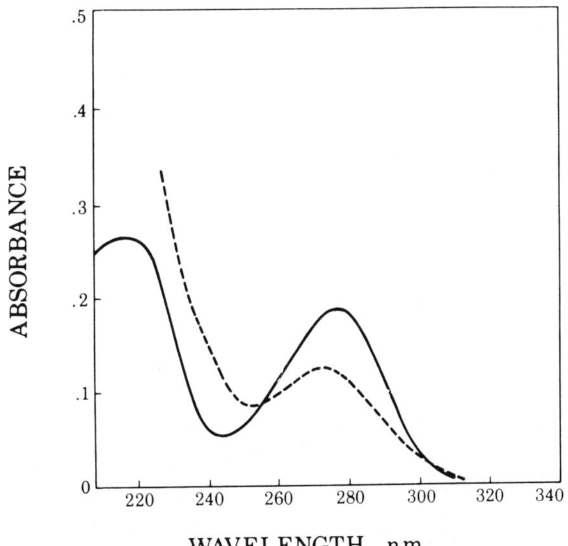
ABSORBANCE
.5
.4
.3
.2
.1
0
220
240
260
280
300
320
340
WAVELENGTH, nm.
O
HN
COOH
HOOC
O
N
O
H3CHC
N–C
O=CHNCH
O
H2NCH
HCOH
OH OH
HOCH
H2COCNH2
O

| No. | Name | Structure | Formula (mol wt) | Melting point °C | $[\alpha]_D$ | Chromatography |
|---|---|---|---|---|---|---|
| 42 | **Polyoxin G** (*246*, *355*) | 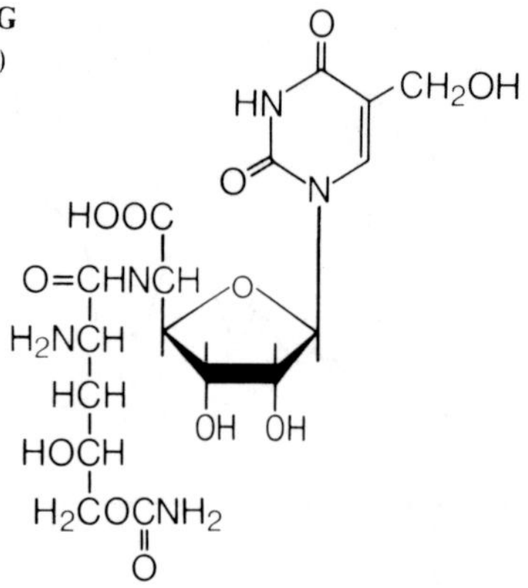 O<br>HN CH$_2$OH<br>O N<br>HOOC<br>O=CHNCH O<br>H$_2$NCH<br>HCH<br>HOCH OH OH<br>H$_2$COCNH$_2$<br>O | $C_{17}H_{25}N_5O_{12}$ (491.41) (*246*) | 190 (*242*) | +37° (c = 1, $H_2O$) (*246*) | Paper (*242*, *243*), Column (*242*, *243*), TLC (*242*, *243*) |

| pK | Purification | Source | Solubility | Toxicity | Miscellaneous |
|---|---|---|---|---|---|
| 3.2, 7.3, 9.3 (*246*) | Aqueous EtOH (*242*) | *Streptomyces cacaoi* var. *asoensis* (*242*) | Soluble in $H_2O$, MeOH and EtOH. | Mice tolerated 500 mg/kg (i.v.). No toxicity against mice or killifish (up to 4 p.p.m.) (10 mg/20 g, i.v.). | uv (*242*, *243*) ir (*242*) and Paper electrophoresis (*243*). |

Biological activity

Phytopathogenic fungi (*242*), antifungal (*243*) and antimicrobial (*243*).

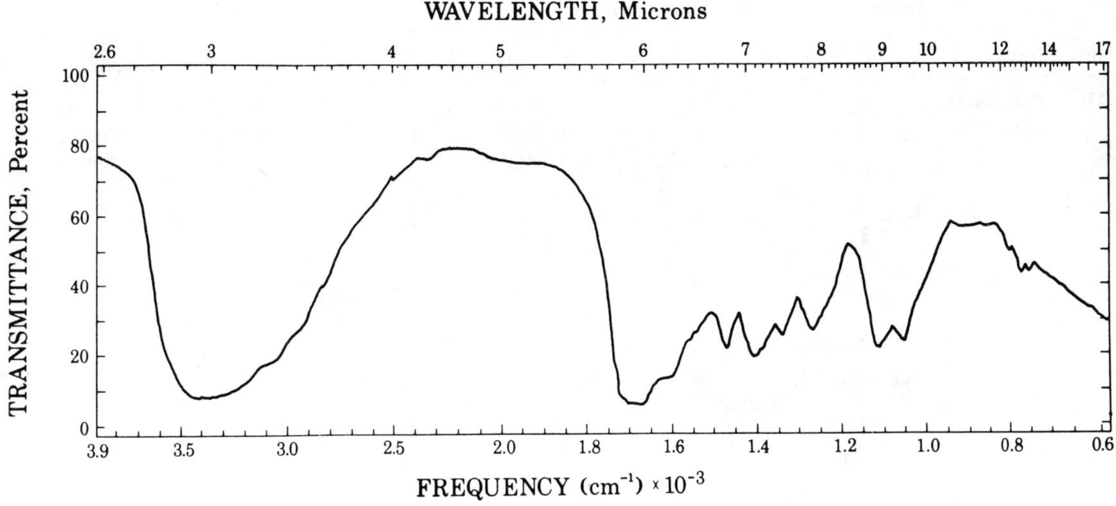


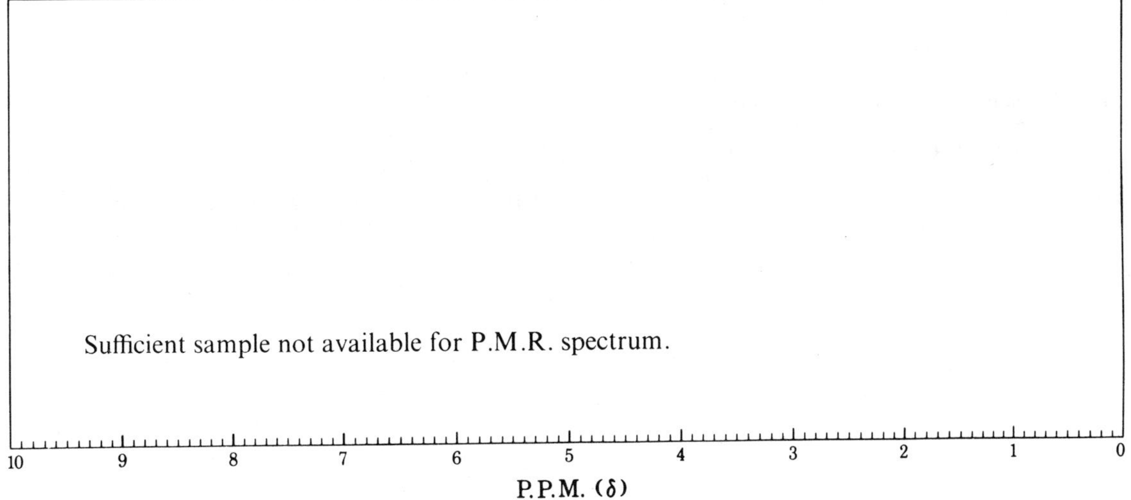


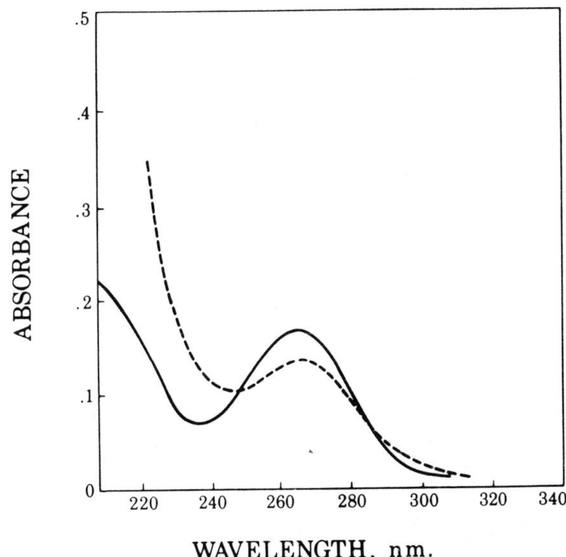


O
HN
CH2OH
O
N
HOOC
O=CHNCH
H2NCH
HCH
OH OH
HOCH
H2COCNH2
O

| No. | Name | Structure | Formula (mol wt) | Melting point °C | $[\alpha]_D^{20}$ | Chromatography |
|---|---|---|---|---|---|---|
| 43 | **Polyoxin H** (*246, 355*) | 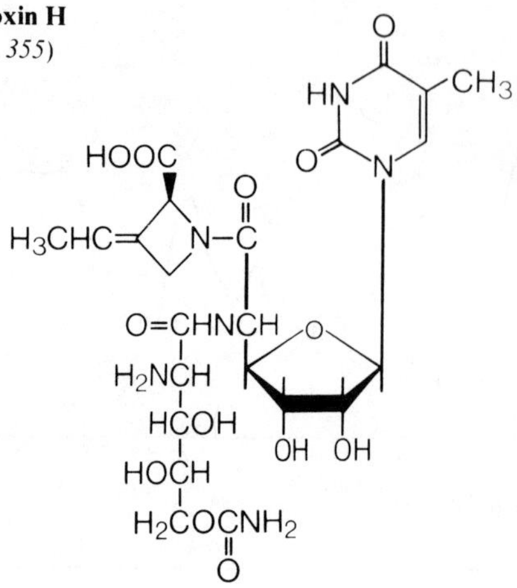 HOOC, $H_3CHC$=(azetidine)N–C(=O)–CH(NH)–C(=O)... structure: O=CHNCH, $H_2NCH$, HCOH, HOCH, $H_2COCNH_2$ (=O); thymine base (HN, O, $CH_3$, N) on ribose (OH OH) | $C_{23}N_{32}N_6O_{13}$ (600.53) (*246*) | — | −38° (c = 1, $H_2O$) (*246*) | TLC (*243*), Paper (*243*), Column (*243*) |

| pK | Purification | Source | Solubility | Toxicity | Miscellaneous |
|---|---|---|---|---|---|
| 3.3, 7.2, 9.4 (*246*) | Aqueous EtOH (*243*) | *Streptomyces cacaoi* var. *asoensis* (*243*) | Soluble in $H_2O$, MeOH and EtOH. | Mice tolerated 500 mg/kg (i.v.). No toxicity against mice or killifish (up to 4 p.p.m.) (10 mg/20 g, i.v.). | uv (*243*), ir (*243*) and Paper electrophoresis (*243*) |

| Biological activity |
|---|
| Antifungal (*243*) and antimicrobial (*243*). |

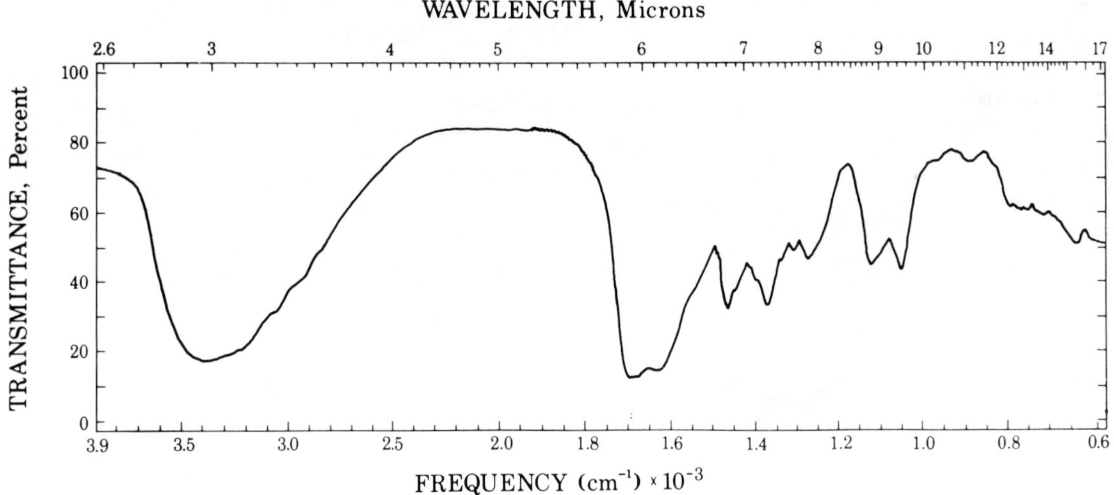


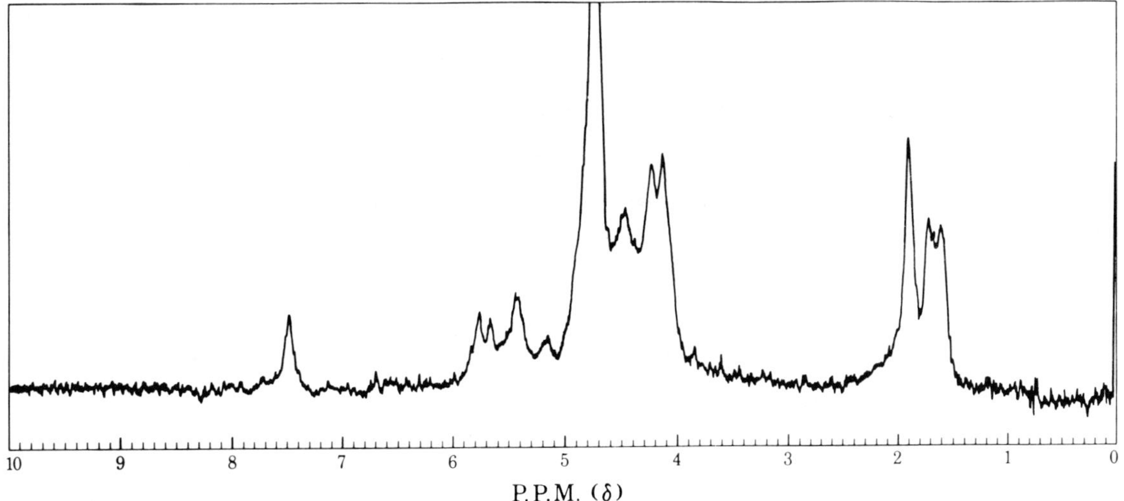


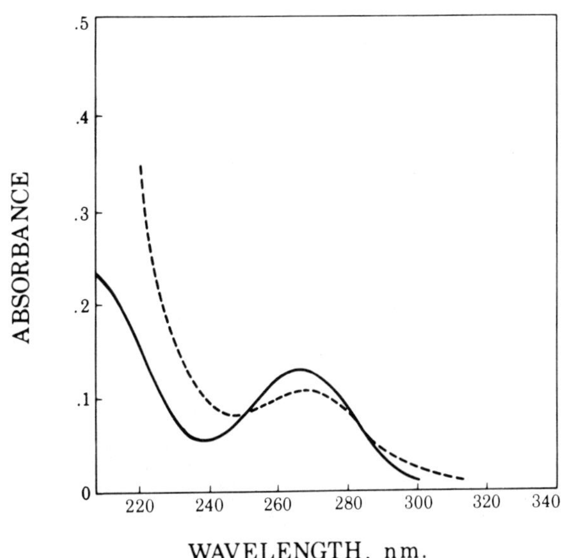


O
HN
CH3
HOOC
O
N
O
H3CHC=
N–C
O=CHNCH
O
H2NCH
HCOH
OH OH
HOCH
H2COCNH2
O

| No. | Name | Structure | Formula (mol wt) | Melting point °C | $[\alpha]_D$ | Chromatography |
|---|---|---|---|---|---|---|
| 44 | **Polyoxin I** (*246, 355*) | 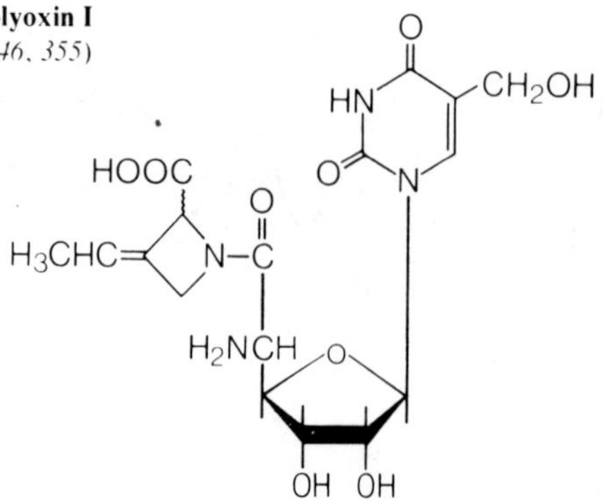 | $C_{17}H_{22}N_4O_9$ (426.4) (*246*) | — | −25° (c = 1, $H_2O$) (*246*) | TLC (*243*), Paper (*243*), Column (*243*) |

| pK | Purification | Source | Solubility | Toxicity | Miscellaneous |
|---|---|---|---|---|---|
| 2.7, 6.1, 9.5 (*246*) | Aqueous EtOH (*243*) | *Streptomyces cacaoi* var. *asoensis* (*243*) | Soluble in $H_2O$, MeOH and EtOH. | Mice tolerated 500 mg/kg (i.v.). No toxicity against mice or killifish (up to 4 p.p.m.) (10 mg/20 g, i.v.). | uv (*243*), ir (*243*), Paper electrophoresis (*243*) |

| Biological activity |
|---|

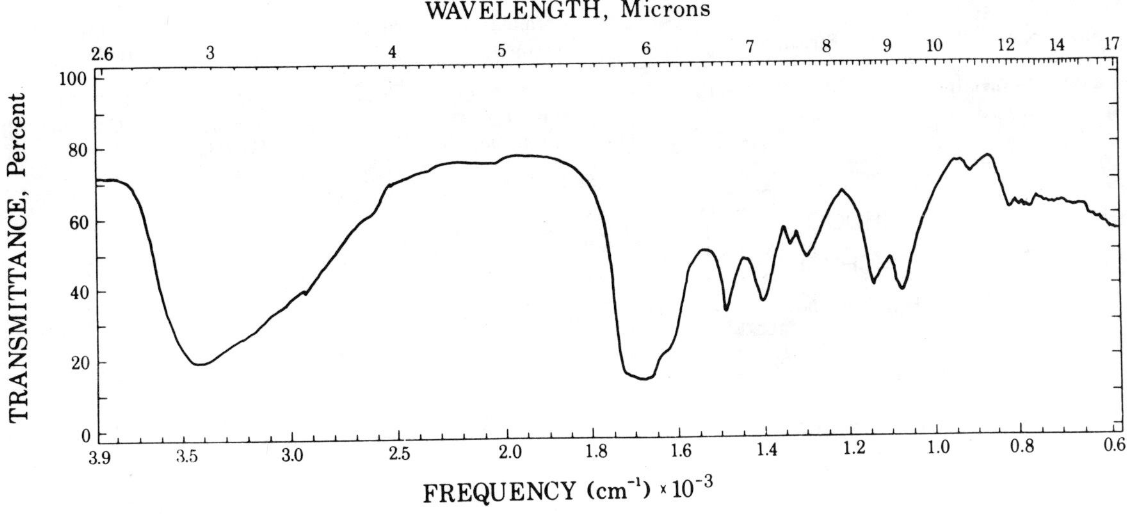
WAVELENGTH, Microns
2.6 3 4 5 6 7 8 9 10 12 14 17
TRANSMITTANCE, Percent
100 80 60 40 20 0
3.9 3.5 3.0 2.5 2.0 1.8 1.6 1.4 1.2 1.0 0.8 0.6
FREQUENCY (cm⁻¹) × 10⁻³

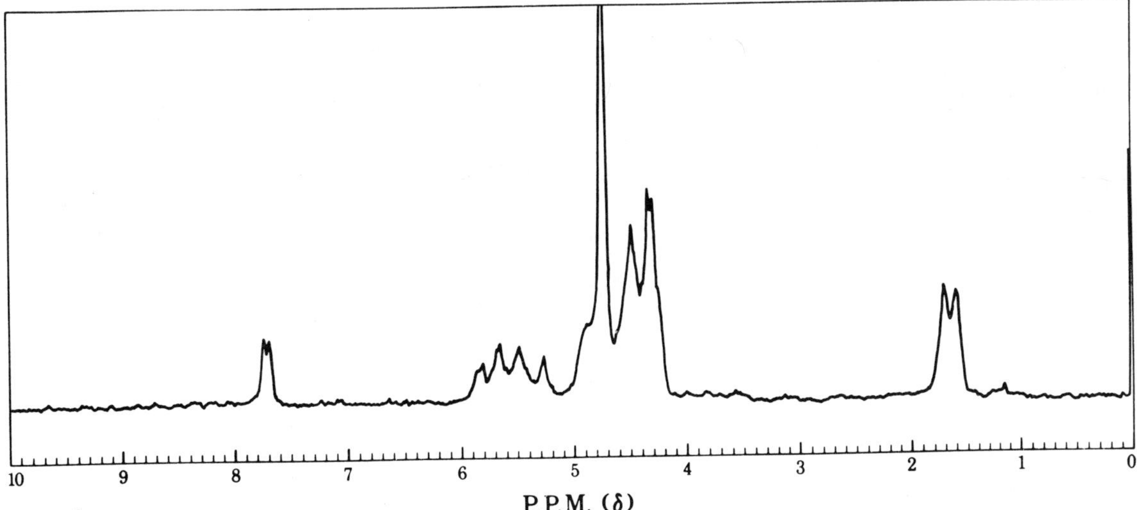
10 9 8 7 6 5 4 3 2 1 0
P.P.M. (δ)

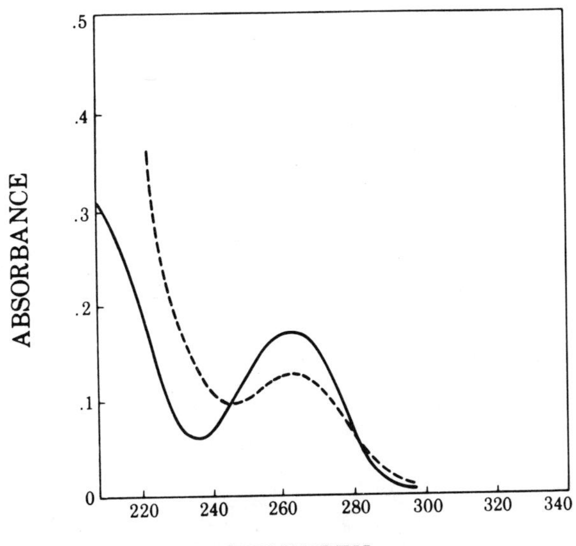
ABSORBANCE
.5 .4 .3 .2 .1 0
220 240 260 280 300 320 340
WAVELENGTH, nm.

| No. | Name | Structure | Formula (mol wt) | Melting point °C | $[\alpha]_D^{20}$ | Chromatog- raphy |
|---|---|---|---|---|---|---|
| 45 | **Polyoxin J** (*246, 355*) | 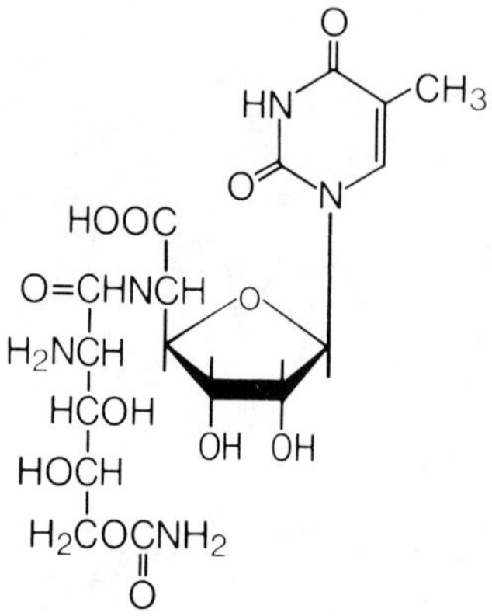 | $C_{17}H_{25}N_5O_{12}$ (491.41) (*244, 246*) | — | +32.0 (c = 1, $H_2O$) (*246*) | TLC (*244*), Column (*244*) |

| pK | Purification | Source | Solubility | Toxicity | Miscellaneous |
|---|---|---|---|---|---|
| 3.0, 7.1, 9.9 (*244, 246*) | Aqueous EtOH | *Streptomyces cacaoi* var. *asoensis* (*244*) | Soluble in $H_2O$, MeOH and EtOH. | Mice tolerated 500 mg/kg (i.v.). | Paper electro- phoresis (*244*), ir (*244*), uv (*244*) |

Biological activity

Tested as antifungal and antimicrobial agents against *Piricularia oryzae*, *Cochliobolus miyabianus*, *Pellicularia sasakii*, *Alternia kikuchiana*, *Sclerotinia cinerera*, *Cladospotium fulvum*, *Corticium rolfsii* and *Guignardia laricina* (*244*).

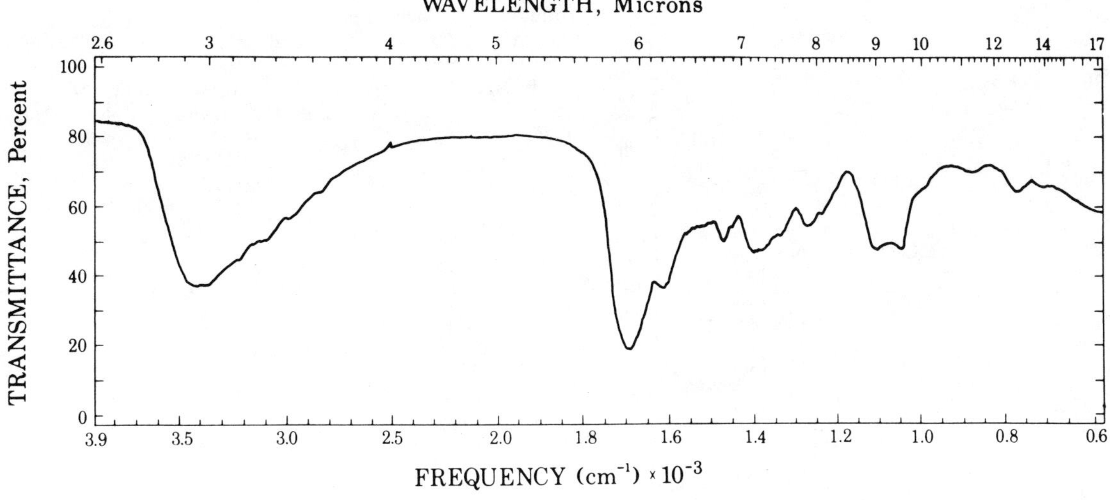

WAVELENGTH, Microns
2.6
3
4
5
6
7
8
9
10
12
14
17
TRANSMITTANCE, Percent
100
80
60
40
20
0
3.9
3.5
3.0
2.5
2.0
1.8
1.6
1.4
1.2
1.0
0.8
0.6
FREQUENCY (cm⁻¹) × 10⁻³

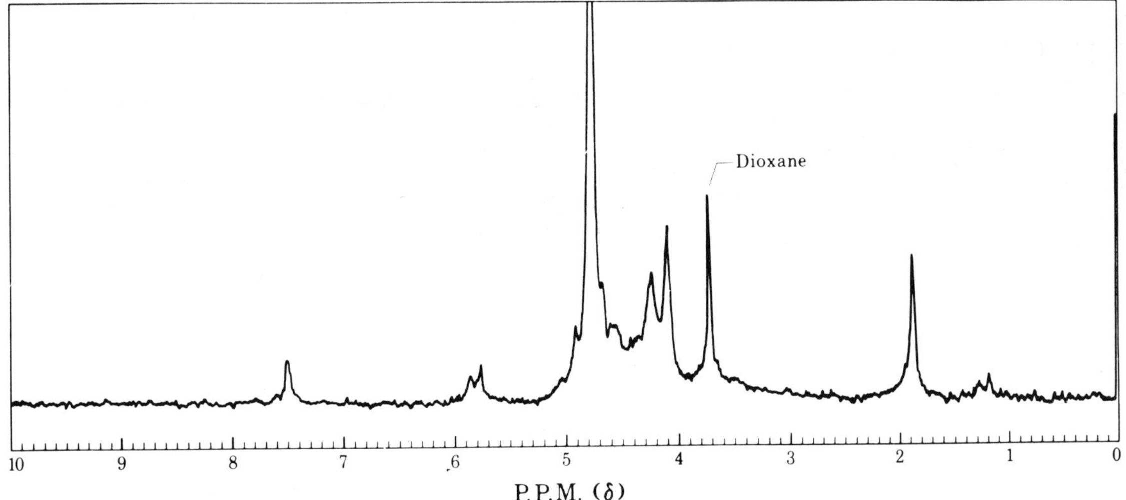

Dioxane
10
9
8
7
6
5
4
3
2
1
0
P.P.M. (δ)

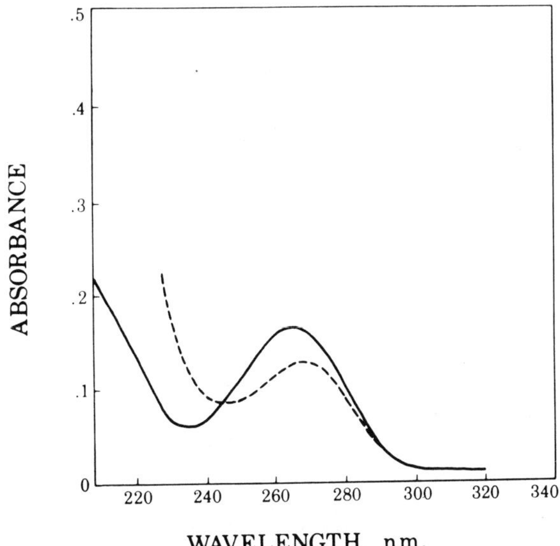

ABSORBANCE
.5
.4
.3
.2
.1
0
220
240
260
280
300
320
340
WAVELENGTH, nm.

O
HN
CH3
O
N
HOOC
O=CHNCH
O
H2NCH
HCOH
OH
OH
HOCH
H2COCNH2
O

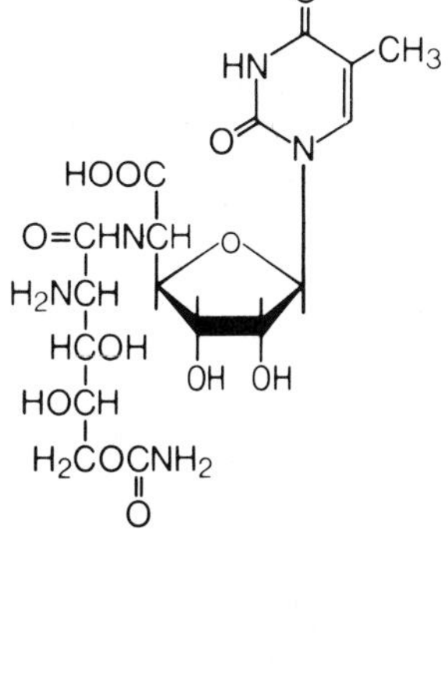

| No. | Name | Structure | Formula (mol wt) | Melting point °C | $[\alpha]_D$ | Chromatography |
|---|---|---|---|---|---|---|
| 46 | **Polyoxin K** (*246, 355*) | 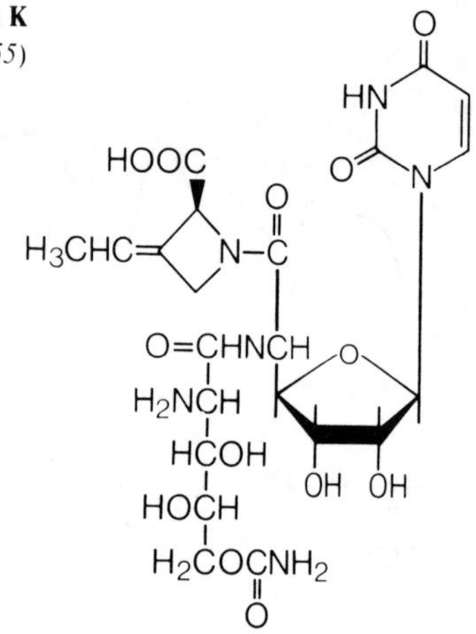 HOOC; $H_3CHC=$ N–C(=O); O=CHNCH; $H_2NCH$; HCOH; HOCH; $H_2COCNH_2$ (=O); HN; O; N; OH OH | $C_{22}H_{30}N_6O_{13}$ (586.53) (*246*) | — | −17° (c = 1, $H_2O$) (*246*), −16.5 (c = 1, $H_2O$) (*244*) | TLC (*244*), Column (*244*) |

| pK | Purification | Source | Solubility | Toxicity | Miscellaneous |
|---|---|---|---|---|---|
| 3.0, 7.2, 9.3 (*246*) | Recrystallized from aqueous EtOH. | *Streptomyces cacaoi* var. *asoensis* (*244*) | Soluble in $H_2O$ and MeOH. | — | ir, paper electrophoresis (*244*) |

Biological activity

Tested as antifungal and antimicrobial agents against *Piricularia oryzae*, *Cochliobolus miyabianus*, *Pellicularia sasakii*, *Alternia kikuchiana*, *Sclerotinia cinerera*, *Cladospotium fulvum*, *Corticium rolfsii* and *Guignardia laricina* (*244*).

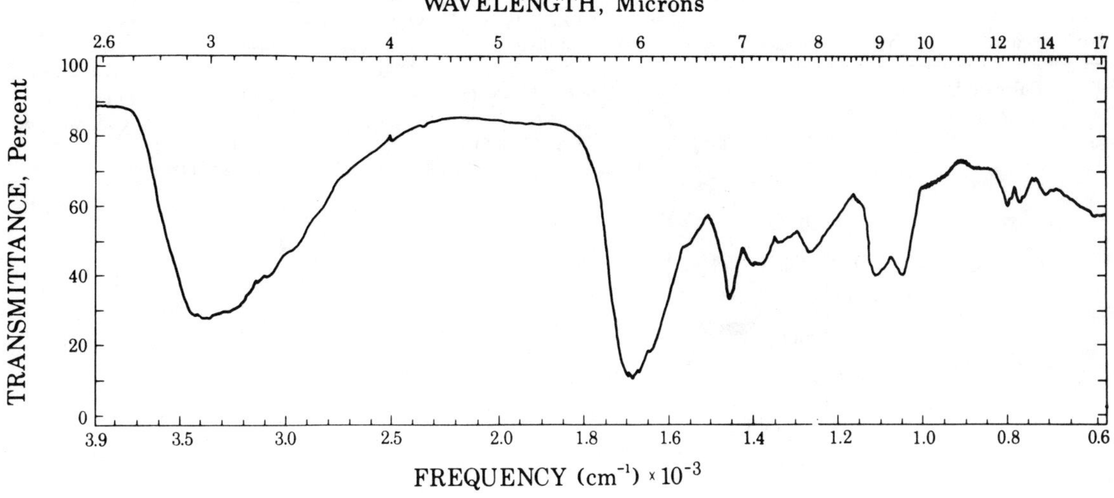


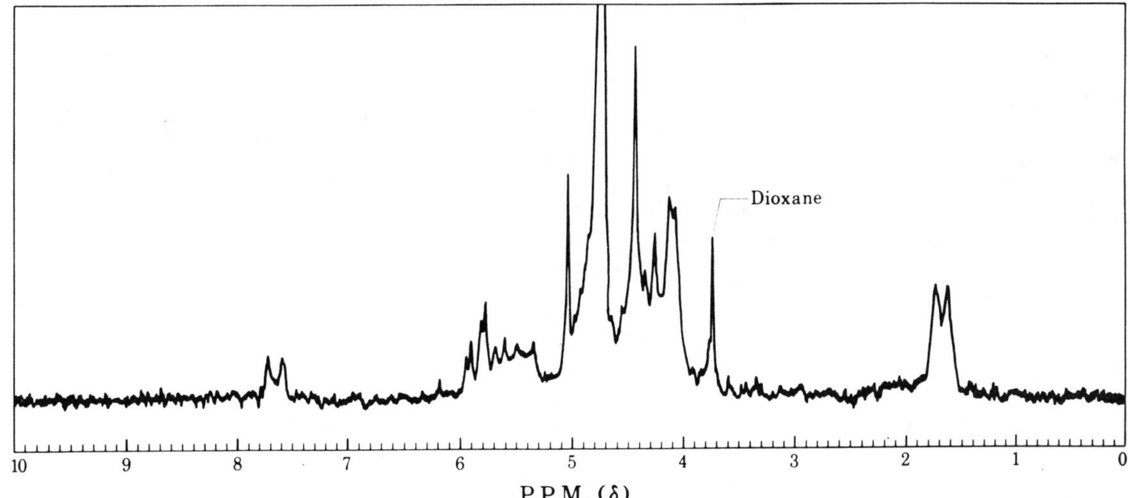


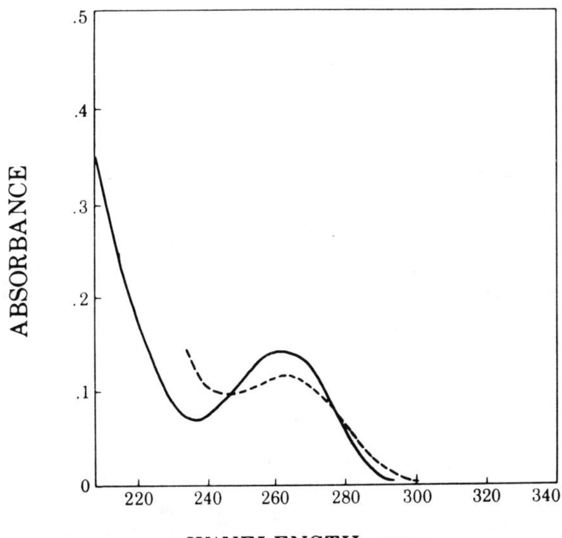


O
HN
O
N
HOOC
O
$H_3CHC$
N—C
O=CHNCH
O
$H_2NCH$
HCOH
OH OH
HOCH
$H_2COCNH_2$
O

| No. | Name | Structure | Formula (mol wt) | Melting point °C | $[\alpha]_D^{20}$ | Chromatography |
|---|---|---|---|---|---|---|
| 47 | **Polyoxin L** (*246, 355*) | 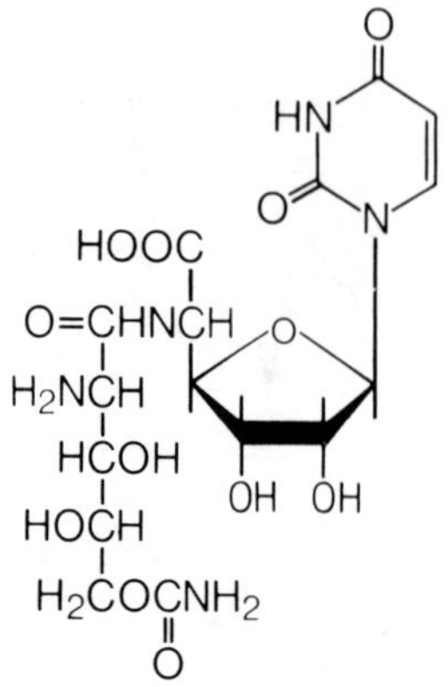 HOOC–CH(NHCHO)... (structure: uracil nucleoside; O=CHNCH, $H_2NCH$, HCOH, HOCH, $H_2COCNH_2$ (=O), OH OH) | $C_{16}H_{23}N_5O_{12}$ (477.40) (*244, 246*) | — | +34° (c = 1, $H_2O$) (*246*), +34.4 (c = 1, $H_2O$) (*244*) | TLC (*244*), Column (*244*) |

| pK | Purification | Source | Solubility | Toxicity | Miscellaneous |
|---|---|---|---|---|---|
| 3.0, 7.1, 9.4 (*246*) | Recrystallized from aqueous alcohol. | *Streptomyces cacaoi* var. *asoensis* (*244*) | Soluble in $H_2O$. | — | ir (*244*), Paper electrophoresis (*244*) |

| Biological activity |
|---|
| Tested as antifungal and antimicrobial agents against *Piricularia oryzae*, *Cochliobolus miyabianus*, *Pellicularia sasakii*, *Alternia kikuchiana*, *Sclerotinia cinerera*, *Cladospotium fulvum*, *Corticium rolfsii*, and *Guignardia laricina* (*244*). |

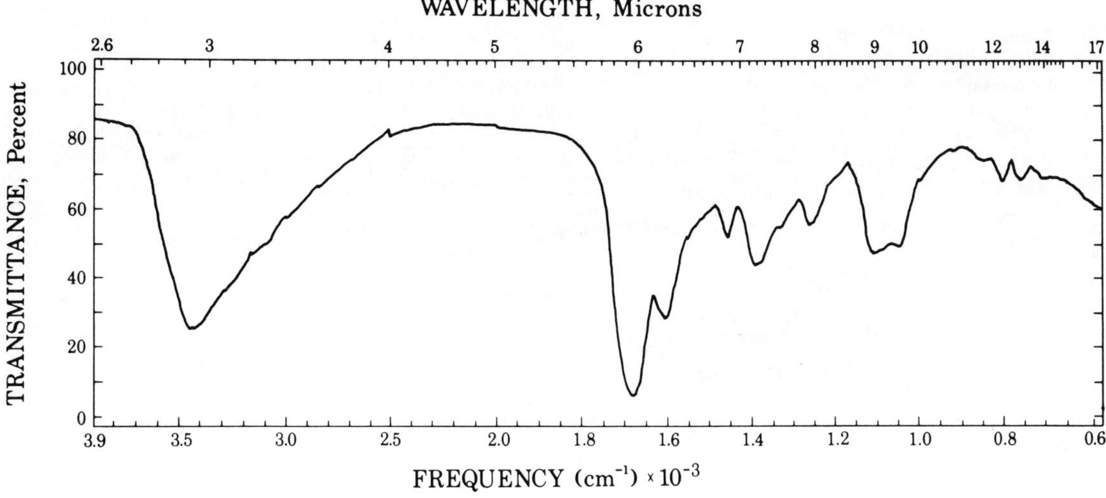


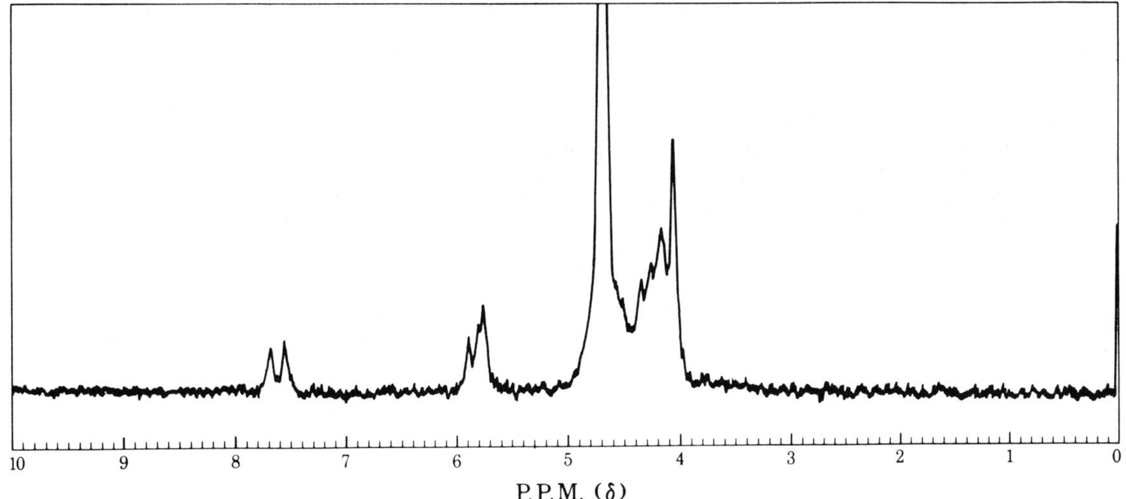


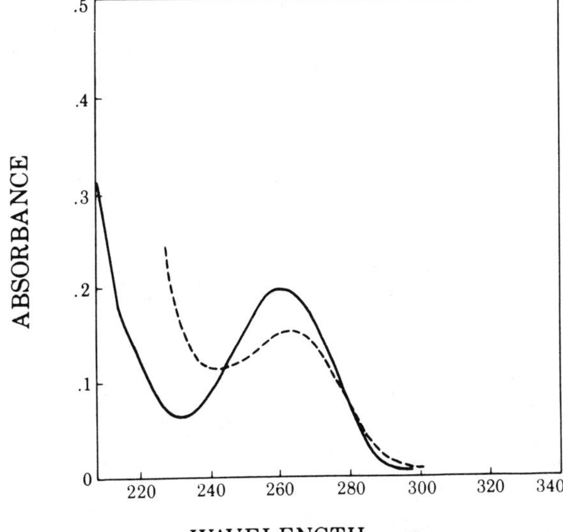


O
HN
O
N
HOOC
O=CHNCH
O
$H_2NCH$
HCOH
OH OH
HOCH
$H_2COCNH_2$
O

| No. | Name | Structure | Formula (mol wt) | Melting point °C | $[\alpha]_D^{25}$ | Chromatography |
|---|---|---|---|---|---|---|
| 48 | **Psicofuranine** (Antibiotic U-9586) (*51*) | 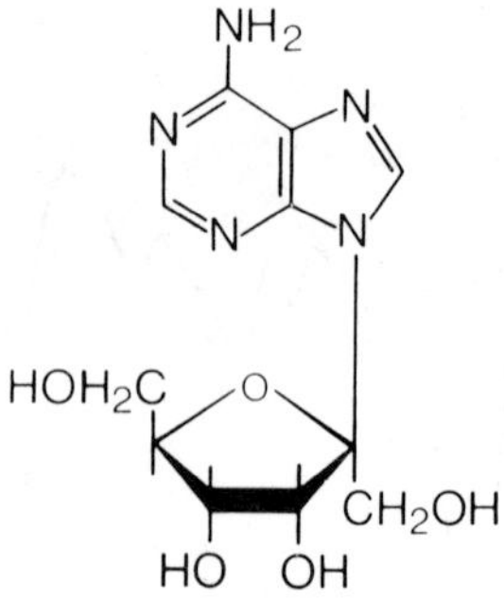 NH2; N; N; N; N; HOH2C; O; CH2OH; HO; OH | $C_{11}H_{15}N_5O_5$ (297.3) | 212–214 dec. | −53.7 (c = 1, DMSO), −68 (c = 1, DMF) | Paper, Column |

| pK | Purification | Source | Solubility | Toxicity | Miscellaneous |
|---|---|---|---|---|---|
| — | Adsorbed on carbon from the filtrate, eluted with $Me_2CO$ and then purified by counter-current distribution. | *Streptomyces hygroscopius* var. *decoyicus* (*249, 250*) | Soluble in DMF, DMSO, $H_2O$, MeOH, EtOH, BuOH and EtOAc. | Inhibition of bacterial systems at 1 to 25 grams per ml. $LD_{50}$ 6.1 to 41 mg/kg (s.c.), 13–68 mg/kg (i.p.) (mice). 10,000 mg/kg (orally) (rats) | uv (*51*), Chemical synthesis (*51*) |

| Biological activity |
|---|
| Active *in vitro* against *E. coli, Salmonella pullorum, Micrococcus pyogenes var. aureus, Micrococcus pyogenes var. albus, Streptococcus haemolyticus, Proteus vulgaris, Pasteurella multocida, Salmonella typhi, Streptococcus faecalis* and *Pseudomonas aeruginosa*. Active *in vivo* (s.c. or orally) against *Micrococcus pyrogenes* var. *aureus, Streptococcus haemolyticus* and *E. coli* in mice (*251*). Active against experimental tumors, Murphy-Sturm lymphosarcoma, Jensen's sarcoma and adenocarcinoma *in vivo* (252). Inhibition of XMP aminase (*253–255*) and PRPP synthesis (*171*). Inhibits PRPP amido transferase activity in intact tumor cells (*175*) and does not act as a substrate for adenosine deaminase (*171, 82, 37*). |

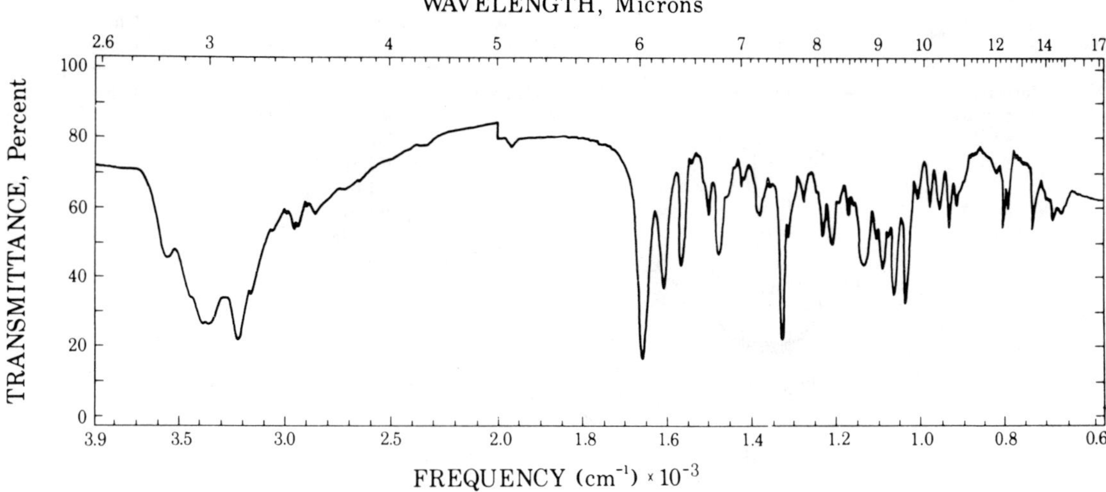


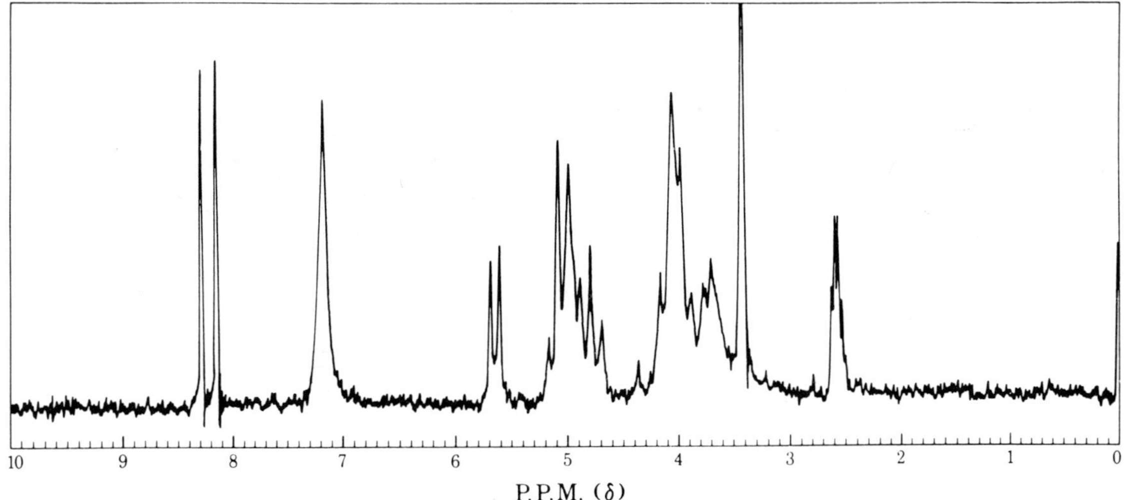


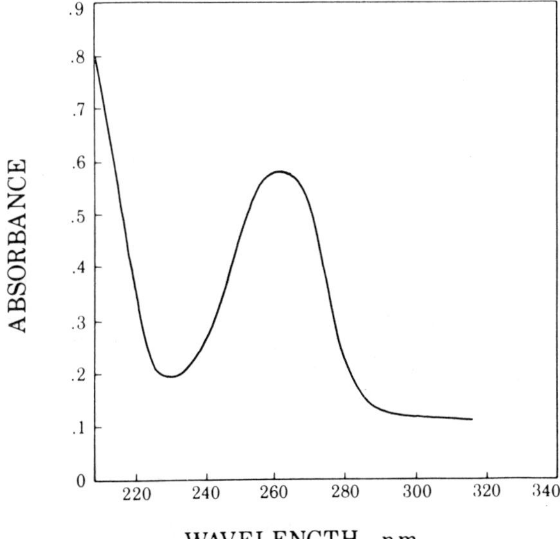


$NH_2$
N
N
N
N
$HOH_2C$
O
$CH_2OH$
HO
OH

| No. | Name | Structure | Formula (mol wt) | Melting point °C | $[\alpha]_D^{25}$ | Chromatography |
|---|---|---|---|---|---|---|
| 49 | **Puromycin (Achromycin)** (*256–260*) | 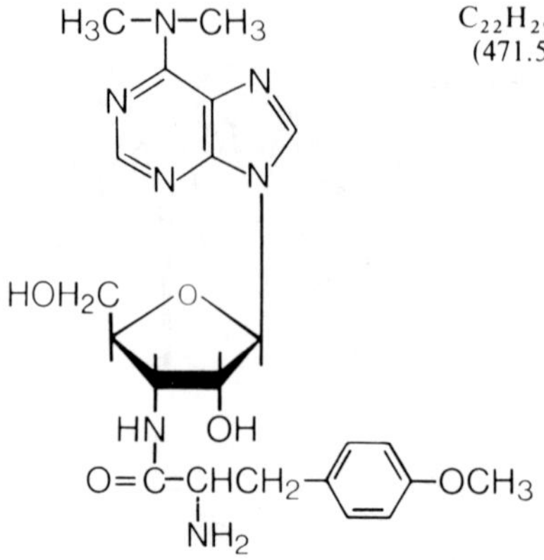 $H_3C-N-CH_3$; N; N; N; N; $HOH_2C$; O; HN; OH; $O=C-CHCH_2-$; $-OCH_3$; $NH_2$ | $C_{22}H_{29}N_7O_5$ (471.52) | 175.5–177 (*256*), 180–187 dec. (sulfate), 146–149 (picrate), 217.5–218 (triacetate) | −11° (EtOH) | Paper |

| pK | Purification | Source | Solubility | Toxicity | Miscellaneous |
|---|---|---|---|---|---|
| 6.8, 7.2 | Recrystallized from $H_2O$. | *Streptomyces alboniger* (*250*) | Sparingly soluble in $H_2O$ and organic solvents. | $LD_{50}$ 335 mg/kg (i.v.) 580 mg/kg (i.p.) (mice) 720 mg/kg (i.p.) and 600 mg/kg (oral) (Guinea pig) (*272*). Maximum tolerated dose in humans 12.5 mg/kg (*276*) | uv (*256, 257*) and neut. equiv (*256, 257*). A report of two possible precursors of puromycin which have been isolated from a commercial sample (*280*). Chemical synthesis (*258*). |

Biological activity

Inhibition of *E. coli* (*270*) and protein synthesis by inhibiting the transfer of leucine from tRNA-leucine into the protein molecule. Active against *Trypanosoma equiperdum*, *Trypanosoma equinum*, *Trypanosoma evansi*, *Trypanosoma rhodosiense*, *Trypanosoma gambiense* and *Trypanosoma congolense* (*273*). Active against *Aspicularis tetraptera* (pinworms), *Syphacia obvelata* and the tapeworm *Hymenolepis nana* var. *fracterna* in mice (*274*). Active (*in vivo*) against *Endamoeba histolytica* in experimental infections (*275*). Significant activity in the treatment of amoebiasis in humans (*276*). Active against sleeping sickness caused by *Trypanosoma gambiense* in humans (*277*). Termination of peptide chain propagation (*278, 279*). Antitumor activity (*281*) and inhibition of poliovirus replication in tissue culture (*282*). Inhibition of replication of influenza virus in tissue culture was effected without destroying all metabolic activity and probably acts by interference with protein synthesis (*283*). Inhibition of cerebral protein synthesis and expression of memory in mice (*284*). It was postulated that swelling of neuronal mitochondria is related to a specific action of puromycin on ribosomal protein synthesis (*284*).

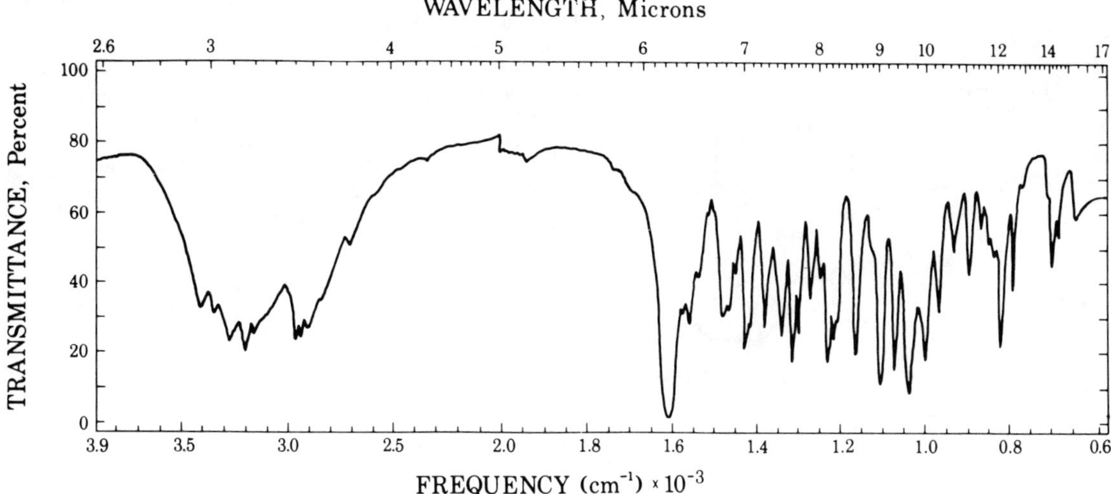


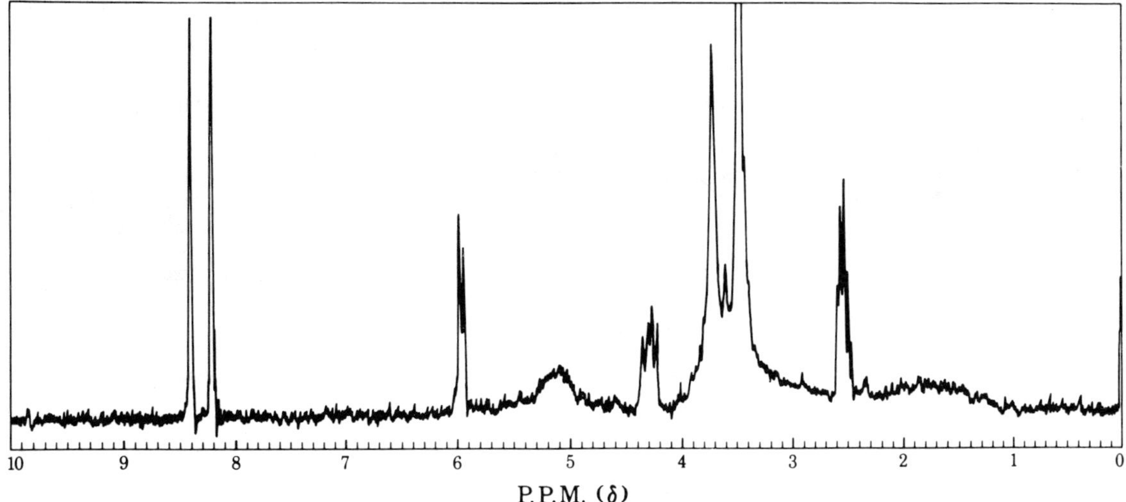


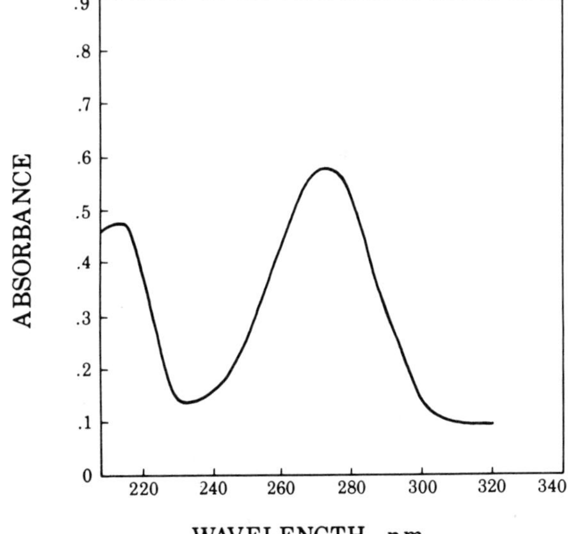


H3C–N–CH3
N
N
N
N
HOH2C
O
HN
OH
O=C–CHCH2–
–OCH3
NH2

| No. | Name | Structure | Formula (mol wt) | Melting point °C | $[\alpha]_D^{20}$ | Chromatography |
|---|---|---|---|---|---|---|
| 50 | **Pyrazomycin** (*285, 287–289*) | 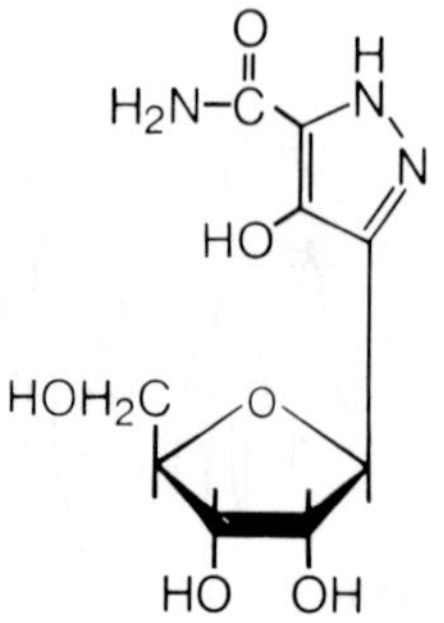 | $C_9H_{13}N_3O_6$ (259.2) (*285, 287*) | 108–113 ($H_2O$) (*285, 287–289*) | −47° (c = 1.25, $H_2O$) (*285, 287, 289*) | Column |

| pK | Purification | Source | Solubility | Toxicity | Miscellaneous |
|---|---|---|---|---|---|
| 6.7 (*285, 287, 289*) | Carbon adsorption and column chromatography (*287–289*). Recrystallized from $H_2O$ (*285, 289*). | *Streptomyces candidus* (*285–289*) | Soluble in $H_2O$, MeOH and EtOH (*287*). | — | uv (*285, 287, 289*), ir (*285, 287, 289*), pmr (*285, 287–289*), mass spectroscopy (*285, 287–289*) |

| Biological activity |
|---|
| Active against vaccinia virus, rhino virus, measles virus, herpes simplex virus, coxsackie B-5 virus (*285*) but was inactive against polio III virus and pseudorabies virus (*285*). Pyrazomycin was found to be a substrate for a nucleoside phosphotransferase and this derivative was a substrate for a 5′-nucleotidase (*285, 288, 289*). The 5′-phosphate derivative also inhibits orotidylic acid decarboxylase (*285*). |

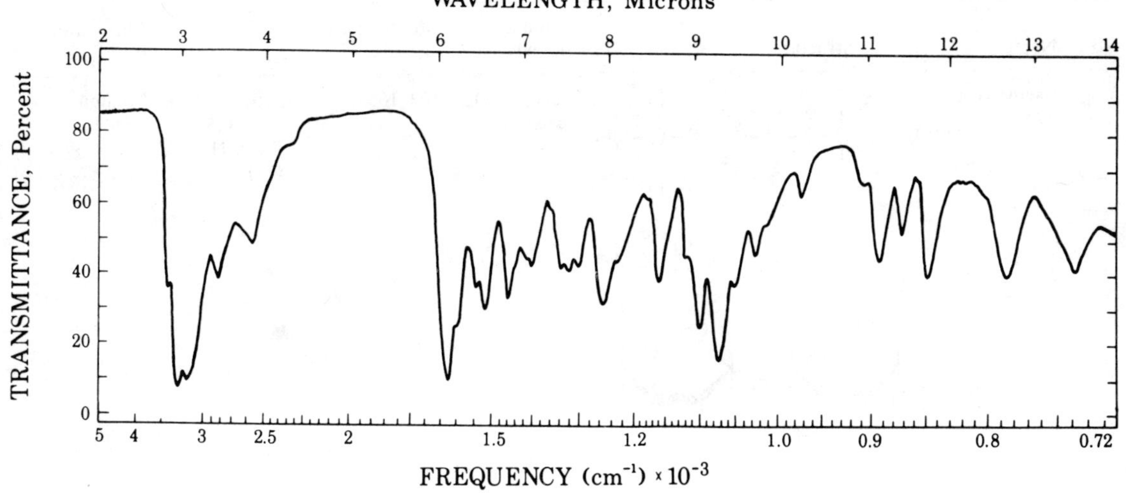
WAVELENGTH, Microns
TRANSMITTANCE, Percent
FREQUENCY (cm⁻¹) × 10⁻³

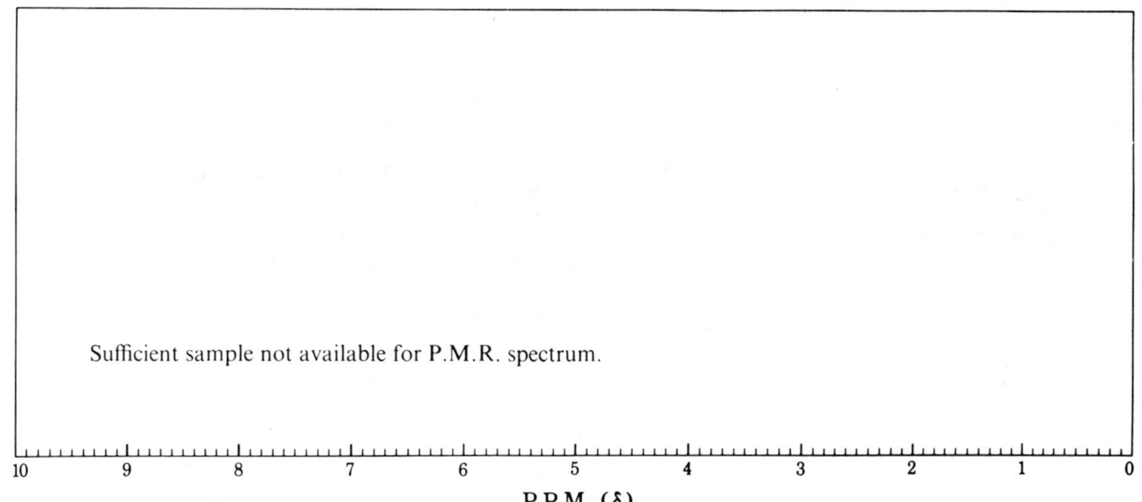
Sufficient sample not available for P.M.R. spectrum.
P.P.M. (δ)

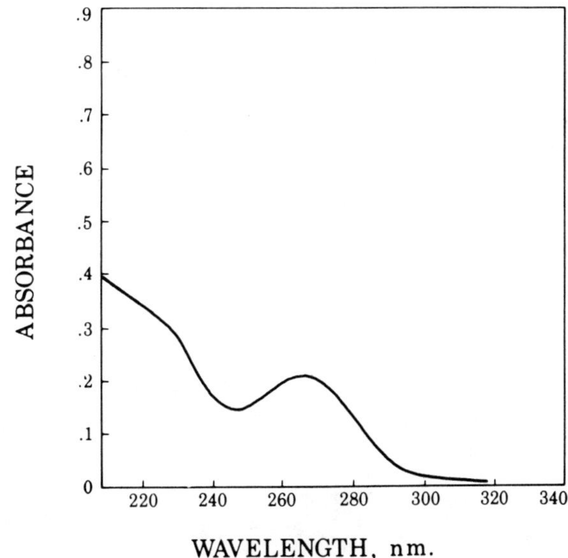
ABSORBANCE
WAVELENGTH, nm.

| No. | Name | Structure | Formula (mol wt) | Melting point °C | $[\alpha]_D^{16}$ | Chromatography |
|---|---|---|---|---|---|---|
| 51 | **Sacromycin** (*13*) | 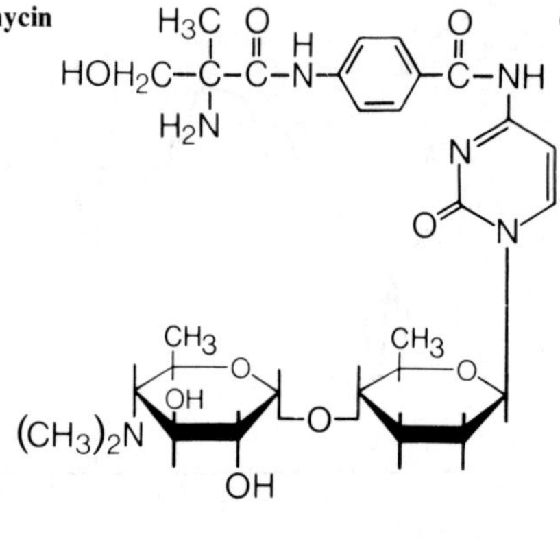 | $C_{29}H_{42}N_6O_9$ (618.69) | 163–165 | $+107.7 \pm 4$ (c = 0.5, 0.1 *N* HCl) (*13*) | Column |

| pK | Purification | Source | Solubility | Toxicity | Miscellaneous |
|---|---|---|---|---|---|
| — | Purified by column (IRC-50) chromatography, eluted, extracted with the HCl salt formed (*13*). Recrystallized from $H_2O$ (*13*). | *Streptomyces sp.* 5223 (*13, 290, 291*) | Soluble in MeOH, EtOH and dilute acid (*13*). Slightly soluble in $H_2O$, BuOH and amyl alcohol. Insoluble in other non-polar solvents (*13*). | An extensive study on toxicity and acute toxicity toward mice by i.v., i.p., i.m. and s.c. was conducted and it was found to have a relatively low toxicity (*13*). | uv (*13*), ir (*13*) |

| Biological activity |
|---|
| Strong inhibition of mycobacteria and gram-positive bacteria (*13, 290, 291*) with complete inhibition of *Staph. aureus* 209P (*13*). Active against human tubercle bacilli (*13*). |

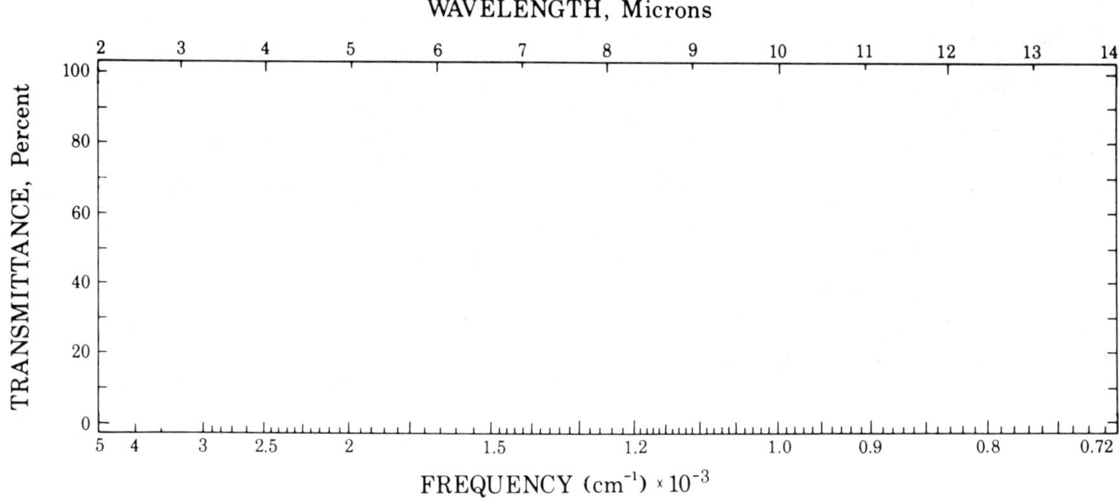
WAVELENGTH, Microns
2
3
4
5
6
7
8
9
10
11
12
13
14
TRANSMITTANCE, Percent
100
80
60
40
20
0
5
4
3
2.5
2
1.5
1.2
1.0
0.9
0.8
0.72
FREQUENCY (cm⁻¹) x 10⁻³

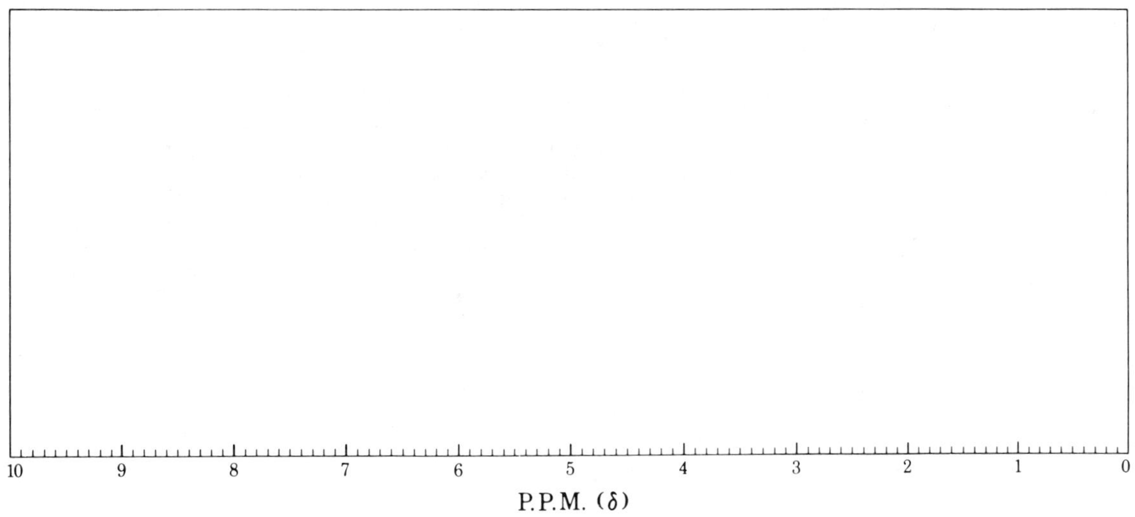
10
9
8
7
6
5
4
3
2
1
0
P.P.M. (δ)

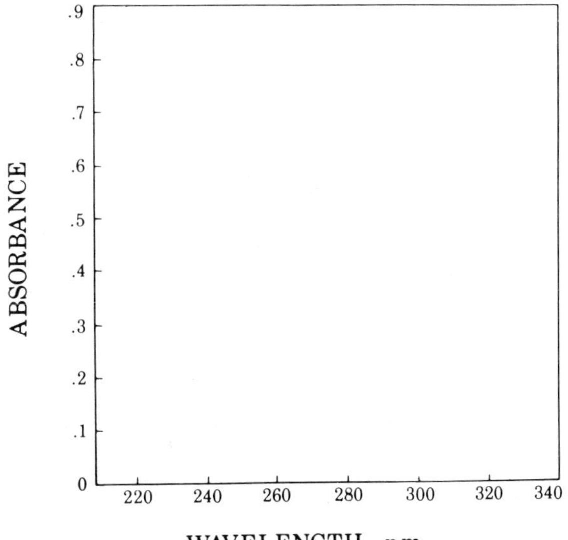
ABSORBANCE
.9
.8
.7
.6
.5
.4
.3
.2
.1
0
220
240
260
280
300
320
340
WAVELENGTH, nm.

| No. | Name | Structure | Formula (mol wt) | Melting point °C | $[\alpha]_D^{26}$ | Chromatography |
|---|---|---|---|---|---|---|
| 52 | **Sangivamycin** (*63, 64*) | 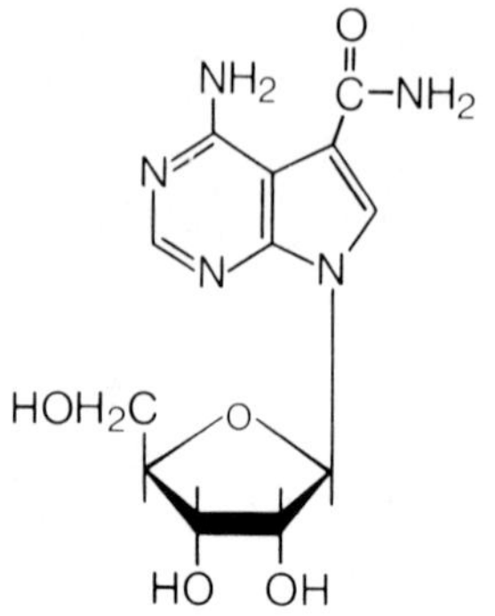 | $C_{12}H_{15}N_5O_5$ (309.3) (*292*) | 260 (*64, 292*) 238 dec. (HCl) (*64, 292*) | −45.7 ± 1.9 (c = 1, 0.1 *N* HCl), −42.2 ± 1.9 (c = 1, 0.1 *N* HCl) (*64, 292*) | TLC (*64, 292*) |

| pK | Purification | Source | Solubility | Toxicity | Miscellaneous |
|---|---|---|---|---|---|
| — | Recrystallized from $H_2O$. | Unidentified species of *Streptomyces* (*292*). | Soluble in $H_2O$, MeOH and dilute acids. Sparingly soluble in EtOH, $Me_2CO$ and BuOH. Insoluble in $CHCl_3$, $CCl_4$, $C_6H_5$ and $Et_2O$. | Initial toxicity study on humans with various types of tumor by i.v. injection (*293*). | ir, pmr, uv (*64, 292*). uv of HCl salt (*64, 292*) and was converted to tubercidin (*64, 292*). Chemical synthesis (*63, 64*). |

| Biological activity |
|---|
| Active against leukemia L-1210 in mice and cytotoxic to HeLa cells in culture (*292*). The nucleotide is incorporated enzymatically in the terminal position of tRNA and then inhibits normal function of tRNA in regards to esterification of amino acids (*294*). Incorporation into RNA in the reaction catalyzed by RNA polymerase from *Micrococcus lysodeikticus* with the use of calf thymus DNA or the copolymer of deoxyadenosine or deoxythymidine [poly d(AT)] as primers (*295*). A study with the nucleotide reductase from *Lactobacillus leichmannii* (*296*) and *E. coli* (*297*) has been reported. |

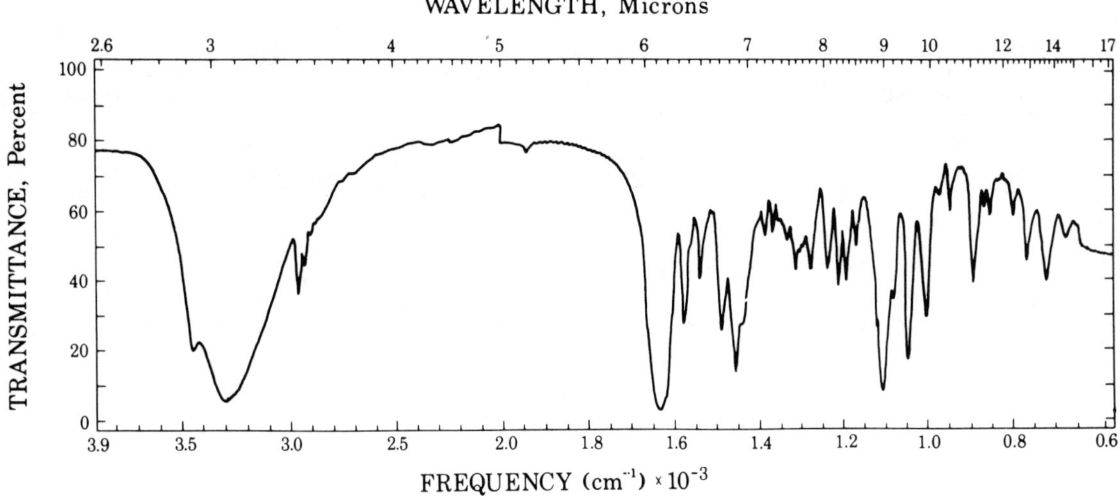


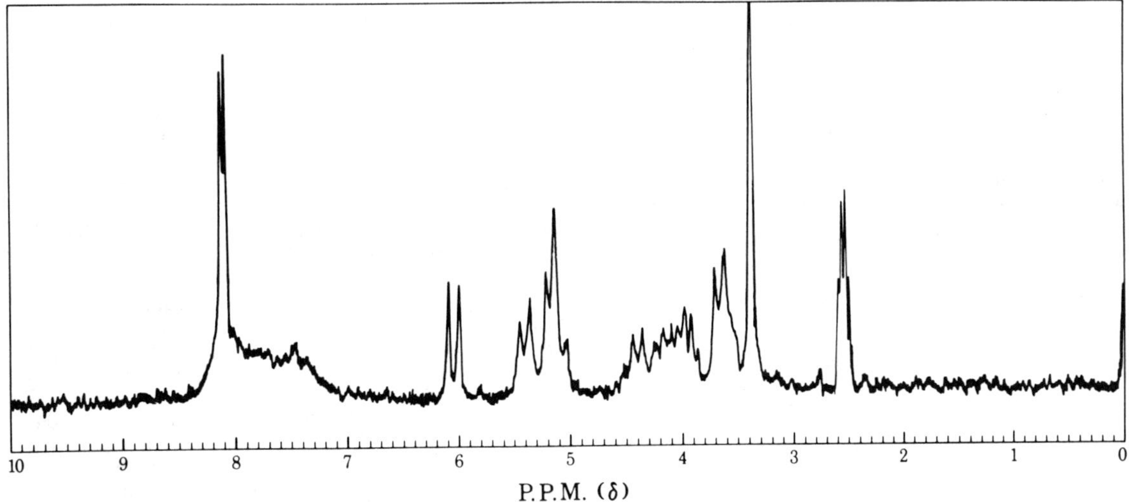


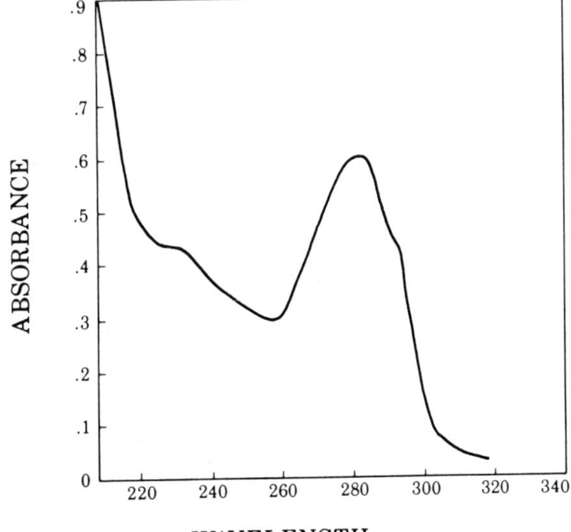


O
$NH_2$ $C-NH_2$
N
N N
$HOH_2C$ O
HO OH

| No. | Name | Structure | Formula (mol wt) | Melting point °C | $[\alpha]_D^{22.5}$ | Chromatography |
|---|---|---|---|---|---|---|
| 53 | **Showdomycin (MSD-125A)** (*308, 298, 302, 304*) | 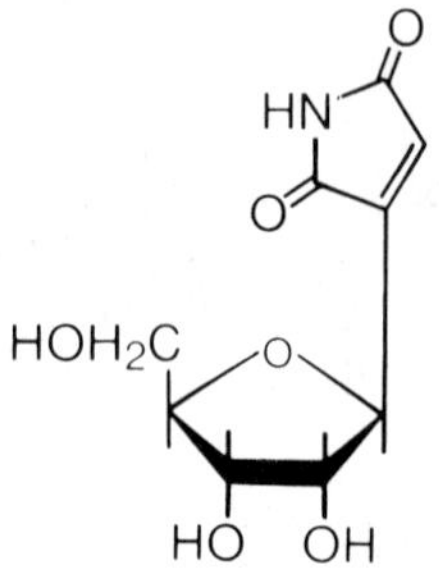 HN, O, O, $HOH_2C$, O, HO, OH | $C_9H_{11}NO_6$ (229.2) (*298, 302, 304*) | 153–154 (*299, 300*), 160–161, 115–116 (triacetate), 140.5–141 (acetonide) (*302*) | +49.9 (c = 1, $H_2O$) (*298–300*), | TLC (*298*), Paper (*303*), Column (*300*) |

| pK | Purification | Source | Solubility | Toxicity | Miscellaneous |
|---|---|---|---|---|---|
| 9.29 (*298*), 9.60 (*302*) | Adsorbed on carbon, extracted, column chromatography (silica gel) and recrystallized from a warm $Me_2CO/C_6H_5$ mixture (*300*). | *Streptomyces showdoensis Nov. sp.* (*299, 300*) | Soluble in $H_2O$ and other polar solvents (*300, 302*). Insoluble in nonpolar solvents (e.g., $Et_2O$, $C_6H_5$ and petroleum ether (*300, 305*). | $LD_{50}$ 25 mg/kg (i.p.), 18 mg/kg (s.c.) and 110 mg/kg (i.v.) in mice (*299*) | uv (*298, 299*) (222 nm, aqueous sol.) (*299*), (222 nm, 95% EtOH) (*302*). Initial absorption at 328 nm in dilute aqueous ammonia followed by a rapid (*298*) loss of uv absorption, ir and pmr (*298*). Hydrogenation furnished the dihydro deriv. (*298*). Resistant toward hydrolysis with conc. HCl. (*302*), X-ray (*305, 302, 306*) and mass spectrometry (*194*). |

Biological activity

Active against *Streptococcus hemolyticus* (*299*) and *Streptococcus pyogenes* (*300*). Antitumor activity against Ehrlich mouse ascites tumor cells *in vivo* and against cultured HeLa cells (*299–301*). It was not a substrate for nucleoside kinase or uridine phosphorylase (*303*). Inhibits UMP kinase and uridine phosphorylase (*303*) with a possible inhibition of orotidylic acid pyrophosphorylase (*303*). The major site of action appears to be the alkylating effect on the sulfhydryl groups of an enzyme (*303, 309*). Inhibition of *E. coli* was reported as well as a study which established that certain ribonucleosides and L-cysteine would reverse this inhibition while ribonucleotides purine and pyrimidine bases were ineffective (*307*). Showdomycin was found to be ineffective in the inhibition of a purine β-ribosidase (*185*). The mode of action of showdomycin on mitochrondrial volume changes appears to be dependent on exposure of thiol groups (*310*).

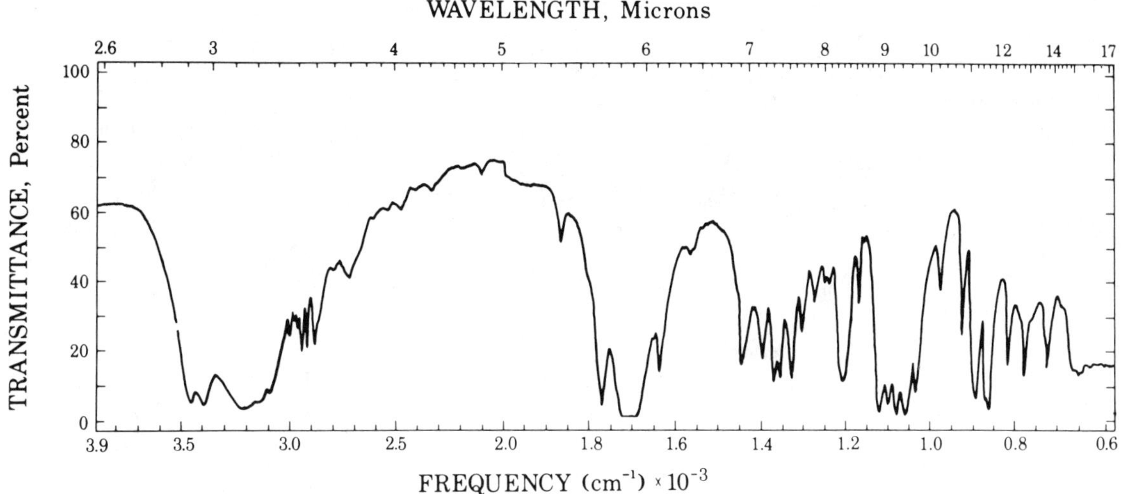


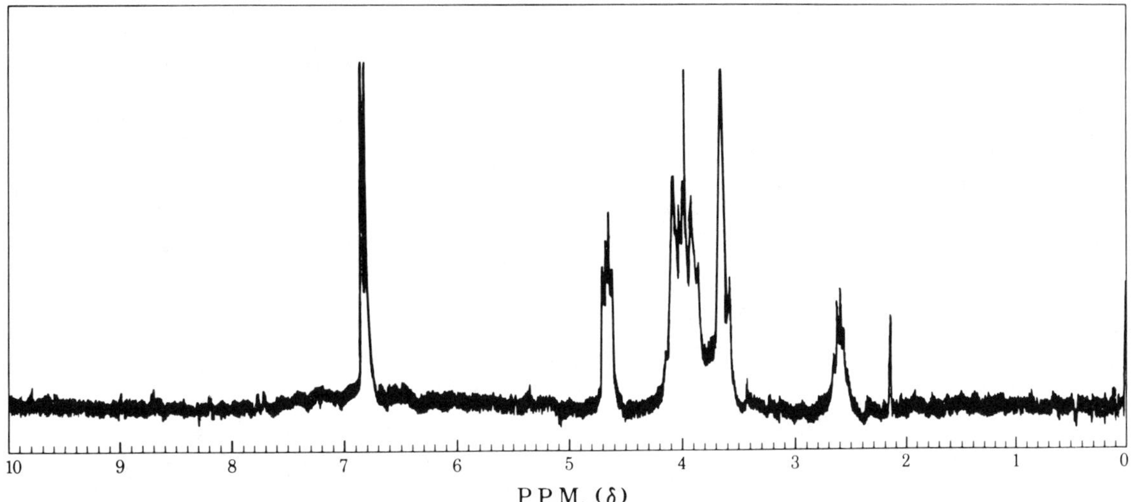


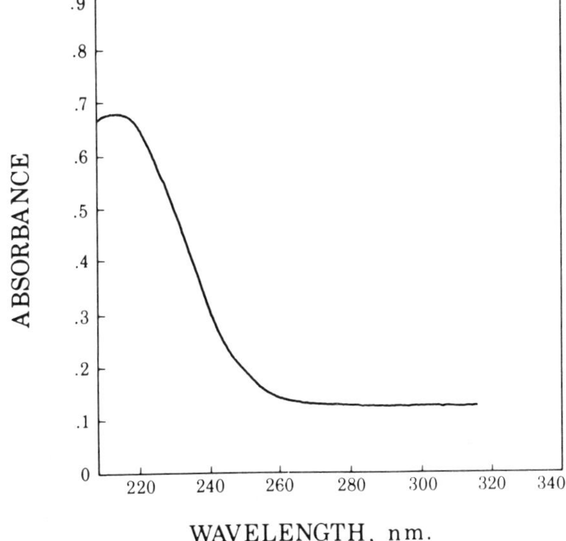


O
HN
O
$HOH_2C$
O
HO OH

| No. | Name | Structure | Formula (mol wt) | Melting point °C | $[\alpha]_D^{25}$ | Chromatography |
|---|---|---|---|---|---|---|
| 54 | **Sparsomycin A** (*313, 314*) | 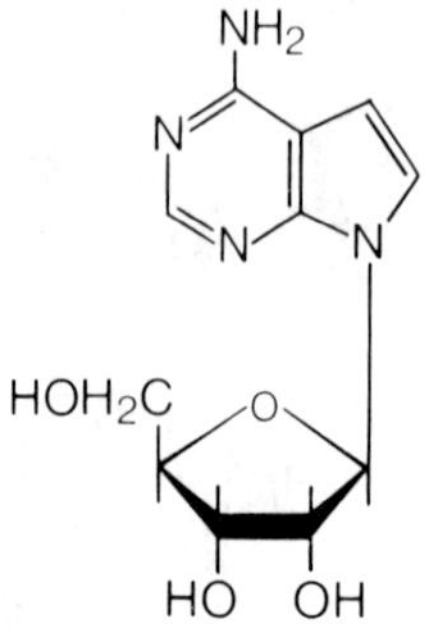 NH2, N, N, N, HOH2C, O, HO, OH | $C_{11}H_{14}N_4O_4$ (266.25) | 247.8–250 (*317*) | −62 (c = 0.718, 0.1 *N* HCl) (*317*) | Partition, Column (*317*) |

| pK | Purification | Source | Solubility | Toxicity | Miscellaneous |
|---|---|---|---|---|---|
| 5–5.2, 5.0 (*317*) | Adsorption on active carbon, eluted, then counter-current distribution (*317, 315*). | *S. sparsogenes* var. *sparsogenes* NRRL 2940 (*313–315, 317*) | Soluble in $H_2O$, MeOH and dilute acids. | — | uv (*317*), ir (*317*) |

| Biological activity |
|---|
| Inhibits bacterial growth and growth of Walker adenocarcinoma in mice (*317*). |

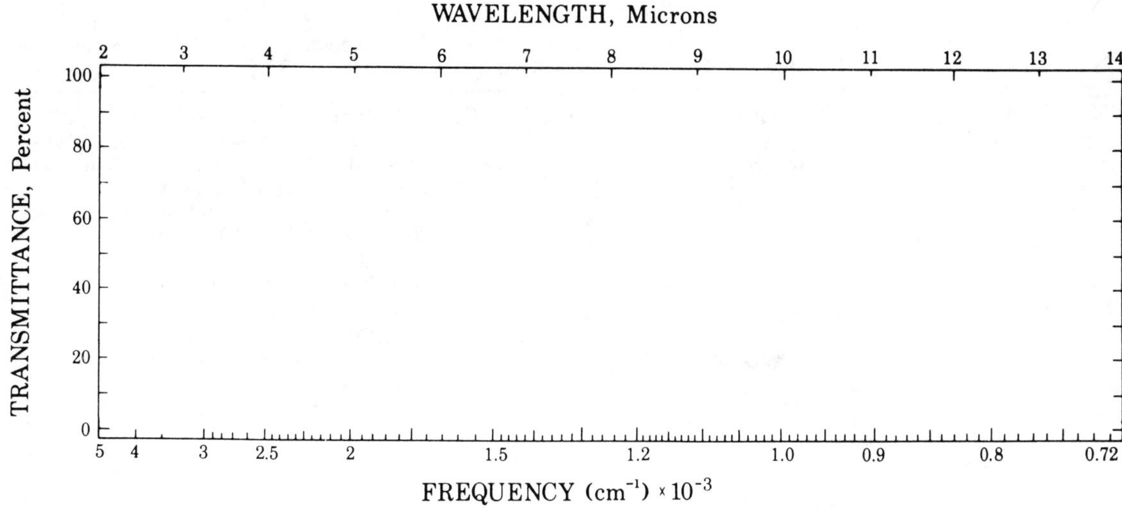
WAVELENGTH, Microns
2
3
4
5
6
7
8
9
10
11
12
13
14
TRANSMITTANCE, Percent
100
80
60
40
20
0
5
4
3
2.5
2
1.5
1.2
1.0
0.9
0.8
0.72
FREQUENCY (cm⁻¹) × 10⁻³

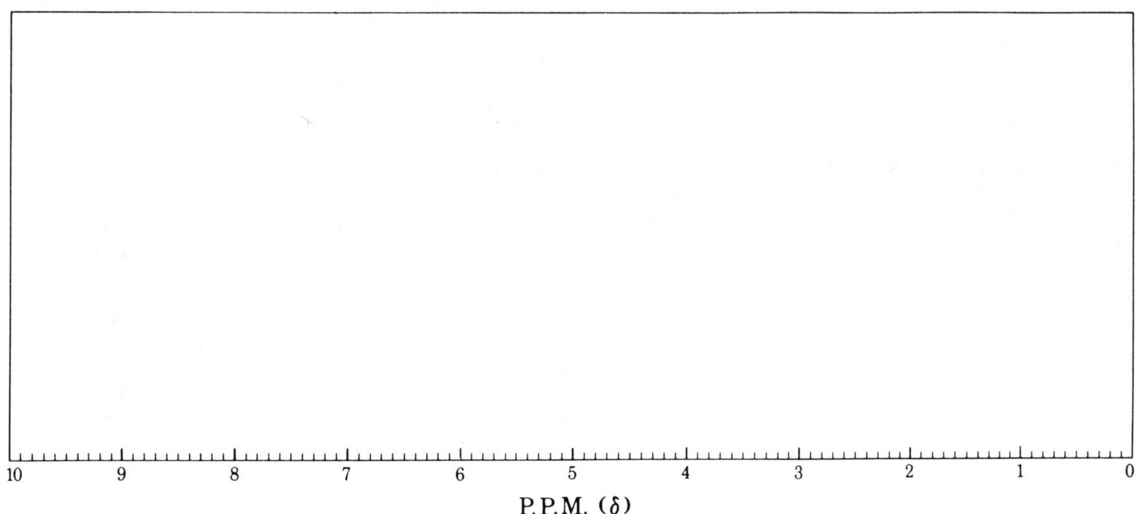
10
9
8
7
6
5
4
3
2
1
0
P.P.M. (δ)

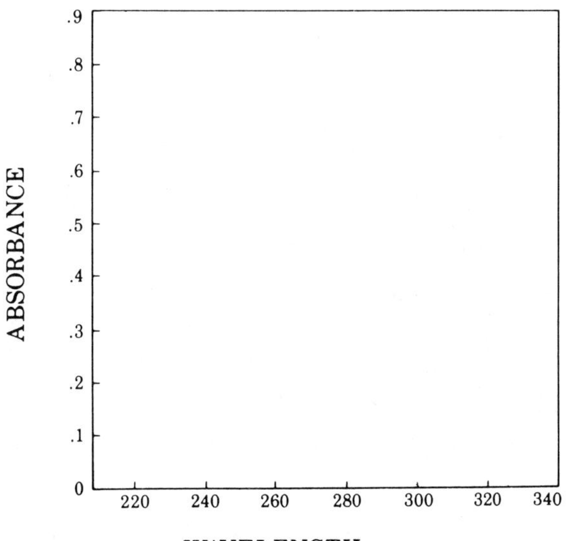
ABSORBANCE
.9
.8
.7
.6
.5
.4
.3
.2
.1
0
220
240
260
280
300
320
340
WAVELENGTH, nm.

| No. | Name | Structure | Formula (mol wt) | Melting point °C | $[\alpha]_D^{17}$ | Chromatography |
|---|---|---|---|---|---|---|
| 55 | **Tubercidin** (*321, 322*) | 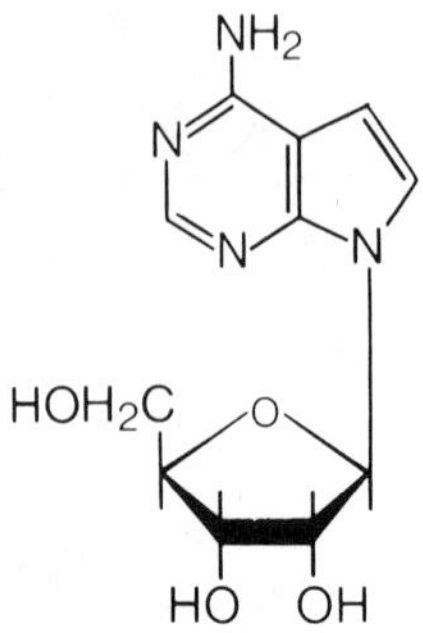 [structure: $NH_2$, N, N, N, $HOH_2C$, O, HO, OH] | $C_{11}H_{14}N_4O_4$ (266.25) (*318, 319*) | 247–248 dec. (*318, 320*), 229–231 (picrate) (*318*), 48–51 (tetraacetate) (*320*), softening at 245 with melting and dec. at 251–263. | −62 (*319*), −67 (*320*) (c = 1, 50% HOAc) (*320*), −59, (0.1 *N* HCl) (*324*), −67 (c = 1, 50% HOAc) (*336*) | Paper (*318, 323, 336*), Column (*335*) |

| pK | Purification | Source | Solubility | Toxicity | Miscellaneous |
|---|---|---|---|---|---|
| 5.2–5.3 (*318, 320, 336*) | Adsorption on carbon and then eluted. Recrystallized from hot $H_2O$ (*318*). Adsorption on Adstar (*335*). | Unidentified Actinomyces strain (*318*). *Streptomyces tubercidicus* (*319–321*). A new strain of *Streptomyces tubercidicus* (*335*). | Insoluble in acetone EtOAc, $CHCl_3$, $C_6H_6$ and pet. ether (*318*). Sparingly soluble in $H_2O$, MeOH and EtOH (*318*). Very soluble in aqueous acid or base (*318*). | $LD_{50}$ 45 mg/kg (i.v.) (mice) (*318, 336*), $LD_{50}$ 35 mg/kg (i.v.) (mice), 1.5 mg/kg (i.v.) (rats) and 25 mg/kg (i.v.) (dogs) (*324, 326*), 40 mg/kg (oral) (mice), 20 mg/kg (oral) (rats) (*324*). Extensive toxicity studies (*324, 326, 338, 339*). | uv (*318, 320, 322, 324, 336*), ir (*318, 320, 336*), cyclonucleoside (*322*) and chemical deamination (*323*). Mass spectrometry (*333*), biosynthesis (*333*), synthesis from the 4-chloro derivative (*340*) and total synthesis (*64, 65*). |

Biological activity

The binding of aminoacyl-sRNA to ribosomes was stimulated by chemically prepared trinucleoside diphosphate analogs containing tubercidin (*328*). A study on the effect of tubercidin on nucleoside uptake in 17 human tumors of various types, *in vitro* (*329*). The 5′-triphosphate derivative of tubercidin was found to function as a substrate for RNA polymerase from *E. coli* (*330*). Inhibits growth of *Streptococcus faecalis* (ATCC 8043) and is incorporated into the nucleic acids of *S. faecalis* (*331, 332*). Incorporation into RNA and DNA and the formation of a DPN analog (nicotinamide-deazadenosine dinucleotide) was reported (*332*). Substrate for mammalian terminal tRNA-C-C-A pyrophosphorylase (*294*). Inhibited *Mycobacterium tuberculosis* BCG (*318*). Inhibited multiplication of NF mouse sarcoma (*318*). Inhibits growth of L-cells and was incorporated into the nucleic acids (*325*). The 2′-deoxy derivative of tubercidin (2′-deoxytubercidin) was isolated, found to inhibit cell growth irreversibly and was incorp. into DNA (*325*). Not a substrate for adenosine deaminase from *Aspergillus oryzae* (*48*). Inhibition of vaccinia virus, mengovirus and Reovirus III (*325*). Inhibition of S180 (ascites), Ehrlich ascites and Jensen sarcoma tumors *in vivo* with equivocal activity against six other tumors and no activity against sixteen other tumors in experimental animals (*326*). A study of the action of ribonucleases $T_1$, $T_2$ and $U_2$ on dinucleoside monophosphates containing Tubercidin (*356*).

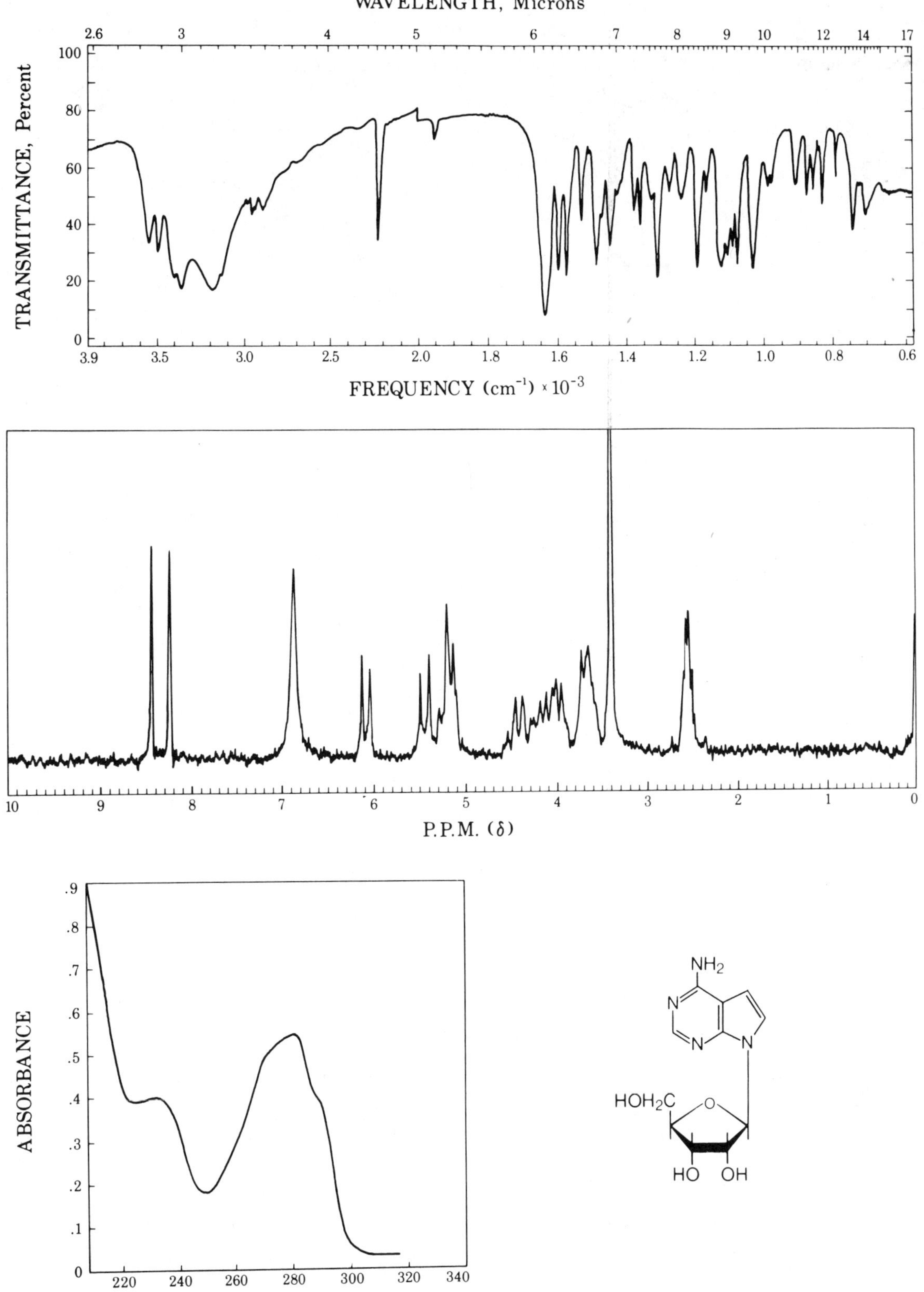

WAVELENGTH, Microns
TRANSMITTANCE, Percent
FREQUENCY (cm⁻¹) × 10⁻³
P.P.M. (δ)
ABSORBANCE
WAVELENGTH, nm.
NH2
N
N
N
HOH2C
O
HO
OH

| No. | Name | Structure | Formula (mol wt) | Melting point °C | $[\alpha]_D^{16}$ | Chromatography |
|---|---|---|---|---|---|---|
| 56 | **Toyocamycin** (*63, 64*) | 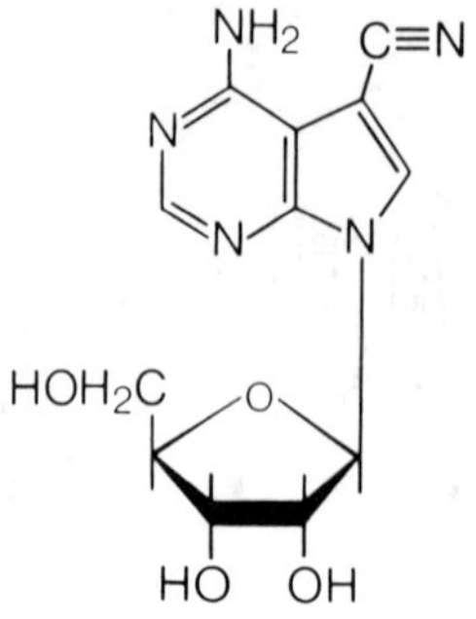 $NH_2$, C≡N, N, N, N, $HOH_2C$, O, HO, OH | $C_{12}H_{13}N_5O_4$ (291.3) (*341–343*) | 243 (*341, 64*), 239–243 ($H_2O$) (*341*), 225–226 dec. (picrate) (*343, 60*), 100 (tetra-acetate) (*343*), 247–250, 243, 237, 100 (acetate) (*60*) | −45.7 (c = 1.05, 0.1 *N* HCl) (*343*) (*60*), −55.6 ± 1.3 (c = 1, 0.1 *N* HCl) (*64*) | Paper (*60*), TLC (*64*) |

| pK | Purification | Source | Solubility | Toxicity | Miscellaneous |
|---|---|---|---|---|---|
| — | Purified by extraction with 80% aqueous acetone and then crystallized from MeOH. Recrystallized from absolute MeOH or acetone (*341*). Recrystallized from $H_2O$ or hot EtOH (*343*). | *Streptomyces toyocaensis* (*341*), *Streptomyces strain* No. 1922 (*343*). | Moderately soluble in MeOH, EtOH, acetone, dioxane and BuOH (*341*). Insoluble in $Et_2O$, EtOAc and $CHCl_3$. Soluble in HOAc and aqueous acids (*343*). | Complete toxicity study (*347*). | uv (*341, 343, 64, 63*), ir (*341, 343, 64, 63*). Chemical deamination (*342, 64*), hydrogenated (*343*), biosynthesis (*344, 357*). Chemical synthesis (*63, 64*). |

Biological activity

Activity against *Candida species* and *Mycobacterium tuberculosis* $H_{37}$ Rv (*341*) and inhibits Nakahara-Fukuoka sarcoma (N-F), *in vivo* (*327*). Not a substrate for adenosine deaminase from *Aspergillis oryzae* (*48*) but the various phosphates were formed by Ehrlich ascites tumor cells (*344*). Incorporation into RNA at internal as well as terminal positions and into DNA presumable as 2′-deoxytoyocamycin (*344*). The nucleotide is incorporated enzymatically in the terminal position of tRNA and then inhibits the normal function of tRNA in regards to esterification of amino acids (*294*). A study on the effect toyocamycin has on RNA synthesis of chick embryo cells which have been infected and not infected with the strain MC29 avian leukosis virus was reported (*345*). Toyocamycin has been found to inhibit the maturation of ribosomal-RNA while allowing normal synthesis of other cellular RNA-specie (*346*). A phase I study with Toyocamycin against several types of tumors (*347*) and a study with nucleotide reductase from *Lactobacillus leichmannii* (*296*) and *E. coli* (*297*) has been reported.

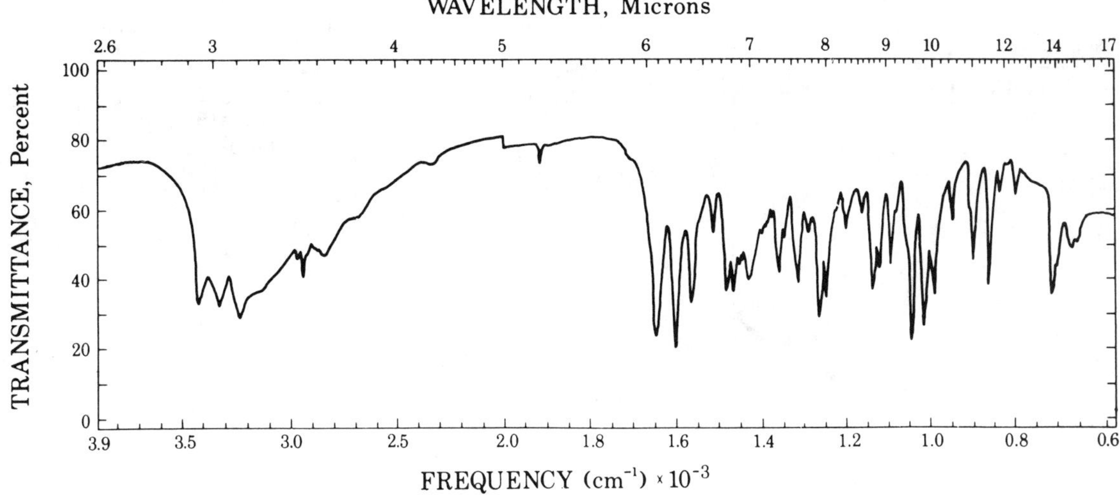
WAVELENGTH, Microns
2.6
3
4
5
6
7
8
9
10
12
14
17
TRANSMITTANCE, Percent
100
80
60
40
20
0
3.9
3.5
3.0
2.5
2.0
1.8
1.6
1.4
1.2
1.0
0.8
0.6
FREQUENCY (cm⁻¹) × 10⁻³

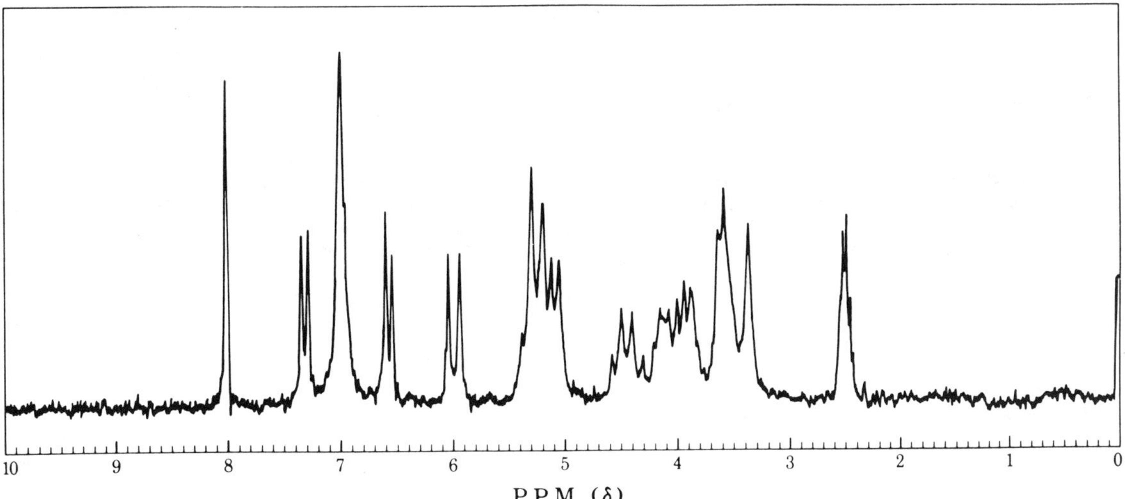
10
9
8
7
6
5
4
3
2
1
0
P.P.M. (δ)

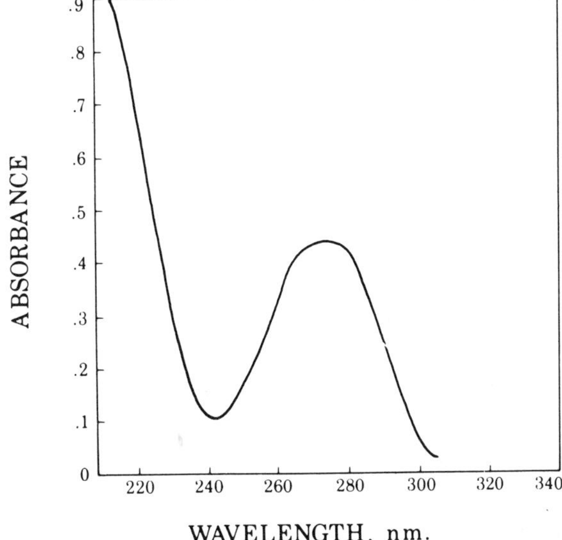
ABSORBANCE
.9
.8
.7
.6
.5
.4
.3
.2
.1
0
220
240
260
280
300
320
340
WAVELENGTH, nm.

| No. | Name | Structure | Formula (mol wt) | Melting point °C | $[\alpha]_D^{15}$ | Chromatography |
|---|---|---|---|---|---|---|
| 57 | **Unamycin B** (*63, 64*) | 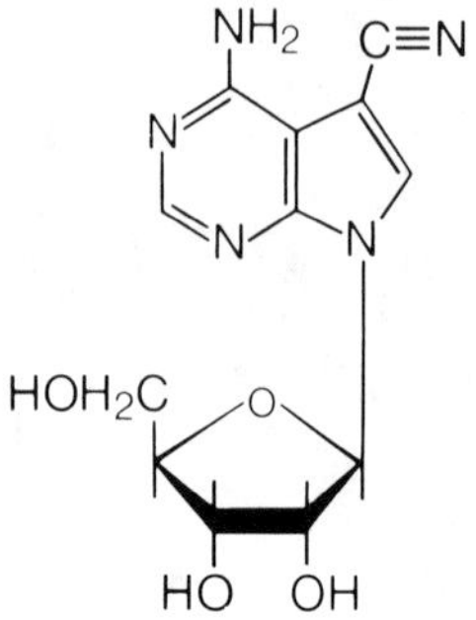 $NH_2$ C≡N N N N $HOH_2C$ O HO OH | $C_{12}H_{13}N_5O_4$ (291.3) | 236–238 dec. (*348*) | −43 (c = 1%, acid/MeOH) (*348*) | Paper (*348*) |

| pK | Purification | Source | Solubility | Toxicity | Miscellaneous |
|---|---|---|---|---|---|
| — | Adsorption on carbon and then elution followed by column chromatography (*348*). | *Streptomyces fungicidicus* (*348*) | Soluble in aqueous acid, acid/alcohols, formamide. Slightly soluble in MeOH, EtOH, BuOH and EtOAc. Insoluble in dioxane, toluene, acetone, ligroin, hexane, benzene and $CCl_4$ (*348*). | $LD_{50}$ 10 mg/kg (i.v.) (mice) and also showed a delayed toxicity (*349*). | uv (*348*), ir (*348*), neut. equiv. (*348*), Chemical synthesis (*63, 64*) |

Biological activity

Weak activity against *B. anthracis* and *Ps. aeruginosa* (*349*) and complete inhibition of *Trichomonas vaginalis* at 100 mcg/ml (*349*). Inhibits *Candida albicans* at 1.5 mcg/ml (1–2 days) and 3 mcg/ml (3 days) (*349*).

| No. | Name | Structure | Formula (mol wt) | Melting point °C | $[\alpha]_D$ | Chromatography |
|---|---|---|---|---|---|---|
| 58 | **Vengicide** (*63, 64*) | 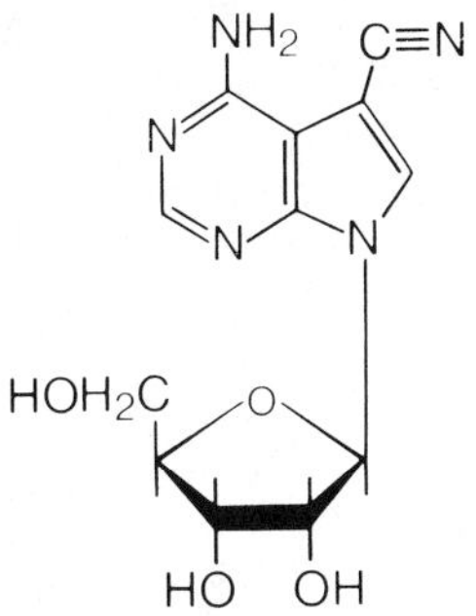 | $C_{24}H_{29}N_{10}O_9$ (*350–352*) $C_{12}H_{13}N_5O_4$ (291.3) (*63, 64*) | 241.5–243 (*350–352*) | −51.6 (0.1 *N* HCl) (*350–352*) | Paper (*350–352*) |

| pK | Purification | Source | Solubility | Toxicity | Miscellaneous |
|---|---|---|---|---|---|
| — | Purified from fermentation broth by adsorption on active carbon, eluted and then recrystallized from $H_2O$. | *Streptomyces vendargensis, nov. spec.* (*350–352*) | Very slight in $H_2O$ at room temperature (*350, 351*). Also very slightly soluble in acetone MeOH, BuOH and insoluble in diethylether (*350, 352*). | — | uv (*350–352*), ir (*350–352*). Chemical synthesis (*63, 64*) |

Biological activity

Active against *Blastomyces sulfureum*, *Blastomyces dermatitides*, *Histophasma capsulatum* and *Candida albicans* (*350–352*).

**Compiled by Leroy B. Townsend and Roland K. Robins.** The authors wish to thank Dr. Kiyoshi Isono for the original spectra (pmr, ir and uv) of the polyoxins, Dr. Robert J. Suhadolnik for generous samples of the nucleoside antibiotics Nos. 1 and 5, Dr. Koert Gerzon for a generous sample of pyrazomycin, Mr. Raymond Panzica, Miss Ritzuko Tokunaga and Mr. Ed Banta for the spectra (pmr, ir and uv) which were determined in our laboratories, and Mrs Jeri Shamy and Miss Linda Dahle for typing the manuscript.

## REFERENCES


1. Suhadolnik, Chassy and Waller, *Biochim. Biophys. Acta, 179,* 258 (1969).
2. Chassy and Suhadolnik, *Biochim. Biophys. Acta, 182,* 316 (1969).
3. Nakazawa, Fumii, Inoue, Hitomi and Miyake, *J. Antibiot. (Tokyo) Ser. B, 7,* 168 (1954).
4. Nakazawa, Fumii, Inoue, Hitomi and Miyake, *Japanese Patent 8, 149* (1955).
5. Tatsuoka, Nakazawa and Inoue, *Yakugaku Zasshi, 75,* 1206 (1955).
6. Tatsuoka, Nakazawa, Inoue and Fujii, *Pharm. Soc. Jap. J., 75,* 1206 (1955); *Chem. Abstr., 50,* 86956 (1956).
7. Tatsuoka *et al., Annu. Rep. Takeda Res. Lab., 13,* 41 (1954).
8. Hanessian and Haskell, *Tetrahedron Lett.,* p. 2451 (1964).
9. Stevens, Nielsen, Blumbergs and Taylor, *J. Amer. Chem. Soc., 86,* 5695 (1964).
10. DeBoer, Caron and Hinman, *J. Amer. Chem. Soc., 75,* 499 (1953).
11. Hinman, Caron and DeBoer, *J. Amer. Chem. Soc., 75,* 5864 (1953).
12. McCormick and Hoehn, *Antibiot. Chemother., 3,* 718 (1953).
13. Hinuma, Kuroya, Yajima, Ishihara, Hamada, Watanabe and Kikuchi, *J. Antibiot. (Tokyo) Ser. A, 8,* 718 (1955).
14. Haskell, Ryder, Frohardt, Fusari, Jakubowski and Bartz, *J. Amer. Chem. Soc., 80,* 743 (1958).
15. Kellogg and Reich, *Antimicrob. Agents Chemother.,* (1969).
16. Burchenal, Yuceoglu, Dagg and Stock, *Proc. Soc. Exp. Biol. Med., 86,* 891 (1954).
17. *British Patent 708, 686* (1954).
18. Bloch and Coutsogeorgopoulos, *Biochemistry, 5,* 3345 (1966).
19. Smith, Lummis and Grady, *Cancer Res., 19,* 847 (1959).
20. Kellogg and Reich, *Abstr. 9th Intersci. Conf. Antimicrob. Agents Chemother.,* No. 140, p. 59 (1969).
21. Sensi, Greco, Gallo and Rolland, *Antibiot. Chemother., 7,* 645 (1957).
22. Parke Davis and Co., *British Patent 707, 332* (1954).
23. Gerber and Lechevalier, *J. Org. Chem., 27,*. 1731 (1962).
24. Guarino and Kredich, *Biochim. Biophys. Acta, 68,* 317 (1963).
25. Lindberg, Klenow and Hansen, *J. Biol. Chem., 242,* 350 (1967).
26. Frederiksen, *Arch. Biochem. Biophys., 113,* 383 (1966).
27. Cory and Suhadolnik, *Biochemistry, 4,* 1729 (1965).
28. Truman and Klenow, *Mol. Pharmacol., 4,* 77 (1968).
29. Shigeura, Boxer, Melon and Sampson, *Biochemistry, 5,* 994 (1966).
30. Shigeura and Sampson, *Nature, 215,* 419 (1967).
31. Gerber, *J. Med. Chem., 7,* 204 (1964).
32. Truman and Frederikson, *Biochim. Biophys. Acta, 182,* 36 (1969).
33. Duggan, *Biochim. Biophys. Acta, 68,* 317 (1963).
34. Pugh, Hubert, Lechevalier and Sulotorovsky, *Antibiot. Chemother., 12,* 310 (1962).
35. Borowsky, Kessner and Recant, *Proc. Soc. Exp. Biol. Med., 97,* 857 (1958).
36. Skipper, Montgomery, Thompson and Schabel, Jr. *Cancer Res., 19,* 425 (1959).
37. Bloch and Nichol, *Antimicrob. Agents Chemother.,* p. 530 (1964).
38. Baker, Schaub and Kissman, *J. Amer. Chem. Soc., 77,* 5911 (1955).
39. Reist and Baker, *J. Org. Chem., 23,* 1083 (1958).
40. Yuntsen, Ohkuma and Ishii, *J. Antibiot. (Tokyo) Ser. A, 9,* 195 (1956).
41. Hoeksema, Slomp and van Tamelen, *Tetrahedron Lett.,* p. 1787 (1964).
42. Yuntsen, *J. Antibiot. (Tokyo) Ser. A, 11,* 233 (1957).
43. Yuntsen, *J. Antibiot. (Tokyo) Ser. A, 11,* 77 (1957).
44. Yuntsen and Yonehara, *Bull. Agr. Chem. Soc. Jap., 21,* 261 (1957).
45. Tanaka, Miyairi and Umezawa, *J. Antibiot. (Tokyo) Ser. A, 13,* 265 (1960).
46. Tanaka, Miyairi and Umezawa, *J. Antibiot. (Tokyo) Ser. A, 14,* 23 (1961).
47. Miyairi, Tanaka and Umezawa, *J. Antibiot. (Tokyo) Ser. A, 14,* 119 (1961).
48. Fukagawa, Sawa, Takeuchi and Umezawa, *J. Antibiot. (Tokyo) Ser. A, 18,* 191 (1965).
49. McCarthy, Jr., Robins and Robins, *J. Amer. Chem. Soc., 90,* 4993 (1968).
50. Hsu, *J. Antibiot. (Tokyo) Ser. A, 11,* 244 (1958).
51. Schroeder and Hoeksema, *J. Amer. Chem. Soc., 81,* 1767 (1959).
52. Beppu, Nose and Arima, *Agr. Biol. Chem., 32,* 197 (1968).
53. Beppu, Nose and Arima, *Agr. Biol. Chem., 32,* 203 (1968).
54. Hanka, *J. Bacteriol., 80,* 30 (1960).
55. Komuro, Nara, Misawa and Kinoshita, *Agr. Biol. Chem., 33,* 230 (1969).
56. Yuntsen, *J. Antibiot. (Tokyo) Ser. A, 11,* (1957).
57. Farkas and Sorm, *Tetrahedron Lett.,* p. 813 (1962); *Collect. Czech. Chem. Commun., 28,* 882 (1963).
58. Schroeder, *U.S. Patent 2,993,039* (1961).
59. Yamamoto, Fujii, Nakazawa, Miyake, Hitomi and Imanishi, *Annu. Rep. Takeda Res. Lab., 16,* 28 (1957).

60. Aszalos, Lemansi, Robison, Davis and Berk, *J. Antibiot. (Tokyo) Ser. A, 19,* 285 (1966).
61. DeBoer and Hinman (Upjohn Co.), *German Patent 926,271* (1955); *Chem. Abstr., 52,* 15848d (1958).
62. Upjohn Co., *British Patent 708,686* (1954); *Chem. Abstr., 48,* 1100A (1954).
63. Kikuchi, *J. Antibiot. (Tokyo) Ser. A, 8,* 145 (1955).
64. Tolman, Robins and Townsend, *J. Amer. Chem. Soc., 90,* 524 (1968).
65. Tolman, Robins and Townsend, *J. Amer. Chem. Soc., 91,* 2102 (1969).
66. Umezawa (Ed.), in *Index of Antibiotics from Actinomycetes,* p. 841. University of Tokyo, Tokyo, Japan (1967).
67. Sensi and Rolland, *British Patent 841,696* (1960).
68. Thomas *et al., French Patent 1,197,629* (1959).
69. Umezawa (Ed.), in *Index of Antibiotics from Actinomycetes,* p. 775. University of Tokyo Press, Tokyo, Japan (1967).
70. Chem. Eng. News, *p. 35 (1968).*
71. Schabel, Jr., *Chemotherapy, 13,* 321 (1968).
72. Privat De Garilke and De Rudder, *C. R. Acad. Sci. (Paris), 259,* 2725 (1964).
73. Freeman, Kuehn and Sultanian, *Ann. N.Y. Acad. Sci., 130,* 330 (1965).
74. Sidwell, Arnett and Dixon, *Abstr. 7th Intersci. Conf. Antimicrob. Agents Chemother.,* No. 64, p. 28–29 (1967).
75. Sidwell, Dixon, Sellers and Schabel, Jr., *Appl. Microbiol., 16,* 370 (1968).
76. Miller, Dixon, Ehrlich, Sloan and McLean, Jr., *Antimicrob. Agents Chemother.,* p. 136 (1968).
77. Sidwell, Dixon, Schabel, Jr. and Kaump, *Antimicrob. Agents Chemother.,* p. 148 (1968).
78. Schardien and Sidwell, *Antimicrob. Agents Chemother.,* p. 155 (1968).
79. Sloan, Miller, Ehrlich, McLean, Jr. and Machamer, *Antimicrob. Agents Chemother.,* p. 161 (1968).
80. Dixon, Sidwell, Miller and Sloan, *Antimicrob. Agents Chemother.,* p. 172 (1968).
81. Kurtz, Fisken, Kaump and Schardien, *Antimicrob. Agents Chemother.,* p. 180 (1968).
82. LePage and Junga, *Cancer Res., 25,* 46 (1965).
83. LePage and Junga, *Cancer Res., 28,* 739 (1963).
84. York and LePage, *Can. J. Biochem., 44,* 331 (1966).
85. York and LePage, *Can. J. Biochem., 44,* 19 (1966).
86. Furth and Cohen, *Cancer Res., 27,* 1528 (1967).
87. Furth and Cohen, *Cancer Res., 28,* 2061 (1968).
88. Miller, Sloan and Silverman, *Abstr. 9th Intersci. Conf. Antimicrob. Agents Chemother.,* No. 23, p. 10 (1969).
89. Cohen, *Prog. Nucl. Acid Res. Mol. Biol., 5,* 1 (1966).
90. Lee, Benitez, Goodman and Baker, *J. Amer. Chem. Soc., 82,* 2648 (1960).
91. Reist, Benitez, Goodman, Baker and Lee, *J. Org. Chem., 25,* 3274 (1962).
92. Glaudemans and Fletcher, Jr., *J. Org. Chem., 28,* 3004 (1963).
93. Lunzmann and Schramm, *Biochim. Biophys. Acta, 169,* 263 (1968).
94. Kusaka, Yamamoto, Shibata, Muro, Kishi and Mizuno, *J. Antibiot. (Tokyo) Ser. A, 21,* 255 (1968).
95. Bennett, Jr., Allan and Hill, *Mol. Pharmacol., 4,* 208 (1968).
96. Kishi, Muroi, Kusaka, Nishikawa, Kikamiya and Mizuno, *Chem. Commun.,* p. 852 (1967).
97. Shealy and Clayton, *J. Amer. Chem. Soc., 91,* 3075 (1969).
98. Hanka, Evans, Mason and Dietz, *Antimicrob. Agents Chemother.,* p. 619 (1966).
99. Bergly and Herr, *Antimicrob. Agents Chemother.,* p. 625 (1966).
100. Sorm and Vesely, *Neoplasma, 11,* 123 (1964).
101. Sorm, Piskalo, Cihak and Vesely, *Experientia (Basel), 20,* 20 (1964).
102. Goldin, Wood, Jr. and Engle, *Cancer Chemother. Rep. Part 2, 1,* 266, 269 (1968).
103. Cihak, Vesely and Sorm, *Biochim. Biophys. Acta, 108,* 516 (1965).
104. Jurovcik, Raska, Sormova and Sorm, *Collect. Czech. Chem. Commun., 31,* 2809 (1966).
105. Vesely, Cihak and Sorm, *Int. J. Cancer, 2,* 639 (1968).
106. Raska, Jurovcik, Fucik, Tykva, Sormova and Sorm, *Collect. Czech. Chem. Commun., 21,* 2808 (1966).
107. Jurovcik, Raska, Sormova and Sorm, *Collect. Czech. Chem. Commun., 30,* 3370 (1965).
108. Cihak, Rykva and Sorm, *Collect. Czech. Chem. Commun., 31,* 3015 (1966).
109. Zadrazil, Ficol, Bartl, Sormova and Sorm, *Biochim. Biophys. Acta, 108,* 70 (1965).
110. Svata, Raska and Sorm, *Experientia (Basel), 22,* 53 (1966).
111. Vesely, Cihak and Sorm, *Biochem. Pharmacol., 17,* 519 (1968).
112. Cihak, Vesely and Sorm, *Biochim. Biophys. Acta, 134,* 486 (1967).
113. Cihak, Vesely and Sorm, *Biochim. Biophys. Acta, 161,* 277 (1968).
114. Vesely, Gostor, Cihak and Sorm, *Z. Naturforsch. Teil B, 24,* 318 (1969).
115. Doskocil and Paces, *Collect. Czech. Chem. Commun., 33,* 4369 (1968).
116. Piskala and Sorm, *Collect. Czech. Chem. Commun., 29,* 2060 (1964).
117. Rao and Renn, *Antimicrob. Agents Chemother.,* p. 77 (1963).
118. *British Patent 707,332* (1954).
119. Haskell, *J. Amer. Chem. Soc., 80,* 747, (1958).
120. Fox, Watanabe and Bloch, *Progr. Nucl. Acid Res. Mol. Biol., 5,* 277 (1966).

121. Takeuchi, Hirayama, Ueda, Sakai and Yonehara, *J. Antibiot. (Tokyo) Ser. A, 11,* 1 (1958).
122. Fukunaga, *Bull. Agr. Chem. Soc. Jap., 19,* 181 (1955); *Chem. Abstr., 50,* 1487C (1956).
123. Misato, Ishii, Asakawa, Okimoto and Fukunaga, *Ann. Phytopathol. Soc. Jap., 26,* 19, 25 (1961).
124. Tanaka, Sakagami, Nishimura, Yamaki and Umezawa, *J. Antibiot. (Tokyo) Ser. A, 14,* 123 (1961).
125. Fox and Watanabe, *Tetrahedron Lett.,* p. 897 (1966).
126. Yonehara and Otake, *Tetrahedron Lett.,* p. 3785 (1966).
127. Yamaguchi, Yamamoto and Tanaka, *J. Biochem., 57,* 667 (1965).
128. Yamaguchi and Tanaka, *J. Biochem., 60,* 632 (1966).
129. Onuma, Nawata and Sito, *Bull. Chem. Soc. Jap., 39,* 1091 (1965).
130. Hirai and Shomomura, *Phytopathology, 55,* 291 (1965).
131. Kitani and Kiso, *Ann. Phytopathol. Soc. Jap., 28,* 293 (1963).
132. Yonehara and Otake, *Antimicrob. Agents Chemother.,* p. 855 (1966).
133. Cunningham, Hutchinson, Manson and Spring, *J. Chem. Soc.,* p. 2299 (1951).
134. Cunningham, Manson, Spring and Hutchinson, *Nature, 166,* 949 (1950).
135. Kacza, Dulaney, Gitterman, Woodruff and Folkers, *Biochem. Biophys. Res. Commun., 14,* 452 (1964).
136. Kacza, Trenner, Arison, Walker and Folkers, *Biochem. Biophys. Res. Commun., 14,* 456 (1964).
137. Frederiksen, Malling and Klenow, *Biochim. Biophys. Acta, 95,* 189 (1965).
138. Rottman and Guarino, *Biochim. Biophys. Acta, 80,* 632, 640 (1964).
139. Jagger, Kredrich and Guarino, *Cancer Res., 21,* 216 (1961).
140. Shigeura and Gordon, *J. Biol. Chem., 240,* 806 (1965).
141. Rich, Meyers, Weinbaum, Cory and Suhadolnik, *Biochim. Biophys. Acta, 95,* 194 (1965).
142. Cory, Suhadolnik, Resnick and Rich, *Biochim. Biophys. Acta, 103,* 646 (1965).
143. Gitterman, Burg, Boxer, Meltz and Hitt, *J. Med. Chem., 8,* 664 (1965).
144. Hanessian, Dejongh and McCloskey, *Biochim. Biophys. Acta, 117,* 480 (1966).
145. Williamson, *Trans. Roy. Soc. Trop. Med. Hyg., 60,* 8 (1966).
146. Chassy and Suhadolnik, *Biochim. Biophys. Acta, 182,* 307 (1969).
147. Lee, Benitz, Anderson, Goodman and Baker, *J. Amer. Chem. Soc., 83,* 1906 (1961).
148. Todd and Ulbricht, *J. Chem. Soc.,* p. 3275 (1960).
149. Walton, Nutt, Jenkins and Holly, *J. Amer. Chem. Soc., 86,* 2952 (1964).
150. Tyrsted and Sartorelli, *Biochim. Biophys. Acta, 155,* 619 (1968).
151. Yonehara and Otake, *Tetrahedron Lett.,* p. 3785 (1966).
152. Yanaka, Sakagami, Nishimura, Tamaki and Umezawa, *J. Antibiot. (Tokyo) Ser. A, 14,* 123 (1961).
153. Niotake, Takeuchi, Endo and Yonehara, *Agr. Biol. Chem., 30,* 126, 132 (1966).
154. Sakagami *et al.,* Unpublished Data in Reference 2 of Reference 152 in this chapter.
155. Tanaka, Nishimura, Miyairi, Sakagami, Hsu, Yonehara and Umezawa, *Gann Suppl., 50,* 7 (1960).
156. Yonehara, Takeuchi, Otake, Endo, Sakagami and Sumuki, *J. Antibiot. (Tokyo) Ser. A, 16,* 195 (1963).
157. *Plant Dis. Rep., 41,* 576 (1967); *Chem. Abstr., 15,* 15865a (1957).
158. Koniuszy, Kaczka, Long and Gray (Merck and Co.), *U.S. Patent 3,137,626* (1964); *Chem. Abstr. 61,* 4925e (1964).
159. Lindnor, Kirkpatrick and Weeks, *Phytopathology, 49,* 802 (1959).
160. Simons, *Phytopathology, 50,* 109 (1960).
161. Wolf, Putter, Fillin and Downing, Jr. (Merck and Co.), *German Patent 1,085,878* (1960).
162. Harries, *J. Econ. Entomol., 54,* 122 (1961).
163. Harries, *J. Econ. Entomol., 56,* 438 (1963).
164. Shanks, Jr. and Chapman, *Virology, 25,* 83 (1965).
165. Harries, *J. Econ. Entomol., 58,* 361 (1965).
166. Wolf, Putter, Fillin, Downing, Jr. and Gillin (Merck and Co.), *U.S. Patent 3,221,008* (1965); *Chem. Abstr., 64,* 6710b (1966).
167. Harries and Wiles, *J. Econ. Entomol., 59,* 694 (1966).
168. Brodasky, *J. Gas Chromatogr., 5,* 311 (1967).
169. DeBoer, Dietz, Johnson, Eble and Hoeksema (Upjohn Co.), *German Patent 1,101,698* (1961); *Chem. Abstr., 55,* 27767h (1961).
170. Bloch and Nichol, *Fed. Proc., 23,* 324 (1964).
171. Bloch and Nichol, *Biochem. Biophys. Res. Commun., 16,* 400 (1964).
172. DeBoer, Dietz, Johnson, Eble and Hoeksema, *French Patent 1,465,395* (1967); *Chem. Abstr., 67,* Pc74261g (1967).
173. DeBoer, Dietz, Johnson, Eble and Hoeksema, *U.S. Patent 3,207,750* (1965); *Chem. Abstr., 63,* 18998f (1965).
174. Hoeksema (Upjohn Co.), *French Patent 1, 376, 307* (1964); *Chem. Abstr., 62,* 14407g (1965).
175. Henderson and Khoo, *J. Biol. Chem., 240,* 3104 (1965).
176. Hori, Ito, Takita, Koyama, Takeuchi and Umezawa, *J. Antibiot. (Tokyo) Ser. A, 17,* 96 (1964).
177. Koyama, Maeda and Umezawa, *Tetrahedron Lett.,* p. 597 (1966).
178. Robins, Townsend, Cassidy, Gerster, Lewis and Miller, *J. Heterocycl. Chem., 3,* 110 (1966).
179. Ishida, Homma, Kumagai, Schimizu, Matsumoto and Izawa, *J. Antibiot. (Tokyo) Ser. A, 20,* 49 (1967).

180. Ishizaka, Takeuchi, Nitta, Koyama, Hori and Umezawa, *J. Antibiot. (Tokyo) Ser. A, 17*, 124 (1964).
181. Aizawa, Hidaka, Otake, Yonehara, Isono, Igarashi and Suzuki, *Agr. Biol. Chem., 29*, 375 (1965).
182. Koyama and Umezawa, *J. Antibiot. (Tokyo) Ser. A, 18*, 175 (1965).
183. Umezawa, Sawa, Fukagawa, Koyama, Murase, Hamada and Takeuchi, *J. Antibiot. (Tokyo) Ser. A, 18*, 178 (1965).
184. Sawa, Fukagawa, Shimauchi, Ito, Hamada, Takeuchi and Umezawa, *J. Antibiot. (Tokyo) Ser. A, 18*, 259 (1965).
185. Reese, Townsend and Parrish, *Arch. Biochem. Biophys., 125*, 175 (1968).
186. Ward, Cerami and Reich, *J. Biol. Chem., 244*, 3243 (1969).
187. Kunimoto, Hori and Umezawa, *J. Antibiot. (Tokyo) Ser. A, 20*, 277 (1967).
188. Sawa, Fukagawa, Homma, Takeuchi and Umezawa, *J. Antibiot. (Tokyo) Ser. A, 20*, 317 (1967).
189. Umezawa, Sawa, Fukagawa, Homma, Ishizuka and Takeuchi, *J. Antibiot. (Tokyo) Ser. A, 20*, 308 (1967).
190. Ishizuka, Sawa, Hori, Takayama, Takeuchi and Umezawa, *J. Antibiot. (Tokyo) Ser. A, 21*, 5 (1968).
191. Ikehara, Murao, Haraka and Nishimura, *Biochim. Biophys. Acta, 174*, 696 (1969).
192. Ward, Reich and Stryer, *J. Biol. Chem., 244*, 1228 (1969).
193. Ikehara, Murao and Nishimura, *Biochim. Biophys. Acta, 182*, 276 (1969).
194. Townsend and Robins, *J. Heterocycl. Chem., 6*, 459 (1969).
195. Otake, Aizawa, Hidaka, Seto and Yonehara, *Agr. Biol. Chem., 29*, 377 (1965).
196. Tsukada, Kunimoto, Hori and Komai, *J. Antibiot. (Tokyo) Ser. A, 22*, 36 (1969).
197. Ishida, Izawa, Homma, Kumagai and Shimizu, *J. Antibiot. (Tokyo) Ser. A, 20*, 129 (1967).
198. Sheen, Kim and Parks, Jr., *Mol. Pharmacol., 4*, 293 (1968).
199. Ishizuka, Sawa, Koyama, Takeuchi and Umezawa, *J. Antibiot. (Tokyo) Ser. A, 21*, 1 (1968).
200. Hori, Wakashiro, Ito, Sawa, Takeuchi and Umezawa, *J. Antibiot. (Tokyo) Ser. A, 21*, 264 (1968).
201. Sawa, Fukagawa, Homma, Wakashiro, Takeuchi and Hori, *J. Antibiot. (Tokyo) Ser. A, 21*, 334 (1968).
202. Kansaki, Higashide, Yamamoto, Shibata, Nakazawa, Iwasaki, Takewaka and Miyake, *J. Antibiot. (Tokyo) Ser. A, 15*, 93 (1962).
203. Fox, Kuwada, Watanabe, Ueda and Whipple, *Antimicrob. Agents Chemother.*, p. 518 (1964).
204. Clark, Jr. and Gunther, *Biochim. Biophys. Acta, 76*, 636 (1963).
205. Casjen and Morris, *Biochim. Biophys. Acta, 108*, 677 (1965).
206. Clark, Jr. and Chang, *J. Biol Chem., 240*, 4734 (1965).
207. Sinohara and Sky-Peck, *Biochem. Biophys. Res. Commun., 18*, 98 (1965).
208. Kredich and Guarino, *J. Biol. Chem., 236*, 3300 (1961).
209. Guarino, Ibershof and Swain, *Biochim. Biophys. Acta, 72*, 62 (1963).
210. Huang, Katagiri and Misato, *J. Antibiot. (Tokyo) Ser. A, 19*, 75 (1966).
211. Guarino and Kredich, *Fed. Proc., 23*, 371 (1964).
212. Umezawa, Takeuchi, Okami and Maeda, *Japanese Patent 10,245* (1964); *Chem. Abstr., 61*, 125921 (1964).
213. Umezawa (Ed.), in *Index of Antibiotics from Actinomycetes*, p. 449. University of Tokyo Press, Tokyo, Japan (1967).
214. Isono and Suzuki, *J. Antibiot. (Tokyo) Ser. A, 13*, 270 (1960).
215. Nakamura, *J. Antibiot. (Tokyo) Ser. A, 14*, 94 (1961).
216. Lofgren, Luning and Hedstrom, *Acta Chem. Scand., 8*, 670 (1954).
217. Lofgren, Takman and Hedstrom, *Sven. Farm. Tidskr., 53*, 321 (1949).
218. Biesele, Slauterback and Margolis, *Cancer, 8*, 87 (1955).
219. Ehrenberg, Hedstrom, Lofgren and Takman, *Sven. Farm. Tidskr., 50*, 645 (1946).
220. Lofgren and Luning, *Acta Chem. Scand., 7*, 215 (1953).
221. Tamm, Folkers and Shunk, *J. Bacteriol., 72*, 59 (1956).
222. Gordon and Brown, *J. Biol. Chem., 220*, 927 (1956).
223. Ballio, *Res. Progr. Org. Biol. Med. Chem. Soc.*, p. 37 (1964).
224. Brown and Weliky, *J. Biol. Chem., 204*, 1019 (1953).
225. Fox, Wempen, Hampton and Doerr, *J. Amer. Chem. Soc., 80*, 1669 (1958).
226. Iwamura and Hashizume, *Tetrahedron Lett.*, p. 643 (1966).
227. Iwamura and Hashizume, *J. Org. Chem., 33*, 1796 (1968).
228. Hampton, Biesele, Moore and Brown, *J. Amer. Chem. Soc., 78*, 5695 (1956).
229. Thomas, Singleton, Lowery, Sharpe, Pruess, Porter, Mowat and Bohonos, *Antibiot. Ann.*, p. 716 (1956–57).
230. Hewitt, Gumble, Taylor and Wallace, *Antibiot. Ann.*, p. 722 (1956–57).
231. Waller, Patrick, Fulmor and Meyer, *J. Amer. Chem. Soc., 79*, 1011 (1957).
232. Patrick and Meyer, *Abstr. 156th Nat. Meet. Amer. Chem. Soc., Atlantic City, New Jersey, September, 1968*, MEDI-024.
233. Morton, Lancaster, VanLear, Fulmor and Meyer, *J. Amer. Chem. Soc., 91*, 1535 (1969).
234. Florini, Bird and Bell, *J. Biol. Chem., 241*, 1091 (1966).
235. Tobie, *J. Parasitol., 43*, 291 (1957).
236. Stephen and Gray, *J. Parasitol., 46*, 509 (1960).
237. Florini, *Antibiotics, 1*, 427 (1967).

238. Thomas, Lowery and Singleton, *U.S. Patent 2,914,525* (1959); *Chem. Abstr., 54,* 5024b (1960).
239. Thomas, Lowery and Singleton (American Cyanamid Co.), *British Patent 815,381* (1959); *Chem. Abstr., 53,* 19312c (1959); *German Patent 1,018,192* (1958).
240. Ishikawa, Niida, Izawa and Fukunaga, *Japanese Patent 13, 792* (1966); *Chem. Abstr., 65* 19272 (1966).
241. Isono, Nagatsu, Kawashima and Suzuki, *Agr. Biol. Chem., 29,* 848 (1965).
242. Suzuki, Isono, Nagatsu, Yamagata, Saski and Hashimoto, *Agr. Biol. Chem., 30,* 817 (1966).
243. Isono, Nagatsu, Kobinata, Saski and Suzuki, *Agr. Biol. Chem., 31* (1967).
244. Isono, Kobinata and Suzuki, *Agr. Biol. Chem., 32,* 792 (1968).
245. Asahi, Sakurai, Isono and Suzuki, *Agr. Biol. Chem., 32,* 1046 (1968).
246. Isono and Suzuki, *Agr. Biol. Chem., 32,* 1193 (1968).
247. Isono and Suzuki, *Tetrahedron Lett.,* p. 203 (1968).
248. Isono and Suzuki, *Tetrahedron Lett.,* p. 1133 (1968).
249. Eble, Hoeksema, Boyack and Savage, *Antibiot. Chemother., 9,* 419 (1959).
250. Varva, Dietz, Churchill, Siminoff and Koepsell, *Antibiot. Chemother., 9,* 427 (1959).
251. Lewis, Reames and Rhuland, *Antibiot. Chemother., 9,* 421 (1959).
252. Evans and Gray, *Antibiot. Chemother., 9,* 675 (1959).
253. Udaka and Moyed, *J. Biol. Chem., 238,* 2797 (1963).
254. Fukuyama and Moyed, *Biochemistry, 3,* 1488 (1964).
255. Fukuyama, *J. Biol. Chem., 241,* 4745 (1966).
256. Porter, Hewitt, Hesseltine, Kraupa, Lowery, Wallace, Bohonos and Williams, *Antibiot. Chemother., 2,* 409 (1952).
257. Fryth, Waller, Hutchings and Williams, *J. Amer. Chem. Soc., 80,* 2736 (1958).
258. Baker, Schaub, Joseph and Williams, *J. Amer. Chem. Soc., 77,* 12 (1955).
259. Waller, Fryth, Hutchings and Williams, *J. Amer. Chem. Soc., 75,* 2025 (1953).
260. The name Achromycin® was reassigned to a tetracycline antibiotic and is no longer used for puromycin.

270. White and White, *Science, 146,* 772 (1964).
271. Yarmolinsky and DeLa Haba, *Proc. Nat. Acad. Sci. U.S.A., 45,* 1721 (1959).
272. Sherman, Taylor and Bond, *Antibiot. Ann.,* p. 754 (1954–1955).
273. Tobie, *Amer. J. Trop. Med. Hyg., 3,* 852 (1954).
274. Gumble, Hewitt, Taylor and Wallace, *Antibiot. Ann.,* p. 260 (1955–1956).
275. Bond, Sherman and Taylor, *Antibiot. Ann.,* p. 751 (1954–1955).
276. Faiguenbaum and Alba, *Biol. Chilena Parasitol., 9,* 94 (1954).
277. Trincao, Franco, Nogueria, Pinto and Muhlpfordt, *Antibiot. Ann.,* p. 596 (1955–1956).
278. Allen and Zamecnik, *Biochim. Biophys. Acta, 55,* 865 (1962).
279. Smith, Traut, Blackburn and Monro, *J. Mol. Biol., 13,* 617 (1965).
280. Pattabiraman and Pogell, *Biochim. Biophys. Acta, 182,* 245 (1969).
281. Sugiura, Stock, Reilly and Schmid, *Cancer Res., 18,* 66 (1958).
282. Levinton, Thoren, Darnell, Jr. and Hooper, *Virology, 16,* 220 (1962).
283. Pilcher and Hobbs, *J. Pharm. Sci., 55,* 119 (1966).
284. Gambetti, Gonatas and Flexner, *Science, 161,* 900 (1968).
285. Gerzon, Private Communication. Eli Lilly and Co., Indianapolis, Indiana.
286. Streightoff, Nelson, Cline, Gerzon, Williams and DeLong, *Abstr. 9th Intersci. Conf. Antimicrob. Agents Chemother.,* No. 18 (1969).
287. Gerzon, Williams, Hoehn, Gorman and DeLong, *Abstr. 2nd Int. Congr. Heterocycl. Chem., Montpellier, France,* C-30, p. 131 (1969).
288. *Chem. Eng. News,* p. 43 (1969).
289. Williams, Gerzon, Hoehn, Gorman and DeLong, *158th Nat. Amer. Chem. Soc. Meeting, New York,* MICRO 38 (1969).
290. Hinuma *et al., 77th Meeting Jap. Antibiot. Res. Ass.* (1954).
291. Kuroya *et al., 27th Gen. Meeting Jap. Bacteriol. Soc.* (1954).
292. Rao, *150th Meeting Amer. Chem. Soc., Atlantic City, N.J.,* p. 24p (1965).
293. Cavins, Hall, Olsen, Khung, Horton, Colsky and Shadduck, *Cancer Chemother. Rep., 51,* 197 (1967).
294. Uretsky, Acs, Reich, Mori and Altwerger, *J. Biol. Chem., 243,* 306 (1968).
295. Suhadolnik, Uematsu and Uematsu, *J. Biol. Chem., 243,* 2761 (1968).
296. Suhadolnik, Finkel and Chassy, *J. Biol. Chem., 243,* 3532 (1968).
297. Chassy and Suhadolnik, *J. Biol. Chem., 243,* 3538 (1968).
298. Darnall, Townsend and Robins, *Proc. Nat. Acad. Sci. U.S.A., 57,* 548 (1967).
299. Nishimura, Mayama, Komatsu, Kato, Shimaoka and Tanaka, *J. Antibiot. (Tokyo) Ser. A, 17,* 148 (1964).
300. Nishimura, *French Patent M2751* (1964); *Chem. Abstr., 62,* 2675b (1965).
301. Matsuura, Shiratori and Katagiri, *J. Antibiot. (Tokyo) Ser. A, 17,* 234 (1965).
302. Nakagawa, Kano, Tsukada and Koyama, *Tetrahedron Lett.,* p. 4105 (1967).
303. Roy-Burman, Roy-Burman and Visser, *Cancer Res., 28,* 1605 (1968).

304. Kano, Nakagawa, Koyama and Tsukada, *Abstr. First Int. Congr. Heterocycl. Chem., Albuquerque, New Mexico,* No. 41 (1967).
305. Tsukada, Nakagawa, Sato, Shiro, Kano and Koyama, *Abstr. 20th Annu. Meeting Chem. Soc. Jap. Ser. 1,* p. 76 (1967).
306. Tsukada, Sato, Shiro and Koyama, *J. Chem. Soc. Sect. B,* p. 843 (1969).
307. Nishimura and Komatsu, *J. Antibiot. (Tokyo) Ser. A, 21,* 250 (1968).
308. An antibiotic substance obtained by Merck and Co. and found to be identical to showdomycin.
309. Hadley, Claybourn and Tschang, *Biochem. Biophys. Res. Commun., 31,* 25 (1968).
310. Hadley, Claybourn and Tschang, *J. Antibiot. (Tokyo) Ser. A, 21,* 575 (1968).
311. Komatsu and Tanaka, *Agr. Biol. Chem., 32,* 1021 (1968).
312. Hadler and Moreau, *Abstr. 9th Intersci. Conf. Antimicrob. Agents Chemother.* No. 136, p. 58 (1969).
313. Owen, Dietz and Camiener, *Antimicrob. Agents Chemother.,* p. 772 (1962).
314. Argoudelis and Herr, *Antimicrob. Agents Chemother.,* p. 780 (1962).
315. Argoudelis *et al., Japanese Patent 27,499* (1964).
316. Upjohn Co., *German Patent 1,188,762* (1962).
317. Upjohn Co., *British Patent 974,541* (1964); *Chem. Abstr., 62,* 5855d (1965).
318. Anzi, Nakamura and Suzuki, *J. Antibiot. (Tokyo) Ser. A, 10,* 201 (1957).
319. Suzuki and Marumo, *J. Antibiot. (Tokyo) Ser. A, 13,* 360 (1960).
320. Suzuki and Marumo, *J. Antibiot. (Tokyo) Ser. A, 14,* 34 (1961).
321. Mizuno, Ikehara, Watanabe and Suzaki, *Chem. Pharm. Bull. (Tokyo), 11,* 1091 (1963).
322. Mizuno, Ikehara, Watanabe, Suzaki and Itoh, *J. Org. Chem., 28,* 3329 (1963).
323. Mizuno, Ikehara, Watanabe and Suzaki, *J. Org. Chem., 28,* 3331 (1963).
324. *Cancer Chemotherapy National Service Center Information Sheet.*
325. Acs, Reich and Mori, *Proc. Nat. Acad. Sci. U.S.A., 52,* 493 (1964).
326. Owen and Smith, *Cancer Chemother Rep., 36,* 19 (1964).
327. Saneyoshi, Tokuzen and Fukuoka, *Gann, 56,* 219 (1965).
328. Ikehara and Oktsuka, *Biochem. Biophys. Res. Commun., 21,* 257 (1965).
329. Wolberg, *Biochem. Pharmacol., 14,* 1921 (1965).
330. Nishimura, Harada and Ikehara, *Biochim. Biophys. Acta, 129,* 301 (1963).
331. Bloch and Nichol, *145th Meeting Amer. Chem. Soc.,* p. 35c (1963).
332. Bloch, Leonard and Nichol, *Biochim. Biophys. Acta, 138,* 10 (1967).
333. Smulson and Suhadolnik, *J. Biol. Chem., 242,* 2872 (1967).
334. Smith, Lummis and Grady, *Cancer Res., 19,* 847 (1959).
335. Shirato, Miyazaki and Suzuki, *J. Ferment. Technol.,* p. 60 (1966).
336. Data compiled by Duvall, *Cancer Chemother. Rep., 30,* 61 (1963).
337. Bennett, Jr., Schnebli, Vail, Allen and Montgomery, *Mol. Pharmacol., 2,* 432 (1966).
338. Smith, Gray, Carlson and Hanze, in *Advances in Enzyme Regulation,* p. 121, Weber, Ed. Pergamon Press, Elmsford, New York (1967).
339. Mihich, Simpson and Mulhern, *Cancer Res., 21,* 116 (1969).
340. Gerster, Carpenter, Robins and Townsend, *J. Med. Chem., 10,* 326 (1967).
341. Nichimura, Katagiri, Sato, Mayama and Shimaoka, *J. Antibiot. (Tokyo) Ser. A, 9,* 60 (1956).
342. Ohkuma, *J. Antibiot. (Tokyo) Ser. A, 13,* 361 (1960).
343. Ohkuma, *J. Antibiot. (Tokyo) Ser. A, 14,* 343 (1961).
344. Suhadolnik and Uematsu, *Fed. Proc., 26,* 855 (1967).
345. Rimzn, Sverak, Langlois, Bonar and Beard, *Cancer Res., 29,* 1707 (1969).
346. Taviten, Uretsky and Acs, *Biochim. Biophys. Acta, 179,* 50 (1969).
347. Wilson, *Cancer Chemother. Rep., 52,* 301 (1968).
348. Matsuoka and Umezawa, *J. Antibiot. (Tokyo) Ser. A, 13,* 114 (1960).
349. Matsuoka, *J. Antibiot. (Tokyo) Ser. A, 13,* 121 (1960).
350. Struyk and Stheeman, *British Patent 764,198* (1956); *Chem. Abstr., 51,* 10009a (1957).
351. Struyk and Stheeman, *Netherlands Patent 109,006* (1954); *Chem. Abstr., 62,* 11114g (1965).
352. Struyk and Stheeman, *Canadian Patent 514,164* (1955).
353. Doskočil and Sörm, *Biochem. Biophys. Res. Commun., 38,* 569 (1970).
354. Fox, Kuwada and Watanabe, *Tetrahedron Lett.,* p. 6029 (1968).
355. Isono, Arahi and Suzaki, *J. Amer. Chem. Soc., 91,* 7490 (1969).
356. Hashimoto, Uchida and Egami, *Biochim. Biophys. Acta, 199,* 535 (1970).
357. Uematsu and Suhadolnik, *Biochemistry, 9,* 1260 (1970).

# SPECTROPHOTOMETRIC CONSTANTS OF RIBONUCLEOTIDES

## TABLE 1
## ULTRAVIOLET ABSORBANCE OF MONONUCLEOTIDES IN 7*M* UREA

| Compound | pH 7.0 (0.05 M Phosphate) | | | | | | | pH 1 (0.1 M HCl) | | | | | | | pH 12 (0.01 M NaOH) | | | | | | |
|---|---|---|---|---|---|---|---|---|---|---|---|---|---|---|---|---|---|---|---|---|---|
| | | | Absorbance ratios[a] | | | | | | | Absorbance ratios[a] | | | | | | | Absorbance ratios[a] | | | | |
| | $\lambda_{min}$ | $\lambda_{max}$ | 240 | 250 | 270 | 280 | 290 | $\lambda_{min}$ | $\lambda_{max}$ | 240 | 250 | 270 | 280 | 290 | $\lambda_{min}$ | $\lambda_{max}$ | 240 | 250 | 270 | 280 | 290 |
| pA | 228 | 261 | 0.357 | 0.739 | 0.727 | 0.206 | | 232 | 258 | 0.395 | 0.795 | 0.762 | 0.277 | | 228 | 260 | 0.371 | 0.742 | 0.730 | 0.205 | |
| pC | 251 | 272.5 | 0.998 | 0.881 | 1.240 | 1.068 | 0.424 | 243 | 281 | 0.342 | 0.456 | 1.795 | 2.333 | 1.949 | 251 | 272.5 | 1.004 | 0.877 | 1.237 | 1.078 | 0.438 |
| pG | 225 | 254 | 0.718 | 1.098 | 0.810 | 0.662 | 0.297 | 229 | 258 | 0.458 | 0.868 | 0.741 | 0.665 | 0.538 | 232 | 258 | 0.513 | 0.863 | 0.969 | 0.650 | 0.133 |
| pU | 231 | 263 | 0.353 | 0.706 | 0.898 | 0.438 | | 231 | 262 | 0.361 | 0.718 | 0.879 | 0.423 | | 242 | 262 | 0.704 | 0.801 | 0.855 | 0.355 | |

## TABLE 2
## ULTRAVIOLET ABSORBANCE OF MONONUCLEOTIDES IN 97% $D_2O$

| Compound | pH 7.0 (0.05 M Phosphate) | | | | | | | pH 1 (0.1 M HCl) | | | | | | | pH 12 (0.01 M NaOH) | | | | | | |
|---|---|---|---|---|---|---|---|---|---|---|---|---|---|---|---|---|---|---|---|---|---|
| | | | Absorbance ratios[a] | | | | | | | Absorbance ratios[a] | | | | | | | Absorbance ratios[a] | | | | |
| | $\lambda_{min}$ | $\lambda_{max}$ | 240 | 250 | 270 | 280 | 290 | $\lambda_{min}$ | $\lambda_{max}$ | 240 | 250 | 270 | 280 | 290 | $\lambda_{min}$ | $\lambda_{max}$ | 240 | 250 | 270 | 280 | 290 |
| pA | 226 | 258 | 0.414 | 0.817 | 0.618 | 0.136 | | 229 | 257 | 0.473 | 0.875 | 0.650 | 0.206 | | 227 | 258 | 0.418 | 0.814 | 0.608 | 0.135 | |
| pC | 249 | 271 | 0.931 | 0.822 | 1.231 | 0.970 | 0.308 | 240 | 279 | 0.220 | 0.430 | 1.706 | 2.028 | 1.482 | 250 | 271 | 0.948 | 0.828 | 1.242 | 0.995 | 0.327 |
| pG | 223 | 252.5 | 0.589 | 1.181 | 0.829 | 0.664 | 0.248 | 227 | 256 | 0.535 | 0.957 | 0.740 | 0.705 | 0.505 | 230 | 264 | 0.543 | 0.883 | 0.951 | 0.549 | |
| pU | 230 | 262 | 0.386 | 0.747 | 0.845 | 0.372 | | 229 | 261 | 0.398 | 0.757 | 0.822 | 0.343 | | 242 | 261 | 0.712 | 0.821 | 0.793 | 0.276 | |

## TABLE 3
## ULTRAVIOLET ABSORBANCE OF MONONUCLEOTIDES IN 90% (V/V) ETHYLENE GLYCOL

| Compound | pH 7.0 (0.05 M Phosphate) | | | | | | | pH 1 (0.1 M HCl) | | | | | | | pH 12 (0.01 M NaOH) | | | | | | |
|---|---|---|---|---|---|---|---|---|---|---|---|---|---|---|---|---|---|---|---|---|---|
| | | | Absorbance ratios[a] | | | | | | | Absorbance ratios[a] | | | | | | | Absorbance ratios[a] | | | | |
| | $\lambda_{min}$ | $\lambda_{max}$ | 240 | 250 | 270 | 280 | 290 | $\lambda_{min}$ | $\lambda_{max}$ | 240 | 250 | 270 | 280 | 290 | $\lambda_{min}$ | $\lambda_{max}$ | 240 | 250 | 270 | 280 | 290 |
| pA | 229 | 260 | 0.374 | 0.743 | 0.743 | 0.251 | — | 233 | 263 | 0.472 | 0.816 | 0.744 | 0.282 | — | 228 | 259 | 0.405 | 0.775 | 0.726 | 0.237 | — |
| pC | 253 | 279 | 1.082 | 0.937 | 1.253 | 1.162 | 0.518 | 243 | 283 | 0.372 | 0.443 | 1.838 | 2.480 | 2.210 | 252 | 274 | 1.040 | 0.909 | 1.267 | 1.165 | 0.560 |
| pG | 224 | 251 | 0.768 | 1.131 | 0.781 | 0.650 | 0.339 | 230 | 257.5 | 0.463 | 0.852 | 0.704 | 0.595 | 0.479 | 226 | 254 | 0.699 | 1.069 | 0.768 | 0.651 | 0.329 |
| pU | 232 | 263 | 0.344 | 0.687 | 0.932 | 0.486 | | 232 | 262 | 0.366 | 0.712 | 0.897 | 0.445 | | 232 | 263 | 0.353 | 0.698 | 0.931 | 0.481 | |

From data of J. L. Hoffman and R. M. Bock. Mononucleoside-5′phosphates obtained from P-L Biochemicals, Milwaukee, Wisconsin. Spectra determined on a Cary 15 spectrometer.

# TABLE 4
# ULTRAVIOLET ABSORBANCE OF MONO– AND OLIGONUCLEOTIDES

| Compound | pH 7 | | | | | | | pH 1 | | | | | | | pH 12 | | | | | | | Slack[b] |
|---|---|---|---|---|---|---|---|---|---|---|---|---|---|---|---|---|---|---|---|---|---|---|
| | | | Absorbance ratios[a] | | | | | | | Absorbance ratios[a] | | | | | | | Absorbance ratios[a] | | | | | |
| | $\lambda_{min}$ | $\lambda_{max}$ | 240 | 250 | 270 | 280 | 290 | $\lambda_{min}$ | $\lambda_{max}$ | 240 | 250 | 270 | 280 | 290 | $\lambda_{min}$ | $\lambda_{max}$ | 240 | 250 | 270 | 280 | 290 | pct. |
| **Ap** | 226 | 259 | 0.41 | 0.79 | 0.65 | 0.14 | 0.00 | 229 | 257 | 0.46 | 0.87 | 0.67 | 0.22 | 0.00 | 227 | 259 | 0.40 | 0.79 | 0.66 | 0.14 | 0.00 | cat.[c] |
| **Cp** | 250 | 271 | 0.94 | 0.85 | 1.17 | 0.90 | 0.28 | 241 | 278 | 0.27 | 0.47 | 1.65 | 1.91 | 1.34 | 225 | 228 | 0.95 | 0.85 | 1.18 | 0.90 | 0.29 | cat. |
| **Gp** | 231 | 264 | 0.81 | 1.17 | 0.82 | 0.68 | 0.26 | 227 | 256 | 0.52 | 0.93 | 0.75 | 0.69 | 0.48 | 225 | 260 | 0.55 | 0.88 | 0.96 | 0.58 | 0.00 | cat. |
| **Up** | 230 | 261 | 0.39 | 0.75 | 0.82 | 0.33 | 0.00 | 229 | 261 | 0.41 | 0.78 | 0.79 | 0.30 | 0.00 | 242 | 261 | 0.75 | 0.83 | 0.79 | 0.27 | 0.00 | cat. |
| **pApA** | 227 | 258 | 0.41 | 0.83 | 0.66 | 0.19 | 0.00 | 229 | 256 | 0.45 | 0.86 | 0.67 | 0.21 | 0.00 | 229 | 258 | 0.42 | 0.82 | 0.66 | 0.18 | 0.00 | 1.0 |
| **ApCp** | 227 | 261 | 0.57 | 0.81 | 0.81 | 0.40 | 0.12 | 233 | 265 | 0.39 | 0.73 | 0.98 | 0.75 | 0.45 | 228 | 261 | 0.59 | 0.81 | 0.82 | 0.41 | 0.12 | 2.3 |
| **ApGp** | 225 | 256 | 0.58 | 0.97 | 0.71 | 0.37 | 0.13 | 228 | 257 | 0.47 | 0.88 | 0.71 | 0.42 | 0.24 | 229 | 259 | 0.46 | 0.83 | 0.79 | 0.35 | 0.00 | 0.9 |
| **ApUp** | 228 | 259 | 0.40 | 0.80 | 0.73 | 0.24 | 0.00 | 230 | 258 | 0.43 | 0.82 | 0.72 | 0.26 | 0.00 | 232 | 259 | 0.53 | 0.81 | 0.71 | 0.20 | 0.00 | 1.8 |
| **CpCp** | 250 | 269 | 0.94 | 0.85 | 1.17 | 0.90 | 0.30 | 241 | 278 | 0.24 | 0.44 | 1.69 | 1.96 | 1.37 | 250 | 270 | 0.95 | 0.85 | 1.19 | 0.92 | 0.33 | 1.3 |
| **CpGp** | 224 | 254 | 0.84 | 1.04 | 0.93 | 0.73 | 0.30 | 233 | 277 | 0.39 | 0.74 | 1.08 | 1.12 | 0.79 | 231 | 250 | 0.69 | 0.86 | 1.04 | 0.72 | 0.20 | 1.8 |
| **GpCp** | 224 | 255 | 0.83 | 1.01 | 0.94 | 0.70 | 0.27 | 233 | 276 | 0.43 | 0.76 | 1.08 | 1.10 | 0.76 | 231 | 267 | 0.71 | 0.88 | 1.04 | 0.70 | 0.17 | 1.7 |
| **GpUp** | 226 | 255 | 0.62 | 0.98 | 0.81 | 0.51 | 0.17 | 229 | 258 | 0.46 | 0.86 | 0.76 | 0.50 | 0.28 | 233 | 261 | 0.64 | 0.87 | 0.90 | 0.47 | 0.00 | 2.8 |
| **UpAp** | 228 | 259 | 0.42 | 0.81 | 0.70 | 0.24 | 0.00 | 229 | 257 | 0.45 | 0.85 | 0.72 | 0.27 | 0.00 | 231 | 259 | 0.54 | 0.83 | 0.70 | 0.21 | 0.00 | 2.1 |
| **UpGp** | 226 | 256 | 0.60 | 0.98 | 0.79 | 0.50 | 0.18 | 229 | 258 | 0.43 | 0.83 | 0.78 | 0.50 | 0.28 | 234 | 261 | 0.61 | 0.85 | 0.89 | 0.48 | 0.00 | 2.9 |
| **UpUp** | 229 | 260 | 0.41 | 0.78 | 0.80 | 0.33 | 0.00 | 230 | 260 | 0.42 | 0.79 | 0.78 | 0.31 | 0.00 | 242 | 260 | 0.79 | 0.85 | 0.78 | 0.29 | 0.00 | 0.4 |
| **pApApA** | 228 | 258 | 0.44 | 0.85 | 0.67 | 0.24 | 0.00 | 230 | 257 | 0.47 | 0.86 | 0.68 | 0.22 | 0.00 | 230 | 258 | 0.40 | 0.82 | 0.68 | 0.24 | 0.00 | 1.6 |
| **ApApCp** | 228 | 258 | 0.55 | 0.85 | 0.76 | 0.39 | 0.11 | 230 | 259 | 0.44 | 0.79 | 0.86 | 0.55 | 0.31 | 232 | 259 | 0.54 | 0.82 | 0.77 | 0.39 | 0.11 | 1.5 |
| **ApApGp** | 226 | 256 | 0.54 | 0.94 | 0.70 | 0.33 | 0.11 | 229 | 257 | 0.47 | 0.88 | 0.70 | 0.35 | 0.17 | 230 | 258 | 0.46 | 0.84 | 0.75 | 0.32 | 0.00 | 1.0 |
| **ApApUp** | 228 | 258 | 0.41 | 0.82 | 0.70 | 0.25 | 0.00 | 229 | 257 | 0.44 | 0.84 | 0.70 | 0.24 | 0.00 | 230 | 258 | 0.49 | 0.83 | 0.70 | 0.22 | 0.00 | 2.1 |
| **ApCpCp** | 229 | 262 | 0.66 | 0.82 | 0.89 | 0.52 | 0.18 | 236 | 270 | 0.37 | 0.67 | 1.16 | 1.04 | 0.69 | 234 | 263 | 0.67 | 0.81 | 0.91 | 0.54 | 0.20 | 2.6 |
| **ApCpGp** | 225 | 257 | 0.65 | 0.94 | 0.80 | 0.49 | 0.19 | 231 | 260 | 0.42 | 0.79 | 0.91 | 0.74 | 0.48 | 230 | 261 | 0.56 | 0.83 | 0.81 | 0.48 | 0.12 | 1.5 |
| **ApGpCp** | 226 | 257 | 0.65 | 0.94 | 0.81 | 0.47 | 0.18 | 231 | 260 | 0.43 | 0.79 | 0.89 | 0.70 | 0.45 | 230 | 261 | 0.56 | 0.83 | 0.87 | 0.47 | 0.12 | 1.8 |
| **ApGpUp** | 227 | 257 | 0.53 | 0.91 | 0.73 | 0.36 | 0.11 | 229 | 257 | 0.45 | 0.85 | 0.73 | 0.39 | 0.18 | 232 | 259 | 0.52 | 0.83 | 0.79 | 0.34 | 0.00 | 1.7 |
| **ApUpGp** | 227 | 257 | 0.51 | 0.90 | 0.73 | 0.36 | 0.11 | 229 | 258 | 0.43 | 0.83 | 0.74 | 0.40 | 0.19 | 231 | 259 | 0.51 | 0.82 | 0.80 | 0.34 | 0.00 | 1.9 |
| **CpApGp** | 226 | 257 | 0.67 | 0.93 | 0.82 | 0.53 | 0.21 | 231 | 260 | 0.42 | 0.79 | 0.91 | 0.74 | 0.48 | 230 | 260 | 0.58 | 0.84 | 0.87 | 0.51 | 0.12 | 0.9 |
| **CpCpGp** | 225 | 257 | 0.85 | 0.98 | 0.99 | 0.77 | 0.33 | 235 | 278 | 0.36 | 0.66 | 1.23 | 1.33 | 0.95 | 232 | 268 | 0.75 | 0.85 | 1.08 | 0.76 | 0.23 | 1.6 |
| **GpApCp** | 226 | 258 | 0.63 | 0.91 | 0.79 | 0.50 | 0.19 | 231 | 259 | 0.44 | 0.80 | 0.89 | 0.71 | 0.46 | 230 | 260 | 0.56 | 0.83 | 0.87 | 0.50 | 0.12 | 1.7 |
| **GpApUp** | 227 | 257 | 0.52 | 0.89 | 0.74 | 0.38 | 0.11 | 229 | 258 | 0.46 | 0.86 | 0.73 | 0.39 | 0.18 | 231 | 259 | 0.53 | 0.83 | 0.80 | 0.36 | 0.00 | 1.8 |
| **GpGpCp** | 224 | 254 | 0.79 | 1.06 | 0.87 | 0.65 | 0.27 | 230 | 260 | 0.45 | 0.82 | 0.92 | 0.87 | 0.61 | 231 | 266 | 0.61 | 0.87 | 1.00 | 0.65 | 0.14 | 2.6 |
| **GpGpUp** | 225 | 254 | 0.68 | 1.05 | 0.80 | 0.55 | 0.21 | 228 | 257 | 0.47 | 0.88 | 0.76 | 0.56 | 0.35 | 233 | 261 | 0.57 | 0.86 | 0.92 | 0.51 | 0.00 | 2.4 |
| **UpApGp** | 226 | 257 | 0.53 | 0.92 | 0.73 | 0.38 | 0.12 | 229 | 258 | 0.44 | 0.85 | 0.74 | 0.40 | 0.19 | 231 | 259 | 0.52 | 0.83 | 0.79 | 0.35 | 0.00 | 2.0 |
| **UpCpCp** | 230 | 264 | 0.72 | 0.83 | 0.99 | 0.65 | 0.20 | 236 | 273 | 0.32 | 0.59 | 1.28 | 1.22 | 0.79 | 248 | 266 | 0.92 | 0.88 | 1.02 | 0.67 | 0.22 | 2.3 |
| **UpUpGp** | 228 | 257 | 0.54 | 0.91 | 0.79 | 0.45 | 0.14 | 229 | 259 | 0.43 | 0.82 | 0.77 | 0.45 | 0.21 | 235 | 260 | 0.67 | 0.86 | 0.87 | 0.43 | 0.00 | 2.5 |
| **pApApApA** | 229 | 257 | 0.43 | 0.85 | 0.68 | 0.25 | 0.00 | 229 | 257 | 0.45 | 0.86 | 0.69 | 0.22 | 0.00 | 230 | 257 | 0.42 | 0.84 | 0.67 | 0.25 | 0.00 | 0.7 |
| **ApApApCp** | 230 | 258 | 0.50 | 0.84 | 0.72 | 0.34 | 0.00 | 232 | 258 | 0.44 | 0.82 | 0.79 | 0.43 | 0.00 | 231 | 258 | 0.51 | 0.83 | 0.73 | 0.35 | 0.00 | 2.1 |
| **ApApApGp** | 227 | 256 | 0.51 | 0.92 | 0.69 | 0.32 | 0.00 | 229 | 257 | 0.46 | 0.87 | 0.69 | 0.31 | 0.13 | 230 | 258 | 0.45 | 0.86 | 0.74 | 0.31 | 0.00 | 0.5 |
| **CpCpCpGp** | 227 | 267 | 0.86 | 0.95 | 1.01 | 0.78 | 0.34 | 236 | 278 | 0.34 | 0.62 | 1.31 | 1.44 | 1.03 | 235 | 268 | 0.78 | 0.86 | 1.09 | 0.79 | 0.27 | 1.3 |
| **UpUpUpGp** | 228 | 258 | 0.49 | 0.85 | 0.80 | 0.41 | 0.00 | 230 | 260 | 0.41 | 0.80 | 0.80 | 0.43 | 0.16 | 237 | 260 | 0.69 | 0.85 | 0.84 | 0.38 | 0.00 | 2.3 |

[a] Absorbance ratios were calculated from optical densities at 240, 250, 270, 280, and 290 nm relative to that at 260 nm. Where optical density was less than 0.1, ratios were not calculated.

[b] Slack = the absolute sum of the catalog mismatch at every wavelength.

[c] cat. = catalog.

Contributed by Jane N. Toal from data in Toal, Rushizky, Pratt and Sober, *Anal. Biochem.*, *23*, 60 (1968). Reproduced by permission of Academic Press, New York.

## TABLE 5
## HYPERCHROMICITY RATIOS OF OLIGONUCLEOTIDES AT DIFFERENT WAVELENGTHS[a]

### Spectrophotometric Constants of Ribonucleotides—(Continued)

### TABLE II. HYPERCHROMICITY RATIOS OF OLIGONUCLEOTIDES AT DIFFERENT WAVELENGTHS (mμ)[a]

| Compound | pH 7 | | | | | | pH 1 | | | | | | pH 12 | | | | | |
|---|---|---|---|---|---|---|---|---|---|---|---|---|---|---|---|---|---|---|
| | 240 | 250 | 260 | 270 | 280 | 290 | 240 | 250 | 260 | 270 | 280 | 290 | 240 | 250 | 260 | 270 | 280 | 290 |
| **pApA** | 1.11 | 1.11 | 1.16 | 1.13 | 0.81 | 0.00 | 1.03 | 1.02 | 1.02 | 1.02 | 1.00 | 0.00 | 1.09 | 1.10 | 1.15 | 1.12 | 0.84 | 0.00 |
| **ApCp** | 1.10 | 1.09 | 1.08 | 1.11 | 1.05 | 0.88 | 1.02 | 1.02 | 1.02 | 1.02 | 1.03 | 1.01 | 1.09 | 1.10 | 1.08 | 1.10 | 1.04 | 0.91 |
| **ApGp** | 1.08 | 1.06 | 1.07 | 1.09 | 1.07 | 0.96 | 1.06 | 1.04 | 1.02 | 1.02 | 1.04 | 1.00 | 1.04 | 1.03 | 1.03 | 1.03 | 0.98 | 0.00 |
| **ApUp** | 1.08 | 1.09 | 1.09 | 1.07 | 0.97 | 0.00 | 1.06 | 1.05 | 1.04 | 1.03 | 1.01 | 0.00 | 1.01 | 1.03 | 1.04 | 1.03 | 0.95 | 0.00 |
| **CpCp** | 1.09 | 1.08 | 1.08 | 1.10 | 1.10 | 0.96 | 1.05 | 1.04 | 1.02 | 1.02 | 1.02 | 1.01 | 1.10 | 1.10 | 1.10 | 1.10 | 1.09 | 0.94 |
| **CpGp** | 1.08 | 1.06 | 1.05 | 1.09 | 1.10 | 0.94 | 1.08 | 1.04 | 1.02 | 1.02 | 1.03 | 1.01 | 1.07 | 1.05 | 1.04 | 1.05 | 1.03 | 0.88 |
| **GpCp** | 1.11 | 1.09 | 1.05 | 1.08 | 1.16 | 1.05 | 1.04 | 1.05 | 1.05 | 1.06 | 1.08 | 1.10 | 1.04 | 1.03 | 1.04 | 1.05 | 1.05 | 0.95 |
| **GpUp** | 1.07 | 1.05 | 1.05 | 1.07 | 1.08 | 0.99 | 1.12 | 1.09 | 1.08 | 1.10 | 1.12 | 1.09 | 1.02 | 1.02 | 1.02 | 1.02 | 1.01 | 0.00 |
| **UpAp** | 1.07 | 1.05 | 1.08 | 1.10 | 0.99 | 0.00 | 1.06 | 1.05 | 1.06 | 1.05 | 0.99 | 0.00 | 1.06 | 1.06 | 1.06 | 1.05 | 0.95 | 0.00 |
| **UpGp** | 1.05 | 1.02 | 1.02 | 1.07 | 1.06 | 0.91 | 1.08 | 1.05 | 1.01 | 1.01 | 1.03 | 1.00 | 1.01 | 1.01 | 1.00 | 1.01 | 0.97 | 0.00 |
| **UpUp** | 0.98 | 1.00 | 1.04 | 1.07 | 1.05 | 0.00 | 1.01 | 1.02 | 1.03 | 1.03 | 0.98 | 0.00 | 0.96 | 0.98 | 1.01 | 1.01 | 0.94 | 0.00 |
| **pApApA** | 1.20 | 1.20 | 1.27 | 1.23 | 0.78 | 0.00 | 1.05 | 1.05 | 1.04 | 1.03 | 1.01 | 0.00 | 1.33 | 1.25 | 1.29 | 1.24 | 0.79 | 0.00 |
| **ApApCp** | 1.20 | 1.19 | 1.22 | 1.22 | 0.99 | 0.85 | 1.04 | 1.04 | 1.04 | 1.04 | 1.04 | 1.02 | 1.21 | 1.21 | 1.23 | 1.22 | 1.00 | 0.90 |
| **ApApGp** | 1.11 | 1.11 | 1.15 | 1.16 | 1.01 | 0.85 | 1.01 | 1.01 | 1.00 | 1.00 | 1.01 | 0.98 | 1.07 | 1.07 | 1.11 | 1.10 | 0.92 | 0.00 |
| **ApApUp** | 1.18 | 1.16 | 1.21 | 1.18 | 0.88 | 0.00 | 1,06 | 1.05 | 1.05 | 1.04 | 0.99 | 0.00 | 1.14 | 1.14 | 1.18 | 1.14 | 0.86 | 0.00 |
| **ApCpCp** | 1.22 | 1.20 | 1.18 | 1.20 | 1.16 | 0.91 | 1.09 | 1.08 | 1.06 | 1.05 | 1.05 | 1.02 | 1.25 | 1.24 | 1.20 | 1.19 | 1.17 | 0.95 |
| **ApCpGp** | 1.14 | 1.11 | 1.12 | 1.16 | 1.13 | 0.90 | 1.06 | 1.06 | 1.05 | 1.05 | 1.05 | 1.01 | 1.11 | 1.10 | 1.09 | 1.10 | 1.05 | 0.85 |
| **ApGpCp** | 1.14 | 1.11 | 1.12 | 1.13 | 1.16 | 1.00 | 1.04 | 1.05 | 1.03 | 1.04 | 1.07 | 1.06 | 1.07 | 1.06 | 1.07 | 1.08 | 1.04 | 0.89 |
| **ApGpUp** | 1.10 | 1.08 | 1.10 | 1.12 | 1.09 | 0.93 | 1.09 | 1.07 | 1.05 | 1.05 | 1.05 | 1.01 | 1.05 | 1.04 | 1.04 | 1.04 | 0.97 | 0.00 |
| **ApUpGp** | 1.11 | 1.09 | 1.10 | 1.12 | 1.10 | 0.93 | 1.10 | 1.08 | 1.05 | 1.04 | 1.04 | 1.00 | 1.04 | 1.04 | 1.04 | 1.04 | 0.97 | 0.00 |
| **CpApGp** | 1.16 | 1.16 | 1.16 | 1.18 | 1.10 | 0.87 | 1.06 | 1.04 | 1.03 | 1.04 | 1.05 | 1.01 | 1.10 | 1.09 | 1.10 | 1.11 | 1.02 | 0.88 |
| **CpCpGp** | 1.15 | 1.12 | 1.10 | 1.15 | 1.15 | 0.89 | 1.05 | 1.05 | 1.03 | 1.03 | 1.04 | 1.01 | 1.09 | 1.08 | 1.07 | 1.08 | 1.08 | 0.89 |
| **GpApCp** | 1.19 | 1.17 | 1.13 | 1.18 | 1.11 | 0.94 | 1.04 | 1.06 | 1.05 | 1.06 | 1.08 | 1.04 | 1.11 | 1.11 | 1.10 | 1.11 | 1.02 | 0.90 |
| **GpApUp** | 1.14 | 1.13 | 1.12 | 1.13 | 1.05 | 0.92 | 1.07 | 1.07 | 1.07 | 1.07 | 1.09 | 1.06 | 1.04 | 1.06 | 1.06 | 1.05 | 0.96 | 0.00 |
| **GpGpCp** | 1.13 | 1.11 | 1.07 | 1.12 | 1.19 | 1.05 | 1.08 | 1.07 | 1.06 | 1.08 | 1.12 | 1.11 | 1.07 | 1.04 | 1.03 | 1.04 | 1.02 | 0.89 |
| **GpGpUp** | 1.09 | 1.08 | 1.07 | 1.11 | 1.11 | 0.97 | 1.12 | 1.09 | 1.08 | 1.09 | 1.11 | 1.07 | 1.07 | 1.03 | 1.02 | 1.01 | 0.99 | 0.00 |
| **UpApGp** | 1.13 | 1.10 | 1.12 | 1.15 | 1.08 | 0.94 | 1.10 | 1.06 | 1.05 | 1.05 | 1.05 | 1.01 | 1.08 | 1.07 | 1.07 | 1.07 | 1.00 | 0.00 |
| **UpCpCp** | 1.09 | 1.07 | 1.09 | 1.14 | 1.14 | 0.91 | 0.99 | 1.02 | 1.03 | 1.04 | 1.05 | 1.02 | 1.02 | 1.03 | 1.07 | 1.11 | 1.10 | 0.90 |
| **UpUpGp** | 1.06 | 1.04 | 1.04 | 1.08 | 1.07 | 0.89 | 1.09 | 1.07 | 1.04 | 1.05 | 1.03 | 0.98 | 1.00 | 1.01 | 1.00 | 1.00 | 0.97 | 0.00 |
| **pApApApA** | 1.29 | 1.28 | 1.36 | 1.31 | 0.76 | 0.00 | 1.09 | 1.07 | 1.06 | 1.03 | 1.01 | 0.00 | 1.32 | 1.29 | 1.38 | 1.36 | 0.80 | 0.00 |
| **ApApApCp** | 1.23 | 1.23 | 1.27 | 1.25 | 0.92 | 0.00 | 1.02 | 1.03 | 1.03 | 1.03 | 1.02 | 0.00 | 1.22 | 1.25 | 1.29 | 1.26 | 0.96 | 0.00 |
| **ApApApGp** | 1.18 | 1.18 | 1.25 | 1.24 | 0.95 | 0.00 | 1.03 | 1.03 | 1.02 | 1.01 | 1.01 | 0.97 | 1.15 | 1.14 | 1.22 | 1.19 | 0.88 | 0.00 |
| **CpCpCpGp** | 1.21 | 1.16 | 1.15 | 1.20 | 1.22 | 0.93 | 1.07 | 1.06 | 1.04 | 1.04 | 1.05 | 1.02 | 1.16 | 1.12 | 1.12 | 1.14 | 1.14 | 0.92 |
| **UpUpUpGp** | 1.08 | 1.06 | 1.07 | 1.10 | 1.07 | 0.00 | 1.10 | 1.06 | 1.05 | 1.05 | 1.03 | 0.97 | 1.00 | 1.02 | 1.03 | 1.02 | 0.98 | 0.00 |

[a] Hyperchromicity ratios were calcualted as the optical density of a hydrolyzed compound divided by the optical density of the corresponding intact compound at the same wavelength. Where the optical density was less than 0.1, ratios were not calculated.

Contributed by Jane N. Toal from data in Toal, Rushizky, Pratt and Sober, *Anal. Biochem., 23,* 60 (1968). Reproduced by permission of Academic Press, New York.

# GENERAL FEATURES OF TRANSFER RIBONUCLEIC ACID STRUCTURE

DR. M. A. Q. SIDDIQUI

A survey of the structural features of twenty-seven transfer ribonucleic acid (tRNA) species of known primary sequences is summarized in the following table and in Figures 1 and 2. Whereas many features are shared by all tRNAs, significant structural differences exist even among tRNAs that are specific for the same amino acid. The base sequences of tRNAs can be arranged to fit into the general cloverleaf structure in spite of the variations in chain lengths (see the table and Figure 1). The amino acid acceptor stem (a) consists of seven base-pairs, with a few exceptions; the T-Ψ-C and anticodon stems (b and d respectively) have five base-pairs each, and the dihydrouridine stem (e) contains either three or four base-pairs. Loops I and II are always seven nucleotides long, whereas loop III has from seven to twelve nucleotides. The extra arm (c) shows considerable variations.

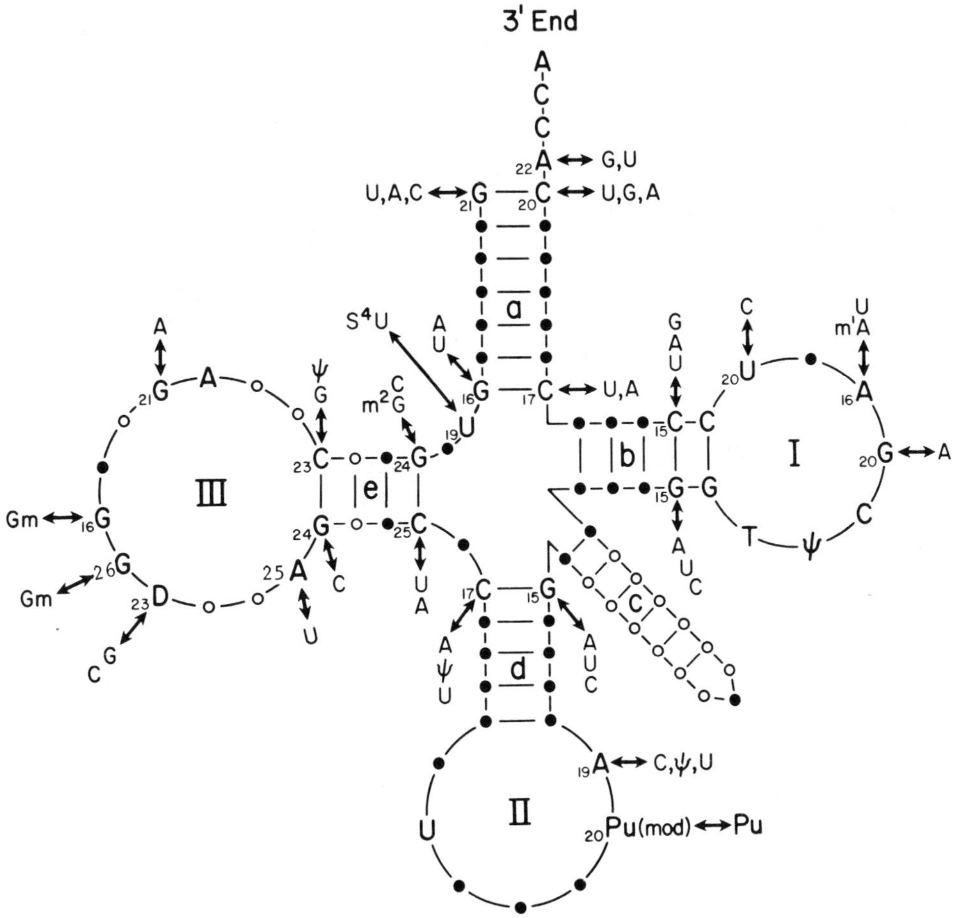

FIGURE 1. General cloverleaf structure of tRNA, with positions of nucleotides common to the known tRNA sequences. •: positions where nucleotide is different in different tRNAs; o: nucleotides that may be absent; I, II, III: T-Ψ-C, anticodon, and dihydrouridine loops respectively; a: amino acid acceptor stem; b: T-Ψ-C stem; c: extra arm; d: anticodon stem; e: dihydrouridine stem; the nucleotides found in most tRNAs, but not in all, are shown by a subscript number preceding the letter, the number denoting the occurrence of the nucleotides out of twenty-seven tRNAs examined; the nucleotides occurring less frequently are indicated by arrows.

An important feature of the tRNA structure is the presence of a large number of minor nucleotides. They are distributed over a large portion of the molecule except for the amino acid acceptor stem (Figure 2). While a majority of minor nucleotides are found at the same sites in all tRNAs, a few , for example pseudouridine (Ψ) and dihydrouridine (D), are located at different positions in the molecule.

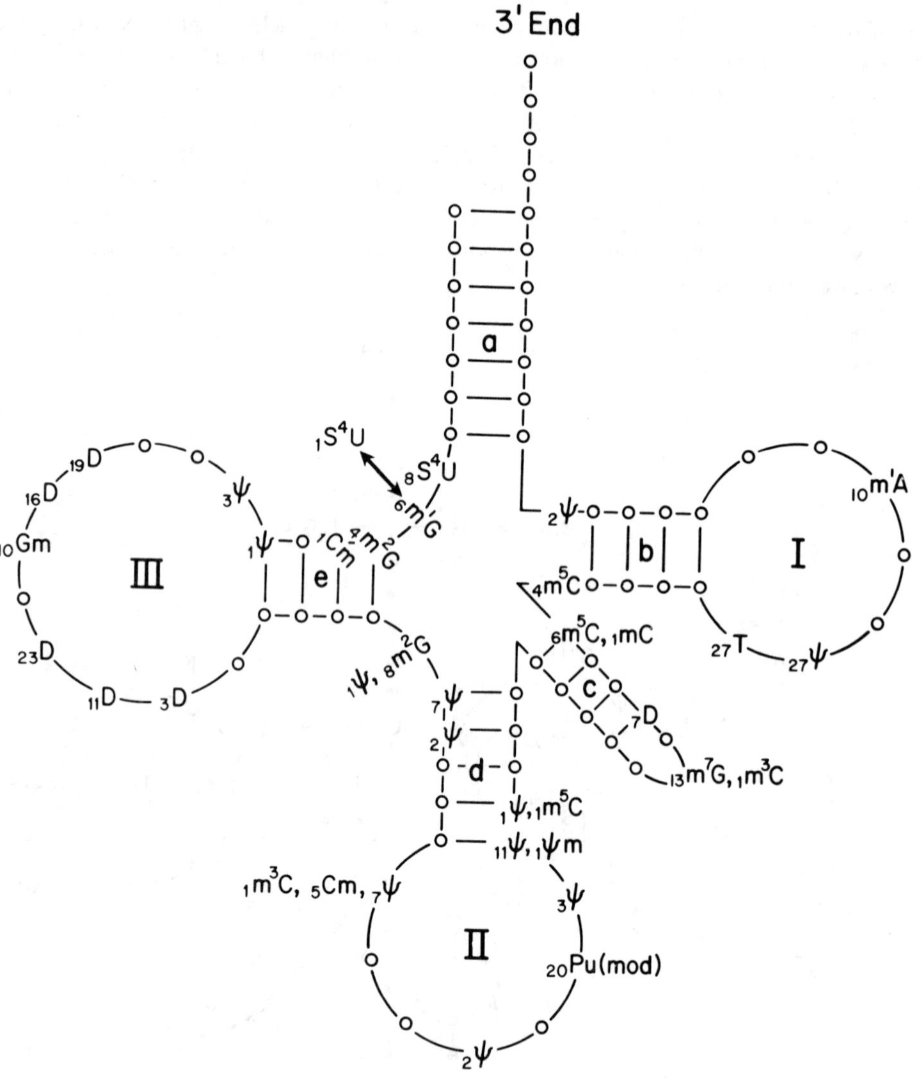

FIGURE 2. Positions and number of minor nucleotides in known tRNA sequences. o: position of nucleotide; other symbols as in Figure 1.

## ABBREVIATIONS AND SYMBOLS

| | | |
|---|---|---|
| A | = | adenosine |
| G | = | guanosine |
| C | = | cytidine |
| U | = | uridine |
| Pu | = | purine |
| I | = | inosine |
| ψ | = | pseudouridine |
| T | = | ribothymidine |
| $m^1A$ | = | 1-methyl-adenosine |
| $m^6A$ | = | $N^6$-methyl-adenosine |
| $m^1G$ | = | 1-methyl-guanosine |
| $m^2G$ | = | $N^2$-methyl-guanosine |
| $m^7G$ | = | 7-methyl-guanosine |
| $m_2^2G$ | = | $N^2$-dimethyl-guanosine |
| $m^3C$ | = | 3-methyl-cytidine |
| $m^5C$ | = | 5-methyl-cytidine |
| $m^1I$ | = | 1-methyl-inosine |
| Gm | = | 2′-O-methyl-guanosine |
| Cm | = | 2′-O-methyl-cytidine |
| Um | = | 2′-O-methyl-uridine |
| Ψm | = | 2′-O-methyl-pseudouridine |
| $i^6A$ | = | $N^6$-isopentenyl-adenosine |
| $ms^2i^6A$ | = | 2′-thiomethyl-$N^6$-isopentenyl-adenosine |
| $t^6A$ | = | N-(purin-6-ylcarbamoyl)-threonine riboside |
| $ac^4C$ | = | $N^4$-acetyl-cytidine |
| D | = | 5,6-dihydrouridine |
| $S^4U$ | = | 4-thiouridine |
| Y | = | modified base with strong fluorescent properties[a] |
| X | = | unknown base |
| * | = | modified base, modification unknown |

### Code for Table Columns

| | | |
|---|---|---|
| (1) | = | Amino Acid Acceptor Stem (a) |
| (2) | = | T-Ψ-C Stems (b) |
| (3) | = | T-Ψ-C Loop (I) |
| (4) | = | Extra Arm (c) |
| (5) | = | Anticodon Stem (d) |
| (6) | = | Anticodon Loop (II) |
| (7) | = | D-Stem (e) |
| (8) | = | D-Loop (III) |
| (A) | = | Number of Base Pairs |
| (B) | = | Number of Bases |
| (C) | = | Number of G-C Pairs |
| (D) | = | Modified Bases |

[a] For structure, see Reference 22.

| Total No. of Bases | (1) | | | (2) | | | (3) | | (4) | | (5) | | | (6) | | (7) | | | (8) | | Modified Bases |
|---|---|---|---|---|---|---|---|---|---|---|---|---|---|---|---|---|---|---|---|---|---|
| | (A) | (C) | (D) | (A) | (C) | (D) | (B) | (D) | (B) | (D) | (A) | (C) | (D) | (B) | (D) | (A) | (C) | (D) | (B) | (D) | |
| **$Ala_1$ (Yeast)[1,2] (Yeast $tRNA^{Ala}_{I}$)** | | | | | | | | | | | | | | | | | | | | | |
| 78 | 6 | 5 | – | 5 | 4 | – | 7 | 1 $\psi$ | 5 | 1 D | 5 | 4 | – | 7 | 1 $\psi$<br>1 $m^1I$ | 4 | 4 | – | 10 | 2 D | 2 $\psi$, 1 $m^1I$,<br>1 $m^2_2G$, 2 D,<br>1 $m^1G$ |
| **Asp (Yeast)[3] (Yeast $tRNA^{Asp}$)** | | | | | | | | | | | | | | | | | | | | | |
| 75 | 7 | 4 | – | 5 | 5 | 1 $m^5C$ | 7 | 1 $\psi$ | 4 | – | 5 | 4 | – | 7 | 1 $\psi$<br>1 $m^1G$ | 4 | 4 | 1 $\psi$ | 8 | 2 D | 3 $\psi$, 1 $m^5C$,<br>1 $m^1G$, 2 D |
| **Gly *(Escherichia coli)*[4] *(E. coli* $tRNA^{Gly}_{III}$)** | | | | | | | | | | | | | | | | | | | | | |
| 76 | 7 | 7 | – | 5 | 3 | – | 7 | 1 $\psi$ | 5 | 1 $m^7G$ | 5 | 4 | – | 7 | – | 4 | 3 | – | 8 | 3 D | 1 $\psi$ |
| **$Leu_1$ *(Escherichia coli)*[5,6] (*E. coli* B $tRNA^{Leu}$ and Leu *E. coli* K12)** | | | | | | | | | | | | | | | | | | | | | |
| 88 | 6 | 5 | – | 5 | 5 | – | 7 | 1 $\psi$ | 15 | – | 5 | 2 | 1 $\psi$ | 7 | 1 $\psi$<br>1 mG | 3 | 3 | – | 11 | 3 D<br>1 Gm | 3 $\psi$, 2 Gm,<br>3 D |
| **$Leu_2$ *(Escherichia coli)*[7] (*E. coli* K12 $tRNA^{Leu}_2$)** | | | | | | | | | | | | | | | | | | | | | |
| 87 | 7 | 5 | – | 5 | 4 | – | 7 | 1 $\psi$ | 15 | – | 5 | 3 | – | 7 | 1 $\psi$<br>1 G* | 3 | 2 | – | 11 | 3 D<br>1 Gm | 2 $\psi$, 1 G*,<br>3 D, 1 Gm |
| **$Leu_3$ (Yeast)[8,9] (Yeast $tRNA^{Leu}_3$)** | | | | | | | | | | | | | | | | | | | | | |
| 85 | 7 | 3 | – | 5 | 1 | – | 7 | 1 $\psi$ | 13 | mC | 5 | 3 | 1 $\psi$ | 7 | 1 $\psi$<br>1 G* | 3 | 3 | G*<br>C* | 11 | 2 D<br>1 Gm | 3 $\psi$, 1 mC,<br>1 $G^+$, 1 G*,<br>1 G*, 1 C*,<br>1 Gm |

| Total No. of Bases | (1) (A) | (1) (C) | (1) (D) | (2) (A) | (2) (C) | (2) (D) | (3) (B) | (3) (D) | (4) (B) | (4) (D) | (5) (A) | (5) (C) | (5) (D) | (6) (B) | (6) (D) | (7) (A) | (7) (C) | (7) (D) | (8) (B) | (8) (D) | Modified Bases |
|---|---|---|---|---|---|---|---|---|---|---|---|---|---|---|---|---|---|---|---|---|---|
| **Leu ($T_4$-Coded)[10] (T4-Bacteriophage-Coded $tRNA^{Leu}$)** | | | | | | | | | | | | | | | | | | | | | |
| 90 | 7 | 7 | – | 5 | 4 | – | 7 | – | 18 | – | 5 | 3 | 1 ψ | 7 | 1 A*<br>1 X | 3 | 2 | – | 10 | 2 D<br>1 Gm | 2 ψ, 1 A*,<br>1 X, 2 D,<br>1 Gm |
| **Ileu *(Escherichia coli)*[11] (*E. coli* $tRNA^{Ileu}$)** | | | | | | | | | | | | | | | | | | | | | |
| 77 | 7 | 4 | – | 5 | 3 | 1 ψ | 7 | 1 ψ | 5 | 1 $m^7G$<br>1 X | 5 | 4 | – | 7 | 1 A* | 4 | 3 | – | 9 | 3 D | 2 ψ, 1 X,<br>1 $m^7G$, 1 A*,<br>3 D |
| **Ileu (Yeast, *Torulopsis utilis*)[12,13] (*T. utilis* $tRNA^{Ileu}$)** | | | | | | | | | | | | | | | | | | | | | |
| 77 | 7 | 5 | – | 5 | 3 | – | 7 | 1 ψ<br>1 $m^1A$ | 5 | 1 $m^5C$<br>1 D | 5 | 3 | 1 ψ | 7 | 1 $t^6A$ | 3 | 3 | – | 11 | 4 D | 1 $m^1A$, 2 ψ,<br>1 $m^5C$, 5 D,<br>1 $t^6A$, 1 $m^2_2G$ |
| **$Met_f$ *(Escherichia coli)*[14-16] (*E. coli* $tRNA^{Met}_f$)** | | | | | | | | | | | | | | | | | | | | | |
| 77 | 6 | 6 | – | 5 | 4 | – | 7 | 1 ψ | 5 | 1 $m^7G$ | 5 | 4 | – | 7 | 1 Cm | 4 | 3 | – | 9 | 1 D | 1 ψ, 1 $m^7G$,<br>1 Cm, 1 D,<br>1 $S^4U$ |
| **$Met_M$ *(Escherichia coli)*[17,18] (*E. coli* $tRNA^{Met}_m$)** | | | | | | | | | | | | | | | | | | | | | |
| 77 | 7 | 5 | – | 4 | 3 | – | 7 | 1 ψ | 5 | 1 $m^7G$<br>1 X | 5 | 2 | 1 ψ | 7 | 1 $C^*$<br>1 A* | 4 | 3 | – | 9 | 3 D<br>1 Gm | 2 ψ, 1 X,<br>1 $m^7G$, 1 A*<br>1 C*, 1 $S^4U$ |
| **Phe *(Escherichia coli)*[19] (*E. coli* $tRNA^{Phe}$)** | | | | | | | | | | | | | | | | | | | | | |
| 76 | 7 | 6 | – | 5 | 3 | – | 7 | 1 ψ | 5 | 1 X<br>1 $m^7G$ | 5 | 4 | 1 ψ | 7 | 1 ψ<br>1 $ms^2i^6A$ | 4 | 3 | – | 8 | 2 D | 3 ψ, 1 X, 1 $m^7G$,<br>1 $ms^2i^6A$, 2 D,<br>1 $S^4U$ |

| Total No. of Bases | (1) (A) | (1) (C) | (1) (D) | (2) (A) | (2) (C) | (2) (D) | (3) (B) | (3) (D) | (4) (B) | (4) (D) | (5) (A) | (5) (C) | (5) (D) | (6) (B) | (6) (D) | (7) (A) | (7) (C) | (7) (D) | (8) (B) | (8) (D) | Modified Bases |
|---|---|---|---|---|---|---|---|---|---|---|---|---|---|---|---|---|---|---|---|---|---|
| **Phe (Yeast)**[20–22] **(Yeast tRNA$^{Phe}$)** | | | | | | | | | | | | | | | | | | | | | |
| 76 | 7 | 3 | – | 5 | 3 | 1 $m^5C$ | 7 | 1 ψ<br>1 $m^1A$ | 5 | 1 $m^7G$ | 5 | 2 | 1 ψ<br>1 $m^5C$ | 7 | 1 Cm<br>1 Gm<br>1 Y | 4 | 3 | 1 $m^2G$ | 8 | 2 D | 1 $m^1A$, 2 ψ,<br>2 $m^5C$, 1 $m^7G$,<br>1 X, 1 Gm,<br>1 $m_2^2G$, 2 D,<br>1 $m^2G$ |
| **Phe (Wheat Germ)**[23–25] **(Wheat Germ tRNA$^{Phe}$)** | | | | | | | | | | | | | | | | | | | | | |
| 76 | 6 | 5 | – | 5 | 4 | – | 7 | 1 ψ<br>1 $m^1A$ | 5 | 1 $m^7G$<br>1 D | 5 | 2 | 2 ψ | 7 | 1 Cm<br>1 Gm<br>1 Y | 4 | 2 | 1 $m^2G$ | 8 | 2 D | 1 $m^1A$, 3 ψ,<br>3 D, 1 $m^7G$,<br>1 Y, 1 Gm,<br>1 Cm, 1 $m_2^2G$,<br>1 $m^2G$ |
| **Ser *(Escherichia coli)***[26] **(*E. coli* tRNA$_2^{Ser}$)** | | | | | | | | | | | | | | | | | | | | | |
| 88 | 7 | 4 | – | 5 | 3 | – | 7 | 1 ψ | 16 | – | 5 | 4 | – | 7 | 1 $ms^2i^6A$<br>1 V<br>1 N | 3 | 3 | – | 11 | 2 D<br>1 $m^2G$ | 1 ψ, 1 $ms^2i^6$,<br>½V, 1 X, 2 D,<br>1 $m^2G$, 1 $S^4U$ |
| **Ser (Yeast)**[27,28] **(Yeast tRNA$_2^{Ser}$, tRNA$_1^{Ser}$)** | | | | | | | | | | | | | | | | | | | | | |
| 85 | 7 | 3 | – | 5 | 4 | – | 7 | 1 ψ | 14 | 1 Um<br>1 $m^5C$ | 5 | 1 | 1 ψ | 7 | 1 ψ<br>1 $i^6A$ | 3 | 3 | 1 $ac^4C$ | 10 | 3 D<br>1 Gm | 3 ψ, 1 $m^5C$,<br>1 Vm, 1 $i^6A$,<br>1 $m_2^2G$, 3 D,<br>1 Gm, 1 $ac^4C$ |
| **Ser (Rat Liver)**[29] **(Rat Liver tRNA$^{Ser}$)** | | | | | | | | | | | | | | | | | | | | | |
| 85 | 7 | 4 | – | 5 | 4 | – | 7 | 1 ψ<br>1 $m^1A$ | 14 | 1 Um<br>1 $m^5C$<br>1 $m^3C$ | 5 | 2 | 1 ψ<br>1 ψm | 7 | 1 $m^3C$<br>1 $i^6A$ | 3 | 3 | 1 $ac^4C$ | 10 | 3 D<br>1 Gm | 1 $m^1A$, 2 ψ,<br>1 $m^5C$, 2 $m^3C$,<br>1 Vm, 1 ψm,<br>1 $i^6A$, 1 $m_2^2G$,<br>3 D, 1 Gm,<br>1 $ac^4C$ |

| Total No. of Bases | (1) (A) | (1) (C) | (1) (D) | (2) (A) | (2) (C) | (2) (D) | (3) (B) | (3) (D) | (4) (B) | (4) (D) | (5) (A) | (5) (C) | (5) (D) | (6) (B) | (6) (D) | (7) (A) | (7) (C) | (7) (D) | (8) (B) | (8) (D) | Modified Bases |
|---|---|---|---|---|---|---|---|---|---|---|---|---|---|---|---|---|---|---|---|---|---|
| **Try *(Escherichia coli)*[30] (*E. coli* su$^+$ ( – su$^-$) tRNA$^{Try}$)** | | | | | | | | | | | | | | | | | | | | | |
| 76 | 7 | 6 | – | 5 | 3 | – | 7 | 1 ψ | 5 | 1 m$^7$G | 5 | 4 | – | 7 | 1 Cm<br>1 ms$^2$i$^6$ A | 4 | 2 | – | 8 | 3 D | 1 ψ, 1 m$^7$G,<br>1 ms$^2$i$^6$ A,<br>1 Cm, 3 D,<br>1 S$^4$U |
| **Try (Yeast)[31] (Yeast tRNA$^{Try}$)** | | | | | | | | | | | | | | | | | | | | | |
| 76 | 7 | 4 | – | 5 | 3 | 1 ψ | 7 | 1 ψ<br>1 m$^1$A | 5 | 1 m$^7$G<br>1 D | 5 | 2 | 3 ψ | 7 | 2 Cm | 4 | 3 | 1 m$^2$G | 7 | 2 D<br>1 Gm | 6 ψ, 1 m$^1$A,<br>3 D, 1 m$^7$G,<br>2 Cm, 1 m$^2$G,<br>1 Gm |
| **Tyr (*Escherichia coli* Su$^+$)[32-37] (*E. coli* [I(su$^-$), su$^+_{III}$, II] tRNA$^{Tyr}$)** | | | | | | | | | | | | | | | | | | | | | |
| 85 | 7 | 6 | – | 5 | 3 | – | 7 | 1 ψ | 13 | – | 5 | 3 | 1 ψ | 7 | 1 ms$^2$i$^6$ A<br>1 G* | 3 | 3 | – | 11 | 1 Gm | 2 ψ, 1 ms$^2$i$^6$ A,<br>1 G*, 1 Gm,<br>2 S$^4$U |
| **Tyr (Yeast, brewer's)[38-40] (Yeast tRNA$^{Tyr}$)** | | | | | | | | | | | | | | | | | | | | | |
| 78 | 7 | 5 | – | 5 | 5 | – | 7 | 1 ψ<br>1 m$^1$A | 5 | 1 D<br>1 m$^5$C | 5 | 2 | 1 ψ | 7 | 1 ψ<br>1 i$^6$A | 3 | 3 | 1 m$^2$G | 12 | 5 D<br>1 Gm | 1 m$^1$A, 3 ψ,<br>1 m$^5$C, 6 D,<br>1 i$^6$A, 1 m$^2_2$G,<br>1 Gm, 1 m$^2$G |
| **Tyr (Yeast, *Torulopsis utilis*)[41] (*T. utilis* tRNA$^{Tyr}$)** | | | | | | | | | | | | | | | | | | | | | |
| 78 | 7 | 5 | – | 5 | 5 | – | 7 | 1 ψ<br>1 m$^1$A | 5 | 1 D<br>1 m$^5$C | 5 | 2 | 2 ψ | 7 | 1 ψ<br>1 i$^6$A | 3 | 3 | 1 m$^2$G | 12 | 5 D<br>1 Gm | 1 m$^1$A, 4 ψ,<br>1 m$^5$C, 6 D,<br>1 i$^6$A, 1 m$^2_2$G,<br>1 Gm, 1 m$^2$G,<br>1 m$^1$G |

| Total No. of Bases | (1) | | | (2) | | | (3) | | (4) | | (5) | | | (6) | | (7) | | | (8) | | Modified Bases |
|---|---|---|---|---|---|---|---|---|---|---|---|---|---|---|---|---|---|---|---|---|---|
| | (A) | (C) | (D) | (A) | (C) | (D) | (B) | (D) | (B) | (D) | (A) | (C) | (D) | (B) | (D) | (A) | (C) | (D) | (B) | (D) | |
| **Val$_1$ *(Escherichia coli)*[42-44] (*E. coli* tRNA$_1^{Val}$)** | | | | | | | | | | | | | | | | | | | | | |
| 76 | 7 | 4 | – | 5 | 4 | – | 7 | 1 $\psi$ | 5 | 1 m$^7$G | 5 | 4 | – | 7 | 1 m$^6$A<br>1 U* | 4 | 3 | – | 8 | 1 D | 1 $\psi$, 1 m$^7$G, 1 m$^6$A, 1 U*, 1 D, 1 S$^4$U |
| **Val$_1$ (Yeast)** | | | | | | | | | | | | | | | | | | | | | |
| 78 | 7 | 3 | – | 5 | 4 | 1 m$^5$C | 7 | 1 $\psi$<br>1 m$^1$A | 5 | 1 m$^7$G<br>1 D | 5 | 3 | 1 $\psi$ | 7 | 1 $\psi$ | 3 | 2 | – | 11 | 3 D<br>1 $\psi$ | 1 m$^1$A, 4 $\psi$, 1 m$^5$C, 1 m$^7$G, 4 D, 1 m$^1$G |
| **Val (Yeast, *Torulopsis utilis*)[45-47] (*T. utilis* tRNA$^{Val}$)** | | | | | | | | | | | | | | | | | | | | | |
| 75 | 7 | 3 | $\theta$ | 5 | 4 | – | 7 | 1 $\psi$<br>1 m$^1$A | 3 | 1 m$^5$C | 5 | 3 | 1 $\psi$ | 7 | 1 $\psi$ | 3 | 2 | – | 11 | 3 D<br>1 $\psi$ | 1 m$^1$A, 4 $\psi$, 1 m$^5$C, 3 D, 1 m$^1$G |
| **Val$_2$ *(Escherichia coli)*[48] (*E. coli* tRNA$_2^{Val}$)** | | | | | | | | | | | | | | | | | | | | | |
| 76 | 7 | 6 | – | 5 | 3 | – | 7 | 1 $\psi$ | 5 | 1 m$^7$G<br>2 X | 5 | 4 | – | 7 | – | 3 | 2 | – | 8 | 3 D | 1 $\psi$, 2 X, 1 m$^7$G, 3 D, 1 S$^4$U |
| **Val$_2$ (Yeast, brewer's)[49] (Yeast tRNA$_2^{Val}$)** | | | | | | | | | | | | | | | | | | | | | |
| 77 | 7 | 3 | – | 5 | 4 | 1 m$^5$C | 7 | 1 $\psi$<br>1 m$^1$A | 5 | 1 m$^7$G<br>1 D | 5 | 3 | 1 $\psi$ | 7 | 1 $\psi$ | 3 | 2 | – | 10 | 3 D<br>1 $\psi$ | 1 m$^1$A, 4 $\psi$, 4 D, 1 m$^7$G, 1 m$^1$G |

## REFERENCES


1. Holley, R. W., Apgar, J., Everett, G. A., Madison, J. T., Marquisee, M., Merrill, S. H., Penswick, J. R., and Zamir, A., *Science, 147,* 1462 (1965).
2. Merril, C. R., *Biopolymers, 6,* 1727 (1968).
3. Keith, G., Gangloff, J., Ebel, J.-P., and Dirheimer, G., *C. R. Acad. Sci. (Paris), 271,* 613 (1970).
4. Squires, C., and Carbon, J., *Fed. Proc., 30,* 1218 (1971).
5. Dube, S. K., Marcker, K. A., and Ydelevich, A., *FEBS Lett., 9,* 168 (1970).
6. Blank, H. U., and Söll, D., *Biochem. Biophys. Res. Commun., 43,* 1192 (1971).
7. Blank, H. U., and Söll, D., *Biochem. Biophys. Res. Commun., 43,* 1192 (1971).
8. Kowalski, S., Yamane, T., and Fresco, J. R., *Science, 172,* 384 (1971).
9. Chang, S. H., and Miller, N., *Fed. Proc., 30,* 1101 (1971).
10. Pinkerton, T. C., Paddock, G., and Abelson, J. N., *Fed. Proc., 30,* 1218 (1971).
11. Yarus, M., and Barrell, B. G., *Biochem. Biophys. Res. Commun., 43,* 729 (1971).
12. Takemura, S., Murakami, M., and Miyazaki, M., *J. Biochem., 65,* 489 (1969).
13. Takemura, S., Murakami, M., and Miyazaki, M., *J. Biochem., 65,* 553 (1969).
14. Dube, S. K., Marcker, K. A., Clark, B. F. C., and Cory, S., *Nature, 218,* 232 (1968).
15. Dube, S. K., Marcker, K. A., Clark, B. F. C., and Cory, S., *Eur. J. Biochem., 8,* 244 (1969).
16. Dube, S. K., and Marcker, K. A., *Eur. J. Biochem., 8,* 256 (1969).
17. Cory, S., Marcker, K. A., Dube, S. K., and Clark, B. F. C., *Nature, 220,* 1039 (1968).
18. Cory, S., and Marcker, K. A., *Eur. J. Biochem., 12,* 177 (1970).
19. Barrell, G. G., and Sanger, F., *FEBS Lett., 3,* 275 (1969).
20. RajBhandary, U. L., Chang, S. H., Stuart, A., Faulkner, R. D., Hoskinson, R. M., and Khorana, H. G., *Proc. Nat. Acad. Sci. U.S.A., 57,* 751 (1967).
21. RajBhandary, U. L., and Simon, H. C., *J. Biol. Chem., 243,* 598 (1968).
22. Nakanishi, K., Furutachi, N., Funamizu, M., Grunberger, D., and Weinstein, I. B., *J. Amer. Chem. Soc., 92,* 7617 (1970).
23. Dudock, B. S., Katz, G., Taylor, E. K., and Holley, R. W., *Proc. Nat. Acad. Sci. U.S.A., 62,* 941 (1969).
24. Katz, G., and Dudock, B. S., *J. Biol. Chem., 244,* 3062 (1969).
25. Dudock, B. S., and Katz, G., *J. Biol. Chem., 244,* 3069 (1969).
26. Ishikura, H., Yamada, Y., and Nishimura, S., *FEBS Lett., 16,* 68 (1971).
27. Zachau, H. G., Dutting, D., and Feldmann, H., *Angew. Chem., 78,* 392 (1966); *Angew. Chem. Int. Ed. Engl., 5,* 422 (1966).
28. Zachau, H. G., Dutting, D., and Feldmann, H., *Hoppe-Seyler's Z. Physiol. Chem., 347,* 212 (1966).
29. Staehelin, M., Rogg, H., Baguley, B. C., Ginsberg, T., and Wehrile, W., *Nature, 219,* 1363 (1968).
30. Hirsh, D., *Nature, 228,* 57 (1970).
31. Keith, G., Roy, A., Ebel, J. P., and Dirheimer, B., *FEBS Lett., 17,* 306 (1971).
32. Goodman, H. M., Abelson, J., Landy, A., Brenner, S., and Smith, J. D., *Nature, 217,* 1019 (1968).
33. Goodman, H. M., Abelson, J., Landy, A., Zadrdazil, S., and Smith, J. D., *Eur. J. Biochem., 13,* 461 (1970).
34. Harada, F., Gross, H. J., Kimura, F., Chang, S. H., Nishimura, S., and RajBhandary, U. L., *Biochem. Biophys. Res. Commun., 33,* 299 (1968).
35. Abelson, J., Brenner, S., Gefter, M., Landy, A., Russell, R., Smith, J. D., and Barnett, L., *FEBS Lett., 3,* 1 (1969).
36. Abelson, J., Gefter, M. L., Barnett, L., Landy, A., Russell, R. L., and Smith, J. D., *J. Mol. Biol., 47,* 15 (1970).
37. Smith, J. D., Barnett, L., Brenner, S., and Russell, R. L., *J. Mol. Biol., 54,* 1 (1970).
38. Madison, J. T., Everett, G. A., and Kung, H., *Science, 153,* 531 (1966).
39. Madison, J. T., Everett, G. A., and Kung, H., *J. Biol. Chem., 242,* 1318 (1967).
40. Madison, J. T., and Kung, H., *J. Biol. Chem., 242,* 1325 (1967).
41. Hashimoto, S., Miyazaki, M., and Takemura, S., *J. Biochem., 65,* 659 (1969).
42. Yaniv, M., and Barrell, B. G., *Nature, 222,* 278 (1969).
43. Harada, F., Kimura, F., and Nishimura, S., *Biochim. Biophys. Acta, 195,* 590 (1969).
44. Murao, K., Saneyoshi, M., Harada, M., and Nishimura, S., *Biochem. Biophys. Res. Commun., 38,* 657 (1970).
45. Takemura, S., Mizutani, T., and Miyazaki, M., *J. Biochem., 63,* 277 (1968).
46. Takemura, S., Mizutani, T., and Miyazaki, M., *J. Biochem., 64,* 827 (1968).
47. Mizutani, T., Miyazaki, M., and Takemura, S., *J. Biochem., 64,* 839 (1968).
48. Yaniv, M., *Nature New Biol., 233,* 118 (1971).
49. Barret, J., Ebel, J. P., and Dirheimer, G., *FEBS Lett., 15,* 286 (1971).

# CONTROLLED PARTIAL HYDROLYSIS OF RNA

ROBERT M. BOCK

Polyribonucleotides of long chain length are readily available as natural materials (viral, ribosomal, messenger, and transfer RNA) and as enzymically synthesized homopolymers and random polymers. Many studies of nucleotide interactions, coding function and nucleotide sequence can be facilitated by controlled partial hydrolysis. Enzymic hydrolysis can be accomplished with endonucleases that cleave specifically at G (RNAse $T_1$), at pyrimidines (pancreatic RNAse) or non-specifically (RNAse $T_2$)[1] to yield mono- and oligonucleotides terminating in a 3′ phosphate or a 2′,3′ cyclic phosphate, depending on the time of digestion. Endonucleases that cleave to yield oligonucleotides with 5′ phosphate termini may be completely non-specific (yeast endonuclease)[2] or relatively specific for cleaving next to G (*Neurospora crassa* endonuclease).[3] The yeast endonuclease has negligible activity on dimers and trimers, so that excellent yields of di- and trinucleotides accumulate.

In this chapter, the parameters that influence the rate and selectivity of chemical hydrolysis of RNA will be considered. Procedures will be given for controlled hydrolysis to oligonucleotides. Acid- and base-catalyzed hydrolyses of RNA in the pH range 1 to 13 proceed via the attack of the 2′ oxygen on the 3′ phosphorus to form a 2′,3′ cyclic phosphate. This cyclic diester can then hydrolyze to form mixed 2′ and 3′ phosphate-ended chains. In the pH range 2 to 11 the cyclic phosphate formation and its hydrolysis show similar pH dependence, but at pH 12 to 13, where the 2′ OH is apparently completely ionized, cyclic ester formation is independent of pH, whereas hydrolysis of the cyclic ester still increases as the pH is raised above 12. Thus, less cyclic ester accumulates during hydrolysis in strongly basic solution.

At low temperatures and moderate to high salt concentrations, RNA bases exhibit considerable "stacking" and hydrogen-bonding interactions. Hydrolysis of the "stacked" nucleotides is at least ten times slower than that of randomly oriented nucleotides. Purines exhibit stronger stacking reactions than pyrimidines, so that in the pH range 7 to 9, clusters of purines resist hydrolysis relative to pyrimidine-rich or mixed-composition regions.[4] At more alkaline pHs and low ionic strengths, electrostatic repulsion strongly influences the relative rates of hydrolysis. Thus, at pH 10 to 12, clusters of A are more stable than clusters of G, and clusters of C are more stable than clusters of U. This is apparently influenced by enolization of G and U, with the charge repulsion of the anionic base residues preventing base stacking interactions. At still higher pH (13 to 14), where each nucleotide residue has a negatively charged 2′ oxygen and a negatively charged phosphodiester, the enolization of the bases causes less relative difference in hydrolysis rates.

Thus pH 9, ionic strength 0.01 and 100°C, or pH 13, ionic strength 0.1 and 80°C are conditions for minimum specificity of hydrolysis; at pH 11, ionic strength 0.05 and 70°C, clusters rich in A are much more stable than other clusters; at pH 9, ionic strength 0.2 and 70°C, purine-rich clusters are stable compared to pyrimidines.

There may be a possibility for selective cleavage by acid catalysis at pH 3 and 1, where selective protonation of C, A and then G could influence the hydrolysis of the phosphodiesters, but the author is not aware of published work on base-selective acid hydrolysis of RNA. Rammler[5] implies that ionic strength is an important variable during acid hydrolysis of poly U, but quantitative rates of bond cleavage were not published.

## CHAIN TERMINI IN PARTIALLY HYDROLYZED RNA

Base-catalyzed hydrolysis of RNA is a two-step reaction. First the chain is cleaved by attack of the diester phosphate by the 2′ hydroxyl, displacing the 5′ esterified portion of the chain and leaving a 2′,3′ cyclic phosphate on the chain terminus. The hydrolytic opening of this cyclic phosphate depends differently on temperature, pH, catalyst and water activity than does the reaction that formed the cyclic ester. Thus one may choose conditions that favor the production of oligonucleotides with cyclic phosphate ends or with phosphomonoester ends.

Under conditions where viral, messenger RNAs or sRNA are inactivated by one or a few breaks per molecule, the ends produced are a free 5′ OH and the 2′,3′ cyclic phosphate ester. Thus phosphatase-released phosphate is usually not a valid measure of the number of breaks.[6] This fact has not been widely appreciated, and numerous authors have incorrectly assumed that their determinations of monoester phosphate were valid assays for the number of cleavage points in RNA. If the partial hydrolysis is conducted at pH 1 to 2 and 20 to 40°C, the cyclic phosphate termini will be hydrolyzed to 2′ and 3′ monoester phosphates in a good yield if the final digest has a chain length of less than 100. The table lists the distribution of cyclic termini under various conditions of hydrolysis.

## CONDITIONS FOR LIMITED RANDOM HYDROLYSIS

The table gives a variety of hydrolysis conditions, which can yield oligonucleotides of the average chain length listed. Experimental and theoretical studies have shown that the mass yield of nucleotide of chain length n is maximized if one cleaves the fraction 1/n of the bonds in a long polynucleotide. The apparent $\Delta H$ of activation for base-catalyzed hydrolysis in the pH range 7 to 11 and ionic strength 0.05 to 0.2 over the temperature range 60 to 100°C is 19 ± 3 Kcal/mole. At temperatures or ionic strengths where order–disorder transitions occur in the polynucleotides, this apparent $\Delta H$ can vary markedly.

The table is presented as a very practical guide for producing oligonucleotide digests. Kinetic constants have been published for hydrolysis conditions less well suited for the practical production of oligonucleotides.[8-11] Mg, Zn, Mn, Cu, and La polyvalent cations all catalyze the hydrolysis of RNA.[12] These catalysts have not been found helpful for enhancing specificity or minimizing side reactions; therefore, one should avoid contamination of the RNA sample with such ions. For hydrolysis to completion under the conditions listed in the table, it is advisable to use ten times the time listed for 50% bond cleavage. Chain cleavage in non-aqueous solvents can provide products with cyclic phosphates at the end of each chain.

### CONDITIONS FOR LIMITED HYDROLYSIS OF RNA

| Polymer | Temp., °C | pH (Buffer) | Bonds Cleaved 1% | 10% | 50% | $P^{=}/(P! + P^{=})$ |
|---|---|---|---|---|---|---|
| Poly A | 100° | 10.0[a] | 150 sec | 26 min | 150 min | 0.1[b] |
| Poly U | 90° | 10.0[a] | 40 sec | 7 min | 46 min | 0.05[b] |
| Poly C | 100° | 10.0[a] | 30 sec | 5 min | 32 min | 0.05[b] |
| RNA | 90° | 7.6[c] | 7 hrs | 3 days | – | – |
| RNA | 100° | 9 (0.01 Γ/2) | – | 17 min | 100 min | 0.1[b] |
| RNA | 70° | 11 (non-random)[c] | – | 20 min | 2 hrs | – |
| RNA | 70° | 9 (non-random)[c] | 30 min | 5 hrs | – | – |
| RNA | 60° | 13 | – | 50 sec | 5 min | – |
| RNA | 27° | 14 (1.0*M* KOH) | 40 sec | 7 min | 40 min | – |
| RNA | 20° | 1 | 10 min | 100 min[e] | 10 hrs | 0.98[b] |
| RNA | 100° | 2 (0.01*M* HCl) | – | 4 min[e] | 50 min | 0.95[b] |
| RNA | 100° | 3 | 4 min | 40 min[e] | – | – |
| Up! | 20° | 1.2 | – | – | 2 min[d] | – |
| Poly A | 100° | 10% Piperidine | – | – | 40 min | 0.95[b] |
| Poly C | 100° | 10% Piperidine | – | – | 10 min | 0.8[b] |

[a] Polymer dissolved in 0.1*M* $NH_4HCO_3$ and adjusted to pH 10.0 with concentrated $NH_4OH$. Data of W. M. Stanley, Jr.
[b] Ratio of monoester to total ends when 10% of the internucleotide bonds have been cleaved.
[c] Buffer of 0.2 ionic strength.
[d] Note that if the average chain length of an alkali-cleaved digest is less than 30, the cyclic phosphate termini can be opened in 20 minutes at pH 1 and 20°C with very little additional cleavage.
[e] There is negligible (less than 1%) 3′ to 2′ migration of the internucleotide diester during the treatment.

Data taken from: *Methods Enzymol., 12,* 218 (1968). Reproduced by permission of Academic Press, New York.

## REFERENCES


1. Rushizky and Sober, *J. Biol. Chem., 239,* 2165 (1964).
2. Nakao, Lee, Halvorson and Bock, *Biochim. Biophys. Acta, 151,* 114, 126 (1968).
3. Linn, *Methods Enzymol., 12,* 247 (1968).
4. Lane and Butler, *Biochim. Biophys. Acta, 33,* 281 (1959).
5. Rammler, Delk and Abell, *Biochemistry, 4,* 2002 (1965).
6. Gordon and Huff, *Biochemistry, 1,* 481 (1962).
7. Stanley, Jr., *Methods Enzymol., 12,* 404 (1968).
8. Bacher and Kauzman, *J. Amer. Chem. Soc., 74,* 3779 (1952).
9. Eigner, Boedtker and Michaels, *Biochim. Biophys. Acta, 52,* 165 (1961).
10. Ginoza, *Nature, 181,* 958 (1958).
11. Stanley and Bock, *Biochemistry, 4,* 1302 (1965).
12. Butzow and Eichhorn, *Biopolymers, 3,* 79 (1965).

# NUCLEOTIDE SEQUENCES OF RIBONUCLEIC ACIDS

DR. B. P. DOCTOR and MARY ANN SODD

## 5S-RIBOSOMAL RIBONUCLEIC ACIDS (5S-rRNA)[a]

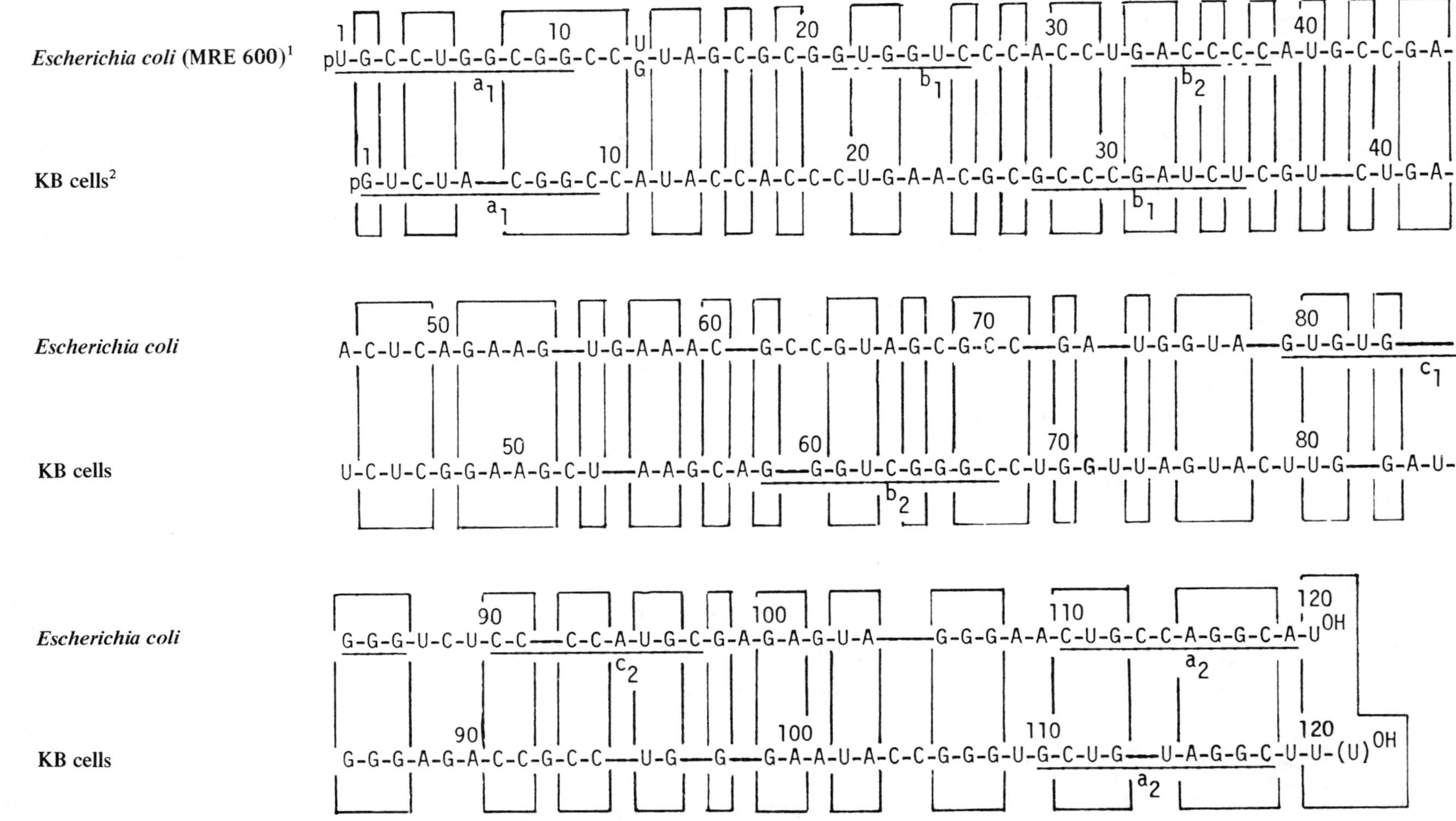

## BACTERIOPHAGE R17 RIBONUCLEIC ACIDS (R17 RNA)[b]

**Coat Protein Cistron**[3]

Primary sequence

```
        -C-A-U-G-G-C-G-U-U-C-G-U-A-C-U-U-A-A-A-U-A-U-G-G-A-A-U-U-A-A-C-U-A-U-U-C-C-A-A-U-U-U-U-C-G-C-U-A-C-G-A-A-C-U-C-C-G-
... Ala | Ala | Try | Arg | Ser | Tyr | Leu | Asn | Met | Glu | Leu | Thr | Ile | Pro | Ile | Phe | Ala | Thr | Asn | Ser | Asp ...
    80                                          90                                          100
```

Secondary structure

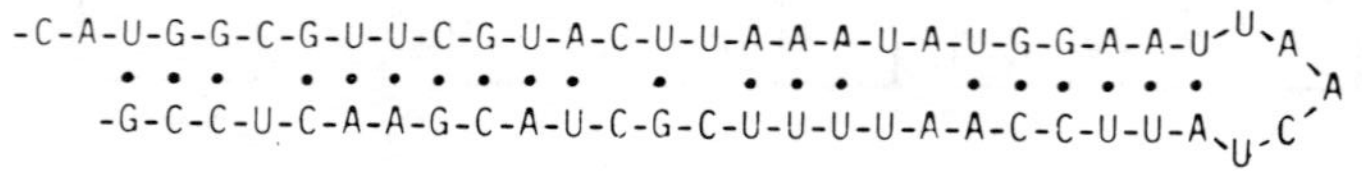

```
-C-A-U-G-G-C-G-U-U-C-G-U-A-C-U-U-A-A-A-U-A-U-G-G-A-A-U-U-A
  •  •  •   •  •  •  •  •  •  •    •     •  •  •    •  •  •  •  •  •   A
   -G-C-C-U-C-A-A-G-C-A-U-C-G-C-U-U-U-U-A-A-C-C-U-U-A-U-C
```

**Ribosome Binding Sites**[4–6]

Coat Protein Initiation Site

Primary sequence

```
A-G-A-G-(C)-C-C-U-C-A-A-C-C-G-G-G-G-U-U-U-G-A-A-G-C-A-U-G-G-C-U-U-C-U-A-A-C-U-U-U-A-C-U-C-A-G-
                                                   | Met_f | Ala | Ser | Asn | Phe | Thr | Gln | ...
```

Secondary structure

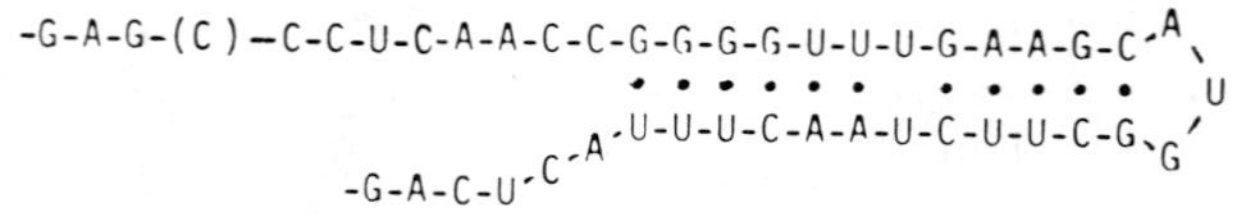

```
-G-A-G-(C)-C-C-U-C-A-A-C-C-G-G-G-G-U-U-U-G-A-A-G-C-A
                 •  •  •  •  •  •    •  •  •  •  •    U
                 A-U-U-U-C-A-A-U-C-U-U-C-G-G
-G-A-C-U-C
```

Synthetase Initiation Site

```
A-A-A-C-A-U-G-A-G-G-A-U-U-A-C-C-C-A-U-G-U-C-G-A-A-G-A-C— A-A-C-A(A,A,G)-
                                   | Met_f | Ser | Lys | Thr | Thr | Lys | ...
```

A Protein Initiation Site

```
A-U-U-C-C-U-A-G-G-A-G-G-U-U-U-G-A-C-C-U-A-U-G-C-G-A-G-C-U-U-U-U-A-G-U-G-
                                       | Met_f | Arg | Ala | Phe | Ser | ...
```

# BACTERIOPHAGE Q$\beta$ RIBONUCLEIC ACID (Q$\beta$ RNA)[b]

## 5′ End (Synthesized *in vitro*)[7]

Primary sequence

```
                   10                  20                   30                  40                  50
pppG-G-G-G-(G)-A-C-C-C-C-C-C-U-U-U-A-G-G-G-G-G-U-C-A-C-(A-C,A-C,C-U-C)-A-G-C-A-G-U-A-C-U-U-C-A-C-U-G-A-G-U-A-U-A-A-G-A-G-G-A-C-

60                  70                  80                   90                  100                 110
A-U-A-U-G-C-C-U-A-A-A-U-U-A-C-C-G-C-G-U-G-G-U-C-U-G-C-G-U-U-U-C-G-G-A-G-C-C-G-A-U-A-A-U-G-A-A-A-U-U-C-U-U-A-A-U-G-A-U-U-

120                 130                 140                  150                 160                 170
U-U-C-A-G-G-A-G-C-U-C-U-G-G-U-U-U-C-C-A-G-A-C-C-U-(U,U,C)-U-A-U-C-G-A-A-U-C-U-U-C-C-G-(A-C-A-C-G,C-A-U-C-C-G,U-G,G)-
```

Secondary structure

```
                         pppG-G-G-G-A-C-C-C-C-C-C-U
                                 • • • • • • •      U
                             A-C-U-G-G-G-G-G-A-U
-C-G-A-(C-U-C,C-A,C-A)-C-
```

## Coat Protein Cistron[8]

Primary sequence

```
Y-A-A-U-U-U-G-A-U-C-A-U-G-G-C-A-A-A-A-U-U-A-G-A-G-A-C-
                    | Met_f | Ala | Lys | Leu | Glu | Thr |...
```

Secondary structure

```
              Y-A-A-U-U-U-G-A-U-C-A
              • • • • •           |
          A-U-U-A-A-A-A-C-G-G-U
-C-A-G-A-G-
```

[a] Homologies are enclosed by solid lines and gaps have been inserted in the sequence to maximize the homology. Areas believed to be involved in base-pairing are underlined, with lower-case letters designating the corresponding regions.

[b] The sequence of amino acids of the bacteriophage proteins is shown under corresponding codon triplets. Centered dots indicate hydrogen bonds.

## REFERENCES


1. Brownlee, Sanger and Barrell, *Nature, 215,* 735 (1967).
2. Forget and Weissmann, *Science, 158,* 1695 (1967).
3. Adams, Jeppeson, Sanger and Barrell, *Nature, 223,* 1009 (1969).
4. Steitz, *Nature, 224,* 957 (1969).
5. Robinson, Frist and Kaesburg, *Science, 166,* 1291 (1969).
6. Jeppeson, Steitz, Gesteland and Spahr, *Nature, 226,* 230 (1970).
7. Billeter, Dahlberg, Goodman, Hindley and Weissmann, *Nature, 224,* 1083 (1969).
8. Hindley and Staples, *Nature, 224,* 964 (1969).

# DNA BASE COMPOSITIONS OF EUKARYOTIC PROTISTS

DR. MANLEY MANDEL

| Organism[a] | G + C, moles %[b] | | | References |
|---|---|---|---|---|
| | From Tm | From Density | Chemical | |
| ***PROTOZOA*** | | | | |
| ***Acanthamoeba castellani*** | — | 56 | — | *21* |
| ***Amoeba proteus*** | — | 66 | — | *21* |
| ***Astasia longa**** | — | 56 | — | *41* |
| ***Blastocrithidia culicis**** | — | 56 | — | *21* |
| ***Colpidium campylum*** | — | 32 | — | *36* |
| ***C. colpidium*** (2) | — | 32.5 | — | *36* |
| ***C. truncatum*** | — | 35 | — | *36* |
| ***Crithidia fasciculata**** (2) | — | 58, 54 | — | *13, 21, 36* |
| ***C. lucilliae*** | — | 57 | — | *36* |
| ***C. oncopelti**** | — | 54 | — | *27, 36* |
| ***Entamoeba invadens*** | — | 23 | — | *21* |
| ***Euglena gracilis**** | | | | |
| **var *bacillaris*** | 50.2 | 48 | 53 | *9, 10, 11, 17, 31, 36, 37* |
| **strain Z** | 50.5 | 49 | 51 | *4* |
| ***Glaucoma chattoni*** | — | 34 | — | *36* |
| ***Leishmania enrietti**** | — | 57 | — | *8* |
| ***L. tarentolae*** | — | 54 | — | *36* |
| ***Naegleria gruberi**** | — | 33 | — | *19* |
| ***Paramecium aurelia*** | — | 29 | — | *36, 39, 44* |
| ***P. caudatum*** | — | — | 28.2 | *12* |
| ***Plasmodium berghei*** | — | — | 40.7 | *51* |
| ***Tetrahymena patula*** | 19 | 25 | 22 | *36* |
| ***T. pyriformis*** | — | — | 25 | *35, 45* |
| ***T. pyriformis* W** | 29 | 30 | — | *36* |
| ***T. pyriformis*** (11) | — | 25 | — | *42, 43* |
| ***T. pyriformis* 3-1, 7-1, 5** | — | 27 | — | *42, 43* |
| ***T. pyriformis* 9-1** | — | 28 | — | *42* |
| ***T. pyriformis* GL*** | — | 29 | — | *42, 43* |
| ***T. pyriformis* E** | — | 31 | — | *42* |
| ***T. pyriformis* ST, 4*** | — | 32 | — | *43* |
| ***T. pyriformis* 9** | — | 30 | — | *43* |
| ***T. rostrata*** | — | 23 | — | *42* |
| ***Trichomonas gallinae*** | 33.2 | 34 | — | *25* |
| ***T. vaginalis*** | 31.2 | 29 | 32 | *25, 29* |
| ***Trypanosoma cruzi*** | — | — | 50.2 | *1, 3* |
| ***T. equiperdum*** | — | 47.9 | — | *32* |
| ***T. gambiense*** | — | — | 45 | *3* |
| ***T. lewisi*** | — | 59 | — | *36* |
| ***ALGAE*** | | | | |
| ***Ankistrodesmus falcatus*** | — | — | 58.7 | *49* |
| ***Chaetoceras decipiens*** | — | — | 39.1 | *49* |
| ***Chara sp.*** | — | — | 49.8 | *49* |
| ***Chlamydomonas angulosa*** | — | 68 | — | *41* |
| ***C. eugametos**** | — | 61 | — | *41* |
| ***C. gyrus*** | — | — | 59.7 | *49* |
| ***C. moewusii*** | — | 60 | — | *41* |
| ***C. reinhardi**** | — | 62, 64, 66 | 65 | *6, 17, 34* |
| ***Chlorella ellipsoidea*** | — | 57 | 59.0 | *6, 15* |
| ***Chlorogonium elongatum**** | — | 54 | — | *41* |
| ***Cricosphaera carterae*** | — | 60.2 | — | *22* |
| ***C. elongata*** | — | 59.2 | — | *22* |
| ***Coccolithus huxleyi*** | — | 64.8 | — | *22* |
| ***Cyclotella cryptica**** | — | 40.8 | — | *23* |
| ***Cylindrotheca fusiformis**** | — | 45.4 | — | *23* |
| ***Cystoseira barbata*** | — | — | 58.8 | *49* |
| ***Dunaliella salina*** | — | — | 60.2 | *49* |

| Organism[a] | G + C, moles %[b] From Tm | From Density | Chemical | References |
|---|---|---|---|---|
| ***ALGAE*—(Continued)** | | | | |
| ***Hydrodictyon reticulatum*** | — | — | 53.5 | *49* |
| ***Isochrysis galbana*** | — | 61 | — | *36* |
| ***Lagerhemia ciliata*** | — | — | 61.1 | *49* |
| ***Monodus subteranneus*** | — | 52 | — | *41* |
| ***Navicula closterium*** | — | 50 | — | *41* |
| ***N. pelliculosa*** | — | 58 | — | *41* |
| ***Nitella sp.*** | — | — | 49.4 | *49* |
| ***Nitzschia angularis**** | — | 46.9 | — | *23* |
| ***Ochromonas danica**** | — | 48 | — | *11* |
| ***Polyides rotundus*** | — | — | 64.3 | *49* |
| ***Polytoma agilis*** | — | 42 | — | *41* |
| ***P. obtusum**** | — | 65 | — | *17* |
| ***P. uvella*** | — | 52 | — | *41* |
| ***Polytomella papillata*** | — | — | 40 | *52* |
| ***Prymnesium parvum*** | — | 58 | — | *36* |
| ***Rhabdonema adriaticum*** | — | — | 36.9 | *49* |
| ***Rhizoclonium japonicum*** | — | — | 53.7 | *49* |
| ***Rhodymenia palmata*** | — | — | 48.9 | *49* |
| ***Scenedesmus accuminatus*** | — | — | 63.8 | *49* |
| ***S. quadricauda*** | — | — | 58.9 | *49* |
| ***Spirogyra sp.*** | — | — | 38.9 | *49* |
| ***Thalassiosira nordenscheldii*** | — | — | 40.3 | *49* |
| ***Ulothrix fimbriata*** | — | 46 | — | *41* |
| ***FUNGI*** | | | | |
| ***Absidia glauca*** | 48 | 44 | — | *40* |
| ***Acrasis rosea*** | — | 36 | — | *24* |
| ***Acytostelium leptosumun*** | — | 36 | — | *24* |
| ***Agaricus bisporus*** | — | — | 44.4 | *47* |
| ***Amanita muscaria*** | — | — | 57.1 | *48* |
| ***A. strobiliformis*** | — | — | 58 | *50* |
| ***Aspergillus niger*** | 52 | 52 | 50.1 | *40, 47* |
| ***A. oryzae* (conidia)** | — | — | 46.0 | *16* |
| ***Blastocladiella emersonii**** | — | 66 | — | *7* |
| ***Botrytis cinerea*** | — | — | 50.9 | *48* |
| ***Bovista sp.*** | — | — | 50.7 | *48* |
| ***Candida pulcherrima*** (3) | 48, 44.5 | 46 | — | *40, 53, 55* |
| ***C. albicans*** (2) | 35.1, 32.5 | — | — | *53, 55* |
| ***C. tropicalis*** (2) | 34.9 | — | — | *53, 57* |
| ***C. claussenii*** | 34.9 | — | — | *53* |
| ***C. stellatoidea*** | 36.7 | — | — | *53* |
| ***C. pelliculosa*** (2) | 36.8, 34.1 | — | — | *53, 54* |
| ***C. truncata*** | 36.9 | — | — | *53* |
| ***C. krusei*** (2) | 39.6, 38 | — | — | *53, 55* |
| ***C. atmosphaerica*** (2) | 39.7, 41.2 | — | — | *53, 57* |
| ***C. parapsilosis*** (2) | 40.4 | — | — | *53, 57* |
| ***C. melinii*** (2) | 40.9, 36.5 | — | — | *53, 55* |
| ***C. pseudotropicalis*** | 41.3 | — | — | *53* |
| ***C. tenuis*** | 44.0 | — | — | *53* |
| ***C. utilis*** (2) | 45.8, 40.2 | — | — | *53, 54* |
| ***C. lipolytica*** | 49.6 | — | — | *53* |
| ***C. brumptii*** | 54.1 | — | — | *53* |
| ***C. catenalata*** | 54.5 | — | — | *53* |
| ***C. zeylanoides*** (3) | 57.6, 51.5 | — | — | *53, 55* |
| ***C. humicola*** | 60 | — | — | *55* |
| ***C. rugosa*** | 47 | — | — | *55* |
| ***C. koshuensis*** | 30 | — | — | *55* |
| ***C. parapsilosis var hokkai*** | 52 | — | — | *55* |
| ***C. punicea*** | 51 | — | — | *55* |
| ***C. gelida*** | 52 | — | — | *55* |
| ***C. diddensii*** | 39.8 | — | — | *57* |
| ***C. oregonensis*** | 48.0 | — | — | *57* |

| Organism[a] | G + C, moles %[b] From Tm | From Density | Chemical | References |
|---|---|---|---|---|
| *FUNGI*—(Continued) | | | | |
| ***Candida pulcherrima*** | 48 | 46 | — | *40* |
| ***Claviceps purpurea*** | — | — | 52.6 | *48* |
| ***Coprinus lagopus*** | — | 52 | — | *24* |
| ***Cryptococcus albidus*** | 55 | 55 | — | *40* |
| ***C. flavus*** | — | 55.0 | — | *56* |
| ***C. gastricus*** (2) | — | 51.0, 65.5 | — | *56* |
| ***C. laurentii var. magnus*** | — | 49.0 | — | *56* |
| ***C. laurentii var. flavescens*** (2) | 56 | 58.0 | — | *55, 56* |
| ***C. melibiosum*** | — | 61.0 | — | *56* |
| ***C. neoformans*** (2) | 46 | 51.5 | — | *55, 56* |
| ***C. skinneri*** | — | 53.0 | — | *56* |
| ***C. terreus*** | — | 59.5 | — | *56* |
| ***C. uniguttulatus*** (2) | — | 51.5, 58.0 | — | *56* |
| ***Debaryomyces globosus*** | 45.1 | — | — | *57* |
| ***D. hansenii*** | 36.6 | — | — | *57* |
| ***D. kloeckeri*** | 40 | 40 | — | *4* |
| ***Dictyostelium discoideum*** | 25 | 22 | 22 | *36, 37* |
| ***D. mucoroides*** | — | 22 | — | *24* |
| ***D. purpureum*** | — | 25 | — | *24* |
| ***Dipodascus uninucleatus*** | — | 43 | — | *40* |
| ***Endomyces reesi*** | 41 | 39 | — | *40* |
| ***Endomycopsis muscicola*** | 34.6 | — | — | *54, 55* |
| ***E. ciferrii*** | 30.5 | — | — | *54* |
| ***E. subpelliculosa*** | 31.0 | — | — | *54* |
| ***E. capsularis*** | 40 | — | — | *55* |
| ***E. fibuligera*** | 37.5 | — | — | *55* |
| ***Fuligo viritans*** | — | — | 33.5 | *48* |
| ***Gelasinospora calospora*** | 53 | 55 | — | *40* |
| ***Hansenula angusta*** (2) | 45.4 | — | — | *54* |
| ***H. minuta*** (2) | 43.9 | — | — | *54* |
| ***H. capsulata*** (2) | 43.3 | — | — | *54, 55* |
| ***H. fabianii*** (4) | 42.7 | — | — | *54* |
| ***H. californica*** (2) | 41.5 | — | — | *54* |
| ***H. wickerhamii*** | 41.7 | — | — | *54* |
| ***H. petersonii*** (2) | 40.1 | — | — | *54* |
| ***H. jadinii*** | 41.0 | — | — | *54* |
| ***H. saturnus*** (2) | 40.5 | — | — | *54* |
| ***H. mrakii*** | 40.2 | — | — | *54* |
| ***H. beijerinckii*** | 40.1 | — | — | *54* |
| ***H. bimundalis*** | 38.4 | — | — | *54* |
| ***H. bimundalis var. americana*** | 40.0 | — | — | *54* |
| ***H. wingei*** | 36.8 | — | — | *54* |
| ***H. canadensis*** (2) | 35.8 | — | — | *54* |
| ***H. holstii*** | 34.1 | — | — | *54* |
| ***H. beckii*** (2) | 33.8 | — | — | *54* |
| ***H. anomala*** (3) | 33.3 | — | — | *54, 55* |
| ***H. schneggii*** (2) | 32.9 | — | — | *54* |
| ***H. silvicola*** (4) | 32.1 | — | — | *54* |
| ***H. subpelliculosa*** | 31.0 | — | — | *54* |
| ***H. ciferrii*** | 30.5 | — | — | *54* |
| ***H. platypodis*** (3) | 30.8 | — | — | *54* |
| ***Helvella esculenta*** | — | — | 50 | *48* |
| ***Labyrinthula sp.*** (2) | — | 58 | — | *22* |
| ***Lichtheimia sp. (Absidia)*** | — | — | 28.5 | *48* |
| ***Lodderomyces elongasporus*** (2) | 39.7 | — | — | *57* |
| ***Lycogala sp.*** | — | — | 42.2 | *48* |
| ***Metschnikowia bicuspidata*** | 44.6 | — | — | *57* |
| ***M. pulcherrima*** (2) | 43.5 | — | — | *57* |
| ***M. pulcherrima*** | 48.3 | — | — | *57* |
| ***M. reukaufii*** | 42.2 | — | — | *57* |
| ***Mucor racemosus*** | — | 38 | — | *40* |
| ***M. rouxianus*** (2) | 41 | 38.5 | — | *40* |

| Organism[a] | G + C, moles %[b] From Tm | From Density | Chemical | References |
|---|---|---|---|---|
| ***FUNGI***—**(Continued)** | | | | |
| ***M. rouxii*** | — | 39 | — | *40* |
| ***M. subtilissimus*** (2) | 39.5 | 39 | — | *40* |
| ***Neurospora crassa*** | 55 | 53 | 54 | *18, 24, 30, 37, 40* |
| ***N. intermedia*** | — | 53 | — | *24* |
| ***N. tetrasperma*** | — | 53 | — | *24* |
| ***Penicillium chrysogenum*** | 52 | 52 | — | *40* |
| ***P. notatum*** | 51 | 53 | — | *40* |
| ***Phycomyces blakesleanus*** | 44 | 39 | 38.7 | *40, 47* |
| ***Physarum polycephalum**** | — | 42.5 | — | *46* |
| ***Pichia membranefaciens*** | 44 | 46 | — | *40* |
| ***Polyporus versicolor*** | — | — | 57 | *50* |
| ***Polysphondylium pallidum*** | — | 29 | — | *24* |
| ***P. violaceum*** | — | 29 | — | *24* |
| ***Protostelium irregularis*** | — | 35 | — | *24* |
| ***Psalliota campestris*** | — | — | 44 | *2* |
| ***Rhizophlyctis rosea*** | — | 44 | — | *22* |
| ***Rhizophydium sp.*** | — | 50.5 | — | *22* |
| ***Rhizopus nigricans*** | — | 49 | — | *24* |
| ***Rhodotorula mucilaginosa*** | 63 | 61 | — | *4* |
| ***R. graminis*** | — | 70.0 | — | *56* |
| ***R. minuta var. minuta*** | — | 53.0 | — | *56* |
| ***R. minuta var. texensis*** (3) | 48 | 53.3 | — | *55, 56* |
| ***R. pallida*** (2) | — | 54.5, 63.5 | — | *56* |
| ***R. glutinis*** | 64 | — | — | *55* |
| ***R. rubra*** | 63.5 | — | — | *55* |
| ***R. slooffii*** | 47.5 | — | — | *55* |
| ***Saccharomyces cereviseae*** | 39.5 | 39 | 40 | *5, 14, 20, 28, 33, 40* |
| ***S. bisporus*** | 44.4 | — | — | *57* |
| ***S. cereviseae var. ellipsoideus*** | 38 | — | 41.0 | *59* |
| ***S. inconspicuus*** | 46.3 | — | — | *57* |
| ***S. lactis*** | 40.0 | — | — | *58* |
| ***S. pombe*** | 43 | — | 44.8 | *59* |
| ***S. rosei*** | 43.9 | — | — | *57* |
| ***S. fragilis*** | 41.5 | 42 | 42.9 | *33, 40* |
| ***S. vini*** | — | 41 | — | *26* |
| ***Sapromyces sp.*** | — | 27 | — | *22* |
| ***Sclerotinia libertiana*** | — | — | 45.5 | *47* |
| ***Schizophyllum commune*** | 58 | 58 | 57.1 | *40, 48* |
| ***Schizosaccharomyces octosporus*** | 42 | 40 | — | *40* |
| ***S. pombe*** | 42 | — | 40.7 | *33, 38* |
| ***Sporobolomyces roseus*** | 50 | 50 | — | *40* |
| ***S. salmonicolor*** | 63 | 63 | — | *40* |
| ***Sporormia sp.*** | 51 | 51 | — | *40* |
| ***Syncephalastrum racemosum*** | 47 | 48 | — | *40* |
| ***Torulopsis stellata*** | 48 | 50 | — | *40* |
| ***Trichothecium roseum*** | — | — | 49.7 | *48* |
| ***Zygorhynchus moelleri*** | 40 | 35 | — | *40* |

[a] The number in parentheses following the name of the organism represents the number of strains analyzed.
[b] Averaged values unless the values were significantly different.
* Satellite DNA present.

Data taken from: Mandel, M., in *Handbook of Biochemistry,* 2nd ed., pp. H-75–H-78, H. A. Sober, Ed. Copyright 1970, The Chemical Rubber Co., Cleveland, Ohio.

## REFERENCES

1. Antonov and Belozerskii, *Dokl. Akad. Nauk SSSR, 138,* 1216 (1961).
2. Belozerskii and Spirin, in *Nucleic Acids,* Vol. 3, p. 147, Chargaff and Davidson, Eds. Academic Press, New York (1960).
3. Bouisset, Breuilland, Carrè, Michel, Larrouy and Raujeva, *C.R. Acad. Sci. (Paris), 262,* 19112 (1966).

4. Brawerman and Eisenstadt, *Biochim. Biophys. Acta, 43,* 374 (1964).
5. Chargaff, in *Nucleic Acids,* Vol. 1, p. 307, Chargaff and Davidson, Eds. Academic Press, New York (1955).
6. Chun, Vaughn and Rich, *J. Mol. Biol., 7,* 130 (1963).
7. Comb, Brown and Katz, *J. Mol. Biol., 8,* 781 (1964).
8. Du Buy, Mattern and Riley, *Science, 147,* 754 (1965).
9. Edelman, Cowan, Epstein and Schiff, *Proc. Nat. Acad. Sci. U.S.A., 52,* 1214 (1964).
10. Edelman, Schiff and Epstein, *J. Mol. Biol., 11,* 769 (1965).
11. Edelman, Schiff and Epstein, Unpublished Data, cited in *Reproduction: Molecular, Subcellular and Cellular,* p. 131, Locke, Ed. Academic Press, New York (1965).
12. Ginzburg (1963), cited in Antonov, *Usp. Sovrem. Biol., 60,* 161 (1965).
13. Guttman and Eisenman, *Nature, 206,* 113 (1965).
14. Huang and Rosenberg, *Anal. Biochem., 16,* 107 (1966).
15. Iwamura and Kuwashima, *Biochim. Biophys. Acta, 82,* 678 (1964).
16. Kogane and Yamagita, *J. Appl. Microbiol., 10,* 61 (1964).
17. Leff, Mandel, Epstein and Schiff, *Biochem. Biophys. Res. Commun., 13,* 126 (1963).
18. Luck and Reich, *Proc. Nat. Acad. Sci. U.S.A., 52,* 931 (1964).
19. Mahler and Fulton, Personal Communication.
20. Mahler and Pereira, *J. Mol. Biol., 5,* 325 (1962).
21. Mandel, in *Chemical Zoology,* Vol. 2, p. 541, Florkin and Scheer, Eds. Academic Press, New York (1967).
22. Mandel, Unpublished Data.
23. Mandel, Lewin and McCarthy, Unpublished Data.
24. Mandel and Dutta, Unpublished Data.
25. Mandel and Honigberg, *J. Protozool., 11,* 114 (1964).
26. Mandel and Mosin, Unpublished Data.
27. Marmur, Cahoon, Shimura and Vogel, *Nature, 197,* 1228 (1963).
28. Marmur and Doty, *J. Mol. Biol., 5,* 109 (1962).
29. Michaels and Treick, *Exp. Parasitol., 12,* 401 (1962).
30. Minagawa, Wagner and Strauss, *Arch. Biochem. Biophys., 80,* 442 (1959).
31. Ray and Hanawalt, *J. Mol. Biol., 11,* 760 (1965).
32. Riou, Pautrizel and Paoletti, *C.R. Acad. Sci. (Paris), 262,* 2376 (1966).
33. Rost and Venner, *Hoppe-Seyler's Z. Physiol. Chem., 339,* 230 (1964).
34. Sager and Ishida, *Proc. Nat. Acad. Sci. U.S.A., 50,* 725 (1963).
35. Scherbaum, *Exp. Cell Res., 13,* 24 (1957).
36. Schildkraut, Mandel, Levisohn, Smith-Sonneborn and Marmur, *Nature, 196,* 795 (1962).
37. Schildkraut, Marmur and Doty, *J. Mol. Biol., 4,* 430 (1962).
38. Serenkov and Kust, *Dokl. Akad. Nauk SSSR, 142,* 1197 (1962).
39. Smith-Sonneborn, Green and Marmur, *Nature, 197,* 385 (1963).
40. Storck, *J. Bacteriol., 91,* 227 (1966).
41. Sueoka, in *The Bacteria,* Vol. 5, p. 9, Gunsalus and Stanier, Eds. Academic Press, New York (1964).
42. Sueoka, *Cold Spring Harbor Symp. Quant. Biol., 26,* 35 (1961).
43. Suyama, *Biochemistry, 5,* 2214 (1966).
44. Suyama and Preer, *Genetics, 52,* 1051 (1965).
45. Swartz, Trautner and Kornberg, *J. Biol. Chem., 237,* 1961 (1962).
46. Szybalski, Personal Communication.
47. Uryson and Belozerskii, *Dokl. Acad. Nauk SSSR, 133,* 708 (1960).
48. Vanyushin, Belozerskii and Bogdanova, *Dokl. Akad. Nauk SSSR, 134,* 1222 (1960).
49. Vanyushin, Belozerskii and Kokurina, *Trans. Moscow Soc. Nat., 24,* 7 (1966).
50. Venner, *Hoppe-Seyler's Z. Physiol. Chem., 333,* 5 (1963).
51. Whitfield, *Aust. J. Biol. Sci., 6,* 234 (1953).
52. Jones and Thompson, *J. Protozool., 10,* 91 (1963).
53. Stenderup and Bak, *J. Gen. Microbiol., 52,* 231 (1968).
54. Nakase and Komagata, *J. Gen. Appl. Bacteriol., 15,* 85 (1969).
55. Nakase and Komagata, *J. Gen. Appl. Bacteriol., 14,* 345 (1968).
56. Storck, Alexopoulos and Phaff, *J. Bacteriol., 98,* 1069 (1969).
57. Meyer and Phaff, *J. Bacteriol., 97,* 52 (1969).
58. Smith, *Ph.D. Dissertation.* University of Wisconsin (1967).
59. Rost and Venner, *Hoppe-Seyler's Z. Physiol. Chem., 339,* 230 (1964).

# DISTRIBUTION OF PURINES AND PYRIMIDINES IN DEOXYRIBONUCLEIC ACIDS

DR. HERMAN S. SHAPIRO

The catalog of deoxyribonucleic acid composition from diverse cellular sources has been organized according to the following scheme.

## PLANT KINGDOM

### 1. Phylum Thallophyta: Subphylum Fungi

Class Basidiomycetes
Class Ascomycetes
Class Phycomycetes
Class Fungi Imperfecti
Class Schizomycetes
Family Groups

### 2. Phylum Thallophyta: Subphylum Algae

Division Cyanophyta (Blue-Green Algae)
Division Rhodophyta (Red Algae)
Division Bacillariophyta (Diatoms)
Division Phaeophyta (Brown Algae)
Division Chlorophyta (Green Algae)
Division Charophyta (Stoneworts)

### 3. Phylum Tracheophyta

Subphylum Lycopsida
Subphylum Sphenopsida
Subphylum Pteropsida
Class Filicineae
Class Gymnospermae
Class Angiospermae

## ANIMAL KINGDOM

### 4. Phylum Protozoa

## VIRUSES

### 5. Bacterial Phages

### 6. Insect Viruses

### 7. Animal Viruses

The arrangement of specimens within each of these groups is in order of increasing molar ratio of adenine + thymine to guanine + cytosine.

Higher plants are listed by trivial names, with the taxonomic designation following in parentheses. Where no precise designation has been indicated in the reference cited, the most common designation for the organism has been assumed by the compiler.

## TABLE 1
## THALLOPHYTA: FUNGI

| Genus Source | Ade[a] | Gua | Cyt | 5Me Cyt | Thy | Molar Ratios[b] Pu/Py | A/T | G/C | (A + T)/(G + C) | Reference |
|---|---|---|---|---|---|---|---|---|---|---|
| ***BASIDIOMYCETES*** | | | | | | | | | | |
| ***Amanita strobiliformis*** | 21.2 | 30.1 | 27.6 | — | 21.0 | 1.06 | 1.01 | 1.09 | 0.73 | *1* |
| ***Polystictus versicolor*** | 22.1 | 28.3 | 28.9 | — | 20.7 | 1.02 | 1.07 | 0.98 | 0.75 | *1* |
| ***Schizophillum commune*** | 21.6 | 27.7 | 29.4 | — | 21.3 | 0.97 | 1.01 | 0.94 | 0.75 | *2* |
| ***Amanita muscaria*** | 21.2 | 28.0 | 29.1 | — | 21.7 | 0.97 | 0.98 | 0.96 | 0.75 | *2* |
| ***Bovista sp.*** | 25.0 | 26.6 | 24.1 | — | 24.3 | 1.07 | 1.03 | 1.10 | 0.97 | *2* |
| ***Agaricus bisporus Lange*** (*A. compestris*) | 28.2 | 22.6 | 21.8 | — | 27.4 | 1.03 | 1.03 | 1.04 | 1.25 | *5* |
| ***ASCOMYCETES*** | | | | | | | | | | |
| ***Neurospora crassa*** | 23.0 | 27.1 | 26.6[c] | — | 23.3 | 1.00 | 0.99 | 1.02 | 0.86 | *4* |
| ***Claviceps purpurea*** | 23.4 | 26.3 | 26.3 | — | 24.0 | 0.99 | 0.98 | 1.00 | 0.90 | *2* |
| ***Aspergillus niger*** | 25.0 | 25.1 | 25.0 | — | 24.9 | 1.00 | 1.00 | 1.00 | 1.00 | *5* |
| ***Gyromitra esculenta*** | 25.0 | 24.9 | 25.1 | — | 25.0 | 1.00 | 1.00 | 0.99 | 1.00 | *2* |
| ***Aspergillus tamarii*** | 26.1 | 24.4 | 23.6 | — | 25.9 | 1.02 | 1.01 | 1.03 | 1.08 | *6* |
| ***Aspergillus oryzae*** conidia, germinating | 27.2 | 22.0 | 25.5 | — | 25.4 | 0.97 | 1.07 | 0.86 | 1.11 | *7* |
| ***Aspergillus oryzae*** conidia, dormant | 28.2 | 20.5 | 24.7 | — | 26.5 | 0.95 | 1.06 | 0.83 | 1.21 | *7* |
| ***Sclerotinia libertiana*** | 27.2 | 22.6 | 22.9 | — | 27.3 | 0.99 | 1.00 | 0.99 | 1.20 | *5* |
| ***Schizosaccharomyces pombe*** | 24.9 | 21.5 | 23.3 | — | 30.3 | 0.87 | 0.82 | 0.92 | 1.23 | *8* |
| ***Saccharomyces fragilis*** | 29.7 | 18.1 | 24.8 | — | 27.3 | 0.92 | 1.09 | 0.73 | 1.33 | *8* |
| ***Saccharomyces cerevisiae*** | 30.1 | 19.3 | 21.7 | — | 28.9 | 0.98 | 1.04 | 0.89 | 1.44 | *8* |
| ***Saccharomyces cerevisiae*** | 31.3 | 18.7 | 17.1 | — | 32.9 | 1.00 | 0.95 | 1.09 | 1.79 | *9* |
| ***Saccharomyces cerevisiae*** | 31.7 | 18.3 | 17.4 | — | 32.6 | 1.00 | 0.97 | 1.05 | 1.80 | *10* |
| ***Saccharomyces cerevisiae*** | 31.2 | 15.6 | 18.5 | — | 34.7 | 0.88 | 0.90 | 0.84 | 1.93 | *11* |
| ***PHYCOMYCETES*** | | | | | | | | | | |
| ***Blastocladiella emersonii*** | 19.3 | 35.6 | 28.6 | — | 16.7 | 1.21 | 1.16 | 1.24 | 0.56 | *12* |
| ***Phycomyces blakesleanus*** + (female) | 30.4 | 19.3 | 19.5 | — | 30.8 | 0.99 | 0.99 | 0.99 | 1.58 | *5* |
| ***Phycomyces blakesleanus*** – (male) | 30.4 | 19.2 | 19.4 | — | 31.0 | 0.98 | 0.98 | 0.99 | 1.59 | *5* |
| ***Lichthemia sp.*** | 34.9 | 15.1 | 13.4 | — | 36.6 | 1.00 | 0.95 | 1.13 | 2.51 | *2* |
| ***FUNGI IMPERFECTI*** | | | | | | | | | | |
| ***Botrytis cinerea*** | 24.4 | 25.0 | 25.9 | — | 24.7 | 0.98 | 0.99 | 0.97 | 0.96 | *2* |
| ***Trichothecium roseum*** | 25.0 | 24.2 | 25.5 | — | 25.3 | 0.97 | 0.99 | 0.95 | 1.01 | *2* |
| ***SCHIZOMYCETES*** | | | | | | | | | | |
| ***Nitrobacteriaceae*** | | | | | | | | | | |
| *Thiobacillus thioparus* | 21.6 | 29.5 | 29.7 | — | 19.2 | 1.05 | 1.13 | 0.99 | 0.69 | *13* |
| *Thiobacillus ferrooxidans* | 20.6 | 30.3 | 28.5 | — | 20.6 | 1.04 | 1.00 | 1.06 | 0.70 | *13* |
| *Nitrosomonas sp.* | 23.3 | 27.3 | 27.0 | — | 22.4 | 1.02 | 1.04 | 1.01 | 0.84 | *14* |
| *Nitrosomonas europaea* | 25.0 | 26.1 | 25.5 | — | 23.4 | 1.04 | 1.07 | 1.02 | 0.94 | *15* |
| ***Pseudomonadaceae*** | | | | | | | | | | |
| *Pseudomonas tabaci* | 16.2 | 33.7 | 33.7 | — | 16.4 | 1.00 | 0.99 | 1.00 | 0.48 | *16* |
| *P. aeruginosa* | 16.8 | 33.0 | 34.0 | — | 16.2 | 0.99 | 1.04 | 0.97 | 0.49 | *17* |
| *P. aeruginosa* | 16.0 | 29.6 | 36.3 | — | 17.8 | 0.84 | 0.90 | 0.82 | 0.51 | *18* |
| *P. aeruginosa*, SI | 17.0 | 34.2 | 31.9 | — | 16.9 | 1.05 | 1.01 | 1.07 | 0.51 | *16* |
| *P. aeruginosa*, 9W | 17.5 | 34.1 | 30.6 | — | 17.8 | 1.07 | 0.98 | 1.11 | 0.55 | *16* |
| *P. aeruginosa*, 22W | 17.8 | 33.1 | 31.1 | — | 17.8 | 1.04 | 1.00 | 1.06 | 0.55 | *16* |
| *P. aeruginosa*, A22 | 18.0 | 33.1 | 30.9 | — | 18.0 | 1.04 | 1.00 | 1.07 | 0.56 | *16* |
| *Xanthomonas oryzae* | 18.2 | 31.8 | 32.6 | — | 17.5 | 1.00 | 1.04 | 0.98 | 0.55 | *84* |
| *Desulfovibrio Gigas* | 17.5 | 30.7 | 33.9 | — | 17.9 | 0.93 | 0.98 | 0.91 | 0.55 | *19* |
| *P. flourescens*, B/S100 | 17.3 | 33.0 | 30.8 | — | 19.0 | 1.01 | 0.91 | 1.07 | 0.57 | *20* |
| *P. flourescens* | 18.2 | 33.0 | 30.0 | — | 18.8 | 1.05 | 0.97 | 1.10 | 0.59 | *16* |
| *Desulfovibrio Hildenborough*, N.C.I.B. 8303 | 18.7 | 32.1 | 30.7 | — | 18.5 | 1.03 | 1.01 | 1.05 | 0.59 | *19* |
| *Desulfovibrio Caret* 41 N.C.I.B. 8393 | 21.4 | 28.9 | 28.5 | — | 21.2 | 1.01 | 1.01 | 1.01 | 0.74 | *19* |
| *Desulfovibrio desulfuricans* (*Vibrio cholinicus*) | 22.4 | 28.6 | 28.2 | — | 20.8 | 1.04 | 1.08 | 1.01 | 0.76 | *19* |
| *Desulfovibrio El Agheila* Z N.C.I.B. 8393 | 21.6 | 26.9 | 29.4 | — | 22.1 | 0.94 | 0.98 | 0.91 | 0.78 | *19* |
| *P. hydrophila* N.R.C. 3426 | 22.2 | 30.4 | 23.8 | — | 23.6 | 1.11 | 0.94 | 1.28 | 0.85 | *21* |
| *Vibrio cholera*, 325 | 25.2 | 24.2 | 24.5 | — | 26.1 | 0.98 | 0.97 | 0.99 | 1.05 | *22* |
| *Vibrio cholera*, 267 | 24.9 | 24.2 | 24.6 | — | 26.3 | 0.96 | 0.95 | 0.98 | 1.05 | *22* |

## TABLE 1 (Continued)
## THALLOPHYTA: FUNGI

| Genus Source | Ade[a] | Gua | Cyt | 5Me Cyt | Thy | Molar Ratios[b] Pu/Py | A/T | G/C | (A+T)/(G+C) | Reference |
|---|---|---|---|---|---|---|---|---|---|---|
| ***Pseudomonadaceae*—(Continued)** | | | | | | | | | | |
| *Vibrio cholera* | 28.8 | 20.0 | 23.3 | — | 27.9 | 0.95 | 1.03 | 0.86 | 1.31 | *16* |
| *Desulfovibrio negrificans*, N.C.I.B. 8395 | 27.7 | 23.1 | 22.3 | — | 26.9 | 1.03 | 1.03 | 1.04 | 1.20 | *19* |
| *Desulfovibrio orientis*, N.C.I.B. 8382 | 30.1 | 20.8 | 21.0 | — | 28.0 | 1.04 | 1.08 | 0.99 | 1.39 | *19* |
| *Pseudococcus citri* (*Xanthomonas citri*) | 34.2 | 16.2 | 17.8 | — | 31.9 | 1.01 | 1.07 | 0.91 | 1.94 | *23* |
| ***Azotobacteriaceae*** | | | | | | | | | | |
| *Azotobacter chroococcum*, 47 opaque form | 20.5 | 28.8 | 28.7 | — | 21.9 | 0.97 | 0.94 | 1.00 | 0.74 | *24* |
| *A. chroococcum*, 47 mucoid form | 20.3 | 29.0 | 28.5 | — | 22.3 | 0.97 | 0.91 | 1.02 | 0.74 | *24* |
| *A. chroococcum*, 54 | 20.6 | 29.2 | 28.0 | — | 22.2 | 0.99 | 0.93 | 1.04 | 0.75 | *25* |
| *A. vinelandii* | 21.9 | 27.5 | 28.3 | — | 22.3 | 0.99 | 0.98 | 0.97 | 0.79 | *25* |
| *A. agile*, 22D | 21.5 | 28.4 | 26.6 | — | 23.7 | 1.00 | 0.91 | 1.07 | 0.82 | *25* |
| ***Rhizobiaceae*** | | | | | | | | | | |
| *Agrobacter tumefaciens* (non-tumorgenic) | 21.2 | 30.6 | 28.1 | — | 20.1 | 1.07 | 1.05 | 1.09 | 0.70 | *16* |
| *Agrobacter tumefaciens* | 21.7 | 30.2 | 28.0 | — | 20.1 | 1.08 | 1.08 | 1.08 | 0.72 | *16* |
| ***Micrococcaceae*** | | | | | | | | | | |
| *Micrococcus lysodeikticus*, ML 1 | 13.8 | 34.5 | 41.1 | — | 10.2 | 0.94 | 1.35 | 0.84 | 0.32 | *26* |
| *M. lysodeikticus*, ML 53-20 | 13.1 | 33.0 | 39.6 | — | 14.2 | 0.86 | 0.92 | 0.83 | 0.38 | *26* |
| *M. lysodeikticus* | 14.4 | 37.3 | 34.6 | — | 13.7 | 1.07 | 1.05 | 1.08 | 0.39 | *16* |
| *Staphylococcus afermentans* 7503 | 12.8 | 36.9 | 37.4 | — | 12.9 | 0.99 | 0.99 | 0.99 | 0.35 | *27* |
| *Staphylococcus afermentans* 7503, mutant 22 | 15.0 | 34.9 | 35.1 | — | 15.0 | 1.00 | 1.00 | 0.99 | 0.43 | *27* |
| *Sarcina lutea* | 13.4 | 37.1 | 37.1 | — | 12.4 | 1.02 | 1.08 | 1.00 | 0.35 | *28* |
| *Sarcina lutea* | 13.6 | 36.4 | 35.6 | — | 14.4 | 1.00 | 0.94 | 1.02 | 0.39 | *17* |
| *Sarcina lutea* | 14.1 | 35.4 | 35.3 | — | 15.2 | 0.98 | 0.93 | 1.00 | 0.41 | *85* |
| *Sarcina lutea* | 18.7 | 32.4 | 31.5 | — | 17.4 | 1.04 | 1.07 | 1.03 | 0.56 | *16* |
| *Sarcina flava* | 15.6 | 33.5 | 35.1 | — | 15.8 | 0.96 | 0.99 | 0.95 | 0.46 | *16* |
| *M. radiodurans*, red-wild type | 17.1 | 32.7 | 33.9 | — | 16.5 | 0.99 | 1.04 | 0.96 | 0.50 | *29* |
| *M. radiodurans*, white, W1 | 17.2 | 32.1 | 34.2 | — | 16.5 | 0.97 | 1.04 | 0.94 | 0.51 | *29* |
| *M. radiodurans* | 17.4 | 33.0 | 33.7 | — | 16.4 | 1.01 | 1.06 | 0.98 | 0.51 | *30* |
| *M. radiodurans* | 19.3 | 30.0 | 30.9 | — | 19.6 | 0.98 | 0.98 | 0.97 | 0.64 | *31* |
| *Micrococcus morrhuae*, 2226 | 17.7 | 31.8 | 31.6 | — | 18.9 | 0.98 | 0.94 | 1.01 | 0.58 | *86* |
| *Micrococcus morrhuae*, 889 | 18.2 | 30.1 | 30.9 | — | 20.8 | 0.94 | 0.88 | 0.97 | 0.64 | *86* |
| *Micrococcus morrhuae*, 2224 | 18.8 | 28.8 | 30.8 | — | 21.6 | 0.91 | 0.87 | 0.94 | 0.68 | *86* |
| *Micrococcus morrhuae*, 537 | 20.2 | 29.2 | 28.6 | — | 22.0 | 0.98 | 0.92 | 1.02 | 0.73 | *86* |
| *Micrococcus morrhuae*, 859 | 21.4 | 26.9 | 30.2 | — | 21.5 | 0.94 | 1.00 | 0.89 | 0.75 | *86* |
| *Sarcina agilis* | 25.6 | 27.0 | 24.1 | — | 23.3 | 1.11 | 1.10 | 1.12 | 0.96 | *85* |
| *Sarcina ureae* | 28.2 | 22.3 | 22.4 | — | 27.1 | 1.02 | 1.04 | 1.00 | 1.24 | *85* |
| *Sarcina ureae*, 1466 | 28.2 | 22.6 | 20.7 | — | 28.5 | 1.03 | 0.99 | 1.09 | 1.31 | *87* |
| *Sarcina ureae*, 1732 | 28.6 | 21.6 | 20.4 | — | 29.4 | 1.01 | 0.97 | 1.06 | 1.38 | *87* |
| *Sarcina ureae*, 380 | 29.7 | 22.1 | 19.1 | — | 29.0 | 1.08 | 1.02 | 1.16 | 1.42 | *87* |
| *Sarcina ureae*, 1743 | 29.6 | 20.5 | 20.2 | — | 29.7 | 1.00 | 1.00 | 1.01 | 1.46 | *87* |
| *Sarcina ureae*, 981 | 28.5 | 20.4 | 19.3 | — | 29.8 | 1.00 | 0.96 | 1.06 | 1.47 | *87* |
| *Sarcina ventriculi* | 29.6 | 20.2 | 21.6 | — | 28.6 | 1.00 | 1.04 | 0.94 | 1.39 | *85* |
| *Staphylococcus aureus*, SA-B | 30.8 | 21.0 | 19.0 | — | 29.2 | 1.07 | 1.05 | 1.11 | 1.50 | *20* |
| *S. aureus*, 209-P | 31.0 | 18.5 | 19.2 | — | 31.2 | 0.98 | 0.99 | 0.96 | 1.65 | *20* |
| *S. pyogenes aureus* | 32.3 | 17.3 | 17.4 | — | 33.0 | 0.98 | 0.98 | 0.99 | 1.88 | *17* |
| *S. aureus*, 209-P | 33.4 | 17.0 | 16.8 | — | 32.8 | 1.02 | 1.02 | 1.01 | 1.96 | *32* |
| *S. aureus*, S44 | 31.8 | 16.1 | 17.4 | — | 34.6 | 0.92 | 0.92 | 0.93 | 1.98 | *33* |
| *S. aureus*, 209 | 33.6 | 16.5 | 15.9 | — | 34.0 | 1.01 | 0.99 | 1.04 | 2.09 | *27* |
| *S. aureus*, 209 (U.V. mutant 2) | 14.4 | 35.8 | 35.2 | — | 14.6 | 1.01 | 0.99 | 1.02 | 0.41 | *27* |
| *Sarcina maxima* | 35.3 | 23.4 | 16.4 | | 24.9 | 1.42 | 1.42 | 1.43 | 1.51 | *85* |
| *S. epidermidis* | 31.0 | 17.5 | 17.6 | — | 33.9 | 0.94 | 0.91 | 0.99 | 1.85 | *20* |
| *M. asaccharolyticus* | 31.2 | 18.8 | 15.3 | — | 34.7 | 1.00 | 0.90 | 1.23 | 1.93 | *16* |
| *Micrococcus halodurans*, 1798 | 33.8 | 15.7 | 16.3 | — | 34.2 | 0.98 | 0.99 | 0.96 | 2.12 | *86* |
| *M. pyogenes*, m320 | 34.2 | 15.6 | 16.4 | — | 33.8 | 0.99 | 1.01 | 0.95 | 2.13 | *16* |
| *M. pyogenes*, Oxford | 35.1 | 15.5 | 16.2 | — | 33.2 | 1.02 | 1.06 | 0.96 | 2.15 | *16* |
| *M. pyogenes*, 145 | 34.4 | 16.2 | 15.2 | — | 34.2 | 1.02 | 1.01 | 1.07 | 2.18 | *16* |
| *M. pyogenes*, 1149 | 34.2 | 16.1 | 15.1 | — | 34.6 | 1.01 | 0.99 | 1.07 | 2.21 | *16* |
| *M. pyogenes*, 1161 | 35.7 | 16.0 | 14.9 | — | 33.4 | 1.07 | 1.07 | 1.07 | 2.24 | *16* |

## TABLE 1 (Continued)
## THALLOPHYTA: FUNGI

| Genus Source | Ade[a] | Gua | Cyt | 5Me Cyt | Thy | Molar Ratios[b] Pu/Py | A/T | G/C | A + T/G + C | Reference |
|---|---|---|---|---|---|---|---|---|---|---|
| ***Neisseriaceae*** | | | | | | | | | | |
| *Neisseriae sicca*, Ne-12 | 24.0 | 26.4 | 25.1 | — | 24.4 | 1.02 | 0.98 | 1.05 | 0.94 | *34* |
| *N. meningitidis*, Ne-15 | 23.5 | 25.7 | 25.6 | — | 25.2 | 0.97 | 0.93 | 1.00 | 0.95 | *34* |
| *N. meningitidis* | 24.6 | 25.5 | 25.0 | — | 24.9 | 1.00 | 0.99 | 1.02 | 0.98 | *16* |
| *N. subflava*, 11076 | 21.9 | 24.1 | 26.4 | — | 27.6 | 0.85 | 0.79 | 0.91 | 0.98 | *34* |
| *N. perflava*, Ne-16 | 26.0 | 25.4 | 24.9 | — | 23.8 | 1.06 | 1.09 | 1.02 | 0.99 | *34* |
| *N. perflava*, Ne-20 | 25.2 | 25.4 | 24.4 | — | 24.9 | 1.03 | 1.01 | 1.04 | 1.01 | *34* |
| *N. flavescens*, 13120 | 23.6 | 25.6 | 24.5 | — | 26.2 | 0.97 | 0.90 | 1.04 | 0.99 | *34* |
| *N. flava*, JJIIA | 25.5 | 25.0 | 24.5 | — | 25.0 | 1.02 | 1.02 | 1.02 | 1.02 | *34* |
| *N. gonorrhoeae* | 25.3 | 25.2 | 24.4 | — | 25.1 | 1.02 | 1.01 | 1.03 | 1.02 | *16* |
| *N. catarrhalis*, Ne-11 | 28.6 | 19.9 | 20.8 | — | 30.6 | 0.94 | 0.93 | 0.96 | 1.45 | *34* |
| *N. catarrhalis*, Ne-13 | 27.9 | 19.1 | 21.0 | — | 32.1 | 0.89 | 0.87 | 0.91 | 1.50 | *34* |
| *Veillonella parvula* | 31.7 | 18.5 | 18.0 | — | 31.8 | 1.01 | 1.00 | 1.03 | 1.74 | *16* |
| ***Lactobacteriaceae*** | | | | | | | | | | |
| *Bifidibacterium bifidum* (*Lactobacillus bifidus*) | 21.9 | 28.3 | 29.3 | — | 20.5 | 1.01 | 1.07 | 0.97 | 0.74 | *16* |
| *Diplococcus pneumoniae* | 30.3 | 21.6 | 18.7 | — | 29.5 | 1.08 | 1.03 | 1.16 | 1.48 | *35* |
| Type III *Pneumococcus* | 29.8 | 20.5 | 18.0 | — | 31.5 | 1.02 | 0.95 | 1.14 | 1.59 | *36* |
| *Lactobacillus acidophilus* | 30.3 | 20.4 | 19.6 | — | 29.7 | 1.03 | 1.02 | 1.04 | 1.50 | *88* |
| Group A *Streptococcus* | 32.5 | 17.1 | 17.6 | — | 32.8 | 0.98 | 0.99 | 0.97 | 1.88 | *37* |
| Group A *Streptococcus*, L form | 32.9 | 16.7 | 17.0 | — | 33.4 | 0.98 | 0.99 | 0.98 | 1.97 | *37* |
| *Strep. faecales*, gr. D | 33.4 | 16.9 | 17.7 | — | 32.0 | 1.01 | 1.04 | 0.95 | 1.89 | *16* |
| *Streptococcus pyogenes* | 30.5 | 19.5 | 20.0 | — | 30.0 | 1.00 | 1.02 | 1.98 | 1.53 | *88* |
| *Strep. zymogenes*, gr. D | 33.1 | 16.2 | 17.4 | — | 33.3 | 0.97 | 0.99 | 0.93 | 1.98 | *16* |
| *Strep. pyog.*, gr. A | 33.4 | 16.6 | 17.0 | — | 33.0 | 1.00 | 1.01 | 0.98 | 1.98 | *16* |
| *Strep. foetidus* | 32.1 | 16.8 | 16.8 | — | 34.3 | 0.96 | 0.94 | 1.00 | 1.98 | *16* |
| *Ramibacterium ramosum* | 35.1 | 14.9 | 15.2 | — | 34.8 | 1.00 | 1.01 | 0.98 | 2.32 | *16* |
| ***Corynebacteriaceae*** | | | | | | | | | | |
| *Diphtheria bacillus*, 11 | 19.8 | 31.1 | 28.9 | — | 20.2 | 1.04 | 1.00 | 1.09 | 0.67 | *38* |
| *Corynebacterium diphtheriae* | 22.5 | 27.2 | 27.3 | — | 23.0 | 0.99 | 0.98 | 1.00 | 0.83 | *17* |
| *Mycobacterium diphtheriae* (*C. diphtheriae*) | 22.5 | 27.2 | 27.3 | — | 23.0 | 0.99 | 0.98 | 1.00 | 0.83 | *39* |
| *Diphtheria bacillus*, PW-8 | 21.4 | 27.8 | 26.6 | — | 24.2 | 0.97 | 0.92 | 1.06 | 0.84 | *38* |
| *C. diphtheriae* | 24.2 | 24.8 | 27.1 | — | 23.9 | 0.96 | 1.01 | 0.92 | 0.93 | *16* |
| *C. parvum* | 21.5 | 29.8 | 28.4 | — | 20.3 | 1.05 | 1.06 | 1.05 | 0.72 | *16* |
| *C. acnes* | 26.3 | 21.1 | 26.8 | — | 25.8 | 0.90 | 1.02 | 0.79 | 1.09 | *16* |
| ***Achromobacteriaceae*** | | | | | | | | | | |
| *Alcaligenes faecalis* | 15.9 | 31.8 | 36.0 | — | 16.2 | 0.91 | 0.98 | 0.88 | 0.47 | *20* |
| *A. faecalis*, 440 | 16.5 | 33.9 | 32.8 | — | 16.8 | 1.02 | 0.98 | 1.03 | 0.50 | *17* |
| *A. faecalis* | 16.5 | 33.9 | 32.9 | — | 16.8 | 1.01 | 0.98 | 1.03 | 0.50 | *40* |
| *Alcaligenes faecalis* | | | | | | | | | | |
| *Achromobacteriaceae* family | 19.3 | 31.1 | 30.6 | — | 19.4 | 1.01 | 0.99 | 1.02 | 0.63 | *41* |
| *Achromobacteriaceae sp.* | 23.6 | 26.9 | 26.3 | — | 23.2 | 1.02 | 1.02 | 1.02 | 0.88 | *42* |
| ***Enterobacteriaceae*** | | | | | | | | | | |
| *Serratia marcescens*, K(κ) | 20.1 | 29.8 | 31.1 | — | 19.0 | 1.00 | 1.05 | 0.96 | 0.64 | *43* |
| *S. marcescens*, CV/R5 | 20.1 | 26.9 | 33.6 | — | 19.4 | 0.89 | 1.04 | 0.80 | 0.65 | *43* |
| *S. marcescens*, EQ | 20.0 | 29.3 | 31.3 | — | 19.4 | 0.97 | 1.03 | 0.94 | 0.65 | *43* |
| *S. marcescens*, HY | 20.0 | 29.4 | 31.4 | — | 19.3 | 0.97 | 1.04 | 0.94 | 0.65 | *43* |
| *S. marcescens*, colorless mutant | 19.7 | 30.2 | 30.0 | — | 20.1 | 1.00 | 0.99 | 1.01 | 0.66 | *44* |
| *S. marcescens*, CN | 20.0 | 29.4 | 31.0 | — | 19.6 | 0.98 | 1.02 | 0.95 | 0.66 | *43* |
| Hybrid, $SC_{67}$ (*E. coli* $K_{12}$ × *S. marcescens*, CV) | 20.1 | 26.9 | 33.4 | — | 19.4 | 0.89 | 1.04 | 0.81 | 0.66 | *43* |
| Hybrid, $SC_{45}$ (*E. coli* $K_{12}$ × *S. marcescens*, CV) | 20.0 | 26.8 | 33.6 | — | 19.6 | 0.88 | 1.02 | 0.80 | 0.66 | *43* |
| Hybrid, $SC_{208}$ (*E. coli* $K_{12}$ × *S. marcescens*, CV) | 20.1 | 26.7 | 33.4 | — | 19.9 | 0.88 | 1.01 | 0.80 | 0.67 | *43* |
| *S. marcesens*, wild N.C.T.C. 1377 | 19.7 | 29.9 | 30.0 | — | 20.3 | 0.99 | 0.97 | 1.00 | 0.67 | *44* |
| *S. marcescens* | 20.4 | 29.4 | 29.6 | — | 20.6 | 0.99 | 0.99 | 0.99 | 0.69 | *45* |
| *S. marcescens* | 21.1 | 29.0 | 29.0 | — | 20.9 | 1.00 | 1.01 | 1.00 | 0.72 | *16* |
| *Aerobacter aerogenes* | 20.9 | 29.7 | 29.1 | — | 20.3 | 1.02 | 1.03 | 1.02 | 0.70 | *46* |
| *A. aerogenes*/Str. | 20.8 | 29.2 | 29.7 | — | 20.3 | 1.00 | 1.02 | 0.98 | 0.70 | *47* |

# TABLE 1 (Continued)
# THALLOPHYTA: FUNGI

| Genus Source | Ade[a] | Gua | Cyt | 5Me Cyt | Thy | Molar Ratios[b] Pu/Py | A/T | G/C | (A+T)/(G+C) | Reference |
|---|---|---|---|---|---|---|---|---|---|---|
| ***Enterobacteriaceae*—(Continued)** | | | | | | | | | | |
| *A. aerogenes* | 21.5 | 29.8 | 28.5 | — | 20.2 | 1.05 | 1.06 | 1.05 | 0.72 | *46* |
| *A. aerogenes* | 21.2 | 28.4 | 29.2 | — | 21.2 | 0.98 | 1.00 | 0.97 | 0.74 | *47* |
| *A. aerogenes* | 20.5 | 29.3 | 27.8 | — | 22.4 | 0.99 | 0.92 | 1.05 | 0.75 | *16* |
| *A. aerogenes* | 21.3 | 28.8 | 28.0 | — | 21.9 | 1.00 | 0.97 | 1.03 | 0.76 | *17* |
| *Bacillus breslau*, 70 (*Salmonella typhimurium*) | 23.1 | 27.6 | 27.8 | — | 21.5 | 1.03 | 1.07 | 0.99 | 0.81 | *48* |
| *Salmonella typhimurium*, 70 | 22.9 | 27.1 | 27.0 | — | 23.0 | 1.00 | 1.00 | 1.00 | 0.85 | *17* |
| *S. typhimurium* | 24.1 | 28.5 | 23.1 | — | 24.2 | 1.11 | 1.00 | 1.23 | 0.94 | *17* |
| *S. typhimurium*, LT-7 | 24.7 | 24.6 | 26.7 | — | 24.1 | 0.97 | 1.02 | 0.92 | 0.95 | *45* |
| *S. typhimurium* | 24.6 | 25.4 | 24.8 | — | 25.2 | 1.00 | 0.98 | 1.02 | 0.99 | *16* |
| *Erwinia carotovora* | 23.3 | 27.1 | 26.9 | — | 22.7 | 1.02 | 1.03 | 1.01 | 0.85 | *17* |
| *Salmonella dysenteriae*, f1.550 (radioresist.) | 28.7 | 21.4 | 21.6 | — | 28.3 | 1.00 | 1.01 | 0.99 | 0.89 | *89* |
| *Salmonella dysenteriae*, f1.550 | 24.9 | 25.0 | 24.8 | — | 25.3 | 1.00 | 0.98 | 1.01 | 1.01 | *89* |
| *E. coli*, S-1-Shch. | 17.3 | 32.5 | 33.0 | — | 17.2 | 0.99 | 1.01 | 0.98 | 0.53 | *89* |
| *Escherichia coli*, $15T^-$ | 23.6 | 27.5 | 26.1 | — | 22.4 | 1.05 | 1.05 | 1.05 | 0.86 | *46* |
| *E. coli*, I | 23.1 | 26.5 | 27.0 | — | 23.4 | 0.98 | 0.98 | 0.99 | 0.87 | *49* |
| *E. coli*, B | 23.8 | 26.8 | 26.3 | — | 23.1 | 1.02 | 1.03 | 1.02 | 0.88 | *50* |
| *E. coli* | 23.8 | 26.0 | 26.4 | — | 23.8 | 0.99 | 1.00 | 0.98 | 0.91 | *48* |
| *E. coli*, B/r | 24.1 | 26.1 | 25.9 | — | 23.9 | 1.01 | 1.01 | 1.01 | 0.92 | *50* |
| *E. coli*, I | 23.9 | 26.0 | 26.2 | — | 23.9 | 1.00 | 1.00 | 0.99 | 0.92 | *17* |
| *E. coli*, B/r | 23.1 | 26.4 | 25.1 | — | 25.0 | 0.99 | 0.92 | 1.05 | 0.93 | *46* |
| *E. coli*, W | 24.4 | 26.3 | 25.5 | — | 23.8 | 1.03 | 1.03 | 1.03 | 0.93 | *51* |
| *E. coli*, I | 24.3 | 25.6 | 25.8 | — | 24.3 | 1.00 | 1.00 | 0.99 | 0.95 | *89* |
| *E. coli*, HfrH | 24.6 | 25.5 | 25.6 | — | 24.3 | 1.00 | 1.01 | 1.00 | 0.96 | *52* |
| *E. coli*, $K_{12}$ | 24.6 | 25.6 | 25.6 | — | 24.3 | 1.01 | 1.01 | 1.00 | 0.96 | *53* |
| *E. coli*, I | 24.6 | 25.5 | 25.6 | — | 24.3 | 1.00 | 1.01 | 1.00 | 0.96 | *54* |
| *E. coli* | 25.0 | 24.5 | 26.3 | — | 24.3 | 0.98 | 1.03 | 0.93 | 0.97 | *55* |
| *E. coli*, $15t^-arg^-$ | 24.8 | 25.3 | 25.4 | — | 24.5 | 1.00 | 1.01 | 1.00 | 0.97 | *45* |
| *E. coli*, B | 24.5 | 24.7 | 25.9 | — | 24.9 | 0.97 | 0.98 | 0.95 | 0.98 | *45* |
| *E. coli*, I | 24.8 | 25.0 | 25.5 | — | 24.8 | 0.99 | 1.00 | 0.98 | 0.98 | *56* |
| *E. coli*, $15t^-arg^-$ | 25.0 | 24.9 | 25.6 | — | 24.4 | 1.00 | 1.02 | 0.97 | 0.98 | *54* |
| *E. coli*, 613 (strepto.-resist.) | 25.2 | 24.9 | 25.1 | — | 24.8 | 1.00 | 1.02 | 0.99 | 1.00 | *89* |
| *E. coli*, Crookes | 26.0 | 22.9 | 27.1 | — | 24.0 | 0.96 | 1.08 | 0.85 | 1.00 | *57* |
| *E. coli* | 24 | 25 | 25 | — | 26 | 0.96 | 0.92 | 1.00 | 1.00 | *58* |
| *E. coli*, $K_{12}$HfrC | 24.9 | 25.0 | 25.1 | — | 25.0 | 1.00 | 1.00 | 1.00 | 1.00 | *43* |
| *E. coli*, SG710 | 25.4 | 24.4 | 25.3 | — | 24.9 | 0.99 | 1.02 | 0.96 | 1.01 | *59* |
| *E. coli*, B/r | 22.9 | 23.9 | 25.7 | — | 27.4 | 0.88 | 0.84 | 0.93 | 1.01 | *60* |
| *E. coli*, K12 Hfr, Thy. | 25.8 | 24.3 | 24.5 | — | 25.4 | 1.00 | 1.02 | 0.99 | 1.05 | *89* |
| *E. coli*, K12 | 26.8 | 24.1 | 24.6 | — | 24.5 | 1.04 | 1.09 | 0.98 | 1.05 | *89* |
| *E. coli*, 613 | 25.7 | 24.2 | 24.6 | — | 25.5 | 1.00 | 1.01 | 0.98 | 1.05 | *89* |
| *E. coli* | 26.8 | 23.2 | 25.0 | — | 25.0 | 1.00 | 1.07 | 0.93 | 1.07 | *18* |
| *E. coli*, K12 Hfr, Thy. (radioresist.) | 31.7 | 19.6 | 18.1 | — | 30.6 | 1.05 | 1.04 | 1.08 | 1.65 | *89* |
| *E. coli*, K12 (radioresist.) | 32.8 | 17.7 | 17.6 | — | 31.9 | 1.02 | 1.03 | 1.01 | 1.83 | *89* |
| *Shigella dysenteriae*, 913 | 23.5 | 26.7 | 26.7 | — | 23.1 | 1.01 | 1.02 | 1.00 | 0.87 | *17* |
| *Salmonella typhosa*, Ty-2 | 23.5 | 26.7 | 26.4 | — | 23.4 | 1.01 | 1.00 | 1.01 | 0.88 | *17* |
| *Salmonella typhosa*, Ty-2 | 23.3 | 26.6 | 26.7 | — | 23.4 | 1.00 | 1.00 | 1.00 | 0.88 | *89* |
| *S. typhosa*, 643 | 24.0 | 26.0 | 25.7 | — | 24.3 | 1.00 | 0.99 | 1.01 | 0.93 | *45* |
| *Salmonella typhosa*, Ty-2 (radioresist.) | 28.8 | 20.8 | 21.0 | — | 29.4 | 0.98 | 0.98 | 0.99 | 1.39 | *89* |
| *Proteus morganii* | 23.7 | 26.3 | 26.7 | — | 23.3 | 1.00 | 1.02 | 0.99 | 0.89 | *17* |
| *Salmonella gallinarum* | 24.8 | 26.1 | 24.1 | — | 25.0 | 1.04 | 0.99 | 1.08 | 0.99 | *16* |
| *S. paratyphi*, A | 24.8 | 24.9 | 25.0 | — | 25.3 | 0.99 | 0.98 | 1.00 | 1.00 | *16* |
| *S. enteritidis*/D4 | 24.4 | 24.7 | 25.2 | — | 25.7 | 0.96 | 0.95 | 0.98 | 1.00 | *16* |
| *S. enteritidis* | 24.9 | 25.0 | 25.0 | — | 25.1 | 1.00 | 0.99 | 1.00 | 1.00 | *16* |
| *Shigella paradysenteriae*, Flexner $Y_6$RO | 24.9 | 24.8 | 24.5 | — | 25.8 | 0.99 | 0.97 | 1.01 | 1.03 | *16* |
| *Shigella paradysenteriae*, Flexner $Y_6$R4 | 25.8 | 25.3 | 23.9 | — | 25.0 | 1.04 | 1.03 | 1.06 | 1.03 | *16* |
| *Proteus mirabilis*, VI | 30.4 | 20.9 | 21.7 | — | 27.0 | 1.05 | 1.13 | 0.96 | 1.35 | *59* |
| *P. mirabilis*, VI/4 | 30.3 | 23.2 | 19.2 | — | 27.3 | 1.15 | 1.11 | 1.21 | 1.36 | *59* |
| *P. mirabilis*, LVI | 30.1 | 20.7 | 19.6 | — | 29.5 | 1.03 | 1.02 | 1.06 | 1.48 | *59* |
| *P. vulgaris* | 30.1 | 19.8 | 20.7 | — | 29.4 | 1.00 | 1.02 | 0.96 | 1.47 | *17* |
| *P. vulgaris* | 32.0 | 18.6 | 17.9 | — | 31.5 | 1.02 | 1.02 | 1.04 | 1.74 | *16* |

## TABLE 1 (Continued)
## THALLOPHYTA: FUNGI

| Genus Source | Ade[a] | Gua | Cyt | 5Me Cyt | Thy | Molar Ratios[b] Pu/Py | A/T | G/C | A+T/G+C | Reference |
|---|---|---|---|---|---|---|---|---|---|---|
| ***Parrobacteriaceae*** | | | | | | | | | | |
| *Fusiformis polymorphus* | 20.7 | 29.9 | 29.0 | — | 20.4 | 1.02 | 1.01 | 1.03 | 0.70 | *16* |
| *Brucella abortus* | 21.0 | 29.0 | 28.9 | — | 21.1 | 1.00 | 1.00 | 1.00 | 0.73 | *17* |
| *Pasteurella pseudotuberculosis*, strain 2613 | 27.6 | 24.0 | 25.2 | — | 23.2 | 1.07 | 1.19 | 0.95 | 1.03 | *61* |
| *P. pseudotuberculosis rodentium* | 26.4 | 23.7 | 24.7 | — | 25.2 | 1.00 | 1.05 | 0.96 | 1.07 | *62* |
| *P. pseudotuberculosis*, 1 | 25.4 | 23.6 | 24.6 | — | 26.4 | 0.96 | 0.96 | 0.96 | 1.07 | *61* |
| *P. pseudotuberculosis*, 162 | 26.2 | 23.4 | 24.3 | — | 26.1 | 0.98 | 1.00 | 0.96 | 1.10 | *61* |
| *P. pestis*, 17 | 26.5 | 23.7 | 25.5 | — | 24.3 | 1.01 | 1.09 | 0.93 | 1.03 | *61* |
| *P. pestis*, EB | 26.0 | 23.4 | 25.4 | — | 25.2 | 0.98 | 1.03 | 0.92 | 1.05 | *61* |
| *P. pestis*, 248 | 26.0 | 23.7 | 24.6 | — | 25.7 | 0.99 | 1.01 | 0.96 | 1.07 | *61* |
| *P. pestis* | 25.9 | 24.1 | 24.3 | — | 25.7 | 1.00 | 1.01 | 0.99 | 1.07 | *62* |
| *P. pestis*, 121 | 26.2 | 23.8 | 23.9 | — | 26.1 | 1.00 | 1.00 | 1.00 | 1.10 | *61* |
| *P. pestis*, 422 | 27.0 | 23.5 | 23.5 | — | 26.0 | 1.02 | 1.04 | 1.00 | 1.13 | *61* |
| *P. pestis*, 1 | 26.1 | 23.1 | 23.9 | — | 26.8 | 0.97 | 0.97 | 0.97 | 1.13 | *61* |
| *P. pestis*, 104 | 26.8 | 23.2 | 23.2 | — | 26.8 | 1.00 | 1.00 | 1.00 | 1.16 | *61* |
| *Ristella insolicitus*, I.S.9 | 28.2 | 23.2 | 20.0 | — | 28.6 | 1.06 | 0.99 | 1.16 | 1.31 | *16* |
| *Ristella insolit.*, N.I.12 | 29.3 | 20.7 | 20.5 | — | 29.5 | 1.00 | 0.99 | 1.01 | 1.43 | *16* |
| *P. suiseptica* | 28.4 | 21.5 | 21.6 | — | 28.5 | 1.00 | 1.00 | 1.00 | 1.32 | *63* |
| *P. boviseptica* | 28.4 | 21.5 | 21.6 | — | 28.5 | 1.00 | 1.00 | 1.00 | 1.32 | *63* |
| *P. boviseptica* | 31.0 | 18.8 | 18.8 | — | 31.4 | 0.99 | 0.99 | 1.00 | 1.66 | *16* |
| *P. aviseptica* | 28.4 | 21.5 | 21.6 | — | 28.5 | 1.00 | 1.00 | 1.00 | 1.32 | *63* |
| *P. aviseptica* | 32.0 | 18.2 | 18.1 | — | 31.7 | 1.01 | 1.01 | 1.01 | 1.75 | *16* |
| *P. tularensis* | 32.4 | 17.6 | 17.1 | — | 32.9 | 1.00 | 0.98 | 1.03 | 1.88 | *17* |
| *Fusiformis fusiformis* | 34.8 | 15.7 | 15.8 | — | 33.7 | 1.02 | 1.03 | 0.99 | 2.17 | *16* |
| *Ristella clostridiform* | 34.4 | 15.8 | 15.4 | — | 34.4 | 1.01 | 1.00 | 1.03 | 2.21 | *16* |
| ***Bacteriaceae*** | | | | | | | | | | |
| *Bacterium paracoli* | 21.1 | 28.9 | 29.1 | — | 21.1 | 1.00 | 1.00 | 0.99 | 0.73 | *27* |
| *Methanobacterium omelianskii* | 28.4 | 22.7 | 21.5 | — | 27.4 | 1.04 | 1.04 | 1.06 | 1.26 | *57* |
| ***Bacillaceae*** | | | | | | | | | | |
| *Bacillus paracoli* mutant 52-1 | 12.4 | 37.8 | 37.6 | — | 12.2 | 1.01 | 1.02 | 1.01 | 0.33 | *27* |
| *B. subtilis*, 168 mutant MK-9 | 17.8 | 32.2 | 31.4 | — | 18.7 | 1.00 | 0.95 | 1.03 | 0.57 | *27* |
| *B. subtilis*, spores | 20.2 | 29.6 | 31.2 | — | 19.0 | 0.99 | 1.06 | 0.95 | 0.64 | *64* |
| *B. subtilis*, veg. cells | 20.4 | 28.2 | 29.0 | — | 22.4 | 0.95 | 0.91 | 0.97 | 0.75 | *64* |
| *B. subtilis*, K(1523) | 27.5 | 22.5 | 22.5 | — | 27.5 | 1.00 | 1.00 | 1.00 | 1.22 | *65* |
| *B. subtilis* (spores) | 28.2 | 21.9 | 21.2 | — | 28.8 | 1.00 | 0.98 | 1.03 | 1.32 | *90* |
| *B. subtilis*, Marburg wild | 28.4 | 21.0 | 21.6 | — | 29.0 | 0.98 | 0.98 | 0.97 | 1.35 | *65* |
| *B. subtilis* | 28.9 | 21.0 | 21.4 | — | 28.7 | 1.00 | 1.01 | 0.98 | 1.36 | *16* |
| *B. subtilis*, 168 | 29.0 | 20.7 | 21.3 | — | 29.0 | 0.99 | 1.00 | 0.97 | 1.38 | *27* |
| *B. subtilis* (veg. cells) | 28.7 | 21.2 | 20.5 | — | 29.3 | 1.00 | 0.98 | 1.03 | 1.39 | *90* |
| *B. stearothermophilus*, 4S | 23.8 | 26.2 | 25.9 | — | 24.2 | 1.00 | 0.98 | 1.01 | 0.92 | *66* |
| *B. stearothermophilus*, 1503-4R | 23.7 | 25.9 | 25.9 | — | 24.3 | 0.99 | 0.98 | 1.00 | 0.93 | *66* |
| *B. stearothermophylicus* | 25.5 | 25.8 | 24.7 | — | 24.1 | 1.05 | 1.06 | 1.04 | 0.98 | *67* |
| *B. circulans*, 1112 | 24.6 | 25.6 | 25.1 | — | 24.6 | 1.01 | 1.00 | 1.02 | 0.97 | *65* |
| *B. circulans*, 1165 | 26.6 | 24.7 | 22.6 | — | 26.1 | 1.05 | 1.02 | 1.09 | 1.11 | *65* |
| *B. brevis*, var. G-B | 27 | 23 | 23 | — | 27 | 1.00 | 1.00 | 1.00 | 1.17 | *68* |
| *B. brevis*, 1031 | 28.4 | 22.4 | 21.6 | — | 27.6 | 1.03 | 1.03 | 1.04 | 1.27 | *65* |
| *B. megaterium*, 203 | 27.6 | 22.9 | 22.4 | — | 27.1 | 1.02 | 1.02 | 1.02 | 1.21 | *65* |
| *B. megaterium* | 31.8 | 19.1 | 18.5 | — | 30.6 | 1.04 | 1.04 | 1.03 | 1.66 | *16* |
| *B. megaterium*, 1166 | 31.1 | 18.3 | 19.3 | — | 31.4 | 0.97 | 0.99 | 0.95 | 1.66 | *65* |
| *B. megaterium*, 1032 | 32.2 | 18.0 | 18.0 | — | 31.8 | 1.01 | 1.01 | 1.00 | 1.78 | *65* |
| *B. aneurinolyticus*, 1077 | 28.9 | 20.8 | 21.4 | — | 28.9 | 0.99 | 1.00 | 0.97 | 1.37 | *65* |
| *B. natto*, 1 (1071) | 29.1 | 20.1 | 21.8 | — | 29.1 | 0.97 | 1.00 | 0.92 | 1.39 | *65* |
| *B. natto*, 8 (1259) | 29.1 | 20.3 | 20.9 | — | 29.7 | 0.98 | 0.98 | 0.97 | 1.43 | *65* |
| *B. natto*, 7 (1191) | 29.0 | 20.6 | 20.6 | — | 29.9 | 0.98 | 0.97 | 1.00 | 1.43 | *65* |
| *Clostridium tetanomorphum* | 30.7 | 20.0 | 19.3 | — | 30.0 | 1.03 | 1.02 | 1.04 | 1.54 | *57* |
| *B. cereus* | 29.9 | 19.6 | 19.5 | — | 31.0 | 0.98 | 0.96 | 1.01 | 1.56 | *69* |
| *B. cereus*, A.T.C.C. 12137 | 32 | 17 | 19 | — | 32 | 0.96 | 1.00 | 0.89 | 1.78 | *70* |
| *B. cereus*, MB-19 | 31.2 | 18.8 | 18.2 | — | 31.8 | 1.00 | 0.98 | 1.03 | 1.70 | *45* |
| *B. cereus*, $p_1$ | 32 | 18 | 18 | — | 32 | 1.00 | 1.00 | 1.00 | 1.78 | *70* |
| *B. cereus*, B.T.C.C. 7587 | 32 | 18 | 17 | — | 33 | 1.00 | 0.97 | 1.06 | 1.86 | *70* |

## TABLE 1 (Continued)
## THALLOPHYTA: FUNGI

| Genus Source | Ade[a] | Gua | Cyt | 5Me Cyt | Thy | Molar Ratios[b] Pu/Py | A/T | G/C | (A + T)/(G + C) | Reference |
|---|---|---|---|---|---|---|---|---|---|---|
| ***Bacillaceae*—(Continued)** | | | | | | | | | | |
| *B. cereus*, $p_2$ | 33 | 17 | 17 | — | 34 | 0.98 | 0.97 | 1.00 | 1.97 | *70* |
| *B. cereus*, $A_{25}$ | 32.3 | 17.9 | 16.6 | — | 33.2 | 1.01 | 0.97 | 1.08 | 1.90 | *16* |
| *B. cereus alesti* | 33.5 | 17.3 | 16.0 | — | 33.2 | 1.03 | 1.01 | 1.08 | 2.00 | *16* |
| *Clostridium butyricum* | 30.5 | 17.8 | 19.6 | — | 32.1 | 0.93 | 0.95 | 0.91 | 1.67 | *57* |
| *B. thuringiensis amer.* | 32.2 | 18.1 | 17.8 | — | 31.9 | 1.01 | 1.01 | 1.02 | 1.79 | *16* |
| *Clostridium bifermentans* | 34.0 | 17.0 | 15.5 | — | 33.5 | 1.04 | 1.01 | 1.10 | 2.08 | *16* |
| *Cl. cylindrosporum* | 33.4 | 14.2 | 18.2 | — | 34.2 | 0.91 | 0.98 | 0.78 | 2.09 | *57* |
| *Cl. valerianicum* | 35.1 | 16.2 | 15.6 | — | 33.1 | 1.05 | 1.06 | 1.04 | 2.14 | *16* |
| *Cl. perfringens* | 34.1 | 15.8 | 15.1 | — | 35.0 | 1.00 | 0.97 | 1.05 | 2.24 | *17* |
| *Cl. pasteurianum* | 34.7 | 14.8 | 16.0 | — | 34.5 | 0.98 | 1.01 | 0.93 | 2.25 | *57* |
| *Plectridium saprogenes* | 34.7 | 16.0 | 14.3 | — | 35.0 | 1.03 | 0.99 | 1.12 | 2.30 | *16* |
| *Cl. acidiurici* | 34.9 | 14.8 | 15.0 | — | 35.4 | 0.99 | 0.99 | 0.99 | 2.36 | *57* |
| *Welchia perfringens* (Fred) | 36.9 | 14.0 | 12.8 | — | 36.3 | 1.04 | 1.02 | 1.09 | 2.73 | *16* |
| ***Thiorhodaceae*** | | | | | | | | | | |
| *Chromatium sp.* | 18.4 | 32.0 | 31.7 | — | 17.9 | 1.02 | 1.03 | 1.01 | 0.57 | *14* |
| ***Athiorhodaceae*** | | | | | | | | | | |
| *Rhodopseudomonas spheroides*, CC1 | 16.1 | 33.1 | 35.3 | — | 15.5 | 0.97 | 1.04 | 0.94 | 0.46 | *45* |
| *Rh. spheroides*, 2.4.1 | 16.1 | 33.9 | 33.8 | — | 16.2 | 1.00 | 0.99 | 1.00 | 0.48 | *45* |
| *Rh. spheroides*, 2.4.1 | 16.2 | 34.3 | 33.4 | — | 16.2 | 1.02 | 1.00 | 1.03 | 0.48 | *71* |
| *Rhodopseudomonas sp.* | 18.2 | 32.0 | 32.0 | — | 17.8 | 1.01 | 1.02 | 1.00 | 0.56 | *13* |
| *Rh. rubrum* | 18.4 | 31.7 | 31.7 | — | 18.2 | 1.00 | 1.01 | 1.00 | 0.58 | *13* |
| ***Chlorobacteriaceae*** | | | | | | | | | | |
| *Chlorobium thiosulphatophilum* | 21.4 | 29.0 | 28.8 | — | 20.8 | 1.02 | 1.03 | 1.02 | 0.73 | *14* |
| ***Mycobacteriaceae*** | | | | | | | | | | |
| *Mycobacterium tuberculosis*, BCG | 16.5 | 34.2 | 33.3 | — | 16.0 | 1.03 | 1.03 | 1.03 | 0.48 | *39* |
| *Myco. tuberculosis*, BCG | 16.5 | 34.0 | 33.3 | — | 16.0 | 1.02 | 1.03 | 1.02 | 0.48 | *17* |
| *Myco. tuberculosis*, BCG (bovine) | 18.3 | 32.7 | 32.3 | — | 16.7 | 1.04 | 1.10 | 1.01 | 0.54 | *72* |
| *Myco. tuberculosis* (bovine) | 17.8 | 29.3 | 33.8 | — | 19.0 | 0.89 | 0.94 | 0.87 | 0.58 | *60* |
| *Myco. tuberculosis* (human strain) | 19.3 | 28.3 | 34.9 | — | 17.5 | 0.91 | 1.10 | 0.81 | 0.58 | *73* |
| *Myco. tuberculosis* (human strain) | 18.0 | 28.5 | 33.5 | — | 20.0 | 0.87 | 0.90 | 0.85 | 0.61 | *60* |
| *Myco. phlei* | 16.3 | 33.8 | 33.5 | — | 16.5 | 1.00 | 0.99 | 1.01 | 0.49 | *55* |
| *Myco. phlei* | 18.0 | 31.6 | 34.8 | — | 15.5 | 0.99 | 1.16 | 0.91 | 0.50 | *73* |
| *Myco. vadosum*, Kras | 20.7 | 29.2 | 28.5 | — | 21.6 | 1.00 | 0.96 | 1.02 | 0.73 | *17* |
| ***Actinomycetaceae*** | | | | | | | | | | |
| *Actinomyces globisporus flaveolus* | 13.8 | 36.3 | 37.2 | — | 12.7 | 1.00 | 1.09 | 0.98 | 0.36 | *39* |
| *A. viridochromogenes* (*Strep. viridochromogenes*) | 13.3 | 36.6 | 37.2 | — | 12.9 | 1.00 | 1.03 | 0.98 | 0.36 | *39* |
| *A. griseus* (*Strep. griseus*) | 13.8 | 35.8 | 37.4 | — | 13.0 | 0.98 | 1.06 | 0.96 | 0.37 | *39* |
| *A. globisporus streptomycini*, Kras | 13.4 | 36.1 | 37.1 | — | 13.4 | 0.98 | 1.00 | 0.97 | 0.37 | *17* |
| *A. globisporus streptomycini* (*Strep. griseus*) | 13.4 | 36.1 | 37.1 | — | 13.4 | 0.98 | 1.00 | 0.97 | 0.37 | *39* |
| *Proactinomyces citreus* (*Nocardia citrea*) | 14.2 | 35.3 | 36.7 | — | 13.8 | 0.98 | 1.03 | 0.96 | 0.39 | *39* |
| *Nocardia rubra* | 14.8 | 36.3 | 35.4 | — | 13.5 | 1.04 | 1.10 | 1.03 | 0.39 | *74* |
| *Nocardia corallina* | 14.5 | 36.0 | 35.0 | — | 14.5 | 1.02 | 1.00 | 1.03 | 0.41 | *74* |
| *Actinosporangium*, sp. N4146 | 14.3 | 35.3 | 35.6 | — | 14.8 | 0.99 | 0.97 | 0.99 | 0.41 | *91* |
| *Microechinospora grisea* | 15.5 | 34.6 | 34.5 | — | 15.4 | 1.00 | 1.01 | 1.00 | 0.45 | *91* |
| *Micropolyspora angiospora* | 17.0 | 32.9 | 33.1 | — | 17.3 | 0.99 | 0.98 | 1.00 | 0.52 | *91* |
| ***Streptomycetaceae*** | | | | | | | | | | |
| *Micromonospora coerulea* (*M. vulgaris*) | 14.3 | 36.2 | 35.6 | — | 13.9 | 1.02 | 1.03 | 1.02 | 0.39 | *39* |
| *Strep. chrysomallus* | 13.8 | 34.8 | 36.0 | — | 15.4 | 0.95 | 0.90 | 0.97 | 0.41 | *74* |
| *Strep. antibioticus*, I | 14.9 | 34.8 | 35.6 | — | 14.7 | 0.99 | 1.01 | 0.98 | 0.42 | *74* |
| *Strep. antibioticus*, II | 14.9 | 36.8 | 33.8 | — | 14.4 | 1.07 | 1.03 | 1.09 | 0.42 | *74* |
| *Strep. parvullus* | 15.4 | 34.6 | 35.6 | — | 14.4 | 1.00 | 1.07 | 0.97 | 0.42 | *74* |
| ***Polyangiaceae*** | | | | | | | | | | |
| *Chondrococcus columnaris*, 2-B58-5b | 29.1 | 22.7 | 20.5 | — | 27.7 | 1.07 | 1.05 | 1.11 | 1.31 | *75* |
| *Chondrococcus columnaris*, 3-Ro62-1 | 28.8 | 23.0 | 20.0 | — | 28.2 | 1.07 | 1.02 | 1.15 | 1.33 | *75* |
| ***Treponemataceae*** | | | | | | | | | | |
| *Treponema pallidum*, Reiter apathogenic | 30.0 | 20.0 | 18.0 | — | 32.0 | 1.00 | 0.94 | 1.11 | 1.63 | *76* |
| ***Rickettsiaceae*** | | | | | | | | | | |
| *Rickettsia burneti* | 29.5 | 22.5 | 22.0 | — | 26.0 | 1.08 | 1.13 | 1.02 | 1.25 | *77* |

## TABLE 1 (Continued)
## THALLOPHYTA: FUNGI

| Genus Source | Ade[a] | Gua | Cyt | 5Me Cyt | Thy | Molar Ratios[b] Pu/Py | A/T | G/C | A + T/G + C | Reference |
|---|---|---|---|---|---|---|---|---|---|---|
| ***Rickettsiaceae*—(Continued)** | | | | | | | | | | |
| *Coxiella burneti* (*Rickettsia*) | 29.1 | 21.9 | 21.0 | — | 27.9 | 1.04 | 1.04 | 1.04 | 1.33 | *92* |
| *Rickettsia prowazeki* | 35.7 | 17.1 | 15.4 | — | 31.8 | 1.12 | 1.12 | 1.10 | 2.08 | *78* |
| ***Borrelomycetaceae*** | | | | | | | | | | |
| *Pleuropneumonia*-like organism, PPLO-07 | 24.8 | 21.1 | 25.8 | — | 28.3 | 0.85 | 0.88 | 0.82 | 1.13 | *79* |
| PPLO, strain 07 | 24.8 | 21.0 | 25.8 | — | 28.3 | 0.85 | 0.88 | 0.81 | 1.13 | *80* |
| *Mycoplasma gallisepticum* (Avian PPLO 5969) | 32.7 | 19.2 | 14.1 | — | 34.0 | 1.08 | 0.96 | 1.36 | 2.00 | *81* |
| *Mycoplasma laidlawii*, 545B | 32.8 | 17.0 | 17.4 | — | 32.8 | 0.99 | 1.00 | 0.98 | 1.91 | *93* |
| *Mycoplasma laidlawii*, PG9 | 33.0 | 16.3 | 17.0 | — | 33.7 | 0.97 | 0.98 | 0.96 | 2.00 | *93* |
| *Mycoplasma laidlawii* | 33.5 | 16.0 | 16.5 | — | 34.0 | 0.98 | 0.99 | 0.97 | 2.08 | *94* |
| *Mycoplasma laidlawii* | 33.4 | 15.8 | 16.7 | — | 34.1 | 0.97 | 0.98 | 0.95 | 2.08 | *94* |
| *Mycoplasma laidlawii*, 544A | 33.2 | 15.7 | 16.8 | — | 34.3 | 0.96 | 0.96 | 0.93 | 2.08 | *93* |
| *Mycoplasma mycoides*, var. mycoides T3 | 35.5 | 15.2 | 14.8 | — | 34.5 | 1.03 | 1.03 | 1.03 | 2.33 | *93* |
| *Mycoplasma mycoides*, capri | 38.0 | 12.5 | 12.7 | — | 36.8 | 1.02 | 1.03 | 0.98 | 2.97 | *94* |
| *Mycoplasma mycoides*, capri | 37.8 | 12.7 | 12.3 | — | 37.2 | 1.02 | 1.02 | 1.03 | 3.00 | *94* |
| *Mycoplasma mycoides*, capri | 38.0 | 12.3 | 12.5 | — | 37.0 | 1.02 | 1.03 | 0.98 | 3.02 | *82* |
| *Mycoplasma mycoides*, capri p.g.3 | 37.9 | 12.1 | 12.6 | — | 37.3 | 1.00 | 1.02 | 0.96 | 3.04 | *83* |

[a] Proportions are expressed in moles of nitrogenous constituents per 100 moles of total constituents.
[b] In preparations containing 5-methylcytosine, the molar ratios include this component with cytosine.
[c] No specific search for methylcytosine was made in this analysis. However, investigations of the DNA of several algae and diatoms do not indicate methylcytosine in lower plants.[3]

Abbreviations: Ade or A = adenine; Gua or G = guanine; Cyt or C = cytosine; 5MeCyt = 5′-methylcytosine; Thy or T = thymine; Pu = sum of purines; Py = sum of pyrimidines.

# TABLE 2
# THALLOPHYTA: ALGAE

| Genus Source | Ade[a] | Gua | Cyt | 5Me Cyt | Thy | Molar Ratios[b] Pu/Py | A/T | G/C | A + T/G + C | Reference |
|---|---|---|---|---|---|---|---|---|---|---|
| ***Cyanophyta*** *(Blue-green Algae)* | | | | | | | | | | |
| *Lingbye aestuarii* | 19.4 | 29.6 | 30.7 | — | 20.0 | 0.97 | 0.97 | 0.96 | 0.65 | *1* |
| *Anacystis nidulans* | 23.5 | 32.5 | 25.5 | — | 18.5 | 1.27 | 1.27 | 1.27 | 0.72 | *1* |
| *Anacystis nidulans* | 23.3 | 29.5 | 24.8 | — | 22.3 | 1.12 | 1.04 | 1.19 | 0.84 | *2* |
| *Mastigocladus laminosus*, Cohn | 25.4 | 26.0 | 24.6 | — | 24.0 | 1.06 | 1.06 | 1.06 | 0.98 | *3* |
| *Coelospherium dubium* | 26.3 | 23.7 | 22.3 | 1.36 | 26.4 | 1.00 | 1.00 | 1.00 | 1.11 | *7* |
| *Aphanizomenon flosaquae* | 29.1 | 22.0 | 22.1 | — | 26.8 | 1.04 | 1.09 | 1.00 | 1.27 | *1* |
| *Aphanizomenon flosaquae* | 30.4 | 19.4 | 17.3 | 1.68 | 31.2 | 0.99 | 0.98 | 1.02 | 1.60 | *7* |
| ***Rhodophyta*** *(Red Algae)* | | | | | | | | | | |
| *Polyides rotundus* | 17.8 | 32.1 | 32.2 | — | 17.8 | 1.00 | 1.00 | 1.00 | 0.55 | *1* |
| *Polyides rotundus* | 18.7 | 30.9 | 27.5 | 3.48 | 19.4 | 0.98 | 0.97 | 1.00 | 0.62 | *7* |
| *Ahnfeltia plicata* | 23.5 | 26.8 | 25.4 | 1.34 | 22.8 | 1.02 | 1.03 | 1.00 | 0.87 | *7* |
| *Rhodymenia palmata* | 25.3 | 24.6 | 24.3 | — | 25.4 | 1.00 | 1.00 | 1.01 | 1.04 | *1* |
| ***Bacillariophyta*** *(Diatoms)* | | | | | | | | | | |
| *Melosira italica* | 26.8 | 22.8 | 20.6 | 2.23 | 27.3 | 0.99 | 0.98 | 1.00 | 1.19 | *7* |
| *Thalassiosira norden* | 29.8 | 20.1 | 20.2 | — | 29.9 | 1.00 | 1.00 | 1.00 | 1.48 | *4* |
| *Chaetoceras decipiens* | 30.8 | 19.9 | 19.2 | — | 30.1 | 1.03 | 1.02 | 1.04 | 1.55 | *4* |
| *Rhabdomena adriaticum* | 31.4 | 18.6 | 18.3 | — | 31.7 | 1.00 | 0.99 | 1.02 | 1.71 | *4* |
| ***Phaeophyta*** *(Brown Algae)* | | | | | | | | | | |
| *Dictiota fasciola* | 19.8 | 29.9 | 26.7 | 2.65 | 20.9 | 0.99 | 0.95 | 1.02 | 0.69 | *7* |
| *Cystoseira barbata* | 20.8 | 29.5 | 29.3 | — | 20.4 | 1.01 | 1.02 | 1.01 | 0.70 | *1* |
| *Phyllophora nervosa* | 21.0 | 28.5 | 26.2 | 2.32 | 21.9 | 0.98 | 0.96 | 1.00 | 0.75 | *7* |
| ***Chlorophyta*** *(Green Algae)* | | | | | | | | | | |
| *Scenedesmus acuminatus* | 18.7 | 32.9 | 30.9 | — | 17.5 | 1.07 | 1.07 | 1.06 | 0.56 | *1* |
| *Stigeoclonium tenue* | 18.4 | 31.3 | 30.0 | 1.55 | 18.4 | 0.99 | 1.00 | 0.99 | 0.59 | *7* |
| *Scenedesmus quadricauda* | 20.2 | 30.8 | 30.2 | — | 18.7 | 1.04 | 1.08 | 1.02 | 0.64 | *1* |
| *Scenedesmus quadricauda* | 20.3 | 30.0 | 26.2 | 3.23 | 20.5 | 1.01 | 0.99 | 1.02 | 0.69 | *7* |
| *Chlamydomonas globosa* | 19.0 | 30.3 | 28.2 | 2.75 | 19.5 | 0.98 | 0.98 | 0.98 | 0.63 | *7* |
| *Lagerchemia cillata* | 19.5 | 30.9 | 30.2 | — | 19.4 | 1.02 | 1.01 | 1.02 | 0.64 | *1* |
| *Lagercheimia ciliata* | 20.2 | 30.5 | 26.5 | 2.93 | 19.7 | 1.03 | 1.02 | 1.04 | 0.67 | *7* |
| *Chlorella vulgaris* | 20.2 | 30.0 | 26.4 | 3.45 | 19.8 | 1.01 | 1.02 | 1.00 | 0.67 | *7* |
| *Chlamydomonas gyrus* | 20.7 | 29.7 | 30.0 | — | 20.5 | 1.00 | 1.01 | 0.99 | 0.69 | *1* |
| *Scenedesmus obliquus* | 20.1 | 29.8 | 26.4 | 3.11 | 20.5 | 1.00 | 0.98 | 1.01 | 0.68 | *7* |
| *Scenedesmus obliquus*, Kütz | 20.6 | 29.7 | 29.4 | — | 20.4 | 1.01 | 1.01 | 1.01 | 0.69 | *5* |
| *Chlorella ellipsoidea* | 20.0 | 29.0 | 30.0 | — | 21.0 | 0.96 | 0.95 | 0.97 | 0.69 | *6* |
| *Chlorella ellipsoidea* | 20.0 | 29.0 | 30.0 | — | 21.0 | 0.96 | 0.95 | 0.97 | 0.69 | *1* |
| *Ankistrodesmus falcatus* | 21.8 | 29.7 | 29.0 | — | 19.5 | 1.06 | 1.12 | 1.02 | 0.70 | *1* |
| *Rhisoclonium laponicum* | 23.0 | 26.7 | 27.1 | — | 23.3 | 0.99 | 0.99 | 0.99 | 0.86 | *1* |
| *Hydrodictyon reticulatum* | 19.4 | 30.6 | 28.7 | 1.42 | 19.7 | 1.00 | 0.99 | 1.02 | 0.64 | *7* |
| *Hydrodictyon reticulatum* | 23.1 | 27.3 | 26.2 | — | 23.4 | 1.02 | 0.99 | 1.04 | 0.87 | *1* |
| *Mougeotia sp.* | 24.0 | 25.4 | 25.4 | — | 25.2 | 0.98 | 0.95 | 1.00 | 0.97 | *1* |
| *Dunaliella salina* | 25.0 | 25.5 | 24.7 | — | 24.8 | 1.02 | 1.01 | 1.03 | 0.99 | *1* |
| *Dunaliella salina* | 24.7 | 25.5 | 23.6 | 1.29 | 24.7 | 1.01 | 1.00 | 1.02 | 0.98 | *7* |
| *Dunaliella viridis* | 24.9 | 25.1 | 23.2 | 1.74 | 24.8 | 1.01 | 1.00 | 1.01 | 0.99 | *7* |
| *Asteromonas gracilis* | 25.1 | 25.0 | 24.1 | 0.93 | 24.9 | 1.00 | 1.01 | 1.00 | 1.00 | *7* |
| *Spirogyra sp.* | 30.7 | 19.2 | 19.8 | — | 30.4 | 0.99 | 1.01 | 0.97 | 1.57 | *1* |
| ***Charophyta*** *(Stoneworts)* | | | | | | | | | | |
| *Chara sp.* | 25.3 | 25.0 | 24.8 | — | 25.2 | 1.01 | 1.00 | 1.01 | 1.01 | *1* |
| *Nitella sp.* | 24.5 | 24.3 | 24.3 | — | 25.7 | 0.98 | 0.95 | 1.00 | 1.03 | *1* |

[a] Proportions are expressed in moles of nitrogenous constituents per 100 moles of total constituents.
[b] In preparations containing 5-methylcytosine, the molar ratios include this component with cytosine.

Abbreviations: Ade or A = adenine; Gua or G = guanine; Cyt or C = cytosine; 5MeCyt = 5′-methylcytosine; Thy or T = thymine; Pu = sum of purines; Py = sum of pyrimidines.

# TABLE 3
# TRACHEOPHYTA

| Genus Source | Ade[a] | Gua | Cyt | 5Me Cyt | Thy | Molar Ratios[b] Pu/Py | A/T | G/C | (A + T)/(G + C) | Reference |
|---|---|---|---|---|---|---|---|---|---|---|
| ***LYCOPSIDA*** | | | | | | | | | | |
| ***Lycopodim clavatum*** | 29.1 | 21.1 | 17.3 | 4.0 | 28.5 | 1.01 | 1.02 | 0.99 | 1.36 | *1* |
| ***SPHENOPSIDA*** | | | | | | | | | | |
| ***Equisetum arvense*** | 29.1 | 21.1 | 18.1 | 2.7 | 29.0 | 1.01 | 1.00 | 1.01 | 1.39 | *1* |
| ***PTEROPSIDA*** | | | | | | | | | | |
| ***Filicineae*** | | | | | | | | | | |
| Bracken fern | 28.3 | 19.8 | 18.2 | 5.6 | 28.4 | 0.92 | 1.00 | 0.83 | 1.30 | *2* |
| *Nephrolepsis sp.* | 28.7 | 21.0 | 19.0 | 2.5 | 28.8 | 0.99 | 1.00 | 0.98 | 1.35 | *1* |
| *Ceratopteris cornuta* | 29.7 | 19.8 | 17.2 | 3.8 | 29.5 | 0.98 | 1.01 | 0.94 | 1.45 | *1* |
| ***Gymnospermae*** | | | | | | | | | | |
| *Pinus sibirica* | 29.2 | 20.8 | 14.6 | 4.9 | 30.5 | 1.00 | 0.96 | 1.07 | 1.48 | *3* |
| *Pinus silvestris* | 30.1 | 19.3 | 16.5 | 3.6 | 30.5 | 0.98 | 0.99 | 0.96 | 1.54 | *1* |
| *Picea excelsa* | 30.7 | 19.5 | 19.6 | — | 30.2 | 1.02 | 1.01 | 1.00 | 1.56 | *1* |
| *Cephalotaxus fortunei* | 31.1 | 18.9 | 15.4 | 3.8 | 30.8 | 1.00 | 1.01 | 0.98 | 1.62 | *1* |
| *Ginkgo biloba* | 31.6 | 17.2 | 17.7 | — | 33.5 | 0.95 | 0.94 | 0.97 | 1.87 | *1* |
| ***Angiospermae*** | | | | | | | | | | |
| Sweet corn, +heterochromatic B chromosomes (*Zea mays*) | 15.1 | 35.0 | 31.0 | 5.3 | 14.5 | 0.99 | 1.04 | 0.96 | 0.42 | *4* |
| Sweet corn, −heterochromatic B chromosomes (*Zea mays*) | 22.1 | 28.2 | 21.1 | 6.1 | 22.8 | 1.01 | 0.97 | 1.04 | 0.81 | *4* |
| *Zea mays* | 25.6 | 24.5 | 17.4 | 7.2 | 25.3 | 1.00 | 1.01 | 1.00 | 1.04 | *1* |
| Corn (*Zea mays*) | 26.8 | 22.8 | 17.0 | 6.2 | 27.2 | 0.98 | 0.99 | 0.98 | 1.17 | *5* |
| *Zea mays*, K64r | 26.8 | 22.8 | 17.0 | 6.2 | 27.2 | 0.98 | 0.99 | 0.98 | 1.17 | *19* |
| Rye (*Secale cereale*) | 27.5 | 19.9 | 20.1 | 10.0 | 24.5 | 0.87 | 0.99 | 0.66 | 1.04 | *2* |
| Rye germ | 27.8 | 22.7 | 16.2 | 5.9 | 27.5 | 1.02 | 1.01 | 1.03 | 1.24 | *6* |
| Rye germ | 28.1 | 21.3 | 16.8 | 5.5 | 28.3 | 0.98 | 0.99 | 0.96 | 1.29 | *7* |
| *Triticum vulgare* | 25.6 | 23.8 | 18.2 | 6.4 | 26.0 | 0.98 | 0.99 | 0.97 | 1.07 | *3* |
| *Triticum vulgare* | 26.3 | 22.9 | 18.8 | 5.8 | 26.2 | 0.97 | 1.00 | 0.93 | 1.11 | *1* |
| Carrot leaf (*Daucus carota*) | 26.7 | 23.2 | 17.3 | 6.0 | 26.8 | 1.00 | 1.00 | 1.00 | 1.15 | *8* |
| Carrot | 26.7 | 23.1 | 17.3 | 5.9 | 26.9 | 0.99 | 0.99 | 1.00 | 1.16 | *8* |
| Carrot, crown gall tumor | 26.8 | 23.1 | 17.3 | 5.9 | 26.9 | 1.00 | 1.00 | 1.00 | 1.16 | *8* |
| Carrot, phloem | 26.9 | 23.0 | 17.2 | 5.9 | 27.0 | 1.00 | 1.00 | 1.00 | 1.17 | *8* |
| *Passiflora coerula* | 27.0 | 23.1 | 16.8 | 6.3 | 26.8 | 1.00 | 1.01 | 1.00 | 1.16 | *1* |
| Wheat germ (*Triticum sativum*) | 25.4 | 22.8 | 18.2 | 5.6 | 28.2 | 0.93 | 0.90 | 0.96 | 1.15 | *2* |
| Wheat germ | 26.5 | 23.5 | 17.2 | 5.8 | 27.0 | 1.00 | 0.98 | 1.02 | 1.15 | *9* |
| Wheat germ | 27.3 | 22.7 | 16.8 | 6.0 | 27.1 | 1.00 | 1.01 | 1.00 | 1.19 | *10* |
| Wheat germ | 27.1 | 20.2 | 19.6 | 5.7 | 27.4 | 0.89 | 0.99 | 0.80 | 1.20 | *11* |
| Wheat germ | 27.2 | 22.6 | 16.6 | 6.2 | 27.4 | 0.99 | 0.99 | 0.99 | 1.20 | *5* |
| Wheat germ | 26.9 | 23.2 | 16.7 | 5.3 | 28.0 | 1.00 | 0.96 | 1.05 | 1.21 | *12* |
| Wheat germ | 28.1 | 21.8 | 16.8 | 5.9 | 27.4 | 1.00 | 1.03 | 0.96 | 1.25 | *13* |
| Wheat germ | 27.6 | 22.0 | 16.9 | 5.6 | 27.9 | 0.98 | 0.99 | 0.98 | 1.25 | *7* |
| Wheat germ | 29.0 | 25.0 | 17.4 | — | 28.6 | 1.17 | 1.01 | 1.44 | 1.36 | *2* |
| Kale (*Brassica oteracea*) | 27.5 | 24.8 | 16.6 | 3.6 | 27.5 | 1.10 | 1.00 | 1.23 | 1.22 | *2* |
| *Styrax obassia* | 27.8 | 22.1 | 17.0 | 5.3 | 27.8 | 1.00 | 1.00 | 0.99 | 1.25 | *1* |
| Rice seedling leaf (*Oryza sativa*) | 28.0 | 22.0 | 18.5 | 3.4 | 28.1 | 1.00 | 1.00 | 1.00 | 1.28 | *14* |
| *Aleurites fordii* | 29.1 | 21.8 | 14.6 | 6.1 | 27.9 | 1.04 | 1.04 | 1.03 | 1.34 | *1* |
| Beet (*Beta vulgaris*) | 27.7 | 25.5 | 12.3 | 4.4 | 30.1 | 1.14 | 0.92 | 1.53 | 1.37 | *2* |
| Barley embryo (*Hordeum vulgare*) | 23.8 | 21.0 | 21.0 | — | 34.3 | 0.81 | 0.70 | 1.00 | 1.38 | *15* |
| Clover (*Trifolium pratense*) | 29.8 | 20.9 | 15.5 | 4.8 | 28.5 | 1.04 | 1.05 | 1.03 | 1.42 | *2* |
| *Brassica oleifera* | 29.1 | 20.5 | 18.7 | 2.0 | 29.7 | 0.98 | 0.98 | 0.99 | 1.43 | *1* |
| *Cordyline australis* | 29.4 | 20.2 | 21.0 | — | 29.4 | 0.98 | 1.00 | 0.96 | 1.43 | *1* |
| *Linum usitatissimum* | 29.6 | 20.1 | 16.8 | 4.0 | 29.5 | 0.99 | 1.00 | 0.97 | 1.44 | *1* |
| *Cucurbita pepo* | 30.2 | 21.0 | 16.1 | 3.7 | 29.0 | 1.04 | 1.04 | 1.06 | 1.44 | *3* |
| *Phaseolus vulgaris* | 29.7 | 20.6 | 14.9 | 5.2 | 29.6 | 1.01 | 1.00 | 1.03 | 1.45 | *3* |
| *Arachis hypogaea* | 29.3 | 20.3 | 14.4 | 6.1 | 29.8 | 0.99 | 0.99 | 0.99 | 1.45 | *3* |
| *Bletia hiacinthiana* | 30.6 | 20.7 | 15.1 | 5.0 | 28.4 | 1.05 | 1.08 | 1.02 | 1.45 | *1* |
| *Papaver somniferum* | 29.6 | 20.6 | 14.8 | 5.3 | 29.8 | 1.01 | 0.99 | 1.04 | 1.46 | *3* |
| Tobacco (*Nicotiana tabacum*) | 29.3 | 23.5 | 12.2 | 4.3 | 30.7 | 1.12 | 0.95 | 1.42 | 1.50 | *16* |
| Tobacco leaf (*Nicotiana tabacum*) | 29.7 | 19.8 | 13.9 | 6.1 | 30.4 | 0.98 | 0.98 | 0.99 | 1.51 | *17* |
| *Cudrania tricuspidata* | 30.5 | 19.7 | 14.3 | 5.5 | 30.0 | 1.01 | 1.02 | 0.99 | 1.53 | *1* |

## TABLE 3 (Continued)
## TRACHEOPHYTA

| Genus Source | Ade[a] | Gua | Cyt | 5Me Cyt | Thy | Molar Ratios[b] Pu/Py | A/T | G/C | A+T/G+C | Reference |
|---|---|---|---|---|---|---|---|---|---|---|
| *Angiospermae*—(Continued) | | | | | | | | | | |
| *Helianthus annuus* | 31.0 | 19.2 | 13.7 | 5.5 | 30.6 | 1.01 | 1.01 | 1.00 | 1.60 | *1* |
| Sunflower (*Helianthus annuus*) | 30.8 | 18.5 | 12.0 | 7.1 | 31.6 | 0.97 | 0.97 | 0.97 | 1.66 | *5* |
| Spinach (*Spinacia oleracea*) | 32.1 | 19.8 | 14.0 | 3.60 | 30.5 | 1.08 | 1.05 | 1.12 | 1.67 | *20* |
| *Allium cepa* | 31.8 | 18.4 | 12.8 | 5.4 | 31.3 | 1.01 | 1.02 | 1.01 | 1.72 | *3* |
| Cotton, diploid (*Gossypium thurberi*) | 31.6 | 18.2 | 14.6 | 3.8 | 31.8 | 0.99 | 0.99 | 0.99 | 1.73 | *18* |
| Cotton, diploid (*G. raimondii*) | 31.6 | 18.2 | 14.6 | 3.8 | 31.8 | 0.99 | 0.99 | 0.99 | 1.73 | *18* |
| Cotton, diploid (*G. herbaceum*) | 32.0 | 17.9 | 13.0 | 5.0 | 32.2 | 0.99 | 0.99 | 0.99 | 1.79 | *18* |
| Cotton, diploid (*G. arboreum*) | 32.1 | 17.8 | 12.8 | 5.0 | 32.2 | 1.00 | 1.00 | 1.00 | 1.81 | *18* |
| Cotton, diploid (*G. armourianum*) | 32.3 | 17.7 | 13.7 | 3.8 | 32.3 | 1.00 | 1.00 | 1.01 | 1.84 | *18* |
| Cotton, diploid (*G. anomalum*) | 32.5 | 17.4 | 12.8 | 4.6 | 32.6 | 1.00 | 1.00 | 1.00 | 1.87 | *18* |
| Cotton, diploid (*G. lobatum*) | 32.7 | 17.2 | 13.6 | 3.6 | 32.8 | 1.00 | 1.00 | 1.00 | 1.90 | *18* |
| Cotton, diploid (*G. gossypoides*) | 32.7 | 17.1 | 13.2 | 4.0 | 33.0 | 0.99 | 0.99 | 0.99 | 1.92 | *18* |
| Cotton, diploid (*G. davidsonii*) | 32.8 | 17.0 | 12.8 | 4.3 | 33.1 | 0.99 | 0.99 | 0.99 | 1.93 | *18* |
| Cotton, diploid (*G. aridum*) | 32.8 | 17.0 | 13.2 | 3.8 | 33.2 | 0.99 | 0.99 | 1.00 | 1.94 | *18* |
| Cotton, diploid (*G. Klotzschianum*) | 33.0 | 17.0 | 12.9 | 4.2 | 33.0 | 1.00 | 1.00 | 0.99 | 1.94 | *18* |
| Cotton, tetraploid (*G. hirsutum*) | 32.8 | 17.0 | 12.8 | 4.6 | 32.9 | 0.99 | 1.00 | 0.98 | 1.91 | *5* |
| Cotton, tetraploid (*G. barbadense*) | 32.9 | 16.8 | 12.5 | 4.6 | 33.1 | 0.99 | 0.99 | 0.98 | 1.95 | *5* |
| Peanut (*Arachis hypogaea*) | 32.1 | 17.6 | 12.3 | 5.7 | 32.2 | 0.99 | 1.00 | 0.98 | 1.80 | *5* |

[a] Proportions are expressed in moles of nitrogenous constituents per 100 moles of total constituents.
[b] In preparations containing 5-methylcytosine, the molar ratios include this component with cytosine.

Abbreviations: Ade or A = adenine; Gua or G = guanine; Cyt or C = cytosine; 5MeCyt = 5′-methylcytosine; Thy or T = thymine; Pu = sum of purines; Py = sum of pyrimidines.

## TABLE 4
## PROTOZOA

| Genus Source | Ade[a] | Gua | Cyt | 5Me Cyt | Thy | Molar Ratios[b] Pu/Py | A/T | G/C | A+T/G+C | Reference |
|---|---|---|---|---|---|---|---|---|---|---|
| ***Euglena gracilis*** | 23.0 | 29.6 | 27.5 | — | 19.9 | 1.11 | 1.16 | 1.08 | 0.75 | *1* |
| ***Euglena gracilis***, strain z | 24.5 | 24.8 | 23.7 | 2.3 | 24.7 | 0.95 | 0.99 | 0.95 | 0.97 | *2* |
| ***Schizotryptanum cruzi*** | 25.3 | 24.7 | 24.7 | — | 25.3 | 1.00 | 1.00 | 1.00 | 1.02 | *3* |
| ***Polytomella papillata*** | 29.6 | 20.6 | 20.8 | — | 29.0 | 1.01 | 1.02 | 0.99 | 1.42 | *4* |
| **Malaria parasite** (*Plasmodium berghii*) | 30.3 | 20.8 | 20.0 | — | 29.0 | 1.04 | 1.04 | 1.04 | 1.45 | *5* |
| ***Tetrahymena pyriformis*** | 35.4 | 14.5 | 14.7 | — | 35.4 | 1.00 | 1.00 | 0.99 | 2.42 | *4* |
| ***Tetrahymena pyriformis*** | 36.6 | 11.5 | 15.0 | — | 36.9 | 0.93 | 0.99 | 0.77 | 2.77 | *6* |

[a] Proportions are expressed in moles of nitrogenous constituents per 100 moles of total constituents.
[b] In preparations containing 5-methylcytosine, the molar ratios include this component with cytosine.

Abbreviations: Ade or A = adenine; Gua or G = guanine; Cyt or C = cytosine; 5MeCyt = 5′-methylcytosine; Thy or T = thymine; Pu = sum of purines; Py = sum of pyrimidines.

## TABLE 5
## BACTERIAL PHAGES

| Phage and Host | Ade[a] | Gua | Cyt | Hm Cyt | Thy | Molar Ratios[b] Pu/Py | A/T | G/C | A + T/G + C | Reference |
|---|---|---|---|---|---|---|---|---|---|---|
| N6, *M. lysodeikticus* | 16.0 | 30.6 | 39.1 | — | 14.6 | 0.87 | 1.10 | 0.78 | 0.44 | *1* |
| N1, *M. lysodeikticus* | 18.0 | 31.9 | 37.2 | — | 13.0 | 0.99 | 1.38 | 0.86 | 0.45 | *1* |
| Phage XP12 (*X. oryzae*) | 16.8 | 33.8 | — | 33.4[c] | 16.0 | 1.02 | 1.05 | 1.01 | 0.49 | *34* |
| Phagus choremis, *Mycobacterium sp.* Jucho | 17.2 | 33.4 | 33.4 | — | 15.9 | 1.03 | 1.08 | 1.00 | 0.50 | *2* |
| MX-1, *Myxococcus xanthus* | 24.1 | 25.0 | 30.8 | — | 20.0 | 0.97 | 1.21 | 0.81 | 0.79 | *3* |
| κ, *S. marcescens* | 23.4 | 25.0 | 28.8 | — | 22.8 | 0.94 | 1.03 | 0.87 | 0.86 | *4* |
| ϕ80, *E. coli* K12 | 23.0 | 27.0 | 26.0 | — | 24.0 | 1.00 | 0.96 | 1.04 | 0.89 | *5* |
| λV virulent, *E. coli* | 21.3 | 22.9 | 27.1 | — | 28.6 | 0.79 | 0.74 | 0.85 | 1.00 | *6* |
| λ temperate | 26.0 | 23.8 | 24.3 | — | 25.8 | 0.99 | 1.01 | 0.98 | 1.08 | *7* |
| $T_3$, *E. coli* | 24.7 | 23.8 | 26.2 | — | 25.3 | 0.94 | 0.98 | 0.91 | 1.00 | *8* |
| $T_3$, DNA | 24.0 | 25.8 | 24.0 | — | 26.2 | 0.99 | 0.92 | 1.08 | 1.01 | *9* |
| P22, *Salmonella* | 25.0 | 25.0 | 25.0 | — | 25.0 | 1.00 | 1.00 | 1.00 | 1.00 | *10* |
| Phage η, *S. marcescens* | 25.5 | 15.8 | 33.5 | — | 25.2 | 0.70 | 1.01 | 0.47 | 1.03 | *4* |
| $T_1$ | 27.0 | 23.0 | 25.0 | — | 25.0 | 1.00 | 1.08 | 0.92 | 1.08 | *11* |
| $T_7$ | 25.8 | 24.4 | 23.3 | — | 26.6 | 1.01 | 0.97 | 1.05 | 1.10 | *12* |
| $T_7$ | 26.0 | 23.8 | 23.6 | — | 26.6 | 0.99 | 0.98 | 1.01 | 1.11 | *13* |
| Phage DD7, *E. coli* C | 27.0 | 22.6 | 22.5 | — | 27.3 | 0.99 | 0.99 | 1.00 | 1.20 | *35* |
| Bacteriophage α, *B. megaterium* | 28.4 | 22.1 | 22.3 | — | 27.4 | 1.02 | 1.04 | 0.99 | 1.26 | *14* |
| ϕX-174, coliphage | 24.6 | 24.1 | 18.5 | — | 32.7 | 0.95 | 0.75 | 1.30 | 1.35 | *15* |
| $A_1$, temperate, *S. typhimurium* | 23.4 | 18.8 | 24.6 | — | 33.3 | 0.73 | 0.70 | 0.76 | 1.31 | *6* |
| Thermophilic TP-84, *B. stearothermophilus* | 27.7 | 22.0 | 22.1 | — | 28.2 | 0.99 | 0.98 | 1.00 | 1.27 | *16* |
| Temperate TP-1C, *B. stearothermophilus* | 28.1 | 21.5 | 21.8 | — | 28.6 | 0.98 | 0.98 | 0.99 | 1.31 | *17* |
| Phage Tϕ3, *B. stearothermophilus* | 29.0 | 19.4 | 20.8 | — | 30.8 | 0.94 | 0.94 | 0.93 | 1.49 | *36* |
| $S_d$, *E. coli* | 28.6 | 21.3 | 21.4 | — | 28.7 | 1.00 | 1.00 | 1.00 | 1.34 | *18* |
| fd, *E. coli* $K_{12}$ | 24.4 | 19.9 | 21.7 | — | 34.1 | 0.79 | 0.72 | 0.92 | 1.41 | *19* |
| E. coliphage fd single strand | 24.4 | 19.9 | 21.7 | — | 34.1 | 0.79 | 0.72 | 0.92 | 1.41 | *20* |
| Phage SPP 1, *B. subtilis* | 28.4 | 21.7 | 20.5 | — | 29.4 | 1.00 | 0.97 | 1.06 | 1.37 | *37* |
| Phage SP 50, *B. subtilis* | 29.5 | 21.2 | 20.5 | — | 28.8 | 1.03 | 1.02 | 1.03 | 1.40 | *38* |
| C2, Myxobacterium, *Chondrococcus columnaris* | 28.6 | 22.9 | 18.4 | — | 30.1 | 1.06 | 0.95 | 1.24 | 1.42 | *21* |
| E. coliphage M13 single strand | 23.3 | 21.1 | 19.8 | — | 35.8 | 0.80 | 0.65 | 1.07 | 1.44 | *20* |
| Bacillus phage S-1 | 30.7 | 20.6 | 20.2 | — | 28.5 | 1.05 | 1.08 | 1.02 | 1.45 | *22* |
| Bacillus phage SP-10 | 30.7 | 20.6 | 19.9 | — | 28.8 | 1.05 | 1.07 | 1.04 | 1.47 | *22* |
| Bacillus phage S-$1_{C13}$ | 31.7 | 19.7 | 20.3 | — | 28.3 | 1.06 | 1.12 | 0.97 | 1.50 | *22* |
| Bacillus phage M2 | 31.6 | 19.0 | 18.4 | — | 31.0 | 1.02 | 1.02 | 1.03 | 1.67 | *22* |
| ϕ3, *B. subtilis* NCTC 3610 | 32.7 | 18.1 | 17.8 | — | 31.4 | 1.03 | 1.04 | 1.02 | 1.79 | *23* |
| $T_5$, DNA | 30.3 | 19.5 | 19.5 | — | 30.8 | 0.99 | 0.98 | 1.00 | 1.57 | *24* |
| $T_5$, *E. coli* | 33.7 | 20.9 | 10.5 | — | 34.9 | 1.20 | 0.97 | 1.99 | 2.18 | *25* |
| $T_4r^+$ | 32.3 | 18.3 | — | 16.3 | 33.1 | 1.02 | 0.98 | 1.12 | 1.89 | *24* |
| $T_4r$ | 32.2 | 18.0 | — | 16.3 | 33.5 | 1.01 | 0.96 | 1.10 | 1.91 | *24* |
| $T_4r^+$ | 32.7 | 18.1 | — | 13.5 | 35.7 | 1.03 | 0.92 | 1.34 | 2.16 | *26* |
| $T_4r^+$ | 33.0 | 18.3 | — | 12.5 | 36.0 | 1.06 | 0.92 | 1.46 | 2.24 | *27* |
| $T_2r$, DNA | 32.4 | 18.3 | — | 17.0 | 32.4 | 1.04 | 1.00 | 1.08 | 1.84 | *24* |
| $T_2r^+$, DNA | 32.5 | 18.2 | — | 16.7 | 32.6 | 1.04 | 1.00 | 1.09 | 1.86 | *24* |
| $T_2r^+$ | 32.0 | 18.0 | — | 16.8 | 33.3 | 1.02 | 0.96 | 1.07 | 1.88 | *24* |
| $T_2r$ | 32.3 | 17.6 | — | 16.7 | 33.4 | 1.01 | 0.97 | 1.05 | 1.91 | *24* |
| $T_2r$ | 32.8 | 16.8 | — | 17.5 | 33.0 | 0.98 | 0.99 | 0.96 | 1.92 | *28* |
| $T_2r^+$ | 31.0 | 18.0 | — | 16.0 | 35.0 | 0.96 | 0.89 | 1.13 | 1.94 | *11* |
| $T_2r$ | 35.0 | 19.0 | — | 15.0 | 31.0 | 1.17 | 1.12 | 1.27 | 1.94 | *11* |
| $T_2$ | 32.0 | 17.0 | — | 16.0 | 36.0 | 0.94 | 0.89 | 1.06 | 2.06 | *29* |
| $T_2r^+$ | 33.2 | 17.9 | — | 13.6 | 35.2 | 1.05 | 0.94 | 1.32 | 2.17 | *26* |
| $T_2$, *E. coli* | 33.2 | 18.2 | — | 12.7 | 35.9 | 1.06 | 0.92 | 1.43 | 2.24 | *25* |
| $T_2r$ | 35.8 | 13.6 | — | 17.2 | 33.2 | 0.98 | 1.08 | 0.79 | 2.24 | *30* |
| $T_2r^+$ | 33.6 | 18.1 | — | 12.7 | 35.6 | 1.07 | 0.94 | 1.43 | 2.25 | *27* |
| $T_6r^+$, DNA | 30.9 | 18.4 | — | 17.4 | 33.3 | 0.97 | 0.93 | 1.06 | 1.79 | *31* |
| $T_6r^+$ | 32.5 | 18.3 | — | 16.7 | 32.5 | 1.03 | 1.00 | 1.10 | 1.86 | *24* |
| $T_6r$ | 32.3 | 17.7 | — | 16.6 | 33.4 | 1.00 | 0.97 | 1.07 | 1.91 | *24* |
| $T_6r^+$ | 32.5 | 17.8 | — | 16.3 | 33.5 | 1.01 | 0.97 | 1.09 | 1.93 | *24* |
| $T_6r^+$ | 33.2 | 17.8 | — | 13.3 | 35.6 | 1.04 | 0.93 | 1.34 | 2.21 | *26* |
| $T_6r^+$ | 34.6 | 13.2 | — | 17.6 | 34.6 | 0.92 | 1.00 | 0.75 | 2.25 | *30* |

## TABLE 5 (Continued)
## BACTERIAL PHAGES

| Phage and Host | Ade[a] | Gua | Cyt | 5Hm Cyt | Thy | Pu/Py | A/T | G/C | A+T/G+C | Reference |
|---|---|---|---|---|---|---|---|---|---|---|
| | | | | | | Molar Ratios[b] | | | | |
| $T_6r^+$ | 33.6 | 18.0 | — | 12.3 | 36.0 | 1.07 | 0.93 | 1.46 | 2.30 | *27* |
| PBS2, *B. subtilis* | 36.2 | 14.0 | 14.3 | — | 35.6[d] | 1.01 | 1.02 | 0.98 | 2.54 | *32* |
| PBS2, *B. subtilis* | 35.9 | 13.4 | 14.7 | — | 35.9[d] | 0.97 | 1.00 | 0.91 | 2.56 | *33* |

[a] Proportions are expressed in moles of nitrogenous constituents per 100 moles of total constituents.
[b] In preparations containing 5-mydroxymethylcytosine, the molar ratios include this component with cytosine.
[c] 5′-Methylcytosine is found in place of cytosine in this preparation.
[d] Deoxyuridylic acid is found in place of thymidylic acid in these DNA preparations.

Abbreviations: Ade or A = adenine; Gua or G = guanine; Cyt or C = cytosine; 5HmCyt = 5′-hydroxymethylcytosine; Thy or T = thymine; Pu = sum of purines; Py = sum of pyrimidines.

## TABLE 6
## INSECT VIRUSES

| Virus Source | Ade[a] | Gua | Cyt | 5Hm Cyt | Thy | Pu/Py | A/T | G/C | A+T/G+C | Reference |
|---|---|---|---|---|---|---|---|---|---|---|
| | | | | | | Molar Ratios[b] | | | | |
| **Polyhedral virus** (Gypsy moth) | 21.2 | 30.5 | 28.3 | — | 20.1 | 1.07 | 1.05 | 1.08 | 0.70 | *1* |
| **Polyhedral virus,** Nun moth (*Lymantria monacha*) | 24.6 | 26.8 | 24.7 | — | 23.8 | 1.06 | 1.03 | 1.09 | 0.94 | *1* |
| **Polyhedral virus** (Spruce budworm) | 24.8 | 26.7 | 24.5 | — | 24.0 | 1.06 | 1.03 | 1.09 | 0.95 | *1* |
| **Polyhedral virus** (*Ptychopoda seriata*) | 26.7 | 24.4 | 23.2 | — | 25.7 | 1.04 | 1.04 | 1.05 | 1.10 | *1* |
| **Polyhedral virus,** Eastern tent caterpillar (*Malacosoma americanum*) | 29.2 | 22.5 | 20.2 | — | 28.0 | 1.07 | 1.04 | 1.11 | 1.34 | *1* |
| **Alfalfa butterfly virus** (*Colias philodice eurytheme*) | 29.9 | 22.4 | 20.1 | — | 27.6 | 1.10 | 1.08 | 1.11 | 1.35 | *1* |
| **Forest tent caterpillar** (*Malacosoma disstria*) | 29.2 | 21.9 | 20.3 | — | 28.5 | 1.05 | 1.02 | 1.08 | 1.37 | *1* |
| **Polyhedral virus** (*Bombyx mori*) | 29.3 | 22.5 | 20.2 | — | 28.0 | 1.07 | 1.05 | 1.11 | 1.34 | *1* |
| **Polyhedral virus** (*Bombyx mori*) | 30.2 | 20.2 | 19.8 | — | 30.0 | 1.01 | 1.01 | 1.02 | 1.51 | *2* |
| **Polyhedral virus** (*Bombyx mori*) | 30.1 | 21.5 | 19.9 | — | 28.5 | 1.07 | 1.06 | 1.08 | 1.42 | *3* |
| **Polyhedral virus** (*Bombyx mori*) | 32.2 | 22.4 | 16.0 | — | 29.4 | 1.20 | 1.10 | 1.40 | 1.60 | *4* |
| **Capsular virus** (*Cacoecia murinana*) | 32.1 | 19.7 | 17.8 | — | 30.5 | 1.07 | 1.05 | 1.11 | 1.67 | *1* |
| **Polyhedral virus** (Pine sawfly) | 32.3 | 19.5 | 17.9 | — | 30.3 | 1.07 | 1.07 | 1.09 | 1.67 | *1* |
| **Capsular virus** (Spruce budworm) | 32.8 | 18.4 | 16.4 | — | 32.4 | 1.05 | 1.01 | 1.12 | 1.87 | *1* |
| **Tipula iridescent virus** (European cranefly) | 35.0 | 15.2 | 15.7 | — | 34.1 | 1.01 | 1.03 | 0.97 | 2.24 | *5* |

[a] Proportions are expressed in moles of nitrogenous constituents per 100 moles of total constituents.
[b] In preparations containing 5-hydroxymethylcytosine, the molar ratios include this component with cytosine.

Abbreviations: Ade or A = adenine; Gua or G = guanine; Cyt or C = cytosine; 5HmCyt = 5′-hydroxymethylcytosine; Thy or T = thymine; Pu = sum of purines; Py = sum of pyrimidines.

# TABLE 7
# ANIMAL VIRUSES

| Virus and Host | Ade[a] | Gua | Cyt | 5Hm Cyt | Thy | Molar Ratios[b] Pu/Py | A/T | G/C | A+T/G+C | Reference |
|---|---|---|---|---|---|---|---|---|---|---|
| **Herpes simplex,** $H_4$ | 13.8 | 37.7 | 35.6 | — | 12.8 | 1.06 | 1.08 | 1.06 | 0.36 | *1* |
| **Herpes virus** | 17.1 | 32.9 | 32.2 | — | 17.6 | 1.00 | 0.97 | 1.02 | 0.53 | *2* |
| **Pseudorabies,** Pr | 13.2 | 37.0 | 36.3 | — | 13.5 | 1.00 | 0.98 | 1.02 | 0.36 | *1* |
| **Adenovirus,** type 4 | 22 | 27 | 30 | — | 20 | 0.98 | 1.10 | 0.90 | 0.74 | *3* |
| **Adenovirus,** type 2 | 22 | 27 | 29 | — | 21 | 0.98 | 1.05 | 0.93 | 0.77 | *3* |
| **Adenovirus 7-E46⁺,** Human | 24.5 | 25.2 | 24.8 | — | 25.5 | 0.99 | 0.96 | 1.02 | 1.00 | *4* |
| **Adenovirus 7-E46⁻,** Human | 25.0 | 24.3 | 24.8 | — | 25.9 | 0.97 | 0.97 | 0.98 | 1.04 | *4* |
| **Adenoassociated virus** | 22.9 | 26.9 | 27.3 | — | 22.8 | 0.99 | 1.00 | 0.99 | 0.84 | *4* |
| **Equine abortus** | 22.1 | 29.2 | 26.8 | — | 22.1 | 1.05 | 1.00 | 1.09 | 0.79 | *5* |
| **Shope papilloma** | 26.6 | 24.5 | 24.2 | — | 24.7 | 1.04 | 1.08 | 1.01 | 1.05 | *6* |
| **Minute virus of mouse,** MVM | 24.2 | 19.2 | 22.1 | — | 34.5 | 0.77 | 0.70 | 0.87 | 1.42 | *9* |
| **Vaccinia virus** | 29.5 | 20.6 | 20.0 | — | 29.9 | 1.01 | 0.99 | 1.03 | 1.46 | *7* |
| **Fowlpox virus,** DNA | 32.3 | 18.0 | 17.2 | — | 32.6 | 1.01 | 0.99 | 1.05 | 1.84 | *8* |
| **Fowlpox virus,** whole | 31.9 | 18.7 | 16.2 | — | 33.3 | 1.02 | 0.96 | 1.15 | 1.87 | *8* |

[a] Proportions are expressed in moles of nitrogenous constituents per 100 moles of total constituents.
[b] In preparations containing 5-hydroxymethylcytosine, the molar ratios include this component with cytosine.

Abbreviations: Ade or A = adenine; Gua or G = guanine; Cyt or C = cytosine; 5HmCyt = 5′-hydroxymethylcytosine; Thy or T = thymine; Pu = sum of purines; Py = sum of pyrimidines.

## REFERENCES

### Thallophyta: Fungi


1. Venner, *Hoppe-Seyler's Z. Physiol. Chem., 333,* 5 (1963).
2. Vanyushin, Belozerskii and Bogdanova, *Dokl. Akad. Nauk SSSR, 134,* 1222 (1960).
3. Low, *Nature, 182,* 1096 (1958).
4. Minagawa, Wagner and Strauss, *Arch. Biochem. Biophys., 80,* 442 (1959).
5. Uryson and Belozerskii, *Dokl. Akad. Nauk SSSR, 132,* 708 (1960).
6. Jones and Thompson, *J. Protozool., 10,* 91 (1963).
7. Kogane and Yanagita, *J. Gen. Appl. Microbiol., 10,* 61 (1964).
8. Rost and Venner, *Hoppe-Seyler's Z. Physiol. Chem., 339,* 230 (1964).
9. Zamenhof and Chargaff, *J. Biol. Chem., 187,* 1 (1950).
10. Vischer, Zamenhof and Chargaff, *J. Biol. Chem., 177,* 429 (1949).
11. Tewari, Votsch, Mahler and Mackler, Personal Communication.
12. Comb, Brown and Katz, *J. Mol. Biol., 8,* 781 (1964).
13. Vanyushin, Kokurina and Belozerskii, *Dokl. Akad. Nauk SSSR, 158,* 722 (1964).
14. Vanyushin and Belozerskii, *Dokl. Akad. Nauk SSSR, 135,* 197 (1960).
15. Anderson, Pramer and Davis, *Biochim. Biophys. Acta, 108,* 155 (1965).
16. Yong Lee, Wahl and Barbu, *Ann. Inst. Pasteur (Paris), 91,* 212 (1956).
17. Spirin, Belozerskii, Shugayeva and Vanyushin, *Biokhimiya, 22,* 744 (1957).
18. Burton, *Biochem. J., 77,* 547 (1960).
19. Sigal, Senez, LeGall and Sebald, *J. Bacteriol., 85,* 1315 (1963).
20. Catlin and Cunningham, *J. Gen. Microbiol., 19,* 522 (1958).
21. Reddi, *Biochim. Biophys. Acta, 15,* 585 (1954).
22. Kutsemakina, *Zh. Mikrobiol. Epidemiol. Immunobiol., 43,* 10 (1966).
23. Loewus, Brown and McLaren, *Nature, 203,* 104 (1964).

24. Belozerskii, Imshenetskii, Zaitseva and Perova, *Mikrobiologiya, 27,* 150 (1958).
25. Zaitseva and Belozerskii, *Mikrobiologiya, 26,* 722 (1957).
26. Scaletti and Naylor, *J. Bacteriol., 78,* 422 (1959).
27. Gause, Loshkareva, Zbarsky and Gause, *Nature, 203,* 598 (1964).
28. Dutta, Jones and Stacey, *J. Gen. Microbiol., 14,* 160 (1956).
29. Moseley and Schein, *Nature, 203,* 1298 (1964).
30. Schein, *Biochem. J., 101,* 647 (1966).
31. Setlow and Duggan, *Biochim. Biophys. Acta, 87,* 664 (1964).
32. Guberniev and Ugoleva, *Dokl. Akad. Nauk SSSR, 133,* 466 (1960).
33. Blobel, *J. Bacteriol., 82,* 425 (1961).
34. Catlin and Cunningham, *J. Gen. Microbiol., 26,* 303 (1961).
35. Mindich and Hotchkiss, *Biochim. Biophys. Acta, 80,* 73 (1964).
36. Daly, Allfrey and Mirsky, *J. Gen. Physiol., 33,* 497 (1950).
37. Panos, *J. Gen. Microbiol., 39,* 131 (1965).
38. Kareva and Filosova, *Akad. Nauk USSR Inst. Biokhim., 38,* 321 (1966).
39. Belozerskii, Shugayeva and Spirin, *Dokl. Akad. Nauk SSSR, 119,* 330 (1958).
40. Spirin, Belozerskii, Shugayeva and Vanyushin, *Biokhimiya, 22,* 744 (1957).
41. Hitawari, *Seikagaku, 31,* 12 (1959).
42. Masui, Iwata, Ishimitsu and Umebayashi, *Biochim. Biophys. Acta, 55,* 384 (1962).
43. Pons, *Biochem. Z., 346,* 26 (1966).
44. Jones and Walker, *J. Gen. Microbiol., 31,* 187 (1963).
45. Rudner, Shapiro and Chargaff, *Biochim. Biophys. Acta, 129,* 85 (1966).
46. Dunn and Smith, *Biochem. J., 68,* 627 (1958).
47. Jones, Marsh and Rizvi, *J. Gen. Microbiol., 17,* 586 (1957).
48. Spirin and Belozerskii, *Biokhimiya, 21,* 768 (1956).
49. Spirin, Belozerskii and Pretel-Martinez, *Dokl. Akad. Nauk SSSR, 111,* 1297 (1956).
50. Harold and Ziporin, *Biochim. Biophys. Acta, 28,* 482 (1958).
51. Chargaff, Schulman and Shapiro, *Nature, 180,* 851 (1957).
52. Rudner, Prokop-Schneider and Chargaff, *Nature, 203,* 479 (1964).
53. Rudner, Rejman and Chargaff, *Proc. Nat. Acad. Sci. U.S.A., 54,* 904 (1965).
54. Rudner, Shapiro and Chargaff, *Nature, 195,* 143 (1962).
55. Kornberg, *Science, 131,* 1503 (1960).
56. Shapiro and Chargaff, *Nature, 188,* 62 (1960).
57. Tonomura, Malkin and Rabinowitz, *J. Bacteriol., 89,* 1438 (1965).
58. Hershey, Dixon and Chase, *J. Gen. Physiol., 36,* 777 (1953).
59. Sarfert and Venner, *Hoppe-Seyler's Z. Physiol. Chem., 340,* 157 (1965).
60. Smith and Wyatt, *Biochem. J., 49,* 144 (1951).
61. Bekker, Kutsemakina and Michaelova, *Dokl. Akad. Nauk SSSR, 142,* 1188 (1962).
62. Bekker and Kutsemakina, *Biokhimiya, 28,* 612 (1963).
63. Bekker and Kutsemakina, *Zh. Mikrobiol. Epidemiol. Immunobiol., 43,* 84 (1966).
64. Masui, Kawasaki and Tabata, *Osaka City Med. J., 6,* 139 (1960).
65. Ikeda, Saito, Miura, Takagi and Aoki, *J. Gen. Appl. Microbiol., 11,* 181 (1965).
66. Welker and Campbell, *J. Bacteriol., 89,* 175 (1965).
67. Evreinova, Bunina and Kuznetsova, *Biokhimiya, 24,* 912 (1959).
68. Stoletov, Glazer and Shestakov, *Nauchn. Dokl. Vyssh. Shk. Biol. Nauki,* p. 179 (1966).
69. Levin, *Biochemistry, 5,* 1618 (1966).
70. Stuy, *J. Bacteriol., 76,* 179 (1958).
71. Haywood, Gray and Chargaff, *Biochim. Biophys. Acta, 61,* 155 (1962).
72. Tsumita and Chargaff, *Biochim. Biophys. Acta, 29,* 568 (1958).
73. Laland, Overend and Webb, *J. Chem. Soc.,* p. 3224 (1952).
74. Zimmer and Venner, *Hoppe-Seyler's Z. Physiol. Chem., 333,* 20 (1963).
75. Kingsbury and Ordal, *J. Bacteriol., 91,* 1327 (1966).
76. Rathlev and Pfau, *Arch. Biochem. Biophys., 106,* 343 (1964).
77. Smith and Stoker, *Brit. J. Exp. Pathol., 32,* 433 (1951).
78. Wyatt and Cohen, *Nature, 170,* 846 (1952).
79. Lynn and Smith, *Ann. N. Y. Acad. Sci., 79,* 493 (1959).
80. Lynn and Smith, *J. Bacteriol., 74,* 811 (1957).
81. Morowitz, Tourtellotte, Guild, Castro, Woese, and Cleverdon, *J. Mol. Biol., 4,* 93 (1962).
82. Jones and Walker, *Nature, 198,* 588 (1963).
83. Jones, Tittensor and Walker, *J. Gen. Microbiol., 40,* 405 (1965).
84. Kuo, Huang and Teng, *J. Mol. Biol., 34,* 373 (1968).
85. Venner, *Acta Biochim. Pol., 14,* 31 (1967).

86. Boháček, Kocur and Martinec, *J. Appl. Bacteriol., 31,* 215 (1968).
87. Boháček, Kocur and Martinec, *Arch. Mikrobiol., 64,* 23 (1968).
88. Jones and Walker, *Arch. Biochem. Biophys., 128,* 579 (1968).
89. Samoilenko and Ivanov, *Bull. Exp. Biol. Med., 63,* 43 (1967).
90. Halvorson, Szulmajster, Cohen and Michelson, *J. Mol. Biol., 28,* 71 (1967).
91. Zyghanov and Krassykova, *Mikrobiologiya, 37,* 969 (1968).
92. Schramek, *Acta Virol., 12,* 18 (1968).
93. Chelton, Jones and Walker, *J. Gen. Microbiol., 50,* 305 (1968).
94. Walker, *Nature, 216,* 711 (1967).

**Thallophyta: Algae**

1. Serenkov, *Izv. Akad. Nauk SSSR Ser. Biol., 27,* 857 (1962).
2. Biswas and Myers, *Nature, 188,* 1029 (1960).
3. Evreinova, Davydova, Sukover and Goryunova, *Dokl. Akad. Nauk SSSR, 137,* 213 (1961).
4. Serenkov and Pakhomova, *Nauchn. Dokl. Vyssh. Shk. Biol. Nauki,* p. 156 (1959).
5. Pakhomova and Serenkov, *Biokhimiya, 28,* 808 (1963).
6. Iwamura and Myers, *Arch. Biochem. Biophys., 84,* 267 (1959).
7. Pakhomova, Zaitseva and Belozerskii, *Dokl. Akad. Nauk SSSR, 182,* 712 (1968).

**Tracheophyta**

1. Vanyushin and Belozerskii, *Dokl. Akad. Nauk SSSR, 129,* 944 (1959).
2. Thomas and Sheratt, *Biochem. J., 62,* 1 (1956).
3. Uryson and Belozerskii, *Dokl. Akad. Nauk SSSR, 125,* 1144 (1959).
4. van Schaik and Pitout, *S. Afr. J. Sci., 62,* 53 (1966).
5. Ergle and Katterman, *Plant Physiol., 36,* 811 (1961).
6. Shapiro and Chargaff, *Biochim. Biophys. Acta, 26,* 608 (1957).
7. Spencer and Chargaff, *Biochim. Biophys. Acta, 68,* 18 (1963).
8. Thomas, *Arch. Biochem. Biophys., 79,* 162 (1959).
9. Wyatt, *Biochem. J., 48,* 584 (1951).
10. Brawerman and Chargaff, *J. Amer. Chem. Soc., 73,* 4052 (1951).
11. Laland, Overend and Webb, *J. Chem. Soc.,* p. 3224 (1952).
12. Hurst, Marko and Butler, *J. Biol. Chem., 204,* 847 (1953).
13. Lipshitz and Chargaff, *Biochim. Biophys. Acta, 19,* 256 (1956).
14. Yoshii, *Nippon Nogei Kagaku Kaishi, 36,* 1 (1962).
15. Chang, *Nature, 198,* 1167 (1963).
16. Tewari and Wildman, *Science, 153,* 1269 (1966).
17. Lyttleton and Petersen, *Biochim. Biophys. Acta, 80,* 391 (1964).
18. Ergle, Katterman and Richard, *Plant Physiol., 39,* 145 (1964).
19. Pitout and Potgieter, *Biochim. Biophys. Acta, 161,* 188 (1968).
20. Bard and Gordon, *Plant Physiol., 44,* 377 (1969).

**Protozoa**

1. Serenkov, *Izv. Akad. Nauk SSSR Ser. Biol., 27,* 857 (1962).
2. Brawerman, Hufnagel and Chargaff, *Biochim. Biophys. Acta, 61,* 340 (1962).
3. Antonov and Belozerskii, *Dokl. Akad. Nauk SSSR, 138,* 1216 (1961).
4. Jones and Thompson, *J. Protozool., 10,* 91 (1963).
5. Whitfeld, *Aust. J. Biol. Sci., 6,* 234 (1953).
6. Scherbaum, *Exp. Cell Res., 13,* 24 (1957).

**Bacterial Phages**

1. Scaletti and Naylor, *J. Bacteriol., 78,* 422 (1959).
2. Tokunaga, Mizuguchi and Murohashi, *J. Bacteriol., 86,* 608 (1963).
3. Burchard and Dworkin, *J. Bacteriol., 91,* 1305 (1966).
4. Pons, *Biochem. Z., 346,* 26 (1966).
5. Yamagishi, Yoshizako and Sato, *Virology, 30,* 29 (1966).
6. Lwoff (citing unpublished data of Smith and Siminowitch), *Bacteriol. Rev., 17,* 269 (1953).
7. Meyer, Mackal, Tao and Evans, Jr., *J. Biol. Chem., 236,* 1141 (1961).
8. Shapiro, Rudner, Miura and Chargaff, *Nature, 205,* 1068 (1965).

9. Kageyama, *Uirusu, 11,* 291 (1961).
10. Zinder (citing unpublished data of Garen and Zinder), *J. Cell. Comp. Physiol., 45 (Suppl. 2),* 23 (1955).
11. Creaser and Taussig, *Virology, 4,* 200 (1957).
12. Lunan and Sinsheimer, *Virology, 2,* 455 (1956).
13. Volkin, Astrachan and Countryman, *Virology, 6,* 545 (1958).
14. Marmur and Cordes, in *Informational Macromolecules (Symposium at Rutgers University, 1962),* p. 79, Vogel, Bryson and Lampen, Eds. Academic Press, New York (1963).
15. Sinsheimer, *J. Mol. Biol., 1,* 43 (1959).
16. Saunders and Campbell, *Biochemistry, 4,* 2836 (1965).
17. Welker and Campbell, *J. Bacteriol., 89,* 175 (1965).
18. Kiselev, Tikhonenko, Kaftanova and Kiselev, *Biokhimiya, 28,* 1065 (1963).
19. Hoffman-Berling, Marvin and Dürwald, *Z. Naturforsch. Teil B, 18,* 876 (1963).
20. Salivar, Tzagoloff and Pratt, *Virology, 24,* 359 (1964).
21. Kingsbury and Ordal, *J. Bacteriol., 91,* 1327 (1966).
22. Ikeda, Saito, Miura, Takagi and Aoki, *J. Gen. Appl. Microbiol., 11,* 181 (1965).
23. Tucker, *Biochem. J., 92,* 58p (1964).
24. Wyatt and Cohen, *Biochem. J., 55,* 774 (1953).
25. Smith and Wyatt, *Biochem. J., 49,* 144 (1951).
26. Wyatt and Cohen, *Nature, 170,* 1072 (1952).
27. Wyatt and Cohen, *Ann. Inst. Pasteur (Paris), 84,* 143 (1953).
28. Kornberg, *Science, 131,* 1503 (1960).
29. Hershey, Dixon and Chase, *J. Gen. Physiol., 36,* 777 (1953).
30. Mayers and Spizizen, *J. Biol. Chem., 210,* 877 (1954).
31. Crampton, Lipshitz and Chargaff, *J. Biol. Chem., 211,* 125 (1954).
32. Takahashi and Marmur, *Biochem. Biophys. Res. Commun., 10,* 289 (1963).
33. Takahashi and Marmur, *Nature, 197,* 794 (1963).
34. Kou, Huang and Teng, *J. Mol. Biol., 34,* 373 (1968).
35. Nikolskaya, Tkatcheva, Vanyushin and Tikhonenko, *Biochim. Biophys. Acta, 155,* 626 (1968).
36. Egbert, *J. Virol., 3,* 528 (1969).
37. Riva, Polsinelli and Falaschi, *J. Mol. Biol., 35,* 347 (1968).
38. Biswal, Kleinschmidt, Spatz and Trautner, *Mol. Gen. Genet., 100,* 39 (1967).


**Insect Viruses**


1. Wyatt, *J. Gen. Physiol., 36,* 201 (1952).
2. Hashinaga, Murakami and Yamafuji, *Enzymologia, 30,* 179 (1966).
3. Onodera, Komano, Himeno and Sakai, *J. Mol. Biol., 13,* 532 (1965).
4. Aizawa and Iida, *J. Insect Pathol., 5,* 344 (1963).
5. Thomas, *Virology, 14,* 240 (1961).


**Animal Viruses**


1. Ben-Porat and Kaplan, *Virology, 16,* 261 (1962).
2. Lando, DeRudder and Privat deGarilhe, *Bull. Soc. Chim. Biol., 47,* 1033 (1965).
3. Green and Piña, *Virology, 20* 199 (1963).
4. Rose, Hoggan and Shatkin, *Proc. Nat. Acad. Sci. U.S.A., 56,* 86 (1966).
5. Darlington and Randall, *Virology, 19,* 322 (1963).
6. Watson and Littlefield, *J. Mol. Biol., 2,* 161 (1960).
7. Wyatt and Cohen, *Biochem. J., 55,* 774 (1953).
8. Randall, Gafford and Darlington, *J. Bacteriol., 83,* 1037 (1962).
9. Crawford, Follett, Burdon and McGeoch, *J. Gen. Virol., 4,* 37 (1969).

# GUANINE-PLUS-CYTOSINE (GC) COMPOSITION OF THE DNA OF BACTERIA, FUNGI, ALGAE AND PROTOZOA

DR. WILLIAM M. NORMORE

The table of bacteria (Table 1) is divided into seven columns. In column 1, the bacteria are listed alphabetically by families, with subdivisions for genera and species; alternate generic or species names are given in parentheses. In general, *Bergey's Manual of Determinative Bacteriology*[1] has been followed for classification, but Skerman's *A Guide to the Identification of the Genera of Bacteria*[2] and original papers were used in placing organisms not listed in *Bergey's Manual.* In column 2, the source of the organism and its strain designation have been given where possible. In columns 3, 4, and 5, the mole percent quanisine-plus-cytosine values are given to the nearest 0.1%. Where ranges of values under any one method are listed, or where there are discrepancies between methods, the original papers should be consulted. The primary references for GC values are given in column A, while column B gives references to hybridization data on the same organisms. An alphabetized index of the genera precedes this table.

The tables for fungi, algae and protozoa (Tables 2, 3, and 4) are similarly arranged, except that column 1 lists the genera in alphabetical order.

The literature survey was concluded in May 1971.

## INDEX TO GENERA


| Genus | Page |
|---|---|
| *Acetobacter* | 664, 665 |
| *Achromobacter* | 587–589 |
| *Acidaminococcus* | 659 |
| *Acinetobacter* | 605, 606 |
| *Actinobacillus* | 606, 607 |
| *Actinomyces* | 593 |
| *Actinoplanes* | 594 |
| *Actinosporangium* | 594 |
| *Aerobacter (Enterobacter)* | 619, 620 |
| *Aerococcus* | 641 |
| *Aeromonas* | 665 |
| *Agrobacterium* | 673–675 |
| *Agromyces* | 593 |
| *Alcaligenes* | 589–591 |
| *Amorphosporangium* | 595 |
| *Ampullariella* | 595 |
| *Anaplasma* | 596 |
| *Arachnia* | 593 |
| *Archangium* | 596 |
| *Arthrobacter* | 612–614 |
| *Azobacter* | 596, 597 |
| *Azotomonas* | 666 |
| *Bacillus* | 598–601 |
| *Bacterium* | 620 |
| *Bacteroides* | 603 |
| *Bdellovibrio* | 676, 677 |
| *Beijerinckia* | 597 |
| *Bifidobacterium* | 593, 630, 631 |
| *Bordetella* | 607 |
| *Borrelia* | 687 |
| *Brevibacterium* | 604, 605 |
| *Brucella* | 607 |
| *Catenabacterium* | 631 |
| *Caulobacter* | 611 |
| *Cellulomonas* | 614 |
| *Chlamydia* | 611 |
| *Chlorobium* | 612 |
| *Chloropseudomonas* | 612 |
| *Chondrococcus* | 658 |
| *Chondromyces* | 663 |
| *Chromatium* | 687 |
| *Chromobacterium* | 675 |
| *Citrobacter* | 620 |
| *Clostridium* | 601–603 |
| *Comamonas* | 677 |
| *Corynebacterium* | 615–618 |
| *Cytophaga* | 619 |
| *Dactylosporangium* | 595 |
| *Derxia* | 597 |
| *Desulfovibrio* | 677, 678 |
| *Desulfotomaculum* | 677 |
| *Diplococcus* | 631 |
| *Erwinia* | 620–625 |
| *Escherichia* | 625, 626 |
| *Flavobacterium* | 592 |
| *Flexibacter* | 619 |
| *Flexibacteria* | 688 |
| *Flexothrix* | 689 |
| *Fusobacterium* | 603 |
| *Gaffkya* | 641 |
| *Gluconobacter* | 666 |
| *Haemophilus* | 607, 608 |
| *Halobacterium* | 666 |
| *Herellea* | 608 |
| *Herpetosiphon* | 689 |
| *Intrasporangium* | 683 |
| *Klebsiella* | 626, 627 |
| *Lactobacillus* | 631–637 |

| Genus | Page |
|---|---|
| *Leptospira* | 687, 688 |
| *Leptotrichia* | 611 |
| *Leuconostoc* | 637 |
| *Leucothrix* | 641 |
| *Listeria* | 618 |
| *Lophomonas* | 666 |
| *Methanobacterium* | 678 |
| *Microbacterium* | 618 |
| *Microbispora (Waksmania)* | 661 |
| *Micrococcus (Staphylococcus)* | 641–646, 648–651 |
| *Microellobosporia* | 595 |
| *Micromonospora* | 651 |
| *Micropolyspora* | 661 |
| *Mima* | 608 |
| *Moraxella* | 608–610 |
| *Mycobacterium* | 652–654 |
| *Mycoplasma* | 654–657 |
| *Mycrocyclus* | 678, 679 |
| *Myxococcus (Myxobacterium)* | 658 |
| *Neisseria* | 659, 660 |
| *Nitrobacter* | 660 |
| *Nitrosomonas* | 661 |
| *Nocardia* | 661, 662 |
| *Oerskovia* | 593 |
| *Paracolobactrum* | 627 |
| *Pasteurella* | 610, 611 |
| *Pediococcus* | 637 |
| *Pectobacterium* | 627, 628 |
| *Planobispora* | 595 |
| *Polyangium* | 663 |
| *Promicromonospora* | 594 |
| *Propionibacterium* | 663, 664 |
| *Prosthecomicrobium* | 689 |
| *Proteus* | 628 |
| *Providencia* | 628 |
| *Pseudomonas* | 666–672 |
| *Pseudonocardia* | 662 |

| Genus | Page |
|---|---|
| *Ramibacterium* | 637 |
| *Rhizobium* | 676 |
| *Rhodopseudomonas* | 596 |
| *Rhodospirillum* | 596 |
| *Rickettsia (Coxiella)* | 676 |
| *Ristella* | 603 |
| *Rothia* | 594 |
| *Salmonella* | 628, 629 |
| *Saprospira* | 682 |
| *Sarcina* | 647, 648 |
| *Serratia* | 629, 630 |
| *Shigella* | 630 |
| *Sorangium* | 676 |
| *Sphaerophorus* | 603 |
| *Sphaerotilus* | 612 |
| *Spirillospora* | 595 |
| *Spirillum* | 679 |
| *Spirochaeta* | 682 |
| *Spirosoma* | 689 |
| *Sporocytophaga* | 659 |
| *Streptobacillus* | 603 |
| *Streptococcus* | 637–640 |
| *Streptomyces* | 683–686 |
| *Streptosporangium* | 595, 596 |
| *Streptoverticillium* | 686 |
| *Thermoactinomyces* | 662 |
| *Thermoactinopolyspora* | 662 |
| *Thermomonospora* | 662 |
| *Thermoplasma* | 658 |
| *Thermus* | 689 |
| *Thiobacillus* | 686, 687 |
| *Treponema* | 688 |
| *Veillonella* | 660 |
| *Vibrio* | 679–682 |
| *Vitreoscilla* | 688 |
| *Wolbachia* | 676 |
| *Xanthomonas* | 672, 673 |

# TABLE 1
# GC COMPOSITION OF THE DNA OF BACTERIA

| | | Mole % Guanosine + Cytosine[b] | | | References[c] | |
|---|---|---|---|---|---|---|
| Organism | Source or Strain[a] | Tm | Chemical Analysis | Buoyant Density | A | B |
| **Achromobacteriaceae** | | | | | | |
| *Achromobacter* | | | | | | |
| *agilis* | NCIB 9986 | 67.6 | | | 171 | |
| *albus* | NCIB 9988 | 61.8 | | | 171 | |
| *anitratus* | ATCC 17912 | 42 | | | 172 | 172 |
| | ATCC 17913 | 43 | | | 172 | 172 |
| | 8, 9 | | | 39.5 | 3 | |
| *aquamarinus* | NCMB 557 | 58.0 | | | 171 | |
| *arsenoxydans* | NCIB 8687 | 59.4 | | | 171 | |
| *butyri* | NCIB 9404 | 60.5 | | | 171 | |
| *citroalcaligenes* | 2723/59, ATCC 17908 | | | 40–41 | 4 | |
| | ATCC 17908 | 42 | | | 172 | 172 |
| *conjunctivae* | ATCC 17905 | 44 | | | 172 | 172 |
| *cycloclastes* | IAM 1013 | 62.1 | | | 171 | |
| | CRM ACY | 64.2 | | | 171 | |
| *fischeri* | | | 40–42 | | 5 | |
| | | | | 44±1 | 228 | |
| *formosus* | NCIB 9987 | 61.5 | | | 171 | |
| *georgiopolitanum* | COC 21 | | | 41±1 | 173 | |
| *haemolysans* | 742/56 | | | 45.0 | 3 | |
| | 2408/57 | | | 41.5 | 3 | |
| | 2181/60, ATCC 17907 | | | 39.5 | 3 | |
| | ATCC 17988 | | | 41.0 | 3 | |
| *haemolyticus* | ATCC 17907 | 43 | | | 172 | 172 |
| | ATCC 17906 | 43 | | | 172 | 172 |
| *halophilus* | AHU 1333 | 61.0 | | | 171 | |
| *hartlebii* | NCIB 8129 | 63.5 | | | 171 | |
| *iophagus* | NCMB 1051 | 64.8 | | | 171 | |
| *liquefaciens* | W. Tulecke | | | 41.0 | 6 | |
| | ATCC 15716 | 44.6 | | | 171 | |
| | NCIB 9989 | 66.0 | | | 171 | |

## TABLE 1 (Continued)
## GC COMPOSITION OF THE DNA OF BACTERIA

| Organism | Source or Strain[a] | Mole % Guanosine + Cytosine[b] Tm | Chemical Analysis | Buoyant Density | References[c] A | B |
|---|---|---|---|---|---|---|
| **Achromobacteriaceae (continued)** | | | | | | |
| *Achromobacter* | | | | | | |
| *lwoffi* | ATCC 17985 | | | 43.0 | 3 | |
| | ATCC 17987 | | | 43.5 | 3 | |
| | 881/57 | | | 41.5 | 3 | |
| | 8858/62 | | | 38.0 | 3 | |
| *metalcaligenes* | ATCC 17905 | 44 | | | 172 | 172 |
| | ATCC 17910 | 46 | | | 172 | 172 |
| *mucosus* | ATCC 17904 | 42 | | | 172 | 172 |
| *turbidus* | AHU 1337 | 55.6 | | | 171 | |
| *venosus* | NCIB 9985 | 61.8 | | | 171 | |
| *viscosus* | NCIB 9408 | 61.2 | | | 171 | |
| species | NCIB 9650 | 55.9 | | | 171 | |
| | Pickett M153 | 56.6 | | | 171 | |
| | Pickett M165 | 56.9 | | | 171 | |
| | Pickett M140 | 58.1 | | | 171 | |
| | Pickett M151 | 58.0 | | | 171 | |
| | Pickett M155 | 58.0 | | | 171 | |
| | Pickett M281 | 68.1 | | | 171 | |
| | Pickett M250 | 67.7 | | | 171 | |
| | NCIB 9387 | 68.2 | | | 171 | |
| | NCMB 1156 | 56.7 | | | 171 | |
| | NCMB 622 | 58.2 | | | 171 | |
| | NCMB 623 | 58.5 | | | 171 | |
| | NCMB 625 | 58.8 | | | 171 | |
| | NCMB 624 | 59.1 | | | 171 | |
| | Pickett M 25 | 62.0 | | | 171 | |
| | Pickett M 22 | 62.0 | | | 171 | |
| | Henderson H 38 | 62.1 | | | 171 | |
| | Pickett M 6 | 62.2 | | | 171 | |
| | NCIB P 18/2 | 62.8 | | | 171 | |
| | NCIB P917/53 | 63.2 | | | 171 | |

# TABLE 1 (Continued)
# GC COMPOSITION OF THE DNA OF BACTERIA

| Organism | Source or Strain[a] | Mole % Guanosine + Cytosine[b] Tm | Chemical Analysis | Buoyant Density | References[c] A | B |
|---|---|---|---|---|---|---|
| **Achromobacteriaceae (continued)** | | | | | | |
| *Achromobacter* | | | | | | |
| species | NCIB P 18/27 | 65.0 | | | 171 | |
| | Pickett M 185 | 66.8 | | | 171 | |
| | N 4-B | 59.4 | 60.8 | | 174 | |
| *Alcaligenes* | | | | | | |
| *denitrificans* | CIP 60.81 | 65.4 | | | 171 | |
| | CIP 60.83 | 66.7 | | | 171 | |
| | CIP 6230 | 68.9 | | | 171 | |
| | CIP X86 | 69.2 | | | 171 | |
| | CIP X73 | 69.8 | | | 171 | |
| *faecalis* | ATCC 8750, NCIB 8156 | 54.8 | | 54.8 | 7 | |
| | 3 strains | | 66.7–69.9 | | 8 | |
| | NCTC 8764 | | | 66.0 | 9 | |
| | NCTC 8769 | | | 63.0 | 9 | |
| | Cs 8 | | | 63.5 | 9 | |
| | Cs 11 | | | 63.0 | 9 | |
| | | 62.0 | | | 10 | |
| | 40 | 58.8 | | | 11 | |
| | 41 | 58.8 | | | 11 | |
| | NCTC 415, ATCC 19018 | 58.9 | | | 11 | |
| | ATCC 8455 | | | 57.1 | 12 | |
| | CIP 6415 | 58.4 | | | 171 | |
| | Pinter 40 | 58.7 | | | 171 | |
| | Pinter 41 | 58.8 | | | 171 | |
| | Delft AF1 | 58.9 | | | 171 | |
| | ATCC 19018 | 58.9 | | | 171 | |
| | NCIB 8156 | 58.9 | | | 171 | |
| | CIP 60.57 | 63.9 | | | 171 | |
| | CIP R40 | 64.9 | | | 171 | |
| | Holding Cs 8 | 65.5 | | | 171 | |
| | Holding Cs 11 | 65.6 | | | 171 | |
| | CIP 62.31 | 65.0 | | | 171 | |
| | CIP R161 | 56.5 | | | 171 | |
| | Lautrop AB 78 | 57.9 | | | 171 | |
| *hemolysans* | ATCC 17988 | 44 | | | 172 | 172 |
| *metalcaligenes* | NCIB 8734 | 60.6 | | | 171 | |
| *odorans* | Lautrop AB 60 | 56.8 | | | 171 | |
| | CCEB 554 | 56.6 | | | 171 | |

## TABLE 1 (Continued)
## GC COMPOSITION OF THE DNA OF BACTERIA

| Organism | Source or Strain[a] | Mole % Guanosine + Cytosine[b] | | | References[c] | |
|---|---|---|---|---|---|---|
| | | Tm | Chemical Analysis | Buoyant Density | A | B |
| **Achromobacteriaceae (continued)** | | | | | | |
| *Alcaligenes* | | | | | | |
| *odorans* | Gilardi 79 | 57.0 | | | 171 | |
| | CCEB 583 | 57.0 | | | 171 | |
| | Gilardi 29 | 57.1 | | | 171 | |
| | CCEB 568 | 57.2 | | | 171 | |
| | Gilardi 115 | 57.4 | | | 171 | |
| | Gilardi 116 | 57.4 | | | 171 | |
| | Gilardi 142 | 57.6 | | | 171 | |
| | CCEB 569 | 57.7 | | | 171 | |
| | Lautrop AB 209 | 57.9 | | | 171 | |
| | Gilardi 117 | 57.9 | | | 171 | |
| | Lautrop AB 373 | 58.1 | | | 171 | |
| | Lautrop AB 1246 | 58.2 | | | 171 | |
| | Mitchell 6 | 58.3 | | | 171 | |
| | Mitchell 1 | 58.4 | | | 171 | |
| | Mitchell 5 | 58.4 | | | 171 | |
| | Lautrop AB 54 | 58.4 | | | 171 | |
| | Lautrop AB 1157 | 58.4 | | | 171 | |
| | Mitchell 4 | 58.5 | | | 171 | |
| | Mitchell 2 | 58.6 | | | 171 | |
| | Mitchell 3 | 58.6 | | | 171 | |
| | Lautrop AB 1472 | 58.6 | | | 171 | |
| species | Lautrop AB 1286 | 57.2 | | | 171 | |
| | Lautrop AB 220 | 57.5 | | | 171 | |
| | Lautrop AB 1289 | 57.8 | | | 171 | |
| | Lautrop AB 1199 | 57.9 | | | 171 | |
| | Lautrop AB 1347 | 58.1 | | | 171 | |
| | Lautrop AB 1194 | 58.0 | | | 171 | |
| | Lautrop AB 1350 | 58.2 | | | 171 | |
| | Lautrop AB 1258 | 58.4 | | | 171 | |
| | Lautrop AB 1367 | 58.4 | | | 171 | |
| | Caselitz AF 61 | 60.0 | | | 171 | |
| | Lautrop AB 374 | 68.6 | | | 171 | |

TABLE 1 (Continued)
GC COMPOSITION OF THE DNA OF BACTERIA

| Organism | Source or Strain[a] | Mole % Guanosine + Cytosine[b] Tm | Chemical Analysis | Buoyant Density | References[c] A | B |
|---|---|---|---|---|---|---|
| **Achromobacteriaceae (continued)** | | | | | | |
| *Alcaligenes* | | | | | | |
| species | Lautrop AB 118 | 64.5 | | | 171 | |
| | Lautrop AB 1416 | 65.6 | | | 171 | |
| | Lautrop AB 52 | 66.4 | | | 171 | |
| | Lautrop AB 53 | 67.4 | | | 171 | |
| | Lautrop AB 1214 | 67.4 | | | 171 | |
| | Lautrop AB 1117 | 67.6 | | | 171 | |
| | Lautrop AB 1013 | 67.6 | | | 171 | |
| | Lautrop AB 104 | 67.8 | | | 171 | |
| | Lautrop AB 1296 | 67.8 | | | 171 | |
| | Lautrop AB 1098 | 68.1 | | | 171 | |
| | Lautrop AB 1475 | 68.3 | | | 171 | |
| | Lautrop AB 107 | 68.4 | | | 171 | |
| | Lautrop AB 1307 | 68.4 | | | 171 | |
| | Lautrop AB 603 | 68.5 | | | 171 | |
| | Lautrop AB 1170 | 68.9 | | | 171 | |
| | Lautrop AB 230 | 68.9 | | | 171 | |
| | Lautrop AB 630 | 47.9 | | | 171 | |
| | Lautrop AB 717 | 53.9 | | | 171 | |
| | Lautrop AB 940 | 58.0 | | | 171 | |
| | Caselitz AF 37 | 61.9 | | | 171 | |
| | Lautrop AB 1295 | 65.2 | | | 171 | |
| | Lautrop AB 1335 | 65.4 | | | 171 | |
| | v. d. Platt B 2 | 66.2 | | | 171 | |
| | Lautrop AB 1391 | 66.6 | | | 171 | |
| | Lautrop AB 1067 | 67.9 | | | 171 | |

## TABLE 1 (Continued)
## GC COMPOSITION OF THE DNA OF BACTERIA

| Organism | Source or Strain[a] | Mole % Guanosine + Cytosine[b] Tm | Chemical Analysis | Buoyant Density | References[c] A | B |
|---|---|---|---|---|---|---|
| **Achromobacteriaceae (continued)** | | | | | | |
| *Flavobacterium* | | | | | | |
| *acidificum* | | | 48–50 | | 5 | |
| | | 48–50 | | | 254 | |
| *aquatile* | ATCC 11947 | 32.0 | | 30.0 | 7, 13 | |
| | | 32–34 | | | 254 | |
| *arborescens* | | | 66–68 | | 5 | |
| | | 66–68 | | | 254 | |
| *buchneri* | A-1 | 42.2 | | | 255 | |
| *capsulatum* | NCIB 9890 | 63 | | | 255 | |
| *dehydrogenans* | | | | 73.5 | 12 | |
| *esteroaromaticum* | | 68–70 | | | 255 | |
| | | | 68–70 | | 5 | |
| *flavescens* | | | 66–68 | | 5 | |
| | | 66–68 | | | 254 | |
| *heparinum* | NCIB 9290 | 42.2 | | | 255 | |
| *meningosepticum* | ATCC 13253 | 36.4 | | | 255 | |
| | ATCC 13254 | 36.4 | | | 255 | |
| | NCTC 10016 | 38.3 | | | 255 | |
| *odoratum* | | | 34–38 | | 5 | |
| | | 34–36 | | | 254 | |
| *pectinovorum* | NCIB 9095 | 32.9 | | | 255 | |
| | NCIB 10021 | 32.7 | | | 255 | |
| *suaveolens* | | | 66–68 | | 5 | |
| | ATCC 958 | 64.5 | | | 7 | |
| | | 66–68 | | | 254 | |
| *vitarumen* | ATCC 10234 | 63.3 | | | 7 | |
| | | | 64–66 | | 5 | |
| species | NCMB 251 | 35.6 | | | 14 | |
| | NCMB 289 | 35.8 | | | 14 | |
| | NCMB 264 | 36.6 | | | 14 | |
| | NCMB 249 | 40.6 | | | 14 | |
| | NCIB 9491 | 51.2 | | | 255 | |
| | NCIB 9942 | 64.4 | | | 255 | |
| | NCIB 9776 | 66.9 | | | 255 | |
| | NCIB 9497 | 73.2 | | | 255 | |
| | NCMB 244 | 62.9 | | | 257 | |

TABLE 1 (Continued)
GC COMPOSITION OF THE DNA OF BACTERIA

| Organism | Source or Strain[a] | Mole % Guanosine + Cytosine[b] Tm | Chemical Analysis | Buoyant Density | References[c] A | B |
|---|---|---|---|---|---|---|
| **Actinomycetaceae** | | | | | | |
| *Actinomyces* | | | | | | |
| *abscessum* | 1852B | | 61.3 | | 15 | |
| *bovis* | | 58.0 | | | 262 | |
| *eriksonii* | | 62.2 | | | 262 | |
| *humiferus* | | 73.0 | | | 263 | |
| *israelii* | | 60.3 | | | 262 | |
| *odontolyticus* | | 59.5–62.0 | | | 262 | |
| *viscosus* | | 59.5 | | | 262 | |
| *Agromyces* | | | | | | |
| *ramosus* | | | | 70.9–71.9 | 265 | |
| *Arachnia* | | | | | | |
| *propionica* | | 66.5 | | | 264 | |
| | | 68.6–69.6 | | | 66 | |
| | | 66.6 | | | 15 | |
| *Bifidobacterium* | | | | | | |
| *bifidum* | | 57.2–60.8 | | | 90 | |
| *Oerskovia* | | | | | | |
| *turbata* | SSIC 891 (Orskov's 27) | 70.5 | | | 266 | |
| | SSIC 891 (Orskov's 27) | | 70.5 | | 205 | |
| | ATCC 12288 | 71.3–78.0 | | | 17, 180 | 181 |
| | | 72 | | | 21 | |
| | IMET 7130 | 70.5 | | | 266 | |
| | IMET 7006 | 71.5 | | | 266 | |
| | IMET 7133 | 72.0 | | | 266 | |
| | IMET 7135 | 70.5 | | | 266 | |
| | SSIC 689 | 71.7 | | | 266 | |
| | SSIC 761 | 75.1 | | | 266 | |
| | SSIC 762 | 71.2 | | | 266 | |
| | 11-49 | 71.2 | | | 266 | |
| | Y-13-3 | 72.4 | | | 266 | |
| species | IMET 7801 | 66.5 | 68.6 | | 205 | 206 |

**TABLE 1 (Continued)**
**GC COMPOSITION OF THE DNA OF BACTERIA**

| Organism | Source or Strain[a] | Mole % Guanosine + Cytosine[b] Tm | Chemical Analysis | Buoyant Density | References[c] A | B |
|---|---|---|---|---|---|---|
| **Actinomycetaceae (continued)** | | | | | | |
| *Promicromonospora* | | | | | | |
| *citrea* | 562 | | 73.0 | | 22 | |
| | USSR-RIA-562 | | 73.5 | | 16 | 16 |
| *Rothia* | | | | | | |
| *dentocariosa* | ATCC 17931 (coccal) | 68.0 | | | 179 | |
| | ATCC 17931 (filamentous) | 69.1 | | | 179 | |
| | CDC X-614a | 65.6 | | | 179 | |
| | CDC X-303 | 65.7 | | | 179 | |
| | CDC X-346 | 65.6 | | | 179 | |
| | RC 27 H. V. Jordan | 68.1 | | | 179 | |
| | RC 29 H. V. Jordan | 65.4 | | | 179 | |
| | RC 30 H. V. Jordan | 66.1 | | | 179 | |
| | RC 37 H. V. Jordan | 65.6 | | | 179 | |
| | RC 44 H. V. Jordan | 65.6 | | | 179 | |
| | H 69 University of Pennsylvania | 65.7 | | | 179 | |
| **Actinoplanaceae** | | | | | | |
| *Actinoplanes* | | | | | | |
| *missouriensis* | CBS J. E. Thiemann | 72 | | | 177 | |
| *philippinensis* | ATCC 12427 | 70.6–76 | 72.1 | | 180, 181 | 182 |
| | CBS J. E. Thiemann | 72 | | | 177 | |
| | 12427 | 73.0 | | | 17 | |
| | P-15 | | | | 16 | 16 |
| *utahensis* | 260 | | 72.6 | | 16 | 16 |
| | CBS J. E. Thiemann | 72 | | | 177 | |
| *Actinosporangium* | | | | | | |
| *violaceum* | USSR-RIA-655 | | 71.6 | | 16 | 16 |
| | 655 | | 70.3 | | 22 | |
| *sp. nov.* | N 4146 | | 70.9 | | 23 | |

# TABLE 1 (Continued)
# GC COMPOSITION OF THE DNA OF BACTERIA

| Organism | Source or Strain[a] | Mole % Guanosine + Cytosine[b] Tm | Chemical Analysis | Buoyant Density | References[c] A | B |
|---|---|---|---|---|---|---|
| **Actinoplanaceae (continued)** | | | | | | |
| *Amorphosporangium* | | | | | | |
| *auranticolor* | 253 | | 71.4 | | 16 | 16 |
| *Ampullariella* | | | | | | |
| *digitata* | 33 | | 72.3 | | 16 | 16 |
| | CBS 19169 J. E. Thiemann | 73 | | | 177 | |
| *Dactylosporangium* | | | | | | |
| *aurantiacum* | D/748, ATCC 23491 | 73 | | | 177 | |
| *thailandense* | D/499, ATCC 23490 | 71 | | | 177 | |
| *Microellobosporia* | | | | | | |
| *cinerea* | 3855 | | 67.6–70.0 | | 16, 22 | 16, 22 |
| *flavea* | 3858 | | 70.3 | | 16 | 16 |
| | 3357 | | 67.9 | | 22 | |
| | ATCC 15332 | 71 | | | 177 | |
| *violacea* | 2732/3 | | 68.4 | | 22, 175 | |
| *Planobispora* | | | | | | |
| *longispora* | Pb/1075, ATCC 23867 | 71 | | | 177 | |
| *parontospora* | B/677, ATCC 23863 | 72 | | | 177 | |
| *rosea* | Pb/1435, ATCC 23866 | 70 | | | 177 | |
| *Spirillospora* | | | | | | |
| *albida* | 761 Couch | 72.9 | 72.9 | | 16, 182 | 16 |
| | CBS J. E. Thiemann | 71 | | | 177 | |
| | 1030 J. E. Thiemann | 72 | | | 177 | |
| *Streptosporangium* | | | | | | |
| *album*[d] | S-16 | | 69.5 | | 16 | 16 |
| *roseum*[d] | 27-b, ATCC 12428 | 68.5 | 70.6 | | 16, 17 | 16 |

## TABLE 1 (Continued)
## GC COMPOSITION OF THE DNA OF BACTERIA

| Organism | Source or Strain[a] | Mole % Guanosine + Cytosine[b] Tm | Chemical Analysis | Buoyant Density | References[c] A | B |
|---|---|---|---|---|---|---|
| **Actinoplanaceae (continued)** | | | | | | |
| *Streptosporangium* | | | | | | |
| *roseum*[d] | | | 72.0 | | 21 | |
| | 27-b J. Couch | 71 | | | 177, 181 | |
| *vulgare* | 765 | | 69.5 | | 22 | |
| species (thermophilic) | 11 | 53.7 | | 49.0 | 24 | |
| **Anaplasmataceae** | | | | | | |
| *Anaplasma* | | | | | | |
| *marginale* | | | 49.9–50.7 | | 25, 26 | |
| **Archangiaceae** | | | | | | |
| *Archangium* | | | | | | |
| *gephyra* | M-18 (University of Windsor) | 68.3 | | | 27 | |
| | M-58 (University of Windsor) | 67.8 | | | 27 | |
| | 2 strains | 67.8–68.3 | | | 318 | |
| **Athiorhodaceae** | | | | | | |
| *Rhodopseudomonas* | | | | | | |
| *palustris* | Taniguchi | | | 66.3 | 28 | |
| *spheroides* | ATCC 14690 (2 bands) | | | Major 71.4, minor 65.3 | 28 | |
| *Rhodospirillum* | | | | | | |
| *rubrum* | | | 63.4 | | 29 | |
| | ATCC 11170 (S-1) | 60.0 | | 66.3–67 | 28, 30, 31 | |
| **Azotobacteriaceae** | | | | | | |
| *Azotobacter* | | | | | | |
| *agilis* | 2 strains | | 54.9–55.1 | | 8 | |
| | SS 1 | 53.5 | | | 32 | |
| | NCIB 8637 | 53.2 | | | 32 | |
| | S | 53.0 | | | 32 | 32 |
| | SS 4 | 52.8 | | | 32 | |
| | VJ | 52.8 | | | 32 | |
| | 9 | 52.5 | | | 32 | 32 |
| | K | 52.5 | | | 32 | 32 |
| | 132 | 52.0 | 50.0 | | 174 | |

## TABLE 1 (Continued)
## GC COMPOSITION OF THE DNA OF BACTERIA

| Organism | Source or Strain[a] | Mole % Guanosine + Cytosine[b] Tm | Chemical Analysis | Buoyant Density | References[c] A | B |
|---|---|---|---|---|---|---|
| **Azotobacteriaceae (continued)** | | | | | | |
| *Azotobacter* | | | | | | |
| *beijerinckii* | NCIB 8948 (ATCC 19360) | 66.2 | | | 32 | |
| | B 2 | 66.2 | | | 32 | 32 |
| | NCIB 9607 | 66.2 | | | 32 | |
| *chroococcum* | 4 strains | | 57.3–57.5 | | 8 | |
| | NCIB 9125 | 66.0 | | | 32 | 32 |
| | A 3 | 64.8 | | | 32 | |
| | 210 | 64.4 | | | 174 | |
| *insignis* | NCIB 9127 | 57.9 | | | 32 | 32 |
| | 2 | 57.7 | | | 32 | |
| | VJ 5 | 57.7 | | | 32 | |
| | L 533 | 56.9 | | | 32 | |
| *macrocytogenes* | NCIB 8200 | 58.6 | | | 32 | 32 |
| | NCIB 9128 | 58.6 | | | 32 | 32 |
| | NCIB 9129 | 58.2 | | | 32 | 32 |
| *miscellum* | ATCC 17962 | 65.6 | | | 33 | |
| *paspali* | 8A | 63.2 | | | 33 | |
| | 22 B | 63.7 | | | 33 | |
| | 15 B | 63.3 | | | 33 | |
| | 23 A | 64.6 | | | 33 | |
| *vinelandii* | 2 strains | 56–56.3 | | | 8 | |
| | ATCC 9104 | 60.0 | | | 30 | |
| | B | 65.9 | | | 174 | |
| | 3 A | 64.9 | | | 174 | |
| | 7492 | 67.0 | 65.0 | | 174 | |
| | KS 4 | 64.7 | | | 174 | |
| | OP | 64.9 | 65.0 | | 174 | |
| | O | 66.5 | 63.5 | | 174 | |
| *Beijerinckia* | | | | | | |
| *derxii* | | 59.1 | | | 32 | 32 |
| *fluminensis* | | 56.2 | | | 32 | 32 |
| *indica* | | 54.7 | | | 32 | |
| *Derxia* | | | | | | |
| *gummosa* | III (ATCC 15995) | 70.4 | | | 32 | 32 |

**TABLE 1 (Continued)**
**GC COMPOSITION OF THE DNA OF BACTERIA**

| Organism | Source or Strain[a] | Mole % Guanosine + Cytosine[b] | | | References[c] | |
|---|---|---|---|---|---|---|
| | | Tm | Chemical Analysis | Buoyant Density | A | B |
| **Bacillaceae** | | | | | | |
| *Bacillus* | | | | | | |
| *alvei* | ATCC 6344 (NCTC 6352) | 32.5–33 | | | 30, 34 | |
| *amyloliquefaciens* | T, P, N, K | 44.4 | | 43.5 | 35 | |
| | F | 44.6 | | 44.9 | 35 | 35 |
| | SB | 44.8 | | 44.9 | 35 | |
| | W | 44.4 | | 44.9 | 35 | |
| | H | 43.4 | | 44.9 | 35 | 183 |
| *aneurinolyticus* | IAM 1077 | | 42.2 | | 36 | |
| *anthracis* | HBA87 | 32.2 | | | 37 | |
| | 30R/Sr | 33.9 | | | 37 | |
| *badius* | NCTC 10333/1 | 50.0 | | | 38 | |
| *brevis* | ATCC 9999 (NCTC 7096) | 42.5–43 | | 45.0 | 30, 31, 34 | |
| | IAM 1031 (ATCC 8185) | | 44.2 | | 36 | |
| | R | | 45.5 | | 29 | |
| | S | | 45.6 | | 29 | |
| | P | | 45.7 | | 29 | |
| *carotarum* | | | | 42.0 | 31 | |
| *cereus* | ATCC 7064 | | 32.1 | | 37 | |
| | A25 | | 34.5 | | 8 | |
| | | | 33.3 | | 8 | |
| | Mutant | | 35.7 | | 8 | |
| | Mutant | | 36.3 | | 8 | |
| | p-1 | | 36.0 | | 8 | |
| | p-2 | | 34.0 | | 8 | |
| | ATCC 7587 | | 35.0 | | 8 | |
| | ATCC 12137 | | 36.0 | | 8 | |
| | MB 19 | 33.0 | 35.0 | 37.0 | 8, 30, 31 34 | |
| | Lamanna | 31.7 | | | 37 | |
| | ATCC 9139 | 32.4 | | | 37 | |
| | NRRL B-569 | 40.1 | | | 37 | |
| | W | 32.1 | | | 37 | |
| | 68 | 32.8 | | | 37 | |
| *cereus* var. *mycoides* | 75a | | | 39.0 | 31 | |
| *circulans* | ATCC 4513 | 35.0 | | | 30, 34 | |
| | IAM 1112 (ATCC 9966) | | 50.1 | | 36 | |
| | IAM 1165 | | 47.3 | | 36 | |

# TABLE 1 (Continued)
# GC COMPOSITION OF THE DNA OF BACTERIA

| Organism | Source or Strain[a] | Mole % Guanosine + Cytosine[b] Tm | Chemical Analysis | Buoyant Density | References[c] A | B |
|---|---|---|---|---|---|---|
| **Bacillaceae (continued)** | | | | | | |
| *Bacillus* | | | | | | |
| *coagulans* | A 22 | | 47.9 | | 39 | |
| | P 22 | | 46.9 | | 39 | |
| | ATCC 10545 | 48.0 | | | 38 | |
| | CCM 843 | 46.8 | | | 38 | |
| | IMAM Renko | 54.8 | | | 38 | |
| *firmus* | | 41.0 | | | 30 | |
| *laterosporus* | ATCC 64 (NCTC 6357) | 40.0 | | | 30, 34 | 34 |
| *lentimorbus* | CCEB 297 | 37.7 | | | 38 | |
| *lentus* | ATCC 10840 | 36.5–37 | | | 30, 34 | |
| *licheniformis* | ATCC 9789 (NRS 243) | 46.0 | 50.7 | 45–46 | 30, 31, 34, 40 | 40 |
| | ATCC 9945a | 43.8 | | | 37 | |
| | Allen | 42.9 | | | 37 | |
| | CD II | 43.0 | | | 37 | |
| | ATCC 9789 | 46.95 | 46.9 | | 184 | |
| | FDO 12 (C. Thorne) | 47.3 | | | 183 | 183 |
| *macerans* | ATCC 7069 | 50.0–50.5 | | 45.0 | 30, 31, 34 | 40, 44 |
| | 7048 | 51.0 | 51.5 | | 174 | |
| | ATCC 8509 | 50.0 | | | 174 | |
| | ATCC 8515 | 49.0 | | | 174 | |
| | ATCC 8518 | 49.0 | | | 174 | |
| *megaterium* | University of Pennsylvania | 37.0 | 37.6 | 38–40 | 8, 30, 31, 40 | 40 |
| | 203 | | 45.5 | | 36 | |
| | IAM 1032 | | 36.0 | | 36 | |
| | IAM 1166 | | 37.5 | | 36 | |
| *megaterium-cereus* | ATCC 14B22 | 34.0 | | | 30, 34 | |
| *natto* | MB 275 | 43.0 | | 44.0 | 30, 34 | |
| *natto (subtilis)* | IAM 1071 (1) | | 42.0 | | 36 | |
| | IAM 1191 (7) | | 41.2 | | 36 | |
| | IAM 1259 (8) | | 41.2 | | 36 | |
| *niger* | ATCC 6454 | 43.0 | | 43.0 | 30, 34 | 40 |
| *pasteurii* | CCM 2056 (ATCC 11859; NCIB 8841; T. Gibson 22) | 41.7 | | | 41 | 42 |

## TABLE 1 (Continued)
## GC COMPOSITION OF THE DNA OF BACTERIA

| Organism | Source or Strain[a] | Mole % Guanosine + Cytosine[b] Tm | Chemical Analysis | Buoyant Density | References[c] A | B |
|---|---|---|---|---|---|---|
| **Bacillaceae (continued)** | | | | | | |
| *Bacillus* | | | | | | |
| *polymyxa* | ATCC 842 | 44.0 | | | 30, 34 | |
| | NRRL B-367 | | | 47–48 | 40 | 40 |
| | 842 | 45.6 | 47.2 | | 174 | |
| | Hino | 51.3 | 53.0 | | 174 | |
| | ATCC 8526 | 43.2 | | | 174 | |
| | ATCC 8527 | 45.1 | | | 174 | |
| | ATCC 813 | 45.1 | | | 174 | |
| *polymyxa asterosporus* | | 45.1 | | | 174 | |
| *popilliae* | CCEB 296 | 41.3 | | | 38 | |
| *pumilus* | ATCC 6631 (NRS 236) | 39.0 | | 40–41 | 30, 34, 40 | 40 |
| | ATCC 6631 | 45.1 | 45.1 | | 184 | |
| | ATCC 7061 | 42.9 | | | 183 | 183 |
| | NCIB 8982 | 43.9 | | | 183 | 183 |
| | ATCC 70 | 41.2 | | | 183 | 183 |
| | BD 2002 (D. Dubnau) | 42.4 | | | 183 | 183 |
| | ATCC 945 | 43.2 | | | 183 | 183 |
| | ATCC 14884 | 43.2 | | | 183 | 183 |
| | ATCC 1 | 42.2 | | | 183 | 183 |
| | NRRL B-3275 | 43.4 | | | 183 | 183 |
| | NRRL B-3275 | | | 41.8 | 183 | 183 |
| | ATCC 945 | | | 41.8 | 183 | 183 |
| | ATCC 14884 | | | 42.9 | 183 | 183 |
| *sphaericus* | ATCC 4525 | 36.5–37 | | | 30,34 | |
| *stearothermophilus* | NCTC 10339/1 | 44.9 | | | 38 | |
| | NCTC 10003/2 | 46.2 | | | 38 | |
| | 194 | 43.5–44 | | 46.0 | 30, 31, 34 | |
| | FJW | 50.97 | 56.0 | | 184 | |
| | 194 | 45.0 | 46.7 | | 184 | |
| | 2184 | 52.9 | 52.2 | | 184 | |
| | 10 | 52.9 | 52.9 | | 184 | |
| *subtilis* | ATCC 4529 | 43.4 | | 42.6 | 35 | |
| | ATCC 7067 | 41.5 | | 42.6 | 35 | |
| | ATCC 9466 | 41.5 | | 41.8 | 35 | |
| | ATCC 6051 | 42.4 | | 41.8 | 35 | |
| | ATCC 6633 | | | 43.0 | 40 | 40 |
| | | | 43.5 | | 43 | |
| | P 1 | 45.9 | | | 37 | |
| | T 7 | 46.1 | | | 37 | |
| | W 23(23-1) | 42.4–46.6 | | 41.8 | 35, 37 | 35, 44 |

## TABLE 1 (Continued)
## GC COMPOSITION OF THE DNA OF BACTERIA

| Organism | Source or Strain[a] | Mole % Guanosine + Cytosine[b] Tm | Chemical Analysis | Buoyant Density | References[c] A | B |
|---|---|---|---|---|---|---|
| **Bacillaceae (continued)** | | | | | | |
| *Bacillus* | | | | | | |
| *subtilis* | 168 I, 168/sr, 168 | 43.6–46.6 | 42.0 | 42.6–44 | 8, 12, 17, 30, 31, 37, 45 | 34, 42, 44 |
| | 168 Mutant MK-9 | 63.3 | 64.0 | | 45 | |
| | 168 Mutant MK-12 | 61.7 | 62.9 | | 45 | |
| | 168 Mutant SC-22 | 65.1 | 65.1 | | 45 | |
| | BQ 2 | 47.7 | | | 46 | |
| | Marburg, wild | | 42.6 | | 36 | |
| | IAM K(1523) | | 45.0 | | 36 | |
| | W 23 (H. R. Burmeister) | 45.6 | | | 183 | 183 |
| | 168 I⁻ (H. R. Burmeister) | 43.9 | | 44.9 | 183 | 183 |
| | ATCC 6633 | 42.9 | | | 183 | 183 |
| *subtilis* var. *aterrimus* | ATCC 7060 | 42.9 | | | 183 | 183 |
| | ATCC 6460 | 42.5–43 | | | 34, 47 | 34 |
| | ATCC 6455 | 43.0 | | | 30 | |
| *subtilis* var. *niger* | Spores | 42.5 | | | 48 | |
| | ATCC 7972 | 43.7 | | | 183 | 183 |
| *thuringiensis* | ATCC 01792 | 33.5–34 | 35.9 | 36.0 | 8, 30, 31, 34 | |
| | CCEB 457 | 40.1 | | | 38 | |
| species | IMAM/BT 5 | 48.6 | | | 38 | |
| | BT/2 | 67.31 | | | 185 | |
| | X1 | 43.2 | 41.5 | | 184 | |
| | IMAM/Cr | 62.7 | | | 38 | |
| | IMAM/TR | 62.4 | | | 38 | |
| | IMAM/O | 62.1 | | | 38 | |
| | IMAM/BT 2 | 67.1 | | | 38 | |
| *Clostridium* | | | | | | |
| *acidiurici* | | 25.3 | 29.8 | 32.2 | 49 | |
| *amylolyticum* | 1 strain | 28 | | | 186 | |
| *barati* | 1 strain | 28 | | | 186 | |
| *bifermentans* | | | 32.5 | | 8 | |

## TABLE 1 (Continued)
## GC COMPOSITION OF THE DNA OF BACTERIA

| Organism | Source or Strain[a] | Mole % Guanosine + Cytosine[b]: Tm | Chemical Analysis | Buoyant Density | References[c] A | B |
|---|---|---|---|---|---|---|
| **Bacillaceae (continued)** | | | | | | |
| *Clostridium* | | | | | | |
| *butylicum* | P | | 32.6 | | 29 | |
| *butylicum (acetobutylicum)* | ATCC 862 | | 38.0 | | 31 | |
| *butyricum* | | 31.5 | 37.4 | | 49 | |
| | (2983) 10 strains | 28 | | | 186 | |
| | (3266) 10 strains | 28 | | | 186 | |
| *chauvoei* | | 26.5 | | 32.0 | 30, 31 | |
| *cylindrosporum* | | 31.7 | 32.4 | | 49 | |
| *kluyveri* | | | 34–36 | 35.0 | 5, 31 | |
| *madisoni* | 16 | 26.5 | | 34.0 | 30, 31 | |
| *multifermentans* | (3266) 5 strains | 28 | | | 186 | |
| | (2983) 1 strain | 28 | | | 186 | |
| *nigrificans* | NCIB 8395 | | | 48.0 | 31 | |
| | 2 strains | | | 45.5 | 50 | |
| | NCIB 8351 | | | 44.7 | 50 | |
| | ATCC 7946(NCA) | | | 44.7 | 50 | |
| | NCIB 8452 | | | 45.6 | 50 | |
| | DL | | | 45.6 | 50 | |
| *paraputrificum* | | 27.0 | | | 21 | |
| *pasteurianum* | | 30.5 | 30.8 | 31.7 | 49 | |
| *perfringens* | 87b | 26.5 | | 32.0 | 30, 31 | |
| | Fred. | | | 26.8 | 8 | |
| | | | | 30.9 | 8 | |
| | 1 strain | 28 | | | 186 | |
| *saccharoperbutylacetonicum* | N1-4 | 31.2 | 27.5–31 | | 51 | |
| | N1-504 | 31.5 | 27.7–31.2 | | 51 | |
| *saprogenes* | | | 30.3 | | 8 | |
| *tartarivorum* | T9-0 | 39.3 | 40.3 | | 187 | |
| *tetani* | 1 | 26.5 | | 34.0 | 30, 31 | |

TABLE 1 (Continued)
GC COMPOSITION OF THE DNA OF BACTERIA

| Organism | Source or Strain[a] | Mole % Guanosine + Cytosine[b] Tm | Chemical Analysis | Buoyant Density | References[c] A | B |
|---|---|---|---|---|---|---|
| **Bacillaceae (continued)** | | | | | | |
| *Clostridium* | | | | | | |
| *tetanomorphum* | | | 39.3 | | 49 | |
| *thermosaccharolyticum* | 3814 | 36.6 | 35.8 | | 187 | |
| *valerianicum* | | | 31.8 | | 8 | |
| species | 1 strain | 29 | | | 186 | |
| **Bacteroidaceae** | | | | | | |
| *Bacteroides* | | | | | | |
| *fragilis* | | 52.7 | | | 52 | |
| *insolitus* | IS 9 | | 43.2 | | 8 | |
| | NL 12 | | 41.2 | | 8 | |
| *melaninogenicus* | 1 | 37.1 | | | 52 | |
| | 2 | 42.6 | | | 52 | |
| *oralis* | 1 | 30.4 | | | 52 | |
| | 2 | 37.1 | | | 52 | |
| | 3 | 33.7 | | | 52 | |
| *ruminicola* | 3 strains } | 35.2–38.2 | | | 52 | |
| *fragilis* | 2 strains } | | | | | |
| *succinogenes* | 3 strains } | 28.5–32.2 | | | 52 | |
| *amylophilus* | 2 strains } | | | | | |
| *Fusobacterium* | | | | | | |
| *fusiformis* | | | 31.5 | | 8 | |
| *polymorphum* | | | 58.9 | | 8 | |
| *Ristella* | | | | | | |
| *clostridiformis* | | | 31.2 | | 8 | |
| *Sphaerophorus* | | | | | | |
| *necrophorus* | | | | 31.0 | 53 | |
| species | | | | 38.0 | 53 | |
| | 1 strain | 27 | | | 186 | |
| *Streptobacillus* | | | | | | |
| *moniliformis* (L-form) | L-1 Rat 30, ATCC 14075 | 23.9±0.1 | | | 130 | |

## TABLE 1 (Continued)
## GC COMPOSITION OF THE DNA OF BACTERIA

| Organism | Source or Strain[a] | Mole % Guanosine + Cytosine[b] Tm | Chemical Analysis | Buoyant Density | References[c] A | B |
|---|---|---|---|---|---|---|
| **Brevibacteriaceae** | | | | | | |
| *Brevibacterium* | | | | | | |
| *acetylicum* | ATCC 954 | 46.8 | | | 191 | |
| | ATCC 953 | 46.6 | | | 191 | |
| *albidum* | IAM 1631 | 70.0 | | | 191 | |
| *ammoniagenes* | ATCC 6872 | 54.6 | | | 191 | |
| | IFM AU-39 | 53.7 | | | 191 | |
| | ATCC 6871 | 54.1 | | | 191 | |
| *citreum* | IAM 1514 | 70.5 | | | 191 | |
| | IAM 1614 | 70.5 | | | 191 | |
| *divaricatum* | NRRL B-2312 | 54.4 | | | 191 | |
| *flavum* | ATCC 14067 (isolate) | 54.1 | | | 191 | |
| *fulvum* | IFM A-34 | 61.7 | | | 191 | |
| *fuscum* | IFM AU-44 | 58.5 | | | 191 | |
| | AJ 1486 Dr.Nakagawa (Osaka University) | 60.7 | | | 191 | |
| | CCEB 227 | 58.3 | | | 191 | |
| *helvolum* | ATCC 11822 | 64.1 | | | 191 | |
| | IAM 1391 | 66.8 | | | 191 | |
| | IAM 1434 | 65.9 | | | 191 | |
| | IAM 1478 | 65.6 | | | 191 | |
| | IAM 1498 | 65.9 | | | 191 | |
| | CCM 193 | 61.7 | | | 191 | |
| *imperiale* | ATCC 8365 | 69.8 | | | 191 | |
| *insectiphilium* | IFM AM-23 | 67.6 | | | 191 | |
| *lactofermentum* | ATCC 13869 (isolate) | 52.7 | | | 191 | |
| | ATCC 13655 (isolate) | 53.7 | | | 191 | |
| *linens* | ATCC 8377 | 62.7 | | | 191 | |
| | ATCC 9172 | 63.4 | | | 191 | |
| | ATCC 9174 | 60.2 | | | 191 | |
| | ATCC 9175 | 60.2 | | 63.3 | 178 | |
| | Isolates | | | | | |
| | R1:13 | | | 66.3 | 178 | |
| | R1:21 | 62.0 | | | 178 | |
| | R1:93 | 49.2 | | | 178 | |

## TABLE 1 (Continued)
## GC COMPOSITION OF THE DNA OF BACTERIA

| Organism | Source or Strain[a] | Mole % Guanosine + Cytosine[b] Tm | Chemical Analysis | Buoyant Density | References[c] A | B |
|---|---|---|---|---|---|---|
| **Brevibacteriaceae (continued)** | | | | | | |
| *Brevibacterium* | | | | | | |
| *linens* | Isolates | | | | | |
| | R1:56 | 59.0 | | | 178 | |
| | R1:98 | | | 67.3 | 178 | |
| | CCM 47 | 62.9 | | | 191 | |
| | IFM AM-27 | 66.8 | | | 191 | |
| | IFM AM-17 | 67.3 | | | 191 | |
| | IFM SC-84 | 67.6 | | | 191 | |
| *lipolyticum* | IAM 1398 | 70.7 | | | 191 | |
| | IAM 1413 | 70.5 | | | 191 | |
| *luteum* | IAM 1623 | 69.8 | | | 191 | |
| *maris* | IFM S-30 | 64.6 | | | 191 | |
| *minutiferula* | IFM AU-27 | 68.3 | | | 191 | |
| *protophormiae* | CCEB 282 | 63.2 | | | 191 | |
| *pusillum* | IAM 1479 | 69.0 | | | 191 | |
| | IAM 1489 | 69.3 | | | 191 | |
| *saperdae* | CCEB 336 | 68.5 | | | 191 | |
| *stationis* | ATCC 14403 | 53.9 | | | 191 | |
| *sulfureum* | IAM 1488 | 62.2 | | | 191 | |
| | IFM AU-38 | 64.4 | | | 191 | |
| | IFM AU-28 | 65.9 | | | 191 | |
| *testaceum* | IAM 1537 | 65.4 | | | 191 | |
| *vitarumen* | ATCC 10234 | 64.4 | | | 191 | |
| **Brucellaceae** | | | | | | |
| *Acinetobacter* | | | | | | |
| *alcaligenes* | ATCC 17923 | 45 | | | 172 | 172 |
| *anitratum* | H 2 | 41 | | | 172 | 172 |
| | type B5W | 41 | | | 172 | 172 |
| *anitratum* | ATCC 15150 | | | | 172 | 172 |
| *(A. calcoaceticus)* | ATCC 15149 | 43 | | | 172 | 172 |
| | ATCC 15151 | 42 | | | 172 | 172 |
| *anitratus* | NCIB 8250 | 42.2 | 38.2 | | 230 | 188 |
| | ATCC 17924 | 45 | | | 172 | 172 |

## TABLE 1 (Continued)
## GC COMPOSITION OF THE DNA OF BACTERIA

| Organism | Source or Strain[a] | Mole % Guanosine + Cytosine[b] Tm | Chemical Analysis | Buoyant Density | References[c] A | B |
|---|---|---|---|---|---|---|
| **Brucellaceae (continued)** | | | | | | |
| *Acinetobacter* | | | | | | |
| *anitratus (Moraxella glucidolytica)* | 32 | 42.4 | | | 11 | |
| | 33 | 42.8 | | | 11 | |
| | 34 | 42.4 | | | 11 | |
| | 35 | 42.2 | | | 11 | |
| | 36 | 42.2 | | | 11 | |
| | NCTC 7844 | 42.1 | | | 11 | |
| | A 4 Lautrop | 40.7 | | | 11 | |
| *lwoffi* | ATCC 9957 | 47.2 | 43.4 | | 229 | 189 |
| | ATCC 17925 | 46 | | | 172 | 172 |
| *(Moraxella) lwoffi* | 6 | 41.3 | | | 11 | |
| | 4 | 42.1 | | | 11 | |
| | 9 | 42.8 | | | 11 | |
| | 28 | 41.6 | | | 11 | |
| | 7 | 42.0 | | | 11 | |
| | 24 | 42.3 | | | 11 | |
| | 1 | 42.2 | | | 11 | |
| | 17 | 42.1 | | | 11 | |
| | 22 | 45.1 | | | 11 | |
| | 26 | 44.6 | | | 11 | |
| | 12 | 44.8 | | | 11 | |
| | 13 | 44.5 | | | 11 | |
| | 27 | 46.8 | | | 11 | |
| | NCTC 5866 | 45.9 | | | 11 | |
| | 11 | 46.9 | | | 11 | |
| | 14 | 46.5 | | | 11 | |
| | 21 | 46.8 | | | 11 | |
| | 10 | 46.6 | | | 11 | |
| | 18 | 46.2 | | | 11 | |
| | 19 | 46.7 | | | 11 | |
| | 16 | 46.6 | | | 11 | |
| | 23 | 46.8 | | | 11 | |
| | 3 | 46.3 | | | 11 | |
| | 15 | 46.6 | | | 11 | |
| | 20 | 46.7 | | | 11 | |
| *winogradskyi* | ATCC 17922 | 42 | | | 172 | 172 |
| *Actinobacillus* | | | | | | |
| *equuli* | VFB 781/64 | 41.8 | | | 54 | |
| | ATCC 13376 | 40.0 | | | 54 | |
| | ATCC 15558 | 40.0 | | | 54 | |
| | ATCC 15560 | 40.7 | | | 54 | |
| | Dorss. 1276/61 | 40.5 | | | 54 | |
| | 9347 | | | 39.5 | 190 | |

TABLE 1 (Continued)
GC COMPOSITION OF THE DNA OF ALGAE

| Organism | Source or Strain[a] | Tm | Chemical Analysis | Buoyant Density | References[c] A | References[c] B |
|---|---|---|---|---|---|---|
| | | Mole % Guanosine + Cytosine[b] | | | | |
| **Brucellaceae (continued)** | | | | | | |
| *Actinobacillus* | | | | | | |
| *lignieresii* | NCTC 4189 | 42.2 | | | 54 | |
| | VFB 238/63 | 42.6 | | | 54 | |
| | VFB 263/63 | 41.8 | | | 54 | |
| | Phill. 493/53 | 41.8 | | | 54 | |
| | Phill. 162 B (4) | 41.8 | | | 54 | |
| | Phill. M 20.1 | 41.0 | | | 54 | |
| | Phill. 97 B (3) | 40.5 | | | 54 | |
| | Phill. 108 B (3) | 41.0 | | | 54 | |
| | Phill. 132 B (3) | 38.7 | | | 54 | |
| | Phill. 173 B (3) | 38.6 | | | 54 | |
| | Phill. 189 B (8) | 39.3 | | | 54 | |
| | 207 | | | 40.0 | 190 | |
| | ATCC 10811 | | | 34.8 | 190 | |
| | ATCC 15557 | | | 38.8 | 190 | |
| *mallei* | 3873 | | | 67.2 | 190 | |
| | 4 | | | 66.8 | 190 | |
| *Bordetella* | | | | | | |
| *bronchiseptica* | NCTC 8761 | 69.5 | | | 11 | |
| | | 66.0 | | | 10 | |
| *pertussis* | 343E/358E | | 67.4 | | 55 | |
| | 124/138/149 | | 67.8 | | 55 | |
| | 154E/196E | | 57.5 | | 55 | |
| *Brucella* | | | | | | |
| *abortus* | 2308 | 57.0 | | | 47 | 47 |
| | 19M | 55.0 | | 56.0 | 30, 31 | |
| | | | 57.9 | | 8 | |
| *melitensis* | 16M | 58.0 | | | 47 | 47 |
| *neotomae* | | 56.0 | | | 10 | |
| | | 57.0 | | | 47 | 47 |
| *ovis* | | 58.0 | | | 47 | 47 |
| *suis* | 1776 | 57.0 | | | 47 | 47 |
| species (beagle organism) | RAA 666 | 56.0 | | | 10 | 10 |
| *Haemophilus* | | | | | | |
| *aegyptius* | | 40.0 | | 39.0 | 30, 31 | |
| *gallinarum* | ATCC 14385 | 41.9 | | | 56 | 56 |

## TABLE 1 (Continued)
## GC COMPOSITION OF THE DNA OF BACTERIA

| Organism | Source or Strain[a] | Mole % Guanosine + Cytosine[b] Tm | Chemical Analysis | Buoyant Density | References[c] A | B |
|---|---|---|---|---|---|---|
| **Brucellaceae (continued)** | | | | | | |
| *Haemophilus* | | | | | | |
| *influenzae* | | | 37.8 | | 8 | |
| | | 39.0 | | 39.0 | 30, 31 | |
| | Rd | 39.8 | | 36.7 | 250 | 207 |
| *parainfluenzae* | | 39.0 | | 39.0 | 30, 31 | |
| *suis* | 3090 | 39.0 | | | 30 | |
| *Herellea* | | | | | | |
| *caseolytica* | ATCC 19002 | 40 | | | 172 | 172 |
| *saponiphilum* | ATCC 19194 | 41 | | | 172 | 172 |
| *vaginicola* | MHD 11/30 | | 39.9 | | 57 | |
| | MHD 305 | | 39.8 | | 57 | |
| | ATCC 9955 | 42 | | | 172 | 172 |
| | ATCC 17961 | 41 | | | 172 | 172 |
| | ATCC 19003 | 42 | | | 172 | 172 |
| | ATCC 19004 | 41 | | | 172 | 172 |
| *Mima* | | | | | | |
| *polymorpha* | MHD 2/10 | | 44.4 | | 57 | |
| | MHD 307 | | 44.4 | | 57 | |
| | ATCC 10973 | | 44.6 | | 57 | |
| | ATCC 9957 | | 43.4 | | 57 | |
| | Raschig | | 42.3 | | 57 | |
| | MHD 12/13 | | 43.8 | | 57 | |
| | ATCC 17959 | 42 | | | 172 | 172 |
| | ATCC 17960 | 47 | | | 172 | 172 |
| *Moraxella* | | | | | | |
| *bovis* | | 41.0 | | | 30 | |
| | NCTC 9426 | | 44.6 | 43.0 | 3, 58 | |
| | NCTC 8561 (ATCC 17947) | | | 42.5 | 3 | |
| | NCTC 9425 (ATCC 17948) | | | 42.5 | 3 | |
| | NCTC 10900 | | | 42.5 | 3 | |
| | ATCC 17949 | 68.5 | | | 171 | |
| *duplex* | ATCC 17975 | 47 | | | 172 | 172 |
| *duplex liquefaciens* | ATCC 17952 | 45 | | | 172 | 172 |
| | ATCC 17952 | 46.8 | 44.3 | | 229 | 189 |

# TABLE 1 (Continued)
# GC COMPOSITION OF THE DNA OF BACTERIA

| Organism | Source or Strain[a] | Mole % Guanosine + Cytosine[b] Tm | Chemical Analysis | Buoyant Density | References[c] A | B |
|---|---|---|---|---|---|---|
| **Brucellaceae (continued)** | | | | | | |
| *Moraxella* | | | | | | |
| *glucidolytica* | ATCC 17978 | 42 | | | 172 | 172 |
| | ATCC 17979 | 42 | | | 172 | 172 |
| *lacunata* | ATCC 19991 | 45.6 | 42.47 | | 229 | 189 |
| | ATCC 10900 | 47.8 | | | 229 | |
| | ATCC 17956 | 48.5 | | | 229 | |
| | ATCC 11748 | | | 42.0 | 3 | |
| *liquefaciens* | | | 40–42 | | 5 | |
| | NCTC 7911 | | 44.3 | 41.5 | 3, 57 | |
| *lwoffi* | IP 55112 | | 39.4 | | 59 | |
| | Vibrio sp. 1, NCIB 8250 | | 38.2 | | 59 | |
| | | 44 | | | 172 | 172 |
| | ATCC 17968 | 45 | | | 172 | 172 |
| | ATCC 17969 | 44 | | | 172 | 172 |
| | ATCC 17976 | | | | 172 | 172 |
| | ATCC 17977 | 43 | | | 172 | 172 |
| | ATCC 17984 | 45 | | | 172 | 172 |
| | ATCC 17985 | 46 | | | 172 | 172 |
| | ATCC 17986 | 44 | | | 172 | 172 |
| | ATCC 17987 | 43 | | | 172 | 172 |
| *lwoffi* (var. *bacteroides*) | ATCC 17985 | 45.9 | | | 229 | |
| *lwoffi* (var. *brevis*) | ATCC 17987 | 47.4 | | | 229 | |
| *lwoffi* (var. *nonliquefaciens*) | 19 (H315) | | | 46–47 | 4 | |
| *nonliquefaciens* | 2770/60 | | | 41.0 | 3 | |
| | 3828/60 | | | 41.0 | 3 | |
| | 826/61 | | | 41.0 | 3 | |
| | 4378/62 | | | 41.0 | 3 | |
| | 4863/62 | | | 41.0 | 3 | |
| | 836/61 (ATCC 19968) | | | 41.0 | 3 | |
| | NCTC 7784 | | | 41.0 | 3 | |
| | 5058/62 | | | 40.0 | 3 | |
| | 13536/62 | | | 40.0 | 3 | |
| | 178/62 | | | 40.0 | 3 | |
| | 5050/62 | | | 41.5 | 3 | |
| | 13385/62 (ATCC 19969) | | | 41.5 | 3 | |
| | 4663/62 (ATCC 19975) | | | 42.0 | 3 | |
| | 752/52 | | | 43.5 | 3 | |

## TABLE 1 (Continued)
## GC COMPOSITION OF THE DNA OF BACTERIA

| Organism | Source or Strain[a] | Mole % Guanosine + Cytosine[b] Tm | Chemical Analysis | Buoyant Density | References[c] A | B |
|---|---|---|---|---|---|---|
| **Brucellaceae (continued)** | | | | | | |
| *Moraxella* | | | | | | |
| *nonliquefaciens* | 11865/52 | | 42.5 | | 57 | |
| | 7146/51 | | 44.7 | | 57 | |
| | 18522/51 | | 44.4 | | 57 | |
| | 19116/51 | | 44.2 | 43.5 | 3, 57 | |
| | ATCC 17953 | 46.2 | | | 229 | |
| | ATCC 17955 | 45 | | | 172 | 172 |
| *osloensis* | D 1 | 46 | | | 172 | 172 |
| | D 2 | 45 | | | 172 | 172 |
| | ATCC 19963 | 46.1 | 44.35 | | 229 | 189 |
| | ATCC 10973 | 48.0 | 44.6 | | 229 | 189 |
| | ATCC 19961 | 48.9 | 44.2 | | 229 | 189 |
| | A 1920 (ATCC 19976) | | | 43.5 | 3 | |
| *tuberculosis* | | | 65.0 | | 208 | |
| species | 5718 | | | 43.5 | 3 | |
| | 5873 | | | 43.5 | 3 | |
| | 8134 | | | 43.5 | 3 | |
| | A 608 | | | 43.5 | 3 | |
| | 5893 | | | 43.0 | 3 | |
| | 8292 | | | 43.0 | 3 | |
| | 9893 | | | 43.0 | 3 | |
| | 9833, 9985, E 1, L 1 | | | 43.0 | 3 | |
| | A 947 (1) | | | 42.5 | 3 | |
| *Pasteurella* | | | | | | |
| *aviseptica* | | | 36.3 | | 8 | |
| *boviseptica* | | | 37.6 | | 8 | |
| *haemolytica* | ATCC 10898 | 40.7 | | | 54 | |
| | VFB 558/243/62 | 40.5 | | | 54 | |
| | VFB 302/63 | 41.2 | | | 54 | |
| | VFB 185/65 | 40.8 | | | 54 | |
| | Bib. F 1 | 40.3 | | | 54 | |
| | Bib. G $7_1$ | 41.5 | | | 54 | |
| | Bib. 927 | 40.0 | | | 54 | |
| | Thomp. 1 C 49 | 41.0 | | | 54 | |
| *pestis* | EV 79 | | | 47.0 | 31 | |
| | AVO 2 | 46.0 | | | 30 | |
| | EV 6 | 46.0 | | | 30 | |
| | 1, 17, 248, 121, 422, 104 (15 strains) | | 48.4 | | 61 | |

# TABLE 1 (Continued)
# GC COMPOSITION OF THE DNA OF BACTERIA

| Organism | Source or Strain[a] | Mole % Guanosine + Cytosine[b] Tm | Chemical Analysis | Buoyant Density | References[c] A | B |
|---|---|---|---|---|---|---|
| **Brucellaceae (continued)** | | | | | | |
| *Pas.eurella* | | | | | | |
| *pseudotuberculosis rodentium* | 1, 162, 2613 | | 48.4 | | 61 | |
| *tularensis* | | 33.0 | 34.7 | 36.0 | 8, 30, 31 | |
| | NIH B-38 (ATCC 6223) | 33.0 | | | 47 | 47 |
| **Caulobacteriaceae** | | | | | | |
| *Caulobacter* | | | | | | |
| *bacteroides* | CB 6 | | | 66.0 | 62 | |
| *crescentus* | KA 3 | | | 62.0 | 62 | |
| | | | | 67 | 209 | |
| *fusiformis* | CB 37 (ATCC 15260) | | | 67.0 | 62 | |
| *henricii* | CB 4 (ATCC 15253) | | | 62.0 | 62 | |
| | KA 1 | | | 65.0 | 62 | |
| *intermedius* | CB-G | | | 65.0 | 62 | |
| *leidyi (Asticcacaulis excentricus)* | KA 4 | | | 55.0 | 62 | |
| *subvibrioides* | CB 79 | | | 67.0 | 62 | |
| *vibrioides* | CB 21 | | | 64.0 | 62 | |
| **Chlamydiaceae** | | | | | | |
| *Chlamydia* | | | | | | |
| *psittaci* | MN | 41.5 | | 41.5 | 63 | 63 |
| | 6 BC | 41.5 | | 41.5 | 63 | 63 |
| *trachomatis* | Peking 2 | 45.0 | | 45.0 | 63 | 63 |
| | MRC 1 | 45.0 | | 45.0 | 63 | 63 |
| | Cal 1 | 45.0 | | 45.0 | 63 | 63 |
| **Chlamydobacteriaceae** | | | | | | |
| *Leptotrichia* | | | | | | |
| *buccalis* | L 11 | 34 | | | 192 | |

## TABLE 1 (Continued)
## GC COMPOSITION OF THE DNA OF BACTERIA

| Organism | Source or Strain[a] | Mole % Guanosine + Cytosine[b] Tm | Chemical Analysis | Buoyant Density | References[c] A | B |
|---|---|---|---|---|---|---|
| **Chlamydobacteriaceae (continued)** | | | | | | |
| *Sphaerotilus* | | | | | | |
| *discophorus* | 35, 36, 41, 42 | | | 69.5 | 64 | |
| *natans* | 6, 16, 18, 21 | | | 70.0 | 64 | |
| **Chlorobacteriaceae** | | | | | | |
| *Chlorobium* | | | | | | |
| *limicola* | Gilroy Hot Spring | 50.0 | | 51.0 | 65 | |
| | Zeulenrode | | | 51.5 | 65 | |
| | Klein-Kalden | 51.0 | | 51.5 | 65 | |
| | Reiershausen | | | 52.0 | 65 | |
| | Moss Landing | 51.0 | | 53.5 | 65 | |
| | PS | 52.0 | | 55.0 | 65 | |
| | Carmel R. 17 | | | 57.0 | 65 | |
| | Federsee | 56.0 | | 58.0 | 65 | |
| | Bennhausen | | | 58.0 | 65 | |
| *thiosulfato-* | | | 57.8 | | 30 | |
| *philium* | Luneberg | | | 52.5 | 65 | |
| | Golf Pond | | | 52.5 | 65 | |
| | Sehestedt | | | 53.5 | 65 | |
| | Berkeley | 56.0 | | 56.5 | 65 | |
| | Carmel R. 24 | | | 57.5 | 65 | |
| | Rothamstedt | | | 57.5 | 65 | |
| | Carmel R. 3 | | | 58.0 | 65 | |
| | Fassajara | | | 58.0 | 65 | |
| *Chloropseudomonas* | | | | | | |
| *ethylicum* | | | 57.6 | | 29 | |
| **Corynebacteriaceae** | | | | | | |
| *Arthrobacter* | | | | | | |
| *atrocyaneus* | CCM 1645 | 69.5 | | | 191 | |
| | ATCC 13752 | | | 70.5 | 178 | |
| | Isolate S1:100 | | | 62.4 | 178 | |
| *aurescens* | CCM 1649 | 61.5 | | | 191 | |
| | ATCC 13344 | | | 59.1 | 178 | |
| *citreus* | CCM 1647 | 62.9 | | | 191 | |
| | AJ 1435 isolate | 62.9 | | | 191 | |
| | ATCC 11642 | 62.9 | | 65.1 | 178 | 191 |
| | AJ 1438 | 64.1 | | | 191 | |

# TABLE 1 (Continued)
# GC COMPOSITION OF THE DNA OF BACTERIA

| Organism | Source or Strain[a] | Mole % Guanosine + Cytosine[b] Tm | Chemical Analysis | Buoyant Density | References[c] A | B |
|---|---|---|---|---|---|---|
| **Corynebacteriaceae (continued)** | | | | | | |
| *Arthrobacter* | | | | | | |
| *crystallopoietes* | NCIB 9499 | | | 62.9 | 178 | |
| | Isolates | | | | | |
| | S2:19 | 64.4 | | | 178 | |
| | S2:53 | | | 65.5 | 178 | |
| *duodecadis* | CBRI 859 | | | 72.5 | 178 | |
| *flavescens* | CBRI 860 | | | 70.3 | 178 | |
| | Isolates | | | | | |
| | S2:55 | 62.0 | | | 178 | |
| | R2:03 | 61.0 | | | 178 | |
| | R2:09 | | | 71.4 | 178 | |
| | R2:39 | 69.2 | | | 178 | |
| | R2:47 | 64.0 | | | 178 | |
| *globiformis* | | | 62–64 | | 5 | |
| | ATCC 8010 | 60.0 | | | 17 | |
| | CBRI 425 | | | 65.1 | 178 | |
| | CBRI 616 | | | 65.1 | 178 | |
| | CBRI 884, ATCC 8010 | 62.0 | | 65.5 | 178 | 191 |
| | CBRI 425 | | | 65.1 | 178 | |
| | CBRI 616 | | | 65.1 | 178 | |
| | CBRI 884, ATCC 8010 | 62.0 | | 65.5 | 178 | |
| *nicotianae* | CBRI 883 | | | 64.0 | 178 | |
| | ATCC 15236 | 60.2 | | | 191 | |
| | CBRI 883 | | | 64.0 | 178 | |
| | ATCC 15236 | 60.2 | | | 191 | |
| *oxydans* | CBRI 899 | | | 62.4 | 178 | |
| | Isolate S2:16 | 64.2 | | | 178 | |
| | CBRI 899 | | | 62.4 | 178 | |
| | Isolate S2:16 | 64.2 | | | 178 | |
| | ATCC 14358 | 62.7 | | 65.9 | 178 | |
| | Isolate S1:13 | | | 64.3 | 178 | |
| | ATCC 14359 | 64.4 | | | 191 | |
| | ATCC 14358 | 62.7 | | 65.9 | 178 | 191 |
| | Isolate S1:13 | | | 64.3 | 178 | 191 |
| | ATCC 14359 | 64.4 | | | 191 | |
| *pascens* | CCM 1653 | 63.7 | | | 191 | |
| | CBRI 858, ATCC 13346 | | | 64.9 | 178 | |
| | CCM 1653 | 63.7 | | | 191 | |
| | CBRI 858, ATCC 13346 | | | 64.9 | 178 | |

## TABLE 1 (Continued)
## GC COMPOSITION OF THE DNA OF BACTERIA

| Organism | Source or Strain[a] | Mole % Guanosine + Cytosine[b] Tm | Chemical Analysis | Buoyant Density | References[c] A | B |
|---|---|---|---|---|---|---|
| **Corynebacteriaceae (continued)** | | | | | | |
| *Arthrobacter* | | | | | | |
| *ramosus* | ATCC 13727 | 62.2 | | 63.1 | 178 | 191 |
| *simplex* | ATCC 6946 | 71.7 | | 74.0 | 178 | 191 |
| *terregens* | ATCC 13345 | | | 68.7 | 178 | |
| *tumescens* | IAM 1447 | 64.1 | | | 191 | |
| | IAM 1458 | 64.6 | | | 191 | |
| | ATCC 6947 | 69.8 | | 72.4 | 178 | 191 |
| | Isolates | | | | | |
| | S1:23 | | | 70.7 | 178 | 191 |
| | S1:55 | | | 69.4 | 178 | 191 |
| | S1:24 | | | 70.7 | 178 | 191 |
| | S1:70 | 64.8 | | | 178 | 191 |
| | S1:47 | 61.0 | | | 178 | 191 |
| *ureafaciens* | ATCC 7562 | 61.7 | | 62.9 | 178 | 191 |
| | IAM 1390 | 62.0 | | | 191 | |
| *variabilis* | CCM 1565 | 70.0 | | | 191 | |
| species | ATCC 12834 | 71.5 | | | 178 | |
| | Isolates | | | | | |
| | S2:26 | | | 71.0 | 178 | |
| | S2:95 | | | 73.3 | 178 | |
| | S2:31 | 71.1 | | | 178 | |
| | S1:19 | 74.0 | | | 178 | |
| | S1:58 | 59.2 | | | 178 | |
| | S1:96 | 68.0 | | | 178 | |
| | 3 C | 61.0 ± 1.5 | | | 193 | |
| *Cellulomonas* | | | | | | |
| *biazotea* | ATCC 486 | 75.0 | | | 17 | |
| | ATCC 486 | 75.6 | | | 266 | |
| *fimi* | ATCC 15724 | 72.0 | | | 191 | |
| | ATCC 484 | 71.7 | | | 191 | |
| | ATCC 8183 | 63.4 | | | 191 | |
| | IMRU 22 | 76.1 | | | 266 | |
| *flavigena* | ATCC 482 | 72.7 | | | 191 | |
| | ATCC 482 | 74.6 | | | 266 | |
| *gelida* | ATCC 488 | 72.4 | | | 191 | |
| | ATCC 488 | 74.2 | | | 266 | |
| *uda* | ATCC 491 | 72.0 | | | 191 | |

**TABLE 1 (Continued)**
**GC COMPOSITION OF THE DNA OF BACTERIA**

| Organism | Source or Strain[a] | Mole % Guanosine + Cytosine[b] Tm | Chemical Analysis | Buoyant Density | References[c] A | B |
|---|---|---|---|---|---|---|
| **Corynebacteriaceae (continued)** | | | | | | |
| *Corynebacterium* | | | | | | |
| *acnes* | 2 strains | | 48.0 | | 8 | |
| *anaerobium* | | | 51.5 | | 66 | |
| | 3130 | | 57.8 | | 67 | |
| *annamensis* | | | 58.0 | | 66 | |
| *aquaticum* | ATCC 14665 | 69.3 | | | 191 | |
| *avidum* | | | 52.0 | | 66 | |
| *bovis* | | | 58.0 | | 66 | |
| | ATCC 7715 | 68.3 | | | 191 | |
| | AJ 1377 | 67.6 | | | 191 | |
| *callunae* | NRRL B-2244 | 51.2 | | | 191 | |
| *cutis commune* | | | 54.0 | | 66 | |
| *diphtheriae* | 3 strains | | 51.9–54.5 | | 66 | |
| | ATCC 11913 | 52.4 | | | 191 | |
| *diphtheroides* | | | 53.0 | | 66 | |
| *enzymicum* | | | 51.5 | | 66 | |
| *equi* | | | 58.5 | | 66 | |
| | ATCC 7698 | 67.6 | | | 39 | |
| | NI | 66.1 | | | 191 | |
| | AJ 1367 (Sanraku Co.) | 65.9 | | | 191 | |
| | AJ 1378 (Sanraku Co.) | 69.5 | | | 191 | |
| | ATCC 6939 | 67.8 | | | 191 | |
| | ATCC 10146 | 67.1 | | | 191 | |
| *fascians* | | | 54.5 | | 66 | |
| | NI | 62.9 | | | 191 | |
| | ATCC 12974 | 67.3 | | | 197 | |
| *flaccumfaciens* | ATCC 6887 | 68.3 | | | 191 | |
| | ATCC 12813 | 74.5 | | | 266 | |
| | ATCC 12813 | 68.5 | | | 191 | |
| *glutamicum* | Type I | | | | | |
| | KY 9002 | | 56.8 | | 68 | |
| | KY 3805 | | 56.8 | | 68 | |
| | KY 3460 | | 57.0 | | 68 | |
| | KY 10122 | | 56.8 | | 68 | |
| | KY 3509 | | 56.8 | | 68 | |

## TABLE 1 (Continued)
## GC COMPOSITION OF THE DNA OF BACTERIA

| Organism | Source or Strain[a] | Mole % Guanosine + Cytosine[b] Tm | Chemical Analysis | Buoyant Density | References[c] A | B |
|---|---|---|---|---|---|---|
| **Corynebacteriaceae (continued)** | | | | | | |
| *Corynebacterium* | | | | | | |
| *glutamicum* | Type I | | | | | |
| | KY 9614 | | 57.0 | | 68 | |
| | KY 10125 | | 56.5 | | 68 | |
| | KY 10124 | | 56.6 | | 68 | |
| | KY 10026 | | 56.9 | | 68 | |
| | KY 10123 | | 56.6 | | 68 | |
| | Type II | | | | | |
| | KY 9606 | | 56.5 | | 68 | |
| | Type III | | | | | |
| | KY 9005 | | 57.5 | | 68 | |
| | Type IV | | | | | |
| | KY 9003 | | 57.7 | | 68 | |
| | Type V | | | | | |
| | KY 10021 | | 56.8 | | 68 | |
| | KY 3807 | | 57.7 | | 68 | |
| | Type VI | | | | | |
| | KY 9026 | | 56.3 | | 68 | |
| | KY 3462 | | 55.6 | | 68 | |
| | Type VII | | | | | |
| | KY 9610 | | 56.2 | | 68 | |
| | Type VIII | | | | | |
| | KY 9594 | | 56.7 | | 68 | |
| | Type IX | | | | | |
| | KY 9024 | | 56.0 | | 68 | |
| | KY 3461 | | 55.0 | | 68 | |
| | KY 10126 | | 56.4 | | 68 | |
| | KY 10127 | | 57.3 | | 68 | |
| | Type X | | | | | |
| | KY 9602 | | 57.0 | | 68 | |
| | Type XI | | | | | |
| | KY 9007 | | 53.6 | | 68 | |
| | KY 3510 | | 53.1 | | 68 | |
| | Type XII | | | | | |
| | KY 9591 | | 65.6 | | 68 | |
| | KY 10101 | | 64.4 | | 68 | |
| | NRRL B2243 | 55.1 | 56.8 | | 194 | 229 |
| | ATCC 13032 | 54.9 | 56.8 | | 194 | 229 |
| | KY 9005 | 54.6 | 57.5 | | 194 | 229 |
| *granulomatis* | | | 51.5 | | 66 | |
| *granulosum* | | | 52.0 | | 66 | |
| *hepatodystrophicans* | | | 52.5 | | 66 | |
| *herculis* | ATCC 13868 | 52.4 | | | 191 | |
| *hoagii* | NI | 65.6 | | | 191 | |
| | ATCC 7005 | 64.6 | | | 191 | |
| | | | 51.5 | | 66 | |

TABLE 1 (Continued)
GC COMPOSITION OF THE DNA OF BACTERIA

| Organism | Source or Strain[a] | Mole % Guanosine + Cytosine[b] Tm | Chemical Analysis | Buoyant Density | References[c] A | B |
|---|---|---|---|---|---|---|
| **Corynebacteriaceae (continued)** | | | | | | |
| *Corynebacterium* | | | | | | |
| *hofmannii* | | | 57.0 | | 66 | |
| *humiferum* | | | 54.5 | | 66 | |
| *hydrocarboclastus* | IAM 1399 | 61.2 | | | 191 | |
| | IAM 1484 | 61.0 | | | 191 | |
| *ilius* | | | 58–60 | | 5 | |
| *insidiosum* | ATCC 10253 | 78.1 | | | 191 | |
| *kutscheri* | | | 58.5 | | 66 | |
| *lacticum* | IMRU 630 | 70.3 | | | 266 | |
| *lilium* | NRRL B-2243 | 52.9 | | | 191 | |
| *liquefasciens* | | | 51.5 | | 66 | |
| *lymphophilum* | | | 54.0 | | 66 | |
| *michiganense* | | | 53.5 | | 66 | |
| | ATCC 7430 | 68.8 | | | 191 | |
| | ATCC 10202 | 70.7 | | | 191 | |
| | ATCC 7429 | 70.2 | | | 191 | |
| | ATCC 4450 | 70.0 | | | 191 | |
| | ATCC 4450 | 67.3 | | | 266 | |
| | PRI 41 | | | 72.5 | 178 | |
| | Isolates | | | | | |
| | S1:97 | | | 40.5 | 178 | |
| | S1:63 | 64.7 | | | 178 | |
| | S2:01 | | | 65.3 | 178 | |
| | S2:76 | 64.8 | | | 178 | |
| *minutissimum* | | | 54.5 | | 66 | |
| *parvum* | | | 58.0 | | 8, 66 | |
| | 936 B | | 54.7 | | 15 | |
| *phocae* | | | 52.0 | | 66 | |
| *poinsettiae* | CCM 1587 | 70.0 | | | 191 | |
| | ATCC 9682 | 72.8 | | | | |
| *pseudodiphthericum* | | | 56–58 | | 5 | |
| | | | 52.5 | | 66 | |
| | IMRU 1076 | 67.5 | | | 266 | |
| *pyogenes* | | | 48.5 | | 66 | |

## TABLE 1 (Continued)
## GC COMPOSITION OF THE DNA OF BACTERIA

| Organism | Source or Strain[a] | Mole % Guanosine + Cytosine[b] | | | References[c] | |
|---|---|---|---|---|---|---|
| | | Tm | Chemical Analysis | Buoyant Density | A | B |
| **Corynebacteriaceae (continued)** | | | | | | |
| *Corynebacterium* | | | | | | |
| *rathayi* | | | 54.0 | | 66 | |
| *renale* | | | 53.0 | | 66 | |
| | 9 | | 55.6 | | 195 | |
| | 35 | | 57.9 | | 195 | |
| | 42 | | 52.6 | | 195 | |
| | 43 | | 53.8 | | 195 | |
| | 45 | | 56.9 | | 195 | |
| | 46 | | 59.8 | | 195 | |
| | 48 | | 53.8 | | 195 | |
| | 77 | | 53.7 | | 195 | |
| | 121 | | 58.1 | | 195 | |
| | RH | | 57.0 | | 195 | |
| | ATCC 10848 | 59.3 | 57.4 | | 195 | 191 |
| *renale cuniculi* | | | 58.0 | | 66 | |
| *sepedonicum* | ATCC 9850 | 69.8 | | | 191 | |
| *vadosum* | | | 57.7 | | 8 | |
| *xerosis* | | | 55.0 | | 66 | |
| | ATCC 9016 | 57.5 | | 59.0 | 30, 31 | |
| | ATCC 373 | 67.3 | | | 191 | |
| | ATCC 7094 | 59.0 | | | 191 | |
| | ATCC 7711 | 68.5 | | | 191 | |
| species | c-85 | | 53.0 | | 66 | |
| | cb-2 | | 53.0 | | 66 | |
| | ct-2 | | 55.5 | | 66 | |
| species (L-form) | ATCC 2381, D-campo B-L | 63.7 ± 0.2 | | | 129 | |
| unidentified | AJ 1958 isolate | 53.7 | | | 191 | |
| | AJ 1490 isolate | 54.1 | | | 191 | |
| | AJ 1548 isolate | 53.9 | | | 191 | |
| | AJ 1562 | 51.2 | | | 191 | |
| | AJ 1549 isolate | 53.4 | | | 191 | |
| | AJ 1488 Dr. Yano (University of Tokyo) | 62.4 | | | 191 | |
| *Listeria* | | | | | | |
| *monocytogenes* | | 38.0 | | 38.0 | 30, 31 | |
| *Microbacterium* | | | | | | |
| *flavum* | ATCC 10340 | 58.3 | | | 191 | |
| *lacticum* | ATCC 8180 | 69.3 | | | 191 | |

## TABLE 1 (Continued)
## GC COMPOSITION OF THE DNA OF BACTERIA

| Organism | Source or Strain[a] | Mole % Guanosine + Cytosine[b]: Tm | Chemical Analysis | Buoyant Density | References[c] A | B |
|---|---|---|---|---|---|---|
| **Cytophagaceae** | | | | | | |
| *Cytophaga* | | | | | | |
| *anitrata* | ATCC 17945 | | | | 172 | 172 |
| | "Vibrio 01" | 40 | | | 172 | 172 |
| *aurantiaca* | ATCC 12208 | | | 42.0 | 69 | |
| *fermentans* | ATCC 12470 | 41.2 | | 39.0 | 69 | |
| *hutchinsonii* | | | | 39.0 | 69 | |
| *johnsonii* | ATCC 17061 (MYX 1.1.1) | | | 33.0 | 69 | |
| | Cook 405 | 34.6 | | 35.0 | 69 | |
| *lwoffi* | ATCC 17946 | | | | 172 | 172 |
| *marinoflava sp. nov.* | ATCC 19326 (NCMB 397) | | | 37.0 | 70, 71 | |
| species | M 3 | 37.1 | | | 318 | |
| | NCMB 292 | 33.6 | | | 319 | |
| | NCMB 11 | 34.2 | | | 319 | |
| | NCMB 296 | 57.3 | | | 317 | |
| | NCIB 8501 | 71 | | | 317 | |
| | NCMB 251 | 35.6 | | | 319 | |
| | NCMB 289 | 35.8 | | | 319 | |
| | NCMB 275 | 35.8 | | | 319 | |
| | NCMB 264 | 36.6 | | | 319 | |
| | NCMB 249 | 40.6 | | | 319 | |
| | NCIB 9497 | | | 67 | 198 | |
| *Flexibacter* | | | | | | |
| *elegans* | | | | 48 | 314 | |
| *rubrum* | | | | 37 | 314 | |
| | | | | 37.0 | 72 | |
| **Enterobacteriaceae** | | | | | | |
| *Aerobacter (Enterobacter)* | | | | | | |
| *aerogenes* | ATCC 13048 | 54.3–54.9 | | 53.1–54.3 | 7, 12, 74 | 73 |
| | ATCC 13882 | 55.6 | | 56.1 | 74 | |
| | 3 strains | 57.8 | | | 75 | |
| | ATCC 14308 | 54.0 | | 54.0 | 7 | |
| | 5 strains | | 55.2–58.9 | | 8 | |
| | MEG | 54.4 | | 54.4 | 7 | |
| | ATCC 211 | 55.5 | | 55.5 | 7 | 73 |
| | ICPB 3151 (banana strain) | | | 56.1 | 74 | |

## TABLE 1 (Continued)
## GC COMPOSITION OF THE DNA OF BACTERIA

| Organism | Source or Strain[a] | Mole % Guanosine + Cytosine[b] Tm | Chemical Analysis | Buoyant Density | References[c] A | B |
|---|---|---|---|---|---|---|
| **Enterobacteriaceae (continued)** | | | | | | |
| *Aerobacter (Enterobacter)* | | | | | | |
| *aerogenes* | Indiana | 55.5 | | 55.5 | 7 | |
| | Balamuth | 56.5 | | 56.5 | 7 | |
| | 1088 (Harvard) | 57.5 | | 57.0 | 30, 31 | |
| | 199 | 59.4 | | | 174 | |
| | DK 32 | 58.6 | | | 174 | |
| | DK 40 | 59.4 | 59.3 | | 174 | |
| | L 1 | 60.4 | | | 174 | |
| | L 10 | 57.9 | | | 174 | |
| | M 5A1 | 58.6 | 57.7 | | 174 | |
| | M 8 | 58.9 | | | 174 | |
| | UW 1 | 60.8 | | | 174 | |
| | CDC 2428-62 | 56.4 | | | 174 | |
| *alvei* | ATCC 23280 | 48.7 | | 48.0 | 74 | |
| *cloacae* | ATCC 13047 | 55.4 | | 54.6 | 74 | |
| | | | 52–54 | | 5 | |
| | CDC 1237-63 | 56.9 | | | 174 | |
| *lipolyticus* | ATCC 14460 | 51.9–53.9 | | 52.6 | 74, 76 | |
| | | | 50–52 | | 5 | |
| species | 1041 | 51.5 | | | 30 | |
| *Bacterium* | | | | | | |
| *paracoli* | ATCC 23280 | 48.0 | | 48.0 | 196 | 197 |
| | ATCC 23281 | 70.5 | | 65.3 | 196 | 197 |
| | | | 57.8 | | 196 | |
| | Mutant 52-1 | | 75.4 | | 196 | |
| *Citrobacter* | | | | | | |
| *freundii* | ATCC 8090 | 52.7 | | 52.6 | 74 | |
| | ATCC 8455 | | | 52.0 | 12 | |
| *Erwinia* | | | | | | |
| *amylovora* | ICPB EA-11 | 53.0 | | 53.6 | 74 | |
| | ICPB EA-131 | | | 54.1 | 74 | |
| | ICPB EA-137 | | | 53.8 | 74 | |
| | ICPB EA-146 | | | 54.1 | 74 | |
| | ICPB EA-162 | | | 54.1 | 74 | |
| | ICPB EA-167 | | | 53.6 | 74 | |
| | ICPB EA-169 | | | 54.1 | 74 | |
| *ananas* | | | 52–54 | | 5 | |
| | ICPB EA-101 | | | 53.6 | 74 | |
| | ICPB EA-133 | | | 53.1 | 74 | |
| | ICPB EA-180 | | | 54.1 | 74 | |
| | ICPB EA-181 | | | 53.6 | 74 | |

TABLE 1 (Continued)
GC COMPOSITION OF THE DNA OF BACTERIA

| Organism | Source or Strain[a] | Mole % Guanosine + Cytosine[b] Tm | Chemical Analysis | Buoyant Density | References[c] A | B |
|---|---|---|---|---|---|---|
| **Enterobacteriaceae (continued)** | | | | | | |
| *Erwinia* | | | | | | |
| *aroideae* | | | 50–52 | | 5 | |
| | ICPB EA-14 | 54.0 | | 53.1 | 74 | |
| | ICPB EA-13 | | | 52.0 | 74 | |
| | ICPB EA-144 | | | 51.0 | 74 | |
| *atroseptica* | ICPB EA-112 | | | 51.3 | 74 | |
| | ICPB EA-153 | | | 53.1 | 74 | |
| *carnegieana* | ICPB EC-186 | | | 51.0 | 74 | |
| | ICPB EC-187 | | | 50.8 | 74 | |
| | ICPB EC-188 | | | 51.7 | 74 | |
| | ICPB EC-189 | | | 51.3 | 74 | |
| | ICPB EC-190 | | | 51.0 | 74 | |
| | ICPB EC-191 | | | 51.0 | 74 | |
| | ICPB EC-221 | | | 51.0 | 74 | |
| | ICPB EC-222 | | | 51.5 | 74 | |
| | ICPB EC-223 | | | 51.0 | 74 | |
| *carotovora* | 17 | | 55.0 | | 66 | |
| | 64 | | 53.5 | | 66 | |
| | 51 | | 55.0 | | 66 | |
| | 58 | | 54.0 | | 66 | |
| | 3a | | 53.5 | | 66 | |
| | | | 54.0 | | 8 | |
| | ATCC 8061 | 51.5 | | 50.0 | 30, 31 | |
| | ICPB EC-138 | 53.4 | | 52.0 | 74 | |
| | ICPB EC-105 | | | 51.0 | 74 | |
| | ICPB EC-131 | | | 52.6 | 74 | |
| | ICPB EC-150 | | | 52.6 | 74 | |
| | ICPB EC-153 | | | 51.7 | 74 | |
| | ICPB EC-208 | | | 52.3 | 74 | |
| *cassavae* | ICPB EC-11 | | | 55.1 | 74 | |
| *chrysanthemi* | | | 52–54 | | 5 | |
| | ICPB EC-16 | 55.6 | | 55.1 | 74 | |
| | ICPB EC-175 | | | 55.1 | 74 | |
| | ICPB EC-183 | | | 56.1 | 74 | |
| | ICPB EC-205 | | | 57.1 | 74 | |
| *cypripedi* | | | 52–54 | | 5 | |
| | ICPB EC-155 | | | 54.1 | 74 | |
| | ICPB EC-160 | | | 54.6 | 74 | |
| *cytolytica* | ICPB EC-207 | | | 56.1 | 74 | |
| *dieffenbachiae* | ICPB ED-102 | | | 55.1 | 74 | |
| | ICPB ED-103 | | | 56.1 | 74 | |

## TABLE 1 (Continued)
## GC COMPOSITION OF THE DNA OF BACTERIA

| Organism | Source or Strain[a] | Mole % Guanosine + Cytosine[b] Tm | Chemical Analysis | Buoyant Density | References[c] A | B |
|---|---|---|---|---|---|---|
| **Enterobacteriaceae (continued)** | | | | | | |
| *Erwinia* | | | | | | |
| *dissolvens* | ICPB ED-106 | 57.1 | | 57.1 | 74 | |
| | ICPB ED-105 | | | 53.6 | 74 | |
| | ICPB ED-109 | | | 57.6 | 74 | |
| | ICPB ED-110 | | | 55.1 | 74 | |
| | ICPB ED-111 | | | 55.8 | 74 | |
| | ICPB ED-112 | | | 57.1 | 74 | |
| | ICPB ED-113 | | | 56.6 | 74 | |
| | ICPB ED-114 | | | 57.1 | 74 | |
| *herbicola (Bacterium herbicola)* | G 138 | 56.1 | | | 77 | |
| | G 139 | 56.2 | | | 77 | |
| | G 140 | 55.5 | | | 77 | |
| | G 141 | 55.5 | | | 77 | |
| | G 142 | 55.9 | | | 77 | |
| | G 143 | 55.4 | | | 77 | |
| | D 11 | 55.1 | | | 77 | |
| *herbicola (lathyri)* | G 144 | 55.0 | | | 77 | |
| | G 145 | 55.8 | | | 77 | |
| | G 157 | 55.9 | | | 77 | |
| | PA | 56.5 | | | 77 | |
| | NCPPB 102 | 55.7 | | | 77 | |
| | G 146 | 58.6 | | | 77 | |
| *herbicola (Bacterium typhi flavum)* | NCTC 9381 | 55.9 | | | 77 | |
| | 708 | 55.8 | | | 77 | |
| | 818 | 56.2 | | | 77 | |
| | 827 | 55.6 | | | 77 | |
| | 892 | 55.2 | | | 77 | |
| | ICPB 3164 | | | 55.1 | 74 | |
| | ICPB 3165 | | | 55.1 | 74 | |
| *herbicola (Bacterium stewartii)* | ICPB SS-11 | | | 55.1 | 74 | |
| | ICPB SS-102 | | | 54.6 | 74 | |
| *herbicola* | ICPB 2553 | | | 55.1 | 74 | |
| | ICPB 2554 | | | 54.1 | 74 | |
| | ICPB 3161 | | | 55.1 | 74 | |
| | ICPB 3162 | | | 53.1 | 74 | |
| | ICPB 3163 | | | 55.1 | 74 | |
| | Muraschi isolate G 150 | 55.7 | | | 77 | |
| | Muraschi isolate G 151 | 55.8 | | | 77 | |
| | Muraschi isolate G 152 | 55.4 | | | 77 | |
| | Muraschi isolate G 153 | 56.1 | | | 77 | |

## TABLE 1 (Continued)
## GC COMPOSITION OF THE DNA OF BACTERIA

| Organism | Source or Strain[a] | Mole % Guanosine + Cytosine[b] | | | References[c] | |
|---|---|---|---|---|---|---|
| | | Tm | Chemical Analysis | Buoyant Density | A | B |
| **Enterobacteriaceae (continued)** | | | | | | |
| *Erwinia* | | | | | | |
| *herbicola (Bacterium stewartii)* | Muraschi isolate G 154 | 56.2 | | | 77 | |
| | Muraschi isolate G 155 | 55.9 | | | 77 | |
| | Muraschi isolate G 156 | 55.5 | | | 77 | |
| | G 150 | 55.9 | | 56.0 | 171 | 199 |
| | G 151 | 56.0 | | 55.1 | 171 | 199 |
| | G 152 | 55.6 | | 56.1 | 171 | 199 |
| *herbicola (ananas)* | BG 1 | 54.4 | | | 77 | |
| | BG 2 | 55.2 | | | 77 | |
| | MF 28 | 54.7 | | | 77 | |
| | Y 1 | 55.0 | | | 77 | |
| *lathyri* | ICPB EL-101 B | | | 55.1 | 74 | |
| | ICPB EL-102 | | | 55.1 | 74 | |
| | ICPB EL-103 | 55.9 | | 55.1 | 74 | |
| | ICPB EL-104 | | | 52.6 | 74 | |
| | ICPB EL-105 | | | 53.1 | 74 | |
| | ICPB EL-106 | | | 57.7 | 74 | |
| | ICPB EL-107 | | | 54.1 | 74 | |
| *mangiferae* | ICPB EM-101 | | | 54.6 | 74 | |
| *maydis* | | | 50–52 | | 5 | |
| *milletiae* | | | 52–54 | | 5 | |
| | JM 1 | 55.8 | | | 77 | |
| | ICPB EM-102 | 55.4 | | 55.1 | 74 | |
| | ICPB EM-113 | | | 54.1 | 74 | |
| | ICPB EM-114 | | | 55.1 | 74 | |
| *nigrificans* | | | 54–56 | | 5 | |
| *nigrifluens* | ICPB EN-104 | 56.6 | | 56.1 | 74 | |
| *nimipressuralis* | | | 52–54 | | 5 | |
| | ICPB EN-1 | 55.4 | | 55.1 | 74 | |
| *oleraceae* | ICPB EO-1 | 51.5 | | 50.5 | 74 | |
| *quercina* | ICPB EQ-101 | | | 55.1 | 74 | |
| | ICPB EQ-102 | | | 54.6 | 74 | |
| *rhapontici* | | | 50–52 | | 5 | |
| | ICPB ER-1 | 51.1 | | 51.0 | 74 | |
| | ICPB ER-102 | | | 53.1 | 74 | |
| | ICPB ER-106 | | | 53.1 | 74 | |

## TABLE 1 (Continued)
## GC COMPOSITION OF THE DNA OF BACTERIA

| Organism | Source or Strain[a] | Mole % Guanosine + Cytosine[b] Tm | Chemical Analysis | Buoyant Density | References[c] A | B |
|---|---|---|---|---|---|---|
| **Enterobacteriaceae (continued)** | | | | | | |
| *Erwinia* | | | | | | |
| *rubrifaciens* | ICPB ER-103 | | | 52.0 | 74 | |
| | ICPB ER-104 | | | 52.6 | 74 | |
| | ICPB ER-105 | | | 52.6 | 74 | |
| *salicis* | ICPB ES-4 | 52.5 | | 51.3 | 74 | |
| | ICPB ES-102 | | | 51.5 | 74 | |
| *solanisapra* | ICPB ES-101 | | | 51.0 | 74 | |
| *tracheiphila* | ICPB ET-5 | | | 52.0 | 74 | |
| | ICPB ET-102 | | | 50.5 | 74 | |
| | ICPB ET-106 | 52.7 | | 50.0 | 74 | |
| *uredovora* | NCPPB 391 | 54.0 | | | 77 | |
| | NCPPB 800 | 53.0 | | | 77 | |
| | NCPPB 802 | 53.7 | | | 77 | |
| | NCPPB 1416 | 54.4 | | | 77 | |
| | NCPPB 802 | 53.9 | | 53.1 | 170 | 199 |
| unclassified | Corn stalk rotters: | | | | | |
| | ICPB EM-107 | | | 53.6 | 74 | |
| | ICPB EM-108 | | | 54.1 | 74 | |
| | ICPB EM-109 | | | 53.1 | 74 | |
| | ICPB EM-110 | | | 53.6 | 74 | |
| | ICPB 111 | | | 53.6 | 74 | |
| | ICPB EM-111 mutant | | | 53.6 | 74 | |
| | ICPB EM-112 | | | 54.1 | 74 | |
| | ICPB EM-209 | | | 53.6 | 74 | |
| | ICPB EM-210 | | | 53.6 | 74 | |
| | ICPB EM-210 mutant | | | 53.6 | 74 | |
| | *Erwinia*-like bacteria: | | | | | |
| | ICPB 2650-651 | | | 55.1 | 74 | |
| | ICPB 2650-653 | | | 54.1 | 74 | |
| | ICPB 2650-656 | | | 55.6 | 74 | |
| | ICPB 2650-658 | | | 53.1 | 74 | |
| | ICPB 2650-661 | | | 55.1 | 74 | |
| | ICPB 2650-662 | | | 54.6 | 74 | |
| | Man (Slotnick) ICPB 2858-A | | | 55.6 | 74 | |
| | Man (Slotnick) ICPB 2858-B | | | 55.1 | 74 | |
| | Man (Muraschi) ICPB 2953 | | | 56.6 | 74 | |
| | Man (Muraschi) ICPB 2956 | | | 54.1 | 74 | |

## TABLE 1 (Continued)
## GC COMPOSITION OF THE DNA OF BACTERIA

| Organism | Source or Strain[a] | Mole % Guanosine + Cytosine[b]: Tm | Chemical Analysis | Buoyant Density | References[c]: A | B |
|---|---|---|---|---|---|---|
| **Enterobacteriaceae (continued)** | | | | | | |
| *Erwinia* | | | | | | |
| unclassified | Man (von Graevenitz) ICPB 2984 | | | 60.2 | 74 | |
| | Man (von Graevenitz) ICPB 2986 | | | 59.7 | 74 | |
| | Man (von Graevenitz) ICPB 2987 | | | 56.1 | 74 | |
| | Man (Ewing) ICPB 2992 | | | 55.1 | 74 | |
| | Man (Ewing) ICPB 3080 | | | 55.6 | 74 | |
| | Leafhopper (Whitcomb) ICPB 2973-L | | | 52.6 | 74 | |
| | Leafhopper (Whitcomb) ICPB 2973-S | | | 53.6 | 74 | |
| | Deer (Muraschi) ICPB 2948 | | | 56.0 | 74 | |
| | Deer (Muraschi) ICPB 2949 | | | 55.1 | 74 | |
| | Deer (Muraschi) ICPB 2950 | | | 56.1 | 74 | |
| *Escherichia* | | | | | | |
| *coli* | | 50.0 | | | 10 | |
| | 15 TAU | | | 51.0 | 12 | |
| | F | | | 51.0 | 12 | |
| | 44B (ATCC 14763) | 50.0 | | 50–51 | 12, 76, 78 | |
| | 0111a | | | 51.0 | 12 | |
| | CA 38 | | | 51.5 | 12 | |
| | W | 50.0 | 51.6 | 50.0 | 8, 30, 40 | 40 |
| | B | 50–52.5 | 50–52.2 | 50–51 | 7, 12, 29, 30, 46, 78, 79, 80 | 16, 46, 47, 58, 73, 79, 81, 82, 83 |
| | C | 50 | 52.2 | 51 | 12, 29, 30 | |
| | 5 strains | 50.5 | | | 75 | |
| | CK | | 52.2 | | 29 | |
| | CCM 1367 | 52.2 | | | 54 | |
| | K 12, K 12 (λ) K 12 HFr 66, K 12 IIIc | 49.8–51.5 | 50.1 | 50–51 | 7, 8, 12, 30, 78 | 44, 60, 73 |
| | ATCC 9723d | 50.0 | | 50.0 | 76, 78 | |
| | ATCC 4157 | 50.0 | | 50.0 | 78 | |
| | T 94M | 49.8 | | 50.0 | 76, 78 | |
| | | | 49.8 | | 8 | |
| | B/r | | 50.3 | | 8 | |

## TABLE 1 (Continued)
## GC COMPOSITION OF THE DNA OF BACTERIA

| Organism | Source or Strain[a] | Mole % Guanosine + Cytosine[b] Tm | Chemical Analysis | Buoyant Density | References[c] A | B |
|---|---|---|---|---|---|---|
| **Enterobacteriaceae (continued)** | | | | | | |
| *Escherichia* | | | | | | |
| *coli* | 5 | 50.5 | | | 75 | |
| | ROW(col-) | | | 50.5 | 78 | |
| | UQ | | 50.5 | | 8 | |
| | ATCC 11775 | 51.7 | | 51.0 | 74, 199 | |
| | 12317(E2) | 50.4 | | 50.7 | 76, 78 | |
| | ROW(17)Str-r | | | 51.0 | 78 | |
| | I | | 52.2 | | 8 | |
| | SM | | 52.5 | | 8 | |
| | ATCC 8739 | 55.0 | 50.0 | | 49 | |
| | 5610-52 | 50.6 | | 50.6 | 31, 76 | |
| | 5611-52 | 50.6 | | 50.6 | 31, 76 | |
| | 5619-52 | 50.0 | | | 30 | |
| | K 12 | 51.5 | | | 183 | 183 |
| | K 12 | 51.95 | | | 194 | 199 |
| | W | | 51.7 | | 199 | 200 |
| | 44 B | | | | 199 | |
| | B | 52.2 | 50.9 | | 199 | 201 |
| | NCIB 8545 | | 50.0 | | 202 | |
| *coli (aurescens)* | ATCC 12814 | 51.3 | | 49.6–51 | 74, 76 | |
| *freundii* | ATCC 8090 | 52.7 | | 52.5 | 199 | |
| *intermedium* | E1-1 | | | 51.0 | 76 | |
| *Klebsiella* | | | | | | |
| *aerogenes* | | | 52–54 | | 5 | |
| *edwardsii* | | | 52–54 | | 5 | |
| *edwardsii* (var. *atlantae*) | ATCC 13887 | 56.8 | | 56.1 | 199 | |
| | ATCC 13886 | 56.6 | | 56.6 | 199 | |
| *edwardsii* sp. *atlantae* | ATCC 13887 | 56.9 | | 56.1 | 74 | |
| *edwardsii* sp. *edwardsii* | ATCC 13886 | 56.5 | | 56.6 | 74 | |
| *enterogenes* | BUMC | | | 57.1 | 12 | |
| *pneumoniae* | BUMC | | | 56.6 | 12 | |
| | 12 strains, including ATCC 10273 and ATCC 9997 | 54.6–55.8 | | | 76 | |
| | ATCC 13883 | 57.6 | | 53.4 | 74 | |
| | | 55.0 | | 56.0 | 30, 31 | |

# TABLE 1 (Continued)
## GC COMPOSITION OF THE DNA OF BACTERIA

| Organism | Source or Strain[a] | Mole % Guanosine + Cytosine[b] | | | References[c] | |
|---|---|---|---|---|---|---|
| | | Tm | Chemical Analysis | Buoyant Density | A | B |
| **Enterobacteriaceae (continued)** | | | | | | |
| *Klebsiella* | | | | | | |
| *pneumoniae* | 8821/6 | 54.2 | | | 76 | 73 |
| | 1 | 60.8 | | | 174 | |
| | 1(anaerogenic strain) | 62.5 | | | 174 | |
| | 10 | 59.8 | | | 174 | |
| | 11 | 61.8 | | | 174 | |
| | 13 | 59.8 | | | 174 | |
| | 17 | 59.2 | 55.8[e] | | 174 | |
| | 17(anaerogenic strain) | 60.8 | | | 174 | |
| | 21 | 58.1 | | | 174 | |
| | 26 | 56.7 | | | 174 | |
| | 39 | 59.8 | 60.4[e] | | 174 | |
| | 69 | 59.4 | | | 174 | |
| | 23 | 56.6 | | | 199 | |
| | ATCC 13883 | 57.6 | | 53.9 | 199 | |
| *rhinoscleromatis* | ATCC 13884 | 57.1 | | 55.6 | 199 | |
| | CDC 3099-63 | 60.4 | | | 174 | |
| | CDC 3101-63 | 60.8 | | | 174 | |
| | CDC 3105-63 | 58.9 | | | 174 | |
| | CDC 3106-63 | 58.9 | | | 174 | |
| | CDC 3107-63 | 60.8 | | | 174 | |
| | | | 54–56 | | 5 | |
| | ATCC 13884 | 57.1 | | 55.6 | 74 | |
| *rubiacearum* | | 55.5 | | | 60 | |
| species | Griffith | 57.9 | 58.7[e] | | 174 | |
| | Silver | 59.8 | 59.8[e] | | 174 | |
| *Paracolobactrum* | | | | | | |
| *aerogenoides* | MKC | 55.0 | | 54.0 | 30, 31 | |
| *(Bacterium) paracoli* | | 54.2 | 57.8 | | 45 | |
| | Mutant 52-1 | 75.5 | 75.4 | | 45 | |
| | Mutant Cu-15 | 54.4 | | | 45 | |
| | Mutant Bu-12 | 54.0 | | | 45 | |
| *Pectobacterium* | | | | | | |
| *aroideae* | ICPB EA-14 | 54.1 | | | 199 | |
| *carotovora* | ICPB EC-138 | 53.4 | | 52.0 | 199 | |
| *chrysanthemi* | ICPB EC-16 | 55.6 | | 55.1 | 199 | |
| *dissolvens* | ICPB ED-106 | 57.1 | | 57.1 | 199 | |

## TABLE 1 (Continued)
## GC COMPOSITION OF THE DNA OF BACTERIA

| Organism | Source or Strain[a] | Mole % Guanosine + Cytosine[b] Tm | Chemical Analysis | Buoyant Density | References[c] A | B |
|---|---|---|---|---|---|---|
| **Enterobacteriaceae (continued)** | | | | | | |
| *Pectobacterium* | | | | | | |
| *nimipressuralis* | ICPB EN-1 | 55.4 | | 55.1 | 199 | |
| *rhapontici* | ICPB ER-1 | 51.0 | | 51.0 | 199 | |
| *Proteus* | | | | | | |
| *anindologenes* | NRRL B-418 | | | 39.8 | 12 | |
| *mirabilis* | 9 | 39.5 | | | 84 | 84 |
| | L-form 9 (ATCC 14168) | 39.5–39.7 | | | 56, 84 | 56, 84 |
| | NRRL B-400 | | | 39.8 | 12 | |
| | 10 strains | 39.3 ± 1.4 | | | 75 | |
| | 35 | 38.0 | | 41.0 | 30, 31 | |
| | ATCC 14273 | 39.5 | | | 56 | 56 |
| *morganii* | | | 53.0 | | 8 | |
| | ATCC 8019 (NCIB 8168) | 51.5 | | 51.0 | 30, 31 | |
| | BUMC | | | 50.0 | 12 | |
| | 10 strains | 50 ± 0.7 | | | 75 | |
| *pseudovaleriae* | NRRL B421 | | | 39.8 | 12 | |
| *rettgeri* | 3478 | 39.5 | | 42.0 | 30, 31 | |
| | 10 strains | 39 ± 1.5 | | | 75 | |
| *vulgaris* | BUMC | | | 39.8 | 12 | |
| | 10 strains | 39.3 ± 1.2 | | | 75 | |
| | 2 strains | | 36.5–40.5 | | 8 | |
| | ATCC 9484 | 37.0 | | 39.0 | 30, 31 | |
| | | | 38–40 | | 5 | |
| | ATCC 4669 | | 38–40 | | 5, 73 | 73 |
| *Providencia* | | | | | | |
| species *(Proteus inconstans)* | | | 40–42 | | 5 | |
| | 29911 | 41.5 ± 0.6 | | | 75 | 60 |
| *Salmonella* | | | | | | |
| *abony* | | | 50–52 | | 5 | |
| *arizona* | PC 145 | 50.0 | | 53.0 | 30, 31 | |
| | 10 strains | 51.0 | | | 75 | |
| *ballerup* | | 51.5 | | | 30 | |
| | Bethesda, 3 strains | 50.7 | | | 75 | |

# TABLE 1 (Continued)
# GC COMPOSITION OF THE DNA OF BACTERIA

| Organism | Source or Strain[a] | Mole % Guanosine + Cytosine[b] Tm | Chemical Analysis | Buoyant Density | References[c] A | B |
|---|---|---|---|---|---|---|
| **Enterobacteriaceae (continued)** | | | | | | |
| *Salmonella* | | | | | | |
| *enteritidis* | 2 strains | | 49.9–50 | | 8 | |
| *gallinarum* | | | 50.2 | | 8 | |
| | University of Massachusetts | | | 52.5 | 12 | |
| *heidelberg* | University of Massachusetts | | | 52.0 | 12 | |
| *paratyphi* | | | 50–52 | | 5 | |
| | A | | 49.9 | | 8 | |
| *pullorum* | | | | 53.1 | 12 | |
| | | 51.0 | | | 85 | |
| *typhi (typhosa)* | 643, ETS 9 | 50.0 | | 51.5–52 | 12, 30, 31 | 60 |
| | T 5501 | | 53.1 | | 8, 29 | |
| *typhi abdominalis* | 319 | | 52.1 | | 8 | |
| *typhimurium* | LT 2Ath C-5(B) | 50.6 | | | 76 | |
| | LT 2Cys(K) | 50.6 | | | 76 | |
| | LT 2 | 51.5 | | 52.5 | 12, 30 | 60 |
| | | | 50.2 | | 8 | |
| | | | 54.0 | | 8 | |
| | 8407 | | 53.8 | | 8 | |
| *Serratia* | | | | | | |
| *indica* | ATCC 4003 | 58–61.7 | | | 76 | |
| *kiliensis* | ATCC 992 | 53.6–56.2 | | | 76, 86 | |
| *marcescens* | 3 strains | 58 | | | 75 | |
| | NIMA | 56.6 | | 59.2 | 12, 76 | |
| | HY | 58 | | 59.7 | 12, 76 | |
| | ATCC 274 (NCTC 1377) | 57.8–61.4 | 59.5 | 59.2 | 12, 76, 86, 87 | |
| | NCTC 1377 mutant | | 59.9 | | 87 | |
| | ATCC 990 | 58.2–62 | | | 76, 86 | |
| | ATCC 990 variant | 58.2–58.5 | | | 76, 86 | |
| | ATCC 13880 | 60.9 | | | 86 | |
| | | 57.5 | | 59.0 | 30, 31 | |
| | ATCC 4261 | 55.2–58.2 | | | 76, 86 | |
| | ATCC 4013 | 56.8–57 | | | 76, 86 | |
| | 293 | 59.9 | | | 79 | 79 |
| | 2 strains | | 58–59.2 | | 8 | |
| | | 58 | | | 10 | |

## TABLE 1 (Continued)
## GC COMPOSITION OF THE DNA OF BACTERIA

| Organism | Source or Strain[a] | Mole % Guanosine + Cytosine[b] Tm | Chemical Analysis | Buoyant Density | References[c] A | B |
|---|---|---|---|---|---|---|
| **Enterobacteriaceae (continued)** | | | | | | |
| *Serratia* | | | | | | |
| *marcescens* | 16 strains | 54.4–58.4 | | | 76, 86 | |
| *marcescens* var. *kiliensis* | ATCC 11111 | 58.8–62.7 | | | 76, 86 | |
| | ATCC 8101 | 58.4–62.2 | | | 76, 86 | |
| *marinorubra* | | 57.2 | | | 88 | |
| | | 59.9 | | | 86 | |
| *plymuthica (plymuthicum)* | ATCC 183 | 57.6–61.2 | | | 76, 86 | |
| *Shigella* | | | | | | |
| *dispar* | | | 50–52 | | 5 | |
| *dysenteriae* | | | 53.4 | | 8 | |
| | 15 | 50.0 | 53.4 | 51.0 | 8, 30, 31, 73 | 73 |
| *flexneri* | BUMC | | | 51.0 | 12 | |
| *paradysenteriae* | Flexner Y6R0 | | 49.3 | | 8 | |
| | Flexner Y6R4 | | 49.2 | | 8 | |
| | | | 50–52 | | 5 | |
| *sonnei* | | | 50–52 | | 5 | |
| | ATCC 11060 | | | 51.0 | 12 | |
| **Lactobacillaceae** | | | | | | |
| *Bifidobacterium* | | | | | | |
| *adolescentis* | 305(Reuter) | | 58.3 | | 15 | |
| | 195(R) | | 58.9 | | 15, 89 | |
| | 319(R) | | 59.2 | | 15, 89 | |
| | 305a(R) | | 60.4 | | 15, 89 | |
| | 298 (R) | | 61.8 | | 15, 89 | |
| *bifidum* | 319(R) | | 61.6 | | 15, 89 | |
| | 28a(R) | | 62.3 | | 15, 89 | |
| | 1620 E(Prévot) | | 62.0 | | 15, 89 | |
| *breve* | 1(R) | | 58.4 | | 15, 89 | |
| | 46(R) | | 62.3 | | 15, 89 | |
| *infantis* | 12(R) | | 60.8 | 57.9 | 15, 89, 90 | |
| *lactensis* | 441 (Dehnert) | | 58.3 | | 15, 89 | |
| | 659(D) | | 58.6 | 57.9 | 15, 89, 90 | |
| | 516/9(D) | | 59.3 | | 15, 89 | |

## TABLE 1 (Continued)
## GC COMPOSITION OF THE DNA OF BACTERIA

| Organism | Source or Strain[a] | Mole % Guanosine + Cytosine[b] Tm | Chemical Analysis | Buoyant Density | References[c] A | B |
|---|---|---|---|---|---|---|
| **Lactobacillaceae (continued)** | | | | | | |
| *Bifidobacterium* | | | | | | |
| *lactensis* | 566/M4(D) | | 59.9 | | 15, 89 | |
| | 426(D) | | 61.1 | | 15, 89 | |
| | 679(D) | | 61.8 | | 15, 89 | |
| | 456(D) | | 63.6 | 57.9 | 15, 89, 90 | |
| | 354(D) | | 64.2 | | 15, 89 | |
| *liberorum* | 29 IES(Negushi) | | 59.3–59.9 | | 15, 89 | |
| | 76e(R) | | 60.3 | | 15, 89 | |
| *longum* | CO(P) | | 57.2 | 57.9 | 15, 89, 90 | |
| | 2921(P) | | 57.2 | | 15, 89 | |
| | 63 BES(N) | | 58.6 | | 15, 89 | |
| | F6 IES(N) | | 60.1 | | 15, 89 | |
| | 3(R) | | 60.2 | | 15, 89 | |
| | 194(R) | | 61.4 | | 15, 89 | |
| *parvulorum* | 17c(R) | | 58.1 | | 15, 89 | |
| | 50(R) | | 59.4 | 57.9 | 15, 89, 90 | |
| *Catenabacterium* | | | | | | |
| *catenaforme* | 1141(P) | | 29.9 | | 15 | |
| | 1871(P) | | 32.7 | | 15 | |
| | 194(R) | | 44.0 | | 15 | |
| | 180(R) | | 46.8 | | 15 | |
| *Diplococcus* | | | | | | |
| *mucosus* | ATCC 17957 | 41 | | | 172 | 172 |
| *pneumoniae* | $R_6$-Ery$^r$-3 | 38.86 | 38.5 | | 213 | |
| | $R_6$-Ery$^r$-3 | 39.0 | | | 245 | |
| | $R_6$-Ery$^r$-3 | | 38.4 | | | 259 |
| *Lactobacillus* | | | | | | |
| *acidophilus* | 505 | | 47.2 | | 39 | |
| | NIRD 1 (NCTC 1723) | | 34.2 | 36.7 ± 0.7 | 90, 91 | |
| | NIRD 3 (ATCC 9857) | | 36.6 | 36.7 ± 0.7 | 90, 91 | |
| | 64N | | 34.9 | 36.7 ± 0.7 | 90, 91 | |
| | 61Z | | 36.7 | 36.7 ± 0.7 | 90, 91 | |
| | 65K | | 36.8 | 36.7 ± 0.7 | 90, 91 | |
| | 63E | | 34.4 | 36.7 ± 0.7 | 90, 91 | |
| | 10FB | 30.1 | | | 92 | |
| | 16FB | 32.6 | | | 92 | |
| | 9FB | 33.0 | | | 92 | |
| | NCIB 8795 | | 40.0 | | 93 | |

**TABLE 1 (Continued)**
**GC COMPOSITION OF THE DNA OF BACTERIA**

| Organism | Source or Strain[a] | Mole % Guanosine + Cytosine[b] Tm | Chemical Analysis | Buoyant Density | References[c] A | B |
|---|---|---|---|---|---|---|
| **Lactobacillaceae (continued)** | | | | | | |
| *Lactobacillus* | | | | | | |
| | | | 38–40 | | 5 | |
| *acidophilus* | Blechman | 39.0 | | 42.0 | 30, 31 | |
| | Farr | 50.2 | | | 210 | |
| *batatas* | | | 47.3 | | 39 | |
| *bifidus* | | | 57.6 | | 8 | |
| *brevis* | NCDO 473 | 45.5 | | | 94, 95, 96, 97 | |
| | ATCC 4006 | 42.0 | | | 95, 96 | |
| | ATCC 8287 | 42.3 | | | 95, 96, 97 | |
| | 1 ken | | 53.1 | | 39 | |
| | NIRD 190 (ATCC 8007) | | 43.2 | 42.7 | 90, 91 | |
| | V-7 (M. E. Sharpe) | | 43.9 | 46 ± 1 | 90, 91 | |
| | OC 737 (M. Rogosa) | | | 46.4 ± 1 | 90 | |
| | F 15 | 45.5 | | | 96 | |
| | F 22 | 42.3 | | | 96 | |
| | S 10 | 42.8 | | | 96 | |
| | F 3 | 44.0 | | | 96 | |
| | F 11 | 44.2 | | | 96 | |
| | S 22 | 44.7 | | | 96 | |
| | 10 SG | 34.5 | | | 92 | |
| | 5 SG | 36.5 | | | 92 | |
| | 7 SG | 36.8 | | | 92 | |
| | 4 SG | 37.0 | | | 92 | |
| | 3 SG | 37.4 | | | 92 | |
| | 4 I | 40.3 | | | 92 | |
| | 3 I | 42.0 | | | 92 | |
| | 1 I | 43.5 | | | 92 | |
| | 38 V | 44.0 | | | 92 | |
| | 18 V | 44.0 | | | 92 | |
| | 19 V | 44.7 | | | 92 | |
| | 40 V | 45.5 | | | 92 | |
| | 28 SB | 42.3 | | | 92 | |
| | 7 I | 42.3 | | | 92 | |
| | 16 | 42.5 | | | 92 | |
| | 20 V | 42.5 | | | 92 | |
| | 2 I | 43.1 | | | 92 | |
| | 5 I | 43.1 | | | 92 | |
| | X1 (NCDO 473) | 45.1 | | | 210 | |
| *buchneri* | BC1 (NCDO 110) | 44.6 | | | 210 | |
| | S 18 | 43.8 | | | 96 | |
| | F 6 | 45.0 | | | 96 | |
| | F 10 | 45.0 | | | 96 | |
| | NIRD 110 (ATCC 4005, NCIB 8007) | 43.9 | 42.6 | 44.8 ± 1.1 | 90, 91, 94, 95, 97 | |

# TABLE 1 (Continued)
# GC COMPOSITION OF THE DNA OF BACTERIA

| Organism | Source or Strain[a] | Mole % Guanosine + Cytosine[b] | | | References[c] | |
|---|---|---|---|---|---|---|
| | | Tm | Chemical Analysis | Buoyant Density | A | B |
| **Lactobacillaceae (continued)** | | | | | | |
| *Lactobacillus* | | | | | | |
| *buchneri* | NIRD 111 (ATCC 9460) | | 42.0 | 44.8 ± 1.1 | 90, 91 | |
| | OC 770 (M. Rogosa) | | | 44.8 ± 1.1 | 90 | |
| *bulgaricus* | | | 45.5 | | 39 | |
| | 1373 | 41.1 | | | 97 | |
| | NIRD 1489 (ATCC 11842) | | 49.4 | 59.3 ± 1 | 90, 91 | |
| | CNRZ 36 | | 48.3 | 50.3 ± 1 | 90, 91 | |
| | ATCC 12278(GA) | 38.3 | | | 210 | |
| *bulgaricus (jugurti)* | ATCC 7993 | 37.3 | | | 210 | |
| *casei* | ATCC 9595 | 45.8 | | | 210 | |
| | OSU | 47.5 | | | 210 | |
| | ATCC 7469 (NCDO 243) | 47.3 | | | 210 | |
| | C5 (NCDO 151) | 47.3 | | | 210 | |
| | 780 | 47.8 | | | 210 | |
| | 356 | 47.5 | | | 210 | |
| | 300 | 47.8 | | | 210 | |
| | 316 | 48.0 | | | 210 | |
| | 512 | | 47.2 | | 39 | |
| | F 14 | 47.0 | | | 96 | |
| | F 12 | 47.0 | | | 96 | |
| | 13 T | 48.0 | | | 96 | |
| *casei* var. *alactosus* | 18 SB | 47.4 | | | 92 | |
| | NCDO 680 | 47.7 | | | 94, 95, 97 | |
| | B 51 | | 49.3 | 46.4 ± 0.8 | 90, 91 | |
| *casei* var. *casei* | 11 LA | 47.0 | | | 92 | |
| | 13 LA | 47.0 | | | 92 | |
| | 12 LA | 46.0 | | | 92 | |
| | 14 LA | 46.0 | | | 92 | |
| | 17 V | 48.0 | | | 92 | |
| | 18 SB | 48.0 | | | 92 | |
| | NIRD 151 | 46–46.3 | 46.2 | 46.4 ± 0.8 | 90, 91, 94, 95, 97 | |
| | NIRD 155 | | 43.4 | 46.4 ± 0.8 | 90, 91 | |
| | NIRD 152 | | 47.0 | 46.4 ± 0.8 | 90, 91 | |
| | 61 BG 3 | | 45.5 | 46.4 ± 0.8 | 90, 91 | |
| | 65 M | | 49.4 | 46.4 ± 0.8 | 90, 91 | |
| *casei* var. *rhamnosus* | 64 H | | 47.4 | 46.4 ± 0.8 | 90, 91 | |
| | NCDO 243 | 47–47.4 | | | 94, 95, 97 | |
| | 7 SB | 47.2 | | | 92 | |

## TABLE 1 (Continued)
## GC COMPOSITION OF THE DNA OF BACTERIA

| Organism | Source or Strain[a] | Mole % Guanosine + Cytosine[b] | | | References[c] | |
|---|---|---|---|---|---|---|
| | | Tm | Chemical Analysis | Buoyant Density | A | B |
| **Lactobacillaceae (continued)** | | | | | | |
| *Lactobacillus* | | | | | | |
| *cellobiosus* | NIRD 927 (ATCC 11740) | 45.7 | 49.8 | 53.1 ± 0.8 | 90, 91, 94, 95, 97 | |
| | NIRD 928 (ATCC 11739) | | 52.5 | 53.1 ± 0.8 | 90, 91 | |
| | G1 (NCDO 927) | 51.2 | | | 210 | |
| *delbrueckii* | ATCC 9649 | 50.0 | | | 210 | |
| | NIRD hors catalogue D6 | | 48.6 | | 91 | |
| | NIRD hors catalogue D5 | 46.9 | | | 91 | |
| | GC 730 (M. Rogosa) | | | 50.0 | 90 | |
| | 1 | | 51.2 | | 39 | |
| | IMAM F-14 | 28.9 | | | 95 | |
| *enzymicum (Corynebacterium)*[f] | | | 39.0 | | 98 | |
| *fermenti* | NIRD 215 (ATCC 9338) | 51.5–51.8 | 49.5 | 53.4 ± 0.5 | 90, 91, 94, 97 | |
| | NIRD 335 | | 50.1 | 53.4 ± 0.5 | 90, 91 | |
| | OC 663 (M. Rogosa) | | | 53.4 ± 0.5 | 90 | |
| | IFO 3229 | | 53.5 | | 39 | |
| | F-1 (CNDO 215) | 51.9 | | | 210 | |
| *fermentum* | | | 52.4 | | 39 | |
| *helveticus* | NIRD 30 | | 36.3 | 39.3 | 90, 91 | |
| | NIRD 261 | | 37.1 | 39.3 | 90, 91 | |
| | NCDO 262 | 31.8 | | | 95, 97 | |
| | OSU | 38.3 | | | 210 | |
| | ATCC 10386 | 37.1 | | | 211 | |
| *homohiochii* | ATCC 15434 | | 46.0 | | 91 | |
| *inulinus (Sporolactobacillus)*[f] | | | 39.3 | | 39 | |
| *japonicus* | | | 45.0 | | 39 | |
| *jugurti* | NIRD 87 (ATCC 521, NCIB 2889) | | 36.5 | 39 ± 0.9 | 90, 91 | |
| | NIRD 99 | | 37.5 | 39 ± 0.9 | 90, 91 | |
| | NIRD 100 (ATCC 10812) | | 37.1 | 39 ± 0.9 | 90, 91 | |

## TABLE 1 (Continued)
## GC COMPOSITION OF THE DNA OF BACTERIA

| Organism | Source or Strain[a] | Mole % Guanosine + Cytosine[b] Tm | Chemical Analysis | Buoyant Density | References[c] A | B |
|---|---|---|---|---|---|---|
| **Lactobacillaceae (continued)** | | | | | | |
| *Lactobacillus* | | | | | | |
| *jugurti* | J-2 (M. E. Sharpe) | | 37.9 | 39 ± 0.9 | | 90, 91 |
| | NIRD 279 | 33.5 | | | | 97 |
| | ISM | 33.5 | | 95 | | |
| | ATCC 521 | 37.1 | | | | 210 |
| *lactis* | 8 C | 36.8 | | | | 92 |
| | NIRD 270, NCTC 7278 (ATCC 8000) | 35.7 | 48.3 | 50.3 ± 1.4 | | 90, 91, 95, 97 |
| | L-1 (M. E. Sharpe) | | 48.2 | 50.3 ± 1.4 | | 90, 91 |
| | 39-A | 49.0 | | | | 210 |
| *leichmannii* | ATCC 4797 | 50.5 | | | | 210 |
| | ATCC 7830 | 50.5 | | | | 210 |
| | 10 FB | 30.1 | | | | 92 |
| | 3 FB | 29.1 | | | | 92 |
| | 2 FB | 29.1 | | | | 92 |
| | 18 FB | 31.8 | | | | 92 |
| | 6 FB | 31.8 | | | | 92 |
| | NIRD 299 (ATCC 4797) | 31.7 | 38.1, 49.4 | 50.8 ± 0.5 | | 39, 90, 91, 95, 97 |
| | NIRD 302 (ATCC 7830) | | 49.2 | 50.8 ± 0.5 | | 90, 91 |
| *plantarum* | NIRD 343 (NCIB 7220) | 43.0 | 43.4 | 45 ± 1 | | 90, 91, 94, 95, 97 |
| | NIRD 773 | | 42.0 | 45 ± 1 | | 90, 91 |
| | P-4 (M.E. Sharpe) | | 44.8 | 45 ± 1 | | 90, 91 |
| | 63 N | | 42.6 | 45 ± 1 | | 90, 91 |
| | 64 L | | 43.1 | 45 ± 1 | | 90, 91 |
| | 61 D | | 43.0 | 45 ± 1 | | 90, 91 |
| | 11 | | 42.9 | | | 39 |
| | | | 44.0 | | | 29 |
| | ATCC 8292 | 43.5 | | | | 92, 95, 96 |
| | ATCC 8041 | 43.8 | | | | 95, 96 |
| | 17 strains | 40–44 | | | | 96 |
| | 53 strains | 38.5–46 | | | | 92 |
| | NCDO 343 | | 43.4 | | | 211 |
| | P-5 (NCDO 343) | 43.9 | | | | 210 |
| | 17-5, ATCC 8014 | 45.1 | | | | 210 |
| *plantarum* (var. *rudensis*) | NIRD 773 | | 42.0 | | | 211 |
| *sake* | | | 42.2 | | | 39 |

## TABLE 1 (Continued)
## GC COMPOSITION OF THE DNA OF BACTERIA

| Organism | Source or Strain[a] | Mole % Guanosine + Cytosine[b] Tm | Chemical Analysis | Buoyant Density | References[c] A | B |
|---|---|---|---|---|---|---|
| **Lactobacillaceae (continued)** | | | | | | |
| *Lactobacillus* | | | | | | |
| *salivarius* | 11-C | 28.3 | | | 92 | |
| | 12-C | 28.4 | | | 92 | |
| | 1-C | 28.9 | | | 92 | |
| | 10-C | 29.1 | | | 92 | |
| | 2-C | 29.4 | | | 92 | |
| | 12-FB | 29.1 | | | 92 | |
| | 6-C | 29.4 | | | 92 | |
| | 7-C | 29.9 | | | 92 | |
| | 3-C | 31.6 | | | 92 | |
| | 4-C | 31.8 | | | 92 | |
| | 9-C | 31.9 | | | 92 | |
| | 5-FB | 31.8 | | | 92 | |
| | 8-FB | 29.4 | | | 92 | |
| | NIRD 929 (ATCC 11742) | 28.2 | 36.6 | 34.7 ± 1.4 | 90, 91, 95, 97 | |
| | 61-AK | | 35.0 | 34.7 ± 1.4 | 90, 91 | |
| | 63-AI | | 33.0 | 34.7 ± 1.4 | 90, 91 | |
| | ATCC 11742 | 32.4 | | | 210 | |
| | 63-AJ | | 33.0 | | 211 | |
| *sili* | IMAM 13T | 48.0 | | | 95 | |
| *thermophilus* | T-1 | | 47.3 | | 39 | |
| *viridescens* | NIRD 403 | | 37.9 | 42.3 | 90, 91 | |
| | NCDO S40 (E3) | | 41.0 | 35.7 | 90, 91 | |
| | ATCC 12706 | 42.7 | | | 210 | |
| *xylosus* | | | 34.9 | | 39 | |
| unclassified | 3 strains | 34.8–35 | | | 96 | |
| unclassified *(Streptobacterium)*[f] | Group I (4 strains) | 40.2–41.1 | | | 92 | |
| | Group II (5 strains) | 44–44.7 | | | 92 | |
| | Group III (1 strain) | 38.8 | | | 92 | |
| | Group IV (1 strain) | 45.0 | | | 92 | |
| | Group V (4 strains) | 38.2–38.9 | | | 92 | |
| | Group VI (10 strains) | 34.5–35.9 | | | 92 | |
| unclassified *(Betabacterium)*[f] | Group I (4 strains) | 34.6–36.5 | | | 92 | |
| | Group II (1 strain) | 41.0 | | | 92 | |

## TABLE 1 (Continued)
## GC COMPOSITION OF THE DNA OF BACTERIA

| Organism | Source or Strain[a] | Mole % Guanosine + Cytosine[b] | | | References[c] | |
|---|---|---|---|---|---|---|
| | | Tm | Chemical Analysis | Buoyant Density | A | B |
| **Lactobacillaceae (continued)** | | | | | | |
| *Lactobacillus* | | | | | | |
| unclassified (*Betabacterium*)[f] | Group III (3 strains) | 35–35.9 | | | | 92 |
| | Group IV (1 strain) | 38.1 | | | | 92 |
| | Group V (1 strain) | 35.4 | | | | 92 |
| | Group VI (8 strains) | 35.5–40.6 | | | | 92 |
| | Group VII (1 strain) | 38.0 | | | | 92 |
| *Leuconostoc* | | | | | | |
| *dextranicum* | | 43.3 | | | | 99 |
| *mesenteroides* | B 07 | | 39.2 | | 39 | |
| | ATCC 12291 | 39.0 | | 42.0 | 30, 31 | |
| | 2 strains | | 40–41.1 | | 39 | |
| *Pediococcus* | | | | | | |
| *cerevisiae* | CCM 833 | | | 44.5 | 86 | |
| | ATCC 8081 | 35.1 | | | 104 | |
| *hennebergi* | | | 38.2 | | 39 | |
| *Ramibacterium* | | | | | | |
| *pleuriticum* | 3282 | | 51.6 | | 89 | |
| *ramosum* | | | 30.1 | | 8 | |
| *Streptococcus* | | | | | | |
| *anaerobius* | Pu 1937 | | 36.2 | | 100 | |
| | 1478-B | | 36.0 | | 100 | |
| | Pu 1076-C | | 35.7 | | 100 | |
| | Pu 10-2 | | 35.5 | | 100 | |
| | Pu 2116-A | | 35.4 | | 100 | |
| | 2711 | | 35.2 | | 100 | |
| | Pu 2010 | | 35.0 | | 100 | |
| | 2923 | | 34.7 | | 100 | |
| | Pr 2464 | | 34.7 | | 100 | |
| | Pu PV | | 34.6 | | 100 | |
| | 2906 A | | 34.5 | | 100 | |
| | Pu 580-D | | 34.3 | | 100 | |
| | Pu 2610 | | 33.7 | | 100 | |
| *bovis* | | | 38–40 | | 5 | |

## TABLE 1 (Continued)
## GC COMPOSITION OF THE DNA OF BACTERIA

| Organism | Source or Strain[a] | Mole % Guanosine + Cytosine[b]: Tm | Chemical Analysis | Buoyant Density | References[c] A | B |
|---|---|---|---|---|---|---|
| **Lactobacillaceae (continued)** | | | | | | |
| *Streptococcus* | | | | | | |
| *cremoris* | | | 38–40 | | 5 | |
| | C-3 | 41.5 | 40.2 | | 212 | |
| *evolutus* | 2668 | | 37.4 | | 100 | |
| | Pr 664 | | 36.9 | | 100 | |
| | 1909-B | | 36.9 | | 100 | |
| | Pr 1099 | | 36.4 | | 100 | |
| | 319-B | | 36.3 | | 100 | |
| *faecalis* | Group D | | 34.6 | | 8 | |
| | 2 strains | | 37–38.6 | | 8 | |
| | | | 34–36 | | 5 | |
| | D-10 | | | 38.0 | 31 | |
| *faecalis* (L-form) | G-KL (T53), ATCC 23240 | 38.2 ± 0.2 | | | 129 | |
| *faecium* | | | 34–36 | | 5 | |
| *faecium* (group D) (var. *durans*) | SF/$A_{kt}$ | 37.72 | | | 213 | |
| | SF/$A_k$ | 38.58 | | | 213 | |
| | SF/O | 36.8 | | | 213 | |
| *foetidus* | 330-B | | 36.2 | | 100 | |
| | 2897 | | 36.0 | | 100 | |
| | M 2661-B | | 36.0 | | 100 | |
| | 1643 | | 35.1 | | 100 | |
| | CZ III | | 34.5 | | 100 | |
| | E 3288 | | 34.0 | | 100 | |
| | 2217 | | 33.4 | | 100 | |
| | E 342-E | | 32.6 | | 100 | |
| | | | 33.6 | | 8 | |
| *intermedius* | 1346-D | | 39.0 | | 100 | |
| | Canada | | 39.0 | | 100 | |
| | 1877 | | 38.2 | | 100 | |
| *lactis* | | | 36.3 | | 39 | |
| *lactis*, *cremoris*, *diacetilactis* | 10 strains | 34.9–37.9 | | | 99 | |
| *lanceolatus* | 74 | | 33.7 | | 100 | |
| | Pu 2759 | | 33.2 | | 100 | |
| | F 580-E | | 31.6 | | 100 | |
| | 873 | | 30.7 | | 100 | |
| *liquefaciens* | | | 34–36 | | 5 | |

TABLE 1 (Continued)
GC COMPOSITION OF THE DNA OF BACTERIA

| Organism | Source or Strain[a] | Mole % Guanosine + Cytosine[b] Tm | Chemical Analysis | Buoyant Density | References[c] A | B |
|---|---|---|---|---|---|---|
| **Lactobacillaceae (continued)** | | | | | | |
| *Streptococcus* | | | | | | |
| *micros* | 783-C | | 31.0 | | 100 | |
| | 1810-A | | 30.5 | | 100 | |
| | 2125 | | 30.3 | | 100 | |
| | 2357-D | | 30.3 | | 100 | |
| | 2204-A | | 30.1 | | 100 | |
| | 2322-A | | 29.8 | | 100 | |
| | 1700-B | | 29.7 | | 100 | |
| | 2261-B | | 29.3 | | 100 | |
| | 2436 | | 29.0 | | 100 | |
| | 783-B | | 28.9 | | 100 | |
| | 712 | | 28.8 | | 100 | |
| | 1685-A | | 28.8 | | 100 | |
| | 1831 | | 28.6 | | 100 | |
| | A 2734-C | | 28.5 | | 100 | |
| | A 1063-B | | 28.4 | | 100 | |
| | Pa 1372 | | 28.3 | | 100 | |
| | 1641-B | | 28.2 | | 100 | |
| | 1936 | | 27.9 | | 100 | |
| | 1970 | | 27.9 | | 100 | |
| *mucosus* | ATCC 17957 | 41 | | | 172 | |
| *mutans* | NCTC 10449 H. V. Jordan | 37.9 | | | 214 | |
| | GS 5 R. J. Gibbons | 37.7 | | | 214 | |
| | Ingbritt B. Krasse | 37.1 | | | 214 | |
| *parvulus* | M 1311 | | 45.9 | | 100 | |
| | 970 | | 44.5 | | 100 | |
| *pneumoniae* | TYP III | | 38.5 | | 8 | |
| *(Diplococcus)* | R 36-A (ATCC 11733) | 39.0 | 42.0 | | 30, 31 | |
| *pneumoniae* | $R_6$-$Ery^r$-3 | 39.0, 38.86 | 38.5, 38.4 | | 213, 245, 251 | |
| *productus* | 2397 | | 44.8 | | 100 | |
| | 2398 | | 44.2 | | 100 | |
| | 2396 | | 43.5 | | 100 | |
| | Lev (T) | | 43.2 | | 100 | |
| *putridus* | M 1782 | | 34.0 | | 100 | |
| | M 1086-A | | 33.9 | | 100 | |
| | 2366 | | 33.1 | | 100 | |
| | 1861 | | 32.8 | | 100 | |
| | 588-E | | 32.6 | | 100 | |
| | 2859 | | 32.5 | | 100 | |
| | 1529 | | 30.8 | | 100 | |

## TABLE 1 (Continued)
## GC COMPOSITION OF THE DNA OF BACTERIA

| Organism | Source or Strain[a] | Mole % Guanosine + Cytosine[b] Tm | Chemical Analysis | Buoyant Density | References[c] A | B |
|---|---|---|---|---|---|---|
| **Lactobacillaceae (continued)** | | | | | | |
| *Streptococcus* | | | | | | |
| *pyogenes* | | | 38–40 | | 5 | |
| | Group A | | 33.6–39.5 | | 8, 93 | |
| *pyogenes* (L-form) | Richards, L., ATCC 19563 | 37.2 ± 0.3 | | | 129 | |
| *salivarius* | I-R 14-Sm[r] | 39.0 | | 42.0 | 30, 31 | |
| *sanguis* | | | 38–40 | | 5 | |
| | 10556 I. L. Shklair | 45.5 | | | 214 | |
| | K-1-R J. M. Tanzer | 45.2 | | | 214 | |
| | SL 1 J. M. Tanzer | 45.1 | | | 214 | |
| | E 49 J. M. Tanzer | 43.7 | | | 214 | |
| | BHT D. D. Zinner | 43.4 | | | 214 | |
| | 167 I. L. Shklair | 43.2 | | | 214 | |
| | FA-1 J. M. Tanzer | 42.2 | | | 214 | |
| *thermophilus* | M 11 | 30.5 | | | 101 | |
| | NIRD 489 | 31.7 | | | 101 | |
| | Pc | 32.3 | | | 101 | |
| | NIRD 573 | 33.5 | | | 101 | |
| | P 6 | 33.5 | | | 101 | |
| | P 1 | 36.0 | | | 101 | |
| | G | 37.2 | | | 101 | |
| | GS | 38.4 | | | 101 | |
| | F 7 | 38.4 | | | 101 | |
| *viridans* | | | 38–40 | | 5 | |
| *zymogenes* | Group D | | 33.6 | | 8 | |
| species | MG 9 | 39.3 | | | 84 | 84 |
| | Group G, D 166-B | | | 41.0 | 31 | |
| | Group A | | 34.5 | | 102 | |
| | Group A (L-form) | | 33.5 | | 102 | |
| | | | 33.5 | | 102 | |
| | Brunol L-form, ATCC 19617 | 37.2 ± 0.1 | | | 129 | |
| species (group H) | Challis | 36.27 | | | 212 | |
| | Challis | | 36.4 | | 215 | |

TABLE 1 (Continued)
GC COMPOSITION OF THE DNA OF BACTERIA

| Organism | Source or Strain[a] | Tm | Chemical Analysis | Buoyant Density | References[c] A | References[c] B |
|---|---|---|---|---|---|---|
| | | Mole % Guanosine + Cytosine[b] | | | | |
| **Leucotrichaceae** | | | | | | |
| *Leucothrix* | | | | | | |
| *mucor* | 1 strain | | | 49.5 | 103 | |
| | 5 strains | | | 49.0 | 103 | |
| | 3 strains | | | 48.5 | 103 | |
| | 1 strain | | | 48.0 | 103 | |
| | 1 strain | | | 46.9 | 103 | |
| | 19 strains, world-wide collection | 48–51 | | | 216 | |
| | 16 strains San Juan Island collection | 48.5–51.0 | | | 216 | |
| | | | 49 | | 267 | |
| **Micrococcaceae** | | | | | | |
| *Aerococcus* | | | | | | |
| *catalasicus* | $C_6$ O. G. Clausen | 39.4 | | | 104 | |
| | $C_9$ O. G. Clausen | 38.9 | | | 104 | |
| *viridans* | $C_2$ O. G. Clausen | 32.0 | | | 104 | |
| | $C_7$ O. G. Clausen | 40.3 | | | 104 | |
| | ATCC 11563 | 38.5 | | | 104 | |
| species | 301 A. C. Baird-Parker | 38.8 | | | 104 | |
| | 302 A. C. Baird-Parker | 37.6 | | | 104 | |
| | 303 A. C. Baird-Parker | 40.9 | | | 104 | |
| *Gaffkya* | | | | | | |
| *homari* | ATCC 10400 (holotype) | 34.0 | | | 105 | |
| *tetragena* (*Staphylococcus* sp.) | ATCC 10875 | 35.0 | | | 105 | |
| *tetragena* (*Micrococcus* sp.) | ATCC 6007 | 64.0 | | | 105 | |
| *Micrococcus* (*Staphylococcus*) | | | | | | |
| (*S.*) *afermentans* | CCM 855 | 73.3 | | | 75 | |
| | NCTC 7563 | 72.8 | | | 106 | |
| | NCTC 7503 | 72.1 | 74.3 | | 45 | |
| | NCTC 7503 (Mutant 44) | 72.1 | 72.8 | | 45 | |

**TABLE 1 (Continued)**
**GC COMPOSITION OF THE DNA OF BACTERIA**

| Organism | Source or Strain[a] | Mole % Guanosine + Cytosine[b] Tm | Chemical Analysis | Buoyant Density | References[c] A | B |
|---|---|---|---|---|---|---|
| **Micrococcaceae (continued)** | | | | | | |
| *Micrococcus (Staphylococcus)* | | | | | | |
| *(S.) afermentans* | NCTC 7503 (Mutant 22) | 67.8 | 70.0 | | 45 | |
| | NCTC 2665 | 72.8–73.3 | | | 106, 107 | |
| | NCTC 7495(A) | 69.0 | | | 105 | |
| | NCTC 7495(B) | 69.0 | | | 105 | |
| *aquivivus* | CCM 316, ATCC 14404 | 47.5–51.0,[g] 51.2 | | | 66, 108 | |
| *(S.) aureus* | Mutant UV-2 | | 71.0 | | 196 | |
| | Mutant UV-16 | | 70.9 | | 196 | |
| | Mutant UV-15 | | 62.9 | | 196 | |
| | NCIB 8625 | | 37.7 | | 189 | |
| | 209 | | 32.4 | | 196 | |
| *calco-aceticus* | ATCC 23055 | | | | 172 | 172 |
| *candidans (S. epidermidis)* | CCM 600 | 36.4 | 37 | | 217 | |
| *cerificans* | | 40 | | | 172 | 172 |
| *conglomeratus* | CCM 2087, CCM 825, CCM 547, ATCC 401 | 64.0–68.5,[g] 63.0–68.9, 69.5 | | | 66, 105, 109 | |
| | CCM 740 | 53.5–58.5,[g] 58.6 | | | 66, 108 | |
| *conglomeratus (M. lactis)* | CCM 2135, ATCC 19102, IAM 1470 | 72.0 | 69 | | 217 | |
| | CCM 2189, OUT 8092 | 70.2 | 70 | | 217 | |
| *conglomeratus (M. luteus)* | CCM 2134, ATCC 19101, IAM 1459 | 71.4 | 68 | | 217 | |
| | CCM 2137, IAM 1480 | 71.2 | 68 | | 217 | |
| | CCM 2136, IAM 1448 | 70.6 | 69 | | 217 | |
| *coraltinus* | ATCC 966 | 69.0 | | | 112 | |
| *cryophilus (Pediococcus cryophilus)* | CCM 900 (ATCC 12226) | 41.5–44.5,[g] 41.3 | | | 66, 108 | |

# TABLE 1 (Continued)
# GC COMPOSITION OF THE DNA OF BACTERIA

| Organism | Source or Strain[a] | Mole % Guanosine + Cytosine[b] Tm | Chemical Analysis | Buoyant Density | References[c] A | B |
|---|---|---|---|---|---|---|
| **Micrococcaceae (continued)** | | | | | | |
| *Micrococcus (Staphylococcus)* | | | | | | |
| *cryophilus* | | | 38–40 | | 5 | |
| *cyaneus* | CCM 856 | 67.0 | | | 110 | |
| | L.A.1.1. | 63.0 | | | 105 | |
| *denitrificans* | CCM 982 (ATCC 13543) | 64.0–65.8 | | | 66, 107, 108 | |
| | CCM 1396 | 66.5–67.0,[g] 66.3 | | | 66, 107, 108 | |
| *eucinetus (Planococcus eucinetus)* | XQ 11(2387) | 40.0 | | | 113 | |
| *eucinetus* | CCM 2388 (XQ 58) | 47.5–48.0,[g] 48.0 | | | 66, 108 | |
| | CCM 2389 (XQ 40) | 49.0–50.5,[g] 50.3 | | | 66, 108 | |
| *flavocyaneus* | ATCC 8673 | 68.0 | | | 105 | |
| | L.A.8.1. (CCM 622) | 68.0 | | | 105 | |
| | L.A.8.2. | 68.0 | | | 105 | |
| | CCM 851 | 71.3 | | | 110 | |
| | CCM 853 | 72.0 | | | 110 | |
| | CCM2147 | 70.8 | | | 105, 110 | |
| *flavocyaneus (luteus)* | CCM 852 (L.E.9) | 64.0–71.3 | | | 105,110 | |
| *flavus* | CCM 210 | 72.8 | | | 107 | |
| | ATCC 10240 | 65.0 | | | 105 | |
| *freudenreichii* | ATCC 407 | 65.0 | | | 105 | |
| | ATCC 407 | | | 58 | 218 | |
| *glutamicus* | ATCC 13032 | 53.4 | | | 191 | |
| *halodenitrificans* | CCM 286 (ATCC 13511) | 65.0–66.0, 65.0–66.0,[g] 62.9–65.0 | 64–66 | | 66, 108, 5 | |
| *infirmus* | ATCC 14396 | 72.8 | | | 113 | |
| *(S.) lactis* | NCTC 1631 | 65.8 | | | 109 | |
| *litoralis (Sarcina morrhuae)* | CCM 2226 | 63.5–65.0,[g] 61.4–65.0 | | | 66, 108, 113 | |

TABLE 1 (Continued)
GC COMPOSITION OF THE DNA OF BACTERIA

| Organism | Source or Strain[a] | Mole % Guanosine + Cytosine[b] Tm | Chemical Analysis | Buoyant Density | References[c] A | B |
|---|---|---|---|---|---|---|
| **Micrococcaceae (continued)** | | | | | | |
| *Micrococcus (Staphylococcus)* | | | | | | |
| *luteus* | CCM 810 | 66.3 | | | 107 | |
| | CCM 132 | 67–71,[g] 71 | | | 66, 108 | |
| | CCM 1674 | 71.0–71.5, 71.5 | | | 66, 108 | |
| | CCM 149 | 73.0–74.5,[g] 74.6 | | | 66, 108 | |
| | CCM 409 | 72.0–75.5,[g] 72.0 | | | 66, 108 | |
| | ATCC 398 Subgroup 3b | 66.3 | | | 112 | |
| | CCM 1048 ATCC 7468, NCIB 8942, *Sarcina flava* | 73.4 | 70 | | 217 | |
| | CCM 166 | 73.4 | 71 | | 217 | |
| | NCTC 7503 | | 74.3 | | 196 | |
| | NCTC 7011 | | 69.0 | | 219 | |
| | Mutant 44 | | 72.8 | | 196 | |
| | Mutant 22 | | 70.0 | | 196 | |
| *luteus (Sarcina lutea)* | CCM 410 (ATCC 272) | 70.0–73.5,[g] 73.7 | | | 66, 108 | |
| | CCM 337 (ATCC 382) | 70.0–72.5,[g] 72.5 | | | 66, 108 | |
| | CCM 310 | 67.5 | | | 109 | |
| *luteus (lysodeikticus)* | CCM 169 (ATCC 4698, NCTC 2665) | 73.3–73.5 | | | 66 | |
| *luteus (ureae)* | CCM 840 ATCC 408 | 71–72,[g] 71.0 | | | 66, 108 | |
| *luteus (M. lactis)* | CCM 268, UEM Sar. 2/49, *Sarcina flava* | 71.8 | 72 | | 217 | |
| *luteus (M. species)* | CCM 310 | 70.3 | 70.0 | | 217 | |
| *lysodeikticus* | CCM 1335 | 72.0 | | | 107 | |
| | 2665 | | 72.0 | | 29 | |
| | | | 71.9 | | 8 | |
| | NRRL B287 | 72.0 | | | 30, 31 | |
| | WRAIR | | | 71.2 | 190 | |
| *maripuniceus* | ATCC 14399 | 72.8 | | | 113 | |

TABLE 1 (Continued)
GC COMPOSITION OF THE DNA OF BACTERIA

| Organism | Source or Strain[a] | Mole % Guanosine + Cytosine[b]: Tm | Chemical Analysis | Buoyant Density | References[c] A | B |
|---|---|---|---|---|---|---|
| **Micrococcaceae (continued)** | | | | | | |
| *Micrococcus (Staphylococcus)* | | | | | | |
| *morrhuae* | CCM 859 | | 57.1 | | 113 | |
| | CCM 537 | | 57.8 | | 113 | |
| | CCM 2224 | | 59.6 | | 113 | |
| | CCM 889 | | 61.0 | | 113 | |
| *mucilaginosus migula* | Pharyngeal isolates 2849/68, S4239/68 | | | 59.0 | 218 | |
| *nishinomiyanensis* (*M.* species) | CCM 2140, OUT 8094 | 67.8 | 68 | | 217 | |
| *(S.) pelagia* | ATCC 14408 | | | | 113 | |
| *perflavus (M. lactis)* | CCM 2139, OUT 8094 | 69.6 | 67 | | 217 | |
| *pulcher (M. lactis)* | CCM 2227, ATCC 15936 | 66.2 | 64 | | 217 | |
| *radiodurans* | CCM 1700 (ATCC 13939) | 60–66,[g] 69.0 | | | 66, 108 | |
| | CCM 1701 | 65.3–65.5 | | | 66, 108 | |
| | Wild type | | 65–68 | | 114 | |
| | $W_1$ | | 66.0 | | 114 | |
| *rhenaus* (var. *miyamizu*) (*M.* species) | CCM 2142, OUT 8100 | 70.5 | 70 | | 217 | |
| *S. roseus (rubens)* | CCM 679 (ATCC 186) | 68–69[g] | | | 66, 108 | |
| | CCM 633 (ATCC 412) | 70.8–72.8, 69–73[g] | | | 66, 108, 109 | |
| | CCM 385 (ATCC 185) | 68–71,[g] 71 | | | 66, 108 | |
| | CCM 560 (ATCC 179) | 70–71,[g] 71 | | | 66, 108 | |
| | CCM 837 (ATCC 416) | 69.7–70.0 | | | 66, 108 | |
| *(S.) roseus* | CCM 146 | 72.8 | | | 75 | |
| | NCTC 7509 | 62.0 | | | 105 | |
| | NCTC 7511 | 75.0 | | | 106 | |
| | NCTC 7512 | 72.8 | | | 106 | |
| *roseus (Sarcina erythromyxa)* | CCM 706 | 72.0 | | | 107 | |

**TABLE 1 (Continued)**
**GC COMPOSITION OF THE DNA OF BACTERIA**

| Organism | Source or Strain[a] | Mole % Guanosine + Cytosine[b] | | | References[c] | |
|---|---|---|---|---|---|---|
| | | Tm | Chemical Analysis | Buoyant Density | A | B |
| **Micrococcaceae (continued)** | | | | | | |
| *Micrococcus (Staphylococcus)* | | | | | | |
| *roseus* | CCM 1679 | 71.1 | 71.0 | | 217 | |
| | CCM 908 | 69.2 | 68.0 | | 217 | |
| | CCM 618 | 71.1 | 70.0 | | 217 | |
| | CCM 347 | 71.5 | 70.0 | | 217 | |
| | CCM 839 | 72.0 | 70.0 | | 217 | |
| | CCM 189 | 72.2 | 70.0 | | 217 | |
| | CCM 562 | 72.8 | 70.0 | | 217 | |
| | CCM 1145 | 73.6 | 71.0 | | 217 | |
| | CCM 570 | 73.8 | 73.0 | | 217 | |
| | CCM 691 | 73.3 | 70.0 | | 217 | |
| | CCM 2390 | 69.0 | 68.5 | | 217 | |
| | CCM 2131 | 68.4 | 68.9 | | 217 | |
| | CCM 1425 | 66.2 | 65.0 | | 217 | |
| *salivaris* (*M.* species) | CCM 2126, ATCC 14344 | 49.8 | 49 | | 217 | |
| *sedentarius* | ATCC 14392 | 69.3 | | | 113 | |
| *sodonensis* | CCM 144 (ATCC 11880) | 71.8 | | | 107 | |
| *varians* | B 2158 | | | 72.4 | 115 | |
| *varians (candidus)* | CCM 1044 | 68.0 | | | 116 | |
| *varians (citreus)* | CCM 1046 | 67.6 | | | 116 | |
| *varians (M. lactis)* | CCM 2132, ATCC 19099, IAM 1392 | 68.9 | 69 | | 217 | |
| | CCM 2133, ATCC 19100, IAM 1404 | 71.2 | 72 | | 217 | |
| *varians (S. lactis)* | CCM 884 (NCTC 7564) | 69.0–69.3 | | | 106, 116 | |
| | CCM 881 (NCTC 1631) | 65.8 | | | 109 | |
| *varians (M. luteus)* | CCM 2253 J. V. Bhat No. 34 | 74.4 | 72 | | 217 | |
| | CCM 418 | 71.5 | 71 | | 217 | |
| species | CCM 836 | 68.3 | | | 108 | |
| | ATCC 8456 | 36.4 | | | 112 | |
| | CCM 678 | 59.6 | | | 112 | |
| species (radioresistant) | | 68.0 | | | 105 | |

# TABLE 1 (Continued)
# GC COMPOSITION OF THE DNA OF BACTERIA

| Organism | Source or Strain[a] | Mole % Guanosine + Cytosine[b] Tm | Chemical Analysis | Buoyant Density | References[c] A | B |
|---|---|---|---|---|---|---|
| **Micrococcaceae (continued)** | | | | | | |
| *Micrococcus (Staphylococcus)* | | | | | | |
| species (*M. roseus*) | CCM 168 | 45.0–49.5,[g] 49.3 | | | 66, 108 | |
| | CCM 1405 | 49.5–50.5,[g] 49.3 | | | 66, 108 | |
| species (*Planococcus*) | NCMB 629 | 39.6 | | | 113 | |
| | NCMB 628 | 42.2 | | | 113 | |
| | NCMB 1491 | 42.2 | | | 113 | |
| *Sarcina* | | | | | | |
| *agilis* | | | 51.1 | | 117 | |
| *aurantiaca* | CCM 836 | 64.0–68.5[g] | | | 66 | |
| | CCM 686 | 68.0 | | | 107 | |
| *citrea* | CCM 248 | 73.3 | | | 107 | |
| *exigua (Micrococcus luteus)* | CCM 1569 | 73.8 | 74 | | 217 | |
| *flava* | | 65.0 | | | 105 | |
| | ATCC 540 | | 70.0 | | 43 | |
| | | | 68.6 | | 8 | |
| | CCM 309 | 73.3 | | | 107 | |
| *lutea* | | | 70.7 | | 117 | |
| | | | 63.9 | | 118 | |
| | | | 72.0 | | 8 | |
| | ATCC 381 | | | 73.5 | 119 | |
| | 26c | 68.0 | | | 30 | |
| | CCM 523 | 73.3 | | | 107 | |
| | ATCC 382 | 72.0 | | | 105 | |
| | 18-FL | | | 71.9 | 115 | |
| *marginata (Micrococcus luteus)* | CCM 265 | 72.8 | | | 107 | |
| *maxima* | | | 39.8 | | 117 | |
| | 11 | | | 28.6 | 119 | |
| | 48 | 29 | 29.5 | | 220 | |
| *oliva (Micrococcus varians)* | CCM 250 | 54.2 | | | 116 | |
| *subflava* | CCM 559 | 72.8 | | | 107 | |
| *ureae* | | | 38–40 | | 5 | |
| | CCM 1743 | 41.5 | 40.7 | | 41 | |

## TABLE 1 (Continued)
## GC COMPOSITION OF THE DNA OF BACTERIA

| Organism | Source or Strain[a] | Mole % Guanosine + Cytosine[b] Tm | Chemical Analysis | Buoyant Density | References[c] A | B |
|---|---|---|---|---|---|---|
| **Micrococcaceae (continued)** | | | | | | |
| *Sarcina* | | | | | | |
| *(Planosarcina)*[f] *ureae* | | | 44.7 | | 117 | |
| *(Sporosarcina)*[f] *ureae* | ATCC 6473 | 43.0 | | | 105 | 42 |
| | CCM 204 | 40.7 | | | 41 | |
| | CCM 634 | 40.0 | | | 41 | |
| | CCM 380 (NRRL B-286) | 41.3 | 41.2 | | 41 | |
| | CCM 752 (CCM 858) | 40.0–40.9 | | | 41, 43 | |
| | CCM 860 (L.E.1.1., NCIB 9150) | 41.3 | | | 41 | |
| | CCM 871 (NCIB 8681) | 40.0 | 41.3 | | 41 | |
| | ATCC 13881 (L.E.1.4., NCIB 9151) | 40.0 | | | 41 | 42 |
| | CCM 981 | 40.0 | 39.7 | | 41 | |
| | CCM 1466 | 40.7 | 43.3 | | 41 | |
| | CCM 1732 | 40.9 | 42.0 | | 41 | |
| *variabilis* | CCM 266 | 72.3 | | | 107 | |
| *ventriculi* | CCM 208 | 69.5 | | | 107 | |
| | | | 41.8 | | 117 | |
| | KV I/VI G | 28 | 30.5 | | 220 | |
| | AL | | | 39.6 | 119 | |
| *(Staphylococcus)* | | | | | | |
| *afermentans* | NCTC 7495 (C) | 32.0 | | | 105 | |
| | NCTC 4819 (A) | 35.0 | | | 105 | |
| | NCTC 4819 (B) | 35.0 | | | 105 | |
| | 7503 | 72.1 | 74.3 | | 45 | |
| | Mutant 44 | 72.1 | 72.8 | | 45 | |
| | Mutant 22 | 67.8 | 70.0 | | 45 | |
| *albus* | Q 12[c] | | | 34.7 | 115 | |
| | 442[c] | | | 35.7 | 115 | |
| | UT 174 | | | 35.7 | 115 | |
| | SX | | | 34.7 | 115 | |
| *(Micrococcus) asaccharolyticus* | | | 34.1 | | 8 | |
| *(Micrococcus) aureus* | SA-B | | 40.0 | | 8 | |

TABLE 1 (Continued)
GC COMPOSITION OF THE DNA OF BACTERIA

| Organism | Source or Strain[a] | Mole % Guanosine + Cytosine[b] Tm | Chemical Analysis | Buoyant Density | References[c] A | B |
|---|---|---|---|---|---|---|
| **Micrococcaceae (continued)** | | | | | | |
| *(Staphylococcus)* | | | | | | |
| *aureus* | | | 32–34 | | 5 | |
| | ATCC 6538P | | 37.5 | | 66 | |
| | ATCC 6538P | | 37.5 | | 66 | |
| | ATCC 14778 | | 31.0 | | 66 | |
| | NCTC 8345 | | 31.5 | | 66 | |
| | NCTC 4163 | 31.2 | | | 106 | |
| | NCTC 8532 | 31.2 | | | 106 | |
| | NCTC 6571 | 30.0 | | | 106 | |
| | NCTC 4136 | 32.8 | | | 106 | |
| | NRRL B313 | 34.0 | | 34.0 | 30, 31 | |
| | | 39.0 | | | 105 | |
| | FDA 209P | | 37.7 | 35.7 | 8, 115 | |
| | SCL I | | | 35.7 | 115 | |
| | 63A 1096 | | | 35.4 | 115 | |
| | 63A | | | 34.7 | 115 | |
| | 16851 | | | 35.7 | 115 | |
| | PS 3C | | | 35.4 | 115 | |
| | 3189 | | | 35.7 | 115 | |
| | BP17 | | | 34.7 | 115 | |
| | CCM 641 | 33.5–35.0 | | | 66 | |
| | CCM 127 | 30–40 | | | 66 | |
| | 909 | | 33.6 | | 29 | |
| | S 26 | 30.7 | | | 109 | |
| | NCTC 8511 | 31.1 | | | 109 | |
| | 23 strains, including ATCC 10831, ATCC 13679, ATCC 13680, NCTC 8511 | 32.4–35.1 | | | 120 | |
| | 209 | 33.2 | 32.4 | | 45 | |
| | Mutant UV 2 | 70.2 | 71.0 | | 45 | |
| | Mutant UV 15 | 70.1 | 69.2 | | 45 | |
| | Mutant UV 16 | 69.8 | 70.9 | | 45 | |
| | Smith, L. (ATCC 19640) | 32.4 ± 0.5 | | | 129 | |
| *aureus*[h] | Mutant UV 2 (ATCC 13679) | 67.6 | | | 120 | |
| *aureus (Micrococcus caseolyticus)* | ATCC 8460 | 35.0 | | | 105 | |
| *aureus (Micrococcus pyogenes)* | 209-P | | 37.7 | | 8 | |

## TABLE 1 (Continued)
## GC COMPOSITION OF THE DNA OF BACTERIA

| Organism | Source or Strain[a] | Mole % Guanosine + Cytosine[b] Tm | Chemical Analysis | Buoyant Density | References[c] A | B |
|---|---|---|---|---|---|---|
| **Micrococcaceae (continued)** | | | | | | |
| *(Staphylococcus)* | | | | | | |
| *aureus (Micrococcus varians)* | CCM 313 | 30.0–32.5,[g] 33.3 | | | 66, 116 | |
| | CCM 529 | 34.0–34.2 | | | 66, 116 | |
| *(M.) candicans* | ATCC 14852 | 30.0 | | | 105 | |
| | | | | 34.7 | 115 | |
| *(M.) caseolyticus* | ATCC 13548 | 35.0 | | | 105 | |
| *citreus (Micrococcus varians)* | CCM 520 (IH 70/48) | 31.8 | | | 109 | |
| *citreus (aureus)* | ATCC 4012 | 36.0 | | | 105 | |
| *epidermidis* | ATCC 155 | 35.0 | | | 105 | |
| | | | 35.1 | | 8 | |
| *epidermidis (Micrococcus cerolyticus)* | CCM 901 (ATCC-12559) | 35.6–36.4 | | | 108,116 | |
| *epidermidis* (*Micrococcus* subgroup 1) | CCM 2435, BP 1652, WO 289a | 36.4 | 38[i] | | 217 | |
| *epidermidis* (*Micrococcus* subgroup 5) | CCM 2436, BP 114, WO 223 | 30.7 | 33[i] | | 217 | |
| *epidermidis* (*Micrococcus* subgroup 6) | CCM 2434, BP 24, WO 216 | 30.7 | 33[i] | | 217 | |
| | CCM 2433, BP 8, WO 215 | 32.0 | 35[i] | | 217 | |
| *euryhalis (Micrococcus varians)* | CCM 315 | 33.8 | | | 116 | |
| | ATCC 14389 (2110) | 33.0 | | | 113 | |
| *(M.) halodurans* | 1789 | | 32.0 | | 113 | |
| *lactis* | NCTC 189 | 32.0 | | | 106 | |
| | NCTC 7944 | 36.5 | | | 106 | |
| | NCTC 1630(A) | 33.0 | | | 105 | |
| | NCTC 1630(B) | 31.0 | | | 105 | |
| *lactis (S. epidermidis)* | CCM 2429, NCTC 1630 | 30.9 | 33[i] | | 217 | |

TABLE 1 (Continued)
GC COMPOSITION OF THE DNA OF BACTERIA

| Organism | Source or Strain[a] | Mole % Guanosine + Cytosine[b] Tm | Chemical Analysis | Buoyant Density | References[c] A | B |
|---|---|---|---|---|---|---|
| **Micrococcaceae (continued)** | | | | | | |
| *(Staphylococcus)* | | | | | | |
| *lactis* (*Micrococcus* species) | CCM 2138, NCTC 7802 | 63.2 | 60[i] | | 217 | |
| | CCM 2430, NCTC 6575 | 65.8 | 68[i] | | 217 | |
| *lactis* *(Micrococcus lactis)* | CCM 2431, NCTC 7565 | 67.1 | 67[i] | | 217 | |
| | CCM 1395, Pohja No. 118 | 70.6 | 72[i] | | 217 | |
| | CCM 1414 Pohja No. 185 | 71.1 | 71[i] | | 217 | |
| | CCM 2432 NCTC 7567 | 71.2 | 70[i] | | 217 | |
| | CCM 1411 Pohja No. 219 | 72.0 | 70[i] | | 217 | |
| | CCM 2141, NCTC 8340 | 72.0 | 70[i] | | 217 | |
| *salivaris* | CCM 2392, CCM 2393 | | | 60.0 | 218 | |
| *saprophyticus* | NCTC 7292 | 31.6 | | | 106 | |
| | NCTC 7612 | 30.8 | | | 106 | |
| | NCTC 901 | 35.5–39.0[g] | | | 66 | |
| *(M.) ureae* | ATCC 13515 (G) | 31.0 | | | 105 | |
| | ATCC 13515 (W) | 31.0 | | | 105 | |
| *(M.) violagabriella* | ATCC 13238 | 30.0 | | | 105 | |
| | | | | 34.7 | 115 | |
| species *(Micrococcus luteus)* | ATCC 398 (W) | 34.0 | | | 105 | |
| **Micromonosporaceae** | | | | | | |
| *Micromonospora* | | | | | | |
| *chalcea* | S | | 72.8 | | 16 | 16 |
| | IPV 751 | 79.1 | | | 162 | |
| *coerulea* | | | 71.8 | | 8 | |
| *fusca* | 472 | | 71.2 | | 22 | |
| species | 401 Virginia Commonwealth University Collection | 73 | | | 177 | |

## TABLE 1 (Continued)
## GC COMPOSITION OF THE DNA OF BACTERIA

| Organism | Source or Strain[a] | Mole % Guanosine + Cytosine[b] | | | References[c] | |
|---|---|---|---|---|---|---|
| | | Tm | Chemical Analysis | Buoyant Density | A | B |
| **Mycobacteriaceae** | | | | | | |
| *Mycobacterium* | | | | | | |
| *avium* | 226 | 68.5 | | | 121 | |
| | P 14 | 69.0 | | | 121 | |
| | ATCC 15769 | 67.4 | | | 121 | |
| *butyricum* | CCM 2242 | 67.7 | | | 221 | |
| | CCM 2067 | 65.1 | | | 221 | |
| | CCM 1693 | 65.3 | | | 221 | |
| *dierenhoferi* | CAMP 7231 | 67.6 | | | 221 | |
| | CAMP 7227 | 66.1 | | | 221 | |
| *flavescens* | 272 | 65.4 | | | 121 | |
| | ATCC 14474 | 64.9 | | | 121 | |
| *fortuitum* | 201 | 65.1 | | | 121 | |
| | ATCC 6841 | 66.1 | | | 121 | |
| | ATCC 6841 | 64.6 | | | 221 | |
| | NCTC 2291 | 66.1 | | | 221 | |
| | Research Institute of T.B., Prague CSSR | 64.3 | | | 221 | |
| *intracellulare* | 294 | 67.3 | | | 121 | |
| | 708 | 68.6 | | | 121 | |
| *kansasii* | 427 | 64.1 | | | 121 | |
| | 448 | 65.3 | | | 121 | |
| | P 1 | 65.7 | | | 121 | |
| | ATCC 12478 | 65.8 | | | 121 | |
| | Veterans' Administration Hospital, Minneapolis, Minnesota | | | 68.5 | 19 | |
| *luteum* | | | 66.4–66.6 | | 29, 122 | |
| *marianum* | 899 | 68.5 | | | 121 | |
| *marinum* | 467 | 64.2 | | | 121 | |
| | 468 | 65.0 | | | 121 | |
| | CAMP 7209 | 64.5 | | | 221 | |
| | Dr. J. Mysak, OHES, Olomouc, CSSR | 63.9 | | | 221 | |
| *pellegrini* | Research Institute of T.B., Prague, CSSR | 63.3 | | | 221 | |

## TABLE 1 (Continued)
## GC COMPOSITION OF THE DNA OF BACTERIA

| Organism | Source or Strain[a] | Mole % Guanosine + Cytosine[b] Tm | Chemical Analysis | Buoyant Density | References[c] A | B |
|---|---|---|---|---|---|---|
| **Mycobacteriaceae (continued)** | | | | | | |
| *Mycobacterium* | | | | | | |
| *phlei* | 762 | 67.4 | | | 121 | |
| | ATCC 19249 | 69.9 | | | 121 | |
| | 5 (Ruth Gordon) | | 69.0 | | 16 | 16 |
| | A. Brodie | 66.0 | 66.4 | 73.0 | 8, 30, 31 | |
| | | | 68.7 | | 8 | |
| | Veterans' Administration Hospital, Minneapolis, Minnesota | | | 68.9 | 19 | |
| | CCM 1889 | 64.7 | | | 221 | |
| | CAMP 5040 | 67.7 | | | 221 | |
| | CCM 111 | 66.6 | | | 221 | |
| *rabinowitsch* | CCM 1690 | 66.3 | | | 221 | |
| *rhodochrous* | 305 | 67.9 | | | 121 | |
| | 355 | 70.0 | | | 121 | |
| | IMRU 502 | 68.7 | | | 266 | |
| *scrofulaceum* | 952 | 68.0 | | | 121 | |
| | ATCC 19981 | 67.9 | | | 121 | |
| *smegmatis* | ATCC 607 | 68.8 | | | 266 | |
| | 763 | 66.7 | | | 121 | |
| | ATCC 14468 | 67.8 | | | 121 | |
| | CCM 2228 | 66.4 | | | 221 | |
| | CCM 2235 | 67.5 | | | 221 | |
| | CCM 2300 | 67.6 | | | 221 | |
| *tuberculosis* | 6 strains | | 62–70.3 | | 8 | |
| | Veterans' Administration Hospital, Minneapolis, Minnesota | | | 64.9 | 19 | |
| | H 37Ra | 66.8 | | | 266 | |
| | H 37Rv | 65.0 | | | 121 | |
| | 731 | 64.3 | | | 121 | |
| *tuberculosis* BCG | Denmark | | 66.3 | | 123 | |
| | France | | 72.7 | | 123 | |
| | Japan | | 67.4 | | 123 | |
| | Brazil | | 67.4 | | 123 | |
| *xenopei* | ATCC 19276 | 65.5 | | | 121 | |
| | 1183 | 64.5 | | | 121 | |
| species (tap-water scotochromogen) | 905 | 65.9 | | | 121 | |
| | ATCC 19277 | 65.4 | | | 121 | |

## TABLE 1 (Continued)
## GC COMPOSITION OF THE DNA OF BACTERIA

| Organism | Source or Strain[a] | Mole % Guanosine + Cytosine[b] | | | References[c] | |
|---|---|---|---|---|---|---|
| | | Tm | Chemical Analysis | Buoyant Density | A | B |
| **Mycobacteriaceae (continued)** | | | | | | |
| *Mycobacterium* | | | | | | |
| species | ATCC 607 | | 67.9 | | 123 | |
| | CCM 1696 | 67.6 | | | 221 | |
| | CCM 2229 | 67.9 | | | 221 | |
| **Mycoplasmataceae** | | | | | | |
| *Mycoplasma* | | | | | | |
| *agalactiae* | Goat milk | 33.6 | | 34.2 | 124, 125 | |
| *arthritidis* | PG 27 | | | 37.0 | 128 | |
| | PG 6 | | | 37.0 | 128 | |
| | PG 6 (Preston), ATCC 19611 | 31.3 ± 0.2 | | | 129 | |
| | H 606 (ATCC 13988) | 30.0–31.9 | | 32.6 | 126 | |
| *arthritidis* (*hominis*, type 2) | Human 39 | 33.7 | | 31.6–36.0 | 126, 128 | |
| | Human 0 7 | | 46.5 | 32.1–36.0 | 126, 127, 128 | |
| *bovigenitalium* | | 33.0 | | | 130 | |
| | Bovine PG 11 (ATCC 14173) | 29.6 | | 28.1 | 126 | |
| | PG 11 (B2), ATCC 19852 | 30.4 ± 0.2 | | | 129 | |
| *bovirhinis* | Bovine PG 43 | | | 24.5 | 126 | |
| | PG 43 (5 M331), ATCC 19884 | 25.4 ± 0.3 | | | 129 | |
| *canis* | | 33.0 | | | 130 | |
| | Canine PG 14 | 28.4 | | 29.1 | 126 | |
| *felis* | CO (ATCC 23391) | 25.2 ± 0.2 | | | 129 | |
| *fermentans* | Human G | | | 27.6 | 125 | |
| | Human G II (ATCC 15474) | 27.8 | | | 126 | |
| | PG 18 (G), ATCC 19989 | 28.7 ± 0.4 | | | 129 | |
| *gallinarum* | Avian PG 16 (ATCC 15319) | 27.5–28.1 | | 26.5 | 56, 126 | 56 |
| | Fowl PG 16, ATCC 19708 | 27.0 ± 0.7 | | | 129 | |
| | | 27.7 | | | 130 | |
| | 54-537 | | | 27.0 | 126 | |

## TABLE 1 (Continued)
## GC COMPOSITION OF THE DNA OF BACTERIA

| Organism | Source or Strain[a] | Mole % Guanosine + Cytosine[b]: Tm | Chemical Analysis | Buoyant Density | References[c]: A | B |
|---|---|---|---|---|---|---|
| **Mycoplasmataceae (continued)** | | | | | | |
| *Mycoplasma* | | | | | | |
| *gallisepticum* | Avian A 5969 | 33.4–35.0 | 33.3 | 33–34 | 30, 31, 126, 131 | |
| | Avian ATCC 15302 (S6) | 32.0–32.7 | | 35.2 | 56, 126 | 56 |
| | Avian 801 | | | 35.7 | 126 | |
| | Avian PG 31 | | | 35.7 | 126 | |
| | PG 31 (X 95), ATCC 19610 | 31.8 ± 0.2 | | | 129 | |
| | 1010 | | | 35.7 | 126 | |
| | 1150 | | | 34.7 | 126 | |
| *gateae* | CS (ATCC 23392) | 28.5 ± 0.1 | | | 129 | |
| *granularum* | Porcine | | | 32.1 | 126 | |
| | Porcine BTS 39 ATCC 19168 | 30.4 | | | 126 | |
| *histotropicum* | Sabin type C, ATCC 23115 | 29.2 ± 0.6 | | | 129 | |
| *hominis*, type 1 | Human 4330 | | | 27.3 | 126 | |
| | Human H 34 (ATCC 15056) | 29.2 | | | 126 | |
| | PG 21 (H 50) (ATCC 23114) | 29.2 ± 0.1 | | | 129 | |
| | Human PPLO 4387 (ATCC 14027) | 28.7 | | | 126 | |
| *hominis*, type 2 | Human ATCC 14152 (Campo) | 31.7–33.9 | | | 126, 130 | |
| *hyorhinis* | Porcine BTS 7 (ATCC 17981) | 27.3 | | | 126 | |
| *iners* | Avian O | | | 29.6 | 126 | |
| | Avian PG 30 | | | 29.1 | 126 | |
| | PG 30 (M) (ATCC 19705) | 29.1 ± 0.2 | | | 129 | |
| *laidlawii* | PG 9 | | 33.3 | | 132 | |
| | PG 8 (Sewage A) (ATCC 23206) | 33.0 ± 0.1 | | | 129 | |
| | PG 9 (Sewage B) (ATCC 23217) | 32.4 ± 0.2 | | | 129 | |
| *laidlawii*, type A | Sewage 11 A (ATCC 14089; 544A) | 30.9–34.6 | 32.5 | 35.5–35.7 | 125, 126, 130, 132, 133 | |

## TABLE 1 (Continued)
## GC COMPOSITION OF THE DNA OF BACTERIA

| Organism | Source or Strain[a] | Mole % Guanosine + Cytosine[b] Tm | Chemical Analysis | Buoyant Density | References[c] A | B |
|---|---|---|---|---|---|---|
| **Mycoplasmataceae (continued)** | | | | | | |
| *Mycoplasma* | | | | | | |
| *laidlawii*, type B | ATCC 14192 (545 B) | 32.3–35.6 | 32.5–34.4 | 33.7 | 126, 130, 132, 134 | 134 |
| *maculosum* | Canine PG 15 | 26.7 | | 29.6 | 126 | |
| *meleagridis* | Avian 529 | | | 28.6 | 126 | |
| | Avian N | | | 28.1 | 126 | |
| *mycoides* var. *capri* | Goat PG 3 | | 24.8–25.0 | | 134, 135, 136 | 134 |
| *mycoides* var. *mycoides* | Bovine | 26.8 | | 26.5 | 124, 125 | |
| | Bovine V 5 | 26.1 | | | 126 | |
| | T 3 | | 30.0 | | 132 | |
| *neurolyticum* | Mouse KSA (ATCC 15099) | 23.0 | | | 126 | |
| | Mouse type A | | | 22.8 | 125 | |
| | Sabin type A (ATCC 19988) | 26.2 ± 0.1 | | | 129 | |
| *orale*, type 1 | Human 823 B (ATCC 15534) | 26.6 | | | 126 | |
| | CH 19299 (ATCC 23714) | 27.8 ± 0.4 | | | 129 | |
| *orale*, type 2 | Human DC 1600 | 26.4 | | | 126 | |
| | CH 20247 (ATCC 23636) | 26.1 ± 0.2 | | | 129 | |
| *pharyngis* | Human LGM (ATCC 19524) | 27.0 | | | 126 | |
| | Patt (ATCC 15544) | 23.9 ± 0.2 | | | 129 | |
| *pneumoniae* | Human Eaton Agent (ATCC 15377) | 38.6–39.3 | | 40.8–41.0 | 84, 124, 125, 126, 133 | |
| | FH (ATCC 15531) | 39.9 ± 0.4 | | | 129 | |
| *pulmonis* | Human KON (ATCC 14267) | 27.5 | | | 126 | |
| | PG 34 (Ash) (ATCC 19612) | 27.9 ± 0.3 | | | 129 | |
| *salivarium* | Human manure | 23.3 | | | 126 | |
| | PG 20 (H 110) (ATCC 23064) | 31.4 ± 0.1 | | | 129 | |
| *spumans* | Canine PG 13 | 28.4 | | 29.1 | 126 | |

## TABLE 1 (Continued)
## GC COMPOSITION OF THE DNA OF BACTERIA

| Organism | Source or Strain[a] | Mole % Guanosine + Cytosine[b]: Tm | Chemical Analysis | Buoyant Density | References[c] A | B |
|---|---|---|---|---|---|---|
| **Mycoplasmataceae (continued)** | | | | | | |
| *Mycoplasma* | | | | | | |
| *synoviae* | Avian 1853 | | | 34.2 | 126 | |
| species | Calf | 23.5–23.6 | | 26.5 | 125, 133 | |
| | Dog PG 24 (C 21, ATCC 23462) | 29.2 ± 0.1 | | | 129 | |
| | Goat kid | 24.0–24.1 | | 25.5 | 125, 133 | |
| | Goat C 30 (KS 1, ATCC 15718) | 28.9 ± 0.2 | | | 129 | |
| | GDL | 27.8 | | | 126 | |
| | Sheep (67-166, ATCC 23243) | 27.2 ± 0.2 | | | 129 | |
| | Human Negroni | 28.3 | | | 126 | |
| | Human (Navel, ATCC 15497) | 26.8 ± 0.4 | | | 129 | |
| | Bovine Donnetta | 32.9 | | | 126 | |
| | Avian 694 | | | 32.1 | 126 | |
| | Avian TU | | | 30.5 | 126 | |
| | Avian C | | | 29.1 | 126 | |
| | Avian PSU 4 | | | 28.6 | 126 | |
| | Avian NY | | | 28.1 | 126 | |
| *Mycoplasma* | | | | | | |
| *columnaris* | LP 8 | 29.8 | | | 255 | |
| | Ek 1 | 35.9 | | | 255 | |
| | NCMB 1038 | 30.3 | | | 255 | |
| *diphtheroid* | D 5 | | | 61.0 | 128 | |
| | Campo D | | | 64.0 | 128 | |
| L-form | D 5-L | | | 64.0 | 128 | |
| | Campo L | | | 64.0 | 128 | |
| species | Avian SA | | | 26.5 | 126 | |
| | Avian 693 | | | 25.5 | 126 | |
| | Avian R 49 | | | 25.0 | 126 | |
| | Avian 695 | | | 24.5 | 126 | |
| | Avian 1805 | | | 24.5 | 126 | |
| | Avian HPR 15 | | | 24.0 | 126 | |
| | Human T strains: | | | | | |
| | 7 (Ford) | 28.2 | | | 137 | |
| | 23 (Ford) | 28.0 | | | 137 | |
| | 27 (Ford) | 28.1 | | | 137 | |
| | 58 (Ford) | 28.5 | | | 137 | |
| | 354 (Ford) | 27.8 | | | 137 | |
| | Pirillo (Ford) | 27.8 | | | 137 | |
| | Cook (Ford) | 27.7 | | | 137 | |
| species (saprophyte) | Bovine PG 10 | | | 34.2 | 126 | |
| | Compost C 15 | | | 32.7 | 126 | |

**TABLE 1 (Continued)**
**GC COMPOSITION OF THE DNA OF BACTERIA**

| Organism | Source or Strain[a] | Mole % Guanosine + Cytosine[b] Tm | Chemical Analysis | Buoyant Density | References[c] A | B |
|---|---|---|---|---|---|---|
| **Mycoplasmataceae (continued)** | | | | | | |
| *Thermoplasma* | | | | | | |
| *acidophila* | ATCC 25905 | | | 25 | 222 | |
| **Myxococcaceae** | | | | | | |
| *Chondrococcus* | | | | | | |
| *coralloides* | University of Windsor strain M 25 | 67.6 | | | 27 | |
| | University of Windsor strain M 2 | 68.1 | | | 27 | |
| | 2 strains | 67.6–68.1 | | | 256 | |
| *Myxococcus (Myxobacterium)* | | | | | | |
| *fulvus* | University of Windsor strain M 17 | 67.4 | | | 27 | |
| | University of Windsor strain M 16 | 67.6 | | | 27 | |
| | Massachusetts | 70.5 | | 71.0 | 69 | |
| | California | 64.8 | | 69.0 | 69 | |
| | M 6 | 70.0 | | | 81 | 81 |
| | 4 strains | 67.4–71 | | | 256 | 258 |
| *virescens* | 5 strains | 67.6–70 | | | 256 | 258 |
| | | | 66–68 | | 5 | |
| | University of Windsor | 67.6 | | | 27 | |
| | New York | 71.0 | | 70.0 | 69 | |
| | California | 70.0 | | 69.0 | 69 | |
| | Wisconsin | 71.2 | | 69.0 | 69 | |
| | Massachusetts | 71.2 | | 68.0 | 69 | |
| *xanthus* | | | 66–68 | | 5 | |
| | University of Windsor strain M 23 | 67.1 | | | 27 | |
| | FB | 70.0 | | | 81 | 81, 138 |
| | | 68.5 | | | 17 | |
| | Massachusetts | 69.3 | | 68.0 | 69 | |
| | Wisconsin | 71.2 | | 70.0 | 69 | |
| | 3 strains | 67.1–70 | | | 256 | 258 |
| species | 2 strains | 69 | | | 258 | |
| | 2 strains | | | 69.0 | 69 | |

TABLE 1 (Continued)
GC COMPOSITION OF THE DNA OF BACTERIA

| Organism | Source or Strain[a] | Mole % Guanosine + Cytosine[b] Tm | Chemical Analysis | Buoyant Density | References[c] A | B |
|---|---|---|---|---|---|---|
| **Myxococcaceae (continued)** | | | | | | |
| *Sporocytophaga* | | | | | | |
| *myxococcoides* | | | 34–36 | | 5 | |
| | California | | | 36.0 | 69 | |
| | Massachusetts | 36.3 | | 36.0 | 69 | |
| | 2 strains | | | 36 | 258 | |
| species | M 5 | 41 | | | 256 | |
| | University of Windsor strain M 5 | 41.0 | | | 27 | |
| **Neisseriaceae** | | | | | | |
| *Acidaminococcus* | | | | | | |
| *fermentans* | VR 2 | | | 55.6 | 139 | |
| | VR 3 | | | 57.1 | 139 | |
| | VR 4 (ATCC 25085) | | | 56.6 | 139 | |
| | VR 5 | | | 56.6 | 139 | |
| | VR 6 | | | 56.1 | 139 | |
| | VR 13 | | | 56.1 | 139 | |
| | VR 14 (ATCC 25088) | | | 56.6 | 139 | |
| | VR 15 | | | 57.4 | 139 | |
| | VR 7 (ATCC 25086) | | | 56.6 | 139 | |
| | VR 8 | | | 56.6 | 139 | |
| | VR 9 | | | 56.6 | 139 | |
| | VR 11 (ATCC 25087) | | | 57.1 | 139 | |
| | VR 12 | | | 56.1 | 139 | |
| *Neisseria* | | | | | | |
| *catarrhalis* | NCTC 4103 | | 45.7 | 42.5 | 3, 140 | |
| | ATCC 8176 | | 42.3 | 42.0 | 3, 140 | |
| | 7269 | | 43.4 | | 140 | |
| | 1 VI A | | 43.4 | | 140 | |
| | 6666 | | 42.6 | | 140 | |
| | 12910/62 | | | 41.0 | 3 | |
| | Ne 11 | 41.0 | 40.7 | 41.0 | 3, 30, 141 | |
| | Ne 13 (11sTr-r) | | 40.1–41.3 | 42.0 | 31, 141 | |
| | ATCC 8193 | 46.7 | 42.25 | | 189 | 229 |
| | ATCC 8176 | 45.5 | | | 189 | 229 |
| | NIH | 41.95 | | 43.9 | 189 | 223 |
| *caviae* | 10293 (ATCC 14659) | | | 44.5 | 3 | |

## TABLE 1 (Continued)
## GC COMPOSITION OF THE DNA OF BACTERIA

| Organism | Source or Strain[a] | Mole % Guanosine + Cytosine[b] Tm | Chemical Analysis | Buoyant Density | References[c] A | B |
|---|---|---|---|---|---|---|
| **Neisseriaceae (continued)** | | | | | | |
| *Neisseria* | | | | | | |
| *caviae* | GP 13 | | 48.6 | | 140 | |
| | GP 11 | | 48.4 | | 140 | |
| | GP 16 | | 48.4 | | 140 | |
| | Q | | 48.4 | | 140 | |
| | R | | 47.7 | | 140 | |
| | GP 8 | | 50.4 | | 140 | |
| | J | | 50.2 | | 140 | |
| | N | | 50.2 | | 140 | |
| | GP 3 | | 50.1 | | 140 | |
| | F | | 49.6 | | 140 | |
| | GP 4 | | 49.3 | | 140 | |
| *cinerea* | 165/61, 159/62 | | | 49.0 | 3 | |
| *flava* | JJ IIA | | 49.5 | | 141 | |
| *flavescens* | ATCC 13120 | 49.0 | 49.2 | 47.0–47.5 | 3, 30, 31, 141 | |
| | ATCC 13115 | | | 47.0 | 3 | |
| | 8263 | | | 46.5 | 3 | |
| *gonorrhoeae* | | | 49.6 | | 8 | |
| | WRAIR 116 | 49.3 | 50.1 | 51.0 | 223 | |
| *meningitidis* | SD 6 | 52.9 | | 51.0 | 223 | |
| | Ne 15 (ATCC 23253) | 51.5 | 51.3 | 50.0 | 30, 31, 141 | |
| | | 51.1 | | 51.1 | 126 | |
| | | | 50.5 | | 8 | |
| *ovis* | 37/59, 917/60 | | | 44.5 | 3 | |
| | 199/55 | | | 45.0 | 3 | |
| *perflava* | 0799 | | 50.5 | | 140 | |
| | N 7 | | 49.2 | | 140 | |
| | Ne 20 | 49.0 | 49.8 | | 30, 141 | |
| | Ne 16 | | 50.3 | 48.0 | 31, 141 | |
| *sicca* | Ne 12 | 49.0 | 51.5 | 51.0 | 30, 31 141 | |
| *subflava* | ATCC 11076 | | 50.5 | | 141 | |
| *winogradskyi* | ATCC 17902 | 42 | | | 172 | 172 |
| *Veillonella* | | | | | | |
| *parvula* | | | 36.5 | | 8 | |
| | | | 34–36 | | 5 | |

# TABLE 1 (Continued)
# GC COMPOSITION OF THE DNA OF BACTERIA

| Organism | Source or Strain[a] | Mole % Guanosine + Cytosine[b] Tm | Chemical Analysis | Buoyant Density | References[c] A | B |
|---|---|---|---|---|---|---|
| **Nitrobacteraceae** | | | | | | |
| *Nitrobacter* | | | | | | |
| *agilis* | | | | 65.0 | 142 | |
| *Nitrosomonas* | | | | | | |
| *europaea* | | | | 52.0 | 142 | |
| | | 50–51 | 51.6 | | 143 | |
| | | 49.6 | | 52.0 | 228 | 224 |
| species | | | 54–56 | | 5 | |
| **Nocardiaceae** | | | | | | |
| *Microbispora (Waksmania)* | | | | | | |
| *rosea* | 567 | | 70.2 | | 22 | |
| | M 20 | | 71.1 | | 16 | 16 |
| | IMRU 3738 | 73.65 | | | 162 | |
| | RIA 567 | | 69.9 | | 175 | |
| *Micropolyspora* | | | | | | |
| *angiospora* | 3479/30 | | 66.0 | | 23 | |
| *brevicatena* | IMRU 1086 W | | 65.8–67.5 | | 16, 22 | 16 |
| *Nocardia* | | | | | | |
| *asteroides* | W1035A (Ruth Gordon) | | 67.8 | | 16 | |
| | 92 | 66.0 | | | 17 | |
| | 96 | 64.0 | 69.0 | | 17 | |
| | IMRU | 77.3 | | 69.4 | 18 | |
| *brasiliensis* | 301 | 65.0 | | | 17 | |
| | 301 | 67.4–68 | | | 179 | 181 |
| *canicruria* | ATCC 11048 (NCTC 8036) | | | 63.6 | 19 | |
| | 57 | | | 62.7 | 19 | |
| *cellulans* | ATCC 12830 | | | 72.7 | 178 | |
| isolates | S2:07 | | | 71.1 | 178 | |
| | S2:20 | 72.7 | | | 178 | |
| | R1:10 | | | 59.0 | 178 | |
| *citreus* | | | 72.0 | | 8 | |

**TABLE 1 (Continued)**
**GC COMPOSITION OF THE DNA OF BACTERIA**

| Organism | Source or Strain[a] | Mole % Guanosine + Cytosine[b] Tm | Chemical Analysis | Buoyant Density | References[c] A | B |
|---|---|---|---|---|---|---|
| **Nocardiaceae (continued)** | | | | | | |
| *Nocardia* | | | | | | |
| *corallina* | ATCC 4273 | | | 68.6 | 19 | 19, 20 |
| | S 5 | | | 68.7 | 19 | 19, 20 |
| | 305 | | | 62.3 | 19 | |
| | Ja 4181 | | 70.95 | | 203 | |
| *erythropolis* | 2 strains (J. N. Adams) | 62 | | | 177 | |
| | | | | 62.5 | 19 | |
| *farcinica* | 2089 | 68.5 | | | 17 | |
| *lutea* | CBS 207 | 69.8 | | | 229 | |
| *opaca* | 765 A (Adams) | | | 63.5 | 204 | |
| | 17a 2 (Yamaguchi) | | 68.1 | | 16 | |
| | 765 A | | | 64.5 | 19 | |
| | ATCC 4276 | | | 68.1 | 19 | 19 |
| *rubra* | JA 4180 | | 71.7 | | 203 | |
| *Pseudonocardia* | | | | | | |
| *thermophila* | ETH 25787 | 79.3 | | | 162 | |
| *Thermoactinomyces* | | | | | | |
| *vulgaris* | | 52.0 | | | 21 | |
| | MP 4 | 53.4–54.4 | | | 163 | |
| | ACTU 3397 | | 72.2 | | 16 | 16 |
| | IPV 702 | 54.1 | | | 162 | |
| | IMRU | 54.8 | | | 162 | |
| *Thermoactinopolyspora* | | | | | | |
| species | 6 | 77.4 | | | 24 | |
| | 10 | 74.1 | | | 24 | |
| *Thermomonospora* | | | | | | |
| species | 5 | 73.9 | | | 24 | |
| | 7 | 43.8 | | 44.0 | 24 | |
| **Polyangiaceae** | | | | | | |
| *Chondromyces* | | | | | | |
| *apiculatus* | | | | 70.0 | 69 | |

TABLE 1 (Continued)
GC COMPOSITION OF THE DNA OF BACTERIA

| Organism | Source or Strain[a] | Mole % Guanosine + Cytosine[b] Tm | Chemical Analysis | Buoyant Density | References[c] A | B |
|---|---|---|---|---|---|---|
| **Polyangiaceae (continued)** | | | | | | |
| *Chondromyces* | | | | | | |
| *apiculatus* | University of Windsor strain M 6 | 69.3 | | | 27 | |
| | 2 strains | 69.3–70 | | | 256 | 258 |
| *brunneus* | 2 strains | 68.7 | | | 256 | |
| | University of Windsor strain M 26 | 68.7 | | | 27 | |
| | University of Windsor strain M 27 | 68.7 | | | 27 | |
| *crocatus* | University of Windsor strain M 38 | 69.6 | | | 27 | |
| | University of Windsor strain M 204 | 69.7 | | | 27 | |
| | 2 strains | 69.6–69.7 | | | 256 | |
| *medius* | 2 strains | 68.5–68.7 | | | 256 | |
| | University of Windsor strain M 15 | 68.7 | | | 27 | |
| | University of Windsor strain M 34 | 68.5 | | | 27 | |
| *Polyangium* | | | | | | |
| *canicruria* | ATCC 11048 | | | 62.2 | 177 | |
| *cellulosum* | | | 68–70 | | 5 | |
| *fuscum* | University of Windsor strain M 29 | 68.5 | | | 27 | |
| | University of Windsor strains M30 and M 31 | 68.3 | | | 27 | |
| **Propionibacteriaceae** | | | | | | |
| *Propionibacterium* | | | | | | |
| *acnes* | (0389) 21 strains | 59 | | | 186 | |
| | 5 strains | 59 | | | 186 | |

## TABLE 1 (Continued)
## GC COMPOSITION OF THE DNA OF BACTERIA

| Organism | Source or Strain[a] | Mole % Guanosine + Cytosine[b] Tm | Chemical Analysis | Buoyant Density | References[c] A | References[c] B |
|---|---|---|---|---|---|---|
| **Propionibacteriaceae (continued)** | | | | | | |
| *Propionibacterium* | | | | | | |
| *adamsonii* | 3 strains | 59 | | | 186 | |
| *anaerobium* | 4 strains | 59 | | | 186 | |
| *arabinosum* | NCIB 5958 | | 70.4 | | 15 | |
| *avidum* | 2 strains | 62 | | | 186 | |
| *freudenreichii* | NCIB 5959 | | 67.1 | | 15 | |
| *jensenii* | (0397) 1 strain | 66 | | | 186 | |
| *pentosaceum* | NCIB 8070 | | 67.8 | | 15 | |
| *shermanii* | | | 68.2 | | 29 | |
| | | | 68.2–68.9 | | 122 | |
| | NCIB 8099 | | 66.9 | | 15 | |
| *technicum* | NCIB 5965 | | 66.4 | | 15 | |
| *thaenii* | NCIB 8072 | | 67.6 | | 15 | |
| *zeae* | NCIB 8100 | | 67.3 | | 15 | |
| **Pseudomonadaceae** | | | | | | |
| *Acetobacter* | | | | | | |
| *aceti* | Ch. 31 | 59.6 | 59.5 | | 144 | |
| *aceti* | P 12 | 64.5 | | | 145 | |
| (*liquefaciens*) | NCIB 9505 | 64.0–64.3 | 65.4 | | 144, 145 | 46 |
| *aceti* (var. *muciparus*) | 5 | 59.5 | | | 144 | |
| *ascendens* | ATCC 9323 | 55.0 | | | 30 | |
| *cerinus* (var. *rosiensis*) | 22 | 55.7 | 56.5 | | 144 | |
| *estunensis* | E | 62.2 | | | 144 | |
| *gluconicum* | 2G | 54.0 | | | 30, 31 | |
| *mesoxydans* | NCIB 8747 | | 61.1 | | 224 | |
| *mesoxydans* (var. *saccharovorans*) | 4 | 60.6 | 61.0 | | 144 | |

TABLE 1 (Continued)
GC COMPOSITION OF THE DNA OF BACTERIA

| Organism | Source or Strain[a] | Mole % Guanosine + Cytosine[b] Tm | Chemical Analysis | Buoyant Density | References[c] A | B |
|---|---|---|---|---|---|---|
| **Pseudomonadaceae (continued)** | | | | | | |
| *Acetobacter* | | | | | | |
| *mobilis* | NCIB 6428 | 58.9 | 58.8 | | 144 | |
| *paradoxus* | P 1 | 55.7 | | | 144 | |
| | P 2 | 55.4 | | | 144 | |
| *pasteurianum* | 11 | 59.8 | | | 144 | |
| *peroxydans* | NCIB 8618 | 62.5 | 61.0 | | 144 | |
| | 3 | 63.4 | | | 144 | |
| | 4 | 63.4 | | | 144 | |
| *rancens* | 23 k1 | 55.4 | | | 144 | |
| | 15 | 56.2 | | | 144 | |
| | NCIB 6428 | | 58.8 | | 225 | |
| *vini acetati* | NCIB 4939 | 57.4 | | | 144 | |
| *xylinoides* | NCIB 4940 | 57.4 | | | 144 | |
| *xylinum* | NCIB 8747 | 60.7 | 61.1 | | 144 | |
| | 25 | 62.8 | | | 144 | |
| *Aeromonas* | | | | | | |
| *formicans* | ATCC 13137 | 58.8 | | | 7 | |
| | | | 58–60 | | 5 | |
| *hydrophila* | | | 54–56 | | 5 | |
| | ATCC 9071 | 54.2–55.7 | | | 7, 76 | 73 |
| | IP R 126 | | 59.0 | | 59 | |
| | IP R 307 | | 59.3 | | 59 | |
| | NCTC 7810 | | 61.4 | | 59 | |
| | NCTC 7812 | | 62.7 | | 59 | |
| | AB 833 | 59.5 | | | 76 | |
| *jamaiscensis* | Caselitz XS 3 | | 62.7 | | 59 | |
| *liquefaciens* | NRRL B 966 | 57.8–59.0 | | | 7, 76 | |
| | ATCC 14715 | 59.0 | | | 7 | |
| | | | 58–59 | | 5 | |
| *punctata* | ATCC 11163 | 56.5 | | 56.6 | 7 | |
| | CCEB 386 | | 60.8 | | 59 | |
| | | | 56–58 | | 5 | |
| *salmonicida* | ATCC 14174 | 61.5 | 58.5 | 226 | 188 | |
| | B6P | 56.6 | | | 227 | |

TABLE 1 (Continued)
GC COMPOSITION OF THE DNA OF BACTERIA

| Organism | Source or Strain[a] | Mole % Guanosine + Cytosine[b] Tm | Chemical Analysis | Buoyant Density | References[c] A | B |
|---|---|---|---|---|---|---|
| **Pseudomonadaceae (continued)** | | | | | | |
| *Azotomonas* | | | | | | |
| *insolita* | J2(3 values) | 57.9–60.9 | 62.0[j] | | 174 | |
| | ATCC 12412 (2 values) | 60.3–60.7 | | | 174 | |
| *Gluconobacter* | | | | | | |
| *oxydans* | 26 | | 61.0 | | 225 | |
| | NCIB 8131 | | 62.1 | | 225 | |
| | SU | | 58.1 | | 225 | |
| | NCIB 4943 | | 61.3 | | 225 | |
| | NCIB 8086 | | 61.0 | | 225 | |
| *Halobacterium* | | | | | | |
| *cutirubrum* | | | | 58–59 | 252 | |
| | | | | 68 | 252 | |
| *salinarium* | | | | 58 | 252 | |
| | | | | 67 | 252 | |
| species | $IE_1$ F. R. Evans | | | 55 | 252 | |
| | $IIG_2$ F. R. Evans | | | 64 | 252 | |
| | $IVA_2$ F. R. Evans | | | 58 | 252 | |
| | $IVA_3$ F. R. Evans | | | 58 | 252 | |
| *Lophomonas* | | | | | | |
| species | Pickett M 284 | 58.3 | | | 171 | |
| | Pickett M 220 | 67.5 | | | 171 | |
| | Pickett M 234 | 61.3 | | | 171 | |
| | Pickett M 242 | 63.1 | | | 171 | |
| | Pickett M 283 | 64.8 | | | 171 | |
| | Pickett M 285 | 67.0 | | | 171 | |
| | Pickett M 286 | 67.2 | | | 171 | |
| *Pseudomonas* | | | | | | |
| *aeruginosa* | NRRL B-23 | 66.0 | | 68.0–68.4 | 30, 31, 146 | 82, 147 |
| | ATCC 8707 | 64.0 | | 67.3 | 7, 146 | |
| | ATCC 8689 | 64.4 | | 67.3 | 7, 146 | |
| | 5 strains | | 64–67 | | 8 | |
| | ICPPB 2020 | | | 67.0 | 146 | |
| | 481 | 66.8 | | | 79 | 79 |
| | Pickett M 147 | 58.1 | | | 171 | |
| | Pickett M 96 | 57.1 | | | 171 | |
| | 17 R. J. Zabransky Mayo Clinic | | | 66.9 | 190 | |

TABLE 1 (Continued)
GC COMPOSITION OF THE DNA OF BACTERIA

| Organism | Source or Strain[a] | Tm | Chemical Analysis | Buoyant Density | References[c] A | References[c] B |
|---|---|---|---|---|---|---|
| | | Mole % Guanosine + Cytosine[b] | | | | |
| **Pseudomonadaceae (continued)** | | | | | | |
| *Pseudomonas* | | | | | | |
| *alcaligenes* | 142 | | | 66.3 | 146 | |
| *aminovorans* | 26 (Dr. Leadbetter) | | | 63.2 | 146 | |
| *atlantica* | | | | 55.0 | 148 | |
| | NCMB 301 | | | 43.5 | 146 | |
| | 506 | 66.5 | | | 79 | 79 |
| | | | 54–56 | | 5 | |
| | | 66.4 | | | 80 | |
| *aureofaciens* | 518 | 62.8 | | | 79 | 79 |
| | 8aB1 | 62.2 ± 1.5 | | 63 ± 1 | 228 | |
| | 8aB1 | 61.8 ± 1.5 | | 63 ± 1 | 228 | |
| *azotocolligans* | NCIB 9391 | 65.6 | | | 33 | |
| *azotogensis* | | 48.1 | | | 32 | 32 |
| *bathycetes* | A2P | 56.0 | | | 227 | |
| | C3M | 57.8 | | | 227 | |
| | C4P | 57.1 | | | 227 | |
| | C1M | 57.4 | | | 227 | |
| | C6P | 57.1 | | | 227 | |
| | C2M #2, ATCC 23597 | | | 57.8 | 227 | |
| | C2M | | | 57.8 | 227 | |
| | A7M | 52.9 | | | 227 | |
| | C5P | 57.0 | | | 227 | |
| | A1M #2 | 57.1 | | | 227 | |
| | C2P | 57.8 | | | 227 | |
| | C3P | 56.6 | | | 227 | |
| | A1P | 56.4 | | | 227 | |
| | A6P | 56.4 | | | 227 | |
| | A5M | 58.5 | | | 227 | |
| | C6M | | | 57.1 | 227 | |
| | B3M | 56.3 | | | 227 | |
| | A6M | 57.6 | | | 227 | |
| | B3P | 57.0 | | | 227 | |
| | A5P | 57.8 | | | 227 | |
| | A1M | 57.8 | | | 227 | |
| | A4M | 56.4 | | | 227 | |
| *chlororaphis* | NRRL B560 | 64.2 | | 64.2 | 7 | |
| | NRRL B977 | 61.0 | | 61.0 | 7 | |
| | 559 | 63.3 | | | 79 | 79 |
| | | | 66–68 | | 5 | |
| *cruciviae* | ATCC 13262 | 36.1 | | 37.3 | 13 | |
| *cuneata* | NCIB 8194 | 60.97 | 62.7 | | 187 | |

## TABLE 1 (Continued)
## GC COMPOSITION OF THE DNA OF BACTERIA

| Organism | Source or Strain[a] | Mole % Guanosine + Cytosine[b] Tm | Chemical Analysis | Buoyant Density | References[c] A | B |
|---|---|---|---|---|---|---|
| **Pseudomonadaceae (continued)** | | | | | | |
| *Pseudomonas* | | | | | | |
| *denitrificans* | ATCC 12133 | 57.5 | | 56.5 | 13 | |
| | CCEB 525 | 62.3 | | | 79, 80 | 79 |
| *desmolytica* | ATCC 15005 | 65.8 | | 64.6 | 13 | |
| *diminuta* | ATCC 13184 | | | 66.3 | 149 | |
| | ATCC 11568 | | | 64.6–67.3 | 13, 149 | |
| | Lautrop AB 102 | | | 65.8 | 149 | |
| | Lautrop AB 236 | | | 66.3 | 149 | |
| | Lautrop AB 265 | | | 67.3 | 149 | |
| | Lautrop AB 328 | | | 66.9 | 149 | |
| | Lautrop AB 359 | | | 66.9 | 149 | |
| | Lautrop AB 1122 | | | 62.2 | 149 | |
| | | | 66–68 | | 5 | |
| | CCEB 513 | 67.3 | | | 79, 80 | 79 |
| | Lautrop AB 1224 | | | 67.3 | 150 | |
| | Lautrop AB 1267 | | | 66.9 | 150 | |
| | Lautrop AB 1268 | | | 66.9 | 150 | |
| | Lautrop AB 1278 | | | 67.3 | 150 | |
| *fluorescens* | Biotype A (12, 126, 186, 392) | | | 60.5 ± 1.1 | 146 | |
| | Biotype B (2, 400, 401, 403, 404, 411) | | | 61.3 ± 1.1 | 146 | |
| | Biotype C (18, 50, 181, 191, 213, 217) | | | 60.6 ± 0.8 | 146 | |
| | Biotype D (30, 31, 32, 35, 388, 389, 390, 391, 393, 394) | | | 63.5 ± 0.9 | 146 | |
| | Biotype E (36, 37, 38, 39, 41, 86) | | | 63.6 ± 0.8 | 146 | |
| | Biotype F (83, 143) | | | 59.4 ± 0.8 | 146 | |
| | Biotype G (33, 34, 267, 269, 271, 272) | | | 60.5 ± 1.0 | 146 | |
| | CCEB 488 | 62.4–62.7 | 59.5 | | 14, 46, 79 | 46, 79 |
| | ATCC 949 | 60.0 | 63.0 | 62.0 | 8, 30, 31 | |
| | ATCC 12121 | 64.0 | | | 7 | |
| | ATCC 10796 | 64.4 | | | 7 | |
| | A 197 | | | 60.0 | 190 | |
| | C7P | 57.8 | | | 227 | |
| | B5P | 58.0 | | | 227 | |
| | A3P | 56.4 | | | 227 | |
| | ATCC 12633 | 61.0 | | 61.0 | 7 | |

## TABLE 1 (Continued)
## GC COMPOSITION OF THE DNA OF BACTERIA

| Organism | Source or Strain[a] | Mole % Guanosine + Cytosine[b] Tm | Chemical Analysis | Buoyant Density | References[c] A | B |
|---|---|---|---|---|---|---|
| **Pseudomonadaceae (continued)** | | | | | | |
| *Pseudomonas* | | | | | | |
| *fluorescens* | ATCC 130341 | 63.4 | | 64.6 | 13 | |
| | B/S 1000 | | 63.8 | | 8 | |
| *fragi* | CCEB 387 | 60.6 | | | 79 | |
| | ATCC 4973 | 62.2 | | 57.5 | 13 | |
| | | | 60–62 | | 5 | |
| *geniculata* | CCEB 338 | 67.7 | | | 14, 79 | 79 |
| | | | 66–68 | | 5 | |
| | NRRL B1603 | | | 60.2 | 146 | |
| | NRRL B2080 | | | 66.3 | 146 | |
| | NRRL B2337 | | | 67.3 | 146 | |
| *iodinum* | 512 | 60.2 | | | 79 | 79 |
| | ATCC 15729 | 60.9 | | 62.6 | 13 | |
| | ATCC 15728 | 61.4 | | 62.6 | 13 | |
| | ATCC 9897 | 63.4 | | 63.6 | 13 | |
| *indoloxidans* | ATCC 9355 | 66.3 | | 64.6 | 13 | |
| *lemoignei* | 443 | | | 58.2 | 146 | |
| *mallei* | NBL 1, 2, 4, 7, 16, 17, 19 | | | 69 ± 1 | 146 | |
| | NCTC 3709 | | | 69.4 | 146 | |
| *maltophilia* | 67, 72, 301, 303 | | | 66.9 ± 0.8 | 146 | |
| | ATCC 13843 | | | 64.6 | 13 | |
| | ATCC 13636 | | | 64.6 | 13 | |
| | ATCC 13637 | | | 64.6 | 13 | |
| | | | 66–68 | | 5 | |
| | NCIB 9203 | 65.6 | | | 229 | |
| *marginalis* | ATCC 10858 | 57.5 | | 58.5 | 13 | |
| *methanica* | (Dr. J. W. Foster) | | | 52.1 | 146 | |
| *multivorans* | 85, 104, 382, 383, 384, 385, 386, 387, 396, 397, 398, 399 | | | 67.6 ± 0.8 | 146 | |
| | ATCC 17616 | | | 67.8 | 190 | |
| *neocistes* | RH 1810 | 64.4 | | 63.6 | 13 | |
| *oleovorans* | | | 60–62 | | 5 | |
| *ovalis* | CCEB 380 | 64.8 | | | 79, 80 | 79 |
| | ATCC 950 | 64.6 | | 64.6 | 13 | |

## TABLE 1 (Continued)
## GC COMPOSITION OF THE DNA OF BACTERIA

| Organism | Source or Strain[a] | Tm | Chemical Analysis | Buoyant Density | References[c] A | References[c] B |
|---|---|---|---|---|---|---|
| | | Mole % Guanosine + Cytosine[b] | | | | |
| **Pseudomonadaceae (continued)** | | | | | | |
| *Pseudomonas* | | | | | | |
| *pavonacae* | CCEB 533 | 45.6 | | | 79 | 79 |
| *phaseolicola* | PM 142 (Dr. M. P. Starr) | | | 58.9 | 146 | |
| | ICPB PM 142 | | | 59.2 | 230 | |
| *piscicida* | 1 strain | | | 43.5 | 151 | |
| | 5 strains | | | 44 | 151 | |
| | 7 strains | | | 44.5 | 151 | |
| | ATCC 15251 | | | 44.5 | 151 | |
| | 4 strains | | | 45 | 151 | |
| | 2 strains | | | 45.5 | 151 | |
| *polycolor* | PP 2 (Dr. M. P. Starr) | | | 67.3 | 146 | |
| *pseudoalcaligenes* | 63, 65, 66, 297, 299, 417 | | | 62.8 ± 0.9 | 146 | |
| | 63 AD (Dr. M. Vernon) | | | 62.2 | 146 | |
| | 63 AD (Dr. M. Vernon) | | | 63.2 | 146 | |
| *pseudomallei* | NBL 111, 113, 114, 117, 121, 123 | | | 69.5 ± 0.7 | 146 | |
| | 295 | | | 69.2 | 190 | |
| | NCTC 1691 | | | 69.2 | 190 | |
| | ATCC 15682 | | | 69.0 | 190 | |
| *putida* | Biotype A 5, 6, 26, 42, 90 | | | 62.5 ± 0.9 | 146 | |
| | Biotype B 53, 96, 98, 110, 153, 157, 158, 167 | | | 60.7 ± 1.1 | 146 | |
| | CCEB 520 | 63.5–63.9 | | | 14, 79, 80 | 79 |
| | 101 | 63.4 | | | 14 | |
| | ATCC 4359 | 58.7 | | 59.6 | 13 | |
| | | | 64–66 | | 5 | |
| | ATCC 12633 | | 63.7 | 63.3 | 229 | |
| | A 3121, CPB 2484 | | | 63.3 | 230 | |
| *putrefaciens* | ATCC 8071 | 45.4 | | 43.4 | 13 | |
| *reptilovora* | ATCC 11252 | 63.9 | | 61.6 | 13 | |
| *rimaefaciens* | PR 107 | 66.5 | | | 79 | 79 |
| *rubescens* | CCEB 519 | 46.1 | | | 79 | 79 |

TABLE 1 (Continued)
GC COMPOSITION OF THE DNA OF BACTERIA

| Organism | Source or Strain[a] | Mole % Guanosine + Cytosine[b] Tm | Chemical Analysis | Buoyant Density | References[c] A | B |
|---|---|---|---|---|---|---|
| **Pseudomonadaceae (continued)** | | | | | | |
| *Pseudomonas* | | | | | | |
| *saccharophila* | | | 68–70 | | 5 | |
| *schuylkilliensis* | NRRL B1104 | | | 60.7 | 146 | |
| | NRRL B1105 | | | 60.9 | 146 | |
| *stanieri* | 220, 223, 224, 225, 228, 320 | | | 62.1 ± 1 | 146 | |
| *stutzeri* | 221, 222, 226, 227, 275, 316, 318, 319, 321, 419 | | | 65 ± 1 | 146 | |
| | | 64.4 | | 64.6 | 13 | |
| | CCEB 522 | | 64.5 | | 79 | 79 |
| | | 64–66 | | | 5 | |
| *syncyanea* | ATCC 9979 | 62.4 | | 61.6 | 13 | |
| *synxantha* | CCEB 607 | 61.0 | | | 79 | 79 |
| *syringae* | 19 P | | 59.9 | | 29 | |
| *tabaci* | | | 67.4 | | 8 | |
| | PT 1 | | | 58.2 | 146 | |
| | | | 66–68 | | 5 | |
| | ICPB PT1 | | | 58.2 | 230 | |
| *taetrolens* | CCEB 381 | 59.8 | | | 79 | 79 |
| *testosteroni* | 15, 16, 25, 27, 28, 78, 79, 138, 139 | | | 61.8 ± 1.1 | 146 | 146 |
| | RH 1104 | 60.7 | | 59.6 | 13 | |
| *vesiculare (Corynebacterium)* | ATCC 11426 | | | 65.8 | 149 | |
| *xanthae* | F 3.0 | | | 69.9 | 146 | |
| species | E 89a | 60.9 | | 59.6 | 13 | |
| | 95 | | | 59.7 | 146 | |
| | 60 | | | 58.7 | 146 | |
| | 298 | | | 64.3 | 146 | |
| | cim 9B | 63.7 ± 1.5 | | 65 ± 1 | 228 | |
| | cim 9A | 63.1 ± 1.5 | | 64 ± 1 | 228 | |
| | cim D | 63.1 ± 1.5 | | 65 ± 1 | 228 | |
| | cim C | 63.2 ± 1.5 | | 65 ± 1 | 228 | |
| | OX Sawyer | 60.9 ± 1.5 | | | 228 | |
| | 115–400 | 56.6 | | | 227 | |
| | 81–4000 | 53.9 | | | 227 | |

## TABLE 1 (Continued)
## GC COMPOSITION OF THE DNA OF BACTERIA

| Organism | Source or Strain[a] | Mole % Guanosine + Cytosine[b]: Tm | Chemical Analysis | Buoyant Density | References[c] A | References[c] B |
|---|---|---|---|---|---|---|
| **Pseudomonadaceae (continued)** | | | | | | |
| *Pseudomanas* | | | | | | |
| species | 83–4000 | 50.0 | | | 227 | |
| | 66–0 | 57.8 | | | 227 | |
| | B2P | 56.1 | | | 227 | |
| | C4M | 56.6 | | | 227 | |
| *Xanthomonas* | | | | | | |
| *alfalfae* | A 121 (ATCC 11765) | 67.3 | | | 79 | 79 |
| *begoniae* | B 3 (ATCC 8718) | 67.1 | | | 14 | 79 |
| | ICPB B 3 | | | 67.4 | 232 | |
| *beticola* | B 109 (ATCC 14047) | 65.7 | | | 79 | 79 |
| *campestris* | C 129 | 68.2 | | | 14 | 79 |
| | ICPB C 129 | | | 68.6 | 232 | |
| *carotae* | ICPB C 104 | | | 67.3 | 232 | |
| | C 104 | 67.0 | | | 14 | 79 |
| | D 230 | 67.3 | | | 14 | |
| *cassavae* | C 110 | 65.9 | | | 79 | 79 |
| *celebensis* | C 144 (ATCC 19046) | 69.2 | | | 79 | 79 |
| *corylina* | C 5 | 66.6 | | | 79 | 79 |
| *geranii* | G 1 | 63.5 | | | 79 | 79 |
| *hederae* | XH 5 | 62.0 | | | 7 | |
| | H 1 | 67.2–67.3 | | | 14, 79 | 79 |
| | ICPB H1 | | | 67.9 | 232 | |
| *hyacinthi* | H 110 | 68.5 | | | 79 | 79 |
| *incanae* | I 3 | 66.3 | | | 79 | 79 |
| *juglandis* | J 107 | 66.3 | | | 14 | 79 |
| | ICPB J 107 | | | 66.6 | 232 | |
| *lactucae* | L 4 | 65.1 | | | 79 | 79 |
| *lespedezae* | L 1 (ATCC 13463) | 66.6 | | | 79 | 79 |
| *muculafoliagardeniae* | M 16 | 67.7 | | | 79 | 79 |

TABLE 1 (Continued)
GC COMPOSITION OF THE DNA OF BACTERIA

| Organism | Source or Strain[a] | Mole % Guanosine + Cytosine[b] Tm | Chemical Analysis | Buoyant Density | References[c] A | B |
|---|---|---|---|---|---|---|
| **Pseudomonadaceae (continued)** | | | | | | |
| *Xanthomonas* | | | | | | |
| *malvacearum* | M 2 | 65.6 | | | 79 | 79 |
| *papavericola* | P 5 (ATCC 14179) | 65.7 | | | 79 | 79 |
| *pelargonii* | J. Koths | 62.4 | | | 7 | |
| | P 121 | 66.5 | | | 14 | 79 |
| | ICPB P121 | | | 66.8 | 232 | |
| *phaseoli* | ICPB P162 | | | 66.4 | 232 | |
| | XP 163 | 63.2 | | | 7 | |
| | XP 8 | 63.2 | | | 7 | |
| | P 162 | 66.0 | | | 14 | 79 |
| *pisi* | XP 171 | 63.8 | | | 7 | |
| *poinsettiaecola* | P 137 (ATCC 11643) | 66.0 | | | 79 | 79 |
| *pruni* | P 10 | 66.8 | | | 79 | 79 |
| *resicatoria* | ICPB V136 | | | 66.6 | 232 | |
| *tamarini* | ICPB T20 | | | 68.2 | 232 | |
| | T 20 | 67.7 | | | 14 | 79 |
| *taraxaci* | T 11 (ATCC 19318) | 64.3 | | | 79 | 79 |
| *translucens* | T 7 | 68.0 | | | 79 | 79 |
| *trifolii* (*Erwina* sp.) | ICPB XT 109 | | | 55.6 | 133 | |
| | LCPB XT 110 | | | 55.6 | 133 | |
| *uredovorus* (*Erwinia* sp.) | ICPB XU 103 | | | 53.1 | 133 | |
| | ICPB XU 104 | | | 53.1 | 133 | |
| *vasculorum* | V 24 | 64.8 | | | 79 | 79 |
| *vesicatoria* | V 136 | 66.4 | | | 14 | 79 |
| *vignicola* | V 118 | 65.7 | | | 79 | 79 |
| **Rhizobiaceae** | | | | | | |
| *Agrobacterium* | | | | | | |
| *ferrugineum* | A 7 | 60.4 | 60 | | 229 | 233 |

## TABLE 1 (Continued)
## GC COMPOSITION OF THE DNA OF BACTERIA

| Organism | Source or Strain[a] | Mole % Guanosine + Cytosine[b] Tm | Chemical Analysis | Buoyant Density | References[c] A | B |
|---|---|---|---|---|---|---|
| **Rhizobiaceae (continued)** | | | | | | |
| *Agrobacterium* | | | | | | |
| *luteum* | A 61 | 58.2 | 57 | | 229 | 233 |
| *radiobacter* | Ra | 62.2 | | 59.5 | 234 | |
| | NCIB 8149 | 60.5 | | | 152 | |
| | ICPPB TR 1 | 61.2 | | | 152 | 46 |
| | Gembloux S 1005 | 61.3 | | | 152 | 46 |
| | ATCC 4718 | 61.8 | | | 152 | 46 |
| | M 2/1 | 62.1 | | | 152 | |
| | ICPPB TR 6 | 62.1 | | | 152 | 46 |
| | L 2/2/1 | 62.5 | | | 152 | |
| *radiobacter (tumefaciens)* | Institut Pasteur RV 3 | 60.9 | | | 152 | |
| *radiobacter* var. *tumefaciens (tumefaciens)* | NCPPB 398 | 59.9 | | | 152 | |
| | Gembloux C | 60.3 | | | 152 | |
| | NCPPB 396 | 60.6 | | | 152 | |
| | ICPPB TT 111 | 60.6 | | | 152 | 46 |
| | ICPPB TT 6 | 60.6 | | | 152 | |
| | ATCC 11.158 | 60.8 | | | 152 | |
| | ATCC 143 | 60.9 | | | 152 | |
| | ATCC 4452 | 60.9 | | | 152 | |
| | ATCC 11.156 | 61.1 | | | 152 | 46 |
| | Gembloux A | 61.1 | | | 152 | |
| | ATCC 11.157 | 61.3 | | | 152 | |
| | Institut Pasteur 42 IV | 61.3 | | | 152 | |
| | Institut Pasteur B 6 | 61.4–61.5 | | | 117, 152 | 46 |
| | NCPPB 4 | 61.4 | | | 152 | 46 |
| | ICPPB TT 9 | 61.5 | | | 152 | |
| | ICPPB TT 107 | 61.8 | | | 152 | |
| | Gembloux tum 6 | 61.5 | | | 152 | |
| | NCPPB 925 | 61.8 | | | 152 | |
| | Institut Pasteur S 1 | 61.8–62 | | | 117, 152 | |
| | ATCC 4720 | 62.1 | | | 152 | 46 |
| | NCPPB 397 | 62.2 | | | 152 | 46 |
| | Delft F | 62.3 | | | 152 | |
| | NCPPB 223 | 61.8 | | | 152 | |
| | NCPPB 794 | 62.3 | | | 152 | |
| | ICPPB TT 133 | 62.5 | | | 152 | |
| | NCPPB 5 | 60.7 | | | 152 | |
| | Delft G | 62.3 | | | 152 | |
| | ICPPB-TR 2 | 59.6 | | | 152 | |

## TABLE 1 (Continued)
## GC COMPOSITION OF THE DNA OF BACTERIA

| Organism | Source or Strain[a] | Mole % Guanosine + Cytosine[b] Tm | Chemical Analysis | Buoyant Density | References[c] A | B |
|---|---|---|---|---|---|---|
| **Rhizobiaceae (continued)** | | | | | | |
| *Agrobacterium* | | | | | | |
| *rhizogenes* | ICPPB TR 101 | 61.1 | | | 152 | 46 |
| | ICPPB TR 107 | 62.3 | | | 152 | 46 |
| | ICPPB TR 7 | 62.8 | | | 152 | 46 |
| | NCIB 8196 | 62.8 | | | 152 | |
| *pseudotsugae* | NCPPB 180 | 67.7 | | | 152 | 46 |
| *psygosphilae* | NCPPB 179 | 56.2 | | | 152 | |
| *sanguineum* | A 91 | 65.2 | 64 | | 229 | 233 |
| *stellulatum* | Stapp 2216 | 66.0 | | | 152 | |
| *tumefaciens* | ATCC 11.095 | 61.1 | | | 152 | |
| | NCPPB 930 | 58.9 | | | 152 | |
| | M 39 | 65.5 | | | 145 | |
| | 2 strains | | 58.2–58.7 | | 8 | |
| | 55 | | 61.6 | | 29 | |
| | S 1 | 58.0 | | 60.2 | 235 | 236 |
| | E 1119.6.1 | 62.2 | | 59.8 | 234 | |
| | A 6 | 61.7 | | 58.9 | 234 | |
| | M 39 | | 62 | 64.8 | 236 | |
| | ATCC 143 | | 58.8 | | 225 | |
| | B 6 | | 60.8 | 59.8 | 237 | |
| | SCA 1 | 60.7 | 59.7 | | 229 | 237 |
| *Chromobacterium* | | | | | | |
| *amethystinum* | ATCC 6915 | | | 62.1 | 190 | |
| *janthinum* | WRAIR | | | 64.5 | 190 | |
| *lividum* | ATCC 12473 | | | 61.9 | 190 | |
| | NCTC 9796 | 65.5 | | | 14 | |
| | NCTC 7150 | 65.4 | | | 14 | |
| | NCIB 9230 | 65.5–65.8 | | | 14 | |
| | GA | 71.4 | | | 14 | |
| | 7 strains | 65.8–66.1 | | | 14 | |
| | MRC MB | 66.1 | | | 46 | 46 |
| | MRC RU | 65.5 | | | 46 | 46 |
| | MRC GA | 71.4 | | | 46 | 46 |
| *violaceum* | NCTC 9373 | 63.4 | | | 14 | |
| | NCTC 9371 | 66.1 | | | 14 | 46 |
| | NCTC 8684 | 67.0 | | | 14 | 46 |
| | NCTC 8685 | 66.9 | | | 14 | 46 |
| | NCTC 9757 | 67.2 | | | 14 | 46 |
| | NCTC 9694 | 67.6 | | | 14 | |
| | ATCC 12472 | | | 64.8 | 190 | |

**TABLE 1 (Continued)**
**GC COMPOSITION OF THE DNA OF BACTERIA**

| Organism | Source or Strain[a] | Mole % Guanosine + Cytosine[b] Tm | Chemical Analysis | Buoyant Density | References[c] A | B |
|---|---|---|---|---|---|---|
| **Rhizobiaceae (continued)** | | | | | | |
| *Rhizobium* | | | | | | |
| *leguminosarum japonicum trifolii* | 41 strains | 57.8–63.8 | | | 153 | |
| *japonicum* | 555 | 61.0 | | 63.0 | 86 | |
| | USDA 311b59 | 64.0 | | | 46 | 46 |
| | 3.2 | 64.8 | | | 46 | 46 |
| *leguminosarum* | USDA 316C10A | 61.4 | | | 46 | 46 |
| | USDA 3F3C1 | 62.5 | | | 46 | 46 |
| *meliloti* | USDA 3DOa30 | 62.3 | | | 46 | 46 |
| | 1.5 | 62.5 | | | 46 | 46 |
| | 441 | | 62.6 | | 29 | |
| species (subpolar flagellation) | 15 strains | 61.5–65.5 | | | 154 | |
| species (peritrichous flagellation) | 20 strains | 59–63 | | | 154 | |
| **Rickettsiaceae** | | | | | | |
| *Rickettsia (Coxiella) burneti* | Paretsky | 42 | | 45 | 30, 31 | |
| | | | 44.5 | | 66 | |
| | | 43.1 | 44.0 | | 170 | |
| | L 35 | 43.4 | 42.9 | | 155 | |
| | Paretsky | | 44.5 | | 253 | |
| *prowazekii* | | | 32.5 | | 66 | |
| *Wolbachia* | | | | | | |
| *persica* | | 30 | | 30 | 63 | |
| **Sorangiaceae** | | | | | | |
| *Sorangium* | | | | | | |
| *cellulosum* | PY | | | 69 | 69 | |
| | PX | | | 69 | 69 | |
| | 1 strain | | | 69 | 258 | |
| **Spirillaceae** | | | | | | |
| *Bdellovibrio* | | | | | | |
| *bacteriovorus* | 100 H-D[k] | | | 50.3 | 230 | |
| | 100 H-I[k] | 50.7 | | 49.5 | 230 | |

# TABLE 1 (Continued)
# GC COMPOSITION OF THE DNA OF BACTERIA

| Organism | Source or Strain[a] | Mole % Guanosine + Cytosine[b] Tm | Chemical Analysis | Buoyant Density | References[c] A | B |
|---|---|---|---|---|---|---|
| **Spirillaceae (continued)** | | | | | | |
| *Bdellovibrio* | | | | | | |
| *bacteriovorus* | 101 H-D[k] | 51.2 | | 50.5 | 230 | |
| | 101 H-I[k] | 50.7 | | 50.0 | 230 | |
| | 109 H-D[k] | | | 50.0 | 230 | |
| | 109 H-I[k] | 50.5 | | 51.5 | 230 | |
| | Sa 109 | | | 49.5 | 230 | |
| | 110 H-D[k] | | | 50.0 | 230 | |
| | 110 H-I[k] | 51.0 | | 50.5 | 230 | |
| | 114 H-D[k] | | | 50.5 | 230 | |
| | 114 H-I[k] | 50.7 | | 50.5 | 230 | |
| | 118 H-D[k] | 51.2 | | 50.0 | 230 | |
| | 118 H-I[k] | 51.2 | | 50.5 | 230 | |
| | 120 H-D[k] | | | 50.0 | 230 | |
| | 120 H-I[k] | 51.0 | | 50.0 | 230 | |
| | 127 H-D[k] | | | 50.5 | 230 | |
| | $X_{tv}$ H-1[k] | 50.7 | | 49.5 | 230 | |
| | OX9-2 H-I[k] | 50.7 | | 50.5 | 230 | |
| | OX9-3 H-I[k] | 50.5 | | 50.5 | 230 | |
| | B H-I[k] | 50.7 | | 50.0 | 230 | |
| | D H-I[k] | 51.2 | | 51.5 | 230 | |
| | E H-I[k] | 51.2 | | 50.0 | 230 | |
| | 233 H-D[k] | | | 50.5 | 230 | |
| | 2484 Se-2 H-I[k] | 49.5 | | 49.5 | 230 | |
| | 2484 Se-3 H-I[k] | 50.7 | | 49.5 | 230 | |
| | A 3.12 H-D[k] | | | 42.9 | 230 | |
| | A 3.12 H-1[k] | 42.7 | | 43.5 | 230 | |
| | 321 H-D[k] | 42.2 | | 42.3 | 230 | |
| | 321 H-I[k] | | | 42.7 | 230 | |
| | (sp, 19) | | | | | |
| *Comamonas* | | | | | | |
| *cyclosites* | NCIB 2581 | 64.6 | 63.9 | | 229 | 238 |
| *neocistes* | NCIB 2582 | 64.9 | 63.6 | | 229 | 238 |
| *terrigena* | NCIB 8193 | 64.1 | 64.6 | | 229 | 238 |
| | ATCC 8461 | | 64.0 | | 239 | |
| *Desulfotomaculum* | | | | | | |
| *nigrificans* | NCIB 8395 | | 45.4 | 48.97 | 240 | 241 |
| *orientis* | NCIB 8382 | | 42.2 | | 240 | |
| *Desulfovibrio* | | | | | | |
| *aestuarii* | NCIB 8380 | | | 55.0 | 31 | |

## TABLE 1 (Continued)
## GC COMPOSITION OF THE DNA OF BACTERIA

| Organism | Source or Strain[a] | Mole % Guanosine + Cytosine[b] Tm | Chemical Analysis | Buoyant Density | References[c] A | B |
|---|---|---|---|---|---|---|
| **Spirillaceae (continued)** | | | | | | |
| *Desulfovibrio* | | | | | | |
| *desulfuricans* | ATCC 7757 | | | 55.3 | 50 | |
| | NCIB 8387 | | | 54.4 | 50 | |
| | NCIB 8307 | | | 54.4 | 50 | |
| | NCIB 8322 | | | 54.4 | 50 | |
| | NCIB 8398 | | | 46.6 | 50 | |
| | NCIB 8403 | 44.1 | | 45.6 | 50 | |
| | NCIB 8365 | | | 45.6 | 50 | |
| | Hildenborough strain 8303, 5 isolates | 58.4–62.8 | | | 174 | |
| | 8303 P | 58.4 | 58.5 | | 174 | |
| | NCIB 8364 | | | 45.6 | 50 | |
| | NCIB 8308 | | | 45.6 | 50 | |
| | NCIB 8303 | 60.6 | 62.8 | 60.2–65 | 31, 50, 156 | |
| | NCIB 8319 | | | 62.1 | 50 | |
| | NCIB 9442 | | | 62.1 | 50 | |
| | NCIB 8311 | | | 61.2 | 50 | |
| | NCIB 8386 | | | 61.2 | 50 | |
| | NCIB 8305 | | | 61.2 | 50 | |
| | NCIB 8302 | | 60.2 | | 50 | |
| | NCIB 8446 | | | 60.2 | 50 | |
| | NCIB 8387 | | | 54.4 | 50 | |
| | NCIB 8401 | | | 60.2 | 50 | |
| | NCIB 9335 | | | 56.3 | 50 | |
| | NCIB 8310 | | | 55.3 | 50 | |
| | NCIB 8312 | | | 55.3 | 50 | |
| | NCIB 8391 | | | 55.3 | 50 | |
| | NCIB 8363 | | | 55.3 | 50 | |
| | NCIB 8388 | | | 55.3 | 50 | |
| | NCIB 8393 | | 57.4 | 55.3 | 50, 156 | |
| *desulfuricans (Vibrio cholinicus)* | ATCC 13541 | | 56.8 | 56.3 | 50, 156 | |
| *orientis* | NCIB 8382 | | 42.2 | 41.7 | 50, 156 | |
| | NCIB 8445 | | | 41.7 | 50 | |
| species | NCIB 8395 | | 45.4 | | 156 | |
| | Gigas | | 64.6 | | 156 | |
| *Methanobacterium* | | | | | | |
| *omelianski* | | | 44.2 | | 49 | |
| *Microcyclus* | | | | | | |
| *aquaticus* | Orskov | | | 67.3 | 157 | |
| | B. Bachmann | | | 66.3 | 157 | |

## TABLE 1 (Continued)
## GC COMPOSITION OF THE DNA OF BACTERIA

| Organism | Source or Strain[a] | Mole % Guanosine + Cytosine[b] Tm | Chemical Analysis | Buoyant Density | References[c] A | B |
|---|---|---|---|---|---|---|
| **Spirillaceae (continued)** | | | | | | |
| *Microcyclus* | | | | | | |
| *flavus* | ATCC 23276 | | | 51.0 | 157 | |
| *major* | BKM B-859 (Gromov) | | | 39.5 | 157 | |
| species | | | 64–66 | | 5 | |
| *Spirillum* | | | | | | |
| *anulus* | | | 54–56 | | 5 | |
| *atlanticum* | NCMB 55 | | 50.7 | | 59 | |
| *itersonii* | | | 54–56 | | 5 | |
| *itersonii* (var. *vulgatum*) | NCIB 9071 | | 57.1 | | 59 | |
| *itersonii* (subsp. *vulgatum*) | ATCC 11331 | 64.0 ± 1.5 | | 65 ± 1 | 228 | |
| *linum* | | | 28–30 | | 59 | |
| *polymorphum* | NCIB 9072 | | 61.3 | | 59 | |
| *serpens* (var. *azotum*) | NCIB 9011 | | 54.0 | | 59 | |
| *serpens* (var. *serpens*) | NCIB 8658 | | 49.2 | | 59 | |
| *serpens* (subsp. *serpens*) | ATCC 11330 | 63.7 ± 1.5 | | 65 ± 1 | 228 | |
| *sinuosum* | NCIB 9010 | | 58.7 | | 59 | |
| species | Veron 63D | | 61.0 | | 59 | |
| *Vibrio* | | | | | | |
| *adaptus* | ATCC 19263 | 63.1 ± 1.5 | | | 228 | |
| *albensis* | ATCC 14547 | 47.6 ± 1.5 | | | 228 | |
| *alcaligenes* | ATCC 14736 | 64.9 ± 1.5 | | | 228 | |
| | NCTC 9239 | | 64.8 | | 59 | |
| *alginolyticus* | | | 44.5 | | 242 | |

## TABLE 1 (Continued)
## GC COMPOSITION OF THE DNA OF BACTERIA

| Organism | Source or Strain[a] | Mole % Guanosine + Cytosine[b] Tm | Chemical Analysis | Buoyant Density | References[c] A | B |
|---|---|---|---|---|---|---|
| **Spirillaceae (continued)** | | | | | | |
| *Vibrio* | | | | | | |
| *anguillarum* | ATCC 14181 | 44.2 ± 1.5 | | | 228 | |
| | NCMB 829 | | 46.0 | | 242 | 242 |
| | NCMB 6 | | 44.4 | | 59 | |
| *bubulus* *(Campylobacter)* | IP 53103 | | 29.5 | | 59 | |
| | IP 5421 | | 29.7 | | 59 | |
| | IP 5422 | | 30.4 | | 59 | |
| | IP 5420 | | 30.9 | | 59 | |
| *cholinicus* | | | 56–58 | | 5 | |
| | | | 56.8 | | 156 | |
| | ATCC 13541 | | 56.3 | | 50 | |
| *cholerae* | NIH 41 | 45.6 | | | 158 | |
| | 35A3, NIH | 47.3 | | | 158 | |
| | C 401 | 46.6 | | | 158 | |
| | C 441 | 46.8 | | | 158 | |
| | VC 9 | 48.3 | | | 158 | |
| | VC 12 | 47.0 | | | 158 | |
| | ATCC 14035 | 46.8 | | | 158 | |
| | P 33/58 | 47.0 | | | 158 | |
| | RH 1094 | 46.1 ± 1.5 | | 48 ± 1 | 228 | |
| | 20A10 | 48.0 ± 1.5 | | 48 ± 1 | 228 | |
| | ATCC 14035 | 46.8 ± 1.5 | | 48 ± 1 | 228 | |
| | NIH 35A3 | 47.3 ± 1.5 | | 48 ± 1 | 228 | |
| *cholerae* (var. *El Tor*) | ATCC 14033 | 46.8 ± 1.5 | | 48 ± 1 | 228 | |
| *comma (cholerae)* | | | 43.3 | | 8 | |
| | ATCC 9459 | 47.0 | 43.3 | 46.0 | 30, 31, 118, 148 | |
| *costicolus* | NCMB 1001 | | 50.2 | | 59 | |
| *cuneatus* | ATCC 6972 (NCIB 8194) | 61.2–65.3 | 62.7 | | 7, 13, 59 | |
| *cyclocistes* | NCIB 2581 | | 63.9 | | 59 | |
| *fetus* *(Campylobacter)* | IP 5396 | | 33.1 | | 59 | |
| | IP 5443 | | 34.4 | | 59 | |
| | IP 5395 | | 35.4 | | 59 | |
| *haloplanktis* | ATCC 14393 | 40.7 ± 1.5 | | | 228 | |
| *ichthyodermis* | NCMB 407 | | 61.0 | | 59 | |
| *marinagilis* | ATCC 14398 | 40.2 ± 1.5 | | | 228 | |

**TABLE 1 (Continued)**
**GC COMPOSITION OF THE DNA OF BACTERIA**

| Organism | Source or Strain[a] | Mole % Guanosine + Cytosine[b] Tm | Chemical Analysis | Buoyant Density | References[c] A | B |
|---|---|---|---|---|---|---|
| **Spirillaceae (continued)** | | | | | | |
| *Vibrio* | | | | | | |
| *marinofulvus* | ATCC 14395 | 40.2 ± 1.5 | | | 228 | |
| *marinopraesens* | ATCC 19648 | 41.0 ± 1.5 | | | 228 | |
| *marinovulgaris* | ATCC 14394 | 40.0 ± 1.5 | | | 228 | |
| *marinus* | ATCC 15381 | | 40.0 | 40.0 | 148 | |
| | ATCC 15382 | | 42.0 | 42.0 | 148 | |
| *metschnikovii* | ATCC 7708 | 47.0 | | 43–43.3 | 7, 13, 148 | |
| | IP A267 | | 45.8 | | 59 | |
| | ATCC 7708 | 47.0 ± 1.5 | | 43 ± 1 | 243 | |
| *neocistes* | NCIB 2582 | | 63.6 | | 59 | |
| | RH 1810 | 64.4 | | 63.6 | 13 | |
| | RH 1810 | 64.4 ± 1.5 | | 64 ± 1 | 243 | |
| *parahaemolyticus* | ATCC 17802 | 45.6 ± 1.5 | | | 228 | |
| | Strain 3 | 44.8 ± 1.5 | | | 228 | |
| | ATCC 17803 | 45.9 ± 1.5 | | | 228 | |
| | A55(05,K 15) | | 46.0 | | 242 | 242 |
| *percolans (Comamonas terrigena)* | ATCC 8461, NCTC 1937, NCIB 8193 | 64–65 | 64.6 | | 7, 9, 59 | |
| | RH 260 | 61.7 | | 62.6 | 13 | |
| *piscium* | NCMB 571 | | 45.7 | | 242 | 242 |
| *ponticus* | ATCC 14391 | 40.7 ± 1.5 | | | 228 | |
| *succinogenes* | | | 49.2 | | 59 | |
| *tyrogenus* | ATCC 7085 | 61.7 | | | 7 | |
| species | El Tor (30 strains) | 46–47 | | 46–47 | 159 | |
| | ATCC 14033 (El Tor) | 46.8 | | | 158 | |
| | HK 1 (El Tor) | 47.3 | | | 158 | |
| | HK 25 (El Tor) | 48.5 | | | 158 | |
| | SLH 29803 (El Tor) | 46.8 | | | 158 | |
| | SLH 30810 (El Tor) | 48.3 | | | 158 | |
| | 2A/62 (El Tor) | 46.8 | | | 158 | |

## TABLE 1 (Continued)
## GC COMPOSITION OF THE DNA OF BACTERIA

| Organism | Source or Strain[a] | Mole % Guanosine + Cytosine[b] Tm | Chemical Analysis | Buoyant Density | References[c] A | B |
|---|---|---|---|---|---|---|
| **Spirillaceae (continued)** | | | | | | |
| *Vibrio* | | | | | | |
| species | 2B/62 (El Tor) | 46.3 | | | 158 | |
| | Bovine R 9 | 33.4 | | | 160 | |
| | Human 662 | 33.3 | | | 160 | |
| | Ovine 751 | 34.7 | | | 160 | |
| | Avian 1802 GB | 32.1 | | | 160 | |
| | Avian 1803 C | 32.4 | | | 160 | |
| | Bovine 655 | 63.5 | | | 160 | |
| | MB 22 | 45.1 | | | 158 | |
| | VL 5 | 44.4 | | | 158 | |
| | 329 | | | 39.0 | 148 | |
| | 822 | | | 43.0 | 148 | |
| | 859 | | | 48.0 | 148 | |
| species (comma-like) | IP 5638 | | 45.7 | | 59 | |
| | NCTC 4711 | 48.3 | 46.8 | | 59, 158 | |
| | NCTC 4715 | 46.3 | 47.7 | | 59, 158 | |
| | NCTC 8042 | | | | 59 | |
| species (non-cholera) | NCTC 4716 | 47.3 | | | 158 | |
| | S 163 | 46.8 | | | 158 | |
| **Spirochaetaceae** | | | | | | |
| *Saprospira* | | | | | | |
| *grandis* | | | | 47.0 | 72 | |
| | | | 44–46 | | 5 | |
| | | | | 47 | 268 | |
| *thermalis* | | | | 37 | 268 | |
| | | | | 37.0 | 72 | |
| *Spirochaeta* | | | | | | |
| *aurantia* | J 1 | | | 66.8 | 161 | |
| *stenostrepta* | Z 1 | | | 60.2 | 161 | |
| *zuelzerae (Treponema)* | ATCC 19044 | | | 56.1 | 161 | |
| species (aerobic, free-living) | Z 4 | | | 59.2 | 161 | |
| species | MB 22 | 45.1 ± 1.5 | | 48 ± 1 | 228 | |
| | NCTC 4711 | | 46.8 | | 242 | 242 |

## TABLE 1 (Continued)
## GC COMPOSITION OF THE DNA OF BACTERIA

| Organism | Source or Strain[a] | Mole % Guanosine + Cytosine[b] Tm | Chemical Analysis | Buoyant Density | References[c] A | B |
|---|---|---|---|---|---|---|
| **Streptomycetaceae** | | | | | | |
| *Intrasporangium* | | | | | | |
| species | 12-17 | 70.7 | | | 266 | |
| *Streptomyces* | | | | | | |
| *albus* | ATCC 618 | 74.3 | 72.3 | 71.4–72.0 | 16, 18, 19 | 16, 19 |
| | 1685 | 73.0 | | | 17 | |
| | G | 75.0 | | 71.0 | 30, 31 | |
| | ATCC 618 | | | 71.4 | 244 | |
| | G | 74 | | | 245 | |
| | IPV 1298 E. Balducci | 72 | | | 177 | |
| *antibioticus I* | JA 3174 | | 70.35 | | 203 | |
| *antibioticus II* | JA 2629 | | 70.67 | | 203 | |
| *argenteolus* | ATCC 11009 | 76.7 | | | 163 | |
| *aureofaciens* | ATCC 10762 | | | 72.4 | 19 | 19 |
| *bobiliae* | ACTU 3310 | | 71.2 | | 16 | 16 |
| | ATCC 3310 | 74.4 | | 70.4 | 18 | |
| *caeruleum* | ATCC 15720 | 72 | 74 | 72.2 | 246 | |
| | ATCC 15812 | | | | 175 | |
| *chrysomallus* | | | 72.0 | | 21 | |
| | JA 1449 | | 70.85 | | 203 | |
| | | | 73.6 | | 203 | |
| *cinnamoneus* | ATCC 11874 | | | 72.1 | 19 | 19 |
| *cinnamoneus (Streptoverticillium)* | | 67.0 | | | 17 | |
| *coelicolor* | 124 | | 70.2 | | 22 | |
| | Muller | | | 72.8 | 19 | 19 |
| | 1945 | | | 72.1 | 19 | 19 |
| | RIA 124 | | 70.4 | | 175 | |
| | Muller | 73 | 76 | 72.8 | 246 | |
| | NRRL B 2419 | 73 | 78 | 72.8 | 146 | |
| *coeruleum (Actinopycnidium)* | USSR RIA 729 | | 70.5–72.5 | | 16, 22 | 16 |
| *diastaticus* | NRRL B 1270 | 77.8 | 70.4 | | 18 | |
| | NRRL B 1270 | 77.8 | | 70.4 | 244 | |

## TABLE 1 (Continued)
## GC COMPOSITION OF THE DNA OF BACTERIA

| Organism | Source or Strain[a] | Mole % Guanosine + Cytosine[b] Tm | Chemical Analysis | Buoyant Density | References[c] A | B |
|---|---|---|---|---|---|---|
| **Streptomycetaceae (continued)** | | | | | | |
| *Streptomyces* | | | | | | |
| *erythraeus* | University of Minnesota | | | 69.8 | 19 | |
| *flaveolus* | ATCC 3319 | 74.8 | | 70.4 | 18 | |
| *fradiae* | 3556 | 72.0 | | | 17 | |
| | ATCC 10745 | 78.5 | | 73.4 | 18 | |
| | Mycelium | | | 73.9 | 164 | |
| | Spores | | | 73.8 | 164 | |
| | Asporogeneous variant mycelium | | | 73.7 | 164 | |
| | MTHU | | 74.5 | | 16 | 16 |
| | IMRU 3535 | | | 71.8 | 19 | 19 |
| | ATCC 10745 | | | 72.4 | 244 | |
| | 3556 | 75.0 | | | 181 | |
| *globisporus* | | | 73.5 | | 8, 176 | |
| *globisporus streptomycini* | | | 73.2 | | 176 | |
| *grisea (Microechinospora)* | p-147 | | 69.1 | | 23 | |
| *griseolus* | IMRU 3325 | | 72.4 | | 16 | |
| | ATCC 3325 | 75.6 | | 70.4 | 18 | |
| *griseus* | ATCC 10137 | 76.8 | 72.1 | 71.4 | 16, 18 | 16 |
| | | | 73.2 | | 8 | |
| | S 104 | 69–69.5 | | | 17, 165 | |
| | E. McCoy | | | 70.4 | 19 | 19 |
| | | 72.0 | | | 81 | |
| | 1945, E. McCoy | 72 | 79 | 71.4 | 246 | |
| | 104 | 70.7–73 | 72 | | 180 | 181 |
| | | | 73 | | 176 | |
| *hachijoensis* | 2725/14 | | 70.9 | | 175 | |
| | H 2609 | | 72.1 | | 16 | 16 |
| | 2725/14 | | 70.7 | | 22 | |
| | 1453/113 | | 70.4 | | 22 | |
| *intermedius* | ATCC 3329 | 73.6 | | 69.4 | 18 | |
| *kentuckensis* | ATCC 12691 | 76.8 | | 69.4 | 18 | |
| *levoris* | 26/1 | | 70.9 | | 175 | |
| | 26/1 | | 70.7 | | 22 | |
| *netropsis* | NRRL 2268 | | 71.5 | | 16 | 16 |

TABLE 1 (Continued)
GC COMPOSITION OF THE DNA OF BACTERIA

| Organism | Source or Strain[a] | Mole % Guanosine + Cytosine[b] Tm | Chemical Analysis | Buoyant Density | References[c] A | B |
|---|---|---|---|---|---|---|
| **Streptomycetaceae (continued)** | | | | | | |
| *Streptomyces* | | | | | | |
| *niveus* | 345 | 67.0 | | | 17 | |
| | 345 | 70 | | | 181 | |
| *parvulus* | JA 4127 | | 70.22 | | 203 | |
| | ATCC 12434 | 77.2 | | 71.4 | 18 | |
| *parvus* | ATCC 12429 | 75.5 | | 70.4 | 18 | |
| *poonensis* | 569 | | 71.9 | | 175 | |
| *poonensis* (*Chainia*) | 569 | | 71.9 | | 22 | |
| *rectus* | IMAM 1c | 79.5 | | | 162 | |
| *rimosus* | ATCC 10970 | | | 70.4 | 19 | 19 |
| *ruber* | IFO 3110 | | 72.8 | | 16 | 16 |
| *rubrireticuli* | ACTU 3631 | | 72.6 | | 16 | 16 |
| *scabies* | L 272V | 72.2 | | | 166 | |
| | I 272A | 72.2 | | | 166 | |
| *streptomycini* | | | 73.2 | | 8 | |
| *thermoviolaceus* (sp. *pingens*) | IMAM 3c | 79.8 | | | 162 | |
| *venezuelae* | S 13 | 71.0 | | 71.4 | 17, 19 | 19, 20 |
| | 86 | | | 71.2 | 19 | 19 |
| | S 13 Virginia Commonwealth University Collection | 72 | | | 177 | |
| | 13 | 70.6–72 | | | 180 | 181 |
| | | 73 | | | 247 | |
| *violaceoruber* | 199 | | | 71.0–71.8 | 17, 19 | 19, 20 |
| | NRRL B 1257 | | | 71.5 | 19 | 19 |
| | 199, University of Minnesota | 72 | 74 | 71.6 | 246 | |
| | NRRL B 1257 | 72 | 73 | 71.8 | 246 | |
| | 307, University of Minnesota | 72 | 74 | 71.6 | 246 | |
| | 3443 R. Gordon Rutgers University | 72 | 74 | 71.4 | 246 | |
| | 3740 R. Gordon Rutgers University | 72 | 75 | 71.4 | 246 | |

## TABLE 1 (Continued)
## GC COMPOSITION OF THE DNA OF BACTERIA

| Organism | Source or Strain[a] | Mole % Guanosine + Cytosine[b] Tm | Chemical Analysis | Buoyant Density | References[c] A | B |
|---|---|---|---|---|---|---|
| **Streptomycetaceae (continued)** | | | | | | |
| *Streptomyces* | | | | | | |
| *violaceoruber* | ATCC 14980 | 72 | 74 | 71.5 | 246 | |
| | 199 | 70.5–74 | | | 180 | 181 |
| | | 72 | | | 247 | |
| *violaceoruber 1* | Sermonti | 72 | 75 | 71.9 | 246 | |
| *violaceus* | ATCC 13760 | 69 | 67 | 69.6 | 246 | |
| *violens* | RIA 565 | | 71.5 | | 182 | |
| | RIA 565 | | 70.9 | | | 175 |
| *violens (Chainia)* | USSR RIA 565 | | 70.9–71.5 | | 16, 22 | 16 |
| *viridis* | ATCC 15386 | 69 | | | 177 | |
| *viridochromogenes* | 45 | 76.0 | 74 | | 181 | 176 |
| | | | 73.5 | 70.3 | 176 | 248 |
| | 93 | 74.0 | | 70.0 | 30, 31 | |
| | | 74.0 | 73.8 | | 8, 17 | |
| | IFO 3113 | | 71.9 | | 16 | 16 |
| *Streptoverticillium* | | | | | | |
| *baldaccii* | IPV 1339 E. Baldacci | 71 | | | 177 | |
| *cinnamoneus* | 1285 | 69.3–70 | | | 180 | 181 |
| *niveus* | | 69.0 | | | 180 | |
| **Thiobacteriaceae** | | | | | | |
| *Thiobacillus* | | | | | | |
| *concretivorus* | NCIB 9514 | | | 51.0 | 142 | |
| | 1 P | | | 51.0 | 142 | |
| | 2 P | | | 52.0 | 142 | |
| *denitrificans* | Baas-Becking (Trudinger) | | | 64.0 | 142 | |
| *ferroxidans* | Aleem | | | 57.0 | 142 | |
| *neopolitanus* | Baas-Becking (Trudinger) | | | 56.0 | 142 | |
| *novellus* | CCM 1077 | 58–59 | | | 66 | |
| | NCIB 9113 | | | 68.0 | 142 | |
| | NCIB 8093 | | | 66.0 | 142 | |

TABLE 1 (Continued)
GC COMPOSITION OF THE DNA OF BACTERIA

| Organism | Source or Strain[a] | Mole % Guanosine + Cytosine[b] Tm | Chemical Analysis | Buoyant Density | References[c] A | B |
|---|---|---|---|---|---|---|
| **Thiobacteriaceae (continued)** | | | | | | |
| *Thiobacillus* | | | | | | |
| *thiocyanoxidans* | NCIB 5177 | | | 63.0 | 142 | |
| *thiooxidans* | NCIB 9112 | | | 52.0 | 142 | |
| | NCIB 8085 | | | 52.0 | 142 | |
| *thioparus* | | 68–70 | | | 5 | |
| | NCIB 8349 | | | 66.0 | 142 | |
| | NCIB 8370 | | | 62.0 | 142 | |
| *trautweinii* | NCIB 9549 | | | 66.0 | 142 | |
| **Thiorhodaceae** | | | | | | |
| *Chromatium* | | | | | | |
| *minutissimum* | | | 63.7 | | 29 | |
| species | D | | | 67.0 | 28 | |
| **Treponemataceae** | | | | | | |
| *Borrelia* | | | | | | |
| *vincentii* | N 9 | | | 46.4 | 161 | |
| *Leptospira* | | | | | | |
| *biflexa* | CDC | | | 38.8 | 161 | 167 |
| Serotype: | | | | | | |
| *andamana* | Correo | 39.1 | | 39.4 | 167 | 167 |
| *patoc* | Patoc 1 | 39.0 | | 38.3 | 167 | 167 |
| *pomona* | RIG | | | 34.7 | 161 | |
| | | | 34–38 | | 5 | |
| *sao-paulo* | Sao Paulo | | | 37.8 | 167 | 167 |
| undetermined | CDC | | | 38.0 | 167 | 167 |
| Pathogenic serotype: | | | | | | |
| *australis* | Ballico | 36.7 | | 35.5 | 167 | 167 |
| *autumnalis* | Akiyama A | 35.5 | | 35.4 | 167 | |
| *ballum* | Mus 127 | | | 39.0 | 167 | |
| *bataviae* | Van Tienen | 36.6 | | 35.3 | 167 | 167 |

**TABLE 1 (Continued)**
**GC COMPOSITION OF THE DNA OF BACTERIA**

| Organism | Source or Strain[a] | Mole % Guanosine + Cytosine[b] Tm | Chemical Analysis | Buoyant Density | References[c] A | B |
|---|---|---|---|---|---|---|
| **Treponemataceae (continued)** | | | | | | |
| *Leptospira* | | | | | | |
| Pathogenic serotype | | | | | | |
| *canicola* | Hond Utrecht | 36.7 | | | 167 | |
| *celledoni* | Celledoni | | | 38.3 | 167 | 167 |
| *copenhageni* | M 20 | 36.6 | | 35.4 | 167 | |
| *hyos* | Mitis Johnson | 40.2 | | | 167 | 167 |
| *icterohaemor-rhagiae* | RGA | 35.5 | | 35.4 | 167 | |
| *javanica* | Veldrat Bataviae 46 | 40.4 | | 39.9 | 167 | 167 |
| | TR 73 | | | 37.7 | 167 | 167 |
| *pomona* | Pomona | 36.0 | | 36.0 | 167 | 167 |
| | Cornelli CB | 36.0 | | 35.4 | 167 | |
| *pyrogenes* | Salinum | | | 34.2 | 167 | |
| species | B 16 | | | 37.2 | 161 | |
| | Turtle strain A 183 | | | 36.0 | 167 | |
| | Turtle strain A 284 | | | 35.0 | 167 | |
| *Treponema* | | | | | | |
| *pallidum* | Reiter | | | 37.8 | 161 | |
| | Nichols | | | 41.3 | 161 | |
| | | | 34–38 | | 5 | |
| **Vitreoscillaceae** | | | | | | |
| *Vitreoscilla* | | | | | | |
| species | | | | 44.0 | 72 | |
| | | | | 44–45 | 267 | |
| | | | | 45.0 | 72 | |
| **Organisms Lacking a Familial Designation** | | | | | | |
| *Flexibacteria* | | | | | | |
| *elegans* | | | | 48.0 | 72 | |

TABLE 1 (Continued)
GC COMPOSITION OF THE DNA OF BACTERIA

| Organism | Source or Strain[a] | Mole % Guanosine + Cytosine[b] Tm | Chemical Analysis | Buoyant Density | References[c] A | B |
|---|---|---|---|---|---|---|
| **Organisms Lacking a Familial Designation (continued)** | | | | | | |
| *Flexothrix* | | | | | | |
| species | | | | 38.0 | 72 | |
| | | | | 38 | 267 | |
| *Herpetosiphon* | | | | | | |
| *aurantiacus* | | | | 48 | 249 | |
| *cohaerens* | II 2 | | | 45 | 249 | |
| *geysericolus* | | | | 48 | 249 | |
| *nigricans* | SS 2 | | | 53 | 249 | |
| *persicus* | T 3 | | | 52 | 249 | |
| *Prosthecomicrobium* | | | | | | |
| *enhydrum* | 9b | | | 65.8 | 168 | |
| *pneumaticum* | 3a | | | 69.4 | 168 | |
| | 3b | | | 69.9 | 168 | |
| *Spirosoma* | | | | | | |
| species | 1, Claus | | | 51.5 | 157 | |
| | 2, Went | | | 51.0 | 157 | |
| *Thermus* | | | | | | |
| *aquaticus* | Y-VII-51D | | | 65.4 | 260 | |
| | Y-VII56-14C | | | 65.4 | 260 | |
| | Y-VII56-3A | | | 65.4 | 260 | |
| | YT-1 | | | 67.4 | 260 | |
| | X-1 | | | 64 | 261 | |
| Unclassified marine and semimarine organisms, and marine isolates of terrestrial types | XG3, XP14, YBC11, XV8, XT8, XT9 | | | 39.0–41.5 | 169 | |
| | 8-2 | | | 59.0 | 169 | |
| | XF1, Y17, T67c, XG4, 3-5, XD1, XA1, XC1, 1C2 | | | 40.0–46.5 | 169 | |
| Unclassified organisms | C5M | 56.8 | | | 227 | |
| | 116-50 | 41.2 | | | 227 | |
| | B1P | 56.1 | | | 227 | |
| | A2M | 56.4 | | | 227 | |
| | XG21 | | | 61.0 | 169 | |

**TABLE 1 (Continued)**
**GC COMPOSITION OF THE DNA OF BACTERIA**

| Organism | Source of Strain[a] | Mole % Guanosine + Gytosine[b]: Tm | Chemical Analysis | Buoyant Density | References[c]: A | B |
|---|---|---|---|---|---|---|
| **Organisms Lacking a Familial Designation (continued)** | | | | | | |
| Unclassified organisms | XR4, XP3, 212, XB2, XL3, X013, 13-11, XD4, 2B1, XL1, 2NG, XH2, XW1 | | | 40.5–48.5 | 169 | |
| | 17L3, 14-2, 1R5, 5L2, 155, 2K10, 1R4 | | | 51.5–60.0 | 169 | |
| | XW14 | | | 52.0 | 169 | |
| | 14-7, 3A5, 2M4, 11-8, 7L7, 3F2, 1G11, 4-5, 2N1, 12-1, 9-13 | | | 49–67 | 169 | |
| | 1F2, 1S1 | | | 59.2–60.5 | 169 | |
| | Snail I | | | 47.5 | 169 | |
| | 2F5, DW286, 3F15, 4-4, DW102, 2F38 | | | 64.5–67.3 | 169 | |

[a] Many of the organisms listed under source or strain have privately designated numbers or letters. Where possible, these organisms have been additionally identified by their designations in national or widely recognized culture collections. The following list is an explanation of the known abbreviations used for those collections:

| | |
|---|---|
| ACTU | Department of Agricultural Chemistry, University of Tokyo, Tokyo, Japan. |
| ATCC | American Type Culture Collection, Washington, D.C., U.S.A. |
| BKM | All Union Collection of Microorganisms, U.S.S.R. |
| BUCM | Baylor University Medical College, Houston, Texas, U.S.A. |
| CBS | Centraalbureau voor Schimmelcultures, Baarn, The Netherlands. |
| CCEB | Culture Collection of Entomogenous Bacteria, Prague, Czechoslovakia. |
| CCM | Czechoslovak Collection of Microorganisms, Brno, Czechoslavakia. |
| CDC | Communicable Disease Center, Atlanta, Georgia, U.S.A. |
| CNRZ | Centre National de la Recherche Zootechnique, Jouy-en-Josas, France. |
| ESB | E. S. Beneke, Biological Research Center, Michigan State University, East Lansing, Michigan, U.S.A. |
| ETH | Eidgenössische Technische Hochschule, Zürich, Switzerland. |
| FDA | Food and Drug Administration, Washington, D.C., U.S.A. |
| Gal. | M. E. Gallegly, Department of Plant Pathology, West Virginia University, Morgantown, West Virginia, U.S.A. |
| IAM | Institute of Applied Microbiology, University of Tokyo, Tokyo, Japan. |
| ICPPB | International Collection of Phytopathogenic Bacteria, University of California, Davis, California, U.S.A. |
| IFO | Institute of Fermentation, Osaka, Japan. |
| IH | Institute of Hygiene, Warsaw, Poland. |
| IMAM | Instituto di Microbiologia Agraria, Milano, Italy. |
| IMRAU | Institute of Microbiology, Rutgers University, New Brunswick, New Jersey, U.S.A. |
| IP (CIP) | Collection Institut Pasteur, Paris, France. |
| IPV | Instituto di Patologia Vegetale, Milano, Italy. |
| ISM | Instituto Sieroterapico Milanese, Milano, Italy. |

| | |
|---|---|
| MHD | Milwaukee Health Department, Milwaukee, Wisconsin, U.S.A. |
| MRC | Medical Research Council (P. H. A. Sneath), Mill Hill, London, England. |
| MTHU | School of Medicine, Tohoku University, Sendai, Japan. |
| NBL | Naval Biological Laboratory, Oakland, California, U.S.A. |
| NCDO | National Collection of Dairy Organisms, Shinfield, Reading, Berkshire, England. |
| NIRD | |
| NCIB | National Collection of Industrial Bacteria, Aberdeen, Scotland. |
| NCMB | National Collection of Marine Bacteria, Aberdeen, Scotland. |
| NCPPB | National Collection of Plant-Pathogenic Bacteria, Harpenden, England. |
| NCTC | National Collection of Type Cultures, Colindale, London, England. |
| NIH | National Institutes of Health, Bethesda, Maryland, U.S.A. |
| NRRL | Northern Utilization Research and Development Division, U.S. Department of Agriculture, Peoria, Illinois, U.S.A. |
| QM | Pioneering Research Center, U.S. Army Laboratory, Nactic, Massachusetts, U.S.A. |
| RH | Dr. Rudolph Hugh, George Washington University, Washington, D.C., U.S.A. |
| RKB | See RSA. |
| RSA | Rancho Santa Ana Botanical Gardens, Claremont, California, U.S.A. |
| SSIC | State Serum Institute, Copenhagen, Denmark. |
| USDA | U.S. Department of Agriculture (L. W. Erdman), Beltsville, Maryland, U.S.A. |
| USSR-RIA | Research Institute for Antibiotics, Moscow, U.S.S.R. |
| UT | University of Texas isolate. |

*b* Mole percent given to the nearest 0.1%.
*c* Column A: primary reference for GC values; column B: reference for hybridization data.
*d* May be homologous with *S. griseüs.*
*e* By chromatography.
*f* Subgenus designation.
*g* Range of values for two methods, Tm and $E_{260}/E_{280}$.
*h* According to the author, this organism may be the same as *Corynebacterium.*
*i* $E_{260}/E_{280}$ at pH 3.
*j* By bromination.
*k* High-GC groups; H-D = host-dependent, H-I = host-independent.

# TABLE 2
# GC COMPOSITION OF THE DNA OF FUNGI

| Organism | Source or Strain[a] | Mole % Guanosine + Cytosine[b] | | | References[c] | |
|---|---|---|---|---|---|---|
| | | Tm | Chemical Analysis | Buoyant Density | A | B |
| *Absidia* | | | | | | |
| *blakesleeana* | QM 6774, NRRL 1304 | | | 52 | 269 | |
| *cylindrospora* | NRRL A12905 | | | 41.5 | 269 | |
| | NRRL A12872 | | | 40.5 | 269 | |
| *glauca* | | 48 | | 44 | 270 | |
| *regneri* | QM 45b | | | 59 | 269 | |
| *spinosa* | NRRL 2797 | | | 41 | 269 | |
| species | | | 39 | | 272 | |
| *Achlya* | | | | | | |
| *ambisexualis* | ESB | | | 54.5 | 269 | |
| *benekei* | ESB | | | 62 | 269 | |
| *flagellata* | ESB | | | 55 | 269 | |
| *klebsiana* | ESB | | | 44.5 | | |
| *oviparvula* | ESB | | | 46.5 | 269 | |
| *Acrasis* | | | | | | |
| *rosea* | | | | 36 | 273 | |
| *Acrothecium* | | | | | | |
| *arenarium* | QM 8024 | | | 45 | 269 | |
| *Actinomucor* | | | | | | |
| *elegans* | UT 199 | | | 40.5 | 269 | |
| | RKB 12 | | | 41.5 | 269 | |
| | RKB 162 | | | 40.5 | | |
| | RKB 349 | | | 40.5 | 269 | |
| | RKB 605 | | | 41.5 | 269 | |
| | RKB 860 | | | 42 | 269 | |
| | RKB 1062 | | | 41 | 269 | |
| | RKB 1702 | | | 43 | 269 | |
| | RKB 1703 | | | 41 | 269 | |
| | RKB 1704 | | | 41 | 269 | |
| | RKB 1705 | | | 41 | 269 | |
| | RKB 1706 | | | 42 | 269 | |
| *Acytostelium* | | | | | | |
| *leptosomun* | | | | 36 | 273 | |

## TABLE 2 (Continued)
## GC COMPOSITION OF THE DNA OF FUNGI

| Organism | Source or Strain[a] | Mole % Guanosine + Cytosine[b] Tm | Chemical Analysis | Buoyant Density | References[c] A | B |
|---|---|---|---|---|---|---|
| *Agaricus* | | | | | | |
| *bisporus* | | | 44.4 | | 274 | |
| *campestris* | | | 44 | | 275 | |
| *Amanita* | | | | | | |
| *muscaria* | | | 57.1 | | 272 | |
| *strobiliformis* | | | 58 | | 276 | |
| *Amauroascus* | | | | | | |
| *verrucosus* | QM 1802M | | | 53 | 269 | |
| | QM 8502 | | | 53 | 269 | |
| *Anixiopsis* | | | | | | |
| *stercoraria* | QM 8503 | | | 50.5 | 269 | |
| | QM 8504 | | | 55 | 269 | |
| *Arachniotus* | | | | | | |
| *flavolutens* | QM 8505 | | | 53 | 269 | |
| | QM 8506 | | | 52.5 | 269 | |
| *reticulatus* | QM 8507 | | | 52 | 269 | |
| *Arthrobotrys* | | | | | | |
| *superba* | QM 1688 | | | 50.5 | 269 | |
| *Aspergillus* | | | | | | |
| *alliaceus* | QM 1885 | | | 52 | 269 | |
| *ambiguus* | QM 8155 | | | 54 | 269 | |
| *aureolus* | QM 1906 | | | 54 | 269 | |
| *avenaceus* | QM 6741 | | | 51 | 269 | |
| *awamori* | QM 6949 | | | 52.5 | 269 | |
| | QM 7397 | | | 52.5 | 269 | |
| *candidus* | QM 1997 | | | 53.5 | 269 | |
| *carbonarius* | QM 331 | | | 54 | 269 | |
| *carbonarius* (*A. fonsecaeus*) | UT 16 | | | 56 | 269 | |

## TABLE 2 (Continued)
## GC COMPOSITION OF THE DNA OF FUNGI

| Organism | Source or Strain[a] | Mole % Guanosine + Cytosine[b] | | | References[c] | |
|---|---|---|---|---|---|---|
| | | Tm | Chemical Analysis | Buoyant Density | A | B |
| *Aspergillus* | | | | | | |
| *clavatus* | UT 10 | | | 55 | 269 | |
| | QM 6884 | | | 52.5 | 269 | |
| *conicus* | QM 7405 | | | 61 | 269 | |
| *fasciculatus* | QM 6950 | | | 51 | 269 | |
| *flavipes* | UT 14 | | | 58 | 269 | |
| *flavus* | UT 15 | | | 51 | 269 | |
| | QM 10e | | | 51.5 | 269 | |
| | QM 7637 | | | 50.5 | 269 | |
| | QM 8190A | | | 50 | 269 | |
| *fumigatus* | QM 1981 | | | 53 | 269 | |
| *giganteus* | UT 19 | | | 52 | 269 | |
| | QM 1970 | | | 52.5 | 269 | |
| | QM 7974 | | | 54.5 | 269 | |
| *japonicus* | QM 332 | | | 54.5 | 269 | |
| *kanagawaensis* | QM 7396 | | | 53 | 269 | |
| *katsuobushi* | QM 8157 | | | 54.5 | 269 | |
| *luchuensis* | QM 5565 | | | 54.5 | 269 | |
| *nidulans* | | | 47 | | 277 | |
| *niger* | | 52 | 50.1 | 52 | 270 | |
| | QM 1999 | | | 52.5 | 269 | |
| | QM 7922 | | | 54 | 269 | |
| | QM 8404 | | | 54 | 269 | |
| | QM 8487 | | | 53.5 | 269 | |
| | QM 8195A | | | 52.5 | 269 | |
| *niger* (var. *cinnamoneus*) | QM 326 | | | 52.5 | 269 | |
| *niveo-glaucus* | UT 6 | | | 54 | 269 | |
| *niveus* | QM 6855 | | | 55 | 269 | |
| *niveus* (var. *bifidus*) | QM 7213 | | | 54 | 269 | |
| *nutans* | QM 8159 | | | 48 | 269 | |
| *ochraceus* | UT 22 | | | 54 | 269 | |
| | QM 58c | | | 52.5 | 269 | |
| | QM 6731 | | | 53.5 | 269 | |

## TABLE 2 (Continued)
## GC COMPOSITION OF THE DNA OF FUNGI

| Organism | Source or Strain[a] | Mole % Guanosine + Cytosine[b] | | | References[c] | |
|---|---|---|---|---|---|---|
| | | Tm | Chemical Analysis | Buoyant Density | A | B |
| *Aspergillus* | | | | | | |
| *oryzae* | QM 1273 | | | 52.5 | 269 | |
| *oryzae* (conidia) | | | 46.0 | | 278 | |
| *parvulus* | QM 7955 | | | 48.5 | 269 | |
| *phoenicis* | QM 329 | | | 52.5 | 269 | |
| *proliferans* | QM 7462 | | | 55 | 269 | |
| *quadricinctus* | QM 6874 | | | 55 | 269 | |
| *quadrilineatus* | QM 7465 | | | 52.5 | 269 | |
| *restrictus* | QM 7305 | | | 52 | 269 | |
| *sydowi* | QM 4d | | | 52.5 | 269 | |
| *tamarii* | QM 506 | | | 49 | 269 | |
| | QM 1223 | | | 49.5 | 269 | |
| | QM 6733 | | | 50.5 | 269 | |
| *terreus* | UT 29 | | | 56.5 | 269 | |
| | QM 1991 | | | 55 | 269 | |
| | QM 1992 | | | 55 | 269 | |
| *terreus* (var. *boedijni*) | QM 7473 | | | 57 | 269 | |
| *terreus* (var. *floccosus*) | QM 7474 | | | 55 | 269 | |
| *unguis* | QM 8f | | | 54 | 269 | |
| | QM 25b | | | 53.5 | 269 | |
| *ustus* | QM 7477 | | | 55 | 269 | |
| *versicolor* | QM 4g | | | 53 | 269 | |
| | QM 432 | | | 51.5 | 269 | |
| *violaceus* | QM 1905 | | | 52 | 269 | |
| *Aureobasidium* | | | | | | |
| *pullulans* | AJ 4981 (salted cuttle-fish) | 51.5 | | | 279 | |

## TABLE 2 (Continued)
## GC COMPOSITION OF THE DNA OF FUNGI

| Organism | Source or Strain[a] | Mole % Guanosine + Cytosine[b] | | | References[c] | |
|---|---|---|---|---|---|---|
| | | Tm | Chemical Analysis | Buoyant Density | A | B |
| *Auxarthon* | | | | | | |
| *brunneum* | QM 8508 | | | 52 | 269 | |
| *californiense* | QM 8509 | | | 50 | 269 | |
| *reticulatum* | QM 8512 | | | 52.5 | 269 | |
| *zuffianum* | QM 8514 | | | 52 | 269 | |
| *Backusia* | | | | | | |
| *terricola* | QM 8602 | | | 52 | 269 | |
| | QM 8603 | | | 52 | 269 | |
| *Basidiobolus* | | | | | | |
| *ranarum* | UT 32 | | | 38 | 269 | |
| *Beauveria* | | | | | | |
| *tenella* | QM 7954 | | | 53 | 269 | |
| *Bispora* | | | | | | |
| *punctata* | QM 7369 | | | 52.5 | 269 | |
| *Bjerkandera* | | | | | | |
| *adusta* | DAOM 22299 | | | 57.5 | 271 | |
| | DAOM 53500 | | | 56 | 271 | |
| *fumosa* | DAOM 10257 | | | 52 | 271 | |
| *Blakesleea* | | | | | | |
| *trispora* | NRRL 2456 | | | 39 | 269 | |
| *Blastocladiella* | | | | | | |
| *emersonii*[d] | | | | 66 | 280 | |
| *Botrytis* | | | | | | |
| *cinerea* | | | 50.9 | | 272 | |
| *Bovista* | | | | | | |
| species | | | 50.7 | | 272 | |
| *Brettanomyces* | | | | | | |
| *bruxellensis* | AJ 5022 (IFO 0677) | 35 | | | 279 | |

TABLE 2 (Continued)
GC COMPOSITION OF THE DNA OF FUNGI

| Organism | Source or Strain[a] | Mole % Guanosine + Cytosine[b] Tm | Chemical Analysis | Buoyant Density | References[c] A | B |
|---|---|---|---|---|---|---|
| *Byssochlamys* | | | | | | |
| *fulva* | QM 6766 | | | 51.5 | 269 | |
| *Candida* | | | | | | |
| *albicans* | AJ 4409 (IFO 1060) | 32.5 | | | 279 | |
| *albicans* (DNA) | CBS 2712 | 35.1 | | | 281 | |
| *atmosphaerica* | NRRL Y-5979 | 41.2 | | | 282 | |
| *atmosphaerica* (DNA) | CSAV 29-50-1 | 39.7 | | | 281 | |
| *brumptii* (DNA) | CBS 564 | 54.1 | | | 281 | |
| *catenulata* (DNA) | CBS 565 | 54.4 | | | 281 | |
| *claussenii* (DNA) | CSAV 29-31-1 | 34.9 | | | 281 | |
| *diddensii* | | 39.8 | | | 282 | |
| *gelida* | AJ 4991 (from Dr. E. DiMenna) | 52 | | | 279 | |
| *humicola* | AJ 4682 (IFO 0753) | 60 | | | 279 | |
| *koshuensis* | AJ 4939 (RIFY WF-87) | 30 | | | 279 | |
| *krusei* | AJ 4421 (IFO 0011) | 38 | | | 279 | |
| *krusei* (DNA) | CBS 573 | 39.6 | | | 281 | |
| *lipolytica* | | 49.5 | | | 283 | |
| *lipolytica* (DNA) | CBS 599 | 49.6 | | | 281 | |
| *melinii* | AJ 4696 (IFO 0747) | 36.5 | | | 279 | |
| *melinii*(DNA) | CBS 661 | 40.9 | | | 281 | |
| *oregonensis* | | 48.0 | | | 282 | |
| *parapsilosis* | CBS 604[e] | 40.0 | | | 282 | |
| *parapsilosis* (DNA) | CBS 604 | 40.8 | | | 281 | |

## TABLE 2 (Continued)
## GC COMPOSITION OF THE DNA OF FUNGI

| Organism | Source or Strain[a] | Mole % Guanosine + Cytosine[b] | | | References[c] | |
|---|---|---|---|---|---|---|
| | | Tm | Chemical Analysis | Buoyant Density | A | B |
| *Candida* | | | | | | |
| *parapsilosis* (var. *hokkai*) | AJ 4611 (IAM 4488) | 52 | | | 279 | |
| *pelliculosa* | AJ 4534 Chinese cabbage | 34.1 | | | 271 | |
| *pelliculosa* (DNA) | CBS 605 | 36.8 | | | 281 | |
| *pseudotropicalis* (DNA) | CBS 607 | 41.3 | | | 281 | |
| *pulcherrima* | | 48 | | 46 | 270 | |
| | AJ 4419 (IFO 0561) | 44.5 | | | 279 | |
| *pulcherrima* (DNA) | CBS 610 | 48.0 | | | 281 | |
| *punicea* | AJ 4764 | 51 | | | 279 | |
| *rugosa* | AJ 4644 (IFO 0750) | 47 | | | 279 | |
| *stellatoidea* (DNA) | CBS 1905 | 35.7 | | | 281 | |
| *tenuis* (DNA) | CBS 615 | 44.0 | | | 281 | |
| *tropicalis* | Sawai | 34.9 | | | 282 | |
| *tropicalis* (DNA) | CBS 120/5 | 34.9 | | | 281 | |
| *truncata* (DNA) | CBS 1899 | 36.9 | | | 281 | |
| *utilis* | AJ 4640 IFO 1086 | 40.2 | | | 279 | |
| *utilis* (DNA) | CBS 621 | 45.8 | | | 281 | |
| *zeylanoides* | AJ 4287 (CBS 2764) | 51.5 | | | 279 | |
| | AJ 4741 | 51.5 | | | 279 | |
| *zeylanoides* (DNA) | CBS 619 | 57.6 | | | 281 | |
| *Ceratocystis* | | | | | | |
| *ulmi* | QM 8426 | | | 53 | 269 | |
| | | | | 56 | 284 | |
| *Cercospora* | | | | | | |
| *salina* | QM 8322 | | | 64.5 | 269 | |

## TABLE 2 (Continued)
## GC COMPOSITION OF THE DNA OF FUNGI

| Organism | Source or Strain[a] | Mole % Guanosine + Cytosine[b] Tm | Chemical Analysis | Buoyant Density | References[c] A | B |
|---|---|---|---|---|---|---|
| *Ceriporiopsis* | | | | | | |
| *placenta* | DAOM QM 1010 | | | 53 | 271 | |
| | DAOM 11663 | | | 54 | 271 | |
| *placenta* (*Poria monticola*) | DAOM 94818 | | | 56.5 | 271 | |
| *Chaetocladium* | | | | | | |
| *brefeldii* | NRRL 2508 | | | 41 | 269 | |
| *Chaetomella* | | | | | | |
| *raphigera* | QM 7359 | | | 53.5 | 269 | |
| *Chaetomium* | | | | | | |
| *bostrychodes* | QM 6711 | | | 59 | 269 | |
| *brasiliense* | QM 623 | | | 51.5 | 269 | |
| *caprinum* | QM 6695 | | | 57.5 | 269 | |
| *causiaeformis* | QM 949 | | | 60 | 269 | |
| *elatum* | QM 606 | | | 56.5 | 269 | |
| *fusisporum* | QM 7960 | | | 49.5 | 269 | |
| *globosum* | QM 6694 | | | 57.5 | 269 | |
| | QM 8199 | | | 57.5 | 269 | |
| | QM 8402 | | | 58 | 269 | |
| | QM 8495A | | | 58 | 269 | |
| | | | | 58 | 284 | |
| *indicum* | QM 46b | | | 58.5 | 269 | |
| | QM 8014 | | | 50 | 269 | |
| *mollipolium* | QM 1007 | | | 57 | 269 | |
| *murorum* | QM 6709 | | | 48.5 | 269 | |
| *subspirilliferum* | QM 8180 | | | 58.5 | 269 | |
| *succineum* | QM 1044 | | | 56.5 | 269 | |
| *tenuissimum* | QM 8178 | | | 57.5 | 269 | |
| *Chloridium* | | | | | | |
| *viride* | QM 1103 | | | 49.5 | 269 | |

## TABLE 2 (Continued)
## GC COMPOSITION OF THE DNA OF FUNGI

| Organism | Source or Strain[a] | Mole % Guanosine + Cytosine[b] | | | References[c] | |
|---|---|---|---|---|---|---|
| | | Tm | Chemical Analysis | Buoyant Density | A | B |
| *Choanephora* | | | | | | |
| *cucurbitarum* | NRRL 2744 | | | 40 | 269 | |
| *Circinella* | | | | | | |
| *linderi* | QM 762, NRRL 2342 | | | 53.5 | 269 | |
| *minor* | QM 6939, NRRL 1453 | | | 53.5 | 269 | |
| *muscae* | QM 629 | | | 36 | 269 | |
| | QM 629M | | | 39.5 | 269 | |
| | QM 7788 | | | 35.5 | 269 | |
| *umbellata* | USDA A-12910 | | | 54.5 | 269 | |
| *Citeromyces* | | | | | | |
| *matritensis* | AJ 4287 (CBS 2764) | 42.5 | | | 279 | |
| *Cladosporium* | | | | | | |
| *cladosporioides* | QM 71d | | | 49 | 269 | |
| | QM 489 | | | 50 | 269 | |
| *herbarum* | QM 3167 | | | 55 | 269 | |
| *resinae* | QM 8598 | | | 54 | 269 | |
| *resinae* f. *avellaneum* | QM 8042 | | | 51.5 | 269 | |
| *Claviceps* | | | | | | |
| *purpurea* | | | 52.6 | | 272 | |
| *Coemansia* | | | | | | |
| *braziliensis* | RSA 77 | | | 50.5 | 269 | |
| *mojavensis* | RSA 71 | | | 54.5 | 269 | |
| *spiralis* | UT 268 | | | 49 | 269 | |
| | RSA 1278 | | | 50.5 | 269 | |
| *Cokeromyces* | | | | | | |
| *poitrassi* | RKB 903 | | | 37 | 269 | |
| | RKB 1095 | | | 36 | 269 | |
| | RKB 1245 | | | 36.5 | 269 | |
| | RKB 1264 | | | 38 | 269 | |
| | RKB 1267 | | | 36.5 | 269 | |

## TABLE 2 (Continued)
## GC COMPOSITION OF THE DNA OF FUNGI

| | | Mole % Guanosine + Cytosine[b] | | | References[c] | |
|---|---|---|---|---|---|---|
| Organism | Source or Strain[a] | Tm | Chemical Analysis | Buoyant Density | A | B |
| *Cokeromyces* | | | | | | |
| *recurvatus* | RSA 1 | | | 31.5 | 269 | |
| | RSA 918 | | | 32 | 269 | |
| *Colletotrichum* | | | | | | |
| *lagenarium* | | | 50 | 53 | 277 | |
| *Coprinus* | | | | | | |
| *lagopus* | | | | 52 | 273 | |
| | | | 52 | 53 | | 277 |
| *Cryptococcus* | | | | | | |
| *albidus* | | 55 | | 55 | 270 | |
| *flavus* (DNA) | CBS 331 | | | 55 | 285 | |
| (*Rhodotorula flava*) | | | 55 | 270 | | |
| *gastricus* | CBS 1927 | | | 51 | 285 | |
| | CBS 2288 | | | 65.5 | 285 | |
| *laurentii* | AJ 4291 (RIFY←CBS 139) | 56 | | | 279 | |
| *laurentii* (var. *magnus*, syn. *Torula heveanensis*) | | | | 49 | 285 | |
| *laurentii* (var. *flavescens*, syn. *Rhodotorula peneaus)* | | | | 58 | 285 | |
| *melibiosum* (syn. *Torulopsis melibiosum)* | | | | 61 | 285 | |
| *neoformans* | AJ 4290 (RIFY←CBS 132) | 46 | | | 279 | |
| | | | | 51.5 | 285 | |
| *skinneri* | | | | 53 | 285 | |
| *terreus* | | | | 59.5 | 285 | |
| *uniguttulatus* | | | | 58 | 285 | |
| *uniguttulatus* (var. *uniguttulatus*, syn. *Cryptococcus neoformans)* | | | | 51.5 | 285 | |

## TABLE 2 (Continued)
## GC COMPOSITION OF THE DNA OF FUNGI

| Organism | Source or Strain[a] | Mole % Guanosine + Cytosine[b] Tm | Chemical Analysis | Buoyant Density | References[c] A | B |
|---|---|---|---|---|---|---|
| *Ctenomyces* | | | | | | |
| *serratus* | QM 8516 | | | 52.5 | 269 | |
| *Cunninghamella* | | | | | | |
| *baineri* | ATCC 6796B | | | 32 | 269 | |
| *bertholletiae* | HLL | | | 27.5 | 269 | |
| *blakesleeana* | NRRL 1368 | | | 31 | 269 | |
| | ATCC 8688b | | | 34 | 269 | |
| *echinulata* | ATCC 8688A | | | 33.5 | 269 | |
| | ATCC 11585A | | | 32.5 | 269 | |
| | ATCC 11585B | | | 32.5 | 269 | |
| | | | | 34 | 284 | |
| *echinulata* (–) | QM 6783, NRRL 1387 | | | 30.5 | 269 | |
| *elegans* | ATCC 6795B | | | 31 | 269 | |
| | ATCC 10025A | | | 31 | 269 | |
| *homothallica* | NRRL 2365 | | | 29.5 | 269 | |
| *verticillata* | ATCC 8983 | | | 31 | 269 | |
| *vesiculosa* | NRRL 3009 | | | 28 | 269 | |
| *Curvularia* | | | | | | |
| *eragrostidis* | QM 7931 | | | 53.5 | 269 | |
| *lunata* | QM 3728 | | | 53 | 269 | |
| *maculans* | QM 666 | | | 53.5 | 269 | |
| | QM 4761 | | | 53.5 | 269 | |
| | QM 6208 | | | 53 | 269 | |
| *siddiquii* | QM 8356 | | | 56 | 269 | |
| *verruciformis* | QM 8326 | | | 52 | 269 | |
| species | QM 8133 | | | 52 | 269 | |
| *Cylindrocephalum* | | | | | | |
| *aureum* | QM 523 | | | 54 | 269 | |
| *Dactylium* | | | | | | |
| *dendroides* | QM 513 | | | 51.5 | 269 | |

TABLE 2 (Continued)
GC COMPOSITION OF THE DNA OF FUNGI

| Organism | Source or Strain[a] | Mole % Guanosine + Cytosine[b] Tm | Chemical Analysis | Buoyant Density | References[c] A | B |
|---|---|---|---|---|---|---|
| *Daedalea* | | | | | | |
| *confragosa* | DAOM 95957 | | | 59.0 | 268 | |
| *confragosa* (*Daedaleopsis confragosa*) | | | | 57.0 | 284 | |
| *quercina* | DAOM F 2278 | | | 55.5 | 269 | |
| *Debaryomyces* | | | | | | |
| *globosus* | CBS 764 | 45.1 | | | 282 | |
| *hansenii* | AJ 4179 (IFO 0023) | 34.5 | | | 279 | |
| | CBS | 36.6 | | | 282 | |
| *kloeckeri* (*hansenii*) | | 40 | | 40 | 270 | |
| *Dictyostelium* | | | | | | |
| *discoiaeum* | | 25 | 22 | 22 | 286 | 287 |
| *mucoroides* | | | | 22 | 273 | |
| *purpureum* | | | | 25 | 273 | |
| *Dictyuchus* | | | | | | |
| *pseudoachlyoides* | ESB | | | 49.5 | 269 | |
| *Dipodascus* | | | | | | |
| *albidus* | AJ 4896 (exudate of tree) | 33 | | | 279 | |
| *uninucleatus* | | | | 43 | 270 | |
| *Dipsacomyces* | | | | | | |
| *acuminosporus* | RSA 1012 | | | 52.5 | 269 | |
| *Eidamella* | | | | | | |
| *deflexa* | QM 8468 | | | 55 | 269 | |
| *Elfvingia* | | | | | | |
| *applanata* | DAOM 10206 | | | 58 | 271 | |

## TABLE 2 (Continued)
## GC COMPOSITION OF THE DNA OF FUNGI

| Organism | Source or Strain[a] | Mole % Guanosine + Cytosine[b] Tm | Chemical Analysis | Buoyant Density | References[c] A | B |
|---|---|---|---|---|---|---|
| *Elfvingia* | | | | | | |
| *applanata* (*Ganoderma*) | DAOM 21775 | | | 59.5 | 271 | |
| *Emericella* | | | | | | |
| *nidulans* | UT 23 | | | 51 | 269 | |
| | UT 24 | | | 54 | 269 | |
| *nidulans* (*Aspergillus nidulans*) | QM 1985 | | | 53 | 269 | |
| *rugulosa* | UT 65-6 | | | 54 | 269 | |
| *variecolor* (*Aspergillus variecolor*) | QM 1910 | | | 53 | 269 | |
| *Emericellopsis* | | | | | | |
| *salmosynnemata* (syn. *Cephalosporium salmosynnemata*) | QM 6889 | | | 53.5 | 269 | |
| *Endomyces* | | | | | | |
| *reesii* | | 41 | | 39 | 270 | |
| *Endomycopsis* | | | | | | |
| *capsularis* | AJ 4271 (IFO 0672) | 40 | | | 279 | |
| *fibuligera* | AJ 4270 (IAM 4247) | 37.5 | | | 279 | |
| *muscicola* | AJ 4272 (moss, M-5-1 (32)) | 34.5 | | | 279 | |
| *Epicoccum* | | | | | | |
| *neglectum* | QM 1070 | | | 51 | 269 | |
| *Eremascus* | | | | | | |
| *fertilis* | QM 6887 | | | 54 | 269 | |
| *Eupenicillium* | | | | | | |
| *baarnense* (*Penicillium baarnense*) | QM 1871 | | | 54.5 | 269 | |

TABLE 2 (Continued)
GC COMPOSITION OF THE DNA OF FUNGI

| Organism | Source or Strain[a] | Mole % Guanosine + Cytosine[b] Tm | Chemical Analysis | Buoyant Density | References[c] A | B |
|---|---|---|---|---|---|---|
| *Eupenicillium* | | | | | | |
| *brefeldianum* (*Penicillium brefeldianum*) | QM 1872 | | | 51 | 269 | |
| *egyptiacum* (*Penicillium egyptiacum*) | QM 7553 | | | 48.5 | 269 | |
| *javanicum* (*Penicillium javanicum*) | UT 135 | | | 51.5 | 269 | |
| | UT 136 | | | 53 | 269 | |
| *Eurotium* | | | | | | |
| *amstelodami* (*Aspergillus amstelodami*) | QM 8405 | | | 56 | 269 | |
| | QM 8486 | | | 54 | 269 | |
| | QM 8486a | | | 52 | 269 | |
| *herbariorum* (*Aspergillus mangini*) | QM 7419 | | | 55 | 269 | |
| *repens* | UT 273 | | | 54.5 | 269 | |
| *rubrum* (*Aspergillus ruber*) | QM 360 | | | 53 | 269 | |
| | QM 1973 | | | 54 | 269 | |
| *Fomes* | | | | | | |
| *fraxineus* | DAOM F2161 | | | 51.0 | 271 | |
| | DAOM F7519 | | | 56.0 | 271 | |
| *fraxinophilus* | DAOM F1931 | | | 54.5 | 271 | |
| *Fomitopsis* (*Fomes*) | | | | | | |
| *pinicola* | DAOM | | | 56.5 | 271 | |
| | DAOM | | | 57 | 271 | |
| | DAOM | | | 57 | 271 | |
| *Fulgio* | | | | | | |
| *viritans* | | | 33.5 | | 272 | |
| *Fusarium* | | | | | | |
| *episphaeria* | QM 7826 | | 48 | 53 | 269 | |
| *moniliforme* | QM 527 | | | 51 | 269 | |
| | QM 1224 | | | 50 | 269 | |

## TABLE 2 (Continued)
## GC COMPOSITION OF THE DNA OF FUNGI

| Organism | Source or Strain[a] | Mole % Guanosine + Cytosine[b] Tm | Chemical Analysis | Buoyant Density | References[c] A | B |
|---|---|---|---|---|---|---|
| *Fusarium* | | | | | | |
| *moniliforme* (var. *minus*) | QM 556 | | | 51 | 269 | |
| *oxysporum* | QM 21c | | | 50.5 | 269 | |
| | QM 47e | | | 50.5 | 269 | |
| *oxysporum* f. *lycopersici* | | | 48 | | 277 | |
| *roseum* | QM 38g | | | 50 | 269 | |
| *sambucum* | QM 7162 | | | 51.5 | 269 | |
| *scirpi* | QM 660 | | | 51.5 | 269 | |
| *solani* | QM 21d | | | 50.5 | 269 | |
| *Ganoderma* | | | | | | |
| *tsugae* | DAOM F569B | | | 54 | 271 | |
| *Gelasinospora* | | | | | | |
| *autosteira* (a) | UT CR-4a | | | 53 | 269 | |
| | QM 7817A | | | 53 | 269 | |
| *autosteira* (A) | UT CR-7A | | | 53 | 269 | |
| | UT CR–10A | | | 53.5 | 269 | |
| *autosteira* | | | | 54 | 284 | |
| *calospora* | | | | 55 | 284 | |
| | | | 52 | | 277 | |
| | | 53 | | 55 | 270 | |
| *cerealis* | | | | 55 | 284 | |
| *tetrasperma* | | | 50 | | 277 | |
| *Geotrichum* | | | | | | |
| *candidum* | AJ 4884 (moss) | 40.5 | | | 279 | |
| *Gilbertella* | | | | | | |
| *persicaria* | NRRL 1546 | | | 40 | 269 | |
| *Gliocladium* | | | | | | |
| *nigrum* | QM 1240 | | | 60 | 269 | |

## TABLE 2 (Continued)
## GC COMPOSITION OF THE DNA OF FUNGI

| Organism | Source or Strain[a] | Mole % Guanosine + Cytosine[b] Tm | Chemical Analysis | Buoyant Density | References[c] A | B |
|---|---|---|---|---|---|---|
| *Gloeophyllum (Lenzites)* | | | | | | |
| *saepiarium* | DAOM 22442 | | | 54 | 271 | |
| | DAOM 22276 | | | 54 | 271 | |
| *Gymnoascus* | | | | | | |
| *reessii* | QM 8517 | | | 54.5 | 269 | |
| | QM 8521 | | | 55 | 269 | |
| *uncinatus* | QM 8590 | | | 51.5 | 269 | |
| *Hanseniaspora* | | | | | | |
| *valbyensis* | AJ 4277 (IFO 0683) | 30 | | | 279 | |
| *Hansenula* | | | | | | |
| *angusta* (*polymorpha*) | AJ 5046 (NRRL Y-2214) | 45.1 | | | 279 | |
| | AJ 4172 (IFO 1071) | 44.9 | | | 279 | |
| *anomala* | AJ 5028 (NRRL Y-366) | 33.9 | | | 279 | |
| | AJ 4160 (Chinese cabbage) | 32.7 | | | 279 | |
| | AJ 4167 (IAM 4213) | 33.4 | | | 279 | |
| *beckii* (*Endomycopsis bispora*) | AJ 5029 (NRRL Y-1482) | 33.7 | | | 279 | |
| | AJ 4173 (IFO 0803) | 33.9 | | | 279 | |
| *beijerinckii* | AJ 5030 | 40.2 | | | 279 | |
| | AJ 5030 (NRRL Y-4818) | 40.2 | | | 279 | |
| *bimundalis* | AJ 5031 | 39.0 | | | 279 | |
| | AJ 5031 | 37.8 | | | 279 | |
| | (NRRL Y-5343) | 37.8 | | | 279 | |
| *bimundalis* (var. *americana*) | AJ 5032 (NRRL Y-2156) | 40.0 | | | 279 | |
| *californica* | AJ 4171 (banana) | 41.5 | | | 279 | |
| | AJ 5033 (NRRL Y-1680) | 41.5 | | | 279 | |

## TABLE 2 (Continued)
## GC COMPOSITION OF THE DNA OF FUNGI

| Organism | Source or Strain[a] | Mole % Guanosine + Cytosine[b] | | | References[c] | |
|---|---|---|---|---|---|---|
| | | Tm | Chemical Analysis | Buoyant Density | A | B |
| *Hansenula* | | | | | | |
| *canadensis* | AJ 5034 (NRRL Y-1888) | 36.1 | | | 279 | |
| | AJ 4177 (IFO 0973) | 35.4 | | | 279 | |
| *capsulata* | AJ 5035 (NRRL Y-1842) | 43.9 | | | 279 | |
| | AJ 4173 (IFO 0721) | 42.7 | | | 279 | |
| | AJ 4174 (IFO 0721) | 42.5 | | | 279 | |
| *ciferrii* | AJ 5036 (NRRL Y-1031) | 30.5 | | | 279 | |
| *fabianii* | AJ 5010 (Dr. K. Kodama, NRRL Y-1871) | 42.4 | | | 279 | |
| | AJ 5037 (NRRL Y-1871) | 41.0 | | | 279 | |
| | AJ 5037 | 41.7 | | | 279 | |
| | AJ 5157 (NRRL Y-1872 | 42.9 | | | 279 | |
| | AJ 5158 (NRRL Y-1873) | 43.2 | | | 279 | |
| | AJ 5159 (NRRL Y-6710) | 43.2 | | | 279 | |
| *holstii* | AJ 5038 (NRRL Y-2155) | 34.1 | | | 279 | |
| *jadinii* | AJ 5039 (NRRL Y-1542) | 41.0 | | | 279 | |
| *minuta* | AJ 4175 (IFO 0975) | 43.2 | | | 279 | |
| | AJ 4175 | 44.1 | | | 279 | |
| | AJ 5040 (NRRL Y-411) | 44.4 | | | 279 | |
| *mrakii* | AJ 5041 (NRRL Y-1364) | 40.2 | | | 279 | |
| *petersonii* | AJ 5042 (NRRL Y-3807) | 41.2 | | | 279 | |
| | AJ 5043 (NRRL Y-3808 | 39.0 | | | 279 | |
| *platypodis* (*Endomycopsis platypodis*) | AJ 5045 (NRRL Y-4169) | 28.5 | | | 279 | |
| | AJ 5045 | 29.3 | | | 279 | |
| | AJ 5066 (CBS 4111) | 33.4 | | | 279 | |

## TABLE 2 (Continued)
## GC COMPOSITION OF THE DNA OF FUNGI

| Organism | Source or Strain[a] | Mole % Guanosine + Cytosine[b] Tm | Chemical Analysis | Buoyant Density | References[c] A | B |
|---|---|---|---|---|---|---|
| *Hansenula* | | | | | | |
| *platypodis* (*Endomycopsis platypodis*) | AJ 5044 (NRRL Y-6732) | 32.0 | | | 279 | |
| *saturunus* | AJ 4176 (IFO 0117) | 39.5 | | | 279 | |
| | AJ 5047 (NRRL Y-1304) | 41.5 | | | 279 | |
| *schneggi* (*Hansenula anomala* var. *schneggi*) | AJ 5028 (NRRL Y-993) | 32.4 | | | 279 | |
| *schneggi* (*Hansenula anomala*) | AJ 4178 (IAM 4269) | 33.4 | | | 279 | |
| *silvicola* | AJ 5160 (NRRL Y-1679 | 32.0 | | | 279 | |
| | AJ 5161 (NRRL Y-2032) | 32.7 | | | 279 | |
| | AJ 5162 (NRRL YB-3450) | 32.4 | | | 279 | |
| | AJ 5048 (NRRL Y-1678) | 31.7 | | | 279 | |
| | AJ 5048 | 31.5 | | | 279 | |
| *subpelliculosa* | AJ 5049 (NRRL Y-1683) | 31.0 | | | 279 | |
| *wickerhamii* | AJ 5050 (NRRL Y-4943) | 41.7 | | | 279 | |
| *wingei* | AJ 5051 (NRRL Y-2340) | 36.8 | | | 279 | |
| *Haplosporangium* | | | | | | |
| *bisporale* | NRRL 2493 | | | 52 | 279 | |
| *Helicoma* | | | | | | |
| *isiola* | QM 760 | | | 54.5 | 269 | |
| *Helicostylium* | | | | | | |
| *piriforme*(+) | RSA 537 | | | 50 | 269 | |
| | RSA 866 | | | 54.5 | 269 | |
| *Helminthosporium* | | | | | | |
| *speciferum* | QM 8535 | | | 53 | 269 | |
| | QM 8536 | | | 69.5 | 269 | |
| | QM 8562 | | | 52.5 | 269 | |
| | QM 8563 | | | 53 | 269 | |

## TABLE 2 (Continued)
## GC COMPOSITION OF THE DNA OF FUNGI

| Organism | Source or Strain[a] | Mole % Guanosine + Cytosine[b]: Tm | Chemical Analysis | Buoyant Density | References[c] A | B |
|---|---|---|---|---|---|---|
| *Helvella* | | | | | | |
| *esculenta* | | | 50 | | 272 | |
| *Humicola* | | | | | | |
| *fuscoatra* | QM 997 | | | 52.5 | 269 | |
| *grisea* (thermophilic) | QM 228 | | | 45.5 | 269 | |
| *Hypocrea* | | | | | | |
| *chlorospora* (*Trichoderma* sp.) | QM 1221 | | | 52 | 269 | |
| *Inonotus* | | | | | | |
| *dryophilus* | F 2034 | | | 50 | 269 | |
| | F 2035 | | | 51 | 269 | |
| *Irpex* | | | | | | |
| *lacteus* | | | | 54 | 269 | |
| | | | | 54.5 | 269 | |
| *Isoachyla* | | | | | | |
| *subterranea* (*I. itoana*) | ESB | | | 61.5 | 269 | |
| *Kloeckera* | | | | | | |
| *apiculata* | AJ 4798 (IFO 0175) | 30 | | | 279 | |
| *Kluyveromyces* | | | | | | |
| *polysporus* | AJ 4278 (IFO 0996) | 31 | | | 279 | |
| *Labyrinthula* | | | | | | |
| species | 2 | | | 58 | 273 | |
| *Laetiporus* | | | | | | |
| *sulphureus* | DAOM F-3474 | | | 54 | 271 | |
| *sulphureus* (*Polyposus*) | DAOM 72246 | | | 56 | 271 | |

# TABLE 2 (Continued)
# GC COMPOSITION OF THE DNA OF FUNGI

| Organism | Source or Strain[a] | Mole % Guanosine + Cytosine[b] | | | References[c] | |
|---|---|---|---|---|---|---|
| | | Tm | Chemical Analysis | Buoyant Density | A | B |
| *Laricifomes (Fomes)* | | | | | | |
| *officinalis* | DAOM 73186 | | | 54.5 | 271 | |
| *Lenzites* | | | | | | |
| *betulina* | DAOM 22291 | | | 59 | 271 | |
| | DAOM 99674 | | | 59 | 271 | |
| *Leptosphaeria* | | | | | | |
| *millefolii* | QM 1285 | | | 56.5 | 269 | |
| *Lichtheimia (Absidia)* | | | | | | |
| species | | | 28.5 | | 272 | |
| *Linderina* | | | | | | |
| *macrospora* | RSA 1724 | | | 56 | 269 | |
| *pennispora* | NRRL A-12619 | | | 42 | 269 | |
| | RSA 3 | | | 30 | 269 | |
| *Lipomyces* | | | | | | |
| *starkeyi* | AJ 4279 (IFO 0687) | 45.5 | | | 279 | |
| *Lodderomyces* | | | | | | |
| *elongasporus* | CBS 2606 | 39.5 | | | 282 | |
| | CBS 2605[e] | 39.8 | | | 282 | |
| *Lycogala* | | | | | | |
| species | | | 42.2 | | 272 | |
| *Meruliopsis (Poria)* | | | | | | |
| *taxicola* | DAOM 21223 | | | 55.5 | 270 | |
| *Metschnikowia* | | | | | | |
| *bicuspidata* | Miami 23-413 | 44.6 | | | 282 | |
| *pulcherrima* | NRRL Y-5941-53 | 42.4 | | | 282 | |
| | | 48.3 | | | 282 | |
| | NRRL YB-2272 | 44.6 | | | 282 | |
| *reukaufii* | | 42.2 | | | 282 | |
| *Monosporium* | | | | | | |
| *apiospermum* | QM 7281 | | | 53.5 | 269 | |

## TABLE 2 (Continued)
## GC COMPOSITION OF THE DNA OF FUNGI

| Organism | Source or Strain[a] | Mole % Guanosine + Cytosine[b] | | | References[c] | |
|---|---|---|---|---|---|---|
| | | Tm | Chemical Analysis | Buoyant Density | A | B |
| *Mortierella* | | | | | | |
| *claussenii* | NRRL 2760 | | | 50 | 269 | |
| *isabellina* | QM 6826, NRRL 1757 | | | 50 | 269 | |
| *minutissima* | NRRL 2591 | | | 49 | 269 | |
| *parvispora* (-) | NRRL 2942 | | | 50.5 | 269 | |
| *Mucor* | | | | | | |
| *ambiguus* | NRRL 1644 | | | 43.5 | 269 | |
| *angulisporus* | NRRL 2657 | | | 46 | 269 | |
| *azygosporus* | NRRL 3068 | | | 46 | 269 | |
| *bacilliformis* | NRRL 2346 | | | 34.5 | 269 | |
| *circinelloides* | NRRL 223 | | | 43 | 269 | |
| *fragilis* | USDA A-12253 | | | 39.5 | 269 | |
| | | | | 39 | 284 | |
| *genevensis* | UT 121 | | | 40 | 269 | |
| *griseo-cyanus* | NRRL 1413 | | | 44 | 269 | |
| *hiemalis* | NRRL 1417 | | | 42.5 | 269 | |
| | NRRL 1419 | | | 43.5 | 269 | |
| *indicus* | NRRL 555 | | | 40 | 269 | |
| *jansseni* | NRRL 2629 | | | 44 | 269 | |
| *mucedo* | NRRL 1425 | | | 29.5 | 269 | |
| *pusillus* | QM 436 | | | 48 | 269 | |
| *racemosus* | QM 79i | | | 41 | 269 | |
| *racemosus* (DNA) | NRRL 1608 | | | 38 | 270 | |
| *ramannianus* | QM 6832, NRRL 1839 | | | 49 | 269 | |
| *recurvatus* | UT 122 | | | 35.5 | 269 | |
| *rouxianus* | | | | 39 | 270 | |
| | | 41 | | 38 | 270 | |

# TABLE 2 (Continued)
# GC COMPOSITION OF THE DNA OF FUNGI

| Organism | Source or Strain[a] | Mole % Guanosine + Cytosine[b] Tm | Chemical Analysis | Buoyant Density | References[c] A | B |
|---|---|---|---|---|---|---|
| *Mucor* | | | | | | |
| *rouxii* | NRRL 1894 | | | 39 | 270 | |
| | | | | 37 | 284 | |
| *subtilissimus* | NRRL 1909 | | | 40 | 269 | |
| | QM 1060 | | | 41.5 | 269 | |
| | | 39 | | 39 | 270 | |
| | | 40 | | 39 | 270 | |
| *Mycotypha* | | | | | | |
| *africana* | RSA 1193 | | | 43 | 269 | |
| *microspora* | UT 266 | | | 47.5 | 269 | |
| | RSA 176 | | | 44.5 | 269 | |
| | RSA 774 | | | 45.5 | 269 | |
| | RSA 775 | | | 42 | 269 | |
| | RSA 1183 | | | 46 | 269 | |
| | RSA 1522 | | | 46 | 269 | |
| | RSA 1559 | | | 43.5 | 269 | |
| *Myrothecium* | | | | | | |
| *inundatum* | QM 206 | | | 53 | 269 | |
| *roridum* | QM 188 | | | 50.5 | 269 | |
| *Myxotrichum* | | | | | | |
| *stipitatum* | QM 8525 | | | 55.5 | 269 | |
| *Naganishia* | | | | | | |
| *globosus* | AJ 4286 (RIFY CH-24) | 47.5 | | | 279 | |
| *Neurospora* | | | | | | |
| *crassa* | | 55 | 54 | 53 | 288 | 287 |
| | | | | 54 | 284 | |
| | | | 54 | | 270 | |
| | | 55 | | 52 | 270 | |
| | | | | 54 | 270 | |
| | | | 53 | 53 | 277 | |
| *intermedia* | | | 52 | 53 | 273 | 277 |
| *sitophila* | | | | 55 | 284 | |
| *tetrasperma* | | | 50 | 53 | 273 | 277 |

## TABLE 2 (Continued)
## GC COMPOSITION OF THE DNA OF FUNGI

| Organism | Source or Strain[a] | Mole % Guanosine + Cytosine[b] | | | References[c] | |
|---|---|---|---|---|---|---|
| | | Tm | Chemical Analysis | Buoyant Density | A | B |
| *Nigrospora* | | | | | | |
| *oryzae* | QM 7977 | | | 42 | 269 | |
| *sphaerica* | QM 1253 | | | 6 | 269 | |
| *Pachysolen* | | | | | | |
| *tannophilus* | AJ 4282 (IFO 1007) | 40 | | | 279 | |
| *Paecilomyces* | | | | | | |
| *varioti* | QM 6764 | | | 50.5 | 269 | |
| | QM 8377 | | | 50.5 | 269 | |
| | QM 8492 | | | 51.5 | 269 | |
| *Penicillium* | | | | | | |
| *abeanum* | QM 8154 | | | 51 | 269 | |
| *adametzi* | QM 1916 | | | 49.5 | 269 | |
| *aeneum* | QM 7290 | | | 53 | 269 | |
| *atramentosum* | QM 7483 | | | 50 | 269 | |
| *braziliense* | QM 7493 | | | 55 | 269 | |
| *brevi-compactum* | QM 7497 | | | 53 | 269 | |
| | QM 8406 | | | 51 | 269 | |
| | QM 8488A | | | 52 | 269 | |
| *capsulatum* | QM 26c | | | 51.5 | 269 | |
| | QM 4869 | | | 52 | 269 | |
| *casei* | QM 7309 | | | 52.5 | 269 | |
| *chrysogenum* | UT 129 | | | 52 | 269 | |
| | QM 941 | | | 51 | 269 | |
| | QM 942 | | | 52 | 269 | |
| | QM 6861 | | | 51.5 | 269 | |
| | QM 7500 | | | 54.5 | 269 | |
| | | 52 | | 52 | 270 | |
| *citreo-viride* | QM 5720 | | | 53 | 269 | |
| *claviforme* | UT 128 | | | 52 | 269 | |
| *claviforme* (*P. silvaticum*) | QM 8040 | | | 50 | 269 | |
| *clavigerum* | QM 1918 | | | 52.5 | 269 | |

TABLE 2 (Continued)
GC COMPOSITION OF THE DNA OF FUNGI

| Organism | Source or Strain[a] | Mole % Guanosine + Cytosine[b] Tm | Chemical Analysis | Buoyant Density | References[c] A | B |
|---|---|---|---|---|---|---|
| *Penicillium* | | | | | | |
| *corylophilum* | QM 7510 | | | 52 | 269 | |
| *cyaneo-fulvum* | QM 7514 | | | 48.5 | 269 | |
| *cyclopium* | QM 8403 | | | 51 | 269 | |
| | QM 8491 | | | 51 | 269 | |
| | QM 8491A | | | 52 | 269 | |
| *daleae* | QM 7551 | | | 52.5 | 269 | |
| *decumbens* | QM 1920 | | | 52 | 269 | |
| *digitatum* | UT 131 | | | 52 | 269 | |
| | UT 132 | | | 52 | 269 | |
| *diversum* | QM 1921 | | | 50 | 269 | |
| *fellutanum* | QM 5716 | | | 53.5 | 269 | |
| | QM 7554 | | | 52.5 | 269 | |
| *frequentans* | UT 133 | | | 51 | 269 | |
| *funiculosum* | QM 28b | | | 49.5 | 269 | |
| | QM 8496A | | | 50 | 269 | |
| | QM 8496B | | | 49.5 | 269 | |
| *granulatum* | QM 6868 | | | 50.5 | 269 | |
| *griseo-azureum* | QM 8150 | | | 51 | 269 | |
| *griseolum* | QM 7523 | | | 53 | 269 | |
| *herquei* | QM 7568 | | | 48.5 | 269 | |
| *humuli* | QM 7570 | | | 51.5 | 269 | |
| *implicatum* | QM 7573 | | | 54 | 269 | |
| *italicum* | UT 134 | | | 51.5 | 269 | |
| *janthinellum* | QM 6865 | | | 53.5 | 269 | |
| | QM 8464 | | | 54 | 269 | |
| *japonicum* | QM 7298 | | | 50.5 | 269 | |
| *jensenii* | QM 7587 | | | 50.5 | 269 | |
| *kojigenum* | QM 7957 | | | 53 | 269 | |
| *lilacino-echinulatum* | QM 7289 | | | 49 | 269 | |

## TABLE 2 (Continued)
## GC COMPOSITION OF THE DNA OF FUNGI

| Organism | Source or Strain[a] | Mole % Guanosine + Cytosine[b] Tm | Chemical Analysis | Buoyant Density | References[c] A | B |
|---|---|---|---|---|---|---|
| *Penicillium* | | | | | | |
| *lilacinum* | QM 4e | | | 61 | 269 | |
| | QM 7592 | | | 59 | 269 | |
| *lividum* | QM 1930 | | | 51 | 269 | |
| *luteo-caeruleum* | QM 8151 | | | 50 | 269 | |
| *martensii* | QM 50a | | | 50 | 269 | |
| *megasporum* | QM 6879 | | | 51 | 269 | |
| *melinii* | QM 1931 | | | 52 | 269 | |
| *namyslowskii* | QM 1932 | | | 55.5 | 269 | |
| *notatum* | UT 140 | | | 52.5 | 269 | |
| | QM 946 | | | 51.5 | 269 | |
| | | 51 | | 53 | 270 | |
| *notatum* (mut. *fulvescens*) | QM 7296 | | | 51.5 | 269 | |
| *ochro-chloron* | QM 7604 | | | 47 | 269 | |
| *olvino-viride* | QM 7605 | | | 51 | 269 | |
| *oxalicum* | QM 7606 | | | 53 | 269 | |
| *phoeniceum* | QM 7608 | | | 48.5 | 269 | |
| *piscarium* | QM 471 | | | 52.5 | 269 | |
| *puberulum* | QM 1556 | | | 52.5 | 269 | |
| | QM 7615 | | | 49 | 269 | |
| *pulvillorum* | QM 1935 | | | 51 | 269 | |
| *purpurogenum* | UT 141 | | | 54.5 | 269 | |
| *purpurogenum* (var. *rubrisclerotium*) | QM 8042 | | | 48.5 | 269 | |
| *raciborskii* | QM 7620 | | | 52 | 269 | |
| *radulatum* | QM 7526 | | | 52 | 269 | |
| *raistrickii* | QM 1936 | | | 50 | 269 | |
| *resedanum* | QM 6966 | | | 50.5 | 269 | |
| *roqueforti* | UT 143 | | | 51 | 269 | |

# TABLE 2 (Continued)
# GC COMPOSITION OF THE DNA OF FUNGI

| Organism | Source or Strain[a] | Mole % Guanosine + Cytosine[b] Tm | Chemical Analysis | Buoyant Density | References[c] A | B |
|---|---|---|---|---|---|---|
| *Penicillium* | | | | | | |
| *rugulosum* | QM 7302 | | | 50.5 | 269 | |
| | QM 7660 | | | 50 | 269 | |
| | QM 7661 | | | 55 | 269 | |
| *sclerotiorum* | UT 227 | | | 50 | 269 | |
| *simplicissimum* | QM 6881 | | | 52 | 269 | |
| *terlikowskii* | QM 7687 | | | 50 | 269 | |
| *variable* | QM 2809 | | | 50 | 269 | |
| *varians* | QM 7691 | | | 49 | 269 | |
| *velutinum* | QM 7686 | | | 52 | 269 | |
| *verruculosum* | QM 3203 | | | 50.5 | 269 | |
| | QM 3698 | | | 49.5 | 269 | |
| | QM 7713 | | | 49 | 269 | |
| | QM 7999 | | | 49 | 269 | |
| *viridicatum* | QM 7595 | | | 54 | 269 | |
| *wentii* | QM 44a | | | 51.5 | 269 | |
| *Pestalotiopsis* | | | | | | |
| species | QM 178 | | | 56 | 269 | |
| *Petalosporus* | | | | | | |
| *anodosus* | QM 85261 | | | 53 | 269 | |
| *Phaeocoriolellus (Lenzites)* | | | | | | |
| *trabeus* | DAOM 22444 | | | 59 | 271 | |
| *Phaeolus (Polyporus)* | | | | | | |
| *schweinitzii* | DAOM 72512 | | | 55 | 271 | |
| *Phellinus* | | | | | | |
| *ferruginosus (Polyporus ferruginosus)* | DAOM 52889 | | | 50 | 271 | |
| *gilvus (Poria gilvus)* | DAOM F1704 | | | 50.5 | 271 | |

## TABLE 2 (Continued)
## GC COMPOSITION OF THE DNA OF FUNGI

| Organism | Source or Strain[a] | Mole % Guanosine + Cytosine[b] | | | References[c] | |
|---|---|---|---|---|---|---|
| | | Tm | Chemical Analysis | Buoyant Density | A | B |
| *Phialophora* | | | | | | |
| *fastigiata* | QM 265 | | | 49.5 | 269 | |
| *lagerbergii* | QM 267 | | | 40.5 | 269 | |
| *verrucosa* | QM 264 | | | 51 | 269 | |
| *Phoma* | | | | | | |
| *pigmentivora* | QM 502 | | | 49 | 269 | |
| *terrestris* | QM 120K (small) | | | 56 | 269 | |
| *Phycomyces* | | | | | | |
| *blakesleeanus* | NRRL 1464 | | | 41 | 269 | |
| | | 44 | 39 | 43 | 270 | |
| *blakesleeanus* (+) | | | 39 | | 274 | |
| | UT 151a | | | 37.5 | 269 | |
| *blakesleeanus* (–) | UT 151b | | | 39.5 | 269 | |
| *Physarum* | | | | | | |
| *polycephalum*[d] | | | | 42.5 | 273 | |
| *Phytophthora* | | | | | | |
| *boehmeriae* | Gal N43 | | | 52.5 | 269 | |
| *cactovorum* | Gal N261 | | | 53.5 | 269 | |
| *calocasiae* | Gal N315 | | | 58 | 269 | |
| *cinnamoni* | Gal N33 | | | 57 | 269 | |
| | Gal N38 | | | 52 | 269 | |
| | Gal N39 | | | 49 | 269 | |
| | Gal N53 | | | 57 | 269 | |
| *cryptogea* | Gal N57 | | | 52 | 269 | |
| *fragariae* | Gal N72 | | | 54 | 269 | |
| *heveae* | Gal N331 | | | 55 | 269 | |
| *infestans* | Gal 63B | | | 54 | 269 | |
| | | 47.5 | | 289 | | |
| *palmivora* | Gal N137 | | | 53 | 269 | |
| *parasitica* | Gal N211 | | | 50.5 | 269 | |

# TABLE 2 (Continued)
# GC COMPOSITION OF THE DNA OF FUNGI

| Organism | Source or Strain[a] | Mole % Guanosine + Cytosine[b] Tm | Chemical Analysis | Buoyant Density | References[c] A | B |
|---|---|---|---|---|---|---|
| *Phytophthora* | | | | | | |
| *parasitica-nicotianae* | Gal N15 | | | 49 | 269 | |
| *Pichia* | | | | | | |
| *kluyveri* | AJ 4426 (banana) | 27 | | | 279 | |
| | AJ 4145 (ATCC 9768) | 26 | | | 279 | |
| *membranaefaciens* | | 44 | | 46 | 270 | |
| | AJ 4110 (IFO 0460) | 40 | | | 279 | |
| *vini* | AJ 4266 (IFO 0795) | 34.5 | | | 279 | |
| *Pilaira* | | | | | | |
| *anomala* | NRRL 2289 | | | 45.5 | 269 | |
| *Piptoporus (Polyporus)* | | | | | | |
| *betulinus* | DAOM 95956 | | | 57 | 271 | |
| | DAOM 100674 | | | 56.5 | 271 | |
| *Pleurotus* | | | | | | |
| *astreatus* | | | | 51.5 | 290 | |
| *Polypaecilum* | | | | | | |
| *insolitum* | QM 7961 | | | 52.5 | 269 | |
| *Polyporus* | | | | | | |
| *balsameus* | DAOM 8407 | | | 56.5 | 271 | |
| *brumalis* | DAOM 11803 | | | 59 | 271 | |
| | DAOM 72515 | | | 58.5 | 271 | |
| *palustris* | DAOM 10250 | | | 53.5 | 271 | |
| | DAOM 10620 | | | 57.5 | 271 | |
| *versicolor* | | | 57 | | 276 | |
| *Polysphondylium* | | | | | | |
| *pallidum* | | | | 29 | 273 | |
| *violaceum* | | | | 29 | 273 | |

## TABLE 2 (Continued)
## GC COMPOSITION OF THE DNA OF FUNGI

| Organism | Source or Strain[a] | Mole % Guanosine + Cytosine[b] Tm | Chemical Analysis | Buoyant Density | References[c] A | B |
|---|---|---|---|---|---|---|
| *Poria* | | | | | | |
| *carbonica* | DAOM 73184 | | | 54.5 | 271 | |
| *cinerascens* | DAOM 21421 | | | 57 | 271 | |
| | DAOM 31968 | | | 57.5 | 271 | |
| *rivulosa* | DAOM 11692 | | | 55.5 | 271 | |
| *sequoiae* | DAOM 8755 | | | 59.5 | 271 | |
| *Protoachlya* | | | | | | |
| *paradoxa* | ESB | | | 60.5 | 269 | |
| *Protostelium* | | | | | | |
| *irregularis* | | | | 35 | 273 | |
| *Pseudeurotium* | | | | | | |
| *multisporum* | QM 7781 | | | 54 | 269 | |
| *zonatum* | QM 8030 | | | 57 | 269 | |
| *Pseudoarachniotus* | | | | | | |
| *citrinus* | QM 8528 | | | 53 | 269 | |
| *reticulatus* | QM 7891 | | | 52.5 | 269 | |
| *Pseudogymnoascus* | | | | | | |
| *roseus* | QM 6969 | | | 50.5 | 269 | |
| *Psalliota* | | | | | | |
| *campestris* | | | 44 | | 275 | |
| *Pycnidiophora* | | | | | | |
| *dispersa* | QM 7827 | | | 54 | 269 | |
| *Pycnoporus* | | | | | | |
| *cinnabarinus* | DAOM 72067 | | | 59 | 271 | |
| *sanguineus* | DAOM 52184 | | | 58.5 | 271 | |
| | DAOM 72096 | | | 59 | 271 | |
| *Pythium* | | | | | | |
| *pulchrum* | ESB | | | 51.5 | 269 | |

## TABLE 2 (Continued)
## GC COMPOSITION OF THE DNA OF FUNGI

| Organism | Source or Strain[a] | Mole % Guanosine + Cytosine[b]: Tm | Chemical Analysis | Buoyant Density | References[c] A | B |
|---|---|---|---|---|---|---|
| *Radiomyces* | | | | | | |
| *embreei* | RKB 914 | | | 46 | 269 | |
| | RKB 984 | | | 50 | 269 | |
| | RKB 985 | | | 46 | 269 | |
| | RKB 1186 | | | 44.5 | 269 | |
| | RKB 1372 | | | 44 | 269 | |
| | RKB 1373 | | | 46 | 269 | |
| | RKB 1374 | | | 46 | 269 | |
| | RKB 1458 | | | 47.5 | 269 | |
| *spectabilis* | UT 242 | | | 47.5 | 269 | |
| | RSA 1620 | | | 44 | 269 | |
| *Rhizophydium* | | | | | | |
| species | | | | 50.5 | 273 | |
| *Rhizophlyctis* | | | | | | |
| *rosea* | | | | 44 | 273 | |
| *Rhizopus* | | | | | | |
| *arrhizus* | UT 62-1 | | | 38 | 261 | |
| | NRRL 1437 | | | 42 | 261 | |
| | NRRL 2542 | | | 38 | 261 | |
| *nigricans* | | | | 49 | 273 | 277 |
| *oligosporus* | NRRL 514 | | | 40.5 | 269 | |
| | NRRL 2549 | | | 38.5 | 269 | |
| | NRRL 2710 | | | 40 | 269 | |
| | NRRL 10-455 | | | 40.5 | 269 | |
| | NRRL A-9848 | | | 40.5 | 269 | |
| | NRRL A-9865 | | | 39.5 | 269 | |
| | NRRL A-9867 | | | 38.5 | 269 | |
| | NRRL A-9868 | | | 40 | 269 | |
| | NRRL A-10.456 | | | 40.5 | 269 | |
| | NRRL A-10.457 | | | 41 | 269 | |
| | NRRL A-10.458 | | | 40.5 | 269 | |
| | NRRL A-11.126 | | | 40 | 269 | |
| *oryzae* | UT 62-18 | | | 37.5 | 269 | |
| | NRRL 1526 | | | 40 | 269 | |
| *Rhodotorula* | | | | | | |
| *glutinis* | AJ 4836 (ATU) | 64 | | | 279 | |
| *graminis* | | | | 70 | 285 | |
| *minuta* (var. *minuta*) | CBS 319 | | | 53 | 285 | |

**TABLE 2 (Continued)**
**GC COMPOSITION OF THE DNA OF FUNGI**

| Organism | Source or Strain[a] | Mole % Guanosine + Cytosine[b] Tm | Chemical Analysis | Buoyant Density | References[c] A | B |
|---|---|---|---|---|---|---|
| *Rhodotorula* | | | | | | |
| *minuta* (var. *texensis*, syn. *Rhodotorula tokyoensis* | CBS 4407 | | | 52.5 | 285 | |
| *minuta* (var. *texensis*, syn. *Rhodotorula texensis*) | CBS 2177 | | | 54 | 285 | |
| *mucilaginosa* | | 63 | | 61 | 270 | |
| *pallida* | CBS 2623 | | | 63.5 | 285 | |
| | CBS 320 | | | 54.5 | 285 | |
| *rubra* | AJ 4837 (ATU IFO 0901) | 63.5 | | | 279 | |
| *slooffii* | AJ 4874 (RIFY CBS 5706) | 47.5 | | | 279 | |
| *texensis* | AJ 4864 (ATU IFO 0710) | 48 | | | 279 | |
| *Saccharomyces* | | | | | | |
| *bisporus* | CBS 702[e] | 44.4 | | | 282 | |
| *carlsbergensis* | NRRL Y-379 | 39 | | | 291 | 291 |
| | | | | 40 | 284 | |
| *cerevisiae* | | 39.5 | 40 | 39 | 245, 270, 292–295 | |
| | NRRL N-123 | 39 | | | 291 | 291 |
| | AJ 4001 (IAM 4512) | 36 | | | 279 | |
| | AJ 4002 (IAM 4274) | 36.5 | | | 279 | |
| | AJ 4005 (ATV) | 36.5 | | | 279 | |
| | AJ 4008 (sea weed) | 36 | | | 279 | |
| | | | | 41 | 284 | |
| | | | | 38 | 284 | |
| | | | | 42 | 284 | |
| | | | | 40 | 284 | |
| | | 38 | | | 270 | |
| | | 41 | | 40 | 270 | |
| | | | 36 | | 270 | |
| | | | | 39 | 270 | |
| | | 37 | | | 282 | |
| | | 36 | | | 282 | |

# TABLE 2 (Continued)
# GC COMPOSITION OF THE DNA OF FUNGI

| Organism | Source or Strain[a] | Mole % Guanosine + Cytosine[b] Tm | Chemical Analysis | Buoyant Density | References[c] A | B |
|---|---|---|---|---|---|---|
| *Saccharomyces* | | | | | | |
| *cerevisiae* (var. *ellipsoideus*) | H 36 | 38.7 | 41.0 | | 295 | |
| *chevalieri* | NRLL Y-1345 | 41 | | | 291 | 291 |
| *dairensis* | NRLL Y-1353 | 37 | | | 291 | 291 |
| *delbrueckii* | AJ 4085 (Chinese cabbage) | 32 | | | 279 | |
| *dobzhanskii* | NRLL Y-1974 | 42 | | | 291 | 291 |
| *exiguus* | AJ 4075 | 30.5 | | | 279 | |
| *florentinus* | AJ 4190 | 38.5 | | | 279 | |
| *fragilis* | | 41.5 | 42.9 | 42 | 270 | 295 |
| | | 42 | | 42 | 270 | |
| | | 41 | | | 282 | |
| | NRLL Y-665 | 41 | | | 291 | 291 |
| | NRLL Y-610 | 41 | | | 291 | 291 |
| *globosus* | NRLL Y-409 | 37 | | | 291 | |
| *inconspicuus* | CBS 3003 | 46.3 | | | 282 | |
| *italicus* | NRLL Y-1373 | 40 | | | 291 | |
| *lactis* | NRLL Y-1140 | 41 | | | 291 | 291 |
| | NRLL Y-1205 | 41 | | | 291 | 291 |
| *marxianus* | NRLL Y-2265 | 40 | | | 291 | 291 |
| *oviformis* | NRLL Y-1356 | 41 | | | 291 | 291 |
| *pastori* | NRLL Y-1603 | 43 | | | 291 | 291 |
| *pombe* | H 28 | 43.2 | 44.8 | | 295 | |
| *rosei* | AJ 4086 (banana) | 40 | | | 279 | |
| | Etchells # Y531 | 43.9 | | | 282 | |
| *rouxii* | NRLL Y-2547 | 40 | | | 291 | 291 |
| *uvarum* | NRLL Y-969 | 40 | | | 291 | 291 |
| *vini* | | | | 41 | 273 | |
| *wickerhamii* | NRRL YB-1411 | 40 | | | 291 | |

## TABLE 2 (Continued)
## GC COMPOSITION OF THE DNA OF FUNGI

| Organism | Source or Strain[a] | Mole % Guanosine + Cytosine[b] | | | References[c] | |
|---|---|---|---|---|---|---|
| | | Tm | Chemical Analysis | Buoyant Density | A | B |
| *Saprolegnia* | | | | | | |
| *ferax* | ESB | | | 49.5 | 269 | |
| *hypogyna* | ESB | | | 55.5 | 269 | |
| *parasitica* | ESB | | | 60.5 | 269 | |
| *Sapromyces* | | | | | | |
| species | | | | 27 | 273 | |
| *Sartorya* | | | | | | |
| *fumigata* (*Aspergillus fischerii*) | UT 13 | | | 54.5 | 269 | |
| *fumigata* (var. *glaber*, rec. as *Aspergillus fischeri* var. *glaber*) | QM 1903 | | | 51.5 | 269 | |
| *Schizophyllum* | | | | | | |
| *commune* | | | | 61 | 284 | |
| | | 58 | | 58 | 270 | |
| | | | 57.1 | | 272 | |
| *Schizosaccharomyces* | | | | | | |
| *octosporus* | | 42 | | 40 | 270 | |
| *pombe* | | 42 | 40.7 | | 295 | 296 |
| *Schwanniomyces* | | | | | | |
| *occidentalis* | AJ 4288 (IAM 4332) | 31.5 | | | 279 | |
| *Sclerotinia* | | | | | | |
| *libertiana* | | | 46 | | 274 | |
| | | | | 45.5 | 274 | |
| *Scopulariopsis* | | | | | | |
| *brevicaulis* | QM 815 | | | 50 | 269 | |
| *brevicaulis* (var. *glabra*) | QM 6875 | | | 52.5 | 269 | |
| *melanospora* | QM 7884 | | | 53 | 269 | |

# TABLE 2 (Continued)
# GC COMPOSITION OF THE DNA OF FUNGI

| Organism | Source or Strain[a] | Mole % Guanosine + Cytosine[b] | | | References[c] | |
|---|---|---|---|---|---|---|
| | | Tm | Chemical Analysis | Buoyant Density | A | B |
| *Scopulariopsis* | | | | | | |
| *repens* | QM 399 | | | 51.5 | 269 | |
| *Sordaria* | | | | | | |
| *humana* | QM 819 | | | 50.5 | 269 | |
| *macrospora* | QM 794 | | | 54 | 269 | |
| | | | | 54 | 284 | |
| *Spadicoides* | | | | | | |
| *xylogenum* | QM 6817 | | | 54 | 269 | |
| *Spondylocladium* | | | | | | |
| *atrovirens* | QM 1793 | | | 52 | 269 | |
| *Spongipellis (Polyporus)* | | | | | | |
| *galactinus* | DAOM 31928 | | | 53 | 271 | |
| | DAOM 53502 | | | 54 | 271 | |
| *Sporobolomyces* | | | | | | |
| *albo-rubescens* (DNA) | CBS 482 | | | 63.0 | 285 | |
| *holsaticus* | CBS 1522 | | | 64.0 | 285 | |
| | CBC 2630 | | | 65.0 | 285 | |
| | CBS 4029 | | | 62.0 | 285 | |
| *holsaticus* (syn. *S. coralliformis*) | CBC 4209 | | | 64.5 | 285 | |
| *odorus* | CBS 483 | | | 65.0 | 285 | |
| *pararoseus* | 2637 | | | 60.5 | 285 | |
| | 4217 | | | 55.0 | 285 | |
| | CBS 484 | | | 51.5 | 285 | |
| *roseus* | CBS 1015 | | | 55.0 | 285 | |
| | CBS 486 | | | 56.0 | 285 | |
| | | 50 | | 50 | 270 | |
| *roseus* (syn. *S. tenius*) | CBS 492 | | | 53.5 | 285 | |
| *roseus* (syn. *S. salmoneus*) | CBS 488 | | | 55.5 | 285 | |

## TABLE 2 (Continued)
## GC COMPOSITION OF THE DNA OF FUNGI

| Organism | Source or Strain[a] | Mole % Guanosine + Cytosine[b] Tm | Chemical Analysis | Buoyant Density | References[c] A | B |
|---|---|---|---|---|---|---|
| *Sporobolomyces* | | | | | | |
| *salmonicolor* | AJ 4878 (RIFY WF 174) | 57.0 | | | 279 | |
| | | 63 | | 63 | 270 | |
| | CBS 490 | | | 63.5 | 285 | |
| | CBS 496 | | | 64.5 | 285 | |
| *Sporormia* | | | | | | |
| *minima* | QM 8592 | | | 53.5 | 269 | |
| species | | 51 | | 51 | 270 | |
| *Sporotrichum* | | | | | | |
| *pruinosum* | QM 168 | | | 58 | 269 | |
| | QM 826 | | | 58 | 269 | |
| *Stachybotrys* | | | | | | |
| *atra* | QM 94d | | | 49 | 269 | |
| | QM 1297 | | | 55.5 | 269 | |
| | QM 8497 | | | 55 | 269 | |
| *Stemphylium* | | | | | | |
| *callistephi* | QM 1326 | | | 53.5 | 269 | |
| species | QM 1484 | | | 57.5 | 269 | |
| *Syncephalastrum* | | | | | | |
| *racemosum* | UT 184 | | | 49 | 269 | |
| | QM 709 | | | 48.5 | 269 | |
| | QM 8011 | | | 48.5 | 269 | |
| | | 47 | | 48 | 270 | |
| *racemosum* (+) | RSA 235 | | | 51 | 269 | |
| | RSA 236 | | | 51.5 | 269 | |
| | RSA 237 | | | 50 | 269 | |
| | RSA 674 | | | 52.5 | 269 | |
| | RSA 702 | | | 51 | 269 | |
| *racemosum*(–) | RSA 24 | | | 50.5 | 269 | |
| | RSA 229 | | | 51 | 269 | |
| | RSA 232 | | | 50 | 269 | |
| | RSA 673 | | | 51.5 | 269 | |
| | RSA 699 | | | 48.5 | 269 | |
| *Talaromyces* | | | | | | |
| *avellaneus* | UT 126 | | | 54.5 | 269 | |

# TABLE 2 (Continued)
# GC COMPOSITION OF THE DNA OF FUNGI

| Organism | Source or Strain[a] | Mole % Guanosine + Cytosine[b]: Tm | Chemical Analysis | Buoyant Density | References[c] A | B |
|---|---|---|---|---|---|---|
| *Talaromyces* | | | | | | |
| *avellaneus (Penicillium avellaneus)* | QM 1849 | | | 52.5 | 269 | |
| | QM 7490 | | | 52 | 269 | |
| *stipitatus* | ATCC 10500 | | | 50 | 269 | |
| | UT 144 | | | 49 | 269 | |
| *vermiculatus* | UT 230 | | | 49 | 269 | |
| *vermiculatus (Penicillium vermiculatus)* | QM 1858 | | | 57 | 269 | |
| *wortmanii* | QM 7322 | | | 50 | 269 | |
| *Thamnidium* | | | | | | |
| *anomalum* | RKB 80 | | | 46 | 269 | |
| | RKB 88 | | | 45.5 | 269 | |
| | RKB 109 | | | 45 | 269 | |
| | RKB 110 | | | 44.5 | 269 | |
| | RKB 169 | | | 44.5 | 269 | |
| | RKB 356 | | | 45 | 269 | |
| | RKB 357 | | | 45 | 269 | |
| | RKB 358 | | | 45.5 | 269 | |
| | RKB 359 | | | 45 | 269 | |
| | RKB 360 | | | 45.5 | 269 | |
| | RKB 361 | | | 44.5 | 269 | |
| | RKB 362 | | | 46 | 269 | |
| *elegans* | NRRL 2467 | | | 53 | 269 | |
| *elegans* (+) | RKB 166 | | | 55 | 269 | |
| | RKB 258 | | | 37 | 269 | |
| | RKB 653 | | | 54 | 269 | |
| *elegans* (–) | RKB 40 | | | 40.5 | 269 | |
| | RKB 74 | | | 55.5 | 269 | |
| | RKB 140 | | | 53.5 | 269 | |
| | RKB 257 | | | 60.5 | 269 | |
| *Thielavia* | | | | | | |
| *sepedonium* | QM 46a | | | 55.5 | 269 | |
| *Thraustotheca* | | | | | | |
| *primoachlya* | ESB | | | 45.5 | 269 | |
| *Torula* | | | | | | |
| *ramosa* | QM 1030 | | | 48.5 | 269 | |

## TABLE 2 (Continued)
## GC COMPOSITION OF THE DNA OF FUNGI

| Organism | Source or Strain[a] | Mole % Guanosine + Cytosine[b] Tm | Chemical Analysis | Buoyant Density | References[c] A | B |
|---|---|---|---|---|---|---|
| *Torulopsis* | | | | | | |
| *aeria* | AJ 4988 (IFO 0377) | 52.5 | | | 279 | |
| *colliculosa* | AJ 4338 (IAM 4426) | 40.5 | | | 279 | |
| *glabrata* | AJ 4359 (NI 7596) | 35.5 | | | 279 | |
| *pinus* | AJ 4989 (IFO 0741) | 34.5 | | | 279 | |
| *stellata* | AJ 4394 | 40.5 | | | 279 | |
| | | 48 | | 50 | 270 | |
| *torresii* | AJ 4937 (IGC 3022) | 48.5 | | | 279 | |
| *Toxotrichum* | | | | | | |
| *cancellatum* | QM 8534 | | | 50 | 269 | |
| *Tremella* | | | | | | |
| *fuciformis* | AJ 5024 | 54.5 | | | 279 | |
| *Trichoderma* | | | | | | |
| *lignorum* | QM 1275 | | | 51 | 269 | |
| *viride* | QM 1512 | | | 49.5 | 269 | |
| *Trichosporon* | | | | | | |
| *behrendii* | AJ 4832 | 32.5 | | | 279 | |
| *cutaneum* | AJ 4815 (RIFY CH-17) | 59.0 | | | 279 | |
| | AJ 4827 | 59.0 | | | 279 | |
| *pullulans* | AJ 4831 | 54.0 | | | 279 | |
| *Trichithecium* | | | | | | |
| *roseum* | QM 102e | | | 54 | 269 | |
| | | | 49.7 | | 272 | |
| *Trigonopsis* | | | | | | |
| *variabilis* | AJ 4876 (IFO 0755) | 44.0 | | | 279 | |

## TABLE 2 (Continued)
## GC COMPOSITION OF THE DNA OF FUNGI

| Organism | Source or Strain[a] | Mole % Guanosine + Cytosine[b] Tm | Chemical Analysis | Buoyant Density | References[c] A | B |
|---|---|---|---|---|---|---|
| *Verticillium* | | | | | | |
| *niviostratum* | QM 5187 | | | 35.5 | 269 | |
| *Wickerhamia* | | | | | | |
| *fluorescens* | AJ 4285 (IFO 1116) | 35 | | | 279 | |
| *Zygorhynchus* | | | | | | |
| *moelleri* | UT 193 | | | 39 | 269 | |
| | | 40 | | 35 | 270 | |

[a] Many of the organisms listed under source or strain have privately designated numbers or letters. Where possible, these organisms have been additionally identified by their designations in national or widely recognized culture collections. An explanation of the known abbreviations used for those collections is given in the footnotes for Table 1.
[b] Mole percent given to the nearest 0.1%.
[c] Column A: primary reference for GC values; column B: reference for hybridization data.
[d] Satellite DNA present.
[e] Type strain.

## TABLE 3
## GC COMPOSITION OF THE DNA OF ALGAE

| Organism | Source or Strain | Mole % Guanosine + Cytosine Tm | Chemical Analysis | Buoyant Density | References |
|---|---|---|---|---|---|
| *Ankistrodesmus* | | | | | |
| *falcatus* | | | 58.7 | | 298 |
| *Chaetoceras* | | | | | |
| *decipiens* | | | 39.1 | | 298 |
| *Chara* | | | | | |
| species | | | 49.8 | | 298 |
| *Chlamydomonas* | | | | | |
| *angulosa* | | | | 68 | 299 |
| *eugametos*[a] | | | | 61 | 299 |

## TABLE 3 (Continued)
## GC COMPOSITION OF THE DNA OF ALGAE

| Organism | Source or Strain | Mole % Guanosine + Cytosine | | | References |
|---|---|---|---|---|---|
| | | Tm | Chemical Analysis | Buoyant Density | |
| *Chlamydomonas* | | | | | |
| *gyrus* | | | 59.7 | | 298 |
| *moewusii* | | | | 60 | 299 |
| *reinhardi*[a] | | | 65 | 62, 64, 66 | 300–302 |
| *Chlorella* | | | | | |
| *ellipsoidea* | | | 59.0 | 57 | 300, 303 |
| *Chlorogonium* | | | | | |
| *elongatum*[a] | | | | 54 | 299 |
| *Coccolithus* | | | | | |
| *huxleyi* | | | | 64.8 | 273 |
| *Cricosphaera* | | | | | |
| *carterae* | | | | 60.2 | 273 |
| *elongata* | | | | 59.2 | 273 |
| *Cyclotella* | | | | | |
| *cryptica*[a] | | | | 40.8 | 273 |
| *Cylindrotheca* | | | | | |
| *fusiformis*[a] | | | | 45.4 | 273 |
| *Cystoseira* | | | | | |
| *barbata* | | | 58.8 | | 298 |
| *Dunaliella* | | | | | |
| *salina* | | | 60.2 | | 298 |
| *Hydrodictyon* | | | | | |
| *reticulatum* | | | 53.5 | | 298 |
| *Isochrysis* | | | | | |
| *galbana* | | | | 61 | 306 |
| *Lagerheimia* | | | | | |
| *ciliata* | | | 61.1 | | 298 |

## TABLE 3 (Continued)
## GC COMPOSITION OF THE DNA OF ALGAE

| Organism | Source or Strain | Mole % Guanosine + Cytosine: Tm | Chemical Analysis | Buoyant Density | References |
|---|---|---|---|---|---|
| *Lyngbya* | | | | | |
| species | 77 Lewin | | | 51 | 267 |
| *Microcoleus* | | | | | |
| *vaginatus* | 6304 M. M. Allen | | | 48 | 267 |
| species | 6401 M. M. Allen | | | 45 | 267 |
| *Monodus* | | | | | |
| *subterraneus* | | | | 52 | 299 |
| *Navicula* | | | | | |
| *closterium* | | | | 50 | 299 |
| *pelliculosa* | | | | 58 | 299 |
| *Nitella* | | | | | |
| species | | | 49.4 | | 298 |
| *Nitzschia* | | | | | |
| *angularis*[a] | | | | 46.9 | 273 |
| *Nodularia* | | | | | |
| *sphaerocarpa* | KFRI 11.1.1 | | | 39 | 267 |
| *Nostoc* | | | | | |
| *punctiform* | IUCC 384 | | | 44 | 267 |
| *Ochromonas* | | | | | |
| *danica*[a] | | | | 48 | 273 |
| *Plectonema* | | | | | |
| *boryanum* | IUCC 581 | | | 48 | 267 |
| | KFRI 925 | | | 48 | 267 |
| *calothricoides* | IUCC 598 | | | 48 | 267 |
| species | 52 R. Lewin | | | 43 | 267 |
| *Polyides* | | | | | |
| *rotundus* | | | 64.3 | | 298 |

## TABLE 3 (Continued)
## GC COMPOSITION OF THE DNA OF ALGAE

| Organism | Source or Strain | Mole % Guanosine + Cytosine | | | References |
|---|---|---|---|---|---|
| | | Tm | Chemical Analysis | Buoyant Density | |
| *Polytoma* | | | | | |
| *agilis* | | | | 42 | 299 |
| *obtusum*[a] | | | | 65 | 301 |
| *uvella* | | | | 52 | 299 |
| *Polytomella* | | | | | |
| *papillata* | | | 40 | | 268 |
| *Prymnesium* | | | | | |
| *parvum* | | | | 58 | 306 |
| *Rhabdonema* | | | | | |
| *adriaticum* | | | 36.9 | | 298 |
| *Rhizoclonium* | | | | | |
| *japonicum* | | | 53.7 | | 298 |
| *Rhodymenia* | | | | | |
| *palmata* | | | 48.9 | | 298 |
| *Scenedesmus* | | | | | |
| *accuminatus* | | | 63.8 | | 298 |
| *quadricauda* | | | 58.9 | | 298 |
| *Spirogyra* | | | | | |
| species | | | 38.9 | | 298 |
| *Thalassiosira* | | | | | |
| *nordenscheldii* | | | 40.3 | | 298 |
| *Ulothrix* | | | | | |
| *fimbriata* | | | | 46 | 299 |

[a] Satellite DNA present.

# TABLE 4
# GC COMPOSITION OF THE DNA OF PROTOZOA

| Organism | Source or Strain | Mole per cent Guanosine + Cytosine | | | References |
|---|---|---|---|---|---|
| | | Tm | Chemical Analysis | Buoyant Density | |
| *Acanthamoeba* | | | | | |
| *castellanii* | | | | 56 | 304 |
| *Amoeba* | | | | | |
| *proteus* | | | | 66 | 304 |
| *Astasia* | | | | | |
| *longa*[a] | | | | 56 | 305 |
| *Blastocrithidia* | | | | | |
| *culicis*[a] | | | | 56 | 304 |
| *Colpidium* | | | | | |
| *campylum* | | | | 32 | 306 |
| *colpidium* (2) | | | | 32.5 | 306 |
| *truncatum* | | | | 35 | 306 |
| *Crithidia* | | | | | |
| *fasciculata*[a] | 2 strains | | | 54, 58 | 304, 306, 307 |
| *lucilliae* | | | | 57 | 306 |
| *onocopetti*[a] | | | | 54 | 306, 308 |
| *Entamoeba* | | | | | |
| *invadens* | | | | 23 | 304 |
| *Euglena* | | | | | |
| *gracilis* | Z | 50.5 | 51 | 49 | 312 |
| *gracilis* (var. *bacillaris*) | | 50.2 | 53 | 48 | 309–311 |
| *Glaucoma* | | | | | |
| *chattoni* | | | | 34 | 304 |
| *Leishmania* | | | | | |
| *enrietti*[a] | | | | 57 | 313 |
| *tarentolae* | | | | 54 | 306 |

**TABLE 4 (Continued)**
**GC COMPOSITION OF THE DNA OF PROTOZOA**

| Organism | Source or Strain | Mole % Guanosine + Cytosine: Tm | Chemical Analysis | Buoyant Density | References |
|---|---|---|---|---|---|
| *Naegleria* | | | | | |
| gruberi[a] | | | | 33 | 314 |
| *Paramecium* | | | | | |
| *aurelia* | | | | 29 | 306, 315, 316 |
| *caudatum* | | | 28.2 | | 317 |
| *Plasmodium* | | | | | |
| *berghei* | | | 40.7 | | 318 |
| *Tetrahymena* | | | | | |
| *patula* | | 19 | 22 | 25 | 306 |
| *pyriformis* | | | 25 | | 319, 320 |
| | W | 29 | | 30 | 306 |
| | (II) | | | 25 | 321, 322 |
| | 3-1, 7-1, 5 | | | 27 | 321, 322 |
| | 9-1 | | | 28 | 321 |
| | GL[a] | | | 29 | 321, 322 |
| | E | | | 31 | 321 |
| | ST, 4[a] | | | 32 | 322 |
| | 9 | | | 30 | 322 |
| *rostrata* | | | | 23 | 321 |
| *Toxoplasma* | | | | | |
| *gondii* | RH strain | | | 53 | 323 |
| *Trichomonas* | | | | | |
| *gallinae* | | 32.2 | | 34 | 324 |
| *vaginalis* | | 31.2 | 32 | 29 | 324, 325 |
| *Trypanosoma* | | | | | |
| *cruzi* | | | 50.2 | | 326, 327 |
| *equiperdum* | | | | 47.9 | 328 |
| *gambiense* | | | 45 | | 327 |
| *lewisi* | | | | 59 | 306 |

[a] Satellite DNA present.

## REFERENCES


1. Breed, Murray and Smith (Eds.), *Bergey's Manual of Determinative Bacteriology,* 7th ed. Bailliere, Tindell and Cox, London, England (1957).
2. Skerman, *A Guide to the Identification of the Genera of Bacteria,* 2nd ed. The Williams and Wilkins Co., Baltimore, Maryland (1967).
3. Bovre, *Acta Pathol. Microbiol. Scand., 69,* 123, (1967).
4. Bauman, Doudoroff and Stanier, *J. Bacteriol., 95,* 1520 (1968).
5. Marmur, Falkow and Mandel, *Annu. Rev. Microbiol., 17,* 329 (1963).
6. Citarella and Colwell, *Can. J. Microbiol., 12,* 418 (1966).
7. Colwell and Mandel, *J. Bacteriol., 87,* 1412 (1964).
8. Belozerskii and Spirin, in *The Nucleic Acids,* Vol. 3, p. 147, Chargaff and Davison, Eds. Academic Press, New York (1960).
9. Leifson and Mandel, *Antonie van Leeuwenhoek J. Microbiol. Serol., 31,* 57 (1966).
10. Hoyer and McCullough, *J. Bacteriol., 96,* 1783 (1968).
11. De Ley, *Antonie van Leeuwenhoek J. Microbiol. Serol., 34,* 109 (1968).
12. Baptist, Shaw and Mandel, *J. Bacteriol., 99,* 180 (1969).
13. Colwell, Citarella and Ryman, *J. Bacteriol., 90,* 1148 (1965).
14. De Ley and Van Muylem, *Antonie van Leeuwenhoek J. Microbiol. Serol., 29,* 344 (1963).
15. Sebald, Gasser and Werner, *Ann. Inst. Pasteur (Paris), 109,* 251 (1965).
16. Yamaguchi, *J. Gen. Appl. Microbiol., 13,* 63 (1967).
17. Jones and Bradley, *Dev. Ind. Microbiol., 5,* 267 (1964).
18. Frontali, Hill and Silvestri, *J. Gen. Microbiol., 38,* 243 (1965).
19. Tewfik and Bradley, *J. Bacteriol., 94,* 1994 (1967).
20. Enquist and Bradley, *Bacteriol. Proc.,* p. G-9 (1968).
21. Fritzche, *Biopolymers, 5,* 863 (1967).
22. Zyghanov, Namestnikova and Krassykova, *Mikrobiologiya, 35,* 92 (1966).
23. Zyghanov and Krassykova, *Mikrobiologiya, 37,* 969 (1968).
24. Craveri, Hill, Manachini and Silvestri, *J. Gen. Microbiol., 41,* 335 (1965).
25. Ellender and Dimopoullos, *Bacteriol. Proc.,* p. GP-47 (1967).
26. Ellender and Dimopoullos, *Proc. Soc. Exp. Biol. Med., 125,* 82 (1967).
27. McCurdy and Wolf, *Can. J. Microbiol., 13,* 1707 (1967).
28. Suyama and Gibson, *Biochem. Biophys. Chem. Commun., 24,* 549 (1966).
29. Vanyushin, Belozerskii, Kokurina and Kadirova, *Nature, 218,* 1066 (1968).
30. Marmur and Doty, *J. Mol. Biol., 5,* 109 (1962).
31. Schildkraut, Marmur and Doty, *J. Mol. Biol., 4,* 430 (1962).
32. De Ley and Park, *Antonie van Leeuwenhoek J. Microbiol. Serol., 32,* 6 (1966).
33. De Ley, *Antonie van Leeuwenhoek J. Microbiol. Serol., 34,* 66 (1968).
34. Marmur, Seaman and Levine, *J. Bacteriol., 85,* 461 (1963).
35. Welker and Campbell, *J. Bacteriol., 94,* 1124 (1967).
36. Ikeda, Saiti, Miura, Takagi and Aoki, *J. Gen. Appl. Microbiol., 11,* 181 (1965).
37. McDonald, Felkner, Turetsky and Matney, *J. Bacteriol., 85,* 1071 (1963).
38. Manachini, Craveri and Guicciardi, *Ann. Microbiol. Enzimol., 18,* 1 (1968).
39. Suzuki, and Kitahara, *J. Gen. Appl. Microbiol., 10,* 305 (1964).
40. Dubnau, Smith, Morell and Marmur, *Proc. Nat. Acad. Sci. U.S.A., 54,* 491 (1965).
41. Boháček, Kocur and Martinec, *Arch. Mikrobiol., 64,* 23 (1968).
42. Herndon and Bott, *J. Bacteriol., 97,* 6 (1969).
43. MacDonald and MacDonald, *Can. J. Microbiol., 8,* 795 (1962).
44. Doi and Igarashi, *J. Bacteriol., 90,* 384 (1965).
45. Gause, Loshkareva, Zbarsky and Gause, *Nature, 203,* 598 (1964).
46. Herberlein, De Ley and Tijtgat, *J. Bacteriol., 94,* 116 (1967).
47. Hoyer and McCullough, *J. Bacteriol., 95,* 444 (1968).
48. Mandel and Rowley, *J. Bacteriol., 85,* 1445 (1963).
49. Tonomura, Malkin and Rabinowitz, *J. Bacteriol., 89,* 1438 (1965).
50. Saunders Campbell and Postgate, *J. Bacteriol., 87,* 1073 (1964).
51. Hongo, Ono, Ogata and Murata, *Agr. Biol. Chem., 30,* 982 (1966).
52. Reddy and Bryant, *Bacteriol. Proc.,* p. G-109 (1967).
53. Dowell, Loper and Hill, *J. Bacteriol., 88,* 1805 (1964).
54. Boháček,and Mraz, *Zentralbl. Bakteriol. Parasitenk. Infektionskr. Hyg. Abt. I Orig., 202,* 438 (1967).
55. Bacon, Overend, Lloyd and Peacocke, *Arch. Biochem. Biophys., 118,* 352 (1967).
56. Rogul, McGee, Wittler and Falkow, *J. Bacteriol., 90,* 1200 (1965).
57. Catlin and Cunningham, *J. Microbiol., 37,* 353 (1964).

58. Brenner, Martin and Hoyer, *J. Bacteriol., 94,* 486 (1967).
59. Sebald and Véron, *Ann. Inst. Pasteur (Paris), 105,* 897 (1963).
60. Brenner, Fanning, Johnson, Citarella and Falkow, *J. Bacteriol., 98,* 637 (1969).
61. Bekker and Kutsemakina, *Biochemistry U.S.S.R., 28,* 612 (1963); English translation, *Biokhimiya, 28,* 498 (1963).
62. Poindexter, *Bacteriol. Rev., 28,* 231 (1964).
63. Kingsbury and Weiss, *J. Bacteriol., 96,* 1421 (1968).
64. Mandel, Johnson and Stokes, *J. Bacteriol., 91,* 1657 (1966).
65. Mandel, Bergendahl and Pfennig, *J. Bacteriol., 89,* 917 (1965).
66. Hill, *J. Gen. Microbiol., 44,* 419 (1966).
67. Mandel, in *Systematic Biology (International Conference Proceedings),* p. 269. National Academy of Sciences, Washington, D.C. (1969).
68. Abe, Takayama and Kinoshita, *J. Gen. Microbiol., 13,* 279 (1967).
69. Mandel and Leadbetter, *J. Bacteriol., 90,* 1795 (1965).
70. Chen, Citarella, Salazar and Colwell, *J. Bacteriol., 91,* 1136 (1966).
71. Colwell, Citarella and Chan, *Can. J. Microbiol., 12,* 1099 (1966).
72. Edelman, Swinton, Schiff, Epstein and Zeldin, *Bacteriol. Rev., 31,* 315 (1967).
73. McCarthy and Bolton, *Proc. Nat. Acad. Sci. U.S.A., 50,* 156 (1963).
74. Starr and Mandel, *Gen. Microbiol., 56,* 113 (1969).
75. Falkow, Ryman and Washington, *J. Bacteriol., 83,* 1318 (1962).
76. Mandel and Rownd, in *Taxonomic Biochemistry and Serology,* p. 585, Leone, Ed. Ronald Press Co., New York (1964).
77. De Ley, *Antonie van Leeuwenhoek J. Microbiol. Serol., 34,* 257 (1968).
78. Rosypal and Rosypalova, *Publ. Fac. Sci. Univ. Jan Ev Purkyne Folia Biol., 7,* 1 (1966).
79. De Ley, Park, Tijtgat and Van Ermengem, *J. Gen, Microbiol., 42,* 43 (1966).
80. De Ley and Friedman, *J. Bacteriol., 89,* 1306 (1965).
81. Johnson and Ordal, *J. Bacteriol., 95,* 893 (1968).
82. Moore and McCarthy, *J. Bacteriol., 94,* 1066 (1967).
83. Ritter and Gerloff, *J. Bacteriol., 92,* 1838 (1966).
84. McGee, Rogul, Falkow and Wittler, *Proc. Nat. Acad. Sci. U.S.A., 54,* 457 (1965).
85. Boháček and Blazicek, *Biophysik, 2,* 233 (1965).
86. Colwell and Mandel, *J. Bacteriol., 89,* 454 (1965).
87. Jones and Walker, *J. Gen. Microbiol., 31,* 187 (1963).
88. Belser *Evolution, 18,* 177 (1964).
89. Werner, Gasser and Sebald, *Zentralbl. Bakteriol. Parasitenk. Infektionskr. Hyg. Abt. I Orig., 198,* 504 (1965).
90. Gasser and Mandel, *J. Bacteriol., 96,* 580 (1968).
91. Gasser and Sebalk, *Ann. Inst. Pasteur (Paris), 110,* 261 (1966).
92. Cantoni, *Boll. Ist. Sieroter. Milan., 45,* 92 (1966).
93. Jones and Walker, *Arch. Biochem. Biophys., 128,* 597 (1968).
94. Cantoni, Hill and Silvestri, *Appl. Microbiol., 13,* 631 (1965).
95. Craveri, Manachini and Cantoni, *Ann. Microbiol. Enzimol., 15,* 209 (1965).
96. Cantoni, Manachini and Craveri, *Ann. Microbiol. Enzimol., 15,* 151 (1965).
97. Cantoni, *Boll. Ist. Sieroter. Milan., 45,* 165 (1966).
98. Mitra, McCleskey and Larson, *Bacteriol. Proc.,* p. G-9 (1965).
99. Knittel, Black, Sandine and Elliker, *Bacteriol. Proc.,* p. G-9 (1965).
100. Romond, Sartory and Malgras, *Ann. Inst. Pasteur (Paris), 111,* 710 (1966).
101. Ottogalli and Galli, *Ann. Microbiol. Enzimol., 17,* 199 (1967).
102. Panos, *J. Gen. Microbiol., 39,* 131 (1965).
103. Brock and Mandel, *J. Bacteriol., 91,* 1659 (1966).
104. Evans and Schultes, *Int. J. Syst. Bacteriol., 19,* 159 (1969).
105. Auletta and Kennedy, *J. Bacteriol., 92,* 28 (1966).
106. Silvestri and Hill, *J. Bacteriol., 90,* 136 (1965).
107. Rosypalová, Boháček and Rosypal, *Antonie van Leeuwenhoek J. Microbiol. Serol., 32,* 192 (1966).
108. Boháček, Kocur and Martinec, *J. Gen. Microbiol., 46,* 369 (1967).
109. Rosypal and Rosypalová, *Publ. Fac. Sci. Univ. Jan Ev Purkyne Brno, 471,* 105 (1966).
110. Rosypalová, Boháček and Rosypal, *Antonie van Leeuwenhoek J. Microbiol. Serol. 32,* 105 (1966).
111. Russell, Herrmann and Dowling, *Biophys. J., 9,* 473 (1969).
112. Jeffries, *Int. J. Syst. Bacteriol., 19,* 183 (1969).
113. Boháček, Kocur and Martinec, *J. Appl. Bacteriol., 31,* 215 (1968).
114. Schein, *Biochem. J., 101,* 647 (1966).
115. Klesius and Schuhardt, *J. Bacteriol., 95,* 739 (1968).
116. Rosypal, Boháček and Rosypalová, *Publ. Fac. Sci. Univ. Jan Ev Purkyne Brno, 471,* 115 (1966).
117. Venner, *Acta Biochim. Pol., 14,* 31 (1967).

118. Lee, Wahl and Barbu, *Ann. Inst. Pasteur (Paris), 91,* 212 (1966).
119. Canale-Parola, Mandel and Kupfer, *Arch. Mikrobiol., 58,* 30 (1967).
120. Garrity, Detrick and Kennedy, *J. Bacteriol., 97,* 557 (1969).
121. Wayne and Gross, *J. Bacteriol., 96,* 1915 (1968).
122. Vorobjeva, Vanushin, Kokurina and Prosvetova, *Mikrobiologiya, 34,* 1003 (1965).
123. Ewa, Sikerska and Priegnitz, *Acta Microbiol.. Pol., 17,* 255 (1968).
124. Neimark and Pène, *Bacteriol. Proc.,* p. M-110 (1965).
125. Neimark, *Ann. N.Y. Acad. Sci., 143,* 31 (1967).
126. Kelton and Mandel, *J. Gen. Microbiol., 56,* 131 (1969).
127. Lynn and Smith, *J. Bacteriol., 74,* 811 (1957).
128. Lynn and Haller, *Antonie van Leeuwenhoek J. Microbiol. Serol., 34,* 249 (1968).
129. Williams, Wittler and Burris, *J. Bacteriol., 99,* 341 (1969).
130. Fox and Folsome, *Bacteriol. Proc.,* p. G-105 (1966).
131. Morowitz, Tourtellotte, Guild, Castro, Woese and Cleverdon, *J. Mol. Biol., 4,* 93 (1962).
132. Chelton, Jones and Walker, *J. Gen. Microbiol., 50,* 305 (1968).
133. Neimark and Pène, *Proc. Soc. Exp. Biol. Med., 118,* 517 (1965).
134. Walker, *Nature, 216,* 711 (1967).
135. Jones and Walker, *Nature, 198,* 588 (1963).
136. Jones, Tittensor and Walker, *J. Gen. Microbiol., 40,* 405 (1965).
137. Bak and Black, *Nature, 219,* 1044 (1968).
138. Johnson and Ordal, *J. Bacteriol., 98,* 319 (1969).
139. Rogosa, *J. Bacteriol., 98,* 756 (1969).
140. La Macchia and Pelczar, *J. Bacteriol., 91,* 514 (1966).
141. Catlin and Cunningham, *J. Gen. Microbiol., 26,* 303 (1961).
142. Jackson, Moriarty and Nicholas, *J. Gen. Microbiol., 53,* 53 (1968).
143. Anderson, Pramer and Davis, *Biochim. Biophys. Acta, 108,* 155 (1965).
144. De Ley and Schell, *J. Gen. Microbiol, 33,* 243 (1963).
145. De Ley, *Antonie van Leeuwenhoek J. Microbiol. Serol., 30,* 281 (1964).
146. Mandel, *J. Gen. Microbiol., 43,* 273 (1966).
147. Mandel, *Annu. Rev. Microbiol., 23,* 700 (1969).
148. Colwell and Mandel, *J. Bacteriol., 88,* 1816 (1964).
149. Ballard, Doudoroff and Stanier, *J. Gen. Microbiol., 53,* 349 (1968).
150. Parejka and Wilson, *J. Bacteriol., 95,* 143 (1968).
151. Mandel, Weeks and Colwell, *J. Bacteriol., 90,* 1492 (1965).
152. De Ley, Bernaerts, Rassel and Guilmot, *J. Gen. Microbiol., 43,* 7 (1966).
153. Wu, Gregory and Hausen, *Bacteriol. Proc.,* p. G-8 (1968).
154. De Ley and Rassel, *J. Gen. Microbiol., 41,* 85 (1965).
155. Schramek, *Acta Virol., 12,* 18 (1968).
156. Sigal, Senez, Le Gall and Sebald, *J. Bacteriol., 85,* 1315 (1963).
157. Claus, Bergendahl and Mandel, *Arch. Mikrobiol., 63,* 26 (1968).
158. Colwell, Adeyemo and Kirtland, *J. Appl. Bacteriol., 31,* 323 (1968).
159. Colwell and Yuter, *Bacteriol. Proc.,* p. G-28 (1965).
160. Basden, Tourtellotte and Tucker, *J. Bacteriol., 95,* 439 (1968).
161. Canale-Parola, Udris and Mandel, *Arch. Mikrobiol., 63,* 385 (1968).
162. Craveri and Manachini, *Ann. Microbiol. Enzimol., 16,* 115 (1966).
163. Craveri and Manachini, *Ann. Microbiol. Enzimol., 16,* 1 (1966).
164. Bednar and Frea, *Bacteriol. Proc.,* p. RT-14 (1967).
165. Germaine and Anderson, *J. Bacteriol., 92,* 662 (1966).
166. Lawrence and Clark, *Can. J. Biochem., 44,* 1685 (1966).
167. Haapala, Rogul, Evans and Alexander, *J. Bacteriol., 98,* 421 (1969).
168. Staley, *J. Bacteriol., 95,* 1921 (1968).
169. Leifson and Mandel, *Int. J. Syst. Bacteriol., 19,* 127 (1969).
170. Shankel, Jungerius and Downs, *Bacteriol. Proc.,* p. G-33 (1965).
171. De Ley, Keysters, Khan-Matsubara and Shewan, *Antonie van Leeuwenhoek J. Microbiol. Serol., 36,* 193 (1970).
172. Johnson, Anderson and Ordal, *J. Bacteriol., 101,* 568 (1970).
173. Colwell, Smith and Chapman, *Can. J. Microbiol., 14,* 165 (1968).
174. Quellette, Burris and Wilson, *Antonie van Leeuwenhoek J. Microbiol. Serol., 35,* 275 (1969).
175. Tsyganov, Namestnikova and Krassikova, *Mikrobiologiya, 35,* 75 (1966).
176. Belozersky, in a publication from the Publishing House of the Academy of Sciences, Moscow, U.S.S.R., p. 194 (1957).
177. Farina and Bradley, *J. Bacteriol., 102,* 30 (1970).
178. Skyring and Quadling, *Can. J. Microbiol., 16,* 95 (1970).

179. Hammond, *J. Bacteriol., 104,* 1024 (1970).
180. Jones and Bradley, *Bacteriol. Proc.,* p. 148 (1963).
181. Jones and Bradley, *Dev. Ind. Microbiol., 5,* 267 (1964).
182. Yagamuchi, *J. Gen. Appl. Microbiol., 13,* 63 (1967).
183. Lovett and Young, *J. Bacteriol., 100,* 658 (1969).
184. Stenesh, Roe and Snyder, *Biochim. Biophys. Acta, 161,* 442 (1968).
185. Guicciardo, Biffi, Manachini, Craveri, Scolastico, Rhindone and Craveri, *Ann. Microbiol. Enzimol., 18,* 191 (1968).
186. Johnson, *Int. J. Syst. Bacteriol., 20,* 421 (1970).
187. Irwin, Unpublished Data, reported by De Ley, *J. Bacteriol., 101,* 738 (1970).
188. Sebald and Veron, *Ann. Inst. Pasteur (Paris), 105,* 897 (1963).
189. Catlin and Cunningham, *J. Gen. Microbiol., 37,* 353 (1964).
190. Rogul, Brendle, Haapala and Alexander, *J. Bacteriol., 101,* 827 (1970).
191. Yamada and Komataga, *J. Gen. Appl. Microbiol., 16,* 215 (1970).
192. Hofstad, *Int. J. Syst. Bacteriol., 20,* 175 (1970).
193. Colwell and Mandel, Unpublished Data.
194. Takayama, Abe and Kinoshita, *J. Agr. Chem. Soc. Jap., 39,* 342 (1965).
195. Kumazawa and Yamagawa, *Jap. J. Vet. Res., 17,* 115 (1969).
196. Gause, Dudnik, Laiko and Netyska, *Science, 157,* 1196 (1964).
197. Rosenkranz and Ellner, *Science, 160,* 893 (1968).
198. Weeks, *J. Appl. Bacteriol., 32,* 13 (1969).
199. Starr and Mandel, Unpublished Data, cited in De Ley, *J. Bacteriol., 101,* 738 (1970).
200. Chargaff, Schulman and Shapiro, *Nature, 180,* 851 (1957).
201. Smith and Wyatt, *Biochem. J., 49,* 144 (1951).
202. Tonomura, Malkin and Rabinowitz, *J. Bacteriol., 89,* 1438 (1965).
203. Zimmer and Venner, *Hoppe Seyler's Z. Physiol. Chem., 333,* 20 (1963).
204. Tewfik and Bradley, *J. Bacteriol., 94,* 1994 (1967).
205. Prausser, *Spisy Prirodoved Fak. Univ. J. E. Purkyne Brno, 475,* 268 (1966).
206. Prausser and Falta, *Z. Allg. Mikrobiol., 8,* 39 (1968).
207. Berns and Thomas, *J. Mol. Biol., 11,* 476 (1965).
208. Wayne and Gross, *J. Bacteriol., 95,* 1481 (1968).
209. Agabian, Keshishian and Shapiro, *Fed. Proc., 29,* 854 (1970).
210. Miller, Sandine and Elliken, *J. Bacteriol., 102,* 278 (1970).
211. Gasser and Sebald, *Ann. Inst. Pasteur (Paris), 110,* 261 (1966).
212. Kittle, Sandine and Elliken, *Bacteriol. Proc.,* p. 41 (1964).
213. Mehta and Hutchinson, *Can. J. Microbiol., 16,* 281 (1970).
214. Coykendall, *Arch. Oral Biol., 15,* 365 (1970).
215. Wang and Hashagen, *J. Mol. Biol., 8,* 333 (1964).
216. Kelly and Brock, *J. Bacteriol., 100,* 14 (1969).
217. Boháček, Kocur and Martinec, *Microbios, 6,* 85 (1970).
218. Bergan, Bovre and Hovig, *Acta Pathol. Microbiol. Scand. Sect. B, 78,* 85 (1970).
219. Rosypalová, Boháček and Rosypal, *Antonie van Leeuwenhoek J. Microbiol. Serol., 32,* 105 (1966).
220. Sarfert and Venner, *Z. Allg. Mikrobiol., 9,* 153 (1969).
221. Slosarek, *Folia Microbiol., 15,* 431 (1970).
222. Darland, Brock, Samsonoff and Conti, *Science, 170,* 1416 (1970).
223. Kingsbury and Weiss, Unpublished Data, cited in De Ley, *J. Bacteriol., 101,* 738 (1970).
224. Jackson, Moriarty and Nicholas, *J. Gen. Microbiol., 53,* 53 (1968).
225. De Ley and Schell, *J. Gen. Microbiol., 33,* 243 (1963).
226. Mandel and Rownd, in *Taxonomic Biochemistry and Serology,* p. 585, Leone, Ed. Ronald Press Co., New York (1964).
227. Hogan and Colwell, *J. Appl. Bacteriol., 32,* 103 (1969).
228. Colwell, *J. Bacteriol., 104,* 410 (1970).
229. De Ley, *J. Bacteriol., 101,* 738 (1970).
230. Siedler, Starr and Mandel, *J. Bacteriol., 100,* 786 (1969).
231. Lee and Boezi, *J. Bacteriol., 92,* 1821 (1969).
232. Friedman and De Ley, *J. Bacteriol., 89,* 95 (1965).
233. Ahrens and Rheinheimer, *Kiel. Meeresforsch., 23,* 127 (1967).
234. Tinbergen, Ph.D. Thesis, State University of Leyden, The Netherlands (1966), cited in De Ley, *J. Bacteriol., 101,* 738 (1970).
235. Mandel, Unpublished Data, cited in De Ley, *J. Bacteriol., 101,* 738 (1970).
236. Sebald, Unpublished Data, cited in De Ley, *J. Bacteriol., 101,* 738 (1964).
237. Wagenbrath, *Flora, 151,* 219 (1961).
238. Sebald and Veron, *Ann. Inst. Pasteur (Paris), 105,* 897 (1963).

239. Leifson and Mandel, *Antonie van Leeuwenhoek J. Microbiol. Serol., 32,* 57 (1966).
240. Sigal, Senez, LeGall and Sebald, *J. Bacteriol., 85,* 1315 (1963).
241. Saunders and Campbell, *J. Bacteriol., 92,* 515 (1966).
242. Hanoaka, Kato and Amano, *Biken J., 12,* 181 (1969).
243. Colwell, Citarella and Ryman, *J. Bacteriol., 90,* 1148 (1965).
244. Frontali, Hill and Silvestri, *J. Gen. Microbiol., 38,* 243 (1965).
245. Marmur and Doty, *J. Mol. Biol., 5,* 109 (1962).
246. Monson, Bradley, Enquist and Cruces, *J. Bacteriol., 99,* 702 (1969).
247. Meinke and Jones, *Bacteriol. Proc.,* p. 155 (1967).
248. Kit, *Nature, 193,* 274 (1962).
249. Lewin, *Can. J. Microbiol., 16,* 517 (1970).
250. De Ley, Unpublished Data, cited in De Ley, *J. Bacteriol., 101,* 738 (1970).
251. Wang and Hutchinson, *J. Mol. Biol., 8,* 333 (1964).
252. Joshi, Guild and Handler, *J. Mol. Biol., 6,* 34 (1963).
253. Wyatt and Cohen, *Nature, 170,* 846 (1952).
254. Marmur, Falkow and Mandel, *Annu. Rev. Microbiol., 17,* 329 (1963).
255. Mitchell, Hendrie, Margaret and Shewan, *J. Appl. Microbiol., 32,* 40 (1969).
256. McCurdy and Wolf, *Can. J. Microbiol., 13,* 1707 (1967).
257. De Ley and van Muylen, *Antonie van Leeuwenhoek J. Microbiol. Serol., 29,* 344 (1963).
258. Mandel and Leadbetter, *J. Bacteriol., 90,* 1795 (1965).
259. Wang and Hutchison, *J. Mol. Biol., 8,* 333 (1964).
260. Brock and Freeze, *J. Bacteriol., 98,* 289 (1969).
261. Ramamley and Hixson, *J. Bacteriol., 103,* 527 (1970).
262. Robertstad, Hester and Gordan, Personal Communication to L. Pine.
263. Gledhill and Casida, *Appl. Microbiol., 18,* 114 (1969).
264. Farina and Bradley, *J. Bacteriol., 102,* 30 (1970).
265. Gledhill and Casida, *Appl. Microbiol., 18,* 340 (1969).
266. Sukapure, Lechevalier, Reber, Higgins, Lechevalier and Prauser, *Appl. Microbiol., 19,* 527 (1970).
267. Edelman, Srointan, Schiff, Epstein and Zeldin, *Bacteriol. Rev., 31,* 315 (1967).
268. Jones and Thompson, *J. Protozool., 10,* 91 (1963).
269. Storck and Alexopoulos, *Bacteriol. Rev., 34,* 126 (1970).
270. Storck, *J. Bacteriol., 91,* 227 (1966).
271. Storck, Nobles and Alexopoulos, *Mycologia, 63,* 38 (1971).
272. Vanyushin, Belozersky and Bogdanova, *Dokl. Acad. Nauk S.S.S.R., 134,* 1222 (1960).
273. Mandel, in *Handbook of Biochemistry,* 2nd ed., p. H-75, Sober, Ed., The Chemical Rubber Co., Cleveland, Ohio (1970).
274. Uryson and Belozersky, *Dokl. Akad. Nauk S.S.S.R., 133,* 708 (1960).
275. Belozersky and Spirin, in *The Nucleic Acids,* Vol. 3, p. 147, Chargaff and Davidson, Eds. Academic Press, New York (1960).
276. Venner, *Z. Physiol. Chem., 333,* 5 (1963).
277. Dutta, Richmann, Woodward and Mandel, *Genetics, 57,* 719 (1967).
278. Kogane and Yamagita, *J. Appl. Microbiol., 10,* 61 (1964).
279. Nakase and Komataga, *J. Gen. Appl. Microbiol., 14,* 345 (1968).
280. Comb, Brown and Katz, *J. Mol. Biol., 8,* 781 (1964).
281. Stenderup and Lethback, *J. Gen. Microbiol., 52,* 231 (1968).
282. Meyer and Phaff, *J. Bacteriol., 97,* 52 (1969).
283. Bak, Christensen and Stenderup, *Nature, 224,* 270 (1969).
284. Villa and Storck, *J. Bacteriol., 96,* 184 (1968).
285. Storck, Alexopoulos and Phaff, *J. Bacteriol., 98,* 1069 (1969).
286. Schildkraut, Mandel, Levisohn, Smith-Sonneborn and Marmur, *Nature, 196,* 795 (1962).
287. Schildkraut, Marmur and Doty, *J. Mol. Biol., 4,* 430 (1962).
288. Luck and Reich, *Proc. Nat. Acad. Sci. U.S.A., 52,* 931 (1964).
289. Clark, Hodgson and Lawrence, *Can. J. Microbiol., 14,* 482 (1967).
290. Storck, Nobles and Alexopoulos, Unpublished Data, cited in Storck and Alexopoulos, *Bacteriol. Rev., 34,* 126 (1970).
291. Bicknell and Douglas, *J. Bacteriol., 101,* 505 (1970).
292. Chargaff in *Nucleic Acids,* Vol. 1, p. 307, Chargaff and Davidson, Eds. Academic Press, New York (1955).
293. Huang and Rosenberg, *Anal. Biochem., 16,* 107 (1966).
294. Mahler and Pereira, *J. Mol. Biol., 5,* 325 (1962).
295. Rost and Venner, *Z. Physiol. Chem. (Hoppe Seyler's), 339,* 230 (1964).
296. Serenkov and Kust, *Dokl. Akad. Nauk S.S.S.R., 142,* 1197 (1962).
297. Storck, *J. Bacteriol., 90,* 1260, (1965).

298. Vanyushin, Belozerskii and Kokurina, *Trans. Moscow Soc. Natur., 24,* 7 (1966).
299. Sueoka, in *The Bacteria,* Vol. 5, p. 9, Gunsalus and Stanier, Eds. Academic Press, New York (1964).
300. Chun, Vaughn and Rich, *J. Mol. Biol., 7,* 130 (1963).
301. Leff, Mandel, Epstein and Schiff, *Biochem. Biophys. Res. Commun., 13,* 126 (1963).
302. Sager and Ishida, *Proc. Nat. Acad. Sci. U.S.A., 50,* 725 (1963).
303. Iwamura and Kuwashima, *Biochem. Biophys. Acta, 82,* 678 (1964).
304. Mandel, in *Chemical Zoology,* Vol. 2, p. 541, Florkin and Scheer, Eds., Academic Press, New York (1967).
305. Sueoka, in *The Bacteria,* Vol. 5, p. 9, (1964).
306. Schildkraut, Mandel, Levisohn, Smith-Sonneborn and Marmur, *Nature, 196,* 795 (1962).
307. Guttman and Eisenman, *Nature, 206,* 113 (1965).
308. Marmur, Cahoon, Shimura and Vogel, *Nature, 197,* 1228 (1963).
309. Edelman, Schiff and Epstein, *J. Mol. Biol., 11,* 769 (1965).
310. Edelman, Cowan, Epstein and Schiff, *Proc. Nat. Acad. Sci. U.S.A., 52,* 1214 (1964).
311. Edelman, Schiff and Epstein, Unpublished Data, cited in Schiff and Epstein, *Reproduction: Molecular, Submolecular and Cellular,* p. 31, Locke, Ed. Academic Press, New York.
312. Braverman and Einstadt, *Biochem. Biophys. Acta, 43,* 374 (1964).
313. DuBuy, Mattern and Riley, *Science, 147,* 754 (1965).
314. Luck and Reich, *Proc. Nat. Acad. Sci. U.S.A., 52,* 931 (1964).
315. Smith-Sonneborn, Green and Marmur, *Nature, 197,* 385 (1963).
316. Suyama and Preer, *Genetics, 52,* 1051 (1965).
317. Ginzburg, cited in Antonov, *Usp. Sovrem. Biol., 60,* 161 (1965).
318. Whitfeld, *Aust. J. Biol. Sci., 6,* 234 (1953).
319. Scherbaum, *Exp. Cell Res., 13,* 24 (1957).
320. Swartz, Trauten and Kornberg, *J. Biol. Chem., 237,* 1961 (1962).
321. Sueoka, *Cold Spring Harbor Symp. Quant. Biol., 26,* 35 (1961).
322. Suyama, *Biochemistry, 5,* 2214 (1966).
323. Neimark and Blaker, *Nature, 216,* 600 (1967).
324. Mandel and Honigberg, *J. Protozool., 11,* 114 (1964).
325. Michaels and Treick, *Exp. Parasitol., 12,* 401 (1962).
326. Antonov and Belozerskii, *Dokl. Akad. Nauk S.S.S.R., 138,* 1216 (1961).
327. Bouisset, Breuilland, Carre-Michel, Larrouy and Raujeva, *C. R. Acad. Sci. (Paris), 262,* 19112 (1966).
328. Riou, Pautrizel and Paoletti, *C. R. Acad. Sci. (Paris), 262,* 2367 (1966).

# THE BASE COMPOSITION OF BACTERIOPHAGE NUCLEIC ACIDS

DR. C. J. ABSHIRE, R. GUAY and J. P. BOULEY

## INTRODUCTION

The nucleic acid base composition of bacteriophages is given in Tables 1 and 2. The GC and ATGC contents are listed irrespective of the method of determination. The phages are listed in the tables by alphabetical order of the host genera (with the exception of the Enterobacteriaceae, which are grouped together because of the numerous cross reactions). The phage names, which follow the host genera, appear in the following order of precedence: Latin alphabet, Greek alphabet, Arabic numeral, Latin numeral. The names of mutant phages are italicized. Concerning the host genera, it should be noted that (a) *Clostridium* includes *Plectridium* and *Welchia,* and (b) *Streptomyces* includes hosts referred to by Russian authors as *Actinomyces.*

Phages with double-stranded DNA include the following: *Pseudomonas* phage PM2, which is cubic and possesses a lipid-containing envelope; a few defective phages or "Killer-particles"; and Phycovirus LPP-1.

The percentage GC for single-stranded nucleic acid-containing phages is not given, because the only method of obtaining it is from the ATGC content.

The tables contain data published up to December 31, 1970.

### Symbols and Abbreviations

| | | | |
|---|---|---|---|
| A | Adenine | 6-MAP | 6-Methylaminopurine |
| C | Cytosine | 5-MC | 5-Methylcytosine |
| G | Guanine | T | Thymine |
| 5-HMC | 5-Hydroxymethylcytosine | U | Uracil |
| 5-HMU | 5-Hydroxymethyluracil | NN | No name |

### TABLE 1
### ATGC OR AUGC CONTENT OF BACTERIOPHAGE NUCLEIC ACIDS

| Host Genus | Phage Designation | A, % | T, % | G, % | C, % | Particulars | References |
|---|---|---|---|---|---|---|---|
| **Double-Stranded DNA** | | | | | | | |
| *Bacillus* | AR9 | 37.3 | 35.0 | 13.8 | 13.9 | U | 122 |
| | Nf | 30.8 | 30.5 | 19.7 | 19.1 | | 107 |
| | *PBS2* | 39.2 | 35.9 | 13.4 | 14.7 | U | 111 |
| | *PBS2* | 36.4 | 35.4 | 14.5 | 13.8 | U | 112 |
| | *PBS2* | 35.5 | 37.3 | 13.6 | 13.7 | U | 3 |
| | SPO1 | 30.2 | 30.7 | 19.2 | 19.9 | 5-HMU | 119 |
| | SP8 | 28.5 | 28.1 | 21.8 | 21.5 | 5-HMU | 3 |
| | *SP8** | 30.3 | 30.8 | 19.2 | 19.6 | 5-HMU | 119 |
| | *SP8r* | 29.7 | 28.7 | 21.9 | 20.7 | 5-HMU | 3 |
| | SP10 | 29.9 | 27.2 | 21.3 | 21.9 | | 82 |
| | SP50 | 29.5 | 28.8 | 21.2 | 20.5 | | 11 |
| | SP50 | 29.4 | 28.8 | 21.2 | 20.4 | | 119 |
| | *SP82* | 29.6 | 27.9 | 21.3 | 21.2 | 5-HMU | 3 |
| | *SP82* | 30.7 | 30.6 | 19.6 | 19.2 | 5-HMU | 119 |
| | SW | 28.2 | 28.2 | 22.5 | 22.3 | 5-HMU | 72 |
| | TP-84 | 27.7 | 28.2 | 22.0 | 22.1 | | 98 |
| | *TP-1C* | 28.1 | 28.6 | 21.5 | 21.8 | | 127 |
| | Tφ3 | 29.0 | 30.8 | 19.4 | 20.8 | | 25 |
| | VX | 28.7 | 28.4 | 21.6 | 21.3 | 5-HMU | 106 |

## TABLE 1 (Continued)
## ATGC OR AUGC CONTENT OF BACTERIOPHAGE NUCLEIC ACIDS

| Host Genus | Phage Designation | A, % | T, % | G, % | C, % | Particulars | References |
|---|---|---|---|---|---|---|---|
| **Double-Stranded DNA (continued)** | | | | | | | |
| *Bacillus (cont.)* | *a* | 28.4 | 27.4 | 22.1 | 22.3 | | 66 |
| | Φe | 30.5 | 30.6 | 18.8 | 20.0 | 5-HMU | 119 |
| | Φe | 30.0 | 28.0 | 20.0 | 20.0 | 5-HMU | 96 |
| | φ3 | 32.7 | 31.4 | 18.1 | 17.8 | | 120 |
| | 2C | 30.5 | 30.3 | 19.5 | 19.6 | 5-HMU | 118 |
| *Caulobacter* | φCbK | 17.4 | 17.4 | 32.7 | 32.5 | | 1 |
| *Chondrococcus* | C2 | 28.8 | 30.3 | 23.0 | 18.5 | | 53 |
| *Clostridium* | HMT | 33.0 | 32.4 | 17.5 | 17.1 | | 45 |
| | HM2 | 32.6 | 32.2 | 17.5 | 17.7 | | 46 |
| | HM3 | 35.1 | 34.9 | 14.6 | 15.4 | | 46 |
| | HM7 | 35.7 | 35.3 | 14.5 | 14.6 | | 46 |
| Enterobacteriaceae | | | | | | | |
| *Klebsiella* | C | 20.0 | 25.4 | 33.1 | 21.5 | | 33 |
| *Escherichia* | C16 | 31.0 | 32.0 | 18.0 | 19.0 | 5-HMC | 49 |
| *Escherichia* | $D_d$-Vi | 32.1 | 32.8 | 19.4 | 16.7 | 5-HMC | 69 |
| *Escherichia* | DD7 | 27.0 | 27.3 | 22.6 | 22.5 | 6-MAP(*), 5-MC(*) | 73 |
| *Klebsiella* | $KL_2$ | 30.8 | 30.8 | 20.5 | 18.0 | 5-HMC | 4 |
| *Klebsiella* | $KL_{12}$ | 28.1 | 27.4 | 21.3 | 23.2 | | 31[a] |
| *Klebsiella* | L1 | 23.6 | 22.6 | 26.9 | 26.9 | | 33 |
| *Klebsiella* | L1 | 23.8 | 22.6 | 27.1 | 26.4 | | 35 |
| *Escherichia* | N15 | 23.0 | 23.0 | 27.0 | 27.0 | | 91 |
| *Escherichia* | N15 | 24.8 | 24.0 | 26.1 | 25.1 | | 31[b] |
| Polyvalent | OX-4 | 31.0 | 34.0 | 18.0 | 17.0 | 5-HMC | 48 |
| *Providencia* | PL26 | 29.6 | 28.8 | 21.0 | 20.6 | | 21 |
| *Salmonella* | P22 | 25.0 | 25.0 | 25.0 | 25.0 | | 109[c] |
| Polyvalent | Sal-3 | 31.0 | 34.0 | 18.0 | 17.0 | 5-HMC | 48 |
| Polyvalent | Sal-4 | 31.0 | 34.0 | 18.0 | 17.0 | 5-HMC | 48 |
| *Escherichia* | $s_d$ | 28.6 | 28.7 | 21.3 | 21.4 | | 113 |
| *Escherichia* | T1 | 27.0 | 25.0 | 23.0 | 25.0 | | 23 |
| *Escherichia* | T2 | 31.5 | 36.0 | 16.5 | 16.3 | 5-HMC | 42 |
| *Escherichia* | T2 | 32.2 | 32.7 | 19.1 | 16.6 | 5-HMC | 69 |
| *Escherichia* | T2 | 33.5 | 30.7 | 18.2 | 17.6 | 5-HMC | 101 |
| *Escherichia* | *T2r* | 34.1 | 32.4 | 16.4 | 16.9 | 5-HMC | 23, 24, 68, 129 |
| *Escherichia* | $T2r^+$ | 32.0 | 33.4 | 18.1 | 16.6 | 5-HMC | 23, 24, 129 |
| *Escherichia* | T3 | 22.8 | 27.8 | 23.5 | 26.1 | | 54 |
| *Escherichia* | *T4r* | 32.2 | 33.5 | 18.0 | 16.3 | 5-HMC | 129 |
| *Escherichia* | $T4r^+$ | 32.3 | 33.1 | 18.3 | 16.3 | 5-HMC | 129 |
| *Escherichia* | T5 | 30.3 | 30.8 | 19.5 | 19.5 | | 129 |
| *Escherichia* | *T6r* | 32.3 | 33.4 | 17.7 | 16.6 | 5-HMC | 129 |
| *Escherichia* | $T6r^+$ | 33.8 | 33.9 | 15.1 | 17.1 | 5-HMC | 68, 129 |
| *Escherichia* | T7 | 25.8 | 26.7 | 24.2 | 23.3 | | 64 |
| *Escherichia* | T7 | 26.0 | 26.6 | 23.6 | 23.8 | | 123 |
| *Serratia* | η | 21.6 | 21.4 | 28.5 | 28.5 | Base X, sugar Y | 89 |
| *Serratia* | κ | 23.4 | 22.8 | 25.0 | 28.8 | | 89 |
| *Escherichia* | λ | 21.3 | 28.6 | 22.9 | 27.1 | | 65 |
| *Escherichia* | λ*dg* | 25.7 | 25.7 | 24.4 | 24.2 | | 50 |
| Polyvalent | χ | 22.4 | 20.2 | 28.6 | 28.8 | | 101 |
| *Klebsiella* | 380 | 21.9 | 22.8 | 27.0 | 28.3 | | 33, 35 |
| *Klebsiella* | 483 | 23.1 | 23.8 | 23.4 | 29.8 | | 35 |
| *Klebsiella* | 490 | 25.0 | 22.4 | 26.1 | 26.5 | | 33, 35 |

TABLE 1 (Continued)
ATGC OR AUGC CONTENT OF BACTERIOPHAGE NUCLEIC ACIDS

| Host Genus | Phage Designation | A, % | T, % | G, % | C, % | Particulars | References |
|---|---|---|---|---|---|---|---|
| **Double-Stranded DNA (continued)** | | | | | | | |
| Enterobacteriaceae (*cont.*) | | | | | | | |
| *Klebsiella* | 632 | 20.8 | 21.5 | 29.0 | 28.8 | | 32 |
| *Klebsiella* | 632 | 20.8 | 21.5 | 28.8 | 28.9 | | 33, 35 |
| *Micrococcus* | N1 | 15.4 | 14.3 | 34.6 | 35.5 | 5-MC(*) | 59 |
| | N1 | 18.0 | 13.0 | 31.9 | 37.2 | 5-MC(*) | 100 |
| | N6 | 16.0 | 14.6 | 30.6 | 39.1 | | 100 |
| *Mycobacterium* | Choremis | 17.2 | 15.9 | 33.4 | 33.4 | | 116 |
| | D4 | 18.0 | 18.0 | 32.0 | 32.0 | | 104 |
| | D28 | 20.0 | 20.0 | 30.0 | 30.0 | | 104 |
| | D29 | 18.0 | 18.1 | 33.9 | 30.0 | | 115 |
| | D29 | 18.0 | 18.0 | 32.0 | 32.0 | | 104 |
| | D32 | 16.0 | 16.0 | 34.0 | 34.0 | | 104 |
| | Leo | 17.6 | 16.9 | 33.6 | 31.9 | | 31[a] |
| | R1 | 17.4 | 14.2 | 33.7 | 34.8 | | 31[a] |
| *Myxococcus* | MX-1 | 24.1 | 20.0 | 25.0 | 30.8 | | 14 |
| *Proteus* | M | 25.7 | 25.4 | 24.6 | 24.3 | | 21 |
| *Pseudomonas* | SD1 | 24.5 | 21.8 | 26.0 | 27.8 | | 105 |
| | SD1 | 24.1 | 23.2 | 25.8 | 27.0 | | 105 |
| | $\phi$-MC | 26.9 | 27.6 | 23.0 | 22.5 | | 19 |
| | $\phi$-2 | 22.5 | 23.0 | 27.5 | 27.0 | | 39 |
| *Streptomyces* | Type II | 16.7 | 15.6 | 32.6 | 35.2 | | 55 |
| *Xanthomonas* | XP12 | 16.8 | 16.0 | 33.8 | 33.4 | 5-MC | 58 |
| **Single-Stranded DNA of Cubic Phages** | | | | | | | |
| Enterobacteriaceae | | | | | | | |
| Polyvalent | $\Phi$R | 23.5 | 33.0 | 24.1 | 19.3 | | 52 |
| Polyvalent | $\Phi$X-174 | 24.6 | 31.5 | 24.9 | 19.0 | | 108 |
| **Single-Stranded DNA of Filamentous Phages** | | | | | | | |
| Enterobacteriaceae | | | | | | | |
| *Escherichia* | AE2 | 26.0 | 32.0 | 21.0 | 21.0 | | 85 |
| *Escherichia* | EC9 | 24.4 | 34.4 | 20.6 | 20.6 | | 16 |
| *Escherichia* | fd | 24.4 | 34.1 | 19.9 | 21.7 | | 44 |
| *Escherichia* | M13 | 23.3 | 35.8 | 21.1 | 19.8 | | 97 |
| *Escherichia* | $\delta$A | 25.7 | 33.1 | 21.2 | 20.0 | | 74 |
| *Xanthomonas* | Xf | 20.6 | 19.3 | 32.6 | 27.5 | | 57 |
| **Single-Stranded RNA of Cubic Phages** | | | | | | | |
| Enterobacteriaceae | | | | | | | |
| *Escherichia* | fr | 24.3 | 23.7 | 27.1 | 24.9 | | 44 |
| *Escherichia* | f2 | 22.1 | 25.1 | 26.8 | 25.9 | | 62 |
| *Escherichia* | MS2 | 22.8 | 25.2 | 27.1 | 24.9 | | 110 |

## TABLE 1 (Continued)
## ATGC OR AUGC CONTENT OF BACTERIOPHAGE NUCLEIC ACIDS

| Host Genus | Phage Designation | A, % | T, % | G, % | C, % | Particulars | References |
|---|---|---|---|---|---|---|---|
| **Single-Stranded RNA of Cubic Phages (continued)** | | | | | | | |
| Enterobacteriaceae (*cont.*) | | | | | | | |
| *Escherichia* | MS2 | 23.3 | 24.8 | 26.4 | 25.5 | | 84 |
| *Escherichia* | Q$\beta$ | 22.1 | 29.1 | 23.7 | 24.7 | | 84 |
| *Escherichia* | R17 | 24.8 | 22.5 | 27.1 | 25.5 | | 86 |
| *Escherichia* | R17 | 23.1 | 25.7 | 26.3 | 24.9 | | 71 |
| *Escherichia* | R17 | 23.0 | 26.0 | 26.0 | 25.0 | | 26 |
| *Escherichia* | R23 | 23.4 | 26.0 | 25.6 | 25.1 | | 126 |
| *Escherichia* | R34 | 23.6 | 26.8 | 23.9 | 25.4 | | 126 |
| *Escherichia* | R40 | 22.4 | 25.5 | 26.9 | 24.9 | | 126 |
| *Escherichia* | ZG | 25.1 | 23.7 | 27.2 | 24.0 | | 10 |
| *Escherichia* | ZIK/1 | 23.6 | 28.3 | 23.9 | 24.2 | | 10 |
| *Escherichia* | ZJ/1 | 24.3 | 28.2 | 23.8 | 23.7 | | 10 |
| *Escherichia* | ZL/3 | 24.8 | 23.4 | 26.9 | 24.9 | | 10 |
| *Escherichia* | ZS/3 | 24.8 | 23.4 | 28.0 | 23.8 | | 10 |
| *Escherichia* | $\alpha$15 | 24.7 | 24.0 | 26.3 | 25.0 | | 10 |
| *Escherichia* | $\beta$ | 23.2 | 24.0 | 27.0 | 25.6 | | 75 |
| *Escherichia* | $\beta$ | 23.3 | 28.0 | 21.6 | 27.2 | | 76 |
| *Escherichia* | $\mu_2$ | 23.8 | 27.1 | 25.4 | 23.7 | | 17 |

(*) Traces.
[a] Reference 3 in the cited reference.
[b] Reference 77 in the cited reference.
[c] Reference *q* in the cited reference: Garen, A., and Zinder, N. D., Unpublished Data, quoted in Zinder, N. D., *J. Cell. Comp. Physiol., 45 (Suppl. 2),* 23 (1955).

## TABLE 2
## GC CONTENT OF BACTERIOPHAGE NUCLEIC ACIDS

| Host Genus | Phage Designation | GC, % | Particulars | References |
|---|---|---|---|---|
| *Bacillus* | A | 35.0–37.0 | | 66, 67 |
| | AR9 | 27.7 | U | 122 |
| | B | 36.0–37.0 | | 66, 67 |
| | C | 37.0 | | 66, 67 |
| | D | 37.0 | | 66, 67 |
| | F | 49.0 | | 125 |
| | G | 40.0–41.0 | | 67 |
| | Nf | 38.8–43.2 | | 107 |
| | PBSH | 43.0 | "Killer particle" | 41 |
| | PBSX | 43.0 | "Killer particle" | 78 |
| | PBS1 | 28.0 | U | 80 |
| | *PBS2* | 27.3–28.3 | U | 3, 111, 112 |
| | SPO1 | 39.1–45.0 | 5-HMU | 10, 80, 81, 119 |
| | SPO2 | 43.0 | | 94 |
| | SPP1 | 43.0 | | 93 |
| | SP3 | 35.0 | | 66, 80 |
| | SP8 | 43.0–43.3 | 5-HMU | 3, 51, 119 |
| | *SP8** | 38.8 | 5-HMU | 119 |
| | *SP8r* | 42.6 | 5-HMU | 3 |
| | SP10 | 42.3–44.0 | | 80, 82 |

## TABLE 2 (Continued)
## GC CONTENT OF BACTERIOPHAGE NUCLEIC ACIDS

| Host Genus | Phage Designation | GC, % | Particulars | References |
|---|---|---|---|---|
| *Bacillus (cont.)* | SP50 | 41.6–44.0 | | 11, 119 |
| | *SP82* | 38.8–45.0 | 5-HMU | 3, 38, 66, 80, 119 |
| | SW | 44.8 | 5-HMU | 72 |
| | t | 41.0–42.0 | | 2 |
| | TP-1 | 42.0 | | 127 |
| | *TP-1C* | 43.3 | | 127 |
| | TP-84 | 41.7–44.1 | | 98, 99 |
| | Tϕ3 | 40.2–44.4 | | 25 |
| | VX | 42.9 | 5-HMU | 106 |
| | α | 41.0–46.0 | | 5, 22, 66 |
| | Φe | 38.8–40.3 | 5-HMU | 80, 95, 96, 119 |
| | ϕ3 | 35.9–36.0 | | 80, 120 |
| | ϕ105 | 43.5 | | 9 |
| | ϕμ-4 | 57.0–60.0 | | 90 |
| | 2C | 39.0–43.0 | 5-HMU | 80, 118, 119 |
| *Bacterium*† | BP1 | 40.7–42.8 | | 121 |
| *Brevibacterium* | AP85 III | 56.6 | | 79 |
| | P4 | 55.3 | | 79 |
| | P465 | 54.0 | | 79 |
| | P468 II | 54.6 | | 79 |
| *Brucella* | F1–F11 (F1, F2, . . . F11) | 45.3–46.7 § | | 15 |
| *Caulobacter* | ϕCbK | 62.3–65.2 | | 1 |
| *Chondrococcus* | C2 | 41.0–41.5 | | 53 |
| *Clostridium* | CT1 | 31.0–32.0 | | 67 |
| | F1 | 39.5 | | 8 |
| | HMT | 34.6 | | 45 |
| | HM2 | 34.6–35.2 | | 46 |
| | HM3 | 30.0–30.5 | | 46 |
| | HM7 | 29.1 | | 46 |
| *Cytophaga* | NCMB 384 | 30.0 | | 18 |
| | NCMB 385 | 30.0 | | 18 |
| Enterobacteriaceae | | | | |
| *Klebsiella* | C | 54.6–58.8 | | 33, 34 |
| *Escherichia* | C16 | 37.0 | 5-HMC | 49 |
| *Escherichia* | Ec15 | 46.4–46.6 | Defective phage | 30 |
| *Escherichia* | $D_d$-Vi | 36.1 | 5-HMC | 69 |
| *Escherichia* | DD7 | 45.1 | 6-MAP(*), 5-MC(*) | 73 |
| *Klebsiella* | KL2 | 38.5 | 5-HMC | 4 |
| *Klebsiella* | KL12 | 44.5 | | 31[a] |
| *Klebsiella* | L1 | 53.5–55.4 | | 33, 34, 35 |
| *Klebsiella* | L2 | 61.5 | | 34 |
| *Escherichia* | Mu1 | 45.7–51.0 | | 117 |
| *Escherichia* | N4 | 48.0 | | 103 |
| *Escherichia* | N15 | 51.2–54.0 | | 31[b], 91 |
| Polyvalent | OX-4 | 35.0 | 5-HMC | 48 |
| Polyvalent | *Plkc* | 46.0 | | 47 |
| Polyvalent | P2 | 50.0 | | 7 |
| Polyvalent | P22 | 46.0–50.0 | | 29, 109[c] |
| *Providencia* | PL25 | 43.2 | | 88 |
| *Providencia* | PL26 | 41.6 | | 21 |

**TABLE 2 (Continued)**
**GC CONTENT OF BACTERIOPHAGE NUCLEIC ACIDS**

| Host Genus | Phage Designation | GC, % | Particulars | References |
|---|---|---|---|---|
| **Enterobacteriaceae (*cont.*)** | | | | |
| Polyvalent | Sal-3 | 35.0 | 5-HMC | 48 |
| Polyvalent | Sal-4 | 35.0 | 5-HMC | 48 |
| *Escherichia* | $s_d$ | 42.7 | | 113 |
| *Escherichia* | T1 | 46.0–48.0 | | 13, 23, 102 |
| *Escherichia* | T2 | 34.1–35.8 | 5-HMC | 43, 69, 101 |
| *Escherichia* | T3 | 49.6–53.0 | | 54, 102 |
| *Escherichia* | T4 | 36.1 | 5-HMC | 79 |
| *Escherichia* | *T4r* | 34.5 | 5-HMC | 129 |
| *Escherichia* | T5 | 39.0–43.0 | | 102, 114, 129 |
| *Escherichia* | *T6r* | 34.4 | 5-HMC | 129 |
| *Escherichia* | $T6r^+$ | 30.8 | 5-HMC | 68 |
| *Escherichia* | T7 | 47.4–51.0 | | 64, 79, 102, 123 |
| *Serratia* | η | 53.9–57.0 | Base X, sugar Y | 89 |
| *Serratia* | κ | 53.8–54.1 | | 89 |
| *Escherichia* | λ | 49.2–51.0 | | 29, 65, 102 |
| *Escherichia* | λ*b2* | 50.5 | | 29 |
| *Escherichia* | λ*dg* | 48.6–49.0 | | 29, 50 |
| *Escherichia* | φN | 50.0 | | 6 |
| *Escherichia* | φ80 | 53.0 | | 29 |
| Polyvalent | χ | 57.0–57.4 | | 101 |
| *Escherichia* | 186 | 52.0 | | 124 |
| *Klebsiella* | 380 | 55.3–55.4 | | 33, 34, 35 |
| *Klebsiella* | 483 | 53.2–55.4 | | 34, 35 |
| *Klebisella* | 490 | 52.6–52.9 | | 33, 34, 35 |
| *Klebsiella* | 632 | 54.1–57.8 | | 32, 33, 34, 35 |
| *Micrococcus* | N1 | 64.0–70.2 | 5-MC(*) | 59, 66, 100, 128 |
| | N6 | 69.7 | | 100 |
| *Mycobacterium* | Choremis | 66.9 | | 116 |
| | C2 | 61.0 | | 70 |
| | D4 | 64.0 | | 104 |
| | D28 | 60.0 | | 104 |
| | D29 | 63.8–64.0 | | 104, 115 |
| | D32 | 68.0 | | 104 |
| | Leo | 65.5 | | 31[a] |
| | R1 | 68.5 | | 31[a] |
| *Myxococcus* | MX-1 | 55.0–56.0 | | 14 |
| *Plectonema* | LPP-1 | 52.7–55.0 | | 37, 63 |
| *Proteus* | M | 48.9 | | 21 |
| | 7/R49 | 47.0 | | 56 |
| | 13vir | 44.1 | | 92 |
| | 107/69 | 39.8 | | 20 |
| *Pseudomonas* | CB3 | 60.4 | | 83 |
| | gh-1 | 57.0 | | 60 |
| | HD2 | 54.5 | | 77 |
| | HD3 | 54.7 | | 77 |
| | HD7 | 53.8 | | 77 |
| | HD11 | 45.7 | | 77 |
| | HD16 | 62.8 | | 77 |
| | HD24 | 54.4 | | 77 |
| | HD44 | 56.5 | | 77 |

## TABLE 2 (Continued)
## GC CONTENT OF BACTERIOPHAGE NUCLEIC ACIDS

| Host Genus | Phage Designation | GC, % | Particulars | References |
|---|---|---|---|---|
| **Pseudomonas (*cont.*)** | HD68 | 53.2 | | 77 |
| | HD95 | 60.5 | | 77 |
| | HD113 | 61.6 | | 77 |
| | PM2 | 42.0–43.0 | | 27, 28 |
| | PX1 | 52.4 | | 83 |
| | PX2 | 68.2 | | 83 |
| | PX3 | 45.0 | | 83 |
| | PX4 | 44.4 | | 83 |
| | PX5 | 39.6 | | 83 |
| | PX7 | 54.6 | | 83 |
| | PX10 | 53.0 | | 83 |
| | PX12 | 55.8 | | 83 |
| | PX14 | 53.8 | | 83 |
| | SD1 | 53.2–53.3 | | 105 |
| | φ-MC | 45.5–46.0 | | 19, 130 |
| | φ-2 | 51.0–55.0 | | 19, 39, 40 |
| | 1X1 | 63.0 | | 66 |
| | 352 | 55.7 | | 77 |
| | 1214 | 59.7 | | 77 |
| *Rhizobium* | PRM1 | 49.0 | | 66 |
| *Saprospira* | NN | 49.5 | Base X | 61 |
| *Staphylococcus* | PA | 31.3 | | 87 |
| | SA | 34.0–35.0 | | 66, 67 |
| | φ131 | 29.9 | | 87 |
| | φ200 | 32.2 | | 87 |
| | 3B | 36.1 | | 87 |
| | 3C | 33.8 | | 87 |
| | 6 | 34.0 | | 87 |
| | 47 | 34.7 | | 87 |
| | 52 | 32.2 | | 87 |
| | 55 | 35.1 | | 87 |
| | 71 | 36.6 | | 87 |
| | 75 | 36.0 | | 87 |
| | 80 | 32.1 | | 87 |
| | 81 | 34.3 | | 87 |
| *Streptococcus* | P3 | 30.0 | | 12 |
| | P9 | 39.0 | | 12 |
| *Streptomyces* | MSP8 | 70.0 | | 66 |
| | Type II | 67.4–68.2 | | 55 |
| *Xanthomonas* | XP5 | 61.5–61.9 | | 36 |
| | XP12 | 67.2 | 5-MC | 58 |

(*) Traces.
† *Bacterium anitratum (B5W).*
§ Range values for F1 to F11.
*a* Reference 3 in the cited reference.
*b* Reference 77 in the cited reference.
*c* Reference *q* in the cited reference: Garen, A., and Zinder, N. D., Unpublished Data, quoted in Zinder, N. D., *J. Cell. Comp. Physiol., 45 (Suppl. 2),* 23 (1955).

## REFERENCES


1. Agabian-Keshishian, N., and Shapiro, L., *J. Virol., 5,* 795 (1970).
2. Aiello, I., Frontali, C., and Tangucci, F., *Ann. Ist. Super. Sanita, 1,* 119 (1965).
3. Alegria, A. H., and Kahan, F. M., *Biochemistry, 7,* 1132 (1968).
4. Anisimova, N. I., Gabrilovich, I. M., Soshina, N. V., and Cherenkevich, S. N., *Biochim. Biophys. Acta, 190,* 225 (1969).
5. Aurisichio, S., in *Procedures in Nucleic Acid Research,* p. 562, G. L. Cantoni and D. R. Davies, Eds. Harper and Row, New York and London, England (1966).
6. de Backer, R., and Bourgaux, P., *Ann. Inst. Pasteur (Paris), 113,* 825 (1967).
7. Bertani, G., Torheim, B., and Laurent, T., *Virology, 32,* 619 (1967).
8. Betz, J. V., *Virology, 36,* 9 (1968).
9. Birdsell, D. C., Hathaway, G. M., and Rutberg, L., *J. Virol., 4,* 264 (1969).
10. Bishop, D. H. L., and Bradley, D. E., *Biochem. J., 95,* 82 (1965).
11. Biswal, N., Kleinschmidt, A. K., Spatz, H. C., and Trautner, T. A., *Mol. Gen. Genet., 100,* 39 (1967).
12. Brock, T. D., Johnson, R. M., and DeVille, W. B., *Virology, 25,* 439 (1965).
13. Brody, E. N., Mackal, R. P., and Evans, E. A., *J. Virol., 1,* 76 (1967).
14. Burchard, R. P., and Dworkin, M., *J. Bacteriol., 91,* 1305 (1966).
15. Calderone, J. G., and Pickett, M. J., *J. Gen. Microbiol., 39,* 1 (1965).
16. Calendi, E., Dettori, R., and Neri, M. G., *G. Microbiol., 14,* 227 (1966).
17. Ceppellini, M., Dettori, R., and Poole, F., *G. Microbiol., 11,* 9 (1963).
18. Chen, P. K., Citarella, R. V., Salazar, O., and Colwell, R. W., *J. Bacteriol., 91,* 1136 (1966).
19. Chow, C. T., and Yamamoto, T., *Can. J. Microbiol., 15,* 1179 (1969).
20. Coetzee, J. N., de Klerk, H. C., and Smit, J. A., *J. Gen. Virol., 1,* 561 (1967).
21. Coetzee, J. N., Smit, J. A., and Prozesky, O. W., *J. Gen. Microbiol., 44,* 167 (1966).
22. Cordes, S., Epstein, H. T., and Marmur, J., *Nature (London), 191,* 1097 (1961).
23. Creaser, E. H., and Taussig, A., *Virology, 4,* 200 (1957).
24. Dunn, D. B., and Smith, J. D., *Biochem. J., 67,* 494 (1957).
25. Egbert, L. N., *J. Virol., 3,* 528 (1969).
26. Enger, M. D., Stubbs, E. A., Mitra, S., and Kaesberg, P., *Proc. Nat. Acad. Sci. U.S.A., 49,* 857 (1963).
27. Espejo, R. T., and Canelo, E. S., *Virology, 34,* 738 (1968).
28. Espejo, R. T., Canelo, E. S., and Sinsheimer, R. L., *Proc. Nat. Acad. Sci. U.S.A., 63,* 1164 (1969).
29. Falkow, S., and Cowie, D. B., *J. Bacteriol., 96,* 777 (1968).
30. Frampton, E. W., and Mandel, M., *J. Virol., 5,* 8 (1970).
31. Gabrilovich, I. M., *Lizogeniya.* Izdatelstvo "Belarusj", Minsk, U.S.S.R. (1970).
32. Gabrilovich, I. M., Anisimova, N. I., Lukelieva, S. I., Polupanov, V. S., and Stefanov, S. B., *Zh. Mikrobiol. Epidemiol. Immunobiol.,* p. 118 (1967).
33. Gabrilovich, I. M., Baturickaya, N. W.; Anisimova, N. I., Kukulianski, A. A., and Polupanov, V. S., *Arch. Immunol. Ther. Exp., 16,* 881 (1968).
34. Gabrilovich, I. M., Lukelieva, S. I., Polupanov, V. S., and Stefanov, S. B., *Arch. Immunol. Ther. Exp., 16,* 870 (1968).
35. Gabrilovich, I. M., Polupanov, V. S., and Anisimova, N. I., *Mol. Biol., 2,* 155 (1968).
36. Ghei, O. K., Eisenstark, A., To, C. M., and Consigli, R. A., *J. Gen. Virol., 3,* 133 (1968).
37. Goldstein, D. A., and Bendet, I. J., *Virology, 32,* 614 (1967).
38. Green, D. M., *J. Mol. Biol., 10,* 438 (1964).
39. Grogan, J. B., and Johnson, E. J., *Virology, 24,* 235 (1964).
40. Grogan, J. B., and Johnson, E. J., *Bacteriol. Proc.,* p. 120 (1964).
41. Haas, M., and Yoshikawa, H., *J. Virol., 3,* 233 (1969).
42. Hershey, A. D., Dixon, J., and Chase, M., *J. Gen. Physiol., 36,* 777 (1953).
43. Hoffmann-Berling, H., Kaerner, H. C., and Knippers, R., *Advan. Virus Res., 12,* 329 (1966).
44. Hoffmann-Berling, H., Marvin, D. A., and Dürwald, H., *Z. Naturforsch. Teil B, 18,* 876 (1963).
45. Hongo, M., Murata, A., and Ogata, S., *Agr. Biol. Chem., 33,* 337 (1969).
46. Hongo, M., Ono, H., Ogata, S., and Murata, A., *Agr. Biol. Chem., 30,* 982 (1966).
47. Ikeda, H., and Tomizawa, J. I., *J. Mol. Biol., 14,* 85 (1965).
48. Jesaitis, M. A., *Fed. Proc., 23,* 319 (1964).
49. Jesaitis, M. A., *Bacteriol. Proc.,* p. 45 (1959).
50. Kaiser, A. D., and Hogness, D. S., *J. Mol. Biol., 2,* 392 (1960).
51. Kallen, R. G., Simon, M., and Marmur, J., *J. Mol. Biol., 5,* 248 (1962).
52. Kay, D., *J. Gen. Microbiol., 27,* 201 (1962).
53. Kingsbury, D. T., and Ordal, E. J., *J. Bacteriol., 91,* 1327 (1966).
54. Knight, C. A., *Advan. Virus Res., 2,* 153 (1954).
55. Kochkina, Z. M., and Rautenstein, Ya. I., *Mikrobiologiya, 36,* 290 (1967).

56. Krizsanovich, K., de Klerk, H. C., and Smit, J. A., *J. Gen. Virol.*, *4*, 437 (1969).
57. Kuo, T. T., Huang, T. C., and Chow, T. Y., *Virology*, *39*, 548 (1969).
58. Kuo, T. T., Huang, T. C., and Teng, M., *J. Mol. Biol.*, *34*, 373 (1968).
59. Lee, C. S., and Davidson, N., *Virology*, *40*, 102 (1970).
60. Lee, L. F., and Boezi, J. A., *J. Bacteriol.*, *92*, 1821 (1966).
61. Lewin, R. A., Crothers, D. M., Correll, D. L., and Reimann, B. E., *Can. J. Microbiol.*, *10*, 75 (1964).
62. Loeb, T., and Zinder, N. D., *Proc. Nat. Acad. Sci. U.S.A.*, *47*, 282 (1961).
63. Luftig, R., and Haselkorn, R., *J. Virol.*, *1*, 344 (1967).
64. Lunan, K. D., and Sinsheimer, R. L., *Virology*, *2*, 455 (1956).
65. Lwoff, A., *Bacteriol. Rev.*, *17*, 269 (1953).
66. Marmur, J., and Cordes, S., in *Informational Macromolecules*, p. 79, H. J. Vogel, V. Bryson and J. O. Lampen, Eds. Academic Press, New York and London, England (1963).
67. Marmur, J., Greenspan, C. M., Palecek, E., Kahan, F. M., Levine, J., and Mandel, M., *Cold Spring Harbor Symp. Quant. Biol.*, *28*, 191 (1963).
68. Mayers, V. L., and Spizizen, J., *J. Biol. Chem.*, *210*, 877 (1954).
69. Mazzarelli, M., Lysenko, A. M., Klimenko, S. M., Tikhonenko, T. I., Rudneva, I. A., and Kokurina, N. K., *Vop. Virusol.*, *12*, 471 (1967).
70. Menezes, J., and Pavilanis, V., *Experentia (Basel)*, *25*, 1112 (1969).
71. Mitra, S., Enger, M. D., and Kaesberg, P., *Proc. Nat. Acad. Sci. U.S.A.*, *50*, 68 (1963).
72. Naroditsky, B. S., Tikhonenko, T. I., and Klimenko, S. M., *Vop. Med. Khim.*, *15*, 476 (1969).
73. Nikolskaya, I. I., Tkatcheva, Z. G., Vanyushin, B. F., and Tikhonenko, T. I., *Biochim. Biophys. Acta*, *155*, 626 (1968).
74. Nishihara, T., and Watanabe, H., *Virus*, *17*, 118 (1967).
75. Nonoyama, M., and Ikeda, Y., *J. Mol. Biol.*, *9*, 763 (1964).
76. Nonoyama, M., Yuki, A., and Ikeda, Y., *J. Gen. Appl. Microbiol.*, *9*, 299 (1963).
77. O'Callaghan, R. J., O'Mara, W., and Grogan, J. B., *Virology*, *37*, 642 (1969).
78. Okamato, K., Mudd, J. A., Mangan, J., Huang, W. M., Subbaiah, T. V., and Marmur, J., *J. Mol. Biol.*, *34*, 413 (1968).
79. Oki, T., and Ogata, K., *Agr. Biol. Chem.*, *32*, 234 (1968).
80. Okubo, S., *Protein (Tokyo)*, *11*, 572 (1966).
81. Okubo, S., Strauss, B., and Stodolsky, M., *Virology*, *24*, 552 (1964).
82. Okubo, S., Stodolsky, M., Bott, K., and Strauss, B., *Proc. Nat. Acad. Sci. U.S.A.*, *50*, 679 (1963).
83. Olsen, R. H., Metcalf, E. S., and Todd, J. K., *J. Virol.*, *2*, 357 (1968).
84. Overby, L. R., Barlow, G. H., Doi, R. H., Jacob, M., and Spiegelman, S., *J. Bacteriol.*, *92*, 739 (1966).
85. Panter, R. A., and Symons, R. H., *Aust. J. Biol. Sci.*, *19*, 565 (1966).
86. Paranchych, W., and Graham, A. F., *J. Cell. Comp. Physiol.*, *60*, 199 (1962).
87. Pillich, J., Hradečná, Z., and Doscočil, J., *Int. J. Radiat. Biol.*, *14*, 299 (1969).
88. Pitout, M. J., and van Rensburg, A. J., *J. Gen. Virol.*, *4*, 615 (1969).
89. Pons, F. W., *Biochem. Z.*, *346*, 26 (1966).
90. Rabussay, D., Zillig, W., and Herrlich, P., *Virology*, *41*, 91 (1970).
91. Ravin, V. K., and Shulga, M. G., *Virology*, *40*, 800 (1970).
92. van Rensburg, A. J., *J. Gen. Virol.*, *5*, 437 (1969).
93. Riva, S., Polsinelli, M., and Falashi, A., *J. Mol. Biol.*, *35*, 347 (1968).
94. Romig, W. R., *Bacteriol. Rev.*, *32*, 349 (1968).
95. Roscoe, D. H., and Tucker, R. G., *Biochem. Biophys. Res. Commun.*, *16*, 106 (1964).
96. Roscoe, D. H., and Tucker, R. G., *Virology*, *29*, 157 (1966).
97. Salivar, W. O., Tzagoloff, H., and Pratt, D., *Virology*, *24*, 359 (1964).
98. Saunders, G. F., and Campbell, L. L., *Biochemistry*, *4*, 2836 (1965).
99. Saunders, G. F., and Campbell, L. L., *J. Bacteriol.*, *91*, 340 (1966).
100. Scaletti, J. V., and Naylor, H. B., *J. Bacteriol.*, *78*, 422 (1959).
101. Schade, S. Z., and Adler, J., *J. Virol.*, *1*, 591 (1967).
102. Schildkraut, C. L., Marmur, J., and Doty, P., *J. Mol. Biol.*, *4*, 430 (1962).
103. Schito, G. C., Rialdi, G., and Pesce, A., *Biochim. Biophys. Acta*, *129*, 491 (1966).
104. Sellers, M. I., and Tokunaga, T., *J. Exp. Med.*, *123*, 327 (1966).
105. Shargool, P. D., and Townsend, E. E., *Can. J. Microbiol.*, *12*, 885 (1966).
106. Shimizu, N., Miura, K., and Aoki, H., *J. Biochem.*, *68*, 265 (1970).
107. Shimizu, N., Miura, K., and Aoki, H., *J. Biochem.*, *68*, 277 (1970).
108. Sinsheimer, R. L., *J. Mol. Biol.*, *1*, 43 (1959).
109. Sinsheimer, R. L., in *The Nucleic Acids*, Vol. 3, p. 187, E. Chargaff and J. N. Davidson, Eds. Academic Press, New York and London, England (1960).
110. Strauss, J. H., and Sinsheimer, R. L., *J. Mol. Biol.*, *7*, 43 (1963).
111. Takahashi, I., and Marmur, J., *Nature (London)*, *197*, 794 (1963).

112. Takahashi, I., and Marmur, J., *Biochem. Biophys. Res. Commun., 10,* 289 (1963).
113. Tikhonenko, T. I., Velikodvorskaya, G. A., and Zemtsova, E. V., *Biokhimiya, 27,* 726 (1962).
114. To, C. M., Eisenstark, A., and Töreci, H., *J. Ultrastruct. Res., 14,* 441 (1966).
115. Tokunaga, T., and Sellers, M. I., *J. Exp. Med., 119,* 139 (1964).
116. Tokunaga, T., Mizuguchi, Y., and Murohashi, T., *J. Bacteriol., 86,* 608 (1963).
117. Torti, F., Barksdale, C., and Abelson, J., *Virology, 41,* 567 (1970).
118. Truffaut, N., *Eur. J. Biochem., 13,* 438 (1970).
119. Truffaut, N., Revet, B., and Soulie, M. O., *Eur. J. Biochem., 15,* 391 (1970).
120. Tucker, R. G., *Biochem. J., 92,* 58 (1964).
121. Twarog, R., and Blouse, L. E., *J. Virol., 2,* 716 (1968).
122. Vanyushin, B. F., Belyaeva, N. N., Kokurina, N. A., Stelmashuk, V. Y., and Tikhonenko, A. S., *J. Mol. Biol., 5,* 724 (1970).
123. Volkin, E., Astrachan, L., and Countryman, J. L., *Virology, 6,* 545 (1958).
124. Wang, J. C., *J. Mol. Biol., 28,* 403 (1967).
125. Watanabe, K., and Szybalski, W., *Jap. J. Microbiol., 11,* 153 (1967).
126. Watanabe, H., and Watanabe, M., *Can. J. Microbiol., 16,* 859 (1970).
127. Welker, N. E., and Campbell, L. L., *J. Bacteriol., 89,* 175 (1965).
128. Wetmur, J. G., Davidson, N., and Scaletti, J. V., *Biochem. Biophys. Res. Commun., 25,* 684 (1966).
129. Wyatt, G. R., and Cohen, S. S., *Biochem. J., 55,* 774 (1953).
130. Yamamoto, T., and Chow, C. T., *Can. J. Microbiol., 14,* 667 (1968).

# VIRAL DNA MOLECULES

DR. LORNE A. MacHATTIE and DR. CHARLES A. THOMAS, JR.

Purified virus particles contain either DNA or RNA, but not both types of nucleic acid. This fact provides a convenient definition of a virus.

Viral DNA molecules are either single polydeoxyribonucleotide chains or Watson-Crick duplexes. All the single-chain molecules that have so far been characterized are circular. The individual component chains of a duplex molecule are usually continuous over the entire duplex length, but certain species contain discontinuities at special points. The base sequences of the linear molecules from a single species may be all identical (unique), or they may be different circular permutations of the same sequence (permuted). In addition, a segment of the base sequence from one end of the molecule may be repeated at its other end (terminal repetition) in identical duplex or complementary single-chain (exposed) form.

In the following table the weight, length, chemical composition, and physical and topological properties of viral DNA molecules are summarized. The list includes only those DNA viruses for which several of these data are available.

## COLUMN HEADINGS, SYMBOLS, AND ABBREVIATIONS

**DNA per Virion.** P = particles counted by plaque assay; M = value based on estimated mass of virion and % DNA or phosphorus; * = value based on phosphorus content per particle, measured by $^{32}P$ star counting; C = value based on electron-microscopic particle count; LS = value obtained by light-scattering.

**S (Sedimentation Coefficient).** R = value based on measurements of relative S in sucrose gradients vs. the DNA of the virus indicated by the subscript (T-series phages are denoted by number only). All other values were obtained by analytic centrifuge.

**Length.** All quoted values were obtained by electron microscopy, using the protein film method.

**MW (Molecular Weight).** E = value obtained by the end-label method (polynucleotide kinase); L = value calculated from length, assuming 192 daltons per Å; L with a subscript = value based on relative length vs. the DNA of the virus indicated by the subscript, whose molecular weight is assumed; LS = value obtained from light-scattering measurements; S = value calculated from the sedimentation constant, assuming linear duplex DNA; R = value based on measurements of relative sedimentation rate in the sucrose gradient vs. the DNA of the virus indicated by the subscript (T-series phages are denoted by number only); B = value calculated from the band width in the CsCl gradient, assuming a calibration factor of x2 in all except the $T_4$, $T_5$ and $T_7$ values.[115]

**G + C Content.** $\rho$ = value calculated from density in CsCl gradient; T = value calculated from temperature. All other values were obtained from base analyses by degradation and chromatography of bases.

**Unusual Bases.** G, C, A, T, and U denote the usual bases; HMC = hydroxymethylcytosine; HMU = hydroxymethyluridine; Y = an unknown strange base; r = replacing.

**Shape.** Lin = linear; ⊙= circular.

**Std. Nicks.** Single-chain breaks occurring at reproducible sites in all molecules of a collection.

**TR (Terminal Repetition).** Exposed = short single-stranded "cohesive sites" at the ends of the molecule.

**Sequence.** Perm = circularly permuted.

**Physical Mapping.** N = nucleotide compositional variation along the DNA molecule, obtained by electron microscopy (subscript m) or by base analysis of fragments from known locations (subscript f); G = genetic markers: spatial distribution in the DNA molecule obtained by electron microscopy of heteroduplex molecules (subscript m) or by genetic characterization of DNA fragments of known size (subscript f).

**Strand Separation.** Single figures give the density difference between the two peaks formed when denatured DNA is banded in a CsCl gradient; OH denotes alkaline conditions; two figures preceded by G indicate the density shifts for the two peaks obtained by banding poly(I·G)- or poly(U·G)-complexed denatured DNA in a CsCl gradient vs. the density of uncomplexed denatured DNA.

## COMPOSITION AND PHYSICAL PROPERTIES OF VIRAL DNA MOLECULES

| Host | Virus | DNA per Virion daltons x $10^{-6}$ | S, Svedbergs | Length, microns | MW daltons x $10^{-6}$ |
|---|---|---|---|---|---|
| 1. *E. coli* | λ | P 30[20] | 33.6[17] | 17;[20,81,135] 15[34] | E, 32;[10] L, 33;[81] S, 31[17] |
| 2. *E. coli* | λb2b5 | | | 13;[81] 14;[135] 12[34] | $L_\lambda$, 26[34,68,135] |
| 3. *E. coli* | φ80 | | | 13.8[117] | $R_\lambda$, 30.2[139] |
| 4. *E. coli* | $T_1$ | | 33.5[14] | 16.0 ± 1[73,14] | $R_{\lambda,7}$, 31;[1,14] $R_7$, 35[15] |
| 5. *E. coli* | $T_2$ | P 160[110] | 63.5 ± 2;[125] 68[35] $R_7$ 57[107] | 56 ± 2[130] | *, 130;[107] L, 118;[130] S, 94–105[115] |
| 6. *E. coli* | $T_4$ | | 61.3 ± 1.3[105] (4.5% > $T_2$)[35] | | B, 105 ± 6; S, 90–100[115] |
| 7. *E. coli* | $T_6$ | | | | |
| 8. *E. coli* | $T_3$ | M 23.0[9] | | 12.2 ± 0.4[72] | L, 23[72] |
| 9. *E. coli* | $T_7$ | *25[107] | 32.0 ± 0.8[125] | 12.5 ± 0.6[12] | LS, 25;[51] *, 25 ± 2;[107] E, 26;[100] B, 23[115] |
| 10. *E. coli* | $T_5$ | *83[107] | $R_7$ 49.2;[106] 53[16] | 38.3 ± 0.9[16] | *, 83;[107] S, 83;[16] B, 64[115] |
| 11. *E. coli* | $T_5$ st0 | *77 ± 2[107] | $R_7$ 48.1[106] | 38.8[2] | $R_7$, 78;[106] *, 77;[107] S, 59[115] |
| 12. *E. coli* | $T_5$ st8 | | $R_7$ 48.1[106] | | $R_5$, 78[106] |
| 13. *E. coli* | $T_5$ st14 | | $R_7$ 48.6[106] | | $R_5$, 80[106] |
| 14. *E. coli* | $T_5$ st20 | | $R_7$ 48.9[106] | | $R_5$, 81[106] |
| 15. *E. coli* | Plkc | | $R_4$ 48, $R_\lambda$ 44[58] | 30;[2] 35[59] | $R_4$, 60[58] |
| 16. *E. coli* and *Salmonella* | χ | P 63[114] | 36[92] | 21.5[114] | $R_\lambda$, 40;[92] L, 42[114] |
| 17. *E. coli* | P2 | | | 13.1[61] | $R_\lambda$, 20-24;[82] $L_\lambda$, 22[61] |
| 18. *E. coli* | 15(defec.) | | | 12.8[76] | $L_\lambda$, 27[76] |
| 19. *E. coli* | φX174 | M, LS 1.6[118] | $S \cdot [\eta]^{1/3} = 23$[39] | 1.77 ± 0.13[42] | LS, 1.7[119] |
| 20. | φX174RF | | $S_{20,w} = 21$, 16[63] | 1.64 ± 0.11[69] | |
| 21. *E. coli* (malee) | fd | | $S \cdot [\eta]^{1/3} = 21.7$[87] | | ~ same φX[87] |
| 22. *E. coli* (male) | M13 | | 24[111] | | |
| 23. *Micrococcus lysodeikticus* | N1 | | $R_7$ 36[136] | 17[136] | S, L, 33 ± 1.5[136] |
| 24. *Salmonella* | P22 | | 31–36[131] | 13.7 ± 0.8[99] | L, 26;[99] $R_{7,\lambda}$, 27[99] |
| 25. *B. subtilis* | PBS1 | | $R_2$ 72[55] | | $R_2$, ~200[55] |
| 26. *B. subtilis* | SPO1 | | 54 ± 2[36] | | |
| 27. *B. subtilis* | SP8 | | 54 ± 2[36] | | S, 69[5] |

## COMPOSITION AND PHYSICAL PROPERTIES OF VIRAL DNA MOLECULES (Continued)

| G + C, mole % | Unusual Bases | Shape | Std. Nicks | No. of Strands | TR | Sequence | Physical Mapping | Strand Separation |
|---|---|---|---|---|---|---|---|---|
| 1. 49[74] | | Lin | No[2] | Duplex | Exposed[53] | Unique[66] | $N_f^-$[121] $N_m$;[60] ,$G_m$[34,135] | G5, 16;[127] OH, 4[54] |
| 2. | | Lin | No[34] | Duplex | Exposed[81] | Unique | | |
| 3. 53[139] | | | | Duplex | Exposed[138] | Unique[138] | | |
| 4. 48[32] | | Lin | No[1] | Duplex | 6.5%[79] | | | G4, 16[71] |
| 5. 35[120] | HMCrC[120] | Lin | No | Duplex | 1–3%[80] | Perm[130] | | G2, 2[71] |
| 6. 34.4[120] | HMCrC[120] | Lin | | Duplex | Yes[80] | Perm[78] | $G_f$[43,89] | G8, 22[50] |
| 7. 34.2[120] | HMCrC[120] | | | Duplex | | | | |
| 8. 49.6[120] | | Lin | No[102] | Duplex | <1%[102] | Unique[102] | | G0, 13[71] |
| 9. 48[137] | | Lin | No[125] | Duplex | 0.7%[102] | Unique[102] | | G2, 18[71] |
| 10. 39[137] | | Lin[16] | Yes[2,16,62] | Duplex | | Unique[133] | | G3, 8[71] |
| 11. | | Lin | Yes[2] | Duplex | | Unique[2] | | |
| 12. | | Lin | Yes[2] | Duplex | | Unique[2] | | |
| 13. | | Lin | Yes[2] | Duplex | | Unique[2] | | |
| 14. | | Lin | | Duplex | | | | |
| 15. ρ, 46[58] | | Lin[58] | No[2] | Duplex | Yes[59] | Perm?[59] | | |
| 16. 57.4[114] | | | | Duplex | Exposed?[114] | | | |
| 17. | | Lin | No[61] | Duplex | Exposed[61;83] | Unique | $N_m$[61] | |
| 18. | | Lin | No[76] | Duplex | 7.4%[76] | Perm[76] | | |
| 19. G, 24.1; C, 18.5; A, 24.6[119] | | ○[41] | | Single | | | | |
| 20. 42[52] | | ○[69] | No[63] | Duplex | | | | |
| 21. G, 19.9; C, 21.7; A, 24.4[86] | | ○[87] | | Single | | | | |
| 22. G, 21; C, 20; A, 23[111] | | | | Single | | | | |
| 23. ρ, 64[136] | | Lin | No[136] | Duplex | Exposed[136] | | | |
| 24. ρ, same $T_7$[131] | | Lin[99] | No[99] | Duplex | 2.4–5%[99] | Perm[99] | | |
| 25. 28[128] | UrT[128] | | | Duplex | | | | |
| 26. 45[90] | HMUrT[90] | | | Duplex | | | | 9[90] |
| 27. 43[84] | HMUrT[84] | | | Duplex | | | | 9[85] |

## COMPOSITION AND PHYSICAL PROPERTIES OF VIRAL DNA MOLECULES (Continued)

| Host | Virus | DNA per Virion daltons x $10^{-6}$ | S, Svedbergs | Length, microns | MW daltons x $10^{-6}$ |
|---|---|---|---|---|---|
| 28. *B. subtilis* | SP82 | P180[44] | $R_7$ 63[45] | | S, 130[44] |
| 29. *B. subtilis* | SP105 | | 48–50[5] | | S, 63[5] |
| 30. *B. subtilis* | $\phi$ 25 | | 64[97] | | |
| 31. *B. subtilis* | SP50 | | $R_2$ 54 ± 2[98] | 49 ± 2[11,98] | $R_{2,3,7}$, 109 ± 6[98] |
| 32. *B. subtilis* | $\phi$1 | | 52[97] | | |
| 33. *B. subtilis* | SP3 | | 69[5] | | S, 150[5] |
| 34. *B. subtilis* | pKc | | 63[97] | | |
| 35. *B. subtilis* | $\phi$ 15 | | 29[97] | | S, 17[97] |
| 36. *B. subtilis* | $\phi$ 105 | | 32[10] | 11.6 ± 0.6[10] | S, 25[10] |
| 37. *B. subtilis* | PBSH(defec.) | | | 4.7 ± 0.2[49] | S, 12; L, 9[49] |
| 38. *B. subtilis* | SPP1 | | 32 ± 1[103] | | S, 25[103] |
| 39. *B. subtilis* | $\phi$ 29 | | 24.5[4] | 5.8 ± 0.1[4] | S, 11.5[4] |
| 40. *Proteus* | 13vir | | | | S, 26.3[134] |
| 41. *Providence* | PL25 | M, 28[93] | 32.5[93] | 12.6 ± 0.6[93] | L, S, 24.3[93] |
| 42. *Pseudomonas* | PB-1 | | | 24[13] | L, 46[13] |
| 43. *Pseudomonas* | PM-2 | P, 17; M7[40] | 128, II20[40] | | S, 5.3[40] |
| 44. *Pseudomonas putida* | gh-1 | | 30.9[75] | | $R_7$, 23[75] |
| 45. *B. stearothermophilus* | TP 84 | | 30[113] | 13.9[113] | S, 22.6[113] |
| 46. *B. stearothermophilus* | T$\phi$ 3 | | 33.6[38] | 12.1[38] | L, 23; S, 28.7[38] |
| 47. *Cyanophyceae* | BGAV, LPP-1 or cyanophage | | 33.5[77] | 13.2 ± 0.5[77] | S, 27[77] |
| 48. *Serratia marcescens* | $\eta$ | P 73[95] | | | |
| 49. *Serratia marcescens* | $\kappa$ | P 120[95] | | | |
| 50. Insect | CIV, SIV, TIV | 145 ± 25[8] | 62[8] | 59[8] | S, L, 130[8] |
| 51. Frog | FV-3 | | | | $R_2$, 130[122] |
| 52. Rat | RV-13 | | | 1.3[104] | L, 1.2?[104] |
| 53. Mouse | MVM | | [25] | 1.2[30] | S, ~1.5[25,30] |
| 54. Mammal | Herpes simplex | C 73[108] | | 53[6] | $R_{vac}$, 110; L, 101[6] |
| 55. Mammal | EAV | C 250[33] | 52[123] | 48[123] | S, 94; L, 92; B, 92[123] |
| 56. Mammal | Pseudorabies | | 44[109] | | S, 68;[109] B, 70[67] |
| 57. Mammal | IBR | | 39[109] | | S, 54[109] |
| 58. Mammal, human | Cytomegalo | | | | B, 64[31] |
| 59. Mammal | LK | | 49[109] | | S, 84[109] |

## COMPOSITION AND PHYSICAL PROPERTIES OF VIRAL DNA MOLECULES (Continued)

| G + C, mole % | Unusual Bases | Shape | Std. Nicks | No. of Strands | TR | Sequence | Physical Mapping | Strand Separation |
|---|---|---|---|---|---|---|---|---|
| 28. | | Lin[45] | | Duplex | | Unique[45] | | 19[44] |
| 29. ρ, T, 44[5] | | | | Duplex | | | | 7[5] |
| 30. | Indicated[97] | | | Duplex | | | | |
| 31. | | Lin[11,98] | No[98] | Duplex | | Unique[98] | | 6[116] |
| 32. ρ, T, 43[97] | | | | Duplex | | | | 5[97] |
| 33. ρ, T, 35[5] | | | | Duplex | | | | 0[5] |
| 34. ρ, 35[97] | | | | Duplex | | | | |
| 35. ρ, T, 35[97] | | | | Duplex | | | | 0[97] |
| 36. ρ, T, 43.5[10] | | | No[10] | Duplex | Exposed?[10] | | | |
| 37. ρ, 43[49] | | Lin | No[49] | Duplex | | | | |
| 38. 43[103] | | | No[103] | Duplex | | | | 12[103] |
| 39. | | Lin | No[4] | Duplex | Exposed[3] | Unique[4] | | |
| 40. T, 44.1[134] | | | | Duplex | | Unique[134] | | |
| 41. T, 43.2[93] | | Lin | | Duplex | | | | |
| 42. | | ○[40] | | Duplex | | | | |
| 43. T, ρ, 43[40] | | Lin | No[40] | Duplex | | | | |
| 44. 57[75] | | | | Duplex | | | | 0[75] |
| 45. 44.1[113] | | | | Duplex | | | | 12[113] |
| 46. 40.3[38] | | Lin[38] | No | Duplex | | | | 8[38] |
| 47. ρ, T, 53[77] | | Lin[77] | | Duplex | | | | |
| 48. (G + Y + C), 57.0[95] | G + Y = C[95] | | | Duplex | | | | |
| 49. 53.8[95] | G < C[95] | | | Duplex | | | | |
| 50. ρ, T, 30 ± 2[8] | | Lin | | Duplex | | | | |
| 51. ρ, T, 56–58[122] | | | | Duplex | | | | |
| 52. | | ○[104] | | Single | | | | |
| 53. 40–43[30] | | Lin[30] | | Single | | | | |
| 54. 68;[108] 70[94] | | | | Duplex | | | | |
| 55. 56[33] | | Lin[123] | | Duplex | | | | |
| 56. ρ, 74[109] | | | | Duplex | | | | |
| 57. ρ, 71[109] | | | | Duplex | | | | |
| 58. ρ, 58[31] | | | | Duplex | | | | |
| 59. ρ, 56[109] | | | | Duplex | | | | |

## COMPOSITION AND PHYSICAL PROPERTIES OF VIRAL DNA MOLECULES (Continued)

| Host | Virus | DNA per Virion daltons x $10^{-6}$ | S, Svedbergs | Length, microns | MW daltons x $10^{-6}$ |
|---|---|---|---|---|---|
| 60. Mammal (pox) | Vaccinia | C 160–180[64,65] | | 80[7] | L, 153; $R_2$, 150[112] |
| 61. Chicken | Fowlpox | 230[126] | | | L, 180–200[56,57] |
| 62. Simian kidney | SV 40 | M 4.0–5.2[88] | 21.2, 16.1[26] | | S, <3.2; B, 2.7[26] |
| 63. Simian | SA-7 | | 30.4[19] | | S, 22[19] |
| 64. Mammal | Polyoma | C 7.5[28] | 21, 15.5[23] | 1.56 ± 0.2[124] | S, 2.9–3.4; B, 2.5[23] |
| 65. Bovine | Papilloma | | | 2.45 ± 0.07[73] | L, 4.9[73] |
| 66. Rabbit | Shope papilloma | M 5[70] | 28, 21[22] | 2.3 ± 0.1[70] | B, 4.2 ;[22] L, 4.4[70] |
| 67. Rabbit | RKV | | I, 8,2; II 5.8[29] | 1.44 ± 0.14[29] | L, 2.8[29] |
| 68. Human | Papilloma | | 28.2, 20.2, 18.0[24] | | S, 5.3; B, 4.0[24] |
| 69. Human | Adeno 2,4 | M 16 ± 4[46,47] | 25[46] | 12 ± 1[48] | S, 23[46] |
| 70. Human | Adeno, 3,21 | | | 11.8 ± 0.8, 13 ± 2[48] | L, 22.7[48] |
| 71. Human | Adeno 12,18 | M 14 ± 4[46,47] | | 11.2 ± 0.7[48] | S, 22.0, 19.6[48] |

## COMPOSITION AND PHYSICAL PROPERTIES OF VIRAL DNA MOLECULES (Continued)

| G + C, mole % | Unusual Bases | Shape | Std. Nicks | No. of Strands | TR | Sequence | Physical Mapping | Strand Separation |
|---|---|---|---|---|---|---|---|---|
| 60. 37[65] | | | | Duplex | Exposed?[7] | | | |
| 61. 35[96] | | | | Duplex | | | | |
| 62. ρ, 41[26] | | ○[26] | No[26] | Duplex | | | | |
| 63. 59.8[19] | | Lin | No[19] | Duplex | | | | |
| 64. ρ, T, 48[21] | | ○[37] | No[37] | Duplex | | | | |
| 65. ρ, 45.5[27] | | ○[73] | | Duplex | | | | |
| 66. ρ, 47[27] | | ○[70] | No[22] | Duplex | | | | |
| 67. ρ, 43[29] | | ○[29] | No | Duplex | | | | |
| 68. ρ, 41[27] | | ○?[24] | No[24] | Duplex | | | | |
| 69. ρ, T, 59[91] | | Lin[48] | | Duplex | | | | |
| 70. ρ, 55; T, 51.6[91] | | Lin[48] | | Duplex | | | | |
| 71. ρ, T, 49[91] | | Lin[48] | | Duplex | | | | |

Data taken from: MacHattie, L. A., and Thomas, C. A., Jr., in *Handbook of Biochemistry,* 2nd ed., pp. H-4–H-7, H. A. Sober, Ed. Copyright 1970, The Chemical Rubber Co., Cleveland, Ohio.

## REFERENCES


1. Abelson and Michalke, Unpublished Data.
2. Abelson and Thomas, *J. Mol. Biol., 18,* 262 (1966).
3. Mosharrafa, Schachtele, Reilly and Anderson, *J. Virol., 6,* 855 (1970).
4. Anderson and Mosharrafa, *J. Virol., 2,* 1185 (1968).
5. Bear, in *Characterization of Phage DNA Infection of Competent Bacillus subtilis (Ph.D. Thesis).* University of California, Los Angeles, California (1966).
6. Becker, Dym and Sarov, *Virology, 36,* 184 (1968).
7. Becker and Sarov, *J. Mol. Biol., 34,* 655 (1968).
8. Bellet and Inman, *J. Mol. Biol., 25,* 425 (1967).
9. Bendet, Schachter and Lauffer, *J. Mol. Biol., 5,* 76 (1962).
10. Birdsell, Hathaway and Rutberg, *J. Virol., 4,* 264 (1969).
11. Biswal, Kleinschmidt, Spatz and Trautner, *Mol. Gen. Genet., 100,* 39 (1967).
12. Bode and Morowitz, *J. Mol. Biol., 23,* 191 (1967).
13. Bradley and Robertson, *J. Gen. Virol., 3,* 247 (1968).
14. Bressler, Kiselev, Maniakov, Mosevitski and Timovski, *Virology, 33,* 1 (1967).
15. Brody, Mackal and Evans, *J. Virol., 1,* 76 (1967).
16. Bujard, *Proc. Nat. Acad. Sci. U.S.A., 62,* 1167 (1969).
17. Burgi and Hershey, *Biophys. J., 3,* 309 (1963).
18. Burgi, Hershey and Ingraham, *Virology, 28,* 11 (1966).
19. Burnett and Harrington, *Proc. Nat. Acad. Sci. U.S.A., 60,* 1023 (1968).
20. Caro, *Virology, 25,* 226 (1965).
21. Crawford, *Virology, 19,* 279 (1963).
22. Crawford, *J. Mol. Biol., 8,* 489 (1964).
23. Crawford, *Virology, 22,* 140 (1964).
24. Crawford, *J. Mol. Biol., 13,* 362 (1965).
25. Crawford, *Virology, 29,* 605 (1966).
26. Crawford and Black, *Virology, 24,* 388 (1964).
27. Crawford and Crawford, *Virology, 21,* 258 (1963).
28. Crawford, Crawford and Watson, *Virology, 18,* 170 (1962).
29. Crawford and Follett, *J. Gen. Virol., 1,* 19 (1967).
30. Crawford, Follett, Burdon and McGeoch, *J. Gen. Virol., 4,* 37 (1969).
31. Crawford and Lee, *Virology, 23,* 105 (1964).
32. Creaser and Taussig, *Virology, 4,* 200 (1957).
33. Darlington and Randall, *Virology, 19,* 322 (1963).
34. Davis and Davidson, *Proc. Nat. Acad. Sci. U.S.A., 60,* 243 (1968).
35. Davison and Freifelder, *Biopolymers, 2,* 15 (1964).
36. Davison, Freifelder and Holloway, *J. Mol. Biol., 8,* 1 (1964).
37. Dulbecco and Vogt, *Proc. Nat. Acad. Sci. U.S.A., 50,* 236 (1963).
38. Egbert, *J. Virol., 3,* 528 (1969).
39. Eigner and Doty, *J. Mol. Biol., 12,* 549 (1965).
40. Espejo and Canelo, *Virology, 34,* 738 (1968); *37,* 495 (1969).
41. Fiers and Sinsheimer, *J. Mol. Biol., 5,* 408, 424 (1962).
42. Freifelder, Kleinschmidt and Sinsheimer, *Science, 146,* 254 (1964).
43. Goldberg, *Proc. Nat. Acad. Sci. U.S.A., 56,* 1457 (1966).
44. Green, *J. Mol. Biol., 10,* 438 (1964).
45. Green, *J. Mol. Biol., 22,* 15 (1966).
46. Green, *Cold Spring Harbor Symp. Quant. Biol., 27,* 219 (1962).
47. Green and Pinã, *Proc. Nat. Acad. Sci. U.S.A., 51,* 1251 (1964).
48. Green, Pinã, Kines, Wensink, MacHattie and Thomas, *Proc. Nat. Acad. Sci. U.S.A., 57,* 1302 (1967).
49. Haas and Yoshikawa, *J. Virol., 3,* 233 (1969).
50. Guha and Szybalski, *Virology, 34,* 608 (1968).
51. Harpst, Krasna and Zimm, *Fed. Proc., 24,* 538 (1965).
52. Hayashi, Hayashi and Spiegelman, *Proc. Nat. Acad. Sci. U.S.A., 50,* 664 (1963).
53. Hershey, Burgi and Ingraham, *Proc. Nat. Acad. Sci. U.S.A., 49,* 748 (1963).
54. Hogness and Simmons, *J. Mol. Biol., 9,* 411 (1964).
55. Hunter, Yamagishi and Takahashi, *J. Virol., 1,* 841 (1967).
56. Gafford and Randall, *J. Mol. Biol., 26,* 303 (1967).
57. Hyde, Randall and Gafford, in *Sixth International Congress for Electron Microscopy,* p. 193. Maruzen Co. Ltd., Tokyo, Japan (1966).
58. Ikeda and Tomizawa, *J. Mol. Biol., 14,* 85 (1965).

59. Ikeda and Tomizawa, *Cold Spring Harbor Symp. Quant. Biol.*, *33*, 791 (1968).
60. Inman, *J. Mol. Biol.*, *28*, 103 (1967).
61. Inman and Bertani, *J. Mol. Biol.*, *44*, 533 (1969).
62. Jacquemin-Sablon and Richardson, *Fed. Proc.*, *27*, 396 (1968).
63. Jansz and Pouwels, *Biochem. Biophys. Res. Commun.*, *18*, 589 (1965).
64. Joklik, *J. Mol. Biol.*, *5*, 265 (1962).
65. Joklik, *Virology*, *18*, 9 (1962).
66. Kaiser, *J. Mol. Biol.*, *4*, 275 (1962).
67. Kaplan and Ben-Porat, *Virology*, *23*, 90 (1964).
68. Kellenberger, Zichichi and Weigle, *Proc. Nat. Acad. Sci. U.S.A.*, *47*, 869 (1961).
69. Kleinschmidt, Burton and Sinsheimer, *Science*, *142*, 961 (1963).
70. Kleinschmidt, Kass, Williams and Knight, *J. Mol. Biol.*, *13*, 749 (1965).
71. Kubinski, Opara-Kubinska and Szybalski, *J. Mol. Biol.*, *20*, 313 (1966).
72. Lang, quoted in Reference 16.
73. Lang, Bujard, Wolff and Russell, *J. Mol. Biol.*, *23*, 163 (1967).
74. Ledinko, *J. Mol. Biol.*, *9*, 834 (1964).
75. Lee and Boezi, *J. Bacteriol.*, *92*, 1821 (1966); Personal Communication.
76. Lee, Davis and Davidson, *J. Mol. Biol.*, *48*, 1 (1970).
77. Luftig and Haselkorn, *J. Virol.*, *1*, 344 (1967).
78. MacHattie, Unpublished Data (1966).
79. MacHattie, Rhoades and Thomas, *J. Mol. Biol.* (in press)
80. MacHattie, Ritchie, Richardson and Thomas, *J. Mol. Biol.*, *23*, 355 (1967).
81. MacHattie and Thomas, *Science*, *144*, 1142 (1964).
82. Mandel, *Mol. Gen. Genet.*, *99*, 88 (1967).
83. Mandel and Berg, *J. Mol. Biol.*, *38*, 137 (1968).
84. Marmur and Cordes, in *Informational Macromolecules*, p. 79, Vogel, Bryson and Lampen, Eds. Academic Press, New York (1963).
85. Marmur and Greenspan, *Science*, *142*, 387 (1963).
86. Marvin and Hoffman-Berling, *Nature*, *197*, 517 (1963).
87. Marvin and Schaller, *J. Mol. Biol.*, *15*, 1 (1966).
88. Mayor, Jameson and Jordan, *Virology*, *19*, 359 (1963).
89. Mosig, *Genetics*, *59*, 137 (1968).
90. Okubo, Strauss and Stodolsky, *Virology*, *24*, 552 (1964).
91. Piña and Green, *Proc. Nat. Acad. Sci. U.S.A.*, *54*, 547 (1965).
92. Pinkerton, Unpublished Observations.
93. Pitout and Van Rensburg, *J. Gen. Virol.*, *4*, 615 (1969).
94. Plummer, Goodheart, Henson, and Bowling, *Virology*, *39*, 134 (1969).
95. Pons, *Biochem. Z.*, *346*, 26 (1966).
96. Randall, Gafford and Darlington, *J. Bacteriol.*, *83*, 1037 (1962).
97. Reilly, in *A Study of the Bacteriophages of Bacillus subtilis and Their Infectious Nucleic Acids (Ph.D. Thesis)*. Western Reserve University, Cleveland, Ohio (1965).
98. Reznikoff and Thomas, *Virology*, *37*, 309 (1969).
99. Rhoades, MacHattie and Thomas, *J. Mol. Biol.*, *37*, 21 (1968).
100. Richardson, *J. Mol. Biol.*, *15*, 49 (1966).
101. Richardson and Weiss, *J. Gen. Physiol.*, *49*, 81 (1966).
102. Ritchie, Thomas, MacHattie and Wensink, *J. Mol. Biol.*, *23*, 365 (1967).
103. Riva, Polsinelli and Falaschi, *J. Mol. Biol.*, *35*, 347 (1968).
104. Robinson and Hetrick, *J. Gen. Virol.*, *4*, 269 (1969).
105. Rosenbloom and Cox, *Biopolymers*, *4*, 747 (1966).
106. Rubenstein and Lanka, *Virology*, *36*, 356 (1968).
107. Rubenstein and Leighton, *Biophys. J.*, *8*, A-80 (1968); *J. Mol. Biol.*, *46*, 313 (1969).
108. Russell and Crawford, *Virology*, *21*, 353 (1963).
109. Russell and Crawford, *Virology*, *22*, 288 (1964).
110. Sadron, in *The Nucleic Acids*, Vol. 3, p. 1, Chargaff and Davidson, Eds. Academic Press, New York (1960).
111. Salivar, Tzagoloff and Pratt, *Virology*, *24*, 359 (1964).
112. Sarov and Becker, *Virology*, *33*, 369 (1967).
113. Saunders and Campbell, *Biochemistry*, *4*, 2836 (1965).
114. Schade and Adler, *J. Virol.*, *1*, 591 (1967).
115. Schmid and Hearst, *J. Mol. Biol.*, *44*, 143 (1969).
116. Sheldrick and Szybalski, *Fed. Proc.*, *25*, 707 (1966).
117. Shinagawa, Hosaka, Yamagishi and Nishi, *Biken J.*, *9*, 3 (1966).
118. Sinsheimer, *J. Mol. Biol.*, *1*, 37 (1959).

119. Sinsheimer, *J. Mol. Biol., 1,* 43 (1959).
120. Sinsheimer, in *The Nucleic Acids,* Vol. 3, p. 194, Chargaff and Davidson, Eds. Academic Press, New York (1960).
121. Skalka, Burgi and Hershey, *J. Mol. Biol., 34,* 1 (1968).
122. Smith and McAuslan, *J. Virol., 4,* 339 (1969).
123. Soehner, Gentry and Randall, *Virology, 26,* 394 (1965).
124. Stoekenius, *Proc. Nat. Acad. Sci. U.S.A., 50,* 737 (1963).
125. Studier, *J. Mol. Biol., 11,* 373 (1965).
126. Szybalski, Erikson, Gentry, Gafford and Randall, *Virology, 19,* 586 (1963).
127. Szybalski, Kubinski and Sheldrick, *Cold Spring Harbor Symp. Quant. Biol., 31,* 123 (1966).
128. Takahashi and Marmur, *Nature, 197,* 794 (1963).
129. Thomas, *J. Gen. Physiol., 49,* 143 (1966).
130. Thomas and MacHattie, *Proc. Nat. Acad. Sci. U.S.A., 52,* 1297 (1964).
131. Thomas and Pinkerton, *J. Mol. Biol., 5,* 356 (1962).
132. Thomas, Pinkerton and Rubenstein, in *Informational Macromolecules,* p. 89, Vogel, Bryson and Lampen, Eds. Academic Press, New York (1963).
133. Thomas and Rubenstein, *Biophys. J., 4,* 93 (1964).
134. Van Rensburg, *J. Gen. Virol., 5,* 437 (1969).
135. Westmoreland, Szybalski and Ris, *Science, 163,* 1343 (1969).
136. Wetmur, Davidson and Scaletti, *Biochem. Biophys. Res. Commun., 25,* 684 (1966).
137. Wyatt and Cohen, *Biochem. J., 55,* 773 (1953).
138. Yamagishi, Nakamura and Ozeki, *Biochem. Biophys. Res. Commun., 20,* 727 (1965).
139. Yamagishi, Yoshizako and Sato, *Virology, 30,* 29 (1966).

# DNA CONTENT PER CELL IN VIRUSES

DR. HERMAN S. SHAPIRO

## TABLE 1
## BACTERIOPHAGES

| Source | DNA/Particle, $\mu g \times 10^{12}$ | Ref. |
|---|---|---|
| a, *Bacillus megaterium* | 67[a] | 1 |
| Sd, *Escherichia coli* | 134–150[a] | 2 |
| | 125–142[a] | 3 |
| $D_4$, *Mycobacterium smegmatis* | 590[a] | 4 |
| $D_{28}$, *Mycobacterium smegmatis* | 530[a] | 4 |
| $D_{29}$, *Mycobacterium smegmatis* | 150[a] | 4 |
| $D_{32}$, *Mycobacterium smegmatis* | 100–330[a] | 4 |
| PLT-22, *Salmonella* | 108[b] | 5 |
| $P_1$, *Salmonella* | 100[a] | 6 |
| $P_{22}$, *Salmonella* | 76[b] | 7 |
| $A_1$, *Salmonella typhimurium* (temperate) | 65[b] | 8 |
| TP-84, *Bacillus stearothermophilus* | 45[a] | 9 |
| T$\phi$3, *Bacillus stearothermophilus* | 38.8–48.0[c] | 40 |
| $\phi$29, *Bacillus subtilis* | 18.9[c] | 41 |
| PBSH, *Bacillus subtilis* 168 | 20[c] | 42 |
| SPO2, *Bacillus subtilis* | 43.4[c] | 43 |
| SPP1, *Bacillus subtilis* | 41.8[c] | 44 |
| 2C, *Bacillus subtilis* 168 | 159–172[c] | 45 |
| SP50, *Bacillus subtilis* | 165[c] | 46 |
| | 167[c] | 47 |
| SP82, *Bacillus subtilis* | 217[a] | 10 |
| $\kappa$, *Serratia marcescens* | 200[c] | 11 |
| n, *Serratia marcescens* | 120[c] | 11 |
| $T_1$, *Escherichia coli* | 70[a] | 12 |
| | 120[a] | 13 |
| | 52.4[c] | 48 |
| | 52–53[c] | 49 |
| | 58[c] | 14 |
| | 76[c] | 15 |
| | 229[c] | 16 |
| $T_2$, *Escherichia coli* | 217[a] | 7 |
| | 217[b] | 17 |
| | 200[c] | 18 |
| | 195[a] | 19 |
| | 167–184[a] | 2 |
| | 230[a] | 12 |
| | 249[a] | 15 |
| | 200[c] | 14 |
| $T_2$r, *Escherichia coli* | 540[a] | 20 |
| | 330[a] | 13 |
| | 522[b] | 16 |
| $T_2r^+$, *Escherichia coli* | 330[a] | 13 |
| | 427[b] | 16 |
| | 306[b] | 21 |
| | 509[b] | 22 |
| $T_2$H, *Escherichia coli* | 260–380[b] | 23 |
| $T_2$L, *Escherichia coli* | 358[b] | 23 |
| $T_3$, *Escherichia coli* | 37.9[c] | 48 |
| | 58[c] | 14 |
| | 98[b] | 15 |
| | 189[b] | 16 |
| $T_4$, *Escherichia coli* | 200[c] | 14 |
| | 303[b] | 23 |
| $T_4$r, *Escherichia coli* | 240[a] | 20 |
| $T_5$, *Escherichia coli* | 180[a] | 12 |
| | 195[b] | 15 |
| | 129–137[a] | 24 |
| | 429[b] | 16 |
| | 58[c] | 14 |
| | 867[b] | 8 |
| | 110[a] | 25 |
| $T_6$, *Escherichia coli* | 423[b] | 26 |
| | 200[c] | 14 |
| $T_6$r, *Escherichia coli* | 360[a] | 20 |
| $T_6r^+$, *Escherichia coli* | 360[a] | 20 |
| | 423[b] | 26 |
| $T_7$, *Escherichia coli* | 184[b] | 16 |
| | 98[b] | 15 |
| | 108[b] | 27 |
| | 58[c] | 14 |
| | 63[a] | 29 |
| | 25[a] | 30 |
| | 50[a] | 28 |
| $\phi$174, *Escherichia coli* | 2.84[a] | 31 |
| $\phi$80, *Escherichia coli* | 49[a] | 32 |
| $\lambda$, wild *Escherichia coli* | 55.1[a] | 33 |
| fd, *Escherichia coli* | 2.17[a] | 34 |
| | 2.29[a] | 35 |
| $\lambda$ dg, *Escherichia coli* | 80[c] | 36 |
| $\lambda$, *Escherichia coli* | 110[c] | 37 |
| | 51.8[a] | 38 |
| $\lambda$, virulent *Escherichia coli* | 130[b] | 8 |
| Virus of *Saprospira grandis* | 45–540[c] | 39 |
| PM-2 | 8.9[c] | 50 |
| gh-1, *Pseudomonas putida* | 37.8[c] | 51 |
| LPP-1, blue-green alga | 41.8[c] | 52 |

[a] Estimated from molecular-weight data cited in the reference and the value of $1.67 \times 10^{-18}$ μg/dalton.
[b] Estimated from the phosphorus content cited in the reference and the value of 9.23% phosphorus in DNA.
[c] Estimated directly.

## TABLE 2
## ANIMAL VIRUSES

| Virus | DNA/Particle, $\mu g \times 10^{12}$ | Ref. | Virus | DNA/Particle, $\mu g \times 10^{12}$ | Ref. |
|---|---|---|---|---|---|
| Nuclear polyhedral virus of | | | Herpes simplex virus | 113.6[b] | 58 |
| *Bombyx mori* | 100[a] | 53 | | 184[a] | 65 |
| *Lymantria dispar* | 100[a] | 53 | Pseudorabies virus | 113.6[b] | 58 |
| *Aporia crataegi* | 100[a] | 53 | | 7.7[b] | 60 |
| Tipula iridescent virus of the | 230[a] | 53 | LK virus | 140.3[b] | 58 |
| European crane fly *Tipula paludosa* | 251[a] | 54 | Bovine rhinotracheitis virus | 90.2[b] | 58 |
| | | | Simian virus 40 | 5.3[b] | 61 |
| Polyoma virus | 5.67[b] | 55 | | 6.7–8.7[b] | 62 |
| Shope papilloma virus | 5.2–5.5[b] | 56 | Adenovirus 5 | 120[a] | 53 |
| Cowpox virus | 27.5[b] | 57 | Adeno-associated virus | 6.0[b] | 63 |
| | 24.6–29.2[a] | 53 | Minute virus of mouse (MVM) | 2.84[a] | 66 |
| Equine abortion virus | 417[b] | 58 | Fowlpox virus (FPV) | 334[a] | 67 |
| | 153.9[b] | 59 | Bovine papilloma virus (BPV) | 8.22[a] | 68 |

[a] Estimated directly.
[b] Estimated from molecular-weight data cited in the reference and the value of $1.67 \times 10^{-18}$ μg/dalton.

Data taken from: Shapiro, H. S., in *Handbook of Biochemistry,* 2nd ed., pp. H-115–H-116, H. A. Sober, Ed. Copyright 1970, The Chemical Rubber Co., Cleveland, Ohio.

### REFERENCES


1. Aurisicchio, Frontali and Graziosi, *Nuovo Cimento Suppl., 25,* 31 (1962).
2. Tikhonenko, *Biokhimiya, 27,* 1015 (1962).
3. Tikhonenko and Zak, *Biokhimiya, 31,* 33 (1966).
4. Sellers and Tokunaga, *J. Exp. Med., 123,* 327 (1966).
5. Zinder, *J. Cell Comp. Physiol. Suppl. 2, 45,* 23 (1955).
6. Ikeda and Tomizawa, *J. Mol. Biol., 14,* 85 (1965).
7. Garen and Zinder, *Virology, 1,* 347 (1955).
8. Lwoff, *Bacteriol. Rev., 17,* 269 (1953).
9. Saunders and Campbell, *J. Bacteriol., 91,* 340 (1966).
10. Green, *J. Mol. Biol., 10,* 438 (1964).
11. Pons, *Biochem. Z., 346,* 26 (1966).
12. Stent, *The Infective Unit, Molecular Biology of Bacterial Viruses.* Charles C Thomas, Springfield, Illinois (1961).
13. Creaser and Taussig, *Virology, 4,* 200 (1957).
14. Allison and Burke, *J. Gen. Microbiol., 27,* 181 (1962).
15. Stent and Fuerst, *J. Gen. Physiol., 38,* 441 (1955).
16. Labaw, *J. Bacteriol., 62,* 169 (1951).
17. Thomas, Jr., *Mezhdunar. Biokhim. Kongr. Moscow 1961 Symp., 1,* 134 (1962).
18. Hershey, *Virology, 1,* 108 (1955).
19. Hershey, Dixon and Chase, *J. Gen. Physiol., 36,* 777 (1953).
20. Cohen and Arbogast, *J. Exp. Med., 91,* 607 (1950).
21. Lesley, French and Graham, *Can. J. Res., 28E,* 281 (1950).
22. Hook, Beard, Taylor, Sharp and Beard, *J. Biol. Chem., 165,* 241 (1946).
23. Hershey, Kamen, Kennedy and Gest, *J. Gen. Physiol., 34,* 305 (1951).
24. Hershey, Burgi and Ingraham, *Biophys. J., 2,* 423 (1962).
25. Frank, Zarnitz and Weidel, *Z. Naturforsch. Teil B, 18,* 281 (1963).
26. Kozloff and Putman, *J. Biol. Chem., 181,* 207 (1949).
27. Volkin, Astrachan and Countryman, *Virology, 6,* 545 (1958).
28. Davison and Freifelder, *J. Mol. Biol., 5,* 635 (1962).
29. Lunan and Sinsheimer, *Virology, 2,* 455 (1956).
30. Davison and Freifelder, *J. Mol. Biol., 5,* 643 (1962).
31. Sinsheimer, *J. Mol. Biol., 1,* 43 (1959).
32. Yamagishi, Yoshizako and Sato, *Virology, 30,* 29 (1966).
33. Caro, *Virology, 25,* 226 (1965).
34. Marvin and Hoffman-Berling, *Nature, 197,* 517 (1963).

35. Hoffman-Berling, Marvin and Dürwald, *Z. Naturforsch. Teil B, 18,* 876 (1963).
36. Kaiser and Hogness, *J. Mol. Biol., 2,* 392 (1960).
37. Sechaud, *Arch. Sci. (Geneva), 13,* 427 (1960).
38. Burgi and Hershey, *Biophys. J., 3,* 309 (1963).
39. Lewin, Crothers, Correll and Reimann, *Can. J. Microbiol., 10,* 75 (1964).
40. Egbert, *J. Virol., 3,* 528 (1969).
41. Anderson and Mosharrafa, *J. Virol., 2,* 1185 (1968).
42. Haas and Yoshikawa, *J. Virol., 3,* 233 (1969).
43. Boice, Eiserling and Romig, *Biochem. Biophys. Res. Commun., 34,* 398 (1969).
44. Riva, Polsinelli and Falaschi, *J. Mol. Biol., 35,* 347 (1968).
45. May, May, Granboulan, Granboulan and Marmur, *Ann. Inst. Pasteur (Paris), 115,* 1029 (1968).
46. Biswal, Kleinschmidt, Spatz and Trautner, *Mol. Gen. Genetics, 100,* 39 (1967).
47. Reznikoff and Thomas, Jr., *Virology, 37,* 309 (1969).
48. Lang, Bujard, Wolff and Russell, *J. Mol. Biol., 23,* 163 (1967).
49. Bresler, Kiselev, Manjakov, Mosevitsky and Timkovsky, *Virology, 33,* 1 (1967).
50. Espejo and Canelo, *Virology, 37,* 495 (1969).
51. Lee and Boezi, *J. Virol., 1,* 1274 (1967).
52. Goldstein and Bendet, *Virology, 32,* 614 (1967).
53. Allison and Burke, *J. Gen. Microbiol., 27,* 181 (1962).
54. Thomas, *Virology, 14,* 240 (1961).
55. Crawford, *Virology, 22,* 149 (1964).
56. Watson and Littlefield, *J. Mol. Biol., 2,* 161 (1960).
57. Joklik, Unpublished Data.
58. Darlington and Randall, *Virology, 19,* 322 (1963).
59. Soehner, Gentry and Randall, *Virology, 26,* 394 (1965).
60. Russell and Crawford, *Virology, 22,* 288 (1964).
61. Ben-Porat and Kaplan, *Virology, 16,* 261 (1962).
62. Crawford and Black, *Virology, 24,* 388 (1964).
63. Mayor, Jamison and Jordon, *Virology, 19,* 359 (1963).
64. Rose, Hoggan and Shatkin, *Proc. Nat. Acad. Sci. U.S.A., 56,* 86 (1966).
65. Becker, Dym and Sarov, *Virology, 36,* 184 (1968).
66. Crawford, Follett, Burdon and McGeoch, *J. Gen. Virol., 4,* 37 (1969).
67. Hyde, Gafford and Randall, *Virology, 33,* 112 (1967).
68. Lang, Bujard, Wolff and Russell, *J. Mol. Biol., 23,* 163 (1967).

# BUOYANT DENSITIES, MELTING TEMPERATURES AND GC CONTENT OF VIRAL DNA

DR. MANLEY MANDEL

| Host | Virus | $\rho_{CsCl}$[a] | T*m*[b] | GC[c] | Comment[d] | Reference |
|---|---|---|---|---|---|---|
| Actinomycete | **φ17** | 1.719 | — | 59 | — | *38* |
| *Agrobacterium radiobacter* | **PR-1001** | 1.712 | — | 53 | 2 Denatured bands | *20* |
| *Bacillus anthracis* | **γ** | 1.696 | — | 36 | — | *38* |
| *B. cereus* | **A** | 1.694 | — | ? | — | — |
| | **B** | 1.695 | — | ? | 2 Denatured bands | *20* |
| | **C** | 1.696 | — | ? | | |
| | **D** | 1.696 | — | ? | — | — |
| *B. licheniformis* | **SP5** | 1.742 | 76.5 | — | Unusual base, 2 denatured bands | *20* |
| *B. megaterium* | **G** | 1.701 | 85.7 | 41 | — | *20* |
| *B. stearothermophilus* | **φμ4** | 1.706 | — | 46 | 2 Denatured bands | *35* |
| | **TP84** | 1.702 | 86.2 | 42 | 2 Denatured bands | *32* |
| | **T3** | — | — | 40.3 | 2 Denatured bands | *40* |
| *B. subtilis* | **φ3** | — | — | 36 | Temperate | *43* |
| | **φ15** | 1.694 | — | 34 | — | *38* |
| | **φ29** | 1.654 | — | 34 | — | *38* |
| | **SP3** | 1.695 | — | 35 | — | *38* |
| | **pKc** | 1.695 | — | 35 | — | *38* |
| | **SPα** | 1.704 | — | 44 | Defective virus | *18, 20, 35* |
| | **SPX** | 1.704 | — | 44 | Defective virus | *18, 20, 35* |
| | **SP50** | 1.704 | — | 44 | 2 Denatured bands | *35, 40, 42* |
| | **φ1** | 1.704 | — | 44 | 2 Denatured bands | *42* |
| | **φ2** | 1.704 | — | 44 | — | *38* |
| | **φ14** | 1.704 | — | 44 | — | *38* |
| | **F** | 1.709 | — | 49 | — | *38* |
| | **SP10** | 1.714 | — | 40.5 | Low T*m*, probably unusual base, density decreases on storage (1.72 → 1.71) | *10, 20, 38* |
| | **PBS1, PBS2** | 1.722 | — | 28 | U replaces T, transducing | *20, 38, 39, 41* |
| | **SP90, SP100 SP70, SP80** | 1.722 | — | — | Identical to PBS1 and 2? | *38* |
| | **1KT, 6a, 6b, 4P, NT** | 1.722 | — | — | U replaces T, transducing may be identical to PBS1 and 2 | *20* |
| | **SP8** | 1.742 | 76.5 | 43 | hmU replaces T, 2 denatured bands | *13, 18, 20, 29, 38* |
| | **SP82, SPO-1, φe, 25, SP60** | 1.742 | — | 43 | hmU replaces T | *29, 35, 38* |
| | **SP6, 7, 9, 13** | 1.743 | 76.5 | 43 | hmU replaces T, 2 denatured bands | *20* |
| | **Vx** | — | — | 38 | May contain unusual base partly replacing T | *8, 10* |
| | **SPP-1** | 1.703 | 88 | 44 | 2 Denatured bands | *48* |
| | **2C** | 1.742 | 78 | 39 | hmU replaces T, 2 denatured bands | *54* |
| *B. tiberius* | **α** | 1.705 | 86.5 | 45 | 2 Denatured bands | *3, 18, 19, 38* |
| *Brucella abortus* | Tbilisi | 1.7089 | — | 48.9 | — | *38* |
| *Clostridium saccharoperbutylacetonicum* | **HM-2** | — | 83.5 | 35 | — | *9* |
| | **HM-3** | — | 81.8 | 30 | — | *9* |
| *Clostridium tetani* | **CT1** | 1.691 | 82.4 | 31.5 | 2 Denatured bands, defective? | *20* |
| *Cytophaga sp.* | **NCMB 384, 385** | 1.691 | 81.5 | 31 | 2 Denatured bands | *1* |
| *Escherichia coli* | **T1** | 1.705 | 89.0 | 48 | — | *16, 34, 36, 38* |
| | **T3** | 1.711 | 90 | 49.6 | — | *19, 34, 36* |
| | **T5** | 1.702 | — | 37 | — | *34, 36, 38* |
| | **T7** | 1.710 | 89.5 | 48 | — | *19, 34, 36, 38* |
| | **T2** | 1.7020 | 83 | 34 | hmC replaces C, glucosylated to various extents. (T2, T4, T6, C16) | *19, 34, 36, 40* |
| | **T4** | 1.7005 | 84 | 34 | | *19, 34, 36* |
| | **T6** | 1.7105 | 84 | 34 | | *19, 34, 36* |
| | **C16** | 1.7103 | — | 34 | | *19, 34, 36* |

| Host | Virus | $\rho_{CsCl}$[a] | $Tm$[b] | GC[c] | Comment[d] | Reference |
|---|---|---|---|---|---|---|
| *Escherichia coli*—(Continued) | | | | | | |
| | **λ** | 1.7093 | 89 | 49 | Two bands in alkaline CsCl, trace of 5-MeC | *14, 36, 38, 42, 44* |
| | **Plkc** | 1.706 | — | 46 | —— | *38* |
| | **M13, f1, fr, fd, AE2** | 1.7224 | — | 41 | Single stranded | *38* |
| | **φR** | 1.7243 | — | 42 | Single stranded | *38* |
| | **φX 174** | 1.725 | — | 42.6 | Single stranded circular G ≠ C | *38, 42* |
| | **φX 174 RF** | 1.708 | — | 42 | Duplex, circular | *42* |
| | **φ15 (TAU⁻)** | 1.7055 | 88 | 46.5 | Defective virus, UV-induced | *17* |
| | **φ80** | 1.710 | — | 51 | Transducing | *45* |
| | **P2** | 1.709 | — | 50 | — | *58* |
| | **N4** | 1.703 | 87.9 | 45 | — | *55, 56* |
| *Micrococcus lyso-deikticus* | **N6** | — | — | 70 | — | *33* |
| | **N1** | 1.724 | — | 65 | — | *20, 33, 59* |
| *Proteus morganii* | **M** | — | — | 48.9 | — | *2* |
| *P. mirabilis* | **13 vir** | — | 87.4 | 44.1 | — | *64* |
| *Providence* | **PL 25** | — | 87.0 | 43.2 | Linear, unique | *65* |
| | **PL 26** | — | — | 41.6 | — | *2* |
| *Pseudomonas aeruginosa* | **1X1** | 1.722 | — | 63 | — | *20* |
| *P. putida* | **gh-1** | 1.716 | — | 57 | — | *66* |
| *Rhizobium meliloti* | **PRM 1** | 1.708 | — | 49 | — | *20* |
| *Salmonella sp.* | **χ** | 1.715 | — | 55, 57.4 | — | *36, 38, 42* |
| *S. typhimurium* | **P22** | 1.710 | — | 50 | — | *34, 36* |
| | **$A_c$** | — | — | 43.4 | Inequality of base pairs, strandedness? | *15* |
| *Serratia marcescens* | **S24V** | 1.714 | 92.5 | 55 | — | *20* |
| | **μ** | — | — | 57 | Unknown base partially replaces G | *42* |
| | **X** | — | — | 53.8 | G less than C, 3% low? | *42* |
| | **t** | 1.715 | — | 56 | Temperate | *17* |
| *Staphylococcus pyogenes* | **SA** | 1.693 | 83.6 | 34 | 2 Denatured bands | *20* |
| | **44A** | 1.687 | — | 27 | — | *38* |
| | **81** | 1.695 | — | 35 | — | *38* |
| | **80** | 1.697 | — | 37 | — | *38* |
| *Streptococcus lactis* | **c2** | 1.696 | — | 37 | Inferred from similarity to *B. cereus* DNA | *67* |
| *Streptomyces griseus* | **MSP8** | 1.729 | — | 70 | — | *20* |
| *Xanthomonas pruni* | **XP5** | 1.721 | — | 62 | — | *20, 38* |
| *X. oryzae* | **Xf** | — | 55 | 60 | Single stranded | *63* |
| Cyanophyceae (*Plectonema boryanum*) | **BGAV LPP-1** | 1.714 | 91.7 | 54 | — | *8, 39* |
| *Bombyx mori* | **Polyhedral** | — | — | 40 | — | *46* |
| Chicken | **Fowl pox** | 1.6945 | 83.5 | 35 | — | *26, 38* |
| Cow | **Allerton** | — | — | 65 | — | *21* |
| | **Cow pox** | — | — | 36 | — | *12* |
| | **IBR** | 1.731 | — | 71.5 | — | *31, 52* |
| | **Mammillitis (BMV)** | — | — | 64 | — | *21* |
| | **Papilloma** | — | — | 45.5 | Circular | *42* |
| Frog | **FV1, FV3** | 1.720 | — | 56, 62 | Density estimate may be high | *22* |
| | **FV3** | 1.720 | 93 | 57 | — | *68* |
| Human | **Adeno 1** | 1.718 | 92.8 | 58.5 | — | *24* |
| | **2** | 1.714 | 92.5 | 55 | — | *24, 27, 23* |
| | **3** | 1.714 | 90.3 | — | Discrepancy not accounted for | *24* |
| | **4** | 1.717 | 92.5 | 57.5 | — | *24* |
| | **5** | 1.717 | 92.6 | 57.5 | — | *24* |
| | **6** | 1.718 | 93.6 | 59 | — | *24* |
| | **7** | 1.711 | 90.3 | 51 | — | *24, 30* |
| | **8** | 1.717 | 90.8 | — | Discrepancy not accounted for | *24* |
| | **9** | 1.720 | 93.4 | 60 | — | *24* |
| | **10** | 1.720 | 93.6 | 60 | — | *24* |
| | **11** | 1.712 | 90.0 | — | Discrepancy not accounted for | *24* |
| | **12** | 1.708 | 89.5 | 49 | — | *24, 27* |
| | **13** | 1.719 | 93.1 | 59 | — | *24* |
| | **14** | 1.715 | 91.0 | — | Discrepancy not accounted for | *24* |
| | **15** | 1.718 | 93.3 | 59 | — | *24* |

| Host | Virus | $\rho_{CsCl}$[a] | Tm[b] | GC[c] | Comment[d] | Reference |
|---|---|---|---|---|---|---|
| Human—(Continued) | | | | | | |
| | **16** | 1.714 | 90.9 | — | Discrepancy not accounted for | *24* |
| | **17** | 1.718 | 93.0 | 58.5 | — | *24* |
| | **18** | 1.708 | 88.8 | 48 | — | *24* |
| | **19** | 1.719 | 93.1 | 59.5 | — | *24* |
| | **20** | 1.719 | 94.2 | 60 | — | *24* |
| | **21** | 1.714 | 90.8 | — | Discrepancy not accounted for | *24* |
| | **22** | 1.718 | 92.9 | 58.5 | — | *24* |
| | **23** | 1.719 | 93.2 | 59 | — | *24* |
| | **24** | 1.719 | 93.7 | 59.5 | — | *24* |
| | **25** | 1.720 | 93.8 | 60 | — | *24* |
| | **26** | 1.719 | 93.7 | 59.5 | — | *24* |
| | **27** | 1.719 | 94.1 | 60 | — | *24* |
| | **28** | 1.719 | 94.1 | 60 | — | *24* |
| | **AAV** | 1.717 | 92 | 54.2 | Buoyant density appears high | *30* |
| | **Cytomegato** | 1.716 | — | 57 | — | *42, 52* |
| | **Herpes simplex 1** | 1.727 | — | 68 | — | *42, 52* |
| | **2** | 1.729 | — | 70 | — | *52* |
| | **Parvovirus H-1** | 1.720 | — | 41.7 | Single stranded | *60* |
| | **Papilloma** | 1.701 | — | 41 | Circular, supercoil | *23* |
| | **Yaba** | 1.692 | 82.5 | 32 | — | *7* |
| Horse | **Equine abortion** | 1.716 | — | 56.5 | — | *37* |
| | **LK (equine herpes)** | 1.717 | — | 56–58 | — | *42, 52* |
| | **Rhinopneumonitis** | 1.714 | — | 54 | — | *23* |
| Mammal | **Vaccinia** | 1.695 | — | 32 | — | *23* |
| | **Laryngotracheitis** | 1.704 | — | 45 | — | *52* |
| Mouse | **Ectromelia (Dohi A)** | — | — | 36 | — | *11* |
| | **MVM** | 1.722 | — | 40.5 | Single stranded | *5, 62* |
| | **Polyoma** | 1.708 | — | 48, 47 | Circular, supercoil | *4, 28, 45* |
| | **Cytomegalo** | 1.717 | — | 58 | — | *52* |
| | | 1.722 | — | 63 | — | *52* |
| Simian | **SV40** | 1.701 | — | 41, 39 | Circular, supercoil | *23, 45* |
| | **SA7** | 1.719 | 93.4 | 60 | — | *53* |
| | **SV11** | — | — | 55 | — | *51* |
| | **SV15** | 1.716 | 92.3 | 57 | — | *50, 53* |
| | **AAV satellite 1** | 1.717 | 92 | 54.2 | Density high | *30* |
| | **4** | 1.716 | 92 | 55 | Single complementary strands in separate virions | *49–51* |
| | **Herpes** | 1.726 | — | 67 | — | *52* |
| Rat | **K (Parvovirus ratt 1)** | 1.704 | 87 | 44 | 2 × $10^6$ daltons | *47* |
| | **RV-13** | 1.72 | — | — | Single stranded | *61* |
| Pig | **Pseudorabies** | 1.732 | — | 74 | — | *38* |
| Rabbit | **Kidney vacuolating** | 1.703 | — | 43 | Circular, supercoil | *6* |
| | **Papilloma, Shope** | 1.711 | 89.5 | 48 | Circular, supercoil | *42, 69* |
| | **Pox** | 1.6959 | — | 36 | — | *11, 38, 40* |
| | **Pseudorabies** | 1.731 | — | 72 | — | *52* |
| | **IBR** | 1.731 | — | 72 | — | *52* |
| | **Herpes, horse 1** | 1.716 | — | 57 | — | *52* |
| Vervet | **Cytomegalo** | 1.710 | — | 51 | — | *52* |
| Dog | **Herpes** | 1.692 | — | 33 | — | *52* |
| Cat | **Herpes** | 1.705 | — | 46 | — | *52* |
| *Galleria Mellonella* | **SIV** | 1.689 | 82 | 31 | — | *57* |
| | **CIV** | 1.686 | 81.3 | 28.7 | — | *57* |
| *Tipula paludosa* | **TIV** | 1.690 | 82.6 | 32.2 | — | *52* |

*a* All buoyant densities are given in g/cm$^3$ relative to *E. coli* DNA at 1.710 g/cm$^3$.
*b* T*m* (°C) in SSC buffer or recalculated to T*m* in that buffer.
*c* Moles %; best estimate from all available data.
*d* HMU = 5-hydroxymethyluracil; 5-MeC = 5-methylcytosine; HMC = hydroxymethylcytosine; A = adenine; T = thymine; G = guanine; C = cytosine; U = uracil.

Data taken from: Mandel, M., in *Handbook of Biochemistry,* 2nd ed., pp. H-9–H-11, H. A. Sober, Ed. Copyright 1970, The Chemical Rubber Co., Cleveland, Ohio.

## REFERENCES


1. Chen, Citarella, Salazar and Colwell, *J. Bacteriol., 91,* 1136 (1966).
2. Coetzee, Smit and Prozesky, *J. Gen. Microbiol., 44,* 167 (1966).
3. Cordes, Epstein and Marmur, *Nature, 191,* 1097 (1961).
4. Crawford, *Virology, 19,* 279 (1963).
5. Crawford, *Virology, 29,* 605 (1966).
6. Crawford and Follett, *J. Gen. Virol., 1,* 19 (1967).
7. Gallagher and Yohn, *Bacteriol. Proc.* (1967).
8. Goldstein and Bendet, *Biophys. Soc. Abstr.,* p. 132 (1967).
9. Hongo, Ono, Ogata and Murata, *Agr. Biol. Chem., 30,* 982 (1966).
10. Ikeda, Saito, Miura, Takagi and Aoki, *J. Gen. Appl. Microbiol., 11,* 181 (1965).
11. Joklik, *Virology, 18,* 9 (1962).
12. Joklik, *J. Mol. Biol., 5,* 265 (1962).
13. Kallen, Simon and Marmur, *J. Mol. Biol., 5,* 248 (1962).
14. Ledinko, *J. Mol. Biol., 9,* 834 (1964).
15. Lwoff, *Bacteriol. Rev., 17,* 269 (1953).
16. Mandel, *J. Mol. Biol., 5,* 435 (1962).
17. Frampton and Mandel, *J. Virol., 5,* 8 (1970).
18. Marmur and Cordes, in *Informational Macromolecules,* p. 79, Vogel, Bryson and Lampen, Eds. Academic Press, New York (1963).
19. Marmur and Doty, *J. Mol. Biol., 5,* 109 (1962).
20. Marmur, Kahan, Riddle and Mandel, in *Acidi Nucleici e Loro Funzione Biologica,* p. 249, Baselli, Ed. Istituto Lombardo – Academia di Scienze e Lettere, Milan, Italy (1964).
21. Martin, Hay, Crawford, Le Bouvier and Crawford, *J. Gen. Microbiol., 45,* 325 (1966).
22. Morris, Spear and Roizman, *Proc. Nat. Acad. Sci. U.S.A., 56,* 1155 (1966).
23. Morrison, Keir, Subak-Sharpe and Crawford, *J. Gen. Virol., 1,* 101 (1967).
24. Piña and Green, *Proc. Nat. Acad. Sci. U.S.A., 54,* 547 (1965).
25. Randall, Gafford and Darlington, *J. Bacteriol., 83,* 1037 (1962).
26. Randall, Gafford, Soehner and Hyde, *J. Bacteriol., 91,* 95 (1966).
27. Rapp, Feldman and Mandel, *J. Bacteriol., 91,* 95 (1966).
28. Ricceri and Cocuzza, *G. Microbiol., 14,* 15 (1966).
29. Roscoe and Tucker, *Virology, 29,* 157 (1966).
30. Rose, Hoggan and Shatkin, *Proc. Nat. Acad. Sci. U.S.A., 56,* 86 (1966).
31. Russell and Crawford, *Virology, 22,* 288 (1964).
32. Saunders and Campbell, *Biochemistry, 4,* 2836 (1965).
33. Scaletti and Naylor, *J. Bacteriol., 78,* 422 (1959).
34. Schildkraut, Marmur and Doty, *J. Mol. Biol., 4,* 430 (1962).
35. Sheldrick and Szybalski, *Fed. Proc., 25,* 207 (1966).
36. Sinsheimer, in *The Nucleic Acids,* Vol. 3, p. 187, Chargaff and Davidson, Eds. Academic Press, New York (1960).
37. Soehner, Gentry and Randall, *Virology, 26,* 394 (1965).
38. Szybalski, *Methods Enzymol., 12,* 330 (1968).
39. Szybalski, Personal Communication.
40. Szybalski, Erickson, Gentry, Gafford and Randall, *Virology, 19,* 586 (1963).
41. Takahashi and Marmur, *Biochem. Biophys. Res. Commun., 10,* 289 (1963).
42. Thomas and MacHattie, *Annu. Rev. Biochem., 36,* 485 (1967).
43. Tucker, *Biochem. J., 92,* 58 (1964).
44. Vinograd, Lebowitz, Radloff, Watson and Laipis, *Proc. Nat. Acad. Sci. U.S.A., 53,* 1104 (1965).
45. Yamagishi, Yoshizako and Sato, *Virology, 30,* 29 (1966).
46. Hashiniga, Murakami and Yamafuji, *Enzymologia, 30,* 179 (1966).
47. May, Niveleau, Berger and Brailovsky, *J. Mol. Biol., 27,* 603 (1967).
48. Riva, Polsinelli and Falaschi, *J. Mol. Biol., 35,* 347 (1968).
49. Mayor, Torikai, Melnick and Mandel, *Science, 166,* 1280 (1969).
50. Parks, Green, Piña and Melnick, *J. Virol., 1,* 980 (1967).
51. Bereczky, Kawaclova and Archetti, *Ann. Sclavo, 10,* 433 (1968).
52. Plummer, Goodheart, Henson and Bowling, *Virology, 39,* 134 (1969).
53. Piña and Green, *Virology, 36,* 321 (1968).
54. May, May, Granboulan, Granboulan and Marmur, *Ann. Inst. Pasteur (Paris), 115,* 1029 (1968).
55. Mandel, Igambi, Bergendahl, Dodson and Scheltgen, *J. Bacteriol., 101,* 333 (1970).
56. Schito, Rialdi and Pesce, *Biochim. Biophys. Acta, 129,* 491 (1966).
57. Bellet and Inman, *J. Mol. Biol., 25,* 425 (1967).
58. Bertani, Torheim and Laurent, *Virology, 32,* 619 (1967).

59. Wetmur, Davidson and Scaletti, *Biochem. Biophys. Res. Commun., 25,* 684 (1966).
60. Usutegui-Gomez, Toolan, Ledinko, Al-Lami and Hopkins, *Virology, 39,* 617 (1969).
61. Robinson and Hetrick, *J. Gen. Virol., 4,* 269 (1969).
62. Crawford, Follett, Burdon and McGeoch, *J. Gen. Virol., 4,* 37 (1969).
63. Kuo, Huang and Chow, *Virology, 39,* 548 (1969).
64. van Rensburg, *J. Gen. Virol., 5,* 437 (1969).
65. Pitout and van Rensburg, *J. Gen. Virol., 4,* 615 (1969).
66. Lee and Boezi, *J. Bacteriol., 92,* 1821 (1966).
67. Anderson and Fraser, *Virology, 35,* 529 (1968).
68. Smith and McAuslan, *J. Virol., 4,* 539 (1969).
69. Watson and Littlefield, *J. Mol. Biol., 2,* 161 (1960).

# DEOXYRIBONUCLEIC ACID CONTENT PER CELL IN FUNGI, ALGAE AND PROTOZOA

DR. HERMAN S. SHAPIRO

## TABLE 1
## FUNGI

| Cell source | DNA/cell $\mu g \times 10^6$ | Reference |
|---|---|---|
| *ASCOMYCETES* | | |
| ***Ophiostoma multiannulatum*** | 0.048 | *1* |
| ***Aspergillus nidulans*** | | |
| Green haploid (per conidium) | $4.38 \times 10^{-8}$ | *2*[a] |
| Green diploid (per conidium) | $10.17 \times 10^{-8}$ | *2*[a] |
| White haploid (per conidium) | $4.57 \times 10^{-8}$ | *2*[a] |
| White diploid (per conidium) | $8.40 \times 10^{-8}$ | *2*[a] |
| ***Aspergillus sojae*** | | |
| Yellow, wild, haploid (per nuclei) | 0.088 | *3* |
| Green diploid (per nuclei) | 0.169 | *3* |
| ***Saccharomyces cerivisiae*** | 0.046 | *4* |
| ***S. cerivisiae,*** haploid | 0.0245 | *5*[a] |
| ***S. cerivisiae,*** diploid | 0.0495 | *5*[a] |
| ***S. cerivisiae,*** triploid | 0.0670 | *5*[a] |
| ***S. cerivisiae,*** tetraploid | 0.102 | *5*[a] |
| ***Neurospora crassa*** (per conidium) | 0.017 | *6* |
| *SCHIZOMYCETES* | | |
| ***Bacillus cereus*** (spores) | 0.0108 | *7*[a] |
| ***B. cereus,*** var. alesti (spores) | 0.0221 | *7*[a] |
| ***B. cereus,*** var. mycoides (spores) | 0.0153 | *7*[a] |
| ***B. cereus,*** var. anthrax (spores) | 0.0122 | *7*[a] |
| ***B. cereus,*** A 30 (spores) | 0 0127 | *7*[a] |
| ***Bacillus cereus,*** B-30-1 (spores) | 0.0127 | *7*[a] |
| ***B. cereus,*** B-30-2 (spores) | 0.0122 | *7*[a] |
| ***B. cereus*** (spores) | 0.00543 | *8* |
| ***B. cereus*** (veg. cells) | 0.0129 | *8* |
| ***B. sotto*** (spores) | 0.0109 | *7*[a] |
| ***B. thuringiensis*** (spores) | 0.0132 | *7*[a] |
| ***B. medusa*** (spores) | 0.0208 | *7*[a] |
| ***B. megaterium*** | 0.025 | *9* |
| ***B. megaterium,*** L (spores) | 0.0106 | *7*[a] |
| ***B. megaterium,*** LM (spores) | 0.0117 | *7*[a] |
| ***B. megaterium,*** Penn (spores) | 0.0103 | *7*[a] |
| ***B. megaterium,*** 350 (spores) | 0.0220 | *7*[a] |
| ***B. megaterium*** (spores) | 0.00405 | *8* |
| ***B. megaterium*** (veg. cells) | 0.0134 | *8* |
| ***B, subtilis*** (spores) | 0.00542 | *7*[a] |
| ***B. subtilis*** (spores) | 0.0051 | *10* |
| ***B. subtilis,*** ws (veg. cells) | 0.0048 | *26* |
| ***B. subtilis,*** ws (spores) | 0.0054 | *26* |
| ***B. subtilis,*** ws (binucl. sporangia) | 0.0097 | *26* |
| ***B. subtilis,*** 168 (veg. cells) | 0.0050 | *26* |
| **Schizomycetes—(Continued)** | | |
| ***B. subtilis,*** 168 (spores) | 0.0058 | *26* |
| ***B. subtilis,*** 168 (binucl. sporangia) | 0.0098 | *26* |
| ***B. subtilis,*** Marburg (veg. cells) | 0.0051 | *26* |
| ***B. subtilis,*** Marburg (spores) | 0.0053 | *26* |
| ***B. subtilis,*** Marburg (binucl. sporangia) | 0.0099 | *26* |
| ***B. subtilis*** | 0.028 | *27* |
| ***B. apiarius*** (spores) | 0.00639 | *7*[a] |
| ***B. stearothermophylicus*** 30° | 0.0203 | *11* |
| ***B. stearothermophylicus*** 44° | 0.0063 | *11* |
| ***B. stearothermophylicus*** 58° | 0.0026 | *11* |
| ***Escherichia coli*** | 0.0047 | *12* |
| ***E. coli*** | 0.009 | *13* |
| ***E. coli,*** B (log phase) | 0.0137 | *14* |
| ***E. coli,*** B (stationary phase) | 0.0078 | *14* |
| ***E. coli,*** B (log phase) | 0.016–0.018 | *15* |
| ***E. coli,*** B (stationary phase) | 0.024–0.026 | *15* |
| ***E. coli,*** B/r (log phase) | 0.0182 | *14* |
| ***Escherichia coli,*** B/r (stationary phase) | 0.0128 | *14* |
| ***E. coli,*** B/r (log phase) | 0.016–0.018 | *15* |
| ***E. coli,*** B/r (stationary phase) | 0.022–0.027 | *15* |
| ***E. coli*** | 0.0267 | *16* |
| ***E. coli***, NCIB 8114 | 0.019 | *27* |
| ***Salmonella typhimurium*** | 0.011 | *17* |
| ***Aerobacter aerogenes*** | 0.002 | *18* |
| ***Aerobacter aerogenes*** | 0.025 | *27* |
| ***Bacterium lactis aerogenes*** | 0.0217 | *18*[a] |
| ***Diplococcus pneumoniae*** | 0.002 | *19* |
| ***D. pneumoniae*** | 0.005 | *20* |
| ***Hemophilus influenzae*** | 0.002–0.0027 | *21* |
| ***H. influenzae*** | 0.002 | *22* |
| ***Mycoplasma gallisepticum*** (Avian PPLO 5969) | $260 \times 10^{-6}$ | *23* |
| ***Mycoplasma pneumoniae*** (Eaton agent PPLO) | $5.2 \times 10^{-6}$ | *24* |
| ***Mycoplasma laidlawii*** (Sewage A PPLO) | $19.0 - 10^{-6}$ | *24* |
| ***M. laidlawii*** (Calif. calf PPLO) | $51.8 \times 10^{-6}$ | *24* |
| ***M. laidlawii*** (Kid strain, goat PPLO) | $51.8 \times 10^{-6}$ | *24* |
| ***Clostridium welchii*** | 0.024 | *25* |
| ***Proteus,*** $P_{18}$ | 0.0184 | *25*[a] |
| ***Proteus,*** $P_{18}$ (L form) | 0.00062 | *25*[a] |
| ***Rh. spheroides***, NCIB 8253 | 0.023 | *27* |
| ***Sarcina lutea*** | 0.020 | *27* |

[a] Content of DNA estimated from the phosphorus analysis cited in the reference and the value of 9.23% phosphorus in DNA.

## TABLE 2
## ALGAE

| Cell source | DNA/cell $\mu g \times 10^6$ | Reference | Cell source | DNA/cell $\mu g \times 10^6$ | Reference |
|---|---|---|---|---|---|
| ***Monochrysis lutheri*** | 0.1 | *1* | ***Thalassiosira fluviatilis*** | 7 | *1* |
| ***Navicula pelliculosa*** | 0.1 | *1* | ***Cachonina niei*** | 10 | *1* |
| ***Skeletonema costatum*** | 0.6 | *1* | ***Ditylum brightwellii*** | 25 | *1* |
| ***Dunaliella tertiolecta*** | 0.6 | *1* | ***Gonyaulax polyhedra*** | 200 | *1* |
| ***Ambphidinium carteri*** | 5 | *1* | ***Anacystis nidulans*** | 0.030 | *2* |
| ***Syracosphaera elongata*** | 5 | *1* | ***Anabaena variabilis*** | 0.036 | *2* |

## TABLE 3
## PROTOZOA

| Cell source | DNA/cell $\mu g \times 10^6$ | Reference | Cell source | DNA/cell $\mu g \times 10^6$ | Reference |
|---|---|---|---|---|---|
| ***Paramecium caudatum*** | 1,530–1,670 | *1* | ***Euglena gracilis,*** bacillaris | 2.52 | *9* |
| **λ particle of *Paramecium aurelia*** | 0.26 | *2* | ***Euglena gracilis,*** R (green) | 2.56 | *9* |
| **Malaria parasite,** *Plasmodium* | 0.59 | *3* | ***Euglena gracilis*** | 2.9 | *10* |
| *berghei* | 0.054 | *4* | ***Euglena gracilis,*** R (colorless) | 3.31–4.00 | *9* |
| ***Tetrahymena pyriformis*** | 13.6 | *5* | ***Euglena gracilis,*** T (green) | 2.70 | *9* |
| ***Tetrahymena pyriformis,*** GL | 15.7 | *6* | ***Euglena gracilis,*** T (colorless) | 4.02–4.30 | *9* |
| ***Urostyla caudata*** | 1,057 | *7* | ***Euglena gracilis*** | 2.3 | *11* |
| ***Euglena gracilis,*** Z | 3.33 | *8*[a] | ***Astasia longa*** | 1.52 | *9* |

[a] Content of DNA estimated from the phosphorus analysis cited in the reference and the value of 9.23% phosphorus in DNA.

Data taken from: Shapiro, H. S., in *Handbook of Biochemistry,* 2nd ed., pp. H-104, H-105, H-107, H. A. Sober, Ed. Copyright 1970, The Chemical Rubber Co., Cleveland, Ohio.

### REFERENCES

#### Fungi

1. von Hofsten, *Physiol. Plant., 16,* 709 (1963).
2. Heagy and Roper, *Nature, 170,* 713 (1952).
3. Ishitani, Uchida and Ikeda, *Exp. Cell Res., 10,* 737 (1956).
4. Williamson and Scopes, *Exp. Cell Res., 24,* 151 (1961).
5. Ogur, Minckler, Lindegren and Lindegren, *Arch. Biochem. Biophys., 40,* 175 (1952).
6. Minagawa, Wagner and Strauss, *Arch. Biochem. Biophys., 80,* 442 (1959).
7. Fitz-James and Young, *J. Bacteriol., 78,* 743 (1959).
8. Hodson and Beck, *J. Bacteriol., 79,* 661 (1960).
9. Spiegelman, Aronson and Fitz-James, *J. Bacteriol., 75,* 102 (1958).
10. Dennis and Wake, *J. Mol. Biol., 15,* 435 (1966).
11. Evreinova, Bunina and Kuznetsova, *Biokhimiya, 24,* 912 (1959).
12. Cairns, *J. Mol. Biol., 4,* 407 (1962).
13. Vendrely, *Ann. Inst. Pasteur (Paris), 94,* 143 (1958).
14. Gillies and Alper, *Biochim. Biophys. Acta, 43,* 182 (1960).
15. Harold and Ziporin, *Biochim. Biophys. Acta, 28,* 482 (1958).
16. Spirin, Skabronskaia and Pretel-Martines, *Mikrobiologiya, 27,* 271 (1958).
17. Lark and Maaløe, *Biochim. Biophys. Acta, 21,* 448 (1956).
18. Caldwell and Hinshelwood, *J. Chem. Soc.,* Part 2, p. 1415 (1950).
19. Fox, *Biochim. Biophys. Acta, 26,* 83 (1957).
20. Lerman and Tolmach, *Biochim. Biophys. Acta, 26,* 68 (1957).

21. Berns and Thomas, Jr., *J. Mol. Biol., 11,* 476 (1965).
22. Zamenhof, Alexander and Leidy, *J. Exp. Med., 98,* 373 (1953).
23. Morowitz, Tourtellotte, Guild, Castro, Woese and Cleverdon, *J. Mol. Biol., 4,* 93 (1962).
24. Neimark and Pène, *Proc. Soc. Exp. Biol. Med., 118,* 517 (1965).
25. Webb, *Science, 118,* 607 (1953).
26. Aubert, Ryter and Schaeffer, *Ann. Inst. Pasteur (Paris), 115,* 989 (1968).
27. Craig, Leach and Carr, *Arch. Mikrobiol., 65,* 218 (1969).

**Algae**

1. Holm-Hansen, *Science, 163,* 87 (1969).
2. Craig, Leach and Carr, *Arch. Mikrobiol., 65,* 218 (1969).

**Protozoa**

1. Gintsburg, *Zh. Obshch. Biol., 22,* 452 (1961).
2. van Wagtendonk and Tanguay, *J. Gen. Microbiol., 33,* 395 (1963).
3. Whitfeld, *Nature, 169,* 751 (1952).
4. Whitfeld, *Aust. J. Biol. Sci., 6,* 234 (1953).
5. Scherbaum, *Exp. Cell Res., 13,* 24 (1957).
6. Scherbaum, Lauderback and Jahn, *Exp. Cell Res., 18,* 150 (1959).
7. Pigon and Edström, *Exp. Cell Res., 16,* 648 (1959).
8. Pogo, Ubero and Pogo, *Exp. Cell Res., 42,* 58 (1966).
9. Neff, *J. Protozool., 7,* 69 (1960).
10. Brawerman, Rebman and Chargaff, *Nature, 187,* 1037 (1960).
11. Edmunds, Jr., *Science, 145,* 266 (1964).

# BUOYANT DENSITIES OF NUCLEIC ACIDS AND POLYNUCLEOTIDES

DR. W. SZYBALSKI and DR. E. H. SZYBALSKI *

## TABLE 1
## SELECTED CELLULAR DNAs[a]

| | Buoyant Density, g/cm$^3$ | | | |
|---|---|---|---|---|
| Source of DNA | $Cs_2SO_4$[b] | CsCl[c] | CsCl[d] | G + C, mole % |
| ***Mycoplasma mycoides* var. *capri*** | 1.4194 | 1.6856 | 1.6791 | 25 |
| ***Spirillum linum*** | 1.4207 | 1.6911 | 1.6846 | 29 |
| ***Clostridium perfringens*** | 1.4212 | 1.6915 | 1.6850 | 31 |
| ***Staphylococcus aureus*** | 1.422 | 1.694 | $1.687_5$ | 33 |
| ***Cytophaga johnsonii*** | 1.4216 | 1.6945 | 1.6880 | 34.5 |
| ***Rhinchosciara angelae*** (Diptera) | 1.422 | 1.695 | $1.688_5$ | 35 |
| ***Bacillus cereus*** | 1.422 | 1.696 | $1.689_5$ | 36 |
| ***Vicia faba*** | 1.422 | 1.696 | $1.689_5$ | 36 |
| ***Proteus vulgaris*** | 1.422 | 1.698 | $1.691_5$ | 38 |
| ***Proteus mirabilis*** | 1.422 | 1.6986 | 1.6921 | 38.5 |
| **Calf thymus** | 1.422 | 1.699 | $1.692_5$ | 39 |
| **Rabbit kidney cells** | 1.422 | 1.699 | $1.692_5$ | 39 |
| **Human cell line D98** | 1.422 | 1.6994 | 1.6929 | 39.5 |
| ***Acinetobacter anitratus*** | 1.423 | $1.699_5$[f] | 1.693 | 39.5[f] |
| **Ehrlich ascites** | 1.423 | 1.700 | $1.693_5$ | 40 |
| **Mouse liver** | 1.423 | 1.700 | $1.693_5$ | 40 |
| **Chicken erythrocytes** | 1.423 | 1.7008 | 1.6943 | 41 |
| ***Neisseria catarrhalis*** | — | 1.701 | $1.694_5$ | 41 |
| ***Acinetobacter haemolysans*** | — | 1.701[f] | $1.694_5$ | 41[f] |
| ***Moraxella nonliquefaciens*** | — | 1.701[f] | $1.694_5$ | 41[f] |
| ***Moraxella bovis*** | — | $1.702_5$[f] | 1.696 | 42.5[f] |
| ***Physarum polycephalum*** | 1.424 | $1.702_5$[e] | 1.696[e] | 42.5 |
| ***Moraxella liquefaciens*** | 1.424 | 1.703[f] | $1.696_5$ | 43[f] |
| ***Moraxella osloensis*** | — | 1.703[f] | $1.696_5$ | 43[f] |
| ***Moraxella phenylpyrouvica*** | — | 1.703[f] | — | 43[f] |
| ***Acinetobacter lwoffi*** | — | 1.703[f] | — | 43[f] |
| **Herring sperm** | 1.424 | 1.703 | $1.696_5$ | 43 |
| ***Bacillus subtilis*** | 1.4240 | 1.7034 | 1.6969 | 43.5 |
| ***Moraxella kingii*** | — | $1.704_5$[f] | 1.698 | 44.5[f] |
| ***Neisseria caviae*** | — | $1.704_5$[f] | 1.698 | 44.5[f] |
| ***Neisseria ovis*** | — | $1.704_5$[f] | 1.698 | 44.5[f] |
| ***Neisseria flavescens*** | — | $1.707_5$[f] | 1.701 | 47.5[f] |
| ***Neisseria cinerea*** | — | 1.709[f] | $1.702_5$ | 49[f] |
| ***Escherichia coli*** | 1.4260 | 1.7100 | 1.7035 | 50 |
| ***Brucella abortus*** | 1.428 | 1.717 | $1.710_5$ | 56 |
| ***Xanthomonas oryzae*** | — | 1.723 | $1.716_5$ | 63 |
| ***Xanthomonas pruni*** | 1.432 | 1.725 | $1.718_5$ | 65 |
| ***Xanthomonas translucens*** | — | 1.7277 | 1.7212 | 67.5 |
| ***Streptomyces chrysomalus*** | 1.4345 | 1.7305 | 1.7240 | 70 |
| ***Micrococcus lysodeikticus*** | 1.435 | 1.731 | $1.724_5$ | 71 |
| ***Streptomyces griseus*** | 1.435 | 1.731 | $1.724_5$ | 71 |
| ***Micrococcus luteus*** | 1.4352 | 1.7311 | 1.7246 | 72 |

[a] The buoyant densities of selected DNAs were determined at 25°C, 44,770 rpm (CsCl) or 31,410 rpm ($Cs_2SO_4$). The densities were abbreviated to the nearest 0.001 (or 0.0005) g/cm$^3$ when only two or three determinations were performed or were measured with an accuracy of ±0.0002 g/cm$^3$.

[b] Measured versus *Escherichia coli* DNA, density 1.4260 g/cm$^3$.

[c] Measured versus *Escherichia coli* DNA, density 1.7100 g/cm$^3$.

[d] Measured versus *Escherichia coli* DNA, density 1.7035 g/cm$^3$.

[e] Second peak at 1.713 g/cm$^3$ or $1.706_5$ g/cm$^3$ respectively.

[f] Bφvre, Fiandt and Szybalski, *Can. J. Microbiol., 15,* 335 (1969). Other data were taken from Szybalski, in *Methods in Enzymology,* Vol. 12, p. 330, Grossman and Moldave, Eds., Academic Press, New York (1968).

*Data previously published in *Handbook of Biochemistry,* 2nd ed., pp. H-13 H-17, H.A. Sober, Ed. Copyright 1970, The Chemical Rubber Co., Cleveland, Ohio.

## TABLE 2
## SELECTED VIRAL DNAs[a]

| Source of DNA (Host/Phage) | Buoyant Density, g/cm³ $Cs_2SO_4$[b] | $CsCl$[c] | $CsCl$[d] | G + C, mole % | Comments |
|---|---|---|---|---|---|
| ***Staphylococcus aureus*** phage 44A | 1.419 | 1.687 | $1.680_5$ | 27 | — |
| ***Staphylococcus aureus*** phage 81 | 1.421 | 1.695 | $1.688_5$ | 35 | — |
| **Fowl pox virus** | 1.4213 | 1.6945 | 1.6880 | 35 | — |
| **Rabbit pox virus** | 1.4217 | 1.6959 | 1.6894 | 36 | — |
| ***Bacillus anthracis*** phage $\gamma$ | 1.422 | 1.696 | $1.689_5$ | 36 | — |
| ***Staphylococcus aureus*** phage 80 | 1.422 | 1.697 | $1.690_5$ | 37 | — |
| ***Escherichia coli*** phage T5 | — | 1.7016 | 1.6951 | 42 | — |
| ***Bacillus tiberius*** phage $\alpha$ | 1.424 | 1.704 | $1.697_5$ | 44 | — |
| **Infectious laryngotracheitis virus** | — | 1.704[e] | — | 45[e] | — |
| ***Escherichia coli*** phage T1 | — | 1.7057 | 1.6992 | 46 | — |
| ***Bacillus stearothermophilus*** phage $\phi\mu4$ | 1.425 | 1.706 | $1.699_5$ | 46 | — |
| ***Brucella abortus*** phage (Tbilisi) | 1.4251 | 1.7089 | 1.7024 | 49 | — |
| ***Escherichia coli*** phages: | | | | | |
| $\lambda dg_{A\text{-}J}$ | — | 1.7083 | 1.7018 | 48 | mol. wt. 23 × $10^6$ |
| $\lambda dg_{L\text{-}J}$ | — | 1.7090 | 1.7025 | 49 | mol. wt. 34 × $10^6$ |
| $\lambda^+$ | 1.426 | 1.7093 | 1.7028 | 49 | mol. wt. 31 × $10^6$ |
| T7 | 1.426 | 1.710 | $1.703_5$ | 50 | — |
| 21 | — | 1.7100 | 1.7035 | 50 | — |
| $\lambda b_2b_5$ | — | 1.7108 | 1.7043 | 51 | mol. wt. 24 × $10^6$ |
| $\lambda cb_2$ | — | 1.7110 | 1.7045 | 51 | mol. wt. 26 × $10^6$ |
| T3 | 1.426 | 1.7105 | 1.7040 | 50 | — |
| $\phi$80 | — | 1.7121 | 1.7056 | 52 | — |
| **Salmonella** phage $\chi$ | 1.428 | 1.715 | $1.708_5$ | 55 | — |
| **Human cytomegalo virus** (CMV) | — | 1.716[e] | — | 57[e] | — |
| **Mouse cytomegalo virus** (CMV) | — | 1.717[e] | — | 58[e] | — |
| **Actinophage** $\phi$17 | 1.429 | 1.719 | $1.712_5$ | 59 | — |
| ***Xanthomonas pruni*** phage XP5 | 1.430 | 1.722 | $1.715_5$ | 62 | — |
| ***Escherichia coli*** phage M13 | 1.4465 | 1.7223 | 1.7158 | 41 | Single stranded DNA |
| ***Escherichia coli*** phage f1 | — | 1.7226 | 1.7161 | 41 | Single-stranded DNA |
| ***Escherichia coli*** phage fr | — | 1.7223 | 1.7158 | 41 | Single-stranded DNA |
| ***Escherichia coli*** phage $\phi$X174 | — | 1.725 | $1.718_5$ | 42 | Single-stranded DNA |
| ***Escherichia coli*** phage $\phi$R | — | 1.7243 | 1.7178 | 42 | Single-stranded DNA |
| **Pseudorabies virus** | 1.436 | 1.732 | $1.725_5$ | 73 | — |
| **Herpes simplex** type 1 | — | 1.727[e] | — | 68[e] | — |
| **Herpes simplex** type 2 | — | 1.729[e] | — | 70[e] | — |
| **Infectious bovine rhinotracheitis virus** | — | 1.731[e] | — | 72[e] | — |
| ***Bacillus subtilis*** phages: | | | | | |
| $\phi$15 | 1.420 | 1.694 | $1.687_5$ | 34 | — |
| $\phi$29 | 1.420 | 1.694 | $1.687_5$ | 34 | — |
| SP3 | 1.421 | 1.695 | $1.688_5$ | 35 | — |
| pKc | 1.421 | 1.695 | $1.688_5$ | 35 | — |
| SP$\alpha$ | 1.424 | 1.703 | $1.696_5$ | 43 | Defective phage |
| SPX | 1.424 | 1.703 | $1.696_5$ | 43 | Defective phage |
| SP50 | 1.424 | 1.704 | $1.697_5$ | 44 | — |
| $\phi$1 | 1.424 | 1.704 | $1.697_5$ | 44 | — |
| $\phi$2 | 1.424 | 1.704 | $1.697_5$ | 44 | — |
| $\phi$14 | 1.424 | 1.704 | $1.697_5$ | 44 | — |
| F | — | 1.709 | $1.702_5$ | 49 | — |
| SP10 | 1.439 | 1.715 to 1.723 | — | 44 | Density decreases upon storage |
| PBS1 | 1.433 | 1.722 | $1.715_5$ | 28 | Uracil replaces thymine |
| PBS2 | 1.433 | 1.722 | $1.715_5$ | 28 | Uracil replaces thymine |
| SP70 | 1.433 | 1.722 | $1.715_5$ | — | Possibly related to PBS1 and 2 |
| SP80 | 1.433 | 1.722 | $1.715_5$ | — | Possibly related to PBS1 and 2 |
| SP90 | 1.433 | 1.722 | $1.715_5$ | — | Possibly related to PBS1 and 2 |
| SP100 | 1.433 | 1.722 | $1.715_5$ | — | Possibly related to PBS1 and 2 |
| SP8 | 1.455 | 1.742 | $1.735_5$ | 43 | Hydroxymethyluracil replaces thymine |
| SP82 | 1.455 | 1.742 | $1.735_5$ | 43 | Hydroxymethyluracil replaces thymine |

## TABLE 2 (Continued)
## SELECTED VIRAL DNAs[a]

| Source of DNA (Host/Phage) | Buoyant Density, g/cm³ $Cs_2SO_4$ [b] | $CsCl$ [c] | $CsCl$ [d] | G + C, mole % | Comments |
|---|---|---|---|---|---|
| *Bacillus subtilus* phages (*cont.*) | | | | | |
| SPO-1 | 1.455 | 1.742 | $1.735_5$ | 43 | Hydroxymethyluracil replaces thymine |
| $\phi$e | 1.455 | 1.743 | $1.736_5$ | 43 | Hydroxymethyluracil replaces thymine |
| 2C | 1.455 | 1.742 | $1.735_5$ | 43 | Hydroxymethyluracil replaces thymine |
| $\phi$25 | 1.455 | 1.742 | $1.735_5$ | 43 | Hydroxymethyluracil replaces thymine |
| SP60 | 1.455 | 1.742 | $1.735_5$ | 43 | Hydroxymethyluracil replaces thymine |

[a] The buoyant densities of selected DNAs were determined at 25°C, 44,770 rpm (CsCl) or 31,410 rpm ($Cs_2SO_4$). The densities were abbreviated to the nearest 0.001 g/cm³ when only two or three determinations were performed or were measured with an accuracy of ±0.0002 g/cm³.
[b] Measured versus *Escherichia coli* DNA, density 1.4260 g/cm³.
[c] Measured versus *Escherichia coli* DNA, density 1.7100 g/cm³.
[d] Measured versus *Escherichia coli* DNA, density 1.7035 g/cm³.
[e] Plummer, Goodheart, Henson and Bowling, *Virology, 39,* 134 (1969). Other data were taken from Szybalski, in *Methods in Enzymology,* Vol. 12, p. 330, Grossman and Moldave, Eds., Academic Press, New York (1968).

## TABLE 3
## T-EVEN COLIPHAGE DNAs[a]

[Glucosylated 5-hydroxymethylcytosine (hmC) replaces cytosine]

| Source of DNA | Buoyant Density, g/cm³ $Cs_2SO_4$ [b] | $CsCl$ [c] | $CsCl$ [d] | G + hmC, mole % | Glucose, $\overline{\text{hmC}}$ mole % | Biose, $\overline{\text{hmC}}$ mole % |
|---|---|---|---|---|---|---|
| T2(o)S[g] | 1.4314 | 1.7060 | 1.6995 | 34 | (0) | — |
| T2gt[f] | — | 1.7057 | 1.6992 | 34 | (0) | — |
| T6(o)[e] | 1.4298 | 1.7052 | 1.6987 | 34 | 14 | 7 |
| T2 | 1.4392 | 1.7020 | 1.6955 | 34 | 80 | 5 |
| T2 × $\overline{\text{T2}}$ | 1.4403 | 1.7018 | 1.6953 | 34 | 85 | 1.5 |
| $\overline{\text{T2}}$ | 1.4430 | 1.7005 | 1.6940 | 34 | 100 | 0 |
| T4 | 1.4430 | 1.7005 | 1.6940 | 34 | 100 | 0 |
| C16 | 1.4478 | 1.7103 | 1.7038 | 34 | 130 | 65 |
| T6 | 1.4510 | 1.7105 | 1.7040 | 34 | 148 | 72 |

[a] Data taken from Erikson and Szybalski, *Virology, 22,* 111 (1964). Reproduced by permission of Academic Press, New York.
[b] Measured versus *Escherichia coli* DNA, density 1.4260 g/cm³, and poly d(A-T), density 1.4240 g/cm³.
[c] Measured versus *Escherichia coli.*
[d] Measured versus *Escherichia coli* DNA, density 1.7035 g/cm³.
[e] Grown in UDPG⁻ host.
[f] Glucose transferase-deficient mutant (Sheldrick, Unpublished Data).
[g] Enzymatically synthesized; glucose-free.

## TABLE 4
## VIRAL AND CELLULAR RNAs IN THE $Cs_2SO_4$ GRADIENT

| Source of RNA | Buoyant Density, $g/cm^3$ | | | References |
|---|---|---|---|---|
| | Single-Stranded | | Double-Stranded | |
| | – | + 1% HCHO | (RF) | |
| **Calf thymus synthetic** | 1.55[a] | — | — | *1* |
| **φx174 coliphage synthetic** | 1.590–1.601[e] | — | — | *2, 3* |
| **MS2 coliphage** | 1.607 | — | 1.609[d] | *4* |
| | 1.626 | — | — | *5, 6* |
| | 1.630 | — | — | *7* |
| **R17 coliphage** | 1.621 | 1.616 | 1.607[b] | *8, 9* |
| | 1.630 | — | 1.606[c] | *10* |
| **fr coliphage** | 1.634 | — | 1.609 | *11* |
| **M12 coliphage** | 1.634 | — | 1.614 | *12* |
| **Polio virus** | 1.63–1.65 | — | 1.60 | *13, 14* |
| | — | — | 1.58 | *15, 9* |
| | — | — | 1.65 | *16* |
| **Wound tumor virus** | — | — | 1.599 | *17* |
| **Reovirus** | — | — | 1.61 | *20* |
| **EMC virus** | 1.63 | — | 1.57 | *21* |
| | 1.69 | — | 1.635 | *22* |
| ***Neurospora crassa*,** ribosomal | 1.637 (ppt)[f] | 1.625 | — | *17* |
| **Brome grass mosaic virus** | 1.631 (ppt) | 1.616 | — | *17* |
| **TMV** (tobacco mosaic virus) | 1.635 (ppt) | 1.627 | — | *8* |
| | 1.640 (ppt) | — | 1.601 | *5* |
| | 1.675 (ppt) | — | 1.620–1.628 | *18* |
| **TMV** (fl⁵U-labeled) | — | 1.652 | — | *8* |
| **TYMV** (turnip yellow mosaic virus) | 1.642–1.65 (ppt) | — | 1.635–1.643 | *18, 19* |
| ***Bacillus subtilis*,** 5s | 1.643 | 1.636 | — | *8* |
| ***Bacillus subtilis*,** 18s | 1.649 | 1.634 | — | *8* |
| ***Bacillus subtilis*,** 23s | 1.653 | 1.637 | — | *8* |
| ***Escherichia coli*,** ribosomal | 1.663 (ppt) | — | — | *10* |
| **Newcastle disease virus** | 1.68 | — | — | *24* |

[a] Calf thymus native DNA, 1425 $g/cm^3$; denatured DNA, 1.451 $g/cm^3$; DNA·RNA synthetic hybrid, 1.490 $g/cm^3$.[1]
[b] + 1% HCHO, 1.606 $g/cm^3$.[17]
[c] R.I. (replicative intermediate), 1.616 $g/cm^3$.[9,10]
[d] In CsCl, 1.868 $g/cm^3$ (51.9% G+C).[23]
[e] φX174 DNA, 1.452 $g/cm^3$; DNA·RNA synthetic hybrid, 1.491 – 1.510 $g/cm^3$.[2,3]
[f] (ppt) indicates formation of an RNA precipitate.[8]

## TABLE 5
## SYNTHETIC POLYNUCLEOTIDES

| Polynucleotide[a] | Buoyant Density, $g/cm^3$ | | | | |
|---|---|---|---|---|---|
| | $Cs_2SO_4$[b] | $Cs_2SO_4$ Alkaline | CsCl[c] | CsCl[d] | CsCl Alkaline |
| rA | 1.570 (*1*) | — | — | — | — |
| rG | 1.693 (*2*), 1.75 (*3*) | — | — | — | — |
| rU | 1.650 (*1*) | — | — | — | — |
| rC | 1.583 (*2*), 1.59 (*3*) 1.63 (*4*) | — | 1.88 (*5*) | — | — |
| dA | 1.379 (*1*) | 1.379 (*1*) | 1.685 (*2*) | 1.628 (*6*) | 1.622 (*7*), 1.623 (*6*) |
| dI | — | 1.45–1.46 (*8*) | — | — | — |
| dG | 1.539 (*2*) | 1.54 (*8*) | — | 1.763 (*6*) | 1.791 (*6*) |
| dT | 1.424 (*1*) | 1.456 (*1*) | — | 1.739 (*6*) | 1.771 (*7*), 1.774 (*6*) |
| dC | 1.42 (*4*) | 1.40–1.41 (*8*) | — | 1.685 (*6*) | 1.722 (*6*) |
| d(br⁵C) | — | 1.73 (*8*) | — | — | — |

## TABLE 5 (Continued)
## SYNTHETIC POLYNUCLEOTIDES

| Polynucleotide[a] | Buoyant Density, g/cm$^3$ | | | | |
|---|---|---|---|---|---|
| | $Cs_2SO_4$[b] | $Cs_2SO_4$ Alkaline | CsCl[c] | CsCl[d] | CsCl Alkaline |
| rA·rU | 1,660 (*1*) | — | — | — | — |
| rA·2rU | 1.702 (*1*) | — | — | — | — |
| rI·rC | 1.62 (*4*) | — | — | — | — |
| rG·rC | 1.66 (*4*), 1.685 (*3*) | — | — | — | — |
| dA·dT | 1.417 (*2*), 1.419 (*9*), 1.432 (*1*) | — | 1.647 (*2*) | $1.637_5$ (*7*) | — |
| dA·2dT | 1.492 (*1*) | — | — | — | — |
| dI·dC | 1.48 (*8*), 1.50 (*4*) | — | — | — | — |
| dG·dC | 1.46 (*8*), 1.467 (*10*) 1.49 (*4*) | — | 1.794 (*10*) 1.796 (*5, 11*) | — | — |
| rA·dT | 1.433 (*1*) | — | — | — | — |
| rA·2dT | 1.519 (*1*) | — | — | — | — |
| (rA·dT)·rU | 1.582 (*1*) | — | — | — | — |
| (rA·rU)·dT | 1.584 (*1*) | — | — | — | — |
| (dA·dT)·rU | 1.525 (*9*), 1.536 (*1*) | — | — | — | — |
| dA·2rU | 1.620 (*1*) | — | — | — | — |
| rI·dC | 1.54 (*4*) | — | — | — | — |
| 2dI·rC | 1.68 (*4*) | — | — | — | — |
| rG·dC | 1.58 (*4*) | — | — | — | — |
| dG·rC | 1.57 (*4*) | — | 1.86 (*5*) | — | — |
| r(A-U)·r(A-U) | 1.614 (*12*) | — | — | — | — |
| r(A-br$^5$U)·r(A-br$^5$U) | 1.695 (*12*) | — | — | — | — |
| d(A-U)·d(A-U) | 1.439 (*17*) | — | 1.718 (*17*) | — | — |
| d(A-hm$^5$U)·d(A-hm$^5$U) | 1.473 (*17*) | — | 1.734 (*17*) | — | — |
| d(A-br$^5$U)·d(A-br$^5$U) | 1.540 (*4, 12*) 1.550 (*16*) | 1.54–1.56 (*8*) | — | — | — |
| d(A-T)·d(A-T) | 1.424 (*6, 10, 13, 17*) 1.426 (*1, 4, 12*) | 1.40 (*8*), 1.416 (*6, 13*) | 1.678 (*10, 17*) 1.679 (*11*) | 1.672 (*6, 7, 9, 13*) | 1.722 (*6, 7, 13*) |
| d(n$^2$A-T)·d(n$^2$A-T) | — | — | 1.718 (*14*) | — | — |
| d(I-C)·d(I-C) | 1.453 (*6, 13*) | 1.436 (*6, 13*) | — | 1.735 (*6, 13*) | 1.766 (*6, 13*) |
| d(G-C)·d(G-C) | 1.448 (*15*) | 1.464 (*6*) | — | 1.741 (*6*) | 1.793 (*6*) |
| d(A-C) | — | — | — | 1.689 (*6, 7*) | 1.684 (*7*), 1.685 (*6*) |
| d(T-C) | 1.460 (*9*) | — | — | — | — |
| d(T-G) | — | — | — | 1.777 (*6, 7*) | 1.826 (*6*), 1.828 (*7*) |
| d(T-C)·d(G-A) | 1.428 (*15*), 1.439 (*2*) 1.466 (*9*) | — | 1.715 (*2*) | 1.711 (*7*) | — |
| d(T-G)·d(C-A) | 1.420 (*2*), 1.422 (*15*) 1.423 (*9*) | — | 1.697 (*2*) | $1.690_5$ (*7*) | — |
| d(T-T-C)·d(G-A-A) | 1.427 (*15*) | — | — | $1.685_5$ (*7*) | — |
| d(T-T-G)·d(C-A-A) | 1.422 (*15*) | — | — | 1.683 (*7*) | — |
| d(T-A-C)·d(G-T-A) | 1.422 (*15*) | — | — | 1.713 (*7*) | — |
| d(A-T-C)·d(G-A-T) | 1.418 (*15*) | — | — | 1.687 (*7*) | — |
| (rA-dU)·(rA-dU) | 1.500 (*12*) | — | — | — | — |
| d(T-C)·d(G-A)·r(U-CH$^+$) | 1.520 (*9*) | — | — | — | — |

[a] The abbreviations for the polynucleotides conform to those proposed by the IUPAC-IUB Commission on Biochemical Nomenclature.

[b] Measured versus *Escherichia coli* DNA, density 1.4260 g/cm$^3$, or d(A-T)•d(A-T), density 1.4240 g/cm$^3$.

[c] Measured versus *Escherichia coli* DNA, density 1.710 g/cm$^3$, or d(A-T)•d(A-T), density 1.678 g/cm$^3$.

[d] Measured versus d(A-T)•d(A-T), density 1.672 g/cm$^3$.

[e] At pH 10.1, 1.46–1.47 g/cm$^3$; at pH 9.4, 1.48 g/cm$^3$.[8]

[f] At pH 11.6, 1.45 g/cm$^3$; at pH 10.9, 1.46 g/cm$^3$.[8]

## REFERENCES


### Viral and Cellular RNAs in the $Cs_2SO_4$ Gradient

1. Warner, Samuels, Abbott and Krakow, *Proc. Nat. Acad. Sci. U.S.A, 49,* 533 (1963).
2. Bassel, Hayashi and Spiegelman, *Proc. Nat. Acad. Sci. U.S.A., 52,* 796 (1964).
3. Sinsheimer and Lawrence, *J. Mol. Biol., 8,* 289 (1964).
4. Shimura, Moses and Nathans, *J. Mol. Biol., 12,* 266 (1965).
5. Burdon, Billeter, Weissmann, Warner, Ochoa and Knight, *Proc. Nat. Acad. Sci. U.S.A., 52,* 768 (1964).
6. Billeter, Weissmann and Warner, *J. Mol. Biol., 17,* 145 (1966).
7. Doi and Spiegelman, *Proc. Nat. Acad. Sci. U.S.A., 49,* 353 (1963).
8. Lozeron and Szybalski, *Biochem. Biophys. Res. Commun., 23,* 612 (1966).
9. Erikson and Franklin *Bacteriol. Rev., 30,* 267 (1966).
10. Erikson, *J. Mol. Biol., 18,* 372 (1966).
11. Kaerner and Hoffmann-Berling, *Z. Naturforsch. Teil B, 19,* 593 (1964).
12. Ammann, Delius and Hofschneider, *J. Mol. Biol., 10,* 557 (1964).
13. Engler and Tolbert, *Virology, 26,* 246 (1965).
14. Bishop, Summers and Levintow, *Proc. Nat. Acad. Sci. U.S.A., 54,* 1273 (1965).
15. Pons, *Virology, 24,* 467 (1964).
16. Baltimore, Becker and Darnell, Jr., *Science, 143,* 1034 (1964).
17. Szybalski, in *Methods in Enzymology,* Vol 12, p. 330, Moldave and Grossman, Eds. Academic Press, New York (1968).
18. Ralph, Matthews, Matus and Mandel, *J. Mol. Biol., 11,* 202 (1965).
19. Mandel, Matthew, Matus and Ralph, *Biochem. Biophys. Res. Commun., 16,* 604 (1964).
20. Shatkin, *Proc. Nat. Acad. Sci. U.S.A., 54,* 1721 (1965).
21. Montagnier and Sanders, *Nature, 199,* 664 (1963).
22. Dalgarno, Martin, Liu and Work, *J. Mol. Biol., 15,* 77 (1966).
23. Kelly, Gould and Sinsheimer, *J. Mol. Biol., 11,* 562 (1965).
24. Kingsbury, *J. Mol. Biol., 18,* 195 (1966).

### Synthetic Polynucleotides

1. Riley, Maling and Chamberlin, *J. Mol. Biol., 20,* 359 (1966).
2. Szybalski, in *Methods in Enzymology,* Vol. 12, p. 330, Grossman and Moldave, Eds. Academic Press, New York (1968).
3. Haselkorn and Fox, *J. Mol. Biol., 13,* 780 (1965).
4. Chamberlin, *Fed. Proc., 24,* 1446 (1965).
5. Schildkraut, Marmur, Fresco and Doty, *J. Biol. Chem., 236,* PC3 (1961).
6. Grant, Shortle and Cantor, *J. Mol. Biol., 54,* 465 (1970).
7. Wells and Blair, *J. Mol. Biol., 27,* 273 (1967).
8. Inman and Baldwin, *J. Mol. Biol., 8,* 452 (1964).
9. Morgan and Wells, *J. Mol. Biol., 37,* 63 (1968).
10. Erikson and Szybalski, *Virology, 22,* 111 (1964).
11. Schildkraut, Marmur and Doty, *J. Mol. Biol., 4,* 430 (1962).
12. Chamberlin, in *Procedures in Nucleic Acid Research,* p. 513, Cantoni and Davies, Eds. Harper and Row, New York (1966).
13. Grant, Harwood and Wells, *J. Amer. Chem. Soc., 90,* 4474 (1968).
14. Cerami, Ward, Reich and Goldberg, *Abstr. Second Int. Biophys. Congr. Vienna,* No. 150 (1966).
15. Wells and Larson, *J. Mol. Biol., 49,* 319 (1970).
16. Wells, Personal Communication.
17. Cassidy, Kahan and Doty, Unpublished Data.

# MELTING TEMPERATURES ($T_m$) OF SYNTHETIC POLYNUCLEOTIDES

DR. E. H. SZYBALSKI and DR. W. SZYBALSKI*

| | $T_m$, °C | | | | | | | |
|---|---|---|---|---|---|---|---|---|
| | Na$^+$ of Molarity | | | | | | | |
| Polynucleotide[a] | 0.01 | 0.1 | 0.15 | 1.0 | Other Solvents[b] | pH[c] | Comment | References |
| **Ribohomopolymers** | | | | | | | | |
| rA | ~45 | ~45 | — | ~45 | — | 7–7.5 | Broad melting curve. Single-stranded, stacked-base conformation at neutral or alkaline pH | *1–9* |
| rA·rA | — | — | >100 | — | 95 | 4 | 0.2 $M$ Na$^+$ | *10–13* |
| | — | — | 85 | 60 | 77 | 4.5 | 0.5 $M$ Na$^+$ | *2, 10, 11, 89* |
| | 80 | 65 | 63 | 40 | 47 | 5 | 0.5 $M$ Na$^+$ | *1, 10, 11, 89* |
| | 60 | 43 | 41 | 20 | 80 | 5.5 | 1 m$M$ Na$^+$ | *3, 10, 11, 89, 90* |
| | 40 | 20 | 19 | — | 59 | 6 | 1 m$M$ Na$^+$ | *10, 11, 89* |
| r(m$^6$A) | — | — | — | — | — | 4–4.8 | No double-helical "acid" form | *13, 14* |
| r(m$^1$A)·r(m$^1$A) | — | — | 85 | — | — | 4 | — | *13* |
| r(he$^6$A) | — | — | — | — | — | 4 | No double-helical "acid" form | *2, 11* |
| r(Am)·r(Am) | — | — | — | — | 59.5 | 5.4 | 0.15 $M$ K$^+$ | *79* |
| rÂ | — | — | — | — | — | — | No double-helical "acid" form | *11, 15* |
| rF | — | — | — | — | — | 7 | Resembles rA at neutral pH | *80* |
| rI·rI·rI | — | — | — | 43 | 35 | 6–7 | 0.5 $M$ Na$^+$. No secondary structure below 0.5 $M$ Na$^+$ | *16–18* |
| r(m$^7$I) | — | — | — | — | — | — | No secondary helical structure | *11, 13* |
| r(m$^1$m$^7$I) | — | — | — | — | — | — | No secondary helical structure | *13* |
| xrG | — | — | — | — | >100 | 7 | 2 m$M$ Na$^+$ | *19, 20* |
| | — | — | — | — | 65 | 6 | $H_2O$ | *20* |
| | — | — | — | — | >100 | 4 | 0.06 $M$ Na$^+$ + 50% ethylene glycol | *21* |
| xr(m$^1$G) | — | — | — | — | 82.5 | 4 | 0.06 $M$ Na$^+$ + 50% ethylene glycol | *21* |
| xr(m$^2$G) | — | — | 70 | — | — | 7 | — | *92* |
| xr(m$^2$m$^2$G) | — | — | 47.5 | — | 32 | 7 | 1.5 m$M$ Na$^+$ | *92* |
| xr(m$^7$G) | — | — | >100 | — | — | 7 | — | *13* |
| r(m$^2$m$^2$m$^7$G) | — | — | — | — | — | 7 | No secondary helical structure | *92* |
| xrX | — | — | 33.4 | — | 25.3 | 7 | 15 m$M$ Na$^+$ | *22, 23* |
| | — | — | 40.6 | — | — | 7 | +0.01 $M$ Mg$^{2+}$ | *22* |
| | — | — | 44.5 | — | — | 5 | — | *22* |
| | — | 23 | — | 35 | 24 | 7.8 | 0.05 $M$ Na$^+$ | *83* |
| | — | 37.5 | — | 56 | 35 | 5.7 | 0.05 $M$ Na$^+$ | *83* |
| rU·rU | — | — | −2.8 | 5.2 | 2.2 | 7 | 0.3 $M$ Na$^+$ | *24* |
| | — | — | 4.9 | — | — | 7 | +0.01 $M$ Mg$^{2+}$ | *23–25* |
| | — | — | — | — | 5.7 | 7 | 0.01 $M$ Mg$^{2+}$ | *26* |
| | — | — | — | — | 8.2 | 7 | 0.1 $M$ Mg$^{2+}$ | *26* |
| | — | — | — | — | 28 | 7 | 10 $\mu M$ spermine | *26* |
| | — | — | — | — | 6.6 | 7 | 0.1 $M$ Mg$^{2+}$ + 0.05 $M$ Na$^+$ | *27* |
| r(io$^5$U)·r(io$^5$U) | — | — | — | — | 21.5 | 7 | 0.1 $M$ Mg$^{2+}$ + 0.05 $M$ Na$^+$ | *27* |
| r(br$^5$U)·r(br$^5$U) | — | — | — | — | 9.4 | 7 | 0.1 $M$ Mg$^{2+}$ + 0.05 $M$ Na$^+$ | *27* |
| r(cl$^5$U)·r(cl$^5$U) | — | — | — | — | <1 | 7 | 0.1 $M$ Mg$^{2+}$ + 0.05 $M$ Na$^+$ | *27* |
| r(fl$^5$U)·r(fl$^5$U) | — | — | — | — | <1 | 7 | 0.1 $M$ Mg$^{2+}$ + 0.05 $M$ Na$^+$ | *27* |
| r(m$^3$U) | — | — | — | — | — | — | No secondary helical structure | *27* |

*Data previously published in *Handbook of Biochemistry,* 2nd ed., pp. H-18–H-23, H. A. Sober, Ed. Copyright 1970, The Chemical Rubber Co., Cleveland, Ohio.

| Polynucleotide[a] | $T_m$, °C — Na$^+$ of Molarity 0.01 | 0.1 | 0.15 | 1.0 | Other Solvents[b] | pH[c] | Comment | Reference |
|---|---|---|---|---|---|---|---|---|
| **Ribohomopolymers—(Continued)** | | | | | | | | |
| r(m$^3$br$^5$U) | — | — | — | — | — | — | No secondary helical structure | *27* |
| r(hm$^5$U) | — | — | — | — | — | — | No secondary helical structure | *28* |
| r(e$^5$U)·r(e$^5$U) | — | — | — | — | −2 | — | 0.01 *M* Mg$^{2+}$ or 0.6 *M* Na$^+$ | *29* |
| rΨ·rΨ | — | — | 60 | — | — | 7 | +0.1 *M* Mg$^{2+}$ | *30* |
| rT·rT | — | 24 | 27.3 | 40 | 35.5 | 7 | 0.01 *M* Mg$^{2+}$ | *26–28, 31* |
| | — | — | — | — | 38 | 7 | 0.1 *M* Mg$^{2+}$ | *26* |
| | — | — | — | — | 51 | 7 | 50 $\mu M$ spermine | *26* |
| | — | — | 40 | — | — | 7 | +0.1 *M* Mg$^{2+}$ | *30* |
| r(m$^3$T) | — | — | — | — | — | 7 | No secondary helical structure | *28* |
| rC | — | ~50 | — | — | — | 7–8 | Broad melting curve. Single-stranded, stacked-base conformation at neutral or alkaline pH | *32–36* |
| rC·rC | — | 73 | — | — | 79 | 3.65 | 0.3 *M* Na$^+$ | *37* |
| | 62.5 | 77 | 79.5 | 82 | 78 | 4 | 0.04 *M* Na$^+$ | *32, 37–39, 85* |
| | — | 82 | — | — | — | 4.2 | — | *40, 41* |
| | — | — | 78 | 72 | 80.5 | 4.5 | 0.04 *M* Na$^+$ | *38, 39, 85* |
| | 75.5 | 69 | 63.3 | 54 | — | 5.0 | — | *36, 38, 39* |
| | | 54 | — | — | — | 5.4 | — | *36* |
| r(br$^5$C)·r(br$^5$C) | — | — | >82 | — | — | 3 | — | *38* |
| | — | — | 65.2 | — | — | 3.5 | — | *38* |
| | — | — | 45.6 | 37.5 | 57.3 | 4 | 0.03 *M* Na$^+$ | *38* |
| r(io$^5$C)·r(io$^5$C) | — | — | 52.8; 90.7 | — | — | 3.5 | 2 structured forms at acid pH | *38* |
| | — | — | 49.5; 76.5 | — | 61.2 | 4 | 0.9 *M* Na$^+$ | *38* |
| | — | — | 41.3; 54.1 | — | — | 4.5 | — | *38* |
| r(m$^5$C)·r(m$^5$C) | — | 79 | — | — | — | 4.2 | — | *40* |
| r(m$^4$C) | — | — | — | — | — | 7–8 | Behaves like rC | *33, 36* |
| | — | — | — | — | — | 4–5.8 | No double-helical "acid" form | *33, 36* |
| r(m$^5$C) | — | — | — | — | — | 7.8 | Broad melting curve. Analogy with rC assumed | *40, 41* |
| r(m$^4$m$^5$C) | — | — | — | — | — | 7.5 | Broad melting curve. Analogy with rC assumed | *41, 85* |
| | — | — | — | — | — | 4 | No double-helical "acid" form | *41* |
| **Ribohomopolymer complexes (purine·purine)** | | | | | | | | |
| rA·rI | — | — | — | — | — | — | No data. Exists at 0.01 *M* Na$^+$ | *11* |
| rA·2rI | — | 39–40.3 | 43.5–44.8 | 57 | 30.2 | 6–7.4 | 0.02 *M* Na$^+$ | *11, 15, 16, 18, 22, 42, 43* |
| | — | — | 53.7 | — | — | 7 | +0.01 *M* Mg$^{2+}$ | *22* |
| rA·2rX | — | — | 59.3; 86.5 | — | — | 7 | — | *22* |
| | — | — | 62.4; 77 | — | — | 7 | +0.01 *M* Mg$^{2+}$ | *22* |
| | — | — | >100 | — | — | 6 | — | *22* |
| | — | 80 | — | — | 84 | 7 | 0.05 *M* Na$^+$ | *83* |
| r(m'A)·rX | — | — | 66.9 | — | — | 7 | — | *22* |
| r(m$^6$A)·rX | — | — | 66.1 | — | — | 7 | — | *22* |
| r(he$^6$A)·rX | — | — | 78.2 | — | — | 7 | — | *22* |
| rÂ·rI | — | — | >100 | — | 89.4 | 7 | 0.02 *M* Na$^+$ | *15, 23* |
| rÂ·2rX | — | — | 75.5 | — | — | 7 | — | *22* |
| 2rI·rX | — | 37 | 38 | — | 51 | 7.2 | 0.7 *M* Na$^+$ | *83* |
| | — | — | 40.2 | — | — | 7 | — | *22* |
| | — | — | 42.5 | — | — | 7 | +0.01 *M* Mg$^{2+}$ | *22* |

| Polynucleotide[a] | $T_m$, °C Na$^+$ of Molarity 0.01 | 0.1 | 0.15 | 1.0 | Other Solvents[b] | pH[c] | Comment | References |
|---|---|---|---|---|---|---|---|---|
| **Ribohomopolymer complexes (purine·pyrimidine)** | | | | | | | | |
| rA·rU | 37 | 56.5 | — | — | 51 | 7 | 0.05 $M$ Na$^+$. Converted to triple-stranded form at molarities above 0.1 | *11, 34, 44–46* |
| | 38–39 | 56–57.5 | — | — | 46.5 | 6.2–7.8 | 25 m$M$ Na$^+$ | *26, 43, 47–50, 86* |
| | — | 56.1 | — | — | 43 | 7 | 1/10 SSC[d] | *8, 51, 52* |
| | — | 67.2 | — | — | — | 7 | +1 m$M$ Mg$^{2+}$ | *51* |
| rA·2rU | 18; 37 | 52.5; 56.5 | — | 80 | 72 | 7 | 0.5 $M$ Na$^+$ | *8, 11, 44–46, 51, 53* |
| | — | 67.7 | — | — | — | 7 | +1 m$M$ Mg$^{2+}$ | *51* |
| rA·2rU—(Continued) | 21; 38 | 51; 57 | 59 | 75 | 27; 41 | 7–7.8 | 17 m$M$ Na$^+$ | *23, 27, 30, 48, 50* |
| rA·r(br$^5$U) | — | 63 | — | — | 69 | 7 | 1/10 SSC[d] | *52, 86* |
| rA·2r(br$^5$U) | — | 92 | 87 | — | — | 7 | +0.02 $M$ Na$_3$ citrate (SSC)[d] | *52, 86* |
| | — | 93 | 95.1 | — | 82 | 7 | 17 m$M$ Na$^+$ | *11, 23, 27, 30* |
| rA·2r(io$^5$U) | — | 90 | 94.8 | — | 81 | 7 | 17 m$M$ Na$^+$ | *11, 23, 27, 30* |
| rA·2r(cl$^5$U) | — | — | 82 | — | — | 7 | +0.02 $M$ Na$_3$ citrate (SSC)[d] | *52* |
| | — | 84 | 86.1 | — | 72 | 7 | 17 m$M$ Na$^+$ | *11, 23, 27, 30* |
| rA·r(fl$^5$U) | — | — | 55 | — | — | 7 | +0.02 $M$ Na citrate (SSC)[d] | *52* |
| | | | | | 40 | 7 | 1/10 SSC[d] | *52* |
| rA·2r(fl$^5$U) | — | — | 49.6 | — | — | 7 | — | *23, 27, 30* |
| rA·2r(ho$^5$U) | — | — | 36; 53 | — | — | 7 | — | *27, 30* |
| rA·r(hm$^5$U) | — | — | 30.5 | — | — | 7 | +0.02 $M$ Na$_3$ citrate (SSC)[d] | *28* |
| rA·2r(e$^5$U) | 47 | 60 | — | 80 | — | 7 | | *29* |
| r(n$^2$A)·rU | — | 87 | — | — | 76 | 7.5 | 0.02 $M$ Na$^+$ | *8, 54* |
| r(n$^2$A)·2rU | — | 53; 87 | — | — | 33; 76 | 7.5 | 0.02 $M$ Na$^+$ | *8* |
| r(m$^6$A)·rU | — | — | — | — | 15 | 7.4 | 0.088 $M$ Na$^+$ | *11, 14* |
| rA·2rΨ | — | — | 58; 82 | — | — | 7 | — | *27, 30* |
| rÂ·rU | — | — | > 100 | — | 88.4 | 7 | 0.02 $M$ Na$^+$ | *15, 23* |
| rA·2rT | 59.7 | 78.2 | 80.5 | — | — | — | — | *55* |
| | — | — | 69.8; 78.3 | — | — | 7 | — | *27, 30* |
| rF·rU | — | — | — | — | 22 | 7.9 | 0.1 $M$ K$^+$ + 0.01 $M$ Tris | *80* |
| rI·rC | 41.5 | 60.2 | — | 75.3 | — | 7.8 | — | *17, 47, 86* |
| | 47 | 62 | 64.5 | — | 56.5 | 7.8 | 0.05 $M$ Na$^+$ | *40, 41, 60* |
| | — | 59 | 61.3–63 | — | 48.3 | 7 | 0.025 $M$ Na$^+$ | *13, 23, 27, 38, 43, 49* |
| | — | — | — | — | 63 | 3 | 0.05 $M$ Na$^+$ | *56* |
| rI·r(io$^5$C) | — | — | 91.2 | — | — | 7 | — | *38* |
| rI·r(br$^5$C) | 66 | 82.8 | 89.2 | — | 90.8 | 7 | 0.4 $M$ Na$^+$ | *13, 23, 27, 38, 86* |
| rI·r(m$^5$C) | 64 | 78 | 80.5 | — | 73.5 | 7.8 | 0.05 $M$ Na$^+$ | *40, 41, 60* |
| r(m$^7$I)·rC | — | — | 39 | — | — | 7 | — | *13* |
| r(m$^7$I)·r(br$^5$C) | — | — | 55.5 | — | — | 7 | — | *13* |
| rG·rC | — | — | — | — | 90 | 7 | 1.5 m$M$ Na$^+$ + 80% methanol | *11, 13, 20* |
| | — | — | — | — | 97 | 7.8 | 1 m$M$ Na$^+$ + 0.1 m$M$ EDTA | *47* |
| | — | — | 78.2 | — | — | 2.5 | — | *11, 82* |
| rG·r(br$^5$C) | — | — | — | — | > 100 | 7 | 1.5 m$M$ Na$^+$ + 80% methanol | *20* |
| | — | — | — | — | > 100 | 2.5 | 15 m$M$ Na$^+$ | *82* |
| rG·r(io$^5$C) | — | — | — | — | > 100 | 2.5 | 15 m$M$ Na$^+$ | *82* |
| r(m$^7$G)·rC | — | — | — | — | 69 | 7 | 1.5 m$M$ Na$^+$ + 80% methanol | *13* |
| r(m$^7$G)·r(br$^5$C) | — | — | — | — | 90 | 7 | 1.5 m$M$ Na$^+$ + 80% methanol | *13* |
| rX·rU | — | — | 49.9 | — | — | 7 | — | *22* |
| | — | — | 54.2 | — | — | 7 | + 0.01 $M$ Mg$^{2+}$ | *22* |
| | — | 46.5 | 48 | — | 43 | 7.8 | 0.05 $M$ Na$^+$ | *83* |
| rX·r(br$^5$U) | — | — | 65.2 | — | — | 7 | — | *22* |
| rX·r(fl$^5$U) | — | 35.5 | 37 | — | 32.5 | 7.8 | 0.05 $M$ Na$^+$ | *83* |

| Polynucleotide[a] | Tm, °C — Na⁺ of Molarity 0.01 | 0.1 | 0.15 | 1.0 | Other Solvents[b] | pH[c] | Comment | References |
|---|---|---|---|---|---|---|---|---|
| **Ribohomopolymer complexes (purine·pyrimidine)—(Continued)** | | | | | | | | |
| rX·r(e⁵U) | — | 46 | 47.5 | — | 43 | 7.8 | 0.05 *M* Na⁺ | *83* |
| rX·rT | — | 57 | 60.5 | — | 54.5 | 7.8 | 0.05 *M* Na⁺ | *83* |
| **Deoxyribohomopolymers** | | | | | | | | |
| dA | ~50 | ~50 | — | ~50 | — | 7–7.5 | Broad melting curve. Analogy with rA assumed | *7, 34, 48, 55* |
| dA·dA[c] | — | ~65 | — | — | — | 4.3 | — | *34, 48, 55, 90* |
| xdI | — | 17 | — | 48 | 38 | — | 0.5 *M* Na⁺ | *17, 57, 58* |
| xdG | — | — | >90 | — | >90 | 7.5–8 | Distilled $H_2O$ | *59* |
| d(ac²G) | — | — | — | — | — | 7.5–8 | No secondary helical structure | *59* |
| dU | — | — | — | — | — | — | No secondary helical structure | *60* |
| dT | — | — | — | — | — | — | No secondary helical structure | *34, 48, 94* |
| dC | — | — | — | — | ~50 | 7.5 | Broad melting curve; SSC[d] | *91* |
| dC·dC | — | 97 | — | — | 87–91 | 5 | 0.4 *M* Na⁺ | *57, 60, 91* |
| | — | 74 | — | — | 66–68 | 6 | 0.4 *M* Na⁺ | *57, 60* |
| | — | 48 | — | — | 32–37 | 7 | 0.4 *M* Na⁺ | *57, 60* |
| d(br⁵C)·d(br⁵C) | — | — | — | — | 33 | 4.7 | 0.4 *M* Na⁺ | *57* |
| d(m⁵C)·d(m⁵C) | — | — | — | — | — | 5–7 | Behaves like dC·dC | *60* |
| d(m⁴m⁵C) | — | — | — | — | — | 5–8.8 | No secondary helical structure | *60* |
| d(e⁴e⁵C) | — | — | — | — | — | 5–8.8 | No secondary helical structure | *60* |
| **Deoxyribohomopolymer complexes (purine·pyrimidine)** | | | | | | | | |
| dA·dT | 48 | 68.5 | — | 87 | 95 | 7.8 | 2 *M* Na⁺; + $10^{-4}$ *M* EDTA | *42, 47* |
| | — | — | 69.5–71.5 | — | 51 | — | SSC; 1/10 SSC[d] | *61* |
| | — | — | — | — | 61.5 | 7 | 0.05 *M* Na⁺ | *34* |
| | 51 | 66 | 75 | — | — | 7.2 | — | *84* |
| | 47.6 | 66.6 | — | 84.9 | — | — | — | *88* |
| dA·2dT | — | 22; 68 | — | 86 | — | 7.8 | — | *48* |
| dA·dU | 35.2 | 55.3 | — | — | 73.9 | — | — | *88* |
| | — | 52 | — | 73 | 43 | 7.8 | 38 m*M* Na⁺ | *93* |
| dA·d(hm⁵U) | 12.0 | 32.3 | — | — | 54.0 | — | — | *88* |
| dI·dC | 27 | 43 | — | 61 | 58 | 6.5–8 | 0.56 *M* Na⁺ | *57, 58, 62* |
| | 27.5 | 46.1–47 | — | 55. | 52 | 7.8 | 1.23 *M* Na⁺ | *17, 47, 60* |
| | — | — | 47.5 | — | 33 | — | SSC; 1/10 SSC[d] | *61* |
| 2dI·dC | — | — | — | 58 | 45; 58 | — | 0.56 *M* Na⁺ | *58* |
| dI·d(br⁵C) | 53 | 72 | — | 92 | 85 | 7.4–8.4 | 0.56 *M* Na⁺ | *57, 58, 62* |
| 2dI·d(br⁵C) | — | 17; 71.5 | — | — | 85 | — | 0.56 *M* Na⁺ | *58* |
| dI·d(m⁵C) | — | 62 | 63 | — | 53 | 7.8 | 0.03 *M* Na⁺ | *60* |
| 2dI·d(m⁵C) | 57; 62 | — | — | 79 | — | 7.8 | — | *60* |
| dG·dC | 82 | >100 | — | — | 64 | 6.4–7.8 | 1 m*M* Na⁺; + $10^{-4}$ *M* EDTA | *47, 62* |
| | 81–83 | — | 106 | — | — | 7–8 | SSC[d], + $10^{-3}$ – $10^{-4}$ EDTA | *59, 63, 64* |
| | 83 | — | — | — | 66 | 6.5 | 2 m*M* Na⁺ | *65* |
| dG·d(br⁵C) | — | — | — | — | >100 | — | 2 m*M* Na⁺ | *65* |
| **Ribohomopolymer·deoxyribohomopolymer hybrids** | | | | | | | | |
| dA·2rX | — | — | — | — | 54 | 7 | 0.2 *M* Na⁺ | *83* |
| rA·dU | — | 55 | — | 68 | 47 | 7.8 | 38 m*M* Na⁺ | *93* |
| rA·dT | 46 | 64.1 | — | 77 | 59 | 7–7.8 | 0.05 *M* Na⁺ | *34, 47, 48* |
| rA·2dT | — | — | — | — | 78 | 7.8 | 3.8 *M* Na⁺ | *48* |
| (rA·dT)·rU | — | — | — | — | 85 | 7.8 | 0.5 *M* Na⁺ | *48* |
| dA·2rU | 15 | 46 | — | 77 | — | 7.8 | — | *48, 55* |
| dA·2r(e⁵U) | 31·5 | 48.2 | 52 | — | — | 7 | — | *55* |
| dA·2rT | 53.2 | 69.2 | 73.2 | — | — | 7 | — | *55* |
| (dA·dT)·rU | — | 28; 68 | — | 73; 84 | 43: 70 | 7.8 | 0.2 *M* Na⁺ | *48, 66* |
| rF·dT | — | — | — | — | 37 | 7.9 | 0.1 *M* K⁺ + 0.01 *M* Tris | *80* |
| rI·dC | 34·8 | 52.3 | — | 64.3 | — | 7.8 | — | *17, 47* |
| dI·rC | 10.1 | 35.4 | — | 52.6 | 49.3 | 7.8 | 0.6 *M* Na⁺ | *17, 47* |
| 2dI·rC | — | — | — | — | 50.5 | 7.8 | 0.8 *M* Na⁺ | *17* |
| rG·dC | — | — | — | — | 90 | 7.8 | 1 m*M* Na⁺ + 0.1 m*M* EDTA | *47* |

| Polynucleotide[a] | $T_m$, °C Na⁺ of Molarity 0.01 | 0.1 | 0.15 | 1.0 | Other Solvents[b] | pH[c] | Comment | References |
|---|---|---|---|---|---|---|---|---|
| **Ribohomopolymer deoxyribohomopolymer hybrids (Continued)** | | | | | | | | |
| dG·rC | — | — | — | — | 83 | — | 1 m*M* $Na^+$ + 0.1 m*M* EDTA | *63* |
| | — | — | — | — | 71 | 7.8 | 1 m*M* $Na^+$ + 0.1 m*M* EDTA | *47* |
| **Copolymers of alternating sequence** | | | | | | | | |
| r(A-U)·r(A-U) | 47.5 | 67 | — | 83 | 27 | 7.5 | 1 m*M* $Na^+$ | *48, 67, 68* |
| | 47 | 62.5 | — | — | 32 | — | 1 m*M* $Na^+$ | *80* |
| | — | 65.1 | — | — | 69 | 6.3–8.1 | 20 m*M* $Na^+$ + 0.5 m*M* $Mg^{2+}$ + 10 m*M* Tris | *69, 70* |
| r(A-br⁵U)·r(A-br⁵U) | 67.3 | 79 | — | — | 53 | 7.5 | 1 m*M* $Na^+$ | *67, 68, 80* |
| r(A-fl⁵U)·r(A-fl⁵U) | — | — | — | — | 32 | — | 1 m*M* $Na^+$ | *80* |
| r(A$s^2$U)·r(A$s^2$U) | — | — | — | — | 68 | 8.1 | 1 m*M* $Na^+$ + 0.5 m*M* $Mg^{2+}$ + 10 m*M* Tris | *69* |
| r(A-Ψ)·r(A-Ψ) | — | — | — | — | 58 | — | 1 m*M* $Na^+$ | *80* |
| r(n²A-U)·r(n²A-U) | 42.5 | 57.5 | — | — | 28 | — | 1 m*M* $Na^+$ | *81* |
| r(n²A-br⁵U)·r(n²A-br⁵U) | | — | — | — | 51 | — | 1 m*M* $Na^+$ | *81* |
| r(n²A-fl⁵U)·r(n²A-fl⁵U) | | — | — | — | 29 | — | 1 m*M* $Na^+$ | *81* |
| r(n²A-T)·r(n²A-T) | 56 | 71 | — | — | 42 | — | 1 m*M* $Na^+$ | *81* |
| r(A-T)·r(A-T) | 60.5 | 76 | — | — | 46 | — | 1 m*M* $Na^+$ | *80* |
| r(F-U)·r(F-U) | — | — | — | — | 33 | — | 1 m*M* $Na^+$ | *80, 81* |
| r(F-br⁵U)·r(F-br⁵U) | — | — | — | — | 63 | — | 1 m*M* $Na^+$ | *80, 81* |
| r(F-fl⁵U)·r(F-fl⁵U) | — | — | — | — | 35 | — | 1 m*M* $Na^+$ | *80, 81* |
| r(F-Ψ)·r(F-Ψ) | — | — | — | — | 32 | — | 1 m*M* $Na^+$ | *80* |
| r(F-T)·r(F-T) | — | — | — | — | 48 | — | 1 m*M* $Na^+$ | *80, 81* |
| d(A-U)·d(A-U) | 37.2 | 58 | — | 72 | — | 7.5 | — | *48, 67, 68* |
| | — | 56.6 | — | — | — | — | — | *88* |
| d(A-br⁵U)·d(A-br⁵U) | 49.5 | 66 | — | 77 | — | 7.5 | — | *62, 67, 68, 71* |
| d(A-hm⁵U)·d(A-hm⁵U) | 38.8 | 55.6 | — | 68.4 | — | — | — | *88* |
| d(A-T)·d(A-T) | 39.9 | 60 | — | 77 | 56 | 7–7.5 | 0.06 *M* $Na^+$ | *48, 57, 62, 67, 68, 71, 72* |
| | 40.7 | 59.6 | — | 74.8 | — | — | — | *88* |
| | 39 | 61 | — | — | 70 | — | 0.2 *M* $Na^+$ | *73* |
| | — | — | 66 | — | 71 | 7 | SSC[d]; 0.5 *M* $Na^+$ | *64, 74* |
| | — | — | — | — | 44 | 7.9 | 0.01 *M* Tris | *75* |
| | — | — | — | — | 55 | 8.1 | 0.02 *M* $Na^+$ + 0.5 m*M* $Mg^{2+}$ + 0.01 *M* Tris | *69* |
| d(A-s⁴T)·d(A-s⁴T) | 35 | 55 | — | — | 63 | — | 0.2 *M* $Na^+$ | *73* |
| d(n²A-T)·d(n²A-T) | — | — | — | — | 66 | 7.9 | 0.01 *M* Tris | *75* |
| d(I-C)·d(I-C) | — | 54 | — | — | 43 | 7.2 | 0.02 *M* $Na^+$ | *76* |
| d(A-G)·d(T-C) | — | — | 84 | — | 68 | 7 | SSC; 1/10 SSC[d] | *77* |
| | — | — | — | — | 90 | 5.8 | 0.2 *M* $K^+$ + 0.01 *M* $Na^+$ + 5 m*M* $Mg^{2+}$ | *78* |
| d(A-C)·d(T-G) | — | — | 91.5 | — | 75 | 7 | SSC; 1/10 SSC[d] | *77* |
| | 69.0 | 85.8 | — | — | 90.2 | — | 0.2 *M* $Na^+$ | *88* |
| d(A-C)·d(U-G) | 71.1 | 87.2 | — | — | 91 | — | 0.2 *M* $Na^+$ | *88* |
| d(A-C)·d(hm⁵U-G) | 66.7 | 83.2 | — | — | 87 | — | 0.2 *M* $Na^+$ | *88* |
| d(A-hm⁵C)·d(T-G) | — | 87.7 | — | — | — | — | — | *88* |
| (rA-dU)·(rA-dU) | 18.4 | — | — | — | — | 7.5 | — | *68* |
| d(T-C)·d(A-G)·r(U-C) | — | — | — | — | 90 | 5.8 | 0.2 *M* $K^+$ + 0.01 *M* $Na^+$ + 5 m*M* $Mg^{2+}$ | *78* |
| DNA (50% G + C) | — | — | 90.5 | — | — | — | + 0.02 *M* $Na_3$ citrate (SSC)[d] | *64* |

[a] The abbreviations for the polynucleotides conform to those proposed by the IUPAC-IUB Commission on Biochemical Nomenclature, except that the prefix "poly" (or subscript *n*) is omitted from each chain to conserve space. Thus, rA•rU and rA•2rU symbolize double- and triple-stranded ribohomopolymer helices respectively, one strand composed of polyadenylate (poly A) and the other strand(s) composed of polyuridylate (poly U); d(A-U)•d(A-U) denotes a double-stranded helix with strictly alternating deoxyadenylate and deoxyuridylate residues on both strands; *x*rG represents a complex of poly rG strands for which the number of strands has not been determined. A = isoadenylate; F = formycin nucleotide; Am = 2′-O-methyladenylate; $^{s}$U = uridine 5′-phosphorothioate. Two melting temperatures separated by a semicolon indicate a two-step thermal dissociation curve.

[b] Solvent specified under Comment.

[c] Unless otherwise specified, the pH is most probably about 7.

[d] SSC = standard saline citrate: 0.15 *M* NaCl–0.02 *M* $Na_3$ citrate.

[e] The existence of a double-helical "acid" form is controversial because of the degradation of poly dA upon heating and aggregation at low pH.

## REFERENCES


1. Barszcz and Shugar, *Acta Biochim. Pol., 11,* 481 (1964).
2. Van Holde, Brahms and Michelson, *J. Mol. Biol., 12,* 726 (1965).
3. Holcomb and Tinoco, *Biopolymers, 3,* 121 (1965).
4. Leng and Felsenfeld, *J. Mol. Biol., 15,* 455 (1966).
5. Applequist and Damle, *J. Amer. Chem. Soc., 88,* 3895 (1966).
6. Eisenberg and Felsenfeld, *J. Mol. Biol., 30,* 17 (1967).
7. Vournakis, Poland and Scheraga, *Biopolymers, 5,* 403 (1967).
8. Felsenfeld and Miles, *Annu. Rev. Biochem., 36,* 407 (1967).
9. Richards, *Eur. J. Biochem., 6,* 88 (1968).
10. Massoulié, *C. R. Acad. Sci., 260,* 5554 (1965).
11. Michelson, Massoulié and Guschlbauer, *Progr. Nucl. Acid Res. Mol. Biol., 6,* 83 (1967).
12. Ts'o, Helmkamp and Sander, *Proc. Nat. Acad. Sci. U.S.A., 48,* 686 (1962).
13. Michelson and Pochon, *Biochim. Biophys. Acta, 114,* 469 (1966).
14. Griffin, Haslam and Reese, *J. Mol. Biol., 10,* 353 (1964).
15. Michelson, Monny, Laursen and Leonard, *Biochim. Biophys. Acta, 119,* 258 (1966).
16. Doty, Boedtker, Fresco, Haselkorn and Litt, *Proc. Nat. Acad. Sci. U.S.A., 45,* 482 (1959).
17. Chamberlin and Patterson, *J. Mol. Biol., 12,* 410 (1965).
18. Sarkar and Yang, *Biochemistry, 4,* 1238 (1965).
19. Fresco and Massoulié, *J. Amer. Chem. Soc., 85,* 1352 (1963).
20. Pochon and Michelson, *Proc. Nat. Acad. Sci. U.S.A., 53,* 1425 (1965).
21. Pochon and Michelson, *Biochim. Biophys. Acta, 145,* 321 (1967).
22. Michelson and Monny, *Biochim. Biophys. Acta, 129,* 460 (1966).
23. Michelson and Monny, *Biochim. Biophys. Acta, 149,* 107 (1967).
24. Michelson and Monny, *Proc. Nat. Acad. Sci. U.S.A., 56,* 1528 (1966).
25. Leng and Michelson, *Biochim. Biophys. Acta, 155,* 91 (1968).
26. Szer, *Acta Biochim. Pol.,* 13, 251 (1966).
27. Massoulié, Michelson and Pochon, *Biochim. Biophys. Acta, 114,* 16 (1966).
28. Scheit, *Biochim. Biophys. Acta, 134,* 17 (1967).
29. Swierkowski and Shugar, *J. Mol. Biol., 47,* 57 (1970).
30. Michelson, *Bull. Soc. Chim. Biol., 47,* 1553 (1965).
31. Swierkowski, Szer and Shugar, *Biochem. Z., 342,* 429 (1965).
32. Fasman, Lindblow and Grossman, *Biochemistry, 3,* 1015 (1964).
33. Brimacombe and Reese, *J. Mol. Biol., 18,* 529 (1966).
34. Ts'o, Rapaport and Bollum, *Biochemistry, 5,* 4153 (1966).
35. Brahms, Maurizot and Michelson, *J. Mol. Biol., 25,* 465 (1967).
36. Brimacombe, *Biochim. Biophys. Acta, 142,* 24 (1967).
37. Akinrimsi, Sander and Ts'o, *Biochemistry, 2,* 340 (1963).
38. Michelson and Monny, *Biochim. Biophys. Acta, 149,* 88 (1967).
39. Guschlbauer, *Proc. Nat. Acad. Sci. U.S.A., 57,* 1441 (1967).
40. Szer and Shugar, *J. Mol. Biol., 17,* 174 (1966).
41. Rabczenko and Szer, *Acta Biochem. Pol., 14,* 369 (1967).
42. Sigler, Davies and Miles, *J. Mol. Biol., 5,* 709 (1962).
43. Gabbay, *Biopolymers, 5,* 727 (1967).
44. Massoulie, *Eur. J. Biochem., 3,* 428 (1968).

45. Blake and Fresco, *J. Mol. Biol., 19,* 145 (1966).
46. Blake, Massoulie and Fresco, *J. Mol. Biol., 30,* 291 (1967).
47. Chamberlin, *Fed. Proc., 24,* 1446 (1965).
48. Riley, Maling and Chamberlin, *J. Mol. Biol., 20,* 359 (1966).
49. Glaser and Gabbay, *Biopolymers, 6,* 243 (1968).
50. Krakauer and Sturtevant, *Biopolymers, 6,* 491 (1968).
51. Stevens and Felsenfeld, *Biopolymers, 2,* 293 (1964).
52. Szer and Shugar, *Acta Biochim. Pol., 10,* 219 (1963).
53. Massoulie, *Eur. J. Biochem., 3,* 439 (1968).
54. Howard, Frazier and Miles, *J. Biol. Chem., 241,* 4293 (1966).
55. Barszcz and Shugar, *Eur. J. Biochem., 5,* 91 (1968).
56. Giannoni and Rich, *Biopolymers, 2,* 399 (1964).
57. Inman, *J. Mol. Biol., 9,* 624 (1964).
58. Inman, *J. Mol. Biol., 10,* 137 (1964).
59. Lefler and Bollum, *J. Biol. Chem., 244,* 594 (1969).
60. Zmudzka, Bollum and Shugar, *Biochemistry, 8,* 3049 (1969).
61. Bollum, in *Procedures in Nucleic Acid Research,* p. 577, Cantoni and Davies, Eds. Harper and Row, New York (1966).
62. Inman and Baldwin, *J. Mol. Biol., 8,* 452 (1964).
63. Schildkraut, Marmur, Fresco and Doty, *J. Biol. Chem., 236,* PC3 (1961).
64. Marmur and Doty, *J. Mol. Biol., 5,* 109 (1962).
65. Radding, Josse and Kornberg, *J. Biol. Chem., 237,* 2869 (1962).
66. Straat, Ts'o and Bollum, *J. Biol. Chem., 244,* 391 (1969).
67. Chamberlin, Baldwin and Berg, *J. Mol. Biol., 7,* 334 (1963).
68. Chamberlin, in *Procedures in Nucleic Acid Research,* p. 513, Cantoni and Davies, Eds. Harper and Row, New York (1966).
69. Matzura and Eckstein, *Eur. J. Biochem., 3,* 448 (1968).
70. Richards and Simpkins, *Eur. J. Biochem., 6,* 93 (1968).
71. Inman and Baldwin, *J. Mol. Biol., 5,* 172 (1962).
72. Scheffler, Elson and Baldwin, *J. Mol. Biol., 36,* 291 (1968).
73. Lezius and Scheit, *Eur. J. Biochem., 3,* 85 (1967).
74. Schachman, Adler, Radding, Lehman and Kornberg, *J. Biol. Chem., 235,* 3242 (1960).
75. Reich, Personal Communication.
76. Grant, Harwood and Wells, *J. Amer. Chem. Soc., 90,* 4474 (1968).
77. Wells, Ohtsuka and Khorana, *J. Mol. Biol., 14,* 221 (1965).
78. Morgan and Wells, *J. Mol. Biol., 37,* 63 (1968).
79. Bobst, Cerutti and Rottman, *J. Amer. Chem. Soc., 91,* 1246 (1969).
80. Ward and Reich, *Proc. Nat. Acad. Sci. U.S.A., 61,* 1494 (1968).
81. Ward and Reich, *J. Biol. Chem., 244,* 1228 (1969).
82. Michelson and Pochon, *Biochim. Biophys. Acta, 174,* 604 (1969).
83. Fikus and Shugar, *Acta Biochim. Pol., 16,* 55 (1969).
84. Tramer, Wierzchowski and Shugar, *Acta Biochim. Pol., 16,* 83 (1969).
85. Zimmer and Szer, *Acta Biochim. Pol., 15,* 339 (1968).
86. Riley and Paul, *J. Mol. Biol. 50,* 439 (1970).
87. Howard, Frazier and Miles, *J. Biol. Chem., 244,* 1291 (1969).
88. Cassidy, Kahan and Doty, Unpublished Data.
89. Guschlbauer and Vetterl, *Fed. Eur. Biochem. Soc. Lett., 4,* 57 (1969).
90. Adler, Grossman and Fasman, *Biochemistry, 8,* 3836 (1969).
91. Adler, Grossman and Fasman, *Proc. Nat. Acad. Sci. U.S.A., 57,* 423 (1967).
92. Pochon and Michelson, *Biochim. Biophys. Acta, 182,* 17 (1969).
93. Zmudzka, Bollum and Shugar, *J. Mol. Biol., 46,* 169 (1969).
94. Cassan and Bollum, *Biochemistry, 8,* 3928 (1969).

# CONTENT[a] OF 6-METHYLAMINOPURINE AND 5-METHYLCYTOSINE IN DNA

DR. HERMAN S. SHAPIRO

| Source of DNA | 6Me Ade[b] | 5Me Cyt[c] | Ref. |
|---|---|---|---|
| Bacteriophage DD7 (*E. coli*) | 0.35 | 0.13 | *1* |
| Bacteriophage Sd (*E. coli*) | <0.10 | <0.10 | *1* |
| Bacteriophage C20 (*Strep. griseus*) | 2.0 | — | *2* |
| *Aerobacter aerogenes* | 0.55 | — | *2* |
| *A. aerogenes* | 0.53 | — | *2* |
| *Agrobacter tumefaciens*, 55 | 0.09 | 0.16 | *3* |
| *Alcaligenes faecalis* | 0.54 | — | *4* |
| *Alcaligenes faecalis* | 0.54 | 0.62 | *3* |
| *Bacillus cereus* | 0.08 | — | *2* |
| *B. brevis*, R | 0.30 | 0.06 | *3* |
| *B. brevis*, S | 0.30 | 0.43 | *3* |
| *B. brevis*, P$^+$ | 0.26 | 0.14 | *3* |
| *Bacterium morganii* | 0.35 | — | *4* |
| *Bact. morganii* | 0.35 | — | *3* |
| Baker's yeast | 1.0 | — | *2* |
| *Blakeslea trispora* | ND[d] | ND[d] | *3* |
| *Brucella abortus* | 0.17 | — | *4* |
| *Brucella abortus* | 0.17 | — | *3* |
| *Chloropseudomonas ethylicum* | 0.13 | — | *3* |
| *Chromatium minutissimum* | 0.05 | — | *3* |
| *Clostridium butylicum*, P | 0.05 | ND[d] | *3* |
| *Corynebacterium diphtheriae* | 0.19 | — | *4* |
| *C. vadosum*, Kras | 0.12 | | *4* |
| *Escherichia coli* | 0.50 | — | *4* |
| *E. coli*, B | 0.48 | ND[d] | *1* |
| *E. coli*, CK | 0.49 | 0.24 | *1* |
| *E. coli*, C | 0.48 | 0.26 | *1* |
| *E. coli*, 15T$^-$ | 0.42 | — | *2* |
| *E. coli*, B/r | 0.42 | — | *2* |
| *Mycobacterium luteum* | 0.05 | 0.10 | *3* |
| *Micrococcus lysodeikticus* | ND[d] | ND[d] | *3* |
| *Propionobacterium shermanii* | 0.04 | 0.52 | *3* |
| *Pseudomonas syringae*, 19P | 0.03 | — | *3* |
| *Rhizobium meliloti*, 441 | 0.11 | — | *3* |
| *Rhodospirillum rubrum* | 0.19 | — | *3* |
| *Salmonella typhosa* | 0.47 | — | *4* |
| *S. typhosa*, Ty-2 | 0.47 | 0.24 | *3* |
| *S. typhosa*, T5501 | 0.43 | 0.27 | *3* |
| *Shigella dysenteriae* | 0.45 | — | *3* |
| *Staphylococcus aureus* | 0.21 | — | *2* |
| *Staph. aureus*, 909 | 0.05 | — | *3* |
| *Staph. albus* | 0.10 | — | *2* |
| *Streptomyces griseus* | 0.28 | — | *2* |
| *Strep. plantarum* | ND[d] | ND[d] | *3* |
| *Ahnfeltia plicata* | 0.19 | 1.34 | *5* |
| *Aphanizomenon flos-aquae* | 0.50 | 1.67 | *5* |
| *Asteromonas gracilis* | 0.40 | 0.93 | *5* |
| *Chlamydomonas globosa* | 0.24 | 2.75 | *5* |
| *Chlorella vulgaris* | 0.28 | 3.45 | *5* |
| *Coelospherium dubium* | 0.42 | 1.35 | *5* |
| *Dictiota fasciola* | 0.13 | 2.65 | *5* |
| *Dunaliella salina* | 0.22 | 1.29 | *5* |
| *Dunaliella viridis* | 0.28 | 1.74 | *5* |
| *Hydrodictyon reticulatum* | 0.15 | 1.42 | *5* |
| *Lagercheimia ciliata* | 0.16 | 2.93 | *5* |
| *Melosira italica* | 0.21 | 2.23 | *5* |
| *Phillophora nervosa* | 0.10 | 2.32 | *5* |
| *Polyides rotundus* | 0.56 | 3.46 | *5* |
| *Scenedesmus obliquus* | 0.19 | 3.11 | *5* |
| *Scenedesmus quadricauda* | 0.19 | 3.23 | *5* |
| *Stigeoclonium tenue* | 0.23 | 1.55 | *5* |
| Broad bean (*Phaseolus vulgaris*) | — | 3.7 | *6* |
| Wheat germ | 0.26 | — | *2* |
| *Suberites domuncula* (sponge) | ND[d] | 1.3 | *7* |
| *Metridium senile* (*coelenterate*) | ND[d] | 0.9 | *7* |
| *Helix pomatia* (*mollusca*) | ND[d] | 0.6 | *7* |
| *Strongylocentrotus intermedius* (echinoderm) | ND[d] | 0.6 | *7* |
| *Cottus*, sp. | ND[d] | 1.7 | *7* |
| *Trichiurus japonicus* | ND[d] | 1.5 | *7* |
| *Spheroides*, sp. | ND[d] | 1.5 | *7* |
| *Oncorhynchus gorbuscha* | ND[d] | 1.6 | *7* |
| *Cyprinus carpio* | ND[d] | 1.2 | *7* |
| *Rana temporaria* | ND[d] | 1.6 | *7* |
| *Testudo horsfieldi* (reptile) | ND[d] | 1.5 | *7* |
| Mouse | ND[d] | 1.1 | *7* |
| Rabbit | ND[d] | 0.9 | *7* |
| Pig | ND[d] | 1.2 | *7* |
| Sheep | ND[d] | 1.1 | *7* |
| Bull | ND[d] | 1.4 | *7* |
| Calf | 0.15 | — | *2* |
| Horse | 0.28 | — | *2* |

[a] Proportions are expressed as moles of nitrogenous constituent per 100 moles of total constituents.
[b] 6MeAde = 6-methylaminopurine.
[c] 5MeCyt = 5-methylcytosine. These individual estimations of 5MeCyt in DNA supplement the data cited elsewhere on the total distribution of purines and pyrimidines in DNA.
[d] ND = not detected; specific searches for these minor components failed to show their presence in the DNA.

## REFERENCES

1. Nikolskaya, Tkacheva, Vanyushin and Tikhonenko, *Biochim. Biophys. Acta, 155,* 626 (1968).
2. Dunn and Smith, *Biochem. J., 68,* 627 (1958).
3. Vanyushin, Belozerskii, Kokurina and Kadirova, *Nature, 218,* 1066 (1968).
4. Vanyushin, Kokurina and Belozerskii, *Dokl. Akad. Nauk SSSR, 161,* 1453 (1965).
5. Pakhomova, Zaitseva and Belozerskii, *Dokl. Akad. Nauk SSSR, 182,* 712 (1968).
6. Baxter and Kirk, *Nature, 222,* 272 (1969).
7. Vanyushin, Tkacheva and Belozerskii, *Nature, 225,* 948 (1970).

# METABOLISM OF NUCLEOSIDES

DR. WILLIAM FIRSHEIN

## METABOLISM OF DEOXYRIBONUCLEOSIDES

The numbers in the pathways refer to the enzymes believed to be responsible for the indicated conversions and serve as a guide to references that can be consulted for additional data. For the desired information regarding the enzymes, refer to the appropriate numbers in the section immediately following the pathway or in the reference guide.

### ABBREVIATIONS

A = adenine
C = cytosine
G = guanine
H = hypoxanthine
T = thymine
U = uracil

dA = deoxyadenosine
dC = deoxycytidine
dG = deoxyguanosine
dI = deoxyinosine
dT = thymidine
dX = deoxyxanthosine

dAMP, dADP, dATP
dGMP, dGDP, dGTP
dCMP, dCDP, dCTP
dUMP, dUDP, dUTP
TMP, TDP, TTP
} mono-, di-, and triphosphates of deoxyribonucleosides

dIMP
dXMP
} monophosphates of deoxyribonucleosides

Ar = adenosine
Cr = cytidine

Gr = guanasine
Ur = uridine

AMP, ADP, ATP
GMP, GDP, GTP
CMP, CDP, CTP
UMP, UDP, UTP
} mono-, di-, and triphosphates of ribonucleosides

acetal. = acetaldehyde
B.am.isobut. = *beta*-amino isobutyric acid
BUr.isobut. = *beta*-ureido isobutyric acid
BUr.prop. = *beta*-ureido propionic acid
$d_iHT$ = dihydrothymine
$d_iHU$ = dihydrouracil
dR1P = deoxyribo-1-phosphate
dR5P = deoxyribo-5-phosphate
$Gl\text{-}3\text{-}PO_4$ = 3-glyceraldehyde phosphate

## Anabolism of Deoxyribonucleosides

### Purines

dG ⇌ (1) dGMP ⇌ (2) dGDP ⇌ (3) dGTP ← (46) GTP ⇌ (35) GDP ⇌ (34) GMP ⇌ (27) Gr

GDP → (42) dGDP

dG ⇌ dX (55, 18); dGMP ⇌ dXMP (19)

dX ⇌ (4) dXMP

dG → (51) dI

dI → (20) dX; dIMP → (21) dXMP

dI ⇌ (5) dIMP

dI ⇌ dA (22, 52); dIMP ⇌ dAMP (53, 23)

dA ⇌ (6) dAMP ⇌ (7) dADP ⇌ (8) dATP ← (47) ATP ⇌ (36) ADP ⇌ (33) AMP ⇌ (28) Ar

ADP → (43) dADP

### Pyrimidines

dC ← (9) dCMP ⇌ (10) dCDP ⇌ (11) dCTP ← (48) CTP ⇌ (37) CDP ⇌ (32) CMP ⇌ (29) Cr

CDP → (44) dCDP; UTP → (41) CTP

dC → (24) dU; dCMP → (25) dUMP; CMP → (40) UMP; Cr → (39) Ur

dU ⇌ (12) dUMP ⇌ (13) dUDP ⇌ (14) dUTP ← (49) UTP ⇌ (38) UDP ⇌ (31) UMP ⇌ (30) Ur

dUTP → (50) dUMP; UDP → (45) dUDP

dUMP → (26) dTMP

dT ⇌ (15) dTMP ⇌ (16) dTDP ⇌ (17) TTP

### The Enzymes Involved

1, 4, 5, 6, 12, 15

Deoxyribonucleoside kinases exist for all deoxyribonucleosides except possibly deoxycytidine. Thus, the origin of cytosine in DNA is probably due to other reactions, notably reductase (see enzymes 42 to 49).

2, 7, 10, 13, 16

Deoxyribonucleotide kinases for all deoxyribonucleotides have been demonstrated and are probably unique enzymes, although ribonucleotides are also phosphorylated and there is evidence at least for enzymes 2 and 7 that they are the same as enzymes 34 and 33 respectively.

| Enzyme | Description |
|---|---|
| 24, 25, 39, 40, 51, 52, 53, 54, 55 | Deaminases. Many specific enzymes at the deoxyribonucleoside or deoxyribonucleotide level. It is not known whether the deaminase is the same for the same base at the nucleotide and nucleoside level (such as dCMP and dC), but there are suggestions that this is the case. |
| 9 | dCMP phosphatase. Since there is probably no dC kinase, the main reaction involves the conversion of dCMP to dC. The back reactions of all the kinases (ribo- and deoxyribo-) also involve phosphatases. There is a phosphatase (dUTP-ase, enzyme 50) in which a pyrophosphate is removed from dUTP to produce dUMP directly, which is then used for TMP synthesis; whether other pyrophosphatases exist for other deoxyribonucleoside triphosphates is not known. |
| 3, 8, 11, 14, 17 | Nucleoside diphosphate kinase. One enzyme may phosphorylate all deoxyribonucleoside and ribonucleoside diphosphates. |
| 18, 19 | dX and/or dXMP aminase (may be the same enzyme) to produce dG and/or dGMP. |
| 20, 21 | dI and/or dIMP dehydrogenase (may be the same enzyme) to produce dX and/or dXMP. |
| 22, 23 | dA and/or dAMP synthetase (may be the same enzyme) for amination of dI and/or dIMP. |
| 26 | Thymidylate synthetase. Probably occurs only at the nucleotide level. |
| 27, 28, 29, 30 | Ribonucleoside kinases. All have been demonstrated for the specific nucleosides, including cytidine kinase. |
| 31, 32, 33, 34 | Ribonucleotide kinases have been demonstrated for all nucleotides, although it is not certain that these kinases are specific for the ribo derivative. At least in two cases the purified kinase that phosphorylates the ribonucleotide will also phosphorylate the deoxyribo derivative (see enzymes 2 and 34, 7 and 33). |
| 35, 36, 37, 38 | Nucleoside diphosphate kinase. May be the same enzyme for ribo- and deoxyribonucleoside diphosphates. |
| 39, 40 | Deaminases for cytidine and cytidylate. |
| 41 | Cytidine triphosphate synthetase. |
| 42, 43, 44, 45 | Ribonucleoside diphosphate reductase. Probably one enzyme exists that converts ribonucleoside diphosphates to deoxyribonucleoside diphosphates. Activity is affected greatly by various deoxyribonucleoside triphosphates. |

| | |
|---|---|
| 46, 47, 48, 49 | Ribonucleoside triphosphate reductase. Similar to the diphosphate reductase, but has thus far been found only in lactobacilli. |
| 50 | dUPTase. Removes pyrophosphate from dUTP to produce dUMP as intermediate in the synthesis of TMP (see enzyme 26). |

**Reference Guide**

| Enzymes | Reference |
|---|---|
| 2, 3, 7, 8, 10, 11, 13, 14, 16, 17, 26 | Bessman, M. J., in *Molecular Genetics,* Part I, p. 1, J. H. Taylor, Ed. Academic Press, New York (1963). |
| 1, 4, 5, 6, 9, 12, 15, 18, 19, 20, 21, 22, 23, 51, 52 | Firshein, W., and Hasselbacher, P., *Biochim. Biophys. Acta, 204,* 60 (1970). |
| 24, 30, 31, 32, 37, 38, 39, 40 | Neuhard, J., and Ingraham, J., *J. Bacteriol., 95,* 2431 (1968). |
| 25 | Siedler, A. J., and Holtz, M. T., *J. Biol. Chem., 238,* 697 (1963). |
| 2, 7, 10, 16 | Firshein, W., *J. Bacteriol., 90,* 327 (1965).<br>Sugino, Y., Teraoka, H., and Shimono, H., *J. Biol. Chem., 241,* 961 (1966). |
| 2, 7, 10, 13, 16, 31, 32, 33, 34 | Canellakis, E. S., Gottesman, M. E., and Kammen, H. O., *Biochim. Biophys. Acta, 39,* 82 (1960). |
| 3, 8, 11, 14, 17 | Berg, P., and Joklik, W. K., *J. Biol. Chem., 210,* 657 (1954). |
| 35, 36, 37, 38 | Okazaki, R., and Kornberg, A., *J. Biol. Chem., 239,* 269 (1964). |
| 41 | Long, C. W., and Pardee, A. B., *J. Biol. Chem., 242,* 4715 (1967). |
| 2, 34 | Oeschger, M. P., and Bessman, M. J., *J. Biol. Chem., 241,* 5452 (1966). |
| 7, 33 | Klenow, H., and Lichtler, E., *Biochim. Biophys. Acta, 23,* 6 (1951). |
| 27, 28, 42, 43, 44, 45, 46, 47, 48, 49, 50 | Reichard, P., and Larson, A., in *Progress in Nucleic Acid Research,* Vol. 7, p. 303, J. N. Davidson and W. E. Cohn, Eds., Academic Press, New York (1967). |
| 26 | Friedkin, M., in *Methods in Enzymology,* Vol. 6, p. 124, S. P. Colowick and N. O. Kaplan, Eds. Academic Press, New York (1963). |

## Catabolism of Deoxyribonucleosides

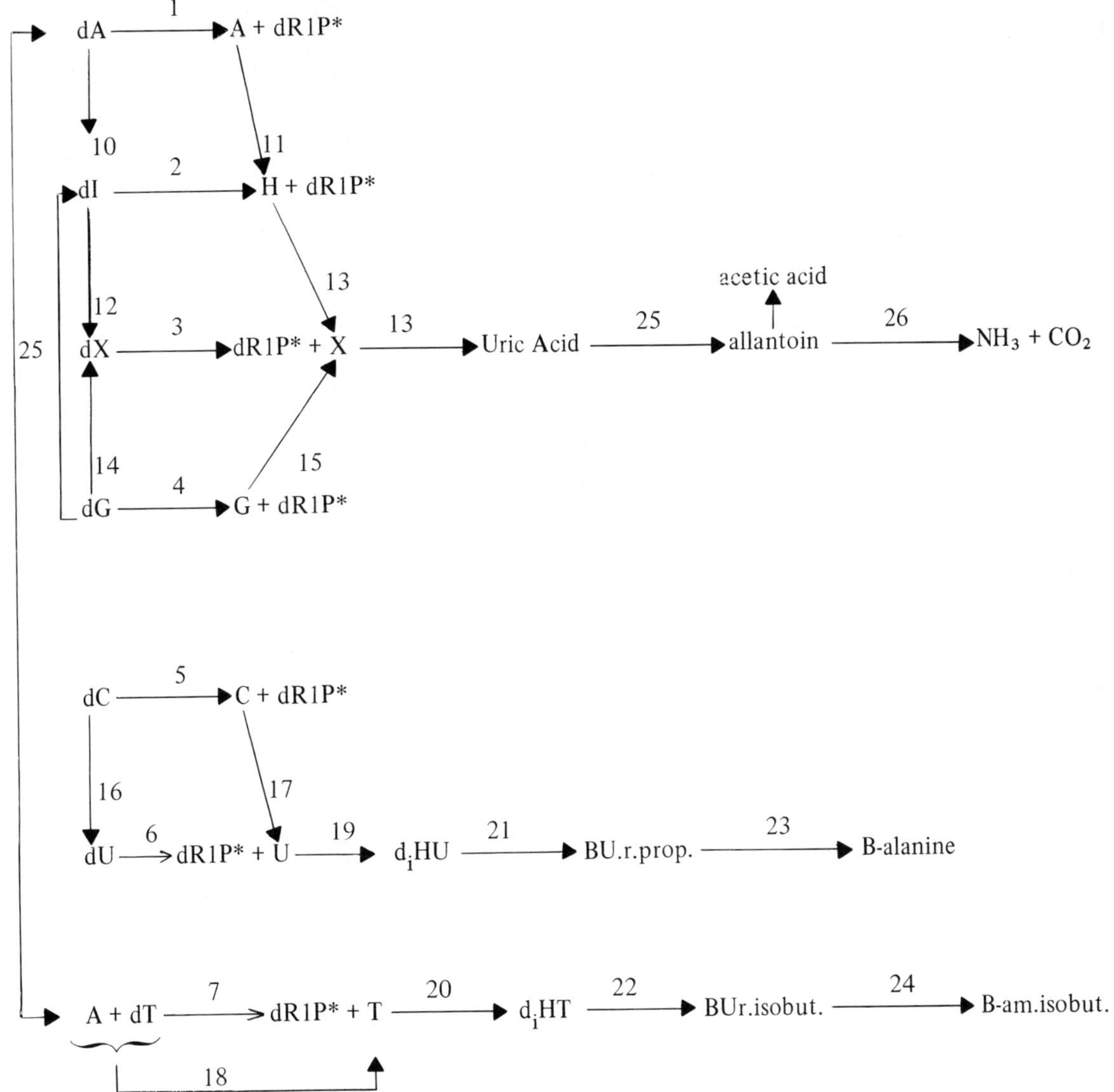


* Catabolism of dR1P is as follows: dR1P $\xrightarrow{8}$ dR5P $\xrightarrow{9}$ Acetal. + G1.3.$PO_4$.

### The Enzymes Involved

| | |
|---|---|
| 1, 2, 3, 4, 5, 6, 7 | Deoxyribonucleoside phosphorylases. Probably unique for each deoxyribonucleoside; adds inorganic phosphorus to deoxyribose after removal from base. |
| 8 | Deoxyribomutase. Converts deoxyribo-1-phosphate to deoxyribo-5-phosphate. |
| 9 | Deoxyriboaldolase. Splits deoxyribo-5-phosphate to acetaldehyde and 3-glyceraldehyde phosphate. |

| Enzymes | Reference |
|---|---|
| 10, 14, 16, 25 | Deoxyribonucleoside deaminases. Probably specific for the specific deoxyribonucleoside shown. |
| 11, 15, 17 | Purine or pyrimidine deaminases unique for the bases shown. It is not known whether the "base" deaminases can deaminate their respective deoxyribonucleosides or vice versa. |
| 12 | dI hydrogenase. |
| 13 | Xanthine oxidase. Converts hypoxanthine to xanthine and xanthine to uric acid. |
| 25 | Uricase. Oxidase that decarboxylates uric acid. |
| 26 | Allantoin is ultimately converted to $CO_2$ + $NH_3$ by several enzymes, including a deaminase, and acetic acid is produced by purine degradation. The pathways shown are taken from those demonstrated in animal tissues as well as in bacteria. |
| 18 | Deoxyribosyltransferase. Converts thymidine to thymine and deoxyadenosine in the presence of adenine. |
| 19, 20 | Dehydrogenases specific for each base. There is no dehydrogenase detected for cytosine, which is first deaminated to uracil (enzyme 17). |
| 21, 22, 23, 24 | Hydrolases involved in the illustrated reactions. It has been reported that barbituric acid as well as urea and malonic acid are intermediates. |

**Reference Guide**

| | |
|---|---|
| 1, 2, 3, 4, 5, 6, 7, 8, 9, 18 | Munch-Petersen, A., *Eur. J. Biochem., 6,* 432 (1968). |
| 10, 12, 14, 16, 25 | Firshein, W., and Hasselbacher, P., *Biochim. Biophys. Acta, 204,* 60 (1970). |
| 11, 13, 17, 19, 23, 25, 26 | Brock, T., in *Biology of Microorganisms,* p. 137. Prentice-Hall, Englewood Cliffs, New Jersey, 1970. |
| 13, 20, 25, 26 | Whitley, H. R., *J. Bacteriol., 63,* 163 (1952). |
| 19, 20, 21, 22, 23, 24 | Wang, T. P., and Lampen, J. O., *J. Biol. Chem., 194,* 775 (1952). |
| 3, 11, 15, 17, 19, 20, 21, 22, 23, 24, 25, 26 | Mahler, H. R., and Cordes, E. H., in *Biological Chemistry,* p. 731. Harper and Row, New York, 1966. |

# METABOLISM OF RIBONUCLEOSIDES

Most of the main reactions involving the synthesis of the immediate precursors of RNA (the ribonucleoside triphosphates) bypass the ribonucleoside level and begin with the synthesis of ribonucleotides, specifically IMP for purines and UMP for pyrimidines. However, almost all microorganisms (not inhibited by organic compounds) can utilize exogenous ribonucleosides and their purine and pyrimidine derivatives in salvage pathways as intermediates in ribonucleotide synthesis.

The numbers in the pathways refer to the enzymes believed to be responsible for the indicated conversions and serve as a guide to references that can be consulted for additional data. For the desired information regarding the enzymes, refer to the appropriate number in the section immediately following the pathway or in the reference guide.

## ABBREVIATIONS

A = adenine
Ar = adenosine
C = cytosine
Cr = cytidine
G = guanine
Gr = guanosine
Hx = hypoxanthine
Ir = inosine
U = uracil
Ur = uridine
X = xanthine
Xr = xanthosine

OMP, IMP, XMP — monophosphates of ribonucleosides

dhOr = dihydroorotic acid
Or = orotic acid

AMP, ADP, ATP; GMP, GDP, GTP; CMP, CDP, CTP; UMP, UDP, UTP — mono-, di-, and triphosphates of ribonucleosides

dADP, dATP; dGDP, dGTP; dCDP, dCTP; dUDP, dUTP — di- and triphosphates of deoxyribonucleosides

acetal. = acetaldehyde
AICR (1 and 2) = 5-amino-4-imidazole carboxamide (succinic) ribonucleotide
AIR = aminoimidazole ribonucleotide
BUr.prop. = *beta*-ureido propionic acid
$d_iHU$ = dihydrouracil
FAICR = foramido-4-imidazole carboxamide ribonucleotide
FGAM = formylglycinamidine ribonucleotide
FGAR = formylglycinamide ribonucleotide
Gl-3-$PO_4$ = 3-glyceraldehyde phosphate
GAR = glycinamide ribonucleotide
PRA = 5-phosphoribosylamine
PRPP = 5-phosphoribosyl pyrophosphate
R - 1 - P = ribose-1-phosphate
R-5-P = ribose-5-phosphate

## Anabolism of Ribonucleosides

### Purines

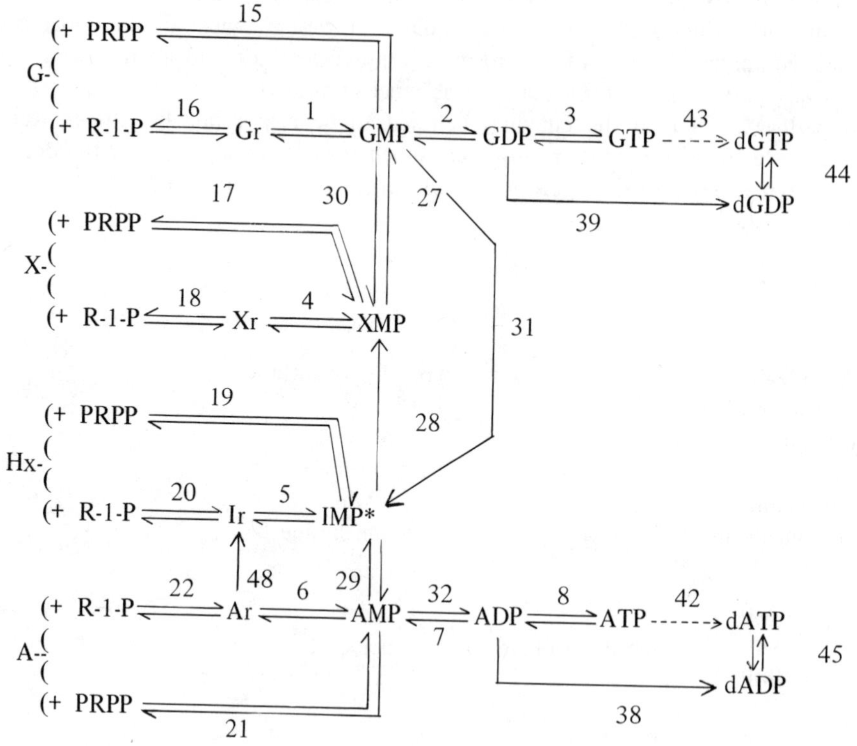

(+ PRPP ⇌ 15 → GMP
G-(
(+ R-1-P ⇌ 16 Gr ⇌ 1 GMP ⇌ 2 GDP ⇌ 3 GTP - - - 43 → dGTP
dGTP ⇅ 44 dGDP
GDP — 39 → dGDP
GMP — 27 — 31 → IMP*
GMP ⇅ 30 XMP

(+ PRPP ⇌ 17 → XMP
X-(
(+ R-1-P ⇌ 18 Xr ⇌ 4 XMP
XMP ← 28 IMP*

(+ PRPP ⇌ 19 → IMP*
Hx-(
(+ R-1-P ⇌ 20 Ir ⇌ 5 IMP*
Ir ← 48 Ar
IMP* ⇅ 29 AMP

(+ R-1-P ⇌ 22 Ar ⇌ 6 AMP ⇌ 32 / 7 ADP ⇌ 8 ATP - - - 42 → dATP
A-(
(+ PRPP ⇌ 21 → AMP
dATP ⇅ 45 dADP
ADP — 38 → dADP

### Pyrimidines

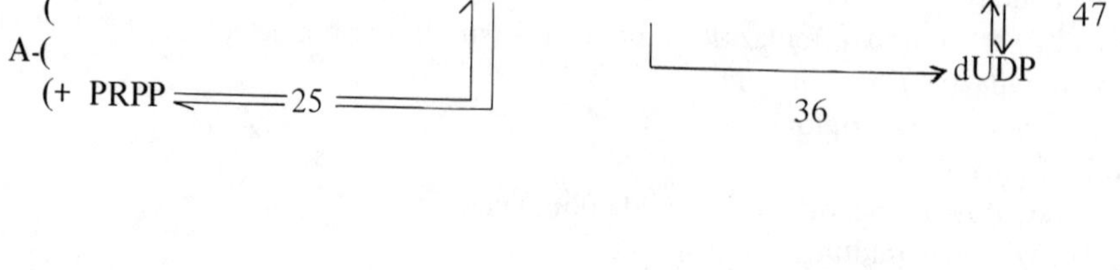

(+ PRPP ⇌ 23 → CMP
C-(
U-(
(+ R-1-P ⇌ 24 Cr ⇌ 9 CMP ⇌ 10 CDP ⇌ 11 CTP - - - 41 → dCTP
CDP — 37 → dCDP
dCDP ⇅ 46 dCTP
Cr ↓ 33 Ur
CMP ↓ 34 UMP*
UTP ↑ 35 CTP

(+ R-1-P ⇌ 26 Ur ⇌ 12 UMP* ⇌ 13 UDP ⇌ 14 UTP - - - 40 → dUTP
A-(
(+ PRPP ⇌ 25 → UMP*
dUTP ⇅ 47 dUDP
UDP — 36 → dUDP

* The major pathways for the synthesis of purine nucleotides are as follows: R-5-p → PRPP → PRA → GAR → FGAR → FGAM → AIR → CAIR → AICR → FAICR → IMP (glutamine is amino donor in three aminations).

* The major pathways for the synthesis of pyrimidine nucleotides are as follows: aspartic acid + carbamyl phosphate → carbamyl aspartic acid → dhOr → Or + PRPP → OMP → UMP (glutamine is amino donor in the synthesis of carbamyl phosphate).

## The Enzymes Involved

| | |
|---|---|
| 1, 4, 5, 6, 9, 12 | Ribonucleoside kinases exist for all ribonucleosides, including cytidine (which is not the case with deoxycytidine). It has not been definitely established that there are unique enzymes. |
| 2, 7, 10, 13 | Ribonucleotide kinases exist for all ribonucleotides, although deoxyribonucleotides may also be phosphorylated by the same kinases. |
| 3, 8, 11, 14, 45, 46, 47, 48 | Nucleoside diphosphate kinase. One enzyme probably phosphorylates all ribo- and deoxyribonucleoside diphosphates. The back reactions for all of the ribonucleoside mono-, di-, and triphosphates are mediated by phosphatases. Whether these are specific for each ribonucleoside (or deoxyribonucleoside) derivative is not known, but it is probable that the enzyme is nonspecific (5′-nucleotidase, for example). |
| 30, 31, 32, 33, 34, 48 | Deaminases. Many specific enzymes, mostly at the ribonucleotide level, which will not deaminate the corresponding ribonucleosides, and vice versa. |
| 27 | Aminase for XMP. Does not aminate Xr. |
| 28 | IMP-dehydrogenase. Does not act on Ir. |
| 29 | AMP synthetase. Does not act on Ar. |
| 35 | Cytidine triphosphate synthetase. |
| 36, 37, 38, 39 | Ribonucleoside diphosphate reductase. One enzyme reduces all ribonucleoside diphosphates. Activity depends greatly on deoxyribonucleoside triphosphates. |
| 31, 40, 42, 43 | Ribonucleoside triphosphate reductase. Similar to diphosphate reductase, but has thus far been found only in lactobacilli. |
| 15, 17, 19, 21, 23, 25 | Ribonucleotide-pyrophosphorylase. At least four to six separate enzymes; however, enzymes 15 and 17 have not been separated from each other. These enzymes, as far as is known, do not exist for deoxyribonucleotides. |
| 16, 18, 20, 22, 24, 26 | Ribonucleoside phosphorylases unique for the various bases. With ribose-1-phosphate, ribonucleosides are formed; with deoxyribose-1-phosphate, deoxyribonucleosides are produced. |

**Reference Guide**

| Enzymes | Reference |
|---|---|
| 2, 7, 10, 13 | Canellakis, E. S., Gottesman, M. E., and Kammen, H. O., *Biochim. Biophys. Acta, 39,* 82 (1966). |
| | Oeschger, M. P., and Bessman, M. J., *J. Biol. Chem., 241,* 5452 (1966). |
| | Sugino, Y., Teraoka, H., and Shimono, H., *J. Biol. Chem., 241,* 961 (1966). |
| 9, 12 | Sköld, O., *J. Biol. Chem., 235,* 3273 (1960). |
| 3, 8, 11, 14, 44, 45, 46, 47 | Bessman, M. J., in *Molecular Genetics,* Part I, p. 1, J. H. Taylor, Ed., Academic Press, New York (1963). |
| 36, 37, 38, 39, 40, 41, 42, 43 | Reichard, P., and Larson, A., in *Progress in Nucleic Acid Research,* Vol. 7, p. 303, J. N. Davidson and W. E. Cohn, Academic Press, New York (1967). |
| 15, 17, 19, 21, 23, 25 | Flaks, J. E., in *Methods in Enzymology,* Vol. 6, p. 136, S. P. Colowick and N. O. Kaplan, Eds., Academic Press, New York (1963). |
| 1, 4, 5, 9, 12, 16, 18, 20, 22, 24, 26 | Mahler, H. R., and Cordes, E. H., in *Biological Chemistry,* 2nd ed., p. 834. Harper and Row, New York (1971). |
| 27, 28, 31 | Magasanik, B., in *Methods in Enzymology,* Vol. 6, p. 106, S. P. Colowick and N. O. Kaplan, Eds., Academic Press, New York (1963). |
| 29 | Lieberman, I., *J. Biol. Chem., 223,* 327 (1956). |
| 33, 34 | Siedler, A. J., and Holtz, M. T., *J. Biol. Chem., 238,* 697 (1963). |
| 35 | Long, C. W., and Pardee, A. B., *J. Biol. Chem., 242,* 4715 (1967). |
| 32 | Kalckar, H. M., *J. Biol. Chem., 167,* 429, 461 (1947). |
| 48 | Wolfenden, R., Tomozawa, Y., and Bamman, B., *Biochemistry, 7,* 3965 (1968). |
| Purine (IMP) | Flaks, J. G., and Lukens, L. W., in *Methods in Enzymology,* Vol. 6, p. 52, S. P. Colowick and N. O. Kaplan, Eds., Academic Press, New York (1963). |
| Pyrimidine (UMP) | Reichard, P., and Sköld, O., in *Methods in Enzymology,* Vol. 6, p. 177, S. P. Colowick and N. O. Kaplan, Eds., Academic Press, New York (1963). |

## Catabolism of Ribonucleosides

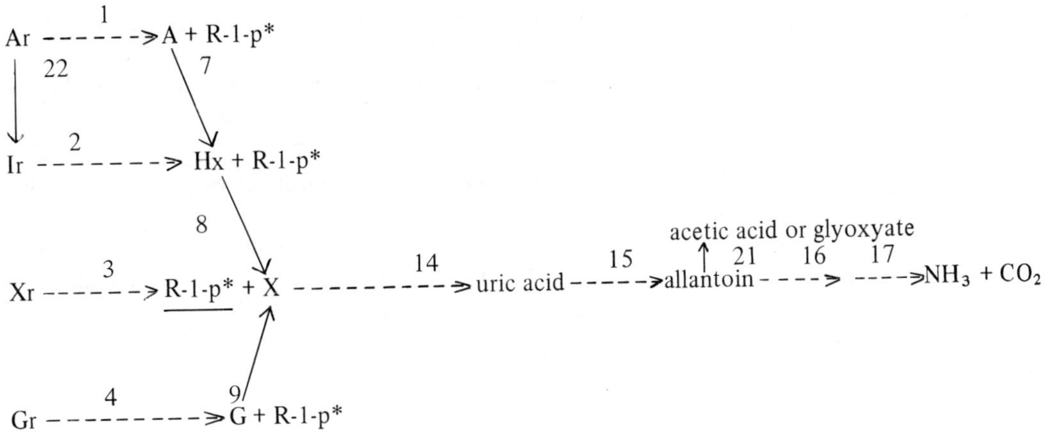


Cr - - - - -(5)- - - ->C + R-1-p*

Cr —(10)→ Ur; C —(11)→ U

Ur - - - - -(6)- - - -> R-1-p* + U - - - -(18)- - - -> diHU- - - - -(19)- - - - ->Bur. prop. - - - -(20)- - - -> B-alanine

* Catabolism of R-1-p is as follows: R-1-p - - -(12)- - - ->R-5-p - - -(13)- - - ->acetal + Gl-3-$PO_4$.

### The Enzymes Involved

| | |
|---|---|
| 1, 2, 3, 4, 5, 6 | Ribonucleoside phosphorylases. Probably unique for each ribonucleoside. Reverse reaction is possible (see anabolism). Each base can also be formed by the reversal of the nucleotide pyrophosphorylase reaction (see anabolism). |
| 7 | Adenase. Specific for adenine deamination. |
| 8, 14 | Xanthine oxidase. Converts hypoxanthine to xanthine and xanthine to uric acid. |
| 15 | Uricase. Oxidase that decarboxylates uric acid. |
| 16, 17, 21 | Allantoin is ultimately converted to $CO_2$ + $NH_3$ by several enzymes including a deaminase, and acetic acid and glyoxylate are produced by bacteria. |
| 9, 10, 11, 22 | Deaminases. Specific for bases and ribonucleosides shown. |
| 12 | Ribomutase. Converts ribose-1-phosphate to ribose-5-phosphate. |

| Enzymes | Reference |
|---|---|
| 13 | Riboaldolase. Splits ribose-5-phosphate to acetaldehyde and 3-glyceraldehyde-phosphate. Enzymes 12 and 13 occur also with deoxyribose; whether the enzymes are the same is not known. |
| 18 | Dehydrogenase specific for uracil. Does not act on cytosine. |
| 19, 20 | Hydrolases involved in the illustrated reactions. It has been reported that barbituric acid as well as urea and malonic acid aldehyde are intermediates. |

**Reference Guide**


| | |
|---|---|
| 1, 2, 3, 4, 5, 6 | Paege, L. M., and Schlenk, F., *Arch. Biochem., 40,* 42 (1952).<br>Mahler, H. R., and Cordes, E. H., in *Biological Chemistry,* 2nd ed., p. 834, Harper and Row, New York (1971). |
| 7 | McElroy, W. D., in *Methods in Enzymology,* Vol. 6, p. 203, S. P. Colowick and N. O. Kaplan, Eds. Academic Press, New York (1963). |
| 8, 14 | Brock, T., in *Biology of Microorganisms,* p. 137, Prentice-Hall, Englewood Cliffs, New Jersey (1970). |
| 10, 11 | Siedler, A. J., and Holtz, M. T., *J. Biol. Chem., 238,* 697 (1963). |
| 22 | Wolfenden, R., Tomozawa, Y., and Bamman, B., *Biochemistry, 7,* 3965 (1968). |
| 9, 15, 16, 17, 18, 19, 20, 21 | Mahler, H. R., and Cordes, E. H., in *Biological Chemistry,* 2nd ed., p. 835, Harper and Row, New York (1971). |
| 12, 13 | Mortlock, R. P., Fossitt, D. D., and Wood, W. A., *Proc. Nat. Acad. Sci. U.S.A., 54,* 572 (1965). |

# MINERALS

# TRACE METALS

DR. E. D. WEINBERG

## PRODUCTS OR STRUCTURES WHOSE YIELD IS STIMULATED OR INHIBITED BY CONCENTRATIONS OF TRACE METALS THAT HAVE NO EFFECT ON THE AMOUNT OF GROWTH

S = stimulated I = inhibited

| Organism | Product or Structure | Metal Concentration x $10^{-5}$ $M^{a}$ | | | | | Reference |
|---|---|---|---|---|---|---|---|
| | | Mn | Fe | Cu | Zn | Co | |
| **Bacteria** | | | | | | | |
| *Bacillus* | | | | | | | |
| *anthracis* | 3,4-Dihydroxybenzoic acid | | S **20** | | | | 1 |
| | Coproporphyrin III | | I **5.0** | | | | 1 |
| | Protective antigen | S **0.5** | | | | | 2 |
| | | I **2.0** | | | | | 2 |
| *licheniformis* | Bacitracin | S 0.07 | | | | | 3 |
| | | I 4.0 | | | | | 3 |
| | Transformants | S **20** | | I **1.0** | | | 4, 5 |
| megaterium | Phage | S **10** | | | | | 6 |
| | Spores | S 0.5 | | S 0.03 | | | 7, 8 |
| *subtilis* | 2,3-Dihydroxybenzoylglycine | | I **0.15** | | | | 9 |
| | Bacillin | S **10** | | | | | 10 |
| | D-Glutamyl polypeptide | S 0.15 | | | | | 11 |
| | Mycobacillin | S 0.6 | S 0.5 | | | | 12 |
| | Subtilin | S 0.5 | | | | | 13 |
| | Transfectants | I **20** | | | | | 14 |
| *Brevibacterium* | | | | | | | |
| *ammoniagenes* | 5′-Inosinic acid | I 0.015 | | | | | 15 |
| *Clostridium* | | | | | | | |
| *acetobutylicum* | Riboflavin | | I 2.5 | | | I 2.5 | 16 |
| *perfringens* | Lecithinase | S **2.0** | I 10 | | S **2.0** | | 17 |
| *tetani* | Neurotoxin | | S 3.0 | | | | 18 |
| *Corynebacterium* | | | | | | | |
| *diphtheriae* | Coproporphyrin | | I 0.75 | | | | 19 |
| | Cytotoxin | | I 0.7 | | | | 20 |

## PRODUCTS OR STRUCTURES WHOSE YIELD IS STIMULATED OR INHIBITED BY CONCENTRATIONS OF TRACE METALS THAT HAVE NO EFFECT ON THE AMOUNT OF GROWTH (Continued)

S = stimulated I = inhibited

| Organism | Product or Structure | Metal Concentration x $10^{-5}$ $M^a$ Mn | Fe | Cu | Zn | Co | Reference |
|---|---|---|---|---|---|---|---|
| **Bacteria (Continued)** | | | | | | | |
| *Escherichia* | | | | | | | |
| *coli* | 2,3-Dihydroxybenzoylserine | | I 0.03 | | | | 21 |
| *Mycobacterium* | | | | | | | |
| *phlei* | Mycobactin | | I **0.3** | | | | 22 |
| *smegmatis* | Salicylic acid | | S **4.0** | | | | 23 |
| *Pseudomonas* | | | | | | | |
| *aeruginosa* | Alkyl-quinolinols | | I 2.0 | | | | 24 |
| | Fluorescin | | I 0.3 | | | | 25 |
| | Pyocyanine | | S 0.3 | | | | 26 |
| *Rhodopseudomonas* | | | | | | | |
| *spheroides* | Bacteriochlorophyll porphyrins | | S 0.2 | | | | 27 |
| *Serratia* | | | | | | | |
| *marcescens* | Pyrryldipyrryl methene | | S 0.3 | | | | 28 |
| | | | I 2.0 | | | | 28 |
| *Shigella* | | | | | | | |
| *shigae* | Neurotoxin | | I 0.6 | | | | 29 |
| *Staphylococcus* | | | | | | | |
| *aureus* | Enterotoxin | | S **10** | | | | 30 |
| **Actinomycetes** | | | | | | | |
| *Actinomyces* sp. | Actinorubin | | S **2.0** | | | | 31 |
| *Streptomyces* sp. | Protease | S **10** | | | | | 32 |
| *antibioticus* | Actinomycin | | S **10** | | S **10** | | 33 |
| *aureofaciens* | Chlortetracycline | | **I 100** | | | | 34 |
| *cinnamonensis* | Monensin | | S 100 | | | | 35 |

## PRODUCTS OR STRUCTURES WHOSE YIELD IS STIMULATED OR INHIBITED BY CONCENTRATIONS OF TRACE METALS THAT HAVE NO EFFECT ON THE AMOUNT OF GROWTH (Continued)

S = stimulated I = inhibited

| Organism | Product or Structure | Metal Concentration x $10^{-5}$ $M^a$ Mn | Fe | Cu | Zn | Co | Reference |
|---|---|---|---|---|---|---|---|
| **Actinomycetes (Continued)** | | | | | | | |
| *Streptomyces* | | | | | | | |
| *fradiae* | Neomycin | I 10 | S 1.0 | | S 0.1 | | 36 |
| | | | I 15 | | I 1.0 | | 36 |
| | Phosphonomycin | | S **4.0** | | | S **34** | 37 |
| *griseus* | Candicidin | | S **4.0** | | S **4.0** | | 38 |
| | Grisein | | S 4.0 | | | | 39 |
| | Streptomycin | | S 1.0 | | S 0.3 | | 40 |
| | | | | | I 20 | | 40 |
| *mediterranei* | Rifamycin | | S **20** | | | | 41 |
| *venezuelae* | Chloramphenicol | | S **2.0** | | S **2.0** | | 42 |
| *verticillatus* | Mitomycin | | S **40** | | | | 43 |
| **Fungi** | | | | | | | |
| *Aspergillus* | | | | | | | |
| *flavus* | Aflatoxin | | | | S 0.5 | | 44 |
| *melleus* | Ochratoxin | | | I **0.12** | S **6.0** | | 45 |
| *niger* | Citric acid | I 0.02 | I 6.0 | | I 2.0 | | 46, 47 |
| | Malformin | S 0.1 | | | I 20 | | 48 |
| | | I 10 | | | | | 48 |
| | Spores | | | | I **100** | | 49 |
| *Candida* | | | | | | | |
| *guilliermondii* | Riboflavin | | I 0.1 | | | | 50 |
| *Claviceps* | | | | | | | |
| *paspali* | Lysergic acid | | | | S 0.5 | | 51 |
| *purpurea* | Ergotamine | | | | S 1.0 | | 52 |
| *Fusarium* | | | | | | | |
| *vasinfectum* | Fusaric acid | | | | S 0.3 | | 53 |
| | | | | | I 0.6 | | 53 |
| *Penicillium* | | | | | | | |
| *chrysogenum* | Penicillin | | S **2.0** | I **1.0** | | | 54 |
| *coronata* | Vesicles | | | | S 12 | | 55 |
| | | | | | I 30 | | 55 |

## PRODUCTS OR STRUCTURES WHOSE YIELD IS STIMULATED OR INHIBITED BY CONCENTRATIONS OF TRACE METALS THAT HAVE NO EFFECT ON THE AMOUNT OF GROWTH (Continued)

S = stimulated I = inhibited

| Organism | Product or Structure | Metal Concentration x $10^{-5}$ *M*[a] Mn | Fe | Cu | Zn | Co | Reference |
|---|---|---|---|---|---|---|---|
| **Fungi (Continued)** | | | | | | | |
| *Penicillium* | | | | | | | |
| *griseofulvum* | Griseofulvin | | | | I **20** | | 56 |
| | Mycelianamide | | I **40** | | | | 57 |
| *notatum* | Penicillin | | | | S 0.1 | | 58 |
| | | | | | I 3.0 | | 58 |
| *rubrum* | Rubratoxin | | S 30 | | S 10 | | 59 |
| *urticae* | 6-Methylsalicylate | | | | I **0.1** | | 60 |
| | Gentisyl alcohol | | I 1.5 | | S **0.1** | | 60 |
| | Patulin | | S 1.5 | | S **0.1** | | 61 |
| *Pythium* | | | | | | | |
| *graminicola* | Oogonia | S **0.03** | | | S **2.0** | | 62 |
| *Ustilago* | | | | | | | |
| *sphaerogena* | Coproporphyrin | | | | S 0.3 | | 63 |
| | Ferrichrome | | | | I 0.2 | S 0.3 | 64 |

[a] Quantities printed in bold type represent 100% end point values; all other quantities represent 50% end point values.

## REFERENCES

1. Chao, K. -C., Hawkins, D., Jr., and Williams, R. P., *Proc. Fed. Amer. Soc. Exp. Biol., 26,* 1532 (1967).
2. Wright, G. G., Hedberg, M. A., and Slein, J. B., *J. Immunol., 72,* 263 (1954).
3. Weinberg, E. D., and Tonnis, S. M., *Appl. Microbiol., 14,* 850 (1966).
4. Thorne, C. B., and Stull, H. B., *J. Bacteriol., 91,* 1012 (1966).
5. Anagnostopoulos, C., and Spizizen, J., *J. Bacteriol., 81,* 741 (1961).
6. Huybers, K., *Ann. Inst. Pasteur (Paris), 84,* 242 (1953).
7. Weinberg, E. D., *Appl. Microbiol., 12,* 436 (1964).
8. Kolodziej, B. J., and Slepecky, R. A., *Nature (London), 194,* 504 (1962).
9. Peters, W. J., and Warren, R. A. J., *J. Bacteriol., 95,* 360, (1968).
10. Foster, J. W., and Woodruff, H. B., *J. Bacteriol., 51,* 363 (1946).
11. Leonard, C. G., Housewright, R. D., and Thorne, C. B., *J. Bacteriol., 76,* 499, (1958).
12. Majumdar, S. K., and Bose, S. K., *J. Bacteriol., 79,* 564 (1960).
13. Jansen, E. F., and Hirschmann, D. J., *Arch. Biochem., 4,* 297 (1944).
14. Bott, K. F., and Wilson, G. A., *Bacteriol. Rev., 32,* 370 (1968).
15. Furuya, A., Shizeo, A., and Kinoshita, S., *Appl. Microbiol., 16,* 981 (1968).
16. Hickey, R. J., *Arch. Biochem., 8,* 439 (1945).
17. Murata, R., Yamamoto, A., Soda, S., and Ito, A., *Jap. J. Med. Sci. Biol., 18,* 189 (1965).
18. Latham, W. C., Bent, D. F., and Levine, L., *Appl. Microbiol., 10,* 146 (1962).
19. Clarke, G. D., *J. Gen. Microbiol., 18,* 698 (1958).
20. Mueller, J. H., *J. Immunol., 42,* 343 (1941).

21. Brot, N., Goodwin, J., and Fales, H., *Biochem. Biophys. Res. Commun., 25,* 454 (1966).
22. Antoine, A. D., and Morrison, N. E., *J. Bacteriol., 95,* 245 (1968).
23. Ratledge, C., and Winder, F. G., *Biochem. J., 84,* 501 (1962).
24. Wensinck, F., van Dalen, A., and Wedema, M., *Antonie van Leeuwenhoek J. Microbiol. Serol., 33,* 73 (1967).
25. Totter, J. R., and Mosely, F. T., *J. Bacteriol., 65,* 45 (1953).
26. Kurachi, M., *Bull. Inst. Chem. Res. Kyoto Univ., 36,* 188 (1958).
27. Lascelles, J., *Biochem. J., 62,* 78 (1956).
28. Waring, W. S., and Werkman, C. H., *Arch. Biochem., 1,* 425 (1943).
29. van Heyningen, W. E., *Brit. J. Exp. Pathol., 36,* 373 (1955).
30. Casman, E. P., *Public Health Rep. (Washington), 73,* 599, (1958).
31. Kelner, A., and Morton, H. E., *J., Bacteriol., 53,* 695 (1947).
32. Mizusawa, K., Ichishawa, E., and Yoshida, F., *Agr. Biol. Chem., 30,* 35 (1966).
33. Katz, E., Pienta, P., and Sivak, A., *Appl. Microbiol., 6,* 236 (1958).
34. Van Dyck, P., and de Somer, P., *Antibiot. Chemother., 2,* 184 (1952).
35. Stark, W. M., Knox, M. C., and Westhead, J. E., *Antimicrob. Agents Chemother.,* p. 353 (1968).
36. Majumdar, M. K., and Majumdar, S. K., *Appl. Microbiol., 13,* 190 (1965).
37. Jackson, M., and Stapley, E. O., *Antimicrob. Agents Chemother.,* p. 291 (1969).
38. Acker, R. F., and Lechevalier, H., *Appl. Microbiol., 2,* 152 (1954).
39. Reynolds, D. M., and Waksman, S. A., *J. Bacteriol., 55,* 739 (1948).
40. Chesters, C. G. C., and Rolinson, G. N., *J. Gen. Microbiol., 5,* 559 (1951).
41. Sensi, P., and Thompson, J. E., in *Progress in Industrial Microbiology*, Vol. 6, p. 21, Hockenhull, D. J. D., ed. Chemical Rubber Co., Cleveland, Ohio (1967).
42. Gallichio, V., Gottlieb, D., and Carter, H. E., *Mycologia, 50,* 490 (1958).
43. Kirsch, E. J., in *Antibiotics*, Vol. 2, p. 66, Gottlieb, D., and Shaw, P. D., eds. Springer Verlag, Berlin, Germany (1967).
44. Mateles, R. I., and Adye, J. C., *Appl. Microbiol., 13,* 208 (1965).
45. Lai, M., Semeniuk, G., and Hesseltine, C. W., *Appl. Microbiol., 19,* 542 (1970).
46. Shu, P., and Johnson, M., *J. Bacteriol., 56,* 577 (1948).
47. Clark, D. S., Ito, K., and Horitsu, H., *Biotechnol. Bioeng., 8,* 465 (1965).
48. Steenbergen, S. M., and Weinberg, E. D., *Growth, 32,* 125 (1968).
49. Foster, J. W., *Bot. Rev., 5,* 207 (1939).
50. Tanner, F. W., Jr., Vojnovick, C., and van Lanen, J. M., *Science, 101,* 180 (1945).
51. Rosazza, J. P., Kelleher, W. J., and Schwarting, A. E., *Appl. Microbiol., 15,* 1270 (1967).
52. Stoll, A., Brack, A., Hofmann, A., and Kobel, H., *U.S. Patent 2,809,920* (1957).
53. Kalyanasundaram, R., and Saraswathi-Devi, L., *Nature (London), 175,* 945 (1955).
54. Koffler, H., Knight, S. G., and Frazier, W. C., *J. Bacteriol., 53,* 115 (1947).
55. Sharp, E. L., and Smith, F. G., *Phytopathology, 42,* 581 (1952).
56. Grove, J. F., in *Antibiotics*, Vol. 2, p. 123, Gottlieb, D., and Shaw, P. D., eds. Springer Verlag, Berlin, Germany (1967).
57. Bayan, A. P., Nager, U. F., and Brown, W. E., *Antimicrob. Agents Chemother.,* p. 669 (1962).
58. Foster, J. W., Woodruff, H. B., and McDaniel, L. E., *J. Bacteriol., 46,* 421 (1943).
59. Hays, A. W., Wyatt, E. P., and King, P. A., *Appl. Microbiol., 20,* 469 (1970).
60. Ehrensvärd, G., *Exp. Cell Res. Suppl., 3,* 102 (1955).
61. Brack, A., *Helv. Chim. Acta, 30,* 1, (1947).
62. Lenny, J. F., and Klemner, H. W., *Nature (London), 209,* 1365 (1966).
63. Komai, H., and Neilands, J. B., *Arch. Biochem. Biophys., 124,* 456 (1968).
64. Komai, H., and Neilands, J. B., *Science, 153,* 751 (1966).

# INORGANIC ENZYME COFACTORS

DR. CLARENCE H. SUELTER

Aluminum, antimony, arsenic, barium, boron, bromine, cadmium, calcium, chromium, cobalt, copper, gallium, iron, lead, lithium, magnesium, manganese, mercury, molybdenum, nickel, potassium, rubidium, selenium, silver, sodium, strontium, tin, titanium, vanadium, and zinc are inorganic elements normally found in biological tissues.[1] While the physiological functions of the majority of these elements are unknown, many are known to serve as cofactors in enzyme catalyzed reactions. There also exist a large number of metalloproteins that do not have a known enzymic function; they may serve as metal storage depots, in the transport of metals, or in detoxification. For those cations important in enzyme catalysis, Williams[2] visualized sodium and potassium as mobile charge carriers, magnesium and calcium as semimobile structure formers and triggers, zinc as a static superacid catalyst, and iron, copper, cobalt, and molybdenum as static redox catalysts.

Of the 840 enzymes known in 1964, 27% have metals built into their structures,[3] require added metals for activity, or are further activated by metal ions. Considerable insight into the role of metals in certain enzymes is now possible because of the application in the past decade of standard enzymological and chemical techniques as well as of specialized physical methods of X-ray crystallography, nuclear magnetic resonance, electron spin resonance, and chemical relaxation.

Enzymes requiring inorganic cofactors have been considered in two main classes, metal-activated enzymes and metalloenzymes. The major difference between metalloenzymes (which have built-in metals) and metal-activated enzymes (to which metals must be added for activity) is a quantitative one.[4] In the latter enzymes, the affinity for the metal is relatively low, but the role of the metal must be the same as in metalloenzymes. A stability constant of $10^8$ $M^{-1}$ has been suggested as a rough dividing line between metalloenzymes and metal-activated enzymes.[5]

In a recent review, Mildvan,[5] after examining the mechanism of reactions catalyzed by metal-requiring enzymes in an attempt to seek generalities that might be applied to other metal-activated enzymes, discussed the participation of cations in terms of three general coordination schemes for the ternary complexes of enzyme, metal, and substrate. For thirty-one enzymes, which were listed according to these tentative coordination schemes as substrate bridge (E-S-M), metal bridge (E-M-S) and enzyme bridge (M-E-S), several generalities emerge. Most, but not all, kinases form enzyme-nucleotide-metal complexes. The important exceptions are those kinases and other phosphotransferase enzymes that utilize pyruvate or phosphoenolpyruvate, which form enzyme-metal bridge complexes. Calcium is often an inhibitor of enzymes that form metal bridge complexes, but an alternative activator of enzymes that form substrate bridge complexes.

Little is known about the detailed role of metals in M-E-S complexes. Presumably it is structural, to stabilize a catalytically active conformation. Enzymes with substrate bridge complexes (E-S-M) are known only for ternary complexes of nucleoside di- and triphosphates. The concept of a metal bridge was initially introduced by Hellerman.[6] Klotz[7] suggested that the role of the metal bridge in metal-activated peptidases[8] was to stabilize the transition state of the peptide substrate. The first direct evidence for a metal bridge complex was obtained for the reaction catalyzed by pyruvate kinase, utilizing nuclear relaxation techniques.[9] For those enzymes requiring more than one inorganic cofactor (see the table below) it is reasonable to conclude that both metals may not necessarily participate through the same type of coordination scheme.

The table lists inorganic cofactors for enzyme-catalyzed reactions, including alternate cofactors, stoichiometry, and source. No claim is made that the listing is complete, especially for enzymes requiring magnesium, since such a compilation would be extensive. It can generally be assumed that all reactions involving di- and triphosphate nucleotides will have a requirement for magnesium. The reader is referred to Vallee and Wacker[10] for a listing of non-enzymic metalloproteins.

| Alternate Cofactor | Enzyme | Stoichiometry, g atoms/mole | Source | Reference |
|---|---|---|---|---|
| **Calcium** | | | | |
| | α-Amylase, E.C. 3.2.1.1 | 1 | Human saliva | 11 |
| | α-Amylase, E.C. 3.2.1.1 | 1–2 | Human saliva | 12 |
| | α-Amylase, E.C. 3.2.1.1 | 1–2 | Porcine pancreas | 13 |
| | α-Amylase, E.C. 3.2.1.1 | 2–3 | *Aspergillus oryzae* | 12, 14 |
| | α-Amylase, E.C. 3.2.1.1 | 3 | *Bacillus subtilis* | 12 |
| | α-Amylase, E.C. 3.2.1.1 | 4 | *Bacillus subtilis* | 15 |
| | Nuclease | 1 | *Staphylococcus aureus* | 16 |
| | Proteinase | | Rat brain | 17, 18 |
| | Proteinase | | Rabbit muscle | 19 |
| | Proteinase | 1–2 | *Pseudomonas aeruginosa* IFO 3080 | 20 |
| **Copper** | | | | |
| | Ascorbate oxidase, E.C. 1.10.3.3 | 8 | *Cucumis sativus* | 21 |
| | Benzylamine oxidase | 1.9–2.3 | Porcine plasma | 22 |
| | Diamine oxidase, E.C. 1.4.3.6 | 2.17 | Porcine kidney | 23 |
| | 3,4-Dihydroxyphenylethylamine β-hydroxylase | 3.64 | Bovine adrenal | 24 |
| | 3,4-Dihydroxyphenylethylamine β-hydroxylase | 5.1 | Bovine adrenal | 25 |
| | Galactose oxidase, E.C. 1.1.3.9 | 1 | *Polyporus circinatus* | 26 |
| | Laccase, E.C. 1.10.3.2 | 4 | *Polyporus versicolar* | 27 |
| | Laccase, E.C. 1.10.3.2 | 4 | *Rhus vernicifera* | 28 |
| | Tyrosinase, E.C. 1.10.3.1 | 1 | *Neurospora crassa* | 29 |
| **Iron** | | | | |
| | Aconitase, E.C. 4.2.1.3 | 2 | Porcine heart | 30 |
| Mn > Fe > Co > Ni > Mg | Arginase, E.C. 3.5.3.1 | | Yeast | 31 |
| | Dihydrocortic dehydrogenase, E.C. 1.3.3.1 | 4 | *Zymobacterium oroticum* | 32, 33 |
| | Metapyrocatechase, E.C. 1.13.1.2 | $Fe^{+3}$ | *Pseudomonas arvilla* | 34 |
| | Pyrocatechase, E.C. 1.13.1.1 | 2 $Fe^{+2}$ | *Pseudomonas arvilla* | 34, 35 |
| **Magnesium** | | | | |
| Mg > Mn | Adenylase kinase, E.C. 2.7.4.3 | 2 | Muscle | 36, 37 |
| Mg = Mn > Co > Ca > Sn > Fe | Arginine kinase, E.C. 2.7.3.3 | | *Homarus vulgaris* (Australian crayfish) | 38 |
| Mn, Fe, Zn | Citrate lyase, E.C. 4.1.3.6 | | *Aerobacter aerogenes* | 39 |
| Mn | Citrate lyase, E.C. 4.1.3.6 | | *Streptococcus diacetilactis* | 40 |
| Mg, Ca, Mn | Creatine kinase, E.C. 2.7.3.2 | 2 | Muscle | 36, 41 |
| | DNA polymerase, E.C. 2.7.7.7 | 1 | *Escherichia coli* | 42 |
| | Enolase, E.C. 4.2.1.11 | 4 | Yeast | 43, 44 |
| Mg > Mn > Ca | Glutamine synthetase, E.C. 6.3.1.2 | 12 | *Escherichia coli* | 45, 46 |
| | Hexokinase, E.C. 2.7.1.1 | | Yeast | 47 |
| Mg >> Mn, Co | Inorganic pyrophosphatase, E.C. 3.6.1.1 | | Yeast | 48 |

| Alternate Cofactor | Enzyme | Stoichiometry, g atoms/mole | Source | Reference |
|---|---|---|---|---|
| **Magnesium (continued)** | | | | |
| Mg > Mn | PEP carboxylase | 6.2 ± 1.5 | Peanut cotyledons | 49, 50 |
| | PEP synthetase | | *Escherichia coli* | 51 |
| Mg > Ni > Co > Mn > Cd > Zn | Phosphoglucomutase, E.C. 2.7.5.1 | 1 | Rabbit muscle | 52, 53 |
| Mg, Mn | 3-Phosphoglyceric acid kinase, E.C. 2.7.2.3 | 1 subunit | Rabbit muscle, yeast | 54, 55, 56 |
| **Magnesium–Monovalent Cations** | | | | |
| Mg > Mn; K > Rb > $NH_4$ | Acetokinase, E.C. 2.7.2.1 | | Rumen microorganism | 57 |
| Mg = Mn = Fe > Co > Ca; K, $NH_4$, Rb, Na, Li | Acetyl CoA synthetase, E.C. 6.2.1.1 | | Bovine heart | 58 |
| | Acetyl CoA synthetase, E.C. 6.2.1.1 | | Rabbit heart | 59 |
| | Acetyl CoA synthetase, E.C. 6.2.1.1 | | Rat heart | 59 |
| K > Rb = $NH_4$ > Cs > Li | Acetyl phosphatase | | Guinea pig kidney cortex | 60 |
| | Adenyl cyclase | | Brain | 61 |
| Mg > Mn | Alanylalanine synthetase, E.C. 6.3.2.4 | | *Streptococcus faecalis* | 62 |
| $NH_4$ > K | Amino acid polymerase | | *Escherichia coli* | 63, 64, 65 |
| K > $NH_4$ > Na | Aspartokinase, E.C. 2.7.2.4 | | *Bacillus polymyxa* | 66 |
| Mg > Mn | Aspartokinase, E.C. 2.7.2.4 | | *Rhodospeudomonas spheroides* | 67 |
| Na, K | ATPase, E.C. 3.6.1.3 | | Human erythrocyte membranes | 68 |
| Na, K | ATPase, E.C. 3.6.1.3 | | Muscle | 69 |
| | Carbamyl phosphatase | | Guinea pig brain microsomes | 70 |
| | Carbamyl phosphate synthetase, E.C. 2.7.2.2 | | *Escherichia coli* | 71 |
| | Carbamyl phosphate synthetase, E.C. 2.7.2.2 | | Rat liver | 72 |
| | Formyltetrahydrofolate synthetase, E.C. 6.3.4.3 | | Chicken liver | 73 |
| $NH_4$ > K > Rb > Cs > Na > Li | Formyltetrahydrofolate synthetase, E.C. 6.3.4.3 | | *Clostridium acidi-urici* | 74 |
| K > $NH_4$ > Rb > Cs > Li > Na | Formyltetrahydrofolate synthetase, E.C. 6.3.4.3 | | *Clostridium cylindrosporum* | 74 |
| Mg >> Mn > Co; $NH_4$ > K > Na | Formyltetrahydrofolate synthetase, E.C. 6.3.4.3 | | Human erythrocytes | 73 |
| $NH_4$ > $K^+$ > Na > Rb > Li > Cs | Formyltetrahydrofolate synthetase, E.C. 6.3.4.3 | | Leukocytes | 75 |
| | Formyltetrahydrofolate synthetase, E.C. 6.3.4.3 | | *Micrococcus aerogenes* | 76 |
| Mg > Mn > Ca | Formyltetrahydrofolate synthetase, E.C. 6.3.4.3 | | Spinach leaves | 77 |
| Rb > K > Na = $NH_4$ > Cs > Li | Fructose-1-phosphate kinase, E.C. 2.7.1.3 | | Bovine liver | 78 |
| | γ-Glutamyl synthetase, E.C. 6.3.2.2 | | *Phaseolus vulgaris* (seedlings) | 79 |
| | γ-Glutamyl synthetase, E.C. 6.3.2.2 | | *Triticum vulgare* (wheat germ) | 80 |
| | Glutathione synthetase, E.C. 6.3.2.3 | | Pigeon liver | 81 |
| | Glutathione synthetase, E.C. 6.3.2.3 | | *Saccharomyces cerevisiae* | 82 |

| Alternate Cofactor | Enzyme | Stoichiometry, g atoms/mole | Source | Reference |
|---|---|---|---|---|
| **Magnesium–Monovalent Cations (continued)** | | | | |
| Mg = Mn > Co | Guanylate kinase, E.C. 2.7.4.8 | | *Escherichia coli* | 83 |
| | Leucyl-RNA synthetase, E.C. 6.1.1.4 | | *Escherichia coli* | 84 |
| $NH_4$ > Na, K | Lysyl-sRNA synthetase, E.C. 6.1.1.6 | | *Escherichia coli* | 85 |
| | D-Malate dehydrogenase (decarboxylating) | | *Escherichia coli* | 86 |
| Rb > K > Cs | L-Malate dehydrogenase (decarboxylating), E.C. 1.1.1.38 | | *Lactobacillus arabinosus* | 87, 88 |
| K = $NH_4$ > Rb > Li > Cs > Na | Methionine adenosyltransferase, E.C. 2.5.1.6 | | Baker's yeast | 89 |
| K > Rb > $NH_4$ > Li > Na > Cs | Methionine adenosyltransferase, E.C. 2.5.1.6 | | Rabbit liver | 89 |
| | NAD synthetase, E.C. 6.3.5.1 | | Rat liver | 90 |
| | NAD synthetase, E.C. 6.3.5.1 | | *Saccharomyces cerevisiae* | 90 |
| $NH_4$ > K | Pantothenate synthetase, E.C. 6.3.2.1 | | *Escherichia coli* | 91 |
| | Peptidase (ribosomal) | | *Escherichia coli* | 92 |
| | Phosphodiesterase, E.C. 3.1.4.1 | | *Escherichia coli* | 93, 94 |
| Mg > Mn | Phosphoenol pyruvate-carboxykinase, E.C. 4.1.1.32 | | *Aspergillus niger* | 95 |
| | Phosphoenol pyruvate-carboxykinase, E.C. 4.1.1.32 | | Porcine liver | 49, 95 |
| K > $NH_4$ | Phosphofructokinase, E.C. 2.7.1.11 | | Rabbit muscle | 96 |
| | Phosphofructokinase, E.C. 2.7.1.11 | | Rat brain | 97 |
| | Phosphofructokinase, E.C. 2.7.1.11 | | Sheep brain | 98 |
| | Phosphofructokinase, E.C. 2.7.1.11 | | Slime mold | 99 |
| K > $NH_4$ | Phosphofructokinase, E.C. 2.7.1.11 | | Yeast | 100 |
| | Phosphoribosyl formylglycine amidine synthetase, E.C. 6.3.5.3 | | Pigeon liver | 101 |
| K, $NH_4$ | Phosphotransacetylase, E.C. 2.3.1.8 | | *Clostridium kluyveri* | 102 |
| Rb > K > Cs | Propionyl-CoA carboxylase, E.C. 6.4.1.3 | | Bovine liver | 103 |
| | Propionyl-CoA carboxylase, E.C. 6.4.1.3 | | Pig heart | 104 |
| K > Rb > Cs > Li | Protein kinase, E.C. 2.7.1.37 | | Brain | 105 |
| | Pyruvate kinase, E.C. 2.7.1.40 | | *Amia colva* | 106 |
| | Pyruvate kinase, E.C. 2.7.1.40 | | *Curcurbita pepo* (seed) | 107 |
| $NH_4$ > K | Pyruvate kinase, E.C. 2.7.1.40 | | Mouse liver | 108 |

| Alternate Cofactor | Enzyme | Stoichiometry, g atoms/mole | Source | Reference |
|---|---|---|---|---|
| **Magnesium–Monovalent Cations (continued)** | | | | |
| Mg > Mn > Co; K > $NH_4$ > Rb > Tl > Cs > Na > Li | Pyruvate kinase, E.C. 2.7.1.40 | 4 Mn, 4 Tl | Muscle | 109, 110 |
| | Pyruvate kinase, E.C. 2.7.1.40 | | Rat brain | 111 |
| Mg > Mn; K > $NH_4$ > Na | Pyruvate kinase, E.C. 2.7.1.40 | | Yeast | 112, 113 |
| | Pyruvate kinase, E.C. 2.7.1.40 | | *Zea mays* (seed) | 114 |
| | Pyruvate phosphate dikinase | | *Bacteroides symbiosus* | 115 |
| | RNA nucleotidyltransferase, E.C. 2.7.7.6 | | *Bacillus stearothermophilus* | 116 |
| | Starch synthetase | | *Glycine max.* (seed) | 117 |
| | Starch synthetase | | *Phaseolus vulgaris* (seed) | 117 |
| | Starch synthetase | | *Pisum sativum* (seed) | 117 |
| | Starch synthetase | | *Solanum tuberosum* (tubers) | 117 |
| | Starch synthetase | | *Triticum aestivum* (seed) | 117 |
| | Starch synthetase | | *Zea mays* (seed) | 117 |
| | Succinyl CoA synthetase, E.C. 6.2.1.5 | | Tobacco leaves | 118 |
| | Tyrosyl-RNA synthetase, E.C. 6.1.1.1 | | Pancreas | 119 |
| K > Rb > $NH_4$ > Cs > Na = Li | Tyrosyl-RNA synthetase, E.C. 6.1.1.1 | | Rat liver | 120 |
| **Manganese** | | | | |
| | Aminopeptidase | | Bovine lens | 121 |
| | Aminopeptidase P | | *Escherichia coli* | 122 |
| | L-Arabinose isomerase, E.C. 5.3.1.3 | | *Escherichia coli* | 123 |
| Mn > Co > Mg > Cu > Ca > Cd | L-Arabinose isomerase, E.C. 5.3.1.3 | | *Lactobacillus gayonii* | 124 |
| Mn > Co, Ni | Arginase, E.C. 3.5.3.1 | 4 | Rat liver | 125 |
| | Histidine ammonia lyase, E.C. 4.3.1.3 | 4 | *Pseudomonas* | 126, 127 |
| Mn > $Fe^{+3}$ > $Fe^{+2}$ > Ca > Co | D-Lyxose isomerase, E.C. 5.3.1a | | *Aerobacter aerogenes* | 128 |
| Mn (tightly bound); Mn or Mg also required | Pyruvate carboxylase, E.C. 6.4.1.1 | 4 | Chicken liver | 129 |
| | Superoxide dismutase | | *Escherichia coli* | 130 |
| | D-Xylose isomerase, E.C. 5.3.1.5 | | *Lactobacillus brevis* | 131 |
| **Manganese–Monovalent Cations** | | | | |
| Mn > Mg > Co >> Ni > Zn > Ca | Aminopeptidase | | *Escherichia coli* | 132 |
| Mn > Mg ~ Fe ~ Zn ~ Co | β-Methyl aspartase, E.C. 4.3.1.2 | | *Bacterium cadaveris* | 133 |
| Mn > Co > Fe > Zn > Cd > Ni | β-Methyl aspartase, E.C. 4.3.1.2 | | *Clostridium tetanomorphum* | 134, 135, 136 |

| Alternate Cofactor | Enzyme | Stoichiometry, g atoms/mole | Source | Reference |
|---|---|---|---|---|
| **Molybdenum** | | | | |
| | Aldehyde oxidase | | Porcine liver | 137 |
| | Aldehyde oxidase | | Rabbit liver | 138 |
| | Nitrate reductase, E.C. 1.2.3.1 | 1 | *Escherichia coli* | 139 |
| | Nitrate reductase, E.C. 1.2.3.1 | 1–2 | *Neurospora crassa* | 140 |
| | Xanthine dehydrogenase | | *Clostridium cylindrosporum* | 141 |
| | Xanthine dehydrogenase | | *Micrococcus lactilyticus* | 32 |
| | Xanthine oxidase, E.C. 1.2.3.2 | | Avian liver | 32 |
| | Xanthine oxidase, E.C. 1.2.3.2 | | Porcine liver | 142 |
| **Monovalent Cations** | | | | |
| K, Rb | Aldehyde dehydrogenase, E.C. 1.2.1.5 | | *Saccharomyces cerevisiae* | 143 |
| K > Rb | Ethanolamine deaminase | | *Clostridium* | 144 |
| | Forminotetrahydrofolate cyclodeaminase | | Rat liver | 145 |
| $NH_4$ > K > Na | Glucose-6-phosphate D-myoinositol 1-phosphate cyclase | | Yeast | 146 |
| $NH_4$ > Rb > K | Glycerol dehydrase | | *Lactobacillus* 208A | 147 |
| | Glycerol dehydrogenase, E.C. 1.1.1.6 | | *Aerobacter aerogenes* | 148 |
| K > Na | Homoserine dehydrogenase, E.C. 1.1.1.3 | | *Escherichia coli* | 149, 150 |
| K > $NH_4$ > Na | 3-Hydroxyisopropyl malate dehydrogenase | | *Salmonella typhimurium* (strain LT-2) | 151 |
| | Inosine monophosphate dehydrogenase, E.C. 1.2.1.14 | | *Aerobacter aerogenes* | 152 |
| K > Na > Rb > $NH_4$ > Li | Methylene tetrahydrofolate dehydrogenase, E.C. 1.5.1.5 | | Baker's yeast | 153 |
| | Phosphoribosyl aminoimidazole carboxamide formyl transferase, E.C. 2.1.2.3 | | Chicken liver, pigeon liver | 154 |
| Rb = $NH_4$ = K = Li > Na | Porphobilinogen synthase, E.C. 4.2.1.24 | | *Rhodopseudomonas spheroides* | 155 |
| $NH_4$ > K > Rb > $CH_3$ $NH_3$ > Na > Cs | Propanediol dehydratase, E.C. 4.2.1.28 | | *Aerobacter aerogenes* | 156 |
| $NH_4$ = K > Na | Serine deaminase, E.C. 4.2.1.13 | | *Escherichia coli* | 157 |
| | Serine deaminase, E.C. 4.2.1.13 | | Rat liver | 158 |
| Rb > Cs > $NH_4$ > K > Na > Li | Tartrate dehydrogenase | | *Pseudomonas putida* | 159 |
| | Tetrahydrofolate dehydrogenase, E.C. 1.5.1.3 | | Carcinoma cells | 160 |
| $NH_4$ > Cs > K > Rb > Na | Tetrahydrofolate dehydrogenase, E.C. 1.5.1.3 | | Guinea pig liver | 160 |
| | Tetrahydrofolate dehydrogenase, E.C. 1.5.1.3 | | Mouse leukemia cells | 160 |
| K > $NH_4$ ⩾ Rb > Li > Na | Threonine deaminase, E.C. 4.2.1.16 | | Sheep liver | 161 |

| Alternate Cofactor | Enzyme | Stoichiometry, g atoms/mole | Source | Reference |
|---|---|---|---|---|
| **Monovalent Cations (continued)** | | | | |
| $NH_4$ > K > Rb > Li > Cs | Threonine deaminase, E.C. 4.2.1.16 | | Yeast | 162 |
| Cs > Rb > K = $NH_4$ > Li > Na | Threonine dehydrogenase | | *Staphylococcus aureus* | 163 |
| | Tryptophanase | | *Escherichia coli* | 164 |
| K > Na | Tryptophane synthase, E.C. 4.2.1.20 | | *Bacillus subtilis* | 165 |
| **Zinc** | | | | |
| | Alcohol dehydrogenase, E.C. 1.1.1.1 | 2 | Human liver | 166 |
| | Alcohol dehydrogenase, E.C. 1.1.1.1 | 4 | Equine liver | 167 |
| | Alcohol dehydrogenase, E.C. 1.1.1.1 | 4 | Yeast | 168 |
| | Alkaline phosphatase, E.C. 3.1.3.1 | 4 | *Escherichia coli* | 169 |
| | Carbonic anhydrase, E.C. 4.2.1.1 | 1 | Bovine and human erythrocytes | 170 |
| Peptidase Co > Zn = Ni >> Mn; esterase Cd > Hg > Zn ⩾ Co > Ni > Pb > Mn | Carboxypeptidase A, E.C. 3.4.2.1 | 1 | Bovine pancreas | 171, 172 |
| | Carboxypeptidase A, E.C. 3.4.2.1 | | Pacific spiny dogfish pancreas and porcine pancreas | 172 |
| Peptidase Co; esterase Cd > Co > Zn | Carboxypeptidase B, E.C. 3.4.2.2 | 1 | Bovine and porcine pancreas | 173 |
| | Dipeptidase | 1 | Mouse ascites tumor | 174 |
| | Dipeptidase | 1 | Porcine kidney | 175 |
| Mg, Mn | Leucine aminopeptidase, E.C. 3.4.1.1 | 4–6 | Porcine kidney | 176 |
| Mg, Mn | Leucine aminopeptidase, E.C. 3.4.1.1 | 8–12 | Bovine lens | 177 |
| | Mannose-6-phosphate isomerase, E.C. 5.3.1.7 | 1 | Yeast | 178 |
| Mg ~ Co > Mn | Protease | 1 | *Serratia* | 179 |
| **Zinc–Calcium** | | | | |
| Apoenzyme (Zn) reactivated by Co + Mn | Neutral proteinase | 1 Zn | *Bacillus subtilis* | 180 |
| Zn apo-activated by Co > Fe > Mn | Protease | | *Bacillus subtilis*, var. *amylosacchariticus* | 180 |
| | Thermolysin | 1 Zn, 4 Ca | *Bacillus thermoproteolyticus* | 181, 182 |
| **Zinc–Cobalt** | | | | |
| | Transcarboxylase | 4 Zn, 2 Co | *Propionibacterium shermanii* | 183 |

| Alternate Cofactor | Enzyme | Stoichiometry, g atoms/mole | Source | Reference |
|---|---|---|---|---|
| **Zinc–Copper** | | | | |
| | Cytocuprein,[184] formerly labeled erythrocuprein, hepatocuprein, and cerebrocuprein, and shown to have superoxide dismutase activity; distribution of superoxide dismutase is reviewed in Reference 185 | 2 Zn, 2 Cu | Human liver, brain, and erythrocytes | 186, 187, 188 |
| **Zinc–Monovalent Cations** | | | | |
| | AMP aminohydrolase, E.C. 3.5.4.6 | | Bovine brain | 189 |
| | AMP aminohydrolase, E.C. 3.5.4.6 | 2 Zn | Rat muscle | 190 |
| | AMP aminohydrolase, E.C. 3.5.4.6 | 2.8 Zn | Rabbit muscle | 191 |
| | Fructose-1,6-diphosphate aldolase (Class II), E.C. 4.1.2.13 | | *Euglena gracilis* | 185 |
| Zn > Co > Fe > Mn > Ni | Fructose-1,6-diphosphate aldolase (Class II), E.C. 4.1.2.13 | 2 Zn | *Saccharomyces cerevisiae* | 192 |
| | Fructose-1,6-diphosphate aldolase (Class II), E.C. 4.1.2.13 | | 25 species of bacteria | 193 |
| | Pyruvate carboxylase, E.C. 6.4.1.1 | 4 Zn | Yeast | 194 |
| | Rhamnulose-1-phosphate aldolase | | *Escherichia coli* | 195 |

Extensive use of reviews in References 3, 5, 10, and 196 was made during the preparation of this table.

## REFERENCES


1. Comar, C. L., and Bronner, F., Eds., *Mineral Metabolism,* Vols. 1 and 2. Academic Press, New York (1960).
2. Williams, R. J. P., *Quart. Rev., 24,* 331 (1970).
3. Dixon, M., and Webb, E. C., Eds., in *Enzymes,* 2nd ed., p. 672. Academic Press, New York (1964).
4. Malmstrom, B. G., and Rosenberg, A., *Advan. Enzymol., 21,* 131 (1959).
5. Mildvan, A. S., in *The Enzymes,* 3rd ed., Vol. 2, p. 446, P. D. Boyer, H. A. Lardy and K. Myrbäck, Eds. Academic Press, New York (1971).
6. Hellerman, L., *Physiol. Rev., 17,* 454 (1937).
7. Klotz, I. M., in *The Mechanism of Enzyme Action,* p. 257. W. D. McElroy and B. Glass, Eds. Johns Hopkins Press, Baltimore, Maryland (1954).
8. Smith, E. L., Davis, N. C., Adams, E., and Spackman, D. H., in *The Mechanism of Enzyme Action,* p. 291, W. D. McElroy and B. Glass, Eds. Johns Hopkins Press, Baltimore, Maryland (1954).

9. Mildvan, A. S., Leigh, J. S., Jr., and Cohn, M., *Biochemistry, 6,* 1805 (1967).
10. Vallee, B. L., and Wacker, W. E. C., in *The Protein,* p. 33, H. Neurath, Ed. Academic Press, New York (1970).
11. Hsiu, J., Fischer, E. H., and Stein, E. A., *Biochemistry, 3,* 61 (1964).
12. Vallee, B. L., Stein, E. A., Sumerwell, W. N., and Fischer, E. H., *J. Biol. Chem., 234,* 2901 (1959).
13. Yamamoto, T., Stein, E. A., and Fischer, E. H., cited in Reference 11.
14. Stein, E. A., Junge, J. M., and Fischer, E. H., *J. Biol. Chem., 235,* 371 (1960).
15. Junge, J. M., Stein, E. A., Neurath, H., and Fischer, E. H., *J. Biol. Chem., 234,* 556 (1959).
16. Cuatrecasas, P., Fuchs, S., and Anfinsen, C. B., *J. Biol. Chem., 242,* 3063 (1967).
17. Drummond, G. I., and Duncan, L., *J. Biol. Chem., 243,* 5532 (1968).
18. Guroff, G., *J. Biol. Chem., 239,* 149 (1964).
19. Huston, R. B., and Krebs, E. G., *Biochemistry, 7,* 2116 (1968).
20. Morihara, K., and Tsuzuki, H., *Biochim. Biophys. Acta, 92,* 351 (1964).
21. Nakamura, T., Makino, N., and Ogura, Y., *J. Biochem., 64,* 189 (1968).
22. Buffoni, F., Della Corte, L., and Knowles, P. F., *Biochem. J., 106,* 575 (1968).
23. Yamada, H., Kumagai, H., Kawasaki, H., Matsui, H., and Ogata, K., *Biochem. Biophys. Res. Commun., 29,* 723 (1967).
24. Friedman, S., and Kaufman, S., *J. Biol. Chem., 241,* 2256 (1966).
25. Blumberg, W. E., Goldstein, M., Lauber, E., and Peisach, J., *Biochim. Biophys. Acta, 99,* 187 (1965).
26. Amaral, D., Bernstein, L., Morse, D., and Horecker, B. L., *J. Biol. Chem., 238,* 2281 (1963).
27. Mosbach, R., *Biochim. Biophys. Acta, 73,* 204 (1963).
28. Reinhammar, B., *Biochim. Biophys. Acta, 205,* 35 (1970).
29. Fling, M., Horowitz, N. H., and Heineman, S. F., *J. Biol. Chem., 238,* 2045 (1963).
30. Villafranca, J. J., and Mildvan, A. S., *J. Biol. Chem., 246,* 772 (1971).
31. Middlehoven, W. J., *Biochim. Biophys. Acta, 191,* 110 (1969).
32. Smith, S. T., Rajagopolan, K. V., and Handler, P., *J. Biol. Chem., 242,* 4108 (1967).
33. Friedmann, H. C., and Vennesland, B., *J. Biol. Chem., 235,* 1526 (1960).
34. Hayaishi, O., *Bacteriol. Rev., 30,* 720 (1966).
35. Kojima, Y., Fujisawa, H., Nakazawa, A., Nakazawa, T., Kanetsuna, H., Taniuchi, H., Zozaki, M., and Hayaishi, O., *J. Biol. Chem., 242,* 3270 (1967).
36. Kuby, S. A., Mahowald, T. A., and Noltman, E. A., *Biochemistry, 1,* 748 (1962).
37. Noda, L., in *The Enzymes,* 2nd ed., p. 143, P. D. Boyer, H. A. Lardy and K. Myrbäck, Eds. Academic Press, New York (1962).
38. Virden, R., Watts, D. C., and Baldwin, E., *Biochem. J., 94,* 536 (1965).
39. Raman, C. S., *Biochim. Biophys. Acta, 52,* 212 (1961).
40. Harvey, R. J., and Collins, E. B., *J. Biol. Chem., 238,* 2648 (1963).
41. Kuby, S. A., and Noltman, E. A., in *The Enzymes,* 2nd ed., Vol. 6, p. 530. P. D. Boyer, H. A. Lardy and K. Myrbäck, Eds. Academic Press, New York (1962).
42. Englund, P. T., Huberman, J. A., Jovin, T. M., and Kornberg, A., *J. Biol. Chem., 244,* 3038 (1969).
43. Wold, F., and Ballou, C. E., *J. Biol. Chem., 227,* 301, 313 (1957).
44. Hanlon, D. P., and Westhead, E. W., *Biochemistry, 8,* 4247 (1969).
45. Denton, M. D., and Ginsburg, A., *Biochemistry, 9,* 617 (1970).
46. Kingdon, H. S., Hubbard, J. S., and Stadtman, E. R., *Biochemistry, 7,* 2136 (1968).
47. Crane, R. K., in *The Enzymes,* 2nd ed., Vol. 6, p. 47, P. D. Boyer, H. A. Lardy and K. Myrbäck, Eds. Academic Press, New York (1962).
48. Kunitz, M., and Robbins, P. W., in *The Enzymes,* 2nd ed., Vol. 5, p. 169, P. D. Boyer, H. A. Lardy and K. Myrbäck, Eds. Academic Press, New York (1961).
49. Millar, R. S., Mildvan, A. S., Chang, H.-C., Easterday, R. L., Maruyama, H., and Lane, M. D., *J. Biol. Chem., 243,* 6030 (1968).
50. Maruyama, H., Easterday, R. L., Chang, H.-C., and Lane, M. D., *J. Biol. Chem., 241,* 4205 (1966).
51. Cooper, R. A., and Kornberg, H. L., *Biochim. Biophys. Acta, 141,* 211 (1967).
52. Ray, W. J., Jr., *J. Biol. Chem., 244,* 3740 (1969).
53. Peck, E. J., Jr., and Ray, W. J., Jr., *J. Biol. Chem., 244,* 3748 (1969).
54. Scopes, R. K., *Biochem. J., 113,* 551 (1969).
55. Larsson-Raznikiewicz, M., and Malmstrom, B. G., *Arch. Biochem. Biophys., 92,* 94 (1961).
56. Caban, C. E., and Hass, L. F., *Fed. Proc., 30,* 1103 (1971).
57. Van Campen, D. R., and Matrone, G., *Biochim. Biophys. Acta, 85,* 410 (1964).
58. Webster, L. T., Jr., *J. Biol. Chem., 241,* 5504 (1966); *242,* 1232 (1967).
59. Von Korff, R. W., *J. Biol. Chem., 203,* 265 (1953).
60. Bader, H., and Sen, A. K., *Biochim. Biophys. Acta, 118,* 116 (1966).
61. Belleau, B., and MacDonald, I. A., Biological Chemistry Abstract No. 81, *Abstracts of the 158th National Meeting of the American Chemical Society, New York, September 1969.* American Chemical Society, Washington, D.C. (1969).

62. Neuhaus, F. C., *J. Biol. Chem., 237*, 778 (1962).
63. Conway, T. W., *Proc. Nat. Acad. Sci. U.S.A., 51*, 1216 (1964).
64. Levine, H., Trindle, M. R., and Moldave, K., *Nature, 211*, 1302 (1966).
65. Maden, B. E. H., and Monro, R. E., *Eur. J. Biochem., 6*, 309 (1968).
66. Paulus, H., and Gray, E., *J. Biol. Chem., 239*, PC4008 (1964).
67. Datta, P., and Prakash, L., *J. Biol. Chem., 241*, 5827 (1966).
68. Bowen, W. J., and Kerwin, T. D., *J. Biol. Chem., 211*, 237 (1954).
69. Ahmed, K., and Judah, J. D., *Biochim. Biophys. Acta, 104*, 112 (1965).
70. Yoshida, H., Izumi, F., and Nagai, K., *Biochim. Biophys. Acta, 120*, 183 (1966).
71. Anderson, P. M., and Meister, A., *Biochemistry, 5*, 3137 (1966).
72. Marshall, M., Metzenberg, R. L., and Cohen, P. P., *J. Biol. Chem., 236*, 2229 (1961).
73. Bertino, J. R., Simmons, B., and Donohue, D. M., *J. Biol. Chem., 237*, 1314 (1962).
74. MacKenzie, R. E., and Rabinowitz, J. C., *J. Biol. Chem., 246*, 3731 (1971).
75. Bertino, J. R., Alenty, A., Gabrio, B. W., and Huennekens, F. M., *Clin. Res., 8*, 206 (1960).
76. Whiteley, H. R., and Huennekens, F. M., *J. Biol. Chem., 237*, 1290 (1962).
77. Hiatt, A. J., *Plant Physiol., 39*, 475 (1964).
78. Parks, R. E., Jr., Ben-Gershom, E., and Lardy, H. A., *J. Biol. Chem., 227*, 231 (1957).
79. Webster, G. C., *Plant Physiol., 28*, 728 (1953).
80. Webster, G. C., and Varner, J. E., *Arch. Biochem. Biophys., 52*, 22 (1954).
81. Snoke, J. E., Yanari, S., and Bloch, K., *J. Biol. Chem., 201*, 573 (1953).
82. Snoke, J. E., *J. Biol. Chem., 213*, 813 (1955).
83. Oeschger, M. P., and Bessman, M. J., *J. Biol. Chem., 241*, 5452 (1966).
84. Yu, C. T., and Hirsch, D., *Biochim. Biophys. Acta, 142*, 149 (1967).
85. Waldenstrom, J., *Eur. J. Biochem., 5*, 239 (1968).
86. Stern, J. R., and Hegre, C. S., *Fed. Proc., 26*, 605 (1967).
87. Lwoff, A., and Ionesco, H., *Compt. Rend., 225*, 77 (1947); *Chem. Abstr., 42*, 5496a (1948).
88. Nossal, P. M., *Biochem. J., 49*, 407 (1951).
89. Mudd, S. H., and Cantoni, G. L., *J. Biol. Chem., 231*, 481 (1958).
90. Preiss, J., and Handler, P., *J. Biol. Chem., 233*, 493 (1958).
91. Maas, W. K., *J. Biol. Chem., 198*, 23 (1952).
92. Tsai, C. S., and Matheson, A. T., *Can. J. Biochem., 43*, 1643 (1965).
93. Singer, M. F., and Tolbert, G., *Science, 145*, 593 (1964).
94. Spahr, P. F., and Schlessinger D., *J. Biol. Chem., 238*, PC2251 (1963).
95. Woronick, C. L., and Johnson, M. J., *J. Biol. Chem., 235*, 9 (1960).
96. Paetkau, V., and Lardy, H. A., *J. Biol. Chem., 242*, 2035 (1967).
97. Muntz, J. A., and Hurwitz, J., *Arch. Biochem. Biophys., 32*, 137 (1951).
98. Lowry, O. H., and Passonneau, J. V., *J: Biol. Chem., 241*, 2268 (1966).
99. Bauman, P., and Wright, B. E., *Biochemistry, 7*, 3653 (1968).
100. Muntz, J. A., *J. Biol. Chem., 171*, 653 (1947).
101. Melnick, I., and Buchanan, J. M., *J. Biol. Chem., 225*, 157 (1957).
102. Stadtman, E. R., *J. Biol. Chem., 196*, 527 (1952).
103. Giorgio, A. J., and Plaut, G. W. E., *Biochim. Biophys. Acta*, 139, 487 (1967).
104. Edwards, J. B., and Keech, D. B., *Biochim. Biophys. Acta, 159*, 167 (1968).
105. Rodnight, R., and Lavin, B. E., *Biochem. J., 93*, 84 (1964).
106. Boyer, P. D., *J. Cell. Comp. Physiol., 42*, 71 (1953).
107. McCollum, R. E., Hageman, R. H., and Tryner, E. H., *Soil Sci., 86*, 324 (1958).
108. Carminatti, H., Jimenez DeAsua, L., Recondo, E., Passeron, S., and Rozengurt, E., *J. Biol. Chem., 243*, 3051 (1968).
109. Kayne, F. J., *Arch. Biochem. Biophys., 143*, 232 (1962).
110. Boyer, P. D., in *The Enzymes*, 2nd ed., Vol. 6, p. 95, P. D. Boyer, H. A. Lardy, and K. Myrback, Eds. Academic Press, New York (1962).
111. Utter, M. F., *J. Biol. Chem., 185*, 499 (1950).
112. Hunsley, J. R., and Suelter, C. H., *J. Biol. Chem., 244*, 4815 (1969).
113. Washio, S., and Mano, Y., *J. Biochem. (Tokyo), 48*, 874 (1960).
114. Miller, G., and Evans, H. J., *Plant Physiol., 32*, 346 (1953).
115. Reeves, R. E., Menzies, R. A., and Hsu, D. S., *J. Biol. Chem., 243*, 5486 (1968).
116. Remold-O'Donnell, E., and Zillig, W., *Eur. J. Biochem., 7*, 318 (1969).
117. Nitsos, R. E., and Evans, H. J., *Plant Physiol., 44*, 1260 (1969).
118. Bush, L. P., *Plant Physiol., 44*, 347 (1969).
119. Schweet, R. S., Arlinghaus, R., Schaeffer, J., and Williamson, A., *Medicine (Baltimore), 43*, 731 (1964).
120. Holley, R. W., Brunngraber, E. F., Saad, F., and Williams, H. H., *J. Biol. Chem., 236*, 197 (1961).
121. Wolff, J. B., and Resnik, R. A., *Biochim. Biophys. Acta, 73*, 588 (1963).

122. Yaron, A., and Meynar, D., *Biochem. Biophys. Res. Commun.*, *32*, 658 (1968).
123. Patrick, J. W., and Lee, N., *J. Biol. Chem.*, *234*, 4312 (1968).
124. Nakamatu, T., and Yamanaka, K., *Biochim. Biophys. Acta*, *178*, 156 (1969).
125. Kolb, H. H., Kolb, H. J., and Greenberg, D. M., *J. Biol. Chem.*, *246*, 395 (1971).
126. Rechler, M. M., *J. Biol. Chem.*, *244*, 551 (1969).
127. Givot, I. L., Mildvan, A. S., and Abeleles, R. H., *Fed. Proc.*, *29*, 531 (1970).
128. Anderson, R. L., and Allison, D. P., *J. Biol. Chem.*, *240*, 2367 (1965).
129. Scrutton, M. C., Utter, M. F., and Mildvan, A. S., *J. Biol. Chem.*, *241*, 3480 (1966).
130. Keele, B. B., Jr., McCord, J. M., and Fridovich, I., *J. Biol. Chem.*, *245*, 6176 (1970).
131. Yamanaka, K., *Biochim. Biophys. Acta*, *151*, 670 (1968).
132. Tsai, C. S., and Matheson, A. T., *Can. J. Biochem.*, *43*, 1643 (1965).
133. Williams, V. R., in *142nd National Meeting of the American Chemical Society, September 1962*, p. 30C. American Chemical Society, Washington, D.C. (1969).
134. Williams, V. R., and Selbin, J., *J. Biol. Chem.*, *239*, 1635 (1964).
135. Barker, H. A., Smyth, R. D., Wilson, R. M., and Weissbach, H., *J. Biol. Chem.*, *234*, 320 (1959).
136. Bright, H. J., *Biochemistry*, *6*, 1191 (1967).
137. Fridovich, I., and Handler, P., *J. Biol. Chem.*, *231*, 899 (1958).
138. Rajagopolan, K. V., Fridovich, I., and Handler, P., *J. Biol. Chem.*, *237*, 922 (1962).
139. Nason, A., *Physiol. Rev.*, *26*, 16 (1962).
140. Garrett, R. H., and Nason, A., *J. Biol. Chem.*, *244*, 2870 (1969).
141. Bradshaw, W. H., and Barker, H. A., *J. Biol. Chem.*, *235*, 3620 (1960).
142. Brumby, P. E., Miller, R. W., and Massey, V., *J. Biol. Chem.*, *240*, 2222 (1965).
143. Milstein, C., and Stoppani, A. O. M., *Biochim. Biophys. Acta*, *28*, 218 (1958).
144. Kaplan, B. H., and Stadtman, E. R., *J. Biol. Chem.*, *243*, 1787 (1968).
145. Tabor, H., and Wyngarden, L., *J. Biol. Chem.*, *234*, 1830 (1959).
146. Chen, I.-W., and Charalampous, F. C., *J. Biol. Chem.*, *240*, 3507 (1965).
147. Smiley, K. L., and Sobolov, M., *Arch. Biochem. Biophys.*, *97*, 538 (1962).
148. Lin, E. C. C., and Magasanik, B., *J. Biol. Chem.*, *235*, 1820 (1960).
149. Barber, E. D., and Bright, H. J., *Proc. Nat. Acad. Sci. U.S.A.*, *60*, 1363 (1968).
150. Patte, J. C., LeBras, G., Loviny, T., and Cohen, G. N., *Biochim. Biophys. Acta*, *67*, 16 (1963).
151. Burns, R. O., Umbarger, H. E., and Gross, S. R., *Biochemistry*, *2*, 1053 (1963).
152. Magasanik, B., Moyed, H. S., and Gehring, L. B., *J. Biol. Chem.*, *226*, 339 (1957).
153. Ramasastri, B. V., Blakley, R. L., *J. Biol. Chem.*, *237*, 1982 (1962).
154. Flaks, J. G., Erwin, M. J., and Buchanan, J. M., *J. Biol. Chem.*, *229*, 603 (1957).
155. Nandi, D. L., Baker-Cohen, K. F., and Shemin, D., *J. Biol. Chem.*, *243*, 1224 (1968).
156. Toraya, T., Sugimoto, Y., Tamao, Y., Shimizu, S., and Fukui, S., *Biochemistry*, *10*, 3475 (1971).
157. Dupourque, D., Newton, W. A., and Snell, E. E., *J. Biol. Chem.*, *241*, 1233 (1966).
158. Pestana, A., and Sols, A., *FEBS Lett.*, *7*, 29 (1970).
159. Kohn, L. D., Packman, P. M., Allen, R. H., and Jakoby, W. B., *J. Biol. Chem.*, *243*, 2479 (1968).
160. Bertino, J. R., *Biochim. Biophys. Acta*, *58*, 377 (1962).
161. Nishimura, J. S., and Greenberg, D. M., *J. Biol. Chem.*, *236*, 2684 (1961).
162. Holzer, H., Cennamo, C., and Boll, M., *Biochem. Biophys. Res. Commun.*, *14*, 487 (1964).
163. Green, M. L., *Biochem. J.*, *91*, 550 (1964).
164. Happold, F. C., and Beechey, R. B., *Biochem. Soc. Symp.*, *15*, 52 (1958).
165. Schwartz, A. K., and Bonner, D. M., *Biochim. Biophys. Acta*, *89*, 337 (1964).
166. von Wartburg, J.-P., Bethune, J. L., and Vallee, B. L., *Biochemistry*, *3*, 1775 (1964).
167. Vallee, B. L., and Hoch, F. L., *J. Biol. Chem.*, *255*, 185 (1957).
168. Vallee, B. L., and Hoch, F. L., *Proc. Nat. Acad. Sci. U.S.A.*, *41*, 327 (1955).
169. Petitclerc, C., Lazdunski, C., Chappelet, D., Moulin, A., and Lazdunski, M., *Eur. J. Biochem.*, *14*, 301 (1970).
170. Davis, R. P., in *The Enzymes*, 2nd ed., Vol. 5, p. 545, P.D. Boyer, H. A. Lardy and K. Myrbäck, Eds. Academic Press, New York (1961).
171. Coleman, J. E., and Vallee, B. L., *J. Biol. Chem.*, *236*, 2244 (1961).
172. Hartshuk, J. A., and Lipscomb, W. N., in *The Enzymes*, 3rd ed., Vol. 3, p. 1, P. D. Boyer, H. A. Lardy and K. Myrbäck, Eds. Academic Press, New York (1971).
173. Folk, J. E., in *The Enzymes*, 3rd ed., Vol. 3, p. 57, P. D. Boyer, H. A. Lardy and K. Myrbäck, Eds. Academic Press, New York (1971).
174. Hayman, S., and Patterson, E. K., *J. Biol. Chem.*, *246*, 660 (1971).
175. Campbell, B. J., Lin, Y. C., Davis, R. V., and Ballew, E., *Biochim. Biophys. Acta*, *118*, 371 (1966).
176. Himmelhoch, S. R., *Arch. Biochem. Biophys.*, *134*, 597 (1969).
177. Fittkau, S., Kettmann, U., and Hanson, H., *J. Labelled Compd.*, *2*, 255 (1966).
178. Gracy, R. W., and Noltman, E. A., *J. Biol. Chem.*, *243*, 4109 (1968).
179. Miyata, K., Tomoda, K., and Isono, M., *Agr. Biol. Chem.*, *35*, 460 (1971).

180. Tsuru, D., Kira, H., Yamamoto, T., and Fukumoto, J., *J. Biol. Chem., 30,* 856 (1966).
181. Feder, J., Garrett, L. R., and Wildi, B. S., *Biochemistry, 10,* 4552 (1971).
182. Latt, S. A., Holmquist, B., and Vallee, B. L., *Biochem. Biophys. Res. Commun., 37,* 333 (1969).
183. Northrop, D. B., and Wood, H. G., *J. Biol. Chem., 244,* 5801 (1969).
184. Carrico, R. J., and Deutsch, H. F., *J. Biol. Chem., 245,* 723 (1969).
185. Rutter, W. J., and Groves, W. E., in *Taxonomic Biochemistry and Serology,* p. 417, C. A. Leone, Ed. Ronald Press, New York (1964).
186. McCord, J. M., and Fridovich, I., *J. Biol. Chem., 244,* 6049 (1969).
187. McCord, J. M., Keele, B. B., Jr., and Fridovich, I., *Proc Nat. Acad. Sci. U.S.A., 68,* 1024 (1971).
188. Carrico, R. J., and Deutsch, H. F., *J. Biol. Chem., 244,* 6087 (1969).
189. Setlow, B., and Lowenstein, J. M., *J. Biol. Chem., 242,* 607 (1967).
190. Raggi, A., Ranieri, M., Taponeco, G., Ronca-Testoni, S., Ronca, G., and Rossi, C. A., *FEBS Lett., 10,* 101 (1970).
191. Zielke, C. L., and Suelter, C. H., *J. Biol. Chem., 242,* 2179 (1971).
192. Rutter, W. J., *Fed. Proc., 23,* 1248 (1964).
193. Lebherz, H. G., and Rutter, W. J., *Biochemistry, 8,* 109 (1969).
194. Scrutton, M. C., and Young, M. R., *Fed. Proc., 29,* 597 (1970).
195. Chiu, T.-H., and Feingold, D. S., *Biochemistry, 8,* 98 (1969).
196. Malkin, R., and Malmstrom, B. G., *Advan. Enzymol., 33,* 177 (1970).

# MICROBIAL IRON TRANSPORT COMPOUNDS

DR. G. C. RODGERS and DR. J. B. NEILANDS

## INTRODUCTION

The iron transport apparatus of aerobic microorganisms consists of (a) low-molecular-weight substances, the iron transport compounds, which coordinate and solubilize ferric ion, and (b) an uptake system for the complexed ion. In the enteric bacteria, the genes for this system are clustered in an "iron operon"; in *Salmonella typhimurium* this is situated at about 20 minutes on the chromosomal map.

An iron transport function has only been proven unequivocally for enterobactin and ferrichrome. However, it is probable that many of the other substances listed in the following tables will be shown to perform the same function. The biosynthesis of the iron transport compounds listed is strongly repressed by the addition of iron to the medium.

In general, the iron transport compounds can be classed as either phenolates, e.g., enterobactin (enterochelin), or hydroxamates, e.g., ferrichrome. Most of them furnish six oxygen atoms, which engage the octahedrally directed bonds of the ferric ion. The stability constant for complex formation is approximately $10^{30}$. Since mainly oxygen atoms are located in the coordination sphere of the iron, the specificity for trivalent iron is relatively high; the metal ion is presumed to be released from the complex following its reduction to the ferrous state.

For a general review of the subject, see Reference 1.

## PHENOLATES

| Compound | Structure | Sources | Reference |
|---|---|---|---|
| Enterobactin (enterochelin) | 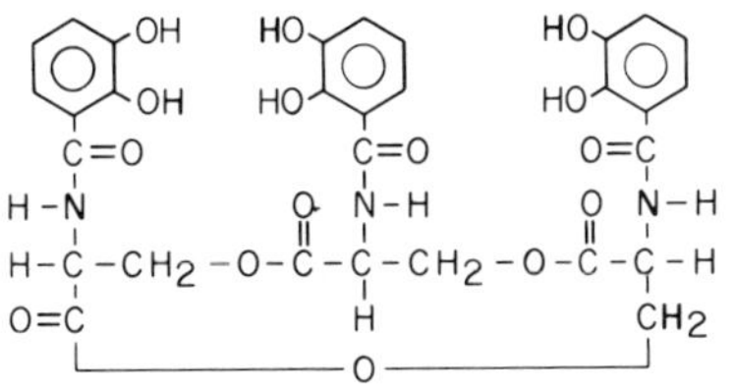 | *Aerobacter aerogenes*<br>*Escherichia coli*<br>*Salmonella typhimurium* | 1<br>1<br>1 |
| 2-N,6-N-Di-(2,4-dihydroxy-benzoyl)-L-lysine | 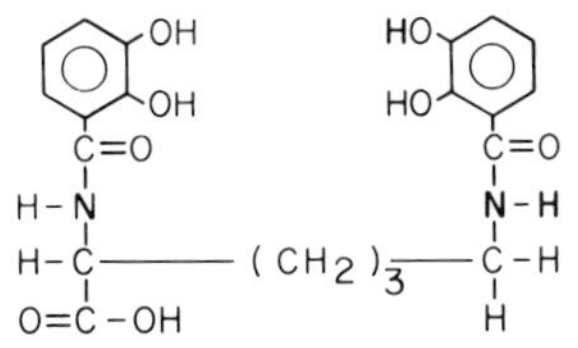 | *Azotobacter vinelandii* | 1 |
| 2,3-Dihydroxy-N-benzoyl-L-serine | 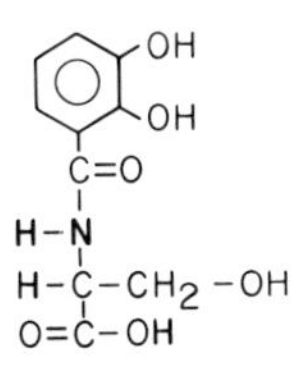 | *Aerobacter aerogenes*<br>*Escherichia coli*<br>*Salmonella typhimurium* | 1<br>1<br>1 |
| 2,3-Dihydroxybenzoylglycine (itoic acid) | 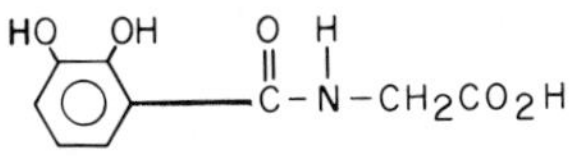 | *Bacillus subtilis* | 1 |

# HYDROXAMATES

## Ferrichrome Family

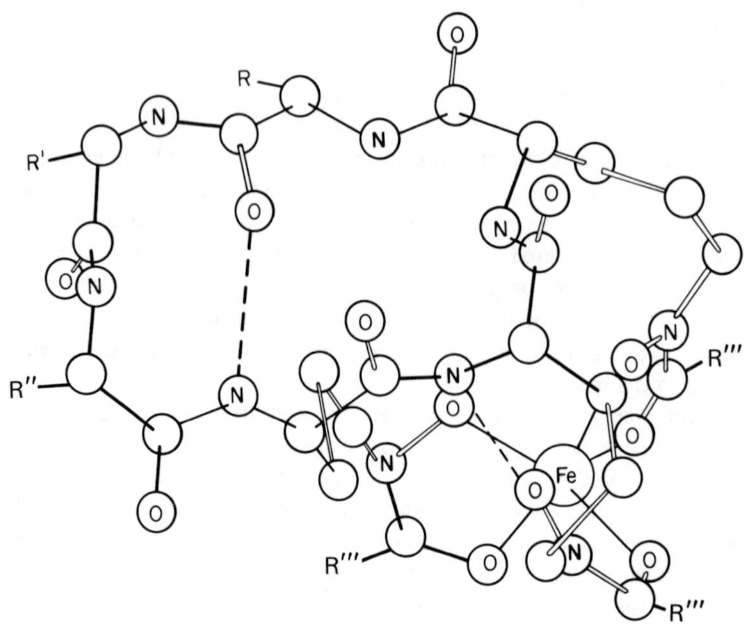


| Compound | Structure | Sources | Reference |
|---|---|---|---|
| Ferrichrome | R=R′=R″=H; R‴=$CH_3$– | Ascomycetes<br>Basidiomycetes<br>Fungi imperfecti<br>including *Aspergillus, Neurospora, Paecilomyces, Penicillium, Spicaria, Ustilago,* and probably *Sphacelotheca* | 1<br>1<br>1 |
| Ferrichrome A | R=R′=HO$CH_2$; R″=H;<br>R‴ = (H)C=C($CH_3$)($CH_2$COOH)<br>(*trans*) | *Ustilago*<br>*maydis*<br>*sphaerogena* | <br>2<br>2 |
| Ferrichrome C | R=R″=H<br>R′=R‴=$CH_3$– (2) | *Cryptococcus*<br>*melibiosum* | <br>1 |
| Ferrichrysin | R=R′=HO$CH_2$–; R″=H;<br>R‴=$CH_3$– | *Aspergillus*<br>*melleus*<br>*terreus* | <br>2<br>2 |
| Ferricrocin | R=R″=H; R′=$CH_2$OH;<br>R‴=$CH_3$– | *Aspergillus*<br>*fumigatus*<br>*humicola*<br>*nidulans*<br>*versicolor* | <br>2<br>2<br>2<br>2 |

| Compound | Structure | Sources | Reference |
|---|---|---|---|
| Ferrirubin | R=R′=$HOCH_2$–; R″=H; 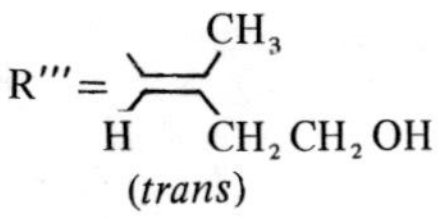 R‴= $\diagdown$C=C$\diagup$ with $CH_3$ and H, $CH_2CH_2OH$ (*trans*) | *Paecilomyces variotі* | 2 |
| | | *Penicillium variable* | 2 |
| | | *Spicaria* | 2 |
| Ferrirhodin | R=R′=$HOCH_2$–; R″=H; 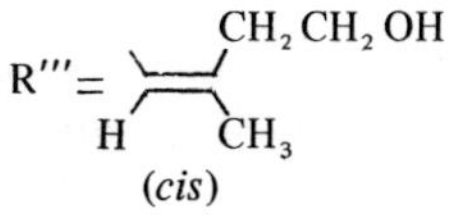 R‴= $\diagdown$C=C$\diagup$ with $CH_2CH_2OH$ and H, $CH_3$ (*cis*) | *Aspergillus nidulans* | 2 |
| | | *versicolor* | 2 |
| Albomycins | | *Actinomyces griseus* | 1 |
| | | *subtropicus* | 2 |
| Form $\delta_2$ | 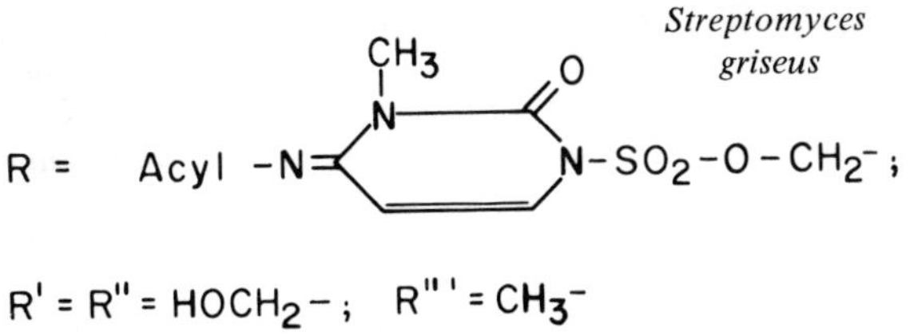 R = Acyl –N= (ring: N–$CH_3$, C=O) N–$SO_2$–O–$CH_2$–; R′ = R″ = $HOCH_2$–; R‴ = $CH_3$– | *Streptomyces griseus* | 1 |
| Form $\epsilon$ | 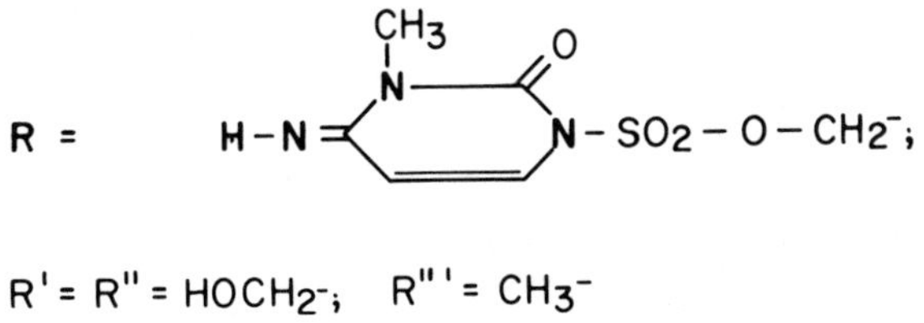 R = H–N= (ring: N–$CH_3$, C=O) N–$SO_2$–O–$CH_2$–; R′ = R″ = $HOCH_2$–; R‴ = $CH_3$– | | |
| Form $\delta_1$ | 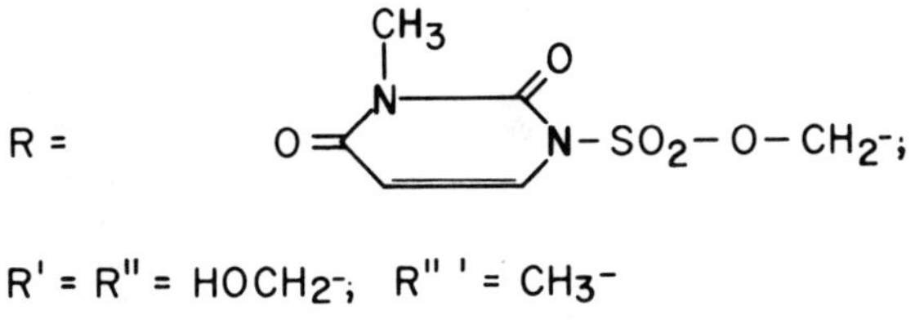 R = O= (ring: N–$CH_3$, C=O) N–$SO_2$–O–$CH_2$–; R′ = R″ = $HOCH_2$–; R‴ = $CH_3$– | | |
| Grisein | May be identical to one of the components of albomycin | *Streptomyces griseus* | 2 |
| Ferribactin | A cyclic decapeptide containing both D- and L-forms of $N^{\delta}$-hydroxyornithine | *Pseudomonas fluorescens* | 1 |
| Sake colorant A | A cyclohexapeptide containing the trihydroxamate of ferrichrome and one residue each of glycine, serine and alanine in the sequence: R = –$CH_2OH$; R′ = R‴ = $CH_3$–; R″ = H (3) | *Aspergillus oryzae* | 4 |

| Compound | Structure | Sources | Reference |
|---|---|---|---|

## Rhodotorulic Acid Family

Compound Structure Sources Reference

| Compound | Structure | Sources | Reference |
|---|---|---|---|
| Rhodotorulic acid | 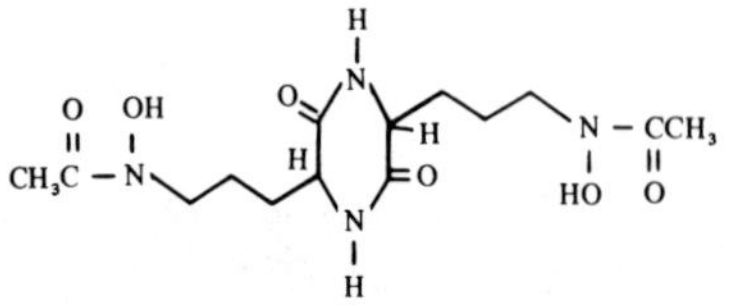 | *Leucosporidium* | |
| | | *scottii* (allotype, mating type a) | 5 |
| | | *Rhodosporidium* | |
| | | *toruloides* (mating type) | 5 |
| | | *Rhodotorula* | |
| | | *glutinis* | |
| | | var. *dairenensis* (type) | 5 |
| | | var. *glutinis* | 5 |
| | | *graminis* (type) | 5 |
| | | *pilimanae* | 5 |
| | | *pilimanae* (type) | 5 |
| | | *rubra* | 5 |
| | | *rubra* (type)[a] | 5 |
| | | *Sporidiobolu* | |
| | | *johnsonii* (type) | 5 |
| | | *ruinenii* (type) | 5 |
| | | *Sporobolomyces* | |
| | | *albo-rubescens* (type) | 5 |
| | | *hispanicus* | 5 |
| | | *pararoseus* (type) | 5 |
| | | *roseus* | 5 |
| Dimerum acid | 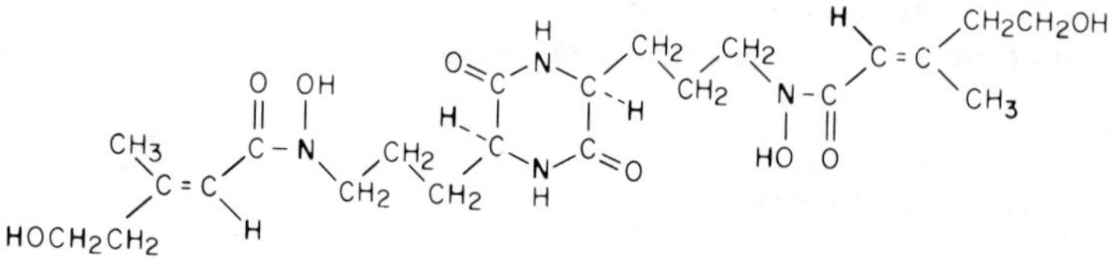 | *Fusarium* | |
| | | *dimerum* | 1 |
| Coprogen (Compound XFe)[b] | 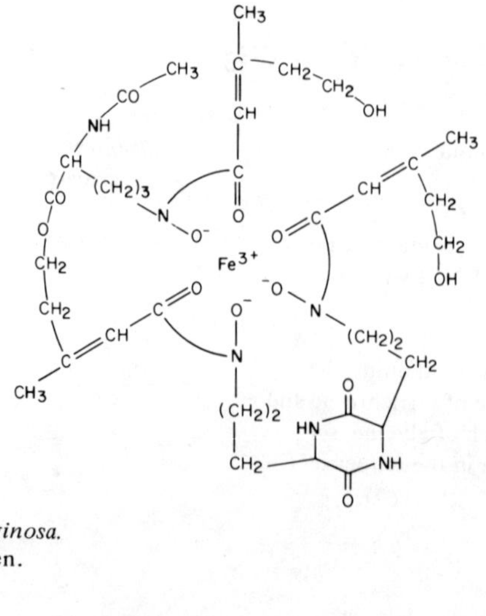 | *Neurospora* | |
| | | *crassa* | 2 |
| | | *Penicillium* | 2 |
| | | *camemberti* | 2 |
| | | *chrysogenum* | 2 |
| | | *notatum* | 2 |
| | | *uriticae* | 2 |

[a] Original designation *R. mucilaginosa.*
[b] Coprogen B = deacetyl coprogen.

## Aerobactin Family

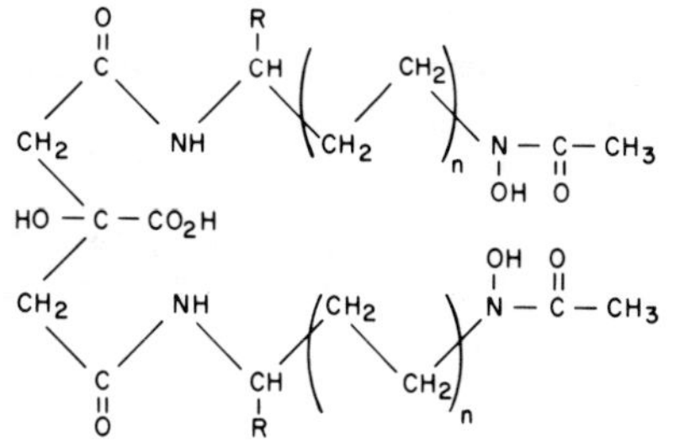

| Compound | Structure | Reference | Sources |
|---|---|---|---|
| Schizokinen | R = H, n = 1 | *Bacillus megaterium* | 6 |
| Terregens factor | R = H, n = 2 | *Arthrobacter pascens* | 8 |
| Aerobactin | R = $CO_2H$, n = 2 | *Aerobacter aerogenes* 62-1 | 1 |

## Mycobactins

The general structure of the ferric mycobactins is shown below. Side chains $R^1$ are alkyl groups having the number of carbon atoms shown.

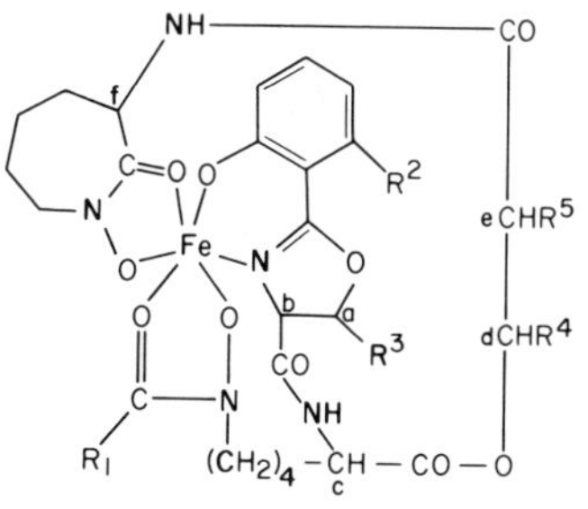

### MYCOBACTIN STRUCTURES

The numbers in the following table show the main types of side chains. Those of greatest abundance are shown in bold type. Asymmetric centers are labeled a to f; lack of an asymmetric center is indicated by the symbol ( ). Blank spaces indicate that the configuration has not been determined.

| Structure | $R^1$ | $R^2$ | $R^3$ | $R^4$ | $R^5$ | a | b | c | d | e | f |
|---|---|---|---|---|---|---|---|---|---|---|---|
| A* | **13**Δ | $CH_3$ | H | $CH_3$ | H | | | | | | |
| F† | **17**,15,13,**11**,9Δ | H | $CH_3$ | $CH_3$ | H | *threo* | | | *S* | ( ) | L |
| H | **19,17**Δ | $CH_3$ | $CH_3$ | $CH_3$ | H | *R* | L | L | *S* | ( ) | L |
| M | 1 | H | $CH_3$ | 18,**17**,16,15‡ | $CH_3$ | | | | *R*§ | *S*§ | |
| N* | 2 | H | $CH_3$ | 18,**17**,16,15‡ | $CH_3$ | | | | | | |
| P | 19,**17**,15*cis*$Δ^1$*n* | $CH_3$ | H | $C_2H_5$ | $CH_3$ | ( ) | L | L | *S* | *R* | L |
| R | **19**Δ | H | H | $C_2H_5$ | $CH_3$ | ( ) | L | L | *R* | *S* | L |
| S | 19,**17,15**,13*cis*Δ | H | H | $CH_3$ | H | ( ) | L | L | *S* | ( ) | L |
| T | 20,**19**,18,17<br>20,**19**,18,17Δ | H | H | $CH_3$ | H | ( ) | | | *R* | ( ) | L |

* Tentative structure.
† Structure inferred from mixture with mycobactin H.
‡ Saturated alkyl groups having the number of carbon atoms shown.
§ Relative configuration at d and e is erythro; absolute configuration is uncertain.

Data taken from: *Bacteriol. Rev., 34,* 107 (1970). Reproduced by permission of the American Society for Microbiology.

## MYCOBACTIN SOURCES

| Structure | Source | Reference | Structure | Source | Reference |
|---|---|---|---|---|---|
| A | *Mycobacterium aurum* | 7 | P | *Mycobacterium phlei* | 7 |
| F | *Mycobacterium fortuitum* | 7 | R | *Mycobacterium terrae* | 7 |
| H | *Mycobacterium thermoresistibile* | 7 | S | *Mycobacterium smegmatis* | 7 |
| M, N | *Mycobacterium marinum* | 7 | T | *Mycobacterium tuberculosis* | 7 |

## Fusarinines

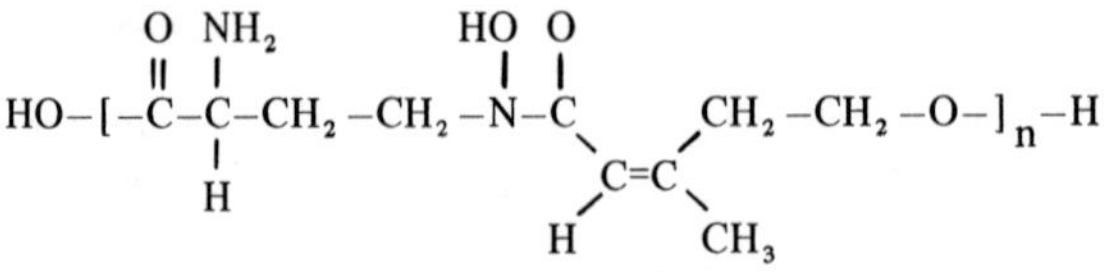

$$HO\text{–}[\text{–}C(=O)\text{–}CH(NH_2)\text{–}CH_2\text{–}CH_2\text{–}N(OH)\text{–}C(=O)\text{–}CH{=}C(CH_3)\text{–}CH_2\text{–}CH_2\text{–}O\text{–}]_n\text{–}H$$

Fusarinines are present in *Fusarium roseum* (ATCC 12822) and in other species of *Fusaria*, as well as in certain species of *Aspergillus, Gibberella,* and *Penicillium.*[1]

| Compound | n | Compound | n |
|---|---|---|---|
| Fusarinine | n = 1 | Fusarinine B | n = 3 |
| Fusarinine A | n = 2 | Fusarinine C | n = 3, cyclo |

## Ferrioxamines

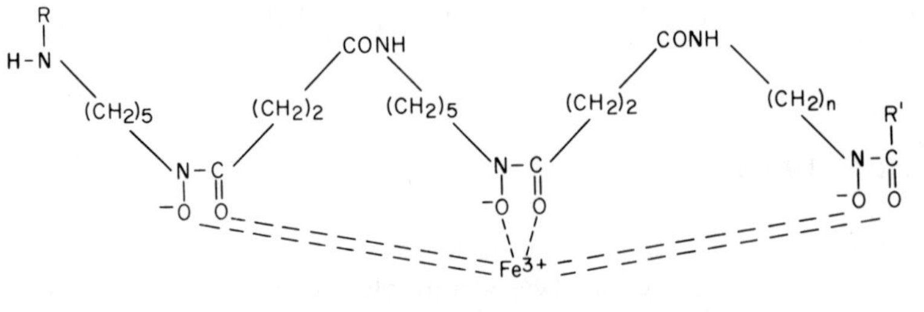

R–NH–$(CH_2)_5$–N(–O$^-$)–C(=O)–$(CH_2)_2$–CONH–$(CH_2)_5$–N(–O$^-$)–C(=O)–$(CH_2)_2$–CONH–$(CH_2)_n$–N(–O$^-$)–C(=O)–R', with the hydroxamate oxygens coordinated to $Fe^{3+}$

**Linear Structure**

HN–$(CH_2)_5$–N(–O$^-$)–C(=O)–$(CH_2)_2$–CONH–$(CH_2)_5$–N(–O$^-$)–C(=O)–$(CH_2)_2$–CONH–$(CH_2)_n$–N(–O$^-$)–C(=O)–$(CH_2)_2$–C=O (ring closed from C=O to HN), with the hydroxamate oxygens coordinated to $Fe^{3+}$

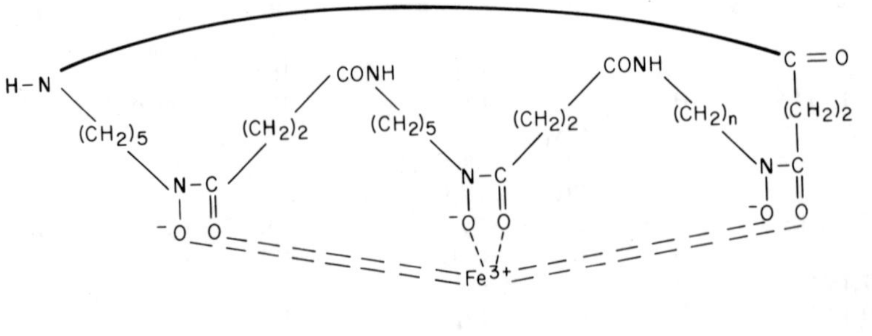

**Cyclic Structure**

Ferrioxamines were found in all actinomycetes that have been investigated for their presence, and in some bacteria.

| Compound | Structure | Sources | Reference |
|---|---|---|---|
| **Linear Ferrioxamines** | | | |
| Ferrioxamine B | R = H; n = 5; R′ = $CH_3$ | *Micromonaspora* | 2 |
| | | *Nocardia* | 2 |
| | | *Streptomyces pilosus* and other *Streptomyces* species | 2 |
| Ferrioxamine $D_1$ | R = $CH_3CO-$; n = 5; R′ = $CH_3$ | | |
| Ferrioxamine G | R = H; n = 5; R′ = $HO_2C(CH_2)_2-$ | | |
| Ferrioxamine $A_1$ | R = H; n = 4; R′ = $HO_2C(CH_2)_2-$ | | |
| Ferrimycin $A_1$ | 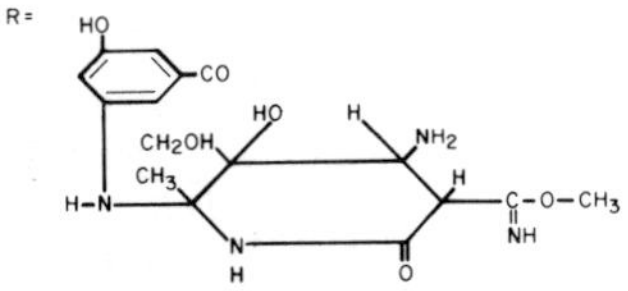 R = [structure: HO, CO, $CH_2OH$, $CH_3$, H–N, HO, H, $NH_2$, H, C(=NH)–O–$CH_3$, N–H, O]; n = 5; R′ = $CH_3-$ | | |
| **Cyclic Ferrioxamines** | | | |
| Ferrioxamine E[a] (Nocardamine) | n = 5 | *Nocardia* | 2 |
| | | *Streptomyces pilosus* | 2 |
| Ferrioxamine $D_2$ | n = 4 | | |
| **Miscellaneous Ferrioxamines** | | | |
| Metabolite C | A derivative of ferrioxamine B in which the terminal amino group is oxidized to COOH | | |
| Ferrioxamine $A_2$ | Contains one residue of acetic acid, two of succinic acid, one of 1-amino-5-hydroxyaminopentane, and two of 1-amino-4-hydroxyaminobutane | | |

## Other Compounds

| Compound | Structure | Sources | Reference |
|---|---|---|---|
| Aspergillic acids | 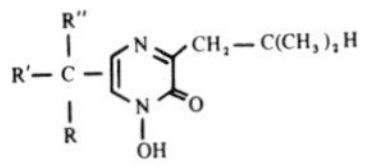 [structure: R″, R′–C–R, pyrazinone ring with N, $CH_2-C(CH_3)_2H$, O, N–OH] | *Aspergillus flavus* | 2 |
| Aspergillic acid | R = H; R′ = $C_2H_5$; R″ = $CH_3$ | *Aspergillus sclerotiorum* | 1 |

[a] Ferrioxamine E and deferrioxamine E are present in *Streptosporangium roseum, Chainia,* and *Streptomyces,*[2] and in *Chromobacterium violaceum.*[1]

| Compound | Structure | Sources | Reference |
|---|---|---|---|
| neo-Aspergillic acid | R = R″ = H; R′ = $C(CH_3)_2H$ | | |
| muta-Aspergillic acid | R = OH, R′ = R″ = $CH_3$ | | |
| hydroxy-Aspergillic acid | R = OH; R′ = $C_2H_5$; R″ = $CH_3$ | | |
| neo-hydroxy-Aspergillic acid | R = OH; R′ = $C(CH_3)_2H$; R″ = H | | |
| Mycelianamide | OH; O; N; $CH_3$; H; O; CH; N; O; OH | *Penicillium griseofulvum* | 2 |
| Pulcherriminic acid (2,5-diisobutyl-3,6-dihydroxypyrazine-1,4-dioxide) | O; HO; N; $CH_2-C(CH_3)_2H$; $H(CH_3)_2C-CH_2$; N; OH; O | *Candida pulcherrima* and related yeasts | 1 |
| | | *Fabospora* | |
| | | *ashbyi* | 1 |
| | | *dobzhanskii* | 1 |
| | | *lactis* | 1 |
| Hadacidin | O OH; ‖ \|; $H-C-N-CH_2-CO_2H$ | *Penicillium aurantioviolaceum* F4070b | 1 |
| | | *frequentans* | 1 |
| Actinonin | O; C; H; N; H; $C_2H_5$; C; O; C; OH; N; H; C; C; C–H; O; $CH_2$; N; $HOH_2C$; $H_3C$; $CH_3$ | *Streptomyces* sp. | 1 |
| 2,4-Dihydroxy-7-methoxy-1,4-bezoxazin-3-one (DIMBOA) | $H_3C-O-$; O; OH; H; O; N; OH | Certain tissues of higher plants, such as corn seedlings and seedlings of higher grasses, lettuce, leaves of tomato, cauliflower, leek and cabbage, and extracts of carrot roots | 1 |
| Ferroverdin | $CH_2=OH-$; $-O-\overset{O}{\overset{\|\|}{C}}-$; OH; N=O; (Deferroferroverdin) | Mycelium of *Streptomyces* sp. | 1 |
| Pyrimine | N; N; H; COO– | *Pseudomonas* GH | 1 |
| | | *roseus fluorescens* | 1 |

## REFERENCES


1. Neilands, J. B., in *Inorganic Biochemistry,* p. 167, G. Eichhorn, Ed. Elsevier, Amsterdam, The Netherlands (1973).
2. Neilands, J. B., *Struct. Bonding, 1,* 59 (1966).
3. Linás, M., Ph.D. Thesis. University of California, Berkeley, California (1971).
4. Tadenuma, M., and Sato, S., *Agr. Biol. Chem., 31,* 1482 (1967).
5. Atkin, C. L., Neilands, J. B., and Phaff, H. J., *J. Bacteriol., 103,* 722 (1970).
6. Mullis, K., Pollack, J. P., and Neilands, J. B., *Biochemistry, 10,* 4894 (1971).
7. Snow, G. A., *Bacteriol. Rev., 34,* 99 (1970).
8. Linke, W. D., Crueger, A., and Dickman, H., *Arch. Mikrobiol., 85,* 44 (1972).

# MISCELLANEOUS INFORMATION

# CELL BREAKAGE

## Bacterial Cell Breakage or Lysis

DR. E. H. COTA-ROBLES and DR. S. M. STEIN

The primary goal of breaking microbial cells in the isolation of cellular components, so that more detailed analysis can be made of the properties of these components. The ultimate goal of such investigations is to develop an understanding of how these components function in an integrated state in the intact cell. Thus, the investigator who wishes to study in detail the permeability properties of a bacterial membrane finds it necessary to isolate the membrane or portions of the membrane in order to make the experimental inquiries desired. Similarly, the investigator wishing to study the organization of ribosomal subunits finds it necessary to disrupt cells.

Occasionally, however, investigators utilize cell breakage to bring a halt to certain biochemical reactions. In particular, bacterial virologists have used cell breakage to isolate viral components, so that a temporal as well as biochemical analysis of viral replication can be undertaken.

In 1960, Marr[a] published a review on the localization of enzymes in bacteria, which succinctly described the principal methods available at that time for effective breakage of bacterial cells. Marr notes that the most frequently used methods were the following: (a) grinding with an abrasive, (b) sonication, (c) ballistic disintegration, and (d) osmotic shock. Amazingly, in this era of technological development few significant innovations have been introduced for disruption of bacterial cells. It is true that the lysozyme-EDTA [(ethylenedinitrilo)tetraacetic acid, edetic acid] method of Repaske with its infinite variations is the most frequently used method currently applied. In particular, EDTA treatment has been found to effect release of periplasmic substances as well as cell wall components. Repaske's system, supplemented with both ionic and non-ionic detergents, has found extensive utility in molecular biology.

The methods cited below include physical, chemical, and biological systems that have been used with some success to disrupt bacterial cells. These methods range from application of explosive decompression to the utilization of direct microbial activity, such as the action of *Bdellovibrio bacteriovorus* to lyse cells. The method of choice is controlled by the experimental goals of the investigator. Generally the investigator must select a method that permits his particular experimental system to be preserved. Thus, for the isolation of a fairly stable cytoplasmic enzyme, such as hexokinase, one need have little worry about inactivation of the enzyme by the use of the methods described. However, when an experimentor desires to investigate a multi-enzyme system, such as the electron-transport system, he must use gentle means to disrupt the cell. Similarly, if an investigator wishes to isolate native undegraded nucleic acids, he must select a method that involves fairly gentle means of disruption.

[a] *The Bacteria,* Vol. 1, pp. 443–468, I. C. Gunsalus and R. Y. Stanier, Eds., Academic Press, New York.

| Method of Lysis or Breakage | Activity Factor(s) | Sensitive Organism(s) | References |
|---|---|---|---|
| **Physical Methods** | | | |
| Ballistic disintegration | | | 1–13 |
| Explosive decompression | | | 14, 15 |
| Grinding | | | 16, 17 |
| Osmotic shock | Sudden dilution with water | Halophiles and halotolerant organisms | 18 |
| | Low salt concentration | *Halobacterium cutirubrum* | 19 |
| | Decreased salt concentration | *Halobacterium salinarum* | 20 |
| | Brief exposure to 1*M* glycerol | *Azotobacter agilia; Rhodospirillum rubrum; Serratia plymuthica* | 21 |
| | Low salt concentration | *Micrococcus halodenitrificans* | 22 |
| Pressure | | | 23–29 |
| Sonication | | | 30–33 |
| | | Actinomycetes | 34–36 |
| Temperature extremes | | Red halophiles | 37 |
| | 25°C | *Escherichia coli; Aerobacter aerogenes* | 38 |
| | Heat treatment | *Escherichia coli; Salmonella pullorum; Pseudomonas fluorescens; P. pyocyanea* | 39 |
| | 25°C | *Staphylococcus aureus* | 40 |
| | Temperatures above optimum | Psychrophiles | 41 |
| **Chemical Methods** | | | |
| Acids | Dichlorophenoxybutyric acid | *Bacillus subtilis* | 42 |
| | Short-chain fatty acids | *Escherichia coli* | 43 |
| | Sodium salts of acetic, butyric, formic and propionic acid | *Escherichia coli* | 43 |
| | *p*-Chloromercurobenzoic acid | *Escherichia coli* | 44 |
| Buffers | Tris® buffer | *Bacillus cereus* | 45 |
| | Sodium acetate buffer | *Escherichia coli* | 45 |
| | Tris® + EDTA | *Vibrio cholerae* | 46 |
| Cations | Addition of monovalent cations; also heat, urea, and ionic detergents | Red halophiles | 47 |
| | Depletion of divalent cations | *Pseudomonas aeruginosa* | 48 |
| | | Marine pseudomonads | 49 |
| Cystamine | Cystamine and closely related derivatives | *Bacillus subtilis* and other *Bacillus* strains | 50, 51 |

| Method of Lysis or Breakage | Activity Factor(s) | Sensitive Organism(s) | References |
|---|---|---|---|
| Detergents | Anionic and non-ionic detergents; also succinylation, phospholipase A, alkaline phosphatase, trypsin, and chymotrypsin | *Mycoplasma laidlawii*, strain B | 52 |
| EDTA | | *Pseudomonas aeruginosa* | 53–55 |
| Glycine | | *Aerobacter aerogenes; Bacillus mesentericus; Escherichia coli* | 56 |
| | | *Escherichia coli* | 57 |
| | | *Mycobacterium smegmatis* | 58 |
| | | *Salmonella typhi* | 59 |
| | | *Vibrio cholerae; V.* El Tor; *Salmonella typhi; S. typhimurium; S. paratyphi* B | 60 |
| Physiological solutions | Ringer's solution | *Bacillus anthracis* | 61, 62 |
| Zeolite, synthetic | | Most bacteria | 63 |
| **Biological Methods** | | | |
| Metabolic disturbance | | | |
| Addition of amino acids | D-Methionine or D-alanine | *Alcaligenes faecalis* | 64 |
| | D-Galactose | *Salmonella enteritidis* | 65, 66 |
| | Glycine, aspartate, or arginine | *Salmonella typhosa* | 67 |
| Antibiotics | Amphomycin | $Gr^+$ bacteria | 68 |
| | Aspartocin | $Gr^+$ bacteria | 69, 70 |
| | Bacitracin | $Gr^+$ bacteria; meningococcus; gonococcus | 71–73 |
| | Bacitracin; also novobiocin, ristocetin, and Vancomycin® | Staphylococci | 74–77 |
| | Cephalosporin C | Wide range | 78, 79 |
| | Circulin® | $Gr^+$ and $Gr^-$ bacteria; $Gr^-$ bacteria are more sensitive than $Gr^+$ bacteria | 80 |
| | D-Cycloserine | *Escherichia coli* | 81 |
| | | *Streptococcus faecalis* | 82 |
| | Megacin | *Bacillus megaterium* | 83 |
| | Penicillin | *Bacillus megaterium* | 84 |
| | | *Escherichia coli* | 85–88 |
| | | Staphylococcus; *Escherichia coli* | 89–96 |
| | Polymyxin | $Gr^-$ bacteria; some $Gr^+$ bacteria | 97, 98 |
| | Pyocin | *Pseudomonas aeruginosa* | 99 |
| | Ristocetin | $Gr^+$ bacteria; mycobacteria | 100, 101 |
| | Subtilin | $Gr^+$ bacteria | 102, 103 |
| | Vancomycin® | $Gr^+$ bacteria especially *Staphylococcus aureus* | 104–107 |

| Method of Lysis or Breakage | Activity Factor(s) | Sensitive Organism(s) | References |
|---|---|---|---|
| Nutritional deficiency | D-Glutamic acid limitation | *Bacillus subtilis* | 108 |
| | Glucose depletion | *Bacillus subtilis* | 109 |
| | Diaminopimelic acid deprivation | Mutant of *Escherichia coli* requiring diaminopimelic acid | 110–113 |
| | N-Acetyl-D-glucosaminidase deficiency | *Lactobacillus bifidus* var. *pennsylvanicus* | 114, 115 |
| | Medium depletion | *Streptococcus faecalis* | 116–119 |
| | L-Lysine deprivation | *Streptococcus faecalis;* vitamin B-deficient mutant of *S. faecalis* | 120, 121 |
| | D-Alanine deprivation | *Streptococcus faecalis;* vitamin B-deficient mutant of *S. faecalis* | 120, 121 |
| Oxygen | Aeration during active growth | Anaerobes | 122 |
| | Halt of aeration and establishment of semiaerobic conditions | *Bacillus megaterium* | 123 |
| | Sudden removal of oxygen from aerated cultures in the log phase | *Bacillus subtilis* | 124, 125 |
| Enzymes | | | |
| Microbial | | | |
| *Aeromonas* | Enzyme from *Aeromonas hydrophila* | *Staphylococcus aureus* | 126 |
| *Bacillus* | Lytic factor induced in *Bacillus cereus* 569 | *Bacillus cereus* 130 | 127 |
| | Filtrate of *Bacillus cereus* | Aerobic spore formers | 128 |
| | Enzyme from spores of *Bacillus cereus* | *Bacillus cereus* | 129 |
| | Enzyme from spores of *Bacillus megaterium* | *Bacillus megaterium* | 130 |
| | Enzyme from sporulating cells of *Bacillus megaterium* | *Bacillus megaterium* | 131 |
| | Autolysin from *Bacillus stearothermophilus* | *Bacillus stearothermophilus* | 132 |
| | Autolysin from *Bacillus subtilis* | *Bacillus subtilis* | 133, 134 |
| | Filtrate of *Bacillus cereus* | *Micrococcus lysodeikticus* | 135 |
| *Chalaropsis* | Enzyme from the fungus *Chalaropsis* | *Staphylococcus aureus* | 136–138 |
| *Escherichia coli* | Enzyme from a defective lysogenic mutant of *Escherichia coli* $K_{12}$ (λ) | $Gr^-$ bacteria; some $Gr^+$ bacteria, i.e., *Bacillus megaterium* and *Staphylococcus aureus* | 139 |
| | Autolysin from *Escherichia coli* | *Escherichia coli* | 140–142 |
| *Flavobacterium* | $L_{11}$ enzyme from *Flavobacterium* | *Staphylococcus aureus; Micrococcus lysodeikticus* | 143 |

| Method of Lysis or Breakage | Activity Factor(s) | Sensitive Organism(s) | References |
|---|---|---|---|
| *Myxobacterium* | Enzyme from *Myxobacterium* | *Arthrobacter crystallopietes; Staphylococcus aureus; Micrococcus lysodeikticus; Sarcina lutea; Rhodospirillum rubrum* | 144 |
| | | *Aerobacter aerogenes; Escherichia coli; Pseudomonas fluorescens; Serratia marcescens; Sarcina lutea; Bacillus subtilis; B. megaterium* | 145–150 |
| | | *Staphylococcus aureus,* strain Copenhagen | 151 |
| *Pseudomonas* | Enzyme from *Pseudomonas* | *Staphylococcus aureus; S. roseus; Gaffkya tetragena; Sarcina lutea* | 152 |
| *Sorangium* | Enzyme from *Sorangium* | *Arthrobacter globiformis; Micrococcus lysodeikticus* | 153, 154 |
| *Staphylococcus* | Enzyme from *Staphylococcus aureus* | *Micrococcus lysodeikticus* | 155, 156 |
| | Autolysin from *Staphylococcus aureus* | *Staphylococcus aureus* | 157, 158 |
| | Lysostaphin from *Staphylococcus staphylolyticus* | Staphylococci | 159 |
| | Enzyme from *Staphylococcus epidermidis* | Staphylococci | 160 |
| *Streptococcus* | Autolysin from *Streptococcus faecalis* | *Streptococcus faecalis* | 161–164 |
| | Muralytic enzyme from Group C *Streptococcus* | Group A *Streptococcus* | 165 |
| *Streptomyces* | Enzyme from *Streptomyces albus* G | *Staphylococcus aureus; Micrococcus lysodeikticus; M. roseus; Corynebacterium diphtheriae* | 166 |
| | | Heat-killed $Gr^-$ species; many living $Gr^+$ species | 167, 168 |
| | R1 enzyme fraction from *Streptomyces albus* G | *Micrococcus radiodurans* | 169 |
| | $L_3$ enzyme from *Streptomyces* | *Corynebacterium diphtheriae* | 170 |
| | Enzyme from *Streptomyces* $F_1$ | $Gr^+$ organisms | 171 |
| Bacteriophage-induced | | | |
| *Bacillus* | | | 172, 173 |
| *Escherichia coli* | | | 174–178 |
| *Klebsiella pneumoniae* | | | 179 |
| *Staphylococcus* | | | 180–182 |
| *Streptococcus* | | | 183–190 |

| Method of Lysis or Breakage | Activity Factor(s) | Sensitive Organism(s) | References |
|---|---|---|---|
| Lysozyme | | Bacteria from stationary-phase cells are less sensitive to lysozyme than actively dividing cells | 191 |
| | Lysozyme | *Bacillus* sp. | 192 |
| | | *Bacillus megaterium* | 193, 194 |
| | | *Bacillus subtilis* | 195, 196 |
| | | *Escherichia coli* | 197 |
| | | *Micrococcus lysodeikticus* | 198–205 |
| | | *Sarcina lutea* | 206–208 |
| | | *Streptococcus faecalis* F24; *S. faecalis* var. *liquefaciens* 31; *S. faecalis* E1 | 209 |
| | | *Sarcina flava* | 210 |
| | Lysozome + EDTA | *Azotobacter vinelandii* | 211 |
| | | *Salmonella typhi* | 212 |
| | | *Escherichia coli* | 213–217 |
| | | *Salmonella paratyphi* B | 218, 219 |
| | Lysozyme + anions | Enterococci | 220 |
| | Lysozyme + Brij®-58 | *Escherichia coli* | 221, 222 |
| | Lysozyme + phagocytin | *Escherichia coli* | 223 |
| | Lysozyme + quaternary ammonia compounds | $Gr^-$ bacteria | 224 |
| | Lysozyme + serum or heat treatment or thioglycolic acid | $Gr^-$ bacteria | 225, 226 |
| | Lysozyme + EDTA or lipase or freezing and thawing | $Gr^-$ bacteria | 227 |
| | Lysozyme + trypsin or trypsin and butanol | $Gr^+$ bacteria | 228 |
| | Lysozyme + EDTA + trypsin | *Lactobacillus casei* | 229 |
| | Polymyxin B sulfate followed by lysozyme | *Neisseria catarrhalis* | 230 |
| | Lysozyme + lipase (steapsin) | *Listeria monocytogenes* | 231 |
| | *n*-Butanol-saturated buffer + lysozyme | *Micrococcus radiodurans* | 232 |
| | Lysozyme + sodium dodecylsulfate | *Micrococcus radiodurans* | 233 |
| | Lysozyme or glycine | *Mycobacterium smegmatis* | 234 |
| | Lysozyme + EDTA + Tris® | *Pseudomonas aeruginosa* | 235 |
| | Lysozyme + EDTA + lack of $Ca^{++}$ | *Spirillum serpens* | 236 |
| | Nafcillin (a semisynthetic penicillin) followed by lysozyme or trypsin | *Staphylococcus aureus* | 237, 238 |
| | Nafcillin, ampicillin, or cloxacillin and cephalothin followed by lysozyme or trypsin | *Staphylococcus aureus* | 239 |
| | Tris® + EDTA + lysozyme or sodium lauryl sulfate | *Vibrio* El Tor | 240, 241 |
| Others | Leucozyme C | *Escherichia coli* | 242 |
| | Phagocytin | *Escherichia coli* | 243 |
| | Anionic and non-ionic detergents; also succinylation, phospholipase A, alkaline phosphatase, trypsin, and chymotrypsin | *Mycoplasma laidlawii*, strain B | 244 |

| Method of Lysis or Breakage | Activity Factor(s) | Sensitive Organism(s) | References |
|---|---|---|---|
| Immune substances | Complement system + lysozyme | Gr$^-$ bacteria that are susceptible to immune bactericidal reaction | 245–249 |
| Direct microbial activity | *Bdellovibrio bacteriovorus* | Gr$^-$ bacteria | 250–254 |

## REFERENCES


1. Cummins, C. S., and Harris, H., *J. Gen. Microbiol., 14,* 583 (1956).
2. Huff, E., Oxley, H., and Silverman, C. S., *J. Bacteriol., 88,* 1155 (1964).
3. King, H. K., and Alexander, H., *J. Gen. Microbiol., 2,* 315 (1948).
4. Mandelstam, J., *Biochem. J., 84,* 294 (1962).
5. Merkenschlager, M., Schlossmann, K., and Kurz, W., *Biochem. Z., 329,* 332 (1957).
6. Mickle, H., *J. R. Microsc. Soc., 68,* 10 (1948).
7. Nossal, P. M., *Aust. J. Exp. Biol. Med. Sci., 31,* 583 (1953).
8. Pickering, B. T., *Biochem. J., 100,* 430 (1966).
9. Sagniez, G., LeCam, M., Madec, Y., and Bernard, S., *Ann. Inst. Pasteur (Paris), 177,* 663 (1969).
10. Salton, M. R. J., and Horne, R. W., *Biochim. Biophys. Acta, 7,* 177 (1951).
11. Sharon, N., and Jeanloz, R. W., *Experientia, 20,* 253 (1962).
12. Shockman, G. D., *Biochim. Biophys. Acta, 59,* 234 (1962).
13. Shockman, G. D., Kalb, J. J., and Toennies, G., *Biochim. Biophys. Acta, 24,* 203 (1957).
14. Foster, J. A., Cowan, R. M., and Maag, T. A., *J. Bacteriol., 83,* 330 (1962).
15. Van Eseltine, W. P., Jones, R. W., and Gilliard, F. E., *Proc. Soc. Exp. Biol. Med., 131,* 1446 (1969).
16. McIlwain, H., *J. Gen. Microbiol., 2,* 288 (1948).
17. Wiggert, W. P., Silverman, M., Utler, M. F., and Werkman, C. H., *Iowa State J. Sci., 14,* 179 (1939).
18. Ingram, M., in *Microbial Ecology,* p. 90, R. E. O. and C. C. Spicer, Eds. Cambridge University Press, New York and London, England (1957).
19. Kushner, D. J., *J. Bacteriol., 87,* 1147 (1964).
20. Mohr, V., and Larsen, H., *J. Gen. Microbiol., 31,* 267 (1963).
21. Robrish, S. A., and Marr, A. G., *Bacteriol. Proc.,* p. 130 (1957).
22. Takahashi, I., and Gibbons, N. E., *Can. J. Microbiol., 5,* 25 (1959).
23. Edebo, L., *J. Biochem. Microbiol. Technol. Eng., 2,* 453 (1960).
24. French, C. S., and Milner, H. W., in *Methods in Enzymology I,* p. 64, S. P. Colowick and N. O. Kaplan, Eds. Academic Press, New York (1955).
25. Hughes, D. E., *Brit. J. Exp. Pathol., 32,* 97 (1950).
26. Milner, H. W., Lawrence, N. S., and French, C. S., *Science, 111,* 633 (1950).
27. Perrine, T. D., Ribi, E., Maki, W., Miller, B., and Oertli, E., *Appl. Microbiol., 10,* 93 (1962).
28. Ribi, E., Perrine, T., List, R., Brown, W., and Goode, G., *Proc. Soc. Exp. Biol. Med., 100,* 647 (1959).
29. Vanderheiden, G. J., Fairchild, A. C., and Jago, E. G. R., *Appl. Microbiol., 19,* 875 (1970).
30. Bosco, G., *J. Infect. Dis., 99,* 270 (1956).
31. Harvey, E. N., *Biol. Bull. (Woods Hole), 59,* 306 (1930).
32. Ikawa, M., and Snell, E. E., *J. Biol. Chem., 235,* 1376 (1960).
33. Marr, A. G., and Cota-Robles, E. H., *J. Bacteriol., 74,* 79 (1957).
34. Becker, B., Lechevalier, M. P., and Lechevalier, H. A., *Appl. Microbiol., 13,* 236 (1965).
35. Roberson, B. S., and Schwab, J. H., *Biochim., Biophys. Acta, 44,* 436 (1960).
36. Salton, M. R. J., *J. Gen. Microbiol., 9,* 512 (1953).
37. Abram, D., and Gibbons, N. E., *Can. J. Microbiol., 7,* 741 (1961).
38. Mitchell, P., and Moyle, J., *Nature, 178,* 993 (1956).
39. Salton, M. R. J., and Horne, R. W., *Biochim. Biophys. Acta, 7,* 19 (1951).
40. Mitchell, P., and Moyle, J., *J. Gen. Microbiol., 16,* 184 (1957).
41. Hagan, P. O., Kushner, D. J., and Gibbons, N. E., *Bacteriol. Proc.,* p. 42 (1964).
42. Roman, M., and Gonzales, C., *Microbiol. Esp., 20,* 63 (1967).
43. Mayo, J. A., and Church, B. D., *Bacteriol. Proc.,* p. 84 (1966).
44. Schaechter, M., and Santomassino, K. A., *J. Bacteriol., 84,* 318 (1962).
45. Mohan, R. R., Kronish, D. P., Pianotti, R. S., Epstein, R. L., and Schwartz, B. S., *J. Bacteriol., 90,* 1355 (1965).
46. Adhikari, P. C., Raychaudhuri, C., and Chatterjee, S. N., *J. Gen. Microbiol., 59,* 91 (1969).

47. Abram, D., and Gibbons, N. E., *Can. J. Microbiol.*, *7*, 741 (1961).
48. Brown, M. R., and Melling, J., *J. Gen. Microbiol.*, *59*, 263 (1969).
49. Buckmire, F. L. A., and MacLeod, R. A., *Can. J. Microbiol.*, *11*, 677 (1965).
50. Weinberg, E. D., *Exp. Cell Res.*, *13*, 175 (1957).
51. Weinberg, E. D., Saz, A. K., and Pilgren, E. Y., *J. Gen. Microbiol.*, *19*, 419 (1958).
52. Smith, P. F., Koostra, W. L., and Mayberry, W. R., *J. Bacteriol.*, *100*, 1166 (1969).
53. Carson, K. J., and Eagon, R. G., *Bacteriol. Proc.*, p. 32 (1964).
54. Gray, G. W., and Wilkinson, S. G., *J. Appl. Bacteriol.*, *28*, 153 (1965).
55. Eagon, R. G., Simmons, G. P., and Carson, K. J., *Can. J. Microbiol.*, *11*, 1041 (1965).
56. Maculla, E. S., and Cowels, P. W., *Science*, *107*, 376 (1948).
57. Gordon, J., Hall, R. A., and Strickland, L. H., *J. Pathol. Bacteriol.*, *64*, 299 (1962).
58. Adamek, L. P., Misson, H., Mohelska, H., and Trinka, L., *Arch. Mikrobiol.*, *69*, 227 (1969).
59. Diena, B. B., Wallace, R., and Greenberg, L., *Can. J. Microbiol.*, *10*, 543 (1964).
60. Jeynes, M. H., *Nature*, *180*, 867 (1957).
61. Stähelin, H., *Schweiz. Z. Allg. Pathol. Bakteriol.*, *17*, 296 (1954).
62. Stähelin, H., *Schweiz. Z. Allg. Pathol. Bakteriol.*, *16*, 111 (1953).
63. Wistreich, G., Lechtman, M. D., Bartholomew, J. W., and Bils, R. F., *Appl. Microbiol.*, *16*, 1269 (1968).
64. Lark, C., and Schichtel, R., *J. Bacteriol.*, *84*, 1241 (1962).
65. Fukasawa, T., and Mikaido, H., *Nature*, *183*, 1131 (1959).
66. Fukasawa, T., and Mikaido, H., *Biochim. Biophys. Acta*, *48*, 470 (1961).
67. Nasier, M. M. R., and Ghatak, S., *Indian J. Microbiol.*, *7*, 91 (1967).
68. Heinemann, B., Kaplan, M. A., Muir, R. D., and Hooper, I. R., *Antibiot. Chemother.*, *3*, 1239 (1953).
69. Shay, A. J., Adam, J., Martin, J. H., Hausmann, W. K., Shu, P., and Bohonos, N., in *Antiobiotics Annual 1959–60*, p. 194, F. Marti-Ibanez, Ed. Antibiotica Inc., New York (1960).
70. Kirsch, E. J., Dornbush, A. C., and Backus, E. K., in *Antibiotics Annual 1959–60*, p. 205, F. Marti-Ibanez, Ed. Antibiotica Inc., New York (1960).
71. Abraham, E. P., in *CIBA Lectures in Microbial Biochemistry*, p. 1. John Wiley and Sons, New York (1957).
72. Abraham, E. P., and Newton, G. G. F., in *CIBA Foundation Symposium on Amino Acids and Peptides with Antimetabolic Activity*, p. 205, G. E. W. Wolstenholme and C. M. O'Connor, Eds. Little, Brown and Co., Boston, Massachusetts (1958).
73. Johnson, B. A., Anker, H., and Meleney, F. L., *Science*, *102*, 376 (1945).
74. Park, J. T., *J. Biol. Chem.*, *194*, 897 (1952).
75. Jordan, D. C., *Biochem. Biophys. Res. Commun.*, *6*, 167 (1961).
76. Reynolds, P. E., *Biochim. Biophys. Acta*, *52*, 403 (1961).
77. Wallas, C. H., and Strominger, J. L., *J. Biol. Chem.*, *238*, 2264 (1963).
78. Newton, G. G. F., and Abraham, E. P., *Biochem. J.*, *62*, 651 (1956).
79. Newton, G. G. F., and Abraham, E. P., *Biochem. J.*, *58*, 103 (1954).
80. Murray, F. J., Tetrault, P. A., Kaufman, O. W., and Koffler, H., *J. Bacteriol.*, *57*, 305 (1949).
81. Ciak, J., and Hahn, F. E., *Antibiot. Chemother.*, *9*, 47 (1959).
82. Shockman, G. D., *Proc. Soc. Exp. Biol. Med.*, *101*, 693 (1959).
83. Ivanovics, G. L., Alfoldi, L., and Nagy, E., *J. Gen. Microbiol.*, *21*, 51 (1959).
84. Fedorova, G. I., *Antibiotiki*, *14*, 880 (1969).
85. Chargaff, E., Schidman, H. M., and Shapiro, H. S., *Nature*, *180*, 851 (1957).
86. Hahn, F. E., and Ciak, J., *Science*, *125*, 119 (1957).
87. Lederberg, J., *Proc. Nat. Acad. Sci. U.S.A.*, *42*, 574 (1956).
88. Liebermeister, K., and Kellenberger, E., *Z. Naturforsch. Teil B.*, *118*, 200 (1956).
89. Mandelstam, J., and Rogers, H. J., *Biochem. J.*, *72*, 654 (1959).
90. Nathenson, S. G., and Strominger, J. L., *J. Pharmacol. Exp. Ther.*, *131*, 1 (1961).
91. Park, J. T., *Biochem. J.*, *70*, 2P (1958).
92. Park, J. T., and Strominger, J. L., *Science*, *125*, 99 (1957).
93. Roberts, J., and Johnson, M. J., *Biochim. Biophys. Acta*, *59*, 458 (1962).
94. Rogers, H. J., and Jeljaszewicz, J., *Biochem. J.*, *84*, 576 (1962).
95. Rogers, H. J., in *Resistance of Bacteria to the Penicillins*, p. 25, CIBA Foundation Group #13, A. V. S. deReuck and M. P. Cameron, Eds. Little, Brown and Co., Boston, Massachusetts (1962).
96. Wylie, E. B., and Johnson, M. J., *Biochim. Biophys. Acta*, *59*, 450 (1962).
97. Newton, B. A., *Bacteriol. Rev.*, *20*, 14 (1956).
98. Newton, B. A., *J. Gen. Microbiol.*, *9*, 54 (1953).
99. Jacob, F., *Ann. Inst. Pasteur (Paris)*, *86*, 149 (1954).
100. Graudy, W. E., Sinclair, A. C., Theriault, R. J., Goldstein, A. W., Rickher, C. J., Warren, H. B., Jr., Oliver, T. J., and Sylvester, J. C., in *Antibiotics Annual 1956–57*, p.680, H. Welch and F. Marti-Ibanez, Eds. Medical Encyclopedia Inc., New York (1957).

101. Philip, J. E., Schenck, J. R., and Hargie, M. P., in *Antibiotics Annual 1956–1957*, p. 699, H. Welch and F. Marti-Ibanez, Eds. Medical Encyclopedia Inc., New York (1957).
102. Bricas, E., and Fromageot, C. L., *Advan. Protein Chem., 4,* 57 (1953).
103. Jansen, E. F., and Hirschmann, D. J., *Arch. Biochem., 4,* 297 (1944).
104. Jordan, D. C., and Inniss, W. E., *Nature, 184,* 1894 (1961).
105. Jordan, D. C., *Biochem. Biophys. Res. Commun., 6,* 167 (1961).
106. McCormick, M. H., Stark, W. M., Pittenger, G. E., Pittenger, R. C., and McGuire, J. M.,, in *Antibiotics Annual 1955–56*, p. 606, H. Welch and F. Marti-Ibanez, Eds. Medical Encyclopedia Inc., New York (1956).
107. Reynolds, P. E., *Biochim. Biophys. Acta, 52,* 403 (1961).
108. Momose, H. J., *Gen. Appl. Microbiol., 7, Suppl. 1,* 359 (1961).
109. Hadjipetrou, L. P., and Stouthamer, A. H., *Antonie van Leeuwenhoeck J. Microbiol. Serol., 29,* 256 (1963).
110. Davis, B. D., and Bauman, N., *Science, 126,* 170 (1957).
111. Davis, B. D., *Nature, 169,* 534 (1952).
112. Meadow, P., Hoare, D. S., and Work, E., *Biochem. J., 66,* 270 (1957).
113. Rhuland, L. E., *J. Bacteriol, 73,* 778 (1957).
114. O'Brien, P. J., Glick, M. G., and Zilliken, F., *Biochim. Biophys. Acta, 37,* 357 (1960).
115. Glick, M. G., Sall, T., Zilliken, F., and Mudd, S., *Biochim. Biophys. Acta, 37,* 361 (1960).
116. Shockman, G. D., Conover, M. J., Kolb, J. J., Phillips, P. M., Riley, L. S., and Toennies, G., *J. Bacteriol., 81,* 36 (1961).
117. Shockman, G. D., *Bacteriol. Rev., 29,* 345 (1965).
118. Shockman, G. D., *Trans. N.Y. Acad. Sci. Ser. II, 26,* 182 (1963).
119. Toennies, G., and Shockman, G. D., *Proc. Int. Congr. Biochem., 4,* 365 (1958).
120. Shockman, G. D., Kolb, J. J., and Toennies, G., *J. Biol. Chem., 230,* 961 (1958).
121. Toennies, G., and Gallant, D. L., *Growth, 13,* 7 (1949).
122. Stolp, H., *Arch. Mikrobiol., 21,* 293 (1955).
123. Kawata, T., Asaki, K., and Takagi, A., *J. Bacteriol., 81,* 160 (1961).
124. Kaufmann, W., and Bauer, K., *J. Gen. Microbiol., 18,* 11 (1958).
125. Nomura, M., and Hosoda, J., *J. Bacteriol., 72,* 573 (1956).
126. Coles, N. W., Gilbo, C. M., and Broad, A. J., *Biochem. J., 111,* 7 (1969).
127. Csuzi, S., and Kramer, M., *Acta Microbiol. Acad. Sci. Hung., 9,* 297 (1962).
128. Norris, J. R., *J. Gen. Microbiol., 16,* 1 (1957).
129. Strange, R. E., and Dark, F. E., *J. Gen. Microbiol., 16,* 236 (1957).
130. Strange, R. E., and Dark, F. E., *J. Gen. Microbiol., 17,* 525 (1957).
131. Dark, F. E., and Strange, R. E., *Nature, 180,* 759 (1957).
132. Welker, N. E., and Campbell, L. L., *Bacteriol. Proc.,* p. 126 (1966).
133. Young, F. E., and Spizizen, J. J., *J. Biol. Chem., 238,* 3126 (1963).
134. Young, F. E., Tipper, D. J., and Strominger, J. L., *J. Biol. Chem., 239,* 3600 (1964).
135. Richmond, M. H., *Biochim. Biophys. Acta, 33,* 78 (1959).
136. Hash, J. H., Wishnick, M., and Miller, P. A., *J. Bacteriol., 87,* 432 (1964).
137. Hash, J. H., *Arch. Biochem. Biophys, 102,* 379 (1963).
138. Tipper, D. J., Strominger, J. L., and Ghuysen, J. G., *Science, 146,* 781 (1964).
139. Jacob, F., and Fuerst, C. R., *J. Gen. Microbiol., 18,* 518 (1958).
140. Pelzer, H., *Z. Naturforsch., Teil B, 18,* 950 (1963).
141. Weidel, W., and Pelzer, H., *Advan. Enzymol., 26,* 193 (1964).
142. Weidel, W., Frank, H., and Leutgeb, W., *J. Gen. Microbiol., 30,* 127 (1963).
143. Kato, K., Kotani, S., Matsubara, T., Kogami, J., Hashimoto, S., Chimori, M., and Kazekawa, I., *Biken J., 5,* 155 (1962).
144. Ensign, J. C., and Wolfe, R. S., *Bacteriol. Proc.,* p. 33 (1964).
145. Kühliwein, H., *Zentralbl. Bakteriol. Parasitenk. Infektionskr. Hyg. Abt. I Orig., 162,* 296 (1955).
146. Noren, B., *Svensk. Bot. Tidskr., 47,* 309 (1953).
147. Noren, B., *Svensk. Bot. Tidskr., 49,* 282 (1955).
148. Noren, B., *Svensk. Bot. Tidskr., 54,* 550 (1960).
149. Noren, B., *Bot. Notis., 113,* 320 (1960).
150. Noren, B., and Raper, K. B., *J. Bacteriol., 84,* 157 (1962).
151. Tipper, D. J., Strominger, J. L., and Ensign, J. C., *Biochemistry, 6,* 906 (1967).
152. Zyskind, J. W., Pattee, P. A., and Lache, M., *Science, 147,* 1458 (1965).
153. Tsai, C. S., Whitaker, D. R., Jurasek, L., and Gillespie, D. C., *Can. J. Biochem., 43,* 1971 (1965).
154. Whitaker, D. R., *Can. J. Biochem., 43,* 1935 (1965).
155. Arvidson, S., Holme, T., and Wadstrom, T., *J. Bacteriol., 104,* 227 (1970).
156. Richmond, M. H., *Biochim. Biophys. Acta, 31,* 564 (1959).
157. Huff, E., Silverman, C. S., Adams, N. J., and Woodruff, S. A., *J. Bacteriol., 103,* 761 (1970).
158. Welsh, M., and Salmon, J., *Ann. Inst. Pasteur (Paris), 79,* 802 (1950).

159. Schindler, C. A., and Schuhardt, V. T., *Proc. Nat. Acad. Sci., U.S.A., 51,* 414 (1964).
160. Suginaka, H., Kashiba, S., Amano, T., Kotani, S., and Imanishi, T., *Biken J., 10,* 109 (1967).
161. Bleiweis, A. S., and Krause, R. M., *J. Exp. Med., 122,* 237 (1965).
162. Conover, M. J., Thompson, J. S., and Shockman, G. D., *Biochem. Biophys. Res. Commun., 23,* 713 (1966).
163. Shockman, G. D., and Cheney, M. C., *J. Bacteriol., 98,* 1199 (1969).
164. Montague, M. D., *Biochim. Biophys. Acta, 86,* 588 (1964).
165. Gooder, H., and Maxted, W. R., *Nature, 182,* 808 (1958).
166. Ghuysen, J. M., Petit, J. F., Munoz, E., and Kato, K., *Fed. Proc., 25,* 410 (1966).
167. McCarty, M., *J. Exp. Med., 96,* 555 (1952).
168. Welsch, M., *Phènomène d'antibiose chez les Actinomycetes,*. J. Duculot, Gembloux, France (1947).
169. Dean, C. J., Feldschreiber, P., and Lett, J. T., *Nature, 209,* 49 (1966).
170. Mori, Y., Kato, K., Matsubara, T., and Kotani, S., *Biken J., 3,* 139 (1960).
171. Munoz, E., Ghuysen, J. M., Leyh-Bouille, M., Petit, J. F., and Tinelli, R., *Biochemistry, 5,* 3091 (1966).
172. Girard, P., and Sertic, V., *C. R. Sèances Soc. Biol. Fil., 118,* 1286 (1935).
173. Murphy, J. S., *Virology, 4,* 563 (1957).
174. Bradley, D. E., *J. Gen. Virol., 3,* 141 (1968).
175. Katz, W. E., and Weidal, W., *Z. Naturforsch. Teil B, 16,* 363 (1961).
176. Panijel, J., and Happert, J., *C. R. Sèances Soc. Biol. Fil., 245,* 240 (1957).
177. Streisinger, G., Mukai, F., Dreyer, W. J., Miller, B., and Horiuchi, S., *Cold Spring Harbor Symp. Quant. Biol., 26,* 5 (1961).
178. Weidal, W., and Katz, W., *Z. Naturforsch. Teil B, 16,* 156 (1961).
179. Humphries, J. C., *J. Bacteriol., 56,* 683 (1948).
180. Gratia, A., and Rhodes, B., *C. R. Sèances Soc. Biol. Fil., 89,* 171 (1923).
181. Ralston, D. J., Baer, B., Lieberman, M., and Krueger, A. P., *J. Gen. Microbiol., 24,* 313 (1961).
182. Ralston, D. J., *Bacteriol. Proc.,* p. 69 (1966).
183. Barkulis, S. S., Smith, C., Boltralik, J. J., and Heymann, H., *J. Biol. Chem., 239,* 4025 (1964).
184. Fox, E. N., *J. Bacteriol., 85,* 536 (1963).
185. Freimer, E. H., *J. Exp. Med., 117,* 377 (1963).
186. Freimer, E. H., Krause, R. M., and McCarty, M., *J. Exp. Med., 110,* 853 (1959).
187. Krause, R. M., *J. Exp. Med., 106,* 365 (1957).
188. Markowitz, A., and Dorfman, A., *J. Biol. Chem., 237,* 273 (1962).
189. Smith, D. G., and Shattok, R. G. E., *J. Gen. Microbiol., 34,* 165 (1964).
190. Zeleznick, L. D., Boltralik, J. J., Barkulis, S. S., Smith, C., and Heymann, H., *Science, 140,* 400 (1963).
191. Chaloupka, J., Kreckova, P., and Rihova, L., *Folia Microbiol., 7,* 269 (1962).
192. Tomcsik, J., and Guex-Holzer, S., *Schweiz. Z. Allg. Pathol. Bacteriol., 15,* 517 (1952).
193. Fedorova, G. I., *Antibiotiki, 14,* 880 (1969).
194. Weibull, C., *J. Bacteriol., 66,* 688 (1953).
195. Mutsaars, W., *Ann. Inst. Pasteur (Paris), 89,* 166 (1955).
196. Wiame, J. M., Storch, R., and Vanderwinkel, E., *Biochim. Biophys. Acta, 18,* 353 (1955).
197. Spizizen, J., *Proc. Nat. Acad. Sci. U.S.A., 43,* 694 (1957).
198. Fedorova, G. I., *Antibiotiki, 14,* 880 (1969).
199. Fleming, A., *Proc. Roy. Soc. London Ser. B Biol. Sci., 93,* 306 (1922).
200. Fleming, A., and Allison, V. D., *Brit. J. Exp. Pathol., 3,* 252 (1922).
201. McQuillen, K., *Biochim. Biophys. Acta, 17,* 382 (1955).
202. Mitchell, P., and Moyle, J., *J. Gen. Microbiol., 15,* 512 (1956).
203. Salton, M. R. J., *Nature, 170,* 746 (1952).
204. Saint-Blancard, J., Chuzel, P., Mathieu, Y., Perrot, J., and Jolles, P., *Biochim. Biophys. Acta, 220,* 300 (1970).
205. Tomcsik, J., and Guex-Holzer, S., *Schweiz. Z. Allg. Pathol. Bacteriol., 15,* 517 (1952).
206. McQuillen, K., *Biochim. Biophys. Acta, 17,* 382 (1955).
207. Mitchell, P., and Moyle, J., *J. Gen. Microbiol., 15,* 512 (1956).
208. Tomcsik, J., and Guex-Holzer, S., *Schweiz. Z. Allg. Pathol. Bacteriol., 15,* 517 (1952).
209. Gooder, H., in *Microbial Protoplasts, Spheroplasts and L-Forms,* p. 40, L. B. Guze, Ed. Williams and Wilkins, Baltimore, Maryland (1968).
210. Colobert, L., and Lenoir, J., *Ann. Inst. Pasteur (Paris), 92,* 74 (1957).
211. Repaske, R., *Biochim. Biophys. Acta, 22,* 189 (1956).
212. Colobert, L., *Ann. Inst. Pasteur (Paris), 95,* 156 (1958).
213. Fraser, D., and Mohler, H. R., *Arch. Biochem. Biophys., 69,* 166 (1957).
214. Mahler, H. R., and Fraser, D., *Biochim. Biophys. Acta, 22,* 197 (1956).
215. Miura, T., and Mizushima, S., *Biochim. Biophys. Acta, 193,* 268 (1969).
216. Repaske, R., *Biochim. Biophys. Acta, 22,* 189 (1956).
217. Rickenberg, H. V., *Biochim. Biophys. Acta, 25,* 206 (1957).
218. Colobert, L., *Ann. Inst. Pasteur (Paris), 95,* 156 (1958).

219. Colobert, L., *C. R. Séances Soc. Biol. Fil., 115,* 1904 (1957).
220. Metcalf, R., and Deibel, R. H., *J. Bacteriol., 99,* 674 (1969).
221. Birdsell, D. C., and Cota-Robles, E. H., *Biochem. Biophys. Res. Commun. 31,* 438 (1968).
222. Godson, G. N., Nigel, G., and Sinsheimer, R. L., *Biochim. Biophys. Acta, 149,* 476 (1967).
223. Zinder, N., and Arndt, W. P., *Proc. Nat. Acad. Sci. U.S.A., 42,* 586 (1956).
224. Ceglowski, W. S., and Lear, S. A., *Appl. Microbiol., 10,* 458 (1962).
225. Gould, G. W., and Hitchins, A. D., *Nature, 197,* 622 (1963).
226. Gould, G. W., and Hitchins, A. D., *J. Gen. Microbiol., 33,* 413 (1963).
227. Kohn, A., *J. Bacteriol., 79,* 697 (1960).
228. Noller, E. C., and Hartsell, S. E., *J. Bacteriol., 81,* 482, 492 (1961).
229. Barker, D. C., and Thorne, K. J. I., *J. Cell Sci., 7,* 755 (1970).
230. Warren, G. H., Gray, J., and Yurchenko, J. A., *J. Bacteriol., 74,* 788 (1957).
231. Ghosh, B. K., and Murray, R. G. E., *J. Bacteriol., 93,* 411 (1967).
232. Driedger, A. A., and Grayston, M. J., *Can. J. Microbiol., 16,* 889 (1970).
233. Kitayama, S., and Matsuyama, A., *Biochem. Biophys. Res. Commun., 33,* 418 (1968).
234. Adamek, L. P., Mison, H., Mohelska, H., and Trnka, L., *Arch. Mikrobiol., 69,* 227 (1969).
235. Repaske, R., *Biochim. Biophys. Acta, 22,* 189 (1956).
236. Murray, R. G. E., in *Microbial Protoplasts, Spheroplasts and L-Forms,* p. 1, L. B. Guze, Ed. Williams and Wilkins, Baltimore, Maryland (1968).
237. Warren, G. H., and Gray, J., *Proc. Soc. Exp. Biol. Med., 114,* 439 (1963).
238. Warren, G. H., and Gray, J., *Proc. Soc. Exp. Biol. Med., 128,* 776 (1968).
239. Warren, G. H., and Gray, J., *Proc. Soc. Exp. Biol. Med., 126,* 15 (1967).
240. Adhikari, P. C., Raychaudhuri, C., and Chatterjee, S. N., *J. Gen. Microbiol., 59,* 91 (1969).
241. Birdsell, D. C., and Cota-Robles, E. H., *Biochem. Biophys. Res. Commun., 31,* 438 (1968).
242. Amano, T., in *Microbial Protoplasts, Spheroplasts and L-Forms,* p. 30, L. B. Guze, Ed. Williams and Wilkins, Baltimore, Maryland (1968).
243. Hirsch, J. G., *J. Exp. Med., 103,* 598 (1956).
244. Smith, P. F., Koostra, W. L., and Mayberry, W. R., *J. Bacteriol., 100,* 1166 (1969).
245. Crombi, L. B., *M. S. Thesis.* University of Minnesota, Minneapolis (1966).
246. Inoue, K., Tanigawa, Y., Takubo, M., Satani, M., and Amano, T., *Biken J., 2,* 1 (1959).
247. Muschel, L. H., in *CIBA Foundation Symposium on Complement,* p. 153, W. Wolstenholme and J. Knight, Eds. Churchill, London, England (1965).
248. Muschel, L. H., in *Microbial Protoplasts, Spheroplasts, and L-Forms,* p. 19, L. B. Guze, Ed. Williams and Wilkins, Baltimore, Maryland (1968).
249. Wilson, L., and Spitznagel, K., Jr., *J. Bacteriol., 96,* 1339 (1968).
250. Stolp, H., and Petzold, H., *Phytopathol. Z., 45,* 364 (1962).
251. Stolp, H., and Starr, M. P., *Bacteriol. Proc.,* p. 47 (1963).
252. Stolp, H., and Starr, M. P., *Antonie Van Leeuwenhoek J. Microbiol. Serol., 29,* 217 (1963).
253. Shilo, M., and Bruff, B., *J. Gen. Microbiol., 40,* 317 (1965).
254. Starr, M. P., and Bargent, N. L., *J. Bacteriol., 91,* 2006 (1966).

## Isolation of Fungal Cell Walls

Dr. Walter J. Nickerson

In general, the preparation of clean cell walls suitable for chemical analysis or electron microscopy entails some form of mechanical agitation of a washed cell mass. Relatively fragile cells may be disrupted in a sonic oscillator,[1] but most yeasts and filamentous fungi require agitation with glass beads. The Mickle disintegrator[2] was one of the first such devices. It operates as an electrically driven "tuning fork", but the cups for cell paste and glass beads are of limited capacity; cooling can be achieved only by operating in a cold room. The Waring blender can accomodate much larger volumes of cell paste and glass beads,[3,4] but again cooling cannot be accomplished efficeintly; long periods (more than one hour) are required for substantial cell breakage.

More recently, the Braun cell homogenizer[5,6,7] has come into widespread usage. It also involves shaking a mixture of glass beads and cell paste, but provision has been made for cooling the mixture during operation. A stream of liquid carbon dioxide (obtained by fitting a tank of carbon dioxide with an adductor tube) cools the shaking vessel. Times for essentially complete cell breakage are short (one to two minutes).

A novel type of disintegrator that employs a smooth spinning disc has recently been described for the preparation of yeast cell walls.[8] A combination of shear forces produced by laminar flow, collision and rolling glass beads causes cell breakage. Little heat is produced in the operation of this device, but breakage times are longer than noted above.

A completely different method for disrupting microorganisms entails the sudden release of high pressures. In the Ribi refrigerated cell fractionator,[7,9,10] a cell suspension is forced through a narrow orifice under a pressure of approximately 35,000 pounds per square inch. Cell walls of yeasts and fungi have been prepared with this apparatus and have been found to possess antigenic activity.

After cells are broken, walls are separated from cellular debris (and glass beads) by centrifugation. Usually a number of centrifugations, alternating water, buffer and denser supporting medium (such as 10% sucrose or mannitol), are required before a preparation is obtained that can be termed an isolated clean cell wall. Light and electron microscopy of stained and unstained preparations are employed to follow the progress of the "cleaning-up" procedure. Infrared spectral analysis[8] has been found to be useful in following the course of purification of yeast cell walls. A decrease in the slope of the lines connecting peaks at 7.1 to 8.1 microns and at 6.05 to 7.1 microns accompanies increasing purification of the wall preparation. The absorption band at 8.1 microns (P = O) is essentially absent in a clean cell wall preparation, indicating the removal of cytoplasmic contamination of nucleic acids and polyphosphates.

## REFERENCES

1. Tokunaga, J., and Bartnicki-Garcia, S., *Arch. Mikrobiol.*, *79*, 293 (1971).
2. Mickle, H., *J. R. Microsc. Soc.*, *68*, 10 (1948).
3. Lamanna, C., and Mallette, M. F., *J. Bacteriol.*, *67*, 503 (1954).
4. Falcone, G., and Nickerson, W. J., *Science*, *124*, 272 (1956).
5. Young, F. E., Spizizen, J., and Crawford, I. P., *J. Biol. Chem.*, *238*, 3119 (1963).
6. Mill, P. J., *J. Gen. Microbiol.*, *44*, 329 (1966).
7. Novaes-Ledieu, M., Jiménez-Martinez, A., and Villanueva, J. R., *J. Gen. Microbiol.*, *47*, 237 (1967).
8. Řeháček, J., Beran, K., and Bičik, V., *Appl. Microbiol.*, *17*, 462 (1969).
9. Ribi, E., Perrine, T., List, R., Brown, W., and Goode, G., *Proc. Soc. Exp. Biol. Med.*, *100*, 647 (1959).
10. Gerhardt, P., and Judge, J. A., *J. Bacteriol.*, *87*, 945 (1964).

# CENTRIFUGATION

DR. C. A. PRICE

## KINDS OF CENTRIFUGATION

Centrifugation may be employed to separate particles on the basis of size (*rate, zone,* or *rate zonal* separation) or density (*isopycnic* or *equilibrium density* separation). In the simplest case a suspension of particles may be resolved by sedimentation into a pellet of particles and the supernatant fluid. Bacteria, for example, are commonly harvested by such simple centrifugation.

A mixture of particles may be partially resolved by *differential* or *fractional* centrifugation, which is essentially a rate separation. In this procedure successively heavier particles are collected by increasing the speed and/or time of centrifugation (Figures 1 and 2). Differential centrifugation remains the most widely used centrifugal technique, but the particles obtained are almost invariably impure and/or recovery is incomplete. One reason is evident from Figure 1: small particles that are initially near the bottom of the tube will be pelleted before all of the larger particles reach the bottom.

*Density gradients* provide much higher resolving power for separating particles according to rate and also offer the possibility of separation according to very small differences in density ($\leqslant 0.001\ gm \cdot cm^{-3}$).[19,25] These two types of density gradient centrifugation (DGC) are illustrated in Figure 3. The capacity of rotors for DGC was increased by about two orders of magnitude for a given speed by Anderson's invention of zonal rotors (Figure 4).[1,2,24] DGC is also employed in analytical rotors,[18] principally for the resolution of mixtures of DNA according to their densities in solutions of cesium salts.

*Continuous-flow* centrifugation is the usual method of choice for harvesting microorganisms from large volumes of culture. Continuous-flow DGC may be required if the particles are sensitive to pelleting, as is the case with many viruses. It has the further advantage of simultaneous harvest and resolution of particles of different densities (Figure 5).[22]

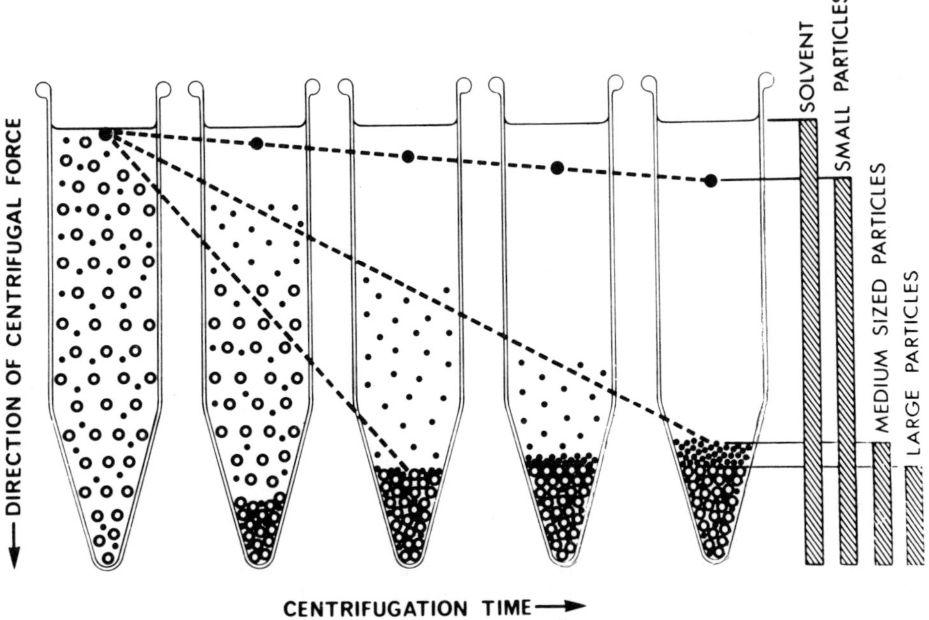


FIGURE 1. Differential centrifugation. In differential centrifugation, large, dense particles sediment more rapidly than small, light particles. Separations can be achieved on the basis of their different rates of sedimentation. (Figure reproduced by courtesy of N. G. Anderson.)

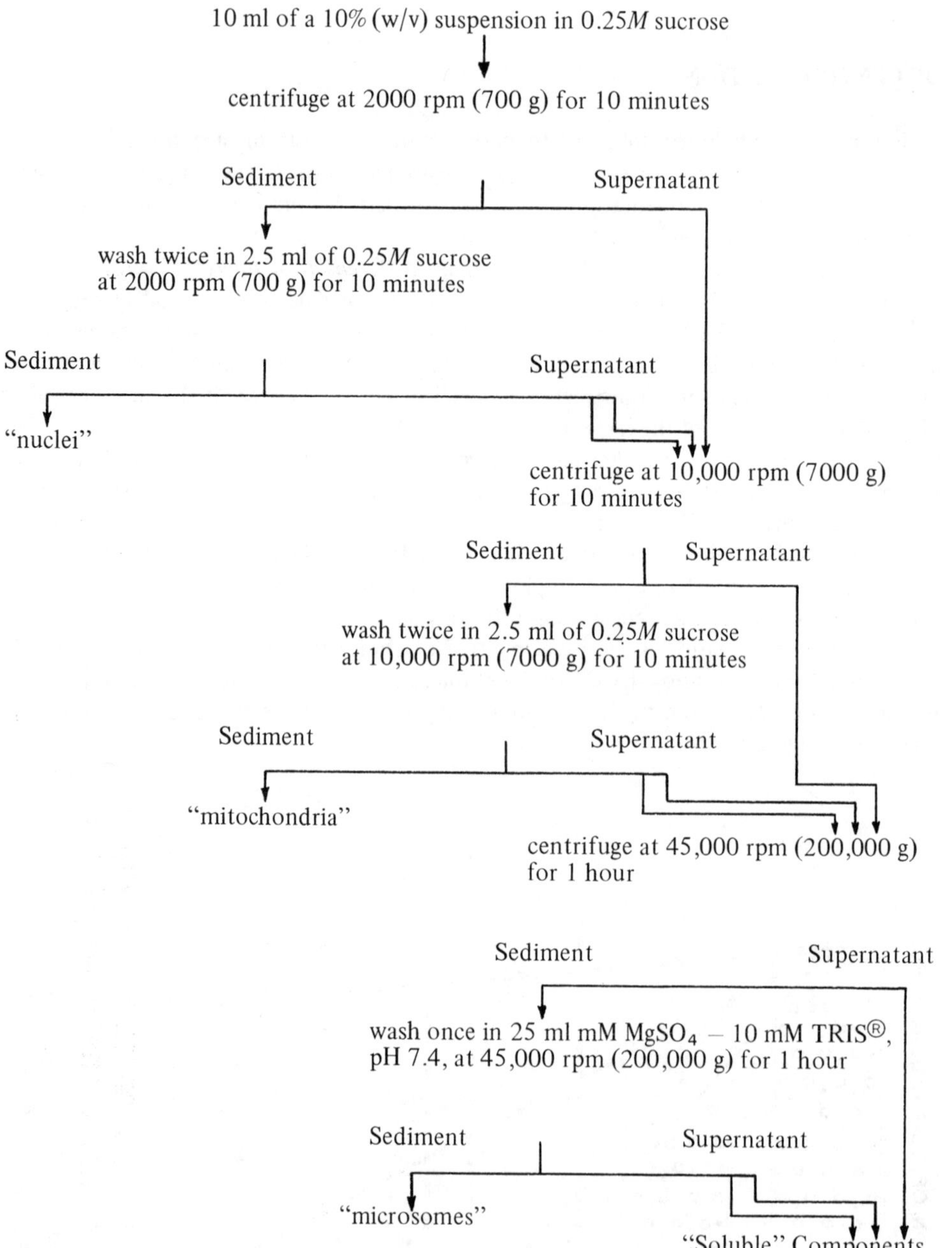


FIGURE 2. Fractionation of cell components by differential centrifugation. This separation scheme, originally designed for liver, has served as a model for most cell and tissue fractionations; it is now used principally as a starting point for the subsequent separation of particulate components by density gradient centrifugation. From: Schneider, W. C., Methods for the Isolation of Particulate Components of the Cell, in *Manometric and Biochemical Techniques,* 5th ed., pp. 196–212, W. W. Umbreit, R. H. Burris and J. F. Stauffer, Eds. (1972). Reproduced by permission of Burgess Publishing Co., Minneapolis, Minnesota.

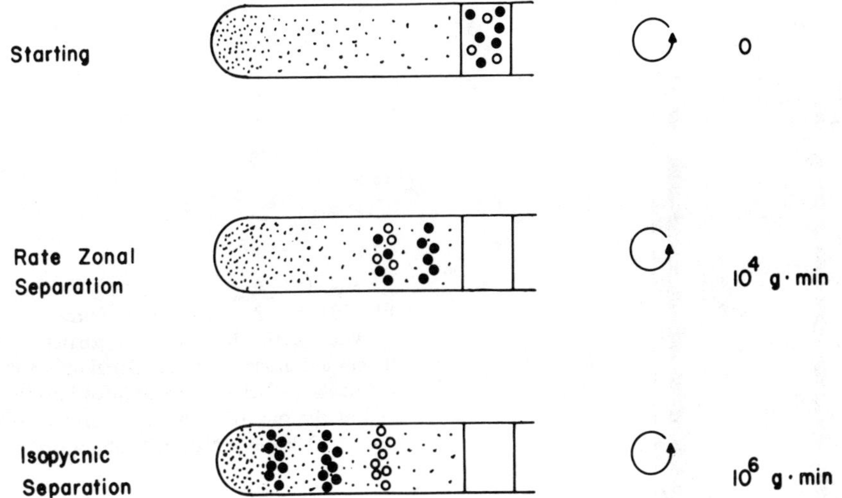


FIGURE 3. Density gradient centrifugation. A mixture of particles is layered over a solution of increasing density; during centrifugation, particles may separate according to differences in size or according to differences in their densities; note that the order of particles in a gradient may shift between rate and isopycnic separation. From: Price, C. A., in *Centrifugation in Density Gradients* (1973). Reproduced by permission of Academic Press, New York.

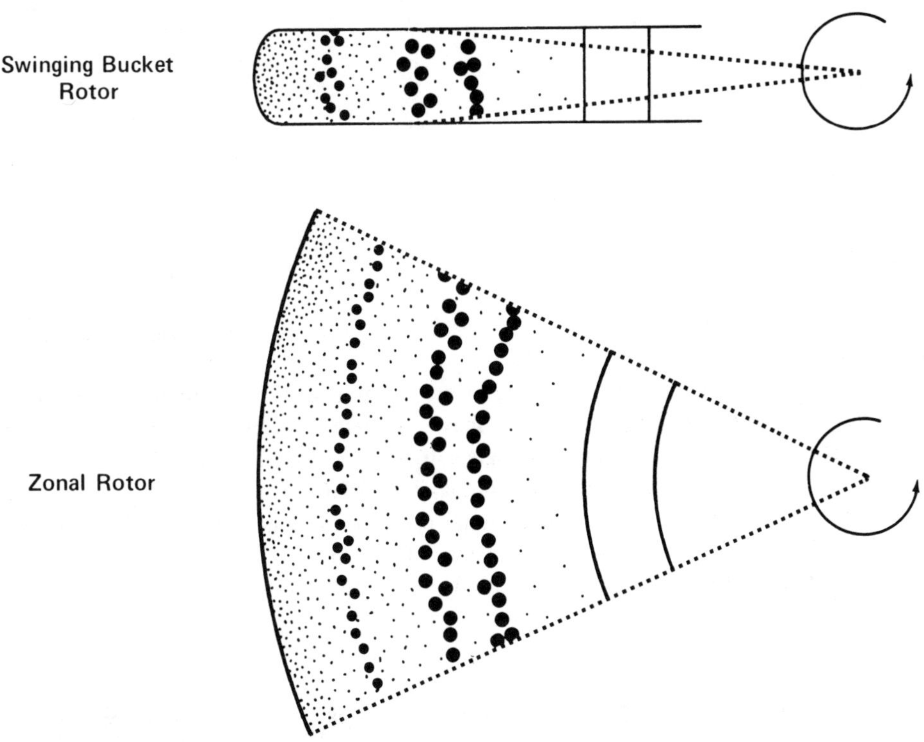


FIGURE 4. Evolution of the zonal rotor. Zonal rotors may be thought of as swinging bucket rotors with the side walls removed; not only is the volume enormously increased, but the phenomenon of "wall effects", in which particles sediment against a lateral wall, is eliminated. From: Price, C. A., in *Centrifugation in Density Gradients* (1973). Reproduced by permission of Academic Press, New York.

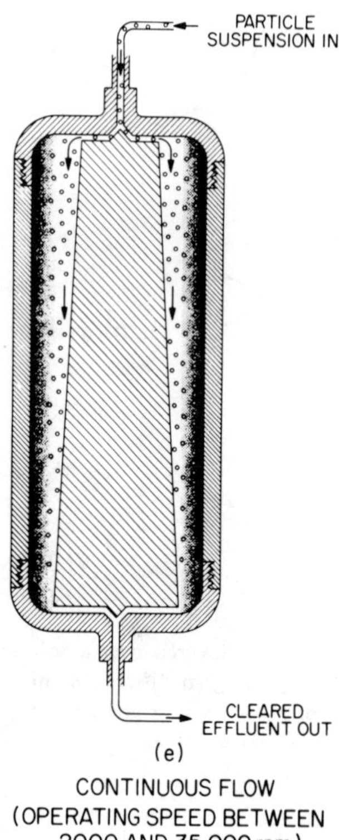


FIGURE 5. A continuous-flow rotor. In this schematic drawing of the K-II rotor, a particle is shown moving downward along a thin annular gradient at the edge of the rotor; the particles are sedimented into the gradient at the end of the run. (Drawing by courtesy of N. G. Anderson and the Oak Ridge National Laboratory.)

## KINDS OF ROTORS

The work horses of the laboratory are *angle rotors*. A representative list of the sizes and speeds commonly available is presented in Table 1.

*Swinging bucket rotors* are preferred over angle rotors for most density gradient work because particles have less tendency to strike the wall, where they may aggregate or stick, but the currently permissible centrifugal fields are only about half of those for angle rotors of the same capacity (Table 2).

*Zonal rotors*, which are favored for most preparative DGC, can be employed in a variety of configurations. Most of the present generation of zonal rotors are immediate descendants of the series of rotors developed by N. G. Anderson's group at Oak Ridge (Table 3).

Some *continuous-flow* configurations of zonal rotors are listed in Table 4.

## TABLE 1
## CHARACTERISTICS OF SOME HIGH-SPEED ANGLE ROTORS IN COMMON USE

| Manufacturer and Designation | Place by Volume | Maximum Speed, rpm | $g_{max}$ | $r_{max}$, cm |
|---|---|---|---|---|
| IEC 959 | 4 x 600 | 6,000 | 7,900 | 19.7 |
| Sorvall GS-3 | 6 x 500 | 9,000 | 13,700 | 15.2 |
| IEC A-28 | 4 x 500 | 14,000 | 28,500 | 13.2 |
| Spinco 15 | 4 x 500 | 15,000 | 35,500 | 14.2 |
| IEC A-54 | 6 x 250 | 20,000 | 54,800 | 12.6 |
| Spinco 19 | 6 x 250 | 19,000 | 53,700 | 13.4 |
| Sorvall GSA | 6 x 250 | 13,000 | 27,500 | 14.6 |
| MSE 59597 | 6 x 250 | 21,000 | 75,000 | 15.3 |
| MSE 59596 | 10 x 100 | 27,500 | 102,000 | 12.1 |
| Spinco 21 | 10 x 94 | 21,000 | 59,000 | 12.0 |
| Spinco 35 | 6 x 94 | 35,000 | 142,800 | 10.5 |
| IEC A-170 | 6 x 90 | 40,000 | 168,300 | 9.4 |
| IEC A-192 | 8 x 50 | 40,000 | 189,600 | 10.7 |
| MSE 59595 | 8 x 50 | 40,000 | 193,000 | 10.9 |
| Sorvall SS-34 | 8 x 50 | 20,000 | 48,200 | 10.8 |
| IEC A-110 | 12 x 40 | 30,000 | 109,900 | 10.9 |
| IEC A-237 | 8 x 40 | 50,000 | 237,300 | 8.5 |
| IEC A-211 | 8 x 40 | 45,000 | 211,200 | 9.3 |
| Spinco 30 | 12 x 38.5 | 30,000 | 105,600 | 10.6 |
| Spinco 60 Ti | 8 x 38.5 | 60,000 | 361,000 | 9.0 |
| Spinco 42.1 | 8 x 38.5 | 42,000 | 195,000 | 9.9 |
| MSE 59593 | 8 x 25 | 60,000 | 379,000 | 9.5 |
| Sorvall SM-24 | 24 x 15 | 20,000 | 49,500 | 11.1 |
| IEC A-168 | 20 x 14 | 40,000 | 168,100 | 9.4 |
| IEC A-269 | 10 x 14 | 55,000 | 268,300 | 7.9 |
| IEC A-321 | 8 x 14 | 60,000 | 319,300 | 7.9 |
| Spinco 75 Ti | 8 x 13.5 | 75,000 | 503,000 | 8.0 |
| Spinco 65 | 8 x 13.5 | 65,000 | 368,400 | 7.8 |
| Spinco 50 Ti | 12 x 13.5 | 50,000 | 226,400 | 8.2 |
| Spinco 40 | 12 x 13.5 | 40,000 | 144,800 | 8.1 |
| Spinco 30.2 | 20 x 10.5 | 30,000 | 94,000 | 9.4 |
| MSE 59592 | 10 x 10 | 65,000 | 407,000 | 8.7 |
| Spinco 50 | 10 x 10 | 50,000 | 198,000 | 7.1 |
| Spinco 40.3 | 18 x 6.5 | 40,000 | 143,000 | 8.0 |
| Spinco 40.2 | 12 x 6.5 | 40,000 | 143,000 | 8.0 |

## TABLE 2
## CHARACTERISTICS OF SOME SWINGING BUCKET ROTORS COMMONLY USED FOR DENSITY GRADIENT CENTRIFUGATION

| Manufacturer and Designation | Place by Volume | Maximum Speed | | | | |
|---|---|---|---|---|---|---|
| | | rpm | $\omega^2$ | $g_{max}$ | $r_{max}$, cm | $\Delta \ell$, cm |
| Sorvall HL 8 | 8 x 100 | 3,700 | $1.50 \times 10^5$ | 3,720 | 24.5 | 10.0 |
| | 16 x 50 | 3,400 | $1.27 \times 10^5$ | 3,080 | 24.0 | 11.6 |
| IEC 269 | 8 x 100 | 3,150 | $1.09 \times 10^5$ | 2,780 | 24.4 | 12.3 |
| | 8 x 50 | 4,100 | $1.84 \times 10^5$ | 3,720 | 19.8 | 9.3 |
| IEC 253 | 12 x 50 | 3,000 | $1.27 \times 10^5$ | 2,400 | 22.7 | 9.3 |
| | 12 x 15 | 3,300 | $1.19 \times 10^5$ | 2,800 | 23.2 | 8.3 |
| MSE 59591 | 3 x 65 | 23,500 | $6.06 \times 10^6$ | 100,000 | 16.2 | 9.8 |
| Spinco SW 25.2 | 3 x 60 | 25,000 | $6.85 \times 10^6$ | 106,900 | 15.4 | 8.9 |
| Sorvall HB-4 | 4 x 50 | 13,000 | $1.85 \times 10^6$ | 27,500 | 14.6 | 9.9 |
| IEC SB-110 | 6 x 40 | 25,000 | $6.85 \times 10^6$ | 110,100 | 15.8 | 8.9 |
| IEC 940 | 4 x 40 | 12,000 | $1.58 \times 10^6$ | 20,800 | 12.9 | 8.9 |
| Spinco SW 27 | 6 x 38.5 | 27,000 | $8.00 \times 10^6$ | 131,000 | 16.2 | 8.9 |
| Spinco SW 25.1 | 3 x 34 | 25,000 | $6.85 \times 10^6$ | 90,000 | 13.0 | 7.6 |
| MSE 59590 | 3 x 23 | 30,000 | $9.87 \times 10^6$ | 129,000 | 12.9 | 7.0 |
| Spinco 27.1 | 6 x 17 | 27,000 | $8.00 \times 10^6$ | 135,000 | 16.7 | 10.2 |
| IEC SB-283 | 6 x 12 | 41,000 | $1.84 \times 10^7$ | 283,000 | 15.1 | 9.6 |
| IEC SB-206 | 6 x 12 | 35,000 | $1.34 \times 10^7$ | 206,000 | 15.1 | 9.6 |
| Spinco SW 40 Ti | 6 x 14 | 40,000 | $1.75 \times 10^7$ | 284,000 | 16.0 | 9.5 |
| Spinco SW 36 | 4 x 13.5 | 36,000 | $1.42 \times 10^7$ | 193,000 | 13.4 | 7.6 |
| Spinco SW 41 Ti | 6 x 13 | 41,000 | $1.84 \times 10^7$ | 286,000 | 15.3 | 8.9 |
| Spinco SW 50.1 | 6 x 5 | 50,000 | $2.74 \times 10^7$ | 300,000 | 10.8 | 5.1 |
| Spinco SW 65L Ti | 3 x 5 | 65,000 | $4.63 \times 10^7$ | 420,000 | 8.9 | 5.1 |
| Spinco SW 50L | 3 x 5 | 50,000 | $2.74 \times 10^7$ | 274,000 | 9.9 | 5.1 |
| MSE 59587 | 3 x 5 | 65,000 | $4.63 \times 10^7$ | 420,000 | 8.9 | 5.1 |
| MSE 59589 | 3 x 5 | 40,000 | $1.75 \times 10^7$ | 178,000 | 9.9 | 5.1 |
| IEC SB-405 | 6 x 4.2 | 60,000 | $3.95 \times 10^7$ | 405,900 | 10.1 | 5.1 |
| Spinco SW 56 Ti | 6 x 4 | 56,000 | $3.44 \times 10^7$ | 408,000 | 11.7 | 6.0 |
| MSE 59588 | 3 x 3 | 50,000 | $2.74 \times 10^7$ | 261,000 | 9.3 | 5.2 |

TABLE 3
CHARACTERISTICS OF CURRENT BATCH-TYPE ZONAL ROTORS

| Manufacturer and Designation | Nearest Oak Ridge Designation | Maximum Speed | | Core Configuration | Gradient Volume, cc | Radius at Edge, cm | $g_{max}$ | Special Characteristics |
|---|---|---|---|---|---|---|---|---|
| | | rpm | $\omega^2$ | | | | | |
| IEC A-12 | A-XII | 4,600 | $2.32 \times 10^5$ | Center-loading | 1300 | 18 | 4,206 | Transparent end caps, largest radius. |
| MSE A | A-XII | 5,000 | $2.74 \times 10^5$ | Center-loading | 1300 | 18 | 4,206 | Transparent end caps, largest radius. |
| IEC Z-15 | – | 8,000 | $7.02 \times 10^5$ | Center-loading | 780 | 12 | 9,000 | Transparent end caps. |
| | | | | MACS, edge-loading | 1005 | 11 | 7,800 | Transparent end caps; MACS = Multiple Alternate Channel Selection; sample may be loaded or unloaded at the center, the edge, or at intermediate radii. |
| MSE HS | – | 10,000 | $1.10 \times 10^6$ | Center-loading | 695 | 10.3 | 11,400 | Transparent end caps. |
| Sorvall SZ-14 | – | 20,000 | $4.39 \times 10^6$ | Reorienting gradient | 1373 | 9.5 | 42,600 | Sample normally loaded while spinning, but unloaded at rest; gradient reorients slowly during deceleration. |
| Spinco JCF-Z | – | 20,000 | $4.39 \times 10^6$ | Center-loading | 1900 | 8.9 | 40,000 | |
| IEC B-29A | B-XV | 35,000 | $1.34 \times 10^7$ | Center-loading | 1670 | 8.9 | 121,100 | |
| | B-XXIX | 35,000 | $1.34 \times 10^7$ | Edge-loading | 1480 | 8.5 | 116,000 | Sample may be loaded or unloaded at the center or the edge. |
| | | 35,000 | $1.34 \times 10^7$ | MACS, semi-batch | 1430 | 8.5 | 116,000 | MACS = Multiple Alternate Sample Selection; repeated volumes of sample may be sedimented into single annular gradient. |
| MSE B-XV Ti | B-XV | 35,000 | $1.34 \times 10^7$ | Center-loading | 1670 | 8.9 | 122,000 | |
| MSE B-XV A1 | B-XV | 25,000 | $6.85 \times 10^6$ | Center-loading | 1670 | 8.9 | 62,000 | |
| Spinco Ti 15 | B-XV | 32,000 | $1.12 \times 10^7$ | Center-loading | 1675 | 8.9 | 101,300 | |
| | B-XXIX | 32,000 | $1.12 \times 10^7$ | Edge-loading | 1350 | 8.4 | 95,600 | Sample may be loaded or unloaded at the center or the edge. |

## TABLE 3 (Continued)
## CHARACTERISTICS OF CURRENT BATCH-TYPE ZONAL ROTORS

| Manufacturer and Designation | Nearest Oak Ridge Designation | Maximum Speed rpm | Maximum Speed $\omega^2$ | Core Configuration | Gradient Volume, cc | Radius at Edge, cm | $g_{max}$ | Special Characteristics |
|---|---|---|---|---|---|---|---|---|
| Spinco Al 15 | B-XV | 22,000 | $5.31 \times 10^6$ | Center-loading | 1675 | 8.9 | 48,100 | |
| | B-XXIX | 22,000 | $5.31 \times 10^6$ | Edge-loading | 1350 | 8.4 | 45,200 | Sample may be loaded or unloaded at the center or the edge. |
| Spinco B-4 | B-IV | 40,000 | $1.75 \times 10^7$ | Center-loading | 1750 | 4.9 | 86,000 | Requires special L-4 drive unit. |
| IEC B-30A | B-XIV | 50,000 | $2.74 \times 10^7$ | Center-loading | 649 | 6.7 | 186,000 | Fastest conventional zonal rotor. |
| | – | 50,000 | $2.74 \times 10^7$ | Edge-loading | 534 | 6.4 | 176,000 | Sample may be loaded or unloaded at the center or the edge. |
| MSE B-XIV Ti | B-XIV | 47,000 | $2.42 \times 10^7$ | Center-loading | 650 | 6.7 | 165,000 | Fastest conventional zonal rotor. |
| MSE B-XIV Al | B-XIV | 35,000 | $1.34 \times 10^7$ | Center-loading | 650 | 6.7 | 91,000 | |
| Spinco Ti 14 | B-XIV | 48,000 | $2.53 \times 10^7$ | Center-loading | 665 | 6.7 | 172,000 | Fastest conventional zonal rotor. |
| | – | 48,000 | $2.53 \times 10^7$ | Edge-loading | 544 | 6.4 | 164,000 | Sample may be loaded or unloaded at the center or the edge. |
| Spinco Al 14 | B-XIV | 35,000 | $1.34 \times 10^7$ | Center-loading | 665 | 6.7 | 91,000 | |
| | – | 35,000 | $1.34 \times 10^7$ | Edge-loading | 544 | 6.4 | 87,000 | Sample may be loaded or unloaded at the center or the edge. |
| Electro-Nucleonics K-5 | K-V | 35,000 | $1.34 \times 10^7$ | Center-loading | 8390 | 6.6 | 90,000 | Alternate core configuration of K-type continuous-flow centrifuge (see Table 4). |
| Electro-Nucleonics RK-5 | – | 35,000 | $1.34 \times 10^7$ | Center-loading | 4180 | 6.6 | 90,000 | Alternate core configuration of RK-type continuous-flow centrifuge (see Table 4). |

## TABLE 4
## CHARACTERISTICS OF CONTINUOUS-FLOW ROTORS

| Manufacturer and Designation | Gradient Volume, cc | Maximum Speed rpm | Maximum Speed $\omega^2$ | Radius at Edge, cm | $g_{max}$ | Flow-Path Length, cm | Drive Unit |
|---|---|---|---|---|---|---|---|
| IEC CF-6 | 515 | 6,000 | $3.95 \times 10^5$ | 14.7 | 6,000 | 92.3 | PR-2, PR-6, PR-6000 |
| Spinco JCF | 400 | 20,000 | $4.39 \times 10^6$ | 8.9 | 40,000 | 8.9 | J-21 |
| Spinco CF-32 Ti | 305 | 32,000 | $1.12 \times 10^7$ | 8.9 | 101,300 | 7.3 | L-series |
| Electro-Nucleonics | | | | | | | |
| K-3 | 3200 | 35,000 | $1.34 \times 10^7$ | 6.6 | 90,000 | 76.2 | All K-series cores are interchangeable in the K-series bowl. |
| K-10 | 8000 | 35,000 | $1.34 \times 10^7$ | 6.6 | 90,000 | 76.2 | |
| K-11 | 380 | 35,000 | $1.34 \times 10^7$ | 6.6 | 90,000 | 76.2 | |
| RK-3 | 1600 | 35,000 | $1.34 \times 10^7$ | 6.6 | 90,000 | 38.1 | All RK-series cores are interchangeable in the RK-series bowl. |
| RK-10 | 3980 | 35,000 | $1.34 \times 10^7$ | 6.6 | 90,000 | 38.1 | |
| RK-11 | 190 | 35,000 | $1.34 \times 10^7$ | 6.6 | 90,000 | 38.1 | |
| J-1 | 780 | 55,000 | $3.32 \times 10^7$ | 4.4 | 150,000 | 38.4 | RK |

## SOME SEDIMENTATION EQUATIONS

The rate of sedimentation along the radial axis $\frac{dr}{dt}$ (equation 1) is proportional to five factors: the sedimentation coefficient of the particle ($S_{20,w}$); the distance from the axis of rotation (r); the square of the angular velocity ($\omega$); the ratio of the viscosity of water ($\eta_w$) to the viscosity of the medium ($\eta_m$); and the ratio of the particle density ($\rho_p$) reduced to that of the medium $\rho_m$ to the particle density reduced to that of water $\rho_w$.

(1) $$\frac{dr}{dt} = S_{20,w}\, r\omega^2 \frac{\eta_w}{\eta_m} \left[\frac{\rho_p - \rho_m}{\rho_p - \rho_w}\right]$$

The sedimentation coefficient of an ideal sphere is a function of the size of the particle, somewhat modified by its density (equation 2).

(2) $$S_{20,w} = \frac{1 - \frac{\rho_m}{\rho_p}}{6\pi a \eta N}$$

For particles other than spherical, the denominator of equation 2 is increased by a *frictional coefficient*. In order to express equation 1 in cgs units, the angular velocity must be expressed in radians$\cdot$sec$^{-1}$ (equation 3).

(3) $$\omega\ (\text{radians sec}^{-1}) = \frac{\text{rpm} \times \pi}{30}$$

It follows from equations 1 and 3 that if a certain separation is achieved in a rotor spun at 10,000 rpm for one hour, the same movement of particles should occur at 20,000 rpm in fifteen minutes. It is appropriate, therefore, to report the conditions of centrifugation in $(\text{rpm})^2(\text{min})$ or, in more fundamental units, radians$^2\cdot$seconds$^{-1}$.

A most precise quantity is the $\int\omega^2\,dt$, which takes into account the changing speed during acceleration and deceleration. This quantity may be estimated or measured by "integral omega-squared-dt" meters obtainable from centrifuge manufacturers. A compilation of $\int \omega^2 dt$ at different speeds and times is presented in Table 5.

The value $r\omega^2$ in equation 1 is the *centrifugal field* $\phi$ (equation 4).

$$\phi = r\omega^2 \quad (4)$$

When r is expressed in cm and $\omega$ is expressed in radians per second, $\phi$ will be expressed in $cm \cdot sec^{-2}$. By dividing this value by 980, one obtains what is popularly known as "gravities" (g), the "relative centrifugal force" (RCF). Alternatively, equation 5 can be applied.

$$RCF = 1.111 \times 10^{-5} \times (r \text{ in cm}) \times (rpm)^2 \quad (5)$$

A nomogram connecting gravity units to radius and speed is shown in Figure 6.

If we refer to Tables 1, 2, and 3, we note that rotors are characterized by their maximum speed ($rpm_{max}$) and their maximum g-value ($g_{max}$), calculated from the value of r at the tip of a tube or the edge of the rotor volume. It has become customary to refer to a centrifugation as having occurred in so many "g-minutes." This is of limited value even in differential centrifugation, especially if the rotor is not identified, since a separation is also determined by the g-value at the top of the tube and the distances that particles must migrate to reach the bottom. In general, it is preferable to report the rotor, the speed, and the time, and to proceed empirically when adapting a known procedure to a rotor of substantially different geometry. Anderson (1968) has shown, however, that differential centrifugation can be treated quantitatively.

The sedimentation coefficient of equations 1 and 2 is usually reported in Svedbergs (equation 6).

$$1 \text{ Svedberg} = 10^{-13} \text{ seconds} \quad (6)$$

The sedimentation coefficients, densities, and frictional coefficients of a variety of biological particles are shown in Table 6.

*Gradient shapes.* The optimization of resolution or capacity in DGC requires a choice of the "shape" of the gradient, meaning the concentration of the solute as a function of the radius or the volume of the gradient.

For *isopycnic separations* only the slope of the gradient is important; one usually chooses gradients that are *linear* with volume (Figure 7). These will be only approximately linear with density.

Homogeneous particles at equilibrium density were shown by Meselson, Stahl and Vinograd[18] to assume a Gaussian distribution with a standard deviation with respect to radius of equation 7.

$$\sigma_r = \sqrt{\frac{R\,T\,\rho_o}{M^* \,(d\rho/dr)_o\, \omega^2 r_o}} \quad (7)$$

For *rate separations* in *swinging buckets*, there are advantages in *isokinetic gradients* (Figure 8),[16,20] in which the distance a particle migrates is proportional to its sedimentation coefficient. The general equation for isokinetic gradients is shown in Equation 8.

$$\frac{r}{\eta_m(r)} \left[\rho_p - \rho_m(r)\right] = \text{constant} \quad (8)$$

Gradient shapes can be approximated by simple exponentials (see *gradient generators* below); Noll[20] and McCarty, Stafford and Brown[17] have computed the starting and limiting solutions for generating isokinetic gradients for the resolution of ribosomes in several common rotors (Tables 7 and 8).

Better resolution can be achieved under special circumstances by the use of *acceleration gradients*,[12] in which faster particles "escape" from slower particles as they migrate into the gradient.

TABLE 5

VALUES OF $\omega^2 \Delta t$ IN RADIANS$^2 \cdot$SECONDS$^{-1}$ FOR DIFFERENT SPEED AND INTERVALS[a]

$$\omega^2 t = \left(\frac{2\pi\ \text{rpm}}{60}\right)^2 t(\text{sec})$$

| rpm | $\omega^2 t$ 10 Minutes 600 Seconds | $\omega^2 t$ 20 Minutes 1,200 Seconds | $\omega^2 t$ 30 Minutes 1,800 Seconds | $\omega^2 t$ 1 Hour 3,600 Seconds | $\omega^2 t$ 2 Hours 7,200 Seconds | $\omega^2 t$ 4 Hours 14,400 Seconds |
|---|---|---|---|---|---|---|
| 500 | $1.6446 \times 10^6$ | $3.2883 \times 10^6$ | $4.9329 \times 10^6$ | $9.8658 \times 10^6$ | $1.9732 \times 10^7$ | $3.9463 \times 10^7$ |
| 1,000 | $6.5785 \times 10^6$ | $1.3153 \times 10^7$ | $1.9732 \times 10^7$ | $3.9463 \times 10^7$ | $7.8926 \times 10^7$ | $1.5785 \times 10^7$ |
| 2,000 | $2.6314 \times 10^7$ | $5.2611 \times 10^7$ | $7.8925 \times 10^7$ | $1.5785 \times 10^8$ | $3.157 \times 10^8$ | $6.314 \times 10^8$ |
| 3,000 | $5.9207 \times 10^7$ | $1.1838 \times 10^8$ | $1.7758 \times 10^8$ | $3.5517 \times 10^8$ | $7.1034 \times 10^8$ | $1.4207 \times 10^9$ |
| 5,000 | $1.6446 \times 10^8$ | $3.2883 \times 10^8$ | $4.932 \times 10^8$ | $9.8658 \times 10^8$ | $1.9732 \times 10^9$ | $3.9463 \times 10^9$ |
| 8,000 | $4.2102 \times 10^8$ | $8.4178 \times 10^8$ | $1.2628 \times 10^9$ | $2.5256 \times 10^9$ | $5.0512 \times 10^9$ | $1.010 \times 10^{10}$ |
| 10,000 | $6.5785 \times 10^8$ | $1.315 \times 10^9$ | $1.9732 \times 10^9$ | $3.9463 \times 10^9$ | $7.8926 \times 10^9$ | $1.5785 \times 10^{10}$ |
| 15,000 | $1.4802 \times 10^9$ | $2.5954 \times 10^9$ | $4.4396 \times 10^9$ | $8.8792 \times 10^9$ | $1.7758 \times 10^{10}$ | $3.5517 \times 10^{10}$ |
| 20,000 | $2.6314 \times 10^9$ | $5.2611 \times 10^9$ | $7.8925 \times 10^9$ | $1.5785 \times 10^{10}$ | $3.157 \times 10^{10}$ | $6.314 \times 10^{10}$ |
| 30,000 | $5.9207 \times 10^9$ | $1.1838 \times 10^{10}$ | $1.7758 \times 10^{10}$ | $3.5517 \times 10^{10}$ | $7.1034 \times 10^{10}$ | $1.4207 \times 10^{11}$ |
| 40,000 | $1.0526 \times 10^{10}$ | $2.1045 \times 10^{10}$ | $3.1583 \times 10^{10}$ | $6.3141 \times 10^{10}$ | $1.2628 \times 10^{11}$ | $2.5256 \times 10^{11}$ |
| 50,000 | $1.6446 \times 10^{10}$ | $3.2883 \times 10^{10}$ | $4.9329 \times 10^{10}$ | $9.8658 \times 10^{10}$ | $1.9732 \times 10^{11}$ | $3.9463 \times 10^{11}$ |
| 55,000 | $1.9901 \times 10^{10}$ | $3.9789 \times 10^{10}$ | $5.969 \times 10^{10}$ | $1.1938 \times 10^{11}$ | $2.3876 \times 10^{11}$ | $4.7752 \times 10^{11}$ |
| 60,000 | $2.3683 \times 10^{10}$ | $4.7352 \times 10^{10}$ | $7.1035 \times 10^{10}$ | $1.4207 \times 10^{11}$ | $2.8414 \times 10^{11}$ | $5.6828 \times 10^{11}$ |

TABLE 5

VALUES OF $\omega^2 \Delta t$ IN RADIANS$^2 \cdot$SECONDS$^{-1}$ FOR DIFFERENT SPEED AND INTERVALS[a]

$$\omega^2 t = \left(\frac{2\pi\ \text{rpm}}{60}\right)^2 t(\text{sec})$$

| rpm | $\omega^2 t$ 6 Hours 21,600 Seconds | $\omega^2 t$ 8 Hours 28,800 Seconds | $\omega^2 t$ 12 Hours 43,200 Seconds | $\omega^2 t$ 18 Hours 64,800 Seconds | $\omega^2 t$ 24 Hours 86,400 Seconds |
|---|---|---|---|---|---|
| 500 | $5.9195 \times 10^7$ | $7.8926 \times 10^7$ | $1.1839 \times 10^8$ | $1.7758 \times 10^8$ | $2.3678 \times 10^8$ |
| 1,000 | $2.3678 \times 10^8$ | $3.157 \times 10^8$ | $4.7356 \times 10^8$ | $7.1033 \times 10^8$ | $9.4711 \times 10^8$ |
| 2,000 | $9.471 \times 10^8$ | $1.2628 \times 10^9$ | $1.8942 \times 10^9$ | $2.8413 \times 10^9$ | $3.7884 \times 10^9$ |
| 3,000 | $2.1310 \times 10^9$ | $2.8414 \times 10^9$ | $4.2620 \times 10^9$ | $6.3931 \times 10^9$ | $8.5241 \times 10^9$ |
| 5,000 | $5.9195 \times 10^9$ | $7.8926 \times 10^9$ | $1.1839 \times 10^{10}$ | $1.7758 \times 10^{10}$ | $2.3678 \times 10^{10}$ |
| 8,000 | $1.5154 \times 10^{10}$ | $2.0205 \times 10^{10}$ | $3.0307 \times 10^{10}$ | $4.5461 \times 10^{10}$ | $6.0614 \times 10^{10}$ |
| 10,000 | $2.3678 \times 10^{10}$ | $3.1570 \times 10^{10}$ | $4.7356 \times 10^{10}$ | $7.1033 \times 10^{10}$ | $9.4711 \times 10^{10}$ |
| 15,000 | $5.3275 \times 10^{10}$ | $7.1034 \times 10^{10}$ | $1.0655 \times 10^{11}$ | $1.5983 \times 10^{11}$ | $2.1310 \times 10^{11}$ |
| 20,000 | $9.471 \times 10^{10}$ | $1.2628 \times 10^{11}$ | $1.8942 \times 10^{11}$ | $2.8413 \times 10^{11}$ | $3.7884 \times 10^{11}$ |
| 30,000 | $2.131 \times 10^{11}$ | $2.8414 \times 10^{11}$ | $4.262 \times 10^{11}$ | $6.3931 \times 10^{11}$ | $8.5241 \times 10^{11}$ |
| 40,000 | $3.7885 \times 10^{11}$ | $5.0513 \times 10^{11}$ | $7.5769 \times 10^{11}$ | $1.1365 \times 10^{12}$ | $1.5154 \times 10^{12}$ |
| 50,000 | $5.9195 \times 10^{11}$ | $7.8926 \times 10^{11}$ | $1.1839 \times 10^{12}$ | $1.7758 \times 10^{12}$ | $2.3678 \times 10^{12}$ |
| 55,000 | $7.1628 \times 10^{11}$ | $9.5504 \times 10^{11}$ | $1.4326 \times 10^{12}$ | $2.1488 \times 10^{12}$ | $2.8651 \times 10^{12}$ |
| 60,000 | $8.5242 \times 10^{11}$ | $1.1366 \times 10^{12}$ | $1.7048 \times 10^{12}$ | $2.5573 \times 10^{12}$ | $3.4097 \times 10^{12}$ |

[a] Calculated by R. E. Canning and W. L. Rasmussan, Molecular Anatomy (MAN) Program, Oak Ridge National Laboratories, Oak Ridge, Tennessee.

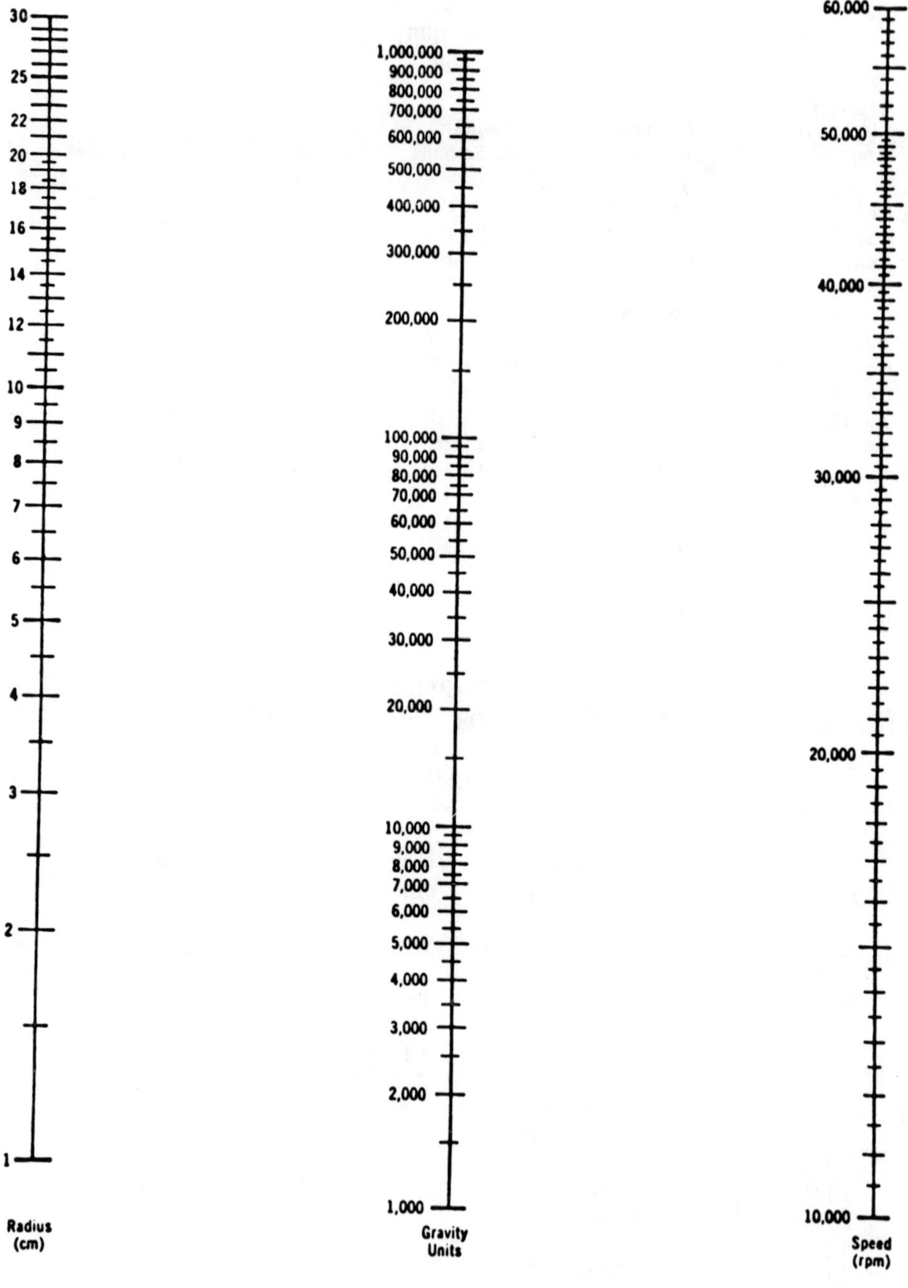


FIGURE 6. Nomogram relating gravity units to rotor speed and radius. A straight line connecting any two columns will intersect the third quantity. (Reproduced by courtesy of International Equipment Company.)

## TABLE 6
## SEDIMENTATION PROPERTIES OF SOME BIOLOGICAL PARTICLES

| Particle | Weight, daltons | Sedimentation Coefficient, Svedbergs | Density, gm·cc$^{-1}$ | Frictional Coefficient, $f/f_o$ |
|---|---|---|---|---|
| Milk lipase[28] | 6,669 | 1.14 | 1.30 | 1.190 |
| Insulin[28] | 24,430 | 1.95 | 1.36 | 1.516 |
| Pig heart lactate dehydrogenase[28] | 109,000 | 6.93 | 1.35 | 1.127 |
| Human fibrinogen[28] | 339,700 | 7.63 | 1.38 | 2.336 |
| *E. coli* "16S" ribosomal RNA[14] | 560,000 | 16.7 | 1.663[a] | |
| Horse liver catalase[28] | 221,600 | 11.20 | 1.40 | 1.246 |
| Apoferritin[28] | 466,900 | 17.60 | 1.34 | 1.141 |
| *E. coli* "30S" ribosomal subunit[9] | 990,000 | 30.6 | 1.72 | |
| Bovine liver glutamate dehydrogenase[28] | 1,015,000 | 26.60 | 1.33 | 1.250 |
| *E. coli* "23S" ribosomal RNA[14] | 1,100,000 | 23 | 1.663[a] | |
| *E. coli* "50S" ribosomal subunit | 1,700,000 | 50.0 | 1.72 | |
| β-Lipoprotein[28] | 2,663,000 | 5.9 | 1.03 | 1.243 |
| *E. coli* "70S" ribosome[9] | 2,690,000 | 69.1 | 1.72 | |
| Turnip yellow mosaic virus[28] | 4,970,000 | 106 | 1.50 | 1.255 |
| Human adenovirus 2 DNA[4] | 23,000,000 | 25 | 1.714[a] | |
| Tobacco mosaic virus[14] | 31,340,000 | 185 | 1.37 | 1.927 |
| Silkworm polyhedral virus[14] | 916,200,000 | 1,871 | 1.30 | 1.515 |
| Lysosomes | | *ca.* 9,000 | 1.18 | |
| Peroxisomes | | *ca.* 20,000 | 1.24 | |
| Mitochondria | | *ca.* 30,000 | 1.18 | |

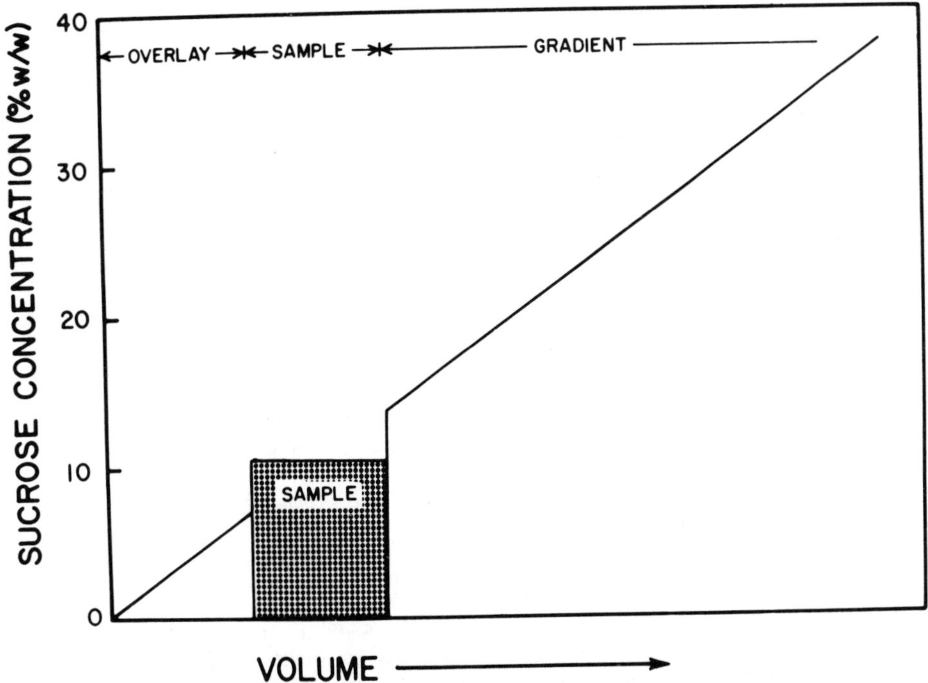


FIGURE 7. Linear gradient. Concentration of the solute is a linear function of volume; density generally will not be exactly proportional to concentration. From: Price, C. A., in *Centrifugation in Density Gradients* (1973). Reproduced by permission of Academic Press, New York.

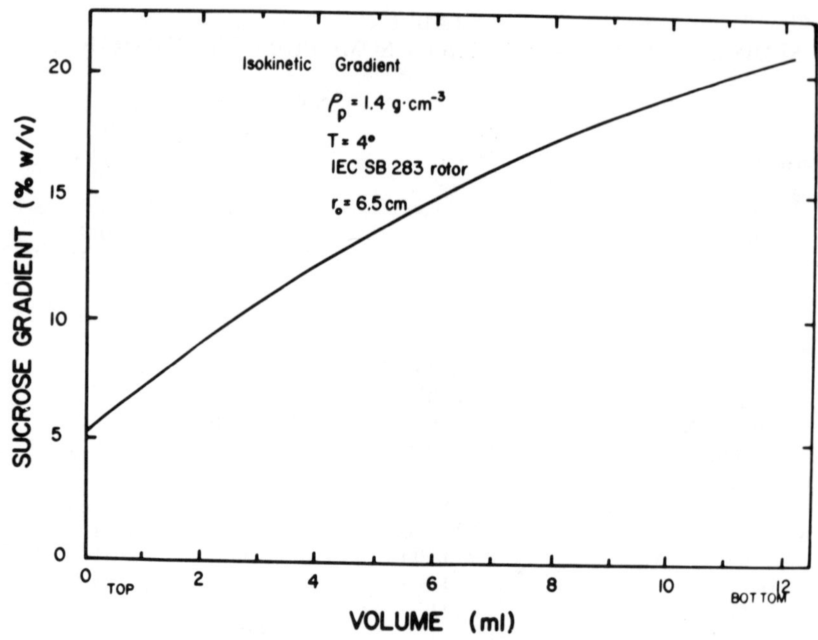


FIGURE 8. Isokinetic gradient. Shown for a particle of density = 1.4 in an IEC SB 283 swinging bucket rotor. Computed by R. A. Bowen (unpublished).

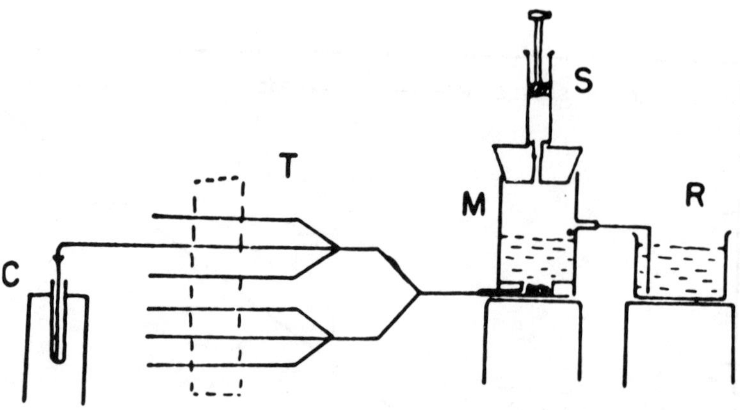


FIGURE 9. Exponential gradient generator. C = centrifuge tube, T = Tygon tubing manifold and pump, M = mixing chamber, S = syringe, R = reservoir. A whole family of exponential shapes may be produced, depending on the volume of the mixing chamber and the concentrations of the starting and limiting solutions (see Figure 10); in general, one should minimize the air space above the fluid in the mixing chamber, since the volume it occupies (and therefore the volume of the mixed fluid) will vary with temperature and uneven flow rates in entering and leaving the chamber. From: McCarty, D. S., Stafford, D., and Brown, O., Resolution and Fractionation of Macromolecules by Isokinetic Sucrose Density Gradient Sedimentation, *Anal. Biochem., 24,* 314 (1968). Reproduced by permission of Academic Press, New York.

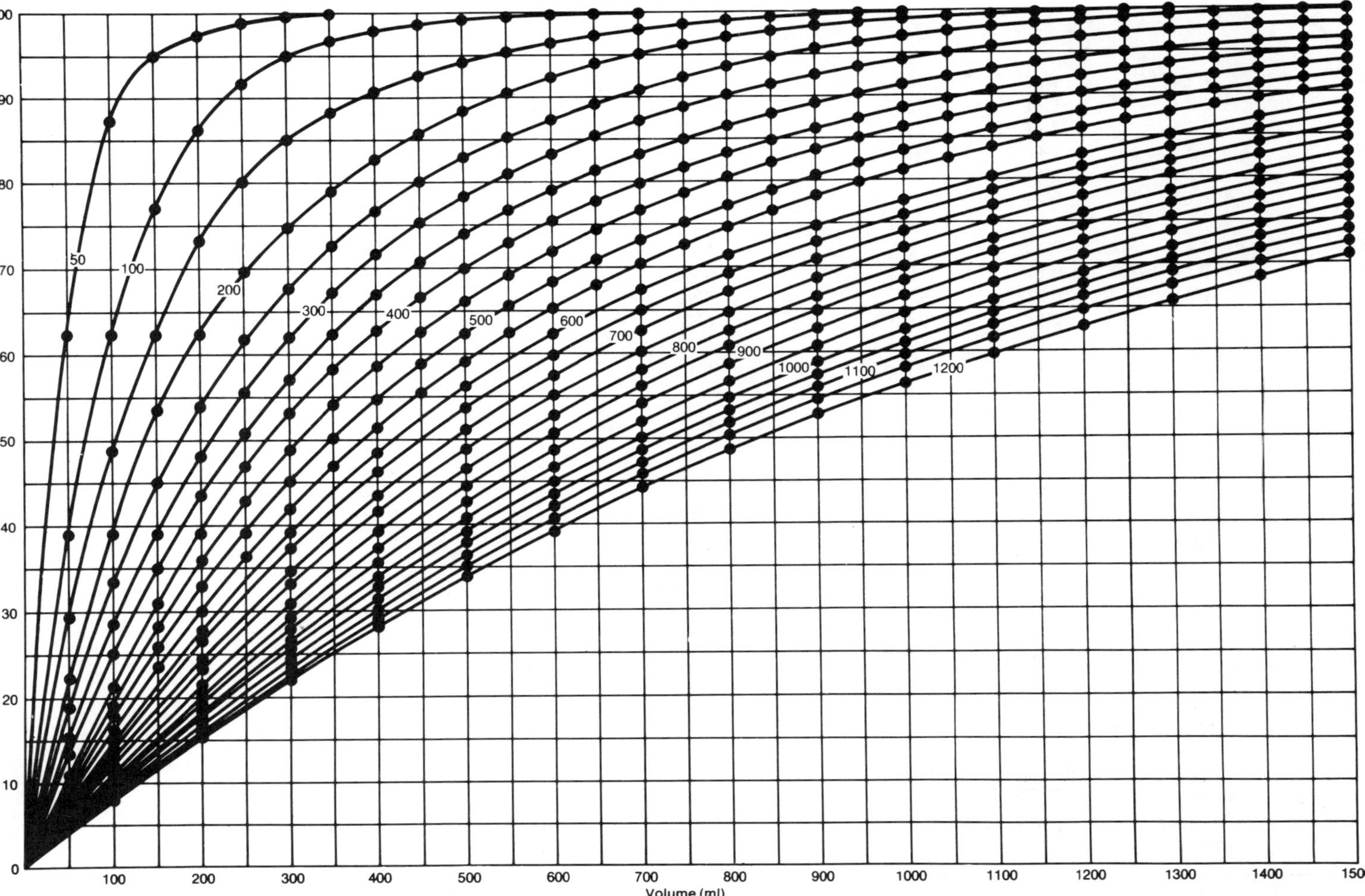


FIGURE 10. Exponential gradient shapes. Consider that the difference in concentration between the starting and limiting solution in the system shown in Figure 9 is $\Delta C$; the ordinate then represents the percentage of $\Delta C$ in the mixing chamber for increasing volumes of the gradient produced (abscissa) and different volumes of the mixing chamber (separately labeled curves). (Drawing by courtesy of the International Equipment Company.)

## TABLE 7
## PARAMETERS FOR ISOKINETIC GRADIENTS IN IEC SB-405 AND SB-283 ROTORS

| Particle Density, g/cm³ | Rotor Temp., °C | SB-405 4.0 ml Gradient Volume 5.34–9.61 cm Radius | | | | SB-283 12.0 ml Gradient Volume 6.74–14.3 cm Radius | | | |
|---|---|---|---|---|---|---|---|---|---|
| | | Concentration, % (w/v) | | Mixing Chamber Volume, ml | Reservoir Concentration, % (w/v) | Concentration, % (w/v) | | Mixing Chamber Volume, ml | Reservoir Concentration, % (w/v) |
| | | Top | Bottom | | | Top | Bottom | | |
| 1.3 | 2 | 0 | 12.3 | 4.30 | 20.4 | 0 | 15.6 | 11.2 | 22.5 |
| 1.3 | 2 | 5 | 16.1 | 4.85 | 24.8 | 5 | 19.4 | 12.5 | 28.4 |
| 1.3 | 2 | 10 | 20.5 | 6.54 | 32.9 | 10 | 23.5 | 13.3 | 32.8 |
| 1.3 | 2 | 15 | 24.9 | 5.36 | 33.9 | 15 | 27.7 | 11.9 | 35.0 |
| 1.3 | 2 | 20 | 29.2 | 5.42 | 37.5 | 20 | 31.8 | 12.7 | 39.4 |
| 1.3 | 2 | 25 | 33.4 | 5.22 | 40.7 | 25 | 35.9 | 12.9 | 43.1 |
| 1.3 | 6 | 0 | 12.5 | 5.73 | 24.8 | 0 | 15.8 | 11.2 | 23.9 |
| 1.3 | 6 | 5 | 16.6 | 4.41 | 24.5 | 5 | 19.9 | 11.4 | 27.9 |
| 1.3 | 6 | 10 | 20.7 | 6.03 | 32.1 | 10 | 23.8 | 13.1 | 32.9 |
| 1.3 | 6 | 15 | 25.1 | 5.69 | 35.1 | 15 | 27.9 | 12.3 | 35.8 |
| 1.3 | 6 | 20 | 29.4 | 5.22 | 37.5 | 20 | 32.0 | 12.0 | 39.0 |
| 1.3 | 6 | 25 | 33.5 | 5.10 | 40.7 | 25 | 36.1 | 12.8 | 43.3 |
| 1.4 | 2 | 0 | 13.3 | 4.19 | 21.6 | 0 | 16.8 | 11.2 | 24.3 |
| 1.4 | 2 | 5 | 17.1 | 4.98 | 26.9 | 5 | 20.7 | 12.8 | 30.8 |
| 1.4 | 2 | 10 | 21.4 | 6.54 | 35.0 | 10 | 24.7 | 13.2 | 34.7 |
| 1.4 | 2 | 15 | 25.9 | 5.27 | 35.5 | 15 | 28.9 | 11.9 | 36.9 |
| 1.4 | 2 | 20 | 30.1 | 5.50 | 39.6 | 20 | 33.0 | 12.6 | 41.2 |
| 1.4 | 2 | 25 | 34.4 | 5.26 | 42.6 | 25 | 37.2 | 13.1 | 45.3 |
| 1.4 | 6 | 0 | 13.4 | 5.38 | 25.6 | 0 | 17.0 | 10.9 | 25.6 |
| 1.4 | 6 | 5 | 17.6 | 4.52 | 26.5 | 5 | 21.2 | 11.6 | 30.2 |
| 1.4 | 6 | 10 | 21.7 | 5.92 | 34.0 | 10 | 25.1 | 13.1 | 35.1 |
| 1.4 | 6 | 15 | 26.1 | 5.61 | 36.8 | 15 | 29.2 | 12.1 | 37.7 |
| 1.4 | 6 | 20 | 30.4 | 5.20 | 39.3 | 20 | 33.2 | 11.8 | 40.7 |
| 1.4 | 6 | 25 | 34.5 | 5.22 | 42.7 | 25 | 37.4 | 13.0 | 45.5 |

### TABLE 7 (Continued)
### PARAMETERS FOR ISOKINETIC GRADIENTS IN IEC SB-405 AND SB-283 ROTORS

| Particle Density, g/cm³ | Rotor Temp., °C | SB-405 4.0 ml Gradient Volume 5.34–9.61 cm Radius | | | | SB-283 12.0 ml Gradient Volume 6.74–14.3 cm Radius | | | |
|---|---|---|---|---|---|---|---|---|---|
| | | Concentration, % (w/v) | | Mixing Chamber Volume, ml | Reservoir Concentration, % (w/v) | Concentration, % (w/v) | | Mixing Chamber Volume, ml | Reservoir Concentration, % (w/v) |
| | | Top | Bottom | | | Top | Bottom | | |
| 1.5 | 2 | 0 | 13.9 | 4.12 | 22.3 | 0 | 17.7 | 10.3 | 25.6 |
| 1.5 | 2 | 5 | 17.7 | 5.08 | 28.3 | 5 | 21.5 | 12.9 | 32.3 |
| 1.5 | 2 | 10 | 22.0 | 6.43 | 36.0 | 10 | 25.5 | 12.9 | 35.7 |
| 1.5 | 2 | 15 | 26.5 | 5.19 | 36.3 | 15 | 29.7 | 11.9 | 38.2 |
| 1.5 | 2 | 20 | 30.7 | 5.49 | 40.7 | 20 | 33.7 | 12.4 | 42.0 |
| 1.5 | 2 | 25 | 35.0 | 5.40 | 44.0 | 25 | 37.9 | 12.9 | 46.2 |
| 1.5 | 6 | 0 | 14.1 | 5.15 | 26.1 | 0 | 17.9 | 11.9 | 26.8 |
| 1.5 | 6 | 5 | 18.3 | 4.57 | 27.8 | 5 | 22.1 | 11.7 | 31.6 |
| 1.5 | 6 | 10 | 22.3 | 5.93 | 35.1 | 10 | 25.9 | 12.9 | 36.3 |
| 1.5 | 6 | 15 | 26.7 | 5.53 | 37.7 | 15 | 30.0 | 11.9 | 38.7 |
| 1.5 | 6 | 20 | 30.9 | 5.14 | 40.2 | 20 | 33.9 | 11.6 | 41.6 |
| 1.5 | 6 | 25 | 35.1 | 5.35 | 44.2 | 25 | 38.1 | 12.9 | 46.6 |
| 1.6 | 2 | 0 | 14.3 | 4.06 | 22.9 | 0 | 18.2 | 10.4 | 26.6 |
| 1.6 | 2 | 5 | 18.2 | 5.14 | 29.3 | 5 | 22.0 | 12.9 | 33.2 |
| 1.6 | 2 | 10 | 22.5 | 6.35 | 36.7 | 10 | 25.9 | 12.8 | 36.3 |
| 1.6 | 2 | 15 | 26.8 | 5.12 | 36.8 | 15 | 30.2 | 11.9 | 38.9 |
| 1.6 | 2 | 20 | 31.1 | 5.46 | 41.3 | 20 | 34.1 | 12.2 | 42.6 |
| 1.6 | 2 | 25 | 35.3 | 5.44 | 44.8 | 25 | 38.3 | 12.7 | 46.8 |
| 1.6 | 6 | 0 | 14.5 | 4.99 | 26.3 | 0 | 18.5 | 10.9 | 27.7 |
| 1.6 | 6 | 5 | 18.7 | 4.59 | 28.6 | 5 | 22.6 | 11.7 | 32.6 |
| 1.6 | 6 | 10 | 22.8 | 5.91 | 35.9 | 10 | 26.4 | 12.8 | 37.0 |
| 1.6 | 6 | 15 | 27.1 | 5.43 | 38.3 | 15 | 30.5 | 11.8 | 39.3 |
| 1.6 | 6 | 20 | 31.3 | 5.09 | 40.7 | 20 | 34.4 | 11.5 | 42.2 |
| 1.6 | 6 | 25 | 35.5 | 5.40 | 45.0 | 25 | 38.5 | 12.9 | 47.3 |

## TABLE 7 (Continued)
## PARAMETERS FOR ISOKINETIC GRADIENTS IN IEC SB-405 AND SB-283 ROTORS

| | | SB-405 4.0 ml Gradient Volume 5.34–9.61 cm Radius | | | | SB-283 12.0 ml Gradient Volume 6.74–14.3 cm Radius | | | |
|---|---|---|---|---|---|---|---|---|---|
| | | Concentration, % (w/v) | | | | Concentration, % (w/v) | | | |
| Particle Density, g/cm³ | Rotor Temp., °C | Top | Bottom | Mixing Chamber Volume, ml | Reservoir Concentration, % (w/v) | Top | Bottom | Mixing Chamber Volume, ml | Reservoir Concentration, % (w/v) |
| 1.7 | 2 | 0 | 14.7 | 4.02 | 23.2 | 0 | 18.7 | 10.4 | 27.3 |
| 1.7 | 2 | 5 | 18.5 | 5.17 | 30.0 | 5 | 22.4 | 12.9 | 33.9 |
| 1.7 | 2 | 10 | 22.8 | 6.28 | 37.1 | 10 | 26.4 | 12.7 | 36.7 |
| 1.7 | 2 | 15 | 27.1 | 5.06 | 37.2 | 15 | 30.5 | 11.8 | 39.4 |
| 1.7 | 2 | 20 | 31.3 | 5.43 | 41.7 | 20 | 34.5 | 12.0 | 42.9 |
| 1.7 | 2 | 25 | 35.6 | 5.45 | 45.3 | 25 | 38.6 | 12.6 | 47.1 |
| 1.7 | 6 | 0 | 14.9 | 4.88 | 26.5 | 0 | 18.9 | 10.9 | 28.3 |
| 1.7 | 6 | 5 | 19.1 | 4.61 | 29.3 | 5 | 23.1 | 11.8 | 33.3 |
| 1.7 | 6 | 10 | 23.1 | 5.89 | 36.5 | 10 | 26.8 | 12.7 | 37.5 |
| 1.7 | 6 | 15 | 27.4 | 5.35 | 38.6 | 15 | 30.9 | 11.7 | 39.7 |
| 1.7 | 6 | 20 | 31.6 | 5.05 | 41.1 | 20 | 34.7 | 11.5 | 42.8 |
| 1.7 | 6 | 25 | 35.7 | 5.41 | 45.5 | 25 | 38.8 | 12.8 | 47.7 |
| 1.8 | 2 | 0 | 14.9 | 3.98 | 23.5 | 0 | 19.0 | 10.4 | 27.8 |
| 1.8 | 2 | 5 | 18.7 | 5.19 | 30.6 | 5 | 22.7 | 12.9 | 34.4 |
| 1.8 | 2 | 10 | 23.0 | 6.23 | 37.4 | 10 | 26.6 | 12.6 | 37.0 |
| 1.8 | 2 | 15 | 27.3 | 5.01 | 37.4 | 15 | 30.8 | 11.8 | 39.7 |
| 1.8 | 2 | 20 | 31.5 | 5.40 | 42.0 | 20 | 34.7 | 12.0 | 43.3 |
| 1.8 | 2 | 25 | 35.8 | 5.44 | 45.7 | 25 | 38.8 | 12.5 | 47.3 |
| 1.8 | 6 | 0 | 15.1 | 4.79 | 26.7 | 0 | 19.3 | 10.9 | 28.8 |
| 1.8 | 6 | 5 | 19.3 | 4.61 | 29.7 | 5 | 23.4 | 11.8 | 33.8 |
| 1.8 | 6 | 10 | 23.3 | 5.86 | 36.9 | 10 | 27.1 | 12.6 | 37.8 |
| 1.8 | 6 | 15 | 27.6 | 5.29 | 38.8 | 15 | 31.1 | 11.6 | 40.0 |
| 1.8 | 6 | 20 | 31.7 | 5.01 | 41.4 | 20 | 34.9 | 11.5 | 43.2 |
| 1.8 | 6 | 25 | 35.9 | 5.42 | 45.9 | 25 | 39.0 | 12.7 | 47.9 |

Note: The gradients are calculated for single gradients of sucrose. To prepare n gradients simultaneously, multiply the volume of the mixing chamber by n. Starting solution is placed in the mixing chamber and limiting solution in the reservoir. The apparatus is shown in Figure 9.

Data taken from: Noll, H., in *Techniques in Protein Biosynthesis,* Vol. 2, pp. 125–127, P. N. Campbell and J. R. Sargent, Eds. (1969). Reproduced by permission of Academic Press, New York.

## TABLE 8
## PARAMETERS FOR ISOKINETIC GRADIENTS IN SOME SPINCO ROTORS

| | | | Rotor SW 39 or 50L 50 ml Delivery Vol. 5.61–9.80 cm | | Rotor SW 25.1 30 ml Delivery Vol. 6.49–12.90 cm | | Rotor SW 25.2 58 ml Delivery Vol. 7.50–15.3 cm | | Rotor SW 25.3 16.5 ml Delivery Vol. 7.55–16.19 cm | |
|---|---|---|---|---|---|---|---|---|---|---|
| Particle Density g/cm$^3$ | Rotor Temp., °C | Concentration, % (w/v), Top | Mixing Chamber Volume, ml | Reservoir Concentration, % (w/v) | Mixing Chamber Volume, ml | Reservoir Concentration, % (w/v) | Mixing Chamber Volume, ml | Reservoir Concentration, % (w/v) | Mixing Chamber Volume, ml | Reservoir Concentration, % (w/v) |
| 1.33 | 5 | 5 | 6.07 | 23.7 | 30.2 | 24.6 | 56.5 | 25.0 | 14.5 | 25.5 |
| 1.33 | 5 | 10 | 6.30 | 27.1 | 31.4 | 28.1 | 58.5 | 28.2 | 14.9 | 28.6 |
| 1.33 | 5 | 15 | 6.50 | 30.4 | 32.2 | 31.2 | 60.2 | 31.5 | 15.3 | 31.8 |
| 1.33 | 20 | 5 | 6.00 | 24.8 | 29.7 | 26.0 | 55.5 | 26.2 | 14.2 | 26.6 |
| 1.33 | 20 | 10 | 6.17 | 28.0 | 30.7 | 29.1 | 56.6 | 30.4 | 14.6 | 29.7 |
| 1.33 | 20 | 15 | 6.37 | 31.4 | 31.6 | 32.3 | 59.1 | 32.5 | 15.1 | 32.9 |
| 1.41 | 5 | 5 | 6.00 | 24.6 | 29.9 | 25.8 | 55.9 | 26.0 | 14.3 | 26.5 |
| 1.41 | 5 | 10 | 6.23 | 28.0 | 31.0 | 29.0 | 57.9 | 29.2 | 14.8 | 26.6 |
| 1.41 | 5 | 15 | 6.47 | 31.4 | 31.9 | 32.3 | 59.6 | 32.5 | 15.2 | 32.8 |
| 1.41 | 20 | 5 | 5.90 | 25.8 | 29.4 | 27.0 | 54.9 | 27.3 | 14.0 | 27.8 |
| 1.41 | 20 | 10 | 6.10 | 29.0 | 30.3 | 30.1 | 55.9 | 31.2 | 14.5 | 30.8 |
| 1.41 | 20 | 15 | 6.30 | 32.4 | 31.4 | 33.3 | 58.6 | 33.6 | 15.0 | 34.0 |
| 1.51 | 5 | 5 | 5.93 | 25.4 | 28.1 | 25.7 | 55.2 | 26.8 | 14.1 | 27.3 |
| 1.51 | 5 | 10 | 6.17 | 28.7 | 30.6 | 29.8 | 57.2 | 30.0 | 14.6 | 30.4 |
| 1.51 | 5 | 15 | 6.37 | 32.0 | 31.6 | 33.0 | 59.0 | 33.2 | 15.1 | 33.6 |
| 1.51 | 20 | 5 | 5.80 | 26.6 | 28.9 | 27.9 | 54.1 | 28.2 | 13.8 | 28.7 |
| 1.51 | 20 | 10 | 6.00 | 29.8 | 29.9 | 31.0 | 54.6 | 32.3 | 14.3 | 31.7 |
| 1.51 | 20 | 15 | 6.23 | 33.1 | 31.0 | 34.2 | 57.9 | 34.4 | 14.8 | 34.8 |
| 1.77 | 5 | 5 | 5.77 | 26.3 | 28.9 | 27.7 | 53.8 | 27.9 | 13.8 | 28.4 |
| 1.77 | 5 | 10 | 6.03 | 29.6 | 29.9 | 30.8 | 55.9 | 31.0 | 14.3 | 31.4 |
| 1.77 | 5 | 15 | 6.23 | 32.9 | 30.9 | 33.9 | 57.7 | 34.1 | 14.8 | 34.5 |
| 1.77 | 20 | 5 | 5.67 | 27.7 | 28.2 | 29.0 | 52.7 | 29.3 | 13.5 | 29.8 |
| 1.77 | 20 | 10 | 5.87 | 30.8 | 29.2 | 32.1 | 54.4 | 32.4 | 14.0 | 32.8 |
| 1.77 | 20 | 15 | 6.10 | 33.1 | 30.3 | 35.0 | 56.7 | 35.4 | 14.5 | 35.9 |

TABLE 8 (Continued)
PARAMETERS FOR ISOKINETIC GRADIENTS IN SOME SPINCO ROTORS

| Particle Density g/cm$^3$ | Rotor Temp., °C | Concentration, % (w/v), Top | Rotor SW 39 or 50L 50 ml Delivery Vol. 5.61–9.80 cm | | Rotor SW 25.1 30 ml Delivery Vol. 6.49–12.90 cm | | Rotor SW 25.2 58 ml Delivery Vol. 7.50–15.3 cm | | Rotor SW 25.3 16.5 ml Delivery Vol. 7.55–16.19 cm | |
|---|---|---|---|---|---|---|---|---|---|---|
| | | | Mixing Chamber Volume, ml | Reservoir Concentration, % (w/v) | Mixing Chamber Volume, ml | Reservoir Concentration, % (w/v) | Mixing Chamber Volume, ml | Reservoir Concentration, % (w/v) | Mixing Chamber Volume, ml | Reservoir Concentration, % (w/v) |
| 1.81 | 5 | 5 | 5.77 | 26.4 | 28.7 | 27.7 | 53.7 | 28.0 | 13.7 | 28.5 |
| 1.81 | 5 | 10 | 6.00 | 29.7 | 29.9 | 30.9 | 55.8 | 31.1 | 14.3 | 31.5 |
| 1.81 | 5 | 15 | 6.23 | 33.0 | 30.8 | 34.0 | 57.6 | 34.2 | 14.7 | 34.6 |
| 1.81 | 20 | 5 | 5.63 | 27.8 | 28.1 | 29.1 | 52.5 | 29.4 | 13.5 | 30.0 |
| 1.81 | 20 | 10 | 5.83 | 30.9 | 29.1 | 32.2 | 52.2 | 29.6 | 13.9 | 32.0 |
| 1.81 | 20 | 15 | 6.07 | 34.2 | 30.2 | 35.3 | 56.5 | 35.5 | 14.5 | 36.0 |
| 1.89 | 5 | 5 | 5.73 | 26.6 | 28.6 | 27.9 | 53.4 | 28.2 | 13.7 | 28.7 |
| 1.89 | 5 | 10 | 5.97 | 29.8 | 29.7 | 31.0 | 55.5 | 31.2 | 14.2 | 31.7 |
| 1.89 | 5 | 15 | 6.20 | 33.1 | 30.7 | 34.1 | 57.3 | 34.4 | 14.6 | 34.8 |
| 1.89 | 20 | 5 | 5.60 | 28.0 | 27.9 | 29.3 | 54.2 | 32.6 | 13.4 | 30.1 |
| 1.89 | 20 | 10 | 5.83 | 31.1 | 28.9 | 32.3 | 54.2 | 32.6 | 13.9 | 33.1 |
| 1.89 | 20 | 15 | 6.07 | 34.3 | 30.1 | 35.5 | 56.3 | 35.7 | 14.4 | 36.1 |

Note: Data for sucrose gradients in Spinco rotors are similar to those given in Table 7.

Data taken from: Noll, H., in *Techniques in Protein Biosynthesis,* Vol. 2, pp. 128–129, P. N. Campbell and J. R. Sargent, Eds. (1969). Reproduced by permission of Academic Press, New York,

In sector-shaped rotors, which include *zonal rotors, linear* gradients are again the most convenient for *isopycnic separation*, but for *rate* separations the twin considerations of capacity and resolution dictate opposing counsels. Both capacity and resolution have been markedly improved by taking advantage of the radius–volume relations and of the phenomenon of *gradient-induced zone narrowing.*[28]

$$\rho_m(r) = \rho_p - \frac{k}{r^n} \tag{9}$$

The largest capacities have been achieved with *hyperbolic* or *Berman gradients* (equation 9; Figure 11),[6] in which gradient-induced zone narrowing and sectorial dilution balance the decrease in capacity due to the falling value of $[\rho_p - \rho_m]$ as the particles migrate into the gradient. A Berman gradient for ribosomal subunits permitted up to 2000 mg to be resolved with greater than 99% purity (Table 9).[9]

For any separation involving large amounts of particles, the sample should be loaded in an inverse gradient.[8] This and other modes of loading particles onto gradients are shown in Figure 12.

Resolution of particles of the same density may be optimized by the use of *equivolumetric gradients,* in which gradient-induced zone narrowing exactly balances sectorial dilution and maintains the volume of particle zones constant (equation 10; Figure 13).[23]

$$\frac{r^2}{\eta_m(r)} [\rho_p - \rho_m(r)] = \text{constant} \tag{10}$$

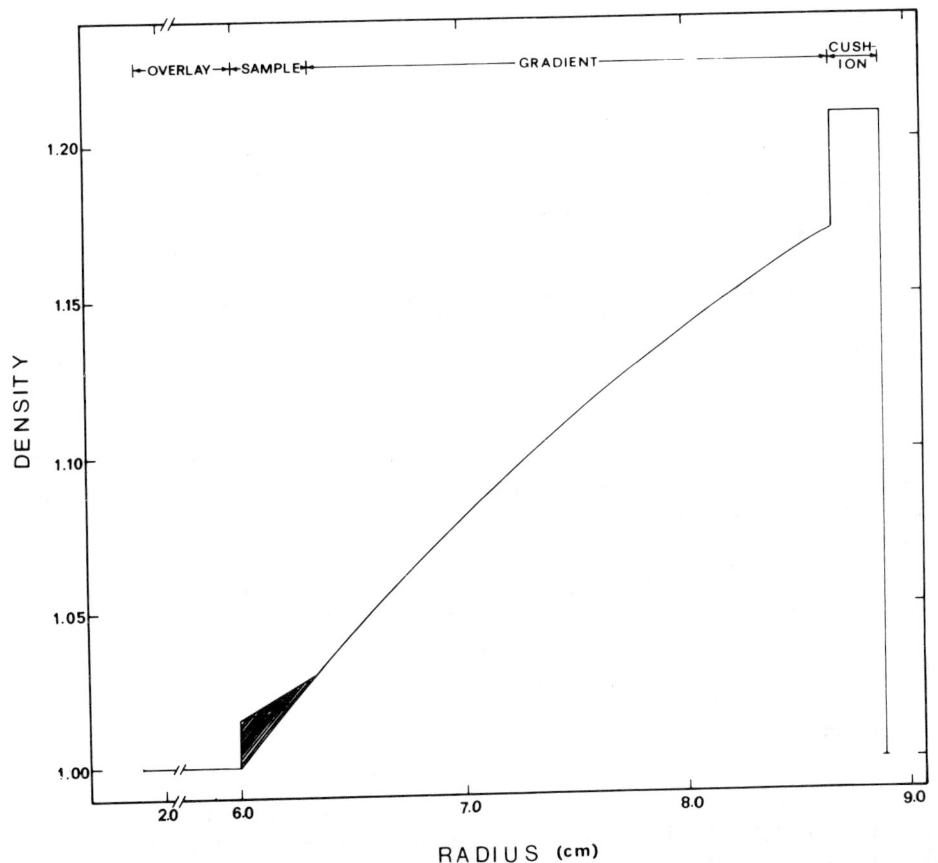


FIGURE 11. Hyperbolic gradient. Shown for ribosomes in a B-XV zonal rotor. From: Eikenberry, E. F., Bickle, T. A., Traut, R. R., and Price, C. A., Separation of Large Quantities of Ribosomal Subunits by Zonal Ultracentifugation, *Eur. J. Biochem.*, *12*, 113 (1970). Reproduced by permission of Springer-Verlag, Berlin, Germany, and New York.

TABLE 9
CONSTRUCTION OF HYPERBOLIC GRADIENT[a] FOR THE SEPARATION OF RIBOSOMES

| Volume, ml | Radius, cm | Density at 5°, g/cm | Sucrose Concentration at 5°, % (w/w) | Computed Factor for Gradient Generator[b] |
|---|---|---|---|---|
| 0–708 | 1.9–6.0 | 1.0 | 0.0 | 0 |
| 750[c] | 6.15 | 1.013[c] | 3.4 | 0.077 |
| 800[c] | 6.32 | 1.028[c] | 7.1 | 0.164 |
| 808 | 6.34 | 1.029 | 7.4 | |
| 850 | 6.49 | 1.042 | 10.4 | 0.244 |
| 900 | 6.66 | 1.055 | 13.4 | 0.318 |
| 950 | 6.82 | 1.067 | 16.1 | 0.387 |
| 1000 | 6.98 | 1.078 | 18.6 | 0.451 |
| 1050 | 7.13 | 1.088 | 20.9 | 0.511 |
| 1100 | 7.28 | 1.098 | 23.0 | 0.567 |
| 1150 | 7.42 | 1.107 | 24.9 | 0.619 |
| 1200 | 7.57 | 1.115 | 26.7 | 0.669 |
| 1250 | 7.71 | 1.124 | 28.5 | 0.711 |
| 1300 | 7.85 | 1.131 | 30.0 | 0.763 |
| 1350 | 7.98 | 1.139 | 31.5 | 0.805 |
| 1400 | 8.12 | 1.146 | 32.9 | 0.846 |
| 1450 | 8.25 | 1.152 | 34.2 | 0.886 |
| 1500 | 8.38 | 1.159 | 35.5 | 0.924 |
| 1550 | 8.52 | 1.165 | 36.8 | 0.962 |
| 1600 | 8.66 | 1.171 | 38.0 | 1.000 |
| 1650 | 8.88 | 1.208 | 45.0 | |

[a] Gradient computed for the separation of particles of 1.56/cc in a B-XV zonal rotor.
[b] Fraction of program cam height for the Beckman Spinco Model 141 gradient pump = volumetric fraction of 38% (w/w) sucrose required to achieve desired density.
[c] These figures computed for the hyperbolic gradient are listed for reference only, since in operation this region is normally occupied by the sample zone.

From: Eikenberry, E. F., Bickle, T. A., Traut, R. R., and Price, C. A., Separation of Large Quantities of Ribosomal Subunits by Zonal Ultracentrifugation, *Eur. J. Biochem.*, *12*, 113 (1970). Reproduced by permission of Springer-Verlag, Berlin, Germany, and New York.

Data for constructing equivolumetric gradients of sucrose at 4° for B-14 and B-30 zonal rotors are given in Tables 10 and 11. First construct a "universal" program for the gradient generator, using Table 10. This table gives the volumetric ratios for mixing starting and limiting solutions as a function of the volume along the gradient. For a variety of particle densities the best single choice for the widest range of conditions is $\rho_p$ = 1.20. Table 11 gives the required final concentrations (at the rotor wall) as a function of the chosen initial concentration (at the sample zone) and the density, $\rho_p$, of the particles being sedimented. For example, if $\rho_p$ = 1.20 g·cm$^{-3}$ and it is desired to have 7.5% (w/w) sucrose at the sample zone, then the starting solution will be 7.5% (w/w) sucrose and the limiting solution will be 31.4% (w/w) sucrose. The sample zone is assumed to be at a standard radius, $r_s$ = 2.36 cm.

## GRADIENT MATERIALS

Sucrose remains the most popular gradient material, but it is not always the most advantageous. Other materials are listed in Table 12. The high densities possible with cesium salts are used for banding nucleic acids and polysaccharides; $D_2O$, Ficoll®, and silica contribute essentially negligible osmotic pressures; glycerol, sorbitol, and Renografin® may be less damaging to membranes than sucrose.

Barber[5] has computed empirical equations for predicting the density and viscosity of sucrose for almost any concentration and temperature (equations 11 and 12, Table 13). These equations have been evaluated by B. S. Bishop from the above reference (Table 14).

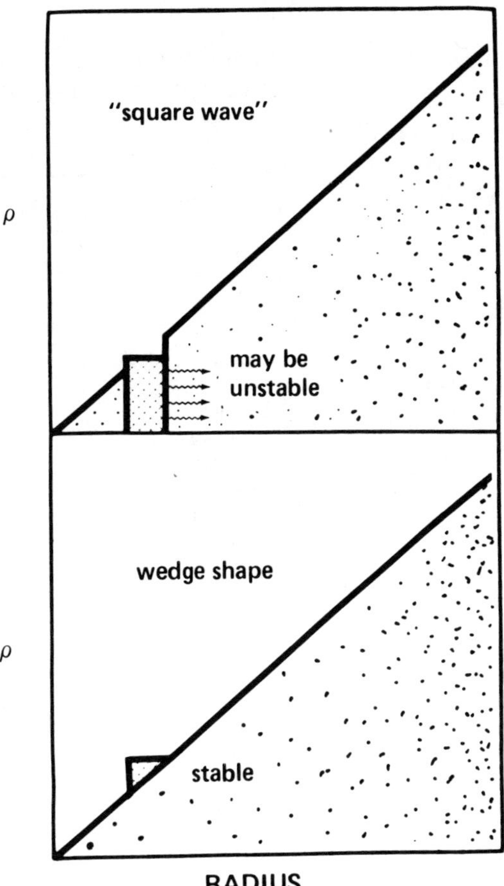


FIGURE 12. Inverse sample gradient. To minimize droplet sedimentation, Britten and Roberts[8] recommended that samples containing a large mass of particles be loaded in an inverse or wedge-shaped gradient; in a conventional rectangular zone, the steep gradient of solute immediately below the zone will cause solute to diffuse into the zone and, if the concentration of particles is too high, cause local density inversions, which become unstable; droplets of the particles then stream into the underlying gradient; in the wedge-shaped zone, the solute gradient extends through the particle zone. From: Price, C. A., Zonal Centrifugation, in *Manometric and Biochemical Techniques,* 5th ed., pp. 213–243, W. W. Umbreit, R. H. Burris and J. F. Stauffer, Eds. (1972). Reproduced by permission of Burgess Publishing Co., Minneapolis, Minnesota.

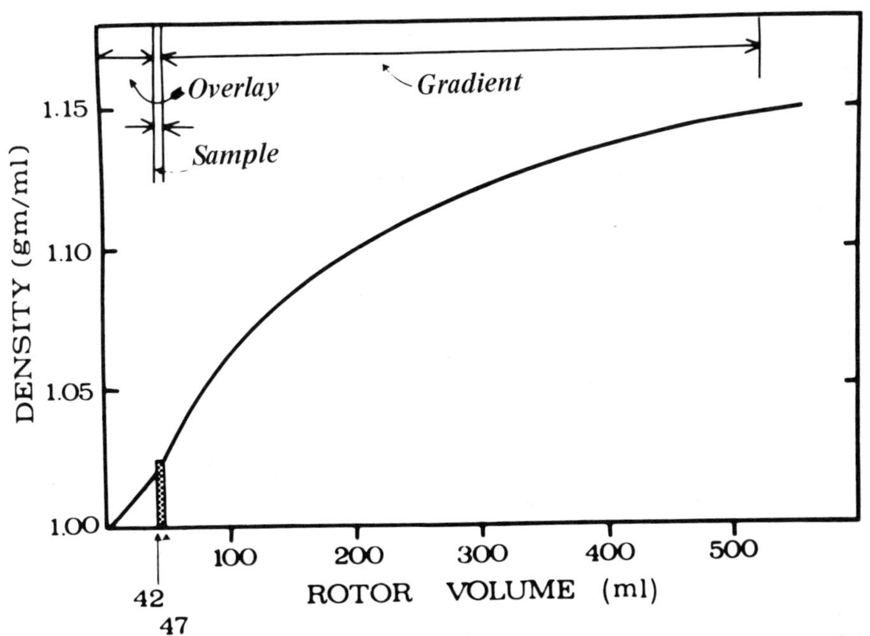


FIGURE 13. Equivolumetric gradient. Shown for ribosomes in a B-30 A zonal rotor. From: Price, C. A., and Hsu, T.-S., The Capacity of Equivolumetric Gradients in Zonal Rotors in the Separation of Ribosomes, *Vierteljahresschr. Naturforsch. Ges. Zürich, 116,* 367 (1971). Reproduced by permission of Verlag Leemann AG., Zürich, Switzerland.

## TABLE 10
## VOLUMETRIC RATIOS OF STARTING AND LIMITING SOLUTION FOR CONSTRUCTING EQUIVOLUMETRIC GRADIENTS OF SUCROSE

| | Volumetric Mixing Fraction | | | |
|---|---|---|---|---|
| Volume, $cm^3$ | $\rho_p$ = 1.10 $g \cdot cm^{-3}$ | $\rho_p$ = 1.15 $g \cdot cm^{-3}$ | $\rho_p$ = 1.20 $g \cdot cm^{-3}$ | $\rho_p$ = 1.40 $g \cdot cm^{-3}$ |
| 0 | 0.0 | 0.0 | 0.0 | 0.0 |
| 25 | 0.191 | 0.177 | 0.170 | 0.167 |
| 50 | 0.325 | 0.303 | 0.291 | 0.286 |
| 75 | 0.426 | 0.398 | 0.385 | 0.377 |
| 100 | 0.506 | 0.475 | 0.460 | 0.451 |
| 125 | 0.570 | 0.538 | 0.522 | 0.512 |
| 150 | 0.623 | 0.591 | 0.575 | 0.565 |
| 175 | 0.669 | 0.637 | 0.621 | 0.610 |
| 200 | 0.708 | 0.677 | 0.661 | 0.651 |
| 225 | 0.742 | 0.713 | 0.697 | 0.686 |
| 250 | 0.772 | 0.744 | 0.729 | 0.719 |
| 275 | 0.799 | 0.772 | 0.758 | 0.748 |
| 300 | 0.823 | 0.798 | 0.784 | 0.775 |
| 325 | 0.844 | 0.821 | 0.809 | 0.800 |
| 350 | 0.864 | 0.843 | 0.831 | 0.823 |
| 375 | 0.881 | 0.863 | 0.852 | 0.844 |
| 400 | 0.898 | 0.881 | 0.871 | 0.864 |
| 425 | 0.913 | 0.898 | 0.889 | 0.883 |
| 450 | 0.926 | 0.913 | 0.906 | 0.901 |
| 475 | 0.939 | 0.928 | 0.922 | 0.917 |
| 500 | 0.951 | 0.942 | 0.936 | 0.933 |
| 525 | 0.962 | 0.955 | 0.951 | 0.948 |
| 550 | 0.973 | 0.967 | 0.964 | 0.962 |
| 575 | 0.982 | 0.979 | 0.976 | 0.975 |
| 600 | 0.991 | 0.990 | 0.988 | 0.988 |
| 625 | 1.000 | 1.000 | 1.000 | 1.000 |

## TABLE 11
## REQUIRED FINAL CONCENTRATIONS FOR CONSTRUCTING EQUIVOLUMETRIC GRADIENTS OF SUCROSE

| | Initial concentration, $c_s$ (w/w) | | | | | | | | | | | | |
|---|---|---|---|---|---|---|---|---|---|---|---|---|---|
| $\rho_p$, $g \cdot cm^{-3}$ | 0.0 | 0.025 | 0.050 | 0.075 | 0.100 | 0.125 | 0.150 | 0.175 | 0.200 | 0.225 | 0.250 | 0.275 | 0.300 |
| 1.10 | .185 | .192 | .198 | .204 | .210 | .216 | .221 | .225 | .229 | .232 | | | |
| 1.15 | .245 | .253 | .261 | .270 | .278 | .286 | .294 | .302 | .309 | .316 | .322 | .327 | .332 |
| 1.20 | .285 | .294 | .304 | .314 | .323 | .334 | .344 | .354 | .364 | .374 | .383 | .392 | .401 |
| 1.25 | .312 | .322 | .332 | .342 | .353 | .364 | .375 | .386 | .398 | .409 | .421 | .432 | .444 |
| 1.30 | .331 | .341 | .351 | .361 | .372 | .383 | .395 | .406 | .418 | .430 | .442 | .455 | .467 |
| 1.35 | .344 | .354 | .364 | .374 | .385 | .396 | .407 | .419 | .431 | .443 | .455 | .468 | .481 |
| 1.40 | .354 | .364 | .373 | .384 | .399 | .405 | .416 | .428 | .440 | .452 | .464 | .477 | .490 |
| 1.45 | .361 | .371 | .381 | .391 | .401 | .412 | .423 | .434 | .446 | .458 | .470 | .483 | .495 |
| 1.50 | .367 | .376 | .386 | .396 | .406 | .417 | .428 | .439 | .451 | .462 | .475 | .487 | .500 |
| 1.55 | .372 | .381 | .391 | .400 | .411 | .421 | .432 | .443 | .454 | .466 | .478 | .490 | |
| 1.60 | .376 | .385 | .394 | .404 | .414 | .424 | .435 | .446 | .457 | .469 | .481 | | |
| 1.70 | .382 | .390 | .400 | .409 | .419 | .429 | .440 | .450 | .462 | .473 | .485 | | |
| 1.80 | .386 | .395 | .404 | .413 | .423 | .433 | .443 | .454 | .465 | .476 | .487 | | |
| 1.90 | .389 | .398 | .407 | .416 | .426 | .435 | .446 | .456 | .467 | .478 | .489 | | |
| 2.00 | .392 | .400 | .409 | .418 | .428 | .437 | .448 | .458 | .469 | .480 | .491 | | |

## TABLE 12
## PHYSICAL PROPERTIES OF SOME GRADIENT MATERIALS[a]

| Substance | Concentration | Stock Solution (stable at 4°C) Density[b] | Viscosity[c] | Refractive Index[d] |
|---|---|---|---|---|
| CsCl | 60% w/w | 1.7900 (20°C) | [e] | 1.4074 |
| $D_2O$ | 100% | 1.105 (20°C)[f] | | 1.33844 |
| Ficoll® | 46.5% w/w | 1.1629 (4°C) | 1020 | 1.3764[g] |
| Glycerol | 100% | 1.2609 (20°C) | 1490 (20°C) | 1.4729 |
| Silica sols[h] | 40.1% w/w | 1.295 (25°C) | 27 (25°C) | |
| Sorbitol | 60% w/w | 1.2584 (4°C) | 102.9 (4°C) | 1.4402 |
| Sucrose | 65% w/w | 1.32600 (4°C) | 56.5 (20°C) | 1.4532 |

[a] Compiled from various sources.
[b] In $g \cdot cc^{-1}$ at the temperature stated.
[c] In centipoise.
[d] At 20°C.
[e] The viscosity of CsCl solutions at O°C as a function of density has been determined by Kaempfer and Meselson (see Reference 12).
[f] The density of CsCl solutions at 25°C has been computed as a function of the refractive index (see Reference 11); $\rho$ = 10.8601 (RI = 13,4974).
[g] A 30% (w/v) solution.
[h] Available under the trade name Ludox® (du Pont de Nemours Chemical Co.).

## TABLE 13
## DENSITY AND VISCOSITY OF SUCROSE AS FUNCTIONS OF CONCENTRATION

**(A)** In the range of 0 to 30° the density of sucrose ($\rho$) may be related to temperature T and weight fraction Y by the relation

$$\textbf{(11)}\ \rho_{T,m} = (B_1 + B_2T + B_3T^2) + (B_4 + B_5T + B_6T^2)Y + (B_7 + B_8T + B_9T^2)Y^2,$$

where $\rho_{T,m}$ = density of a sucrose solution, T = temperature (°C), Y = weight fraction sucrose, and the B's are constants.

| Constant | Value* |
|---|---|
| $B_1$ | 1.0003698 |
| $B_2$ | $3.9680504 \times 10^{-5}$ |
| $B_3$ | $-5.8513271 \times 10^{-6}$ |
| $B_4$ | 0.38982371 |
| $B_5$ | $-1.0578919 \times 10^{-3}$ |
| $B_6$ | $1.2392833 \times 10^{-5}$ |
| $B_7$ | 0.17097594 |
| $B_8$ | $4.7530081 \times 10^{-4}$ |
| $B_9$ | $-8.9239737 \times 10^{-6}$ |

* Values are given to eight figures for machine calculations; use of the first five figures would be sufficient for hand calculations.

**(B)** The viscosity of sucrose ($\eta$) between 0° and 80° can be expressed as a fraction of temperature T and the mole fraction y.

$$\textbf{(12)}\qquad \log \eta_{T,m} = A + \frac{B}{T + C}$$

## TABLE 13 (Continued)
## DENSITY AND VISCOSITY OF SUCROSE AS FUNCTIONS OF CONCENTRATION

The mole fraction y is related to the weight fraction Y and the molecular weights of sucrose S and water W by the relation

$$y = \frac{Y/S}{Y/S + (1-Y)/W}$$

The constants A and B are calculated from y by the relation

$$A = D_0 + D_1 y + D_2 y^2 + D_3 y^3 \ldots D_\eta y^\eta$$

| Coefficients† | Range of Equation, weight %: 0 to 48 | 48 to 75 |
|---|---|---|
| $D_0$ | -1.5018327 | -1.0803314 |
| $D_1$ | 9.4112153 | $-2.0003484 \times 10^1$ |
| $D_2$ | $-1.1435741 \times 10^3$ | $4.6066898 \times 10^2$ |
| $D_3$ | $1.0504137 \times 10^5$ | $-5.9517023 \times 10^3$ |
| $D_4$ | $-4.6927102 \times 10^6$ | $3.5627216 \times 10^4$ |
| $D_5$ | $1.0323349 \times 10^8$ | $-7.8542145 \times 10^4$ |
| $D_6$ | $-1.1028981 \times 10^9$ | |
| $D_7$ | $4.5921911 \times 10^9$ | |

† Coefficient subscript indicates the exponent of the composition by which the coefficient is to be multiplied.

$$B = E_0 + E_1 y + E_2 y^2 + E_3 y^3 \ldots E_\eta y^\eta$$

| Coefficients‡ | Range of Equation, weight %: 0 to 48 | 48 to 75 |
|---|---|---|
| $E_0$ | $2.1169907 \times 10^2$ | $1.3975568 \times 10^2$ |
| $E_1$ | $1.6077073 \times 10^3$ | $6.6747329 \times 10^3$ |
| $E_2$ | $1.6911611 \times 10^5$ | $-7.8716105 \times 10^4$ |
| $E_3$ | $-1.4184371 \times 10^7$ | $9.0967578 \times 10^5$ |
| $E_4$ | $6.0654775 \times 10^8$ | $-5.5380830 \times 10^6$ |
| $E_5$ | $-1.2985834 \times 10^{10}$ | $1.2451219 \times 10^7$ |
| $E_6$ | $1.3532907 \times 10^{11}$ | |
| $E_7$ | $-5.4970416 \times 10^{11}$ | |

‡ Coefficient subscript indicates the exponent of the composition by which the coefficient is to be multiplied.

The value of C is related to the weight fraction y plus three additional constants.

$$C = G_1 - G_2 \left[1 + \left(\frac{y}{G_3}\right)^2\right]^{1/2},$$

where $G_1 = 146.06635$, $G_2 = 25.251728$, and $G_3 = 0.070674842$.

From: Barber, E. J., The Development of Zonal Centrifuges, in *National Cancer Institute Monograph No. 21*, pp. 219–239, N. G. Anderson, Ed. (1966). Reproduced by permission of the National Cancer Institute, Bethesda, Maryland.

# TABLE 14
## DENSITY AND VISCOSITY OF SUCROSE AT DIFFERENT TEMPERATURES†

Temperature = 0.0°C

| % Sucrose, w/w | Density, g/cc | Viscosity, cP | % Sucrose, w/w | Density, g/cc | Viscosity, cP |
|---|---|---|---|---|---|
| 1.0 | 1.0043 | 1.8306 | 36.0 | 1.1629 | 10.3146 |
| 2.0 | 1.0082 | 1.8843 | 37.0 | 1.1680 | 11.1955 |
| 3.0 | 1.0122 | 1.9414 | 38.0 | 1.1732 | 12.1880 |
| 4.0 | 1.0162 | 2.0021 | 39.0 | 1.1784 | 13.3100 |
| 5.0 | 1.0203 | 2.0666 | 40.0 | 1.1837 | 14.5831 |
| 6.0 | 1.0244 | 2.1352 | 41.0 | 1.1889 | 16.0331 |
| 7.0 | 1.0285 | 2.2083 | 42.0 | 1.1943 | 17.6905 |
| 8.0 | 1.0326 | 2.2862 | 43.0 | 1.1996 | 19.5917 |
| 9.0 | 1.0368 | 2.3692 | 44.0 | 1.2050 | 21.7808 |
| 10.0 | 1.0411 | 2.4578 | 45.0 | 1.2104 | 24.3104 |
| 11.0 | 1.0453 | 2.5523 | 46.0 | 1.2159 | 27.2445 |
| 12.0 | 1.0496 | 2.6533 | 47.0 | 1.2214 | 30.6598 |
| 13.0 | 1.0539 | 2.7614 | 48.0 | 1.2269 | 34.5693 |
| 14.0 | 1.0583 | 2.8771 | 49.0 | 1.2324 | 39.2342 |
| 15.0 | 1.0627 | 3.0010 | 50.0 | 1.2380 | 44.7427 |
| 16.0 | 1.0671 | 3.1340 | 51.0 | 1.2436 | 51.2860 |
| 17.0 | 1.0716 | 3.2768 | 52.0 | 1.2493 | 59.1069 |
| 18.0 | 1.0761 | 3.4304 | 53.0 | 1.2550 | 68.5183 |
| 19.0 | 1.0806 | 3.5958 | 54.0 | 1.2607 | 79.9240 |
| 20.0 | 1.0852 | 3.7742 | 55.0 | 1.2665 | 93.8504 |
| 21.0 | 1.0898 | 3.9668 | 56.0 | 1.2723 | 110.9897 |
| 22.0 | 1.0944 | 4.1751 | 57.0 | 1.2781 | 132.2682 |
| 23.0 | 1.0991 | 4.4008 | 58.0 | 1.2840 | 158.9211 |
| 24.0 | 1.1038 | 4.6457 | 59.0 | 1.2899 | 192.6349 |
| 25.0 | 1.1085 | 4.9121 | 60.0 | 1.2958 | 235.7207 |
| 26.0 | 1.1133 | 5.2023 | 61.0 | 1.3018 | 291.3948 |
| 27.0 | 1.1181 | 5.5192 | 62.0 | 1.3078 | 364.1926 |
| 28.0 | 1.1229 | 5.8660 | 63.0 | 1.3138 | 460.5754 |
| 29.0 | 1.1278 | 6.2464 | 64.0 | 1.3199 | 589.9290 |
| 30.0 | 1.1327 | 6.6646 | 65.0 | 1.3260 | 766.0437 |
| 31.0 | 1.1376 | 7.1257 | 66.0 | 1.3321 | 1009.5676 |
| 32.0 | 1.1426 | 7.6355 | 67.0 | 1.3383 | 1351.8850 |
| 33.0 | 1.1476 | 8.2006 | 68.0 | 1.3445 | 1841.7961 |
| 34.0 | 1.1527 | 8.8291 | 69.0 | 1.3507 | 2556.5417 |
| 35.0 | 1.1578 | 9.5301 | 70.0 | 1.3570 | 3621.1423 |

Temperature = 5.0°C

| % Sucrose, w/w | Density, g/cc | Viscosity, cP | % Sucrose, w/w | Density, g/cc | Viscosity, cP |
|---|---|---|---|---|---|
| 1.0 | 1.0043 | 1.5583 | 36.0 | 1.1614 | 8.2356 |
| 2.0 | 1.0082 | 1.6028 | 37.0 | 1.1665 | 8.9053 |
| 3.0 | 1.0121 | 1.6500 | 38.0 | 1.1717 | 9.6560 |
| 4.0 | 1.0161 | 1.7000 | 39.0 | 1.1768 | 10.5003 |
| 5.0 | 1.0201 | 1.7530 | 40.0 | 1.1821 | 11.4535 |
| 6.0 | 1.0241 | 1.8094 | 41.0 | 1.1873 | 12.5331 |
| 7.0 | 1.0282 | 1.8692 | 42.0 | 1.1926 | 13.7605 |
| 8.0 | 1.0323 | 1.9329 | 43.0 | 1.1979 | 15.1605 |
| 9.0 | 1.0365 | 2.0006 | 44.0 | 1.2033 | 16.7632 |
| 10.0 | 1.0406 | 2.0728 | 45.0 | 1.2087 | 18.6042 |
| 11.0 | 1.0448 | 2.1497 | 46.0 | 1.2141 | 20.7267 |
| 12.0 | 1.0491 | 2.2318 | 47.0 | 1.2195 | 23.1819 |
| 13.0 | 1.0534 | 2.3194 | 48.0 | 1.2250 | 25.9727 |
| 14.0 | 1.0577 | 2.4131 | 49.0 | 1.2306 | 29.2818 |
| 15.0 | 1.0620 | 2.5133 | 50.0 | 1.2361 | 33.1611 |
| 16.0 | 1.0664 | 2.6207 | 51.0 | 1.2417 | 37.7337 |
| 17.0 | 1.0708 | 2.7358 | 52.0 | 1.2474 | 43.1553 |
| 18.0 | 1.0753 | 2.8594 | 53.0 | 1.2530 | 49.6241 |
| 19.0 | 1.0798 | 2.9923 | 54.0 | 1.2587 | 57.3936 |
| 20.0 | 1.0843 | 3.1354 | 55.0 | 1.2645 | 66.7908 |
| 21.0 | 1.0889 | 3.2896 | 56.0 | 1.2702 | 78.2404 |
| 22.0 | 1.0935 | 3.4561 | 57.0 | 1.2760 | 92.3043 |
| 23.0 | 1.0981 | 3.6361 | 58.0 | 1.2819 | 109.7226 |
| 24.0 | 1.1028 | 3.8310 | 59.0 | 1.2877 | 131.4927 |
| 25.0 | 1.1075 | 4.0425 | 60.0 | 1.2937 | 158.9610 |
| 26.0 | 1.1122 | 4.2723 | 61.0 | 1.2996 | 193.9751 |
| 27.0 | 1.1169 | 4.5226 | 62.0 | 1.3056 | 239.0989 |
| 28.0 | 1.1217 | 4.7957 | 63.0 | 1.3116 | 297.9216 |
| 29.0 | 1.1266 | 5.0943 | 64.0 | 1.3176 | 375.5669 |
| 30.0 | 1.1315 | 5.4216 | 65.0 | 1.3237 | 479.4153 |
| 31.0 | 1.1364 | 5.7811 | 66.0 | 1.3298 | 620.2959 |
| 32.0 | 1.1413 | 6.1771 | 67.0 | 1.3360 | 814.3062 |
| 33.0 | 1.1463 | 6.6144 | 68.0 | 1.3422 | 1085.8899 |
| 34.0 | 1.1513 | 7.0987 | 69.0 | 1.3484 | 1472.7625 |
| 35.0 | 1.1563 | 7.6365 | 70.0 | 1.3546 | 2034.3164 |

## TABLE 14 (Continued)
## DENSITY AND VISCOSITY OF SUCROSE AT DIFFERENT TEMPERATURES

**Temperature = 10.0°C**

| % Sucrose, w/w | Density, g/cc | Viscosity, cP | % Sucrose, w/w | Density, g/cc | Viscosity, cP |
|---|---|---|---|---|---|
| 1.0 | 1.0040 | 1.3430 | 36.0 | 1.1598 | 6.6916 |
| 2.0 | 1.0079 | 1.3804 | 37.0 | 1.1649 | 7.2107 |
| 3.0 | 1.0118 | 1.4199 | 38.0 | 1.1700 | 7.7899 |
| 4.0 | 1.0157 | 1.4617 | 39.0 | 1.1752 | 8.4383 |
| 5.0 | 1.0196 | 1.5059 | 40.0 | 1.1803 | 9.1667 |
| 6.0 | 1.0236 | 1.5528 | 41.0 | 1.1856 | 9.9878 |
| 7.0 | 1.0277 | 1.6025 | 42.0 | 1.1908 | 10.9164 |
| 8.0 | 1.0317 | 1.6553 | 43.0 | 1.1961 | 11.9701 |
| 9.0 | 1.0358 | 1.7114 | 44.0 | 1.2014 | 13.1700 |
| 10.0 | 1.0400 | 1.7711 | 45.0 | 1.2068 | 14.5409 |
| 11.0 | 1.0441 | 1.8345 | 46.0 | 1.2122 | 16.1127 |
| 12.0 | 1.0484 | 1.9022 | 47.0 | 1.2176 | 17.9204 |
| 13.0 | 1.0526 | 1.9743 | 48.0 | 1.2231 | 19.9623 |
| 14.0 | 1.0569 | 2.0513 | 49.0 | 1.2286 | 22.3695 |
| 15.0 | 1.0612 | 2.1336 | 50.0 | 1.2341 | 25.1728 |
| 16.0 | 1.0655 | 2.2216 | 51.0 | 1.2397 | 28.4541 |
| 17.0 | 1.0699 | 2.3159 | 52.0 | 1.2453 | 32.3162 |
| 18.0 | 1.0743 | 2.4170 | 53.0 | 1.2509 | 36.8887 |
| 19.0 | 1.0788 | 2.5254 | 54.0 | 1.2566 | 42.3360 |
| 20.0 | 1.0833 | 2.6420 | 55.0 | 1.2623 | 48.8679 |
| 21.0 | 1.0878 | 2.7675 | 56.0 | 1.2681 | 56.7541 |
| 22.0 | 1.0923 | 2.9027 | 57.0 | 1.2739 | 66.3478 |
| 23.0 | 1.0969 | 3.0486 | 58.0 | 1.2797 | 78.1086 |
| 24.0 | 1.1016 | 3.2063 | 59.0 | 1.2855 | 92.6489 |
| 25.0 | 1.1062 | 3.3771 | 60.0 | 1.2914 | 110.7842 |
| 26.0 | 1.1109 | 3.5622 | 61.0 | 1.2973 | 133.6190 |
| 27.0 | 1.1157 | 3.7632 | 62.0 | 1.3033 | 162.6641 |
| 28.0 | 1.1204 | 3.9820 | 63.0 | 1.3093 | 200.0011 |
| 29.0 | 1.1252 | 4.2206 | 64.0 | 1.3153 | 248.5555 |
| 30.0 | 1.1301 | 4.4812 | 65.0 | 1.3214 | 312.4668 |
| 31.0 | 1.1349 | 4.7666 | 66.0 | 1.3275 | 397.6987 |
| 32.0 | 1.1398 | 5.0798 | 67.0 | 1.3336 | 512.9407 |
| 33.0 | 1.1448 | 5.4245 | 68.0 | 1.3398 | 671.1133 |
| 34.0 | 1.1498 | 5.8047 | 69.0 | 1.3460 | 891.7009 |
| 35.0 | 1.1548 | 6.2252 | 70.0 | 1.3522 | 1204.6399 |

**Temperature = 15.0°C**

| % Sucrose, w/w | Density, g/cc | Viscosity, cP | % Sucrose, w/w | Density, g/cc | Viscosity, cP |
|---|---|---|---|---|---|
| 1.0 | 1.0034 | 1.1702 | 36.0 | 1.1581 | 5.5221 |
| 2.0 | 1.0073 | 1.2019 | 37.0 | 1.1631 | 5.9315 |
| 3.0 | 1.0111 | 1.2354 | 38.0 | 1.1682 | 6.3864 |
| 4.0 | 1.0150 | 1.2708 | 39.0 | 1.1734 | 6.8934 |
| 5.0 | 1.0189 | 1.3082 | 40.0 | 1.1785 | 7.4604 |
| 6.0 | 1.0229 | 1.3477 | 41.0 | 1.1837 | 8.0965 |
| 7.0 | 1.0269 | 1.3895 | 42.0 | 1.1889 | 8.8126 |
| 8.0 | 1.0309 | 1.4339 | 43.0 | 1.1942 | 9.6213 |
| 9.0 | 1.0350 | 1.4809 | 44.0 | 1.1995 | 10.5377 |
| 10.0 | 1.0391 | 1.5309 | 45.0 | 1.2048 | 11.5795 |
| 11.0 | 1.0432 | 1.5840 | 46.0 | 1.2102 | 12.7679 |
| 12.0 | 1.0474 | 1.6405 | 47.0 | 1.2156 | 14.1276 |
| 13.0 | 1.0516 | 1.7007 | 48.0 | 1.2211 | 15.6545 |
| 14.0 | 1.0558 | 1.7648 | 49.0 | 1.2265 | 17.4452 |
| 15.0 | 1.0601 | 1.8333 | 50.0 | 1.2320 | 19.5179 |
| 16.0 | 1.0644 | 1.9064 | 51.0 | 1.2376 | 21.9286 |
| 17.0 | 1.0688 | 1.9847 | 52.0 | 1.2432 | 24.7470 |
| 18.0 | 1.0732 | 2.0685 | 53.0 | 1.2488 | 28.0603 |
| 19.0 | 1.0776 | 2.1582 | 54.0 | 1.2544 | 31.9782 |
| 20.0 | 1.0820 | 2.2546 | 55.0 | 1.2601 | 36.6392 |
| 21.0 | 1.0865 | 2.3582 | 56.0 | 1.2658 | 42.2200 |
| 22.0 | 1.0911 | 2.4696 | 57.0 | 1.2716 | 48.9497 |
| 23.0 | 1.0956 | 2.5896 | 58.0 | 1.2774 | 57.1228 |
| 24.0 | 1.1002 | 2.7191 | 59.0 | 1.2832 | 67.1281 |
| 25.0 | 1.1048 | 2.8590 | 60.0 | 1.2891 | 79.4768 |
| 26.0 | 1.1095 | 3.0103 | 61.0 | 1.2950 | 94.8528 |
| 27.0 | 1.1142 | 3.1743 | 62.0 | 1.3009 | 114.1797 |
| 28.0 | 1.1189 | 3.3523 | 63.0 | 1.3069 | 138.7120 |
| 29.0 | 1.1237 | 3.5458 | 64.0 | 1.3129 | 170.1869 |
| 30.0 | 1.1285 | 3.7567 | 65.0 | 1.3189 | 211.0244 |
| 31.0 | 1.1334 | 3.9869 | 66.0 | 1.3250 | 264.6538 |
| 32.0 | 1.1382 | 4.2387 | 67.0 | 1.3311 | 335.9805 |
| 33.0 | 1.1431 | 4.5148 | 68.0 | 1.3373 | 432.1655 |
| 34.0 | 1.1481 | 4.8183 | 69.0 | 1.3434 | 563.7854 |
| 35.0 | 1.1531 | 5.1527 | 70.0 | 1.3497 | 746.7412 |

## TABLE 14 (Continued)
## DENSITY AND VISCOSITY OF SUCROSE AT DIFFERENT TEMPERATURES

**Temperature = 20.0°C**

| % Sucrose, w/w | Density, g/cc | Viscosity, cP | % Sucrose, w/w | Density, g/cc | Viscosity, cP |
|---|---|---|---|---|---|
| 1.0 | 1.0026 | 1.0296 | 36.0 | 1.1563 | 4.6205 |
| 2.0 | 1.0064 | 1.0569 | 37.0 | 1.1613 | 4.9485 |
| 3.0 | 1.0102 | 1.0856 | 38.0 | 1.1663 | 5.3115 |
| 4.0 | 1.0140 | 1.1159 | 39.0 | 1.1714 | 5.7144 |
| 5.0 | 1.0179 | 1.1478 | 40.0 | 1.1766 | 6.1630 |
| 6.0 | 1.0219 | 1.1815 | 41.0 | 1.1817 | 6.6641 |
| 7.0 | 1.0258 | 1.2171 | 42.0 | 1.1870 | 7.2259 |
| 8.0 | 1.0298 | 1.2548 | 43.0 | 1.1922 | 7.8574 |
| 9.0 | 1.0339 | 1.2947 | 44.0 | 1.1975 | 8.5698 |
| 10.0 | 1.0380 | 1.3371 | 45.0 | 1.2028 | 9.3760 |
| 11.0 | 1.0421 | 1.3820 | 46.0 | 1.2081 | 10.2914 |
| 12.0 | 1.0462 | 1.4297 | 47.0 | 1.2135 | 11.3338 |
| 13.0 | 1.0504 | 1.4806 | 48.0 | 1.2189 | 12.4979 |
| 14.0 | 1.0546 | 1.5347 | 49.0 | 1.2244 | 13.8568 |
| 15.0 | 1.0588 | 1.5923 | 50.0 | 1.2299 | 15.4209 |
| 16.0 | 1.0631 | 1.6539 | 51.0 | 1.2354 | 17.2296 |
| 17.0 | 1.0675 | 1.7196 | 52.0 | 1.2409 | 19.3311 |
| 18.0 | 1.0718 | 1.7899 | 53.0 | 1.2465 | 21.7855 |
| 19.0 | 1.0762 | 1.8652 | 54.0 | 1.2522 | 24.6681 |
| 20.0 | 1.0806 | 1.9459 | 55.0 | 1.2578 | 28.0727 |
| 21.0 | 1.0851 | 2.0325 | 56.0 | 1.2635 | 32.1183 |
| 22.0 | 1.0896 | 2.1255 | 57.0 | 1.2693 | 36.9575 |
| 23.0 | 1.0941 | 2.2255 | 58.0 | 1.2750 | 42.7847 |
| 24.0 | 1.0987 | 2.3332 | 59.0 | 1.2808 | 49.8539 |
| 25.0 | 1.1033 | 2.4493 | 60.0 | 1.2867 | 58.4957 |
| 26.0 | 1.1079 | 2.5747 | 61.0 | 1.2926 | 69.1469 |
| 27.0 | 1.1126 | 2.7103 | 62.0 | 1.2985 | 82.3910 |
| 28.0 | 1.1173 | 2.8571 | 63.0 | 1.3044 | 99.0098 |
| 29.0 | 1.1221 | 3.0164 | 64.0 | 1.3104 | 120.0721 |
| 30.0 | 1.1268 | 3.1894 | 65.0 | 1.3164 | 147.0460 |
| 31.0 | 1.1316 | 3.3777 | 66.0 | 1.3225 | 181.9787 |
| 32.0 | 1.1365 | 3.5831 | 67.0 | 1.3286 | 227.7530 |
| 33.0 | 1.1414 | 3.8076 | 68.0 | 1.3347 | 288.5073 |
| 34.0 | 1.1463 | 4.0535 | 69.0 | 1.3408 | 370.2395 |
| 35.0 | 1.1513 | 4.3234 | 70.0 | 1.3470 | 481.7917 |

**Temperature = 25.0°C**

| % Sucrose, w/w | Density, g/cc | Viscosity, cP | % Sucrose, w/w | Density, g/cc | Viscosity, cP |
|---|---|---|---|---|---|
| 1.0 | 1.0014 | 0.9139 | 36.0 | 1.1543 | 3.9143 |
| 2.0 | 1.0052 | 0.9376 | 37.0 | 1.1593 | 4.1808 |
| 3.0 | 1.0090 | 0.9625 | 38.0 | 1.1643 | 4.4746 |
| 4.0 | 1.0128 | 0.9886 | 39.0 | 1.1694 | 4.7994 |
| 5.0 | 1.0167 | 1.0162 | 40.0 | 1.1745 | 5.1596 |
| 6.0 | 1.0206 | 1.0452 | 41.0 | 1.1797 | 5.5604 |
| 7.0 | 1.0245 | 1.0758 | 42.0 | 1.1848 | 6.0079 |
| 8.0 | 1.0285 | 1.1082 | 43.0 | 1.1901 | 6,5088 |
| 9.0 | 1.0325 | 1.1424 | 44.0 | 1.1953 | 7.0715 |
| 10.0 | 1.0366 | 1.1787 | 45.0 | 1.2006 | 7.7057 |
| 11.0 | 1.0407 | 1.2171 | 46.0 | 1.2059 | 8.4226 |
| 12.0 | 1.0448 | 1.2579 | 47.0 | 1.2113 | 9.2354 |
| 13.0 | 1.0489 | 1.3013 | 48.0 | 1.2167 | 10.1386 |
| 14.0 | 1.0531 | 1.3474 | 49.0 | 1.2221 | 11.1884 |
| 15.0 | 1.0574 | 1.3965 | 50.0 | 1.2276 | 12.3906 |
| 16.0 | 1.0616 | 1.4488 | 51.0 | 1.2331 | 13.7732 |
| 17.0 | 1.0659 | 1.5047 | 52.0 | 1.2386 | 15.3705 |
| 18.0 | 1.0702 | 1.5644 | 53.0 | 1.2442 | 17.2251 |
| 19.0 | 1.0746 | 1.6282 | 54.0 | 1.2498 | 19.3894 |
| 20.0 | 1.0790 | 1.6965 | 55.0 | 1.2554 | 21.9288 |
| 21.0 | 1.0835 | 1.7697 | 56.0 | 1.2611 | 24.9251 |
| 22.0 | 1.0879 | 1.8483 | 57.0 | 1.2668 | 28.4828 |
| 23.0 | 1.0924 | 1.9326 | 58.0 | 1.2726 | 32.7334 |
| 24.0 | 1.0970 | 2.0233 | 59.0 | 1.2784 | 37.8474 |
| 25.0 | 1.1016 | 2.1209 | 60.0 | 1.2842 | 44.0442 |
| 26.0 | 1.1062 | 2.2260 | 61.0 | 1.2901 | 51.6111 |
| 27.0 | 1.1108 | 2.3395 | 62.0 | 1.2959 | 60.9275 |
| 28.0 | 1.1155 | 2.4621 | 63.0 | 1.3019 | 72.4955 |
| 29.0 | 1.1202 | 2.5948 | 64.0 | 1.3078 | 86.9939 |
| 30.0 | 1.1250 | 2.7386 | 65.0 | 1.3138 | 105.3420 |
| 31.0 | 1.1298 | 2.8946 | 66.0 | 1.3199 | 128.8052 |
| 32.0 | 1.1346 | 3.0643 | 67.0 | 1.3259 | 159.1388 |
| 33.0 | 1.1395 | 3.2493 | 68.0 | 1.3320 | 198.8241 |
| 34.0 | 1.1444 | 3.4512 | 69.0 | 1.3382 | 251.3983 |
| 35.0 | 1.1493 | 3.6721 | 70.0 | 1.3444 | 321.9836 |

† Programmed by Barbara S. Bishop, MAN Program, Oak Ridge National Laboratories, Oak Ridge, Tennessee, from Reference 5.

For the preparation of CsCl solutions of a desired density ($\rho^o$) from stock solutions of a known density ($\rho_c$), the equations 13 and 14* are used. The volume of water ($v_w$) to be added to a volume of stock solution ($v_c$) is given in equation 13.

(13)
$$v_w = v_c \left[\frac{\rho_c - \rho^o}{\rho^o - 0.997}\right]$$

Alternatively, the volume of water required to prepare 1 ml of solution of density $\rho^o$ ($v_w{}^1$) is calculated as follows:

(14)
$$v_w{}^1 = \frac{\rho_c - \rho^o}{\rho_c - 0.997}$$

The concentrations of CsCl solutions are commonly determined by their index of refraction; the empirical relation is expressed in equation 15.[11,32]

(15)
$$\rho^{25.0°} = a\eta_D{}^{25} - b,$$

where $\eta_D{}^{25°}$ is the refractive index at 25° determined with the sodium D line and a and b are empirical constants. Values for a and b for a number of salts are shown in Table 15.

Ludlum and Warner[14] give the following formula for density, $\rho_{25°}$, of $Cs_2SO_4$ (equation 16):

(16)
$$\rho_{25°} = 0.9954 + 11.1066\,(\eta - \eta°) + 26.4460\,(\eta - \eta°)^2,$$

where $\eta$ is the refractive index of the solution and $\eta°$ that of water.

The densities, viscosities, and refractive indices of Ficoll® are shown in Figures 14, 15, and 16 respectively.

## TABLE 15
## COEFFICIENTS FOR CALCULATION OF DENSITY FROM REFRACTIVE INDEXES OF SOLUTIONS AT 25°C

| Solute | Coefficients a | Coefficients b | Density Range |
|---|---|---|---|
| $Cs_2SO_4$ | 12.1200 | 15.1662 | 1.15–1.40 |
| $Cs_2SO_4$ | 13.6986 | 17.3233 | 1.40–1.70 |
| CsBr | 9.9667 | 12.2876 | 1.25–1.35 |
| CsCl | 10.8601 | 13.4974 | 1.25–1.90 |
| Cs acetate | 10.7527 | 13.4247 | 1.80–2.05 |
| Cs formate | 13.7363 | 17.4286 | 1.72–1.82 |
| KBr | 6.4786 | 7.6431 | 1.10–1.35 |
| RbBr | 9.1750 | 11.2410 | 1.15–1.65 |

Data taken from: Vinograd, J., and Hearst, J. E., Equilibrium Sedimentation of Macromolecules and Viruses in a Density Gradient, *Fortschr. Chem. Org. Naturst., 20,* 372 (1962). Reproduced by permission of Springer-Verlag, New York.

* See MAN for acknowledgments.

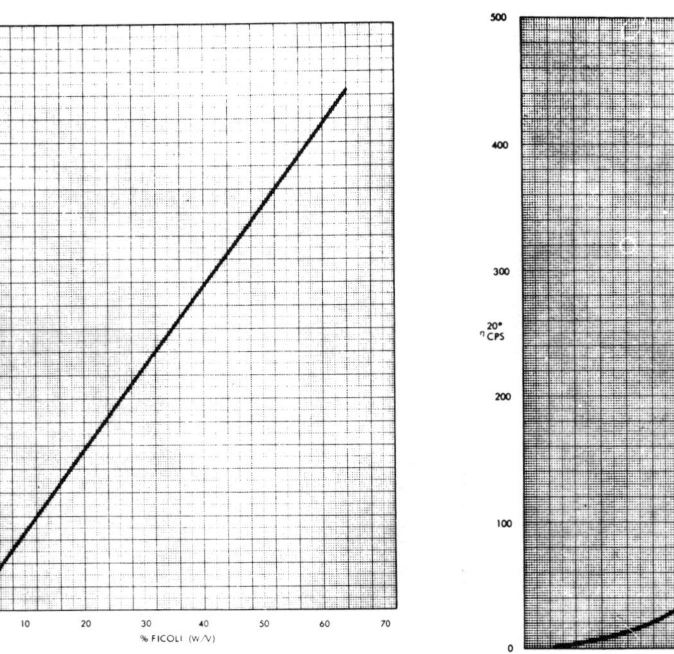


FIGURE 14. Density of Ficoll® as a function of concentration. (Reproduced by courtesy of Pharmacia Fine Chemicals, Inc., Piscataway, New Jersey.) For additional density data refer to Pretlow, T. G., Boone, C. W., Shrager, R. I., and Weiss, G. H., *Anal. Biochem.*, *29*, 230 (1969).

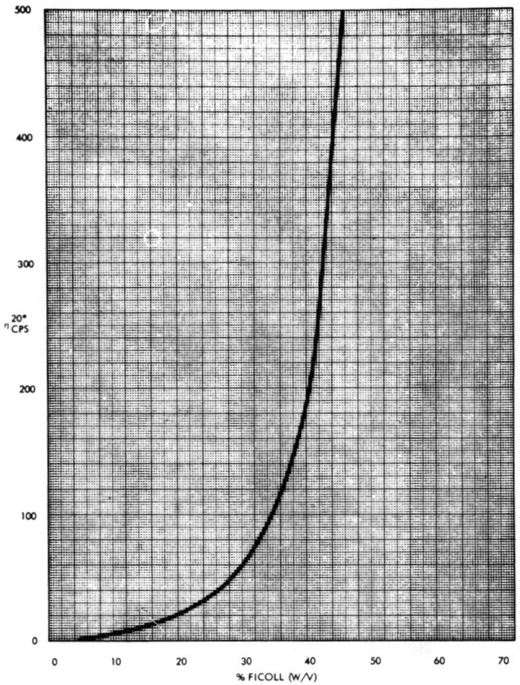

FIGURE 15. Viscosity of Ficoll® as a function of concentration. (Reproduced by courtesy of Pharmacia Fine Chemicals, Inc., Piscataway, New Jersey.) For additional viscosity data refer to Pretlow, T. G., Boone, C. W., Shrager, R. I., and Weiss, G. H., *Anal. Biochem.*, *29*, 230 (1969).

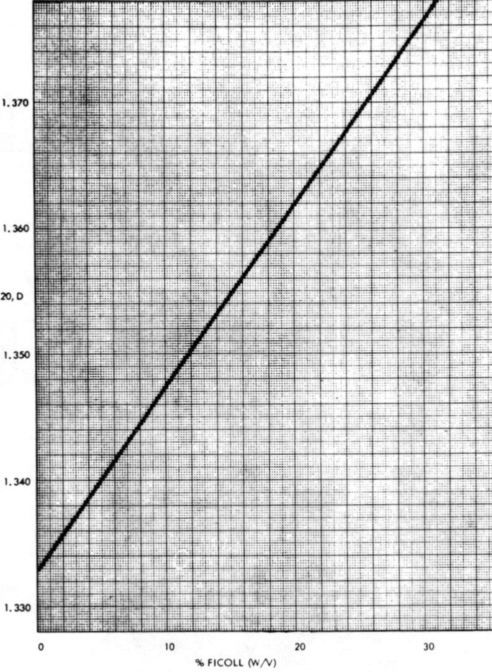


FIGURE 16. Refractive index of Ficoll® as a function of concentration. (Reproduced by courtesy of Pharmacia Fine Chemicals, Inc., Piscataway, New Jersey.)

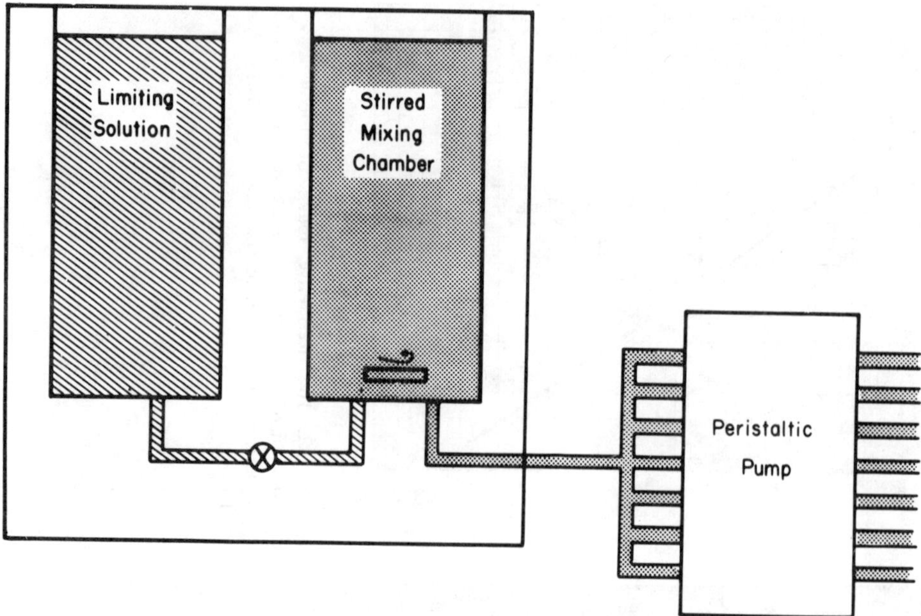


FIGURE 17. Two-cylinder generator. In its common version with two identical cylinders, this generator produces linear gradients; in order to split the gradient into two or more equal gradients, a peristaltic or proportioning pump pulls the output from the mixing cylinder. From: Price, C. A., in *Centrifugation in Density Gradients* (1973). Reproduced by permission of Academic Press, New York.

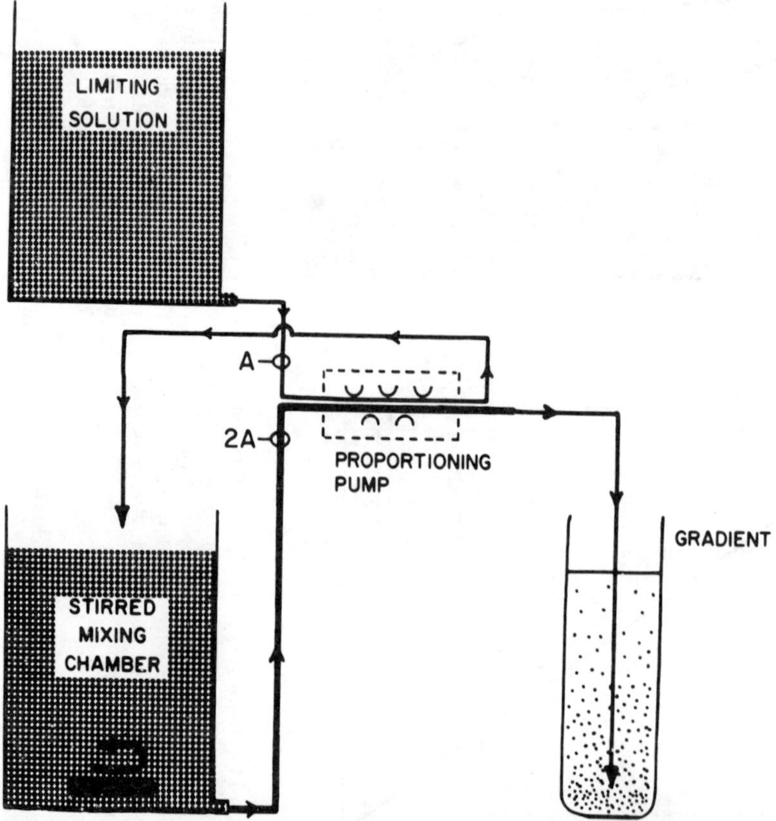


FIGURE 18. Proportioning-pump generator. This system also produces linear gradients; as in a two-cylinder generator, the mixing chamber is charged with a volume of starting solution equal to one half the final volume of the gradient; the cross-sectional areas of the two pieces of tubing (A and 2A) passing through the proportioning pump must have a relation of 1:2, as shown. From: Price, C. A., in *Centrifugation in Density Gradients* (1973). Reproduced by permission of Academic Press, New York.

## GRADIENT GENERATORS

Linear gradients are easily constructed with two-cylinder devices[7,13] (Figure 17) or proportioning pumps (Figure 18). Exponential gradients, which can be made to approximate isokinetic, hyperbolic, and equivolumetric gradients, can be constructed with equally simple mixing chambers (see Figure 9). With a given volume of the mixing chamber, one may obtain a whole family of exponential gradients by varying the ratios of initial and reservoir concentrations (see Figure 10). It is generally more accurate, and certainly more convenient, to split the output of a large gradient maker than to prepare multiple, separate small gradients, but the separate channels must be *pulled* with some type of peristaltic pump (see Figure 17), not pushed.

For linear gradients prepared from highly viscous components, such as 50% (w/w) sucrose, a positive-displacement pump is essential. A small device may be constructed inexpensively with plastic syringes and a dual-syringe pump (Figure 19). Larger pumps, either partly or fully programmable, are widely available (Table 16), but none have yet proven to be fully satisfactory for zonal rotors. A simple and effective fixed-volume generator for exponential gradients has been designed (Figure 20).

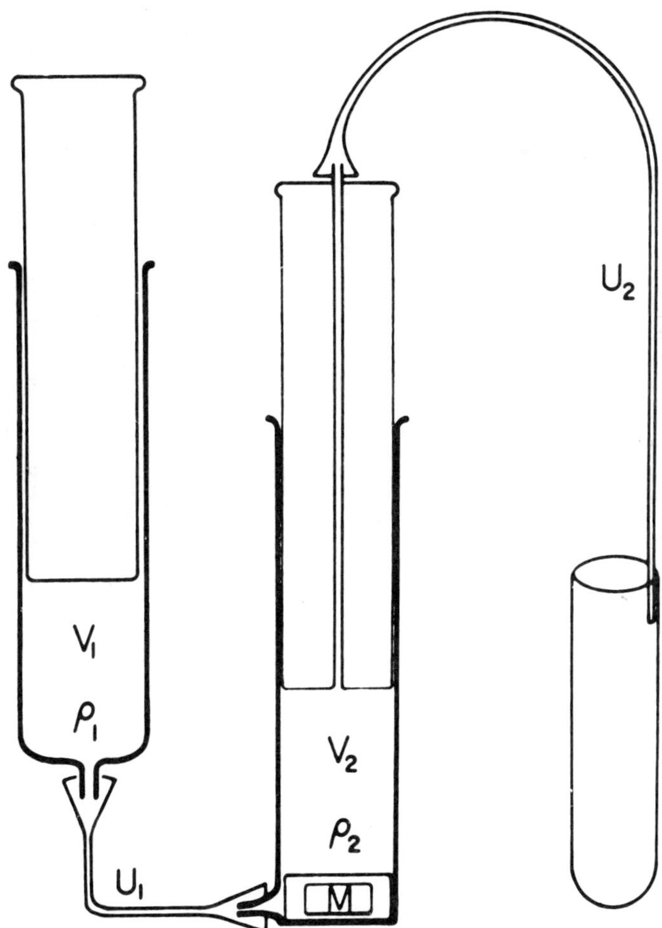


FIGURE 19. A simple syringe-driven gradient generator. One may fashion a small gradient suitable for relatively viscous solutions by connecting two plastic syringes in a motor-driven syringe pump; if both syringes are driven, a linear gradient is produced; if the mixing chamber is stationary, an exponential gradient is produced. From: Moore, D. H. (Ed.), in *Physical Techniques in Biological Research,* 2nd ed., Vol. 2, Part B, p. 297 (1969). Reproduced by permission of Academic Press, Inc.

## TABLE 16
## CHARACTERISTICS OF COMMERCIALLY AVAILABLE GRADIENT GENERATORS

| Manufacturer and Designation | Kinds of Gradients Delivered | | | Maximum Volume, cc | Maximum Pumping Rate, cc/min | Suitability with Viscous Materials | | Ease of Sterilization |
|---|---|---|---|---|---|---|---|---|
| | Linear | Exponential | Fully Programmable | | | | | |
| IEC No. 3651 | + | + | | 2000 | 70 | +++ | | ++ |
| IEC No. 3660 | | + | | *a* | 100 | ++ | | +++ |
| ISCO Dialagrad 380 | | | *b* | ∞ | 60 | ++ | | + |
| ISCO Dialagrad 382 | | | *b* | ∞ | 10 | ++ | | + |
| ISCO 570 | + | + | | 80 | 16 | +++ | | + |
| ISOLAB Gradipore | | | + | 1500 | *c* | – | | – |
| LKB GM-1 | +[e] | | | 500 | *c* | + | | |
| LKB Ultragrad | | | + | ∞ | *d* | *d* | + | + |
| MSE Fixed Profile | +[e] | | | 2000 | *c* | + | | +++ |
| MSE Automatic Variable | | | + | ∞ | | | | +++ |
| Sorvall | + | + | | | *c* | | | |
| Spinco 141 | | | + | 6000 | 66 | – | | – |

[a] Depends on selection of mixing chamber. Commercial version of the system is shown in Figure 20.
[b] Gradient is determined by smoothed fit to eleven specified points along the gradient.
[c] Depends on gravity or choice of auxiliary pump.
[d] Pumps suitable for sucrose have not been fully described.
[e] Very little experience has been reported on the use of these for density gradient centrifugation.

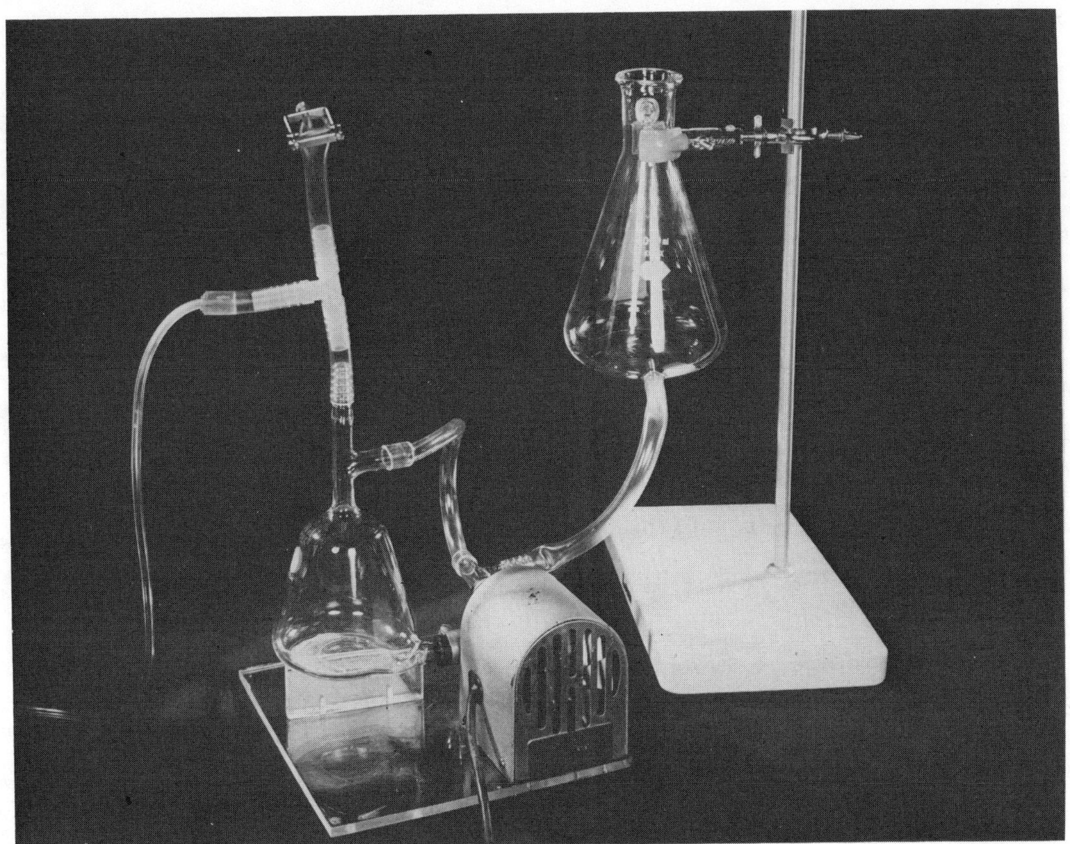

FIGURE 20. An exponential-gradient generator suitable for zonal rotors. Anderson and Rutenberg[4] described this inexpensive generator, which is suitable for relatively large volumes of viscous solutions; mixing is accomplished by circulating the fluid between the pump and the mixing chamber by way of a tangential port. From: Anderson, N. G., and Rutenberg, E., *Anal. Biochem., 21,* 260 (1967). Reproduced by permission of Academic Press, New York.

## REFERENCES


1. Anderson, N. G., Zonal Centrifuges and Other Separation Systems, *Science, 154,* 103 (1966).
2. Anderson, N. G. (Ed.), Zonal Centrifugation, in *National Cancer Institute Monograph No. 21.* National Cancer Institute, Bethesda, Maryland (1966).
3. Anderson, N. G., Analytical Techniques for Cell Fraction, VIII, Analytical Differential Centrifugation in Angle-Head Rotors, *Anal. Biochem., 23,* 72 (1968).
4. Anderson, N. G., and Rutenberg, E., Analytical Techniques for Cell Fractions, VII, A Simple Gradient-Forming Apparatus, *Anal. Biochem., 21,* 259 (1967).
5. Barber, E. J., The Development of Zonal Centrifuges, in *National Cancer Institute Monograph No. 21,* pp. 219–239, N. G. Anderson, Ed. National Cancer Institute, Bethesda, Maryland (1966).
6. Berman, A. S., Theory of Centrifugation: Miscellaneous Studies, in *National Cancer Institute Monograph No. 21,* pp. 41–76, N. G. Anderson, Ed. National Cancer Institute, Bethesda, Maryland (1966).
7. Bock, R. M., and Ling, N. S., Devices for Gradient Elution Chromatography, *Anal. Chem., 26,* 1543 (1954).
8. Britten, R. J., and Roberts, R. B., High-Resolution Density Gradient Sedimentation Analysis, *Science, 131,* 32 (1960).
9. Eikenberry, E. F., Bickle, T. A., Traut, R. R., and Price, C. A., Separation of Large Quantities of Ribosomal Subunits by Zonal Ultracentrifugation, *Eur. J. Biochem., 12,* 113 (1970).
10. Hill, W. E., Rosetti, G. P., and Van Holde, K. E., Physical Studies on *Escherichia coli, J. Mol. Biol., 44,* 263 (1969).
11. Ifft, J. B., Voet, D. H., and Vinograd, J., The Determination of Density Distributions and Density Gradients in Binary Solutions at Equilibrium in the Ultracentrifuge, *J. Chem. Phys., 65,* 1138 (1961).

12. Kaempfer, R., and Meselson, M., Sedimentation Velocity Analysis In Accelerating Gradients, in *Methods in Enzymology,* Vol. 20, pp. 520–528, K. Moldave and L. Grossman, Eds. Academic Press, New York (1971).
13. Lakshmanan, T. K., and Lieberman, S., An Improved Method of Gradient Elution Chromatography and Its Application To the Separation of Urinary Ketosteroids, *Arch. Biochem. Biophys., 53,* 258 (1954).
14. Ludlum, D. B., and Warner, R. C., Equilibrium Centrifugation in Cesium Sulfate Solutions, *J. Biol. Chem., 240,* 2961 (1965).
15. MacHattie, L. A., and Thomas, C. A., Viral DNA Molecules, in *Handbook of Biochemistry,* 2nd ed., pp. H-3–H-8, H. A. Sober, Ed. The Chemical Rubber Co., Cleveland, Ohio (1970).
16. Martin, R. G., and Ames, B. N., A Method for Determining the Sedimentation Behavior of Enzymes: Application to Protein Mixtures, *J. Biol. Chem., 236,* 1372 (1961).
17. McCarty, D. S., Stafford, D., and Brown, O., Resolution and Fractionation of Macromolecules by Isokinetic Sucrose Density Gradient Sedimentation, *Anal. Biochem., 24,* 314 (1968).
18. Meselson, M., Stahl, F. W., and Vinograd, J., Equilibrium Sedimentation of Macromolecules in Density Gradients, *Proc. Nat. Acad. Sci. U.S.A., 43,* 581 (1957).
19. Moore, D. H. (Ed.), Gradient Centrifugation, in *Physical Techniques in Biological Research,* 2nd ed., Vol. 2, Part B, pp. 285–314. Academic Press, New York (1969).
20. Noll, H., Characterization of Macromolecules by Constant Velocity Sedimentation, *Nature, 215,* 360 (1967).
21. Noll, H., Polysomes: Analysis of Structure and Function, in *Techniques in Protein Biosynthesis,* Vol. 2, pp. 101–179, P. N. Campbell and J. R. Sargent, Eds. Academic Press, London, England (1969).
22. Perardi, T. E., Leffler, R. A. A., and Anderson, N. G., K-Series Centrifuges, II, Performance of the K-II Rotor, *Anal. Biochem., 32,* 495 (1969).
23. Pollack, M. S., and Price, C. A., Equivolumetric Gradients for Zonal Rotors: Separation of Ribosomes, *Anal. Biochem., 42,* 38 (1971).
24. Price, C. A., Zonal Centrifugation, in *Manometric and Biochemical Techniques,* 5th ed., pp. 213–234, W. W. Umbreit, R. H. Burris and J. F. Stauffer, Eds. Burgess Publishing Co., Minneapolis, Minnesota (1972).
25. Price, C. A., Centrifugation in Density Gradients, Academic Press, New York (1973).
26. Price, C. A., and Hsu, T.-S., The Capacity of Equivolumetric Gradients in Zonal Rotors in the Separation of Ribosomes, *Vierteljahresschr. Naturforsch. Ges. Zürich, 116,* 367 (1971).
27. Schneider, W. C., Methods for the Isolation of Particulate Components of the Cell, in *Manometric and Biochemical Techniques,* 5th ed., pp. 196–212, W. W. Umbreit, R. H. Burris and J. F. Stauffer, Eds. Burgess Publishing Co., Minneapolis, Minnesota (1972).
28. Schumaker, V. N., Limiting Law for Boundary Spreading in Zone Centrifugation, *Sep. Sci., 1,* 409 (1966).
29. Smith, M. H., Peptides and Proteins, in *Handbook of Biochemistry,* 2nd ed., pp. C-3–C-35, H. A. Sober, Ed. The Chemical Rubber Co., Cleveland, Ohio (1970).
30. Tissières, A., Watson, J. D., Schlessinger, D., and Hollingworth, B. R., Ribonucleoprotein Particles from *E. coli, J. Mol. Biol., 1,* 221 (1959).
31. Vinograd, J., Sedimentation Equilibrium in a Buoyant Density Gradient, in *Methods in Enzymology,* Vol. 6, pp. 854–870, S. P. Colowick and N. O. Kaplan, Eds. Academic Press, New York (1963).
32. Vinograd, J., and Hearst, J. E., Equilibrium Sedimentation of Macromolecules and Viruses in a Density Gradient, *Fortschr. Chem. Org. Naturst., 20,* 372 (1962).

# TABLE OF ATOMIC WEIGHTS

For the sake of completeness all known elements are included in the list. Several of those more recently discovered are represented only by the unstable isotopes. The value in parenthesis in the atomic weight column is, in each case, the mass number of the most stable isotope.**

| Name | Symbol | At. No. | International atomic weight | | Valence |
|---|---|---|---|---|---|
| | | | 1966 | 1959 | |
| Actinium | Ac | 89 | ........ | (227) | ......... |
| Aluminum | Al | 13 | 26.9815 | 26.98 | 3 |
| Americium | Am | 95 | ........ | (243) | 3, 4, 5, 6 |
| Antimony, stibium | Sb | 51 | 121.75 | 121.76 | 3, 5 |
| Argon | Ar | 18 | 39.948 | 39.944 | 0 |
| Arsenic | As | 33 | 74.9216 | 74.92 | 3, 5 |
| Astatine | At | 85 | ........ | (210) | 1, 3, 5, 7 |
| Barium | Ba | 56 | 137.34 | 137.36 | 2 |
| Berkelium | Bk | 97 | ........ | (247) | 3, 4 |
| Beryllium | Be | 4 | 9.0122 | 9.013 | 2 |
| Bismuth | Bi | 83 | 208.980 | 208.99 | 3, 5 |
| Boron | B | 5 | 10.811 | 10.82 | 3 |
| Bromine | Br | 35 | 79.904[1] | 79.916 | 1, 3, 5, 7 |
| Cadmium | Cd | 48 | 112.40 | 112.41 | 2 |
| Calcium | Ca | 20 | 40.08 | 40.08 | 2 |
| Californium | Cf | 98 | ........ | (251) | ......... |
| Carbon | C | 6 | 12.01115 | 12.011 | 2, 4 |
| Cerium | Ce | 58 | 140.12 | 140.13 | 3, 4 |
| Cesium | Cs | 55 | 132.905 | 132.91 | 1 |
| Chlorine | Cl | 17 | 35.453 | 35.457 | 1, 3, 5, 7 |
| Chromium | Cr | 24 | 51.996 | 52.01 | 2, 3, 6 |
| Cobalt | Co | 27 | 58.9332 | 58.94 | 2, 3 |
| Columbium, see *Niobium* | | | | | |
| Copper | Cu | 29 | 63.546[2] | 63.54 | 1, 2 |
| Curium | Cm | 96 | ........ | (247) | 3 |
| Dysprosium | Dy | 66 | 162.50 | 162.51 | 3 |
| Einsteinium | Es | 99 | ........ | (254) | ......... |
| Erbium | Er | 68 | 167.26 | 167.27 | 3 |
| Europium | Eu | 63 | 151.96 | 152.0 | 2, 3 |
| Fermium | Fm | 100 | ........ | (257) | ......... |
| Fluorine | F | 9 | 18.9984 | 19.00 | 1 |
| Francium | Fr | 87 | ........ | (223) | 1 |
| Gadolinium | Gd | 64 | 157.25 | 157.26 | 3 |
| Gallium | Ga | 31 | 69.72 | 69.72 | 2, 3 |
| Germanium | Ge | 32 | 72.59 | 72.60 | 4 |
| Gold, aurum | Au | 79 | 196.967 | 197.0 | 1, 3 |
| Hafnium | Hf | 72 | 178.49 | 178.50 | 4 |
| Helium | He | 2 | 4.0026 | 4.003 | 0 |
| Holmium | Ho | 67 | 164.930 | 164.94 | 3 |
| Hydrogen | H | 1 | 1.00797 | 1.0080 | 1 |
| Indium | In | 49 | 114.82 | 114.82 | 3 |
| Iodine | I | 53 | 126.9044 | 126.91 | 1, 3, 5, 7 |
| Iridium | Ir | 77 | 192.2 | 192.2 | 3, 4 |
| Iron, ferrum | Fe | 26 | 55.847 | 55.85 | 2, 3 |
| Krypton | Kr | 36 | 83.80 | 83.80 | 0 |
| Lanthanum | La | 57 | 138.91 | 138.92 | 3 |
| Lawrencium | Lr | 103 | (257) | | |
| Lead, plumbum | Pb | 82 | 207.19 | 207.21 | 2, 4 |
| Lithium | Li | 3 | 6.939 | 6.940 | 1 |
| Lutetium | Lu | 71 | 174.97 | 174.99 | 3 |
| Magnesium | Mg | 12 | 24.312 | 24.32 | 2 |
| Manganese | Mn | 25 | 54.9380 | 54.94 | 2, 3, 4, 6, 7 |
| Mendelevium | Md | 101 | ........ | (256) | ......... |
| Mercury, hydrargyrum | Hg | 80 | 200.59 | 200.61 | 1, 2 |
| Molybdenum | Mo | 42 | 95.94 | 95.95 | 3. 4. 6 |
| Neodymium | Nd | 60 | 144.24 | 144.27 | 3 |
| Neon | Ne | 10 | 20.183 | 20.183 | 0 |
| Neptunium | Np | 93 | ........ | (237) | 4, 5, 6 |
| Nickel | Ni | 28 | 58.71 | 58.71 | 2, 3 |
| Niobium (columbium) | Nb | 41 | 92.906 | 92.91 | 3, 5 |
| Nitrogen | N | 7 | 14.0067 | 14.008 | 3, 5 |
| Nobelium | No | 102 | ........ | (254) | ......... |
| Osmium | Os | 76 | 190.2 | 190.2 | 2, 3, 4, 8 |
| Oxygen | O | 8 | 15.9994 | 16.000 | 2 |
| Palladium | Pd | 46 | 106.4 | 106.4 | 2, 4, 6 |
| Phosphorus | P | 15 | 30.9738 | 30.975 | 3, 5 |
| Platinum | Pt | 78 | 195.09 | 195.09 | 2, 4 |
| Plutonium | Pu | 94 | ........ | (244) | 3, 4, 5, 6 |
| Polonium | Po | 84 | ........ | (209) | ......... |
| Potassium, kalium | K | 19 | 39.102 | 39.100 | 1 |
| Praseodymium | Pr | 59 | 140.907 | 140.92 | 3 |
| Promethium | Pm | 61 | ........ | (145) | 3 |
| Protactinium | Pa | 91 | ........ | (231) | ......... |
| Radium | Ra | 88 | ........ | (226) | 2 |
| Radon | Rn | 86 | ........ | (222) | 0 |
| Rhenium | Re | 75 | 186.2 | 186.22 | ......... |
| Rhodium | Rh | 45 | 102.905 | 102.91 | 3 |
| Rubidium | Rb | 37 | 85.47 | 85.48 | 1 |
| Ruthenium | Ru | 44 | 101.07 | 101.1 | 3, 4, 6, 8 |
| Samarium | Sm | 62 | 150.35 | 150.35 | 2, 3 |
| Scandium | Sc | 21 | 44.956 | 44.96 | 3 |
| Selenium | Se | 34 | 78.96 | 78.96 | 2, 4, 6 |
| Silicon | Si | 14 | 28.086 | 28.09 | 4 |
| Silver, argentum | Ag | 47 | 107.868[3] | 107.873 | 1 |
| Sodium, natrium | Na | 11 | 22.9898 | 22.991 | 1 |
| Strontium | Sr | 38 | 87.62 | 87.63 | 2 |
| Sulfur | S | 16 | 32.064 | 32.066* | 2, 4, 6 |
| Tantalum | Ta | 73 | 180.948 | 180.95 | 5 |
| Technetium | Tc | 43 | ........ | (97) | 6, 7 |
| Tellurium | Te | 52 | 127.60 | 127.61 | 2, 4, 6 |
| Terbium | Tb | 65 | 158.924 | 158.93 | 3 |
| Thallium | Tl | 81 | 204.37 | 204.39 | 1, 3 |
| Thorium | Th | 90 | 232.038 | (232) | 4 |
| Thulium | Tm | 69 | 168.934 | 168.94 | 3 |
| Tin, stannum | Sn | 50 | 118.69 | 118.70 | 2, 4 |
| Titanium | Ti | 22 | 47.90 | 47.90 | 3, 4 |
| Tungsten (wolfram) | W | 74 | 183.85 | 183.86 | 6 |
| Uranium | U | 92 | 238.03 | 238.07 | 4, 6 |
| Vanadium | V | 23 | 50.942 | 50.95 | 3, 5 |
| Xenon | Xe | 54 | 131.30 | 131.30 | 0 |
| Ytterbium | Yb | 70 | 173.04 | 173.04 | 2, 3 |
| Yttrium | Y | 39 | 88.905 | 88.91 | 3 |
| Zinc | Zn | 30 | 65.37 | 65.38 | 2 |
| Zirconium | Zr | 40 | 91.22 | 91.22 | 4 |

* Because of natural variations in the relative abundances of the isotopes of sulfur the atomic weight of this element has a range of ±0.003.

** The 1959 atomic weights are based on O = 16.000 whereas those of 1966 are based on the isotope $C^{12}$.

1. ±0.002
2. ±0.001
3. ±0.001

# PERIODIC TABLE OF THE ELEMENTS

KEY TO CHART: Atomic Number → **50** ; Symbol → Sn ; Atomic Weight → 118.69 ; ← Oxidation States +2, +4 ; ← Electron Configuration –18–18–4

| 1a | 2a | 3b | 4b | 5b | 6b | 7b | 8 | 8 | 8 | 1b | 2b | 3a | 4a | 5a | 6a | 7a | 0 | Orbit |
|---|---|---|---|---|---|---|---|---|---|---|---|---|---|---|---|---|---|---|
| **1** H +1 −1<br>1.00797<br>1 | | | | | | | | | | | | | | | | | **2** He 0<br>4.0026<br>2 | K |
| **3** Li +1<br>6.939<br>2–1 | **4** Be +2<br>9.0122<br>2–2 | | | | | | | | | | | **5** B +3<br>10.811<br>2–3 | **6** C +2 +4 −4<br>12.01115<br>2–4 | **7** N +1 +2 +3 +4 +5 −1 −2 −3<br>14.0067<br>2–5 | **8** O −2<br>15.9994<br>2–6 | **9** F −1<br>18.9984<br>2–7 | **10** Ne 0<br>20.183<br>2–8 | K–L |
| **11** Na +1<br>22.9898<br>2–8–1 | **12** Mg +2<br>24.312<br>2–8–2 | Transition Elements | | | | | Group 8 | | | Transition Elements | | **13** Al +3<br>26.9815<br>2–8–3 | **14** Si +2 +4 −4<br>28.086<br>2–8–4 | **15** P +3 +5 −3<br>30.9738<br>2–8–5 | **16** S +4 +6 −2<br>32.064<br>2–8–6 | **17** Cl +1 +5 +7 −1<br>35.453<br>2–8–7 | **18** Ar 0<br>39.948<br>2–8–8 | K–L–M |
| **19** K +1<br>39.102<br>–8–8–1 | **20** Ca +2<br>40.08<br>–8–8–2 | **21** Sc +3<br>44.956<br>–8–9–2 | **22** Ti +2 +3 +4<br>47.90<br>–8–10–2 | **23** V +2 +3 +4 +5<br>50.942<br>–8–11–2 | **24** Cr +2 +3 +6<br>51.996<br>–8–13–1 | **25** Mn +2 +3 +4 +7<br>54.9380<br>–8–13–2 | **26** Fe +2 +3<br>55.847<br>–8–14–2 | **27** Co +2 +3<br>58.9332<br>–8–15–2 | **28** Ni +2 +3<br>58.71<br>–8–16–2 | **29** Cu +1 +2<br>63.546<br>–8–18–1 | **30** Zn +2<br>65.37<br>–8–18–2 | **31** Ga +3<br>69.72<br>–8–18–3 | **32** Ge +2 +4<br>72.59<br>–8–18–4 | **33** As +3 +5 −3<br>74.9216<br>–8–18–5 | **34** Se +4 +6 −2<br>78.96<br>–8–18–6 | **35** Br +1 +5 −1<br>79.904<br>–8–18–7 | **36** Kr 0<br>83.80<br>–8–18–8 | –L–M–N |
| **37** Rb +1<br>85.47<br>–18–8–1 | **38** Sr +2<br>87.62<br>–18–8–2 | **39** Y +3<br>88.905<br>–18–9–2 | **40** Zr +4<br>91.22<br>–18–10–2 | **41** Nb +3 +5<br>92.906<br>–18–12–1 | **42** Mo +6<br>95.94<br>–18–13–1 | **43** Tc +4 +6 +7<br>(97)<br>–18–13–2 | **44** Ru +3<br>101.07<br>–18–15–1 | **45** Rh +3<br>102.905<br>–18–16–1 | **46** Pd +2 +4<br>106.4<br>18–18–0 | **47** Ag +1<br>107.868<br>–18–18–1 | **48** Cd +2<br>112.40<br>–18–18–2 | **49** In +3<br>114.82<br>–18–18–3 | **50** Sn +2 +4<br>118.69<br>–18–18–4 | **51** Sb +3 +5 −3<br>121.75<br>–18–18–5 | **52** Te +4 +6 −2<br>127.60<br>–18–18–6 | **53** I +1 +5 +7 −1<br>126.9044<br>–18–18–7 | **54** Xe 0<br>131.30<br>–18–18–8 | –M–N–O |
| **55** Cs +1<br>132.905<br>–18–8–1 | **56** Ba +2<br>137.34<br>–18–8–2 | **57*** La +3<br>138.91<br>–18–9–2 | **72** Hf +4<br>178.49<br>–32–10–2 | **73** Ta +5<br>180.948<br>–32–11–2 | **74** W +6<br>183.85<br>–32–12–2 | **75** Re +4 +6 +7<br>186.2<br>–32–13–2 | **76** Os +3 +4<br>190.2<br>–32–14–2 | **77** Ir +3 +4<br>192.2<br>–32–15–2 | **78** Pt +2 +4<br>195.09<br>–32–16–2 | **79** Au +1 +3<br>196.967<br>32–18–1 | **80** Hg +1 +2<br>200.59<br>–32–18–2 | **81** Tl +1 +3<br>204.37<br>–32–18–3 | **82** Pb +2 +4<br>207.19<br>–32–18–4 | **83** Bi +3 +5<br>208.980<br>–32–18–5 | **84** Po +2 +4<br>(209)<br>–32–18–6 | **85** At<br>(210)<br>–32–18–7 | **86** Rn 0<br>(222)<br>–32–18–8 | –N–O–P |
| **87** Fr +1<br>(223)<br>–18–8–1 | **88** Ra +2<br>(226)<br>–18–8–2 | **89**** Ac +3<br>(227)<br>–18–9–2 | | | | | | | | | | | | | | | | –O–P–Q |

| | | | | | | | | | | | | | | | | Orbit |
|---|---|---|---|---|---|---|---|---|---|---|---|---|---|---|---|---|
| *Lanthanides | **58** Ce +3 +4<br>140.12<br>–20–8–2 | **59** Pr +3<br>140.907<br>–21–8–2 | **60** Nd +3<br>144.24<br>–22 8 2 | **61** Pm +3<br>(145)<br>–23–8–2 | **62** Sm +2 +3<br>150.35<br>–24–8–2 | **63** Eu +2 +3<br>151.96<br>–25–8–2 | **64** Gd +3<br>157.25<br>–25–9–2 | **65** Tb +3<br>158.924<br>–27–8–2 | **66** Dy +3<br>162.50<br>–28–8–2 | **67** Ho +3<br>164.930<br>29–8–2 | **68** Er +3<br>167.26<br>–30–8–2 | **69** Tm +3<br>168.934<br>–31–8–2 | **70** Yb +2 +3<br>173.04<br>–32–8–2 | **71** Lu +3<br>174.97<br>–32–9–2 | | –N–O–P |
| **Actinides | **90** Th +4<br>(232)<br>–18–10–2 | **91** Pa +5 +4<br>(231)<br>–20–9–2 | **92** U +3 +4 +5 +6<br>(238)<br>–21–9–2 | **93** Np +3 +4 +5 +6<br>(237)<br>–22–9–2 | **94** Pu +3 +4 +5 +6<br>(244)<br>–24–8–2 | **95** Am +3 +4 +5 +6<br>(243)<br>–25–8–2 | **96** Cm +3<br>(247)<br>–25–9–2 | **97** Bk +3 +4<br>(247)<br>–27–8–2 | **98** Cf +3<br>(251)<br>–28–8–2 | **99** Es<br>(254)<br>–29–8–2 | **100** Fm<br>(257)<br>–30–8–2 | **101** Md<br>(256)<br>–31–8–2 | **102** No<br>(254)<br>–32–8–2 | **103** Lw<br>–32–9–2 | **104** —<br>–32–10 2 | –O–P–Q |

Numbers in parentheses are mass numbers of most stable isotope of that element.

# DISSOCIATION AND IONIZATION CONSTANTS

## DISSOCIATION CONSTANTS OF ORGANIC BASES IN AQUEOUS SOLUTIONS

| Compound | °C | Step | $pK_a$ | $K_a$ |
|---|---|---|---|---|
| Acetamide | 25 | | 0.63 | $2.34 \times 10^{-1}$ |
| Acridine | 20 | | 5.58 | $2.63 \times 10^{-6}$ |
| α-Alanine | 25 | | 2.345 | $4.52 \times 10^{-3}$ |
| Alanine | | | | |
| glycyl | 25 | | 3.153 | $7.03 \times 10^{-4}$ |
| methoxy-(DL) | 25 | | 2.037 | $9.18 \times 10^{-3}$ |
| phenyl | 25 | 2 | 9.19 | $6.61 \times 10^{-10}$ |
| Allothreonine | 25 | 1 | 2.108 | $7.80 \times 10^{-3}$ |
| | 25 | 2 | 9.096 | $8.02 \times 10^{-10}$ |
| *n*-Amylamine | 25 | | 10.63 | $2.34 \times 10^{-11}$ |
| Aniline | 25 | | 4.63 | $2.34 \times 10^{-5}$ |
| *n*-allyl | 25 | | 4.17 | $6.76 \times 10^{-5}$ |
| 4-(*p*-aminobenzoyl) | 25 | 1 | 2.932 | $1.17 \times 10^{-3}$ |
| 4-benzyl | 25 | | 2.17 | $6.76 \times 10^{-3}$ |
| 2-bromo | 25 | | 2.53 | $2.95 \times 10^{-3}$ |
| 3-bromo | 25 | | 3.58 | $2.63 \times 10^{-4}$ |
| 4-bromo | 25 | | 3.86 | $1.38 \times 10^{-4}$ |
| 4-bromo-N,N-dimethyl | 25 | | 4.232 | $5.86 \times 10^{-5}$ |
| *o*-chloro | 25 | | 2.65 | $2.24 \times 10^{-3}$ |
| *m*-chloro | 25 | | 3.46 | $3.47 \times 10^{-4}$ |
| *p*-chloro | 25 | | 4.15 | $7.08 \times 10^{-5}$ |
| 3-chloro-N,N-dimethyl | 20 | | 3.837 | $1.46 \times 10^{-4}$ |
| 4-chloro-N,N-dimethyl | 20 | | 4.395 | $4.03 \times 10^{-5}$ |
| 3,5-dibromo | 25 | | 2.34 | $4.57 \times 10^{-3}$ |
| 2,4-dichloro | 22 | | 2.05 | $8.91 \times 10^{-3}$ |
| N,N-diethyl | 22 | | 6.61 | $2.46 \times 10^{-7}$ |
| N,N-dimethyl | 25 | | 5.15 | $7.08 \times 10^{-6}$ |
| N,N-dimethyl-3-nitro | 25 | | 2.626 | $2.37 \times 10^{-3}$ |
| N-ethyl | 24 | | 5.12 | $7.59 \times 10^{-6}$ |
| 2-fluoro | 25 | | 3.20 | $6.31 \times 10^{-4}$ |
| 3-fluoro | 25 | | 3.50 | $3.16 \times 10^{-4}$ |
| 4-fluoro | 25 | | 4.65 | $2.24 \times 10^{-5}$ |
| 2-iodo | 25 | | 2.60 | $2.51 \times 10^{-3}$ |
| N-methyl | 25 | | 4.848 | $1.41 \times 10^{-5}$ |
| 4-methylthio | 25 | | 4.35 | $4.46 \times 10^{-5}$ |
| 3-nitro | 25 | | 2.466 | $3.42 \times 10^{-3}$ |
| 4-nitro | 25 | | 1.00 | $1.00 \times 10^{-1}$ |
| 2-sulfonic acid | 25 | 2 | 2.459 | $3.47 \times 10^{-3}$ |
| 3-sulfonic acid | 25 | 2 | 3.738 | $1.82 \times 10^{-4}$ |
| 4-sulfonic acid | 25 | 2 | 3.227 | $5.92 \times 10^{-4}$ |
| *o*-Anisidine | 25 | | 4.52 | $3.02 \times 10^{-5}$ |
| *m*-Anisidine | 25 | | 4.23 | $5.89 \times 10^{-5}$ |
| *p*-Anisidine | 25 | | 5.34 | $4.57 \times 10^{-6}$ |
| Arginine | 25 | 1 | 1.8217 | $1.5 \times 10^{-2}$ |
| | 25 | 2 | 8.9936 | $1.01 \times 10^{-9}$ |
| Asparagine | 20 | 1 | 2.213 | $6.12 \times 10^{-3}$ |
| | 20 | 2 | 8.85 | $1.41 \times 10^{-9}$ |
| glycyl | 25 | 1 | 2.942 | $1.14 \times 10^{-3}$ |
| | 18 | 2 | 8.44 | $3.63 \times 10^{-9}$ |
| DL-Aspartic acid | 1 | 1 | 2.122 | $7.55 \times 10^{-3}$ |
| | 1 | 2 | 4.006 | $1.00 \times 10^{-4}$ |
| Acetidine (trimethylimidine) | 25 | | 11.29 | $5.12 \times 10^{-12}$ |
| Aziridine | 25 | | 8.01 | $9.77 \times 10^{-9}$ |
| Benzene | | | | |
| 4-aminoazo | 25 | | 2.82 | $1.51 \times 10^{-3}$ |
| 2-aminoethyl (β-phenylamine) | 25 | | 9.84 | $1.45 \times 10^{-10}$ |
| 4-dimethylaminoazo | 25 | | 3.226 | $5.94 \times 10^{-4}$ |

## DISSOCIATION CONSTANTS OF ORGANIC BASES IN AQUEOUS SOLUTIONS (Continued)

| Compound | °C | Step | $pK_a$ | $K_a$ |
|---|---|---|---|---|
| Benzidine | 30 | 1 | 4.66 | $2.19 \times 10^{-5}$ |
| | 30 | 2 | 3.57 | $2.69 \times 10^{-4}$ |
| Benzimidazole | 25 | | 5.532 | $2.94 \times 10^{-6}$ |
| 2-ethyl | 25 | | 6.18 | $6.61 \times 10^{-7}$ |
| 2-methyl | 25 | | 6.19 | $6.46 \times 10^{-7}$ |
| 2-phenyl | 25 | 1 | 5.23 | $5.89 \times 10^{-6}$ |
| Benzoic acid | | | | |
| 2-amino (anthranilic acid) | 25 | 1 | 2.108 | $7.80 \times 10^{-3}$ |
| | 25 | 2 | 4.946 | $1.13 \times 10^{-5}$ |
| 4-amino | 25 | 1 | 2.501 | $3.15 \times 10^{-3}$ |
| | 25 | 2 | 4.874 | $1.33 \times 10^{-5}$ |
| Benzylamine | 25 | | 9.33 | $4.67 \times 10^{-10}$ |
| Betaine | 0 | | 1.83 | $1.48 \times 10^{-2}$ |
| Biphenyl | | | | |
| 2-amino | 22 | | 3.82 | $1.51 \times 10^{-4}$ |
| *trans*-Bornylamine | 25 | | 10.17 | $6.76 \times 10^{-11}$ |
| Brucine | 25 | 1 | 8.28 | $5.24 \times 10^{-9}$ |
| Butane | | | | |
| 1-amino-3-methyl | 25 | | 10.60 | $2.51 \times 10^{-11}$ |
| 2-amino-2-methyl | 19 | | 10.85 | $1.41 \times 10^{-11}$ |
| 1,4-diamino (putrescine) | 10 | 1 | 11.15 | $7.08 \times 10^{-12}$ |
| *n*-Butylamine | 20 | | 10.77 | $1.69 \times 10^{-11}$ |
| *t*-Butylamine | 18 | | 10.83 | $1.48 \times 10^{-11}$ |
| Butyric acid | | | | |
| 4-amino | 25 | 1 | 4.0312 | $9.31 \times 10^{-5}$ |
| | 25 | 2 | 10.5557 | $2.78 \times 10^{-11}$ |
| *n*-Butyric acid | | | | |
| glycyl-2-amino | 25 | 1 | 3.1546 | $7.01 \times 10^{-4}$ |
| Cacodylic acid | 25 | 1 | 1.57 | $2.69 \times 10^{-2}$ |
| | | 2 | 6.27 | $5.37 \times 10^{-7}$ |
| β-Chlortriethylammonium | 25 | | 8.80 | $1.59 \times 10^{-9}$ |
| Cinnoline | 20 | | 2.37 | $4.27 \times 10^{-3}$ |
| Codeine | 25 | | 8.21 | $6.15 \times 10^{-9}$ |
| Cyclohexaneamine | | | | |
| *n*-butyl | 25 | | 11.23 | $5.89 \times 10^{-12}$ |
| Cyclohexylamine | 24 | | 10.66 | $2.19 \times 10^{-11}$ |
| Cystine | 30 | 1 | 1.90 | $1.25 \times 10^{-2}$ |
| | 30 | 2 | 8.24 | $5.76 \times 10^{-9}$ |
| *n*-Decylamine | 25 | | 10.64 | $2.29 \times 10^{-11}$ |
| Diethylamine | 40 | | 10.489 | $3.24 \times 10^{-11}$ |
| Diisobutylamine | 21 | | 10.489 | $1.23 \times 10^{-11}$ |
| Diisopropylamine | 28.5 | | 10.96 | $1.09 \times 10^{-11}$ |
| Dimethylamine | 25 | | 10.732 | $1.85 \times 10^{-11}$ |
| *n*-Diphenylamine | 25 | | 0.79 | $1.62 \times 10^{-1}$ |
| *n*-Dodecaneamine (laurylamine) | 25 | | 10.63 | $2.35 \times 10^{-11}$ |
| *d*-Ephedrine | 10 | | 10.139 | $7.26 \times 10^{-11}$ |
| *l*-Ephedrine | 10 | | 9.958 | $1.10 \times 10^{-10}$ |
| Ethane | | | | |
| 1-amino-3-methoxy | 10 | | 9.89 | $1.29 \times 10^{-10}$ |
| 1,2-bismethylamino | 25 | 1 | 10.40 | $3.98 \times 10^{-11}$ |
| | 25 | 2 | 8.26 | $5.50 \times 10^{-9}$ |
| Ethanol | | | | |
| 2-amino | 25 | | 9.50 | $3.16 \times 10^{-10}$ |
| Ethylamine | 20 | | 10.807 | $1.56 \times 10^{-11}$ |
| Ethylenediamine | 0 | 1 | 10.712 | $1.94 \times 10^{-11}$ |
| | 0 | 2 | 7.564 | $2.73 \times 10^{-8}$ |

## DISSOCIATION CONSTANTS OF ORGANIC BASES IN AQUEOUS SOLUTIONS (Continued)

| Compound | °C | Step | $pK_a$ | $K_a$ |
|---|---|---|---|---|
| *l*-Glutamic acid | 25 | 1 | 2.13 | $7.41 \times 10^{-3}$ |
| | 25 | 2 | 4.31 | $4.90 \times 10^{-5}$ |
| Glutamic acid | | | | |
| *a*-monoethyl | 25 | 1 | 3.846 | $1.42 \times 10^{-4}$ |
| | 25 | 2 | 7.838 | $1.45 \times 10^{-8}$ |
| *l*-Glutamine | | | 9.28 | $5.25 \times 10^{-10}$ |
| *l*-Glutathione | 25 | 2 | 3.59 | $2.57 \times 10^{-4}$ |
| Glycine | 25 | 1 | 2.3503 | $4.46 \times 10^{-3}$ |
| | 25 | 2 | 9.7796 | $1.68 \times 10^{-10}$ |
| *n*-acetyl | 25 | | 3.6698 | $2.14 \times 10^{-4}$ |
| dimethyl | 5 | | 10.3371 | $4.60 \times 10^{-11}$ |
| glycyl | 25 | | 3.1397 | $7.25 \times 10^{-4}$ |
| glycylglycyl | 25 | 1 | 3.225 | $5.96 \times 10^{-4}$ |
| | 25 | 2 | 8.090 | $8.13 \times 10^{-9}$ |
| leucyl | 25 | 1 | 3.25 | $5.62 \times 10^{-4}$ |
| | 25 | 2 | 8.28 | $5.25 \times 10^{-9}$ |
| methyl (sarcosine) | 25 | 1 | 2.21 | $6.16 \times 10^{-3}$ |
| | 25 | 2 | 10.12 | $7.58 \times 10^{-11}$ |
| phenyl | 25 | 1 | 1.83 | $1.48 \times 10^{-2}$ |
| | 25 | 2 | 4.39 | $4.07 \times 10^{-5}$ |
| N,*n*-propyl | 25 | 1 | 2.35 | $4.46 \times 10^{-3}$ |
| | 25 | 2 | 10.19 | $6.46 \times 10^{-11}$ |
| tetraglycyl | 20 | 1 | 3.10 | $7.94 \times 10^{-4}$ |
| | 20 | 2 | 8.02 | $9.55 \times 10^{-9}$ |
| Glycylserine | 25 | 1 | 2.9808 | $1.04 \times 10^{-3}$ |
| | 25 | 2 | 8.38 | $4.17 \times 10^{-9}$ |
| Heptadecaneamine | 25 | | 10.63 | $2.35 \times 10^{-11}$ |
| Heptane | | | | |
| 1-amino | 25 | | 10.66 | $2.19 \times 10^{-11}$ |
| 2-amino | 19 | | 10.88 | $1.58 \times 10^{-11}$ |
| 2-methylamino | 17 | | 10.99 | $1.02 \times 10^{-11}$ |
| Hexadecaneamine | 25 | | 10.61 | $2.46 \times 10^{-11}$ |
| Hexamethylenediamine | 0 | 1 | 111.857 | $1.39 \times 10^{-12}$ |
| | 0 | 2 | 0.762 | $1.73 \times 10^{-11}$ |
| Hexanoic acid | | | | |
| 6-amino | 25 | 1 | 4.373 | $4.23 \times 10^{-5}$ |
| | 25 | 2 | 10.804 | $1.57 \times 10^{-11}$ |
| *n*-Hexylamine | 25 | | 10.56 | $2.75 \times 10^{-11}$ |
| *dl*-Histidine | 25 | 1 | 1.80 | $1.58 \times 10^{-2}$ |
| | 25 | 2 | 6.04 | $9.12 \times 10^{-7}$ |
| | 25 | 3 | 9.33 | $4.67 \times 10^{-10}$ |
| Histidine | | | | |
| *β*-alanyl (carnosine) | 20 | 1 | 2.73 | $1.86 \times 10^{-3}$ |
| | 20 | 2 | 6.87 | $1.35 \times 10^{-7}$ |
| | 20 | 3 | 9.73 | $1.48 \times 10^{-10}$ |
| Imidazol | 25 | | 6.953 | $1.11 \times 10^{-7}$ |
| 2,4-dimethyl | 25 | | 8.359 | $5.50 \times 10^{-9}$ |
| 1-methyl (oxalmethyline) | 25 | | 6.95 | $1.12 \times 10^{-7}$ |
| Indane | | | | |
| 1-amino (*d*-1-hydrindamine) | 22.5 | | 9.21 | $6.17 \times 10^{-10}$ |
| Isobutyric acid | | | | |
| 2-amino | 25 | 1 | 2.357 | $4.30 \times 10^{-3}$ |
| | 25 | 2 | 10.205 | $6.23 \times 10^{-11}$ |
| Isoleucine | 25 | 1 | 2.318 | $4.81 \times 10^{-3}$ |
| | 25 | 2 | 9.758 | $1.74 \times 10^{-10}$ |
| Isoquinoline (leucoline) | 20 | | 5.42 | $3.80 \times 10^{-6}$ |
| 1-amino | 20 | | 7.59 | $2.57 \times 10^{-8}$ |
| 7-hydroxy | 20 | 1 | 5.68 | $2.09 \times 10^{-6}$ |
| | 20 | 2 | 8.90 | $1.26 \times 10^{-9}$ |

## DISSOCIATION CONSTANTS OF ORGANIC BASES IN AQUEOUS SOLUTIONS (Continued)

| Compound | °C | Step | $pK_a$ | $K_a$ |
|---|---|---|---|---|
| L-Leucine | 25 | 1 | 2.328 | $4.70 \times 10^{-3}$ |
| | 25 | 2 | 9.744 | $1.80 \times 10^{-10}$ |
| Leucine | | | | |
| glycyl | 25 | | 3.18 | $6.61 \times 10^{-4}$ |
| Methionine | 25 | 1 | 2.22 | $6.02 \times 10^{-3}$ |
| | 25 | 2 | 9.27 | $5.37 \times 10^{-10}$ |
| Methylamine | 25 | | 10.657 | $2.70 \times 10^{-11}$ |
| Morphine | 25 | | 8.21 | $6.16 \times 10^{-9}$ |
| Morpholine | 25 | | 8.33 | $4.67 \times 10^{-9}$ |
| Naphthalene | | | | |
| 1-amino-6-hydroxy | 25 | | 3.97 | $1.07 \times 10^{-4}$ |
| dimethylamino | 25 | | 4.566 | $2.72 \times 10^{-5}$ |
| $\alpha$-Naphthylamine | 25 | | 3.92 | $1.20 \times 10^{-4}$ |
| *n*-methyl | 27 | | 3.67 | $2.13 \times 10^{-4}$ |
| $\beta$-Naphthylamine | 25 | | 4.16 | $6.92 \times 10^{-5}$ |
| *cis*-Neobornylamine | 25 | | 10.01 | $9.77 \times 10^{-11}$ |
| Nicotine | 25 | 1 | 8.02 | $9.55 \times 10^{-9}$ |
| | 25 | 2 | 3.12 | $7.59 \times 10^{-4}$ |
| *n*-Nonylamine | 25 | | 10.64 | $2.29 \times 10^{-11}$ |
| Norleucine | 25 | | 2.335 | $4.62 \times 10^{-3}$ |
| Octadecaneamine | 25 | | 10.60 | $2.51 \times 10^{-11}$ |
| Octylamine | 25 | | 10.65 | $2.24 \times 10^{-11}$ |
| Ornithine | 25 | 1 | 1.705 | $1.97 \times 10^{-2}$ |
| | 25 | 2 | 8.690 | $2.04 \times 10^{-9}$ |
| Papaverine | 25 | | 6.40 | $3.98 \times 10^{-7}$ |
| Pentane | | | | |
| 3-amino | 17 | | 10.59 | $2.57 \times 10^{-11}$ |
| 3-amino-3-methyl | 16 | | 11.01 | $9.77 \times 10^{-12}$ |
| *n*-Pentadecylamine | 25 | | 10.61 | $2.46 \times 10^{-11}$ |
| Pentanoic acid | | | | |
| 5-amino (valeric acid) | 25 | 1 | 4.270 | $5.37 \times 10^{-5}$ |
| | 25 | 2 | 10.766 | $1.71 \times 10^{-11}$ |
| Perimidine | 20 | | 6.35 | $4.47 \times 10^{-7}$ |
| Phenanthridine | 20 | | 5.58 | $2.63 \times 10^{-6}$ |
| 1,10-Phenanthroline | 25 | | 4.84 | $1.44 \times 10^{-5}$ |
| *o*-Phenetidine (2-ethoxyaniline) | 28 | | 4.43 | $3.72 \times 10^{-5}$ |
| *m*-Phenetidine (3-ethoxyaniline) | 25 | | 4.18 | $6.60 \times 10^{-5}$ |
| *p*-Phenetidine (4-ethoxyaniline) | 28 | | 5.20 | $6.31 \times 10^{-6}$ |
| $\alpha$-Picoline | 20 | | 5.97 | $1.07 \times 10^{-6}$ |
| $\beta$-Picoline | 20 | | 5.68 | $2.09 \times 10^{-6}$ |
| $\gamma$-Picoline | 20 | | 6.02 | $9.55 \times 10^{-7}$ |
| Pilocarpine | 30 | | 6.87 | $1.35 \times 10^{-7}$ |
| Piperazine | 23.5 | 1 | 9.83 | $1.48 \times 10^{-10}$ |
| | 23.5 | 2 | 5.56 | $2.76 \times 10^{-6}$ |
| 2,5-dimethyl (*trans*-) | 25 | 1 | 9.66 | $2.19 \times 10^{-10}$ |
| | 25 | 2 | 5.20 | $6.31 \times 10^{-6}$ |
| Piperidine | 25 | | 11.123 | $7.53 \times 10^{-12}$ |
| 3-acetyl | 25 | | 3.18 | $6.61 \times 10^{-4}$ |
| 1-*n*-butyl | 23 | | 10.47 | $3.39 \times 10^{-11}$ |
| 1,2-dimethyl | 25 | | 10.22 | $6.03 \times 10^{-11}$ |
| 1-ethyl | 23 | | 10.45 | $3.55 \times 10^{-11}$ |
| 1-methyl | 25 | | 10.08 | $8.32 \times 10^{-11}$ |
| 2,2,6,6-tetramethyl | 25 | | 11.07 | $8.51 \times 10^{-12}$ |
| 2,2,4-trimethyl | 30 | | 11.04 | $9.12 \times 10^{-12}$ |

## DISSOCIATION CONSTANTS OF ORGANIC BASES IN AQUEOUS SOLUTIONS (Continued)

| Compound | °C | Step | $pK_a$ | $K_a$ |
|---|---|---|---|---|
| Proline | 25 | 1 | 1.952 | $1.11 \times 10^{-10}$ |
| | 25 | 2 | 10.640 | $2.29 \times 10^{-11}$ |
| hydroxy | 25 | 1 | 1.818 | $1.52 \times 10^{-2}$ |
| | 25 | 2 | 9.662 | $2.18 \times 10^{-10}$ |
| Propane | | | | |
| 1-amino-2,2-dimethyl | 25 | | 10.15 | $7.08 \times 10^{-11}$ |
| 1,2-diamino | 25 | 1 | 9.82 | $1.52 \times 10^{-10}$ |
| | 25 | 2 | 6.61 | $2.46 \times 10^{-7}$ |
| 1,3-diamino | 10 | 1 | 10.94 | $1.15 \times 10^{-11}$ |
| | 10 | 2 | 9.03 | $9.33 \times 10^{-10}$ |
| 1,2,3-triamino | 20 | 1 | 9.59 | $2.57 \times 10^{-10}$ |
| | 20 | 2 | 7.95 | $1.12 \times 10^{-8}$ |
| Propanoic acid | | | | |
| 3-amino (β-alanine) | 25 | 1 | 3.551 | $2.81 \times 10^{-4}$ |
| | 25 | 2 | 10.238 | $5.78 \times 10^{-11}$ |
| Propylamine | 20 | | 10.708 | $1.96 \times 10^{-11}$ |
| Pteridine | 20 | | 4.05 | $8.91 \times 10^{-5}$ |
| 2-amino-4,6-dihydroxy | 20 | 2 | 6.59 | $2.57 \times 10^{-7}$ |
| | 20 | 3 | 9.31 | $4.90 \times 10^{-10}$ |
| 2-amino-4-hydroxy | 20 | 1 | 2.27 | $5.37 \times 10^{-3}$ |
| | 20 | 2 | 7.96 | $1.10 \times 10^{-8}$ |
| 6-chloro | 20 | | 3.68 | $2.09 \times 10^{-4}$ |
| 6-hydroxy-4-methyl | 20 | 1 | 4.08 | $8.32 \times 10^{-5}$ |
| | 20 | 2 | 6.41 | $3.89 \times 10^{-7}$ |
| Purine | 20 | 1 | 2.30 | $5.01 \times 10^{-3}$ |
| | 20 | 2 | 8.96 | $1.10 \times 10^{-9}$ |
| 6-amino (adenine) | 25 | 1 | 4.12 | $7.59 \times 10^{-5}$ |
| | 25 | 2 | 9.83 | $1.48 \times 10^{-10}$ |
| 2-dimethylamino | 20 | 1 | 4.00 | $1.00 \times 10^{-4}$ |
| | 20 | 2 | 10.24 | $5.75 \times 10^{-11}$ |
| 8-hydroxy | 20 | 1 | 2.56 | $2.75 \times 10^{-3}$ |
| | 20 | 2 | 8.26 | $9.49 \times 10^{-9}$ |
| Pyrazine | 27 | | 0.65 | $2.24 \times 10^{-1}$ |
| 2-methyl | 27 | | 1.45 | $3.54 \times 10^{-2}$ |
| methylamino | 25 | | 3.39 | $4.07 \times 10^{-4}$ |
| Pyridazine | 20 | | 2.24 | $5.76 \times 10^{-3}$ |
| Pyridine | 25 | | 5.25 | $5.62 \times 10^{-6}$ |
| 2-aldoxime | 20 | 1 | 3.59 | $2.57 \times 10^{-4}$ |
| | 20 | 2 | 10.18 | $6.61 \times 10^{-11}$ |
| 2-amino | 20 | | 6.82 | $1.51 \times 10^{-7}$ |
| 4-amino | 25 | | 9.1141 | $7.69 \times 10^{-10}$ |
| 2-benzyl | 25 | | 5.13 | $7.41 \times 10^{-6}$ |
| 3-bromo | 25 | | 2.84 | $1.45 \times 10^{-3}$ |
| 3-chloro | 25 | | 2.84 | $1.45 \times 10^{-3}$ |
| 2,5-diamino | 20 | | 6.48 | $3.31 \times 10^{-7}$ |
| 2,3-dimethyl (2,3-lutidine) | 25 | | 6.57 | $2.69 \times 10^{-7}$ |
| 2,4-dimethyl (2,4-lutidine) | 25 | | 6.99 | $1.02 \times 10^{-7}$ |
| 3,5-dimethyl (3,5-lutidine) | 25 | | 6.15 | $7.08 \times 10^{-7}$ |
| 2-ethyl | 25 | | 5.89 | $1.28 \times 10^{-6}$ |
| 2-formyl | 20 | | 3.80 | $1.59 \times 10^{-4}$ |
| 2-hydroxy (2-pyridol) | 20 | 1 | 0.75 | $9.82 \times 10^{-1}$ |
| | 20 | 2 | 11.65 | $2.24 \times 10^{-12}$ |
| 4-hydroxy | 20 | 1 | 3.20 | $6.31 \times 10^{-4}$ |
| | 20 | 2 | 11.12 | $7.59 \times 10^{-12}$ |
| methoxy | 25 | | 6.47 | $3.30 \times 10^{-7}$ |
| 4-methylamino | 20 | | 9.65 | $2.24 \times 10^{-10}$ |
| 2,4,6-trimethyl | 25 | | 7.43 | $3.72 \times 10^{-8}$ |

## DISSOCIATION CONSTANTS OF ORGANIC BASES IN AQUEOUS SOLUTIONS (Continued)

| Compound | °C | Step | $pK_a$ | $K_a$ |
|---|---|---|---|---|
| Pyrimidine | | | | |
| 2-amino | 20 | | 3.45 | $3.54 \times 10^{-4}$ |
| 2-amino-4,6-dimethyl | 20 | | 4.82 | $1.51 \times 10^{-5}$ |
| 2-amino-5-nitro | 20 | | 0.35 | $4.46 \times 10^{-1}$ |
| Pyrrolidine | 25 | | 11.27 | $5.37 \times 10^{-12}$ |
| 1,2-dimethyl | 26 | | 10.20 | $6.31 \times 10^{-11}$ |
| *n*-methyl | 25 | | 10.32 | $4.79 \times 10^{-11}$ |
| Quinazoline | 20 | | 3.43 | $3.72 \times 10^{-4}$ |
| 5-hydroxy | 20 | 1 | 3.62 | $2.40 \times 10^{-4}$ |
| | 20 | 2 | 7.41 | $3.89 \times 10^{-8}$ |
| Quinine | 25 | 1 | 8.52 | $3.02 \times 10^{-9}$ |
| | 25 | 2 | 4.13 | $7.41 \times 10^{-5}$ |
| Quinoline | 20 | | 4.90 | $1.25 \times 10^{-5}$ |
| 3-amino | 20 | | 4.91 | $1.23 \times 10^{-5}$ |
| 3-bromo | 25 | | 2.69 | $2.04 \times 10^{-3}$ |
| 8-carboxy | 25 | | 1.82 | $1.51 \times 10^{-2}$ |
| 3-hydroxy (3-quinolinol) | 20 | 1 | 4.28 | $5.25 \times 10^{-5}$ |
| | 20 | 2 | 8.08 | $8.32 \times 10^{-9}$ |
| 8-hydroxy (8-quinolinol) | 20 | 1 | 5.017 | $1.21 \times 10^{-6}$ |
| | 25 | 2 | 9.812 | $1.54 \times 10^{-10}$ |
| 8-hydroxy-5-sulfo | 25 | 1 | 4.112 | $7.73 \times 10^{-5}$ |
| | 25 | 2 | 8.757 | $1.75 \times 10^{-9}$ |
| 6-methoxy | 20 | | 5.03 | $9.33 \times 10^{-6}$ |
| 2-methyl (quinaldine) | 20 | | 5.83 | $1.48 \times 10^{-6}$ |
| 4-methyl (lepidine) | 20 | | 5.67 | $2.14 \times 10^{-6}$ |
| 5-methyl | 20 | | 5.20 | $6.31 \times 10^{-6}$ |
| Quinoxaline (quinazine) | 20 | | 0.56 | $3.63 \times 10^{-1}$ |
| Serine (2-amino-3-hydroxypropanoic acid) | 25 | 1 | 2.186 | $5.49 \times 10^{-3}$ |
| | 25 | 2 | 9.208 | $6.19 \times 10^{-10}$ |
| Strychnine | 25 | | 8.26 | $5.49 \times 10^{-9}$ |
| Taurine (2-aminoethane sulfonic acid) | 25 | 2 | 9.0614 | $8.69 \times 10^{-10}$ |
| Tetradecaneamine (myristilamine) | 25 | | 10.62 | $2.40 \times 10^{-11}$ |
| Thiazole | 20 | | 2.44 | $3.63 \times 10^{-3}$ |
| 2-amino | 20 | | 5.36 | $4.36 \times 10^{-5}$ |
| Threonine | 25 | 1 | 2.088 | $8.16 \times 10^{-3}$ |
| | 25 | 2 | 9.10 | $7.94 \times 10^{-10}$ |
| *o*-Toluidine | 25 | | 4.44 | $3.63 \times 10^{-5}$ |
| *m*-Toluidine | 25 | | 4.73 | 1.86 x 10 ;5 |
| *p*-Toluidine | 25 | | 5.08 | $8.32 \times 10^{-6}$ |
| 1,3,5-Triazine | | | | |
| 2,4,6-triamino | 25 | | 5.00 | $1.00 \times 10^{-5}$ |
| Tridecaneamine | 25 | | 10.63 | $2.35 \times 10^{-11}$ |
| Triethylamine | 18 | | 11.01 | $9.77 \times 10^{-12}$ |
| Trimethylamine | 25 | | 9.81 | $1.55 \times 10^{-10}$ |
| Tryptophan | 25 | 1 | 2.43 | $3.72 \times 10^{-3}$ |
| | 25 | 2 | 9.44 | $3.63 \times 10^{-10}$ |
| Tyrosine | 25 | 2 | 9.11 | $7.76 \times 10^{-10}$ |
| | 25 | 3 | 10.13 | $7.41 \times 10^{-11}$ |
| amide | 25 | | 7.33 | $4.68 \times 10^{-8}$ |
| Urea | 21 | | 0.10 | $7.94 \times 10^{-1}$ |
| Valine | 25 | 1 | 2.286 | $5.17 \times 10^{-3}$ |
| | 25 | 2 | 9.719 | $1.91 \times 10^{-10}$ |

## DISSOCIATION CONSTANTS OF INORGANIC BASES IN AQUEOUS SOLUTIONS

| Compound | °C | Step | $pK_b$ | $K_b$ |
|---|---|---|---|---|
| Ammonium hydroxide | 25 | | 4.75 | $1.79 \times 10^{-5}$ |
| Arsenous oxide | 25 | | 3.96 | $1.1 \times 10^{-4}$ |
| Beryllium hydroxide | 25 | 2 | 10.30 | $5 \times 10^{-11}$ |
| Calcium hydroxide | 25 | 1 | 2.43 | $3.74 \times 10^{-3}$ |
| | 30 | 2 | 1.40 | $4.0 \times 10^{-2}$ |
| Deuterammonium hydroxide | 25 | | 4.96 | $1.1 \times 10^{-5}$ |
| Hydrazine | 20 | | 5.77 | $1.7 \times 10^{-6}$ |
| Hydroxylamine | 20 | | 7.97 | $1.07 \times 10^{-8}$ |
| Lead hydroxide | 25 | | 3.02 | $9.6 \times 10^{-4}$ |
| Silver hydroxide | 25 | | 3.96 | $1.1 \times 10^{-4}$ |
| Zinc hydroxide | 25 | | 3.02 | $9.6 \times 10^{-4}$ |

[a]Approximately 0.1–0.01 *N*

## DISSOCIATION CONSTANTS OF ORGANIC ACIDS IN AQUEOUS SOLUTIONS

| Compound | °C | Step | $pK$ | $K$ |
|---|---|---|---|---|
| Acetic acid | 25 | | 4.75 | $1.76 \times 10^{-5}$ |
| Acetoacetic acid | 18 | | 3.58 | $2.62 \times 10^{-4}$ |
| Acrylic acid | 25 | | 4.25 | $5.6 \times 10^{-5}$ |
| Adipamic acid | 25 | | 4.63 | $2.35 \times 10^{-5}$ |
| Adipic acid | 25 | 1 | 4.43 | $3.71 \times 10^{-5}$ |
| | 25 | 2 | 4.41 | $3.87 \times 10^{-5}$ |
| *d*-Alanine | 25 | | 9.87 | $1.35 \times 10^{-10}$ |
| Allantoin | 25 | | 8.96 | $1.10 \times 10^{-9}$ |
| Alloxanic acid | 25 | | 6.64 | $2.3 \times 10^{-7}$ |
| *α*-Aminoacetic acid (Glycine) | 25 | | 9.78 | $1.67 \times 10^{-10}$ |
| *o*-Aminobenzoic acid | 25 | | 6.97 | $1.07 \times 10^{-7}$ |
| *m*-Aminobenzoic acid | 25 | | 4.78 | $1.67 \times 10^{-5}$ |
| *p*-Aminobenzoic acid | 25 | | 4.92 | $1.2 \times 10^{-5}$ |
| *o*-Aminobenzosulfonic acid | 25 | | 2.48 | $3.3 \times 10^{-3}$ |
| *m*-Aminobenzosulfonic acid | 25 | | 3.73 | $1.85 \times 10^{-4}$ |
| *p*-Aminobenzosulfonic acid | 25 | | 3.24 | $5.81 \times 10^{-4}$ |
| Anisic acid | 25 | | 4.47 | $3.38 \times 10^{-5}$ |
| *o*-*β*-Anisylpropionic acid | 25 | | 4.80 | $1.59 \times 10^{-5}$ |
| *m*-*β*-Anisylpropionic acid | 25 | | 4.65 | $2.24 \times 10^{-5}$ |
| *p*-*β*-Anisylpropionic acid | 25 | | 4.69 | $2.04 \times 10^{-5}$ |
| Ascorbic acid | 24 | 1 | 4.10 | $7.94 \times 10^{-5}$ |
| | 16 | 2 | 11.79 | $1.62 \times 10^{-12}$ |
| DL-Aspartic acid | 25 | 1 | 3.86 | $1.38 \times 10^{-4}$ |
| | 25 | 2 | 9.82 | $1.51 \times 10^{-10}$ |
| Barbituric acid | 25 | | 4.01 | $9.8 \times 10^{-5}$ |
| Benzoic acid | 25 | | 4.19 | $6.46 \times 10^{-5}$ |
| Benzosulfonic acid | 25 | | 0.70 | $2 \times 10^{-1}$ |
| Bromoacetic acid | 25 | | 2.69 | $2.05 \times 10^{-3}$ |
| *o*-Bromobenzoic acid | 25 | | 2.84 | $1.45 \times 10^{-3}$ |
| *m*-Bromobenzoic acid | 25 | | 3.86 | $1.37 \times 10^{-4}$ |
| *n*-Butyric acid | 20 | | 4.81 | $1.54 \times 10^{-5}$ |
| *iso*-Butyric acid | 18 | | 4.84 | $1.44 \times 10^{-5}$ |
| Cacodylic acid | 25 | | 6.19 | $6.4 \times 10^{-7}$ |

## DISSOCIATION CONSTANTS OF ORGANIC ACIDS IN AQUEOUS SOLUTIONS (Continued)

| Compound | °C | Step | *pK* | *K* |
|---|---|---|---|---|
| *n*-Caproic acid | 18 | | 4.83 | $1.43 \times 10^{-5}$ |
| *iso*-Caproic acid | 18 | | 4.84 | $1.46 \times 10^{-5}$ |
| Chloroacetic acid | 25 | | 2.85 | $1.40 \times 10^{-3}$ |
| *o*-Chlorobenzoic acid | 25 | | 2.92 | $1.20 \times 10^{-3}$ |
| *m*-Chlorobenzoic acid | 25 | | 3.82 | $1.5 \times 10^{-4}$ |
| *p*-Chlorobenzoic acid | 25 | | 3.98 | $1.04 \times 10^{-4}$ |
| α-Chlorobutyric acid | RT[a] | | 2.86 | $1.39 \times 10^{-3}$ |
| β-Chlorobutyric acid | RT | | 4.05 | $8.9 \times 10^{-5}$ |
| γ-Chlorobutyric acid | RT | | 4.52 | $3.0 \times 10^{-5}$ |
| *o*-Chlorocinnamic acid | 25 | | 4.23 | $5.89 \times 10^{-5}$ |
| *m*-Chlorocinnamic acid | 25 | | 4.29 | $5.13 \times 10^{-5}$ |
| *p*-Chlorocinnamic acid | 25 | | 4.41 | $3.89 \times 10^{-5}$ |
| *o*-Chlorophenoxyacetic acid | 25 | | 3.05 | $8.91 \times 10^{-4}$ |
| *m*-Chlorophenoxyacetic acid | 25 | | 3.07 | $8.51 \times 10^{-4}$ |
| *p*-Chlorophenoxyacetic acid | 25 | | 3.10 | $7.94 \times 10^{-4}$ |
| *o*-Chlorophenylacetic acid | 25 | | 4.07 | $1.18 \times 10^{-5}$ |
| *m*-Chlorophenylacetic acid | 25 | | 4.14 | $7.25 \times 10^{-5}$ |
| *p*-Chlorophenylacetic acid | 25 | | 4.19 | $6.46 \times 10^{-5}$ |
| β-(*o*-Chlorophenyl) propionic acid | 25 | | 4.58 | $2.63 \times 10^{-4}$ |
| β-(*m*-Chlorophenyl) propionic acid | 25 | | 4.59 | $2.57 \times 10^{-5}$ |
| β-(*p*-Chlorophenyl) propionic acid | 25 | | 4.61 | $2.46 \times 10^{-5}$ |
| α-Chloropropionic acid | 25 | | 2.83 | $1.47 \times 10^{-3}$ |
| β-Chloropropionic acid | 25 | | 3.98 | $1.04 \times 10^{-4}$ |
| *cis*-Cinnamic acid | 25 | | 3.89 | $1.3 \times 10^{-4}$ |
| *trans*-Cinnamic acid | 25 | | 4.44 | $3.65 \times 10^{-5}$ |
| Citric acid | 18 | 1 | 3.08 | $8.4 \times 10^{-4}$ |
| | 18 | 2 | 4.74 | $1.8 \times 10^{-5}$ |
| | 18 | 3 | 5.40 | $4.0 \times 10^{-6}$ |
| *o*-Cresol | 25 | | 10.20 | $6.3 \times 10^{-11}$ |
| *m*-Cresol | 25 | | 10.01 | $9.8 \times 10^{-11}$ |
| *p*-Cresol | 25 | | 10.17 | $6.7 \times 10^{-11}$ |
| *trans*-Crotonic acid | 25 | | 4.69 | $2.03 \times 10^{-5}$ |
| Cyanoacetic acid | 25 | | 2.45 | $3.65 \times 10^{-3}$ |
| γ-Cyanobutyric acid | 25 | | 2.42 | $3.80 \times 10^{-3}$ |
| *o*-Cyanophenoxyacetic acid | 25 | | 2.98 | $1.05 \times 10^{-3}$ |
| *m*-Cyanophenoxyacetic acid | 25 | | 3.03 | $9.33 \times 10^{-4}$ |
| *p*-Cyanophenoxyacetic acid | 25 | | 2.93 | $1.18 \times 10^{-3}$ |
| Cyanopropionic acid | 25 | | 2.44 | $3.6 \times 10^{-3}$ |
| Cyclohexane-1: | 25 | 1 | 3.45 | $3.55 \times 10^{-4}$ |
| 1-dicarboxylic acid | 25 | 2 | 6.11 | $7.76 \times 10^{-7}$ |
| Cyclopropane-1: | 25 | 1 | 1.82 | $1.51 \times 10^{-2}$ |
| 1-dicarboxylic acid | 25 | 2 | 7.43 | $3.72 \times 10^{-8}$ |
| DL-Cysteine | 30 | 1 | 8.14 | $7.25 \times 10^{-9}$ |
| | 30 | 2 | 10.34 | $4.6 \times 10^{-11}$ |
| L-Cystine | 25 | 1 | 7.85 | $1.4 \times 10^{-8}$ |
| | 25 | 2 | 9.85 | $1.4 \times 10^{-10}$ |
| Deuteroacetic acid (in $D_2O$) | 25 | | 5.25 | $5.5 \times 10^{-6}$ |
| Dichloroacetic acid | 25 | | 1.48 | $3.32 \times 10^{-2}$ |
| Dichloroacetylacetic acid | ? | | 2.11 | $7.8 \times 10^{-3}$ |
| 2,3-Dichlorphenol | 25 | | 7.44 | $3.6 \times 10^{-8}$ |
| 2,2-Dihydroxybenzoic acid | 25 | | 2.94 | $1.14 \times 10^{-3}$ |
| 2,5-Dihydroxybenzoic acid | 25 | | 2.97 | $1.08 \times 10^{-3}$ |
| 3,4-Dihydroxybenzoic acid | 25 | | 4.48 | $3.3 \times 10^{-5}$ |
| 3,5-Dihydroxybenzoic acid | 25 | | 4.04 | $9.1 \times 10^{-5}$ |
| Dihydroxymalic acid | 25 | | 1.92 | $1.2 \times 10^{-2}$ |
| Dihydroxytartaric acid | 25 | | 1.92 | $1.2 \times 10^{-2}$ |
| Dimethylglycine | 25 | | 9.89 | $1.3 \times 10^{-10}$ |
| Dimethylmalic acid | 25 | 1 | 3.17 | $6.83 \times 10^{-4}$ |
| | 25 | 2 | 6.06 | $8.72 \times 10^{-7}$ |

[a]RT = room temperature.

## DISSOCIATION CONSTANTS OF ORGANIC ACIDS IN AQUEOUS SOLUTIONS (Continued)

| Compound | °C | Step | *pK* | *K* |
|---|---|---|---|---|
| Dimethylmalonic acid | 25 | | 3.15 | $7.08 \times 10^{-4}$ |
| Dinicotinic acid | 25 | | 2.80 | $1.6 \times 10^{-3}$ |
| 2,4-Dinitrophenol | 15 | | 3.96 | $1.1 \times 10^{-4}$ |
| 3,6-Dinitrophenol | 15 | | 5.15 | $7.1 \times 10^{-6}$ |
| Diphenylacetic acid | 25 | | 3.94 | $1.15 \times 10^{-4}$ |
| Ethylbenzoic acid | 25 | | 4.35 | $4.47 \times 10^{-5}$ |
| Ethylphenylacetic acid | 25 | | 4.37 | $4.27 \times 10^{-5}$ |
| Fluorobenzoic acid | 17 | | 2.90 | $1.25 \times 10^{-3}$ |
| Formic acid | 20 | | 3.75 | $1.77 \times 10^{-4}$ |
| *trans*-Fumaric acid | 18 | 1 | 3.03 | $9.30 \times 10^{-4}$ |
| | 18 | 2 | 4.44 | $3.62 \times 10^{-5}$ |
| Furancarboxylic acid | 25 | | 3.15 | $7.1 \times 10^{-4}$ |
| Furoic acid | 25 | | 3.17 | $6.76 \times 10^{-4}$ |
| Gallic acid | 25 | | 4.41 | $3.9 \times 10^{-5}$ |
| Glutaramic acid | 25 | | 4.60 | $3.98 \times 10^{-5}$ |
| Glutaric acid | 25 | 1 | 4.34 | $4.58 \times 10^{-5}$ |
| | 25 | 2 | 5.41 | $3.89 \times 10^{-6}$ |
| Glycerol | 25 | | 14.15 | $7 \times 10^{-15}$ |
| Glycine | 25 | | 9.87 | $1.67 \times 10^{-10}$ |
| Glycol | 25 | | 14.22 | $6 \times 10^{-15}$ |
| Glycollic acid | 25 | | 3.83 | $1.48 \times 10^{-4}$ |
| Heptanoic acid | 25 | | 4.89 | $1.28 \times 10^{-5}$ |
| Hexahydrobenzoic acid | 25 | | 4.90 | $1.26 \times 10^{-5}$ |
| Hexanoic acid | 25 | | 4.88 | $1.31 \times 10^{-5}$ |
| Hippuric acid | 25 | | 3.80 | $1.57 \times 10^{-4}$ |
| Histidine | 25 | | 9.17 | $6.7 \times 10^{-10}$ |
| Hydroquinone | 20 | | 10.35 | $4.5 \times 10^{-11}$ |
| *o*-Hydroxybenzoic acid | 19 | 1 | 2.97 | $1.07 \times 10^{-3}$ |
| | 18 | 2 | 13.40 | $4 \times 10^{-14}$ |
| *m*-Hydroxybenzoic acid | 19 | 1 | 4.06 | $8.7 \times 10^{-5}$ |
| | 19 | 2 | 9.92 | $1.2 \times 10^{-10}$ |
| *p*-Hydroxybenzoic acid | 19 | 1 | 4.48 | $3.3 \times 10^{-5}$ |
| | 19 | 2 | 9.32 | $4.8 \times 10^{-10}$ |
| β-Hydroxybutyric acid | 25 | | 4.70 | $2 \times 10^{-5}$ |
| γ-Hydroxybutyric acid | 25 | | 4.72 | $1.9 \times 10^{-5}$ |
| β-Hydroxypropionic acid | 25 | | 4.51 | $3.1 \times 10^{-5}$ |
| γ-Hydroxyquinoline | 20 | | 9.51 | $3.1 \times 10^{-10}$ |
| Iodoacetic acid | 25 | | 3.12 | $7.5 \times 10^{-4}$ |
| *o*-Iodobenzoic acid | 25 | | 2.85 | $1.4 \times 10^{-3}$ |
| *m*-Iodobenzoic acid | 25 | | 3.80 | $1.6 \times 10^{-4}$ |
| Itaconic acid | 25 | 1 | 3.85 | $1.40 \times 10^{-4}$ |
| | 25 | 2 | 5.45 | $3.56 \times 10^{-6}$ |
| Lactic acid | 100 | | 3.08 | $8.4 \times 10^{-4}$ |
| Lutidinic acid | 25 | | 2.15 | $7.0 \times 10^{-3}$ |
| Lysine | 25 | | 10.53 | $2.95 \times 10^{-11}$ |
| Maleic acid | 25 | 1 | 1.83 | $1.42 \times 10^{-2}$ |
| | 25 | 2 | 6.07 | $8.57 \times 10^{-7}$ |
| Malic acid | 25 | 1 | 3.40 | $3.9 \times 10^{-4}$ |
| | 25 | 2 | 5.11 | $7.8 \times 10^{-6}$ |
| Malonic acid | 25 | 1 | 2.83 | $1.49 \times 10^{-3}$ |
| | 25 | 2 | 5.69 | $2.03 \times 10^{-6}$ |
| DL-Mandelic acid | 25 | | 3.85 | $1.4 \times 10^{-4}$ |

## DISSOCIATION CONSTANTS OF ORGANIC ACIDS IN AQUEOUS SOLUTIONS (Continued)

| Compound | °C | Step | *pK* | *K* |
|---|---|---|---|---|
| Mesaconic acid | 25 | 1 | 3.09 | $8.22 \times 10^{-4}$ |
| | 25 | 2 | 4.75 | $1.78 \times 10^{-5}$ |
| Mesitylenic acid | 25 | | 4.32 | $4.8 \times 10^{-5}$ |
| Methyl-*o*-aminobenzoic acid | 25 | | 5.34 | $4.6 \times 10^{-6}$ |
| Methyl-*m*-aminobenzoic acid | 25 | | 5.10 | $8 \times 10^{-6}$ |
| Methyl-*p*-aminobenzoic acid | 25 | | 5.04 | $9.2 \times 10^{-6}$ |
| *o*-Methylcinnamic acid | 25 | | 4.50 | $3.16 \times 10^{-5}$ |
| *m*-Methylcinnamic acid | 25 | | 4.44 | $3.63 \times 10^{-5}$ |
| *p*-Methylcinnamic acid | 25 | | 4.56 | $2.76 \times 10^{-5}$ |
| β-Methylglutaric acid | 25 | | 4.24 | $5.75 \times 10^{-5}$ |
| *n*-Methylglycine | 18 | | 9.92 | $1.2 \times 10^{-10}$ |
| Methylmalonic acid | 25 | | 3.07 | $1.17 \times 10^{-4}$ |
| Methylsuccinic acid | 25 | 1 | 4.13 | $7.4 \times 10^{-5}$ |
| | 25 | 2 | 5.64 | $2.3 \times 10^{-6}$ |
| *o*-Monochlorophenol | 25 | | 8.49 | $3.2 \times 10^{-9}$ |
| *m*-Monochlorophenol | 25 | | 8.85 | $1.4 \times 10^{-9}$ |
| *p*-Monochlorophenol | 25 | | 9.18 | $6.6 \times 10^{-10}$ |
| Naphthalenesulfonic acid | 25 | | 0.57 | $2.7 \times 10^{-1}$ |
| α-Naphthoic acid | 25 | | 3.70 | $2 \times 10^{-4}$ |
| β-Naphthoic acid | 25 | | 4.17 | $6.8 \times 10^{-5}$ |
| α-Naphthol | 25 | | 9.34 | $4.6 \times 10^{-10}$ |
| β-Naphthol | 25 | | 9.51 | $3.1 \times 10^{-10}$ |
| Nitrobenzene | 0 | | 3.98 | $1.05 \times 10^{-4}$ |
| *o*-Nitrobenzoic acid | 18 | | 2.16 | $6.95 \times 10^{-3}$ |
| *m*-Nitrobenzoic acid | 25 | | 3.47 | $3.4 \times 10^{-4}$ |
| *p*-Nitrobenzoic acid | 25 | | 3.41 | $3.93 \times 10^{-4}$ |
| *o*-Nitrophenol | 25 | | 7.17 | $6.8 \times 10^{-8}$ |
| *m*-Nitrophenol | 25 | | 8.28 | $5.3 \times 10^{-9}$ |
| *p*-Nitrophenol | 25 | | 7.15 | $7 \times 10^{-8}$ |
| *o*-Nitrophenylacetic acid | 25 | | 4.00 | $1.00 \times 10^{-5}$ |
| *m*-Nitrophenylacetic acid | 25 | | 3.97 | $1.07 \times 10^{-4}$ |
| *p*-Nitrophenylacetic acetic | 25 | | 3.85 | $1.41 \times 10^{-4}$ |
| *o*-β-Nitrophenylpropionic acid | 25 | | 4.50 | $3.16 \times 10^{-5}$ |
| *p*-β-Nitrophenylpropionic acid | 25 | | 4.47 | $3.39 \times 10^{-5}$ |
| Nonanic acid | 25 | | 4.96 | $1.09 \times 10^{-5}$ |
| Octanoic acid | 25 | | 4.89 | $1.28 \times 10^{-5}$ |
| Oxalic acid | 25 | 1 | 1.23 | $5.90 \times 10^{-2}$ |
| | 25 | 2 | 4.19 | $6.40 \times 10^{-5}$ |
| Phenol | 20 | | 9.89 | $1.28 \times 10^{-10}$ |
| Phenylacetic acid | 18 | | 4.28 | $5.2 \times 10^{-5}$ |
| *o*-Phenylbenzoic acid | 25 | | 3.46 | $3.47 \times 10^{-4}$ |
| γ-Phenylbutyric acid | 25 | | 4.76 | $1.74 \times 10^{-5}$ |
| α-Phenylpropionic acid | 25 | | 4.64 | $2.27 \times 10^{-5}$ |
| β-Phenylpropionic acid | 25 | | 4.37 | $4.25 \times 10^{-5}$ |
| *o*-Phthalic acid | 25 | 1 | 2.89 | $1.3 \times 10^{-3}$ |
| | 25 | 2 | 5.51 | $3.9 \times 10^{-6}$ |
| *m*-Phthalic acid | 25 | 1 | 3.54 | $2.9 \times 10^{-4}$ |
| | 18 | 2 | 4.60 | $2.5 \times 10^{-5}$ |
| *p*-Phthalic acid | 25 | 1 | 3.51 | $3.1 \times 10^{-4}$ |
| | 16 | 2 | 4.82 | $1.5 \times 10^{-5}$ |
| Picric acid | 25 | | 0.38 | $4.2 \times 10^{-1}$ |
| Pimelic acid | 25 | | 4.71 | $3.09 \times 10^{-5}$ |
| Propionic acid | 25 | | 4.87 | $1.34 \times 10^{-5}$ |
| *iso*-Propylbenzoic acid | 25 | | 4.40 | $3.98 \times 10^{-5}$ |
| 2-Pyridinecarboxylic acid | 25 | | 5.52 | $3 \times 10^{-6}$ |
| 3-Pyridinecarboxylic acid | 25 | | 4.85 | $1.4 \times 10^{-5}$ |

## DISSOCIATION CONSTANTS OF ORGANIC ACIDS IN AQUEOUS SOLUTIONS (Continued)

| Compound | °C | Step | *pK* | *K* |
|---|---|---|---|---|
| 4-Pyridinecarboxylic acid | 25 | | 4.96 | $1.1 \times 10^{-5}$ |
| Pyrocatechol | 20 | | 9.85 | $1.4 \times 10^{-10}$ |
| Quinolinic acid | 25 | | 2.52 | $3 \times 10^{-3}$ |
| Resorcinol | 25 | | 9.81 | $1.55 \times 10^{-10}$ |
| Saccharin | 18 | | 11.68 | $2.1 \times 10^{-12}$ |
| Suberic acid | 25 | | 4.52 | $2.99 \times 10^{-5}$ |
| Succinic acid | 25 | 1 | 4.16 | $6.89 \times 10^{-5}$ |
| | 25 | 2 | 5.61 | $2.47 \times 10^{-6}$ |
| Sulfanilic acid | 25 | | 3.23 | $5.9 \times 10^{-4}$ |
| *a*-Tartaric acid | 25 | 1 | 2.98 | $1.04 \times 10^{-3}$ |
| | 25 | 2 | 4.34 | $4.55 \times 10^{-5}$ |
| *meso*-Tartaric acid | 25 | 1 | 3.22 | $6 \times 10^{-4}$ |
| | 25 | 2 | 4.82 | $1.53 \times 10^{-5}$ |
| Theobromine | 18 | | 7.89 | $1.3 \times 10^{-8}$ |
| Terephthalic acid | 25 | | 3.51 | $3.1 \times 10^{-4}$ |
| Thioacetic acid | 25 | | 3.33 | $4.7 \times 10^{-4}$ |
| Thiophenecarboxylic acid | 25 | | 3.48 | $3.3 \times 10^{-4}$ |
| *o*-Toluic acid | 25 | | 3.91 | $1.22 \times 10^{-4}$ |
| *m*-Toluic acid | 25 | | 4.27 | $5.32 \times 10^{-5}$ |
| *p*-Toluic acid | 25 | | 4.36 | $4.33 \times 10^{-5}$ |
| Trichloroacetic acid | 25 | | 0.70 | $2 \times 10^{-1}$ |
| Trichlorophenol | 25 | | 6.00 | $1 \times 10^{-6}$ |
| 2,4,6-Trihydroxybenzoic acid | 25 | | 1.68 | $2.1 \times 10^{-2}$ |
| Trimethylacetic acid | 18 | | 5.03 | $9.4 \times 10^{-6}$ |
| 2,4,6-Trinitrophenol | 25 | | 0.38 | $4.2 \times 10^{-1}$ |
| Tryptophan | 25 | | 9.38 | $4.2 \times 10^{-10}$ |
| Tyrosine | 17 | | 8.40 | $3.98 \times 10^{-9}$ |
| Uric acid | 12 | | 3.89 | $1.3 \times 10^{-4}$ |
| *n*-Valeric acid | 18 | | 4.82 | $1.51 \times 10^{-5}$ |
| *iso*-Valeric acid | 25 | | 4.77 | $1.7 \times 10^{-5}$ |
| Veronal | 25 | | 7.43 | $3.7 \times 10^{-8}$ |
| Vinylacetic acid | 25 | | 4.34 | $4.57 \times 10^{-5}$ |
| Xanthine | 40 | | 9.91 | $1.24 \times 10^{-10}$ |

## DISSOCIATION CONSTANTS OF INORGANIC ACIDS IN AQUEOUS SOLUTIONS[a]

| Compound | °C | Step | *pK* | *K* |
|---|---|---|---|---|
| Arsenic acid | 18 | 1 | 2.25 | $5.62 \times 10^{-3}$ |
| | 18 | 2 | 6.77 | $1.70 \times 10^{-3}$ |
| | 18 | 3 | 11.60 | $3.95 \times 10^{-12}$ |
| Arsenious acid | 25 | | 9.23 | $6 \times 10^{-10}$ |
| *o*-Boric acid | 20 | 1 | 9.14 | $7.3 \times 10^{-10}$ |
| | 20 | 2 | 12.74 | $1.8 \times 10^{-13}$ |
| | 20 | 3 | 13.80 | $1.6 \times 10^{-14}$ |
| Carbonic acid | 25 | 1 | 6.37 | $4.30 \times 10^{-7}$ |
| | 25 | 2 | 10.25 | $5.61 \times 10^{-11}$ |

[a]Approximately 0.1–0.01 *N*.

## DISSOCIATION CONSTANTS OF INORGANIC ACIDS IN AQUEOUS SOLUTIONS (Continued)

| Compound | °C | Step | *pK* | *K* |
|---|---|---|---|---|
| Chromic acid | 25 | 1 | 0.74 | $1.8 \times 10^{-1}$ |
| | 25 | 2 | 6.49 | $3.20 \times 10^{-7}$ |
| Germanic acid | 25 | 1 | 8.59 | $2.6 \times 10^{-9}$ |
| | 25 | 2 | 12.72 | $1.9 \times 10^{-13}$ |
| Hydrocyanic acid | 25 | | 9.31 | $4.93 \times 10^{-10}$ |
| Hydrofluoric acid | 25 | | 3.45 | $3.53 \times 10^{-4}$ |
| Hydrogen sulfide | 18 | 1 | 7.04 | $9.1 \times 10^{-3}$ |
| | 18 | 2 | 11.96 | $1.1 \times 10^{-12}$ |
| Hydrogen peroxide | 25 | | 11.62 | $2.4 \times 10^{-12}$ |
| Hypobromous acid | 25 | | 8.69 | $2.06 \times 10^{-9}$ |
| Hypochlorous acid | 18 | | 7.53 | $2.95 \times 10^{-8}$ |
| Hypoiodous acid | 25 | | 10.64 | $2.3 \times 10^{-11}$ |
| Iodic acid | 25 | | 0.77 | $1.69 \times 10^{-1}$ |
| Nitrous acid | 12.5 | | 3.37 | $4.6 \times 10^{-4}$ |
| Periodic acid | 25 | | 1.64 | $2.3 \times 10^{-2}$ |
| *o*-Phosphoric acid | 25 | 1 | 2.12 | $7.52 \times 10^{-3}$ |
| | 25 | 2 | 7.21 | $6.23 \times 10^{-8}$ |
| | 18 | 3 | 12.67 | $2.2 \times 10^{-13}$ |
| Phosphorous acid | 18 | 1 | 2.00 | $1.0 \times 10^{-2}$ |
| | 18 | 2 | 6.59 | $2.6 \times 10^{-7}$ |
| Pyrophosphoric acid | 18 | 1 | 0.85 | $1.4 \times 10^{-1}$ |
| | 18 | 2 | 1.49 | $3.2 \times 10^{-2}$ |
| | 18 | 3 | 5.77 | $1.7 \times 10^{-6}$ |
| | 18 | 4 | 8.22 | $6 \times 10^{-9}$ |
| Selenic acid | 25 | 2 | 1.92 | $1.2 \times 10^{-2}$ |
| Selenous acid | 25 | 1 | 2.46 | $3.5 \times 10^{-3}$ |
| | 25 | 2 | 7.31 | $5 \times 10^{-8}$ |
| *o*-Silicic acid | 30 | 1 | 9.66 | $2.2 \times 10^{-10}$ |
| | 30 | 2 | 11.70 | $2 \times 10^{-12}$ |
| | 30 | 3 | 12.00 | $1 \times 10^{-12}$ |
| | 30 | 4 | 12.00 | $1 \times 10^{-12}$ |
| *m*-Silicic acid | RT[b] | 1 | 9.70 | $2 \times 10^{-10}$ |
| | RT | 2 | 12.00 | $1 \times 10^{-12}$ |
| Sulfuric acid | 25 | 2 | 1.92 | $1.20 \times 10^{-2}$ |
| Sulfurous acid | 18 | 1 | 1.81 | $1.54 \times 10^{-2}$ |
| | 18 | 2 | 6.91 | $1.02 \times 10^{-7}$ |
| Telluric acid | 18 | 1 | 7.68 | $2.09 \times 10^{-8}$ |
| | 18 | 2 | 11.29 | $6.46 \times 10^{-12}$ |
| Tellurous acid | 25 | 1 | 2.48 | $3 \times 10^{-3}$ |
| | 25 | 2 | 7.70 | $2 \times 10^{-8}$ |
| Tetraboric acid | 25 | 1 | 4.00 | $\sim 10^{-4}$ |
| | 25 | 1 | 9.00 | $\sim 10^{-9}$ |

[b]RT = room temperature

## DISSOCIATION CONSTANTS ($K_b$) OF AQUEOUS AMMONIA FROM 0 TO 50°C[a]

| °C | $pK_b$ | $K_b$ | °C | $pK_b$ | $K_b$ | °C | $pK_b$ | $K_b$ |
|---|---|---|---|---|---|---|---|---|
| 0 | 4.862 | $1.374 \times 10^{-5}$ | 20 | 4.767 | $1.710 \times 10^{-5}$ | 35 | 4.733 | $1.849 \times 10^{-5}$ |
| 5 | 4.830 | $1.479 \times 10^{-5}$ | 25 | 4.751 | $1.774 \times 10^{-5}$ | 40 | 4.730 | $1.862 \times 10^{-5}$ |
| 10 | 4.804 | $1.570 \times 10^{-5}$ | 30 | 4.740 | $1.820 \times 10^{-5}$ | 45 | 4.726 | $1.879 \times 10^{-5}$ |
| 15 | 4.782 | $1.652 \times 10^{-5}$ | | | | 50 | 4.723 | $1.892 \times 10^{-5}$ |

[a] Values of $K_b$ are accurate to ±0.005; determined by emf method and taken from Bates, R. G. and Pinching, G. D., *J. Amer. Chem. Soc.*, *72*, 1393 (1950).

## IONIZATION CONSTANTS FOR WATER

| °C | $-\log_{10} K_w$ | °C | $-\log_{10} K_w$ |
|---|---|---|---|
| 0 | 14.9435 | 30 | 13.8330 |
| 5 | 14.7338 | 35 | 13.6801 |
| 10 | 14.5346 | 40 | 13.5348 |
| 15 | 14.3463 | 45 | 13.3960 |
| 20 | 14.1669 | 50 | 13.2617 |
| 24 | 14.0000 | 55 | 13.1369 |
| 25 | 13.9965 | 60 | 13.0171 |

## IONIZATION CONSTANTS FOR DEUTERIUM OXIDE FROM 10 TO 50°C

| °C | $pK_m$ | $pK_c$ | °C | $pK_m$ | $pK_c$ |
|---|---|---|---|---|---|
| 10 | 15.526 | 15.439 | 30 | 14.784 | 14.699 |
| 20 | 15.136 | 15.049 | 40 | 14.468 | 14.385 |
| 25 | 14.955 | 14.869 | 50 | 14.182 | 14.103 |

Note:

The subscript *m* indicates values on the molal scale; the subscript *c* indicates values on the molar scale. Data taken from *NBS Technical Note 400.*

## IONIZATION CONSTANTS OF ACIDS IN WATER AT VARIOUS TEMPERATURES

| Tabular Value Equals | 0°C | 5°C | 10°C | 15°C | 20°C | 25°C | 30°C | 35°C | 40°C | 45°C | 50°C |
|---|---|---|---|---|---|---|---|---|---|---|---|
| **Acetic Acid** | | | | | | | | | | | |
| $K_A \times 10^5$ | 1.657 | 1.700 | 1.729 | 1.745 | 1.753 | 1.754 | 1.750 | 1.728 | 1.703 | 1.670 | 1.633 |
| **Boric Acid** | | | | | | | | | | | |
| $K_A \times 10^{10}$ | | 3.63 | 4.17 | 4.72 | 5.26 | 5.79 | 6.34 | 6.86 | 7.38 | | 8.32 |
| ***n*-Butyric Acid** | | | | | | | | | | | |
| $K_A \times 10^5$ | 1.563 | 1.574 | 1.576 | 1.569 | 1.542 | 1.515 | 1.484 | 1.439 | 1.395 | 1.347 | 1.302 |
| **Carbonic Acid** | | | | | | | | | | | |
| $K_{1A} \times 10^7$ | 2.64 | 3.04 | 3.44 | 3.81 | 4.16 | 4.45 | 4.71 | 4.90 | 5.04 | 5.13 | 5.19 |
| **Chloroacetic Acid** | | | | | | | | | | | |
| $K_A \times 10^3$ | 1.528 | | 1.488 | | | 1.379 | | | 1.230 | | |
| **Citric Acid** | | | | | | | | | | | |
| $K_{1A} \times 10^4$ | 6.03 | 6.31 | 6.69 | 6.92 | 7.21 | 7.45 | 7.66 | 7.78 | 7.96 | 7.99 | 8.04 |
| $K_{2A} \times 10^5$ | 1.45 | 1.54 | 1.60 | 1.65 | 1.70 | 1.73 | 1.76 | 1.77 | 1.78 | 1.76 | 1.75 |
| $K_{3A} \times 10^7$ | 4.05 | 4.11 | 4.14 | 4.13 | 4.09 | 4.02 | 3.99 | 3.78 | 3.69 | 3.45 | 3.28 |

## IONIZATION CONSTANTS OF ACIDS IN WATER AT VARIOUS TEMPERATURES (Continued)

| Tabular Value Equals | 0°C | 5°C | 10°C | 15°C | 20°C | 25°C | 30°C | 35°C | 40°C | 45°C | 50°C |
|---|---|---|---|---|---|---|---|---|---|---|---|
| **Formic Acid** | | | | | | | | | | | |
| $K_A \times 10^4$ | 1.638 | 1.691 | 1.728 | 1.749 | 1.765 | 1.772 | 1.768 | 1.747 | 1.716 | 1.685 | 1.650 |
| **Glycine** | | | | | | | | | | | |
| $K_{1A} \times 10^7$ | | 3.82 | 3.99 | 4.17 | 4.32 | 4.46 | 4.57 | 4.66 | 4.73 | 4.77 | 4.79 |
| **Glycollic Acid** | | | | | | | | | | | |
| $K_A \times 10^4$ | 1.334 | | | | | 1.475 | | | | | 1.415 |
| **Lactic Acid** | | | | | | | | | | | |
| $K_A \times 10^4$ | 1.287 | | | | | 1.374 | | | | | 1.270 |
| **Malonic Acid** | | | | | | | | | | | |
| $K_{2A} \times 10^6$ | 2.140 | 2.165 | 2.152 | 2.124 | 2.076 | 2.014 | 1.948 | 1.863 | 1.768 | 1.670 | 1.575 |
| **Oxalic Acid** | | | | | | | | | | | |
| $K_{2A} \times 10^5$ | 5.91 | 5.82 | 5.70 | 5.55 | 5.40 | 5.18 | 4.92 | 4.67 | 4.41 | 4.09 | 3.83 |
| **Phenolsulfonic Acid** | | | | | | | | | | | |
| $K_{2A} \times 10^{10}$ | 4.45 | 5.20 | 6.03 | 6.92 | 7.85 | 8.85 | 9.89 | 10.94 | 12.00 | 13.09 | 14.16 |
| **Phosphoric Acid** | | | | | | | | | | | |
| $K_{1A} \times 10^3$ | 8.968 | | | | | 7.516 | | | | | 5.495 |
| $K_{2A} \times 10^8$ | 4.85 | 5.24 | 5.57 | 5.89 | 6.12 | 6.34 | 6.46 | 6.53 | 6.58 | 6.59 | 6.55 |
| **Propionic Acid** | | | | | | | | | | | |
| $K_A \times 10^5$ | 1.274 | 1.305 | 1.326 | 1.336 | 1.338 | 1.336 | 1.326 | 1.310 | 1.280 | 1.257 | 1.229 |

**Notes:**

1. All values are on the molar scale.
2. Reproducibility between various workers is about ± (0.01–0.02) x $10^5$.

## COMPOSITION AND PROPERTIES OF SOME INORGANIC ACIDS AND BASES SUPPLIED AS CONCENTRATED AQUEOUS SOLUTIONS

| Compound | Composition, weight % | | Formula Weight | Molarity | Specific Gravity |
|---|---|---|---|---|---|
| Acetic acid | 99–100 | $CH_3COOH$ | 60.05 | 17.5 | 1.05 |
| Ammonium hydroxide | 28–30 | $NH_4OH$ | 35.05 | 7.4 | 0.90 |
| Hydriodic acid | 47–47.5 | HI | 127.91 | 5.5 | 1.5 |
| Hydrobromic acid | 47–49 | HBr | 80.93 | 9.0 | 1.5 |
| Hydrochloric acid | 36.5–38 | HCl | 36.46 | 12.0 | 1.18 |
| Hydrofluoric acid | 48–51 | HF | 20.01 | 28.9 | 1.17 |
| Phosphoric acid | 85 | $H_3PO_4$ | 98.00 | 14.7 | 1.7 |
| Sulfuric acid | 95–98 | $H_2SO_4$ | 98.08 | 18.0 | 1.84 |

# VISCOSITY AND DENSITY OF SUCROSE IN WATER

| Sucrose, % | Density,[a] g/ml | Viscosity,[b] cp | Sucrose, % | Density,[a] g/ml | Viscosity,[b] cp | Sucrose, % | Density,[a] g/ml | Viscosity[b] cp |
|---|---|---|---|---|---|---|---|---|
| **0°C** | | | | | | | | |
| 0 | 1.0004 | 1.780 | 24 | 1.1037 | 4.646 | 48 | 1.2269 | 34.57 |
| 1 | 1.0043 | 1.830 | 25 | 1.1085 | 4.912 | 49 | 1.2324 | 39.23 |
| 2 | 1.0082 | 1.884 | 26 | 1.1133 | 5.202 | 50 | 1.2380 | 44.74 |
| 3 | 1.0122 | 1.941 | 27 | 1.1181 | 5.519 | 51 | 1.2436 | 51.29 |
| 4 | 1.0162 | 2.002 | 28 | 1.1229 | 5.866 | 52 | 1.2493 | 59.11 |
| 5 | 1.0203 | 2.066 | 29 | 1.1278 | 6.246 | 53 | 1.2550 | 68.52 |
| 6 | 1.0244 | 2.135 | 30 | 1.1327 | 6.665 | 54 | 1.2607 | 79.92 |
| 7 | 1.0285 | 2.208 | 31 | 1.1376 | 7.126 | 55 | 1.2665 | 93.85 |
| 8 | 1.0326 | 2.286 | 32 | 1.1426 | 7.635 | 56 | 1.2723 | 111.0 |
| 9 | 1.0368 | 2.369 | 33 | 1.1476 | 8.201 | 57 | 1.2781 | 132.3 |
| 10 | 1.0411 | 2.458 | 34 | 1.1527 | 8.829 | 58 | 1.2840 | 158.9 |
| 11 | 1.0453 | 2.552 | 35 | 1.1578 | 9.530 | 59 | 1.2899 | 192.6 |
| 12 | 1.0496 | 2.653 | 36 | 1.1629 | 10.31 | 60 | 1.2958 | 235.7 |
| 13 | 1.0539 | 2.761 | 37 | 1.1680 | 11.20 | 61 | 1.3018 | 291.4 |
| 14 | 1.0583 | 2.877 | 38 | 1.1732 | 12.19 | 62 | 1.3078 | 364.2 |
| 15 | 1.0627 | 3.001 | 39 | 1.1784 | 13.31 | 63 | 1.3138 | 460.6 |
| 16 | 1.0671 | 3.134 | 40 | 1.1836 | 14.58 | 64 | 1.3199 | 589.9 |
| 17 | 1.0716 | 3.277 | 41 | 1.1889 | 16.03 | 65 | 1.3260 | 766.0 |
| 18 | 1.0760 | 3.430 | 42 | 1.1942 | 17.69 | 66 | 1.3321 | 1010. |
| 19 | 1.0806 | 3.596 | 43 | 1.1996 | 19.59 | 67 | 1.3383 | 1352. |
| 20 | 1.0852 | 3.774 | 44 | 1.2050 | 21.78 | 68 | 1.3445 | 1842. |
| 21 | 1.0898 | 3.967 | 45 | 1.2104 | 24.31 | 69 | 1.3507 | 2556. |
| 22 | 1.0944 | 4.175 | 46 | 1.2159 | 27.24 | 70 | 1.3570 | 3621. |
| 23 | 1.0991 | 4.401 | 47 | 1.2213 | 30.66 | | | |
| **5°C** | | | | | | | | |
| 0 | 1.0004 | 1.516 | 24 | 1.1027 | 3.831 | 48 | 1.2250 | 25.97 |
| 1 | 1.0043 | 1.558 | 25 | 1.1074 | 4.042 | 49 | 1.2306 | 29.28 |
| 2 | 1.0082 | 1.603 | 26 | 1.1122 | 4.272 | 50 | 1.2361 | 33.16 |
| 3 | 1.0121 | 1.650 | 27 | 1.1169 | 4.523 | 51 | 1.2417 | 37.73 |
| 4 | 1.0161 | 1.700 | 28 | 1.1218 | 4.796 | 52 | 1.2474 | 43.16 |
| 5 | 1.0201 | 1.753 | 29 | 1.1266 | 5.094 | 53 | 1.2530 | 49.62 |
| 6 | 1.0241 | 1.809 | 30 | 1.1315 | 5.422 | 54 | 1.2587 | 57.39 |
| 7 | 1.0282 | 1.869 | 31 | 1.1364 | 5.781 | 55 | 1.2645 | 66.79 |
| 8 | 1.0323 | 1.933 | 32 | 1.1413 | 6.177 | 56 | 1.2702 | 78.24 |
| 9 | 1.0365 | 2.001 | 33 | 1.1463 | 6.614 | 57 | 1.2760 | 92.30 |
| 10 | 1.0406 | 2.073 | 34 | 1.1513 | 7.099 | 58 | 1.2819 | 109.7 |
| 11 | 1.0448 | 2.150 | 35 | 1.1563 | 7.637 | 59 | 1.2877 | 131.4 |
| 12 | 1.0491 | 2.232 | 36 | 1.1614 | 8.236 | 60 | 1.2936 | 158.9 |
| 13 | 1.0534 | 2.319 | 37 | 1.1665 | 8.905 | 61 | 1.2996 | 194.0 |
| 14 | 1.0577 | 2.413 | 38 | 1.1717 | 9.656 | 62 | 1.3056 | 239.1 |
| 15 | 1.0620 | 2.513 | 39 | 1.1768 | 10.50 | 63 | 1.3116 | 297.9 |
| 16 | 1.0664 | 2.621 | 40 | 1.1820 | 11.45 | 64 | 1.3176 | 375.6 |
| 17 | 1.0708 | 2.736 | 41 | 1.1873 | 12.53 | 65 | 1.3237 | 479.4 |
| 18 | 1.0753 | 2.859 | 42 | 1.1926 | 13.76 | 66 | 1.3298 | 620.3 |
| 19 | 1.0798 | 2.992 | 43 | 1.1979 | 15.16 | 67 | 1.3360 | 814.3 |
| 20 | 1.0843 | 3.135 | 44 | 1.2033 | 16.76 | 68 | 1.3422 | 1086. |
| 21 | 1.0889 | 3.290 | 45 | 1.2087 | 18.60 | 69 | 1.3484 | 1473. |
| 22 | 1.0934 | 3.456 | 46 | 1.2140 | 20.73 | 70 | 1.3546 | 2034. |
| 23 | 1.0981 | 3.636 | 47 | 1.2195 | 23.18 | | | |

**10°C**

| Sucrose, % | Density,[a] g/ml | Viscosity,[b] cp | Sucrose, % | Density,[a] g/ml | Viscosity,[b] cp | Sucrose, % | Density,[a] g/ml | Viscosity,[b] cp |
|---|---|---|---|---|---|---|---|---|
| 0 | 1.0002 | 1.308 | 24 | 1.1016 | 3.206 | 48 | 1.2231 | 19.96 |
| 1 | 1.0040 | 1.343 | 25 | 1.1062 | 3.377 | 49 | 1.2286 | 22.37 |
| 2 | 1.0079 | 1.380 | 26 | 1.1109 | 3.562 | 50 | 1.2341 | 25.17 |
| 3 | 1.0118 | 1.420 | 27 | 1.1157 | 3.763 | 51 | 1.2397 | 28.45 |
| 4 | 1.0157 | 1.462 | 28 | 1.1204 | 3.982 | 52 | 1.2453 | 32.32 |
| 5 | 1.0196 | 1.506 | 29 | 1.1252 | 4.220 | 53 | 1.2510 | 36.89 |
| 6 | 1.0236 | 1.553 | 30 | 1.1300 | 4.481 | 54 | 1.2566 | 42.34 |
| 7 | 1.0277 | 1.603 | 31 | 1.1349 | 4.767 | 55 | 1.2623 | 48.87 |
| 8 | 1.0317 | 1.655 | 32 | 1.1398 | 5.080 | 56 | 1.2681 | 56.75 |
| 9 | 1.0358 | 1.711 | 33 | 1.1448 | 5.424 | 57 | 1.2739 | 66.35 |
| 10 | 1.0400 | 1.771 | 34 | 1.1498 | 5.805 | 58 | 1.2797 | 78.11 |
| 11 | 1.0442 | 1.835 | 35 | 1.1548 | 6.225 | 59 | 1.2855 | 92.65 |
| 12 | 1.0484 | 1.902 | 36 | 1.1598 | 6.692 | 60 | 1.2914 | 110.8 |
| 13 | 1.0526 | 1.974 | 37 | 1.1649 | 7.211 | 61 | 1.2973 | 133.6 |
| 14 | 1.0569 | 2.051 | 38 | 1.1700 | 7.790 | 62 | 1.3033 | 162.7 |
| 15 | 1.0612 | 2.134 | 39 | 1.1752 | 8.438 | 63 | 1.3093 | 200.0 |
| 16 | 1.0655 | 2.222 | 40 | 1.1803 | 9.167 | 64 | 1.3153 | 248.6 |
| 17 | 1.0699 | 2.316 | 41 | 1.1856 | 9.988 | 65 | 1.3214 | 312.5 |
| 18 | 1.0743 | 2.417 | 42 | 1.1908 | 10.92 | 66 | 1.3275 | 397.7 |
| 19 | 1.0788 | 2.525 | 43 | 1.1961 | 11.97 | 67 | 1.3336 | 512.9 |
| 20 | 1.0833 | 2.642 | 44 | 1.2014 | 13.17 | 68 | 1.3398 | 671.1 |
| 21 | 1.0878 | 2.767 | 45 | 1.2068 | 14.54 | 69 | 1.3460 | 891.7 |
| 22 | 1.0924 | 2.903 | 46 | 1.2122 | 16.11 | 70 | 1.3522 | 1205. |
| 23 | 1.0969 | 3.049 | 47 | 1.2176 | 17.92 | | | |

**15°C**

| Sucrose, % | Density,[a] g/ml | Viscosity,[b] cp | Sucrose, % | Density,[a] g/ml | Viscosity,[b] cp | Sucrose, % | Density,[a] g/ml | Viscosity,[b] cp |
|---|---|---|---|---|---|---|---|---|
| 0 | 0.9996 | 1.140 | 24 | 1.1002 | 2.719 | 48 | 1.2211 | 15.65 |
| 1 | 1.0034 | 1.170 | 25 | 1.1048 | 2.859 | 49 | 1.2265 | 17.44 |
| 2 | 1.0073 | 1.202 | 26 | 1.1095 | 3.010 | 50 | 1.2320 | 19.52 |
| 3 | 1.0111 | 1.235 | 27 | 1.1142 | 3.174 | 51 | 1.2376 | 21.93 |
| 4 | 1.0150 | 1.271 | 28 | 1.1189 | 3.352 | 52 | 1.2432 | 24.75 |
| 5 | 1.0189 | 1.308 | 29 | 1.1237 | 3.546 | 53 | 1.2488 | 28.06 |
| 6 | 1.0229 | 1.348 | 30 | 1.1285 | 3.757 | 54 | 1.2544 | 31.98 |
| 7 | 1.0269 | 1.390 | 31 | 1.1334 | 3.987 | 55 | 1.2601 | 36.64 |
| 8 | 1.0309 | 1.434 | 32 | 1.1382 | 4.239 | 56 | 1.2658 | 42.22 |
| 9 | 1.0350 | 1.481 | 33 | 1.1432 | 4.515 | 57 | 1.2716 | 48.95 |
| 10 | 1.0391 | 1.531 | 34 | 1.1481 | 4.818 | 58 | 1.2774 | 57.12 |
| 11 | 1.0432 | 1.584 | 35 | 1.1531 | 5.153 | 59 | 1.2832 | 67.12 |
| 12 | 1.0474 | 1.640 | 36 | 1.1581 | 5.522 | 60 | 1.2891 | 79.48 |
| 13 | 1.0516 | 1.701 | 37 | 1.1632 | 5.932 | 61 | 1.2950 | 94.85 |
| 14 | 1.0558 | 1.765 | 38 | 1.1682 | 6.386 | 62 | 1.3009 | 114.2 |
| 15 | 1.0601 | 1.833 | 39 | 1.1734 | 6.894 | 63 | 1.3069 | 138.7 |
| 16 | 1.0644 | 1.906 | 40 | 1.1785 | 7.461 | 64 | 1.3129 | 170.2 |
| 17 | 1.0688 | 1.985 | 41 | 1.1837 | 8.097 | 65 | 1.3189 | 211.0 |
| 18 | 1.0732 | 2.068 | 42 | 1.1889 | 8.813 | 66 | 1.3250 | 264.6 |
| 19 | 1.0776 | 2.158 | 43 | 1.1942 | 9.621 | 67 | 1.3311 | 336.0 |
| 20 | 1.0820 | 2.255 | 44 | 1.1995 | 10.54 | 68 | 1.3373 | 432.2 |
| 21 | 1.0865 | 2.358 | 45 | 1.2048 | 11.58 | 69 | 1.3434 | 563.8 |
| 22 | 1.0910 | 2.470 | 46 | 1.2102 | 12.77 | 70 | 1.3497 | 746.7 |
| 23 | 1.0956 | 2.590 | 47 | 1.2156 | 14.13 | | | |

| Sucrose, % | Density,[a] g/ml | Viscosity,[b] cp | Sucrose, % | Density,[a] g/ml | Viscosity,[b] cp | Sucrose, % | Density,[a] g/ml | Viscosity,[b] cp |
|---|---|---|---|---|---|---|---|---|
| **20°C** | | | | | | | | |
| 0 | 0.9988 | 1.004 | 24 | 1.0987 | 2.333 | 48 | 1.2189 | 12.50 |
| 1 | 1.0026 | 1.030 | 25 | 1.1033 | 2.449 | 49 | 1.2244 | 13.86 |
| 2 | 1.0064 | 1.057 | 26 | 1.1079 | 2.575 | 50 | 1.2299 | 15.42 |
| 3 | 1.0102 | 1.086 | 27 | 1.1126 | 2.710 | 51 | 1.2354 | 17.23 |
| 4 | 1.0140 | 1.116 | 28 | 1.1173 | 2.857 | 52 | 1.2409 | 19.33 |
| 5 | 1.0179 | 1.148 | 29 | 1.1220 | 3.016 | 53 | 1.2465 | 21.79 |
| 6 | 1.0219 | 1.181 | 30 | 1.1268 | 3.189 | 54 | 1.2522 | 24.67 |
| 7 | 1.0258 | 1.217 | 31 | 1.1316 | 3.378 | 55 | 1.2578 | 28.07 |
| 8 | 1.0298 | 1.255 | 32 | 1.1365 | 3.583 | 56 | 1.2635 | 32.11 |
| 9 | 1.0339 | 1.295 | 33 | 1.1414 | 3.808 | 57 | 1.2693 | 36.96 |
| 10 | 1.0380 | 1.337 | 34 | 1.1463 | 4.053 | 58 | 1.2750 | 42.78 |
| 11 | 1.0421 | 1.382 | 35 | 1.1513 | 4.323 | 59 | 1.2808 | 49.85 |
| 12 | 1.0462 | 1.430 | 36 | 1.1563 | 4.621 | 60 | 1.2867 | 58.50 |
| 13 | 1.0504 | 1.480 | 37 | 1.1613 | 4.948 | 61 | 1.2926 | 69.15 |
| 14 | 1.0546 | 1.535 | 38 | 1.1663 | 5.311 | 62 | 1.2985 | 82.39 |
| 15 | 1.0588 | 1.592 | 39 | 1.1714 | 5.714 | 63 | 1.3044 | 99.01 |
| 16 | 1.0631 | 1.654 | 40 | 1.1766 | 6.163 | 64 | 1.3104 | 120.1 |
| 17 | 1.0674 | 1.720 | 41 | 1.1817 | 6.664 | 65 | 1.3164 | 147.0 |
| 18 | 1.0718 | 1.790 | 42 | 1.1870 | 7.226 | 66 | 1.3225 | 182.0 |
| 19 | 1.0762 | 1.865 | 43 | 1.1922 | 7.857 | 67 | 1.3286 | 227.8 |
| 20 | 1.0806 | 1.946 | 44 | 1.1975 | 8.570 | 68 | 1.3347 | 288.5 |
| 21 | 1.0851 | 2.032 | 45 | 1.2028 | 9.376 | 69 | 1.3408 | 370.2 |
| 22 | 1.0896 | 2.125 | 46 | 1.2081 | 10.29 | 70 | 1.3470 | 481.8 |
| 23 | 1.0941 | 2.225 | 47 | 1.2135 | 11.33 | | | |
| **25°C** | | | | | | | | |
| 0 | 0.9977 | 0.8913 | 24 | 1.0970 | 2.023 | 48 | 1.2167 | 10.14 |
| 1 | 1.0014 | 0.9139 | 25 | 1.1016 | 2.121 | 49 | 1.2221 | 11.19 |
| 2 | 1.0052 | 0.9376 | 26 | 1.1062 | 2.226 | 50 | 1.2276 | 12.39 |
| 3 | 1.0090 | 0.9625 | 27 | 1.1108 | 2.339 | 51 | 1.2331 | 13.77 |
| 4 | 1.0128 | 0.9886 | 28 | 1.1155 | 2.462 | 52 | 1.2386 | 15.37 |
| 5 | 1.0167 | 1.016 | 29 | 1.1202 | 2.595 | 53 | 1.2442 | 17.22 |
| 6 | 1.0206 | 1.045 | 30 | 1.1250 | 2.739 | 54 | 1.2498 | 19.39 |
| 7 | 1.0246 | 1.076 | 31 | 1.1298 | 2.895 | 55 | 1.2554 | 21.93 |
| 8 | 1.0285 | 1.108 | 32 | 1.1346 | 3.064 | 56 | 1.2611 | 24.92 |
| 9 | 1.0325 | 1.142 | 33 | 1.1395 | 3.249 | 57 | 1.2668 | 28.48 |
| 10 | 1.0366 | 1.179 | 34 | 1.1444 | 3.451 | 58 | 1.2726 | 32.73 |
| 11 | 1.0407 | 1.217 | 35 | 1.1493 | 3.672 | 59 | 1.2784 | 37.85 |
| 12 | 1.0448 | 1.258 | 36 | 1.1543 | 3.914 | 60 | 1.2842 | 44.04 |
| 13 | 1.0489 | 1.301 | 37 | 1.1593 | 4.181 | 61 | 1.2901 | 51.61 |
| 14 | 1.0531 | 1.347 | 38 | 1.1643 | 4.475 | 62 | 1.2959 | 60.93 |
| 15 | 1.0574 | 1.396 | 39 | 1.1694 | 4.799 | 63 | 1.3019 | 72.50 |
| 16 | 1.0616 | 1.449 | 40 | 1.1745 | 5.160 | 64 | 1.3078 | 86.99 |
| 17 | 1.0659 | 1.505 | 41 | 1.1797 | 5.560 | 65 | 1.3138 | 105.3 |
| 18 | 1.0702 | 1.564 | 42 | 1.1848 | 6.008 | 66 | 1.3199 | 128.8 |
| 19 | 1.0746 | 1.628 | 43 | 1.1901 | 6.509 | 67 | 1.3259 | 159.1 |
| 20 | 1.0790 | 1.696 | 44 | 1.1953 | 7.072 | 68 | 1.3320 | 198.8 |
| 21 | 1.0835 | 1.770 | 45 | 1.2006 | 7.706 | 69 | 1.3382 | 251.4 |
| 22 | 1.0879 | 1.848 | 46 | 1.2059 | 8.423 | 70 | 1.3444 | 322.0 |
| 23 | 1.0924 | 1.933 | 47 | 1.2113 | 9.236 | | | |

30°C

| Sucrose, % | Density,[a] g/ml | Viscosity,[b] cp | Sucrose, % | Density,[a] g/ml | Viscosity,[b] cp | Sucrose, % | Density,[a] g/ml | Viscosity,[b] cp |
|---|---|---|---|---|---|---|---|---|
| 0 | 0.9963 | 0.7978 | 24 | 1.0951 | 1.771 | 48 | 1.2144 | 8.344 |
| 1 | 1.0000 | 0.8176 | 25 | 1.0997 | 1.854 | 49 | 1.2198 | 9.168 |
| 2 | 1.0038 | 0.8384 | 26 | 1.1043 | 1.943 | 50 | 1.2252 | 10.10 |
| 3 | 1.0075 | 0.8601 | 27 | 1.1089 | 2.039 | 51 | 1.2307 | 11.18 |
| 4 | 1.0113 | 0.8830 | 28 | 1.1136 | 2.143 | 52 | 1.2362 | 12.42 |
| 5 | 1.0152 | 0.9069 | 29 | 1.1183 | 2.255 | 53 | 1.2418 | 13.84 |
| 6 | 1.0191 | 0.9322 | 30 | 1.1230 | 2.376 | 54 | 1.2474 | 15.50 |
| 7 | 1.0230 | 0.9588 | 31 | 1.1278 | 2.506 | 55 | 1.2530 | 17.43 |
| 8 | 1.0270 | 0.9868 | 32 | 1.1326 | 2.648 | 56 | 1.2586 | 19.69 |
| 9 | 1.0310 | 1.016 | 33 | 1.1374 | 2.802 | 57 | 1.2643 | 22.36 |
| 10 | 1.0350 | 1.048 | 34 | 1.1423 | 2.970 | 58 | 1.2701 | 25.52 |
| 11 | 1.0391 | 1.081 | 35 | 1.1472 | 3.153 | 59 | 1.2758 | 29.30 |
| 12 | 1.0432 | 1.116 | 36 | 1.1522 | 3.353 | 60 | 1.2816 | 33.84 |
| 13 | 1.0473 | 1.154 | 37 | 1.1572 | 3.572 | 61 | 1.2875 | 39.34 |
| 14 | 1.0515 | 1.193 | 38 | 1.1622 | 3.813 | 62 | 1.2933 | 46.05 |
| 15 | 1.0557 | 1.235 | 39 | 1.1672 | 4.079 | 63 | 1.2992 | 54.30 |
| 16 | 1.0599 | 1.280 | 40 | 1.1723 | 4.372 | 64 | 1.3052 | 64.53 |
| 17 | 1.0642 | 1.328 | 41 | 1.1775 | 4.697 | 65 | 1.3112 | 77.35 |
| 18 | 1.0685 | 1.380 | 42 | 1.1826 | 5.058 | 66 | 1.3172 | 93.54 |
| 19 | 1.0728 | 1.434 | 43 | 1.1878 | 5.461 | 67 | 1.3232 | 114.2 |
| 20 | 1.0772 | 1.493 | 44 | 1.1931 | 5.912 | 68 | 1.3293 | 140.9 |
| 21 | 1.0816 | 1.555 | 45 | 1.1983 | 6.418 | 69 | 1.3354 | 175.8 |
| 22 | 1.0861 | 1.622 | 46 | 1.2036 | 6.988 | 70 | 1.3416 | 222.0 |
| 23 | 1.0906 | 1.694 | 47 | 1.2090 | 7.632 | | | |

[a] Original data were stated to a precision of about 1 part in 10,000; maximum deviation from the original data is 7 parts in 10,000.

[b] Precision of the original data was between 1 part in 1,000 and 1 part in 10,000; maximum deviation from the original data is 4 parts in 1,000 in the range covered by this table.

Data are based on equations developed by Barber, E. J., *J. Nat. Cancer Inst. Monograph, 21,* 219 (1966). Compiled by Anderson, N. G., *Handbook of Biochemistry,* 2nd ed., pp. J-288–J-291, Sober, H. A., ed., Chemical Rubber Co., Cleveland, Ohio (1970). Reproduced with the author's permission.

# DENSITY OF CESIUM CHLORIDE SOLUTIONS AS A FUNCTION OF REFRACTIVE INDEX AT 25°C[a]

| Refractive Index, Sodium D Line | Density, g/cc | Refractive Index, Sodium D Line | Density, g/cc | Refractive Index, Sodium D Line | Density, g/cc |
|---|---|---|---|---|---|
| 1.34400 | 1.09857 | 1.35020 | 1.16591 | 1.35640 | 1.23324 |
| 1.34410 | 1.09966 | 1.35030 | 1.16699 | 1.35650 | 1.23433 |
| 1.34420 | 1.10075 | 1.35040 | 1.16808 | 1.35660 | 1.23541 |
| 1.34430 | 1.10183 | 1.35050 | 1.16917 | 1.35670 | 1.23650 |
| 1.34440 | 1.10292 | 1.35060 | 1.17025 | 1.35680 | 1.23758 |
| 1.34450 | 1.10400 | 1.35070 | 1.17134 | 1.35690 | 1.23867 |
| 1.34460 | 1.10509 | 1.35080 | 1.17242 | 1.35700 | 1.23976 |
| 1.34470 | 1.10618 | 1.35090 | 1.17351 | 1.35710 | 1.24084 |
| 1.34480 | 1.10726 | 1.35100 | 1.17460 | 1.35720 | 1.24193 |
| 1.34490 | 1.10835 | 1.35110 | 1.17568 | 1.35730 | 1.24301 |
| 1.34500 | 1.10943 | 1.35120 | 1.17677 | 1.35740 | 1.24410 |
| 1.34510 | 1.11052 | 1.35130 | 1.17785 | 1.35750 | 1.24519 |
| 1.34520 | 1.11161 | 1.35140 | 1.17894 | 1.35760 | 1.24627 |
| 1.34530 | 1.11269 | 1.35150 | 1.18003 | 1.35770 | 1.24736 |
| 1.34540 | 1.11378 | 1.35160 | 1.18111 | 1.35780 | 1.24844 |
| 1.34550 | 1.11486 | 1.35170 | 1.18220 | 1.35790 | 1.24953 |
| 1.34560 | 1.11595 | 1.35180 | 1.18328 | 1.35800 | 1.25062 |
| 1.34570 | 1.11704 | 1.35190 | 1.18437 | 1.35810 | 1.25170 |
| 1.34580 | 1.11812 | 1.35200 | 1.18546 | 1.35820 | 1.25279 |
| 1.34590 | 1.11921 | 1.35210 | 1.18654 | 1.35830 | 1.25387 |
| 1.34600 | 1.12029 | 1.35220 | 1.18763 | 1.35840 | 1.25496 |
| 1.34610 | 1.12138 | 1.35230 | 1.18871 | 1.35850 | 1.25605 |
| 1.34620 | 1.12247 | 1.35240 | 1.18980 | 1.35860 | 1.25713 |
| 1.34630 | 1.12355 | 1.35250 | 1.19089 | 1.35870 | 1.25822 |
| 1.34640 | 1.12464 | 1.35260 | 1.19197 | 1.35880 | 1.25930 |
| 1.34650 | 1.12572 | 1.35270 | 1.19306 | 1.35890 | 1.26039 |
| 1.34660 | 1.12681 | 1.35280 | 1.19414 | 1.35900 | 1.26148 |
| 1.34670 | 1.12790 | 1.35290 | 1.19523 | 1.35910 | 1.26256 |
| 1.34680 | 1.12898 | 1.35300 | 1.19632 | 1.35920 | 1.26365 |
| 1.34690 | 1.13007 | 1.35310 | 1.19740 | 1.35930 | 1.26473 |
| 1.34700 | 1.13115 | 1.35320 | 1.19849 | 1.35940 | 1.26582 |
| 1.34710 | 1.13224 | 1.35330 | 1.19957 | 1.35950 | 1.26691 |
| 1.34720 | 1.13333 | 1.35340 | 1.20066 | 1.35960 | 1.26799 |
| 1.34730 | 1.13441 | 1.35350 | 1.20175 | 1.35970 | 1.26908 |
| 1.34740 | 1.13550 | 1.35360 | 1.20283 | 1.35980 | 1.27016 |
| 1.34750 | 1.13658 | 1.35370 | 1.20392 | 1.35990 | 1.27125 |
| 1.34760 | 1.13767 | 1.35380 | 1.20500 | 1.36000 | 1.27234 |
| 1.34770 | 1.13876 | 1.35390 | 1.20609 | 1.36010 | 1.27342 |
| 1.34780 | 1.13984 | 1.35400 | 1.20718 | 1.36020 | 1.27451 |
| 1.34790 | 1.14093 | 1.35410 | 1.20826 | 1.36030 | 1.27559 |
| 1.34800 | 1.14201 | 1.35420 | 1.20935 | 1.36040 | 1.27668 |
| 1.34810 | 1.14310 | 1.35430 | 1.21043 | 1.36050 | 1.27777 |
| 1.34820 | 1.14419 | 1.35440 | 1.21152 | 1.36060 | 1.27885 |
| 1.34830 | 1.14527 | 1.35450 | 1.21261 | 1.36070 | 1.27994 |
| 1.34840 | 1.14636 | 1.35460 | 1.21369 | 1.36080 | 1.28102 |
| 1.34850 | 1.14744 | 1.35470 | 1.21478 | 1.36090 | 1.28211 |
| 1.34860 | 1.14853 | 1.35480 | 1.21586 | 1.36100 | 1.28320 |
| 1.34870 | 1.14962 | 1.35490 | 1.21695 | 1.36110 | 1.28428 |
| 1.34880 | 1.15070 | 1.35500 | 1.21804 | 1.36120 | 1.28537 |
| 1.34890 | 1.15179 | 1.35510 | 1.21912 | 1.36130 | 1.28645 |
| 1.34900 | 1.15287 | 1.35520 | 1.22021 | 1.36140 | 1.28754 |
| 1.34910 | 1.15396 | 1.35530 | 1.22129 | 1.36150 | 1.28863 |
| 1.34920 | 1.15505 | 1.35540 | 1.22238 | 1.36160 | 1.28971 |
| 1.34930 | 1.15613 | 1.35550 | 1.22347 | 1.36170 | 1.29080 |
| 1.34940 | 1.15722 | 1.35560 | 1.22455 | 1.36180 | 1.29188 |
| 1.34950 | 1.15830 | 1.35570 | 1.22564 | 1.36190 | 1.29297 |
| 1.34960 | 1.15939 | 1.35580 | 1.22672 | 1.36200 | 1.29406 |
| 1.34970 | 1.16048 | 1.35590 | 1.22781 | 1.36210 | 1.29514 |
| 1.34980 | 1.16156 | 1.35600 | 1.22890 | 1.36220 | 1.29623 |
| 1.34990 | 1.16265 | 1.35610 | 1.22998 | 1.36230 | 1.29731 |
| 1.35000 | 1.16373 | 1.35620 | 1.23107 | 1.36240 | 1.29840 |
| 1.35010 | 1.16482 | 1.35630 | 1.23215 | 1.36250 | 1.2994[illegible] |

| Refractive Index, Sodium D Line | Density, g/cc |
|---|---|
| 1.36260 | 1.30057 |
| 1.36270 | 1.30166 |
| 1.36280 | 1.30274 |
| 1.36290 | 1.30383 |
| 1.36300 | 1.30492 |
| 1.36310 | 1.30600 |
| 1.36320 | 1.30709 |
| 1.36330 | 1.30817 |
| 1.36340 | 1.30926 |
| 1.36350 | 1.31035 |
| 1.36360 | 1.31143 |
| 1.36370 | 1.31252 |
| 1.36380 | 1.31360 |
| 1.36390 | 1.31469 |
| 1.36400 | 1.31578 |
| 1.36410 | 1.31686 |
| 1.36420 | 1.31795 |
| 1.36430 | 1.31903 |
| 1.36440 | 1.32012 |
| 1.36450 | 1.32121 |
| 1.36460 | 1.32229 |
| 1.36470 | 1.32338 |
| 1.36480 | 1.32446 |
| 1.36490 | 1.32555 |
| 1.36500 | 1.32664 |
| 1.36510 | 1.32772 |
| 1.36520 | 1.32881 |
| 1.36530 | 1.32989 |
| 1.36540 | 1.33098 |
| 1.36550 | 1.33207 |
| 1.36560 | 1.33315 |
| 1.36570 | 1.33424 |
| 1.36580 | 1.33532 |
| 1.36590 | 1.33641 |
| 1.36600 | 1.33750 |
| 1.36610 | 1.33858 |
| 1.36620 | 1.33967 |
| 1.36630 | 1.34075 |
| 1.36640 | 1.34184 |
| 1.36650 | 1.34293 |
| 1.36660 | 1.34401 |
| 1.36670 | 1.34510 |
| 1.36680 | 1.34618 |
| 1.36690 | 1.34727 |
| 1.36700 | 1.34836 |
| 1.36710 | 1.34944 |
| 1.36720 | 1.35053 |
| 1.36730 | 1.35161 |
| 1.36740 | 1.35270 |
| 1.36750 | 1.35379 |
| 1.36760 | 1.35487 |
| 1.36770 | 1.35596 |
| 1.36780 | 1.35704 |
| 1.36790 | 1.35813 |
| 1.36800 | 1.35922 |
| 1.36810 | 1.36030 |
| 1.36820 | 1.36139 |
| 1.36830 | 1.36247 |
| 1.36840 | 1.36356 |
| 1.36850 | 1.36465 |
| 1.36860 | 1.36573 |
| 1.36870 | 1.36682 |
| 1.36880 | 1.36790 |
| 1.36890 | 1.36899 |
| 1.36900 | 1.37008 |
| 1.36910 | 1.37116 |
| 1.36920 | 1.37225 |
| 1.36930 | 1.37333 |
| 1.36940 | 1.37442 |
| 1.36950 | 1.37551 |
| 1.36960 | 1.37659 |
| 1.36970 | 1.37768 |
| 1.36980 | 1.37876 |
| 1.36990 | 1.37985 |
| 1.37000 | 1.38094 |
| 1.37010 | 1.38202 |
| 1.37020 | 1.38311 |
| 1.37030 | 1.38420 |
| 1.37040 | 1.38528 |
| 1·37050 | 1.38637 |
| 1.37060 | 1.38745 |
| 1.37070 | 1.38854 |
| 1.37080 | 1.38963 |
| 1.37090 | 1.39071 |
| 1.37100 | 1.39180 |
| 1.37110 | 1.39288 |
| 1.37120 | 1.39397 |
| 1.37130 | 1.39506 |
| 1.37140 | 1.39614 |
| 1.37150 | 1.39723 |
| 1.37160 | 1.39831 |
| 1.37170 | 1.39940 |
| 1.37180 | 1.40049 |
| 1.37190 | 1.40157 |
| 1.37200 | 1.40266 |
| 1.37210 | 1.40374 |
| 1.37220 | 1.40483 |
| 1.37230 | 1.40592 |
| 1.37240 | 1.40700 |
| 1.37250 | 1.40809 |
| 1.37260 | 1.40917 |
| 1.37270 | 1.41026 |
| 1.37280 | 1.41135 |
| 1.37290 | 1.41243 |
| 1.37300 | 1.41352 |
| 1.37310 | 1.41460 |
| 1.37320 | 1.41569 |
| 1.37330 | 1.41678 |
| 1.37340 | 1.41786 |
| 1.37350 | 1.41895 |
| 1.37360 | 1.42003 |
| 1.37370 | 1.42112 |
| 1.37380 | 1.42221 |
| 1.37390 | 1.42329 |
| 1.37400 | 1.42438 |
| 1.37410 | 1.42546 |
| 1.37420 | 1.42655 |
| 1.37430 | 1.42764 |
| 1.37440 | 1.42872 |
| 1.37450 | 1.42981 |
| 1.37460 | 1.43089 |
| 1.37470 | 1.43198 |
| 1.37480 | 1.43307 |
| 1.37490 | 1.43415 |
| 1.37500 | 1.43524 |
| 1.37510 | 1.43632 |
| 1.37520 | 1.43741 |
| 1.37530 | 1.43850 |
| 1.37540 | 1.43958 |
| 1.37550 | 1.44067 |
| 1.37560 | 1.44175 |
| 1.37570 | 1.44284 |
| 1.37580 | 1.44393 |
| 1.37590 | 1.44501 |
| 1.37600 | 1.44610 |
| 1.37610 | 1.44718 |
| 1.37620 | 1.44827 |
| 1.37630 | 1.44936 |
| 1.37640 | 1.45044 |
| 1.37650 | 1.45153 |
| 1.37660 | 1.45261 |
| 1.37670 | 1.45370 |
| 1.37680 | 1.45479 |
| 1.37690 | 1.45587 |
| 1.37700 | 1.45696 |
| 1.37710 | 1.45804 |
| 1.37720 | 1.45913 |
| 1.37730 | 1.46022 |
| 1.37740 | 1.46130 |
| 1.37750 | 1.46239 |
| 1.37760 | 1.46347 |
| 1.37770 | 1.46456 |
| 1.37780 | 1.46565 |
| 1.37790 | 1.46673 |
| 1.37800 | 1.46782 |
| 1.37810 | 1.46890 |
| 1.37820 | 1.46999 |
| 1.37830 | 1.47108 |
| 1.37840 | 1.47216 |
| 1.37850 | 1.47325 |
| 1.37860 | 1.47433 |
| 1.37870 | 1.47542 |
| 1.37880 | 1.47651 |
| 1.37890 | 1.47759 |
| 1.37900 | 1.47868 |
| 1.37910 | 1.47976 |
| 1.37920 | 1.48085 |
| 1.37930 | 1.48194 |
| 1.37940 | 1.48302 |
| 1.37950 | 1.48411 |
| 1.37960 | 1.48519 |
| 1.37970 | 1.48628 |
| 1.37980 | 1.48737 |
| 1.37990 | 1.48845 |
| 1.38000 | 1.48954 |
| 1.38010 | 1.49062 |
| 1.38020 | 1.49171 |
| 1.38030 | 1.49280 |
| 1.38040 | 1.49388 |
| 1.38050 | 1.49497 |
| 1.38060 | 1.49605 |
| 1.38070 | 1.49714 |
| 1.38080 | 1.49823 |
| 1.38090 | 1.49931 |
| 1.38100 | 1.50040 |
| 1.38110 | 1.50148 |
| 1.38120 | 1.50257 |
| 1.38130 | 1.50366 |
| 1.38140 | 1.50474 |
| 1.38150 | 1.50583 |
| 1.38160 | 1.50691 |
| 1.38170 | 1.50800 |
| 1.38180 | 1.50909 |
| 1.38190 | 1.51017 |

| Refractive Index, Sodium D Line | Density, g/cc | Refractive Index, Sodium D Line | Density, g/cc | Refractive Index, Sodium D Line | Density, g/cc |
|---|---|---|---|---|---|
| 1.38200 | 1.51126 | 1.38850 | 1.58185 | 1.39500 | 1.65244 |
| 1.38210 | 1.51234 | 1.38860 | 1.58293 | 1.39510 | 1.65353 |
| 1.38220 | 1.51343 | 1.38870 | 1.58402 | 1.39520 | 1.65461 |
| 1.38230 | 1.51452 | 1.38880 | 1.58511 | 1.39530 | 1.65570 |
| 1.38240 | 1.51560 | 1.38890 | 1.58619 | 1.39540 | 1.65678 |
| 1.38250 | 1.51669 | 1.38900 | 1.58728 | 1.39550 | 1.65787 |
| 1.38260 | 1.51777 | 1.38910 | 1.58836 | 1.39560 | 1.65896 |
| 1.38270 | 1.51886 | 1.38920 | 1.58945 | 1.39570 | 1.66004 |
| 1.38280 | 1.51995 | 1.38930 | 1.59054 | 1.39580 | 1.66113 |
| 1.38290 | 1.52103 | 1.38940 | 1.59162 | 1.39590 | 1.66221 |
| 1.38300 | 1.52212 | 1.38950 | 1.59271 | 1.39600 | 1.66330 |
| 1.38310 | 1.52320 | 1.38960 | 1.59379 | 1.39610 | 1.66439 |
| 1.38320 | 1.52429 | 1.38970 | 1.59488 | 1.39620 | 1.66547 |
| 1.38330 | 1.52538 | 1.38980 | 1.59597 | 1.39630 | 1.66656 |
| 1.38340 | 1.52646 | 1.38990 | 1.59705 | 1.39640 | 1.66764 |
| 1.38350 | 1.52755 | 1.39000 | 1.59814 | 1·39650 | 1.66873 |
| 1.38360 | 1.52863 | 1.39010 | 1.59923 | 1.39660 | 1.66982 |
| 1.38370 | 1.52972 | 1.39020 | 1.60031 | 1.39670 | 1.67090 |
| 1.38380 | 1.53081 | 1.39030 | 1.60140 | 1.39680 | 1.67199 |
| 1.38390 | 1.53189 | 1.39040 | 1.60248 | 1.39690 | 1.67307 |
| 1.38400 | 1.53298 | 1.39050 | 1.60357 | 1.39700 | 1.67416 |
| 1.38410 | 1.53406 | 1.39060 | 1.60466 | 1.39710 | 1.67525 |
| 1.38420 | 1.53515 | 1.39070 | 1.60574 | 1.39720 | 1.67633 |
| 1.38430 | 1.53624 | 1.39080 | 1.60683 | 1.39730 | 1.67742 |
| 1.38440 | 1.53732 | 1.39090 | 1.60791 | 1.39740 | 1.67850 |
| 1.38450 | 1.53841 | 1.39100 | 1.60900 | 1.39750 | 1.67959 |
| 1.38460 | 1.53949 | 1.39110 | 1.61009 | 1.39760 | 1.68068 |
| 1.38470 | 1.54058 | 1.39120 | 1.61117 | 1.39770 | 1.68176 |
| 1.38480 | 1.54167 | 1.39130 | 1.61226 | 1.39780 | 1.68285 |
| 1.38490 | 1.54275 | 1.39140 | 1.61334 | 1.39790 | 1.68393 |
| 1.38500 | 1.54384 | 1.39150 | 1.61443 | 1.39800 | 1.68502 |
| 1.38510 | 1.54492 | 1.39160 | 1.61552 | 1.39810 | 1.68611 |
| 1.38520 | 1.54601 | 1.39170 | 1.61660 | 1.39820 | 1.68719 |
| 1.38530 | 1.54710 | 1.39180 | 1.61769 | 1.39830 | 1.68828 |
| 1.38540 | 1.54818 | 1.39190 | 1.61877 | 1.39840 | 1.68936 |
| 1.38550 | 1.54927 | 1.39200 | 1.61986 | 1.39850 | 1.69045 |
| 1.38560 | 1.55035 | 1.39210 | 1.62095 | 1.39860 | 1.69154 |
| 1.38570 | 1.55144 | 1.39220 | 1.62203 | 1.39870 | 1.69262 |
| 1.38580 | 1.55253 | 1.39230 | 1.62312 | 1.39880 | 1.69371 |
| 1.38590 | 1.55361 | 1.39240 | 1.62420 | 1.39890 | 1.69479 |
| 1.38600 | 1.55470 | 1.39250 | 1.62529 | 1.39900 | 1.69588 |
| 1.38610 | 1.55578 | 1.39260 | 1.62638 | 1.39910 | 1.69697 |
| 1.38620 | 1.55687 | 1.39270 | 1.62746 | 1.39920 | 1.69805 |
| 1.38630 | 1.55796 | 1.39280 | 1.62855 | 1.39930 | 1.69914 |
| 1.38640 | 1.55904 | 1.39290 | 1.62963 | 1.39940 | 1.70022 |
| 1.38650 | 1.56013 | 1.39300 | 1.63072 | 1.39950 | 1.70131 |
| 1.38660 | 1.56121 | 1.39310 | 1.63181 | 1.39960 | 1.70240 |
| 1.38670 | 1.56230 | 1.39320 | 1.63289 | 1.39970 | 1.70348 |
| 1.38680 | 1.56339 | 1.39330 | 1.63398 | 1.39980 | 1.70457 |
| 1.38690 | 1.56447 | 1.39340 | 1.63506 | 1.39990 | 1.70565 |
| 1.38700 | 1.56556 | 1.39350 | 1.63615 | 1.40000 | 1.70674 |
| 1.38710 | 1.56664 | 1.39360 | 1.63724 | 1.40010 | 1.70783 |
| 1.38720 | 1.56773 | 1.39370 | 1.63832 | 1.40020 | 1.70891 |
| 1.38730 | 1.56882 | 1.39380 | 1.63941 | 1.40030 | 1.71000 |
| 1.38740 | 1.56990 | 1.39390 | 1.64049 | 1.40040 | 1.71108 |
| 1.38750 | 1.57099 | 1.39400 | 1.64158 | 1.40050 | 1.71217 |
| 1.38760 | 1.57207 | 1.39410 | 1.64267 | 1.40060 | 1.71326 |
| 1.38770 | 1.57316 | 1.39420 | 1.64375 | 1.40070 | 1.71434 |
| 1.38780 | 1.57425 | 1.39430 | 1.64484 | 1.40080 | 1.71543 |
| 1.38790 | 1.57533 | 1.39440 | 1.64592 | 1.40090 | 1.71651 |
| 1.38800 | 1.57642 | 1.39450 | 1.64701 | 1.40100 | 1.71760 |
| 1.38810 | 1.57750 | 1.39460 | 1.64810 | 1.40110 | 1.71869 |
| 1.38820 | 1.57859 | 1.39470 | 1.64918 | 1.40120 | 1.71977 |
| 1.38830 | 1.57968 | 1.39480 | 1.65027 | 1.40130 | 1.72086 |
| 1.38840 | 1.58076 | 1.39490 | 1.65135 | 1.40140 | 1.72194 |

| Refractive Index, Sodium D Line | Density, g/cc | Refractive Index, Sodium D Line | Density, g/cc | Refractive Index, Sodium D Line | Density, g/cc |
|---|---|---|---|---|---|
| 1.40150 | 1.72303 | 1.40800 | 1.79362 | 1.41450 | 1.86421 |
| 1.40160 | 1.72412 | 1.40810 | 1.79471 | 1.41460 | 1.86530 |
| 1.40170 | 1.72520 | 1.40820 | 1.79579 | 1.41470 | 1.86638 |
| 1.40180 | 1.72629 | 1.40830 | 1.79688 | 1.41480 | 1.86747 |
| 1.40190 | 1.72737 | 1.40840 | 1.79796 | 1.41490 | 1.86856 |
| 1.40200 | 1.72846 | 1.40850 | 1.79905 | 1.41500 | 1.86964 |
| 1.40210 | 1.72955 | 1.40860 | 1.80014 | 1.41510 | 1.87073 |
| 1.40220 | 1.73063 | 1.40870 | 1.80122 | 1.41520 | 1.87181 |
| 1.40230 | 1.73172 | 1.40880 | 1.80231 | 1.41530 | 1.87290 |
| 1.40240 | 1.73280 | 1.40890 | 1.80339 | 1.41540 | 1.87399 |
| 1.40250 | 1.73389 | 1.40900 | 1.80448 | 1.41550 | 1.87507 |
| 1.40260 | 1.73498 | 1.40910 | 1.80557 | 1.41560 | 1.87616 |
| 1.40270 | 1.73606 | 1.40920 | 1.80665 | 1.41570 | 1.87724 |
| 1.40280 | 1.73715 | 1.40930 | 1.80774 | 1.41580 | 1.87833 |
| 1.40290 | 1.73823 | 1.40940 | 1.80882 | 1.41590 | 1.87942 |
| 1.40300 | 1.73932 | 1.40950 | 1.80991 | 1.41600 | 1.88050 |
| 1.40310 | 1.74041 | 1.40960 | 1.81100 | 1.41610 | 1.88159 |
| 1.40320 | 1.74149 | 1.40970 | 1.81208 | 1.41620 | 1.88267 |
| 1.40330 | 1.74258 | 1.40980 | 1.81317 | 1.41630 | 1.88376 |
| 1.40340 | 1.74366 | 1.40990 | 1.81425 | 1.41640 | 1.88485 |
| 1.40350 | 1.74475 | 1.41000 | 1.81534 | 1.41650 | 1.88593 |
| 1.40360 | 1.74584 | 1.41010 | 1.81643 | 1.41660 | 1.88702 |
| 1.40370 | 1.74692 | 1.41020 | 1.81751 | 1.41670 | 1.88810 |
| 1.40380 | 1.74801 | 1.41030 | 1.81860 | 1.41680 | 1.88919 |
| 1.40390 | 1.74909 | 1.41040 | 1.81969 | 1.41690 | 1.89028 |
| 1.40400 | 1.75018 | 1.41050 | 1.82077 | 1.41700 | 1.89136 |
| 1.40410 | 1.75127 | 1.41060 | 1.82186 | 1.41710 | 1.89245 |
| 1.40420 | 1.75235 | 1.41070 | 1.82294 | 1.41720 | 1.89353 |
| 1.40430 | 1.75344 | 1.41080 | 1.82403 | 1.41730 | 1.89462 |
| 1.40440 | 1.75452 | 1.41090 | 1.82512 | 1.41740 | 1.89571 |
| 1.40450 | 1.75561 | 1.41100 | 1.82620 | 1.41750 | 1.89679 |
| 1.40460 | 1.75670 | 1.41110 | 1.82729 | 1.41760 | 1.89788 |
| 1.40470 | 1.75778 | 1.41120 | 1.82837 | 1.41770 | 1.89896 |
| 1.40480 | 1.75887 | 1.41130 | 1.82946 | 1.41780 | 1.90005 |
| 1.40490 | 1.75995 | 1.41140 | 1.83055 | 1.41790 | 1.90114 |
| 1.40500 | 1.76104 | 1.41150 | 1.83163 | 1.41800 | 1.90222 |
| 1.40510 | 1.76213 | 1.41160 | 1.83272 | 1.41810 | 1.90331 |
| 1.40520 | 1.76321 | 1.41170 | 1.83380 | 1.41820 | 1.90439 |
| 1.40530 | 1.76430 | 1.41180 | 1.83489 | 1.41830 | 1.90548 |
| 1.40540 | 1.76538 | 1.41190 | 1.83598 | 1.41840 | 1.90657 |
| 1.40550 | 1.76647 | 1.41200 | 1.83706 | 1.41850 | 1.90765 |
| 1.40560 | 1.76756 | 1.41210 | 1.83815 | 1.41860 | 1.90874 |
| 1.40570 | 1.76864 | 1.41220 | 1.83923 | 1.41870 | 1.90982 |
| 1.40580 | 1.76973 | 1.41230 | 1.84032 | 1.41880 | 1.91091 |
| 1.40590 | 1.77081 | 1.41240 | 1.84141 | 1.41890 | 1.91200 |
| 1.40600 | 1.77190 | 1.41250 | 1.84249 | 1.41900 | 1.91308 |
| 1.40610 | 1.77299 | 1.41260 | 1.84358 | 1.41910 | 1.91417 |
| 1.40620 | 1.77407 | 1.41270 | 1.84466 | 1.41920 | 1.91525 |
| 1.40630 | 1.77516 | 1.41280 | 1.84575 | 1.41930 | 1.91634 |
| 1.40640 | 1.77624 | 1.41290 | 1.84684 | 1.41940 | 1.91743 |
| 1.40650 | 1.77733 | 1.41300 | 1.84792 | 1.41950 | 1.91851 |
| 1.40660 | 1.77842 | 1.41310 | 1.84901 | 1.41960 | 1.91960 |
| 1.40670 | 1.77950 | 1.41320 | 1.85009 | 1.41970 | 1.92068 |
| 1.40680 | 1.78059 | 1.41330 | 1.85118 | 1.41980 | 1.92177 |
| 1.40690 | 1.78167 | 1.41340 | 1.85227 | 1.41990 | 1.92286 |
| 1.40700 | 1.78276 | 1.41350 | 1.85335 | 1.42000 | 1.92394 |
| 1.40710 | 1.78385 | 1.41360 | 1.85444 | 1.42010 | 1.92503 |
| 1.40720 | 1.78493 | 1.41370 | 1.85552 | 1.42020 | 1.92611 |
| 1.40730 | 1.78602 | 1.41380 | 1.85661 | 1.42030 | 1.92720 |
| 1.40740 | 1.78710 | 1.41390 | 1.85770 | 1.42040 | 1.92829 |
| 1.40750 | 1.78819 | 1.41400 | 1.85878 | 1.42050 | 1.92937 |
| 1.40760 | 1.78928 | 1.41410 | 1.85987 | 1.42060 | 1.93046 |
| 1.40770 | 1.79036 | 1.41420 | 1.86095 | 1.42070 | 1.93154 |
| 1.40780 | 1.79145 | 1.41430 | 1.86204 | 1.42080 | 1.93263 |
| 1.40790 | 1.79253 | 1.41440 | 1.86313 | 1.42090 | 1.93372 |

| Refractive Index, Sodium D Line | Density, g/cc | Refractive Index, Sodium D Line | Density, g/cc | Refractive Index, Sodium D Line | Density, g/cc |
|---|---|---|---|---|---|
| 1.42100 | 1.93480 | 1.42120 | 1.93697 | 1.42140 | 1.93915 |
| 1.42110 | 1.93589 | 1.42130 | 1.93806 | | |

[a] Data were calculated from the following equation of Ifft, Voet and Vinograd, *J. Phys. Chem.*, *65*, 1138 (1961):

$$\rho^{25^\circ}C = (10.8601 \times R.I.) - 13.4974.$$

where R.I. is the refractive index (sodium D line, 25°C).

Table compiled by Anderson, N. G., and Anderson, N. L., *Handbook of Biochemistry,* 2nd ed., pp. J-292–J-296, Sober, H. A., ed., The Chemical Rubber Co., Cleveland, Ohio (1970). Reproduced with the authors' permission.

# BUFFER SOLUTIONS, pH INDICATORS, AND FLUORESCENT INDICATORS

## Buffer Solutions

DR. R. A. ROBINSON

### OPERATIONAL DEFINITIONS OF pH

The operational definition of pH is:

$$pH = pH(s) + E/k,$$

where E is the e.m.f. of the cell

$$H_2 \mid \text{Solution, pH} \mid \text{Saturated KCl} \mid \text{Solution, pH(s)} \mid H_2,$$

the half-cell on the left containing the solution whose pH is being measured and that on the right a standard buffer mixture of known pH; k = 2.303 RT/F, where R is the gas constant, T the temperature (°K), and F the value of the faraday.

Alternatively, the cell

Glass electrode | Solution, pH | Saturated calomel electrode

can be used, the glass electrode being calibrated by a standard buffer mixture or, if possible, two standard buffer mixtures whose pH values lie on either side of that of the solution being measured. Suitable standard buffer mixtures are:

0.05*M* potassium hydrogen phthalate (pH = 4.008 at 25°C)
0.025*M* potassium dihydrogen phosphate
0.025*M* disodium hydrogen phosphate (pH = 6.865 at 25°C)
0.01*M* borax (pH = 9.180 at 25°C)

For most purposes pH can be equated to $-\log_{10}\gamma_{H^+}m_{H^+}$, i.e., to the negative logarithm of the hydrogen ion activity. There is a small difference between those two quantities if pH > 9.2 or pH < 4.0, given by:

$$-\log\gamma_{H^+}m_{H^+} = pH + 0.014(pH - 9.2) \text{ for } pH > 9.2$$

$$-\log\gamma_{H^+}m_{H^+} = pH + 0.009(4.0 - pH) \text{ for } pH < 4.0$$

It should be noted that in the table listing the round values of pH at 25°C the value for $-\log\gamma_{H^+}m_{H^+}$ is quoted rather than pH, when there is a difference between these two values.

### REFERENCES

1. Bates, R. G., *Electrometric pH Determinations: Theory and Practice*. Wiley and Sons, New York (1954).
2. Robinson, R. A., and Stokes, R. H., *Electrolyte Solutions*, 2nd ed. Butterworths, London, England; Academic Press, New York (1959).
3. Bates, R. G., *J. Res. Nat. Bur. Stand.*, *66A*, 179 (1962).

**National Bureau of Standards:**

4. Bates, R. G., and Acree, S. F., *Research, 34,* 373 (1945).
5. Hamer, W. J., Pinching, C. D., and Acree, S. F., *Research, 36,* 47 (1946).
6. Manor, G. G., DeLollis, N. J., Lindwall, P. W., and Acree, S. F., *Research, 36,* 543 (1946).
7. Bates, R. G., *Research, 39,* 411 (1947).
8. Bates, R. G., Bower, V. E., Miller, R. G., and Smith, E. R., *Research, 47,* 433 (1951).
9. Bower, V. E., Bates, R. G., and Smith, E. R., *Research, 51,* 189 (1953).
10. Bower, V. E., and Bates, R. G., *Research, 55,* 197 (1955).
11. Bower, V. E., and Smith, R. E., *Research, 56,* 305 (1956).
12. Bower, V. E., and Bates, R. G., *Research, 59,* 261 (1956).
13. Bates, R. G., and Bower, V. E., *Anal. Chem., 28,* 1322 (1956).

## SOLUTIONS GIVING ROUND VALUES OF pH AT 25°C

### Code for Solutions

A: 25 ml of 0.2*M* potassium chloride + *x* ml of 0.2*M* hydrochloric acid
B: 50 ml of 0.1*M* potassium hydrogen phthalate + *x* ml of 0.1*M* hydrochloric acid
C: 50 ml of 0.1*M* potassium hydrogen phthalate + *x* ml of 0.1*M* sodium hydroxide
D: 50 ml of 0.1*M* potassium dihydrogen phosphate + *x* ml of 0.1*M* sodium hydroxide
E: 50 ml of 0.1*M* tris(hydroxymethyl)aminomethane + *x* ml of 0.1*M* hydrochloric acid
F: 50 ml of 0.025*M* borax + *x* ml of 0.1*M* hydrochloric acid
G: 50 ml of 0.025*M* borax + *x* ml of 0.1*M* sodium hydroxide
H: 50 ml of 0.05*M* sodium bicarbonate + *x* ml of 0.1*M* sodium hydroxide
I: 50 ml of 0.05*M* disodium hydrogen phosphate + *x* ml of 0.1*M* sodium hydroxide
J: 25 ml of 0.2*M* potassium chloride + *x* ml of 0.2*M* sodium hydroxide

| Solution A | | Solution B | | Solution C | | Solution D | | Solution E | |
|---|---|---|---|---|---|---|---|---|---|
| pH | *x* | pH | *x* | pH | *x* | pH | *x* | pH | *x* |
| 1.00 | 67.0 | 2.20 | 49.5 | 4.10 | 1.3 | 5.80 | 3.6 | 7.00 | 46.6 |
| 1.10 | 52.8 | 2.30 | 45.8 | 4.20 | 3.0 | 5.90 | 4.6 | 7.10 | 45.7 |
| 1.20 | 42.5 | 2.40 | 42.2 | 4.30 | 4.7 | 6.00 | 5.6 | 7.20 | 44.7 |
| 1.30 | 33.6 | 2.50 | 38.8 | 4.40 | 6.6 | 6.10 | 6.8 | 7.30 | 43.3 |
| 1.40 | 26.6 | 2.60 | 35.4 | 4.50 | 8.7 | 6.20 | 8.1 | 7.40 | 42.0 |
| 1.50 | 20.7 | 2.70 | 32.1 | 4.60 | 11.1 | 6.30 | 9.7 | 7.50 | 40.3 |
| 1.60 | 16.2 | 2.80 | 28.9 | 4.70 | 13.6 | 6.40 | 11.6 | 7.60 | 38.5 |
| 1.70 | 13.0 | 2.90 | 25.7 | 4.80 | 16.5 | 6.50 | 13.9 | 7.70 | 36.6 |
| 1.80 | 10.2 | 3.00 | 22.3 | 4.90 | 19.4 | 6.60 | 16.4 | 7.80 | 34.5 |
| 1.90 | 8.1 | 3.10 | 18.8 | 5.00 | 22.6 | 6.70 | 19.3 | 7.90 | 32.0 |
| 2.00 | 6.5 | 3.20 | 15.7 | 5.10 | 25.5 | 6.80 | 22.4 | 8.00 | 29.2 |
| 2.10 | 5.1 | 3.30 | 12.9 | 5.20 | 28.8 | 6.90 | 25.9 | 8.10 | 26.2 |
| 2.20 | 3.9 | 3.40 | 10.4 | 5.30 | 31.6 | 7.00 | 29.1 | 8.20 | 22.9 |
| | | 3.50 | 8.2 | 5.40 | 34.1 | 7.10 | 32.1 | 8.30 | 19.9 |
| | | 3.60 | 6.3 | 5.50 | 36.6 | 7.20 | 34.7 | 8.40 | 17.2 |
| | | 3.70 | 4.5 | 5.60 | 38.8 | 7.30 | 37.0 | 8.50 | 14.7 |
| | | 3.80 | 2.9 | 5.70 | 40.6 | 7.40 | 39.1 | 8.60 | 12.2 |
| | | 3.90 | 1.4 | 5.80 | 42.3 | 7.50 | 41.1 | 8.70 | 10.3 |
| | | 4.00 | 0.1 | 5.90 | 43.7 | 7.60 | 42.8 | 8.80 | 8.5 |
| | | | | | | 7.70 | 44.2 | 8.90 | 7.0 |
| | | | | | | 7.80 | 45.3 | 9.00 | 5.7 |
| | | | | | | 7.90 | 46.1 | | |
| | | | | | | 8.00 | 46.7 | | |

| Solution F pH | Solution F x | Solution G pH | Solution G x | Solution H pH | Solution H x | Solution I pH | Solution I x | Solution J pH | Solution J x |
|---|---|---|---|---|---|---|---|---|---|
| 8.00 | 20.5 | 9.20 | 0.9 | 9.60 | 5.0 | 10.90 | 3.3 | 12.00 | 6.0 |
| 8.10 | 19.7 | 9.30 | 3.6 | 9.70 | 6.2 | 11.00 | 4.1 | 12.10 | 8.0 |
| 8.20 | 18.8 | 9.40 | 6.2 | 9.80 | 7.6 | 11.10 | 5.1 | 12.20 | 10.2 |
| 8.30 | 17.7 | 9.50 | 8.8 | 9.90 | 9.1 | 11.20 | 6.3 | 12.30 | 12.8 |
| 8.40 | 16.6 | 9.60 | 11.1 | 10.00 | 10.7 | 11.30 | 7.6 | 12.40 | 16.2 |
| 8.50 | 15.2 | 9.70 | 13.1 | 10.10 | 12.2 | 11.40 | 9.1 | 12.50 | 20.4 |
| 8.60 | 13.5 | 9.80 | 15.0 | 10.20 | 13.8 | 11.50 | 11.1 | 12.60 | 25.6 |
| 8.70 | 11.6 | 9.90 | 16.7 | 10.30 | 15.2 | 11.60 | 13.5 | 12.70 | 32.2 |
| 8.80 | 9.6 | 10.00 | 18.3 | 10.40 | 16.5 | 11.70 | 16.2 | 12.80 | 41.2 |
| 8.90 | 7.1 | 10.10 | 19.5 | 10.50 | 17.8 | 11.80 | 19.4 | 12.90 | 53.0 |
| 9.00 | 4.6 | 10.20 | 20.5 | 10.60 | 19.1 | 11.90 | 23.0 | 13.00 | 66.0 |
| 9.10 | 2.0 | 10.30 | 21.3 | 10.70 | 20.2 | 12.00 | 26.9 | | |
| | | 10.40 | 22.1 | 10.80 | 21.2 | | | | |
| | | 10.50 | 22.7 | 10.90 | 22.0 | | | | |
| | | 10.60 | 23.3 | 11.00 | 22.7 | | | | |
| | | 10.70 | 23.8 | | | | | | |
| | | 10.80 | 24.25 | | | | | | |

Data taken from: Robinson, R. A., and Stokes, R. H., *Electrolyte Solutions,* 2nd ed. (1959). Reproduced by permission of Butterworths, London, England.

## PROPERTIES OF STANDARD AQUEOUS BUFFER SOLUTIONS AT 25°C

| Buffer Substance | Molality, $m$ | Weight of Salt in Air per Liter Solution | Density, g/ml | Molarity, $M$ | Dilution Value, $\Delta pH_{1/2}$ | $\Delta pH_S$ [a] | Buffer Value, equiv. per pH | Temperature Coefficient ($dpH_S/dt$), units per °C |
|---|---|---|---|---|---|---|---|---|
| **Tetroxalate Solution** | | | | | | | | |
| $KH_3(C_2O_4)_2 \cdot 2H_2O$ | 0.05 | 12.61 | 1.0032 | 0.04962 | +0.186 | −0.0028 | 0.070 | +0.001 |
| **Tartrate Solution** | | | | | | | | |
| $KHC_4H_4O_6$, sat. | 0.0341 | – | 1.0036 | 0.034 | +0.049 | −0.003 | 0.027 | −0.0014 |
| **Phthalate Solution** | | | | | | | | |
| $KHC_8O_4H_4$ | 0.05 | 10.12 | 1.0017 | 0.04958 | +0.052 | −0.0009 | 0.016 | +0.0012 |
| **Phosphate Solution** | | | | | | | | |
| $KH_2PO_4$ + | 0.025 | 3.39 | 1.0028[b] | 0.0249 | +0.080[b] | −0.0006[b] | 0.029[b] | −0.0028[b] |
| $Na_2HPO_4$ | 0.025 | 3.53 | | 0.0249 | | | | |
| $KH_2PO_4$ + | 0.008695 | 1.179 | 1.0020[b] | 0.008665 | +0.07[c] | −0.0005[b] | 0.016[b] | −0.0028[b] |
| $Na_2HPO_4$ | 0.03043 | 4.30 | | 0.03032 | | | | |
| **Borax Solution** | | | | | | | | |
| $Na_2B_4O_7 \cdot 10H_2O$ | 0.01 | 3.80 | 0.9996 | 0.009971 | +0.01 | −0.0001 | 0.020 | −0.0082 |
| **Calcium Hydroxide Solution** | | | | | | | | |
| $Ca(OH)_2$, sat. | 0.0203 | – | 0.9991 | 0.02025 | −0.28 | +0.0014 | 0.09 | −0.033 |

[a] $\Delta pH_S = pH_S(M \text{ molar solution}) - pH_S(m \text{ molal solution})$.
[b] For mixture at given concentration.
[c] Calculated value for mixture at given concentration.

## pH VALUES OF STANDARD SOLUTIONS AT TEMPERATURES FROM 0 to 95°C

| Temperature, °C | Tetroxalate, 0.05*m* | Tartrate, 0.0341*m*, saturated at 25°C | Phthalate, 0.05*m* | Phosphate[a] | Phosphate[b] | Borax, 0.01*m* | Calcium Hydroxide, saturated at 25°C |
|---|---|---|---|---|---|---|---|
| 0 | 1.666 | – | 4.003 | 6.984 | 7.534 | 9.464 | 13.423 |
| 5 | 1.668 | – | 3.999 | 6.951 | 7.500 | 9.395 | 13.207 |
| 10 | 1.670 | – | 3.998 | 6.923 | 7.472 | 9.332 | 13.003 |
| 15 | 1.672 | – | 3.999 | 6.900 | 7.448 | 9.276 | 12.810 |
| 20 | 1.675 | – | 4.002 | 6.881 | 7.429 | 9.225 | 12.627 |
| 25 | 1.679 | 3.557 | 4.008 | 6.865 | 7.413 | 9.180 | 12.454 |
| 30 | 1.683 | 3.552 | 4.015 | 6.853 | 7.400 | 9.139 | 12.289 |
| 35 | 1.688 | 3.549 | 4.024 | 6.844 | 7.389 | 9.102 | 12.133 |
| 38 | 1.691 | 3.548 | 4.030 | 6.840 | 7.384 | 9.081 | 12.043 |
| 40 | 1.694 | 3.547 | 4.035 | 6.838 | 7.380 | 9.068 | 11.984 |
| 45 | 1.700 | 3.547 | 4.047 | 6.834 | 7.373 | 9.038 | 11.841 |
| 50 | 1.707 | 3.549 | 4.060 | 6.833 | 7.367 | 9.011 | 11.705 |
| 55 | 1.715 | 3.554 | 4.075 | 6.834 | – | 8.985 | 11.574 |
| 60 | 1.723 | 3.560 | 4.091 | 6.836 | – | 8.962 | 11.449 |
| 70 | 1.743 | 3.580 | 4.126 | 6.845 | – | 8.921 | – |
| 80 | 1.766 | 3.609 | 4.164 | 6.859 | – | 8.885 | – |
| 90 | 1.792 | 3.650 | 4.205 | 6.877 | – | 8.850 | – |
| 95 | 1.806 | 3.674 | 4.227 | 6.886 | – | 8.833 | – |

[a] Solution of 0.025*m* $KH_2PO_4$ and 0.025*m* $Na_2HPO_4$.
[b] Solution of 0.008695*m* $KH_2PO_4$ and 0.03043*m* $Na_2HPO_4$.

## APPROXIMATE pH VALUES OF SOME ACIDS, BASES, BIOLOGIC MATERIALS AND FOODS

All values in the following table are based on measurements made at 25°C and are rounded off to the nearest tenth.

| Substance | pH | Substance | pH |
|---|---|---|---|
| **Acids** | | | |
| Acetic acid, *N* | 2.4 | Hydrochloric acid, 0.01*N* | 2.0 |
| Acetic acid, 0.1*N* | 2.9 | Hydrocyanic acid, 0.1*N* | 5.1 |
| Acetic acid, 0.01*N* | 3.4 | Hydrogen sulfide, 0.1*N* | 4.1 |
| Alum, 0.1*N* | 3.2 | Lactic acid, 0.1*N* | 2.4 |
| Arsenious acid, saturated | 5.0 | Malic acid, 0.1*N* | 2.2 |
| Benzoic acid, 0.01*N* | 3.1 | Orthophosphoric acid, 0.1*N* | 1.5 |
| Boric acid, 0.1*N* | 5.2 | Oxalic acid, 0.1*N* | 1.6 |
| Carbonic acid, saturated | 3.8 | Sulfuric acid, *N* | 0.3 |
| Citric acid, 0.1*N* | 2.2 | Sulfuric acid, 0.1*N* | 1.2 |
| Formic acid, 0.1*N* | 2.3 | Sulfuric acid, 0.01*N* | 2.1 |
| Hydrochloric acid, *N* | 0.1 | Sulfurous acid, 0.1*N* | 1.5 |
| Hydrochloric acid, 0.1*N* | 1.1 | Tartaric acid, 0.1*N* | 2.2 |
| **Bases** | | | |
| Ammonia, *N* | 11.6 | Potassium hydroxide, 0.1*N* | 13.0 |
| Ammonia, 0.1*N* | 11.1 | Potassium hydroxide, 0.01*N* | 12.0 |
| Ammonia, 0.01*N* | 10.6 | Sodium bicarbonate, 0.1*N* | 8.4 |
| Borax, 0.1*N* | 9.2 | Sodium carbonate, 0.1*N* | 11.6 |
| Calcium carbonate, saturated | 9.4 | Sodium hydroxide, *N* | 14.0 |
| Ferrous hydroxide, saturated | 9.5 | Sodium hydroxide, 0.1*N* | 13.0 |
| Lime, saturated | 12.4 | Sodium hydroxide, 0.01*N* | 12.0 |
| Magnesia, saturated | 10.5 | Sodium metasilicate, 0.1*N* | 12.6 |
| Potassium cyanide, 0.1*N* | 11.0 | Sodium sesquicarbonate, 0.1*M* | 10.1 |
| Potassium hydroxide, *N* | 14.0 | Trisodium phosphate, 0.1*N* | 12.0 |

| Substance | pH | Substance | pH |
|---|---|---|---|
| **Biologic Materials** | | | |
| Bile, human | 6.8–7.0 | Gastric contents, human | 1.0–3.0 |
| Blood, plasma, human | 7.3–7.5 | Milk, human | 6.6–7.6 |
| Blood, whole, dog | 6.9–7.2 | Saliva, human | 6.5–7.5 |
| Duodenal contents, human | 4.8–8.2 | Spinal fluid, human | 7.3–7.5 |
| Feces, human | 4.6–8.4 | Urine, human | 4.8–8.4 |
| **Foods** | | | |
| Apples | 2.9–3.3 | Milk, cows | 6.3–6.6 |
| Apricots | 3.6–4.0 | Olives | 3.6–3.8 |
| Asparagus | 5.4–5.8 | Oranges | 3.0–4.0 |
| Bananas | 4.5–4.7 | Oysters | 6.1–6.6 |
| Beans | 5.0–6.0 | Peaches | 3.4–3.6 |
| Beers | 4.0–5.0 | Pears | 3.6–4.0 |
| Beets | 4.9–5.5 | Peas | 5.8–6.4 |
| Blackberries | 3.2–3.6 | Pickles, dill | 3.2–3.6 |
| Bread, white | 5.0–6.0 | Pickles, sour | 3.0–3.4 |
| Butter | 6.1–6.4 | Pimento | 4.6–5.2 |
| Cabbage | 5.2–5.4 | Plums | 2.8–3.0 |
| Carrots | 4.9–5.3 | Potatoes | 5.6–6.0 |
| Cheeses | 4.8–6.4 | Pumpkin | 4.8–5.2 |
| Cherries | 3.2–4.0 | Raspberries | 3.2–3.6 |
| Cider | 2.9–3.3 | Rhubarb | 3.1–3.2 |
| Corn | 6.0–6.5 | Salmon | 6.1–6.3 |
| Crackers | 6.5–8.5 | Sauerkraut | 3.4–3.6 |
| Dates | 6.2–6.4 | Shrimp | 6.8–7.0 |
| Eggs, fresh white | 7.6–8.0 | Soft drinks | 2.0–4.0 |
| Flour, wheat | 5.5–6.5 | Spinach | 5.1–5.7 |
| Gooseberries | 2.8–3.0 | Squash | 5.0–5.4 |
| Grapefruit | 3.0–3.3 | Strawberries | 3.0–3.5 |
| Grapes | 3.5–4.5 | Sweet potatoes | 5.3–5.6 |
| Hominy (lye) | 6.8–8.0 | Tomatoes | 4.0–4.4 |
| Jams, fruit | 3.5–4.0 | Tuna | 5.9–6.1 |
| Jellies, fruit | 2.8–3.4 | Turnips | 5.2–5.6 |
| Lemons | 2.2–2.4 | Vinegar | 2.4–3.4 |
| Limes | 1.8–2.0 | Water, drinking | 6.5–8.0 |
| Maple syrup | 6.5–7.0 | Wines | 2.8–3.8 |

Data from: *Modern pH and Chlorine Control,* Taylor Chemicals, Inc. Reproduced by permission of the publishers.

## pH Indicators

| Approximate pH Range | Indicator | Color Change | Preparation |
|---|---|---|---|
| 0.0–1.0 | Cresol red [*o*-cresolsulfonphthalein] | Red to yellow | 0.1 g in 26.2 ml of 0.01*N* sodium hydroxide + 223.8 ml of water |
| 0.0–1.6 | Methyl violet | Yellow to blue | 0.01–0.05% in water |
| 0.0–1.8 | Crystal violet | Yellow to blue | 0.02% in water |
| 0.0–2.4 | Ethyl violet | Yellow to blue | 0.1 g in 50 ml of methanol + 50 ml of water |

| Approximate pH Range | Indicator | Color Change | Preparation |
|---|---|---|---|
| 0.2–1.8 | Malachite green | Yellow to blue-green | Water |
| | Methyl green | Yellow to blue | 0.1% in water |
| | 2-(*p*-Dimethylaminophenylazo)pyridine | Yellow to blue | 0.1% in ethanol |
| 0.4–1.8 | Cresol red [*o*-cresolsulfonphthalein] | Yellow to red | 0.1 g in 26.2 ml of 0.01*N* sodium hydroxide + 223.8 ml of water |
| 1.0–2.2 | Quinaldine red | Colorless to red | 1% in ethanol |
| 1.0–3.0 | Paramethyl red [*p*-(*p*-dimethylamino-phenylazo)benzoic acid, sodium salt] | Red to yellow | Ethanol |
| 1.2–2.4 | Metanil yellow [*m*-(*p*-anilinophenylazo)-benzenesulfonic acid, sodium salt] | Red to yellow | 0.01% in water |
| 1.2–2.6 | 4-Phenylazodiphenylamine | Red to yellow | 0.01 g in 1 ml of *N* hydro-chloric acid + 50 ml of ethanol + 49 ml of water |
| 1.2–2.8 | Thymol blue [thymolsulfonphthalein] | Red to yellow | 0.1 g in 21.5 ml of 0.01*N* sodium hydroxide + 229.5 ml of water |
| | Metacresol purple [*m*-cresolsulfon-phthalein] | Red to yellow | 0.1 g in 26.2 ml of 0.01*N* sodium hydroxide + 223.8 ml of water |
| 1.4–2.8 | Orange IV [*p*-(*p*-anilinophenylazo)-benzenesulfonic acid, sodium salt] | Red to yellow | 0.01% in water |
| | 4-*o*-Tolylazo-*o*-toluidine | Orange to yellow | Water |
| 2.2–3.6 | Erythrosine, disodium salt | Orange to red | 0.1% in water |
| 2.2–4.2 | Benzopurpurine 48 | Violet to red | 0.1% in water |
| 2.6–4.8 | N,N-Dimethyl-*p*-(*m*-tolylazo)aniline | Red to yellow | 0.1% in water |
| 2.8–4.0 | 2,4-Dinitrophenol | Colorless to yellow | Saturated water solution |
| 2.8–4.4 | *p*-Dimethylaminoazobenzene [N,N-di-methyl-*p*-phenylazoaniline] | Red to yellow | 0.1 g in 90 ml of ethanol + 10 ml of water |
| 3.0–4.0 | 4,4′-Bis(2-amino-1-naphthylazo)-2,2′-stilbenedisulfonic acid | Purple to red | 0.1 g in 5.9 ml of 0.05*N* sodium hydroxide + 94.1 ml of water |
| 3.0–4.2 | Tetrabromophenolphthaleinethyl ester, potassium salt | Yellow to blue | 0.1% in ethanol |
| 3.0–4.6 | Bromophenol blue [3′,3″,5′,5″-tetra-bromophenolsulfonphthalein] | Yellow to blue | 0.1 g in 14.9 ml of 0.01*N* sodium hydroxide + 235.1 ml of water |
| 3.0–5.0 | Congo red | Blue to red | 0.1% in water |
| 3.2–4.2 | Methyl orange–xylene cyanole solution | Purple to green | Ready for use |
| 3.2–4.4 | Methyl orange | Red to yellow | 0.01% in water |

| Approximate pH Range | Indicator | Color Change | Preparation |
|---|---|---|---|
| 3.4–4.8 | Ethyl orange | Red to yellow | 0.05–0.2% in water or aqueous ethanol solution |
| 3.5–4.8 | 4-(4-Dimethylamino-1-naphthylazo)-3-methoxybenzenesulfonic acid | Violet to yellow | 0.1% in 60% ethanol |
| 3.8–5.4 | Bromocresol green [3′,3″,5′,5″-tetrabromo-*m*-cresolsulfonphthalein] | Yellow to blue | 0.1 g in 14.3 ml of 0.01*N* sodium hydroxide + 235.7 ml of water |
| 3.8–6.4 | Resazurin | Orange to violet | Water |
| 4.0–5.6 | 4-Phenylazo-1-naphthylamine | Red to yellow | 0.1% in ethanol |
| 4.0–5.8 | Ethyl red | Colorless to red | 0.1 g in 50 ml of methanol + 50 ml of water |
| 4.4–5.6 | 2-(*p*-Dimethylaminophenylazo)pyridine | Red to yellow | 0.1% in ethanol |
| 4.4–5.8 | 4-(*p*-Ethoxyphenylazo)-*m*-phenylenediamine monohydrochloride | Orange to yellow | 0.1% in water |
| 4.4–6.2 | Lacmoid | Red to blue | 0.2% in ethanol |
| 4.6–6.0 | Alizarin red S | Yellow to red | Dilute solution in water |
| 4.8–6.0 | Methyl red | Red to yellow | 0.02 g in 60 ml of ethanol + 40 ml of water |
| 4.8–6.6 | Propyl red | Red to yellow | Ethanol |
| 5.2–6.8 | Bromocresol purple [5′,5″-dibromo-*o*-cresolsulfonphthalein] | Yellow to purple | 0.1 g in 18.5 ml of 0.01*N* sodium hydroxide + 231.5 ml of water |
| | Chlorophenol red [3′,3″-dichlorophenolsulfonphthalein] | Yellow to red | 0.1 g in 23.6 ml of 0.01*N* sodium hydroxide + 226.4 ml of water |
| 5.4–6.6 | *p*-Nitrophenol | Colorless to yellow | 0.1% in water |
| 5.6–7.2 | Alizarin | Yellow to red | 0.1% in methanol |
| 6.0–7.0 | 2-(2,4-Dinitrophenylazo)-1-naphthol-3,6-disulfonic acid, disodium salt | Yellow to blue | 0.1% in water |
| 6.0–7.6 | Bromothymol blue [3′,3″-dibromothymolsulfonphthalein] | Yellow to blue | 0.1 g in 16 ml of 0.01*N* sodium hydroxide + 234 ml of water |
| 6.4–8.0 | *m*-Dinitrobenzoylene urea [6,8-dinitro-2,4-(1H)-quinazolinedione] | Colorless to yellow | 25 g in 115 ml of *M* sodium hydroxide + 234 ml of sodium chloride in 100 ml of water |
| 6.6–7.8 | Brilliant yellow | Yellow to orange | 1% in water |
| 6.6–8.0 | Phenol red [phenolsulfonphthalein] | Yellow to red | 0.1 g in 28.2 ml of 0.01*N* sodium hydroxide + 221.8 ml of water |

| Approximate pH Range | Indicator | Color Change | Preparation |
|---|---|---|---|
| 6.8–8.0 | Neutral red | Red to amber | 0.01 g in 50 ml of ethanol + 50 ml of water |
| 6.8–8.6 | *m*-Nitrophenol | Colorless to amber | 0.3% in water |
| 7.0–8.8 | Cresol red [*o*-cresolsulfonphthalein] | Yellow to red | 0.1 g in 26.2 ml of 0.01*N* sodium hydroxide + 223.8 ml of water |
| 7.4–8.6 | Curcumin | Yellow to red | Ethanol |
| 7.4–9.0 | Metacresol purple [*m*-cresolsulfon-phthalein] | Yellow to purple | 0.1 g in 26.2 ml of 0.01*N* sodium hydroxide + 223.8 ml of water |
| 8.0–9.0 | 4,4′-Bis(4-amino-1-naphthylazo)-2,2′-stilbenedisulfonic acid | Blue to red | 0.1 g in 5.9 ml of 0.05*N* sodium hydroxide + 94.1 ml of water |
| 8.0–9.6 | Thymol blue [thymolsulfonphthalein] | Yellow to blue | 0.1 g in 21.5 ml of 0.01*N* sodium hydroxide + 229.5 ml of water |
| 8.2–9.8 | *o*-Cresolphthalein | Colorless to red | 0.04% in ethanol |
| 8.2–10.0 | *p*-Naphtholbenzene | Orange to blue | 1% in dilute alkali |
| | Phenolphthalein | Colorless to pink | 0.05 g in 50 ml of ethanol + 50 ml of water |
| 8.4–9.6 | Ethyl-bis(2,4-dimethylphenyl)acetate | Colorless to blue | Saturated solution in 50% acetone alcohol |
| 9.4–10.6 | Thymolphthalein | Colorless to blue | 0.04 g in 50 ml of ethanol + 50 ml of water |
| 10.1–12.0 | Alizarin yellow R [5-(*p*-nitrophenylazo)-salicylic acid, sodium salt] | Yellow to red | 0.01% in water |
| 11.0–12.4 | Alizarin | Red to purple | 0.1% in methanol |
| 11.4–12.6 | *p*-(2,4-Dihydroxyphenylazo)benzene-sulfonic acid, sodium salt | Yellow to orange | 0.1% in water |
| 11.4–13.0 | 5,5′-Indigodisulfonic acid, sodium salt | Blue to yellow | Water |
| 11.5–13.0 | 2,4,6-Trinitrotoluene | Colorless to orange | 0.1–0.5% in ethanol |
| 12.0–14.0 | 1,3,5-Trinitrobenzene | Colorless to orange | 0.1–0.5% in ethanol |
| 12.2–13.2 | Clayton yellow | Yellow to amber | 0.1% in water |

# Fluorescent Indicators

DR. JACK DEMENT

Fluorescent indicators are substances that show definite changes in fluorescence with change in pH. Some fluorescent materials are not suitable for indicators, since their change in fluorescence is too gradual. Fluorescent indicators find greatest utility in the titration of opaque, highly turbid or deeply colored solutions. A long-wavelength ultraviolet ("black light") lamp in a dimly lighted room provides the best environment for titrations involving fluorescent indicators, although bright daylight is sometimes sufficient to evoke a response in the bright-green, yellow or orange fluorescent indicators. Titrations are carried out in nonfluorescent glassware. All glassware should be checked before use, to be certain that it does not fluoresce in the wavelength of the light involved in the titration. The meniscus of the liquid in the burette can be followed when a few particles of an insoluble fluorescent solid are dropped onto its surface.

The indicators in the following table are arranged by approximate pH range covered. In some of the dyestuffs the end point may vary slightly with the source or manufacturer.

| | | From | | To | |
|---|---|---|---|---|---|
| Indicator | C.I. | pH | Color | pH | Color |
| **pH 0 to 2** | | | | | |
| Benzoflavine | – | 0.3 | yellow fl. | 1.7 | green fl. |
| 3,6-Dioxyphthalimide | – | 0 | blue fl. | 2.4 | green fl. |
| Eosine YS | 768 | 0 | yellow | 3.0 | yellow fl. |
| Erythrosine | 772 | 0 | yellow | 3.6 | yellow fl. |
| Esculin | – | 1.5 | colorless | 2 | blue fl. |
| 4-Ethoxyacridone | – | 1.2 | green fl. | 3.2 | blue fl. |
| 3,6-Tetramethyldiaminooxanthone | – | 1.2 | green fl. | 3.4 | blue fl. |
| **pH 2 to 4** | | | | | |
| Chromotropic acid | – | 3.5 | colorless | 4.5 | blue fl. |
| Fluorescein | 766 | 4 | colorless | 4.5 | green fl. |
| Magdala red | – | 3.0 | purple | 4.0 | fl. |
| α-Naphthylamine | – | 3.4 | colorless | 4.8 | blue fl. |
| β-Naphthylamine | – | 2.8 | colorless | 4.4 | violet fl. |
| Phloxine | 774 | 3.4 | colorless | 5.0 | bright-yellow fl. |
| Salicylic acid | – | 2.5 | colorless | 3.5 | blue fl. |
| **pH 4 to 6** | | | | | |
| Acridine | 788 | 4.9 | green fl. | 5.1 | violet |
| Dichlorofluorescein | – | 4.0 | colorless | 5.0 | green fl. |
| 3,6-Dioxyanthone | – | 5.4 | colorless | 7.6 | blue-violet fl. |
| Erythrosine | 722 | 4.0 | colorless | 4.5 | yellow-green fl. |
| β-Methylesculetin | – | 4.0 | colorless | 6.2 | blue fl. |
| Neville-Winther acid | – | 6.0 | colorless | 6.5 | blue fl. |
| Quininic acid | – | 4.0 | yellow | 5.0 | blue fl. |
| Quinine (first end point) | – | 5.0 | blue fl. | 6.1 | violet fl. |
| Resorufin | – | 4.4 | yellow fl. | 6.4 | weak-orange fl. |
| **pH 6 to 8** | | | | | |
| Acid R phosphine | – | | (claimed for pH range 6.0–7.0) | | |
| Brilliant diazol yellow | – | 6.5 | colorless | 7.5 | violet fl. |
| Cleves acid | – | 6.5 | colorless | 7.5 | green fl. |
| Coumaric acid | – | 7.2 | colorless | 9.0 | green fl. |

| Indicator | C.I. | From pH | From Color | To pH | To Color |
|---|---|---|---|---|---|
| **pH 6 to 8 (continued)** | | | | | |
| 3,6-Dioxyphthalic dinitrile | – | 5.8 | blue fl. | 8.2 | green fl. |
| Magnesium 8-hydroxyquinolate | – | 6.5 | colorless | 7.5 | golden fl. |
| β-Methylumbelliferone | – | 7.0 | colorless | 7.5 | blue fl. |
| 1-Naphthol-4-sulfonic acid | – | 6.0 | colorless | 6.5 | blue fl. |
| Orcinaurine | – | 6.5 | colorless | 8.0 | green fl. |
| Patent phosphine | 789 | (for pH range 6.0–7.0, green–yellow fl.) | | | |
| Thioflavine | 816 | (for pH range 6.5–7.0, yellow fl.) | | | |
| Umbelliferone | – | 6.5 | colorless | 7.6 | blue fl. |
| **pH 8 to 10** | | | | | |
| Acridine orange | 788 | 8.4 | orange | 10.4 | green fl. |
| Ethoxyphenylnaphthostilbazonium chloride | – | 9 | green fl. | 11 | non-fl. |
| G salt | – | 9.0 | dull-blue fl. | 9.5 | bright-blue fl. |
| Naphthazol derivatives | – | 8.2 | colorless | 10.0 | yellow or green fl. |
| α-Naphthionic acid | – | 9 | blue fl. | 11 | green fl. |
| 2-Naphthol-3,6-disulfonic acid | – | 9.5 | dark-blue fl. | | light-blue fl. at higher pH |
| β-Naphthol | – | 8.6 | colorless | | blue fl. at higher pH |
| α-Naphtholsulfonic acid | – | 8.0 | dark-blue fl. | 9.0 | bright-violet fl. |
| 1,4-Naphtholsulfonic acid | – | 8.2 | dark-blue fl. | | light-blue fl. at higher pH |
| Orcinsulfonphthalein | – | 8.6 | yellow | 10.0 | fl. |
| Quinine (second end point) | – | 9.5 | violet fl. | 10.0 | colorless |
| R salt | – | 9.0 | dull-blue fl. | 9.5 | bright-blue fl. |
| Sodium 1-naphthol-sulfonate | – | 9.0 | dark-blue fl. | 10.0 | bright-violet fl. |
| **pH 10 to 12** | | | | | |
| Coumarin | – | 9.8 | deep-green fl. | 12 | light-green fl. |
| Eosine BN | 771 | 10.5 | colorless | 14.0 | yellow fl. |
| Papaverine (permanganate oxidized) | – | 9.5 | yellow fl. | 11.0 | blue fl. |
| Schaffer's salt | – | 5.0 | violet fl. | 11.0 | green-blue fl. |
| SS acid (sodium salt) | – | 10.0 | violet fl. | 12.0 | yellow |
| **pH 12 to 14** | | | | | |
| Cotarnine | – | 12.0 | yellow fl. | 13.0 | white fl. |
| α-Naphthionic acid | – | 12 | blue fl. | 13 | green fl. |
| β-Naphthionic acid | – | 12 | blue fl. | 13 | violet fl. |

## ANALYSES OF MEDIA CONSTITUENTS

### TABLE 1
### TRACE METAL ANALYSIS OF DIFCO NUTRIENT BROTH, BATCH 490560

| Element | X-Ray Fluorescence | Emission Spectroscopy | Neutron Activation Analysis | Best Estimate |
|---|---|---|---|---|
| **Major Components**[a] | | | | |
| Cl | – | – | 1.86 | 1.86 |
| Na | – | >10 | 1.65 | 1.65 |
| K | – | 1 to 10 | <1 | 1 (?) |
| P | – | >10 | <3.5 | <3.5 |
| **Minor Components**[b] | | | | |
| Al | – | 10 to 100 | <66 | 50 (?) |
| Sb | <10 | – | – | <10 |
| As | <10 | – | – | <10 |
| Ba | <10 | <10 | <3,400 | <10 |
| Br | 23 | – | 100 | 20–100 |
| Cd | <10 | – | <3,000 | <10 |
| Ca | – | 10 to 100 | <11,000 | 50 (?) |
| Cs | – | 10 to 100 | 2 | 2 |
| Cr | – | – | <31 | <31 |
| Co | <10 | – | <0.4 | <0.4 |
| Cu | – | <10 | <18 | <10 |
| I | <10 | – | <10,000 | <10 |
| Fe | – | 10 to 100 | <200 | 50(?) |
| Pb | <10 | – | – | <10 |
| Li | – | <10 | – | <10 |
| Mg | – | 100 to 1,000 | <2,000 | 500 (?) |
| Mn | – | – | <5 | <5 |
| Hg | – | – | <22 | <22 |
| Mo | <10 | – | – | <10 |
| Ni | <10 | – | <20,000 | <10 |
| Rb | 132 | 100 to 1,000 | 152 | 140 |
| Si | – | 100 to 1,000 | <13,000 | 500 (?) |
| Sr | <10 | – | – | <10 |
| Sn | <10 | – | <85 | <10 |
| Ti | <10 | – | – | <10 |
| Zn | <10 | – | <3 | <3 |
| Zr | <10 | – | – | <10 |

[a] Results for major components are given in percentages.
[b] Results for minor components are given in parts per million (ppm).

Data taken from: Kempner, E. S., *Appl. Microbiol., 15,* 1525 (1967). Reproduced by permission of the American Society for Microbiology, Washington, D.C.

## TABLE 2
## MINOR ELEMENT COMPOSITION OF YEAST EXTRACTS (DIFCO)

| Ash, % | Element, μg/g dry weight | | | | | | | | | | | | | | | | | |
|---|---|---|---|---|---|---|---|---|---|---|---|---|---|---|---|---|---|---|
| | Al | Ba | Cd | Co | Cr | Cu | Fe | Ga | Mg | Mn | Mo | Ni | Pb | Sn | Sr | Ti | V | Zn |
| **Batch No. 385062** | | | | | | | | | | | | | | | | | | |
| 12.25 | 3.0 | 1.1 | 1.2 | 6.1 | 10.7 | 91.7 | 121 | 0.01 | 1160 | 3.2 | 3.7 | 6.3 | 10.5 | 0.18 | 1.0 | 3.2 | 36.3 | 104.0 |
| **Batch No. 406870** | | | | | | | | | | | | | | | | | | |
| 12.71 | 2.1 | 1.0 | 1.3 | 1.0 | 9.4 | 53.5 | 156 | 0.13 | 1580 | 2.0 | 2.6 | 30.8 | 4.6 | 0.03 | 1.0 | 1.4 | 31.2 | 58.6 |
| **Batch No. 416490** | | | | | | | | | | | | | | | | | | |
| 12.84 | 3.0 | 1.3 | 2.0 | 4.0 | 17.4 | 68.7 | 155 | 0.20 | 1540 | 2.3 | 8.0 | 32.9 | 4.2 | 0.14 | 1.4 | 2.9 | 66.1 | 104.0 |
| **Batch No. 419285** | | | | | | | | | | | | | | | | | | |
| 14.24 | 3.8 | 1.7 | 1.8 | 4.3 | 11.0 | 41.6 | 135 | 0.02 | 980 | 1.4 | 6.3 | 9.0 | 2.6 | 0.05 | 1.2 | 2.9 | 45.8 | 57.3 |
| **Batch No. 421072** | | | | | | | | | | | | | | | | | | |
| 12.33 | 3.5 | 1.3 | 1.3 | 2.0 | 11.6 | 101.0 | 185 | 0.07 | 1100 | 2.4 | 9.1 | 11.9 | 12.0 | 0.04 | 0.84 | 4.8 | 39.1 | 46.2 |
| **Mean** | | | | | | | | | | | | | | | | | | |
| 12.87 | 31 | 1.3 | 1.5 | 3.5 | 12.0 | 71.3 | 150 | 0.09 | 1270 | 2.3 | 5.9 | 18.2 | 6.8 | 0.09 | 1.1 | 3.0 | 43.7 | 74.0 |
| **Coefficient of Variation (%) of Individual Values** | | | | | | | | | | | | | | | | | | |
| 6.2 | 21 | 21 | 23 | 58 | 26 | 35 | 16 | 93 | 21 | 29 | 47 | 70 | 62 | 77 | 20 | 40 | 31 | 38 |

Data taken from: Grant, C. L., and Pramer, D., *J. Bacteriol., 81*, 870 (1962). Reproduced by permission of the American Society for Microbiology, Washington, D.C.

TABLE 3
APPROXIMATE COMPOSITION OF BBL PEPTONES AND YEAST EXTRACT

| Catalog Number<br>Name<br>Source<br>Hydrolysis | 11842<br>Acidicase®<br>Peptone<br>Casein<br>HCl | 11869<br>Gelysate®<br>Peptone<br>Gelatin<br>Pancreatic | 11879<br>Lactalysate®<br>Peptone<br>Lactalbumin<br>Pancreatic | 11892<br>Myosate®<br>Peptone<br>Cardiac<br>Muscle<br>Pancreatic | 11905<br>Phytone®<br>Peptone<br>Soy<br>Meal<br>Papaic | 11918<br>Thiotone®<br>Peptone<br>Animal<br>Tissues<br>Peptic | 11920<br>Trypticase®<br>Peptone<br>Casein<br>Pancreatic | 11928<br>Yeast Extract<br>Yeast Cells<br>Autolytic<br>Enzymes |
|---|---|---|---|---|---|---|---|---|
| Nitrogen, % | | | | | | | | |
| Total | 8.0 | 16 | 11.9 | 12.3 | 9.2 | 12.8 | 11.7 | 10.3 |
| Amino | 6.4 | 1 | 6.9 | 4.8 | 1.8 | 3.2 | 3.5 | 5.5 |
| NaCl | 37.2 | 0.6 | 1.5 | 1.2 | 4.4 | 2.3 | 0.5 | 0.5 |
| Ca | 0.05 | Trace | 0.19 | 0.002 | 0.05 | 0.05 | 0.35 | 0.06 |
| Fe | 0.0045 | 0.006 | 0.056 | 0.016 | 0.02 | 0.02 | 0.03 | 0.20 |
| K | 0.4 | 0.06 | 0.32 | 2.0 | 3.99 | 1.4 | 0.24 | 3.4 |
| Mg | 0.003 | 0.001 | 0.005 | 0.06 | 0.19 | 0.05 | 0.03 | 0.07 |
| P | 0.32 | 0.07 | 0.34 | 0.63 | 0.38 | 0.6 | 0.65 | 1.16 |
| S | 0.066 | 0.35 | 0.43 | 1.0 | 0.39 | 1.0 | 0.73 | |
| Carbohydrates, % | 0.0 | 0.0 | + | + | 37 | + | 0.0 | 16.6 |
| Amino Acids, % | | | | | | | | |
| Arginine | 1.4 | 8.0 | 3.1 | 4.7 | 4.6 | 5.0 | 2.6 | 3.5 |
| Aspartic Acid | 3.7 | 3.9 | 8.2 | 7.0 | 5.8 | 5.9 | 5.1 | |
| Cystine | 0.3 | 0.1 | 2.1 | 0.4 | 0.5 | 0.6 | 0.3 | 1.6 |
| Glycine | 1.0 | 20.7 | 1.7 | 5.2 | 2.8 | 9.3 | 1.8 | |
| Glutamic Acid | 14.2 | 9.1 | 10.6 | 12.9 | 9.3 | 10.0 | 17.0 | |
| Histidine | 0.7 | 0.9 | 1.9 | 1.6 | 1.6 | 1.8 | 2.4 | 1.5 |
| Isoleucine | 2.7 | 1.6 | 5.4 | 4.0 | 2.5 | 3.3 | 5.0 | 4.7 |
| Leucine | 3.5 | 2.8 | 10.9 | 5.8 | 3.2 | 6.0 | 7.1 | 6.4 |
| Lysine | 3.7 | 4.3 | 10.0 | 5.2 | 3.6 | 5.5 | 5.3 | 6.5 |
| Methionine | 1.7 | 1.0 | 2.5 | 1.8 | 0.6 | 1.6 | 2.4 | 2.0 |
| Phenylalanine | 0.7 | 1.9 | 3.4 | 3.2 | 3.6 | 3.3 | 3.8 | 3.5 |
| Proline | 4.0 | 14.0 | 6.3 | 4.9 | 3.4 | 7.3 | 11.5 | |
| Threonine | 2.5 | 1.8 | 3.8 | 3.0 | 1.8 | 3.5 | 3.5 | 3.3 |
| Tryptophan | 0.0 | 0.2 | 2.0 | 1.2 | 0.7 | 0.7 | 0.9 | 1.0 |
| Tyrosine | 3.1 | 0.8 | 3.3 | 1.2 | 1.9 | 2.1 | 2.3 | 4.0 |
| Valine | 4.1 | 2.2 | 4.1 | 4.0 | 2.0 | 4.6 | 5.6 | 4.8 |

## TABLE 3 (Continued)

## APPROXIMATE COMPOSITION OF BBL PEPTONES AND YEAST EXTRACT

| Catalog Number<br>Name<br>Source<br>Hydrolysis | 11842<br>Acidicase®<br>Peptone<br>Casein<br>HCl | 11869<br>Gelysate®<br>Peptone<br>Gelatin<br>Pancreatic | 11879<br>Lactalysate®<br>Peptone<br>Lactalbumin<br>Pancreatic | 11892<br>Myosate®<br>Peptone<br>Cardiac<br>Muscle<br>Pancreatic | 11905<br>Phytone®<br>Peptone<br>Soy<br>Meal<br>Papaic | 11918<br>Thiotone®<br>Peptone<br>Animal<br>Tissues<br>Peptic | 11920<br>Trypticase®<br>Peptone<br>Casein<br>Pancreatic | 11928<br>Yeast Extract<br>Yeast Cells<br>Autolytic<br>Enzymes |
|---|---|---|---|---|---|---|---|---|
| Vitamins, μg/g | | | | | | | | |
| Biotin | 0.018 | 0.014 | 0.062 | 0.10 | 0.35 | 0.18 | 0.083 | 4 |
| Choline | 0.0 | 0.0 | 1.15 | 4,250 | 3.05 | 1,980 | 0.0 | 2,000 |
| Cyanocobalamin | 0.00006 | 0.13 | 0.0039 | 0.37 | 0.00115 | 0.5 | 0.45 | 0.0 |
| Folic Acid | 0.0057 | 0.052 | 0.038 | 1.06 | 0.81 | 1.17 | 0.67 | 20 |
| Niacin | 0.10 | 2 | 6.3 | 390 | 33 | 212 | 8 | 400 |
| Pantothenic Acid | 0.26 | 1.76 | 4.0 | 1.76 | 7.6 | 8.9 | 2.2 | 100 |
| Pyridoxine | 0.024 | 1.7 | 0.7 | 11.9 | 4.0 | 3.2 | 0.06 | 30 |
| Riboflavin | 0.10 | 4 | 0.69 | 37 | 4.3 | 19 | 5.75 | 50 |
| Thiamine | 0.105 | 160 | 0.44 | 96 | 1.9 | + | ,+ | 100 |
| PABA | | 0.164 | 0.08 | | 5.4 | 0.66 | 0.21 | 24 |

Note: Blank spaces indicate the absence of data, not absence of the component.

Data taken from: *BBL Manual of Products and Laboratory Procedures*, 5th ed., p. 163 (1968). Reproduced by permission of BBL Division of BioQuest, Cockeysville, Maryland.

## TABLE 4
## TYPICAL ANALYSES OF BACTO-PEPTONES

| Constituent, % | Bacto-Peptone | Proteose Peptone (Difco) | Bacto-Tryptone | Bacto-Tryptose | Neo-peptone (Difco) | Bacto-Protone |
|---|---|---|---|---|---|---|
| Total nitrogen | 16.16 | 14.37 | 13.14 | 13.76 | 14.33 | 15.41 |
| Primary proteose N | 0.06 | 0.60 | 0.20 | 0.40 | 0.46 | 5.36 |
| Secondary proteose N | 0.68 | 4.03 | 1.63 | 2.83 | 3.03 | 7.60 |
| Peptone N | 15.38 | 9.74 | 11.29 | 10.52 | 10.72 | 2.40 |
| Ammonia N | 0.04 | 0.00 | 0.02 | 0.01 | 0.12 | 0.05 |
| Free amino N (Van Slyke) | 3.20 | 2.66 | 4.73 | 3.70 | 2.82 | 1.86 |
| Amide N | 0.49 | 0.94 | 1.11 | 1.03 | 1.23 | 0.00 |
| Mono-amino N | 9.42 | 7.61 | 7.31 | 7.46 | 7.56 | 0.00 |
| Di-amino N | 4.07 | 4.51 | 3.45 | 3.98 | 4.43 | 0.00 |
| Tryptophan | 0.29 | 0.51 | 0.77 | 0.64 | 0.73 | 1.03 |
| Tyrosine | 0.98 | 2.51 | 4.39 | 3.45 | 4.72 | 2.99 |
| Cystine (Sullivan) | 0.22 | 0.56 | 0.19 | 0.38 | 0.39 | 0.27 |
| Organic sulfur | 0.33 | 0.60 | 0.53 | 0.57 | 0.63 | 0.45 |
| Inorganic sulfur | 0.29 | 0.04 | 0.04 | 0.04 | 0.09 | 0.16 |
| Phosphorus | 0.22 | 0.47 | 0.97 | 0.72 | 0.19 | 0.27 |
| Chlorine | 0.27 | 3.95 | 0.29 | 2.77 | 0.84 | 0.38 |
| Sodium | 1.08 | 2.84 | 2.69 | 2.77 | 0.45 | 0.30 |
| Potassium | 0.22 | 0.70 | 0.30 | 0.50 | 0.85 | 0.06 |
| Calcium | 0.058 | 0.137 | 0.96 | 0.117 | 0.198 | 0.263 |
| Magnesium | 0.056 | 0.118 | 0.045 | 0.082 | 0.051 | 0.057 |
| Manganese | 0.0000 | 0.0002 | 0.0000 | 0.0001 | 0.0000 | 0.0000 |
| Iron | 0.0033 | 0.0056 | 0.0104 | 0.0080 | 0.0041 | 0.0023 |
| Ash | 3.53 | 9.61 | 7.28 | 8.45 | 3.90 | 2.50 |
| Ether-soluble extract | 0.37 | 0.32 | 0.30 | 0.31 | 0.30 | 0.31 |
| Reaction pH[a] | 7.0 | 6.8 | 7.2 | 7.3 | 6.8 | 6.7 |

[a] 1% solution in distilled water after autoclaving for 15 minutes at 121°C.

Data taken from: *Difco Manual of Dehydrated Culture Media and Reagents,* 9th ed., p. 265. Difco Laboratories, Detroit, Michigan (1969).

## TABLE 5
## ANALYSIS OF SOY PROTEIN HYDROLYSATE

| Representative Analysis | | Nitrogen Distribution | |
|---|---|---|---|
| Protein | 67.07% | Proteoses | 54% |
| Ash | 15.28% | Peptones | 45% |
| NFE | 11.20% | Amino N | 1% |
| Moisture | 6.45% | | |

### Amino Acid Profile

| | | | |
|---|---|---|---|
| Arginine | 6.6% | Threonine | 3.5% |
| Cystine | 1.7% | Tryptophan | 1.2% |
| Histidine | 2.6% | Valine | 4.7% |
| Isoleucine | 5.4% | Tyrosine | 3.6% |
| Leucine | 7.2% | Serine | 3.8% |
| Lysine | 6.1% | Glutamic acid | 16.5% |
| Methionine | 1.5% | Alanine | 3.0% |
| Phenylalanine | 4.7% | Proline | 4.5% |

Data furnished by J. H. DeLamar and Son, Inc., Chicago, Illinois.

# TABLE 6
# ANALYSES OF REPRESENTATIVE SOY PRODUCTS

| Constituent | Powdered Soya | | Meat Packers Lo-Fat Soy Flour | Extracted Soy Flour | | Lecithinated Soy Flour | | Extracted Soy Grits[a] | | |
|---|---|---|---|---|---|---|---|---|---|---|
| | Hi-Fat I | Lo-Fat I | F | I-200 | F-200 | 5% | 15% | I | F | FF |
| **General Analysis** | | | | | | | | | | |
| Moisture, % | 6.0 | 6.0 | 6.0 | 6.0 | 6.0 | 7.0 | 6.0 | 10.0 | 10.0 | 11.0 |
| Protein (N x 6.25) | 46.0 | 52.5 | 52.5 | 53.0 | 53.0 | 51.0 | 45.5 | 53.0 | 53.0 | 53.0 |
| Protein solubility, %[b] | 60.0 | 60.0 | 20.0 | 60.0 | 20.0 | 60.0 | 60.0 | 55.0 | 20.0 | 20.0 |
| Fat, % | 14.5 | 4.0 | 4.5 | 0.6 | 0.6 | 6.5 | 15.5 | 0.6 | 0.6 | 0.6 |
| Carbohydrates, % | 28.0 | 31.5 | 31.0 | 33.9 | 33.9 | 29.5 | 28.0 | 29.9 | 29.9 | 29.9 |
| Fiber, %[c] | 2.5 | 2.5 | 2.5 | 2.5 | 2.5 | 2.5 | 2.5 | 2.5 | 2.5 | 2.5 |
| Ash, % | 5.5 | 6.0 | 6.0 | 6.5 | 6.5 | 6.0 | 5.5 | 6.5 | 6.5 | 6.5 |
| Water absorption, % | 250 | 250 | 250 | 275 | 275 | 300 | 250 | 350 | 300 | 290 |
| Urease activity[d] | 25.0 | 25.0 | 2.0 | 25.0 | 2.0 | 25.0 | 25.0 | 25.0 | 2.0 | 0.5 |
| **Vitamin Content per Pound** | | | | | | | | | | |
| Vitamin A, I.U. | 204 | 110 | 100 | 71 | 70 | 107 | 204 | 70 | 69 | 70 |
| Vitamin $B_1$, mg | 5.37 | 6.04 | 6.00 | 6.13 | 6.06 | 5.82 | 5.37 | 6.00 | 5.93 | 6.00 |
| Riboflavin, mg | 1.53 | 1.36 | 1.40 | 1.58 | 1.58 | 1.46 | 1.53 | 1.58 | 1.58 | 1.58 |
| Pantothenic acid, mg | 10.5 | 11.8 | 11.8 | 12.0 | 11.8 | 11.5 | 10.5 | 11.7 | 11.6 | 11.6 |
| Niacin, mg | 10.1 | 11.3 | 11.5 | 11.5 | 11.4 | 10.9 | 10.1 | 11.3 | 11.1 | 11.2 |
| Pyridoxine, mg | 3.4 | 3.7 | 3.7 | 3.8 | 3.8 | 3.6 | 3.4 | 3.8 | 3.8 | 3.8 |
| Biotin, mg | 0.29 | 0.26 | 0.30 | 0.30 | 0.30 | 0.27 | 0.29 | 0.30 | 0.30 | 0.30 |
| Inositol, mg | 1518 | 1706 | 1700 | 1630 | 1712 | 2011 | 1518 | 1704 | 1684 | 1700 |
| Choline, mg | 1171 | 1308 | 1308 | 1330 | 1317 | 1970 | 1171 | 1298 | 1285 | 1290 |
| **Minerals** | | | | | | | | | | |
| Calcium, % | 0.33 | 0.38 | 0.38 | 0.38 | 0.37 | 0.36 | 0.33 | 0.37 | 0.37 | 0.37 |
| Phosphorus | 0.68 | 0.76 | 0.76 | 0.78 | 0.77 | 0.85 | 0.68 | 0.76 | 0.75 | 0.85 |
| Magnesium, % | 0.25 | 0.28 | 0.28 | 0.28 | 0.28 | 0.27 | 0.25 | 0.27 | 0.27 | 0.28 |
| Manganese, % | 0.004 | 0.004 | 0.004 | 0.004 | 0.004 | 0.004 | 0.004 | 0.004 | 0.004 | 0.004 |
| Potassium, % | 1.64 | 1.84 | 1.83 | 1.87 | 1.85 | 1.78 | 1.64 | 1.83 | 1.81 | 1.83 |
| Iron, % | 0.009 | 0.010 | 0.010 | 0.010 | 0.010 | 0.010 | 0.009 | 0.010 | 0.010 | 0.010 |
| Copper, % | 0.001 | 0.001 | 0.001 | 0.001 | 0.001 | 0.001 | 0.001 | 0.001 | 0.001 | 0.001 |

[a] Approximate screen size: fine, 64 (I); medium, 28–64 (I); coarse, 8–20 (F).
[b] In water.
[c] Included in total carbohydrates.
[d] cc of $N/10$ HCl per gram of product.

Data furnished by A. E. Staley Manufacturing Co., Decatur, Illinois.

# TABLE 7
# ANALYSIS OF COTTON SEED FLOUR (PHARMAMEDIA)

## Typical Analysis, %

| | |
|---|---|
| Total solids | 95.30 |
| Protein (N x 6.25) | 56.19 |
| Amino nitrogen (lysine as standard) | 4.67 |
| Ammonia nitrogen | 1.27 |
| NFE (carbohydrates) | 24.13 |
| Reducing sugars | 1.19 |
| Non-reducing sugars | 1.16 |
| Fat (linoleic, oleic, and palmitic fatty acids) | 5.53 |
| Ash | 6.51 |
| Fiber | 2.94 |
| Moisture | 4.70 |
| Gossypol (a yellow pigment) | 0.07 |
| pH (aqueous solution) | 6 |

## Solubles, %[a]

| | |
|---|---|
| Total solubles | 32.56 |
| Soluble amino nitrogen | 1.65 |
| Soluble phosphorus | 0.52 |
| Soluble iron | 0.0015 |
| Soluble magnesium | 0.363 |

## Amino Acids, %[b]

| | | | |
|---|---|---|---|
| Lysine | 3.04 | Aspartic acid | 7.54 |
| Leucine | 5.54 | Serine | 4.00 |
| Isoleucine | 2.87 | Glutamic acid | 17.39 |
| Threonine | 2.90 | Proline | 3.85 |
| Valine | 3.99 | Glycine | 3.72 |
| Phenylalanine | 4.91 | Alanine | 3.78 |
| Tryptophan | 0.55 | Tyrosine | 3.10 |
| Methionine | 1.65 | Histidine | 2.16 |
| Cystine | 1.43 | Arginine | 7.12 |

## Minerals, ppm

| | | | |
|---|---|---|---|
| Calcium | 2,360 | Sulfates | 780 |
| Chlorides | 486 | Magnesium | 6,800 |
| Phosphorus | 12,200 | Potassium | 14,300 |
| Iron | 103 | Sodium | 175 |

## Vitamins, mg/lb

| | | | |
|---|---|---|---|
| Carotene | 0.17 | Pantothenic acid | 19.6 |
| Total tocopherols | 8.2 | Choline | 1,470 |
| Ascorbic acid | 20.7 | Pyridoxine | 5.4 |
| Thiamine | 7.1 | Biotin | 0.23 |
| Riboflavine | 4.9 | Folic acid | 0.84 |
| Niacin | 27.1 | Inositol | 4,900 |

[a] All values were determined relative to 1 g of Pharmamedia, which was autoclaved in 100 ml of water for 15 minutes at 120°C.

[b] Moore-Stein technique, calculated on 16% nitrogen basis; Pharmamedia are based on 9.10% nitrogen.

Data furnished by Traders, Protein Division, Fort Worth, Texas.

## TABLE 8
## GENERAL ANALYSIS OF CORN STEEP LIQUOR

| Determination | Percent | Determination | Percent |
|---|---|---|---|
| Water[a] | 45–55 | Lactic acid[a] | 5–15 |
| Total Kjeldahl N[a] | 2.7–4.5* | Ash[a] | 9–10 |
| Van Slyke amino N[b] | 1.0–1.8* | Volatile acids (as acetic acid)[d] | 0.1–0.3 |
| Volatile N[c] | 0.15–0.40 | $SO_2$[d] | 0.009–0.015 |
| Free reducing sugar (as glucose)[a] | 0.1–11.0 | | |

* The amino-*N*/total Kjeldahl-N ratio ranges from 0.30 to 0.50.
[a] Based on determinations made on 1,000 different lots.
[b] Based on determinations made on 50 different lots.
[c] Based on determinations made on 15 different lots.
[d] Based on determinations made on 10 different lots.

Data taken from: Liggett, R. W., and Koffler, H., *Bacteriol. Rev., 12,* 302 (1948). Reproduced by permission of the American Society for Microbiology, Washington, D.C.

## TABLE 9
## COMPOSITION OF THE ASH OF CORN STEEP LIQUOR

| Element | Percent of Dry Matter I[a] | Percent of Dry Matter II[b] | Element | Percent of Dry Matter I[a] | Percent of Dry Matter II[b] |
|---|---|---|---|---|---|
| Al | | 0.032 | Mn | 0.004 | 0.012 |
| Ca | 0.5–1.5 | | Mo | | 0.0006 |
| Cd | | 0.0029 | P | 2.0–3.0 | 2.75 |
| Cu | 0–0.001 | 0.0033 | K | 1.0–2.0 | |
| Fe | 0.01–0.05 | 0.052 | S | 0.34 | |
| Pb | | 0.055 | Zn | 0.005 | 0.0005 |
| Mg | 0.5–1.0 | 1.05 | | | |

[a] I = range of quantitative determinations made on ten different lots.
[b] II = data on a single lot.[1]

*Note:* A semiquantitative spectroscopic analysis on a single lot[2] indicates the presence of the following elements: Al, As, B, Ca, Cr, Co, Cu, Fe, Pb, Li, Mg, Mn, Ni, P, K, Si, Ag, Sn, W, and Zn; the following metals were not detected: Sb, Be, Bi, Cd, Cb, Ce, Au, La, Hg, Pt, Sr, Ta, Ti, V, and Zr. In another spectroscopic analysis, also on a single lot, the workers[3] found the following elements present: Ba, Pb, Mo, Ni, Rb, Sn, V, and Zn; Cr, Co, Li, and Ag were absent.

Data taken from: Liggett, R. W., and Koffler, H., *Bacteriol. Rev., 12,* 303 (1948). Reproduced by permission of the American Society for Microbiology, Washington, D.C.

### REFERENCES

1. Perlman, D., Thesis, Penicillin Production in Submerged Culture. University of Wisconsin, Madison, Wisconsin (1945).
2. Koffler, H., Knight, S. G., and Frazier, W. C., *J. Bacteriol., 53,* 115 (1947).
3. Cook, R. O., Tullock, W. J., Brown, M. B., and Brodie, J., *Biochem. J., 39,* 314 (1945).

# A NOMOGRAM FOR AMMONIUM SULFATE SOLUTIONS

In the purification of enzymes and other proteins it is frequently necessary to fractionate with ammonium sulfate by collecting the fraction that is precipitated on passing from one given percentage of saturation to another. One is then faced with the problem of calculating how much solid ammonium sulfate must be added to a given volume of a solution of a known degree of saturation to bring it to a certain desired degree of saturation.

Such a calculation from physical tables is distinctly laborious, due to the form in which the data are given. The main difficulty is caused by the fact that, owing to change of volume, the degree of saturation is far from being proportional to the amount of the salt added to a given volume of water.

The nomogram shown below, which is self-explanatory, avoids the necessity for calculation. A straight line through the initial saturation and the desired saturation gives the amount of solid $(NH_4)_2SO_4$ to be added to one liter of the solution; a line from this point passing through the volume of the solution gives the amount required. An example is indicated by the dotted lines. This shows that 500 ml of a 0.25-saturated solution requires the addition of 193 g of solid ammonium sulfate to bring it to 0.8 saturation. It must be noted that the final scale reads downwards. Full saturation is taken as being given by adding 760 g to one liter of water, i.e., as saturation at room temperature.

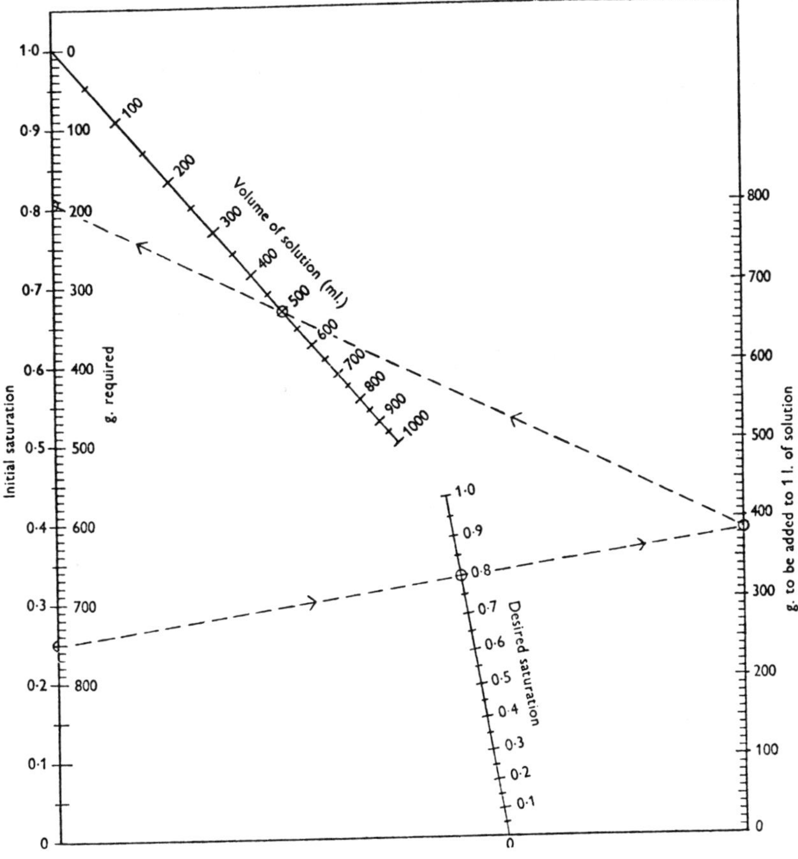


Taken from: Dixon, M., *Biochem. J., 54,* 457–458 (1953). Reproduced by permission of The Biochemical Society, London, England.

## RECIPROCALS OF NUMBERS

For given values of N, the following table contains the corresponding values of 1000/N, N + 273.15* and 1000/N + 273.15.

| N | 1000/N | N + 273.15 | 1000/N + 273.15 |
|---|---|---|---|
| −273.15 | −3.66099 | 0.00 | |
| .14 | 6113 | .01 | 100000.000 |
| .13 | 6126 | .02 | 50000.000 |
| .12 | 6139 | .03 | 33333.333 |
| .11 | 6153 | .04 | 25000.000 |
| .10 | 6166 | .05 | 20000.000 |
| .09 | 6180 | .06 | 16666.667 |
| .08 | 6193 | .07 | 14285.714 |
| .07 | 6206 | .08 | 12500.000 |
| .06 | 6220 | .09 | 11111.111 |
| .05 | 6233 | .10 | 10000.000 |
| .04 | 6247 | .11 | 9090.909 |
| .03 | 6260 | .12 | 8333.333 |
| .02 | 6274 | .13 | 7692.308 |
| .01 | 6287 | .14 | 7142.857 |
| −273.00 | 6300 | .15 | 6666.667 |
| −272.99 | 6314 | .16 | 6250.000 |
| .98 | 6327 | .17 | 5882.353 |
| .97 | 6341 | .18 | 5555.556 |
| .96 | 6354 | .19 | 5263.158 |
| .95 | 6367 | .20 | 5000.000 |
| .94 | 6381 | .21 | 4761.905 |
| .93 | 6394 | .22 | 4545.455 |
| .92 | 6408 | .23 | 4347.826 |
| .91 | 6421 | .24 | 4166.667 |
| .90 | 6435 | .25 | 4000.000 |
| .89 | 6448 | .26 | 3846.154 |
| .88 | 6461 | .27 | 3703.704 |
| .87 | 6475 | .28 | 3571.429 |
| .86 | 6488 | .29 | 3448.276 |
| .85 | 6502 | .30 | 3333.333 |
| .84 | 6515 | .31 | 3225.806 |
| .83 | 6529 | .32 | 3125.000 |
| .82 | 6542 | .33 | 3030.303 |
| .81 | 6555 | .34 | 2941.176 |
| .80 | 6569 | .35 | 2857.143 |
| .79 | 6582 | .36 | 2777.778 |
| .78 | 6596 | .37 | 2702.703 |
| .77 | 6609 | .38 | 2631.579 |
| .76 | 6623 | .39 | 2564.103 |
| .75 | 6636 | .40 | 2500.000 |
| .74 | 6650 | .41 | 2439.024 |
| .73 | 6663 | .42 | 2380.952 |
| .72 | 6676 | .43 | 2325.581 |
| .71 | 6690 | .44 | 2272.727 |
| .70 | 6703 | .45 | 2222.222 |
| .69 | 6717 | .46 | 2173.913 |
| .68 | 6730 | .47 | 2127.660 |
| .67 | 6744 | .48 | 2083.333 |
| .66 | 6757 | .49 | 2040.816 |
| .65 | 6771 | .50 | 2000.000 |
| .64 | 6784 | .51 | 1960.784 |
| .63 | 6797 | .52 | 1923.077 |
| .62 | 6811 | .53 | 1886.792 |
| .61 | 6824 | .54 | 1851.852 |
| .60 | 6838 | .55 | 1818.182 |
| .59 | 6851 | .56 | 1785.714 |
| .58 | 6865 | .57 | 1754.386 |
| .57 | 6878 | .58 | 1724.138 |
| .56 | 6892 | .59 | 1694.915 |
| .55 | 6905 | .60 | 1666.667 |
| .54 | 6919 | .61 | 1639.344 |
| .53 | 6932 | .62 | 1612.903 |
| .52 | 6946 | .63 | 1587.302 |
| .51 | 6959 | .64 | 1562.500 |
| .50 | 6972 | .65 | 1538.462 |
| .49 | 6986 | .66 | 1515.152 |
| .48 | 6999 | .67 | 1492.537 |
| .47 | 7013 | .68 | 1470.588 |
| .46 | 7026 | .69 | 1449.275 |
| .45 | 7040 | .70 | 1428.571 |

| N | 1000/N | N + 273.15 | 1000/N + 273.15 |
|---|---|---|---|
| −272.44 | −3.67053 | 0.71 | 1408.451 |
| .43 | 7067 | .72 | 1388.889 |
| .42 | 7080 | .73 | 1369.863 |
| .41 | 7094 | .74 | 1351.351 |
| .40 | 7107 | .75 | 1333.333 |
| .39 | 7121 | .76 | 1315.789 |
| .38 | 7134 | .77 | 1298.701 |
| .37 | 7148 | .78 | 1282.051 |
| .36 | 7161 | .79 | 1265.823 |
| .35 | 7175 | .80 | 1250.000 |
| .34 | 7188 | .81 | 1234.568 |
| .33 | 7202 | .82 | 1219.512 |
| .32 | 7215 | .83 | 1204.819 |
| .31 | 7229 | .84 | 1190.476 |
| .30 | 7242 | .85 | 1176.471 |
| .29 | 7255 | .86 | 1162.791 |
| .28 | 7269 | .87 | 1149.425 |
| .27 | 7282 | .88 | 1136.364 |
| .26 | 7296 | .89 | 1123.596 |
| .25 | 7309 | .90 | 1111.111 |
| .24 | 7323 | .91 | 1098.901 |
| .23 | 7336 | .92 | 1086.957 |
| .22 | 7350 | .93 | 1075.269 |
| .21 | 7363 | .94 | 1063.830 |
| .20 | 7377 | .95 | 1052.632 |
| .19 | 7390 | .96 | 1041.667 |
| .18 | 7404 | .97 | 1030.928 |
| .17 | 7417 | .98 | 1020.408 |
| .16 | 7431 | .99 | 1010.101 |
| .15 | 7444 | 1.00 | 1000.000 |
| .14 | 7458 | 1.01 | 990.099 |
| .13 | 7471 | .02 | 980.392 |
| .12 | 7485 | .03 | 970.874 |
| .11 | 7498 | .04 | 961.538 |
| .10 | 7512 | .05 | 952.381 |
| .09 | 7525 | .06 | 943.396 |
| .08 | 7539 | .07 | 934.579 |
| .07 | 7552 | .08 | 925.926 |
| .06 | 7566 | .09 | 917.431 |
| .05 | 7579 | .10 | 909.091 |
| .04 | 7593 | .11 | 900.901 |
| .03 | 7607 | .12 | 892.857 |
| .02 | 7620 | .13 | 884.956 |
| .01 | 7634 | .14 | 877.193 |
| −272.00 | 7647 | .15 | 869.565 |
| −271.99 | 7661 | .16 | 862.069 |
| .98 | 7674 | .17 | 854.701 |
| .97 | 7688 | .18 | 847.458 |
| .96 | 7701 | .19 | 840.336 |
| .95 | 7715 | .20 | 833.333 |
| .94 | 7728 | .21 | 826.446 |
| .93 | 7742 | .22 | 819.672 |
| .92 | 7755 | .23 | 813.008 |
| .91 | 7769 | .24 | 806.452 |
| .90 | 7782 | .25 | 800.000 |
| .89 | 7796 | .26 | 793.651 |
| .88 | 7809 | .27 | 787.402 |
| .87 | 7823 | .28 | 781.250 |
| .86 | 7836 | .29 | 775.194 |
| .85 | 7850 | .30 | 769.231 |
| .84 | 7863 | .31 | 763.359 |
| .83 | 7877 | .32 | 757.576 |
| .82 | 7891 | .33 | 751.880 |
| .81 | 7904 | .34 | 746.269 |
| .80 | 7918 | .35 | 740.741 |
| .79 | 7931 | .36 | 735.294 |
| .78 | 7945 | .37 | 729.927 |
| .77 | 7958 | .38 | 724.638 |
| .76 | 7972 | .39 | 719.424 |
| .75 | 7985 | .40 | 714.286 |
| .74 | 7999 | .41 | 709.220 |

* For conversion from the Kelvin temperature scale to the Celsius temperature scale, N may be read as °C and N + 273.15 as °K.

| N | 1000/N | N + 273.15 | 1000/N + 273.15 |
|---|---|---|---|
| −271.73 | −3.68012 | 1.42 | 704.225 |
| .72 | 8026 | .43 | 699.301 |
| .71 | 8039 | .44 | 694.444 |
| .70 | 8053 | .45 | 689.655 |
| .69 | 8067 | .46 | 684.932 |
| .68 | 8080 | .47 | 680.272 |
| .67 | 8094 | .48 | 675.676 |
| .66 | 8107 | .49 | 671.141 |
| .65 | 8121 | .50 | 666.667 |
| .64 | 8134 | .51 | 662.252 |
| .63 | 8148 | .52 | 657.895 |
| .62 | 8161 | .53 | 653.595 |
| .61 | 8175 | .54 | 649.351 |
| .60 | 8188 | .55 | 645.161 |
| .59 | 8202 | .56 | 641.026 |
| .58 | 8216 | .57 | 636.943 |
| .57 | 8229 | .58 | 632.911 |
| .56 | 8243 | .59 | 628.931 |
| .55 | 8256 | .60 | 625.000 |
| .54 | 8270 | .61 | 621.118 |
| .53 | 8283 | .62 | 617.284 |
| .52 | 8297 | .63 | 613.497 |
| .51 | 8311 | .64 | 609.756 |
| .50 | 8324 | .65 | 606.061 |
| .49 | 8338 | .66 | 602.410 |
| .48 | 8351 | .67 | 598.802 |
| .47 | 8365 | .68 | 595.238 |
| .46 | 8378 | .69 | 591.716 |
| .45 | 8392 | .70 | 588.235 |
| .44 | 8406 | .71 | 584.795 |
| .43 | 8419 | .72 | 581.395 |
| .42 | 8433 | .73 | 578.035 |
| .41 | 8446 | .74 | 574.713 |
| .40 | 8460 | .75 | 571.429 |
| .39 | 8473 | .76 | 568.182 |
| .38 | 8487 | .77 | 564.972 |
| .37 | 8501 | .78 | 561.798 |
| .36 | 8514 | .79 | 558.659 |
| .35 | 8528 | .80 | 555.556 |
| .34 | 8541 | .81 | 552.486 |
| .33 | 8555 | .82 | 549.451 |
| .32 | 8568 | .83 | 546.448 |
| .31 | 8582 | .84 | 543.478 |
| .30 | 8596 | .85 | 540.541 |
| .29 | 8609 | .86 | 537.634 |
| .28 | 8623 | .87 | 534.759 |
| .27 | 8636 | .88 | 531.915 |
| .26 | 8650 | .89 | 529.101 |
| .25 | 8664 | .90 | 526.316 |
| .24 | 8677 | .91 | 523.560 |
| .23 | 8691 | .92 | 520.833 |
| .22 | 8704 | .93 | 518.135 |
| .21 | 8718 | .94 | 515.464 |
| .20 | 8732 | .95 | 512.821 |
| .19 | 8745 | .96 | 510.204 |
| .18 | 8759 | .97 | 507.614 |
| .17 | 8772 | .98 | 505.051 |
| .16 | 8786 | 1.99 | 502.513 |
| .15 | 8800 | 2.00 | 500.000 |
| .14 | 8813 | .01 | 497.512 |
| .13 | 8827 | .02 | 495.050 |
| .12 | 8840 | .03 | 492.611 |
| .11 | 8854 | .04 | 490.196 |
| .10 | 8868 | .05 | 487.805 |
| .09 | 8881 | .06 | 485.437 |
| .08 | 8895 | .07 | 483.092 |
| .07 | 8908 | .08 | 480.769 |
| .06 | 8922 | .09 | 478.469 |
| .05 | 8936 | .10 | 476.190 |
| .04 | 8949 | .11 | 473.934 |
| .03 | 8963 | .12 | 471.698 |
| .02 | 8976 | .13 | 469.484 |
| .01 | 8990 | .14 | 467.290 |
| −271.00 | 9004 | .15 | 465.116 |

| N | 1000/N | N + 273.15 | 1000/N + 273.1 |
|---|---|---|---|
| −270.99 | 9017 | .16 | 462.963 |
| .98 | 9031 | .17 | 460.829 |
| .97 | 9045 | .18 | 458.716 |
| .96 | 9058 | .19 | 456.621 |
| .95 | 9072 | .20 | 454.545 |
| .94 | 9085 | .21 | 452.489 |
| .93 | 9099 | .22 | 450.450 |
| .92 | 9113 | .23 | 448.430 |
| .91 | 9126 | .24 | 446.429 |
| .90 | 9140 | .25 | 444.444 |
| .89 | 9154 | .26 | 442.478 |
| .88 | 9167 | .27 | 440.529 |
| .87 | 9181 | .28 | 438.596 |
| .86 | 9194 | .29 | 436.681 |
| .85 | 9208 | .30 | 434.783 |
| .84 | 9222 | .31 | 432.900 |
| .83 | 9235 | .32 | 431.034 |
| .82 | 9249 | .33 | 428.185 |
| .81 | 9263 | .34 | 427.350 |
| .80 | 9276 | .35 | 425.532 |
| .79 | 9290 | .36 | 423.729 |
| .78 | 9303 | .37 | 421.941 |
| .77 | 9317 | .38 | 420.168 |
| .76 | 9331 | .39 | 418.410 |
| .75 | 9344 | .40 | 416.667 |
| .74 | 9358 | .41 | 414.938 |
| .73 | 9372 | .42 | 413.223 |
| .72 | 9385 | .43 | 411.523 |
| .71 | 9399 | .44 | 409.836 |
| .70 | 9413 | .45 | 408.163 |
| .69 | 9426 | .46 | 406.504 |
| .68 | 9440 | .47 | 404.858 |
| .67 | 9454 | .48 | 403.226 |
| .66 | 9467 | .49 | 401.606 |
| .65 | 9481 | .50 | 400.000 |
| .64 | 9495 | .51 | 398.406 |
| .63 | 9508 | .52 | 396.825 |
| .62 | 9522 | .53 | 395.257 |
| .61 | 9535 | .54 | 393.701 |
| .60 | 9549 | .55 | 392.157 |
| .59 | 9563 | .56 | 390.625 |
| .58 | 9576 | .57 | 389.105 |
| .57 | 9590 | .58 | 387.597 |
| .56 | 9604 | .59 | 386.100 |
| .55 | 9617 | .60 | 384.615 |
| .54 | 9631 | .61 | 383.142 |
| .53 | 9645 | .62 | 381.679 |
| .52 | 9658 | .63 | 380.228 |
| .51 | 9672 | .64 | 378.788 |
| .50 | 9686 | .65 | 377.358 |
| .49 | 9699 | .66 | 375.940 |
| .48 | 9713 | .67 | 374.532 |
| .47 | 9727 | .68 | 373.134 |
| .46 | 9740 | .69 | 371.747 |
| .45 | 9754 | .70 | 370.370 |
| .44 | 9768 | .71 | 369.004 |
| .43 | 9781 | .72 | 367.647 |
| .42 | 9795 | .73 | 366.300 |
| .41 | 9809 | .74 | 364.964 |
| .40 | 9822 | .75 | 363.636 |
| .39 | 9836 | .76 | 362.319 |
| .38 | 9850 | .77 | 361.011 |
| .37 | 9864 | .78 | 359.712 |
| .36 | 9877 | .79 | 358.423 |
| .35 | 9891 | .80 | 357.143 |
| .34 | 9905 | .81 | 355.872 |
| .33 | 9918 | .82 | 354.610 |
| .32 | 9932 | .83 | 353.357 |
| .31 | 9946 | .84 | 352.113 |
| .30 | 9959 | .85 | 350.877 |
| .29 | 9973 | .86 | 349.650 |
| .28 | 9987 | .87 | 348.432 |
| .27 | −3.70000 | .88 | 347.222 |
| .26 | 0014 | .89 | 346.021 |

| N | 1000/N | N + 273.15 | 1000/N + 273.15 |
|---|---|---|---|
| 270.25 | −3.70028 | 2.90 | 344.828 |
| .24 | 0041 | .91 | 343.643 |
| .23 | 0055 | .92 | 342.466 |
| .22 | 0069 | .93 | 341.297 |
| .21 | 0083 | .94 | 340.136 |
| .20 | 0096 | .95 | 338.983 |
| .19 | 0110 | .96 | 337.838 |
| .18 | 0124 | .97 | 336.700 |
| .17 | 0137 | .98 | 335.570 |
| .16 | 0151 | 2.99 | 334.448 |
| .15 | 0165 | 3.00 | 333.333 |
| .14 | 0178 | 3.01 | 332.226 |
| .13 | 0192 | .02 | 331.126 |
| .12 | 0206 | .03 | 330.033 |
| .11 | 0220 | .04 | 328.947 |
| .10 | 0233 | .05 | 327.869 |
| .09 | 0257 | .06 | 326.797 |
| .08 | 0261 | .07 | 325.733 |
| .07 | 0274 | .08 | 324.675 |
| .06 | 0288 | .09 | 323.625 |
| .05 | 0302 | .10 | 322.581 |
| .04 | 0316 | .11 | 321.543 |
| .03 | 0329 | .12 | 320.513 |
| .02 | 0343 | .13 | 319.489 |
| .01 | 0357 | .14 | 318.471 |
| −270.00 | 0370 | .15 | 317.460 |
| −269.99 | 0384 | .16 | 316.456 |
| .98 | 0398 | .17 | 315.457 |
| .97 | 0412 | .18 | 314.465 |
| .96 | 0425 | .19 | 313.480 |
| .95 | 0439 | .20 | 312.500 |
| .94 | 0453 | .21 | 311.526 |
| .93 | 0466 | .22 | 310.559 |
| .92 | 0480 | .23 | 309.598 |
| .91 | 0494 | .24 | 308.642 |
| .90 | 0508 | .25 | 307.692 |
| .89 | 0521 | .26 | 306.748 |
| .88 | 0535 | .27 | 305.810 |
| .87 | 0549 | .28 | 304.878 |
| .86 | 0563 | .29 | 303.951 |
| .85 | 0576 | .30 | 303.030 |
| .84 | 0590 | .31 | 302.115 |
| .83 | 0604 | .32 | 301.205 |
| .82 | 0617 | .33 | 300.300 |
| .81 | 0631 | .34 | 299.401 |
| .80 | 0645 | .35 | 298.507 |
| .79 | 0659 | .36 | 297.619 |
| .78 | 0672 | .37 | 296.736 |
| .77 | 0686 | .38 | 295.858 |
| .76 | 0700 | .39 | 294.985 |
| .75 | 0714 | .40 | 294.118 |
| .74 | 0727 | .41 | 293.255 |
| .73 | 0741 | .42 | 292.398 |
| .72 | 0755 | .43 | 291.545 |
| .71 | 0769 | .44 | 290.698 |
| .70 | 0782 | .45 | 289.855 |
| .69 | 0796 | .46 | 289.017 |
| .68 | 0810 | .47 | 288.184 |
| .67 | 0824 | .48 | 287.356 |
| .66 | 0837 | .49 | 286.533 |
| .65 | 0851 | .50 | 285.714 |
| .64 | 0865 | .51 | 284.900 |
| .63 | 0879 | .52 | 284.091 |
| .62 | 0892 | .53 | 283.286 |
| .61 | 0906 | .54 | 282.486 |
| .60 | 0920 | .55 | 281.690 |
| .59 | 0934 | .56 | 280.899 |
| .58 | 0947 | .57 | 280.112 |
| .57 | 0961 | .58 | 279.330 |
| .56 | 0975 | .59 | 278.552 |
| .55 | 0989 | .60 | 277.778 |
| .54 | 1002 | .61 | 277.008 |
| .53 | 1016 | .62 | 276.243 |
| .52 | 1030 | .63 | 275.482 |

| N | 1000/N | N + 273.15 | 1000/N + 273.15 |
|---|---|---|---|
| −269.51 | −3.71044 | 3.64 | 274.725 |
| .50 | 1058 | .65 | 273.973 |
| .49 | 1071 | .66 | 273.224 |
| .48 | 1085 | .67 | 272.480 |
| .47 | 1099 | .68 | 271.739 |
| .46 | 1112 | .69 | 271.003 |
| .45 | 1126 | .70 | 270.270 |
| .44 | 1140 | .71 | 269.542 |
| .43 | 1154 | .72 | 268.817 |
| .42 | 1168 | .73 | 268.097 |
| .41 | 1181 | .74 | 267.380 |
| .40 | 1195 | .75 | 266.667 |
| .39 | 1209 | .76 | 265.957 |
| .38 | 1223 | .77 | 265.252 |
| .37 | 1237 | .78 | 264.550 |
| .36 | 1250 | .79 | 263.852 |
| .35 | 1264 | .80 | 263.158 |
| .34 | 1278 | .81 | 262.467 |
| .33 | 1292 | .82 | 261.780 |
| .32 | 1306 | .83 | 261.097 |
| .31 | 1319 | .84 | 260.417 |
| .30 | 1330 | .85 | 259.740 |
| .29 | 1347 | .86 | 259.067 |
| .28 | 1361 | .87 | 258.398 |
| .27 | 1374 | .88 | 257.732 |
| .26 | 1388 | .89 | 257.069 |
| .25 | 1402 | .90 | 256.410 |
| .24 | 1416 | .91 | 255.754 |
| .23 | 1430 | .92 | 255.102 |
| .22 | 1443 | .93 | 254.453 |
| .21 | 1457 | .94 | 253.807 |
| .20 | 1471 | .95 | 253.165 |
| .19 | 1485 | .96 | 252.525 |
| .18 | 1499 | .97 | 251.889 |
| .17 | 1512 | .98 | 251.256 |
| .16 | 1526 | 3.99 | 250.627 |
| .15 | 1540 | 4.00 | 250.000 |
| .14 | 1554 | 4.01 | 249.377 |
| .13 | 1568 | .02 | 248.756 |
| .12 | 1581 | .03 | 248.139 |
| .11 | 1595 | .04 | 247.525 |
| .10 | 1609 | .05 | 246.914 |
| .09 | 1623 | .06 | 246.305 |
| .08 | 1637 | .07 | 245.700 |
| .07 | 1650 | .08 | 245.098 |
| .06 | 1664 | .09 | 244.499 |
| .05 | 1678 | .10 | 243.902 |
| .04 | 1692 | .11 | 243.309 |
| .03 | 1706 | .12 | 242.718 |
| .02 | 1720 | .13 | 242.131 |
| .01 | 1733 | .14 | 241.546 |
| −269.00 | 1747 | .15 | 240.964 |
| −268.99 | 1761 | .16 | 240.385 |
| .98 | 1775 | .17 | 239.808 |
| .97 | 1789 | .18 | 239.234 |
| .96 | 1802 | .19 | 238.663 |
| .95 | 1816 | .20 | 238.095 |
| .94 | 1830 | .21 | 237.530 |
| .93 | 1844 | .22 | 236.967 |
| .92 | 1858 | .23 | 236.407 |
| .91 | 1872 | .24 | 235.849 |
| .90 | 1885 | .25 | 235.294 |
| .89 | 1899 | .26 | 234.742 |
| .88 | 1913 | .27 | 234.192 |
| .87 | 1927 | .28 | 233.645 |
| .86 | 1941 | .29 | 233.100 |
| .85 | 1955 | .30 | 232.558 |
| .84 | 1968 | .31 | 232.019 |
| .83 | 1982 | .32 | 231.481 |
| .82 | 1996 | .33 | 230.947 |
| .81 | 2010 | .34 | 230.415 |
| .80 | 2024 | .35 | 229.885 |
| .79 | 2038 | .36 | 229.358 |
| .78 | 2051 | .37 | 228.833 |

| N | 1000/N | N + 273.15 | 1000/N + 273.15 |
|---|---|---|---|
| −268.77 | −3.72065 | 4.38 | 228.311 |
| .76 | 2079 | .39 | 227.790 |
| .75 | 2093 | .40 | 227.273 |
| .74 | 2107 | .41 | 226.757 |
| .73 | 2121 | .42 | 226.244 |
| .72 | 2135 | .43 | 225.734 |
| .71 | 2148 | .44 | 225.225 |
| .70 | 2162 | .45 | 224.719 |
| .69 | 2176 | .46 | 224.215 |
| .68 | 2190 | .47 | 223.714 |
| .67 | 2204 | .48 | 223.214 |
| .66 | 2218 | .49 | 222.717 |
| .65 | 2232 | .50 | 222.222 |
| .64 | 2245 | .51 | 221.729 |
| .63 | 2259 | .52 | 221.239 |
| .62 | 2273 | .53 | 220.751 |
| .61 | 2287 | .54 | 220.264 |
| .60 | 2301 | .55 | 219.780 |
| .59 | 2315 | .56 | 219.298 |
| .58 | 2329 | .57 | 218.818 |
| .57 | 2342 | .58 | 218.341 |
| .56 | 2356 | .59 | 217.865 |
| .55 | 2370 | .60 | 217.391 |
| .54 | 2384 | .61 | 216.920 |
| .53 | 2398 | .62 | 216.450 |
| .52 | 2412 | .63 | 215.983 |
| .51 | 2426 | .64 | 215.517 |
| .50 | 2439 | .65 | 215.054 |
| .49 | 2453 | .66 | 214.592 |
| .48 | 2467 | .67 | 214.133 |
| .47 | 2481 | .68 | 213.675 |
| .46 | 2495 | .69 | 213.220 |
| .45 | 2509 | .70 | 212.766 |
| .44 | 2523 | .71 | 212.314 |
| .43 | 2537 | .72 | 211.864 |
| .42 | 2550 | .73 | 211.416 |
| .41 | 2564 | .74 | 210.970 |
| .40 | 2578 | .75 | 210.526 |
| .39 | 2592 | .76 | 210.084 |
| .38 | 2606 | .77 | 209.644 |
| .37 | 2620 | .78 | 209.205 |
| .36 | 2634 | .79 | 208.768 |
| .35 | 2648 | .80 | 208.333 |
| .34 | 2662 | .81 | 207.900 |
| .33 | 2675 | .82 | 207.469 |
| .32 | 2689 | .83 | 207.039 |
| .31 | 2703 | .84 | 206.612 |
| .30 | 2717 | .85 | 206.186 |
| .29 | 2731 | .86 | 205.761 |
| .28 | 2745 | .87 | 205.339 |
| .27 | 2759 | .88 | 204.918 |
| .26 | 2773 | .89 | 204.499 |
| .25 | 2787 | .90 | 204.082 |
| .24 | 2800 | .91 | 203.666 |
| .23 | 2814 | .92 | 203.252 |
| .22 | 2828 | .93 | 202.840 |
| .21 | 2842 | .94 | 202.429 |
| .20 | 2856 | .95 | 202.020 |
| .19 | 2870 | .96 | 201.613 |
| .18 | 2884 | .97 | 201.207 |
| .17 | 2898 | .98 | 200.803 |
| .16 | 2912 | 4.99 | 200.401 |
| .15 | 2926 | 5.00 | 200.000 |
| .14 | 2940 | 5.01 | 199.601 |
| .13 | 2953 | .02 | 199.203 |
| .12 | 2967 | .03 | 198.807 |
| .11 | 2981 | .04 | 198.413 |
| .10 | 2995 | .05 | 198.020 |
| .09 | 3009 | .06 | 197.628 |
| .08 | 3023 | .07 | 197.239 |
| .07 | 3037 | .08 | 196.850 |
| .06 | 3051 | .09 | 196.464 |
| .05 | 3065 | .10 | 196.078 |
| .04 | 3079 | .11 | 195.695 |

| N | 1000/N | N + 273.15 | 1000/N + 273. |
|---|---|---|---|
| −268.03 | −3.73093 | 5.12 | 195.313 |
| .02 | 3106 | .13 | 194.932 |
| .01 | 3120 | .14 | 194.553 |
| −268.00 | 3134 | .15 | 194.175 |
| −267.99 | 3148 | .16 | 193.798 |
| .98 | 3162 | .17 | 193.424 |
| .97 | 3176 | .18 | 193.050 |
| .96 | 3190 | .19 | 192.678 |
| .95 | 3204 | .20 | 192.308 |
| .94 | 3218 | .21 | 191.939 |
| .93 | 3232 | .22 | 191.571 |
| .92 | 3246 | .23 | 191.205 |
| .91 | 3260 | .24 | 190.840 |
| .90 | 3274 | .25 | 190.476 |
| .89 | 3288 | .26 | 190.114 |
| .88 | 3301 | .27 | 189.753 |
| .87 | 3315 | .28 | 189.394 |
| .86 | 3329 | .29 | 189.036 |
| .85 | 3343 | .30 | 188.679 |
| .84 | 3357 | .31 | 188.324 |
| .83 | 3371 | .32 | 187.970 |
| .82 | 3385 | .33 | 187.617 |
| .81 | 3399 | .34 | 187.266 |
| .80 | 3413 | .35 | 186.916 |
| .79 | 3427 | .36 | 186.567 |
| .78 | 3441 | .37 | 186.220 |
| .77 | 3455 | .38 | 185.874 |
| .76 | 3469 | .39 | 185.529 |
| .75 | 3483 | .40 | 185.185 |
| .74 | 3497 | .41 | 184.843 |
| .73 | 3511 | .42 | 184.502 |
| .72 | 3525 | .43 | 184.162 |
| .71 | 3539 | .44 | 183.824 |
| .70 | 3552 | .45 | 183.486 |
| .69 | 3566 | .46 | 183.150 |
| .68 | 3580 | .47 | 182.815 |
| .67 | 3594 | .48 | 182.482 |
| .66 | 3608 | .49 | 182.149 |
| .65 | 3622 | .50 | 181.818 |
| .64 | 3636 | .51 | 181.488 |
| .63 | 3650 | .52 | 181.159 |
| .62 | 3664 | .53 | 180.832 |
| .61 | 3678 | .54 | 180.505 |
| .60 | 3692 | .55 | 180.180 |
| .59 | 3706 | .56 | 179.856 |
| .58 | 3720 | .57 | 179.533 |
| .57 | 3734 | .58 | 179.211 |
| .56 | 3748 | .59 | 178.891 |
| .55 | 3762 | .60 | 178.571 |
| .54 | 3776 | .61 | 178.253 |
| .53 | 3790 | .62 | 177.936 |
| .52 | 3804 | .63 | 177.620 |
| .51 | 3818 | .64 | 177.305 |
| .50 | 3832 | .65 | 176.991 |
| .49 | 3846 | .66 | 176.678 |
| .48 | 3860 | .67 | 176.367 |
| .47 | 3874 | .68 | 176.056 |
| .46 | 3888 | .69 | 175.747 |
| .45 | 3902 | .70 | 175.439 |
| .44 | 3916 | .71 | 175.131 |
| .43 | 3930 | .72 | 174.825 |
| .42 | 3944 | .73 | 174.520 |
| .41 | 3958 | .74 | 174.216 |
| .40 | 3972 | .75 | 173.913 |
| .39 | 3986 | .76 | 173.611 |
| .38 | 4000 | .77 | 173.310 |
| .37 | 4014 | .78 | 173.010 |
| .36 | 4028 | .79 | 172.712 |
| .35 | 4042 | .80 | 172.414 |
| .34 | 4056 | .81 | 172.117 |
| .33 | 4070 | .82 | 171.821 |
| .32 | 4083 | .83 | 171.527 |
| .31 | 4097 | .84 | 171.233 |
| .30 | 4111 | .85 | 170.940 |

| N | 1000/N | N + 273.15 | 1000/N + 273.15 |
|---|---|---|---|
| -267.29 | −3.74125 | 5.86 | 170.648 |
| .28 | 4139 | .87 | 170.358 |
| .27 | 4153 | .88 | 170.068 |
| .26 | 4167 | .89 | 169.779 |
| .25 | 4181 | .90 | 169.492 |
| .24 | 4195 | .91 | 169.205 |
| .23 | 4209 | .92 | 168.919 |
| .22 | 4223 | .93 | 168.634 |
| .21 | 4237 | .94 | 168.350 |
| .20 | 4251 | .95 | 168.067 |
| .19 | 4266 | .96 | 167.785 |
| .18 | 4280 | .97 | 167.504 |
| .17 | 4294 | .98 | 167.224 |
| .16 | 4308 | 5.99 | 166.945 |
| .15 | 4322 | 6.00 | 166.667 |
| .14 | 4336 | .01 | 166.389 |
| .13 | 4350 | .02 | 166.113 |
| .12 | 4364 | .03 | 165.837 |
| .11 | 4378 | .04 | 165.563 |
| .10 | 4392 | .05 | 165.289 |
| .09 | 4406 | .06 | 165.017 |
| .08 | 4420 | .07 | 164.745 |
| .07 | 4434 | .08 | 164.474 |
| .06 | 4448 | .09 | 164.204 |
| .05 | 4462 | .10 | 163.934 |
| .04 | 4476 | .11 | 163.666 |
| .03 | 4490 | .12 | 163.399 |
| .02 | 4504 | .13 | 163.132 |
| .01 | 4518 | .14 | 162.866 |
| −267.00 | 4532 | .15 | 162.602 |
| −266.99 | 4546 | .16 | 162.338 |
| .98 | 4560 | .17 | 162.075 |
| .97 | 4574 | .18 | 161.812 |
| .96 | 4588 | .19 | 161.551 |
| .95 | 4602 | .20 | 161.290 |
| .94 | 4616 | .21 | 161.031 |
| .93 | 4630 | .22 | 160.772 |
| .92 | 4644 | .23 | 160.514 |
| .91 | 4658 | .24 | 160.256 |
| .90 | 4672 | .25 | 160.000 |
| .89 | 4686 | .26 | 159.744 |
| .88 | 4700 | .27 | 159.490 |
| .87 | 4714 | .28 | 159.236 |
| .86 | 4728 | .29 | 158.983 |
| .85 | 4742 | .30 | 158.730 |
| .84 | 4756 | .31 | 158.479 |
| .83 | 4770 | .32 | 158.228 |
| .82 | 4784 | .33 | 157.978 |
| .81 | 4799 | .34 | 157.729 |
| .80 | 4813 | .35 | 157.480 |
| .79 | 4827 | .36 | 157.233 |
| .78 | 4841 | .37 | 156.986 |
| .77 | 4855 | .38 | 156.740 |
| .76 | 4869 | .39 | 156.495 |
| .75 | 4883 | .40 | 156.250 |
| .74 | 4897 | .41 | 156.006 |
| .73 | 4911 | .42 | 155.763 |
| .72 | 4925 | .43 | 155.521 |
| .71 | 4939 | .44 | 155.280 |
| .70 | 4953 | .45 | 155.039 |
| .69 | 4967 | .46 | 154.799 |
| .68 | 4981 | .47 | 154.560 |
| .67 | 4995 | .48 | 154.321 |
| .66 | 5009 | .49 | 154.083 |
| .65 | 5023 | .50 | 153.846 |
| .64 | 5038 | .51 | 153.610 |
| .63 | 5052 | .52 | 153.374 |
| .62 | 5066 | .53 | 153.139 |
| .61 | 5080 | .54 | 152.905 |
| .60 | 5094 | .55 | 152.672 |
| .59 | 5108 | .56 | 152.439 |
| .58 | 5122 | .57 | 152.207 |
| .57 | 5136 | .58 | 151.976 |
| .56 | 5150 | .59 | 151.745 |

| N | 1000/N | N + 273.15 | 1000/N + 273.15 |
|---|---|---|---|
| −266.55 | −3.75164 | 6.60 | 151.515 |
| .54 | 5178 | .61 | 151.286 |
| .53 | 5192 | .62 | 151.057 |
| .52 | 5206 | .63 | 150.830 |
| .51 | 5220 | .64 | 150.602 |
| .50 | 5235 | .65 | 150.376 |
| .49 | 5249 | .66 | 150.150 |
| .48 | 5263 | .67 | 149.925 |
| .47 | 5277 | .68 | 149.701 |
| .46 | 5291 | .69 | 149.477 |
| .45 | 5305 | .70 | 149.254 |
| .44 | 5319 | .71 | 149.031 |
| .43 | 5333 | .72 | 148.810 |
| .42 | 5347 | .73 | 148.588 |
| .41 | 5361 | .74 | 148.368 |
| .40 | 5375 | .75 | 148.148 |
| .39 | 5389 | .76 | 147.929 |
| .38 | 5404 | .77 | 147.710 |
| .37 | 5418 | .78 | 147.493 |
| .36 | 5432 | .79 | 147.275 |
| .35 | 5446 | .80 | 147.059 |
| .34 | 5460 | .81 | 146.843 |
| .33 | 5474 | .82 | 146.628 |
| .32 | 5488 | .83 | 146.413 |
| .31 | 5502 | .84 | 146.199 |
| .30 | 5516 | .85 | 145.985 |
| .29 | 5530 | .86 | 145.773 |
| .28 | 5545 | .87 | 145.560 |
| .27 | 5559 | .88 | 145.349 |
| .26 | 5573 | .89 | 145.138 |
| .25 | 5587 | .90 | 144.928 |
| .24 | 5601 | .91 | 144.718 |
| .23 | 5615 | .92 | 144.509 |
| .22 | 5629 | .93 | 144.300 |
| .21 | 5643 | .94 | 144.092 |
| .20 | 5657 | .95 | 143.885 |
| .19 | 5672 | .96 | 143.678 |
| .18 | 5686 | .97 | 143.472 |
| .17 | 5700 | .98 | 143.266 |
| .16 | 5714 | 6.99 | 143.062 |
| .15 | 5728 | 7.00 | 142.857 |
| .14 | 5742 | .01 | 142.653 |
| .13 | 5756 | .02 | 142.450 |
| .12 | 5770 | .03 | 142.248 |
| .11 | 5784 | .04 | 142.045 |
| .10 | 5799 | .05 | 141.844 |
| .09 | 5813 | .06 | 141.643 |
| .08 | 5827 | .07 | 141.443 |
| .07 | 5841 | .08 | 141.243 |
| .06 | 5855 | .09 | 141.044 |
| .05 | 5869 | .10 | 140.845 |
| .04 | 5883 | .11 | 140.647 |
| .03 | 5897 | .12 | 140.449 |
| .02 | 5912 | .13 | 140.252 |
| .01 | 5926 | .14 | 140.056 |
| −266.00 | 5940 | .15 | 139.860 |
| −265.99 | 5954 | .16 | 139.665 |
| .98 | 5968 | .17 | 139.470 |
| .97 | 5982 | .18 | 139.276 |
| .96 | 5996 | .19 | 139.082 |
| .95 | 6011 | .20 | 138.889 |
| .94 | 6025 | .21 | 138.696 |
| .93 | 6039 | .22 | 138.504 |
| .92 | 6053 | .23 | 138.313 |
| .91 | 6067 | .24 | 138.122 |
| .90 | 6081 | .25 | 137.931 |
| .89 | 6095 | .26 | 137.741 |
| .88 | 6110 | .27 | 137.552 |
| .87 | 6124 | .28 | 137.363 |
| .86 | 6138 | .29 | 137.174 |
| .85 | 6152 | .30 | 136.986 |
| .84 | 6166 | .31 | 136.799 |
| .83 | 6180 | .32 | 136.612 |
| .82 | 6194 | .33 | 136.426 |

| N | 1000/N | N + 273.15 | 1000/N + 273.15 |
|---|---|---|---|
| −265.81 | −3.76209 | 7.34 | 136.240 |
| .80 | 6223 | .35 | 136.054 |
| .79 | 6237 | .36 | 135.870 |
| .78 | 6251 | .37 | 135.685 |
| .77 | 6265 | .38 | 135.501 |
| .76 | 6279 | .39 | 135.318 |
| .75 | 6294 | .40 | 135.135 |
| .74 | 6308 | .41 | 134.953 |
| .73 | 6322 | .42 | 134.771 |
| .72 | 6336 | .43 | 134.590 |
| .71 | 6350 | .44 | 134.409 |
| .70 | 6364 | .45 | 134.228 |
| .69 | 6378 | .46 | 134.048 |
| .68 | 6393 | .47 | 133.869 |
| .67 | 6407 | .48 | 133.690 |
| .66 | 6421 | .49 | 133.511 |
| .65 | 6435 | .50 | 133.333 |
| .64 | 6449 | .51 | 133.156 |
| .63 | 6463 | .52 | 132.979 |
| .62 | 6478 | .53 | 132.802 |
| .61 | 6492 | .54 | 132.626 |
| .60 | 6506 | .55 | 132.450 |
| .59 | 6520 | .56 | 132.275 |
| .58 | 6534 | .57 | 132.100 |
| .57 | 6549 | .58 | 131.926 |
| .56 | 6563 | .59 | 131.752 |
| .55 | 6577 | .60 | 131.579 |
| .54 | 6591 | .61 | 131.406 |
| .53 | 6605 | .62 | 131.234 |
| .52 | 6619 | .63 | 131.062 |
| .51 | 6634 | .64 | 130.890 |
| .50 | 6648 | .65 | 130.719 |
| .49 | 6662 | .66 | 130.548 |
| .48 | 6676 | .67 | 130.378 |
| .47 | 6690 | .68 | 130.208 |
| .46 | 6705 | .69 | 130.039 |
| .45 | 6719 | .70 | 129.870 |
| .44 | 6733 | .71 | 129.702 |
| .43 | 6747 | .72 | 129.534 |
| .42 | 6761 | .73 | 129.366 |
| .41 | 6776 | .74 | 129.199 |
| .40 | 6790 | .75 | 129.032 |
| .39 | 6804 | .76 | 128.866 |
| .38 | 6818 | .77 | 128.700 |
| .37 | 6832 | .78 | 128.535 |
| .36 | 6847 | .79 | 128.370 |
| .35 | 6861 | .80 | 128.205 |
| .34 | 6875 | .81 | 128.041 |
| .33 | 6890 | .82 | 127.877 |
| .32 | 6903 | .83 | 127.714 |
| .31 | 6918 | .84 | 127.551 |
| .30 | 6932 | .85 | 127.389 |
| .29 | 6946 | .86 | 127.226 |
| .28 | 6960 | .87 | 127.065 |
| .27 | 6974 | .88 | 126.904 |
| .26 | 6989 | .89 | 126.743 |
| .25 | 7003 | .90 | 126.582 |
| .24 | 7017 | .91 | 126.422 |
| .23 | 7031 | .92 | 126.263 |
| .22 | 7045 | .93 | 126.103 |
| .21 | 7060 | .94 | 125.945 |
| .20 | 7074 | .95 | 125.786 |
| .19 | 7088 | .96 | 125.628 |
| .18 | 7102 | .97 | 125.471 |
| .17 | 7117 | .98 | 125.313 |
| .16 | 7131 | 7.99 | 125.156 |
| .15 | 7145 | 8.00 | 125.000 |
| .14 | 7159 | .01 | 124.844 |
| .13 | 7173 | .02 | 124.688 |
| .12 | 7188 | .03 | 124.533 |
| .11 | 7202 | .04 | 124.378 |
| .10 | 7216 | .05 | 124.224 |
| .09 | 7223 | .06 | 124.069 |
| .08 | 7245 | .07 | 123.916 |

| N | 1000/N | N + 273.15 | 1000/N + 273. |
|---|---|---|---|
| −265.07 | −3.77259 | 8.08 | 123.762 |
| .06 | 7273 | .09 | 123.609 |
| .05 | 7287 | .10 | 123.457 |
| .04 | 7302 | .11 | 123.305 |
| .03 | 7316 | .12 | 123.153 |
| .02 | 7330 | .13 | 123.001 |
| .01 | 7344 | .14 | 122.850 |
| −265.00 | 7358 | .15 | 122.699 |
| −264.99 | 7373 | .16 | 122.549 |
| .98 | 7387 | .17 | 122.399 |
| .97 | 7401 | .18 | 122.249 |
| .96 | 7415 | .19 | 122.100 |
| .95 | 7430 | .20 | 121.951 |
| .94 | 7444 | .21 | 121.803 |
| .93 | 7458 | .22 | 121.655 |
| .92 | 7472 | .23 | 121.507 |
| .91 | 7487 | .24 | 121.359 |
| .90 | 7501 | .25 | 121.212 |
| .89 | 7515 | .26 | 121.065 |
| .88 | 7529 | .27 | 120.919 |
| .87 | 7544 | .28 | 120.773 |
| .86 | 7558 | .29 | 120.627 |
| .85 | 7572 | .30 | 120.482 |
| .84 | 7586 | .31 | 120.337 |
| .83 | 7601 | .32 | 120.192 |
| .82 | 7615 | .33 | 120.048 |
| .81 | 7629 | .34 | 119.904 |
| .80 | 7644 | .35 | 119.760 |
| .79 | 7658 | .36 | 119.617 |
| .78 | 7672 | .37 | 119.474 |
| .77 | 7686 | .38 | 119.332 |
| .76 | 7701 | .39 | 119.190 |
| .75 | 7715 | .40 | 119.048 |
| .74 | 7729 | .41 | 118.906 |
| .73 | 7743 | .42 | 118.765 |
| .72 | 7758 | .43 | 118.624 |
| .71 | 7772 | .44 | 118.483 |
| .70 | 7786 | .45 | 118.343 |
| .69 | 7800 | .46 | 118.203 |
| .68 | 7815 | .47 | 118.064 |
| .67 | 7829 | .48 | 117.925 |
| .66 | 7843 | .49 | 117.786 |
| .65 | 7858 | .50 | 117.647 |
| .64 | 7872 | .51 | 117.509 |
| .63 | 7886 | .52 | 117.371 |
| .62 | 7900 | .53 | 117.233 |
| .61 | 7915 | .54 | 117.096 |
| .60 | 7929 | .55 | 116.959 |
| .59 | 7943 | .56 | 116.822 |
| .58 | 7958 | .57 | 116.686 |
| .57 | 7972 | .58 | 116.550 |
| .56 | 7986 | .59 | 116.414 |
| .55 | 8000 | .60 | 116.279 |
| .54 | 8015 | .61 | 116.144 |
| .53 | 8029 | .62 | 116.009 |
| .52 | 8043 | .63 | 115.875 |
| .51 | 8058 | .64 | 115.741 |
| .50 | 8072 | .65 | 115.607 |
| .49 | 8086 | .66 | 115.473 |
| .48 | 8100 | .67 | 115.340 |
| .47 | 8115 | .68 | 115.207 |
| .46 | 8129 | .69 | 115.075 |
| .45 | 8143 | .70 | 114.943 |
| .44 | 8158 | .71 | 114.811 |
| .43 | 8172 | .72 | 114.679 |
| .42 | 8186 | .73 | 114.548 |
| .41 | 8201 | .74 | 114.416 |
| .40 | 8215 | .75 | 114.286 |
| .39 | 8229 | .76 | 114.155 |
| .38 | 8243 | .77 | 114.025 |
| .37 | 8258 | .78 | 113.895 |
| .36 | 8272 | .79 | 113.766 |
| .35 | 8286 | .80 | 113.636 |
| .34 | 8301 | .81 | 113.507 |

| N | 1000/N | N + 273.15 | 1000/N + 273.15 |
|---|---|---|---|
| 264.33 | −3.78315 | 8.82 | 113.379 |
| .32 | 8329 | .83 | 113.250 |
| .31 | 8344 | .84 | 113.122 |
| .30 | 8358 | .85 | 112.994 |
| .29 | 8372 | .86 | 112.867 |
| .28 | 8387 | .87 | 112.740 |
| .27 | 8401 | .88 | 112.613 |
| .26 | 8415 | .89 | 112.486 |
| .25 | 8430 | .90 | 112.360 |
| .24 | 8444 | .91 | 112.233 |
| .23 | 8458 | .92 | 112.108 |
| .22 | 8472 | .93 | 111.982 |
| .21 | 8487 | .94 | 111.857 |
| .20 | 8501 | .95 | 111.732 |
| .19 | 8515 | .96 | 111.607 |
| .18 | 8530 | .97 | 111.483 |
| .17 | 8544 | .98 | 111.359 |
| .16 | 8558 | 8.99 | 111.235 |
| .15 | 8573 | 9.00 | 111.111 |
| .14 | 8587 | .01 | 110.988 |
| .13 | 8601 | .02 | 110.865 |
| .12 | 8616 | .03 | 110.742 |
| .11 | 8630 | .04 | 110.619 |
| .10 | 8644 | .05 | 110.497 |
| .09 | 8659 | .06 | 110.375 |
| .08 | 8673 | .07 | 110.254 |
| .07 | 8687 | .08 | 110.132 |
| .06 | 8702 | .09 | 110.011 |
| .05 | 8716 | .10 | 109.890 |
| .04 | 8730 | .11 | 109.769 |
| .03 | 8745 | .12 | 109.649 |
| .02 | 8759 | .13 | 109.529 |
| .01 | 8774 | .14 | 109.409 |
| 264.00 | 8788 | .15 | 109.290 |
| 263.99 | 8802 | .16 | 109.170 |
| .98 | 8817 | .17 | 109.051 |
| .97 | 8831 | .18 | 108.932 |
| .96 | 8845 | .19 | 108.814 |
| .95 | 8860 | .20 | 108.696 |
| .94 | 8874 | .21 | 108.578 |
| .93 | 8888 | .22 | 108.460 |
| .92 | 8903 | .23 | 108.342 |
| .91 | 8917 | .24 | 108.225 |
| .90 | 8931 | .25 | 108.108 |
| .89 | 8946 | .26 | 107.991 |
| .88 | 8960 | .27 | 107.875 |
| .87 | 8974 | .28 | 107.759 |
| .86 | 8989 | .29 | 107.643 |
| .85 | 9003 | .30 | 107.527 |
| .84 | 9018 | .31 | 107.411 |
| .83 | 9032 | .32 | 107.296 |
| .82 | 9046 | .33 | 107.181 |
| .81 | 9061 | .34 | 107.066 |
| .80 | 9075 | .35 | 106.952 |
| .79 | 9089 | .36 | 106.838 |
| .78 | 9104 | .37 | 106.724 |
| .77 | 9118 | .38 | 106.610 |
| .76 | 9133 | .39 | 106.496 |
| .75 | 9147 | .40 | 106.383 |
| .74 | 9161 | .41 | 106.270 |
| .73 | 9176 | .42 | 106.157 |
| .72 | 9190 | .43 | 106.045 |
| .71 | 9204 | .44 | 105.932 |
| .70 | 9219 | .45 | 105.820 |
| .69 | 9233 | .46 | 105.708 |
| .68 | 9248 | .47 | 105.597 |
| .67 | 9262 | .48 | 105.485 |
| .66 | 9276 | .49 | 105.374 |
| .65 | 9291 | .50 | 105.263 |
| .64 | 9305 | .51 | 105.152 |
| .63 | 9320 | .52 | 105.042 |
| .62 | 9334 | .53 | 104.932 |
| .61 | 9348 | .54 | 104.822 |
| .60 | 9363 | .55 | 104.712 |

| N | 1000/N | N + 273.15 | 1000/N + 273.15 |
|---|---|---|---|
| −263.59 | −3.79377 | 9.56 | 104.603 |
| .58 | 9391 | .57 | 104.493 |
| .57 | 9406 | .58 | 104.384 |
| .56 | 9420 | .59 | 104.275 |
| .55 | 9435 | .60 | 104.167 |
| .54 | 9449 | .61 | 104.058 |
| .53 | 9463 | .62 | 103.950 |
| .52 | 9478 | .63 | 103.842 |
| .51 | 9492 | .64 | 103.734 |
| .50 | 9507 | .65 | 103.627 |
| .49 | 9521 | .66 | 103.520 |
| .48 | 9535 | .67 | 103.413 |
| .47 | 9550 | .68 | 103.306 |
| .46 | 9564 | .69 | 103.199 |
| .45 | 9579 | .70 | 103.093 |
| .44 | 9593 | .71 | 102.987 |
| .43 | 9607 | .72 | 102.881 |
| .42 | 9622 | .73 | 102.775 |
| .41 | 9636 | .74 | 102.669 |
| .40 | 9651 | .75 | 102.564 |
| .39 | 9665 | .76 | 102.459 |
| .38 | 9680 | .77 | 102.354 |
| .37 | 9694 | .78 | 102.249 |
| .36 | 9708 | .79 | 102.145 |
| .35 | 9723 | .80 | 102.041 |
| .34 | 9737 | .81 | 101.937 |
| .33 | 9752 | .82 | 101.833 |
| .32 | 9766 | .83 | 101.729 |
| .31 | 9780 | .84 | 101.626 |
| .30 | 9795 | .85 | 101.523 |
| .29 | 9809 | .86 | 101.420 |
| .28 | 9824 | .87 | 101.317 |
| .27 | 9838 | .88 | 101.215 |
| .26 | 9853 | .89 | 101.112 |
| .25 | 9867 | .90 | 101.010 |
| .24 | 9881 | .91 | 100.908 |
| .23 | 9896 | .92 | 100.806 |
| .22 | 9910 | .93 | 100.705 |
| .21 | 9925 | .94 | 100.604 |
| .20 | 9939 | .95 | 100.503 |
| .19 | 9954 | .96 | 100.402 |
| .18 | 9968 | .97 | 100.301 |
| .17 | 9983 | .98 | 100.200 |
| .16 | 9997 | .99 | 100.100 |
| .15 | −3.80011 | 10.00 | 100.000 |
| .14 | 0026 | .01 | 99.9001 |
| .13 | 0040 | .02 | 99.8004 |
| .12 | 0055 | .03 | 99.7009 |
| .11 | 0069 | .04 | 99.6016 |
| .10 | 0084 | .05 | 99.5025 |
| .09 | 0098 | .06 | 99.4036 |
| .08 | 0113 | .07 | 99.3049 |
| .07 | 0127 | .08 | 99.2063 |
| .06 | 0141 | .09 | 99.1080 |
| .05 | 0156 | .10 | 99.0099 |
| .04 | 0170 | .11 | 98.9120 |
| .03 | 0185 | .12 | 98.8142 |
| .02 | 0199 | .13 | 98.7167 |
| .01 | 0214 | .14 | 98.6193 |
| 263 | 3.80228 | 10.15 | 98.52216 |
| 262 | 3.81679 | 11.15 | 89.68610 |
| 261 | 3.83142 | 12.15 | 82.30453 |
| 260 | 3.84615 | 13.15 | 76.04563 |
| 259 | 3.86100 | 14.15 | 70.67138 |
| 258 | 3.87597 | 15.15 | 66.00660 |
| 257 | 3.89105 | 16.15 | 61.91950 |
| 256 | 3.90625 | 17.15 | 58.30904 |
| 255 | 3.92157 | 18.15 | 55.09642 |
| 254 | 3.93701 | 19.15 | 52.21932 |
| 253 | 3.95257 | 20.15 | 49.62779 |
| 252 | 3.96825 | 21.15 | 47.28132 |
| 251 | 3.98406 | 22.15 | 45.14673 |
| 250 | 4.00000 | 23.15 | 43.19654 |
| 249 | 4.01606 | 24.15 | 41.40787 |

| N | 1000/N | N + 273.15 | 1000/N + 273.15 |
|---|---|---|---|
| −248 | −4.03226 | 25.15 | 39.76143 |
| 247 | 4.04858 | 26.15 | 38.24091 |
| 246 | 4.06504 | 27.15 | 36.83241 |
| 245 | 4.08163 | 28.15 | 35.52398 |
| 244 | 4.09836 | 29.15 | 34.30532 |
| 243 | 4.11523 | 30.15 | 33.16750 |
| 242 | 4.13223 | 31.15 | 32.10273 |
| 241 | 4.14938 | 32.15 | 31.10420 |
| 240 | 4.16667 | 33.15 | 30.16591 |
| 239 | 4.18410 | 34.15 | 29.28258 |
| 238 | 4.20168 | 35.15 | 28.44950 |
| 237 | 4.21941 | 36.15 | 27.66252 |
| 236 | 4.23729 | 37.15 | 26.91790 |
| 235 | 4.25532 | 38.15 | 26.21232 |
| 234 | 4.27350 | 39.15 | 25.54278 |
| 233 | 4.29185 | 40.15 | 24.90660 |
| 232 | 4.31034 | 41.15 | 24.30134 |
| 231 | 4.32900 | 42.15 | 23.72479 |
| 230 | 4.34783 | 43.15 | 23.17497 |
| 229 | 4.36681 | 44.15 | 22.65006 |
| 228 | 4.38597 | 45.15 | 22.14839 |
| 227 | 4.40529 | 46.15 | 21.66847 |
| 226 | 4.42478 | 47.15 | 21.20891 |
| 225 | 4.44444 | 48.15 | 20.76843 |
| 224 | 4.46429 | 49.15 | 20.34588 |
| 223 | 4.48430 | 50.15 | 19.94018 |
| 222 | 4.50450 | 51.15 | 19.55034 |
| 221 | 4.52489 | 52.15 | 19.17546 |
| 220 | 4.54545 | 53.15 | 18.81468 |
| 219 | 4.56621 | 54.15 | 18.46722 |
| 218 | 4.58716 | 55.15 | 18.13237 |
| 217 | 4.60829 | 56.15 | 17.80944 |
| 216 | 4.62963 | 57.15 | 17.49781 |
| 215 | 4.65116 | 58.15 | 17.19690 |
| 214 | 4.67290 | 59.15 | 16.90617 |
| 213 | 4.69484 | 60.15 | 16.62510 |
| 212 | 4.71698 | 61.15 | 16.35323 |
| 211 | 4.73934 | 62.15 | 16.09010 |
| 210 | 4.76190 | 63.15 | 15.83531 |
| 209 | 4.78469 | 64.15 | 15.58846 |
| 208 | 4.80769 | 65.15 | 15.34919 |
| 207 | 4.83092 | 66.15 | 15.11716 |
| 206 | 4.85437 | 67.15 | 14.89203 |
| 205 | 4.87805 | 68.15 | 14.67351 |
| 204 | 4.90196 | 69.15 | 14.46132 |
| 203 | 4.92611 | 70.15 | 14.25517 |
| 202 | 4.95050 | 71.15 | 14.05481 |
| 201 | 4.97512 | 72.15 | 13.86001 |
| 200 | 5.00000 | 73.15 | 13.67054 |
| 199 | 5.02513 | 74.15 | 13.48618 |
| 198 | 5.05051 | 75.15 | 13.30672 |
| 197 | 5.07614 | 76.15 | 13.13198 |
| 196 | 5.10204 | 77.15 | 12.96176 |
| 195 | 5.12821 | 78.15 | 12.79591 |
| 194 | 5.15464 | 79.15 | 12.63424 |
| 193 | 5.18135 | 80.15 | 12.47661 |
| 192 | 5.20833 | 81.15 | 12.32286 |
| 191 | 5.23560 | 82.15 | 12.17285 |
| 190 | 5.26316 | 83.15 | 12.02646 |
| 189 | 5.29101 | 84.15 | 11.88354 |
| 188 | 5.31915 | 85.15 | 11.74398 |
| 187 | 5.34759 | 86.15 | 11.60766 |
| 186 | 5.37634 | 87.15 | 11.47447 |
| 185 | 5.40541 | 88.15 | 11.34430 |
| 184 | 5.43478 | 89.15 | 11.21705 |
| 183 | 5.46448 | 90.15 | 11.09262 |
| 182 | 5.49451 | 91.15 | 10.97093 |
| 181 | 5.52486 | 92.15 | 10.85187 |
| 180 | 5.55556 | 93.15 | 10.73537 |
| 179 | 5.58659 | 94.15 | 10.62135 |
| 178 | 5.61798 | 95.15 | 10.50972 |
| 177 | 5.64972 | 96.15 | 10.40042 |
| 176 | 5.68182 | 97.15 | 10.29336 |
| 175 | 5.71429 | 98.15 | 10.18849 |

| N | 1000/N | N + 273.15 | 1000/N + 273 |
|---|---|---|---|
| −174 | −5.74713 | 99.15 | 10.08573 |
| 173 | 5.78035 | 100.15 | 9.98502 |
| 172 | 5.81395 | 101.15 | 9.88631 |
| 171 | 5.84795 | 102.15 | 9.78956 |
| 170 | 5.88235 | 103.15 | 9.69462 |
| 169 | 5.91716 | 104.15 | 9.60154 |
| 168 | 5.95238 | 105.15 | 9.51022 |
| 167 | 5.98802 | 106.15 | 9.42063 |
| 166 | 6.02410 | 107.15 | 9.33271 |
| 165 | 6.06061 | 108.15 | 9.24642 |
| 164 | 6.09756 | 109.15 | 9.16170 |
| 163 | 6.13497 | 110.15 | 9.07853 |
| 162 | 6.17284 | 111.15 | 8.99685 |
| 161 | 6.21118 | 112.15 | 8.91663 |
| 160 | 6.25000 | 113.15 | 8.83783 |
| 159 | 6.28931 | 114.15 | 8.76040 |
| 158 | 6.32911 | 115.15 | 8.68432 |
| 157 | 6.36943 | 116.15 | 8.60956 |
| 156 | 6.41026 | 117.15 | 8.53606 |
| 155 | 6.45161 | 118.15 | 8.46382 |
| 154 | 6.49351 | 119.15 | 8.39278 |
| 153 | 6.53595 | 120.15 | 8.32293 |
| 152 | 6.57895 | 121.15 | 8.25423 |
| 151 | 6.62252 | 122.15 | 8.18666 |
| 150 | 6.66667 | 123.15 | 8.12018 |
| 149 | 6.71141 | 124.15 | 8.05477 |
| 148 | 6.75676 | 125.15 | 7.99041 |
| 147 | 6.80272 | 126.15 | 7.92707 |
| 146 | 6.84932 | 127.15 | 7.86473 |
| 145 | 6.89655 | 128.15 | 7.80336 |
| 144 | 6.94444 | 129.15 | 7.74293 |
| 143 | 6.99301 | 130.15 | 7.68344 |
| 142 | 7.04225 | 131.15 | 7.62486 |
| 141 | 7.09220 | 132.15 | 7.56716 |
| 140 | 7.14286 | 133.15 | 7.51033 |
| 139 | 7.19424 | 134.15 | 7.45434 |
| 138 | 7.24638 | 135.15 | 7.39919 |
| 137 | 7.29927 | 136.15 | 7.34484 |
| 136 | 7.35294 | 137.15 | 7.29129 |
| 135 | 7.40741 | 138.15 | 7.23851 |
| 134 | 7.46269 | 139.15 | 7.18649 |
| 133 | 7.51880 | 140.15 | 7.13521 |
| 132 | 7.57576 | 141.15 | 7.08466 |
| 131 | 7.63359 | 142.15 | 7.03482 |
| 130 | 7.69231 | 143.15 | 6.98568 |
| 129 | 7.75194 | 144.15 | 6.93722 |
| 128 | 7.81250 | 145.15 | 6.88942 |
| 127 | 7.87402 | 146.15 | 6.84229 |
| 126 | 7.93651 | 147.15 | 6.79579 |
| 125 | 8.00000 | 148.15 | 6.74992 |
| 124 | 8.06452 | 149.15 | 6.70466 |
| 123 | 8.13008 | 150.15 | 6.66001 |
| 122 | 8.19672 | 151.15 | 6.61594 |
| 121 | 8.26446 | 152.15 | 6.57246 |
| 120 | 8.33333 | 153.15 | 6.52955 |
| 119 | 8.40336 | 154.15 | 6.48719 |
| 118 | 8.47478 | 155.15 | 6.44538 |
| 117 | 8.54701 | 156.15 | 6.40410 |
| 116 | 8.62069 | 157.15 | 6.36345 |
| 115 | 8.69565 | 158.15 | 6.32311 |
| 114 | 8.77193 | 159.15 | 6.28338 |
| 113 | 8.84956 | 160.15 | 6.24415 |
| 112 | 8.92857 | 161.15 | 6.20540 |
| 111 | 9.00901 | 162.15 | 6.16713 |
| 110 | 9.09091 | 163.15 | 6.12933 |
| 109 | 9.17431 | 164.15 | 6.09199 |
| 108 | 9.25926 | 165.15 | 6.05510 |
| 107 | 9.34579 | 166.15 | 6.01866 |
| 106 | 9.43396 | 167.15 | 5.98265 |
| 105 | 9.52381 | 168.15 | 5.94707 |
| 104 | 9.61538 | 169.15 | 5.91191 |
| 103 | 9.70874 | 170.15 | 5.87717 |
| 102 | 9.80392 | 171.15 | 5.84283 |
| 101 | 9.90099 | 172.15 | 5.80889 |

| N | 1000/N | N + 273.15 | 1000/N + 273.15 |
|---|---|---|---|
| −100 | −10.00000 | 173.15 | 5.77524 |
| 99 | 10.10101 | 174.15 | 5.74218 |
| 98 | 10.20408 | 175.15 | 5.70939 |
| 97 | 10.30928 | 176.15 | 5.67698 |
| 96 | 10.41667 | 177.15 | 5.64493 |
| 95 | 10.52632 | 178.15 | 5.61325 |
| 94 | 10.63830 | 179.15 | 5.58191 |
| 93 | 10.75269 | 180.15 | 5.55093 |
| 92 | 10.86957 | 181.15 | 5.52029 |
| 91 | 10.98901 | 182.15 | 5.48999 |
| 90 | 11.11111 | 183.15 | 5.46001 |
| 89 | 11.23596 | 184.15 | 5.43035 |
| 88 | 11.36364 | 185.15 | 5.40103 |
| 87 | 11.49425 | 186.15 | 5.37201 |
| 86 | 11.62791 | 187.15 | 5.34331 |
| 85 | 11.76471 | 188.15 | 5.31491 |
| 84 | 11.90476 | 189.15 | 5.28681 |
| 83 | 12.04819 | 190.15 | 5.25901 |
| 82 | 12.19512 | 191.15 | 5.23149 |
| 81 | 12.34568 | 192.15 | 5.20427 |
| 80 | 12.50000 | 193.15 | 5.17732 |
| 79 | 12.65823 | 194.15 | 5.15066 |
| 78 | 12.82051 | 195.15 | 5.12426 |
| 77 | 12.98701 | 196.15 | 5.09814 |
| 76 | 13.15789 | 197.15 | 5.07228 |
| 75 | 13.33333 | 198.15 | 5.04668 |
| 74 | 13.51351 | 199.15 | 5.02134 |
| 73 | 13.69863 | 200.15 | 4.99625 |
| 72 | 13.88889 | 201.15 | 4.97141 |
| 71 | 14.08451 | 202.15 | 4.94682 |
| 70 | 14.28571 | 203.15 | 4.92247 |
| 69 | 14.49275 | 204.15 | 4.89836 |
| 68 | 14.70588 | 205.15 | 4.87448 |
| 67 | 14.92537 | 206.15 | 4.85084 |
| 66 | 15.15152 | 207.15 | 4.82742 |
| 65 | 15.38462 | 208.15 | 4.80423 |
| 64 | 15.62500 | 209.15 | 4.78126 |
| 63 | 15.87302 | 210.15 | 4.75851 |
| 62 | 16.12903 | 211.15 | 4.73597 |
| 61 | 16.39344 | 212.15 | 4.71365 |
| 60 | 16.66667 | 213.15 | 4.69153 |
| 59 | 16.94915 | 214.15 | 4.66962 |
| 58 | 17.24138 | 215.15 | 4.64792 |
| 57 | 17.54386 | 216.15 | 4.62642 |
| 56 | 17.85714 | 217.15 | 4.60511 |
| 55 | 18.18182 | 218.15 | 4.58400 |
| 54 | 18.51852 | 219.15 | 4.56308 |
| 53 | 18.86792 | 220.15 | 4.54236 |
| 52 | 19.23077 | 221.15 | 4.52182 |
| 51 | 19.60784 | 222.15 | 4.50146 |
| 50 | 20.00000 | 223.15 | 4.48129 |
| 49 | 20.40816 | 224.15 | 4.46130 |
| 48 | 20.83333 | 225.15 | 4.44148 |
| 47 | 21.27660 | 226.15 | 4.42184 |
| 46 | 21.73913 | 227.15 | 4.40238 |
| 45 | 22.22222 | 228.15 | 4.38308 |
| 44 | 22.72727 | 229.15 | 4.36395 |
| 43 | 23.25581 | 230.15 | 4.34499 |
| 42 | 23.80952 | 231.15 | 4.32620 |
| 41 | 24.39024 | 232.15 | 4.30756 |
| 40 | 25.00000 | 233.15 | 4.28908 |
| 39 | 25.64103 | 234.15 | 4.27077 |
| 38 | 26.31579 | 235.15 | 4.25260 |
| 37 | 27.02703 | 236.15 | 4.23460 |
| 36 | 27.77778 | 237.15 | 4.21674 |
| 35 | 28.57143 | 238.15 | 4.19903 |
| 34 | 29.41176 | 239.15 | 4.18148 |
| 33 | 30.30303 | 240.15 | 4.16406 |
| 32 | 31.25000 | 241.15 | 4.14680 |
| 31 | 32.25806 | 242.15 | 4.12967 |
| 30 | 33.33333 | 243.15 | 4.11269 |
| 29 | 34.48276 | 244.15 | 4.09584 |
| 28 | 35.71429 | 245.15 | 4.07914 |
| 27 | 37.03704 | 246.15 | 4.06256 |

| N | 1000/N | N + 273.15 | 1000/N + 273.15 |
|---|---|---|---|
| −26 | −38.46154 | 247.15 | 4.04613 |
| 25 | 40.00000 | 248.15 | 4.02982 |
| 24 | 41.66667 | 249.15 | 4.01365 |
| 23 | 43.47826 | 250.15 | 3.99760 |
| 22 | 45.45455 | 251.15 | 3.98168 |
| 21 | 47.61905 | 252.15 | 3.96589 |
| 20 | 50.00000 | 253.15 | 3.95023 |
| 19 | 52.63158 | 254.15 | 3.93468 |
| 18 | 55.55556 | 255.15 | 3.91926 |
| 17 | 58.82353 | 256.15 | 3.90396 |
| 16 | 62.50000 | 257.15 | 3.88878 |
| 15 | 66.66667 | 258.15 | 3.87372 |
| 14 | 71.42857 | 259.15 | 3.85877 |
| 13 | 76.92308 | 260.15 | 3.84394 |
| 12 | 83.33333 | 261.15 | 3.82922 |
| 11 | 90.90909 | 262.15 | 3.81461 |
| 10 | 100.00000 | 263.15 | 3.80011 |
| 9 | 111.11111 | 264.15 | 3.78573 |
| 8 | 125.00000 | 265.15 | 3.77145 |
| 7 | 142.85714 | 266.15 | 3.75728 |
| 6 | 166.66667 | 267.15 | 3.74322 |
| 5 | 200.00000 | 268.15 | 3.72926 |
| 4 | 250.00000 | 269.15 | 3.71540 |
| 3 | 333.33333 | 270.15 | 3.70165 |
| 2 | 500.00000 | 271.15 | 3.68780 |
| 1 | 1000.00000 | 272.15 | 3.67444 |
| 0 | .......... | 273.15 | 3.66099 |
| + 1 | 1000.00000 | 274.15 | 3.64764 |
| 2 | 500.00000 | 275.15 | 3.63438 |
| 3 | 333.33333 | 276.15 | 3.62122 |
| 4 | 250.00000 | 277.15 | 3.60815 |
| 5 | 200.00000 | 278.15 | 3.59518 |
| 6 | 166.66667 | 279.15 | 3.58230 |
| 7 | 142.85714 | 280.15 | 3.56952 |
| 8 | 125.00000 | 281.15 | 3.55682 |
| 9 | 111.11111 | 282.15 | 3.54421 |
| 10 | 100.00000 | 283.15 | 3.53170 |
| 11 | 90.90909 | 284.15 | 3.51927 |
| 12 | 83.33333 | 285.15 | 3.50693 |
| 13 | 76.92308 | 286.15 | 3.49467 |
| 14 | 71.42857 | 287.15 | 3.48250 |
| 15 | 66.66667 | 288.15 | 3.47041 |
| 16 | 62.50000 | 289.15 | 3.45841 |
| 17 | 58.82353 | 290.15 | 3.44649 |
| 18 | 55.55556 | 291.15 | 3.43466 |
| 19 | 52.63158 | 292.15 | 3.42290 |
| 20 | 50.00000 | 293.15 | 3.41122 |
| 21 | 47.61905 | 294.15 | 3.39963 |
| 22 | 45.45455 | 295.15 | 3.38811 |
| 23 | 43.47826 | 296.15 | 3.37667 |
| 24 | 41.66667 | 297.15 | 3.36530 |
| 25 | 40.00000 | 298.15 | 3.35402 |
| 26 | 38.46154 | 299.15 | 3.34280 |
| 27 | 37.03704 | 300.15 | 3.33167 |
| 28 | 35.71429 | 301.15 | 3.32060 |
| 29 | 34.48276 | 302.15 | 3.30961 |
| 30 | 33.33333 | 303.15 | 3.29870 |
| 31 | 32.25806 | 304.15 | 3.28785 |
| 32 | 31.25000 | 305.15 | 3.27708 |
| 33 | 30.30303 | 306.15 | 3.26637 |
| 34 | 29.41176 | 307.15 | 3.25574 |
| 35 | 28.57143 | 308.15 | 3.24517 |
| 36 | 27.77778 | 309.15 | 3.23468 |
| 37 | 27.02703 | 310.15 | 3.22425 |
| 38 | 26.31579 | 311.15 | 3.21388 |
| 39 | 25.64103 | 312.15 | 3.20359 |
| 40 | 25.00000 | 313.15 | 3.19336 |
| 41 | 24.39024 | 314.15 | 3.18319 |
| 42 | 23.80952 | 315.15 | 3.17309 |
| 43 | 23.25581 | 316.15 | 3.16306 |
| 44 | 22.72727 | 317.15 | 3.15308 |
| 45 | 22.22222 | 318.15 | 3.14317 |
| 46 | 21.73913 | 319.15 | 3.13332 |
| 47 | 21.27660 | 320.15 | 3.12354 |

| N | 1000/N | N + 273.15 | 1000/N + 273.15 |
|---|---|---|---|
| +48 | +20.83333 | 321.15 | 3.11381 |
| 49 | 20.40816 | 322.15 | 3.10414 |
| 50 | 20.00000 | 323.15 | 3.09454 |
| 51 | 19.60784 | 324.15 | 3.08499 |
| 52 | 19.23077 | 325.15 | 3.07550 |
| 53 | 18.86792 | 326.15 | 3.06607 |
| 54 | 18.51852 | 327.15 | 3.05670 |
| 55 | 18.18182 | 328.15 | 3.04739 |
| 56 | 17.85714 | 329.15 | 3.03813 |
| 57 | 17.54386 | 330.15 | 3.02893 |
| 58 | 17.24138 | 331.15 | 3.01978 |
| 59 | 16.94915 | 332.15 | 3.01069 |
| 60 | 16.66667 | 333.15 | 3.00165 |
| 61 | 16.39344 | 334.15 | 2.99267 |
| 62 | 16.12903 | 335.15 | 2.98374 |
| 63 | 15.87302 | 336.15 | 2.97486 |
| 64 | 15.62500 | 337.15 | 2.96604 |
| 65 | 15.38462 | 338.15 | 2.95727 |
| 66 | 15.15152 | 339.15 | 2.94855 |
| 67 | 14.92537 | 340.15 | 2.93988 |
| 68 | 14.70588 | 341.15 | 2.93126 |
| 69 | 14.49275 | 342.15 | 2.92269 |
| 70 | 14.28571 | 343.15 | 2.91418 |
| 71 | 14.08451 | 344.15 | 2.90571 |
| 72 | 13.88889 | 345.15 | 2.89729 |
| 73 | 13.69863 | 346.15 | 2.88892 |
| 74 | 13.51351 | 347.15 | 2.88060 |
| 75 | 13.33333 | 348.15 | 2.87233 |
| 76 | 13.15789 | 349.15 | 2.86410 |
| 77 | 12.98701 | 350.15 | 2.85592 |
| 78 | 12.82051 | 351.15 | 2.84779 |
| 79 | 12.65823 | 352.15 | 2.83970 |
| 80 | 12.50000 | 353.15 | 2.83166 |
| 81 | 12.34568 | 354.15 | 2.82366 |
| 82 | 12.19512 | 355.15 | 2.81571 |
| 83 | 12.04819 | 356.15 | 2.80781 |
| 84 | 11.90476 | 357.15 | 2.79994 |
| 85 | 11.76471 | 358.15 | 2.79213 |
| 86 | 11.62791 | 359.15 | 2.78435 |
| 87 | 11.49425 | 360.15 | 2.77662 |
| 88 | 11.36364 | 361.15 | 2.76893 |
| 89 | 11.23596 | 362.15 | 2.76129 |
| 90 | 11.11111 | 363.15 | 2.75368 |
| 91 | 10.98901 | 364.15 | 2.74612 |
| 92 | 10.86957 | 365.15 | 2.73860 |
| 93 | 10.75269 | 366.15 | 2.73112 |
| 94 | 10.63830 | 367.15 | 2.72368 |
| 95 | 10.52632 | 368.15 | 2.71628 |
| 96 | 10.41667 | 369.15 | 2.70893 |
| 97 | 10.30928 | 370.15 | 2.70161 |
| 98 | 10.20408 | 371.15 | 2.69433 |
| 99 | 10.10101 | 372.15 | 2.68709 |
| 100 | 10.00000 | 373.15 | 2.67989 |
| 101 | 9.90099 | 374.15 | 2.67272 |
| 102 | 9.80392 | 375.15 | 2.66560 |
| 103 | 9.70874 | 376.15 | 2.65851 |
| 104 | 9.61538 | 377.15 | 2.65146 |
| 105 | 9.52381 | 378.15 | 2.64445 |
| 106 | 9.43396 | 379.15 | 2.63748 |
| 107 | 9.34579 | 380.15 | 2.63054 |
| 108 | 9.25926 | 381.15 | 2.62364 |
| 109 | 9.17431 | 382.15 | 2.61677 |
| 110 | 9.09091 | 383.15 | 2.60994 |
| 111 | 9.00901 | 384.15 | 2.60315 |
| 112 | 8.92857 | 385.15 | 2.59639 |
| 113 | 8.84956 | 386.15 | 2.58967 |
| 114 | 8.77193 | 387.15 | 2.58298 |
| 115 | 8.69565 | 388.15 | 2.57632 |
| 116 | 8.62069 | 389.15 | 2.56970 |
| 117 | 8.54701 | 390.15 | 2.56312 |
| 118 | 8.47458 | 391.15 | 2.55656 |
| 119 | 8.40336 | 392.15 | 2.55004 |
| 120 | 8.33333 | 393.15 | 2.54356 |
| 121 | 8.26446 | 394.15 | 2.53711 |

| N | 1000/N | N + 273.15 | 1000/N + 273.1 |
|---|---|---|---|
| +122 | +8.19672 | 395.15 | 2.53068 |
| 123 | 8.13008 | 396.15 | 2.52430 |
| 124 | 8.06452 | 397.15 | 2.51794 |
| 125 | 8.00000 | 398.15 | 2.51162 |
| 126 | 7.93651 | 399.15 | 2.50532 |
| 127 | 7.87402 | 400.15 | 2.49906 |
| 128 | 7.81250 | 401.15 | 2.49283 |
| 129 | 7.75194 | 402.15 | 2.48663 |
| 130 | 7.69231 | 403.15 | 2.48047 |
| 131 | 7.63359 | 404.15 | 2.47433 |
| 132 | 7.57576 | 405.15 | 2.46822 |
| 133 | 7.51880 | 406.15 | 2.46214 |
| 134 | 7.46269 | 407.15 | 2.45610 |
| 135 | 7.40741 | 408.15 | 2.45008 |
| 136 | 7.35294 | 409.15 | 2.44409 |
| 137 | 7.29927 | 410.15 | 2.43813 |
| 138 | 7.24638 | 411.15 | 2.43220 |
| 139 | 7.19424 | 412.15 | 2.42630 |
| 140 | 7.14286 | 413.15 | 2.42043 |
| 141 | 7.09220 | 414.15 | 2.41458 |
| 142 | 7.04225 | 415.15 | 2.40877 |
| 143 | 6.99301 | 416.15 | 2.40298 |
| 144 | 6.94444 | 417.15 | 2.39722 |
| 145 | 6.89655 | 418.15 | 2.39149 |
| 146 | 6.84932 | 419.15 | 2.38578 |
| 147 | 6.80272 | 420.15 | 2.38010 |
| 148 | 6.75676 | 421.15 | 2.37445 |
| 149 | 6.71141 | 422.15 | 2.36883 |
| 150 | 6.66667 | 423.15 | 2.36323 |
| 151 | 6.62252 | 424.15 | 2.35766 |
| 152 | 6.57895 | 425.15 | 2.35211 |
| 153 | 6.53595 | 426.15 | 2.34659 |
| 154 | 6.49351 | 427.15 | 2.34110 |
| 155 | 6.45161 | 428.15 | 2.33563 |
| 156 | 6.41026 | 429.15 | 2.33019 |
| 157 | 6.36943 | 430.15 | 2.32477 |
| 158 | 6.32911 | 431.15 | 2.31938 |
| 159 | 6.28931 | 432.15 | 2.31401 |
| 160 | 6.25000 | 433.15 | 2.30867 |
| 161 | 6.21118 | 434.15 | 2.30335 |
| 162 | 6.17284 | 435.15 | 2.29806 |
| 163 | 6.13497 | 436.15 | 2.29279 |
| 164 | 6.09756 | 437.15 | 2.28754 |
| 165 | 6.06061 | 438.15 | 2.28232 |
| 166 | 6.02410 | 439.15 | 2.27713 |
| 167 | 5.98802 | 440.15 | 2.27195 |
| 168 | 5.95238 | 441.15 | 2.26680 |
| 169 | 5.91716 | 442.15 | 2.26168 |
| 170 | 5.88235 | 443.15 | 2.25657 |
| 171 | 5.84795 | 444.15 | 2.25149 |
| 172 | 5.81395 | 445.15 | 2.24643 |
| 173 | 5.78035 | 446.15 | 2.24140 |
| 174 | 5.74713 | 447.15 | 2.23639 |
| 175 | 5.71429 | 448.15 | 2.23140 |
| 176 | 5.68182 | 449.15 | 2.22643 |
| 177 | 5.64972 | 450.15 | 2.22148 |
| 178 | 5.61798 | 451.15 | 2.21656 |
| 179 | 5.58659 | 452.15 | 2.21166 |
| 180 | 5.55556 | 453.15 | 2.20677 |
| 181 | 5.52486 | 454.15 | 2.20192 |
| 182 | 5.49451 | 455.15 | 2.19708 |
| 183 | 5.46448 | 456.15 | 2.19226 |
| 184 | 5.43478 | 457.15 | 2.18747 |
| 185 | 5.40541 | 458.15 | 2.18269 |
| 186 | 5.37634 | 459.15 | 2.17794 |
| 187 | 5.34759 | 460.15 | 2.17320 |
| 188 | 5.31915 | 461.15 | 2.16849 |
| 189 | 5.29101 | 462.15 | 2.16380 |
| 190 | 5.26316 | 463.15 | 2.15913 |
| 191 | 5.23560 | 464.15 | 2.15448 |
| 192 | 5.20833 | 465.15 | 2.14984 |
| 193 | 5.18135 | 466.15 | 2.14523 |
| 194 | 5.15464 | 467.15 | 2.14064 |
| 195 | 5.12821 | 468.15 | 2.13607 |

| N | 1000/N | N + 273.15 | 1000/N + 273.15 |
|---|---|---|---|
| +196 | 5.10204 | 469.15 | 2.13151 |
| 197 | 5.07614 | 470.15 | 2.12698 |
| 198 | 5.05051 | 471.15 | 2.12247 |
| 199 | 5.02513 | 472.15 | 2.11797 |
| 200 | 5.00000 | 473.15 | 2.11349 |
| 201 | 4.97512 | 474.15 | 2.10904 |
| 202 | 4.95050 | 475.15 | 2.10460 |
| 203 | 4.92611 | 476.15 | 2.10018 |
| 204 | 4.90196 | 477.15 | 2.09578 |
| 205 | 4.87805 | 478.15 | 2.09139 |
| 206 | 4.85437 | 479.15 | 2.08703 |
| 207 | 4.83092 | 480.15 | 2.08268 |
| 208 | 4.80769 | 481.15 | 2.07835 |
| 209 | 4.78469 | 482.15 | 2.07404 |
| 210 | 4.76190 | 483.15 | 2.06975 |
| 211 | 4.73934 | 484.15 | 2.06548 |
| 212 | 4.71698 | 485.15 | 2.06122 |
| 213 | 4.69484 | 486.15 | 2.05698 |
| 214 | 4.67290 | 487.15 | 2.05276 |
| 215 | 4.65116 | 488.15 | 2.04855 |
| 216 | 4.62963 | 489.15 | 2.04436 |
| 217 | 4.60829 | 490.15 | 2.04019 |
| 218 | 4.58716 | 491.15 | 2.03604 |
| 219 | 4.56621 | 492.15 | 2.03190 |
| 220 | 4.54545 | 493.15 | 2.02778 |
| 221 | 4.52489 | 494.15 | 2.02368 |
| 222 | 4.50450 | 495.15 | 2.01959 |
| 223 | 4.48430 | 496.15 | 2.01552 |
| 224 | 4.46429 | 497.15 | 2.01147 |
| 225 | 4.44444 | 498.15 | 2.00743 |
| 226 | 4.42478 | 499.15 | 2.00341 |
| 227 | 4.40529 | 500.15 | 1.99940 |
| 228 | 4.38596 | 501.15 | 1.99541 |
| 229 | 4.36681 | 502.15 | 1.99144 |
| 230 | 4.34783 | 503.15 | 1.98748 |
| 231 | 4.32900 | 504.15 | 1.98354 |
| 232 | 4.31034 | 505.15 | 1.97961 |
| 233 | 4.29185 | 506.15 | 1.97570 |
| 234 | 4.27350 | 507.15 | 1.97180 |
| 235 | 4.25532 | 508.15 | 1.96792 |
| 236 | 4.23729 | 509.15 | 1.96406 |
| 237 | 4.21941 | 510.15 | 1.96021 |
| 238 | 4.20168 | 511.15 | 1.95637 |
| 239 | 4.18410 | 512.15 | 1.95255 |
| 240 | 4.16667 | 513.15 | 1.94875 |
| 241 | 4.14938 | 514.15 | 1.94496 |
| 242 | 4.13223 | 515.15 | 1.94118 |
| 243 | 4.11523 | 516.15 | 1.93742 |
| 244 | 4.09836 | 517.15 | 1.93367 |
| 245 | 4.08163 | 518.15 | 1.92994 |
| 246 | 4.06504 | 519.15 | 1.92623 |
| 247 | 4.04858 | 520.15 | 1.92252 |
| 248 | 4.03226 | 521.15 | 1.91883 |
| 249 | 4.01606 | 522.15 | 1.91516 |
| 250 | 4.00000 | 523.15 | 1.91150 |
| 251 | 3.98406 | 524.15 | 1.90785 |
| 252 | 3.96825 | 525.15 | 1.90422 |
| 253 | 3.95257 | 526.15 | 1.90060 |
| 254 | 3.93701 | 527.15 | 1.89699 |
| 255 | 3.92157 | 528.15 | 1.89340 |
| 256 | 3.90625 | 529.15 | 1.88982 |
| 257 | 3.89105 | 530.15 | 1.88626 |
| 258 | 3.87597 | 531.15 | 1.88271 |
| 259 | 3.86100 | 532.15 | 1.87917 |
| 260 | 3.84615 | 533.15 | 1.87564 |
| 261 | 3.83142 | 534.15 | 1.87213 |
| 262 | 3.81679 | 535.15 | 1.86863 |
| 263 | 3.80228 | 536.15 | 1.86515 |
| 264 | 3.78788 | 537.15 | 1.86168 |
| 265 | 3.77358 | 538.15 | 1.85822 |
| 266 | 3.75940 | 539.15 | 1.85477 |
| 267 | 3.74532 | 540.15 | 1.85134 |
| 268 | 3.73134 | 541.15 | 1.84792 |
| 269 | 3.71747 | 542.15 | 1.84451 |

| N | 1000/N | N + 273.15 | 1000/N + 273.15 |
|---|---|---|---|
| +270 | 3.70370 | 543.15 | 1.84111 |
| 271 | 3.69004 | 544.15 | 1.83773 |
| 272 | 3.67647 | 545.15 | 1.83436 |
| 273 | 3.66300 | 546.15 | 1.83100 |
| 274 | 3.64964 | 547.15 | 1.82765 |
| 275 | 3.63636 | 548.15 | 1.82432 |
| 276 | 3.62319 | 549.15 | 1.82100 |
| 277 | 3.61011 | 550.15 | 1.81769 |
| 278 | 3.59712 | 551.15 | 1.81439 |
| 279 | 3.58423 | 552.15 | 1.81110 |
| 280 | 3.57143 | 553.15 | 1.80783 |
| 281 | 3.55872 | 554.15 | 1.80457 |
| 282 | 3.54610 | 555.15 | 1.80131 |
| 283 | 3.53357 | 556.15 | 1.79808 |
| 284 | 3.52113 | 557.15 | 1.79485 |
| 285 | 3.50877 | 558.15 | 1.79163 |
| 286 | 3.49650 | 559.15 | 1.78843 |
| 287 | 3.48432 | 560.15 | 1.78524 |
| 288 | 3.47222 | 561.15 | 1.78205 |
| 289 | 3.46021 | 562.15 | 1.77888 |
| 290 | 3.44828 | 563.15 | 1.77573 |
| 291 | 3.43643 | 564.15 | 1.77258 |
| 292 | 3.42466 | 565.15 | 1.76944 |
| 293 | 3.41297 | 566.15 | 1.76632 |
| 294 | 3.40136 | 567.15 | 1.76320 |
| 295 | 3.38983 | 568.15 | 1.76010 |
| 296 | 3.37838 | 569.15 | 1.75701 |
| 297 | 3.36700 | 570.15 | 1.75392 |
| 298 | 3.35570 | 571.15 | 1.75085 |
| 299 | 3.34448 | 572.15 | 1.74779 |
| 300 | 3.33333 | 573.15 | 1.74474 |
| 301 | 3.32226 | 574.15 | 1.74171 |
| 302 | 3.31126 | 575.15 | 1.73868 |
| 303 | 3.30033 | 576.15 | 1.73566 |
| 304 | 3.28947 | 577.15 | 1.73265 |
| 305 | 3.27869 | 578.15 | 1.72965 |
| 306 | 3.26797 | 579.15 | 1.72667 |
| 307 | 3.25733 | 580.15 | 1.72369 |
| 308 | 3.24675 | 581.15 | 1.72073 |
| 309 | 3.23625 | 582.15 | 1.71777 |
| 310 | 3.22581 | 583.15 | 1.71482 |
| 311 | 3.21543 | 584.15 | 1.71189 |
| 312 | 3.20513 | 585.15 | 1.70896 |
| 313 | 3.19489 | 586.15 | 1.70605 |
| 314 | 3.18471 | 587.15 | 1.70314 |
| 315 | 3.17460 | 588.15 | 1.70025 |
| 316 | 3.16456 | 589.15 | 1.69736 |
| 317 | 3.15457 | 590.15 | 1.69448 |
| 318 | 3.14465 | 591.15 | 1.69162 |
| 319 | 3.13480 | 592.15 | 1.68876 |
| 320 | 3.12500 | 593.15 | 1.68591 |
| 321 | 3.11526 | 594.15 | 1.68308 |
| 322 | 3.10559 | 595.15 | 1.68025 |
| 323 | 3.09598 | 596.15 | 1.67743 |
| 324 | 3.08642 | 597.15 | 1.67462 |
| 325 | 3.07692 | 598.15 | 1.67182 |
| 326 | 3.06748 | 599.15 | 1.66903 |
| 327 | 3.05810 | 600.15 | 1.66625 |
| 328 | 3.04878 | 601.15 | 1.66348 |
| 329 | 3.03951 | 602.15 | 1.66072 |
| 330 | 3.03030 | 603.15 | 1.65796 |
| 331 | 3.02115 | 604.15 | 1.65522 |
| 332 | 3.01205 | 605.15 | 1.65248 |
| 333 | 3.00300 | 606.15 | 1.64976 |
| 334 | 2.99401 | 607.15 | 1.64704 |
| 335 | 2.98507 | 608.15 | 1.64433 |
| 336 | 2.97619 | 609.15 | 1.64163 |
| 337 | 2.96736 | 610.15 | 1.63894 |
| 338 | 2.95858 | 611.15 | 1.63626 |
| 339 | 2.94985 | 612.15 | 1.63359 |
| 340 | 2.94118 | 613.15 | 1.63092 |
| 341 | 2.93255 | 614.15 | 1.62827 |
| 342 | 2.92398 | 615.15 | 1.62562 |
| 343 | 2.91545 | 616.15 | 1.62298 |

| N | 1000/N | N + 273.15 | 1000/N + 273.15 |
|---|---|---|---|
| +344 | 2.90698 | 617.15 | 1.62035 |
| 345 | 2.89855 | 618.15 | 1.61773 |
| 346 | 2.89017 | 619.15 | 1.61512 |
| 347 | 2.88184 | 620.15 | 1.61251 |
| 348 | 2.87356 | 621.15 | 1.60992 |
| 349 | 2.86533 | 622.15 | 1.60733 |
| 350 | 2.85714 | 623.15 | 1.60475 |
| 351 | 2.84900 | 624.15 | 1.60218 |
| 352 | 2.84091 | 625.15 | 1.59962 |
| 353 | 2.83286 | 626.15 | 1.59706 |
| 354 | 2.82486 | 627.15 | 1.59451 |
| 355 | 2.81690 | 628.15 | 1.59198 |
| 356 | 2.80899 | 629.15 | 1.58945 |
| 357 | 2.80112 | 630.15 | 1.58692 |
| 358 | 2.79330 | 631.15 | 1.58441 |
| 359 | 2.78552 | 632.15 | 1.58190 |
| 360 | 2.77778 | 633.15 | 1.57940 |
| 361 | 2.77008 | 634.15 | 1.57691 |
| 362 | 2.76243 | 635.15 | 1.57443 |
| 363 | 2.75482 | 636.15 | 1.57196 |
| 364 | 2.74725 | 637.15 | 1.56949 |
| 365 | 2.73973 | 638.15 | 1.56703 |
| 366 | 2.73224 | 639.15 | 1.56458 |
| 367 | 2.72480 | 640.15 | 1.56213 |
| 368 | 2.71739 | 641.15 | 1.55970 |
| 369 | 2.71003 | 642.15 | 1.55727 |
| 370 | 2.70270 | 643.15 | 1.55485 |
| 371 | 2.69542 | 644.15 | 1.55243 |
| 372 | 2.68817 | 645.15 | 1.55003 |
| 373 | 2.68097 | 646.15 | 1.54763 |
| 374 | 2.67380 | 647.15 | 1.54524 |
| 375 | 2.66667 | 648.15 | 1.54285 |
| 376 | 2.65957 | 649.15 | 1.54048 |
| 377 | 2.65252 | 650.15 | 1.53811 |
| 378 | 2.64550 | 651.15 | 1.53574 |
| 379 | 2.63852 | 652.15 | 1.53339 |
| 380 | 2.63158 | 653.15 | 1.53104 |
| 381 | 2.62467 | 654.15 | 1.52870 |
| 382 | 2.61780 | 655.15 | 1.52637 |
| 383 | 2.61097 | 656.15 | 1.52404 |
| 384 | 2.60417 | 657.15 | 1.52172 |
| 385 | 2.59740 | 658.15 | 1.51941 |
| 386 | 2.59067 | 659.15 | 1.51711 |
| 387 | 2.58398 | 660.15 | 1.51481 |
| 388 | 2.57732 | 661.15 | 1.51252 |
| 389 | 2.57069 | 662.15 | 1.51023 |
| 390 | 2.56410 | 663.15 | 1.50795 |
| 391 | 2.55754 | 664.15 | 1.50568 |
| 392 | 2.55102 | 665.15 | 1.50342 |
| 393 | 2.54453 | 666.15 | 1.50116 |
| 394 | 2.53807 | 667.15 | 1.49891 |
| 395 | 2.53165 | 668.15 | 1.49667 |
| 396 | 2.52525 | 669.15 | 1.49443 |
| 397 | 2.51889 | 670.15 | 1.49220 |
| 398 | 2.51256 | 671.15 | 1.48998 |
| 399 | 2.50627 | 672.15 | 1.48776 |
| 400 | 2.50000 | 673.15 | 1.48555 |
| 401 | 2.49377 | 674.15 | 1.48335 |
| 402 | 2.48756 | 675.15 | 1.48115 |
| 403 | 2.48139 | 676.15 | 1.47896 |
| 404 | 2.47525 | 677.15 | 1.47678 |
| 405 | 2.46914 | 678.15 | 1.47460 |
| 406 | 2.46305 | 679.15 | 1.47243 |
| 407 | 2.45700 | 680.15 | 1.47026 |
| 408 | 2.45098 | 681.15 | 1.46811 |
| 409 | 2.44499 | 682.15 | 1.46595 |
| 410 | 2.43902 | 683.15 | 1.46381 |
| 411 | 2.43309 | 684.15 | 1.46167 |
| 412 | 2.42718 | 685.15 | 1.45953 |
| 413 | 2.42131 | 686.15 | 1.45741 |
| 414 | 2.41546 | 687.15 | 1.45529 |
| 415 | 2.40964 | 688.15 | 1.45317 |
| 416 | 2.40385 | 689.15 | 1.45106 |
| 417 | 2.39808 | 690.15 | 1.44896 |

| N | 1000/N | N + 273.15 | 1000/N + 273.15 |
|---|---|---|---|
| +418 | 2.39234 | 691.15 | 1.44686 |
| 419 | 2.38663 | 692.15 | 1.44477 |
| 420 | 2.38095 | 693.15 | 1.44269 |
| 421 | 2.37530 | 694.15 | 1.44061 |
| 422 | 2.36967 | 695.15 | 1.43854 |
| 423 | 2.36407 | 696.15 | 1.43647 |
| 424 | 2.35849 | 697.15 | 1.43441 |
| 425 | 2.35294 | 698.15 | 1.43236 |
| 426 | 2.34742 | 699.15 | 1.43031 |
| 427 | 2.34192 | 700.15 | 1.42827 |
| 428 | 2.33645 | 701.15 | 1.42623 |
| 429 | 2.33100 | 702.15 | 1.42420 |
| 430 | 2.32558 | 703.15 | 1.42217 |
| 431 | 2.32019 | 704.15 | 1.42015 |
| 432 | 2.31481 | 705.15 | 1.41814 |
| 433 | 2.30947 | 706.15 | 1.41613 |
| 434 | 2.30415 | 707.15 | 1.41413 |
| 435 | 2.29885 | 708.15 | 1.41213 |
| 436 | 2.29358 | 709.15 | 1.41014 |
| 437 | 2.28833 | 710.15 | 1.40815 |
| 438 | 2.28311 | 711.15 | 1.40617 |
| 439 | 2.27790 | 712.15 | 1.40420 |
| 440 | 2.27273 | 713.15 | 1.40223 |
| 441 | 2.26757 | 714.15 | 1.40027 |
| 442 | 2.26244 | 715.15 | 1.39831 |
| 443 | 2.25734 | 716.15 | 1.39636 |
| 444 | 2.25225 | 717.15 | 1.39441 |
| 445 | 2.24719 | 718.15 | 1.39247 |
| 446 | 2.24215 | 719.15 | 1.39053 |
| 447 | 2.23714 | 720.15 | 1.38860 |
| 448 | 2.23214 | 721.15 | 1.38667 |
| 449 | 2.22717 | 722.15 | 1.38475 |
| 450 | 2.22222 | 723.15 | 1.38284 |
| 451 | 2.21729 | 724.15 | 1.38093 |
| 452 | 2.21239 | 725.15 | 1.37903 |
| 453 | 2.20751 | 726.15 | 1.37713 |
| 454 | 2.20264 | 727.15 | 1.37523 |
| 455 | 2.19780 | 728.15 | 1.37334 |
| 456 | 2.19298 | 729.15 | 1.37146 |
| 457 | 2.18818 | 730.15 | 1.36958 |
| 458 | 2.18341 | 731.15 | 1.36771 |
| 459 | 2.17865 | 732.15 | 1.36584 |
| 460 | 2.17391 | 733.15 | 1.36398 |
| 461 | 2.16920 | 734.15 | 1.36212 |
| 462 | 2.16450 | 735.15 | 1.36027 |
| 463 | 2.15983 | 736.15 | 1.35842 |
| 464 | 2.15517 | 737.15 | 1.35658 |
| 465 | 2.15054 | 738.15 | 1.35474 |
| 466 | 2.14592 | 739.15 | 1.35291 |
| 467 | 2.14133 | 740.15 | 1.35108 |
| 468 | 2.13675 | 741.15 | 1.34925 |
| 469 | 2.13220 | 742.15 | 1.34744 |
| 470 | 2.12766 | 743.15 | 1.34562 |
| 471 | 2.12314 | 744.15 | 1.34382 |
| 472 | 2.11864 | 745.15 | 1.34201 |
| 473 | 2.11416 | 746.15 | 1.34021 |
| 474 | 2.10970 | 747.15 | 1.33842 |
| 475 | 2.10526 | 748.15 | 1.33663 |
| 476 | 2.10084 | 749.15 | 1.33485 |
| 477 | 2.09644 | 750.15 | 1.33307 |
| 478 | 2.09205 | 751.15 | 1.33129 |
| 479 | 2.08768 | 752.15 | 1.32952 |
| 480 | 2.08333 | 753.15 | 1.32776 |
| 481 | 2.07900 | 754.15 | 1.32600 |
| 482 | 2.07469 | 755.15 | 1.32424 |
| 483 | 2.07039 | 756.15 | 1.32249 |
| 484 | 2.06612 | 757.15 | 1.32074 |
| 485 | 2.06186 | 758.15 | 1.31900 |
| 486 | 2.05761 | 759.15 | 1.31726 |
| 487 | 2.05339 | 760.15 | 1.31553 |
| 488 | 2.04918 | 761.15 | 1.31380 |
| 489 | 2.04499 | 762.15 | 1.31208 |
| 490 | 2.04082 | 763.15 | 1.31036 |
| 491 | 2.03666 | 764.15 | 1.30864 |

| N | 1000/N | N + 273.15 | 1000/N + 273.15 |
|---|---|---|---|
| +492 | 2.03252 | 765.15 | 1.30693 |
| 493 | 2.02840 | 766.15 | 1.30523 |
| 494 | 2.02429 | 767.15 | 1.30353 |
| 495 | 2.02020 | 768.15 | 1.30183 |
| 496 | 2.01613 | 769.15 | 1.30014 |
| 497 | 2.01207 | 770.15 | 1.29845 |
| 498 | 2.00803 | 771.15 | 1.29676 |
| 499 | 2.00401 | 772.15 | 1.29509 |
| 500 | 2.00000 | 773.15 | 1.29341 |
| 501 | 1.99601 | 774.15 | 1.29174 |
| 502 | 1.99203 | 775.15 | 1.29007 |
| 503 | 1.98807 | 776.15 | 1.28841 |
| 504 | 1.98413 | 777.15 | 1.28675 |
| 505 | 1.98020 | 778.15 | 1.28510 |
| 506 | 1.97627 | 779.15 | 1.28345 |
| 507 | 1.97239 | 780.15 | 1.28180 |
| 508 | 1.96850 | 781.15 | 1.28016 |
| 509 | 1.96464 | 782.15 | 1.27853 |
| 510 | 1.96078 | 783.15 | 1.27689 |
| 511 | 1.95695 | 784.15 | 1.27527 |
| 512 | 1.95313 | 785.15 | 1.27364 |
| 513 | 1.94932 | 786.15 | 1.27202 |
| 514 | 1.94553 | 787.15 | 1.27041 |
| 515 | 1.94175 | 788.15 | 1.26879 |
| 516 | 1.93798 | 789.15 | 1.26719 |
| 517 | 1.93424 | 790.15 | 1.26558 |
| 518 | 1.93050 | 791.15 | 1.26398 |
| 519 | 1.92678 | 792.15 | 1.26239 |
| 520 | 1.92308 | 793.15 | 1.26080 |
| 521 | 1.91939 | 794.15 | 1.25921 |
| 522 | 1.91571 | 795.15 | 1.25762 |
| 523 | 1.91205 | 796.15 | 1.25604 |
| 524 | 1.90840 | 797.15 | 1.25447 |
| 525 | 1.90476 | 798.15 | 1.25290 |
| 526 | 1.90114 | 799.15 | 1.25133 |
| 527 | 1.89753 | 800.15 | 1.24977 |
| 528 | 1.89394 | 801.15 | 1.24821 |
| 529 | 1.89036 | 802.15 | 1.24665 |
| 530 | 1.88679 | 803.15 | 1.24510 |
| 531 | 1.88324 | 804.15 | 1.24355 |
| 532 | 1.87970 | 805.15 | 1.24200 |
| 533 | 1.87617 | 806.15 | 1.24046 |
| 534 | 1.87266 | 807.15 | 1.23893 |
| 535 | 1.86916 | 808.15 | 1.23739 |
| 536 | 1.86567 | 809.15 | 1.23586 |
| 537 | 1.86220 | 810.15 | 1.23434 |
| 538 | 1.85874 | 811.15 | 1.23282 |
| 539 | 1.85529 | 812.15 | 1.23130 |
| 540 | 1.85185 | 813.15 | 1.22979 |
| 541 | 1.84843 | 814.15 | 1.22827 |
| 542 | 1.84502 | 815.15 | 1.22677 |
| 543 | 1.84162 | 816.15 | 1.22526 |
| 544 | 1.83824 | 817.15 | 1.22377 |
| 545 | 1.83486 | 818.15 | 1.22227 |
| 546 | 1.83150 | 819.15 | 1.22078 |
| 547 | 1.82815 | 820.15 | 1.21929 |
| 548 | 1.82482 | 821.15 | 1.21780 |
| 549 | 1.82149 | 822.15 | 1.21632 |
| 550 | 1.81818 | 823.15 | 1.21485 |
| 551 | 1.81488 | 824.15 | 1.21337 |
| 552 | 1.81159 | 825.15 | 1.21190 |
| 553 | 1.80832 | 826.15 | 1.21043 |
| 554 | 1.80505 | 827.15 | 1.20897 |
| 555 | 1.80180 | 828.15 | 1.20751 |
| 556 | 1.79856 | 829.15 | 1.20605 |
| 557 | 1.79533 | 830.15 | 1.20460 |
| 558 | 1.79211 | 831.15 | 1.20315 |
| 559 | 1.78891 | 832.15 | 1.20171 |
| 560 | 1.78571 | 833.15 | 1.20026 |
| 561 | 1.78253 | 834.15 | 1.19883 |
| 562 | 1.77936 | 835.15 | 1.19739 |
| 563 | 1.77620 | 836.15 | 1.19596 |
| 564 | 1.77305 | 837.15 | 1.19453 |
| 565 | 1.76991 | 838.15 | 1.19310 |

| N | 1000/N | N + 273.15 | 1000/N + 273.15 |
|---|---|---|---|
| +566 | 1.76678 | 839.15 | 1.19168 |
| 567 | 1.76367 | 840.15 | 1.19026 |
| 568 | 1.76056 | 841.15 | 1.18885 |
| 569 | 1.75747 | 842.15 | 1.18744 |
| 570 | 1.75439 | 843.15 | 1.18603 |
| 571 | 1.75131 | 844.15 | 1.18462 |
| 572 | 1.74825 | 845.15 | 1.18322 |
| 573 | 1.74520 | 846.15 | 1.18182 |
| 574 | 1.74216 | 847.15 | 1.18043 |
| 575 | 1.73913 | 848.15 | 1.17904 |
| 576 | 1.73611 | 849.15 | 1.17765 |
| 577 | 1.73310 | 850.15 | 1.17626 |
| 578 | 1.73010 | 851.15 | 1.17488 |
| 579 | 1.72712 | 852.15 | 1.17350 |
| 580 | 1.72414 | 853.15 | 1.17213 |
| 581 | 1.72117 | 854.15 | 1.17075 |
| 582 | 1.71821 | 855.15 | 1.16939 |
| 583 | 1.71527 | 856.15 | 1.16802 |
| 584 | 1.71233 | 857.15 | 1.16666 |
| 585 | 1.70940 | 858.15 | 1.16530 |
| 586 | 1.70648 | 859.15 | 1.16394 |
| 587 | 1.70358 | 860.15 | 1.16259 |
| 588 | 1.70068 | 861.15 | 1.16124 |
| 589 | 1.69779 | 862.15 | 1.15989 |
| 590 | 1.69492 | 863.15 | 1.15855 |
| 591 | 1.69205 | 864.15 | 1.15721 |
| 592 | 1.68919 | 865.15 | 1.15587 |
| 593 | 1.68634 | 866.15 | 1.15453 |
| 594 | 1.68350 | 867.15 | 1.15320 |
| 595 | 1.68067 | 868.15 | 1.15187 |
| 596 | 1.67785 | 869.15 | 1.15055 |
| 597 | 1.67504 | 870.15 | 1.14923 |
| 598 | 1.67224 | 871.15 | 1.14791 |
| 599 | 1.66945 | 872.15 | 1.14659 |
| 600 | 1.66667 | 873.15 | 1.14528 |
| 601 | 1.66389 | 874.15 | 1.14397 |
| 602 | 1.66113 | 875.15 | 1.14266 |
| 603 | 1.65837 | 876.15 | 1.14136 |
| 604 | 1.65563 | 877.15 | 1.14006 |
| 605 | 1.65289 | 878.15 | 1.13876 |
| 606 | 1.65017 | 879.15 | 1.13746 |
| 607 | 1.64745 | 880.15 | 1.13617 |
| 608 | 1.64474 | 881.15 | 1.13488 |
| 609 | 1.64204 | 882.15 | 1.13359 |
| 610 | 1.63934 | 883.15 | 1.13231 |
| 611 | 1.63666 | 884.15 | 1.13103 |
| 612 | 1.63399 | 885.15 | 1.12975 |
| 613 | 1.63132 | 886.15 | 1.12848 |
| 614 | 1.62866 | 887.15 | 1.12721 |
| 615 | 1.62602 | 888.15 | 1.12594 |
| 616 | 1.62338 | 889.15 | 1.12467 |
| 617 | 1.62075 | 890.15 | 1.12341 |
| 618 | 1.61812 | 891.15 | 1.12215 |
| 619 | 1.61551 | 892.15 | 1.12089 |
| 620 | 1.61290 | 893.15 | 1.11963 |
| 621 | 1.61031 | 894.15 | 1.11838 |
| 622 | 1.60772 | 895.15 | 1.11713 |
| 623 | 1.60514 | 896.15 | 1.11588 |
| 624 | 1.60256 | 897.15 | 1.11464 |
| 625 | 1.60000 | 898.15 | 1.11340 |
| 626 | 1.59744 | 899.15 | 1.11216 |
| 627 | 1.59490 | 900.15 | 1.11093 |
| 628 | 1.59236 | 901.15 | 1.10969 |
| 629 | 1.58983 | 902.15 | 1.10846 |
| 630 | 1.58730 | 903.15 | 1.10724 |
| 631 | 1.58479 | 904.15 | 1.10601 |
| 632 | 1.58228 | 905.15 | 1.10479 |
| 633 | 1.57978 | 906.15 | 1.10357 |
| 634 | 1.57729 | 907.15 | 1.10235 |
| 635 | 1.57480 | 908.15 | 1.10114 |
| 636 | 1.57233 | 909.15 | 1.09993 |
| 637 | 1.56986 | 910.15 | 1.09872 |
| 638 | 1.56740 | 911.15 | 1.09751 |
| 639 | 1.56495 | 912.15 | 1.09631 |

| N | 1000/N | N + 273.15 | 1000/N + 273.15 |
|---|---|---|---|
| +640 | 1.56250 | 913.15 | 1.09511 |
| 641 | 1.56006 | 914.15 | 1.09391 |
| 642 | 1.55763 | 915.15 | 1.09272 |
| 643 | 1.55521 | 916.15 | 1.09152 |
| 644 | 1.55280 | 917.15 | 1.09033 |
| 645 | 1.55039 | 918.15 | 1.08915 |
| 646 | 1.54799 | 919.15 | 1.08796 |
| 647 | 1.54560 | 920.15 | 1.08678 |
| 648 | 1.54321 | 921.15 | 1.08560 |
| 649 | 1.54083 | 922.15 | 1.08442 |
| 650 | 1.53846 | 923.15 | 1.08325 |
| 651 | 1.53610 | 924.15 | 1.08208 |
| 652 | 1.53374 | 925.15 | 1.08091 |
| 653 | 1.53139 | 926.15 | 1.07974 |
| 654 | 1.52905 | 927.15 | 1.07857 |
| 655 | 1.52672 | 928.15 | 1.07741 |
| 656 | 1.52439 | 929.15 | 1.07625 |
| 657 | 1.52207 | 930.15 | 1.07510 |
| 658 | 1.51976 | 931.15 | 1.07394 |
| 659 | 1.51745 | 932.15 | 1.07279 |
| 660 | 1.51515 | 933.15 | 1.07164 |
| 661 | 1.51286 | 934.15 | 1.07049 |
| 662 | 1.51057 | 935.15 | 1.06935 |
| 663 | 1.50830 | 936.15 | 1.06820 |
| 664 | 1.50602 | 937.15 | 1.06707 |
| 665 | 1.50376 | 938.15 | 1.06593 |
| 666 | 1.50150 | 939.15 | 1.06479 |
| 667 | 1.49925 | 940.15 | 1.06366 |
| 668 | 1.49701 | 941.15 | 1.06253 |
| 669 | 1.49477 | 942.15 | 1.06140 |
| 670 | 1.49254 | 943.15 | 1.06028 |
| 671 | 1.49031 | 944.15 | 1.05915 |
| 672 | 1.48810 | 945.15 | 1.05803 |
| 673 | 1.48588 | 946.15 | 1.05691 |
| 674 | 1.48368 | 947.15 | 1.05580 |
| 675 | 1.48148 | 948.15 | 1.05469 |
| 676 | 1.47929 | 949.15 | 1.05357 |
| 677 | 1.47710 | 950.15 | 1.05247 |
| 678 | 1.47493 | 951.15 | 1.05136 |
| 679 | 1.47275 | 952.15 | 1.05025 |
| 680 | 1.47059 | 953.15 | 1.04915 |
| 681 | 1.46843 | 954.15 | 1.04805 |
| 682 | 1.46628 | 955.15 | 1.04696 |
| 683 | 1.46413 | 956.15 | 1.04586 |
| 684 | 1.46199 | 957.15 | 1.04477 |
| 685 | 1.45985 | 958.15 | 1.04368 |
| 686 | 1.45773 | 959.15 | 1.04259 |
| 687 | 1.45560 | 960.15 | 1.04150 |
| 688 | 1.45349 | 961.15 | 1.04042 |
| 689 | 1.45138 | 962.15 | 1.03934 |
| 690 | 1.44928 | 963.15 | 1.03826 |
| 691 | 1.44718 | 964.15 | 1.03718 |
| 692 | 1.44509 | 965.15 | 1.03611 |
| 693 | 1.44300 | 966.15 | 1.03504 |
| 694 | 1.44092 | 967.15 | 1.03397 |
| 695 | 1.43885 | 968.15 | 1.03290 |
| 696 | 1.43678 | 969.15 | 1.03183 |
| 697 | 1.43472 | 970.15 | 1.03077 |
| 698 | 1.43266 | 971.15 | 1.02971 |
| 699 | 1.43062 | 972.15 | 1.02865 |
| 700 | 1.42857 | 973.15 | 1.02759 |
| 701 | 1.42653 | 974.15 | 1.02654 |
| 702 | 1.42450 | 975.15 | 1.02548 |
| 703 | 1.42248 | 976.15 | 1.02443 |
| 704 | 1.42045 | 977.15 | 1.02338 |
| 705 | 1.41844 | 978.15 | 1.02234 |
| 706 | 1.41643 | 979.15 | 1.02129 |
| 707 | 1.41443 | 980.15 | 1.02025 |
| 708 | 1.41243 | 981.15 | 1.01921 |
| 709 | 1.41044 | 982.15 | 1.01817 |
| 710 | 1.40845 | 983.15 | 1.01714 |
| 711 | 1.40647 | 984.15 | 1.01611 |
| 712 | 1.40449 | 985.15 | 1.01507 |
| 713 | 1.40252 | 986.15 | 1.01404 |

| N | 1000/N | N + 273.15 | 1000/N + 273.15 |
|---|---|---|---|
| +714 | 1.40056 | 987.15 | 1.01302 |
| 715 | 1.39860 | 988.15 | 1.01199 |
| 716 | 1.39665 | 989.15 | 1.01097 |
| 717 | 1.39470 | 990.15 | 1.00995 |
| 718 | 1.39276 | 991.15 | 1.00893 |
| 719 | 1.39082 | 992.15 | 1.00791 |
| 720 | 1.38889 | 993.15 | 1.00690 |
| 721 | 1.38696 | 994.15 | 1.00588 |
| 722 | 1.38504 | 995.15 | 1.00487 |
| 723 | 1.38313 | 996.15 | 1.00386 |
| 724 | 1.38122 | 997.15 | 1.00286 |
| 725 | 1.37931 | 998.15 | 1.00185 |
| 726 | 1.37741 | 999.15 | 1.00085 |
| 727 | 1.37552 | 1000.15 | .99985 |
| 728 | 1.37363 | 1001.15 | .99885 |
| 729 | 1.37174 | 1002.15 | .99785 |
| 730 | 1.36986 | 1003.15 | .99686 |
| 731 | 1.36799 | 1004.15 | .99587 |
| 732 | 1.36612 | 1005.15 | .99488 |
| 733 | 1.36426 | 1006.15 | .99389 |
| 734 | 1.36240 | 1007.15 | .99290 |
| 735 | 1.36054 | 1008.15 | .99192 |
| 736 | 1.35870 | 1009.15 | .99093 |
| 737 | 1.35685 | 1010.15 | .98995 |
| 738 | 1.35501 | 1011.15 | .98897 |
| 739 | 1.35318 | 1012.15 | .98800 |
| 740 | 1.35135 | 1013.15 | .98702 |
| 741 | 1.34953 | 1014.15 | .98605 |
| 742 | 1.34771 | 1015.15 | .98508 |
| 743 | 1.34590 | 1016.15 | .98411 |
| 744 | 1.34409 | 1017.15 | .98314 |
| 745 | 1.34228 | 1018.15 | .98217 |
| 746 | 1.34048 | 1019.15 | .98121 |
| 747 | 1.33869 | 1020.15 | .98025 |
| 748 | 1.33690 | 1021.15 | .97929 |
| 749 | 1.33511 | 1022.15 | .97833 |
| 750 | 1.33333 | 1023.15 | .97737 |
| 751 | 1.33156 | 1024.15 | .97642 |
| 752 | 1.32979 | 1025.15 | .97547 |
| 753 | 1.32802 | 1026.15 | .97452 |
| 754 | 1.32626 | 1027.15 | .97357 |
| 755 | 1.32450 | 1028.15 | .97262 |
| 756 | 1.32275 | 1029.15 | .97168 |
| 757 | 1.32100 | 1030.15 | .97073 |
| 758 | 1.31926 | 1031.15 | .96979 |
| 759 | 1.31752 | 1032.15 | .96885 |
| 760 | 1.31579 | 1033.15 | .96791 |
| 761 | 1.31406 | 1034.15 | .96698 |
| 762 | 1.31234 | 1035.15 | .96604 |
| 763 | 1.31062 | 1036.15 | .96511 |
| 764 | 1.30890 | 1037.15 | .96418 |
| 765 | 1.30719 | 1038.15 | .96325 |
| 766 | 1.30548 | 1039.15 | .96232 |
| 767 | 1.30378 | 1040.15 | .96140 |
| 768 | 1.30208 | 1041.15 | .96048 |
| 769 | 1.30039 | 1042.15 | .95955 |
| 770 | 1.29870 | 1043.15 | .95863 |
| 771 | 1.29702 | 1044.15 | .95772 |
| 772 | 1.29534 | 1045.15 | .95680 |
| 773 | 1.29366 | 1046.15 | .95589 |
| 774 | 1.29199 | 1047.15 | .95497 |
| 775 | 1.29032 | 1048.15 | .95406 |
| 776 | 1.28866 | 1049.15 | .95315 |
| 777 | 1.28700 | 1050.15 | .95224 |
| 778 | 1.28535 | 1051.15 | .95134 |
| 779 | 1.28370 | 1052.15 | .95043 |
| 780 | 1.28205 | 1053.15 | .94953 |
| 781 | 1.28041 | 1054.15 | .94863 |
| 782 | 1.27877 | 1055.15 | .94773 |
| 783 | 1.27714 | 1056.15 | .94684 |
| 784 | 1.27551 | 1057.15 | .94594 |
| 785 | 1.27389 | 1058.15 | .94505 |
| 786 | 1.27226 | 1059.15 | .94415 |
| 787 | 1.27065 | 1060.15 | .94326 |

| N | 1000/N | N + 273.15 | 1000/N + 273.15 |
|---|---|---|---|
| +788 | 1.26904 | 1061.15 | .94237 |
| 789 | 1.26743 | 1062.15 | .94149 |
| 790 | 1.26582 | 1063.15 | .94060 |
| 791 | 1.26422 | 1064.15 | .93972 |
| 792 | 1.26263 | 1065.15 | .93883 |
| 793 | 1.26103 | 1066.15 | .93795 |
| 794 | 1.25945 | 1067.15 | .93708 |
| 795 | 1.25786 | 1068.15 | .93620 |
| 796 | 1.25628 | 1069.15 | .93532 |
| 797 | 1.25471 | 1070.15 | .93445 |
| 798 | 1.25313 | 1071.15 | .93358 |
| 799 | 1.25156 | 1072.15 | .93271 |
| 800 | 1.25000 | 1073.15 | .93184 |
| 801 | 1.24844 | 1074.15 | .93097 |
| 802 | 1.24688 | 1075.15 | .93010 |
| 803 | 1.24533 | 1076.15 | .92924 |
| 804 | 1.24378 | 1077.15 | .92838 |
| 805 | 1.24224 | 1078.15 | .92751 |
| 806 | 1.24069 | 1079.15 | .92666 |
| 807 | 1.23916 | 1080.15 | .92580 |
| 808 | 1.23762 | 1081.15 | .92494 |
| 809 | 1.23609 | 1082.15 | .92409 |
| 810 | 1.23457 | 1083.15 | .92323 |
| 811 | 1.23305 | 1084.15 | .92238 |
| 812 | 1.23153 | 1085.15 | .92153 |
| 813 | 1.23001 | 1086.15 | .92068 |
| 814 | 1.22850 | 1087.15 | .91984 |
| 815 | 1.22699 | 1088.15 | .91899 |
| 816 | 1.22549 | 1089.15 | .91815 |
| 817 | 1.22399 | 1090.15 | .91730 |
| 818 | 1.22249 | 1091.15 | .91646 |
| 819 | 1.22100 | 1092.15 | .91563 |
| 820 | 1.21951 | 1093.15 | .91479 |
| 821 | 1.21803 | 1094.15 | .91395 |
| 822 | 1.21655 | 1095.15 | .91312 |
| 823 | 1.21507 | 1096.15 | .91228 |
| 824 | 1.21359 | 1097.15 | .91145 |
| 825 | 1.21212 | 1098.15 | .91062 |
| 826 | 1.21065 | 1099.15 | .90979 |
| 827 | 1.20919 | 1100.15 | .90897 |
| 828 | 1.20773 | 1101.15 | .90814 |
| 829 | 1.20627 | 1102.15 | .90732 |
| 830 | 1.20482 | 1103.15 | .90650 |
| 831 | 1.20337 | 1104.15 | .90567 |
| 832 | 1.20192 | 1105.15 | .90485 |
| 833 | 1.20048 | 1106.15 | .90404 |
| 834 | 1.19904 | 1107.15 | .90322 |
| 835 | 1.19760 | 1108.15 | .90240 |
| 836 | 1.19617 | 1109.15 | .90159 |
| 837 | 1.19474 | 1110.15 | .90078 |
| 838 | 1.19332 | 1111.15 | .89997 |
| 839 | 1.19190 | 1112.15 | .89916 |
| 840 | 1.19048 | 1113.15 | .89835 |
| 841 | 1.18906 | 1114.15 | .89755 |
| 842 | 1.18765 | 1115.15 | .89674 |
| 843 | 1.18624 | 1116.15 | .89594 |
| 844 | 1.18483 | 1117.15 | .89513 |
| 845 | 1.18343 | 1118.15 | .89433 |
| 846 | 1.18203 | 1119.15 | .89354 |
| 847 | 1.18064 | 1120.15 | .89274 |
| 848 | 1.17925 | 1121.15 | .89194 |
| 849 | 1.17786 | 1122.15 | .89115 |
| 850 | 1.17647 | 1123.15 | .89035 |
| 851 | 1.17509 | 1124.15 | .88956 |
| 852 | 1.17371 | 1125.15 | .88877 |
| 853 | 1.17233 | 1126.15 | .88798 |
| 854 | 1.17096 | 1127.15 | .88719 |
| 855 | 1.16959 | 1128.15 | .88641 |
| 856 | 1.16822 | 1129.15 | .88562 |
| 857 | 1.16686 | 1130.15 | .88484 |
| 858 | 1.16550 | 1131.15 | .88406 |
| 859 | 1.16414 | 1132.15 | .88328 |
| 860 | 1.16279 | 1133.15 | .88250 |
| 861 | 1.16144 | 1134.15 | .88172 |

| N | 1000/N | N + 273.15 | 1000/N + 273.15 |
|---|---|---|---|
| +862 | 1.16009 | 1135.15 | .88094 |
| 863 | 1.15875 | 1136.15 | .88017 |
| 864 | 1.15741 | 1137.15 | .87939 |
| 865 | 1.15607 | 1138.15 | .87862 |
| 866 | 1.15473 | 1139.15 | .87785 |
| 867 | 1.15340 | 1140.15 | .87708 |
| 868 | 1.15207 | 1141.15 | .87631 |
| 869 | 1.15075 | 1142.15 | .87554 |
| 870 | 1.14943 | 1143.15 | .87478 |
| 871 | 1.14811 | 1144.15 | .87401 |
| 872 | 1.14679 | 1145.15 | .87325 |
| 873 | 1.14548 | 1146.15 | .87249 |
| 874 | 1.14416 | 1147.15 | .87173 |
| 875 | 1.14286 | 1148.15 | .87097 |
| 876 | 1.14155 | 1149.15 | .87021 |
| 877 | 1.14025 | 1150.15 | .86945 |
| 878 | 1.13895 | 1151.15 | .86870 |
| 879 | 1.13766 | 1152.15 | .86794 |
| 880 | 1.13636 | 1153.15 | .86719 |
| 881 | 1.13507 | 1154.15 | .86644 |
| 882 | 1.13379 | 1155.15 | .86569 |
| 883 | 1.13250 | 1156.15 | .86494 |
| 884 | 1.13122 | 1157.15 | .86419 |
| 885 | 1.12994 | 1158.15 | .86345 |
| 886 | 1.12867 | 1159.15 | .86270 |
| 887 | 1.12740 | 1160.15 | .86196 |
| 888 | 1.12613 | 1161.15 | .86122 |
| 889 | 1.12486 | 1162.15 | .86047 |
| 890 | 1.12360 | 1163.15 | .85973 |
| 891 | 1.12233 | 1164.15 | .85900 |
| 892 | 1.12108 | 1165.15 | .85826 |
| 893 | 1.11982 | 1166.15 | .85752 |
| 894 | 1.11857 | 1167.15 | .85679 |
| 895 | 1.11732 | 1168.15 | .85605 |
| 896 | 1.11607 | 1169.15 | .85532 |
| 897 | 1.11483 | 1170.15 | .85459 |
| 898 | 1.11359 | 1171.15 | .85386 |
| 899 | 1.11235 | 1172.15 | .85313 |
| 900 | 1.11111 | 1173.15 | .85241 |
| 901 | 1.10988 | 1174.15 | .85168 |
| 902 | 1.10865 | 1175.15 | .85096 |
| 903 | 1.10742 | 1176.15 | .85023 |
| 904 | 1.10619 | 1177.15 | .84951 |
| 905 | 1.10497 | 1178.15 | .84879 |
| 906 | 1.10375 | 1179.15 | .84807 |
| 907 | 1.10254 | 1180.15 | .84735 |
| 908 | 1.10132 | 1181.15 | .84663 |
| 909 | 1.10011 | 1182.15 | .84592 |
| 910 | 1.09890 | 1183.15 | .84520 |
| 911 | 1.09769 | 1184.15 | .84449 |
| 912 | 1.09649 | 1185.15 | .84378 |
| 913 | 1.09529 | 1186.15 | .84306 |
| 914 | 1.09409 | 1187.15 | .84235 |
| 915 | 1.09290 | 1188.15 | .84164 |
| 916 | 1.09170 | 1189.15 | .84094 |
| 917 | 1.09051 | 1190.15 | .84023 |
| 918 | 1.08932 | 1191.15 | .83952 |
| 919 | 1.08814 | 1192.15 | .83882 |
| 920 | 1.08696 | 1193.15 | .83812 |
| 921 | 1.08578 | 1194.15 | .83742 |
| 922 | 1.08460 | 1195.15 | .83672 |
| 923 | 1.08342 | 1196.15 | .83602 |
| 924 | 1.08225 | 1197.15 | .83532 |
| 925 | 1.08108 | 1198.15 | .83462 |
| 926 | 1.07991 | 1199.15 | .83392 |
| 927 | 1.07875 | 1200.15 | .83323 |
| 928 | 1.07759 | 1201.15 | .83254 |
| 929 | 1.07643 | 1202.15 | .83184 |
| 930 | 1.07527 | 1203.15 | .83115 |
| 931 | 1.07411 | 1204.15 | .83046 |
| 932 | 1.07296 | 1205.15 | .82977 |
| 933 | 1.07181 | 1206.15 | .82908 |
| 934 | 1.07066 | 1207.15 | .82840 |
| 935 | 1.06952 | 1208.15 | .82771 |

| N | 1000/N | N + 273.15 | 1000/N + 273.15 |
|---|---|---|---|
| +936 | 1.06838 | 1209.15 | .82703 |
| 937 | 1.06724 | 1210.15 | .82634 |
| 938 | 1.06610 | 1211.15 | .82566 |
| 939 | 1.06496 | 1212.15 | .82498 |
| 940 | 1.06383 | 1213.15 | .82430 |
| 941 | 1.06270 | 1214.15 | .82362 |
| 942 | 1.06157 | 1215.15 | .82294 |
| 943 | 1.06045 | 1216.15 | .82227 |
| 944 | 1.05932 | 1217.15 | .82159 |
| 945 | 1.05820 | 1218.15 | .82092 |
| 946 | 1.05708 | 1219.15 | .82024 |
| 947 | 1.05597 | 1220.15 | .81957 |
| 948 | 1.05485 | 1221.15 | .81890 |
| 949 | 1.05374 | 1222.15 | .81823 |
| 950 | 1.05263 | 1223.15 | .81756 |
| 951 | 1.05152 | 1224.15 | .81689 |
| 952 | 1.05042 | 1225.15 | .81623 |
| 953 | 1.04932 | 1226.15 | .81556 |
| 954 | 1.04822 | 1227.15 | .81490 |
| 955 | 1.04712 | 1228.15 | .81423 |
| 956 | 1.04603 | 1229.15 | .81357 |
| 957 | 1.04493 | 1230.15 | .81291 |
| 958 | 1.04384 | 1231.15 | .81225 |
| 959 | 1.04275 | 1232.15 | .81159 |
| 960 | 1.04167 | 1233.15 | .81093 |
| 961 | 1.04058 | 1234.15 | .81027 |
| 962 | 1.03950 | 1235.15 | .80962 |
| 963 | 1.03842 | 1236.15 | .80896 |
| 964 | 1.03734 | 1237.15 | .80831 |
| 965 | 1.03627 | 1238.15 | .80766 |
| 966 | 1.03520 | 1239.15 | .80700 |
| 967 | 1.03413 | 1240.15 | .80635 |
| 968 | 1.03306 | 1241.15 | .80570 |
| 969 | 1.03199 | 1242.15 | .80506 |
| 970 | 1.03093 | 1243.15 | .80441 |
| 971 | 1.02987 | 1244.15 | .80376 |
| 972 | 1.02881 | 1245.15 | .80312 |
| 973 | 1.02775 | 1246.15 | .80247 |
| 974 | 1.02669 | 1247.15 | .80183 |
| 975 | 1.02564 | 1248.15 | .80119 |
| 976 | 1.02459 | 1249.15 | .80054 |
| 977 | 1.02354 | 1250.15 | .79990 |
| 978 | 1.02249 | 1251.15 | .79926 |
| 979 | 1.02145 | 1252.15 | .79863 |
| 980 | 1.02041 | 1253.15 | .79799 |
| 981 | 1.01937 | 1254.15 | .79735 |
| 982 | 1.01833 | 1255.15 | .79672 |
| 983 | 1.01729 | 1256.15 | .79608 |
| 984 | 1.01626 | 1257.15 | .79545 |
| 985 | 1.01523 | 1258.15 | .79482 |
| 986 | 1.01420 | 1259.15 | .79419 |
| 987 | 1.01317 | 1260.15 | .79356 |
| 988 | 1.01215 | 1261.15 | .79293 |
| 989 | 1.01112 | 1262.15 | .79230 |
| 990 | 1.01010 | 1263.15 | .79167 |
| 991 | 1.00908 | 1264.15 | .79105 |
| 992 | 1.00806 | 1265.15 | .79042 |
| 993 | 1.00705 | 1266.15 | .78980 |
| 994 | 1.00604 | 1267.15 | .78917 |
| 995 | 1.00503 | 1268.15 | .78855 |
| 996 | 1.00402 | 1269.15 | .78793 |
| 997 | 1.00301 | 1270.15 | .78731 |
| 998 | 1.00200 | 1271.15 | .78669 |
| 999 | 1.00100 | 1272.15 | .78607 |
| 1000 | 1.00000 | 1273.15 | .78545 |
| 1005 | .99502 | 1278.15 | .78238 |
| 1010 | .99010 | 1283.15 | .77933 |
| 1015 | .98522 | 1288.15 | .77631 |
| 1020 | .98039 | 1293.15 | .77331 |
| 1025 | .97561 | 1298.15 | .77033 |
| 1030 | .97087 | 1303.15 | .76737 |
| 1035 | .96618 | 1308.15 | .76444 |
| 1040 | .96154 | 1313.15 | .76153 |
| 1045 | .95694 | 1318.15 | .75864 |

| N | 1000/N | N + 273.15 | 1000/N + 273.15 |
|---|---|---|---|
| +1050 | .95238 | 1323.15 | .75577 |
| 1055 | .94787 | 1328.15 | .75293 |
| 1060 | .94340 | 1333.15 | .75010 |
| 1065 | .93897 | 1338.15 | .74730 |
| 1070 | .93458 | 1343.15 | .74452 |
| 1075 | .93023 | 1348.15 | .74176 |
| 1080 | .92593 | 1353.15 | .73902 |
| 1085 | .92166 | 1358.15 | .73630 |
| 1090 | .91743 | 1363.15 | .73359 |
| 1095 | .91324 | 1368.15 | .73091 |
| 1100 | .90909 | 1373.15 | .72825 |
| 1105 | .90498 | 1378.15 | .72561 |
| 1110 | .90090 | 1383.15 | .72299 |
| 1115 | .89686 | 1388.15 | .72038 |
| 1120 | .89286 | 1393.15 | .71780 |
| 1125 | .88889 | 1398.15 | .71523 |
| 1130 | .88496 | 1403.15 | .71268 |
| 1135 | .88106 | 1408.15 | .71015 |
| 1140 | .87719 | 1413.15 | .70764 |
| 1145 | .87336 | 1418.15 | .70514 |
| 1150 | .86957 | 1423.15 | .70267 |
| 1155 | .86580 | 1428.15 | .70021 |
| 1160 | .86207 | 1433.15 | .69776 |
| 1165 | .85837 | 1438.15 | .69534 |
| 1170 | .85470 | 1443.15 | .69293 |
| 1175 | .85106 | 1448.15 | .69054 |
| 1180 | .84746 | 1453.15 | .68816 |
| 1185 | .84388 | 1458.15 | .68580 |
| 1190 | .84034 | 1463.15 | .68346 |
| 1195 | .83682 | 1468.15 | .68113 |
| 1200 | .83333 | 1473.15 | .67882 |
| 1205 | .82988 | 1478.15 | .67652 |
| 1210 | .82645 | 1483.15 | .67424 |
| 1215 | .82305 | 1488.15 | .67198 |
| 1220 | .81967 | 1493.15 | .66973 |
| 1225 | .81633 | 1498.15 | .66749 |
| 1230 | .81301 | 1503.15 | .66527 |
| 1235 | .80972 | 1508.15 | .66306 |
| 1240 | .80645 | 1513.15 | .66087 |
| 1245 | .80321 | 1518.15 | .65870 |
| 1250 | .80000 | 1523.15 | .65653 |
| 1255 | .79681 | 1528.15 | .65439 |
| 1260 | .79365 | 1533.15 | .65225 |
| 1265 | .79051 | 1538.15 | .65013 |
| 1270 | .78740 | 1543.15 | .64803 |
| 1275 | .78431 | 1548.15 | .64593 |
| 1280 | .78125 | 1553.15 | .64385 |
| 1285 | .77821 | 1558.15 | .64179 |
| 1290 | .77519 | 1563.15 | .63973 |
| 1295 | .77220 | 1568.15 | .63769 |
| 1300 | .76923 | 1573.15 | .63567 |
| 1305 | .76628 | 1578.15 | .63365 |
| 1310 | .76336 | 1583.15 | .63165 |
| 1315 | .76046 | 1588.15 | .62966 |
| 1320 | .75758 | 1593.15 | .62769 |
| 1325 | .75472 | 1598.15 | .62572 |
| 1330 | .75188 | 1603.15 | .62377 |
| 1335 | .74906 | 1608.15 | .62183 |
| 1340 | .74627 | 1613.15 | .61991 |
| 1345 | .74349 | 1618.15 | .61799 |
| 1350 | .74074 | 1623.15 | .61609 |
| 1355 | .73801 | 1628.15 | .61419 |
| 1360 | .73529 | 1633.15 | .61231 |
| 1365 | .73260 | 1638.15 | .61044 |
| 1370 | .72993 | 1643.15 | .60859 |
| 1375 | .72727 | 1648.15 | .60674 |
| 1380 | .72464 | 1653.15 | .60491 |
| 1385 | .72202 | 1658.15 | .60308 |
| 1390 | .71942 | 1663.15 | .60127 |
| 1395 | .71685 | 1668.15 | .59947 |
| 1400 | .71429 | 1673.15 | .59768 |
| 1405 | .71174 | 1678.15 | .59589 |
| 1410 | .70922 | 1683.15 | .59412 |
| 1415 | .70671 | 1688.15 | .59236 |

| N | 1000/N | N + 273.15 | 1000/N + 273.15 |
|---|---|---|---|
| +1420 | .70423 | 1693.15 | .59062 |
| 1425 | .70175 | 1698.15 | .58888 |
| 1430 | .69930 | 1703.15 | .58715 |
| 1435 | .69686 | 1708.15 | .58543 |
| 1440 | .69444 | 1713.15 | .58372 |
| 1445 | .69204 | 1718.15 | .58202 |
| 1450 | .68966 | 1723.15 | .58033 |
| 1455 | .68729 | 1728.15 | .57865 |
| 1460 | .68493 | 1733.15 | .57698 |
| 1465 | .68259 | 1738.15 | .57532 |
| 1470 | .68027 | 1743.15 | .57367 |
| 1475 | .67797 | 1748.15 | .57203 |
| 1480 | .67568 | 1753.15 | .57040 |
| 1485 | .67340 | 1758.15 | .56878 |
| 1490 | .67114 | 1763.15 | .56717 |
| 1495 | .66890 | 1768.15 | .56556 |
| 1500 | .66667 | 1773.15 | .56397 |
| 1505 | .66445 | 1778.15 | .56238 |
| 1510 | .66225 | 1783.15 | .56081 |
| 1515 | .66007 | 1788.15 | .55924 |
| 1520 | .65789 | 1793.15 | .55768 |
| 1525 | .65574 | 1798.15 | .55613 |
| 1530 | .65359 | 1803.15 | .55459 |
| 1535 | .65147 | 1808.15 | .55305 |
| 1540 | .64935 | 1813.15 | .55153 |
| 1545 | .64725 | 1818.15 | .55001 |
| 1550 | .64516 | 1823.15 | .54850 |
| 1555 | .64309 | 1828.15 | .54700 |
| 1560 | .64103 | 1833.15 | .54551 |
| 1565 | .63898 | 1838.15 | .54403 |
| 1570 | .63694 | 1843.15 | .54255 |
| 1575 | .63492 | 1848.15 | .54108 |
| 1580 | .63291 | 1853.15 | .53962 |
| 1585 | .63091 | 1858.15 | .53817 |
| 1590 | .62893 | 1863.15 | .53673 |
| 1595 | .62696 | 1868.15 | .53529 |
| 1600 | .62500 | 1873.15 | .53386 |
| 1605 | .62305 | 1878.15 | .53244 |
| 1610 | .62112 | 1883.15 | .53103 |
| 1615 | .61920 | 1888.15 | .52962 |
| 1620 | .61728 | 1893.15 | .52822 |
| 1625 | .61538 | 1898.15 | .52683 |
| 1630 | .61350 | 1903.15 | .52544 |
| 1635 | .61162 | 1908.15 | .52407 |
| 1640 | .60976 | 1913.15 | .52270 |
| 1645 | .60790 | 1918.15 | .52134 |
| 1650 | .60606 | 1923.15 | .51998 |
| 1655 | .60423 | 1928.15 | .51863 |
| 1660 | .60241 | 1933.15 | .51729 |
| 1665 | .60060 | 1938.15 | .51596 |
| 1670 | .59880 | 1943.15 | .51463 |
| 1675 | .59701 | 1948.15 | .51331 |
| 1680 | .59524 | 1953.15 | .51199 |
| 1685 | .59347 | 1958.15 | .51069 |
| 1690 | .59172 | 1963.15 | .50939 |
| 1695 | .58997 | 1968.15 | .50809 |
| 1700 | .58824 | 1973.15 | .50680 |
| 1705 | .58651 | 1978.15 | .50552 |
| 1710 | .58480 | 1983.15 | .50425 |
| 1715 | .58309 | 1988.15 | .50298 |
| 1720 | .58140 | 1993.15 | .50172 |
| 1725 | .57971 | 1998.15 | .50046 |
| 1730 | .57803 | 2003.15 | .49921 |
| 1735 | .57637 | 2008.15 | .49797 |
| 1740 | .57471 | 2013.15 | .49673 |
| 1745 | .57307 | 2018.15 | .49550 |
| 1750 | .57143 | 2023.15 | .49428 |
| 1755 | .56980 | 2028.15 | .49306 |
| 1760 | .56818 | 2033.15 | .49185 |
| 1765 | .56657 | 2038.15 | .49064 |
| 1770 | .56497 | 2043.15 | .48944 |
| 1775 | .56338 | 2048.15 | .48825 |
| 1780 | .56180 | 2053.15 | .48706 |
| 1785 | .56022 | 2058.15 | .48587 |

| N | 1000/N | N + 273.15 | 1000/N + 273.15 |
|---|---|---|---|
| +1790 | .55866 | 2063.15 | .48470 |
| 1795 | .55710 | 2068.15 | .48352 |
| 1800 | .55556 | 2073.15 | .48236 |
| 1805 | .55402 | 2078.15 | .48120 |
| 1810 | .55249 | 2083.15 | .48004 |
| 1815 | .55096 | 2088.15 | .47889 |
| 1820 | .54945 | 2093.15 | .47775 |
| 1825 | .54795 | 2098.15 | .47661 |
| 1830 | .54645 | 2103.15 | .47548 |
| 1835 | .54496 | 2108.15 | .47435 |
| 1840 | .54348 | 2113.15 | .47323 |
| 1845 | .54201 | 2118.15 | .47211 |
| 1850 | .54054 | 2123.15 | .47100 |
| 1855 | .53908 | 2128.15 | .46989 |
| 1860 | .53763 | 2133.15 | .46879 |
| 1865 | .53619 | 2138.15 | .46769 |
| 1870 | .53476 | 2143.15 | .46660 |
| 1875 | .53333 | 2148.15 | .46552 |
| 1880 | .53191 | 2153.15 | .46444 |
| 1885 | .53050 | 2158.15 | .46336 |
| 1890 | .52910 | 2163.15 | .46229 |
| 1895 | .52770 | 2168.15 | .46122 |
| 1900 | .52632 | 2173.15 | .46016 |
| 1905 | .52493 | 2178.15 | .45911 |
| 1910 | .52356 | 2183.15 | .45805 |
| 1915 | .52219 | 2188.15 | .45701 |
| 1920 | .52083 | 2193.15 | .45597 |
| 1925 | .51948 | 2198.15 | .45493 |
| 1930 | .51813 | 2203.15 | .45390 |
| 1935 | .51680 | 2208.15 | .45287 |
| 1940 | .51546 | 2213.15 | .45184 |
| 1945 | .51414 | 2218.15 | .45083 |
| 1950 | .51282 | 2223.15 | .44981 |
| 1955 | .51151 | 2228.15 | .44880 |
| 1960 | .51020 | 2233.15 | .44780 |
| 1965 | .50891 | 2238.15 | .44680 |
| 1970 | .50761 | 2243.15 | .44580 |
| 1975 | .50633 | 2248.15 | .44481 |
| 1980 | .50505 | 2253.15 | .44382 |
| 1985 | .50378 | 2258.15 | .44284 |
| 1990 | .50251 | 2263.15 | .44186 |
| 1995 | .50125 | 2268.15 | .44089 |
| 2000 | .50000 | 2273.15 | .43992 |
| 2005 | .49875 | 2278.15 | .43895 |
| 2010 | .49751 | 2283.15 | .43799 |
| 2015 | .49628 | 2288.15 | .43703 |
| 2020 | .49505 | 2293.15 | .43608 |
| 2025 | .49383 | 2298.15 | .43513 |
| 2030 | .49261 | 2303.15 | .43419 |
| 2035 | .49140 | 2308.15 | .43325 |
| 2040 | .49020 | 2313.15 | .43231 |
| 2045 | .48900 | 2318.15 | .43138 |
| 2050 | .48780 | 2323.15 | .43045 |
| 2055 | .48662 | 2328.15 | .42953 |
| 2060 | .48544 | 2333.15 | .42861 |
| 2065 | .48426 | 2338.15 | .42769 |
| 2070 | .48309 | 2343.15 | .42678 |
| 2075 | .48193 | 2348.15 | .42587 |
| 2080 | .48077 | 2353.15 | .42496 |
| 2085 | .47962 | 2358.15 | .42406 |
| 2090 | .47847 | 2363.15 | .42316 |
| 2095 | .47733 | 2368.15 | .42227 |
| 2100 | .47619 | 2373.15 | .42138 |
| 2105 | .47506 | 2378.15 | .42049 |
| 2110 | .47393 | 2383.15 | .41961 |
| 2115 | .47281 | 2388.15 | .41873 |
| 2120 | .47170 | 2393.15 | .41786 |
| 2125 | .47059 | 2398.15 | .41699 |
| 2130 | .46948 | 2403.15 | .41612 |
| 2135 | .46838 | 2408.15 | .41526 |
| 2140 | .46729 | 2413.15 | .41440 |
| 2145 | .46620 | 2418.15 | .41354 |
| 2150 | .46512 | 2423.15 | .41269 |
| 2155 | .46404 | 2428.15 | .41184 |

| N | 1000/N | N + 273.15 | 1000/N + 273.15 |
|---|---|---|---|
| +2160 | .46296 | 2433.15 | .41099 |
| 2165 | .46189 | 2438.15 | .41015 |
| 2170 | .46083 | 2443.15 | .40931 |
| 2175 | .45977 | 2448.15 | .40847 |
| 2180 | .45872 | 2453.15 | .40764 |
| 2185 | .45767 | 2458.15 | .40681 |
| 2190 | .45662 | 2463.15 | .40598 |
| 2195 | .45558 | 2468.15 | .40516 |
| 2200 | .45455 | 2473.15 | .40434 |
| 2205 | .45351 | 2478.15 | .40353 |
| 2210 | .45249 | 2483.15 | .40271 |
| 2215 | .45147 | 2488.15 | .40191 |
| 2220 | .45045 | 2493.15 | .40110 |
| 2225 | .44944 | 2498.15 | .40030 |
| 2230 | .44843 | 2503.15 | .39950 |
| 2235 | .44743 | 2508.15 | .39870 |
| 2240 | .44643 | 2513.15 | .39791 |
| 2245 | .44543 | 2518.15 | .39712 |
| 2250 | .44444 | 2523.15 | .39633 |
| 2255 | .44346 | 2528.15 | .39555 |
| 2260 | .44248 | 2533.15 | .39477 |
| 2265 | .44150 | 2538.15 | .39399 |
| 2270 | .44053 | 2543.15 | .39321 |
| 2275 | .43956 | 2548.15 | .39244 |
| 2280 | .43860 | 2553.15 | .39167 |
| 2285 | .43764 | 2558.15 | .39091 |
| 2290 | .43668 | 2563.15 | .39014 |
| 2295 | .43573 | 2568.15 | .38939 |
| 2300 | .43478 | 2573.15 | .38863 |
| 2305 | .43384 | 2578.15 | .38788 |
| 2310 | .43290 | 2583.15 | .38712 |
| 2315 | .43197 | 2588.15 | .38638 |
| 2320 | .43103 | 2593.15 | .38563 |
| 2325 | .43011 | 2598.15 | .38489 |
| 2330 | .42918 | 2603.15 | .38415 |
| 2335 | .42827 | 2608.15 | .38341 |
| 2340 | .42735 | 2613.15 | .38268 |
| 2345 | .42644 | 2618.15 | .38195 |
| 2350 | .42553 | 2623.15 | .38122 |
| 2355 | .42463 | 2628.15 | .38050 |
| 2360 | .42373 | 2633.15 | .37977 |
| 2365 | .42283 | 2638.15 | .37905 |
| 2370 | .42194 | 2643.15 | .37834 |
| 2375 | .42105 | 2648.15 | .37762 |
| 2380 | .42017 | 2653.15 | .37691 |
| 2385 | .41929 | 2658.15 | .37620 |
| 2390 | .41841 | 2663.15 | .37550 |
| 2395 | .41754 | 2668.15 | .37479 |
| 2400 | .41667 | 2673.15 | .37409 |
| 2405 | .41580 | 2678.15 | .37339 |
| 2410 | .41494 | 2683.15 | .37270 |
| 2415 | .41408 | 2688.15 | .37200 |
| 2420 | .41322 | 2693.15 | .37131 |
| 2425 | .41237 | 2698.15 | .37062 |
| 2430 | .41152 | 2703.15 | .36994 |
| 2435 | .41068 | 2708.15 | .36926 |
| 2440 | .40984 | 2713.15 | .36858 |
| 2445 | .40900 | 2718.15 | .36790 |
| 2450 | .40816 | 2723.15 | .36722 |
| 2455 | .40733 | 2728.15 | .36655 |
| 2460 | .40650 | 2733.15 | .36588 |
| 2465 | .40568 | 2738.15 | .36521 |
| 2470 | .40486 | 2743.15 | .36454 |
| 2475 | .40404 | 2748.15 | .36388 |
| 2480 | .40323 | 2753.15 | .36322 |
| 2485 | .40241 | 2758.15 | .36256 |
| 2490 | .40161 | 2763.15 | .36191 |
| 2495 | .40080 | 2768.15 | .36125 |
| 2500 | .40000 | 2773.15 | .36060 |
| 2505 | .39920 | 2778.15 | .35995 |
| 2510 | .39841 | 2783.15 | .35931 |
| 2515 | .39761 | 2788.15 | .35866 |
| 2520 | .39683 | 2793.15 | .35802 |
| 2525 | .39604 | 2798.15 | .35738 |

| N | 1000/N | N + 273.15 | 1000/N + 273.15 |
|---|---|---|---|
| +2530 | .39526 | 2803.15 | .35674 |
| 2535 | .39448 | 2808.15 | .35611 |
| 2540 | .39370 | 2813.15 | .35547 |
| 2545 | .39293 | 2818.15 | .35484 |
| 2550 | .39216 | 2823.15 | .35421 |
| 2555 | .39139 | 2828.15 | .35359 |
| 2560 | .39063 | 2833.15 | .35296 |
| 2565 | .38986 | 2838.15 | .35234 |
| 2570 | .38911 | 2843.15 | .35172 |
| 2575 | .38835 | 2848.15 | .35111 |
| 2580 | .38760 | 2853.15 | .35049 |
| 2585 | .38685 | 2858.15 | .34988 |
| 2590 | .38610 | 2863.15 | .34927 |
| 2595 | .38536 | 2868.15 | .34866 |
| 2600 | .38462 | 2873.15 | .34805 |
| 2605 | .38388 | 2878.15 | .34745 |
| 2610 | .38314 | 2883.15 | .34684 |
| 2615 | .38241 | 2888.15 | .34624 |
| 2620 | .38168 | 2893.15 | .34564 |
| 2625 | .38095 | 2898.15 | .34505 |
| 2630 | .38023 | 2903.15 | .34445 |
| 2635 | .37951 | 2908.15 | .34386 |
| 2640 | .37879 | 2913.15 | .34327 |
| 2645 | .37807 | 2918.15 | .34268 |
| 2650 | .37736 | 2923.15 | .34210 |
| 2655 | .37665 | 2928.15 | .34151 |
| 2660 | .37594 | 2933.15 | .34093 |
| 2665 | .37523 | 2938.15 | .34035 |
| 2670 | .37453 | 2943.15 | .33977 |
| 2675 | .37383 | 2948.15 | .33920 |
| 2680 | .37313 | 2953.15 | .33862 |
| 2685 | .37244 | 2958.15 | .33805 |
| 2690 | .37175 | 2963.15 | .33748 |
| 2695 | .37106 | 2968.15 | .33691 |
| 2700 | .37037 | 2973.15 | .33634 |
| 2705 | .36969 | 2978.15 | .33578 |
| 2710 | .36900 | 2983.15 | .33522 |
| 2715 | .36832 | 2988.15 | .33466 |
| 2720 | .36765 | 2993.15 | .33410 |
| 2725 | .36697 | 2998.15 | .33354 |
| 2730 | .36630 | 3003.15 | .33298 |
| 2735 | .36563 | 3008.15 | .33243 |
| 2740 | .36496 | 3013.15 | .33188 |
| 2745 | .36430 | 3018.15 | .33133 |
| 2750 | .36364 | 3023.15 | .33078 |
| 2755 | .36298 | 3028.15 | .33023 |
| 2760 | .36232 | 3033.15 | .32969 |
| 2765 | .36166 | 3038.15 | .32915 |
| 2770 | .36101 | 3043.15 | .32861 |
| 2775 | .36036 | 3048.15 | .32807 |
| 2780 | .35971 | 3053.15 | .32753 |
| 2785 | .35907 | 3058.15 | .32700 |
| 2790 | .35842 | 3063.15 | .32646 |
| 2795 | .35778 | 3068.15 | .32593 |
| 2800 | .35714 | 3073.15 | .32540 |
| 2805 | .35651 | 3078.15 | .32487 |
| 2810 | .35587 | 3083.15 | .32434 |
| 2815 | .35524 | 3088.15 | .32382 |
| 2820 | .35461 | 3093.15 | .32330 |
| 2825 | .35398 | 3098.15 | .32277 |
| 2830 | .35336 | 3103.15 | .32225 |
| 2835 | .35273 | 3108.15 | .32173 |
| 2840 | .35211 | 3113.15 | .32122 |
| 2845 | .35149 | 3118.15 | .32070 |
| 2850 | .35088 | 3123.15 | .32019 |
| 2855 | .35026 | 3128.15 | .31968 |
| 2860 | .34965 | 3133.15 | .31917 |
| 2865 | .34904 | 3138.15 | .31866 |
| 2870 | .34843 | 3143.15 | .31815 |
| 2875 | .34783 | 3148.15 | .31765 |
| 2880 | .34722 | 3153.15 | .31714 |
| 2885 | .34662 | 3158.15 | .31664 |
| 2890 | .34602 | 3163.15 | .31614 |
| 2895 | .34542 | 3168.15 | .31564 |

| N | 1000/N | N + 273.15 | 1000/N + 273.15 |
|---|---|---|---|
| +2900 | .34483 | 3173.15 | .31514 |
| 2905 | .34423 | 3178.15 | .31465 |
| 2910 | .34364 | 3183.15 | .31415 |
| 2915 | .34305 | 3188.15 | .31366 |
| 2920 | .34247 | 3193.15 | .31317 |
| 2925 | .34188 | 3198.15 | .31268 |
| 2930 | .34130 | 3203.15 | .31219 |
| 2935 | .34072 | 3208.15 | .31171 |
| 2940 | .34014 | 3213.15 | .31122 |
| 2945 | .33956 | 3218.15 | .31074 |
| 2950 | .33898 | 3223.15 | .31026 |
| 2955 | .33841 | 3228.15 | .30977 |
| 2960 | .33784 | 3233.15 | .30930 |
| 2965 | .33727 | 3238.15 | .30882 |
| 2970 | .33670 | 3243.15 | .30834 |
| 2975 | .33613 | 3248.15 | .30787 |
| 2980 | .33557 | 3253.15 | .30739 |
| 2985 | .33501 | 3258.15 | .30692 |
| 2990 | .33445 | 3263.15 | .30645 |
| 2995 | .33389 | 3268.15 | .30598 |
| 3000 | .33333 | 3273.15 | .30552 |
| 3005 | .33278 | 3278.15 | .30505 |
| 3010 | .33223 | 3283.15 | .30459 |
| 3015 | .33167 | 3288.15 | .30412 |
| 3020 | .33113 | 3293.15 | .30366 |
| 3025 | .33058 | 3298.15 | .30320 |
| 3030 | .33003 | 3303.15 | .30274 |
| 3035 | .32949 | 3308.15 | .30228 |
| 3040 | .32895 | 3313.15 | .30183 |
| 3045 | .32841 | 3318.15 | .30137 |
| 3050 | .32787 | 3323.15 | .30092 |
| 3055 | .32733 | 3328.15 | .30047 |
| 3060 | .32680 | 3333.15 | .30002 |
| 3065 | .32626 | 3338.15 | .29957 |
| 3070 | .32573 | 3343.15 | .29912 |
| 3075 | .32520 | 3348.15 | .29867 |
| 3080 | .32468 | 3353.15 | .29823 |
| 3085 | .32415 | 3358.15 | .29778 |
| 3090 | .32362 | 3363.15 | .29734 |
| 3095 | .32310 | 3368.15 | .29690 |
| 3100 | .32258 | 3373.15 | .29646 |
| 3105 | .32206 | 3378.15 | .29602 |
| 3110 | .32154 | 3383.15 | .29558 |
| 3115 | .32103 | 3388.15 | .29515 |
| 3120 | .32051 | 3393.15 | .29471 |
| 3125 | .32000 | 3398.15 | .29428 |
| 3130 | .31949 | 3403.15 | .29385 |
| 3135 | .31898 | 3408.15 | .29341 |
| 3140 | .31847 | 3413.15 | .29298 |
| 3145 | .31797 | 3418.15 | .29256 |
| 3150 | .31746 | 3423.15 | .29213 |
| 3155 | .31696 | 3428.15 | .29170 |
| 3160 | .31646 | 3433.15 | .29128 |
| 3165 | .31596 | 3438.15 | .29085 |
| 3170 | .31546 | 3443.15 | .29043 |
| 3175 | .31496 | 3448.15 | .29001 |
| 3180 | .31447 | 3453.15 | .28959 |
| 3185 | .31397 | 3458.15 | .28917 |
| 3190 | .31348 | 3463.15 | .28875 |
| 3195 | .31299 | 3468.15 | .28834 |
| 3200 | .31250 | 3473.15 | .28792 |
| 3205 | .31201 | 3478.15 | .28751 |
| 3210 | .31153 | 3483.15 | .28710 |
| 3215 | .31104 | 3488.15 | .28668 |
| 3220 | .31056 | 3493.15 | .28627 |
| 3225 | .31008 | 3498.15 | .28587 |
| 3230 | .30960 | 3503.15 | .28546 |
| 3235 | .30912 | 3508.15 | .28505 |
| 3240 | .30864 | 3513.15 | .28464 |
| 3245 | .30817 | 3518.15 | .28424 |
| 3250 | .30769 | 3523.15 | .28384 |
| 3255 | .30722 | 3528.15 | .28343 |
| 3260 | .30675 | 3533.15 | .28303 |
| 3265 | .30628 | 3538.15 | .28263 |

| N | 1000/N | N + 273.15 | 1000/N + 273.15 |
|---|---|---|---|
| +3270 | .30581 | 3543.15 | .28223 |
| 3275 | .30534 | 3548.15 | .28184 |
| 3280 | .30488 | 3553.15 | .28144 |
| 3285 | .30441 | 3558.15 | .28104 |
| 3290 | .30395 | 3563.15 | .28065 |
| 3295 | .30349 | 3568.15 | .28026 |
| 3300 | .30303 | 3573.15 | .27987 |
| 3305 | .30257 | 3578.15 | .27947 |
| 3310 | .30211 | 3583.15 | .27908 |
| 3315 | .30166 | 3588.15 | .27870 |
| 3320 | .30120 | 3593.15 | .27831 |
| 3325 | .30075 | 3598.15 | .27792 |
| 3330 | .30030 | 3603.15 | .27753 |
| 3335 | .29985 | 3608.15 | .27715 |
| 3340 | .29940 | 3613.15 | .27677 |
| 3345 | .29895 | 3618.15 | .27638 |
| 3350 | .29851 | 3623.15 | .27600 |
| 3355 | .29806 | 3628.15 | .27562 |
| 3360 | .29762 | 3633.15 | .27524 |
| 3365 | .29718 | 3638.15 | .27486 |
| 3370 | .29674 | 3643.15 | .27449 |
| 3375 | .29630 | 3648.15 | .27411 |
| 3380 | .29586 | 3653.15 | .27374 |
| 3385 | .29542 | 3658.15 | .27336 |
| 3390 | .29499 | 3663.15 | .27299 |
| 3395 | .29455 | 3668.15 | .27262 |
| 3400 | .29412 | 3673.15 | .27225 |
| 3405 | .29369 | 3678.15 | .27188 |
| 3410 | .29326 | 3683.15 | .27151 |
| 3415 | .29283 | 3688.15 | .27114 |
| 3420 | .29240 | 3693.15 | .27077 |
| 3425 | .29197 | 3698.15 | .27041 |
| 3430 | .29155 | 3703.15 | .27004 |
| 3435 | .29112 | 3708.15 | .26968 |
| 3440 | .29070 | 3713.15 | .26931 |
| 3445 | .29028 | 3718.15 | .26895 |
| 3450 | .28986 | 3723.15 | .26859 |
| 3455 | .28944 | 3728.15 | .26823 |
| 3460 | .28902 | 3733.15 | .26787 |
| 3465 | .28860 | 3738.15 | .26751 |
| 3470 | .28818 | 3743.15 | .26715 |
| 3475 | .28777 | 3748.15 | .26680 |
| 3480 | .28736 | 3753.15 | .26644 |
| 3485 | .28694 | 3758.15 | .26609 |
| 3490 | .28653 | 3763.15 | .26573 |
| 3495 | .28612 | 3768.15 | .26538 |
| 3500 | .28571 | 3773.15 | .26503 |
| 3505 | .28531 | 3778.15 | .26468 |
| 3510 | .28490 | 3783.15 | .26433 |
| 3515 | .28450 | 3788.15 | .26398 |
| 3520 | .28409 | 3793.15 | .26363 |
| 3525 | .28369 | 3798.15 | .26329 |
| 3530 | .28329 | 3803.15 | .26294 |
| 3535 | .28289 | 3808.15 | .26259 |
| 3540 | .28249 | 3813.15 | .26225 |
| 3545 | .28209 | 3818.15 | .26191 |
| 3550 | .28169 | 3823.15 | .26156 |
| 3555 | .28129 | 3828.15 | .26122 |
| 3560 | .28090 | 3833.15 | .26088 |
| 3565 | .28050 | 3838.15 | .26054 |
| 3570 | .28011 | 3843.15 | .26020 |
| 3575 | .27972 | 3848.15 | .25987 |
| 3580 | .27933 | 3853.15 | .25953 |
| 3585 | .27894 | 3858.15 | .25919 |
| 3590 | .27855 | 3863.15 | .25886 |
| 3595 | .27816 | 3868.15 | .25852 |
| 3600 | .27778 | 3873.15 | .25819 |
| 3605 | .27739 | 3878.15 | .25785 |
| 3610 | .27701 | 3883.15 | .25752 |
| 3615 | .27663 | 3888.15 | .25719 |
| 3620 | .27624 | 3893.15 | .25686 |
| 3625 | .27586 | 3898.15 | .25653 |
| 3630 | .27548 | 3903.15 | .25620 |
| 3635 | .27510 | 3908.15 | .25588 |

| N | 1000/N | N + 273.15 | 1000/N + 273.15 |
|---|---|---|---|
| +3640 | .27473 | 3913.15 | .25555 |
| 3645 | .27435 | 3918.15 | .25522 |
| 3650 | .27397 | 3923.15 | .25490 |
| 3655 | .27360 | 3928.15 | .25457 |
| 3660 | .27322 | 3933.15 | .25425 |
| 3665 | .27285 | 3938.15 | .25393 |
| 3670 | .27248 | 3943.15 | .25360 |
| 3675 | .27211 | 3948.15 | .25328 |
| 3680 | .27174 | 3953.15 | .25296 |
| 3685 | .27137 | 3958.15 | .25264 |
| 3690 | .27100 | 3963.15 | .25232 |
| 3695 | .27064 | 3968.15 | .25201 |
| 3700 | .27027 | 3973.15 | .25169 |
| 3705 | .26991 | 3978.15 | .25137 |
| 3710 | .26954 | 3983.15 | .25106 |
| 3715 | .26918 | 3988.15 | .25074 |
| 3720 | .26882 | 3993.15 | .25043 |
| 3725 | .26846 | 3998.15 | .25012 |
| 3730 | .26810 | 4003.15 | .24980 |
| 3735 | .26774 | 4008.15 | .24949 |
| 3740 | .26738 | 4013.15 | .24918 |
| 3745 | .26702 | 4018.15 | .24887 |
| 3750 | .26667 | 4023.15 | .24856 |
| 3755 | .26631 | 4028.15 | .24825 |
| 3760 | .26596 | 4033.15 | .24795 |
| 3765 | .26560 | 4038.15 | .24764 |
| 3770 | .26525 | 4043.15 | .24733 |
| 3775 | .26490 | 4048.15 | .24703 |
| 3780 | .26455 | 4053.15 | .24672 |
| 3785 | .26420 | 4058.15 | .24642 |
| 3790 | .26385 | 4063.15 | .24611 |
| 3795 | .26350 | 4068.15 | .24581 |
| 3800 | .26316 | 4073.15 | .24551 |
| 3805 | .26281 | 4078.15 | .24521 |
| 3810 | .26247 | 4083.15 | .24491 |
| 3815 | .26212 | 4088.15 | .24461 |
| 3820 | .26178 | 4093.15 | .24431 |
| 3825 | .26144 | 4098.15 | .24401 |
| 3830 | .26110 | 4103.15 | .24372 |
| 3835 | .26076 | 4108.15 | .24342 |
| 3840 | .26042 | 4113.15 | .24312 |
| 3845 | .26008 | 4118.15 | .24283 |
| 3850 | .25974 | 4123.15 | .24253 |
| 3855 | .25940 | 4128.15 | .24224 |
| 3860 | .25907 | 4133.15 | .24195 |
| 3865 | .25873 | 4138.15 | .24165 |
| 3870 | .25840 | 4143.15 | .24136 |
| 3875 | .25806 | 4148.15 | .24107 |
| 3880 | .25773 | 4153.15 | .24078 |
| 3885 | .25740 | 4158.15 | .24049 |
| 3890 | .25707 | 4163.15 | .24020 |
| 3895 | .25674 | 4168.15 | .23991 |
| 3900 | .25641 | 4173.15 | .23963 |
| 3905 | .25608 | 4178.15 | .23934 |
| 3910 | .25575 | 4183.15 | .23905 |
| 3915 | .25543 | 4188.15 | .23877 |
| 3920 | .25510 | 4193.15 | .23848 |
| 3925 | .25478 | 4198.15 | .23820 |
| 3930 | .25445 | 4203.15 | .23792 |
| 3935 | .25413 | 4208.15 | .23763 |
| 3940 | .25381 | 4213.15 | .23735 |
| 3945 | .25349 | 4218.15 | .23707 |
| 3950 | .25316 | 4223.15 | .23679 |
| 3955 | .25284 | 4228.15 | .23651 |
| 3960 | .25253 | 4233.15 | .23623 |
| 3965 | .25221 | 4238.15 | .23595 |
| 3970 | .25189 | 4243.15 | .23567 |
| 3975 | .25157 | 4248.15 | .23540 |
| 3980 | .25126 | 4253.15 | .23512 |
| 3985 | .25094 | 4258.15 | .23484 |
| 3990 | .25063 | 4263.15 | .23457 |
| 3995 | .25031 | 4268.15 | .23429 |
| 4000 | .25000 | 4273.15 | .23402 |
| 4005 | .24969 | 4278.15 | .23375 |

| N | 1000/N | N + 273.15 | 1000/N + 273.15 |
|---|---|---|---|
| +4010 | .24938 | 4283.15 | .23347 |
| 4015 | .24907 | 4288.15 | .23320 |
| 4020 | .24876 | 4293.15 | .23293 |
| 4025 | .24845 | 4298.15 | .23266 |
| 4030 | .24814 | 4303.15 | .23239 |
| 4035 | .24783 | 4308.15 | .23212 |
| 4040 | .24752 | 4313.15 | .23185 |
| 4045 | .24722 | 4318.15 | .23158 |
| 4050 | .24691 | 4323.15 | .23131 |
| 4055 | .24661 | 4328.15 | .23105 |
| 4060 | .24631 | 4333.15 | .23078 |
| 4065 | .24600 | 4338.15 | .23051 |
| 4070 | .24570 | 4343.15 | .23025 |
| 4075 | .24540 | 4348.15 | .22998 |
| 4080 | .24510 | 4353.15 | .22972 |
| 4085 | .24480 | 4358.15 | .22946 |
| 4090 | .24450 | 4363.15 | .22919 |
| 4095 | .24420 | 4368.15 | .22893 |
| 4100 | .24390 | 4373.15 | .22867 |
| 4105 | .24361 | 4378.15 | .22841 |
| 4110 | .24331 | 4383.15 | .22815 |
| 4115 | .24301 | 4388.15 | .22789 |
| 4120 | .24272 | 4393.15 | .22763 |
| 4125 | .24242 | 4398.15 | .22737 |
| 4130 | .24213 | 4403.15 | .22711 |
| 4135 | .24184 | 4408.15 | .22685 |
| 4140 | .24155 | 4413.15 | .22660 |
| 4145 | .24125 | 4418.15 | .22634 |
| 4150 | .24096 | 4423.15 | .22608 |
| 4155 | .24067 | 4428.15 | .22583 |
| 4160 | .24038 | 4433.15 | .22557 |
| 4165 | .24010 | 4438.15 | .22532 |
| 4170 | .23981 | 4443.15 | .22507 |
| 4175 | .23952 | 4448.15 | .22481 |
| 4180 | .23923 | 4453.15 | .22456 |
| 4185 | .23895 | 4458.15 | .22431 |
| 4190 | .23866 | 4463.15 | .22406 |
| 4195 | .23838 | 4468.15 | .22381 |
| 4200 | .23810 | 4473.15 | .22356 |
| 4205 | .23781 | 4478.15 | .22331 |
| 4210 | .23753 | 4483.15 | .22306 |
| 4215 | .23725 | 4488.15 | .22281 |
| 4220 | .23697 | 4493.15 | .22256 |
| 4225 | .23669 | 4498.15 | .22231 |
| 4230 | .23641 | 4503.15 | .22207 |
| 4235 | .23613 | 4508.15 | .22182 |
| 4240 | .23585 | 4513.15 | .22157 |
| 4245 | .23557 | 4518.15 | .22133 |
| 4250 | .23529 | 4523.15 | .22108 |
| 4255 | .23502 | 4528.15 | .22084 |
| 4260 | .23474 | 4533.15 | .22060 |
| 4265 | .23447 | 4538.15 | .22035 |
| 4270 | .23419 | 4543.15 | .22011 |
| 4275 | .23392 | 4548.15 | .21987 |
| 4280 | .23364 | 4553.15 | .21963 |
| 4285 | .23337 | 4558.15 | .21939 |
| 4290 | .23310 | 4563.15 | .21915 |
| 4295 | .23283 | 4568.15 | .21891 |
| 4300 | .23256 | 4573.15 | .21867 |
| 4305 | .23229 | 4578.15 | .21843 |
| 4310 | .23202 | 4583.15 | .21819 |
| 4315 | .23175 | 4588.15 | .21795 |
| 4320 | .23148 | 4593.15 | .21772 |
| 4325 | .23121 | 4598.15 | .21748 |
| 4330 | .23095 | 4603.15 | .21724 |
| 4335 | .23068 | 4608.15 | .21701 |
| 4340 | .23041 | 4613.15 | .21677 |
| 4345 | .23015 | 4618.15 | .21654 |
| 4350 | .22989 | 4623.15 | .21630 |
| 4355 | .22962 | 4628.15 | .21607 |
| 4360 | .22936 | 4633.15 | .21584 |
| 4365 | .22910 | 4638.15 | .21560 |
| 4370 | .22883 | 4643.15 | .21537 |
| 4375 | .22857 | 4648.15 | .21514 |

| N | 1000/N | N + 273.15 | 1000/N + 273.15 |
|---|---|---|---|
| +4380 | .22831 | 4653.15 | .21491 |
| 4385 | .22805 | 4658.15 | .21468 |
| 4390 | .22779 | 4663.15 | .21445 |
| 4395 | .22753 | 4668.15 | .21422 |
| 4400 | .22727 | 4673.15 | .21399 |
| 4405 | .22701 | 4678.15 | .21376 |
| 4410 | .22676 | 4683.15 | .21353 |
| 4415 | .22650 | 4688.15 | .21330 |
| 4420 | .22624 | 4693.15 | .21308 |
| 4425 | .22599 | 4698.15 | .21285 |
| 4430 | .22573 | 4703.15 | .21262 |
| 4435 | .22548 | 4708.15 | .21240 |
| 4440 | .22523 | 4713.15 | .21217 |
| 4445 | .22497 | 4718.15 | .21195 |
| 4450 | .22472 | 4723.15 | .21172 |
| 4455 | .22447 | 4728.15 | .21150 |
| 4460 | .22422 | 4733.15 | .21128 |
| 4465 | .22396 | 4738.15 | .21105 |
| 4470 | .22371 | 4743.15 | .21083 |
| 4475 | .22346 | 4748.15 | .21061 |
| 4480 | .22321 | 4753.15 | .21039 |
| 4485 | .22297 | 4758.15 | .21017 |
| 4490 | .22272 | 4763.15 | .20995 |
| 4495 | .22247 | 4768.15 | .20972 |
| 4500 | .22222 | 4773.15 | .20951 |
| 4505 | .22198 | 4778.15 | .20929 |
| 4510 | .22173 | 4783.15 | .20907 |
| 4515 | .22148 | 4788.15 | .20885 |
| 4520 | .22124 | 4793.15 | .20863 |
| 4525 | .22099 | 4798.15 | .20841 |
| 4530 | .22075 | 4803.15 | .20820 |
| 4535 | .22051 | 4808.15 | .20798 |
| 4540 | .22026 | 4813.15 | .20776 |
| 4545 | .22002 | 4818.15 | .20755 |
| 4550 | .21978 | 4823.15 | .20733 |
| 4555 | .21954 | 4828.15 | .20712 |
| 4560 | .21930 | 4833.15 | .20690 |
| 4565 | .21906 | 4838.15 | .20669 |
| 4570 | .21882 | 4843.15 | .20648 |
| 4575 | .21858 | 4848.15 | .20626 |
| 4580 | .21834 | 4853.15 | .20605 |
| 4585 | .21810 | 4858.15 | .20584 |
| 4590 | .21786 | 4863.15 | .20563 |
| 4595 | .21763 | 4868.15 | .20542 |
| 4600 | .21739 | 4873.15 | .20521 |
| 4605 | .21716 | 4878.15 | .20500 |
| 4610 | .21692 | 4883.15 | .20479 |
| 4615 | .21668 | 4888.15 | .20458 |
| 4620 | .21645 | 4893.15 | .20437 |
| 4625 | .21622 | 4898.15 | .20416 |
| 4630 | .21598 | 4903.15 | .20395 |
| 4635 | .21575 | 4908.15 | .20374 |
| 4640 | .21552 | 4913.15 | .20354 |
| 4645 | .21529 | 4918.15 | .20333 |
| 4650 | .21505 | 4923.15 | .20312 |
| 4655 | .21482 | 4928.15 | .20292 |
| 4660 | .21459 | 4933.15 | .20271 |
| 4665 | .21436 | 4938.15 | .20250 |
| 4670 | .21413 | 4943.15 | .20230 |
| 4675 | .21390 | 4948.15 | .20210 |
| 4680 | .21368 | 4953.15 | .20189 |
| 4685 | .21345 | 4958.15 | .20169 |
| 4690 | .21322 | 4963.15 | .20148 |
| 4695 | .21299 | 4968.15 | .20128 |
| 4700 | .21277 | 4973.15 | .20108 |
| 4705 | .21254 | 4978.15 | .20088 |
| 4710 | .21231 | 4983.15 | .20068 |
| 4715 | .21209 | 4988.15 | .20048 |
| 4720 | .21186 | 4993.15 | .20027 |
| 4725 | .21164 | 4998.15 | .20007 |
| 4730 | .21142 | 5003.15 | .19987 |
| 4735 | .21119 | 5008.15 | .19967 |
| 4740 | .21097 | 5013.15 | .19948 |
| 4745 | .21075 | 5018.15 | .19928 |

| N | 1000/N | N + 273.15 | 1000/N + 273.15 |
|---|---|---|---|
| +4750 | .21053 | 5023.15 | .19908 |
| 4755 | .21030 | 5028.15 | .19888 |
| 4760 | .21008 | 5033.15 | .19868 |
| 4765 | .20986 | 5038.15 | .19849 |
| 4770 | .20964 | 5043.15 | .19829 |
| 4775 | .20942 | 5048.15 | .19809 |
| 4780 | .20921 | 5053.15 | .19790 |
| 4785 | .20899 | 5058.15 | .19770 |
| 4790 | .20877 | 5063.15 | .19751 |
| 4795 | .20855 | 5068.15 | .19731 |
| 4800 | .20833 | 5073.15 | .19712 |
| 4805 | .20812 | 5078.15 | .19692 |
| 4810 | .20790 | 5083.15 | .19673 |
| 4815 | .20768 | 5088.15 | .19654 |
| 4820 | .20747 | 5093.15 | .19634 |
| 4825 | .20725 | 5098.15 | .19615 |
| 4830 | .20704 | 5103.15 | .19596 |
| 4835 | .20683 | 5108.15 | .19577 |
| 4840 | .20661 | 5113.15 | .19557 |
| 4845 | .20640 | 5118.15 | .19538 |
| 4850 | .20619 | 5123.15 | .19519 |
| 4855 | .20597 | 5128.15 | .19500 |
| 4860 | .20576 | 5133.15 | .19481 |
| 4865 | .20555 | 5138.15 | .19462 |
| 4870 | .20534 | 5143.15 | .19443 |
| 4875 | .20513 | 5148.15 | .19424 |
| 4880 | .20492 | 5153.15 | .19406 |
| 4885 | .20471 | 5158.15 | .19387 |
| 4890 | .20450 | 5163.15 | .19368 |
| 4895 | .20429 | 5168.15 | .19349 |
| 4900 | .20408 | 5173.15 | .19331 |
| 4905 | .20387 | 5178.15 | .19312 |
| 4910 | .20367 | 5183.15 | .19293 |
| 4915 | .20346 | 5188.15 | .19275 |
| 4920 | .20325 | 5193.15 | .19256 |
| 4925 | .20305 | 5198.15 | .19238 |
| 4930 | .20284 | 5203.15 | .19219 |
| 4935 | .20263 | 5208.15 | .19201 |
| 4940 | .20243 | 5213.15 | .19182 |
| 4945 | .20222 | 5218.15 | .19164 |
| 4950 | .20202 | 5223.15 | .19146 |
| 4955 | .20182 | 5228.15 | .19127 |
| 4960 | .20161 | 5233.15 | .19109 |
| 4965 | .20141 | 5238.15 | .19091 |
| 4970 | .20121 | 5243.15 | .19073 |
| 4975 | .20101 | 5248.15 | .19054 |
| 4980 | .20080 | 5253.15 | .19036 |
| 4985 | .20060 | 5258.15 | .19018 |
| 4990 | .20040 | 5263.15 | .19000 |
| 4995 | .20020 | 5268.15 | .18982 |
| 5000 | .20000 | 5273.15 | .18964 |
| 5005 | .19980 | 5278.15 | .18946 |
| 5010 | .19960 | 5283.15 | .18928 |
| 5015 | .19940 | 5288.15 | .18910 |
| 5020 | .19920 | 5293.15 | .18892 |
| 5025 | .19900 | 5298.15 | .18875 |
| 5030 | .19881 | 5303.15 | .18857 |
| 5035 | .19861 | 5308.15 | .18839 |
| 5040 | .19841 | 5313.15 | .18821 |
| 5045 | .19822 | 5318.15 | .18804 |
| 5050 | .19802 | 5323.15 | .18786 |
| 5055 | .19782 | 5328.15 | .18768 |
| 5060 | .19763 | 5333.15 | .18751 |
| 5065 | .19743 | 5338.15 | .18733 |
| 5070 | .19724 | 5343.15 | .18716 |
| 5075 | .19704 | 5348.15 | .18698 |
| 5080 | .19685 | 5353.15 | .18681 |
| 5085 | .19666 | 5358.15 | .18663 |
| 5090 | .19646 | 5363.15 | .18646 |
| 5095 | .19627 | 5368.15 | .18628 |
| 5100 | .19608 | 5373.15 | .18611 |
| 5105 | .19589 | 5378.15 | .18594 |
| 5110 | .19569 | 5383.15 | .18576 |
| 5115 | .19550 | 5388.15 | .18559 |

| N | 1000/N | N + 273.15 | 1000/N + 273.15 |
|---|---|---|---|
| +5120 | .19531 | 5393.15 | .18542 |
| 5125 | .19512 | 5398.15 | .18525 |
| 5130 | .19493 | 5403.15 | .18508 |
| 5135 | .19474 | 5408.15 | .18491 |
| 5140 | .19455 | 5413.15 | .18474 |
| 5145 | .19436 | 5418.15 | .18456 |
| 5150 | .19417 | 5423.15 | .18439 |
| 5155 | .19399 | 5428.15 | .18422 |
| 5160 | .19380 | 5433.15 | .18406 |
| 5165 | .19361 | 5438.15 | .18389 |
| 5170 | .19342 | 5443.15 | .18372 |
| 5175 | .19324 | 5448.15 | .18355 |
| 5180 | .19305 | 5453.15 | .18338 |
| 5185 | .19286 | 5458.15 | .18321 |
| 5190 | .19268 | 5463.15 | .18304 |
| 5195 | .19249 | 5468.15 | .18288 |
| 5200 | .19231 | 5473.15 | .18271 |
| 5205 | .19212 | 5478.15 | .18254 |
| 5210 | .19194 | 5483.15 | .18238 |
| 5215 | .19175 | 5488.15 | .18221 |
| 5220 | .19157 | 5493.15 | .18204 |
| 5225 | .19139 | 5498.15 | .18188 |
| 5230 | .19120 | 5503.15 | .18171 |
| 5235 | .19102 | 5508.15 | .18155 |
| 5240 | .19084 | 5513.15 | .18138 |
| 5245 | .19066 | 5518.15 | .18122 |
| 5250 | .19048 | 5523.15 | .18106 |
| 5255 | .19029 | 5528.15 | .18089 |
| 5260 | .19011 | 5533.15 | .18073 |
| 5265 | .18993 | 5538.15 | .18057 |
| 5270 | .18975 | 5543.15 | .18040 |
| 5275 | .18957 | 5548.15 | .18024 |
| 5280 | .18939 | 5553.15 | .18008 |
| 5285 | .18921 | 5558.15 | .17992 |
| 5290 | .18904 | 5563.15 | .17975 |
| 5295 | .18886 | 5568.15 | .17959 |
| 5300 | .18868 | 5573.15 | .17943 |
| 5305 | .18850 | 5578.15 | .17927 |
| 5310 | .18832 | 5583.15 | .17911 |
| 5315 | .18815 | 5588.15 | .17895 |
| 5320 | .18797 | 5593.15 | .17879 |
| 5325 | .18779 | 5598.15 | .17863 |
| 5330 | .18762 | 5603.15 | .17847 |
| 5335 | .18744 | 5608.15 | .17831 |
| 5340 | .18727 | 5613.15 | .17815 |
| 5345 | .18709 | 5618.15 | .17799 |
| 5350 | .18692 | 5623.15 | .17784 |
| 5355 | .18674 | 5628.15 | .17768 |
| 5360 | .18657 | 5633.15 | .17752 |
| 5365 | .18639 | 5638.15 | .17736 |
| 5370 | .18622 | 5643.15 | .17721 |
| 5375 | .18605 | 5648.15 | .17705 |
| 5380 | .18587 | 5653.15 | .17689 |
| 5385 | .18570 | 5658.15 | .17674 |
| 5390 | .18553 | 5663.15 | .17658 |
| 5395 | .18536 | 5668.15 | .17642 |
| 5400 | .18519 | 5673.15 | .17627 |
| 5405 | .18501 | 5678.15 | .17611 |
| 5410 | .18484 | 5683.15 | .17596 |
| 5415 | .18467 | 5688.15 | .17580 |
| 5420 | .18450 | 5693.15 | .17565 |
| 5425 | .18433 | 5698.15 | .17550 |
| 5430 | .18416 | 5703.15 | .17534 |
| 5435 | .18399 | 5708.15 | .17519 |
| 5440 | .18382 | 5713.15 | .17503 |
| 5445 | .18365 | 5718.15 | .17488 |
| 5450 | .18349 | 5723.15 | .17473 |
| 5455 | .18332 | 5728.15 | .17458 |
| 5460 | .18315 | 5733.15 | .17442 |
| 5465 | .18298 | 5738.15 | .17427 |
| 5470 | .18282 | 5743.15 | .17412 |
| 5475 | .18265 | 5748.15 | .17397 |
| 5480 | .18248 | 5753.15 | .17382 |
| 5485 | .18232 | 5758.15 | .17367 |

| N | 1000/N | N + 273.15 | 1000/N + 273.15 |
|---|---|---|---|
| +5490 | .18215 | 5763.15 | .17352 |
| 5495 | .18198 | 5768.15 | .17337 |
| 5500 | .18182 | 5773.15 | .17322 |
| 5505 | .18165 | 5778.15 | .17307 |
| 5510 | .18149 | 5783.15 | .17292 |
| 5515 | .18132 | 5788.15 | .17277 |
| 5520 | .18116 | 5793.15 | .17262 |
| 5525 | .18100 | 5798.15 | .17247 |
| 5530 | .18083 | 5803.15 | .17232 |
| 5535 | .18067 | 5808.15 | .17217 |
| 5540 | .18051 | 5813.15 | .17202 |
| 5545 | .18034 | 5818.15 | .17188 |
| 5550 | .18018 | 5823.15 | .17173 |
| 5555 | .18002 | 5828.15 | .17158 |
| 5560 | .17986 | 5833.15 | .17143 |
| 5565 | .17969 | 5838.15 | .17129 |
| 5570 | .17953 | 5843.15 | .17114 |
| 5575 | .17937 | 5848.15 | .17099 |
| 5580 | .17921 | 5853.15 | .17085 |
| 5585 | .17905 | 5858.15 | .17070 |
| 5590 | .17889 | 5863.15 | .17056 |
| 5595 | .17873 | 5868.15 | .17041 |
| 5600 | .17857 | 5873.15 | .17027 |
| 5605 | .17841 | 5878.15 | .17012 |
| 5610 | .17825 | 5883.15 | .16998 |
| 5615 | .17809 | 5888.15 | .16983 |
| 5620 | .17794 | 5893.15 | .16969 |
| 5625 | .17778 | 5898.15 | .16954 |
| 5630 | .17762 | 5903.15 | .16940 |
| 5635 | .17746 | 5908.15 | .16926 |
| 5640 | .17730 | 5913.15 | .16911 |
| 5645 | .17715 | 5918.15 | .16897 |
| 5650 | .17699 | 5923.15 | .16883 |
| 5655 | .17683 | 5928.15 | .16869 |
| 5660 | .17668 | 5933.15 | .16854 |
| 5665 | .17652 | 5938.15 | .16840 |
| 5670 | .17637 | 5943.15 | .16826 |
| 5675 | .17621 | 5948.15 | .16812 |
| 5680 | .17606 | 5953.15 | .16798 |
| 5685 | .17590 | 5958.15 | .16784 |
| 5690 | .17575 | 5963.15 | .16770 |
| 5695 | .17559 | 5968.15 | .16756 |
| 5700 | .17544 | 5973.15 | .16742 |
| 5705 | .17528 | 5978.15 | .16728 |
| 5710 | .17513 | 5983.15 | .16714 |
| 5715 | .17498 | 5988.15 | .16700 |
| 5720 | .17483 | 5993.15 | .16686 |
| 5725 | .17467 | 5998.15 | .16672 |
| 5730 | .17452 | 6003.15 | .16658 |
| 5735 | .17437 | 6008.15 | .16644 |
| 5740 | .17422 | 6013.15 | .16630 |
| 5745 | .17406 | 6018.15 | .16616 |
| 5750 | .17391 | 6023.15 | .16603 |
| 5755 | .17376 | 6028.15 | .16589 |
| 5760 | .17361 | 6033.15 | .16575 |
| 5765 | .17346 | 6038.15 | .16561 |
| 5770 | .17331 | 6043.15 | .16548 |
| 5775 | .17316 | 6048.15 | .16534 |
| 5780 | .17301 | 6053.15 | .16520 |
| 5785 | .17286 | 6058.15 | .16507 |
| 5790 | .17271 | 6063.15 | .16493 |
| 5795 | .17256 | 6068.15 | .16479 |
| 5800 | .17241 | 6073.15 | .16466 |
| 5805 | .17227 | 6078.15 | .16452 |
| 5810 | .17212 | 6083.15 | .16439 |
| 5815 | .17197 | 6088.15 | .16425 |
| 5820 | .17182 | 6093.15 | .16412 |
| 5825 | .17167 | 6098.15 | .16398 |
| 5830 | .17153 | 6103.15 | .16385 |
| 5835 | .17138 | 6108.15 | .16372 |
| 5840 | .17123 | 6113.15 | .16358 |
| 5845 | .17109 | 6118.15 | .16345 |
| 5850 | .17094 | 6123.15 | .16331 |
| 5855 | .17079 | 6128.15 | .16318 |

| N | 1000/N | N + 273.15 | 1000/N + 273.15 |
|---|---|---|---|
| +5860 | .17065 | 6133.15 | .16305 |
| 5865 | .17050 | 6138.15 | .16292 |
| 5870 | .17036 | 6143.15 | .16278 |
| 5875 | .17021 | 6148.15 | .16265 |
| 5880 | .17007 | 6153.15 | .16252 |
| 5885 | .16992 | 6158.15 | .16239 |
| 5890 | .16978 | 6163.15 | .16225 |
| 5895 | .16964 | 6168.15 | .16212 |
| 5900 | .16949 | 6173.15 | .16199 |
| 5905 | .16935 | 6178.15 | .16186 |
| 5910 | .16920 | 6183.15 | .16173 |
| 5915 | .16906 | 6188.15 | .16160 |
| 5920 | .16892 | 6193.15 | .16147 |
| 5925 | .16878 | 6198.15 | .16134 |
| 5930 | .16863 | 6203.15 | .16121 |
| 5935 | .16849 | 6208.15 | .16108 |
| 5940 | .16835 | 6213.15 | .16095 |
| 5945 | .16821 | 6218.15 | .16082 |
| 5950 | .16807 | 6223.15 | .16069 |
| 5955 | .16793 | 6228.15 | .16056 |
| 5960 | .16779 | 6233.15 | .16043 |
| 5965 | .16764 | 6238.15 | .16030 |
| 5970 | .16750 | 6243.15 | .16018 |
| 5975 | .16736 | 6248.15 | .16005 |
| 5980 | .16722 | 6253.15 | .15992 |
| 5985 | .16708 | 6258.15 | .15979 |
| 5990 | .16694 | 6263.15 | .15966 |
| 5995 | .16681 | 6268.15 | .15954 |
| 6000 | .16667 | 6273.15 | .15941 |
| 6005 | .16653 | 6278.15 | .15928 |
| 6010 | .16639 | 6283.15 | .15916 |
| 6015 | .16625 | 6288.15 | .15903 |
| 6020 | .16611 | 6293.15 | .15890 |
| 6025 | .16598 | 6298.15 | .15878 |
| 6030 | .16584 | 6303.15 | .15865 |
| 6035 | .16570 | 6308.15 | .15853 |
| 6040 | .16556 | 6313.15 | .15840 |
| 6045 | .16543 | 6318.15 | .15827 |
| 6050 | .16529 | 6323.15 | .15815 |
| 6055 | .16515 | 6328.15 | .15802 |
| 6060 | .16502 | 6333.15 | .15790 |
| 6065 | .16488 | 6338.15 | .15777 |
| 6070 | .16474 | 6343.15 | .15765 |
| 6075 | .16461 | 6348.15 | .15753 |
| 6080 | .16447 | 6353.15 | .15740 |
| 6085 | .16434 | 6358.15 | .15728 |
| 6090 | .16420 | 6363.15 | .15715 |
| 6095 | .16407 | 6368.15 | .15703 |
| 6100 | .16393 | 6373.15 | .15691 |
| 6105 | .16380 | 6378.15 | .15679 |
| 6110 | .16367 | 6383.15 | .15666 |
| 6115 | .16353 | 6388.15 | .15654 |
| 6120 | .16340 | 6393.15 | .15642 |
| 6125 | .16327 | 6398.15 | .15630 |
| 6130 | .16313 | 6403.15 | .15617 |
| 6135 | .16300 | 6408.15 | .15605 |
| 6140 | .16287 | 6413.15 | .15593 |
| 6145 | .16273 | 6418.15 | .15581 |
| 6150 | .16260 | 6423.15 | .15569 |
| 6155 | .16247 | 6428.15 | .15557 |
| 6160 | .16234 | 6433.15 | .15544 |
| 6165 | .16221 | 6438.15 | .15532 |
| 6170 | .16207 | 6443.15 | .15520 |
| 6175 | .16194 | 6448.15 | .15508 |
| 6180 | .16181 | 6453.15 | .15496 |
| 6185 | .16168 | 6458.15 | .15484 |
| 6190 | .16155 | 6463.15 | .15472 |
| 6195 | .16142 | 6468.15 | .15460 |
| 6200 | .16129 | 6473.15 | .15448 |
| 6205 | .16116 | 6478.15 | .15437 |
| 6210 | .16103 | 6483.15 | .15425 |
| 6215 | .16090 | 6488.15 | .15413 |
| 6220 | .16077 | 6493.15 | .15401 |
| 6225 | .16064 | 6498.15 | .15389 |

| N | 1000/N | N + 273.15 | 1000/N + 273.15 |
|---|---|---|---|
| +6230 | .16051 | 6503.15 | .15377 |
| 6235 | .16038 | 6508.15 | .15365 |
| 6240 | .16026 | 6513.15 | .15354 |
| 6245 | .16013 | 6518.15 | .15342 |
| 6250 | .16000 | 6523.15 | .15330 |
| 6255 | .15987 | 6528.15 | .15318 |
| 6260 | .15974 | 6533.15 | .15307 |
| 6265 | .15962 | 6538.15 | .15295 |
| 6270 | .15949 | 6543.15 | .15283 |
| 6275 | .15936 | 6548.15 | .15271 |
| 6280 | .15924 | 6553.15 | .15260 |
| 6285 | .15911 | 6558.15 | .15248 |
| 6290 | .15898 | 6563.15 | .15237 |
| 6295 | .15886 | 6568.15 | .15225 |
| 6300 | .15873 | 6573.15 | .15213 |
| 6305 | .15860 | 6578.15 | .15202 |
| 6310 | .15848 | 6583.15 | .15190 |
| 6315 | .15835 | 6588.15 | .15179 |
| 6320 | .15823 | 6593.15 | .15167 |
| 6325 | .15810 | 6598.15 | .15156 |
| 6330 | .15798 | 6603.15 | .15144 |
| 6335 | .15785 | 6608.15 | .15133 |
| 6340 | .15773 | 6613.15 | .15121 |
| 6345 | .15760 | 6618.15 | .15110 |
| 6350 | .15748 | 6623.15 | .15099 |
| 6355 | .15736 | 6628.15 | .15087 |
| 6360 | .15723 | 6633.15 | .15076 |
| 6365 | .15711 | 6638.15 | .15064 |
| 6370 | .15699 | 6643.15 | .15053 |
| 6375 | .15686 | 6648.15 | .15042 |
| 6380 | .15674 | 6653.15 | .15030 |
| 6385 | .15662 | 6658.15 | .15019 |
| 6390 | .15649 | 6663.15 | .15008 |
| 6395 | .15637 | 6668.15 | .14997 |
| 6400 | .15625 | 6673.15 | .14985 |
| 6405 | .15613 | 6678.15 | .14974 |
| 6410 | .15601 | 6683.15 | .14963 |
| 6415 | .15588 | 6688.15 | .14952 |
| 6420 | .15576 | 6693.15 | .14941 |
| 6425 | .15564 | 6698.15 | .14929 |
| 6430 | .15552 | 6703.15 | .14918 |
| 6435 | .15540 | 6708.15 | .14907 |
| 6440 | .15528 | 6713.15 | .14896 |
| 6445 | .15516 | 6718.15 | .14885 |
| 6450 | .15504 | 6723.15 | .14874 |
| 6455 | .15492 | 6728.15 | .14863 |
| 6460 | .15480 | 6733.15 | .14852 |
| 6465 | .15468 | 6738.15 | .14841 |
| 6470 | .15456 | 6743.15 | .14830 |
| 6475 | .15444 | 6748.15 | .14819 |
| 6480 | .15432 | 6753.15 | .14808 |
| 6485 | .15420 | 6758.15 | .14797 |
| 6490 | .15408 | 6763.15 | .14786 |
| 6495 | .15396 | 6768.15 | .14775 |
| 6500 | .15385 | 6773.15 | .14764 |
| 6505 | .15373 | 6778.15 | .14753 |
| 6510 | .15361 | 6783.15 | .14742 |
| 6515 | .15349 | 6788.15 | .14732 |
| 6520 | .15337 | 6793.15 | .14721 |
| 6525 | .15326 | 6798.15 | .14710 |
| 6530 | .15314 | 6803.15 | .14699 |
| 6535 | .15302 | 6808.15 | .14688 |
| 6540 | .15291 | 6813.15 | .14677 |
| 6545 | .15279 | 6818.15 | .14667 |
| 6550 | .15267 | 6823.15 | .14656 |
| 6555 | .15256 | 6828.15 | .14645 |
| 6560 | .15244 | 6833.15 | .14635 |
| 6565 | .15232 | 6838.15 | .14624 |
| 6570 | .15221 | 6843.15 | .14613 |
| 6575 | .15209 | 6848.15 | .14602 |
| 6580 | .15198 | 6853.15 | .14592 |
| 6585 | .15186 | 6858.15 | .14581 |
| 6590 | .15175 | 6863.15 | .14571 |
| 6595 | .15163 | 6868.15 | .14560 |

| N | 1000/N | N + 273.15 | 1000/N + 273.15 |
|---|---|---|---|
| +6600 | .15152 | 6873.15 | .14549 |
| 6605 | .15140 | 6878.15 | .14539 |
| 6610 | .15129 | 6883.15 | .14528 |
| 6615 | .15117 | 6888.15 | .14518 |
| 6620 | .15106 | 6893.15 | .14507 |
| 6625 | .15094 | 6898.15 | .14497 |
| 6630 | .15083 | 6903.15 | .14486 |
| 6635 | .15072 | 6908.15 | .14476 |
| 6640 | .15060 | 6913.15 | .14465 |
| 6645 | .15049 | 6918.15 | .14455 |
| 6650 | .15038 | 6923.15 | .14444 |
| 6655 | .15026 | 6928.15 | .14434 |
| 6660 | .15015 | 6933.15 | .14423 |
| 6665 | .15004 | 6938.15 | .14413 |
| 6670 | .14993 | 6943.15 | .14403 |
| 6675 | .14981 | 6948.15 | .14392 |
| 6680 | .14970 | 6953.15 | .14382 |
| 6685 | .14959 | 6958.15 | .14372 |
| 6690 | .14948 | 6963.15 | .14361 |
| 6695 | .14937 | 6968.15 | .14351 |
| 6700 | .14925 | 6973.15 | .14341 |
| 6705 | .14914 | 6978.15 | .14330 |
| 6710 | .14903 | 6983.15 | 14320 |
| 6715 | .14892 | 6988.15 | .14310 |
| 6720 | .14881 | 6993.15 | .14300 |
| 6725 | .14870 | 6998.15 | .14289 |
| 6730 | .14859 | 7003.15 | .14279 |
| 6735 | .14848 | 7008.15 | .14269 |
| 6740 | .14837 | 7013.15 | .14259 |
| 6745 | .14826 | 7018.15 | .14249 |
| 6750 | .14815 | 7023.15 | .14239 |
| 6755 | .14804 | 7028.15 | .14228 |
| 6760 | .14793 | 7033.15 | .14218 |
| 6765 | .14782 | 7038.15 | .14208 |
| 6770 | .14771 | 7043.15 | .14198 |
| 6775 | .14760 | 7048.15 | .14188 |
| 6780 | .14749 | 7053.15 | .14178 |
| 6785 | .14738 | 7058.15 | .14168 |
| 6790 | .14728 | 7063.15 | .14158 |
| 6795 | .14717 | 7068.15 | .14148 |
| 6800 | .14706 | 7073.15 | .14138 |

| N | 1000/N | N + 273.15 | 1000/N + 273.15 |
|---|---|---|---|
| 6805 | .14695 | 7078.15 | .14128 |
| 6810 | .14684 | 7083.15 | .14118 |
| 6815 | .14674 | 7088.15 | .14108 |
| 6820 | .14663 | 7093.15 | .14098 |
| 6825 | .14652 | 7098.15 | .14088 |
| 6830 | .14641 | 7103.15 | .14078 |
| 6835 | .14631 | 7108.15 | .14068 |
| 6840 | .14620 | 7113.15 | .14058 |
| 6845 | .14609 | 7118.15 | .14049 |
| 6850 | .14599 | 7123.15 | .14039 |
| 6855 | .14588 | 7128.15 | .14029 |
| 6860 | .14577 | 7133.15 | .14019 |
| 6865 | .14567 | 7138.15 | .14009 |
| 6870 | .14556 | 7143.15 | .13999 |
| 6875 | .14545 | 7148.15 | .13990 |
| 6880 | .14535 | 7153.15 | .13980 |
| 6885 | .14524 | 7158.15 | .13970 |
| 6890 | .14514 | 7163.15 | .13960 |
| 6895 | .14503 | 7168.15 | .13951 |
| 6900 | .14493 | 7173.15 | .13941 |
| 6905 | .14482 | 7178.15 | .13931 |
| 6910 | .14472 | 7183.15 | .13921 |
| 6915 | .14461 | 7188.15 | .13912 |
| 6920 | .14451 | 7193.15 | .13902 |
| 6925 | .14440 | 7198.15 | .13892 |
| 6930 | .14430 | 7203.15 | .13883 |
| 6935 | .14420 | 7208.15 | .13873 |
| 6940 | .14409 | 7213.15 | .13864 |
| 6945 | .14399 | 7218.15 | .13854 |
| 6950 | .14388 | 7223.15 | .13844 |
| 6955 | .14378 | 7228.15 | .13835 |
| 6960 | .14368 | 7233.15 | .13825 |
| 6965 | .14358 | 7238.15 | .13816 |
| 6970 | .14347 | 7243.15 | .13806 |
| 6975 | .14337 | 7248.15 | .13797 |
| 6980 | .14327 | 7253.15 | .13787 |
| 6985 | .14316 | 7258.15 | .13778 |
| 6990 | .14306 | 7263.15 | .13768 |
| 6995 | .14296 | 7268.15 | .13759 |
| 7000 | .14286 | 7273.15 | .13749 |

# SQUARES, SQUARE ROOTS, CUBES, AND CUBE ROOTS

$\sqrt{100n} = 10\sqrt{n}$; $\sqrt{1000n} = 10\sqrt{10n}$; $\sqrt{\frac{1}{10}n} = \frac{1}{10}\sqrt{10n}$; $\sqrt{\frac{1}{100}n} = \frac{1}{10}\sqrt{n}$, $\sqrt{\frac{1}{1000}n} = \frac{1}{100}\sqrt{10n}$; $\sqrt[3]{1000n} = 10\sqrt[3]{n}$; $\sqrt[3]{10{,}000n} = 10\sqrt[3]{10n}$; $\sqrt[3]{100{,}000n} = 10\sqrt[3]{100n}$; $\sqrt[3]{\frac{1}{10}n} = \frac{1}{10}\sqrt[3]{100n}$; $\sqrt[3]{\frac{1}{100}n} = \frac{1}{10}\sqrt[3]{10n}$; $\sqrt[3]{\frac{1}{1000}n} = \frac{1}{10}\sqrt[3]{n}$.

| $n$ | $n^2$ | $\sqrt{n}$ | $\sqrt{10n}$ | $n^3$ | $\sqrt[3]{n}$ | $\sqrt[3]{10n}$ | $\sqrt[3]{100n}$ |
|---|---|---|---|---|---|---|---|
| 1 | 1 | 1.000 000 | 3.162 278 | 1 | 1.000 000 | 2.154 435 | 4.641 589 |
| 2 | 4 | 1.414 214 | 4.472 136 | 8 | 1.259 921 | 2.714 418 | 5.848 035 |
| 3 | 9 | 1.732 051 | 5.477 226 | 27 | 1.442 250 | 3.107 233 | 6.694 330 |
| 4 | 16 | 2.000 000 | 6.324 555 | 64 | 1.587 401 | 3.419 952 | 7.368 063 |
| 5 | 25 | 2.236 068 | 7.071 068 | 125 | 1.709 976 | 3.684 031 | 7.937 005 |
| 6 | 36 | 2.449 490 | 7.745 967 | 216 | 1.817 121 | 3.914 868 | 8.434 327 |
| 7 | 49 | 2.645 751 | 8.366 600 | 343 | 1.912 931 | 4.121 285 | 8.879 040 |
| 8 | 64 | 2.828 427 | 8.944 272 | 512 | 2.000 000 | 4.308 869 | 9.283 178 |
| 9 | 81 | 3.000 000 | 9.486 833 | 729 | 2.080 084 | 4.481 405 | 9.654 894 |
| **10** | 100 | 3.162 278 | 10.00000 | 1 000 | 2.154 435 | 4.641 589 | 10.00000 |
| 11 | 121 | 3.316 625 | 10.48809 | 1 331 | 2.223 980 | 4.791 420 | 10.32280 |
| 12 | 144 | 3.464 102 | 10.95445 | 1 728 | 2.289 428 | 4.932 424 | 10.62659 |
| 13 | 169 | 3.605 551 | 11.40175 | 2 197 | 2.351 335 | 5.065 797 | 10.91393 |
| 14 | 196 | 3.741 657 | 11.83216 | 2 744 | 2.410 142 | 5.192 494 | 11.18689 |
| 15 | 225 | 3.872 983 | 12.24745 | 3 375 | 2.466 212 | 5.313 293 | 11.44714 |
| 16 | 256 | 4.000 000 | 12.64911 | 4 096 | 2.519 842 | 5.428 835 | 11.69607 |
| 17 | 289 | 4.123 106 | 13.03840 | 4 913 | 2.571 282 | 5.539 658 | 11.93483 |
| 18 | 324 | 4.242 641 | 13.41641 | 5 832 | 2.620 741 | 5.646 216 | 12.16440 |
| 19 | 361 | 4.358 899 | 13.78405 | 6 859 | 2.668 402 | 5.748 897 | 12.38562 |
| **20** | 400 | 4.472 136 | 14.14214 | 8 000 | 2.714 418 | 5.848 035 | 12.59921 |
| 21 | 441 | 4.582 576 | 14.49138 | 9 261 | 2.758 924 | 5.943 922 | 12.80579 |
| 22 | 484 | 4.690 416 | 14.83240 | 10 648 | 2.802 039 | 6.036 811 | 13.00591 |
| 23 | 529 | 4.795 832 | 15.16575 | 12 167 | 2.843 867 | 6.126 926 | 13.20006 |
| 24 | 576 | 4.898 979 | 15.49193 | 13 824 | 2.884 499 | 6.214 465 | 13.38866 |
| 25 | 625 | 5.000 000 | 15.81139 | 15 625 | 2.924 018 | 6.299 605 | 13.57209 |
| 26 | 676 | 5.099 020 | 16.12452 | 17 576 | 2.962 496 | 6.382 504 | 13.75069 |
| 27 | 729 | 5.196 152 | 16.43168 | 19 683 | 3.000 000 | 6.463 304 | 13.92477 |
| 28 | 784 | 5.291 503 | 16.73320 | 21 952 | 3.036 589 | 6.542 133 | 14.09460 |
| 29 | 841 | 5.385 165 | 17.02939 | 24 389 | 3.072 317 | 6.619 106 | 14.26043 |
| **30** | 900 | 5.477 226 | 17.32051 | 27 000 | 3.107 233 | 6.694 330 | 14.42250 |
| 31 | 961 | 5.567 764 | 17.60682 | 29 791 | 3.141 381 | 6.767 899 | 14.58100 |
| 32 | 1 024 | 5.656 854 | 17.88854 | 32 768 | 3.174 802 | 6.839 904 | 14.73613 |
| 33 | 1 089 | 5.744 563 | 18.16590 | 35 937 | 3.207 534 | 6.910 423 | 14.88806 |
| 34 | 1 156 | 5.830 952 | 18.43909 | 39 304 | 3.239 612 | 6.979 532 | 15.03695 |
| 35 | 1 225 | 5.916 080 | 18.70829 | 42 875 | 3.271 066 | 7.047 299 | 15.18294 |
| 36 | 1 296 | 6.000 000 | 18.97367 | 46 656 | 3.301 927 | 7.113 787 | 15.32619 |
| 37 | 1 369 | 6.082 763 | 19.23538 | 50 653 | 3.332 222 | 7.179 054 | 15.46680 |
| 38 | 1 444 | 6.164 414 | 19.49359 | 54 872 | 3.361 975 | 7.243 156 | 15.60491 |
| 39 | 1 521 | 6.244 998 | 19.74842 | 59 319 | 3.391 211 | 7.306 144 | 15.74061 |
| **40** | 1 600 | 6.324 555 | 20.00000 | 64 000 | 3.419 952 | 7.368 063 | 15.87401 |
| 41 | 1 681 | 6.403 124 | 20.24846 | 68 921 | 3.448 217 | 7.428 959 | 16.00521 |
| 42 | 1 764 | 6.480 741 | 20.49390 | 74 088 | 3.476 027 | 7.488 872 | 16.13429 |
| 43 | 1 849 | 6.557 439 | 20.73644 | 79 507 | 3.503 398 | 7.547 842 | 16.26133 |
| 44 | 1 936 | 6.633 250 | 20.97618 | 85 184 | 3.530 348 | 7.605 905 | 16.38643 |
| 45 | 2 025 | 6.708 204 | 21.21320 | 91 125 | 3.556 893 | 7.663 094 | 16.50964 |
| 46 | 2 116 | 6.782 330 | 21.44761 | 97 336 | 3.583 048 | 7.719 443 | 16.63103 |
| 47 | 2 209 | 6.855 655 | 21.67948 | 103 823 | 3.608 826 | 7.774 980 | 16.75069 |
| 48 | 2 304 | 6.928 203 | 21.90890 | 110 592 | 3.634 241 | 7.829 735 | 16.86865 |
| 49 | 2 401 | 7.000 000 | 22.13594 | 117 649 | 3.659 306 | 7.883 735 | 16.98499 |
| **50** | 2 500 | 7.071 068 | 22.36068 | 125 000 | 3.684 031 | 7.937 005 | 17.09976 |

| $n$ | $n^2$ | $\sqrt{n}$ | $\sqrt{10n}$ | $n^3$ | $\sqrt[3]{n}$ | $\sqrt[3]{10n}$ | $\sqrt[3]{100n}$ |
|---|---|---|---|---|---|---|---|
| **50** | 2 500 | 7.071 068 | 22.36068 | 125 000 | 3.684 031 | 7.937 005 | 17.09976 |
| 51 | 2 601 | 7.141 428 | 22.58318 | 132 651 | 3.708 430 | 7.989 570 | 17.21301 |
| 52 | 2 704 | 7.211 103 | 22.80351 | 140 608 | 3.732 511 | 8.041 452 | 17.32478 |
| 53 | 2 809 | 7.280 110 | 23.02173 | 148 877 | 3.756 286 | 8.092 672 | 17.43513 |
| 54 | 2 916 | 7.348 469 | 23.23790 | 157 464 | 3.779 763 | 8.143 253 | 17.54411 |
| 55 | 3 025 | 7.416 198 | 23.45208 | 166 375 | 3.802 952 | 8.193 213 | 17.65174 |
| 56 | 3 136 | 7.483 315 | 23.66432 | 175 616 | 3.825 862 | 8.242 571 | 17.75808 |
| 57 | 3 249 | 7.549 834 | 23.87467 | 185 193 | 3.848 501 | 8.291 344 | 17.86316 |
| 58 | 3 364 | 7.615 773 | 24.08319 | 195 112 | 3.870 877 | 8.339 551 | 17.96702 |
| 59 | 3 481 | 7.681 146 | 24.28992 | 205 379 | 3.892 996 | 8.387 207 | 18.06969 |
| **60** | 3 600 | 7.745 967 | 24.49490 | 216 000 | 3.914 868 | 8.434 327 | 18.17121 |
| 61 | 3 721 | 7.810 250 | 24.69818 | 226 981 | 3.936 497 | 8.480 926 | 18.27160 |
| 62 | 3 844 | 7.874 008 | 24.89980 | 238 328 | 3.957 892 | 8.527 019 | 18.37091 |
| 63 | 3 969 | 7.937 254 | 25.09980 | 250 047 | 3.979 057 | 8.572 619 | 18.46915 |
| 64 | 4 096 | 8.000 000 | 25.29822 | 262 144 | 4.000 000 | 8.617 739 | 18.56636 |
| 65 | 4 225 | 8.062 258 | 25.49510 | 274 625 | 4.020 726 | 8.662 391 | 18.66256 |
| 66 | 4 356 | 8.124 038 | 25.69047 | 287 496 | 4.041 240 | 8.706 588 | 18.75777 |
| 67 | 4 489 | 8.185 353 | 25.88436 | 300 763 | 4.061 548 | 8.750 340 | 18.85204 |
| 68 | 4 624 | 8.246 211 | 26.07681 | 314 432 | 4.081 655 | 8.793 659 | 18.94536 |
| 69 | 4 761 | 8.306 624 | 26.26785 | 328 509 | 4.101 566 | 8.836 556 | 19.03778 |
| **70** | 4 900 | 8.366 600 | 26.45751 | 343 000 | 4.121 285 | 8.879 040 | 19.12931 |
| 71 | 5 041 | 8.426 150 | 26.64583 | 357 911 | 4.140 818 | 8.921 121 | 19.21997 |
| 72 | 5 184 | 8.485 281 | 26.83282 | 373 248 | 4.160 168 | 8.962 809 | 19.30979 |
| 73 | 5 329 | 8.544 004 | 27.01851 | 389 017 | 4.179 339 | 9.004 113 | 19.39877 |
| 74 | 5 476 | 8.602 325 | 27.20294 | 405 224 | 4.198 336 | 9.045 042 | 19.48695 |
| 75 | 5 625 | 8.660 254 | 27.38613 | 421 875 | 4.217 163 | 9.085 603 | 19.57434 |
| 76 | 5 776 | 8.717 798 | 27.56810 | 438 976 | 4.235 824 | 9.125 805 | 19.66095 |
| 77 | 5 929 | 8.774 964 | 27.74887 | 456 533 | 4.254 321 | 9.165 656 | 19.74681 |
| 78 | 6 084 | 8.831 761 | 27.92848 | 474 552 | 4.272 659 | 9.205 164 | 19.83192 |
| 79 | 6 241 | 8.888 194 | 28.10694 | 493 039 | 4.290 840 | 9.244 335 | 19.91632 |
| **80** | 6 400 | 8.944 272 | 28.28427 | 512 000 | 4.308 869 | 9.283 178 | 20.00000 |
| 81 | 6 561 | 9.000 000 | 28.46050 | 531 441 | 4.326 749 | 9.321 698 | 20.08299 |
| 82 | 6 724 | 9.055 385 | 28.63564 | 551 368 | 4.344 481 | 9.359 902 | 20.16530 |
| 83 | 6 889 | 9.110 434 | 28.80972 | 571 787 | 4.362 071 | 9.397 796 | 20.24694 |
| 84 | 7 056 | 9.165 151 | 28.98275 | 592 704 | 4.379 519 | 9.435 388 | 20.32793 |
| 85 | 7 225 | 9.219 544 | 29.15476 | 614 125 | 4.396 830 | 9.472 682 | 20.40828 |
| 86 | 7 396 | 9.273 618 | 29.32576 | 636 056 | 4.414 005 | 9.509 685 | 20.48800 |
| 87 | 7 569 | 9.327 379 | 29.49576 | 658 503 | 4.431 048 | 9.546 403 | 20.56710 |
| 88 | 7 744 | 9.380 832 | 29.66479 | 681 472 | 4.447 960 | 9.582 840 | 20.64560 |
| 89 | 7 921 | 9.433 981 | 29.83287 | 704 969 | 4.464 745 | 9.619 002 | 20.72351 |
| **90** | 8 100 | 9.486 833 | 30.00000 | 729 000 | 4.481 405 | 9.654 894 | 20.80084 |
| 91 | 8 281 | 9.539 392 | 30.16621 | 753 571 | 4.497 941 | 9.690 521 | 20.87759 |
| 92 | 8 464 | 9.591 663 | 30.33150 | 778 688 | 4.514 357 | 9.725 888 | 20.95379 |
| 93 | 8 649 | 9.643 651 | 30.49590 | 804 357 | 4.530 655 | 9.761 000 | 21.02944 |
| 94 | 8 836 | 9.695 360 | 30.65942 | 830 584 | 4.546 836 | 9.795 861 | 21.10454 |
| 95 | 9 025 | 9.746 794 | 30.82207 | 857 375 | 4.562 903 | 9.830 476 | 21.17912 |
| 96 | 9 216 | 9.797 959 | 30.98387 | 884 736 | 4.578 857 | 9.864 848 | 21.25317 |
| 97 | 9 409 | 9.848 858 | 31.14482 | 912 673 | 4.594 701 | 9.898 983 | 21.32671 |
| 98 | 9 604 | 9.899 495 | 31.30495 | 941 192 | 4.610 436 | 9.932 884 | 21.39975 |
| 99 | 9 801 | 9.949 874 | 31.46427 | 970 299 | 4.626 065 | 9.966 555 | 21.47229 |
| **100** | 10 000 | 10.00000 | 31.62278 | 1 000 000 | 4.641 589 | 10.00000 | 21.54435 |

| $n$ | $n^2$ | $\sqrt{n}$ | $\sqrt{10n}$ | $n^3$ | $\sqrt[3]{n}$ | $\sqrt[3]{10n}$ | $\sqrt[3]{100n}$ |
|---|---|---|---|---|---|---|---|
| **100** | 10 000 | 10.00000 | 31.62278 | 1 000 000 | 4.641 589 | 10.00000 | 21.54435 |
| 101 | 10 201 | 10.04988 | 31.78050 | 1 030 301 | 4.657 010 | 10.03322 | 21.61592 |
| 102 | 10 404 | 10.09950 | 31.93744 | 1 061 208 | 4.672 329 | 10.06623 | 21.68703 |
| 103 | 10 609 | 10.14889 | 32.09361 | 1 092 727 | 4.687 548 | 10.09902 | 21.75767 |
| 104 | 10 816 | 10.19804 | 32.24903 | 1 124 864 | 4.702 669 | 10.13159 | 21.82786 |
| 105 | 11 025 | 10.24695 | 32.40370 | 1 157 625 | 4.717 694 | 10.16396 | 21.89760 |
| 106 | 11 236 | 10.29563 | 32.55764 | 1 191 016 | 4.732 623 | 10.19613 | 21.96689 |
| 107 | 11 449 | 10.34408 | 32.71085 | 1 225 043 | 4.747 459 | 10.22809 | 22.03575 |
| 108 | 11 664 | 10.39230 | 32.86335 | 1 259 712 | 4.762 203 | 10.25986 | 22.10419 |
| 109 | 11 881 | 10.44031 | 33.01515 | 1 295 029 | 4.776 856 | 10.29142 | 22.17220 |
| **110** | 12 100 | 10.48809 | 33.16625 | 1 331 000 | 4.791 420 | 10.32280 | 22.23980 |
| 111 | 12 321 | 10.53565 | 33.31666 | 1 367 631 | 4.805 896 | 10.35399 | 22.30699 |
| 112 | 12 544 | 10.58301 | 33.46640 | 1 404 928 | 4.820 285 | 10.38499 | 22.37378 |
| 113 | 12 769 | 10.63015 | 33.61547 | 1 442 897 | 4.834 588 | 10.41580 | 22.44017 |
| 114 | 12 996 | 10.67708 | 33.76389 | 1 481 544 | 4.848 808 | 10.44644 | 22.50617 |
| 115 | 13 225 | 10.72381 | 33.91165 | 1 520 875 | 4.862 944 | 10.47690 | 22.57179 |
| 116 | 13 456 | 10.77033 | 34.05877 | 1 560 896 | 4.876 999 | 10.50718 | 22.63702 |
| 117 | 13 689 | 10.81665 | 34.20526 | 1 601 613 | 4.890 973 | 10.53728 | 22.70189 |
| 118 | 13 924 | 10.86278 | 34.35113 | 1 643 032 | 4.904 868 | 10.56722 | 22.76638 |
| 119 | 14 161 | 10.90871 | 34.49638 | 1 685 159 | 4.918 685 | 10.59699 | 22.83051 |
| **120** | 14 400 | 10.95445 | 34.64102 | 1 728 000 | 4.932 424 | 10.62659 | 22.89428 |
| 121 | 14 641 | 11.00000 | 34.78505 | 1 771 561 | 4.946 087 | 10.65602 | 22.95770 |
| 122 | 14 884 | 11.04536 | 34.92850 | 1 815 848 | 4.959 676 | 10.68530 | 23.02078 |
| 123 | 15 129 | 11.09054 | 35.07136 | 1 860 867 | 4.973 190 | 10.71441 | 23.08350 |
| 124 | 15 376 | 11.13553 | 35.21363 | 1 906 624 | 4.986 631 | 10.74337 | 23.14589 |
| 125 | 15 625 | 11.18034 | 35.35534 | 1 953 125 | 5.000 000 | 10.77217 | 23.20794 |
| 126 | 15 876 | 11.22497 | 35.49648 | 2 000 376 | 5.013 298 | 10.80082 | 23.26967 |
| 127 | 16 129 | 11.26943 | 35.63706 | 2 048 383 | 5.026 526 | 10.82932 | 23.33107 |
| 128 | 16 384 | 11.31371 | 35.77709 | 2 097 152 | 5.039 684 | 10.85767 | 23.39214 |
| 129 | 16 641 | 11.35782 | 35.91657 | 2 146 689 | 5.052 774 | 10.88587 | 23.45290 |
| **130** | 16 900 | 11.40175 | 36.05551 | 2 197 000 | 5.065 797 | 10.91393 | 23.51335 |
| 131 | 17 161 | 11.44552 | 36.19392 | 2 248 091 | 5.078 753 | 10.94184 | 23.57348 |
| 132 | 17 424 | 11.48913 | 36.33180 | 2 299 968 | 5.091 643 | 10.96961 | 23.63332 |
| 133 | 17 689 | 11.53256 | 36.46917 | 2 352 637 | 5.104 469 | 10.99724 | 23.69285 |
| 134 | 17 956 | 11.57584 | 36.60601 | 2 406 104 | 5.117 230 | 11.02474 | 23.75208 |
| 135 | 18 225 | 11.61895 | 36.74235 | 2 460 375 | 5.129 928 | 11.05209 | 23.81102 |
| 136 | 18 496 | 11.66190 | 36.87818 | 2 515 456 | 5.142 563 | 11.07932 | 23.86966 |
| 137 | 18 769 | 11.70470 | 37.01351 | 2 571 353 | 5.155 137 | 11.10641 | 23.92803 |
| 138 | 19 044 | 11.74734 | 37.14835 | 2 628 072 | 5.167 649 | 11.13336 | 23.98610 |
| 139 | 19 321 | 11.78983 | 37.28270 | 2 685 619 | 5.180 101 | 11.16019 | 24.04390 |
| **140** | 19 600 | 11.83216 | 37.41657 | 2 744 000 | 5.192 494 | 11.18689 | 24.10142 |
| 141 | 19 881 | 11.87434 | 37.54997 | 2 803 221 | 5.204 828 | 11.21346 | 24.15867 |
| 142 | 20 164 | 11.91638 | 37.68289 | 2 863 288 | 5.217 103 | 11.23991 | 24.21565 |
| 143 | 20 449 | 11.95826 | 37.81534 | 2 924 207 | 5.229 322 | 11.26623 | 24.27236 |
| 144 | 20 736 | 12.00000 | 37.94733 | 2 985 984 | 5.241 483 | 11.29243 | 24.32881 |
| 145 | 21 025 | 12.04159 | 38.07887 | 3 048 625 | 5.253 588 | 11.31851 | 24.38499 |
| 146 | 21 316 | 12.08305 | 38.20995 | 3 112 136 | 5.265 637 | 11.34447 | 24.44092 |
| 147 | 21 609 | 12.12436 | 38.34058 | 3 176 523 | 5.277 632 | 11.37031 | 24.49660 |
| 148 | 21 904 | 12.16553 | 38.47077 | 3 241 792 | 5.289 572 | 11.39604 | 24.55202 |
| 149 | 22 201 | 12.20656 | 38.60052 | 3 307 949 | 5.301 459 | 11.42165 | 24.60719 |
| **150** | 22.500 | 12.24745 | 38.72983 | 3 375 000 | 5.313 293 | 11.44714 | 24.66212 |

| $n$ | $n^2$ | $\sqrt{n}$ | $\sqrt{10n}$ | $n^3$ | $\sqrt[3]{n}$ | $\sqrt[3]{10n}$ | $\sqrt[3]{100n}$ |
|---|---|---|---|---|---|---|---|
| **150** | 22 500 | 12.24745 | 38.72983 | 3 375 000 | 5.313 293 | 11.44714 | 24.66212 |
| 151 | 22 801 | 12.28821 | 38.85872 | 3 442 951 | 5.325 074 | 11.47252 | 24.71680 |
| 152 | 23 104 | 12.32883 | 38.98718 | 3 511 808 | 5.336 803 | 11.49779 | 24.77125 |
| 153 | 23 409 | 12.36932 | 39.11521 | 3 581 577 | 5.348 481 | 11.52295 | 24.82545 |
| 154 | 23 716 | 12.40967 | 39.24283 | 3 652 264 | 5.360 108 | 11.54800 | 24.87942 |
| 155 | 24 025 | 12.44990 | 39.37004 | 3 723 875 | 5.371 685 | 11.57295 | 24.93315 |
| 156 | 24 336 | 12.49000 | 39.49684 | 3 796 416 | 5.383 213 | 11.59778 | 24.98666 |
| 157 | 24 649 | 12.52996 | 39.62323 | 3 869 893 | 5.394 691 | 11.62251 | 25.03994 |
| 158 | 24 964 | 12.56981 | 39.74921 | 3 944 312 | 5.406 120 | 11.64713 | 25.09299 |
| 159 | 25 281 | 12.60952 | 39.87480 | 4 019 679 | 5.417 502 | 11.67165 | 25.14581 |
| **160** | 25 600 | 12.64911 | 40.00000 | 4 096 000 | 5.428 835 | 11.69607 | 25.19842 |
| 161 | 25 921 | 12.68858 | 40.12481 | 4 173 281 | 5.440 122 | 11.72039 | 25.25081 |
| 162 | 26 244 | 12.72792 | 40.24922 | 4 251 528 | 5.451 362 | 11.74460 | 25.30298 |
| 163 | 26 569 | 12.76715 | 40.37326 | 4 330 747 | 5.462 556 | 11.76872 | 25.35494 |
| 164 | 26 896 | 12.80625 | 40.49691 | 4 410 944 | 5.473 704 | 11.79274 | 25.40668 |
| 165 | 27 225 | 12.84523 | 40.62019 | 4 492 125 | 5.484 807 | 11.81666 | 25.45822 |
| 166 | 27 556 | 12.88410 | 40.74310 | 4 574 296 | 5.495 865 | 11.84048 | 25.50954 |
| 167 | 27 889 | 12.92285 | 40.86563 | 4 657 463 | 5.506 878 | 11.86421 | 25.56067 |
| 168 | 28 224 | 12.96148 | 40.98780 | 4 741 632 | 5.517 848 | 11.88784 | 25.61158 |
| 169 | 28 561 | 13.00000 | 41.10961 | 4 826 809 | 5.528 775 | 11.91138 | 25.66230 |
| **170** | 28 900 | 13.03840 | 41.23106 | 4 913 000 | 5.539 658 | 11.93483 | 25.71282 |
| 171 | 29 241 | 13.07670 | 41.35215 | 5 000 211 | 5.550 499 | 11.95819 | 25.76313 |
| 172 | 29 584 | 13.11488 | 41.47288 | 5 088 448 | 5.561 298 | 11.98145 | 25.81326 |
| 173 | 29 929 | 13.15295 | 41.59327 | 5 177 717 | 5.572 055 | 12.00463 | 25.86319 |
| 174 | 30 276 | 13.19091 | 41.71331 | 5 268 024 | 5.582 770 | 12.02771 | 25.91292 |
| 175 | 30 625 | 13.22876 | 41.83300 | 5 359 375 | 5.593 445 | 12.05071 | 25.96247 |
| 176 | 30 976 | 13.26650 | 41.95235 | 5 451 776 | 5.604 079 | 12.07362 | 26.01183 |
| 177 | 31 329 | 13.30413 | 42.07137 | 5 545 233 | 5.614 672 | 12.09645 | 26.06100 |
| 178 | 31 684 | 13.34166 | 42.19005 | 5 639 752 | 5.625 226 | 12.11918 | 26.10999 |
| 179 | 32 041 | 13.37909 | 42.30839 | 5 735 339 | 5.635 741 | 12.14184 | 26.15879 |
| **180** | 32 400 | 13.41641 | 42.42641 | 5 832 000 | 5.646 216 | 12.16440 | 26.20741 |
| 181 | 32 761 | 13.45362 | 42.54409 | 5 929 741 | 5.656 653 | 12.18689 | 26.25586 |
| 182 | 33 124 | 13.49074 | 42.66146 | 6 028 568 | 5.667 051 | 12.20929 | 26.30412 |
| 183 | 33 489 | 13.52775 | 42.77850 | 6 128 487 | 5.677 411 | 12.23161 | 26.35221 |
| 184 | 33 856 | 13.56466 | 42.89522 | 6 229 504 | 5.687 734 | 12.25385 | 26.40012 |
| 185 | 34 225 | 13.60147 | 43.01163 | 6 331 625 | 5.698 019 | 12.27601 | 26.44786 |
| 186 | 34 596 | 13.63818 | 43.12772 | 6 434 856 | 5.708 267 | 12.29809 | 26.49543 |
| 187 | 34 969 | 13.67479 | 43.24350 | 6 539 203 | 5.718 479 | 12.32009 | 26.54283 |
| 188 | 35 344 | 13.71131 | 43.35897 | 6 644 672 | 5.728 654 | 12.34201 | 26.59006 |
| 189 | 35 721 | 13.74773 | 43.47413 | 6 751 269 | 5.738 794 | 12.36386 | 26.63712 |
| **190** | 36 100 | 13.78405 | 43.58899 | 6 859 000 | 5.748 897 | 12.38562 | 26.68402 |
| 191 | 36 481 | 13.82027 | 43.70355 | 6 967 871 | 5.758 965 | 12.40731 | 26.73075 |
| 192 | 36 864 | 13.85641 | 43.81780 | 7 077 888 | 5.768 998 | 12.42893 | 26.77732 |
| 193 | 37 249 | 13.89244 | 43.93177 | 7 189 057 | 5.778 997 | 12.45047 | 26.82373 |
| 194 | 37 636 | 13.92839 | 44.04543 | 7 301 384 | 5.788 960 | 12.47194 | 26.86997 |
| 195 | 38 025 | 13.96424 | 44.15880 | 7 414 875 | 5.798 890 | 12.49333 | 26.91606 |
| 196 | 38 416 | 14.00000 | 44.27189 | 7 529 536 | 5.808 786 | 12.51465 | 26.96199 |
| 197 | 38 809 | 14.03567 | 44.38468 | 7 645 373 | 5.818 648 | 12.53590 | 27.00777 |
| 198 | 39 204 | 14.07125 | 44.49719 | 7 762 392 | 5.828 477 | 12.55707 | 27.05339 |
| 199 | 39 601 | 14.10674 | 44.60942 | 7 880 599 | 5.838 272 | 12.57818 | 27.09886 |
| **200** | 40 000 | 14.14214 | 44.72136 | 8 000 000 | 5.848 035 | 12.59921 | 27.14418 |

| $n$ | $n^2$ | $\sqrt{n}$ | $\sqrt{10n}$ | $n^3$ | $\sqrt[3]{n}$ | $\sqrt[3]{10n}$ | $\sqrt[3]{100n}$ |
|---|---|---|---|---|---|---|---|
| **200** | 40 000 | 14.14214 | 44.72136 | 8 000 000 | 5.848 035 | 12.59921 | 27.14418 |
| 201 | 40 401 | 14.17745 | 44.83302 | 8 120 601 | 5.857 766 | 12.62017 | 27.18934 |
| 202 | 40 804 | 14.21267 | 44.94441 | 8 242 408 | 5.867 464 | 12.64107 | 27.23436 |
| 203 | 41 209 | 14.24781 | 45.05552 | 8 365 427 | 5.877 131 | 12.66189 | 27.27922 |
| 204 | 41 616 | 14.28286 | 45.16636 | 8 489 664 | 5.886 765 | 12.68265 | 27.32394 |
| 205 | 42 025 | 14.31782 | 45.27693 | 8 615 125 | 5.896 369 | 12.70334 | 27.36852 |
| 206 | 42 436 | 14.35270 | 45.38722 | 8 741 816 | 5.905 941 | 12.72396 | 27.41295 |
| 207 | 42 849 | 14.38749 | 45.49725 | 8 869 743 | 5.915 482 | 12.74452 | 27.45723 |
| 208 | 43 264 | 14.42221 | 45.60702 | 8 998 912 | 5.924 992 | 12.76501 | 27.50138 |
| 209 | 43 681 | 14.45683 | 45.71652 | 9 129 329 | 5.934 472 | 12.78543 | 27.54538 |
| **210** | 44 100 | 14.49138 | 45.82576 | 9 261 000 | 5.943 922 | 12.80579 | 27.58924 |
| 211 | 44 521 | 14.52584 | 45.93474 | 9 393 931 | 5.953 342 | 12.82609 | 27.63296 |
| 212 | 44 944 | 14.56022 | 46.04346 | 9 528 128 | 5.962 732 | 12.84632 | 27.67655 |
| 213 | 45 369 | 14.59452 | 46.15192 | 9 663 597 | 5.972 093 | 12.86648 | 27.72000 |
| 214 | 45 796 | 14.62874 | 46.26013 | 9 800 344 | 5.981 424 | 12.88659 | 27.76331 |
| 215 | 46 225 | 14.66288 | 46.36809 | 9 938 375 | 5.990 726 | 12.90663 | 27.80649 |
| 216 | 46 656 | 14.69694 | 46.47580 | 10 077 696 | 6.000 000 | 12.92661 | 27.84953 |
| 217 | 47 089 | 14.73092 | 46.58326 | 10 218 313 | 6.009 245 | 12.94653 | 27.89244 |
| 218 | 47 524 | 14.76482 | 46.69047 | 10 360 232 | 6.018 462 | 12.96638 | 27.93522 |
| 219 | 47 961 | 14.79865 | 46.79744 | 10 503 459 | 6.027 650 | 12.98618 | 27.97787 |
| **220** | 48 400 | 14.83240 | 46.90416 | 10 648 000 | 6.036 811 | 13.00591 | 28.02039 |
| 221 | 48 841 | 14.86607 | 47.01064 | 10 793 861 | 6.045 944 | 13.02559 | 28.06278 |
| 222 | 49 284 | 14.89966 | 47.11688 | 10 941 048 | 6.055 049 | 13.04521 | 28.10505 |
| 223 | 49 729 | 14.93318 | 47.22288 | 11 089 567 | 6.064 127 | 13.06477 | 28.14718 |
| 224 | 50 176 | 14.96663 | 47.32864 | 11 239 424 | 6.073 178 | 13.08427 | 28.18919 |
| 225 | 50 625 | 15.00000 | 47.43416 | 11 390 625 | 6.082 202 | 13.10371 | 28.23108 |
| 226 | 51 076 | 15.03330 | 47.53946 | 11 543 176 | 6.091 199 | 13.12309 | 28.27284 |
| 227 | 51 529 | 15.06652 | 47.64452 | 11 697 083 | 6.100 170 | 13.14242 | 28.31448 |
| 228 | 51 984 | 15.09967 | 47.74935 | 11 852 352 | 6.109 115 | 13.16169 | 28.35600 |
| 229 | 52 441 | 15.13275 | 47.85394 | 12 008 989 | 6.118 033 | 13.18090 | 28.39739 |
| **230** | 52 900 | 15.16575 | 47.95832 | 12 167 000 | 6.126 926 | 13.20006 | 28.43867 |
| 231 | 53 361 | 15.19868 | 48.06246 | 12 326 391 | 6.135 792 | 13.21916 | 28.47983 |
| 232 | 53 824 | 15.23155 | 48.16638 | 12 487 168 | 6.144 634 | 13.23821 | 28.52086 |
| 233 | 54 289 | 15.26434 | 48.27007 | 12 649 337 | 6.153 449 | 13.25721 | 28.56178 |
| 234 | 54 756 | 15.29706 | 48.37355 | 12 812 904 | 6.162 240 | 13.27614 | 28.60259 |
| 235 | 55 225 | 15.32971 | 48.47680 | 12 977 875 | 6.171 006 | 13.29503 | 28.64327 |
| 236 | 55 696 | 15.36229 | 48.57983 | 13 144 256 | 6.179 747 | 13.31386 | 28.68384 |
| 237 | 56 169 | 15.39480 | 48.68265 | 13 312 053 | 6.188 463 | 13.33264 | 28.72430 |
| 238 | 56 644 | 15.42725 | 48.78524 | 13 481 272 | 6.197 154 | 13.35136 | 28.76464 |
| 239 | 57 121 | 15.45962 | 48.88763 | 13 651 919 | 6.205 822 | 13.37004 | 28.80487 |
| **240** | 57 600 | 15.49193 | 48.98979 | 13 824 000 | 6.214 465 | 13.38866 | 28.84499 |
| 241 | 58 081 | 15.52417 | 49.09175 | 13 997 521 | 6.223 084 | 13.40723 | 28.88500 |
| 242 | 58 564 | 15.55635 | 49.19350 | 14 172 488 | 6.231 680 | 13.42575 | 28.92489 |
| 243 | 59 049 | 15.58846 | 49.29503 | 14 348 907 | 6.240 251 | 13.44421 | 28.96468 |
| 244 | 59 536 | 15.62050 | 49.39636 | 14 526 784 | 6.248 800 | 13.46263 | 29.00436 |
| 245 | 60 025 | 15.65248 | 49.49747 | 14 706 125 | 6.257 325 | 13.48100 | 29.04393 |
| 246 | 60 516 | 15.68439 | 49.59839 | 14 886 936 | 6.265 827 | 13.49931 | 29.08339 |
| 247 | 61 009 | 15.71623 | 49.69909 | 15 069 223 | 6.274 305 | 13.51758 | 29.12275 |
| 248 | 61 504 | 15.74802 | 49.79960 | 15 252 992 | 6.282 761 | 13.53580 | 29.16199 |
| 249 | 62 001 | 15.77973 | 49.89990 | 15 438 249 | 6.291 195 | 13.55397 | 29.20114 |
| **250** | 62 500 | 15.81139 | 50.00000 | 15 625 000 | 6.299 605 | 13.57209 | 29.24018 |

| $n$ | $n^2$ | $\sqrt{n}$ | $\sqrt{10n}$ | $n^3$ | $\sqrt[3]{n}$ | $\sqrt[3]{10n}$ | $\sqrt[3]{100n}$ |
|---|---|---|---|---|---|---|---|
| **250** | 62 500 | 15.81139 | 50.00000 | 15 625 000 | 6.299 605 | 13.57209 | 29.24018 |
| 251 | 63 001 | 15.84298 | 50.09990 | 15 813 251 | 6.307 994 | 13.59016 | 29.27911 |
| 252 | 63 504 | 15.87451 | 50.19960 | 16 003 008 | 6.316 360 | 13.60818 | 29.31794 |
| 253 | 64 009 | 15.90597 | 50.29911 | 16 194 277 | 6.324 704 | 13.62616 | 29.35667 |
| 254 | 64 516 | 15.93738 | 50.39841 | 16 387 064 | 6.333 026 | 13.64409 | 29.39530 |
| 255 | 65 025 | 15.96872 | 50.49752 | 16 581 375 | 6.341 326 | 13.66197 | 29.43383 |
| 256 | 65 536 | 16.00000 | 50.59644 | 16 777 216 | 6.349 604 | 13.67981 | 29.47225 |
| 257 | 66 049 | 16.03122 | 50.69517 | 16 974 593 | 6.357 861 | 13.69760 | 29.51058 |
| 258 | 66 564 | 16.06238 | 50.79370 | 17 173 512 | 6.366 097 | 13.71534 | 29.54880 |
| 259 | 67 081 | 16.09348 | 50.89204 | 17 373 979 | 6.374 311 | 13.73304 | 29.58693 |
| **260** | 67 600 | 16.12452 | 50.99020 | 17 576 000 | 6.382 504 | 13.75069 | 29.62496 |
| 261 | 68 121 | 16.15549 | 51.08816 | 17 779 581 | 6.390 677 | 13.76830 | 29.66289 |
| 262 | 68 644 | 16.18641 | 51.18594 | 17 984 728 | 6.398 828 | 13.78586 | 29.70073 |
| 263 | 69 169 | 16.21727 | 51.28353 | 18 191 447 | 6.406 959 | 13.80337 | 29.73847 |
| 264 | 69 696 | 16.24808 | 51.38093 | 18 399 744 | 6.415 069 | 13.82085 | 29.77611 |
| 265 | 70 225 | 16.27882 | 51.47815 | 18 609 625 | 6.423 158 | 13.83828 | 29.81366 |
| 266 | 70 756 | 16.30951 | 51.57519 | 18 821 096 | 6.431 228 | 13.85566 | 29.85111 |
| 267 | 71 289 | 16.34013 | 51.67204 | 19 034 163 | 6.439 277 | 13.87300 | 29.88847 |
| 268 | 71 824 | 16.37071 | 51.76872 | 19 248 832 | 6.447 306 | 13.89030 | 29.92574 |
| 269 | 72 361 | 16.40122 | 51.86521 | 19 465 109 | 6.455 315 | 13.90755 | 29.96292 |
| **270** | 72 900 | 16.43168 | 51.96152 | 19 683 000 | 6.463 304 | 13.92477 | 30.00000 |
| 271 | 73 441 | 16.46208 | 52.05766 | 19 902 511 | 6.471 274 | 13.94194 | 30.03699 |
| 272 | 73 984 | 16.49242 | 52.15362 | 20 123 648 | 6.479 224 | 13.95906 | 30.07389 |
| 273 | 74 529 | 16.52271 | 52.24940 | 20 346 417 | 6.487 154 | 13.97615 | 30.11070 |
| 274 | 75 076 | 16.55295 | 52.34501 | 20 570 824 | 6.495 065 | 13.99319 | 30.14742 |
| 275 | 75 625 | 16.58312 | 52.44044 | 20 796 875 | 6.502 957 | 14.01020 | 30.18405 |
| 276 | 76 176 | 16.61325 | 52.53570 | 21 024 576 | 6.510 830 | 14.02716 | 30.22060 |
| 277 | 76 729 | 16.64332 | 52.63079 | 21 253 933 | 6.518 684 | 14.04408 | 30.25705 |
| 278 | 77 284 | 16.67333 | 52.72571 | 21 484 952 | 6.526 519 | 14.06096 | 30.29342 |
| 279 | 77 841 | 16.70329 | 52.82045 | 21 717 639 | 6.534 335 | 14.07780 | 30.32970 |
| **280** | 78 400 | 16.73320 | 52.91503 | 21 952 000 | 6.542 133 | 14.09460 | 30.36589 |
| 281 | 78 961 | 16.76305 | 53.00943 | 22 188 041 | 6.549 912 | 14.11136 | 30.40200 |
| 282 | 79 524 | 16.79286 | 53.10367 | 22 425 768 | 6.557 672 | 14.12808 | 30.43802 |
| 283 | 80 089 | 16.82260 | 53.19774 | 22 665 187 | 6.565 414 | 14.14476 | 30.47395 |
| 284 | 80 656 | 16.85230 | 53.29165 | 22 906 304 | 6.573 138 | 14.16140 | 30.50981 |
| 285 | 81 225 | 16.88194 | 53.38539 | 23 149 125 | 6.580 844 | 14.17800 | 30.54557 |
| 286 | 81 796 | 16.91153 | 53.47897 | 23 393 656 | 6.588 532 | 14.19456 | 30.58126 |
| 287 | 82 369 | 16.94107 | 53.57238 | 23 639 903 | 6.596 202 | 14.21109 | 30.61686 |
| 288 | 82 944 | 16.97056 | 53.66563 | 23 887 872 | 6.603 854 | 14.22757 | 30.65238 |
| 289 | 83 521 | 17.00000 | 53.75872 | 24 137 569 | 6.611 489 | 14.24402 | 30.68781 |
| **290** | 84 100 | 17.02939 | 53.85165 | 24 389 000 | 6.619 106 | 14.26043 | 30.72317 |
| 291 | 84 681 | 17.05872 | 53.94442 | 24 642 171 | 6.626 705 | 14.27680 | 30.75844 |
| 292 | 85 264 | 17.08801 | 54.03702 | 24 897 088 | 6.634 287 | 14.29314 | 30.79363 |
| 293 | 85 849 | 17.11724 | 54.12947 | 25 153 757 | 6.641 852 | 14.30944 | 30.82875 |
| 294 | 86 436 | 17.14643 | 54.22177 | 25 412 184 | 6.649 400 | 14.32570 | 30.86378 |
| 295 | 87 025 | 17.17556 | 54.31390 | 25 672 375 | 6.656 930 | 14.34192 | 30.89873 |
| 296 | 87 616 | 17.20465 | 54.40588 | 25 934 336 | 6.664 444 | 14.35811 | 30.93361 |
| 297 | 88 209 | 17.23369 | 54.49771 | 26 198 073 | 6.671 940 | 14.37426 | 30.96840 |
| 298 | 88 804 | 17.26268 | 54.58938 | 26 463 592 | 6.679 420 | 14.39037 | 31.00312 |
| 299 | 89 401 | 17.29162 | 54.68089 | 26 730 899 | 6.686 883 | 14.40645 | 31.03776 |
| **300** | 90 000 | 17.32051 | 54.77226 | 27 000 000 | 6.694 330 | 14.42250 | 31.07233 |

| n | $n^2$ | $\sqrt{n}$ | $\sqrt{10n}$ | $n^3$ | $\sqrt[3]{n}$ | $\sqrt[3]{10n}$ | $\sqrt[3]{100n}$ |
|---|---|---|---|---|---|---|---|
| **300** | 90 000 | 17.32051 | 54.77226 | 27 000 000 | 6.694 330 | 14.42250 | 31.07233 |
| 301 | 90 601 | 17.34935 | 54.86347 | 27 270 901 | 6.701 759 | 14.43850 | 31.10681 |
| 302 | 91 204 | 17.37815 | 54.95453 | 27 543 608 | 6.709 173 | 14.45447 | 31.14122 |
| 303 | 91 809 | 17.40690 | 55.04544 | 27 818 127 | 6.716 570 | 14.47041 | 31.17556 |
| 304 | 92 416 | 17.43560 | 55.13620 | 28 094 464 | 6.723 951 | 14.48631 | 31.20982 |
| 305 | 93 025 | 17.46425 | 55.22681 | 28 372 625 | 6.731 315 | 14.50218 | 31.24400 |
| 306 | 93 636 | 17.49286 | 55.31727 | 28 652 616 | 6.738 664 | 14.51801 | 31.27811 |
| 307 | 94 249 | 17.52142 | 55.40758 | 28 934 443 | 6.745 997 | 14.53381 | 31.31214 |
| 308 | 94 864 | 17.54993 | 55.49775 | 29 218 112 | 6.753 313 | 14.54957 | 31.34610 |
| 309 | 95 481 | 17.57840 | 55.58777 | 29 503 629 | 6.760 614 | 14.56530 | 31.37999 |
| **310** | 96 100 | 17.60682 | 55.67764 | 29 791 000 | 6.767 899 | 14.58100 | 31.41381 |
| 311 | 96 721 | 17.63519 | 55.76737 | 30 080 231 | 6.775 169 | 14.59666 | 31.44755 |
| 312 | 97 344 | 17.66352 | 55.85696 | 30 371 328 | 6.782 423 | 14.61229 | 31.48122 |
| 313 | 97 969 | 17.69181 | 55.94640 | 30 664 297 | 6.789 661 | 14.62788 | 31.51482 |
| 314 | 98 596 | 17.72005 | 56.03570 | 30 959 144 | 6.796 884 | 14.64344 | 31.54834 |
| 315 | 99 225 | 17.74824 | 56.12486 | 31 255 875 | 6.804 092 | 14.65897 | 31.58180 |
| 316 | 99 856 | 17.77639 | 56.21388 | 31 554 496 | 6.811 285 | 14.67447 | 31.61518 |
| 317 | 100 489 | 17.80449 | 56.30275 | 31 855 013 | 6.818 462 | 14.68993 | 31.64850 |
| 318 | 101 124 | 17.83255 | 56.39149 | 32 157 432 | 6.825 624 | 14.70536 | 31.68174 |
| 319 | 101 761 | 17.86057 | 56.48008 | 32 461 759 | 6.832 771 | 14.72076 | 31.71492 |
| **320** | 102 400 | 17.88854 | 56.56854 | 32 768 000 | 6.839 904 | 14.73613 | 31.74802 |
| 321 | 103 041 | 17.91647 | 56.65686 | 33 076 161 | 6.847 021 | 14.75146 | 31.78106 |
| 322 | 103 684 | 17.94436 | 56.74504 | 33 386 248 | 6.854 124 | 14.76676 | 31.81403 |
| 323 | 104 329 | 17.97220 | 56.83309 | 33 698 267 | 6.861 212 | 14.78203 | 31.84693 |
| 324 | 104 976 | 18.00000 | 56.92100 | 34 012 224 | 6.868 285 | 14.79727 | 31.87976 |
| 325 | 105 625 | 18.02776 | 57.00877 | 34 328 125 | 6.875 344 | 14.81248 | 31.91252 |
| 326 | 106 276 | 18.05547 | 57.09641 | 34 645 976 | 6.882 389 | 14.82766 | 31.94522 |
| 327 | 106 929 | 18.08314 | 57.18391 | 34 965 783 | 6.889 419 | 14.84280 | 31.97785 |
| 328 | 107 584 | 18.11077 | 57.27128 | 35 287 552 | 6.896 434 | 14.85792 | 32.01041 |
| 329 | 108 241 | 18.13836 | 57.35852 | 35 611 289 | 6.903 436 | 14.87300 | 32.04291 |
| **330** | 108 900 | 18.16590 | 57.44563 | 35 937 000 | 6.910 423 | 14.88806 | 32.07534 |
| 331 | 109 561 | 18.19341 | 57.53260 | 36 264 691 | 6.917 396 | 14.90308 | 32.10771 |
| 332 | 110 224 | 18.22087 | 57.61944 | 36 594 368 | 6.924 356 | 14.91807 | 32.14001 |
| 333 | 110 889 | 18.24829 | 57.70615 | 36 926 037 | 6.931 301 | 14.93303 | 32.17225 |
| 334 | 111 556 | 18.27567 | 57.79273 | 37 259 704 | 6.938 232 | 14.94797 | 32.20442 |
| 335 | 112 225 | 18.30301 | 57.87918 | 37 595 375 | 6.945 150 | 14.96287 | 32.23653 |
| 336 | 112 896 | 18.33030 | 57.96551 | 37 933 056 | 6.952 053 | 14.97774 | 32.26857 |
| 337 | 113 569 | 18.35756 | 58.05170 | 38 272 753 | 6.958 943 | 14.99259 | 32.30055 |
| 338 | 114 244 | 18.38478 | 58.13777 | 38 614 472 | 6.965 820 | 15.00740 | 32.33247 |
| 339 | 114 921 | 18.41195 | 58.22371 | 38 958 219 | 6.972 683 | 15.02219 | 32.36433 |
| **340** | 115 600 | 18.43909 | 58.30952 | 39 304 000 | 6.979 532 | 15.03695 | 32.39612 |
| 341 | 116 281 | 18.46619 | 58.39521 | 39 651 821 | 6.986 368 | 15.05167 | 32.42785 |
| 342 | 116 964 | 18.49324 | 58.48077 | 40 001 688 | 6.993 191 | 15.06637 | 32.45952 |
| 343 | 117 649 | 18.52026 | 58.56620 | 40 353 607 | 7.000 000 | 15.08104 | 32.49112 |
| 344 | 118 336 | 18.54724 | 58.65151 | 40 707 584 | 7.006 796 | 15.09568 | 32.52267 |
| 345 | 119 025 | 18.57418 | 58.73670 | 41 063 625 | 7.013 579 | 15.11030 | 32.55415 |
| 346 | 119 716 | 18.60108 | 58.82176 | 41 421 736 | 7.020 349 | 15.12488 | 32.58557 |
| 347 | 120 409 | 18.62794 | 58.90671 | 41 781 923 | 7.027 106 | 15.13944 | 32.61694 |
| 348 | 121 104 | 18.65476 | 58.99152 | 42 144 192 | 7.033 850 | 15.15397 | 32.64824 |
| 349 | 121 801 | 18.68154 | 59.07622 | 42 508 549 | 7.040 581 | 15.16847 | 32.67948 |
| **350** | 122 500 | 18.70829 | 59.16080 | 42 875 000 | 7.047 299 | 15.18294 | 32.71066 |

| $n$ | $n^2$ | $\sqrt{n}$ | $\sqrt{10n}$ | $n^3$ | $\sqrt[3]{n}$ | $\sqrt[3]{10n}$ | $\sqrt[3]{100n}$ |
|---|---|---|---|---|---|---|---|
| **350** | 122 500 | 18.70829 | 59.16080 | 42 875 000 | 7.047 299 | 15.18294 | 32.71066 |
| 351 | 123 201 | 18.73499 | 59.24525 | 43 243 551 | 7.054 004 | 15.19739 | 32.74179 |
| 352 | 123 904 | 18.76166 | 59.32959 | 43 614 208 | 7.060 697 | 15.21181 | 32.77285 |
| 353 | 124 609 | 18.78829 | 59.41380 | 43 986 977 | 7.067 377 | 15.22620 | 32.80386 |
| 354 | 125 316 | 18.81489 | 59.49790 | 44 361 864 | 7.074 044 | 15.24057 | 32.83480 |
| 355 | 126 025 | 18.84144 | 59.58188 | 44 738 875 | 7.080 699 | 15.25490 | 32.86569 |
| 356 | 126 736 | 18.86796 | 59.66574 | 45 118 016 | 7.087 341 | 15.26921 | 32.89652 |
| 357 | 127 449 | 18.89444 | 59.74948 | 45 499 293 | 7.093 971 | 15.28350 | 32.92730 |
| 358 | 128 164 | 18.92089 | 59.83310 | 45 882 712 | 7.100 588 | 15.29775 | 32.95801 |
| 359 | 128 881 | 18.94730 | 59.91661 | 46 268 279 | 7.107 194 | 15.31198 | 32.98867 |
| **360** | 129 600 | 18.97367 | 60.00000 | 46 656 000 | 7.113 787 | 15.32619 | 33.01927 |
| 361 | 130 321 | 19.00000 | 60.08328 | 47 045 881 | 7.120 367 | 15.34037 | 33.04982 |
| 362 | 131 044 | 19.02630 | 60.16644 | 47 437 928 | 7.126 936 | 15.35452 | 33.08031 |
| 363 | 131 769 | 19.05256 | 60.24948 | 47 832 147 | 7.133 492 | 15.36864 | 33.11074 |
| 364 | 132 496 | 19.07878 | 60.33241 | 48 228 544 | 7.140 037 | 15.38274 | 33.14112 |
| 365 | 133 225 | 19.10497 | 60.41523 | 48 627 125 | 7.146 569 | 15.39682 | 33.17144 |
| 366 | 133 956 | 19.13113 | 60.49793 | 49 027 896 | 7.153 090 | 15.41087 | 33.20170 |
| 367 | 134 689 | 19.15724 | 60.58052 | 49 430 863 | 7.159 599 | 15.42489 | 33.23191 |
| 368 | 135 424 | 19.18333 | 60.66300 | 49 836 032 | 7.166 096 | 15.43889 | 33.26207 |
| 369 | 136 161 | 19.20937 | 60.74537 | 50 243 409 | 7.172 581 | 15.45286 | 33.29217 |
| **370** | 136 900 | 19.23538 | 60.82763 | 50 653 000 | 7.179 054 | 15.46680 | 33.32222 |
| 371 | 137 641 | 19.26136 | 60.90977 | 51 064 811 | 7.185 516 | 15.48073 | 33.35221 |
| 372 | 138 384 | 19.28730 | 60.99180 | 51 478 848 | 7.191 966 | 15.49462 | 33.38215 |
| 373 | 139 129 | 19.31321 | 61.07373 | 51 895 117 | 7.198 405 | 15.50849 | 33.41204 |
| 374 | 139 876 | 19.33908 | 61.15554 | 52 313 624 | 7.204 832 | 15.52234 | 33.44187 |
| 375 | 140 625 | 19.36492 | 61.23724 | 52 734 375 | 7.211 248 | 15.53616 | 33.47165 |
| 376 | 141 376 | 19.39072 | 61.31884 | 53 157 376 | 7.217 652 | 15.54996 | 33.50137 |
| 377 | 142 129 | 19.41649 | 61.40033 | 53 582 633 | 7.224 045 | 15.56373 | 33.53105 |
| 378 | 142 884 | 19.44222 | 61.48170 | 54 010 152 | 7.230 427 | 15.57748 | 33.56067 |
| 379 | 143 641 | 19.46792 | 61.56298 | 54 439 939 | 7.236 797 | 15.59121 | 33.59024 |
| **380** | 144 400 | 19.49359 | 61.64414 | 54 872 000 | 7.243 156 | 15.60491 | 33.61975 |
| 381 | 145 161 | 19.51922 | 61.72520 | 55 306 341 | 7.249 505 | 15.61858 | 33.64922 |
| 382 | 145 924 | 19.54482 | 61.80615 | 55 742 968 | 7.255 842 | 15.63224 | 33.67863 |
| 383 | 146 689 | 19.57039 | 61.88699 | 56 181 887 | 7.262 167 | 15.64587 | 33.70800 |
| 384 | 147 456 | 19.59592 | 61.96773 | 56 623 104 | 7.268 482 | 15.65947 | 33.73731 |
| 385 | 148 225 | 19.62142 | 62.04837 | 57 066 625 | 7.274 786 | 15.67305 | 33.76657 |
| 386 | 148 996 | 19.64688 | 62.12890 | 57 512 456 | 7.281 079 | 15.68661 | 33.79578 |
| 387 | 149 769 | 19.67232 | 62.20932 | 57 960 603 | 7.287 362 | 15.70014 | 33.82494 |
| 388 | 150 544 | 19.69772 | 62.28965 | 58 411 072 | 7.293 633 | 15.71366 | 33.85405 |
| 389 | 151 321 | 19.72308 | 62.36986 | 58 863 869 | 7.299 894 | 15.72714 | 33.88310 |
| **390** | 152 100 | 19.74842 | 62.44998 | 59 319 000 | 7.306 144 | 15.74061 | 33.91211 |
| 391 | 152 881 | 19.77372 | 62.52999 | 59 776 471 | 7.312 383 | 15.75405 | 33.94107 |
| 392 | 153 664 | 19.79899 | 62.60990 | 60 236 288 | 7.318 611 | 15.76747 | 33.96999 |
| 393 | 154 449 | 19.82423 | 62.68971 | 60 698 457 | 7.324 829 | 15.78087 | 33.99885 |
| 394 | 155 236 | 19.84943 | 62.76942 | 61 162 984 | 7.331 037 | 15.79424 | 34.02766 |
| 395 | 156 025 | 19.87461 | 62.84903 | 61 629 875 | 7.337 234 | 15.80759 | 34.05642 |
| 396 | 156 816 | 19.89975 | 62.92853 | 62 099 136 | 7.343 420 | 15.82092 | 34.08514 |
| 397 | 157 609 | 19.92486 | 63.00794 | 62 570 773 | 7.349 597 | 15.83423 | 34.11381 |
| 398 | 158 404 | 19.94994 | 63.08724 | 63 044 792 | 7.355 762 | 15.84751 | 34.14242 |
| 399 | 159 201 | 19.97498 | 63.16645 | 63 521 199 | 7.361 918 | 15.86077 | 34.17100 |
| **400** | 160 000 | 20.00000 | 63.24555 | 64 000 000 | 7.368 063 | 15.87401 | 34.19952 |

| $n$ | $n^2$ | $\sqrt{n}$ | $\sqrt{10n}$ | $n^3$ | $\sqrt[3]{n}$ | $\sqrt[3]{10n}$ | $\sqrt[3]{100n}$ |
|---|---|---|---|---|---|---|---|
| **400** | 160 000 | 20.00000 | 63.24555 | 64 000 000 | 7.368 063 | 15.87401 | 34.19952 |
| **401** | 160 801 | 20.02498 | 63.32456 | 64 481 201 | 7.374 198 | 15.88723 | 34.22799 |
| **402** | 161 604 | 20.04994 | 63.40347 | 64 964 808 | 7.380 323 | 15.90042 | 34.25642 |
| **403** | 162 409 | 20.07486 | 63.48228 | 65 450 827 | 7.386 437 | 15.91360 | 34.28480 |
| **404** | 163 216 | 20.09975 | 63.56099 | 65 939 264 | 7.392 542 | 15.92675 | 34.31314 |
| **405** | 164 025 | 20.12461 | 63.63961 | 66 430 125 | 7.398 636 | 15.93988 | 34.34143 |
| **406** | 164 836 | 20.14944 | 63.71813 | 66 923 416 | 7.404 721 | 15.95299 | 34.36967 |
| **407** | 165 649 | 20.17424 | 63.79655 | 67 419 143 | 7.410 795 | 15.96607 | 34.39786 |
| **408** | 166 464 | 20.19901 | 63.87488 | 67 917 312 | 7.416 860 | 15.97914 | 34.42601 |
| **409** | 167 281 | 20.22375 | 63.95311 | 68 417 929 | 7.422 914 | 15.99218 | 34.45412 |
| **410** | 168 100 | 20.24846 | 64.03124 | 68 921 000 | 7.428 959 | 16.00521 | 34.48217 |
| **411** | 168 921 | 20.27313 | 64.10928 | 69 426 531 | 7.434 994 | 16.01821 | 34.51018 |
| **412** | 169 744 | 20.29778 | 64.18723 | 69 934 528 | 7.441 019 | 16.03119 | 34.53815 |
| **413** | 170 569 | 20.32240 | 64.26508 | 70 444 997 | 7.447 034 | 16.04415 | 34.56607 |
| **414** | 171 396 | 20.34699 | 64.34283 | 70 957 944 | 7.453 040 | 16.05709 | 34.59395 |
| **415** | 172 225 | 20.37155 | 64.42049 | 71 473 375 | 7.459 036 | 16.07001 | 34.62178 |
| **416** | 173 056 | 20.39608 | 64.49806 | 71 991 296 | 7.465 022 | 16.08290 | 34.64956 |
| **417** | 173 889 | 20.42058 | 64.57554 | 72 511 713 | 7.470 999 | 16.09578 | 34.67731 |
| **418** | 174 724 | 20.44505 | 64.65292 | 73 034 632 | 7.476 966 | 16.10864 | 34.70500 |
| **419** | 175 561 | 20.46949 | 64.73021 | 73 560 059 | 7.482 924 | 16.12147 | 34.73266 |
| **420** | 176 400 | 20.49390 | 64.80741 | 74 088 000 | 7.488 872 | 16.13429 | 34.76027 |
| **421** | 177 241 | 20.51828 | 64.88451 | 74 618 461 | 7.494 811 | 16.14708 | 34.78783 |
| **422** | 178 084 | 20.54264 | 64.96153 | 75 151 448 | 7.500 741 | 16.15986 | 34.81535 |
| **423** | 178 929 | 20.56696 | 65.03845 | 75 686 967 | 7.506 661 | 16.17261 | 34.84283 |
| **424** | 179 776 | 20.59126 | 65.11528 | 76 225 024 | 7.512 572 | 16.18534 | 34.87027 |
| **425** | 180 625 | 20.61553 | 65.19202 | 76 765 625 | 7.518 473 | 16.19806 | 34.89766 |
| **426** | 181 476 | 20.63977 | 65.26868 | 77 308 776 | 7.524 365 | 16.21075 | 34.92501 |
| **427** | 182 329 | 20.66398 | 65.34524 | 77 854 483 | 7.530 248 | 16.22343 | 34.95232 |
| **428** | 183 184 | 20.68816 | 65.42171 | 78 402 752 | 7.536 122 | 16.23608 | 34.97958 |
| **429** | 184 041 | 20.71232 | 65.49809 | 78 953 589 | 7.541 987 | 16.24872 | 35.00680 |
| **430** | 184 900 | 20.73644 | 65.57439 | 79 507 000 | 7.547 842 | 16.26133 | 35.03398 |
| **431** | 185 761 | 20.76054 | 65.65059 | 80 062 991 | 7.553 689 | 16.27393 | 35.06112 |
| **432** | 186 624 | 20.78461 | 65.72671 | 80 621 568 | 7.559 526 | 16.28651 | 35.08821 |
| **433** | 187 489 | 20.80865 | 65.80274 | 81 182 737 | 7.565 355 | 16.29906 | 35.11527 |
| **434** | 188 356 | 20.83267 | 65.87868 | 81 746 504 | 7.571 174 | 16.31160 | 35.14228 |
| **435** | 189 225 | 20.85665 | 65.95453 | 82 312 875 | 7.576 985 | 16.32412 | 35.16925 |
| **436** | 190 096 | 20.88061 | 66.03030 | 82 881 856 | 7.582 787 | 16.33662 | 35.19618 |
| **437** | 190 969 | 20.90454 | 66.10598 | 83 453 453 | 7.588 579 | 16.34910 | 35.22307 |
| **438** | 191 844 | 20.92845 | 66.18157 | 84 027 672 | 7.594 363 | 16.36156 | 35.24991 |
| **439** | 192 721 | 20.95233 | 66.25708 | 84 604 519 | 7.600 139 | 16.37400 | 35.27672 |
| **440** | 193 600 | 20.97618 | 66.33250 | 85 184 000 | 7.605 905 | 16.38643 | 35.30348 |
| **441** | 194 481 | 21.00000 | 66.40783 | 85 766 121 | 7.611 663 | 16.39883 | 35.33021 |
| **442** | 195 364 | 21.02380 | 66.48308 | 86 350 888 | 7.617 412 | 16.41122 | 35.35689 |
| **443** | 196 249 | 21.04757 | 66.55825 | 86 938 307 | 7.623 152 | 16.42358 | 35.38354 |
| **444** | 197 136 | 21.07131 | 66.63332 | 87 528 384 | 7.628 884 | 16.43593 | 35.41014 |
| **445** | 198 025 | 21.09502 | 66.70832 | 88 121 125 | 7.634 607 | 16.44826 | 35.43671 |
| **446** | 198 916 | 21.11871 | 66.78323 | 88 716 536 | 7.640 321 | 16.46057 | 35.46323 |
| **447** | 199 809 | 21.14237 | 66.85806 | 89 314 623 | 7.646 027 | 16.47287 | 35.48971 |
| **448** | 200 704 | 21.16601 | 66.93280 | 89 915 392 | 7.651 725 | 16.48514 | 35.51616 |
| **449** | 201 601 | 21.18962 | 67.00746 | 90 518 849 | 7.657 414 | 16.49740 | 35.54257 |
| **450** | 202 500 | 21.21320 | 67.08204 | 91 125 000 | 7.663 094 | 16.50964 | 35.56893 |

| $n$ | $n^2$ | $\sqrt{n}$ | $\sqrt{10n}$ | $n^3$ | $\sqrt[3]{n}$ | $\sqrt[3]{10n}$ | $\sqrt[3]{100n}$ |
|---|---|---|---|---|---|---|---|
| **450** | 202 500 | 21.21320 | 67.08204 | 91 125 000 | 7.663 094 | 16.50964 | 35.56893 |
| 451 | 203 401 | 21.23676 | 67.15653 | 91 733 851 | 7.668 766 | 16.52186 | 35.59526 |
| 452 | 204 304 | 21.26029 | 67.23095 | 92 345 408 | 7.674 430 | 16.53406 | 35.62155 |
| 453 | 205 209 | 21.28380 | 67.30527 | 92 959 677 | 7.680 086 | 16.54624 | 35.64780 |
| 454 | 206 116 | 21.30728 | 67.37952 | 93 576 664 | 7.685 733 | 16.55841 | 35.67401 |
| 455 | 207 025 | 21.33073 | 67.45369 | 94 196 375 | 7.691 372 | 16.57056 | 35.70018 |
| 456 | 207 936 | 21.35416 | 67.52777 | 94 818 816 | 7.697 002 | 16.58269 | 35.72632 |
| 457 | 208 849 | 21.37756 | 67.60178 | 95 443 993 | 7.702 625 | 16.59480 | 35.75242 |
| 458 | 209 764 | 21.40093 | 67.67570 | 96 071 912 | 7.708 239 | 16.60690 | 35.77848 |
| 459 | 210 681 | 21.42429 | 67.74954 | 96 702 579 | 7.713 845 | 16.61897 | 35.80450 |
| **460** | 211 600 | 21.44761 | 67.82330 | 97 336 000 | 7.719 443 | 16.63103 | 35.83048 |
| 461 | 212 521 | 21.47091 | 67.89698 | 97 972 181 | 7.725 032 | 16.64308 | 35.85642 |
| 462 | 213 444 | 21.49419 | 67.97058 | 98 611 128 | 7.730 614 | 16.65510 | 35.88233 |
| 463 | 214 369 | 21.51743 | 68.04410 | 99 252 847 | 7.736 188 | 16.66711 | 35.90820 |
| 464 | 215 296 | 21.54066 | 68.11755 | 99 897 344 | 7.741 753 | 16.67910 | 35.93404 |
| 465 | 216 225 | 21.56386 | 68.19091 | 100 544 625 | 7.747 311 | 16.69108 | 35.95983 |
| 466 | 217 156 | 21.58703 | 68.26419 | 101 194 696 | 7.752 861 | 16.70303 | 35.98559 |
| 467 | 218 089 | 21.61018 | 68.33740 | 101 847 563 | 7.758 402 | 16.71497 | 36.01131 |
| 468 | 219 024 | 21.63331 | 68.41053 | 102 503 232 | 7.763 936 | 16.72689 | 36.03700 |
| 469 | 219 961 | 21.65641 | 68.48357 | 103 161 709 | 7.769 462 | 16.73880 | 36.06265 |
| **470** | 220 900 | 21.67948 | 68.55655 | 103 823 000 | 7.774 980 | 16.75069 | 36.08826 |
| 471 | 221 841 | 21.70253 | 68.62944 | 104 487 111 | 7.780 490 | 16.76256 | 36.11384 |
| 472 | 222 784 | 21.72556 | 68.70226 | 105 154 048 | 7.785 993 | 16.77441 | 36.13938 |
| 473 | 223 729 | 21.74856 | 68.77500 | 105 823 817 | 7.791 488 | 16.78625 | 36.16488 |
| 474 | 224 676 | 21.77154 | 68.84766 | 106 496 424 | 7.796 975 | 16.79807 | 36.19035 |
| 475 | 225 625 | 21.79449 | 68.92024 | 107 171 875 | 7.802 454 | 16.80988 | 36.21578 |
| 476 | 226 576 | 21.81742 | 68.99275 | 107 850 176 | 7.807 925 | 16.82167 | 36.24118 |
| 477 | 227 529 | 21.84033 | 69.06519 | 108 531 333 | 7.813 389 | 16.83344 | 36.26654 |
| 478 | 228 484 | 21.86321 | 69 13754 | 109 215 352 | 7.818 846 | 16.84519 | 36.29187 |
| 479 | 229 441 | 21.88607 | 69.20983 | 109 902 239 | 7.824 294 | 16.85693 | 36.31716 |
| **480** | 230 400 | 21.90890 | 69 28203 | 110 592 000 | 7.829 735 | 16.86865 | 36.34241 |
| 481 | 231 361 | 21.93171 | 69.35416 | 111 284 641 | 7.835 169 | 16.88036 | 36.36763 |
| 482 | 232 324 | 21.95450 | 69.42622 | 111 980 168 | 7.840 595 | 16.89205 | 36.39282 |
| 483 | 233 289 | 21.97726 | 69 49820 | 112 678 587 | 7.846 013 | 16.90372 | 36.41797 |
| 484 | 234 256 | 22.00000 | 69.57011 | 113 379 904 | 7.851 424 | 16.91538 | 36.44308 |
| 485 | 235 225 | 22.02272 | 69.64194 | 114 084 125 | 7.856 828 | 16.92702 | 36.46817 |
| 486 | 236 196 | 22.04541 | 69.71370 | 114 791 256 | 7.862 224 | 16.93865 | 36.49321 |
| 487 | 237 169 | 22.06808 | 69.78539 | 115 501 303 | 7.867 613 | 16.95026 | 36.51822 |
| 488 | 238 144 | 22.09072 | 69.85700 | 116 214 272 | 7.872 994 | 16.96185 | 36.54320 |
| 489 | 239 121 | 22.11334 | 69.92853 | 116 930 169 | 7.878 368 | 16.97343 | 36.56815 |
| **490** | 240 100 | 22.13594 | 70.00000 | 117 649 000 | 7.883 735 | 16.98499 | 36.59306 |
| 491 | 241 081 | 22.15852 | 70.07139 | 118 370 771 | 7.889 095 | 16.99654 | 36.61793 |
| 492 | 242 064 | 22.18107 | 70.14271 | 119 095 488 | 7.894 447 | 17.00807 | 36.64278 |
| 493 | 243 049 | 22.20360 | 70.21396 | 119 823 157 | 7.899 792 | 17.01959 | 36.66758 |
| 494 | 244 036 | 22.22611 | 70.28513 | 120 553 784 | 7.905 129 | 17.03108 | 36.69236 |
| 495 | 245 025 | 22.24860 | 70.35624 | 121 287 375 | 7.910 460 | 17.04257 | 36.71710 |
| 496 | 246 016 | 22.27106 | 70.42727 | 122 023 936 | 7.915 783 | 17.05404 | 36.74181 |
| 497 | 247 009 | 22.29350 | 70.49823 | 122 763 473 | 7.921 099 | 17.06549 | 36.76649 |
| 498 | 248 004 | 22.31591 | 70.56912 | 123 505 992 | 7.926 408 | 17.07693 | 36.79113 |
| 499 | 249 001 | 22.33831 | 70.63993 | 124 251 499 | 7.931 710 | 17.08835 | 36.81574 |
| **500** | 250 000 | 22.36068 | 70.71068 | 125 000 000 | 7.937 005 | 17.09976 | 36.84031 |

| n | n² | √n | √10n | n³ | ∛n | ∛10n | ∛100n |
|---|---|---|---|---|---|---|---|
| **500** | 250 000 | 22.36068 | 70.71068 | 125 000 000 | 7.937 005 | 17.09976 | 36.84031 |
| **501** | 251 001 | 22.38303 | 70.78135 | 125 751 501 | 7.942 293 | 17.11115 | 36.86486 |
| **502** | 252 004 | 22.40536 | 70.85196 | 126 506 008 | 7.947 574 | 17.12253 | 36.88937 |
| **503** | 253 009 | 22.42766 | 70.92249 | 127 263 527 | 7.952 848 | 17.13389 | 36.91385 |
| **504** | 254 016 | 22.44994 | 70.99296 | 128 024 064 | 7.958 114 | 17.14524 | 36.93830 |
| **505** | 255 025 | 22.47221 | 71.06335 | 128 787 625 | 7.963 374 | 17.15657 | 36.96271 |
| **506** | 256 036 | 22.49444 | 71.13368 | 129 554 216 | 7.968 627 | 17.16789 | 36.98709 |
| **507** | 257 049 | 22.51666 | 71.20393 | 130 323 843 | 7.973 873 | 17.17919 | 37.01144 |
| **508** | 258 064 | 22.53886 | 71.27412 | 131 096 512 | 7.979 112 | 17.19048 | 37.03576 |
| **509** | 259 081 | 22.56103 | 71.34424 | 131 872 229 | 7.984 344 | 17.20175 | 37.06004 |
| **510** | 260 100 | 22.58318 | 71.41428 | 132 651 000 | 7.989 570 | 17.21301 | 37.08430 |
| **511** | 261 121 | 22.60531 | 71.48426 | 133 432 831 | 7.994 788 | 17.22425 | 37.10852 |
| **512** | 262 144 | 22.62742 | 71.55418 | 134 217 728 | 8.000 000 | 17.23548 | 37.13271 |
| **513** | 263 169 | 22.64950 | 71.62402 | 135 005 697 | 8.005 205 | 17.24669 | 37.15687 |
| **514** | 264 196 | 22.67157 | 71.69379 | 135 796 744 | 8.010 403 | 17.25789 | 37.18100 |
| **515** | 265 225 | 22.69361 | 71.76350 | 136 590 875 | 8.015 595 | 17.26908 | 37.20509 |
| **516** | 266 256 | 22.71563 | 71.83314 | 137 388 096 | 8.020 779 | 17.28025 | 37.22916 |
| **517** | 267 289 | 22.73763 | 71.90271 | 138 188 413 | 8.025 957 | 17.29140 | 37.25319 |
| **518** | 268 324 | 22.75961 | 71.97222 | 138 991 832 | 8.031 129 | 17.30254 | 37.27720 |
| **519** | 269 361 | 22.78157 | 72.04165 | 139 798 359 | 8.036 293 | 17.31367 | 37.30117 |
| **520** | 270 400 | 22.80351 | 72.11103 | 140 608 000 | 8.041 452 | 17.32478 | 37.32511 |
| **521** | 271 441 | 22.82542 | 72.18033 | 141 420 761 | 8.046 603 | 17.33588 | 37.34902 |
| **522** | 272 484 | 22.84732 | 72.24957 | 142 236 648 | 8.051 748 | 17.34696 | 37.37290 |
| **523** | 273 529 | 22.86919 | 72.31874 | 143 055 667 | 8.056 886 | 17.35804 | 37.39675 |
| **524** | 274 576 | 22.89105 | 72.38784 | 143 877 824 | 8.062 018 | 17.36909 | 37.42057 |
| **525** | 275 625 | 22.91288 | 72.45688 | 144 703 125 | 8.067 143 | 17.38013 | 37.44436 |
| **526** | 276 676 | 22.93469 | 72.52586 | 145 531 576 | 8.072 262 | 17.39116 | 37.46812 |
| **527** | 277 729 | 22.95648 | 72.59477 | 146 363 183 | 8.077 374 | 17.40218 | 37.49185 |
| **528** | 278 784 | 22.97825 | 72.66361 | 147 197 952 | 8.082 480 | 17.41318 | 37.51555 |
| **529** | 279 841 | 23.00000 | 72.73239 | 148 035 889 | 8.087 579 | 17.42416 | 37.53922 |
| **530** | 280 900 | 23.02173 | 72.80110 | 148 877 000 | 8.092 672 | 17.43513 | 37.56286 |
| **531** | 281 961 | 23.04344 | 72.86975 | 149 721 291 | 8.097 759 | 17.44609 | 37.58647 |
| **532** | 283 024 | 23.06513 | 72.93833 | 150 568 768 | 8.102 839 | 17.45704 | 37.61005 |
| **533** | 284 089 | 23.08679 | 73.00685 | 151 419 437 | 8.107 913 | 17.46797 | 37.63360 |
| **534** | 285 156 | 23.10844 | 73.07530 | 152 273 304 | 8.112 980 | 17.47889 | 37.65712 |
| **535** | 286 225 | 23.13007 | 73.14369 | 153 130 375 | 8.118 041 | 17.48979 | 37.68061 |
| **536** | 287 296 | 23.15167 | 73.21202 | 153 990 656 | 8.123 096 | 17.50068 | 37.70407 |
| **537** | 288 369 | 23.17326 | 73.28028 | 154 854 153 | 8.128 145 | 17.51156 | 37.72751 |
| **538** | 289 444 | 23.19483 | 73.34848 | 155 720 872 | 8.133 187 | 17.52242 | 37.75091 |
| **539** | 290 521 | 23.21637 | 73.41662 | 156 590 819 | 8.138 223 | 17.53327 | 37.77429 |
| **540** | 291 600 | 23.23790 | 73.48469 | 157 464 000 | 8.143 253 | 17.54411 | 37.79763 |
| **541** | 292 681 | 23.25941 | 73.55270 | 158 340 421 | 8.148 276 | 17.55493 | 37.82095 |
| **542** | 293 764 | 23.28089 | 73.62065 | 159 220 088 | 8.153 294 | 17.56574 | 37.84424 |
| **543** | 294 849 | 23.30236 | 73.68853 | 160 103 007 | 8.158 305 | 17.57654 | 37.86750 |
| **544** | 295 936 | 23.32381 | 73.75636 | 160 989 184 | 8.163 310 | 17.58732 | 37.89073 |
| **545** | 297 025 | 23.34524 | 73.82412 | 161 878 625 | 8.168 309 | 17.59809 | 37.91393 |
| **546** | 298 116 | 23.36664 | 73.89181 | 162 771 336 | 8.173 302 | 17.60885 | 37.93711 |
| **547** | 299 209 | 23.38803 | 73.95945 | 163 667 323 | 8.178 289 | 17.61959 | 37.96025 |
| **548** | 300 304 | 23.40940 | 74.02702 | 164 566 592 | 8.183 269 | 17.63032 | 37.98337 |
| **549** | 301 401 | 23.43075 | 74.09453 | 165 469 149 | 8.188 244 | 17.64104 | 38.00646 |
| **550** | 302 500 | 23.45208 | 74.16198 | 166 375 000 | 8.193 213 | 17.65174 | 38.02952 |

| $n$ | $n^2$ | $\sqrt{n}$ | $\sqrt{10n}$ | $n^3$ | $\sqrt[3]{n}$ | $\sqrt[3]{10n}$ | $\sqrt[3]{100n}$ |
|---|---|---|---|---|---|---|---|
| **550** | 302 500 | 23.45208 | 74.16198 | 166 375 000 | 8.193 213 | 17.65174 | 38.02952 |
| 551 | 303 601 | 23.47339 | 74.22937 | 167 284 151 | 8.198 175 | 17.66243 | 38.05256 |
| 552 | 304 704 | 23.49468 | 74.29670 | 168 196 608 | 8.203 132 | 17.67311 | 38.07557 |
| 553 | 305 809 | 23.51595 | 74.36397 | 169 112 377 | 8.208 082 | 17.68378 | 38.09854 |
| 554 | 306 916 | 23.53720 | 74.43118 | 170 031 464 | 8.213 027 | 17.69443 | 38.12149 |
| 555 | 308 025 | 23.55844 | 74.49832 | 170 953 875 | 8.217 966 | 17.70507 | 38.14442 |
| 556 | 309 136 | 23.57965 | 74.56541 | 171 879 616 | 8.222 899 | 17.71570 | 38.16731 |
| 557 | 310 249 | 23.60085 | 74.63243 | 172 808 693 | 8.227 825 | 17.72631 | 38.19018 |
| 558 | 311 364 | 23.62202 | 74.69940 | 173 741 112 | 8.232 746 | 17.73691 | 38.21302 |
| 559 | 312 481 | 23.64318 | 74.76630 | 174 676 879 | 8.237 661 | 17.74750 | 38.23584 |
| **560** | 313 600 | 23.66432 | 74.83315 | 175 616 000 | 8.242 571 | 17.75808 | 38.25862 |
| 561 | 314 721 | 23.68544 | 74.89993 | 176 558 481 | 8.247 474 | 17.76864 | 38.28138 |
| 562 | 315 844 | 23.70654 | 74.96666 | 177 504 328 | 8.252 372 | 17.77920 | 38.30412 |
| 563 | 316 969 | 23.72762 | 75.03333 | 178 453 547 | 8.257 263 | 17.78973 | 38.32682 |
| 564 | 318 096 | 23.74868 | 75.09993 | 179 406 144 | 8.262 149 | 17.80026 | 38.34950 |
| 565 | 319 225 | 23.76973 | 75.16648 | 180 362 029 | 8.267 029 | 17.81077 | 38.37215 |
| 566 | 320 356 | 23.79075 | 75.23297 | 181 321 496 | 8.271 904 | 17.82128 | 38.39478 |
| 567 | 321 489 | 23.81176 | 75.29940 | 182 284 263 | 8.276 773 | 17.83177 | 38.41737 |
| 568 | 322 624 | 23.83275 | 75.36577 | 183 250 432 | 8.281 635 | 17.84224 | 38.43995 |
| 569 | 323 761 | 23.85372 | 75.43209 | 184 220 009 | 8.286 493 | 17.85271 | 38.46249 |
| **570** | 324 900 | 23.87467 | 75.49834 | 185 193 000 | 8.291 344 | 17.86316 | 38.48501 |
| 571 | 326 041 | 23.89561 | 75.56454 | 186 169 411 | 8.296 190 | 17.87360 | 38.50750 |
| 572 | 327 184 | 23.91652 | 75.63068 | 187 149 248 | 8.301 031 | 17.88403 | 38.52997 |
| 573 | 328 329 | 23.93742 | 75.69676 | 188 132 517 | 8.305 865 | 17.89444 | 38.55241 |
| 574 | 329 476 | 23.95830 | 75.76279 | 189 119 224 | 8.310 694 | 17.90485 | 38.57482 |
| 575 | 330 625 | 23.97916 | 75.82875 | 190 109 375 | 8.315 517 | 17.91524 | 38.59721 |
| 576 | 331 776 | 24.00000 | 75.89466 | 191 102 976 | 8.320 335 | 17.92562 | 38.61958 |
| 577 | 332 929 | 24.02082 | 75.96052 | 192 100 033 | 8.325 148 | 17.93599 | 38.64191 |
| 578 | 334 084 | 24.04163 | 76.02631 | 193 100 552 | 8.329 954 | 17.94634 | 38.66422 |
| 579 | 335 241 | 24.06242 | 76.09205 | 194 104 539 | 8.334 755 | 17.95669 | 38.68651 |
| **580** | 336 400 | 24.08319 | 76.15773 | 195 112 000 | 8.339 551 | 17.96702 | 38.70877 |
| 581 | 337 561 | 24.10394 | 76.22336 | 196 122 941 | 8.344 341 | 17.97734 | 38.73100 |
| 582 | 338 724 | 24.12468 | 76.28892 | 197 137 368 | 8.349 126 | 17.98765 | 38.75321 |
| 583 | 339 889 | 24.14539 | 76.35444 | 198 155 287 | 8.353 905 | 17.99794 | 38.77539 |
| 584 | 341 056 | 24.16609 | 76.41989 | 199 176 704 | 8.358 678 | 18.00823 | 38.79755 |
| 585 | 342 225 | 24.18677 | 76.48529 | 200 201 625 | 8.363 447 | 18.01850 | 38.81968 |
| 586 | 343 396 | 24.20744 | 76.55064 | 201 230 056 | 8.368 209 | 18.02876 | 38.84179 |
| 587 | 344 569 | 24.22808 | 76.61593 | 202 262 003 | 8.372 967 | 18.03901 | 38.86387 |
| 588 | 345 744 | 24.24871 | 76.68116 | 203 297 472 | 8.377 719 | 18.04925 | 38.88593 |
| 589 | 346 921 | 24.26932 | 76.74634 | 204 336 469 | 8.382 465 | 18.05947 | 38.90796 |
| **590** | 348 100 | 24.28992 | 76.81146 | 205 379 000 | 8.387 207 | 18.06969 | 38.92996 |
| 591 | 349 281 | 24.31049 | 76.87652 | 206 425 071 | 8.391 942 | 18.07989 | 38.95195 |
| 592 | 350 464 | 24.33105 | 76.94154 | 207 474 688 | 8.396 673 | 18.09008 | 38.97390 |
| 593 | 351 649 | 24.35159 | 77.00649 | 208 527 857 | 8.401 398 | 18.10026 | 38.99584 |
| 594 | 352 836 | 24.37212 | 77.07140 | 209 584 584 | 8.406 118 | 18.11043 | 39.01774 |
| 595 | 354 025 | 24.39262 | 77.13624 | 210 644 875 | 8.410 833 | 18.12059 | 39.03963 |
| 596 | 355 216 | 24.41311 | 77.20104 | 211 708 736 | 8.415 542 | 18.13074 | 39.06149 |
| 597 | 356 409 | 24.43358 | 77.26578 | 212 776 173 | 8.420 246 | 18.14087 | 39.08332 |
| 598 | 357 604 | 24.45404 | 77.33046 | 213 847 192 | 8.424 945 | 18.15099 | 39.10513 |
| 599 | 358 801 | 24.47448 | 77.39509 | 214 921 799 | 8.429 638 | 18.16111 | 39.12692 |
| **600** | 360 000 | 24.49490 | 77.45967 | 216 000 000 | 8.434 327 | 18.17121 | 39.14868 |

| $n$ | $n^2$ | $\sqrt{n}$ | $\sqrt{10n}$ | $n^3$ | $\sqrt[3]{n}$ | $\sqrt[3]{10n}$ | $\sqrt[3]{100n}$ |
|---|---|---|---|---|---|---|---|
| **600** | 360 000 | 24.49490 | 77.45967 | 216 000 000 | 8.434 327 | 18.17121 | 39.14868 |
| 601 | 361 201 | 24.51530 | 77.52419 | 217 081 801 | 8.439 010 | 18.18130 | 39.17041 |
| 602 | 362 404 | 24.53569 | 77.58866 | 218 167 208 | 8.443 688 | 18.19137 | 39.19213 |
| 603 | 363 609 | 24.55606 | 77.65307 | 219 256 227 | 8.448 361 | 18.20144 | 39.21382 |
| 604 | 364 816 | 24.57641 | 77.71744 | 220 348 864 | 8.453 028 | 18.21150 | 39.23548 |
| 605 | 366 025 | 24.59675 | 77.78175 | 221 445 125 | 8.457 691 | 18.22154 | 39.25712 |
| 606 | 367 236 | 24.61707 | 77.84600 | 222 545 016 | 8.462 348 | 18.23158 | 39.27874 |
| 607 | 368 449 | 24.63737 | 77.91020 | 223 648 543 | 8.467 000 | 18.24160 | 39.30033 |
| 608 | 369 664 | 24.65766 | 77.97435 | 224 755 712 | 8.471 647 | 18.25161 | 39.32190 |
| 609 | 370 881 | 24.67793 | 78.03845 | 225 866 529 | 8.476 289 | 18.26161 | 39.34345 |
| **610** | 372 100 | 24.69818 | 78.10250 | 226 981 000 | 8.480 926 | 18.27160 | 39.36497 |
| 611 | 373 321 | 24.71841 | 78.16649 | 228 099 131 | 8.485 558 | 18.28158 | 39.38647 |
| 612 | 374 544 | 24.73863 | 78.23043 | 229 220 928 | 8.490 185 | 18.29155 | 39.40795 |
| 613 | 375 769 | 24.75884 | 78.29432 | 230 346 397 | 8.494 807 | 18.30151 | 39.42940 |
| 614 | 376 996 | 24.77902 | 78.35815 | 231 475 544 | 8.499 423 | 18.31145 | 39.45083 |
| 615 | 378 225 | 24.79919 | 78.42194 | 232 608 375 | 8.504 035 | 18.32139 | 39.47223 |
| 616 | 379 456 | 24.81935 | 78.48567 | 233 744 896 | 8.508 642 | 18.33131 | 39.49362 |
| 617 | 380 689 | 24.83948 | 78.54935 | 234 885 113 | 8.513 243 | 18.34123 | 39.51498 |
| 618 | 381 924 | 24.85961 | 78.61298 | 236 029 032 | 8.517 840 | 18.35113 | 39.53631 |
| 619 | 383 161 | 24.87971 | 78.67655 | 237 176 659 | 8.522 432 | 18.36102 | 39.55763 |
| **620** | 384 400 | 24.89980 | 78.74008 | 238 328 000 | 8.527 019 | 18.37091 | 39.57892 |
| 621 | 385 641 | 24.91987 | 78.80355 | 239 483 061 | 8.531 601 | 18.38078 | 39.60018 |
| 622 | 386 884 | 24.93993 | 78.86698 | 240 641 848 | 8.536 178 | 18.39064 | 39.62143 |
| 623 | 388 129 | 24.95997 | 78.93035 | 241 804 367 | 8.540 750 | 18.40049 | 39.64265 |
| 624 | 389 376 | 24.97999 | 78.99367 | 242 970 624 | 8.545 317 | 18.41033 | 39.66385 |
| 625 | 390 625 | 25.00000 | 79.05694 | 244 140 625 | 8.549 880 | 18.42016 | 39.68503 |
| 626 | 391 876 | 25.01999 | 79.12016 | 245 314 376 | 8.554 437 | 18.42998 | 39.70618 |
| 627 | 393 129 | 25.03997 | 79.18333 | 246 491 883 | 8.558 990 | 18.43978 | 39.72731 |
| 628 | 394 384 | 25.05993 | 79.24645 | 247 673 152 | 8.563 538 | 18.44958 | 39.74842 |
| 629 | 395 641 | 25.07987 | 79.30952 | 248 858 189 | 8.568 081 | 18.45937 | 39.76951 |
| **630** | 396 900 | 25.09980 | 79.37254 | 250 047 000 | 8.572 619 | 18.46915 | 39.79057 |
| 631 | 398 161 | 25.11971 | 79.43551 | 251 239 591 | 8.577 152 | 18.47891 | 39.81161 |
| 632 | 399 424 | 25.13961 | 79.49843 | 252 435 968 | 8.581 681 | 18.48867 | 39.83263 |
| 633 | 400 689 | 25.15949 | 79.56130 | 253 636 137 | 8.586 205 | 18.49842 | 39.85363 |
| 634 | 401 956 | 25.17936 | 79.62412 | 254 840 104 | 8.590 724 | 18.50815 | 39.87461 |
| 635 | 403 225 | 25.19921 | 79.68689 | 256 047 875 | 8.595 238 | 18.51788 | 39.89556 |
| 636 | 404 496 | 25.21904 | 79.74961 | 257 259 456 | 8.599 748 | 18.52759 | 39.91649 |
| 637 | 405 769 | 25.23886 | 79.81228 | 258 474 853 | 8.604 252 | 18.53730 | 39.93740 |
| 638 | 407 044 | 25.25866 | 79.87490 | 259 694 072 | 8.608 753 | 18.54700 | 39.95829 |
| 639 | 408 321 | 25.27845 | 79.93748 | 260 917 119 | 8.613 248 | 18.55668 | 39.97916 |
| **640** | 409 600 | 25.29822 | 80.00000 | 262 144 000 | 8.617 739 | 18.56636 | 40.00000 |
| 641 | 410 881 | 25.31798 | 80.06248 | 263 374 721 | 8.622 225 | 18.57602 | 40.02082 |
| 642 | 412 164 | 25.33772 | 80.12490 | 264 609 288 | 8.626 706 | 18.58568 | 40.04162 |
| 643 | 413 449 | 25.35744 | 80.18728 | 265 847 707 | 8.631 183 | 18.59532 | 40.06240 |
| 644 | 414 736 | 25 37716 | 80.24961 | 267 089 984 | 8.635 655 | 18.60495 | 40.08316 |
| 645 | 416 025 | 25.39685 | 80.31189 | 268 336 125 | 8.640 123 | 18.61458 | 40.10390 |
| 646 | 417 316 | 25.41653 | 80.37413 | 269 586 136 | 8.644 585 | 18.62419 | 40.12461 |
| 647 | 418 609 | 25.43619 | 80.43631 | 270 840 023 | 8.649 044 | 18.63380 | 40.14530 |
| 648 | 419 904 | 25.45584 | 80.49845 | 272 097 792 | 8.653 497 | 18.64340 | 40.16598 |
| 649 | 421 201 | 25.47548 | 80.56054 | 273 359 449 | 8.657 947 | 18.65298 | 40.18663 |
| **650** | 422 500 | 25.49510 | 80.62258 | 274 625 000 | 8.662 391 | 18.66256 | 40.20726 |

| $n$ | $n^2$ | $\sqrt{n}$ | $\sqrt{10n}$ | $n^3$ | $\sqrt[3]{n}$ | $\sqrt[3]{10n}$ | $\sqrt[3]{100n}$ |
|---|---|---|---|---|---|---|---|
| **650** | 422 500 | 25 49510 | 80.62258 | 274 625 000 | 8.662 391 | 18.66256 | 40.20726 |
| 651 | 423 801 | 25.51470 | 80.68457 | 275 894 451 | 8.666 831 | 18.67212 | 40.22787 |
| 652 | 425 104 | 25.53429 | 80.74652 | 277 167 808 | 8.671 266 | 18.68168 | 40.24845 |
| 653 | 426 409 | 25.55386 | 80.80842 | 278 445 077 | 8.675 697 | 18.69122 | 40.26902 |
| 654 | 427 716 | 25.57342 | 80.87027 | 279 726 264 | 8.680 124 | 18.70076 | 40.28957 |
| 655 | 429 025 | 25 59297 | 80.93207 | 281 011 375 | 8.684 546 | 18.71029 | 40.31009 |
| 656 | 430 336 | 25.61250 | 80.99383 | 282 300 416 | 8.688 963 | 18.71980 | 40.33059 |
| 657 | 431 649 | 25.63201 | 81.05554 | 283 593 393 | 8.693 376 | 18.72931 | 40.35108 |
| 658 | 432 964 | 25.65151 | 81.11720 | 284 890 312 | 8.697 784 | 18.73881 | 40.37154 |
| 659 | 434.281 | 25.67100 | 81.17881 | 286 191 179 | 8.702 188 | 18.74830 | 40.39198 |
| **660** | 435 600 | 25.69047 | 81.24038 | 287 496 000 | 8.706 588 | 18.75777 | 40.41240 |
| 661 | 436 921 | 25.70992 | 81.30191 | 288 804 781 | 8.710 983 | 18.76724 | 40.43280 |
| 662 | 438 244 | 25.72936 | 81.36338 | 290 117 528 | 8.715 373 | 18.77670 | 40.45318 |
| 663 | 439 569 | 25.74879 | 81.42481 | 291 434 247 | 8.719 760 | 18.78615 | 40.47354 |
| 664 | 440 896 | 25.76820 | 81.48620 | 292 754 944 | 8.724 141 | 18.79559 | 40.49388 |
| 665 | 442 225 | 25.78759 | 81.54753 | 294 079 625 | 8.728 519 | 18.80502 | 40.51420 |
| 666 | 443 556 | 25.80698 | 81.60882 | 295 408 296 | 8.732 892 | 18.81444 | 40.53449 |
| 667 | 444 889 | 25.82634 | 81.67007 | 296 740 963 | 8.737 260 | 18.82386 | 40.55477 |
| 668 | 446 224 | 25.84570 | 81.73127 | 298 077 632 | 8.741 625 | 18.83326 | 40.57503 |
| 669 | 447 561 | 25.86503 | 81.79242 | 299 418 309 | 8.745 985 | 18.84265 | 40.59526 |
| **670** | 448 900 | 25.88436 | 81.85353 | 300 763 000 | 8.750 340 | 18.85204 | 40.61548 |
| 671 | 450 241 | 25.90367 | 81.91459 | 302 111 711 | 8.754 691 | 18.86141 | 40.63568 |
| 672 | 451 584 | 25.92296 | 81.97561 | 303 464 448 | 8.759 038 | 18.87078 | 40.65585 |
| 673 | 452 929 | 25.94224 | 82.03658 | 304 821 217 | 8.763 381 | 18.88013 | 40.67601 |
| 674 | 454 276 | 25.96151 | 82.09750 | 306 182 024 | 8.767 719 | 18.88948 | 40.69615 |
| 675 | 455 625 | 25.98076 | 82.15838 | 307 546 875 | 8.772 053 | 18.89882 | 40.71626 |
| 676 | 456 976 | 26.00000 | 82.21922 | 308 915 776 | 8.776 383 | 18.90814 | 40.73636 |
| 677 | 458 329 | 26.01922 | 82.28001 | 310 288 733 | 8.780 708 | 18.91746 | 40.75644 |
| 678 | 459 684 | 26.03843 | 82.34076 | 311 665 752 | 8.785 030 | 18.92677 | 40.77650 |
| 679 | 461 041 | 26.05763 | 82.40146 | 313 046 839 | 8.789 347 | 18.93607 | 40.79653 |
| **680** | 462 400 | 26.07681 | 82.46211 | 314 432 000 | 8.793 659 | 18.94536 | 40.81655 |
| 681 | 463 761 | 26.09598 | 82.52272 | 315 821 241 | 8.797 968 | 18.95465 | 40.83655 |
| 682 | 465 124 | 26.11513 | 82.58329 | 317 214 568 | 8.802 272 | 18.96392 | 40.85653 |
| 683 | 466 489 | 26.13427 | 82.64381 | 318 611 987 | 8.806 572 | 18.97318 | 40.87649 |
| 684 | 467 856 | 26.15339 | 82.70429 | 320 013 504 | 8.810 868 | 18.98244 | 40.89643 |
| 685 | 469 225 | 26.17250 | 82.76473 | 321 419 125 | 8.815 160 | 18.99169 | 40.91635 |
| 686 | 470 596 | 26.19160 | 82.82512 | 322 828 856 | 8.819 447 | 19.00092 | 40.93625 |
| 687 | 471 969 | 26.21068 | 82.88546 | 324 242 703 | 8.823 731 | 19.01015 | 40.95613 |
| 688 | 473 344 | 26.22975 | 82.94577 | 325 660 672 | 8.828 010 | 19.01937 | 40.97599 |
| 689 | 474 721 | 26.24881 | 83.00602 | 327 082 769 | 8.832 285 | 19.02858 | 40.99584 |
| **690** | 476 100 | 26.26785 | 83.06624 | 328 509 000 | 8.836 556 | 19.03778 | 41.01566 |
| 691 | 477 481 | 26.28688 | 83.12641 | 329 939 371 | 8.840 823 | 19.04698 | 41.03546 |
| 692 | 478 864 | 26.30589 | 83.18654 | 331 373 888 | 8.845 085 | 19.05616 | 41.05525 |
| 693 | 480 249 | 26.32489 | 83.24662 | 332 812 557 | 8.849 344 | 19.06533 | 41.07502 |
| 694 | 481 636 | 26.34388 | 83.30666 | 334 255 384 | 8.853 599 | 19.07450 | 41.09476 |
| 695 | 483 025 | 26.36285 | 83.36666 | 335 702 375 | 8.857 849 | 19.08366 | 41.11449 |
| 696 | 484 416 | 26.38181 | 83.42661 | 337 153 536 | 8.862 095 | 19.09281 | 41.13420 |
| 697 | 485 809 | 26.40076 | 83.48653 | 338 608 873 | 8.866 338 | 19.10195 | 41.15389 |
| 698 | 487 204 | 26.41969 | 83.54639 | 340 068 392 | 8.870 576 | 19.11108 | 41.17357 |
| 699 | 488 601 | 26.43861 | 83.60622 | 341 532 099 | 8.874 810 | 19.12020 | 41.19322 |
| **700** | 490 000 | 26.45751 | 83.66600 | 343 000 000 | 8.879 040 | 19.12931 | 41.21285 |

| n | $n^2$ | $\sqrt{n}$ | $\sqrt{10n}$ | $n^3$ | $\sqrt[3]{n}$ | $\sqrt[3]{10n}$ | $\sqrt[3]{100n}$ |
|---|---|---|---|---|---|---|---|
| **700** | 490 000 | 26.45751 | 83.66600 | 343 000 000 | 8.879 040 | 19.12931 | 41.21285 |
| 701 | 491 401 | 26.47640 | 83.72574 | 344 472 101 | 8.883 266 | 19.13842 | 41.23247 |
| 702 | 492 804 | 26.49528 | 83.78544 | 345 948 408 | 8.887 488 | 19.14751 | 41.25207 |
| 703 | 494 209 | 26.51415 | 83.84510 | 347 428 927 | 8.891 706 | 19.15660 | 41.27164 |
| 704 | 495 616 | 26.53300 | 83.90471 | 348 913 664 | 8.895 920 | 19.16568 | 41.29120 |
| 705 | 497 025 | 26.55184 | 83.96428 | 350 402 625 | 8.900 130 | 19.17475 | 41.31075 |
| 706 | 498 436 | 26.57066 | 84.02381 | 351 895 816 | 8.904 337 | 19.18381 | 41.33027 |
| 707 | 499 849 | 26.58947 | 84.08329 | 353 393 243 | 8.908 539 | 19.19286 | 41.34977 |
| 708 | 501 264 | 26.60827 | 84.14274 | 354 894 912 | 8.912 737 | 19.20191 | 41.36926 |
| 709 | 502 681 | 26.62705 | 84.20214 | 356 400 829 | 8.916 931 | 19.21095 | 41.38873 |
| **710** | 504 100 | 26.64583 | 84.26150 | 357 911 000 | 8.921 121 | 19.21997 | 41.40818 |
| 711 | 505 521 | 26.66458 | 84.32082 | 359 425 431 | 8.925 308 | 19.22899 | 41.42761 |
| 712 | 506 944 | 26.68333 | 84.38009 | 360 944 128 | 8.929 490 | 19.23800 | 41.44702 |
| 713 | 508 369 | 26.70206 | 84.43933 | 362 467 097 | 8.933 669 | 19.24701 | 41.46642 |
| 714 | 509 796 | 26.72078 | 84.49852 | 363 994 344 | 8.937 843 | 19.25600 | 41.48579 |
| 715 | 511 225 | 26.73948 | 84.55767 | 365 525 875 | 8.942 014 | 19.26499 | 41.50515 |
| 716 | 512 656 | 26.75818 | 84.61678 | 367 061 696 | 8.946 181 | 19.27396 | 41.52449 |
| 717 | 514 089 | 26.77686 | 84.67585 | 368 601 813 | 8.950 344 | 19.28293 | 41.54382 |
| 718 | 515 524 | 26.79552 | 84.73488 | 370 146 232 | 8.954 503 | 19.29189 | 41.56312 |
| 719 | 516 961 | 26.81418 | 84.79387 | 371 694 959 | 8.958 658 | 19.30084 | 41.58241 |
| **720** | 518 400 | 26.83282 | 84.85281 | 373 248 000 | 8.962 809 | 19.30979 | 41.60168 |
| 721 | 519 841 | 26.85144 | 84.91172 | 374 805 361 | 8.966 957 | 19.31872 | 41.62093 |
| 722 | 521 284 | 26.87006 | 84.97058 | 376 367 048 | 8.971 101 | 19.32765 | 41.64016 |
| 723 | 522 729 | 26.88866 | 85.02941 | 377 933 067 | 8.975 241 | 19.33657 | 41.65938 |
| 724 | 524 176 | 26.90725 | 85.08819 | 379 503 424 | 8.979 377 | 19.34548 | 41.67857 |
| 725 | 525 625 | 26.92582 | 85.14693 | 381 078 125 | 8.983 509 | 19.35438 | 41.69775 |
| 726 | 527 076 | 26.94439 | 85.20563 | 382 657 176 | 8.987 637 | 19.36328 | 41.71692 |
| 727 | 528 529 | 26.96294 | 85.26429 | 384 240 583 | 8.991 762 | 19.37216 | 41.73606 |
| 728 | 529 984 | 26.98148 | 85.32292 | 385 828 352 | 8.995 883 | 19.38104 | 41.75519 |
| 729 | 531 441 | 27.00000 | 85.38150 | 387 420 489 | 9.000 000 | 19.38991 | 41.77430 |
| **730** | 532 900 | 27.01851 | 85.44004 | 389 017 000 | 9.004 113 | 19.39877 | 41.79339 |
| 731 | 534 361 | 27.03701 | 85.49854 | 390 617 891 | 9.008 223 | 19.40763 | 41.81247 |
| 732 | 535 824 | 27.05550 | 85.55700 | 392 223 168 | 9.012 329 | 19.41647 | 41.83152 |
| 733 | 537 289 | 27.07397 | 85.61542 | 393 832 837 | 9.016 431 | 19.42531 | 41.85056 |
| 734 | 538 756 | 27.09243 | 85.67380 | 395 446 904 | 9.020 529 | 19.43414 | 41.86959 |
| 735 | 540 225 | 27.11088 | 85.73214 | 397 065 375 | 9.024 624 | 19.44296 | 41.88859 |
| 736 | 541 696 | 27.12932 | 85.79044 | 398 688 256 | 9.028 715 | 19.45178 | 41.90758 |
| 737 | 543 169 | 27.14774 | 85.84870 | 400 315 553 | 9.032 802 | 19.46058 | 41.92655 |
| 738 | 544 644 | 27.16616 | 85.90693 | 401 947 272 | 9.036 886 | 19.46938 | 41.94551 |
| 739 | 546 121 | 27.18455 | 85.96511 | 403 583 419 | 9.040 966 | 19.47817 | 41.96444 |
| 740 | 547 600 | 27.20294 | 86.02325 | 405 224 000 | 9.045 042 | 19.48695 | 41.98336 |
| 741 | 549 081 | 27.22132 | 86.08136 | 406 869 021 | 9.049 114 | 19.49573 | 42.00227 |
| 742 | 550 564 | 27.23968 | 86.13942 | 408 518 488 | 9.053 183 | 19.50449 | 42.02115 |
| 743 | 552 049 | 27.25803 | 86.19745 | 410 172 407 | 9.057 248 | 19.51325 | 42.04002 |
| 744 | 553 536 | 27.27636 | 86.25543 | 411 830 784 | 9.061 310 | 19.52200 | 42.05887 |
| 745 | 555 025 | 27.29469 | 86.31338 | 413 493 625 | 9.065 368 | 19.53074 | 42.07771 |
| 746 | 556 516 | 27.31300 | 86.37129 | 415 160 936 | 9.069 422 | 19.53948 | 42.09653 |
| 747 | 558 009 | 27.33130 | 86.42916 | 416 832 723 | 9.073 473 | 19.54820 | 42.11533 |
| 748 | 559 504 | 27.34959 | 86.48699 | 418 508 992 | 9.077 520 | 19.55692 | 42.13411 |
| 749 | 561 001 | 27.36786 | 86.54479 | 420 189 749 | 9.081 563 | 19.56563 | 42.15288 |
| **750** | 562 500 | 27.38613 | 86.60254 | 421 875 000 | 9.085 603 | 19.57434 | 42.17163 |

| $n$ | $n^2$ | $\sqrt{n}$ | $\sqrt{10n}$ | $n^3$ | $\sqrt[3]{n}$ | $\sqrt[3]{10n}$ | $\sqrt[3]{100n}$ |
|---|---|---|---|---|---|---|---|
| **750** | 562 500 | 27.38613 | 86.60254 | 421 875 000 | 9.085 603 | 19.57434 | 42.17163 |
| 751 | 564 001 | 27.40438 | 86.66026 | 423 564 751 | 9.089 639 | 19.58303 | 42.19037 |
| 752 | 565 504 | 27.42262 | 86.71793 | 425 259 008 | 9.093 672 | 19.59172 | 42.20909 |
| 753 | 567 009 | 27.44085 | 86.77557 | 426 957 777 | 9.097 701 | 19.60040 | 42.22779 |
| 754 | 568 516 | 27.45906 | 86.83317 | 428 661 064 | 9.101 727 | 19.60908 | 42.24647 |
| 755 | 570 025 | 27.47726 | 86.89074 | 430 368 875 | 9.105 748 | 19.61774 | 42.26514 |
| 756 | 571 536 | 27.49545 | 86.94826 | 432 081 216 | 9.109 767 | 19.62640 | 42.28379 |
| 757 | 573 049 | 27.51363 | 87.00575 | 433 798 093 | 9.113 782 | 19.63505 | 42.30243 |
| 758 | 574 564 | 27.53180 | 87.06320 | 435 519 512 | 9.117 793 | 19.64369 | 42.32105 |
| 759 | 576 081 | 27.54995 | 87.12061 | 437 245 479 | 9.121 801 | 19.65232 | 42.33965 |
| **760** | 577 600 | 27.56810 | 87.17798 | 438 976 000 | 9.125 805 | 19.66095 | 42.35824 |
| 761 | 579 121 | 27.58623 | 87.23531 | 440 711 081 | 9.129 806 | 19.66957 | 42.37681 |
| 762 | 580 644 | 27.60435 | 87.29261 | 442 450 728 | 9.133 803 | 19.67818 | 42.39536 |
| 763 | 582 169 | 27.62245 | 87.34987 | 444 194 947 | 9.137 797 | 19.68679 | 42.41390 |
| 764 | 583 696 | 27.64055 | 87.40709 | 445 943 744 | 9.141 787 | 19.69538 | 42.43242 |
| 765 | 585 225 | 27.65863 | 87.46428 | 447 697 125 | 9.145 774 | 19.70397 | 42.45092 |
| 766 | 586 756 | 27.67671 | 87.52143 | 449 455 096 | 9.149 758 | 19.71256 | 42.46941 |
| 767 | 588 289 | 27.69476 | 87.57854 | 451 217 663 | 9.153 738 | 19.72113 | 42.48789 |
| 768 | 589 824 | 27.71281 | 87.63561 | 452 984 832 | 9.157 714 | 19.72970 | 42.50634 |
| 769 | 591 361 | 27.73085 | 87.69265 | 454 756 609 | 9.161 687 | 19.73826 | 42.52478 |
| **770** | 592 900 | 27.74887 | 87.74964 | 456 533 000 | 9.165 656 | 19.74681 | 42.54321 |
| 771 | 594 441 | 27.76689 | 87.80661 | 458 314 011 | 9.169 623 | 19.75535 | 42.56162 |
| 772 | 595 984 | 27.78489 | 87.86353 | 460 099 648 | 9.173 585 | 19.76389 | 42.58001 |
| 773 | 597 529 | 27.80288 | 87.92042 | 461 889 917 | 9.177 544 | 19.77242 | 42.59839 |
| 774 | 599 076 | 27.82086 | 87.97727 | 463 684 824 | 9.181 500 | 19.78094 | 42.61675 |
| 775 | 600 625 | 27.83882 | 88.03408 | 465 484 375 | 9.185 453 | 19.78946 | 42.63509 |
| 776 | 602 176 | 27.85678 | 88.09086 | 467 288 576 | 9.189 402 | 19.79797 | 42.65342 |
| 777 | 603 729 | 27.87472 | 88.14760 | 469 097 433 | 9.193 347 | 19.80647 | 42.67174 |
| 778 | 605 284 | 27.89265 | 88.20431 | 470 910 952 | 9.197 290 | 19.81496 | 42.69004 |
| 779 | 606 841 | 27.91057 | 88.26098 | 472 729 139 | 9.201 229 | 19.82345 | 42.70832 |
| **780** | 608 400 | 27.92848 | 88.31761 | 474 552 000 | 9.205 164 | 19.83192 | 42.72659 |
| 781 | 609 961 | 27.94638 | 88.37420 | 476 379 541 | 9.209 096 | 19.84040 | 42.74484 |
| 782 | 611 524 | 27.96426 | 88.43076 | 478 211 768 | 9.213 025 | 19.84886 | 42.76307 |
| 783 | 613 089 | 27.98214 | 88.48729 | 480 048 687 | 9.216 950 | 19.85732 | 42.78129 |
| 784 | 614 656 | 28.00000 | 88.54377 | 481 890 304 | 9.220 873 | 19.86577 | 42.79950 |
| 785 | 616 225 | 28.01785 | 88.60023 | 483 736 625 | 9.224 791 | 19.87421 | 42.81769 |
| 786 | 617 796 | 28.03569 | 88.65664 | 485 587 656 | 9.228 707 | 19.88265 | 42.83586 |
| 787 | 619 369 | 28.05352 | 88.71302 | 487 443 403 | 9.232 619 | 19.89107 | 42.85402 |
| 788 | 620 944 | 28.07134 | 88.76936 | 489 303 872 | 9.236 528 | 19.89950 | 42.87216 |
| 789 | 622 521 | 28.08914 | 88.82567 | 491 169 069 | 9.240 433 | 19.90791 | 42.89029 |
| **790** | 624 100 | 28.10694 | 88.88194 | 493 039 000 | 9.244 335 | 19.91632 | 42.90840 |
| 791 | 625 681 | 28.12472 | 88.93818 | 494 913 671 | 9.248 234 | 19.92472 | 42.92650 |
| 792 | 627 264 | 28.14249 | 88.99438 | 496 793 088 | 9.252 130 | 19.93311 | 42.94458 |
| 793 | 628 849 | 28.16026 | 89.05055 | 498 677 257 | 9.256 022 | 19.94150 | 42.96265 |
| 794 | 630 436 | 28.17801 | 89.10668 | 500 566 184 | 9.259 911 | 19.94987 | 42.98070 |
| 795 | 632 025 | 28.19574 | 89.16277 | 502 459 875 | 9.263 797 | 19.95825 | 42.99874 |
| 796 | 633 616 | 28.21347 | 89.21883 | 504 358 336 | 9.267 680 | 19.96661 | 43.01676 |
| 797 | 635 209 | 28.23119 | 89.27486 | 506 261 573 | 9.271 559 | 19.97497 | 43.03477 |
| 798 | 636 804 | 28.24889 | 89.33085 | 508 169 592 | 9.275 435 | 19.98332 | 43.05276 |
| 799 | 638 401 | 28.26659 | 89.38680 | 510 082 399 | 9.279 308 | 19.99166 | 43.07073 |
| **800** | 640 000 | 28.28427 | 89.44272 | 512 000 000 | 9.283 178 | 20.00000 | 43.08869 |

| $n$ | $n^2$ | $\sqrt{n}$ | $\sqrt{10n}$ | $n^3$ | $\sqrt[3]{n}$ | $\sqrt[3]{10n}$ | $\sqrt[3]{100n}$ |
|---|---|---|---|---|---|---|---|
| **800** | 640 000 | 28.28427 | 89.44272 | 512 000 000 | 9.283 178 | 20.00000 | 43.08869 |
| 801 | 641 601 | 28.30194 | 89.49860 | 513 922 401 | 9.287 044 | 20.00833 | 43.10664 |
| 802 | 643 204 | 28.31960 | 89.55445 | 515 849 608 | 9.290 907 | 20.01665 | 43.12457 |
| 803 | 644 809 | 28.33725 | 89.61027 | 517 781 627 | 9.294 767 | 20.02497 | 43.14249 |
| 804 | 646 416 | 28.35489 | 89.66605 | 519 718 464 | 9.298 624 | 20.03328 | 43.16039 |
| 805 | 648 025 | 28.37252 | 89.72179 | 521 660 125 | 9.302 477 | 20.04158 | 43.17828 |
| 806 | 649 636 | 28.39014 | 89.77750 | 523 606 616 | 9.306 328 | 20.04988 | 43.19615 |
| 807 | 651 249 | 28.40775 | 89.83318 | 525 557 943 | 9.310 175 | 20.05816 | 43.21400 |
| 808 | 652 864 | 28.42534 | 89.88882 | 527 514 112 | 9.314 019 | 20.06645 | 43.23185 |
| 809 | 654 481 | 28.44293 | 89.94443 | 529 475 129 | 9.317 860 | 20.07472 | 43.24967 |
| **810** | 656 100 | 28.46050 | 90.00000 | 531 441 000 | 9.321 698 | 20.08299 | 43.26749 |
| 811 | 657 721 | 28.47806 | 90.05554 | 533 411 731 | 9.325 532 | 20.09125 | 43.28529 |
| 812 | 659 344 | 28.49561 | 90.11104 | 535 387 328 | 9.329 363 | 20.09950 | 43.30307 |
| 813 | 660 969 | 28.51315 | 90.16651 | 537 367 797 | 9.333 192 | 20.10775 | 43.32084 |
| 814 | 662 596 | 28.53069 | 90.22195 | 539 353 144 | 9.337 017 | 20.11599 | 43.33859 |
| 815 | 664 225 | 28.54820 | 90.27735 | 541 343 375 | 9.340 839 | 20.12423 | 43.35633 |
| 816 | 665 856 | 28.56571 | 90.33272 | 543 338 496 | 9.344 657 | 20.13245 | 43.37406 |
| 817 | 667 489 | 28.58321 | 90.38805 | 545 338 513 | 9.348 473 | 20.14067 | 43.39177 |
| 818 | 669 124 | 28.60070 | 90.44335 | 547 343 432 | 9.352 286 | 20.14889 | 43.40947 |
| 819 | 670 761 | 28.61818 | 90.49862 | 549 353 259 | 9.356 095 | 20.15710 | 43.42715 |
| **820** | 672 400 | 28.63564 | 90.55385 | 551 368 000 | 9.359 902 | 20.16530 | 43.44481 |
| 821 | 674 041 | 28.65310 | 90.60905 | 553 387 661 | 9.363 705 | 20.17349 | 43.46247 |
| 822 | 675 684 | 28.67054 | 90.66422 | 555 412 248 | 9.367 505 | 20.18168 | 43.48011 |
| 823 | 677 329 | 28.68798 | 90.71935 | 557 441 767 | 9.371 302 | 20.18986 | 43.49773 |
| 824 | 678 976 | 28.70540 | 90.77445 | 559 476 224 | 9.375 096 | 20.19803 | 43.51534 |
| 825 | 680 625 | 28.72281 | 90.82951 | 561 515 625 | 9.378 887 | 20.20620 | 43.53294 |
| 826 | 682 276 | 28.74022 | 90.88454 | 563 559 976 | 9.382 675 | 20.21436 | 43.55052 |
| 827 | 683 929 | 28.75761 | 90.93954 | 565 609 283 | 9.386 460 | 20.22252 | 43.56809 |
| 828 | 685 584 | 28.77499 | 90.99451 | 567 663 552 | 9.390 242 | 20.23066 | 43.58564 |
| 829 | 687 241 | 28.79236 | 91.04944 | 569 722 789 | 9.394 021 | 20.23880 | 43.60318 |
| **830** | 688 900 | 28.80972 | 91.10434 | 571 787 000 | 9.397 796 | 20.24694 | 43.62071 |
| 831 | 690 561 | 28.82707 | 91.15920 | 573 856 191 | 9.401 569 | 20.25507 | 43.63822 |
| 832 | 692 224 | 28.84441 | 91.21403 | 575 930 368 | 9.405 339 | 20.26319 | 43.65572 |
| 833 | 693 889 | 28.86174 | 91.26883 | 578 009 537 | 9.409 105 | 20.27130 | 43.67320 |
| 834 | 695 556 | 28.87906 | 91.32360 | 580 093 704 | 9.412 869 | 20.27941 | 43.69067 |
| 835 | 697 225 | 28.89637 | 91.37833 | 582 182 875 | 9.416 630 | 20.28751 | 43.70812 |
| 836 | 698 896 | 28.91366 | 91.43304 | 584 277 056 | 9.420 387 | 20.29561 | 43.72556 |
| 837 | 700 569 | 28.93095 | 91.48770 | 586 376 253 | 9.424 142 | 20.30370 | 43.74299 |
| 838 | 702 244 | 28.94823 | 91.54234 | 588 480 472 | 9.427 894 | 20.31178 | 43.76041 |
| 839 | 703 921 | 28 96550 | 91.59694 | 590 589 719 | 9.431 642 | 20.31986 | 43.77781 |
| **840** | 705 600 | 28.98275 | 91.65151 | 592 704 000 | 9.435 388 | 20.32793 | 43.79519 |
| 841 | 707 281 | 29.00000 | 91.70605 | 594 823 321 | 9.439 131 | 29.33599 | 43.81256 |
| 842 | 708 964 | 29.01724 | 91.76056 | 596 947 688 | 9.442 870 | 20.34405 | 43.82992 |
| 843 | 710 649 | 29.03446 | 91.81503 | 599 077 107 | 9.446 607 | 20.35210 | 43.84727 |
| 844 | 712 336 | 29.05168 | 91.86947 | 601 211 584 | 9.450 341 | 20.36014 | 43.86460 |
| 845 | 714 025 | 29.06888 | 91.92388 | 603 351 125 | 9.454 072 | 20.36818 | 43.88191 |
| 846 | 715 716 | 29.08608 | 91.97826 | 605 495 736 | 9.457 800 | 20.37621 | 43.89922 |
| 847 | 717 409 | 29.10326 | 92.03260 | 607 645 423 | 9.461 525 | 20.38424 | 43.91651 |
| 848 | 719 104 | 29.12044 | 92.08692 | 609 800 192 | 9.465 247 | 20.39226 | 43.93378 |
| 849 | 720 801 | 29.13760 | 92.14120 | 611 960 049 | 9.468 966 | 20.40027 | 43.95105 |
| **850** | 722 500 | 29.15476 | 92.19544 | 614 125 000 | 9.472 682 | 20.40828 | 43.96830 |

| $n$ | $n^2$ | $\sqrt{n}$ | $\sqrt{10n}$ | $n^3$ | $\sqrt[3]{n}$ | $\sqrt[3]{10n}$ | $\sqrt[3]{100n}$ |
|---|---|---|---|---|---|---|---|
| **850** | 722 500 | 29.15476 | 92.19544 | 614 125 000 | 9.472 682 | 20.40828 | 43.96830 |
| 851 | 724 201 | 29.17190 | 92.24966 | 616 295 051 | 9.476 396 | 20.41628 | 43.98553 |
| 852 | 725 904 | 29.18904 | 92.30385 | 618 470 208 | 9.480 106 | 20.42427 | 44.00275 |
| 853 | 727 609 | 29.20616 | 92.35800 | 620 650 477 | 9.483 814 | 20.43226 | 44.01996 |
| 854 | 729 316 | 29.22328 | 92.41212 | 622 835 864 | 9.487 518 | 20.44024 | 44.03716 |
| 855 | 731 025 | 29.24038 | 92.46621 | 625 026 375 | 9.491 220 | 20.44821 | 44.05434 |
| 856 | 732 736 | 29.25748 | 92.52027 | 627 222 016 | 9.494 919 | 20.45618 | 44.07151 |
| 857 | 734 449 | 29.27456 | 92.57429 | 629 422 793 | 9.498 615 | 20.46415 | 44.08866 |
| 858 | 736 164 | 29.29164 | 92.62829 | 631 628 712 | 9.502 308 | 20.47210 | 44.10581 |
| 859 | 737 881 | 29.30870 | 92.68225 | 633 839 779 | 9.505 998 | 20.48005 | 44.12293 |
| **860** | 739 600 | 29.32576 | 92.73618 | 636 056 000 | 9.509 685 | 20.48800 | 44.14005 |
| 861 | 741 321 | 29.34280 | 92.79009 | 638 277 381 | 9.513 370 | 20.49593 | 44.15715 |
| 862 | 743 044 | 29.35984 | 92.84396 | 640 503 928 | 9.517 052 | 20.50387 | 44.17424 |
| 863 | 744 769 | 29.37686 | 92.89779 | 642 735 647 | 9.520 730 | 20.51179 | 44.19132 |
| 864 | 746 496 | 29.39388 | 92.95160 | 644 972 544 | 9.524 406 | 20.51971 | 44.20838 |
| 865 | 748 225 | 29.41088 | 93.00538 | 647 214 625 | 9.528 079 | 20.52762 | 44.22543 |
| 866 | 749 956 | 29.42788 | 93.05912 | 649 461 896 | 9.531 750 | 20.53553 | 44.24246 |
| 867 | 751 689 | 29.44486 | 93.11283 | 651 714 363 | 9.535 417 | 20.54343 | 44.25949 |
| 868 | 753 424 | 29.46184 | 93.16652 | 653 972 032 | 9.539 082 | 20.55133 | 44.27650 |
| 869 | 755 161 | 29.47881 | 93.22017 | 656 234 909 | 9.542 744 | 20.55922 | 44.29349 |
| **870** | 756 900 | 29.49576 | 93.27379 | 658 503 000 | 9.546 403 | 20.56710 | 44.31048 |
| 871 | 758 641 | 29.51271 | 93.32738 | 660 776 311 | 9.550 059 | 20.57498 | 44.32745 |
| 872 | 760 384 | 29.52965 | 93.38094 | 663 054 848 | 9.553 712 | 20.58285 | 44.34440 |
| 873 | 762 129 | 29.54657 | 93.43447 | 665 338 617 | 9.557 363 | 20.59071 | 44.36135 |
| 874 | 763 876 | 29.56349 | 93.48797 | 667 627 624 | 9.561 011 | 20.59857 | 44.37828 |
| 875 | 765 625 | 29.58040 | 93.54143 | 669 921 875 | 9.564 656 | 20.60643 | 44.39520 |
| 876 | 767 376 | 29.59730 | 93.59487 | 672 221 376 | 9.568 298 | 20.61427 | 44.41211 |
| 877 | 769 129 | 29.61419 | 93.64828 | 674 526 133 | 9.571 938 | 20.62211 | 44.42900 |
| 878 | 770 884 | 29.63106 | 93.70165 | 676 836 152 | 9.575 574 | 20.62995 | 44.44588 |
| 879 | 772 641 | 29.64793 | 93.75500 | 679 151 439 | 9.579 208 | 20.63778 | 44.46275 |
| **880** | 774 400 | 29.66479 | 93.80832 | 681 472 000 | 9.582 840 | 20.64560 | 44.47960 |
| 881 | 776 161 | 29.68164 | 93.86160 | 683 797 841 | 9.586 468 | 20.65342 | 44.49644 |
| 882 | 777 924 | 29.69848 | 93.91486 | 686 128 968 | 9.590 094 | 20.66123 | 44.51327 |
| 883 | 779 689 | 29.71532 | 93.96808 | 688 465 387 | 9.593 717 | 20.66904 | 44.53009 |
| 884 | 781 456 | 29.73214 | 94.02127 | 690 807 104 | 9.597 337 | 20.67684 | 44.54689 |
| 885 | 783 225 | 29.74895 | 94.07444 | 693 154 125 | 9.600 955 | 20.68463 | 44.56368 |
| 886 | 784 996 | 29.76575 | 94.12757 | 695 506 456 | 9.604 570 | 20.69242 | 44.58046 |
| 887 | 786 769 | 29.78255 | 94.18068 | 697 864 103 | 9.608 182 | 20.70020 | 44.59723 |
| 888 | 788 544 | 29.79933 | 94.23375 | 700 227 072 | 9.611 791 | 20.70798 | 44.61398 |
| 889 | 790 321 | 29.81610 | 94.28680 | 702 595 369 | 9.615 398 | 20.71575 | 44.63072 |
| **890** | 792 100 | 29.83287 | 94.33981 | 704 969 000 | 9.619 002 | 20.72351 | 44.64745 |
| 891 | 793 881 | 29.84962 | 94.39280 | 707 347 971 | 9.622 603 | 20.73127 | 44.66417 |
| 892 | 795 664 | 29.86637 | 94.44575 | 709 732 288 | 9.626 202 | 20.73902 | 44.68087 |
| 893 | 797 449 | 29.88311 | 94.49868 | 712 121 957 | 9.629 797 | 20.74677 | 44.69756 |
| 894 | 799 236 | 29.89983 | 94.55157 | 714 516 984 | 9.633 391 | 20.75451 | 44.71424 |
| 895 | 801 025 | 29.91655 | 94.60444 | 716 917 375 | 9.636 981 | 20.76225 | 44.73090 |
| 896 | 802 816 | 29.93326 | 94.65728 | 719 323 136 | 9.640 569 | 20.76998 | 44.74756 |
| 897 | 804 609 | 29.94996 | 94.71008 | 721 734 273 | 9.644 154 | 20.77770 | 44.76420 |
| 898 | 806 404 | 29.96665 | 94.76286 | 724 150 792 | 9.647 737 | 20.78542 | 44.78083 |
| 899 | 808 201 | 29.98333 | 94.81561 | 726 572 699 | 9.651 317 | 20.79313 | 44.79744 |
| **900** | 810 000 | 30.00000 | 94.86833 | 729 000 000 | 9.654 894 | 20.80084 | 44.81405 |

| $n$ | $n^2$ | $\sqrt{n}$ | $\sqrt{10n}$ | $n^3$ | $\sqrt[3]{n}$ | $\sqrt[3]{10n}$ | $\sqrt[3]{100n}$ |
|---|---|---|---|---|---|---|---|
| **900** | 810 000 | 30.00000 | 94.86833 | 729 000 000 | 9.654 894 | 20.80084 | 44.81405 |
| 901 | 811 801 | 30.01666 | 94.92102 | 731 432 701 | 9.658 468 | 20.80854 | 44.83064 |
| 902 | 813 604 | 30.03331 | 94.97368 | 733 870 808 | 9.662 040 | 20.81623 | 44.84722 |
| 903 | 815 409 | 30.04996 | 95.02631 | 736 314 327 | 9.665 610 | 20.82392 | 44.86379 |
| 904 | 817 216 | 30.06659 | 95.07891 | 738 763 264 | 9.669 176 | 20.83161 | 44.88034 |
| 905 | 819 025 | 30.08322 | 95.13149 | 741 217 625 | 9.672 740 | 20.83929 | 44.89688 |
| 906 | 820 836 | 30.09983 | 95.18403 | 743 677 416 | 9.676 302 | 20.84696 | 44.91341 |
| 907 | 822 649 | 30.11644 | 95.23655 | 746 142 643 | 9.679 860 | 20.85463 | 44.92993 |
| 908 | 824 464 | 30.13304 | 95.28903 | 748 613 312 | 9.683 417 | 20.86229 | 44.94644 |
| 909 | 826 281 | 30.14963 | 95.34149 | 751 089 429 | 9.686 970 | 20.86994 | 44.96293 |
| **910** | 828 100 | 30.16621 | 95.39392 | 753 571 000 | 9.690 521 | 20.87759 | 44.97941 |
| 911 | 829 921 | 30.18278 | 95.44632 | 756 058 031 | 9.694 069 | 20.88524 | 44.99588 |
| 912 | 831 744 | 30.19934 | 95.49869 | 758 550 528 | 9.697 615 | 20.89288 | 45.01234 |
| 913 | 833 569 | 30.21589 | 95.55103 | 761 048 497 | 9.701 158 | 20.90051 | 45.02879 |
| 914 | 835 396 | 30.23243 | 95.60335 | 763 551 944 | 9.704 699 | 20.90814 | 45.04522 |
| 915 | 837 225 | 30.24897 | 95.65563 | 766 060 875 | 9.708 237 | 20.91576 | 45.06164 |
| 916 | 839 056 | 30.26549 | 95.70789 | 768 575 296 | 9.711 772 | 20.92338 | 45.07805 |
| 917 | 840 889 | 30.28201 | 95.76012 | 771 095 213 | 9.715 305 | 20.93099 | 45.09445 |
| 918 | 842 724 | 30.29851 | 95.81232 | 773 620 632 | 9.718 835 | 20.93860 | 45.11084 |
| 919 | 844 561 | 30.31501 | 95.86449 | 776 151 559 | 9.722 363 | 20.94620 | 45.12721 |
| **920** | 846 400 | 30.33150 | 95.91663 | 778 688 000 | 9.725 888 | 20.95379 | 45.14357 |
| 921 | 848 241 | 30.34798 | 95.96874 | 781 229 961 | 9.729 411 | 20.96138 | 45.15992 |
| 922 | 850 084 | 30.36445 | 96.02083 | 783 777 448 | 9.732 931 | 20.96896 | 45.17626 |
| 923 | 851 929 | 30.38092 | 96.07289 | 786 330 467 | 9.736 448 | 20.97654 | 45.19259 |
| 924 | 853 776 | 30.39737 | 96.12492 | 788 889 024 | 9.739 963 | 20.98411 | 45.20891 |
| 925 | 855 625 | 30.41381 | 96.17692 | 791 453 125 | 9.743 476 | 20.99168 | 45.22521 |
| 926 | 857 476 | 30.43025 | 96.22889 | 794 022 776 | 9.746 986 | 20.99924 | 45.24150 |
| 927 | 859 329 | 30.44667 | 96.28084 | 796 597 983 | 9.750 493 | 21.00680 | 45.25778 |
| 928 | 861 184 | 30.46309 | 96.33276 | 799 178 752 | 9.753 998 | 21.01435 | 45.27405 |
| 929 | 863 041 | 30.47950 | 96.38465 | 801 765 089 | 9.757 500 | 21.02190 | 45.29030 |
| **930** | 864 900 | 30.49590 | 96.43651 | 804 357 000 | 9.761 000 | 21.02944 | 45.30655 |
| 931 | 866 761 | 30.51229 | 96.48834 | 806 954 491 | 9.764 497 | 21.03697 | 45.32278 |
| 932 | 868 624 | 30.52868 | 96.54015 | 809 557 568 | 9.767 992 | 21.04450 | 45.33900 |
| 933 | 870 489 | 30.54505 | 96.59193 | 812 166 237 | 9.771 485 | 21.05203 | 45.35521 |
| 934 | 872 356 | 30.56141 | 96.64368 | 814 780 504 | 9.774 974 | 21.05954 | 45.37141 |
| 935 | 874 225 | 30.57777 | 96.69540 | 817 400 375 | 9.778 462 | 21.06706 | 45.38760 |
| 936 | 876 096 | 30.59412 | 96.74709 | 820 025 856 | 9.781 946 | 21.07456 | 45.40377 |
| 937 | 877 969 | 30.61046 | 96.79876 | 822 656 953 | 9.785 429 | 21.08207 | 45.41994 |
| 938 | 879 844 | 30.62679 | 96.85040 | 825 293 672 | 9.788 909 | 21.08956 | 45.43609 |
| 939 | 881 721 | 30.64311 | 96.90201 | 827 936 019 | 9.792 386 | 21.09706 | 45.45223 |
| **940** | 883 600 | 30.65942 | 96.95360 | 830 584 000 | 9.795 861 | 21.10454 | 45.46836 |
| 941 | 885 481 | 30.67572 | 97.00515 | 833 237 621 | 9.799 334 | 21.11202 | 45.48448 |
| 942 | 887 364 | 30.69202 | 97.05668 | 835 896 888 | 9.802 804 | 21.11950 | 45.50058 |
| 943 | 889 249 | 30.70831 | 97.10819 | 838 561 807 | 9.806 271 | 21.12697 | 45.51668 |
| 944 | 891 136 | 30.72458 | 97.15966 | 841 232 384 | 9.809 736 | 21.13444 | 45.53276 |
| 945 | 893 025 | 30.74085 | 97.21111 | 843 908 625 | 9.813 199 | 21.14190 | 45.54883 |
| 946 | 894 916 | 30.75711 | 97.26253 | 846 590 536 | 9.816 659 | 21.14935 | 45.56490 |
| 947 | 896 809 | 30.77337 | 97.31393 | 849 278 123 | 9.820 117 | 21.15680 | 45.58095 |
| 948 | 898.704 | 30.78961 | 97.36529 | 851 971 392 | 9.823 572 | 21.16424 | 45.59698 |
| 949 | 900 601 | 30.80584 | 97.41663 | 854 670 349 | 9.827 025 | 21.17168 | 45.61301 |
| **950** | 902 500 | 30.82207 | 97.46794 | 857 375 000 | 9.830 476 | 21.17912 | 45.62903 |

| $n$ | $n^2$ | $\sqrt{n}$ | $\sqrt{10n}$ | $n^3$ | $\sqrt[3]{n}$ | $\sqrt[3]{10n}$ | $\sqrt[3]{100n}$ |
|---|---|---|---|---|---|---|---|
| **950** | 902 500 | 30.82207 | 97.46794 | 857 375 000 | 9.830 476 | 21.17912 | 45.62903 |
| 951 | 904 401 | 30.83829 | 97.51923 | 860 085 351 | 9.833 924 | 21.18655 | 45.64503 |
| 952 | 906 304 | 30.85450 | 97.57049 | 862 801 408 | 9.837 369 | 21.19397 | 45.66102 |
| 953 | 908 209 | 30.87070 | 97.62172 | 865 523 177 | 9.840 813 | 21.20139 | 45.67701 |
| 954 | 910 116 | 30.88689 | 97.67292 | 868 250 664 | 9.844 254 | 21.20880 | 45.69298 |
| 955 | 912 025 | 30.90307 | 97.72410 | 870 983 875 | 9.847 692 | 21.21621 | 45.70894 |
| 956 | 913 936 | 30.91925 | 97.77525 | 873 722 816 | 9.851 128 | 21.22361 | 45.72489 |
| 957 | 915 849 | 30.93542 | 97.82638 | 876 467 493 | 9.854 562 | 21.23101 | 45.74082 |
| 958 | 917 764 | 30.95158 | 97.87747 | 879 217 912 | 9.857 993 | 21.23840 | 45.75675 |
| 959 | 919 681 | 30.96773 | 97.92855 | 881 974 079 | 9.861 422 | 21.24579 | 45.77267 |
| **960** | 921 600 | 30.98387 | 97.97959 | 884 736 000 | 9.864 848 | 21.25317 | 45.78857 |
| 961 | 923 521 | 31.00000 | 98.03061 | 887 503 681 | 9.868 272 | 21.26055 | 45.80446 |
| 962 | 925 444 | 31.01612 | 98.08160 | 890 277 128 | 9.871 694 | 21.26792 | 45.82035 |
| 963 | 927 369 | 31.03224 | 98.13256 | 893 056 347 | 9.875 113 | 21.27529 | 45.83622 |
| 964 | 929 296 | 31.04835 | 98.18350 | 895 841 344 | 9.878 530 | 21.28265 | 45.85208 |
| 965 | 931 225 | 31.06445 | 98.23441 | 898 632 125 | 9.881 945 | 21.29001 | 45.86793 |
| 966 | 933 156 | 31.08054 | 98.28530 | 901 428 696 | 9.885 357 | 21.29736 | 45.88376 |
| 967 | 935 089 | 31.09662 | 98.33616 | 904 231 063 | 9.888 767 | 21.30470 | 45.89959 |
| 968 | 937 024 | 31.11270 | 98.38699 | 907 039 232 | 9.892 175 | 21.31204 | 45.91541 |
| 969 | 938 961 | 31.12876 | 98.43780 | 909 853 209 | 9.895 580 | 21.31938 | 45.93121 |
| **970** | 940 900 | 31.14482 | 98.48858 | 912 673 000 | 9.898 983 | 21.32671 | 45.94701 |
| 971 | 942 841 | 31.16087 | 98.53933 | 915 498 611 | 9.902 384 | 21.33404 | 45.96279 |
| 972 | 944 784 | 31.17691 | 98.59006 | 918 330 048 | 9.905 782 | 21.34136 | 45.97857 |
| 973 | 946 729 | 31.19295 | 98.64076 | 921 167 317 | 9.909 178 | 21.34868 | 45.99433 |
| 974 | 948 676 | 31.20897 | 98.69144 | 924 010 424 | 9.912 571 | 21.35599 | 46.01008 |
| 975 | 950 625 | 31.22499 | 98.74209 | 926 859 375 | 9.915 962 | 21.36329 | 46.02582 |
| 976 | 952 576 | 31.24100 | 98.79271 | 929 714 176 | 9.919 351 | 21.37059 | 46.04155 |
| 977 | 954 529 | 31.25700 | 98.84331 | 932 574 833 | 9.922 738 | 21.37789 | 46.05727 |
| 978 | 956 484 | 31.27299 | 98.89388 | 935 441 352 | 9.926 122 | 21.38518 | 46.07298 |
| 979 | 958 441 | 31.28898 | 98.94443 | 938 313 739 | 9.929 504 | 21.39247 | 46.08868 |
| **980** | 960 400 | 31.30495 | 98.99495 | 941 192 000 | 9.932 884 | 21.39975 | 46.10436 |
| 981 | 962 361 | 31.32092 | 99.04544 | 944 076 141 | 9.936 261 | 21.40703 | 46.12004 |
| 982 | 964 324 | 31.33688 | 99.09591 | 946 966 168 | 9.939 636 | 21.41430 | 46.13571 |
| 983 | 966 289 | 31.35283 | 99.14636 | 949 862 087 | 9.943 009 | 21.42156 | 46.15136 |
| 984 | 968 256 | 31.36877 | 99.19677 | 952 763 904 | 9.946 380 | 21.42883 | 46.16700 |
| 985 | 970 225 | 31.38471 | 99.24717 | 955 671 625 | 9.949 748 | 21.43608 | 46.18264 |
| 986 | 972 196 | 31.40064 | 99.29753 | 958 585 256 | 9.953 114 | 21.44333 | 46.19826 |
| 987 | 974 169 | 31.41656 | 99.34787 | 961 504 803 | 9.956 478 | 21.45058 | 46.21387 |
| 988 | 976 144 | 31.43247 | 99.39819 | 964 430 272 | 9.959 839 | 21.45782 | 46.22948 |
| 989 | 978 121 | 31.44837 | 99.44848 | 967 361 669 | 9.963 198 | 21.46506 | 46.24507 |
| **990** | 980 100 | 31.46427 | 99.49874 | 970 299 000 | 9.966 555 | 21.47229 | 46.26065 |
| 991 | 982 081 | 31.48015 | 99.54898 | 973 242 271 | 9.969 910 | 21.47952 | 46.27622 |
| 992 | 984 064 | 31.49603 | 99.59920 | 976 191 488 | 9.973 262 | 21.48674 | 46.29178 |
| 993 | 986 049 | 31.51190 | 99.64939 | 979 146 657 | 9.976 612 | 21.49396 | 46.30733 |
| 994 | 988 036 | 31.52777 | 99.69955 | 982 107 784 | 9.979 960 | 21.50117 | 46.32287 |
| 995 | 990 025 | 31.54362 | 99.74969 | 985 074 875 | 9.983 305 | 21.50838 | 46.33840 |
| 996 | 992 016 | 31.55947 | 99.79980 | 988 047 936 | 9.986 649 | 21.51558 | 46.35392 |
| 997 | 994 009 | 31.57531 | 99.84989 | 991 026 973 | 9.989 990 | 21.52278 | 46.36943 |
| 998 | 996 004 | 31.59114 | 99.89995 | 994 011 992 | 9.993 329 | 21.52997 | 46.38492 |
| 999 | 998 001 | 31.60696 | 99.94999 | 997 002 999 | 9.996 666 | 21.53716 | 46.40041 |
| **1000** | 1 000 000 | 31.62278 | 100.00000 | 1 000 000 000 | 10.000 000 | 21.54435 | 46.41589 |

## DISTRIBUTION OF $t$

| $n$ | Probability .9 | .8 | .7 | .6 | .5 | .4 | .3 | .2 | .1 | .05 | .02 | .01 | .001 |
|---|---|---|---|---|---|---|---|---|---|---|---|---|---|
| 1 | .158 | .325 | .510 | .727 | 1.000 | 1.376 | 1.963 | 3.078 | 6.314 | 12.706 | 31.821 | 63.657 | 636.619 |
| 2 | .142 | .289 | .445 | .617 | .816 | 1.061 | 1.386 | 1.886 | 2.920 | 4.303 | 6.965 | 9.925 | 31.598 |
| 3 | .137 | .277 | .424 | .584 | .765 | .978 | 1.250 | 1.638 | 2.353 | 3.182 | 4.541 | 5.841 | 12.924 |
| 4 | .134 | .271 | .414 | .569 | .741 | .941 | 1.190 | 1.533 | 2.132 | 2.776 | 3.747 | 4.604 | 8.610 |
| 5 | .132 | .267 | .408 | .559 | .727 | .920 | 1.156 | 1.476 | 2.015 | 2.571 | 3.365 | 4.032 | 6.869 |
| 6 | .131 | .265 | .404 | .553 | .718 | .906 | 1.134 | 1.440 | 1.943 | 2.447 | 3.143 | 3.707 | 5.959 |
| 7 | .130 | .263 | .402 | .549 | .711 | .896 | 1.119 | 1.415 | 1.895 | 2.365 | 2.998 | 3.499 | 5.408 |
| 8 | .130 | .262 | .399 | .546 | .706 | .889 | 1.108 | 1.397 | 1.860 | 2.306 | 2.896 | 3.355 | 5.041 |
| 9 | .129 | .261 | .398 | .543 | .703 | .883 | 1.100 | 1.383 | 1.833 | 2.262 | 2.821 | 3.250 | 4.781 |
| 10 | .129 | .260 | .397 | .542 | .700 | .879 | 1.093 | 1.372 | 1.812 | 2.228 | 2.764 | 3.169 | 4.587 |
| 11 | .129 | .260 | .396 | .540 | .697 | .876 | 1.088 | 1.363 | 1.796 | 2.201 | 2.718 | 3.106 | 4.437 |
| 12 | .128 | .259 | .395 | .539 | .695 | .873 | 1.083 | 1.356 | 1.782 | 2.179 | 2.681 | 3.055 | 4.318 |
| 13 | .128 | .259 | .394 | .538 | .694 | .870 | 1.079 | 1.350 | 1.77 | 2.160 | 2.650 | 3.012 | 4.221 |
| 14 | .128 | .258 | .393 | .537 | .692 | .868 | 1.076 | 1.345 | 1.761 | 2.145 | 2.624 | 2.977 | 4.140 |
| 15 | .128 | .258 | .393 | .536 | .691 | .866 | 1.074 | 1.341 | 1.753 | 2.131 | 2.602 | 2.947 | 4.073 |
| 16 | .128 | .258 | .392 | .535 | .690 | .865 | 1.071 | 1.337 | 1.746 | 2.120 | 2.583 | 2.921 | 4.015 |
| 17 | .128 | .257 | .392 | .534 | .689 | .863 | 1.069 | 1.333 | 1.740 | 2.110 | 2.567 | 2.898 | 3.965 |
| 18 | .127 | .257 | .392 | .534 | .688 | .862 | 1.067 | 1.330 | 1.734 | 2.101 | 2.552 | 2.878 | 3.922 |
| 19 | .127 | .257 | .391 | .533 | .688 | .861 | 1.066 | 1.328 | 1.729 | 2.093 | 2.539 | 2.861 | 3.883 |
| 20 | .127 | .257 | .391 | .533 | .687 | .860 | 1.064 | 1.325 | 1.725 | 2.086 | 2.528 | 2.845 | 3.850 |
| 21 | .127 | .257 | .391 | .532 | .686 | .859 | 1.063 | 1.323 | 1.721 | 2.080 | 2.518 | 2.831 | 3.819 |
| 22 | .127 | .256 | .390 | .532 | .686 | .858 | 1.061 | 1.321 | 1.717 | 2.074 | 2.508 | 2.819 | 3.792 |
| 23 | .127 | .256 | .390 | .532 | .685 | .858 | 1.060 | 1.319 | 1.714 | 2.069 | 2.500 | 2.807 | 3.767 |
| 24 | .127 | .256 | .390 | .531 | .685 | .857 | 1.059 | 1.318 | 1.711 | 2.064 | 2.492 | 2.797 | 3.745 |
| 25 | .127 | .256 | .390 | .531 | .684 | .856 | 1.058 | 1.316 | 1.708 | 2.060 | 2.485 | 2.787 | 3.725 |
| 26 | .127 | .256 | .390 | .532 | .684 | .856 | 1.058 | 1.315 | 1.706 | 2.056 | 2.479 | 2.779 | 3.707 |
| 27 | .127 | .256 | .389 | .531 | .684 | .855 | 1.057 | 1.314 | 1.703 | 2.052 | 2.473 | 2.771 | 3.690 |
| 28 | .127 | .256 | .389 | .530 | .683 | .855 | 1.056 | 1.313 | 1.701 | 2.048 | 2.467 | 2.763 | 3.674 |
| 29 | .127 | .256 | .389 | .530 | .683 | .854 | 1.055 | 1.311 | 1.699 | 2.045 | 2.462 | 2.756 | 3.659 |
| 30 | .127 | .256 | .389 | .530 | .683 | .854 | 1.055 | 1.310 | 1.697 | 2.042 | 2.457 | 2.750 | 3.646 |
| 40 | .126 | .255 | .388 | .529 | .681 | .851 | 1.050 | 1.303 | 1.684 | 2.021 | 2.423 | 2.704 | 3.551 |
| 60 | .126 | .254 | .387 | .527 | .679 | .848 | 1.046 | 1.296 | 1.671 | 2.000 | 2.390 | 2.660 | 3.460 |
| 120 | .126 | .254 | .386 | .526 | .677 | .845 | 1.041 | 1.289 | 1.658 | 1.980 | 2.358 | 2.617 | 3.373 |
| $\infty$ | .126 | .253 | .385 | .524 | .674 | .842 | 1.036 | 1.282 | 1.645 | 1.960 | 2.326 | 2.576 | 3.291 |

Data taken from: Fisher, R. A. and Yates, F., *Statistical Tables*, 4th ed., p. 40, Hafner Publishing Co., New York (1953). Reproduced by permission of Oliver and Boyd, Edinburgh, Scotland.

# DISTRIBUTION OF $x^2$

| $n$ | Probability .99 | .98 | .95 | .90 | .80 | .70 | .50 | .30 | .20 | .10 | .05 | .02 | .01 | .001 |
|---|---|---|---|---|---|---|---|---|---|---|---|---|---|---|
| 1 | $.0^3157$ | $.0^3628$ | .00393 | .0158 | .0642 | .148 | .455 | 1.074 | 1.642 | 2.706 | 3.841 | 5.412 | 6.635 | 10.827 |
| 2 | .0201 | .0404 | .103 | .211 | .446 | .713 | 1.386 | 2.408 | 3.219 | 4.605 | 5.991 | 7.824 | 9.210 | 13.815 |
| 3 | .115 | .185 | .352 | .584 | 1.005 | 1.424 | 2.366 | 3.665 | 4.642 | 6.251 | 7.815 | 9.837 | 11.345 | 26.266 |
| 4 | .297 | .429 | .711 | 1.064 | 1.649 | 2.195 | 3.357 | 4.878 | 5.989 | 7.779 | 9.488 | 11.668 | 13.277 | 18.467 |
| 5 | .554 | .752 | 1.145 | 1.610 | 2.343 | 3.000 | 4.351 | 6.064 | 7.289 | 9.236 | 11.070 | 13.388 | 15.086 | 20.515 |
| 6 | .872 | 1.134 | 1.635 | 2.204 | 3.070 | 3.828 | 5.348 | 7.231 | 8.558 | 10.645 | 12.592 | 15.033 | 16.812 | 22.457 |
| 7 | 1.239 | 1.564 | 2.167 | 2.833 | 3.822 | 4.671 | 6.346 | 8.383 | 9.803 | 12.017 | 14.067 | 16.622 | 18.475 | 24.322 |
| 8 | 1.646 | 2.032 | 2.733 | 3.490 | 4.594 | 5.527 | 7.344 | 9.524 | 11.030 | 13.362 | 15.507 | 18.168 | 20.090 | 26.125 |
| 9 | 2.088 | 2.532 | 3.325 | 4.168 | 5.380 | 6.393 | 8.343 | 10.656 | 12.242 | 14.684 | 16.919 | 19.679 | 21.666 | 27.877 |
| 10 | 2.558 | 3.059 | 3.940 | 4.865 | 6.179 | 7.267 | 9.342 | 11.781 | 13.442 | 15.987 | 18.307 | 21.161 | 23.209 | 29.588 |
| 11 | 3.053 | 3.609 | 4.575 | 5.578 | 6.989 | 8.148 | 10.341 | 12.899 | 14.631 | 17.275 | 19.675 | 22.618 | 24.725 | 31.264 |
| 12 | 3.571 | 4.178 | 5.226 | 6.304 | 7.807 | 9.034 | 11.340 | 14.011 | 15.812 | 18.549 | 21.026 | 24.054 | 26.217 | 32.909 |
| 13 | 4.107 | 4.765 | 5.892 | 7.042 | 8.634 | 9.926 | 12.340 | 15.119 | 16.985 | 19.812 | 22.362 | 25.472 | 27.688 | 34.528 |
| 14 | 4.660 | 5.368 | 6.571 | 7.790 | 9.467 | 10.821 | 13.339 | 16.222 | 18.151 | 21.064 | 23.685 | 26.873 | 29.141 | 36.123 |
| 15 | 5.229 | 5.985 | 7.261 | 8.547 | 10.307 | 11.721 | 14.339 | 17.322 | 19.311 | 22.307 | 24.996 | 28.259 | 30.578 | 37.697 |
| 16 | 5.812 | 6.614 | 7.962 | 9.312 | 11.152 | 12.624 | 15.338 | 18.418 | 20.465 | 23.542 | 26.296 | 29.633 | 32.000 | 39.252 |
| 17 | 6.408 | 7.255 | 8.672 | 10.085 | 12.002 | 13.531 | 16.338 | 19.511 | 21.615 | 24.769 | 27.587 | 30.995 | 33.409 | 40.790 |
| 18 | 7.015 | 7.906 | 9.390 | 10.865 | 12.857 | 14.440 | 17.338 | 20.601 | 22.760 | 25.989 | 28.869 | 32.346 | 34.805 | 42.312 |
| 19 | 7.633 | 8.567 | 10.117 | 11.651 | 13.716 | 15.352 | 18.338 | 21.689 | 23.900 | 27.204 | 30.144 | 33.687 | 36.191 | 43.820 |
| 20 | 8.260 | 9.237 | 10.851 | 12.443 | 14.578 | 16.266 | 19.337 | 22.775 | 25.038 | 28.412 | 31.410 | 35.020 | 37.566 | 45.315 |
| 21 | 8.897 | 9.915 | 11.591 | 13.240 | 15.445 | 17.182 | 20.337 | 23.858 | 26.171 | 29.615 | 32.671 | 36.343 | 38.932 | 46.797 |
| 22 | 9.542 | 10.600 | 12.338 | 14.041 | 16.314 | 18.101 | 21.337 | 24.939 | 27.301 | 30.813 | 33.924 | 37.659 | 40.289 | 48.268 |
| 23 | 10.196 | 11.293 | 13.091 | 14.848 | 17.187 | 19.021 | 22.337 | 26.018 | 28.429 | 32.007 | 35.172 | 38.968 | 41.638 | 49.728 |
| 24 | 10.856 | 11.992 | 13.848 | 15.659 | 18.062 | 19.943 | 23.337 | 27.096 | 29.553 | 33.196 | 36.415 | 40.270 | 42.980 | 51.179 |
| 25 | 11.524 | 12.697 | 14.611 | 16.473 | 18.940 | 20.867 | 24.337 | 28.172 | 30.675 | 34.382 | 37.652 | 41.566 | 44.314 | 52.620 |
| 26 | 12.198 | 13.409 | 15.379 | 17.292 | 19.820 | 21.792 | 25.336 | 29.246 | 31.795 | 35.563 | 38.885 | 42.856 | 45.642 | 54.052 |
| 27 | 12.879 | 14.125 | 16.151 | 18.114 | 20.703 | 22.719 | 26.336 | 30.310 | 32.912 | 36.741 | 40.113 | 44.140 | 46.963 | 55.476 |
| 28 | 13.565 | 14.847 | 16.928 | 18.939 | 21.588 | 23.647 | 27.336 | 31.391 | 34.027 | 37.916 | 41.337 | 45.419 | 48.278 | 56.893 |
| 29 | 14.256 | 15.574 | 17.708 | 19.768 | 22.475 | 24.577 | 28.336 | 32.461 | 35.139 | 39.087 | 42.557 | 46.693 | 49.588 | 58.302 |
| 30 | 14.953 | 16.306 | 18.493 | 20.599 | 23.364 | 25.508 | 29.336 | 33.530 | 36.250 | 40.256 | 43.773 | 47.962 | 50.892 | 59.703 |

Note: For larger values of $n$, the expression $\sqrt{2x^2} - \sqrt{2n-1}$ may be used as a normal deviate with unit variance, remembering that the probability for $x^2$ corresponds with that of a single tail of the normal curve.

Data taken from: Fisher, R. A. and Yates, F., *Statistical Tables,* 4th ed., p. 41, Hafner Publishing Co., New York (1953).

# RANDOM PERMUTATIONS OF TWENTY NUMBERS

| | 0 | 1 | 2 | 3 | 4 | 5 | 6 | 7 | 8 | 9 |
|---|---|---|---|---|---|---|---|---|---|---|
| 0 | 1 15 6 8 5 | 2 13 7 4 10 | 11 9 10 1 6 | 10 2 12 13 9 | 7 6 10 0 1 | 16 19 14 8 18 | 1 15 3 10 17 | 3 16 0 15 9 | 14 19 12 17 10 | 13 17 11 12 14 |
| | 10 17 12 11 9 | 18 19 9 3 15 | 8 5 14 2 16 | 0 14 7 4 8 | 19 8 11 17 5 | 0 12 13 3 4 | 18 7 2 8 16 | 17 8 6 14 7 | 11 13 7 5 3 | 10 4 8 6 9 |
| | 2 16 18 0 3 | 11 1 6 12 0 | 0 7 15 4 17 | 5 15 6 3 19 | 12 16 9 2 4 | 6 10 2 5 7 | 0 11 5 12 9 | 5 12 1 10 13 | 0 2 6 8 18 | 1 0 3 18 16 |
| | 14 7 4 13 19 | 17 5 14 8 16 | 3 18 13 19 12 | 16 1 17 11 18 | 13 15 18 3 14 | 1 17 9 15 11 | 14 19 13 6 4 | 11 4 19 18 2 | 1 16 9 4 15 | 2 7 19 5 15 |
| 1 | 11 19 2 15 1 | 3 16 13 7 12 | 3 15 5 0 11 | 12 9 16 10 1 | 2 16 6 18 19 | 11 14 10 13 5 | 4 14 3 10 1 | 6 4 19 14 17 | 18 17 9 19 6 | 0 13 18 16 2 |
| | 7 14 5 4 3 | 6 18 8 2 19 | 7 12 13 18 4 | 6 13 2 19 5 | 0 7 10 15 3 | 0 17 4 9 7 | 8 11 18 9 6 | 16 7 3 11 2 | 5 14 13 1 4 | 11 6 12 14 17 |
| | 16 9 18 0 12 | 9 1 11 10 5 | 8 10 17 6 16 | 18 4 0 8 7 | 8 14 5 9 13 | 8 12 16 6 15 | 16 2 15 12 13 | 10 8 0 9 5 | 7 15 12 3 2 | 8 15 1 19 10 |
| | 8 13 10 17 6 | 0 17 14 15 4 | 9 1 14 19 2 | 11 17 14 15 3 | 1 11 4 12 17 | 3 18 19 2 1 | 19 7 17 5 0 | 1 18 15 12 13 | 0 10 8 11 16 | 7 5 4 9 3 |
| 2 | 18 8 11 9 15 | 19 13 4 1 11 | 18 6 3 17 11 | 3 1 14 12 0 | 2 17 0 7 19 | 1 2 10 9 12 | 17 7 4 3 19 | 13 0 9 5 17 | 1 12 9 3 16 | 0 11 8 4 12 |
| | 4 5 14 3 0 | 2 0 7 16 18 | 4 14 9 10 16 | 8 2 10 13 18 | 14 9 12 11 8 | 6 11 15 0 17 | 10 0 13 11 14 | 15 11 1 12 7 | 5 17 0 7 13 | 7 2 15 3 18 |
| | 7 10 16 12 6 | 5 17 6 8 14 | 1 12 2 8 19 | 17 7 11 16 4 | 6 3 16 13 5 | 16 8 4 18 5 | 8 9 5 2 18 | 10 14 4 18 16 | 18 4 19 11 15 | 10 13 6 16 5 |
| | 19 2 1 13 17 | 9 10 12 15 3 | 15 5 0 13 7 | 5 15 9 19 6 | 15 10 1 4 18 | 13 3 14 19 7 | 16 12 6 15 1 | 3 2 19 6 8 | 8 6 2 10 14 | 9 17 19 1 14 |
| 3 | 4 0 1 12 7 | 12 9 0 5 19 | 15 16 2 5 19 | 5 14 16 11 6 | 11 7 5 12 2 | 15 11 8 5 3 | 18 14 8 19 15 | 8 16 19 0 4 | 13 8 5 1 14 | 1 11 3 17 7 |
| | 3 14 5 15 6 | 7 3 4 6 1 | 7 8 3 4 12 | 15 13 3 1 17 | 13 19 15 14 3 | 1 9 18 0 14 | 6 4 3 7 1 | 13 5 11 14 6 | 18 11 2 4 12 | 9 16 0 6 5 |
| | 8 17 11 19 10 | 8 13 2 18 16 | 6 1 0 17 10 | 19 2 7 8 18 | 9 10 8 6 1 | 4 12 16 6 17 | 0 13 12 2 10 | 1 18 17 7 12 | 10 9 3 15 6 | 4 18 8 15 2 |
| | 9 18 16 13 2 | 11 17 14 10 15 | 9 13 18 11 14 | 12 9 0 10 4 | 16 0 4 18 17 | 19 10 2 7 13 | 17 9 16 5 11 | 15 10 3 2 9 | 17 16 0 7 19 | 13 14 12 19 10 |
| 4 | 0 2 5 1 18 | 14 16 19 7 8 | 14 6 8 1 19 | 5 2 3 14 13 | 18 9 2 12 16 | 0 1 18 4 15 | 13 14 0 2 11 | 0 14 5 12 13 | 0 6 18 7 16 | 7 9 19 11 5 |
| | 15 6 14 12 8 | 1 11 9 3 4 | 10 7 12 13 5 | 18 9 0 12 16 | 3 10 7 8 13 | 17 8 13 6 16 | 3 19 18 1 6 | 9 16 7 6 4 | 8 11 1 10 15 | 18 16 2 12 6 |
| | 13 16 17 3 19 | 10 6 5 17 13 | 17 0 15 18 4 | 10 6 15 17 7 | 15 14 17 0 6 | 14 9 12 19 3 | 12 10 4 15 17 | 3 19 18 10 1 | 3 17 5 2 13 | 14 13 1 17 0 |
| | 10 9 7 4 11 | 12 15 2 18 0 | 3 9 11 16 2 | 8 19 1 11 4 | 11 19 4 1 5 | 11 5 2 10 7 | 5 9 16 7 8 | 2 17 15 11 8 | 14 4 9 12 19 | 3 15 4 10 8 |
| 5 | 6 2 8 4 14 | 7 16 6 2 1 | 14 8 3 19 4 | 7 16 11 19 17 | 2 11 0 1 7 | 10 4 12 5 19 | 4 13 11 9 2 | 1 6 19 16 11 | 5 3 11 9 0 | 3 6 7 9 1 |
| | 18 1 10 16 9 | 5 8 17 0 12 | 9 15 0 11 5 | 5 14 18 9 3 | 19 13 18 8 6 | 8 7 14 2 18 | 10 15 16 17 6 | 17 5 14 9 13 | 2 7 12 15 10 | 11 16 4 12 10 |
| | 7 15 5 3 12 | 9 18 10 3 11 | 1 6 2 18 7 | 0 10 4 8 12 | 9 15 3 17 16 | 0 3 9 15 13 | 19 8 14 3 7 | 10 18 8 2 15 | 8 6 19 1 14 | 17 8 2 13 19 |
| | 13 0 11 19 17 | 15 19 4 14 13 | 13 16 17 10 12 | 15 13 1 6 2 | 4 12 5 10 14 | 6 17 1 16 11 | 18 0 1 12 5 | 3 0 7 12 4 | 16 17 13 4 18 | 0 5 14 18 15 |
| 6 | 19 13 17 0 9 | 8 7 10 5 4 | 12 15 5 17 9 | 17 19 0 14 11 | 4 10 7 19 2 | 17 10 3 0 14 | 19 9 4 17 1 | 11 7 17 3 10 | 5 18 1 13 8 | 13 10 18 16 19 |
| | 3 5 1 7 14 | 9 1 15 6 12 | 6 1 18 14 4 | 3 2 18 8 16 | 0 16 3 8 6 | 2 9 18 19 1 | 10 14 15 18 3 | 14 1 2 19 6 | 11 0 6 3 19 | 1 11 14 8 4 |
| | 8 6 15 10 4 | 13 18 19 2 11 | 2 16 19 7 11 | 4 10 6 12 1 | 14 11 17 18 15 | 8 15 6 11 12 | 6 16 5 13 7 | 15 0 5 9 8 | 17 9 16 10 15 | 0 6 5 9 17 |
| | 2 11 12 18 16 | 14 17 16 0 3 | 0 8 13 3 10 | 7 15 5 9 13 | 12 9 13 5 1 | 4 5 13 16 7 | 0 8 11 12 2 | 12 16 13 4 18 | 4 14 12 2 7 | 15 2 12 3 7 |
| 7 | 3 1 17 13 19 | 14 1 12 6 8 | 16 12 6 7 1 | 2 18 3 8 6 | 11 1 6 8 9 | 6 1 10 11 15 | 2 11 10 9 8 | 17 7 6 15 19 | 10 0 14 3 19 | 1 3 13 9 14 |
| | 5 18 12 15 2 | 18 17 2 3 5 | 15 5 18 0 17 | 4 7 14 0 10 | 18 12 15 0 19 | 8 7 2 14 4 | 1 13 18 3 16 | 18 0 16 4 3 | 13 18 1 8 2 | 6 12 4 18 11 |
| | 0 9 16 4 10 | 15 19 13 16 4 | 8 10 19 3 11 | 19 1 9 11 13 | 13 2 10 14 5 | 13 17 0 9 19 | 12 14 17 6 19 | 10 2 9 5 12 | 12 11 15 9 7 | 16 8 2 17 7 |
| | 11 8 14 7 6 | 0 7 10 9 11 | 2 4 9 13 14 | 15 12 16 5 17 | 4 3 7 16 17 | 16 3 12 18 5 | 5 15 7 4 0 | 14 8 1 11 13 | 4 5 16 6 17 | 0 19 15 5 10 |
| 8 | 19 16 6 8 15 | 10 11 9 3 12 | 13 6 4 8 5 | 13 7 6 1 2 | 12 15 13 5 3 | 10 15 0 11 9 | 19 18 7 11 17 | 13 14 10 6 18 | 18 10 17 7 1 | 8 10 6 18 11 |
| | 2 10 4 14 18 | 14 2 4 17 0 | 1 2 9 14 11 | 8 9 0 12 11 | 4 9 18 16 7 | 3 16 13 5 2 | 4 5 0 16 2 | 2 12 4 1 3 | 12 13 4 8 9 | 9 17 13 19 1 |
| | 9 0 12 5 13 | 15 5 16 18 6 | 3 12 0 15 7 | 19 14 5 4 3 | 2 10 0 17 1 | 18 19 14 4 12 | 12 1 6 10 3 | 17 19 9 16 15 | 14 16 3 15 5 | 15 14 4 16 3 |
| | 11 1 7 3 17 | 19 7 13 8 1 | 16 10 17 18 19 | 15 17 10 18 16 | 6 11 19 8 14 | 17 6 1 7 8 | 13 14 15 9 8 | 11 8 7 0 5 | 2 0 19 11 6 | 2 5 12 0 7 |
| 9 | 2 7 18 0 17 | 5 6 13 15 16 | 4 15 9 3 10 | 13 18 19 4 3 | 9 17 15 11 10 | 13 10 11 3 14 | 6 13 1 19 12 | 9 16 2 8 6 | 13 2 1 9 5 | 3 15 10 5 1 |
| | 3 11 1 16 15 | 7 9 19 10 4 | 13 8 6 14 12 | 9 16 14 5 0 | 6 1 4 16 7 | 7 9 0 12 5 | 4 7 15 8 3 | 19 14 5 15 10 | 8 6 14 4 16 | 16 8 12 13 0 |
| | 13 19 12 6 8 | 3 1 12 18 8 | 0 16 7 5 2 | 17 1 11 15 10 | 19 5 14 3 0 | 18 19 17 8 6 | 2 18 14 16 5 | 1 4 13 18 11 | 0 18 11 17 12 | 7 9 4 2 11 |
| | 9 4 10 14 5 | 17 11 2 0 14 | 18 17 11 1 19 | 2 6 8 12 7 | 12 2 8 13 18 | 2 16 15 4 1 | 9 0 17 11 10 | 3 17 12 7 0 | 19 15 10 3 7 | 18 6 17 14 19 |

| — | 0 | 1 | 2 | 3 | 4 | 5 | 6 | 7 | 8 | 9 |
|---|---|---|---|---|---|---|---|---|---|---|
| 10 | 17 7 3 13 12 | 17 1 12 3 2 | 2 9 16 4 6 | 3 4 10 13 16 | 6 17 13 19 0 | 7 13 17 11 1 | 14 10 16 4 7 | 10 12 15 17 5 | 15 2 3 12 1 | 0 9 11 8 6 |
| | 16 15 1 10 14 | 13 10 14 15 16 | 3 5 13 10 17 | 14 0 18 5 9 | 12 15 18 9 3 | 0 4 14 6 19 | 8 11 13 19 15 | 2 9 4 1 8 | 13 4 10 18 19 | 7 18 15 12 1 |
| | 2 19 8 4 5 | 5 6 7 9 19 | 18 8 7 12 1 | 17 15 12 11 6 | 14 2 1 7 4 | 8 5 2 16 18 | 0 9 18 6 3 | 13 0 14 16 7 | 16 9 7 8 0 | 19 13 10 17 5 |
| | 0 9 11 6 18 | 18 0 11 8 4 | 15 0 11 14 19 | 7 1 19 2 8 | 10 16 5 8 11 | 12 10 15 3 9 | 12 1 5 2 17 | 11 6 19 18 3 | 14 17 11 5 6 | 14 16 2 4 3 |
| 11 | 16 1 17 15 11 | 6 16 1 18 17 | 8 16 11 3 9 | 4 10 15 16 12 | 6 13 8 12 1 | 1 7 2 9 4 | 14 3 4 1 10 | 10 16 0 15 3 | 1 19 8 6 13 | 0 14 18 11 5 |
| | 8 6 14 7 10 | 19 9 10 3 4 | 12 0 4 2 13 | 1 9 8 6 5 | 9 17 5 10 11 | 10 0 17 19 3 | 2 0 12 17 19 | 7 8 19 13 17 | 17 3 16 15 18 | 3 19 13 12 4 |
| | 9 0 3 5 4 | 13 0 12 11 2 | 6 5 19 14 7 | 18 7 14 13 0 | 18 14 7 4 0 | 8 16 13 5 11 | 15 9 16 7 18 | 5 1 9 2 4 | 4 5 14 0 12 | 10 7 2 15 1 |
| | 19 18 2 12 13 | 7 8 5 14 15 | 10 18 1 15 17 | 3 19 11 17 2 | 3 19 15 2 16 | 12 6 14 15 18 | 11 13 5 6 8 | 18 14 12 6 11 | 2 10 7 11 9 | 6 16 9 17 8 |
| 12 | 4 13 10 18 14 | 6 10 8 11 9 | 3 7 4 2 14 | 15 17 12 6 0 | 8 0 13 12 18 | 0 16 6 4 18 | 17 11 6 16 15 | 1 19 0 4 8 | 17 19 5 8 16 | 18 19 17 6 10 |
| | 2 5 0 1 11 | 16 13 5 0 3 | 1 5 13 18 15 | 14 4 16 19 5 | 2 3 19 9 5 | 14 5 1 2 3 | 13 1 18 7 19 | 16 6 18 14 17 | 3 7 18 4 0 | 3 1 0 11 16 |
| | 19 7 3 16 6 | 12 2 14 4 7 | 0 9 10 11 17 | 3 11 13 7 2 | 7 11 14 10 15 | 15 13 19 7 12 | 5 3 9 2 12 | 2 12 13 7 9 | 13 1 14 6 11 | 2 13 5 4 8 |
| | 15 12 9 8 17 | 17 18 1 15 19 | 19 16 6 12 8 | 10 9 8 18 1 | 6 16 1 17 4 | 11 8 17 9 10 | 14 4 0 8 10 | 15 5 3 11 10 | 12 15 2 10 9 | 7 14 9 12 15 |
| 13 | 11 1 5 0 14 | 0 7 17 19 3 | 16 2 18 15 1 | 10 0 1 17 11 | 19 6 18 5 16 | 15 11 5 14 8 | 11 14 15 18 5 | 16 18 7 17 11 | 7 19 12 2 13 | 8 13 4 0 3 |
| | 19 9 2 4 8 | 11 12 13 10 9 | 9 4 5 3 6 | 16 19 7 6 3 | 10 15 13 11 7 | 10 1 12 3 19 | 13 7 3 8 10 | 13 9 10 19 1 | 10 15 3 1 17 | 5 10 7 9 17 |
| | 17 10 6 12 15 | 5 15 6 8 1 | 8 13 17 14 7 | 15 8 5 4 2 | 12 0 14 1 4 | 4 18 7 13 0 | 16 4 1 0 19 | 0 12 8 6 14 | 9 16 5 4 14 | 15 11 14 2 16 |
| | 16 13 7 18 3 | 14 18 2 4 16 | 11 19 10 0 12 | 14 18 13 9 12 | 3 9 8 2 17 | 9 6 17 16 2 | 12 6 9 2 17 | 5 4 2 15 3 | 0 6 8 18 11 | 6 19 18 1 12 |
| 14 | 18 5 12 3 6 | 11 16 1 8 3 | 16 6 0 8 18 | 13 17 14 15 0 | 9 5 16 4 18 | 2 7 13 3 12 | 8 0 3 15 10 | 8 7 5 15 10 | 1 15 0 13 11 | 18 2 5 6 19 |
| | 1 2 11 15 7 | 18 14 19 9 13 | 17 5 7 19 10 | 8 12 2 16 11 | 19 1 11 17 10 | 10 17 5 11 8 | 16 6 2 14 9 | 1 9 4 6 0 | 8 17 3 10 12 | 8 13 9 10 4 |
| | 14 4 10 9 17 | 12 2 5 17 4 | 4 9 12 3 2 | 9 18 6 4 10 | 12 14 2 13 6 | 14 9 19 16 6 | 17 18 4 19 11 | 11 19 13 17 2 | 5 4 6 18 7 | 12 15 7 3 1 |
| | 16 8 0 13 19 | 7 0 10 6 15 | 14 13 1 15 11 | 19 3 7 1 5 | 0 7 8 3 15 | 4 15 18 1 0 | 1 13 12 5 7 | 12 16 18 14 3 | 19 14 9 16 2 | 16 0 17 14 11 |
| 15 | 5 1 11 7 10 | 2 14 4 6 18 | 18 4 17 16 5 | 15 0 1 14 4 | 14 13 16 7 12 | 19 5 15 7 9 | 19 8 3 1 12 | 14 6 8 7 18 | 7 17 14 13 4 | 18 10 15 16 0 |
| | 14 16 6 15 12 | 8 10 1 13 19 | 9 6 1 19 11 | 8 6 2 10 18 | 6 2 8 10 19 | 17 18 13 11 16 | 4 17 13 18 2 | 9 5 19 17 2 | 12 8 19 1 3 | 13 14 17 19 9 |
| | 17 3 4 2 0 | 17 16 12 3 11 | 7 2 13 15 8 | 12 9 7 16 3 | 3 1 11 9 18 | 14 3 6 10 1 | 10 16 9 7 0 | 1 4 12 11 16 | 0 10 5 11 2 | 6 2 3 12 4 |
| | 18 13 9 19 8 | 15 0 7 5 9 | 14 3 12 0 10 | 5 13 11 17 19 | 17 5 15 4 0 | 4 12 8 0 2 | 14 15 6 11 5 | 15 0 13 10 3 | 9 16 6 15 18 | 11 8 1 7 5 |
| 16 | 15 14 0 13 3 | 19 4 12 3 6 | 6 15 9 10 12 | 8 7 0 17 15 | 12 18 17 3 5 | 7 14 0 2 3 | 12 9 14 13 7 | 13 1 11 19 4 | 18 3 4 0 1 | 13 19 3 7 4 |
| | 4 17 2 11 16 | 14 11 13 18 17 | 4 19 14 7 5 | 10 13 1 2 6 | 15 19 14 7 16 | 1 11 9 5 16 | 2 1 11 15 0 | 2 16 5 9 3 | 8 2 14 17 7 | 16 10 9 14 15 |
| | 10 18 12 6 8 | 0 7 10 1 8 | 17 3 16 2 18 | 19 4 9 12 11 | 8 4 13 6 11 | 17 13 10 12 18 | 16 3 19 17 4 | 15 7 10 8 18 | 15 13 6 19 12 | 6 2 11 8 18 |
| | 7 9 1 19 5 | 9 2 16 5 15 | 13 8 1 0 11 | 3 18 16 5 14 | 1 2 9 0 10 | 4 19 8 6 15 | 18 5 8 6 10 | 0 17 6 12 14 | 10 16 9 11 5 | 1 0 5 17 12 |
| 17 | 4 7 18 3 8 | 1 15 16 18 13 | 2 7 12 4 0 | 14 12 15 17 0 | 16 17 6 5 11 | 7 2 4 1 9 | 16 8 14 6 19 | 2 16 9 6 12 | 4 16 2 7 18 | 19 3 6 2 16 |
| | 13 15 5 6 0 | 2 8 6 10 3 | 9 15 19 13 5 | 8 18 4 2 16 | 10 4 15 7 13 | 15 19 11 16 14 | 2 5 11 7 0 | 8 15 10 14 18 | 0 17 15 10 6 | 14 15 12 5 4 |
| | 10 2 9 17 12 | 17 5 19 4 0 | 18 16 10 8 6 | 13 1 10 7 3 | 14 1 3 2 0 | 0 10 18 17 13 | 15 17 4 3 9 | 5 13 19 1 4 | 1 12 5 14 19 | 9 7 11 10 8 |
| | 11 14 1 19 16 | 11 9 7 14 12 | 3 11 17 1 14 | 9 19 6 5 11 | 18 19 8 9 12 | 5 8 12 6 3 | 1 12 10 18 13 | 7 3 0 11 17 | 3 13 9 8 11 | 18 13 17 1 0 |
| 18 | 7 10 16 12 1 | 19 0 16 14 8 | 14 11 12 2 5 | 4 0 15 5 9 | 11 18 17 4 2 | 5 14 11 16 1 | 5 3 16 2 0 | 16 7 6 8 1 | 0 13 4 11 14 | 14 9 16 15 1 |
| | 18 8 14 3 19 | 2 9 18 10 4 | 15 1 0 6 18 | 7 18 13 17 1 | 15 0 12 5 8 | 18 17 0 10 15 | 1 13 10 19 7 | 4 15 14 17 10 | 15 9 6 8 1 | 13 7 8 11 17 |
| | 13 17 9 5 15 | 5 6 13 1 11 | 8 16 13 10 3 | 19 10 14 11 2 | 14 7 19 1 10 | 2 13 6 8 9 | 18 14 12 17 6 | 9 12 19 0 11 | 3 17 19 10 2 | 5 18 3 10 12 |
| | 2 6 4 11 0 | 3 12 7 17 15 | 19 4 7 9 17 | 12 8 3 16 6 | 13 16 6 9 3 | 4 12 7 19 3 | 11 9 8 15 4 | 18 2 13 3 5 | 5 18 12 7 16 | 6 4 0 2 19 |
| 19 | 4 1 10 18 13 | 17 15 9 19 4 | 7 4 19 12 16 | 6 7 9 13 3 | 4 2 19 3 0 | 12 7 6 19 0 | 12 14 7 19 16 | 8 12 10 3 19 | 5 6 3 16 10 | 1 2 6 9 5 |
| | 0 8 14 16 12 | 7 18 12 3 14 | 5 0 9 3 10 | 0 12 8 17 18 | 9 13 7 14 5 | 2 14 8 16 18 | 4 8 18 10 2 | 11 4 6 5 13 | 19 1 4 12 13 | 17 19 15 14 13 |
| | 9 15 7 3 17 | 1 8 13 2 5 | 2 14 8 17 11 | 16 1 2 4 5 | 15 12 17 8 6 | 5 13 11 17 4 | 13 3 0 5 6 | 2 16 9 18 0 | 17 2 0 15 9 | 10 18 3 12 7 |
| | 2 6 5 19 11 | 0 10 6 16 11 | 18 13 15 1 6 | 15 10 19 14 11 | 10 1 18 16 11 | 9 15 10 3 1 | 17 11 1 9 15 | 1 15 17 7 14 | 18 8 7 11 14 | 0 8 4 16 11 |

Data from: Fisher, R. A., and Yates, F., *Statistical Tables,* 4th ed., pp. 122–123, Hafner Publishing Co., New York (1953). Reproduced by permission of Oliver and Boyd, Edinburgh, Scotland.

# COMPONENTS OF ATMOSPHERIC AIR*

| Constituent | Content by Volume, % | Content by Volume, ppm |
|---|---|---|
| $N_2$ | 78.084 ± 0.004 | |
| $O_2$ | 20.946 ± 0.002 | |
| $CO_2$ | 0.033 ± 0.001 | |
| A | 0.934 ± 0.001 | |
| Ne | | 18.18 ± 0.04 |
| He | | 5.24 ± 0.004 |
| Kr | | 1.14 ± 0.01 |
| Xe | | 0.087 ± 0.001 |
| $H_2$ | | 0.5 |
| $CH_4$ | | 2 |
| $N_2O$ | | 0.5 ± 0.1 |

*Exclusive of water vapor.

# CONSTANT HUMIDITY

The following table shows the percent humidity and the aqueous tension at the given temperature within a closed space when an excess of the substance indicated is in contact with a saturated aqueous solution of the given solid phase.

| Solid Phase | °C | Percent Humidity | Aqueous Tension, mm Hg |
|---|---|---|---|
| $H_3PO_4 \cdot ½H_2O$ | 24 | 9 | 1.99 |
| $KC_2H_3O_2$ | 168 | 13 | 738 |
| $LiCl \cdot H_2O$ | 20 | 15 | 2.60 |
| $KC_2H_3O_2$ | 20 | 20 | 3.47 |
| KF | 100 | 22.9 | 174 |
| NaBr | 100 | 22.9 | 174 |
| NaCl, $KNO_3$ and $NaNO_3$ | 16.39 | 30.49 | 4.23 |
| $CaCl_2 \cdot 6H_2O$ | 24.5 | 31 | 7.08 |
| $CaCl_2 \cdot 6H_2O$ | 20 | 32.3 | 5.61 |
| $CaCl_2 \cdot 6H_2O$ | 18.5 | 35 | 5.54 |
| $CrO_3$ | 20 | 35 | 6.08 |
| $CaCl_2 \cdot 6H_2O$ | 10 | 38 | 3.47 |
| $CaCl_2 \cdot 6H_2O$ | 5 | 39.8 | 2.59 |
| $Zn(NO_3)_2 \cdot 6H_2O$ | 20 | 42 | 7.29 |
| $K_2CO_3 \cdot 2H_2O$ | 24.5 | 43 | 9.82 |
| $K_2CO_3 \cdot 2H_2O$ | 18.5 | 44 | 6.96 |
| $KNO_2$ | 20 | 45 | 7.81 |
| KCNS | 20 | 47 | 8.16 |
| NaI | 100 | 50.4 | 383 |
| $Ca(NO_3)_2 \cdot 4H_2O$ | 24.5 | 51 | 11.6 |
| $NaHSO_4 \cdot H_2O$ | 20 | 52 | 9.03 |
| $Na_2Cr_2O_7 \cdot 2H_2O$ | 20 | 52 | 9.03 |
| $Mg(NO_3)_2 \cdot 6H_2O$ | 24.5 | 52 | 11.9 |
| $NaClO_3$ | 100 | 54 | 410 |
| $Ca(NO_3)_2 \cdot 4H_2O$ | 18.5 | 56 | 8.86 |
| $Mg(NO_3)_2 \cdot 6H_2O$ | 18.5 | 56 | 8.86 |
| KI | 100 | 56.2 | 427 |
| $NaBr \cdot 2H_2O$ | 20 | 58 | 10.1 |
| $Mg(C_2H_3O_2)_2 \cdot 4H_2O$ | 20 | 65 | 11.3 |
| $NaNO_2$ | 20 | 66 | 11.5 |
| $NH_4Cl$ and $KNO_3$ | 30 | 68.6 | 21.6 |
| KBr | 100 | 69.2 | 526 |
| $NH_4Cl$ and $KNO_3$ | 25 | 71.2 | 16.7 |
| TlCl | 100.1 | 99.7 | 761 |
| $NH_4Cl$ and $KNO_3$ | 20 | 72.6 | 12.6 |
| $NaClO_3$ | 20 | 75 | 13.0 |
| $(NH_4)_2SO_4$ | 108 | 75 | 754 |
| $NaC_2H_3O_2 \cdot 3H_2O$ | 20 | 76 | 13.2 |
| $H_2C_2O_4 \cdot 2H_2O$ | 20 | 76 | 13.2 |
| $NH_4Cl$ | 30 | 77.5 | 24.4 |
| $Na_2S_2O_3 \cdot 5H_2O$ | 20 | 78 | 13.5 |
| $NH_4Cl$ | 25 | 79.3 | 18.6 |
| $NH_4Cl$ | 20 | 79.5 | 13.8 |
| $(NH_4)_2SO_4$ | 20 | 81 | 14.1 |
| $(NH_4)_2SO_4$ | 25 | 81.1 | 19.1 |
| $(NH_4)_2SO_4$ | 30 | 81.1 | 25.6 |
| KBr | 20 | 84 | 14.6 |
| $Tl_2SO_4$ | 104.7 | 84.8 | 768 |
| $KHSO_2$ | 20 | 86 | 14.9 |
| $Na_2CO_3 \cdot 10H_2O$ | 24.5 | 87 | 20.9 |
| $BaCl_2 \cdot 2H_2O$ | 24.5 | 88 | 20.1 |
| $K_2CrO_4$ | 20 | 88 | 15.3 |
| $Pb(NO_3)_2$ | 103.5 | 88.4 | 760 |
| $ZnSO_4 \cdot 7H_2O$ | 20 | 90 | 15.6 |
| $Na_2CO_3 \cdot 10H_2O$ | 18.5 | 92 | 14.6 |
| $NaBrO_3$ | 20 | 92 | 16.0 |
| $K_2HPO_4$ | 20 | 92 | 16.0 |
| $NH_4H_2PO_4$ | 30 | 92.9 | 29.3 |
| $NH_4H_2PO_4$ | 25 | 93 | 21.9 |
| $Na_2SO_4 \cdot 10H_2O$ | 20 | 93 | 16.1 |
| $NH_4H_2PO_4$ | 20 | 93.1 | 16.2 |
| $ZnSO_4 \cdot 7H_2O$ | 5 | 94.7 | 6.10 |
| $Na_2SO_3 \cdot 7H_2O$ | 20 | 95 | 16.5 |
| $Na_2HPO_4 \cdot 12H_2O$ | 20 | 95 | 16.5 |
| NaF | 100 | 96.6 | 734 |
| $Pb(NO_3)_2$ | 20 | 98 | 17.0 |
| $CuSO_4 \cdot 5H_2O$ | 20 | 98 | 17.0 |
| $TlNO_3$ | 100.3 | 98.7 | 759 |

# CONSTANT HUMIDITY WITH SULFURIC ACID SOLUTIONS

The following table gives the relative humidity and the pressure of aqueous vapor of air in equilibrium conditions above aqueous solutions of sulfuric acid.

| Density of Acid Solution | Percent Sulfuric Acid | Percent Relative Humidity | Vapor Pressure at 20°C |
|---|---|---|---|
| 1.00 | 1 | 100.0 | 17.4 |
| 1.05 | 8 | 97.5 | 17.0 |
| 1.10 | 15 | 93.9 | 16.3 |
| 1.15 | 21 | 88.8 | 15.4 |
| 1.20 | 28 | 80.5 | 14.0 |
| 1.25 | 34 | 70.4 | 12.2 |
| 1.30 | 40 | 58.3 | 10.1 |
| 1.35 | 45 | 47.2 | 8.3 |
| 1.40 | 51 | 37.1 | 6.5 |
| 1.50 | 60 | 18.8 | 3.3 |
| 1.60 | 69 | 8.5 | 1.5 |
| 1.70 | 78 | 3.2 | 0.6 |

## CONVERSION OF TRANSPARENCY TO OPTICAL DENSITY

Transparency of a layer of material is defined as the ratio of the intensity of the transmitted light to that of the incident light. Opacity is the reciprocal of the transparency. Optical density is the common logarithm of the opacity. Thus,

$$\text{Transparency} = \frac{I_t}{I_i},$$

$$\text{Opacity} = \frac{1}{\text{Transparency}} = \frac{I_i}{I_t},$$

$$\text{Optical density} = \log_{10} \frac{I_i}{I_t},$$

where $I_i$ = intensity of incident light and $I_t$ = intensity of transmitted light.

| Transparency | Density | Transparency | Density | Transparency | Density | Transparency | Density |
|---|---|---|---|---|---|---|---|
| 0.000 | – | .035 | 1.456 | .070 | 1.155 | .105 | .9788 |
| .001 | 3.000 | .036 | 1.444 | .071 | 1.149 | .106 | .9747 |
| .002 | 2.699 | .037 | 1.432 | .072 | 1.143 | .107 | .9706 |
| .003 | 2.523 | .038 | 1.420 | .073 | 1.137 | .108 | .9666 |
| .004 | 2.398 | .039 | 1.409 | .074 | 1.131 | .109 | .9626 |
| .005 | 2.301 | .040 | 1.398 | .075 | 1.125 | .110 | .9586 |
| .006 | 2.222 | .041 | 1.387 | .076 | 1.119 | .111 | .9547 |
| .007 | 2.155 | .042 | 1.377 | .077 | 1.114 | .112 | .9508 |
| .008 | 2.097 | .043 | 1.367 | .078 | 1.108 | .113 | .9469 |
| .009 | 2.046 | .044 | 1.357 | .079 | 1.102 | .114 | .9431 |
| .010 | 2.000 | .045 | 1.347 | .080 | 1.097 | .115 | .9393 |
| .011 | 1.959 | .046 | 1.337 | .081 | 1.092 | .116 | .9356 |
| .012 | 1.921 | .047 | 1.328 | .082 | 1.086 | .117 | .9318 |
| .013 | 1.886 | .048 | 1.319 | .083 | 1.081 | .118 | .9281 |
| .014 | 1.854 | .049 | 1.310 | .084 | 1.076 | .119 | .9244 |
| .015 | 1.824 | .050 | 1.301 | .085 | 1.071 | .120 | .9208 |
| .016 | 1.796 | .051 | 1.292 | .086 | 1.066 | .121 | .9172 |
| .017 | 1.770 | .052 | 1.284 | .087 | 1.060 | .122 | .9137 |
| .018 | 1.745 | .053 | 1.276 | .088 | 1.055 | .123 | .9101 |
| .019 | 1.721 | .054 | 1.268 | .089 | 1.051 | .124 | .9066 |
| .020 | 1.699 | .055 | 1.260 | .090 | 1.046 | .125 | .9031 |
| .021 | 1.678 | .056 | 1.252 | .091 | 1.041 | .126 | .8996 |
| .022 | 1.658 | .057 | 1.244 | .092 | 1.036 | .127 | .8962 |
| .023 | 1.638 | .058 | 1.237 | .093 | 1.032 | .128 | .8928 |
| .024 | 1.620 | .059 | 1.229 | .094 | 1.027 | .129 | .8894 |
| .025 | 1.602 | .060 | 1.222 | .095 | 1.022 | .130 | .8861 |
| .026 | 1.585 | .061 | 1.215 | .096 | 1.018 | .131 | .8827 |
| .027 | 1.569 | .062 | 1.208 | .097 | 1.013 | .132 | .8794 |
| .028 | 1.553 | .063 | 1.201 | .098 | 1.009 | .133 | .8761 |
| .029 | 1.538 | .064 | 1.194 | .099 | 1.004 | .134 | .8729 |
| .030 | 1.523 | .065 | 1.187 | .100 | 1.000 | .135 | .8697 |
| .031 | 1.509 | .066 | 1.180 | .101 | .9957 | .136 | .8665 |
| .032 | 1.495 | .067 | 1.174 | .102 | .9914 | .137 | .8633 |
| .033 | 1.482 | .068 | 1.168 | .103 | .9872 | .138 | .8601 |
| .034 | 1.469 | .069 | 1.161 | .104 | .9830 | .139 | .8570 |

| Transparency | Density | Transparency | Density | Transparency | Density | Transparency | Density |
|---|---|---|---|---|---|---|---|
| .140 | .8539 | .190 | .7212 | .240 | .6198 | .290 | .5376 |
| .141 | .8508 | .191 | .7190 | .241 | .6180 | .291 | .5361 |
| .142 | .8477 | .192 | .7167 | .242 | .6162 | .292 | .5346 |
| .143 | .8447 | .193 | .7144 | .243 | .6144 | .293 | .5331 |
| .144 | .8416 | .194 | .7122 | .244 | .6126 | .294 | .5317 |
| .145 | .8386 | .195 | .7100 | .245 | .6108 | .295 | .5302 |
| .146 | .8356 | .196 | .7077 | .246 | .6091 | .296 | .5287 |
| .147 | .8327 | .197 | .7055 | .247 | .6073 | .297 | .5272 |
| .148 | .8297 | .198 | .7033 | .248 | .6056 | .298 | .5258 |
| .149 | .8268 | .199 | .7011 | .249 | .6038 | .299 | .5243 |
| .150 | .8239 | .200 | .6990 | .250 | .6021 | .300 | .5229 |
| .151 | .8210 | .201 | .6968 | .251 | .6003 | .301 | .5215 |
| .152 | .8182 | .202 | .6946 | .252 | .5986 | .302 | .5200 |
| .153 | .8153 | .203 | .6925 | .253 | .5969 | .303 | .5186 |
| .154 | .8125 | .204 | .6904 | .254 | .5952 | .304 | .5171 |
| .155 | .8097 | .205 | .6882 | .255 | .5935 | .305 | .5157 |
| .156 | .8069 | .206 | .6861 | .256 | .5918 | .306 | .5143 |
| .157 | .8041 | .207 | .6840 | .257 | .5901 | .307 | .5128 |
| .158 | .8013 | .208 | .6819 | .258 | .5884 | .308 | .5114 |
| .159 | .7986 | .209 | .6799 | .259 | .5867 | .309 | .5100 |
| .160 | .7959 | .210 | .6778 | .260 | .5850 | .310 | .5086 |
| .161 | .7932 | .211 | .6757 | .261 | .5834 | .311 | .5072 |
| .162 | .7905 | .212 | .6737 | .262 | .5817 | .312 | .5058 |
| .163 | .7878 | .213 | .6716 | .263 | .5800 | .313 | .5045 |
| .164 | .7852 | .214 | .6696 | .264 | .5784 | .314 | .5031 |
| .165 | .7825 | .215 | .6676 | .265 | .5768 | .315 | .5017 |
| .166 | .7799 | .216 | .6655 | .266 | .5751 | .316 | .5003 |
| .167 | .7773 | .217 | .6636 | .267 | .5735 | .317 | .4989 |
| .168 | .7747 | .218 | .6615 | .268 | .5719 | .318 | .4976 |
| .169 | .7721 | .219 | .6596 | .269 | .5702 | .319 | .4962 |
| .170 | .7696 | .220 | .6576 | .270 | .5686 | .320 | .4949 |
| .171 | .7670 | .221 | .6556 | .271 | .5670 | .321 | .4935 |
| .172 | .7645 | .222 | .6536 | .272 | .5654 | .322 | .4921 |
| .173 | .7620 | .223 | .6517 | .273 | .5638 | .323 | .4908 |
| .174 | .7594 | .224 | .6498 | .274 | .5622 | .324 | .4895 |
| .175 | .7570 | .225 | .6478 | .275 | .5607 | .325 | .4881 |
| .176 | .7545 | .226 | .6459 | .276 | .5591 | .326 | .4868 |
| .177 | .7520 | .227 | .6440 | .277 | .5575 | .327 | .4855 |
| .178 | .7496 | .228 | .6421 | .278 | .5560 | .328 | .4841 |
| .179 | .7471 | .229 | .6402 | .279 | .5544 | .329 | .4828 |
| .180 | .7447 | .230 | .6383 | .280 | .5528 | .330 | .4815 |
| .181 | .7423 | .231 | .6364 | .281 | .5513 | .331 | .4802 |
| .182 | .7399 | .232 | .6345 | .282 | .5498 | .332 | .4789 |
| .183 | .7375 | .233 | .6326 | .283 | .5482 | .333 | .4776 |
| .184 | .7352 | .234 | .6308 | .284 | .5467 | .334 | .4763 |
| .185 | .7328 | .235 | .6289 | .285 | .5452 | .335 | .4750 |
| .186 | .7305 | .236 | .6271 | .286 | .5436 | .336 | .4737 |
| .187 | .7282 | .237 | .6253 | .287 | .5421 | .337 | .4724 |
| .188 | .7258 | .238 | .6234 | .288 | .5406 | .338 | .4711 |
| .189 | .7235 | .239 | .6216 | .289 | .5391 | .339 | .4698 |

| Transparency | Density | Transparency | Density | Transparency | Density | Transparency | Density |
|---|---|---|---|---|---|---|---|
| .340 | .4685 | .390 | .4089 | .440 | .3565 | .490 | .3098 |
| .341 | .4673 | .391 | .4078 | .441 | .3556 | .491 | .3089 |
| .342 | .4660 | .392 | .4067 | .442 | .3546 | .492 | .3080 |
| .343 | .4647 | .393 | .4056 | .443 | .3536 | .493 | .3072 |
| .344 | .4634 | .394 | .4045 | .444 | .3526 | .494 | .3063 |
| .345 | .4622 | .395 | .4034 | .445 | .3516 | .495 | .3054 |
| .346 | .4609 | .396 | .4023 | .446 | .3507 | .496 | .3045 |
| .347 | .4597 | .397 | .4012 | .447 | .3497 | .497 | .3036 |
| .348 | .4584 | .398 | .4001 | .448 | .3487 | .498 | .3028 |
| .349 | .4572 | .399 | .3990 | .449 | .3478 | .499 | .3019 |
| .350 | .4559 | .400 | .3979 | .450 | .3468 | .500 | .3010 |
| .351 | .4547 | .401 | .3969 | .451 | .3458 | .501 | .3002 |
| .352 | .4535 | .402 | .3958 | .452 | .3449 | .502 | .2993 |
| .353 | .4522 | .403 | .3947 | .453 | .3439 | .503 | .2984 |
| .354 | .4510 | .404 | .3936 | .454 | .3429 | .504 | .2975 |
| .355 | .4498 | .405 | .3925 | .455 | .3420 | .505 | .2967 |
| .356 | .4486 | .406 | .3915 | .456 | .3410 | .506 | .2959 |
| .357 | .4473 | .407 | .3904 | .457 | .3401 | .507 | .2950 |
| .358 | .4461 | .408 | .3893 | .458 | .3391 | .508 | .2941 |
| .359 | .4449 | .409 | .3883 | .459 | .3382 | .509 | .2933 |
| .360 | .4437 | .410 | .3872 | .460 | .3372 | .510 | .2924 |
| .361 | .4425 | .411 | .3862 | .461 | .3363 | .511 | .2916 |
| .362 | .4413 | .412 | .3851 | .462 | .3354 | .512 | .2907 |
| .363 | .4401 | .413 | .3840 | .463 | .3344 | .513 | .2899 |
| .364 | .4389 | .414 | .3830 | .464 | .3335 | .514 | .2890 |
| .365 | .4377 | .415 | .3819 | .465 | .3325 | .515 | .2882 |
| .366 | .4365 | .416 | .3809 | .466 | .3316 | .516 | .2873 |
| .367 | .4353 | .417 | .3799 | .467 | .3307 | .517 | .2865 |
| .368 | .4342 | .418 | .3788 | .468 | .3298 | .518 | .2857 |
| .369 | .4330 | .419 | .3778 | .469 | .3288 | .519 | .2848 |
| .370 | .4318 | .420 | .3768 | .470 | .3279 | .520 | .2840 |
| .371 | .4306 | .421 | .3757 | .471 | .3270 | .521 | .2831 |
| .372 | .4295 | .422 | .3747 | .472 | .3260 | .522 | .2823 |
| .373 | .4283 | .423 | .3737 | .473 | .3251 | .523 | .2815 |
| .374 | .4271 | .424 | .3726 | .474 | .3242 | .524 | .2807 |
| .375 | .4260 | .425 | .3716 | .475 | .3233 | .525 | .2798 |
| .376 | .4248 | .426 | .3706 | .476 | .3224 | .526 | .2790 |
| .377 | .4237 | .427 | .3696 | .477 | .3215 | .527 | .2782 |
| .378 | .4225 | .428 | .3685 | .478 | .3206 | .528 | .2774 |
| .379 | .4214 | .429 | .3675 | .479 | .3197 | .529 | .2766 |
| .380 | .4202 | .430 | .3665 | .480 | .3188 | .530 | .2757 |
| .381 | .4191 | .431 | .3655 | .481 | .3179 | .531 | .2749 |
| .382 | .4179 | .432 | .3645 | .482 | .3170 | .532 | .2741 |
| .383 | .4168 | .433 | .3635 | .483 | .3161 | .533 | .2733 |
| .384 | .4157 | .434 | .3625 | .484 | .3152 | .534 | .2725 |
| .385 | .4145 | .435 | .3615 | .485 | .3143 | .535 | .2717 |
| .386 | .4134 | .436 | .3605 | .486 | .3134 | .536 | .2708 |
| .387 | .4123 | .437 | .3595 | .487 | .3125 | .537 | .2700 |
| .388 | .4112 | .438 | .3585 | .488 | .3116 | .538 | .2692 |
| .389 | .4101 | .439 | .3575 | .489 | .3107 | .539 | .2684 |

| Transparency | Density | Transparency | Density | Transparency | Density | Transparency | Density |
|---|---|---|---|---|---|---|---|
| .540 | .2676 | .590 | .2291 | .640 | .1938 | .690 | .1612 |
| .541 | .2668 | .591 | .2284 | .641 | .1932 | .691 | .1605 |
| .542 | .2660 | .592 | .2277 | .642 | .1925 | .692 | .1599 |
| .543 | .2652 | .593 | .2269 | .643 | .1918 | .693 | .1593 |
| .544 | .2644 | .594 | .2262 | .644 | .1911 | .694 | .1586 |
| .545 | .2636 | .595 | .2255 | .645 | .1904 | .695 | .1580 |
| .546 | .2628 | .596 | .2248 | .646 | .1898 | .696 | .1574 |
| .547 | .2620 | .597 | .2240 | .647 | .1891 | .697 | .1568 |
| .548 | .2612 | .598 | .2233 | .648 | .1884 | .698 | .1562 |
| .549 | .2604 | .599 | .2226 | .649 | .1877 | .699 | .1555 |
| .550 | .2596 | .600 | .2219 | .650 | .1871 | .700 | .1549 |
| .551 | .2589 | .601 | .2211 | .651 | .1864 | .701 | .1543 |
| .552 | .2581 | .602 | .2204 | .652 | .1857 | .702 | .1537 |
| .553 | .2573 | .603 | .2197 | .653 | .1851 | .703 | .1531 |
| .554 | .2565 | .604 | .2190 | .654 | .1844 | .704 | .1524 |
| .555 | .2557 | .605 | .2182 | .655 | .1838 | .705 | .1518 |
| .556 | .2549 | .606 | .2175 | .656 | .1831 | .706 | .1512 |
| .557 | .2541 | .607 | .2168 | .657 | .1824 | .707 | .1506 |
| .558 | .2534 | .608 | .2161 | .658 | .1818 | .708 | .1500 |
| .559 | .2526 | .609 | .2154 | .659 | .1811 | .709 | .1493 |
| .560 | .2518 | .610 | .2147 | .660 | .1805 | .710 | .1487 |
| .561 | .2510 | .611 | .2140 | .661 | .1798 | .711 | .1481 |
| .562 | .2503 | .612 | .2132 | .662 | .1791 | .712 | .1475 |
| .563 | .2495 | .613 | .2125 | .663 | .1785 | .713 | .1469 |
| .564 | .2487 | .614 | .2118 | .664 | .1778 | .714 | .1463 |
| .565 | .2479 | .615 | .2111 | .665 | .1772 | .715 | .1457 |
| .566 | .2472 | .616 | .2104 | .666 | .1765 | .716 | .1451 |
| .567 | .2464 | .617 | .2097 | .667 | .1759 | .717 | .1445 |
| .568 | .2457 | .618 | .2090 | .668 | .1752 | .718 | .1439 |
| .569 | .2449 | .619 | .2083 | .669 | .1746 | .719 | .1433 |
| .570 | .2441 | .620 | .2076 | .670 | .1739 | .720 | .1427 |
| .571 | .2434 | .621 | .2069 | .671 | .1733 | .721 | .1421 |
| .572 | .2426 | .622 | .2062 | .672 | .1726 | .722 | .1415 |
| .573 | .2418 | .623 | .2055 | .673 | .1720 | .723 | .1409 |
| .574 | .2411 | .624 | .2048 | .674 | .1713 | .724 | .1403 |
| .575 | .2403 | .625 | .2041 | .675 | .1707 | .725 | .1397 |
| .576 | .2396 | .626 | .2034 | .676 | .1701 | .726 | .1391 |
| .577 | .2388 | .627 | .2027 | .677 | .1694 | .727 | .1385 |
| .578 | .2381 | .628 | .2020 | .678 | .1688 | .728 | .1379 |
| .579 | .2373 | .629 | .2013 | .679 | .1681 | .729 | .1373 |
| .580 | .2366 | .630 | .2007 | .680 | .1675 | .730 | .1367 |
| .581 | .2358 | .631 | .2000 | .681 | .1668 | .731 | .1361 |
| .582 | .2351 | .632 | .1993 | .682 | .1662 | .732 | .1355 |
| .583 | .2343 | .633 | .1986 | .683 | .1655 | .733 | .1349 |
| .584 | .2336 | .634 | .1979 | .684 | .1649 | .734 | .1343 |
| .585 | .2328 | .635 | .1972 | .685 | .1643 | .735 | .1337 |
| .586 | .2321 | .636 | .1965 | .686 | .1637 | .736 | .1331 |
| .587 | .2314 | .637 | .1959 | .687 | .1630 | .737 | .1325 |
| .588 | .2306 | .638 | .1952 | .688 | .1624 | .738 | .1319 |
| .589 | .2299 | .639 | .1945 | .689 | .1618 | .739 | .1314 |

| Transparency | Density | Transparency | Density | Transparency | Density | Transparency | Density |
|---|---|---|---|---|---|---|---|
| .740 | .1308 | .790 | .1024 | .840 | .0757 | .890 | .0506 |
| .741 | .1302 | .791 | .1018 | .841 | .0752 | .891 | .0501 |
| .742 | .1296 | .792 | .1013 | .842 | .0747 | .892 | .0496 |
| .743 | .1290 | .793 | .1007 | .843 | .0742 | .893 | .0491 |
| .744 | .1284 | .794 | .1002 | .844 | .0736 | .894 | .0487 |
| .745 | .1278 | .795 | .0996 | .845 | .0731 | .895 | .0482 |
| .746 | .1273 | .796 | .0991 | .846 | .0726 | .896 | .0477 |
| .747 | .1267 | .797 | .0985 | .847 | .0721 | .897 | .0472 |
| .748 | .1261 | .798 | .0980 | .848 | .0716 | .898 | .0467 |
| .749 | .1255 | .799 | .0975 | .849 | .0711 | .899 | .0462 |
| .750 | .1249 | .800 | .0969 | .850 | .0706 | .900 | .0458 |
| .751 | .1244 | .801 | .0964 | .851 | .0701 | .901 | .0453 |
| .752 | .1238 | .802 | .0958 | .852 | .0696 | .902 | .0448 |
| .753 | .1232 | .803 | .0953 | .853 | .0690 | .903 | .0443 |
| .754 | .1226 | .804 | .0948 | .854 | .0685 | .904 | .0438 |
| .755 | .1221 | .805 | .0942 | .855 | .0680 | .905 | .0434 |
| .756 | .1215 | .806 | .0937 | .856 | .0675 | .906 | .0429 |
| .757 | .1209 | .807 | .0931 | .857 | .0670 | .907 | .0424 |
| .758 | .1203 | .808 | .0926 | .858 | .0665 | .908 | .0419 |
| .759 | .1198 | .809 | .0921 | .859 | .0660 | .909 | .0414 |
| .760 | .1192 | .810 | .0915 | .860 | .0655 | .910 | .0410 |
| .761 | .1186 | .811 | .0910 | .861 | .0650 | .911 | .0405 |
| .762 | .1180 | .812 | .0904 | .862 | .0645 | .912 | .0400 |
| .763 | .1175 | .813 | .0899 | .863 | .0640 | .913 | .0395 |
| .764 | .1169 | .814 | .0894 | .864 | .0635 | .914 | .0391 |
| .765 | .1163 | .815 | .0888 | .865 | .0630 | .915 | .0386 |
| .766 | .1158 | .816 | .0883 | .866 | .0625 | .916 | .0381 |
| .767 | .1152 | .817 | .0878 | .867 | .0620 | .917 | .0376 |
| .768 | .1146 | .818 | .0872 | .868 | .0615 | .918 | .0371 |
| .769 | .1141 | .819 | .0867 | .869 | .0610 | .919 | .0367 |
| .770 | .1135 | .820 | .0862 | .870 | .0605 | .920 | .0362 |
| .771 | .1129 | .821 | .0856 | .871 | .0600 | .921 | .0357 |
| .772 | .1124 | .822 | .0851 | .872 | .0595 | .922 | .0353 |
| .773 | .1118 | .823 | .0846 | .873 | .0590 | .923 | .0348 |
| .774 | .1113 | .824 | .0841 | .874 | .0585 | .924 | .0343 |
| .775 | .1107 | .825 | .0835 | .875 | .0580 | .925 | .0339 |
| .776 | .1102 | .826 | .0830 | .876 | .0575 | .926 | .0334 |
| .777 | .1096 | .827 | .0825 | .877 | .0570 | .927 | .0329 |
| .778 | .1090 | .828 | .0820 | .878 | .0565 | .928 | .0325 |
| .779 | .1085 | .829 | .0815 | .879 | .0560 | .929 | .0320 |
| .780 | .1079 | .830 | .0809 | .880 | .0555 | .930 | .0315 |
| .781 | .1073 | .831 | .0804 | .881 | .0550 | .931 | .0310 |
| .782 | .1068 | .832 | .0799 | .882 | .0545 | .932 | .0306 |
| .783 | .1062 | .833 | .0794 | .883 | .0540 | .933 | .0301 |
| .784 | .1057 | .834 | .0788 | .884 | .0535 | .934 | .0296 |
| .785 | .1051 | .835 | .0783 | .885 | .0530 | .935 | .0292 |
| .786 | .1046 | .836 | .0778 | .886 | .0526 | .936 | .0287 |
| .787 | .1040 | .837 | .0773 | .887 | .0521 | .937 | .0282 |
| .788 | .1035 | .838 | .0767 | .888 | .0516 | .938 | .0278 |
| .789 | .1029 | .839 | .0762 | .889 | .0511 | .939 | .0273 |

| Transparency | Density | Transparency | Density | Transparency | Density | Transparency | Density |
|---|---|---|---|---|---|---|---|
| .940 | .0269 | .955 | .0200 | .970 | .0132 | .985 | .0066 |
| .941 | .0264 | .956 | .0195 | .971 | .0128 | .986 | .0061 |
| .942 | .0260 | .957 | .0191 | .972 | .0123 | .987 | .0057 |
| .943 | .0255 | .958 | .0186 | .973 | .0119 | .988 | .0052 |
| .944 | .0250 | .959 | .0182 | .974 | .0114 | .989 | .0048 |
| .945 | .0246 | .960 | .0177 | .975 | .0110 | .990 | .0044 |
| .946 | .0241 | .961 | .0173 | .976 | .0106 | .991 | .0039 |
| .947 | .0237 | .962 | .0168 | .977 | .0101 | .992 | .0035 |
| .948 | .0232 | .963 | .0164 | .978 | .0097 | .993 | .0030 |
| .949 | .0227 | .964 | .0159 | .979 | .0092 | .994 | .0026 |
| .950 | .0223 | .965 | .0155 | .980 | .0088 | .995 | .0022 |
| .951 | .0218 | .966 | .0150 | .981 | .0083 | .996 | .0017 |
| .952 | .0214 | .967 | .0146 | .982 | .0079 | .997 | .0013 |
| .953 | .0209 | .968 | .0141 | .983 | .0074 | .998 | .0009 |
| .954 | .0204 | .969 | .0137 | .984 | .0070 | .999 | .0004 |
| | | | | | | 1.000 | .0000 |

Data taken from: *Handbook of Chemistry and Physics,* 53rd ed., pp. E-235–E-237, R. C. Weast, Ed. Copyright 1972, The Chemical Rubber Co., Cleveland, Ohio.

# PROPERTIES OF VARIOUS LABORATORY MATERIALS

## Plastics

### PHYSICAL PROPERTIES

TABLE 1
SOME PROPERTIES OF SELECTED FORMULATIONS

| Clarity or % Light Transmission | Refractive Index | Specific Gravity | Specific Heat | Thermal Conductivity | Dielectric Constant, 60 Cycles | Water Absorption, % |
|---|---|---|---|---|---|---|
| **Cellulose Acetate** | | | | | | |
| Translucent to opaque | 1.46–1.50 | 1.24–1.34 | 0.3–0.42 | 4–8 | 3.5–7.5 | 1.9–6.5 |
| **Methyl Methacrylate** | | | | | | |
| 90–92 | 1.49 | 1.17–120 | 0.35 | 4.6 | 3.5–4.5 | 0.3–0.4 |
| **Nylon (Polyamide)** | | | | | | |
| Transparent to opaque | 1.53 | 1.09–1.14 | 0.4 | 5.2–5.8 | 4.1–4.6 | 0.4–3.3 |
| **Polyethylene, High-Density** | | | | | | |
| Translucent to opaque | 1.54 | 0.941–0.965 | 0.55 | 11–12.4 | 2.25–2.35 | <0.010 |
| **Polyethylene, Regular** | | | | | | |
| Translucent to opaque | 1.5 | 0.910–0.925 | 0.55 | 8.0 | 2.25–2.35 | <0.015 |
| **Polypropylene** | | | | | | |
| Transparent to opaque | 1.49 | 0.90–0.91 | 0.46 | 3.3 | | <0.01 |
| **Polystyrene, High-Impact** | | | | | | |
| Translucent to opaque | | 0.98–1.10 | 0.32–0.35 | 1.0–3.0 | 2.45–5.75 | 0.1–0.3 |
| **Polystyrene, Regular** | | | | | | |
| 88–99 | 1.59–1.60 | 1.04–1.065 | 0.32 | 2.4–3.3 | 2.45–2.65 | 0.03–0.05 |
| **Teflon* (Polytetrafluoroethylene)** | | | | | | |
| Opaque | 1.35 | 2.1–2.2 | 0.25 | 6 | 2.0 | 0.005 |
| **Vinyl, Rigid** | | | | | | |
| Transparent to opaque | 1.52–1.55 | 1.35–1.45 | 0.2–0.28 | 3.0–7 | 3.2–3.6 | 0.07–0.4 |

*Trade name, E.I. duPont de Nemours and Company.

Data reproduced by courtesy of Bel-Art Products, Pequannock, New Jersey.

Note: The above table shows averages and ranges. Many properties vary with manufacturer, formulation and testing laboratory. Most of the data given here were obtained from tests conducted according to ASTM standards.

## TABLE 2
## SOME PROPERTIES OF OTHER SELECTED FORMULATIONS

| Temperature Limit, °C | Specific Gravity | Tensile Strength, psi | Brittleness Temperature, °C | Water Absorption, % | Flexibility | Transparency | Relative $O_2$ Permeability | Autoclavable |
|---|---|---|---|---|---|---|---|---|
| **Noryl** | | | | | | | | |
| 135 | 1.06 | 9,600 | | 0.07 | Rigid | Opaque | | Yes |
| **Polyallomer** | | | | | | | | |
| 130 | 0.90 | 2,900 | − 40 | <0.02 | Slight | Translucent | 0.20 | Yes |
| **Polycarbonate** | | | | | | | | |
| 135 | 1.20 | 8,000 | −135 | <0.35 | Rigid | Clear | 0.15 | Yes |
| **Polyethylene, Conventional** | | | | | | | | |
| 80 | 0.92 | 2,000 | −100 | <0.01 | Excellent | Translucent | 0.40 | No |
| **Polyethylene, Linear** | | | | | | | | |
| 120 | 0.95 | 4,000 | −100 | <0.01 | Rigid | Opaque | 0.08 | With caution |
| **Polymethylpentene** | | | | | | | | |
| 175 | 0.83 | 4,000 | | <0.01 | Rigid | Clear | 2.0 | Yes |
| **Polypropylene** | | | | | | | | |
| 135 | 0.90 | 5,000 | 0 | <0.02 | Rigid | Translucent | 0.11 | Yes |
| **Polystyrene, General Purpose** | | | | | | | | |
| 70 | 1.07 | 6,000 | Normally somewhat brittle at room temperatures | 0.05 | Rigid | Clear | 0.11 | No |
| **Polyvinyl Chloride** | | | | | | | | |
| 70 | 1.34 | 6,500 | − 30 | 0.06 | Rigid | Clear | 0.01 | No |
| **Styrene–Acrylonitrile** | | | | | | | | |
| 95 | 1.07 | 11,000 | − 25 | 0.23 | Rigid | Clear | 0.03 | No |
| **Teflon* FEP** | | | | | | | | |
| 205 | 2.15 | 3,000 | −270 | <0.01 | Excellent | Translucent | 0.59 | Yes |

* Trade name, E.I. duPont de Nemours and Company.

Data reproduced by courtesy of Nalge Company, Division of Ritter Pfaudler Corporation, Rochester, New York.

## CHEMICAL RESISTANCE

The data presented in Table 3 were obtained from tests performed in accordance with ASTM D543-56T procedures. The plastics used in these tests were as free from internal and external stresses as possible, so that any attack was due solely to chemical action.

Because the plastics–chemicals relationship is subject to so many variables, this table should be used only as a guide to expected behavior and should not be interpreted as recommending contact of any plastic with a given reagent without testing under actual conditions of use.

Some effects of chemical attack are: color change due to oxidation, extraction or absorption; change in strength and flexibility; change in surface appearance, such as pitting, crazing, cracking, blistering, frosting or softening; dimensional changes due to distortion, swelling or shrinking; and weight changes due to absorption and extraction.

Various factors can retard chemical attack considerably. For example, extra wall thickness, carbon filling, and intermittent use enable polyethylene and polypropylene to resist even chemicals rated as "not recommended" for years.

### Code

Resins:
- CPE = conventional polyethylene
- LPE = linear polyethylene
- PA = polyallomer
- PP = polypropylene
- TPX* = polymethylpentene
- FEP = Teflon† FEP
- PC = polycarbonate
- SA = styrene–acrylonitrile
- PVC = polyvinyl chloride

Ratings:
- E = excellent – 30 days of constant exposure will cause no damage; may even tolerate exposure for years
- G = good – little or no damage after 30 days of constant exposure
- F = fair – some signs of attack after 7 days of constant exposure
- N = not recommended – noticeable signs of attack occur within minutes to hours after exposure; however, actual failure might take years

Note: In the table, the first letter represents the rating at room temperature and the second letter the rating at 52°C.

* Trade name, Imperial Chemical Industries Limited.
† Trade name, E. I. duPont de Nemours and Company.

### TABLE 3
### RESISTANCE GUIDE FOR SPECIFIC CHEMICAL COMPOUNDS

| Chemical Compound | CPE | LPE | PA | PP | TPX | FEP | PC | SA | PVC |
|---|---|---|---|---|---|---|---|---|---|
| Acetaldehyde | GN | GF | GN | GN | GN | EE | FN | NN | GN |
| Acetamide, sat. | EE | EE | EE | EE | EE | EE | NN | EE | NN |
| Acetic acid, 5% | EE | EE | EE | EE | EE | EE | EG | EE | EE |
| Acetic acid, 50% | EE | EE | EE | EE | EE | EE | EG | EF | EG |
| Acetone | EE | EE | EE | EE | EE | EE | NN | NN | FN |
| Adipic acid | EG | EE | EG | EE | EE | EE | EE | EE | EG |
| Alanine | EE | EE | EE | EE | EE | EE | NN | EE | NN |
| Allyl alcohol | EE | EE | EE | EE | EG | EE | EG | NN | GF |
| Aluminum hydroxide | EG | EE | EG | EG | EG | EE | FN | EG | EG |
| Aluminum salts | EE | EE | EE | EE | EE | EE | EG | EE | EE |
| Ammonia | EE | EE | EE | EE | EE | EE | NN | EG | EG |
| Ammonium acetate, sat. | EE | EE | EE | EE | EE | EE | EE | EE | EE |
| Ammonium glycolate | EG | EE | EG | EG | EG | EE | GF | EE | EE |
| Ammonium hydroxide, 5% | EE | EE | EE | EE | EE | EE | FN | EE | EE |
| Ammonium hydroxide, conc. | EG | EE | EG | EG | EG | EE | NN | EG | EG |

## TABLE 3 (Continued)
## RESISTANCE GUIDE FOR SPECIFIC CHEMICAL COMPOUNDS

| Chemical Compound | CPE | LPE | PA | PP | TPX | FEP | PC | SA | PVC |
|---|---|---|---|---|---|---|---|---|---|
| Ammonium oxalate | EG | EE | EG | EG | EG | EE | EE | EE | EE |
| Ammonium salts | EE | EE | EE | EE | EE | EE | EG | EE | EG |
| *n*-Amyl acetate | GF | EG | GF | GF | GF | EE | NN | NN | FN |
| Amyl chloride | NN | FN | NN | NN | NN | EE | NN | NN | NN |
| Antimony salts | EE | EE | EE | EE | EE | EE | EE | EE | EE |
| Arsenic salts | EE | EE | EE | EE | EE | EE | EE | EE | EE |
| Aspirin, powder | EE | EE | EE | EE | EE | EE | EE | EE | EE |
| Barium salts | EE | EE | EE | EE | EE | EE | EE | EE | EG |
| Benzaldehyde | EG | EE | EG | EG | EG | EE | FN | NN | NN |
| Benzene | FN | GG | GF | GF | GF | EE | NN | NN | NN |
| Benzoic acid, sat. | EE | EE | EG | EG | EG | EE | EG | EG | EG |
| Benzyl acetate | EG | EE | EG | EG | EG | EE | FN | NN | FN |
| Benzyl alcohol | NN | FN | NN | NN | NN | EE | GF | NN | GF |
| Bismuth salts | EE | EE | EE | EE | EE | EE | EE | EE | EE |
| Boric acid, 10% | EE | EE | EE | EE | EE | EE | EE | EE | EE |
| Boric acid, sat. | EE | EE | EE | EE | EE | EE | EE | EE | EE |
| Boron salts | EE | EE | EE | EE | EE | EE | EE | EE | EE |
| Brine | EE | EE | EE | EE | EE | EE | EE | EE | EE |
| Bromine | NN | FN | NN | NN | NN | EE | FN | NN | GN |
| Bromobenzene | NN | FN | NN | NN | NN | EE | NN | NN | NN |
| Butadiene | NN | FN | NN | NN | NN | EE | NN | NN | FN |
| *n*-Butyl acetate | GF | EG | GF | GF | GF | EE | NN | NN | NN |
| *n*-Butyl alcohol | EE | EE | EE | EE | EG | EE | GF | GN | GF |
| *sec*-Butyl alcohol | EG | EE | EG | EG | EG | EE | GF | GF | GG |
| *tert*-Butyl alcohol | EG | EE | EG | EG | EG | EE | GF | EG | EG |
| Butyric acid | NN | FN | NN | NN | NN | EE | FN | NN | GN |
| Cadmium salts | EE | EE | EE | EE | EE | EE | EE | EE | EE |
| Calcium hydroxide, conc. | EE | EE | EE | EE | EE | EE | NN | EE | EE |
| Calcium salts | EE | EE | EE | EE | EE | EE | EG | EE | EE |
| Carbazole | EE | EE | EE | EE | EE | EE | NN | EE | NN |
| Carbon dioxide | EE | EE | EE | EE | EE | EE | EG | EE | FF |
| Carbon monoxide | EE | EE | EE | EE | EE | EE | EE | EE | EE |
| Carbon tetrachloride | FN | GF | GF | GF | NN | EE | NN | GN | GF |
| Castor oil | EE | EE | EE | EE | EE | EE | EE | EE | EE |
| Cedarwood oil | NN | FN | NN | NN | NN | EE | GF | GG | FN |
| Cellosolve* acetate | EG | EE | EG | EG | EG | EE | FN | NN | FN |
| Cesium salts | EE | EE | EE | EE | EE | EE | EE | EE | EE |
| Chlorine, 10% in air | GN | EF | GN | GN | GN | EE | EG | GG | EE |
| Chlorine, 10%, moist | GN | GF | GN | GN | GN | EE | GF | NN | EG |
| Chloroacetic acid | EE | EE | EG | EG | EG | EE | FN | NN | FN |
| *p*-Chloroacetophenone | EE | EE | EE | EE | EE | EE | NN | NN | NN |
| Chloroform | FN | GF | GF | GF | FN | EE | NN | NN | NN |
| Chromic acid, 10% | EE | EE | EE | EE | EE | EE | EG | EG | EG |
| Chromic acid, 50% | EE | EE | EG | EG | EG | EE | EG | GF | EF |
| Cinnamon oil | NN | FN | NN | NN | NN | EE | GF | NN | NN |

## TABLE 3 (Continued)
## RESISTANCE GUIDE FOR SPECIFIC CHEMICAL COMPOUNDS

| Chemical Compound | CPE | LPE | PA | PP | TPX | FEP | PC | SA | PVC |
|---|---|---|---|---|---|---|---|---|---|
| Citric acid | EE | EE | EE | EE | EE | EE | EG | EE | GG |
| Citric acid, crystals | EE | EE | EE | EE | EE | EE | EE | EE | EG |
| Coconut oil | EE | EE | EG | EE | EG | EE | EE | EE | GF |
| Cupric salts | EE | EE | EE | EE | EE | EE | EE | EE | EE |
| Cuprous salts | EE | EE | EE | EE | EE | EE | EE | EE | EE |
| Cyclohexane | GF | EG | GF | GF | NN | EE | EG | NN | GF |
| Decalin‡ | GF | EG | GF | GF | FN | EE | NN | NN | EG |
| *o*-Dichlorobenzene | FN | FF | FN | FN | FN | EE | NN | NN | GN |
| *p*-Dichlorobenzene | FN | GF | EF | EF | GF | EE | NN | NN | NN |
| Diethyl benzene | NN | FN | NN | NN | NN | EE | FN | NN | NN |
| Diethyl ether | NN | FN | NN | NN | NN | EE | NN | NN | FN |
| Diethyl ketone | GF | GG | GG | GG | GF | EE | NN | NN | NN |
| Diethyl malonate | EE | EE | EE | EE | EG | EE | FN | NN | GN |
| Diethylene glycol | EE | EE | EE | EE | EE | EE | GF | EE | FN |
| Diethylene glycol ethyl ether | EE | EE | EE | EE | EE | EE | FN | NN | FN |
| Dimethyl formamide | EE | EE | EE | EE | EE | EE | NN | NN | FN |
| Dimethyl sulfoxide | EE | EE | EE | EE | EE | EE | NN | NN | NN |
| 1,4-Dioxane | GF | GG | GF | GF | GF | EE | GF | NN | FN |
| Dipropylene glycol | EE | EE | EE | EE | EE | EE | GF | EE | GF |
| Ethyl acetate | EE | EE | EG | EE | EG | EE | NN | NN | FN |
| Ethyl acetate, 85–88% | EE | EE | EE | EE | EE | EE | FN | NN | FN |
| Ethyl alcohol | EG | EE | EG | EG | EG | EE | EG | EN | EG |
| Ethyl alcohol, 40% | EG | EE | EG | EG | EG | EE | EG | EN | EE |
| Ethyl benzene | FN | GF | FN | FN | FN | EE | NN | NN | NN |
| Ethyl benzoate | FF | GG | GF | GF | GF | EE | NN | NN | NN |
| Ethyl butyrate | GN | GF | GN | GN | FN | EE | NN | NN | NN |
| Ethyl chloride, liquid | FN | FF | FN | FN | FN | EE | NN | NN | NN |
| Ethyl cyanoacetate | EE | EE | EE | EE | EE | EE | FN | NN | FN |
| Ethyl lactate | EE | EE | EE | EE | EE | EE | FN | NN | FN |
| Ethylene chloride | GN | GF | FN | FN | NN | EE | NN | NN | NN |
| Ethylene glycol | EE | EE | EE | EE | EE | EE | GF | EE | EE |
| Ethylene glycol methyl ether | EE | EE | EE | EE | EE | EE | FN | NN | FN |
| Ethylene oxide | FF | GF | FF | FF | FN | EE | FN | NN | FN |
| Ferric salts | EE | EE | EE | EE | EE | EE | EE | EE | EE |
| Ferrous salts | EE | EE | EE | EE | EE | EE | EE | EE | EE |
| Fluorine | FN | GN | FN | FN | FN | EG | GF | NN | EG |
| Formaldehyde, 10% | EE | EE | EE | EE | EG | EE | EG | EG | GF |
| Formaldehyde, 40% | EG | EE | EG | EG | EG | EE | EG | EG | GF |
| Formic acid, 3% | EG | EE | EG | EG | EG | EE | EG | EG | GF |
| Formic acid, 50% | EG | EE | EG | EG | EG | EE | EG | FF | GF |
| Formic acid, 98–100% | EG | EE | EG | EG | EF | EE | EF | FN | FN |
| Gasoline, aviation | FN | GF | GF | GF | GF | EE | FN | EF | NN |
| Gasoline, ethyl | FN | GF | GF | GF | GF | EE | FF | EE | EN |
| Gasoline, regular | FN | GG | GF | GF | GF | EE | FF | EE | GN |
| Gasoline, white | FN | GF | GF | GF | GF | EE | FN | EE | NN |
| Glacial acetic acid | EG | EE | EG | EG | EG | EE | GF | NN | EG |
| Glycerine | EE | EE | EE | EE | EE | EE | EE | EE | EE |
| *n*-Heptane | FN | GF | FF | FF | FF | EE | EG | EE | FN |
| 2-Heptyl | EG | EE | EG | EG | EG | EE | EG | EG | EG |
| Hexane | NN | GF | FF | EF | FF | EE | FN | EE | GN |
| Hydrochloric acid, 1–5% | EE | EE | EG | EE | EG | EE | EE | EE | EE |

## TABLE 3 (Continued)
## RESISTANCE GUIDE FOR SPECIFIC CHEMICAL COMPOUNDS

| Chemical Compound | CPE | LPE | PA | PP | TPX | FEP | PC | SA | PVC |
|---|---|---|---|---|---|---|---|---|---|
| Hydrochloric acid, 20% | EE | EE | EG | EE | EG | EE | EG | EE | EG |
| Hydrochloric acid, 35% | EE | EE | EG | EG | EG | EE | EG | EG | GF |
| Hydrofluoric acid, 4% | EG | EE | EG | EG | EG | EE | EE | EG | GF |
| Hydrofluoric acid, 48% | EE | EE | EE | EE | EE | EE | EG | GF | GF |
| Hydrogen | EE | EE | EE | EE | EE | EE | EE | EE | EE |
| Hydrogen peroxide, 3% | EE | EE | EE | EE | EE | EE | EE | EE | EE |
| Hydrogen peroxide, 30% | EG | EE | EG | EG | EG | EE | EE | EE | EE |
| Hydrogen peroxide, 90% | EG | EE | EG | EG | EG | EE | EE | EE | EG |
| Isobutyl alcohol | EE | EE | EE | EE | EG | EE | EG | EG | EG |
| Isopropyl acetate | GF | EG | GF | GF | GF | EE | NN | NN | NN |
| Isopropyl alcohol | EE | EE | EE | EE | EE | EE | EE | EF | EG |
| Isopropyl benzene | FN | GF | NN | FN | NN | EE | NN | NN | NN |
| Kerosene | FN | GG | GF | GF | GF | EE | GF | EE | EE |
| Lactic acid, 3% | EG | EE | EG | EG | EG | EE | EG | EE | GF |
| Lactic acid, 85% | EE | EE | EG | EG | EG | EE | EG | EE | GF |
| Lead salts | EE | EE | EE | EE | EE | EE | EE | EE | EE |
| Lithium salts | EE | EE | EE | EE | EE | EE | GF | EE | EE |
| Magnesium salts | EE | EE | EE | EE | EE | EE | EG | EE | EE |
| Mercuric salts | EE | EE | EE | EE | EE | EE | EE | EE | EE |
| Mercurous salts | EE | EE | EE | EE | EE | EE | EE | EE | EE |
| Methoxyethyl oleate | EG | EE | EG | EG | EG | EE | FN | NN | NN |
| Methyl alcohol | EE | EE | EE | EE | EE | EE | FN | NN | EF |
| Methyl ethyl ketone | EG | EE | EG | EG | EF | EE | NN | NN | NN |
| Methyl isobutyl ketone | GF | EG | GF | GF | FF | EE | NN | NN | NN |
| Methyl propyl ketone | GF | EG | GF | GF | FF | EE | NN | NN | NN |
| Mineral oil | GN | EE | EG | EE | EG | EE | EG | EE | EG |
| Nickel salts | EE | EE | EE | EE | EE | EE | EE | EE | EE |
| Nitric acid, 1% | EE | EE | EE | EE | EE | EE | EG | EG | EG |
| Nitric acid, 10% | EE | EE | EE | EE | EE | EE | EG | EG | EG |
| Nitric acid, 50% | EG | EE | EG | EG | EG | EE | GF | GF | GF |
| Nitric acid, 70% | EN | EF | GN | EN | GN | EE | FN | NN | FN |
| Nitrobenzene | NN | FN | NN | NN | NN | EE | NN | NN | NN |
| Nitrosyl salts | NN | NN | NN | NN | NN | EE | NN | NN | NN |
| *n*-Octane | EE | EE | EE | EE | EE | EE | GF | NN | FN |
| Orange oil | FN | GF | GF | GF | FF | EE | FF | EE | FN |
| Ozone | EG | EE | EG | EG | EE | EE | EG | EE | EG |
| Perchloric acid | EE | EE | EE | EE | EG | EE | EG | EG | EF |
| Phosphoric acid, 1–5% | EE | EE | EE | EE | EE | EE | EE | EE | EE |
| Phosphoric acid, 85% | EE | EE | EG | EG | EG | EE | EG | EG | EG |
| Phosphorus salts | EE | EE | EE | EE | EE | EE | EE | EE | EE |
| Pine oil | GN | EG | EG | EG | GF | EE | GF | FN | FN |
| Potassium hydroxide, 1% | EE | EE | EE | EE | EE | EE | FN | EE | EE |
| Potassium hydroxide, conc. | EE | EE | EE | EE | EE | EE | NN | EG | EG |
| Potassium salts | EE | EE | EE | EE | EE | EE | EG | EE | EE |
| Propane gas | NN | FN | NN | NN | NN | EE | FN | EE | EG |
| Propylene glycol | EE | EE | EE | EE | EE | EE | GF | EE | FN |
| Propylene oxide | EG | EE | EG | EG | EG | EE | GF | NN | FN |
| Resorcinol, sat. | EE | EE | EE | EE | EE | EE | GF | EE | FN |
| Resorcinol, 5% | EE | EE | EE | EE | EE | EE | GF | EE | GN |
| Salicylaldehyde | EG | EE | EG | EG | EG | EE | GF | NN | FN |

## TABLE 3 (Continued)
## RESISTANCE GUIDE FOR SPECIFIC CHEMICAL COMPOUNDS

| Chemical Compound | CPE | LPE | PA | PP | TPX | FEP | PC | SA | PVC |
|---|---|---|---|---|---|---|---|---|---|
| Salicylic acid, powder | EE | EE | EE | EE | EG | EE | EG | EE | GF |
| Salicylic acid, sat. | EE | EE | EE | EE | EE | EE | EG | EE | GF |
| Silver acetate | EE | EE | EE | EE | EE | EE | EG | EE | GG |
| Silver salts | EG | EE | EG | EG | EE | EE | EE | EE | EG |
| Sodium acetate, sat. | EE | EE | EE | EE | EE | EE | EG | EE | GF |
| Sodium benzoate, powder | EE | EE | EE | EE | EE | EE | EG | EE | EE |
| Sodium benzoate, sat. | EE | EE | EE | EE | EE | EE | EE | EE | EE |
| Sodium hydroxide, 1% | EE | EE | EE | EE | EE | EE | FN | EE | EE |
| Sodium hydroxide, 50% | EE | EE | EE | EE | EE | EE | NN | EG | EG |
| Sodium hydroxide, sat. | EE | EE | EE | EE | EE | EE | NN | EG | EG |
| Sodium salts | EE | EE | EE | EE | EE | EE | EG | EE | EE |
| Stannic salts | EE | EE | EE | EE | EE | EE | EE | EE | EG |
| Stannous salts | EE | EE | EE | EE | EE | EE | EE | EE | EG |
| Stearic, crystals | EE | EE | EE | EE | EE | EE | EG | EE | EG |
| Sulfuric acid, 1–6% | EE | EE | EE | EE | EE | EE | EE | EE | EG |
| Sulfuric acid, 20% | EE | EE | EG | EG | EG | EE | EG | EG | EG |
| Sulfuric acid, 60% | EG | EE | EG | EG | EG | EE | EG | EG | GF |
| Sulfuric acid, 98% | EG | EE | EE | EE | EE | EE | GF | NN | NN |
| Sulfur dioxide, dry | EE | EE | EE | EE | EE | EE | EG | GF | EG |
| Sulfur dioxide, liquid, 46 psi | NN | FN | NN | NN | NN | EE | GN | NN | FN |
| Sulfur dioxide, wet | EE | EE | EE | EE | EE | EE | EG | GF | EG |
| Sulfur salts | FN | GF | FN | FN | FN | EE | FN | NN | NN |
| Tartaric acid, powder | EE | EE | EE | EE | EE | EE | EG | EE | EG |
| Tartaric acid, sat. | EE | EE | EE | EE | EE | EE | EG | EE | EG |
| Tetrahydrofuran | FN | GF | GF | GF | FF | EE | NN | NN | NN |
| Thionyl salts | NN | NN | NN | NN | NN | EE | NN | NN | NN |
| Titanium salts | EE | EE | EE | EE | EE | EE | EE | EE | EE |
| Toluene | FN | GG | GF | GF | FF | EE | FN | NN | FN |
| Tributyl citrate | GF | EG | GF | GF | GF | EE | NN | NN | FN |
| Trichloroethane | NN | FN | NN | NN | NN | EE | NN | NN | NN |
| Triethylene glycol | EE | EE | EE | EE | EE | EE | EG | EE | GF |
| Tripropylene glycol | EE | EE | EE | EE | EE | EE | EG | EE | GF |
| Turkey red oil | EE | EE | EE | EE | EE | EE | EG | EE | EG |
| Turpentine | FN | GG | GF | GF | FF | EE | FN | GF | GF |
| Undecyl alcohol | EF | EG | EG | EG | EG | EE | GF | EE | EF |
| Urea | EE | EE | EE | EE | EG | EE | NN | EE | GN |
| Vinylidene chloride | NN | FN | NN | NN | NN | EE | NN | NN | NN |
| Water | EE | EE | EE | EE | EE | EE | EE | EE | EE |
| Xylene | GN | GF | FN | FN | FN | EE | NN | NN | NN |
| Zinc salts | EE | EE | EE | EE | EE | EE | EE | EE | EE |
| Zinc stearate | EE | EE | EE | EE | EE | EE | EE | EE | EG |

* Trade name, Union Carbide Corporation.
† Trade name, E. I. duPont de Nemours and Company.
‡ Trade name, The Dow Chemical Company.

Data reproduced by courtesy of Nalge Company, Division of Ritter Pfaudler Corporation, Rochester, New York.

The ratings given in Table 4 apply to most compounds of the classes indicated. Specific materials should be tested under conditions of actual use, if any doubt exists.

Because so many factors can affect the chemical resistance of a given product, this table should be used only as a guide. For example, the combination of compounds of two or more classes may cause an undesirable chemical effect. Other factors affecting chemical resistance include temperature, pressure and other stresses, length of exposure, and concentration of the chemical. As the maximum useful temperature of the plastic is approached, resistance to chemical attack decreases.

**Code**

Resins:
- CPE = conventional polyethylene
- LPE = linear polyethylene
- PA = polyallomer
- PP = polypropylene
- TPX = polymethylpentene
- FEP = Teflon FEP
- PC = polycarbonate
- NO = noryl
- PS = general-purpose polystyrene
- SA = styrene–acrylonitrile
- PVC = polyvinyl chloride

Ratings:
- E = excellent – long exposures (up to 1 year) at room temperature have no effect
- G = good – short exposures (less than 24 hours) at room temperature cause no damage
- F = fair – short exposures at room temperature cause little or no damage under unstressed conditions
- N = not recommended – short exposures may cause permanent damage

**TABLE 4**
**RESISTANCE GUIDE FOR COMMON SOLVENTS AT ROOM TEMPERATURE**

| Class of Substances | CPE | LPE | PA | PP | TPX | FEP | PC | NO | PS | SA | PVC |
|---|---|---|---|---|---|---|---|---|---|---|---|
| Acids, inorganic | E | E | E | E | E | E | E | G | N | E | G |
| Acids, organic | E | E | E | E | E | E | G | E | G | E | G |
| Alcohols | E | E | E | E | E | E | G | E | G | G | G |
| Aldehydes | G | G | G | G | G | E | F | G | N | F | F |
| Amines | G | G | G | G | G | E | N | F | G | G | N |
| Bases | E | E | E | E | E | E | N | E | G | E | E |
| Dimethyl sulfoxide | E | E | E | E | E | E | N | E | N | N | N |
| Esters | E | E | E | E | E | E | N | F | N | N | F |
| Ethers | G | G | G | G | G | E | F | N | F | N | F |
| Foods | E | E | E | E | E | E | E | G | E | G | G |
| Glycols | E | E | E | E | E | E | G | E | G | G | F |
| Hydrocarbons, aliphatic | G | G | G | G | G | E | F | G | N | E | F |
| Hydrocarbons, aromatic | G | G | G | G | F | E | N | N | N | N | N |
| Hydrocarbons, halogenated | G | G | G | G | F | E | N | N | N | N | N |
| Ketones | G | G | G | G | G | E | N | N | N | N | N |
| Mineral oil | E | E | E | E | E | E | E | E | G | G | E |
| Oils, essential | G | G | G | G | G | E | G | F | N | F | N |
| Oils, lubricating | G | E | E | E | E | E | G | E | G | G | E |
| Oils, vegetable | E | E | E | E | E | E | E | E | G | E | E |
| Proteins, unhydrolyzed | E | E | E | E | E | E | E | G | G | E | G |
| Salts | E | E | E | E | E | E | E | E | E | E | E |
| Silicones | G | E | E | E | E | E | E | E | G | G | G |
| Water | E | E | E | E | E | E | E | E | E | E | E |

Data reproduced by courtesy of Nalge Company, Division of Ritter Pfaudler Corporation, Rochester, New York.

Table 5 is based on information extracted from current literature and is intended to be used solely as a guide in selecting the proper tube material. The user is urged to make preliminary tests under actual conditions of use.

**Code**

Resins:
- CN = cellulose nitrate
- PA = polyallomers
- PC = polycarbonate
- PE = polyethylene
- PP = polypropylene

Ratings:
- E = excellent
- G = good
- F = fair
- N = not recommended

## TABLE 5
## RESISTANCE GUIDE FOR SELECTION OF PLASTIC ULTRACENTRIFUGE TUBES

| Chemical Compound | CN | PA | PC | PE | PP |
|---|---|---|---|---|---|
| Acetaldehyde | N | G | N | G | G |
| Acetic acid, 10% | E | E | E | E | E |
| Acetic acid, glacial | N | E | N | E | E |
| Acetic anhydride | | | N | N | E |
| Acetone | N | E | N | G | F |
| Acetophenone | | G | N | G | G |
| Aluminum salts | E | E | E | G | E |
| Ammonium salts | E | E | E | E | |
| Ammonium hydroxide, 10% | N | E | N | E | E |
| Ammonium hydroxide, conc. | N | E | N | E | E |
| Amyl acetate | N | F | N | F | F |
| Amyl alcohol | N | G | E | G | F |
| Amyl chloride | | | | F | |
| Aniline | | | | G | N |
| Aqua regia | N | | N | | N |
| Asphalt | E | | E | E | |
| Barium salts | | E | E | E | E |
| Benzaldehyde | | G | F | F | G |
| Benzene | E | N | N | N | N |
| Benzoic acid | E | E | E | E | E |
| Benzyl alcohol | E | E | N | E | E |
| Borax | E | E | E | E | |
| Boric acid | E | E | E | E | E |
| Butyl acetate | N | E | E | E | G |
| Butyl alcohol | N | E | E | E | E |
| Calcium salts | E | E | G | E | E |
| Carbon disulfide | | | | | N |
| Carbon tetrachloride | E | N | N | F | F |
| Chlorobenzene | N | N | N | F | F |
| Chlorine, aq. sol. | | | E | F | G |
| Chloroform | F | N | N | N | F |
| Chlorosulfonic acid | | N | N | N | F |

| Chemical Compound | CN | PA | PC | PE | PP |
|---|---|---|---|---|---|
| Chromic acid | N | G | E | G | G |
| Chromic acid, 30% | N | G | G | G | G |
| Citric acid | E | E | E | E | E |
| Copper salts | | E | E | G | E |
| Cottonseed oil | E | E | E | E | E |
| Creosote | E | E | | E | E |
| Cresol | | | N | | |
| Cyclohexanol | | G | N | F | G |
| Cyclohexanone | | G | N | N | F |
| Decalin* | | | | N | |
| Dibutyl phthalate | | | | F | |
| Diethyl ether | N | N | N | G | G |
| Diethyl ketone | N | N | N | G | G |
| Dioxane | | F | N | N | F |
| Ethyl acetate | N | F | N | E | F |
| Ethyl alcohol, 50% | E | E | E | E | E |
| Ethyl alcohol, 95% | N | E | E | E | E |
| Ethylene chlorohydrin | N | F | N | F | F |
| Ethylene dichloride | N | F | N | F | F |
| Ethylene glycol | E | E | G | E | E |
| Ferric salts | | E | E | E | E |
| Ferrous salts | | E | E | E | E |
| Formaldehyde, 40% | E | E | E | E | E |
| Formic acid | E | E | F | E | E |
| Gasoline | N | N | F | N | N |
| Glycerin | E | E | E | E | E |
| Hydrochloric acid, 10% | E | E | E | E | E |
| Hydrochloric acid, conc. | N | E | F | E | G |
| Hydrocyanic acid | E | E | E | E | E |
| Hydrofluoric acid, 10% | G | E | G | E | E |
| Hydrofluoric acid, 50% | N | E | N | E | E |
| Hydrogen peroxide, 3% | E | E | E | G | G |

* Trade name, E. I. duPont de Nemours and Company.

## TABLE 5 (Continued)
## RESISTANCE GUIDE FOR SELECTION OF PLASTIC ULTRACENTRIFUGE TUBES

| Chemical Compound | CN | PA | PC | PE | PP |
|---|---|---|---|---|---|
| Hydrogen sulfite, aq. sol. | E | E | E | E | E |
| Iodine, tincture | | E | F | G | G |
| Isopropyl alcohol | N | E | E | G | |
| Kerosene | E | N | E | N | N |
| Lacquer | N | | N | G | |
| Lactic acid, 20% | | E | E | F | E |
| Linseed oil | E | E | E | G | E |
| Magnesium salts | E | E | E | E | E |
| Malic acid | | E | E | G | E |
| Manganese salts | | E | E | E | E |
| Mercury salts | E | E | E | E | E |
| Methyl alcohol, abs. | N | E | N | G | G |
| Methyl ethyl ketone | N | E | N | F | F |
| Methylene chloride | N | F | N | F | F |
| Nickel salts | E | E | E | E | E |
| Nitric acid, 10% | E | E | E | G | E |
| Nitric acid, conc. | N | E | N | N | F |
| Nitrobenzene | | F | N | N | F |
| Oils, vegetable | E | E | E | G | E |
| Oleic acid | E | E | E | G | |
| Oxalic acid | E | E | E | E | E |
| Palmitic acid | | E | E | G | G |
| Perchloric acid | | E | N | E | E |
| Phenol | | E | N | G | N |
| Phosphoric acid, 10% | E | E | E | E | E |
| Phosphoric acid, conc. | F | E | G | G | F |
| Picric acid | | | | G | |
| Potassium salts | E | E | E | E | E |
| Potassium hydroxide, 5% | E | E | N | E | E |
| Potassium hydroxide, conc. | E | E | N | E | E |
| Sodium salts | E | E | F | E | E |
| Sodium hydroxide, 1% | E | E | N | G | E |
| Sodium hydroxide, 10% | N | E | N | G | G |
| Sodium hydroxide, conc. | N | E | N | G | F |
| Stearic acid | | E | F | G | E |
| Sucrose solution | E | E | E | E | E |
| Sulfuric acid, 10% | E | E | N | E | E |
| Sulfuric acid, conc. | N | E | E | N | E |
| Sulfurous acid | | | | G | E |
| Tannic acid | F | G | | G | E |
| Tartaric acid | | E | | E | E |
| Tetrahydrofuran | | | | N | |
| Thiopen | | | | N | |
| Toluene | E | N | N | G | N |
| Trichloroethylene | | N | N | N | N |
| Turpentine | | F | N | N | F |
| Urea | E | | E | E | E |
| Xylene | N | N | N | N | N |
| Zinc salts | E | E | E | E | E |

Data from: *Iscotables,* 3rd ed., p. 30, 1970, by permission of the publishers, Instrumentation Specialties Company, Lincoln, Nebraska.

## LABORATORY PERFORMANCE

## TABLE 6
## PERFORMANCE CHARACTERISTICS OF PLASTICS IN LABORATORY USE

| General Characteristics | Clarity | Autoclave Results | Heat Distortion Point | Burning Rate | Gas Permeability of Thin-Wall Products | | |
|---|---|---|---|---|---|---|---|
| | | | | | $O_2$ | $N_2$ | $CO_2$ |
| **ABS (Acrylonitrile–Butadiene–Styrene)** | | | | | | | |
| Very tough; somewhat flexible | Opaque | Melts | 77–93°C<br>170–200°F | Slow | | | |
| **Cellulose Acetate** | | | | | | | |
| Tough; somewhat flexible | Clear | Melts | 43–90°C<br>110–194°F | Slow | Very low | Very low | High |

## TABLE 6 (Continued)
## PERFORMANCE CHARACTERISTICS OF PLASTICS IN LABORATORY USE

| General Characteristics | Clarity | Autoclave Results | Heat Distortion Point | Burning Rate | Gas Permeability of Thin-Wall Products $O_2$ | $N_2$ | $CO_2$ |
|---|---|---|---|---|---|---|---|
| **Cellulose Nitrate** | | | | | | | |
| Tough; fairly clear | Clear | Melts | 60–71°C<br>140–160°F | Fast (explosive) | | | |
| **Methyl Methacrylate** | | | | | | | |
| Finest optical qualities; easily fabricated | Clear | Melts | 71–88°C<br>160–190°F | Slow | Very high | Very low | |
| **Nylon** | | | | | | | |
| Tough; heat resistant; machinable; high-moisture vapor transmission | Opaque | OK | 150–180°C<br>300–356°F | Self-extinguishing | Very low | Very low | |
| **Polycarbonate** | | | | | | | |
| Very tough; inert; high-temperature resistant | Clear | OK | 138–143°C<br>280–290°F | Self-extinguishing | Very low | Very low | Low |
| **Polyethylene** | | | | | | | |
| Biologically inert; high chemical resistance | Opaque | Melts | 40–49°C<br>105–120°F | Slow | High | Low | Very high |
| **Polypropylene** | | | | | | | |
| Biologically inert; high chemical resistance; exceptional toughness | Translucent | Withstands several cycles | 121°C<br>250°F | Slow | High | Low | Very high |
| **Polypropylene Film** | | | | | | | |
| Popular as film material | Clear | OK | 126°C<br>260°F | Slow | High | Low | Very high |
| **Polystyrene** | | | | | | | |
| Biologically inert; hard; excellent optical qualities | Clear | Melts | 64–80°C<br>147–175°F | Slow | Low | Very low | High |
| **Polystyrene, High-Impact** | | | | | | | |
| Rubber content improves strength of styrene | Opaque | Melts | 64–90°C<br>147–195°F | Slow | | | |
| **Polyvinyl Chloride, Plasticized** | | | | | | | |
| Inert; tough; high chemical resistance | Clear | Melts | 43–80°C<br>110–175°F | Self-extinguishing | Low | | High |

TABLE 6 (Continued)
PERFORMANCE CHARACTERISTICS OF PLASTICS IN LABORATORY USE

| General Characteristics | Clarity | Autoclave Results | Heat Distortion Point | Burning Rate | Gas Permeability of Thin-Wall Products $O_2$ | $N_2$ | $CO_2$ |
|---|---|---|---|---|---|---|---|
| **PTE (Teflon*)** | | | | | | | |
| Biologically inert; chemically inert; high-heat resistant; slippery surface | Opaque | OK | 121°C<br>250°F | None | | | |
| **Styrene–Acrylonitrile** | | | | | | | |
| Stronger than polystyrene | Clear | Melts | 90–93°C<br>195–200° F | Slow | Very low | Very low | Low |
| **Thermosetting Polyester Film** | | | | | | | |
| Popular as film material | Clear | OK | 121°C | Self-extinguishing | Very low | Very low | Very low |
| **Vinyl Chloride** | | | | | | | |
| Popular as film material | Clear | Melts | 54–66°C<br>130–150°F | Self-extinguishing | Low | | High |

* Trade name, E. I. duPont de Nemours and Company.

Data taken from: Thermoplastic Properties Chart, *Falcon Plastics Catalog,* p. 24, by permission of copyright owners, Falcon Plastics, Division of BioQuest, Oxnard, California.

TABLE 7
EFFECTS OF LABORATORY REAGENTS IN ROUTINE STORAGE AND CONTACT PERIODS

| Weak Acids | Strong Acids | Weak Alkalies | Strong Alkalies | Organic Solvents |
|---|---|---|---|---|
| **ABS (Acrylonitrile–Butadiene–Styrene)** | | | | |
| None | Oxidizing acids attack | None | None | Soluble in ketones, esters and chloroform |
| **Cellulose Acetate** | | | | |
| Slight | Decomposes | Slight | Decomposes | Softens in alcohol; soluble in ketones and esters |
| **Cellulose Nitrate** | | | | |
| Slight | Decomposes | Slight | Decomposes | Softens in alcohol; slightly affected by hydrocarbons; soluble in ketones and esters |
| **Methyl Methacrylate** | | | | |
| Slight | Oxidizing acids attack | Slight | Slight | Soluble in ketones, esters and aromatic hydrocarbons |
| **Nylon** | | | | |
| Resistant | Attacked | None | None | Resistant |

## TABLE 7 (Continued)
## EFFECTS OF LABORATORY REAGENTS IN ROUTINE STORAGE AND CONTACT PERIODS

| Weak Acids | Strong Acids | Weak Alkalies | Strong Alkalies | Organic Solvents |
|---|---|---|---|---|
| **Polycarbonate** | | | | |
| None | Very resistant | None | Slowly attacked | Soluble in chlorinated hydrocarbons; partly soluble in aromatics |
| **Polyethylene** | | | | |
| Resistant | Oxidizing acids attack | Resistant | Resistant | Resistant |
| **Polypropylene** | | | | |
| Resistant | Oxidizing acids attack | None | Very resistant | Resistant below 175°F |
| **Polypropylene Film** | | | | |
| Resistant | Oxidizing acids | None | Very resistant | Resistant below 175°F |
| **Polystyrene** | | | | |
| None | Oxidizing acids attack | None | None | Soluble in aromatic chlorinated hydrocarbons |
| **Polystyrene, High-Impact** | | | | |
| None | Oxidizing acids attack | None | None | Soluble in aromatic chlorinated hydrocarbons |
| **Polyvinyl Chloride, Plasticized** | | | | |
| None | None | None | None | Soluble in ketones and esters |
| **PTE (Teflon*)** | | | | |
| None | None | None | None | Resistant |
| **Styrene–Acrylonitrile** | | | | |
| None | Oxidizing acids attack | None | None | Soluble in ketones, esters and chlorinated hydrocarbons |
| **Thermosetting Polyester Film** | | | | |
| Resistant | Resistant | None | None | Good to excellent resistance |
| **Vinyl Chloride** | | | | |
| None | Resistant | Resistant | Resistant | Slight; resistant to hydrocarbons, ketones, etc. |

* Trade name, E. I. duPont de Nemours and Company.

Data taken from: Thermoplastic Properties Chart, *Falcon Plastics Catalog,* p. 24, by permission of copyright owners, Falcon Plastics, Division of BioQuest, Oxnard, California.

## Commercial Rubbers

### CHEMICAL EFFECTS OF COMMON SOLVENTS ON VARIOUS COMMERCIAL RUBBERS

**Code**

NP = neoprene
TH = thiokol
EP = ethylene propylene
SL = silicone
SP = styrene–butadiene (Buna-S)
FS = fluorosilicone

PU = polyurethane
PA = polyacrylate
AB = acrylonitrile–butadiene (Buna-N)
VT = viton
BP = isobutylene–isoprene (Butyl)

| Solvent | Percent Volume Change | | | | | | | | | | |
|---|---|---|---|---|---|---|---|---|---|---|---|
| | NP | TH | EP | SL | SB | FS | PU | PA | AB | VT | BP |
| **Immersion for 70 Hours at Room Temperature** | | | | | | | | | | | |
| Acetone | +31 | +7 | +2 | +18 | +18 | +205S | +87 | +201 | +125 | +165 | +2 |
| Aniline | +51 | +325 | +1 | +2 | +24 | −4 | +261 | +272 | +225 | +1 | +1 |
| Benzene | +168 | +113 | +82 | +70 | +23 | +23 | +109 | +214 | +155 | +10 | +92 |
| Butyl acetate | +107 | +34 | +18 | +78 | +99 | +175 | +100 | +175 | +116 | +294 | +27 |
| Butyl carbitol | +42 | +15 | +1 | +4 | +18 | +5 | +81 | +84 | +40 | +67 | NC |
| Butyl Cellosolve* | +31 | +7 | +2 | +100 | +23 | +7 | +72 | +89 | +38 | +8 | +1 |
| Carbon disulfide | +163 | +51 | +135 | +48 | +191 | +20 | +42 | +51 | +56 | +1 | +139 |
| Carbon tetrachloride | +142 | +53 | +152 | +103 | +207 | +12 | +76 | +214 | +81 | +1 | +173 |
| Chloroform | +191 | +311 | +113 | +99 | +217 | +36 | +263 | +296 | +270 | +9 | +139 |
| Cyclohexane | +73 | +3 | +148 | +102 | +143 | +15 | +21 | +24 | +11 | NC | +195 |
| Diphenyl (Dowtherm[†] A) | +185 | +150 | +16 | NC | +174 | +4 | +110 | +375 | +145 | NC | +10 |
| Ethyl ether | +64 | +12 | +58 | +101 | +82 | +60 | +36 | +99 | +44 | +97 | +51 |
| Formaldehyde | +2 | +3 | +1 | +2 | +2 | +2 | +9 | +13 | +2 | +1 | NC |
| Isooctane (ASTM fuel A) | +14 | NC | +103 | +93 | +52 | +3 | NC | −2 | +2 | NC | +112 |
| Sodium hydroxide, concentrated | +1 | NC | NC | NC | −2 | NC | +2 | NC | +1 | NC | NC |
| Toluene | +159 | +81 | +132 | +104 | +202 | +23 | +99 | +179 | +141 | +18 | +180 |
| Toluene, 30% (ASTM fuel B) | +65 | +10 | +128 | +109 | +112 | +24 | +12 | +39 | +31 | +1 | +145 |
| Trichloroethylene | +187 | +150 | +163 | +98 | +212 | +26 | +152 | +248 | +178 | +4 | +178 |
| Turpentine | +60 | −4 | +163 | +98 | +153 | +13 | +21 | +22 | +9 | +1 | +182 |

## CHEMICAL EFFECTS OF COMMON SOLVENTS ON VARIOUS COMMERCIAL RUBBERS (Continued)

| Solvent | Percent Volume Change | | | | | | | | | | |
|---|---|---|---|---|---|---|---|---|---|---|---|
| | NP | TH | EP | SL | SB | FS | PU | PA | AB | VT | BP |
| **Immersion for 70 Hours at 158°F** | | | | | | | | | | | |
| Acetic acid, concentrated | +16 | +23 | +16 | +3 | +19 | +12 | +172 | +154 | +55 | +112 | +1 |
| Chromic acid, 10% | +3 | D | −3 | NC | D | −1 | S | +6 | D | NC | NC |
| Cresylic acid | +11 | +260 | +1 | +4 | +32 | +2 | +404 | +266 | +200 | NC | NC |
| Diacetone alcohol | +27 | +35 | −1 | +6 | +11 | +40 | +131 | +190 | +119 | +218 | +1 |
| Furfural alcohol | +37 | D | NC | +3 | +13 | +14 | +153 | +215 | +163 | +80 | +1 |
| Hexyl alcohol | +14 | +14 | +12 | +27 | +17 | +7 | +77 | +104 | +31 | +5 | +10 |
| Hydrochloric acid, concentrated | +2 | D | +2 | +12S | +3 | +5 | +50S | +4 | +7 | +4 | NC |
| Methyl alcohol | +8 | +15 | +1 | NC | +2 | +8 | +28 | +140 | +18 | +68 | −2 |
| Nitric acid, concentrated | D | D | +2 | +25S | D | +2 | D | +52 | D | +5 | +4S |
| Soap solution, concentrated Castile | +6 | +5 | +2 | +4 | +5 | +1 | −1 | +33 | +4 | +1 | NC |
| Sulfuric acid, concentrated | D | D | +7 | D | D | D | D | D | D | NC | S |
| Sulfuric acid, 50% | NC | D | −1 | NC | +3 | NC | +36 | +3 | +1 | NC | +1 |
| **Immersion for 70 Hours at 212°F** | | | | | | | | | | | |
| ASTM oil No. 1 | +8 | −2 | +83 | +7 | +44 | NC | −8 | +2 | −2 | NC | +104 |
| ATMS oil No. 3 | +68 | +5 | +120 | +23 | +105 | +3 | −34 | +16 | +15 | +1 | +217 |
| Brake fluid (Wagner) | +13 | +23 | NC | +8 | +10 | D | D | +166 | +47 | +50S | +3 |
| Castor oil | +10 | −3 | −2 | +2 | +4 | NC | +19 | +11 | +6 | NC | NC |
| Detergent solution, concentrated Surf‡ | +7 | +11 | +22 | +2 | +4 | +3 | +15 | +31 | +3 | +2 | NC |
| Ethylene glycol | +7 | +10 | +2 | +3 | +4 | +1 | +8 | +37 | +3 | +1 | −1 |
| Linseed oil | +40 | +1 | +22 | +3 | +56 | NC | NC | +4 | +5 | NC | +11 |
| Salt solution, concentrated NaCl | NC | +1 | +1 | NC | NC | +1 | +1 | +2 | −1 | +16 | NC |
| Tributyl phosphate | +138 | +78 | +15 | +36 | +72 | D | +234 | +158 | +134 | +430S | +18 |
| Water | +7 | +45 | +6 | +3 | +10 | NC | +15 | +23 | +5 | NC | +1 |

## CHEMICAL EFFECTS OF COMMON SOLVENTS ON VARIOUS COMMERCIAL RUBBERS (Continued)

| Solvent | Percent Volume Change | | | | | | | | | | |
|---|---|---|---|---|---|---|---|---|---|---|---|
| | NP | TH | EP | SL | SB | FS | PU | PA | AB | VT | BP |
| **Immersion for 70 Hours at 300°F** | | | | | | | | | | | |
| Dibasic ester (MIL-L-7808B) | +87 | −8 | +81 | +16 | D | +8 | S | +36 | +27 | +10 | S |
| Diphenyl (Dowtherm A) | D | D | +77 | +15 | D | +9 | D | D | D | +13 | D |
| Hypoid (Parapoid†† 10-C) | S | S | +124 | +16S | +154 | D | D | +30 | +9H | D | D |
| Silicate ester (OS-45) | +34 | S | +36 | +35 | D | +5 | −6 | +8 | +3 | +4 | D |
| Silicone oil (DC-200) | −2 | S | −5 | +17 | −3 | NC | −9 | −3 | −4 | NC | −5 |

* Trade name, Union Carbide Corporation.

† Trade name, The Dow Chemical Company.

‡ Trade name, Lever Bros. Co.

†† Trade name, Enjay Co., Inc.

Data reproduced by courtesy of Precision Rubber Products Corporation, Lebanon, Tennessee.

Note: D = disintegration; H = excessive hardening; S = excessive softening and loss of physical properties; NC = no change.

# MISCIBILITY OF ORGANIC SOLVENT PAIRS

The classifications for Tables 1 and 2 were made at 20°C in the following manner. One-milliliter portions of each solvent comprising a pair were shaken together for approximately one minute. If no interfacial meniscus was observed after the contents of the tube were allowed to settle, the solvent pair was considered to be miscible (M). If a meniscus was observed without apparent change in the volume of either solvent, the pair was regarded as immiscible (I). This classification is a qualitative one, since solvent pairs may exhibit varied degrees of partial miscibility while existing as separate phases. If an obvious change occurred in the volume of each solvent but a meniscus was present, the pair was classified as partially miscible (S). The designation R indicates that the two solvents reacted.

The classifications for Table 3 were made in a similar manner, but using 5-ml portions of each solvent. In this table the symbol "Is" is used where solvent pairs show a pronounced degree of partial miscibility.

## TABLE 1

| Compound number | Compounds | Acetone | Isoamyl acetate | *n*-Amyl cyanide | Benzene | Benzyl ether | 2-Bromoethyl acetate | Chloroform | Cinnamaldehyde | Di-*n*-amylamine | Di-*n*-butyl carbonate | Diethylacetic acid | Diethylenetriamine | Diethyl formamide | Diisobutyl ketone | Diisopropylamine | Di-*n*-propyl aniline | Ethyl alcohol | Ethyl benzoate | Ethyl ether | Ethyl phenylacetate | Heptadecanol[a] | 3-Heptanol | *n*-Heptyl acetate | *n*-Hexyl ether | Methyl isopropyl ketone | 4-Methyl-*n*-valeric acid | *o*-Phenetidine | Sulfuric acid (concd.) | Tetradecanol[a] | Tri-*n*-butyl phosphate | Triethylene glycol | Triethylenetetramine | 2,6,8-Trimethyl 4-nonanone | Compound number |
|---|---|---|---|---|---|---|---|---|---|---|---|---|---|---|---|---|---|---|---|---|---|---|---|---|---|---|---|---|---|---|---|---|---|---|---|
| 1 | Acetone | .. | M | M | M | M | M | M | M | M | M | M | M | M | M | M | M | M | M | M | M | M | M | M | M | M | M | M | R | M | M | M | M | M | 1 |
| 2 | Isoamyl acetate | M | .. | M | M | M | M | M | M | M | M | M | M | M | M | M | M | M | M | M | M | M | M | M | M | M | M | M | R | M | M | I | M | M | 2 |
| 3 | *n*-Amyl cyanide | M | M | .. | M | M | M | M | M | M | M | M | M | M | M | M | M | M | M | M | M | M | M | M | M | M | M | M | R | M | M | M | M | M | 3 |
| 4 | Benzene | M | M | M | .. | M | S | M | M | M | M | M | M | M | M | M | M | M | M | M | M | M | M | M | M | M | M | M | I | M | M | S | M | M | 4 |
| 5 | Benzyl ether | M | M | M | M | .. | M | M | M | M | M | M | M | M | M | M | M | M | M | M | M | M | M | M | M | M | M | M | R | M | M | I | M | M | 5 |
| 6 | 2-Bromoethyl acetate | M | M | M | S | M | .. | M | M | R | M | M | R | R | M | R | M | M | M | M | M | M | M | M | S | M | M | M | R | M | M | M | R | M | 6 |
| 7 | Chloroform | M | M | M | M | M | M | .. | M | M | M | M | M | M | M | M | M | M | M | M | M | M | M | M | M | M | M | M | I | M | M | M | M | M | 7 |
| 8 | Cinnamaldehyde | M | M | M | M | M | M | M | .. | M | M | M | R | M | M | M | M | M | M | M | M | M | M | M | M | M | M | R | R | M | M | M | R | M | 8 |
| 9 | Di-*n*-amylamine | M | M | M | M | M | R | M | M | .. | M | R | I | R | M | M | M | M | M | M | M | M | M | M | M | M | M | M | R | M | M | S | I | M | 9 |
| 10 | Di-*n*-butyl carbonate | M | M | M | M | M | M | M | M | M | .. | M | M | M | M | M | M | M | M | M | M | M | M | M | M | M | M | M | R | M | M | I | I | M | 10 |
| 11 | Diethylacetic acid | M | M | M | M | M | M | M | M | R | M | .. | R | M | M | R | M | M | M | M | M | M | M | M | M | M | M | M | R | M | M | M | R | M | 11 |
| 12 | Diethylenetriamine | M | M | M | M | M | R | M | R | I | M | R | .. | R | M | M | M | M | M | M | M | M | M | R | I | R | R | M | R | M | M | M | M | I | 12 |
| 13 | Diethyl formamide | M | M | M | M | M | R | M | M | R | M | M | R | .. | M | R | M | M | M | M | M | M | M | M | I | M | M | M | R | M | M | M | R | M | 13 |
| 14 | Diisobutyl ketone | M | M | M | M | M | M | M | M | M | M | M | M | M | .. | M | M | M | M | M | M | M | M | M | M | M | M | M | R | M | M | I | M | M | 14 |
| 15 | Diisopropylamine | M | M | M | M | M | R | M | M | M | M | R | M | R | M | .. | M | M | M | M | M | M | M | M | M | M | R | M | R | M | M | M | M | M | 15 |
| 16 | Di-*n*-propylaniline | M | M | M | M | M | M | M | M | M | M | M | M | M | M | M | .. | M | M | M | M | M | M | M | M | M | M | M | R | M | M | I | M | M | 16 |
| 17 | Ethyl alcohol | M | M | M | M | M | M | M | M | M | M | M | M | M | M | M | M | .. | M | M | M | M | M | M | M | M | M | M | M | M | M | M | M | M | 17 |
| 18 | Ethyl benzoate | M | M | M | M | M | M | M | M | M | M | M | M | M | M | M | M | M | .. | M | M | M | M | M | M | M | M | M | M | M | M | M | M | M | 18 |
| 19 | Ethyl ether | M | M | M | M | M | M | M | M | M | M | M | M | M | M | M | M | M | M | .. | M | M | M | M | M | M | M | M | R | M | M | I | M | M | 19 |
| 20 | Ethyl phenylacetate | M | M | M | M | M | M | M | M | M | M | M | M | M | M | M | M | M | M | M | .. | M | M | M | M | M | M | M | R | M | M | M | M | M | 20 |
| 21 | Heptadecanol[a] | M | M | M | M | M | M | M | M | M | M | M | M | M | M | M | M | M | M | M | M | .. | M | M | M | M | M | M | R | M | M | I | M | M | 21 |
| 22 | 3-Heptanol | M | M | M | M | M | M | M | M | M | M | M | M | M | M | M | M | M | M | M | M | M | .. | M | M | M | M | M | R | M | M | M | M | M | 22 |
| 23 | *n*-Heptyl acetate | M | M | M | M | M | M | M | M | M | M | M | R | M | M | M | M | M | M | M | M | M | M | .. | M | M | M | M | R | M | M | I | R | M | 23 |
| 24 | *n*-Hexyl ether | M | M | M | M | M | S | M | M | M | M | M | I | I | M | M | M | M | M | M | M | M | M | M | .. | M | M | M | R | M | M | I | I | M | 24 |
| 25 | Methyl isopropyl ketone | M | M | M | M | M | M | M | M | M | M | M | R | M | M | M | M | M | M | M | M | M | M | M | M | .. | M | M | R | M | M | M | R | M | 25 |
| 26 | 4-Methyl-*n*-valeric acid | M | M | M | M | M | M | M | M | M | M | M | R | M | M | R | M | M | M | M | M | M | M | M | M | M | .. | M | R | M | M | M | R | M | 26 |
| 27 | *o*-Phenetidine | M | M | M | M | M | M | M | R | M | M | M | M | M | M | M | M | M | M | M | M | M | M | M | M | M | M | .. | R | M | M | M | M | M | 27 |
| 28 | Sulfuric acid (concd.) | R | R | R | I | R | R | I | R | R | R | R | R | R | R | R | R | M | M | R | R | R | R | R | R | R | R | R | .. | R | R | R | R | R | 28 |
| 29 | Tetradecanol[a] | M | M | M | M | M | M | M | M | M | M | M | M | M | M | M | M | M | M | M | M | M | M | M | M | M | M | M | R | .. | M | I | M | M | 29 |
| 30 | Tri-*n*-butyl phosphate | M | M | M | M | M | M | M | M | M | M | M | M | M | M | M | M | M | M | M | M | M | M | M | M | M | M | M | R | M | .. | M | M | M | 30 |
| 31 | Triethylene glycol | M | I | M | S | I | M | M | M | S | I | M | M | M | I | M | I | M | M | I | M | I | M | I | I | M | M | M | R | I | M | .. | M | I | 31 |
| 32 | Triethylenetetramine | M | M | M | M | M | R | M | R | I | I | R | M | R | M | M | M | M | M | M | M | M | M | R | I | R | R | M | R | M | M | M | .. | I | 32 |
| 33 | 2,6,8-Trimethyl 4-nonanone | M | M | M | M | M | M | M | M | M | M | M | I | M | M | M | M | M | M | M | M | M | M | M | M | M | M | M | R | M | M | I | I | .. | 33 |

[a] Trademark, Union Carbide Corporation.

Data from: Jackson, W. M., and Drury, J. S., *Ind. Eng. Chem., 51,* 1491 (1959). Reproduced by permission of the American Chemical Society, Washington, D.C.

## TABLE 2

| Compound number | Compounds | Acetone | Isoamyl acetate | *n*-Amyl cyanide | Anisaldehyde | Benzene | Benzyl ether | Chloroform | *o*-Cresol | Diisobutyl ketone | Diethylacetic acid | Diethyl formamide | Di-*n*-propyl aniline | Ethyl alcohol | Ethyl ether | 3-Heptanol | *n*-Heptyl acetate | *n*-Hexyl ether | α-Methylbenzylamine | α-Methylbenzyldiethanolamine | α-Methylbenzyldimethylamine | α-Methylbenzylethanolamine | 2-Methyl-5-ethylpyridine | Methyl isopropyl ketone | 4-Methyl-*n*-valeric acid | *o*-Phenetidine | 2-Phenylethylamine | Isopropanolamine | Pyridine | Salicylaldehyde | Tetradecanol[a] | Tri-*n*-butyl phosphate | Triethylenetetramine | 2,6,8-Trimethyl 4-nonanone |
|---|---|---|---|---|---|---|---|---|---|---|---|---|---|---|---|---|---|---|---|---|---|---|---|---|---|---|---|---|---|---|---|---|---|---|
| 1 | 1,3-Butylene glycol | M | I | M | I | I | I | M | M | I | M | M | I | M | S | M | I | I | M | M | M | M | M | M | M | M | M | M | M | M | M | M | M | I |
| 2 | 2,3-Butylene glycol | M | M | M | M | S | I | M | M | M | M | M | I | M | M | M | M | I | M | M | M | M | M | M | M | M | M | M | M | M | M | M | M | I |
| 3 | 2-Chloroethanol | M | M | M | M | M | M | M | M | M | M | M | M | M | M | M | M | M | R | M | M | M | R | M | M | M | R | R | M | M | M | M | M | M |
| 4 | 3-Chloro-1,2-propanediol | M | M | M | M | I | M | M | M | M | M | M | I | M | M | M | M | I | R | M | M | M | M | M | M | M | R | R | M | M | S | M | R | S |
| 5 | Dibutyl hydrogen phosphite | M | M | M | M | M | M | M | M | M | M | M | M | M | M | M | M | M | M | M | M | M | M | M | M | M | M | M | M | M | M | M | M | M |
| 6 | Diethylene glycol dibutyl ether | M | M | M | M | M | M | M | M | M | M | M | M | M | M | M | M | M | R | M | S | M | M | M | M | M | R | R | M | M | M | M | R | M |
| 7 | Diethylene glycol diethyl ether | M | M | M | M | M | M | M | M | M | M | M | M | M | M | M | M | M | M | M | M | M | M | M | M | M | M | M | M | M | M | M | M | M |
| 8 | Diethylene glycol monobutyl ether | M | M | M | M | M | M | M | M | M | M | M | M | M | M | M | M | M | M | M | M | M | M | M | M | M | M | M | M | M | M | M | M | M |
| 9 | Diethylene glycol monoethyl ether | M | M | M | M | M | M | M | M | M | M | M | M | M | M | M | M | I | M | M | M | M | M | M | M | M | M | M | M | M | M | M | M | M |
| 10 | Diethylene glycol monomethyl ether | M | M | M | M | M | M | M | M | M | M | M | M | M | M | M | M | I | M | M | M | M | M | M | M | M | M | M | M | M | M | M | M | M |
| 11 | Dipropylene glycol | M | M | M | M | M | M | M | M | M | M | M | M | M | M | M | M | I | M | M | M | M | M | M | M | M | M | M | M | M | M | M | M | M |
| 12 | Ethylene diacetate | M | M | M | M | M | M | M | M | M | M | M | M | M | M | M | M | I | M | M | M | M | M | M | M | M | M | M | M | M | M | M | M | M |
| 13 | Ethylene glycol | M | I | I | I | I | I | S | M | I | M | M | I | M | I | M | I | I | M | M | M | M | M | I | M | M | M | M | M | I | I | S | M | I |
| 14 | Ethyl glycol ethylbutyl ether | M | M | M | M | M | M | M | M | M | M | M | M | M | M | M | M | M | M | M | M | M | M | M | M | M | M | M | M | M | M | M | M | M |
| 15 | Ethylene glycol monobutyl ether | M | M | M | M | M | M | M | M | M | M | M | M | M | M | M | M | M | M | M | M | M | M | M | M | M | M | M | M | M | M | M | M | M |
| 16 | Ethylene glycol monoethyl ether | M | M | M | M | M | M | M | M | M | M | M | M | M | M | M | M | M | M | M | M | M | M | M | M | M | M | M | M | M | M | M | M | M |
| 17 | Ethylene glycol monomethyl ether | M | M | M | M | M | M | M | M | M | M | M | M | M | M | M | M | M | M | M | M | M | M | M | M | M | M | M | M | M | M | M | M | M |
| 18 | Ethylene glycol monophenyl ether | M | M | M | M | M | M | M | M | M | M | M | M | M | M | M | M | M | M | M | M | M | M | M | M | M | M | M | M | M | M | M | M | M |
| 19 | Glycerol | I | I | I | I | I | I | I | M | I | I | M | I | M | I | I | I | I | M | M | I | M | M | I | I | I | M | M | M | I | I | I | M | I |
| 20 | 1,2-Propanediol | M | M | M | M | I | I | M | M | I | M | M | I | M | S | M | I | I | M | M | M | M | M | M | M | M | M | M | M | M | M | M | M | I |
| 21 | 1,3-Propanediol | M | I | I | I | I | I | M | M | I | M | M | I | M | I | M | I | I | M | M | M | M | M | M | M | M | M | M | M | I | S | M | M | I |
| 22 | Triethylene glycol | M | I | M | M | S | I | M | M | I | M | M | I | M | I | M | I | I | M | M | M | M | M | M | M | M | M | M | M | M | I | M | M | I |
| 23 | Triethyl phosphate | M | M | M | M | M | M | M | M | M | M | M | M | M | M | M | M | M | M | M | M | M | M | M | M | M | M | M | M | M | M | M | M | M |
| 24 | Trimethylene chlorohydrin | M | M | M | M | M | M | M | M | M | M | M | M | M | M | M | M | M | R | M | M | M | R | M | M | M | R | R | M | M | M | M | R | M |

[a] Trademark, Union Carbide Corporation.

Data from: Jackson, W. M., and Drury, J. S., *Ind. Eng. Chem.*, *51*, 1492–1493 (1959). Reproduced by permission of the American Chemical Society, Washington, D.C.

# TABLE 3

| Compound number | Compounds | Acetone | Acetyl acetone | 2-Amino-2-methyl-1-propanol | Aniline | Benzaldehyde | Benzene | Benzin | Benzyl alcohol | Butyl acetate | Butyl alcohol | *n*-Butyl ether | Capryl alcohol | Carbon tetrachloride | Diacetone alcohol | Diethanolamine | Diethyl cellosolve | Diethyl ether | Dimethylaniline | Ethyl alcohol | Ethyl benzoate | Ethylene glycol | 2-Ethylhexanol | Formamide | Furfuryl alcohol | Glycerol | Hydroxyethyl-ethylenediamine | Isoamyl alcohol | Methyl isobutyl ketone | Nitromethane | Dibutoxytetra-ethylene glycol | Pyridine | Triethanolamine | Trimethylene glycol |
|---|---|---|---|---|---|---|---|---|---|---|---|---|---|---|---|---|---|---|---|---|---|---|---|---|---|---|---|---|---|---|---|---|---|---|
| 1 | Acetone | .. | M | M | .. | M | M | M | M | M | M | M | M | M | M | M | M | M | M | M | M | M | M | M | M | I | M | M | M | M | M | M | .. | M |
| 2 | Acetyl acetone | M | .. | R | .. | M | M | M | M | M | M | M | M | M | M | R | M | M | M | M | M | M | M | M | M | I | R | M | M | M | M | M | .. | M |
| 3 | Adiponitrile | M | M | M | M | .. | M | .. | M | M | M | I | .. | I | M | .. | M | I | M | M | M | I | I | M | M | I | M | I | M | .. | M | M | M | M |
| 4 | 2-Amino-2-methyl-1-propanol | M | R | .. | .. | M | M | I | M | M | M | Is | M | M | R | M | M | M | M | M | M | M | M | M | M | M | M | M | M | M | M | M | .. | M |
| 5 | Benzaldehyde | M | M | M | .. | .. | M | M | M | M | M | M | M | M | M | I | M | M | M | M | M | Is | M | M | M | Is | R | M | M | M | M | M | .. | M |
| 6 | Benzene | M | M | M | .. | M | .. | M | M | M | M | M | M | M | Is | I | M | M | M | M | M | I | M | I | M | I | I | M | M | I | M | M | .. | I |
| 7 | Benzin | M | M | I | .. | M | M | .. | I | M | M | M | M | M | I | I | M | M | M | M | M | M | M | I | I | I | Is | M | M | M | M | M | .. | I |
| 8 | Benzonitrile | M | M | M | M | .. | M | .. | M | M | M | M | .. | M | M | .. | M | M | M | M | M | I | M | I | M | I | M | M | M | .. | M | M | M | I |
| 9 | Benzothiazole | M | M | M | M | .. | M | .. | M | M | M | M | .. | M | M | .. | M | M | M | M | M | M | M | I | M | I | M | M | M | .. | M | M | M | M |
| 10 | Benzyl alcohol | M | M | M | .. | M | M | I | .. | M | M | M | M | M | M | M | M | M | M | M | M | M | M | M | M | M | M | M | M | M | M | M | .. | M |
| 11 | Benzyl mercaptan | M | M | I | M | .. | M | .. | M | M | M | M | .. | M | M | .. | M | M | M | M | M | I | M | I | M | I | M | M | M | .. | M | M | R | I |
| 12 | Butyl acetate | M | M | M | .. | M | M | M | M | .. | M | M | M | M | M | I | M | M | M | M | M | Is | M | I | M | I | I | M | M | M | M | M | .. | Is |
| 13 | Butyl alcohol | M | M | M | .. | M | M | M | M | M | .. | M | M | M | M | M | M | M | M | M | M | M | M | M | M | M | M | M | M | M | M | M | .. | M |
| 14 | *n*-Butyl ether | M | M | Is | .. | M | M | M | M | M | M | .. | M | M | M | I | M | M | M | M | M | I | M | I | M | I | I | M | M | I | M | M | .. | I |
| 15 | Capryl alcohol | M | M | M | .. | M | M | M | M | M | M | M | .. | M | M | M | M | M | M | M | M | M | M | I | M | I | M | M | M | Is | M | M | .. | M |
| 16 | Carbon tetrachloride | M | M | M | .. | M | M | M | M | M | M | M | M | .. | Is | I | M | M | M | M | M | I | M | I | M | I | I | M | M | M | M | M | .. | I |
| 17 | Diacetone alcohol | M | M | R | .. | M | Is | I | M | M | M | M | M | Is | .. | M | M | M | M | M | M | M | M | M | M | I | R | M | M | M | M | M | .. | M |
| 18 | Diethanolamine | M | R | M | .. | I | I | I | M | I | M | I | M | I | M | .. | I | I | Is | M | M | M | M | M | M | M | M | M | I | I | I | M | .. | M |
| 19 | Diethyl Cellosolve | M | M | M | .. | M | M | M | M | M | M | M | M | M | M | I | .. | M | M | M | M | M | M | M | M | I | I | M | M | M | M | M | .. | M |
| 20 | Diethyl ether | M | M | M | .. | M | M | M | M | M | M | M | M | M | M | I | M | .. | M | M | M | I | M | I | M | I | I | M | M | M | M | M | .. | I |
| 21 | Dimethylaniline | M | M | M | .. | M | M | M | M | M | M | M | M | M | M | Is | M | M | .. | M | M | I | M | I | M | I | I | M | M | M | M | M | .. | I |
| 22 | Di-*N*-propylaniline | M | M | I | M | .. | M | .. | M | M | M | M | .. | M | I | .. | M | M | M | M | M | I | M | I | M | I | I | M | M | .. | M | M | I | I |
| 23 | Ethyl alcohol | M | M | M | .. | M | M | M | M | M | M | M | M | M | M | M | M | M | M | .. | M | M | M | M | M | M | M | M | M | M | M | M | .. | M |
| 24 | Ethyl benzoate | M | M | M | .. | M | M | M | M | M | M | M | M | M | M | M | M | M | M | M | .. | I | M | I | M | I | M | M | M | M | M | M | .. | Is |
| 25 | Ethyl isothiocyanate | M | M | R | M | .. | M | .. | M | M | M | M | .. | M | M | .. | M | M | M | M | M | I | M | I | M | I | R | M | M | .. | M | M | M | I |
| 26 | Ethyl thiocyanate | M | M | M | M | .. | M | .. | M | M | M | M | .. | M | M | .. | M | M | M | M | M | I | M | I | M | I | M | M | M | .. | M | M | M | I |
| 27 | Ethylene glycol | M | M | M | .. | Is | I | M | M | Is | M | I | M | I | M | M | M | I | I | M | I | .. | M | M | M | M | M | M | I | I | M | M | .. | M |
| 28 | 2-Ethylhexanol | M | M | M | .. | M | M | M | M | M | M | M | M | M | M | M | M | M | M | M | M | M | .. | I | M | I | M | M | M | I | M | M | .. | M |
| 29 | Formamide | M | M | M | .. | M | I | I | M | I | M | I | i | I | M | M | M | I | I | M | I | M | I | .. | M | M | M | M | Is | M | M | M | .. | M |
| 30 | Furfuryl alcohol | M | M | M | .. | M | M | I | M | M | M | M | M | M | M | M | M | M | M | M | M | M | M | M | .. | M | M | M | M | M | M | M | .. | M |
| 31 | Glycerol | I | I | M | .. | I | I | I | M | I | M | I | I | I | I | M | I | I | I | M | I | M | I | M | M | .. | M | I | I | I | I | M | .. | M |
| 32 | Hydroxyethyl-ethylenediamine | M | R | M | .. | R | I | Is | M | I | M | I | M | I | R | M | I | I | I | M | M | M | M | M | M | M | .. | M | M | M | M | M | .. | M |
| 33 | Isoamyl alcohol | M | M | M | .. | M | M | M | M | M | M | M | M | M | M | M | M | M | M | M | M | M | M | M | M | I | M | .. | M | M | M | M | .. | M |
| 34 | Isoamyl sulfide | M | M | I | M | .. | M | .. | M | M | M | M | .. | M | I | .. | M | M | M | M | M | I | M | I | I | I | I | M | M | .. | M | M | I | I |
| 35 | Isobutyl mercaptan | M | M | M | M | .. | M | .. | M | M | M | M | .. | M | M | .. | M | M | M | M | M | I | M | I | M | I | I | M | M | .. | M | M | R | R |
| 36 | Methyl disulfide | M | M | M | M | .. | M | .. | M | M | M | M | .. | M | M | .. | M | M | M | M | M | I | M | I | M | I | I | M | M | . | M | M | I | R |
| 37 | Methyl isobutyl ketone | M | M | M | .. | M | M | M | M | M | M | M | M | M | M | I | M | M | M | M | M | I | M | Is | M | I | M | M | .. | M | M | M | .. | I |
| 38 | Nitromethane | M | M | M | .. | M | I | M | M | M | M | I | Is | M | M | I | M | M | M | M | M | I | I | M | M | I | M | M | M | .. | M | M | .. | I |
| 39 | Dibutoxytetra-ethylene glycol | M | M | M | .. | M | M | M | M | M | M | M | M | M | M | I | M | M | M | M | M | M | M | M | M | I | M | M | M | M | .. | M | .. | M |
| 40 | Pyridine | M | M | M | .. | M | M | M | M | M | M | M | M | M | M | M | M | M | M | M | M | M | M | M | M | M | M | M | M | M | M | .. | .. | M |
| 41 | Tri-*n*-butylamine | M | M | I | I | .. | M | .. | M | M | M | M | .. | M | I | .. | M | M | M | M | M | I | M | I | M | I | I | M | M | .. | M | M | I | I |
| 42 | Trimethylene glycol | M | M | M | .. | M | I | I | M | Is | M | I | M | I | M | M | M | I | I | M | Is | M | M | M | M | M | M | M | I | I | M | M | .. | .. |

Data from: Drury, J. S., *Ind. Eng. Chem.*, *44*, 2744 (1952). Reproduced by permission of the American Chemical Society, Washington, D.C.

## REFRACTIVE INDICES OF LIQUIDS

The compounds in the following table are listed in the order of increasing indices. Unless otherwise indicated by a superscript, the indices given are for 20°C.

| Liquid | $n_D$ |
|---|---|
| Methanol | $1.3276^{25}$ |
| Acetaldehyde | 1.3316 |
| Water | 1.3330 |
| Methylene chloride | $1.3348^{15}$ |
| Acetonitrile | $1.34596^{16.5}$ |
| Diethyl ether | $1.3497^{24.8}$ |
| Dimethylamine | $1.350^{17}$ |
| *n*-Pentane | 1.3577 |
| Acetone | $1.35886^{19.4}$ |
| Methyl acetate | 1.35935 |
| *n*-Propyl nitrite | 1.3613 |
| Ethyl alcohol, anhydrous | $1.36242^{18.35}$ |
| Dimethyl carbonate | 1.3687 |
| Propionitrile | $1.36888^{14.6}$ |
| Ethyl propyl ether | 1.36948 |
| Formic acid | 1.37137 |
| Isobutyl nitrite | $1.37151^{22.1}$ |
| Acetic acid | 1.37182 |
| Ethyl acetate | $1.37216^{18.9}$ |
| Isobutyraldehyde | 1.37302 |
| *n*-Hexane | 1.37536 |
| Isopropylamine | $1.37698^{15.4}$ |
| Isopropyl alcohol | 1.37757 |
| Methyl propionate | $1.37767^{18.5}$ |
| Butylamine, tertiary | $1.37940^{18}$ |
| Propyl ether | 1.3807 |
| Ethyl methyl ketone | $1.38071^{15.9}$ |
| Ethyl borate | 1.381 |
| Butyronitrile | $1.3816^{24}$ |
| Nitromethane | 1.3818 |
| Acetal | 1.3819 |
| Butyraldehyde | 1.38433 |
| *n*-Propyl acetate | 1.38438 |
| Ethyl nitrate | $1.38484^{21.5}$ |
| Ethyl carbonate | 1.38456 |
| *n*-Amyl nitrite | 1.38506 |
| *n*-Propyl alcohol | 1.38543 |
| *n*-Heptane | $1.3855^{25}$ |
| Butyl chloride, tertiary | $1.3869^{18}$ |
| Isoamyl nitrite | $1.38708^{20.7}$ |
| Diethylamine | $1.38730^{18}$ |
| Propionic acid | $1.38736^{19.9}$ |
| Butyl alcohol, tertiary | 1.38779 |
| Isopropyl methyl ketone | $1.38788^{16}$ |
| *n*-Propyl chloride | 1.3886 |
| Methyl propyl ketone | $1.38946^{20.2}$ |
| Acetyl chloride | 1.38976 |
| *n*-Propylamine | $1.39006^{16.6}$ |
| Nitroethane | $1.39007^{24.3}$ |
| Isovaleraldehyde | 1.3902 |
| Isobutyl acetate | $1.3907^{19}$ |
| Valeronitrile | 1.3909 |
| Acrylonitrile | 1.393 |
| Ethyl butyrate | $1.39302^{18}$ |
| Diacetyl | $1.3933^{18}$ |
| *n*-Propyl propionate | 1.3935 |
| Butyl alcohol, secondary | 1.397 |
| *n*-Propyl nitrate | 1.3972 |
| *n*-Octane | 1.3975 |
| Isobutyl alcohol | $1.3976^{15}$ |
| Isobutylamine | $1.39878^{17}$ |
| Ethyl propyl ketone | $1.39899^{22}$ |
| Butyric acid | 1.39906 |
| *n*-Butyl alcohol | 1.3993 |
| Acrolein | 1.39975 |
| Ethyl *n*-butyrate | 1.400 |
| Methacrylonitrile | 1.4002 |
| Triethylamine | 1.40032 |
| *n*-Butylamine | 1.401 |
| Isobutyl chloride | $1.4010^{15}$ |
| Amyl acetate | 1.4012 |
| *n*-Butyl nitrate | $1.40130^{23.2}$ |
| Isobutyl nitrate | $1.40130^{23.3}$ |
| *n*-Butyl chloride | 1.4015 |
| 1-Nitropropane | 1.4015 |
| Isoamyl acetate | $1.40170^{17.9}$ |
| Isovaleric acid | $1.40178^{22.4}$ |
| Methyl cellosolve | 1.4028 |
| Propionic anhydride | 1.4038 |
| Cyclopentane | 1.4039 |
| Tetrahydrofuran | $1.4040^{25}$ |
| Dipropylamine | $1.40455^{19.5}$ |
| Paraldehyde | 1.40486 |
| *n*-Butyl *n*-butyrate | 1.4049 |
| Propionyl chloride | 1.40507 |
| Amyl alcohol, tertiary | 1.4052 |
| Isoamyl alcohol | 1.4075 |
| Isoamyl ether | 1.408 |
| Valeric acid | 1.4086 |
| Diisobutylamine | 1.40934 |
| *pri*-*n*-Amyl alcohol | 1.40994 |
| Diethyl oxalate | 1.41011 |
| Isoamyl chloride | 1.4103 |
| *n*-Amyl chloride | $1.4119^{18}$ |
| *n*-Decane | 1.41203 |
| Isoamyl nitrate | $1.4122^{22}$ |
| Tetramethylethylene | 1.4128 |
| Isovaleryl chloride | $1.41361^{24.3}$ |
| Cyclopentanol | 1.41530 |
| Valeryl chloride | 1.41555 |
| *n*-Hexyl methyl ketone | 1.41613 |
| *n*-Hexanol | $1.4162^{25}$ |
| 1,1-Dichloroethane | 1.41655 |
| Diethyl acetic acid | $1.41788^{10}$ |
| Undecane | 1.4184 |
| Allylamine | $1.41943^{22}$ |
| Methyl urethan | $1.4200^{18.9}$ |
| Diethyl succinate | 1.42007 |
| Octyl acetate | 1.4204 |
| Ethyl acetoacetate | $1.42092^{16.6}$ |
| Dibutyl ketone | $1.421^{15}$ |
| Furan | 1.42157 |

| Liquid | $n_D$ |
|---|---|
| Diethyl ketone | $1.3939^{16.6}$ |
| Butyl alcohol, secondary | 1.3949 |
| Butylamine, secondary | $1.39501^{16.7}$ |
| Butyl acetate | 1.3951 |
| Valeraldehyde | 1.3952 |
| Butyl chloride, secondary | $1.3953^{25}$ |
| Butyl methyl ketone | $1.39694^{17.4}$ |
| *n*-Heptyl alcohol | 1.42410 |
| Isopropyl bromide | 1.42508 |
| Triisobutylamine | $1.42519^{17.3}$ |
| Succinaldehyde | 1.4254 |
| Vinyl acetic acid | $1.4257^{15}$ |
| Cyclohexane | 1.4264 |
| Methyl carbitol | $1.4264^{27}$ |
| Butyl bromide, tertiary | 1.428 |
| 1,1-Dichloropropane | 1.42887 |
| Heptyl methyl carbinol | $1.4290^{25}$ |
| *n*-Octylamine | 1.430 |
| Ethylene glycol | 1.4311 |
| Nonanoic acid | 1.4330 |
| Undecanal | 1.4334 |
| *n*-Propyl bromide | 1.43414 |
| Butyl bromide, secondary | $1.4344^{25}$ |
| Glycerol tributyrate | 1.4359 |
| Diethyl malate | 1.4362 |
| Cyclopentanone | 1.4366 |
| Crotonaldehyde | $1.43838^{17.3}$ |
| 1,2-Dichloropropane | 1.4388 |
| Isobutyl bromide | $1.4391^{15}$ |
| Tetranitromethane | $1.43976^{16.9}$ |
| *n*-Butyl bromide | 1.4398 |
| Trimethylene glycol | 1.4398 |
| Isoamyl bromide | 1.4412 |
| Lactic acid (*dl*) | 1.4414 |
| Tridecane | $1.4419^{16.8}$ |
| Epichlorohydrin | $1.44195^{11.5}$ |
| Dimethyl malate | 1.4425 |
| Mesityl oxide | $1.4425^{22}$ |
| Cycloheptane | 1.4440 |
| 1,2-Dichloroethane | 1.4443 |
| Cyclohexene | $1.44507^{22}$ |
| Formamide | $1.44530^{22.7}$ |
| Tetradecane | 1.4459 |
| 1,3-Dichloropropane | 1.4469 |
| Diethylene glycol | 1.4475 |
| Chloroform | 1.4476 |
| *sym.*-Dichloroethylene | $1.4490^{15}$ |
| *n*-Octyl bromide | $1.4503^{25}$ |
| Trichloroethyl acetate | 1.45068 |
| Cyclohexanone | 1.4507 |
| Piperidine | 1.4534 |
| Ethanolamine | 1.4539 |
| Ethylenediamine, anhyd. | $1.4540^{26}$ |
| Trichloroethylene | $1.4556^{25}$ |
| Chloral | 1.45572 |
| *d*-Citronellol | 1.4566 |
| *sym.*-Dichloroethyl ether | 1.4570 |
| Rose oil | $1.4570–1.4630^{30}$ |
| Eucalyptus oil | 1.4580–1.4700 |
| Eucalyptol | $1.4584^{15}$ |
| Lavender oil | 1.4590–1.4700 |
| Peppermint oil | 1.4590–1.4650 |
| Chloropicrin | $1.46075^{23}$ |
| Coriander oil | 1.4620–1.4720 |
| *dl*-Bornyl acetate | $1.4623^{22}$ |
| Peanut oil | $1.4625–1.4645^{40}$ |
| Acrylic acid | 1.4224 |
| Cyclopentene | 1.42246 |
| Ethyl chloracetate | 1.42274 |
| Diisoamylamine | $1.42289^{21}$ |
| 1,4-Dioxane | 1.4232 |
| *m* Heptylamine | 1.424 |
| Ethyl bromide | 1.4241 |
| Dichloroacetic acid | $1.4659^{22}$ |
| Turpentine oil | 1.4680–1.4780 |
| Myristica oil (West Indian) | 1.4690–1.4760 |
| Decalin (*trans.*) | $1.46994^{18}$ |
| 1,1,2-Trichloroethane | 1.4711 |
| Orange oil | 1.4723–1.4737 |
| Bitter orange oil | 1.4725–1.4755 |
| Glycerin | 1.4729 |
| Lemon oil | 1.4738–1.4755 |
| Chenopodium oil | 1.4740–1.4790 |
| Myristica oil (East Indian) | 1.4740–1.4880 |
| Crotonic anhydride | 1.47446 |
| Dwarf pine needle oil | 1.4750–1.4800 |
| Poppy seed oil | 1.4766–1.4774 |
| Pine oil | 1.4780–1.4820 |
| Geraniol | 1.4798 |
| Nitroglycerin | $1.482^{18.6}$ |
| Decalin (*cis*) | 1.4828 |
| Caraway oil | 1.4840–1.4880 |
| Spearmint oil | 1.4840–1.4910 |
| Furfuryl alcohol | $1.4850^{25}$ |
| Triethanolamine | 1.4852 |
| Geranial | 1.48752 |
| Toluene | $1.4893^{24}$ |
| Furfuryl chloride | 1.4941 |
| *sym.*-Tetrachloroethane | 1.4942 |
| Cumene | $1.4947^{15}$ |
| Thyme oil | 1.4950–1.5050 |
| *m*-Xylene | 1.4973 |
| α-Ionone | $1.49842^{22.3}$ |
| Benzene | 1.50142 |
| Diethyl phthalate | 1.5019 |
| *o*-Cymene | $1.50206^{16.1}$ |
| Isopropyl iodide | 1.5026 |
| 2-Picoline | $1.50293^{16.7}$ |
| Phenyl acetate | 1.503 |
| Pyrrole | 1.5035 |
| 3-Picoline | $1.50432^{24}$ |
| Butyl phenyl ether | 1.5046 |
| *n*-Propyl iodide | 1.50508 |
| *o*-Xylene | 1.50545 |
| Tetrachloroethylene | 1.50547 |
| Isoamyl salicylate | 1.506 |
| Bay oil | 1.5070–1.5160 |
| Phenetole | 1.5076 |
| Pyridine | $1.50919^{21}$ |
| Ethyl isothiocyanate | 1.5134 |
| Isovalerophenone | $1.51385^{15.8}$ |
| Dimethyl phthalate | $1.51546^{20.8}$ |
| Anisole | 1.51791 |
| Methyl benzoate | $1.51810^{16}$ |
| Isopropyl phenyl ketone | $1.51919^{16.6}$ |
| β-Ionone | $1.51977^{18.9}$ |
| *p*-Chlorotoluene | $1.5199^{19}$ |
| Phenyl propyl ketone | $1.52016^{18.3}$ |
| Propylene dibromide | 1.5203 |
| *m*-Chlorotoluene | $1.5214^{19}$ |
| Carvacrol | 1.52295 |
| Benzyl acetate | 1.5232 |

| Liquid | $n_D$ |
|---|---|
| Cardamom oil | 1.4630–1.4660 |
| Carbon tetrachloride | $1.46305^{15}$ |
| *d*-Fenchone | $1.4636^{18}$ |
| Phytol | 1.46380 |
| Rosemary oil | 1.4640–1.4760 |
| Bergamot oil | 1.4650–1.4675 |
| Allyl bromide | 1.46545 |
| *dl*-Pinene | 1.4658 |
| *o*-Tolunitrile | $1.52720^{23.1}$ |
| Thiophene | 1.5287 |
| *o*-Chlorotoluene | 1.5288 |
| Benzonitrile | 1.52892 |
| Methyl iodide | $1.5293^{21}$ |
| Acetophenone | $1.53418^{19}$ |
| Methyl salicylate | 1.5369 |
| Safrol | 1.5383 |
| Indan | $1.53877^{16.4}$ |
| Benzyl alcohol | 1.53956 |
| *m*-Cresol | 1.5398 |
| Benzylamine | 1.5401 |
| *m*-Tolualdehyde | $1.54068^{21.4}$ |
| *N,N*-Diethylaniline | $1.54105^{22.3}$ |
| Benzyl chloride | $1.5415^{15}$ |
| Eugenol | $1.5416^{19.4}$ |
| *m*-Nitrotoluene | $1.5425^{30}$ |
| *sym.*-Dibromoethylene | 1.5428 |
| Carvacrylamine | $1.543^{19}$ |
| Styrene | $1.54344^{17}$ |
| *m*-Dichlorobenzene | $1.54570^{20.9}$ |
| Tetralin | 1.54614 |
| *p*-Tolualdehyde | $1.54693^{16.6}$ |
| *o*-Chlorophenol | $1.5473^{40}$ |
| *o*-Nitrotuluene | $1.54739^{20.4}$ |
| *o*-Tolualdehyde | $1.54852^{19}$ |
| *o*-Dichlorobenzene | $1.5518^{22}$ |
| Nitrobenzene | 1.55291 |
| Anise oil | 1.5530–1.5600 |
| *N*-Ethylaniline | $1.55558^{20.3}$ |
| Nicotine | $1.52392^{22.4}$ |
| Phenylethyl alcohol | 1.5240 |
| Chlorobenzene | 1.5248 |
| Sassafras oil | 1.5250–1.5350 |
| Ethyl salicylate | $1.52511^{14.4}$ |
| Furfural | 1.52608 |
| Clove oil | 1.5270–1.5350 |
| Pimenta oil | 1.5270–1.5400 |
| *N,N*-Dimethylaniline | 1.55819 |
| Bromobenzene | $1.5625^{15}$ |
| Benzyl benzoate | $1.5681^{21}$ |
| *N*-Methylaniline | $1.57021^{21.2}$ |
| Phthalyl chloride | $1.57099^{15.5}$ |
| Indene | $1.57107^{12.7}$ |
| *m*-Toluidine | $1.5711^{22}$ |
| *o*-Toluidine | 1.57276 |
| Salicylaldehyde | $1.57358^{19.7}$ |
| Isoeugenol | $1.5739^{19}$ |
| Dibenzylamine | $1.57432^{22}$ |
| *o*-Anisidine | 1.57536 |
| Anisaldehyde | $1.5764^{13}$ |
| Aniline | 1.5863 |
| *o*-Chloroaniline | 1.5895 |
| *m*-Chloroaniline | $1.59424^{20.7}$ |
| Boromoform | $1.5980^{19}$ |
| Cinnamon oil | 1.6020–1.6135 |
| Phenylhydrazine | $1.6081^{20.3}$ |
| *m*-Dibromobenzene | $1.6083^{17.5}$ |
| *o*-Dibromobenzene | $1.6117^{17.5}$ |
| Iodobenzene | $1.62145^{18.5}$ |
| Isoquinoline | $1.62233^{25.1}$ |
| Quinoline | $1.62450^{24.9}$ |
| Carbon disulfide | $1.62950^{18}$ |
| *sym.*-Tetrabromoethane | 1.638 |
| Phenyl isothiocyanate | $1.64918^{23.1}$ |
| 1-Naphthaldehyde | $1.65464^{19.3}$ |
| 1-Bromonaphthalene | $1.65876^{19.4}$ |
| Methylene iodide | $1.7425^{15}$ |

Data taken from: *Merck Index,* 7th ed., pp. 1530–1532 (1960). Reproduced by permission of Merck and Co., Inc., Rahway, New Jersey.

# SPECIFIC GRAVITY OF LIQUIDS

Specific gravity and density are not identical, although the abbreviation "d" is frequently used to designate specific gravity. Specific gravity and density are numerically equal when water is the standard of reference for specific gravity and g/ml is the unit designation for density.

The numerical value for specific gravity is usually written with a superscript (indicating the temperature of the liquid) and a subscript (indicating the temperature of the liquid to which it is referred); for example, $d^{25}_{4}$ 1.724, or sp gr $1.724^{25}_{4}$. When superscript and subscript are omitted, the value given represents the specific gravity at 20°C referred to water at 4°C. Unless otherwise specified, the standard of reference is understood to be water.

Since the density of water is highest at 4°C, the specific gravity of a liquid with reference to water will be higher at all temperatures other than 4°C. To obtain the specific gravity with reference to water at the same temperature as the liquid, multiply the specific gravity of $^{15}_{4}$, $^{20}_{4}$, or $^{25}_{4}$ by 1.001, 1.002, or 1.003 respectively.

The liquids in the following table are listed in the order of increasing specific gravities.

| Liquid | Specific Gravity |
|---|---|
| *n*-Pentane | 0.626 |
| *n*-Hexane | 0.660 |
| 1-Butyne | $0.668^{0}_{4}$ |
| Dimethylamine | $0.680^{0}_{4}$ |
| Isoprene | 0.681 |
| *n*-Heptane | 0.684 |
| 2-Butyne | $0.688^{25}$ |
| 1,5-Hexadiene | 0.688 |
| Isopropylamine | $0.694^{15}_{4}$ |
| Butylamine, tertiary | 0.696 |
| Triethylboron | $0.696^{23}$ |
| Ethylamine | $0.706^{0}_{4}$ |
| Diethylamine | $0.711^{18}_{4}$ |
| 2,4-Hexadiene | 0.711 |
| Diethyl ether | 0.713 |
| *n*-Nonane | 0.716 |
| Triethylamine | $0.723^{25}_{4}$ |
| Butylamine, secondary | 0.724 |
| Isopropyl ether | 0.726 |
| Ethyl methyl ether | $0.726^{0}_{4}$ |
| 2,4-Heptadiene | $0.733^{21.5}_{4}$ |
| Isobutylamine | $0.724^{25}_{4}$ |
| Propyl ether | 0.736 |
| Methyl propyl ether | 0.738 |
| Dipropylamine | 0.738 |
| Ethyl *n*-propyl ether | 0.739 |
| *n*-Butylamine | 0.740 |
| Undecane | 0.741 |
| *N,N*-Dimethylamylamine | 0.743 |
| Ethyl isopropyl ether | $0.745^{0}_{4}$ |
| Isoamylamine | 0.751 |
| Cyclopentane | 0.751 |
| Butyl ethyl ether | 0.752 |
| Isohexylamine | $0.758^{25}_{4}$ |
| Isobutyl ether | $0.761^{15}_{4}$ |
| Allylamine | 0.761 |
| *n*-Amylamine | 0.761 |
| Butyl methyl ether | $0.764^{0}_{4}$ |
| Allyl ether ether | 0.765 |
| *n*-Dodecane | $0.766^{0}_{4}$ |
| Dibutylamine | 0.767 |
| *n*-Butyl ether | $0.769^{20}_{20}$ |
| Cyclopentene | 0.774 |
| *n*-Heptylamine | 0.777 |
| Cyclohexane | 0.778 |
| Methanol, anhydrous | 0.791 |
| Acetone | 0.792 |
| Isobutyraldehyde | 0.794 |
| Acrylonitrile | 0.797 |
| Ethyl alcohol, anhydrous | $0.798^{15.56}_{15.56}$ |
| Valeronitrile | 0.801 |
| Isovaleraldehyde | $0.803^{17}_{4}$ |
| *n*-Propyl alcohol | 0.804 |
| Allyl ether | $0.805^{18}_{0}$ |
| Ethyl methyl ketone | 0.805 |
| Isobutyl alcohol | $0.806^{15}_{4}$ |
| Propionaldehyde | 0.807 |
| Butyl alcohol, secondary | 0.808 |
| Amyl alcohol, tertiary | 0.809 |
| Methyl propyl ketone | 0.809 |
| *n*-Butyl alcohol | 0.810 |
| Cycloheptane | 0.810 |
| Cyclohexene | 0.810 |
| Isoamyl alcohol | $0.813^{15}_{4}$ |
| Ethyl propyl ketone | $0.813^{21.8}_{4}$ |
| *pri-n*-Amyl alcohol | 0.814 |
| Heptyl ether | $0.815^{0}_{4}$ |
| Diethyl ketone | $0.816^{19}_{4}$ |
| Ethyl alcohol, 95 per cent | $0.816^{15.56}_{15.56}$ |
| Butyraldehyde | 0.817 |
| Dipropyl ketone | 0.817 |
| Ethyl butyl ketone | 0.818 |
| *n*-Hexyl methyl ketone | 0.819 |
| 1-Hexanol | 0.819 |
| 3-Hexanol | 0.819 |
| Isoamyl alcohol, secondary | 0.819 |
| Pinacolin | $0.821^{0}_{4}$ |
| Amyl methyl ketone | $0.822^{15}_{4}$ |
| Cycloheptene | 0.823 |
| *n*-Octyl alcohol | 0.825 |
| 2-Undecanone | 0.826 |
| Light Liquid Petrolatum | $0.828–0.880^{25}_{25}$ |
| 2-Hexanol | $0.829^{0}_{4}$ |
| *n*-Decyl alcohol | 0.829 |
| *n*-Undecylaldehyde | 0.830 |
| Butyl methyl ketone | $0.830^{0}_{4}$ |
| 1-Undecanol | $0.833^{23}_{4}$ |
| Acrolein | 0.841 |
| Orange oil | $0.842–0.846^{25}_{25}$ |
| Bitter orange oil | $0.845–0.851^{25}_{25}$ |

| Liquid | Specific Gravity |
|---|---|
| *n*-Octylamine | $0.779^{20}_{20}$ |
| Isoamyl ether | $0.781^{15}_{15}$ |
| Propionitrile | 0.783 |
| Acetonitrile | $0.783^{25}_{25}$ |
| *n*-Butyl ether | $0.784^{0}_{4}$ |
| Isopropyl alcohol | 0.785 |
| Isovaleronitrile | 0.788 |
| Butyl alcohol, tertiary | 0.789 |
| Mesityl oxide | 0.854 |
| Myristica oil | $0.854–0.910^{25}_{25}$ |
| *p*-Cymene | 0.857 |
| *dl*-Pinene | 0.858 |
| Isopropyl chloride | 0.859 |
| 2-Diethylaminoethanol | $0.860^{25}_{25}$ |
| Liquid Petrolatum | $0.860–0.905^{25}_{25}$ |
| Piperidine | 0.861 |
| Cumene | 0.863 |
| Coriander oil | $0.863–0.875^{25}_{25}$ |
| Orange flower oil | $0.863–0.880^{25}_{25}$ |
| Phytol | $0.864^{0}_{4}$ |
| *m*-Xylene | 0.864 |
| Toluene | 0.866 |
| Ethyl benzene | 0.867 |
| *m*-Cymene | 0.870 |
| Isoamyl acetate | $0.870^{25}_{4}$ |
| Isopropyl acetate | 0.870 |
| Isobutyl nitrite | $0.870^{20}_{20}$ |
| Butyl chloride, secondary | 0.871 |
| Octyl acetate | $0.873^{20}_{20}$ |
| Isobutyl acetate | 0.875 |
| Isoamyl nitrite | $0.875^{25}_{25}$ |
| Bergamot oil | $0.875–0.880^{25}_{25}$ |
| Lavender oil | $0.875–0.888^{25}_{25}$ |
| *o*-Cymene | 0.876 |
| Benzene | $0.879^{15}_{4}$ |
| Amyl acetate | $0.879^{20}_{20}$ |
| Geraniol | $0.881^{16}_{4}$ |
| *n*-Amyl chloride | 0.883 |
| Isobutyl chloride | $0.883^{15}$ |
| *n*-Butyl chloride | 0.884 |
| Pine needle oil | $0.884–0.886^{15}_{15}$ |
| Citronella oil | $0.885–0.912^{25}_{25}$ |
| 2-Dimethylaminoethanol | 0.887 |
| *n*-Propyl acetate | 0.887 |
| 1-Menthol | $0.890^{15}_{15}$ |
| Propyl chloride | $0.890^{20}_{20}$ |
| Isoamyl chloride | 0.893 |
| Rosemary oil | $0.894–0.912^{25}_{25}$ |
| Oleic acid | $0.895^{18}_{4}$ |
| Isodurene | $0.896^{0}_{4}$ |
| Peppermint oil | $0.896–0.908^{25}_{25}$ |
| *o*-Xylene | 0.897 |
| Ethyl nitrite | $0.900^{15}_{15}$ |
| Caraway oil | $0.900–0.910^{25}_{25}$ |
| Ethyl acetate | 0.902 |
| Linoleic acid | 0.903 |
| Cubeb oil | $0.905–0.925^{25}_{25}$ |
| Eucalyptus oil | $0.905–0.925^{25}_{25}$ |
| Diethyl Carbitol | 0.907 |
| Styrene | 0.907 |
| Undecylenic acid | $0.908^{25}_{4}$ |
| Olive oil | $0.910–0.915^{25}_{25}$ |
| Expressed almond oil | $0.910–0.915^{25}_{25}$ |
| Persic oil | $0.910–0.923^{25}_{25}$ |
| Thyme oil | $0.910–0.935^{25}_{25}$ |
| *n*-Butyl nitrite | $0.911^{0}_{4}$ |
| Peanut oil | $0.912–0.920^{25}_{25}$ |

| Liquid | Specific Gravity |
|---|---|
| Butyl chloride, tertiary | $0.847^{15}_{4}$ |
| Rose oil | $0.848–0.863^{30}_{15}$ |
| Lemon oil | $0.849–0.855^{25}_{25}$ |
| Amyl ether ketone | $0.850^{0}_{4}$ |
| *n*-Amyl nitrite | 0.853 |
| Rectified turpentine oil | $0.853–0.862^{25}_{25}$ |
| Dwarf pine needle oil | $0.853–0.871^{25}_{25}$ |
| Allyl alcohol | 0.854 |
| Coconut oil | $0.918–0.923^{25}_{25}$ |
| Cod liver oil | $0.918–0.927^{25}_{25}$ |
| Halibut liver oil | $0.920–0.930^{25}_{25}$ |
| Eucalyptol | $0.921–0.923^{25}_{25}$ |
| Ethyl formate | $0.924^{25}_{4}$ |
| Soya oil | $0.924–0.927^{15}_{15}$ |
| Linseed oil | $0.925–0.935^{25}_{25}$ |
| Pine oil | $0.927–0.940^{25}_{25}$ |
| Methyl acetate | 0.928 |
| Cellosolve | 0.930 |
| Ionone | $0.933–0.937^{25}_{25}$ |
| *N,N*-Diethylaniline | 0.935 |
| Furan | 0.937 |
| Allyl chloride | 0.938 |
| Valeric acid | 0.942 |
| Castor oil | $0.945–0.965^{25}_{25}$ |
| Cyclohexanone | 0.948 |
| Pyrrole | 0.948 |
| Cyclopentanone | 0.948 |
| Cyclopentanol | 0.949 |
| Isobutyric acid | 0.949 |
| 2-Picoline | $0.950^{15}_{4}$ |
| Chenopodium oil | $0.950–0.980^{25}_{25}$ |
| Myrcia oil | $0.950–0.990^{25}_{25}$ |
| Cycloheptanone | 0.951 |
| Fennel oil | $0.953–0.973^{25}_{25}$ |
| Dimethylaniline | 0.956 |
| 4-Picoline | $0.957^{15}_{4}$ |
| 3-Picoline | $0.961^{15}_{4}$ |
| Indan | 0.965 |
| Methyl cellosolve | 0.966 |
| Phenetole | 0.967 |
| Vitamin $K_1$ | $0.967^{25}_{25}$ |
| Tetralin | 0.970 |
| Carvacrol | 0.976 |
| Pyridine | $0.978^{25}_{4}$ |
| Anise oil | $0.978–0.988^{25}_{25}$ |
| Ethyl urethan | 0.981 |
| Benzylamine | $0.983^{19}_{4}$ |
| Benzyl acetone | $0.989^{23}_{17}$ |
| *m*-Toluidine | 0.989 |
| Carbitol | 0.990 |
| Dimethyl glyoxal | $0.990^{15}_{15}$ |
| Isoamyl benzoate | $0.993^{19}_{4}$ |
| Paraldehyde | 0.994 |
| Anisole | 0.995 |
| Isoamyl nitrate | $0.996^{22}_{4}$ |
| Morpholine | 0.999 |
| Water | $0.9970^{0}_{4}$ |
| Water | $0.9999^{20}_{20}$ |
| Water | $1.0000^{4.08}_{4.08}$ |
| Isobutyl benzoate | $1.002^{15}_{4}$ |
| *o*-Toluidine | 1.004 |
| Indene | 1.006 |
| Nicotine | 1.009 |
| Benzonitrile | $1.010^{15}_{15}$ |
| Hydrazine | $1.011^{15}_{4}$ |
| Ethanolamine | 1.018 |
| Pimenta oil | $10.018–1.048^{25}_{25}$ |

| Liquid | Specific Gravity |
|---|---|
| Mustard oil | $0.914–0.916^{15}_{15}$ |
| Corn oil | $0.914–0.921^{25}_{25}$ |
| Methyl propionate | 0.915 |
| Glycerin trioleate | 0.915 |
| Cottonseed oil | $0.915–0.921^{25}_{25}$ |
| Sesame oil | $0.916–0.921^{25}_{25}$ |
| Spearmint oil | $0.917–0.934^{25}_{25}$ |
| Cardamom oil | $0.917–0.947^{25}_{25}$ |
| Glycerol tributyrate | 1.035 |
| Propylene glycol | $1.036^{25}_{4}$ |
| Phlorol | $1.037^{12}$ |
| Butyl nitrate, secondary | $1.038^{0}_{4}$ |
| Bitter almond oil | $1.038–1.060^{25}_{25}$ |
| Clove oil | $1.038–1.060^{25}_{25}$ |
| Ethyl succinate | 1.040 |
| Benzyl ether | 1.043 |
| Benzyl alcohol | $1.045^{25}_{4}$ |
| Cinnamon oil | $1.045–1.063^{25}_{25}$ |
| *o*-Cresol | 1.047 |
| *n*-Butyl phthalate | 1.047 |
| *n*-Butyl nitrate | $1.048^{0}_{4}$ |
| Acetic acid (glacial) | $1.049^{25}_{25}$ |
| Benzaldehyde | $1.050^{15}_{4}$ |
| Ethyl benzoate | $1.051^{15}_{4}$ |
| Ethyl malonate | 1.055 |
| Benzyl acetate | $1.057^{16}_{4}$ |
| Allyl benzoate | $1.058^{15}_{15}$ |
| *n*-Propyl nitrate | 1.058 |
| Succinaldehyde | 1.064 |
| Thiophene | 1.064 |
| Methyl carbonate | $1.065^{17}_{4}$ |
| Eugenol | 1.066 |
| *p*-Chlorotoluene | 1.070 |
| *m*-Chlorotoluene | 1.072 |
| Diethyl maleate | $1.074^{15}_{15}$ |
| Benzofuran | $1.078^{15}_{15}$ |
| *o*-Chlorotoluene | 1.082 |
| Acetic anhydride | $1.087^{15}_{4}$ |
| *o*-Anisidine | 1.092 |
| Methyl benzoate | $1.094^{15}_{4}$ |
| Quinoline | 1.095 |
| *m*-Anisidine | 1.096 |
| Diethanolamine | 1.097 |
| Benzyl chloride | $1.103^{18}_{4}$ |
| Aldol | 1.103 |
| Acetyl chloride | 1.105 |
| Ethyl nitrate | 1.105 |
| Chlorobenzene | 1.107 |
| Polyethylene Glycol 400 | $1.110–1.140^{25}_{25}$ |
| Cinnamaldehyde | $1.112^{15}_{4}$ |
| Benzyl benzoate | $1.118^{25}_{4}$ |
| Diethylene glycol | $1.118^{20}_{20}$ |
| Anisaldehyde | 1.123 |
| Diethyl phthalate | $1.123^{25}_{4}$ |
| Triethanolamine | 1.124 |
| Polyethylene Glycol 300 | $1.124–1.130^{25}_{25}$ |
| Furfuryl alcohol | 1.130 |
| Nitromethane | 1.130 |
| Formamide | 1.134 |
| Ethyl salicylate | $1.136^{15}_{4}$ |
| *m*-Nitrotoluene | 1.157 |
| Ethyl chloroacetate | 1.159 |
| Furfural | 1.160 |
| Glycerol triacetate | 1.161 |
| *o*-Nitrotoluene | 1.163 |
| Salicylaldehyde | 1.167 |

| Liquid | Specific Gravity |
|---|---|
| Aniline | 1.022 |
| Sparteine | 1.023 |
| Phenylethyl alcohol | $1.024^{15}_{4}$ |
| Dibenzylamine | 1.026 |
| Chloroacetal | $1.026^{16}_{4}$ |
| Acetophenone | $1.033^{15}_{15}$ |
| 1,4-Dioxane | 1.034 |
| *m*-Cresol | 1.034 |
| Methyl salicylate | $1.184^{25}_{25}$ |
| Dimethyl phthalate | $1.189^{25}_{25}$ |
| Nitrobenzene | $1.205^{15}_{4}$ |
| Isoamyl bromide | $1.210^{15}_{4}$ |
| Benzoyl chloride | $1.219^{15}_{15}$ |
| *sym.*-Dichloroethyl ether | 1.222 |
| Butyl bromide, tertiary | 1.222 |
| Formic acid | $1.226^{15}_{4}$ |
| Methyl chloroacetate | $1.238^{20}_{20}$ |
| Amyl bromide | $1.246^{0}_{4}$ |
| Lactic acid (*dl*) | $1.249^{15}_{4}$ |
| *uns.*-Ethylene dichloride | 1.252 |
| Butyl bromide, secondary | 1.258 |
| Glycerol | 1.260 |
| Carbon disulfide | 1.263 |
| *n*-Butyl bromide | $1.269^{25}_{4}$ |
| Isobutyl bromide | $1.272^{15}_{4}$ |
| *sym.*-Dichloroethylene | $1.291^{15}_{4}$ |
| *o*-Dichlorobenzene | $1.307^{20}_{20}$ |
| Isopropyl bromide | 1.310 |
| Ethylsulfuric acid | $1.316^{17}_{4}$ |
| Methylene chloride | $1.335^{15}_{4}$ |
| *n*-Propyl bromide | 1.353 |
| *m*-Xylyl bromide | $1.371^{23}_{4}$ |
| Benzotrichloride | $1.380^{15}_{4}$ |
| Ethyl trichloroacetate | 1.383 |
| Allyl bromide | 1.398 |
| Ethyl bromide | 1.430 |
| Benzyl bromide | $1.438^{22}_{0}$ |
| Hydrogen peroxide, anhydrous | $1.465^{0}_{4}$ |
| Trichloroethylene | 1.465 |
| Chloroform | $1.498^{15}_{4}$ |
| Bromobenzene | $1.499^{15}_{15}$ |
| Chloral | 1.512 |
| Trichloroethanol | $1.550^{20}_{20}$ |
| Dichloroacetic acid | 1.563 |
| Benzoyl bromide | $1.570^{15}_{4}$ |
| Glycerophosphoric acid | $1.590^{19}_{4}$ |
| Nitroglycerin | $1.592^{25}_{4}$ |
| Carbon tetrachloride | 1.595 |
| Tetrachloroethane | 1.600 |
| Tetrachloroethylene | $1.631^{15}_{4}$ |
| Chloropicrin | 1.651 |
| Diphosgene | $1.653^{14}_{4}$ |
| Thionyl chloride | $1.655^{10.4}$ |
| Acetyl bromide | $1.663^{16}_{4}$ |
| Isopropyl iodide | 1.703 |
| Ethyl iodide | 1.933 |
| Ethylene bromide | $2.170^{25}_{4}$ |
| Ethylene dibromide | $2.172^{25}_{25}$ |
| Methyl iodide | 2.251 |
| Bromal | $2.300^{15}_{4}$ |
| Methylene bromide | 2.495 |
| Bromoform | 2.890 |
| Tetrabromoethane | 2.964 |
| Methylene iodide | 3.325 |
| Mercury | 13.546 |

Data taken from: *Merck Index,* 7th ed., pp. 1532–1535 (1960). Reproduced by permission of Merck and Co., Inc., Rahway, New Jersey.

# INFRARED CORRELATION[a]

## CHART 1

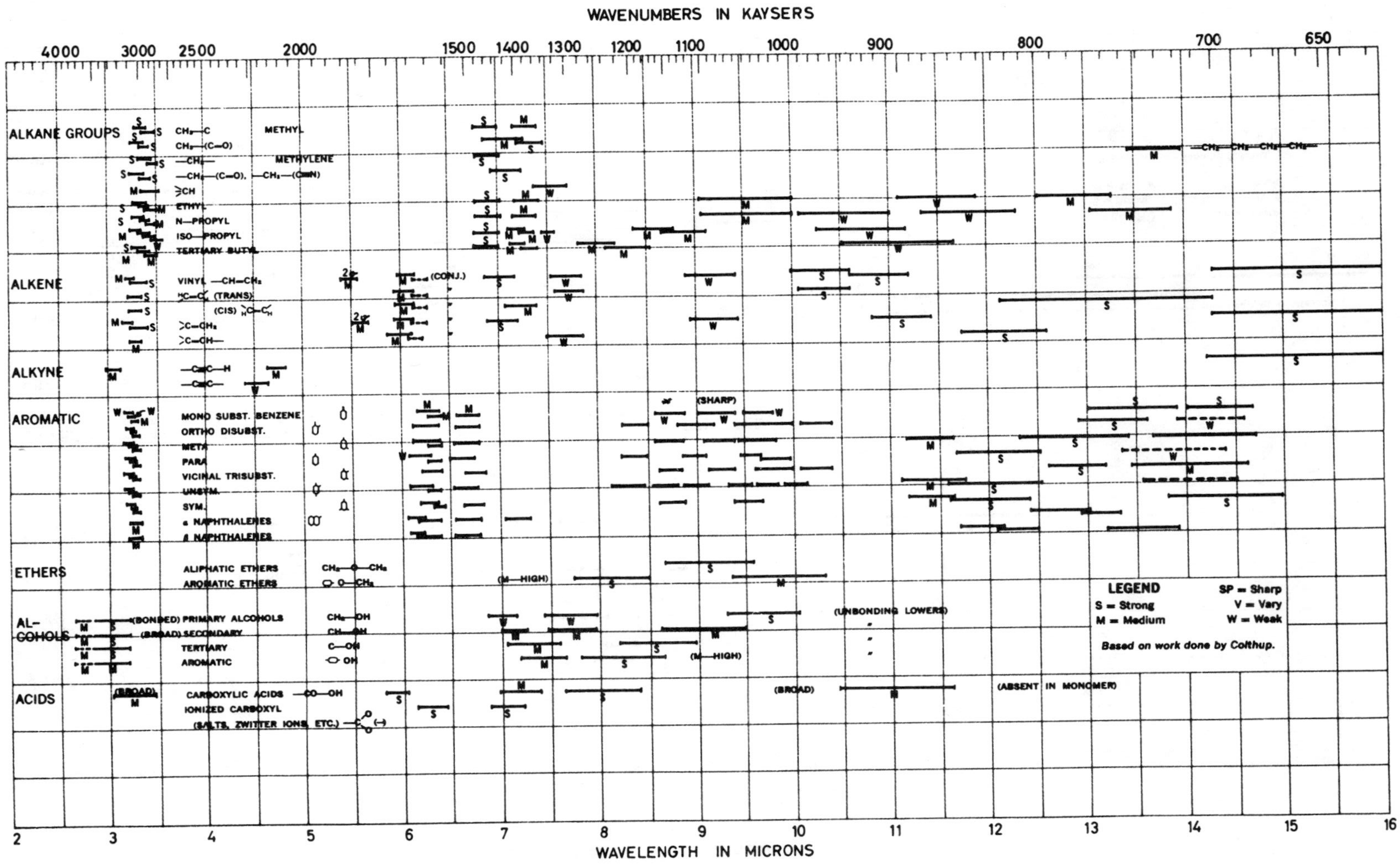

WAVENUMBERS IN KAYSERS

ALKANE GROUPS: $CH_3$—C; $CH_3$—(C=O) (METHYL); —$CH_2$—; —$CH_2$—(C=O), —$CH_2$—(C≡N) (METHYLENE); ⋛CH; ETHYL; N—PROPYL; ISO—PROPYL; TERTIARY BUTYL; —$CH_2$—$CH_2$—$CH_2$—$CH_2$—

ALKENE: VINYL —CH=$CH_2$; (TRANS); (CIS); >C=$CH_2$; >C=CH—; (CONJ.)

ALKYNE: —C≡C—H; —C≡C—

AROMATIC: MONO SUBST. BENZENE; ORTHO DISUBST.; META; PARA; VICINAL TRISUBST.; UNSYM.; SYM.; α NAPHTHALENES; β NAPHTHALENES; (SHARP)

ETHERS: ALIPHATIC ETHERS $CH_3$—O—$CH_3$; AROMATIC ETHERS; (M—HIGH)

ALCOHOLS: (BONDED) PRIMARY ALCOHOLS $CH_2$—OH; (BROAD) SECONDARY CH—OH; TERTIARY C—OH; AROMATIC; (UNBONDING LOWERS); (M—HIGH)

ACIDS: (BROAD); CARBOXYLIC ACIDS —CO—OH; IONIZED CARBOXYL (SALTS, ZWITTER IONS, ETC.); (BROAD); (ABSENT IN MONOMER)

LEGEND
S = Strong
M = Medium
SP = Sharp
V = Vary
W = Weak

Based on work done by Colthup.

WAVELENGTH IN MICRONS

[a] Prepared from information by Beckman Instruments.

CHART 1 (Continued)

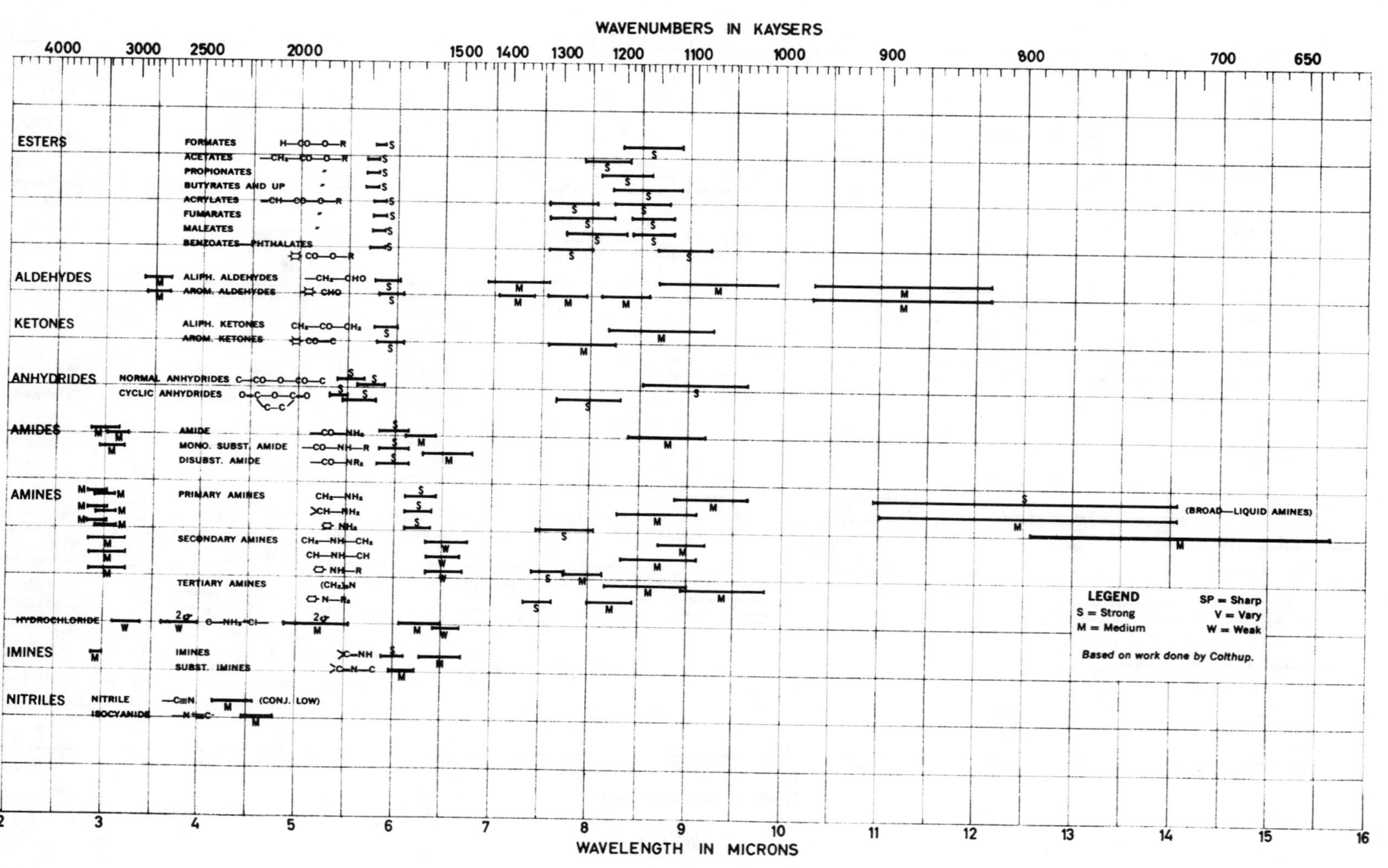

CHART 1 (Continued)

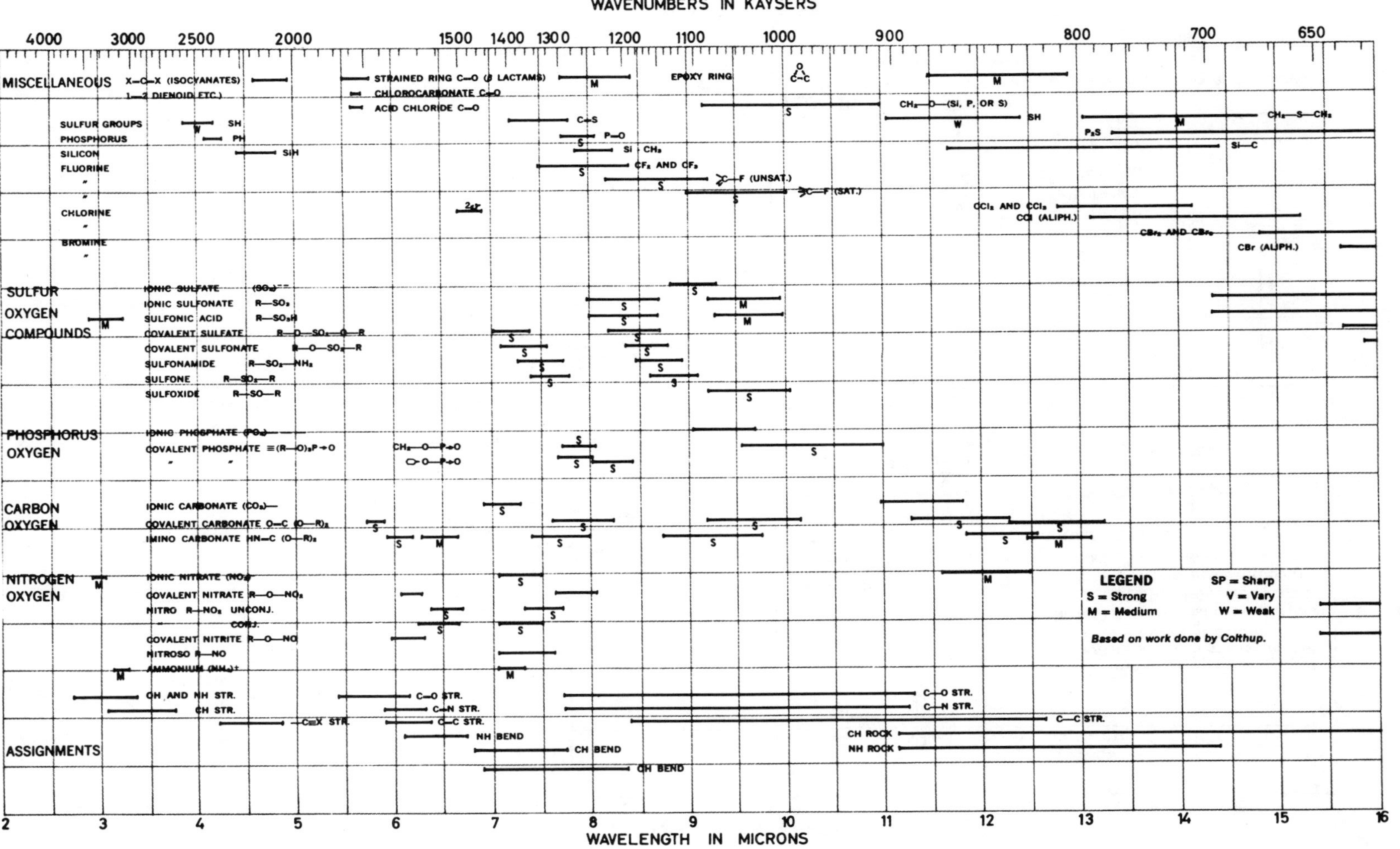

# CHART 2[b]

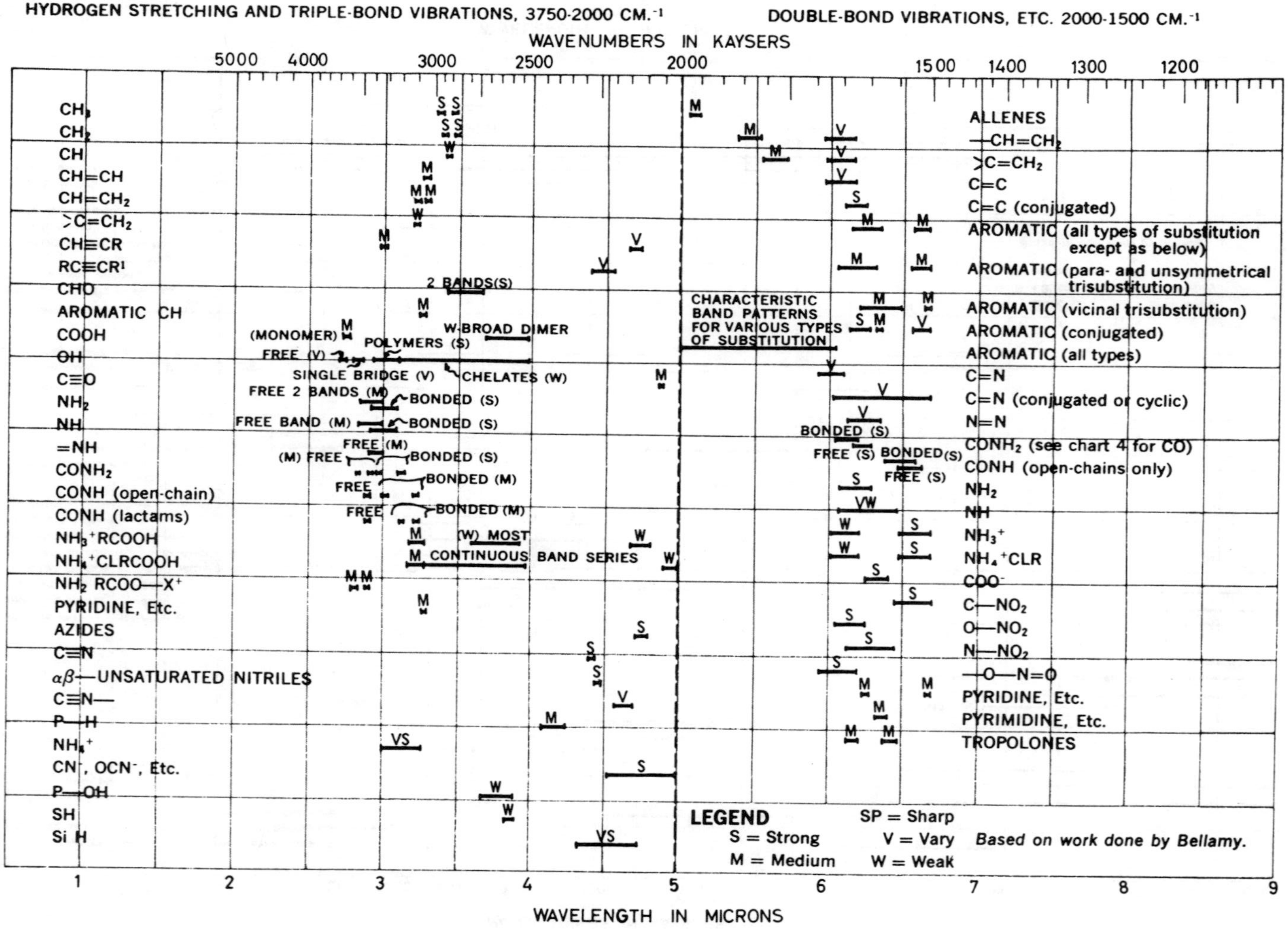

HYDROGEN STRETCHING AND TRIPLE-BOND VIBRATIONS, 3750-2000 CM.$^{-1}$

DOUBLE-BOND VIBRATIONS, ETC. 2000-1500 CM.$^{-1}$

WAVENUMBERS IN KAYSERS

5000 4000 3000 2500 2000 1500 1400 1300 1200

$CH_3$
$CH_2$
CH
CH=CH
$CH=CH_2$
$>C=CH_2$
CH≡CR
RC≡CR[1]
CHO
AROMATIC CH
COOH
OH
C≡O
$NH_2$
NH
=NH
$CONH_2$
CONH (open-chain)
CONH (lactams)
$NH_3^+RCOOH$
$NH_4^+CLRCOOH$
$NH_2$ RCOO—$X^+$
PYRIDINE, Etc.
AZIDES
C≡N
αβ—UNSATURATED NITRILES
C≡N—
P—H
$NH_4^+$
$CN^-$, $OCN^-$, Etc.
P—OH
SH
Si H

2 BANDS (S)
(MONOMER) M
W-BROAD DIMER
POLYMERS (S)
FREE (V)
SINGLE BRIDGE (V)
CHELATES (W)
FREE 2 BANDS (M)
BONDED (S)
FREE BAND (M)
BONDED (S)
FREE (M)
(M) FREE
BONDED (S)
FREE
BONDED (M)
FREE
BONDED (M)
(W) MOST
M CONTINUOUS BAND SERIES

CHARACTERISTIC BAND PATTERNS FOR VARIOUS TYPES OF SUBSTITUTION
BONDED (S)
FREE (S) BONDED (S)
FREE (S)

ALLENES
—CH=$CH_2$
$>C=CH_2$
C=C
C=C (conjugated)
AROMATIC (all types of substitution except as below)
AROMATIC (para- and unsymmetrical trisubstitution)
AROMATIC (vicinal trisubstitution)
AROMATIC (conjugated)
AROMATIC (all types)
C=N
C=N (conjugated or cyclic)
N=N
$CONH_2$ (see chart 4 for CO)
CONH (open-chains only)
$NH_2$
NH
$NH_3^+$
$NH_4^+CLR$
$COO^-$
C—$NO_2$
O—$NO_2$
N—$NO_2$
—O—N=O
PYRIDINE, Etc.
PYRIMIDINE, Etc.
TROPOLONES

**LEGEND**
S = Strong
M = Medium
SP = Sharp
V = Vary
W = Weak

*Based on work done by Bellamy.*

1 2 3 4 5 6 7 8 9

WAVELENGTH IN MICRONS

[b] This chart presents some information regarding structure, double-bond vibrations, hydrogen stretching and triple-bond vibrations.

## CHART 3[c]

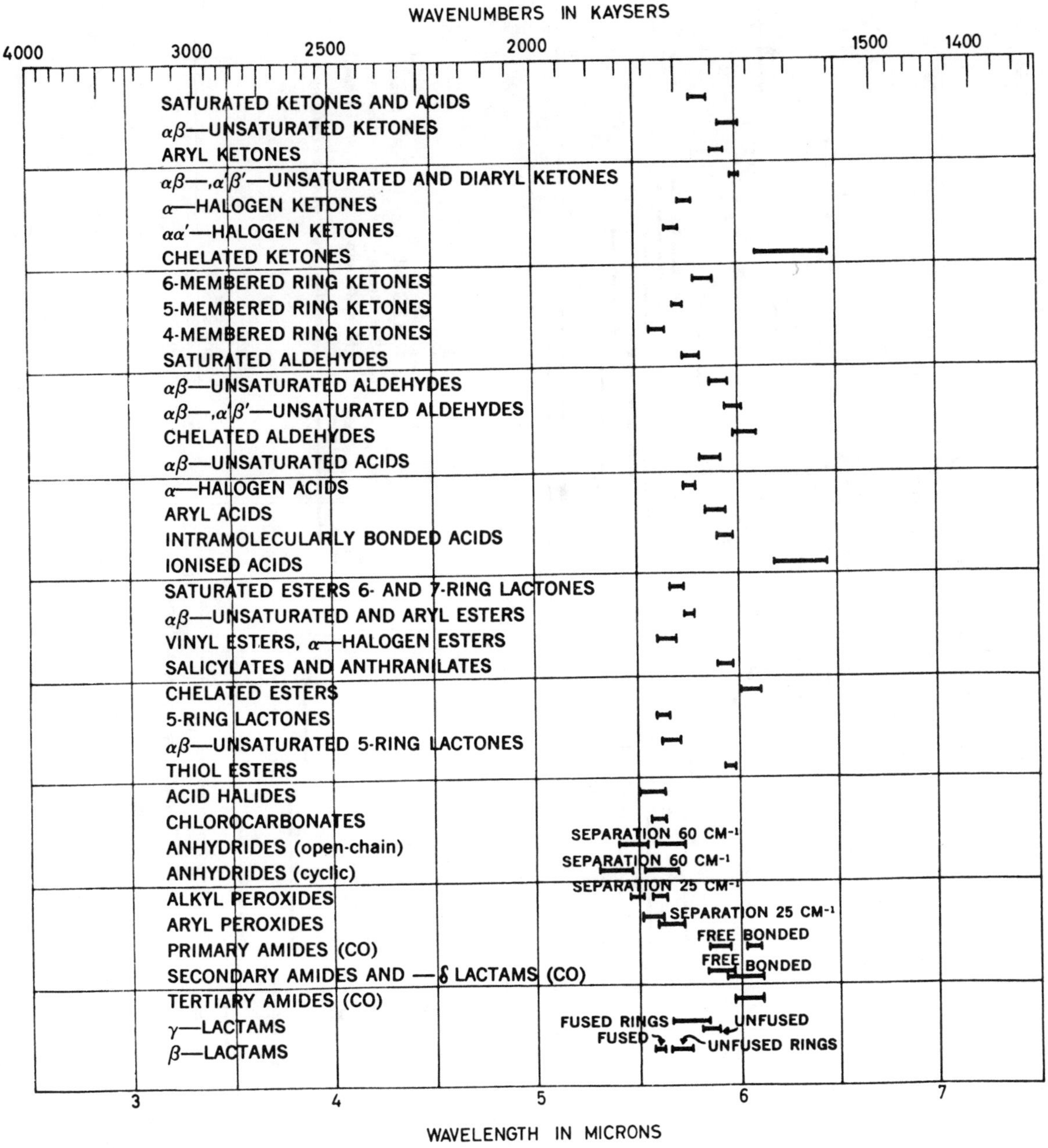

WAVENUMBERS IN KAYSERS

4000 3000 2500 2000 1500 1400

SATURATED KETONES AND ACIDS
αβ—UNSATURATED KETONES
ARYL KETONES
αβ—,α'β'—UNSATURATED AND DIARYL KETONES
α—HALOGEN KETONES
αα'—HALOGEN KETONES
CHELATED KETONES
6-MEMBERED RING KETONES
5-MEMBERED RING KETONES
4-MEMBERED RING KETONES
SATURATED ALDEHYDES
αβ—UNSATURATED ALDEHYDES
αβ—,α'β'—UNSATURATED ALDEHYDES
CHELATED ALDEHYDES
αβ—UNSATURATED ACIDS
α—HALOGEN ACIDS
ARYL ACIDS
INTRAMOLECULARLY BONDED ACIDS
IONISED ACIDS
SATURATED ESTERS 6- AND 7-RING LACTONES
αβ—UNSATURATED AND ARYL ESTERS
VINYL ESTERS, α—HALOGEN ESTERS
SALICYLATES AND ANTHRANILATES
CHELATED ESTERS
5-RING LACTONES
αβ—UNSATURATED 5-RING LACTONES
THIOL ESTERS
ACID HALIDES
CHLOROCARBONATES
ANHYDRIDES (open-chain) SEPARATION 60 CM⁻¹
ANHYDRIDES (cyclic) SEPARATION 60 CM⁻¹
ALKYL PEROXIDES SEPARATION 25 CM⁻¹
ARYL PEROXIDES SEPARATION 25 CM⁻¹
PRIMARY AMIDES (CO) FREE BONDED
SECONDARY AMIDES AND —δ LACTAMS (CO) FREE BONDED
TERTIARY AMIDES (CO)
γ—LACTAMS FUSED RINGS UNFUSED
β—LACTAMS FUSED UNFUSED RINGS

3 4 5 6 7

WAVENUMBERS IN MICRONS

WAVELENGTH IN MICRONS

[c] This chart presents some correlations between structure and the carbonyl vibrations of some classes of organic compounds. In all cases the absorption bands are strong and fall within the range from 1,900 to 1,500 $cm^{-1}$.

# CHART 4[d]

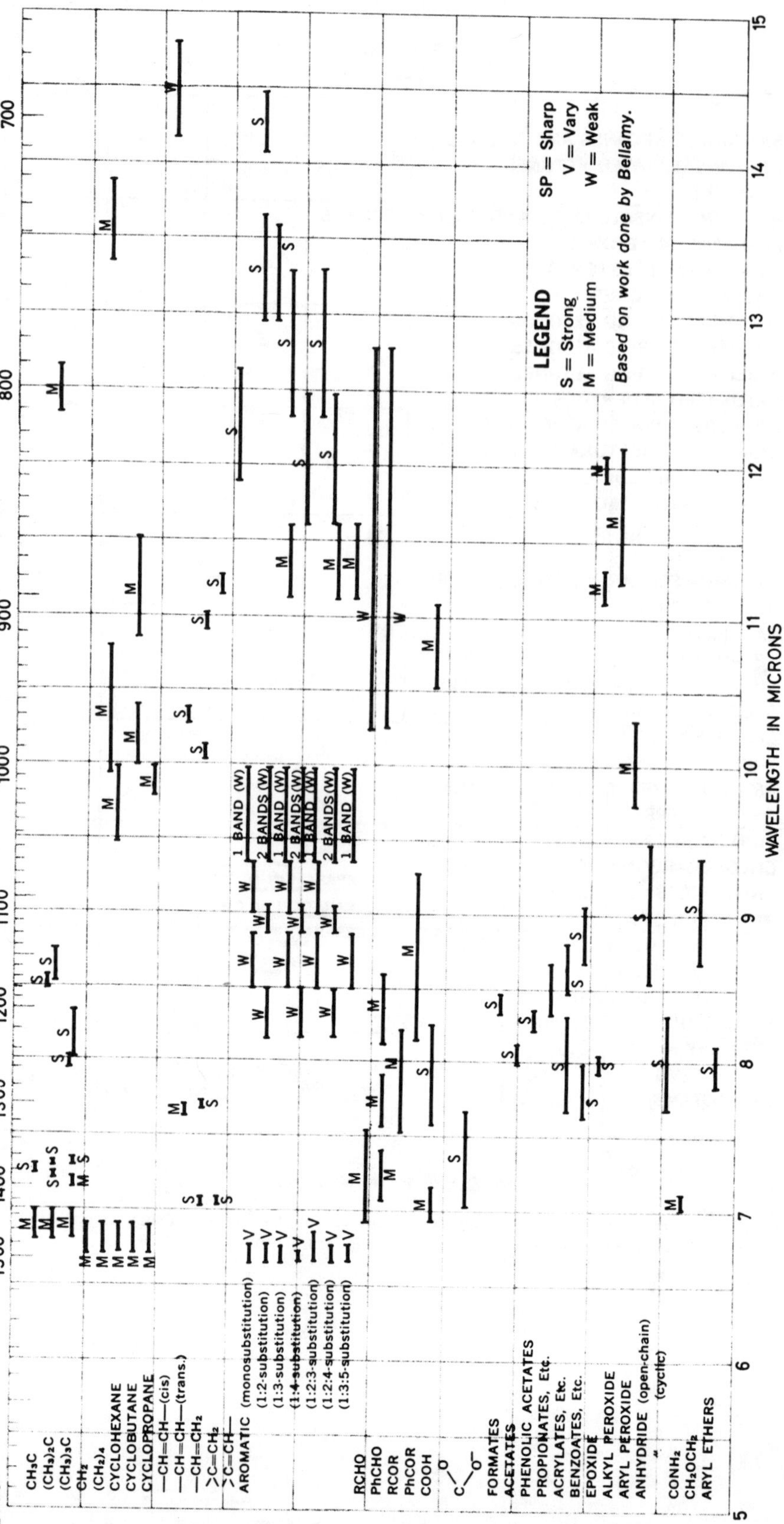

[d] This chart presents some correlations between structure and single-bond vibrations for a number of classes of compounds that have absorption between 1,500 and 650 cm$^{-1}$.

# CHART 4 (Continued)

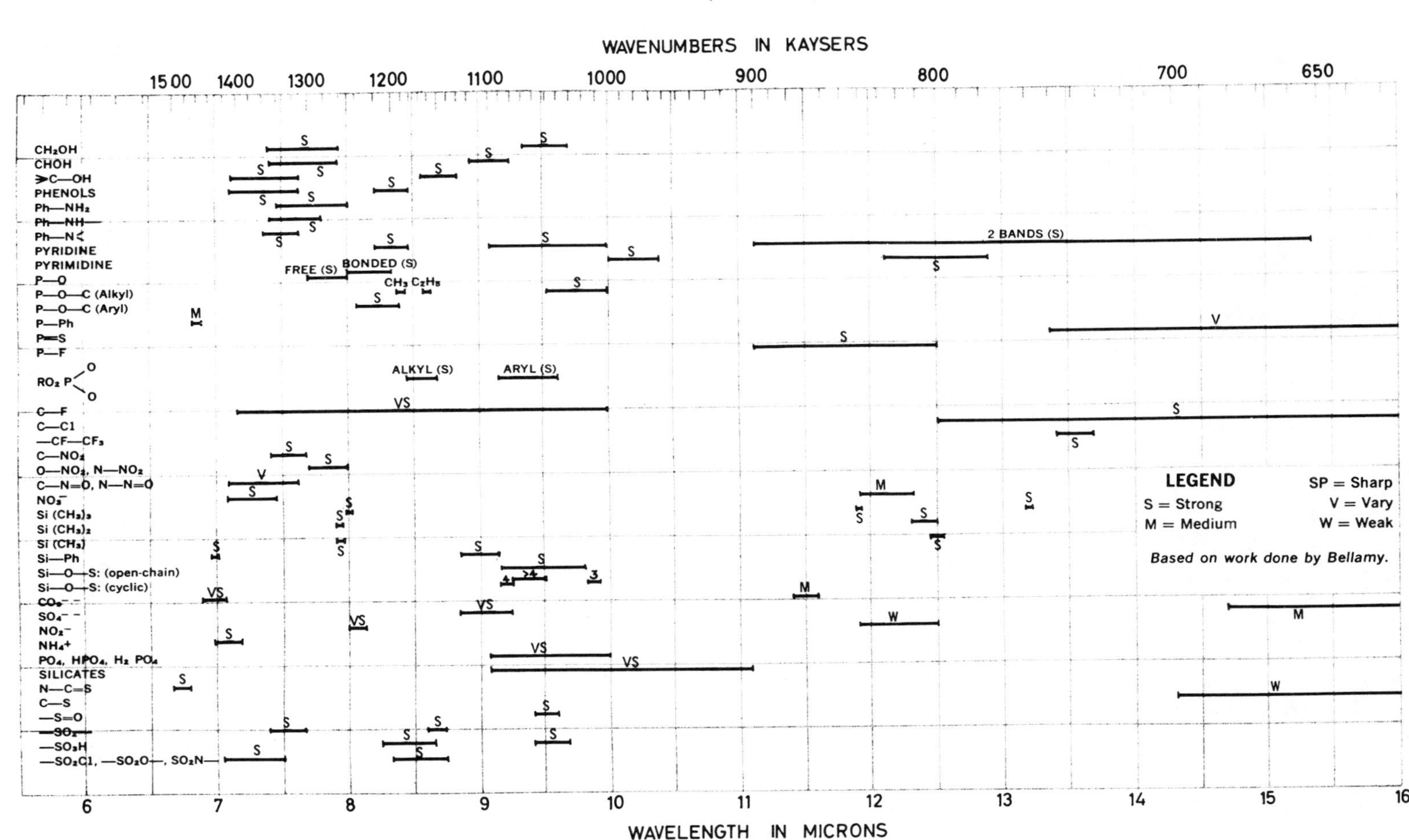

# FAR-INFRARED CORRELATION

## Vibrational Frequency Correlation[a]

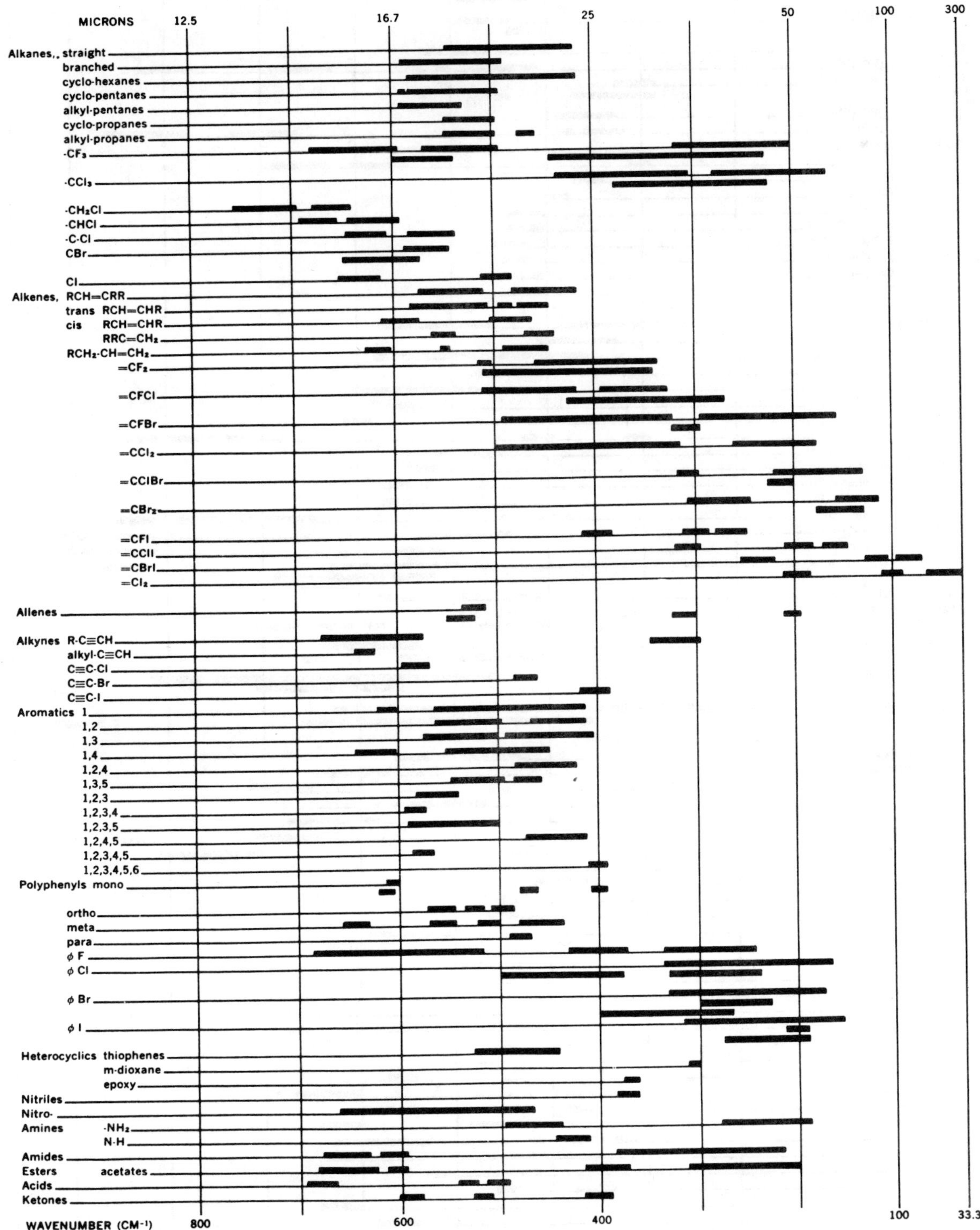


[a] Based on evidence by James E. Stewart of Beckman Instruments. Because research in this area is still continuing, the data in this chart are not all-inclusive.

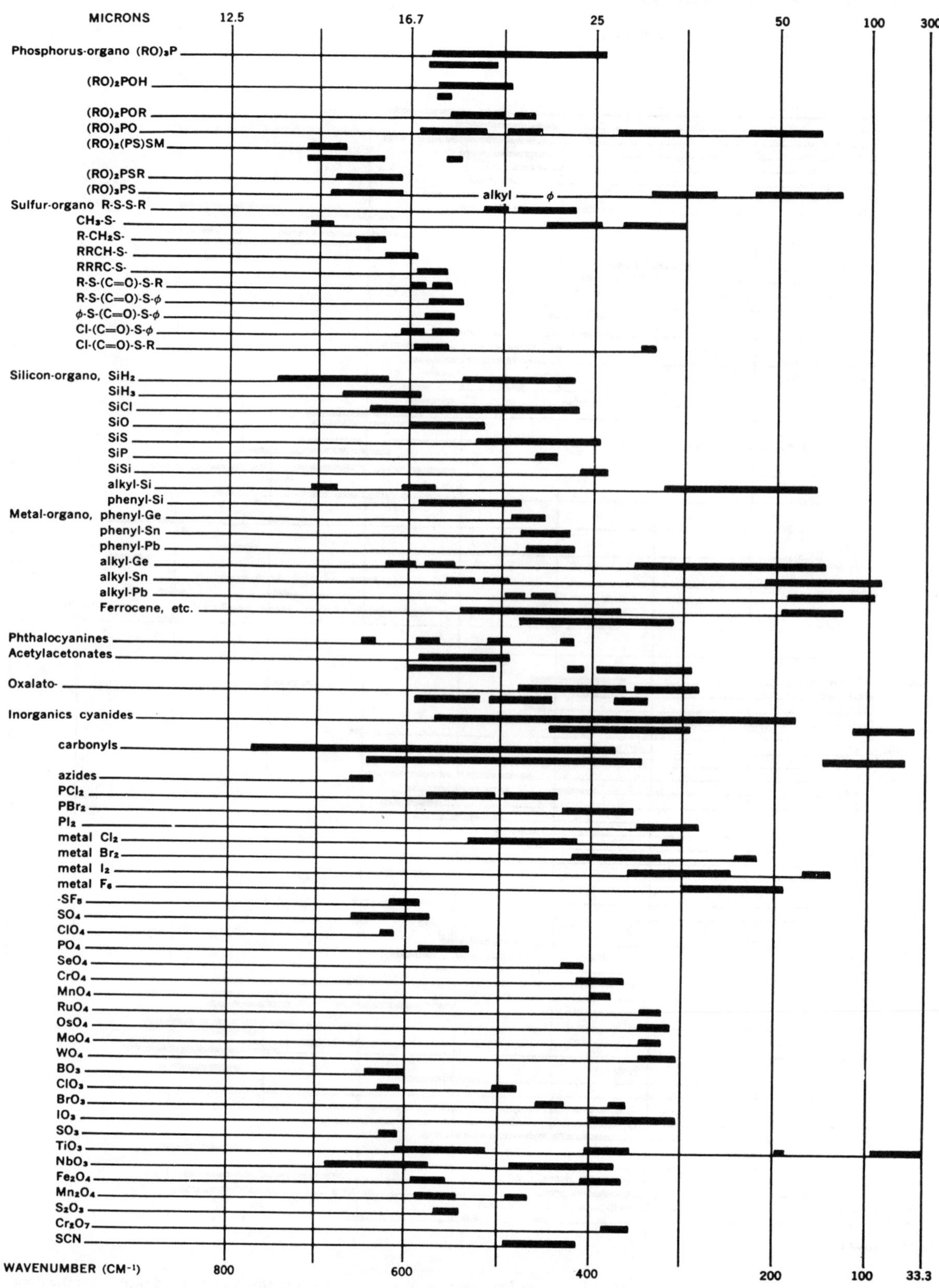
MICRONS
12.5
16.7
25
50
100
300
Phosphorus-organo (RO)₃P
(RO)₂POH
(RO)₂POR
(RO)₃PO
(RO)₂(PS)SM
(RO)₂PSR
(RO)₃PS
alkyl
ϕ
Sulfur-organo R-S-S-R
CH₃-S-
R-CH₂S-
RRCH-S-
RRRC-S-
R-S-(C=O)-S-R
R-S-(C=O)-S-ϕ
ϕ-S-(C=O)-S-ϕ
Cl-(C=O)-S-ϕ
Cl-(C=O)-S-R
Silicon-organo, SiH₂
SiH₃
SiCl
SiO
SiS
SiP
SiSi
alkyl-Si
phenyl-Si
Metal-organo, phenyl-Ge
phenyl-Sn
phenyl-Pb
alkyl-Ge
alkyl-Sn
alkyl-Pb
Ferrocene, etc.
Phthalocyanines
Acetylacetonates
Oxalato-
Inorganics cyanides
carbonyls
azides
PCl₂
PBr₂
PI₂
metal Cl₂
metal Br₂
metal I₂
metal F₆
-SF₅
SO₄
ClO₄
PO₄
SeO₄
CrO₄
MnO₄
RuO₄
OsO₄
MoO₄
WO₄
BO₃
ClO₃
BrO₃
IO₃
SO₃
TiO₃
NbO₃
Fe₂O₄
Mn₂O₄
S₂O₃
Cr₂O₇
SCN
WAVENUMBER (CM⁻¹)
800
600
400
200
100
33.3

## Characteristic Frequencies Between ~ 700 and 300 $cm^{-1}$

FREEMAN F. BENTLEY, LEE D. SMITHSON and ADELE L. ROZEK

This chart summarizes the characteristic frequencies known to occur between approximately 700 and $300^{-1}$. Those who anticipate using this region of the spectrum should consult *Infrared Spectra and Characteristic Frequencies between ~ 700 and 300 $cm^{-1}$* by Interscience Publishers, a division of John Wiley and Sons, Inc., for a complete discussion of the characteristic frequencies summarized in this chart, a large collection of infrared spectra (700 and 300 $cm^{-1}$) of most of the common organic and inorganic compounds, and an extensive bibliography of references to infrared data below ~ 700 $cm^{-1}$.

In this chart the black horizontal bars indicate the range of the spectrum in which the characteristic frequencies have been observed to occur in the compounds investigated. The number of compounds investigated is given immediately to the right of the names or structures of the compounds. Obviously those characteristic frequency ranges based upon a limited number of compounds should be used with caution.

The letters above the bars indicate the relative intensities of the absorption bands. These intensities are based upon the strongest band in the spectra (700 and 300 $cm^{-1}$) of specific classes of compounds investigated, and they cannot be compared accurately with the intensities given for other classes.

When known, the specific vibration giving rise to the characteristic frequency is printed in abbreviated form immediately to the right of the bar indicating the frequency range, except when lack of space prevents this. When there can be no ambiguity, this information may be printed in other available spaces. In doubtful cases, arrows are used for clarification.

Naturally, the characteristic frequencies vary in their specificity and analytical value. The user is, therefore, cautioned to use this chart with some reserve. After reviewing this chart, the reader should be aware that there are many characteristic frequencies in the 700 and 300 $cm^{-1}$ region. Used cautiously, this chart can be of considerable value in the elucidation of structures of unknown compounds.

It is important to emphasize that the region of the infrared between ~ 700 and 300 $cm^{-1}$ should be used in conjunction with the more conventional 5,000 and 700 $cm^{-1}$ region. Much of the value of the 700 and 300 $cm^{-1}$ region can only be realized after interpreting the spectrum between 5,000 and 700 $cm^{-1}$.

The following symbols and abbreviations are used:

| **Symbol or Abbreviation** | |
|---|---|
| $\alpha$CCC | In-plane bending of benzene ring |
| Antisym. | Antisymmetrical (Asymmetrical) |
| ~ | Approximately |
| $\beta$ | In-plane bending of ring substituent bond |
| $\delta$ | In-plane bending |
| $\gamma$ | Out-of-plane bending |
| i.p. | In-plane |
| m | Medium |
| $\nu$ | Stretching |
| $\nu_s$ | Symmetrical stretching |
| $\nu_{as}$ | Antisymmetrical stretching |
| o.p. | Out-of-plane |
| $\parallel$ | Parallel |
| $\perp$ | Perpendicular |
| $\phi$ | Phenyl |
| $\phi$CC | Out-of-plane bending of aromatic ring |
| r | Rocking |
| s | Strong |
| sh | Shoulder |
| Sym. | Symmetrical |
| v | Variable |
| w | Weak |
| "X" sensitive | An aromatic vibrational mode whose frequency position is greatly dependent on the nature of the substituent. |

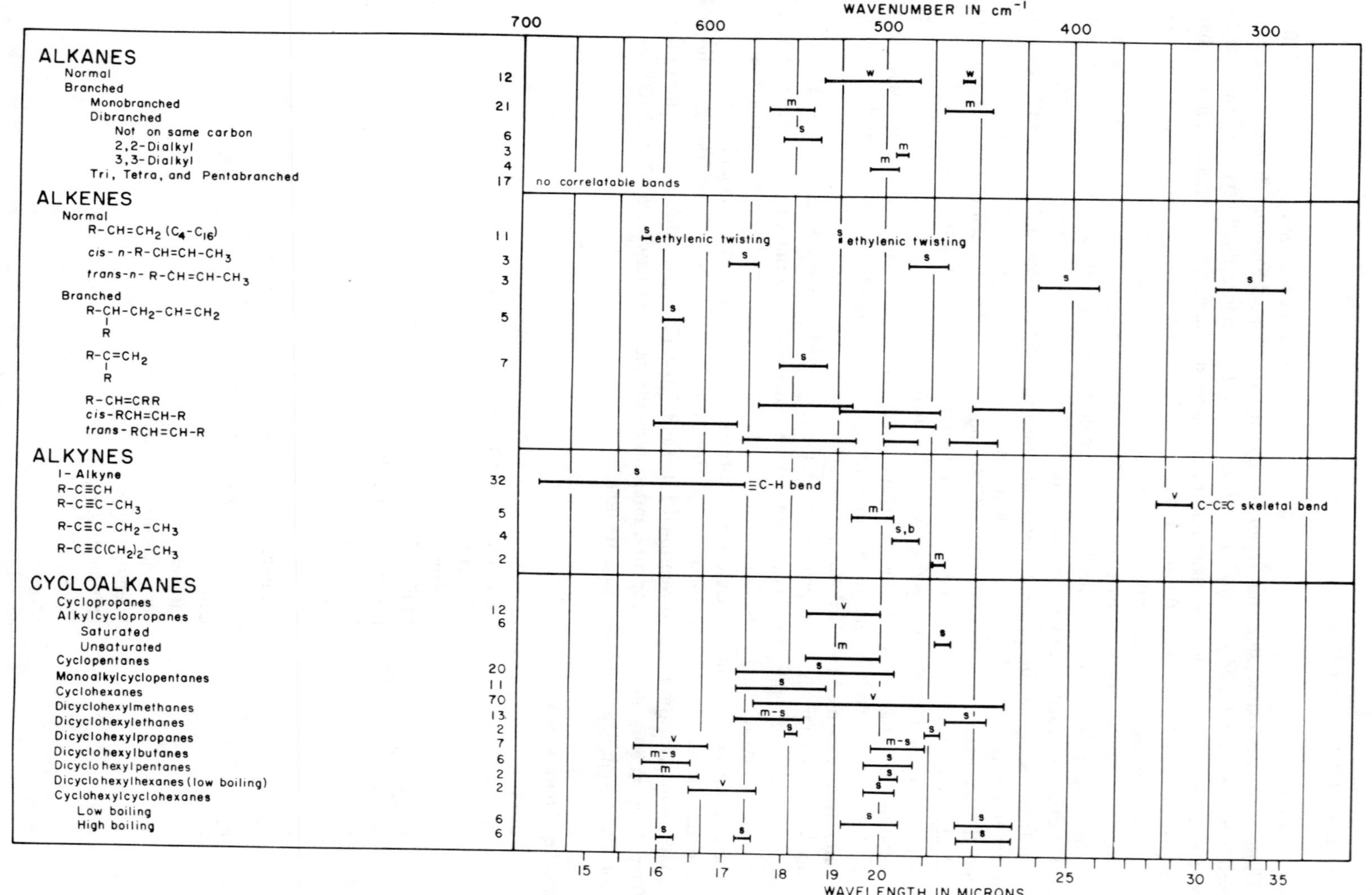
WAVENUMBER IN cm$^{-1}$
700
600
500
400
300
ALKANES
Normal
Branched
Monobranched
Dibranched
Not on same carbon
2,2-Dialkyl
3,3-Dialkyl
Tri, Tetra, and Pentabranched
12
21
6
3
4
17
no correlatable bands
ALKENES
Normal
R-CH=CH$_2$ (C$_4$-C$_{16}$)
*cis*-*n*-R-CH=CH-CH$_3$
*trans*-*n*-R-CH=CH-CH$_3$
Branched
R-CH-CH$_2$-CH=CH$_2$
R
R-C=CH$_2$
R
R-CH=CRR
*cis*-RCH=CH-R
*trans*-RCH=CH-R
11
3
3
5
7
ethylenic twisting
ethylenic twisting
ALKYNES
1-Alkyne
R-C≡CH
R-C≡C-CH$_3$
R-C≡C-CH$_2$-CH$_3$
R-C≡C(CH$_2$)$_2$-CH$_3$
32
5
4
2
≡C-H bend
C-C≡C skeletal bend
CYCLOALKANES
Cyclopropanes
Alkylcyclopropanes
Saturated
Unsaturated
Cyclopentanes
Monoalkylcyclopentanes
Cyclohexanes
Dicyclohexylmethanes
Dicyclohexylethanes
Dicyclohexylpropanes
Dicyclohexylbutanes
Dicyclohexylpentanes
Dicyclohexylhexanes (low boiling)
Cyclohexylcyclohexanes
Low boiling
High boiling
12
6
20
11
70
13
2
7
6
2
2
6
6
15
16
17
18
19
20
25
30
35
WAVELENGTH IN MICRONS

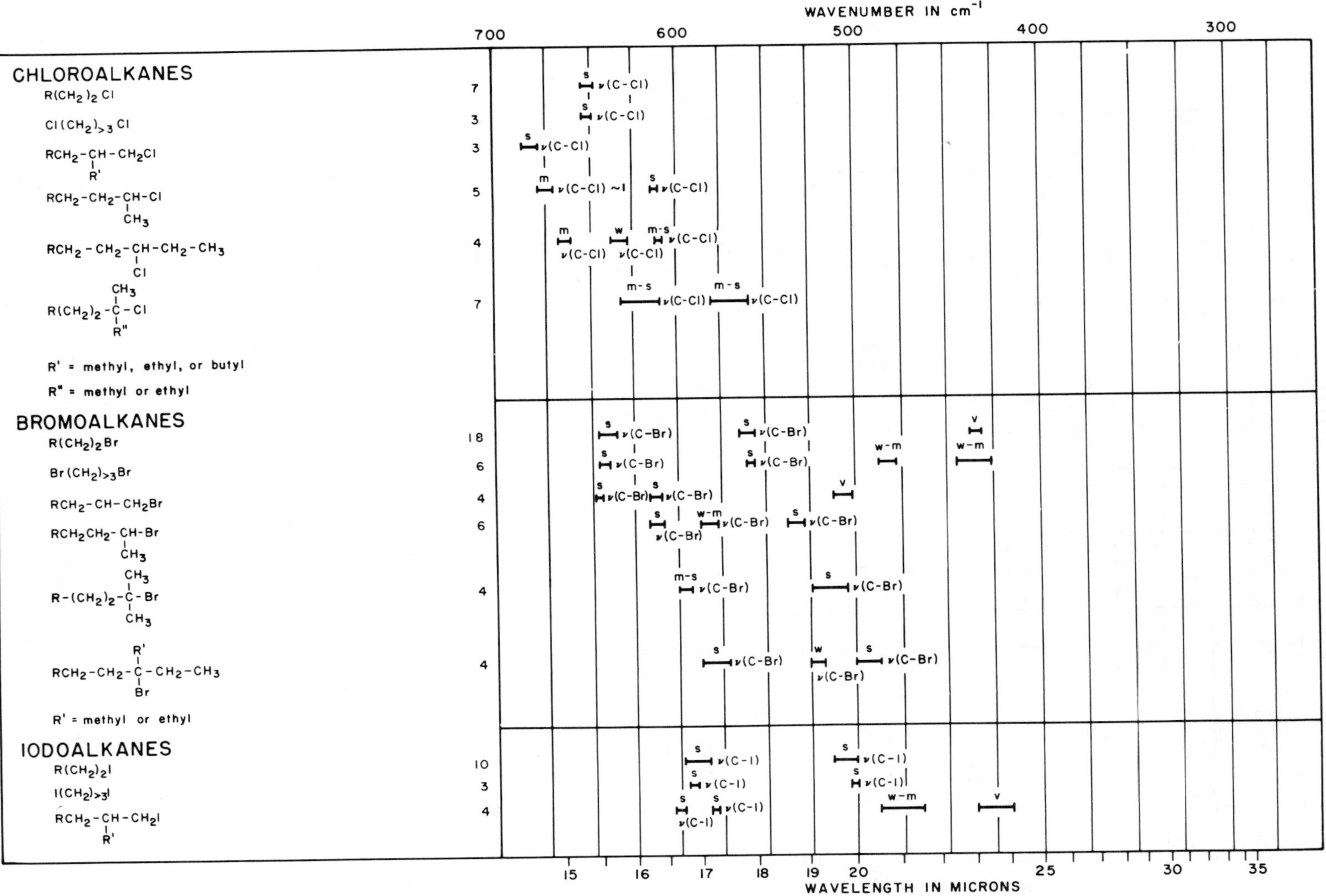
WAVENUMBER IN cm⁻¹
700
600
500
400
300
CHLOROALKANES
R(CH2)2Cl
7
s ν(C-Cl)
Cl(CH2)>3Cl
3
s ν(C-Cl)
RCH2-CH-CH2Cl
R'
3
s ν(C-Cl)
RCH2-CH2-CH-Cl
CH3
5
m ν(C-Cl) ~I
s ν(C-Cl)
RCH2-CH2-CH-CH2-CH3
Cl
4
m ν(C-Cl)
w ν(C-Cl)
m-s ν(C-Cl)
CH3
R(CH2)2-C-Cl
R"
7
m-s ν(C-Cl)
m-s ν(C-Cl)
R' = methyl, ethyl, or butyl
R" = methyl or ethyl
BROMOALKANES
R(CH2)2Br
18
s ν(C-Br)
s ν(C-Br)
v
w-m
Br(CH2)>3Br
6
s ν(C-Br)
s ν(C-Br)
w-m
RCH2-CH-CH2Br
4
s ν(C-Br)
s ν(C-Br)
v
RCH2CH2-CH-Br
CH3
6
s ν(C-Br)
w-m ν(C-Br)
s ν(C-Br)
CH3
R-(CH2)2-C-Br
CH3
4
m-s ν(C-Br)
s ν(C-Br)
R'
RCH2-CH2-C-CH2-CH3
Br
4
s ν(C-Br)
w ν(C-Br)
s ν(C-Br)
R' = methyl or ethyl
IODOALKANES
R(CH2)2I
10
s ν(C-I)
s ν(C-I)
I(CH2)>3I
3
s ν(C-I)
s ν(C-I)
RCH2-CH-CH2I
R'
4
s ν(C-I)
s ν(C-I)
w-m
v
15
16
17
18
19
20
25
30
35
WAVELENGTH IN MICRONS

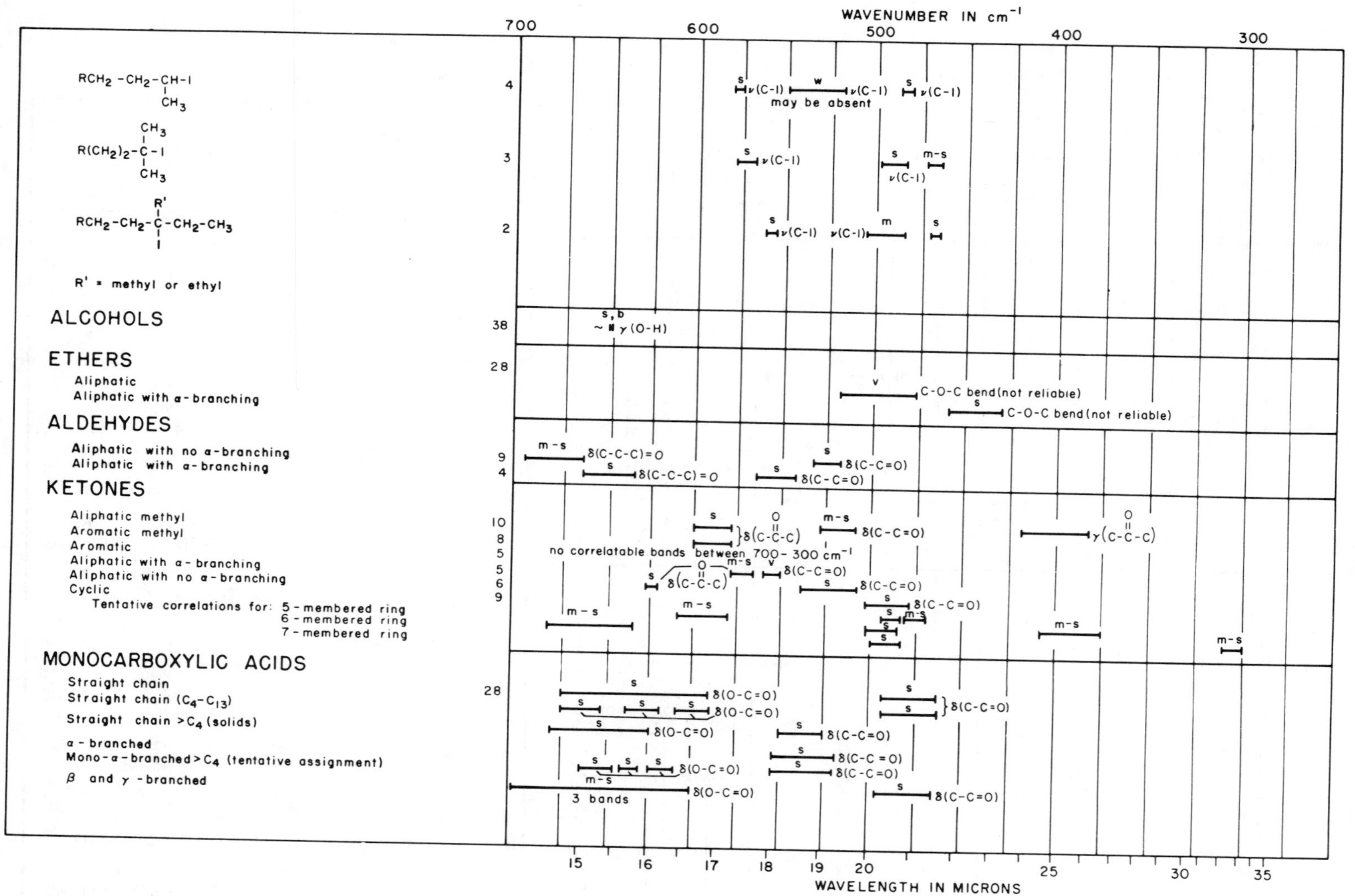

WAVENUMBER IN cm⁻¹
700 600 500 400 300
RCH2-CH2-CH-I CH3
R(CH2)2-C-I CH3 CH3
RCH2-CH2-C-CH2-CH3 R' I
R' = methyl or ethyl
ALCOHOLS
ETHERS
Aliphatic
Aliphatic with α-branching
ALDEHYDES
Aliphatic with no α-branching
Aliphatic with α-branching
KETONES
Aliphatic methyl
Aromatic methyl
Aromatic
Aliphatic with α-branching
Aliphatic with no α-branching
Cyclic
Tentative correlations for: 5-membered ring
6-membered ring
7-membered ring
MONOCARBOXYLIC ACIDS
Straight chain
Straight chain (C4-C13)
Straight chain >C4 (solids)
α-branched
Mono-α-branched >C4 (tentative assignment)
β and γ-branched
ν(C-I)
may be absent
γ(O-H)
C-O-C bend (not reliable)
δ(C-C-C)=O
δ(C-C=O)
no correlatable bands between 700-300 cm⁻¹
δ(O-C=O)
3 bands
15 16 17 18 19 20 25 30 35
WAVELENGTH IN MICRONS

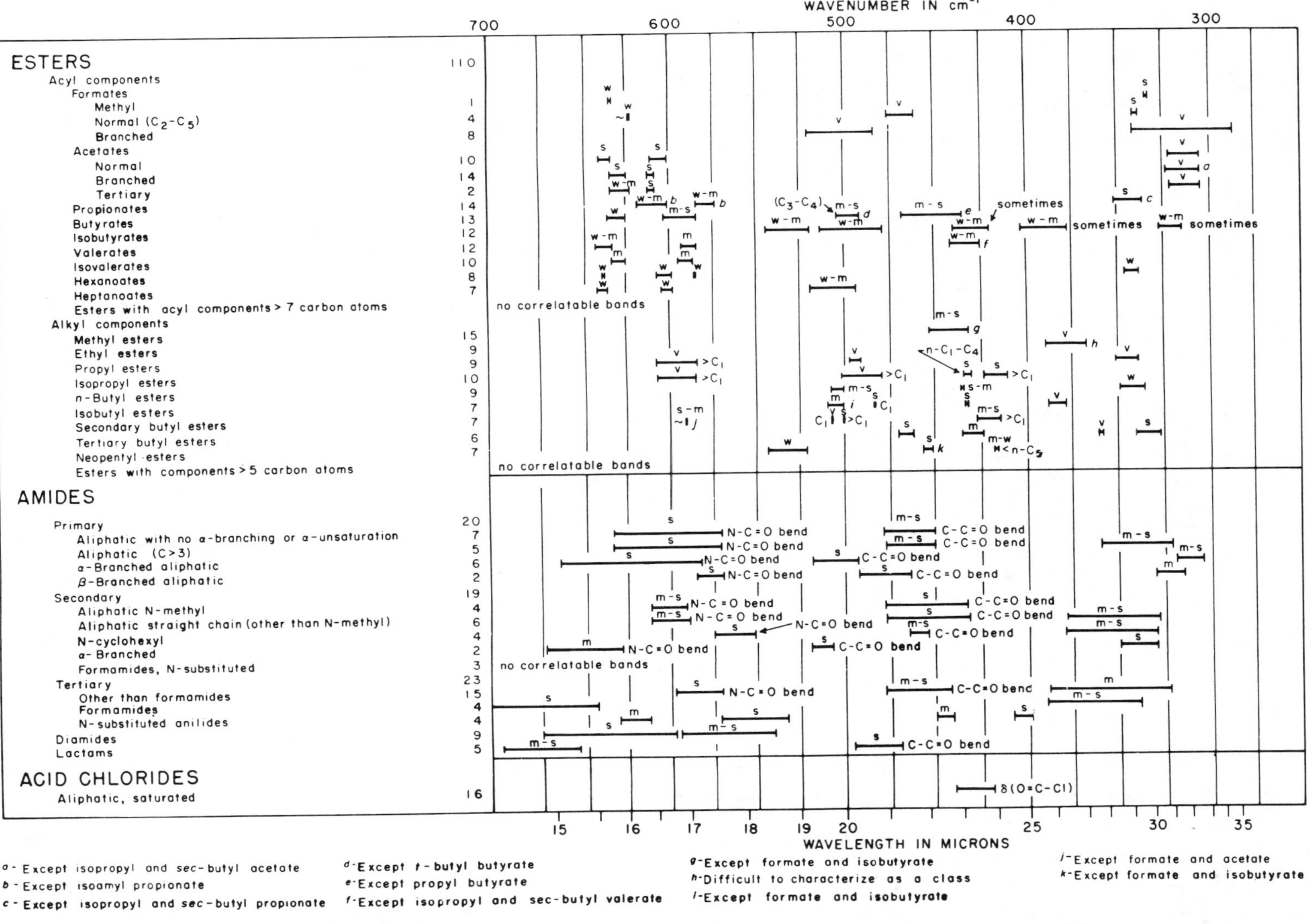


a - Except isopropyl and *sec*-butyl acetate
b - Except isoamyl propionate
c - Except isopropyl and *sec*-butyl propionate
d - Except *t*-butyl butyrate
e - Except propyl butyrate
f - Except isopropyl and sec-butyl valerate
g - Except formate and isobutyrate
h - Difficult to characterize as a class
i - Except formate and isobutyrate
j - Except formate and acetate
k - Except formate and isobutyrate

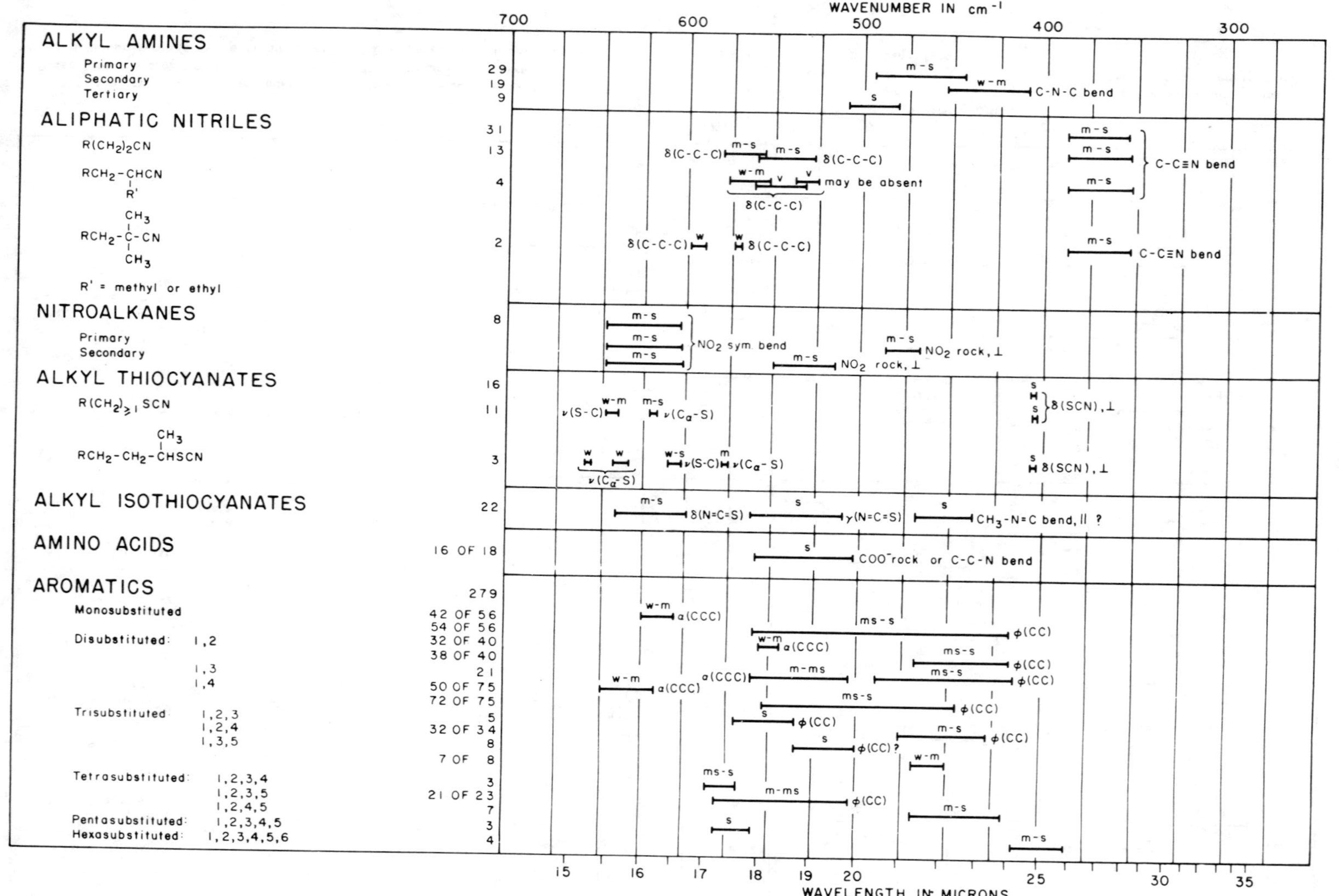
WAVENUMBER IN cm⁻¹
700
600
500
400
300
ALKYL AMINES
Primary
Secondary
Tertiary
ALIPHATIC NITRILES
R(CH2)2CN
RCH2-CHCN
R'
RCH2-C(CH3)2-CN
R' = methyl or ethyl
NITROALKANES
Primary
Secondary
ALKYL THIOCYANATES
R(CH2)≥1 SCN
RCH2-CH2-CH(CH3)SCN
ALKYL ISOTHIOCYANATES
AMINO ACIDS
AROMATICS
Monosubstituted
Disubstituted: 1,2
1,3
1,4
Trisubstituted 1,2,3
1,2,4
1,3,5
Tetrasubstituted: 1,2,3,4
1,2,3,5
1,2,4,5
Pentasubstituted: 1,2,3,4,5
Hexasubstituted: 1,2,3,4,5,6
29
19
9
31
13
4
2
8
16
11
3
22
16 OF 18
279
42 OF 56
54 OF 56
32 OF 40
38 OF 40
21
50 OF 75
72 OF 75
5
32 OF 34
8
7 OF 8
3
21 OF 23
7
3
4
m-s
w-m
C-N-C bend
s
δ(C-C-C)
δ(C-C-C)
may be absent
C-C≡N bend
NO2 sym. bend
NO2 rock, ⊥
ν(S-C)
ν(Cα-S)
δ(SCN), ⊥
δ(N=C=S)
γ(N=C=S)
CH3-N=C bend, ∥ ?
COO⁻ rock or C-C-N bend
α(CCC)
φ(CC)
φ(CC) ?
15
16
17
18
19
20
25
30
35
WAVELENGTH IN MICRONS

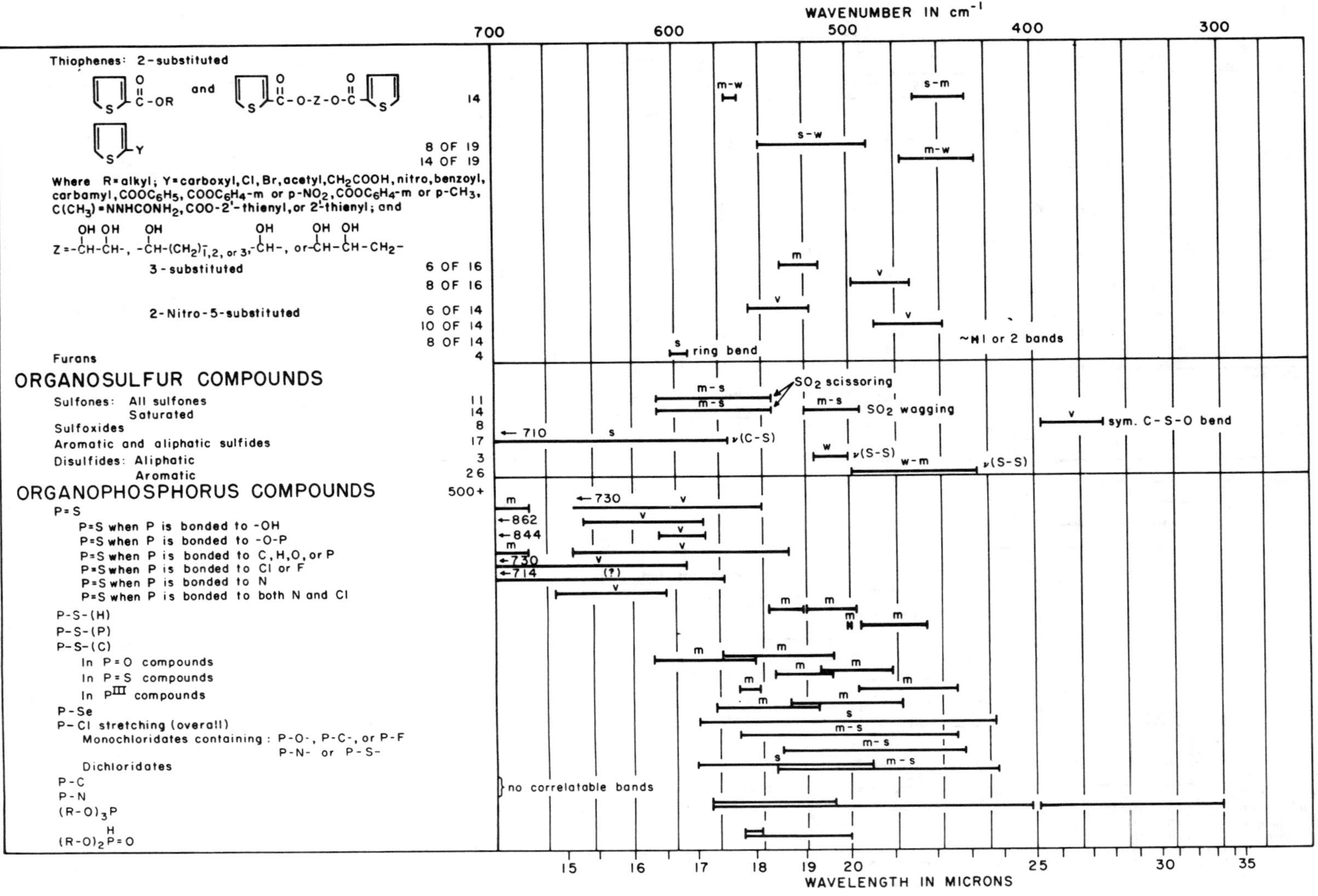
WAVENUMBER IN cm⁻¹
700
600
500
400
300
Thiophenes: 2-substituted
C-OR and C-O-Z-O-C
14
Y
8 OF 19
14 OF 19
Where R=alkyl; Y=carboxyl, Cl, Br, acetyl, CH2COOH, nitro, benzoyl, carbamyl, COOC6H5, COOC6H4-m or p-NO2, COOC6H4-m or p-CH3, C(CH3)=NNHCONH2, COO-2'-thienyl, or 2'-thienyl; and
Z=-CH(OH)-CH(OH)-, -CH(OH)-(CH2)1,2, or 3-CH(OH)-, or -CH(OH)-CH(OH)-CH2-
3-substituted
6 OF 16
8 OF 16
2-Nitro-5-substituted
6 OF 14
10 OF 14
8 OF 14
Furans
4
ORGANOSULFUR COMPOUNDS
Sulfones: All sulfones
11
Saturated
14
Sulfoxides
8
Aromatic and aliphatic sulfides
17
Disulfides: Aliphatic
3
Aromatic
26
ORGANOPHOSPHORUS COMPOUNDS
500+
P=S
P=S when P is bonded to -OH
P=S when P is bonded to -O-P
P=S when P is bonded to C, H, O, or P
P=S when P is bonded to Cl or F
P=S when P is bonded to N
P=S when P is bonded to both N and Cl
P-S-(H)
P-S-(P)
P-S-(C)
In P=O compounds
In P=S compounds
In PIII compounds
P-Se
P-Cl stretching (overall)
Monochloridates containing: P-O-, P-C-, or P-F
P-N- or P-S-
Dichloridates
P-C
P-N
(R-O)3P
(R-O)2P(H)=O
m-w
s-m
s-w
m-w
m
v
v
v
s ring bend
~1 or 2 bands
m-s
SO2 scissoring
m-s
m-s
SO2 wagging
v
sym. C-S-O bend
← 710
s
ν(C-S)
w
ν(S-S)
w-m
ν(S-S)
m
← 730
v
← 862
v
← 844
v
m
v
v
← 730
v
← 714
(?)
v
m
m
m
m
m
m
m
m
m
m
m
m
m
m
s
m-s
m-s
s
m-s
no correlatable bands
15
16
17
18
19
20
25
30
35
WAVELENGTH IN MICRONS

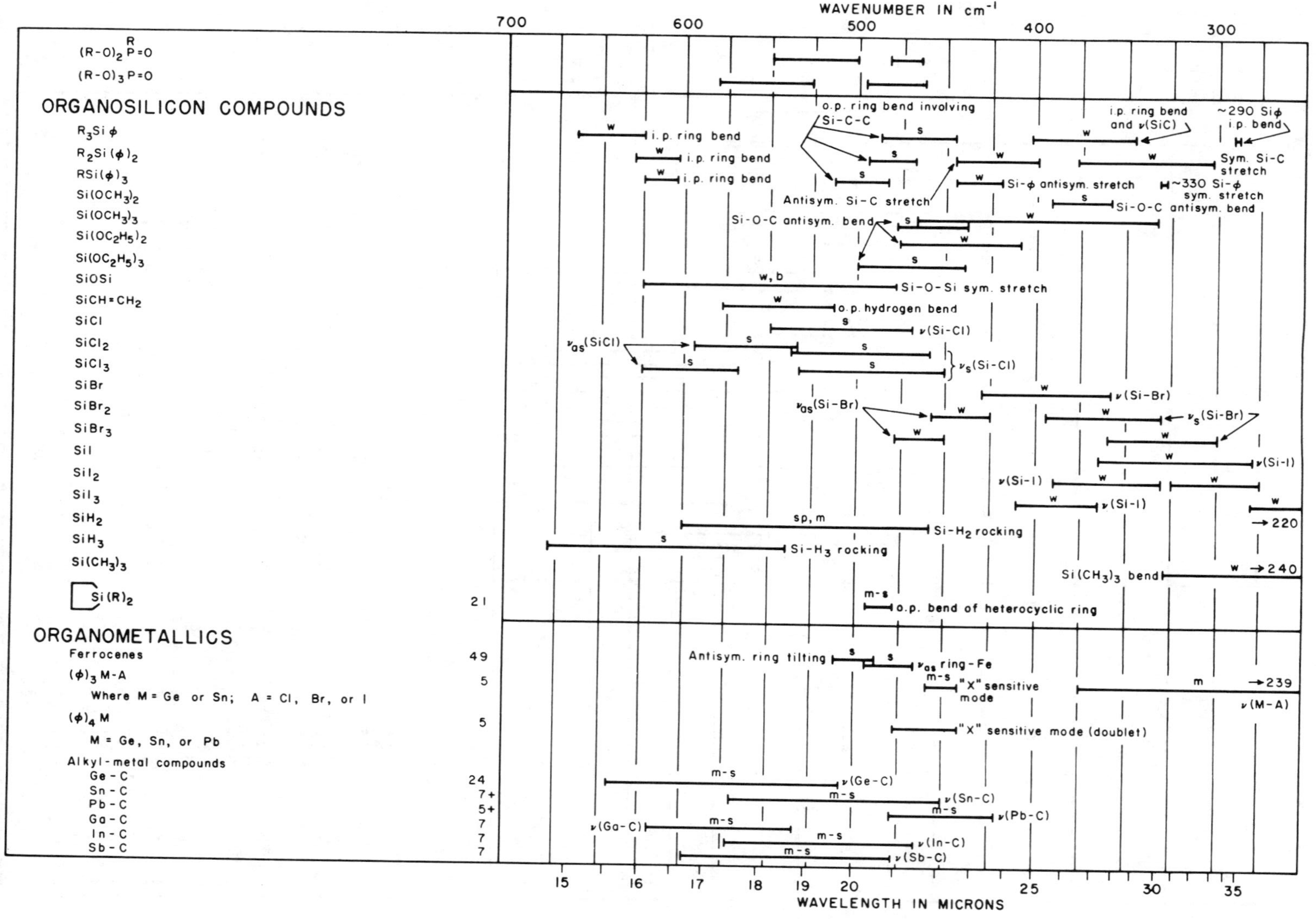
WAVENUMBER IN cm⁻¹
700
600
500
400
300
(R-O)2 P=O (R)
(R-O)3 P=O
ORGANOSILICON COMPOUNDS
R3Si φ
R2Si(φ)2
RSi(φ)3
Si(OCH3)2
Si(OCH3)3
Si(OC2H5)2
Si(OC2H5)3
SiOSi
SiCH=CH2
SiCl
SiCl2
SiCl3
SiBr
SiBr2
SiBr3
SiI
SiI2
SiI3
SiH2
SiH3
Si(CH3)3
Si(R)2
21
ORGANOMETALLICS
Ferrocenes
49
(φ)3 M-A
5
Where M = Ge or Sn; A = Cl, Br, or I
(φ)4 M
5
M = Ge, Sn, or Pb
Alkyl-metal compounds
Ge-C
24
Sn-C
7+
Pb-C
5+
Ga-C
7
In-C
7
Sb-C
7
o.p. ring bend involving Si-C-C
i.p. ring bend
i.p. ring bend
i.p. ring bend
i.p. ring bend and ν(SiC)
~290 Siφ i.p. bend
Sym. Si-C stretch
Si-φ antisym. stretch
~330 Si-φ sym. stretch
Antisym. Si-C stretch
Si-O-C antisym. bend
Si-O-C antisym. bend
Si-O-Si sym. stretch
o.p. hydrogen bend
ν(Si-Cl)
νas(SiCl)
νs(Si-Cl)
ν(Si-Br)
νas(Si-Br)
νs(Si-Br)
ν(Si-I)
ν(Si-I)
ν(Si-I)
→220
Si-H2 rocking
Si-H3 rocking
Si(CH3)3 bend
→240
o.p. bend of heterocyclic ring
Antisym. ring tilting
νas ring-Fe
"X" sensitive mode
→239
ν(M-A)
"X" sensitive mode (doublet)
ν(Ge-C)
ν(Sn-C)
ν(Pb-C)
ν(Ga-C)
ν(In-C)
ν(Sb-C)
w
s
m
m-s
w, b
sp, m
15
16
17
18
19
20
25
30
35
WAVELENGTH IN MICRONS

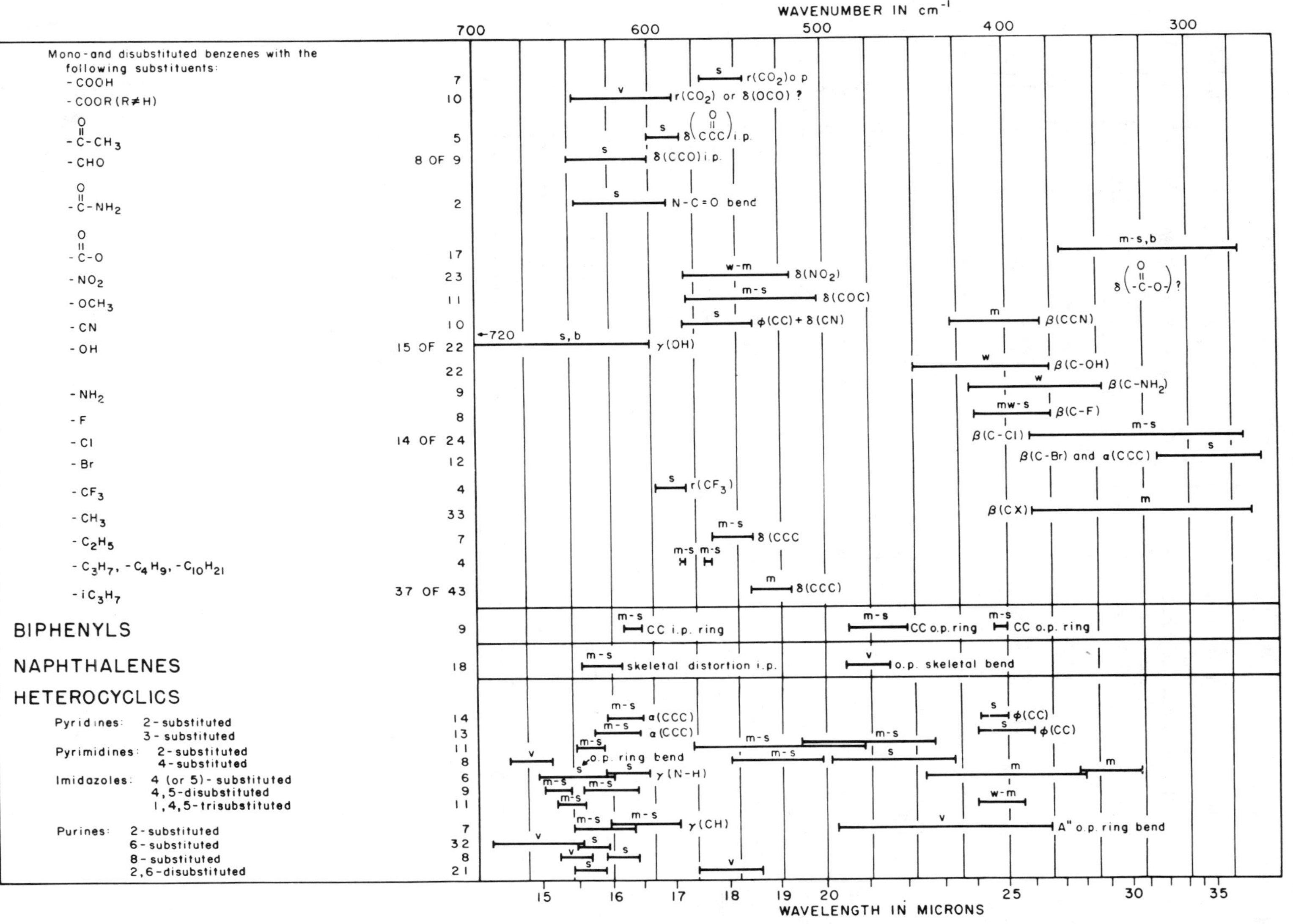
WAVENUMBER IN $cm^{-1}$
700
600
500
400
300
Mono-and disubstituted benzenes with the following substituents:
-COOH 7
s r($CO_2$) o.p.
-COOR (R≠H) 10
v r($CO_2$) or δ(OCO) ?
-C(=O)-$CH_3$ 5
s δ(C(=O)CC) i.p.
-CHO 8 OF 9
s δ(CCO) i.p.
-C(=O)-$NH_2$ 2
s N-C=O bend
-C(=O)-O 17
m-s,b
δ(-C(=O)-O-) ?
-$NO_2$ 23
w-m δ($NO_2$)
-$OCH_3$ 11
m-s δ(COC)
-CN 10
s φ(CC)+δ(CN)
m β(CCN)
-OH 15 OF 22
←720 s,b γ(OH)
22
w β(C-OH)
-$NH_2$ 9
w β(C-$NH_2$)
-F 8
mw-s β(C-F)
-Cl 14 OF 24
β(C-Cl) m-s
-Br 12
β(C-Br) and α(CCC) s
-$CF_3$ 4
s r($CF_3$)
-$CH_3$ 33
β(CX) m
-$C_2H_5$ 7
m-s δ(CCC
-$C_3H_7$, -$C_4H_9$, -$C_{10}H_{21}$ 4
m-s m-s
-i$C_3H_7$ 37 OF 43
m δ(CCC)
BIPHENYLS 9
m-s CC i.p. ring
m-s CC o.p. ring
m-s CC o.p. ring
NAPHTHALENES 18
m-s skeletal distortion i.p.
v o.p. skeletal bend
HETEROCYCLICS
Pyridines: 2-substituted 14
3-substituted 13
m-s α(CCC)
m-s α(CCC)
s φ(CC)
s φ(CC)
Pyrimidines: 2-substituted 11
4-substituted 8
m-s
o.p. ring bend
v
m-s
m-s
m-s
s
Imidazoles: 4 (or 5)-substituted 6
4,5-disubstituted 9
1,4,5-trisubstituted 11
s
s γ(N-H)
m-s
m-s
m-s
m
m
w-m
Purines: 2-substituted 7
6-substituted 32
8-substituted 8
2,6-disubstituted 21
m-s
m-s γ(CH)
v A" o.p. ring bend
v
s
v
s
s
v
15
16
17
18
19
20
25
30
35
WAVELENGTH IN MICRONS

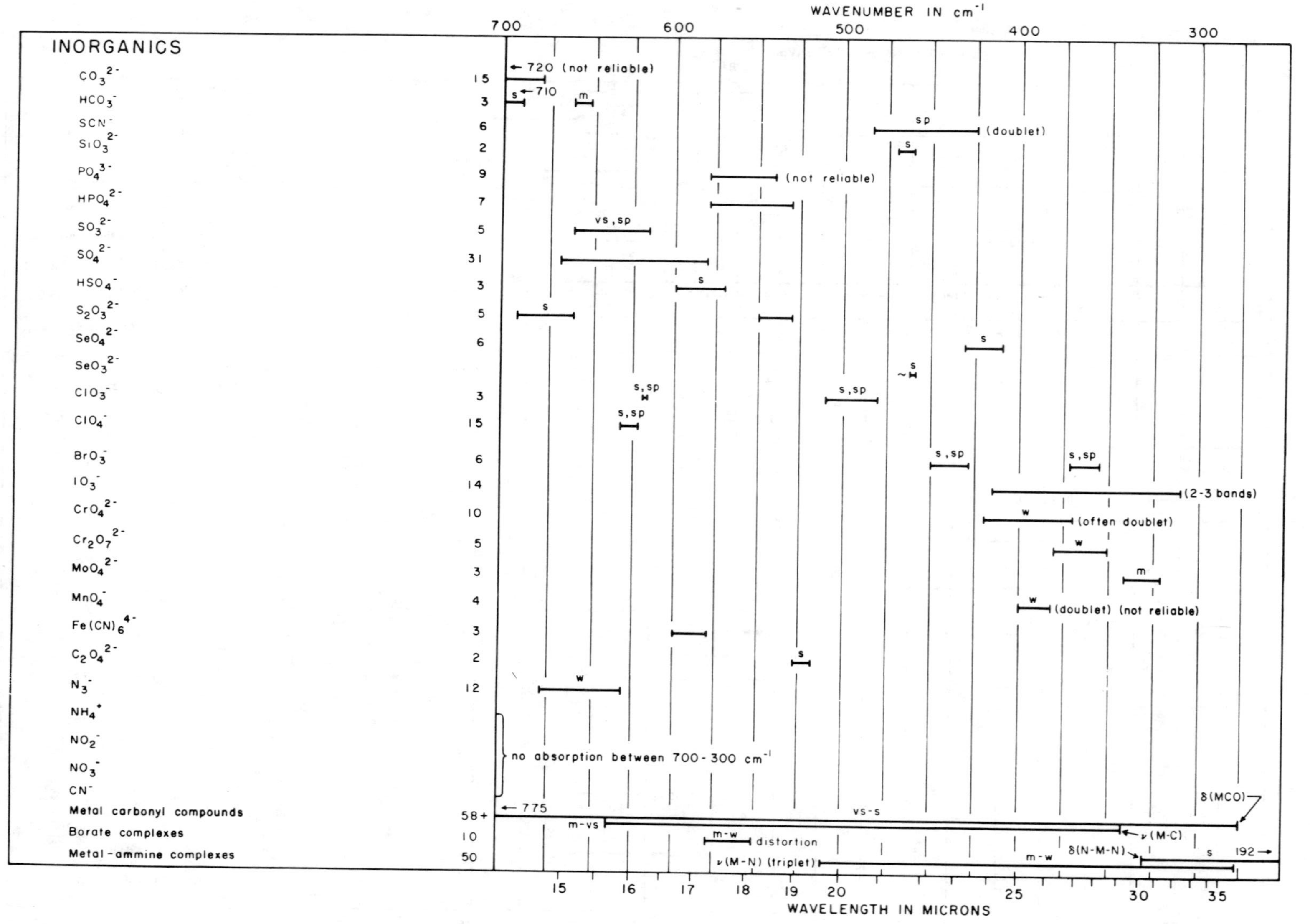
WAVENUMBER IN cm⁻¹
700
600
500
400
300
INORGANICS
CO3 2- 15
HCO3 - 3
SCN - 6
SiO3 2- 2
PO4 3- 9
HPO4 2- 7
SO3 2- 5
SO4 2- 31
HSO4 - 3
S2O3 2- 5
SeO4 2- 6
SeO3 2-
ClO3 - 3
ClO4 - 15
BrO3 - 6
IO3 - 14
CrO4 2- 10
Cr2O7 2- 5
MoO4 2- 3
MnO4 - 4
Fe(CN)6 4- 3
C2O4 2- 2
N3 - 12
NH4 +
NO2 -
NO3 -
CN -
Metal carbonyl compounds 58+
Borate complexes 10
Metal-ammine complexes 50
← 720 (not reliable)
s ← 710
m
sp (doublet)
s
(not reliable)
vs, sp
s
s
s
~s
s, sp
s, sp
s, sp
s, sp
s, sp
(2-3 bands)
w (often doublet)
w
m
w (doublet) (not reliable)
s
w
no absorption between 700-300 cm⁻¹
← 775
vs-s
δ(MCO)
m-vs
ν(M-C)
m-w distortion
δ(N-M-N)
m-w
s
192 →
ν(M-N) (triplet)
15
16
17
18
19
20
25
30
35
WAVELENGTH IN MICRONS

# CHARACTERISTIC NMR SPECTRAL POSITIONS FOR HYDROGEN IN ORGANIC STRUCTURES

This chart is useful for quick qualitative determination of proton spectrum lines by providing a tabulation of line positions obtained with tetramethylsilane as an internal reference. The listing has been kept as simple as possible for this purpose. The proton spectrum lines are arranged according to the chemical shift relative to tetramethylsilane and are given in values of $\tau$ and $\sigma$. The purpose of this table is to supplement tables available in standard references and to summarize information available in the literature.

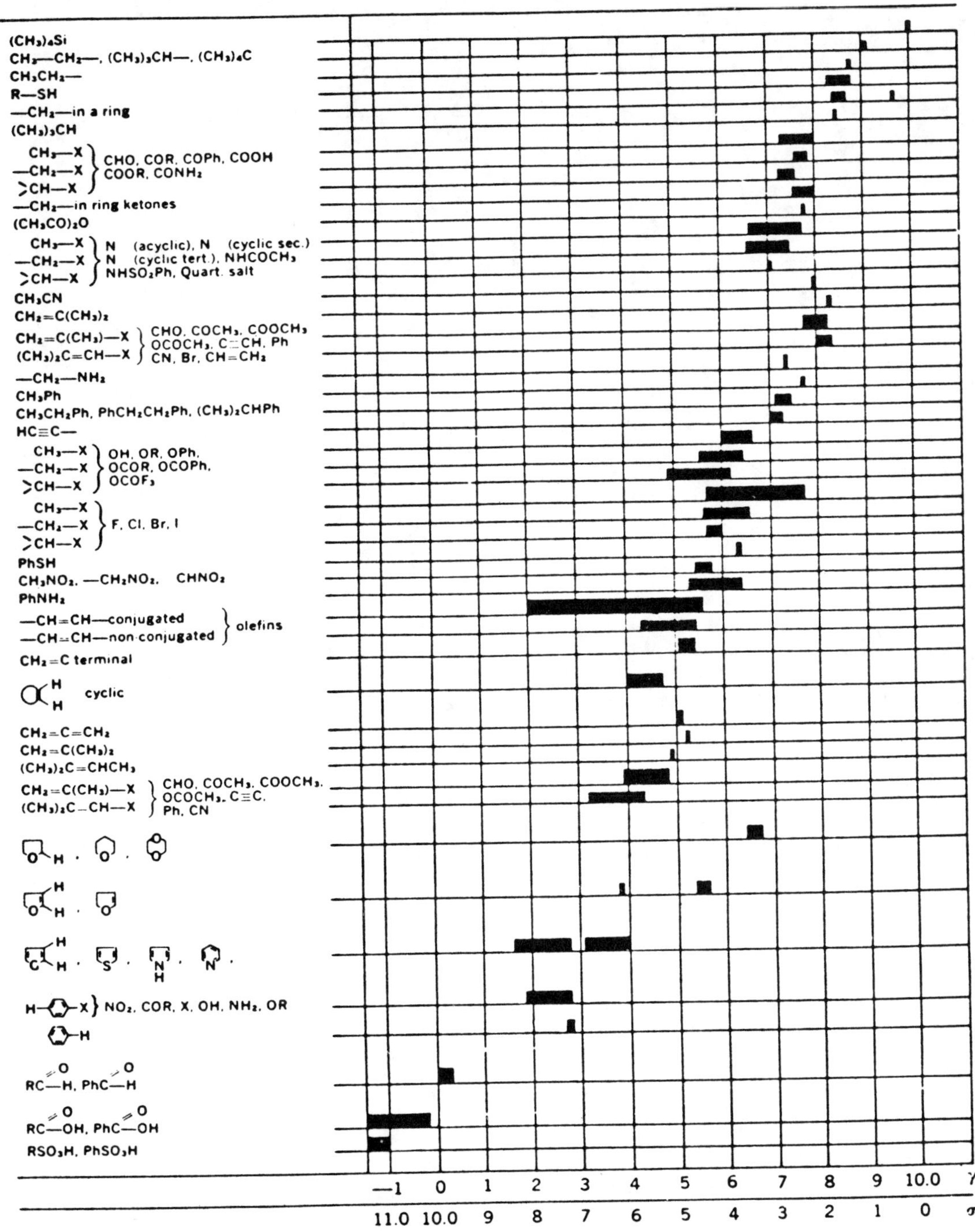


Data taken from: *J. Chem. Educ.*, *41*, 38 (1964). Reproduced by permission of the American Chemical Society, Washington, D.C.

# INDEX

## A

Acetamide, dissociation constant in aqueous solution, 885
Acetic acid, 248, 249, 891, 897, 898, 912
  approximate pH values at 25°C, 912
  chemical and physical characteristics, 249,
  composition and properties of concentrated aqueous solution, 898
  dissociation constant in aqueous solution, 891
  ionization constants in water, 897
Acetidine, dissociation constant in aqueous solution, 885
Acetoacetic acid, dissociation constant in aqueous solution, 891
N[4]-Acetylcytidine, 356, 429, 548
  in tRNAs, 548
  natural occurrence, 429
  physical constants, 356
  spectral properties, 356
N[4]-Acetylcytosine, physical and spectral data, 337
N-Acetyl-D-galactosamine, structure, 90
N-Acetylglucosamine, content in fungal cell walls, 202–209
N-Acetyl-D-glucosamine, structure, 90
N-Acetyl-D-mannosamine, structure, 90
Acridine, dissociation constant in aqueous solution, 885
Acrylic acid, dissociation constant in aqueous solution, 891
Actinomycetes, effects of trace metals, 806, 807
Actinonin, structure and source, 830
Adenine, 332, 741–743
  content in bacteriophage nucleic acids, 741–743
  physical constants, 332
  spectral properties, 332
Adenosine, 347, 543, 547
  in tRNAs, 543, 547
  physical constants, 347
  spectral properties, 347
Adipamic acid, dissociation constant in aqueous solution, 891
Adipic acid, dissociation constant in aqueous solution, 891
Aerobactin, structure and source, 827
Air components, 979
Alanine, 4, 5, 11, 12, 16–24, 53, 57, 65–68, 75, 77, 210, 212, 546, 891, 923, 925
  abbreviation, 4
  antagonists, D-configuration, 57
  content in fungal cell walls, 210, 212
  dissociation constant in aqueous solution, D-configuration, 891
  distribution in selected proteins, 16–24
  far-ultraviolet absorption spectra, 67, 68
  in media composition, 923, 925
  ionization constants, 11, 12
    at 25°C, DL-configuration, 11
    in aqueous ethanol solution, 11
    in aqueous formaldehyde solutions, DL-configuration, 12
  isoelectric point, DL-configuration, 11
  α-ketoacid analogs, 65
  molecular weight, 4
  solubility in water, L- and DL-configurations, 53
Alanine (*cont.*)
  specific rotations, 75, 77
    D-configuration, 77
    L-configuration, 75, 77
  structural features, 546
  structure, 5
α-Alanine, 57, 885
  antagonists, 57
  dissociation constant in aqueous solution, 885
β-Alanine, 8, 57, 65
  antagonists, 57
  α-keto acid analogs, 65
  structure, 8
Alanine compounds, dissociation constants in aqueous solutions, 885
L-Albizziine, specific rotation, 75
Albomycins, structures and sources, 825
Aldaric acids, properties, 105
Alditols, properties, 97, 98
Aldonic acids, properties, 101–104
Aldoses, natural, properties, 106–119
Algae, 147–162, 561, 562, 575, 729–732, 772
  distribution of purines and pyrimidines in DNA, 575
  DNA base compositions, 561, 562
  DNA content per cell, 772
  GC composition of DNA, 729–732
  polysaccharides, 147–162
    apochromatic groups, 162
    Bacillariophyta, 151–153
    Chlorophyta, 151–156
    Chrysophyta, 155
    Cryptophyta, 150
    Cyanophyta, 147, 148
    Euglenophyta, 154
    Phaeophyta, 156, 157
    Pyrrophyta, 150, 151
    Rhodophyta, 149
    Xanthophyta, 156
Allantoin, dissociation constant in aqueous solution, 891
Allomycin, selected properties, 440
Allothreonine, dissociation constants in aqueous solutions, 885
Allylglycine, specific rotation, L-configuration, 76
Alum, approximate pH value at 25°C, 912
Aluminum, in media composition, 919, 920, 926
Amicetin, selected properties, 441
Amicetin B, selected properties, 442
Amides, far-ultraviolet absorption spectra, 68
Amino acids, 3–81, 210–213, 220, 921, 923–925
  abbreviations, 4
  absorbance values of aromatic amino acids in neutral, alkaline and acid solutions, 71–73
  antagonists, 57–59
  distribution in selected proteins, 16–24
  far-ultraviolet absorption spectra, 67, 68
  in bacterial endotoxins, 220
  in fungal cell walls, 210–213
  in media composition, 921, 923–925
  ionization constants, 11, 12
    at 25°C, 11
    in aqueous ethanol solutions, 11

Amino acids (*cont.*)
ionization constants (*cont.*)
in aqueous formaldehyde solutions, 12
isoelectric points, 11
α-keto acid analogs, 65, 66
molecular weights, 4
optical isomers, 9–12
sequences in microbial proteins, 31–50
solubilities in water, 53–56
specific rotations, 75–81
structures, 5–8
ultraviolet absorption characteristics of N-acetyl methyl esters, 69
α-Amino adipic acid, 7, 65, 75
α-keto acid analogs, 65
specific rotation, L-configuration, 75
structure, 7
Amino alditols, properties, 100
Aminobenzoic acid, 891, 922
dissociation constants in aqueous solutions, 891
in media composition, 922
α-Aminobutyric acid, 7, 65, 75
α-keto acid analogs, 65
specific rotations, L-configuration, 75
structure, 7
L-α-Aminocapriylic acid, specific rotations, 76
α-Aminoheptylic acid, 65, 76
α-keto acid analogs, 65
specific rotations, L-configuration, 76
α-Amino-ε-hydroxycaproic acid, 65, 76
α-ketoacid analogs, 65
specific rotations, L-configuration, 76
α-Aminoisobutyric acid, structure, 8
β-Aminoisobutyric acid, specific rotation, 75
α-Aminophenylacetic acid, 65, 76
α-keto acid analogs, 65
specific rotations, L-configuration, 76
Amino sugar acids, properties, 105, 106
Amino sugars, natural, properties, 123–132
Ammonia, 897, 912
approximate pH values at 25°C, 912
dissociation constants, aqueous solution, 897
Ammonium hydroxide, 891, 898
composition and properties of concentrated aqueous solution, 898
dissociation constant in aqueous solution, 891
Ammonium sulfate solutions, nomogram for degree of saturation, 927
*n*-Amylamine, dissociation constant in aqueous solution, 885
Angustamycin A, selected properties, 446, 447
Angustamycin C, selected properties, 448
Aniline, dissociation constant in aqueous solution, 885
Anisic acid, dissociation constant in aqueous solution, 891
Anisidine, dissociation constants in aqueous solutions, 885
Antibiotics, 8, 437–531
amino acid derivatives, structures, 8
nucleosides, selected properties, 437–531
Antigen O, 190, 192–194
in bacterial cell walls, 190, 192–194
extraction, 190, 193
Antigen O (*cont.*)
in bacterial cell walls (*cont.*)
lipid components, 193
protein components, 193
variation in lipopolysaccharide structure, 190
Antimony, in media composition, 919
Arabinose, 90, 147–152, 155–162, 202–209
content in algae, 148–151, 155–162
content in fungal cell walls, 202–209
structure, L-configuration, 90
Arachidic acid, chemical and physical characteristics, 248, 249
Arachidonic acid, chemical and physical characteristics, 252, 253
Arginine, 4, 7, 11, 12, 16–24, 57, 65, 67, 68, 75, 77, 210, 212, 885, 921, 923, 925
abbreviation, 4
antagonists, 57
content in fungal cell walls, 210, 212
dissociation constants in aqueous solutions, 885
distribution in selected proteins, 16–24
far-ultraviolet absorption spectra, 67, 68
in media composition, 921, 923, 925
ionization constants, 11, 12
at 25°C, L-configuration, 11
in aqueous ethanol solutions, 11
in aqueous formaldehyde solutions, L-configuration, 12
isoelectric point, L-configuration, 11
α-keto acid analogs, 65
molecular weight, 4
specific rotations, L-configuration, 75, 77
structure, 7
Argon, content in atmospheric air, 979
Aristeromycin, selected properties, 458
Arsenic, in media composition, 919
Arsenic acid, dissociation constants in aqueous solutions, 895
Arsenious acid, dissociation constant in aqueous solution, 895
Arsenous oxide, dissociation constant in aqueous solution, 891
Ascorbic acid, 891, 925
dissociation constants in aqueous solutions, 891
in media composition, 925
Acosterol, properties, 270
Ash content in media composition, 923–926
Asparagine, 4, 5, 57, 65, 75, 210, 212, 885
abbreviation, 4
antagonist, 57
content in fungal cell walls, 210, 212
dissociation constants in aqueous solutions, 885
α-keto acid analogs, 65
molecular weight, 4
specific rotation, L-configuration, 75
structure, 5
Aspartic acid, 4, 6, 11, 16–24, 53, 57, 65, 67, 75, 77, 546, 885, 891, 921, 925
abbreviation, 4
antagonists, 57
dissociation constants in aqueous solutions, DL-configuration, 885, 891

Aspartic acid (*cont.*)
distribution in selected proteins, 16–24
far-ultraviolet absorption spectrum, 67
in media composition, 921, 925
ionization constants, 11, 12
at 25°C, L-configuration, 11
in aqueous ethanol solutions, 11
in aqueous formaldehyde solutions, L-configuration, 12
isoelectric point, L-configuration, 11
α-keto acid analogs, 65
molecular weight, 4
solubility in water, L- and DL-configurations, 53
specific rotations, 75, 77
D-configuration, 77
L-configuration, 75, 77
structural features, 546
structure, 6
Aspergillic acid, structures and sources, 829, 830
Atomic weights of elements, 881
5-Azacytidine, selected properties, 460, 461
Azaserine, structure, 8

## B

Bacteria, 167–196, 215–227, 277–299, 307–314, 587–689, 805, 806, 833–839
cell breakage or lysis, 833–839
biological methods, 835–839
chemical methods, 834, 835
physical methods, 834
cell wall composition, 167–196, 215–227
differences between Gram-positive and Gram-negative organisms, 169, 170
lipids, 189
lipopolysaccharides, 215–227
lipopolysaccharide–protein complexes, 190–194
peptidoglycan, 169, 171–182
polysaccharides, acidic, 182, 183
polysaccharides, neutral, 185–189
proteins, 189, 190
teichoic acids, 182–185
cell wall synthesis, 194–196
effects of antibiotics, 196
effects of trace metals, 805, 806
fatty acid content, 277–299
GC composition of DNA, 587–689
Achromobacteriaceae, 587–592
Actinomycetaceae, 593, 594–596
Archangiaceae, 596
Athiorhodaceae, 596
Azotobacteriaceae, 596, 597
Bacillaceae, 598–603
Bacteroidaceae, 603
Brevibacteriaceae, 604, 605
Brucellaceae, 605–611
Caulobacteriaceae, 611
Chlamydiaceae, 611
Chlamydobacteriaceae, 611, 612
Chlorobacteriaceae, 612
Bacteria (*cont.*)
GC composition of DNA (*cont.*)
Corynebacteriaceae, 612–618
Cytophagaceae, 619
Enterobacteriaceae, 619–630
Lactobacillaceae, 630–640
Leucotrichaceae, 641
Mycobacteriaceae, 652–654
Micrococcaceae, 641–651
Micromonosporaceae, 651
Mycoplasmataceae, 654–658
Myxococcaceae, 658, 659
Neisseriaceae, 659, 660
Nitrobacteraceae, 661
Nocardiaceae, 661, 662
Polyangiaceae, 662, 663
Propionibacteriaceae, 663, 664
Pseudomonadaceae, 664–673
Rhizobiaceae, 673–676
Rickettsiaceae, 676
Sorangiaceae, 676
Spirillaceae, 676–682
Spirochaetaceae, 682
Streptomycetaceae, 683–686
Thiobacteriaceae, 686, 687
Thiorhodaceae, 687
Treponemataceae, 687, 688
Vitreoscillaceae, 688
lipid content, 307–314
Bamicetin, selected properties, 462
Barbituric acid, dissociation constant in aqueous solution, 891
Barium, in media composition, 919, 920
Base composition, 561–564, 741–747
deoxyribonucleic acids of eukaryotic protists, 561–564
nucleic acids of bacteriophages, 741–747
ATGC or AUGC content, 741–744
GC content, 744–747
Benzene compounds, dissociation constants in aqueous solutions, 885
Benzidine, dissociation constants in aqueous solutions, 886
Benzimidazole, dissociation constant in aqueous solution, 886
Benzoic acid, 891, 912
approximate pH value at 25°C, 912
dissociation constant in aqueous solution, 891
Benzylamine, dissociation constant in aqueous solution, 886
Betaine, dissociation constant in aqueous solution, 886
Biotin, in media composition, 922, 924, 925
Blasticidin B, selected properties, 464, 465
Borax solution, 911, 912
pH values at temperatures from 0 to 95°C, 912
properties at 25°C, 911
Boric acid, 897, 912
approximate pH value at 25°C, 912
ionization constants in water, 897
Bromine, in media composition, 919
Brucine, dissociation constant in aqueous solution, 886
Buffer solutions, 909–913

Buffer solutions (*cont.*)
approximate pH values of some acids, bases, biologic materials and foods, 912, 913
operational definitions of pH, 909
pH values of standard solutions at temperatures from 0 to 95°C, 912
properties of standard aqueous buffer solutions at 25°C, 911
solutions giving round values of pH at 25°C, 910, 911
Buoyant densities, 765–767, 775–779
selected cellular DNAs, 775
selected viral DNAs, 776, 777
synthetic polynucleotides, 778, 779
T-even coliphage DNAs, 777
viral and cellular RNAs in the $Cs_2SO_4$ gradient, 778
viral DNA, 765–767
Butane compounds, dissociation constants in aqueous solutions, 886
Butylamine, dissociation constants in aqueous solutions, 886
Butyric acid, 248, 249, 891
chemical and physical characteristics, 248, 249
dissociation constants in aqueous solutions, 891

## C

Cacodylic acid, dissociation constants in aqueous solutions, 886, 891
Cadmium, in media composition, 919, 920, 926
Calcium, 812, 919, 921, 923–926
in enzyme-catalyzed reactions, 812
in media composition, 919, 921, 923–926
Calcium carbonate, saturated, approximate pH value at 25°C, 912
Calcium hydroxide, dissociation constants in aqueous solutions, 891
Calcium hydroxide solution, 911, 912
pH values at temperatures from 0 to 95°C, 912
properties at 25°C, 911
Capric acid, chemical and physical characteristics, 248, 249
Caproic acid, 248, 249, 892
chemical and physical characteristics, 248, 249
dissociation constants in aqueous solutions, 892
Caproleic acid, chemical and physical characteristics, 248, 249
Caprylic acid, chemical and physical characteristics, 248, 249
Carbohydrates, 224, 923–925
in media composition, 923–925
metabolic effects of bacterial endotoxins, 224
Carbon dioxide content in atmospheric air, 979
Carbonic acid, 895, 897, 912
approximate pH value of saturated solution at 25°C, 912
dissociation constants in aqueous solutions, 895
ionization constants in water, 897
5-Carboxymethyluridine, 370, 431
natural occurrence, 431
physical constants, 370
spectral properties, 370
Cardiolipin, structure, 261
Carotene, in media composition, 925
Cell breakage, 833–839, 844
bacterial, 833–839
biological methods, 835–839
chemical methods, 834, 835
physical methods, 834
fungal, 844
Cell wall composition, 167–196, 201–213
bacterial, 167–196
differences between Gram-positive and Gram-negative organisms, 169, 170
lipids, 189
lipopolysaccharide–protein complexes, 190–194
peptidoglycan, 169, 171–182
polysaccharides, acidic, 182, 183
polysaccharides, neutral, 185–189
proteins, 189, 190
teichoic acids, 182–185
fungal, 201–213
amino acids and protein, 210, 210–213
monosaccharides, 207–209
Cell wall synthesis, bacterial, 194–196
effects of antibiotics, 196
Centrifugation, 845–879
gradient generators, 876–879
characteristics of commercially available models, 878
gradient materials, 866, 869–875
coefficients for calculation of density from refractive indexes of solutions at 25°C, 874
density and viscosity of sucrose as functions of concentration, 869, 870
density and viscosity of sucrose at different temperatures, 871–873
physical properties of some gradient materials, 869
rotors, 848–853
batch-type zonal, 851, 852
continuous-flow, 853
high-speed angle, 849
swinging-bucket, 850
sedimentation equations, 853–866
equivolumetric gradients of sucrose, 868
hyperbolic gradient for the separation of ribosomes, 866
isokinetic gradients, 860–864
sedimentation properties of some biological particles, 857
values for different speeds and intervals, 855
techniques, 845–848
Cesium, in media composition, 919
Cesium chloride solutions, density as a function of refractive index, 903–907
Chloramphenicol, structure, 8
Chlorides, in media composition, 925
Chlorine, in media composition, 919, 923
Chloroacetic acid, 892, 897
dissociation constant in aqueous solution, 892
ionization constants in water, 897
Chlorobenzoic acid, dissociation constants in aqueous solutions, 892
Chlorobutyric acid, dissociation constants in aqueous solutions, 892

Chlorocinnamic acid, dissociation constants in aqueous solutions, 892
Chlorophenoxyacetic acid, dissociation constants in aqueous solutions, 892
Chlorophenylacetic acid, dissociation constants in aqueous solutions, 892
Chloropropionic acid, dissociation constants in aqueous solutions, 892
Cholestanol, properties, 265, 268
Cholesterol, properties, 265
Choline, in media composition, 924, 925
Chondrillasterol, properties, 271
Chromic acid, dissociation constants in aqueous solutions, 896
Chromium, in media composition, 919, 920
Cinnamic acid, dissociation constants in aqueous solutions, 892
Cinnoline, dissociation constant in aqueous solution, 886
Citric acid, 892, 897, 912
  approximate pH value at 25°C, 912
  dissociation constants in aqueous solutions, 892
  ionization constants in water, 897
Citrulline, 7, 65, 75, 78
  α-keto acid analogs, 65
  specific rotations, L-configuration, 75, 78
  structure, 7
Cladinose, structure, 91
Coat proteins, amino acid sequences, 31, 32
Cobalt, 805–808, 919, 920
  effects of trace amounts, 805–808
    in Actinomycetes, 806, 807
    in bacteria, 805, 806
    in fungi, 807, 808
  in media composition, 919, 920
Codeine, dissociation constant in aqueous solution, 886
Constant-humidity data, 981
Conversion, transparency to optical density, 983–988
Copper, 805–808, 812, 919, 920, 924, 926
  effects of trace amounts, 805–808
    in Actinomycetes, 806, 807
    in bacteria, 805, 806
    in fungi, 807, 808
  in media composition, 919, 920, 924, 926
  in enzyme-catalyzed reactions, 812
Coprogen, structure and sources, 826
Coprostanol, properties, 265, 268
Corbisterol, properties, 271
Cord factor, structure, 319
Corrinic acid, structure, 318
Corynolic acid, structure, 318
Corynomycolenic acid, structure, 318
Corynomycolic acid, structure, 318
Cresol, dissociation constants in aqueous solutions, 892
Crotonic acid, chemical and physical characteristics, 248, 249
Cubes and cube roots of numbers, 953–972
Cyanocobalamin, in media composition, 922
Cyanopropionic acid, dissociation constant in aqueous solution, 892
β-Cyclohexylalanine, α-ketoacid analogs, 65
Cyclohexylglycine, α-keto acid analogs, 65
  specific rotation, D-configuration, 75
Cycloserine, 8, 75
  structure, 8
Cystathionine, 7, 75, 76
  specific rotation, L-allo isomer, 76
  specific rotation, L-configuration, 75
  structure, 7
Cysteic acid, α-keto analogs, 65
Cysteine, 4, 6, 16–24, 57, 67, 68, 75, 210, 212, 892
  abbreviation, 4
  antagonist, 57
  content in fungal cell walls, 210, 212
  dissociation constants in aqueous solutions, DL-configuration, 892
  distribution in selected proteins, 16–24
  far-ultraviolet absorption spectra, 67, 68
  molecular weight, 4
  specific rotations, L-configuration, 75
  structure, 6
Cystine, 4, 6, 11, 68, 69, 78, 886, 892, 921, 923, 925
  abbreviation, 4
  dissociation constants in aqueous solutions, 886
  dissociation constants in aqueous solutions, L-configuration, 892
  far-ultraviolet absorption spectra, 68
  in media composition, 921, 923, 925
  ionization constants at 25°C, L-configuration, 11
  isoelectric point, L-configuration, 11
  molecular weight, 4
  solubility in water, L-configuration, 53
  specific rotations, 75, 78
    D-configuration, 78
    L-configuration, 75, 78
  structure, 6
  ultraviolet absorption characteristics of N-acetyl methyl ester, 69
Cytidine, 355, 543, 546, 547
  in tRNAs, 543, 546, 547
  physical constants, 355
  spectral properties, 355
Cytomycin, selected properties, 468
Cytosine, 336, 741–743
  content in bacteriophage nucleic acids, 741–743
  physical constants, 336
  spectral properties, 336
Cytovirin, selected properties, 469

## D

Decoyinine, selected properties, 470
24-Dehydrocholestadione-3β-ol, properties, 265
7-Dehydrocholesterol, properties, 265
22-Dehydrocholesterol, properties, 266
Dehydroergosterol, properties, 268
14-Dehydroergosterol, properties, 268
24-Dehydroergosterol, properties, 268
Deoxyadenosine, physical and spectral data, 377
Deoxycytidine, physical and spectral data, 378
Deoxyguanosine, physical and spectral data, 380
Deoxyribonucleic acids, 561–564, 567–580, 585–734, 741–743, 751–756, 761, 762, 765–767, 771, 772, 775–777, 789

Deoxyribonucleic acids (*cont.*)
ATGC or AUGC content, 741–743
double-stranded DNA of bacteriophages, 741–743
single-stranded DNA of cubic phages, 743
single-stranded DNA of filamentous phages, 743
buoyant densities, 775–777
selected cellular DNAs, 775
selected viral DNAs, 776, 777
T-even coliphage DNAs, 777
composition and physical properties of molecules found in viruses, 751–756
content per cell, 761, 762, 771, 772
in algae, 772
in animal viruses, 762
in bacteriophages, 761
in fungi, 771
in protozoa, 772
distribution of purines and pyrimidines, 567–580
in algae, 575
in animal viruses, 580
in bacterial phages, 578, 579
in fungi, 568–574
in insect viruses, 579
in protozoa, 577
in tracheophytes, 576, 577
GC composition, 561–564, 585–734
in algae, 729–732
in bacteria, 587–689
in eukaryotic protists, 561–564
in fungi, 692–729
in protozoa, 733, 734
5-methylcytosine content, 789
6-methylaminopurine content, 789
properties of viral DNA, 765–767
Deoxyribonucleosides, 377–382, 791–796
metabolism, 791–796
anabolism, 792–794
catabolism, 795, 796
physical constants, 377–382
spectral properties, 377–382
Deoxyribonucleotides, physical and spectral data, 410–419
Deoxyuridine, physical and spectral data, 381
Deuterium oxide, ionization constants, 897
α, γ-Diaminobutyric acid, L-configuration, specific rotations, 75
Diaminopimelic acid, in bacterial cell walls, 182
α, ε-Diaminopimelic acid, 8, 57, 75
antagonists, 57
specific rotations, L-configuration, 75
structure, 8
α, β-Diaminopropionic acid, L-configuration, specific rotations, 75
Dichloroacetic acid, dissociation constant in aqueous solution, 892
Dichloroacetylacetic acid, dissociation constant in aqueous solution, 892
2,3-Dichlorphenol, dissociation constant in aqueous solution, 892
Dicholesterylether, properties, 273
Diethylamine, dissociation constant in aqueous solution, 886
5-Dihydroergosterol, properties, 269
22-Dihydroergosterol, properties, 269
Dihydrositosterol, properties, 273
Dihydrouracil, physical and spectral data, 345
Dihydrouridine, physical and spectral data, 372
5,6-Dihydrouridine, 431, 543, 544, 546–550
in tRNAs, 543, 544, 546–550
natural occurrence, 431
11,12-Dihydroxyarachidic acid, chemical and physical properties, 254, 255
Dihydroxybenzoic acid, dissociation constants in aqueous solutions, 892
Dihydroxypentadecyclic acid, chemical and physical characteristics, 254, 255
9,10-Dihydroxystearic acid, chemical and physical characteristics, 254, 255
Diiodotyrosine, L-configuration, 11, 54, 75, 78
ionization constants at 25°C, 11
isoelectric point, 11
solubility in water, 54
specific rotations, 75, 78
Diisobutylamine, dissociation constant in aqueous solution, 886
Diisopropylamine, dissociation constant in aqueous solution, 886
Dimerum acid, structure and source, 826
$N^6$,$N^6$-Dimethyladenine, physical and spectral data, 335
$N^6$,$N^6$-Dimethyladenosine, 352, 427
natural occurrence, 427
physical constants, 352
spectral properties, 352
Dimethylamine, dissociation constant in aqueous solution, 886
$N^4$,$O^2$-Dimethylcytidine, 359, 428
natural occurrence, 428
physical constants, 359
spectral properties, 359
$N^2$,$N^2$-Dimethylguanine, physical and spectral data, 339
$N^2$-Dimethylguanosine, in tRNAs, 546–549
$N^2$,$N^2$-Dimethylguanosine, 361, 429
natural occurrence, 429
physical constants, 361
spectral properties, 361
*n*-Diphenylamine, dissociation constant in aqueous solution, 886
Diphosphatidyl glycerol, *see* Cardiolipin
Diphtheric acid, structure, 318
Dissociation constants, 855–897
aqueous ammonia, 897
inorganic acids in aqueous solutions, 895, 896
inorganic bases in aqueous solutions, 891
organic acids in aqueous solutions, 891–895
organic bases in aqueous solutions, 885–891
Distribution of $t$, 973
Distribution of $x^2$, 975
2,4-Dithiouridine, physical and spectral data, 366
DNA, *see* Deoxyribonucleic acid(s)

## E

Elaidic acid, chemical and physical characteristics, 250, 251
Elements, 881, 883, 919, 920, 926
  atomic weights, 881
  in media composition, 919–923, 926
  periodic table, 883
Endotoxins, *see* Lipopolysaccharides
Enterobactin, structure and sources, 823
Enzymes, 38–45, 83–86, 224
  amino acid sequences, 38–45
  metabolic effects of bacterial endotoxins, 224
  numbering and classification, 83–86
    hydrolases, 85
    isomerases, 85, 86
    ligases, 86
    lyases, 85
    oxidoreductases, 84
    transferases, 84, 85
Enzyme cofactors, inorganic, in enzyme-catalyzed reactions, 811–818
  copper, 812
  iron, 812
  magnesium, 812, 813
  magnesium–monovalent cations, 813–815
  manganese, 815
  manganese–monovalent cations, 815
  molybdenum, 816
  monovalent cations
  zinc, 817
  zinc–calcium, 817
  zinc–cobalt, 817
  zinc–copper, 818
  zinc–monovalent cations, 818
*d*-Ephedrine, dissociation constants in aqueous solutions, 886
Epicoprostanol, properties, 268
Episterol, properties, 269
Epoxystearic acid, chemical and physical characteristics, 254, 255
Ergostanol, properties, 271
Ergosterol, properties, 265, 269
Erucic acid, chemical and physical data, 250, 251
Ethane compounds, dissociation constants in aqueous solutions, 886
Ethionine, 65, 76
  *α*-keto acid analogs, 65
  specific rotations, L-configuration, 76
Ethylamine, dissociation constant in aqueous solution, 886
Ethylenediamine, dissociation constants in aqueous solutions, 886
Eukaryotic cytochromes *c*, amino acid sequences, 33–36

## F

Far-infrared correlation, 1025–1036
  characteristic frequencies between ~700 and 300 cm$^{-1}$, 1027–1036
  vibrational frequency correlation, 1025, 1026
Fats, in media composition, 924, 925
Fatty acids, 248–259, 275, 277–306, 925
  chemical and physical characteristics, 248–259
    branched-chain fatty acids, 256, 257
    cyclo fatty acids, 254–257
    epoxy fatty acids, 248, 249
    hydroxyalkanoic acids, 252–255
    hydroxy-unsaturated fatty acids, 256–259
    keto fatty acid, 254, 255
    saturated fatty acids, 248, 249
    unsaturated fatty acids, dienoic, 250, 251
    unsaturated fatty acids, hexaenoic, 252, 253
    unsaturated fatty acids, monoethenoic, 248–251
    unsaturated fatty acids, pentaenoic, 252, 253
    unsaturated fatty acids, tetranoic, 252, 253
    unsaturated fatty acids, trienoic, 250–253
  composition in specific microorganisms, 277–306
    bacteria, 277–299
    fungi, 302–306
    viruses, 299, 300
    yeasts, 301, 302
  in media composition, 925
Fecosterol, properties, 270
Ferredoxins, amino acid sequences, 46, 47
Ferribactin, structure and source, 825
Ferrichromes, structures and sources, 824
Ferrichrysin, structure and sources, 824
Ferricrocin, structure and sources, 824
Ferrioxamines, structures and sources, 828, 829
Ferrirhodin, structure and sources, 825
Ferrirubin, structure and sources, 825
Ferroverdin, structure and source, 830
Fiber content in media composition, 924, 925
Fluorescent indicators, 917, 918
Folic acid, in media composition, 922, 925
Formic acid, 248, 249, 893, 898, 912
  approximate pH value at 25°C, 912
  chemical and physical characteristics, 248, 249
  dissociation constant in aqueous solution, 893
  ionization constants in water, 898
Formycin, selected properties, 472, 473
Formycin B, selected properties, 474, 475
Fructose, content in algae, 160, 161
Fucose, 90, 147, 148, 151–153, 156–158, 202–209
  content in algae, 147, 148, 151–153, 156–158
  content in fungal cell walls, 202–209
  structure, D-configuration, 90
Fucosterol, properties, 270
Fungi, 201–213, 302–306, 316, 317, 562–564, 568–574, 692–729, 771, 807, 808, 844
  cell wall composition, 201–213
    amino acids and protein, 210–213
    monosaccharides, 202–209
  distribution of purines and pyrimidines in DNA, 568–574
  DNA content per cell, 771
  effects of trace metals, 807, 808
  fatty acid content, 302–306
  GC composition of DNA, 562–564, 692–729
  isolation of cell walls, 844
  lipid content, 316, 317
Fungisterol, properties, 269

Fusarinines, structures and sources, 828

## G

Gallic acid, dissociation constant in aqueous solution, 893
Gadoleic acid, chemical and physical characteristics, 250, 251
Galactosamine, content in fungal cell walls, 202–209
Galactose, 90, 147–153, 155–162, 202–209
  content in algae, 147–153, 155–162
  content in fungal cell wals, 202–209
  structure, D-configuration, 90
Galacturonic acid, 90, 148, 162
  content in algae, 148, 162
  structure, D-configuration, 90
Gallium, in media composition, 920
GC composition, 585–734
  algal DNA, 729–732
  bacterial DNA, 587–689
    Achromobacteriaceae, 587–592
    Actinomycetaceae, 593, 594
    Actinoplanaceae, 594–596
    Anaplasmataceae, 596
    Archangiaceae, 596
    Athiorhodaceae, 596
    Azotobacteriaceae, 596, 597
    Bacillaceae, 598–603
    Bacteroidaceae, 603
    Brevibacteriaceae, 604, 605
    Brucellaceae, 605–611
    Caulobacteriaceae, 611
    Chlamydiaceae, 611
    Chlamydobacteriaceae, 611, 612
    Chlorobacteriaceae, 612
    Corynebacteriaceae, 612–618
    Cytophagaceae, 619
    Enterobacteriaceae, 619–630
    Lactobacillaceae, 630–640
    Leucotrichaceae, 641
    Micrococcaceae, 641–651
    Micromonosporaceae, 651
    Mycobacteriaceae, 652–654
    Mycoplasmataceae, 654–658
    Myxococcaceae, 658, 659
    Neisseriaceae, 659, 660
    Nitrobacteraceae, 661
    Nocardiaceae, 661, 662
    Polyangiaceae, 662, 663
    Propionibacteriaceae, 663, 664
    Pseudomonadaceae, 664–673
    Rhizobiaceae, 673–676
    Ricketssiaceae, 676
    Sorangiaceae, 676
    Spirillaceae, 676–682
    Spirochaetaceae, 682
    Streptomycetaceae, 683–686
    Thiobacteriaceae, 686, 687
    Thiorhodaceae, 687
    Treponemataceae, 687, 688
    Vitreoscillaceae, 688
  fungal DNA, 692–729
GC composition (*cont.*)
  protozoal DNA, 733, 734
GC content, 561–564, 765–767
  in deoxyribonucleic acids of eukaryotic protists, 561–564
  in viral DNA, 765–767
Germanic acid, dissociation constants in aqueous solution, 896
Gheddic acid, chemical and physical characteristics, 248, 249
Glucosamine, content in fungal cell walls, 202–209
Glucose, 90, 147–153, 155–162, 202–209
  content in algae, 147–153, 155–162
  content in fungal cell walls, 202–209
  structure, D-configuration, 90
Glucuronic acid, 90, 148, 149, 153, 155, 158, 160, 161, 202–209
  content in algae, 148, 149, 153, 155, 158, 160, 161
  content in fungal cell walls, 202–209
  structure, D-configuration, 90
Glutamic acid, 4, 6, 11, 12, 16–24, 54, 57, 65, 67, 75, 78, 210, 212, 887, 921, 923, 925
  abbreviation, 4
  antagonists, 57
  content in fungal cell walls, 210, 212
  dissociation constants in aqueous solutions, 887
  distribution in selected proteins, 16–24
  far-ultraviolet absorption spectra, 67
  in media composition, 921, 923, 925
  ionization constants, 11, 12
    at 25°C, L-configuration, 11
    in aqueous ethanol solutions, 11
    in aqueous formaldehyde solutions, L-configuration, 12
  isoelectric point, L-configuration, 11
  α-keto acid analogs, 65
  molecular weight, 4
  solubility in water, L- and DL-configurations, 54
  specific rotations, 75, 78
    D-configuration, 78
    L-configuration, 75, 78
  structure, 6
Glutamine, 4, 5, 57, 65, 75, 78, 887
  abbreviation, 4
  antagonists, 57
  dissociation constant in aqueous solution, 887
  α-keto acid analogs, 65
  molecular weight, 4
  specific rotations, L-configuration, 75, 78
  structure, 5
Glutaramic acid, dissociation constant in aqueous solution, 893
Glutaric acid, dissociation constants in aqueous solutions, 893
Glycerol, dissociation constant in aqueous solution, 893
Glycine, 4, 5, 11, 12, 16–24, 54, 57, 65, 210, 213, 546, 887, 893, 898, 921, 925
  abbreviation, 4
  antagonist, 57
  content in fungal cell walls, 210, 212
  dissociation constants in aqueous solutions, 887, 893
  distribution in selected proteins, 16–24

Glycine (*cont.*)
in media composition, 921, 925
ionization constants, 11, 12, 898
at 25°C, 11
in aqueous ethanol solutions, 11
in aqueous formaldehyde solutions, 12
in water, 898
isoelectric point, 11
α-keto acid analogs, 65
molecular weight, 4
solubility in water, 54
structural features, 546
structure, 5
Glycogen, occurrence, 93
Glycol, dissociation constant in aqueous solution, 893
Glycollic acid, 893, 898
dissociation constant in aqueous solution, 893
ionization constants in water, 898
Glycylserine, dissociation constants in aqueous solutions, 887
Gradient generators, 876–879
characteristics of commercially available models, 878
Gradient materials, 866, 869–875
coefficients for calculation of density from refractive indexes of solutions at 25°C, 874
density and viscosity of sucrose, 869–873
at different temperature, 871–873
as functions of concentration, 869, 870
physical properties, 869
Gradients, 860–864, 866, 868
equivolumetric, sucrose, 868
hyperbolic, separation of ribosomes, 866
isokinetic, 860–864
parameters for IEC rotors, 860–862
parameters for Spinco rotors, 863, 864
Grisein, structure and source, 825
Guanine, 338, 741–743
content in bacteriophage nucleic acids, 741–743
physical constants, 338
spectral properties, 338
Guanosine, 360, 543, 546, 549
in tRNAs, 543, 546, 549
physical constants, 360
spectral properties, 360

## H

Hadacidin, structure and sources, 830
Haliclonasterol, properties, 270
Helium content in atmospheric air, 979
Heptane compounds, dissociation constants in aqueous solutions, 887
Heptanoic acid, dissociation constant in aqueous solution, 983
Heptylic acid, chemical and physical characteristics, 248, 249
Hexanoic acid, dissociation constant in aqueous solution, 893
Hexosamine, content in fungal cell walls, 202–209
Hippuric acid, dissociation constant in aqueous solution, 893
Hippuric acid (*cont.*)
isoelectric point, 11
Histidine, 4, 7, 11, 12, 16–24, 57, 65, 67, 68, 75, 78, 210, 212, 887, 893, 921, 923, 925
abbreviation, 4
antagonists, 57
content in fungal cell walls, 210, 212
dissociation constants in aqueous solutions, 887, 893
distribution in selected proteins, 16–24
far-ultraviolet absorption spectra, 67, 68
in media composition, 921, 923, 925
ionization constants, 11, 12
at 25°C, L-configuration, 11
in aqueous ethanol solutions, 11
in aqueous formaldehyde solutions, L-configuration, 12
α-keto acid analogs, 65
molecular weight, 4
specific rotations, 75, 78
D-configuration, 78
L-configuration, 75, 78
structure, 7
Homocystine, L-configuration, specific rotation, 76
Homoglutamine, L-configuration specific rotations, 76
Homolanthionine, L-configuration, specific rotation, 76
Homoserine, L-configuration, specific rotation, 75
Hydrazine, dissociation constant in aqueous solution, 891
Hydriodic acid, composition and properties of concentrated aqueous solution, 898
Hydrobromic acid, composition and properties of concentrated aqueous solution, 898
Hydrochloric acid, 898, 912
approximate pH values at 25°C, 912
composition and properties of concentrated aqueous solution, 898
Hydrocyanic acid, 896, 912
approximate pH value at 25°C, 912
dissociation constant in aqueous solution, 896
Hydrofluoric acid, 896, 898
composition and properties of concentrated aqueous solution, 898
dissociation constant in aqueous solution, 896
Hydrogen, 979, 1037
characteristic NMR spectral positions in organic structures, 1037
content in atmospheric air, 979
Hydrogen peroxide, dissociation constant in aqueous solution, 896
Hydrogen sulfide, 896, 912
approximate pH value at 25°C, 912
dissociation constants in aqueous solutions, 896
Hydroquinone, dissociation constant in aqueous solution, 893
2-Hydroxyadenine, physical and spectral data, 333
2-Hydroxyadenosine, physical and spectral data, 349
Hydroxybenzoic acid, dissociation constants in aqueous solutions, 893
Hydroxybutyric acid, dissociation constants in aqueous solutions, 893
Hydroxycholesterol, properties, 267
Hydroxylamine, dissociation constant in aqueous solution, 891

Hydroxylauric acid, chemical and physical characteristics, 252, 253
$N^6$-(*cis*-4-Hydroxy-3-methylbut-2-enyl)adenosine, 351, 428
  natural occurrence, 428
  physical constants, 351
  spectral properties, 351
Hydroxymyristic acid, chemical and physical characteristics, 252, 253
Hydroxypalmitic acid, chemical and physical characteristics, 252–255
Hydroxyproline, 6, 11, 12, 54, 75, 78, 79, 211, 213, 889
  content in fungal cell walls, 211, 213
  dissociation constants in aqueous solutions, 889
  ionization constants, L-configuration, 11, 12
    at 25°C, 11
    in aqueous formaldehyde solution, 12
  isoelectric point, L-configuration, 11
  solubility in water, L-configuration, 54
  specific rotations, 75, 78, 79
    D-configuration, 79
    L-configuration, 75, 78
  structure, 6
Hydroxylysine, 7, 75, 76
  specific rotations, L-configuration, 75, 76
  structure, 7
β-Hydroxypropionic acid, dissociation constant in aqueous solution, 893
γ-Hydroxyquinoline, dissociation constant in aqueous solution, 893
Hydroxystearic acid, chemical and physical characteristics, 254, 255
5-Hydroxyuracil, physical and spectral data, 344
5-Hydroxyuridine, 371, 430
  natural occurrence, 430
  physical constants, 371
  spectral properties, 371
Hypobromous acid, dissociation constant in aqueous solution, 896
Hypochlorous acid, dissociation constant in aqueous solution, 896
Hypoiodous acid, dissociation constant in aqueous solution, 896
Hypoxanthine, physical and spectral data, 340

## I

Infrared correlation charts, 1017–1024
Inosamines, properties, 100
Inosine, 363, 430
  natural occurrence, 430
  physical constants, 363
  spectral properties, 363
Inositols, 98, 99, 924, 925
  in media composition, 924, 925
  properties, 98, 99
Inososes, properties, 99
Iodic acid, dissociation constant in aqueous solution, 96
Iodine, in media composition, 919
Iodoacetic acid, dissociation constant in aqueous solution, 893
Iodobenzoic acid, dissociation constants in aqueous solutions, 893
Ionization constants, 897, 898
  acids in water, 897
  deuterium oxide, 897
  water, 897
Iron, 805–808, 812, 919–921, 923–926
  effects of trace amounts, 805–808
    in Actinomycetes, 806, 807
    in bacteria, 805, 806
    in fungi, 807, 808
  in enzyme-catalyzed reactions, 812
  in media composition, 919–921, 923–926
Iron-transport compounds, 823–830
  hydroxamates, 824–830
    aerobactin family, 827
    ferrichrome family, 824, 825
    ferrioxamines, 828, 829
    fusarinines, 828
    mycobactins, 827, 828
    rhodotorulic acid family, 826
  phenolates, 823
Isoarachidic acid, chemical and physical characteristics, 256, 257
Isobarbituridine, *see* 5-Hydroxyuridine
Isobehenic acid, chemical and physical characteristics, 258, 259
Isocerotic acid, chemical and physical characteristics, 258, 259
Isocrotonic acid, chemical and physical characteristics, 248, 249
Isohydrosorbic acid, chemical and physical characteristics, 248, 249
Isolauric acid, chemical and physical characteristics, 256, 257
Isoleucine, 4, 5, 11, 16–24, 54, 58, 65, 75, 76, 79, 210, 212, 547, 887, 921, 923, 925
  abbreviation, 4
  antagonists, 58
  content in fungal cell walls, 210, 212
  dissociation constants in aqueous solutions, 887
  distribution in selected proteins, 16–24
  in media composition, 921, 923, 925
  ionization constants, 11
  isoelectric point, DL-configuration, 11
  α-keto acid analogs, 65
  molecular weight, 4
  solubility in water, DL-configuration, 54
  specific rotations, 75, 76, 79
    D-configuration, 79
    L-configuration, 75, 76, 79
  structural features, 547
  structure, 5
Isolignoceric acid, chemical and physical characteristics, 258, 259
Isomontanic acid, chemical and physical characteristics, 258, 259
Isomyristic acid, chemical and physical characteristics, 256, 257
Isopalmitic acid, chemical and physical characteristics, 256, 257

Isopentadecylic acid, chemical and physical characteristics, 256, 257
Isoquinoline, dissociation constants in aqueous solutions, 887
Isostearic acid, chemical and physical characteristics, 256, 257
Isoundecylic acid, chemical and physical characteristics, 256, 257
Isovaleric acid, chemical and physical characteristics, 256, 257
Isovaline, L-configuration, specific rotations, 76
Itaconic acid, dissociation constants in aqueous solutions, 893

### J

Jalapinolic acid, chemical and physical characteristics, 254, 255
Juniperic acid, chemical and physical characteristics, 254, 255

### K

α-Kainic acid, L-configuration, specific rotations, 75
Kamlolenic acid, chemical and physical characteristics, 256, 257
7-Ketocholesterol, properties, 267
Ketoses, natural, occurrence, 120–122
Krypton content in atmospheric air, 979

### L

Lacceroic acid, chemical and physical characteristics, 248, 249
Lactarinic acid, chemical and physical characteristics, 254, 255
Lactic acid, 893, 898, 912, 926
  approximate pH value at 25°C, 912
  dissociation constant in aqueous solution, 893
  in media composition, 926
  ionization constants in water, 898
Lactobacillic acid, chemical and physical characteristics, 254, 255
Lanthionine, 7, 76
  specific rotation, L-configuration, 76
  structure, 7
Lathosterol, properties, 267
Lauric acid, chemical and physical characteristics, 248, 249
Lauroleic acid, chemical and physical characteristics, 248, 249
Lead, in media composition, 919, 920, 926
Leucine, 4, 5, 11, 12, 16–24, 54, 58, 65, 67, 68, 75, 79, 210, 212, 546, 547, 888, 921, 923, 925
  abbreviation, 4
  antagonists, 58
  content in fungal cell walls, 210, 212
  dissociation constants in aqueous solution, L-configuration, 888
Leucine (*cont.*)
  distribution in selected proteins, 16–24
  far-ultraviolet absorption spectra, 67, 68
  in media composition, 921, 923, 925
  ionization constants, 11, 12
    at 25°C, DL-configuration, 11
    in aqueous formaldehyde, L- and DL-configurations, 12
  isoelectric point, DL-configuration, 11
  α-keto acid analogs, 65
  molecular weight, 4
  solubility in water, L- and DL-configurations, 54
  specific rotations, 75, 79
    D-configuration, 79
    L-configuration, 75, 79
  structural features, 546, 547
  structure, 5
Levulinic acid, chemical and physical characteristics, 254, 255
Licanic acid, chemical and physical characteristics, 254, 255
Lignoceric acid, chemical and physical characteristics, 248, 249
Linderic acid, chemical and physical characteristics, 248, 249
α-Linoleic acid, chemical and physical characteristics, 250, 251
α-Linolenic acid, chemical and physical characteristics, 250, 251
γ-Linolenic acid, chemical and physical characteristics, 250, 251
Lipids, 189, 216, 219, 224, 245–247, 261–263, 275, 276, 307–317, 324, 325
  composition in specific microorganisms, 307–317
    bacteria, 307–314
    fungi, 316, 317
    viruses, 314–316
    yeasts, 316
  in bacterial cell walls, 189, 193
    Gram-negative organisms, 193
    Gram-positive organisms, 189
  metabolic effects of bacterial endotoxins, 224
  moiety in lipopolysaccharides, 216, 219
  nomenclature, 245–247
  proton chemical shifts, 324, 325
  structures of types found in microorganisms, 261–263
Lipopolysaccharides, 190–193, 215–227
  assay methods, 220
  biological properties, 220–227
    detoxification, 222, 226
    effects on blood, 221, 223, 224
    effects on metabolism, 221, 224, 225
    effects on the endocrine system, 221, 224
    effects on the reticuloendothelial system, 224, 225
    effects on the vascular system, 221, 225
    endotoxin tolerance, 222, 226
    fever, 221, 223
    immunological phenomena, 222, 225, 226
    non-specific resistance to infection, 222, 226
    pathology, 221, 227
    Shwartzman phenomenon, 222, 227
  cellular localization, 215, 216

Lipopolysaccharides (*cont.*)
extraction and purification methods, 216, 217
in bacterial cell walls, 190–193
sugar constituents, 190–192
variation in structure, 190
structure, 216–221
amino acids, 220
lipid moiety, 216, 219
O-polysaccharide, 220
R-core, 216–219
Liquids, 1009–1011, 1013–1015
refractive indices, 1009–1011
specific gravity, 1013–1015
Lithium, in media composition, 919
LPS, *see* Lipopolysaccharides
Lysine, 4, 7, 11, 12, 16–24, 58, 65, 67, 68, 75, 79, 211, 213, 893, 921, 923, 925
abbreviation, 4
antagonists, 58
content in fungal cell walls, 211, 213
dissociation constant in aqueous solution, 893
distribution in selected proteins, 16–24
far-ultraviolet absorption spectra, 67, 68
in media composition, 921, 923, 925
ionization constants, 11, 12
at 25°C, L-configuration, 11
in aqueous ethanol solutions, 11
in aqueous formaldehyde solutions, L-configuration, 12
isoelectric point, L-configuration, 11
α-keto acid analogs, 65
molecular weight, 4
specific rotations, 75, 79
D-configuration, 79
L-configuration, 75, 79
structure, 7
Lyxose, D-configuration, structure, 90

## M

Magnesium, 812, 813, 919–921, 923–926
in enzyme-catalyzed reactions, 812, 813
in media composition, 919–921, 923–926
Magnesium–monovalent cations, in enzyme-catalyzed reactions, 813–815
Maleic acid, dissociation constants in aqueous solutions, 893
Malic acid, 893, 912
approximate pH value at 25°C, 912
dissociation constants in aqueous solutions, 893
Malonic acid, 893, 898
dissociation constants in aqueous solutions, 893
ionization constants in water, 898
Manganese, 805–808, 815, 919, 920, 923, 926
effects of trace amounts, 805–808
in Actinomycetes, 806, 807
in bacteria, 805, 806
in fungi, 807, 808
in enzyme-catalyzed reactions, 815
in media composition, 919, 920, 923, 926
Manganese–monovalent cations, in enzyme-catalyzed reactions, 815
Mannose, 90, 147–149, 151–153, 155–162, 202–209
content in algae, 147–149, 151–153, 155–162
content in fungal cell walls, 202–209
structure, D-configuration, 90
Mannuronic acid, D-configuration, structure, 90
Margaric acid, chemical and physical characteristics, 248, 249
Media constituents, 919–926
bacto-peptones, 923
corn steep liquor, 926
cotton seed flour, 925
nutrient broth, DIFCO, trace metals, 919
peptones, BBL, 921, 922
soy products, representative, 924
soy protein hydrolysate, 923
yeast extracts, 920–922
approximate composition, BBL, 921, 922
minor element composition, 920
Melissic acid, chemical and physical characteristics, 248, 249
Mercury, in media composition, 919
Metals, 805–808, 919–923
in media composition, 919–923
trace amounts, 805–808
effects in Actinomycetes, 806, 807
effects in bacteria, 805, 806
effects in fungi, 807, 808
in DIFCO nutrient broth, 919
Methane content in atmospheric air, 979
Methionine, 4, 6, 11, 16–24, 55, 58, 65, 75, 79, 80, 211, 213, 547, 888, 921, 923, 925
abbreviation, 4
antagonists, 58
content in fungal cell walls, 211, 213
dissociation constants in aqueous solutions, 888
distribution in selected proteins, 16–24
far-ultraviolet absorption spectra, 67, 68
ionization constants at 25°C, D-configuration, 11
isoelectric point, DL-configuration, 11
α-keto acid analogs, 65
in media composition, 921, 923, 925
molecular weight, 4
solubility in water, L- and DL-configurations, 55
specific rotations, 75, 79, 80
D-configuration, 80
L-configuration, 75, 79
structural features, 547
structure, 6
Methionine sulfone, α-keto acid analogs, 65
Methionine sulfoxide, structure, 7
Methyladenine, physical and spectral data, 332–335
Methyladenosine, 347, 349, 350, 353, 427, 428, 543, 544, 547–550
in tRNAs, 543, 544, 547–550
natural occurrence, 427, 428
physical constants, 347, 349, 350, 353
spectral properties, 347, 349, 350, 353
Methylamine, dissociation constant in aqueous solution, 888

6-Methylaminopurine, content in DNA, 789
Methylcytidine, 356–358, 428, 429, 546–550
  in tRNAs, 546–550
  natural occurrence, 428, 429
  physical constants, 356–358
  spectral properties, 356–358
Methylcytosine, physical and spectral data, 336, 337
5-Methylcytosine, content in DNA, 789
24-Methylenecholesterol, properties, 269
Methyl esters, unsaturated, 323, 326, 327
  mass spectra, 326, 327
  NMR spectra, 323
N-Methylglycine, structure, 7
Methylguanine, physical and spectral data, 339, 340
Methylguanosine, 360–363, 429, 430, 546–550
  in tRNAs, 546–550
  natural occurrence, 429, 430
  physical constants, 360–363
  spectral properties, 360–363
Methylhistidine, structures, 8
Methylhypoxanthine, physical and spectral data, 340, 341
1-Methylinosine, 364, 430, 546
  in tRNAs, 546
  natural occurrence, 430
  physical constants, 364
  spectral properties, 364
Methyl linoleate, 323, 326, 327
  mass spectrum, 326, 327
  NMR spectrum, 323
Methyl oleate, 323, 326
  mass spectrum, 326
  NMR spectrum, 323
2′-O-Methylpseudouridine, physical and spectral data, 375
α-Methylserine, specific rotations, 76
Methyluracil, physical and spectral data, 342
Methyluridine, 367, 368, 430
  natural occurrence, 430
  physical constants, 367, 368
  spectral properties, 367, 368
2′-O-Methyluridine, 371, 430, 548
  in tRNAs, 548
  natural occurrence, 430
  physical constants, 371
  spectral properties, 371
7-Methylxanthine, physical and spectral data, 346
Mevalonic acid, chemical and physical characteristics, 252, 253
Minerals, 224, 225, 919–921, 923–926
  in media composition, 919–921, 923–926
  metabolic effects of bacterial endotoxins, 224, 225
Miscibility of organic solvents, 1005–1007
Molybdenum, 816, 919, 920, 926
  in enzyme-catalyzed reactions, 816
  in media composition, 919, 920, 926
Monochlorphenol, dissociation constants in aqueous solutions, 894
Mononucleotides, ultraviolet absorbance, 539, 540
Monosaccharides, 89–92, 147–162, 201–209
  anomers, 91
  branched-chain sugars, 91
  classification, 89
Monosaccharides (*cont.*)
  constituents of algal polysaccharides, 147–162
  in fungal cell walls, 201–209
  glyceraldehyde configurations, 91
  pyranose ring projections, 92
  structures, 90
Monovalent cations, in enzyme-catalyzed reactions, 816, 817
Montanic acid, chemical and physical characteristics, 248, 249
Moroctic acid, chemical and physical characteristics, 252, 253
Morphine, dissociation constant in aqueous solution, 888
Morpholine, dissociation constant in aqueous solution, 888
Muramic acid, in bacterial cell walls, 182
Mycarose, structure, 91
Mycelianamide, structure and source, 830
Mycobactins, 827, 828
  sources, 828
  structures, 827
Mycoceranic acid, chemical and physical characteristics, 258, 259
Mycocerosic acid, structure, 318
Mycolic acids, structure, 318
Mycolipenic acid, structure, 318
Myristic acid, chemical and physical characteristics, 248, 249
Myristoleic acid, chemical and physical characteristics, 248, 249

## N

Naphthalene compounds, dissociation constants in aqueous solutions, 888
Naphthoic acid, dissociation constants in aqueous solutions, 894
Naphthol, dissociation constants in aqueous solutions, 894
Naphthylamine, dissociation constants in aqueous solutions, 888
Nariterashin, selected properties, 480
Naritheracin, selected properties, 480
Nebularine, 364, 482, 483
  physical constants, 364
  selected properties, 482, 483
  spectral properties, 364
Neon content in atmospheric air, 979
Neospongosterol, properties, 270
Nervonic acid, chemical and physical characteristics, 250, 251
Niacin, in media composition, 922, 924, 925
Nickel, in media composition, 919, 920
Nicotine, dissociation constants in aqueous solutions, 888
Nisinic acid, chemical and physical characteristics, 252, 253
Nitroarginine, α-keto acid analogs, 65
Nitrobenzene, dissociation constant in aqueous solution, 894
Nitrobenzoic acid, dissociation constants in aqueous solutions, 894

Nitrogen, 921, 923, 925, 926, 979
content in atmospheric air, 979
in media composition, 921, 923, 925, 926
Nitrophenol, dissociation constants in aqueous solutions, 894
Nitrous acid, dissociation constant in aqueous solution, 896
Nitrous oxide content in atmospheric air, 979
NMR spectral positions for hydrogen in organic structures, 1037
Norleucine, 65, 888
dissociation constant in aqueous solution, 888
α-keto acid analogs, 65
Norleucine, 55, 76
solubility in water, DL-configuration, 55
specific rotations, L-configuration, 76
Norvaline, 65, 76
α-keto acid analogs, 65
specific rotations, L-configuration, 76
Nucleic acids, 329–802
base composition, 561–564, 741–747
DNA of bacteriophages, 741–747
DNA of eukaryotic protists, 561–564
RNA of bacteriophages, 743–747
buoyant densities, 765–767, 775–779
cellular DNAs, 775
synthetic polynucleotides, 779
T-even coliphage DNAs, 777
viral DNAs, 765–767, 776, 777
viral and cellular RNAs in the $Cs_2SO_4$ gradient, 778
controlled partial hydrolysis of RNA, 553, 554
deoxyribonucleic acids, 561–777, 789
base composition of bacteriophages, 741–747
base composition of eukaryotic protists, 561–564
buoyant densities, viral, 765–767
composition and physical properties of molecules in viruses, 751–756
content of 6-methylaminopurine and 5-methylcytosine, 789
content per cell in algae, 772
content per cell in fungi, 771
content per cell in protozoa, 772
content per cell in viruses, 761, 762
distribution of purines and pyrimidines, 567–580
GC compositions, 585–734, 765–767
melting temperatures, viral, 765–767
distribution of purines and pyrimidines in DNA, 567–580
algae, 575–577
animal viruses, 580
bacteriophages, 578, 579
fungi, 568–574
insect viruses, 579
protozoa, 577
GC composition of DNA, 585–734, 765–767
algae, 729–732
bacteria, 587–690
fungi, 692–729
protozoa, 733, 734
viruses, 765–767
melting temperatures, 765–767, 781–785
synthetic polynucleotides, 781–785
Nucleic acids (*cont.*)
Melting temperatures (*cont.*)
viral DNA, 765–767
metabolism of nucleosides, 791–802
deoxyribonucleosides, 791–796
ribonucleosides, 797–802
nucleosides, 347–382, 437–531, 791–802
antibiotics, 437–531
metabolism, 791–802
physical constants, 347–382
spectral properties, 347–382
nucleosides, modified, natural occurrence, 427–431
nucleotides, 383–419, 539–541, 557–559, 779, 781–785
buoyant densities of synthetic polynucleotides, 779
melting temperatures of synthetic polynucleotides, 781–785
physical constants, 383–419
sequences in ribonucleic acids, 557–559
spectral properties, 383–419
spectrophotometric constants of ribonucleotides, 539–541
purines, 332–346, 567–580
distribution in DNA, 567–580
physical constants, 332–346
spectral properties, 332–346
pyrimidines, 332–346, 567–580
distribution in DNA, 567–580
physical constants, 332–346
spectral properties, 332–346
ribonucleic acids, 543–559, 743–747
base composition of bacteriophages, 743–747
controlled partial hydrolysis, 553, 554
nucleotide sequences, 557–559
structural features of tRNA, 543–550
transfer ribonucleic acid structure, 543–550
viral DNA molecules, composition and physical properties, 751–756
Nucleocidin, selected properties, 484
Nucleoside antibiotics, selected properties, 437–531
Nucleosides, 347–382, 791–802
metabolism, 791–802
deoxyribonucleosides, 791–796
ribonucleosides, 797–802
physical constants, 377–382
deoxyribonucleosides, 377–382
ribonucleosides, 347–376
spectral properties, 377–382
deoxyribonucleosides, 377–382
ribonucleosides, 347–376
Nucleosides, modified, natural occurrence, 427–433
Nucleotides, 383–419, 539–541, 557–559, 779, 781–785
buoyant densities of synthetic polynucleotides, 779
melting temperatures of synthetic polynucleotides, 781–785
physical constants, 383–419
deoxyribonucleotides, 410–419
ribonucleotides, 383–409
sequences in ribonucleic acids, 557–559
spectral properties, 383–419
deoxyribonucleotides, 410–419

Nucleotides (*cont.*)
spectral properties (*cont.*)
ribonucleotides, 383–409
spectrophotometric constants of ribonucleotides, 539–541
Numbers, 929–952, 953–972, 977, 978
cubes and cube roots, 953–972
random permutations, 977, 978
reciprocals, 929–952
squares and square roots, 953–972

## O

O-antigen, *see* Antigen O
Obtusilic acid, chemical and physical characteristics, 248–249
Octanoic acid, dissociation constant in aqueous solution, 894
Octylamine, dissociation constant in aqueous solution, 888
Oleic acid, chemical and physical characteristics, 250, 251
Oligonucleotides, 540, 541
hyperchromicity ratios at different wavelengths, 541
ultraviolet absorbance, 540
Optical density, conversion from transparency, 983–988
Ornithine, 8, 58, 66, 76, 888
antagonists, 58
dissociation constants in aqueous solutions, 888
$\alpha$-keto acid analogs, 66
specific rotations, L-configuration, 76
structure, 8
Orotic acid, physical and spectral data, 345
Orotidine, physical and spectral data, 373
Orthophosphoric acid, approximate pH value at 25°C, 912
Oxalic acid, 894, 898, 912
approximate pH value at 25°C, 912
dissociation constants in aqueous solutions, 894
ionization constants in water, 898
Oxygen content in atmospheric air, 979
Oyamycin, selected properties, 485

## P

Palmitic acid, chemical and physical characteristics, 248, 249
Palmitoleic acid, chemical and physical characteristics, 248, 249
Palysterol, properties, 272
Pantothenic acid, in media composition, 922
Papaverine, dissociation constant in aqueous solution, 888
Paratose, D-configuration, structure, 90
Parinaric acid, chemical and physical characteristics, 252, 253
Pelargonic acid, chemical and physical characteristics, 248, 249
Penicillamine, D-configuration, specific rotation, 76
Pentadecylic acid, chemical and physical characteristics, 248, 249
Pentahomoserine, $\alpha$-keto acid analogs, 66
Pentane compounds, dissociation constants in aqueous solutions, 888
Peptidoglycan, in bacterial cell walls, 169, 171–182
Perimidine, dissociation constant in aqueous solution, 888
Periodic acid, dissociation constant in aqueous solution, 896
Periodic table of the elements, 883
Phenanthridine, dissociation constant in aqueous solution, 888
1, 10-Phenanthroline, dissociation constant in aqueous solution, 888
Phenetidine, dissociation constants in aqueous solutions, 888
Phenol, dissociation constant in aqueous solution, 894
Phenolsulfonic acid, ionization constants in water, 898
Phenylacetic acid, dissociation constant in aqueous solution, 894
Phenylalanine, 4, 6, 11, 12, 16–24, 55, 58, 59, 66–69, 72, 73, 75, 80, 211, 213, 547, 548, 921, 923, 925
abbreviation, 4
antagonists, 58, 59
difference of acid versus neutral spectra, 73
difference of alkaline versus neutral spectra, 72
distribution in selected proteins, 16–24
content in fungal cell walls, 211, 213
far-ultraviolet absorption spectra, 67, 68
in media composition, 921, 923, 925
ionization constants, 11, 12
at 25°C, DL-configuration, 11
in aqueous formaldehyde solutions, L- and DL-configuration, 12
isoelectric point, DL-configuration, 11
$\alpha$-keto acid analogs, 66
molecular absorbance values in neutral and alkaline solutions, 72
molecular weight, 4
solubility in water, L- and DL-configurations, 55
specific rotations, 75, 80
D-configuration, 80
L-configuration, 75, 80
structural features, 547, 548
structure, 6
ultraviolet absorption characteristics of N-acetyl methyl ester, 69
ultraviolet absorption spectrum at pH 6, 69
Phenylpropionic acid, dissociation constants in aqueous solutions, 894
Phenylserine, L-configuration, 8, 76
specific rotations, 76
structure, 8
pH indicators, 913–916
Phosphate solution, 911, 912
pH values at temperatures from 0 to 95°C, 912
properties at 25°C, 911
Phosphatidic acid, structure, 261
Phosphatidyl choline, structure, 261
Phosphatidyl ethanolamine, structure, 261
Phosphatidyl glycerol, structure, 261
Phosphatidyl inositol, structure, 261
Phosphatidyl serine, structure, 261

Phosphoric acid, 896, 898
composition and properties of concentrated aqueous solution, 898
dissociation constants in aqueous solutions, 896
ionization constants in water, 898
Phosphorous acid, dissociation constants in aqueous solutions, 896
Phosphorus, in media composition, 919, 921, 923–926
Phosphoserine, structure, 7
Phthalate solution, 911, 912
pH values at temperatures from 0 to 95°C, 912
properties at 25°C, 911
Phthalic acid, dissociation constants in aqueous solutions, 894
Picoline, dissociation constants in aqueous solutions, 888
Picric acid, dissociation constant in aqueous solution, 894
Pilacetin, selected properties, 486
Pilocarpine, dissociation constant in aqueous solution, 888
Pimelic acid, dissociation constant in aqueous solution, 894
Pipecolic acid, 65, 76
α-keto acid analogs, 65
specific rotation, L-configuration, 76
Piperazine, dissociation constants in aqueous solutions, 888
Piperidine, dissociation constant in aqueous solution, 888
Plasmalogen, structure, 262
Plastics, 989–1001
chemical resistance, 991–998
common solvents at room temperature, 996, 997
specific compounds, 991–996
ultracentrifuge tubes, 997, 998
effects of laboratory reagents, 1000, 1001
laboratory performance, 998–1000
physical properties, 989, 990
Poly-β-hydroxybutyrate, structure, 263
Polynucleotides, synthetic, 778, 779, 781–785
buoyant densities, 778, 779
melting temperatures, 781–785
copolymers of alternating sequence, 785
deoxyribohomopolymers, 784
deoxyribohomopolymer complexes, purine–pyrimidine, 784
ribohomopolymers, 781, 782
ribohomopolymer complexes, purine–purine, 782
ribohomopolymer complexes, purine–pyrimidine, 783, 784
ribohomopolymer-deoxyribohomopolymer hybrids, 784
Polyoxins, selected properties, 488–511
Polysaccharides, 92–96, 147–162, 182–189, 220
algal, 147–162
extracellular, 93–96
in bacterial cell walls, 182–189
O-antigenic specificity, 220
storage materials, 92
Poriferasterol, properties, 271
Potassium, in media composition, 919, 921, 923–926
Potassium cyanide, approximate pH value at 25°C, 912
Potassium hydroxide, approximate pH values at 25°C, 912
Proline, 4, 6, 11, 12, 16–24, 55, 59, 66–68, 75, 80, 211, 213, 889, 921, 923, 925
abbreviation, 4
antagonists, 59
content in fungal cell walls, 211, 213
dissociation constants in aqueous solutions, 889
distribution in selected proteins, 16–24
far-ultraviolet absorption spectra, 67, 68
in media composition, 921, 923, 925
ionization constants, 11, 12
at 25°C, L-configuration, 11
in aqueous ethanol solutions, 11
in aqueous formaldehyde solutions, L-configuration, 12
isoelectric point, L-configuration, 11
isoelectric point, L-configuration, 11
α-keto acid analogs, 66
molecular weight, 4
solubility in water, L-configuration, 55
specific rotations, 75, 80
D-configuration, 80
L-configuration, 75, 80
structure, 6
Propane compounds, dissociation constants in aqueous solutions, 889
Propionic acid, 248, 249, 894, 898
chemical and physical characteristics, 248, 249
dissociation constant in aqueous solution, 894
ionization constants on water, 898
Propylamine, dissociation constant in aqueous solutions, 889
Proteins, 16–24, 31–50, 189, 190, 193, 210, 212, 224, 923–925
amino acid composition, 16–24
amino acid sequences, 31–50
coat proteins, 31, 32
enzymes, 38–45
eukaryotic cytochromes *c*, 33–36
ferredoxins, 46, 47
miscellaneous, 48–50
in bacterial cell walls, 189, 190, 193
in fungal cell walls, 210, 212
in media composition, 923–925
metabolic effects of bacterial endotoxins, 224
Protists, eukaryotic, DNA base composition, 561–564
algae, 561, 562
fungi, 562–564
protozoa, 561
Protozoa, 561, 577, 733, 734
distribution of purines and pyrimidines in DNA, 577
DNA base compositions, 561
DNA content per cell, 772
GC composition of DNA, 733, 734
Pseudouridine, 373–375, 543, 544, 546–550
in tRNAs, 543, 544, 546–550
physical constants, 373–375
spectral properties, 373–375
Pteridine, dissociation constant in aqueous solution, 889
Pulcherriminic acid, structure and sources, 830
Purines, 331–376, 543, 544, 567–580, 889
in tRNAs, 543, 544

Purines (*cont.*)
dissociation constant in aqueous solution, 889
distribution in DNA, 567–580
of algae, 575
of animal viruses, 580
of bacterial phages, 578, 579
of fungi, 568–574
of insect viruses, 579
of protozoa, 577
of tracheophytes, 576, 577
physical constants, 331–376
spectral properties, 331–376
Puromycin, selected properties, 514, 515
Pyrazine, dissociation constant in aqueous solution, 889
Pyrazomycin, selected properties, 516, 517
Pyridine, dissociation constant in aqueous solution, 889
Pyridizine, dissociation constant in aqueous solution, 889
Pyridoxine, in media composition, 922, 924, 925
Pyrimidine compounds, dissociation constants in aqueous solution, 890
Pyrimidines, 331–346, 567–580, 890
distribution in DNA, 567–580
of algae, 575
of animal viruses, 580
of bacterial phages, 578, 579
of fungi, 568–574
of insect viruses, 579
of protozoa, 577
of tracheophytes, 576, 577
physical constants, 331–346
spectral properties, 331–346
Pyrimine, structure and source, 830
Pyrocatechol, dissociation constant in aqueous solution, 895
Pyrophosphorous acid, dissociation constants in aqueous solutions, 896
Pyrrolidine, dissociation constant in aqueous solution, 890

## Q

Quinazoline, dissociation constant in aqueous solution, 890
Quinine, dissociation constants in aqueous solutions, 890
Quinoline, dissociation constant in aqueous solution, 890
Quinolinic acid, dissociation constant in aqueous solution, 895
Quinovose, D-configuration, structure, 90

## R

Random permutations of twenty numbers, 977, 978
Reciprocals of numbers, 929–952
Refractive indices of liquids, 1009–1011
Resorcinol, dissociation constant in aqueous solution, 895
Rhamnolipid, structure, 263
Rhamnose, 90, 147–153, 155–162, 202–209
content in algae, 147–153, 155–162
content in fungal cell walls, 202–209
structure, D-configuration, 90
Rhodotorulic acid, structure and sources, 826
Riboflavin, in media composition, 922, 924, 925
Ribonucleic acids, 543–550, 553, 554, 557–559, 743, 744, 778
ATGC or AUGC content in cubic phages, 743, 744
buoyant densities of viral and cellular RNAs in the $Cs_2SO_4$ gradient, 778
controlled partial hydrolysis, 553, 554
nucleotide sequences, 557–559
bacteriophage Qβ ribonucleic acid, 559
bacteriophage R17 ribonucleic acids, 558
5S-ribosomal ribonucleic acids, 557
structural features of tRNAs, 543–550
Ribonucleosides, 347–376, 797–802
metabolism, 797–802
anabolism, 798–800
catabolism, 801, 802
physical constants, 347–376
spectral properties, 347–376
Ribonucleotides, 383–409, 539–541
physical constants, 383–409
spectral properties, 383–409
spectrophotometric constants, 539–541
Ribose, 148, 155, 156, 202–209
content in algae, 148, 155, 156
content in fungal cell walls, 202–209
Ribothymidine, in tRNAs, 543, 544
Ricinoleic acid, chemical and physical characteristics, 256, 257
RNA, *see* Ribonucleic acids
Rubbers, commercial, chemical effects of common solvents, 1002–1004
Rubidium, in media composition, 919

## S

Sabinic acid, chemical and physical characteristics, 252, 253
Saccharin, dissociation constant in aqueous solution, 895
Sacromycin, selected properties, 518
Sake colorant A, structure and source, 825
Sangivamycin, selected properties, 520, 521
Sarcosine, structure, 7
Sargasterol, properties, 271
Schizokinen, structure and source, 827
Sedimentation equations, 853–866
equivolumetric gradients of sucrose, 868
hyperbolic gradient for the separation of ribosomes, 866
isokinetic gradients, 860–864
parameters for IEC rotors, 860–862
parameters for Spinco rotors, 863, 864
sedimentation properties of some biological particles, 857
values for different speeds and intervals, 855
Selenic acid, dissociation constant in aqueous solution, 896
Selenous acid, dissociation constants in aqueous solutions, 896
Serine, 4, 5, 11, 12, 16–24, 55, 59, 66–68, 75, 80, 211, 213, 548, 890, 923, 925
abbreviation, 4

Serine (*cont.*)
antagonists, 59
content in fungal cell walls, 211, 213
dissociation constants in aqueous solutions, 890
distribution in selected proteins, 16–24
far-ultraviolet absorption spectra, 67, 68
in media composition, 923, 925
ionization constants, DL-configuration, 11, 12
at 25°C, 11
in aqueous formaldehyde solutions, 12
isoelectric point, DL-configuration, 11
α-keto acid analogs, 66
molecular weight, 4
solubility in water, L- and DL-configurations, 55
specific rotations, 75, 80
D-configuration, 80
L-configuration, 75, 80
structural features, 548
structure, 5
Showdomycin, selected properties, 522, 523
Shwartzman phenomenon, 222
Silicic acid, dissociation constants in aqueous solutions, 896
Silicon, in media composition, 919
Silver hydroxide, dissociation constant in aqueous solution, 891
β-Sitosterol, properties, 266, 271
γ-Sitosterol, properties, 272
Sodium, in media composition, 919, 923, 925
Sodium carbonate, approximate pH value at 25°C, 912
Sodium chloride, in media composition, 921
Sodium hydroxide, approximate pH value at 25°C, 912
Solvents, miscibility, 1005–1007
Sorbic acid, chemical and physical characteristics, 250, 251
Sparsomycin, selected properties, 524
Specific gravity of liquids, 1013–1015
Sphingomyelin, structure, 262
α-Spinasterol, properties, 272
Squares and square roots of numbers, 953–972
Starch, occurrence, 92
Stearic acid, chemical and physical characteristics, 248, 249
Sterculic acid, chemical and physical characteristics, 254, 255
Sterols, properties, 265–273
Stigmasterol, properties, 265, 272
Stillingic acid, chemical and physical characteristics, 250, 251
Streptose, structure, 91
Strontium, in media composition, 919, 920
Strychnine, dissociation constant in aqueous solution, 890
Succinic acid, dissociation constants in aqueous solutions, 895
Sucrose, 93, 868–873, 899–902
construction of equivolumetric gradients, 868
density and viscosity, 869–873, 899–902
as functions of concentration, 869, 870
at different temperatures, 871–873
in water, 899–902
occurrence, 93
Sugar, neutral, content in fungal cell walls, 202–209
Sulfanilic acid, dissociation constant in aqueous solution, 895
Sulfates, in media composition, 925
Sulfur, in media composition, 921, 923, 926
Sulfur dioxide, in media composition, 926
Sulfuric acid, 896, 898, 912
approximate pH values at 25°C, 912
composition and properties of concentrated aqueous solution, 898
dissociation constant in aqueous solution, 896
Sulfurous acid, approximate pH value at 25°C, 912

## T

Tartaric acid, 895, 912
approximate pH value at 25°C, 912
dissociation constants in aqueous solutions, 895
Tartrate solution, 911, 912
pH values at temperatures from 0 to 95°C, 912
properties at 25°C, 911
Taurine, 7, 55, 890
dissociation constant in aqueous solution, 890
solubility in water, 55
structure, 7
Teichoic acids, 182–185
as antigens, 185
location and chemical nature in lactobacilli, 185
extraction from bacterial cell walls, 183, 184
in bacterial cell walls, 182–185
Telluric acid, dissociation constants in aqueous solutions, 896
Terregens factor, structure and sources, 827
Tetroxalate solution, 911, 912
pH values at temperatures from 0 to 95°C, 912
properties at 25°C, 911
Theanine, L-configuration, specific rotation, 76
Theobromine, dissociation constant in aqueous solution, 895
Thiamine, in media composition, 922, 924, 925
Thiazole, dissociation constant in aqueous solution, 890
2-Thiocytidine, 355, 429
natural occurrence, 429
physical constants, 355
spectral constants, 355
2-Thio-5(N-methylaminomethyl)uridine, 369, 431
natural occurrence, 431
physical constants, 369
spectral properties, 369
4-Thiouracil, physical and spectral data, 342
2-Thiouridine, physical and spectral data, 365
4-Thiouridine, 367, 430, 543, 544, 547–550
in tRNAs, 543, 544, 547–550
natural occurrence, 430
physical constants, 367
spectral properties, 367
Threonine, 4, 5, 16–24, 59, 66–68, 75, 76, 80, 211, 213, 890, 921, 923, 925
abbreviation, 4
antagonists, 59

Threonine (*cont.*)
content in fungal cell walls, 211, 213
dissociation constants in aqueous solutions, 890
distribution in selected proteins, 16–24
far-ultraviolet absorption spectra, 67, 68
in media composition, 921, 923, 925
α-keto acid analogs, 66
molecular weight, 4
specific rotation, 75, 76, 80
D-configuration, 80
L-configuration, 75, 76, 80
structure, 5
Thymidine, physical and spectral data, 382
Thymine, 343, 741–743
content in bacteriophage nucleic acids, 741–743
physical constants, 343
spectral properties, 343
Thyroxine, 59, 76, 81
antagonist, 59
specific rotations, L-configuration, 76, 81
Tin, in media composition, 919, 920
Titanium, in media composition, 919, 920
Tocopherols, in media composition, 925
Toluic acid, dissociation constants in aqueous solutions, 895
Toluidine, dissociation constants in aqueous solutions, 890
Toyocamycin, selected properties, 528, 529
Trace metals, 805–808, 919
effects in Acinomycetes, 806, 807
effects in bacteria, 805, 806
effects in fungi, 807, 808
in DIFCO nutrient broth, 919
Tracheophytes, distribution of purines and pyrimidines in DNA, 576, 577
Transfer ribonucleic acids, structural features, 543–550
Transparency, conversion to optical density, 983–988
Trichloroacetic acid, dissociation constant in aqueous solution, 895
Trichlorophenol, dissociation constant in aqueous solution, 895
Tridecylic acid, chemical and physical characteristics, 248, 249
Triethylamine, dissociation constant in aqueous solution, 890
Trimethylamine, dissociation constant in aqueous solution, 890
$N^2$,$N^2$,7-Trimethylguanosine, 362, 429
natural occurrence, 429
physical constants, 362
spectral properties, 362
Trisodium phosphate, approximate pH value at 25°C, 912
Tryptophan, 4, 6, 11, 12, 16–24, 55, 59, 66–69, 71–73, 75, 81, 549, 890, 895, 921, 923, 925
abbreviation, 4
antagonists, 59
difference of acid versus neutral spectra, 73
difference of alkaline versus neutral spectra, 72
dissociation constants in aqueous solutions, 890, 895
distribution in selected proteins, 16–24
far-ultraviolet absorption spectra, 67, 68
in media composition, 921, 923, 925
Tryptophan (*cont.*)
ionization constants, L-configuration, 11, 12
at 25°C, 11
in aqueous formaldehyde solutions, 12
isoelectric point, L-configuration, 11
α-keto acid analogs, 66
molecular absorbance values in neutral and alkaline solutions, 71
molecular weight, 4
solubility in water, L-configuration, 55
specific rotations, 75, 81
D-configuration, 81
L-configuration, 75, 81
structural features, 549
structure, 6
ultraviolet absorption characteristics of N-acetyl methyl ester, 69
ultraviolet absorption spectrum at pH 6, 69
Tubercidin, selected properties, 526, 527
Tuberculostearic acid, 256, 257, 318
chemical and physical characteristics, 256, 257
structure, 318
Tyrosine, 4, 6, 11, 12, 16–24, 56, 59, 66–69, 71–73, 75, 81, 211, 213, 549, 890, 895, 921, 923, 925
abbreviation, 4
antagonists, 59
content in fungal cell walls, 211, 213
difference of acid versus neutral spectra, 73
difference of alkaline versus neutral spectra, 72
dissociation constants in aqueous solutions, 890, 895
distribution in selected proteins, 16–24
far-ultraviolet absorption spectra, 67, 68
in media composition, 921, 923, 925
ionization constants, L-configuration, 11, 12
at 25°C, 11
in aqueous formaldehyde solutions, 12
isoelectric point, L-configuration, 11
α-keto acid analogs, 66
molecular absorbance values in neutral and alkaline solutions, 71
molecular weight, 4
solubility in water, L-configuration, 56
specific rotations, 75, 81
D-configuration, 81
L-configuration, 75, 81
structural features, 549
structure, 6
ultraviolet absorption characteristics of N-acetyl methyl ester, 69
ultraviolet absorption spectrum at pH 6, 69
Tyvelose, D-configuration, structure, 90

## U

Unamycin B, selected properties, 530
Undecylic acid, chemical and physical characteristics, 248, 249
Uracil, physical and spectral data, 341
Urea, dissociation constant in aqueous solution, 890
Uric acid, dissociation constant in aqueous solution, 895

Uridine, 365, 543, 550
in tRNAs, 543, 550
physical constants, 365
spectral properties, 365

## V

Vaccenic acid, chemical and physical characteristics, 250, 251
Valeric acid, 248, 249, 895
chemical and physical characteristics, 248, 249
dissociation constants in aqueous solutions, 895
Valine, 4, 5, 11, 12, 16–24, 56, 59, 66, 75, 81, 211, 213, 550, 890, 921, 923, 925
abbreviation, 4
antagonists, 59
content in fungal cell walls, 211, 213
dissociation constants in aqueous solutions, 890
distribution in selected proteins, 16–24
in media composition, 921, 923, 925
ionization constants, 11, 12
at 25°C, DL-configuration, 11
in aqueous ethanol solutions, 11
in aqueous formaldehyde solutions, DL-configuration, 12
isoelectric point, DL-configuration, 11
α-keto acid analogs, 66
molecular weight, 4
solubility in water, L- and DL-configurations, 56
specific rotation, 75, 81
D-configuration, 81
L-configuration, 75, 81
structural features, 550
structure, 5
Vanadium, in media composition, 920
Vengicide, selected properties, 531
Viruses, 299, 300, 314–316, 578–580, 741–747, 751–756, 761, 762, 765–767, 776–778
base composition of nucleic acids in bacteriophages, 741–747
ATGC or AUGC content, 741–744
GC content, 744–747
buoyant densities of nucleic acids, 776–778
RNAs in the $Cs_2SO_4$ gradient, 778
selected DNAs, 776, 777
T-even coliphage DNAs, 777
distribution of purines and pyrimidines in DNA, 578–580
animal viruses, 580
bacterial phages, 578, 579
insect viruses, 579
DNA content per cell, 761, 762
in animal viruses, 762
Viruses (*cont.*)
in bacteriophages, 761
fatty acid content, 299, 300
lipid content, 314–316
molecular composition and physical properties of DNA, 751–756
properties of DNA, 765–767
Vitamins, in media composition, 922, 924, 925

## W

Water, ionization constants, 897

## X

Xanthine, 345, 895
dissociation constant in aqueous solution, 895
physical constants, 345
spectral properties, 345
Xanthosine, physical and spectral data, 376
Xenon content in atmospheric air, 979
Xylose, 90, 148–153, 155–162, 202–209
content in algae, 148–153, 155–162
content in fungal cell walls, 202–209
structure, D-configuration, 90

## Y

Yeasts, 301, 302, 316
fatty acid content, 301, 302
lipid content, 316

## Z

Zeatin, physical and spectral data, 346
Zinc, 805–808, 817, 919, 920, 926
effects of trace amounts, 805–808
in Actinomycetes, 806, 807
in bacteria, 805, 806
in fungi, 807, 808
in enzyme-catalyzed reactions, 817
in media composition, 919, 920, 926
Zinc–calcium, in enzyme-catalyzed reactions, 817
Zinc–cobalt, in enzyme-catalyzed reactions, 817
Zinc–copper, in enzyme-catalyzed reactions, 818
Zinc–monovalent cations, in enzyme-catalyzed reactions, 818
Zirconium, in media composition, 919
Zymosterol, properties, 266